CORROSION

Volume 1

Metal/Environment Reactions

Edited by

L. L. Shreir, PhD, CChem, FRIC, FIM, FICorrT, FIMF, OBE
R. A. Jarman, MSc, PhD, CEng, MIEE, FIW
G. T. Burstein, MSc, PhD, MA

Butterworth-Heinemann Ltd
Linacre House, Jordan Hill, Oxford OX2 8DP

A member of the Reed Elsevier group

OXFORD   LONDON   BOSTON
MUNICH   NEW DELHI   SINGAPORE   SYDNEY
TOKYO   TORONTO   WELLINGTON

First published 1963
Second edition 1976
Third edition 1994

© The several contributors listed on pages xvii–xxii, 1963, 1976, 1994

All rights reserved. No part of this publication
may be reproduced in any material form (including
photocopying or storing in any medium by electronic
means and whether or not transiently or incidentally
to some other use of this publication) without the
written permission of the copyright holder except in
accordance with the provisions of the Copyright,
Designs and Patents Act 1988 or under the terms of a
licence issued by the Copyright Licensing Agency Ltd,
90 Tottenham Court Road, London, England W1P 9HE.
Applications for the copyright holder's written permission
to reproduce any part of this publication should be addressed
to the publishers

British Library Cataloguing in Publication Data

Corrosion – 3Rev. ed
 I. Shreir, L. L.   II. Jarman, R. A.
 III. Burstein, G. T.
 620.1623

ISBN 0 7506 1077 8 (for both volumes)

Library of Congress Cataloguing in Publication Data

Corrosion/edited by L. L. Shreir, R. A. Jarman, G. T. Burstein.
  p.   cm.
  Includes bibliographical references and index.
  Contents: v. 1. Metal/evironment reactions – v. 2. Corrosion control.
  ISBN 0 7506 1077 8
  1. Corrosion and anti-corrosives.   I. Shreir, L. L.
  II. Jarman, R. A.   III. Burstein, G. T.
  TA462.C6513   1993
  620.1′1223–dc20   93-13859
                                                    CIP

Printed and bound in Great Britain by Hartnolls, Bodmin

CONTENTS

Volume 1. Metal/Environment Reactions

L. L. Shreir, OBE
Preface to the third edition
Preface to the first edition
List of contributors

1. Principles of Corrosion and Oxidation

1.1   Basic Concepts of Corrosion

1.1A   Appendix – Classification of Corrosion Processes

1.2   Nature of Films, Scales and Corrosion Products on Metals

1.3   Effects of Metallurgical Structure on Corrosion

1.4   Corrosion in Aqueous Solutions

1.5   Passivity and Localised Corrosion

1.6   Localised Corrosion

1.7   Bimetallic Corrosion

1.8   Lattice Defects in Metal Oxides

1.9   Continuous Oxide Films

1.10   Discontinuous Oxide Films

1.11   Erosion Corrosion

2. Environments

2.1   Effect of Concentration, Velocity and Temperature

2.2   The Atmosphere

2.3   Natural Waters

2.4   Sea Waters

2.5   Soil in the Corrosion Process

2.6   The Microbiology of Corrosion

2.7   Chemicals

2.8   Corrosion by Foodstuffs

2.9   Mechanisms of Liquid-metal Corrosion

2.10 Corrosion in Fused Salts

2.11 Corrosion Prevention in Lubricant Systems

2.12 Corrosion in the Oral Cavity

2.13 Surgical Implants

3. Ferrous Metals and Alloys

3.1   Iron and Steel

3.2   Low-alloy Steels

3.3   Stainless Steels

3.4   Corrosion Resistance of Maraging Steels

3.5   Nickel–Iron Alloys

3.6   Cast Iron

3.7   High-nickel Cast Irons

3.8   High-chromium Cast Irons

3.9   Silicon–Iron Alloys

3.10 Amorphous (Ferrous and Non-Ferrous) Alloys

4. Non-Ferrous Metals and Alloys

4.1 Aluminium and Aluminium Alloys

4.2 Copper and Copper Alloys

4.3 Lead and Lead Alloys

4.4 Magnesium and Magnesium Alloys

4.5 Nickel and Nickel Alloys

4.6 Tin and Tin Alloys

4.7 Zinc and Zinc Alloys

5. Rarer Metals

15.1 Beryllium

5.2 Molybdenum

5.3 Niobium

5.4 Titanium and Zirconium

5.5 Tantalum

5.6 Uranium

5.7 Tungsten

6. The Noble Metals

6.1 The Noble Metals

7. High-Temperature Corrosion

7.1 Environments

7.2 The Oxidation Resistance of Low-alloy Steels

7.3 High-temperature Corrosion of Cast Iron

7.4   High-alloy Steels

7.5   Nickel and its Alloys

7.6   Thermodynamics and Kinetics of Gas–Metal Systems

8. Effect of Mechanical Factors on Corrosion

8.1   Mechanisms of Stress-corrosion Cracking

8.2   Stress-corrosion Cracking of Ferritic Steels

8.3   Stress-corrosion Cracking of Stainless Steels

8.4   Stress-corrosion Cracking of High-tensile Steels

8.5   Stress-corrosion Cracking of Titanium, Magnesium and Aluminium Alloys

8.6   Corrosion Fatigue

8.7   Fretting Corrosion

8.8   Cavitation Damage

8.9   Outline of Fracture Mechanics

8.10   Stress-corrosion Test Methods

8.10A Appendix – Stresses in Bent Specimens

Volume 2. Corrosion Control

Introduction to Volume 2

9. Design and Economic Aspects of Corrosion

9.1 Economic Aspects of Corrosion

9.2 Corrosion Control in Chemical and Petrochemical Plant

9.3 Design for Prevention of Corrosion in Buildings and Structures

9.4 Design in Marine and Offshore Engineering

9.5 Design in Relation to Welding and Joining

9.5A Appendix – Terms Commonly Used in Joining

10. Cathodic and Anodic Protection

10.1 Principles of Cathodic Protection

10.2 Sacrificial Anodes

10.3 Impressed-current Anodes

10.4 Practical Applications of Cathodic Protection

10.5 Stray-current Corrosion

10.6 Cathodic-protection Interaction

10.7 Cathodic-protection Instruments

10.8 Anodic Protection

11. Pretreatment and Design for Metal Finishing

11.1 Pretreatment Prior to Applying Coatings

11.2 Pickling in Acid

11.3 Chemical and Electrolytic Polishing

11.4 Design for Corrosion Protection by Electroplated Coatings

11.5 Design for Corrosion Protection by Paint Coatings

12. Methods of Applying Metallic Coatings

12.1 Electroplating

12.2 Principles of Applying Coatings by Hot Dipping

12.3 Principles of Applying Coatings by Diffusion

12.4 Principles of Applying Coatings by Metal Spraying

12.5 Miscellaneous Methods of Applying Metallic Coatings

13. Protection by Metallic Coatings

13.1 The Protective Action of Metallic Coatings

13.2 Aluminium Coatings

13.3 Cadmium Coatings

13.4 Zinc Coatings

13.5 Tin and Tin Alloy Coatings

13.6 Copper and Copper Alloy Coatings

13.7 Nickel Coatings

13.8 Chromium Coatings

13.9 Noble Metal Coatings

14. Protection by Paint Coatings

14.1 Paint Application Methods

14.2	Paint Formulation	
14.3	The Mechanism of the Protective Action of Paints	
14.4	Paint Failure	
14.5	Paint Finishes for Industrial Applications	
14.6	Paint Finishes for Structural Steel for Atmospheric Exposure	
14.7	Paint Finishes for Marine Application	
14.8	Protective Coatings for Underground Use	
14.9	Synthetic Resins	
14.10	Glossary of Paint Terms	

15. Chemical Conversion Coatings

15.1 Coatings Produced by Anodic Oxidation

15.2 Phosphate Coatings

15.3 Chromate Treatments

16. Miscellaneous Coatings

16.1 Vitreous Enamel Coatings

16.2 Thermoplastics

16.3 Temporary Protectives

17. Conditioning the Environment

17.1 Conditioning the Atmosphere to Reduce Corrosion

17.2 Corrosion Inhibition: Principles and Practice

17.3 The Mechanism of Corrosion Prevention by Inhibitors

17.4 Boiler and Feed-water Treatment

18. Non-Metallic Materials

18.1   Carbon

18.2   Glass and Glass-ceramics

18.3   Vitreous Silica

18.4   Glass Linings and Coatings

18.5   Stoneware

18.6   Plastics and Reinforced Plastics

18.7   Rubber and Synthetic Elastomers

18.8   Corrosion of Metals by Plastics

18.9   Wood

18.10 The Corrosion of Metals by Wood

19. Corrosion Testing, Monitoring and Inspection

19.1   Corrosion Testing

19.1A Appendix – Removal of Corrosion Products

19.1B Appendix – Standards for Corrosion Testing

19.2   The Potentiostat and its Applications to Corrosion Studies

19.3   Corrosion Monitoring and Inspection

19.4   Inspection of Paints and Painting Operations

20. Electrochemistry and Metallurgy Relevant to Corrosion

20.1   Outline of Electrochemistry

20.2   Outline of Chemical Thermodynamics

20.3   The Potential Difference at a Metal/Solution Interface

20.4   Outline of Structural Metallurgy Relevant to Corrosion

21. Useful Information

21.1 Tables

21.2 Glossary of Terms

21.3 Symbols and Abbreviations

21.4 Calculations Illustrating the Economics of Corrosion

Index

L. L. SHREIR, OBE 1914-1992

Lionel Louis Shreir OBE, died on 5th November 1992 after a lifetime devoted to the science and technology of corrosion and education. His industrial career spanned a period of 19 years, from 1929 to 1948, during which time he was employed by the Mond Nickel Company, Baker Platinum Ltd and Plessey Ltd. At the same time he continued his higher education on a part time basis at the Chelsea and Battersea Polytechnics and Sir John Cass College in London. In 1948, he joined the staff of Battersea Polytechnic, subsequently renamed the University of Surrey, eventually attaining the position of Reader in Corrosion. In 1962 Lionel became Head of Metallurgy of the Sir John Cass College (now London Guildhall University), a post he enjoyed by greatly expanding the Department, its research and general reputation until his retirement in 1979.

Lionel's contribution to corrosion was outstanding. In addition to his encyclopedic work *Corrosion*, the present edition being dedicated to his memory, he was author of more than 70 papers and was editor of *Corrosion Science* for many years. He was engaged as a consultant to a number of organisations up to the time of his death whilst his research initiatives covered many fields, including hydrogen in metals, anodic oxidation and electrodeposition. He was the third recipient of the U. R. Evans Award in 1978 and was awarded the OBE in 1982 in recognition of his services to corrosion. In this context, one of his most notable activities was to advise on the protection of the Thames Barrier. In the past year a Lionel Shreir Award was awarded for the first time by the Institute of Corrosion Science and Technology.

Regardless of his achievements Lionel was kind, modest and a very caring man. He will be affectionately remembered for his boundless energy and infectious enthusiasm by his peers, colleagues, friends and the countless past students privileged to have made his acquaintance during what was a remarkably active life.

R A JARMAN

PREFACE TO THE THIRD EDITION

The huge success of the first two editions of *Corrosion* has inevitably created the demand for a third edition. Corrosion science and technology, like most of the physical sciences, has progressed and advanced significantly in the seventeen years since the second edition was published. Such knowledge requires transferral from the laboratory and the journal literature to the wider audience: the student, the teacher, the engineer, the metallurgist and workers in other fields who require knowledge and understanding of the interactions of materials with their environments. The previous two editions, the fruits of Lionel Shreir's hard labours, have fulfilled this multiple role admirably and the new editors hope that this new edition will continue to do so. The fact that Lionel worked so hard on producing the third edition but did not live to see its publication, is a personal and deeply poignant sorrow for us, as it must be for the many readers of *Corrosion* who knew and respected him as scientist and friend.

The ever-increasing research into corrosion, and the knowledge that this produces is driven to a small part by the corrosion scientist him- or her-self in seeking a detailed understanding of the intricacies of the interfacial processes driving corrosion and passivation. Such a self-fulfilling drive cannot of itself however, be indefinitely sustainable, despite the fascination that this science engenders, since research is costly. Such advances are led primarily by the continuing need to predict, control and prevent corrosion as an engineering imperative. Corrosion science, multidisciplinary in itself, is probably unique in crossing the borders of almost all the technologies: environmental stability of all components of those technologies remains a prime requirement for their success. New technologies, new engineering practices, new materials and new processes can succeed only if the behaviour of their components with the environment is satisfactory, and predictably so. The eighties and nineties, and beyond, see a further need to underpin research and development into corrosion and protection – the growing awareness of the necessity for conservation, of materials and of energy, the so-called green issues. Most materials and components made from them require large energy resources to produce; clearly the quest for longevity and reliability of structures is a significant and worthy contribution towards conserving energy and materials, quite additional to minimising the heavy cost of corrosion failures.

As with the second edition, the new volumes have been revised according to the general format and structure of their antecedents. Some sections have been completely rewritten to bring them up to date, while others have been altered and extended. New sections have been included to cover areas not previously treated. The incorporation of new authors to carry out such revisions and additions is the inevitable consequence of the fact that thirty years have elapsed since *Corrosion* first appeared. The multiplicity of authors for the new edition leads (as with previous editions) to a variety of styles of writing and variation in treatment and emphasis of subject matter. One hopes this is beneficial to the work in providing a broader cross-section of corrosion science and technology as a whole: it is for the reader and casual user to judge. One hopes too, that the third edition remains a tribute to the man who initiated *Corrosion*.

GTB
Cambridge

PREFACE TO THE FIRST EDITION

The enormous scope of the subject of corrosion follows from the definition which has been adopted in the present work. Corrosion will include all reactions at a metal/environment interface irrespective of whether the reaction is beneficial or detrimental to the metal concerned — no distinction is made between chemical or electropolishing of a metal in an acid and the adventitious deterioration of metal plant by acid attack. It follows, therefore, that a comprehensive work on the subject of corrosion should include an account of batteries, electrorefining, chemical machining, chemical and electrochemical polishing, etc.

The fact remains, nonetheless, that the environmental reaction of a metal used as a construction material is the most important type of corrosion reaction, and the one of most concern to the engineer. The technological and economic consequences of the wastage of metals by corrosion are now fully appreciated, and figures have been published which show the enormous financial losses, both to the individual organisation and to the economy of the country as a whole, resulting from the deterioration of metals. The need for conserving metals has been publicised by Dr U. R. Evans, Dr J. C. Hudson, Mr T. H. Turner, Professor H. H. Uhlig, Dr W. H. J. Vernon, and others, and the 'corrosion consciousness' which prevails today is largely due to their efforts.

In the light of what has been said above, little further explanation of the implications of the title of the present work is required. Its treatment of the subject of corrosion will centre round the *control* of the environmental interactions of metals and alloys used as materials of construction.

The effective control of corrosion reactions must be based on an understanding of the mechanism of such reactions and on the application of this knowledge to practical problems. The work, regarded as a whole, represents an attempt, therefore, to present the subject of corrosion as a synthesis of corrosion science and corrosion engineering. Thus in the planning of the content an attempt has been made to strike a suitable balance between the primarily scientific and the primarily practical aspects, and so the nature of individual sections ranges from the fundamental and theoretical to the essentially practical.

It is hoped that this approach has resulted in a work that will be of some value to the student, the corrosion worker, and the engineer in the field of corrosion.

Corrosion represents the joint effort of over 100 authors, all of whom have been free, within the necessary limitations of length, to express their own views. Grateful acknowledgements are made to the individual authors from Great Britain, the United States, and Canada for their valuable and enthusiastic co-operation.

The task of the editor in finding suitable authors for various topics was considerably lightened by the fact that the majority of corrosion specialists in this country belong either to the Corrosion Group of the Society of Chemical Industry or to the Institute of Metal Finishing, and acknowledgements are made to Mr S. C. Britton (then Secretary, Corrosion Group) and to Dr S. Wernick and Mr I. S. Hallows (Hon. Secretary and Assistant Editor respectively of the I.M.F.). The editor wishes to express his appreciation of the considerable assistance received from Dr E. C. Rhodes and Dr G. L. J. Bailey of INCO (Mond) in providing authors from this organisation.

The editor also acknowledges with pleasure the encouragement and assistance he has received from Mr L. W. Derry (Head of Department of Metallurgy) and Dr D. M. A. Leggett (Principal) of the Battersea College of Technology, and from Dr A. M. Ward (Principal) of Sir John Cass College.

Throughout the course of this work the content and subject matter have been discussed with various workers in the field, and the editor would like to take this opportunity to thank Mr S. C. Britton, Professor C. W. Davies, Dr T. P. Hoar, Dr E. C. Potter, and others for their advice and constructive criticism.

He would also like to take this opportunity to express his appreciation to Dr U. R. Evans and Dr W. H. J. Vernon for assistance given when he first contemplated entering the field of corrosion, and for their encouragement and advice in connection with the present work.

Finally, grateful acknowledgements are made by the editor to Mr T. F. Saunders and Mrs N. E. Orna, M.A., of George Newnes (Technical Books) for their kind co-operation at all stages of the work.

L.L.S.

1963

CONTRIBUTORS

K G Adamson*, AMCST, LIM
Development Officer,
Magnesium Elektron Ltd, Manchester

D F Aitken*, MA
British Rubber Manufacturers' Association (Formerly)

J C B Alcock*, ARCS, DSc, PhD, CChem, FRIC
Professor and Chairman,
Dept. of Metallurgy and Materials Science, University of Toronto, Canada

M D Allen, CEng, MIM, MICorrST
Penspen, London

D Ames*, MICorr
Associate, Spencer and Partners, Consulting Engineers

K F Anderson*
Morganite Carbon Ltd. (Formerly)

B Angell
Section Head, Corrosion and Protection, Defence Research Agency, Poole, Dorset

J E Antill*, PhD, BSc
Head, Chemical Metallurgy Group, Materials Development Division, UKAEA, Harwell

V Ashworth
Global Corrosion Consultants Ltd Shifnal, Shropshire

D J Astley, BSc, ARCS, PhD, PGCE
Formerly Senior Technical Officer IMI Research and Development, Birmingham

J C Bailey*, BSc, FIM
Formerly Deputy Director (Technical), Aluminium Federation

W E Ballard*, CChem, FRIC, FIM
Consultant,
Formerly Managing Director, Metallisation Ltd.

*Contributor to earlier editions

T A Banfield*, PhD, DIC, ARCS, CChem, FRIC, FICorr, FTSC
Deputy Manager,
Group Research Laboratory,
Berger Jenson and Nicholson Ltd.

P J Barnes, BSc, MRSC, CChem, ATSC
Consultant

G E Barrett*, BSc, ARIC, PhD
Technical Director, Plastic Coatings Ltd.

E W Beale*, ARIC
Senior Scientific Officer,
Materials Quality Assurance Directorate, Ministry of Defence

J Bentley*, BSc, DipChemEng, CEng, MIChemE
Principal Chemist, Wastes Division Directorate General, Water Engineering, Dept. of the Environment

W Betteridge*, DSc, FInstP, FIM
Consultant,
Formerly of International Nickel Ltd.

P J Boden*, PhD, CEng, CChem, FRSC, MIM, FICorr
Senior Lecturer, Dept. of Metallurgy and Materials Science,
Nottingham University

C J L Booker*, BSc, PhD, ARIC, FICorr
Formerly Senior Lecturer in Corrosion Science,
Dept. of Metallurgy and Materials, City of London Polytechnic

J W L F Brand*, MITE, TEng(CEI), MICorr
Divisional Manager,
Corrosion Control Division,
Corrosion and Welding Engineering Ltd.

CONTRIBUTORS

C F Britton*, LRIC, MICorr
*Product Manager, Instruments,
Magnachem (UK) Ltd.*

S C Britton*, MA, CChem, FRIC, FIM,
IMF, FICorr
Tin Research Institute (Retired)

J A Brydson*, FPRI, ANCRT
Technical Consultant

T R Bullett, BSc, CPhys, FInstP, FICorrST

W Bullough*, BSc, ARIC
*Principal Research Officer,
BSC Research Centre,
Strip Mills Division*

G T Burstein, MSc, PhD, MA
*Affiliated Lecturer,
Department of Materials Science and
Metallurgy
University of Cambridge*

V E Carter*, FICorr, FIMF
*Corrosion and metal finishing consultant,
Formerly of BNFRA*

J E Castle*, BSc, PhD, CChem, FRSC
FICorr
*Head of Department,
University of Surrey, Guildford*

K A Chandler*, BSc, ARSM, FICorr
Head, Corrosion Advice Bureau, BSC

B Chatterjee*, PhD, AIM, ARIC
*Research Scientist,
British Aluminium Co.*

A R L Chivers*, MA
*Senior Technical Officer,
Zinc Development Association,
London*

M Clarke*, BSc, PhD, DSc, CChem, FRIC,
FIM, FICorr, FIMF
*Consultant, formerly Principal Lecturer,
Dept. of Metallurgy and Materials,
City of London Polytechnic*

R J Clarke*, MA, CEng, FIChemE, FIFST
*Hon. Visiting Lecturer in Food
Engineering,
Queen Elizabeth College, London*

H G Cole*, BSc, FIMF, FICorr
*Principal Scientific Officer, Ministry of
Defence (Procurement Executive)*

H H Collins*, BSc, CChem, FRIC
*Superintendent, Chemistry Research,
Stanton & Staveley/BSC*

J Congleton, BSc, PhD, FIM, CEng
*Senior Lecturer
Department of Mechanical, Materials and
Manufacturing Engineering University of
Newcastle upon Tyne*

J B Cotton*, CChem, AMCT, ARIC,
FICorr
Industrial Consultant

R A Coltis
*Senior Lecturer in Corrosion Science and
Engineering, Corrosion and Protection
Centre, UMIST*

R N Cox, BSc, CEng, MIM
*Building Research Establishment,
Garston, Watford*

G W Currer, CEng, MIEE, MICorrST
Consultant

D P Dautovich*, MSc, PhD
*Corrosion Engineer, Research Division,
Ontario Hydro, Canada*

K Julyan Day*, FICorr, FTSC, MBIM
Anti-Corrosion Consultant

J Dodd*, BSc, FIM, FIBF
*Metallurgical Consultant
Dodd and Associates, Colorado, USA*

P D Donovan*, MSc, ARIC, FIM
*Principal Scientific Officer,
Ministry of Defence*

C W Drane*, BSc, CChem, FRIC
*Technical Manager, Water Specialities
and Services,
Industrial Chemicals Division, Albright and
Wilson Ltd.*

F G Dunkley*, FICorr
*Consultant,
Formerly of British Rail, Derby*

E J Easterbrook*, BSc(Eng), ARSM,
AMIMM, MIM
*Formerly Principal Lecturer,
Dept. of Metallurgy and Materials,
City of London Polytechnic*

J Edwards, BSc, PhD
*The Paint Research Association,
Teddington, Middlesex*

T E Evans*, BSc, ARIC, FICorr
*Principal Technologist,
International Nickel Ltd., Birmingham*

D Eyre, BSc, MSc, PhD, MICorr
*Principal Corrosion Engineer
Spencer and Partners, London*

J A von Fraunhofer, PhD, MSc, FRSC
*Director, Laboratory of Molecular and
Materials Science, School of Dentistry,
University of Louisville, Kentucky*

P C Frost,
*Senior Research Scientist,
Cookson Group plc
Yarnton, Oxfordshire*

CONTRIBUTORS

D Fyfe*, MA, PhD
*Senior Marketing Engineer,
Chemetics Ltd,
Montreal, Canada*

D R Gabe, BSc, MMet, PhD, CEng, FIM, FICorr
*Institute of Polymer Technology and
Materials Engineering, Loughborough*

P J Gay*, BSc, FTSC, FICorr
Consultant

J S Gerrard*, AMIEE
*Formerly Joint General Manager,
Metal and Pipeline Endurance Ltd.*

P T Gilbert*, BSc, PhD, CChem, FRIC, FIM, FIMarE, FICorr, CEng
*Metallurgical Consultant
Formerly BCIRA*

E W F Gillham*, MSc, PhD, CChem, FRIC, CEng, FInstF
*Consultant in Land Restoration,
Formerly of CEGB, Midland Region*

D J Godfrey*, BSc, PhD
*Principal Scientific Officer,
Admiralty Materials Laboratory,
Poole, Dorset*

V R Gray*, BA, MA, PhD, FIWSc, MNZIC
*Senior Scientist, Chemistry Division,
DSIR, New Zealand*

T B Grimley*, BSc, PhD
*Reader in Theoretical Chemistry,
Dept. of Inorganic, Physical and
Industrial Chemistry,
University of Liverpool*

B H Hanson, BSc
*Consultant
Formerly IMI Titanium*

J O Harris*, PhD
*Professor of Bacteriology,
Kansas State University*

S J Harris, MSc, PhD, CEng, FIM, FIMF
*Department of Metallurgy and Materials
Science,
University of Nottingham*

A C Hart*, DTech, BSc, CChem, MRSC, FIMF
*Managing Director,
Hart Coating Technology*

A A B Harvey*, MSc, CChem, FRIC, FIMF
*Senior Technical Officer,
British Standards Institution*

K Hashimoto, DSc
*Professor
Institute for Materials Research
Tohoku University, Sendai, Japan*

M Hess*, CChem, FRIC, FIMF, FTSC
Member of the Association of Consulting
Scientists
*Principal of Manfred Hess,
Consulting Chemist and Paint Technologist*

G L Higgins*, Bsc, MIMF
Chemetall Ltd, Aylesbury

D K Hill*, DSc, PhD, FSGT
*Formerly Technical Manager,
British Indestructo Glass Ltd. (Retired)*

J Hines*, MA, PhD
*Formerly North West Region Materials
Group Manager, ICI Ltd.*

R A E Hooper, BMech, FIM, CEng, FICorr
*Group Technical Manager,
Authur Lee & Sons plc, Sheffield*

H Howarth*, AMICorr, AMet
*Investigator, Production Metallurgy Section,
Special Steels Division, BSC*

J C Hudson*, DSc, DIC, ARCS, FIM
*Consultant,
Formerly of BISRA*

D E Hughes*, MA, DSc, PhD, CBiol, FIBiol
*Professor Emeritus (Microbiology)
University of Wales*

M L Hughes*,
*Consultant,
Formerly of BISRA*

R S Hullcoop
*Ray Hullcoop and Associates, High
Wycombe*

D Inman*, BSc, ARCS, FRIC, PhD, DSc, DIC, MIMM
*Reader in Chemical Metallurgy,
Nuffield Fellow in Extraction Metallurgy,
Dept. of Metallurgy and Materials Science,
Imperial College, London*

R A Jarman*, MSc, PhD, CEng, MIEE, MWeldE, FIM
*Consultant, formerly
School of Engineering,
University of Greenwich*

D A Jones*, BS, MS, PhD (Rensselaer Polytechnic Inst.)
*Senior Research Engineer,
Research Laboratory,
U.S. Steel Corporation, Penn., USA*

L Kenworthy*, MSc, ARCS, CChem, FRIC, FIM, FICorrT
*Consultant,
Formerly Navy Dept. (Ministry of Defence)*

B T Kelly, MSc, ChP, InstP
Consultant Physicist

E G King, CEng, BSc, MIM, MIWeld
Consultant

G N King, MSc
Department of Metallurgy
University of Nottingham

D G Kingerley, MSc, BSc, CChem, MRSC, CEng, MInstE, FICorr
Dept of Materials Engineering and Materials Design, University of Nottingham

D Kirkwood, PhD
Senior Lecturer
School of Mechanical and Offshore Engineering
The Robert Gordon University, Aberdeen

F LaQue*, BSc, LLD, Past President, Nat. Assoc. Corrosion Eng., Am. Soc. Test and Mat., Electrochemical Society
Senior Lecturer, Scripps Institution of Oceanography, University of California

D N Layton*, PhD, MSc, ARCS, DIC, MInstP, FIMF
Managing Director,
Fredk. Mountford (Birmingham) Ltd.

EurIng **M F Leclerc**, BSc, PhD, MIM, CEng
Technical Executive
Biomet Ltd

D A Lewis*, BSc(Eng), FICorr
Partner of Spencer and Partners,
Consulting Engineers

E L Littauer*, BSc, PhD, MIM, AMIMM
Manager, Electrochemistry and Environmental Sciences,
Lockheed Missiles and Space Co.,
California, USA

G O Lloyd*, BSc
Principal Scientific Officer,
Division of Materials Applications,
National Physical Laboratory

N A Lockington*, MA, PhD, ARIC, FIM
Metallurgist, Director,
The Chrome-Alloying Co. Ltd.

G L Long*, PhD, CChem, FRIC
Principal Scientific Officer,
Energy Technology Support Unit,
UKAEA, Harwell

W A Luce, BMetEng
Retired

P Lydon
Roxby Engineering International Ltd, Kent

W B Mackay*, BSc, AIMF
Director, B.K.L. Alloys Ltd.

J Mackowiak, BSc, PhD, CEng, MIM
Retired Senior Lecturer, University of Surrey, Guildford
Consultant in high temperature corrosion

C A May*, MSc, PhD
Lecturer, School of Engineering,
University of Greenwich

J E O Mayne*, DSc, ARCS, DIC, CChem, FRIC, FICorr
Dept. of Materials Science and Metallurgy
University of Cambridge

P McIntyre, BSc, PhD, CEng, FIM
Technology Consultant
National Power plc,
Research and Engineering,
Swindon, Wiltshire

A D Mercer*, BSc
Principal Scientific Officer
National Physical Laboratory

N S C Millar*, CChem, FRIC, FICeram, MICorr, MBIM
General Manager, Thermovitrine Ltd.

W G O'Donnell, BSc, MSc, CChem, MRSC, APRI
Plascoat Systems Ltd, Surrey

J W Oldfield, BSc, PhD
Managing Director
Cortest Laboratories Ltd, Sheffield

R J Oliphant, BA, MSc, PhD, AWIEM
Technical Specialist
WRc plc, Swindon

D S Oliver*, BSc, PhD, FIM, FInstP
Group Director of Research and Development,
Pilkington Bros. Ltd.

M W O'Reilly*, Dip Tech, LRIC
Decorative Paints Market Team Leader,
ICIPaints Division

S Orman*, BSc, PhD, FICorr, CChem, FRIC
Senior Principal Scientific Officer,
AWRE, Aldermaston

R N Parkins*, BSc, PhD, DSc, FIM
Professor and former Head of Department,
Metallurgy and Engineering Materials,
University of Newcastle upon Tyne

A W Pearson*, MIM
Research Division, British Aluminium Co. Ltd.

J S Picard, D-è-Sc
Research Director,
Centre National de la Recherche Scientifique, Laboratoire d'ELectroshime Analytique et Appliquée,
Ecole Nationale Supériere de Chimie de Paris, France

CONTRIBUTORS

L W Pinder*, BSc, MICorr
Research Officer
PowerGen

L Pinion, BSc
Consultant

R Pinner*, BSc, FICorr, FIMF
Consultant

J S Pitman, FIM
Research Scientist
Servicised Ltd

P Poole*, BSc, PhD
Senior Scientific Officer,
Royal Aircraft Establishment

F C Porter*, MA, FIM, FICorr, FIMF
Zinc Development Association, London

B S Poulson, BSc, PhD, FICorr
Chief Technologist (Joining and Surface Engineering)
International Research and Development Ltd.

J K Prall*, PhD, BSc, CChem, FRIC
Section Manager,
Unilever Research Laboratory

J T Pringle
ICI Paints, Slough

B A Proctor, DSc, FIP
Formerly Manager, Fibres and Glass,
Pilkington Group Research

R P M Procter*, MA, PhD, CEng, FIM, FICorr
Vice Principal
The University of Manchester Institute of Science and Technology, UMIST

E F Redknap*, BSc
Retired

F H Reid*, BSc, CChem, FRIC, FIMF
Consultant,
Formerly of International Nickel Ltd.

J A Richardson, BSc, PhD, MIM, MRSC, CChem, CEng
Manager, Materials, ICI Engineering,
Cleveland

M O W Richardson, BTech, PhD, CChem, FRSC, FPRI IPTME
Loughborough University of Technology

R G Robson*, BSc(Eng), MIEE
Chartered Engineer

D van Rooyen*, BSc, PhD
Advisory Scientist, Westinghouse Bettis Atomic Power Labs., USA

EurIng C E D Rowe, BSc, CEng, MIM
Manager, Technical Services/Quality Assurance, Climax Special Metals Fabrications Ltd, Brentwood

J C Rowlands*, FICorr
Defence Research Agency, Holton Heath

J Sadowska-Mazur*, MSc (Gdansk)
Formerly Research Assistant, City of London Polytechnic

F W Salt*, BSc, PhD, FIMF
Head, Basic Surface Studies Section
BSC Research Centre,
Strip Mills Division

S R J Sauders, BSc, PhD, DIC
National Physical Laboratory

M J Schofield, PhD, MICorr
Technical Manager,
Cortest Laboratories Ltd.

I R Scholes, BSc, CChem, FRSC, FICorr
Formerly Manager,
IMI Research and Development
Wilton, Birmingham

B A Scott*, ARCS, BSc, PhD, CChem, FRIC
Deputy Information Officer, Group Technical Information Service,
British Aluminium Co. Ltd.

P M Scott, BSc, PhD
Framatome, Paris

J C Scully*, MA, PhD, CEng, FIM, FICorr
Senior Teaching Fellow,
School of Materials,
University of Leeds

H J Sharp*, PhD, MSc, CChem, FIM, FPRI
Director of International Associates

R E Shaw*, BSc, FIM, FIMF
ICI Paints Division (Retired)

P G Sheasby, BSc, FIMF
Alcan International Ltd, Banbury

L Sherwood
Formerly Global Corrosion Consultants Ltd
Shifnal, Shropshire

G S Shipley*, FICeram
Technical Advisor, Hathernware Ltd.

N R Short, BSc, PhD
Department of Civil Engineering,
Aston University

L L Shreir*, PhD, CChem, FRIC, FIM, FICorr, FIMF
Former Head, Dept. of Metallurgy and Materials, City of London Polytechnic

H Silman*, BSc, CEng, CChem, FRIC, FIChemE, FIM, FIMF
Consultant

E W Skerrey*, BSc, CChem, FRIC, FICorrT, AIM
Assistant Manager, Application Technology Department, Research Division, British Aluminium Co. Ltd.

R A Smith*, BSc, PhD, FIM
*Manager,
Research Laboratory,
International Nickel Ltd.*

J F Stanners*, BSc, FICorr
*Head of Corrosion Research,
Inter-Services Laboratory, BSC*

T N Tate
DRA Swynnerton, Defence Research Agency, Nr Stone, Staffs

W H Tatton*, ARIC, FIMF, FTSC
*Technical Officer,
British Standards Institution*

D S Tawil, BSc
*Technical Marketing Manager,
Magnesium Elektron Inc,
Lakehurst, New Jersey*

J G N Thomas*, BSc, PhD, ARCS, DIC
*Formerly Corrosion Section, Division of Materials Applications,
National Physical Laboratory*

A W Thorley, BSc, AIM
*Consultant
Formerly UK Atomic Energy Authority*

J E Truman*, AMet
*Consultant metallurgist
Jessop Saville Ltd, Sheffield*

S Turgoose, MA, PhD, MICorrST
Lecturer in Corrosion Science and Engineering, UMIST

G P A Turner, MA
Formerly Industrial Paints Research Manager, ICI Paints, Slough

R Walker*, BSc, DipEd, MSc, MSc(Eng), PhD
Lecturer, Metallurgy and Materials Technology Dept, University of Surrey, Guildford

G W Walkiden*, BSc, CChem, FRIC, MIM
Consultant, Ever Ready Central Laboratories

J R Walters*
Consultant to British Post Office

R B Waterhouse*, MA, PhD, FIM, FICorr
*Reader in Metallurgy,
Dept. of Metallurgy and Materials Science, University of Nottingham*

K O Watkins*, FIM, FICorr
Corrosion Advice Bureau, BSC

S A Watson*, BSc, PhD, CChem, FRIC, FIMF
*Senior Development Officer,
International Nickel Ltd.*

H C Wesson*, MA, BSc, CChem, FRIC
*Formerly Technical Manager,
Lead Development Association (Retired)*

E E White*, CChem, FRIC, FIM, CEng, MIMM, FCS, MIInfSc, FICorrT, FIMF
*Consultant
Inter-Services Laboratory, BSC*

N R Whitehouse, BSc PhD
*The Paint Research Association,
Teddington, Middlesex*

C Wilson
Escol Products Ltd, Huntingdon

R W Wilson*, MA, PhD, CEng, FICorr, FIM
*Senior Consultant
CAPCIS, Manchester*

P A Woods*, BSc
HM Inspector of Factories

K H R Wright*, BSc, PhD, MInstP
*Senior Principal Scientific Officer,
Materials Group, National Engineering Laboratory*

1 PRINCIPLES OF CORROSION AND OXIDATION

1.1	Basic Concepts of Corrosion	**1**:3
1.1A	Appendix—Classification of Corrosion Processes	**1**:16
1.2	Nature of Films, Scales and Corrosion Products on Metals	**1**:22
1.3	Effects of Metallurgical Structure on Corrosion	**1**:36
1.4	Corrosion in Aqueous Solutions	**1**:55
1.5	Passivity and Localised Corrosion	**1**:118
1.6	Localised Corrosion	**1**:151
1.7	Bimetallic Corrosion	**1**:213
1.8	Lattice Defects in Metal Oxides	**1**:244
1.9	Continuous Oxide Films	**1**:254
1.10	Discontinuous Oxide Films	**1**:268
1.11	Erosion Corrosion	**1**:293

1.1 Basic Concepts of Corrosion

Modern technology has at its disposal a wide range of constructional materials — metals and alloys, plastics, rubber, ceramics, composites, wood, etc. and the selection of an appropriate material for a given application is the important responsibility of the design engineer. No general rules govern the choice of a particular material for a specific purpose, and a logical decision involves a consideration of the relevant properties, ease of fabrication, availability, relative costs, etc. of a variety of materials; frequently the ultimate decision is determined by economics rather than by properties, and ideally the material selected should be the cheapest possible that has adequate properties to fulfil the specific function.

Where metals are involved, mechanical, physical and *chemical* properties must be considered, and in this connection it should be observed that whereas mechanical and physical properties can be expressed in terms of constants, the chemical properties of a given metal are dependent entirely on the precise environmental conditions prevailing during service. The relative importance of mechanical, physical and chemical properties will depend in any given case on the application of the metal. For example, for railway lines elasticity, tensile strength, hardness and abrasion resistance will be of major importance, whereas electrical conductivity will be of primary significance in electrical transmission. In the case of heat-exchanger tubes, good thermal conductivity is necessary, but this may be outweighed in certain environments by chemical properties in relation to the aggressiveness of the two fluids involved — thus although the thermal conductivity of copper is superior to that of aluminium brass or the cupronickels, the alloys are preferred when high velocity sea water is used as the coolant, since copper has very poor chemical properties under these conditions.

While a metal or alloy may be selected largely on the basis of its mechanical or physical properties, the fact remains that there are very few applications where the effect of the interaction of a metal with its environment can be completely ignored, although the importance of this interaction will be of varying significance according to circumstances; for example, the slow uniform wastage of steel of massive cross section (such as railway lines or sleepers) is of far less importance than the rapid perforation of a buried steel pipe or the sudden failure of a vital stressed steel component in sodium hydroxide solution.

The effect of the metal/environment interaction on the environment itself is frequently more important than the actual deterioration of the metal (*see* Section 2.7). For instance, lead pipes cannot be used for conveying plumbosolvent waters, since a level of lead > 0.1 p.p.m. is toxic; similarly, galvanised steel may not be used for certain foodstuffs owing to the toxicity of zinc salts (*see* Section 2.8). In many chemical processes selection of a particular metal may be determined by the need to avoid contamination of the environment by traces of metallic impurities that would affect colour or taste of products or catalyse undesirable reactions; thus copper and copper alloys cannot be used in soap manufacture, since traces of copper ions result in colouration and rancidification of the soap. In these circumstances it will be essential to use unreactive and relatively expensive metals, even though the environment would not result in the rapid deterioration of cheaper metals such as mild steel. A further possibility is that contamination of the environment by metals' ions due to the corrosion of one metal can result in the enhanced corrosion of another when the two are in contact with the same environment. Thus the slow uniform corrosion of copper by a cuprosolvent domestic water may not be particularly deleterious to copper plumbing, but it can result in the rapid pitting and consequent perforation of galvanised steel and aluminium that subsequently comes into contact with the copper-containing water (Sections 4.1, 4.2 and 4.7).

Finally, it is necessary to point out that for a number of applications metals are selected in preference to other materials because of their visual appearance, and for this reason it is essential that brightness and reflectivity are retained during exposure to the atmosphere; stainless steel is now widely used for architectural purposes, and for outdoor exposure the surface must remain bright and rust-free without periodic cleaning (Section 3.3). On the other hand, the slow-weathering steels, which react with the constituents of the atmosphere to form an adherent uniform coating of rust, are now being used for cladding buildings (Section 3.2), in spite of the fact that a rusty surface is usually regarded as aesthetically unpleasant.

The interaction of a metal or alloy (or a non-metallic material) with its environment is clearly of vital importance in the performance of materials of construction, and the fact that the present work is largely confined to a detailed consideration of such interactions could create the impression that this was the sole factor of importance in materials selection. This, of course, is not the case although it is probably true to say that this factor is the one that is the most neglected by the design engineer.

Definitions of Corrosion

In the case of non-metallic materials, the term *corrosion* invariably refers to their-deterioration from chemical causes, but ·a similar concept is not necessarily applicable to metals. Many authorities[1] consider that the term *metallic corrosion* embraces all interactions of a metal or alloy (solid or liquid) with its environment, irrespective of whether this is deliberate and beneficial or adventitious and deleterious. Thus this definition of corrosion, which for convenience will be referred to as the *transformation* definition, will include, for example, the deliberate anodic dissolution of zinc in cathodic protection and electroplating as well as the spontaneous gradual wastage of zinc roofing sheet resulting from atmospheric exposure.

On the other hand, *corrosion* has been defined[2] as 'the undesirable deterioration' of a metal or alloy, i.e. an interaction of the metal with its environment that adversely affects those properties of the metal that are to be preserved. This definition — which will be referred to as the *deterioration* definition — is also applicable to non-metallic materials such as glass, concrete, etc. and embodies the concept that corrosion is always deleterious. However, the restriction of the definition to undesirable chemical reactions of a metal results in anomalies which will become apparent from a consideration of the following examples.

Steel, when exposed to an industrial atmosphere, reacts to form the reaction product rust, of approximate composition $Fe_2O_3 \cdot H_2O$, which being loosely adherent does not form a protective barrier that isolates the metal from the environment; the reaction thus proceeds at an approximately linear rate until the metal is completely consumed. Copper, on the other hand forms an adherent green patina, corresponding approximately with bronchantite, $CuSO_4 \cdot 3Cu(OH)_2$, which is protective and isolates the metal from the atmosphere. Copper roofs installed 200 years ago are still performing satisfactorily, and it is apparent that the formation of bronchantite is not deleterious to the function of copper as roofing material — indeed, in this particular application it is considered to enhance the appearance of the roof, although a similar patina formed on copper water pipes would be aesthetically objectionable.

The rapid dissolution of a vessel constructed of titanium in hot 40% H_2SO_4 with the formation of Ti^{4+} aquo cations conforms with both definitions of corrosion, but if the potential of the metal is raised (anodic protection) a thin adherent protective film of anatase, TiO_2, is formed, which isolates the metal from the acid so that the rate of corrosion is enormously decreased. The formation of this very thin oxide film on titanium, like that of the relatively thick bronchantite film on copper, clearly conforms with the *transformation* definition of corrosion, but not with the *deterioration* definition, since in these examples the rate and extent of the reaction is not significantly detrimental to the metal concerned. Again, magnesium, zinc or aluminium is deliberately sacrificed when these metals are used for the cathodic protection of steel structures, but as these metals are clearly not required to be maintained as such, their consumption in this particular application cannot, according to the *deterioration*, be regarded as corrosion. Furthermore, corrosion reactions are used to advantage in technological processes such as pickling, etching, chemical and electrochemical polishing and machining, etc.

The examples already discussed lead to the conclusion that any reaction of a metal with its environment must be regarded as a corrosion process irrespective of the extent of the reaction or of the rates of the initial and subsequent stages of the reaction. It is not illogical, therefore, to regard *passivity*, in which the reaction product forms a very thin protective film that controls rate of the reaction at an acceptable level, as a limiting case of a corrosion reaction. Thus both the rapid dissolution of *active* titanium in 40% H_2SO_4 and the slow dissolution of *passive* titanium in that acid must be regarded as corrosion processes, even though the latter will not be detrimental to the metal during the anticipated life of the vessel.

It follows that in deciding whether the corrosion reaction is detrimental to a metal in a given application, the precise form of attack on the metal (general, intergranular, etc.), the nature of the reaction products (protective or non-protective), the velocity and extent of the reaction and the location of the corrosion reaction must all be taken into account. In addition, due consideration must be given to the effect of the corrosion reaction on the environment itself. Thus corrosion reactions are not always detrimental, and our ability to use highly reactive metals such as aluminium, titanium, etc. in aggressive environments is due to a limited initial corrosion reaction, which results in the formation of a rate-controlling corrosion product. Expressions such as 'preventing corrosion', 'combating corrosion' or even 'fighting corrosion' are misleading; with the majority of metals corrosion cannot be avoided and 'corrosion control' rather than 'prevention' is the desired goal. The implication of 'control' in this context is that (*a*) neither the form, nor the extent, nor the rate of the corrosion reaction must be detrimental to the metal used as a constructional material for a specific purpose, and (*b*) for certain applications the corrosion reaction must not result in contamination of the environment. The scope of corrosion control is considered in more detail in the *Introduction to Volume 2*, but it is relevant to mention here that it must involve a consideration of materials, availability, fabrication, protective methods and economics in relation to the specific function of the metal and its anticipated life. At one extreme corrosion control in certain environments may be effected by the use of thick sections of mild steel without any protective system, at the other the environmental conditions prevailing may necessitate the use of platinum.

The scope of the term 'corrosion' is continually being extended, and Fontana and Staehle have stated[3] that *'corrosion* will include the reaction of metals, glasses, ionic solids, polymeric solids and composites with environments that embrace liquid metals, gases, non-aqueous electrolytes and other non-aqueous solutions'.

Vermilyea, who has defined corrosion as a process in which atoms or molecules are removed one at a time, considers that evaporation of a metal into vacuum should come within the scope of the term, since atomically it is similar to other corrosion processes[4].

Evans[5] considers that corrosion may be regarded as a branch of chemical thermodynamics or kinetics, as the outcome of electron affinities of metals and non-metals, as short-circuited electrochemical cells, or as the demolition of the crystal structure of a metal.

These considerations lead to the conclusion that there is probably a need for two definitions of corrosion, which depend upon the approach adopted:

1. Definition of corrosion in the context of Corrosion Science: the reaction of a solid with its environment.
2. Definition of corrosion in the context of Corrosion Engineering: the reaction of an engineering constructional metal (material) with its environment with a consequent deterioration in properties of the metal (material).

Methods of Approach to Corrosion Phenomena

The effective use of metals as materials of construction must be based on an understanding of their physical, mechanical and chemical properties. These last, as pointed out earlier, cannot be divorced from the environmental conditions prevailing. Any fundamental approach to the phenomena of corrosion must therefore involve consideration of the structural features of the metal, the nature of the environment and the reactions that occur at the metal/environment interface. The more important factors involved may be summarised as follows:

1. *Metal*—composition, detailed atomic structure, microscopic and macroscopic heterogeneities, stress (tensile, compressive, cyclic), etc.
2. *Environment*—chemical nature, concentrations of reactive species and deleterious impurities, pressure, temperature, velocity, impingement, etc.
3. *Metal/environment interface*—kinetics of metal oxidation and dissolution, kinetics of reduction of species in solution; nature and location of corrosion products; film growth and film dissolution, etc.

From these considerations it is evident that the detailed mechanism of metallic corrosion is highly complex and that an understanding of the various phenomena will involve many branches of the pure and applied sciences, e.g. metal physics, physical metallurgy, the various branches of chemistry, bacteriology, etc. although the emphasis may vary with the particular system under consideration. Thus in stress-corrosion cracking (*see* Section 8.1) emphasis may be placed on the detailed metallurgical structure in relation to crack propagation resulting from the conjoint action of corrosion at localised areas and mechanical tearing, while in underground corrosion the emphasis may be on the mechanism of bacterial action in relation to the kinetics of the overall corrosion reaction (*see* Section 2.6).

Although the mechanism of corrosion is highly complex the actual control of the majority of corrosion reactions can be effected by the application of relatively simple concepts. Indeed, the Committee on Corrosion and Protection[6] concluded that 'better dissemination of existing knowledge' was the most important single factor that would be instrumental in decreasing the enormous cost of corrosion in the U.K.

Corrosion as a Chemical Reaction at a Metal/Environment Interface

As a first approach to the principles which govern the behaviour of metals in specific environments it is preferable for simplicity to disregard the detailed structure of the metal and to consider corrosion as a heterogeneous chemical reaction which occurs at a metal/non-metal interface and which involves the metal itself as one of the reactants (cf. catalysis). Corrosion can be expressed, therefore, by the simple chemical reaction:

$$aA + bB = cC + dD \qquad \ldots(1.1)$$

where A is the metal and B the non-metal reactant (or reactants) and C and D the products of the reaction. The non-metallic reactants are frequently referred to as *the environment* although it should be observed that in a complex environment the major constituents may play a very subsidiary role in the reaction. Thus in the 'atmospheric' corrosion of steel, although nitrogen constitutes approximately 75% of the atmosphere, its effect, compared with that of moisture, oxygen, sulphur dioxide, solid particles, etc. can be disregarded (in the high-temperature reaction of titanium with the atmosphere, on the other hand, nitrogen is a significant factor).

One of the reaction products (say, C) will be an oxidised form of the metal, and D will be a reduced form of the non-metal — C is usually referred to as the *corrosion product*, although the term could apply equally to D. In its simplest form, reaction 1.1 becomes

$$aA + bB = cC \quad \ldots(1.2)$$

e.g. $\quad\quad\quad\quad 4Fe + 3O_2 = 2Fe_2O_3$

where the reaction product can be regarded either as an oxidised form of the metal or as the reduced form of the non-metal. Reactions of this type which do not involve water or aqueous solutions are referred to as 'dry' corrosion reactions.

The corresponding reaction in aqueous solution is referred to as a 'wet' corrosion reaction, and the overall reaction (which actually occurs by a series of intermediate steps) can be expressed as

$$4Fe + 2H_2O + 3O_2 = 2Fe_2O_3 \cdot H_2O \quad \ldots(1.3)$$

Thus in all corrosion reactions one (or more) of the reaction products will be an oxidised form of the metal, aquo cations (e.g. Fe^{2+} (aq.), Fe^{3+} (aq.)), aquo anions (e.g. $HFeO_2^-$(aq.), FeO_4^{2-} (aq.)), or solid compounds (e.g. $Fe(OH)_2$, Fe_3O_4, $Fe_3O_4 \cdot H_2O$, $Fe_2O_3 \cdot H_2O$), while the other reaction product (or products) will be the reduced form of the non-metal. Corrosion may be regarded, therefore, as a *heterogeneous redox reaction at a metal/non-metal interface in which the metal is oxidised and the non-metal is reduced.* In the interaction of a metal with a specific non-metal (or non-metals) under specific environmental conditions, the chemical nature of the non-metal, the chemical and physical properties of the reaction products, and the environmental conditions (temperature, pressure, velocity, viscosity, etc.) will clearly be important in determining the form, extent and rate of the reaction.

Environment

Environments are considered in detail in Chapter 2, but some examples of the behaviour of normally reactive and non-reactive metals in simple chemical solutions will be considered here to illustrate the fact that corrosion is dependent on the nature of the environment; the thermodynamics of the systems and the kinetic factors involved are considered in Sections 1.4 and 1.9.

Gold is stable in most strong reducing acids, whereas iron corrodes rapidly, yet finely divided gold can be quickly dissolved in oxygenated cyanide solutions which may be contained in steel tanks. A mixture of caustic soda and sodium nitrate can be fused in an iron or nickel crucible, whereas this melt would have a disastrous effect on a platinum crucible.

Copper is relatively resistant to dilute sulphuric acid but will corrode if oxygen or oxidising agents are present in the acid, whereas austenitic stainless steels are stable in this acid only if oxygen or other oxidising agents are present. Iron will corrode rapidly in oxygenated water but extremely slowly if all oxygen is removed; if, however, oxygen is brought rapidly and simultaneously to all parts of the metal surface the rate will become very slow, owing to the formation of a protective oxide film. Lead will dissolve rapidly in nitric acid, more slowly in hydrochloric acid, and very slowly in sulphuric acid. These examples show that the corrosion behaviour of a metal cannot be divorced from the specific environmental conditions prevailing, which determine the rate, extent (after a given period of time) and form of the corrosion process.

Metal

Heterogeneities associated with a metal have been classified in Table 1.1 as atomic (*see* Fig. 1.1), microscopic (visible under an optical microscope), and macroscopic, and their effects are considered in various sections of the present work. It is relevant to observe, however, that the detailed mechanism of all aspects of corrosion, e.g. the passage of a metallic cation from the lattice to the solution, specific effects of ions and species in solution in accelerating or inhibiting corrosion or causing stress-corrosion cracking, etc. must involve a consideration of the detailed atomic structure of the metal or alloy.

The corrosion behaviour of different constituents of an alloy is well known, since the etching techniques used in metallography are essentially corrosion processes which take advantage of the different corrosion rates of phases as a means of identification, e.g. the grain boundaries are usually etched more rapidly than the rest of the grain owing to the greater reactivity of the disarrayed metal (*see* Sections 1.3 and 20.4).

Table 1.1 Heterogeneities in metals

1. *Atomic* (as classified by Ehrlich and Turnbull[7], see Fig. 1.1).
   (*a*) Sites within a given surface layer ('normal' sites); these vary according to the particular crystal plane (Fig. 1.2).
   (*b*) Sites at edges of partially complete layers.
   (*c*) Point defects in the surface layer: vacanies (molecules missing in surface layer), kink sites (molecules missing at edge of layer), molecules adsorbed on top of complete layer.
   (*d*) Disordered molecules at point of emergence of dislocations (screw or edge) in metal surface.

2. *Microscopic*
   (*a*) Grain boundaries—usually, but not invariably, more reactive than grain interior.
   (*b*) Phases—metallic (single metals, solid solutions, intermetallic compounds), non-metallic, metal compounds, impurities, etc.—heterogeneities due to thermal or mechanical causes.

3. *Macroscopic*
   (*a*) Grain boundaries.
   (*b*) Discontinuities on metal surface—cut edges, scratches, discontinuities in oxide films (or other chemical films) or in applied metallic or non-metallic coatings.
   (*c*) Bimetallic couples of dissimilar metals.
   (*d*) Geometrical factors—general design, crevices, contact with non-metallic materials. etc.

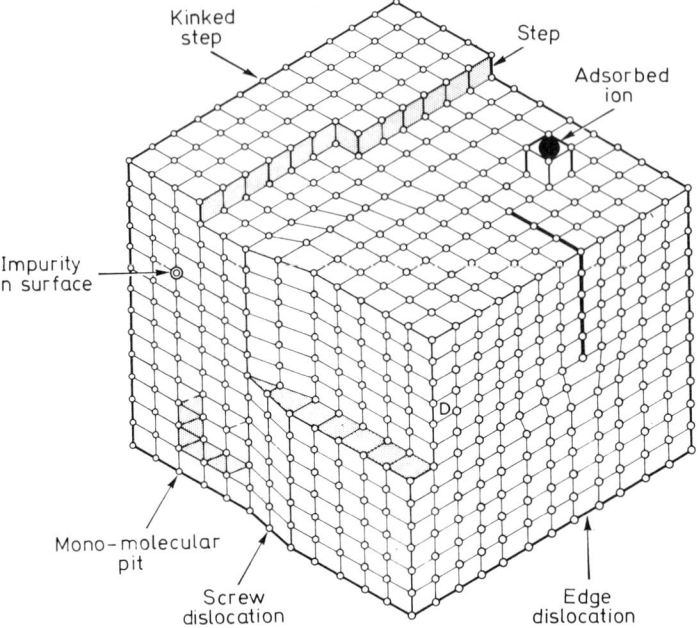

Fig.1.1 Surface imperfections in a crystal (after Erlich and Turnbull[7])

Fig.1.2 Hard-sphere model of face-centred cubic (f.c.c.) lattice showing various types of sites. Numbers denote Miller indices of atom places and the different shadings correspond to differences in the number of nearest neighbours (courtesy Erlich and Turnbull[7])

BASIC CONCEPTS OF CORROSION

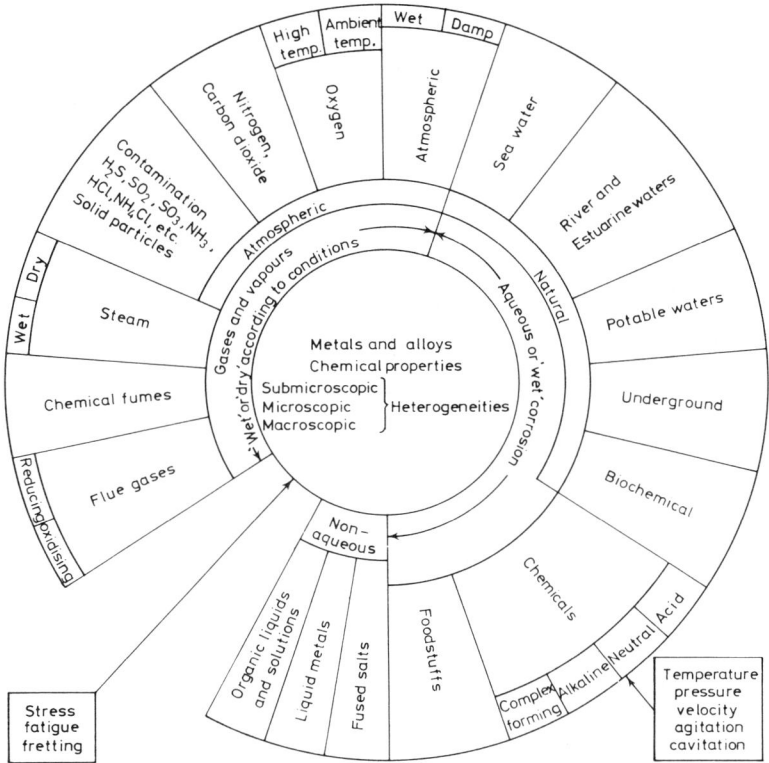

Fig.1.3 Environments in corrosion

Macroscopic heterogeneities, e.g. crevices, discontinuities in surface films, bimetallic contacts etc. will have a pronounced effect on the location and the kinetics of the corrosion reaction and are considered in various sections throughout this work. Practical environments are shown schematically in Fig. 1.3, which also serves to emphasise the relationship between the detailed structure of the metal, the environment, and external factors such as stress, fatigue, velocity, impingement, etc.

Types of Corrosion

Corrosion can affect the metal in a variety of ways which depend on its nature and the precise environmental conditions prevailing, and a broad classification of the various forms of corrosion in which five major types have been identified, is presented in Table 1.2. Thus an 18Cr–8Ni stainless steel will corrode uniformly during polishing, active dissolution or passivation, but will corrode locally during intergranular attack, crevice corrosion or pitting; in certain circumstances selective attack along an 'active path' in conjunction with a tensile stress may lead to a transgranular fracture. Types of corrosion are dealt with in more detail in Appendix 1.1A.

Table 1.2 Types of corrosion

Type	Characteristic	Examples
1. Uniform (or almost uniform)	All areas of the metal corrode at the same (or similar) rate	Oxidation and tarnishing; active dissolution in acids; anodic oxidation and passivity; chemical and electrochemical polishing; atmospheric and immersed corrosion in certain cases
2. Localised	Certain areas of the metal surface corrode at higher rates than others due to 'heterogeneities' in the metal, the environment or in the geometry of the structure as a whole. Attack can range from being slightly localised to pitting	Crevice corrosion; filiform corrosion; deposit attack; bimetallic corrosion; intergranular corrosion; weld decay
3. Pitting	Highly localised attack at specific areas resulting in small pits that penetrate into the metal and may lead to perforation	Pitting of passive metals such as the stainless steels, aluminium alloys, etc., in the presence of specific ions, e.g. Cl^- ions
4. Selective dissolution	One component of an alloy (usually the most active) is selectively removed from an alloy	Dezincification; dealuminification; graphitisation
5. Conjoint action of corrosion and a mechanical factor	Localised attack or fracture due to the synergistic action of a mechanical factor and corrosion	Erosion–corrosion, fretting corrosion, impingement attack, cavitation damage; stress corrosion cracking, hydrogen cracking, corrosion fatigue

Ideally, the metal selected, or the protective system applied to the metal, should be such that no corrosion occurs at all, but this is seldom technologically or economically feasible. It is necessary, therefore, to tolerate a rate and a form of corrosion that will not be significantly detrimental to the properties of the metal during its anticipated life. Thus, providing the corrosion rate is known, the slow uniform corrosion of a metal can frequently be allowed for in the design of the structure; for example, in the case of a metal that shows an active/passive transition the rate of corrosion in the passive region is usually acceptable whereas the rate in the active region is not. It follows that certain forms of corrosion can be tolerated and that corrosion control is possible, providing that the rate and form of the corrosion reaction are predictable and can be allowed for in the design of the structure.

Pitting is regarded as one of the most insidious forms of corrosion, since it frequently leads to perforation and to a consequent corrosion failure. In other cases pitting may result in loss of appearance, which is of major importance when the metal concerned is used for decorative architectural purposes. However, aluminium saucepans that have been in service for some time are invariably pitted, although the pits seldom penetrate the metal, i.e. the saucepan remains functional and the pitted appearance is of no significance in that particular application.

These considerations lead to the conclusion that the relationship between corrosion and deterioration of properties of a metal is highly complex, and involves a consideration of a variety of factors such as the rate and form of corrosion and the specific function of the metal concerned; certain forms of corrosion such as uniform attack can be tolerated, whereas others such as pitting and stress corrosion cracking that ultimately lead to complete loss of function, cannot.

The implications of the terms *predictable* and *unpredictable* used in the context of corrosion require further consideration, since they are clearly dependent on the knowledge and expertise of the engineer, designer or corrosion designer who takes the decision on the metal or alloy to be used, or the procedure to be adopted, to control corrosion in a specific environmental situation. On this basis a corrosion failure (i.e. failure of the function of the metal due to corrosion within a period that is significantly less than the anticipated life of the structure) may be the result of one or more of the following possibilities:

1. *Predictable.* (*a*) The knowledge and technology are available but have not been utilised by the designer; this category includes a wide variety of design features such as the wrong choice of materials, introduction of crevices and bimetallic contacts etc., and is the most frequent cause of corrosion failures. (*b*) The knowledge and technology are available, but have not been applied for economic reasons; e.g. inadequate pretreatment of steel prior to painting and the use of unprotected mild steel for silencers and exhaust systems of cars.

2. *Unpredictable.* (*a*) The design has been based on specific environmental conditions, which have subsequently changed during the operation of the process; in this connection it should be noted that small changes in the chemical nature of the environment, temperature, pressure and velocity may lead to significant changes in the corrosion rate and form: the catastrophic oxidation and failure of steel bolts in nuclear reactors in the U.K. resulting from an increase in the temperature of the carbon dioxide is an example of an unpredictable failure due to a change in environmental conditions. (*b*) There is insufficient knowledge and experience of the metal, alloy or the environment to predict with certainty that failure will not occur; examples could be quoted of new alloys that have been subjected to an extensive series of carefully planned corrosion tests, but have failed in service.

Professor M. Fontana[8] has made the statement that "Virtually all premature corrosion failures these days occur for reasons which were already well known and these failures can be prevented". It is apparent from this statement, and from the conclusions reached by the Committee on Corrosion and Protection, that category 1 is responsible for the majority of incidents of corrosion failure that could have been avoided if those responsible were better informed on the hazards of corrosion and on the methods that should have been used to control it.

Principles of Corrosion

It has been stated that metallic corrosion is an art rather than a science and that, at present, insufficient knowledge is available to predict with any certainty how a particular metal or alloy will behave in a specific environment[4]. It should be appreciated that the decision to use a particular metal or alloy in preference to others in a given environment or to employ a particular protective system is based usually on previous experience and empirical testing (*see* Chapter 19) rather than on the application of scientific knowledge — the technology of corrosion is without doubt in advance of corrosion science and many of the phenomena of corrosion are not fully understood. Thus the phenomena of passivity which was first observed by Faraday in 1836 is still a subject of controversy, the specific effect of certain anions in causing stress-corrosion cracking of certain alloy systems is not fully understood, and dezincification of brasses can be prevented by additions of arsenic (or other elements such as antimony or phosphorus) but no adequate theory has been submitted to explain the action of these elements (*see* Section 4.2).

An understanding of the basic principles of the science of metallic corrosion is clearly vital for corrosion control, and as knowledge of the subject advances the application of scientific principle rather than an empirical approach may be used for such purposes as the selection of corrosion inhibitors, formulation of corrosion-resisting alloys, etc.

Terminology

The classification given in Table 1.2 is based on the various forms that corrosion may take, but the terminology used in describing corrosion phenomena frequently places emphasis on the environment or cause of attack rather than the form of attack. Thus the broad classification of corrosion reactions into 'wet' or 'dry' is now generally accepted, and the nature of the process is frequently made more specific by the use of an adjective that indicates type or environment, e.g. concentration — cell corrosion, crevice corrosion, bimetallic corrosion and atmospheric corrosion,

Table 1.3 Terminology in corrosion

Type of attack	Environmental	Cause of attack	Mechanical factors	Corrosion product
general (uniform)	wet*	concentration cell	stress	rusting
localised	dry	bimetallic cell	fretting	tarnishing
pitting (or intense)	atmospheric	active-passive cell	fatigue	scaling
intergranular	immersed	stray current (electrolysis)	cavitation	green rot
transgranular	underground	hydrogen evolution	erosion	tin pest
selective	sea water	oxygen absoption	impingement	
parting	chemical	impingement		
catastrophic	fused-salt	hydrogen embrittlement		
layer	flue-gas	caustic embrittlement		
filiform	biochemical			
	bacterial			
	high-temperature			
	liquid-metal			

*See Appendix to this section.

high-temperature corrosion, sea-water corrosion, etc. Alternatively, the phenomenon is described in terms of the corrosion product itself — tarnishing, rusting, green rot. The terminology used in corrosion is given in Table 1.3 and is considered in more detail in Appendix 1.1A.

L. L. SHREIR

REFERENCES

1. Hoar, T. P., *J. Appl. Chem.*, **11**, 121 (1961); Vernon, W. H. J., *The Conservation of Natural Resources*, Instn. of Civil Engrs., London, 105 (1957); Potter, E. C., *Electrochemistry*, Cleaver-Hume, London, 231 (1956)
2. Uhlig, H. H. (Ed.), *The Corrosion Handbook*, Wiley, New York and Chapman and Hall, London (1948); Uhlig, H. H., *Corrosion and Corrosion Control*, Wiley, New York (1971); Fontana, M. G. and Greene, N. D., *Corrosion Engineering*, McGraw-Hill (1967)
3. Fontana, M. G. and Staehle, R. W., *Advances in Corrosion Science and Technology*, Plenum Press, New York (1990)
4. Vermilyea, D. A., *Proc. 1st International Congress on Metallic Corrosion*, London, 1961, Butterworths, London, 62 (1962)
5. Evans, U. R., *The Corrosion and Oxidation of Metals*, Arnold, London, 12 (1960)
6. *Report of the Committee on Corrosion and Protection*, Department of Trade and Industry, H.M.S.O. (1971)
7. Ehrlich, G. and Turnbull, D., *Physical Metallurgy of Stress Corrosion Fracture*, Interscience, New York and London, 47 (1959)
8. Fontana, M. G., *Corrosion*, **27**, 129 (1971)

1.1A Appendix — Classification of Corrosion Processes

Existing Classifications

A logical and scientific classification of corrosion processes, although desirable, is by no means simple, owing to the enormous variety of corrosive environments and the diversity of corrosion reactions, but the broad classification of corrosion reactions into 'wet' or 'dry' is now generally accepted, and the terms are in common use. The term 'wet' includes all reactions in which an aqueous solution is involved in the reaction mechanism; implicit in the term 'dry' is the absence of water or an aqueous solution.

These terms are evidently ambiguous; for example, it is not always clear whether 'wet' is confined to aqueous solutions — the 'wetting' of solids by mercury indicates that liquid-metal corrosion should be classified as 'wet'. Even if the term is restricted to aqueous solutions, the difficulty arises that the mechanism of growth of magnetite scale during the reaction of the interior of a boiler drum with dilute caustic soda at high temperatures and pressures is best interpreted in terms of a 'dry' corrosion process. Similar considerations apply to the reactions of aluminium and zirconium with high-temperature water.

Considering oxidation as a typical 'dry' reaction it follows from Fig. 1.A1*a* that at the *interfaces*:

$$M \rightleftharpoons (M^{z+}\circ/O) + z(e\circ/O)$$

where $M^{z+}\circ$ is an interstitial metal ion, $e\circ$ an interstitial electron and /O indicates the metal/oxide interface (Section 1.8).

If the metal dissolves to enter a vacant site, then

$$M \rightleftharpoons (M^{z+}\square/O + ze\square/O)$$

where $M^{z+}\square$ represents a cation vacancy and $e\square$ a positive hole.

At the gas/oxide interface the O_2 gas ionises

$$(\tfrac{1}{2}O_2/\text{ads.}) + 2(e/X) \rightleftharpoons (O^{2-}/\text{ads.})$$

where /X indicates the gas/oxide interface.

By definition, these *interfaces* can be considered as anodes and cathodes respectively.

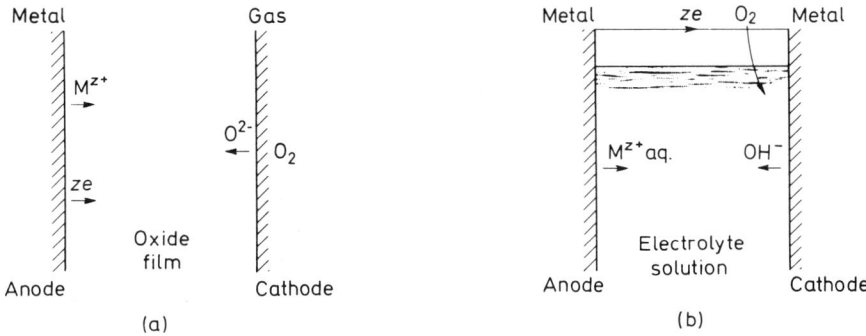

Fig.1.A1 Anodes and cathodes in corrosion processes. (a) 'Dry' corrosion and (a) 'wet' corrosion

The corresponding 'wet' corrosion half-reactions (Fig. 1.A1b) are:

$$[M^{z+} + ze]_{lattice} + H_2O \rightarrow M^{z+}aq.$$

and
$$O_2 + 2H_2O + 4e = 4OH^-$$

or
$$O_2 + 4H^+ + 4e = 2H_2O$$

'Dry' Corrosion

These are generally metal/gas or metal/vapour reactions involving non-metals such as oxygen, halogens, hydrogen sulphide, sulphur vapour, etc. and oxidation, scaling and tarnishing are the more important forms. A characteristic of these reactions is that the initial oxidation of the metal, reduction of the non-metal, and formation of compound must occur at one and the same place at the metal/non-metal interface. Should the compound be volatile or discontinuous, further interaction at the interface (or through a thin film of constant thickness) is possible and in most cases the reaction rate will tend to remain constant with time (linear law). If the film is continuous it will present a barrier to the reactants and further interaction will necessitate passage of the reactants through the film by (a) diffusion of the non-metal or (a) diffusion and migration of ions of the reactants. The detailed mechanisms of these reactions are considered in Sections 1.8–1.10, but it is appropriate to observe that the formation of a continuous film of reactant product at a metal/non-metal interface will result in a growth rate which, when the film becomes sufficiently thick to be rate determining, decreases as the film thickens, i.e. parabolic, logarithmic, asymptotic, cubic, etc.

'Wet' Corrosion

In 'wet' corrosion the oxidation of the metal and reduction of a species in solution (electron acceptor or oxidising agent) occur at different areas on the metal surface with consequent electron transfer through the metal from the anode (metal oxidised) to the cathode (electron acceptor reduced); the thermodynamically stable phases formed at the metal/solution interface may be solid compounds or hydrated ions (cations or anions) which may be transported away from the interface by processes such as migration, diffusion and convection (natural or forced). Under these circumstances the reactants will not be separated by a barrier and the rate law will tend to be linear. Subsequent reaction with the solution may result in the formation of a stable solid phase, but as this will form away from the interface it will not be protective – the thermodynamically stable oxide can affect the kinetics of the reaction only if it forms a film or precipitates on the metal surface (*see* Sections 1.4 and 1.5).

Further points which distinguish 'wet' from 'dry' corrosion are:

1. In 'wet' corrosion the metal ions are hydrated – the hydration energy of most metal ions is very large and thus facilitates ionisation (*see* Section 1.9).
2. In 'wet' corrosion ionisation of oxygen to hydroxyl must involve the hydronium ion or water.
3. In 'dry' corrosion the direct ionisation of oxygen occurs.

Corrosion in Organic Solvents

Corrosion reactions in aggressive organic solvents are becoming a more frequent occurrence owing to developments in the chemical and petrochemical industries, and these reactions can lead to the deterioration of the metal and to undesirable changes in the solvent. This aspect of corrosion has recently been the subject of an extensive review by Heitz[1] who has considered the mechanisms of the reactions, the similarities between corrosion in organic solvents and in aqueous solutions, the methods of study and the occurrence of the phenomenon in industrial processes.

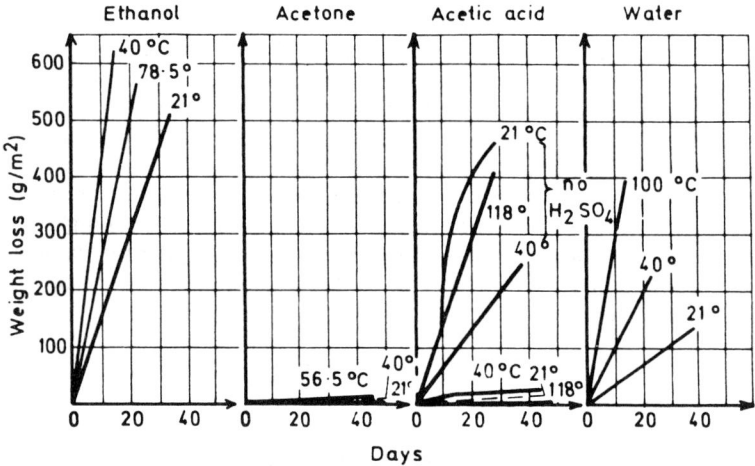

Fig.1.A2 Corrosion of nickel in different solvents containing 0.05 wt.% H_2SO_4 at various temperatures (after Heitz[1])

Figure 1.A2 shows the weight loss against time curve for nickel in various solvents containing 0·05 wt.% H_2SO_4 at various temperatures, and illustrates the unpredictable nature of corrosion in organic solvents. Thus the corrosion rates in ethanol are far greater than those in the aqueous acid whereas in acetone the rate is practically zero; even more surprising is the fact that in acetic acid the addition of 0·05%H2SO4 actually decreases the corrosion rate.

Heitz classifies corrosion reactions in organic solvents into

1. Electrochemical reactions, which follow a similar mechanism to those in aqueous solution.
2. Chemical reactions, which involve direct charge transfer between the metal atom in the lattice of the metal and the oxidising species.

In the case of electrochemical reactions the partial anodic reaction results in the formation of a solvated metal cation $M^{z+}_{solv.}$, a charged or uncharged metal complex MX^- or a solid compound MX_z, where X is a halogen ion, organic acid anion, etc.

The cathodic partial reactions are as follows:

(a) Reduction of a solvated proton to H_2 gas

$$H^+_{solv.} + e \rightarrow \tfrac{1}{2}H_2$$

(b) Reduction of acidic hydrogen of a proton donor

$$HA + e \rightarrow \tfrac{1}{2}H_2 + A^-$$

where A^- is a carboxylic acid anion, alcoholate ion, etc.

(c) Reduction of an oxidising gas Y

$$Y_z + zme \rightarrow zY^{m-}$$

where Y can be O_2, Cl_2, F_2, Br_2, O_3, N_2O_4, etc.

(d) Reduction of oxidising ions such as Fe^{3+}, Cu^{2+}, MnO_4^-, ClO_3^-, etc. It is evident from the above that in many systems the reaction of a metal with an organic solvent follows a mechanism that is similar to the electrochemical mechanism of corrosion in aqueous solution.

Non-electrochemical processes may be represented by the general equation

$$M + 2C_xH_yX_z \rightarrow MX_2 + C_{2x}H_{2y}X_{2z-2}$$

where X is a halogen and M is a divalent metal, e.g. the Grignard reaction

$$Mg + CH_3Cl \rightarrow CH_3MgCl$$

A further type of chemical process, which is analogous to high-temperature corrosion, is the reaction of metals with organic sulphur compounds, which follow the equation

$$2M + 2RSH \rightarrow 2MS + H_2 + R_2$$

Heitz quotes a number of case studies of corrosion of metals in organic solvents and concludes that the phenomenology indicates no specific differences from that experienced in aqueous corrosion. Thus general corrosion, pitting, crevice corrosion, intergranular corrosion, erosion-corrosion cracking, hydrogen embrittlement, etc. can all occur in organic solvents. The methods of control also follow that used for corrosion in aqueous

1:20 APPENDIX – CLASSIFICATION OF CORROSION PROCESSES

solutions, although there are certain differences. Thus cathodic and anodic protection are seriously limited by the resistivity of the solvent, and paint coatings deteriorate rapidly in contact with the solvent.

Suggested Classification and Nomenclature

On a basis of the preceding discussion, the classification and nomenclature outlined in Table 1.Al is suggested as a possible alternative to the accepted classification of corrosion reactions into 'wet' and 'dry'.

Table 1.1 Classification of spontaneous corrosion reactions

1. *Film-free Chemical Interaction*
   (a) Metal/gas – oxide or compound volatile (e.g. reaction of molybdenum with oxygen, reaction of iron or aluminium with chlorine).
   (b) Metal/liquid.
   Reactions of solid metals with liquid metals (e.g. dissolution of aluminium in mercury) Dissolution of metal in their fused halides (e.g. lead in lead chloride).
   Dissolution of metals in non-aqueous solutions (e.g. reaction of aluminium with carbon tetrachloride).
2. *Electrochemical*
   (a) *Inseparable anode/cathode type (insep. A/C)*
   Reactions with aqueous solutions. Uniform dissolution or corrosion of metals in acid, alkaline or neutral solutions (e.g. dissolution of zinc in hydrochloric acid or in caustic soda solution; general corrosion of zinc in water or during atmospheric exposure). Reactions with non-aqueous solution (e.g. dissolution of copper in a solution of ammonium acetate and bromine in alcohol).
   Reactions with fused salts.
   (b) *Separable anode/cathode type (sep. A/C)*
   All reactions of metals in aqueous or non-aqueous solutions or in fused salts where one area of the metal surface is predominantly anodic and the other is predominantly cathodic so that the sites are physically identifiable.
   (c) *Interfacial anode/cathode type in which the metal surface is filmed*
      (i) Metal/gas and metal/vapour reactions
         All reactions in which charge is transported through a film of reaction product on the metal surface – the film may or may not be rate determining (e.g. parabolic, logarithmic, asymptotic, etc. or linear growth laws, respectively).
      (ii) Metal/solution reactions
         All reactions involving the uniform formation and growth of a film of reaction product (e.g. reaction of metals with high-temperature water, reaction of copper with sulphur dissolved in carbon disulphide).

It is considered that the main types of corrosion reactions can be classified as follows:

1. Film-free chemical interaction in which there is direct chemical reaction of a metal with its environment. The metal remains film-free and there is no transport of charge.
2. Electrochemical reactions which involve transfer of charge across an interface. These electrochemical reactions can be further subdivided into:
   (a) Inseparable anode/cathode type (insep. A/C). The anodes and cathodes cannot be distinguished by experimental methods although their presence is postulated by theory, i.e. the *uniform*

dissolution of metals in acid*, alkaline or neutral aqueous solutions, in non-aqueous solution, or in fused salts.

(b) Separable anode/cathode type (sep. A/C). Certain areas of the metal can be distinguished experimentally as predominantly anodic or cathodic, although the distances of separation of these areas may be as small as fractions of a millimetre. In these reactions there will be a macroscopic flow of charge through the metal.

(c) Interfacial anode/cathode type (interfacial A/C). One entire interface will be the anode and the other will be the cathode. Thus in Fig. 1.A1a the metal/metal oxide interface might be regarded as the anode and the metal/oxygen interface as the cathode.

It is apparent that, in general, 2(a) and 2(b) include corrosion reactions which are normally classified as 'wet', while 2(c) includes those which are normally classified as 'dry'.

The terminology suggested can be illustrated by reference to the corrosion behaviour of iron:

1. Reaction of iron with oxygen at room temperature or with oxygen or water at high temperatures — interfacial A/C type.
2. Reaction of iron with oxygenated water or with reducing acids — inseparable A/C type.
3. Reaction of iron containing a discontinuous magnetite scale with oxygenated water, crevice corrosion, water-line attack, 'long-line' corrosion of buried iron pipes, etc. — separable A/C type.

Although it is realised that this classification and terminology has certain limitations, it represents a preliminary attempt to provide a more rational classification of corrosion processes than that based on 'wet' and 'dry'.

Acknowledgement

Grateful thanks are due to Dr. W. B. Jepson, Dr. M. Pryor and Mr. J. N. Wanklyn for helpful discussions during the preparation of this Appendix.

L. L. SHREIR

REFERENCES

1. Heitz, E., 'Corrosion of Metals in Organic Solvents', *Advances in Corrosion Science and Technology* (ed. M. G. Fontana and R. W. Staehle), Vol. 4, Plenum Press, 149 (1974)

---

*Dr. Pryor considers that in certain cases of uniform dissolution of metals in acids (e.g. Al in hydrochloric or sulphuric acid) or alkalis a thin film of oxide is present on the metal surface — the film is not rate-determining but its presence would indicate that reactions of this type should be classified under 2 (c).

1.2 Nature of Films, Scales and Corrosion Products on Metals

The study of corrosion is essentially the study of the nature of the metal reaction products (corrosion products) and of their influence on the reaction rate. It is evident that the behaviour of metals and alloys in most practical environments is highly dependent on the solubility, structure, thickness, adhesion, etc. of the solid metal compounds that form during a corrosion reaction. These may be formed naturally by reaction with their environment (during processing of the metal and/or during subsequent exposure) or as a result of some deliberate pretreatment process that is used to produce thicker films or to modify the nature of existing films. The importance of these solid reaction products is due to the fact that they frequently form a kinetic barrier that isolates the metal from its environment and thus controls the rate of the reaction; the protection afforded to the metal will, of course, depend on the physical and chemical properties outlined above.

In general, reaction products (films*, scales and corrosion products) may be formed under the following environmental conditions.

(a) Direct reaction with a gas (O_2, CO_2, CO, $H_2 + O_2$, H_2S, etc.) at temperatures that range from ambient to very high (1 000–2 000°C).

(b) Direct reaction with an aqueous solution with the formation of a thin invisible film (passivation) or of a thick visible corrosion product (protective or non-protective).

(c) By the deliberate formation of thick oxide films (e.g. anodising) at elevated potentials or by changing the nature of existing films by chemical treatments (e.g. chromating or phosphating).

For example, in a dry atmosphere a reactive metal such as aluminium may carry a natural protective oxide film of only some 3 nm thickness, while for increased corrosion resistance aluminium may be anodised to give a coating 10^4 times thicker (*see* Section 15.1). However, thickness alone does not provide a criterion of protection; and although a thick protective layer of millscale is formed on iron and steel during processing it is not continuous owing to spalling, and the attack on the exposed substrate at the discontinuities is far greater than if the surface was bare. Thus the kinetics of attack

---

*The distinction between a film and scale is not well defined, but it is usual to use the former when referring to a *thin* continuous layer of reaction product (visible or invisible) whilst the latter is normally used for *thick* high-temperature layer (always visible).

will be related to a variety of other factors such as composition, structure, continuity, adhesion to the substrate, cohesion, mechanical properties, etc. of the film or scale of reaction products.

This section describes in general terms the variation in the nature of very thin films originating in the initial reaction of a metal with its environment and their progression to the thicker overgrowths that control the kinetics. Recent developments in instrumental techniques have led to significant advances in the characterisation of these film- and scale-forming systems, and a summary of the experimental approaches available is provided at the end of the section. It is appropriate to consider first the products of reaction formed by a gaseous oxidising atmosphere and then to proceed to a consideration of the effect of water and aqueous systems.

Initial Surface Reaction States

The application of ultra-high vacuum techniques to low-energy electron diffraction (L.E.E.D.) studies of very clean metal surfaces in low-pressure oxidising and sulphidising atmospheres over a range of temperatures above ambient has provided detailed information on the initial states of interaction[1,2]. The following sequence of events is generally observed in the case of exposure to oxygen:

1. Rapid physical adsorption of molecular oxygen.
2. Chemisorption of atomic oxygen to form a partial or complete monolayer.
3. Further chemisorption of atomic oxygen into a second layer and/or further physical adsorption of O_2.

In Stage 2 a distinct structural modification to an expanded lattice at submonolayer coverages has been observed on nickel, indicating that the oxygen ions become progressively incorporated into the metal lattice. These two-dimensional crystals then gradually transform into a three-dimensional nickel oxide lattice as more oxygen becomes incorporated. Subsequent exposure to high-temperature conditions (>1 000°C) has confirmed the extreme stability of the Stage 2 state.

Similarly, under low-temperature conditions (<25°C) three stages have been recognised and defined[3] as follows:

1. Physical adsorption of oxygen resulting in the formation of one or more monolayers of oxide and requiring no activation energy.
2. Electron tunnelling through the stable oxide film to the adsorbed oxygen which sets up a potential and causes ion drift, thus resulting in logarithmic oxide growth.
3. Film rearrangement resulting in the formation of oxide subgrain and grain boundaries; these paths of easy ion migration promote the formation of oxide 'islands' and result in an increase in the growth rate of the oxide.

Oxide films formed at low temperatures are initially continuous and amorphous, but may undergo local crystallisation with the incorporation of the oxide 'islands', a process that is facilitated by water, heat, high electric fields and mechanical stress[4].

Thin-Film Region

Studies of thermally grown oxides in the thin-film region (<100 nm) have revealed[5] on single crystal substrates interesting details of epitaxy, stress generation, mosaic structure and film topography, and oxidation rate anisotropic behaviour. Mismatch between the oxide lattice and the metal substrate gives rise to stresses which may find relief in the generation of mosaic structures consisting of small crystallites (5-100 nm diameter) whose lattices are slightly twisted or tilted with respect to one another. Their boundaries represent potential paths of easy diffusion through the oxide.

The uniformity of film thickness is dependent upon temperature and pressure. The nucleation rate rises with pressure, such that at pressures above atmospheric the high rate of nucleation can lead to comparatively uniform oxide films, while increase in temperature reduces the density of oxide nuclei, and results in non-uniformity. Subsequently, lateral growth of nuclei over the surface is faster than the rate of thickening until uniform coverage is attained, when the consolidated film grows as a continuous layer[2].

Growth of oxide nuclei may also be accompanied by the appearance of whiskers and platelets under certain conditions[6]. It has been demonstrated that oxidation of iron in air at about 200°C initially leads to nuclei of Fe_3O_4 developing to form a porous layer. Over this homogeneous oxide layer, nuclei of α-Fe_2O_3 appear and spread over the Fe_3O_4, but no γ-Fe_2O_3 is observed. After 30 days whiskers of α-Fe_2O_3 appear, ultimately reaching a length of 1 μm. At higher temperatures, too, whiskers of α-Fe_2O_3 appear and subsequently develop into crystallographic platelets. In general, products of this nature occur as fine features developing from otherwise protective films.

Scale-Forming Situations

In considering film growth at higher temperatures, a changeover to diffusion control, which is dependent on concentration gradients, tends to give rise to parabolic and paralinear kinetics as substantial scales form at thicknesses of 1-100 μm or more. This is the area of vital concern in the development and application of engineering alloys for high-temperature resistance, and is in distinct contrast to the thin-film régime. Nevertheless, the initial state of the metal surface can still influence subsequent oxidation behaviour. Thus, different oxidation patterns may be observed depending upon whether the surface is electropolished, hydrogen reduced, mechanically abraded or cathodically pretreated. When metals of variable valency become subjected to the oxidising potential gradient across the scale, a duplex or multiple series of layers forms. The classical case of iron oxidised above 600°C has been well established[6], and it has been shown that the system consists of $Fe/FeO/Fe_3O_4/Fe_2O_3/O_2$. In these situations film thickening occurs by transport of cations, anions, vacancies and electrons across the various phase boundaries, which is possible owing to the non-stoichiometric composition of the various coexistent oxides (*see* Sections 1.8, 1.9 and 7.2).

A rather different situation arises when mild steel is exposed to liquid water or dilute sodium hydroxide at 300-360°C. Here a duplex Fe_3O_4 scale is formed, consisting of an inner adherent protective film in contact with

NATURE OF FILMS, SCALES AND CORROSION PRODUCTS ON METALS 1:25

an outer poorly adherent crystalline layer of magnetite (*see* Section 1.10). In alloy systems the course of events is complicated by such factors[7,8] as:

1. The affinity of the component metals for each other and for the non-metal.
2. The diffusion rates of atoms in the alloy and of ions in the compounds.
3. The mutual solubilities of the products present in the oxidation layers.
4. The formation of ternary compounds, e.g. spinels (*see* Table 1.4).
5. The relative volumes of the various phases.

In practice, thermal cycling rather than isothermal conditions more frequently occurs, leading to a deviation from steady state thermodynamic conditions and introducing kinetic modifications. Lattice expansion and contraction, the development of stresses and the production of voids at the alloy-oxide interface, as well as temperature-induced compositional changes, can all give rise to further complications. The resulting loss of scale adhesion and spalling may lead to *breakaway oxidation*[9,11] in which linear oxidation replaces parabolic oxidation (*see* Section 1.10).

Examination of the structural consequences of these complex interacting factors is now being elucidated in considerable detail by systematic application of electron optical and X-ray analysis techniques[9], as well as by a range of other methods[10].

In certain systems the oxidation reactions may lead to a particularly protective single phase being formed at the surface, e.g. magnetite (Fe_3O_4) in the case of iron and steel and γ-Al_2O_3 in the case of aluminium. The 'spinel' ($MgAl_2O_4$) lattice is important in relation to the protection it affords to alloys used at high temperatures, and such structures often occur with a continuously varying stoichiometry as a 'double oxide' phase, which may provide an effective kinetic barrier to the oxidation process (e.g. $NiO \cdot Cr_2O_3$ spinel in Cr-Ni-Fe alloys). Some examples are given in Table 1.4.

The spinel structure is of especial significance in the corrosion behaviour of iron and alloy steels both at high temperatures and in aqueous environments. Its crystallographic unit cell can be represented as $8XY_2O_4$ (or $X_8Y_{16}O_{32}$) in which the valencies of the metal ions X and Y may be (*a*) X^{II}, Y^{III}; (*b*) X^{IV}, Y^{II} (giving rise to the so-called '2-3 spinels' or the '4-2 spinels'; and (*c*) X^{VI}, Y^{I}. The structure is based on a cell containing 32 oxygen atoms in a close-packed cubic arrangement. This provides for the incorporation of the X atoms in eight equivalent tetrahedral sites and the Y atoms in 16 equivalent octahedral sites. 'Inverse' spinels follow a different arrangement, represented by $Y(XY)O_4$, in which half of the Y atoms are located tetrahedrally, while the remaining Y atoms together with the X atoms are randomly arranged among the 16 octahedral positions. More generally, some spinels exist with a fraction λ of Y cations in tetrahedral sites where $0 > \lambda < \frac{1}{2}$.

Table 1.4 Spinel phases encountered in alloy oxidation

n-type	$MgFe_2O_4$	$ZnCo_2O_4$
	$NiFe_2O_4$	$MgAl_2O_4$
	$ZnFe_2O_4$	$ZnAl_2O_4$
p-type	$MgCr_2O_4$	$ZnCr_2O_4$
	$FeCr_2O_4$	$CoAl_2O_4$
	$CoCr_2O_4$	$NiAl_2O_4$

It should be noted that single metal oxides such as Fe_3O_4 and Co_3O_4 are inverse spinels, while Mn_3O_4 is a normal spinel. The spinel structure is prominent in the oxides on iron and aluminium[18,19]. The oxides M_2O_3 (and also the hydroxides and oxy-hydroxides $M(OH)_3$ and $MO \cdot OH$) exist in the α and γ forms. Corundum and haematite represent the isostructural α forms, while the γ forms have cubic spinel-like structures deficient in metal ions. For example, in γ-Fe_2O_3 there are only $21\frac{1}{3}$ Fe^{3+} ions per unit cell of $32 O^{2-}$ ions, and these are randomly distributed among the eight tetrahedral and 16 octahedral 'available' sites. In magnetite, represented as $Fe^{3+}(Fe^{2+} Fe^{3+})O_4$, one third of the cations are Fe^{2+} and continuous interchange of electrons between Fe^{2+} and Fe^{3+} ions in the 16-fold positions accounts for its extremely high electronic conductivity. Careful oxidation of Fe_3O_4 yields γ-Fe_2O_3, which may be converted back into Fe_3O_4 by heating *in vacuo* at 250°C. Because wüstite (FeO) ideally has the NaCl-type structure (f.c.c. anion lattice), with four Fe^{2+} and four O^{2-} ions per unit cell, deviations from stoichiometry lead to not every octahedral site being filled in the metal deficient lattice (e.g. at 570°C $Fe_{0.93}O$ contains cation vacancies and compensating Fe^{3+} ions). At lower temperatures disproportionation occurs:

$$4FeO \rightleftharpoons \alpha\text{-}Fe + Fe_3O_4$$

Therefore the relationship between these interconvertible structures originates from a cubic anion lattice of $32 O^{2-}$ ions in the cell. With 32 Fe^{2+} ions in the octahedral holes stoichiometric FeO is formed. Replacement of a number of Fe^{2+} ions with two-thirds of their number of Fe^{3+} ions maintains electrical neutrality but provides non-stoichiometric $Fe_{1-x}O$. Continual replacement in this way to leave 24 Fe atoms in the cubic cell produces Fe_3O_4, and further exchange to an average of $21\frac{1}{3}$ Fe^{3+} ions leads to γ-Fe_2O_3

$$Fe_{1-x}O \rightarrow Fe_3O_4 \rightarrow \gamma\text{-}Fe_2O_3$$

In actual oxidation, the cubic anion lattice becomes extended by the addition of new layers of close-packed O^{2-} ions into which Fe atoms migrate to give rise to the appropriate stable structures.

The defect γ-structures may be stabilised by the presence of Li^+ or H^+ ions (e.g. $LiFe_5O_8$). Cation diffusion rates in these and other lattices developed on metal surfaces play an important rôle in governing corrosion behaviour.

Surface Reaction Products Formed in Aqueous Environments

Whereas a film formed in dry air consists essentially of an anhydrous oxide and may reach a thickness of 3 nm, in the presence of water (ranging from condensed films deposited from humid atmospheres to bulk aqueous phases) further thickening occurs as partial hydration increases the electron tunnelling conductivity[3]. Other components in contaminated atmospheres may become incorporated (e.g. H_2S, SO_2, CO_2, Cl^-), as described in Sections 2.2 and 3.1.

Films may thus range from thin transparent oxides (passive films on Al, Cr, Ti and Fe–Cr alloys), or thin visible sulphides (on Cu and Ag) to thicker

Table 1.5 Variations in the nature and thickness of the product formed on aluminium under different conditions

Formation conditions	Nature of oxide film	Thickness (nm)
Dry air or O_2	Amorphous Al_2O_3	1-2
Humid atmosphere	AlOOH + $Al_2O_3.3H_2O$	50-100
Boiling water	AlOOH (or $Al_2O_3.H_2O$	500-2000
Chemical conversion	AlOOH + anions of solution	1000-5000
Anodic oxidation (barrier films)	Amorphous + crystalline Al_2O_3 + anions of solution	1000-3000

'visible films, which may be compact, adherent and protective (anodic oxide films on Al and Ti, $PbSO_4$ films on Pb, etc.) or bulky, poorly adherent and non-protective (rust on steel, 'white rust' on Zn). In some cases, fairly precise limits can be placed on the nature and thickness of the products formed under different conditions, as with aluminium illustrated in Table 1.5. In other cases, the undesirable wastage of the basis metal (e.g. the rusting of steel) is of more significance than the thickness of the corrosion product, although the nature of the latter may provide information useful in interpreting the mechanism of its formation.

Thus in industrial atmospheres the presence of $FeSO_4.4H_2O$ has been identified in combination with α- and γ- FeO.OH, and the two latter incorporate free water in excess of the composition $Fe_2O_3.H_2O$. Furthermore, although some of the corrosion product may be adherent, most of it is not[12] (Sections 3.1 and 3.2).

In the fully immersed situation where the corrosion product is produced by a secondary reaction such as $M^{2+} + 2H_2O \rightarrow M(OH)_2 + 2H^+$, as in the case of iron or zinc in dilute aqueous aerated chloride solutions, the sites of the anodic and cathodic processes are separated, and widely so in the partially immersed condition. Thus OH^- ions are formed at the cathode and $M^{2+}_{aq.}$ ions at the anode, giving rise to dispersed $M(OH)_2$ where they meet and react; under these circumstances the corrosion product cannot influence the kinetics. If chloride or sulphate is present, a basic compound $M_x(OH)_y(X)_z$ may form whose range of stability will depend upon the concentration of the anion pX and the pH of the solution; diagrams with axes pX and pH have been constructed that show the range of stability of these basic compounds. In the case of iron, the $Fe(OH)_2$ formed initially is subsequently oxidised to yellow FeO(OH) or $Fe_2O_3.H_2O$, or in low oxygen conditions black Fe_3O_4 is formed containing green reduced corrosion products. Vertical surfaces allow ready detachment of the products formed, while they may settle on a horizontally corroding surface and provide some blanketing action, restraining access of oxygen to the surface. Precise identification of the products and a knowledge of the pH at their location on the surface may provide information on the conditions of formation[13].

Thin Passive Films

In considering passivity and passivation (Sections 1.4 and 1.5), the nature of the surface product (the passivating film) entering into the process between the curve for active dissolution and that for the onset of film breakdown or oxygen evolution, assumes considerable significance.

As the system passes from the active to the passive state the initial interaction depends on the composition of the aqueous phase[14]. An initial chemisorbed state on Fe, Cr and Ni has been postulated in which the adsorbed oxygen is abstracted from the water molecules[2]. This has features in common with the metal/gaseous oxygen interaction mentioned previously. With increase in anodic potential a distinct 'phase' oxide or other film substance emerges at thicknesses of 1-4 nm. Increase in the anodic potential may lead to the sequence

$$M \rightarrow \underset{\text{monolayer}}{M-OH} \rightarrow \underset{\text{multilayer}}{M(OH)_2} \rightarrow \underset{\text{phase oxide}}{MO}$$

which has been suggested for Ni in acid solutions, and Cd and Zn in alkaline solutions. On the other hand, Fe in strong H_2SO_4 first forms a layer of $FeSO_4$ crystals, which at higher potentials is replaced by an Fe_2O_3 film, the normal product formed during anodic polarisation in dilute acid[15]. In near-neutral solutions the passive film on Fe (2-6 nm thick) has been characterised as the so-called cubic oxide γ-Fe_2O_3 overlying a thin film of Fe_3O_4 on the metal surface[16].

The nature of γ-Fe_2O_3 in passive films is very significant and has been reviewed in detail[17]. Here again a spinel structure is prominent (derived from magnetite). Its structure is considered to be cation defective with protons (H^+) progressively replacing Fe^{2+} ions in the Fe_3O_4 spinel, and leading to a continuous series of solid solutions of which Fe_3O_4 and Fe_2O_3 are the end products. In some cases an HFe_5O_8 composition is indicated in which some Fe^{2+} ions have been replaced by protons. The implication of this mechanism of replacement of Fe^{2+} ions is that water is incorporated into the passive film by a process of oxidative hydrolysis of the initial Fe_3O_4 substrate as the potential of the metal is progressively raised.

An important feature of such films is their low ionic conductivity that restricts cation transport through the film substance. Electronic semi-conduction, however, permits other electrode processes (oxidation of H_2O to O_2) to take place at the surface without further significant film growth. At elevated anodic potentials adsorption and entry of anions, particularly chloride ions, may lead to instability and breakdown of these protective films (Sections 1.5 and 1.6).

Thick Anodic Films

Where the electronic conductivity of the film substance is low, as in the case of the 'valve' metals (Al, Nb, Ta, Zr, Ti), an increase in anode potential gives rise to a high electric field across the passive layer. Under these circumstances ion transport occurs and film growth continues to several hundred volts with thicknesses rising to hundreds of nanometres. At low voltages an amorphous or microcrystalline 'barrier' oxide is formed, which may recrystallise thermally or by the action of a high field to γ-Al_2O_3, β-Ta_2O_5 or TiO_2, etc. A 'mosaic' structure has been attributed to these amorphous films[17] to account for their high field conduction properties. In the case of a valve metal with variable valency a number of anodic oxides may form over a range of anodic potential, e.g. Ti in strongly oxidising conditions gives TiO_2, while anodic passivation at lower potentials leads to Ti_2O_3, 3-4 TiO_2, or even Ti_3O_5. Furthermore, different structural modifications can be produced depending on the precise conditions of formation. For example, with Al[20] and Ti[21] high temperatures and high formation voltage tend to favour crystalline modifications as compared with the more commonly observed amorphous oxides. While, in general, anodic films produced represent those expected from thermodynamic data, significant free-energy gradients may exist across the film substance. Such situations may lead to complex geometrical arrays of different compounds as shown by Burbank[22]

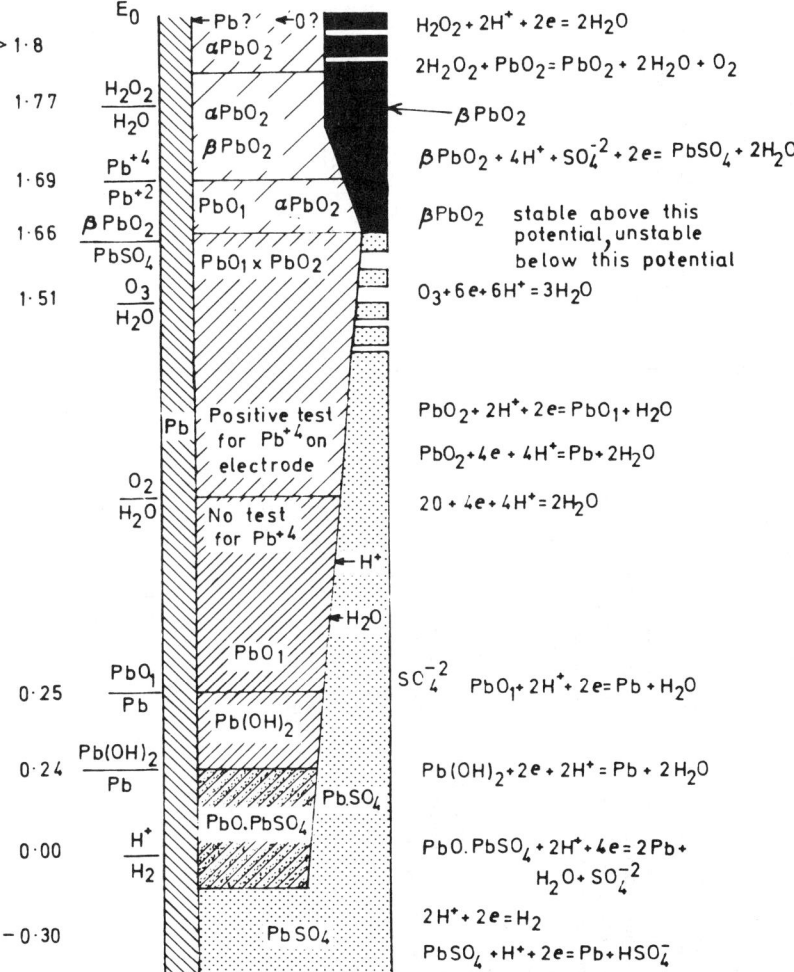

Fig. 1.4 Schematic representation of the reaction products formed on lead in sulphuric acid and their distribution over a range of anodic potentials (after Burbank[22])

Table 1.6 Schematic representation of experimental techniques and their range of application (extended from the table of Wood[10])

Technique (electrochemical)	Technique	Thickness range (approximate) of technique	Technique
Cyclic voltammetry (adsorption, monolayers)	Ellipsometry (kinetics)	1 nm — ADSORBED LAYERS, VERY THIN FILMS AND NUCLEI	Field ion microscopy, scanning tunnelling microscopy (morphology; analysis, etc.)
Potentiodynamic polarisation (passivation, activation)	Electrometric reduction (kinetics; thickness)		L.E.E.D. (structure)
Cathodic reduction (thickness)	Interference colours and spectrophotometry (kinetics; thickness)		Auger electron spectroscopy
Frequency response analysis (electrical properties, heterogeneity)	A.C. impedance (thickness; conduction mechanisms and profiles; compactness; crystallinity)		X-ray photo-electron spectroscopy } (composition, thickness)
Chronopotentiometry (kinetics)	Electrical methods (kinetics; thickness)		Secondary ion mass spectrometry
Chronoamperometry (kinetics)	Manometric and volumetric methods (kinetics)		Ion scattering spectroscopy
Photoelectrochemical methods (electronic properties, heterogeneity)	Thermogravimetry (kinetics from very thin films to thick scales; stoichiometry)		Electron diffraction (structure; epitaxy; grain size; preferred orientation)
	Electrical conductivity of oxides and allied methods (defect structures; conduction mechanisms; transport numbers)	200 nm — FILMS	Transmission electron microscopy (topography; local thickness; substructure — all thickness ranges) STEM/EDX
	Radioactive tracers and allied methods (kinetics; self diffusion; markers)	Stress measurements	Scanning electron microscopy (thickness; topography; porosity; barrier layers; fracture sections)
	Inert markers (transport mechanisms)	Adhesion	Energy dispersive X-ray analysis (EDX)
	Gas adsorption (surface area)	Stress/strain characteristics	Electron probe microanalysis (composition; diffusion profiles; local thickness)
		1 μm — Creep	X-ray absorption spectroscopy } (composition, structure)
		Hardness (oxide mechanical properties; oxygen solution in metal)	X-ray absorption fine structure
		100 μm — Thermal cycling tests	

SCALES

- Surface-enhanced Raman spectroscopy (chemistry)
- Laser microprobe mass spectrometry (composition)
- X-ray fluorescence analysis (composition; thickness)
- X-ray diffraction (structure; grain size; preferred orientation; stress)
- Scanning laser microscopy
- Optical microscopy (local thickness; topography; nucleation; general morphology; internal oxidation)
- I.R. spectroscopy (specialised analysis and applications)
- Spectrographic analysis (trace element analysis)
- Chemical analysis (analysis; stoichiometry)
- Vacuum fusion analysis (oxygen solubility in metal)

for anodic films of lead in sulphuric acid (Fig. 1.4) in which it can be seen that the nature and thickness of the oxidation products are highly dependent on the anodic potential.

In the particular case of aluminium in acid electrolytes, an initially formed thin barrier film breaks down to give a porous coating which can be grown to a considerable thickness (Table 1.5). The voltage remains low as the porous anodic coating continues to thicken. Significant amounts of the acid anion (SO_4^{2-}, PO_4^{3-}, CrO_4^{2-}) may be incorporated into the oxides so produced, together with protons to provide a degree of hydration (*see* Section 15.1). These features can significantly influence the structure and properties of the coatings obtained.

Techniques of Examination

This limited survey has indicated the wide range of chemical compounds, particularly oxides, which may be formed on a metal surface as a result of a corrosion process. The nature of such films and scales needs to be carefully characterised. Fortunately, a wide spectrum of experimental techniques is now available to provide such valuable information, and others are under development. A convenient summary is provided in Table 1.6.

In this scheme the nature of the surface product is arbitrarily divided into (*a*) adsorbed layers, very thin films and nuclei (1-200 nm thickness); (*b*) thin films (200 nm-1 μm), and (*c*) scales (above 1 μm). The principal techniques are located as appropriately as possible to indicate their areas of useful application. The spectrum thus ranges from the regime of very clean metal surfaces to grossly thick scales which may result from exposure to industrial oxidising atmospheres. Initial interaction may be studied by field-ion or field-emission spectroscopy and low energy electron diffraction, after which time the kinetics of the growth process may be followed by such techniques as ellipsometry, thermogravimetry, or electrometric reduction, while the structure may be examined by electron microscopy, electron diffraction or X-ray microanalysis. Stoichiometric and defect characteristics may be examined by a number of electrical methods. As the thickness approaches scale dimensions less sensitive techniques become applicable. Information on stress distribution, hardness, porosity, adhesion as well as thermal cycling characteristics also become accessible. Chemical analysis, scanning electron microscopy, X-ray diffraction techniques and gas adsorption data may provide further information on the composition, structure and porosity of thick scales, while electron probe microanalysis permits detailed examination of the concentration profiles across specimen sections. Many of these techniques are equally applicable to films formed under aqueous electrochemical conditions.

Recent Developments

In recent years the number of techniques available for analysis of metal surfaces has proliferated greatly[23-30]. Many of the new methods are ultra-high vacuum (UHV) techniques suitable for analyses of films ranging in thickness from a single monolayer to around a micrometre[23-28]. These techniques are still being improved and updated and many of them have attained a high degree of accuracy and sensitivity. Most noteworthy and probably most widely spread are X-ray photoelectron spectroscopy (XPS) and Auger electron spectroscopy (AES). These highly sensitive UHV techniques provide quantitative chemical analyses of surfaces and are sensitive to even sub-monolayer levels of atoms. They are sensitive to all atoms except hydrogen (and helium for AES). Even here, XPS can be used to provide some information on the presence of H^+ in oxide films by analysis of the oxygen signal. AES has the great advantage over XPS of being highly spatially resolved, enabling chemical 'maps' to be generated; these show the distribution of elements across the surface. XPS, although less spatially resolved (recent developments of the technique have improved this significantly), has the advantage over AES of being sensitive to the chemical state of the atoms; the technique can distinguish readily atoms in different oxidation states. Both techniques can be used to generate depth-profiles of the composition. Secondary ion mass spectrometry (SIMS) and ion scattering spectroscopy (ISS) fulfil a similar function to AES and XPS. They are less widely available, but can be used to great sensitivity (sub-monolayer up to around a micrometre, with depth profiling) and can be used for elemental mapping. To date, they are less quantitative than AES and XPS.

The composition of surface films can be determined as a function of depth using these UHV techniques. Such depth profiles are usually provided by sequential removal and analysis of layers of the surface films, removal being achieved by sputtering with an ionized noble gas beam. XPS can alternatively achieve a depth/composition profile by angular resolution, a non-destructive technique, successful for films up to the escape depth of the photoelectrons, typically around 1 to 3 nm in thickness. The technique finds widespread use in the analysis of the very thin passivating films formed electrolytically on metals such as stainless steels, for which it is very powerful indeed. These UHV methods generally provide *ex-situ* analyses, that is to say, the surface must be removed from the environment in which the film was formed and transferred to a UHV chamber; some features of the surface films may be altered by the analytical technique itself, particularly with very thin films which are formed electrochemically. The same is true of laser microprobe mass spectrometry (LAMMS), a very rapid method of producing a spot elemental analysis of a surface to a depth of around a micrometre, but not yet fully quantitative. LAMMS operates by transient ablation of the surface with an intense focused laser beam, and issues a mass spectrum of the ablated fragments. Because AES uses a primary electron beam as a probe, the technique can be more destructive to the surface than XPS, which employs a beam of soft X-rays.

Several UHV techniques which have been developed have not found such wide use in corrosion analysis, despite potential applicability. Ultraviolet photoelectron spectroscopy (UPS) is one of these, operating in a similar fashion to XPS (but using an ultraviolet excitation), and probing the valence electrons, rather than the core electrons of the atoms. Because the energies of the valence electrons are so very sensitive to the precise state of the atom, the technique is in principle very informative; however exactly this high sensitivity renders the data difficult to interpret, particularly as a routine analytical procedure. By and large the techniques which find application in corrosion are those which are relatively easy to use and easy to interpret.

Electrical characteristics of surface films formed electrochemically can be analysed using frequency response analysis (FRA) (sometimes called electrochemical impedance spectroscopy, or EIS)[23-25, 29, 30, 31]. This technique is capable of detecting separate components of films by resolving their separate resistance and capacitances *in situ*, for which most other electrochemical techniques are blind. The method has found wide application in the analysis of the passive state. It is also widely used to yield useful information on the state of applied surface coatings, such as paints.

Measurement of photocurrents generated by illuminating the surface while it is polarized in solution is increasingly being used to probe electronic properties of surface films generated electrochemically[23, 24, 30]. By focusing the light source and scanning the probe over the electrochemically polarised surface, this technique can be used to yield a photocurrent map of the surface. Other *in situ* measurements employing illumination of thin surface films generated electrochemically also yield characteristic information on passivating oxide films; these include ellipsometry, infrared spectroscopy and surface-enhanced Raman spectroscopy (SERS)[23, 24, 30, 31].

The very new techniques of scanning tunnelling microscopy (STM) and atomic force microscopy (AFM) have yet to establish themselves in the field of corrosion science. These techniques are capable of revealing surface structure to atomic resolution, and are totally undamaging to the surface. They can be used in principle in any environment *in situ*, even under polarization within an electrolyte. Their application to date has been chiefly to clean metal surfaces and surfaces carrying single monolayers of adsorbed material, rendering examination of the adsorption of inhibitors possible. They will indubitably find use in passive film analysis.

C. J. L. BOOKER
G. T. BURSTEIN

REFERENCES

1. Benard, J., 'Adsorption of Oxidant and Oxide Nucleation', in *Oxidation of Metals and Alloys*, Seminar, 1970; American Society for Metals, Ohio, 1 (1971)
2. Uhlig, H. H., Proceedings of the Third International Congress on Metallic Corrosion, Moscow, 1966, Vol. 1, 25 (1969); also *Corros. Sci.*, 7, 325 (1967)
3. Fehlner, F. P. and Mott, N. F., 'Oxidation in the Thin-film Range', as Reference 1, 37 (1971)
4. Fehlner, F. P. and Mott, N. F., *Oxid. Metals*, 2, 59 (1970)
5. Cathcart, J. V., 'The Structure and Properties of Thin Oxide Films', as Reference 1, 17 (1971)
6. Grauer, R. and Feitknecht, W., *Corrosion Sci.*, 6, 301 (1966)
7. Hauffe, K., *Metalloberflache*, 8, 97 (1954)
8. Kubaschewski, O. and Hopkins, B. E., *Oxidation of Metals and Alloys*, Butterworths, London, 114 (1967)
9. Wood, G. C., 'The Structures of Thick Scales on Alloys', as Reference 1, 201 (1971)
10. Wood, G. C., in *Techniques in Metals Research*, Rapp, R. A. (Assoc. Ed.), Vol. 4, Interscience, New York, 494 (1970)
11. Douglas, D. L., 'Exfoliation and the Mechanical Properties of Scales', as Reference 1, 137 (1971)

12. Evans, U. R., *Corrosion and Oxidation of Metals*, First Supplementary Volume, Arnold, London, 194 (1968)
13. Feitknecht, W. and Keller, G., *Z. Anorg. Chem.*, **262**, 61 (1950); Feitknecht, W., Weidman, H. and Haberli, E., *Helv. Chim. Acta.*, **26**, 1911 (1943), **32**, 2294 (1949) and **33**, 922 (1950)
14. Brusic, V., 'Passivation and Passivity', in *Oxides and Oxide Films* (Ed. Diggle, J. W.), Marcell Dekker, New York (1972)
15. Evans, U. R., as Reference 12, 98 (1968)
16. Bloom, M. C. and Goldberg, L., *Corros. Sci.*, **5**, 623 (1965)
17. Dignam, M. J., as Reference 14, 91 (1972)
18. Wells, A. F., *Structural Inorganic Chemistry*, 3rd edn, Clarendon Press, Oxford (1962)
19. Greenwood, N. N., *Ionic Crystals, Lattice Defects and Nonstoichiometry*, Butterworths, London, 92, 101 (1968)
20. Diggle, J. W., Downie, T. C. and Goulding,. C. W., *Chem. Rev.*, **69**, 365 (1969)
21. Aladjem, A., *J. Mat. Sci.*, **8**, 688 (1973)
22. Burbank, J., *J. Electrochem. Soc.*, **106**, 369 (1959)
23. Froment, M. (Ed.), *Passivity of Metals and Semiconductors*, Elsevier, Amsterdam, (1983)
24. MacDougall, B. R., Alwitt, R. S. and Ramanarayanan, T. A. (Eds.), *Oxide Films on Metals and Alloys*, Proceedings, **92-22**, The Electrochemical Society, Pennington, New Jersey (1992)
25. McCafferty, E. and Brodd, R. J. (Eds.), *Surfaces, Inhibition and Passivation*, Proceedings, **86-7**, The Electrochemical Society, Pennington, New Jersey (1986)
26. Rapp, R. A. (Ed.), *High Temperature Corrosion*, NACE, Houston, Texas (1983)
27. Bennett, M. J. and Lorimer, G. W. (Eds.), *Microscopy of Oxidation*, Institute of Metals, London (1991)
28. Augustynski, J. and Balsenc, L., in *Modern Aspects of Electrochemistry*, No. 13, (Eds. Conway, B. E. and Bockris, J. O'M.), 251, Plenum Press, New York (1979)
29. Macdonald, D. D. and McKubre, M. C. H., in *Modern Aspects of Electrochemistry*, **14**, (Eds. Bockris, J. O'M., Conway, B. E. and White, R. E.), 61, Plenum Press, New York (1982)
30. Ferreira, M. S. G. and Melendres, C. A. (Eds.), *Electrochemical and Optical Techniques for the Study and Monitoring of Metallic Corrosion*, NATO ASI series, Kluwer Academic Publishers, Dordrecht (1991)
31. Efrima, S., in *Modern Aspects of Electrochemistry*, **16**, (Eds. Conway, B. E., White, R. E. and Bockris, J. O'M.). 253, Plenum Press, New York (1985)

1.3 Effects of Metallurgical Structure on Corrosion*

The objective of this section is to show by means of specific examples how the various crystalline defects and structural features described in Section 20.4 can affect the form, location and kinetics of the corrosion of metals and alloys.

Effect of Crystal Defects on Corrosion—General Considerations

Before considering specific examples it is appropriate to note that there are, in principle, two quite distinct ways in which crystal defects can affect corrosion behaviour.

Firstly, they might be expected to have an effect when corrosion occurs under conditions of active (film-free) anodic dissolution and is not limited by the diffusion of oxygen or some other species in the environment. However, if the rate of active dissolution is controlled by the rate of oxygen diffusion, or if, in general terms, the rate-controlling process does not take place at the metal surface, the effect of crystal defects might be expected to be minimal.

Secondly, crystal defects might be expected to affect the corrosion behaviour of metals which owe their corrosion resistance to the presence of thin passive or thick protective films on their surface. The crystal defects and structural features discussed in Section 20.4 might, in principle, be expected to affect the thickness, strength, adhesion, porosity, composition, solubility, etc. of these surface films, and hence, in turn, the corrosion behaviour of the filmed metal surfaces. Clearly, this is the more common situation in practice.

Finally, it should be noted that in both cases the effect of crystal defects and microstructural features must, in general, be to tend to make the corrosion less uniform and more localised.

---

* The basic concepts of physical metallurgy are considered in Section 20.4, which should be regarded by those who are not conversant with the subject as an introduction to this section. Some of the diagrams referred to here will be found in Section 20.4.

Active Dissolution and Crystal Defects—Energy Considerations

The crystal defects described in Section 20.4 are all regions of higher energy than the adjacent perfect crystal lattice; they are therefore all inherently more chemically active and hence are potential sites for preferential attack under conditions of active dissolution. This preferential attack is, however, masked in highly aggressive environments, when there is very rapid dissolution and severe general corrosion, and it is not observed when the corrosion rate is controlled either by oxygen diffusion or some other process not occurring at the metal/environment interface. Furthermore, although the energy associated with the various defects may be quite large in metallurgical terms, when converted to a potential difference it is quite small in electrochemical terms, being not more than a few millivolts, at the most.

Etching of Single Crystals and Polycrystals

There is no evidence that any particular crystal structure is more readily corroded than any other. For example, the difference in the corrosion behaviour of austenitic and ferritic stainless steels is, of course, due to compositional rather than structural differences.

Using single crystals it has been shown that different low-index crystal faces (*see* Section 20) exhibit different corrosion rates. However, the relative corrosion rate of the different faces varies with the environment and these structural effects are of little practical significance. On the other hand, the fact that polycrystal grains of different crystallographic orientation may corrode at different rates, is of some importance.

A freshly polished metal surface appears quite featureless even when viewed at high magnification, while on etching different grains are attacked to differing degrees, as shown in Fig. 20.28 (bottom). The surface of grain B in Fig. 20.36a probably corresponds to a low-index low-energy plane while the surfaces of grains A and C correspond to high-index high-energy planes. In fact, the surfaces of grains A and C actually consist of low-index terraces separated by ledges with kinks in them. Since dissolution occurs most readily from kinks and ledges, owing to the lower co-ordination number of atoms at such sites, grains A and C will be attacked more rapidly than grain B, as illustrated schematically in Fig. 1.5a. It must be emphasised, however, that this is primarily a laboratory effect, albeit an important one. In practice, preferential corrosion of grains of a particular crystallographic orientation is not generally a problem. One possible exception to this is the etching of coarse-grained brass door-handles by sweaty hands!

Dislocations, Etch Pits and the Effect of Cold Work on Corrosion

Preferential corrosion or attack at many other types of crystal defect may also be best illustrated during the etching of metallographically polished

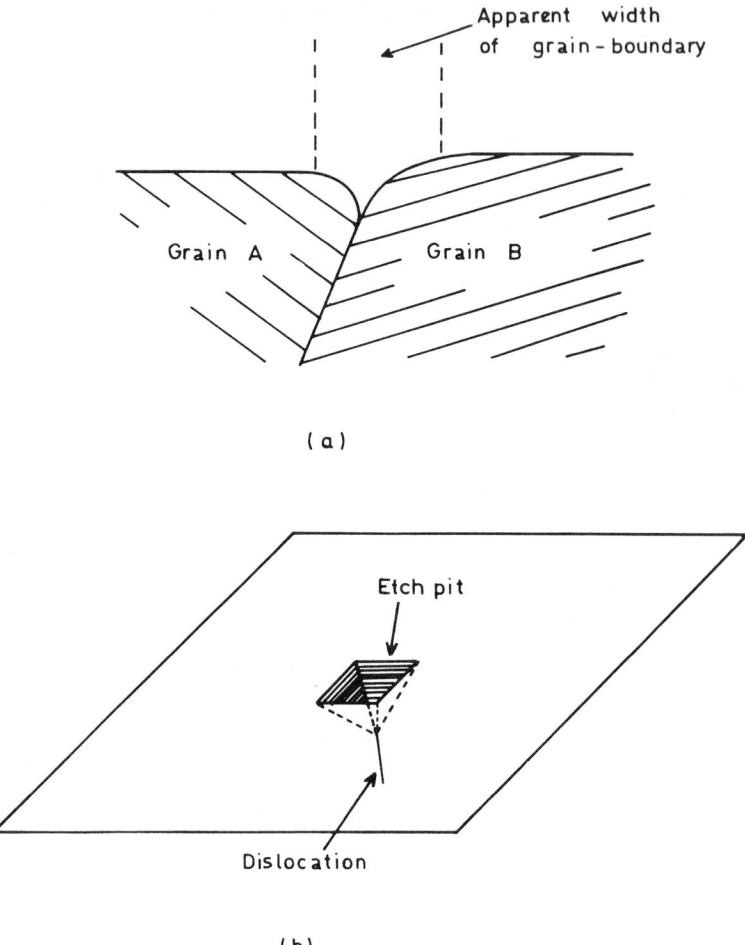

Fig. 1.5(a) Grain boundary intersecting an etched metallographic surface and (b) etch pit at a dislocation intersecting an etched metallographic surface

surfaces. Thus emergent dislocations intersecting metal surfaces may be revealed by the use of appropriate etchants. There is preferential attack at each dislocation, and small crystallographic etch pits are produced[1] as shown schematically in Fig. 1.5b. However, this effect is again of little practical significance, and the development of etch pits is used primarily as a research technique in the study of dislocations. Moreover, the technique must be used with some caution with metals, since there is often doubt as to whether there is a one-to-one correlation between etch pits and emergent dislocations, and because etch pits can also develop at other defects in surface films. Finally, there are many instances where it is thought that etch pits are produced only as a result of segregation of impurities to the dislocations. 'Clean' dislocations may not result in etch pits.

Potentially of somewhat greater practical significance is the effect of cold work on the corrosion of metals. When an annealed material is heavily cold worked, something of the order of 8–80 kJ/kg mol of energy may be stored in the material, as a result of the increased dislocation density, etc. This energy difference is, however, only equivalent to a potential difference of a few millivolts or so between the annealed and the cold-worked material. There is thus at the most only a small difference in the driving force for corrosion in the two cases. However, it is possible that the kinetics of the various anodic and cathodic processes could nevertheless be quite different on annealed and cold-worked surfaces; this would also result in annealed and cold-worked metals exhibiting significantly different corrosion rates. Certainly it has been experimentally observed that cold work markedly increases the corrosion rate of steel and aluminium in acids. The interpretation of this effect is, however, still not clear. Several authors suggest that the increased corrosion rate is due to the increased dislocation density *per se*, possibly as a result of an increased number of kink sites on the surface increasing the anodic exchange current density. On the other hand Foroulis and Uhlig[2] suggest that the increased corrosion rate is due to the segregation of carbon and nitrogen to dislocations, and that the cathodic (hydrogen evolution) reaction is kinetically easier at these sites; this is supported by their observation that cold work does not increase the corrosion rate of high-purity iron.

In natural waters, cold-worked commercial carbon steels of the same composition corrode at more or less the same rate as annealed steels, presumably because the corrosion rate in this case is controlled by the diffusion of oxygen. Unprotected carbon steels are sometimes exposed to natural waters, and it is this latter situation which is of greater practical importance than the behaviour of steels in acids, since steels should never be used in these environments unless they are protected.

Etching of Grain Boundaries and Intergranular Corrosion

During metallographic etching, twin and grain boundaries are preferentially attacked, as is apparent in Fig. 9.25 (bottom). Shallow grooves develop at these boundaries, and they therefore appear, in the microscope, as dark lines of finite width, as illustrated schematically in Fig. 1.5a. The best experimental evidence available indicates that even the grain boundaries in very high purity metals are slightly grooved by appropriate etchants. This is due to the grain boundaries being inherently more active than the adjacent crystal lattice, as implied by the energy associated with grain boundaries in metals. However, the grain boundaries in impure metals and alloys are generally much more readily etched, primarily as a result of segregation to them of the impurities and alloying additions. In this context it is important to note that grain-boundary regions may be preferentially attacked either because segregation makes them more base or because segregation makes them more noble; in the latter case the grain boundary itself acts as a local cathode, and the region immediately adjacent to the grain boundary is preferentially attacked. The subject of segregation and preferential attack at grain boundaries has been reviewed by Aust and Iwao[3].

Again it must be emphasised that preferential etching of twin and grain boundaries is predominantly a laboratory effect. There are no practical instances of significant corrosion problems resulting from the preferential attack of twin boundaries. In practice, grain-boundary effects in metals and alloys are *usually* of little or no consequence in the corrosion of metals. Severe intergranular corrosion (in the absence of tensile stress) is generally observed as a practical problem only when there is very gross segregation or solute depletion at grain boundaries, or, in certain instances, when there is marked intergranular precipitation, as discussed below.

Intergranular Corrosion of Austenitic Stainless Steels (Section 3.3)

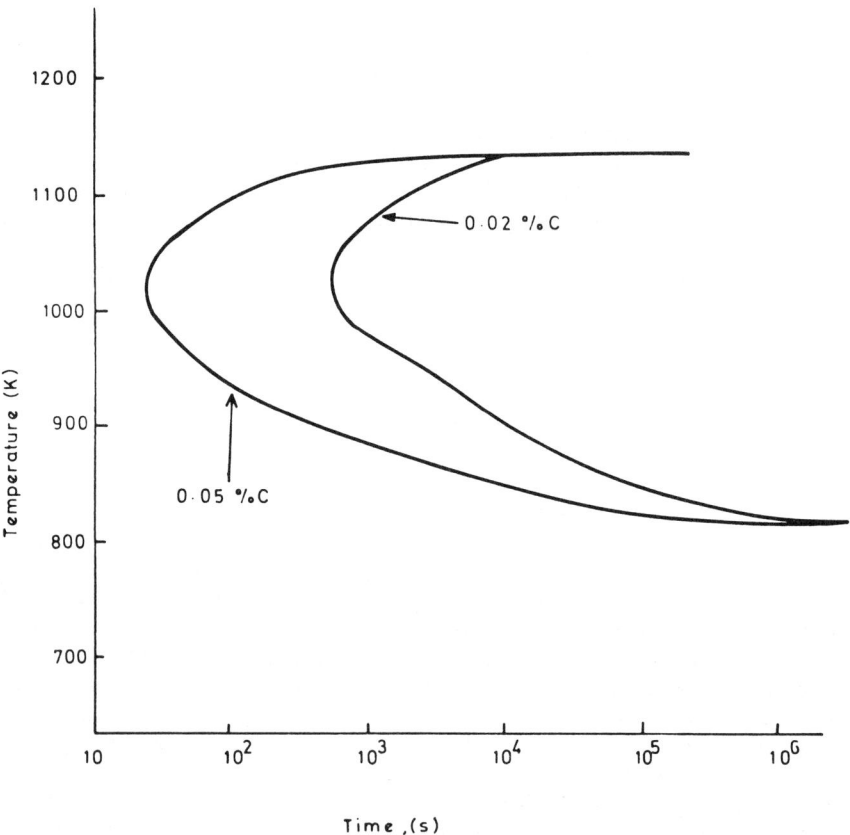

Fig. 1.6 Curves of the effect of temperature on the time required to sensitise two austenitic stainless steels of different carbon content

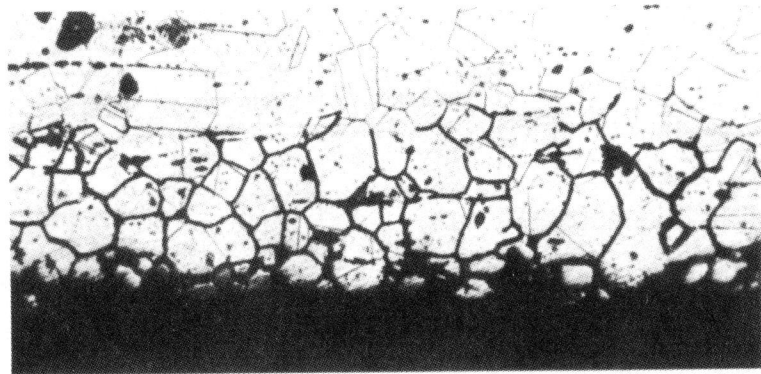

Fig. 1.7 Light micrograph showing intergranular corrosion of a sensitised austenitic stainless steel; × 200

As is well known, certain austenitic stainless steels may be 'sensitised' by certain heat treatments and made highly susceptible to intergranular corrosion. Sensitisation occurs when the alloys are held in, or slowly cooled through, the temperature range 1120–820 K. Quenching through the critical temperature range does not result in sensitisation. The degree of sensitisation, and therefore the susceptibility to intergranular corrosion, depend critically on the time at temperature, the temperature within the critical range or on the cooling rate through the critical temperature range, as well as on alloy composition, in particular the carbon content. These effects are illustrated schematically in Fig. 1.6. The intergranular corrosion, an example of which is shown in Fig. 1.7, is observed in a wide variety of environments in which austenitic stainless steels would normally be expected to have good corrosion resistance.

The generally accepted mechanism for sensitisation and the resultant intergranular corrosion was first proposed by Bain, et al.[4] and is basically as follows. During sensitisation, thin feathery precipitates of a chromium-rich carbide ($M_{23}C_6$ where $M = Fe_{0.2-0.3}Cr_{0.8-0.7}$) nucleate and grow in the austenite grain boundaries. These carbide particles, which can only be seen using electron microscopy, are only stable below about 1120 K; at higher temperatures they do not form, or if already present, tend to dissolve. On the other hand, below about 820 K, the diffusion rate of chromium in steels is too low for precipitation of the chromium-rich carbide to occur within a practical time scale. During precipitation of the carbide, which contains 70–80 wt.% Cr, the austenite matrix adjacent to the grain boundaries becomes depleted of chromium. In particular, the chromium level in these regions falls below the approximately 12% Cr required in solid solution to confer corrosion resistance, i.e. to permit the formation of a complete and protective passive film on the steel surface. The regions adjacent to the grain boundaries are therefore no longer passive and hence corrode preferentially. This mechanism is illustrated schematically in Fig. 1.8. The preferential attack of the non-passive chromium-depleted regions adjacent to the grain boundaries will be accelerated by the fact that these regions will be less noble than both the carbide precipitates in the grain boundary and the passive grain

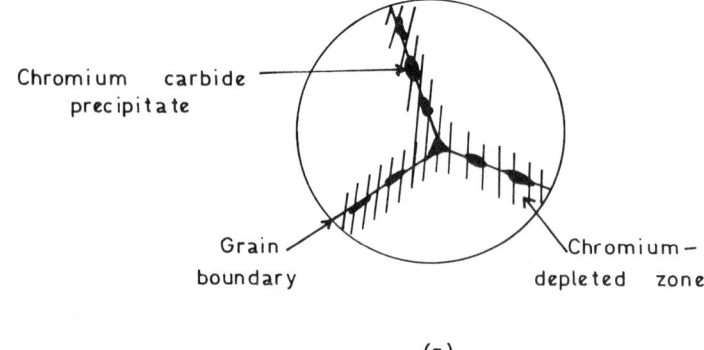

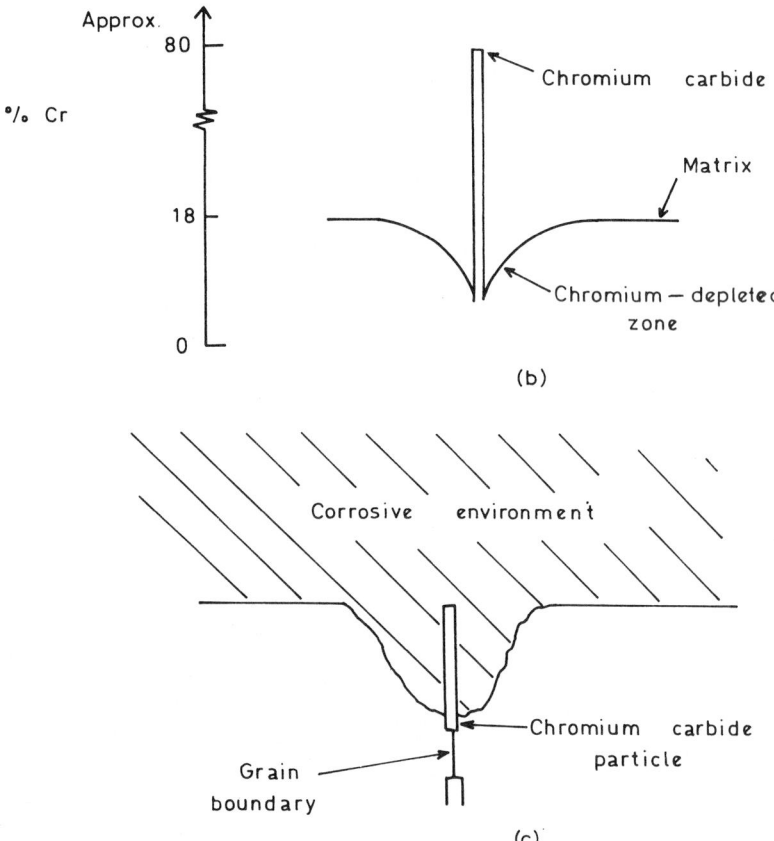

Fig. 1.8(a) Intergranular precipitation of chromium carbide particles in a sensitised austenitic stainless steel and the consequent chromium-depleted zones adjacent to the grain boundaries, (b) variation of the chromium content across a grain boundary in a sensitised austenitic stainless steel (18Cr) and (c) intergranular corrosion of a sensitised austenitic stainless steel

interiors. This effect is, of course, further exacerbated by the unfavourable high cathode/anode area ratio. It should be emphasised, however, that there will generally be little or no potential difference between the carbide precipitates and the passive grain interiors. Intergranular corrosion is therefore only observed when there is chromium depletion of the grain-boundary regions and not when there is carbide precipitation without chromium depletion.

In practice, three methods are available for preventing sensitisation and intergranular corrosion of austenitic stainless steels:

1. Quenching through the critical temperature range (if necessary after heat treating well above 1120 K to dissolve any existing chromium carbides).
2. The use of very low carbon (usually <0·03%) austenitic stainless steels, in which intergranular carbide precipitation does not occur within practical time scales.
3. The use of stabilised austenitic stainless steels to which small amounts (usually <1%) of strong carbide formers such as titanium or niobium have been added. The carbides of these elements form preferentially to chromium carbide at temperatures above the critical temperature range; chromium depletion of the grain-boundary regions is therefore not observed after slow cooling through that range.

In principle it is also possible to eliminate the effects of sensitisation by prolonged heat treatment within the critical temperature range to allow diffusion of chromium from the grain interiors to level out and eliminate the region of chromium depletion adjacent to the grain boundaries. In practice, however, the times involved (many hundreds of hours) are too long.

A more detailed treatment of sensitisation of austenitic stainless steels, of intergranular corrosion of austenitic stainless steels without sensitisation, and of sensitisation and intergranular corrosion of ferritic stainless steels and high-nickel alloys, is given by Cowan and Tedmon[5].

Weld Decay and Knife-Line Attack

A particularly important manifestation of sensitisation and intergranular corrosion of austenitic stainless steels is the phenomenon of weld decay, which is illustrated schematically in Fig. 1.9. During the welding of austenitic stainless steels that are potentially susceptible to sensitisation (i.e. steels that are not stabilised or low-carbon), a band of material in the heat-affected zone adjacent to the fusion zone of the weld is held within the critical temperature range and becomes sensitised (*B* in Fig. 1.9). Metal nearer the fusion zone is held above the upper limit of the critical temperature range (*A* in Fig. 1.9) while metal further away does not reach the lower limit of the critical temperature range (*C* in Fig. 1.9). An intermediate band of material parallel to, but away from, the fusion zone is therefore susceptible to severe intergranular corrosion, as shown in Fig. 1.9*b*. Weld decay is more severe with gas welding than with electric arc welding, and with thick plates rather than thin sheets, owing basically to different temperature-time profiles (*see also* Section 9.5).

A somewhat similar phenomenon is knife-line attack which may be observed after welding titanium or niobium stabilised austenitic stainless steels. In this case there is a very narrow band of severe intergranular attack along the interface between the parent metal and the fusion zone. During welding, the parent metal immediately adjacent to the fusion zone is heated to just below the melting point and both chromium carbides and niobium or titanium carbides dissolve completely. On cooling rapidly, the conditions are such that when relatively thin sections are welded, neither chromium carbide nor niobium or titanium carbide have time to precipitate. If the weld is now

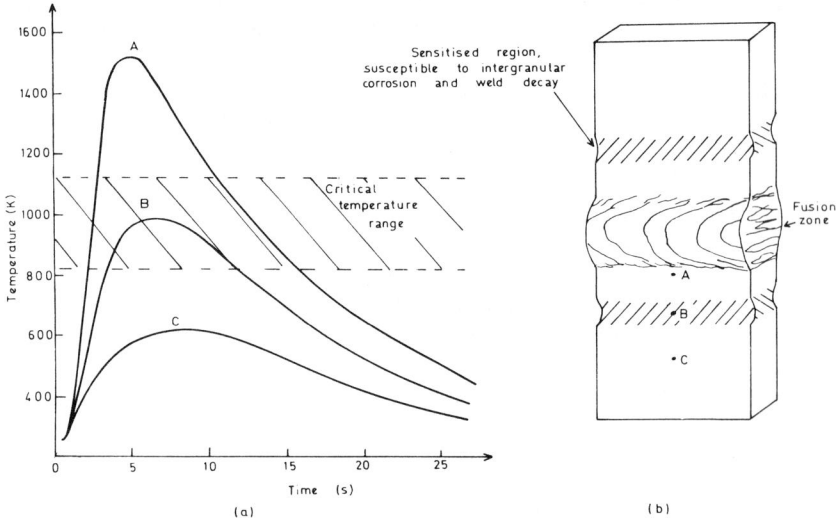

Fig. 1.9(*a*) Curves showing the variation of temperature with time at various points adjacent to a fusion weld in an austenitic stainless steel and (*b*) weld decay in an unstabilised austenitic stainless steel

heated to within the critical temperature range (e.g. to relieve residual stresses or during multi-pass welding) chromium carbide will precipitate rapidly but the temperature will be too low to precipitate niobium or titanium carbide (other than very slowly). The narrow band of material which during welding reached a temperature high enough to dissolve the stabilising carbides (>1 500 K) thus becomes sensitised and susceptible to intergranular corrosion. The remedy is to heat the fabricated structure or component to about 1 340 K after welding so that the chromium carbide dissolves and titanium (or niobium) carbide forms; following this solution treatment the rate of cooling is not important.

Intergranular Corrosion of Aluminium Alloys

A number of aluminium alloys may also, depending on their metallurgical structure, be susceptible to severe intergranular corrosion. The alloys concerned are primarily the precipitation-hardenable Al–Cu and Al–Zn–Mg based alloys and the work-hardenable Al–Mg alloys containing more than 3% Mg. Since there is much less inherent difference in the corrosion resistance of the grain-boundary regions and the grain centres than there is in the case of sensitised austenitic stainless steels, the mechanism of the intergranular corrosion in these alloys is primarily electrochemical, involving local cell action between grain-boundary precipitates and the adjacent matrix.

Aluminium alloys in which intergranular precipitation is not observed (e.g. commercial purity Al and Al–Mn alloys) or in which there is little or no potential difference between the matrix and any intergranular precipitates (e.g. balanced Al–Mg–Si alloys with Mg_2Si intergranular precipitates) are generally not markedly susceptible to severe intergranular corrosion. On the other hand, aluminium alloys in which the intergranular precipitates are markedly more noble than the matrix phase (e.g. Al–Cu base alloys with $CuAl_2$ intergranular precipitates), or alloys in which the precipitates are markedly more base (e.g. Al–Mg alloys and Al–Zn–Mg base alloys with Mg_2Al_3 and $MgZn_2$ intergranular precipitates, respectively) may be susceptible to severe intergranular corrosion. The latter precipitates corrode preferentially, while the former stimulate preferential corrosion of the adjacent matrix. The degree of susceptibility to intergranular attack depends on the nature, amount, size, distribution, etc. of the *inter*-granular precipitates (and to a lesser extent of the *intra*granular precipitates), and hence on the heat treatment of the alloy (*see* Figs. 20.31 and 20.34). In general, the precipitation-hardenable alloys are more likely to be susceptible to intergranular corrosion when aged to peak hardness and less likely to be susceptible in the overaged condition. In the work-hardenable Al–Mg alloys the tendency to intergranular precipitation (and hence to intergranular corrosion) increases with increasing Mg contents, with increasing cold work and with increased ageing times at temperatures below about 400 K. Chloride-containing environments, in particular, are liable to cause severe intergranular corrosion of susceptible aluminium alloys.

Intergranular Corrosion in Other Alloy Systems

A number of other alloy systems may also be susceptible to intergranular corrosion. For example, zinc die-casting alloys containing aluminium may be susceptible to intergranular attack in steam- and chloride-containing environments. Stray currents often result in intergranular corrosion of lead cable sheaths. These instances are, however, relatively unimportant compared to the intergranular corrosion of sensitised stainless steels and, to a lesser extent, to intergranular corrosion of intermediate- and high-strength aluminium alloys.

Effect of Grain Structure on Corrosion

The grain structure of alloys, as well as intergranular precipitation, can also markedly affect their corrosion behaviour. For example, the corrosion resistance of certain wrought metals may be less on surfaces perpendicular to the hot-or cold-working direction than on surfaces parallel to this direction. Typically there may be severe localised corrosion starting on the faces perpendicular to the working direction and proceeding into the metal in the working direction, while the surfaces parallel to the working direction remain relatively unattacked. Such *end-grain* attack, which is basically the result of the grain structure being elongated in the working direction, has been observed in austenitic stainless steels, titanium alloys and mild steel.

Layer Corrosion

The most marked effect of grain structure on corrosion is observed in wrought aluminium alloys. These alloys generally do not recrystallise during heat treatment after rolling, extrusion, etc. mainly because their grain boundaries are pinned by inclusions; they therefore exhibit the elongated pancake-shaped grain structure shown in Fig. 20.34. As a result of this structure, these alloys may be susceptible to *exfoliation* (also known as *layer* or *lamellar*) corrosion. The attack proceeds along a number of narrow planar paths (usually but not necessarily intergranular) parallel to the working direction. The corrosion products formed force the layers apart and cause the metal to swell and, in severe instances, to disintegrate into separate sheets of metal (i.e. to exfoliate). Exfoliation is most common and severe in Al–Cu, Al–Zn–Mg and Al–Mg based alloys, but mild exfoliation also occurs in Al–Mg–Si alloys. Since the exfoliation is normally intergranular it is clear that exfoliation and intergranular corrosion are associated, and exfoliation is usually affected by intergranular precipitation and hence by heat treatment. However, aluminium alloys that are susceptible to intergranular attack will not be susceptible to exfoliation corrosion if they have an equiaxed grain structure. Transgranular exfoliation is thought to be the result of segregation in the original ingot persisting in the wrought alloy.

Stress-corrosion Cracking

Grain structure also affects the stress-corrosion behaviour of high-strength age-hardenable aluminium alloys. Cracking in these alloys is always exclusively intergranular. When they are stressed in the short transverse direction (*a* in Fig. 1.10) their highly elongated, pancake-shaped grain structure ensures that an easy path for crack propagation is readily available. On the other hand, when stressed in the long-transverse or the longitudinal direction (*b* and *c* respectively in Fig. 1.10) the possible intergranular crack paths are clearly complex and difficult. Many high-strength aluminium alloys are therefore quite susceptible to stress-corrosion cracking when stressed in the short-transverse direction but quite resistant or immune when stressed in the long-transverse or longitudinal directions. This result is of considerable

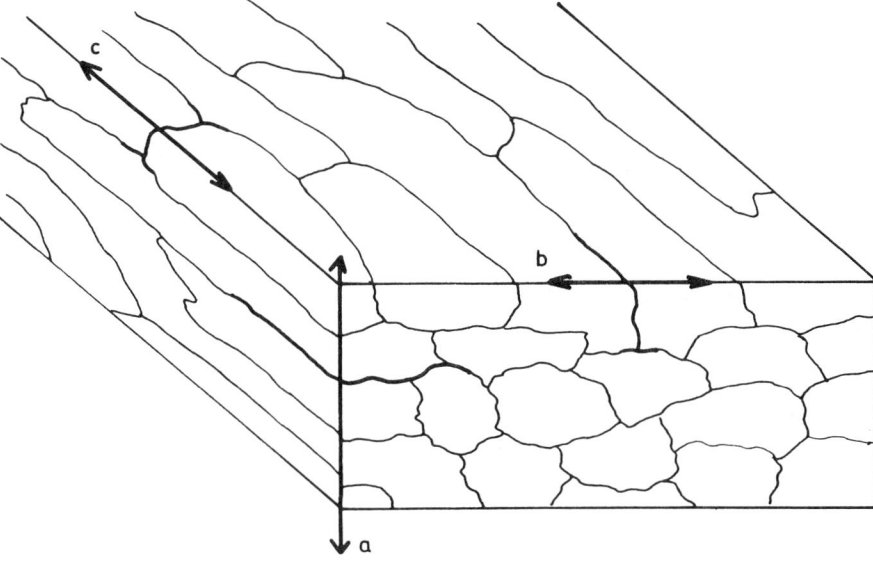

Fig. 1.10 Grain structure of a wrought high-strength precipitation-hardening aluminium alloy showing potential crack growth paths

practical importance: high-strength aluminium alloys can often be used in sheet form (when the short-transverse tensile stresses are generally negligible) in tempers in which they cannot normally be used in thick sections of forgings (when the tensile stresses in the short-transverse direction may be high). The distinction is, for example, between an aircraft's skin and its wing spars; with a susceptible alloy stress-corrosion cracking is likely to be a problem in the latter instance, but less so in the former.

Corrosion of Impure Metals and Single-phase Alloys

Many of the forms of corrosion already discussed have been caused or affected not only by metallurgical structure but also by the segregation of impurities or alloying additions to dislocations, grain boundaries, precipitates, etc. However, the presence of impurities or alloying elements in homogeneous solid solution can also markedly affect corrosion behaviour, without any segregation effects. In the context of this section it is the deleterious effects of soluble impurities and alloying additions that are relevant, rather than the beneficial effects, such as the addition of chromium and nickel to iron to produce stainless steels, or the addition of nickel and small concentrations of iron to copper to give cupronickels.

As is well known, high-purity zinc corrodes much less rapidly in dilute acids than commercial purity material; in the latter instance, impurities (particularly copper and iron) are exposed on the surface of the zinc to give local cathodes with low hydrogen overpotentials; this result is of practical significance only in the use of zinc for sacrificial anodes in cathodic protection or for anodes in dry cells. In neutral environments, where the cathodic reaction is oxygen reduction, there is very little difference in the corrosion rates of pure and impure zinc.

In contrast, the selective dissolution or leaching-out by corrosion of one component of a single-phase alloy is of considerable practical importance. The most common example of this phenomenon, which is also referred to as 'parting', is dezincification, i.e. the selective removal of zinc from brass (*see* Section 1.6). Similar phenomena are observed in other binary copper-base alloys, notably Cu–Al, as well as in other alloy systems.

Corrosion and Selective Dissolution in Two-phase Alloys

In principle the selective dissolution of the less noble component of a single-phase alloy would perhaps be expected and is in fact observed (dezincification of an α-brass, etc.) even though the details of the mechanism by which it occurs is not yet fully understood. In contrast, the preferential attack of the less noble phase of a two-phase alloy is not only expected and observed — the mechanism by which it occurs in practice is also quite clear. Selective dissolution of the more active phase of a two-phase alloy is best exemplified by the graphitic corrosion (or *graphitisation*) of grey cast iron.

Graphitisation

Cast irons, although common, are in fact quite complex alloys. The iron–carbon phase diagram exhibits a eutectic reaction at 1 420 K and 4·3 wt.%C (*see* Fig. 20.44). One product of this eutectic reaction is always austenite; however, depending on the cooling rate and the composition of the alloy, the other product may be cementite or graphite. The graphite may be in the form of flakes which are all interconnected (although they appear separate on a

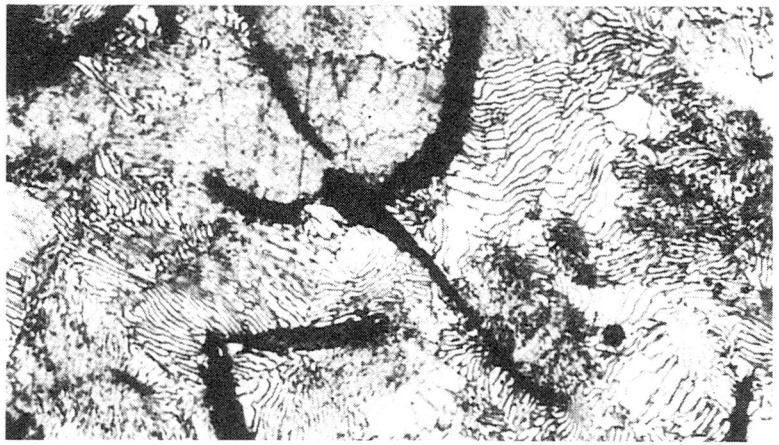

Fig. 1.11 Light micrograph of the microstructure of a pearlitic grey cast iron; × 720

metallographic section), or spheroids, which are all separate. The resultant alloys are known as white, grey and spheroidal graphite cast irons, respectively. During subsequent cooling the austenite may transform to ferrite, pearlite or martensite, or it may, in the case of high-nickel irons, be retained to room temperature. The microstructure of a grey cast iron with a predominantly pearlitic matrix is shown in Fig. 1.11.

Graphite is more noble than any of the other phases in cast iron and is a very good cathode material; highly effective galvanic cells therefore exist between the graphite and the surrounding, less noble matrix. In grey cast irons, the matrix therefore corrodes preferentially, leaving behind a network of interconnected graphite flakes which is very porous and weak. The attack is often not readily apparent on superficial inspection. White cast irons are not susceptible to graphitisation since they contain no graphite; spheroidal graphite cast irons are also not susceptible to graphitisation, since although they do contain graphite it is in the form of discrete spheroids which have a limited effect, instead of the interconnected graphite flakes in grey cast iron. Thus not only the existence but also the distribution of a cathodic phase is important (*see* also Section 3.6).

Influence of Structure on Surface Films—Pitting Corrosion

Metals which owe their good corrosion resistance to the presence of thin, passive or protective surface films may be susceptible to pitting attack when the surface film breaks down locally and does not reform. Thus stainless steels, mild steels, aluminium alloys, and nickel and copper-base alloys (as well as many other less common alloys) may all be susceptible to pitting attack under certain environmental conditions, and pitting corrosion provides an excellent example of the way in which crystal defects of various kinds can affect the integrity of surface films and hence corrosion behaviour.

In general, pitting corrosion may be divided into two stages, pit initiation and pit propagation. During pit initiation the passive film breaks down and does not reform. During pit propagation, the small active sites formed during the initiation stage propagate, often very rapidly, to form pits. The most recent ideas on the mechanism of pit initiation and propagation are dealt with in some detail in Reference 6. The propagation of pits is relatively well understood and is comparatively insensitive to the structure of the metal (*see* Sections 1.5 and 1.6).

On the other hand, pit initiation which is the necessary precursor to propagation, is less well understood but is probably far more dependent on metallurgical structure. A detailed discussion of pit initiation is beyond the scope of this section. The two most widely accepted models are, however, as follows. Heine, *et al.*[7] suggest that pit initiation on aluminium alloys occurs when chloride ions penetrate the passive oxide film by diffusion via lattice defects. McBee and Kruger[8] indicate that this mechanism may also be applicable to pit initiation on iron. On the other hand, Evans[9] has suggested that a pit initiates at a point on the surface where the rate of metal dissolution is momentarily high, with the result that more aggressive anions are attracted to the point and produce a local environment that is favourable to further dissolution, i.e. an autocatalytic process similar to that operative in pit propagation. This view has recently found increasing support, since there is now evidence that pits initiate at flaws or discontinuities in the passive film which result from mechanical, geometrical or compositional inhomogeneities in the metal surface[10,11]. The latter model, in particular, predicts a strong influence of metallurgical structure on the integrity of the passive film and hence on susceptibility to pitting corrosion.

In practice many metallurgical factors do appear to affect pitting corrosion. For example, severe cold work increases the pitting susceptibility of austenitic stainless steels, while molybdenum and nitrogen alloying additions, in particular, reduce it. Pitting is less likely to occur on smooth, polished surfaces than on rough, etched, ground or machined surfaces. Austenitic stainless steels are more susceptible to pitting if they have been held briefly in the sensitising temperature range. Pure aluminium is much more resistant to pitting than impure metal and alloys, particularly those containing copper. In general, the more homogeneous a metal surface the better is the resistance of passive films on that surface to pitting. In austenitic stainless steels, pits have been observed to initiate at grain boundaries and also at certain sulphide inclusions. These effects are all evidence of the fact that crystal defects and metallurgical structure and composition affect the thickness, strength, solubility, porosity, etc. of passive films, and hence the susceptibility of those films to localised breakdown and pitting.

Effect of Mechanical Stresses on Corrosion

The presence of stresses does not usually affect the general corrosion behaviour of metals and alloys to any very significant extent. However, two extremely important forms of localised corrosion may occur when metals are simultaneously exposed to stress and a corrosive environment. Metals subjected simultaneously to alternating stresses and any corrosive environment may be subject to corrosion fatigue, while certain alloys exposed simultaneously to tensile stresses and fairly specific environmental conditions may fail by stress-corrosion cracking. Other sections (*see* Chapter 8) deal specifically with the mechanism and phenomenology of corrosion fatigue and stress-corrosion cracking of various alloy systems, and it is not the intention to duplicate that material in this section. However, the susceptibility of many alloys to stress-corrosion cracking is determined not only by the presence of tensile stresses and specific environmental conditions, but also by the metallurgical structure of the metals. These instances will be discussed briefly, by way of further examples of the effect of structure on the corrosion of metals.

Stress-corrosion Cracking of Copper-base Alloys

Single-phase α-brasses are susceptible to stress-corrosion cracking in the presence of moist ammonia vapour or certain ammonium compounds[12]. Here the predominant metallurgical variable is alloy composition, and in practice brasses containing less than 10–15% Zn seldom fail by stress corrosion; above about 15% Zn, the stress-corrosion susceptibility increases with zinc content. Other structural factors are secondary: cold-worked brass, in practice, is more likely than annealed material to fail by stress-corrosion crocking, but this is probably only a reflection of the fact that the residual stresses are likely to be higher in cold-worked than in annealed alloys. The stress-corrosion susceptibility of α-brasses increases with increasing grain size. There is also evidence that decreasing stacking-fault energy results in a transition from inter- to transgranular cracking in a number of binary copper-base alloys[13] and that the presence of order (*see* Section 20.4) increases the susceptibility of certain complex copper-base alloys[14]. It must be emphasised, however, that these are only secondary factors which merely tend to increase or decrease fairly marginally the stress-corrosion susceptibility.

Stress-corrosion Cracking of Aluminium-base Alloys

In contrast to brasses, metallurgical structure plays a predominant rôle in determining the susceptibility of high-strength aluminium alloys to stress-corrosion cracking in the presence of tensile stresses and moist chloride-containing environments. Under these conditions these alloys may vary from highly susceptible, to practically immune, to intergranular stress-corrosion cracking, depending on their microstructure, as determined by heat treatment[15] (*see* Section 8.5).

The effect of grain shape on the stress-corrosion behaviour of aluminium alloys has already been discussed. The effect of heat treatment on the stress-corrosion susceptibility of high-strength precipitation-hardenable Al–Zn–Mg alloys is illustrated schematically in Fig. 1.12. In the solution-heat-treated and quenched condition, these alloys are very resistant to stress-corrosion cracking but they are also too weak to be of much use in this condition. On ageing, the alloys become progressively stronger (*see* Section 20.4), but also increasingly susceptible to stress corrosion, as shown in Fig. 1.12. Maximum stress-corrosion susceptibility is observed in the intermediate-strength, under-aged condition; thereafter the alloys become increasingly more resistant to stress-corrosion cracking. Thus, as shown in Fig. 1.12 the highest-strength, peak-aged condition is moderately susceptible to stress corrosion, while the intermediate-strength, over-aged condition is relatively resistant. In practice, therefore, there is a choice between maximum strength alloys with moderate stress-corrosion susceptibility and somewhat lower strength alloys with little stress-corrosion susceptibility. As suggested above, the former condition or temper might be selected for thin sheet applications while the latter heat treatment would be specified for thick sections. There is some doubt as to whether the effect of heat treatment on the stress-corrosion susceptibility of precipitation-hardenable aluminium alloys results from variations in the precipitate-free zone width and the intergranular precipitate morphology, or from variations in the interaction of intragranular precipitates with dislocations. The effect of microstructure on the stress-corrosion susceptibility of Al–Cu, although again very substantial, is somewhat less straightforward than in the case of Al–Zn–Mg alloys[15].

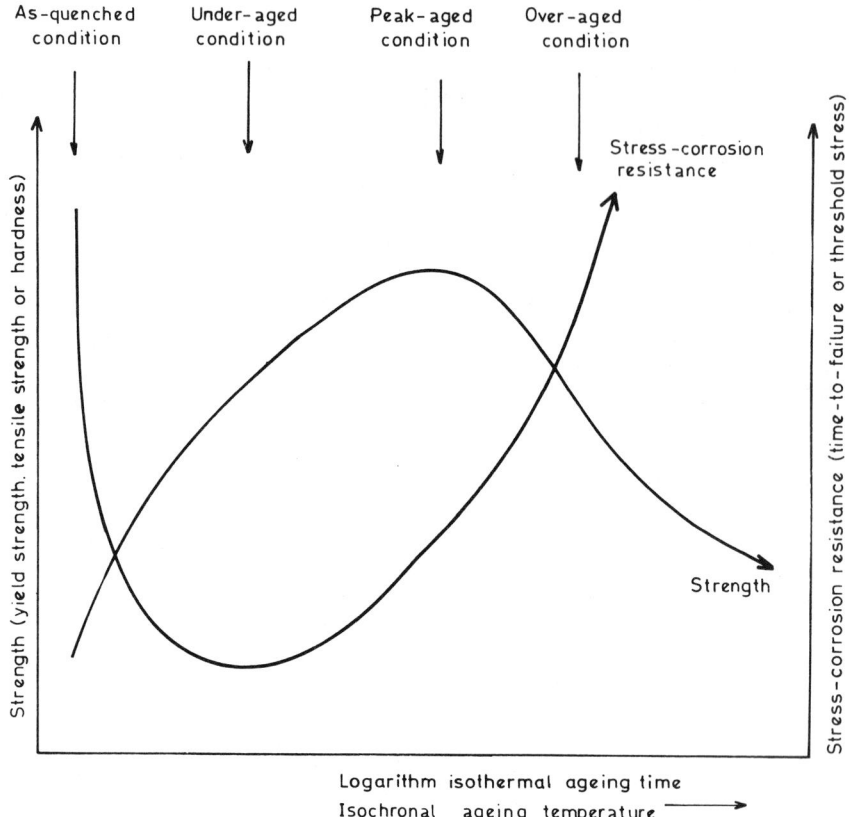

Fig. 1.12 Curves showing the relationship between strength, stress-corrosion susceptibility and heat treatment for a high-strength precipitation-hardening aluminium alloy

Microstructure also plays a predominant rôle in determining the stress-corrosion susceptibility of the work-hardenable Al–Mg alloys. The Al–Mg system, like the Al–Cu system, exhibits decreasing solubility with decreasing temperature, and on ageing a solution-heat-treated and quenched alloy, precipitation of Mg_2Al_3 is observed. However, the alloy is not strengthened by this precipitation, as it occurs either as very coarse, widely dispersed intragranular precipitates (which do not interact with dislocations), or as a more or less continuous intergranular film. These alloys can therefore only be strengthened by cold working. Nevertheless, the precipitate morphology controls the stress-corrosion susceptibility; alloys which exhibit continuous films of intergranular precipitate are highly susceptible to stress-corrosion cracking (and to exfoliation and intergranular corrosion), while those which exhibit coarse intragranular precipitates (or no precipitation at all) are generally much less susceptible or resistant. The higher the magnesium level, the greater the degree of cold work, and the lower the ageing temperature, the more likely is the formation of a continuous intergranular precipitate and therefore the greater is the potential stress-corrosion susceptibility.

Stress-corrosion Cracking of Steels

High-strength low-alloy quenched and tempered steels (i.e. steels with yield strengths greater than about 900 MN/m^2) may be susceptible to stress-corrosion cracking in the presence of moisture. The crack path is usually intergranular with respect to the prior-austenite grain boundaries, and the mechanism of cracking is generally accepted as involving some form of hydrogen embrittlement. The major metallurgical variable in this instance of environmentally induced cracking is the strength level—the stronger the steel the greater is its susceptibility. However, at constant strength level, steels with martensitic structures are considerably more susceptible to cracking than steels with bainitic structures (*see* Section 20.4). Again, at constant strength level, it has been shown that the crack growth rate decreases and the time-to-failure increases as the prior-austenite grain size is reduced[16].

In practice, by far the most common case of stress corrosion is that occurring when austenitic stainless steels are simultaneously exposed to tensile stresses and hot, aqueous, aerated, chloride-containing environments. In this case the major variable is alloy composition and structure; virtually all austenitic stainless steels are more or less susceptible to stress-corrosion cracking in these environments, while ferritic and ferritic/austenitic stainless steels are highly resistant or immune.

Stress-corrosion cracking of all types of steels formed the topic of a recent conference[17], the proceedings of which deal in some detail with the effect of structure on the stress-corrosion susceptibility of these alloys.

Conclusions

This discussion on the relationships between structure and corrosion should not be taken as exhaustive. For example, the stress-corrosion cracking behaviour of titanium-base alloys in a variety of environments is affected to differing degrees by the microstructure of the alloys[18]. Again, the effect of the changes in structure produced by welding on the corrosion behaviour of metals and alloys is of great practical importance[19]. This topic has been considered above in relation to stainless steels, but it is also of considerable importance in the welding, brazing and soldering of other alloy systems. In certain instances, the corrosion resistance of a weld is markedly affected by the structure of the weld metal and the adjacent heat-affected zone. Further examples of the effect of metallurgical structure on corrosion phenomena are provided by (*a*) the possible rôle of emergent dislocations and slip-steps in the mechanism of stress-corrosion cracking of austenitic stainless steels, (*b*) by the rôle of δ-ferrite in the corrosion of certain austenitic stainless steels, (*c*) by the rôle of local spheroidisation in 'ringworm' corrosion of mild steel and (*d*) by the rôle of manganese sulphide and other inclusions in the pitting of mild steel.

Since corrosion is essentially a reaction between a metal and its environment, the very significant effect of crystal defects and metallurgical structure on certain corrosion phenomena is to be expected. It is no more possible to neglect the metallurgical aspects of a corrosion problem than it is to overlook the environmental and electrochemical factors.

R. P. M. PROCTER

REFERENCES

1. Ives, M. B., in *Proc. U. R. Evans Internat. Conf. on Localised Corrosion*, N.A.C.E. (Houston)
2. Foroulis, Z. A. and Uhlig, H. H., *J. Electrochem. Soc.*, **111**, 522 (1964)
3. Aust, K. and Iwao, O., in *Proc. U. R. Evans Internat. Conf. on Localised Corrosion*, N.A.C.E. (Houston)
4. Bain, E. C., Aborn, R. H. and Rutherford, J. B., *Trans. Amer. Soc. Steel Treating*, **21**, 481 (1933)
5. Cowan, R. L. and Tedmon, C. S., in *Advances in Corrosion Science and Technology*, Vol. III, Plenum Press, New York (1973)
6. *Proc. U. R. Evans Internat. Conf. on Localised Corrosion*, N.A.C.E. (Houston)
7. Heine, M. A., Keir, D. S. and Pryor, M. J., *J. Electrochem. Soc.*, **112**, 29 (1965)
8. McBee, C. L. and Kruger, J., in *Proc. U. R. Evans Internat. Conf. on Localised Corrosion*, N.A.C.E. (Houston)
9. Evans, U. R., *Corrosion*, **7**, 238 (1951)
10. Richardson, J. A. and Wood, G. C., *Corr. Sci.*, **10**, 313 (1970)
11. Ashworth, V., Boden, P. J., Leach, J. S. Ll. and Nehru, A. J., *Corr. Sci.*, **10**, 481 (1970)
12. Pugh, E. N., Craig, J. V. and Sedriks, A. J., in *Proc. Internat. Conf. on Fundamental Aspects of Stress-corrosion Cracking*, N.A.C.E., Houston (1969)
13. Ohtani, N. and Dodd, R. A., *Corrosion*, **21**, 161 (1965)
14. Popplewell, J. M., Procter, R. P. M. and Ford, J. A., *Corr. Sci.*, **12**, 193 (1972)
15. Speidel, M. O. and Hyatt, M. V., in *Advances in Corrosion Science and Technology*, Vol. II, Plenum Press, New York (1972)
16. Procter, R. P. M. and Paxton, H. W., *Trans. A.S.M.*, **62**, 989 (1969)
17. *Proc. Internat. Conf. on Stress-corrosion Cracking and Hydrogen Embrittlement of Iron Base Alloys*, N.A.C.E. (Houston)
18. Blackburn, M. J., Feeney, J. A. and Beck, T. R., in *Advances in Corrosion Science and Technology*, Vol. III, Plenum Press, New York (1973)
19. Lancaster, J. F., *Metallurgy of Welding, Brazing and Soldering*, George Allen and Unwin, London (1970)

1.4 Corrosion in Aqueous Solutions

In this section the interaction of a metal with its aqueous environment will be considered from the viewpoint of thermodynamics and electrode kinetics, and in order to simplify the discussion it will be assumed that the metal is a homogeneous continuum, and no account will be taken of submicroscopic, microscopic and macroscopic heterogeneities, which are dealt with elsewhere (*see* Sections 1.3 and 20.4). Furthermore, emphasis will be placed on uniform corrosion since localised attack is considered in Section 1.6.

Aqueous environments will range from very thin condensed films of moisture to bulk solutions, and will include natural environments such as the atmosphere, natural waters, soils, body fluids, etc. as well as chemicals and food products. However, since environments are dealt with fully in Chapter 2, this discussion will be confined to simple chemical solutions, whose behaviour can be more readily interpreted in terms of fundamental physico-chemical principles, and additional factors will have to be considered in interpreting the behaviour of metals in more complex environments. For example, iron will corrode rapidly in oxygenated water, but only very slowly when oxygen is absent; however, in an anaerobic water containing sulphate-reducing bacteria, rapid corrosion occurs, and the mechanism of the process clearly involves the specific action of the bacteria (*see* Section 2.6).

All corrosion reactions in aqueous solutions are characterised by the following features:

1. The electrified interface between the metal and the electrolyte solution (the metal surface may be film-free or partially or completely covered with films or corrosion products).
2. Transfer of positive charge from the metal to the solution with consequent oxidation of the metal to a higher valency state.
3. Transfer of positive charge from the solution to the metal with consequent reduction of a species in solution (an electron acceptor) to a lower valency state.
4. Transfer of charge through the solution and corroding metal.

It follows that corrosion is an electrochemical reaction in which the metal itself is a reactant and is oxidised (loss of electrons) to a higher valency state, whilst another reactant, an electron acceptor, in solution is reduced (gain of electrons) to a lower valency state. This may be regarded as a concise expression of the 'electrochemical mechanism of corrosion'.

Thermodynamics and Kinetics of Corrosion Reactions

Thermodynamics provides a means of predicting the *equilibrium state* of a system of specified components, but provides no information on the detailed course of the reaction nor of the rate at which the system proceeds to equilibrium. With the aid of suitable catalysts the reaction may be made to proceed by different mechanisms and at different rates, but the final position of equilibrium, which can be predicted unequivocally by thermodynamics, remains unchanged. At elevated temperatures reactions proceed to equilibrium far more quickly than at ambient temperatures, and for this reason metal/gas equilibria at elevated temperatures can be defined entirely in terms of the thermodynamics of the systems under consideration (*see* Section 7.6). The position is frequently quite different with systems at ambient temperatures, since although a reaction may have a very pronounced tendency to proceed in a given direction this may be entirely nullified, by the very slow rate at which it proceeds to its equilibrium state. Thus although the interaction of hydrogen and oxygen to form water at 25°C is accompanied by a large decrease in the standard free energy (*see* Section 20.4)

$$H_2(g.) + \tfrac{1}{2}O_2(g.) \rightarrow H_2O(g.) \qquad \Delta G^{\ominus}_{298} = -237 \text{ kJ}$$

the rate of reaction is so slow that the reaction may be regarded as not occurring at all; however, in the presence of a catalyst, e.g. platinum black, the reaction proceeds instantaneously with explosive violence. It follows that although the change in free energy ΔG provides a quantitative measure of the *tendency* of a reaction to proceed in a given direction, it gives no information on the rate of the reaction.

It is indeed fortunate that the *rates* of the majority of corrosion reactions are very slow, since it is this factor that permits metals to be used as constructional materials. Considering metals in contact with oxygen at room temperatures, it follows from the standard free energies of formation of metal oxides that, with the exception of gold, the oxide of the metal is the thermodynamically stable state; the free energy decreases with temperature but the common metals are thermodynamically stable in an oxygen atmosphere only at very elevated temperatures (Fig. 7.55, Section 7.6). However, in practice, highly reactive metals like aluminium, magnesium, tantalum, niobium, etc. are relatively stable in oxygen whereas other reactive metals such as the alkali and alkaline-earth metals oxidise rapidly and completely. In these examples, the kinetics of the reaction is determined by the nature of the oxide film formed on the metal surface, which forms a protective barrier in the former group of metals and a non-protective one in the latter. Nevertheless, it should be noted that when the surface area is very large compared to the volume, the thermodynamic tendency for oxidation will outweigh kinetics, and metals such as aluminium, magnesium, titanium and iron in the form of very fine powders, can ignite spontaneously at ambient temperatures.

Most metals (copper, silver, gold, mercury and the platinum metals are the exceptions) are found in nature in the combined form as minerals, which are then reduced to the metal by the expenditure of sufficient energy (chemical, electrical or thermal) to reverse the natural spontaneous reaction. It follows that most metals are thermodynamically unstable and will tend to revert back to the thermodynamically stable combined state when exposed to the atmosphere, but the rate at which the reaction proceeds may be so slow that for all practical purposes the metal is stable. It is of some significance that very reactive metals such as tantalum, niobium and titanium, which have high affinities for oxygen and are reduced from their oxides with the expenditure of considerable chemical energy, are highly stable in the majority of environments; in fact, it is unlikely that these metals will ever revert back to oxides during atmospheric exposure owing to the highly protective nature of the thin oxide film. However, similar considerations do not apply to the less reactive metals iron and mild steel, since the oxide formed on these metals (rust) is not so protective as that formed on certain of the more thermodynamically reactive metals.

These examples show that the thermodynamic tendency of metals to corrode is frequently outweighed by kinetic factors that control the rate of the process so that the final position of equilibrium, i.e. complete conversion to corrosion product, is attained only slowly or not at all. Although the formation of oxide films, or films of other corrosion products, are of vital importance in controlling the rate of a corrosion reaction in aqueous solution, it must be emphasised that in certain circumstances the activation energy of the process or the rate of diffusion of species to and from the metal surface may be more significant than film formation. Thus the slow rate of corrosion of pure zinc in sulphuric acid, in which the oxide film is thermodynamically unstable, is dependent on the high activation energy needed for the reduction of hydrogen ions to hydrogen gas. In neutral solutions, on the other hand, diffusion of oxygen to the metal surface may be the rate-determining factor.

The magnitude of ΔG, the free energy change, of a specific corrosion reaction provides a measure of the spontaneity of the reaction and of the extent to which it will proceed before equilibrium is attained; if $\Delta G \ll 0$

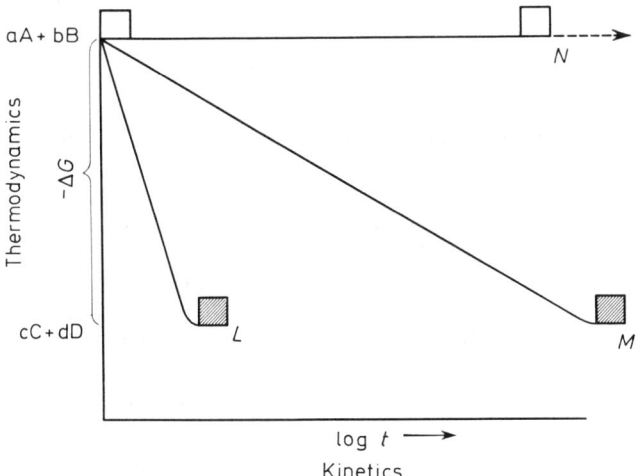

Fig. 1.13 Block on inclined plane illustrating the relationship between the thermodynamic tendency ($-\Delta G$) of a corrosion reaction $aA + bB \rightarrow cC + dD$ and the rate of conversion of metal A to corrosion products C and D

(negative) the tendency for a metal to react with species in solution will be high, but whether or not the reaction proceeds to any extent will depend on kinetic factors; if $\Delta G > 0$ (positive) the metal is stable and no further consideration need be given to kinetics; if $\Delta G = 0$ the system is at equilibrium and will not proceed in either direction. The relationship between the free energy decrease $-\Delta G$ and the rate of the reaction

$$aA + bB = cC + dD$$

where A is the metal, B is a reactant in solution, and C and D are the reaction (or corrosion) products, is illustrated in Fig. 1.13, which shows that although the thermodynamic tendency is the same for L, M and N the rates are different. This illustration may be exemplified by regarding L as an unprotected steel pipe buried in a corrosive soil, M the pipe protected by a bituminous coating and N the pipe coated and also cathodically protected; in each case the free energy change for the corrosion reaction

$$2Fe + H_2O + \tfrac{3}{2}O_2 \rightarrow Fe_2O_3.H_2O$$

is the same, but the rate at which the reaction proceeds to equilibrium (complete conversion of iron to $Fe_2O_2.H_2O$) varies significantly.

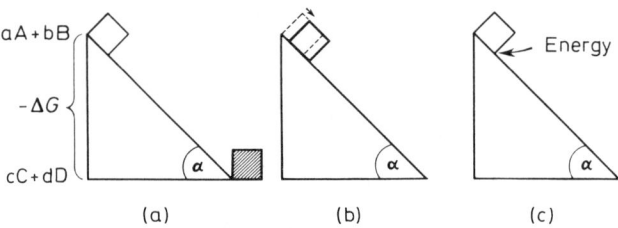

Fig. 1.14 Block on inclined plane illustrating the significance of (a) corrosion, (b) passivity and (c) immunity

Figure 1.14 illustrates the concepts implicit in the terms *corrosion*, *passivity* and *immunity*, which are of particular significance in both the thermodynamic and kinetic approach to corrosion reactions. In this analogy the metal A is in the form of a cube that rests on the top of an inclined plane and is in contact with the environment B. The slope of the plane α may be regarded as a measure of the tendency of the metal to slide down the plane with progressive formation of corrosion product ($-\Delta G$) until equilibrium is attained at the bottom of the plane when all the metal is converted to corrosion product. The rate at which the block slides down the plane, which in this analogy corresponds to the rate of the reaction, will depend both on α and on the frictional resistance between the metal, or the metal filmed with corrosion product, and the plane. In Fig. 1.14a the frictional resistance of both metal and corrosion product is assumed to be low, and the metal proceeds rapidly down the plane with a correspondingly high corrosion rate. In Fig. 1.14b the corrosion product is assumed to form a continuous adherent film on the metal surface that has a high frictional resistance; it follows that progress of the block down the plane will be decreased to a very low value as soon as a thin film of oxide is formed, and although some corrosion will occur it will not affect the bulk properties of the metal block significantly.

In this analogy Fig. 1.14a represents a corrosion rate that could not be tolerated in practice, whereas Fig. 1.14b represents the phenomenon of *passivity* in which the rate of corrosion is controlled at an acceptable level by virtue of the kinetic barrier of corrosion product. It should be noted, however, that the metal in the passive state is still thermodynamically unstable, and any environmental factor that results in the removal of the oxide film, either locally or generally, will result in rapid corrosion; a characteristic of passive metals is that in certain environmental situations localised breakdown of the oxide film occurs with consequent pitting of the exposed metal, the filmed surface surrounding the pit remaining unattacked. Finally, no corrosion will occur if the block can be maintained at the top of the plane by a continuous supply of energy of a magnitude that is equal to or greater than the energy liberated when the reaction proceeds spontaneously. The analogy in Fig. 1.14c represents the phenomenon of *immunity* that forms the basis of the important practical method of corrosion control known as *cathodic protection* in which the potential of the metal is depressed sufficiently to prevent it oxidising (*see* Chapter 10).

Thermodynamic Approach to Corrosion Reactions*

The standard electrode potentials[1] $E^\ominus$, or the standard chemical potentials $\mu^\ominus$, may be used to calculate the free energy decrease $-\Delta G$ and the equilibrium constant K of a corrosion reaction (*see* Appendix 9.1A). Any corrosion reaction in aqueous solution must involve oxidation of the metal and reduction of a species in solution (an electron acceptor) with consequent electron transfer between the two reactants. Thus the corrosion of zinc ($E^\ominus_{Zn^{2+}/Zn} = -0.76$ V) in a reducing acid of pH = 4 ($a_{H^+} = 10^{-4}$) may be represented by the reaction:

$$Zn + 2H^+ (a_{H^+} = 10^{-4}) = Zn^{2+} + H_2(p_{H_2} = 1) \quad \ldots(1.4)$$

which consists of the two half reactions

$$Zn = Zn^{2+}(aq.) + 2e \,(\text{oxidation of } Zn \to Zn^{2+}) \quad \ldots(1.5)$$

and

$$2H^+ + 2e = H_2 \,(\text{reduction of } H^+ \to H_2) \quad \ldots(1.6)$$

Since $\Delta G = -zFE$ and since 1 F (F is a Faraday) $\approx 96\,500$ C

$$\Delta G_{Zn^{2+}/Zn} = -2 \times 96\,500 \times -(-0.76)\,\text{J mol}^{-1}\,Zn \quad \ldots(1.7)$$

and

$$\Delta G_{H^+/H_2} = -2 \times 96\,500 \times (-4 \times 0.59)\,\text{J mol}^{-1}\,H_2 \quad \ldots(1.8)$$

and for the overall reaction

$$\Delta G_{\text{reaction}} = -2 \times 96\,500 \times (0.76 - 0.24) = -100.3\,\text{kJ mol}^{-1}\,Zn$$

which shows that the reaction of zinc with hydrogen ions at $a_{H^+} = 10^{-4}$ g ion/l is spontaneous and proceeds in the direction as written in equation 1.4. However, in spite of the large free-energy decrease for equation 1.4 the rate of corrosion, which in this particular system is controlled by the rate of hydrogen evolution (equation 1.6), is extremely slow.

(At this point it should be noted that the free energies of half reactions can be added algebraically to evaluate $\Delta G_{\text{reaction}}$ and similar considerations

*Thermodynamic data are given in Chapter 21, Tables 21.5 and 21.6.

apply to the evaluation of E_{reaction}, providing the number of electrons z in the two half reactions and in the complete reaction are the same (*see* example just given). However, should z differ, then it is essential to sum the *free energies* in order to evaluate E_{reaction}, and this can be illustrated by calculating the value of $E^{\ominus}_{\text{Cu}^{2+}/\text{Cu}^+}$ from $E^{\ominus}_{\text{Cu}^{2+}/\text{Cu}} = 0 \cdot 346$ V, and $E^{\ominus}_{\text{Cu}^+/\text{Cu}} = 0 \cdot 522$ V:

$$\text{Cu}^{2+} + 2e = \text{Cu}, \; -\Delta G^{\ominus} = 2\,\text{F} \times 0 \cdot 346 = 0 \cdot 792\,\text{F} \quad \ldots(1.9)$$

$$\text{Cu}^+ + e = \text{Cu}, \; -\Delta G^{\ominus} = 1\,\text{F} \times 0 \cdot 522 = 0 \cdot 522\,\text{F} \quad \ldots(1.10)$$

Subtracting equation 1.10 from equation 1.9

$$\text{Cu}^{2+} + e = \text{Cu}^+, \; -\Delta G^{\ominus} = (0 \cdot 792\,\text{F} - 0 \cdot 522\,\text{F}) = 0 \cdot 170\,\text{F}$$

$$\therefore \; 1\,\text{F}\,E^{\ominus}_{\text{Cu}^{2+}/\text{Cu}^+} = 1\,\text{F} \times 0 \cdot 170, \text{ and } E^{\ominus}_{\text{Cu}^{2+}/\text{Cu}^+} = 0 \cdot 17\,\text{V}$$

Direct subtraction of the standard electrode potential would have given the incorrect value $E^{\ominus}_{\text{Cu}^{2+}/\text{Cu}} = 0 \cdot 346 - 0 \cdot 522 = -0 \cdot 176$ V.)

In this example of the corrosion of zinc in a reducing acid of pH = 4, the corrosion product is Zn^{2+} (aq.), but at higher pHs the thermodynamically stable phase will be Zn(OH)_2 and the equilibrium activity of Zn^{2+} will be governed by the solubility product of Zn(OH)_2 and the pH of the solution; at still higher pHs ZnO_2^- anions will become the stable phase and both Zn^{2+} and Zn(OH)_2 will become unstable. However, a similar thermodynamic approach may be adopted to that shown in this example.

Table 1.7 Half reactions involving the oxidation of a metal in aqueous solutions

Equation number	Half reaction
1	$M \rightarrow M^{z+}$(aq.) $+ ze$ Oxidation to acquo cations
2(a)	$M + z\text{H}_2\text{O} \rightarrow M(\text{OH})_z + z\text{H}^+ + ze$ ⎫ Oxidation to
or 2(b)	$M + z\text{OH}^- \rightarrow M(\text{OH})_z + ze$ ⎭ metal hydroxide
3(a)	$M + z\text{H}_2\text{O} \rightarrow MO_z^{z-}$(aq.) $+ 2z\text{H}^+ + ze$ ⎫ Oxidation to
or 3(b)	$M + z\text{OH}^- \rightarrow MO_z^{z-}$(aq.) $+ z\text{H}^+ + ze$ ⎭ aquo anions

Table 1.7 shows typical half reactions for the oxidation of a metal M in aqueous solutions with the formation of aquo cations, solid hydroxides or aquo anions[2]. The equilibrium potential for each half reaction can be evaluated from the chemical potentials of the species involved (*see* Appendix 9.1A) and it should be noted that there is no difference thermodynamically between equations 2(*a*) and 2(*b*) nor between 3(*a*) and 3(*b*) when account is taken of the chemical potentials of the different species involved.

In view of the importance of the hydronium ion, H_3O^+, and dissolved oxygen as electron acceptors in corrosion reactions, some values of the redox potentials E and chemical potentials μ for the equilibria

$$2\text{H}^+ + 2e = \text{H}_2, \; E = 0 \cdot 00 - 0 \cdot 059\text{pH} - 0 \cdot 030 \log p_{\text{H}_2} \quad \ldots(1.11)$$

and $\tfrac{1}{2}\text{O}_2 + \text{H}_2\text{O} + 2e = 2\text{OH}^-, \; E = 1 \cdot 23 - 0 \cdot 059\text{pH} + 0 \cdot 15 \log p_{\text{O}_2}$
$$\ldots(1.12)$$

at 25°C are given in Table 1.8. The following should be noted:

1. Dissolved oxygen has a higher redox potential than the hydrogen ion at all values of pH, i.e. it is a more powerful oxidant.

Table 1.8 Potentials of the $H^+/\frac{1}{2}H_2$ and O_2/OH^- electrodes ($p_{H_2} = p_{O_2} = 1$ atm*)

Equilibrium	Activity	pH	E (V)	$\mu = 2 \times 96\,500\,E$ kJ
$2H^+ + 2e = H_2$	$a_{H^+} = 1$	0	0·00	0·00
(see equation 1.11)	$a_{H^+} = 10^{-7}$	7	−0·414	−79·89
	$a_{H^+} = 10^{-14}$	14	−0·828	−159·8
$\frac{1}{2}O_2 + H_2O + 2e = 2OH^-$	$a_{OH^-} = 1$	14	0·401	77·4
(see equation 1.12)	$a_{OH^-} = 10^{-7}$	7	0·815	157·3
	$a_{OH^-} = 10^{-14}$	0	1·229	237·5

* 1 atm = 1·013 × 10⁵ Pa (Pascal).

2. In both equilibria E decreases with increase in pH.
3. A plot of E vs. pH for each equilibrium gives a linear curve of slope 0·059 (see curves ℓ and m in Fig. 1.15 (bottom)).
4. A decrease in p_{O_2} will displace curve m in Fig. 1.15 (bottom) in the negative direction, whereas a decrease in p_{H_2} will displace curve ℓ in the positive direction (equations 1.11 and 1.12).

Table 1.9 shows the values of $\Delta G^\ominus$, K and the equilibrium activities of metal cations and pressures (fugacities) of hydrogen gas for the reaction of typical metals with a reducing acid ($a_{H^+} = 1$) to form metal cations M^{z+} (aq.). Equilibria such as Au^{3+}/Au, Ag^+/Ag and Cu^{2+}/Cu have redox potentials $>0\cdot0$ V, and the metal (the reduced species of the equilibrium) may be regarded as being thermodynamically stable in a reducing acid, since the equilibrium activities of metal cations are negligible; thus in these systems no further consideration of the rates of the reaction are necessary. On the other hand, the metals M of M^{z+}/M equilibria that are negative to the H^+/H_2 equilibrium are thermodynamically unstable, and the high values of the equilibrium activities of M^{z+} indicate that the reaction will proceed to completion, although at a rate that cannot be predicted from thermodynamic consideration alone. Thus pure zinc corrodes slowly in hydrochloric acid, more quickly if the zinc contains impurities of lead and more quickly still when the impurities are copper; in each of these three cases the thermodynamic tendency and the final position of equilibrium (see Table 1.9) are the same, although the rates at which equilibrium is achieved are markedly different. (Note that each equilibrium must be written so that $-\Delta G^\ominus > 0$ (positive), i.e. in the direction in which it proceeds spontaneously so that $E^\ominus_{\text{reaction}}$ is always positive although $E^\ominus_{M^{z+}/M}$ may be either positive or negative.)

Table 1.9 Thermodynamics of the reaction of metals with acid solutions ($a_{H^+} = 1$, $E^\ominus_{H^+/\frac{1}{2}H_2} = 0\cdot00$ V)

Reaction	$E^\ominus_{M^{z+}/M}$ (V)	$E^\ominus_{\text{reaction}}$ (V)	$-\Delta G^\ominus$ (kJ)	K	$a_{M^{z+}}$ (when $p_{H_2} = 1$)	p_{H_2} (when $a_{M^{z+}} = 1$)
$\frac{2}{3}Au^{3+} + H_2 = 2H^+ + \frac{2}{3}Au$	1·50	1·50	289·4	$7\cdot2 \times 10^{50}$	$8\cdot5 \times 10^{-77}$	$1\cdot4 \times 10^{-51}$
$2Ag^+ + H_2 = 2H^+ + 2Ag$	0·79	0·79	152·4	$6\cdot0 \times 10^{26}$	$4\cdot1 \times 10^{-14}$	$1\cdot7 \times 10^{-27}$
$Cu^{2+} + H_2 = 2H^+ + Cu$	0·337	0·337	65·0	$2\cdot6 \times 10^{11}$	$3\cdot8 \times 10^{-12}$	$3\cdot8 \times 10^{-12}$
$Sn + 2H^+ = H_2 + Sn^{2+}$	−0·136	0·136	26·2	$4\cdot0 \times 10^4$	$4\cdot0 \times 10^4$	$4\cdot0 \times 10^4$
$Fe + 2H^+ = H_2 + Fe^{2+}$	−0·440	0·440	84·9	$8\cdot1 \times 10^{14}$	$8\cdot1 \times 10^{14}$	$8\cdot1 \times 10^{14}$
$Zn + 2H^+ = H_2 + Zn^{2+}$	−0·763	0·763	147·2	$7\cdot2 \times 10^{25}$	$7\cdot2 \times 10^{25}$	$7\cdot2 \times 10^{25}$
$\frac{2}{3}Al + 2H^+ = H_2 + \frac{2}{3}Al^{3+}$	−1·66	1·66	320·3	$1\cdot9 \times 10^{56}$	$2\cdot6 \times 10^{84}$	$1\cdot9 \times 10^{56}$
$Mg + 2H^+ = H_2 + Mg^{2+}$	−2·37	2·37	457·3	$2\cdot2 \times 10^{80}$	$2\cdot2 \times 10^{80}$	$2\cdot2 \times 10^{80}$

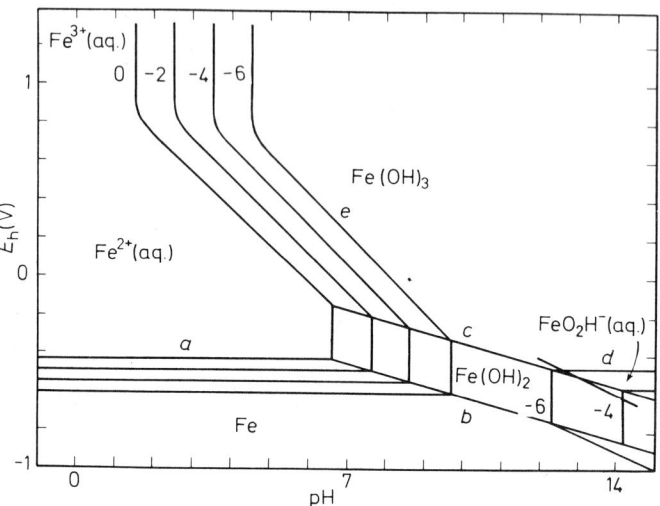

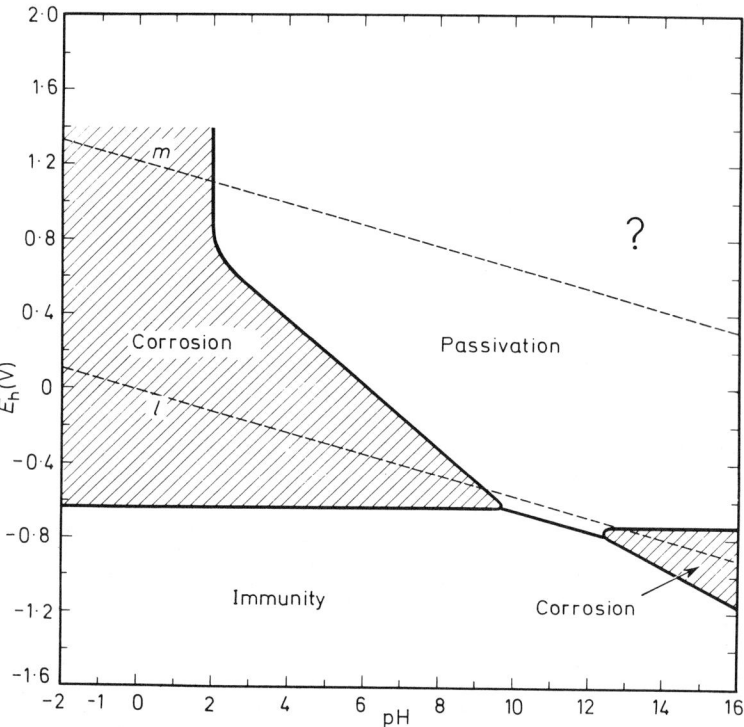

Fig. 1.15(top) Equilibrium potential–pH diagram for the Fe–H$_2$O system showing the zones of stability of cations, anions and solid hydroxides (after Deltombe and Pourbaix[7]) and (bottom) simplified version showing zones of corrosion, immunity and passivity (curve l is the H$_2$O/H$_2$ equilibrium at $p_{H_2} = 1$ and curve m is the O$_2$/H$_2$O equilibrium at $p_{O_2} = 1$)

Table 1.10 Thermodynamics of reactions of noble metals with acids and alkaline solutions containing oxygen ($p_{O_2} = 1$ atm)

Reaction	pH	$E^{\ominus}_{M^{z+}/M}$ (V)	$E^{\ominus}_{reaction}$ (V)	$-\Delta G^{\ominus}$ (kJ mol^{-1})	K	$a_{M^{z+}}$ (when $p_{O_2}=1$)	p_{O_2} (when $a_{M^{z+}}=1$)
$2Au^+ + H_2O = 2Au + \frac{1}{2}O_2 + 2H^+$	0	1·68	0·45	86·8	$1·7 \times 10^{15}$	$2·4 \times 10^{-8}$	$2·9 \times 10^{30}$
$2Ag + \frac{1}{2}O_2 + 2H^+ = 2Ag^+ + H_2O$	0	0·79	0·44	84·9	$8·1 \times 10^{14}$	$2·8 \times 10^{7}$	$1·5 \times 10^{-30}$
$2Cu + \frac{1}{2}O_2 + 2H^+ = 2Cu^+ + H_2O$	0	0·521	0·71	137·0	$1·1 \times 10^{20}$	$1·0 \times 10^{10}$	$8·2 \times 10^{-41}$
$2Au^+ + 2OH^- = \frac{1}{2}O_2 + H_2O + 2Au$	14	1·68	1·28	247·0	$2·4 \times 10^{43}$	$2·0 \times 10^{-22}$	$5·8 \times 10^{86}$
$2Ag^+ + 2OH^- = \frac{1}{2}O_2 + H_2O + 2Ag$	14	0·79	0·39	75·2	$1·6 \times 10^{13}$	$2·5 \times 10^{-7}$	$2·6 \times 10^{26}$
$2Cu^+ + 2OH^- = \frac{1}{2}O_2 + H_2O + 2Cu$	14	0·521	0·12	23·1	$1·2 \times 10^{4}$	$1·0 \times 10^{-2}$	$1·4 \times 10^{8}$

Table 1.10 shows the effect of raising the redox potential of the acid by means of dissolved oxygen ($p_{O_2} = 1$ atm), and it can be seen that the noble metals (metals more positive than 0·0 V) silver and copper are now thermodynamically unstable, and due consideration must be given therefore to the factors that control the rates of the reaction. Oxygen has been selected in this example because of its omnipresence, but similar considerations would apply to other oxidants of high redox potential, e.g. nitric acid, ferric salts, hydrogen peroxide, perchlorate, persulphate, etc. In this connection it is of relevance to draw attention to the compositions and redox potentials of the various etching solutions used for specific metals and alloys in metallography, which reveal the structure by a controlled corrosion reaction during which different structural features corrode at different rates.

These examples show how the tendency of a corrosion reaction to proceed in the direction of oxidation of metal to M^{z+}(aq.) may be increased by increasing the redox potential of the solution, i.e. by increasing $-\Delta G$ of the reaction. In the data given in Tables 1.9 and 1.10 it has been assumed that the metal is oxidised solely to M^{z+}(aq.) and that no complex ions have formed, a situation that is more rare than usual. Thus copper is stable in dilute hydrochloric acid (Table 1.9) but will corrode in the hot concentrated acid owing to the formation of $CuCl_2^-$ complexes; it will also form complexes with a variety of chemicals, including NH_3, CN^-, etc. Ferrous ions form complexes with ammonium salts, sodium ethylene diamine tetraacetates or concentrated solutions of sodium polyphosphates[3], and both Fe^{2+} and Fe^{3+} form complexes in chloride solutions. Table 1.10 also provides thermodynamic data for the reaction of silver, gold and copper with oxygenated alkaline solutions, and from the low equilibrium activity of Ag^+ and Au^+ it can be assumed that these metals are stable in that environment. However, in the presence of CN^- anions, cyano-complexes are formed and there is an equilibrium between M^+ and $M(CN)_2^-$ that is defined by the instability constant (Table 1.11)[4], and although the a_{M^+} (which is dependent upon the redox potential and pH of the solution and is unaffected by CN^-) is very low, the $a_{M(CN)_2^-}$ is significantly high. Thus addition of CN^- to an oxygenated sodium hydroxide solution will result in the corrosion of gold, and this reaction is utilised in extractive metallurgy for the leaching of gold from its ores by means of an oxygenated solution of sodium cyanide.

The data given in Tables 1.9 and 1.10 have been based on the assumption that metal cations are the sole species formed, but at higher pH values oxides, hydrated oxides or hydroxides may be formed, and the relevant half reactions will be of the form shown in equations 2(a) and 2(b) (Table 1.7). In these circumstances the $a_{M^{z+}}$ will be governed by the solubility product of the solid compound and the pH of the solution. At higher pH values the solid compound may become unstable with respect to metal anions (equations 3(a) and 3(b), Table 1.7), and metals like aluminium, zinc, tin and lead, which form amphoteric oxides, corrode in alkaline solutions. It is evident, therefore, that the equilibrium between a metal and an aqueous solution is far more complex than that illustrated in Tables 1.9 and 1.10. Nevertheless, as will be discussed subsequently, a similar thermodynamic approach is possible.

Finally, it is necessary to observe that the values of activities and fugacities calculated are thermodynamic quantities that cannot always be realised in practice, e.g. very high activities of metal ions cannot be attained because of solubility consideration and very low activities have no physical significance.

Table 1.11 Equilibrium between M^+ and $M(CN)_2^-$ in cyanide solutions ($a_{OH^-} = 1$ (pH = 14); $a_{CN^-} = 1$; $p_{O_2} = 1$)

Equilibrium	Instability constant	a_{M^+}	$a_{M(CN)_2^-}$
$Au(CN)_2^- = Au^+ + 2CN^-$	10^{-39}	2×10^{-22}	10^{17}
$Ag(CN)_2^- = Ag^+ + 2CN^-$	10^{-20}	$2 \cdot 5 \times 10^{-7}$	10^{13}
$Cu(CN)_2^- = Cu^+ + 2CN^-$	10^{-16}	$9 \cdot 1 \times 10^{-8}$	10^9

Potential–pH Equilibrium Diagrams

Pourbaix and his co-workers[5,6] have calculated the phases at equilibrium for M/H_2O systems at 25°C from the chemical potentials of the species involved in the equilibria, and have expressed the data in the form of *equilibrium diagrams* having pH vs. E, the *equilibrium potential* (vs. S.H.E.) as ordinates. These diagrams, which are analogous to the composition-temperature diagrams of alloy systems, provide a thermodynamic basis for the study of corrosion reactions, although, as emphasised by Pourbaix, their limitations in relation to practical problems must be appreciated. Since these diagrams are referred to throughout this work, some emphasis is given to their significance in this section; reference should be made to the numerous publications of Pourbaix and his co-workers for a fuller account of the subject.

It should be emphasised that potential–pH diagrams can also be constructed from experimental E_p–I curves, where E_p is the *polarised potential* and I the current[6]. These diagrams, which are of more direct practical significance than the equilibrium potential–pH equilibrium diagrams constructed from thermodynamic data, show how a metal in a natural environment (e.g. iron in water of given chloride ion concentration) may give rise to general corrosion, pitting, perfect or imperfect passivity, or to immunity, depending on the pH and potential (*see* Section 1.6, Fig. 1.56).

Construction of Potential–pH Diagrams

Pourbaix has classified the various equilibria that occur in aqueous solution into homogeneous and heterogeneous, and has subdivided them according to whether the equilibria involve electrons and/or hydrogen ions. The general equation for a half reaction is

$$aA + mH^+ + ze^- = bB + cH_2O \qquad \ldots(1.13)$$

which shows that the equilibrium between the reactant A and product B depends on a_{H^+} (pH) and the electrode or redox potential E; if neither hydrogen ions nor electrons are involved then the equilibrium is independent of pH and E. From the Nernst equation (*see* page 20.69), and substituting for $-\log a_{H^+} = $ pH

$$E = E^\ominus - 0\cdot 059 \frac{m}{z}\text{pH} - \frac{0\cdot 059}{z}\log \frac{a_B^b}{a_A^a} \qquad \ldots(1.14)$$

It follows from equation 1.14 that for any constant ratio of a_B^b/a_A^a the E vs. pH relationship will be linear with a slope $-0\cdot 059m/z$, and that when $a_B^b = a_A^a = 1$ the intercept of the curve on the E axis (i.e. pH = 0) will be $E^\ominus$, the standard equilibrium potential, which by definition is the potential when the species involved in the equilibrium are at unit activity.

Pourbaix has evaluated all possible equilibria between a metal M and H_2O (*see* Table 1.7) and has consolidated the data into a single potential–pH diagram, which provides a pictorial summary of the anions and cations (nature and activity) and solid oxides (hydroxides, hydrated oxides and oxides) that are at equilibrium at any given pH and potential; a similar approach has been adopted for certain M–H_2O–X systems where X is a non-metal, e.g. Cl^-, CN^-, CO_2, SO_4^{2-}, PO_4^{3-}, etc. at a defined concentration. These diagrams give the activities of the metal cations and anions at any specified E and pH, and in order to define corrosion in terms of an equilibrium activity, Pourbaix has selected the arbitrary value of 10^{-6} g ion/l, i.e. corrosion of a metal is defined in terms of the pH and potential that give an equilibrium activity of metal cations or anions $> 10^{-6}$ g ion/l; conversely, passivity and immunity are defined in terms of an equilibrium activity of $< 10^{-6}$ g ion/l. (Note that g ion/l is used here because this is the unit used by Pourbaix; in the S.I. the relative activity is dimensionless.)

Fe/H$_2$O system Figure 1.15 (top) is a simplified version[7] of the Fe–H$_2$O potential–pH equilibrium diagram [the region of stability of magnetite (Fe$_3$O$_4$) is not included] and it is instructive to consider some of the more important equilibria involved:

Curve a, $Fe^{2+} + 2e = Fe$; $E = -0\cdot 440 + 0\cdot 030 \log a_{Fe^{2+}}$ $\qquad \ldots(1.15)$

Curve b, $Fe(OH)_2 + 2H^+ + 2e = Fe + 2H_2O$;

$$E = -0\cdot 047 - 0\cdot 059\text{pH} \qquad \ldots(1.16)$$

Curve c, $Fe(OH)_3 + H^+ + e = Fe(OH)_2 + H_2O$;
$$E = 0\cdot271 - 0\cdot059 pH \qquad \ldots(1.17)$$
Curve d, $Fe(OH)_3 + e = FeO_2H^- + H_2O$;
$$E = -0\cdot81 - 0\cdot059 \log a_{FeO_2H^-} \qquad \ldots(1.18)$$
Curve e, $Fe(OH)_3 + 3H^+ + e = Fe^{2+} + 3H_2O$;
$$E = 1\cdot060 - 0\cdot177 pH - 0\cdot059 \log a_{Fe^{2+}} \qquad \ldots(1.19)$$

Since equation 1.15 does not involve H^+ it is pH independent, and variation of $a_{Fe^{2+}}$ will result in a series of curves parallel to the pH axis that extend across the diagram until the pH is sufficiently high to reduce the $a_{Fe^{2+}}$ to $<10^{-6}$ g ion/l by formation of $Fe(OH)_2$. The relevant equilibrium is:

$$Fe^{2+} + 2H_2O = Fe(OH)_2 + 2H^+$$

for which values of $\mu^{\ominus}$ (J) for the species involved are Fe^{2+}(aq.), $-84\,760$; H_2O, $-236\,700$; $Fe(OH)_2$, $-482\,000$, and since

$$\log K = \frac{-\Sigma \nu_i \mu_i^{\ominus}}{2\cdot303\,RT} = \frac{-(\Sigma \nu_i \mu_i^{\ominus}{}_{(products)} - \Sigma \nu_i \mu_i^{\ominus}{}_{(reactants)})}{2\cdot303\,RT} \qquad \ldots(1.20a)$$

and at 25°C

$$\log K = \frac{[-482\,000 - 2(-236\,700) - (-84\,760)]}{2\cdot303 \times 8\cdot315 \times 298\cdot16} = -13\cdot37 \qquad \ldots(1.20b)$$

Since $\quad K = \dfrac{a_{H^+}^2}{a_{Fe^{2+}}}$, then $\log K = \log a_{H^+}^2 - \log a_{Fe^{2+}} \qquad \ldots(1.21)$

and $\qquad \log a_{Fe^{2+}} = 13\cdot27 - 2pH \qquad \ldots(1.22)$

Thus at $a_{Fe^{2+}} = 10^{-6}$ the equilibrium pH will be 9·7 and the $a_{Fe^{2+}}$ will be $<10^{-6}$ at any higher pH value; it follows that the formation of a new solid phase $Fe(OH)_2$ at a sufficiently high pH must limit the zone of corrosion as defined by Pourbaix.

Whereas lowering the potential results in a decrease in $a_{Fe^{2+}}$, the converse applies when the potential is raised. However, this increase in activity is again limited by the formation of a solid phase. Thus curve e of Fig. 1.15 (top) gives the equilibrium between $Fe(OH)_3$ and Fe^{2+} at any predetermined activity of the latter in the range $10^0 - 10^{-6}$. At $a_{Fe^{2+}} = 10^{-6}$ g-ion/l, $E = [1\cdot06 + (-6 \times 0\cdot059)] - 0\cdot177 pH$ which defines the boundary between corrosion and passivity at high potentials (equation 1.19).

At high pH values and low potentials, Fe, Fe^{2+}, Fe^{3+}, $Fe(OH)_2$ and $Fe(OH)_3$, etc. will be thermodynamically unstable with respect to FeO_2H^- and a further limited zone of corrosion will appear on the right-hand side of the diagram.

Significance of Zones in Potential–pH Diagrams

The above outline of the method adopted in the construction of the potential–pH diagram of the $Fe-H_2O$ system serves to illustrate the essentially thermodynamic nature of diagrams of this type, which therefore cannot provide any information on the rates of corrosion processes. However, on the basis of certain assumptions that have no thermodynamic significance, it is possible to separate the Fe–H$_2$O diagram into the following zones [Fig. 1.15 (bottom)]

Corrosion: activity of Fe^{2+}, Fe^{3+}(aq.) or FeO$_2^-$(aq.) $> 10^{-6}$ g ion/l (0·06 p.p.m. of Fe in the case of the cations).

Passivity: Fe(OH)$_2$, Fe$_3$O$_4$ or Fe(OH)$_3$ in equilibrium with metal ions at an activity $< 10^{-6}$ g ion/l.

Immunity: Fe metal in equilibrium with Fe^{2+}(aq.) or FeO$_2$H$^-$(aq.) at an activity $< 10^{-6}$ g ion/l.

It should be noted that Fig. 1.15 (top) is based entirely on thermodynamic data and is therefore correctly described as an equilibrium diagram, since it shows the phases (nature and activity) that exist at equilibrium. However, the concepts implicit in the terms *corrosion*, *immunity* and *passivity* lie outside the realm of thermodynamics, and, for example, passivity involves both thermodynamic and kinetic concepts; it follows that Fig. 1.15 (bottom) cannot be regarded as a true equilibrium diagram, although it is based on one that has been constructed entirely from thermodynamic data.

In Fig. 1.15 (bottom) curves ℓ and m show the potential–pH relationships for the reversible hydrogen and oxygen electrodes at $p_{H_2} = p_{O_2} = 1$ atm respectively. Within the area confined by the curves ℓ and m, H$_2$O is thermodynamically stable and $p_{H_2} < 1$ and $p_{O_2} < 1$; whereas below ℓ and above m, $p_{H_2} > 1$ atm, and $p_{O_2} > 1$ atm, respectively (*see* equations 1.11 and 1.12). Thus the diagram shows the solid phases of iron, the activities of metal ions and the pressures of hydrogen and oxygen gas that are at equilibrium at any given potential and pH when pure iron reacts with pure water.

This can be illustrated by considering the changes that will *tend* to occur when iron with a coating of rust (Fe$_2$O$_3$.H$_2$O) is immersed in oxygenated water at pHs and potentials that correspond with the various zones in the Fe–H$_2$O diagram[6].

Immunity Any Fe$_2$O$_3$ on the surface (or any Fe^{2+} in solution) will be reduced to metal, and $a_{Fe^{2+}} < 10^{-6}$ g ion/l; water will be reduced to hydrogen and $p_{H_2} > 1$ atm; any dissolved oxygen present will be reduced to OH$^-$, and $p_{O_2} \ll 1$ atm.

Corrosion (in acid solutions) At low potentials iron will be oxidised to Fe^{2+} and Fe$_2$O$_3$ reduced to Fe^{2+}, and the $a_{Fe^{2+}}$ will be $> 10^{-6}$ g ion/l; water will be reduced to hydrogen or remain stable, depending upon whether E is below or above curve ℓ. At high potentials iron will be oxidised to Fe^{3+} and Fe$_2$O$_3$ will dissolve to form Fe^{3+} ($E^{\ominus}_{Fe^{3+}/Fe^{2+}} = 0·76$ V); water will be stable or will be oxidised to oxygen, depending upon whether E is below or above curve m, respectively.

Passivation According to Fig. 1.15 (top) all the Fe will be converted to Fe$_2$O$_3$, whilst the rust originally present will be unaffected. According to Fig. 1.15 (bottom) the rust will be unaffected, whilst the iron surface exposed to the solution through pores in the rust will be passivated by a protective film of Fe$_2$O$_3$. Water will be stable except at high potentials where it will be oxidised to O$_2$.

Thus the *tendency* for an electrochemical reaction at a metal/solution interface to proceed in a given direction may be defined in terms of the relative values of the actual electrode potential E (experimentally determined and expressed with reference to the S.H.E.) and the reversible or equilibrium potential E_r (calculated from $E^{\ominus}$ and the activities of the species involved in the equilibrium).

When $E > E_r$ the reaction can only proceed in the direction of oxidation.
When $E < E_r$ the reaction can only proceed in the direction of reduction.
When $E = E_r$ the reaction is at equilibrium.

This can be summarised by the relationship

$$(E - E_r)I \geqslant 0 \qquad \ldots(1.23)$$

in which I, the reaction current, is regarded as positive in the case of oxidation and negative in the case of reduction. The parameter $(E - E_r)$, which may be positive, negative or zero, is termed the *overpotential* or *affinity*, and gives the tendency for the reaction to proceed in the direction of oxidation or reduction or to be at equilibrium, respectively. However, the precise magnitude of I will depend upon kinetic factors, which will be considered subsequently.

Advantages and Limitations of Diagrams

Although the zones of corrosion, immunity and passivity are clearly of fundamental importance in corrosion science it must be emphasised again that they have serious limitations in the solution of practical problems, and can lead to unfortunate misconceptions unless they are interpreted with caution. Nevertheless, Pourbaix and his co-workers, and others, have shown that these diagrams used in conjunction with E–i curves for the systems under consideration can provide diagrams that are of direct practical use to the corrosion engineer. It is therefore relevant to consider the advantages and limitations of the equilibrium potential–pH diagrams.

The M–H_2O diagrams present the equilibria at various pHs and potentials between the metal, metal ions and solid oxides and hydroxides for systems in which the only reactants are metal, water, and hydrogen and hydroxyl ions; a situation that is extremely unlikely to prevail in real solutions that usually contain a variety of electrolytes and non-electrolytes. Thus a solution of pH 1 may be prepared from either hydrochloric, sulphuric, nitric or perchloric acids, and in each case a different anion will be introduced into the solution with the consequent possibility of the formation of species other than those predicted in the M–H_2O system. In general, anions that form soluble complexes will tend to extend the zones of corrosion, whereas anions that form insoluble compounds will tend to extend the zone of passivity. However, provided the relevant thermodynamic data are available, the effect of these anions can be incorporated into the diagram, and diagrams of the type M–H_2O–X are available in Cebelcor reports and in the published literature.

The effect of anions on the zones of corrosion and passivation can be exemplified by a comparison of the Pb–H_2O and Pb–H_2O–SO_4^{2-} equilibrium diagrams (*see* Section 4.3, Figs. 4.13 and 4.14) and it can be seen that in the presence of SO_4^{2-} the corrosion zone corresponding with stability of

Pb^{2+} is completely replaced by $PbSO_4$ so that passivation is possible in the acid region owing to the thermodynamic stability of $PbSO_4$.

Similar considerations apply to the potential E which can be varied by means of an auxiliary electrode and an external source of e.m.f., or by varying the redox potential of the solution, and in the case of the latter a given redox potential may be achieved by using different oxidants. Thus at pH 7 and $E = -0.44$ V, iron will be in the zone of corrosion ($a_{Fe^{2+}} = 1$), and the potential could be raised into the passive region by either dissolved oxygen, potassium chromate or potassium perchlorate. However, the effects produced will depend upon a variety of factors, and whereas passivation can be achieved if chromate is present in sufficient concentration, it may cause pitting at lower concentrations. Perchlorate will tend to cause pitting, and dissolved oxygen can result in localised attack and will passivate iron only if it is brought rapidly and simultaneously to all parts of the metal surface.

A further serious limitation is that diagrams evaluated from thermodynamic data at 25°C have little relevance in high-temperature aqueous corrosion, but it is now possible to construct diagrams[8,9] that are applicable at elevated temperatures from data obtained at 25° (*see* Section 2.1).

Pourbaix[6] has studied the behaviour of iron in city water (Brussels) at

Table 1.12 State environments and iron in Brussels water; see also Fig. 1.16 (after Pourbaix[6])

Experiment	Sample number	Solution		pH	E_H (V)	State of metal*	Gas
a	1	H_2O distilled		8·1	−0·486	●	—
	2	NaCl	1 g/l	6·9	−0·445	●	—
	3	H_2SO_4	1 g/l	2·3	−0·351	●	H_2
	4	$NaHSO_4$	1 g/l	6·4	−0·372	●	—
	5	NaOH	1 g/l	11·2	+0·026	○	—
	6	K_2CrO_4	1 g/l	8·5	+0·235	○	—
	7	K_2CrO_4 + NaCl	1 g/l	8·6	−0·200	◐	—
	8	$KMnO_4$	0·2 g/l	6·7	−0·460	●	—
	9	$KMnO_4$	1 g/l	7·1	+0·900	○	—
	10	H_2O_2	0·3 g/l	5·7	−0·200	●	—
	11	H_2O_2	3·0 g/l	3·4	+0·720	○	O_2
	12	Brussels city water		7·0	−0·450	●	—
b	13	NaOH 40 g/l degassed		13·7	−0·810	●	H_2
c	14	city water-iron-copper		7·5	−0·445	●	—
	15	city water-iron-zinc		7·5	−0·690	○	H_2
	16	city water-iron-magnesium		7·5	−0·910	○	H_2
	17	city water-iron-platinum		7·5	−0·444	●	—
c′	14′	city water-iron-copper		7·8	−0·385	●	—
	15′	city water-iron-zinc		7·7	−0·690	○	H_2
	16′	city water-iron-magnesium		8·7	−0·495	○	H_2
	17′	city water-iron-platinum		—	—	●	—
d	18	$NaHCO_3$ 0·1 M Pole −		8·4	−0·860	●	H_2
	19	$NaHCO_3$ 0·1 M Pole +		8·4	−0·350	●	—
	20	$NaHCO_3$ 0·1 M Pole −		8·4	−0·885	○	H_2
	21	$NaHCO_3$ 0·1 M Pole −		8·4	+1·380	○	O_2
e	22	$NaHCO_3$ 0·1 M Pole −		8·4	−0·500	●	—
	23	$NaHCO_3$ 0·1 M Pole +		8·4	+1·550	○	O_2
	24	$NaHCO_3$ 0·1 M Pole −		8·4	−1·000	○	H_2
	25	$NaHCO_3$ 0·1 M Pole +		8·4	+1·550	○	O_2

* ● General corrosion; ◐ local corrosion; ○ absence of corrosion.

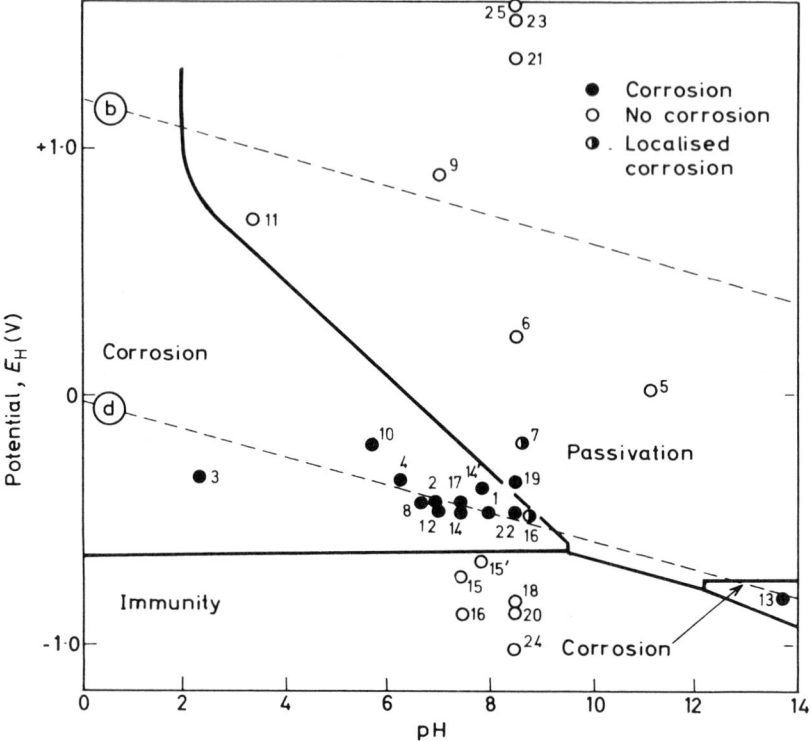

Fig. 1.16 Potential-pH diagram for the Fe-H$_2$O system in which results obtained for the behaviour of iron in Brussels water have been inserted (*see* Table 1.12) (after Pourbaix[6])

various pHs and potentials (Table 1.12); the study of the former was effected by adding acids and alkalis and the latter by applying an external e.m.f., by coupling the iron to either more positive (Cu, Pt) or more negative metals (e.g. Zn, Mg), or by adding oxidants (e.g. K$_2$CrO$_4$, KMnO$_4$, H$_2$O$_2$). The corrosion rate of the iron has been determined and results have been inserted in the potential-pH diagram for Fe-H$_2$O (Fig. 1.16), and it can be seen that in this particular water there is good agreement between the predictions of the diagram and the corrosion behaviour of the iron. However, it does not follow that this correlation would necessarily apply to all fresh waters or to sea-water.

Zones of Corrosion

The Zn-H$_2$O (Fig. 1.17) diagram[10] shows that extensive corrosion zones exist at both low and high pH values (compare the very restricted corrosion zone in the Fe-H$_2$O diagram at high pHs); similar zones in the region of low and high pH are obtained with other amphoteric metals such as aluminium, lead and tin. The diagram for Zn-H$_2$O predicts with some accuracy the behaviour of the metal in practice, where it has been established that zinc corrodes rapidly outside the range pH 6–12·5 but is passive within

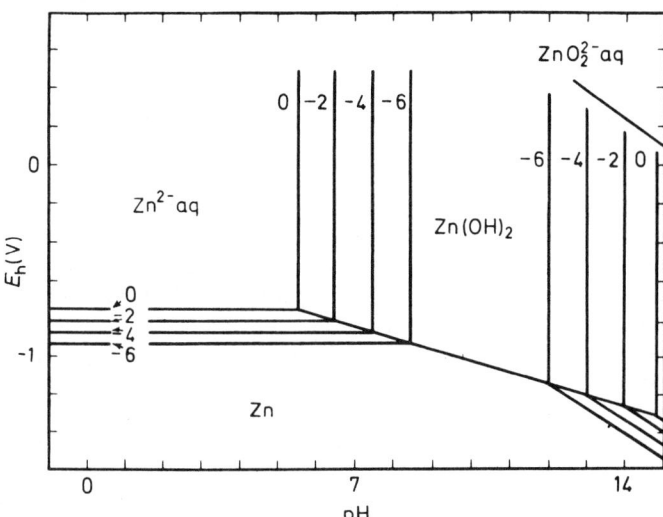

Fig. 1.17 Simplified potential-pH diagram for the Zn/H$_2$O system (after Delahay, Pourbaix and Van Rysselberghe[10])

it. On the other hand, iron in sodium hydroxide at 25°C at potentials and pHs corresponding with the triangle of corrosion on the right-hand side of Fig. 1.15 [FeO$_2$H$^-$(aq.) stable] shows little evidence of attack, although the presence of this zone does explain the phenomenon of caustic cracking. However, it should be noted that caustic cracking normally occurs in alkaline waters at elevated temperatures and that temperature will have a marked effect on both the thermodynamics and kinetics of the reaction (*see* Section 2.1).

Finally, it should be noted that although the arbitrary activity of 10^{-6} g ion/l represents a very low concentration of metal ions it could be significant in certain circumstances, e.g. lead at that concentration would render potable water toxic. It should also be noted that if the equilibrium is continuously disturbed, e.g. by a flowing solution, significant amounts of metal will corrode even at an equilibrium activity of 10^{-6} g ion/l.

Zone of Immunity

The region of immunity [Fig. 1.15 (bottom)] illustrates how corrosion may be controlled by lowering the potential of the metal, and this zone provides the thermodynamic explanation of the important practical method of cathodic protection (Section 11.1). In the case of iron in near-neutral solutions the potential $E = -0\cdot62$ V for immunity corresponds approximately with the practical criterion adopted for cathodically protecting the metal in most environments, i.e. $-0\cdot52$ to $-0\cdot62$ V (*vs.* S.H.E.). It should be observed, however, that the diagram provides no information on the rate of charge transfer (the current) required to depress the potential into the region of immunity, which is the same ($< -0\cdot62$ V) at all values of pH below 9·8. Consideration of curve ℓ for the H$_2$/H$_2$O equilibrium shows that as the pH decreases the thermodynamic tendency of water to become reduced to hydrogen increases, and although theoretically cathodic protection is feasible in the acid region it would be economically impracticable, since hydrogen evolution at a very high rate would occur at the potentials required to achieve immunity. Thus cathodic protection is normally confined to near-neutral solutions.

In the case of the Al–H_2O diagram system (Section 4.1, Fig. 4.41) immunity in the near-neutral region can be achieved only at potentials < -1.82 V, which cannot be attained in practice owing to the hydrogen evolution reaction, which is the thermodynamically preferred process. However, owing to the presence of a surface oxide film the potential of aluminium in practical environments is far more positive than the reversible potential, e.g. the corrosion potential in sea-water is -0.55 V compared to $E^{\ominus}_{Al^{3+}/Al} = -1.7$ V, and cathodic protection may be achieved in practice by making the potential 100 mV more negative than the corrosion potential. This is because aluminium in neutral chloride-containing environments corrodes by pitting, and the criterion of cathodic protection is thus the critical pitting potential (see Sections 1.5 and 1.6) and not the zone of immunity of the potential–pH diagram, and similar considerations apply to the cathodic protection of stainless steels.

The Al–H_2O diagram does show, however, the danger that may arise due to an increase in pH when the metal is cathodically protected in near-neutral solutions; indeed, the possibility of alkaline corrosion has seriously limited the use of cathodic protection for aluminium structures.

Zone of Passivity

Although thermodynamics can predict the region of pH and potential in which solid oxides, hydroxides and other compounds are stable, it can provide no other information; thus on the basis of these considerations alone a metal in the passive region should be completely converted to a solid compound by reacting with water with a consequent loss of properties.

Implicit in the concept of passivity is the assumption that the solid compound forms a kinetic barrier between the reactants so that further interaction becomes very slow. Whether this occurs in practice will depend on the position of formation of the oxide (an oxide produced by the sequence $M \rightarrow M^{2+}$(aq.) $\rightarrow M_2O_3 \cdot H_2O$ is likely to precipitate away from the metal surface owing to the mobility of the M^{2+}(aq.) ion and to be non-protective, and an oxide produced directly, e.g. $M \rightarrow M_xO_y$ is likely to form on the metal surface and to be protective), the adhesion of the oxide to the metal, the solubility of the oxide, its cohesion, crystal form, etc. Thus iron in a neutral chloride solution maintained in the region of passivation by dissolved oxygen will corrode owing to the fact that the hydrated oxide $Fe_2O_3 \cdot H_2O$ precipitates away from the metal surface and is therefore non-protective. Similarly, metals such as magnesium, aluminium and zinc, which according to the relevant potential–pH diagram are all passive in near-neutral solutions at elevated potentials, can be used as sacrificial anodes in sea-water, since the presence of the chloride ion precludes passivation; in fact in this particular application it is essential to ensure that the metal does not passivate, and in the case of aluminium and zinc, additions of mercury may be used to prevent the formation of a protective film thus facilitating uniform corrosion (*see* Section 10.2).

A number of metals and alloys can be passivated in the acid region at elevated potentials, although this phenomenon is not evident from the pH–potential diagrams, which give the impression that the metal will corrode. Thus iron is passivated in fuming nitric acid, and aluminium in nitric acid at concentrations $> \approx 70\%$; iron, nickel and cobalt can be passivated in sulphuric acid by raising the potential by applying an external source of e.m.f. The reason for this behaviour is that although the passive zone is based on the thermodynamic stability of solid compounds it is possible for these compounds to exist as metastable phases outside the regions defined by thermodynamic data. Under these circumstances the rate of corrosion of the metal will be controlled by the rate of transport of metal cations through the film and by the dissolution of the oxide in the solution (*see* Section 1.4).

The fact that oxides can exist as metastable phases is illustrated by the Ni–H$_2$O diagram (Fig. 1.18) in which the curves for the various oxides of nickel have been extrapolated into the acid region of Ni^{2+} stability, and this diagram emphasises the fact that nickel can be passivated outside the region of thermodynamic stability of the oxides[11].

The converse situation occurs when complexants are present, and this can be exemplified by the Cu–H$_2$O–NH$_3$ system (Fig. 1.19)[12] in which the zones

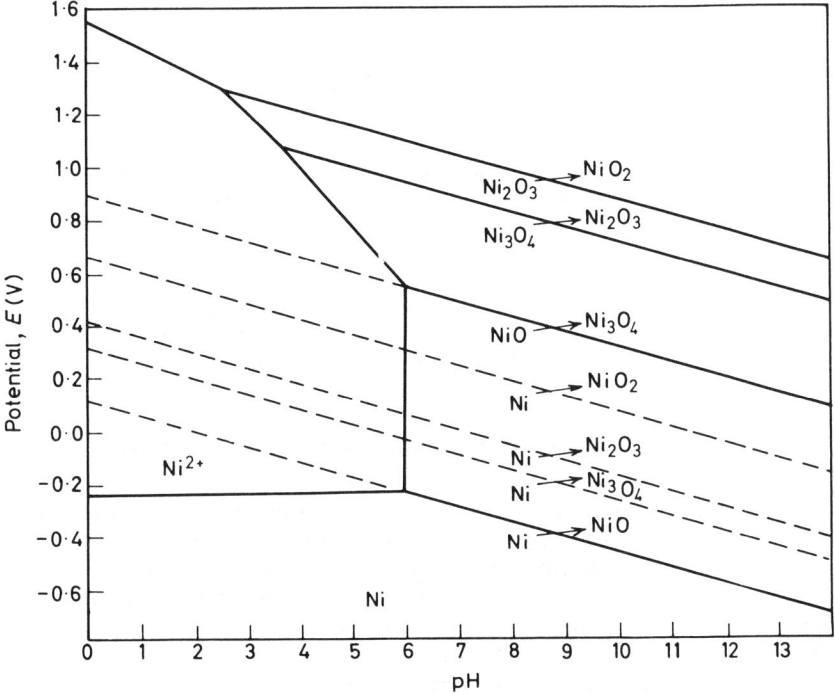

Fig. 1.18 Modified potential–pH diagram for the Ni–H$_2$O system; the curves showing the stability of the nickel oxides have been extrapolated into the acid region to indicate the formation of metastable oxides (after De Gromoboy and Shreir[11])

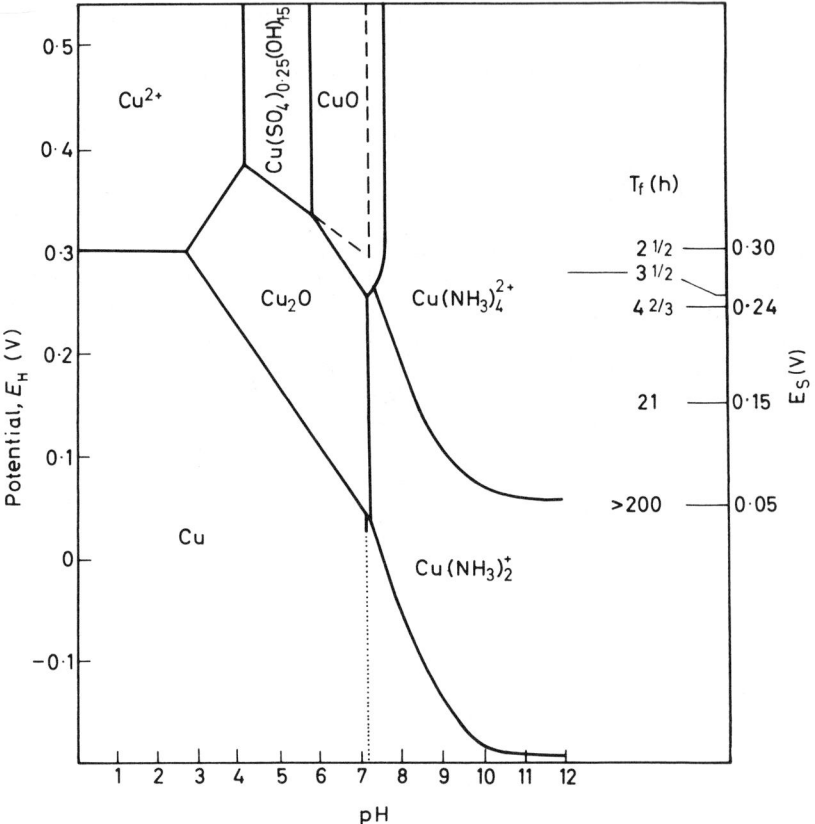

Fig. 1.19 Potential–pH diagram for copper in solutions containing Cu^{2+} and $(NH_4)_2SO_4$ (after Mattson[12]) with superimposed times to fracture T_f of direct-loaded α-brass wires held at various potentials in the solution of pH 7·2; the specimen without external polarisation had $T_f = 3\frac{1}{2}$ h (after Hoar and Booker[13])

of passivation due to the stability of Cu_2O and CuO are confined to a narrow region of pH, whereas in the $Cu-H_2O$ water system passivation can be achieved at all pHs $> \approx 6\cdot 8$ (Section 4.2, Fig. 4.10).

Importance of Potential–pH Diagrams in Corrosion Science

Since the remainder of this section will be confined to a consideration of rates of corrosion reactions, it is appropriate to conclude this review of potential–pH diagrams with an assessment of their significance in corrosion science. In this connection, it is relevant to consider the composition–temperature equilibrium phase diagrams of alloy systems, which provide the foundation for the study of the structure of metals and alloys, although it is recognised that these diagrams have serious limitations owing to the fact that many commercial alloys are not in a state of equilibrium. This can be exemplified by the Fe–C equilibrium diagram, in which ferrite (α-Fe) and graphite are the phases at equilibrium; however, a variety of metastable phases such as cementite and martensite can be formed by suitable heat treatments. The Fe-graphite equilibrium diagram predicts that a liquid alloy containing 3% C when cooled to ambient temperatures will consist of ferrite and graphite, but if the alloy is cooled rapidly graphite formation is suppressed, resulting in a hard, brittle white iron composed of cementite and ferrite. Subsequent annealing at 870°C will cause the system to tend to the equilibrium state, and the cementite will decompose slowly with the formation of the stable phase graphite. There are numerous examples that could be quoted to show that commercial alloys contain metastable phases that do not conform with the predictions of the relevant equilibrium diagram, but the usefulness of these diagrams is not disputed. However, in the case of steels the kinetics of the isothermal transformation of austenite can be presented in the form of transformation-time-temperature (TTT) diagrams which show the phases that form at different temperatures after a given time (the axes are transformation temperature *vs.* the logarithm of the time of transformation). These TTT diagrams (*see* Section 20.4) may be regarded as being analogous to the E-log I diagrams, since the former provide information on the isothermal rate of mass transfer, whilst the latter provide information on both charge transfer and mass transfer.

Thus the potential-pH diagrams and the E-I diagrams may be regarded as complementary in the study of corrosion phenomena and in the solution of corrosion problems.

A survey of the literature (*see* pages 1.114 to 1.117) shows that numerous workers in the field of corrosion have used potential-pH diagrams in order to throw more light on the mechanism of a corrosion process. As an example, some consideration will be given to the stress corrosion of α-brass, which also serves to illustrate diagrams of the type M-H_2O-X, where in this particular case X is NH_3. Mattsson[12] constructed an equilibrium potential-pH diagram for copper in solutions of $(NH_4)_2SO_4$ at various pHs (Fig. 1.19), and also studied experimentally the rate of cracking of brass in dilute solutions containing NH_3, Cu^{2+} ions and SO_4^{2-} ions at various pH values of from 2 to 11. It was found that intergranular cracking occurred mainly in the range pH 6·3-7·7, and was most rapid in the range pH 7·1-7·3. The diagram shows that, thermodynamically, Cu_2O formation becomes increasingly easy up to pH 7·3, and although previous workers in this field had observed that a black film of corrosion product (subsequently identified as Cu_2O) accompanied cracking, it was regarded as of little significance. Mattsson suggested that the cuprous oxide stimulated cracking, and this was confirmed by the subsequent work of Hoar and Booker[13] who studied the time to failure (T_f) of stressed brass wires at various pHs and potentials (E_s) corresponding with significant zones in the potential-pH diagram. They found that rapid fracture occurred in the pH range 7·1-7·3, and that at fracture the potential of the brass had risen from 0·15 to 0·25 V, corresponding with the anodic formation of Cu_2O from Cu. At potentials below −0·05 V, at which Cu_2O cannot form, cracking was arrested. It is not appropriate here to consider the precise rôle of Cu_2O in the cracking process, but this example does serve to illustrate the usefulness of potential-pH diagrams in providing information that can assist in establishing the mechanism of corrosion phenomena[13-15].

Electrochemical Mechanism of Corrosion

The rate (or kinetics) and form of a corrosion reaction will be affected by a variety of factors associated with the metal and the metal surface (which can range from a planar outer surface to the surface within pits or fine cracks), and the environment. Thus heterogeneities in a metal (*see* Section 1.3) may have a marked effect on the kinetics of a reaction without affecting the thermodynamics of the system; there is no reason to believe that a perfect single crystal of pure zinc completely free from lattic defects (a hypothetical concept) would not corrode when immersed in hydrochloric acid, but it would probably corrode at a significantly slower rate than polycrystalline pure zinc, although there is no thermodynamic difference between these two forms of zinc. Furthermore, although heavy metal impurities in zinc will affect the rate of reaction they cannot alter the final position of equilibrium.

The essential features of the electrochemical mechanism of corrosion were outlined at the beginning of the section, and it is now necessary to consider the factors that control the rate of corrosion of a single metal in more detail. However, before doing so it is helpful to examine the charge transfer processes that occur at the two separable electrodes of a well-defined electrochemical cell in order to show that since the two half reactions constituting the overall reaction are *interdependent*, their rates and extents will be equal.

Electrochemical and Electrolytic Cells

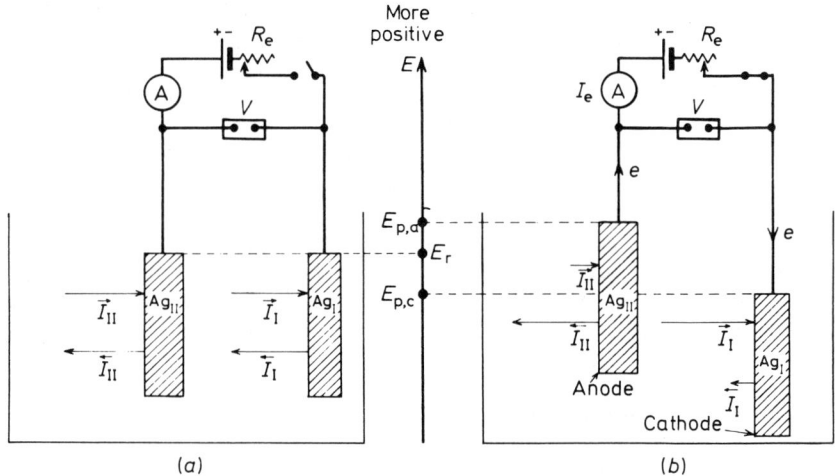

Fig. 1.20 Cell consisting of two reversible Ag^+/Ag electrodes (Ag in $AgNO_3$ solution). The rate and direction of charge transfer is indicated by the length and arrow-head as follows: gain of electrons by $Ag^+ + e \rightarrow Ag \longrightarrow$; loss of electrons by $Ag \rightarrow Ag^+ + e \longleftarrow$. (*a*) Both electrodes at equilibrium and (*b*) electrodes polarised by an external source of e.m.f.; the position of the electrodes in the vertical direction indicates the potential change. (V, high-impedance voltmeter; A, ammeter; R, variable resistance)

An electrochemical cell is a device by means of which the enthalpy (or heat content) of a spontaneous chemical reaction is converted into electrical energy; conversely, an electrolytic cell is a device in which electrical energy is used to bring about a chemical change with a consequent increase in the enthalpy of the system. Both types of cells are characterised by the fact that during their operation charge transfer takes place at one electrode in a direction that leads to the oxidation of either the electrode or of a species in solution, whilst the converse process of reduction occurs at the other electrode.

For simplicity a cell consisting of two identical electrodes of silver immersed in silver nitrate solution will be considered first (Fig. 1.20a), i.e. $Ag_I/AgNO_3/Ag_{II}$. On open circuit each electrode will be at equilibrium, and the rate of transfer of silver ions from the metal lattice to the solution and from the solution to the metal lattice will be equal, i.e. the electrodes will be in a state of dynamic equilibrium. The rate of charge transfer, which may be regarded as either the rate of transfer of silver cations (positive charge) in one direction, or the transfer of electrons (negative charge) in the opposite direction, in an electrochemical reaction is the current I, so that for the equilibrium at electrode I

$$\overrightarrow{I}_{Ag,I} = \overleftarrow{I}_{Ag,I} = I_{0,I} \qquad \ldots(1.24)$$

In this equation $I_{0,I}$ is the equilibrium exchange current, and the arrow convention adopted is that $\overrightarrow{I}_{Ag}$ represents the rate of cathodic reduction

$$Ag^+(aq.) + e \rightarrow Ag(l) \qquad \ldots(1.25)$$

and $\overleftarrow{I}_{Ag}$ represent the rate of anodic oxidation

$$Ag(l) \rightarrow Ag^+(aq.) + e \qquad \ldots(1.26)$$

If the areas of the electrodes are assumed to be $1\,cm^2$, and taking the equilibrium exchange current density i_0 for the Ag^+/Ag equilibrium to be $10^{-2}\,A\,cm^{-2}$, then I_0 will be $10^{-2}\,A$, which is a very high rate of charge transfer. A similar situation will prevail at electrode II, and rates of exchange of silver ions and the potential will be the same as for electrode I.

It is apparent from this that since the rates of the cathodic and anodic processes at each electrode are equal, there will be no net transfer of charge; in fact, with this particular cell, consisting of two identical electrodes in the same electrolyte solution, a similar situation would prevail even if the electrodes were short-circuited, since there is no tendency for a spontaneous reaction to occur, i.e. the system is at equilibrium and $\Delta G = 0$.

Consider now the transfer of electrons from electrode II to electrode I by means of an external source of e.m.f. and a variable resistance (Fig. 1.20b). Prior to this transfer the electrodes are both at equilibrium, and the equilibrium potentials of the metal/solution interfaces will therefore be the same, i.e. $E_I = E_{II} = E_r$, where E_r is the reversible or equilibrium potential. When transfer of electrons at a slow rate is made to take place by means of the external e.m.f., the equilibrium is disturbed and the rates of the charge transfer processes become unequal. At electrode I, $\overrightarrow{I}_{Ag,I} > \overleftarrow{I}_{Ag,I}$, and there is now a net *cathodic* reaction (equation 1.25). At electrode II, $\overleftarrow{I}_{Ag,II} > \overrightarrow{I}_{Ag,II}$, and there is now a net *anodic* reaction (equation 1.26).

The rate of transfer of electrons in the external circuit I_e, which is the rate actually measured by the ammeter, is the difference between rates of the dominant or forward reaction and the subsidiary or reverse reaction at each electrode, and it follows that

$$I_e = \overrightarrow{I}_{Ag,I} - \overleftarrow{I}_{Ag,I} = \overleftarrow{I}_{Ag,II} - \overrightarrow{I}_{Ag,II} \qquad \ldots(1.27)$$

By definition, electrode II at which oxidation is the predominant reaction is the *anode*, whereas electrode I at which reduction is the predominant reaction is the *cathode*. It is apparent that the removal of electrons from Ag_{II} will result in the potential of its interface becoming more positive, whilst the concomitant supply of electrons to the interface of Ag_I will make its potential become more negative than the equilibrium potential:

$$E_{p,c} < E_r < E_{p,a} \qquad \ldots(1.28)$$

where $E_{p,c}$ and $E_{p,a}$ are the polarised potentials of the cathode and anode, respectively.

If the resistance in the external circuit is decreased sufficiently so that

$$E_{p,c} \ll E_r \ll E_{p,a} \qquad \ldots(1.29)$$

the rates of the reverse subsidiary reactions may be disregarded and

$$\overrightarrow{I}_{Ag,c} = \overleftarrow{I}_{Ag,a} = I_e \qquad \ldots(1.30)$$

Concentration Cells

In the previous example of an electrolytic cell the two electrodes were immersed in the same solution of silver nitrate, and the system was therefore thermodynamically at equilibrium. However, if the activities of Ag^+ at the electrodes differ, the system is unstable, and charge transfer will occur in a direction that tends to equalise the activities, and equilibrium is achieved only when they are equal.

Figure 1.21 shows a reversible concentration cell in which $a_{Ag,I} > a_{Ag,II}$, and for the equilibrium $Ag^+ + e \rightleftharpoons Ag$ the e.m.f. of the cell will be

$$E_{r,cell} = \frac{RT}{zF} \ell n \frac{a_{Ag,I}}{a_{Ag,II}} = 0 \cdot 059 \log \frac{a_{Ag,I}}{a_{Ag,II}} \text{ at } 25°C \qquad \ldots(1.31)$$

in which $E_{r,cell}$ must be positive since $a_{Ag,I} > a_{Ag,II}$. If $a_{Ag,I} = 1 \cdot 0$ and $a_{Ag,II} = 0 \cdot 1$, then $E_{r,cell} = 0 \cdot 059$ V and electrode I will be the cathode [$E_r = (0 \cdot 79 + 0 \cdot 006)$V] and electrode II the anode [$E_r = (0 \cdot 079 - 0 \cdot 060)$V]. If the cell operates spontaneously, charge transfer takes place until $a_{Ag,I} = a_{Ag,II}$ when ΔG becomes zero, i.e. the system is at equilibrium.

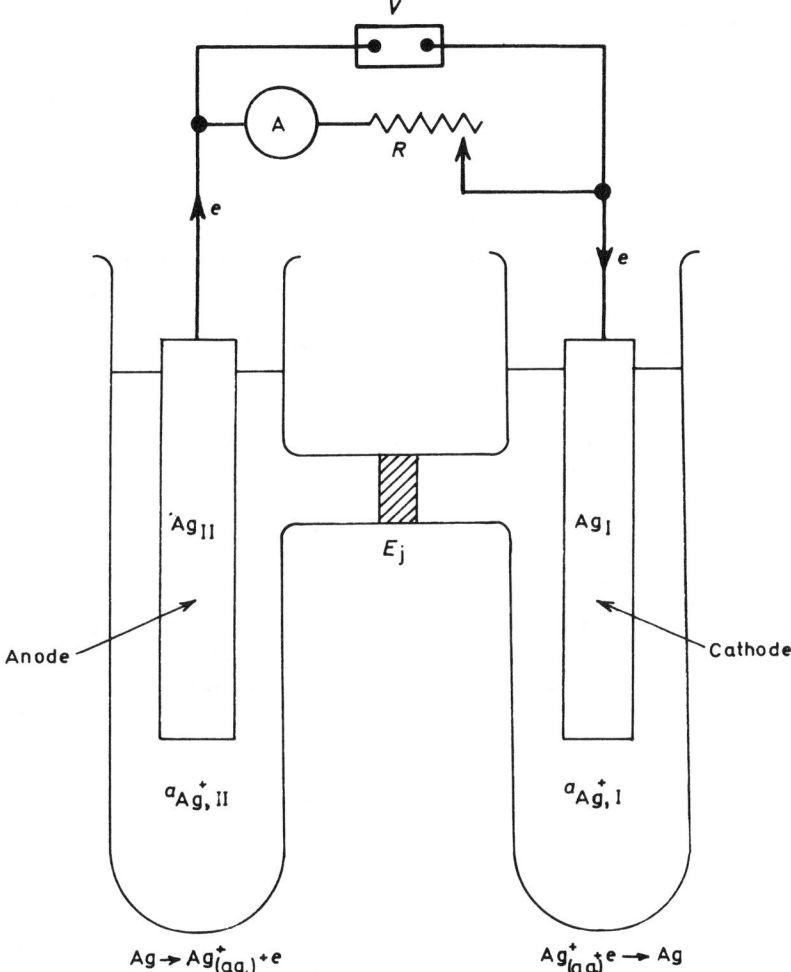

Fig. 1.21 Concentration cell in which $a_{Ag^+, II} < a_{Ag^+, I}$ so that charge transfer occurs spontaneously and proceeds until the activities are equal (E_j is the liquid junction potential at the sintered glass plug that is used to minimise mixing of the two solutions)

Rates and Extents

The current I is the rate of charge transfer, and a rate of $1 A \equiv 1 Cs^{-1}$. The charge on the electron is $1 \cdot 602\,0 \times 10^{-19}$ C, and since 1 mol of an element contains $6 \cdot 023\,5 \times 10^{23}$ atoms (Avogadro's number) the cathodic reduction of a univalent metal ion M^+ will require $6 \cdot 023\,5 \times 10^{23} \times 1 \cdot 602\,0 \times 10^{-19}$ = 96 494 C. This statement is essentially Faraday's law and 1 faraday = 96 494 C ≈ 96 500 C.

In the more general case of the anodic oxidation or cathodic reduction of a species by change transfer, zF coulombs will be required for 1 mol where

z is the number of electrons required to carry out one act of the electron transfer process. Thus the *rate* of charge transfer of an electrochemical reaction is given by

$$k_e = \frac{I}{zF} \quad (\text{mols s}^{-1}) \quad \ldots(1.32)$$

or

$$k_e = \frac{IM}{zF} \quad (\text{kg s}^{-1}) \quad \ldots(1.33)$$

where M is the molar mass (kg mol^{-1}) of the species involved. In these examples of cells the areas of the electrodes have been disregarded, but in electrode kinetics it is the rate per unit area of the electrode that is of significance. The rate per unit area is the current density i, and

$$i = IS^{-1} \quad \ldots(1.34)$$

where S is the area, and i therefore has units of A cm^{-2}, A m^{-2}, mA cm^{-2}, etc. Thus equation 1.33 can be expressed in terms of a rate per unit area

$$k_e = \frac{iM}{zF} \quad (\text{kg cm}^{-2}\,\text{s}^{-1}) \quad \ldots(1.35)$$

where i is in A cm^{-2}. (Note that although the metre is the recommended S.I. unit, the centimetre is so widely used that it will be retained in certain units in this book.)

The *extent* of an electrochemical reaction is the quantity of charge Q (coulombs) transferred in a given time t, and

$$Q = It \quad \ldots(1.36)$$

where t is the time (s). Since 1 mol of charge transfer requires zF coulombs

$$Q = \frac{It}{zF} \quad (\text{mol}), \quad \text{or } Q = \frac{it}{zF} \quad (\text{mol cm}^{-2}) \quad \ldots(1.37)$$

alternatively, $Q = \dfrac{ItM}{zF}$ (kg), or $Q = \dfrac{itM}{zF}$ (kg cm^{-2}) $\quad \ldots(1.38)$

It is apparent (Fig. 1.21) that at potentials removed from the equilibrium potential (*see* equation 1.30) the rate of charge transfer of (*a*) silver cations from the metal to the solution (anodic reaction), (*b*) silver aquo cations from the solution to the metal (cathodic reaction) and (*c*) electrons through the metallic circuit from anode to cathode, are equal, so that any one may be used to evaluate the rates of the others. The rate is most conveniently determined from the rate of transfer of electrons in the metallic circuit (the current I) by means of an ammeter, and if I is maintained constant it can also be used to evaluate the extent. A more precise method of determining the quantity of charge transferred is the coulometer, in which the extent of a single well-defined reaction is determined accurately, e.g. by the quantity of metal electrodeposited, by the volume of gas evolved, etc. The reaction $Ag^+(aq.) + e = Ag$ is utilised in the silver coulometer, and provides one of the most accurate methods of determining the extent of charge transfer.

Partial Cathodic and Anodic Reactions

The reversible cells described are characterised by the fact that the same charge transfer process $Ag^+(aq.) \rightleftharpoons Ag(l)$ takes place at each electrode. However, this type of cell is not typical, and the fact that the exchange processes usually differ can be exemplified by considering an electrolytic cell consisting of two platinum electrodes immersed in a solution of deoxygenated Na_2SO_4, in which water is decomposed into hydrogen and oxygen

$$H_2O \rightarrow H_2 + \tfrac{1}{2}O_2$$

by the interdependent reactions

$$2H_2O + 2e = H_2 + 2OH^- \quad \text{(cathode)} \quad \ldots(1.39)$$

and

$$H_2O = \tfrac{1}{2}O_2 + 2H^+ + 2e \quad \text{(anode)} \quad \ldots(1.40)$$

In contrast to the cell consisting of two reversible Ag^+/Ag electrodes, charge transfer cannot occur until the e.m.f. of the equivalent reversible cell is exceeded ($> 1 \cdot 23$ V)*, and in view of the high overpotential for oxygen evolution the rate will not be significant until the e.m.f. is more than about $1 \cdot 6$ V. If the deoxygenated Na_2SO_4 solution is replaced by an oxygenated sodium chloride solution, the following *additional* reactions are also possible

$$O_2 + 2H_2O + 2e = 2OH^- \quad \text{(cathode)} \quad \ldots(1.41)$$

$$2Cl^- \rightarrow Cl_2 + 2e \quad \text{(anode)} \quad \ldots(1.42)$$

It is not appropriate here to consider the kinetics of the various electrode reactions, which in the case of the oxygenated NaCl solution will depend upon the potentials of the electrodes, the pH of the solution, activity of chloride ions, etc. The significant points to note are that (*a*) an anode or cathode can support more than one electrode process and (*b*) the sum of the rates of the *partial cathodic reactions* must equal the sum of the rates of the *partial anodic reactions*. Since there are four exchange processes (equations 1.39–1.42) there will be eight partial reactions, but if the reverse reactions are regarded as occurring at an insignificant rate then

$$\vec{I}_{O_2} + \vec{I}_{H_2} = \overleftarrow{I}_{Cl_2} + \overleftarrow{I}_{O_2} = I_e \quad \ldots(1.43)$$

This leads to the fundamental concept that irrespective of the number of electrode processes or whether they occur on one or more than one electrode surface

$$\Sigma I_c = \Sigma I_a \quad (C\,s^{-1}) \quad \ldots(1.44)$$

i.e. during charge transfer the sum of the partial cathodic currents must equal the sum of the partial anodic current.

From Faraday's law

$$\Sigma \frac{I_c}{zF} = \Sigma \frac{I_a}{zF} \quad (mol\,s^{-1}) \quad \ldots(1.45)$$

which means that the sum of the cathodic processes must equal the sum of the anodic processes with respect to both the *rate* at any instant of time and the *extent* after any period of time. This law applies equally to these cells,

*The value of $1 \cdot 23$ V follows from Table 1.8 and Fig. 1.15 (bottom).

Electrode Area and Current Density

It is now necessary to consider equation 1.45 in terms of the rates per unit area of the electrode surfaces, which may be equal or unequal depending on circumstances.

When the area of the cathode equals the areas of the anode, equation 1.44 is applicable, and I can be replaced by i, the current density:

$$\Sigma i_c = \Sigma i_a \qquad (\text{C cm}^{-2}\text{s}^{-1}) \qquad \ldots(1.46)$$

and

$$\sum \frac{i_c}{zF} = \sum \frac{i_c}{zF} \qquad (\text{mol cm}^{-2}\text{s}^{-1}) \qquad \ldots(1.47)$$

which may be regarded as the criterion for the *uniform corrosion* of a single metal. However, if the area of the anode is smaller than that of the cathode, then $i_a > i_c$ and

$$\Sigma i_a > \Sigma i_c \qquad \ldots(1.48)$$

whereas if the cathode area is smaller, $i_c > i_a$ and

$$\Sigma i_c > \Sigma i_a \qquad \ldots(1.49)$$

and it follows that the criterion for *localised attack* (or pitting) will be

$$\frac{\Sigma i_a}{\Sigma i_c} > 1 \qquad \ldots(1.50)$$

and that the greater the magnitude of this ratio the more intense will be the attack. Equation 1.50 is referred to as the pitting ratio. The current density i requires a knowledge of both the current I and the area of the electrode S, and the latter is seldom equal to the geometrical or superficial area. In fact, it is possible to distinguish the following types of areas in relation to electrode processes that take place on metal surfaces: *geometrical*, *true*, *active* and *effective*.

In uniform corrosion the *superficial* or *geometrical* area of the metal is used to evaluate both the anodic and cathodic current density, although it might appear to be more logical to take half of that area. However, surfaces are seldom smooth and the true surface area may be twice to three times that of the geometrical area (a cleaved crystal face or an electropolished single crystal would have a true surface area that approximates to its superficial area). It follows, therefore, that the true current density is smaller than the superficial current density, but whether the area used for calculating i_a and i_c is taken as either equal to or half of the superficial area, is unimportant compared to the fact that they are equal. Furthermore, the number of metal atoms at the metal surface that are active in the processes of metal dissolution or cathodic reduction are usually far less than the total number available. Thus Hoar and Notman[16] have shown during the anodic dissolution of nickel that the current density at any given potential is increased by a factor of 10 by cold working the annealed nickel. Similar considerations apply to the hydrogen evolution reaction, in which the coverage of the atoms on a metal surface with adsorbed hydrogen atoms can range from very small to about 100%, depending on the mechanism of the reaction (*see* Section 9.1). Thus the active surface area (the number of atomic sites that participate in the electrode process) may be appreciably less than the total number of sites available.

Finally, it is important to point out that although in localised corrosion the anodic and cathodic areas are physically distinguishable, it does not follow that the total geometrical areas available are actually involved in the charge transfer process. Thus in the corrosion of two dissimilar metals in contact (bimetallic corrosion) the metal of more positive potential (the predominantly cathodic area of the bimetallic couple) may have a very much larger area than that of the predominantly anodic metal, but only the area adjacent to the anode may be effective as a cathode. In fact in a solution of high resistivity the effective areas of both metals will not extend appreciably from the interface of contact. Thus the effective areas of the anodic and cathodic sites may be much smaller than their geometrical areas.

It follows from equation 1.45 that the corrosion rate of a metal can be evaluated from the rate of the cathodic process, since the two are faradaically equivalent; thus either the rate of hydrogen evolution or of oxygen reduction may be used to determine the corrosion rate, providing no other cathodic process occurs. If the anodic and cathodic sites are physically separable the rate of transfer of charge (the current) from one to the other can also be used, as, for example, in evaluating the effects produced by coupling two dissimilar metals. There are a number of examples quoted in the literature where this has been achieved, and reference should be made to the early work of Evans[17] who determined the current and the rate of anodic dissolution in a number of systems in which the anodes and cathodes were physically separable.

More recently, Fontana and Greene[18] measured the current between a pit in stainless steel and the surrounding metal; the pit was allowed to form, and cut out from the surrounding metal (the cathode), its edge was insulated and it was then replaced in the hole with a suitable connection for measuring the current flow between the pit and the surrounding metal. These workers showed that under certain conditions i_a was about a thousand times i_c.

When the anodic and cathodic sites are inseparable the corrosion current cannot be determined directly by an ammeter, but it can be evaluated electrochemically by the *linear polarisation* technique (*see* Sections 19.1–19.3).

Electrochemical Cells and Corrosion Cells

One of the most well-known electrochemical cells that is used for the conversion of chemical energy into electrical energy is the Daniell cell

$$Zn\,|\,ZnSO_4(aq.)\,|\,CuSO_4(aq.)\,|\,Cu \qquad \ldots(1.51)$$

in which the spontaneous reaction

$$Cu^{2+}(aq.) + Zn \rightarrow Zn^{2+}(aq.) + Cu \qquad \ldots(1.52)$$

takes place. The half reactions that constitute the overall reaction are

$$Zn \rightarrow Zn^{2+}(aq.) + 2e \quad \text{(anodic reaction)}$$
$$Cu^{2+}(aq.) + 2e \rightarrow Cu \quad \text{(cathodic reaction)}$$

and electrons are transferred from the zinc to the copper through the metallic circuit.

During the operation of the cell (or during the direct interaction of zinc metal and cupric ions in a beaker) the zinc is oxidised to Zn^{2+} and corrodes, and the Daniell cell has been widely used to illustrate the electrochemical mechanism of corrosion. This analogy between the Daniell cell and a corrosion cell is perhaps unfortunate, since it tends to create the impression that corrosion occurs only when two dissimilar metals are placed in contact and that the electrodes are always physically separable. Furthermore, although reduction of Cu^{2+}(aq.) does occur in certain corrosion reactions it is of less importance than reduction of H_3O^+ ions or dissolved oxygen.

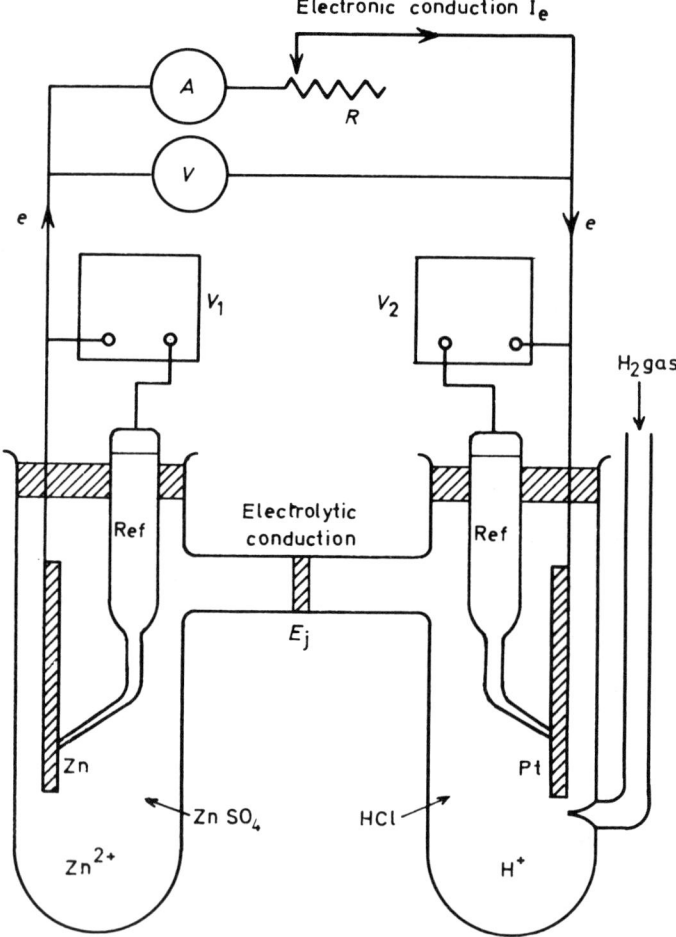

Fig. 1.22 Spontaneous corrosion of zinc in acid illustrated by the reversible cell $Zn|Zn^{2+}|H_3O^+, H_2|Pt$. The individual potentials of the electrodes are determined by a reference electrode (Ref) and a Luggin capillary to minimise the *IR* drop in the solution

For these reasons a somewhat different approach will be adopted here, and an attempt will be made to show how a corrosion reaction may be represented by a well-defined reversible electrochemical cell, although again there are a number of difficulties. Consider the corrosion of metallic zinc in a reducing acid

$$Zn + 2H_3O^+ \rightarrow Zn^{2+}(aq.) + H_2 + H_2O \quad \ldots(1.53)$$

which occurs spontaneously when zinc is immersed in hydrochloric acid; the dissolution of the zinc is usually quite uniform so that there is no means by which anodic and cathodic sites can be identified physically. The half reactions involved are

$$Zn \rightarrow Zn^{2+} + 2e \quad \text{(anodic reaction)} \quad \ldots(1.54)$$

$$2H^+ + 2e \rightarrow H_2 \quad \text{(cathodic reaction)} \quad \ldots(1.55)$$

and it can be seen although the anodic reaction is the same as that in the Daniell cell the cathodic reaction is different. It will be assumed that when the zinc corrodes, randomly dispersed atoms on the surface form the anodic and cathodic sites, and that equations 1.54 and 1.55 can proceed with charge transfer through the zinc. Since the zinc corrodes uniformly the total anodic and cathodic areas must be equal to one another, and this electrochemical reaction, in which the anodic and cathodic sites are inseparable, could be represented by the reversible cell

$$Zn\,|\,Zn^{2+}\,|\,H_3O^+, H_2\,|\,Pt \quad \ldots(1.56)$$

consisting of a reversible Zn^{2+}/Zn electrode and a reversible hydrogen electrode (Fig. 1.22). There will of course be a liquid junction (indicated by the line) and a corresponding liquid junction potential, but the latter will be disregarded for the purpose of the present discussion. This cell clearly does not represent what actually occurs during the corrosion of zinc, and an obvious objection that can be raised is that during corrosion the hydrogen evolution reaction (equation 1.55) occurs in a zinc surface and not on one of platinised platinum. Nevertheless, a reversible cell of this type does serve as a convenient starting point.

By means of a resistance in the circuit the spontaneous corrosion reaction can be made to proceed at a predetermined rate, and the rate can be measured by means of an ammeter A. At the same time the potentials of the individual electrodes can be measured by means of a suitable reference electrode, a Luggin capillary and high-impedance voltmeters V_1 and V_2. At equilibrium there is no net transfer of charge ($I_c = I_a = 0$), and the e.m.f. of the cell is a maximum and equals the difference between the reversible potentials of the two electrodes

$$E_{r,\,cell} = E_{r,\,c} - E_{r,\,a} \quad \ldots(1.57)$$

where $E_{r,\,cell}$ is the reversible e.m.f. of the cell, and $E_{r,\,c}$ and $E_{r,\,a}$ are the reversible potential of the cathode and an anode, respectively.

The driving force of the reaction is the free energy change ΔG which is related to the reversible or equilibrium e.m.f. of the cell by the relationship

$$\Delta G = -zFE_{r,\,cell} \quad \ldots(1.58)$$

and, as emphasised in Section 20.2, both ΔG and $E_{r,\text{cell}}$ are thermodynamic quantities that provide a means of evaluating the equilibrium constant K, and hence the activities of the reactants and products when the reaction comes to equilibrium (Table 1.9).

If now the resistance in the external circuit is decreased slightly the reaction will proceed at a finite rate, and the electrodes constituting the cell will become mutually polarised and displaced from their equilibrium values, i.e. the polarised potential of the anode (Zn^{2+}/Zn) will become more positive, whilst that of the cathode ($2H^+/H_2$) will become more negative (Fig. 1.23).

The displacement of the potential of an electrode from its reversible value is the *overpotential* η, and

$$\eta = E_p - E_r \qquad \ldots(1.59)$$

where E_p is the polarised potential and E_r the reversible or equilibrium potential. Since $E_{p,c} < E_{r,c}$ (more negative)

$$\eta_c = E_{p,c} - E_{r,c} < o \qquad \ldots(1.60)$$

and the cathode overpotential η_c is always *negative*, although $E_{p,c}$ may be positive or negative depending on the sign of E_r and the magnitude of η_c. (If the potential of a Cu^{2+}/Cu electrode, where $E_r = 0 \cdot 34\,V$, is polarised cathodically to $0 \cdot 32\,V$, then $\eta_c = -0 \cdot 020\,V$; if the same procedure is

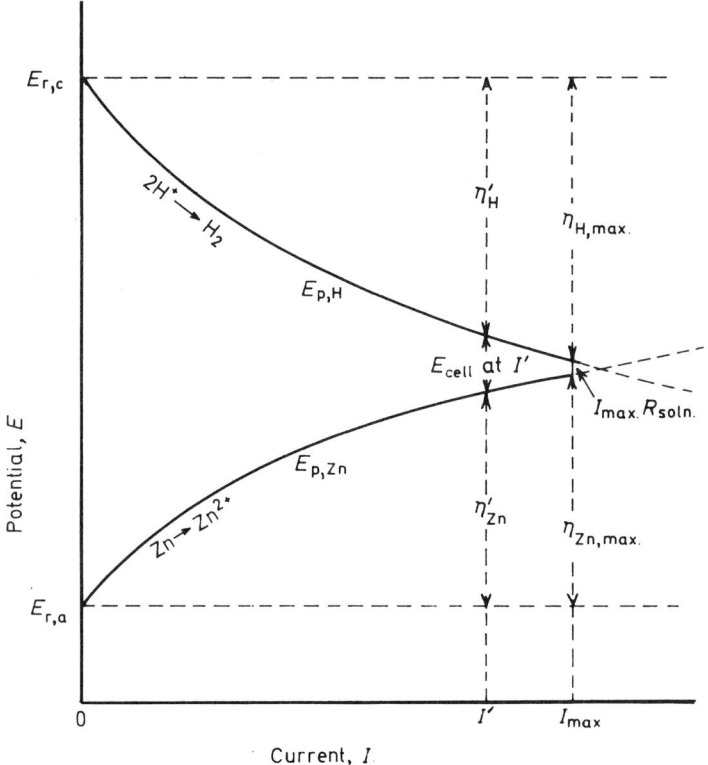

Fig. 1.23 E-I curves for the corrosion of zinc (see Fig. 1.22) showing the relationship between E_r, E_p and η for the cathodic and anodic half reactions

adopted with a Zn^{2+}/Zn electrode where $E_r = -0.76\,V$, E_p will be $-0.78\,V$, but η_c will still be $-0.020\,V$.) Conversely, $E_{p,a} > E_{r,a}$ (more positive), and since

$$\eta_a = E_{p,a} - E_{r,c} > 0 \qquad \ldots(1.61)$$

the anode overpotential is always *positive*.

It should be noted that whereas E is always relative to a specified reference electrode this will not apply to the overpotential (*see* equation 1.59).

As the rate of charge transfer is increased by decreasing the resistance R_e in the circuit, the magnitudes of η_c and η_a increase thus decreasing the magnitude of the polarised e.m.f. of the cell $E_{p,\text{cell}}$. It follows from Fig. 1.23 that for any given rate of charge transfer I

$$E_{p,\text{cell}} = E_{r,\text{cell}} - (\eta_a + \eta_c + IR_{\text{soln.}}) \qquad \ldots(1.62)$$

where η_a and η_c are the *magnitudes* of the overpotentials (the negative sign for η_c must be omitted) corresponding to the rate I, and $R_{\text{soln.}}$ is the electrolytic resistance of the solution.

Since $\qquad E_{p,\text{cell}} = IR_e$

$$I = \frac{E_{r,\text{cell}} - (\eta_a + \eta_c + IR_{\text{soln.}})}{R_e} \qquad \ldots(1.63)$$

which shows that for any given value of R_e the rate of the process I increases with

(*a*) Increase in the magnitude of the reversible e.m.f. of the cell.
(*b*) Decrease in the magnitudes of the anode and cathode potentials.
(*c*) Decrease in the electrolytic resistivity of the solution.

Thus, irrespective of $E_{r,\text{cell}}$, a thermodynamic parameter, the rate will be controlled by the irreversibility of the reaction, which is reflected in the magnitudes of the anode and cathode overpotentials.

If the two electrodes are short-circuited $R_e \to 0$, and $IR_e \to 0$, and $E_{p,\text{cell}}$ will attain its minimum value. If the conductivity is very high and $E_{p,\text{cell}}$ is small enough to be disregarded it follows from equation 1.62 that

$$E_{r,\text{cell}} = \eta_a + \eta_c + IR_{\text{soln.}} \qquad \ldots(1.64)$$

and $\qquad\qquad I = \dfrac{E_{r,\text{cell}} - (\eta_a + \eta_c)}{R_{\text{soln.}}} \qquad \ldots(1.65)$

Equations 1.62–1.65 apply when the anodes and cathodes are separable so that the rate of transfer of charge can be measured by means of an ammeter in the metallic circuit. If $R_{\text{soln.}}$ is significant, then $E_{p,c} > E_{p,a}$, and $E_{p,\text{cell}} > 0$; if $R_{\text{soln.}}$ is very small $E_{p,c} \sim E_{p,a}$ and $E_{p,\text{cell}} \to 0$, but η_c will not necessarily be equal to η_a.

It is now appropriate to apply the above considerations of the operation of a well-defined electrochemical cell to the uniform corrosion of a metal in a solution of high conductivity, and under these circumstances both IR_e and $IR_{\text{soln.}}$ may be regarded as negligible. Thus $E_{p,\text{cell}}$ will tend to zero, and $E_{p,c}$ will tend to be equal to $E_{p,a}$ (within 1–2 mV)

$$\therefore \qquad E_{p,c} = E_{p,a} = E_{\text{corr.}} \qquad \ldots(1.66)$$

where $E_{corr.}$ is the *corrosion potential*, and from equation 1.64

$$E_{r,\,cell} = \eta_c + \eta_a \qquad \ldots(1.67)$$

The above considerations show that the rate of a corrosion reaction is dependent on both the thermodynamic parameter $E_{r,\,cell}$ and the kinetic parameters η_a and η_c. It is also apparent that (*a*) the potential actually measured when corrosion reaction occurs on a metal surface is *mixed, compromise* or *corrosion* potential whose magnitude depends on $E_{r,\,cell}$ and on the $E_{p,\,c}-I$ and $E_{p,\,a}-I$ relationships, and (*b*) direct measurement of I_e is not possible when the electrodes are inseparable.

Overpotentials [19, 20]

The various types of overpotentials are dealt with in more detail in Section 9.1 but it is appropriate here to outline the significant factors in relation to their importance in controlling the rate of corrosion reaction.

Activation overpotential η_A For any given electrode process under specified conditions, charge transfer at a finite rate will involve an activation overpotential η_A, which provides the activation energy required for the reactant to surmount the energy barrier that exists between the energy states of the reactant and product. Some reactions are kinetically easy (e.g. $Ag^+(aq.) + e \rightarrow Ag$) and thus require only a small activation overpotential, whilst others (e.g. $H_3O^+ + e \rightarrow \frac{1}{2}H_2$ on metals such as Hg, Pb and Zn) are kinetically difficult and high activation overpotentials are required. Most electrode processes involve more than one step; one of them is usually more sluggish than the others and is thus rate determining, and the activation energy is required, therefore, to maintain the rate of the rate-determining step (r.d.s.), since the other steps may be regarded as being at equilibrium. The activation energy $E^\ddagger$ is given by

$$E^\ddagger = zF\eta_A \qquad \ldots(1.68)$$

where $E^\ddagger$ is in joules per mole and z is the number of electrons involved in one act of the rate-determining step.

The activation overpotential, and hence the activation energy, varies exponentially with the rate of charge transfer per unit area of electrode surface, as defined by the well-known Tafel equation

$$\eta_A = a + b \log i \qquad \ldots(1.69)$$

where i is the current density, and a and b are the Tafel constants which vary with the nature of the electrode process and with the nature of the solution. Thus η_A will be linearly related to $\log i$ at overpotentials greater than $0 \cdot 010\,V$, and the position and slope of the curve will be dependent on the magnitudes of a and b, which are in turn dependent on the equilibrium exchange current density i_0, the transfer coefficient α and the number of electrons z involved in one act of the rate-determining step. The Tafel equation for a cathodic process can be expressed (*see* Section 20.1) in the form

$$\eta_{A,\,c} = \frac{RT}{\alpha zF}\ln i_0 - \frac{RT}{\alpha zF}\ln i_c \qquad \ldots(1.70)$$

and since $2 \cdot 303\, RT/F \ln x = 0 \cdot 059 \log x$ at 25°C

$$\eta_{A,c} = \frac{0 \cdot 059}{\alpha_c z} \log i_0 - \frac{0 \cdot 059}{\alpha_c z} \log i_c \text{ at } 25°C \qquad \ldots(1.71)$$

where $\eta_{A,c}$ is the activation overpotential of the cathodic process. Similarly, the activation overpotential of an anodic process is given by

$$\eta_{A,a} = -\frac{0 \cdot 059}{\alpha_a z} \log i_0 + \frac{0 \cdot 059}{\alpha_a z} \log i_a \text{ at } 25°C \qquad \ldots(1.72)$$

It is evident from these expressions that since in the Tafel region i (the current density actually determined) must be greater than i_0 (the equilibrium exchange current density), the signs of the overpotentials will conform to equations 1.60 and 1.61, i.e. $\eta_{A,c}$ will be negative and $\eta_{A,a}$ will be positive.

Furthermore, the smaller the magnitude of i_0 the greater the magnitude of η_A and the lower the rate of the electrode process at any given polarised

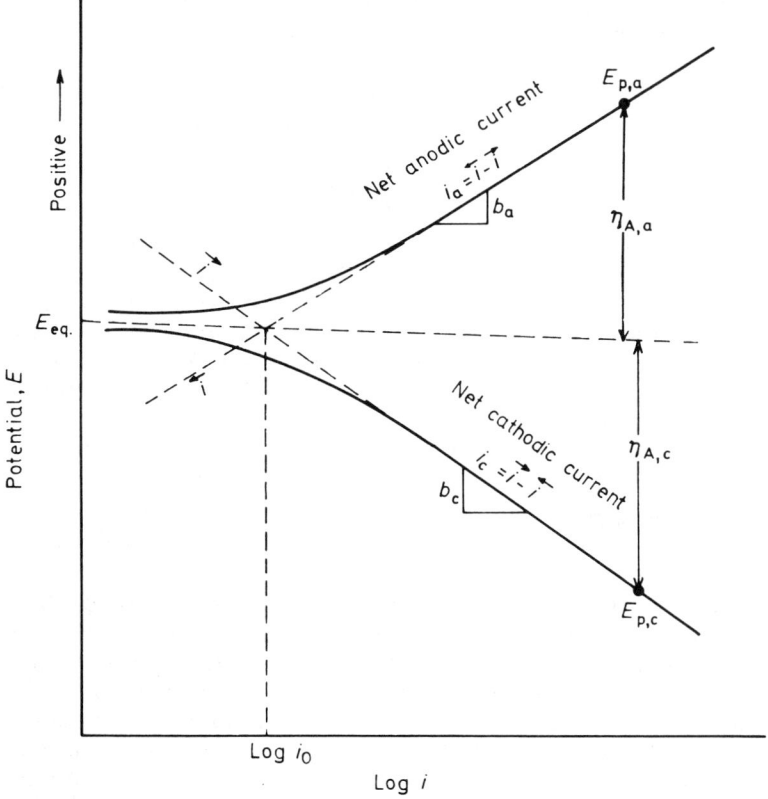

Fig. 1.24 Tafel lines for a single exchange process. The following should be noted: (a) linear E–log i curves are obtained only at overpotentials greater than $0 \cdot 052$ V (at less than $0 \cdot 052$ V E vs. i is linear); (b) the extrapolated anodic and cathodic E–log i curves intersect at i_0 the equilibrium exchange current density; and (c) i_a and i_c the anodic and cathodic current densities actually measured at the differences between $\overleftarrow{i}$ and $\overrightarrow{i}$, and $\overrightarrow{i}$ and $\overleftarrow{i}$, respectively

potential E_p. Thus the equilibrium exchange current density i_0 is the most significant parameter in controlling the rate of a corrosion process in which one (or both) of the electrode processes involve an appreciable activation energy. Figure 1.24 shows the cathodic and anodic Tafel lines for a single exchange process at an electrode, in which i_a and i_c are the anodic and cathodic current densities actually measured.

Transport (diffusion and concentration) overpotentials η_T Previous considerations have been confined to the kinetics of charge transfer but the rate of an electrode reaction will also depend on mass transfer, i.e. the rate at which the reactant is transported to the surface of the electrode and the rate at which the product is transported away from the electrode. Transport through the solution to and from the metal surface occurs by diffusion, ionic migration (transport of electrical charge through the solution) and convection, and of these diffusion through the thin static layer of solution adjacent to the metal surface, the diffusion layer δ, is usually of the greatest significance. However, this is not always the case in practical systems, particularly where dissolved oxygen is the cathodic reactant, and in certain circumstances the rate of diffusion through the bulk solution to the metal/solution interface may be rate determining.

The limiting current density (the maximum possible rate/unit area under the conditions prevailing) for a cathodic process is given by

$$i_L = \frac{DzFc}{\delta(1 - n_+)} \times 10^{-3} \qquad \ldots(1.73)$$

where i_L is the limiting current density (A cm^{-2}), z is the number of electrons required for one step of the electrode process involving 1 mol of the cathode reactant, D is the diffusion coefficient, c is the concentration of the reactant (mol dm^{-3}), δ is the thickness of the diffusion layer (cm), and n_+ the transport number of the cation; the term $(1 - n_+)$ can be neglected if ions other than the species involved in the electrode process are responsible for ionic migration.

The relation between transport overpotential and current density for a cathodic reaction is given by

$$\eta_T = \frac{RT}{zF} \ln\left(1 - \frac{i}{i_L}\right) = \frac{0 \cdot 059}{z} \log\left(1 - \frac{i}{i_L}\right) \text{ at } 25°C \quad \ldots(1.74)$$

and it is evident that the smaller i_L the greater the magnitude of the overpotential due to transport. Unlike activation overpotential, transport overpotential is not controlled by the kinetics of charge transfer, and the magnitude of η_T will be the same for any cation (providing z, D_i and c are the same) and any metal surface. Thus the rate-controlling parameter in transport overpotential is i_L, and it will be seen that any factor in equation 1.73 that causes i_L to increase will result in an increase in the corrosion rate, providing the latter is solely determined by the kinetics of the cathodic process. Figure 1.25a shows the relationship between η and log i when the rate is controlled solely by transport, and Fig. 1.25b shows the relationship when both transport and activated charge transfer are involved. It should be noted that whereas in transport overpotential z is the number of electrons involved in one act of the reaction, in activation overpotential z is the number of electrons involved in one act of the rate-determining step.

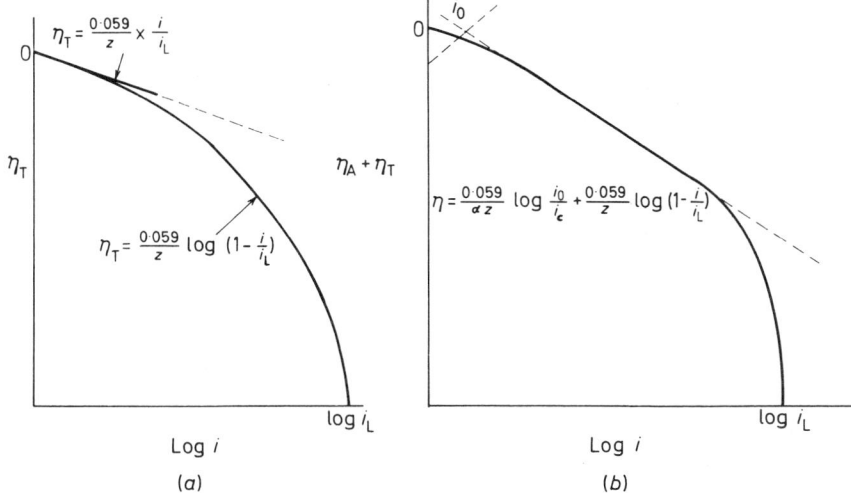

Fig. 1.25 η vs. log i curves for a cathodic reaction (a) when the rate is solely controlled by transport and (b) when both transport and activated charge transfer are rate determining. (Derivations of the relationships are provided in Section 1.9)

Resistance overpotential η_R Since in corrosion the resistance of the metallic path for charge transfer is negligible, resistance overpotential η_R is determined by factors associated with the solution or with the metal surface. Thus resistance overpotential may be defined as

$$\eta_R = I(R_{soln.} + R_f) \qquad \ldots(1.75)$$

where $R_{soln.}$ is the electrical resistance of the solution, which is dependent on the electrical resistivity (Ω cm) of the solution and the geometry of the corroding system, and R_f is the resistance produced by films or coatings formed on or applied to the surface of the sites. Thus, in addition to the resistivity of the solution, any insulating film deposited either at the cathodic or anodic sites that restricts or completely blocks contact between the metal and the solution will increase the resistance overpotential, although the resistivity of the solution is unaffected. This applies particularly to the deposition of $CaCO_3$ [and $Mg(OH)_2$] at the cathodic sites during corrosion in hard waters due to the increase in pH produced by the cathodic process, and since the anodic and cathodic sites are usually close together the calcareous scale will also block the anodic sites, and thus decrease the corrosion rate.

Similar considerations also apply to the dielectric films formed on the metal surface during anodising, and, for example, in the case of the valve metals (Al, Ti, Ta, Nb, etc.) IR drops of hundreds of volts may be produced by the anodic oxide film formed on the metal surfaces. Paint films applied to a metal surface also exert resistance control (*see* Section 14.3).

All these types of polarisation will be present to a greater or lesser extent in most corrosion reactions, but if one is more significant than the others it will control the rate of the reaction. This leads to a classification of corrosion reactions according to whether the cathodic or anodic reaction is rate determining (cathodic control or anodic control), which can be made even more specific by including such terms as 'activation', 'transport' and 'resistance'. Thus the slow corrosion of zinc in solutions of reducing acids is controlled by the high activation energy required for the hydrogen evolution reaction (cathodic activation control), whereas the rapid corrosion of the metal in concentrated sodium hydroxide is controlled by transport of OH^- and ZnO_2^- to and away from the metal/solution interface, respectively (anodic transport control).

Graphical Methods of Expressing Corrosion Rates

The graphical method of showing how the corrosion rate $I_{corr.}$ is dependent on the extent of the polarisation of the anodic and cathodic reactions constituting the corrosion reaction was due originally to Evans[17,22] who used

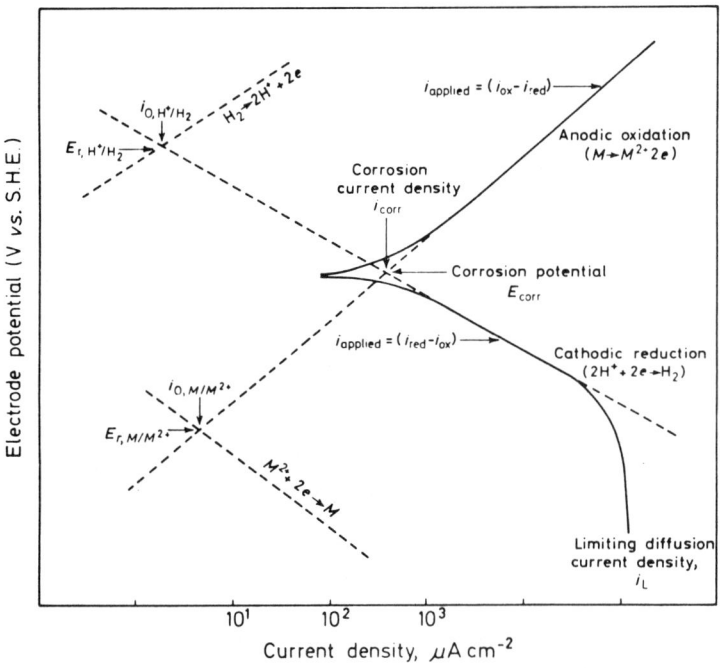

Fig. 1.26 E vs. log i curves for the corrosion of a metal in a reducing acid in which there are two exchange processes (c.f. Fig. 1.24) involving oxidation of $M \rightarrow M^{2+}$ are reduction of $H^+ \rightarrow H_2$. Note that (a) the reverse reactions for exchange process are negligible at potentials removed from E_r, (b) the potential actually measured is the corrosion potential $E_{corr.}$, which is mixed potential, and (c) the E vs. $i_{appl.}$ curves (where $i_{appl.}$ is the applied current density) when extrapolated intersect at $E_{corr.}$

the co-ordinates E and I to illustrate how the electrochemical mechanism of corrosion could be applied to a variety of corroding systems. In these 'Evans' diagrams, both the cathodic and anodic partial reactions constituting the overall corrosion reaction are presented as linear E–I curves that converge and intersect at a point, which defines the corrosion potential $E_{corr.}$ and the corrosion current $I_{corr.}$. Figure 1.26 shows the E log i curves for the two half-reactions involved in the corrosion of a metal in an acid. Comparison should be made with Fig. 1.24 for a *single* exchange process, and it should be note that at significantly high overpotentials the reverse reaction for each half-reaction may be neglected.

A typical Evans diagrams for the corrosion of a single metal is illustrated in Fig. 1.26a (compare with Fig. 1.23 for two separable electrodes), and it can be seen that the E_c–I and E_a–I curves are drawn as straight lines that intersect at a point that defines $E_{corr.}$ and $I_{corr.}$ (it is assumed that the resistance for the solution is negligible). $E_{corr.}$ can of course be determined by means of a reference electrode, but since the anodic and cathodic sites are inseparable direct determination of $I_{corr.}$ by means of an ammeter is not

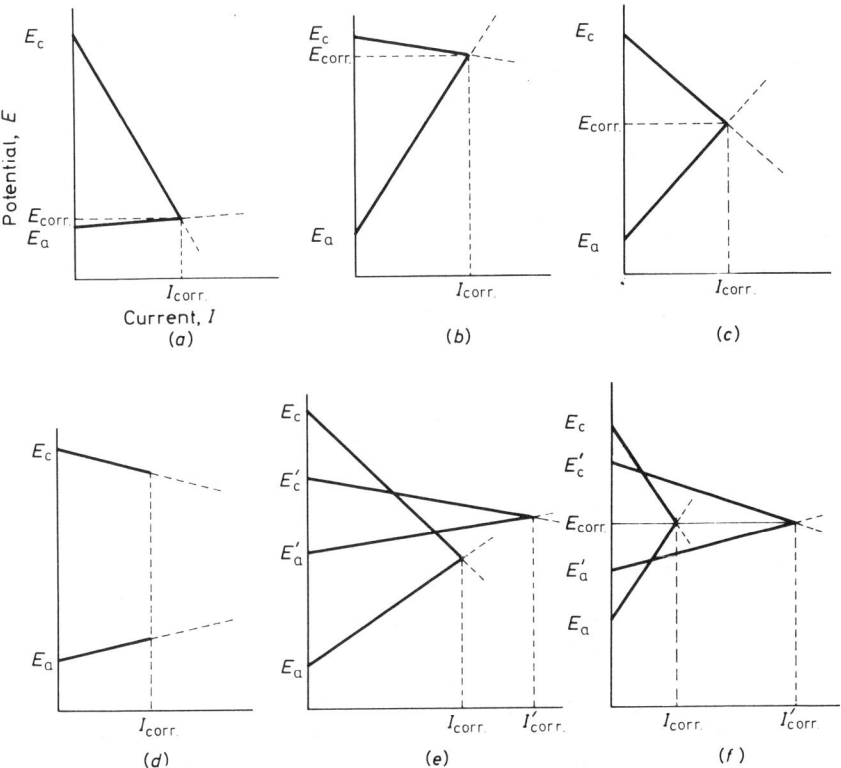

Fig. 1.27 Evans diagrams illustrating (*a*) cathodic control, (*b*) anodic control, (*c*) mixed control, (*d*) resistance control, (*e*) how a reaction with a higher thermodynamic tendency ($E_{r, cell}$) may result in a smaller corrosion rate than one with a lower thermodynamic tendency and (*f*) how $E_{corr.}$ gives no indication of the corrosion rate

possible, and indirect methods must be used (e.g. weight loss and the application of Faraday's law).

The equilibrium potentials $E_{r,c}$ and $E_{r,a}$ can be calculated from the standard electrode potentials of the H^+/H_2 and M/M^{z+} equilibria taking into account the pH and $a_{M^{z+}}$; although the pH may be determined an arbitrary value must be used for the activity of metal ions, and $a_{M^{z+}} = 1$ is not unreasonable when the metal is corroding actively, since it is the activity in the diffusion layer rather than that in the bulk solution that is significant. From these data it is possible to construct an Evans diagram for the corrosion of a single metal in an acid solution, and a similar approach may be adopted when dissolved O_2 or another oxidant is the cathode reactant.

Figures 1.27a to d show how the Evans diagram can be used to illustrate how the rate may be controlled by either the polarisation of one or both of the partial reactions (cathodic, anodic or mixed control) constituting corrosion reaction, or by the resistivity of the solution or films on the metal surface (resistance control). Figures 1.27e and f illustrate how kinetic factors may be more significant than the thermodynamic tendency ($E_{r,\text{cell}}$) and how $E_{\text{corr.}}$ provides no information on the corrosion rate.

The Evans diagram has been used for illustrating various types of corrosion phenomena ranging from the uniform corrosion of a single metal to the enhanced corrosion of one metal when it is coupled to another (bimetallic corrosion), and since the diagram can include only the predominant cathodic and anodic reactions all others are regarded as negligible. Thus if zinc is coupled to iron and the couple is immersed in an oxygenated neutral solution there are at least four possible exchange processes (eight half-reactions), but for the purpose of the Evans diagram only the reduction of dissolved oxygen on the iron surface and the oxidation of $Zn \rightarrow Zn^{2+}$ need to be considered. This tends to create the erroneous impression that each metal sustains only one electrode reaction, whereas in reality the more anodic metal may support a cathodic reaction, although it is predominantly anodic, and the converse applies to the cathodic metal.

In the Evans diagram the curves show the E_p–I relationship, whereas it is evident from previous consideration that E_p and η are functions of the current density i. In the case of a single metal $I_c = I_a$, and since $S_c = S_a$, $i_c = i_a$. However, this is not possible when the anodic and cathodic areas are not equal, and Fig. 1.28 shows how bimetallic corrosion of two dissimilar metals can be represented by an Evans-type diagram. It can be seen that although the curves intersect at $E_{\text{corr.}}$, which must be determined by placing the reference electrode at some distance from the couple, the magnitudes of i_c and i_a are different; it is also evident that the more dangerous bimetallic situation is when S_c is large and S_a is small (see Section 1.7).

Over the years the original Evans diagrams have been modified by various workers[23] who have replaced the linear E–I curves by curves that provide a more fundamental representation of the electrode kinetics of the anodic and cathodic processes constituting a corrosion reaction (see Fig. 1.26). This has been possible partly by the application of electrochemical theory and partly by the development of newer experimental techniques. Thus the cathodic curve is plotted so that it shows whether activation-controlled charge transfer (equation 1.70) or mass transfer (equation 1.74) is rate determining. In addition, the potentiostat (see Section 20.2) has provided

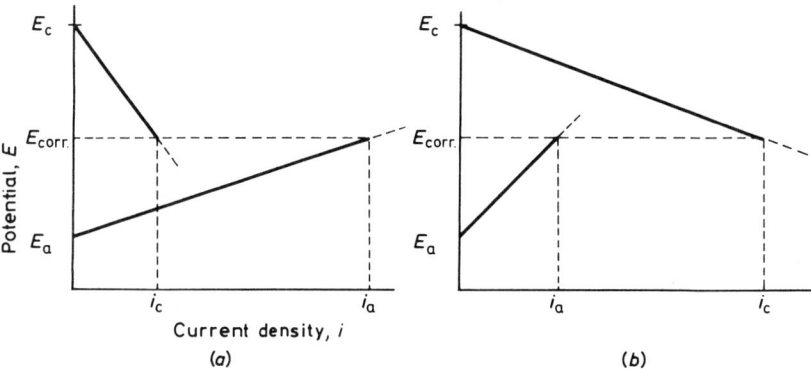

Fig. 1.28 Evans diagram illustrating a corrosion process (e.g. a bimetallic couple) in which the area of the cathode is not equal to that of the anode. (a) $S_c > S_a$ so that $i_c < i_a$ and (b) $S_a > S_c$ so that $i_a < i_c$

a powerful tool for studying the detailed shape of the anodic curve of metals that show an active–passive transition, which has meant that the linear anodic E–I curve used originally has been replaced by the characteristic discontinuous potentiostatic curve. Nevertheless, all these modifications are based on the original concepts of U.R. Evans whose 'Evans diagrams' provided a major step forward in our understanding of the electrochemical mechanism of corrosion. In conclusion it is appropriate to mention that whereas in the Evans diagrams both the anodic and cathodic currents are drawn on the same side of the E axis (i.e. both positive) many workers (particularly Pourbaix and his co-workers) adopt the approach originally devised by Wagner and Traud[24], in which the cathodic curve is taken as negative and drawn on the left-hand side of the E axis whilst the converse applies to the anodic curve (Fig. 1.29).

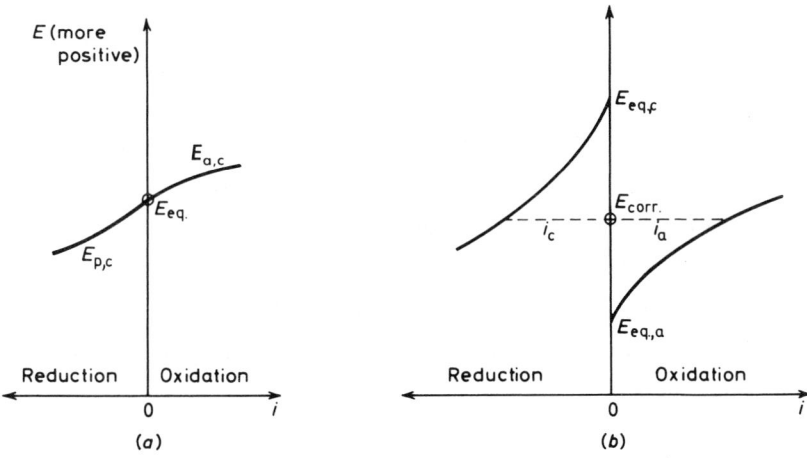

Fig. 1.29 Wagner–Traud method of representing (a) a single reversible reaction and (b) a corrosion reaction (note that $E_{corr.}$ is the potential when $i_c = i_a$)

Cathodic Reactions in Corrosion

General Considerations

It follows from the electrochemical mechanism of corrosion that the rates of the anodic and cathodic reactions are interdependent, and that either or both may control the rate of the corrosion reaction. It is also evident from thermodynamic considerations (Tables 1.9 and 1.10) that for a species in solution to act as an electron acceptor its redox potential must be more positive than that of the M^{z+}/M equilibrium or of any other equilibrium involving an oxidised form of the metal.

The hydrogen evolution reaction (h.e.r.) and the oxygen reduction reaction (equations 1.11 and 1.12) are the two most important cathodic processes in the corrosion of metals, and this is due to the fact that hydrogen ions and water molecules are invariably present in aqueous solution, and since most aqueous solutions are in contact with the atmosphere, dissolved oxygen molecules will normally be present.

In the complete absence of oxygen, or any other oxidising species, the h.e.r. will be the only cathodic process possible, and if the anodic reaction is only slightly polarised the rate will be determined by the kinetics of the h.e.r. on the particular metal under consideration (cathodic control). However, when dissolved oxygen is present both cathodic reactions will be possible, and the rate of the corrosion reaction will depend upon a variety of factors such as the reversible potential of the metal/metal ion system, the pH of the solution, the concentration of oxygen, the kinetics of the h.e.r. and the oxygen reduction reaction on the metal under consideration, temperature, etc. In general, the contribution made by the h.e.r. will increase in significance with decrease in pH, but this too will depend upon the nature of the metal and metal oxide. Thus metals like zinc and aluminium, whose oxides are amphoteric, are thermodynamically unstable in alkaline solutions (*see* Fig. 1.17) and will react with water at high pHs with consequent hydrogen evolution and formation of metal anions. In this connection it should be noted that in neutral or alkaline solutions the activity of H_3O^+ is too low for it to participate in the h.e.r., and under these circumstances the water molecule will act as the electron acceptor

$$H_3O^+ + e \rightarrow \tfrac{1}{2}H_2 + H_2O \quad \text{(acid solutions)} \quad \ldots(1.76)$$

$$H_2O + e \rightarrow \tfrac{1}{2}H_2 + OH^- \quad \text{(neutral and alkaline solutions)} \quad \ldots(1.77)$$

and for the oxygen reduction reaction

$$\tfrac{1}{2}O_2 + 2H_3O^+ + 2e \rightarrow 3H_2O \quad \text{(acid solutions)} \quad \ldots(1.78)$$

$$\tfrac{1}{2}O_2 + H_2O + 2e \rightarrow 2OH^- \quad \text{(neutral and alkaline solutions)} \quad \ldots(1.79)$$

It should also be noted that both reactions will result in an increase in pH in the diffusion layer.

The Hydrogen Evolution Reaction (H.E.R.)[19, 20]

Although the h.e.r. involves transport of H_3O^+ ions (or H_2O molecules) to the metal surface by diffusion and migration, the activation energy for charge transfer is usually of the greater significance, and a corrosion reaction in which the h.e.r. is the cathodic process is frequently controlled by the activation overpotential of the latter.

If it is assumed that the transfer coefficient $\alpha = 0 \cdot 5$, and taking $z = 1$, equation 1.70 becomes

$$\eta_{A,H} = 0 \cdot 12 \log i_0 - 0 \cdot 12 \log i_c \qquad \ldots(1.80)$$

which is identical with the original Tafel equation[25], since i_0 is a constant for a given metal and for given conditions of the solution (*see* Chapter 21.1, Table 21.12, for values of i_0). Thus for activation-controlled transfer the significant parameter is the equilibrium exchange current density i_0, and the smaller i_0 the smaller i_c at a given overpotential. For a corrosion reaction in which both the anodic and cathodic reactions are under activation control, then

for cathodic control $i_{0,c} \ll i_{0,a}$ and $E_{\text{corr.}} \to E_{r,a}$

and

for anodic control $i_{0,a} \ll i_{0,c}$ and $E_{\text{corr.}} \to E_{r,c}$

Figure 1.30 gives examples of single metals corroding in a highly conducting acid in which both the anodic and cathodic reactions are assumed to be under activation control, and it can be seen that at $E_{\text{corr.}}$

$$E_{p,c} = E_{p,a} = E_{\text{corr.}} \qquad \ldots(1.81)$$

and

$$i_c = i_a = i_{\text{corr.}} \qquad \ldots(1.82)$$

but $\eta_c \neq \eta_c$

In order to evaluate $i_{\text{corr.}}$ from the Tafel equation it must be expressed in terms of E_p. By definition

$$\eta = E_p - E_r$$

and from equations 1.81 and 1.82 for the cathodic reaction

$$E_{\text{corr.}} = E_{r,c} + \frac{0 \cdot 059}{\alpha_c z_c} \log \frac{i_{0,c}}{i_{\text{corr.}}} \qquad \ldots(1.83)$$

and for the anodic reaction

$$E_{\text{corr.}} = E_{r,a} + \frac{0 \cdot 059}{\alpha_a z_a} \log \frac{i_{\text{corr.}}}{i_{0,a}} \qquad \ldots(1.84)$$

By equating equations 1.83 and 1.84 $i_{\text{corr.}}$ can be evaluated, providing the Tafel parameters are known, and $E_{\text{corr.}}$ can then be calculated from either of these two equations. Alternatively, by replacing $(E_{r,c} - E_{r,a})$ by $E_{r,\text{cell}}$

$$E_{r,\text{cell}} = \frac{0 \cdot 059}{\alpha_a z_a} \log \frac{i_{\text{corr.}}}{i_{0,a}} + \frac{0 \cdot 059}{\alpha_c z_c} \log \frac{i_{\text{corr.}}}{i_{0,c}} \qquad \ldots(1.85)$$

which provides another convenient method of calculating $i_{\text{corr.}}$.

It can also be shown[19] by means of the above equations, and by assuming that $z_a = z_c = 1$, that

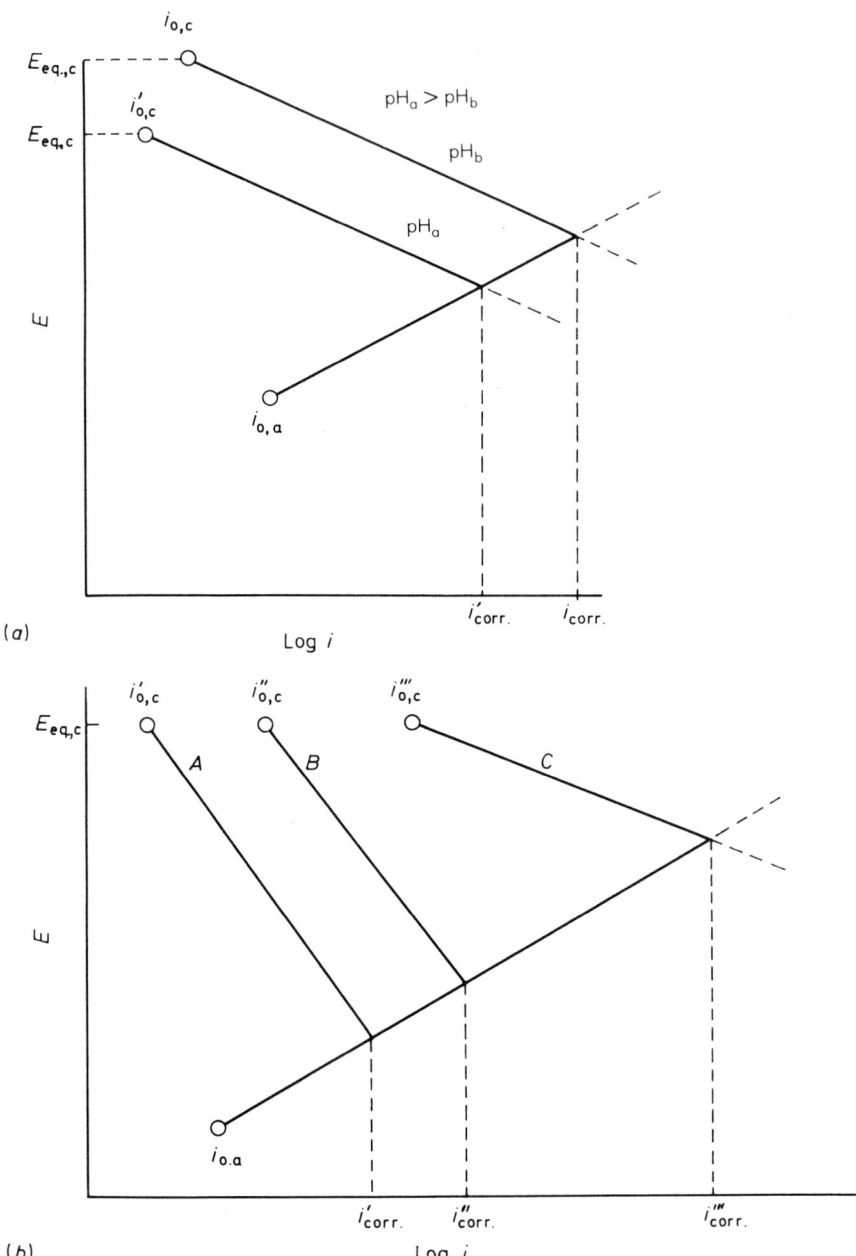

Fig. 1.30 Corrosion of a metal in an acid in which both metal dissolution and hydrogen evolution are under activation control so that the $E.\log i$ curves are linear. (a) Effect of pH on $E_{eq.}$ and $i_{0,H}$; an increase in pH (decrease in a_{H^+}) lowers $E_{eq.,c}$ and decreases $i_{0,c}$. (b) Effect of impurities with different exchange current densities for the h.e.r. on the corrosion rate of a metal (the metal could be regarded as zinc and curve B as representing the h.e.r. on the pure metal; curve A could represent amalgamation with a consequent decrease in the corrosion rate, and curve C the increase in rate produced by impurities of platinum

$$i_{\text{corr.}} = i_{0,a}^{\alpha_c/\alpha_t} \times i_{0,c}^{\alpha_a/\alpha_t} \times \exp\left(\frac{\alpha_a \alpha_c}{\alpha_t} \cdot \frac{FE_{r,\text{cell}}}{RT}\right) \quad \ldots(1.86)$$

where for convenience $(\alpha_a + \alpha_c)$ has been denoted by α_t.
If it is assumed that $\alpha_a = \alpha_c = 0\cdot 5$ then equation 1.86 simplifies to

$$i_{\text{corr.}} \approx i_{0,a}^{\frac{1}{2}} \times i_{0,c}^{\frac{1}{2}} \times \exp\left(\frac{1}{4} \cdot \frac{FE_{r,\text{cell}}}{RT}\right) \quad \ldots(1.87)$$

It can also be shown that

$$E_{\text{corr.}} \approx \frac{RT}{\alpha_t F}\ell n \frac{i_{0,c}}{i_{0,a}} + \left(\frac{\alpha_a}{\alpha_t} \times E_{r,a}\right) + \left(\frac{\alpha_c}{\alpha_t} \times E_{r,c}\right) \quad \ldots(1.88)$$

which if α_a and α_c are taken as $0\cdot 5$ simplifies to

$$E_{\text{corr.}} \approx \frac{RT}{F}\ell n \frac{i_{0,c}}{i_{0,a}} + \tfrac{1}{2}(E_{r,a} + E_{r,c}) \quad \ldots(1.89)$$

Equations 1.83 and 1.84, or the equations derived from them (1.85 to 1.89), may be used to calculate $i_{\text{corr.}}$ and $E_{\text{corr.}}$, providing the various parameters involved are known. The equations also serve to illustrate how $i_{\text{corr.}}$ and $E_{\text{corr.}}$ depend upon a thermodynamic factor ($E_{r,\text{cell}}$, or $E_{r,c}$ and $E_{r,a}$) and the kinetic factors α and i_0 for each of the half reactions that constitute the corrosion reaction.

Values of i_0 for the hydrogen evolution reaction, oxygen reduction reaction, metal deposition, etc. are given in Tables 21.12–21.17 in Section 21 and it is important to note that they are dependent on the nature of the metal and the solution. Thus in the case of the hydrogen evolution reaction the magnitude of i_0 decreases with increase in pH (equation 1.90) and since E_r also decreases (Fig. 1.30a) it follows that a decrease in rate is to be expected on both kinetic and thermodynamic grounds. Furthermore, as the pH increases, transport of hydrogen ions to the metal surface will become increasingly important, and diffusion rather than charge transfer may become rate determining.

Oxygen Reduction

Most aqueous solutions (ranging from bulk natural water and chemical solutions to thin condensed films of moisture) will be in contact with the atmosphere and will contain dissolved oxygen, which can act as a cathode reactant. The saturated solubility of oxygen in pure water at 25°C is only about 10^{-3} mol dm^{-3}, and the solubility decreases significantly with increase in temperature and slightly with concentration of dissolved salts (see Table 21.20 in Section 21 for oxygen solubilities). On the other hand, the concentration of H_3O^+ in acid solutions, which is given by the pH, is high, and since this ion has a high rate of diffusion its rate of reduction is normally controlled by the activation energy for electron transfer. Furthermore, the vigorous evolution of hydrogen that occurs during corrosion facilitates transport, so that diffusion is not a significant factor in controlling the rate of the reaction except at very high current densities. As the pH in acid solutions increases the h.e.r. becomes kinetically more difficult, and requires a higher over-potential. It has been shown[20] that the exchange current density decreases with increase in pH:

$$i_{0,\text{Hss}} = i^0_{0,\text{H}} \times c^{0.5}_{\text{H}_3\text{O}^+} \qquad \ldots(1.90)$$

in which $i^0_{0,\text{H}}$ is the exchange current density at $c_{\text{H}^+} = 1$.

The situation is different, however, in near-neutral or alkaline solutions in which the concentration of H_3O^+ will be small ($< 10^{-7}\,\text{mol dm}^{-3}$), and in these solutions the water molecule will act as the electron acceptor, and although diffusion occurs rapidly its reduction is kinetically more difficult than that of H_3O^+, and will therefore require a higher activation overpotential.

The relationship between E and pH for the H^+/H_2 and $\text{O}_2/\text{H}_2\text{O}$ equilibria have been given in Table 1.8, and it is evident that thermodynamically dissolved oxygen is a far more powerful electron acceptor than H_3O^+ at all pHs. Thus, in the case of the corrosion of $\text{Fe} \rightarrow \text{Fe}^{2+}$ ($E^{\ominus}_{\text{Fe}^{2+}/\text{Fe}} = -0.44\,\text{V}$) in an oxygen-free neutral solution the thermodynamic tendency for corrosion will be very small ($E_{\text{reaction}} \approx 0.02\,\text{V}$) compared with the thermodynamic tendency in oxygenated water ($E_{\text{reaction}} \approx 1.2\,\text{V}$). Similar consideration will not, of course, apply to the more negative metals such as magnesium ($E^{\ominus}_{\text{Mg}^{2+}/\text{Mg}} = -2.1\,\text{V}$).

From these considerations it follows that, in general, oxygen reduction will be more significant than hydrogen evolution in near-neutral solutions, and that in the case of the former, transport of oxygen to the metal surface will be more significant than activation-controlled electron transfer. A further important factor is that in near-neutral solutions solid corrosion products will be thermodynamically stable and will affect the corrosion rate either by passivating the metal or by forming a barrier that hinders transport of oxygen to the metal surface. For these reasons corrosion rates in acid solutions are usually much higher than in neutral solution.

Transport of Oxygen

Before electron transfer can occur the oxygen in the atmosphere must be transported to the metal/solution interface, and this involves the following steps[21]:

(a) Transfer of oxygen across the atmosphere/solution interface.
(b) Transport through the solution (by diffusion and by natural and forced convection) to the diffusion layer.
(c) Transport across the static solution at the metal/solution interface (the diffusion layer δ) by diffusion.

As far as (a) is concerned the greater the surface area of the interface (as may be produced by agitating the solution) the greater the transport of oxygen to the solution. On the other hand, a restricted interface (e.g. a volumetric graduated flask filled to the graduation mark) will decrease oxygen transfer, a factor which must be taken into account in corrosion testing. In the case of (b) oxygen will be transported through the solution by natural convection (e.g. thermal gradients, evaporation at surfaces) or by forced convection (e.g. agitation of the solution, movement of the metal), but these are fairly rapid compared with (c), which is normally rate determining; however, (b) could be significant when a metal is immersed at depth in a static solution.

It follows from this that the limiting current density i_L is the most significant parameter in a corrosion reaction in which oxygen is the cathodic reactant, and that any factor that increases i_L will increase the corrosion rate, since at $E_{corr.}$

$$i_L = i_{corr.} \quad \quad \quad \ldots (1.91)$$

For the reaction $O_2 + 2H_2O + 4e = 4OH^-$, and taking the concentration of oxygen to be 10^{-3} mol dm^{-3}, the diffusion coefficient to be 10^{-5} cm^2 s^{-1} and δ to be 0·05 cm in an unstirred solution, it can be calculated from equation 1.73 that $i_L \approx 80\,\mu$ A cm^{-2}; for a vigorously stirred solution $\delta \approx 0.001$ cm and $i_L \approx 4$ mA cm^{-2}. Thus in this particular case the corrosion rate due to oxygen reduction can range from about 0·08 to 4mA cm^{-2}, depending on the degree of agitation of the solution.

Figure 1.31a to c shows how an increase in the concentration of dissolved oxygen or an increase in velocity increases i_L and thereby increases $i_{corr.}$. It has been shown in equation 1.73 that i_L increases with the concentration of oxygen and temperature, and with decrease in thickness of the diffusion layer, and similar considerations apply to $i_{corr.}$. Thus Uhlig, Triadis and Stern[26] found that the corrosion rate of mild steel in slowly moving water at

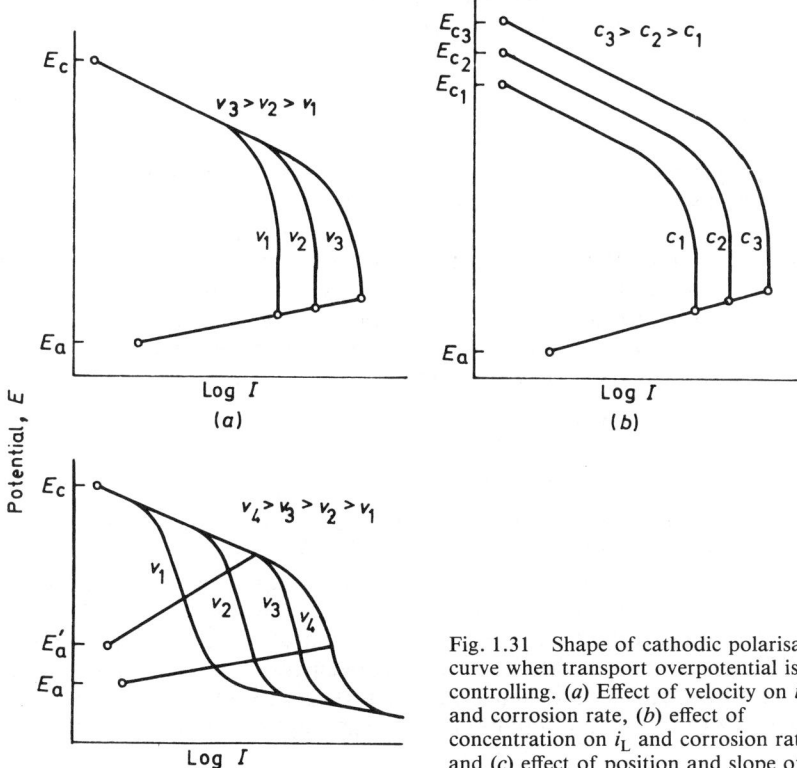

Fig. 1.31 Shape of cathodic polarisation curve when transport overpotential is rate controlling. (a) Effect of velocity on i_L and corrosion rate, (b) effect of concentration on i_L and corrosion rate and (c) effect of position and slope of anodic curve (after Stern[23])

25°C containing 165 p.p.m. of $CaCl_2$ increased linearly with increase in the concentration of dissolved oxygen. However, in distilled water it is possible to passivate pure iron, providing oxygen is brought rapidly and simultaneously to all parts of the metal surface; this was demonstrated first by Evans in his 'whirling ring experiment' in which a steel ring was supported on a glass rod that rotated eccentrically[27]. A similar effect was observed by Uhlig, Triadis and Stern[26] who showed that although initially the corrosion rate of iron in distilled water increased rapidly with concentration of oxygen, it started to decrease when a critical concentration of oxygen (12 ml /l) was attained; this critical value increased with concentration of dissolved salts and temperature, and decreased with increase in velocity and pH. Thus under certain circumstances the rate of reduction of oxygen may be sufficient to exceed the critical current density required for passivation so that even mild steel in oxygenated distilled water will become passive. This is considered further in later paragraphs.

The effect of temperature is complex since there are two conflicting factors, (a) a decrease in the oxygen concentration which results in a decrease in $i_{corr.}$ and (b) an increase in the diffusion coefficient that increases about 3% per degree K rise in temperature. In a closed system from which oxygen cannot escape there is a linear increase in rate with temperature that corresponds with the increase in the diffusion coefficient. However, in an open system although the rate follows that for the closed system initially, the rate starts to decrease at about 70°C due to the decrease in oxygen solubility, which at that temperature becomes more significant than the increase in the diffusion coefficient[28] (*see* Section 2.1).

Mechanism

The mechanism of the oxygen reduction reaction[29] is by no means as fully understood as the h.e.r., and a major experimental difficulty is that in acid solutions (pH = 0) $E^{\ominus}_{O_2/H_2O} = 1 \cdot 23$, which means that oxygen will start to be reduced at potentials at which most metals anodically dissolve. For this reason accurate data on kinetics is available only for the platinum metals. In the case of an iridium electrode at which oxygen reduction is relatively rapid, a number of reaction sequences have been proposed, of which the most acceptable appear to be the following[30]:

1. $O_2 + 2M \rightleftharpoons 2MO$
2. $MO + H^+ + e \xrightarrow{r.d.s.} MOH$
3. $MOH + H^+ + e \rightleftharpoons M + H_2O$

where step 2 is the rate-determining step. The exchange current densities for oxygen reduction on the platinum metals (Pt, Rh, Ir) in acid solutions are about 10^{-10} A cm^{-2}, and it is even smaller for iron, i.e. about 10^{-14} A cm^{-2}.

It has been emphasised that the oxygen reduction reaction is diffusion controlled, and it might be thought that the nature of the metal surface is unimportant compared with the effect of concentration, velocity and temperature that all affect i_L and hence $i_{corr.}$. However, in near-neutral solutions the surface of most metals will be coated (partially or completely) with either thin invisible films or with thick visible corrosion products, and the latter in particular will impede electron transfer; thus metals like platinum, gold, silver and copper are more efficient cathodes for oxygen reduction than metals like steel, zinc or lead (*see* Section 1.7).

Simultaneous Cathodic Reactions

Corrosion reactions involving two simultaneous cathodic processes have already been referred to, and it is now appropriate to consider the graphical method of representing the corrosion rate. It should be noted that although the simultaneous reduction of H_3O^+ and dissolved oxygen occurs frequently this does not exhaust the possibilities, and reactions such as $Fe^{3+} \rightarrow Fe^{2+}$, $Cu^{2+} \rightarrow Cu$, $Cl_2 \rightarrow Cl^-$ may accompany either or both of the above reactions. Similar considerations apply to electrolytic processes such as cathodic protection using an impressed current system in which simultaneous reduction of H_3O^+ ions and dissolved oxygen (and sometimes reduction of dissolved Cl_2 and ClO^- produced by the anodically generated Cl_2) occurs at the surface of the metal protected.

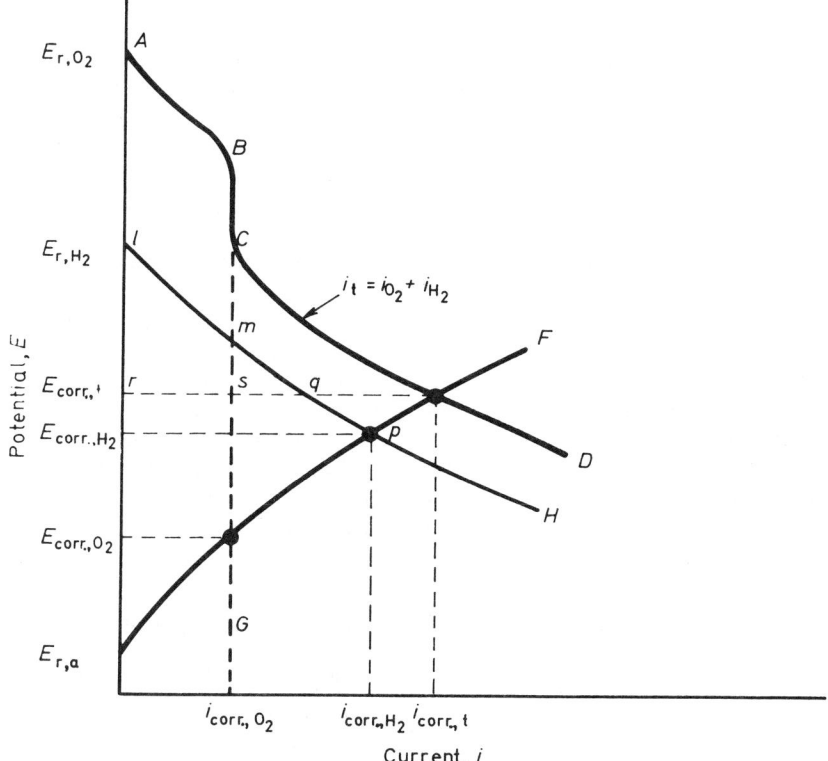

Fig. 1.32 *E–i* curve for the simultaneous cathodic reduction of H_3O^+ (curve *lmqpH*) and dissolved oxygen (*ABCG*) which give the combined curve *ABCD*; $E_{r,a} F$ is the anodic curve for $M \rightarrow M^{z+}$

If the anodic process is assumed to be solely $M \to M^{z+}$ then for the corrosion of a single metal

$$i_{\text{corr.}} = \Sigma i_c = i_{H_2} + i_x = i_t \qquad \ldots(1.92)$$

where i_x is the partial cathodic current density produced by an oxidant in solution and i_t is the total current density.

Figure 1.32 shows the E–i curves for a metal corroding in an acid in which both dissolved oxygen and H_3O^+ ions act as cathode reactants (note that in order to illustrate the summation of the partial currents to give the total current $i_t = i_{H_2} + i_{O_2}$, i rather than log i has been used). The total cathodic curve $ABCD$ is the sum of the partial cathodic currents, and it can be seen that the corrosion potential due to the total cathodic current is more positive than either $E_{\text{corr.},H_2}$ or $E_{\text{corr.},O_2}$. However, whereas in the case of oxygen reduction the rate remains constant once i_L is attained, this does not apply to the h.e.r., which means that the presence of O_2 will decrease the rate of hydrogen evolution from p to q. At $E_{\text{corr.},t}$ the relative rate i_{O_2}/i_{H_2} will be in the ratio rs/rq with the h.e.r. dominating. The relative contributions to the corrosion rate made by dissolved oxygen and H_3O^+ ions will depend on the position of each of the curves; thus if the anodic curve passed through the cathodic curves at their point of intersection m, then $i_{\text{corr.},H_2} = i_{\text{corr.},O_2}$ and their contributions to the total corrosion rate would be equal.

The situation will of course be different if the potential is maintained constant, e.g. in cathodic protection. For example, if the potential was maintained constant at, say, $E_{\text{corr.},t}$ in an oxygen-free solution the rate due to the h.e.r. would be rq which would be unaffected if the solution was then aerated. However, since the corrosion rate would then increase, the current required to maintain the potential constant would have to be increased.

Table 1.13 shows the effect of dissolved oxygen on the corrosion rate of mild steel in oxygenated and oxygen-free dilute acids[31], and it can be seen that the ratio varies with the nature and concentration of the acid; oxygen in $1\cdot 2N$ HNO_3 as would be anticipated has little effect since this acid is readily reduced cathodically. In general, the higher the concentration of the acid the lower the contribution made by dissolved oxygen (*see* results for HCl), and this appears to be due to more rapid hydrogen evolution in the more concentrated acid that tends to screen the surface of the metal from the dissolved oxygen. Uhlig[32] quotes examples, showing how traces of oxygen in dilute H_2SO_4 or substantial amounts in the more concentrated acid can act as an inhibitor. Thus zone-refined iron corroded at 680 mg dm^{-2} d^{-1} in oxygen-free $0\cdot 05$ M H_2SO_4, but at only 415 mg dm^{-2} d^{-1} when the acid was aerated.

Table 1.13 Effect of dissolved oxygen on the corrosion rate (mm y^{-1}) of mild steel in acids*

Acid	Oxygen-containing (O.C)	Oxygen-free (O.F.)	Ratio O.C./O.F.
6% acetic	13·97	0·15	87
6% H_2SO_4	9·14	0·76	12
4% HCl	12·19	0·79	16
0·04% HCl	9·91	0·14	71
1·2% HNO_3	46·23	39·88	1·2

*Data after Whitman and Russell[31].

In case of a pure Fe-9·2 Co alloy in 0·05 M H_2SO_4, in which the corrosion rate is high, the rate was found to increase when the oxygenated acid was deoxygenated. These examples show that the role of oxygen in corrosion reactions is far more complex than would appear from the kinetic curves illustrated above.

Anodic *E–i* Curves

It is now appropriate to consider the kinetics of the anodic reaction with particular reference to the phenomenon of passivity, but since the mechanism is dealt with in detail in Section 1.5 this discussion will place the emphasis on the anodic *E–i* curves.

It has been assumed that the anodic curve for $M \rightarrow M^{z+}$(aq.) conforms with the Tafel relationship, and that E–log i is linear throughout the range of potentials under consideration. It follows, therefore, that charge transfer rather than mass transfer is rate determining, and that the linearity of the E–log i curve will be maintained until transport of metal ions away from the surface becomes significant. It is not proposed to consider metal dissolution in detail here, but it is appropriate to illustrate the complexity of the process by considering the anodic dissolution of iron:

$$Fe \rightarrow Fe^{2+}(aq.) + 2e \qquad \ldots(1.93)$$

This follows the Tafel relationship with a Tafel slope of $2/3\, RT/F = 0·04$ V. Since direct transfer of two electrons as shown in equation 1.92 is highly unlikely, it would appear that the mechanism might involve a two-step process in which one step is rate determining, say

$$Fe \xrightarrow{r.d.s.} Fe^+ + e \qquad \ldots(1.94)$$

$$Fe^+ \rightleftharpoons Fe^{2+} + 2e \qquad \ldots(1.95)$$

However, Despic and Drazic[33] found that the reaction rate was dependent on the concentration of both Fe^{2+} and OH^-, and that

$$i_{0,Fe} = i^0_{0,Fe} \times c_{Fe^{2+}} \times c_{OH^-} \qquad \ldots(1.96)$$

and although the former dependence was expected the latter was not. This observation that the rate increased with increase in hydroxyl concentration (increase in pH), together with a number of other diagnostic criteria, has lead to the view that the most probable mechanism involves the following sequence of steps:

$$Fe + H_2O \rightleftharpoons FeOH + H^+ + e \qquad \ldots(1.97)$$

$$FeOH \xrightarrow{r.d.s.} FeOH^+ + e \qquad \ldots(1.98)$$

$$FeOH^+ + H^+ \rightleftharpoons Fe^{2+} + H_2O \qquad \ldots(1.99)$$

It must be emphasised that although the rate of anodic dissolution of iron increases with increase in pH this will not necessarily apply to the corrosion rate which will be dependent on a number of other factors, e.g. the thermodynamics and kinetics of the cathodic reaction, film formation, etc.

Active–Passive Transitions

On the basis of previous considerations it is evident that the rate of an activation-controlled electrode reaction should increase with increase in potential according to the Tafel relationship. However, in certain metal (alloy)/electrolyte solutions the rate of corrosion decreases to a comparatively low value when the potential is raised above a critical value, and the metal is said to have become *passive*. It must be emphasised again that the passivation of a metal or alloy is dependent on its nature and on the nature of the solution; iron can be passivated only under specific environmental conditions whereas tantalum is passive in most environments.

Although passivity is normally the result of a corrosion reaction involving both a cathodic and anodic reaction, it is studied more easily by making the whole of the metal surface anodic by means of an external source of e.m.f. and a counter-electrode. In this connection it should be noted that there are a number of procedures that can be used to determine the anodic E–i relationship.

1. Galvanostatically—the current density is maintained constant at a predetermined value and the steady-state potential is measured.
2. Potentiostatically—the potential of the metal/solution interface is maintained constant at a predetermined value and the steady-state current density is measured.

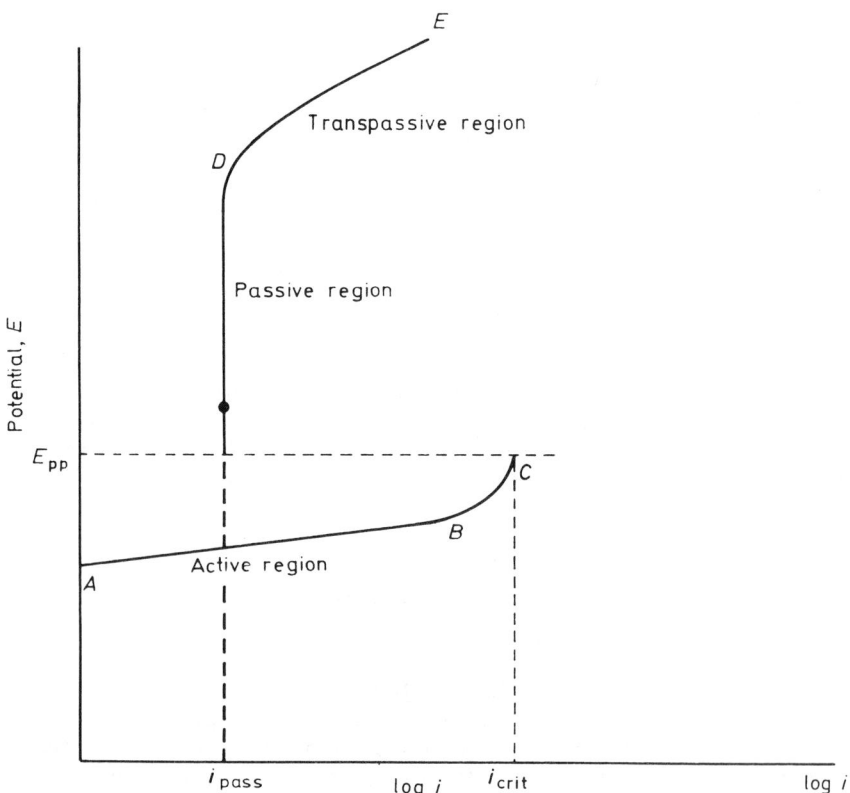

Fig. 1.33 Regions of the potentiostatically determined anodic E–i curve for a metal that shows an active–passive transition

The former method requires a constant-current power source whilst the latter require a potentiostat[34] (*see* Section 19.2).

Each of these two procedures can be varied by proceeding from a low to a high current density (or potential) or from a high to a low current density (or potential); the former is referred to as *forward* polarisation and the latter as *reverse* polarisation. Furthermore, there are a number of variations of the potentiostatic technique, and in the potentiokinetic method the potential of the electrode is made to vary continuously at a predetermined rate, the current being monitored on a recorder; in the pulse method the electrode is given a pulse of potential and the current transient is determined by means of an oscilloscope.

Both the galvanostatic and potentiostatic method have their own particular spheres of application, and it is not always advantageous to reject the former in favour of the latter, although there is an increasing tendency to do so. Nevertheless, the potentiostatic method does have a distinct advantage in studies of passivity, since it is capable of defining more precisely the potential and current density at which the transition from the active (charge transfer controlled $M \rightarrow M^{z+}$) to the passive state takes place; this is far more difficult to achieve with the galvanostatic method owing to the sudden decrease in current (equivalent to an increase in resistance) that accompanies the onset of passivation (*see* Fig. 1.39). A further advantage of the potentiostatic method is that most corrosion processes take place under conditions of constant redox potential rather than constant current so that results obtained potentiostatically are more directly applicable to practical situations.

The typical features of a metal/solution system that exhibits an active to passive transition is shown in Fig. 1.33, which represents diagrammatically the potentiostatically determined anodic E–log i curve for iron in H_2SO_4. Initially, the curve conforms to the Tafel equation and curve AB, which is referred to as the *active region*, corresponds with the reaction Fe → Fe^{2+}(aq). At B there is a departure from linearity that becomes more pronounced as the potential is increased, and at a potential C the current decreases to a very small value. The current density and potential at which the transition occurs are referred to as the *critical current density* $i_{crit.}$, and the *passivation potential* E_{pp}, respectively. In this connection it should be noted that whereas E_{pp} is determined from the active to passive transition, the Flade potential[35] E_F is determined from the passive to active transition

Table 1.14 Approximate Flade potentials of some metals at pH 0*

Metal	Flade potential (V, *vs.* S.H.E.)	Comparison with reversible potential of the equilibrium specified
Au	1·36	(Approximates to Au_2O_3/Au)
Pt	0·87	(Equivalent to PtO/Pt)
Fe	0·58	(0·62 V > Fe_2O_3/Fe)
Ag	0·40	(Equivalent to Ag_2O/Ag)
Ni	0·36	(0·24 V > NiO/Ni)
Cr	−0·22	(0·54 V > Cr_2O_3/Cr)
Ti	−0·24	(0·94 V > TiO_2/Ti)

*Data after West[37].

and does not therefore include extraneous factors such as the *IR* drop through the insulating film, a precursor of true passivity, or pH changes that occur in the base of the pores in the film. For these reasons E_{pp} is somewhat more positive (about 0·001–0·03 V) than E_F, although for most purposes they may be regarded as being approximately the same (see Section 1.5). The Flade potential E_F was found by Franck[36] to be a linear function of pH, and the general equation can be expressed in the following form:

$$E_F = E_F^{\ominus} - 0.059\, pH \qquad \ldots(1.100)$$

in which $E_F^{\ominus}$ is the Flade potential at pH 0. This equation, which shows that E_F becomes more negative with decrease in pH, is applicable to iron, nickel and chromium–iron alloys.

The two significant parameters in passivation are E_{pp} and $i_{crit.}$, and although they are evaluated by means of the potentiostatically determined anodic *E–i* curve they are equally applicable to chemical passivation in which the redox potential and kinetics of the cathodic reaction determines the potential of the metal/solution interface. Typical values of E_F in acid solutions at pH 0 are given in Table 1.14[37], and it is evident that whereas metals like chromium and titanium, which have negative passivation potentials, may be passivated by the hydrogen ion in acid solutions ($E_{H^+/H_2}^{\ominus} = 0.00$ V at pH 0) the other metals will require solutions of higher redox potentials. For passivation two conditions must be satisfied, (*a*) the redox potential of the solution must be more positive than E_{pp} and (*b*) the rate of the cathodic reaction must be greater than $i_{crit.}$. This is illustrated by the passivation of iron ($E_F = 0.58$ V) in nitric acid, an oxidising acid of very high redox potential (about 1·1 V) and high limiting current density. Iron can be passivated chemically in acid solutions only by means of the powerful oxidising agent fuming nitric acid, but in reducing acids (e.g. H_2SO_4) it can be passivated by raising its potential into the passive region by means of an external e.m.f. and a counter-electrode—a process that is known as *anodic protection*. However, if iron is alloyed with metals that passivate more readily (smaller $i_{crit.}$ and more negative E_F) such as chromium and nickel, the ease of passivation of these two metals is imparted to the alloy. Table 1.15[38] shows how $i_{crit.}$, E_F and the quantity of charge required to fully passivate the alloys all decrease with increase in chromium and chromium plus

Table 1.15 Critical current densities ($i_{crit.}$), Flade potentials (E_F) and charge required for passivation of Fe–Cr alloys in 10% H_2SO_4*

Composition (wt.% Cr)	$i_{crit.}$ (mA cm^{-2})	E_F (V, vs. S.H.E.)	Charge (mC cm^{-2})
0	1 000	0·58	1 600
2·8	360	0·58	620
6·7	340	0·35	70
9·5	27	0·15	14
12	27	0·01	15
14	19	−0·03	9
16	12	−0·02	9
18	11	+0·10	8
18 + 8% Ni	2	−0·10	8

*Data after Olivier[38] (see also Ref. 37).

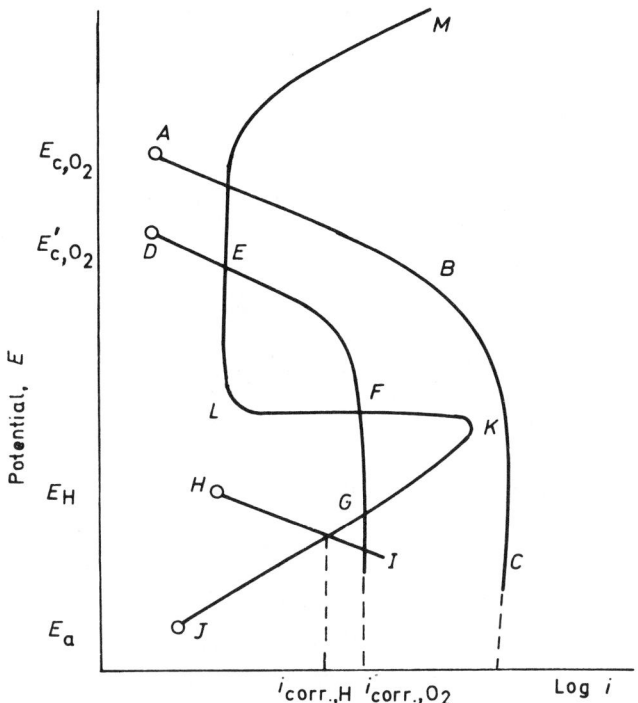

Fig. 1.34 Corrosion and passivation of Fe-18Cr-8Ni stainless steel. Potentiostatic anodic curve *JKLM*; hydrogen evolution reaction, curve *HI*; low concentration of dissolved oxygen, curve *DEFG*; high concentration of dissolved oxygen, curve *ABC* (Section 3.3)

nickel additions Figure 1.34 shows diagrammatically how Fe-18Cr-8Ni stainless steel in sulphuric acid will corrode when the solution is oxygen free, since neither of the two criteria for passivation is satisfied. The presence of a small amount of oxygen in the solution (curve *DEFG*) could worsen the situation, since $i_L < i_{crit.}$. On the other hand an ample supply of oxygen could result in complete passivation. Figures 1.35[39] and 1.36[40] are experimentally determined anodic curves for a ferritic steel (Type 430, 18Cr) and for an Fe-18Cr-8Ni austenitic steel, respectively, and it can be seen how $i_{crit.}$ varies with alloy composition and with concentration of the sulphuric acid. The ferritic stainless steels in acid solutions cannot be passivated by additions of dissolved oxygen, since i_L for oxygen reduction is insufficiently large, but passivation could be achieved if another oxidant (HNO_3, Fe^{3+}, Cu^{2+}) with a high limiting current density was present in the acid.

A further possibility in the passivation of a metal in reducing acid is to change the kinetics of the h.e.r. by alloying the metal with a noble metal that has a higher exchange current density than that of the metal to be passivated. This was achieved first by Tomashov[41] who alloyed Fe-18Cr-8Ni stainless steel with Pt, Pd or Cu. Subsequently, Stern and Wissenberg[42] applied the same principle to Ti, a metal that has a low $i_{crit.}$ in hydrochloric acid as well as in sulphuric acid, and the addition of 0·1-0·2% palladium to the titanium enables passivation to take place even in boiling 10% HCl.

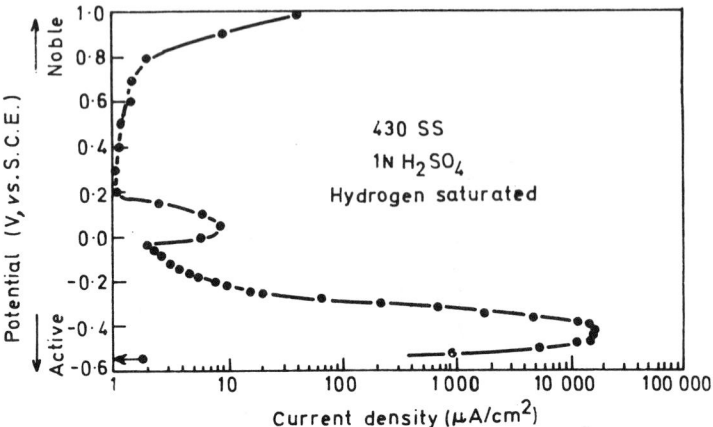

Fig. 1.35 Potentiostatic anodic curve[34] for an 18Cr ferritic stainless steel (type 430) in H_2SO_4 showing the high value of $i_{crit.}$ (after Greene[39])

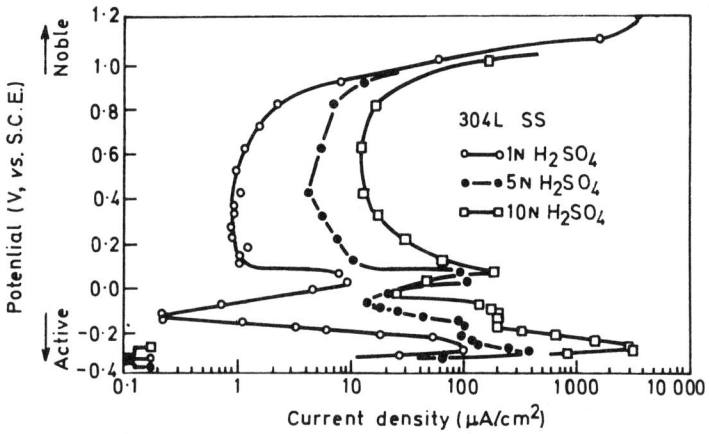

Fig. 1.36 Potentiostatic anodic curve[40] for an Fe-18Cr-8Ni austenitic stainless steel (type 304L) in various concentrations of H_2SO_4. (Note how both $i_{crit.}$ and $i_{pass.}$ increase with concentration of H_2SO_4 and how the passive region ranges from about 0·1 to 0·8 V vs. S.C.E.). (After Fontana and Greene[40])

Figure 1.37 shows diagrammatically how the kinetics of the h.e.r. are altered by the presence of noble-metal alloying additions.

It is not appropriate here to consider the mechanism of passivation (*see* Section 1.5), but it is apparent from Fig. 1.33 that the transition from the active to the passive state must be associated with a fundamental change in the nature of the metal surface, and it is now the generally accepted view that passivity is due to the formation of a very thin solid film of metal corrosion product, usually of oxide, on the metal surface. In this connection it should be noted that metal oxides are thermodynamically unstable in acid solutions so that the oxide formed during passivation must be regarded as a *metastable* form of the oxide that is stable at a higher pH (Fig. 1.18). It follows that the

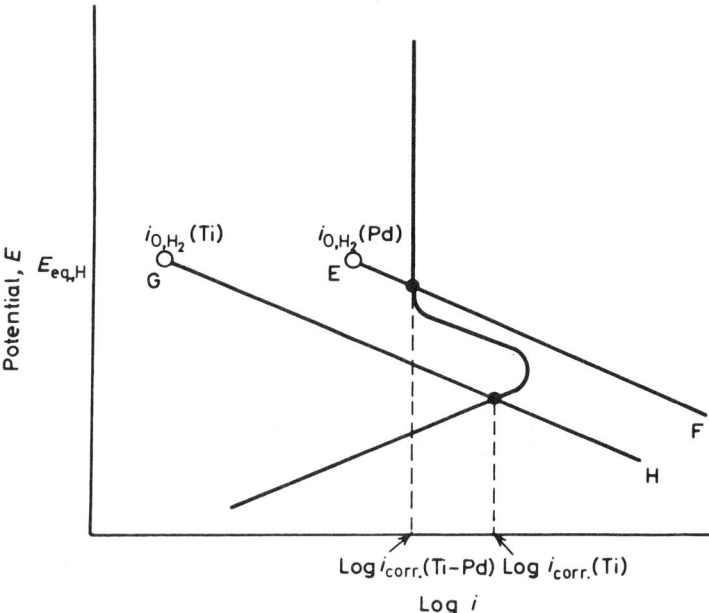

Fig. 1.37 Passivation of titanium in hot 10% HCl resulting from the alloying addition of 0·2% Pd. Curve *EF*, kinetics of the h.e.r. on Pd; curve *GH*, kinetics of the h.e.r. on the titanium surface

passive oxide will tend to dissolve slowly in the acid solution to form stable metal cations; in the case of iron in sulphuric acid the oxide dissolves to form Fe^{3+}(aq.) indicating that the passive oxide is Fe_2O_3. Thus in the passive state the corrosion rate is decreased to a low value by the formation of a rate controlling metastable oxide on the surface of the metal.

The passive current must be associated with an electrode process, which at the low potentials prevailing during chemical passivation (<1 V) cannot be due to the oxidation of OH^- or H_2O molecules to oxygen, with transfer of electrons from the solution to the metal through the intervening oxide. It follows that it must be an ionic current, and since ionic currents in oxide lattices require high fields (about 10^6 V cm^{-1}) the thickness of the oxide cannot exceed 10 nm. Since the potential is constant the field must be constant, which means that the thickness of the oxide must remain constant, and this is possible only if its rate of formation $i_{pass.}$ equals its rate of dissolution in the acid. Thus in the passive state the interface between the metal and the passivating metal oxide progresses slowly into the metal, which confirms the view expressed earlier that passivity may be regarded as a limiting case of a corrosion reaction.

As the potential is raised the passive current density $i_{pass.}$ remains reasonably constant until *D* (Fig. 1.33) when it starts to increase along curve *DE*, which is referred to as the *transpassive* region of the potentiostatic *E–i* curve. Depending on the nature of the metal/solution system and the potential, the transpassive region may be associated with gas evolution (oxygen, chlorine, etc.) or corrosion of the metal, or both reactions may proceed simultaneously.

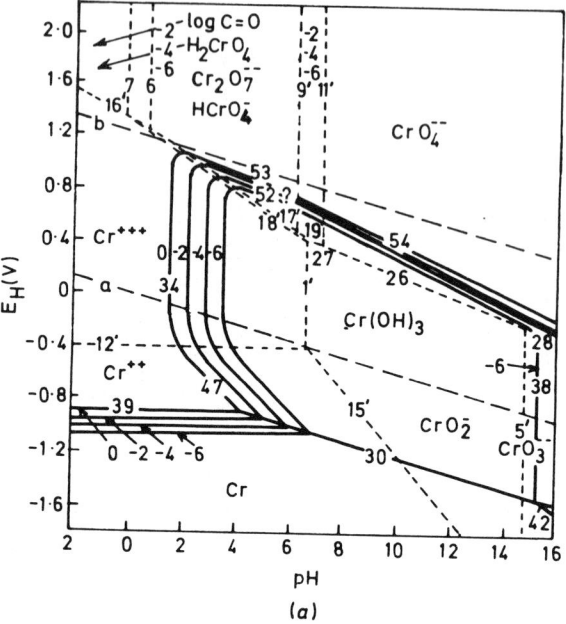

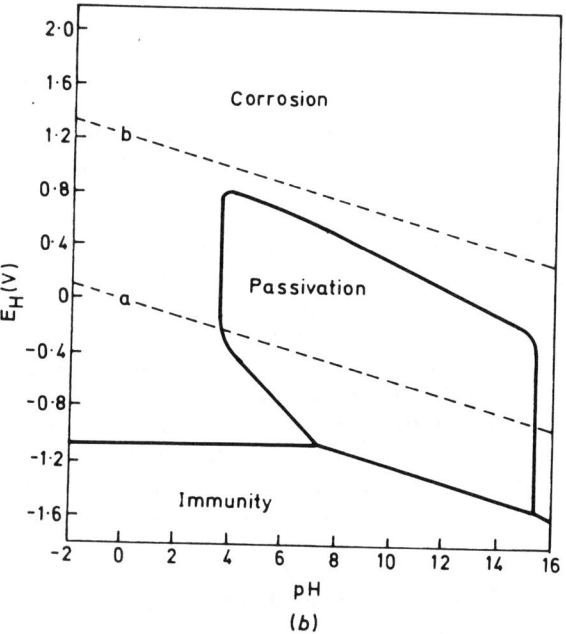

Fig. 1.38(a) Equilibrium potential–pH diagram for the Cr–H$_2$O system and (b) potential–pH diagram showing zones of corrosion, passivity and immunity (after Pourbaix[6])

In the case of chromium in 1 N H_2SO_4 transpassivity occurs at about 1·1 V (below the potential for oxygen evolution, since the equilibrium potential in acid solutions at pH 0 is 1·23 V and oxygen evolution requires an appreciable overpotential) and is associated with oxidation of chromium to dichromate anions:

$$2Cr + 7H_2O \rightarrow Cr_2O_7^{2-} + 14H^+ + 12e \qquad \ldots(1.101)$$

A similar situation prevails in the case of chromium-containing alloys such as the stainless steels, which may corrode at high potentials due to the formation of $Cr_2O_7^{2-}$ in the transpassive region; for this reason Fe-18Cr-8Ni stainless steel corrodes at a significant rate in hot concentrated HNO_3, and may corrode in reducing acids during anodic protection if the potential of the metal is allowed to become too positive.

Figures 1.38a and b are potential–pH equilibrium diagrams for the $Cr-H_2O$ equilibrium, and it can be seen that passivation occurs at a relatively negative potential over a wide range of pH, although once again the impression is created that the metal cannot be passivated in strongly acid solutions. However, at elevated potentials passivation by Cr_2O_3 gives way to the formation of soluble anions ($Cr_2O_7^{2-}$ or CrO_4^{2-} depending upon pH) and this is associated with an increase in the corrosion rate as shown by the transpassive region of the potentiostatic E–log i curve. Oxygen evolution occurs initially when iron in sulphuric acid is in the transpassive region, but Franck and Weil[43] have shown that at potentials greater than 1·65 V corrosion of iron with the formation of Fe^{3+} and oxygen evolution occur simultaneously; in alkaline solutions iron is normally passive, but at elevated potentials it may corrode with the formation of FeO_4^{2-}.

On the other hand, metals such as Ta, Nb, Ti, Zr, Al, etc. (the 'valve' metals[44]) do not exhibit transpassive behaviour, and in appropriate electrolyte solutions film growth at high fields rather than corrosion and/or oxygen evolution is the predominant reaction; thus aluminium can be anodised to 500 V or more in an ammonium borate buffer; titanium can be anodised to about 400 V in formic acid and tantalum can be anodised to high voltages in most acids, including hydrochloric acid.

L. L. SHREIR

REFERENCES

1. Latimer, W. M., *Oxidation Materials*, Prentice Hall, New Jersey (1952)
2. Hoar, T. P., 'The Anodic Behaviour of Metals', in *Modern Aspects of Electrochemistry*, No. 2, Ed. by J. O'M. Bockris, Butterworths, London (1959)
3. Uhlig, H. H., *J. Electrochem. Soc.*, **109**, 11C (1962)
4. Sillen, L. C. and Martell, A. E., *Stability Constants of Metal-Ion Complexes*, Chemical Society, London (1962)
5. Pourbaix, M., *Atlas of Potential/pH Diagrams*, Pergamon, Oxford (1962) and CEBELCOR (Centre Belge D'etude de la Corrosion) Rapports Techniques. (*See* 'List of Reports' published by CEBELCOR)
6. Pourbaix, M., *Lectures on Electrochemical Corrosion*, Plenum Press, New York, London (1973)
7. Deltombe, E. and Pourbaix, M., *CEBELCO R Rapport Techniques*, **7** (1954) and Proceedings of 6th Meeting CITCE, Butterworths, London, 118 (1954)

8. Ashworth, V. and Boden, P. J., *Corros. Sci.*, **10**, 709 (1970)
9. Townsend, H. E., *Corros. Sci.*, **10** No. 5, 343 (1970)
10. Delahay, P., Pourbaix, M. and Van Rysselberghe, P., *Proceedings*, 2nd Meeting of CITCE, Tamburino, Milan, 118 (1950)
11. De Gromoboy, T. S. and Shreir, L. L., *Electrochim. Acta.*, **3**, 895 (1966)
12. Mattsson, E., *Electrochim. Acta.*, **3**, 279 (1961)
13. Hoar, T. P. and Booker, C. J. L., *Corros. Sci.*, **5**, 821 (1965)
14. Booker, C. J. L., *Proc. Int. Conf. Fundamental Aspects of Stress Corrosion Cracking* (ed. by R. W. Staehle), N.A.C.E., Houston, 178 (1969)
15. Hoar, T. P. and Rothwell, G. P., *Electrochim. Acta.*, **15**, 1037 (1970)
16. Hoar, T. P. and Notman, G. K., *Trans. Inst. Met. Finish.*, **39**, 166 (1962)
17. Evans, U. R., *The Corrosion and Oxidation of Metals*, Arnold, London (1961)
18. Greene, M. D. and Fontana, M. G., *Corrosion*, **15**, 41 (1959) and **15**, 55 (1959)
19. Bockris, J. O'M. and Drazic, D., *Electro-chemical Science*, Taylor and Francis, London (1972)
20. Bockris, J. O'M. and Reddy, A. K. N., *Modern Electrochemistry*, Plenum Press, 1970 (published by Macdonald and Co., London, 1970)
21. Tomashov, N. D., *Theory of Corrosion and Protection of Metals*, Collier-Macmillan Ltd., London (1966)
22. Evans, U. R., *J. Franklin Inst.*, **208**, 52 (1929)
23. Stern, M., *J. Electrochem. Soc.*, **104**, 56 (1957); ibid., *Corrosion*, **13**, 755 (1957); ibid., **14**, 329t (1958); Bonhoeffer, K. F., *Corrosion*, **11**, 304t (1955); and Edeleanu, C., *Nature*, **173**, 739(1954)
24. Wagner, C. and Traud, W., *Z. Elektrochem.*, **44**, 391–402 (1938)
25. Tafel, J., *Z. Physik. Chem.*, **50A**, 641 (1905); Frumkin, A. N., 'Hydrogen Overvoltage and Adsorption Phenomena' in P. Delahey and C. Tobias's *Advances in Electrochemistry and Electrochemical Engineering*, Vol. 1 (1961); ibid., Vol. 3 (1964); and Vetter, K. J., *Electrochemical Kinetics*, Academic Press, Inc., New York (1967)
26. Uhlig, H., Triadis, D. and Stern, M., *J. Electrochem. Soc.*, **102**, 59 (1955)
27. See Ref. 17, p. 102–103
28. Speller, F., *Corrosion, Causes and Prevention*, McGraw-Hill, New York (1951)
29. Hoare, J. P., *The Electrochemistry of Oxygen*, Interscience, New York (1962)
30. Damjanovic, A., 'The Kinetics of the Mechanism of the Oxygen Reduction', in Bockris, J. O'M. and Conway, B. E. (editors), *Modern Aspects of Electrochemistry*, No. 5, Plenum Press, New York (1969)
31. Whitman, W. and Russell, R., *Ind. Eng. Chem.*, **17**, 348 (1925)
32. Uhlig, H., *Corrosion and Corrosion Control*, Wiley, 103 (1972)
33. Despic, A. R., *Electrochim. Acta.*, **4**, 325 (1961); Drazic, D. M., *Electrochim. Acta.*, **7**, 293 (1962). (*See* also Ref. 20, p. 1080)
34. von Fraunhofer, J. A. and Banks, C. H., *Potentiostat and its Applications*, Butterworths, London (1972)
35. Flade, F., *Z. Physik. Chem.*, **76**, 513 (1911)
36. Franck, U. F., *Z. Naturforschung*, **4A**, 378 (1949)
37. West, J. F., *Electrodeposition and Corrosion Processes*, van Nostrand Reinhold Co. Ltd., London, 90 (1970)
38. Olivier, R., *6th CITCE Meeting*, Poitiers, 1954, Butterworths, London, 314 (1955)
39. Greene, N. D., *Corrosion*, **18**, 136t (1962)
40. Fontana, M. G. and Greene, N. D., *Corrosion Engineering*, McGraw-Hill, 337 (1967)
41. Tomashov, N. D., *Corrosion*, **14**, 299t (1958)
42. Stern, M. and Wissenberg, H., *J. Electrochem. Soc.*, **106**, 759 (1959)
43. Franck, U. F. and Weil, K. G., *Z. Elektrochem.*, **56**, 814 (1952)
44. Young, L., *Anodic Oxide Films*, Academic Press, London (1961)

BIBLIOGRAPHY

Electrochemistry

Bockris, J. O'M. and Reddy, A. K. N., *Modern Electrochemistry*, Macdonald, London (1971)
Bockris, J. O'M. and Drazic, D., *Electro-chemical Science*, Taylor and Francis, London (1972)
Davies, C. W., *Electrochemistry*, Newnes, London (1967)

Delahey, P. and Tobias, C. W. (Eds.), *Advances in Electrochemistry and Electrochemical Engineering*, Vols. 1-9, John Wiley (1972)
Delahey, P., *New Instrumental Methods in Electrochemistry*, Interscience, London (1954)
Potter, E. C., *Electrochemistry, Principles and Applications*, Cleaver-Hume, London (1961)

Textbooks and reference books on corrosion
Chilton, J. P., *Principles of Metallic Corrosion*, 2nd edition, Royal Institute of Chemistry, London (1969)
Evans, U. R., *The Corrosion and Oxidation of Metals*, Arnold, London (1961); 1st Supplementary Volume (1968)
Evans, U. R., *An Introduction to Metallic Corrosion*, Arnold, London (1963)
Fontana, M. G. and Greene, N. D., *Corrosion Engineering*, McGraw-Hill, New York (1967)
Fontana, M. G. and Staehle, R. W., *Advances in Corrosion Science and Technology*, Plenum Press, New York [Vol. 1 (1970); Vol. 2 (1971); Vol. 3 (1973); Vol. 4 (1974)]
La Que, F. and Copson, H. R., *Corrosion Resistance of Metals and Alloys*, van Nostrand-Reinhold, New York (1963)
McKay, R. and Worthington, R., *Corrosion Resistance of Metals and Alloys*, 2nd edition, Reinhold, New York (1963)
Pourbaix, M., *Atlas of Electrochemical Equilibria in Aqueous Solutions*, Pergamon Press, Oxford
Pourbaix, M., *Lectures on Electrochemical Corrosion*, Plenum Press, New York–London (1973)
Rabold, E., *Corrosion Guide*, Elsevier (1968)
Scully, J. C., *The Fundamentals of Corrosion*, Pergamon Press, Oxford (1966)
Tomashov, N. D., *Theory of Corrosion and Protection of Metals*, Macmillan, New York (1966)
Uhlig, H. H. (Ed.), *Corrosion Handbook*, John Wiley, New York (1955)
Uhlig, H. H., *Corrosion and Corrosion Control*, John Wiley, New York (1967)
West, J. M., *Electrodeposition and Corrosion Processes*, van Nostrand-Reinhold, London (1971)

Books on specialised topics
Ailor, W. H., *Handbook on Corrosion Testing and Evaluation*, John Wiley, New York (1971)
Bregman, J. I., *Corrosion Inhibitors*, Collier-Macmillan Co., New York (1963)
Chapman, F. A., *Corrosion Testing Procedures*, Butler and Tanner, London (1964)
Kofsted, P., *High Temperature Oxidation of Metals*, John Wiley, New York (1965)
Kubaschewski, O. and Hopkins, B. F., *Oxidation of Metals and Alloys*, Butterworths, London (1972)
Logan, H. I., *The Stress Corrosion of Metals*, John Wiley, New York (1968)
Miller, J. D. A. (Ed.), *Microbial Aspects of Metallurgy*, Medical and Technical Publishing Co., London (1971)
Morgan, J. H., *Cathodic Protection*, Leonard Hill, London (1959)
Parker, M. E., *Pipeline Corrosion and Cathodic Protection*, Gulf Publishing Co., Houston, Texas
Rhodin, T. N. (Ed.), *Physical Metallurgy of Stress Corrosion Fracture*, Interscience, New York (1959)
Scully, J. C. (Ed.), *The Theory of Stress Corrosion Cracking in Alloys*, NATO, Brussels (1971)

Potential-pH diagrams in research
Markovic, T., 'Ein Thermodynamisches Kriterium für den Kathodischen Korrosionsschutz des Eisens mit Magnesium als Opferanode', *Corros. Sci.*, **2**, 51 (1962)
Clerbois, F. and Massart, J., 'Comportement Anodique des Metaux en Milieu Aqueux – I Utilisation des Courbes Potentiostatiques à la Determination des Mecanismes de Corrosion', *Corros. Sci.*, **2**, 1 (1962)
Nagayama, M. and Okamoto, G., 'The Anodic Behaviour of Passive Iron in Chromic Acid-Chromate Solutions', *Corros. Sci.*, **2**, 203 (1962)
Uusitalo, E. and Heinanen, J., 'Corrosion of Steel in Soft Waters', *Corros. Sci.*, **2**, 281 (1962)
Littlewood, R., 'Diagrammatic Representation of the Thermodynamics of Metal-fused Chloride Systems', *J. Electrochem. Soc.*, **109**, 525 (1962)
Bouet, J. and Brenet, J. P., 'Contribution a l'Etude du Diagramme Tension/pH du Fer en Milieux Sulfures', *Corros. Sci.*, **3**, 51 (1963)

Horvath, J. and Novak, M., 'Potential/pH Equilibrium Diagrams of Some Me-S-H$_2$O Ternary Systems and Their Interpretation from the Point of View of Metallic Corrosion', *Corros. Sci.*, **4**, 159 (1964)

Ruetschi, P. and Angstadt, R. T., 'Anodic Oxidation of Lead at Constant Potential', *J. Electrochem. Soc.*, **111**, 1323 (1964)

Richaud, H.,'Application des Courbes de Polarisation a l'Etude des Inhibiteurs de l'Aluminium, Notamment en presence de Chlorures', *Corros. Sci.*, **4**, 191 (1964)

Pourbaix, M., 'A Comparative Review of Electrochemical Methods of Assessing Corrosion and the Behavior in Practice of Corrodible Material', *Corros. Sci.*, **5**, 67 (1965)

Pourbaix, M. and Vandervelden, F., 'Intentiostatic and Potentiostatic Methods, Their Use to Predetermine the Circumstances for Corrosion or Non-corrosion of Metals and Alloys', *Corros. Sci.*, **5**, 81 (1965)

Horváth, J. and Hackl, L., 'Check of the Potential/pH Equilibrium Diagrams of Different Metal-Sulphur-Water Ternary Systems by Intermittent Galvanostatic Polarisation Method', *Corros. Sci.*, **5**, 525 (1965)

Johnson, H. E. and Leja, J., 'On the Potential/pH Diagrams of the Cu-NH$_3$-H$_2$O and Zn-NH$_3$-H$_2$O Systems', *J. Electrochem. Soc.*, **112**, 638 (1965)

Letowski, F. and Niemiec, J., 'On the Potential-pH Diagrams of the Cu-NH$_3$-H$_2$O and Zn-NH$_3$-H$_2$O Systems', Discussion, *J. Electrochem. Soc.*, **113**, 629 (1966)

Verink, E. D., 'Simplified Procedure for Constructing Pourbaix Diagrams', *Corros. Sci.*, **23**, 371 (1967)

Katoh, M., 'Influence of Chelating Agent (Citric Acid) and F$^-$ on Corrosion of Aluminium', *Corros. Sci.*, **8**, 423 (1968)

Bartonicek, R. and Lukasovská, M., 'A Potential-pH Diagram for the System Cu-NH$_3$-Cl$^-$-H$_2$O', *Corros. Sci.*, **9**, 35 (1969)

Pourbaix, M., 'Recent Applications of Electrode Potential Measurements in the Thermodynamics and Kinetics of Corrosion of Metals', *Corros.*, **25**, 267 (1969)

de Nora, O., Gallone, P., Traini, C. and Meneghini, G., 'On the Mechanism of Anodic Chlorate Oxidation', *J. Electrochem. Soc.*, **116**, 147 (1969)

Lennox, T. J., 'Limitations on the Use of Pourbaix Diagrams to Predict De-alloying or Other Corrosion Characteristics', Discussion, *Corrosion*, **26**, 397 (1970)

Verink, E. D. and Parrish, P. A., 'Use of Pourbaix Diagrams in Predicting the Susceptibility to De-alloying Phenomena', *Corrosion*, **26**, 214 (1970)

Townsend, H. E., 'Potential-pH Diagrams at Elevated Temperature for the System Fe-H$_2$O', *Corros. Sci.*, **10**, 543 (1970)

Ashworth, V. and Boden, P. J., 'Potential-pH Diagrams at Elevated Temperatures', *Corros. Sci.*, **10**, 709 (1970

Verink, E. D. and Pourbaix, M., 'Use of Electrochemical Hysteresis Techniques in Developing Alloys for Saline Exposures', *Corrosion*, **27**, 495 (1971)

Pourbaix, A., 'Characteristics of Localised Corrosion of Steel in Chloride Solutions', *Corrosion*, **27**, 449 (1971)

Cron, C. J., Payer, J. H. and Staehle, R. W., 'Dissolution Behaviour of Fe-Fe$_3$C Structures as a Function of pH, Potential and Anion; an Electron Microscopic Study', *Corrosion*, **27**, 1 (1971)

Cowan, R. L. and Staehle, R. W., 'The Thermodynamics and Electrode Kinetic Behaviour of Nickel in Acid Solution in the Temperature Range 25-300°C', *J. Electrochem. Soc.*, **118**, 557 (1971)

Brook, P. A., 'A Computer Method of Calculating Potential-pH Diagrams', *Corros. Sci.*, **11**, 389 (1971)

Bardal, E., 'pH and Potential Measurements on Mild Steel and Cast Iron During Periodic Cathodic Polarisation at 20°C and 90°C', *Corros. Sci.*, **11**, 371 (1971)

Bartoníček, R. and Lukasovská, M., 'A Completed Potential-pH Diagram for the System pH-NH$_3$-Cl$^-$-H$_2$O', *Corros. Sci.*, **11**, 111 (1971)

Zehr, S. W., 'A Study of the Intergranular Cracking of U−7·5 wt. %, Nb−2·5 wt. % Zr (Mulberry) Alloy in Aqueous Chloride Solutions', *Corrosion*, **28**, 196 (1972)

Leidheiser, H. and Kissinger, R., 'Chemical Analysis of the Liquid Within a Propagating Stress Corrosion Crack in 70:30 Brass Immersed in Concentrated NH$_4$OH', *Corrosion*, **28**, 218 (1972)

Heidersbach, R. H. and Verink, E. D., 'The Dezincification of α and β-brasses', *Corrosion*, **28**, 397 (1972)

Macdonald, D. D. and Butler, P., 'The Thermodynamics of the Al–H_2O System at Elevated Temperatures', *Corros. Sci.*, **13**, 259 (1973)

Misawa, T., 'The Thermodynamics Consideration for the Fe–H_2O System at 25°C', *Corros. Sci.*, **13**, 659 (1973)

Cowan, R. L. and Kaznoff, A. I., 'Electrochemical Measurements of Corrosion Processes in a Boiling Water Nuclear Reactor', *Corrosion*, **29**, 123 (1973)

1.5 Passivity and Localised Corrosion

Introduction

Passivity is a state of low corrosion rate brought about under a high anodic driving force, or potential, by the presence of an interfacial solid film, usually an oxide. The phenomenon has been defined in a number of ways[1,2], most of which are similar in meaning. Although from the more literal sense of the word, passivity could include metals immune from corrosion (such as gold in water), in the parlance of corrosion science, this is specifically excluded. Passive metals are thermodynamically unstable[3]: they possess a kinetic stability, which is engendered by a solid interfacial film, and without which corrosion would occur. The oxide film, itself formed anodically through a mechanism very similar to that of the corrosion process, stifles the rate of further oxidation (corrosion and further passivation) by forming a barrier between the metal surface and its environment. So many metallic structures and components depend for their stability on the state of passivity. From an engineering point of view the phenomenon of passivity is a remarkable one, remarkable because of the extremely thin film required to procure passivity. The typical passivating oxide film on many metals is only some 1–10 nm in thickness, and is produced by oxidation of the surface to a depth measured in monolayers of atoms. Without the passivating oxide film many metallic structures would corrode at a very fast rate, phenomenological evidence for this being provided by the very fast rates at which localised corrosion (such as pitting corrosion or stress-corrosion cracking) can propagate when passivity is disrupted and regeneration of the passivating oxide film cannot take place. This thin oxide film is the stabilising barrier which separates the potentially active metal from the environment for structures smaller than a hypodermic needle and larger than an aircraft. Not surprisingly, passivity and passivation have been examined extensively in the past many decades using a large range of techniques; despite the great deal of information and knowledge that has thereby been generated, there are many questions that remain to be answered before a full understanding of the subject is achieved. For example, ever since the advent of stainless steel, it has been realised that this important class of materials owes its stainlessness to the surface oxide film, and even now, despite the large amount of research, using a wide range of *in situ* and *ex situ* techniques, the exact nature of this oxide film remains to be resolved. Passivity of iron has been long recognised[4-6]; the modern theories concerning the passivating oxide film are primarily due to Evans[7-10].

Passivity of a metal lies in contrast to its activity, in which the metal corrodes freely under an anodic driving force. The passive state is well illustrated by reference to a classical polarisation curve prepared potentiostatically or potentiodynamically (Figure 1.39). As the potential is raised

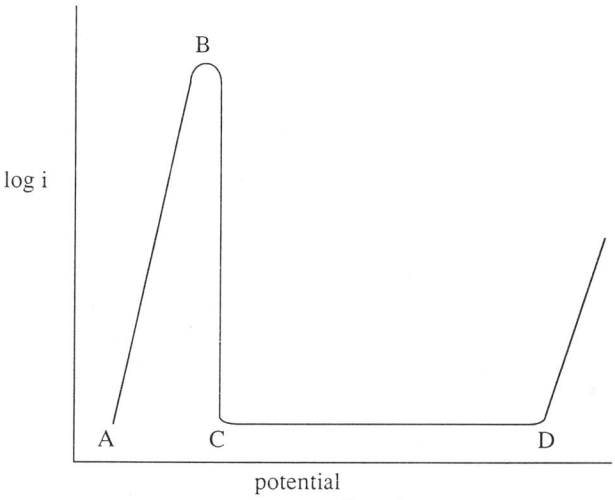

Fig. 1.39 Schematic anodic polarisation curve for a metal. Region AB describes active dissolution of the metal. BC is the active/passive transition, with passivation commencing at B. Passivation is complete only at potentials higher than C. The metal is passive over the range CD

(in the anodic or positive sense) above the equilibrium potential between the metal and its dissolved ions, so the driving force towards oxidation increases and the rate of dissolution increases, in the classical case representing an exponential rise of current with potential according to the Tafel equation[11]. When the potential is high enough, a dramatic reduction in the dissolution rate occurs, and the rate of dissolution remains low with further increase in potential. This latter state is the state of passivity, and the minimum potential at which the low oxidation rate exists is the passivation potential. It is a feature of the passive state that the oxidation rate is independent of potential, or nearly so[12]. This feature does not define the state of passivity uniquely however, since diffusion-controlled dissolution reactions are also independent of potential, and passivity is not due to dissolution under diffusion-control. Some surface films too, while displaying all the qualitative properties of passivity with respect to the polarisation curve, in fact allow a high corrosion current to pass, and cannot be regarded as passivating. An extreme example of this is the phenomenon of electropolishing, where the rate of dissolution is controlled by migration of ions through a viscous salt film layer and the rate is independent of potential, but the current density is high[12-15]. It is a prime requirement of passivity that the oxidation rate is small; how small depends entirely on the application. The state of passivity is never perfect, and a passive metal always corrodes at a finite rate, albeit that this rate may be very low. The surface film forms a barrier between the metal and its environment, which retards further oxidation. Raising the potential in the passive state, while increasing the driving force towards oxidation, serves to thicken the surface oxide film, thereby increasing the barrier towards further oxidation. It is this increase in the barrier film thickness with increase in potential which generates the potential-independence of the oxidation rate as long as the metal remains passive. The phenomenon is used in high-voltage anodising of some metals, where highly resistive oxide films can be grown to thicknesses of several hundred nanometres at high voltages.

The above description of passivity refers to passivity stimulated by an externally applied potential. Passivity of real structures is however, generally achieved in the absence of externally applied power (although it can be stimulated by application of an anodic voltage, the practice of anodic protection[21]). In the absence of an applied potential the sink for the electrons generated by the anodic oxidation of the metal is a cathodic reaction which occurs on the passive metal surface itself. The cathodic reaction thereby provides the driving force. In aqueous systems this is generally the reduction of dissolved oxygen or the reduction of water to hydrogen (or both simultaneously, depending upon the potential involved) occurring on the passive oxide surface, as:

$$O_2 + 2H_2O + 4e^- \rightarrow 4OH^- \qquad \ldots(1.102)$$

or

$$2H^+ + 2e^- \rightarrow H_2 \qquad \ldots(1.103)$$

These same cathodic reactions also fuel the corrosion of metals if corrosion (rather than passivity) is the dominant process. It is to be noted that both cathodic reactions involve an increase in pH; the anodic site correspondingly tends to develop a decreased pH. These changes in pH over the anodic and cathodic sites can be important since they can lead to destabilisation of the passive state. The state of passivity involving a cathodic reaction on the passive surface is sometimes referred to as 'chemical passivity', as opposed to 'anodic passivity' or 'electrochemical passivity' (stimulated by an applied potential). The two phenomena are however very similar: both are electrochemical in nature. The only mechanistic difference lies in the fact that under chemical passivity the transferred electrons must pass *through* the passivating oxide film from the metal atoms to the cathodic reactant, since the anodic reaction (metal oxidation) occurs at the metal/film interface, and the cathodic reaction at the film/electrolyte interface. Under externally applied anodic stimulation, this is not the case, the electrons being passed around an external circuit to a separate cathode. Electrons pass fairly readily through most passivating oxide films, even those of high resistivity, because they are so thin. Other cathodic reactants may also behave as passivating agents. Their role is variable depending on their nature. Thus for example, chromates added to the aqueous environment raise the potential of the metal by accelerating the cathodic reaction. If the potential is thereby raised into the passive regime, the metal is passivated essentially by the added chromate, which then functions as a corrosion inhibitor. The reduced chromate also aids passivity by depositing Cr(III) into the passivating film. The earlier concept that oxygen itself is required as an adsorbed layer for passivation is of course, not tenable, since passivity can be achieved readily with alternative oxidizing agents, or with an applied potential in the absence of dissolved oxygen (or any other oxidizing agent): passivity is due to a surface film, usually an oxide, as a distinct solid phase and usually of thickness greater than one monolayer. The oxide ions required to form the film come from water molecules, and the role of dissolved oxygen is merely that of an oxidising agent.

Other corrosion inhibitors also enhance passivity without electrochemical reduction, by depositing insoluble oxidation products into the passivating film. For example benzoate ions cause deposition of ferric benzoate into the oxide, but do not provide any cathodic reaction.

The phenomenon of chemical passivity can be considered by inspecting the polarisation curves for the separate anodic and cathodic reactions, shown in Figures 1.40 and 1.41. The two forms of passivating agent (that giving accelerated cathodic reaction and that providing insoluble reaction product) are both illustrated. Using simply the fact that the anodic and cathodic reactions must occur at an equal rate, the point of intersection of these two kinetic lines in Figures 1.40 and 1.41 describes the condition of the metal. To achieve chemical passivity, the potential must be sufficiently high to overcome the active loop (otherwise the oxidising agent accelerates corrosion), and it is easy to see that there must therefore be sufficient cathodic reactant of sufficiently high driving force.

The process of formation of a passivating oxide film is an anodic one; the driving force for its formation is raised by raising the potential anodically

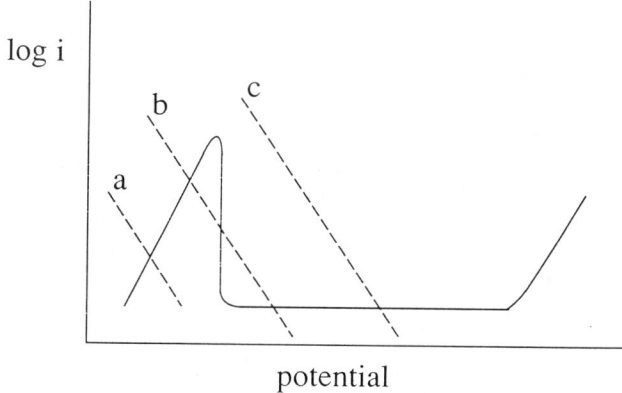

Fig. 1.40 Schematic anodic polarisation curve for a passivatable metal (solid line), shown together with three alternative cathodic reactions (broken line). Open-circuit corrosion potentials are determined by the intersection between the anodic and cathodic reaction rates. Cathode a intersects the anodic curve in the active region and the metal corrodes. Cathode b intersects at three possible points for which the metal may actively corrode or passivate, but passivity could be unstable. Only cathode c provides stable passivity. The lines a, b and c respectively could represent different cathodic reactions of increasing oxidizing power, or they could represent the same oxidizing agent at increasing concentration.

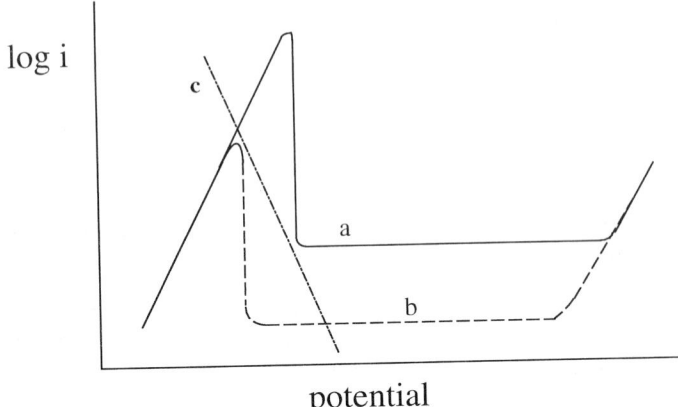

Fig. 1.41 Schematic anodic polarisation curves for a passivatable metal showing the effect of a passivating agent that has no specific cathodic action, but forms a sparingly soluble salt with the metal cation. a: without the passivating agent. b: with the passivating agent. The passive current density, the active/passive transition and the critical current density are all lowered in b. The effect of the cathodic reaction c, is to render the metal active in case a, and passive in case b

across the metal/environment interface. Thus for example, the driving force for formation of a passivating oxide film on iron by the reaction.

$$2Fe + 3H_2O \rightarrow Fe_2O_3 + 6H^+ + 6e^- \qquad \ldots(1.104)$$

is raised as the potential is raised above the equilibrium potential between the metal and its oxide. Raising the potential however, also raises the driving force towards oxidation of the metal quite generally and thus increases the propensity towards accelerated corrosion. The two processes of accelerated corrosion and passivation are therefore generally in competition, and the balance between the two can be quite delicate. Examples abound, both in the laboratory and in the field, of passivity failure, leading frequently to rapid and sustained localised corrosion. Mechanisms describing the origins of stress-corrosion cracking and corrosion-fatigue inherently involve rupture of the passivating oxide film, and the consequently rapid reactions that may ensue on the bared metal surface. The basis on which such forms of localised corrosion occur are well understood. In its passive state the metal is thermodynamically very unstable. The conditions of the metal, the solution and the potential dictate its state of passivity. If passivity is lost by whatever mechanism (for example by oxide film rupture or through reaction of local heterogeneities), the metal reacts anodically at high potential, and therefore at high rate, with the cathodic area occurring chiefly on the surrounding undamaged passive surface. One consequence of such anodic reaction is the transport of ions to carry the current, and the generation of acidity through hydrolysis of the reaction products[26-30]. (Note that the reaction above describing ferric oxide formation on iron itself generates H^+.) If these processes generate a local solution of sufficient aggressiveness, then localized corrosion will ensue: the rate will be fast because of the high potential. If the local solution developed is not sufficiently aggressive or the reaction rate not high enough (because the potential is too low) the site repassivates (i.e. regenerates the passivating oxide film). These phenomena are dealt with in more detail below.

Passivity normally exists within a well defined potential range, below which the metal may activate and corrode, and above which it may transpassivate and corrode. The potential range is characteristic of the metal/environment system, and critically, of the available cathodic reaction. Because all anodic processes occurring through the passivating film, including film growth itself, are controlled by the *ionic* conductance of the film[12,31], a low ionic conductivity to metal cations as well as oxide ions is beneficial to passivity. Oxides of high ionic conductivity would grow to be thick, and thick oxide films are more likely to be mechanically unstable, because they are more crystalline, and more defective. The passivating film thickness is also controlled by the rate of dissolution of the oxide into the electrolyte. The *electron* conductivity of the oxide is of less consequence. Metals such as aluminium and tantalum, whose oxides are of low electronic conductivity, can be as passive as iron and copper, whose oxides are highly electronically conductive. Although passivity in open circuit requires electron passage through the oxide, this is rarely a limiting step. Consideration of the variables associated with each system is thus critical in determining the state of passivity.

Determination of the Passive Corrosion Rate

If a metal were perfectly passive and the solubility of its ions were zero, the passive corrosion rate would be zero but it would take a long time to reach that state. This is never the case, and passivity always implies an ultimate steady state corrosion rate. It is generally important to know that rate. This is determined as the passive current density, i_{pass}, through polarisation experiments. The rate of penetration in the passive state is then $dx/dt = Mi_{pass}/zF\rho$ where M is the atomic weight of the metal, z is the electron number, F the Faraday constant and ρ the density of the metal. Polarization curves such as that shown in Figure 1.39 are commonly measured by potential sweep measurements (at constant sweep rate, dE/dt); such curves also yield the passive range of potential. The value achieved however, depends on the imposed potential sweep rate because at any potential the steady state value of i_{pass} takes some time to be reached. This time to reach a steady state, often overlooked or ignored in laboratory studies of passivity, can be measured in minutes, hours, or even days. Fast sweep rates may therefore yield an erroneously high value. The passive current density is thus more accurately obtained at constant potential. A further feature, which can yield an erroneously low value of i_{pass}, is the presence of a cathodic reactant, such as oxygen. To overcome this, measurements in the laboratory are made in deaerated solution; full deaeration is however, not possible. The lower the true passive current density, the greater the inaccuracy in the measured value. Measurement of i_{pass} at constant potential does not resolve this inaccuracy. The error can be recognised by measurement using a rotating disc or cylinder electrode. Variation in the rotation rate then alters the rate of transport of oxygen, and is observed as an *apparent* reduction in i_{pass} as the rotation rate increases. What passive corrosion rate constitutes a viably passive structure is very much open to question and dependent on the application. One of the most demanding states of passivity is that required by metallic surgical implant materials, where even small passive corrosion rates of metals, (undamaging to the metal) which show toxicity in ionic form and whose ions are not excreted at the appropriate rate, may accumulate to significantly toxic levels. It is nevertheless quite common experience for the passive corrosion rate to be so low, that a well engineered structure corroding in its passive state, is visibly unchanged after very many years service.

The Polarisation Curve

It is worthwhile to consider the general form of the anodic polarisation curve obtained when the potential of a metal is scanned in deaerated solution. Figure 1.42 shows a series of schematic curves which describe the onset of passivity (as shown in Figure 1.39) as well as the behaviour of the passive state itself. Such curves can only be determined potentiostatically or potentiodynamically; galvanostatic measurement of potential cannot provide the polarisation curve of a passivating metal because the potential is a three-valued function of the current density. At low anodic potential (region AB) in solutions where the metal ion has appreciable solubility, the metal dissolves actively with the current rising exponentially with increase in potential (see figure 1.39). Passivation commences at B and is complete at C; region BC comprises the active/passive transition. The passive current than continues to flow with further increase in potential, approximately independently of potential, until point D, F, H or J depending upon the properties of the passivating film with respect to the electrolyte solution.

In solutions of a sufficient concentration of aggressive ions, such as chloride, premature failure of passivity occurs with many metals (such as aluminium, iron, nickel and their alloys) at point D by the nucleation of pitting corrosion. Failure of passivity is localised, and the current rises with further increase in potential (region DE) as the pits grow and more pits form. Note that the current density is usually obtained by dividing the measured current by the surface area of the specimen under test. When localised corrosion, such as pitting occurs, the actively dissolving area is very much smaller than this (otherwise the corrosion could not be termed localised). The current density measured as such does not thus represent the *local* corrosion rate, or the penetration rate, which is in fact much faster. The rise in mean current density in region DE of Figure 1.42 is not necessarily an exponential function of the potential. The region carries an increasing number of pits with increase in potential as well as their growing surface area with time. This region also shows significant noise in the measured current.

In the absence of aggressive anions, the passivating oxide film may itself begin to be oxidised at point F to produce a soluble anion, and passivity again fails, this time by oxidative dissolution. The current rises with further increase in potential (region FG) as the dissolution rate rises. This occurs for example with chromium and manganese by oxidation to CrO_4^{2-} and MnO_4^{2-} or MnO_4^- respectively. If the oxide film is stable, and is also a

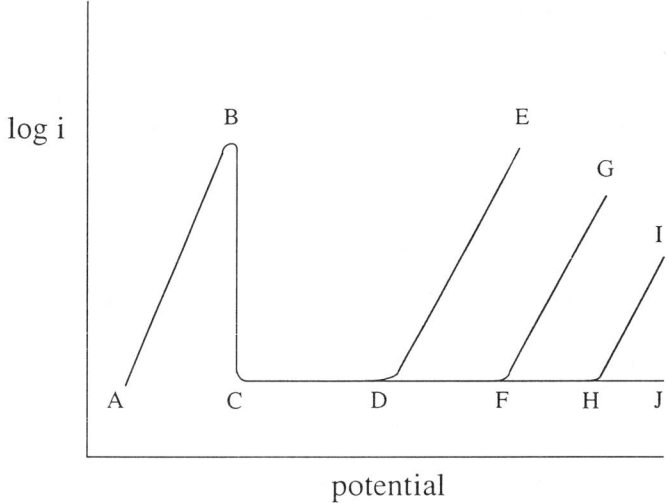

Fig. 1.42 Schematic anodic polarisation curve of a metal showing possible behaviour of the passive region. AB is the active region and BC the active/passive transition. The metal is passive at potentials more positive than C. If the solution contains aggressive anions such as chloride, passivity may break down at D (the pitting potential) and the current rises with further increase in potential (DE) as corrosion pits propagate and more nucleate. The region DE is not necessarily linear. If pitting agents are absent, but the passivating film can be transpassively dissolved, then transpassive corrosion commences at F, increasing in rate with increasing potential (FG). If the passivating film is chemically and electrochemically stable, and is conductive to electrons, oxygen evolution commences at H, and increases in rate to I. The metal remains passive. If the film is stable and insulating to electrons, oxide film growth continues with further increase in potential (HJ) and the metal remains passive

good electron conductor, such as occurs with iron and nickel in neutral or moderately alkaline solutions, then point H marks the potential at which the solution itself begins to be oxidised. In aqueous solution this is the oxidation of water to oxygen, for which the standard electrode potential is 1.228 V(SHE) at 25°C. The rise of current with increase in potential in region HI then simply describes the increasing rate of oxygen evolution. Passivity here is not disturbed (unless the oxide film is damaged mechanically by the oxygen bubbles). This process can only occur if the oxide film is an electron conductor, since the oxidation of water requires the withdrawal of electrons from the water molecules through the oxide to the metal: oxidation of water occurs at the film/electrolyte interface.

Finally, if the film is stable in solution, and is a poor electron conductor, such as occurs on aluminium and tantalum in halide-free environments, oxygen evolution is not possible. Raising the potential beyond point H in Figure 1.42 causes the film to grow further, with the current continuing to be approximately independent of potential (region HJ). It is under these conditions that high-voltage anodising can be adopted to produce relatively thick oxide films, as described briefly below.

Anodising Under High Applied Voltage

For metal/environment couples where the passivating oxide film does not break down, does not oxidatively dissolve, and does not conduct electrons readily, the only mechanism for relaxing the increasing electric field as the potential is increased is the thickening of the oxide film. The process occurs by ion conduction through the film. In fact, if the potential of aluminium immersed in a suitable anodising electrolyte is raised suddenly to a high value, transient evolution of oxygen bubbles is observed. The oxide is of course a dielectric, and the sudden potential jump imposes dielectric breakdown with conduction of electrons because of the increased electric field. The effect is rapidly quenched as the barrier oxide film grows simultaneously, relaxing the electric field.

In many cases passivating and highly electrically insulating oxide films can be grown to thicknesses of several hundred nanometres, and even micrometres; such films can be observed readily in the scanning electron microscope[16-20,31]. The applied anodic voltages can be in the region of several tens, and even hundreds of volts, and can be applied as a d.c. current or voltage, or as an a.c. voltage, depending on the required properties of the oxide. The most widely anodised metal industrially is aluminium. Although a.c. anodising of aluminium imposes a cathodic half-cycle for each anodic half-cycle, oxide film reduction does not occur, contrary to what might be expected. This is because the oxide film on aluminium is very irreducible. In the appropriate electrolyte the passivating oxide film can be induced to be overlaid with a much thicker microporous oxide layer. Electrolytes which induce formation of a porous overlayer tend to be the more aggressive ones, for example aluminium in sulphuric or phosphoric acid. Inevitably, the porous layer is underlaid with a non-porous barrier layer (otherwise the high applied voltage would induce vigorous oxygen evolution in those areas lacking the electron-insulating oxide). The barrier layer is thus named because it carries most of the electric field and controls the film growth current; the barrier layer is the passivating component. Its thickness increases linearly with the applied voltage, consistent with the behaviour of other passivating oxide films. This duplex oxide film formed on aluminium finds wide technological application. Anodising of titanium to appropriate oxide film thicknesses produces vigorous interference colours, a phenomenon exploited in the ingenious art of Pedeferri[32].

Mechanism and Kinetics of Passivation

The basic mechanism of passivation is easy to understand. When the metal atoms of a fresh metal surface are oxidised (under a suitable driving force) two alternative processes occur. They may enter the solution phase as solvated metal ions, passing across the electrical double layer, or they may remain on the surface to form a new solid phase, the passivating film. The former case is active corrosion, with metal ions passing freely into solution via adsorbed intermediates. In many real corrosion cases, the metal ions, despite dissolving, are in fact not very soluble, or are not transported away from the vicinity of the surface very quickly, and may consequently still deposit on the surface as an oxide or insoluble salt. Often these cases are not passivating because the insoluble corrosion product is deposited from the solution phase as a poorly adherent, loose, crystalline substance. Formation of rust on steel is one such example; blue and green deposits on corroded copper plumbing are another. Oxide films which are passivating are most commonly formed when they are generated *directly* on the metal surface without the constituent metal ions passing first into solution (e.g.[33]). While the precise structures of passivating films remain contentious[34, 35], they are known to be coherent and continuous, necessarily so since any lack of continuity would provide paths for active dissolution leading to localised corrosion.

Formation of the first layer (a monolayer) of passivating oxide film on a denuded metal surface occurs very simply by the loss of protons from the adsorbed intermediate oxidation products, such intermediates being common to both dissolution and passivation processes[36-44]. Thus for example, the first oxidative step in the anodic oxidation of nickel is the formation of the unstable adsorbed intermediate NiOH by

$$Ni + H_2O \rightarrow NiOH_{ads} + H^+ + e^- \qquad \ldots(1.105)$$

Similar initial reactions occur on many metals such as iron and cobalt. This intermediate can now react further in one of two ways. Oxidation and *protonation* of the intermediate to Ni(II) leads to dissolved nickel ions (active corrosion) which are unable to passivate the metal:

$$NiOH_{ads} + H^+ \rightarrow Ni^{2+}_{aq} + H_2O + e^- \qquad \ldots(1.106)$$

On the other hand, oxidation and *deprotonation* of the same intermediate leads directly to passivation by:

$$NiOH_{ads} \rightarrow NiO_f + H^+ + e^- \qquad \ldots(1.107)$$

where the subscript ads represents the adsorbed intermediate, aq represents the aquated metal ion in solution, and f represents the oxide film. These reactions are able to proceed because the solvent, water, is effectively an infinite source, and sink, for H^+. The intermediate $NiOH_{ads}$ is common to both the dissolution and passivation reactions. It can be regarded as an hydroxide of Ni(I) or an adsorbed OH radical; the distinction is probably not important to the case of passivation. Because of subsequent further reaction, it is better considered as the compound of univalent nickel. This intermediate does not exist as a bulk species. Its existence as an intermediate is undoubtedly due to the fact that it is adsorbed to the metal surface on one side, and carries the aqueous double layer on the other. Which of reactions 1.106 and 1.107 occurs preferentially depends critically on the solution composition, its pH, the potential, and of course, temperature. It is apparent that the nickel atoms involved in going from Ni to NiO via steps (1.105) and (1.107) never leave the metal surface. Formation of NiO (or the hydrated form, $Ni(OH)_2$) by precipitation from the dissolved nickel ions formed in reaction (1.106) does not lead directly to passivity. For iron, formation of oxides from dissolved Fe^{2+} via precipitation from solution leads to non-protective rust. (Brown rusts involve further oxidation of Fe^{2+} by dissolved oxygen). Such precipitated oxides are not usually passivating. The passivating oxide formed by reaction (1.107) can only occur when the equilibrium potential between the Ni and NiO is reached or exceeded. Higher oxides of nickel may also impart passivity once the appropriately higher equilibrium potential is exceeded[33]. Clearly, such oxides form preferentially in solutions where the oxide is effectively insoluble: for many metals these are solutions of neutral and mildly alkaline pH. Nevertheless, in acidic solutions where the oxide is thermodynamically unstable relative to the dissolved metal cation, passivation can still occur (and does so on many metals, including nickel) as so well exhibited by chromium. Cr(III) oxide is thermodynamically unstable in acidic systems[3], but the metal is still very passive in sulphuric acid (at the appropriate potentials); the implication is that the dissolution of the passivating oxide into acidic solutions, which must inevitably occur, is very slow, and the rate of corrosion of chromium in its passive state is undoubtedly controlled by this dissolution rate. (It is chiefly chromium (III) which gives rise to passivation of alloys containing sufficient quantities of Cr, such as stainless steels and nickel/chromium alloys.)

There are in fact many possible further steps that could be included in the basic mechanisms described above, for example, involving adsorption processes for H_2O and OH^-, and interactions among the adsorbed species. The most widely studied is the reaction of iron, but broadly similar steps are encountered in the anodic oxidation of many metals. These several mechanisms still include the same basic steps as described above[36-44].

The first monolayer is generally formed before any significant film thickening occurs, because of the very strong dependence of the film growth rate on the electric field across the reacting interface. A mechanism of the type shown in equations 1.105 and 1.107 does not necessarily involve an oxide nucleation step. Rather, it is a random oxidation of the exposed surface metal atoms until the first monolayer of oxide is complete[43-46]. Such a model could account for the suggestion that passive films may in fact be amorphous, with no definable crystal structure. Alternatively for some metals, an initial nucleation step may occur with two-dimensional spreading of the oxide over the surface to produce expanding circular 'pancakes' of oxide which link up when they impinge on one another. This mechanism, which involves different kinetics, has not however, been found in repassivation experiments involving mechanically stripped transition metal surfaces[43-46]. Growth of the first monolayer of the passivating oxide already causes significant reduction in the rate of oxidation of the metal, but in general, the process does not stop there.

Once the metal surface has become covered with its first monolayer of the oxide film, the metal is then separated from its environment by that oxide, and film thickening ensues: metal cations or oxide anions (or both) are transported through the existing oxide film, thus relaxing the electric field which drives the mobile ions, until a steady state is reached. Ion migration (or vacancy migration) under the electric field controls the film growth rate[12, 31, 47-50]. As the film thickens at constant potential, the relaxing electric field across it reduces the rate of field-assisted ion migration, and the rate of film growth declines. The steady state is reached when the rate of film thickening equals the rate of film dissolution into the environment, giving a net constant-rate dissolution of metal. Different metals display different passive corrosion rates and it is the numerical value of this rate which dictates whether the metal can be regarded as sufficiently passive for specific engineering purposes. Highly passive metals usually carry very thin oxides, and the electric field sustained by the overall metal/oxide-film/electrolyte interphase region is then very high. For example at an overpotential of 0.2 V and an oxide film thickness of 2 nm the mean electric field is *ca.* 10^8 V m^{-1}. The precise distribution of this electric field, namely that fraction lying across the film itself and that lying across the two phase interfaces is not known, although several models exist[49-53]. Similarly, the identity of the transporting ions is often not known. Although bulk oxides can often be classified as to whether they are cation-vacant, anion-vacant or contain excess ions interstitially, thus dictating the mode of ionic charge transport, passive films are so thin, that properties of bulk oxides may not necessarily be applicable. The effect is strikingly realised when it is considered that many passivating oxide films are often no thicker than one or two unit cells of the equivalent bulk oxide, and their transport properties may consequently differ from those of the bulk material. Only in the model of Macdonald *et al.*[52,53] is such distinction drawn for the corrosion and passivation characteristics, where the model predicts that the transport of metal cations across the oxide from the metal interface to the electrolyte interface causes dissolution, and the transport of oxide ions in the reverse direction causes film growth.

Derivation of the rate of oxide film growth proceeds as follows. When the metal carrying its oxide film is at its equilibrium potential, no overpotential exists across the oxide, and the rate at which ions move through the oxide from the metal to the electrolyte equals the rate at which they move back again. Expressed as current densities, the forward rate is i_f and the backward rate i_b; these rates are equal at equilibrium. Ions moving through the film do so via vacancies (anionic or cationic, leading to anionic or cationic migration), or interstitially. They thereby encounter an energy barrier which they must surmount in order to move at all: the width of this energy barrier is the distance between anionic or cationic sites in the oxide lattice if the mechanism is by vacancy migration, or the distance between interstitial sites if the mobile ions move interstitially. This distance is $2a$. The height of the energy barrier is the activation energy, ΔG^*. Thus, at equilibrium

$$i_f = i_b = zFk \exp\left(-\frac{\Delta G^*}{RT}\right) \qquad \ldots(1.108)$$

where k is the rate constant, z is the charge number, F is the Faraday constant, R the gas constant and T the temperature. Equation 1.108 is an expression of the Arrhenius equation.

We now apply an anodic overpotential η, for film growth, and for simplicity, we regard this overpotential as being distributed entirely across the oxide film. If the oxide film has thickness x, then the electric field across the film is $d\eta/dx$, which equals η/x if the overpotential lies linearly across the oxide film. The potential drop across the distance $2a$ (a single energy barrier) is then $2a\eta/x$. The activation energy in the forward direction (i.e. with the electric field) is thereby reduced by an amount $a\eta zF/x$, and that in the backward direction (i.e. against the electric field) increased by an amount $a\eta zF/x$. Thus, under the applied overpotential η

$$i_f = zFk\exp\left(-\frac{(\Delta G^* - a\eta zF/x)}{RT}\right)$$

$$= zFk\exp\left(-\frac{\Delta G^*}{RT}\right)\exp\left(\frac{a\eta zF}{xRT}\right) \quad \ldots(1.109)$$

Similarly in the backward direction

$$i_b = zFk\exp\left(-\frac{(\Delta G^* + a\eta zF/x)}{RT}\right)$$

$$= zFk\exp\left(-\frac{\Delta G^*}{RT}\right)\exp\left(-\frac{a\eta zF}{xRT}\right) \quad \ldots(1.110)$$

The following parameters are constants (at constant temperature) and are simplified to

$$A = zFk\exp\left(-\frac{\Delta G^*}{RT}\right) \quad \ldots(1.111)$$

and

$$B = \frac{azF}{RT} \quad \ldots(1.112)$$

The net current density, i, in the forward direction (i.e. that producing film growth) is

$$i = i_f - i_b \quad \ldots(1.113)$$

and

$$i = A\exp\left(\frac{B\eta}{x}\right) - A\exp\left(-\frac{B\eta}{x}\right)$$

$$= 2A\sinh\left(\frac{B\eta}{x}\right) \quad \ldots(1.114)$$

This is the general expression for film growth under an electric field. The same basic relationship can be derived if the forward and reverse rate constants, k, are regarded as different, and the forward and reverse activation energies, ΔG^* are correspondingly different: these parameters are equilibrium parameters, and are both incorporated into the constant A. The parameters A and B are constants for a particular oxide: A has units of current density (A m^{-2}) and B has units of reciprocal electric field (m V^{-1}). Equation 1.114 has two limiting approximations.

(a) When the electric field, η/x, is large (high η and/or small film thickness, x), $2\sinh y = \exp y$, and

$$i = A\exp\left(\frac{B\eta}{x}\right) \quad \ldots(1.115)$$

This is commonly known as the high field equation. It is of similar form to the Tafel equation for activation controlled electrochemical reactions with the additional feature that the potential driving the reaction lies across an oxide film of variable thickness, x, rather than across the double layer (of constant thickness). The kinetics of oxide film growth according to this equation *approximate* to (but are not identical to) inverse logarithmic film growth[48,54,55]. The passivity of most passive metals is associated with a high electric field since most passivating films are very thin. A high electric field implies either a thin passivating film (x is small) or a high overpotential (η is large) or both.

(b) When the electric field is small (i.e. η is small and/or x is large), $\sinh y = y$, and

$$i = \frac{2AB\eta}{x} \qquad \ldots(1.116)$$

Equation 1.116 is ohmic ($i \propto \eta$ for constant film thickness): the term $x/2AB$ can be regarded as the film resistance. The equation is identical to parabolic film growth, for which the film thickens with the square root of time at constant potential.

Because passivating films are generally very thin, passivation is normally a high field process and equation 1.115 can replace equation 1.114. The same relationships can be applied to high voltage anodising of metals such as aluminium and titanium. However, under these conditions, the oxide sometimes assumes a duplex structure, with an inner barrier layer and an outer porous layer: for these, the film thickness x, relates to that of the barrier film only, since it is largely this part which carries the applied voltage. Notice that the relationships make no assumption about the identity of the mobile ion, and are equally applicable whether the anion or the cation is mobile within the film lattice causing film growth. Because the anodic current in the passive state is approximately independent of potential in the steady state, the above kinetic equations imply that the thickness of the passivating oxide film increases linearly with increase in the overpotential, and therefore with the applied potential. This has been confirmed by ellipsometric analysis of the oxide film.

Because the film growth rate depends so strongly on the electric field across it (equation 1.115), separation of the anodic and cathodic sites for metals in open circuit is of little consequence, provided film growth is the exclusive reaction. Thus if one site is anodic, and an adjacent site cathodic, film thickening on the anodic site itself causes the two sites to swap roles so that the film on the former cathodic site also thickens correspondingly. Thus the anodic and cathodic sites of the stably passive metal 'dance' over the surface. If however, permanent separation of sites can occur, as for example, where the anodic site has restricted access to the cathodic component in the electrolyte (as in a crevise), then breakdown of passivity and associated corrosion can follow.

The kinetics outlined above, first observed empirically by Güntherschulze and Betz[56], were modelled by Verwey[47] with the rate-controlling energy barrier being that between to adjacent cation sites within the oxide film. The same basic form can be derived if the rate-controlling energy barrier is that between a metal atom on the metal surface and an adjacent cation site in the film[12,48]. The rate is then limited by ion injection into the film rather than ion migration through the film. It should also be noted that the above kinetic relationships can be modified to take account of various other processes occurring during anodic oxide film growth, such as space charges within the film and charging of the phase interfaces[49,50,57].

The point defect model of Macdonald et al.[53] draws a distinction between cation and anion mobility, with mobile cations leading to metal dissolution and mobile anions (oxide ions) leading to passivity: this model has the inherent attraction of incorporating a mechanism for metal dissolution. The procedure leads to direct logarithmic film growth (at constant potential), as does the place exchange mechanism of Sato and Cohen[58]. Apart from the work of Macdonald[53], the passive state dissolution rate has been studied far less than the oxide film growth rate, and it is not really known why the dissolution rate of the passivating oxide should be so slow. The process of oxide dissolution has been reviewed[59] and is complex. The complexity of the dissolution has been exemplified by Nii[60], who showed that the rate of dissolution of nickel oxide particles (with no metal substrate) in aqueous solution is very strongly dependent on trace quantities (for concentrations down to ca. 10^{-6} M) of ionic impurities in the solution. The rate of dissolution of oxides is influenced by the polarisation of the oxides, a notion clearly very important to the passivity of metals.

The kinetics outlined above refer strictly to single crystals, with no metallurgical defects, such as grain boundaries. Polycrystalline metals and alloys contain many such defects, and their passivation is expected to be different, especially when the defects may contain high concentrations of impurity atoms, such as sulphur segregated in the grain boundaries in steels. The kinetics of passive oxide film growth, as measured electrochemically, may well be dominated by the passivation of the grain faces because of their overwhelmingly large surface area. However, the longer term stability of the passive state may become dominated by the grain boundaries (as in the case where intergranular corrosion occurs) or by inclusions, precipitates or other microscopic defects. This feature is very important in consideration of measured passive corrosion rates.

Determination of the kinetics of passivation can be carried out potentiostatically provided any previously formed oxide film, such as that formed by prior exposure to air is first removed. Such removal can be achieved by reduction at low potential for fairly noble metals, but cannot be readily achieved for the relatively base metals such as chromium, aluminium or titanium. Some metals, such as iron, become charged with hydrogen during cathodic reduction of their oxide films, a process predicted by the thermodynamics of the two processes. Many alloys may undergo changes in surface composition and structure during applied electrochemical reduction, and subsequent reoxidation kinetics may not be representative of the true alloy surface. Under these circumstances the electrochemical kinetics of oxide film growth can only be determined if the preformed oxide film is stripped away mechanically, by cutting with an inert scribe, such as diamond, sapphire or boron nitride[61-68]. Such techniques then allow measurement of the kinetics of passivation through potentiostatic current transients or through galvanostatic potential transients. Passivation in open circuit is also amenable to such examination[68].

Thermodynamics of Passivity

The thermodynamics of passivity, primarily due to Pourbaix[3], are usefully considered by examining the equilibria between metals and their oxides in contact with water. Before doing so however, it is worth bearing a number a points in mind. First, the passivating oxide film is in many cases so thin, that its thermodynamics may not be identical to that of bulk oxides. This is already implicit in the discussion of the adsorbed NiOH intermediate in nickel oxidation above. The matter is of particular importance when it is realised that the two faces of the oxide film, that in contact with the metal and that in contact with the environment, are so very different. By its nature, the passive film must therefore be regarded as extremely anisotropic in the sense that its dimensions are very much larger in the two directions parallel to the metal surface than the third dimension, perpendicular to the metal surface. Second, the electric field across any existing passivating oxide is very large in the direction perpendicular to the metal surface and insignificant in the parallel direction, adding to its anisotropy. Finally, in the case of engineering alloys such as stainless steels, it is frequently difficult to ascribe the passivating oxide film to a known stoichiometry or structure, and the use of thermodynamic data of known bulk oxides is open to question. This problem is clearly less pronounced for pure metals. Nevertheless, consideration of the thermodynamics of standard bulk metal oxides can still provide surprisingly good agreement with observation, and can often be used effectively and predictively.

As with all determinations of thermodynamic stability, we commence by defining all stable phases possible, and their standard chemical potentials. For most metals there are many such phases, including oxides, hydroxides and dissolved ions. For brevity here, only the minimum number of phases is considered. The simplest system is a metal M, which can oxidise to form a stable dissolved product, M^{2+} (corrosion), or to form a stable oxide MO (passivation). In aqueous environments three equilibria can thereby be defined:

$$M^{2+} + 2e^- \rightarrow M \qquad \ldots(1.117)$$

$$MO + 2H^+ + 2e^- \rightarrow M + H_2O \qquad \ldots(1.118)$$

$$M^{2+} + H_2O \rightarrow MO + 2H^+ \qquad \ldots(1.119)$$

The equilibrium potentials (E_0) for reactions 1.117 and 1.118 are then given respectively by the Nernst equation:

$$E_0 = E^0 - \frac{RT}{2F} \ln \frac{(M)}{(M^{2+})} \qquad \ldots(1.120)$$

$$E_0 = E^0 - \frac{RT}{2F} \ln \frac{(M)(H_2O)}{(MO)(H^+)^2} \qquad \ldots(1.121)$$

where the brackets represent the activities of the components and E^0 is the standard equilibrium potential (or standard electrode potential) for the reaction. The value of E^0 is readily calculated from the standard chemical potentials of the relevant components. By defining the standard states of the metal, the oxide, and the solvent as the pure components (unit mole fraction), the only solution variables are the M^{2+} and H^+ activities, provided the metal, the oxide and the water are pure. Thus, in decadic logs, equations 1.120 and 1.121 respectively are

$$E_0 = E^0 + \frac{2.303\,RT}{2F} \log(M^{2+}) \qquad \ldots(1.122)$$

and

$$E_0 = E^0 - \frac{2.303\,RT}{F} \text{pH} \qquad \ldots(1.123)$$

For the non-electrochemical reaction 1.119, the equilibrium constant (K) for the reaction is written as

$$\text{pH} = -\tfrac{1}{2}\log K - \tfrac{1}{2}\log(M^{2+}) \qquad \ldots(1.124)$$

The thermodynamic equilibria are then represented for fixed dissolved metal ion activity as a plot of E_0 against the pH of the solution, as shown in Figure 1.43, noting that equation 1.122 is independent of pH and equation 1.124 is independent of potential. The figure shows that oxide stability is achieved at potentials greater than E_0 for reaction 1.123 and in solutions of pH greater than that described by equation 1.124. If the potential is greater than E_0 for reaction 1.122 and the pH more acidic than that given by equation 1.124 the metal corrodes to give M^{2+}. If the potential is low the metal is immune to oxidation. The diagram thus describes the *thermodynamically stable* phases of the metal in terms of regions of passivity, corrosion and immunity as a function of E and pH. The positions of the lines representing equations 1.122 and 1.124 depend of course on the value of the M^{2+} activity selected. It is common (but not necessary) to adopt a concentration of 10^{-6} M for this purpose, but the issue is critical in determining the relative thermodynamic stabilities of the dissolved component and the oxide. The fact that only equilibria with water are considered in the example specifies that any dissolved components present in solution have no effect on the reactions of the metal and its components. Reactive ions such as Cl^- dissolved in the solution modify the equilibria by their own free energy considerations. Although more complicated, such systems can nevertheless be thermodynamically calculated if the formation constants of the associated complexes and compounds can be defined.

The line representing equation 1.123 is shown broken for pH more acidic than equation 1.124. The oxide is not thermodynamically stable in this region of pH, and for true thermodynamic equilibrium should not be shown at all. Many metals however, still show a passive region in this state, provided the potential is higher than that given by equation 1.123. The origin of this is of course, kinetic. If an oxide film does form, and if it is very slow to dissolve (it must dissolve since the oxide is unstable relative to the dissolved ion) then passivation ensues. The state has sometimes been referred to as 'metastable passivity'[33]. While it might reasonably be rationalised that passivity is brought about by a very low solubility of the dissolved metal ion, thereby effectively preventing metal dissolution through formation of a sparingly soluble oxide layer, this is not a necessary condition for passivity.

Certainly a thermodynamically stable oxide layer is more likely to generate passivity. However, the existence of the metastable passive state implies that an oxide film may (and in many cases does) still form in solutions in which the oxides are very soluble. This occurs for example, on nickel, aluminium and stainless steel, although the passive corrosion rate in some systems can be quite high. What is required for passivity is the *rapid* formation of the oxide film and its *slow* dissolution, or at least the slow dissolution of metal ions *through* the film. The potential must, of course be high enough for oxide formation to be thermodynamically possible. With these criteria, it is easily understood that a low passive current density requires a low conductivity of ions (but not necessarily of electrons) within the oxide.

It is not possible to predict thermodynamically *whether* or not a metastable passivating film forms in acidic solutions. Thus nickel, cobalt and aluminium exhibit metastable passivity in acidic solution (although not necessarily practicably so); cadmium does not. Metals which display metastable passivity, particularly in acidic solutions, often show polarisation curves whose potential ranges can be correlated with the thermodynamic potentials of the potential/pH (Pourbaix) diagram (Figure 1.43). Thus for example, nickel in sulphuric acid shows an active loop between the potentials represented by lines a and b in the Pourbaix diagram; as the potential is increased further, passivity commences when the potential exceeds approximately that described by equation 1.123, observed as an active/passive transition. Because metastable passivity is a kinetic phenomenon, however, a significant overpotential may be required to produce it. Where more than one oxide may exist, it is not necessarily that associated with the lowest metal oxidation state that provides the best state of passivity. On thermodynamic grounds alone, one predicts the classic form of the anodic polarisation curve, in which the passive range of potential is preceded (at lower potential) by an active loop as shown in Figures 1.39–1.42. From the point of view of corrosion engineering for acidic systems, metastable passivity is of the utmost importance. Most structural metals designed for acidic environments are used in this state of metastable passivity. These are often chromium-containing alloys (stainless steels and nickel-based alloys), since chromium shows extremely good metastable passivity in acidic environments.

It is worth emphasising too, that the position of those lines representing equilibria with the dissolved species, M^{2+}, depend critically on the solubility of the ion, which is a continuous function of pH. For example, iron in moderately alkaline solution is expected to be very passive[3], and so it is in borate solutions (in the absence of aggressive ions). However, the anodic polarization curve still shows a small active loop at low potential.[69]

The form of Figure 1.43 is common among many metals in solutions of acidic to neutral pH of non-complexing anions. Some metals such as aluminium and zinc, whose oxides are amphoteric, lose their passivity in alkaline solutions, a feature reflected in the potential/pH diagram. This is likely to arise from the rapid rate at which the oxide is attacked by the solution, rather than from direct attack on the metal, although at low potential, active dissolution is predicted thermodynamically[3]. The reader is referred to the classical work of Pourbaix[3] for a full treatment of potential/pH diagrams of pure metals in equilibrium with water.

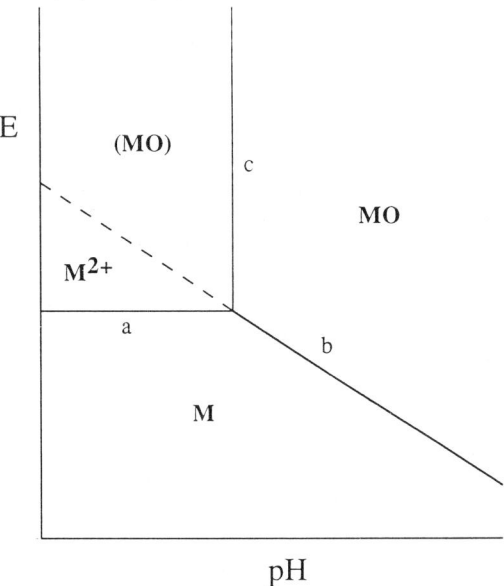

Fig. 1.43 Schematic potential/pH diagram for a metal M in equilibrium with water in the absence of complexing species. Line a represents equations 1.117 and 1.122. Line b represents equations 1.118 and 1.123. Line c represents equations 1.119 and 1.124. The stable phases are marked in bold. The metastable phase is in parentheses. The broken line is an extrapolation of equation 1.123 and indicates possible metastable passivity

Chemical Passivity and Electrochemical Passivity

Passivity is provided and maintained by the appropriate cathodic reaction such as reaction 1.102 or 1.103: it is a requirement of this reaction that the potential is held within the passive range. This may be achieved by the cathodic reactant in the solution, such as oxygen, or it may be applied from an external power supply. In the absence of further oxidizing agents only reaction 1.102 or 1.103 can provide and maintain passivity. For more noble metals reduction of water is not possible, and passivity can be maintained only by reaction 1.102. These metals are, of course, immune from corrosion (and passivation) in the absence of dissolved oxygen. Base metals such as aluminium, when bared in solution (these metals inevitably carry a prior air-formed oxide film), are rendered passive by reduction of water because the initial potential is so low[61,62,68]. Passivity can then be *maintained* by reduction of oxygen because the potential rises significantly during passivation. Passivating agents in solution may be divided broadly into two groups. Those which provide the cathodic reactant are oxidising agents and raise the open-circuit potential of the metal into the passive regime, but only if employed to sufficient concentration. The process is shown schematically in Figure 1.40. In insufficient concentration such oxidising agents raise the potential further into the active region, thereby accelerating corrosion. Such behaviour is well typified by the long-known reaction between iron and nitric acid. In concentrated nitric acid iron is passive, but if the acid is diluted, very powerful corrosion occurs with vigorous evolution of nitrogen oxides. This process was reported long ago by Keir[5] and Faraday[6]. Many oxidising agents fall into this category, such as chromates, molybdates, oxygen, and hydrogen peroxide acting on steels. Some oxidising agents become aggressive in reduced form, and cannot be regarded as passivating agents; these include for example, the chlorate anion which is cathodically reduced to chloride on iron. The chloride anion can initiate pitting corrosion, particularly at high potential, and the high potential is itself brought about by that same cathodic reaction.

Other passivating agents, such as phosphates[70-73] and benzoates[74] (for steels) do not fall into this category and they have no oxidising action. Their role in enhancing passivity is to aid formation of the passive film itself under whatever cathode is present in solution. In a polarisation curve this would be observed as a decrease in the passive current density, often coupled with a decrease in the potential of the active/passive transition (see Figure 1.41). These chemical passivating agents act by forming highly insoluble compounds with the dissolving metal (such as ferric phosphates and benzoates). The salts are precipitated over areas of the metal surface which may become active. Presumably a passivating oxide is then formed beneath the deposited metal salt, since the metal is then no longer capable of rapid dissolution, although this has not been firmly established. The formation of azelaic acid from linseed oil[75,76] allows precipitation of sparingly soluble ferric azelate as a passivating agent in linseed oil-based paints. Again, the metal presumably passivates by oxide film formation beneath the deposited azelate. For these passivating agents to be effective, an oxidising agent, such as oxygen must of course, be present.

Chromate ions fall into both categories, namely as both an oxidising agent and a source of insoluble film[77,78]. Their oxidising power raises the potential into the passive region for a number of metal systems. The reduced form is Cr(III), either as Cr_2O_3 or as a spinel with the metal oxide of the substrate, which is deposited onto the metal surface; this is itself a passivating agent and acts as such for many metals. These prime processes render chromates such effective passivating agents and are responsible for the use of chromates as inhibitive pigments in metal priming paints (often as zinc chromate).

Other forms of similar chemical passivating agent act where no oxide is formed at all. These must be mentioned here, although their action is not truly passivating. Although the dissolution-precipitation mechanism for oxide formation on metals like iron does not generally lead to passivation, some salt films formed by precipitation of dissolved ions from solution may give rise to significant passivity. Thus silver in acidic chloride solutions forms a very stable insoluble layer of AgCl, which reduces substantially the corrosion rate of the metal. The chloride is formed, perhaps only partially, by precipitation from solution. As such the AgCl layer is thick relative to the normally formed oxide film on metals such as aluminium or stainless steel, and results in visible tarnishing of the metal. Although the corrosion rate is thereby rendered low the metal is not strictly passive, and indeed, tarnishing of silver by silver chloride is not only unsightly—it is slowly damaging. The behaviour of lead in sulphuric acid is similar because of the low solubility of lead sulphate.

Passivity can also be enhanced by incorporating components that catalyse the cathodic reaction, thereby raising the potential. By alloying titanium with small quantities of platinum or palladium, or by implantation of these elements into the surface, the cathodic hydrogen evolution reaction is thus accelerated, and titanium becomes anodically protected in appropriate environments[79]. Interestingly, this simple form of anodic protection does not involve an external power supply. Platinum and palladium function as good hydrogen catalysts — they raise the exchange current density for hydrogen evolution. A similar mechanism has been proposed for the action of lead-based pigments in priming paints[75].

Passivation in Non-Aqueous Liquids

Far less is known of the corrosion and passivation of metals in non-aqueous solvents and solutions[80], and the topic can be mentioned only briefly here. The prime requirement for passivity of metals in water in the formation of an oxide film by anodic reaction with oxygen obtained from the water. This is not always available in organic solvents. Thus polarization of nickel in methanol containing sulphuric acid[81] shows no passive region at all in solutions containing less than 2 wt-% water, but shows an increasingly large passive potential range, and lower passive current density, as the water content is increased above 2 wt-%. Passive oxide formation arises from reaction with the water component. Reaction of the metal with the methanol component to produce the oxide is clearly not possible. Some metals can react with the methanol component of methanolic solutions, but such reaction produces the non-passivating methoxide rather than the oxide and results in corrosion. The effect is seen clearly by the reaction of titanium with alkaline methanol solutions[67]. Generally, the effect of chloride in inducing pitting corrosion and stress-corrosion cracking in metals such as titanium is increased substantially as the water content of the methanol solution is reduced, demonstrating the destabilizing effect of the reduced water activity on the stability of the passive state. Both passivation and corrosion can be difficult to anticipate in non-aqueous liquids in the absence of laboratory tests.

Passivity of Alloys

While the same basic mechanisms for passivity of pure metals also applies to alloys, the processes involved in the passivation of alloys have an added complexity. In many cases only one component of the alloy has the property of being passive in a particular environment. Alloys such as stainless steels, which contain highly passive components (chromium in this case), owe their corrosion resistance to the surface enrichment of the passivating component[82] Thus stainless steels resist corrosion in many acidic systems (where iron or carbon steel would be poorly passive or not passive at all) by a passivating oxide film containing Cr predominantly as Cr(III). Surface analytical techniques such as Auger electron and X-ray photoelectron spectroscopies reveal substantial enrichment of chromium in the passivating oxide film on these alloys[34, 35, 83-87]. There are only two ways by which this enrichment can arise. Either the other component (iron in the case of binary iron/chromium stainless steels) is left behind at the metal surface when the chromium oxidises, or it is dissolved during passivation. In aqueous corrosion, particularly at ambient temperatures, it is undoubtedly dissolution of the iron component that causes chromium enrichment of passive stainless steel surfaces[66]. This contrasts gaseous oxidation of the same class of metals at high temperature, where chromium-rich oxide films are also formed, but iron cannot dissolve in the environment, and diffusion in the metal phase is much more rapid. Classically, a minimum amount of chromium is required in the alloy before chromium enrichment in the oxide film can be achieved: for stainless steels this is determined to be around 12–13 wt-%. Steels containing less than this amount of chromium are not regarded as stainless. The basis of this is as follows. In environments corrosive towards iron but passivating towards chromium (such as sulphuric acid), iron dissolves selectively from the surface of the alloy, and so the surface concentration of chromium increases to a level where the basic chromium-rich oxide film can cover the surface completely: the metal is then passive. This film contains significant quantities of Cr(III), but iron ions are also found in its structure, generally at an decreasing level towards the film/electrolyte interface. Steels containing greater than about 12–13% chromium show a dramatic improvement in their passivating properties in comparison with those of lower chromium content[22]. Analyses of such oxide films are performed using *ex-situ* surface analytical techniques, such as Auger electron spectroscopy, X-ray photoelectron spectroscopy and secondary ion mass spectrometry coupled with some form of depth profiling. Most such analyses performed to date have been carried out on surfaces exposed to the solution for relatively little time (seconds or minutes), despite the fact that steady state passivity of stainless steels takes much longer to achieve. Whether after a sufficiently long period of exposure the oxide film on stainless steel is composed entirely of chromium oxides is not known.

The ease with which stainless steels can passivate then increases with the level of chromium within the alloy and so materials with higher chromium content are more passive (i.e. conduct a lower passive current density) and passivate more readily (i.e. the critical current density is lower and the active/passive transition is lower in potential). They are also passive in more aggressive solutions: the pitting potential is higher.

The precise mechanism by which surface enrichment of chromium through dissolution of iron in stainless steels occurs is still under some debate[66, 88-90]. For alloys which are single phase solid solutions, the phenomenon is readily understood if the passivating element is the major component of the alloy. The corrodible element then simply dissolves from the metal surface enriching the surface with the passivating element which then completes its passivating oxide. However, far less than 50% chromium is required to achieve passivity. Examination of the repassivation of stainless steels shows evidence of dissolution of the corrodible components from the film as it grows[66]. Excess iron ions in the film are dissolved and replaced by chromium ions from the metal, progressively enriching the film with chromium until passivity is reached. However, the process may not be so simple. Recent ideas of percolation through binary alloys involve dissolution of individual and clusters of iron atoms from the alloy surface[88-90]. The residual chromium atoms are able to diffuse over the surface, progressively uncovering iron atoms and iron atom clusters. These are dissolved until there is sufficient chromium on the surface to link up into a continuous passivating network. The threshold concentration of Cr in the alloy (*ca.* 12%) is then determined as the percolation threshold below which interlinked networks of iron atom clusters exist through the alloy and it is no longer stainless. Above this threshold composition, continuous percolation clusters do not exist and the alloy is passive and stainless. The attraction of this model is that its concepts can be extended to the nucleation of pitting corrosion of stainless steels[91] (see below). The distinction between the two modes of iron dissolution and associated chromium enrichment is however, a fine one.

Stainless steels show another interesting feature. At high potentials passivation by an oxide rich in Cr(III) is lost since the Cr(III) component of the oxide is transpassively dissolved to chromate ions (line FG in Figure 1.42). In solutions in which iron oxides are stable at high potential (mildly alkaline) and the metal does not undergo pitting corrosion, the resulting iron oxide can then provide some passivity: the roles of the two components of the alloy are thus reversed. If the non-chromium components of the steel are also not passive at high potentials, the metal corrodes transpassively. Very powerful oxidising agents can bring about transpassive corrosion of stainless steel. Under normal conditions of use in dilute aqueous solutions however, these high potentials would not be achieved, and chromium provides the prime passivity of these materials.

For stainless steels to remain passive, the level of chromium defined by the bulk composition must be maintained throughout the entire metal matrix. If local regions exist within the alloy microstructure where the chromium level is depleted, these become susceptible to localised corrosion, and increasingly so as the chromium content locally is reduced. Ultimately if the chromium content is reduced locally below *ca.* 12%, this region no longer carries the composition of a stainless steel, and corrodes as a non-stainless alloy steel. However, any reduction at all in the local chromium concentration of the alloy matrix is deleterious to passivity, since these alloys become progressively less passive as the chromium concentration is decreased. Such conditions are found under sensitising heat treatments, commonly in or around welds (the heat-affected zone). In these regions, at the appropriate temperature, chromium carbides precipitate in the grain boundaries depleting the matrix around the carbides of chromium[22]. The grain boundaries may then activate in service under conditions where the grain matrix is passive.

In multiple phase alloys the situation is different. The passivity of each phase is of importance, but there can also be a significant synergism between the behaviour of adjacent phases. The presence of copper-rich precipitates in aluminium alloys containing the element accelerates the cathodic reaction locally because the exchange current density for the cathodic reaction on the relatively noble precipitate is higher than it is on the surrounding aluminium matrix. In passivating electrolytes and the absence of pitting agents such as chloride, passivity is thus enhanced because of the increased potential. However, in the presence of chloride, the potential may be raised into the pitting regime, and passivity then becomes degraded. Aluminium alloys containing more noble alloying elements such as copper or iron are generally less resistant to pitting corrosion than pure aluminium.

Compositions of Passivating Oxide Films

The composition, structure and even thickness of passivating oxide films on metals are in fact extremely difficult to determine, chiefly because these oxide films are so thin, and because they are 'held together' by the compositions of the interfaces (with the metal and with the electrolyte) as well as by the high electric field they carry. Many methods of determination are *ex situ* techniques, requiring removal of the specimen from the electric field as well as from the electrolyte. These methods include Auger electron spectroscopy (AES), X-ray photoelectron spectroscopy (XPS) and secondary ion mass spectrometry (SIMS), as well as X-ray diffraction (XRD)[34,35]. A range of *in situ* methods, such as ellipsometry, frequency response analysis and photocurrent spectroscopy are also in use to provide information on the passive film[34,35]. These techniques however, give more indirect information; they do not present compositional or structural information directly, but probe other properties of the passive interface, from which more detailed understanding must be inferred. The new techniques of scanning tunnelling and atomic force microscopies, which can be used *in situ*, may yield some information in the future.

The thickness of the oxide film is most readily determined by cathodic reduction, provided the film is cathodically reducible and the reduction is 100% current efficient. The cathodic charge involved then determines the film thickness for films of known composition. The method is really only suited to more noble metals, and for most structural metals is not relevant. Film thicknesses determined by ellipsometry have been used successfully, but assumptions of the refractive index must be made. The passivating film on iron is variously determined to be Fe_3O_4, γ-Fe_2O_3 or FeOOH, depending on the conditions under which it is formed[3,34,35]. The thickness of the film formed in borate buffer solutions is up to *ca.* 4–5 nm, the value depending on the potential at which it is formed. The thickness is a linear function of potential. There is now significant evidence that the Fe(III) film which passivates iron contains some Fe(IV)[92], a phenomenon which may account for the sometimes necessarily high overpotential required to passivate the metal. Surface analyses of passive films on iron, as well as other metals such as aluminium and stainless steel, show that the outer surfaces are usually hydrated, equivalent to hydroxide ions forming part of the outer structure in place of oxide ions[93-95]. This is of course, only to be expected from film growth in aqueous solution. The inner regions of the films tend to be less hydrated or not hydrated at all. The passivating oxide on aluminium is reckoned to be amorphous, but crystallises with increasing thickness during anodising.

The incorporation of anions from the electrolyte, such as borate and carbonate, into the oxide has also been shown to occur on iron and cobalt[96], such anions being restricted to the outer layers of the film. Attempts to find incorporation of chloride into passive iron surfaces from chloride-containing electrolytes have met with contradictory results, and it is still not known whether or not chloride incorporation is a precursor step to pitting corrosion[34, 35, 97-100]. Neither has this mechanistically important but vexed question been answered unambiguously for other metals.

The oxide film composition on alloys is inevitably more complex; some discussion of chromium-containing alloys is given above. Generally, and not surprisingly, the components of the alloy are to be found in the passivating oxide film (e.g.[100]), but the role that these play in determining passivity and its breakdown is far more open to question. Stainless steels show the presence of iron, chromium and nickel in the oxide, (and molybdenum for alloys containing the element) with elemental ratios which depend on pH and potential. By various *ex situ* techniques, chromium is found in (III), (IV) and (VI) oxidation states in varying proportions[34, 35, 84-87, 100]. The reasons why changes in the oxide film composition should improve passivity and resistance to breakdown remain speculative.

Breakdown of Passivity and Pitting

Probably less surprising than the phenomenon and characteristics of passivity itself, is the breakdown of passivity, and the very high rate of localised corrosion that can ensue. The passivating oxide film is extremely thin, and its existence and stability depend critically on the composition of the environment and the very high electric field it carries. It is in essence fragile, chemically, electrochemically and mechanically, but the fragility depends on the metal and the environment as a system. Mechanical breakdown of the oxide film arises when the material is under tensile stress and can give rise to stress-corrosion cracking or corrosion fatigue. Mechanical breakdown can also arise from an erosive or cavitative action from the environment. These processes generate a fresh metal surface, the reaction of which can be two-fold. In suitably quiescent environments the bare metal surface repassivates, and the electrochemical reaction with the environment causes no significant damage, although a continuing erosive action does lead to loss of material through wear. During corrosion fatigue too, the regrowth of the passivating film over the surface can help wedge open an incipient crack during the compressive half-cycle, allowing mechanical crack growth.

Chemical breakdown of passivity occurs naturally enough if the potential is lowered into the active regime (region AB in Figure 1.42). This can occur if the cathodic reactant supporting passivity, such as oxygen dissolved in the electrolyte, is depleted, as occurs locally in crevices or occluded regimes of the structure. Chemical breakdown also arises at high potential in the presence of aggressive ions, usually anions, which interact with the metal enhancing dissolution (region DE in Figure 1.42). The most common aggressive anion is chloride, aggressive to very many passive metals. In such cases, absence of the aggressive ion would impose passivity at the same potential: the process is thus essentially different from crevice corrosion.

Localised corrosion can occur if the passivating element of an alloy is locally depleted from the matrix. This may occur in the sensitized regions of stainless steels for example, as outlined above. Other alloy components too, may be deleterious towards the passivity of alloys. For example, the presence of significant quantities of sulphur as an impurity in ferrous and nickel based alloys is generally degrading towards passivity[30, 35, 101, 102]. Sulphur becomes incorporated into the oxide film or accumulates at the metal/oxide interface; the element retards formation of the oxide film and catalyses dissolution of iron and nickel, often remaining adsorbed on the dissolving metal surface. In extreme cases, a sulphide film forms on the metal surface, which is non-protective. Appropriate heat treatment of such alloys allows segregation of the sulphur to the grain boundaries or precipitation of sulphides, both leading to very high local concentrations of sulphur and premature failure of passivity at these sites.

Where the environment is *potentially* corrosive however (i.e. contains aggressive ions), even though the overall surface remains passive, corrosion can advance into the bare metal surface. The question is, how can this happen? In stress-corrosion cracking attack occurs chiefly around the active/passive transition (BC in Figure 1.42), where oxide film growth still occurs, but the state of passivity is potentially unstable, because of the proximity in potential of the active region (AB in Figure 1.42). Dissolution occurs simultaneously with repassivation (regrowth of the oxide film). A small ohmic resistance in the electrolyte between the external cathodic area and the anodic incipient crack tip may be sufficient to lower the potential of the crack tip into the active regime. Alternatively, a small increase in acidity developed at the crack tip as a consequence of anodic reaction may enlarge the active loop a little by raising the potential of the active/passive transition locally. Some anodic reactions at the crack tip may themselves cause local brittleness, such as dealloying or the diffusion of cathodically reduced hydrogen in atomic form into the metal ahead of the crack tip. These processes allow repeated rupture at the crack tip, even if it has already repassivated, because of the embrittled local metal matrix. Stress-corrosion cracking also occurs near the passive/active transition at higher potential when such a transition is available (near point D in Figure 1.42). In this case chloride undoubtedly enhances the ability of an incipient crack to propagate by enhancing dissolution at the crack tip, even though the pitting potential (the potential of point D) is not necessarily exceeded.

There are many forms of pitting corrosion (the formation of holes by corrosive action). The various forms have been reviewed by Szklarska-Smialowska[103]. Sometimes activation of a steel brought about by lowering the potential from the passive potential regime into the active regime produces etch pits, a form of uneven corrosion associated with different crystal planes within the metal. This process is less sensitive to the type of anion present, and more associated with the pH changes that occur during activation in unbuffered solutions. The pitting process considered here is a process involving the formation of small holes in an otherwise passive surface as a consequence of the aggressive action of a component of the electrolyte. It is a form of localized corrosion which involves loss of passivity locally, but does not require a stress at all (although it may be enhanced by a stress). It usually occurs at high potential, where one might expect passivity to dominate. In pitting corrosion, passivity is lost or ruptured through the chemical or electrochemical action of a specific component of the environment. The most commonly encountered pitting agent in aqueous electrolytes is chloride, although some other anions may also induce pitting. Notably, these are the other halides. Sulphur oxy-anions such as thiosulphate[30,104], and indeed sulphide ions, are conducive towards pitting corrosion in ferrous materials and nickel-based materials. It is of some consequence that most anions capable of causing pitting in particular metals also induce stress-corrosion cracking of those same metals, although the converse is not necessarily true. Many metals undergo pitting corrosion in chloride solutions of sufficient concentration; these include steels and stainless steels, nickel, copper and aluminium, and their alloys, and many more. A few metals are resistant to chloride-induced pitting, including chromium, titanium and tantalum, but that is not to say they are necessarily immune to pitting, only that they are resistant. Some other anions may also induce pitting but outside the normally encountered range of corrosion potentials, and are therefore not regarded as specific pitting agents. For example, nitrate ions, which fairly passivating towards aluminium under normal potentials, induce pitting in dilute nitric acid solutions at high potentials. Such potentials are only accessible through application of an external voltage source.

Precisely how these anions nucleate pitting in a passive metal surface is still largely unsolved, although a range of models exists. Pits in chloride solutions are often rare events[30,105]. In stainless steels, pits can be very widely spaced apart, and most of the surface is passive. Pit propagation rates are however, very fast: a propagation rate equivalent to a current density of $10-100 \text{ kA m}^{-2}$ is common[30,105]. In chloride solutions pitting is characterised by a minimum potential, called the pitting potential[103] (point D in Figure 1.42). Only above the pitting potential is pitting observed to occur. The metal is in essence passive below this potential, and this is a commonly used design criterion. However, detailed examination of the passive region (i.e. above the passivation potential and below the pitting potential, the region CD in Figure 1.42) of steels and stainless steels shows that the passive current is more noisy in chloride solutions than it is in the absence of chloride[30,105-107]. The origin of this noise is the nucleation and propagation of 'metastable' pits (also sometimes called unstable pits). Metastable pits are those which grow for a short period and then die through repassivation[30,105-110]. These are observed both below and above the pitting potential. Although these metastable pits do not generally cause significant damage to the metal, apart from a degree of microscopic surface roughening, they are evidence that passivity is not fully stable in the presence of chloride ions, even below the pitting potential. For these materials at least, the pitting potential (sometimes called the breakdown potential, erroneously since breakdown of passivity occurs at lower potentials for ferrous materials) is not a pit nucleation potential: it can only be regarded as the minimum potential for *stable* pitting. Pits which are stable propagate almost indefinitely, either until total penetration of the metal, or until they are large enough to be washed out by the bulk electrolyte. Such noise in the current can be observed throughout the passive range of potential if the experiment is carried out carefully enough[105].

The pitting potential (point D in Figure 1.42) is sometimes treated as a type of equilibrium potential. There are empirical reasons for this. The pitting potential is often observed to decrease linearly with log (Cl$^-$) (the chloride activity or concentration), giving an apparently Nernstian form[103].

However, observation of metastable pitting at potentials well below the pitting potential probably precludes this notion. In addition, if the potential sweep direction is reversed at point E, the polarization curve shows strong hysteresis. The negatively moving sweep lies at higher currents, and the potential at which the pitting metal repassivates is much lower than point D. (The point where the metal repassivates in the negative sweep is sometimes called the repassivation potential.) It must also be noted that the pitting potential itself depends on the prior surface treatment of the metal, such as the surface roughness[111]. These facts are inconsistent with the pitting potential describing some simple electrochemical equilibrium. The pitting potential is nevertheless an extremely useful engineering parameter.

Pitting corrosion requires an incubation period or induction time to nucleate[112,113]. Even after initial nucleation, further events are sporadic in time. Several possible mechanisms of nucleation have been put forward. Complexing of metal cations on the surface of the film has been proposed to lead to local dissolution of the oxide[112], the site then receding to the metal surface. Alternatively, chloride migration through the oxide film may lead to its accumulation as the metal chloride at the metal film interface. This, if it occurs, could well lead to mechanical failure of the film by bursting, since metal chlorides are of greater molar volume than the corresponding oxides. The notion is appealing since it provides for a high local chloride concentration at the metal/film interface once the film is ruptured, providing an immediate pit propagation path. Film rupture could instead be induced by condensation of vacancies at the metal/film interface[53]. The idea of mechanical film rupture carries the added appeal that pitting events on iron microelectrodes observed potentiostatically show an initiating very fast current spike[106,114]. The event is microscopically violent. Recent observations with stainless steel microelectrodes show that pits here are also initiated with a violent current jump[115,116]. The magnitudes of the events are very small, generally up to some hundreds of picoamperes, indicative not of a low reaction rate, but of the microscopic size of the nucleation event. Interestingly, many of these nucleation events show no propagation stage at all, not even into a metastable pit, but die immediately after nucleation. The implication is that while pit growth may be a rare event over the metal surface, the nucleation event may in fact be rather common. The same nucleation mechanism may not of course be applicable to all metals which pit in chloride solution, although current fluctuations are observed from aluminium in chloride solutions below the pitting potential. It has also been suggested that passive films may always undergo continuous breakdown/repair events on a microscopic scale, even in solutions which are fully passivating. This interesting idea takes the nucleation event away from the properties of the aggressive chloride anion. Chloride would then serve merely to propagate pits from these nuclei, and prevent their repassivation.

Steels and stainless steels show preferential nucleation of pits at inclusions, most notably sulphide inclusions[30,118]. Other sulphur-rich regions in ferrous and nickel-based alloys may also lead to premature failure. It has been shown that accumulation of sulphur on the surface of these materials retards passivity and enhances dissolution of the metal. These effects occur in any solution in which the metal shows an active region and they are also preferential pitting sites in the presence of chloride. A recent notion[91] for pit nucleation in stainless steels suggests that iron-rich clusters to be found randomly distributed in the metal structure are active when they become exposed to the electrolyte, since they do not carry sufficient chromium to allow their passivation. These are the pit nucleation sites. They become progressively exposed to the solution through the normal passive dissolution rate. This model cannot of course be generally applicable to pitting corrosion since it cannot account for pit nucleation in single component metals, such as iron and aluminium. In fact, probably any site on the metal surface capable of depassivation for any reason, is a potential pit nucleation site. For it to develop into a pit simply requires an anodic reaction capable of procuring a more aggressive anolyte, as outlined below.

The fact that amorphous metals show a higher resistance to pitting corrosion than their polycrystalline counterparts can offer more than one explanation. Such amorphous metals made from iron, chromium and nickel to simulate a stainless steel show significantly high pitting potentials[118-122]. These alloys are usually made with significant quantities of non-metallic element additions such as boron in order to retard crystallization during quenching from the melt. Such elements, which are found in the oxide film[123], may aid passivity and inhibit pit nucleation by aiding amorphisation of the passive film or by their own specific chemical action. More probably however, it is the absence of metallurgical defects such as grain boundaries and inclusions which inhibits pit nucleation.

Once nucleated, anodic reaction at the incipient pit then requires a corresponding cathodic reaction on the surrounding passive surface if the metal is pitting in open circuit, or at the counter electrode if pitting is carried out potentiostatically. Anodic reaction generates cations by dissolution, and H^+ by hydrolysis of the dissolved cations. Acidity is also generated if the metal repassivates. Both the metal cations and H^+ require neutralisation, accomplished by ingress of anions. If chloride is the only available anion, its local concentration increases; the locally higher chloride concentration enhances local metal dissolution and prevents repassivation[124]. The enhanced dissolution then draws more chloride into the now propagating pit, enhancing dissolution even further. Growth of the pit is in essence being fuelled by its own reaction products: it is a feed-back mechanism. The process is sometimes termed autocatalytic. Pit growth is further supported by the higher local pH. The pit anolyte becomes a saturated (or near-saturated) solution of the metal chloride, and highly acidic. Some 60–70% of saturation of the metal chloride is required to prevent repassivation of stainless steel. It follows from this, that the presence of other anions in solution, in addition to chloride, but which are not aggressive (e.g. SO_4^{2-}) would reduce the ability of the metal to undergo pitting. Both ions are then transportable into the pit anolyte and it is correspondingly more difficult to build up sufficient chloride to establish propagation: this is found in practice[125,126].

It is apparent that such a mechanism depends strongly on the transport properties of the electrolyte components into and out of the pit. In stainless steels pit propagation has been shown to be controlled by diffusion of ions between the pit interior and the external electrolyte[105,126]. The behaviour of the pit site depends on its degree of occlusion: pitting is easier on rougher surfaces because of the greater occlusion of sites. Surfaces of rougher finish usually display a lower pitting potential. More occluded sites can more readily retain the dissolved cations required to draw chloride into the pit anolyte.

In the early stages of pit propagation on stainless steel in chloride solution, when the pit is small, its depth alone is insufficient to act as a diffusion barrier. Diffusion is then restricted by a perforated cover which exists over the pit mouth; the perforation arises from the initiating event, and the cover exists because of undermining of the surrounding passive film as the pit expands. This perforated cover is critical to survival of the pit through the so-called metastable period of growth. *Small* ruptures in the cover during the metastable growth period enhance the propagation rate without terminating the pit. However, if this cover is *totally* lost by rupture before a critical stage in the growth of the pit, the anolyte is washed out (the outward diffusion of cations becomes fast) and the pit repassivates (hence the term metastable). Pit growth does not then achieve stability. The critical parameter here is the product of the pit growth current density and its radius[105,126]: this pit stability product must reach a value of *ca.* 3 mA cm^{-1} for pit growth on stainless steel to stabilise. Once the critical pit stability product is achieved, the pit can propagate without its cover. The pit depth alone is then a sufficient diffusion barrier to ensure the appropriate high chloride concentration within the anolyte, and pit growth has achieved stability. Corrosion pits in chloride solution thus pass through three consecutive steps of growth. The nucleation event is followed by metastable pit growth. Growth is metastable because the pit depth alone is an insufficient barrier to diffusion and may terminate at any point by rupture of the sustaining cover, resulting in repassivation. If the pit grows to a stage where the pit depth alone is sufficient to maintain the diffusion barrier, it enters the third stage and its growth becomes stable. Most pits die at the nucleation stage, and show no propagation. A few survive to become metastable; of those propagating metastably, only a few achieve stability.

It is worth mentioning that although pitting corrosion is generally undesirable, the process does in fact have its uses. Aluminium sheet used for the preparation of printers' lithographic sheet is subject to a preliminary treatment involving a highly controlled electrochemically induced pitting process in electrolytes based on either hydrochloric acid or nitric acid[127]. This procedure, termed electrograining, is designed to roughen the surface on a microscopic scale prior to anodising. Pitting corrosion may also be induced into the nickel anodes used in nickel electroplating baths containing chloride in order to prevent their passivation and allow continuous replenishment of the nickel ions in the electrolyte as they are removed.

G.T. BURSTEIN

REFERENCES

1. Wagner, C., *Corros. Sci.*, **5**, 751 (1965)
2. Brusic, V., *Oxides and Oxide films*, **1**, Ed. J. W. Diggle, Marcel Dekker, New York (1972)
3. Pourbaix, M., *Atlas of Electrochemical Equilibria in Aqueous Solutions*, Pergamon/CEBELCOR, Oxford (1966)
4. Tomashov, N. D. and Chernova, G. P., *Passivity and Protection of Metals*, 9, Plenum Press, New York (1967)

5. Keir, J., *Philos. Trans. Roy. Soc. Lond.*, **80**, 359 (1790)
6. Faraday, M., *Experimental Researches in Electricity*, **2**, University of London, London (1844)
7. Evans, U. R., *Industr. Engng. Chem.*, **17**, 363 (1925)
8. Evans, U. R., *The Corrosion and Oxidation of Metals*, Ch. 7, Arnold, London (1960)
9. Evans, U. R., *The Corrosion and Oxidation of Metals: First Supplementary Volume*, Ch. 7, Arnold, London (1968)
10. Evans, U. R., *The Corrosion and Oxidation of Metals: Second Supplementary Volume*, Ch. 7, Arnold, London (1976)
11. Tafel, J., *Zeit. Physik. Chem.*, **50A**, 641 (1905)
12. Hoar, T. P., in *Modern Aspects of Electrochemistry*, Ed. J. O'M. Bockris, **2**, 262, Butterworths, London, (1959)
13. Hoar, T. P. and Mowatt, J. A. S., *Nature*, **165**, 64 (1950)
14. Hoar, T. P., Mears, D. C. and Rothwell, G. P., *Corros. Sci.*, **5**, 279 (1965)
15. Hoar, T. P. and Mears, D. C., *Proc. Roy. Soc. Lond.*, **A294**, 486 (1966)
16. Hoar, T. P. and Mott, N. F., *J. Phys. Chem. Solids*, 9, 97 (1959)
17. Tajima, S., in *Advances in Corrosion Science and Engineering*, Eds. M. G. Fontana and R. W. Staehle, **1**, Plenum, New York (1970)
18. Wood, G. C., in *Oxides and Oxide films*, **2**, Ed. J. W. Diggle, Marcel Dekker, New York (1972)
19. Furneaux, R. C., Thompson, G. E. and Wood, G. C., *Corros. Sci.*, **18**, 853 (1978)
20. Shimuzu, K., Kobayashi, K., Thompson, G. E. and Wood, G. C., *Philos. Mag.* A, **66**, 643 (1992)
21. Edeleanu, C. and Gibson, J. G., *Chem. Ind.*, 301 (1961)
22. Uhlig, H. H., *Corrosion and Corrosion Control*, 2nd ed, Ch. 5, John Wiley and Sons, New York (1971)
23. Staehle, R. W., Forty, A. J. and van Rooyen, D. (Eds), *Fundamental Aspects of Stress Corrosion Cracking*, NACE, Houston, Texas (1969)
24. Scully, J. C. (Ed.), *The Theory of Stress Corrosion Cracking in Alloys*, NATO, Brussels, (1971)
25. *Stress Corrosion Cracking and Hydrogen Embrittlement of Iron Base Alloys*, NACE, Houston, Texas, (1975)
26. Turnbull, A. (Ed.), *Corrosion Chemistry in Pits, Crevices and Cracks*, HMSO, London (1987)
27. Kolotyrkin, J. M., *Corrosion*, **19**, 261t (1963)
28. Wranglen, G., *Corros. Sci.*, **14**, 331 (1974)
29. Staehle, R. W., Brown, B. F., Kruger, J. and Agrawal, A. (Eds), *Localized Corrosion*, NACE, Houston, Texas (1974)
30. Isaacs, H. S., Bertocci, U., Kruger, J. and Smialowska, S. (Eds), *Advances in Localized Corrosion*, NACE, Houston, Texas (1990)
31. Young, L., *Anodic Oxide Films*, Academic Press, London (1961)
32. Pedeferri, P., *Imagination on Titanium*, Cooperativa Libraria Universitaria del Politecnico, Milan (1987)
33. de Gromoboy, T. S. and Shreir, L. L., *Electrochim. Acta*, **11**, 895 (1966)
34. Froment, M. (Ed.), *Passivity of Metals and Semiconductors*, Elsevier, Amsterdam (1983)
35. MacDougall, B. R., Alwitt, R. S. and Ramanarayanan, T. A. (Eds), *Oxide Films on Metals and Alloys*, Proceedings, **92-22**, The Electrochemical Society, Pennington, New Jersey (1992)
36. Kabanov, B., Burstein, R. and Frumkin, A., *Disc. Farad. Soc.*, **1**, 259 (1947)
37. Heusler, K. E., *Zeit. Elektrochem.*, **62**, 582 (1958)
38. Bockris, J. O'M., Drazic, D. and Despic, A., *Electrochim. Acta*, **4**, 325 (1961)
39. Weissmantel, C., Schwabe, K. and Hecht, G., *Werkst Korros.*, **12**, 353 (1961)
40. Sato, N. and Okamoto, G., *J. Electrochem. Soc.*, **111**, 197 (1964)
41. Bockris, J. O'M. and Reddy, A. K. N., *Modern Electrochemistry*, **2**, Macdonald, London (1970)
42. Bessone, J., Karakaya, L., Lorbeer, P. and Lorenz, W. J., *Electrochim. Acta.* **22**, 1147 (1977)
43. Burstein, G. T. and Ashley, G. W., *Corrosion*, **39**, 241 (1983)
44. Misra, R. D. K. and Burstein, G. T., *Corros. Sci.*, **24**, 305 (1984)
45. Burstein, G. T. and Newman, R. C., *Electrochim. Acta*, **25**, 1009 (1980)
46. Burstein, G. T. and Newman, R. C., *J. Electrochem. Soc.*, **128**, 2270 (1981)

47. Verwey, E. J. W., *Physica*, **2**, 1059 (1935)
48. Cabrera, N. and Mott, N. F., *Repts. Prog. Phys.*, **12**, 163 (1948/49)
49. Dignam, M. J., *Oxides and Oxide films*, **1**, Ed. J. W. Diggle, 91, Marcel Dekker, New York (1972)
50. Fromhold, A. T., *Oxides and Oxide Films*, **3**, Eds. J. W. Diggle and A. K. Vijh, 1, Marcel Dekker, New York (1976)
51. Kirchheim, R., *Electrochim. Acta*, **32**, 1619 (1987)
52. Davenport, A. J. and Brustein, G. T., *J. Electrochem. Soc.*, **137**, 1496 (1990)
53. Chao, C. Y., Lin, L. F. and Macdonald, D. D., *J. Electrochem. Soc.*, **128**, 1187, 1194 (1981)
54. Ghez, R., *J. Chem. Phys.*, **56**, 1838 (1973)
55. Burstein, G. T. and Davenport, A. J., *J. Electrochem. Soc.*, **136**, 936 (1989)
56. Güntherschultze, A. and Betz, H., *Zeit. Phys.*, **91**, 70, **92**, 367 (1934)
57. Dewald, J. F., *J. Phys. Chem Solids*, **2**, 55 (1957)
58. Sato, N. and Cohen, M., *J. Electrochem. Soc.*, **111**, 512, 624 (1964)
59. Diggle, J. W., *Oxides and Oxide films*, **2**, (Ed. J. W. Diggle), 280, Marcel Dekker, New York (1973)
60. Nii, K., *Corros. Sci.*, **10**, 571 (1970)
61. Hagyard, T. and Williams, J. R., *Trans. Farad. Soc.*, **57**, 2288 (1961)
62. Hagyard, T. and Earl, W. B., *J. Electrochem. Soc.*, **114**, 694 (1967)
63. Beck, T. R., *Corrosion*, **30**, 408 (1974)
64. Burstein, G. T. and Ashley, G. W. A., *Corrosion*, **40**, 110 (1984)
65. Burstein, G. T. and Marshall, P. I., *Corros. Sci.*, **23**, 125 (1983)
66. Burstein, G. T. and Marshall, P. I., *Corros. Sci.*, **24**, 449 (1984)
67. Burstein, G. T. and Whillock, G. O. H., *J. Electrochem. Soc.*, **136**, 1313, 1320 (1989)
68. Burstein, G. T. and Cinderey, R. J., *Corros. Sci.*, **32**, 1195 (1991)
69. Sato, N. and Kudo, K., *Electrochim. Acta*, **16**, 477, (1971)
70. Mayne, J. E. O. and Pryor, M. J., *J. Chem. Soc.*, 1831 (1949)
71. Cohen, M., *J. Phys. Chem.*, **56**, 415 (1952)
72. Mellors, G. W., Cohen, M. and Beck, A. F., *J. Electrochem. Soc.*, **105**, 332 (1952)
73. Mayne, J. E. O. and Menter, J. W., *J. Chem. Soc.*, 103 (1954)
74. Brasher, D. M. and Mercer, A. D., *Br. Corros. J.*, **3**, 120 (1968)
75. Appleby, A. J. and Mayne, J. E. O., *J. Oil Col. Chem. Assoc.*, **50**, 897 (1967)
76. Mayne, J. E. O. and Page, C. L., *Br. Corros. J.*, **7**, 111, 115 (1972)
77. Hoar, T. P. and Evans, U. R., *J. Chem. Soc.*, 2476 (1932)
78. Szklarska-Smialowska, Z. and Staehle, R. W., *J. Electrochem. Soc.*, **121**, 1146 (1974)
79. Stern, M. and Wissenberg, H., *J. Electrochem. Soc.*, **106**, 755, 759 (1959)
80. Heitz, E.; in *Advances in Corrosion Science and Technology*, **4**, Eds. M. G. Fontana and R. W. Staehle, 149, Plenum Press, New York (1974)
81. Mazza, F., *Werkst. Korros.*, **20**, 199 (1969)
82. Vernon, W. H. J., Wormwell, F. and Nurse, T. J., *J. Iron Steel Inst.*, **150**, 81 (1944)
83. Cahoon, J. R, and Bandy, R., *Corrosion*, **38**, 299 (1982)
84. Olefjord, I., *Mater. Sci. Engng.*, **42**, 161 (1980)
85. Olefjord, I. and Elfstrom, B. O., *Corrosion*, **38**, 46 (1982)
86. Castle, J. E. and Clayton, C. R., *Corros. Sci.*, **17**, 7 (1977)
87. Asami, K., Hashimoto, K. and Shimodaira, S., *Corros. Sci.*, **16**, 387 (1976)
88. Sieradzki, K. and Newman, R. C., *J. Electrochem Soc.*, **133**, 1979 (1986)
89. Song, Q., Newman, R. C., Cottis, R. A. and Sieradzki, K., *J. Electrochem. Soc.*, **137**, 435 (1990)
90. Song, Q., Newman, R. C., Cottis, R. A. and Sieradzki, K., *Corros. Sci.*, **31**, 621 (1990)
91. Williams, D. E., Newman, R. C., Song, Q. and Kelly, R. G., *Nature*, **350**, 216 (1991)
92. Chin, Y-T. and Cahan, B. D., *J. Electrochem. Soc.*, **139**, 2432 (1992)
93. Okamoto, G. *Corros. Sci.*, **13**, 471 (1973)
94. McBee, C. L. and Kruger, J., *Electrochim. Acta*, **17**, 1337 (1977)
95. Bessone, J. B., Salinas, D. R., Mayer, C. E., Ebert, M. and Lorenz, W. J., *Electrochim. Acta*, **37**, 2283 (1992)
96. Burstein, G. T. and Davies, D. H., *Corros. Sci.*, **20**, 989, 1143 (1980)
97. Goetz, R., MacDougall, B. J. and Graham, M. J., *Electrochim. Acta*, **31** 1299 (1986)
98. Mitrovic-Stepanovic, V., MacDougall, B. J. and Graham, M. J., *Corros. Sci.*, **27**, 239 (1987)
99. Landolt, D., Mischler, S, Vogel, A. and Mathieu, H. J., *Corros. Sci.*, **31**, 431 (1990)

100. Mischler, S., Vogel, A., Mathieu, H. J. and Landolt, D., *Corros. Sci.*, **32**, 925 (1991)
101. Marcus, P., *C. R. Acad. Sci. Paris.*, Sér II, **305**, 675 (1987)
102. Marcus, P. and Moscatelli, M., *Mém. Etud. Sci. Rev. Métall.*, **85**, 561 (1988)
103. Szklarska-Smialowska, Z., *Pitting Corrosion of Metals*, NACE, Houston, Texas, (1986)
104. Newman, R. C., *Corrosion*, **41**, 450 (1985)
105. Pistorius, P. C. and Burstein, G. T., *Phil. Trans. Roy. Soc. Lond.* series A, **341**, 531 (1992)
106. Bertocci, U., Koike, M., Leigh, S., Qiu, F. and Yang, G., *J. Electrochem. Soc.*, **133**, 1782 (1986)
107. Isaacs, H. S., *Corros. Sci.*, **29**, 313 (1989)
108. Forchhammer, P. and Engell, H. J., *Werkst. Korros.*, **20**, 1 (1969)
109. Williams, D. E., Westcott, C. and Fleischmann, M., *J. Electrochem. Soc.*, **132**, 1796, 1804 (1985)
110. Stockert, L. and Böhni, H., *Mater. Sci. Forum*, **44**, **45**, 313 (1989)
111. Coates, G. E., *Mater. Perf.*, **29**(8), 61 (1990)
112. Hoar, T. P. and Jacob, W. R., *Nature*, **216**, 1299 (1967)
113. Heusler, K. E. and Fischer, L., *Werkst. Korros.*, **27**, 551 (1976)
114. Bertocci, U. and Yang-Xiang, Y., *J. Electrochem. Soc.*, **131**, 1011, (1984)
115. Riley, A. M., Wells, D. B. and Williams, D. E., *Corros. Sci.*, **32**, 1307 (1991)
116. Burstein, G. T. and Mattin, S. P., *Philos. Mag. Letters*, **66**, 127 (1992)
117. Szklarska-Smialowska, Z., *Corrosion*, **28**, 388 (1972)
118. Masumoto, T. and Hashimoto, K., *Ann. Rev. Mater. Sci.*, **8**, 215 (1978)
119. Hashimoto, K., *Amorphous Metallic Alloys*, Ed. F. E. Luborsky, 471, Butterworths, London (1983)
120. Diegle, R. B., Sorensen, N. R., Tsuru, T. and Latanison, R. M., *Treatise on Materials Science and Technology*, **23**, Ed. J. C. Scully, 59, Academic Press, London (1983)
122. Wislawska, M. and Janik-Czakor, M., *Brit. Corros. J.*, **20**, 36 (1985)
123. Burstein, G. T., *Corrosion*, **37**, 549 (1981)
124. Galvele, J. R., *J. Electrochem. Soc.*, **123**, 464 (1976)
125. Leckie, H. P. and Uhlig, H. H., *J. Electrochem. Soc.*, **115**, 1262 (1966)
126. Pistorius, P. C. and Burstein, G. T., *Corros. Sci.*, **33**, 1885 (1992)
127. Laevers, P., Terryn, H. and Vereecken, J. *Trans. Inst. Met. Finish.*, **70**, 105 (1992)

1.6 Localised Corrosion

The various types of localised corrosion have been enumerated in Table 1.2 in Section 1.1, and many of them are dealt with in some detail in other sections of this volume. For this reason this section will be confined to a consideration of the factors that give rise to crevice corrosion, filiform corrosion, pitting, selective leaching and erosion-corrosion and of the mechanisms of these forms of localised attack.

Corrosion can range from the highly uniform (chemical or electrochemical polishing) to the highly localised such as occurs during pitting, intergranular attack and stress-corrosion cracking. Uniform, or near-uniform, corrosion without doubt accounts for the greater proportion of metal deterioration in terms of both mass of metal converted to corrosion products and cost. However, although detrimental it is at least predictable on the basis of laboratory and field testing, so that allowance can be made for it in the design of a structure (*see* Section 9.2). In completely uniform corrosion the anodic and cathodic sites are physically inseparable, and although this type of corrosion occurs in many practical situations (e.g. chemical and electrochemical polishing, passivity) corrosion is more usually near-uniform, particularly in natural environments. Thus the appearance of the rust on steel that has been exposed to the atmosphere for some time may give the impression that attack has been quite uniform, but removal of the rust with an inhibited acid will reveal that the surface is undulating, indicating that although the whole surface has corroded the rates at different areas of the surface are not uniform. This is due to a variety of factors associated with the structure of the metal, the nature of the surface, deposits of dirt and corrosion products, etc.

Localised corrosion, on the other hand, is far less predictable, and this applies particularly to pitting in which the location of the pits on the metal surface and their distribution and size will depend upon the precise structure of the metal and environmental conditions prevailing. In pitting, the rate of corrosion into the metal is significantly greater than that parallel to the metal surface, but a precise distinction between highly localised attack and pitting is largely arbitrary. Champion[1] distinguishes between 'semi-local' corrosion and pitting by defining the latter as a pit geometry in which the ratio of average width to average depth of the attacked area is 4:1 or less; however, a ratio of 1:1 is a generally more acceptable definition of a pit. Pitting is a particularly insidious form of corrosion, since although the extent of the reaction (moles corroded) is very small the fact that the attack is predominantly into the metal results in rapid perforation when thin metal sections are involved with consequent leakage of the fluid contained in the structure. Pitting is usually regarded as more detrimental than near-uniform corrosion, but this is not always the case, and Hoar[2] has pointed out that whereas uniform corrosion of a bearing could lead to misalignment and seizure, pitting could be beneficial in providing pockets for the lubricant. Figure 1.44 shows the various types of attack[3] that can range from completely uniform to highly localised and Fig. 1.45 how a pit can initiate a stress-corrosion crack[4] thus transforming one type of localised attack into another.

Although the form of pits and their rate of propagation into the metal is far less predictable than near-uniform attack, their location is frequently determined by the metallurgical structure of the metal and the geometry of the system. Thus in certain alloys (*see* Section 1.3) attack is localised at grain boundaries leading to intergranular corrosion, in which the loss of metal may be far greater than that predicted by Faraday's law since grains of the

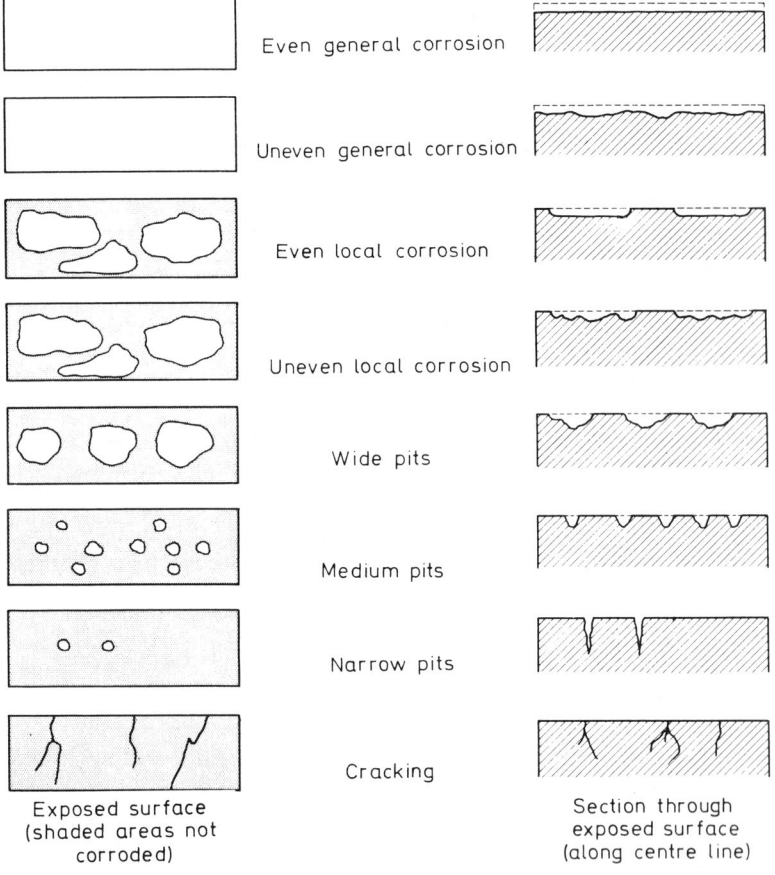

Fig. 1.44 Different types of corrosion (after Greene and Fontana[3])

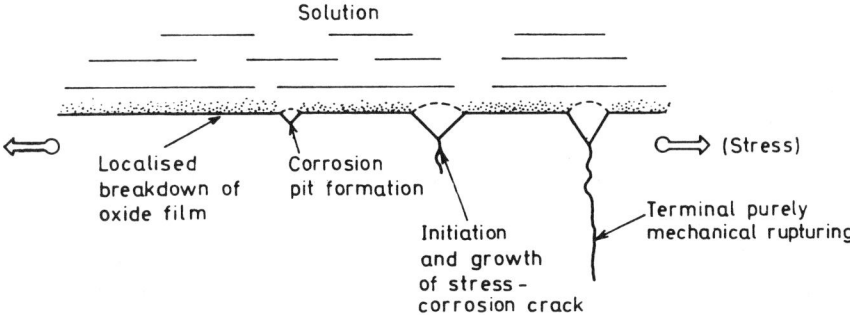

Fig. 1.45 Breakdown of oxide film leading to a pit and crack when a high-strength steel is subjected to a tensile stress in a chloride solution (after Brown[4])

metal may become detached from the substrate with a consequent rapid decrease in the cross-section leading to cracking and fracture. Similar considerations apply to localised attack resulting from dissimilar metals in contact (bimetallic corrosion), crevices, impingement of high velocity waters, etc. in which the area subjected to attack is usually predictable from the geometry of the system. On the other hand, the location of pits on metals

Table 1.16 Heterogeneities that lead to localised attack*

Heterogeneity	Area of metal that is predominantly anodic
METAL AND METAL SURFACE	
Differences in metallurgical structure	Grain boundaries, more reactive phases (solid solutions, intermetallic compounds, etc.)
Differences in metallurgical condition due to thermal or mechanical treatment	Cold-worker areas adjacent to annealed areas, metal subjected to external stress anodic to unstressed metal
Discontinuities in conducting oxide film or scale or discontinuities in applied metallic coatings that are cathodic to the substrate	Exoised area of substrate metal. In the case of passive metals defects in the passive film result in an active-passive cell with intense localised attack on the active area
Crevices or deposits on a metal surface or any other geometrical configuration that results in differences in the concentration of the cathodic reactant	The area of the metal in contact with the lower concentration of the cathode reactant, although there are exceptions to this rule
Dissimilar metals in contact (bimetallic corrosion)	The metal with the more negative corrosion potential in the environmental conditions prevailing (note that the standard electrode potentials are seldom applicable and the galvanic series can be misleading)
ENVIRONMENT	
Differences in aeration or in the concentration of other cathode reactants	Metal area in contact with the lower concentration
Differences in temperature	Metal area in contact with the higher temperature solution
Differences in velocity	Metal in contact with solutions of higher velocity
Differences in pH or salt concentration	Metal in contact with the solution of lower pH or higher salt concentration

*The table provides a general indication of the area that is likely to be anodic, but it must be emphasised that there are many situations in which the heterogeneity will have no effect or where the converse to the above may occur.

that depend upon passivity for their low corrosion rate is frequently unpredictable, and the pits will tend to be randomly dispersed on a metal surface that is apparently quite uniform.

Although corrosion is due to the thermodynamic instability of a metal in a specific environment, and although in many metal/environment systems attack will tend to be uniform, there are a variety of factors associated with the metal, the environment and the geometry of the system that may result in the attack being localised.

Table 1.16 gives examples of heterogeneities and geometrical factors that lead to localised attack, but it must be emphasised that they must be regarded as general guidelines, and whether or not the heterogeneity leads to localised attack will depend on the environmental conditions prevailing; stressed areas of a metal may be predominantly anodic compared with unstressed areas in a mild environment, but will have very little effect if the environment is an aggressive acid; a sensitised stainless steel will suffer enhanced intergranular attack in nitric acid, but will be unaffected if the environment is non-aggressive, e.g. milk (*see* Section 1.3).

Principles

Unfortunately, there is no general theory that will explain all the forms of localised attack that occur with the variety of metal/environment systems encountered in practice, e.g. the mechanism of the pitting of stainless steels in Cl^--containing solutions is quite different from the dezincification of brass in a fresh natural water. Nevertheless, many of the following factors play an important part in most forms of localised attack:

1. The cathode/anode area relationship.
2. Differential aeration.
3. pH changes at the cathodic and anodic sites.
4. Corrosion products (films) present initially on the metal surface and those formed during the corrosion reaction.

Anodic and Cathodic Current Densities

In Section 1.4 it was pointed out that a fundamental principle of corrosion is that the sum of the rates of the cathodic reactions must equal the sum of the rates of the anodic reactions, irrespective of whether the attack is uniform or localised:

$$\Sigma I_a = \Sigma I_c \qquad \ldots(1.125)$$

Furthermore, if attack is *uniform*, and assuming that there is only a single predominantly anodic and cathodic reaction, then

$$\frac{I_a}{S_a} = \frac{I_c}{S_c} \quad \text{or,} \quad i_a = i_c \qquad \ldots(1.126)$$

since the area of the cathode S_c equals the area of the anode S_a. On the other hand, if attack is localised $S_a < S_c$ and

$$i_a > i_c \quad \text{or,} \quad \frac{i_a}{i_c} > 1$$

and the larger the ratio $i_a : i_c$ the more intense the attack.

Thus localised attack usually involves a corrosion cell consisting of a large cathodic area and a small anodic area, and since I_a must equal I_c, the effect will become more pronounced the higher the rate of the cathodic process and the larger the effective area of the cathode.

This principle may be exemplified by considering the corrosion of steel in an oxygenated water, in which attack will be uniform providing the surface is free from a magnetite scale (millscale) and the geometry of the system is such that the concentration of dissolved oxygen is the same on all parts of the metal surface (Fig. 1.46a). If, however, a discontinuous film of magnetite Fe_3O_4 (a stable oxide of high electronic conductivity) is present on the surface, attack will be localised at the discontinuity owing to the large cathode area/small anode area relationship (Fig. 1.46b), and the more rapidly the oxygen arrives at the surface of the millscale the more rapid is the attack on the exposed steel (Fig. 1.46c).

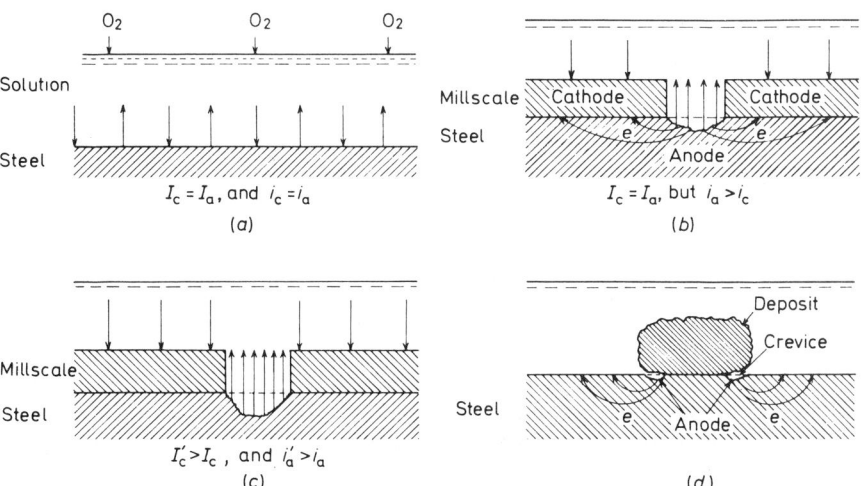

Fig. 1.46 Localised attack due to discontinuity in millscale or a deposit at a steel surface. (a) Uniform corrosion, (b) localised attack at a discontinuity in millscale, (c) increase in rate due to increase in the supply of oxygen and (d) crevice formed by deposit. (Arrows pointing downwards represents the cathodic current, and arrows pointing upwards the anodic current)

A similar effect can be produced if a crevice is present in the steel, since the geometry of the system is such that whereas oxygen can diffuse readily to the metal surface outside the crevice it can only gain access to the metal within the crevice through its very narrow mouth (Fig. 1.46d), and the large cathode: anode area ratio leads to localised attack of the metal within the crevice.

From these two examples, which as will be seen subsequently, present a very oversimplified picture of the actual situation, it is evident that macroheterogeneities can lead to localised attack by forming a large cathode/small anode corrosion cell. For localised attack to proceed, an ample and continuous supply of the electron acceptor (dissolved oxygen in the example, but other species such as the H^+ ion and Cu^{2+} can act in a similar manner) must be present at the cathode surface, and the anodic reaction must not be stifled by the formation of protective films of corrosion products. In general, localised attack is more prevalent in near-neutral solutions in which dissolved oxygen is the cathode reactant; thus in a strongly acid solution the millscale would be removed by reductive dissolution (*see* Section 11.2) and attack would become uniform.

Localised attack can, however, occur on a surface of metal that is apparently uniform, and this occurs particularly with the highly passive metals that depend on a thin invisible protective film of oxide for their corrosion resistance. In such cases submicroscopic defects in the passive film may form the sites at which pits are initiated, thus giving rise to a situation similar to that shown in Fig. 1.46.

Differential Aeration

It can be seen from Table 1.16 that differences in composition (nature or concentration) of the environment can lead to localised attack, and in Section 1.4 it was shown how differences in the activity of silver ions can give rise to a reversible concentration cell in which the silver electrodes in contact with the solution containing the lower and higher concentration of Ag^+ ions are the anode and cathode, respectively. Concentration cells of this type are rare in practice, but can occur during the corrosion of copper and copper alloys.

However, most solutions are in contact with atmospheric oxygen, and geometrical situations can arise where transport of oxygen through the solution by convection (natural or forced) and diffusion to one part of the metal occurs rapidly, whereas it is slow or even negligible at another (Fig. 1.46*d*). Under these circumstances the concentration of oxygen will be higher at one part of the metal surface and lower at the other, and this situation can lead to localised attack. Evans[5,6] in his Differential Aeration Principle stated that any geometrical factor that results in a higher concentration of oxygen at one part of a metal surface and a lower concentration (or zero concentration) at another will result in the former becoming the cathode and the latter the anode of the corrosion cell, with consequent localised attack. The evidence for this principle was based on a cell divided into two components by a porous diaphragm in which two steel electrodes were immersed in KCl solution (Fig. 1.47). Oxygen was bubbled into one compartment, and it was shown from the direction of the transfer of the electrons that the steel electrode in this compartment became the cathode, whilst the steel in the nonoxygenated compartment became the anode of the differential aeration cell.

At first sight it might appear that differential aeration could be explained in terms of a reversible oxygen concentration cell, for which

$$E = \frac{0 \cdot 059}{4} \log \frac{c_{II}}{c_{I}} \qquad \ldots (1.127)$$

where c_{II} and c_{I} are the concentrations of oxygen at the aerated and non-aerated electrodes, respectively. However, since a reversible oxygen electrode is extremely difficult to achieve in practice, and since the potentials are irreversible corrosion potentials involving at least two electrode reactions, this explanation is not valid.

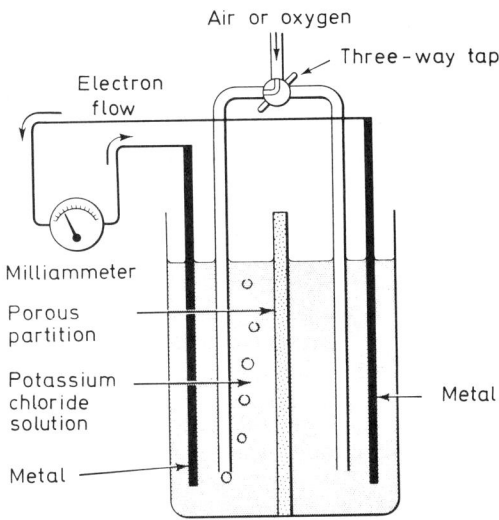

Fig. 1.47 Early form of the Evans 'differential aeration cell'. (Courtesy U. R. Evans)

In Section 1.4 (*see* Fig. 1.31*b*) it has been shown that when a corrosion reaction is controlled by the rate of oxygen diffusion, both the rate of corrosion and the corrosion potential increase with i_L, the limiting current density, i.e.

$$i_L = i_{corr} = k c_{O_2} \qquad \ldots (1.128)$$

Thus if the two pieces of steel in the differential aeration cell are not connected together the aerated steel will have a higher corrosion rate and more positive potential than the non-aerated steel. On short-circuiting the two pieces of steel, electrons will be transferred from the more negative to the more positive electrode (aerated steel), and this will result in an increase in the corrosion rate of the former and decrease in the rate of the latter, the potentials becoming equal, i.e. the potential of the aerated electrode becomes more negative and that of the non-aerated electrode more positive (*see* Fig. 1.62). On the other hand, if the two electrodes are connected through a high resistance, the non-aerated electrode will corrode at a lower rate than the aerated electrode.

It is evident from this description of the operation of the differential aeration cell that other factors must be involved if the differential aeration principle is to operate in the manner described by Evans, and this involves a consideration of the pH at the interfaces, the nature of the electrolyte, the anodic reaction and ohmic resistance effects.

Pourbaix[7,8] has shown by means of *E-i* diagrams (Wagner and Traud type) that the differential principle is applicable only at certain pH values, and the situation when the pH is the same in *both* the aerated and non-aerated zones is as follows:

1. pH < 8; differential aeration results in a small increase in the electrode potentials of both zones with a consequent increase in both corrosion rates.
2. pH ≈ 8 − 10; the differential aeration cell operates normally, and although the potential of the aerated zone increases, its corrosion rate decreases owing to passivation; the potential and corrosion rate of the non-aerated zone increases; a high current flows from the non-aerated to the aerated zone.
3. pH ≈ 10 − 13; aeration causes a significant increase in the potentials of both zones, without appreciable current flow.

It is now appreciated that for the differential aeration principle to apply the aerated zone must become passive, and that whether this occurs or not will depend on the initial pH of the solution, the change in pH produced by the OH^- ions resulting from the oxygen-reduction reaction and on the potential of the electrode. Similar conclusions were drawn by Bianchi[9] during a study of the corrosion of zinc, and it was observed that differential aeration did not occur when a buffered solution was used, since the pH at the aerated zone was unable to increase to a value sufficient to cause passivation; under these circumstances attack occurred at the water line. Schaschl and Marsh[10] showed that differences in the concentration of NaCl may result in differences in oxygen concentration, the higher the salt concentration the lower the solubility of oxygen so that these areas become the anodic zones of the differential aeration cell. Irrespective of salt concentration, however, the predominantly cathodic zone was always the one where the oxygen concentration was highest, and the cell current was found to be proportional to the difference in oxygen concentration in the solutions in the two compartments. They also point out that although initially the non-aerated zone may not be completely oxygen-free, any oxygen present is rapidly depleted in oxidising the $Fe(OH)_2$ to $Fe(OH)_3$.

Many of the studies of differential aeration have been carried out in the laboratory using a differential aeration cell containing sodium chloride solution, and measuring the current flowing in the external circuit when the electrodes are short-circuited. Although the results obtained in this way are useful in studying the mechanism of differential aeration they are sometimes misleading, and in certain situations the more highly aerated zone (the cathode of the macrocouple) corrodes at a higher rate. Tomashov[11] found that the less aerated zones of steel piles driven into the sea bed corroded at a slower rate than the more aerated zones in spite of the fact that the former was the anode and the latter the cathode of the macro-cell. He points out that this situation can arise where an increase in aeration has little effect on the anodic reaction, but markedly decreases the cathode polarisation of the local-action cells. This situation applies to sea-water in which the high concentration of Cl^- ions prevents anodic passivation of the steel, and under these circumstances the more highly aerated zone will corrode at a higher rate. A further factor that must be taken into account is the electrical resistance of the metal separating the two, since if this is significant the two zones will not be short-circuited, a situation that applies in the case of the steel piling[11].

Examples of differential aeration Before considering the factors that complicate localised attack it is relevant to refer to some of the classical work of U. R. Evans[6] on differential aeration. Perhaps the simplest example, although in corrosion science nothing is as simple as it first appears, is the distribution of anodic and cathodic sites produced when a drop of an electrolyte solution (e.g. NaCl or Na_2SO_4) is placed on a horizontal steel surface. Assuming that no dissolved oxygen is present initially and that mass transfer occurs solely by diffusion, it is apparent that transport of oxygen from the atmosphere to the surface of the metal by diffusion will occur most rapidly through the thin layer of solution at the periphery of the drop and most slowly at the centre of the drop. Thus the metal surface adjacent to the periphery will become the cathode of the differential aeration cell whilst the metal in the interior will become the anode (Fig. 1.48a and b). The OH^- ions formed by the cathodic reduction of dissolved oxygen and the Fe^{2+} ions formed by anodic dissolution will be brought together by migration and diffusion and will combine to form $Fe(OH)_2$ that is rapidly oxidised by the dissolved oxygen to rust, $Fe_2O_3.H_2O$. If the electrolyte contains a small

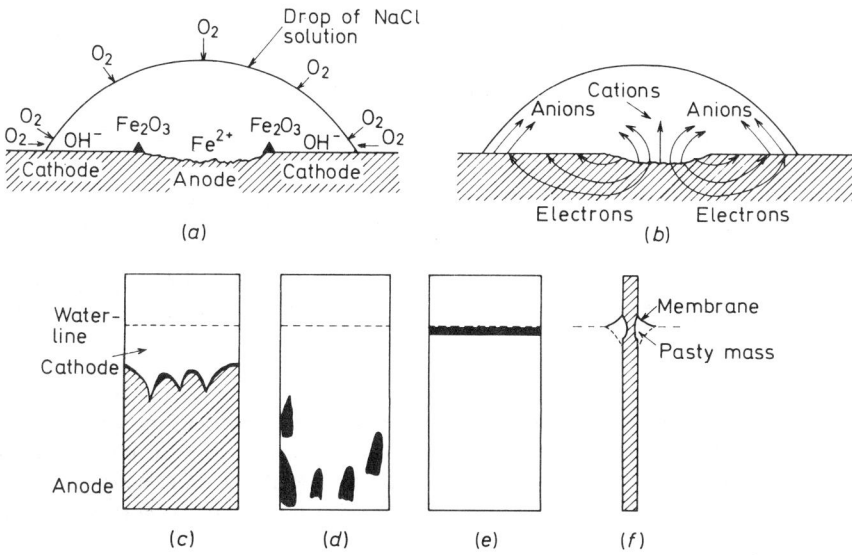

Fig. 1.48 Examples of differential aeration cells (a) and (b) Differential aeration cells formed by the geometry of a drop of NaCl solution on a steel surface; (c) differential aeration cells formed by the geometry of a vertical steel plate partly immersed in a NaCl solution. Increasing concentrations of Na_2CO_3 decrease the anodic area (d) until at a sufficient concentration attack is confined to the water line (e); (f) shows the membrane of corrosion products formed at water line (after Evans[6])

concentration of ferroxyl indicator (potassium ferricyanide plus phenolphthalein) the Fe^{2+} ions formed at the anodic area are revealed by the formation of Prussian blue and the OH^- ions formed at the cathodic area by the pink colour of phenolphthalein in alkaline solutions.

It should be noted that since the rust is formed at a position in between the anodic and cathodic sites it will not influence the kinetics of the corrosion reaction.

A similar situation arises when a vertical metal plate is partly immersed in an electrolyte solution (Fig. 1.48c), and owing to differential aeration the upper area of the plate will become cathodic and the lower area anodic. With time the anodic area extends upwards owing to the mixing of the anolyte and catholyte by convection and by the neutralisation of the alkali by absorption of atmospheric carbon dioxide.

Severe attack frequently occurs at a water-line, which in practice can range from structural steel partly immersed in a natural water to a lacquered tin can used for containing emulsion paint. This can be illustrated by adding increasing amounts of sodium carbonate to a sodium chloride solution in which a steel plate is partly immersed (Fig. 1.48c, d and e). With increase in concentration of the inhibitor, attack decreases and becomes confined to the water-line. The attack at the water-line is intense and is characterised by a triangular pasty mass of corrosion products bounded on the upper surface by a dark-brown membrane that follows the contour of the water-line. The mechanism of water-line attack is not clear, but it is likely that the membrane of corrosion products results in the formation of an *occluded cell*, in which the anolyte and catholyte are prevented from mixing. These occluded cells are discussed in more detail subsequently.

pH Changes During Localised Attack

The majority of the different types of localised attack dealt with in this section occur in near-neutral solutions containing dissolved oxygen, and the pH changes that occur at the anodic and cathodic sites are of fundamental importance when the reaction products are prevented from mixing by the geometry of the system.

It has been shown in Section 1.4 that cathodic reduction of dissolved oxygen may result in an increase in pH of the solution in the vicinity of the metal

$$O_2 + H_2O + 2e = 2OH^- \qquad \ldots(1.129)$$

and a similar consideration applies if the cathodic reaction involves hydrogen evolution. It is also evident that this increase in pH and the presence of dissolved oxygen will favour passivation of the metal surface so that its potential will increase and it will become the cathode of the macro-corrosion cell.

At the anodic site the metal will form metal cations

$$M \rightarrow M^{2+} + 2e \qquad \ldots(1.130)$$

and if it is assumed that the metal hydroxide is the thermodynamically stable phase in the solution under consideration, the metal ions will be hydrolysed by water with the formation of H^+ ions

$$M^{2+} + 2H_2O \rightarrow M(OH)_2 + 2H^+ \qquad \ldots(1.131)$$

Thus if the cathodic and anodic sites are separated from one another by the geometry of the system and if the solution is relatively stagnant the pH of the anolyte will decrease whereas that of the catholyte will increase.

It should be noted that whereas a completely soluble hydroxide (e.g. NaOH) will give a solution of high pH in which the pH will increase with concentration of the hydroxide, the pH of a solution of a sparingly soluble hydroxide will depend upon the equilibrium constant for hydrolysis and the activity of metal ions.

The equilibrium between the ions of a metal M_i and its hydroxide (or hydrated oxide) M_0 due to hydrolysis can be written in the form[12]:

$$M_i + cH_2O \rightarrow M_0 + mH^+ \qquad \ldots(1.132)$$

and since

$$\log K = \frac{-\Sigma \nu_i \mu_i^{\ominus}}{5710}$$

then

$$\log K = \frac{-[(\mu_{M_0}^{\ominus} + m\mu_{H^+}^{\ominus}) - (\mu_{M_i}^{\ominus} + c\mu_{H_2O}^{\ominus})]}{5710} \qquad \ldots(1.133)$$

Thus for the equilibrium

$$Fe^{3+} + 3H_2O = Fe(OH)_3 + 3H^+ \qquad \ldots(1.134)$$

$$\log K = \log \frac{a_{H^+}^3}{a_{Fe^{3+}}} \qquad \ldots(1.135)$$

and $\log K$ can be evaluated as -4.84 from the standard chemical potentials of the species involved in the equilibrium. Thus from equation 1.135

$$pH = \tfrac{1}{3}(4.84 - \log a_{Fe^{3+}}) \qquad \ldots(1.136)$$

and, for example, when $a_{Fe^{3+}} = 10^{-2}$ the pH will be $2 \cdot 3$.

Similar considerations will apply to other metal hydroxides, and Table 1.17[13] gives the hydrolysis reactions and the equilibrium pHs for metal ions

Table 1.7 Hydrolysis reactions and equilibrium pHs for metal ions of significance in acid formation in an occluded cell in stainless steels*

Hydrolysis reaction	Equilibrium pH
$Fe^{2+} + 2H_2O \rightleftharpoons Fe(OH)_2 + 2H^+$	$pH = 6 \cdot 64 - \tfrac{1}{2} \log a_{Fe^{2+}}$
$Cr^{3+} + 3H_2O \rightleftharpoons Cr(OH)_3 + 3H^+$	$pH = 1 \cdot 53 - \tfrac{1}{3} \log a_{Cr^{3+}}$
$Mi^{2+} + 2H_2O \rightleftharpoons Ni(OH)_2 + 2H^+$	$pH = 6 \cdot 1 - \tfrac{1}{2} \log a_{Ni^{2+}}$
$Mo^{3+} + 2H_2O \rightleftharpoons MoO_2 + 4H^+ + e$	$pH^{\dagger} = (0 \cdot 311 - 0 \cdot 059 \log a_{Mo^{3+}} - E^{\ominus})/0 \cdot 236$
$Mn^{2+} + 2H_2O \rightleftharpoons Mn(OH)_2 + 2H^+$	$pH = 7 \cdot 66 - \tfrac{1}{2} \log a_{mn^2}$

*Data after Pourbaix[13].
†$E^{\ominus} = -0 \cdot 20 \, V$ (vs. S.C.E.).

that are of significance in acid formation during the corrosion of stainless steels, and it is evident that metal ions like Cr^{3+} and Mo^{3+} will have the greatest effect in the pH decrease. However, this treatment is valid only if the metal forms simple cations, and if complex ions are formed the pH may be even lower than that predicted from equilibria of the type given in equation 1.132.

Cells of this type, in which the anolyte and catholyte are separated from one another, are referred to as *occluded* cells, and they may be formed by

(a) the original geometry by the system (crevices), (b) by a corrosion reaction that results in a pit or crack, or (c) by the formation of a membrane of insoluble corrosion products that form over the anodic site. In these occluded cells the pH of the solution at the cathodic sites will tend to increase and the pH at the anodic sites to decrease, thus stimulating attack at the latter.

The development of acidity within an occluded cell is by no means a new concept, and it was used by Hoar[14] as early as 1947 in his 'Acid Theory of Pitting' to explain the pitting of passive metals in solutions containing Cl^- ions. According to Hoar the Cl^- ions migrate to the anodic sites and the metal ions at these sites hydrolyse with the formation of HCl, a strong acid that inhibits the formation of a protective film of oxide or hydroxide. Edeleanu and Evans[15] followed the pH changes when aluminium was made anodic in Cl^- solutions and found that the pH decreased from 8·8 to 5·3.

However, it is only in recent years that attempts have been made to determine the actual pH within an occluded cell. Thus Brown et al.[16] determined the pH of the solution exuding from a propagating crack in a high-strength steel immersed in neutral NaCl (using indicator paper, an Sb/Sb_2O_3 electrode and a glass electrode) and found that it decreased to about pH 3. Wilde[17] studied hydrogen permeation through a 12% Cr martensitic steel membrane by means of the Devanathan permeation cell 19.2 and showed that in 3% NaCl permeation occurred when the cathode side of the membrane was not cathodically polarised, indicating the formation of acid within the pits. Both these workers were concerned with stress-corrosion cracking of the alloys being studied, and both concluded that the mechanism of failure of the alloys was due to hydrogen absorption and hydrogen cracking resulting from the formation of acid.

Suzuki, Yamake and Kitamura[13] determined the pHs, chloride ion concentrations, metal ion concentrations and the potentials of artificial pits in Fe, Cr, Ni and Mo, and in three austenitic stainless steels during anodic polarisation in 0·5 N NaCl at 70°C. In the case of the pure metals the pH values were found to be lower than those calculated from the metal ion concentrations (Table 1.17), and the experimentally determined pHs were as follows:

Fe 4·71; Cr 0·09; Ni 2·93; Mo 0·020

indicating that chromium and molybdenum when present in an alloy would have the greatest effect in lowering the pH within the pit.

The results for the steels are given in Table 1.18.

Table 1.18 pH, chloride ion concentration and potential for artificial pits in stainless steels*

	Type 304L (Fe-18Cr-10Ni-0·13Mo)	Type 316L (Fe-17Cr-13·7Ni-2·4Mo)	Fe-18Cr-16Ni-5Mo
pH	approx. 0·6 to 0·80	approx. 0·06 to 0·17	approx. −0·13 to 0·08
Cl^- conc. (N)	3·87	6·47	6·20
Steady-state potential (V, vs. S.C.E.)	−0·25 to −0·021	−0·02 to −0·22	−0·18 to −0·20

*Data after Suzuki, Yamake and Kitamura[13].

The fact that the pH values of the pure metals were lower than the theoretical values was attributed to the formation of hydroxy-chloro complexes of the metal and to the high chloride ion concentration in the pit, and the results highlight the very pronounced decrease in pH that can occur in an occluded cell, particularly when the alloy contains high concentrations of chromium and molybdenum. They also showed that migration of chloride ions into the solution in the pit can result in a 7-12-fold increase in concentration, and that the potential in the pit is in the active region.

Similar results have been obtained by Bogar and Fujii[18] using an artificial crevice formed between the steel and an acrylic plastic for a range of Fe-Cr alloys immersed in 3·5% NaCl. Initially the pH corresponded with the bulk solution, but after a short time it started to decrease and attained a steady-value after a few hours. The pH decreased with increase in chromium content of the alloy and attained a minimum value of 1·8 with the Fe-25Cr alloy, and higher chromium contents produced no further decrease. It was concluded that hydrolysis of Fe^{2+} determined the pH when the chromium content was less than 1%, hydrolysis of both Fe^{3+} and Cr^{3+} was responsible in the range 1·15% chromium and that Cr^{3+} dominated when the chromium content was above 15%. These results indicate that in any occluded cell the pH will be significantly lower than that of the bulk solution and the chloride ion concentration will be significantly higher. In the case of the passive metals the potential of the metal within the occluded cell will be in the active region, whereas the freely exposed surface will be in the passive region, with a consequent difference in potential between these two areas.

Pickering and Frankenthal[18a] have determined the potentials of the bottom of a pit and the passive surface of the freely exposed metal during the pitting of pure iron and certain stainless steels, and concluded that the metal within the pit is in the active state. It was established during potentiostatic studies that the bottom of the pit could have a potential as low as about $-0·2$ V whereas that of the passive surface could be as high as 1·4 V, so that the *IR* drop between these two surfaces could be as much as 1·6 V. They consider that this is due to the formation of bubbles of hydrogen gas at the bottom of the pit, which constricts the current flow with a consequent high *IR* drop, and results in a high anodic current density at that area which is independent of the resistivity of the solution. On the other hand, Franck[18b] came to the conclusion that the large *IR* drop was due to a layer of solution of high resistivity at the pit/solution interface that was caused by the inability of the corrosion products to diffuse rapidly away from the surface. Pickering and Frankenthal[18a] consider that the fluctuations in the corrosion current observed during these studies was due to the movement of gas bubbles that caused a variation in the *IR* drop between the passive cathode and active anode within the pit; they also consider that a similar mechanism involving hydrogen bubbles is applicable to crevice corrosion.

The importance of occluded cells cannot be overemphasised, and Brown[19] considers that pitting, crevice corrosion, intergranular attack, filiform corrosion and hydrogen cracking are characterised by local acidification due to hydrolysis of metal ions, and that this phenomenon is of major significance in the overall mechanism.

Crevice Corrosion

Intense localised corrosion, ranging from small pits to extensive corrosion over the whole surface, can occur within narrow crevices that may be formed by:

1. The geometry of the structure, e.g. riveted plates, welded fabrications, threaded joints.
2. Contact of the metal with non-metallic solids, e.g. plastics, rubber, glass.
3. Deposits of sand, dirt or permeable corrosion products on the metal surface (a type of crevice corrosion that is referred to as *deposit attack*).

Different types of crevices are shown in Fig. 1.49, and reference should also be made to Section 19.2 for other examples. The phenomenon is referred to as *crevice* corrosion, and is characterised by a geometrical configuration in which the cathode reactant (usually dissolved oxygen) can

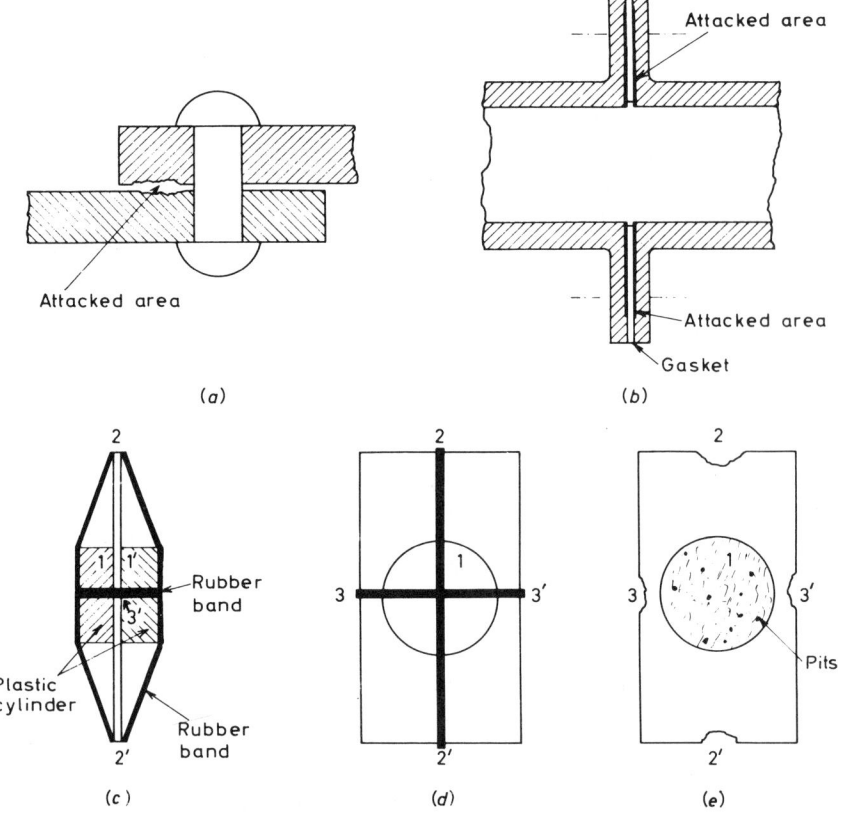

Fig. 1.49 Crevice corrosion. (*a*) Crevice resulting from the joining of two plates of steel and (*b*) crevice due to a gasket between two flanged pipes; (*c*), (*d*) and (*e*) show the method used by Streicher[21] to test different steels for their propensity to crevice corrosion; two plastic cylinders are held onto the sheet of metal by two rubber bands of the same size, giving rise to three different types of crevices in duplicate

readily gain access by convection (natural and forced) and diffusion to the metal surface outside the crevice, whereas access to the layer of the stagnant solution within the crevice is far more difficult and can be achieved only by diffusion through the narrow mouth of the crevice. This leads to a difference in concentration of the cathode reactant at the two surfaces, and for this reason it is sometimes referred to as *concentration cell* corrosion. Fibrous materials (such as gaskets for joints) that draw the solution into the crevice formed between the gasket and the metal by capillary action are particularly conducive to crevice corrosion, and the pitting of stainless steels is invariably stimulated by the presence of a crevice or a deposit (dirt, sand, marine growths, etc.).

Crevice corrosion, which can range from near-uniform attack to pitting at the metal surface within the crevice, can occur with a variety of metal and alloys ranging from the noble metals silver and copper to the very electronegative metals such as aluminium and titanium, and it is particularly prone to occur with metals and alloys that rely on passivity for their corrosion resistance. The environment can be any aggressive solution (acid or neutral), including natural waters, but solutions containing chloride ions are the most conducive to crevice corrosion. The crevice must be wide enough to permit entry of the solution, but sufficiently narrow to maintain a stagnant zone of solution within the crevice so that entry of the cathode reactant and removal of the reaction products is very slow and occurs only by diffusion and migration (if the species are charged); crevice widths are of the order of about 0·025 to 0·1 mm. In this connection it should be noted that macro-crevices can also result in enhanced corrosion during atmospheric exposure, but this is due to the fact that dirt and salts accumulate in the crevice and retain moisture so that the metal is maintained in contact with a wet poultice containing electrolytes.

Most cases of crevice corrosion take place in near-neutral solutions in which dissolved oxygen is the cathode reactant, but in the case of copper and copper alloys crevice corrosion can occur owing to differences in the concentration of Cu^{2+} ions; however, in the latter the mechanism appears to be different, since attack takes place at the exposed surface close to the crevice and not within the crevice; in fact, the inside of the crevice may actually be cathodic and copper deposition is sometimes observed, particularly in the Cu-Ni alloys. Similar considerations apply in acid solutions in which the hydrogen ion is the cathode reactant, and again attack occurs at the exposed surface close to the crevice.

Stainless steels are particularly prone to crevice corrosion, and even the Fe-18Cr-8Ni-3Mo type of austenitic stainless steel, which is highly resistant to pitting when the surface is free from crevices, is susceptible although initiation of attack may take 1-2 years[20].

Streicher[21] has studied various stainless steels, Inconels, Hastelloys and pure metals such as Ni, Mo, Ti, Ta and Nb using an artificial crevice (Fig. 1.49c to e) that gives rise to three different types of crevice conditions. In order to accelerate attack a 10% $FeCl_3$ solution (pH 1·6) at 50°C was used and attack was found to occur within 24 h on all susceptible alloys, whereas it could take up to 4 months when the surface was free from a crevice. A stainless steel (SP-2; Fe-18Cr-1ONi-2·5Mo-2·5Si) was found to be most resistant to attack, whereas all other stainless steels (including the molybdenum-containing alloys) failed.

Mechanism

At first sight the mechanism of crevice corrosion appears to be simply the formation of a differential aeration cell in which the freely exposed metal outside the crevice is predominantly cathodic whilst the metal within the crevice is predominantly or solely anodic; the large cathode current acts on the small anodic area thus resulting in intense attack. However, although differential aeration plays an important rôle in the mechanism, the situation in reality is far more complex, owing to the formation of acid within the crevice.

It is appropriate to consider first the crevice corrosion of mild steel in oxygenated neutral sodium chloride, and then to consider systems in which the metal is readily passivated. Initially, the whole surface will be in contact with a solution containing oxygen so that attack, with oxygen reduction providing the cathodic process, occurs on both the freely exposed surface and the surface within the crevice (Fig. 1.50). However, whereas the freely exposed surface will be accessible to dissolved oxygen by convection and diffusion, access of oxygen to the solution within the crevice can occur only

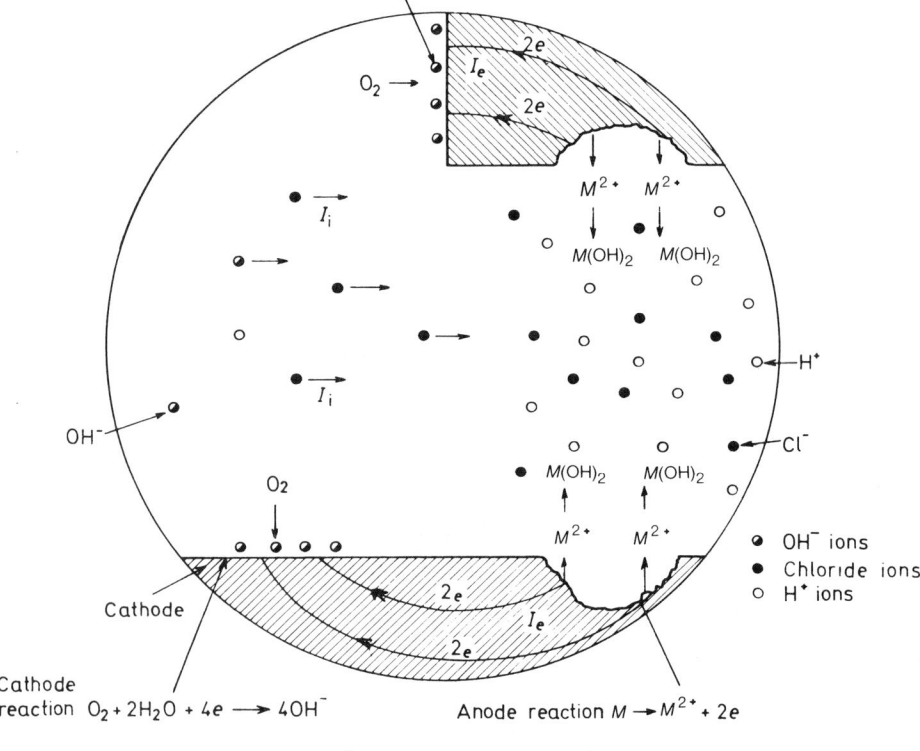

Fig. 1.50 Mechanism of crevice corrosion at, for example, the steel joint shown in Fig. 1.49*a* when immersed in a chloride solution after Reference 22

by diffusion; furthermore, it will be rapidly removed from the solution outside the crevice by cathodic reduction on the freely exposed surface and by reaction with Fe(OH)$_2$ at the mouth of the crevice. Thus under these conditions the oxygen concentration within the crevice will rapidly become negligible, and oxygen reduction within the crevice will cease.

The cathodic reduction of oxygen on the large area of freely exposed surface will result in anodic dissolution of the metal within the crevice, and the excess of positive charges (metal cations) formed within the solution will result in the migration of Cl$^-$ and OH$^-$ from the bulk solution to maintain charge balance (see Fig. 19.39); since the contribution made by the very low concentration of OH$^-$ ions to ionic conduction is small it will not affect the pH significantly. Hydrolysis of the metal chloride then occurs with the formation of Fe(OH)$_2$, which is not protective, and H$^+$ ions, and the pH falls to about 3; the presence of the high chloride concentration (3-10 times that of the bulk solution) and H$^+$ ions prevents passivation and facilitates anodic dissolution. As the anodic dissolution becomes kinetically easier, the potential of the freely exposed surface becomes more negative, i.e. it becomes partially cathodically protected by the electrons liberated by the anodic reaction[22]. This sequence of events results in an *autocatalytic* process, and although there is frequently an incubation period before attack commences it subsequently proceeds at a rate that increases with time. This is due to the gradual increase in the concentration of M^{z+} ions in the solution within the crevice, which stimulates migration of Cl$^-$ ions, and the presence of these ions in the crevice then stimulates more rapid anodic dissolution. Thus this cycle of events causes the anodic reaction to proceed at a rate that increases with time.

In the case of the stainless steels, or other readily passivated metals, the rapid reduction of dissolved oxygen on the freely exposed surface will be sufficient to exceed the critical current density so that the metal will become passive with a potential greater than E_{pp}, whereas the metal within the crevice will be active with a potential less than E_{pp}. The passivation of the freely exposed surface will be facilitated by the rise in pH resulting from oxygen reduction, whilst passivation within the crevice will be impeded by the high concentration of Cl$^-$ ions (which increases the critical current density for passivation) and by the H$^+$ ions (which increases the passivation potential E_{pp}, see Section 1.4).

In recent years there have been a number of investigations of crevice corrosion in which a variety of geometrical configurations have been used to simulate the conditions prevailing within a narrow crevice. Rosenfeld and Marshakov[23] used a cylinder of steel surrounded by an annular cylinder of plastic of slightly greater internal diameter than the external diameter of the metal. They also used a disc of glass, which contained a Luggin capillary, separated by a predetermined distance from a disc of the steel to determine the potential of a metal within a crevice, and found that in the case of stainless steel the difference in potential between the freely exposed surface and the surface within the crevice was as high as 600 mV. Similar studies have been made by Karlberg and Wranglen[24] who have determined the E-log i relationships of the freely exposed metal and of the metal within the crevice. They consider that the stages in the mechanism of crevice corrosion of a 13% Cr steel in neutral NaCl are as follows:

1. All areas of the steel are passive, although i_{pass} is comparatively high.
2. Thickening of the passive film on all surfaces occurs and i_{pass} decreases; a small current now flows from the crevice to the outer surface, but this is so small that the surfaces are practically equipotential.
3. The difference in oxygen concentration is now large and the potential of the metal within the crevice is now more negative than the freely exposed metal; the predominant reaction in the crevice is anodic dissolution resulting in a high concentration of Fe^{2+} and Cr^{3+} ions.
4. Hydrolysis of Fe^{2+} and Cr^{3+} results in a lowering of the pH; the cell current increases with consequent increased migration of Cl^- into the crevice.
5. Since E_{pp} of the metal within the crevice is raised by the decrease in pH, the metal is active whereas the outer surface is passive, and this represents the active propagation stage of crevice corrosion; there is a significant IR drop between the two zones.

In conclusion it is appropriate to refer to titanium, a highly passive metal that when first used industrially was thought to be resistant to crevice corrosion. Although this is the case at ambient temperatures it is attacked by concentrated solutions of Cl^-, I^-, Br^- and SO_4^{2-} at elevated temperatures of greater than 95°C. Figure 1.51 shows crevice corrosion at the gasket/metal interface on the flange of a titanium pipe used for hot hypochlorite solutions. Takamura[25], on the basis of studies of freely exposed surfaces of titanium and surfaces within an artificial crevice in concentrated neutral and acid chloride solutions ($NaCl$, $AlCl_3$, $ZnCl_2$, $MgCl_2$, $CaCl_2$, NH_4Cl), concluded that attack within a crevice was due to acid formation by hydrolysis reactions of the type

$$Ti \rightarrow Ti^{2+} + 2e$$
$$Ti^{2+} + 2H_2O \rightarrow TiO_2 + 4H^+ + 4e$$

Under these circumstances the metal's surface within the crevice became active and it corroded with the formation of a yellowish-white corrosion product that was identified as being mainly rutile TiO_2. On the other hand, a Ti-0·13Pd alloy was found to be immune from crevice corrosion, since the presence of the palladium facilitated passivation of the metal surfaces forming the crevice.

Fig. 1.51 Crevice corrosion resulting from the crevice produced between the gasket and the flange of a titanium pipe used for conveying a hot hypochlorite solution. The attacked areas are coated with a hard deposit of titanium oxides, whilst the unattacked area of metal outside the crevice remains bright

Griess[26] has observed crevice corrosion of titanium in hot concentrated solutions of Cl^-, SO_4^{2-} I^- ions, and considers that the formation of acid within the crevice is the major factor in the mechanism. He points out that at room temperature $Ti(OH)_3$ precipitates at pH 3, and $Ti(OH)_4$ at pH 0·7, and that at elevated temperatures and at the high concentrations of Cl^- ions that prevail within a crevice the activity of hydrogen ions could be even greater than that indicated by the equilibrium pH values at ambient temperatures. Alloys that remain passive in acid solutions of the same pH as that developed within a crevice should be more immune to crevice attack than pure titanium, and this appears to be the case with alloys containing 0·2% Pd, 2% Mo or 2% Ni.

Controlling Crevice Corrosion

An outline of the methods of controlling crevice corrosion is provided here, and for further details reference should be made to Chapter 9, and particularly to Sections 9.2 and 9.5. The obvious method of controlling crevice corrosion is to avoid crevices in the design of the structure and to avoid geometrical conditions that lead to the formation of deposits on the metal surface. Thus welded butt joints should be used in preference to riveted or bolted joints. Alternatively, if crevices cannot be avoided in design they should be sealed by welding, soldering, or by the use of caulking compounds.

Solids should be removed from process liquors and vessels should be designed so that deposition of solids does not occur, e.g. by avoiding sharp corners and stagnant areas by providing facilities for complete drainage of the process liquor.

The interface between a gasket and a metal flange forms a very effective crevice, particularly if the gasket is composed of an absorbent material, but this can be minimised by using non-absorbent materials such as p.t.f.e.

Metals and alloys vary in their ability to resist crevice corrosion, and this applies particularly to those that rely on passivity for their resistance to corrosion. Titanium and high-nickel alloys such as the Inconels and Hastelloys are amongst the most resistant, but even these will be attacked under highly aggressive environmental conditions.

Filiform Corrosion

Filiform corrosion[27] is characterised by the formation of a network of threadlike filaments of corrosion products on the surface of a metal coated with a transparent lacquer or a paint film, as a result of exposure to a humid atmosphere. This phenomenon first attracted attention because of its formation on lacquered steel, and for this reason it is sometimes referred to as *underfilm* corrosion, but although it is most readily observed under a transparent lacquer it can also occur under an opaque paint film or on a bare metal surface. Filiform corrosion has been observed on steel, zinc, magnesium and aluminium coated with lacquers and paints, and with aluminium foil coated with paper. Surface treatment of the metal by phosphating or chromating lessens the tendency for filiform corrosion to occur, but it is not completely effective in preventing it. The damage caused by corrosion is very slight, and the major objection to this form of attack is its detrimental effect on the appearance of the metal, particularly when the coating is transparent.

Filiform corrosion on steel covered with a transparent lacquer, which in common with other organic finishes is permeable to oxygen and water, and takes the form of red threads with a green or blue head (Fig. 1.52). Propagation of the thread occurs via the active head, the body or tail left behind being inactive. The threads are $0 \cdot 1$–$0 \cdot 5$ mm in width and propagate at $0 \cdot 4$ mm d^{-1} in random directions, but a feature of their propagation is that they never intersect. If a propagating head approaches an inactive tail it becomes deflected in such a way that the angle of approach equals the angle of deflection, and if two propagating heads meet propagation ceases[20].

The main factor in causing filiform corrosion is the relative humidity of the atmosphere, and if this is below 65% (the critical relative humidity for the atmospheric corrosion of most metals, see Section 2.2) it will not occur. As the relative humidity increases the thickness of the filaments increases; at 65–80% relative humidity they are very thin, at 80–95% relative humidity they are much wider and at approximately 95% relative humidity they broaden sufficiently to form blisters.

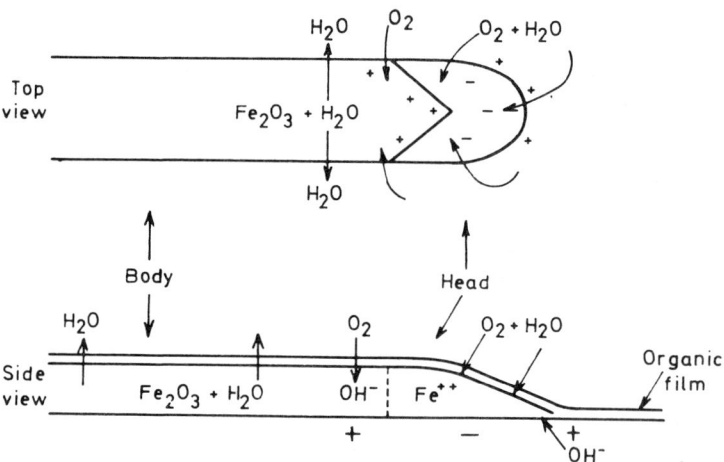

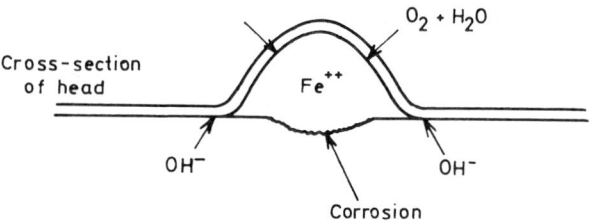

Fig. 1.52 Mechanism of filiform corrosion showing how atmospheric oxygen and water enter the active head through the film (lacquer) and how water leaves through the inactive tail. This results in a high concentration of oxygen at the 'V'-shaped interface between the tail and the head, and to a differential aeration cell (after Uhlig[20])

The mechanism of filiform corrosion[28-30] is highly complex, but it has many of the features of crevice corrosion in that differential aeration and acid formation have an important rôle in the propagation of the active head. Slabaugh and Grutheer[29] found that the blue-green active head was composed of a concentration solution of Fe^{2+}, which is supplied with moisture from the atmosphere by osmosis through the film of lacquer. In the tail the concentration of Fe^{2+} is kept low by the precipitation of $Fe_2O_3.H_2O$ (rust) and water is removed by osmotic action, i.e. atmospheric water enters at the head and is removed at the tail, which therefore tends to dry out. Oxygen also diffuses through the film, but owing to lateral diffusion it attains a higher concentration at the 'V'-shaped interface between the head and the tail (although it is also high at the periphery of the head) and a lower concentration in the centre of the head. Thus a differential aeration cell is set up, the active head forming an occluded cell in which the metal corrodes anodically at the centre and forward portions of the head, and it is possible that owing to hydrolysis acid formation also occurs at this area. Hydroxyl ions formed at the periphery of the head play a rôle in attacking the film at the interface and thus weakening the film/metal bond. Behind the 'V'-shaped interface the metal for a portion of the length of the filament acts as a relatively large cathode that stimulates activity at the metal within the propagating head. Figure 1.52 provides a schematic picture of the mechanism of filiform corrosion[20].

The obvious remedy is to maintain the relative humidity below 65%, but this is not always practicable, and at present no completely satisfactory remedy has been found.

Pitting Corrosion (see also Section 1.5)

Introduction

Pitting may be defined as a limiting case of localised attack in which only small areas of the metal surface are attacked whilst the remainder is largely unaffected, and this definition is applicable irrespective of the mechanism involved; dezincification, crevice corrosion and impingement attack can all result in pitting, although the mechanisms of these three processes are quite different.

Pitting is insidious, since it frequently results in perforation of the metal with consequent leakage of fluid, but where massive sections of metal are involved its effect is usually of little significance, although it can lead to a deterioration in surface appearance, e.g. in stainless steel used for architectural purposes. Furthermore, although pits may initiate and propagate to a certain depth they frequently become inactive before penetrating the cross-section of the metal — an observation that can be confirmed by examining an old aluminium saucepan.

Whereas many forms of localised attack, including pitting, can often be ascribed to a well-defined heterogeneity associated with the metal/environment system (e.g. a crevice, a discontinuity in millscale, an area impinged upon by a high velocity solution), pitting can occur in a system that is apparently free from heterogeneities, providing the solution contains an

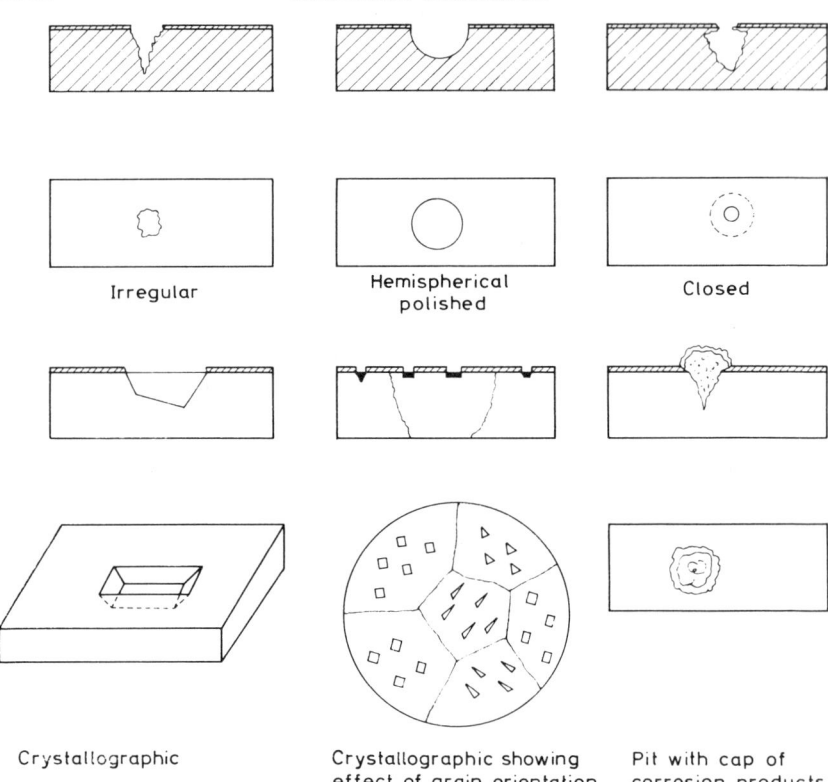

Fig. 1.53 Some different types of pits

aggressive anion such as Cl$^-$. It should be emphasised that whereas crevice corrosion can take place in most corrosive electrolyte solutions, pitting on a freely exposed surface will take place only if certain anions are present in the solution. Thus an 18/8 stainless steel free from crevices will pit in a stagnant solution of sodium chloride and the pit sites will be random and unpredictable; if a crevice is present the pits will be located within the crevice and the rate of pitting will be greater than at the freely exposed surface. On the other hand, the steel will not pit if immersed in a solution of, for example, sodium sulphate.

Crevice corrosion and pitting have a number of features in common, and it has been stated that pitting may be regarded as crevice corrosion in which the pit forms its own crevice; however, whereas a macroscopic heterogeneity determines the site of attack in crevice corrosion, the sites of attack in pitting are determined by microscopic or sub-microscopic features in the passive film (*see* Sections 1.3 and 1.5).

Figure 1.53 shows diagrammatically various types of pits that can range from hemispherical with a polished surface, in which crystallographic etching has been completely suppressed, to crystallographic pits whose sides are composed of the crystal planes that corrode at the slowest rate. Pits formed on Ni during anodic polarisation in an acetic acid-acetate buffer of pH 4·6 are shown in Fig. 1.54.

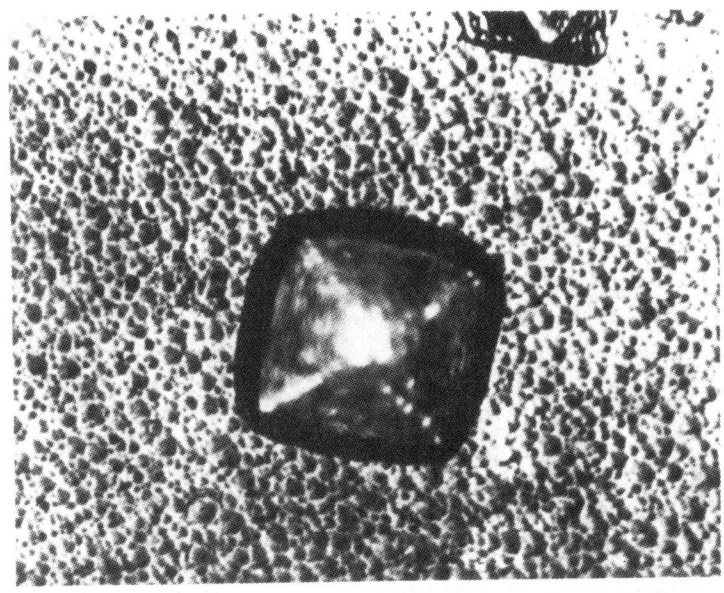

Fig. 1.54 Pits formed during anodic polarisation in an acetic acid-acetate buffer of pH 4·6 containing thiourea or NaCl. (a) General attack and formation of crystallographic pits on nickel in the buffer + 10^{-5}M thiourea ($\times$ 200), (b) crystallograhic pits formed in the buffer +10^{-6} M

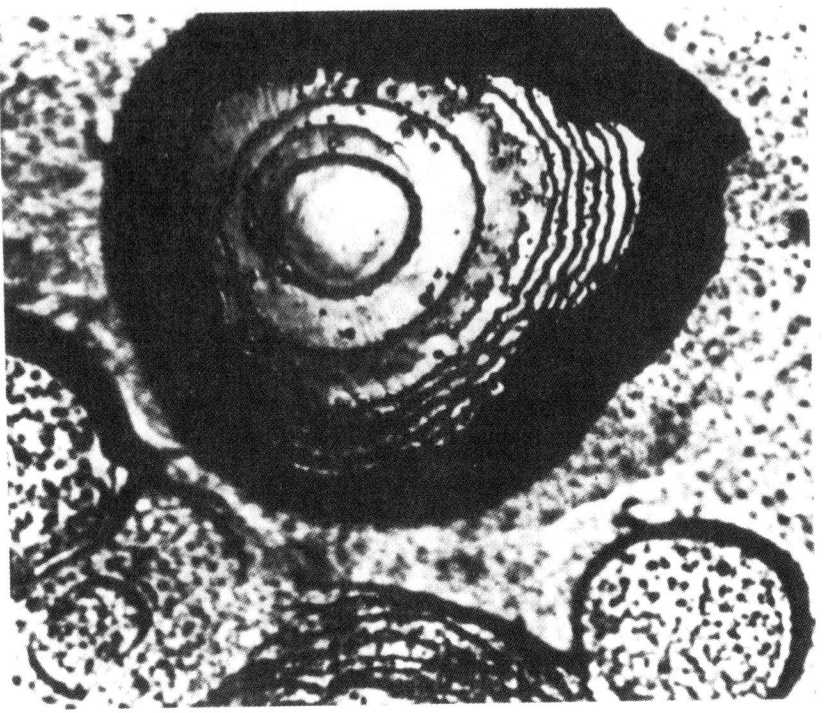

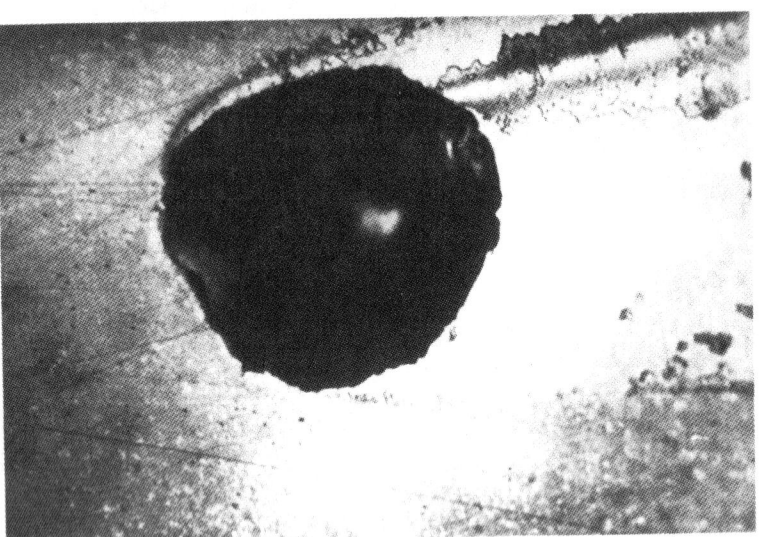

thiourea (× 400), (c) hemispherical spiral-shaped pits formed on nickel in the buffer $+10^{-4}$ M thiourea (× 400) and (d) hemispherical pit formed on Fe-12Cr in the buffer $+ 0.5$ M NaCl (× 200). (After P. Surendra, PhD thesis, University of London, 1971)

Many of the factors that have been considered above in relation to crevice corrosion are equally applicable to pitting, e.g. the large cathode/small anode area relationship, differences in concentration of the cathode reactant, auto-catalytic formation of acid within the occluded cell (*see* Fig. 1.49), etc. but there are certain distinct differences that must be considered.

Pitting can occur with a number of metals and alloys, and as a rough guide the more expensive the alloy the lower its propensity for pitting! Thus Fe–17Cr stainless steel is more susceptible to pitting in chloride solutions than Fe–18Cr–8Ni, which in turn is more susceptible than the Fe–18Cr–8Ni–3Mo steel, and titanium is superior to all of them. However, it is important to avoid sweeping generalisations regarding the propensity of an alloy to pitting or of the pitting or inhibitive properties of anions, and each system must be considered individually. Thus the pitting of certain stainless steels will occur in solutions containing Cl^- ions, Br^- (but not I^- or F^-) ions, hypochlorite or thiosulphate anions, but pitting will tend to be suppressed by the presence of oxyanions such as NO_3^- or SO_4^{2-}. On the other hand, copper pits in natural fresh waters containing a very small concentration of Cl^- and a significantly higher concentration of SO_4^{2-} ions; in waters with a higher Cl^- concentration attack is far less localised[31].

Pitting can occur over a range of pHs and potentials, but in all cases the potential of the metal must be within the passive region so that the major part of the surface is passive. Furthermore, the redox potential of the solution must exceed a certain critical value for pits to initiate. In near-neutral solution this potential is achieved by the presence of dissolved oxygen, but oxidising metal cations of a higher redox potential than the critical potential are more conducive to rapid pitting than dissolved oxygen, e.g. $FeCl_3$, $CUCl_2$ and $HgCl_2$. For this reason $FeCl_3$ is widely used in testing alloys for their resistance to pitting. Stagnant conditions that prevent mixing of the alkaline catholyte and acid anolyte will favour pitting, whilst any movement of the solution will tend to suppress pitting. It follows that a crevice or the entrapment of a solution in an inactive part of a plant will be conducive to pitting, whereas movement of the solution will lower the pitting propensity. Conversely, a pit may initiate, but the local concentration of Cl^- and H^+ ions that are required for continued propagation may be removed by convection currents so that after a short time the pit becomes inactive.

Pitting, Breakdown or Rupture Potential

A characteristic of the pitting of a number of metals that rely on passivity for their corrosion resistance is the *breakdown* potential E_b (which is also referred to as the *critical pitting* or *rupture* potential), which may be regarded as the most negative potential to cause the initiation and propagation of one or more pits, or as the most positive potential that results in a decrease in current due to passivation of the entire surface. The breakdown potential for a given metal will, of course, vary with the nature of the solution, but it will also be dependent on the method used for its determination (anodic polarisation using potentiostatic, potentiokinetic or galvanostatic techniques or chemical solutions of high redox potential such as $FeCl_3$) and on the duration of the test. It cannot, therefore, be regarded as a precise parameter.

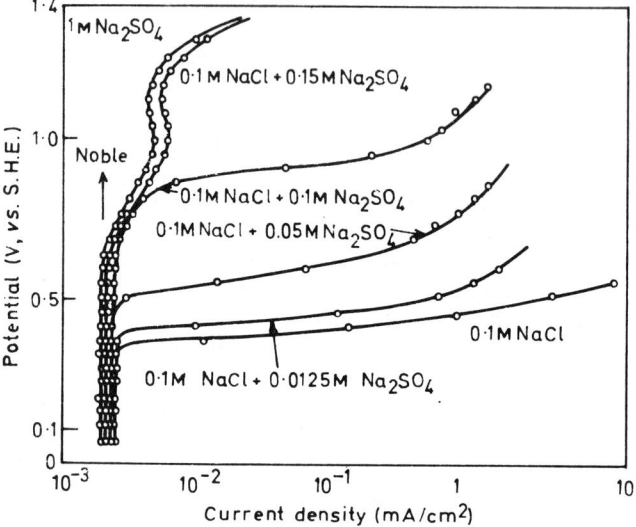

Fig. 1.55 Breakdown potentials of Fe-18Cr-8Ni stainless steel in 0·1 M NaCl plus various concentrations of SO_4^{2-} ions during potentiostatic anodic polarisation (after Leckie and Uhlig[32])

Figure 1.55 shows potentiostatic anodic curves obtained by Leckie and Uhlig[32] for Fe–18Cr–8Ni stainless steel in 0·1 M NaCl containing various concentrations of SO_4^{2-} ions, and it can be seen that in all solutions containing less than 0·15 M Na_2SO_4 there is a sudden increase in current at a certain potential, named the *critical pitting* or *breakdown* potential, within the passive region. This potential is a manifestation of pitting, and it can be seen that it becomes more positive with increasing concentration of SO_4^{2-} ions and is suppressed at concentrations at and above 0·15M Na_2SO_4.

Table 1.19 gives some critical pitting potentials collated by Uhlig[20] for different metals in 0·1 N NaCl at 25°C, in which in most cases the potential was first increased incrementally and held constant at each potential for 5 min in order to obtain an approximate value of the critical pitting potential. This may be followed by a separate series of experiments in which fresh specimens are held at potentials close to E_b for 12 h or more and the surface

Table 1.19 Critical pitting potentials in 0·1 N NaCl at 25°C[20]

Metal	Potential (V, vs. S.H.E.)
Al	−0·37
Ni	0·28
Zr	0·46
Fe-18Cr-8Ni stainless steel	0·26
Fe-30Cr	0·62
Cr	> 1·0
Ti	> 1·0 (1 N NaCl)
	approx. 1·0 (1 N NaCl at 200°C)

*Approx. 8–12V in sea-water.

Table 1.20 Minimum concentration of chloride ions for initiation of pitting[33]

Metal	Concentration of chloride ions (N)
Fe	0·0003
Fe–5·6Cr	0·017
Fe–11·6Cr	0·069
Fe–20Cr	0·1
Fe–18·6Cr–9·9Ni	0·1
Fe–24·5Cr	1·0
Fe–29·4Cr	1·0

Table 1.21 Effect of molybdenum additions to Fe–15Cr–13Ni stainless steel on the critical pitting potential E_b^* (mV, vs. S.H.E.)

Molybdenum content (wt. %)	E_b in 0·1 N NaCl at 25°C	E_b in 0·1 N NaBr at 25°C
0·0	280	540
0·42	300	525
0·92	340	—
1·4	400	510
1·8	535	—
2·4	722	467

*Data after V. Hospadaruk and J. V. Petrocelli, *J. Electrochem. Soc.*, **113**, 878 (1962) and N. D. Tomashov, G. P. Chernova and O. N. Marcova, *Corrosion*, **20**, 166t (1964).

observed with lower-power microscope to detect pit formation. Pitting will occur in practice if the redox potential of the solution is more positive than the potentiostically determined critical pitting potential, and providing the conditions (concentration of chloride ions, temperature, etc.) are the same. It follows from Table 1.19 that chromium and titanium will not pit in oxygenated 0·1 N NaCl at 25°C, but titanium will pit in boiling saturated $CaCl_2$ solution, owing to the fact that E_b becomes more negative with increase in temperature and chloride ion concentration; it will also pit in 0·1 N NaCl at 25°C if the potential is raised by an external source of e.m.f. to about 10 V. A measure of the tendency of metals to pit can also be obtained from the minimum concentration of chloride ions that are required to initiate pitting, and Stolica[33] has used this method to study a range of Cr–Fe alloys in 1 N H_2SO_4 containing different concentrations of chloride ions. He showed that whereas iron pitted when the concentration of chloride ions was as small as 3×10^{-5} N, concentrations as high as 1 N were required to cause pitting of Fe–29Cr alloy (Table 1.20). Table 1.21 shows how additions of Mo increase the critical potential of a 15Cr–13Ni stainless steel in chloride and bromide solutions.

Nature of Solution and Temperature

The critical pitting potential will, in general, decrease with the concentration of chloride ions, but will increase with the concentration of inhibiting oxyanions such as OH^-, SO_4^{2-}, NO_3^- CrO_4^{2-}, etc. Thus whereas Fe–18Cr–8Ni stainless steel will pit in 10% $FeCl_3$ in a matter of hours, pitting is completely arrested if 3% $NaNO_3$ present in the solution owing to the fact that the presence of the NO_3^- increases the breakdown potential above that of the redox potential of the Fe^{3+}/Fe^{2+} equilibrium[34].

Uhlig et al. have established the following relationships between E_b (V, vs. S.H.E.) a_{Cl^-} for Fe-18Cr-8Ni stainless steel[32,35]

$$E_b = -0.088 \log a_{Cl^-} + 0.168 \qquad \ldots(1.137)$$

and for aluminium[36]

$$E_b = -0.124 \log a_{Cl^-} - 0.0504 \qquad \ldots(1.138)$$

Uhlig et al. have also determined the minimum activity of inhibiting oxyanions $a_{inh.}$ to inhibit the pitting of Fe-18Cr-8Ni stainless steel[32,35] and aluminium[36], and found that equations of the type

$$\log a_{Cl^-} = A \log a_{inh.} + B \qquad \ldots(1.139)$$

are obeyed in which A and B are constants that depend on the metal and the nature of the inhibiting anion. The order of the effectives of oxyanions in inhibiting pitting was found to be as follows:

for Fe-18Cr-8Ni: $\quad OH^- > NO_3^- > \bar{A}c > SO_4^{2-} > ClO_4^-$

for aluminium: $\quad NO_3^- > CrO_4^- > \bar{A}c > $ benzoate $> SO_4^{2-}$

where $\bar{A}c$ signifies the acetate anion.

The effect of pH appears to be controversial. Some workers find a slight increase in E_b with increase in pH in the acid region, whilst others report that there is practically no change. In the alkaline region, however, E_b becomes significantly more positive with increase in pH owing to the passivating ability of the OH^- ion.

An increase in temperature significantly decreases E_b, an observation that is frequently neglected with unfortunate consequences. An interesting example was observed recently in which Fe-18Cr-8Ni stainless-steel steam-heated pans used for the manufacture of synthetic cream containing a small concentration of sodium chloride were found to pit after 3-4 weeks. The cream was manufactured at 70°C, but the pan was heated with superheated steam, and on removal of the cream by an outlet at the bottom of the pan the residue of cream on the sides of the pan was subjected to temperatures well above 70°C, with consequent pitting as a result of the small amount of salt present in the cream. The obvious solution to this problem was to use an Fe-18Cr-10Ni-3Mo stainless steel, which is more resistant to pitting attack.

Induction Period for Pitting

It is apparent that the critical pitting potential for a given alloy depends on the concentration of chloride ions, on the concentration of inhibiting anions in the solution and on the temperature of the solution. Unfortunately, the situation is complicated further by the fact that there is an induction period for the onset of pitting, which means that the pitting propensity of an alloy cannot be assessed precisely on the basis of potentiostatic determination of short duration. The induction time τ will decrease with increase in potential and with increase in chloride ion concentration, and in connection with the latter Stolica[37] obtained the following relationship:

For pure iron $\quad \tau = 21 \cdot 4 \, c_{Cl^-} \, \text{min}^{-1}$ ...(1.140)

For Fe-5·6Cr $\quad \tau = 1 \cdot 54 \, (c_{Cl^-} - 0 \cdot 02) \text{min}^{-1}$ ...(1.141)

For Fe-11·6Cr $\quad \tau = 2 \cdot 3 \, (c_{Cl^-} - 0 \cdot 069) \text{min}^{-1}$ ...(1.142)

Mutually Protective Effect

Pits seldom form in close proximity to one another and it would appear that the area of passivated metal, which acts as the cathode for the local cell, is protected by the anodic dissolution of metal within the pit — a phenomenon that is referred to as the *mutually protective effect* (*see* Section 1.5).

Protection Potential

It has been pointed out in Section 1.3 that although the *equilibrium* potential-pH diagrams are based solely on thermodynamic data it is possible to construct practical potential-pH diagrams from experimentally determined $E - i$ curves. Figure 1.56a shows the potentiokinetic $E - i$ curves for Armco iron in chloride-free solutions of different pHs obtained by Pourbaix[38], in which general corrosion occurs below the passivation potential P and above the region of immunity, and it can be seen that it is possible to construct a *practical* potential-pH diagram from these curves showing the zones of immunity, general corrosion and passivity that will prevail under various conditions of pH and potential. However, if the same procedure is adopted with solutions of different pH containing 10^{-2} mol dm^{-3} of chloride ion, a sudden increase in current occurs when the potential is raised to a value r, the breakdown potential E_b, at which pits are initiated and give rise to an anodic current (Fig. 1.56b). If after attaining E_b the potential is now lowered, the curve is not retraced (electrochemical hysteresis) and will intersect the i axis ($i = 0$) at a potential p at which neither anodic oxidation nor cathodic reduction can occur, i.e. pitting is arrested. The potential p is referred to as the *protection* potential E_p against pitting, since at and below E_p the metal, will not pit and the whole surface will remain passive. E_p is always more negative than E_b, and whereas pitting will occur on a pit-free surface above E_b, it will occur only in the range of potentials between E_p and E_b if the surface is already pitted, i.e. between E_p and E_b prior pits will continue to propagate, but initiation of new ones will not be possible.

Pourbaix, on the basis of the breakdown potential E_b and the protection potential E_p, distinguishes between the following states of a metal surface, which have been incorporated in the potential-pH diagram shown in Fig. 1.56b:

1. Perfect passivation. The potential-pH region between the passivation potential E_{pp} and the protection potential E_p, in which pits are not

initiated nor do they propagate if already present owing to passivation.
2. Imperfect passivation. The potential-pH region between E_p and E_b, in which pits already present can continue to propagate.
3. Pitting region. The potential-pH region above E_b, at which pits can both initiate and propagate.

The electrochemical hysteresis method just described is now becoming widely used to characterise the pitting propensity of alloys, and it has been used by Verink, and Pourbaix[39] to study the behaviour of a range of Fe-Cr alloys (0·5-24·9% Cr), Cu-10Ni and Cu-10Ni-1Fe alloys in solutions of different pH and chloride ion concentrations. However, Wilde[39b] has shown that E_p is not a unique parameter since it varies in magnitude with the amount of localised attack produced during anode polarisation. This is discussed more fully in Chapter 19 when considering testing for crevice corrosion and pitting.

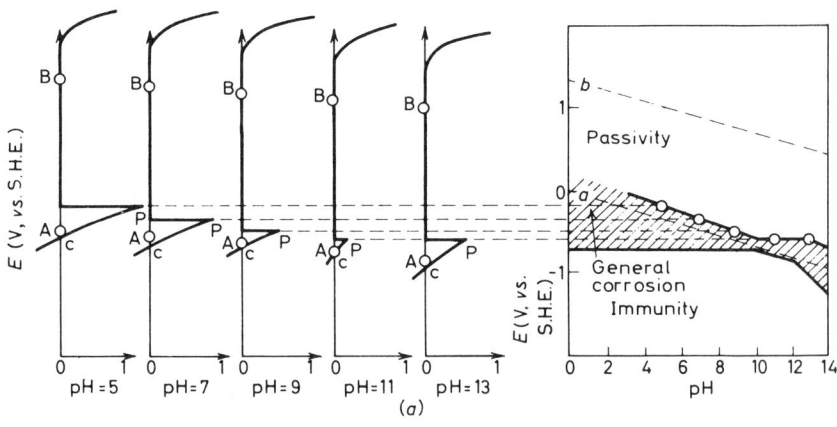

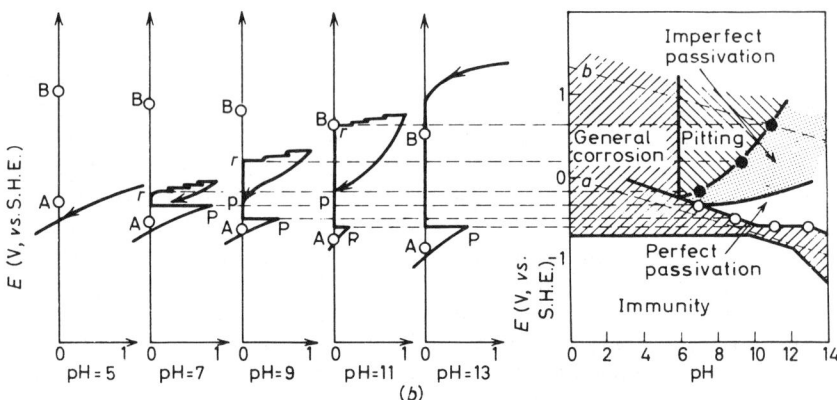

Fig. 1.56(a) $E-i$ curves and experimental potential-pH diagram for Armco iron in chloride-free solutions of different pHs (A is the unpolarised potential and P the passivation potential) and (b) $E-i$ curves and experimental potential-pH diagram for Armco iron in solutions of different pHs containing 10^{-2} mol dm^{-3} of chloride ion (r is the rupture potential and p the protection potential). (After Pourbaix[38, 39])

Mechanism

In view of the fact that there are two opposing views on the mechanism of passivity it is not surprising that a similar situation prevails concerning the mechanism of breakdown of passivity. The solid film theory of passivity and breakdown of passivity is dealt with in some detail in Section 1.5, so that it is appropriate here to discuss briefly the views based on the adsorption theory.

Uhlig and Gilman[34], and Kolotyrkin[40] explain the breakdown of passivity in terms of competitive adsorption between chloride ions in solution and the adsorbed monolayer of oxygen on the surface of the metal. According to Uhlig[20,34] although the metal has a greater affinity for oxygen than chloride ions, adsorption of the latter will be favoured by an increase in the potential until a value is reached where the adsorbed oxygen at specific sites is replaced by chloride ions, which catalyse anodic dissolution. Thus the induction period required for pitting, the decrease in the breakdown potential with increase in chloride ion concentration and the increase when passivating anions are also present in solution, are all explained in terms of competitive adsorption between chloride ions (and the passivating oxyanions) and the adsorbed oxygen on the metal surface.

Vermilyea[41] has adopted a thermodynamic approach to pitting, and considers that the critical pitting potential is the potential at which the metal salt of the aggressive ion (e.g. $AlCl_3$) is in equilibrium with metal oxide (e.g. Al_2O_3). On the basis of this theory the critical pitting potential should decrease by 0.059 V per decade increase in chloride ion concentration. Vermilyea's theory successfully predicts the values of the critical potentials for Al, Mg, Fe and Ni, but in the case of Zr, Ti and Ta there are large discrepancies.

An extensive review of pitting corrosion has been compiled by Szklarska-Smialowski[42], and the reader is recommended to consult this publication for further details.

Control of Pitting

Since stagnant conditions favour pitting attack it follows that it will be stimulated by the presence of crevices, deposits and by stagnant volumes of solutions, and these should therefore be avoided. In fact, many of the precautions enumerated to avoid crevice corrosion apply equally to pitting, However, the best procedure is to use an alloy that is resistant to pitting under the environmental conditions prevailing, and reference should be made to the recent paper by Streicher[21] who has examined the resistance to pitting and crevice corrosion of a range of metals including stainless steels, Inconels, Hastelloys and titanium.

Pitting of Carbon Steels

At the beginning of this section a simple explanation was provided of the localised attack that can occur when steel with a discontinuous coating of

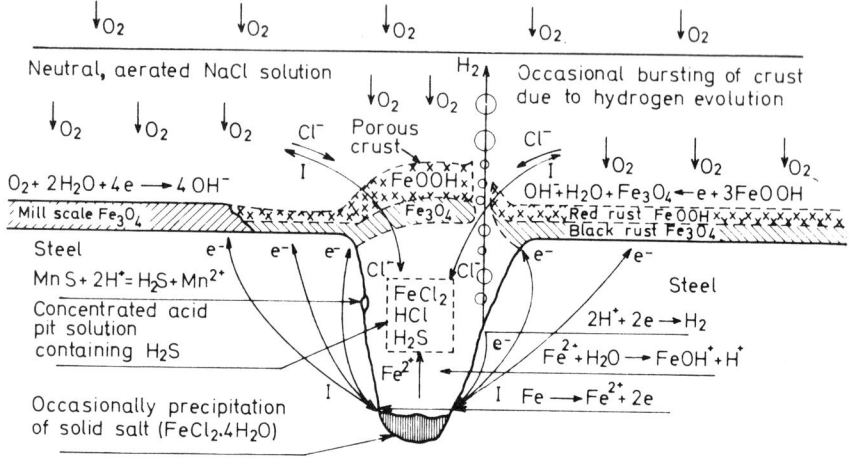

Fig. 1.57 Electrochemical reactions that occur when a pit is initiated at sulphide inclusion in a carbon steel (after Wranglen[43])

magnetic is exposed to an oxygenated water (Fig. 1.46). However, the situation is far more complex than that described, and is characterised by the formation of a hemispherical membrane of corrosion products (a tubercle, cap or mound) that inhibits diffusion of dissolved oxygen to the metal beneath, thus resulting in an occluded cell.

Wranglen[43] considers that sulphide inclusions are responsible for the initiation of attack in both carbon steels and stainless steels, and on this basis he has provided a detailed exposition of the pitting of a carbon steel at an inclusion of MnS when the steel is immersed in an oxygenated chloride solution (Fig. 1.57). The reactions of significance are given in the diagram, but certain features of the mechanism are of interest since they illustrate the complexity of the process.

Pit interior Within the pit the primary anodic reaction is

$$Fe \rightarrow Fe^{2+} + 2e \qquad \ldots(1.143)$$

which is followed by hydrolysis and the generation of H^+

$$Fe^{2+} + H_2O \rightarrow FeOH^+ + H^+ \qquad \ldots(1.144)$$

The decrease in pH results in dissolution of some of the MnS

$$MnS + 2H^+ \rightarrow H_2S + Mn^{2+} \qquad \ldots(1.145)$$

thus providing S^{2-} and HS^- that stimulate attack by decreasing the activation overpotential for the dissolution of Fe (and Ni).

The electrons released are partly accepted by dissolved oxygen at the surface millscale and partly by the H^+, with the consequent formation of H_2 gas. The concentration of chloride ions within the pit will increase owing to migration, and this too will stimulate dissolution.

Pit mouth A membrane of magnetite (Fe_3O_4) and rust (FeOOH) is formed, which prevents the intermingling of the acid anolyte and alkaline catholyte, by the following steps:

Oxidation of $FeOH^+$ and Fe^{2+} by dissolved oxygen occurs

$$2FeOH^+ + \tfrac{1}{2}O_2 + 2H^+ \rightarrow 2FeOH^{2+} + H_2O \qquad \ldots(1.146)$$

$$2Fe^{2+} + \tfrac{1}{2}O_2 + 2H^+ \rightarrow 2Fe^{3+} + H_2O \qquad \ldots(1.147)$$

followed by hydrolysis of the reaction products

$$FeOH^{2+} + H_2O \rightarrow Fe(OH)_2^+ + H^+ \qquad \ldots(1.148)$$

$$Fe^{3+} + H_2O \rightarrow FeOH^{2+} + H^+ \qquad \ldots(1.149)$$

and the precipitation of magnetite and rust

$$2FeOH^{2+} + Fe^{2+} + 2H_2O \rightarrow Fe_3O_4 + 6H^+ \qquad \ldots(1.150)$$

$$Fe(OH)_2^+ + OH^- \rightarrow FeOOH + H_2O \qquad \ldots(1.151)$$

Outside the pit Reduction of dissolved oxygen

$$O_2 + 2H_2O + 4e \rightarrow 4OH^- \qquad \ldots(1.152)$$

and reduction of rust to magnetite

$$3FeOOH + e \rightarrow Fe_3O_4 + H_2O + OH^- \qquad \ldots(1.153)$$

This area will be passivated by the increase in pH due to the cathodically produced OH^- ions, and partially cathodically protected by the electrons liberated by the anodic processes within the pit. The tubercle thus results in an occluded cell with the consequent acidification of the anodic sites. Wranglen considers that in view of the fact that crystals of $FeCl_2 \cdot 4H_2O$ are sometimes observed at the bottom of a pit the solution within the pit is a saturated solution of that salt, and that this will correspond with an equilibrium pH of about 3·5.

It is also of interest to note that Wranglen considers that the decrease in the corrosion rate of steel in the atmosphere and the pitting rate in acid and neutral solution brought about by small alloying additions of copper is due to the formation of Cu_2S, which reduces the activity of the HS^- and S^{2-} ions to a very low value so that they do not catalyse anodic dissolution, and a similar mechanism was put forward by Fyfe *et al.*[44] to explain the corrosion resistance of copper-containing steels when exposed to industrial atmospheres.

Pitting of Aluminium

The pitting of aluminium in chloride-containing waters follows a similar mechanism to that of steels (Fig. 1.58), and again the characteristic feature of the process is the formation of acid within the occluded cell[45]. The passivating film of Al_2O_3 surrounding the pit acts as the cathode, but its effectiveness in reducing dissolved oxygen is significantly enhanced if copper is either deposited on the surface or enters the lattice of the Al_2O_3, and it is well known that the pitting of aluminium occurs rapidly when the water contains a trace of copper ions (*see* Section 4.1). Similar considerations apply to intermetallic phases such as $FeAl_3$ and $CuAl_2$, which can increase the kinetics of oxygen reduction.

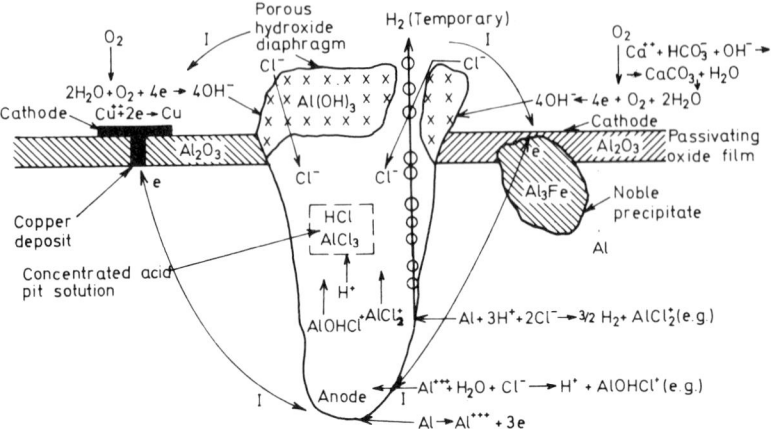

Fig. 1.58 Pit on aluminium showing how the rate of pitting may be facilitated by an intermetallic phase (Al_3Fe) or by a deposit of copper (after Wranglen[45])

Pitting of Copper

Previous considerations of pitting have been largely confined to metals and alloys that have a strong tendency to passivate, but since the pitting of copper has a number of unusual features it is appropriate to consider it in some detail. Reference to the potential–pH diagram for the Cu–H_2O (Section 4.2) system shows that in neutral solutions at the potentials encountered in oxygenated waters the stable form of copper is Cu_2O, and the corrosion resistance of copper thus depends upon whether or not the Cu_2O forms a protective film.

Copper and its alloys in certain fresh waters give rise to a form of localised attack that is referred to as *nodular* pitting in which the attacked areas are covered by small mounds or nodules composed of corrosion products and of $CaCO_3$ precipitated from the water. This is a serious problem in view of the extensive use of copper pipes and tanks for water supplies, and in aggressive water these may perforate in a relatively short time.

Figure 1.59 shows the type of pit and corrosion products[46] that form on copper pipes used for hard (or moderately hard) well waters, and this type of attack is most prevalent when the pipe is used for conveying cold water. The pit interior is almost invariably covered with solid Cu_2Cl_2, and across the mouth of the pit there is a thin membrane of Cu_2O containing one or more holes; this membrane is supported on the underside by a more substantial layer of coarsely crystalline Cu_2O formed by hydrolysis of Cu_2Cl_2. Above the Cu_2O membrane there is a roughly hemispherical mound of $CaCO_3$ containing insoluble copper salts, mainly basic carbonate and chloride. According to Campbell[47,48] it is possible to distinguish two types of pits and those described above are most prevalent in waters used in the USA, Belgium, Holland and the UK. Another type of pit occurs in certain soft water areas (mainly in Sweden and Germany), but only when the temperature is above 60°C; these pits are of a smaller cross-section than those obtained in hard waters, and contain a very hard crystalline Cu_2O

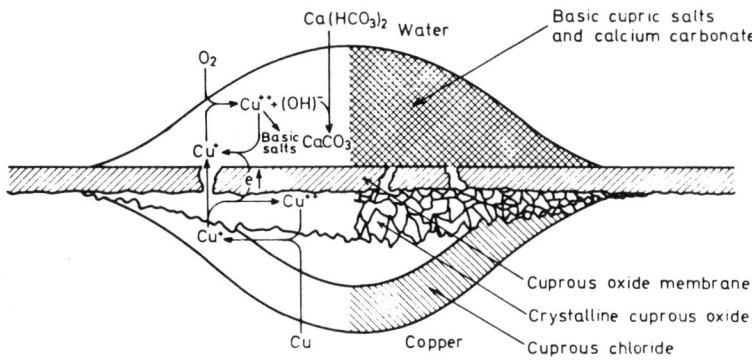

Fig. 1.59 Pit formed on a copper surface (protected by a film of Cu_2O) in a hard water (after Lucey[46])

that may be capped by small black or greenish-black mounds of Cu_2O and basic copper sulphate, but often no mound of corrosion product is produced. Subsequent considerations will be confined to the type of pit shown in Fig. 1.59.

Campbell[49-51] during an investigation of pitting of copper tubes in the UK showed that pitting only occurred in deep well supplies of very low organic content, whereas river or lake waters that contain organic impurities did not give rise to pitting. He suggested that these non-aggressive waters contained an organic substance that facilitated the formation of a protective layer of Cu_2O, whereas in its absence the Cu_2O formed a loose, coarsely crystalline, non-protective deposit. The inhibitor has not been identified, but it has been shown to be a negatively charged colloidal substance with acidic properties, which can be detected by a whitish-blue fluorescence when the water is exposed to ultra-violet radiation; it has been suggested that it is possibly an unsaturated delta lactone.

Campbell also showed that a high proportion of copper pipes that failed in the aggressive cold waters contained a carbon deposit that was formed by the breakdown of the drawing lubricant residue during bright annealing. A glassy thin film of Cu_2O can also form during bright annealing, and initially it was considered that both the carbon and the glassy Cu_2O could act as a cathode, and thus stimulate attack at discontinuities in the film and lead to pitting.

Devroey and Depommier[52] in Belgium, and Campbell have found, however, in further investigations that glassy cuprous oxide scales formed in tubes during annealing do not cause pitting corrosion but become converted in service to a dull protective oxide. The glassy cuprous oxide found in tubes that had failed in service were shown to have been formed by the action of the water and to be associated with the presence of very thin carbon films produced in the tubes during manufacture.

The compositions of waters that give rise to pitting have been the subject of numerous investigations, and studies by Obrecht et al.[31] of the pitting of copper tubes in the USA have shown that it is invariably a cold water phenomenon that occurs with hard well waters. These waters contained over 5 p.p.m. of dissolved CO_2, 10–40 p.p.m. being typical, a pH in the range

7·0-7·8, 10-12 p.p.m. of dissolved O_2 and an SO_4^{2-}/Cl^- ratio of 3-4:1. Lucey[53] has analysed about 120 waters in the UK, whose behaviour in relation to the pitting of copper is well established, and has constructed a nomogram that provides a means of predicting pitting propensity from an analysis of the water. It would appear that an increase in the SO_4^{2-} or Na^+ ion concentration or an increase in the concentration of dissolved oxygen increases the pitting propensity, whereas the converse applies to the Cl^- or NO_3^- ion concentration and pH. Thus whereas the pitting of stainless steels is favoured by a high $Cl^-:SO_4^{2-}$ ratio (see Fig. 1.55) the pitting of copper is favoured by a high $SO_4^{2-}:Cl^-$ ratio. A high pH reduces the pitting propensity, and many surface waters contain 'humic acid' that enables more calcium bicarbonate to be held in solution than the equilibrium value, thus resulting in a high and stable pH.

Mechanism

The mechanism of pitting is highly complex, and reference should be made to the original papers for further details. However, it is of interest to consider certain views on this subject, since some of them introduce new concepts.

May[54] was the first to stress the important rôle of Cu_2Cl_2 within the pits on the mechanism, and he considered that it acted as a screen that prevented dissolved oxygen gaining access to the bottom of the pit thus preventing the formation of a protective Cu_2O film; the low solubility of Cu_2Cl_2 also maintained the activity of copper ions at a low value and thus facilitated anodic dissolution of the copper.

Pourbaix[12,55] has constructed equilibrium potential-pH diagrams for the $Cu-Cl-H_2O$ system and for the $Cu-Cl-CO_2-SO_3-H_2O$ system using concentrations of ions that correspond to those actually present in Brussels water. This work has shown that at the bottom of a pit the phases Cu, Cu_2O and Cu_2Cl_2 are at equilibrium in a solution containing 246 p.p.m. of Cl^- and 270 p.p.m. of Cu^{2+} only at a potential of 270 mV (vs. S.H.E.) and pH 3·5. The potential of the metal outside the pit in a water of pH 7-8 depends on the concentration of oxygen, the overpotential for its reduction on the passive surface film, etc. but it is normally about 300 mV, which is only 30-70 mV higher than the potential within the pits. Under these circumstances the pits do not grow appreciably. However, if a carbon film is present on the surface of the copper, a differential aeration cell will be set up, and the potential of the surface will increase this in turn will increase the potential of the surface of the interior of the pit above the equilibrium value of 270 mV. Under these circumstances the equilibrium will be disturbed and the copper within the pit will corrode rapidly forming Cu^{2+}. This mechanism thus involves many of the conventional features of pitting, i.e. differential aeration, a large cathode:anode surface area ratio and the development of acidity within the pit by hydrolysis of Cu_2Cl_2 that prevents a protective film of Cu_2O from forming.

Lucey[46] examined a number of examples of pitting of copper pipes and tanks from hard water districts, and found that there was no more calcium carbonate scale deposited around the pits than on other parts of the metal surface. There was, however, a large amount of $CaCO_3$ in the mound immediately above the pit, and this suggested that the reduction of dissolved oxygen takes place immediately above the pit and not on the surrounding surface. He also showed that the Cu_2O membrane could act as thin bipolar electrode, the upper surface acting as a cathode and the under surface as an anode. Thus the Cu_2Cl_2 produced within the pit is anodically oxidised to Cu^{2+}, and this ion can then attack the copper within the pits by the disproportionation reaction

$$Cu^{2+} Cu \rightarrow 2Cu^{+} \qquad \ldots(1.154)$$

The principal cathodic reaction on the upper surface of the membrane is the reduction of Cu_{2+} that is formed by the reaction of Cu^{+} with dissolved oxygen in the water; these Cu^{+} ions are provided partly from the diffusion through the pores in the oxide membrane from within the pit and partly from those produced by cathodic reduction (equation 1.154). Lucey's theory thus rejects the conventional large cathode:small anode relationship that is invoked to explain localised attack, and this concept of an electronically conducting membrane has also been used by Evans[56] to explain localised attack on steel due to a discontinuous film of magnetite.

Selective Leaching or De-alloying

In certain alloys and under certain environmental conditions selective removal of one metal (the most electrochemically active) can occur resulting in either localised attack, with the consequent possibility of perforation (plug type), or in a more uniform attack (layer type) that results in a weakening of the strength of the component. Although the selective removal of metals such as Al, Fe, Co, Ni and Cr from their alloys is known, the most prevalent form of de-alloying is the selective removal of zinc from the brasses — a phenomenon that is known as *dezincification*.

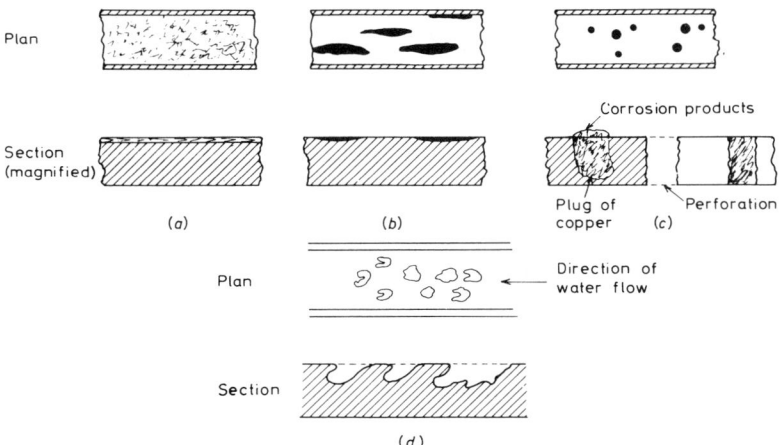

Fig. 1.60 Dezincification and impingement attack of copper-alloy tubes. (*a*) Uniform layer dezincification of a brass, (*b*) banded dezincification of a brass, (*c*) plug-type dezincification and (*d*) impingement attack

In this connection it is of interest to refer to the parting of Ag–Au alloys in nitric acid in which the silver can be completely removed from the alloy providing the ratio of Ag:Au is greater than 2·5:1; at somewhat lower ratios separation is incomplete and at low ratios no separation is possible and the alloy is unattacked by the acid. Similarly, dezincification does not occur when the zinc in a brass is less than 15%, and red brass (15% Zn) has only a slight tendency to dezincify, whereas the α-brasses (Cu–30Zn) are highly susceptible and the β-brasses (Cu–40Zn) even more susceptible; low alloying additions of Al and Mn in brasses appear to have very little effect in reducing dezincification.

Dezincification is readily apparent, since the yellow colour of the brass is replaced by the characteristic red of copper, which may take the form of small plugs or of layers that in some cases can extend over the whole of the surface (Fig. 1.60). In plug-type dezincification a mechanically weak, porous residue of copper is produced, which may remain *in situ* or become removed by the pressure of water, leading to a perforation. In the layer type the transformation of the alloy into a mechanically weak layer of copper results in loss of strength, and failure may occur by splitting when the metal is subjected to water pressure or to external stress.

Dezincification can occur over a wide range of pH, although the pH appears to affect the form of attack. Thus the layer type is favoured when the environment is acid and the brass has a high zinc content, whilst the plug type is more prevalent when the environment is neutral, slightly acid or alkaline and the zinc content is relatively low. In both cases dezincification is favoured by stagnant conditions (compare erosion-corrosion that becomes prevalent at high velocities), by the presence of chloride ions and by the formation of porous scales and deposits that lead to stagnant crevice conditions.

Dezincification of α-brass can be readily prevented by suitable alloying additions, and this was achieved first by adding 1% Sn. However, elements such as As, Sb and P are more effective, and alloying additions of 0·02–0·06% As are widely used for this purpose. Unfortunately, no alloying element has been found that prevents the dezincification of the two-phase $\alpha\beta$-brasses, which are more susceptible than the α-brasses, and their use must be avoided under environmental conditions that are conducive to dezincification.

Mechanism

Two theories have been proposed to explain dezincification, but since both have considerable support the precise mechanism remains unresolved. One theory proposes that the zinc is selectively leached from the alloy leaving a porous residue of metallic copper *in situ* (*cf.* parting of Ag–Au alloys), whilst the other proposes that the whole of brass dissolves and that the copper immediately redeposits at sites close to where the brass was dissolved.

There is considerable metallographic and electrochemical evidence in support of each theory and it is of interest to note that two of the most authoritative works on corrosion appear to support opposite views — Uhlig[20] favours the selective dissolution of zinc theory, whereas Fontana and Greene[22] favour the dissolution-precipitation theory.

Metallographic studies by Polushkin and Shuldener[57] revealed that the copper residue in a dezincified α-brass contained twins and residual grain boundaries that resembled those present in the parent metal, which appears to support the dissolution of zinc theory. However, other workers claim that twins found in the copper residue are similar to those found in electrolytically deposited copper. Horton[58] used time-lapse colour photomicrography to observe the changes that took place during the dezincification of an α-brass in 3% NaCl at 50°C. The brass was cold worked and annealed to produce a coarse-grained structure with a number of annealing twins, but none of these features was reproduced in the copper, indicating that the whole alloy dissolved.

Any selective dissolution mechanism must explain how the zinc within the alloy diffuses to the surface at which the reaction takes place. Pickering and Wagner[59] used X-ray and electron diffraction to study the selective dissolution of copper from Cu–Au alloys, and subsequently a similar approach was used by Pickering[60] to study the selective dissolution of zinc from ε- and γ-brasses (zinc-rich alloys) during anodic polarisation in a variety of electrolyte solutions. Pickering found that the partially dissolved alloys gave rise to new intermediate and terminal phases having a higher copper concentration than the original alloys. It was concluded that volume diffusion of zinc could occur via the vacancies (mono- and di-) created at the surface by anodic dissolution; the diffusion coefficient for di-vacancies in copper at 25°C was calculated to be $1 \cdot 3 \times 10^{-12}\,\text{cm}^2\,\text{s}^{-1}$.

Lucey[61] concludes from his electrochemical studies that dezincification involves anodic dissolution of both copper and zinc followed by the cathodic deposition of copper, and on this basis he has explained why arsenic is capable of inhibiting dezincification of α-brass but not of αβ-brass.

When dezincification occurs in service the brass dissolves anodically and this reaction is electrochemically balanced by the reduction of dissolved oxygen present in the water at the surface of the brass. Both the copper and zinc constituents of the brass dissolve, but the copper is not stable in solution at the potential of dezincifying brass and is rapidly reduced back to metallic copper. Once the attack becomes established, therefore, two cathodic sites exist—the first at the surface of the metal, at which dissolved oxygen is reduced, and a second situated close to the advancing front of the anodic attack where the copper ions produced during the anodic reaction are reduced to form the porous mass of copper which is characteristic of dezincification. The second cathodic reaction can only be sufficient to balance electrochemically the anodic dissolution of the copper of the brass, and without the support of the reduction of oxygen on the outer face (which balances dissolution of the zinc) the attack cannot continue.

The potentials of film-free α-brass and β-brass in solutions comparable to those existing inside the alloy at the advancing front of attack were found to be $-0 \cdot 38$ V and $-0 \cdot 56$ V (vs. S.H.E.), respectively. It was also established, taking into account the activities of copper ions in equilibrium with the sparingly soluble corrosion product Cu_2Cl_2, that whereas Cu^{2+} ions can be reduced to copper at $-0 \cdot 16$ V the reduction of Cu^+ ions is possible only at potentials more negative than $-0 \cdot 41$ V. Thus whereas the β-phase of an αβ-brass can reduce both Cu^{2+} and Cu^+ ions, the α-brass can reduce only the Cu^{2+} ion.

Lucey points out that although arsenic can prevent dezincification of an α-brass it cannot prevent pitting, during which Cu_2Cl_2 is formed and is subsequently hydrolysed or oxidised to the secondary corrosion product Cu_2O. Thus during dezincification the advancing front of attack results in the formation of Cu_2Cl_2, and the Cu^+ ions can be readily reduced to metallic copper by the very negative β-phase, but not by the α-phase. The fact that α-brasses dezincify is explained by the formation of Cu^{2+} by the disproportionation of Cu^+ ions (formed from Cu_2Cl_2)

$$2Cu^+ \rightleftharpoons Cu^{2+} + Cu \qquad \ldots(1.155)$$

which are readily reduced directly to copper by the α-phase. Under these circumstances dezincification will proceed as long as Cu^{2+} ions are present at the advancing front. If, however, metallic arsenic is deposited at the advancing front in preference to copper metal, the Cu^{2+} ions are reduced to Cu^+ according to the following cycle:

$$3Cu^2 + As \rightarrow 3Cu^+ + As^{3+}$$

$$3Cu + As^{3+} \rightarrow 3Cu^+ + As$$

in which arsenic metal is regenerated. Thus dezincification of α-brass is prevented by the presence of a very small quantity of arsenic, and slow hydrolysis of Cu_2Cl_2 to Cu_2O takes place.

Erosion—Corrosion

The effect of movement of the solution or of the metal on the rate and form of corrosion is complex, and on the basis of previous considerations (*see also* Sections 1.4 and 2.1) the situation can be summarised as follows:

1. Increase in velocity may increase the rate by bringing the cathode reactant more rapidly to the surface of the metal thus decreasing cathodic polarisation, and by removing metal ions thus decreasing anodic polarisation.
2. Increase in velocity may decrease the rate by bringing the cathode reactant to the surface at a rate that exceeds $i_{crit.}$, thus causing passivation.
3. Decrease in velocity will favour all forms of localised attack in which an occluded cell is involved in the mechanism, and will also favour selective dissolution of alloys that are susceptible to this form of attack.

However, movement at appreciable rates can result in another form of attack that is brought about by the conjoint action of erosion and corrosion; hence the term *erosion–corrosion* that includes all forms of accelerated attack in which protective films, and even the metal surface itself, are removed by the abrasive action of movement of a fluid (gas or liquid) at high velocity. In general, the higher the velocity the more abrasive the solution.

Erosion–corrosion in the widest sense of the term will include impingement attack, cavitation damage and fretting corrosion, but since the latter two are dealt with in separate sections (*see* Sections 8.7 and 8.8) they will not be considered here.

The most significant effect of erosion–corrosion is the constant removal of protective films (which may range from thick visible films of corrosion products to the thin invisible passivating films) from the metal's surface, thus resulting in localised attack at the areas at which the film is removed. This can be caused by movement at high velocities, and will be particularly prone to occur if the solution contains solid particles (e.g. insoluble salts, sand and silt) that have an abrasive action. Impingement attack is a form of erosion–corrosion in which a turbulent stream of water containing entangled air bubbles and solid particles hits a metal surface, disrupts the protective film and thus results in pitting.

In addition to the mechanical damage of the protective film, velocity or movement will also bring the cathode reactant more rapidly to the metal surface thus decreasing cathode polarisation.

The ability of a metal or alloy to withstand erosion–corrosion depends upon the nature of the environment which can range from a natural water to a concentrated acid, so that it is difficult to make sweeping generalisations and each system must be considered individually. In most systems the rate of attack increases with velocity, and metals and alloys that have an acceptable corrosion rate in static solutions may corrode rapidly when the solution (or the metal) is moving at a high velocity. Table 1.22[22] shows that up to about 120 cm s^{-1} the effect is small, but at about 820 cm s^{-1} there is a rapid increase in the corrosion rate. However, there are exceptions to this rule, and Fontana and Greene[22] point out that whereas the corrosion rate of aluminium in fuming nitric acid at 42°C increases with velocity, the converse applies to type 347 stainless steel owing to the different corrosion mechanisms involved. Thus aluminium is protected by films of aluminium nitrate and aluminium oxide and since the former is removed at intermediate velocities (up to 100 cm s^{-1}) and the latter at higher velocities, the corrosion

Table 1.22 Corrosion of metals by sea-water moving at different velocities (g m^{-2} d^{-1} × 10^2)*

Metal	Velocity		
	30·5 cm s^{-1} (1 ft s^{-1})†	122 cm s^{-1} (4 ft s^{-1})‡	823 cm s^{-1} (27 ft s^{-1})§
Carbon steel	3·4	7·2	25·4
Cast iron	4·5	—	27·0
Silicon bronze	0·1	0·2	34·3
Admiralty brass	0·2	2·0	17·0
Hydraulic bronze	0·4	0·1	33·9
G bronze	0·7	0·2	28·0
Aluminium bronze (10% Al)	0·5	—	23·6
Aluminium brass	0·2	—	10·5
Cu–10Ni–0·8Fe	0·5	—	9·9
Cu–30Ni–0·05Fe	0·2	—	19·9
Cu–30Ni–0–5Fe	<0·1	<0·1	3·9
Monel	<0·1	<0·1	0·4
Stainless steel (*316*)	0·1	0	<0·1
Hastelloy C	<0·1	—	0·3
Titanium	0	—	0

*Data after Fontana and Greene[22]; source International Nickel Co.
† Immersed in tidal current.
‡ Immersed in sea-water flume.
§ Attached to immersed rotating disc.

Fig. 1.61 Erosion–corrosion. Impeller fabricated from Cu-10Sn-2Zn alloy used for pumping a slurry containing 5-10% H_2SO_4 after six months' service. The blades of the impeller were extensively corroded. Note the directional form of the attack

rate increases with increase in velocity. In the case of the stainless steel in stagnant fuming nitric acid the metal is attacked autocatalytically by the nitrous acid formed by cathodic reduction of the nitric acid, and if the nitrous acid is swept away the rate decreases.

This example of aluminium illustrates the importance of the protective film, and films that are hard, dense and adherent will provide better protection than those that are loosely adherent or that are brittle and therefore crack and spall when the metal is subjected to stress. The ability of the metal to reform a protective film is highly important and metals like titanium and tantalum that are readily passivated are more resistant to erosion–corrosion than copper, brass, lead and some of the stainless steels. There is some evidence that the hardness of a metal is a significant factor in resistance to erosion–corrosion, but since alloying to increase hardness will also affect the chemical properties of the alloy it is difficult to separate these two factors. Thus although copper is highly susceptible to impingement attack its resistance increases with increase in zinc content, with a corresponding increase in hardness. However, the increase in resistance to attack is due to the formation of a more protective film rather than to an increase in hardness.

Attack by erosion–corrosion may be uniform or localised, and in the case of the latter it is characterised by the formation of grooves or rounded depressions that are smooth and free from corrosion products. Frequently the areas of localised attack follow a pattern that is indicative of the direction of movement of the metal or the environment (Fig. 1.61).

Impingement Attack

Impingement attack, as is implicit in the name, is a form of erosion–corrosion in which the solution strikes the metal surface at a high velocity—a situation that can occur at bends, tees and sudden changes in section in a water system; at steam-turbine blades; and at the surface of tubes in which debris or corrosion products have accumulated. Thus in the majority of cases of impingement attack a geometrical feature of the system results in turbulence at one or more parts of the metal surface. Attack can occur in this way with a solution that is free from gas bubbles or suspended solids, but it is markedly accentuated when they are present.

Air bubbles entangled in cooling waters are particularly detrimental, and their importance was stressed by Bengough and May[62] in early studies of condenser-tube alloys, in which they used a 'jet impingement test' (*see* Section 19.1) to simulate conditions that prevail in practice. These workers considered that the almost static boundary layer of water adjacent to the metal surface (*see* Section 2.1) prevents the impingement of rapidly moving water. However, if the water contains air bubbles that have a diameter that is greater than a certain critical value (the thickness of the boundary layer) the bubbles on striking the boundary layer are subjected to differential forces that disrupt them and the boundary layer, thus enabling the water to impinge on the metal surface at the point of bubble disruption with consequent removal of the protective film. On the other hand, Campbell and Carter[63] consider that attack is due to highly turbulent flow at the attacked area, which can be caused by air-free water at comparatively high speeds and at lower speeds if air bubbles of a certain size are present in the water. A similar turbulent condition is produced by particles of foreign debris, by corrosion products on the tube wall and by marine crustacea lodged within the tube. The turbulent water breaks through the boundary layer and causes continuous breakdown of the protective film and removal of metal ions so that anode polarisation is small. The conducting Cu_2O film on the metal surrounding the impinged area provides a large efficient cathode for the cathodic reduction of the dissolved oxygen that is brought rapidly to its surface, and under these circumstances attack is rapid and intense. The pits formed during the impingement of water containing entangled air bubbles are of a smooth rounded shape, and frequently resemble the imprints of tiny horseshoes (as if the horse was walking up-stream) indicating the direction of the flow (Figs. 1.60*d* and 4.11).

The Role of Iron in the Resistance of Aluminium Brass to Impingement Attack

The nature of the protective film on the surface of a metal in relation to its resistance to erosion–corrosion has already been emphasised, and it is generally accepted that alloying additions that improve resistance to this form of attack do so by modifying the nature of the film. This applies to the addition of 2% Al to brass to form aluminium-brass and to the addition of 1–2% Fe to the cupro-nickels, and in both cases the alloying addition results in a pronounced increase in resistance to impingement attack. However, since impingement attack involves the characteristic large cathode/small anode relationship its control could also be effected by making the cathodic reaction more difficult, and in this connection it is of interest to consider the rôle of iron in relation to the resistance of aluminium-brass to impingement attack.

It was established in the 1920s by chemical analysis that certain aluminium-brass condenser tubes that were found to have a remarkably high resistance to impingement attack were characterised by the presence of iron in the protective film, which had entered the film owing to the presence of iron corrosion products in the water. Subsequently, when it became common practice to cathodically protect the steel boxes, deliberate injections of small amounts of $FeSO_4 \cdot 7H_2O$ were made to the water to maintain the source of iron, and this procedure is now common practice when aluminium-brass tubes are used, particularly when the cooling water is polluted. North and Pryor[64] found that when a solution of sodium chloride containing a small concentration of Fe^{2+} ions was electrolysed between copper electrodes the anode became coated with a film of Cu_2O whilst the cathode became coated with a dark-brown film that was identified as lepidocrocite γ-FeOOH, which was formed in the alkaline catholyte by the reaction

$$2Fe^{2+} + 4OH^- + \tfrac{1}{2}O_2 \rightarrow 2FeOOH + H_2O \quad \ldots(1.156)$$

According to these workers the FeOOH is present in the water as a positively charged colloid that migrates and adheres to the negatively charged cathode forming a non-conducting film that impedes electron transfer between the copper and the dissolved oxygen. A similar situation would prevail in practice, so that the rôle of the iron is to control the kinetics of the cathodic reaction rather than that of the anodic reaction. Studies of a 2·5% Fe-Cu alloy, in which only 0·1% of the iron is in solid solution, indicated that the iron corrodes by galvanic action and that the FeOOH is deposited initially around the iron particles where the cathodic current density is a maximum, but it subsequently spreads out over the whole surface.

Gasparini et al.[65] have emphasised the importance of the zeta-potentials of the filmed metal's surface and the colloidal FeOOH in relation to the mechanism, and consider that the latter has a positive zeta-potential only in demineralised water. In natural waters the zeta potential at pHs at or greater than 7·5 is negative, and since the Cu_2O film has a positive zeta potential that extends into the solution during water flow (streaming potential) the FeOOH deposits on the Cu_2O, thus forming a non-conducting film. Although North and Pryor[64], and Gasperini et al.[65] differ as to the nature of the charge on the FeOOH they are in accord as to the mechanism involved.

Control of Corrosion–Erosion

The most important method for controlling corrosion–erosion is the use of materials that are resistant to this form of attack, and further information can be obtained by consulting the sections that are devoted to metals and alloys (*see* particularly Section 4.2).

The geometry of the system is a further factor that enables control to be effected, and in general, reducing the velocity of the solution and ensuring that flow is laminar and not turbulent will reduce the tendency for attack by erosion–corrosion. Thus the pipe diameter should be as large as possible, consistent with other considerations; bends should have a large radius and inlet and outlets should be streamlined so that there is not a sudden change in section.

As far as the environment is concerned the main factor is to remove solid particles by settling and filtration and to remove marine crustacea by screening.

L. L SHREIR

REFERENCES

1. Champion, F. A., *Corrosion Testing Procedures*, Chapman and Hall, London (1962)
2. Hoar, T. P., *J. Soc. Chem. Ind.*, **69**, 356 (1950)
3. Greene, N. D. and Fontana, M. G., *Corrosion*, **15**, 25t (1959)
4. Brown, B. F., in *The Theory of Stress Corrosion Cracking of Alloys* (ed. J. C. Scully), NATO, Brussels, 187 (1971)
5. Evans, U. R., *Industr. Engng. Chem.*, **17**, 363, 370 (1925)
6. Evans, U. R., *The Corrosion and Oxidation of Metals*, Arnold, London (1961)
7. Pourbaix, M., *Rapports Technique*, No. 84, CEBELCOR (1960)
8. Pourbaix, M., *J. Electrochem. Soc.*, **101**, 217C (1954)
9. Bianchi, G., *Corrosion*, **14**, 245t (1958)
10. Schaschl, E. and Marsh, G. A., *Corrosion*, **16**, 461t (1960)
11. Tomashov, N. D., *Theory of Corrosion and Protection of Metals*, Collier-Macmillan, London, 470 (1966)
12. Pourbaix, M., *Lectures on Electrochemical Corrosion*, Plenum Press, New York–London (1973)
13. Suzuki, T., Yamake, M. and Kitamura, Y., *Corrosion*, **29**, 18 (1973)
14. Hoar, T. P., *Disc. Faraday Soc.*, No. 1, 299 (1947)
15. Edeleanu, C. and Evans, U. R., *Trans. Faraday Soc.*, **47**, 1121 (1951)
16. Brown, B. F., Fujii, C. T. and Dahlberg, E. P., *J. Electrochem. Soc.*, **116**, 218 (1969) and Smith, J. A., Peterson, M. H. and Brown, B. F., *Corrosion*, **26**, 539 (1970)
17. Wilde, B. E., *Corrosion*, **27**, 32 (1971)
18. Bogar, F. D. and Fujii, C. T., N.R.L. Report 7690, Washington (1974)
18a. Pickering, H. W. and Frankenthal, R. P., *J. Electrochem. Soc.*, **119**, 1297 (1972)
18b. Franck, U. F., *Proc. 1st International Congress on Metallic Corrosion*, Butterworths, London (1962)
19. Brown, B. F., *Corrosion*, **26**, 249 (1970)
20. Uhlig, H. H., *Corrosion and Corrosion Control*, John Wiley, New York (1971)
21. Streicher, M. A., *Corrosion*, **30**, 77 (1974)
22. Fontana, M. G. and Greene, N. D., *Corrosion Engineering*, McGraw-Hill, New York (1967)
23. Rosenfeld, I. L. and Marshakov, I. K., *Corrosion*, **20**, 115t (1964)
24. Karlberg, G. and Wranglen, G., *Corros. Sci.*, **11**, 499 (1971)
25. Takamura, A., *Corrosion*, **23**, 306 (1967)
26. Griess, J. C., *Corrosion*, **24**, 96 (1968)
27. Sharmon, C., *Nature*, **153**, 621 (1944) and *Chem. Ind.*, **46**, 1126 (1952)
28. Van Loo, M., Laiderman, D. and Bruhn, R., *Corrosion*, **9**, 277 (1953)
29. Slabaugh, W. and Grutheer, M., *Ind. Eng. Chem.*, **46**, 1014 (1954)
30. Kaesche, H., *Werkst. Korros.*, **10**, 688 (1959)
31. Obrecht, M. F., Sartor, W. E. and Keyes, J. M., *Proc. 4th International Congress on Metallic Corrosion* (1969)
32. Leckie, H. and Uhlig, H. H., *J. Electrochem. Soc.*, **113**, 1262 (1966)
33. Stolica, N. D., *Corros. Sci.*, **9**, 455 (1969)
34. Uhlig, H. H. and Gilman, J., *Corrosion*, **19**, 261t (1963)
35. Uhlig, H. H. and Gilman, J., *Z. Phys. Chem.*, **226**, 127 (1964)
36. Bohni, H. and Uhlig, H. H., *J. Electrochem. Soc.*, **116**, 906 (1969)
37. Stolica, N. D., *Corros. Sci.*, **9**, 205 (1969)
38. Pourbaix, M., *Corrosion*, **26**, 431 (1970)
39. Verink, E. D. and Pourbaix, M., *Corrosion*, **27**, 495 (1971)
39b. Wilde, B. E., *Corrosion*, **28**, 283 (1972)
40. Kolotyrkin, Y., *Corrosion*, **19**, 261t (1963)
41. Vermilyea, D., *J. Electrochem. Soc.*, **118**, 529 (1971)
42. Szklarska-Smialowski, Z., *Corrosion*, **27**, 223 (1971)

43. Wranglen, G., *Corros. Sci.*, **14**, 331 (1974)
44. Fyfe, D., Shanahan, C. E. A. and Shreir, L. L., *Corros. Sci.*, **10**, 817 (1971)
45. Wranglen, G., *Introduction to Corrosion and Protection of Metals*, Butler and Tanner, London (1972)
46. Lucey, V. F., *Br. Corr. J.*, **2**, 175 (1967)
47. Campbell, H. S., Miscellaneous Publications MP 574, Aug. (1972) and MP 568, Feb. (1972), British Non-ferrous Research Association
48. Campbell, H. S., *Water Treatment and Examination*, **20**, 11 (1971)
49. Campbell, H. S., *J. Inst. Metals*, **77**, 345 (1950)
50. Campbell, H. S., *J. Appl. Chem.*, **4**, 633 (1954)
51. Campbell, H. S., *Proc. 2nd International Congress on Metallic Corrosion*, 237 (1966)
52. Devroey, P. and Depommier, C., *Visseries et Tritileries Reunies S.A.*, Madrelen, Belgium, Unpublished Report (1962)
53. Lucey, V. F., BNFRA Research Report No. 2, June, 1692 (1968)
54. May, R., *J. Inst. Metals*, **32**, 65 (1953)
55. Pourbaix, M., Van Muylder, J. and Van Laer, J., *CEBELCOR Rapport Technique*, No. 125 (1965) and *Corros. Sci.*, **7**, 795 (1967); Pourbaix, M., *Corrosion*, **25**, 267 (1969)
56. Evans, U. R., *Corros. Sci.*, **9**, 813 (1969)
57. Polushkin, E. and Shuldener, H., *Trans. Amer. Inst. Mining Met. Eng.*, **161**, 214 (1945)
58. Horton, R. M., *Corrosion*, **26**, 160 (1970)
59. Pickering, H. and Wagner, C., *J. Electrochem. Soc.*, **114**, 698 (1967)
60. Pickering, H., *J. Electrochem. Soc.*, **117**, 8 (1970)
61. Lucey, V. F., *Br. Corr. J.*, **1**, 9 (1965) and **1**, 52 (1965)
62. Bengough, G. D. and May, R., *J. Inst. Met.*, **32**, 81 (1924)
63. Campbell, H. S. and Carter, V. E., *J. Inst. Met.*, **90**, 362 (1961)
64. North, R. F. and Pryor, M. J., *Corros. Sci.*, **8**, 149 (1968) and **9**, 509 (1969)
65. Gasparini, R., Rocca, C. D. and Ioannilli, E., *Corros, Sci.*, **10**, 157 (1970)

BIBLIOGRAPHY*†

Differential aeration
Shatalov, A. Ya. and Santsova, V. P., 'Effect of the Solution Composition of Zinc During Uneven Aeration', *Sb. Tr. Voroneshsk Otd. Vses. Khim Obshehestva*, **2**, 89 (1959); *C.A.*, **59**, 12478g
Skorchelletti, V. V. and Golubeva, N. K., 'Effect of Nonequilibrium Aeration on the Rate of Corrosion of Iron', *Zh. Prikl. Khim.*, **35**, 1570 (1962); *C.A.*, **57**, 12198a
Ruther, W. E. and Hart, R. K., 'Influence of Oxygen on High Temperature Aqueous Corrosion of Iron', *Corrosion*, **19**, 127t (1963)
Ulanovskii, I. B., Korovin, Yu. M. and Sevastyanov, V. F., 'Effect of Hydrogen Sulfide on the Electrode Potential of Carbon Steel', *Zh. Prikl. Khim.*, **37**, 1736 (1964); *C.A.*, **61**, 11607d
Golubeva, N. K. and Skorchelletti, V. V., 'The Effect of Differential Aeration on the Rate of Corrosion of Iron', *Corros. Sci.*, **5**, 203 (1965)
Gladysheva, V. P. and Shatalov, A. Ya., 'Effect of Hydrogen-ion Concentration on the Performance of Differentially Aerated Couples', *Izv. Vysshikh. Uchebn. Zavedenii, Khim. i Khim. Tekhnol.*, **9**, 48 (1966); *C.A.*, **65**, 5002a
Grubitsch, H., 'The Area of Capture Principle with Oxygen Corrosion in Electrolytes', *Werkstoffe Korrosion*, **17**, 679 (1966); *C.A.*, 19674f
Saraby-Reintjes, A., 'Differential Aeration Cell', *J. Electro. Chem. Interfacial Electrochem.*, **37**, 357 (1972); *C.A.*, **77**, 55484f
Gladysheva, V. P. and Shatalov, A. Ya., 'Electrochemical Principles of the Operation of Differential Aeration Couples', *Izv. Vyssh. Ucheb. Zavad., Khim. i Khim. Tekhnol.*, **16**, 1508 (1973); *C.A.*, **80**, 43386z

Note: C.A. means *Chemical Abstracts*.
*Acknowledgements are made to Mrs J. Sadowska-Mazur for preparing this bibliography on localised attack.

†**Bibliography to 1976.**
Since the second edition of *Corrosion* was published a vast body of literature on localised corrosion has arisen. This is most easily accessed through computer library sources. Older references, however, may be harder to trace in this way, and so are presented here.

Gladysheva, V. P. and Shatalov, A. Ya., 'Corrosion of Metals During Differential Aeration', Part 2 'Work of Differential Aeration Pairs During Varied Oxygen Concentration Proportions at the Cathode and Anode', *Izv. Vyssh. Ucheb. Zaved., Khim. i Khim. Tekhnol.*, **16**, 1663 (1973); *C.A.*, **80**, 55218j

Richard, A. and Nicolaides, G., 'Differential Aeration Corrosion of Passivating Metal Under a Moist Film of Locally Variable Thickness', *J. Electrochem. Soc.*, **121**, 183 (1974)

Crevice corrosion

Anderson, W. K. and McGoff, M. J., 'Corrosion of Zircaloy in Crevices Under Nucleate Boiling Conditions', *US At. Energy Comm. Kapl.*, 2203 (1962); *C.A.*, **57**, 3173

Smirnova, T. N., Ganina, N. V., Kolotygina, V. B. and Kazakova, V. M., 'Crevice Corrosion on Objects Made of the Alloy AMr6', *Zashchita Metal ot Karrozii, Sb. Kuibyshev*, 27 (1965); *C.A.*, **65**,1908d

Ulanovskii, I. B., 'Conditions for Cathodic Protection of Stainless Steels in Crevices', *Zashchita Metal*, **1**, 643 (1965); *C.A.*, **64**, 10751

Holmes, D. R. and Mann, G. M. W., 'A Critical Survey of Possible Factors Contributing to Internal Boiler Corrosion', *Corrosion*, **21**, 370 (1965)

Ruskol, Yu. S. and Klinov, I. Ya., 'Crevice Corrosion of Ti-alloys in Acids', *Khim. i Neft. Mashinostr.*, 28 (1966); *C.A.*, **65**, 16488a

France Jr., W. D. and Greene Jr., N. D., 'Passivation of Crevices During Anodic Protection', *Corrosion*, **24**, 247 (1968)

Swann, P. R., 'Mechanism of Corrosion Tunnelling with Special Reference to Cu_3Au', *Corrosion*, **25**, 147 (1969)

Landrum, R. J., 'Designing for Corrosion Resistance', *Chem. Eng.*, **76**, 118 (1969)

Lizlovs, E. A., 'Polarization Cell for Potentiostatic Crevice Corrosion Testing', *J. Electrochem., Soc.*, **117**, 256C, Atlantic City Meeting (1970)

Lizlovs, E. A., 'Polarization Cell for Potentiostatic Crevice Corrosion', *J. Electrochem. Soc.*, **117**, 1335 (1970)

Vermilyea, D. A. and Tedmon Jr., C. S., 'A Simple Crevice Corrosion Theory', *J. Electrochem. Soc.*, **117**, 437 (1970)

Isasi, J. A. and Metzger, M., 'Structure-dependent Corrosion of High-purity Al in HNO_3 and H_2SO_4', *Corros. Sci.*, **11**, 631 (1971)

Karlberg, G. and Wranglen, G., 'On the Mechanism of Crevice Corrosion of Stainless Cr Steels', *Corrosion Sci.*, **11**, 499 (1971)

Vincentini, B., Sinigaglia, D. and Taccani, G., 'Crevice Corrosion: Calculation of the Voltage and Current Distribution Along the Crevice', *Werks. Korros.*, **22**, 916 (1971)

Chen, C. M., Beck, F. H. and Fontana, M. G., 'Crevice Effect During Polarisation of Ti-8Al-IMo-1V Alloy in Aqueous and Methanol Environment', *Corrosion*, **27**, 234 (1971)

Korovin, Yu. M. and Ulanovskii, I. B., 'Corrosion in the Crevice of Stainless Steel Alloyed with Copper', *Zashch. Metal*, **8**, 425 (1972); *C.A.*, **77**, 121420k

Richman, R. B., 'Device for Determining Crevice Corrosion by Polarisation Techniques', U.S./633,099(Cl.324/71C; G.01n) (1972); *C.A.*, **76**, 120896n

Covington, L. C., 'Rôle of Multivalent Metal Ions in Suppressing Crevice Corrosion of Ti', *2nd Titanium Sc. Technol. Proc. Int. Conf., 1972* (1973); *C.A.*, **76**, 73409b

Peterson, M. H. and Lennox, T. J., 'A Study of Cathodic Polarisation and pH Changes in Metal Crevices', *Corrosion*, **29**, 406 (1973)

Bates, J. F., 'Cathodic Protection to Prevent Crevice Corrosion of Stainless Steel in Halide Media', *Corrosion*, **29**, 28 (1973)

Bowers, J. M., 'Crevice Corrosion of a 70-30 Chromium Modified Cupronickel and its Relations to the Experimental Pourbaix Diagram', US Nat. Tech. Inform. Serv., AD Rep. No. 762596 (1973). Avail. NTIS from Govt. Rep. Announce. (US), **76**, 84 (1973); *C.A.*, **79**, 148936m

Crolet, J. L. and Defranoux, J. M., 'Calculation of the Incubation Period for Crevice Corrosion of Stainless Steel', *Centre Belge Étude Corros. Rapport Tech.*, **122**, 210 3/1 to 210 3/10 (1973); *C.A.*, **79**, 72821x

Korovin, Yu. M. and Ulanovskii, I. B., 'Sea Water Crevice Corrosion of Ni, Some of its Alloys and Stainless Steel', *Zashch. Metal*, **9**, 309 (1973)

Crolet, J. and Defranoux, J.M., 'Calcul du Temps d'Incubation de la Corrosion Caverneuse des Aciers Inoxydables', *Corros. Sci.*, **13**, 575 (1973)

Isaacs, H. S., 'The Behaviour of Resistive Layers in the Localised Corrosion of Stainless Steel', *J. Electrochem. Soc.*, **120**, 1456 (1973)

Degerbeck, J., 'On Accelerated Pitting and Crevice Corrosion Tests', *J. Electrochem. Soc.*, **120**, 175 (1973)

McCafferty, E., 'Electrochemical Behaviour of Iron with Crevices in Nearly Neutral Chloride Solutions', *J. Electrochem. Soc.*, **120**, 234C (1973)

Ambrose, J. R. and Kruger, J., 'Crevice Corrosion of 304 Stainless Steel in 1·0 N NaCl Solution', *J. Electrochem. Soc.*, **120**, 234C (1973)

Occluded corrosion cell — electrochemical changes at the crack tip

Barth, C. F., Steigerwaid, E. A. and Troiano, A. R., 'Hydrogen Permeability and Delayed Failure of Polarised Martensitic Steels', *Corrosion*, **25**, 353 (1969)

Rhodes, P. R., 'Mechanism of Chloride Stress-corrosion Cracking of Austenitic Stainless Steel', *Corrosion*, **25**, 462 (1969)

Sandoz, G., Fujii, C. T. and Brown, B. F., 'Solution Chemistry Within Stress-corrosion Cracks in Alloy Steels', *Corros. Sci.*, **10**, 839 (1970)

Smith, J. A., Peterson, M. H. and Brown, B. F., 'Electrochemical Conditions at the Tip of an Advancing Stress-corrosion Crack in AISI 4340 Steel', *Corrosion*, **26**, 539 (1970)

Brown, B. F., 'Concept of the Occluded Corrosion Cell', *Corrosion*, **26**, 249 (1970)

Sedriks, A. J., Green, J. A. S. and Novak, D. L., 'On the Chemistry of the Solution at Tips of Stress-corrosion Cracks in Aluminium Alloys', *Corrosion*, **27**, 199 (1971)

Vermilyea, D. A., 'A Theory for the Propagation of Stress-corrosion Cracking in Metals', *J. Electrochem. Soc.*, **119**, 405 (1972)

Wilde, B. E. and Kim, C. D., 'The Rôle of Hydrogen in the Mechanism of Stress-corrosion Cracking of Austenitic Stainless Steel in Hot Chloride Media', *Corrosion*, **28**, 350 (1972)

Lin, F. and Hochman, R. F., 'Electrochemical Study of Stress-corrosion Cracking of Ti 8-1-1 Alloy and NaCl Solutions', *Corrosion*, **28**, 182 (1972)

Brignold, G. J., 'Electrochemical Aspects of Stress Corrosion of Steels in Alkaline Solutions', *Corrosion*, **28**, 307 (1972)

Gray, H. R., 'Ion and Laser Microprobes Applied to the Measurement of Corrosion-Produced Hydrogen on a Microscopic Scale', *Corrosion*, **28**, 47 (1972)

Smith, T., 'A Capillary Model for Stress-corrosion Cracking of Metals in Fluid Media, *Corros. Sci.*, **12**, 45 (1972)

Gabrielli, C., Keddam, M. and Takenouti, H., 'Progression of a Hole at the Surface of Passive Iron', *C.R. Acad. Sci. Ser. C.*, **277**, 743 (1973); *C.A.*, 43389c

Gal-Or, L., Raz, Y. and Yahalom, J., 'Mathematical Characterisation of Corrosion Currents in Local Electrolytic Cells', *J. Electrochem. Soc.*, **120**, 598 (1973)

Suzuki, T., Yambe, M. and Kitomura, Y., 'Composition of an Anolyte Within Pit Anode of Austenitic Stainless Steel in Chloride Solution', *Corrosion*, **29**, 18 (1973)

Pourbaix, M., 'Electrochemical Aspects of Stress-corrosion Cracking', *CEBELCOR Rapp. Tech.*, **118**(RT 199) (1971); *C.A.*, 76, 12065e

Wilde, B. E., 'Mechanism of Cracking of High-strength Martensitic Stainless Steel in NaCl Solutions', *Corrosion*, **27**, 326 (1971)

Intergranular attack

Levin, I. A. and Maksimova, G. F., 'Effect of Cold Work on the Tendency Towards Intergranular Corrosion of Type 18-8 Stainless Steel, *Khim Mashinostroenie*, **5**, 35 (1961); *C.A.*, **56**, 3219

Gerasimov, V. V. and Alexandrova, V. N., 'Intercrystalline Corrosion of Steels of the 1 Kh 18N9T Type in Distilled Water', *Metalloved i Term. Obrabotka Metal*, **2**, 53 (1962); *C.A.*, **56**, 13900f

Kuzub, V. S., Tsinmar, A. I., Kuzub, L. G. and Dolotova, T. S., 'Intercrystalline Corrosion of Stainless Steel in Concentrated HNO_3', *Zh. Prikl. Khim.*, **35**, 2794 (1962); *C.A.*, **58**, 9873e

Aylward, J. R. and Whitener, E. M., 'Dissolution of Zr in HCl plus CH_3OH', *J. Electrochem. Soc.*, **109**, 87 (1962)

Steverding, B., 'Intergranular Sulfur Corrosion in Missile Thrust Chamber Nickel Tubes', *Corrosion*, **18**, 433t (1962)

Rowlands, J. C., 'Electrochemical Aspects of Preferential Phase Corrosion in Complex Alloys', *Corros. Sci.*, **2**, 89 (1962)

Younger, R. N., Baker, R. G. and Littlewood, R., 'The Relationship Between Microstructure and Intercrystalline Corrosion in an 18Cr-12Ni-1Nb Austenitic Steel', *Corros. Sci.*, **2**, 157 (1962)

Greenblatt, J. H., 'Initial Corrosion of Al Alloys in High Temperature Water', *Corrosion*, **19**, 295t (1963)

Parker, E. A., 'Roping and Intergranular Corrosion of 430 Stainless Steel', *Am. Soc. Testing Mater.*, Spec. Tech. Pub. No. 369, 348 (1963); *C.A.*, **63**, 14476a

Resek, F., 'Intergranular Corrosion of C-Cr, Cr-Ni, Cr-Ni-Mn and Other Steels', *Zashchita Mater.*, **11**, 379 (1963); *C.A.*, **60**, 15518e

Voeltzel, J., Henry, G., Plateau, J. and Crussard, C., 'Precipitation and Intercrystalline Corrosion in a Stainless Steel of the 18-8Ti Type', *Mem. Sci. Rev Met.*, **60**, 879 (1963); *C.A.*, **60**, 8956c

Cihal, V., 'Explanation of Intercrystalline Corrosion of Welded, Stabilised Cr-Ni Stainless Steels', *Bergakademie*, **15**, 23 (1963); *C.A.*, **58**, 13523d

Fleitman, A. H., Romano, A. J. and Kiermut, C. J., 'Corrosion of Carbon Steel by High Temperature Liquid Mercury', *J. Electrochem. Soc.*, **110**, 964 (1963)

Haynie, F. H. and Ketcham, S. J., 'Electrochemical Behaviour of Al Alloys Susceptible to Intergranular Corrosion. Electrode Kinetics of Oxide-covered Al', *Corrosion*, **19**, 403t (1963)

Ketcham, S. J. and Haynie, F. H., 'Electrochemical Behaviour of Al Alloys Susceptible to Intergranular Corrosion. Effect of Cooling Rate on Structure and Electrochemical Behaviour in 202A Al Alloy', *Corrosion*, **19**, 242t (1963)

Tingley, I. I., 'Corrosion Resistance of Type 329 Stainless Steel in HNO_3 containing Chloride Ion', *Corrosion*, **19**, 408t (1963)

Stickler, R. and Vinckier, A., 'Electron Microscope Investigation of the Intergranular Corrosion Fracture Surfaces in a Sensitised Austenitic Stainless Steel', *Corros. Sci.*, **3**, 1 (1963)

von Schwenk, W. and Buhler, H.-E., 'Beoboshtungen an einem Kornzerfallsen falligen Austenitischen Cr-Ni-Stahl im Aktivzustand', *Corros. Sci.*, **3**, 145 (1963)

Streicher, M. A., 'Relationship of Heat Treatment and Microstructure to Corrosion Resistance in Wrought Ni-Cr-Mo Alloys', *Corrosion*, **19**, 272t (1963)

Levin, I. A. and Volikova, I. G., 'The Accelerated Method for Testing the Stability of a Single-phase Corrosion Resistant Steel to Intercrystalline Corrosion', *Zavodsk Lab.*, **30**, 816 (1964)

Hua, P.-T., Shen, H.-S., Chou, C. C. and Chung, T. T., 'Relation Between Intercrystalline Corrosion and Electrode Potential of Austenitic Stainless Steel', *K'O Hsueh Tung Pao*, **2**, 56 (1964); *C.A.*, **60**, 15516h

Baumel, A., 'Deutung der Ursachen der Interkristallinen Korrosion von Nichtrostended Stahlen in Zusammenhang mit der Chromverormungstheorie', *Corros. Sci.*, **4**, 89 (1964)

Mieluch, J. and Smialowski, M., 'The Behaviour of Grain Boundaries in Iron During Anodic Polarisation in Ammonium Nitrate Solution', *Corros. Sci.*, **4**, 237 (1964)

Streicher, M. A., 'Effect of Heat Treatment, Composition and Microstructure on Corrosion of 18Cr-8Ni-Ti Stainless Steel in Acids', *Corrosion*, **20**, 57t (1964)

Hua, P.-T., Hsin, K.-H., Sun, C. and Hsi, S.-Y., 'An Electrochemical Method for the Estimation of Intergranular Corrosion Susceptibility of Duralumin', *Chung Kuo K'O Hsueh Yuan Ying Yung Hua Hsueh Yen Chiu So Chi K'an*, **14**, 78 (1965); *C.A.*, **64**, 9233e

Singrorell, G. and D'Angelo, M., 'Intergranular Corrosion of Cr-Ni Austenitic Steels by Potentiodynamic Polarisation Curves', *Chim. Ind.*, **47**, 1215 (1965); *C.A.*, **64**, 6216b

Nichols, R. and Restoker, W., 'Intergranular Corrosion Penetration in an Age-hardenable Al Alloy', *J. Electrochem. Soc.*, **112**, 108 (1965)

Doshi, C. P. and Austin, W. W., 'Effect of Grain Size on Carbide Precipitation and Intergranular Corrosion in AISI Type 201 Stainless Steel', *Corrosion*, **21**, 332 (1965)

Armijo, J. S., 'Intergranular Corrosion of Nonsensitised Austenitic Stainless Steels—Part I: Environmental Variables', *Corrosion*, **21**, 235 (1965)

Lynes, W., 'Intergranular Corrosion of Alpha-brass and Some Effects of Stress', *Corrosion*, **21**, 125 (1965)

Baeumel, A. and Tramposch, O., 'Investigation of Intercrystalline Grain Boundary Corrosion of Austenitic Manganese-chromium Steels by Water and Aqueous Salt Solutions', *Werkst. Korros.*, **17**, 110 (1966); *C.A.*, **64**, 13847d

Myers, J. R., Crow, W. B., Beck, F. H. and Saxer, R. K., 'Observation on the Anodic Behaviour of Nickel and Chromium Surface Topography and Temperature Effect', *Corrosion*, **22**, 32 (1966)

Fleitman, A., Romano, J. and Klamut, C. J., 'Boiling Mercury Corrosion of Certain Refractory Metals and Stainless Steels from 593–703°C', *Corrosion*, **22**, 137 (1966)

Samans, C. H., Meyer, A. R. and Tisinai, G. F., 'Sensitisation and Thermal Stabilisation of Ni-Cr-Mo-Fe-W Alloy', *Corrosion*, **22**, 336 (1966)

Osozawa, K., Bohenkamp, K. and Engell, H. J., 'Potentiostatic Study on the Intergranular Corrosion of an Austenitic Cr-Ni Stainless-steel', *Corros. Sci.*, **6**, 421 (1966)

Coriou, M., Grail, L., Mahieu, C. and Pelas, M., 'Sensitivity to Stress Corrosion and Intergranular Attack of High-nickel Austenitic Alloys', *Corrosion*, **22**, 280 (1966)

Gellings, P. J. and de Jongh, M. A., 'Grain Boundary Oxidation and the Chromium-depletion Theory of Intercrystalline Corrosion of Austenitic Stainless Steels', *Corros. Sci.*, **7**, 413 (1967)

Armijo, J. S., 'Impurity Adsorption and Intergranular Corrosion of Austenitic Stainless Steel in Boiling HNO_3-$K_2Cr_2O_7$ Solutions', *Corros. Sci.*, **7**, 143 (1967)

Raymond, E. L., 'Mechanism of Sensitisation and Stabilisation of Incoloy Nickel-Iron-Chromium Alloy 825', *Corrosion*, **24**, 180 (1968)

Haneman, R. E. and Aust, K. T., 'Solute Clustering and Intergranular Corrosion', *Scr. Met.*, **2**, 235 (1968); *C.A.*, **67**, 63432d

Nikitin, V. I., 'Intercrystalline Corrosion by Liquid Metals by Dissolution Process', *Izv. Akad. Nauk. SSSR, Metal*, **1**, 213 (1968); *C.A.*, **67**, 98156g

Desestret, A., Epelboin, I., Froment, M. and Guiraldenq, P., 'Sur la Comparison des Facies d'Attaque Electrolytique et Thermique d'Aciers Inoxydables Austenitiques a Teneur Variable en Si', *Corros. Sci.*, **8**, 225 (1968)

France, W. D. and Greene, N. D., 'Predicting the Intergranular Corrosion of Austenitic Stainless Steels', *Corros. Sci.*, **8**, 9 (1968)

Butler, G., Ison, H. C. K. and Stretton, P., Short Communication – 'Catastrophic Grain Boundary Corrosion in a Cr-Fe Alloy', *Corros. Sci.*, **8**, 281 (1968)

Bond, A. P. and Lizlovs, E. A., 'Intergranular Corrosion of Ferritic Stainless Steel', *J. Electrochem. Soc.*, **115**, 233C (1968)

Armijo, J. S., 'Intergranular Corrosion of Nonsensitised Austenitic Stainless Steel', *Corrosion*, **24**, 24 (1968)

Ward, C. T., Mathis, D. L. and Staehle, R. W., 'Intergranular Attack of Sensitised Austenitic Stainless Steel by Water Containing Fluoride Ions', *Corrosion*, **25**, 394 (1969)

Hill, B. and Trueb, L. F., 'Resistance of Explosion-bonded Stainless Steel Clads to Intergranular Corrosion and Stress Corrosion Cracking', *Corrosion*, **25**, 23 (1969)

Gaivele, J. R. and De Micheli, S. M., 'Mechanism of Intergranular Corrosion of Al-Cu Alloys', *Proc. 4th Int. Congr. Metal. Corros.*, 439 (1969); *C.A.*, **77**, 134210s

Bond, A. P. and Lizlovs, E. A., 'Intergranular Corrosion of Ferritic Stainless Steels', *J. Electrochem. Soc.*, **116**, 1305 (1969)

Lev, V. Z., Klinov, I. Ya. and Shapiro, M. B., 'Potentiostatic Method for Determining the Intercrystalline Corrosion Tendency of Cast Alloy EP 375', *Tr. Mosk. Inst. Khim. Mashinostr.*, **2**, 245 (1970); *C.A.*, **77**, 82803r

Hajto, N. and Vrabely, E., 'Comparison of Tests for Intercrystalline Corrosion Tendency', *Banyasz. Kohasz. Lapok. Kohasz.*, **103**, 385 (1970); *C.A.*, **77**, 48934n

Berge, P., 'Corrosion Intergranulaire d'un Alliage du Type Inconel 600 dans la Vapeur d'Eau', *Corros. Sci.*, **10**, 185 (1970)

Gaivele, J. R. and de Micheli, S. M., 'Mechanism of Intergranular Corrosion of Al-Cu Alloys', *Corros. Sci.*, **10**, 795 (1970)

Cihal, V. and Kasova, I., 'Relation Between Carbide Precipitation and Intercrystalline Corrosion of Stainless Steel', *Corros. Sci.*, **10**, 875 (1970)

France, W. D. and Greene, N. D., 'Some Effects of Experimental Procedures on Controlled Potential Corrosion Tests of Sensitised Austenitic Stainless Steel', *Corros. Sci.*, **10**, 379 (1970)

Tedmon, C. S. (Jr.), 'Intergranular Corrosion of Austenitic Stainless Steel', *J. Electrochem. Soc.*, **118**, 192 (1971)

Hodges, R. J., 'Intergranular Corrosion in High Purity Ferritic Stainless Steel. Isothermal Time-Temp. Sensitisation Measurements', *Corrosion*, **27**, 164 (1971)

Hodges, R. J., 'Intergranular Corrosion in High Purity Ferritic Stainless Steel. Effect of Cooling Rate and Alloy Composition', *Corrosion*, **27**, 119 (1971)

Tedmon, C. S. (Jr.) and Vermilyea, D. A., 'Carbide Sensitisation and Intergranular Corrosion of Ni Base Alloys', *Corrosion*, **27**, 376 (1971)

Gulyaev, A. P. and Tokareva, T. B., 'Intergranular Corrosion of Austenitic Chromium-Nickel Steels', *Inst. Khim. Mashinostr.*, **38**, 3 (1971); *C.A.*, **79**, 44784u

Tedmon, C. S. (Jr.), Vermilyea, D. A. and Broecker, D. E., 'Effect of Cold Work on Intergranular Corrosion of Sensitised Stainless Steel', *Corrosion*, **27**, 104 (1971)

Henthorne, M. and De Bold, T. A., 'Intergranular Corrosion Resistance', *Corrosion*, **27**, 255 (1971)

Streicher, M. A. – Letter to the Editor, *Corros. Sci.*, **11**, 275 (1971) – Reply to Authors to 'Some Effects of Experimental Procedures', Short Communication by France, W. D. and Greene, N. D., *Corros. Sci.*, **10**, 379 (1970)

Wilson, F. G., 'Mechanism of Intergranular Corrosion of Austenitic Stainless Steel', *Br. Corros. J.*, **6**, 100 (1971)

Gizhermo, R. and Khristo, E., 'Effect of the Deoxidation Method on the Intercrystalline-corrosion Tendency of Cr–Ni Austenitic Steels', *Metalurgiye*, **5**, 17 (1972); *C.A.*, **80**, 98898y

Joshi, A. and Stein, D. F., 'Chemistry of Grain Boundaries and its Relation to Intergranular Corrosion of Austenitic Stainless Steel', *Corrosion*, **28**, 321 (1972)

Heinz, K., 'Susceptibility of Materials to Intergranular Corrosion', *Prakt. Metallogr.*, **9**, 441 (1972); *C.A.*, **77**, 167771d

Gegelova, N. B., Mudzhiri, Y. N., Knyazheva, V. M. and Topchiashvili, L. I., 'Determination of Intercrystalline Corrosion Tendency of Stainless Steel by Potentiodynamic Method', *Zashch. Metal.*, **8**, 420 (1972); *C.A.*, **77**, 121422n

van der Horst, J. M. A., 'Grain Boundary Attack of 316L Stainless Steel in Ammonia-rich Environments', *J. Electrochem. Soc.*, **119**, 216C (1972)

Froment, M. and Vignaud, C., 'Study of Intergranular Corrosion of Aluminium Using Bicrystals', *J. Electrochem. Soc.*, **119**, 216C (1972)

Steiner, A., 'Effect of Phosphorous Level on General and Intercrystalline Corrosion in 18/8 Type Steel', *Pr. Inst. Hutn.*, **24**, 255 (1972); *C.A.*, **80**, 6185n

Anisimova, M. S. and Chikurova, A. A., 'Effects of Heat-treatment Conditions and Structure of Kh18N9TL Cast Steel on its Susceptibility to Intercrystalline Corrosion', *Optimiz Met. Protsessov*, **6**, 163 (1972)

Zehr, S. W., 'A Study of the Intergranular Cracking of U-7·5wt.%Nb-2·5wt.%Zr (Mulberry) Alloy in Aqueous Chloride Solutions', *Corrosion*, **28**, 196 (1972)

Ballard, D. B., Bennett, L. H. and Swartzendruber, L. J., 'Intergranular Embrittlement of Cu–Pd Alloys in Sea-water', *Corrosion*, **28**, 368 (1972)

Westbrook, J. H., 'The Rôle of Precipitation and Segregation at Grain Boundaries in Corrosion Failure', *J. Electrochem. Soc.*, **119**, 216C (1972)

Lewis, E. C., 'Grain Boundary Corrosion of Sensitised Type 304 Stainless Steel by a Noble-to-active Polarisation Scan', *J. Electrochem. Soc.*, **119**, 219C (1972)

Frankenthal, R. P. and Pickering, H. W., 'Intergranular Corrosion of Ferritic Stainless Steels', *J. Electrochem. Soc.*, **119**, 216C (1972)

Cihal, V. and Jezek, J., 'Corrosion of Stainless Steel in the Immediate Vicinity of the Weld Metal', *Brit. Corr. J.*, **7**, 76 (1972)

Pakhomova, N. A. and Levin, I. A., 'Effect of Grain Size on the Inter-crystalline Corrosion of Stainless Steel Type 18-8', *Zashch. Metal*, **9**, 676 (1973); *C.A.*, **80**, 86207j

Brown, M. H. and Kirchner, R. W., 'Corrosion of High Alloy Weldments: 1. Sensitisation of Wrought High Nickel Alloys', *Corrosion*, **29**, 470 (1973)

Osozawa, K., 'Metallurgical Aspects of Intergranular Corrosion of Stainless Steel', *Boshoku Gijutsu*, **22**, 267 (1973); *C.A.*, **80**, 85981V

Mikhovskii, M., Dzhambazova, L. and Marinova, I., 'Determining the Extent of Inter-crystalline Corrosion by Measuring Barkhausen Noise', *Tekh. Misul.*, **10**, 101 (1973); *C.A.*, **80**, 103010j

Levente, B., 'Effect of Stabilising Elements on the Susceptibility to Intercrystalline Corrosion of Heat and Corrosion Resistant Austenitic Steels', *Gepgyartastechnologia*, **13**, 528 (1973); *C.A.*, **80**, 102971m

Gruetzner, G., 'Sensitising a Cr–Ni–N_2 Austenitic Stainless Steel to Intergranular Corrosion by Dichromium Nitride Precipitation', *Cent. Doc. Siderurg, Circ. Inform. Tech.*, **30**, 1165 (1973); *C.A.*, **80**, 39816x

Van der Horst, J. M. A., 'Grain Boundary Attack of *3116L* Stainless Steel in Ammonia-rich Environments', *J. Electrochem. Soc.*, **120**, 512 (1973)

Frankenthal, R. P. and Pickering, W. H., 'Intergranular Corrosion of a Ferritic Stainless Steel', *J. Electrochem. Soc.*, **120**, 23 (1973)

Dzhambasova, L., Mikhovskii, M. and Ivanov, D., 'Acoustic Study of Intergranular Corrosion in Non-ferrous Alloys', *Tekh. Misul.*, **10**, 115 (1973)

Freid, M. Kh. and Suprunov, V. A., 'Effect of Secondary Phases on the Corrosion Behaviour of Cr–Ni–Ti Steel Kh 18N 10T', *Izv. Vyssh. Ucheb. Zaved. Khim., Khim. Tekhnol.*, **16**, 1037 (1973); *C.A.*, **79**, 111101j

Sandor, P., 'Resistance to Intercrystalline Corrosion in Stainless Steel Weld Metal', *Anticorros. Methods Mater.*, **20**, 3 (1973); *C.A.*, **79**, 56661q

Savkina, L. Ya., Lazareva, N. A. and Feldgandler, E. G., 'Intercrystalline Corrosion Tendency of Low-carbon Cr–Ni Steel', *Metalloved, Term. Obrab. Metal.*, **55** (1973); *C.A.*, **79**, 34366V

Hodge, F. G., 'Effect of Ageing on the Anodic Behaviour of Ni–Cr–Mo Alloys', *Corrosion*, **29**, 375 (1973)

Cowan, R. L. and Tedmon, C. S., 'Intergranular Corrosion of Fe–Ni–Cr Alloys', in *Advances* in *Corrosion Science and Technology*, **3** (ed. M. G. Fontana and R. W. Staehle), Plenum Press (1973)

Streicher, M. A., 'The Rôle of Carbon, Nitrogen and Heat Treatment in the Dissolution of Fe–Cr Alloys in Acids', *Corrosion*, **29**, 337 (1973)

Selective dissolution

Desai, M. N., Talati, J. D. and Trivedi, A. K. M., 'Dezincification of Brasses', *J. Indian Chem. Soc.*, **38**, 565 (1961); *C.A.*, **56**, 11334e

Bumbulis, J. and Graydon, W. F., 'Dissolution of Brass in H_2SO_4 Solutions', *J. Electrochem. Soc.*, **109**, 1130 (1962)

Crennell, J. T. and Sawyer, L. J. E., 'Cathodic Protection Against Dezincification', *J. Appl. Chem.*, **12**, 170 (1962)

Marshakov, I. K. and Bogdanov, V. P., 'Mechanism of the Selective Corrosion of Cu–Zn Alloys', *Zh. Fiz. Khim.*, **37**, 2767 (1963); *C.A.*, **60**, 7758q

Tanabe, Z., 'The Relation Between Dealuminisation and Anodic Polarisation of Aluminium Bronze', *Sumitomo Keikinzoku Giho*, **4**, 59 (1963); *C.A.*, **62**, 223b

Gerischer, H., 'Anodic Behaviour of Noble Metal Alloys and Resistance Limits in the Corrosion of Such Alloys', *Korrosion*, **16**, 21 (1963); *C.A.*, **61**, 14180g

Joseph, G. and Arce, M. T., 'Brass Dezincification', *Rev. Indian*, **2**, 60 (1963); *C.A.*, **61**, 11693c

Hashimoto, K., Ogava, S. and Shimodaira, S., 'Dezincification of α-brass', *Trans. Japan Inst. Metals*, **4**, 42 (1963); *C.A.*, **60**, 2589g

Piatti, L. and Grour, R., 'Selective Corrosion of Cu Alloys', *Korrosion*, **14**, 551 (1963); *C.A.*, **59**, 8436g

Robinson, F. P. A. and Sholit, M., 'The Dezincification of Brass', *Corrosion Technol.*, **11**, 11 (1964); *C.A.*, **61**, 2785f

Tanabe, Z., 'Dealuminisation of Cu–Al Alloys and Polarisation Measurements', *Boshoku Gijutsu*, **14**, 56 (1965); *C.A.*, **63**, 3971d

Turner, M. E. ., 'Further Studies on the Influence of Water Composition on the Dezincification of Duplex Brass Fittings', *Proc. Soc. Water Treat. Exam.*, **14**, 81 (1965); *C.A.*, **63**, 15984e

Frade, G. and Lacombe, P., 'Influence of Intergranular Diffusion in the Kinetics of Dezincification of 70/30 Brasses Without Additions at 450–900°C', *Compt. Rend.*, **260**, 5022 (1965); *C.A.*, **63**, 5323c

Lucey, V. F., 'The Mechanism of Dezincification and the Effect of Arsenic, Part 1', *Brit. Corrosion J.*, **1**, 9 (1965–66)

Lucey, V. F., 'The Mechanism of Dezincification and the Effect of Arsenic, Part 2', *Brit. Corrosion J.*, **1**, 51 (1965–66)

Rowlands, J. C., 'Preferential Phase Corrosion of Naval Brass in Sea-water', *2nd Int. Congr. Metal Corrosion*, 1963, New York City (1966); *C.A.*, **65**, 10282h

Joseph, G. and Arce, M. T., 'Contribution to the Study of Brass Dezincification', *Corros. Sci.*, **7**, 597 (1967)

Sugawara, H. and Ebiko, H., 'Dezincification of Brass', *Corros. Sci.*, **7**, 513 (1967)

Wilde, B. E. and Teterin, G. A., 'Anodic Dissolution of Copper–Zinc Alloys in Alkaline Solutions', *Brit. Corrosion J.*, **2**, 125 (1967)

Ugiansky, G. M. and Ellinger, G. A., 'Corrosion of Monel 400 in High CO_2 Well Water', *Corrosion*, **24**, 134 (1968)

Feller, H. G., 'Untersuchungen zur Selectiven Korrosion mit der Ringscheibenelektrode im System Cu–Zn in Schwefelsaure', *Corros. Sci.*, **8**, 259 (1968)

Pickering, H. W. and Byrne, P. J., 'Partial Currents During Anodic Dissolution of Cu–Zn Alloys at Constant Potential', *J. Electrochem. Soc.*, **116**, 1492 (1968)

Heidersbach, R., 'Clarification of the Mechanism of De-alloying Phenomenon', *Corrosion*, **24**, 171 (1968)

Langenegger, E. E. and Robinson, F. P. A., 'Effect of the Polarisation Technique on Dezincification Rates and Physical Structure of Dezincified Zones', *Corrosion*, **24**, 411 (1968)

Brooks, W. B., 'Discussion of the De-alloying Phenomenon', *Corrosion*, **24**, 171 (1968)

Pickering, H. W., 'Volume Diffusion During Anodic Dissolution of a Binary Alloy', *J. Electrochem. Soc.*, **115**, 143 (1968)

Pickering, H. W., 'Preferential Anodic Dissolution in Binary Alloys', *Proc. Conf. Fundam. Aspects Stress Corros. Cracking*, 159, N.A.C.E. (1969)

Langenegger, E. E. and Robinson, F. P. A., 'A Study of Mechanism of Dezincification of Brasses', *Corrosion*, **25**, 59 (1969)

Langenegger, E. E. and Robinson, F. P. A., 'The Rôle of As in Preventing Dezincification of α-brass', *Corrosion*, **25**, 137 (1969)
Horton, R. M., 'New Metallographic Evidence for Dezincification of Brass by Redeposition of Cu', *Corrosion*, **26**, 160 (1970)
Verink, E. D. (Jr.) and Parrish, P. A., 'Use of Pourbaix Diagrams in Predicting the Susceptibility to De-alloying Phenomena', *Corrosion*, **26**, 214 (1970)
Mattsson, E., 'Mechanism of Exfoliation (Layer Corrosion) of Al-5%Zn-1%Mg', *Brit. Corrosion J.*, **6**, 73 (1971)
Gal-Or, L., Zanker, L. and Yahalom, J., 'Preferential Dissolution of Inclusions in Hastelloy C', *Corros. Sci.*, **11**, 107 (1971)
Vernik, E. D. (Jr.) and Heindersbach, R. H. (Jr.), 'Evaluation of the Tendency for Dealloying in Metal Systems', *Amer. Soc. Test. Mater., Spec. Tech. Publ.*, **516**, 303 (1972); *C.A.*, **80**, 102944e
Leidheiser, H. and Kissinger, R., 'Chemical Analysis of the Liquid Within Propagatory Stress-corrosion Cracks in 70:30 Brass Immersed in Concentrated Ammonium Hydroxide', *Corrosion*, **28**, 218 (1972)
Langenegger, E. E. and Callaghan, B. G., 'Use of an Empirical Potential Shift Technique for Predicting Dezincification Rates of αβ-brasses in Chloride Media', *Corrosion*, **28**, 245 (1972)
Heidersbach, R. H. and Vernik, E. D., 'The Dezincification of α and β Brasses', *Corrosion*, **28**, 397 (1972)
Revie, R. W. and Uhlig, H. H., 'Further Evidence Regarding the Dezincification Mechanism', *Corros. Sci.*, **12**, 669 (1972)
Nielsen, K. and Rislund, E., 'Comparative Study of Dezincification Tests', *Br. Corrosion J.*, **8**, 106 (1973)

Pitting (papers published in English)
Robinson, F. P. A., 'Pitting Corrosion; Cause, Effect, Detection and Prevention', *Corrosion Technol.*, **7**, 237, 266 (1960)
Kolotyrkin, Y. M., 'Electrochemical Behaviour of Metals During Anodic and Chemical Passivation in Electrolytic Solutions', *1st Intern. Congr. Metallic Corrosion*, Butterworths, London, 10 (1962)
Clarke, F. E. and Ristaino, A. J., 'New Clues in the Boiler Tube Pitting Puzzle', *1st Intern. Congr. Metallic Corrosion*, Butterworths, London, 403 (1962)
Kimmel, A. L., Holingshand, W. R. and Shea, E. P., 'Effect of Dissolved Oxygen on Pitting of Boiler Tubes', *Am. Soc. Mech. Engrs.*, Paper No. 61-WA-285
Colborne, G. F., Allen, A. R. and Thunaes, A., 'Crash Repair Program to Control Serious Pitting Corrosion on Process Tanks', *Corrosion*, **17**, 20 (1961)
Flint, G. N. and Melbourne, S. H., 'The Corrosion of Decorative Ni and Cr Coatings—a Metallographic and Potential Study', *Trans. Inst. Met. Finishing*, **38**, 35 (1961)
Menasce, S., 'Corrosion of 18-5 Stainless Steel in Dilute HCl', *Rev. Met.*, **58**, 951 (1961)
Doyle, D. P. and Godard, H. P., 'Influence of Cl$^-$ Additions to Water on the Corrosion Behaviour of Al', *Oil Can.*, 14 No. 33, 42 (1962); *C.A.*, **57**, 8330h
Fraser, W. A. and Bloom, M. C., 'Corrosion of Mild Steel in 40% NaOH Solution at 316°C', *Corrosion*, **18**, 163t (1962)
Bloom, M. C., Fraser, W. A. and Krulfeld, M., 'Corrosion of Steel in Concentrated LiOH Solution at 316°C', *Corrosion*, **18**, 401t (1962)
Videm, K., 'Pitting Corrosion of Aluminium in Contact with Stainless Steel', *Proc. Conf. on Corrosion Reactor Mater.*, Salzburg, Austria, **1**, 391 (1962); *C.A.*, **60**, 1412g
Lyon, D. H., Salva, S. J. and Shaw, B. C., Etch Pits in Germanium: Detection and Effects', *J. Electrochem. Soc.*, **110**, 184c (1963)
Bertocci, U., Hulett, L. D. and Jenkins, L. H., 'On the Formation of Electrochemical Etch Pits on the (111) Face of Copper', *J. Electrochem. Soc.*, **110**, 1190 (1963)
Weed, R. D., 'Dendritic Pitting of Stellite by Alkaline Permanganate Solutions', *J. Electrochem. Soc.*, **110**, 178c (1963)
Fraser, W. A. and Bloom, M. C., 'Corrosion of Mild Steel in 15%NaOH at 316°C', *J. Electrochem. Soc.*, **110**, 177C (1963)
Pourbaix, M., Klimczak-Mathieu, Martens, C. and Meunier, J., 'Potentiokinetic and Corrosimetric Investigations of the Corrosion Behaviour of Alloy Steels', *Corros. Sci.*, **3**, 239 (1963)
Lefrancois, P. A. and Hoyt, W. B., 'Chemical Thermodynamics of High Temperature Reactions in Metal Testing Corrosion', *Corrosion*, **19**, 360t (1963)

Wesley, W. A., 'Pitting of Decorative Nickel-Chromium Coatings', *Tech. Proc. Am. Electroplaters Soc.*, **50**, 9 (1963)

Bloom, M. C., Newport, G. N. and Fraser, W. A., 'The Growth and Breakdown of Protective Films in High-temperature Aqueous Solutions 15% NaOH at 316°C', *J. Electrochem. Soc.*, **111**, 1343 (1964)

Dugdale, I. and Cotton, J. B., 'The Anodic Polarisation of Titanium in Halide Solutions', *Corros. Sci.*, **4**, 397 (1964)

Herman, R. S., 'Radiographic-Photographic Method for Measuring Depth and Distribution of Pitting', *Corrosion*, **20**, 361 (1964)

Tomashov, N. D., Chernova, G. P. and Marcova, O. N., 'Effect of Supplementary Alloying Elements on Pitting Corrosion Susceptibility of 18Cr-14Ni Stainless Steel', *Corrosion*, **20**, 166 (1964)

Schwenk, W., 'Theory of Stainless Steel Pitting', *Corrosion*, **20**, 129 (1964)

Murray, G. A. W., 'Artificial Pits for Quantitative Studies of Corrosion of Aluminium Alloys in Natural Waters', *Corrosion*, **20**, 329 (1964)

Koehler, E. L. and Evans, S., 'Pitting and Uniform Corrosion of Aluminium by pH 3·5 Citrate Buffer Solution', *J. Electrochem. Soc.*, **111**, 17 (1964)

Uhlig, H. H. and Gilman, J. R., 'Pitting of 18-8 Stainless Steel in Ferric Chloride Inhibited by Nitrates', *Corrosion*, **20**, 289 (1964)

Fisher, A. O., 'Magnesium Anodes Control Pitting in an Inhibited Circulating Cooling Water System', *Mater. Protect.*, **3**, 64 (1964)

Pouillard, E. and Finckbohner, A., 'Removal of Surface Chromium in Relation to Pitting Corrosion of Stainless Steel', *Dev. Met.*, **62**, 721 (1965); *C.A.*, **63**, 14476b

Hoar, T. P., Mears, D. C. and Rothwell, G. P., 'The Relationships Between Anodic Passivity, Brightening and Pitting', *Corros. Sci.*, **5**, 279 (1965)

Piggott, A. R., Leckie, H. and Shreir, L. L., 'Anodic Polarisation of Titanium in HCOOH – I: Anodic Behaviour of Titanium in Relation to Anodising Conditions', *Corros. Sci.*, **5**, 165 (1965)

Heine, M. A., Keir, D. S. and Pryor, M. J., 'The Specific Effects of Chloride and Sulphate Ions on Oxide Covered Aluminium', *J. Electrochem. Soc.*, **112**, 24 (1965)

Freeman, L. I. and Kolotyrkin, Ya. M., 'Pitting Corrosion of Iron by Perchlorate Ions', *Corros. Sci.*, **5**, 199 (1965)

Greene, N. D. and Judd, G., 'Relation Between Anodic Dissolution and Resistance to Pitting Corrosion', *Corrosion*, **21**, 15 (1965)

Bond, A. P., Bolling, G. F. and Damian, H. A., 'Influence of Microsegregation on Pitting Corrosion in High-purity Aluminium', *J. Electrochem. Soc.*, **112**, 178c (1965)

Pickering, H. W. and Frankenthal, R. P., 'A Transmission Electron Microscope Study of the Breakdown of Passivity on Fe-24% Cr', *J. Electrochem. Soc.*, **112**, 761 (1965)

Board, P. W. and Elbourne, R. G. P., 'Pitting Corrosion in Plain Cans Containing Acid Foods', *Food Technol.*, **19**, 1571 (1965); *C.A.*, **64**, 2661c

Voogel, P., 'Pit Corrosion of Internally Painted Cargo and Cargo/Ballast Tanks of Ocean-going Crude Oil Tankers', *J. Oil Colour Chemists Assoc.*, **48**, 597 (1965)

White, J. H., Yaniv, A. E. and Schick, H., 'The Corrosion of Metals in the Water of the Dead Sea', *Corros. Sci.*, **6**, 447 (1966)

Novakovskii, V. M. and Sorokina, A. N., 'Model Study of Chloride Pitting in 18-8 Stainless Steel', *Corros. Sci.*, **61**, 227 (1966)

Kato, T. R. and Price, D. H., 'Effect of Environmental Conditions on the Pitting of Uranium', *AEC*, Accession No. 25486, Dept. No. NLCO-957 (1966); *C.A.*, **65**, 18069e

Davies, D. E. and Lotlikat, M. M., 'Passivation and Pitting Characteristics of Zinc', *Brit. Corrosion J.*, **1**, 149 (1966)

Bond, A. P., Bolling, G. F., Damian, H. A. and Biloni, H., 'Microsegregation and the Tendency for Pitting Corrosion in High-purity Aluminium', *J. Electrochem. Soc.*, **113**, 773 (1966)

Camp, E. K., 'Pitting Corrosion of Titanium in Aqua Regia', *J. Electrochem. Soc.*, **113**, 204c (1966)

Craig, H. L., 'Rotating Disc Study of the Pitting of Aluminium in Alkaline Chloride Solutions', *J. Electrochem. Soc.*, **113**, 203c (1966)

Hospadaruk, V. and Petrocelli, J. V., 'The Pitting Potential of Stainless Steel in Chloride Solutions', *J. Electrochem. Soc.*, **113**, 878 (1966)

Steigerwald, R. F., 'Effect of Chromium Content on Pitting Behaviour of Fe-Cr Alloys', *Corrosion*, **22**, 107 (1966)

Lebet, R. and Piotrowski, A., 'Resistance to Pitting of Types *202* and *321* Steels to Sulphuric Acid and Sodium Chloride Solutions', *Corrosion*, **22**, 257 (1966)

Spaepen, G. J. and Fevery-de Meyer, M. J., 'Electrochemical Corrosion Experiments at Temperatures About 100°C', *Corros. Sci.*, **7**, 405 (1967)

Bond, A. P. and Lizlovs, E. A., 'Anodic Polarisation of Some Ferritic Stainless Steels in Chloride Media', *J. Electrochem. Soc.*, **114**, 199c (1967)

Joseph, G. and Perret, R., 'Inhibitor Action on the Corrosion of Iron in Salt-water Solution', *Corros. Sci.*, **7**, 553 (1967)

Frazer, M. J. and Langstaff, R. D., 'Influence of Cupric Ions on the Behaviour of Surface-active Agents Towards Aluminium', *Brit. Corrosion J.*, **2**, 11 (1967)

Szklarska-Smialowska, Z. and Janik-Czachor, H., 'Pitting Corrosion of 13Cr–Fe Alloy in Na_2SO_4 Solutions Containing Chloride Ions', *Corros. Sci.*, **7**, 65 (1967)

Rosenfeld, I. L. and Danilov, I. S., 'Electrochemical Aspects of Pitting Corrosion', *Corros. Sci.*, **7**, 129 (1967)

Hoar, T. P., 'The Production and Breakdown of the Passivity of Metals', *Corros. Sci.*, **7**, 341 (1967)

Butler, G. and Beynon, J. G., 'The Corrosion of Mild Steel in Boiling Salt Solutions', *Corros. Sci.*, **7**, 385 (1967)

Beck, T. R., 'Pitting Corrosion of Titanium in Halide Solutions', *J. Electrochem. Soc.*, **114**, 201c (1967)

Frankenthal, R. P., 'The Effect of Surface Preparation on Pitting and Anodic Dissolution of Iron–Chromium Alloys', *J. Electrochem. Soc.*, **114**, 201c (1967)

Horvath, J. and Uhlig, H., 'Metallurgical Factors Affecting the Critical Potential for Pitting Corrosion of Cr–Fe–Ni Alloys', *J. Electrochem. Soc.*, **114**, 201c (1967)

Murray, G. A. W., 'Rôle of Iron in Aluminium on the Initiation of Pitting in Water', *Br. Corros. J.*, **2**, 216 (1967)

Penn, J. H. and Murray, G. A. W., 'Effect of Acidic Gelatinous Materials on Pitting Corrosion of Aluminium Hollow-ware', *Br. Corros. J.*, **2**, 193 (1967)

Lucey, V. F., 'Mechanism of Pitting Corrosion of Copper in Supply Waters', *Br. Corros. J.*, **2**, 175 (1967)

Finley, H. F., 'An Extreme-value Statistical Analysis of Maximum Pit Depths and Time to First Perforation', *Corrosion*, **23**, 83 (1967)

Mears, D. C. and Rothwell, G. P., 'Effects of Probe Position on Potentiostatic Control During the Breakdown of Passivity', *J. Electrochem. Soc.*, **115**, 36 (1968)

Frankenthal, R. P., 'The Effect of Surface Preparation and of Deformation on the Pitting and Anodic Dissolution of Fe–Cr Alloys', *Corros. Sci.*, **8**, 491 (1968)

Bond, A. P. and Lizlovs, E. A., 'Anodic Polarisation of Austenitic Stainless Steel in Chloride Media', *J. Electrochem. Soc.*, **115**, 1130 (1968)

Szummer, A., Szklarska-Smialowska, Z. and Janik-Czachor, M., 'Electron Microprobe Study of the Corrosion Pits Nucleation of Fe-16Cr Single Crystals', *Corros. Sci.*, **8**, 827 (1968)

Ijzermans, A. B. and van der Krogt, A. J., 'Pitting Corrosion of an Austenitic Cr–Ni Stainless Steel in H_2SO_4 Containing H_2S', *Corros. Sci.*, **8**, 679 (1968)

Horvath, J. and Uhlig, H. H., 'Critical Potentials for Pitting Corrosion of Ni, Cr–Ni, Cr–Fe and Related Stainless Steels', *J. Electrochem. Soc.*, **115**, 791 (1968)

Black, J. R., 'Etch Pit Formation in Silicon at Al–Si Contacts Due to the Transport of Silicon in Aluminium by Momentum Exchange with Conducting Electrons', *J. Electrochem. Soc.*, **115**, 242c (1968)

Posey, F. A., 'Pitting and Crevice Corrosion of Titanium and its Alloys', *J. Electrochem. Soc.*, **115**, 231c (1968)

Mattson, E. and Fredriksson, A. M., 'Pitting Corrosion in Copper Tubes — Cause of Corrosion and Counter Measures', *Br. Corros. J.*, **3**, 246 (1968)

Arora, G. P., Isasi, J. A. and Metzger, M., 'Protective Films and Structure Dependent Corrosion in Aluminium', *Corrosion*, **25**, 445 (1969)

Pourbaix, M., 'Recent Applications of Electrode Potential Measurements in the Thermodynamics and Kinetics of Corrosion of Metals', *Corrosion*, **25**, 267 (1969)

Lizlovs, E. A. and Bond, A. P., 'Anodic Polarisation of Some Ferritic Stainless Steels in Chloride Media', *J. Electrochem. Soc.*, **116**, 574 (1969)

Smialowski, M., Szklarska-Smialowska, Z., Rychcik, M. and Szummer, A., 'Effect of Sulphide Inclusions in a Commercial Stainless Steel on the Nucleation of Corrosion Pits', *Corros. Sci.*, **9**, 123 (1969)

Rajagopalan, K. S., Venu, K. and Viswanathan, M., 'Activation of Passivated Steel in Borax Solution', *Corros. Sci.*, **9**, 169 (1969)

Jones, D. A. and Greene, N. D., 'Electrochemical Detection of Localised Corrosion', *Corrosion*, **25**, 367 (1969)

Rossum, J. R., 'Prediction of Pitting Rates in Ferrous Metals from Soil Parameters', *J. Amer. Water Works Ass.*, **61**, 305 (1969); *C.A.*, **71**, 35404c

Szklarska-Smialowska, Z. and Czachor-Janik, M., 'Electrochemical Investigation of the Nucleation and Propagation of Pits in Iron-Chromium Alloys', *Br. Corros. J.*, **4**, 138 (1969)

Jones, R. L., 'Potter-Mann and Bloom Iron Oxide Films Grown on Mild Steel in 280C 5 N NaOH: Their Significance in Pitting', *Corrosion*, **29**, 133 (1969)

Obrecht, M. F., Sastor, W. E. and Keyes, J. M., 'Integrated Design of Field Test Panel Pilot Unit for Investigating Pitting Corrosion of Copper Water Tube by Potable Water Supplies', *Proc. 4th Int. Congr. Met. Corr.*, 1969, 576 (1972)

Joseph, G., Perret, R. and Spacek, J., 'Pitting Corrosion of Copper Tubings', *Proc. 4th Int. Congr. Metal. Corros.*, 1969, 817 (1972)

Bohni, H. and Uhlig, H. H.'Environmental Factors Affecting the Critical Pitting Potential of Aluminium', *J. Electrochem. Soc.*, **116**, 906 (1969)

Hodge, F. G. and Wilde, B. E., 'Effect of Chloride Ion on the Anodic Dissolution Kinetics of Cr-Ni Binary Alloys in Dilute H_2SO_4', *Corrosion*, **26**, 146 (1970)

Foley, R. T., 'Rôle of the Chloride Ion in Iron Corrosion', *Corrosion*, **26**, 58 (1970)

Richardson, J. A. and Wood, G. C., 'A Study of the Pitting Corrosion of Aluminium by Scanning Electron Microscopy', *Corros. Sci.*, **10**, 313 (1970)

France, W. D. (Jr.) and Greene, N. D., 'Comparison of Chemically and Electrolytically Induced Pitting Corrosion', *Corrosion*, **26**, 1 (1970)

Brigham, R. J., 'Discussion on Chemically and Electrolytically Induced Pitting', *Corrosion*, **26**, 200 (1970)

Abdul Azim, A. A., 'Pitting Corrosion of Lead in $H_2SO_4 + HClO_4$', *Corros. Sci.*, **10**, 421 (1970)

Szklarska-Smialowska, Z., 'Electron Microprobe Study of the Effect of Sulphide Inclusions on the Nucleation of Corrosion Pits in Stainless Steels', *Br. Corros. J.*, **5**, 159 (1970)

Weinstein, M. and Speirs, K., 'Mechanisms of Chloride-activated Pitting Corrosion of Martensitic Stainless Steels', *J. Electrochem. Soc.*, **117**, 256 (1970)

Wilde, B. E. and Williams, E., 'On the Correspondence Between Electrochemical and Chemical Accelerated Pitting Corrosion Tests', *J. Electrochem. Soc.*, **117**, 775 (1970)

Bianchi, G., Cerguetti, A., Mazza, F. and Torchio, S., 'Chemical Etching and Pitting of Stainless Steels', *Corros. Sci.*, **10**, 19 (1970)

Ijzermans, A. B., 'Pitting Corrosion and Intergranular Attack of Austenitic Cr-Ni Stainless Steels in NaSCN', *Corros. Sci.*, **10**, 607 (1970)

Ashworth, V., Boden, P. J., Leach, J. S. Ll. and Nehru, A. Y., 'On the Cl^- Breakdown of Passive Films on Mild Steel', *Corros. Sci.*, 10, 481 (1970)

Payer, J. H. and Staehle, R. W., 'Localised Attack on Metal Surfaces', *Corros. Fatigue: Chem. Mech. Microstruct.*, 1971, 211 (1972); *C.A.*, **80**, 40010

Vijh, A. K., 'A Possible Interpretation of the Influence of Chloride Ions on the Anodic Behaviour of Some Metals', *Corros. Sci.*, **11**, 161 (1971)

Forchhammer, P. and Engell, H. J., 'Investigations of the Corrosion of Stainless Steels by Potentiokinetic Measurements', *Corros. Sci.*, **11**, 49 (1971)

Jackson, R. P. and van Rooyen, D., 'Electrochemical Evaluation of Resistance of Stainless Alloys to Chloride Media', *Corrosion*, **27**, 203 (1971)

Pourbaix, A., 'Characteristics of Localised Corrosion of Steel in Chloride Solutions', *Corrosion*, **27**, 449 (1971)

Tokuda, T. and Ives, M. B., 'Pitting Corrosion of Nickel', *Corros. Sci.*, **11**, 297 (1971)

Rajagopalan, K. S. and Venu, K., 'Activation of Passivated Steel in Na_2CrO_4 Containing NaCl', *Corrosion*, **27**, 506 (1971)

Szklarska-Smialowska, Z., 'Effect of the Ratio of Cl^-/SO_4^{2-} in Solution on the Pitting Corrosion of Nickel', *Corros. Sci.*, **11**, 209 (1971)

Vermilyea, D. A., 'Concerning the Critical Pitting Potential', *J. Electrochem. Soc.*, **118**, 529 (1971)

Wilde, B. E. and Williams, E., 'The Relevance of Accelerated Electrochemical Pitting Tests to the Long Term Pitting and Crevice Corrosion Behaviour of Stainless Steels in Marine Environments', *J. Electrochem. Soc.*, **118**, 1056 (1971)

Fraker, A. C. and Ruff, A. W. (Jr.), 'Observations of Hot Saline Water Corrosion of Aluminium Alloys', *Corrosion*, **27**, 151 (1971)

Janik-Czachor, M., 'Electrochemical and Microscopic Study of Pitting Corrosion of Ultra-pure Iron', *Br. Corros. J.*, **6**, 57 (1971)

Butler, G., Ison, H. C. K. and Mercer, A. D., 'Some Important Aspects of Corrosion in Central Heating Systems', *Br. Corros. J.*, **6**, 31 (1971)

Johnson, W. K., 'Recent Development in Pitting Corrosion of Aluminium', *Br. Corros. J.*, **6**, 200 (1971)

Bond, A. P., 'Effects of Molybdenum on the Pitting Potentials of 18%Cr Ferritic Stainless Steels at Various Temperatures', *J. Electrochem. Soc.*, **118**, 208c (1971)

Mansfeld, F., 'The Effect of Water on Passivity and Pitting of Titanium in Solutions of Methanol and Hydrogen Chloride', *J. Electrochem. Soc.*, **118**, 1412 (1971)

Takuda, T. and Ives, M. B., 'Aggressive Ion Accessibility and the Morphology of Corrosion Pits', *J. Electrochem. Soc.*, **118**, 1404 (1971)

Szklarska-Smialowska, Z. and Janik-Czachor, M., 'The Analysis of Electrochemical Methods for the Determination of Characteristic Potentials of Pitting Corrosion', *Corros. Sci.*, **11**, 901 (1971)

Pryor, M. J., contribution to a discussion on 'A Study of the Pitting Corrosion of Aluminium by Scanning Electron Microscopy', by J. A. Richardson and G. C. Ward, *Corros. Sci.*, **11**, 463 (1971)

Zahavi, J. and Metzger, M., 'Electron Microscope Study of Breakdown and Repair of Anodic Films on Aluminium', *J. Electrochem. Soc.*, **119**, 1479 (1972)

Wilde, B. E., 'A Critical Appraisal of Some Popular Laboratory Electrochemical Tests for Predicting the Localised Corrosion Resistance of Stainless Alloys in Sea-water', *Corrosion*, **28**, 283 (1972)

Suzuki, T. and Kitamura, Y., 'Critical Potential for Growth of Localised Corrosion of Stainless Steel in Chloride Media', *Corrosion*, **28**, 1 (1972)

Rarey, C. R. and Aronson, A. H., 'Pitting Corrosion of Sensitised Ferritic Stainless Steel', *Corrosion*, **28**, 255 (1972)

Brigham, R. J., 'Pitting of Molybdenum Bearing Austenitic Stainless Steel', *Corrosion*, **28**, 177 (1972)

Vernik, E. D., Lee, T. S. and Cusumano, R. L., 'Influence of Prior Electrochemical History on the Propagation of Localised Corrosion', *Corrosion*, **28**, 348 (1972)

Szklarska-Smialowska, Z., 'Influence of Sulfide Inclusions on the Pitting Corrosion of Steels', *Corrosion*, **28**, 388 (1972)

Lizlovs, E. A. and Bond, A. P., 'Polarisation Behaviour of High-purity 13 and 18% Chromium Stainless Steels', *J. Electrochem. Soc.*, **119**, 219c (1972)

Bogar, F. D. and Foley, R. T., 'The Influence of Chloride Ion on the Pitting of Aluminium', *J. Electrochem. Soc.*, **119**, 462 (1972)

Beck, J. R., Mahaffey, D. W. and Olsen, J. H. 'Pitting and Deposits With an Organic Fluid by Electrolysis and by Fluid Flow', *J. Electrochem. Soc.*, **119**, 155 (1972)

Kato, M., Inoue, T. and Sato, O., 'New Microautoradiography Using Auger Electrons in the Study of the Structure and Pitting Corrosion of Aluminium', *Proc. Symp. Nucl. Tech. Basic Metal Ind.*, 1972, 541 (1973)

Seys, A. A. and van Haute, A. A., 'Pitting Potential Measurements by the Static Potential Band Method', *Philipp. Geogr. J.*, **16**, 107 (1972); *C.A.*, **79**, 99726g

Pourbaix, M., 'Significance of Protection Potential in Pitting, Intergranular Corrosion and Stress-corrosion Cracking', *J. Less-common Metals*, **28**, 51 (1972)

Abdul Azim, A. A. and Afifi, S. E., 'Electrochemical Behaviour of Lead and Some Selected Pb–Sb Alloys in $H_2SO_4 + HClO_4$', *Corros. Sci.*, **12**, 603 (1972)

Isaacs, H. S. and Kissel, G., 'Surface Preparation and Pit Propagation in Stainless Steels', *J. Electrochem. Soc.*, **119**, 1628 (1972)

Pickering, H. W. and Frankenthal, R. P., I: electrochemical studies 'On the Mechanism of Localised Corrosion of Iron and Stainless Steel', *J. Electrochem. Soc.*, **119**, 1297 (1972)

Frankenthal, R. P. and Pickering, W. H., II: morphological studies 'On the Mechanism of Localised Corrosion of Iron and Stainless Steel', *J. Electrochem. Soc.*, **119**, 1304 (1972)

Zahavi, J. and Metzger, M., 'On the Rôle of Grain Boundaries in Film Breakdown and Pitting of Aluminium', *J. Electrochem. Soc.*, **119**, 218c (1972)

Szklarska-Smialowska, Z., 'The Kinetics of Pit Growth on Nickel in Solutions with Different Cl^-/SO_4^{2-} Ratios', *Corros. Sci.*, **12**, 527 (1972)

Bianchi, G., Corguetti, A., Mazza, F. and Torchio, S., 'Electronic Properties of Oxide Films and Pitting Susceptibility of Type *304* Stainless Steel', *Corros. Sci.*, **12**, 495 (1972)

Szklarska-Smialowska, Z. and Mankowski, J., 'Effect of Temperature on the Kinetics of

Development of Pits in Stainless Steel in 0·5 N NaCl + 0·1 N H_2SO_4 Solution', *Corros. Sci.*, **12**, 925 (1972)

Vijh, A. K., 'The Influence of Solid State Cohesion of Metals on Their Pitting Potentials', *Corros. Sci.*, **12**, 935 (1972)

Thomas, J. G. N. and Tiller, A. K., 'Formation and Breakdown of Surface Films on Copper in $NaHCO_3$ and NaCl. Effects of Temperature and pH', *Br. Corr. J.*, **7**, 263 (1972)

Butler, G., Stretton, P. and Beynon, J. G., 'Initiation and Growth of Pits on High-purity Iron and its Alloys with Chromium and Copper in Neutral Chloride Solutions', *Br. Corr. J.*, **7**, 168 (1972)

Janik-Czachor, M., Szummer, A. and Szklarska-Smialowska, Z., 'Effect of Sulphur and Manganese on Nucleation of Corrosion Pits in Iron', *Br. Corr. J.*, **7**, 90 (1972)

Board, P. H., Holland, R. V. and Steele, R. J., 'Prediction of Pitting Corrosion in Tinplate from Capacitance Measurements', *Br. Corr. J.*, **7**, 87 (1972)

Lucey, V. F., 'Developments Leading to the Present Understanding of the Mechanism of Pitting Corrosion of Copper', *Br. Corr. J.*, **7**, 36 (1972)

Maskell, R. V., 'Pitting Potentials', AECL(Rep.) 4159, Atomic Energy Can. Ltd., 17 (1972)

Beck, T. R., 'One-dimensional Pits in Titanium', *J. Electrochem. Soc.*, **119**, 218c (1972)

Clark, M. B. and Zupancic, 'The Electrochemical Corrosion of Magnesium', *J. Electrochem. Soc.*, **119**, 218c (1972)

Richardson, J. A. and Wood, G. C., 'The Interpretation of Impedance Charges on Oxide-coated Aluminium Produced by Immersion in Inhibitive and Corrosive Aqueous Media', *J. Electrochem. Soc.*, **120**, 193 (1973)

Mao, K. W. and Hoare, J. P., 'The Anodic Dissolution of Mild Steel in Solutions Containing Both Cl^- and NO^-_3 Ions', *Corros. Sci.*, **13**, 799 (1973)

Broli, A. and Holtan, H., 'Use of Potentiokinetic Methods for the Determination of Characteristic Potentials for Pitting Corrosion of Aluminium in a Deaerated Solution of 3% NaCl', *Corros. Sci.*, **13**, 237 (1973)

Rozgonyi, G. A. and Lizuka, T., 'Etch Pit Studies of GaP Liquid Phase Epitoxial Layers', *J. Electrochem. Soc.*, **120**, 673 (1973)

Bond, A. P., 'Effects of Molybdenum on the Pitting Potentials of Ferritic Stainless Steel at Various Temperatures', *J. Electrochem. Soc.*, **120**, 603 (1973)

Kato, M., Inoue, T., Goto, K., Ito, G. and Shimizu, Y., 'Effect of Dissolved Oxidising Agents and Inhibitors on Pitting Corrosion of Aluminium in Water', *Aluminium*, **49**, 289 (1973)

Richardson, J. A. and Godwin, A. W., 'Localised Corrosion of Stainless Steel During Food Processing', *Br. Corros. J.*, **8**, 259 (1973)

Bonewitz, R. A., 'An Electrochemical Evaluation of 1100, 5052 and 6063 Aluminium Alloys for Desalination', *Corrosion*, **29**, 215 (1973)

Datha, M. and Landolt, D., 'Stoichiometry of Anodic Nickel Dissolution in NaCl and $NaClO_3$ Under Active and Transpassive Conditions', *Corros. Sci.*, **13**, 187 (1973)

Mansfeld, F., 'Passivity and Pitting of Al, Ni, Ti and Stainless Steel in $CH_3OH + H_2SO_4$', *J. Electrochem. Soc.*, **120**, 188 (1973)

Yahalom, J., Ives, L. K. and Kruger, J., 'On the Nature of Films Over Corrosion Pits in Stainless Steel', *J. Electrochem. Soc.*, **120**, 384 (1973)

Beck, T. R., 'Pitting of Titanium – 1: Titanium Foil Experiments', *J. Electrochem. Soc.*, **120**, 1310 (1973)

Beck, T. R., 'Pitting of Titanium – 2: One-dimensional Pit Experiments', *J. Electrochem. Soc.*, **120**, 1317 (1973)

Brigham, R. J. and Tozer, E. W., 'Temperature as a Pitting Criterion', *Corrosion*, **29**, 33 (1973)

Bruce, S., 'Specialist Steels Combat Corrosion by Chloride-containing Cooling Water', *Process Eng.*, **88** (1973); *C.A.*, **80**, 6176k

Cornwell, F. J., Wildsmith, G. and Gilbert, P. T., 'Pitting Corrosion in Copper Tubes in Cold Water Service', *Br. Corros. J.*, **8**, 202 (1973)

Broli, A., Holtan, H. and Midjo, M., 'Use of Potentiokinetic and Potentiostatic Methods for the Determination of Characteristic Potentials for Pitting Corrosion of an Fe–Cr Alloy', *Br. Corros. J.*, **8**, 173 (1973)

Retief, R., 'Corrosion of Condenser Tubes Due to Carbonaceous Film on the Surface', *Br. Corros. J.*, **8**, 264 (1973)

Seys, A. A. and Van Haute, A. A., 'Pitting Potential Measurements by Means of the Static Potential Band Method', *Corrosion*, **29**, 329 (1973)

Zamin, M. and Ives, M. B., 'Effect of Chloride Ion Concentration on the Anodic Dissolution Behaviour of Nickel', *Corrosion*, **29**, 319 (1973)

Uhlig, H. H., 'Distinguishing Characteristics of Pitting and Crevice Corrosion', *Mater. Prot. Performance*, **12**, 42 (1973); *C.A.*, **78**, 105248a
Herbsleb, G. and Schwenk, W., 'Electrochemical Investigations of Pitting Corrosion', *Corros. Sci.*, **13**, 739 (1973)
Vijh, A. K., 'The Pitting Potentials of Metals: the Case of Titanium', *Corros. Sci.*, **13**, 805 (1973)
Fraker, A. C., 'Effect of Solution pH on the Saline Water Corrosion of Titanium Alloys', *Proc. 2nd Int. Conf. Titanium Sci. Technol.*, 1972 (1973)
Tatsuya, K. and Shuichi, F., 'Pitting Corrosion of Titanium in High-temperature Halide Solutions', *Proc. 2nd Int. Conf. Titanium Sci. Technol.*, **4**, 2383 (1973)
Mao, K. W. and Chin, D. T., 'Anodic Behaviour of Mild Steel in $NaClO_3$ at High Current Densities', *J. Electrochem. Soc.*, **121**, 191 (1974)
Zahavi, J. and Metzger, M., 'Effect of Chloride on Growth on Anodic Film', *J. Electrochem. Soc.*, **121**, 268 (1974)

Pitting (papers published in foreign languages)
Kutznelnigg, 'Pitting, the Typical Corrosion of Nickel Coatings', *Korrosion*, **13**, 64 (1960)
Schmeken, H., 'Electrochemical Measurements on Corrosion of Zinc with Regard to Pitting Possibilities', *Korrosion*, **13**, 65 (1960)
Sverepa, O., 'Pitting Corrosion of Aluminium in Water', *Korose Ochrana Mater.*, 1 (1961); *C.A.*, **57**, 12237h
Videm, K., 'Pitting of Aluminium in Contact with Stainless Steel', *Tek. Ukeblad*, **108**, 775 (1961); *C.A.*, **57**, 10912h
Sverepa, O. and Svobodny, Z., 'Corrosion Resistance of Aluminium Storing Vessels', *Chem. Prumysl.*, **11**, 644 (1961); *C.A.*, **56**, 7030b
Coriou, H., 'Les Problemes de Corrosion Aqueuse dans le Domaine de l'Energic Nucleaire', *Corros. Sci.*, **1**, 132 (1961)
Rozenfeld, I. L, and Danilov, I. S., 'The Mechanism of Pitting Corrosion of Stainless Steel', *Dokl. Akad. Nauk. SSSR*, **147**, 1417-19 (1962); *C.A.*, **58**, 8773e
Pourbaix, M., 'Potentiokinetic and Corrosimetric Studies on the Behaviour of Alloy Steels', CEBELCOR Rappt. Tech. No. 120, 1 (1962); *C.A.*, **58**, 13431b
Sakiyama, K. and Fuyimoto, M., 'Effects of Free Chlorine on the Pitting Corrosion of Gas Cooler Tubes', *Boshoku Gijutsu*, **11**, 299 (1962); *C.A.*, **58**, 5257b
Kaesche, H., 'Uniform Dissolution and Pitting of Aluminium Electrodes', *Z. Physik. Chem.*, **34**, 87 (1962); *C.A.*, **58**, 41546
Gillman, V. A. and Kolotyrkin, Ya. M., 'The Mechanism of Zirconium Pitting in Halide Solutions', *Dokl. Akad. Nauk. SSSR*, **143**, 640 (1962); *C.A.*, **57**, 3184a
Videm, K., 'Pitting Corrosion of Aluminium in Contact with Stainless Steel', Kjeller Rept. KR-18, 16 (1962); *C.A.*, **57**, 8315a
Prazak, M. and Cihal, V., 'Die Potentiostatische Untersuchung des Einflusses Einiger Leigierungselemente auf die Electrochemischem und Korrosions-eigenschaften Nischtrostender Stahle', *Corros. Sci.*, **2**, 71 (1962)
Vanleugengaghe, C., Klimczak-Mathieu, L., Meunier, J. and Pourbaix, M., 'Influence de Chlorures et d'Oxyolants sur le Compartement d'Aciers Inoxydables en Solution Bicarbonique', *Corros. Sci.*, **2**, 29 (1962)
Prazak, M., Tousek, F. and Spanily, V., 'Rôle of Anion Adsorption During Pitting Corrosion and Corrosion Cracking of Metals', *Zashch. Metal.*, **5**, 371 (1962); *C.A.*, **71**, 97654t
Rozenfeld, I. L. and Danilov, I. S,, 'Pitting Tendency of Stainless Steels', *Tr. Vses. Mezhvuz. Nauchn. Konf. po Vopr. Bor'by s Korrozici* (Moscow: Gos. Zzd. Neft i Gorno-Toplivn Prom.), 18 (1962); *C.A.*, **60**, 3806f
Schwenk, W., 'Diskussion uber die Ursache und Stabilitat der Lochfrasskorrosion in Zusammenlang mit dem 'Alles-oder-Nichts' Gesetz der Passivitat und Einen Kinetichen Modell', *Corros. Sci.*, **3**, 107 (1963)
Defranoux, J. M., 'Methodes Electrochimiques d'Etude du Compartement des Aciers Inoxydables en Milieu Chlorure', *Corros. Sci.*, **3**, 75 (1963)
Harzbecker, H., 'Pitting Corrosion in Steam Generators', *Energietechnic*, **13**, 502 (1963); *C.A.*, **64**, 9434a
Tomashov, N. D., Chernova, G. P. and Markova, O. N., 'Effect of Alloying Elements on the Tendency of Stainless Chromium–Nickel Steels to Pitting Corrosion', *Sa. Korroziya Metal i Splavov*, 73 (1963); *C.A.*, **60**, 3805b

Freiman, L. I. and Kolotyrkin, Ya. M., 'Pitting Corrosion of Iron by Perchlorate Ions', *Dokl. Akad. Nauk. SSSR*, **153**, 886 (1963); *C.A.*, **60**, 7732e

Roedeker, W. and Friche, W., 'An Experimental Investigation into the Pitting of Various Types of Steel Plate', *Blech*, **11**, 302 (1964); *C.A.*, **61**, 9230f

Rozenfeld, I. L. and Danilov, I. S., 'Pitting Corrosion of Passivated Stainless Steel', *Z. Physik. Chem.*, **226**, 257 (1964); *C.A.*, **61**, 13005d

Uhlig, H. H. and Gilman, J. R., 'Inhibition of Pitting Corrosion of Stainless Steel 18/8 in Iron(III) Chloride Solutions by Nitrates', *Z. Physik. Chem.*, **226**, 127 (1964); *C.A.*, **61**, 9231c

Fisher, W. R., 'Pitting Corrosion, Especially of Titanium. 1: Corrosion Studies', *Techn. Mitt. Krupp, Forschber*, **22**, 65 (1964); *C.A.*, **62**, 15859c

Kazinachei, B. Ya., Balashova, N. N. and Shuvalova, M. A., 'Control of Anti-pitting Additives in Nickel Bath', *Sb. Statei Vses. Zaochn. Politechn. Ins.*, **32**, 52 (1964); *C.A.*, **62**, 15859

Storchai, E. I. and Turkovskaya, A. V., 'The Pitting Corrosion of Aluminium Alloys', *Zashch. Metal*, **1**, 293 (1965); *C.A.*, **63**, 7999c

Schwenk, W., 'Diskussion der Eigenschaften des Lochfrass-potentials Nichtnostender Stahle in Zusemmenhang mit Seinen Bestimmungsmethoden', *Corros. Sci.*, **5**, 245 (1965)

Blauchet, 'The Pitting Attack of Magnesium and its Alloys', *Comm. Energie At. Rappt. CEA-R*, 2815 (1965); *C.A.*, **65**, 1761e

Morioka, S., Sawanda, Y. and Shisbara, K., 'Effect of SO_2 Gas on Pitting of Austenitic Stainless Steel', *Boshoku Gijutsu*, **14**, 535 (1965); *C.A.*, **64**, 19086h

Fokin, M. N., Kurtepov, M. M. and Bochkareva, E. F., 'Investigation of Pitting and Crevice Corrosion of Stainless Steels in Sea-water', *Korroziya Metal i Splavov*, **2**, 35 (1965); *C.A.*, **64**, 122820

Waight, F. H., 'Fatigue Pitting in Gears. Programme, Results and Discussion of the Results', *Erdoel Kohle*, **18**, 976 (1965) and **18**, 899 (1965); *C.A.*, **64**, 10877d

Blanchet, J. and Coriou, H., 'Electrochemical Methods for the Study of Corrosion, Experimental Techniques and Industrial Applications, Electrochemical Study of Corrosion by Pitting of Magnesium', CEBELCOR Rappt. Tech. No. 135, 5 (1965); *C.A.*, **64**, 9233a

Herbsleb, G., 'The Inhibition of Chloride Corrosion (Pitting) on Chemical-resistant Steels by Nitrate, Sulfate and Chromate Ions, and the Methods for Determining the Pit Corrosion Potential', *Werkst. Korros.*, **16**, 929 (1965); *C.A.*, **64**, 4702c

Vetter, K. J., 'Thermodynamic Theory of Corrosion Pitting', *Ber. Bunsenges. Physik. Chem.*, **69**, 683 (1965); *C.A.*, **64**, 3018f

van Muylder, J., Pourbaix, M. and van Lear, P., 'Electrochemical Characteristics of Pits from the Corrosion of Copper in the Presence of H_2O and Aqueous Chloride Solutions', CEBELCOR Rappt. Tech. No. 127, 26 (1965); *C.A.*, **64**, 3018c

Szklarska-Smialowska, Z., 'Pitting Corrosion of Iron in Ammonium Nitrate Solution Containing Chloride Ions', *Bull. Acad. Polon. Sci., Ser. Ser: Chim.*, **13**, 221 (1965); *C.A.*, **63**, 6662h

Golovina, G. V., Florianovich, G. M. and Kolotyrkin, Ya. M., 'Kinetics of the Initial Stages of Activation of Iron-Chromium Alloys by Halogen Ions', *Elektrokhimiya*, **1**, 12 (1965); *C.A.*, **63**, 302h

Van Lear, P., Van Muylden, J., de Zoubov, N. and Pourbaix, M., 'Accelerated Methods for Estimating the Risk of Pitting of Copper in the Presence of Water. Application to the Study of the Influence of a Copper Contact with Another Metal or with Graphite', CEBELCOR Rappt. Tech. No. 128, May (1965)

Tousek, J., Cihal, V. and Prazak, M., 'Die Einwirkung von Halogenionen auf die Korrosion der durch Molybdan Modifizurten Chrom-Nickel-Stahle', *Corros. Sci.*, **6**, 105 (1966)

Hubner, W., 'Pitting Corrosion of Passivated Metals', *Svensk Kem. Tidstr.*, **78**, 321 (1966); *C.A.*, **66**, 25356w

Hatch, G. B., 'Maximum Self-generated Anodic Current Density as an Inhibitor Pitting Index', Ill. State Water Surv., Circ. No. 91, 24 (1966); *C.A.*, **66**, 81814f

Herbsleb, G., 'Pitting Corrosion on Metals with Electron-conductive Passive Layers', *Werkst. Korros.*, **17**, 649 (1966); *C.A.*, **66**, 5337m

Freiman, L. I. and Kolotyrkin, Ya. M., 'Pitting Corrosion of Aluminium in Solutions of Sodium Perchlorate and Perchloric Acid', *Zashch. Metal*, **2**, 488 (1966); *C.A.*, **65**, 19674d

Novakovskii, V. M. and Sorokina, A. N., 'Comparative Electrochemistry of Stress Corrosion and Pitting of Stainless Steels in Chloride Solutions', *Zashch. Metal*, **2**, 416 (1966); *C.A.*, **65**, 18152g

Willert, H., 'Investigations of the Corrosion Resistance of a 13%Cr Steel in Aqueous Salt Solutions', *Neue Huette*, **11** 72 (1966); *C.A.*, **65**, 14954b

Tousek, J., 'Passivity of Nickel', *Collection Czech. Chem. Commun.*, **31**, 3083 (1966); *C.A.*, **65**, 10116h

Rozenfel'd, I. L. and Danilov, I. S., 'Electrochemistry of Pitting Corrosion. 1: Pit Formation During Spontaneous Dissolution of Stainless Steels', *Zashch. Metal.*, **2**, 134 (1966); *C.A.*, **65**, 10114a

Bond, A. P., Bolling, G. F. and Domian, H. A., 'Microsegregation and the Tendency for Pitting Corrosion in High-purity Aluminium', *Ber. Bunsenges. Physik. Chem.*, **70**, 773 (1966); *C.A.*, **65**, 8510b

Kuzub, V. S. and Kachanov, V. A., 'Possibility of Anodic Protection of Stainless Steels from Chloride Pitting in Dilute Nitric Acid', *Zashch. Metal.*, **2**, 358 (1966); *C.A.*, **65**, 6716f

Rickert, H. and Holzaepfel, G., 'The Coexistence of Active and Passive Zones on an Electrode Surface', *Werkst. Korros.*, **17**, 376 (1966); *C.A.*, **65**, 3322f

Pourbaix, M., 'Effect of Chlorides on the Behaviour of Metals and Alloys in the Presence of Aqueous Solutions', *Symp. Coupling Basic Appl. Corross. Res. Dialogue*, 67 (1966); *C.A.*, **72**, 50238r

Pourbaix, M., Van Muylder, J. and Van Laer, P., 'Sur la Tension d'Electrode du Cuivre en Presence d'Eau de Bruxelles; Influence de la Lumière et des Conditions de Circulation de l'Eau', *Corros. Sci.*, **7**, 795 (1967)

Takenori, N., Norio, S., Tatsuo, I. and Okamoto, G., 'Electrochemical Test for Pitting Corrosion in Stainless Steels', *Hakkaido Daigaku Kogakubu Kenkyu Hokoku*, **44**, 1 (1967); *C.A.*, **70**, 16534h

Izaki, T. and Arai, K., 'Pitting Corrosion of Aluminium and its Alloys in Neutral Solutions', *Nippon Kinzoku Gakkaishi*, **31**, 1023 (1967); *C.A.*, **68**, 26190r

Herbsleb, G. and Engell, H. J., 'Pitting Corrosion of Passive Iron in Sulphuric Acid Containing Chloride Ions', *Werkst. Korros.*, **17**, 365 (1966); *C.A.*, **67**, 104517y

Herbsleb, G. and Schwenk, W., 'A Pitting Corrosion Indicator Test on Chromium and Cr–Ni Steels in Solutions Containing Chloride and Bromide', *Werkst. Korros.*, **18**, 685 (1967); *C.A.*, **67**, 87126a

Defranoux, J. M., 'Sur le Compartement des Aciers Inoxydables en Presence d'Eau de Mor Froide et Chaude', *Corros. Sci.*, **8**, 245 (1968)

Garz, I., Worch, H. and Schatt, W., 'Untersuchungen uber die Anodische Auflosung und die Lochfrapkorrosion von Nickel-Einkristallelektrochen', *Corros. Sci.*, **9**, 71 (1969)

Riskin, I. V. and Turkovskaya, A. V., 'Pitting Corrosion of Kh 18N10T Steel on a Rotating Disk Electrode', *Zashch. Metal*, **5**, 443 (1969); *C.A.*, 97645r

Graefen, H., 'Electrochemical Investigation of Localised Corrosion', *Ind. Chim. Belge*, **32**, 422 (1967); *C.A.*, **70**, 111038h

Kuznetsov, V. V. and Verzhbitskaya, L. V., 'Mechanism of Pitting Corrosion of Carbon Steel in Fresh Water', *Uch. Zap. Perm. Univ.*, **229**, 131 (1970); *C.A.*, 120661g

Yoshihiro, H., 'Influence of Crystal Structure on Corrosion', *Denki Kagoku*, **9**, 691 (1970); *C.A.*, **74**, 60040a

Vetter, K. J. and Strehblow, H. H., 'Formation and Shape of Pitting Corrosion Pits in Iron and Theoretical Conclusions on Pitting Corrosion', *Ber. Bunsenges Phys. Chem.*, **74**, 1024 (1970); *C.A.*, **74**, 8865j

Uhlig, H. H., 'Modern Look at Pitting Corrosion', *Boshoku Gijutsu*, **19**, 171 (1970); *C.A.*, **74**, 8835z

Parlapanskii, M. and Mutafchiev, I., 'Antipitting Action of Some Surface-active Agents', *God. Vissh. Khimikotekhnol Inst.*, **16**, 263 (1971); *C.A.*, **79**, 37944t

von Baeckmann, W., Gisen, F. and Funk, D., 'Breaks Caused by Corrosion Pitting', *Gesammelte Ber. Betr. Forch Ruhrgas AG.*, **19**, 25 (1971); *C.A.*, **76**, 120662h

Schatt, W. and Worch, H., 'Nahere Charakterisierung der bei Nickel-Einkristallen zum Lochfrass Fuhrenden Versetzungen', *Corros. Sci.*, **11**, 623 (1971)

Tinck, M., Petitjean, J. P., Van der Poosten, P. H. and Blave, A., 'Multipotentiostat for the Rapid Determination of the Pitting Potential of Stainless Steel', *Cent. Belge Éttude Corros. Rapp. Tech.*, **122**, 210 (1972); *C.A.*, **79**, 99730d

Kato, K. and Kono, T., 'Effect of Alloying Elements on Pitting Corrosion Resistance of 17%Cr–13%Ni–3%Mo (SOS316) Austenitic Stainless Steel', *Denki Seiko*, **43**, 229 (1972); *C.A.*, **79**, 34374w

Grassiani, M., 'Pit Corrosion of Austenitic Stainless Steels in Artificial Sea-water', *Trib. CEBEDEAU*, **25**, 515 (1972); *C.A.*, **79**, 8531a

Tousek, J., 'Die Aktivierende Wirkung der Nitrationen auf die Geschwiadigkeit der Metallauflosung im Passiven Zustand', *Corros. Sci.*, **12**, 799 (1972)

Tousek, J., 'Die Kinetik der Lochfrasskorrosion von Metallen', *Corros. Sci.*, **12**, 1 (1972)

Yashihiro, H., 'Electrochemistry of Pitting Corrosion', *Boshoku Gijutsu*, **21**, 503 (1972); *C.A.*, **78**, 143032h

Cherepakhova, G. L. and Shreider, A. V., 'Effect of the Ions of Cooling Waters on the Pitting Corrosion of Al-Mg Alloy', *Zh. Prik. Khim.*, **45**, 1958 (1972)

Tousek, J., 'Kinetics of the Pitting Corrosion of Iron and Nickel', *Czech. Chem. Commun.*, **37**, 1454 (1972); *C.A.*, **77**, 55490e

Tousek, J., 'Pitting Corrosion Mechanism', *Werkst. Korros.*, **23**, 109 (1972); *C.A.*, 120666n

Pourbaix, M., 'Significance of Protective Potential During Pitting Corrosion, Intergranular Corrosion, and Stress Cracking Corrosion', *Rev. Roum. Chim.*, **17**, 239 (1972); *C.A.*, **76**, 120554z

Boehni, H., 'Pit Corrosion of Metallic Materials and Detection Methods', *Schweiz. Arch. Angar. Wiss. Tech.*, **36**, 41 (1970); *C.A.*, **72**, 106498v

Tousek, J., 'Untersuchungen uber den Lochfrass von Ni und Fe der Durch Sulfationen Hervongerufen Wird', *Corros. Sci.*, **12**, 15 (1972)

Freiman, L. I., Lap Le, M. and Raskin, G. S., 'Rôle of Local Changes in Solution Composition During Pitting on Iron', *Zashch. Metal*, **9**, 680 (1973); *C.A.*, 43390w

Inoue, T. and Kato, M., 'Pitting Corrosion of Aluminium Using Macro- and Microautoradiography', *Keikinzoku*, **23**, 78 (1973); *C.A.*, **79**, 121179e

Brigham, R. J. and Tozer, E. W., 'Effect of Microstructure on Pitting of Stainless Steel', *Can. Met. Quart.*, **12**, 171 (1973); *C.A.*, **79**, 1174576

Brennert, S. and Eklund, G., 'Pitting Corrosion of Stainless Steel. Propagation Mechanism', *Scan. J. Met.*, **2**, 269 (1973); *C.A.*, **79**, 138952z

Sato, E., Tamura, T. and Okabe, T., 'Aluminium Anode for Cathodic Protection. VII: Pitting and Protection Potentials of Aluminium in NaCl Solution (Effects of pH)', *Kinzoku Hyomen Gijutsu*, **24**, 139 (1973); *C.A.*, **79**, 142225g

Ciolac, S., Vasilescu, E. and Stretcu, M., 'Pitting Corrosion of Metals. II: Characteristics of the Pitting Process and Methods of Study and Protection', *Stud. Cercet. Chim.*, **21**, 669 (1973); *C.A.*, **79**, 99678t

Ciolac, S., Vasilescu, E., Stretcu, M. and Joachimescu, O., 'Pitting Corrosion of Metals. Pitting–Passivity Correlation', *Stud. Cercet. Chim.*, **21**, 597 (1973)

Herbsleb, G. and Schwenk, W., 'Flow Dependence of the Pitting Corrosion of Cr-Ni Steel in NaCl Solution. 2: Tests with Ultrasonics', *Werkst. Korros.*, **24**, 267 (1973); *C.A.*, **79**, 56638n

El Din Shams, A. M., Bodran, M. M. and Khalil, S. E., 'Corrosion Behaviour of Manganese-containing Stainless Steel. 3: Their Susceptibility Towards Pitting Corrosion', *Werkst. Korros.*, **24**, 290 (1973); *C.A.*, **79**, 56642j

Imoi, H., Saito, Y., Kobayashi, M. and Fujiyama, S., 'Pitting-corrosion-resistant Chromium Stainless Steel', *Japan Kokai 7300*, 221 (1973); *C.A.*, **79**, 22569a

Sato, E., Tamura, T. and Okabe, T., 'Aluminium Anode for Cathodic Protection. 7: Pitting and Corrosion Potentials for Gallium in Sodium Chloride Solutions', *Kinzoku Hyomen Gijutsu*, **24**, 82 (1973); *C.A.*, **79**, 12792d

Freiman, L. I., Lap Le, M. and Kolotyrkin, J. M., 'Depassivation and Repassivation of Iron and Iron Alloys in Halide Solutions', *Z. Phys. Chem.*, **252**, 76 (1973); *C.A.*, **79**, 8530z

Freiman, L. I. and Lap Le, M., 'Limiting Potentials Which Determine the Resistance of Metals to Pitting Corrosion', *Z. Phys. Chem.*, **252**, 65 (1973); *C.A.*, **78**, 154238f

Janik-Czachor, M., 'Pitting Corrosion of Zinc Single Crystals', *Bull. Acad. Pol. Sci., Ser. Sci. Chim.*, **21**, 159 (1973); *C.A.*, **78**, 154240a

Herbsleb, G. and Schwenk, W., 'Electrochemical Study of the Pitting Corrosion of Corrosion-resistant Steels', *Werkst. Korros.*, **24**, 763 (1973); *C.A.*, **88**, 21970h

Poatsch, W., 'Optical Studies of the Mechanism of Pitting Corrosion', *Ber. Bunsenges Phys. Chem.*, **77**, 895 (1973); *C.A.*, **80**, 90201v

Zhurin, A. I., Kosmynin, A. I. and Vlasenko, O. B., 'Corrosion of Aluminium Cathodes During the Electrodeposition of Zinc', *Izv. Vyssh. Ucheb. Zaved. Tsvet. Met.*, **16**, 71 (1973); *C.A.*, **80**, 77471p

Worch, H., Garz, I. and Schott, W., 'Influence of Surface Structure and Chemical Composition on the Pitting Corrosion of Nickel Single Crystals', *Werkst. Korros.*, **24**, 872 (1973); *C.A.*, **80**, 90185

1.7 Bimetallic Corrosion

This section is concerned with the kinetics of reactions occurring when dissimilar metals are in direct electrical contact in corrosive solutions or atmospheres. Under these conditions, enhanced and aggressive corrosion of the more negative member of the bimetallic couple can be experienced together with partial or complete cathodic protection of the more positive metal*. In many cases the flow of current $I_{galv.}$ resulting from the bimetallic contact is intense and much greater than that experienced from variations of oxygen content over the surface of a single metal (differential aeration) or from variations of metal-ion concentration at different locations on a single metal (concentration-cell corrosion). Bimetallic corrosion is qualitatively well understood, but since it is a highly complex process, it has proved difficult to handle its kinetics in a quantitative fashion. Furthermore, it must be emphasised that metal surfaces are seldom unfilmed so that the kinetics of electrode reactions will be affected by the film on the metal surface.

General Theory

The factors that are of importance in the enhancement of the corrosion rate of one metal when it is in direct electronic contact with another (*cf.* cathodic protection where contact is by conducting wire) are as follows.

1. The *corrosion potentials* of the metals M_A and M_B forming the couple under the environmental conditions that prevail in practice.
2. The nature and kinetics of the cathodic reaction at the surface of the more positive metal and the nature and kinetics of the anodic reaction at the surface of the more negative metal.
3. The relative areas of M_A and M_B (*see* page **1.82**).
4. The nature and the conductivity of the electrolyte solution.

The corrosion potentials of the two metals in the environment under consideration will determine the direction of the transfer of electrons, but will provide no information on the rate of electron transfer, i.e. the magnitude of the galvanic current $I_{galv.}$. Thus if $E_{corr.,A}$ is more positive than $E_{corr.,B}$ the transfer of electrons will be from M_B to M_A with a consequent increase in the corrosion potential (more positive) of M_B and a decrease in that of M_A; the corrosion rate of M_B will consequently increase and the corrosion rate of M_A will decrease compared with the rates when the metals

*This is utilised in the cathodic protection of metals using sacrificial anodes (*see* Section 10.2).

are uncoupled. However, it must be emphasised that the difference in the corrosion potentials provides no information on the kinetics of bimetallic corrosion, a fact that may be illustrated by considering the couples Mg–Hg and Mg–Pt immersed in sea-water. The corrosion potentials (vs. S.H.E.) of the metals under consideration are Mg $\approx -1\cdot 0$ V, Pt $\approx 0\cdot 0$ V and Hg $\approx 0\cdot 0$ V, so that the e.m.f.s are the same, i.e. about $1\cdot 0$ V. However, whereas coupling platinum to magnesium results in a pronounced increase in the corrosion rate of the latter, the effect produced by coupling magnesium and mercury is insignificant. The reason for this apparent anomaly is that magnesium in sea-water corrodes by the overall reaction

$$Mg + 2H_2O \rightarrow Mg(OH)_2 + H_2$$

in which the cathodic reaction is reduction of water to hydrogen gas. The rate of corrosion is thus determined by the rate of the hydrogen evolution reaction which occurs with kinetic ease at a platinum surface ($i_0 \approx 10^{-3}$ Acm^{-2}), but with extreme difficulty at a mercury surface ($i_0 \approx 10^{-11}$ Acm^{-2}). Thus the platinum provides an additional cathode at which hydrogen evolution is kinetically easy, and since $I_c = I_a$ (see Section 1.4) the corrosion rate of the magnesium is increased.

Graphic estimation of the corrosion rate and corrosion potential of a metal immersed in a corrosive high-conductivity electrolyte, from the intersection of the polarisation curves for the appropriate anodic and cathodic reactions, has been proposed and explained by several authorities[1,2]. These polarisation curves can be further used to illustrate the effect of imposing additional anodic or cathodic potentials on to a corroding metal (see also Sections 1.4 and 10.1).

Figure 1.62a shows the effect of depressing the potential of a metal corroding in a solution of high electrolytic conductivity. As the potential is progressively depressed, the relation between potential and total cathodic current is defined by the curve CBD, which is a typical cathodic polarisation curve for reduction of dissolved oxygen, of which the section CB is the cathodic polarisation curve for local-cell corrosion in the absence of an applied potential. As the potential is depressed along the curve BD by the increasing polarising current yz, $y'z'$, $y''z''$, the corrosion current (rate) of the metal is given by the progressively decreasing horizontal intercept of the anodic polarisation curve, i.e. xy, $x'y'$, $x''y''$, etc. When the potential of the metal is depressed below A, complete cathodic protection results and the original anodic sites now support the cathodic reduction of oxygen, and of metal ions, as shown by curve AE. At potentials between B and A, partial cathodic protection is obtained, with the corrosion rates decreasing as the potentials approach A.

Figure 1.62b shows the result of raising the potential of a corroding metal. As the potential is raised above B, the current/potential relationship is defined by the line BD, the continuation of the local cell anodic polarisation curve, AB. The corrosion rate of an anodically polarised metal can very seldom be related quantitatively by Faraday's law to the *external* current flowing, $I_{galv.}$[3,4]. Instead, the measured corrosion rate will usually exceed that rate calculated from the external current flow by an amount that will be designated as the local-cell corrosion rate of the anode, i.e. xy, $x'y'$, $x''y''$, which are the horizontal current intercepts, at the appropriate potential, of the cathodic polarisation curve CB (Fig. 1.62b). Thus the higher the potential of the anodically polarised metal, the smaller will be the local-cell corrosion rate of the anode and the more closely will the total corrosion rate of the anode approach that calculated from the flow of current in the external circuit. Since the slope of the cathodic polarisation curve is usually much greater than that of the anodic polarisation curve, it follows that the local-cell corrosion rate should be relatively little affected by variations of galvanic *current*.

When two dissimilar metals of different corrosion potentials are coupled in the same solution, the more positive metal will serve as a cathode, its

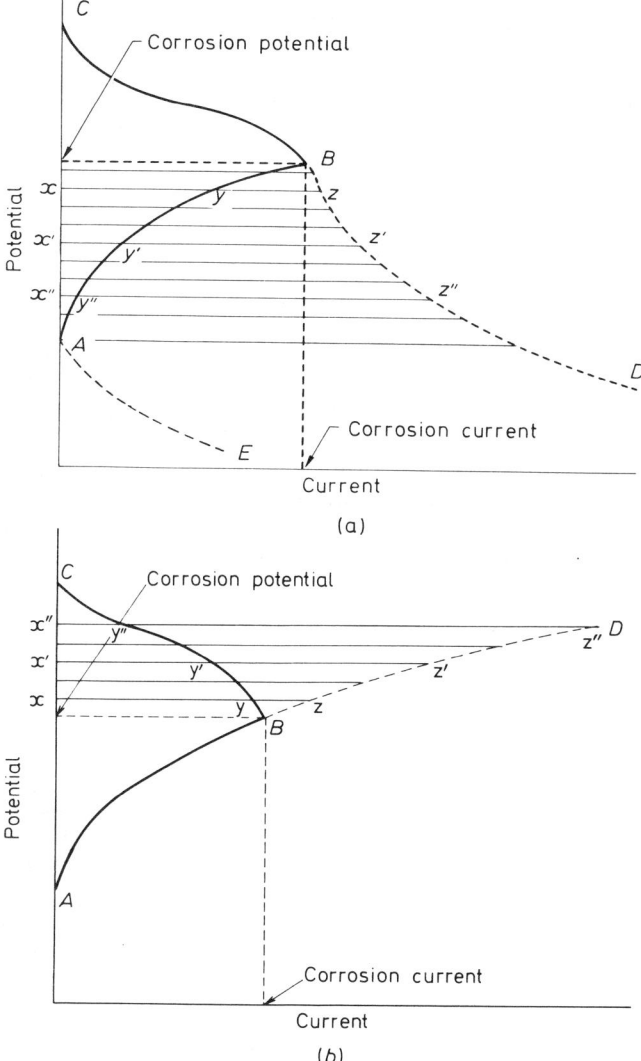

Fig. 1.62 Potential/current curves for a metal polarised (*a*) cathodically and (*b*) anodically. The horizontal intercepts xy, $x'y'$, $x''y''$ with AB and CB represent the local cell currents respectively, and yz, $y'z'$, $y''z''$ the externally applied currents (cathodic and anodic). In bimetallic corrosion yz, $y'z'$, etc. will be the galvanic current $I_{\text{galv.}}$ flowing from M_B to M_A (see page 1.192)

characteristic uncoupled corrosion rate in the same solution being partly or completely retarded as shown in Fig. 1.62a. The more negative metal will suffer enhanced corrosion, its total corrosion rate being given by the sum of the rate due to the flow of current in the external circuit and the potential-dependent local-cell corrosion rate as shown in Fig. 1.62b. If the areas of the two metals are comparatively small and if the solution has a high conductivity the effect of the transfer of electrons $I_{galv.}$ will be to make the whole surface equipotential, but this will not be so if the areas are large or if the solution has a high resistivity. In this connection (see Section 1.4) it should be emphasised that the effective areas of the metals constituting the bimetallic couple are not necessarily the same as the geometrical areas, and in a solution of high resistivity only the areas of the metals adjacent to their interface will be significant in relation to the magnitude of $I_{galv.}$ and the anodic current density at the more negative metal of the couple. For a discussion on mathematical modelling of bimetallic corrosion distribution, the reader is referred to the section entitled Distribution of Bimetallic Corrosion in Real Systems, p. 1:238.

Potentials

It is evident that the standard electrode potentials of M^{z+}/M equilibria, as given in the e.m.f. series of metals, are quite irrelevant in predicting the effect of coupling two dissimilar metals. The galvanic series of metals (see Table 21.8) in which metals are listed according to their corrosion potentials in chloride solutions, are more helpful, but these too can be misleading when other environments are involved, and when data are not available in tables or in the literature it is necessary to determine the corrosion potentials of the metals under the environmental conditions that prevail in practice. Furthermore, the magnitude of the galvanic current is dependent only in part on the difference in corrosion potentials, being also strongly influenced by the slope of the cathodic polarisation curve of the more positive metal, and/or that of the anodic polarisation curve of the less positive metal. Films of oxide or insoluble reaction products on the anode or the cathode also introduce

Table 1.23 Weight loss of iron for Fe/M bimetallic couples in 1% NaCl*

M	Weight loss of iron (mg)	Weight loss of M (mg)	Difference in standard electrode potentials[6] V†
Copper	183·1	0·0	+0·785
Nickel	181·1	0·7	+0·19
Tin	171·1	2·5	+0·30
Lead	183·2	3·6	+0·31
Aluminium	9·8	105·9	−1·23
Cadmium	0·4	307·9	+0·04
Zinc	0·4	688·0	−0·32
Magnesium	0·0	3104·0	−1·90

* Weight loss data after Bauer and Vogel[5].
† The + sign indicates that iron has a more negative standard electrode potential than the second metal.

important kinetic and resistive effects which have a major influence on the magnitude of the galvanic current flow and which are not accounted for by standard electrode potentials. Some of the foregoing factors are illustrated by Table 1.23 which contains early information developed by Bauer and Vogel[5] on the weight losses of iron and various dissimilar metals coupled together in 1% sodium chloride solution containing dissolved oxygen.

Even discounting the case of aluminium, which is usually covered by protective oxide films, it is evident from Table 1.23 that the quantitative connection between the galvanic corrosion rate of the more active member of the couple and the difference of reversible potentials of the two metals, is non-existent.

This situation has encouraged the empirical approach of using the difference in *corrosion* potentials of dissimilar metals in a single solution (usually sea-water or sodium chloride) as a means of assessing the anticipated intensity of galvanic corrosion of the less positive metal in the same solution. This approach has sometimes been carried to the extreme of setting a maximum value of $0 \cdot 25$ V difference in corrosion potential in sea-water as representing the safe limit for elimination of greatly aggravated corrosion of the less positive member of the couple. Such a criterion clearly has little or no technical significance, not only because it ignores polarisation effects, but also because irreversible corrosion potentials cannot be subtracted in the fashion of reversible half-cell potentials (*see* page **1.59**). The fallacy of employing corrosion potential difference as a means of quantifying or even predicting the severity of bimetallic corrosion has been emphasised by a number of authors[7-16].

Table 1.24 Grouping of compatible metals and alloys

Group I	Magnesium and its alloys*
Group II	Cadmium, zinc and aluminium, and their alloys
Group III	Iron, lead and tin, and their alloys (except stainless steels)
Group IV	Copper, chromium, nickel, silver, gold, platinum, titanium, cobalt and rhodium, and their alloys; stainless steels; graphite

*The studies of Bothwell[79] would suggest that super-purity aluminium could be placed in Group I.

Galvanic corrosion rates can, of course, be reliably obtained by direct measurement using zero-resistance ammetry or weight-loss determinations (*see* below) or can be estimated in various electrolytes if the corrosion potentials of the dissimilar metals and the slopes of the appropriate anodic and cathodic polarisation curves are known[4, 11, 12, 16, 17]. Unfortunately, the slopes of the anodic and cathodic polarisation curves are strongly dependent on the particular metal and on the composition of the electrolyte. Non-stable oxide or reaction-product films, which may either thin or thicken during galvanic corrosion, also impose variable resistive effects which make the practical prediction of galvanic corrosion rates uncertain. Accordingly, it is hardly surprising that a combination of practical experience and less frequent quantitative investigation often dictates the choice of compatible metals in different environments. Promisel and Mustin (Table 1.24[18]) and Evans and Rance (Table 1.25[19]) have published reliable classifications of compatible and incompatible metals. The type of information in Table 1.25 has now

Table 1.25 Degree of corrosion at bimetallic contacts*

Metal considered	Contact metal 1 Gold, platinum, rhodium, silver	2 Monel, Inconel, nickel-molybdenum alloys	3 Cupronickels silver solder, aluminium bronzes, tin bronzes, gunmetals	4 Copper brasses, 'nickel silvers'	5 Nickel	6 Lead, tin and soft solders	7 Steel and cast iron
1. Gold, platinum, rhodium, silver	—	A	A	A	A	A	A
2. Monel, Inconel, nickel-molybdenum alloys	B	—	A	A	A	A	A
3. Cupronickels, silver solder, aluminium bronzes, tin bronzes, gun metals	C(k)	B or C	—	A	A	A	A
4. Copper, brasses, 'nickel silvers'	C(k)	B or C	B or C(g)	—	B or C	B or C(p)	A
5. Nickel	C	B	A	A	—	A	A
6. Lead, tin and soft solders	C	B or C(t)	B or C(q)	B or C(q)	B	—	A or C(r)
7. Steel and cast iron (a) (f) (w)	C	C	C	C	C(k)	C(k)	—
8. Cadmium (u)	C	C	C	C	C	B	C
9. Zinc (u)	C	C	C	C	C	B	C
10. Magnesium and magnesium alloys (chromated) (b) (a)	D	D	D	D	D	C	D
11. Austenitic Fe-18Cr-8Ni	A	A	A	A	A	A	A
12. Stainless steel Fe-18Cr-2Ni	C	A or C(s)	A or C(s)	A or C(s)	A	A	A
13. 13% Cr	C	C	C	C	B or C	A	A
14. Chromium	A	A	A	A	A	A	A
15. Titanium	A	A	A	A	A	A	A
16. Aluminium and aluminium alloys (n) (a) (w)	D	C	D(e)	D(e)	C(k)	B or C	B or C

*Based on data provided by members of the I.S.M.R.C. Corrosion and Electrodeposition Committee and others, and arranged by Mrs. V. E. Rance.
A. The corrosion of the 'metal considered' is not increased by the 'contact metal'.
B. The corrosion of the 'metal considered' may be slightly increased by the 'contact metal'.
C. The corrosion of the 'metal considered' may be markedly increased by the 'contact metal'. (Acceleration is likely to occur only when the metal becomes wet by moisture containing an electrolyte, e.g. salt, acid, combustion products. In ships, acceleration may be expected to occur under in-board conditions, since salinity and condensation are frequently present. Under less severe conditions the acceleration may be slight or negligible.)
D. When moisture is present, this combination is inadvisable, even in mild conditions, without adequate protective measures.

8	9	10	11	12	13	14	15	16
Cadmium	Zinc	Magnesium and magnesium alloys (chromated)	Stainless Austenitic Fe–18Cr–2Ni	Fe–18Cr–2Ni	13% Cr	Chromium	Titanium	Aluminium and aluminium alloys
A	A	A	A	A	A	A	A	A
A	A	A	A	A	A	A	A(x)	A
A	A	A	B or C	B	A	B or C	B or C	A(e)
A	A	A	B or C	B or C	A	B or C	B or C	A(e)
A	A	A	B or C	B or C	A	B or C	B or C	A
A	A or C(r)	A	B or C	B or C	B or C	B or C	B or C	A
A(m)	A(m) (l)	A	C	C	C	C(k)	C	A(m)
—	A	A	C	C	C	C	C	B
B	—	A	C	C	C	C	C	C(j)
B or C	B or C	—	C	C	C	C	C	B or C(c)
A	A	A	(v)	A	A	A	A	A
A	A	A	A	(v)	A	A	(o)	A
A	A	A	C	C	(v)	C	C	A
A	A	A	A	A	A	—	A	A
A	A	A	A	A	A	A	—	A
A	A	A(c) (h)	B or C	B or C	B or C	B or C(d)	C	(v)

been further refined and extended by a committee meeting under the auspices of the British Standards Institute[20]. Reference 20 gives detailed guidance on bimetallic corrosion effects for both aqueous and atmospheric conditions. The most commonly considered electrolyte is sea-water because highly aggravated corrosion of less noble metals commonly occurs in this environment.

Degree of Corrosion at Bimetallic Contacts

Table 1.25 is reproduced from *Corrosion and its Prevention at Bimetallic Contacts* (H.M.S.O., London, 1958) by permission of the Director of Publications. The reader should note that it is recommended that the Table be used only in conjunction with the Introduction to the original publication.

Notes to Table 1.25

(a) The exposure of iron, steel, magnesium alloys and unclad aluminium–copper alloys in an unprotected condition in corrosive environments should be avoided whenever possible even in the absence of bimetallic contact.

(b) The behaviour of magnesium alloys in bimetallic contacts is particularly influenced by the environment, depending especially on whether an electrolyte can collect and remain as a bridge across the contact. The behaviour indicated in the table refers to fairly severe conditions. Under conditions of total immersion or the equivalent, magnesium alloys should be electrically insulated from other metals. In less severe conditions complete insulation is not necessary, but steel, brass and copper parts should be galvanised or cadmium plated, and jointing compound (D.T.D. 369A) used during assembly. Under conditions of good ventilation and drainage, contacts classified as *D* have given satisfactory service, e.g. brass and steel push-fit and cast-in inserts in magnesium castings.

(c) Where contact between magnesium alloys and aluminium alloys is necessary, adverse galvanic effects will be minimised by using aluminium alloys containing little or no copper (0·1% max.).

(d) If in contact with thin (decorative) chromium plate, the symbol is *C*, but with thick plating (as used for wear resistance) the symbol is *B*.

(e) When contacts between copper or copper-rich materials and aluminium alloys cannot be avoided, a much higher degree of protection against corrosion is obtained by first plating the copper-rich material with tin or nickel and then with cadmium, than by applying a coating of cadmium of similar thickness.

(f) The corrosion of mild steel may sometimes be increased by coupling with cast iron, especially when the exposed area of the mild steel is small compared with the cast iron.

(g) Instances may arise in which corrosion of copper or brasses may be accelerated by contact with bronzes or gunmetals, e.g. the corrosion of copper sea-water-carrying pipelines may be accelerated by contact with gunmetal valves, etc.

(h) When magnesium corrodes in sea-water or certain other electrolytes, alkali formed at the aluminium cathode may attack the aluminium.

(j) When it is not practicable to use other more suitable methods of protection e.g. spraying with aluminium, zinc may be useful for the protection of steel in contact with aluminium, despite the accelerated attack upon the coating.

(k) This statement should not necessarily discourage the use of the 'contact metal' as a coating for the 'metal considered', provided that continuity is good; under abrasive conditions, however, even a good coating may become discontinuous.

(l) In most supply waters at temperatures above about 60°C, zinc may accelerate the corrosion of steel.

(m) In these cases the 'contact metal' may provide an excellent protective coating for the 'metal considered', the latter usually being electrochemically protected at gaps in the coating.

(*n*) When aluminium is alloyed with appreciable amounts of copper it becomes more noble and when alloyed with appreciable amounts of zinc or magnesium it becomes less noble. These remarks apply to bimetallic contacts and not to inherent corrosion resistance. Such effects are mainly of interest when the aluminium alloys are connected with each other.

(*o*) No data available.

(*p*) In some immersed conditions, the corrosion of copper or brass may be seriously accelerated at pores or defects in tin coatings.

(*q*) In some immersed conditions there may be serious acceleration of the corrosion of soldered seams in copper or copper alloys.

(*r*) When exposed to the atmosphere in contact with steel or galvanised steel, lead can be rapidly corroded with formation of PbO at narrow crevices where the access of air is restricted.

(*s*) Serious acceleration of corrosion of Fe-18Cr-2Ni stainless steel in contact with copper or nickel alloys may occur at crevices where the oxygen supply is low.

(*t*) Normally the corrosion of lead-tin soldered seams is not significantly increased by their contact with the nickel-base alloys, but under a few immersed conditions the seams may suffer enhanced corrosion.

(*u*) The corrosion product on zinc is, in certain circumstances, more voluminous and less adherent than that on cadmium. Where this is known to be the case, it should be borne in mind in making a choice between these two metals.

(*v*) These joints are liable to corrosion in crevices where these are not filled with jointing compound (*see* para. 8 of the Introduction).

(*w*) Corrosion products from iron or steel reaching aluminium, or corrosion products from aluminium reaching iron or steel, may sometime cause serious local corrosion through oxygen screening or in other ways, even when the total destruction of metal is finished.

(*x*) The corrosion of Monel can be increased under immersed sea-water conditions.

The realisation that the current flowing between the two members of a bimetallic couple, rather than their potential difference, provides a more realistic measure of the degree of hazard involved, is inherent from the discussion on polarisation curves already presented earlier in this chapter. Because, however, the traditional method of measurement of current flow necessarily involves a series resistance insertion between the coupled members, this affects the behaviour of the couple, and valid galvanic-current-flow measurement has only become possible with the development of zero-resistance ammeters and potentiostats. Although the first zero-resistance ammeter was developed in 1951[21] they did not come into more general use until about 1970[22]. Fully automatic balancing zero-impedance ammeters are now available commercially and their use has enabled a more accurate assessment of the degree of galvanic hazard to be made. Using this technique, a fresh assessment has been made of the galvanic hazard involved when a range of materials of construction used in marine engineering are coupled in the presence of sea-water[8,9]. A summary of some of the results of this investigation is provided in Table 1.26, in which the degree of acceleration of corrosion is expressed by means of a numerical factor which should be used with the uncoupled rate of corrosion of each metal or alloy in sea-water, which is obtained under the same experimental conditions. Data shown in the table, and those given in References 8 and 9, are based on both short-term (100 hours) zero-resistance ammetry[8,9] and long-term (*ca.* one year) weight-loss studies[8,9,23-25]. Account has therefore been made of the effects of long-term film deposition and local-cell corrosion. Even such a sophisticated treatment does not produce a faultless assessment of the hazard in every circumstance. In practice, uncertainty is almost inevitable in systems subject to varying speeds of liquid flow[26], and where anyway deep crevices may exist between coupled metals. In addition, uncertainty

Table 1.26 Total corrosion rate accelerration factors due to dissimilar metal coupling at 1:1 area ratio in flowing sea-water (mean flow rate about 1-2 m/s)

Coupled metal (wrought form)	Uncoupled corrosion rate* (mm/y)	Acceleration factor† due to dissimilar metal coupling with:	
		Titanium	Mild steel
Zinc	0·05	4	10
SIC aluminium	0·008	30	60
Mild steel	0·15	2	1
Lead	0·01	3	3 MS
2% aluminium brass	0·01	3	3 MS
10% aluminium bronze	0·02	3	3 MS
Nickel aluminium bronze	0·015	1	3 MS
Copper	0·03	6	3 MS
90/10 cupro-nickel (1% Fe)	0·02	3	3 MS
Monel 400	0·005	2	3 MS
Stainless steel type 316	0·005	2	3 MS

*Corrosion rates relate to general corrosion only and are average rates obtained over about one year's exposure.
†Acceleration factors quoted for coupled metal corroding unless indicated by suffix (MS) for mild steel corroding.

will attach to the cathode to anode area ratio in a real system (see Distribution of Bimetallic Corrosion in Real Systems, p. 1.238).

Bimetallic Corrosion in Aqueous Solutions

Dissolved Oxygen

Most cases of practical bimetallic corrosion in solutions occur under conditions when the solution contains dissolved oxygen. Accordingly, the primary cathodic reaction is the reduction of dissolved oxygen

$$O_2 + 4H^+ + 4e \rightarrow 2H_2O$$

at the surface of the cathodic metal. The maximum rate at which dissolved oxygen, at a given bulk concentration and pH, can be cathodically reduced is clearly limited by the rate at which the oxygen can diffuse from the bulk electrolyte to the surface of the cathode. Accordingly, in static solutions a maximum concentration and temperature dependent diffusion rate exists, which often limits the galvanic current, and the rate of bimetallic corrosion is then under cathodic diffusion control. This situation is quite common in practical bimetallic corrosion. Although i_L should be independent of the nature of the metal (c.f. the hydrogen evolution reaction) it will be seen that the noble metals are far more efficient cathodes than metals filmed with unreducible corrosion products.

It also follows that if the solution is stirred the rate of arrival of oxygen at the cathode will be increased. This will result in a corresponding increase in the rate of bimetallic corrosion as is shown in Fig. 1.63 for the aluminium-mild steel couple in stirred 1·0 N NaCl solution[27]. The increase in galvanic corrosion rate will be in the inverse relation to the slope of the anodic polarisation curve of the more negative metal, provided that the cathodic reaction is not totally diffusion controlled.

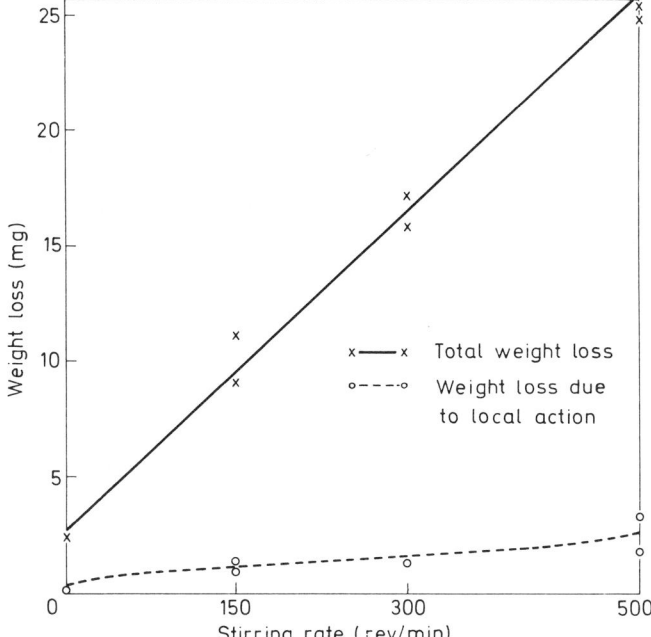

Fig. 1.63 Effect of stirring rate on the aluminium–mild steel couple in 1·0 N NaCl at 25°C. Equal areas (20 cm^2) of aluminium and steel coupled together for 24 h (after Pryor and Keir[27])

The rate at which dissolved oxygen can be reduced under fixed experimental conditions is also strongly dependent on the nature of the surface of the cathodic member of the couple. If the cathode carries no oxide films, or if any pre-existing oxide films are reduced by the flow of galvanic current, reduction of dissolved oxygen takes place readily; it follows that the noble metals Pt, Au, Ag and Cu will be efficient cathodes in bimetallic corrosion involving oxygen reduction as the cathodic reaction. If, however, the reduction of oxygen takes place on a stable oxide film on the cathode, the reaction often tends to be much more polarised. Figure 1.64 shows polarisation curves for zinc anodes coupled to copper and aluminium cathodes in 1·0 N NaCl. Any pre-existing oxide films on the copper will be cathodically reduced at very low current densities, as shown by Dyess and Miley[28], as follows:

$$Cu_2O + 2H^+ + 2e \rightarrow 2Cu + H_2O$$

and/or

$$CuO + 2H^+ + 2e \rightarrow Cu + H_2O$$

Accordingly, a bare copper surface is soon presented to the electrolyte. Aluminium, on the other hand, maintains an oxide-covered surface under these conditions, and it is evident from Fig. 1.64, which is constructed from the work of Pryor and Keir[27,29], that the reduction of dissolved oxygen is highly polarised and severely limits the galvanic current flow. Aluminium is

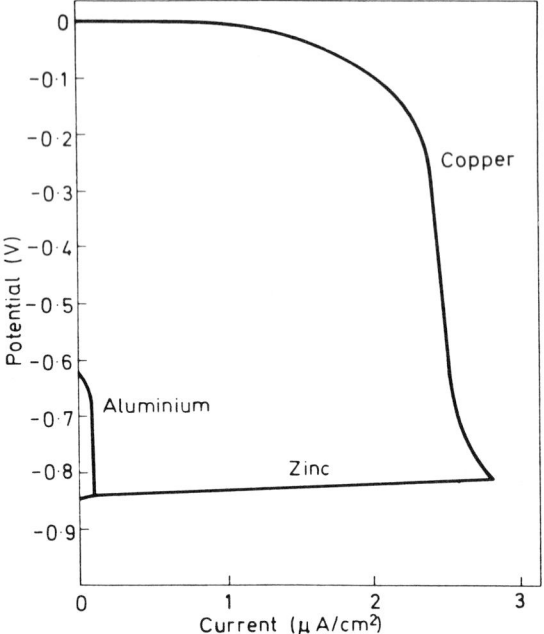

Fig. 1.64 Polarisation curves for zinc anodes coupled to aluminium and copper cathodes in 1·0 N NaCl at 25°C (potentials vs. S.H.e.) (after Pryor and Kier[27,29])

an extreme example of an oxide-covered cathode because the electronic resistance of the natural oxide film is so high that it permits only small isolated areas to act as effective cathodes at low current density[29].

The existence of oxide films of varying resistance on the surface of a cathode results in a corresponding variation of galvanic current flow. This

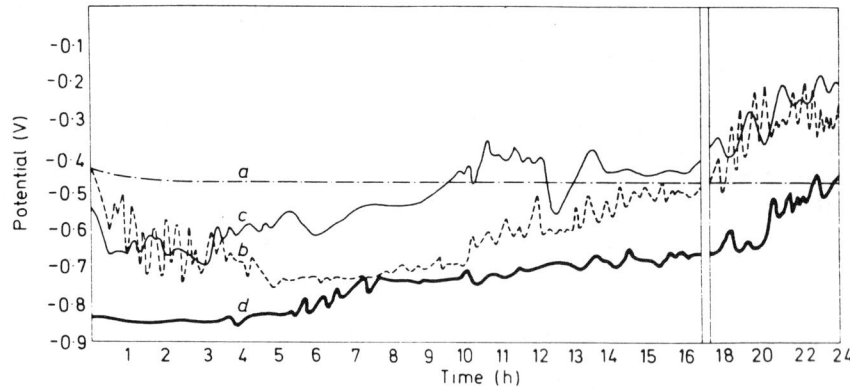

Fig. 1.65 Potential/time curves for cathodically polarised and unpolarised iron in 1·0 N NaCl at 25°C. Curve a unpolarised Fe; curve b Fe polarised cathodically at 20 μA/cm^2, pH = 6·2; curve c Fe polarised cathodically at 20 μA/cm^2, pH = 8·9; and curve d Fe polarised cathodically at 40 μA/cm^2, pH = 6·2 (potentials vs. S.H.E.) (after Pryor[30])

situation is far more common than is generally believed, since it is frequently observed on steel cathodes at current densities of less than $40 \, \mu A/cm^2$. Figure 1.65 shows the variation of potential with time of mild-steel cathodes polarised cathodically at constant impressed current densities of 20 and $40 \, \mu A/cm^2$ in $1 \cdot 0 \, N$ sodium chloride solution[30]. The potential fluctuates rapidly but generally moves in the noble direction with increasing time to values of around $-0 \cdot 2 \, V$ (S.H.E.). This behaviour has been attributed to an overall thickening of the original air-formed oxide film due to a combination of locally increased pH at the cathode and the presence of dissolved oxygen[31]. The potential fluctuations in the active direction have been attributed to sporadic film breakdown by the large concentration of chloride ions.

Similar effects to those shown in Fig. 1.65 have been observed by Pryor and Keir[27] in certain bimetallic couples having steel cathodes. Figure 1.66

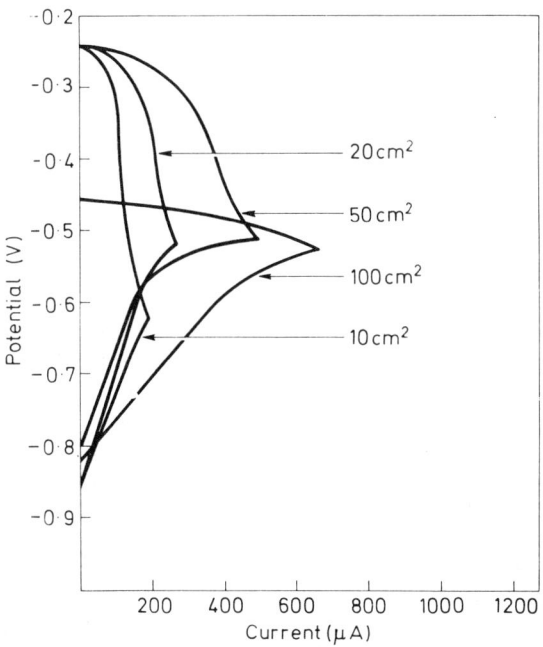

Fig. 1.66 Effect of area of steel cathode on the polarisation curves of the Al-Fe couple in $1 \cdot 0 \, N$ NaCl at $25°C$ using $100 \, cm^2$ of aluminium (potential *vs*. S.H.E.) (after Pryor and Keir[27])

shows anodic and cathodic polarisation curves for the aluminium-mild steel couple in $1 \cdot 0 \, N$ NaCl. The polarisation curves were determined by shorting different areas of mild-steel cathode to $100 \, cm^2$ of aluminium for 24 h, rapidly introducing a variable and increasing external resistance into the circuit, and measuring the potential of both electrodes. Figure 1.66 shows that the transient open-circuit potentials of the steel cathodes are ennobled far above the normal value of corrosion potential (around $-0 \cdot 45 \, V$) in this solution, particularly when smaller cathodes are used. In fact, maximum values of around $-0 \cdot 25 \, V$ are obtained, which are similar to the values obtained by impressed current methods (Fig. 1.65). Since the closed-circuit potential of the couples in Fig. 1.66 remains remarkably constant, the galvanic current at zero external resistance shows marked fluctuations with time, somewhat similar to, but less extreme than, the potential fluctuations in Fig. 1.65. Comparable behaviour has also been observed with the zinc-mild steel couple in sodium chloride solution[27], but the open-circuit potentials of the cathodes are less ennobled and the current fluctuations are more gradual.

The electrochemical effects of slowly and erratically thickening oxide films on iron cathodes are, of course, eliminated when the film is destroyed by reductive dissolution[32] and the iron is maintained in the film-free condition. Such conditions are obtained when iron is coupled to uncontrolled magnesium anodes in high-conductivity electrolytes and when iron is coupled to aluminium in high-conductivity solutions of pH less than 4·0 or more than 12·0[33]. In these cases, the primary cathodic reaction (after reduction of the oxide film) is the evolution of hydrogen.

In sea-water, the increase of pH adjacent to the surface of cathodes brought about by the reduction of oxygen leads to the deposition of films of calcium carbonate and magnesium hydroxide[34,35]. Such film deposition often results in a gradual decrease in the rate of galvanic corrosion of the more negative members of couples immersed in sea-water.

Catchment Area Principle

Practical instances of bimetallic corrosion usually involve electrolytes containing dissolved oxygen. Where the galvanic current is limited by the rate of diffusion of dissolved oxygen to the cathode, the galvanic corrosion rate will be under cathodic control. Whitman and Russell[36], in their classic investigation, showed that, under these conditions, the galvanic corrosion rate war, directly proportional to the area of the cathodic metal and independent of the area of the anodic metal. This principle was later referred to as the *catchment area* principle. It follows from the work of Whitman and Russell that the intensity (i.e. the anodic current density) of galvanic corrosion of an anode coupled to a cathode of constant area is inversely proportional to the anodic area. This principle is of major practical importance in view of its obvious use in minimising galvanic corrosion by proper selection of the relative areas of cathodic and anodic metals.

The catchment area principle is well illustrated by experiments on the Al–Fe and Zn–Fe couples shown in Figs. 1.67 and 1.68[27]. In Fig. 1.67 the area of aluminium and zinc was held constant and the area of steel cathode varied. Both the number of coulombs flowing between the two metals and the anodic weight losses increase with increasing cathodic area, and the current flow is approximately proportional to the cathodic area of each couple. Figure 1.68 shows the effect of holding the steel cathode area constant (100 cm^2) and varying the area of the aluminium anode. The weight loss of aluminium (the dotted line) increases slightly with increase in the anodic area exposed rather than remaining constant as predicted by the catchment area principle. Similar results have been obtained with zinc anodes of various areas coupled to steel cathodes[27]. In part this is due to an increase in the

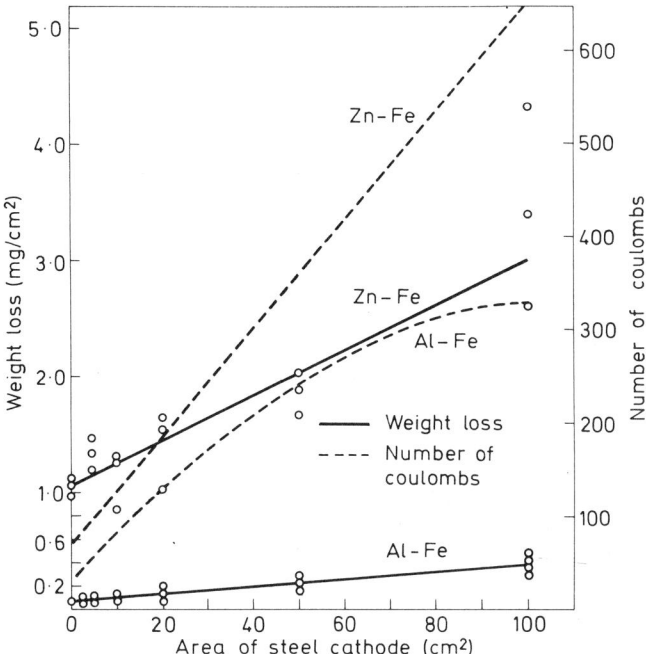

Fig. 1.67 Effect of mild-steel cathode on the weight loss of aluminium and zinc anodes (100 cm² area) and on the number of coulombs flowing in the aluminium–mild steel and zinc–mild steel couples over a 96 h period in 1·0 N NaCl at 25°C (after Pryor and Keir[27])

local action corrosion with increasing anodic area[27]. Furthermore, the anodic polarisation curves for both aluminium and zinc have small but finite slopes and so the couple potential decreases slightly with increasing anodic area (decrease in polarisation of the anode metal) thereby increasing the proportion of the weight loss due to local action (Fig. 1.62b). However, the *intensity* of corrosion (solid line) is very much as predicted by the catchment area principle and increases greatly with decreasing anodic area.

For further discussion of cathode to anode area ratio effects see References[8, 9, 11, 15, 20, 37 and 38], and also refer to the section entitled Distribution of Bimetallic Corrosion in Real Systems, p. 1.238.

Hydrogen Evolution

When the predominant cathodic reaction is reduction of dissolved oxygen the magnitude of $I_{galv.}$ will often be determined by the rate of diffusion of oxygen to the surface of the more positive metal constituting the bimetallic couple, and providing the metal surface is free from films or corrosion products the rate will be independent of the nature of the cathode. The position is quite different when the predominant cathodic reaction is hydrogen evolution, since under these circumstances charge transfer rather than diffusion is rate determining, and different metals will have different catalytic activities for the hydrogen evolution reaction.

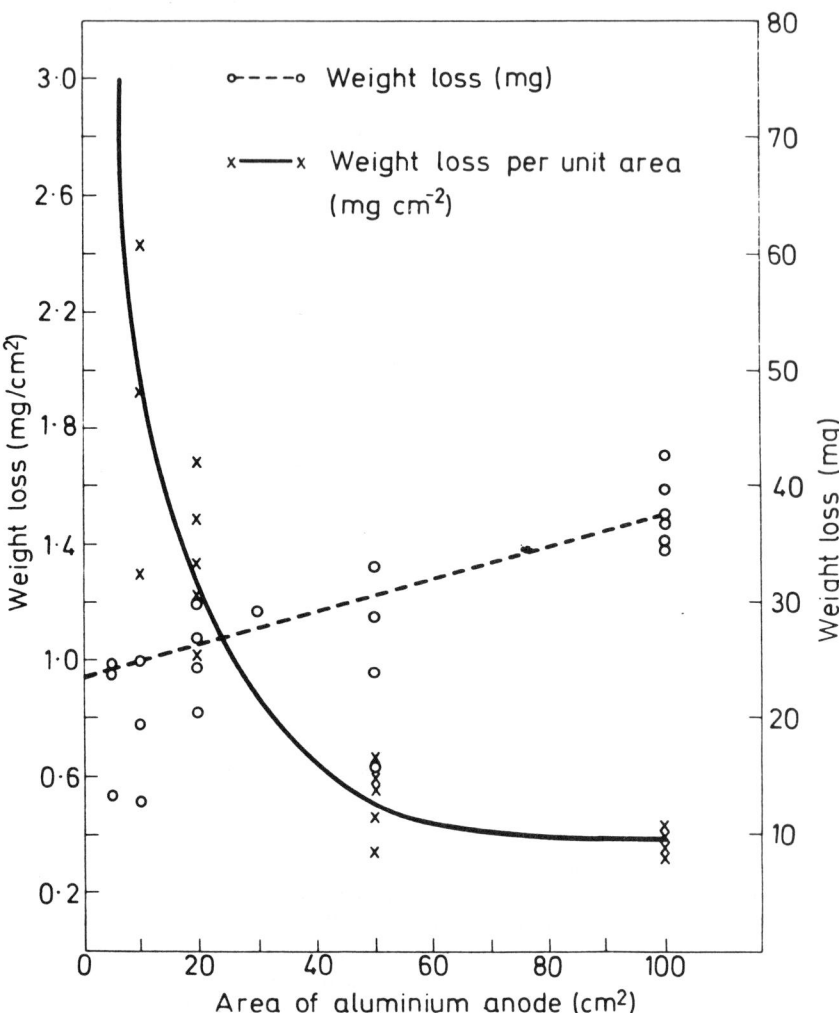

Fig. 1.68 Effect of anode area on the total corrosion and on the intensity of corrosion of aluminium coupled with mild-steel cathodes (100 cm² area) over a 96 h period in 1·0 N NaCl at 25°C (after Pryor and Keir[27])

If it is assumed that anodic polarisation of the more negative metal of the couple is insignificant, then it can be shown[39] from the Tafel relationship for the hydrogen evolution reaction that

$$\log i_{\text{galv.}} = \frac{-E_{\text{corr.}} - 0.059\,\text{pH}}{b_c} + \log \frac{S_c}{S_a} i_{0,c}$$

where $E_{\text{corr.}}$ is the corrosion potential of the couple (*vs.* S.H.E.) determined at some distance away from the couple to obtain the average value, b_c is the Tafel slope, and $i_{0,c}$ the exchange current density for the hydrogen evolution reaction at the surface of the more positive metal, and S_c and S_a are the areas of the cathode and anode metal, respectively. In this equation it is assumed that the potential of the more positive metal at which hydrogen evolution occurs is that of a reversible hydrogen electrode at 1 atm pressure, i.e. $E_H = 0.00 - 0.059 \text{pH}$.

Electrolyte Composition

The composition of the electrolyte has, of course, a major influence on the magnitude and often the direction of the galvanic current flow. The majority of these effects can be rationalised from the effect of the electrolyte on the corrosion potential and, more particularly, the slopes of the appropriate anodic and cathodic polarisation curves. The electrolyte does, however, have one further and important effect on the *distribution* of galvanic corrosion on the anodic metal[40]. When the electrolyte has a high conductivity, such as would be experienced in sea-water, the galvanic corrosion on the less noble metal is distributed in a comparatively uniform fashion and the total weight loss is high. As the electrolyte conductivity decreases, galvanic corrosion of the less noble metal becomes concentrated primarily around the bimetallic junction and the total weight loss of the anode decreases. The intensity of the galvanic corrosion in the interfacial region is relatively little influenced by electrolytic conductivity since it is generally controlled at near to a maximum value by polarisation effects, although it is usually somewhat greater in electrolytes of lower conductivity[41]. It is also implicit that the catchment area principle is not applicable in very low conductivity solutions. The effect of the resistivity of the electrolyte solution on the distribution of attack is illustrated in Fig. 1.69. Methods for quantifying the distribution of galvanic corrosion are described in the section entitled Bimetallic Corrosion in Real Systems, p. **1.238**.

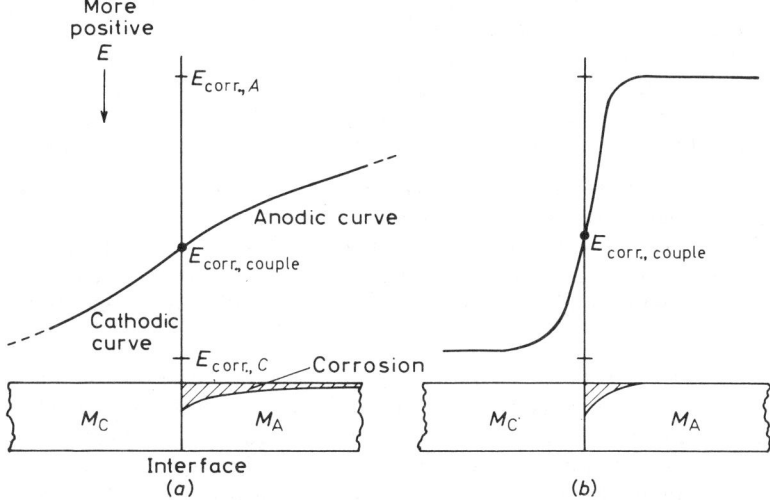

Fig. 1.69 Effect of resistivity of solution on the distribution of corrosion on the more negative metal M_A of a bimetallic couple. (*a*) Solution of very low resistivity and (*b*) solution of very high resistivity. Note that when the resisitivity is high the effective areas of the cathodic and anodic metals are confined to the interface between the two metals

Published work relating to bimetallic corrosion in sodium chloride solution is reported in References[3-5, 7, 17, 27, 29, 30 and 42-44]; in sea-water in References[8, 9, 12-14, 16, 23-25 and 44-47]; in fresh waters in References[10, 16, 42 and 48]; in mineral acids in References[11, 38, 39 and 49]; in water/glycol mixtures in Reference[50] and in the oral cavity in Reference[51].

A large proportion of these references contain quantitative data on bimetallic corrosion obtained either by electrochemical methods or by weight-loss determination. References giving useful quantitative data may be classified in terms of the metal couples studied, as follows: References[8-10, 12, 14 and 25] each deals with a wide range of metal couples; References[7, 13, 16, 42-45 and 50] deal specifically with couples including aluminium alloys; References[17 and 23] deal specifically with couples including titanium; References[24 and 49] deal specifically with couples including mild steel; Reference[38] deals specifically with galvanic corrosion between phases in ferritic/austenitic stainless steels and Reference[11] deals with couples made up from stainless steels, nickel alloys, titanium or graphite.

Note that Reference[11] draws attention to the possibility of an increase of anodic polarisation of the more negative member of a couple leading to a decrease in galvanic corrosion rate. There can also be a risk of increased corrosion of the more positive member of a couple. Both these features can arise as a result of active/passive transition effects on certain metals in certain environments.

Data relating to the galvanic effects of graphite contained in composite materials exposed to sodium chloride solution are to be found in References[52 and 53].

Bimetallic Corrosion in Corrosive Atmospheres

There are many special factors controlling atmospheric bimetallic corrosion that entitle it to separate treatment. The 'electrolyte' in atmospheric corrosion consists of a thin condensed film of moisture containing any soluble contaminants in the atmosphere such as acid fumes from industrial atmospheres and chlorides from marine atmospheres. This type of electrolyte has two characteristics which are summarised in a paper by Rosenfel'd[54].

1. The cathodic reduction of dissolved oxygen contained in a thin condensed electrolyte layer is much less severely polarised than the reduction of dissolved oxygen in the corresponding bulk electrolyte.
2. The electrolytic resistance of the condensed layer parallel to the surface of the metal is high even when it contains dissolved acid fumes or chlorides.

It follows from (1) that the more negative metal, at a bimetallic junction, can be subject to more aggravated attack because of this lowered cathodic polarisation. In part, this may result from the greater ease of replenishment of dissolved oxygen under conditions where the ratio of surface area to electrolyte volume is very high. Rosenfel'd has also produced evidence to show that rapid convective mixing in the condensed layer, under conditions of lowered relative humidity which permit rapid evaporation, further hastens the arrival of dissolved oxygen at the cathode and results in an additional decrease of cathodic polarisation and a corresponding increase in galvanic corrosion rate. Atmospheric galvanic corrosion rates are defined mainly by contact with a less polarised cathode, but also, to a lesser degree, by the nature of the dissolved salts in the condensed electrolyte layer. Rosenfel'd[54] showed that anions in condensed electrolyte layers do not necessarily have the same effect on bimetallic corrosion rates as is observed under conditions of total immersion in the bulk electrolyte. His experiments demonstrate that sulphates produce more intense atmospheric galvanic corrosion of iron than chlorides and that the latter anion is without specific harmful effects on iron. This contention is not universally supported by atmospheric bimetallic corrosion tests on couples with iron anodes. The high electrolytic resistance of the thin condensed electrolyte layer parallel to the bimetallic surface has a controlling influence on the distribution of corrosion on the anode. The galvanic attack will normally be highly localised and qualitatively similar to that experienced under conditions of total immersion in very low conductivity electrolytes. Indeed it is quite rare to find atmospheric galvanic corrosion extending much further than 12-25 mm from the bimetallic junction. The effective cathodic area of the more noble metal tends to be limited by similar considerations (*see* Fig. 1.69*b*). It is, accordingly, evident that the catchment area principle is not applicable in the strict sense to atmospheric bimetallic corrosion since only small and approximately *equal* areas of the dissimilar metals in the immediate vicinity of a bimetallic junction interact electrochemically; thus although the geometrical area of the cathode metal may be large, its effective area is very limited. It should also be noted that whereas under immersed conditions the rate of bimetallic corrosion often remains reasonably constant, it could be variable when the couple is exposed to the atmosphere owing to the variation in atmospheric conditions, i.e. periodic wetting and drying of the surface. Rust forming on unprotected mild steel lowers the dew point thus promoting wetting conditions[16].

Further information on atmospheric bimetallic corrosion can be found in References [15, 20, 24, 25 and 55 to 61].

Bimetallic Corrosion in Other Environments

Information on bimetallic corrosion for buried metals may be obtained in References [15, 62 and 63]; for embedded metals in References [15 and 64], and for non-aqueous liquids in References [15, 65] (liquid fluorine) and [66] (liquid ammonia).

Variable Polarity

Instances where the normal polarity of a galvanic couple becomes reversed during service have given rise to severe practical corrosion problems and have been the subject of a great deal of investigation. Schikorr[67] was probably the first to note that the polarity of the zinc–iron couple could reverse, with iron becoming anodic to the zinc in certain supply waters when the temperature was raised above 60°C. Subsequently this problem was widely investigated by Hoxeng and Prutton[68], Kenworthy and Smith[69], Schuldener and Lehrman[70] and many others in view of its major importance relative to the severe corrosion of galvanised-steel domestic hot-water heaters[71]. It has been confirmed that the polarity reversal occurs at elevated temperature, in excess of about 60°C, and is aided by the presence of bicarbonates and nitrates in the water[68]. It has also been shown that the presence of sulphates and/or chlorides in the water decreases the probability of the polarity reversal. Schuldener and Lehrman[72] showed that under conditions of temperature and electrolyte composition which favour the zinc-steel polarity reversal, the reversal occurs more rapidly when the area of steel exposed to the electrolyte is increased. This observation suggests that galvanised steel that has been extensively pre-corroded, with zinc as the anode, may be more susceptible to subsequent polarity reversal under the appropriate conditions. Furthermore the grain size of hot-dipped galvanised coatings may well influence the rate of polarity reversal, since Caplan and Sereda[73] showed that voids of up to 0·5 mm in diameter and extending at least as far as the Fe-Zn intermetallic layers, exist in these coatings where three zinc 'spangles' meet at a point. The size of the flaws was observed to increase with increasing thickness of the base steel.

This polarity reversal has sometimes been attributed[74] to the formation, at the higher temperatures involved, of an electronically conducting zinc oxide film rather than the normally observed zinc hydroxide of poor electronic conductivity which is stable at room temperature. Recent work by Glass and Ashworth indicates that, subsequent to polarity reversal, zinc does not promote significant galvanic corrosion on steel, but rather that the principal damage results from failure of zinc to cathodically protect the steel[75]. It is generally less well realised that the aluminium-iron couple also undergoes a polarity reversal, like the zinc-iron couple, at elevated temperatures and under somewhat similar conditions[76]. The form of the corrosion product on aluminium also changes with increasing temperature from bayerite (β-Al_2O_3.$3H_2O$, monoclinic) to boehmite (γ-AlO.OH, monoclinic) to boehmite (γ-AlO.OH, orthorhombic), but since the boehmite has a very low electronic conductivity, the intensity of galvanic attack on the steel is low. Polarity reversal has also been observed for cadmium-iron[77] and lead-iron couples[78].

The relative polarities of iron and tin are considered in Sections 4.6 and 14.5 and will not be discussed here.

The variable polarity of the aluminium-iron couple has already been noted. It is also known that it is virtually impossible to protect commercial aluminium alloys cathodically by coupling them to magnesium in a high-conductivity electrolyte such as sea-water, since the development of a high local pH at the aluminium cathode results in cathodic corrosion. Some interesting results have been reported by Bothwell[79] concerning the compatibility of super-purity >99·99% Al aluminium with the magnesium alloys AZ31B (3% Al, 1% Zn, 0·4% Mn) and AZ3IA (3% Al, 1% Zn, 0·4% Mn, 0·15% Ca) in sodium chloride and sea-water. Here the higher electronic resistance of the natural oxide film on super-purity aluminium appears to limit the current flow and the local rise of pH at the cathode, thereby preventing normal aggressive cathodic corrosion.

Akimov[80] was one of the earliest investigators to report that the pitting of aluminium and aluminium alloys in sea-water could be prevented by coupling to zinc. He also drew attention to the occurrence of a tenacious black film on the aluminium. Since then additional observations have been made on the beneficial effects of coupling zinc to various aluminium alloys in chloride solutions. For instance, zinc will often stop the stress-corrosion cracking of susceptible Al–Mg alloys. The protective action of zinc is, however, somewhat unreliable and strongly dependent on the method of surface preparation of the aluminium alloy[81].

Experiments carried out by Keir, van Rooyen and Pryor have clarified the variable behaviour of the Al–Zn couple in chloride solutions. Using high-purity aluminium and zinc electrodes of equal size coupled together in sodium chloride solution, it was found that zinc is initially anodic to aluminium but that within one day the polarity of the couple reverses and remains as such subsequently (Fig. 1.70). This reversal in polarity appears to be due to the accumulation of Zn^{2+} in solution. Accordingly, with decrease in distance between the electrodes, and in solution-volume: electrode-area ratio, the polarity reversal occurs much more rapidly. The accumulation of Zn^{2+} in solution depresses the potential of the aluminium from an initial value of about -0.5 V to a final open-circuit value of about -1.0 V (vs. S.H.E.). The corrosion rates of both the aluminium and the zinc electrodes are greater than in the absence of bimetallic contact, but the corrosion of the aluminium is changed from the characteristic pitting, usually observed in nearly neutral chlorides, to a desirable mild uniform attack. The polarity reversal is not

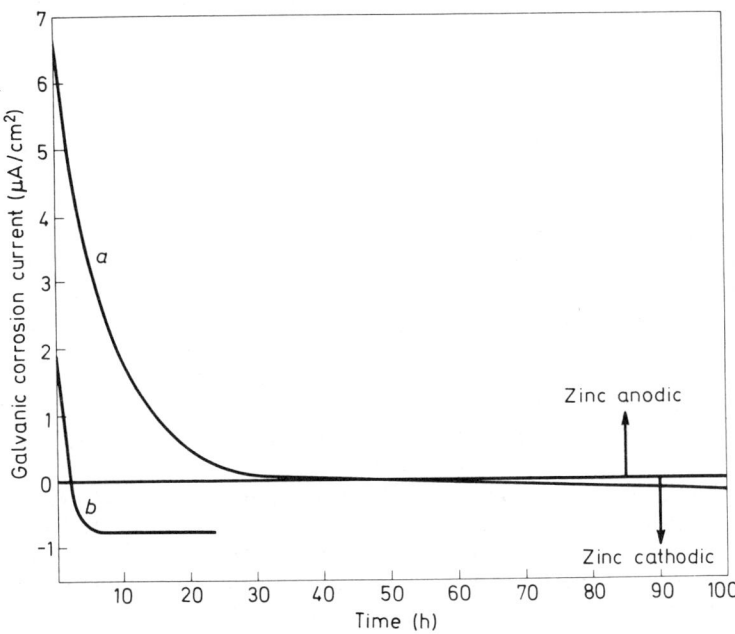

Fig. 1.70 Polarity reversal of the Al–Zn couple in 1.0 N sodium chloride at 25°C. Curve *a* aluminium and zinc electrodes 150 mm apart; 16 ml of solution per square centimetre of electrode; curve *b* aluminium and zinc electrodes 20 mm apart, 10 ml of solution per square centimetre of electrode

detectable by potential measurements alone, since the potential of the couple remains constant, to within a few millivolts at a value of around $-0 \cdot 83$ V. When the aluminium is cathodic, polarisation curves indicate that the galvanic corrosion is under cathodic control; after the polarity reversal, the much weaker galvanic corrosion appears to be under anodic control.

It remains to be determined whether the previous experiments[80, 81], which have been interpreted as confirming the cathodic protection of aluminium by zinc, can be truly interpreted in this fashion or whether they are due to the accumulation of Zn^{2+} in the electrolyte. Under laboratory conditions, and under some practical conditions in stagnant solutions or in recirculating systems, the latter explanation is quite likely.

Two dissimilar metals, such as iron and aluminium, may cause aggravated corrosion effects even if they are not in electrical contact. This subject is, however, outside the scope of this section, and has been treated in detail elsewhere[82, 83]. Heavy metal ions, such as copper ions, are particularly liable to produce galvanic effects by redeposition on a less noble metal; the phenomenon is discussed in Sections 4.1, 4.2 and 9.3.

Protective Measures

Under Conditions of Total Immersion

Protective measures against bimetallic corrosion should ideally start before the particular installation or equipment is built[15]. Reference should be made to tables showing compatibility of metals, alloys and non-metallic materials (*see* Table 1.25) and to the literature. However, it must be emphasised that the environment obviously plays a most important rôle in bimetallic corrosion, and that there are a number of situations in which apparently incompatible materials in contact can be used without adverse effects.

Assuming that some incompatible materials of construction must be used, much can be done in the way of initial design to minimise future problems. Under conditions of total immersion in high-conductivity electrolytes containing dissolved oxygen, the catchment area principle is fundamental in minimising galvanic corrosion problems. Reduction of the area of the more noble metal together with use of the maximum area of the less noble metal gives a combination of small galvanic current and minimum intensity of attack on the less noble metal. This approach becomes less effective as the conductivity of the electrolyte decreases.

In simple equipment, galvanic corrosion can be eliminated by complete electrical insulation. Bushings, washers and pipe fittings of nylon or Teflon have become quite popular for this purpose despite their obvious limitations in mechanical properties (*see* Section 9.5). Particular examples of the use of this type of fitting include insulating domestic aluminium or galvanised-iron hot-water heaters and tanks from black iron or copper plumbing. Electrical insulation often tends to be ineffective in complicated equipment on account of the numerous other electronically conducting paths that may exist[84]. It is always well to check supposedly electrically isolated metal components with an ohmmeter in order to confirm that the desired electrical isolation has, in fact, been achieved.

Increasing the electrolytic resistance of the solution path is a possible method of reducing the galvanic corrosion rate. Little significant practical benefit accompanies this approach in high-conductivity electrolytes since severe galvanic corrosion can exist at locations which are several metres distant from the actual bimetallic joint[84]. This approach, accordingly, has merit mainly in low-conductivity electrolytes such as certain supply waters, and in the case of atmospheric galvanic corrosion.

Deaeration has occasionally been used as a means of controlling bimetallic corrosion under conditions of total immersion, and this method of control can be used successfully, if physical conditions permit, provided that the less noble metal is not sufficiently electrochemically active to permit rapid evolution of hydrogen at the more noble metal, as is observed, for instance, in many bimetallic couples involving magnesium anodes.

Metallic coatings have been widely and successfully used as a means of alleviating many bimetallic corrosion problems both under conditions of total immersion and in corrosive atmospheres. If, for instance, aluminium and steel must be jointed together in sea-water, the galvanic corrosion can be largely eliminated by aluminising the steel either by hot dipping or by flame spraying, as is more popular in Europe. Both zinc and cadmium are also fairly compatible with aluminium and so the steel may be protected with thin coatings of these metals without incurring the risk of aggravated galvanic corrosion; cadmium plating has even been applied to stainless steel for this purpose. The use of dissimilar metallic coatings eliminates bimetallic corrosion only if the coating is initially free from voids and remains so in service, a circumstance seldom realised in practice. Metallic coatings on the steel that contain or develop voids still reduce the galvanic corrosion rate (because of the smaller area of steel exposed) provided that they are anodic to the substrate.

Cathodic protection with a sacrificial anode that is less noble than either member of the couple is frequently used to reduce the severity of bimetallic corrosion, particularly that resulting from the use of bronze propellers in steel ship hulls.

Paint coatings also receive extensive practical use for protecting against galvanic corrosion in atmospheres and under conditions of total immersion. The best practice, where feasible, calls for complete painting of both members of the bimetallic couple. If only one member of the couple can be painted, the *cathodic* metal should receive this treatment; since paint coatings are seldom free from holidays, painting the cathode will reduce the total cathodic area and hence the galvanic corrosion rate. Painting the anodic metal alone represents bad practice under conditions of total immersion in a high-conductivity electrolyte, because the original cathode area is undiminished, and corrosion will then take place at holidays or damaged areas in the coating on the anodic metal at a high intensity[85]. For optimum protection against galvanic corrosion, repainting should be carried out on a regular schedule since the protection afforded by most paints can be rather limited in duration.

The use of soluble inhibitors as a means of controlling bimetallic corrosion presents many technical problems. Apart from the fact that this method is limited in applicability to recirculating systems, efficient anodic inhibitors, such as chromates, are frequently quite specific in their action and so certain bimetallic couples, such as the Al–Cu couple in chloride solutions[40], are extremely difficult to control by a single anodic inhibitor. Accordingly, other methods of treating bimetallic corrosion, as already described, are often preferred, with inhibition being relegated to special applications such as automotive cooling systems where these previous methods are either not feasible or economical. Inhibition of automotive cooling systems is generally achieved by highly complex mixtures of inhibitors, often involving combinations of borax, nitrates and organic adsorption inhibitors, and even then complete success is not always attained (*see* Section 17.2). Various inhibitors were compared in tests by Brunoro *et al.*[86].

Advice on the reduction of bimetallic corrosion at welded and brazed joints can be found in Reference[87]. The use of replaceable wastage pieces to take up the bimetallic corrosion in various systems is proposed in References[84 and 85].

In Corrosive Atmospheres

Bimetallic corrosion in atmospheres is confined to the area of the less noble metal in the vicinity of the bimetallic joint, owing to the high electrolytic resistance of the condensed electrolyte film. Electrolytic resistance considerations limit the effective anodic and cathodic areas to approximately equal size and therefore prevent alleviation of atmospheric galvanic corrosion through strict application of the catchment area principle.

With this exception, many of the methods already described for protecting against bimetallic corrosion under conditions of total immersion may be similarly used for preventing atmospheric galvanic corrosion. These include selection of compatible metals, metallic coatings and painting. It is, however, more common to use the principle of increasing the resistance of the solution path for preventing galvanic corrosion. Since the solution-path resistance is already high, the additional means that are required to increase resistance further are simple and generally inexpensive. In many cases, taping the immediate joint area with mastic tapes, with or without chromate impregnation, will suffice, provided that the whole of the bimetallic contact is covered to a distance of about 25 mm from the junction and on either side. Vulcanising a rubber or Neoprene ball around small joints has also been used very satisfactorily. While these methods protect a joint against atmospheric galvanic corrosion, it can hardly be overemphasised that they are *not* applicable to protecting against bimetallic corrosion where the same joint is totally immersed in a high-conductivity electrolyte.

Some Beneficial Effects of Galvanic Coupling

So far this section has been primarily concerned with the harmful aspects of bimetallic corrosion, in which the less noble member of the couple is subjected to attack of unusual severity. It is, however, implicit that bimetallic corrosion can be beneficial in that it will usually reduce or prevent corrosion of the more noble metal. Refer to Sections 11.2 and 11.4 for further details. Another very beneficial aspect of bimetallic corrosion is power generation from chemical cells, but this subject is outside the scope of this section.

The principles of bimetallic corrosion have, in addition, been used in an elegant fashion for the development of highly corrosion-resistant alloys.

Draley and Ruther[88] observed that commercial-purity aluminium (1100 alloy) was subject to catastrophic intergranular corrosion in distilled water above 200°C with the corrosion rate increasing very rapidly with temperature. In most cases enhanced attack occurred at grain boundaries and around second-phase stringers. Draley and Ruther showed that the rapid intergranular disintegration of the aluminium was associated with the entry of hydrogen into the metal from cathodic sites. They proposed that, if alternate cathodic sites of lower hydrogen overpotential could be provided, the hydrogen would have a much better chance of being evolved as harmless bubbles instead of entering the aluminium and causing intergranular disintegration. This hypothesis was confirmed by adding 5 p.p.m. of Ni^{2+} to the distilled water. The small amounts of nickel plating-out on the aluminium surface were sufficient to protect against catastrophic corrosion of 1100 alloy at 275°C, although the 1100 was subject to slow and uniform attack. Subsequently, 1100 alloy specimens were electroless nickel-plated and found to resist catastrophic corrosion for 80 days at 315°C. Finally Draley and Ruther alloyed small amounts of nickel (0·5% or more) with commercially pure aluminium and obtained consistent protection against catastrophic corrosion at temperatures of up to 350°C. The nickel, being largely insoluble in aluminium, exists primarily as $NiAl_3$ constituent which evidently possesses a low hydrogen overpotential and protects against catastrophic corrosion. Its action is augmented by the simultaneous presence of iron in the alloy, the improved corrosion resistance probably being due to the presence of an Al–Ni–Fe constituent[89]. A commercial aluminium alloy designated as 8001 (1% Ni, 0·6% Fe) now exists for high-temperature water service.

It has long been known that alloys such as austenitic stainless steel and metals such as titanium, while exhibiting passive behaviour in mildly or strongly oxidising solutions, often suffer active corrosion at a high rate in reducing acids. Tomashov et al.[90,91] were among the first to point out the possibility of improving the corrosion resistance of stainless steel, chromium and titanium by increasing the stability of the passive state with small alloying additions of noble metals such as silver, palladium and platinum. This work was extended for titanium in reducing-type acids by Stern and Wissenberg[92] who primarily investigated the effect of platinum and palladium. The principle by which noble alloying additions are effective in improving the corrosion resistance of titanium is illustrated in Fig. 1.71 which is taken from the work of Stern and Wissenberg. In a reducing acid where hydrogen is evolved from a titanium surface, the exchange current is relatively small. It may be increased and the cathodic Tafel slope decreased by providing local noble-metal cathodes. The intercept of the cathodic and the anodic polarisation curves is shifted in the more noble direction and, if the shift is large enough to raise the mixed potential into the passive potential region in Fig. 1.71, essentially passive behaviour in reducing acids such as boiling HCl can result (see Section 5.4). Similar improvements in the corrosion resistance of chromium in sulphuric and hydrochloric acids have been found by Greene, Bishop and Stern[93] to accompany alloying with small amounts of rhodium, palladium or osmium. These noble-metal alloying

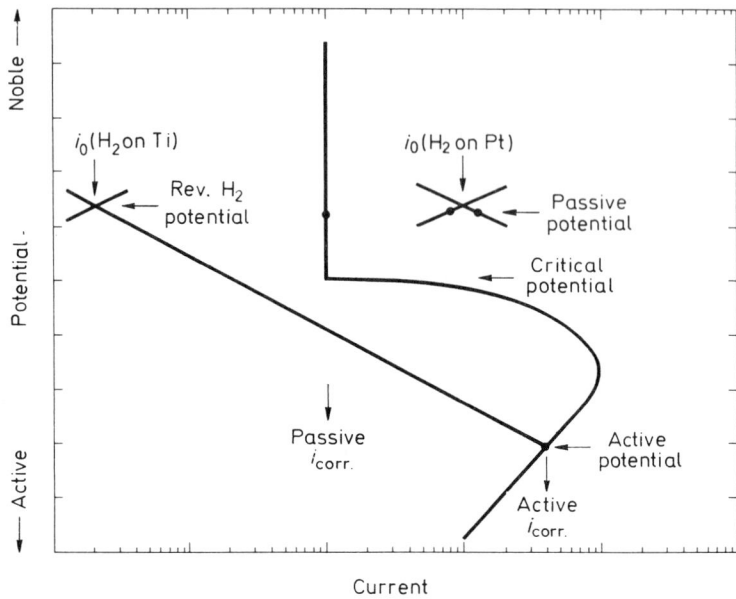

Fig. 1.71 How alloying with a noble metal produces a passive mixed potential and a marked reduction in corrosion rate (after Stern and Wissenberg[92])

additions have the further advantage of not significantly reducing the normally excellent corrosion resistance of chromium in oxidising acids such as nitric acid, whereas elements such as platinum, iridium and ruthenium confer excellent corrosion resistance on chromium in non-oxidising acids, but increase its corrosion rate in oxidising acids.

It is evident from the foregoing that the principles of bimetallic corrosion are being applied in a progressively more widespread and successful fashion to the development of alloys of maximum corrosion resistance.

Distribution of Bimetallic Corrosion in Real Systems

The influence of electrolyte conductivity on the distribution of bimetallic corrosion has already been described in qualitative terms earlier in the section (*see* Fig. 1.69). For a bimetal couple of small size, one could expect an approximately even distribution of corrosion in an electrolyte of high conductivity, such as sea-water. In real systems, such as heat-exchangers,[94] steam condensers, pumps, pipework and off-shore rigs, the almost inevitable presence of a mix of different metals leads to the development of galvanic corrosion which is unevenly distributed, even in electrolytes of high conductivity, because of large system dimensions. Furthermore, the effective cathode to anode area ratio in real mixed-metal systems will differ from the geometrical area ratio. Indeed, there could well be doubt as to which metals will be cathodic and which anodic in complex systems. Thus laboratory-derived data relating to simple bimetal couples made up from small electrode samples cannot be used to give an accurate indication of even the maximum corrosion rates of the more negative metals in a real system[9]. Maximum corrosion rates of the more negative metals in a real system[9]. Maximum corrosion rates calculated on the basis of geometrical area ratio can be typically two or three times too low.

The quantification of the probable extent and magnitude of bimetallic corrosion for a new system at the design stage, for it is at this stage that remedial actions, such as the provision of wastage pieces of thicker material, can most readily be made[34, 35], is difficult to achieve by means of a corrosion evaluation using a-scaled-down model of the full-scale prototype design. Such a procedure is time consuming and also presents grave problems associated with necessary scaling of the conductivity of the test electrolyte by dilution which can itself have an effect upon the anodic and cathodic reactions[95]. An alternative to setting up scaled-down tests is to mathematically model the corrosive processes for the full-sized system in order to predict the distribution of bimetallic corrosion, using cathodic and anodic polarisation curves relating to fresh or filmed metal, as desired. Such modelling also, of course, enables the prediction of the extent of cathodic protection on the more positive metals in the system.

As the corrosion rate, inclusive of local-cell corrosion, of a metal is related to electrode potential, usually by means of the Tafel equation and, of course, Faraday's second law of electrolysis, a necessary precursor to corrosion rate calculation is the assessment of electrode potential distribution on each metal in a system. In the absence of significant concentration variations in the electrolyte,[96] a condition certainly satisfied in most practical sea-water systems, the exact prediction of electrode potential distribution at a given time involves the solution of the Laplace equation for the electrostatic potential (P) in the electrolyte at the position given by the three spatial coordinates (x, y, z).

$$\frac{\partial^2 P}{\partial x^2} + \frac{\partial^2 P}{\partial y^2} + \frac{\partial^2 P}{\partial z^2} = 0$$

The solution of the Laplace equation is not trivial even for relatively simple geometries and analytical solutions are usually not possible. Series solutions have been obtained for simple geometries assuming linear polarisation kinetics[97-101]. More complex electrode kinetics and/or geometries have been dealt with by various numerical methods of solution such as finite difference[102, 103], finite element[104, 105] and boundary element.[41, 106]

The numerical approaches to the solution of the Laplace equation usually demand access to minicomputers with fast processing capabilities. Numerical methods of this sort are essential when the electrolyte is unconfined, as for an off-shore rig or a submarine hull. However, where the electrolyte is confined, as within essentially cylindrical equipment such as pipework and heat-exchangers, or for restricted electrolyte depths, a simpler modelling procedure may be adopted in the case of electrolytes of good conductivity, such as sea-water[34, 35]. This simpler procedure enables computation to be carried out on small, desk-top microcomputers.

For electrolytes of low resistivity, it can be shown that the electrode potential distribution within cylindrical equipment is often very closely approximated to by neglecting the radial potential variation i.e. by assuming current flows only axially[34, 35, 107, 108]. Astley[35] has demonstrated that sea-water systems with diameters of up to at least 500 mm can be examined making a 'unidirectional current flow' assumption.

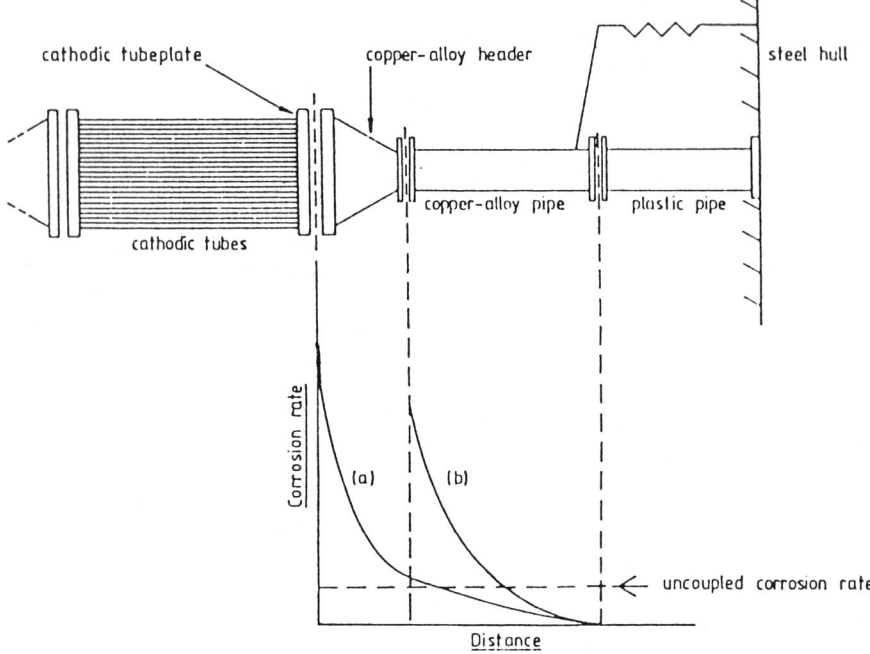

Fig. 1.72 Schematic representation of calculated bimetallic corrosion distribution; (a) connected header, (b) electrically-insulated heater

The differential equation to be solved for a cylindrical system assuming unidirectional current flow conditions apply is[34,35]

$$\frac{d^2E}{dx^2} = \frac{2\rho i}{r}$$

where E is the electrode potential at distance, x, along the system, ρ is the electrolyte resistivity, i is the local surface current density at distance, x, and r is the system radius. The solution of this equation depends upon the potential dependence of i, i.e. the form of the cathodic or anodic polarisation curve for each metal in the system. Analytical solutions have been derived for linear polarisation kinetics[34,107,109-111], for Tafel conditions[34,112], for Butler-Volmer conditions[9,35,107,110,111,113,114] and for combinations of linear polarisation kinetics with a potential-independent current density[34].

Astley has used unidirectional current flow analysis to assess bimetallic corrosion and cathodic protection distribution within a number of real seawater systems[35,115]. Thus, design-stage analysis has been made of the bimetallic corrosion distribution within a marine heat-exchanger system having a cathodic metal tube-bundle, two conical bronze headers and two seven-metre long, 350 mm dia cupronickel feed-pipes each connected to a steel hull by lengths of plastic piping.[35,115] Mathematical modelling for this system revealed the likely magnitude of corrosion rates within the headers and feed-pipes, both before and after electrical insulation of the headers at their flanges, and also quantified any possible decrease in bimetallic corrosion that could arise due to electrical connection of the steel hull to the feed-pipes. The schematic corrosion rate distributions are shown in Figure 1.72. Design decisions with respect to header insulation and the employment of wastage pieces were thereby facilitated. Quantitative confirmation of the predicted corrosion pattern in this system was subsequently obtained in pilot-scale trials.

The development of mathematical modelling techniques is proving to be a significant advance in the assessment of the bimetallic corrosion hazard in real systems.

M. J. PRYOR
D. J. ASTLEY

REFERENCES

1. Evans, U. R., *The Corrosion and Oxidation of Metals*, Arnold, London (1961)
2. Gatty, O. and Spooner, E. C., *Electrode Potential Behaviour of Corroding Metals in Aqueous Solutions*, Oxford University Press, London (1938)
3. Mansfeld, F., *Corrosion*, **29**, 403 (1973)
4. Baboian, R., Paper 58, 'Corrosion 85', N.A.C.E., Houston (1985)
5. Bauer, O. and Vogel, O., *Mitt. MatPrüfAmt. Inst. Metallforsch. Berl.*, **36**, 114 (1918)
6. Latimer, W. M., *Oxidation States of the Elements and Their Potentials in Aqueous Solutions*, Prentice-Hall, New York, 294 (1950)
7. Mansfeld, F., *Corrosion*, **30**, 343 (1974)
8. Scholes, I. R., Astley, D. J. and Rowlands, J. C., Sixth European Congress on Metallic Corrosion, SCI, London, 161 (1977)
9. Astley, D. J. and Rowlands, J. C., *Br. Corr. J.*, **20**, 90 (1985)
10. Linder, M. and Mattson, E., Seventh Scandinavian Corrosion Congress, Norway, 19 (1975)
11. Davis, G. O., Kolts, J. and Sridhar, N., *Corrosion*, **42**, 329 (1986)
12. Hack, H. P. and Scully, J. R., *Corrosion*, **42**, 79 (1986)
13. Lee, H. Y., Son, U. T. and Kim, S. J., *J. Korean Inst. Met.*, **20**, 31 (1982)
14. Kuron, D., Kilian, R. and Grafen, H., *Z. Werkstofftech*, **11**, 382 (1980)
15. *Guides to Practice in Corrosion Control*, No 14; 'Bimetallic Corrosion', Dept. of Industry (1982)
16. Reboul, M. C., *Corrosion*, **35**, 423 (1979)
17. Shalaby, L. A., *Corrosion Science*, **11**, 767 (1971)
18. Promisel, N. E. and Mustin, G. S., *Corrosion*, **7**, 339 (1951)
19. Evans, U. R. and Rance, V. E., *Corrosion and Its Prevention at Bimetallic Contacts*, H.M.S.O., London (1958)
20. 'Commentary on corrosion at bimetallic contacts and its alleviation', British Standards Institute, PD6484 (1979), confirmed August (1984)
21. Godard, H. P., *Corrosion*, **7**, 93 (1951)
22. Lauer, G. and Mansfeld, F., *Corrosion*, **26**, 504 (1970)
23. Cotton, J. B. and Downing, B. P., *Trans. Inst. Mar. Eng.*, **69**, 311 (1957)
24. Johnson, K. E. and Abbott, J. S., *Br. Corr. J.*, **9**, 171 (1974)
25. Southwell, C. R., Bultman, J. D. and Alexander, A. L., *Materials Performance*, **15**, 9 (1976)
26. Danek, G. J., 'The effect of seawater velocity on the corrosion behaviour of metals', *Naval Engineers Journal*, No. 763, (1966)
27. Pryor, M. J. and Keir, D. S., *J. Electrochem. Soc.*, **104**, 269 (1957)
28. Dyess, J. B. and Miley, H. A., *Trans. Amer. Inst. Min. (Metall.) Engrs.*, **133**, 239 (1939)
29. Pryor, M. J. and Keir, D. S., *J. Electrochem. Soc.*, **102**, 605 (1955)
30. Pryor, M. J., *Nature, Lond.*, **178**, 1245 (1956)
31. Mayne, J. E. O., Menter, J. W. and Pryor, M. J., *J. Chem. Soc.*, 1831 (1949)
32. Evans, U. R., *J. Chem. Soc.*, 478 (1930)
33. Pryor, M. J. and Keir, D. S., *J. Electrochem. Soc.*, **105**, 629 (1958)

34. Astley, D. J., *Corrosion Science*, **23**, 801 (1983)
35. Astley, D. J., *Galvanic Corrosion* (Ed. H. P. Hack) ASTM STP 978 pp. 53-78 (1988)
36. Whitman, W. G. and Russell, R. P., *Industr. Engng. Chem.*, **16**, 276 (1924)
37. Mansfeld, F., *Corrosion*, **27**, 436 (1971)
38. Yau, Y. H. and Streicher, M. A., *Corrosion*, **43**, 366 (1987)
39. Uhlig, H. H., *Corrosion and Corrosion Control*, Wiley, New York (1971)
40. Eldridge, G. G. and Mears, R. B., *Industr. Engng. Chem.*, **37**, 736 (1945)
41. Bardal, E., Johnsen, R. and Per Olav Gartland, *Corrosion*, **40**, 628 (1984)
42. Mansfeld, F., *Corrosion Science*, **15**, 183 (1975)
43. Mansfeld, F., *Corrosion Science*, **15**, 239 (1975)
44. Mansfeld, F., and Kenkel, J. V., *Corrosion*, **33**, 376 (1977)
45. Lennox, T. J., Peterson, M. H., Smith, J. A. and Groover, R. E., *Materials Performance*, **13**, 31 (1974)
46. Robson, D. N. C., Section 2.3 of 'Corrosion and Marine Growth on Offshore Structures', 69, J. Wiley & Sons, (1984)
47. Wei, M. W., *Corrosion*, **23**, 261 (1967)
48. Venczel, J. and Wranglen, G., *Corrosion Science*, **7**, 461 (1967)
49. Wranglen, G. and Inam Khokar, M., *Corrosion Science*, **9**, 439 (1969)
50. Monticelli, C., Brunoro, G., Trabanelli, G. and Frignani, A., *Werkst. Korros.*, **38**, 83 (1987)
51. Marek, M., 'Corrosion of Dental Materials', *Encyclopedia of Materials Science and Engineering*, Vol. 2, Pergamon Press, 896 (1986)
52. Trzaskoma, P. P., *Corrosion*, **42**, 609 (1986)
53. Belluci, F. D., Martino, A. and Liberti, C., *J. Appl. Electrochemistry*, **16**, 15 (1986)
54. Rosenfel'd, N., *Proceedings of the First International Congress on Metallic Corrosion, London, 1961*, Butterworths, London, 243 (1962)
55. 'Determination of bimetallic corrosion in outdoor exposure tests', BS 6682 (1986)
56. Kucera, V. and Mattson, E., *Atmospheric Corrosion of Bimetallic Structures*, ex *Atmospheric Corrosion*, 561, J. Wiley and Sons, (1982)
57. Pelensky, M. A., Jaworski, J. J. and Gallaccio, A., *ASTM STP* **646**, 58 (1978)
58. Baboian, R., *ASTM STP* **646**, 17 (1978)
59. 'Protection of Electrical Power Equipment against Climatic Conditions', BS CP 1014 (1963)
60. Godard, H. P., *Materials Protection*, **2**, 40 (1963)
61. Latimer, K. G., 2nd Inter. Congress Metal Corrosion, 780 New York City, (1966)
62. Escalante, E. and Gerhold, W. F., ASTM 'Field and Laboratory Studies', 81 (1976)
63. Schick, G. and Mitchell, D. A., ASTM 'Field and Laboratory Studies', 69 (1976)
64. Vrable, J. B., *Materials Performance*, **21**, 51 (1982)
65. Toy, S. M., English, W. D. and Crane, W. E., *Corrosion*, **24**, 418 (1968)
66. Jones, D. A. and Wilde, B. E., *Corrosion*, **33**, 46 (1977)
67. Schikorr, G., *Trans. Electrochem. Soc.*, **76**, 247 (1939)
68. Hoxeng, R. B. and Prutton, C. F., *Corrosion*, **5**, 330 (1949)
69. Kenworthy, L. and Smith, M. D., *J. Inst. Met.*, **70**, 463 (1944)
70. Schuldener, H. L. and Lehrman, L., *J. American Wat. Wks. Assoc.*, **49**, 1432 (1957)
71. Cohen, M., Thomas, W. R. and Sereda, P. J., *Canad. J. Technol.*, **29**, 435 (1951)
72. Schuldener, H. L. and Lehrman, L., *Corrosion*, **14**, 545t (1958)
73. Caplan, D. and Sereda, P. J., *Canad. J. Technol.*, **31**, 172 (1953)
74. Gilbert, P. T., *J. Electrochem. Soc.*, **99**, 18 (1952)
75. Glass, G. K. and Ashworth, V., *Corrosion Science*, **25**, 971 (1985)
76. Gabe, D. R. and El Hassan, A. M., *Br. Corr. J.*, **21**, 185 (1986)
77. Zanker, L. and Yahalom, J., *Corrosion Science*, **9**, 157 (1969)
78. Gouda, V. K., Shalaby, L. A. and Abdul Azim, A. A., *Br. Corr. J.*, **8**, 81 (1973)
79. Bothwell, M. R., *J. Electrochem. Soc.*, **106**, 1014, 1019 (1959)
80. Akimov, G., *Korros. Metallsch.*, **6**, 84 (1930)
81. Edeleanu, C. and Evans, U. R., *Trans. Faraday Soc.*, **47**, 1121 (1951)
82. Bird, C. E. and Evans, U. R., *Corr. Technol.*, **3**, 279 (1956)
83. See for instance Evans, U. R., *The Corrosion and Oxidation of Metals*, Arnold, London, 205-6 (1960)
84. Gilbert, P. T., 'Considerations arising from the use of dissimilar metals in seawater piping systems', 5th International Congress on Marine Corrosion and Fouling, Barcelona (1980)
85. Rowe, L. C., *Automotive Engineering*, **82**, 40 (1974)

86. Brunoro, G., Zucchi, F. and Zucchini, M., *Mater. Chem.*, **5**, 135 (1980)
87. Jarman, R. A. and Shreir, L. L., *Welding and Metal Fabrication*, 444 (1987)
88. Draley, J. E. and Ruther, W. E., *Corrosion*, **12**, 480t (1956); *J. Electrochem. Soc.*, **104**, 329 (1957)
89. Phillips, H. W. L., *J. Inst. Met.*, **69**, 275 (1943)
90. Tomashov, N. D. and Chernova, G. P., *C. R. Acad. Sci. U.R.S.S.*, **89**, 121 (1953)
91. Tomashov, N. D., Altovsky, R. M. and Arakelov, A. G., *C. R. Acad. Sci. U.R.S.S.*, **121**, 885 (1958)
92. Stern, M. and Wissenberg, H., *J. Electrochem. Soc.*, **106**, 755, 759 (1959)
93. Greene, N. D., Bishop, C. R. and Stern, M., *J. Electrochem. Soc.*, **108**, 836 (1961)
94. Gehring, G. A., Kuester, C. K. and Maurer, J. R., Paper 80, 'Corrosion 80', N.A.C.E., Houston (1980)
95. Agar, J. N. and Hoar, T. P., *Discuss. Faraday Soc.*, **1**, 158 (1947)
96. Newman, J., *Electrochemical Systems*, Prentice-Hall (1973)
97. Waber, J. T. and Ruth, J. M., Los Alamos Lab. Microfich LA-1993 (1956)
98. McCafferty, E., *Corrosion Science*, **16**, 183 (1976)
99. McCafferty, E., *J. Electrochem. Soc.*, **124**, 1869 (1977)
100. Melville, P. H., *J. Electrochem. Soc.*, **126**, 2081 (1979)
101. Melville, P. H., *J. Electrochem. Soc.*, **127**, 864 (1980)
102. Doig, P. and Flewitt, P. E. J., *J. Electrochem. Soc.*, **126**, 2057 (1979)
103. Munn, R. S. and Clark, J. H., Paper 74, 'Corrosion 83', N.A.C.E., Houston (1983)
104. Helle, H. P. E., Beck, G. H. M. and Ligtelijn, J. Th., *Corrosion*, **37**, 522 (1981)
105. Forrest, A. W., Fu, J. W. and Bicicchi, R. T., Paper 150, 'Corrosion 80', N.A.C.E., Houston (1980)
106. Danson, D. J. and Warne, M. A., Paper 211, 'Corrosion 83', N.A.C.E., Houston (1983)
107. Frumkin, A. N., *Zh. fiz. Khim.*, **23**, 1477 (1949)
108. de Levie, R. in *Advances in Electrochemistry and Electrochemical Engineering*, Vol. 6, Ed. Delahay, P., Interscience (1967)
109. Sato, S. and Yamauchi, S., Sumitomo, *Light Metal Techn. Rep.*, **17**, 24 (1976)
110. Chizmadzhev, Yu. A., Markin, V. S., Tarasevich, M. R. and Chirkov, Yu. G., *Macrokinetics of Processes in Porous Media*, Nauka, Moscow (1971)
111. Reingeverts, M. D., Parputs, I. V. and Sukhotin, A. M., *Soviet Electrochemistry*, **16**, 35 (1980)
112. Mueller, W. A., *J. Electrochem. Soc.*, **110**, 698 (1963)
113. Posey, F. A., *J. Electrochem. Soc.*, **111**, 1173 (1964)
114. Alkire, R. and Mirarefi, A. A., *J. Electrochem. Soc.*, **120**, 1507 (1973)
115. Astley, D. J., 'Prediction of Galvanic Corrosion in Marine Heat-exchangers', Institute of Metals Conference, Bristol (1986)

1.8 Lattice Defects in Metal Oxides

When a metal oxide is in contact with one of its components (metal or oxygen), the condition for thermodynamic equilibrium cannot, in general, be satisfied unless the crystal is non-stoichiometric, i.e. unless it contains an excess of one of the two components. The reason for this is that although energy must be expended in incorporating the excess component, the entropy of the system increases extremely rapidly at first, and then more slowly as the non-stoichiometry increases. Thus the equilibrium condition, namely that the free energy of the system is a minimum, is satisfied only for some finite degree of non-stoichiometry. The thermodynamic functions are shown schematically as functions of α, the degree of non-stoichiometry, in Fig. 1.72. We note at this stage that it is sufficient to discuss the equilibrium between an oxide and oxygen gas because the other case (equilibrium between the oxide and the metal) is then covered by putting the oxygen pressure equal to the dissociation pressure of the oxide.

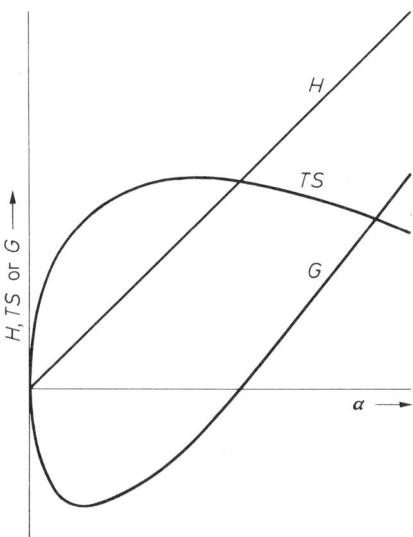

Fig. 1.72 Thermodynamic functions G, H and TS as functions of α, the degree of non-stoichiometry. G is the free enthalpy, H the enthalpy and S entropy

There are two basic questions which can be decided only by experiments. First, we must know whether the metal or the oxygen is present in excess, and second, we must know how the excess component is incorporated in the oxide lattice. In connection with the latter question we have to remember that a non-stoichiometric crystal remains electrically neutral (except in narrow regions near the surfaces), so that if the excess component is present in the crystal as ions, lattice defects with charges of opposite sign must necessarily be present also (*see* Figs. 1.77 and 1.78). The most important defect structures will be discussed in this section.

The presence in an oxide of an excess of one component provides a mechanism for the transport of material. This transport mechanism, which is vital in understanding the formation of a continuous oxide film on a metal, is also discussed in this section. An important feature here is that an excess of one component may provide a transport mechanism, not for itself, but for the other component.

p-type Oxides

Cu_2O contains excess oxygen which is taken up in such a way as to build up new layers of the oxide. Thus, the excess oxygen is present as O^{2-} ions on their normal lattice sites. To form oxygen ions, electrons are needed, and to build up new layers of the crystal, Cu^+ ions are needed. Both species, electrons and Cu^+, ions, are supplied from the interior of the crystal. Thus excess oxygen is incorporated in Cu_2O by forming vacant cation sites ($Cu^+\square$), and vacant electron levels ($e\square$) in the crystal. The vacant electron levels are called *positive holes*, and in Cu_2O they may be pictured as Cu^{2+} ions at normal cation sites in the lattice. The cation vacancies are negative charges in the Cu_2O lattice, while the positive holes are positive charges. Consequently, they attract each other, and at low temperatures may stick together. We shall assume that the temperature is high enough for this trapping of positive holes by cation vacancies to be ignored. The chemical reaction giving non-stoichiometric Cu_2O may then be written

$$\tfrac{1}{2}O_2 \rightarrow Cu_2O + 2(Cu^+\square) + 2(e\square) \qquad \ldots(1.157)$$

four cation vacancies and four positive holes being produced by every oxygen molecule absorbed. The non-stoichiometric oxide conducts electricity with the movement of positive holes, and because of this the conductivity is said to be *p*-type (positive carriers).

If n is the concentration of defects (cation vacancies or positive holes) at equilibrium, then, applying the law of mass action to equation 1.157

$$n = p^{1/8}K^{1/4} \qquad \ldots(1.158)$$

where p is the oxygen pressure, and K the equilibrium constant for the reaction. The formula for K can be found only from statistical mechanics, and a simple calculation is instructive. The important quantity which we need to estimate is the entropy change when one molecule of oxygen is absorbed, with the formation of four defects in the lattice. This entropy change may be divided into two parts. The first is the entropy change, ΔS_1, when one molecule is absorbed, with the formation of defects at specified lattice points; the second is the entropy change, ΔS_2, when we allow for the fact that the defects may be formed anywhere in the crystal. ΔS_1 is independent of the existing defect concentration, and, in addition, is almost certainly negative because it involves the loss of translational and rotational degrees of freedom of the oxygen molecule. ΔS_2 is positive and depends on the existing defect concentration. To calculate ΔS_2 we proceed as follows. Let there be N cation sites in the crystal and n defects. The number of arrangements, P, of the cation vacancies on the lattice sites is

$$P = N!/n!(N-n)! \qquad \ldots(1.159)$$

which contributes an entropy $S_c = k\ell nP$, where k is Boltzmann's constant. Taking logarithms in equation 1.159, using Stirling's approximation

$$(\ln x! \approx x\ln x - x \text{ as } x \to \infty)$$

and putting $n/N = \alpha$ we find

$$S_c = -Nk[(1-\alpha)\ln(1-\alpha) + \alpha\ln\alpha] \qquad \ldots(1.160)$$

We note that S_c is positive and goes through a maximum as α increases. If the positive holes were localised on the cations, they would give an entropy contribution S_p exactly equal to S_c. The positive holes have, however, considerable mobility (see below), and are perhaps best treated as an ideal gas consisting of particles of effective mass m. In this case[1]

$$S_p = Nk\alpha[\tfrac{3}{2}\ln(\phi\Omega^{2/3}) + \tfrac{5}{2} - \ln\alpha + \ln 2] \qquad \ldots(1.161)$$

where $\phi = 2\pi mkT/h^2$, h being Planck's constant. Ω is the volume of the oxide per metal ion, and the term $Nk\alpha\ln 2$ takes account of electron spin. We note that S_p is positive but shows no maximum. Adding equations 1.160 and 1.161, and differentiating with respect to α, we obtain ΔS_2. Remembering that $\alpha \ll 1$, we find

$$\Delta S_2 = k[6\ln(\phi\Omega^{2/3}) + 6 + 4\ln 2 - 8\ln\alpha]$$

showing that ΔS_2 is large and positive for small values of α.

ΔS, the entropy change in equation 1.157, is given by

$$\Delta S = \tfrac{1}{2}(\Delta S_1 + \Delta S_2)$$

Since ΔS_1 is independent of α, we see that ΔS becomes positive and increases without limit as $\alpha \to 0$. Thus, although the enthalpy change ΔH in equation 1.157 may be large and positive, the equilibrium condition $\Delta G = \Delta H - T\Delta S = 0$ is satisfied for some value of α different from zero except at $T = 0$. This proves that Cu_2O is non-stoichiometric at any temperature above the absolute zero.

Regarding ΔS_1, the simplest assumption we can make is that it comes entirely from the loss of translational and rotational degrees of freedom of the oxygen molecule when it is absorbed into the crystal. A standard calculation then gives[1] (the electronic ground state of O_2 is a triplet)

$$\Delta S_1 = -k[25 + (7/2)\ln(T/298) - \ln(p/\text{atm}) - \ln 3]$$

and equating ΔH and $T\Delta S$ to find the equilibrium condition we obtain

$$\alpha = (9\cdot 3 \times 10^{-2})(298/T)^{7/16}(48\phi^6\Omega^4)^{1/8}(p/\text{atm})^{1/8}\exp(-\Delta H/4kT)$$

Comparison with equation 1.158 would give the formula for K, but because of the approximate nature of the formula for ΔS_1 we cannot rely on the factor $(9 \cdot 3 \times 10^{-2})(298/T)^{7/16}$.

Other p-type oxides are NiO, FeO, CoO, Ag_2O, MnO and SnO. For NiO the reaction giving non-stoichiometry is

$$\tfrac{1}{2}O_2 \rightarrow NiO + (Ni^{2+}\square) + 2(e\square)$$

and the concentration of positive holes is twice that of the cation vacancies. From the mass-action formula, the concentration of positive holes, n_p, is given by

$$n_p = 2^{1/3} p^{1/6} K^{1/3}$$

For all p-type oxides, the defect concentration, and hence the electrical conductivity, increases with the oxygen pressure.

n-type Oxides

ZnO contains excess metal which is accommodated interstitially, i.e. at positions in the lattice which are unoccupied in the perfect crystal. The process by which ZnO in oxygen gas acquires excess metal may be pictured as follows*. The outer layers of the crystal are removed, oxygen is evolved, and zinc atoms go into interstitial positions in the oxide. We represent interstitial zinc by $(Zn\bigcirc)$. However, the interstitial zinc atoms may ionise to give $(Zn^+ \bigcirc)$ or even $(Zn^{2+} \bigcirc)$. The extra electrons produced in this way must occupy electron levels which would be vacant in the perfect crystal. We represent them by the symbol $(e\bigcirc)$, and refer to them as free electrons. They can be pictured as Zn^+ ions at normal cation sites. We see therefore that three reactions can be written, each giving non-stoichiometric ZnO:

$$ZnO \rightarrow \tfrac{1}{2}O_2 + (Zn\bigcirc) \qquad \ldots(1.162)$$

$$ZnO \rightarrow \tfrac{1}{2}O_2 + (Zn^+ \bigcirc) + (e\bigcirc) \qquad \ldots(1.163)$$

$$ZnO \rightarrow \tfrac{1}{2}O_2 + (Zn^{2+} \bigcirc) + 2(e\bigcirc) \qquad \ldots(1.164)$$

The exact situation with ZnO is not altogether clear. Under most experimental conditions it seems that equation 1.163 is the important reaction, but equation 1.164 cannot be ignored at high temperatures. Applying the mass-action formula to equation 1.163 we have for n, the concentration of defects (interstitial Zn^+ ions or free electrons)

$$n = p^{-1/4} K^{1/2}$$

where K is the equilibrium constant of the reaction. The non-stoichiometric oxide conducts electricity with the movement of free electrons, and because of this the conductivity is said to be n-type (negative carriers).

Other n-type oxides are TiO_2, CdO, Al_2O_3 and V_2O_5. For TiO_2, the reaction giving the non-stoichiometry is[2]

$$\tfrac{1}{2}TiO_2 \rightarrow \tfrac{1}{2}O_2 + (O^{2-}\square) + 2(e\bigcirc)$$

*The mechanism is of no consequence in determining the equilibrium situation.

and the defects are anion vacancies and free electrons (Ti^{3+} ions). From the mass-action formula, the concentration of free electrons n_e is given by*

$$n_e = 2^{1/3} p^{-1/6} K^{1/3}$$

For all n-type oxides, the defect concentration, and hence the electrical conductivity, decreases with the oxygen pressure.

Motion of Lattice Defects

Lattice defects do not occupy fixed positions in the crystal. In Cu_2O for example, a positive hole moves about because an electron from a neighbouring Cu^+ ion can fall into the vacant level which the positive hole represents [Fig. 1.73 (top)]. Likewise, the cation vacancy moves when a neighbouring cation jumps into the vacant site [Fig. 1.73 (bottom)]. We expect the positive hole to be the more mobile because its motion is an electronic process, not an ionic one. This expectation is confirmed by measurements of the defect mobilities. The mobility, v, of a charged defect is the drift velocity acquired by the defect in an applied electric field of unit strength. It is related to the conductivity, σ, by $\sigma = nev$, where n is the concentration and e the charge of the defect, and to the diffusion coefficient D by Einstein's relation $v/D = e/kT$ (see Section 1.9). For Cu_2O at 1000°C, Wagner and Hammen[3] found for the hole mobility $v_p \approx 10 \, cm^2 \, s^{-1} \, V^{-1}$, and for the mobility of cation vacancies, $v_c \approx 5 \times 10^{-4} \, cm^2 \, s^{-1} \, V^{-1}$. The temperature dependencies of v_p and v_c are, however, fundamentally different from one another, as with decreasing temperature v_c falls but v_p increases. At room temperature[4], v_p is about $80 \, cm^2 \, s^{-1} \, V^{-1}$.

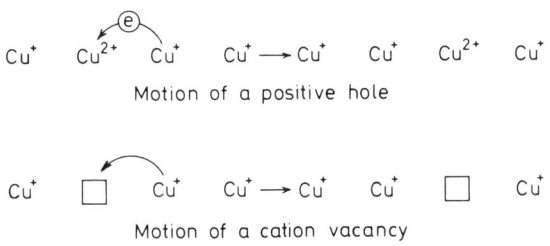

Fig. 1.73 Motion of positive holes and cation vacancies

The reason for this can be seen as follows. In a perfect crystal with the ions held fixed, a positive hole would move about like a free particle with a mass m depending on the nature of the crystal. In an applied electric field, the hole would be uniformly accelerated, and a mobility could not be defined. The existence of a mobility in a real crystal derives from the fact that the uniform acceleration is continually disturbed by deviations from a perfect lattice structure. Among such deviations, the thermal motions of the ions, and in particular, the longitudinal polarisation vibrations, are most important in obstructing the uniform acceleration of the hole. Since the amplitude of the lattice vibrations increases with temperature, we see how the mobility of a

---

*This assumes that anion vacancies are present to a significant extent only in the non-stoichiometric oxide, which is not in fact true for TiO_2.

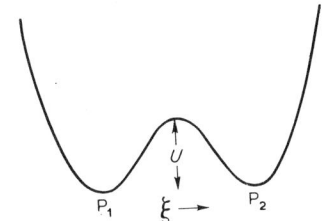

Fig. 1.74 Energy of the crystal as a function of ξ, the co-ordinate of the moving cation

hole decreases as the temperature is raised. On the other hand, a cation vacancy can move only by virtue of the thermal motions of the neighbouring cations so that the vacancy becomes more mobile as the temperature is raised. A parallel argument can of course be given for *n*-type oxides which contain free electrons and either interstitial cations or anion vacancies.

When a metal oxidises to form a continuous oxide film, the transport of material through the oxide is a necessary step in the oxidation process. From what has already been said, it is clear that the slow process will almost certainly be the transport of ions (vacancies) not electrons (positive holes). Let us therefore examine the temperature dependence of the diffusion coefficient (and hence the mobility) of a cation vacancy. Let q be the probability per second that a cation next to a vacancy jumps into the vacancy. When this happens, the vacancy moves a distance a equal to the cation-cation distance. The diffusion coefficient D_c is by definition given by $D_c = qa^2$. Hence we need a formula for q. In Fig. 1.74 we show the energy of the lattice as a function of ξ, the co-ordination of the moving cation. When the cation is at P_1, the vacancy is at P_2, and vice versa. Between P_1 and P_2 there is a symmetrical barrier of height U, and the cation can pass from P_1 to P_2 only if it acquires energy greater than U. This energy is of course supplied by the lattice vibrations. The probability that the ion has this energy is proportional to the usual Boltzmann factor, $\exp(-U/kT)$, so we can write

$$q = q_0 \exp(-U/kT)$$

q_0, which has dimensions of s^{-1}, can be calculated only by a detailed analysis of the jumping process, but this elementary consideration is sufficient to establish the characteristic formulae

$$\left. \begin{array}{ll} D_c = D_o \exp(-U/kT) & D_o = q_o a^2 \\ v_c = v_o \exp(-U/kT) & v_o = D_o e/kT = q_o a^2 e/kT \end{array} \right\} \quad \ldots(1.165)$$

showing the strong temperature dependencies of D_c and v_c. The *self-diffusion coefficient* of Cu$^+$, in Cu$_2$O is, of course, $(n/N)D_c$. Naturally, we can also derive relations like equation 1.165 for interstitial ions.

Finally we notice that in the *p*-type oxides Cu$_2$O and NiO, the presence of excess oxygen actually provides, through the formation of cation vacancies, a transport mechanism for the metal, while in an *n*-type oxide like TiO$_2$, the excess metal, by forming anion vacancies, provides a transport mechanism for oxygen. With *n*-type oxides like ZnO and Al$_2$O$_3$, where the excess metal is accommodated interstitially, a transport mechanism is, of course, provided for the excess component itself.

Defect Clustering

The point defects discussed above exist as such, only at low enough concentrations (perhaps up to 0.5% depending on the material). At higher concentrations the point defects can aggregate into clusters, or become ordered, or can be eliminated by forming 2- or 3-dimensional defects such as shear planes, or voids. In grossly non-stoichiometric oxides such as reduced TiO_2, VO_2, CrO_2, WO_3 and MoO_3, simple point defects may be in the minority.

Shear-plane formation is shown schematically in the diagram below. A structure (a) with aligned oxygen vacancies shears to eliminate these vacancies in favour of an extended planar defect in the cation lattice (b). This process decreases the oxygen to metal ratio in the defect, which is therefore positively charged as were the original oxygen vacancies; the compensating negative charge is still present as reduced metal ions, as Ti^{2+} or Ti^{3+} in non-stoichiometric TiO_2 for example. Shear structures of this sort are observed routinely in high resolution transmission electron microscopy[5].

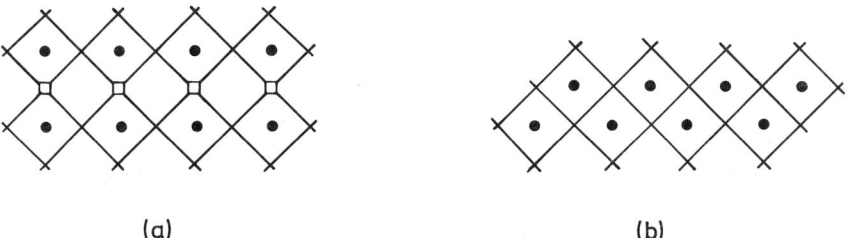

Schematic illustration of shear-plane formation. Structure (*a*) with aligned oxygen vacancies □ shears to eliminate these vacancies in favour of an extended planar defect in the cation lattice as in (*b*). ● cations; oxygen ions are at the mesh intersections

Defect clusters are well-established in transition-metal oxides having the sodium chloride lattice, in oxygen-excess FeO for example[6]. The basic defect unit is almost certainly an interstitial Fe^{3+} ion, electrostatically stabilised by having all four nearest cation sites vacant, i.e. the unit is the tetrahedral complex $(4(Fe^{2+} \square) - Fe^{3+} \bigcirc]^{5-}$, but linear aggregates of two or three such units are apparently even more stable[7]. The most important consequence of this defect clustering for metal oxidation is that the cation self-diffusion coefficient is reduced because the point defects are bound in these larger, and consequently less mobile clusters. Clusters are in dynamic equilibrium with point defects, and therefore defect clusters *can* move by 'dissolving' one one side, and 'growing' on another. However the role of defect clusters in mass transport remains obscure, and even where point defects are the minority species, they may nevertheless, provide the mass transport.

Determination and Characterisation of Defect Structures

The presence of small concentrations of point defects changes the density of a crystal, and four values of the density can be calculated, depending on whether the deviation from stoichiometry is due to interstitials, or vacancies, and whether they are on the metal or oxygen sublattices.

Consider the stoichiometric oxide MO_n. If the unit cell of volume Ω contains z molecules, the density is

$$z(M + nO)/N_A\Omega$$

where M and O are the atomic weights, and N_A is Avagadro's number. For a non-stoichiometric oxide MO_{n+x} with interstitial oxygen, the density is

$$z(M + nO + xO)/N_A\Omega$$

while with metal vacancies it is

$$zn\left(\frac{M}{n+x} + O\right)/N_A\Omega.$$

If the non-stoichiometry is due to excess metal, and we write the formula as $M_{1+x}O_n$, the two values of the density are

$$z(M + xM + nO)/N_A\Omega$$

for interstitial metal, and

$$z\left(M + \frac{n}{1+x}O\right)/N_A\Omega$$

for oxygen vacancies. Thus, experimental determination of the density by pyknometric methods, and of the unit cell volume by X-ray measurements of the lattice parameters, for a series of oxides with different oxygen/metal ratios, should yield precise information on the nature of the point defects causing the non-stoichiometry. An example is such a study is that[8] on UO_{n+x} which establishes interstitial oxygen as the defect for small x. For large x ($x \sim 0.15$), clusters of oxygen interstitials *and vacancies* are formed[9].

To get more information on point defects, in particular their state of ionisation (interstitial atoms or ions, doubly or singly charged oxygen vacancies etc.), it is in principle only necessary to measure, as a function of the oxygen pressure, some property of the oxide (its weight, its electrical conductivity, the metal or oxygen self-diffusion coefficients for example) which is directly proportional to the defect concentration. This is clear from the discussions of non-stoichiometric Cu_2O, NiO, and ZnO above. Unfortunately, the number of cases where a clear conclusion can be reached is small because, although the relation

$$\text{Measured Property} \propto p^{1/n} \qquad \text{(a)}$$

where p is the oxygen pressure, is frequently observed, n is often found to be non-integral, and moreover, its value depends on the ranges of temperature and oxygen pressure over which the measurements are made. NiO and CoO are good examples, where values of n between 4 and 6 are found[10]. But the explanation appears to be simple: vacant cation sites and positive holes are incompletely dissociated, i.e. the singly charged vacancy ($Ni^{2+}\square$, $\square$e) is present. If this defect were the dominant species, the mass-action formula applied to

$$\tfrac{1}{2}O_2 \rightarrow NiO + (Ni^{2+}\,\square,\, e\square) + (e\square) \qquad \text{(b)}$$

would give for n_p, the concentration of free positive holes

$$n_p = p^{1/4} K^{1/2} \qquad \text{(c)}$$

instead of the $p^{1/6}$ oxygen-pressure dependence given above for completely dissociated species. Values between 1/4 and 1/6 are therefore easy to understand.

Investigations based on equation (a) are *indirect*. Direct structural studies using diffraction techniques (X-ray or neutron), or electron microscopy, while they cannot detect the low concentrations of defects present in NiO or CoO, are indispensible to the study of grossly non-stoichiometric oxides like FeO, TiO_2, WO_3 etc., and particularly electron microscopes with a point-to-point resolution of about 0.2 nm are widely used. The first direct observation of a point defect (actually a complex of two interstitial metal atoms, and two oxygen atoms in $Nb_{12}O_{29}$) was made[11] using electron microscopy.

Nomenclature

A standard nomenclature for point defects has evolved. It is given below.

Defect	Standard notation	Notation in this section
'Free' Electron	e'	$(e\bigcirc)$
Positive hole	$h^\cdot$	$(e\square)$
Metal vacancy	V_M'', V_M'	$(M^{2+}\,\square)$, $(M^+\,\square)$ or $(M^{2+}\,\square,\,e\square)$
Oxygen vacancy	$V_{\ddot{O}}$	$(O^{2-}\,\square)$
Singly charged oxygen vacancy	$V_{\dot{O}}$	—
Interstitial metal	M_i, M_i'', M_i'	$(M\bigcirc)$, $(M^{2+}\bigcirc)$, $(M^+\bigcirc)$
Interstitial oxygen	O_i''	—

T. B. GRIMLEY

REFERENCES

1. See for example Rushbrooke, G. S., *Introduction to Statistical Mechanics*, Oxford University Press, 135 (1949)
2. Breckenridge, R. G. and Hosler, W. R., *Phys. Rev.*, **91**, 793 (1953)
3. Wagner, C. and Hammen, H., *Z. Phys. Chem.*, **B40**, 197 (1938)
4. See for example Brattain, W. H., *Rev. Mod. Phys.*, **23**, 203 (1951)
5. Iijima, S., *J. Solid State Chem.*, **14**, 52 (1975)
6. Koch, F. and Cohen, J. B., *Acta Crystallogr.*, Sect. B **25**, 275 (1969)
7. Catlow, C. R. A. and Fender, B. E. F., *J. Phys.* **C8**, 3267 (1975)
8. Lynds, L., Young, W. A., Mohl, J. S., and Libowitz, G. G., in *Nonstoichiometric Compounds*, American Chemical Society, Washington, D.C., 58 (1963)
9. Catlow, C. R. A., in *Nonstoichiometric Oxides*, Ed. O. Toft Sørensen, Academic Press, New York, 61 (1981)
10. Tallon, N. M., in *Defects and Transport in Oxides*, Eds M. S. Seltzer and R. I. Jaffee, Plenum Press, New York, 239 (1974)
11. Iijima, S. and Allpress, J. G., *J. Solid State Chem.*, **7**, 94 (1973)

BIBLIOGRAPHY

Greenwood, N. N., *Ionic Crystals, Lattice Defects and Nonstoichiometry*, Butterworths, London (1968)

Seltzer, M. S. and Jaffee, R. I. (Eds), *Defects and Transport in Oxides*, Plenum Press, New York (1974)

Sørensen, O. T. (Ed.), *Nonstoichiometric Oxides*, Academic Press, New York (1981)

Kröger, T. A. and Vink, H. J., *Solid State Phys.*, **3**, 310 (1956)

Van Gool, W., *Principles of Defect Chemistry of Crystalline Solids*, Academic Press, London and New York (1966)

1.9 Continuous Oxide Films

When a solid metal is attacked by oxygen gas, the product of the reaction is the metal oxide which, if it is not volatile, builds up as a surface layer on the metal. The oxide layer may be protective or non-protective. A non-protective layer does not inhibit the continued access of oxygen to the unchanged metal; the rate of growth of such an oxide layer is independent of its thickness X and the law of growth is $dX/dt = k_1$. On integration this gives the linear law

$$X - X_0 = k_1 t$$

and k_1 is the *linear rate constant*. Discontinuous films are considered in Section 1.10.

A protective oxide layer forms a continuous barrier between the reactants (oxygen and metal), which inhibits the reaction. The simplest assumption that can be made about the effectiveness of this barrier is that its protecting power is directly proportional to its thickness. Mathematically, $dX/dt = k_2/X$, which on integration gives the parabolic law,

$$X - X_0^2 = 2k_2 t$$

where k_2 is the *parabolic rate constant*. This law is obeyed for the high-temperature oxidation of many metals.

Since a protective layer separates the primary reactants, further growth of the layer involves two essential steps: (*a*) surface reactions at the metal/oxide and oxide/oxygen interfaces and (*b*) transport of material through the oxide. The rate of growth will be controlled by whichever of these two steps is the slower. If the layer is thin enough, a surface reaction must control the rate. Indeed, during the initial stages of oxide formation step (*b*) is absent. As the layer thickens, step (*b*) may be distinguished as a definite process which ultimately controls the rate. When this stage is reached, the surface reactions, by which both metal and oxygen are incorporated into the oxide, proceed to near equilibrium. The products of these reactions are then transported through the oxide by a driving force derived from the free energy change in the oxidation reaction. The first problem is to decide the nature of these reaction products.

Surface Reactions

Metal oxides are not normally stoichiometric, although the non-stoichiometry may be too small to be detected by ordinary methods of analysis (Section 1.8). Cuprous oxide, for example, absorbs oxygen to build up new layers of the oxide with the formation of vacant cation sites ($Cu^+ \square$) and positive holes ($e\square$).

The reaction of Cu_2O with oxygen may be written,

$$\tfrac{1}{2}O_2 \rightarrow Cu_2O + 2(Cu^+ \square) + 2(e\square) \qquad \ldots(1.166)$$

two cation vacancies and two positive holes being formed for each oxygen atom absorbed. If $n(Cu^+ \square)$ and $n(e\square)$ are the concentrations of cation vacancies and positive holes when thermodynamic equilibrium is established, the usual mass action formula is

$$n^2(Cu^+ \square)n^2(e\square) = p^{1/2}K \qquad \ldots(1.167)$$

where p is the oxygen pressure and K the equilibrium constant. Except in narrow regions near the surfaces where space charges can exist, the oxide must be electrically neutral. In the interior, therefore,

$$n(Cu^+ \square) = n(e\square) = n$$

say, and

$$n = p^{1/8}K^{1/4} \qquad \ldots(1.168)$$

The electrical conductivity is proportional to n. Equation 1.168 therefore predicts an electrical conductivity varying as $p^{1/8}$. Experimental results show proportionality to $p^{1/7}$, and this discrepancy is probably due to incomplete disorder of cation vacancies and positive holes. An effect of this sort (deviation from ideal thermodynamic behaviour) is not allowed for in the simple mass action formula of equation 1.167.

Consider now the system Cu/Cu_2O in oxygen gas at a pressure p_X (X signifies the oxide/oxygen interface in Fig. 1.75). Ignoring space charges, n_X the equilibrium concentration of cation vacancies or positive holes at the Cu_2O/O_2 interface, is given by

$$n_X = p_X^{1/8}K^{1/4} \qquad \ldots(1.169)$$

and n_0 their equilibrium concentration at the Cu/Cu_2O interface, by

$$n_0 = p_0^{1/8}K^{1/4} \qquad \ldots(1.170)$$

In equation 1.170 p_0 is the dissociation pressure of Cu_2O (oxygen pressure for the equilibrium $2Cu + \tfrac{1}{2}O_2 \rightleftharpoons Cu_2O$). If $p_X > p_0$, then $n_X > n_0$, and the oxidation proceeds with cation vacancies and positive holes being created at the Cu_2O/O_2 interface and moving inwards to be destroyed at the Cu/Cu_2O interface. A similar situation exists whenever a p-type oxide is formed, for example with Ni/NiO.

As an example of a different type of oxide, we may consider ZnO. This oxide evolves oxygen and forms cations in interstitial positions ($Zn^+ \circ$) or ($Zn^{2+} \circ$), and free electrons ($e\circ$). If the interstitial zinc ions are only singly charged, the reaction describing the non-stoichiometry may be written

$$ZnO \rightarrow \tfrac{1}{2}O_2 + (Zn^+ \circ) + (e\circ)$$

Using the simple mass action formula, the equations corresponding to 1.169 and 1.170 are

$$n_x = p_x^{-1/4} K^{1/2}$$
$$n_0 = p_0^{-1/4} K^{1/2}$$
...(1.171)

with p_0 equal to the dissociation pressure of ZnO. If $p_x > p_0$ then $n_x < n_0$, and the oxidation proceeds with interstitial cations and free electrons moving outwards from the Zn/ZnO interface. A similar situation exists whenever the oxidation product is an n-type oxide. The system Al/Al$_2$O$_3$ is another example.

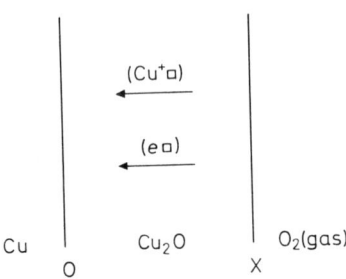

Fig. 1.75 Movement of defects in the oxidation of copper

Wagner's Theory of the Parabolic Law

For definiteness, the oxidation of copper to copper(I) oxide may be considered. Our picture of the process is that cation vacancies and positive holes formed at the Cu$_2$O/O$_2$ interface by equation, 1.166 are transported to the Cu/Cu$_2$O interface where they are destroyed by copper dissolving in the non-stoichiometric oxide. We require an expression for the rate of oxidation.

We denote by x the distance from the metal surface, and by $n_c(x)$ and $n_p(x)$ the concentrations of cation vancancies and positive holes in the oxide. Let v_c and v_p be their mobilities, and D_c and D_p their diffusion coefficients. Let $F(x)$ be the electrostatic field in the oxide. J_c, the flux of cation vacancies (number crossing unit area per second), will be expressed by

$$J_c = -D_c \frac{dn_c}{dx} - n_c v_c F \qquad ...(1.172)$$

and J_p, the flux of positive holes, by

$$J_p = -D_p \frac{dn_p}{dx} + n_p v_p F \qquad ...(1.173)$$

In equation 1.172 the first term on the right-hand side is the flux due to a concentration gradient, the second is that due to the electric field. It should be noted that cation vacancies are negatively charged carriers of electricity. The terms in equation 1.173 have similar meanings; positive holes are positively charged carriers. If the system were at equilibrium we would have $J_c = J_p = 0$ and the field F would be negligible except in narrow regions near the two interfaces, but when oxidation is proceeding this is not generally the case.

In a steady state of oxidation $J_p = J_c = J$, say. Eliminating F from equations 1.172 and 1.173 and using Einstein's relation (see below) $v/D = e/kT$, where k is Boltzmann's constant and T the absolute temperature,

$$J = -(kT/e)[n_c v_c n_p v_p/(n_c v_c + n_p v_p)]\frac{d}{dx}\ln(n_c n_p) \quad \ldots(1.174)$$

The negative sign means that cation vacancies and positive holes move inwards, i.e. in the negative direction of x. For Cu_2O, the positive holes are much more mobile than the cation vacancies, and we can assume that $n_p v_p \gg n_c v_c$. The oxidation flux is then

$$J = -(kT/e)n_c v_c \frac{d}{dx}\ln(n_c n_p) \quad \ldots(1.175)$$

This equation can be obtained in another way which may be more instructive. Assume that the slow step in the oxidation is the transport of cation vacancies. The positive holes may then be considered to take up their equilibrium distribution, defined by Boltzmann's equation

$$n_p = n_p(0)\exp(-eV/kT) \quad \ldots(1.176)$$

Here n_p is the concentration of positive holes at any point in the oxide where the electrical potential is V, and $n_p(0)$ is their concentration at the Cu/Cu_2O interface where $V = 0$. Differentiating equation 1.176, we obtain an equation for the electric field, namely

$$F = -\frac{dV}{dx} = (kT/e)\frac{d}{dx}\ln n_p \quad \ldots(1.177)$$

Substituting this in equation 1.172 and using Einstein's relation between the mobility and the diffusion coefficient of cation vacancies, we obtain equation 1.175 for the oxidation flux J_c. This derivation shows that equation 1.175 is valid if all processes involving positive holes (and therefore electrons) are so fast that the oxidation flux can be carried without significantly disturbing their equilibrium distribution. We note at this stage that to derive Einstein's relation it is only necessary to compare equation 1.176 with the equation obtained by integrating equation 1.173 for $J_p = 0$. The integration gives

$$n_p = n_p(0)\exp(-v_p V/D_p)$$

and this establishes the result $v_p/D_p = e/kT$. The corresponding result for cation vacancies is derived similarly.

Returning now to equation 1.175, we cannot in general proceed without knowing how n_c and n_p vary through the oxide. However, if the oxide layer is thick enough, the situation is simple, for we can assume that electrical neutrality is preserved in the oxide except in narrow regions near the two interfaces where there are space charges. Thus we have $n_c = n_p = n$ say, except in the space-charge regions which, however, we can neglect if the oxide layer is thick enough. Then equation 1.175 gives

$$J = -2(kT/e)v_c \frac{dn}{dx} \quad \ldots(1.178)$$

However, J must be independent of x, so this equation can be integrated to give

$$JX = -2(kT/e)v_c(n_X - n_0) \qquad \ldots(1.179)$$

where X is the thickness of the oxide layer, and n_X and n_0 are the concentrations of cation vacancies or positive holes at the two interfaces. These concentrations are of course given by equations 1.169 and 1.170.

Our picture of the transport process in these thick oxide layers is that there is a uniform concentration gradient of defects (cation vacancies and positive holes) across the layer. But it is important to notice that the oxidation flux is exactly twice that to be expected if diffusion alone were responsible for the transport of cation vacancies. The reason for this is, of course, that the more mobile positive holes set up an electric field which assists the transport of the slower-moving cation vacancies.

If Ω is the volume of the oxide per metal atom, the rate of growth, dX/dt, is equal to $|J|\Omega$. Thus from equation 1.179 we derive the parabolic law

$$\left. \begin{array}{l} dX/dt = k_2/X \\ k_2 = 2D_c\Omega(n_X - n_0) \end{array} \right\} \qquad \ldots(1.180)$$

This formula for k_2 can be cast into another form by using equations 1.169 and 1.170. We note first that in these latter equations $K^{1/4}$ is the concentration of defects in Cu_2O at 1 atm pressure of oxygen, so that $(K^{1/4}D_c\Omega)$ is the self-diffusion coefficient of Cu^+ in Cu_2O at this oxygen pressure. Call this self-diffusion coefficient D_c^0, then

$$k_2 = 2D_c^0(p_X^{1/8} - p_0^{1/8}) \qquad \ldots(1.181)$$

This equation shows that if $p_X \gg p_0$, k_2 depends upon the oxygen pressure and the temperature in the same way as the self-diffusion coefficient of Cu^+ in Cu_2O. Regarding the temperature dependence, self-diffusion coefficients in solids depend exponentially on the temperature (*see* Section 1.8), i.e.

$$D_c^0 = A\exp(-Q/kT)$$

and Q is the *activation energy* for self-diffusion. Thus for the temperature dependence of k_2 we have

$$k_2 = B\exp(-Q/kT)$$

and the activation energy for the oxidation reaction should be the same as that for self-diffusion of cations in the oxide.

The above account of the oxidation of Cu to Cu_2O is a simplified version of the more general theory developed by Wagner[1,2]. Cu_2O is a *p*-type oxide. As an example of a system where an *n*-type oxide is formed, we shall consider the oxidation of Zn to ZnO. Here Zn dissolves in the oxide at the Zn/ZnO interface to give interstitial cations and free electrons. These defects cross to the ZnO/O_2 interface and react with oxygen to build up new layers of the oxide. The slow step in the oxidation is the transport of interstitial cations, and if these are singly charged we still have equation 1.179 for the flux, except that v_c is the mobility of interstitial cations and n_X and n_0 are given by equation 1.171. We note further that J is positive in this case because defects move outwards. The expression for the parabolic rate constant corresponding to equation 1.181 is

$$k_2 = 2D_c^0(p_0^{-1/4} - p_X^{-1/4})$$

and if $p_X \gg p_0$, the oxidation rate should be *independent* of the oxygen pressure. This pressure independence should be noted since the self-diffusion coefficient of Zn in ZnO does depend on the oxygen pressure. Further, the temperature dependence of k_2 is *not* simply that of the self-diffusion coefficient, since p_0, the dissociation pressure of ZnO, is also temperature dependent. Because of this the activation energy for the oxidation reaction is less than that for self-diffusion by half the heat of formation of ZnO.

Thin Oxide Films

Wagner's theory of the parabolic law involves the following assumptions:
1. The oxide layer is compact and adherent.
2. The slow step is the transport of material through the oxide layer.
3. The layer is so thick that the space-charge regions at the two interfaces are unimportant and the oxide can be regarded as electrically neutral.

In the early stages of oxidation when the oxide layer is thin, it is clear that assumption (3) must be invalid. The limiting simple case when the layer is so thin that space charges can be neglected because they are small compared with the surface charges has been considered by Mott[3]. All other assumptions are the same as in Wagner's theory. In the oxidations of Cu, for example, we assume that electronic equilibrium is established in the system $Cu/Cu_2O/O_2$. This sets up an electrical potential difference across the oxide layer because electrons are transferred from the metal to form oxygen ions adsorbed on the outer surface of the oxide. If the surface charge formed in this way is large compared with the space charge in the oxide, the electric field is uniform and equal to V/X where V is the potential drop across the oxide. In a thin oxide layer this field may be very large. For example, a potential difference of 1 V gives a field of 10^5 V/cm in a layer 100 nm thick. Now the flux of cation vacancies (and hence the oxidation flux) is given by equation 1.172, and if F is as large as this, a significant oxidation rate is to be expected even at ordinary temperatures where the diffusion coefficient is very small. For the system Cu/Cu_2O the theory gives a cubic law of growth $(dX/dt = k_3/X^2)$ at ordinary temperatures, and a parabolic law at high temperatures. The parabolic rate constant is, however, entirely different from that in Wagner's theory. When the oxidation product is an *n*-type oxide like ZnO or Al_2O_3, the law of growth is parabolic both at ordinary temperatures and at high temperatures. The two rate constants are different, and both differ from that in Wagner's theory. For further details, the original papers[3–6] should be consulted.

Very Thin Oxide Layers

Many metals oxidise rapidly at first when exposed to oxygen at sufficiently low temperatures, but after a few minutes, when a very thin oxide layer has been formed, the reaction virtually ceases. Oxide layers formed in this way are about 5 nm thick. Aluminium and chromium are well-known examples, showing this sort of behaviour at room temperature. A theory of the effect has been proposed by Mott[3,4].

For definiteness consider the system Al/Al$_2$O$_3$. It is assumed as before that electronic equilibrium is established so that there is a field in the oxide associated with the presence of oxygen ions adsorbed on the outer surface. In a very thin layer this field will be enormous (about 10^6 V/cm if $X = 10$ nm), and we cannot assume, as we did in equation 1.172, that the contribution which this field makes to the flux of cations is simply proportional to the field. An investigation of the transport process in such strong fields (see below) shows that the flux increases exponentially with the field, and because of this cations are transported through the oxide much more rapidly than would be expected on the basis of equation 1.172. It seems unlikely therefore that cation transport can be the slow step in the reaction, and the rate will be controlled instead by a surface reaction. For the system Al/Al$_2$O$_3$, the slow step is probably that by which Al^{3+} ions enter interstitial positions in the oxide at the Al/Al$_2$O$_3$ interface. The rate of this process is also influenced by the strong electric field. The potential energy diagram for an ion leaving the metal and entering the oxide is shown in Fig. 1.76. P represents an ion in the metal surface and I_1, I_2, are interstitial positions in the

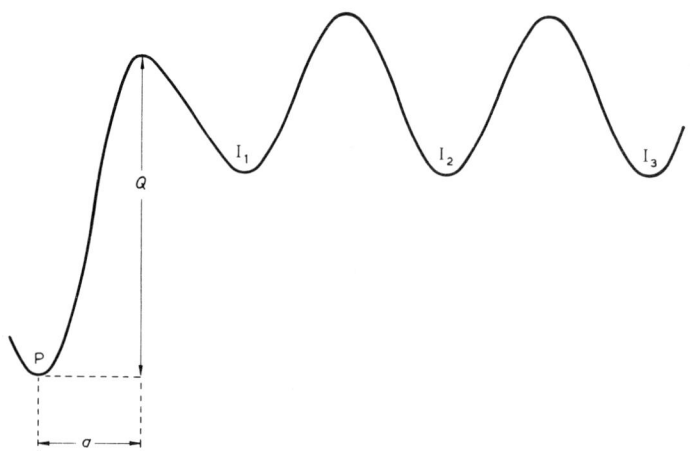

Fig. 1.76 Potential energy of an interstitial ion near the metal/oxide interface

oxide. The height of the barrier between P and I_1 is Q when the field is zero. The field lowers this barrier by an amount $zeaF$ where ze is the charge on the ion and a the distance between P and the top of the barrier. The probability per second that an ion jumps from P to I_1 in the absence of a field is $v \exp(-Q/kT)$, where v is the vibration frequency of an ion at P. In a field F, this probability is increased by the factor $\exp(zeaF/kT)$. Hence if $\mathscr{R}$ is the rate at which ions can enter the oxide when there is no field, the rate in the field is*

$$\mathscr{R} \exp(zeaF/kT)$$

If this process determines the oxidation rate, then with $F = V/X$, the law of growth is

---

* An argument similar to this establishes the exponential relation between the cation flux *through* the oxide and the field F.

$$dX/dt = \Omega \mathscr{R} \exp(\lambda/X)$$
$$\lambda = zeaV/kT$$
...(1.182)

If $V = 1$ V, $a = 0.25$ nm, and $z = 3$, $\lambda = 30$ nm at 300 K, so that for a film 1 nm thick, the field increases the rate of growth by a factor of about 10^{13}. The term in the growth law due to the field, namely $\exp(\lambda/X)$, is large only when X is small. Because of this a thin oxide film can form even at low temperatures where $\mathscr{R}$, the ordinary rate of entry of ions into the oxide, is negligible. As the film thickens, the factor $\exp(\lambda/X)$ decreases rapidly, and the rate of growth soon falls to such a low value that, for practical purposes, oxidation has ended.

Effects of Alloying

An important aspect of any theory of the oxidation of a pure metal is that it enables us to see how the protective power of the oxide layer can be altered by the introduction of alloying constituents into the metal. According to Wagner's theory, the parabolic rate constant for the system Ni/NiO for example depends upon the concentration of cation vacancies in the oxide in equilibrium with oxygen gas. If this concentration can be reduced, the oxidation rate is reduced. Now this can be done if cations of lower valency than Ni^{2+} can be got into the oxide (Fig. 1.77). Suppose, for example, that a little Li is added to the Ni. Each Li^+ ion which replaces Ni^{2+} is a negative

O^{2-}	Ni^{2+}	O^{2-}	Ni^{3+}	O^{2-}	Ni^{2+}	
Ni^{2+}	O^{2-}	Ni^{2+}	O^{2-}	Ni^{2+}	O^{2-}	NiO
O^{2-}	☐	O^{2-}	Ni^{3+}	O^{2-}	Ni^{2+}	
Ni^{2+}	O^{2-}	Ni^{2+}	O^{2-}	Ni^{2+}	O^{2-}	
O^{2-}	Li^+	O^{2-}	Ni^{3+}	O^{2-}	Ni^{2+}	
Ni^{2+}	O^{2-}	Ni^{2+}	O^{2-}	Ni^{2+}	O^{2-}	$NiO + Li_2O$
O^{2-}	Li^+	O^{2-}	Ni^{3+}	O^{2-}	Ni^{2+}	
Ni^{2+}	O^{2-}	Ni^{2+}	O^{2-}	Ni^{2+}	O^{2-}	
O^{2-}	Ni^{2+}	O^{2-}	Ni^{3+}	O^{2-}	Ni^{2+}	
Cr^{3+}	O^{2-}	Ni^{2+}	O^{2-}	Ni^{2+}	O^{2-}	$NiO + Cr_2O_3$
O^{2-}	☐	O^{2-}	Ni^{3+}	O^{2-}	Ni^{2+}	
Ni^{2+}	O^{2-}	Cr^{3+}	O^{2-}	☐	O^{2-}	

Fig. 1.77 Effects of Li_2O and Cr_2O_3 on the defect structure of NiO

charge in the NiO lattice. To preserve electrical neutrality, one positive hole ($e\square$) must be created for each Li^+ ion introduced. But the product $n(Ni^{2+}\square) n(e\square)$ is fixed by the reaction governing the non-stoichiometry of NiO. Hence $n(Ni^{2+}\square)$ falls and the oxidation rate is reduced. By a similar argument, an alloying constituent of higher valency than Ni^{2+} (Cr^{3+} for example) which enters the oxide layer in place of Ni^{2+} increases the oxidation rate.

When the oxidation product is an n-type oxide like ZnO, the conditions are reversed (Fig. 1.78). If a monovalent ion like Li^+ enters the oxide layer in place of Zn^{2+} one free electron ($e\circ$) is destroyed. But the product $n(Zn^+\circ)n(e\circ)$ is fixed by the reaction governing the non-stoichiometry of ZnO. Hence $n(Zn^+\circ)$, the concentration of interstitial Zn^+ ions, increases, and the oxidation rate, which depends upon the concentration of these ions in the oxide in equilibrium with metallic Zn, increases.

This simple account of the effect of alloying constituents is valid only if the second metal shares in the oxide formation by dissolving freely in the oxide of the basis metal. Further, the second metal should be present in the oxide layer in such low concentrations that it can be regarded as an impurity in the oxide of the basis metal. If the alloying constituent is insoluble in the oxide of the original metal, or if a new phase, for example a spinel, is formed, the discussion fails. The spinel $NiCr_2O_4$ is in fact formed in the oxidation of Ni–Cr alloys when the Cr content is high enough, and the oxidation rate is decreased, not increased as we would expect from the simple discussion above.

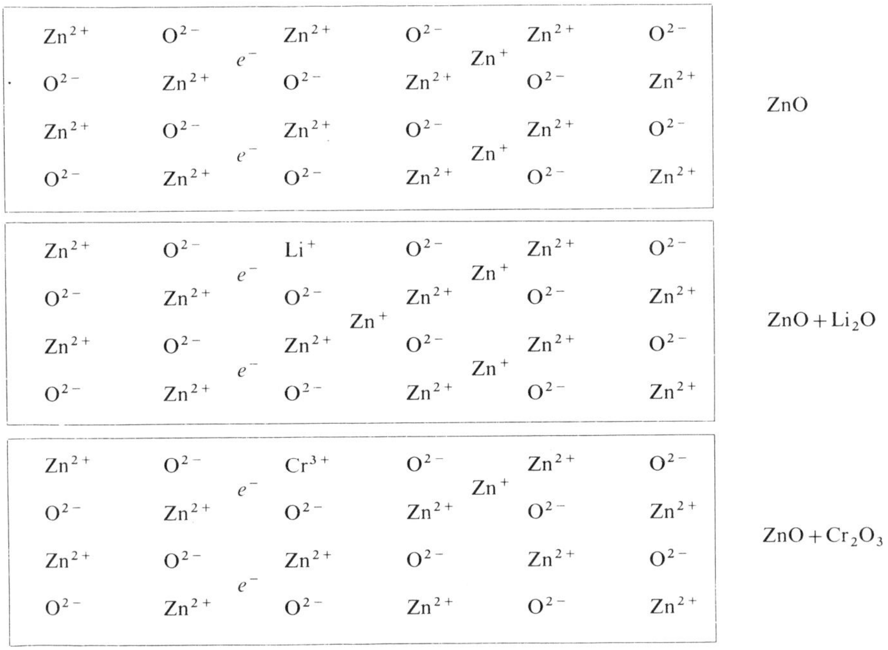

Fig. 1.78 Effects of Li_2O and Cr_2O_3 on the defect structure of ZnO

CONTINUOUS OXIDE FILMS

To examine the situation with alloys in a little more detail, the Cu–Ni alloys will first be considered. Here the mutual solubility of the two oxides NiO and Cu_2O can probably be neglected, and these are the only two possible oxidation products. Assume for simplicity that the alloy is thermodynamically ideal, and let x_{Cu} and x_{Ni} be the mole fractions in the alloy. Consider the reactions

$$2Cu + \tfrac{1}{2}O_2 \rightarrow Cu_2O$$

$$Ni + \tfrac{1}{2}O_2 \rightarrow NiO$$

whereby Cu_2O and NiO are formed by oxidation of the *alloy*. The equilibrium conditions are

$$\left.\begin{array}{l} x_{Cu}^2 p^{1/2} = p^{1/2}(Cu_2O) \\ x_{Ni} p^{1/2} = p^{1/2}(NiO) \end{array}\right\} \qquad \ldots(1.183)$$

where $p(Cu_2O)$ and $p(NiO)$ are the dissociation pressures of the two oxides, and p is the effective oxygen pressure at the alloy surface. The two relations of equation 1.183 are illustrated in Fig. 1.79 by plotting p against x_{Ni}. Note that $p(NiO) < p(Cu_2O)$ and $x_{Ni} + x_{Cu} = 1$. The two curves intersect at one value of x_{Ni}, and this defines the alloy composition for which Cu_2O and NiO can co-exist on the surface. If the Ni content is higher than this critical amount, only NiO is stable; for lower Ni contents only Cu_2O is stable. If therefore the diffusion coefficients of Cu and Ni in the alloy were very large so that the composition of the alloy in the surface region did not change during oxidation, the situation would be simple, and the oxidation product would be either Cu_2O or NiO except at one critical alloy composition where

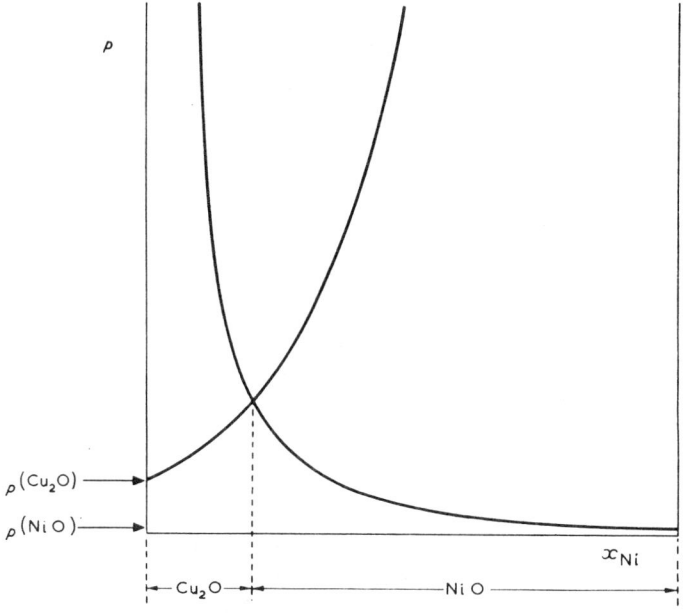

Fig. 1.79 Surface oxides on Cu–Ni alloys

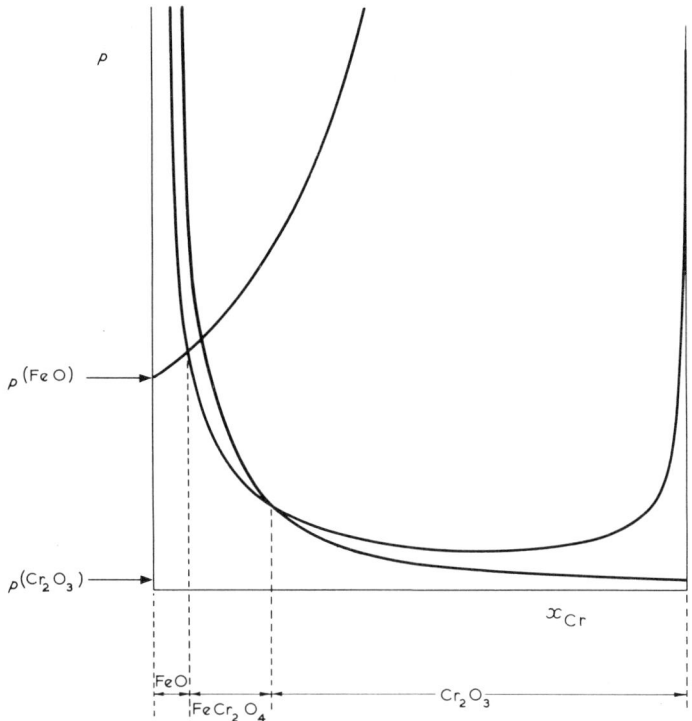

Fig. 1.80 Surface oxides on Fe–Cr alloys

both oxides would be formed. But the diffusion coefficients in the alloy are not usually large enough for this simple analysis to apply, and if we start with a bulk alloy composition in the region where we would expect only NiO to be formed then, as oxidation proceeds, the surface region of the alloy is depleted of Ni. Thus Cu diffuses inwards and Ni has to be supplied from the interior of the alloy. The composition of the alloy at the surface necessary to maintain the supply of Ni into the oxide layer may be such that the Ni content is below the critical concentration for which only NiO is formed. If this is so, both Cu_2O and NiO are formed as oxidation products; if not, we still get only NiO. The effect of finite diffusion rates in the alloy is therefore that the critical *bulk* alloy composition for the exclusive formation of NiO is pushed to a higher Ni content than we would expect from elementary consideration. Similarly, the minimum Cu content for the exclusive formation of Cu_2O is pushed higher. There is therefore a *range* of compositions of the bulk alloy in which both Cu_2O and NiO are formed together.

This analysis shows that if the oxides of the two components of a binary alloy are mutually insoluble, and if one of the components has a much greater affinity for oxygen than the other, then the oxide of the baser metal will be formed exclusively even though it is present in the alloy in only a small amount. It seems that the importance of beryllium as an alloying constituent can be explained in this way. It has a high affinity for oxygen [$p(BeO) \approx 10^{-30}$ atm at 1000°C] and also forms a highly protective oxide layer. The oxidation resistance of metals more stable to oxygen than Be, but which normally oxidise faster, should therefore be improved by the addition of Be, provided that the oxide of the basis metal is not soluble in BeO. The addition of Be to Cu is an example.

Another example which can be argued on the same lines is that of Cr-Fe alloys. This is more complicated, and for simplicity we may assume that only FeO could be formed as the oxidation product of Fe. In addition, Cr_2O_3 and the spinel $FeCr_2O_4$ can be formed. We expect the dissociation pressures to be in the order $p(FeO) > p(FeCr_2O_4) > p(Cr_2O_3)$ and Fig. 1.80 may be constructed showing two critical Cr contents of the alloy. Below the first only FeO is formed, between the first and the second only $FeCr_2O_4$, and above the second only Cr_2O_3. The existence of finite diffusion rates in the alloy will, of course, smear out these divisions. There is, however, the possibility that by adding Cr to Fe, either the spinel $FeCr_2O_4$ or the single oxide Cr_2O_3 is formed exclusively as the oxidation product. A Cr_2O_3 layer is certainly protective. A spinel layer will be protective if the diffusion coefficient for Fe^{2+} (or Fe^{3+}) in the spinel is lower than that in the oxides of iron. We note that the protective layer (either Cr_2O_3 or $FeCr_2O_4$) is formed next to the alloy. Beyond this there will almost certainly be another layer composed mainly of the oxides of iron. This portion is without influence on the protective properties of Cr.

Experimental Techniques

Oxidation is followed by measuring the gain in weight of the specimen with time. An electrostatic field applied across the growing oxide enhances or reduces the oxidation rate according to the polarity of the field, and the charge on the moving species. The movement, or lack of it, of an inert marker placed on the metal prior to oxidation indicates whether the oxide grows by metal moving outwards, or oxygen moving inwards (*see* Section 1.10). Techniques of modern surface science (Auger Electron Spectroscopy (AES), Secondary Ion Mass Spectrometry (SIMS), X-ray Photoelectron Spectroscopy (XPS), Ion Scattering Spectroscopy (ISS), for example) are used to determine the composition, and the thickness of tarnish films. Three examples must suffice. Firstly, ion scattering has been used[7] to analyse air-formed films on Fe-Cr alloys. Incident ^{20}Ne or ^{3}He ions with energies in the range $1.3\,keV$ scattered at 90°, when energy-analysed, have peaks for each element in the surface of the film, and since the incident beam sputters the surface, a depth profile is also obtained. As expected from the discussion above, at the oxide/air interface the Cr/Fe ratio is low, as is the metal/oxygen ratio, and the Cr/Fe ratio increases going into the oxide. But an unexpected finding is that the latter ratio peaks a short distance into the oxide. No explanation of this has been given. A second example of the use of surface-sensitive analytical techniques is the investigation[8] using AES, and argon-ion sputtering, of the composition, and thickness of the films formed on Ni in air in various relative humidities. The findings of this work will be mentioned below. Angularly resolved XPS, unlike depth profiling by sputtering, is non-destructive. Photoelectrons from the metal, and from its different oxides, are identified by their chemical shifts. Those originating from the metal are attenuated by the oxide film, and the current $I(X, \theta)$, measured at an angle θ to the normal for a film of thickness X is

$$I(X, \theta) \propto \exp(-X/\lambda_{ox}\cos\theta)$$

where λ_{ox} is the electron mean free path in the oxide. Oxide thickness up to around 20 nm depending on the system, can be measured in this way, and in addition, by varying θ, the uniformity of the film can be checked. Measurements using different XPS lines (when they are present) enables the kinetics of multi-layer film growth to be determined. XPS studies of the oxidation of Nb at 300 K have been published[9]. Nb_2O_5 grows on a thin layer ($\sim$ .5 nm) of NbO_x (x < 1) to a limiting thickness of about 6 nm in several days, with the growth law

$$dX/dt = 2\Omega\mathscr{R}\sinh(\lambda/X)$$

which is (1.182) modified to include the back-reaction.

Atmospheres other than Pure Oxygen

Atmospheres containing H_2O vapour, or SO_2 are technically very important, but much more fundamental work is needed, and there is space here for only elementary considerations.

Consider Ni exposed to O_2/H_2O vapour mixtures. Possible oxidation products are NiO and $Ni(OH)_2$, but the large molar volume of $Ni(OH)_2$, (24 cm^3 compared with that of Ni, 6.6 cm^3) means that the hydroxide is not likely to form as a continuous film. From thermodynamic data, $Ni(OH)_2$ is the stable species in pure water vapour, and in all O_2/H_2O vapour mixtures in which O_2 is present in measurable quantities, and certainly if the partial pressure of O_2 is greater than the dissociation pressure of NiO. But the actual reaction product is determined by kinetics, not by thermodynamics, and because the mechanism of hydroxide formation is more complex than oxide formation, $Ni(OH)_2$ is only expected to form in the later stages of the oxidation at the NiO/gas interface. As it does so, cation vacancies are formed in the oxide according to

$$H_2O(g) + \tfrac{1}{2}O_2(g) \rightarrow Ni(OH)_2 + (Ni^{2+}\square) + 2(e\square)$$

which lowers the degree of protection afforded by the NiO film[8].

Similar considerations apply to O_2/SO_2 mixtures. Taking Cu as an example, thermodynamic data show that in the presence of SO_2, the sulphate $CuSO_4$ is the stable species even when the partial pressure of O_2 is as low as the dissociation pressure of Cu_2O. Even so, for kinetic reasons, Cu_2O should form first, and be converted to $CuSO_4$ at the Cu_2O/gas interface in the later stages of the reaction. Because the volume of the sulphate per metal atom is so much larger than that of the oxide it replaces, the sulphate is unlikely to be continuous. Furthermore, its growth creates defects in the Cu_2O film:

$$SO_2(g) + O_2(g) \rightarrow CuSO_4 + (Cu^+\square) + (e\square)$$

which lowers its protective power.

The situation with Ni in O_2/SO_2 mixtures is different. When the partial pressure of O_2 is as low as the dissociation pressure of NiO, the sulphide NiS, not the sulphate, is the stable species. Consequently, in the presence of NiO, which for kinetic reasons is expected to form in preference to the sulphate, NiS forms at the Ni/NiO interface. It is observed[10,11] that the sulphide forms as a thin layer between the metal, and the oxide, and also grows into the oxide as a network providing an easy path for the transport of Ni to the solid/gas interface. No quantitative theory of oxidation leading to this film morphology, which is observed[12] also with Co, exists. Film growth is initially linear indicating that a surface reaction controls the rate, but in the later stages of the reaction, film growth obeys a parabolic law because the transport of Ni in the sulphide network controls the rate.

T. B. GRIMLEY

REFERENCES

1. Wagner, C., *Z. Phys. Chem. B.*, **21**, 25 (1933)
2. Wagner, C., *Z. Phys. Chem. B.*, **32**, 47 (1936)
3. Mott, N. F., *J. Chim. Phys.*, **44**, 172 (1947)
4. Cabrera, N. and Mott, N. F., *Rep. Progr. Phys.*, **12**, 163 (1948-49)
5. Grimley, T. B. and Trapnell, B. M. W., *Proc. Roy. Soc. A.*, **234**, 405 (1956)
6. Hauffe, K. and Schottky, W., *Halbleiterprobleme*, **5**, 203 (1960)
7. Frankenthal, R. P. and Malm, D. L., *J. Electrochem. Soc.* **123**, 186 (1976)
8. Kulpa, S. H. and Frankenthal, R. P., *J. Electrochem. Soc.* **124**, 1588 (1977)
9. Grundner, M. and Halbritter, J., *Surface Sci.* **136**, 144 (1984)
10. Luthra, K. L. and Worrell, W. L., *Metall. Trans.* **94**, 1055 (1978)
11. Luthra, K. L. and Worrell, W. L., *Metall. Trans.* **10A** 621 (1979)
12. Jacobson, N. S. and Worrell, W. L., *J. Electrochem. Soc.* **131**, 1182 (1984)

BIBLIOGRAPHY

Kofstad, P., *High-Temperature Oxidation of Metals*, John Wiley, New York (1966).
Lawless, K. R., "The Oxidation of Metals", Rep. Prog. Phys. **37**, 231 (1974)
Birks, N. and Meier, G. H., *Introduction to High Temperature Oxidation of Metals*, Edward Arnold, London (1983)
Kubaschewski, O. and Hopkins, B. E., *Oxidation of Metals and Alloys*, Butterworths, London (1962)
Hauffe, K., *Oxydation von Metallen und Metallegierungen*, Springer, Berlin (1956)
Garner, W. E. (Ed.), *Chemistry of the Solid State*, Butterworths, London (1955)
Bénard, J. (Ed.), *L'Oxydation des Méteaux*, Vol. 1, Gauthier-Villars et Cie, Paris (1962)

1.10 Discontinuous Oxide Films

The Applicability of Rate Laws

Section 1.9 showed that as long as an oxide layer remains adherent and continuous it can be expected to increase in thickness in conformity with one of a number of possible rate laws. This qualification of continuity is most important; the direct access of oxidant to the metal by way of pores and cracks inevitably means an increase in oxidation rate, and often in a manner in which the lower rate is not regained. In common with other phase change reactions the volume of the solid phase alters during the course of oxidation; it is the manner in which this change is accommodated which frequently determines whether the oxide will develop discontinuities. It is found, for example, that oxidation behaviour depends not only on time and temperature but also on specimen geometry, oxide strength and plasticity or even on specific environmental interactions such as volatilisation or dissolution.

The models derived for continuous oxide layers remain valuable when porous oxides are formed; they provide a frame of reference against which deviations may be examined and give a basis for understanding the factors governing the location of new oxide. In many cases, however, the experimentally derived rate 'laws' no longer have a unique interpretation. For example, the linear rate law relating the thickness of oxide, x, to the time, t

$$x = k/t \qquad \ldots(1.184)$$

can describe a situation in which an oxide regularly fails when it reaches a critical thickness[1], or one in which the oxide volatilises[2] at a uniform rate. Similarly, the logarithmic rate law

$$x = k_a \log(1 + k_b t) \qquad \ldots(1.185)$$

has been shown to describe many differing situations, including rate control by chemisorption[3], by oxide nucleation[4], or by cavity production[5].

In certain circumstances even the parabolic rate law may be observed under conditions in which the oxide is porous and permeated by the oxidising environment[6]. In these cases it has been shown that it is diffusion of one or other of the reactants through the fluid phase which is rate controlling. More usually however the porous oxide is thought to grow on the surface of a lower oxide which is itself growing at a parabolic rate. The overall rate of growth is then said to be *paralinear*[7,8] and may be described by the sum of linear and parabolic relationships (*see* equations 1.197 and 1.198).

The Volume Change on Oxidation of a Metal

The formation of discontinuities, particularly the grosser forms of pores and cracks, in an oxide layer is often attributable to the mass flows and volume changes occasioned by oxidation. As can be seen from Table 1.27 it is usual

Table 1.27 Metals which form porous oxides in dry oxygen

Linear oxidation Type 1 Curve (Fig. 1.89)		Paralinear oxidation Type 2 Curve (Fig. 1.89)	
Metal	Oxide/metal volume ratio (r)	Metal	Oxide/metal volume ratio (r)
Mg	0·81	Ce	1·22
Ca	0·64	La	1·1
Nb (400°)	2·49	Mo	3·24
		W	3·35
		U	1·94
		Nb (450°)	2·49
		Ta	2·54
		Ti	1·73
		Zr	1·56
		Hf	1·60
		Th	1·35

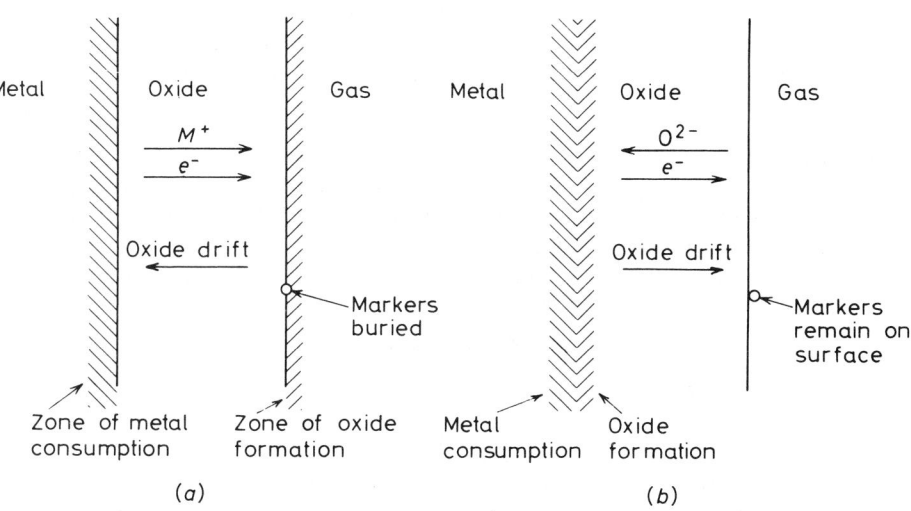

Fig. 1.81 Oxidation of flat surfaces. (a) When cations diffuse the initially formed oxide drifts towards the metal; (b) when anions diffuse the oxide drifts in the opposite direction

for the volume of an oxide to differ greatly from that of the equivalent amount of metal. It is in fact remarkable that such a phase change can occur in so many cases of oxidation without oxide disruption. Moreover, oxidation reactions are rarely topotactic but take place by transport of one of the reactants to a plane of reaction. It is thus not only the net volume change which must be considered, but also the local volume changes due to oxide creation in some zones and metal consumption in others[9].

Fortunately the oxidation of many metals takes place by the diffusion of the metal cation[10]. This flux is outwards through the oxide layer, and the work of adhesion[11] enables the loss of metal to be compensated for by a drift of the oxide towards the metal (Fig. 1.81). Thus the stresses set up in the maintenance of oxide/metal contact are compressive and, as such, can be more readily withstood by most oxides. Nevertheless, it is these general movements of the oxide scale which are ultimately responsible for discontinuities in the majority of cases and it is appropriate to discuss transport-induced flows before proceeding any further.

Mass Transport in Growing Oxide Layers

Oxide Drift

In the very early stages of oxidation the oxide layer is discontinuous; both kinetic[12] and electron microscope[13-15] studies have shown that oxidation commences by the lateral extension of discrete oxide nuclei. It is only once these interlace that the direction of mass transport becomes of importance. In the majority of cases the metal then diffuses across the oxide layer in the form of cations and electrons (cationic diffusion), or as with the heavy metal oxides, oxygen may diffuse as O^{2-} ions with a flow of electrons in the reverse direction (anionic diffusion). The number of metals oxidising by both cationic and anionic diffusion is believed to be small, since a favourable energy of activation for one ion generally means an unfavourable value for the other[10].

A consequence of single-ion diffusion is that the mass movement must be compensated for by an opposing drift (relative to a fixed point deep in the metal) of the existing oxide layer if oxidation is not to be stifled by lack of one of the reactants. The effect may be illustrated by reference to a metal surface of infinite extent (Fig. 1.81).

When cations are the diffusing species (Fig. 1.81a), metal is consumed either by solution in the oxide as interstitial cations and electrons

$$M = M^+ + e^- \qquad \ldots(1.186)$$

or by reaction with metal ion vacancies in the oxide

$$M + M^+\square + \oplus = \text{null} \qquad \ldots(1.187)$$

where as is conventional the right-hand side of the equation represents the excess or deficiency of ions in the stoichiometric oxide, and $M^+\square$ and $\oplus$ represent the vacant sites in the ionic lattice and positive holes in the full band, respectively. Since oxide is formed at the gas/oxide interface, oxide–metal contact is only maintained by translation of the initial oxide through the volume occupied by that part of the metal which has been oxidised.

In systems in which anionic diffusion prevails (Fig. 1.81*b*), metal is consumed by direct reaction to form the diffusing oxygen species

$$M \rightarrow MO + O^{2-}\square + 2e^- \qquad \ldots(1.188)$$

The oxygen vacancies then diffuse to the gas interface where they are annihilated by reaction with adsorbed oxygen. The important point, however, is that metal is consumed and oxide formed in the same reaction zone. The oxide drift has thus only to accommodate the *net* volume difference between the metal and its equivalent amount of oxide. In theory this net volume change could represent an increase or a decrease in the volume of the system, but in practice all metal oxides in which anionic diffusion predominates have a lower metal density than that of the original metal. There is thus a net expansion and the oxide drift is away from the metal.

Oxide movements are determined by the positioning of 'inert' markers on the surface of the oxide[16-18]. At various intervals of time their position can be observed relative to, say, the centreline of the metal as seen in metallographic cross-section. In the case of cation diffusion the metal–interface–marker distance remains constant and the marker moves towards the centreline; when the anion diffuses, the marker moves away from both the metal–oxide interface and the centreline of the metal. In the more usual observation the position of the marker is determined relative to the oxide/gas interface. It can be appreciated from Fig. 1.81 that when anions diffuse the marker remains on the surface, but when cations move the marker translates at a rate equivalent to the total amount of new oxide formed. Bruckman[19] recently has re-emphasised the care that is necessary in the interpretation of marker movements in the oxidation of lower to higher oxides.

Oxidation of Non-planar Surfaces

Oxide movements on plane surfaces, such as those just described, do not create stress; stress will arise however, when the oxide movement is constrained by the presence of a corner, or when the metal is curved, so that there is a progressive strain on the lateral dimensions of the oxide. Since oxides are brittle the appearance of tensional stress can be expected to lead to brittle failure; examples are given in Figs. 1.82 and 1.83.

When convex surfaces oxidise by cationic diffusion compressive stresses occur within the oxide along planes parallel to the metal surfaces and reach a maximum at the metal/oxide interface. Frequently oxides are able to flow plastically under compression at moderate temperatures and the stress is relieved. Hales[20], for example, has proved this to occur by the density of dislocations present in a nickel oxide layer. Sometimes however, relief is abrupt, giving rise to oxide distortions of the type illustrated in Fig. 1.82. An example of oxide buckling observed during the oxidation at 500° of a Cu–10Ni alloy is shown in Fig. 1.84; features of this kind are not unusual and have been followed in the oxidation of pure iron using a hot stage and oxygen blanket in the scanning electron microscope[21].

The compressive forces may be contained by adhesion and cohesion in the oxide–metal system; experiments designed to reveal the presence and

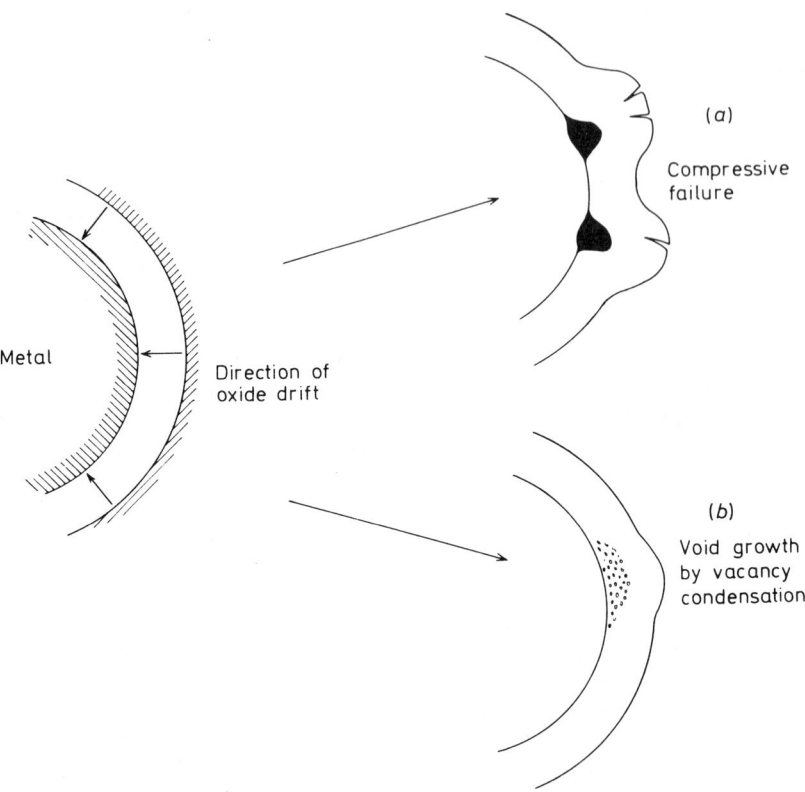

Fig. 1.82 Oxidation of a convex surface by cation diffusion; the compressive stress in the initially formed oxide may lead to (*a*) failure by buckling or to (*b*) void precipitation

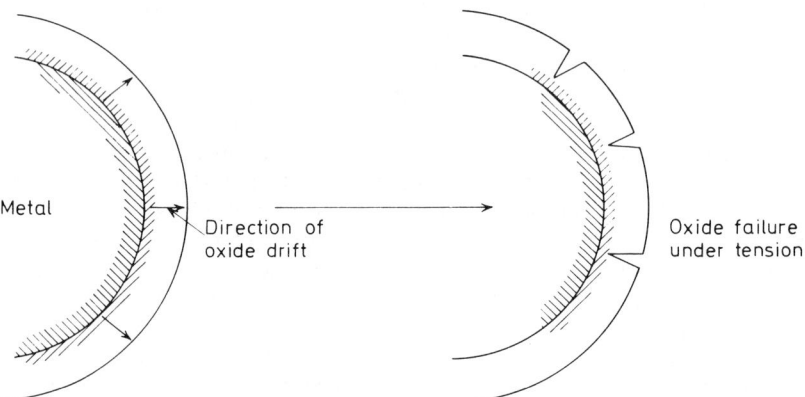

Fig. 1.83 Oxidation of a convex surface by anion diffusion; the outward translation of the oxide gives tensile cracking in the initially formed oxide

magnitude of such stresses in *continuous* films will be described later.

Anionic diffusion in the oxidation of a convex surface creates a situation which is the reverse of that just described. The oxide is in tension along planes parallel to the surface and fracture may be expected to occur readily in perpendicular directions and starting from the gas/metal interface. Although very thin films may have resistance to fracture[22], thick films frequently acquire the morphology shown in Fig. 1.83.

Concave surfaces are of industrial importance, in relation to the internal surface of bores, holes and pipes, but are not found on typical solid testpieces and have received much less discussion. The stress patterns will tend to be the opposite of those found on convex surfaces; for example, an oxide growing by cation diffusion should be in tension at the metal interface. Bruce and Hancock[23] have discussed the oxidation of curved surfaces and show how the time to adhesive failure of the oxide can be predicted if its mechanical properties are known.

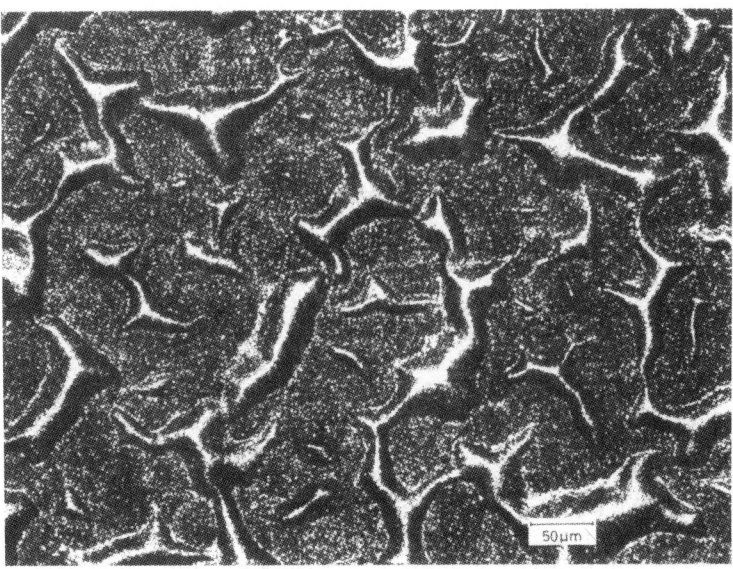

Fig. 1.84 Surface of a Cu–10Ni alloy after oxidation in oxygen at 500°C, showing blistering, probably associated with CuO formation over voids at the metal/oxide interface (courtesy Central Electricity Research Laboratories)

The potential influence of shape on the correct design of laboratory testpieces has been discussed in detail by Romanski[24]. Samples of iron in the form of discs, cylinders, plates or parallelepipeds, and of a wide range of areas, were sulphidised under controlled conditions. The parabolic rate constant could be expressed in terms of the area A of the samples by

$$k_p = A/(a + bA) \qquad \ldots(1.189)$$

at any given temperature. In this expression the constant a depended on the geometry of the specimen and hence on scale rigidity, and b on specimen purity, and thus probably on scale plasticity. The form of this equation was confirmed with the systems Ni-S, Cu-S and Cu-O. Romanski concluded that plate or disc specimens with an area of 30-40 cm^2 were required to obtain rates approaching the maximum for infinite surfaces.

Oxide Drift and Pore Formation

Clearly geometric constraints have an important influence on the formation of oxide layers. There are, however, many reported examples[25-28] of the development of porosity at the oxide-metal interface on plane surfaces well away from corners. This phenomenon is generally ascribed to the suitability of the interface as a sink for lattice vacancies generated during oxidation; the point was lucidly discussed by Vermilyea[29]. Voids of this type may range in size from a few nanometres, as described by Boggs[26] for the oxidation of tin, through the crypt-like cavities in the beautiful micrographs of Howes[30], to the complete separation witnessed in the oxidation of iron by Pheil[31] and described in a paper which also recorded the first use of inert markers. Birchenall has also given interesting observations on the behaviour of large cavities[32]. Such porosity is more commonly associated with cation diffusion although Pemsler[33] has suggested that condensation of *oxygen* vacancies at the metal-oxide interface may be the cause of the voids observed during the oxidation of zirconium. The work of Cohen and his group at Ottawa concerning the oxidation of pure irons[34-39] and more recently, of nickel[40] has made a notable contribution to this field. In particular, they have shown how the dislocation network of the underlying metal is effective in removing vacancies from the metal-oxide interface and thus preventing void formation. The effect is less marked in the case of nickel where the more important effect of cold work in the metal is to yield an oxide structure which is consistent with high diffusion rates. It seems likely that the major difference between the oxidation of these two metals lies in the plasticity of their oxide layers. Bruce and Hancock[41] have detected repetitive oxide cracking during the oxidation of iron by use of a vibrating testpiece, whereas oxide cracking did not occur when nickel testpieces were used. This is in good accord with the behaviour of dislocations in nickel *oxide* during oxidation; Hales[20] has shown that these become optimised to permit maximum creep rates and moreover that the preferred orientation and columnar nature of the oxide also derives from this selection principle. The work is related very nicely to the analysis by Harris and Masters[11] of the work of metal/oxide adhesion.

When voids do form the first formed oxide may lose contact with the retreating metal surface; the way is then open for the metal consumption zone[9] to be filled with a secondary growth which is usually fine grained and porous[42]. Often the inner layer appears to have formed from the inception of oxidation since the inner-outer layer interface retains the shape and dimensions of the original metal surface[43-45]. More often, however, the balance between inner and outer layer formation alters from point to point, with the inner layer being favoured at corners and edges. On alloys the inner layer may consist of particles of the more noble metal oxide which, as in the case of the copper-nickel alloys[45], provide a spaceframe supporting the outer layer. Duplex layers of this nature are sometimes seen on pure metals, the classic example being the case of nickel. Sartell and Li[46] have shown from diffraction measurements that the two layers on nickel differ in their lattice constants, the outer being in compression. They envisaged the outer oxide to grow by anionic diffusion and the inner by cation movement. It now seems likely that the inner layer grows in oxygen formed by decomposition of the outer layer; the case has been well argued by Bruckman[9] and supported by the work of Douglass[42]. What is not so clear is how the outer layer on pure metals continues to grow since it demands the movement of both metal and oxygen across the inner layer when they already have sufficient chemical potential to react. The problem does not arise when alloys oxidise since the inner layer can be traversed at potentials below those at which the species react. Mrowec and Webber[43] have made numerous observations on systems of this kind and neatly summarise their findings in a discussion of a paper by Kofstad and Hed[47] on cobalt–chromium alloys.

There are a number of factors which will trigger void formation during oxidation: Cohen and his co-workers[34-39] have shown that annealing of the metal (to remove dislocations) will do so, whilst Douglass[42], and Wulf, Carter and Wallwork[48] have shown that the presence of a continuous film of a more noble alloying element is just as effective. Tuck et al.[49] consider that the extra adhesion of the oxide on iron when this metal is heated at 950°C in oxygen containing water vapour is due to the beneficial effect of hydrogen on oxide plasticity. There are many similar observations in the literature.

The Influence of Voids on Oxidation Kinetics

Voids as Diffusion Barriers

It was Evans[5,50] who first suggested that cavities could act as diffusion barriers and derived the logarithmic rate law from the progressive nucleation of voids. Boggs[51], and several co-workers[26,27] proved the Evans mechanism to apply to the oxidation of tin over the temperature range 150–220°C in a series of papers illustrating the combined use of electron microscopy and kinetic measurements. The logarithmic rate law also describes the oxidation of a number of other metals in the thin-film range, e.g. Mg[52], Cu[53] and Ni[54], but there is as yet no evidence that the Evans mechanism applies to these cases. Possibly pores have not been sought with the same rigour as that used by Boggs. However, Douglass[42] used a rearrangement of a method first used by Tylecote[55] to show that void condensation causes a three-fold reduction in the parabolic rate constant for a dilute chromium–nickel alloy at 800°C; he compared the rates of oxidation of two cylinders, one of them being plugged at either end to exclude the atmosphere from the interior. Voids did not form at the metal–oxide interface on the plugged cylinder since the interior then acted as a sink for diffusing vacancies.

Probably the most comprehensive measurements of the effect of voids on rates are those of Cohen[34-39] and his school. They have published data on the oxidation of pure irons for a wide temperature range and for oxygen pressures ranging from $1 \cdot 3 \times 10^{-4} \, N/m^2$ to $100 \, kN/m^2$. The interactions between void formation and oxygen uptake are complex but only at pressures below $1 \cdot 3 \times 10^{-2} \, N/m^2$ do voids have no effect. Some of their results are summarised in Fig. 1.85; over the pressure range $1 \cdot 3 \times 10^{-2} \, N/m^2$ to

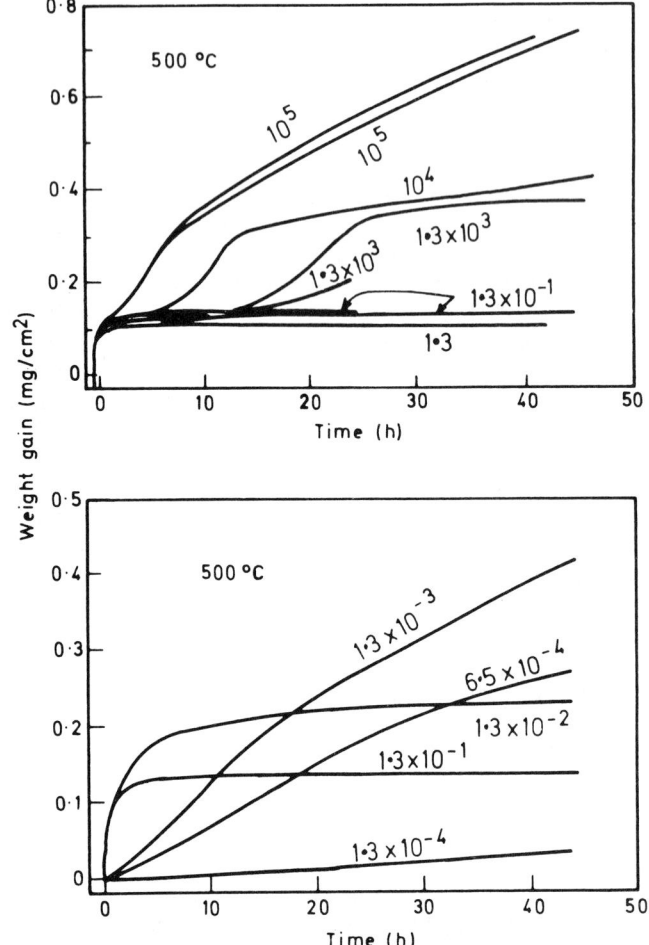

Fig. 1.85 Oxidation of high-purity iron in oxygen at differing pressures. All figures on curves are in N m^{-2}. At $1\cdot3 \times 10^{-4}$ to $1\cdot3 \times 10^{-3}$ N/m^2 torr the rate is controlled by the impact of molecular oxygen; at $1\cdot3 \times 10^{-2}$ N/m^2 torr the initial rate of oxidation is sufficiently high to give void precipitation and the rate decreases with pressure increasing to $1\cdot3$ N/m^2; at pressures greater than this the crack-heal mechanism becomes operative and the rate again increases with pressure (after Hussey and Cohen[38])

$1\cdot3$ N/m^2 the rate decreases as vacancies start arriving at the interface more rapidly than they can be assimilated. Hussey and Cohen[37,38] have evidence that the voids form in the Fe$_3$O$_4$ layer before the outer Fe$_2$O$_3$ is nucleated. However, this rapidly takes place once the supply of iron is stifled by the voids, oxide blisters and cracks, and oxygen ingress causes the rate to accelerate again. The curves indicate that blistering is a repetitive process and this is confirmed by micrographs from the hot-stage scanning electron microscope (Fig. 1.86). It is a salutary fact that the material showing the most protective behaviour (Fig. 1.85) has an oxide which is already partly separated from the metal.

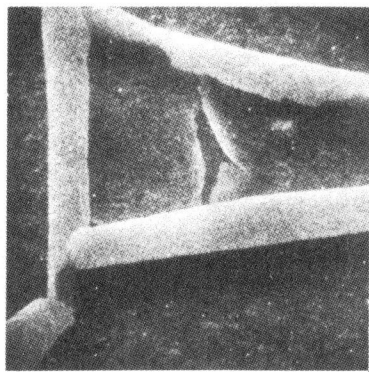 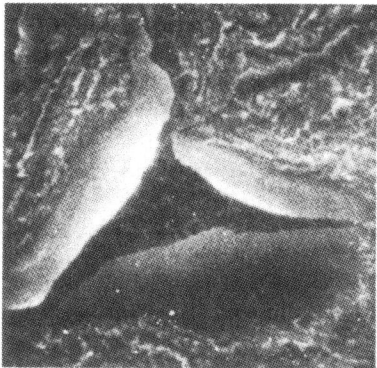

Fig. 1.86 'Stills' from a scanning electron microscope study by time-lapse photography of iron oxidation showing the results of the crack-heal mechanism. Left, 1 mm ≡ 1 μm; right, 1 mm ≡ 0·5 μm (courtesy Central Electricity Research Laboratories)

Notwithstanding the large amount of work on pure iron and binary alloys, it remains difficult to translate the results to commercially useful steels. It is believed, on the one hand, that effusion of carbon monoxide can cause non-healing fissures in the scale[56], and on the other, that silicon creates self-healing layers at the metal interface[57].

The Importance of Voids in 'Short-circuit' Diffusion

Several authors[58-63] have suggested that in some systems voids, far from acting as diffusion barriers, may actually assist transport by permitting a dissociation–recombination mechanism. The presence of elements which could give rise to carrier molecules, e.g. carbon or hydrogen[60,62], and thus to the behaviour illustrated in Fig. 1.87, would particularly favour this mechanism. The oxidant side of the pore functions as a sink for vacancies diffusing from the oxide/gas interface by a reaction which yields gas of sufficiently high chemical potential to oxidise the metal side of the pore. The vacancies created by this reaction then travel to the metal/oxide interface where they are accommodated by plastic flow, or they may form additional voids by the mechanisms already discussed. The reaction sequence at the various interfaces (Fig. 1.87b) for the oxidation of iron (prior to the formation of Fe_2O_3) would be

at A $\quad 2O_2 = Fe_3O_4 + Fe^{\frac{8}{3}+} \square + 8e$ ...(1.190)

at B $\quad Fe_3O_4 + 3Fe^{\frac{8}{3}+} \square + 8e + 4CO = 4CO_2$ ...(1.191)

at C $\quad 4CO_2 = Fe_3O_4 + 3Fe^{\frac{8}{3}+} \square + 8e + 4CO$ ...(1.192)

at D $\quad 3Fe^{\frac{8}{3}+} \square + 8e + 3Fe_{metal} = $ null ...(1.193)

Notice that oxide is utilised by the reaction at interface B at the same rate as it is formed at A, so that the void effectively moves through the growing oxide with the distance AB remaining constant. It may be recalled that a truly inert marker placed at the point *B* in the *continuous* oxide would be expected to remain fixed with reference to the point *D* (the metal–oxide interface). The distinction is important since voids involved in the mechanism in this way remain in a string across the oxide and can thus co-operate to move the oxidant towards the metal at a high rate[64]. What is not so easy to explain is why, when this mechanism is thought to apply, the oxide/metal and metal/oxide interfaces remain flat; the implication is that the abnormally high flux carried by the porous regions redistributes, at the gas/oxide interface, to the benefit of the whole surface. Gibbs[65] has shown that as long as surface diffusion is fast this mechanism gives rise to an altered parabolic rate law. Smeltzer[66], however, has argued that short-circuited routes may be progressively lost as the oxide increases in thickness; thus there is a transition from a short-circuit diffusion process at short times to the usual parabolic dependence at long times.

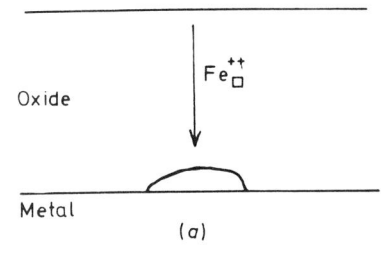

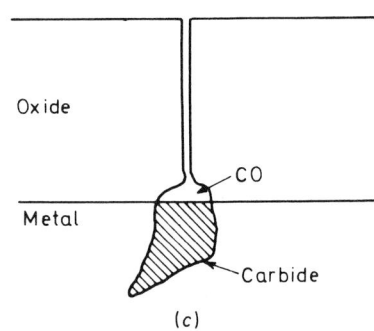

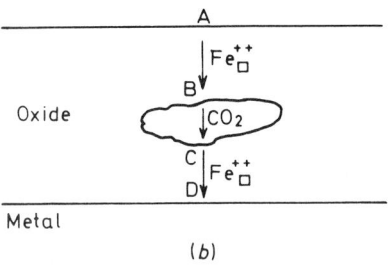

Fig. 1.87 Voids at an oxide/metal interface. They may grow by (*a*) condensation of vacancies from the metal as well as the local oxide; they impede transport as shown in Fig. 1.85. However, voids which become filled with a carrier gas (*b*) may act as short-circuit paths by the reactions at *A*, *B*, *C* and *D*, given in the text (after Birks and Rickert[58]). Alternatively (*c*) gaseous by-products of oxidation may maintain fissures in the oxide (after Boggs and Kachik[56])

All voids described so far have been formed at, or released from, the metal–oxide interface. Birchenal[67] has discussed the formation and growth of voids within the oxide scale by condensation of vacancies. For this to occur anions must be removed in some way and he has suggested creep and slip in the anion lattice as possible mechanisms. More recently, Cox[64] has used a sensitive porosimeter as well as special metallographic techniques to study strings of voids formed in zirconium oxide as a result of recrystallisation—the work is discussed in more detail below. Although Birchenal's model predicted greater deviations from the parabolic rate law than were in fact found, the phenomenon of pore growth within the scale seems to be real and it is as well to keep the implications for anion movement in mind.

Oxide Cracking and Void Cavities

As the area of individual cavities at the metal–oxide interface increases there is an increased probability that the oxide will crack and permit oxygen access to a large area of unprotected metal. Such cracking is then reflected in the overall kinetics and the rate curve takes the form shown by the curves in Fig. 1.85 for pressures greater than $1 \cdot 3$ kN/m^2; it is known to occur with copper[68,69], iron[38] and iron–chromium alloys[70,71] over certain temperature ranges. The oxide formed has become known as a Pfeil-type porous oxide[31]. When failure occurs rather infrequently the rate curve appears as a succession of parabolas, each having as origin the critical values of time and thickness (t_c and x_c) at which the previous film failed. The net rate of oxidation is thus close to that given by the linear rate law with the value of the constant k_l equal to x_c/t_c, or

$$k_1 = (x_c/k_p)^{\frac{1}{2}} \qquad \ldots(1.194)$$

In some cases the number of oxide layers can be related directly to the number of breaks in the curve and there is then no doubt that the acceleration derives from repetitive stress-induced oxide cracking.

Growth Laws of Oxidation

Oxidation at a Linear Rate

Metals which oxidise at a linear rate can follow two types of oxidation curve; there are those (Type 1) for which the transition to the linear rate is far more abrupt and irreversible than that associated with Pfeil-type behaviour. The oxidation curve (Fig. 1.88) shows that in the region of the point A, i.e. at 'breakaway', the rate alters from a value which is initially very small to one which is both large and which no longer decreases with time, e.g. Cathcart *et al.*[72] have shown that sodium will oxidise protectively in pure oxygen, forming a film with a limiting thickness of about 150 nm at 25°C, whereas in air breakaway occurs and a crust many centimetres thick may be formed[73]. The situation is similar with magnesium; at 575°C, and prior to breakaway, the protective oxide is transparent although after this event the rate may be so large that the metal takes fire[1]. Not all rate transitions are as large in magnitude; that of steel in the CO_2 coolant of nuclear reactors leads to a linear rate of about 25 μm/y[74], and this, because of the inaccessibility of the steel, has necessitated temperature reductions in operating reactors, consequently with a considerable loss in power output.

The second type of behaviour (Fig. 1.89) is much closer to that which one might predict from the regular cracking of successive oxide layers, i.e. the rate decreases to a constant value. Often the oxide–metal volume ratio (Table 1.27) is much greater than unity, and oxidation occurs by oxygen transport in the continuous oxide; in some examples the data can be fitted by the paralinear rate law, which is considered later. Destructive oxidation of this type is shown by many metals such as molybdenum, tungsten and tantalum which would otherwise have excellent properties for use at high temperatures.

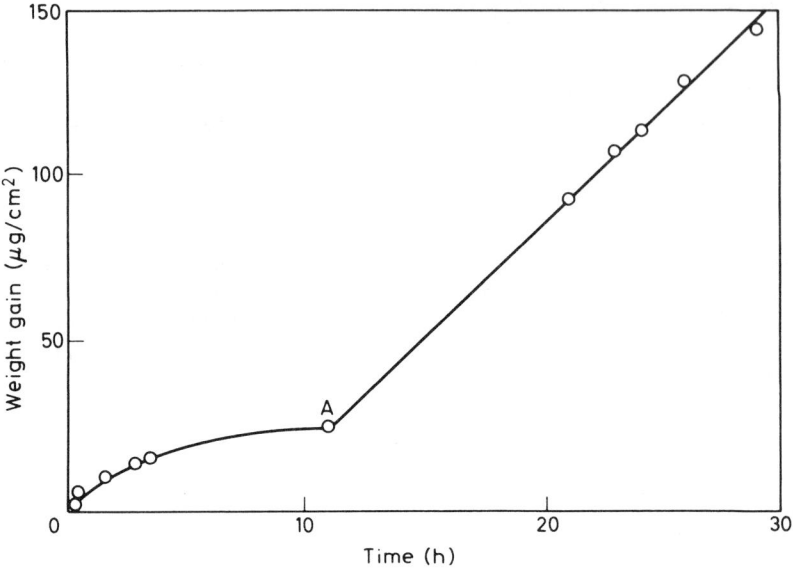

Fig. 1.88 Early stages of oxidation of magnesium at 525°C, but at a lower pressure of 13 kN/m² than the example in Fig. 1.89 illustrating breakaway

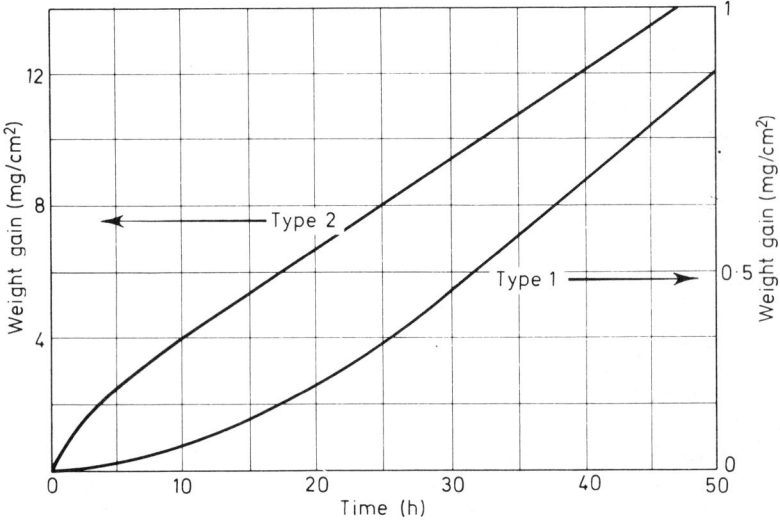

Fig. 1.89 Oxidation of magnesium at 500°C illustrating the increase in rate to the constant value (Type 1) (after Gregg and Jepson[1]) and the oxidation of tungsten at 700°C (after Webb, Norton and Wagner[88]) illustrating the decrease in rate to the constant value (Type 2)

Probably the only feature common to the mechanism of oxidation of the two groups is that, because of crack or pore formation in the continuous oxide, the rate of transport of oxygen in a molecular form has increased to the point where a phase-boundary reaction has assumed rate control. In accord with this interpretation there is frequently a marked reduction in the activation energy for oxidation when non-protective oxidation commences.

The mechanism and the theories of linear oxidation must be discussed with reference to specific examples. This will shortly be done, but it will be helpful to return first of all to the theories of Pilling and Bedworth.

The Pilling–Bedworth Theory

Pilling and Bedworth[68] made the earliest attempt to classify and interpret the oxidation behaviour of metals, and since they were working before any marker experiments had been carried out, they assumed that the predominating transport was always of oxygen to the metal. They consequently argued that the oxide layer would be continuous if its volume was greater than that of the metal consumed by its production. Conversely, if the oxide was less in volume than that of the equivalent mass of metal, the layer would be cellular and porous so that gaseous oxygen would reach the metal surface. As experimental evidence has accumulated it has become clear that this simple rule does not fit the facts; and understandably so since the original premise concerning oxygen transport is not generally true. In particular it is now known that all the metals with oxide: metal volume ratios of less than unity, e.g. Na_2O: Na = $0 \cdot 55$[72], MgO: Mg = $0 \cdot 81$[52] and CaO: Ca = $0 \cdot 64$[75], form oxides which remain thin and continuous indefinitely at temperatures below critical values t_c which are characteristic of the metals and which are often close to their melting points (for Na, $t_c = 48°C$; for Mg, $t_c = 550°C$). As Table 1.25 shows, the oxide ratios of metals which do exhibit linear oxidation of one form or another cover the whole range of possible values so that the rule gives no guidance at all in this respect. In some circumstances the rule is found to be in accord with the sign of the stress set up in the metal during oxidation so that some authors have felt that some residual anion diffusion must occur, notwithstanding the relative values of self-diffusion coefficients. This point will be returned to later.

The Mechanism of Breakaway Corrosion

Over the years, breakaway has become very strongly associated with stress-induced oxide cracking, especially following the work of Pilling and Bedworth, but such a proposition is unwarranted as a generalisation and is difficult to prove for any specific case. The only general feature of breakaway is the very fine state of subdivision of the porous oxide[76]. This is usually beyond the resolving power of even the scanning electron microscope[77]. The coarse cracks which are seen in many optical micrographs are almost certainly secondary features associated with the proximity of the reaction zone to the metal–oxide interface and the subsequent drift of the oxide scale. The factors suggesting that the development of fine-grained porosity in oxides is not merely a stress-assisted reticulation type of failure are (*a*) the long induction periods, when stresses should be relieved rather than developed; (*b*) the major rôle of gas phase impurities (examples are water in the case of sodium[72] and beryllium[78], hydrocarbons with magnesium[52] and CO with steel in CO_2[44]); (c) the promotion of breakaway of steel[79] in high-temperature water or CO_2[44], and of zirconium[80] by layers of platinum; and (d) the curious fact that metals which happily formed a protective oxide when they first contacted oxygen refused to do so when the process was repeated.

Much of the difficulty in demonstrating the mechanism of breakaway in a particular case arises from the thinness of the reaction zone and its location at the metal–oxide interface. Workers must consider (a) whether the oxide is cracked or merely recrystallised[81]; (b) whether the oxide now results from direct molecular reaction, or whether a barrier layer remains[82]; (c) whether the inception of a side reaction (e.g. $2CO \rightarrow CO_2 + C$)[44] caused failure; or (d) whether a new transport process, chemical transport or volatilisation[83], has become possible. In developing these mechanisms both arguments and experimental technique require considerable sophistication. As a few examples one may cite the use of density and specific surface-area measurements as routine[75]; of porosimetry by a variety of methods[64]; of optical microscopy[84,85], electron microscopy[21] and X-ray diffraction[86] at reaction temperature; of tracer[44], electric field[87] and stress measurements. Excellent metallographic sectioning is taken for granted in this field of research.

As has been intimated certain breakaway reactions are of great technological importance and a correspondingly large amount of research has been carried out on these, but as yet no consensus has obtained for the mechanism of linear oxidation of Type 1 (Fig. 1.89) for any one of the metals. The papers of Cox[64,87] on the oxidation of zirconium and its alloys are, however, well worth study; the work included the development of a mercury porosimeter sensitive to pores of about 10 nm in size and the investigation of electron transport in the pre-breakaway oxide layer. Cox concludes that it is electron transport which is rate controlling in the early stages and that breakaway is the recrystallisation of the oxide, induced by a tensional stress, which creates continuous porosity[81] by void condensation at grain boundaries. Intermetallic particles, which in zirconium alloys are associated with easy routes for electron transport in the pre-breakaway film, appear to be located at the base of pores after breakaway. The porosimetry measurements indicate that the pores have a diameter of less than 10 nm and this correlates with the size of pore-like features seen in replicas by transmission electron microscopy. The model received additional support from hot-stage X-ray diffraction[86] which showed recrystallisation to a columnar structure to be concurrent with the rate transition. Bradhurst and Heuer[85] prefer to turn the mechanism around, arguing that recrystallisation occurs as a result of cracking in the traditional way; they have confirmed that there is stress relief in the oxide–metal system at the time of breakaway and have evidence from the hot stage microscope of crack-like features moving across the surface. The viewpoints of both schools are summarised in letters to the *Journal of Nuclear Materials*[81,85].

The discovery by Fiegna and Weisgerber[80] that noble metals are able to catalyse the breakaway corrosion of zirconium has not been built into either of the main theories. Antill *et al.*[44] have also found it difficult to explain their similar observation for the oxidation of steel by CO_2. Reactor grade CO_2 contains both water and CO as impurities; CO is also produced by the reaction

$$3Fe + 4CO_2 = Fe_3O_4 + 4CO \qquad \ldots(1.195)$$

and Antill's thesis is that it is the disproportionation of the carbon monoxide (the Boudouard reaction)

$$2CO = CO_2 + C \qquad \ldots(1.196)$$

which causes fragmentation of the protective oxide. However, in this system the action of the noble metal appears to be associated with the water impurity.

Water as an impurity is known to promote the breakaway corrosion of a number of metals; in addition to iron in CO_2 the effect has been reported for magnesium (hydrocarbons have more effect on the oxidation of this metal), beryllium, zirconium and sodium. In the latter case water is known to convert the oxide to deliquescent NaOH but acceleration of beryllium oxidation probably results from hydride formation and mechanical damage to the oxide.

The Mechanisms of Paralinear Oxidation

Some metals oxidise at a rate which decreases, rather than increases (Type 2 in Fig. 1.89). Cerium behaves in this fashion at temperatures between 40°C and 130°C, and Loriers[7,8] has suggested that the curve derives from the competition between the two oxides Ce_2O_3 and CeO_2. It was proposed that the inner layer Ce_2O_3 was continuous and grew under diffusion control but transformed at a constant rate to an outer layer of CeO_2. That is, if we write y and z as the thickness of the inner and outer layers respectively, then

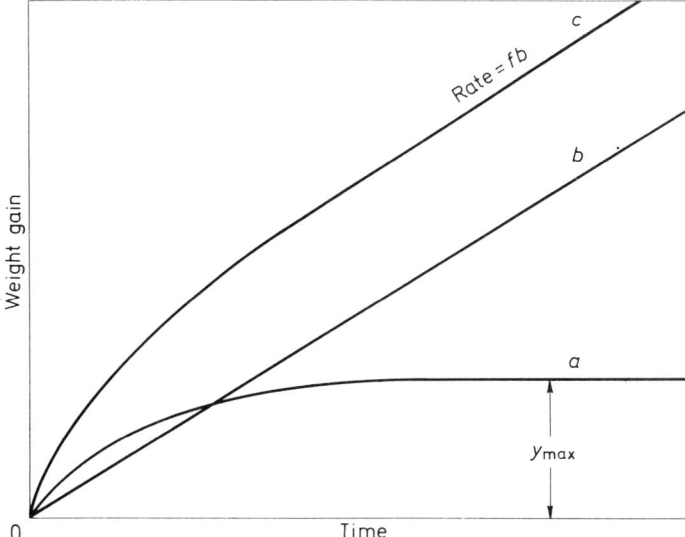

Fig. 1.90 Kinetic interpertation of paralinear oxidation. Curves a and b correspond to the growth of the inner compact layer and the outer porous layer, respectively; curve c represents the total weight and is the algebraic sum of curves a and b. Note that as oxidation proceeds, y tends to a limiting value y_{max} (curve a) and the overall rate of oxidation tends to a constant value fb

$$dy/dt = (a/y) - b \qquad \ldots(1.197)$$

and

$$dz/dt = fb \qquad \ldots(1.198)$$

where f is the ratio of the oxygen content per gramme atom in the outer layer to that in the inner layer, a is one half the parabolic rate constant for diffusion in the inner layer, and fb is the linear rate constant.

The integrated forms of equations 1.197 and 1.198 are illustrated in Fig. 1.90 together with that for the total weight gain w which is given by the sum of y and z. It is evident that as oxidation proceeds the rate of thickening of the inner layer progressively decreases and its thickness tends to a limiting value, i.e.

$$y_{\text{max.}} = a/b \qquad \ldots(1.199)$$

so that the overall rate of oxidation tends to the constant value

$$dw/dt = fb \qquad \ldots(1.200)$$

The particular interest in this form of oxidation stems from the fact that the important group of metals Nb, Ta, Mo and W show similar behaviour although it is only with tungsten that the kinetics are strictly paralinear. The model was first applied to the oxidation of tungsten by Webb, Norton and Wagner[88] for the temperature range 700–1000°C. They were able to demonstrate the presence of a barrier film (probably a metastable modification of one of the intermediate oxide phases of tungsten such as W_4O_{11}) between the metal and the outer layer of porous tungstic oxide. Subsequently Jepson and Aylmore[89] measured the amount of porous oxide formed on tungsten oxidised at temperatures in the range 750–800°C by krypton sorption and by metallographic methods. They discovered that the porous oxide did not form at a constant rate as required by the paralinear model; instead its rate of formation decreased with time with kinetics similar to those of weight gain vs. time. They concluded that the weight of combined oxygen in the barrier layer must be very small. Kellet and Rodgers[90] have since thrown even more doubt on the concept of a barrier layer by their finding that the black oxide 'barrier layer' and the yellow porous oxide have the same chemical composition.

Oxidation in the Presence of Subscales

The aforementioned inconsistencies between the paralinear model and actual observations point to the possibility that there is a different mechanism altogether. The common feature of these metals, and their distinction from cerium, is their facility for dissolving oxygen. The relationship between this process and an oxidation rate which changes from parabolic to a linear value was first established by Wallwork and Jenkins[91] from work on the oxidation of titanium. These authors were able to determine the oxygen distribution in the metal phase by microhardness traverses across metallographic sections; comparison of the results with the oxidation kinetics showed that the rate became linear when the metal surface reached oxygen saturation, at a composition of about Ti_3O. It was thought that the porous layer of TiO_2 was formed by exfoliation of metal layers but remained in good contact since (a) when the oxygen demand of the metal fell, cations diffused into, and discoloured, the oxide and (b) if the oxidising atmosphere was removed, the oxide redissolved.

The general thesis of this work was supported by the work of Osthagen and Kofstad[92] with zirconium in oxygen at 800°C, who found that the rate became linear at a surface composition of Zr_6O, and by that of Smeltzer et al.[93]. The former authors related the rate change to the poor influence of the suboxide (which is believed to be volatile) on the interface adhesion. Pemsler[94], in an important series of papers, has developed the theme of oxidation of prior-formed zirconium–oxygen alloys but, curiously, was unable to reproduce the linear portion of the oxidation curve. He found the rate to remain parabolic at all temperatures up to 1300°C and with a rational rate constant which was independent of the degree of oxygen saturation. Pemsler developed a sensitive technique for the determination of the oxygen profile in the metal and showed ordered alloys to exist at the compositions ZrO_x, where x takes, in turn, the values 0·16, 0·21, 0·26 and 0·32.

Kofstad[95] had earlier suggested that the initial protective period in the oxidation of niobium (this metal belongs to the group which shows Type 1 behaviour) is to be interpreted in terms of the dissolution of oxygen and the formation of a suboxide. Breakaway then corresponds to the nucleation of the pentoxide Nb_2O_5 on the surface. Cathcart[96] and his co-workers observed blister-like cracks in the oxide at breakaway, which is to be expected since Nb_2O_5 grows by anion diffusion and has an oxide: metal volume ratio of 2·49:1. Later work has confirmed this and it thus seems possible that there is no barrier layer, merely an interaction with the underlying metal[97], at least for certain conditions of temperature and oxygen pressure.

Theories of the oxidation of tantalum in the presence of suboxide have been developed by Stringer[98]. By means of single-crystal studies he has been able to show that a rate anisotropy stems from the orientation of the suboxide which is precipitated in the form of thin plates. Their influence on the oxidation rate is least when they lie parallel to the metal interface, since the stresses set up by their oxidation to the pentoxide are most easily accommodated. By contrast, when the plates are at 45° to the surface, complex stresses are established which create characteristic chevron markings and cracks in the oxide. The cracks in this case follow lines of pores generated by oxidation of the plates. This behaviour is also found with niobium, but surprisingly, these pores are not formed when Ta–Nb alloys are oxidised[99], and the rate anisotropy disappears. However, the rate remains linear; it seems that this is another case in which molecular oxygen travels by submicroscopic routes.

The Role of Metal Dissolution or Volatilisation in the Formation of Porous Oxides

As we have seen, a consequence of the formation of porous oxide is that the rate-controlling step reverts to that of a phase boundary reaction and therefore becomes independent of the oxide thickness. When these circumstances are such that the porous oxide becomes unusually thick, or if the oxidising medium is unusually dense, a form of the parabolic rate law may be re-established. In this case the relevant diffusion coefficient is that of the transported component in the fluid phase permeating the oxide layer. Perhaps the best known system in which the apparently paradoxical association of porous oxide formation with parabolic kinetics is observed is that in which iron or mild steel reacts with water or alkaline solutions at temperatures within the region of 300°C. At this temperature water acts as an oxidising agent, even in the absence of electrochemical coupling, and very hard compact layers of oxide are formed. This oxide scale is found to consist of two layers, both magnetite, and marker experiments have been used to establish the important fact that the inner layer occupies exactly the volume of the metal consumed by oxidation[100]. It follows from these observations that the inner layer of oxide supports counter diffusing and nearly equivalent fluxes of iron and oxygen and, by inference from the observed parabolic rate law, one of these is rate controlling. Potter[101] considered the inner layer to be continuous and to grow by oxygen ion diffusion; however the rate of oxygen transport during corrosion is 10^8 times greater than would be expected to occur by diffusion in a continuous oxide film, and it is difficult to account for the equivalence of the ion fluxes over a wide range of temperature and solution composition[102].

The fact that the outer layer was porous was known to Potter and Mann, but the possibility that the inner layer was porous was first discussed by Field et al.[103] who adduced evidence that the oxide had only 90% of its bulk theoretical density. It was subsequently shown that the inner layer consists of individual crystallites of $0 \cdot 1$–$0 \cdot 2\,\mu m$ in size[104] and that the porosity between them was interconnected[6]. The knowledge that molecular water penetrated to the reaction zone close to the metal surface enabled Castle and Masterson[6] to construct a model in which the dissolution of the metal matrix represented a competing reaction with the growth of oxide nuclei. When dissolution is sufficiently fast the growing oxide nuclei can be undermined before a continuous stable film of oxide is formed. Since dissolution will also expose new sites for oxide nucleation the process can be repeated indefinitely so long as the resolved iron is removed from the vicinity of the metal surface. It is this efflux of iron in soluble form from the metal surface which becomes the rate-controlling transport reaction during the oxidation of steel in high-temperature water. The model is useful since it correctly predicts the dependence of the reaction rate on the solution pH[6,105]. This model, or variants[106] of it, is also able to explain the behaviour of aluminium in high-temperature water[107], of steels in molten salts[108] and of non-ferrous metals and alloys in low-temperature aqueous solutions[109,110]. There is also evidence that the transport of metals in volatile form, across similarly porous oxide, may be an important feature of oxidation in steam[83] and in special circumstances where the vapour pressure of the metal is high[82].

The more important cases of oxide volatilisation occur in the platinum metals[111] and with the refractory metals[2] at high temperatures. In these systems, unlike the aforementioned, it is the higher valence oxide which is the more volatile so that at sufficiently high temperature the metal may be oxide free. Gulbransen[2] has shown that the rate of oxidation is then con-

Stresses in Oxide Layers

Much of the earlier part of this chapter has dealt with the observable effects of stress in an oxide layer. Oxide buckling or even failure, recrystallisation, the promotion of columnar grain growth or whiskers, and, above all else, the creation of non-protective oxide, have all been attributed to stress. It has been shown (page **1**.270) how mass transport in oxides leads to oxide drift and thus, on finite and non-planar surfaces, to a change of shape with consequent generation of stress. Other mechanisms of stress generation are: epitaxial strain, i.e. the development of oxide layers with an orientation which permits some correspondence between the lattice parameter of the metal sublattice in the oxide and that of the metal; oxide formation in cracks or grain boundaries; the formation of higher oxides; oxygen solution in the metal, with or without subscale formation; and shrinkage of the metal by assimilation of vacancies from the metal oxide interface. The experimental evidence relating to these forms of stress generation has been reviewed by Stringer[112] in an article which draws together work from wide-ranging fields.

In order to understand the more catastrophic effects of oxide stress its magnitude must be determined in oxide layers which are still continuous. Several methods have been evolved to permit this measurement including low-energy electron diffraction (L.E.E.D.)[113] for the very early stages of oxide growth, X-ray diffraction-line broadening[114] and techniques in which the strain developed by the oxide layer when grown under conditions which permit stress relief is measured by bending or extension of the metal substrate[112, 115, 116]. This latter class encompasses a variety of geometrically designed metal testpieces. The method has become known as the *flexure* or Stoney[117] method and is worthy of special comment particularly since it is the method most widely adopted for measurements on thick film.

The Flexure Method

The method makes use of the tendency of a metal foil to bend when it is oxidised unilaterally, i.e. on one side only, and has been developed from the method used by Stoney to measure stresses in electrodeposits, as long ago as 1909. There are important problems in the translation of the technique to oxidation studies. Firstly, it is difficult to completely arrest oxidation on the 'inert' side of the testpiece. The problems of diffusion across a barrier layer of electrodeposited or vapour-deposited metal for a long time restricted its use to low temperatures, e.g. Ta at 350–550°C[118], Nb at 425°C[118] and Cu at 200–400°C, but recently Pawel and Cathcart have reported the use of Al–3Au evaporated layers which, on uranium alloys, will enable temperatures of up to 800°C to be used[119]. Secondly, there is an interpretive difficulty since although there are two stress systems, acting at right angles in the plane of the oxide, a strip (or even a disc) generally bends in a plane perpendicular to only one of them[120].

Because of the experimental difficulty other workers have circumvented the problem in various ways. In one of the first demonstrations of stress, Evans[121] examined the flexure of detached oxides at room temperature whilst soon after this Dankov and Churaev[122] examined the stresses present at reaction temperature by oxidation of layers of the metals evaporated onto mica substrates. Their technique has been criticised[29] on the grounds that the major stresses would refer to oxide formation in the pores of the evaporated layer—and also that the thin films oxidised may unduly reflect epitaxial stresses. Engell and Wever[123] oxidised both sides of a spiral of iron at 700°C and relied on the different lengths of the inward- and outward-facing surfaces of the helix to provide a differential force. This technique would be unduly influenced by rate differences on either side of the helix—especially as derived from the curvature itself[112]. Bradhurst and Heuer[124] also oxidised both sides of the testpiece and obtained their results from the change in curvature at room temperature when the oxide was removed from one face. The method was used to determine stresses in the oxide on Zircaltoy 2 and zirconium at 500–700°C and is successful only if the oxide does not spall on cooling and if the correction for differential thermal expansion can be experimentally determined. Bradhurst and Heuer were successful and showed that stresses steadily increased until relieved on the Zircaltoy 2 at the point of breakaway. Yet another technique has been utilised by Appleby and Tylecote[125] who protected one side of disc-shaped specimens by reducing atmospheres while the upper side was oxidised at temperatures up to 950°C.

The interpretive difficulty has been discussed in detail by Pawel[120] and by Morton[126] in papers which appeared almost simultaneously. Both authors use arguments to show that the simple formula of bending-beam theory utilised by Stoney

$$\sigma = \frac{Et^2}{6rd} \qquad \ldots(1.201)$$

where E is Young's modulus, t is the thickness of metal, r the radius of curvature, d the oxide thickness and σ the stress, is suitable only as a first approximation.

Since there is an isotropic growth stress in the plane of the oxide it is necessary to consider the two principal stresses σ_x and σ_y given by

$$\sigma_x = \frac{E(\varepsilon_x + \nu\varepsilon_y)}{1 - \nu^2} \qquad \ldots(1.202)$$

for biaxial plane stress in a plate, and

$$\sigma_y = \frac{E(\nu\varepsilon_x + \varepsilon_y)}{1 - \nu^2} \qquad \ldots(1.203)$$

where ε_x and ε_y are the strains, x is the long axis and y the short axis of the testpiece, respectively, and ν is the Poisson's ratio. If the testpiece is free to bend in both directions under these stresses then $\varepsilon_x = \varepsilon_y$ and

$$\sigma_x = \frac{Et^2}{(1 - \nu)6rd} \qquad \ldots(1.204)$$

More usually however the initiation of bending on a plane perpendicular to y (say) and through x will render the testpiece sufficiently rigid to preclude bending in the orthogonal plane through y. As Morton points out this mode of deformation must inevitably occur in the helical testpieces used in the spiral contractometer. An appropriate formula is then $\varepsilon_y \approx 0$ and

$$\sigma_x = \frac{Et^2}{(1-\nu^2)6rd} \qquad \ldots(1.205)$$

although Timoshenko's[127] early analysis shows that it is this relationship which reverts to the Stoney formula for suitably shaped specimens.

The rigidity of the y axis prevents the development of spherical surfaces for all but very small displacements. Morton suggests that the limit is reached when the displacement is equal to the metal thickness. This condition was satisfied in the high-temperature studies of Appleby and Tylecote[125] and spherical doming of the disc specimen occurred. When the oxide is not very thin compared with the metal both the moduli for oxide and metal must be considered. Stringer[112], in his excellent review of stress generation and relief in oxide layers, quotes a corrected formula, originally due to Brenner and Senderoff[128]

$$\sigma = E_{ox.}t\frac{(t+d)}{6rd} - \frac{(E_{ox.} - E_{met.})t^3}{6rd(t+d)} \qquad \ldots(1.206)$$

(which omits the Poisson's ratio correction) and quotes data for magnesium which show that the simple formula would be in error by more than 50% when the oxide layer had a thickness of 10% of that of the metal. It is perhaps apposite to remind the reader that these corrections are pertinent to films which are very thick compared with those formed at breakaway in the Type 1 oxidation. Magnesium undergoes breakaway when the oxide has maintained a stable thickness of about 5×10^{-7} m for several hundred hours. Either on the Pilling–Bedworth model or the oxide drift argument the film would contain a biaxial stress of the order of only about 2 MN/m^2 tension or compression even if there has been no stress relief by plastic flow. The oxide should be well able to support this. In fact it is occasionally possible to evaporate magnesium leaving a near-perfect box of transparent oxide which shows no sign whatsoever of having been in strained conditions — yet eventually such magnesium will go into breakaway.

As reliable stress measurements become extended to greater temperatures the extent to which growth stresses are dissipated by plastic flow of the oxide becomes more apparent — the values of around 700 MN/m^2 measured for the oxidation of tantalum[118], niobium[118, 129] and Zircalloy 2[124] must be contrasted with the value of 70 MN/m^2 (no breakaway) found for the oxidation of zirconium at 700–900°C[124] and 869 kN/m^2 for copper at this same temperature[125]. The stress measured for the formation of cuprous oxide at 900°C was only 371 kN/m$^{2(125)}$. The very low stress calculated for the oxidation of copper at 700°C is, however, surprising since the crack-heal mechanism of oxidation is known to be operative at slightly lower temperatures, and also in view of the direct relation between stress in the oxide scale and the formation of whiskers[130]. They are, however, in reasonable accord with the X-ray strain values measured by Homma and Issike[131] for the oxidation of copper at 500 and 600°C and the very high values are undoubtedly associated with oxygen solution and oxide wedging in the metal.

Stress Relief by Metal Creep

The fact that uniaxially oxidised metals bend suggests that thin foils should stretch when biaxially oxidised. This has been frequently observed[112] and moreover thin-walled tubing may both decrease in diameter[132] and increase in length[133,134] under the compressive loading of oxidation. Such a loading on the metal may seriously reduce its effective tensile strength[135] without relieving much of the compressive stress in the oxide, and introduces a dependence of creep behaviour on the oxygen potential of an environment[136]. Although outside the scope of this section, the paper on anodic oxidation of loaded aluminium wires by Leach and Neufeld[137] gives an indication of the probable depth of this field, in which little work has yet been carried out.

J. E. CASTLE

REFERENCES

1. Gregg, S. J. and Jepson, W. B., *J. Inst. Met.*, **87**, 187 (1958-59)
2. Gulbransen, E. A., *Corrosion*, **26**, 1 (1970)
3. Landsberg, P. T., *J. Chem. Phys.*, **23**, 1079 (1955)
4. Vernon, W., Akeroyd, E. and Stroud, E., *J. Inst. Met.*, **65**, 301 (1939)
5. Evans, U. R., *Trans. Electrochem. Soc.*, **91**, 547 (1947)
6. Castle, J. E. and Masterson, H. G., *Corros. Sci.*, **6**, 93 (1966)
7. Lorriers, J., *C.R. Acad, Sci., Paris*, **229**, 547 (1949)
8. Lorriers, J., *C.R. Acad. Sci., Paris*, **231**, 522 (1950)
9. Bruckman, A., *Corros. Sci.*, **7**, 51 (1967)
10. Harrop, P. J., *J. Mats. Sci.*, **3**, 206 (1968)
11. Masters, B. C. and Harris, J. E., *Proc. Roy. Soc.*, **292A**, 240 (1966)
12. Boggs, W. E., Kachik, R. H. and Pellissier, G. E., *J. Electrochem. Soc.*, **112**, 539 (1965)
13. Taylor, M. E., Holmes, E. and Boden, P. J., *Corros. Sci.*, **9**, 683 (1969)
14. Pignocco, A. J. and Pellissier, G. E., *J. Electrochem. Soc.*, **112**, 1188 (1965)
15. Uhlig, H. H., *Corros. Sci.*, **7**, 325 (1967)
16. Sacks, K., *Metallurgia*, **54**, 11 (1956)
17. Mrowec, S. and Weber, T., *Acta Met.*, **8**, 819 (1960)
18. Jorgenson, P. J., *J. Chem. Phys.*, **37**, 874 (1962)
19. Bruckman, A. and Simkovich, G., *Corros. Sci.*, **12**, 595 (1972)
20. Hales, R., *Corros. Sci.*, **12**, 555 (1972)
21. Castle, J. E. and Hunt, M. R., to be published in *Corros. Sci.*
22. Pashley, D. W., *Advances Phys.*, **5**, 173 (1956)
23. Bruce, D. and Hancock, P., *J. Iron and Steel Inst.*, **208**, 1021 (1970)
24. Romanski, J., *Corros. Sci.*, **8**, 67 (2 papers) (1968)
25. Tylecote, R. F., *J. Iron Steel Inst.*, **195**, 380 (1960)
26. Boggs, W. E., Kachick, R. H. and Pellisier, G. E., *J. Electrochem. Soc.*, 1086 (1961)
27. Boggs, W. E., Trozzo, P. S. and Pellisier, G. E., ibid., 13 (1961)
28. Caplan, D. and Cohen, M., *Corrosion*, **15**, 14lt (1969)
29. Vermilyea, D. A., *Acta Met.*, **5**, 492 (1957)
30. Howes, V. R., *Corros. Sci.*, **10**, 99 (1970)
31. Pfeil, L. B., *J. Iron and Steel Inst.*, **119**, 501 (1929)
32. Juenker, D. W., Meussener, R. A. and Birchenall, C. E., *Corrosion*, **14**, 57 (1958)
33. Pemsler, J., *J. Electrochem. Soc.*, **112**, 477 (1965)
34. Caplan, D. and Cohen, M., *Corros. Sci.*, **6**, 521 (1966)
35. Caplan, D. and Cohen, M., *Corros. Sci.*, **7**, 725 (1967)
36. Caplan, D., Graham, M. J. and Cohen, M., *Corros. Sci.*, **10**, 1 (1970)
37. Hussey, R. J. and Cohen, M. J., *Corros. Sci.*, **11**, 699 (1971)
38. Hussey, R. J. and Cohen, M. J., *Corros. Sci.*, **11**, 713 (1971)
39. Caplan, D., Sproule, G. I. and Hussey, R. J., *Corros. Sci.*, **10**, 9 (1971)

40. Caplan, D., Graham, M. J. and Cohen, M., *J. Electrochem. Soc.*, **119**, 1205 (1972)
41. Bruce, D. and Hancock, P., *J. Inst. Met.*, **97**, 140 (1969)
42. Douglass, D. L., *Corros. Sci.*, **8**, 665 (1968)
43. Mrowec, S. and Weber, T., *J. Electrochem. Soc.*, **117**, 1531 (1970)
44. Antill, J. E., Peakall, K. A. and Warburton, J. B., *Corros. Sci.*, **8**, 689 (1968)
45. Whittle, D. P. and Wood, G. G., *Corros. Sci.*, **8**, 295 (1968)
46. Sartell, J. A. and Li, C. H., *J. Inst. Met.*, **90**, 92 (1961)
47. Kofstad, P. N. and Hed, A. Z., *J. Electrochem. Soc.*, **116**, 1542 (1969)
48. Wulf, G. L., Carter, J. J. and Wallwork, G. R., *Corros. Sci.*, **9**, 689 (1969)
49. Tuck, C. W., Odgers, M. and Sacks, K., *Corros. Sci.*, **9**, 271 (1969)
50. Evans, U. R., *The Corrosion and Oxidation of Metals*, Arnold, London (1961)
51. Boggs, W. E., *J. Electrochem. Soc.*, **108**, 124 (1961)
52. Castle, J. E., Gregg, S. J. and Jepson, W. B., *J. Electrochem. Soc.*, **109**, 1018 (1962)
53. Uhlig, H. H., *Acta Met.*, **4**, 541 (1956)
54. Uhlig, H. H., Pickett, J. and MacNairn, J., *Acta Met.*, **7**, 111 (1959)
55. Tylecote, R. F. and Michett, T. E., *J. Iron and Steel Inst.*, **196**, 445 (1960)
56. Boggs, W. E. and Kachik, R. H., *J. Electrochem. Soc.*, **116**, 424 (1969)
57. Wood, G. C., Richardson, J. A., Hobby, M. C. and Bouttead, J., *Corros. Sci.*, **9**, 655 (1969)
58. Birks, N. and Rickert, H., *J. Inst. Met.*, **91**, 30 (1962)
59. Wood, G. C., Wright, J. G. and Fergusson, J. M., *Corros. Sci.*, **5**, 645 (1965)
60. Fuji, C. T. and Meussner, R. A., *J. Electrochem. Soc.*, **114**, 435 (1967)
61. Mrowec, S., *Corros. Sci.*, **7**, 563 (1967)
62. Antill, J. E., Peakall, K. A. and Warburton, J. B., *Corros. Sci.*, **8**, 689 (1968)
63. Birks, N., *British Corros. J.*, **3**, 56 (1968)
64. Cox, B., *J. Nuc. Mats.*, **27**, 1 (1968)
65. Gibbs, G. B., *Corrosion Sci.*, **7**, 165 (1967)
66. Smeltzer, W. W., Haering, R. R. and Kirkaldy, J. S., *Acta. Met.*, **9**, 880 (1961)
67. Birchenal, C. E., *J. Electrochem. Soc.*, **103**, 619 (1956)
68. Pilling, N. B. and Bedworth, R. E., *J. Inst. Met.*, **29**, 529 (1923)
69. Tylecote, R. F., *J. Inst. Met.*, **81**, 681 (1953)
70. Caplan, D. and Cohen, M., *J. Metals, Trans. Amer. Inst. Min. (Metall.) Engrs.*, **203**, 336 (1955)
71. Mortimer, D. and Post, M. L., *Corros. Sci.*, **8**, 499 (1968)
72. Cathcart, J. V., Hall, C. C. and Smith, G. P., *Acta. Met.*, **5**, 249 (1957)
73. Howland, W. H. and Epstein, L. F., *Ind. Eng. Chem.*, **49**, 1931 (1957)
74. Antill, J. E., Campbell, C. B., Goodison, D., Jepson, W. B. and Stevens, C. G., *Proc. 3rd Conf. Peaceful Uses of Atomic Energy*, **9**, 523 (1964)
75. Gregg, S. J. and Jepson, W. B., *J. Chem. Soc.*, 712 (1960)
76. Aylmore, D. W., Gregg, J. J. and Jepson, W. B., *J. Electrochem. Soc.*, **106**, 1010 (1959)
77. Castle, J. E. and Wood, C. G., *Scanning Electron Microscopy* (Ed. O. Johari), I.I.T. Research Inst., Chicago, 39 (1968)
78. Aylmore, D. W., Gregg, S. J. and Jepson, W. B., *J. Nuc. Mats.*, **3**, 190 (1961)
79. Castle, J. E. and Mann, G. M. W., *Corros. Sci.*, **6** (1966)
80. Fiegna, A. and Weisgerber, P., *J. Electrochem. Soc.*, **115**, 369 (1968)
81. Cox, B., *J. Nuc. Mats.*, **41**, 96 (1971)
82. Dupre, B. and Shreiff, R., *J. Nuc. Mats.*, **42**, 260 (1972)
83. Surman, P. L. and Castle, J. E., *Corros. Sci.*, **9**, 771 (1969)
84. Fern, F. H. and Antill, J. E., *Corros. Sci.*, **10**, 649 (1970)
85. Bradhurst, D. H. and Heuer, P. M., *J. Nuc. Mats.*, **41**, 101 (1971)
86. Roy, C. and David, G., *J. Nuc. Mats.*, **37**, 71 (1970)
87. Cox, B., *J. Nuc. Mats.*, **31**, 48 (1969)
88. Webb, W. W., Norton, J. T. and Wagner, C., *J. Electrochem. Soc.*, **103**, 107 (1956)
89. Jepson, W. B. and Aylmore, D. W., *J. Electrochem. Soc.*, **108**, 942 (1961)
90. Kellett, E. A. and Rodgers, S. E., *J. Electrochem. Soc.*, **110**, 503 (1965)
91. Wallwork, G. R. and Jenkins, N. E., *J. Electrochem. Soc.*, **106**, 10 (1959)
92. Osthagen, K. and Kofstad, P., *J. Electrochem. Soc.*, **109**, 204 (1962)
93. Smeltzer, W. W., *Can. Met. Quart.*, **11**, 41 (1962)
94. Pemsler, J., *J. Electrochem. Soc.*, **111**, 383 (1964)
95. Kofstad, P., *Proc. 1st Int. Cong. Met. Corrosion, London, 1961*, Butterworths, London, 181 (1962)

96. Cathcart, J. V., Campbell, J. J. and Smith, G. P., *J. Electrochem. Soc.*, **105**, 442 (1958)
97. Weirich, L. J. and Larsen, W. L., *J. Electrochem. Soc.*, **119**, 465 (1972)
98. Stringer, J., *J. Less Common Metals*, **12**, 301 (1967)
99. Dooley, R. B. and Stringer, J., *J. Less Common Metals*, **24**, 139 (1971)
100. Potter, E. C. and Mann, G. M. W., *Proc. 1st Int. Cong. Met. Corrn.*, London, 1961, Butterworths, London, 417 (1962)
101. Potter, E. C., *Mitt. V.G.B.*, **76**, 19 (1962)
102. Castle, J. E. and Surman, P. L., *J. Phys. Chem.*, **71**, 4255 (1967)
103. Field, L. M., Stanley, R. C., Adams, A. M. and Holmes, D. R., *Proc. 2nd Int. Cong. Met. Corrosion*, N.A.C.E., New York (1963)
104. Harrison, P. L., Holmes, D. R. and Teore, P., V.G.B. Conference on Feed-water Treatment, Essen (1965)
105. Potter, E. C. and Mann, G. M. W., *B. Corrosion J.*, **1**, 26 (1965)
106. Bignold, G. J., Garnsey, R. and Mann, G. M. W., *Corros. Sci.*, **12**, 325 (1972)
107. Castle, J. E., unpublished work
108. Holmes, D. R., Discussion of Paper by A. U. Seybolt, *4th Int. Cong. on Met. Corros.*, Amsterdam, 560 (1969)
109. Vermilyea, D. A. and Vedder, W., *Trans. Farad. Soc.*, **66**, 2644 (1970)
110. Green, J. A. S., Mengelberg, H. D. and Yolhen, H. T., *J. Electrochem. Soc.*, **117**, 433 (1970)
111. Betteridge, W. and Rhys, D. W., *1st Int. Cong. Met. Corros.*, Butterworths, London, 186 (1962)
112. Stringer, J., *Corros. Sci.*, **10**, 513 (1970)
113. Sickafus, E. N. and Bonzel, H. P., *Recent Progress in Surface Science IV*, A.P., New York and London (1971)
114. Swank, T. F. and Lawless, K. R., *Advances in X-ray Analysis*, **10**, Plenum Press, New York, 234 (1966)
115. Pawel, R. E., Cathcart, J. V. and Campbell, J. J., *J. Electrochem. Soc.*, **110**, 551 (1963)
116. Jaenicke, W., Leistikow, S. and Städder, J., *J. Electrochem. Soc.*, **111**, 1031 (1964)
117. Stoney, G. C., *Proc. Roy. Soc., Lond.*, **A82**, 172 (1909)
118. Pawel, R. E. and Campbell, J. J., *Acta Met.*, **14**, 1827 (1966)
119. Pawel, R. E. and Cathcart, J. V., *J. Electrochem. Soc.*, **118**, 1776 (1971)
120. Pawel, R. E., *J. Electrochem. Soc.*, **116**, 1144 (1969)
121. Evans, U. R., *Inst. Met. Symp. on Stresses in Metals*, 219 (1947)
122. Dankov, D. D. and Churaev, P. V., *Dokl. Akad. Nauk.* SSSR, **73**, 1221 (1950)
123. Engell, H. and Wever, F., *Acta Metall.*, **5**, 695 (1957)
124. Bradhurst, D. H. and Heuer, P. M., *J. Nuc. Mats.*, **37**, 35 (1970)
125. Appleby, W. K. and Tylecote, R. F., *Corros. Sci.*, **10**, 325 (1970)
126. Morton, V. M., *Corros. Sci.*, **9**, 261 (1969)
127. Timoshenko, S., *Mech. Engrs.*, **45**, 259 (1923)
128. Brenner, A. and Senderoff, S., *J. Res. Natn. Bur. Stand.*, **42**, 105 (1949)
129. Weirich, L. J. and Larsen, W. L., *J. Electrochem. Soc.*, **119**, 465 (1972)
130. Sartell, J. A., Stokes, R. J., Bendel, S. H., Johnson, T. L. and Li, C. H., *Trans. A.I.M.M.E.*, **215**, 211 (1959)
131. Homma, T. and Issiki, S., *Acta Met.*, **12**, 1092 (1964)
132. Moore, W. J., *J. Chem. Phys.*, **21**, 1117 (1953)
133. Hoden, J. D., Knight, C. J. and Thomas, M. W., *Br. Corros. J.*, **3**, 47 (1968)
134. Buresch, F. L. and Bollenrath, F., *J. Nuc. Mats.*, **248**, 270 (1967)
135. Weirich, L. J. and Larsen, W. L., *J. Electrochem. Soc.*, **119**, 472 (1972)
136. Knights, C. E. and Perkins, R., *J. Nuc. Mats.*, **36**, 180 (1970)
137. Leach, J. S. L. and Neufeld, P., *Proc. Br. Ceramic Soc.*, 49 (1966)

1.11 Erosion Corrosion

The majority of corrosion problems involve situations where the environment moves, this may increase or decrease the rate of corrosion. It is only cases where both anodic and cathodic processes are activation controlled that will be unaffected by the relative movement between the surface and the environment. As well as influencing the rate of existing processes solution movement can introduce different rate controlling steps. With the exception of cavitation, flow induced corrosion problems are generally termed erosion corrosion, encompassing flow enhanced dissolution and impingement attack. The fluid can be aqueous or gaseous, single or multiphase. The actual corrosion component of the attack will be variable[1] but we can conceptually imagine a spectrum of processes from those where loss is due to corrosion to those where mechanical loss dominates as shown in Table 1.28.

Table 1.28 Spectrum of erosion corrosion process (Poulson[1])

Dissolution dominant

Flow thins protective film to equilibrium thickness which is a function of both mass transfer rate and growth kinetics. Erosion corrosion rate is controlled by the dissolution rate of the protective film.

Film is locally removed by dissolution, surface shear stress or particle/bubble impact; but it can repassivate. Erosion corrosion rate is a function of the frequency of film removal, bare metal dissolution rate and subsequent repassivation rate.

Film is removed and does not reform. Erosion corrosion rate is the rate the bare metal can dissolve.

Film is removed and underlying metal surface is mechanically damaged which contributes to overall metal loss i.e. erosion corrosion rate is equal to bare metal dissolution rate plus possibly synergistic effect of mechanical damage.

Film is removed and mechanical damage to underlying metal is the dominant damage mechanism.

Mechanical damage dominant

Important features of erosion corrosion are that it usually occurs in situations where the metal is normally covered by a protective film. Accelerated attack is then caused by the rate of dissolution of the film increasing, the film dissolving away or being damaged mechanically. Usually surfaces that have undergone erosion corrosion where dissolution dominates have characteristic features that are variously described as scallops, or horseshoe shaped pits. However not surprisingly the surface morphology will vary with the damage mechanism as shown in Figure 1.91.

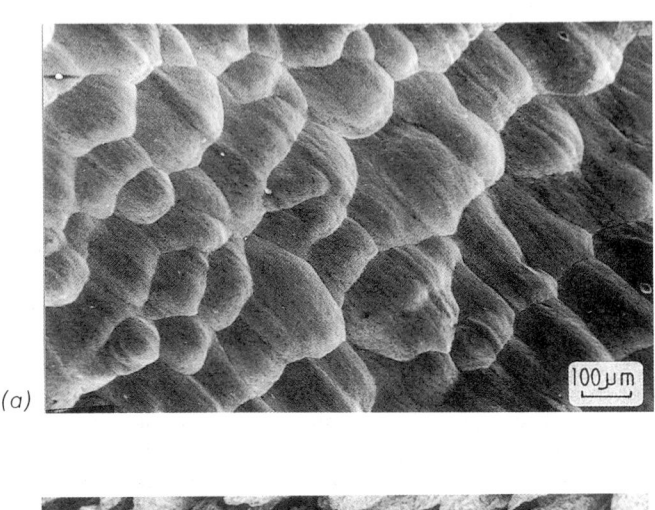

Fig. 1.91 Variation in possible erosion corrosion surface morphology (a) single phase attack of mild steel in water (b) droplet erosion corrosion of 12% Cr steel

Table 1.29 tabulates most known examples of erosion corrosion problems occuring in aqueous systems. Historically, erosion corrosion first became a problem with the copper alloy (70%Cu 29%Zn 1%Sn) condensers of naval ships[2]. Erosion corrosion of copper alloys has been an ongoing problem since then. The other major problem areas are (a) power plants where steels are exposed to water or water/steam mixtures in the temperature range 90°–280°C[3] (b) the oil and gas industry where steels are exposed to various liquid, gas, and sometimes solids combinations containing carbon dioxide.

Table 1.29 Examples of occurrences of erosion corrosion

Material	Environment	Example
Copper alloys	Seawater	Condenser tubing. Piping. Pump impellers.
Carbon steels	Water, water/steam. Sulphuric acid. Carbonate solution. Coal/water. Water/CO$_2$/oil/sand. Two phase sulphide environments.	Boiler tubes. Process plant piping. Process plant piping. Coal slurry transportation. Oil production pipes. Oil refining equipment.
12%Cr steel	Water droplets in steam.	Turbine blades.
Austenitic stainless steels	Sulphuric acid slurry.	Process plant pump impeller
Aluminium alloys	Water. Concentrated fuming HNO$_3$.	Nuclear fuel cans. Process plant.
Lead	Sulphuric acid.	Sulphuric acid manufacturer.

Hydrodynamic Parameters

It has long been known that increasing the velocity and the presence of bends, welds or other features promoting increased turbulence increases the risk of erosion corrosion. For copper alloys both laboratory data and practical experience suggest that for a given tube diameter there will be a critical velocity above which erosion corrosion could become a problem. This critical velocity will be a function of the environment and will be higher for the more resistant alloys[4] (Table 1.30). However, knowing the critical velocity in 25 mm diameter tubing does not allow its prediction in 200 cm diameter pipe or other geometries. What is needed is the identification of the rate controlling hydrodynamic parameter.

For copper alloys Efird[5] proposed that a critical shear stress (*see* Table 1.31) at the surface could be obtained which corresponded with the removal

Table 1.30 Critical velocity of copper alloys on seawater (from reference[4])

Alloy	Composition	Critical velocity in 25 mm dia. tube ms^{-1}	Critical shear stress Nm^{-2}
Cupro Nickel with Cr	83 Cu 16·5 Ni 0·5 Cr 0·75 Fe	9	297
70-30 Cupro Nickel	70 Cu 30 Ni 0·6 Fe	4·5-4·6	48
90-10 Cupro Nickel	88·7 Cu 10 Ni 1·3 Fe	3-3·6	43
Aluminium Bronzes	91 Cu 7 Al 2·5 Fe 0·5 Si	2·7	—
Arsenical Al Brass	77·5 Cu 20·5 Zn 2 Al 0·06 As	2·4	19
Inhibited Admiralty Metal	71 Cu 28 Zn 1 Sn 0·06 As	1·2-1·8	—
Low Si Bronze	98·5 Cu 1·5 Si	0·9	—
Phosphorus deoxidised Copper	99·9 Cu 0·02 P	0·6-0·9	9·6

Table 1.31 Hydrodynamic parameters relevant to erosion corrosion

---

V is fluid velocity, γ is fluid kinematic viscosity, μ is fluid dynamic viscosity
ρ is fluid density, D is diffusion coefficient, ΔP is pressure drop
x is distance along tube, ΔC is concentration driving force
K is mass transfer coefficient, J is mass rate of reaction
Re is the Reynolds number Vd/γ or $\rho Vd/\mu$
Sh is the Sherwood number Kd/D
Sc is the Schmid Number γ/D

The surface shear stress τ is a consequence of the velocity difference between the metal surface and the fluid velocity. For tubular geometries it can be obtained from pressure drop measurements or calculated:

$$\tau = \Delta P/4(x/d) = f(\rho V^2/2)$$

Where f is known the Fanning friction factor which is a function of Re and the roughness of the tube relative to its diameter. For practical systems f varies from 0.002–0.01. Very little information on local values of τ in other geometries.

For any diffusion controlled reaction

$$J = K\Delta C$$

K is normally obtained from non-dimensional correlations between Sh, Re and Sc of the form

$$Sh = a Re^x Sc^y$$

It can be obtained from the available literature or measured experimentally. If the erosion corrosion rate (*ECR*) is directly proportional to the mass transfer rate:

$$ECR = \frac{D}{d} a Re^x Sc^y \Delta CP$$

For situations controlled by anodic dissolution of a film $P = 1/\text{density of metal}$, but if the corrosion is controlled by the cathodic reaction $P = 1/\text{density of metal} \times nc\,Ma/na\,Mc$ where n and M are the number of electrons and the molecular masses of anodic and cathodic reactants.

---

of protective corrosion films. For a single geometry (plate parallel to tube flow) Efird obtained critical shear stresses for a variety of copper alloys (Table 1.30). However the ideas was not tested to see if the concept was applicable to other geometries. Syrett has pointed out that the critical shear stress values obtained seem remarkably low to mechanically disrupt a surface layer and that it was more likely that the films were being dissolved away by the increased rate of mass transfer to and from the surface. This suggestion was in fact made earlier by Lush[7] *et al.* who used the shear stress to characterise the rate of mass transfer at the surface. However the mass transfer rate is normally specified by the mass transfer coefficient (*see* Table 1.31). This is more readily obtained experimentally and can be obtained in the form of a non-dimensional correlation, for many geometries of interest from the available literature. Early applications of the mass transfer concept to corrosion include those of King[8], Levich[9] and Ross[10] but it is not clear who first applied it to erosion corrosion. The erosion corrosion of carbon steels in concentrated sulphuric acid was investigated by Ellison and Schmeal[11] who showed that the corrosion rate could be predicted from available mass transfer correlations and the required solution parameters.

However the expected simple linear relationship between mass transfer and erosion corrosion does not always hold for a variety of reasons which include:

1. *Change in rate controlling step*
   If the film is removed above a certain mass transfer rate[12] there will be a sudden increase in the erosion corrosion rate which will tend to rise to the rate the bare metal can dissolve.
2. *Spatial separation of anodic and cathodic areas*
   Corrosion involves both anodic and cathodic reactions and the localisation of these can lead to lack of conformity with mass transfer expectations[13].
3. *Series control*
   A number of workers[7,14] have suggested that there are situations in which two processes in series control the erosion corrosion rate, for example diffusion plus partial activation control, leading to a lower dependency on mass transfer than expected.
4. *Coupled reactions*
   Bignold[14] has postulated that increasing flow increases both mass transfer and by lowering the free corrosion potential the oxide solubility. This would lead to a higher dependency on mass transfer than expected.
5. *Inappropriate mass transfer values*
   Most mass transfer data is for smooth surfaces[15]. It is well established that as a surface suffers erosion corrosion it will roughen and the mass transfer rate will increase. It has been suggested[16] that when this occurs the roughness becomes more important than the original shape in controlling the mass transfer.

Examples of some of these effects and the resulting mass transfer erosion corrosion behaviour are shown in Figure 1.92.

Material/Environmental Influences

Since the formation nature and breakdown of protective surface films depends on both material and environmental parameters such influences on erosion corrosion will be discussed together. Particular attention will be paid to the copper/seawater and carbon steel/water (steam) systems.

Briefly the important developments in copper alloys with respect to their erosion corrosion behaviour in seawater have been:

(a) Aluminium brasses which are widely used as condenser materials.
(b) The 90-10 and 70-30 cupro nickels which are used extensively in seawater systems where their improved resistance to erosion corrosion and anti-fouling characteristics are valuable.
(c) The important role of controlled iron additions to both the aluminium brasses and the cupro-nickels[18] (Figure 1.93) in improving the resistance to erosion corrosion. There is evidence that maximum benefits are obtained if all the iron is in solid solution.
(d) The dramatic role of chromium (Figure 1.94) in increasing the erosion corrosion behaviour of the cupro nickels[19] particularly those with nickel contents greater than 16%. Again heat treatment effects are important and rapid cooling is required to keep the Cr in solid solution for maximum resistance.

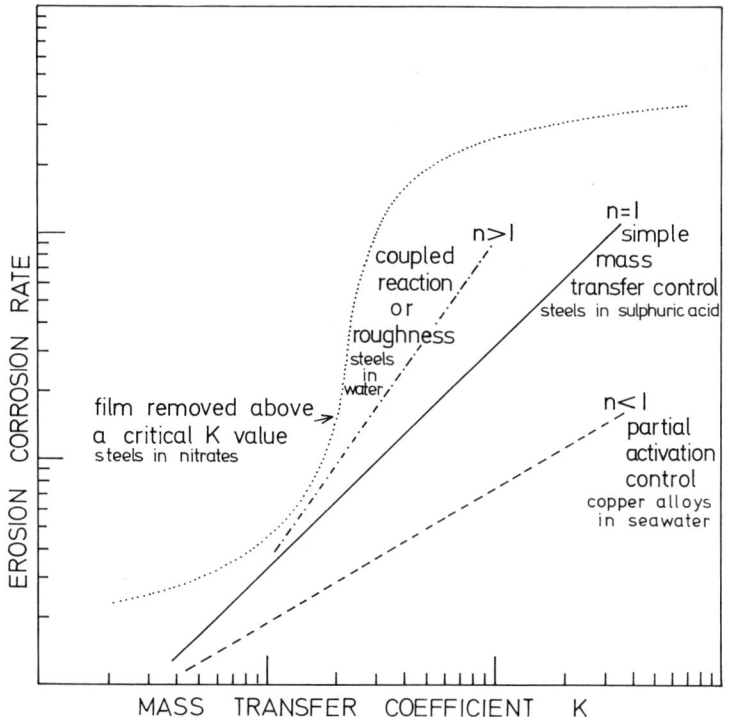

Fig. 1.92 Possible relationships between erosion corrosion and mass transfer (after Poulson[1])

The relative erosion corrosion resistance of these alloys has been previously given in Table 1.30.

Environmentally the important detrimental role of sulphide ions from pollution or the decay of organic matter, has been highlighted in recent years[20] (Figure 1.95). The additions of ferrous sulphate has proved extremely beneficial even in the presence of sulphides[21] (Figure 1.95). It is worth indicating that surface pretreatments, using mains water, maintaining flow or draining after use, have all been suggested as means to prevent problems during preservice testing when the risks of sulphide problems is highest.

There is no general agreement on the mechanism of these various effects. For example the beneficial effects of iron additions have been attributed to it: increasing the amount of nickel in the corrosion film[22], increasing the electronic or ionic resistivity of the film and impeding the anodic[23] or cathodic[24] reactions or improving the mechanical strength of the film[25]. Similarly the detrimental effects of sulphides have been suggested to result from: the oxygen reaction being catalysed[26], the evolution of hydrogen being promoted, copper sulphide being precipitated thus lowering the corrosion potential[27] or copper sulphide being incorporated into the film and reducing its mechanical properties[28].

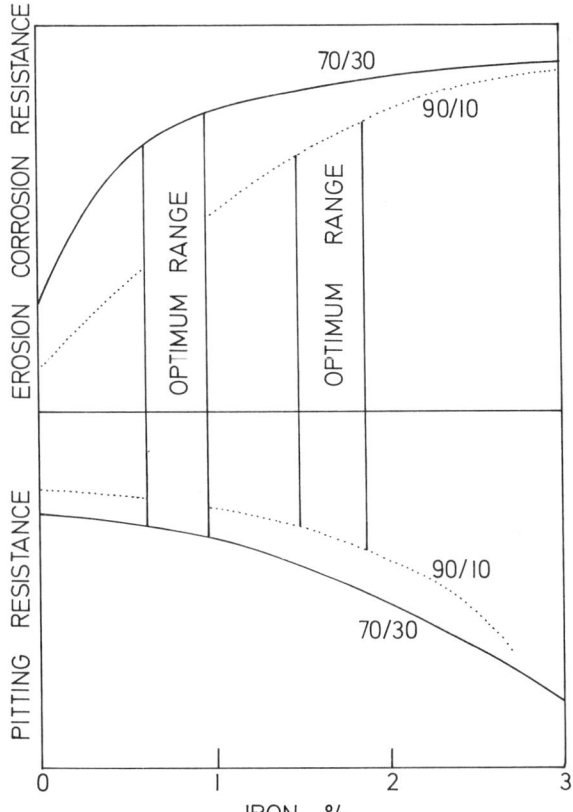

Fig. 1.93 Schematic diagram showing beneficial effect of iron on erosion corrosion of cupro nickel (after Pearson[18])

For carbon steels in water, or water/steam mixtures at temperatures in the range 90°C–300°C the most important material property is the chromium content[29] (Figure 1.96). In single phase flow additions of 1%Cr have prevented erosion corrosion under both laboratory tests and service exposures, and levels of Cr as low as 0.1% can have substantial benefits. In two-phase flow it is generally thought that the conditions are more arduous and the 2.25CrlMo is often specified but if the velocity is high enough even the 12Cr steel used in turbine blades suffers problems. This is probably an example where mechanical effects predominate.

Environmentally the most important variables are pH, oxygen content and temperature of the water (Figure 1.96). In single phase conditions both high pH and additions of low levels of oxygen have been used to prevent erosion corrosion[30]. However, because of partitioning effects between water and steam this is more difficult to achieve in two-phase flow. Although additions of morpholine[31] or AMP[32] (2-amino-2-methyl-propan-1-ol) have been successfully used to control pH.

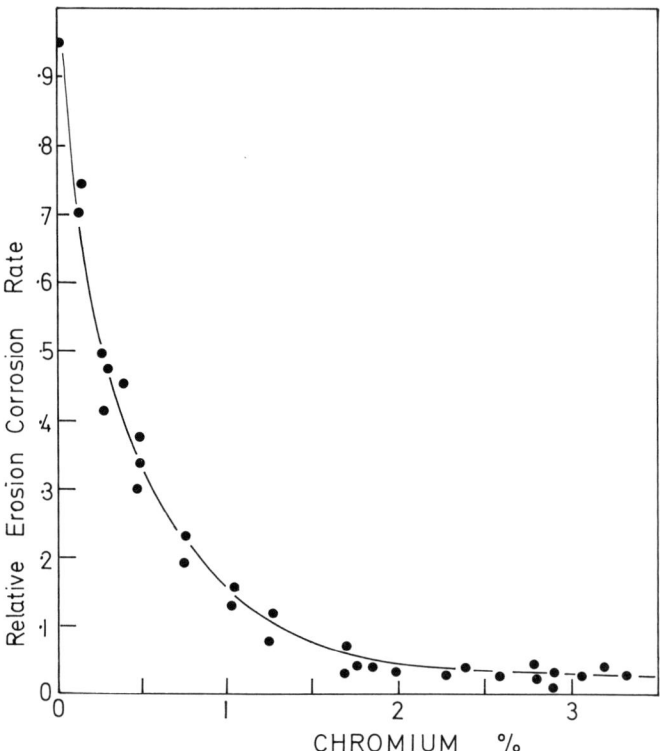

Fig. 1.94 Effect of chromium additions on the erosion corrosion of 70/30 cupro nickel in seawater (after Anderson[19])

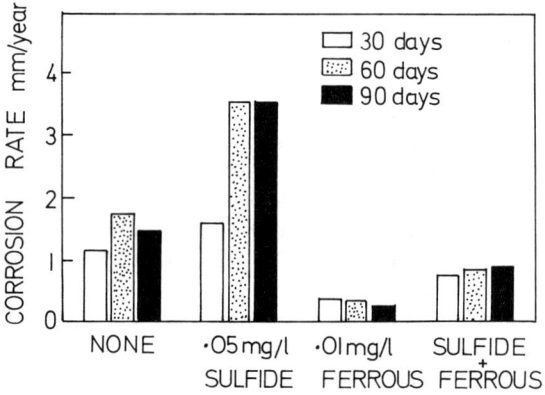

Fig. 1.95 Effect of sulphide and ferrous additions on the behaviour of 70/30 cupro nickel in seawater (after Hack[21])

Mechanistically chromium additions have been shown[33] to significantly enrich (10X) in the magnetite oxide layer and, it has been suggested that this lowers its solubility. Additions of small amounts of oxygen[34] to the water, increases the metal's potential and promotes the formation of haematite

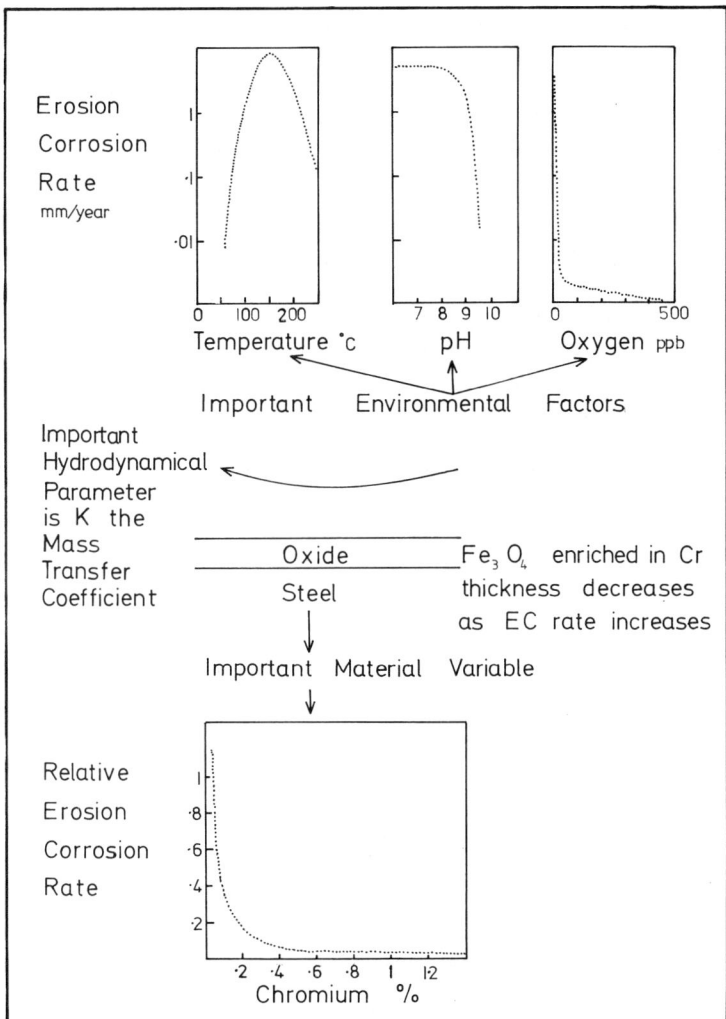

Fig. 1.96 Summary of factors controlling the erosion corrosion of steels in water

which also appears less soluble. The effects of both temperature and pH are thought to be related to the solubility of magnetite[14]; these effects are incorporated into Figure 1.96.

Predicting the Occurrence of Erosion Corrosion

Basically there are two approaches to predicting the occurrence of erosion corrosion. Practical or experience based methods typified by Keller's approach[36] for carbon steels in wet steam. Keller developed an equation that related the erosion corrosion rate as a function of temperature, steam quality, velocity and geometric factor. In recent years this approach has been improved to account for material and environmental variables by a number of organisations[35,37]; and computer programs are available[37]. The second approach[14,17] is based on the generation of high precision erosion corrosion data from controlled laboratory tests. This data is then used to define the variable relationship between erosion corrosion rate and mass transfer under the range of environmental and material variables of interest.

Preventing the Occurrence of Erosion Corrosion

Erosion corrosion may be considered at the design stage when a wider range of options is available to prevent its occurrence than if it occurs after the item of plant has been built. Various options and examples for these two situations are summarised in Table 1.32.

Table 1.32 Preventing the occurrence of erosion corrosion

Solution	Existing plant	New plant
Reduce velocity	Yes, but economic penalty	Yes, by increasing flow area
Ensure favourable geometry: Limit weld penetrations, tightness of bends, steps at flanges etc.	Not usually possible	Good practice to minimise problem
Use thicker material	Yes when replacing a failure but more resistant material better option	Yes, but resistant material better option
Use different material	Yes when replacing failure if compatible	Usually possible
Use of insert	Sometimes e.g. tube inlets but must ensure no steps introduced	
Coatings either organic or metallic	Yes, if access possible, must ensure coating doesn't detach	Might be cheaper solution than more expensive material
Modifying the environment	Usually possible but costs can be excessive in once through systems depending on additive	If suitable material not available
Apply anodic or cathodic protection	Yes, if possible and throwing power sufficient	

B. S. POULSON

REFERENCES

1. Poulson, B. *Plant Corrosion, Predictions of Materials Performance*, eds. J. E. Strutt and J. Nickolls, Ellis Horwood, Chichester, (1987)
2. Viscount J. R. Jellicoe, The Grand Fleet 1914-1916, Cassell
3. Proc of Specialists Meeting held at 'Les Renardiers' May 1982 on Erosion Corrosion of Steels in High Temperature Water and Wet Steam, Eds P. Berge and E. Khan, Electricité de France Paris (1983)

4. *ASM Metals Handbook*, **13**, 624, ASM Metals Park, Ohio (1981)
5. Efird, K. D. *Corrosion* **33**, 3 (1976)
6. Syrett, B. C. *Corrosion* **32**, 242 (1975)
7. Lush, P. A. *et al.*, Proc 6th European Cong. on Corrosion, London (1977)
8. King, C. V. In *Surface Chemistry of Metals and Semi Conductors*, Ed H. C. Gates, 357, J. Wiley, New York (1959)
9. Levich Physiochemical Hydrodynamics, Prentice Hall, New Jersey (1962)
10. Ross, T. R. *et al.*, *Jrn. Electrochem. Soc.*, **113**, 334 (1986)
11. Ellison, B. T. and Schmeal, W. R. *Jrn. Electrochem. Soc.*, **125**, 521 (1978)
12. Coney, M. Private Communication (1982)
13. La Que, F. L. *Corrosion*, **13**, 3034 (1957)
14. Bignold, G. J. *et al.*, *Proc. of UK Corr.*, **83**, Inst. Corrosion Sci. and Tech., Birmingham (1983)
15. Poulson, B. *Corrosion Sci.*, **23**, 391 (1983)
16. Poulson, B. *Corrosion Sci.*, **30**, 743 (1990)
17. Bouchacort, M. *Proc. of UK Corr.*, **88**, Inst. Corrosion Sci. and Tech., Birmingham (1988)
18. Pearson, C. *British Corrosion Journal*, **7** (1972)
19. Anderson D. B. and Badio, F. A. *Trans ASME*, **95** (1973)
20. Gilbert, P. T. *Trans. Institute of Marine Engineers* **66**, (1954)
21. Hack, H. P. and Gudes, J. P. *Corrosion*, **79**, paper 234.
22. Efird, K. D. *Corrosion*, **33**, 347 (1977)
23. Bailey G. L. J. *Jrn. Inst. of Metals*, **79**, 243 (1951)
24. Kato, C. *et al.*, *J. Electrochem. Soc.*, **127**, 1890 (1980)
25. Castle, J. E. and Parvizi, M. S. *Corros. Prev. Control*, Feb., (1986)
26. Eiselstein L. E. *et al.*, *Corros. Sci.*, **23**, 223 (1983)
27. Syrett B. S. and Wing, S. S. *Corrosion*, **36**, 23 (1980)
28. Little, B., Wagner, P. and Mensfeld, F. *Int. Met. Reviews*, **36**, 253 (1991)
29. Duereux, J. (in Reference 3)
30. Heitmann, H. G. and Kestner, W. (in Reference 3).
31. Gill, G. M. *et al.* (in Reference 3)
32. Penfold, D. P. *et al.*, Conf. on Water Chemistry of Nuclear Systems, 4, NES, Bournemouth (1986)
33. Woolsey, I. Private communication (1983)
34. Penfold, D. *et al.*, *Nuclear Energy*, **25**, 257 (1986)
35. Heitmann, H. G., Kastner, W. *VGB Krafwerk Stechrick*, **64**, 452, (1984)
36. Keller, H. *VGB Kraftwekstechnik*, (1974)
37. Chexal, B. *et al.* CHECWORKS[TM] Integrated Corrosion Software, *EPRI Journal*, Jul/Aug, (1992)

2 ENVIRONMENTS

2.1	Effect of Concentration, Velocity and Temperature	**2**:3
2.2	The Atmosphere	**2**:31
2.3	Natural Waters	**2**:43
2.4	Sea Water	**2**:60
2.5	Soil in the Corrosion Process	**2**:73
2.6	The Microbiology of Corrosion	**2**:87
2.7	Chemicals	**2**:99
2.8	Corrosion by Foodstuffs	**2**:114
2.9	Mechanisms of Liquid-metal Corrosion	**2**:120
2.10	Corrosion in Fused Salts	**2**:130
2.11	Corrosion Prevention in Lubricant Systems	**2**:143
2.12	Corrosion in the Oral Cavity	**2**:155
2.13	Surgical Implants	**2**:164

2.1 Effect of Concentration, Velocity and Temperature

In Chapter 1 the electrochemical mechanism of corrosion was considered in detail and it was shown how the kinetics of cathodic and anodic partial reactions control the rate of overall corrosion reaction. In this section the effects of environmental factors such as concentration, velocity and temperature will be considered on the assumption that either the anodic or cathodic reaction, but not both, is rate controlling. Thus if a metal is corroding under cathodic control it is apparent that the velocity of the solution will be more significant when diffusion of the cathodic reactant is rate controlling, although temperature may still have an effect. On the other hand if the cathodic process requires a high activation energy, temperature will have the most significant effect.

The effects of concentration, velocity and temperature are complex and it will become evident that these factors can frequently outweigh the thermodynamic and kinetic considerations detailed in Section 1.4. Thus it has been demonstrated in Chapter 1 that an increase in hydrogen ion concentration will raise the redox potential of the aqueous solution with a consequent increase in rate. On the other hand, an increase in the rate of the cathodic process may cause a decrease in rate when the metal shows an active/passive transition. However, in complex environmental situations these considerations do not always apply, particularly when the metals are subjected to certain conditions of high velocity and temperature.

The Effect of Anion Concentration on the Rate of Corrosion

The numerous metals and alloys used in practice show such a wide variation in response to various anions in acid and alkaline solutions that common features are difficult to discern and a basis for predicting corrosion behaviour is not very apparent.

Although Pourbaix (potential–pH) diagrams (Sections 1.4 and 7.6) have led to a greater under-standing of the changes in the corrosion behaviour of a metal due to a change in pH, they are less instructive about the behaviour of alloys and about the maxima in the corrosion rate that frequently occurs with increasing concentration of acid. For instance, Fe–18Cr–8Ni stainless steel has a low rate of corrosion in dilute sulphuric acid which increases with increase in concentration, to a maximum, followed by a decreasing rate on further increase of concentration (Fig. 2.1). Recently thermochemical data has become available so that potential-pH diagrams[1-4] have been calculated for anions such as SO_4^{2-}, Cl^-, citrate, S^{2-}, etc. However, there is no method

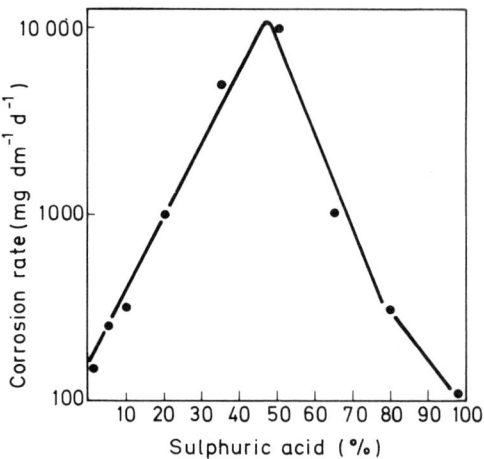

Fig. 2.1 Corrosion rate of Fe-18Cr-8Ni as a function of sulphuric acid concentration (20°C)

of calculating the relationships for alloys, although present studies on alloy dissolution[5,6] suggest that for some simple alloy systems this may be possible since dissolution of each phase appears to be a simple additive mechanism.

Anodic Dissolution under Film-free Conditions

With regard to the anodic dissolution under film-free conditions in which the metal does not exhibit passivity, and neglecting the accompanying cathodic process, it is now generally accepted that the mechanism of active dissolution for many metals results from hydroxyl ion adsorption[7-9], and the sequence of steps for iron are as follows:

$$Fe + H_2O \rightleftharpoons Fe(OH)^-_{ads.} + H^+$$
$$Fe(OH)^-_{ads.} \rightleftharpoons Fe(OH)_{ads.} + e$$
$$Fe(OH)_{ads.} \rightleftharpoons Fe^{2+} + OH^- + e$$

where $Fe(OH)^-_{ads.}$ signifies adsorption of OH^- on the Fe surface.

The most important outcome of this theory is that the rate of dissolution should be potentially greater as the pH increases, which is in conflict with simple concepts of corrosion kinetics. However, the theory has been proved to be applicable to many systems, and Bonhoeffer and Heusler[8] found that iron in sulphuric acid corroded at a greater rate with increase in pH, whilst Kabanov *et al.*[9] found that it corroded faster in alkaline solution than in acid solution for the same electrode potential.

Since the hydroxyl anion is involved in the mechanism given before, the implication is that other anions may also take part in the dissolution process, and that the effect of various chemicals may be interpreted in the light of the effect of each anion species. Most studies have been in solutions of sulphuric and hydrochloric acids and typically the reaction postulated for active dissolution in the presence of sulphuric acid is:

$$\mathrm{Fe(OH)_{ads.} + SO_4^{2-} \rightarrow FeSO_4 + OH^- + \mathit{e}},$$

followed by dissociation to Fe^{2+} and SO_4^{2-} aquo ions.

Studies of other metals in sulphuric acid[10], hydrochloric acid plus sodium chloride[11] and formic acid strongly support direct anion participation in the dissolution process. In general, it appears that whilst OH^- ions (and water molecules) have the largest accelerating effect on the rate of corrosion, other anions are also effective and this explains why some strong acids are more aggressive than others, in that they have different abilities to compete with hydroxyl ions in the dissolution process. This effect of different anions in increasing the rate of dissolution manifests itself as an increase in the exchange current density, as shown by Bockris et al.[7] who gave the following series of anions in order of acceleration of dissolution:

$$NO_3^- < CH_3COO^- < Cl^- < SO_4^{2-} < ClO_4^-$$

However, in the pH range 1-4, the effect of the OH^- ion predominates to such an extent that corrosion rates are similar in the presence of many other anions at concentrations less than 0.1 M. Since an adsorption process is involved in the mechanism, the corrosion rate in the pH range 1-4 may be represented by the Freundlich equation:

$$\text{Corrosion rate} = KC_{OH^-}^n$$

where C_{OH^-} is the concentration of OH^- ions and n is a small integer, often = 2. At higher pHs and concentrations of anions the rate of corrosion can be markedly reduced by either (*a*) precipitation and crystallisation on the surface of corrosion products, or (*b*) adsorption of specific anions that cover the surface and decrease adsorption of OH^-, i.e. competitive adsorption.

An important example of (*a*) is mild steel which may be used for containing concentrated sulphuric acid (greater than 70% H_2SO_4) because of the process of sulphation[12] or in the case of 10 M phosphoric acid[13] because of phosphatisation by ferrous phosphate; in each case the salt crystallises on the metal surface forming a mechanical barrier. Under these circumstances, and providing the salt layer is not disturbed by mechanical scraping or by flow of the solution, the corrosion rate will decrease to a low level. An example of (*b*) is the decrease in the corrosion rate of iron in dilute sulphuric acid caused by halide ions, e.g. I^- ions lower the rate by 95% in 1 N H_2SO_4, perhaps as a consequence of the high polarisability of halide anions[14].

An increase followed by a decrease in corrosion rate at a certain critical concentration is a commonly observed phenomenon for many metals and alloys. If the anion concentration at which the decrease takes place is high, then the anion species is deemed to be aggressive, but if low the anion is referred to as inhibitive. A considerable amount of experimental work in relation to the effect of specific ions on corrosion has been carried out on mild steel[15-17] and zinc[18].

Film-forming Conditions

The corrosion rate of many important metals and alloys is controlled by the formation of a passive film, and the thermodynamics and kinetics of their formation and breakdown are dealt with in Section 1.2.

The dissolution of passive films, and hence the corrosion rate, is controlled by a chemical activation step. In contrast to the enhancement of the rate of dissolution by OH^- ions under film-free conditions, the rate of dissolution of the passive film is increased by increasing the H^+ ion concentration, and the rate of corrosion in film-forming conditions such as near-neutral solutions follows the empirical Freundlich adsorption isotherm:

$$\text{Corrosion rate} = KC_{H+}^n$$

where n is an integer and C_{H+} is the concentration of hydrogen ions. It has been observed that in general, rates are controlled mainly by this equation, but the nature and concentration of anions do have an effect. Many anions such as Cl^- appear to be capable of causing pitting and breakdown of the film as the concentration increases. As the concentration in the bulk solution increases, corrosion products precipitate in the pits and blocking occurs with a subsequent reduction in dissolution rate.

For both film-free and film-forming conditions a decrease in corrosion rate is observed as the concentration of the anion increases. For some anions the maximum in the corrosion rate may be attained at low concentrations depending on the species and concentration (Fig. 2.2). One form of inhibition

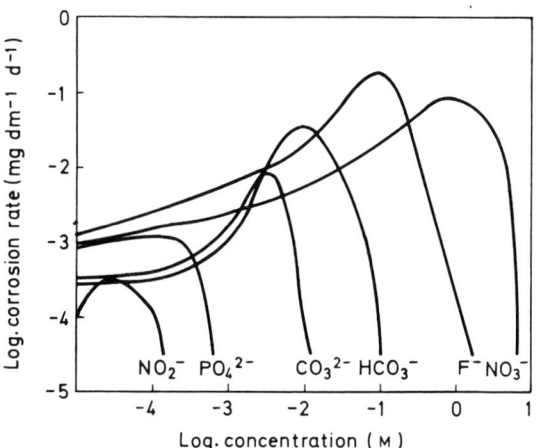

Fig. 2.2 Effect of increasing anion concentration on corrosion rate of mild steel in sodium salt solutions (after Brasher)

based on this effect may be achieved by adding an anion type that reaches its maximum corrosion rate at low concentration to a solution containing a more aggressive anion. Legault et al.[19] have studied this subject in detail and show that a Freundlich-type equation may still apply:

$$\text{Corrosion rate} = a + b \frac{C_{\text{agg.}}}{C_{\text{ihb.}}}$$

where a and b are constants, $C_{\text{agg.}}$ is the concentration of anions which stimulate corrosion and $C_{\text{ihb.}}$ is the concentration of anion type which precipitates and blocks the surface.

Effect of Solution Velocity on the Rate of Dissolution

Since the effect of concentration has been shown to lead usefully to salt passivation in many metal/anion systems the flow rate will markedly affect precipitation and subsequent corrosion behaviour. There have been many investigations of the influence of flow on the rate of corrosion, but lack of an awareness of hydrodynamic parameters has led to many experiments of questionable validity. In these circumstances it is not surprising that results from service failures do not correlate with laboratory tests. Some of these difficulties arise from the incomplete understanding of the theory of mass transport that still exists, especially in concentrated solution, although methods are now becoming available that are leading to more accurate prediction of corrosion rates in flowing systems. The most successful application of hydrodynamic theory to date has been for metals dissolving under essentially film-free conditions for a process that is unambiguously controlled by the arrival of the reactants at the surface, i.e. when the activation process is very fast compared to diffusion. For this reason it is customary to decide for each corrosion process which of the four main processes predominates and controls the corrosion rate. These four processes arise from the two electrochemical reactions, anodic and cathodic, and whether activation (act.) or concentration (conc.) overpotential is the dominant process. Thus, of the four possibilities the rate of mass transport will be involved in at least three of them as follows:

Anode: act.   act.   conc.   conc.
Cathode: act.   conc.   act.   conc.

In contrast, the temperature may only become important in the case of perhaps one combination since this parameter will have the greatest influence on activation-controlled processes. This arises because corrosion processes controlled by concentration overpotential have a limiting diffusion current, which in many cases imposes a maximum value on the corrosion rate even when activation polarisation is decreased (see Section 20.1). When concentration overpotential predominates then the limiting current density will give a good estimate of the corrosion rate. Since the limiting current density is determined by the flow-rate it should be possible to predict the changes in corrosion rate if the relationship between flow and concentration overpotential is known. If the limiting current density of the corrosion cell exceeds the

Concentration Overpotential at an Anode

There is an important difference between anode and cathode concentration overpotential. In the former, where anions directly participate in the dissolution process, ions accumulate at the surface and the rate is governed by metal ions moving out to the bulk solution; in the latter the rate is controlled by cathodic reactants (hydrogen ions, dissolved oxygen) moving towards the surface. In the case of cathode processes the limiting diffusion current is due to the depletion of ions as a result of the high rate of reaction, but in anodic processes no such limit is possible. There is, however, a type of rate-limiting behaviour when the solution next to the surface becomes saturated and crystallisation on the surface occurs. Solution flow will stimulate both electrode reactions by providing fresh solution with more ions for the cathodic process and with more water molecules to dilute the saturated solution formed at the anode. The rate of corrosion when concentration overpotential is controlling is governed by the diffusion of ions and the length of the path between concentration of ions in the bulk solution and at the surface. Fick's first law* can be applied to this situation and the rate/unit area, in terms of the current density i, can be described by:

$$i = \frac{zFD(a_s - a_b)}{\delta(1 - t)}$$

where z = number of electrons in the electrode process,
F = Faraday constant (96 485 C/mol),
D = diffusion coefficient (m²/s),
a_s = activity of ions at the surface (mol/m³),

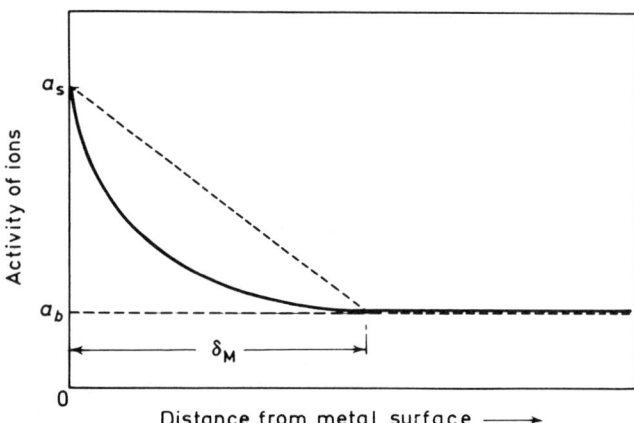

Fig. 2.3 Distribution of ions during anodic polarisation, showing the arbitrary value used for diffusion path length

*See Section 20.1 for a more detailed derivation.

a_b = activity of ions in the bulk solution (mol/m^3),
δ = path length of diffusing ions (m) and
t = transport number of all the ions involved in the electrode process; in high concentrations of electrolytes, NaCl, Na$_2$SO$_4$, HCl, etc. the charge is carried by ions not involved in the corrosion process and may be assumed to be unity.

The rate is a maximum when $a_s = 0$ for a cathodic process and when it reaches saturation for an anodic process. For isothermal conditions D is a constant, and at the limiting current $(a_s - a_b)$ is a constant, and in this case the controlling factor is δ the path length for diffusion. Thus the variation of δ with solution flow rate determines the corrosion rate.

The change in activity along the diffusion path length is unlikely to be linear and from diffusion theory the distribution of ions is most likely to be that given in Fig. 2.3. The distribution shows the difficulty of choosing a value for δ and a compromise value is used, δ_M, which is the effective diffusion layer thickness when the diffusion gradient is assumed to be linear. The effect of changing velocity on corrosion processes can best be understood through the factors that change δ_M. Fortunately, there are many practical examples of corrosion where the rate is wholly controlled by either the anode or cathode concentration overpotential so that the parameters that control the effective layer thickness should be known. This is best done by the use of hydrodynamic theory.

Application of Hydrodynamics

It is important at the outset to define more closely the effective film thickness δ_M. In any electrolyte solution in contact with a metal surface there is a static layer of solution next to the surface whose thickness will decrease as the solution velocity increases. The way in which this velocity changes the hydrodynamic thickness (δ_H) is complex and depends on such factors as viscosity, geometry, temperature and surface roughness*.

It is also necessary to separate *laminar* flow when a stagnant layer of well defined thickness δ_H is formed, from *turbulent* flow when values of δ_H are very low and when flow towards and away from the surface is complex. The analogy of mass transport with heat transfer[20,21] has led to successful methods of regarding the mass transport interaction with fluid flow, since the behaviour of heat is in many ways similar to mass transport depending as it does on a driving force, i.e. the heat gradient may be regarded as analogous to the concentration gradient.

Relationships between δ_H and δ_M have been established for certain geometries, e.g. for a rotating disc Levich[21] has found that for laminar flow $\delta_H \approx 5\delta_M$. The mathematical proofs of these relationships are not appropriate here, but a useful non-mathematical account of the application of hydrodynamic theory to mass transport has been given by King[22]. The most important variables are the main stream solution velocity U, the characteristic length L (diameter in the case of a rotating cylinder) and the kinematic

---

*Reference 20 gives definitions pertaining to the various diffusion layers for the special case of electrochemical mass transfer.

viscosity v (m²/s). For application to mass transport such as cathodic reduction or anodic dissolution, which are dependent on ion or molecular (dissolved oxygen) transport, the variables are the diffusion coefficient, D, and the activity difference between surface and main solution, Δa (mol/m³).

Using the mathematical technique of dimensionless group analysis, the rate of mass transport (R_M) in terms of moles per unit area per unit time can be shown to be a function of these variables, which when grouped together can be related to the rate by a power term. For many systems under laminar flow conditions it has been shown that the following relationship holds:

$$\frac{R_M L}{D \Delta a} = K \left(\frac{UL}{v}\right)^x \left(\frac{v}{D}\right)^y \qquad \ldots(2.1)$$

where K is a constant, and x and y are exponents that are very often $\frac{1}{2}$ and $\frac{1}{3}$ respectively. The dimensionless groups in equation 2.1 are referred to as follows:

$$\frac{UL}{v} = (Re) \text{ (Reynolds No.)}; \frac{v}{D} = (Sc) \text{ (Schmidt No.)};$$

$$\frac{R_M L}{D \Delta a} = (Sh) \text{ (Sherwoood No.)};$$

where U is the main stream velocity (m/s), D is the diffusion coefficient (m²/s), v is the kinematic viscosity (m²/s) and L is the characteristic length (m). The application of this equation is only useful if:

1. The relationship between δ_H and δ_M is known.
2. Concentration has a power exponent of unity, i.e. conforms to a first order reaction.
3. Dissolution is uniform (etching); otherwise, for rough surfaces such as pitting, turbulent flow regimes may occur even at low solution velocities.

The rate R_M can be converted to the limiting current density i_L by Faraday's law, so that from physical measurements on the solution it is possible to calculate $i_{corr.}$, since it is of the same order as i_L.

It has not been possible to calculate constant K or exponents x and y directly from theory, except in one case, so that they have to be determined by experiment. The geometry of a system, i.e. flat plates, stirrers, pipes, etc. have a large effect in determining the magnitude of the constants and by using reactions whose parameters are well established for one geometry it has been possible to gather data for many other systems, by using the known reactions in other geometries.

In the case of the one system that can be predicted from theory, i.e. the rotating-disc electrode (radius r), this is proving to be a useful tool for understanding the effects of flow on corrosion reactions. The equation can be rearranged to give the limiting current density i_L, and velocity ÷ the characteristic length U/L can be interpreted as the angular velocity ω. The equation developed by Levich[21] by substituting in equation 2.1 is:

$$i_L = 0 \cdot 0062 zF \left(\frac{U}{L}\right) \left(\frac{UL}{v}\right)^{\frac{1}{2}} \left(\frac{v}{D}\right)^{\frac{1}{3}} D\bar{r}\Delta a \qquad \ldots(2.2)$$

EFFECT OF CONCENTRATION, VELOCITY AND TEMPERATURE

Substituting for UL by ω and for the limiting case for the cathodic reduction process when $a_s = 0$, the activity term Δa is then equal to a_b (the concentration of ions in the bulk solution), and then

$$i_L = 0\cdot 0062 z F \omega^{\frac{1}{2}} v^{-\frac{1}{6}} D^{\frac{2}{3}} a_b r^{-1} \qquad \ldots(2.3)$$

Zembura[23] has made specific use of the rotating disc for investigation of the effect of flow on corrosion reactions. This work has shown that it is possible to determine the type of control (activation or concentration polarisation) of zinc dissolving in $0\cdot 1$ N Na_2SO_4 (de-aerated), which followed closely the predicted increase in hydrogen ion reduction as the flow rate increased, and proved that in this example

$$i_{\text{corr.}} \approx i_L$$

Although the rotating disc is useful in understanding the mechanism of corrosion it is necessary to evaluate flow rates in the turbulent region which characterises the effect of velocity in real systems. King[22] collected data for rotating cylinders, which give turbulent flow regimes*, even at low rotation speeds. For many systems he found that several data fitted the relationship:

Table 2.1 Corrosion rates from hydrodynamic parameters for pipes and annuli

Flow regime	Appropriate equation	Ref. No.
Pipes		
1. Laminar flow $(Re) > 2000$	$\dfrac{i_L}{zFD\Delta a} = 1\cdot 614 \left[\dfrac{1}{d}\right] (Re)^{0\cdot 33} (Sc)^{0\cdot 33} \left[\dfrac{d}{L}\right]^{0\cdot 33}$ only if $(Re)(Sc)\left[\dfrac{d}{L}\right] < 8$; d = diameter	25
2. Turbulent flow $(Re) < 5000$	$\dfrac{i_L}{zFD\Delta a} = 0\cdot 276 \left[\dfrac{1}{d}\right] (Re)^{0\cdot 58} (Sc)^{0\cdot 33} \left[\dfrac{d}{L}\right]^{0\cdot 33}$	25
3. Turbulent flow $(Re) > 5000$	$\dfrac{i_L}{zFD\Delta a} = 0\cdot 023 \left[\dfrac{1}{d}\right] (Re)^{0\cdot 8} (Sc)^{0\cdot 33}$	56
Annuli	$\dfrac{i_L}{zFD\Delta a} = K\left[\dfrac{1}{d_e}\right] (Re)^{0\cdot 33} (Sc)^{0\cdot 33} \left[\dfrac{d_e}{L}\right]^{0\cdot 33}$,	
1. Laminar flow $(Re) > 2000$	radius ratio $= \dfrac{\text{core radius}}{\text{outer radius}}$ When radius ratio $= 0\cdot 25$, $K = 1\cdot 8$ and when radius ratio $= 0\cdot 125$, $K = 2\cdot 03$; $d_e =$ annular equivalent diameter $= d_2 - d_1$	27
2. Turbulent flow	for radius ratio $0\cdot 5$: $\dfrac{i_L}{zFD\Delta a} = 0\cdot 276 \left[\dfrac{1}{d_e}\right] (Re)^{0\cdot 58} (Sc)^{0\cdot 33} \left[\dfrac{d_e}{L}\right]^{0\cdot 33}$	27

*For rotating cylinders the exponent x for Reynolds number is very often unity for turbulent flow, and therefore L may be included in the constant term for a particular geometry of cynlinder.

$$i_L = 0{\cdot}017zFa_b U \left(\frac{D}{v}\right)^{0{\cdot}83}$$

where i_L is in A/m^2, a_b is the activity in the bulk solution and the equation assumes that the reaction is under cathodic control.

The corrosion rates in pipes and annuli are very important, but for these geometries the effect must be evaluated experimentally, since the theory cannot predict the relationships with any accuracy. Although the importance of the interaction of flow rate and corrosion has long been appreciated, there have been very few studies where hydrodynamic parameters have been considered or even measured and controlled[24,25]. However, there has been much work where these factors have been taken into consideration for corroding systems[26,27], and the work of Ross and Wragg[27] contains a valuable review of previous work. A great deal of work has been done in the general field of electrochemistry which may be applicable to corroding electrodes when concentration polarisation of one of the half reactions predominates. Table 2.1 gives some useful equations for pipes and annuli.

Correlations Between Flow Rate in Rotating Discs, Cylinders and Smooth Tubes

The rotating disc and rotating cylinder have been successfully applied in the laboratory to study the effect of flow on corrosion rates and are much easier to use than actual pipelines and other real geometries. The results of these tests can now be correlated to geometries likely to be found in pipes, pumps, bends, etc. in plant by use of dimensionless group analysis. There-

Table 2.2 Correlation between rotating disc, rotating cylinder and smooth tubes

where W = rotation speed in r.p.m.
V = flow rate in tubes, m s^{-1}
v = kinematic viscosity m^2 s^{-1}
Rd = disc radius, m
Rc = cylinder radius, m

Rotating disc (laminar flow): $W = \dfrac{V}{0{\cdot}0066\,(Rd)^{0{\cdot}3}}$ (61)

Rotating cylinder (turbulent flow): $W = \dfrac{V^{0{\cdot}9} + \text{const.}}{0{\cdot}026\,(Rc)^{1{\cdot}6}}$ (61)

Examples of rotating cylinder

Steel: 94 to 98% H_2SO_4, 60°C, $W = \dfrac{V^2 + 0{\cdot}0025}{0{\cdot}5\,v}$ (62)

Copper: 0·1 M HCl + 198 gl^{-1} of Fe^{3+} at 30°C, $W = \dfrac{V^2 + 0{\cdot}004}{0{\cdot}25\,v}$ (62)

Lead: 2 M NaOH + 0·1 M NaNO$_3$, 40°C, $W = \dfrac{V^2 + 0{\cdot}88}{0{\cdot}36\,v}$ (62)

Erosion Corrosion Rates at Jets, Nozzles, Orifices and Other Flow Expansions

When corrosion rates are mainly dependent on diffusion, especially of dissolved oxygen, carbon dioxide and/or hydrogen ions in weak acid solutions, then it is possible to relate corrosion rates in terms of hydrodynamic parameters, i.e. erosion corrosion. This makes corrosion allowance in design, at the drawing-board stage, a possibility. Engineers, who already carry out similar calculations to estimate the dimensions of pipelines and other flow systems, can use these concepts to predict erosion rates. Such calculations could be used as a guide to selection of materials or inhibitor type when more realistic estimations are made of the rate of corrosion damage. Thus, effective corrosion control might therefore be achieved by a larger pipe size, longer bends, more sophisticated 'tee' junctions, and slower pump speeds as an alternative to the more formal methods of corrosion control which generally are more costly.

For many cooling waters, including seawater and also drinking water, where corrosion rates are 70 to 100% of the limiting diffusion current, the use of dimensionless group analysis can then be applied.

Suitable equations have been given for pipelines in Fig. 2.4 and these may be compared with the equation for impinging jets and nozzles or orifices. A more detailed review of this and other hydrodynamic relationships are given by B. Poulson[57].

Impinging Jet or Nozzle

There are many examples of increased corrosion at or near nozzles and jets and this is a recurring problem requiring frequent replacement and maintenance. The use of hydrodynamics and, in particular, dimensional group analysis, can show the most important parameters and can indicate the comparative rates of corrosion.

In the case of jets and nozzles, the general pattern is a stagnant area beneath the jet and an area adjacent which suffers increased flow rate and therefore corrosion. The essential parameters are shown in Fig. 2.4 and for turbulent flow the Sherwood No. (Sh), according to Chin and Tsang[61], becomes:

$$Sh = 1 \cdot 12 \, Re^{0.5} \, Sc^{0.33} \, (H/d)^{-0.054}$$

where H = height of the jet from a flat plate
d = diameter of the jet stream
x = diameter of jet area

This equation applies for turbulent flow when Re is between 4000 and 16 000; x/d between 0·1 and 1·0. For H, d and x, see Fig. 2.4.

This may be compared with fully developed turbulent flow along a flat sheet or tube when:

2:14 EFFECT OF CONCENTRATION, VELOCITY AND TEMPERATURE

Geometry	Diagram	Re	Sh
Rotating disc	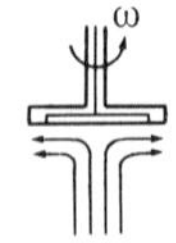	$Re = \dfrac{\omega y^2}{\gamma}$ $Re_c = 1.7 - 3.5 \times 10^5$	$\overline{Sh}_L = 0.6205\, Re^{0.5}\, Sc^{0.33}$ $\overline{Sh}_r = 0.0078\, Re^{0.9}\, Sc^{0.33}$
Rotating cylinder	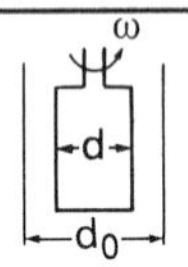	$Re = \dfrac{dV}{\gamma} = \dfrac{\omega d^2}{2\gamma}$ $Re_c \simeq 200$	$\overline{Sh}_{smooth} = 0.079\, Re^{0.7}\, Sc^{0.356}$ $\overline{Sh}_{rough} = (1.25 + 5.76\, \log_{10} d/\varepsilon)^{0.7}\, Re\, Sc^{0.356}$
Impinging jet	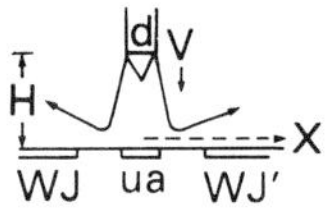	$Re = \dfrac{dV}{\gamma}$ $Re_c \simeq 2000$	$\overline{Sh}_{UA} = 1.12\, Re^{0.5}\, Sc^{0.33}\, (H/d)^{0.051}$ $\overline{Sh}_{WJ} = 0.65\, Re^{0.84} \left(\dfrac{x}{d}\right)^{1.2}$
Nozzle or orifice	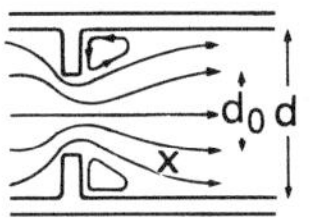	$Re_o = \dfrac{d_o V_o}{\gamma}$ $Re_c \simeq 500$	$\overline{Sh}_{max} = 0.276\, Re^{0.66}\, Sc^{0.33}$ $\dfrac{Sh_x}{Sh_{fd}} = 1 + A_x \left[1 + B_x \left(\dfrac{Re^{0.66}}{0.0165\, Re^{0.66}} - 21\right)\right]$
Tube	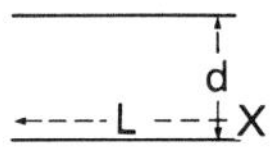	$Re = \dfrac{dV}{\gamma}$ $Re_c \simeq 2000$	$\overline{Sh}_L = 1.614\, (Re\, Sc\, d/L)^{0.33}$ $\overline{Sh}_T = 0.276\, Re^{0.563}\, Sc^{0.33}\, (d/L)^{0.33}$ $\overline{Sh}_{Fdf} = 0.0165\, Re^{0.86}\, Sc^{0.33}$

Fig. 2.4

$$Sh = 0\cdot 023\, Re^{0\cdot 8}\, Sc^{0\cdot 33}$$

This may be compared with the corrosion rate increase at an orifice of the same diameter ratio, given that

$$Sh_o = 0\cdot 276\, Re^{0\cdot 66}\, Sc^{0\cdot 33}$$

$$Re = \frac{Ud}{v}$$

The zone beyond the orifice is likely to be corroding faster than for the material at the orifice.

Assuming $H/d = 5$ then

$$\left(\frac{H}{d}\right)^{-0\cdot 054} = 0\cdot 91$$

and since $Sc^{0\cdot 33}$ is common, then the increase in the Sherwood No. (*Sh*) as a result of the jet, is proportional to the increase in corrosion rate and an approximate estimate can be made from the ratio:

$$1\cdot 12\, Re^{0\cdot 5} : 0\cdot 023\, Re^{0\cdot 8}$$

For example, when $Re = 5000$, the ratio equals 79:21, and therefore an increase in corrosion rate of about four times is found as a result of the jet.

Angled Jets

If jets are inclined at angles other than 90° to the flat plate, the overall mass transport appears to be unchanged except that the stagnation area is nearer to the down side of the jet[59].

Orifices and Other Restrictions in Flow

In practice, deliberate changes in flow are necessary to proportion quantities of flow into various systems, and to determine flow rate by various measuring devices by restrictions, e.g. Venturi meters and rotameters.

Accidental restrictions in flow, causing considerable premature failure, have included wrongly sized packing gaskets in pipe and tube joints, welding ferrules and baffling, in heat exchangers, all of which could have been allowed for at the design stage if hydrodynamic relationships had been applied.

It has been suggested[60] that (see Fig. 2.4):

$$Sh_{max} = 0 \cdot 276 \, Re^{0 \cdot 66} \, Sc^{0 \cdot 33} \left(\frac{d}{d_0}\right)$$

where d = diameter of the pipe
d_0 = diameter of the orifice.

The effect of an orifice in a tube on corrosion rate compared with the rate in a smooth tube can be calculated from:

$$\frac{Sh_0}{Sh_t} = \frac{0 \cdot 276}{0 \cdot 023} \cdot \frac{Re^{0 \cdot 66}}{Re^{0 \cdot 8}} \cdot \frac{d}{d_0}$$

Where Sh_0 is for the turbulance at x, and Sh_t is for turbulent flow in a smooth pipe.

If the Reynolds No. is 100 000 and the ratio of the diameter of the tube and orifice is 2, then the increase in corrosion rate at x is:

$$\frac{0 \cdot 276}{0 \cdot 023} \cdot \frac{(100\,000)^{0 \cdot 66}}{(100\,000)^{0 \cdot 8}} \cdot 2 \simeq 5 \text{ times}$$

Temperature and Flow Rate

For diffusion controlled corrosion reactions e.g. dissolved oxygen reduction, and the effect of temperature which increases diffusion rates, then by substituting viscosity and the diffusion coefficients at appropriate temperatures into the Reynolds No. and Schmidt No., changes in corrosion rate can be calculated.

For example, Oldfield and Todd[63] have confirmed that for mild steel in seawater then the rate of corrosion can be predicted from:

$$\text{C.R. in mm/yr} = 0 \cdot 0117 \cdot Co_2 U^{0 \cdot 9} \frac{D^{0 \cdot 75}}{\nu}$$

when Co_2 = concentration of oxygen in ppb
U = flow rate in cm/sec
D = diffusion coefficient, cm^2/sec
ν = kinematic viscosity, cm^2/sec

By substituting the appropriate values for viscosity and diffusion at various temperatures, they found that corrosion rates could be calculated which were confirmed by experiment. The corrosion rates represent maxima, and in real systems, corrosion products, scale and fouling would reduce these values often by 50%. The equation was useful in predicting the worst effects of changing the flow and temperature. The method assumes that the corrosion rate is the same as the limiting diffusion of oxygen; at least initially this seems correct.

Effect of Flow During Conditions that Lead to Film Formation

For many corroding systems the rate is controlled by the presence of a film of corrosion products. This may range from the relatively thin films on metals such as aluminium and magnesium to massive deposits found on corroding iron pipes. Both situations have been considered from the view-point of hydrodynamic theory. In the case of magnesium corroding in hydrochloric acid, Marangozis[28] has derived a relationship that includes chemical dissolution of $Mg(OH)_2$ as a controlling factor, the rate being controlled by diffusion of OH^- ions. For rotating cylinders the rate of dissolution followed the relationship

$$R_{Mg^{2+}} = 2 \cdot 10^7 (K_{OH^-})(C_{Mg^{2+}})$$

where K_{OH^-} is the mass transport coefficient of hydroxyl ions, $C_{Mg^{2+}}$ is derived from the solubility product of $Mg(OH)_2$ and the units are gm^{-2}d^{-1}. K_{OH^-} can be calculated from purely physical data, thus

$$K_{OH^-} = K^0_{OH^-} \left[\frac{(Sc)^{0 \cdot 33}_{H^+}}{(Sc)^{0 \cdot 33}_{OH^-}} + \frac{(Sc)^{0 \cdot 66}_{OH^-}}{(Sc)^{0 \cdot 66}_{H^+}} \right] \frac{(C_{H^+})}{(C_{OH^-})}$$

where (Sc) = Schmidt number (ν/D),
(C_{H^+}) = concentration (or activity) in main solution,
(C_{OH^-}) = concentration (or activity) at the surface which may be obtained from the solubility product of $Mg(OH)_2$, and
$K^0_{OH^-}$ = the mass transport coefficient of hydroxyl ions *without* chemical reaction and is obtained from:

$$K^0_{OH^-} = \left[0 \cdot 143 (Sc)^{0 \cdot 33}_{OH^-} \left(\frac{Uh}{\nu} \right)^{0 \cdot 7} \left(\frac{D}{r+h} \right) \right],$$

where U = peripheral velocity of the rotating cylinder,
h = distance between cylinder and vessel wall,
r = radius of cylinder, and
D = diffusion coefficient of OH^- ions.

The equations are valid up to about 0·1 M HCl, but at high rates surface roughening gives errors as the flow regime changes to turbulent flow. The analysis of the hydrodynamic situation when a porous film, several micrometres in thickness, is formed by precipitation of corrosion products has been made by Mahato et al.[29]. The importance of this work is that it takes into account the varying rate as this layer begins to thicken. The system they considered was the internal rusting of steel pipes by natural water. The presence of corrosion product gave rise to turbulent flow, even at low flow rate and the characteristic length L was the internal pipe diameter which decreased with time as corrosion product accumulated. Thus the dimensional group theory was modified to accommodate the 'unsteady state' mass transport.

The details of this calculation are given by Mahato, and the final result was:

$$\text{Weight loss} = 52 \cdot 1 (Re)^{0.54} [(0 \cdot 96t + 0 \cdot 31)^{0.5} - 0.56]$$

where the weight loss is in $gm^{-2}d^{-1}$ and the time t is in hours.

An important assumption was that the solution was dilute (in this case natural water of approximately 100 p.p.m. total dissolved solids) since there are difficulties in applying mass transport equations for certain situations in concentrated electrolyte solution, where a knowledge of activities is uncertain and this can lead to large errors.

The above work is important, since many practical corrosion systems involve a thick but porous film of corrosion products, e.g. rusting, sulphatising, tuberculation and atmospheric corrosion, and the approach may lead to a more valid corrosion testing technique for these situations.

Some Effects of Temperature on Corrosion Reactions

In contrast to the influence of velocity, whose primary effect is to increase the corrosion rates of electrode processes that are controlled by the diffusion of reactants, temperature changes have the greatest effect when the rate determining step is the activation process. In general, if diffusion rates are doubled for a certain increase in temperature, activation processes may be increased by 10–100 times, depending on the magnitude of the activation energy.

Bearing in mind the importance of the rate determining process and because of the complex situation in corrosion reactions of having two electrode processes, the effect of temperature is best illustrated by reference to specific situations.

Cathode Reactions

Hydrogen evolution process In de-aerated solutions when this process is under activation control the main effect of increasing the temperature is to increase the exchange current. Typical examples of the magnitude of this change have been given by Conway et al.[30] who found that for nickel the exchange current increased from approximately 10^{-2} A/m^2 to $1 \cdot 0$ A/m^2 when the temperature changed from 10 to 75°C and the activation energy was about 59 kJ/mol. Thus the rate of corrosion would be increased by at least 100 times if the anode process was unaffected by the temperature increase, whilst for control by concentration polarisation, the diffusion coefficient for hydrogen ions would increase perhaps only twice over the same temperature range.

Dissolved oxygen reduction process Corrosion processes governed by this cathode reaction might be expected to be wholly controlled by concentration polarisation because of the low solubility of oxygen, especially in concentrated salt solution. The effect of temperature increase is complex in that the diffusivity of oxygen molecules increases, but solubility decreases. Data are scarce for these effects but the net mass transport of oxygen should increase with temperature[31] until a maximum is reached (estimated at about 80°C) when the concentration falls as the boiling point is approached. Thus the corrosion rate should attain a maximum at 80°C and then decrease with further increase in temperature.

A striking example of the interaction of solution velocity and concentration is given by Zembura[23] who found that for copper in aerated $0 \cdot 1$ N H_2SO_4, the controlling process was the oxygen reduction reaction and that up to 50°C, the 'slow step' is the activation process for that reaction. At 75°C the process is now controlled by diffusion, and increasing solution velocity has a large effect on the corrosion rate (Fig. 2.5), but little effect at temperatures below 50°C. This study shows how unwise it is to separate these various

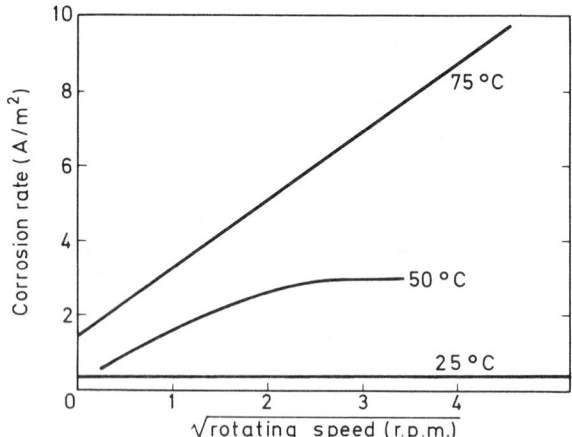

Fig. 2.5 Corrosion rate of a copper rotating disc in air-saturated $0 \cdot 1$ N H_2SO_4 at various temperatures (after Zembura[23])

Anode Reactions

It is convenient to consider three stages of anode polarisation with regard to temperature effects, (*a*) under film-free conditions, (*b*) under film-forming conditions and (*c*) at the active-passive transition.

Film-free conditions It has been observed for many metals that the magnitude of $i_{crit.}$ (see Section 1.4) increases with temperature and that the activation energy for dissolution is low, suggestive of a diffusion-limited anode process when the migration of corrosion products away from the surface is rate controlling. Some examples of the value of the activation energy for this process are given in Table 2.4.

These data have been obtained by anodic polarisation work and might therefore be more relevant when cathodic reduction of oxygen takes place that can increase the corrosion potential to high positive values.

Table 2.4 Activation energies for critical currents ($i_{crit.}$) for passivation

Temperature range	Activation energy (kJ/mol)	System	Reference No.
25–98°C	19·2	Ni in 0·05 N H_2SO_4	32
25–75°C	23·4	Ni in 0·12 N H_2SO_4	32
25–100°C	37·6	Cr-Ni-Mn steels in H_2SO_4	33
25–98°C	46·0	Ti in HCl	34
4–30°C	41·8	Fe in 1 N H_2SO_4	35

The relationship between $i_{crit.}$ and temperature can be described by an Arrhenius type of equation:

$$\log i_{crit.} = A - \frac{E^{\ddagger}}{RT} \qquad \ldots(2.4)$$

where A is a constant and $E^{\ddagger}$ is the activation energy. A more useful form is:

$$\log i_{crit.} = \log i_{crit.}^{T} - \frac{E^{\ddagger}}{RT} + K$$

where $\log i_{crit.}^{T}$ is the value at, say, 25°C.

For many metals the critical current density for passivation ($i_{crit.}$) increases with increasing pH of the solution:

$$\log i_{crit.} = \log i_{crit.^0} + K\text{pH} \qquad \ldots(2.5)$$

where K is a constant depending on the metal or alloy and $\log i_{crit.^0}$ is the value when pH = 0.

This relationship is in accordance with the hydroxyl adsorption theory where polarisation is decreased with increasing hydroxyl concentration

2:20 EFFECT OF CONCENTRATION, VELOCITY AND TEMPERATURE

(unless film formation interferes). Equations 2.4 and 2.5 may be combined to determine the magnitude of i_{crit} at various temperatures and pH. Since control is due mainly to the diffusion of anodic products from the surface, changes in flow would be expected to have a large effect, and should be considered when designing anodic protections systems.

Film formation In a few instances the temperature dependence of i_p, the passive current, has been observed and high activation energies (46 to −84 kJ/mol) have been obtained, indicating a large increase in rate as the temperature increases (see Table 2.5).

Table 2.5 Activation energies of passivation current densities, i_p

Temperature range	Activation energy (kJ/mol)	System	Reference No.
25–60°C	60	Sn in 1 N NaOH	37
25–60°C	49	Sn in 0·1 N NaOH	37
4–25°C	54	Fe in 1 N H_2SO_4	35
25–100°C	78	Ni in 1 N H_2SO_4	36
25–75°C	56	Cu in 0·1 N H_2SO_4	23
60–100°C	50	Inconel 825	38
	46	Corronel 230	38
25–60°C	48	Cr-Mn-Ni steels	33
10–25°C	75	Ti in 10 N HCl	39
25–70°C	92	Ti in 10 N HCl	39

Thus the rate of change of i_p under activation control is much faster than i_{crit} which is under diffusion control, and for the same condition of solution velocity the two rates could become equal at some common temperature, i.e. $i_{\text{crit}} = i_p$, and there is *no* active–passive transition. For many of the systems given in the table this temperature is about 100°C. Above this temperature the measured activation energy is lower and diffusion control is established.

In practice the danger of aerated systems becomes apparent when the temperature is above a certain minimum, for there is no passive film formation, and it is clear that anodic protection cannot be effective in these circumstances.

The dissolution of passive films is, in the main, controlled by a chemical activation step in contrast to film-free conditions at i_{crit}. Many protective anodic films are oxides and hydroxides whose dissolution depends upon the hydrogen ion concentration, and the rate follows a Freundlich adsorption equation:

$$\log i_p = \log k + n \log C^{H+} \qquad \ldots(2.6)$$

where k is constant and n is related to the valency of the cations; $n = 0.5$, 0.33 and 0.25 for monovalent, divalent and trivalent ions, respectively. On rearranging equation 2.6 a more useful form is obtained:

$$\log i_p = \log i_p^0 - n\text{pH}^*$$

where $\log i_p^0$ is the current density (at constant temperature) when pH = 0. This equation can be combined with the temperature relationship assuming the pH is constant:

$$\log i_p = \log i_{pT} - \frac{-E^{\ddagger}}{2 \cdot 3RT} + K$$

where i_{pT} is the passivation current at some standard temperature, e.g. 25°C.

Hence
$$\log i_p = K_1 - n\text{pH} + K_2 \left(\frac{1}{T}\right) \qquad \ldots(2.7)$$

This assumes that the pH does not change over the temperature range considered. However, corrections are small for pH values of from 0 to 5, but significant in the range 6–9 as shown in Fig. 2.6.

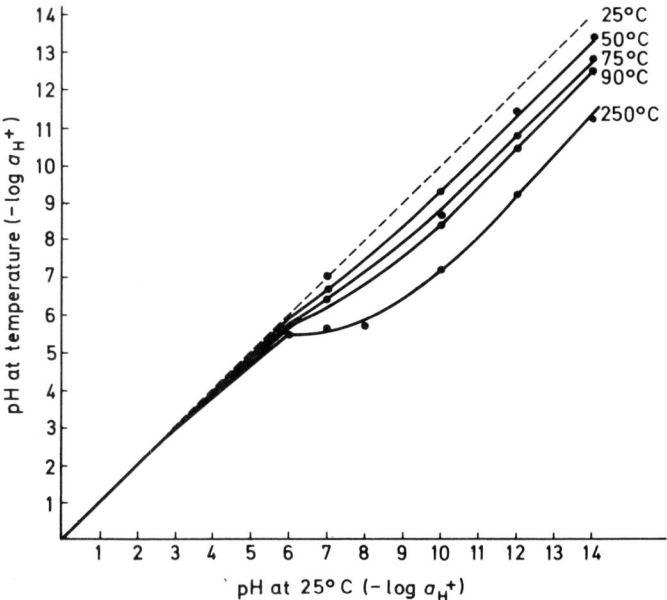

Fig. 2.6 Change in $-\log a_{H^+}$ with temperature[54]

Active–passive transition It has been shown that i_p, the current required to maintain a passive film, increases with temperature at a much greater rate than the critical current for passivation as a result of an activation-controlled process. At some temperature i_p will exceed $i_{crit.}$ and *no* active–passive transition will be observed, and more important no protection by a passive film is possible because of the high rate of dissolution. At this stage the slow process becomes the diffusion of reactants and control of the rate is

* Note that the pH dependence is opposite to that for $i_{crit.}$ which increases as pH increases.

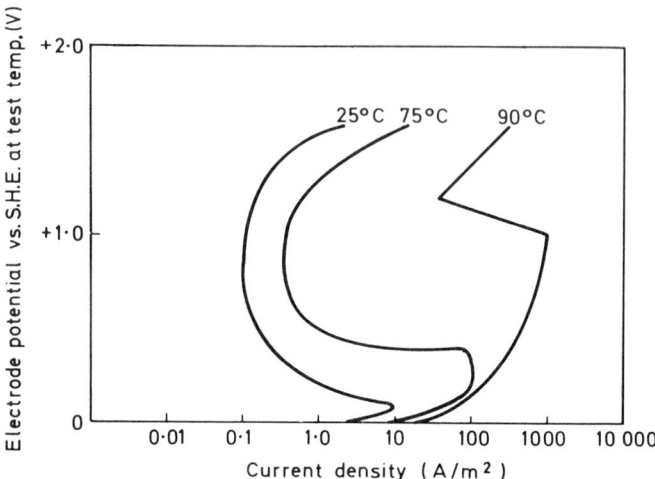

Fig. 2.7 Anodic polarisation of nickel at various temperatures, in 0·05 N H_2SO_4 + 0·05 N K_2SO_4, pH 1·3 (after Cowan and Staehle[32])

sensitive to solution flow changes. The temperature at which this occurs has only been determined in a few cases and an example is given in Fig. 2.7. It must be emphasised that the rate of dissolution of the passive film is also determined in many cases by the pH of the solution, and for some metals and alloys this effect probably occurs at low temperatures below the boiling point of aqueous solutions when under acidic conditions.

Some metals and alloys have low rates of film dissolution (low i_p) even in solutions of very low pH, e.g. chromium and its alloys, and titanium. In these cases the value of i_p is quite low, and although it increases as the temperature increases, a maximum is reached when the solution boils. The maximum current is below $i_{crit.}$ and breakdown does not occur. However, in certain alloys, e.g. Cr–Fe, the protective film may change in composition on increasing the anode potential to give oxides that are more soluble at low pH and are therefore more susceptible to temperature increases. This occurs in the presence of cathode reactants such as chromic acid which allow polarisation of the anode.

For many metals and alloys the determination of i_p is complex, and its magnitude is governed by many factors such as surface finish, rate of formation, alloying constituents, and the presence of those anions, such as halides, that promote localised breakdown. In many instances the attack on passive films by halide ions shows a temperature and concentration dependence similar to the effect of hydrogen ions, i.e. the rate of film dissolution increases with concentration in accordance with a Freundlich adsorption relationship

$$\text{Rate} = KC^n$$

where C is the concentration of halide, and the effect of temperature indicates a diffusion-controlled process.

Thermodynamic Considerations

Potential–pH diagrams have been calculated for most metals and many non-metal systems by Pourbaix[40] and others for temperatures at 25°C. In addition, theoretical relationships have been established[41,42] which enable the diagrams to be calculated at other temperatures.

The diagram for Fe–H_2O has been calculated by two different methods[43,44] and gives similar results. The most interesting feature of the diagrams is that they predict that for pH 13.0 (determined at 25°C) the thermodynamic tendency for corrosion increases markedly when the temperature is raised above about 75°C.

Another important result is the effect of temperature on the pH scale. This change cannot be calculated from fundamental principles, but it is possible to calculate the change in activity of hydrogen ion as a function of temperature. In general, as the activity changes as a function of the dissociation constant of water, there is very little change at high concentrations of hydrogen ions, but significant changes do occur from the neutral to the alkaline region (Fig. 2.6).

Potential–pH diagrams may be made more useful if the solution pH at 25°C is given on the axis (since this would be measured in practice) and the diagram drawn for the temperature of interest. Because of the non-linear relationship, the various lines become curved as shown in Fig. 2.8. Experimental verifications of these types of diagram are few, but Cowan and Staehle[34] have studied the behaviour of nickel in sulphate solutions at pH values in the acidic region. Their main conclusions were that at 300°C in the pH range 5–10, the corrosion of nickel should be thermodynamically impossible, and in the range pH 2–5 and 10–13 the driving force for corrosion should be small in oxygen-free solutions. These observations confirm subsequent experimental tests.

The calculations of these diagrams are complicated, but computer methods are now available[45], and using these techniques the effect of temperature on potential–pH diagrams has been calculated for many metals[46]. Their usefulness, however, awaits further experimental evidence to confirm the predictions.

Heat Transfer

The discussion to date has assumed isothermal conditions, but in many practical situations corrosion reactions have to be considered when the electrode is acting as a cooler or a heater.

From various studies[47,48] it is becoming clear that in spite of a heat flux, the overriding parameter is the temperature at the interface between the metal electrode and the solution, which has an effect on diffusion coefficients and viscosity. If the variations of these parameters with temperature are known, then i_L (and i_{corr}) can be calculated from the hydrodynamic equations.

When film-forming reactions occur and activation control is the rate-determining factor then the interfacial temperature again will determine the extent of corrosion.

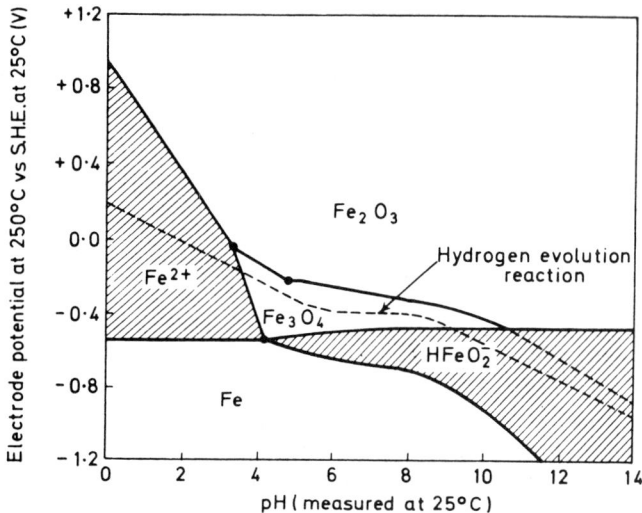

Fig. 2.8 Potential–pH diagram calculated for Fe–H$_2$O system at 250°C. The pH scale refers to the solution measured at 25°C and then raised to 250°C (after Ashworth[55])

The work of Porter *et al.*[48] has shown that for copper in phosphoric acid the interfacial temperature was the main factor, and furthermore this was the case for positive or negative heat flux. Activation energies were determined for this system; they indicated that concentration polarisation was the rate-determining process, and by adjustment of the diffusion coefficient and viscosity for the temperature at the interface and the application of dimensional group analysis it was found that:

$$i_L = 43 \cdot 3(C_S - C_B)D^{\frac{2}{3}} v^{-\frac{1}{6}}$$

where C_S = surface concentration of ions,
C_B = bulk concentration of ions,
D = diffusion coefficient and
v = kinematic viscosity.

This equation provides a means of predicting i_L (A/m^2), from which an approximation of $i_{corr.}$ can be made from physical measurements alone.

Boiling Heat Transfer

This represents a special case of high-level turbulence at a surface by the formation of steam and the possibility of the concentration of ions as water evaporates into the steam bubbles[49, 50]. For those metals and alloys in a particular environment that allow diffusion-controlled corrosion processes, rates will be very high except in the case where dissolved gases such as oxygen are the main cathodic reactant. Under these circumstances gases will be expelled into the steam and are not available for reaction. However, under conditions of sub-cooled forced circulation, when cool solution is continually approaching the hot metal surface, the dissolved oxygen appears to be effective[51] and cathodic processes are stimulated. When the activation step is rate controlling the boiling temperature represents a maximum in the rate. Only small changes (of about 10°C) are possible for metal temperatures to exceed the boiling point because of film boiling, when steam effectively covers the whole surface and corrosion rates become negligible. There is a danger of metal damage by a rapid rise in temperature (Fig. 2.9) when the 'cooling' action of evaporation at the surface is prevented, a situation that is obviously to be avoided in the design. When activation processes are in control the small temperature rise allowable above the equilibrium boiling point may increase rates of dissolution by two orders of magnitude[51].

Butler and Ison[49] have suggested that variation in corrosion rate can be influenced by surface roughness, which allows a large number of nuclei for steam bubble formation. In these circumstances they have suggested that concentration of ions in solution next to the surface will be greater, and their observations on corrosion damage indicate that the steam bubbles may provide crevices or at least enhanced conditions for dissolution at the triple interface (solution/metal/steam).

Thermogalvanic Corrosion

It is impossible to design heat exchangers where all surfaces are isothermal and in many cases such differences are required by the design. For instance, a steam cooler may have a de-superheating zone, a condensing zone and a liquid cooling zone on the same metal tube, but at different positions along its length (Fig. 2.10). The question arises as to whether such temperature differences on the same metal surface in contact with the same electrolyte

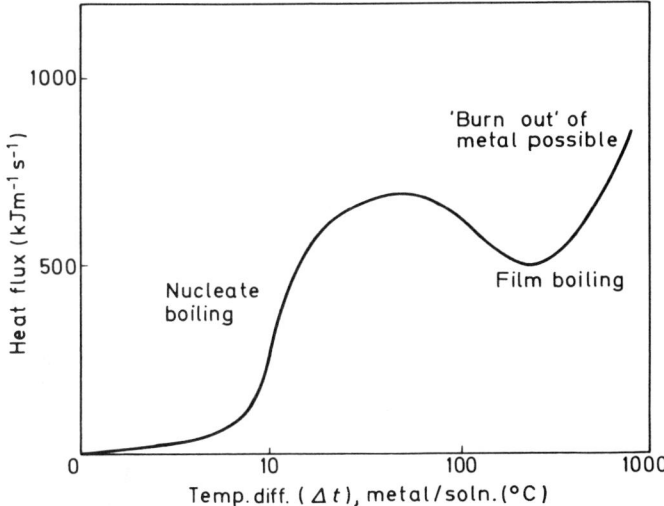

Fig. 2.9 Typical boiling heat-transfer characteristics; Δt is the temperature difference between the solution and the metal surface

2:26 EFFECT OF CONCENTRATION, VELOCITY AND TEMPERATURE

solution (on the cooling side), can have sufficient electrode potential differences to give rise to a galvanic cell, i.e. a thermal galvanic cell. Electrode potentials change with temperature, but as shown previously, temperature changes may also affect the kinetics of dissolution, especially activation-controlled processes. The main rôle of thermogalvanic cells is in polarising existing electrode processes, which, depending on other aspects of the environment, may accelerate or decelerate corrosion.

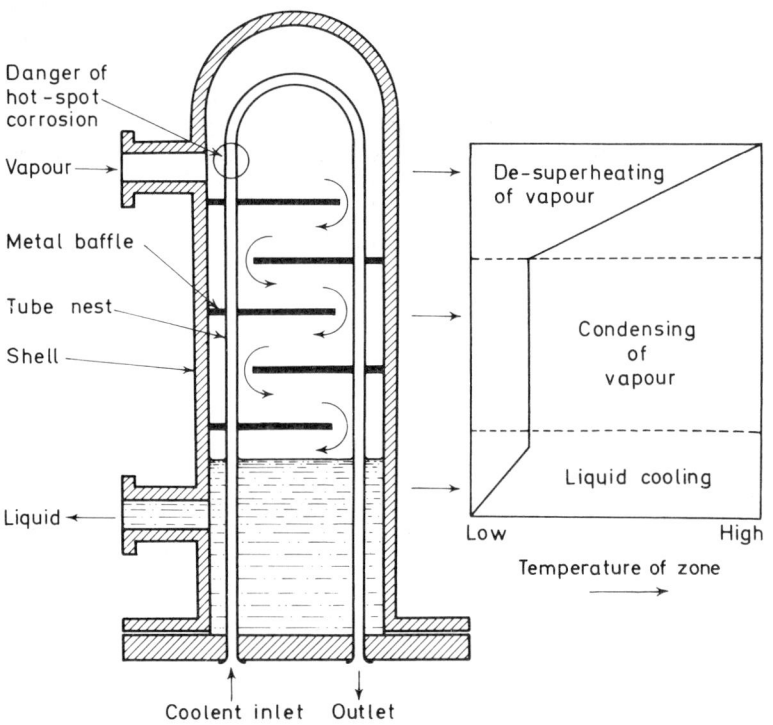

Fig. 2.10 Temperature distribution in a typical heat exchanger

Origin and Magnitude of Thermogalvanic Potentials

The e.m.f. of a thermogalvanic cell is the result of four main effects[42]: (a) electrode temperature, (b) thermal liquid junction potential, (c) metallic thermocouple and (d) thermal diffusion gradient or Soret.

The driving force of a thermogalvanic corrosion cell is therefore the e.m.f. attributable to these four effects, but modified by anodic and cathodic polarisation of the metal electrodes as a result of local action corrosion processes.

In practical systems, (c) and (d) are often very small especially on the same metal surface when solution flow occurs by convection or forced circulation. In neutral solutions, (b) may be small but is somewhat larger in acid solution. On this basis several workers have determined effect (a) as a guide to the subsequent behaviour of a thermogalvanic cell. The main usefulness of such a calculation is to decide whether a hot anode or a hot cathode is produced. In many corroding systems, a large cathode area to anode area is detrimental, because of the many situations where corrosion is controlled by diffusion of reactants to the cathode. Such a situation exists at the entrance to a heat exchanger producing a hot zone, and if this is anodic to the larger area of cooler metal then a thermogalvanic cell is set up, having a potentially enhanced corrosion rate[52, 53].

It should be noted that the simple Nernst equation *cannot* be used since the standard electrode potential $E^\ominus$ is markedly temperature dependent. By means of irreversible thermodynamics[42] equations have been computed to calculate these potentials and are in good agreement with experimentally determined results.

In general, temperature coefficients of electrode potential are in the range $\pm 0 \cdot 1$ to $2 \, mV/°C$ and in many practical systems temperature differences rarely exceed 75°C, so that the driving force for thermogalvanic corrosion is small and would be subject mainly to resistance control. However, in many instances the temperature change also decreases polarisation (see Fig. 2.11) so that if the resistance of the solution is not high severe attack can ensue. Because of the resistance effect attack is confined to a small area of largest temperature gradient, leading to deep notches at the edge of the heated zone, i.e. the dangerous situation of a small anode and large cathode.

Active–passive Transitions

Whilst temperature coefficients suggest modest potential differences, these calculations do not take into account the large potential changes that can occur when thermal effects allow transition from active to passive states.

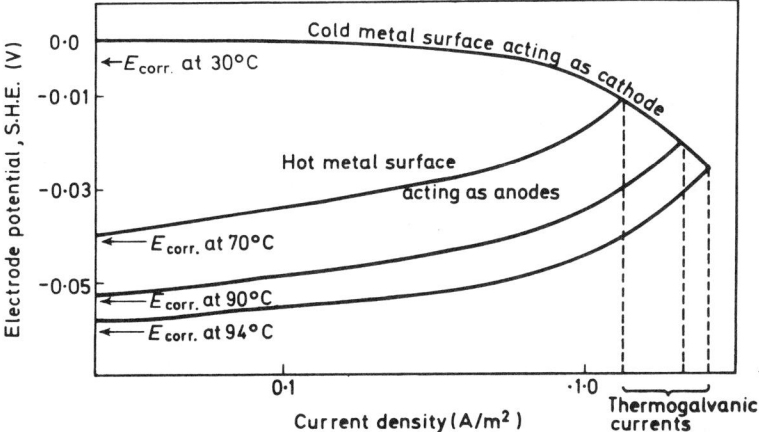

Fig. 2.11 Influence of temperature on the anodic polarisation of copper in aerated 3% NaCl solution[51]. $E_{corr.}$ is the corrosion potential of the hot metal when not in contact with the cold metal

Potentials become more positive (a hot cathode) as a result of thickening of the passive film. Such changes have been observed[45] (Fig. 2.12), and moreover, in the presence of aggressive anions, when the thermal effects allow the change to passivity, then overall general corrosion is changed to deep pitting on the electrode surface. Thermogalvanic coupling would enhance this effect.

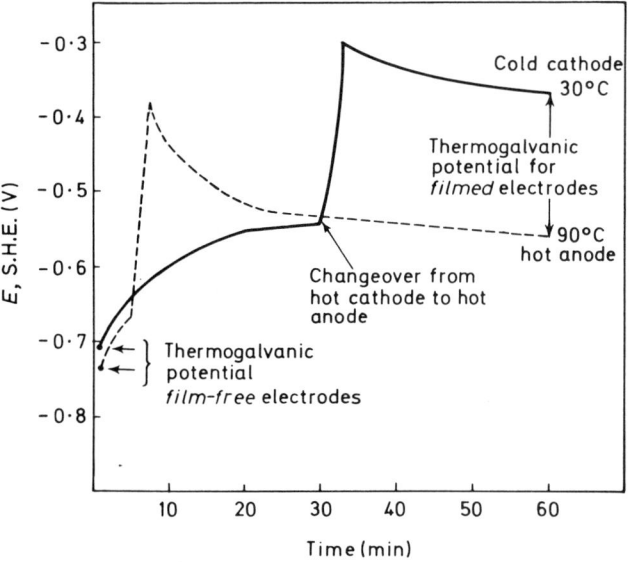

Fig. 2.12 Potential-time curves on mild steel in sodium borate/hydrochloric acid buffer solutions, pH 7·60, oxygen-saturated solution (after Ashworth[55])

These observations show that kinetic factors can outweigh thermodynamic effects and the situation of the mutual polarisation of two electrodes in a corrosion cell leads to either negative or positive temperature coefficients. Prediction of thermogalvanic action from consideration of the anode process alone can therefore be misleading. Thermogalvanic corrosion rates may be low in most circumstances but they are persistent for long periods, existing as long as temperature differences exist, i.e. the operating period of plant. They represent a dormant situation that can accelerate corrosion if the environment changes, e.g. high conductivity, and increase in aggressive ion concentration.

P. J. BODEN

REFERENCES

1. Pourbaix, M., *Electrochim. Acta*, **12**, 184 (1967) and Ness, P., *Electrochim. Acta*, **12**, 161 (1967)
2. Pourbaix, M., *Corrosion*, **25**, 267 (1967)
3. Mattson, E., *Electrochim. Acta*, **3**, 279 (1961)
4. Horvath, J. and Hackl, L., *Corr. Sci.*, **5**, 528–538 (1965)

5. Greene, N. D., *J. Electrochem. Soc.*, **107**,457 (1960) and France, W. D. and Greene, N. D., *Corrosion*, **24**, 403 (1968)
6. Cron, C. J., Payer, J. H. and Staehle, R. W., *Corrosion*, **27**, 1 (1971)
7. Bockris, J. O'M., Drazic, D. and Despic, A. R., *Electrochim. Acta*, **4**, 325 (1961)
8. Bonhoeffer, K. F. and Heusler, K. E., *Z. Phys. Chem.*, N.F., **8**, 390 (1956)
9. Kabanov, B. N. and Leikis, D. I., *Dokl. Akad Nauk. SSSR*, **58**, 1 685 (1947)
10. Florianovich, G. M., Sokolova, L. A. and Kolotyrkin, Ya. M., *Electrochim. Acta*, **12**, 897 (1967)
11. Oakes, G. and West, J. M., *Brit. Corr. J.*, **4**, 66 (1969)
12. Hines, J. G., *Electrochim. Acta*, **10**, 225 (1965)
13. Florianovich, G. M., Kolotyrkin, Ya. M. and Kononova, M. D., *Proceedings of the 4th International Congress on Metallic Corrosion*, Amsterdam, N.A.C.E. (1969)
14. Ammar, I., Darwish, S. and Etman, M., *Electrochim. Acta*, **12**, 485 (1967)
15. Brasher, D. M. and Mercer, A. D., *Brit. Corr. J.*, **3**,120 (1968)
16. Brasher, D. M., Reichenberg, D. and Mercer, A. D., *Brit. Corr. J.*, **3**,144 (1968)
17. Brasher, D. M., *Brit. Corr. J.*, **4**, 122 (1969)
18. Gouda, V. K., Khedr, M. G. A. and Am Sham El Din, *Corr. Sci.*, **7**, 221 (1967)
19. Legault, R. A., Mori, S. and Leckie, H. P., *Corrosion*, **26** (1970)
20. Ibi, N., *Electrochim. Acta*, **1**, 117 (1959)
21. Levich, V. G., *Physicochemical Hydrodynamics*, Prentice-Hall Inc. (1962)
22. King, C. V., *Surface Chemistry of Metals and Semiconductors*, editor H. C. Gatos, John Wiley, New York, 357 (1959)
23. Zembura, Z., *Corr. Sci.*, **8**, 703 (1968)
24. Ross, T. K. and Hitchen, B. P. L., *Corros. Sci.*, **1**, 65 (1961)
25. Ross, T. K., Wood, G. C. and Mahmud, I., *J. Electrochem. Soc.*, **113**, 334 (1966)
26. Van Shaw, P., Reiss, L. P. and Hanratty, T. J., *Am. Inst. Chem. Engrs. J.*, **9**, 362 (1963)
27. Ross, T. K. and Wragg, A. A., *Electrochim. Acta*, **10**, 1 093 (1965)
28. Marangozis, J., *Corrosion*, **24**, 255 (1968)
29. Mahato, B. K., Steward, F. R. and Shemilt, L. W., *Corr. Sci.*, **10**, 737 (1968)
30. Conway, B. E., Beatty, E. M. and DeMaine, P. A. D., *Electrochime Acta*, **7**, 39 (1962)
31. Speller, F., *Corrosion*, McGraw-Hill, London (1951)
32. Cowan, R. L. and Staehle, R. W., *J. Electrochem. Soc.*, **118**, 557 (1971)
33. Murgulescu, I. G. and Radovici, O., *Proceedings of the 1st Cong. On Metallic Corrosion*, Butterworths, London, 109 (1961)
34. Griess, J. C., *Corrosion*, **24**, 97 (1968)
35. Finley, T. C. and Myers, J. R., *Corrosion*, **26**, 544 (1970)
36. Okomoto, G. O. and Kobayashi, H., *Z. Electrochem.*, **62**, 755 (1958)
37. Pugh, M., Warner, D. Gabe, *Corr. Sci.*, **7**, 807 (1967)
38. Robinson, F. P. A. and Golante, L., *Proc. 2nd Int. Cong. on Metallic Corrosion*, N.A.C.E., New York (1963)
39. Kucera, V., Novak, P., Franz, F. and Koritta, J., *Korrozija Zaščita Titana*, G.N.T.I.M.L., Moscow (1964)
40. Pourbaix, M., *Atlas of Electrochemical Equilibria in Aqueous Solutions*, Pergamon Press, Oxford (1966)
41. Criss, C. M. and Cobble, J. W., *J. Am. Chem. Soc.*, **86**, 5 390 (1964)
42. de Bethune, A. J., Licht, T. S. and Swendeman, N., *J. Electrochem. Soc.*, **106**, 616 (1959)
43. Townsend, H. E., *Corr. Sci.*, **10**, 343 (1970)
44. Ashworth, V. and Boden, P. J., *Corr. Sci.*, **10**, 709 (1970)
45. Brook, P. A., *Corr. Sci.*, **11**, 389 (1971)
46. Brook, P. A., *Corr. Sci.*, **12** (1972)
47. Ross, T. K., *Brit. Corr. J.*, **2**, 131 (1967)*
48. Porter, D. T., Donimirska, M. and Wall, R ., *Corr. Sci.*, **8**, 833 (1968)
49. Butler, G. and Ison, H. C. K., *Proc. 1st Int. Cong. on Metallic Corrosion*, Butterworths, London (1961)
50. Freeborn, J. and Lewis, D., *J. Mech. Eng. Soc.*, **4**, 46 (1962)
51. Boden, P. J., *Corr. Sci.*, **11**, 353 (1971)
52. Breckon, G. and Gilbert, P. T., *Proc. 1st Int. Cong. on Metallic Corrosion*, Butterworths, London (1961)
53. Bem, R. S. and Campbell, H. S., ibid.
54. Ashworth, V. and Boden, P. J., *J. Electrochem. Soc.*, **119**, 6 (1972)
55. Ashworth, V. and Boden, P. J., *Corr. Sci.*, **14**, 209 (1974)

56. Chilton, T. H. and Colburn, A. P., *Ind. Eng. Chem.*, **26**, 1183 (1934)
57. Poulson, B., *Corr. Sci.*, **23**, 391 (1983)
58. Chin, D. T., Tsang, C. H., *J. Electrochem. Soc.*, **125**, 1461 (1978)
59. Martin, H., 'Advances in Heat Transfer', **13**, 1, Academic Press New York (1977)
60. Tagg. D. J., Pattrick, M. A., Wragg, A. A., *Trans. I. Chem. E.*, **57**, 176 (1979)
61. Heitz, E., *Werkstoffe und Korrosion*, **15**, 63 (1964)
62. Poluboyartseva, L. A. *et al.*, *J. Appl. Chem.*, **36**, 1210 (1963)
63. Oldfield, J. and Todd, P., *Desalination*, **31**, 365 (1979)

* Detailed review presented at symposium on 'Corrosion under Heat Transfer in Liquid Media', reported in *Br. Corr. J.*, **2** (1967)

2.2 The Atmosphere

Metals are more frequently exposed to the atmosphere than to any other corrosive environment. Atmospheric corrosion is also the oldest corrosion problem known to mankind, yet even today it is not fully understood. The principal reason for this paradox lies in the complexity of the variables which determine the kinetics of the corrosion reactions. Thus, corrosion rates vary from place to place, from hour to hour and from season to season. Equally important, this complexity makes meaningful results from laboratory experiments very difficult to obtain.

However, the object of this section is to outline the principles which govern atmospheric corrosion, and the emphasis is placed on metals whose atmospheric corrosion is of economic importance. These include iron and steel, zinc, copper, lead, aluminium and chromium.

Classification of Atmospheric Corrosion

Atmospheric corrosion can be conveniently classified as follows:

(*a*) Dry oxidation.
(*b*) Damp corrosion.
(*c*) Wet corrosion.

Dry Oxidation

This takes place in the atmosphere with all metals that have a negative free energy of oxide formation. Gold does not oxidise and this property is utilised in the coating of electronic components where even the thinnest layers of corrosion product cannot be tolerated. For metals forming non-porous oxides (alkali metals are an exception) the films rapidly reach a limiting thickness since ion diffusion through the oxide lattice is extremely slow at ambient temperatures, and at the limiting thickness, the oxide films on metals are invisible. For example, those on iron are typically 30 Å thick. For certain metals and alloys these films are so fault-free or rapidly self-healing that they confer remarkable protection on the substrate, e.g. stainless steel, titanium and chromium.

The tarnishing of copper and silver in dry air containing traces of hydrogen sulphide (Table 2.6) is another example of film growth by lattice diffusion at ambient temperatures. In these cases defects in the sulphide lattice enable the films to grow to visible thicknesses with the consequent formation of tarnish films which are aesthetically objectionable and may have a significant effect on the behaviour of the metals in particular applications, e.g. electrical contacts.

Table 2.6 Typical concentration of atmospheric impurities

Impurity	Typical concentrations ($\mu g/m^3$)
Sulphur dioxide[*,1]	Industrial region: winter 350, summer 100
	Rural region: winter 100, summer 40
Sulphur trioxide	Approximately 1% of the sulphur dioxide content
Hydrogen sulphide[2]	Industrial region: 1·5–90 ⎫
	Urban region: 0·5–1·7 ⎬ values measured
	Rural region: 0·15–0·45 ⎭ in the spring
Ammonia[3]	Industrial region: 4·8
	Rural region: 2·1
Chloride[3]	Industrial inland: winter 8·2, summer 2·7
(air sampled)	Rural coastal: annual average 5·4
Chloride[3]	Industrial inland: winter 7·9, summer 5·3
(rainfall sampled)	Rural coastal: winter 57, summer 18
	(these values in mg/l)
Smoke particles[1]	Industrial region: winter 250, summer 100
	Rural region: winter 60, summer 15

* There are two methods that are commonly used for estimating sulphur dioxide:
(*a*) Lead peroxide 'candle' method. The weight gain, caused by lead sulphate formation as sulphur dioxide reacts with a specified surface area of lead peroxide paste, is measured.
(*b*) Hydrogen peroxide titrimetric method. A known volume ofair is pumped through a weak hydrogen peroxide solution in which the sulphur dioxide is oxidised to sulphuric acid. The acid content is estimated by titration. In Ref. 1 the second method was used, the air first being filtered to yield an estimate of particulate matter.

However, in this section emphasis is placed upon *damp* and *wet* atmospheric corrosion which are characterised by the presence of a thin, invisible film of electrolyte solution on the metal surface (damp type) or by visible deposits of dew, rain, sea-spray, etc. (wet type). In these categories may be placed the rusting of iron and steel (both types involved), 'white rusting' of zinc (wet type) and the formation of patinae on copper and its alloys (both types).

The corrosion products may be soluble or insoluble. If insoluble, they usually reduce the rate of corrosion by isolating the substrate from the corrosive environment. Less commonly, they may stimulate corrosion by offering little physical protection while retaining moisture in contact with the metal surface for longer periods.

Soluble corrosion products may increase corrosion rates in two ways. Firstly, they may increase the conductivity of the electrolyte solution and thereby decrease 'internal resistance' of the corrosion cells. Secondly, they may act hygroscopically to form solutions at humidities at and above that in equilibrium with the saturated solution (Table 2.7). The 'fogging' of nickel in SO_2-containing atmospheres, due to the formation of hygroscopic nickel sulphate, exemplifies this type of behaviour. However, whether the corrosion products are soluble or insoluble, protective or non-protective, the

Table 2.7 Relative humidities of air in equilibrium with saturated salt solutions at 20°C[4-7]

Salt in solution	r.h. (%)	Salt in solution	r.h. (%)
$CuSO_4 \cdot 5H_2O$	98	$NaCl$	76
K_2SO_4	98	$CuCl_2 \cdot 2H_2O$	68
Na_2SO_4	93	$FeCl_2$	56
$Na_2CO_3 \cdot 10H_2O$	92	$NiCl_2$	54
$FeSO_4 \cdot 7H_2O$	92	$K_2CO_3 \cdot 2H_2O$	44
$ZnSO_4 \cdot 7H_2O$	90	$MgCl_2 \cdot 6H_2O$	34
$3CdSO_4 \cdot 8H_2O$	89	$CaCl_2 \cdot 6H_2O$	32
KCl	86	$ZnCl_2 \cdot xH_2O$	10
$(NH_4)_2SO_4$	81	NH_4Cl	80

corrosive atmosphere experienced by the substrate (often referred to as the 'micro-environment') is modified from the macro-environment experienced by a bare substrate. For this reason, corrosion rates are rarely constant for extended periods of atmospheric exposure.

Composition of the Atmosphere

Nominal Composition

The composition given in Table 2.8 is global and, for most components, is reasonably constant for all locations, but the water vapour content will obviously vary according to the climatic region, season of the year, time of the day, etc. However, only oxygen, carbon dioxide and water vapour need to be considered in the context of atmospheric corrosion.

Carbon dioxide was once thought essential for the rusting of ferrous metals (viz. the carbonic acid theory of rusting) but is now considered of relatively minor importance[8,9]. However, basic zinc carbonate is frequently found in the corrosion products of zinc and small amounts of siderite ($FeCO_3$) are found in ferrous rusts.

Table 2.8 Approximate constitution[10] of the atmosphere at 10°C and 100 kN/m^2 (excluding impurities)

Constituents	g/m^3	Weight (%)	Constituents	mg/m^3	p.p.m. by weight
Air	1 172	100	Neon	14	12
Nitrogen	879	75	Krypton	4	3
Oxygen	269	23	Helium	0·8	0·7
Argon	15	1·26	Xenon	0·5	0·4
Water vapour	8	0·70	Hydrogen	0·05	0·04
Carbon dioxide	0·5	0·04			

Water vapour is essential to the formation of an electrolyte solution which will support the electrochemical corrosion reactions, and its concentration in the atmosphere is usually expressed in terms of the *relative humidity* (r.h.).

This is defined as the percentage ratio of the water vapour pressure in the atmosphere compared to that which would saturate the atmosphere at the same temperature. Alternatively, the difference in temperature between the ambient atmosphere and that to which it would have to be cooled before moisture condensed from it, is also used as a measure of moisture content. This difference in temperature is called the *dew point depression*. The actual temperature at which condensation takes place is known as the *dew point*. The relative humidity is then expressed as:

$$\text{r.h.} = \frac{\text{Saturated vapour pressure of } H_2O \text{ at the dew point}}{\text{Saturated vapour pressure of } H_2O \text{ at ambient temp.}} \times 100\%.$$

Oxygen from the atmosphere, dissolved in the electrolyte solution provides the cathode reactant in the corrosion process. Since the electrolyte solution is in the form of thin films or droplets, diffusion of oxygen from the atmosphere/electrolyte solution interface to the solution/metal interface is rapid. Moreover, convection currents within these thin films of solution may play a part in further decreasing concentration polarisation of this cathodic process[11]. Oxygen may also oxidise soluble corrosion products to less soluble ones which form more or less protective barriers to further corrosion, e.g. the oxidation of ferrous species to the less soluble ferric forms in the rusting of iron and steel.

Atmospheric Contaminants

In a sense this subdivision of the composition of the atmosphere is arbitrary since some of the so-called contaminants are derived partly or wholly from natural sources. However, in that their concentrations vary appreciably within very narrow geographical limits, they may be distinguished from the contents of Table 2.8 (with the possible exception of water vapour). Table 2.6 lists those contaminants which are important from a corrosion standpoint. Excluded are contaminants found only in very specific locations, e.g. in the vicinity of a chemical works. The concentrations given are intended only to indicate general levels in the usual classification of environments and not to define a particular environment.

Sulphur oxides These (SO_2 is the most frequently encountered oxide) are powerful stimulators of atmospheric corrosion, and for steel and particularly zinc the correlation between the level of SO_2 pollution and corrosion rates is good[12-14]. However, in severe marine environments, notably in the case of zinc, the chloride contamination may have a higher correlation coefficient than SO_2.

The SO_2 in the atmosphere is derived from two sources. Firstly, from the aerial oxidation of H_2S produced naturally (see later) and secondly from the combustion of sulphur-containing fuels. In industrialised countries the second source predominates, but on a global scale only about one-fifth of the total sulphur pollution is derived from human activity. In 1969, the total sulphur emission, expressed in terms of SO_2, from burnt fuel in the UK was $6 \cdot 06 \times 10^6$ tons. In densely populated countries sulphur pollution levels are very much related to the domestic heating cycle, and in the UK maximum pollution levels are reached in January/February and the minimum usually occurs in August[1]. This cyclic pattern is closely reflected by corrosion rate variations[15,16], corrosion being heaviest in the winter months despite lower average temperatures.

A more detailed consideration of the rôle of SO_2 as a corrosion stimulator will be given later.

Hydrogen sulphide This is produced by the putrefaction of organic sulphur compounds or by the action of sulphate-reducing bacteria in anaerobic conditions (e.g. in polluted river estuaries). It is fairly rapidly oxidised to SO_2 and concentrations are considerably lower than those of SO_2^2 (Table 2.6). Nevertheless it is responsible for the tarnishing of copper and silver at normal atmospheric concentrations.

Nitrogen compounds These also arise from both natural and synthetic sources. Thus ammonia is formed in the atmosphere during electrical storms, but increases in the ammonium ion concentration in rainfall over Europe in recent years are attributed to increased use of artificial fertilisers. Ammonium compounds in solution may increase the wettability of a metal[17] and the action of ammonia and its compounds in causing 'season cracking', a type of stress-corrosion cracking of cold-worked brass, is well documented.

Saline particles These are of two main types. The first is ammonium sulphate formed in heavily industrialised areas where appreciable concentrations of ammonia and SO_3 or of H_2SO_4 aerosol co-exist. It is a strong stimulator of the initiation of corrosion, being hygroscopic and acidic. The second is marine salt, mainly sodium chloride but quite appreciable quantities of potassium, magnesium and calcium ions are analysed in rainfall[3]. Chlorides are also produced in industrial areas and for the UK the fall-off in concentration of marine salt with distance from the sea is partially masked by chloride produced by the industrial regions in the centre of the country[3]. Chlorides are also hygroscopic and the chloride ion is highly aggressive to some metals, e.g. stainless steel.

Other airborne particles These are also divisible into two groups. Firstly, the inert non-absorbent particles, usually siliceous, which can only affect corrosion by facilitating differential aeration processes at points of contact.

Secondly, absorbent particles such as charcoal and soot are intrinsically inert but have surfaces or infrastructures that adsorb SO_2, and by either co-adsorption of water vapour or condensation of water within the structure, catalyse the formation of a corrosive acid electrolyte solution. 'Dirt' with soot assists the formation of patinae on copper and its alloys by retaining soluble corrosion products long enough for them to be converted to protective, insoluble basic salts.

Other Atmospheric Variables

Temperature This may be more or less of an important factor, depending on the metal considered. For example, while zinc is characterised by a very low positive temperature coefficient of corrosion rate[19], steel has a high positive coefficient[15,19]. The rate of drying of electrolyte solution from the metal surface, directly into the atmosphere or through layers of corrosion product, is strongly temperature dependent. In these contexts the metal *surface* temperature is probably more important than ambient temperature although the latter obviously strongly influences the former. However, many other factors will affect the metal temperature, including the thermal capacity of the metal structure, its orientation with respect to the sun, the intensity of sunlight, the reflectivity of the metal surface or its corrosion products, wind velocity and direction, the thermal insulating properties of insoluble corrosion products, and so on. The prevailing wind direction is also an important factor in relation to increases in corrosion rates to be expected from the proximity of large industrial plants producing appreciable concentrations of potentially corrosive pollutants.

Electrolyte Solution Formation

Wetness of a metal surface The *time of wetness* of the metal surface is an exceedingly complex, composite variable. It determines the duration of the electrochemical corrosion process. Firstly it involves a consideration of all the means by which an electrolyte solution can form in contact with the metal surface. Secondly, the conditions under which this solution is stable with respect to the ambient atmosphere must be considered, and finally the rate of evaporation of the solution when atmospheric conditions change to make its existence unstable. Attempts have been made to measure directly the time of wetness[18], but these have tended to use metals forming non-bulky corrosion products (see Section 20.1). The literature is very sparse on the rôle of insoluble corrosion products in extending the time of wetness, but considerable differences in moisture desorption rates are found for rusted steels of slightly differing alloy content, e.g. mild steel and Cor-Ten.

Critical relative humidity The primary value of the critical relative humidity denotes that humidity below which no corrosion of the metal in question takes place. However, it is important to know whether this refers to a clean metal surface or one covered with corrosion products. In the latter case a secondary critical humidity is usually found at which the rate of corrosion increases markedly[8]. This is attributed to the hygroscopic nature of the corrosion product (see later). In the case of iron and steel it appears that there may even be a tertiary critical humidity[15]. Thus at about 60% r.h. rusting commences at a very slow rate (primary value)[20]; at 75-80% r.h. there is a sharp increase in corrosion rate probably attributable to capillary condensation of moisture within the rust[8,21]. At 90% r.h. there is a further increase in rusting rate[15] corresponding to the vapour pressure of saturated ferrous sulphate solution[5], ferrous sulphate being identifiable in rust as crystalline agglomerates[16]. The primary critical r.h. for uncorroded metal surfaces seems to be virtually the same for all metals, but the secondary values vary quite widely.

Moisture precipitation Apart from wetting by sea-spray, moisture may either be deposited on a surface by rainfall or dew formation. For a known ambient humidity the dew point can be calculated, using the expression given previously, from standard tables giving the saturated vapour pressure of

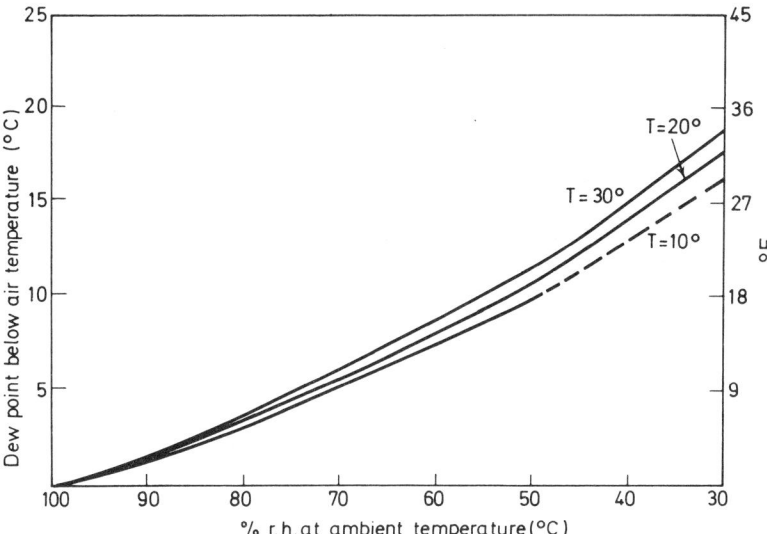

Fig. 2.13 Dew point depression below ambient temperature as a function of the relative humidity of the ambient atmosphere over a range of temperature

water at various temperatures (e.g. Handbook of the Chemical Rubber Company). However, Fig. 2.13 sets out these relationships graphically and from a knowledge of the ambient relative humidity and ambient temperature, the dew point depression may be read off.

Since gaseous pollution, particularly of SO_2, tends to be concentrated near ground level, dew can be considerably more acid than rain which forms at higher altitudes. Moreover, dew can, unlike rain, wet completely sheltered surfaces. Thin sheets of metal which closely follow changes in ambient temperature are more likely to have dew formed on them than more massive sections of higher thermal capacity which will cool more slowly. 'White rusting' of galvanised sheeting is usually attributable to dew formation in poorly ventilated conditions.

Rainfall, besides wetting the metal surface, can be beneficial in leaching otherwise deleterious soluble species and this can result in marked decreases in corrosion rate[15,16]. A recent survey of rainfall analyses for Europe has shown that, with the exception of the UK, the acidity and sulphate content of rainfall markedly increased in the period 1956 to 1966, pH values having fallen by 0·05 to 0·10 units per ann[22]. The exception of the UK may be due to anti-pollution measures introduced in this period. However, even in the UK a pH of 4 is not uncommon for rainfall in industrial areas. The significance of electrolyte solution pH will be discussed in the context of corrosion mechanisms. The remaining cases of electrolyte formation are those in which it exists in equilibrium with air at a relative humidity below 100%.

Capillary condensation The vapour pressure above a concave meniscus of water is less than that in equilibrium with a plane water surface. It is therefore possible for moisture to condense in narrow capillaries from an atmosphere of less than 100% r.h.

The relative lowering of the saturated vapour pressure of water is described by the Thomson equation:

$$p = p_0 e^{-2\sigma M/dRTr}$$

where p and p_0 are the saturated vapour pressures above a concave meniscus of radius r, and a plane surface, respectively; σ is the surface tension of the liquid at an absolute temperature T, d its density and M its molecular weight; and R is the gas constant. Thus, as the value of r decreases (r can be approximately equated to the radius of the capillary concerned) so the relative humidity at which condensation takes place within the capillary also decreases (Table 2.9).

Table 2.9 Capillary radii for condensation at given humidities

Capillary radius (Å)	Relative humidity for condensation
360	98
94	90
47	80
30	70
21	60
15	50

This concept may be invoked to account for electrolyte formation in microcracks in a metal surface or in the re-entrant angle formed by a dust particle and the metal surface. More importantly, it can also explain electrolyte formation in the pores of corrosion product and hence the secondary critical humidity discussed earlier. Ferric oxide gel is known to exhibit capillary condensation characteristic[23] and pore sizes deduced from measurements of its adsorptive capacity[23] are of the right order of magnitude to explain a secondary critical relative humidity ≈70% for rusted steel[20].

Chemical condensation This occurs when soluble corrosion products or atmospheric contaminants are present on the metal surface. When the humidity exceeds that in equilibrium with a saturated solution of the soluble species, a solution, initially saturated, is formed until equilibrium is established with the ambient humidity. The contaminants have already been detailed and of the corrosion products, obviously sulphates, chlorides and carbonates are most important in this context. However, in some cases there is a lack of reliable data on the vapour pressure exerted by saturated solutions of likely corrosion products. The useful data was summarised in Table 2.7.

In practice, however, the soluble components are often contained in a matrix of insoluble product and formation of electrolyte by both capillary and chemical condensation may occur in the same humidity range.

Adsorbed electrolyte layers In this case the water molecules are bound to the metal surface by Van der Waals' forces. It is estimated that at 55% r.h. the film on polished iron is about 15 molecular layers thick, increasing to 90 molecular layers at just below 100% r.h.[4]. Such films are capable of supporting electrochemical corrosion processes and these have been studied. As the humidity is reduced below 100% and the moisture layers become thinner, polarisation of the cathodic and particularly the anodic process rapidly becomes enormous and corrosion virtually ceases below about 60% r.h.[4].

The Rôle of Sulphur Dioxide in Atmospheric Corrosion

Sulphur dioxide plays such an important rôle in the corrosion of metals in the atmospheres of industrialised countries that detailed consideration of its action seems justified. For all metals SO_2 appears to be selectively adsorbed from the atmosphere, less so for aluminium than for other metals, and for rusty steel it is almost quantitatively adsorbed even from dry air at $0°C$[15]. Under humid conditions sulphuric acid is formed, the oxidation of SO_2 to SO_3 being catalysed by metals and by metallic oxides.

For some non-ferrous metals (copper, lead, nickel) the attack by sulphuric acid is probably direct with the formation of sulphates. Lead sulphate is barely soluble and gives good protection. Nickel and copper sulphates are deliquescent but are gradually converted (if not leached away) into insoluble basic sulphates[15], e.g. $Cu\{Cu(OH)_2\}_3SO_4$, and the metals are thus protected after a period of active corrosion. For zinc and cadmium the sulphur acids probably act by dissolution of the protective basic carbonate film[15]. This reforms, consuming metal in the process, redissolves, and so on. Zinc and cadmium sulphates are formed in polluted winter conditions whereas in the purer atmospheres of the summer the corrosion products include considerable amounts of oxide and basic carbonate[15].

Thus for non-ferrous metals, SO_2 is consumed in the corrosion reactions whereas in the rusting of iron and steel it is believed[15,24] that ferrous sulphate is hydrolysed to form oxides and that the sulphuric acid is regenerated. Sulphur dioxide thus acts as a catalyst such that one SO_4^{2-} ion can catalyse the dissolution of more than 100 atoms of iron before it is removed by leaching, spalling of rust or the formation of basic sulphate[24]. These reactions can be summarised as follows:

$$Fe + SO_2 + O_2 \rightleftharpoons FeSO_4$$
$$4FeSO_4 + O_2 + 6H_2O \rightleftharpoons 4FeOOH + 4H_2SO_4$$
$$4H_2SO_4 + 4Fe + O_2 \rightleftharpoons 4FeSO_4 + 4H_2O, \textit{et seq.}$$

Rosenfel'd[11,25] considers that SO_2 can act as a depolariser of the cathodic process. However, this effect has only been demonstrated with much higher levels of SO_2 (0·5%) than are found in the atmosphere (Table 2.4) and the importance of this action of SO_2 has yet to be proved for practical environments. However, SO_2 is 1 300 times more soluble than O_2 in water[11] and therefore its concentration in solution may be considerably greater than would be expected from partial pressure considerations. This high solubility would make it a more effective cathode reactant than dissolved oxygen even though its concentration in the atmosphere is comparatively small.

Electrochemistry of Atmospheric Corrosion

This has already been touched upon in several of the previous paragraphs. Russian workers have extensively examined the electrochemistry of corrosion under thin moisture films and the reader is referred to the work of Rosenfel'd, Tomashov, Klark and co-workers for fuller details[4,11,25,26]. It has been found that the corrosion rate reaches a maximum when the moisture film is around 150 μm thick.

The cathodic process in atmospheric corrosion is often stated to be oxygen reduction, and indeed in many cases the evidence is that this is so[27,28], i.e.

$$O_2 + 2H_2O + 4e^- \rightleftharpoons 4OH^-$$

Kaesche[9] considers that proton reduction may also play a rôle in polluted environments where the pH of the electrolyte is likely to be low. This would be particularly likely in the case of iron if the Schikorr mechanism, involving the presence of sulphuric acid, did in fact operate. However, Russian work[4,11] has shown that oxygen depolarisation is many times more efficient in thin moisture films than in bulk solutions and therefore proton reduction may not be important in affecting corrosion rates.

In the rusting of iron and steel, Evans[29] considers that the anodic reaction of

$$Fe \rightleftharpoons Fe^{2+} + 2e^-$$

is balanced by the cathodic reduction of ferric rust to magnetite under wet conditions when access of oxygen is limited:

$$4Fe_2O_3 + Fe^{2+} + 2e^- \rightleftharpoons 3Fe_3O_4$$

As the rust dries and is permeated by oxygen, magnetite is reoxidised to rust with a net gain of $0 \cdot 5 Fe_2O_3$:

$$3Fe_3O_4 + 0 \cdot 75 O_2 \rightleftharpoons 4 \cdot 5 Fe_2O_3$$

There is considerable evidence that under certain conditions this mechanism may be operative.

Effect of Corrosion Products on Corrosion Rates

The change in corrosion rate with time varies markedly for different metals due to the differing degrees of protection conferred by the corrosion products.

Lead, aluminium and copper corrode initially but eventually form completely protective films[4]. Nickel in urban atmospheres does not form a completely protective film, the corrosion/time curve being nearly parabolic[4]. The corrosion rate of zinc appears to become linear after an initial period of decreasing corrosion rate[4].

The behaviour of steel depends very much on the alloying elements present for any given environment. Thus the decrease in corrosion rate with time for mild steel is very much slower than for a low-alloy steel. This can be attributed to the much more compact nature of the rust formed on the latter type of steel and this is clearly illustrated in Figs. 2.14(a) and (b).

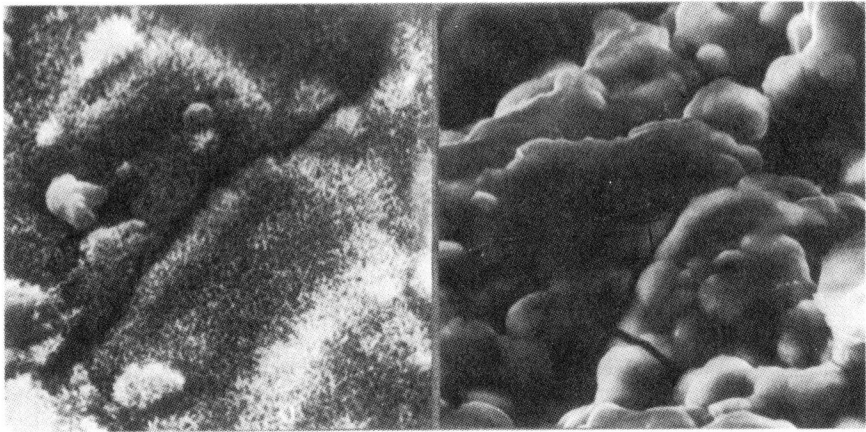

Fig. 2.14 Surface textures of rust on (a) mild steel and (b) Cor-Ten steel exposed in an industrial atmosphere for 2·5 years [(a) × 1 260 and (b) × 1 320]

Weather conditions at the time of initial exposure of zinc and steel have a large influence on the protective nature of the initial corrosion products[15]. This can still be detected some months after initial exposure. Finally, rust on steel contains a proportion of ferrous sulphate which increases with increase in SO_2 pollution of the atmosphere. The effect of this on corrosion rate is so strong that mild steel transferred from an industrial atmosphere to a rural one corrodes for some months as though it was still exposed to the industrial environment[24].

Conclusion

In a section of this brevity it is impossible to cover all aspects of the 'atmosphere'. There are therefore gaps concerning such topics as corrosion at sub-zero temperatures, effect of surface orientation and inclination on corrosion rates and the influence of organic vapours on metallic corrosion. Neither has it been possible to describe the intensive efforts now being made to monitor continuously atmospheric variables such as 'time of wetness' with a view to predicting the long-term corrosion behaviour of metals in a particular area without resorting to long-term trials.

D. FYFE

REFERENCES

1. *Investigation of Air Pollution, April 1968 to March 1969 — National Survey, Smoke and Sulphur Dioxide*, Min. Tech., Warren Spring Laboratory, Stevenage, Herts.
2. Smith, A. F. *et al.*, *J. Appl. Chem.*, **11**, 317 (1961)
3. Stevenson, C. M., *Q. J. Roy. Meteorol. Soc.*, **94**, 56 (1968)

4. Tomashov, N. D., *Theory of Corrosion and Protection of Metals*, Macmillan, New York (1966)
5. Schikorr, G., *Werk. Korr.*, **18**, 514 (1967)
6. Young, J. F., *J. Appl. Chem.*, **17**, 241 (1967)
7. O'Brien, F. E. M., *J. Sci. Instrum.*, **25**, 73 (1948)
8. Vernon, W. H. J., *Trans. Far. Soc.*, **31**, 1 668 (1935)
9. Kaesche, H., *Werk. Korr.*, **15**, 379 (1964) and B.I.S.I.T.S. No. 5 271
10. Meetham, A. R., *Atmospheric Pollution: its Origins and Prevention*, Pergamon, London (1956) and Stern, A. C., (Ed.), *Air Pollution*, Academic Press, New York, 2nd edn. (1968)
11. Rosenfel'd, I. L., *Proc. 1st Int. Corros. Cong. (London)*, Butterworths (1962)
12. Hudson, J. C. and Stanners, J. F., *J. Appl. Chem.*, **3**, 86 (1953)
13. Chandler, K. A. and Kilcullen, M. B., *Br. Corros. J.*, **3**, 87 (1968)
14. Haynie, F. H. and Upham, J. B., *Mats. Prot.*, **9** No. 8, 35 (1970)
15. Schikorr, G., *Werk. Korr.*, **15**, 457 (1964) and B.I.S.I.T.S. No. 3 947
16. Schwarz, H., ibid., **16**, 93 (1965) and B.I.S.I.T.S. No. 4269
17. Ross, T. K. and Callaghan, B. G., *Nature*, **211**, 25 (1966)
18. Sereda, P. J., *Ind. Engng. Chem.*, **53**, 157 (1960)
19. Stanners, J. F., *Br. Corros. J.*, **5**, 117 (1970)
20. Skorchelletti, V. V. and Tukachinsky, S. E., *J. Appl. Chem. (USSR)*, **28**, 615 (1955)
21. Barton, K. and Bartonova, Z., *Werk. Korr.*, **21** No. 2, 85 (1970) and B.I.S.I.T.S. No. 8 349
22. Persson, G., *Acidity and Concentration of Sulphate in Precipitation Over Europe*, Report, Swedish National Nature Conservancy Office, December (1968)
23. Broad, D. W. and Foster, A. G., *J. Chem. Soc.*, 446 (1946)
24. Schikorr, G., *Werk. Korr.*, **14**, 69 (1963)
25. Rosenfei'd, I. L. and Zhigalova, K., *Corrosion of Metals and Alloys* (Ed. by C. Booker), Metallurgizdat, Moscow
26. Klark, G. B., *et al.*, ibid.
27. Brokskii, A. I., *Zhur. Fiz. Khim.*, **30**, 676 (1956)
28. Roikh, I. L., ibid., **32**, 1137 (1958)
29. Evans, U. R., *Trans. Inst. Met. Fin.*, **37**, 1 (1960)

2.3 Natural Waters

Introduction

Metals immersed or partly immersed in water tend to corrode because of their thermodynamic instability. Natural waters contain dissolved solids and gases and sometimes colloidal or suspended matter; all these may affect the corrosive properties of the water in relation to the metals with which it is in contact. The effect may be either one of stimulation or one of suppression, and it may affect either the cathodic or the anodic reaction; more rarely there may be a general blanketing effect. Some metals form a natural protective film in water and the corrosiveness of the water to these metals depends on whether or not the dissolved materials it contains assist in the maintenance of a self-healing film.

The metals most commonly used for water systems are iron and steel. These metals often have some sort of applied protective coating; galvanised steel, for example, relies on a thin layer of zinc, which is anodic to the steel except at high temperatures. Many systems, however, contain a wide variety of other metals and the effect of various water constituents on these must be considered. The more usual are copper, brasses, bronzes, lead, aluminium, stainless steel and solder.

The passage of a natural water through a pipe may modify the composition of the water and hence its corrosive properties. Consumption of constituents which in the circumstances may be corrosion inducing — e.g. oxygen or carbon dioxide, may reduce the water's corrosive properties. Dissolution of a metal into water may, on the other hand, make it more corrosive. An example of this is the attack of some waters on copper and the subsequent increased pitting corrosion of less noble metals such as iron, galvanised steel and aluminium. It has been suggested that this enhanced pitting is caused by the redeposition of minute quantities of copper on the less noble metal thus setting up numerous bimetallic corrosion cells[1].

Failure of the metal can be the most important effect of a corrosive water, but other effects may arise from small concentrations of metallic ion produced by corrosion. A natural water passed through a lead pipe may contain a toxic concentration of that metal; with copper there is a greater tolerance from the toxicity point of view but staining of fabrics and sanitary fittings may be objectionable. With iron, similarly, discoloration of the water may be unpleasant and may cause damage to materials being processed.

Constituents or Impurities of Water

The concentrations of various substances in water in dissolved, colloidal or suspended form are relatively low but vary considerably; for example, a hardness of 300–400 p.p.m. (as $CaCO_3$) is sometimes tolerated in public supplies, whereas dissolved iron to the extent of 1 mg/litre would be unacceptable.

In treated water for high-pressure boilers or where radiation effects are important, as in some nuclear projects, impurities are measured in very small units (e.g. μg/litre or p.p. 10^9), but for most purposes it is convenient to express results in mg/litre. In water analysis, determinations (except occasionally for dissolved gases) are made on a weight/volume basis but some analysts still express results in terms of parts per million (p.p.m.). The difference between mg/litre and p.p.m. is small and for practical purposes the two units are interchangeable. For some calculations, the use of milliequivalents per litre or equivalents per million (e.p.m.) has advantages but has not found much application. Hardness, whatever the constituent salts, is usually expressed as p.p.m. $CaCO_3$ (see Table 2.10).

Table 2.10 Units of measurement and of hardness

Units of Measurement – Conversion Factors

Milligrams per litre (mg/ℓ) = parts per million (p.p.m.)
Part per 100 000 = 10 mg/litre
Grains per Imperial gallon = 14·25 mg/litre.
Grains per US gallon = 17·1 mg/litre.

Hardness Units

Parts per million (mg/ℓ) as $CaCO_3$.
Degree French = parts per 100 000 as $CaCO_3$ (= 10 mg/litre $CaCO_3$).
Degree Clark, English or British = grains per Imperial gallon or p.p. 70 000 as $CaCO_3$ (= 14·25 mg/litre $CaCO_3$).
Degree German = parts per 100 000 as CaCO (= 17·8 mg/litre $CaCO_3$).

Water analysis for drinking-water supplies is concerned mainly with pollution and bacteriological tests. For industrial supplies a mineral analysis is of more interest. Table 2.11 includes a typical selection and gives some indication of the wide range that can be found.

The important constituents can be classified as follows:

1. Dissolved gases (oxygen, nitrogen, carbon dioxide, ammonia, sulphurous gases).
2. Mineral constituents, including hardness salts, sodium salts (chloride, sulphate, nitrate, bicarbonate, etc.), salts of heavy metals, and silica.
3. Organic matter, including that of both animal and vegetable origin, oil, trade waste (including agricultural) constituents and synthetic detergents.
4. Microbiological forms, including various types of algae and slime-forming bacteria.

NATURAL WATERS 2:45

Table 2.11 Typical water analyses (results in mg/litre)

	Very soft lake water	Moderately soft surface water	Slightly hard river water	Moderately hard river water	Hard borehole water (chalk formation)	Slightly hard borehole water containing sodium bicarbonate	Very hard underground water
pH value	6·3	6·8	7·4	7·5	7·1	8·3	7·1
Alkalinity to methyl orange (CaCO$_3$)	2	38	90	180	250	278	470
Total hardness (CaCO$_3$)	10	53	120	230	340	70	559
Calcium hardness (CaCO$_3$)	5	36	85	210	298	40	451
Sulphate (SO$_4$)	6	20	39	50	17	109	463
Chloride (Cl)	5	11	24	21	4	94	149
Silica (SiO$_2$)	trace	0·3	3	4	7	12	6
Dissolved solids	33	88	185	332	400	620	1670

Dissolved Gases

Of the dissolved gases occurring in water, oxygen occupies a special position as it stimulates the corrosion reaction. Carbon dioxide is scarcely less important; this constituent must, however, be considered in relation to other constituents, especially calcium hardness.

Nitrogen is present with oxygen although the ratio is not the same as in air. It has little importance in connection with corrosion, but can be a nuisance if changes in physical conditions bring about its release from solution.

Other gases which are occasionally present usually arise from pollution. Ammonia, which in various forms may be present in waste waters, attacks copper and copper alloys; its presence in estuarine waters is one of the main causes of condenser-tube corrosion.

Hydrogen sulphide and sulphur dioxide are also usually the result of pollution; sometimes they are produced by the interaction of two contaminants, but sometimes bacterial action may be contributory. Both gases may initiate or accelerate corrosion of most metals.

The significance of small concentrations of these and other impurities in high-pressure steam-boiler feed water is discussed in Section 17.4.

Oxygen Dissolved oxygen is probably the most significant constituent affecting corrosion, its importance lying in the fact that it is the most important cathodic depolariser in neutral solutions. Other depolarisers also occur, but as oxygen is an almost universal constituent of natural waters its importance will readily be understood.

In surface waters, the oxygen concentration approximates to saturation, but in the presence of green algae supersaturation may occur. Underground waters are more variable in oxygen content and some waters containing ferrous bicarbonate are oxygen-free. Contact with air, however, usually gives rise to an oxygen concentration similar to the figures in Table 2.12, which are for distilled water. The solubility is slightly less in the presence of dissolved solids, but this effect is not very significant in natural waters containing less than 1 000 p.p.m. dissolved solids.

Table 2.12 Solubility of oxygen in distilled water

Temperature (°C)	Oxygen content of air-saturated water	
	mg/litre	ml/litre
0	14·16	9·90
5	12·37	8·65
10	10·92	7·64
15	9·76	6·83
20	8·84	6·18
25	8·11	5·67

For some applications, notably feed-water treatment for high-pressure boilers, removal of oxygen is essential. For most industrial purposes, however, de-aeration is not applicable, since the water used is in continuous contact with air, from which it would rapidly take up more oxygen. Attention must therefore be given to creating conditions under which oxygen will stifle rather than stimulate corrosion.

It has been shown that pure distilled water is least corrosive when fully aerated and that some inhibitors function better in the presence of oxygen[2]. In these cases oxygen acts as a passivator of the anodic areas of the corrosion cells. These facts do not, however, modify the foregoing statements on the significance of oxygen in waters as used in practice.

A major difficulty in applying corrosion inhibitors is that the oxygen content of the water may not be the same at all points. For example, in a thin layer of water between a flake of scale (or almost any other foreign body) and the metal on which it is lying the oxygen can be depleted; the difference in oxygen content between the body of water and the stagnant water will then set up a corrosion current which is difficult to suppress. Rather similar conditions occur at the water line of a vessel containing water with air above it. Even if the water is conditioned to prevent corrosion under submerged conditions, the protection may not extend to the water line, especially if the water has a high dissolved-solids content. Fluctuation in water level extends the area of localised attack.

Carbon dioxide and calcium carbonate The effect of carbon dioxide is closely linked with the bicarbonate content. Normal carbonates are rarely found in natural waters but sodium bicarbonate is found in some underground supplies. Calcium bicarbonate is the most important, but magnesium bicarbonate may be present in smaller quantities; in general, it may be regarded as having properties similar to those of the calcium compound except that on decomposition by heat it deposits magnesium hydroxide whereas calcium bicarbonate precipitates the carbonate.

Calcium bicarbonate requires excess carbon dioxide in solution to stabilise it; the necessary concentration depends on the other constituents of the water and the temperature.

The concentrations of carbon dioxide in water can be classified as follows:

1. The amount required to produce carbonate.
2. The amount required to convert carbonate to bicarbonate.
3. The amount required to keep the calcium bicarbonate in solution.
4. Any excess over that accounted for in types 1, 2 and 3.

With insufficient carbon dioxide of type 3 (and none of type 4) the water will be supersaturated with calcium carbonate and a slight increase in pH (at the local cathodes) will tend to cause its precipitation. If the deposit is continuous and adherent the metal surface may become isolated from the water and hence protected from corrosion. If type 4 carbon dioxide is present there can be no deposition of calcium carbonate and old deposits will be dissolved; there cannot therefore be any protection by calcium carbonate scale.

The mathematical relationship between carbon dioxide, calcium bicarbonate and calcium carbonate has been studied by several workers, including Langelier[3,4]. The simpler form of his equation is

$$pH_s = pCa + pAlk + (pK_2 - pK_s) \text{ at constant temperature}$$

where pH_s = saturation pH value,
pCa = negative logarithm of the calcium concentration expressed as p.p.m. $CaCO_3$,
$pAlk$ = negative logarithm of the alkalinity to methyl orange expressed in p.p.m. of equivalent $CaCO_3$,
pK_2 = ionisation constant of HCO_3^-, $\left[\dfrac{[H^+][CO_3^{2-}]}{[HCO_3^-]}\right]$ and
pK_s = solubility product of $CaCO_3$.

This simple formula does not apply to pH values over 9·0, and high salinities affect its accuracy. The term $(pK_2 - pK_s)$ is a function of temperature and ionic strength (dissolved solids). In an analysis of a given water at a constant temperature much useful information can be obtained from the equation.

The saturation index of a water (S.I.) is defined as:

$$\text{S.I.} = pH - pH_s$$

where pH is the actual pH of the water. If the saturation index is positive the water will be supersaturated with calcium carbonate whereas if it is negative the water will be aggressive to calcium carbonate. Graphical forms of the expression are of most practical value and that devised by Powell, Bacon and Lill[5] is shown in Fig. 2.15. These authors have also developed the method so that it can be applied to provide a water which has a constant saturation index over a fair temperature range; this is mainly of interest to operators of industrial cooling systems[6]. The corresponding formula for the magnesium hydroxide equilibrium is:

$$pH_s(Mg) = \tfrac{1}{2}(pMg - pK_{s(Mg)}) + pK_w$$

is incorporated in the Langelier diagram[4].

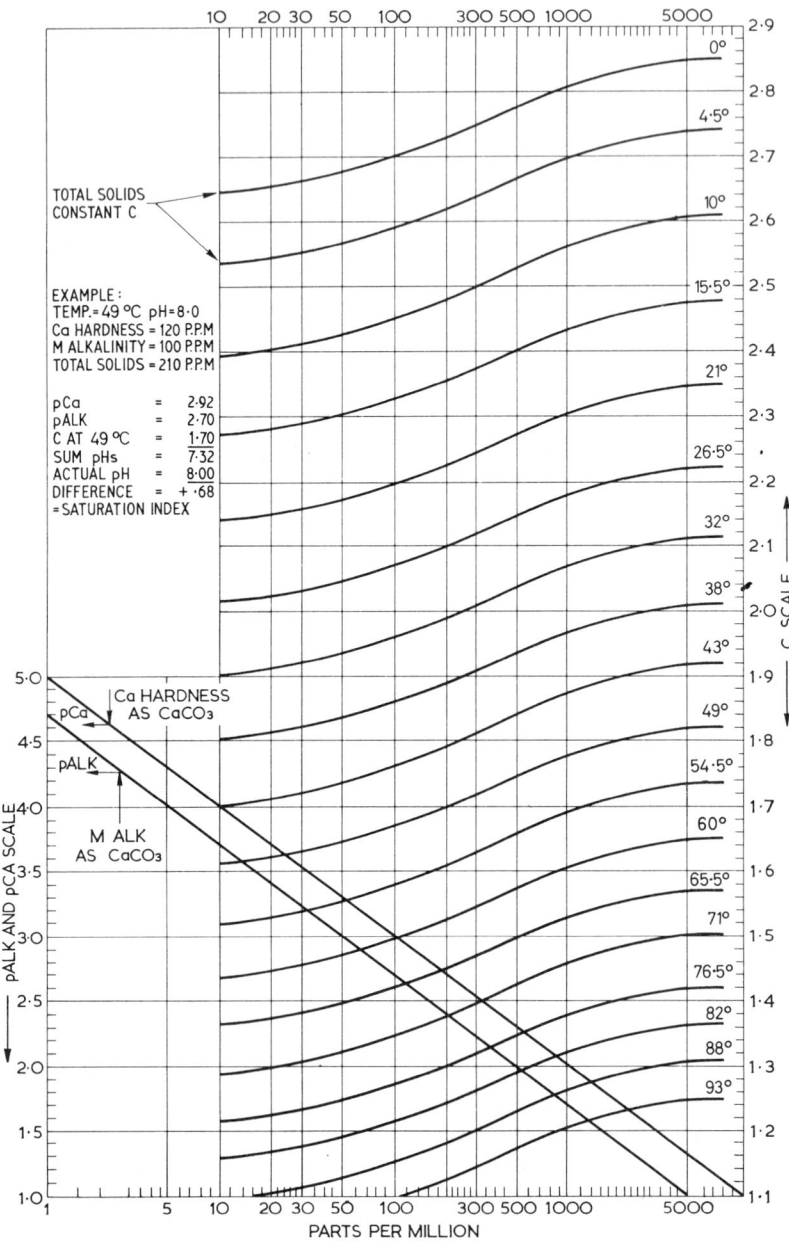

Fig. 2.15 Langelier saturation index chart (after the American Chemical Society).

A distinction must be made between a thick layer of deposit — whether of calcium carbonate or of other material — and a protective layer. The ideal protection in fact consists of layers of negligible thickness which do not impede water or heat flow and which are self-healing. This is difficult to achieve with natural waters. A water which is exactly in equilibrium in respect of calcium carbonate is normally corrosive to steel (unless it contains natural inhibitors of other types) because it has no power to form a calcium carbonate deposit. Supersaturated waters on the other hand, unless suitably treated, will form a substantial scale, but whether this inhibits corrosion or not depends on adherence to the metal and porosity. The degree of protection afforded by calcium carbonate has been studied by McCauley[7].

The carbon dioxide content (i.e. of types 3 and 4) can be ascertained from the pH of the water and its alkalinity by a formula devised by Tillmans[8]

$$\text{pH} = \log \left(\frac{\text{alkalinity} \times 0.203 \times 10^7}{\text{free CO}_2} \right)$$

where alkalinity (expressed as $CaCO_3$) and free CO_2 are in p.p.m.

The actual figure is, however, of value only in relation to calcium carbonate content and for calculation of alkali additions for pH corrections. A graphical form is included in the Langelier diagram[4].

The significance of carbon dioxide in corrosion is also discussed in some detail by Simmonds[9].

Mineral Constituents

Hardness salts   The hardness figures for natural supplies are very varied but most natural supplies in the U.K. fall into well-defined groups. The most important of these are:

1. Upland waters of low hardness, as supplied to most towns in Scotland, Wales and the North of England.
2. Hard underground waters, mainly in the East and South of England, mostly from chalk, sandstone or limestone strata.

A few supplies are intermediate in composition. Many of them are derived from river sources and vary according to season.

The usual classification of water by hardness (Thresh, Beale and Suckling) is as follows:

<50 p.p.m. $CaCO_3$	soft
50–100 p.p.m. $CaCO_3$	moderately soft
100–150 p.p.m. $CaCO_3$	slightly hard
150–250 p.p.m. $CaCO_3$	moderately hard
250–350 p.p.m. $CaCO_3$	hard
>350 p.p.m. $CaCO_3$	very hard.

The corrosive properties of natural waters are governed by many factors and cannot be related to hardness alone, but the following trends are apparent:

1. Soft upland waters are aggressive to most metals, their behaviour depending to some extent on pH values as discussed on p. 2.53. They are inevitably unsaturated with respect to calcium carbonate and it is not usually practicable to modify the carbonate equilibrium to make them non-aggressive.

2. Very hard waters are usually not very aggressive provided that they are supersaturated with calcium carbonate. Underground waters with a low pH value and high carbon dioxide content are, however, aggressive unless corrective treatment is applied.
3. Waters of intermediate hardness frequently contain fair amounts of other constituents and there is often a tendency for the scale to be loosely attached, permitting corrosion to occur irregularly underneath. In most waters the bicarbonate content is less than the hardness, but a few natural waters are known where the reverse is the case. These waters have been partially softened by the zeolite process which occurs underground, and then contain sodium bicarbonate which, together with the high concentration of chloride and other minerals, may accelerate attack.

Dissolved mineral salts The principal ions found in water are calcium, magnesium, sodium, bicarbonate, sulphate, chloride and nitrate. A few parts per million of iron or manganese may sometimes be present and there may be traces of potassium salts, whose behaviour is very similar to that of sodium salts. From the corrosion point of view the small quantities of other acid radicals present, e.g. nitrite, phosphate, iodide, bromide and fluoride, have little significance. Larger concentrations of some of these ions, notably nitrite and phosphate, may act as corrosion inhibitors, but the small quantities present in natural waters will have little effect. Some of the minor constituents have other beneficial or harmful effects, e.g. there is an optimum concentration of fluoride for control of dental caries and very low iodide or high nitrate concentrations are objectionable on medical grounds.

Chlorides have probably received the most study in relation to their effect on corrosion. Like other ions, they increase the electrical conductivity of the water so that the flow of corrosion currents will be facilitated. They also reduce the effectiveness of natural protective films, which may be permeable to small ions; the effect of chloride on stainless steel is an extreme example but a similar effect is noted to a lesser degree with other metals. Turner[10] has observed that the meringue dezincification of duplex brasses is affected by the chloride/bicarbonate hardness ratio.

Nitrate is very similar in its effects to chloride but is usually present in much smaller concentrations.

Sulphate in general appears to behave very similarly; Hatch and Rice have shown that small concentrations in distilled water increase corrosion more than similar concentrations of chloride[11]. In practice, high-sulphate waters may attack concrete, and the performance of some inhibitors appears to be adversely affected by the presence of sulphate. Sulphates have also a special rôle in bacterial corrosion under anaerobic conditions. Both sulphates and nitrates are acceptable in low-pressure boiler feed water as they are believed to be of value in controlling caustic cracking.

Conventional combinations Salts of strong acids and alkalis are, of course, almost completely ionised in dilute solutions. For some purposes, however, it is convenient to regard the ions as being in combination, and various systems of 'conventional combinations' have been developed. In Britain, the system most used takes the metals and acid radicals in the order shown in Table 2.13 (after Thresh, Beale and Suckling). For example, if the amount

Table 2.13 Conventional combinations*†

Ferrous carbonate	56 Fe ≡	60 CO_3	≡ 116 $FeCO_3$
Calcium carbonate	40 Ca ≡	60 CO_3	≡ 100 $CaCO_3$
Calcium sulphate	40 Ca ≡	96 SO_4	≡ 136 $CaSO_4$
Calcium chloride‡	40 Ca ≡	71 Cl	≡ 111 $CaCl_2$
Calcium nitrate‡	40 Ca ≡	124 NO_3	≡ 164 $Ca(NO_3)_2$
Magnesium carbonate	24 Mg ≡	60 CO_3	≡ 84 $MgCO_3$
Magnesium sulphate	24 Mg ≡	96 CO_4	≡ 120 $MgSO_4$
Magnesium chloride‡	24 Mg ≡	71 Cl	≡ 95 $MgCl_2$
Magnesium nitrate‡	24 Mg ≡	124 NO_3	≡ 148 $Mg(NO_3)_2$
Sodium carbonate	46 Na ≡	60 CO_3	≡ 106 Na_2CO_3
Sodium sulphate	46 Na ≡	96 SO_4	≡ 142 Na_2CO_4
Sodium chloride‡	46 Na ≡	71 Cl	≡ 117 NaCl
Sodium nitrate‡	46 Na ≡	124 NO_3	≡ 170 $NaNO_3$
Sodium silicate§	46 Na ≡	136 Si_2O_5	≡ 182 $Na_2Si_2O_5$

* If potassium is present in significant quantities and is determined it is usually inserted in the conventional combination table after magnesium and before sodium (78 K ≡ 138 K_2CO_3 ≡ 174 K_2SO_4 ≡ 149 KCl ≡ 202 KNO_3).
† The figures incorporated in the table are 'equivalent to $CaCO_3$' or double the chemical equivalent weights. If it is desired to express analytical figures as milli-equivalents per litre (e.p.m.), the concentrations in mg/litre (p.p.m.) must be divided by half the table figure (e.g. 150 p.p.m. $CaCO_3$ ≡ 3 e.p.m.; 96 p.p.m. Mg ≡ 8·0 e.p.m.). If this procedure is adopted for both anions and cations, the totals of each should be identical.
‡ Sometimes chlorides and nitrates are taken in reversed order throughout but this is rarely of much significance.
§ The sodium silicates present are variable in composition; the formula given above must be regarded as a 'rough average'.

of bicarbonate ion is more than the sum of the equivalents of ferrous iron and calcium the presence of magnesium bicarbonate, and possibly sodium bicarbonate, is postulated. If, however, the bicarbonate content is less than the calcium equivalent the water is assumed to contain calcium sulphate.

The significance of conventional combinations arises largely in two classes of supply: (*a*) those in which the method indicates sodium bicarbonate to be present and (*b*) those similarly found to contain magnesium chloride or calcium chloride.

Waters containing sodium bicarbonate are derived from underground sources where zeolitic materials are present. They occur in various parts of Britain but the best-known group comprises those in central London and other parts of the Thames valley where the water, originating in the chalk, is modified by passage through zeolitic rock. Although some softening occurs, the hardness may still be appreciable; the waters also have high chloride concentrations and usually a fair amount of carbon dioxide. (A typical analysis is given in Table 2.11.) They are among the most difficult corrosive waters to deal with unless the bicarbonates and excess carbon dioxide are removed. Alkali additions are rarely effective and conditions are far from ideal for the effective use of inhibitors.

The importance of magnesium chloride has probably been exaggerated. There is little doubt that it can act as a catalyst in corrosion reactions by hydrolysing to form hydrochloric acid, being then regenerated by reaction between ferrous chloride and magnesium hydroxide. There is, however, little evidence that this reaction takes place in cold- or hot-water systems, and it is probably confined to steam boilers where it might be a cause of corrosive attack underneath scale deposits; it does not constitute a problem in a properly conditioned boiler water.

Brief reference has already been made to iron- and manganese-bearing waters. The small amount of deposit formed from these waters is not likely to have much effect on corrosion although there is always a possibility that attack will occur under sludge deposits. Most iron-bearing waters contain substantial amounts of carbon dioxide which may be troublesome. Manganese-bearing waters may be of a similar type but they sometimes contain complex organic compounds of manganese for which special treatment may be needed. Manganese deposits have been associated with type II copper pitting corrosion.

Another mineral constituent of water is silica, present both as a colloidal suspension and dissolved in the form of silicates. The concentration varies very widely and, as silicates are sometimes applied as corrosion inhibitors, it might be thought that the silica content would affect the corrosive properties of a water. In general, the effect appears to be trivial; the fact that silicate inhibitors are used in waters with a high initial silica content suggests that the form in which silica is present is important.

Organic Matter

The types of organic matter in supplies are very diverse and may be present in suspension, or in colloidal or true solution. It is largely decaying vegetable matter but there are many other possible sources including run-off from fields and domestic and industrial wastes. An increasing volume of literature is appearing on organic pollution but the significance of this in relation to corrosion receives little, if any, attention. A comprehensive account of all the possible constituents is beyond the scope of this section but it may be useful to consider the effects of some types of organic matter.

In the first place, there is the masking effect of deposits which may result from suspended matter thrown down on to hot surfaces or at areas where velocity is reduced. They may also form from material coming out of true or colloidal solution. A partially covered surface is always liable to attack. In an aerated water, the distribution of oxygen will be uneven so that corrosion currents will be set up by the cells so produced, corrosion normally occuring at the points where the oxygen content is lower. In waters free from oxygen, other 'differentials' may result in corrosion cells. Another aspect, especially in systems where de-aerated water might be used, is that deposits may lead to over-heating and failures of a different kind, e.g. bursting of boiler tubes.

Among contaminants one of the most objectionable is oil, especially in systems where water is strongly heated. A relatively small amount of oil on a heating surface can produce very rapid failures. An indirect effect of oil, or other contaminants which form films on the water surface, is that the film isolates the water from air so that in polluted water anaerobic conditions may develop with the encouragement of objectionable bacterial activity.

Some of the worst corrosive effects in soft waters are attributed to a rather wide group of organic acids abstracted from peat and mosses, sometimes called *peaty* acids. Such waters have low pH values and are often discoloured. They affect ferrous metals appreciably and also attack lead and copper. Corrosion control, either of steel or of copper, is rarely achieved solely by pH correction of such waters.

Organic Growths

Natural waters may contain organic growths of various kinds, including algae and slime-forming bacteria, which may have a direct or indirect effect on corrosion.

The effect may be of two main types: (*a*) the masking effect of living or dead organisms which is little different from that of other materials, and (*b*) the effect of alterations in composition brought about by the organisms. Algae, for example, may remove carbon dioxide and produce oxygen, while other organisms often consume oxygen. Under anaerobic conditions the sulphate-reducing bacteria produce sulphide, and hence hydrogen sulphide, from dissolved sulphate, with disastrous effects. This mechanism is often responsible for external attack on cast iron mains in waterlogged soil, but is not unknown in hot-water systems where the temperature in stagnant branches may promote the development of such organisms (see Sections 2.5 and 2.6).

The voluminous deposits associated with iron bacteria, although objectionable in other ways, rarely have much effect on corrosion as they form only over a long period and the alteration in water composition is negligible.

pH of Water

Reference has previously been made to pH in connection with calcium carbonate, but it has also a more general significance. The pH of natural waters is, in fact, rarely outside the fairly narrow range of 4.5 to 8.5. High values, at which corrosion of steel may be suppressed, and low values, at which gaseous hydrogen evolution occurs, are not often found in natural waters.

According to weight-loss measurements, steel corrodes at approximately the same rate throughout the range of pH found in natural waters. The form which the corrosion takes is, however, affected by the pH. At values between 7·5 and 9·0 there is a tendency for the corrosion products to adhere in a hard crusty deposit[12]. Sometimes there are separate 'tubercles', but these are more usually joined up to form a more or less continuous layer. Attack under the deposit is, however, usually irregular. At lower pH values, adherent corrosion products are not so evident although a very hard form of deposit is sometimes seen in pipes which have been in service for some years. Loss of head due to scaling of a pipe is more commonly found in the higher pH range; at lower pH values, 'red water' complaints, arising from corrosion products in suspension, are more common. Inhibitors may reduce the amount of corrosion, but if inhibition is not complete the type of attack is unaltered. For this reason, it is difficult to prevent corrosion in the tuberculating range as a small amount of attack produces an adherent corrosion product which puts a barrier between the inhibitor and the metal.

Cast iron behaves in a manner similar to steel at alkaline pH values but at low pH values it is subject to graphitisation.

Copper is affected to a marked extent by pH value. In acid waters, slight corrosion occurs and the small amount of copper in solution causes green staining of fabrics and sanitary ware. In addition redeposition of copper on aluminium or galvanised surfaces sets up corrosion cells resulting in pitting of these metals. In most waters the critical pH value is about 7·0 but in soft water containing organic acids it may be higher. The 'pitting' corrosion of copper is independent of the general nature of the water but occurs only when (a) certain carbonaceous or oxide films occur on the metal surface and (b) when the water does not contain a natural organic inhibitor* which is, in fact, present in many supplies[13,14] (see also Sections 1.6 and 4.2).

Lead is affected by carbonate content, pH value and mineral constituents. With soft waters the simplest method of control is usually to increase the pH value by adding alkali.

Zinc coatings on steel (galvanised) are attacked in the same way as iron, but usually more slowly. Very alkaline waters are usually aggressive to zinc and will often remove galvanised coatings; the corrosion products consist of basic zinc carbonate or other basic compounds and may take the form of a thick creamy deposit or hard abrasive particles.

Rates of Flow

The effect of deposits has been referred to in relation to organic matter. Oxygen depletion can, of course, also occur under other types of deposit. Water velocity plays some part here, as with a good flow deposition is less likely.

Apart from this effect, increased velocity usually increases corrosion rates by removing corrosion products which otherwise might stifle the anodic reaction and, by providing more oxygen, may stimulate the cathodic reaction[15].

Temperature

The effect of temperature is complex. At very high temperatures such effects as reversal of polarities, as in the Zn/Fe couples of a galvanised surface, may be produced, and, where there are temperature gradients, corrosion cells may be set up. The more general effects may, however, be summarised as follows: (a) the velocity of corrosion reactions is greater at increased temperatures, (b) temperature changes may affect solubility of corrosion products or shift the position of such equilibria as that existing between calcium carbonate and carbon dioxide, (c) gases are less soluble at increased temperature, an effect which is, however, partly offset by greater diffusion rates and (d) modification of pH value. This last effect is bound up with the previous two and is mainly of importance in affecting the form and distribution of corrosion products.

The overall effect is that corrosion is usually more rapid at higher temperatures, the corrosion products being often more objectionable in nature. There are, however, exceptions to this generalisation and the increased rate

---

* The chemical nature of the inhibitor is not known

Assessing the Corrosivity of Natural Waters from their Chemical Analysis

Although the Langelier index is probably the most frequently quoted measure of a water's corrosivity, it is at best a not very reliable guide. All that the index can do, and all that its author claimed for it[16], is to provide an indication of a water's thermodynamic tendency to precipitate calcium carbonate. It cannot indicate if sufficient material will be deposited to completely cover all exposed metal surfaces; consequently a very soft water can have a strongly positive index but still be corrosive. Similarly the index cannot take into account if the precipitate will be in the appropriate physical form, i.e. a semi-amorphous 'egg-shell' like deposit that spreads uniformly over all the exposed surfaces rather than forming isolated crystals at a limited number of nucleation sites. The egg-shell type of deposit has been shown to be associated with the presence of organic material which affects the growth mechanism of the calcium carbonate crystals[17]. Where a substantial and stable deposit is produced on a metal surface, this is an effective anti-corrosion barrier and forms the basis of a chemical treatment to protect water pipes[18]. However, the conditions required for such a process are not likely to arise with any natural waters.

As well as the conventional chemical parameters generally useful in gauging a water's corrosivity e.g. pH, chloride, sulphate etc., various ratios of ions have been found to be significant for particular problems. Thus an increase in the corrosion rate of iron occurs when the chloride:carbonate ratio exceeds 3:1[19] and attack of the dezincification prone brasses arises when the chloride to carbonate hardness ratio exceeds 1:3[20]. More recently the aggressivity of the chloride ion to galvanically coupled lead or tin-lead solders has been found to be suppressed when the sulphate:chloride ratio exceeds 2:1[21]. The most spectacular example of this approach is that involving six parameters, pH, chloride, sulphate, nitrate, sodium ion and dissolved oxygen, that have to be taken into account when calculating the propensity of waters to support type I copper pitting[22]. However such examples, which require a computer program to carry them out conveniently and provide semi-quantitative answers, are unfortunately rare. More usually only very limited correlations can be made between water composition and corrosivity, and even where no multiple ion effects are involved, the response to a change in one parameter may be difficult to model mathematically e.g. the corrosion of iron which passes through a maximum between pH 7·5–8 in some natural waters[23].

Considerations such as these can lead to unexpected problems where waters are mixed, either at a treatment works or in a tidal zone within a distribution network into which two sources are fed separately. Within the author's experience, problems of an erosion attack on copper pipe have occurred at fittings, especially where the ends of the copper tube have been belled out to meet the requirements of the bye-laws for underground pipe[24], with mixtures of waters that were satisfactory when supplied separately.

Calculations to determine if the mixing of the supplies had produced an increase in free CO_2, the most likely explanation of this effect, proved negative.

More recently, attempts have been made to correlate mathematically the chemical composition of natural waters and their aggressivity to iron by direct measurements on corrosion coupons[25] or pipe samples removed from distribution systems[26]. This work has been of limited success, either producing a mathematical best fit only for the particular data set examined or very general trends. The particular interest to the water supply industry of the corrosivity of natural waters to cast iron has led to the development of a simple corrosion rig for the direct measurement of corrosion rates[27]. The results obtained using this rig has suggested an aggressivity classification of waters by source type i.e.

Source type	Corrosivity to iron (mdd)
hard aerated borehole	typically 5
lowland river derived	typically 15
soft upland	typically 40

All these rates are, of course, quite low and the problem with corrosion of cast iron water mains is its effect on water quality rather than deterioration of the asset.

The corrosion rig has been used to study the effect of inhibitors e.g. silicate and phosphate commonly used to overcome problems with iron. This has revealed that these 'inhibitors' hardly affect the long-term corrosion rate, indeed in certain circumstances they may actually increase it. They produce their effect by stabilising the corrosion product developed, thereby preventing the water quality deterioration which is the real complaint[27].

The above catalogue of difficulties, in relating the aggressivity of natural waters to their chemical composition, arises precisely because of the low corrosion rates that are usually found with most metals. Under such circumstances, water composition is only one of many factors that determine the rate of attack. The other factors include flow regime, temperature and the conditions under which the initial corrosion product is laid down. The best summary of the behaviour of metals commonly used in natural waters is still that produced by Campbell for the Society of Water Treatment and Examination[28].

Recent Developments

EC Directive relating to the quality of water intended for human consumption

The significance of this directive[29], from the corrosion point of view, is that for the first time legally enforceable limits for the concentrations of toxic metals in drinking water have been defined. This has greatly increased the importance of contamination as a consequence of corrosion, as opposed to simple mechanical failure, and has required a reassessment of the suitability of various metals and alloys traditionally used in the supply of water for domestic purposes.

Chief among these has been the use of lead, mainly for service pipes but also for header tanks in parts of Scotland, and in tin-lead solders for capillary joints of copper tube. For both materials corrosion rates are low in the range of drinking waters supplied. However, the limit for lead set in the drinking water standard is so very low, $0 \cdot 05$ mg ℓ^{-1} in a running water sample, that it can be readily exceeded, especially in large plumbing systems where the water can have significant residence times.

Because of these considerations, no new lead pipe is being installed. Also, as a result of extensive research[30], the contamination from pipe already in place is being controlled by reducing the solubility of the lead corrosion product in the water concerned. For soft waters (carbonate hardness of < 50 mg $CaCO_3$ ℓ^{-1}) this is readily achievable by increasing the pH to $8-8 \cdot 5$[31]. Thus the pH values of $6 \cdot 3$ and $6 \cdot 8$, given above in Table 2.11 as typical of soft water, are no longer typical for treated supplies. For waters with higher carbonate hardness values, where contamination problems can arise because of the formation the soluble lead carbonate ion pair complex[32], raising the pH is much less effective. For these waters orthophosphate additions are made which converts the corrosion product to a lead phosphate complex with a sufficiently low solubility.

Because of the galvanic interaction with the copper, lead contamination from tin-lead soldered joints, or lead-copper pipe junctions, cannot be controlled by reducing the solubility of the corrosion product layer. Galvanic coupling does not simply increase the rate of corrosion, it also increases the corrosivity of the water by converting chloride, which otherwise acts as a corrosion inhibitor for lead, into an aggressive ion[21]. Although unacceptable contamination only arises where soldered joints are made badly, problems have occurred in practice, especially in large buildings with long water residence times such as hospitals and schools where particularly vulnerable members of society are exposed. Given that practically equivalent non-lead solders are already available, it has now become policy not to use tin-lead alloys in contact with potable water and the appropriate British Standard has been amended accordingly[33].

A British Standards 'draft for development' has been developed[34] which defines a test procedure to determine the potential of metals to contaminate drinking water in contravention of the requirements of the EC Directive. Although primarily meant for new materials, traditional plumbing alloys will also have to be shown to be satisfactory.

Organics

There is an increasing tendency to treat drinking waters to remove organic material. This is to minimise the formation of haloforms, produced when the water is chlorinated, which have health implications[35]. Organics are known to affect certain corrosion processes, e.g. type I copper pitting and the formation of protective corrosion product layers. However, the outcome of this development is difficult to predict as not all the organic material present is removed.

Nitrates

There has been an increasing level of nitrate contamination of borehole supplies in the east of England, because of the use of agricultural fertilisers since the Second World War[36]. Nitrates are known to exacerbate certain corrosion processes e.g. at soldered joints; however the maximum value allowed for this ion by the EC drinking water directive (50 mg NO_3 ℓ^{-1}) should limit its significance.

C. W. DRANE
R. J. OLIPHANT

REFERENCES

1. Kenworthy, L., *J. Inst. Met.*, **69**, 67 (1943)
2. Uhlig, H. H., Triadis, D. N. and Stern, M., *J. Electrochem. Soc.*, **102**, (1955)
3. Langelier, W. F., *J. Amer. Wat. Wks. Ass.*, **28**, 1500 (1936)
4. Langelier, W. F., *J. Amer. Wat. Wks. Ass.*, **38**, 169 (1946)
5. Powell, S. T., Bacon, H. W. and Lill, J. R., *Industr. Engng. Chem.*, **37**, 842 (1945)
6. Powell, S. T., Bacon, H. W. and Lill, S. R., *Industr. Engng. Chem.*, **40**, 435 (1948)
7. McCauley, R. F. and Abdullah, M. O., *J. Amer. Wat. Wks. Ass.*, **50**, 1419 (1958)
8. Tillmans, J., *Die Chemische Untersuchung von Wasser und Abwasser*, Verlage von Willhelm Knapp, Saale (1936)
9. Simmonds, M. A., 'Carbon Dioxide in Domestic Water Supplies', *Proc. Soc. Water Treatment and Examination*, **12**, 4, 197 (1963) and **13**, 1, 40 (1964)
10. Turner, M. E. D., ibid., **10**, 2, 162 (1961) and **14**, 2, 81 (1965)
11. Hatch, G. B. and Rice, Owen, *J. Amer. Wat. Wks. Ass.*, **51**, 719 (1959)
12. Hatch, G. B. and Rice, Owen, *Industr. Engng. Chem.*, **37**, 710 (1945)
13. Gilbert, P. T., 'Dissolution by Fresh Waters of Copper from Copper Pipes', *Proc. Soc. Water Treatment and Examination*, **15**, 3, 165 (1966)
14. Lucey, V. F., *Br. Corros. J.*, **2**, 175 (1967)
15. Eliassen, R., Perada, C., Romeo, A. S. and Skrinde, R. T., *J. Amer. Wat. Wks. Ass.*, **48**, 1005 (1956)
16. Langelier, W. F., *Amer. Wat. Wks. Ass.*, **38**, 169 (1949)
17. Campbel, H. S., Turner, M. E. D., *Jour Inst. Wat. Eng. & Sci.*, **37**, 1, 55 (1983)
18. Hasson, D., Karman, M., 5th. International Conference on the Internal and External Protection of Pipes, Innsbruck, Austria, conference sponsored by BHRA, Cranfield, England (October 1983)
19. Lawson, T. E., 'Corrosion by Domestic Water', *Bull 59*, Ill. State Water Survey, Urbana (1975)
20. Turner, M. E. D., *Jour Soc. Wat. Treat. & Exam.*, **10**, 2, 162 (1961)
21. Oliphant, R. J., Water Research Centre Report ER125 E (November 1983)
22. Lucey, V. F., BNFMRA Research Report No A.1838 (December 1972)
23. Larson, T. E., Skold, R. V., *Corros.*, **14**, 6, 43 (1958)
24. Guide to the application and interpretation of the model water byelaws (1986 Edition), Ellis Harwood Limited, Publishers, Byelaw 52, 120.
25. Singley, J. E., *J. Amer. Wat. Wks. Ass.*, **73**, 579 (1981)
26. Oliphant, R. J., *Assoc. Wat. Offices Jour*, **23**, 3, 29, (1987)
27. Williams, S. M., Ainsworth, R. G., and Elvidge, A. F., 'A method of assessing the corrosivity of water towards iron', Source document 3, Water Mains Rehabilitation Manual, Water Research Centre/Water Authorities Association (1986)
28. Campbell, H. S., *J. Soc. Wat Treat & Exam.*, **20**, 1, 11 (1971)
29. Council Directive relating to the quality of water intended for human consumption, Official Journal of the European communities No. L 229, 11 (August 1980)
30. Seminar 'Lead in Drinking Water', Lorch Foundation, Lane End, High Wycombe, Organised by the Water Research Centre, (March 1981)
31. Gregory, R. *ibid*. Paper 16
32. Hunt, T. E and Jackson, P. J., *ibid*. Paper 9

33. BS 219:1977 Specification for soft solders (as amended 1987)
34. BS DD 201 (1991)
35. Fawell, J. K., Fielding, M. and Ridgway, J. W., *J. Institute of Water and Environmental Management* **1**, 1, 61 (1987)
36. Beresford, S. A. A., *International Journal of Epidemiology*, **14**, 1, 57 (1985)

BIBLIOGRAPHY

Holden, W. S. (Ed.), *Water Treatment and Examination*, Churchill, London (1970)

Tarzwell, C. M., Proc. 29th International Water Conference, Engineers' Society of Western Pennsylvania, 1 (1968)

Corrosion of Iron and Steel by Industrial Waters and its Prevention, Special Report No. 41, I.S.I., London (1949)

Manual of British Water Engineering Practice, Vols. I to III. Published for the Institution of Water Engineers, Heffer, Cambridge, 4th edn. (1969)

Manual on Industrial Water and Industrial Waste Water, American Society for Testing Materials, Philadelphia (1962)

Water Quality and Treatment, American Water Works Association, New York, 3rd edn. (1971)

Water Quality Criteria, Resources Agency of California, State Water Quality Control Board, 2nd edn. (1963)

Water Quality Criteria, American Society for Testing and Materials, S.T.P. No. 416 (1967)

Water—1968, Chemical Engineering Progress, Symposium Series, American Institute of Chemical Engineers (1968)

2.4 Sea Water

Sea water is the only electrolyte containing a relatively high concentration of salts that occurs commonly in nature, covering as it does over two-thirds of the earth's surface. It is both the most familiar and one of the most severe of natural corrosive agents.

Chemical Composition

Ocean sea water is roughly equivalent in strength to a $3\frac{1}{2}$% w/v solution of sodium chloride, but it has a much more complex composition, embodying a number of major constituents, and traces at least of almost all naturally occurring elements. For convenience, however, the concentration of salts in any sample of sea water is expressed in terms of the chloride content, either as *chlorinity* or as *salinity*. Both these units are again subject to arbitrary definition and do not conform simply to the chemical composition.

Chlorinity When a sample of sea water is titrated with silver nitrate, bromides and iodides, as well as chlorides are precipitated. In calculating the chlorinity (Cl), the entire halogen content is taken as chloride, and chlorinity is defined as the weight in grams of silver required for precipitation of total halogen content per kilogram of sea water, multiplied by 0·328 533. (Chlorinity is always expressed as parts per thousand, using the symbol ‰.)

Salinity This term is intended to denote the total proportion of dissolved salts in sea water. As it is inconvenient to determine directly, it is normally derived from the chlorinity, defined and determined as above, using the empirical relationship:

$$\text{Salinity} = 1 \cdot 80655 \times \text{chlorinity}$$

Like chlorinity, it is expressed in parts per thousand.

Constancy of composition The validity of these arbitrary conversions depends on the constancy of the ratios of the various dissolved salts. It is a remarkable and important fact that, except where there is gross dilution or contamination, the relative proportions of the major constituents of sea water are practically constant all over the world.

Table 2.14[1] gives the composition of sea water of 19 parts per thousand chlorinity.

Table 2.14 Major constituents of sea water
(parts per thousand)
(Chlorinity = 19‰, density at 20°C = 1·024 3)

Chloride (Cl^-)	18·979 9
Sulphate (SO_4^{2-})	2·648 6
Bicarbonate (HCO_3^-)	0·139 7
Bromide (Br^-)	0·064 6
Fluoride (F^-)	0·001 3
Boric acid (H_3BO_3)	0·026 0
Sodium (Na^+)	10·556 1
Magnesium (Mg^{2+})	1·272 0
Calcium (Ca^{2+})	0·400 1
Potassium (K^+)	0·380 0
Strontium (Sr^{2+})	0·013 3

(Data courtesy Prentice-Hall, Inc., USA.)

Variations of salinity In the major oceans the salinity of sea water does not vary widely, lying in general between 33 and 37 parts per thousand, a figure of 35 parts per thousand, equivalent to 19·4 parts per thousand chlorinity is commonly taken as the average for 'open-sea' water.

Local conditions may modify this profoundly in special areas. In the Arctic and Antarctic, and where there is dilution by large rivers, the salinity may be considerably less, and it may vary greatly according to season. Salinity is well below normal in the Baltic, and may fall nearly to zero at the head of the Gulf of Bothnia. In enclosed seas like the Mediterranean, Black Sea and Red Sea, on the other hand, where there is rapid evaporation, salinity may reach 40 parts per thousand. The total salt content of the inland Dead Sea is 260 g/kg compared to 37 g/kg for the Atlantic Ocean.

Minor constituents Sea water contains a multitude of organic and inorganic molecules some of which form metallic complexes which even in trace amounts can significantly affect the corrosion mechanism. Trace metallic complexes also play an important rôle in determining the physiology of biological organisms whose presence in sea water can exert considerable control over corrosion reactions. The presence of such complexing agents in sea water could explain the difficulty of simulating the natural product for corrosion research investigations in the laboratory. (See Section 20.1 for compositions of artificial sea waters.)

Variability of Seawater Vertical sections through seawater showing the distribution of temperature, salinity, and oxygen for the Pacific Ocean and Western Atlantic Ocean are shown in Figures 21.3 and 21.4. The global variability of natural seawater and its effects on corrosion have been reviewed[26], in particular with respect to seasonal variation of temperature, salinity, oxygen and pH in the Pacific surface water. Data is also given on the depth profiles for temperature, salinity, oxygen and pH at various sites around North America.

Similar information on temperature, dissolved oxygen, salinity and density has been published for the Northern North Atlantic[27], and in more detail for the seas around the British Isles[28]; the latter also includes hydrographic data on the contents of dissolved metals (Zn, Ni, Cu, Cd, Hg, Mn) and nutrient cations (phosphate, nitrate).

Data from extensive trials investigating the effect of depth on corrosion of materials in the Pacific have now been published[29].

Physical Properties

Density The density of sea water is, of course, related to its salinity (or chlorinity). If ρ_0 is the density of sea water at 0° C in g/ml, σ_0 is defined as $(\rho_0 - 1)1\,000$ and the relationship between density and chlorinity is given by the equation

$$\sigma_0 = -0\cdot 069 + 1\cdot 470\,8\,\text{Cl} - 0\cdot 001\,570\,\text{Cl}^2 + 0\cdot 000\,039\,8\,\text{Cl}^3$$

Since, however, density is affected to a considerable degree by temperature, and since its accurate measurement demands special apparatus and great care, it is not a reliable measure of the 'strength' of sea water.

Electrical Conductivity This is often a convenient and accurate measurement of salinity or chlorinity. Here, too, there is considerable variation with temperature, so that simultaneous observation of temperature is essential. Figure 2.16[3] shows the relationship between conductivity and chlorinity at various temperatures.

Temperature The surface temperature of sea water ranges between about $-2°$ C and 35°C, while the temperature of a shallow surface layer may run even higher. A general picture of the variation with geographical location is given by Table 2.15[4].

Seasonal variations are associated not only directly with the elevation of the sun, but also with changes of surface currents depending on the prevailing winds. The annual variation is generally quite small in the tropics and greatest in the temperate zones, where it may amount to about 10° C.

In general, water at great depths in the oceans is not subject to temperature fluctuations and even in the tropics seldom exceeds 10° C.

The 'freezing point' of sea water, defined as the temperature at which ice crystals begin to form, is $-2°$ C.

Dissolved Gases

Dissolved oxygen is a very important factor in the corrosion of metals immersed in sea water. Because of its biological significance, a vast amount

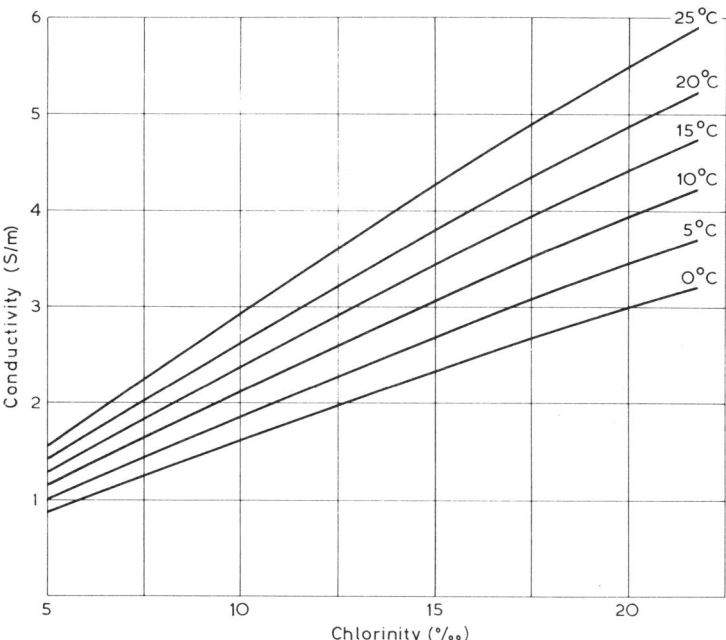

Fig. 2.16 Relationship between conductivity and chlorinity of sea water (after Prentice-Hall, Inc., USA). Note that $S = \Omega^{-1}$

Table 2.15 Average surface temperature of the oceans between parallels of latitude (°C)

North latitude	Atlantic ocean	Indian ocean	Pacific ocean	South latitude	Atlantic ocean	Indian ocean	Pacific ocean
70°–60°	5·60	—	—	70°–60°	−1·30	−1·50	−1·30
60°–50°	8·66	—	5·74	60°–50°	1·76	1·63	5·00
50°–40°	13·16	—	9·99	50°–40°	8·68	8·67	11·16
40°–30°	20·40	—	18·62	40°–30°	16·90	17·00	16·98
30°–20°	24·16	26·14	23·38	30°–20°	21·20	22·53	21·53
20°–10°	25·81	27·23	26·42	20°–10°	23·16	25·85	25·11
10°– 0°	26·66	27·88	27·20	10°– 0°	25·18	27·41	26·01

(Data courtesy Prentice-Hall, Inc., USA.)

of information about its variation in ocean masses has been collected, but insufficient detail is available about the coastal and harbour waters which are of most importance in the corrosion of fixed structures.

Sea water of normal salinity, in equilibrium with the atmosphere, has the following oxygen contents (compare Table 2.14):

Temperature (°C)	−2	0	5	10	15	20	30
Dissolved oxygen (ml/l)	8·52	8·08	7·16	6·44	5·86	5·38	5·42

The dissolved oxygen content of *surface* oceanic water is mainly determined by its biological history; it always tends, by solution from the air, towards saturation values. Estuarial water may be grossly deficient in oxygen; this results in the rapid multiplication of anaerobic bacteria, and in extreme cases the rate of corrosion may be controlled by the bacteria rather than by dissolved oxygen (Section 2.6).

It has been suggested[2] that the oxygen content of the deep water in the Atlantic ocean is high due to the southward flow of the cold oxygen-saturated water through the funnel of the north Atlantic, but in the Pacific ocean the oxygen content decreases with depth due to negligible water flow through the Bering Strait.

pH

Natural sea water is well buffered and normally lies between 8·1 and 8·3 but may fall to 7·0 in stagnant basins with the formation of hydrogen sulphide produced by anaerobic bacteria.

For the solubility in seawater of oxygen, nitrogen and carbon dioxide at various temperatures and chlorinities refer to Tables 21.21 and 21.22.

The freezing point, temperature of maximum density, osmotic pressure and specific heat for seawater of various salinities are given in Table 21.23.

Potentials of Metals in Sea Water

An important factor in the corrosion of a metal in sea water is its electrical potential. This is of course especially the case when two or more electrically connected metals are immersed in a single system.

The 'open-circuit' potential of most metals in sea water is not a constant and varies with the oxygen content, water velocity, temperature and metallurgical and surface condition of the metal.

In static air-free sea water the potential of iron or steel reaches a steady-state value of $-0·75$ V ($vs.$ S.C.E., $E = 0·246$ V) which should be compared with the more noble potential of $-0·61$ V observed under conditions of high velocity and aeration (Table 2.16). This potential of $-0·75$ V for iron in sea water is important in the practice of cathodic protection.

The values in Table 2.16 show how the potentials obtained under service conditions differ from the standard electrode potentials which are frequently calculated from thermodynamic data. Thus aluminium, which is normally coated with an oxide film, has a more noble value than the equilibrium potential $E^{\ominus}_{Al^{3+}/Al} = -1·66$ V $vs.$ S.H.E. and similar considerations apply to 'passive' stainless steel (see Chapter 21).

Although Table 2.16 shows which metal of a couple will be the anode and will thus corrode more rapidly, little information regarding the corrosion current, and hence the corrosion rate, can be obtained from the e.m.f. of the cell. The kinetics of the corrosion reaction will be determined by the rates of the electrode processes and the corrosion rates of the anode of the couple will depend on the rate of reduction of hydrogen ions or dissolved oxygen at the cathode metal (Section 1.4).

Table 2.16 Potentials of metals in aerated moving sea water (Potentials are *negative* to the S.C.E., $E = 0.246$ V)

Metal	Potential (V)	Reference from which figures taken
Magnesium	1·5	5
Zinc	1·03	6
Aluminium	0·79	6
Cadmium	0·7	5
Steel	0·61	6
Lead	0·5	5
Solder (50/50)	0·45	5
Tin	0·42	5
Naval brass	0·30	5
Copper	0·28	5
Aluminium brass	0·27	5
Gun metal	0·26	5
Cupro-nickel 90/10	0·26	5
Cupro-nickel 80/20	0·25	5
Cupro-nickel 70/30	0·25	5
Nickel	0·14	5
Silver	0·13	6
Titanium	0·10	7
Stainless steel 18/8 (passive)	0·08	6
Stainless steel 18/8 (active)	0·53	6

Table 2.17 Effect of exposure period on corrosion rate of mild steel, copper and aluminium[5]

Exposure time (months)	Average corrosion rate for period (mm/y)		
	Steel	Copper	Aluminium
1	0·33	—	—
2	0·25	—	—
3	0·19	—	—
6	0·15	—	—
12	0·13	0·034	0·004 3
24	0·11	0·019	0·002 1
48	0·11	0·018	0·0017

Corrosion Rates

Ferrous Metals

Ferrous metals, of which steel is technically the most important, have a remarkably steady rate of corrosion when fully immersed in sea water. The corrosion of mild steel is very rapid initially but falls off gradually over several months to a fairly steady rate (Table 2.17).

In extended exposure periods of up to 16 years in tropical sea water, Southwell and Alexander[8] obtained an average corrosion rate for steel of 0·18 mm/y in the first year, falling off to a constant rate after 4 years at 0·025 mm/y. They also quote pitting rates as 1 mm/y in the first year falling off dramatically over the second to fourth years and ultimately continuing at a rate comparable with the average rate of penetration giving an average rate for exposure for 16 years of 0·08 mm/y. However, the pitting rate is generally quoted as several orders of magnitude greater than the average rate of penetration, with values of 0·25 to 0·4 mm/y[9].

A comprehensive table of corrosion rates in sea water has been compiled by LaQue[10]. This appears to show no obvious dependence of corrosion rates on the geographical location of the testing site, and few of the rates depart widely from an average of 0·11 mm/y. It is suggested that a figure of 0·13 mm/y may be taken as a reasonable estimate of the expected rate of corrosion of steel or iron continuously immersed in sea water under natural conditions, in any part of the world.

The theory has been advanced that the rapid growth of marine fouling in the tropics may provide a protective shield which counteracts the effect of the greater activity of the hotter water, and LaQue[11] has pointed out that in flowing sea water, when no fouling organisms became attached to small fully immersed specimens, corrosion of steel at 11° C proceeded at 0·18 mm/y compared with 0·36 mm/y at 21° C. This increase corresponds with what would be expected from chemical kinetics, where the rate of reaction is approximately doubled for a rise of 10° C.

It is significant that most of the data from which a remarkable uniformity of attack is deduced are derived from small isolated panels. This is the most convenient form of specimen for measurements of corrosion rates by loss of weight; but it eliminates the important effect of galvanic currents passing between remote parts of a large structure. It is believed that the experience of civil engineers and other users would not support the conclusion suggested by panel tests that corrosion is no faster in tropical than in temperate waters.

Ambler and Bain[12] found that isolated panels exposed in half-tide conditions are normally more rapidly corroded than those fully immersed, a factor of 2 to 4 being not unusual, but in commercial ports the presence of oil contamination may greatly reduce half-tide corrosion by filming the metal surface.

Humble[13] investigated the corrosion of coupled and uncoupled steel plates distributed in a vertical line extending above high-water and below low-water levels and gives a diagram showing the corrosion profile of steel piling in sea water, based on five years' exposure at Kure Beach. This shows two maxima, one in the 'splash zone' above high-tide level, and the other just below low-tide level. In the tidal zone, between these, there is a minimum corrosion rate.

The explanation of this pattern is that the well-aerated areas in the tidal zone become strongly cathodic while the metal just below water becomes anodic. This distribution is in striking contrast to the results quoted by Ambler and Bain[12].

It is generally agreed that steel composition within the range practical for ship plate has little influence on the corrosion rate in sea water[14-17].

Owing to the laborious task of obtaining corrosion rates from gravimetric measurements, data for the effect of exposure time on corrosion rates have been very limited. However, with the more recent use of polarisation resistance measurements it would appear that in the absence of macro-biofouling settlement the depth of penetration rate for mild steel in the Channel, North Sea and North Atlantic varies with time according to the relationship:

$$d = 0.126\, t^{0.92}$$

where d is the average depth of corrosion penetration (mm) occurring in time t (years). This expression agrees fairly well with experience gained from the wreck of the *Holland I* where the average corrosion of the steel hull had occurred to a depth of 6 mm in 70 years[30]. The presence of biofouling settlement may, however, considerably reduce the depth of corrosion, as for instance in the case of artifacts jettisoned by Captain Cook on the Coral Reef in 1770, where coral formation had almost completely protected cast and wrought iron for 200 years[31].

The presence of shell fouling affects the corrosion of steel structures in the intertidal zone where it has been found that the rust formed consists of irregular layers or iron oxides and lime, the latter accounting for up to 15% by weight of the corrosion product[32]. The corrosion rate of mild steel in UK waters for the full immersion and intertidal zone is typically 0.08 mm/y compared with 0.1 to 0.25 mm/y in the splash zone according to the strength of wave action. Above the splash zone corrosion diminishes rapidly to 0.05–0.1 mm/y[32].

Non-ferrous Metals

Many of the common non-ferrous metals corrode relatively slowly in still or slowly-moving sea water. Typical figures are given in Table 2.18.

The effect of exposure time on the corrosion of copper and aluminium is illustrated in Table 2.17. The results quoted by Southwell, Hummer and

Table 2.18 Corrosion rates of non-ferrous metals and alloys in sea water

Material	Corrosion rate (mm/y)
Copper (*a*) full immersion	0·003 8
(*b*) half-tide	0·002 5
Brass (Cu–10 to 35 Zn)	0·004 5
Aluminium brass (Cu–22Zn–2Al)	0·002 0
Admiralty brass (Cu–29Zn–1Sn)	0·004 6
Gun-metal (Cu–10Sn–2Zn)	0·002 5
Phosphor bronze	0·002 5
Aluminium bronze (95Cu–4Al)	0·003 8
Copper-nickel-iron (Cu–5Ni–1Fe)	0·003 8
Cupro-nickel (70Cu–30Ni)	0·001 3
Nickel	0·002 5
Monel	0·002 5
Aluminium (99.8%)	0·000 38
(98%)	0·000 76
(5 Mg)	0·000 30
Lead	0·001 0
Zinc	0·001 8

Alexander[18-19] for corrosion rates of copper and aluminium in tropical waters compared with those obtained around the British Isles suggests that the corrosion rate increases by a factor of two for every 10° C rise in temperature.

The effect of alloying additions on the marine corrosion properties of non-ferrous metals can be very significant, and for copper-based alloys has been comprehensively reviewed by Bradley[20].

For comprehensive reviews of published marine corrosion data refer to references 29 and 33.

Crevice Corrosion

Stainless steel, and aluminium and its alloys, derive their excellent corrosion resistance from the self-repairing protective oxide film which renders them passive. Repair of the film depends on access of oxygen, and in crevices this is often inadequate, with the result that the metal in a crevice becomes 'active'. As the fully exposed areas, usually relatively large, act as a cathode, rapid and sometimes disastrous corrosion may result. The use of stainless steel under sea water needs the greatest care to avoid this trouble, and as a rule one of the resistant copper alloys is the better choice;. the danger from built-in crevices may be foreseen and avoided by careful design, but crevice corrosion also occurs behind marine fouling organisms or other deposits which it may not be possible to prevent. (See Section 1.6.)

In recent years the mechanism of crevice has been mathematically modelled and a more thorough understanding of the corrosion processes has been evolved[34-36]. From such mathematical modelling it is feasible to predict critical crevice dimensions to avoid crevice corrosion determined with relatively simple electrochemical measurements on any particular stainless steel.

Effect of Depth

Little scientific examination of the deterioration of materials at depth has been undertaken except that by the US Naval Civil Engineering Laboratory and Naval Research, Laboratory. The results of this work were reported by Reinhart in 1966 and more recently the work has been reviewed by Kirk[21]. Typical corrosion data for a selection of metals exposed in the Pacific Ocean at several sites and for different times are shown in Table 2.19 and are compared with results obtained in surface waters at Wrightsville Beach by International Nickel Inc.

The general indication of the results in this table is that the corrosion rates of non-ferrous metals increase with depth in spite of lower temperatures and lower oxygen concentrations than at the surface. It was noted in the paper by Kirk[21] that the results at depth were typical of the variation of performance of these materials experienced on numerous occasions in surface sea water. A notable exception was for aluminium alloys of the 5000 (Al–Mg) and 6000 (Al–Mg–Si) series which had good resistance to corrosion

Table 2.19 Effect of depth on the corrosion rate of some metals and alloys

Material	Rate of metal penetration at various exposure depths (mm/y)					Form of attack
	0 m	704 m	1 600 m	1 700 m	2 050 m	
Zinc	0·015	0·058	0·018	0·091	0·150	General
Mild steel	0·127	0·043	0·023	0·020	0·058	General
Aluminium alloy 5052*	—	—	—	>0·576	—	Pitting
'G' bronze†	0·008	0·005	0·008	0·018	0·008	General
Cu-10Ni	0·008	0·020	0·018	0·015	0·015	General
Cu-30Ni	0·005	0·023	0·015	0·025	0·030	General
Stainless steel (type 410)	—	1·270	1·270	1·270	1·270	Pitting
Incoloy 825	—	Slight	Slight	0	Slight	Pitting
Stainless steel (type 316)	—	0	0·025	0	Slight	Pitting
Monel 400	—	0·035	>0·035	0·038	>0·092	Pitting
Exposure conditions						
Temperature (°C)	5-30	7·2	2·5	2·3	2·7	
Oxygen concentration (p.p.m.)	5-10	0·6	1·8	2·8	1·7	

* Al-4Mg alloy. † Admiralty Gunmetal (Cu-10Sn-2Zn, BS 1400 G1).

in shallow waters, but were found to suffer very severe crevice corrosion in deep sea water. Interpreting these data into practical terms it would seem that there is little justification for expecting lower corrosion rates with increased depth in spite of changes in the sea-water chemistry, and corrosion rates would be expected to show the normal variations encountered in surface waters.

Although in the deep-ocean corrosion tests the oxygen concentrations were considerably lower than at the surface, it is perhaps not too surprising that this does not lead to reduction in corrosion since water movement brings a fresh supply of oxygen to the corroding metal surface. In desalination investigations it has been shown that even one part per million of dissolved oxygen in sea water can sustain a corrosion reaction on some materials[22]. In terms of applications for deep-water engineering it is probably the deterioration of materials by the combined action of mechanical loading and corrosion such as stress corrosion and corrosion fatigue which is of major concern.

Drisco and Brouillett[23] have examined a number of protective coatings on mild steel and compared their performance at 2·1 km with that at shallow depth. They concluded that with thick coatings of over 0·3 mm there was negligible difference, but with thinner coatings there was some loss of protection at holidays under deep immersion conditions. An exception was with soft coatings such as asphalt and coal tar which performed better at depth owing to their susceptibility to damage by certain marine fouling organisms such as barnacles at the surface, whereas such species were not encountered at depth.

See reference 29 for further information.

Effect of Water Speed

Hardly any quantitative results on the effect of movement on corrosion of steel are available. Water movement can markedly affect the corrosion process in controlling the rate of transport of reactants to the corrosion site, and the removal of the corrosion reaction products.

A curve is given by LaQue[10] which indicates that the corrosion rates are approximately as follows:

Water speed (m/s)	0	1·5	3	5	7
Corrosion rate (mm/y)	0·13	0·5	0·74	0·86	0·89

This is presumably an estimated average curve, as no numerical data are quoted, and it may be assumed to refer to bare steel. This conclusion is not supported by the results of Volkening[24], whose main interest was in the effect of chlorination and who shows that although corrosion increased with velocity of chlorinated sea water, when plain sea water was used velocity had little effect. There can be no doubt that painting will very much reduce the effect of water speed, as also will marine fouling or slime.

Excessive corrosion rates are commonly observed on those parts of a ship's hull which are exposed to high and turbulent flow of water, e.g. leading edges of rudders and shaft-brackets. The pitting found in these places is stimulated by selective local damage to paint films and possibly also by the proximity of a bronze propeller. The contribution of high water speed to this accelerated corrosion cannot be separately assessed. An indirect relation between water speed and corrosion arises from the fact that marine fouling organisms (in particular shell-fouling) do not settle if the water speed is more than about 1·5–2 m/s. Fouling may somewhat restrict general corrosion by its shielding effect, but may also cause crevice corrosion.

Impingement Attack

This form of attack, especially as affecting copper alloys in sea water, has been widely studied since the pioneer work of Bengough and May[25]. Impingement attack of sea water pipe and heat exchanger systems is considered in Sections 1.6 and 4.2. In such engineering systems the water flow is invariably turbulent and the thickness of the laminar boundary layer is an important factor in controlling localised corrosion.

At very high water speeds *cavitation-damage* (Section 8.8) is sustained by any metal; high-speed bronze propellers, for instance, may suffer seriously. This form of attack is mainly mechanical, although an element of true corrosion may be present, and is not specifically associated with sea water.

With respect to general corrosion, once a surface film is formed the rate of corrosion is essentially determined by the ionic concentration gradient across the film. Consequently the corrosion rate tends to be independent of water flow rate across the corroding surface. However, under impingement conditions where the surface film is unable to form or is removed due to the shear stress created by the flow, the corrosion rate is theoretically velocity (V) dependent and is proportional to the power $V^{1/2}$ for laminar flow and

$V^{2/3}$ under turbulent flow[37]. Under cavitation conditions the loss of metal in addition to corrosion may be mechanically induced and the velocity dependence has a higher power relationship[38,39] with values between V^3 and V^7, a popular mean value being V^6. It has been shown that impingement resistance is not just a simple property of a material with respect to turbulent flow but is dependent on polarisation characteristics under flow conditions[40] and hence is very susceptible to the bimetallic effect of coupling to more noble materials.

J. C. ROWLANDS

REFERENCES

1. Sverdrup, H. V., Johnson, M. W. and Fleming, R. H., *The Oceans*, Prentice-Hall, N.Y., 173 (1942)
2. Compton, K. G., *Corrosion*, **26**, 448 (1970)
3. As Reference 1, but p. 72
4. As Reference 1, but p. 127
5. Central Dockyard Laboratory, Portsmouth (unpublished)
6. LaQue, F. L., *Proc. Amer. Soc. Test. Mater.*, **51**, 541 (1951)
7. Cotton, J. B. and Downing, B. P., *Trans. Inst. Mar. Engrs.*, **69**, 311 (1959)
8. Southwell C. R. and Alexander A. L., *Materials Protection*, **9**, 14 (1970)
9. Fink, F. W., *Corrosion of Metals in Sea Water*, Battelle Memorial Inst., PB 171 344 (1960)
10. LaQue, F. L., *The Corrosion Handbook*, (Ed. H. H. Uhlig) Wiley, New York; Chapman and Hall, London, (2nd edition) 391 (1948)
11. LaQue, F. L., *Corrosion*, **6**, 162 (1958)
12. Ambler, H. R. and Bain, A. J., *J. Appl. Chem.*, **5**, 437 (1955)
13. Humble, H. A., *Corrosion*, **5**, 292 (1949)
14. Hudson, J. C., *J. Iron Steel Inst.*, **166**, 123 (1950)
15. Boudot, H. and Chaudron, G., *Rev. Métall.*, **43**, 1 (1946)
16. Forgeson B. W., Southwell, C. R. and Alexander, A. L., *Corrosion*, **16**, 105t (1960)
17. Evans, U. R. and Rance, V., *Corrosion and its Prevention at Bimetallic Contacts*, H.M.S.O., 3rd edn. (1963)
18. Southwell, C. R., Hummer, C. W. and Alexander, A. L., *Materials Protection*, **4**, 30 (1965)
19. Southwell, C. R., Alexander, A. L. and Hummer, C. W., *Materials Protection*, **7**, 41 (1968)
20. Bradley, J. N., *Inst. Metals Metallurgical Review* (1971)
21. Kirk, W. W., *Proc. Workshop Conf. on High Pressure Aquarian Systems*, N.A.C.E. (1971)
22. Schreiber, C. F., Osborn, O. and Coley, F. H., *Mat. Prot.*, **7**, 24 (1968)
23. Drisco, R. W. and Brouillett, C. V., *Mat. Prot.*, **32** (1966)
24. Volkening, V. B., *Corrosion*, **6**, 123 (1950)
25. Bengough, G. D. and May, R., *J. Inst. Met.*, **32**, 204 (1924)
26. Dexter, S. C. and Culberson, C., *Mat Perf.* **19**, 16 (1980)
27. Dietrich, G., *Atlas of the Hydrography of the Northern North Atlantic*, Pub. Conseil International Power l'Exploration de la Mer. Service Hydrographique, Charlottenlund Slot, Denmark (1969)
28. Min. of Agriculture, Fisheries and Food, *Atlas of the Seas Around the British Isles*, HMSO ISBN 0 907545 00 9 (1981)
29. Schumacher, M., *Seawater Corrosion Handbook*, Noyes Data Corporation, Park Ridge, USA (1979)
30. Elliott, S. *Metal Construction*, **16**, 20 (1984)
31. Knuckey, P. J., Private Communication (1984)
32. Morley, J., *I Corr ST Bulletin*, **19**, 2 (1981)
33. Katz, W., *Corrosion Data Sheets–Seawater*, DECHEMA, Frankfurt/Main (1976)
34. Oldfield, J. W. and Sutton, W. H., *Br Corros J*, **13**, 13 (1978)

35. Oldfield, J. W. and Sutton, W. H., *Br Corr J*, **13**, 104 (1978)
36. Kain, R. M. and Lee, T. S., 5th Int Conf Marine Corrosion and Fouling, Barcelona (1980)
37. Lush, P. A. *et al.*, Eurocor, **77**, *Soc Chem Ind*, 137 (1977)
38. Hutton, S. P., 2nd Int Conf on Cavitation, Edinburgh, I Mech E, **41** (1983)
39. Pylaev, N. I. and Sonikov, A. A., *Energomastinostroenie*, **12**, 4 (1972)
40. Rowlands, J. C. and Angell, B. A., UK Corr, 83, *I Corr ST*, **133** (1983)

2.5 Soil in the Corrosion Process

Introduction

Soil has been defined in many ways, often depending upon the particular interests of the person proposing the definition. In discussion of the soil as an environmental factor in corrosion, no strict definitions or limitations will be applied; rather, the complex interaction of all earthen materials will come within the scope of the discussion. It is obvious only a general approach to the topic can be given, and no attempt will be made to give full and detailed information on any single facet of the topic.

Soil is distinguished by the complex nature of its composition and of its interaction with other environmental factors. No two soils are exactly alike, and extremes of structure, composition and corrosive activity are found in different soils. Climatic factors of rainfall, temperature, air movement and sunlight can cause marked alterations in soil properties which relate directly to the rates at which corrosion will take place on metals buried in these soils.

Soil Genesis

The condition of any soil represents a stage in the changing process of soil evolution. Soils develop, mature and change with the passage of time. Whereas the time required for a true soil to develop from the parent rock of the earth may be thousands of years, rapid changes can result in a few years when soils are cultivated, irrigated, or otherwise subjected to man's manipulation. The type of soil that develops from the parent material will depend upon the various physical, chemical and biological factors of the environment.

The weathering process which eventually reduces the rock of the parent material to the inorganic constituents of soil comprises both physical and chemical changes. Size reduction from rocks to the colloidal state depends not only upon the mechanical action of natural forces but also on chemical solubilisation of certain minerals, action of plant roots, and the effects of organic substances formed by biological activity.

Interrelated with change in particle size and changes in type and kind of soil minerals present, organic matter is formed and accumulates as an integral part of the soil. Organic-matter content varies from practically none in sands to almost 100%, as exemplified by peat formations. The amount of organic matter present thus reflects the interaction of all environmental factors influencing chemical and biological activity. Whether the percentage of organic matter increases or decreases depends then upon the relation between the rate at which it is being formed by growth, death and accumulation of plant material, and that at which the microbiological activities within the soil are causing the decomposition of the complex organic molecules.

Moisture must be considered of primary importance in soil formation, in weathering, and in all of the changes taking place within the soil. The types of soil that form depend to a great extent upon the rainfall situation. Too little rainfall will prevent development of plant and animal life with their soil-building action. Too much moisture has a similar effect in preventing normal soil formation.

Closely associated with rainfall and climate is the acid or basic reaction which develops as a soil matures. When rainfall is high, water percolates through the soil, dissolving the soluble components, and leaching out alkaline minerals of the weathering rock. This happens whether a soil is developing from a naturally acidic or a naturally alkaline parent material. The end result is a shift in reaction to an acid condition. The degree to which this acidity develops depends upon many factors such as the parent minerals, biological activity, and temperature, related to the moisture situation. Should the loss of water from a soil be mainly by surface evaporation (as in arid regions), the dissolved salts tend to accumulate near the surface and alkaline conditions usually develop.

Although conditions of high rainfall and moderate to warm temperatures usually lead to an overall decrease in organic matter (particularly in cultivated soils), exceptions occur when the amount of water is great enough to prevent the adequate aeration necessary for maximum microbial activity. Swampy areas with peat and muck soils are the result. In a parallel manner, low temperatures of sub-polar regions slow down decomposition of organic materials and again highly organic soils develop.

The Corrosion Process in Soil

Although the soil as a corrosive environment is probably of greater complexity than any other environment, it is possible to make some generalisations regarding soil types and corrosion. It is necessary to emphasise that corrosion in soils is extremely variable and can range from the rapid to the negligible. This can be illustrated by the fact that buried pipes have become perforated within one year, while archaeological specimens of ancient iron have probably remained in the soil for hundreds of years without significant attack.

Corrosion in soil is aqueous, and the mechanism is electrochemical (see Section 1.4), but the conditions in the soil can range from 'atmospheric' to completely immersed (Sections 2.2 and 2.3). Which conditions prevail depends on the compactness of the soil and the water or moisture content. Moisture retained within a soil under field dry conditions is largely held within the capillaries and pores of the soil. Soil moisture is extremely significant in this connection, and a dry sandy soil will, in general, be less corrosive than a wet clay.

Although the mechanism will be essentially electrochemical, there are many characteristic features of soil as a corrosive environment which will be considered subsequently; it can, however, be stated here that the actual corrosiveness of a soil will depend upon an interaction between rainfall, climate and soil reaction.

A characteristic feature of the soil is its heterogeneity. Thus variation in soil composition or structure can result in different environments acting on different parts of the same metal surface, and this can give rise to differing electrical potentials at the metal/soil interface. This will result in the establishment of predominantly cathodic or predominantly anodic areas, and the consequent passage of charge through the metal and through the soil. Differences in oxygen concentration (differential aeration), or differences in acidity or salt concentrations may thus give rise to corrosion cells. The distance of the separation of the anodic and cathodic areas can range from very small to miles ('long-line' corrosion).

The conductivity of the soil is important as it is evident from the electrochemical mechanism of corrosion that this can be rate-controlling; a high conductivity will be conducive of a high corrosion rate. In addition, the conductivity of the soil is important for 'stray-current corrosion' (see Section 10.5), and for cathodic protection (Chapter 10).

Properties of Soils Related to Corrosion

Soil Texture and Structure

Soils are commonly named and classified according to the general size range of their particulate matter. Thus sandy, silt and clay types derive their names from the predominant size range of inorganic constituents. Particles between 0·07 and about 2 mm are classed as sands. Silt particles range from 0·005 mm to 0·07, and clay particle size ranges from 0·005 mm mean diameter down to colloidal matter.

The proportion of the three size groups will determine many of the properties of the soil. Although a number of systems have been used to classify soils as to texture, the one shown in Fig. 2.17 represents commonly used terminology for various proportions of sand, silt and clay.

Since soils contain organic matter, moisture, gases and living organisms as well as mineral particles, it is apparent that the relative size range does not determine the whole nature of the soil structure. In fact most soils consist of aggregates of particles within a matrix of organic and inorganic colloidal matter rather than separate individual particles. This aggregation gives a crumb-like structure to the soil, and leads to friability, more ready penetration of moisture, greater aeration, less erosion by water and wind, and generally greater biological activity. The loss of the aggregated structure can occur as the result of mechanical action, or by chemical alteration such as excess alkali accumulation. Destruction of the structure or 'puddling' greatly alters the physical nature of the soil.

Mention should be made of the soil profile (section through soil showing various layers) because it is important to recognise that the soil's surface

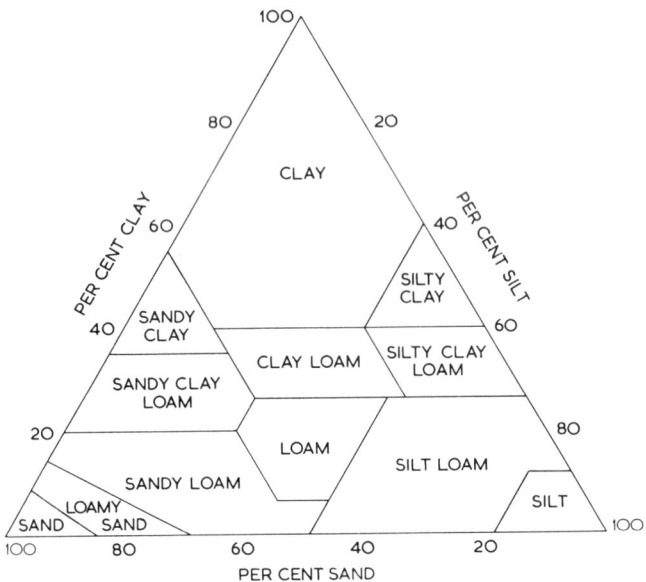

Fig. 2.17 Proportions of sand, silt and clay making up the various groups of soils classified on the basis of particle size.

gives a very poor indication of the underlying strata. Pipe-lines are buried several feet below surface soils and corrosion surveys based on surface observations give little information as to the actual environs of the pipe when buried.

The Clay Fraction

Clays make up the most important inorganic constituents of soil. They consist of various minerals depending on the mineral composition of the parent material, and on the type and degree of weathering. Often clays may be grouped in a family series, depending upon the weathered condition, as, for example, montmorillonite → illite → kaolinite. Weathering of montmorillonite causes loss of potassium and magnesium which alters the crystalline structure, and eventually kaolinite results. In this example (and also for other clay mineral groups) marked changes occur in the physical properties of a soil as clay minerals undergo the weathering process.

Montmorillonite clays absorb water readily, swell greatly and confer highly plastic properties to a soil. Thus soil stress (Section 14.8) occurs most frequently in these soils and less commonly in predominantly kaolinitic types. Similarly, a soil high in bentonite will show more aggressive corrosion than a soil with a comparable percentage of kaolinite. A chalky soil usually shows low corrosion rates. Clay mineralogy and the relation of clays to corrosion deserves attention from corrosion engineers. Many important relationships are not fully understood and there is need for extensive research in this area.

Aeration and Oxygen Diffusion

The pore space of a soil may contain either water or a gaseous atmosphere. Thus the aeration of a soil is directly related to the amount of pore space present and to the water content. Soils of fine texture due to a high clay content contain more closely packed particles and have less pore capacity for gaseous diffusion than an open-type soil such as sand.

Oxygen content of soil atmosphere is of special interest in corrosion. It is generally assumed that the gases of the upper layers of soil are similar in composition to the atmosphere above the soil, except for a higher carbon dioxide content. Relatively few data are available showing oxygen content of soils at depths of interest to the corrosion engineer. Judging by the fact that plant roots require oxygen to penetrate a soil, however, it may be assumed that soil gases at depths of 6 m or more contain significant amounts of oxygen.

Diffusion of gases into soil is enhanced by a number of climatic factors. Temperature changes from day to night conditions cause expansion and contraction of the surface-soil gases. Variation in barometric pressure has a bellows-like effect on gaseous diffusion. To illustrate the magnitude of this diffusion rate on a large scale, it may be recalled that air within the more than 43 km of underground passages of the Carlsbad Caverns in New Mexico undergoes a complete change each day, despite the fact that the single opening of these caverns to the surface is only a metre or so in diameter.

Biological activity within the soil tends to decrease the oxygen content and replace the oxygen with gases from metabolic activity, such as carbon dioxide. Most biological activity occurs in the upper 150 mm of soil, and it is in this region that diffusion would be most rapid. Factors which tend to increase microbial respiration, such as the addition of large amounts of readily decomposed organic matter, or factors which decrease diffusion rates (water saturation) will lead to development of anaerobic conditions within the soil. The significant microbiological relationships to corrosion under both aerobic and anaerobic situations are discussed in Section 2.6.

Water Relations

No corrosion occurs in a completely dry environment. In soil, water is needed for ionisation of the oxidised state at the metal surface. Water is also needed for ionisation of soil electrolytes, thus completing the circuit for flow of a current maintaining corrosive activity. Apart from its participation in the fundamental corrosion process, water markedly influences most of the other factors relating to corrosion in soils. Its rôle in weathering and soil genesis has already been mentioned.

Types of Soil Moisture

1. *Free ground water*. At some depth below the surface, water is constantly present. This distance to the water table may vary from a few metres to hundreds of metres, depending upon the geological formations present.

Only a small amount of the metal used in underground service is present in the ground water zone. Such structures as well casings and under-river pipelines are surrounded by ground water. The corrosion conditions in such a situation are essentially those of an aqueous environment.

2. *Gravitational water.* Water entering soil at the surface from rainfall or some other source moves downward. This gravitational water will flow at a rate governed largely by the physical structure regulating the pore space at various zones in the soil profile. An impervious layer of clay, a 'puddled' soil, or other layers of material resistant to water passage may act as an effective barrier to the gravitational water and cause zones of water accumulation and saturation. This is often the situation in highland swamp and bog formation. Usually gravitational water percolates rapidly to the level of the permanent ground water.

3. *Capillary water.* Most soils contain considerable amounts of water held in the capillary spaces of the silt and clay particles. The actual amount present depends upon the soil type and weather conditions. Capillary moisture represents the important reservoir of water in soil which supplies the needs of plants and animals living in or on the soil. Only a portion of capillary water is available to plants. 'Moisture-holding capacity' of a soil is a term applied to the ability of a soil to hold water present in the form of capillary water. It is obvious that the moisture-holding capacity of a clay is much greater than that of a sandy type soil. Likewise, the degree of corrosion occurring in soil will be related to its moisture-holding capacity, although the complexities of the relationships do not allow any quantitative or predictive applications of the present state of knowledge.

Significance of Fluctuations in Water Content

Except for zones below the level of permanent ground water where the environment is water-saturated, and for zones of dry surface sand, continual variation may be expected to occur in the water content of soils. This is usually dependent on rainfall, snow, flooding and such climatic influences, though irrigation practices in many agricultural areas influence water content and hence the corrosion rates.

Water losses from the soil represent the sum of downward movement of gravitational water and surface losses by evaporation. Man's activities, other than drainage procedures or long-term water use from pumps in industrial areas, do not usually influence the downward movement of water. On the other hand, agricultural practices have a great effect on surface evaporation losses.

As mentioned earlier, there is an inverse relationship between water volumes and oxygen concentration in soil. As soils dry, conditions become more aerobic and oxygen diffusion rates become higher. The wet-dry or anaerobic-aerobic alternation, either temporal or spatial, leads to higher corrosion rates than would be obtained within a constant environment. Oxygen-concentration-cell formation is enhanced. This same fluctuation in water and air relations also leads to greater variation in biological activity within the soil.

Chemical Properties of Soils

Soil reaction (pH) The relationship between the environment and development of acid or alkaline conditions in soil has been discussed with respect to formation of soils from the parent rock materials. Soil acidity comes in part by the formation of carbonic acid from carbon dioxide of biological origin and water. Other acidic development may come from acid residues of weathering, shifts in mineral types, loss of alkaline or basic earth elements by leaching, formation of organic or inorganic acids by microbial activity, plant root secretions, and man-made pollution of the soil, especially by industrial wastes.

As with other factors, no direct statements can be made relating the reaction of a soil to its corrosive properties. Extremely acid soils (pH 4·0 and lower) can cause rapid corrosion of bare metals of most types. This degree of acidity is not common, being limited to certain-bog soils and soils made acid by large accumulations of acidic plant materials such as needles in a coniferous forest. Most soils range from pH5·0 to pH8·O, and corrosion rates are apt to depend on many other environmental factors rather than soil reaction *per se*. The 45-year study of underground corrosion conducted by the United States Bureau of Standards[1] included study of the effect of soils of varying pH on different metals, and extensive data were reported.

Soluble salts of the soil Water in the soil should most properly be considered as the solvent for salts of the soil; the result being the soil solution. In temperate climates and moderate rainfall areas, the soil solution is relatively dilute, with total dissolved salts ranging from 80 to 1 500 p.p.m.[2]. Regions of extensive rainfall show lower concentrations of soluble salts as the result of leaching action. Conversely, soils in arid regions are usually quite high in salts as these salts are carried to the surface layers of the soil by water movement due to surface evaporation.

Generally, the most common cations in the soil solution are potassium, sodium, magnesium and calcium. Alkali soils are high in sodium and potassium, while calcareous soils contain predominantly magnesium and calcium. Salts of all four of these elements tend to accelerate metallic corrosion by the mechanisms mentioned. The alkaline earth elements, calcium and magnesium, however, tend to form insoluble oxides and carbonates in non-acid conditions. These insoluble precipitates may result in a protective layer on the metal surface and reduced corrosive activity.

The anionic portions of the soil solution play a rôle of equal importance to the cations. The anions function in the manner outlined for cations in conductivity and concentration-cell action, and have an additional action if they react with the metal cation and form insoluble salts. Thus, if the metal is lead and the predominant anion is sulphate, a layer of insoluble lead sulphate may precipitate on the metal surface and form an effective barrier against further loss of metal.

Another important relationship between the salts of the soil and corrosion has to do with biological activity. Since the growth of plants and microorganisms depends upon the proper inorganic mineral nutrients, the action of these forms of life varies with the mineral content of the soil. While many of the possible indirect effects, such as the role of various nitrogenous materials in bacterial growth and corrosion, have not as yet been studied, one well-documented situation is known. This relationship of sulphur and sulphates to bacterial activity in corrosion is fully discussed in Section 2.6.

The salts content of soils may be markedly altered by man's activities. The effect of cathodic protection will be discussed later in this section. Fertiliser use, particularly the heavy doses used in lawn care, introduces many chemicals into the soil. Industrial wastes, salt brines from petroleum production, thawing salts on walks and roads, weed-killing salts at the base of metal structures, and many other situations could be cited as examples of alteration of the soil solution. In tidal areas or in soils near extensive salt deposits, depletion of fresh ground-water supplies has resulted in a flow of brackish or salty sea water into these soils, causing increased corrosion.

The Environment of the Pipe-line Ditch

A comprehensive study of the soil and microbial situation in the backfilled zone of pipe-line ditches has shown a number of significant facts[3-6]. The results of over a thousand bell-hole studies along operating oil and gas pipe-lines in widely separated geographical areas of the United States has led to the conclusion that the pipe-line ditch represents a marked disturbance of the

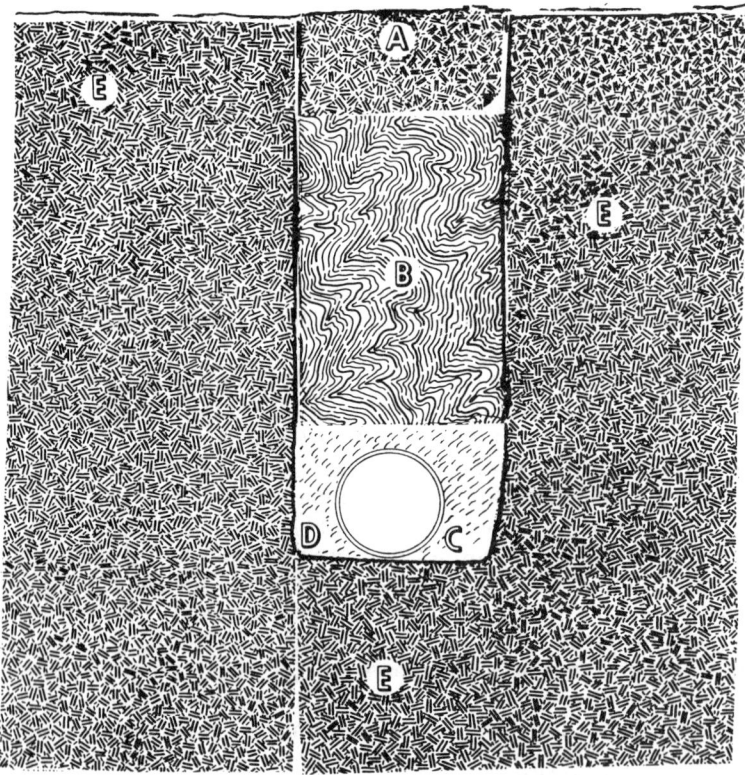

Fig. 2.18 Cross-section of soil and backfill areas surrounding underground pipe-lines

natural soil situation. Figure 2.18 indicates the general zones of interest. The operation of ditching and back-filling has resulted in a zone above the pipe (*B*) which never settles but remains less compact than the undisturbed soil. In this zone, water may penetrate and leave more rapidly. Aeration is more efficient, as shown by the presence of strictly aerobic bacteria in abundant population.

The bottom of the pipe-line ditch (*C* and *D*, Fig. 2.18) has a higher moisture content than undisturbed soil at comparable depths. Many instances of free water at the ditch bottom were reported. Differences at the surface between backfill (A) and soil of the right-of-way (*E*) were less and tended to decrease with passage of time. Conclusive evidence was obtained indicating a greatly increased activity of bacteria in the backfill zone. Some of this may have resulted from the mixing of surface and sub-surface soils during ditching and back-filling. High populations of bacteria adjacent to the organic matter of coating or coating and wrapped-in systems on the external surface of the pipe indicated that these organic compounds served as an available food supply. The presence of hydrocarbon-utilising bacteria was a common finding particularly when the soil was in contact with asphalt protective coatings.

These concepts of the altered soil situation in the pipe-line ditch have important implications to the corrosion process. The increased aeration, the high moisture of the ditch bottom, presence of organic matter in coatings, and high microbial populations all lead to greatly increased possibilities for the development of heterogeneity and the formation of zones differing in oxidation-reduction potentials. The action of moisture and micro-organisms on asphaltic coatings, unbonding of the coating[6], and formation of cathodic and anodic areas on metal surfaces are all directly related to the disturbed environment of the pipe-line ditch.

Cathodic Protection and Soil Properties

The modern procedure to minimise corrosion losses on underground structures is to use protective coatings between the metal and soil and to apply cathodic protection to the metal structure (see Chapter 11). In this situation, soils influence the operation in a somewhat different manner than is the case with unprotected bare metal. A soil with moderately high salts content (low resistivity) is desirable for the location of the anodes. If the impressed potential is from a sacrificial metal, the effective potential and current available will depend upon soil properties such as pH, soluble salts and moisture present. When rectifiers are used as the source of the cathodic potential, soils of low electrical resistance are desirable for the location of the anode beds. A protective coating free from holidays and of uniformly high insulation value causes the electrical conducting properties of the soil to become of less significance in relation to corrosion rates (Section 15.8).

Effect of cathodic protection on soils Long-term application of an electrical potential to the metal structure with resulting flow of electrical current through the soil has two noticeable effects, the magnitude of which will be in proportion to the time and amount of current passing through the soil.

The most commonly observed effect of current flow is the development of alkaline conditions at the cathode. On bare metal this alkaline zone may exist only at the metal surface and may often reach pH values of 10 to 12. When the soil solution contains appreciable calcium or magnesium these cations usually form a layer of carbonate or hydroxide at the cathodic area. On coated lines the cations usually move to holidays or breaks in the coating. On failing asphalt or asphalt mastic type coatings, masses of precipitated calcium and magnesium often form nodules or tubercles several centimetres in diameter.

The existence of an electrical potential causes not only cation and anion movement but also migration of moisture toward the cathode. This movement of water (electroendosmosis) is due to the asymmetrical nature of the polar groups of the water molecule. In arid regions water leaving the anode area may cause the soil surrounding the anodes to become so dry that proper current densities cannot be maintained along the line. To alleviate this, some pipe-line companies have had to transport water into desert areas to re-moisten anode beds.

Moisture films are frequently found under unbonded protective coatings of asphalt and plastic tapes. The nature and origin of this water is still unknown but is of great interest because of its relationship to bond failure, microbial utilisation of asphalt and hydrocarbons, and efficiency of cathodic protection[6].

Long-line currents As the result of the use of protective coatings and cathodic protection, present-day pipe-lines are usually constructed of welded joints and the line forms a continuous conductor rather than a series of insulated sections. This situation led to the finding of the so-called long-line currents. Often low currents of medium to high voltage have been observed. The cause and significance of this phenomenon is not known. Theories as to the origin of these currents are:

1. *Pick-up of stray current (a.c. or d.c.)* (Section 10.5). Decreased use of d.c. in many areas has led to less possibilities of pick-up of direct current from utilities, mines, etc. The importance of grounded a.c. systems has been discounted, but Waters[7] has shown that alternating currents can accelerate corrosion. Furthermore the rectifying effects of oxide films, clay minerals and other soil factors are not understood.

2. *Current induction due to earth's rotation.* The long lines act as conductors, and variations in magnetic flux could cause induced currents. A few studies have shown long-line current activity to be greater at high activity of the aurora borealis, which is known to be related to earth's magnetism. The existence of extremely long-length waves of electromagnetic radiation[8] gives another possibility.

3. *Atmospheric lightning.* Discharges of static electricity in the various forms of lightning represent high potentials of extremely short duration. The dissipation of this potential through the earth's crust may well be the origin of the long-line currents.

4. *Differences in soil potential.* Since pipe-lines pass through zones of aerated and unaerated soil and possibilities for electrolytic-cell formation are great, the observed currents may have resulted from soil dissimilarities.

Those interested in long-line currents are referred to the publications of Gish[9], Gill and Rogers[10] and Mudd[11].

Methods Proposed for Evaluation of Soil Corrosivity

Because corrosion rates of metals buried in soil show extreme variation, a test procedure to indicate the expected corrosion activity of a given soil would be extremely valuable. The discussion on the heterogeneity of soils, however, indicates the complex nature of the situation and thus also suggests that the likelihood of success of a single survey-type procedure would be slight. Many types of tests have been suggested. Certain ones are in use by corrosion engineers, and others remain to receive further study. The various types of survey vary from the observations of Denison and Ewing[12] that corrosivity of Ohio soils could be related to colour and texture, to complex laboratory testing equipment. It is obvious that a useful test procedure should be relatively rapid and capable of field use, show small changes in environmental relations, and give results predicting relative corrosion rates. In the paragraphs which follow, only the general nature of the test will be discussed, and the reader is referred to Sections 10.4 and 10.7.

Soil resistivity The rôle of soil in the electrical circuitry of corrosion is now apparent. Thus the conductivity of the soil represents an important parameter. Soil resistivity has probably been more widely used than any other test procedure. Opinions of experts vary somewhat as to the actual values in terms of ohm centimetres which relate to metal-loss rates. The extended study of the US Bureau of Standards[1] presents a mass of data with soil-resistivity values given. A weakness of the resistivity procedure is that it neither indicates variations in aeration and pH of the soil, nor microbial activity in terms of coating deterioration or corrosion under anaerobic conditions. Furthermore, as shown by Costanzo[13], rainfall fluctuations markedly affect readings. Despite its short comings, however, this procedure represents a valuable survey method. Scott[14] points out the value of multiple data and the statistical nature of the resistivity readings as related to corrosion rates (see also Chapter 10).

Oxidation-reduction potential Because of the interest in bacterial corrosion under anaerobic conditions, the oxidation-reduction situation in the soil was suggested as an indication of expected corrosion rates. The work of Starkey and Wight[15], McVey[16], and others led to the development and testing of the so-called *redox* probe. The probe with platinum electrodes and copper sulphate reference cells has been described as difficult to clean. Hence, results are difficult to reproduce. At the present time this procedure does not seem adapted to use in field tests. Of more importance is the fact that the data obtained by the redox method simply indicate anaerobic situations in the soil. Such data would be effective in predicting anaerobic corrosion by sulphate-reducing bacteria, but would fail to give any information regarding other types of corrosion.

Electrolytic method This procedure is also known as the Williams Corfield test[17]. It is based on loss of metal from iron electrodes buried in a water-saturated soil through which current from a 6-V battery is passed. It does not reflect field conditions and depends upon soil conductance under saturated conditions.

Polarisation-curve procedures The Denison[18] method is to measure the current at various degrees of polarisation of metal in soil in a special cell. While this test is considered quite accurate, it has the disadvantage that the measurements are made in the laboratory and cannot be made in the field.

Combination electrical methods Tomashov and Mikhailovsky[19] describe a method developed in the Soviet Union. This test is essentially a combination of resistivity measurement and polarisation rates on iron electrodes in soil *in situ*. The usefulness and value of this procedure has not as yet been determined by practical application by corrosion engineers. The development of this combination test does, however, represent an attempt to integrate some of the complex factors controlling corrosion rates in soil. Much more research on these factors and methods of measurement should in the future enable the corrosion engineer to evaluate soil properties with respect to application of corrosion-alleviating operations.

Polarisation-resistance method The polarisation-resistance method (see Section 20.1) has been used for determining corrosion rates of metals buried underground.

Soil Corrosivity Assessment

The development of soil corrosivity assessment techniques has largely been due to the pipeline industry's requirements for better corrosion risk assessment and the reduction of pipeline failures. Corrosion in soil is a complex process and over the years several parameters have been identified as having a significant effect on the corrosion rate in a given soil.

Measurement of some of these parameters identifies the risk of a particular type of corrosion, for example pH measurements assess the risk of acid attack and redox potential measurements is used to assess the suitability of the soil for microbiological corrosion, a low redox potential indicates that the soil is anaerobic and favourable for the life cycle of anaerobic bacteria such as to sulphate-reducing bacteria. Other measurements are more general, resistivity measurements being the most widely quoted. However, as yet no single parameter has been identified which can confidently be expected to assess the corrosion risk of a given soil. It is therefore common practice to measure several parameters and make an assessment from the results.

Most of the accepted corrosivity assessment techniques have been outline above. Some of the techniques are used widely, others are more controversial. However, it must be accepted that even with a combination of available techniques no corrosivity assessment survey will accurately predict the corrosion rate for metals in every soil.

Table 2.20 Soil Corrosivity Assessment Technique from the German Gas and Water Works Engineers' Association Standard (DVGW GW9)

Item	Measured Value	Marks
Soil composition	Calcareous, marly limestone Sandy marl, not stratified sand	+2
	Loam, sandy loam (loam content 75% or less) marly loam, sandy claysoil (silt content 75% or less)	0
	Clay, marly clay, humus,	−2
	Peat, thick loam, marshy soil	−4
Ground-water level at buried position	None	0
	Exist	−1
	Vary	−2
Resistivity	10,000 ohm. cm or more	0
	10,000–5,000	−1
	5,000–2,300	−2
	2,300–1,000	−3
	1,000 or less	−4
Moisture content	20% or less	0
	20% or more	−1
pH	6 or more	0
	6 or less	−2
Sulphide and hydrogen sulphide	None	0
	Trace	−2
	Exist	−4
Carbonate	5% or more	+2
	5–1	+1
	1 or less	0
Chloride	100 mg/kg or less	0
	100 mg/kg more	−1
Sulphate	200 mg/kg or less	0
	200–500	−1
	500–1,000	−2
	1,000 or more	−3
Cinder and coke	None	0
	Exist	−4

Soil is regarded as non-corrosive if the total of the above is 0 or higher; Slightly corrosive if 0 to −4; corrosive if −5 to −10 and very corrosive if −10 or less.

Corrosion when it occurs can take many different forms from general uniform attack to pitting corrosion. In a given situation some forms of corrosion are more deleterious to the metal structures than others. Pitting corrosion, although overall weight loss is small, is more likely to lead to early failure of a pipeline than uniform corrosion, with a considerably higher overall weight loss.

Although certain conditions very often lead to a particular type of attack, attempts to categorise soil corrosion in this way cannot be made on a general basis and most corrosivity assessment techniques categorise the soil as reacting to bare steel or iron in one of four ways:

Non aggressive
Mildly aggressive
Aggressive
Very aggressive

Varying degrees of emphasis can be placed on certain parameters and this results in a variety of techniques. This preference for some tests over others is due to a number of reasons including:

1. The metal under consideration, although usually iron or steel, may vary in its resistance to attack.
2. Some industries are interested only in certain aspects of soil assessment and do not require a detailed assessment.
2. Some of these tests are not applicable to soils encountered in that industry.
3. The responsible engineer does not have confidence in some of the techniques.

Perhaps the most widely known measurement technique is that adopted by the West German Gas Industry[20] and developed by Steinrath[21] for buried pipework. This assigns a value (See Table 2.20) to each parameter measured; the summation of these values determines the corrosivity of the soil. The parameters measured are shown in Table 2.20. Although this technique was developed for the pipeline industry it can be used with some success for general soil corrosivity assessment.

J. O. HARRIS
D. EYRE

REFERENCES

1. Romanoff, M., *Underground Corrosion*, Nat. Bur. Stand., Circular No. 579, Washington (1957)
2. Russell, E. J., *Soil Conditions and Plant Growth*, Longmans, London (1932)
3. Harris, J. O., *Kansas Agric. Exp. Sta. Tech. Bull.*, No. 102, Manhattan (1959)
4. Harris, J. O., *Corrosion*, **16**, 149 (1960)
5. Harris, J. O., *Proceedings*, Sixth Ann. Appalachian Underground Corrosion Short Course, West Virginia Univ., Morgantown, 198 (1961)
6. Harris, J. O., *Proc. Pacific Cst. Gas Ass.*, **52**, 109 (1961)
7. Waters, F. O., *Mater. Protect.*, **1** No. 3, 26 (1962)
8. Heirtzler, J. R., *Sci. Amer.*, **206**, 128 (1962)
9. Gish, O. H., *Sci. Mon., N. Y.*, **32**, 5 (1930)
10. Gill, S. and Rogers, W., *Physics*, **1**, 194 (1931)
11. Mudd, O. C., *Oil Gas J.*, **38**, 48 (1939)
12. Denison, I. A. and Ewing, S. P., *Soil Sci.*, **40**, 287 (1935)
13. Costanzo, F. E., *Corrosion*, **14**, 363 (1958)
14. Scott, G. N., *Corrosion*, **14**, 396 (1958)
15. Starkey, R. L. and Wight, K. M., *Anaerobic Corrosion of Iron in Soil*, Amer. Gas Assoc., New York (1945)
16. McVey, R. E., Proceedings of the Fifth Ann. Appalachian Underground Corrosion Short Course, West Virginia Univ., Morgantown, 23 (1960)
17. Corfield, G., *West. Gas.*, **7**, 123 (1930)
18. Denison, I. A., *Nat. Bur. Stand. J.*, **17**, 363 (1936)
19. Tomashov, N. D. and Mikhailovsky, Y. N., *Corrosion*, **15**, 77 (1959)
20. German Gas and Water Works Engineers' Association Standard, *Merkblatt für die Beurteilung der Korrosionsgefährdung von Eisen und Stahl im Erdboden*, DVGW GW9, Frankfurt, DVGW (1971)
21. Steinrath, H., *Untersuchungsmethoden Zur Beurteilung der Aggressivität von Boden*, Frankfurt, DVGW (1966)

2.6 The Microbiology of Corrosion

The role of microbes in the corrosion of metals is due to the chemical activities (metabolism) associated with the microbial growth and reproduction[1]. Under favourable growth conditions doubling times of 60–120 min are common. By reason of such rapid growth the onset of changes may be sudden, and even when apparently supressed by mechanical or chemical cleaning often return because a residual low number of living organisms rapidly grow again when favourable conditions are restored[2]. These characteristics are typical of widespread biodeterioration caused by microbes in all industries of which corrosion is a special case. With a few exceptions such as synthetic polymers, all materials including natural products such as cotton, wood, rubber and oils, and man-made materials such as concrete, complex organic chemicals and metals, can be attacked. Rarely a single microbial species is involved, but usually biodeterioration, including corrosion, results from an association of a number of different microbes. For instance, a rapid growth of an aerobic organism may so deplete oxygen that strictly anaerobic sulphate-reducers associated with cathodic depolarisation then appear, and metallic corrosion results. In many years the complicated associations in such microbial ecosystems have become increasingly recognised. Their corrosive effects on metals can be attributed to the removal of electrons from the metal and formation of a corrosion products by:

(a) Direct chemical action of metabolic products such as sulphuric acid, inorganic or organic sulphides and chelating agents such as organic acids.
(b) Cathodic depolarisation associated with anaerobic growth.
(c) Changes in oxygen potential, salt concentration, pH, etc. which establish local electrochemical cells.
(d) Removal of corrosion inhibitors (oxidation of nitrite or amines) or protective coatings (bitumen on buried pipes).
(e) The presence of the biomass itself or residues of biomass such as hygroscopic salt deposits from cells burnt-on in annealing.

It must be stressed that the identification of the causative organism(s) may be extremely difficult since it often depends on the quantitative determination of numbers of each microbial type in a complex mixture together with an assessment of its chemical and physical activities in that particular environment.

Although bacteria may predominate, moulds, yeasts and protozoa may be associated with bacteria, or, under some conditions, may either cause corrosion by themselves, or modify it drastically.

Although many of the effects of microbes on metal are associated with growth this is not necessarily so because a biomass once established may cease to increase but continue its chemical activities often at an accelerated rate, once the controls on growth are relaxed.

Methods of protecting materials against microbial corrosion include:

(a) Coatings, particularly of resistant synthetic polymers or paints containing inhibiting salts (e.g. Cu^{2+}, Cr^{3+} and Zn^{2+}).
(b) Controlled dosing with appropriate biocides.
(c) Changes in environmental conditions unfavourable to microbial growth, e.g. removal of water from lubricating or fuel oils, good industrial housekeeping, temperature changes.
(d) Designs based on fundamental knowledge of microbial ecology. This implies co-operation between engineers and biologists aimed at reducing infection and maintaining unfavourable conditions for microbial growth.

Acid Corrosion

Massive and rapid corrosion of metal, concrete and limestones under aerobic conditions is usually caused by the action of sulphuric acid formed by the oxidation of sulphur or sulphide by members of the genus Thiobacillus[3]. The majority of this genus grow by assimilating CO_2 at the expenditure of energy produced by the oxidation of sulphur, sulphite, thiocyanate and tri- or tetrathionate; some strains are sensitive to low concentrations of H_2S. The oxidation of sulphur may produce a concentration of up to 1-2% H_2SO_4 and it is this that produces corrosion; these organisms are also exploited for ore leaching[4] and in the biological treatment of coke oven effluents[5]. Thiobacillus thio-oxidans commonly occurs in soil and water and is to be suspected where corrosion is associated with very low pH in the immediate vicinity of the metal[6]. It may be isolated on acid media and reliably estimated by plate counting on a solid medium.

Acid production and corrosion associated with pyritic deposits is caused by Ferrobacillus ferro-oxidans[7].

Detection T. thio-oxidans is best detected by the strongly acid conditions it generates in a mineral salt solution on which sulphur is floating.

Prepare and sterilise by steaming a medium of $(NH_4)_2SO_4$, 0.2 to 0.4 g; KH_2PO_4, 3 to 4 g; $CaCl_2$, 0.25 g; $MgSO_4$, 0.5 g; $FeSO_4$, 10 mg; sulphur 10 g; tap water to 1000 ml; pH $5+/-0.3$ (some authorities recommend a trace metal mixture in place of the ferrous salt). Add 1 ml or 1 g of material to be tested to 100 ml of medium in a conical flask and incubate in air at 30°C. From four days to two weeks the pH of samples should be measured at intervals. An abrupt drop to 2.5 or lower indicates growth of T. thio-oxidans. Little turbidity appears in the medium; under a microscope the sulphur particles are seen to be surrounded by motile stubby Gram-negative rods.

Estimation[8] *The above medium is reinforced with 10 g/l of thiocyanate, sulphur is omitted and it is prepared as pour plates by the addition of 3% agar. Organisms other than Thiobacilli will grow from spread samples, but the Thiobacilli are easily distinguished by sulphur haloes (see Fig. 2.19).*

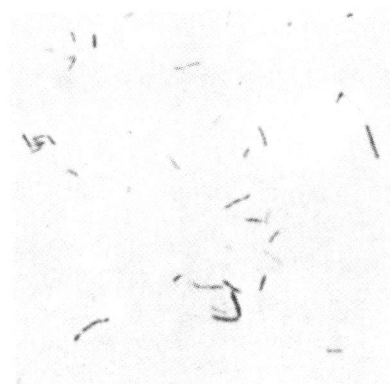

Fig. 2.19 *Thiobacillus thio-oxidans* (NCIB 8 342). Usually stubby rods, but a few elongated forms can be seen (these are most common in old cultures), × 260 (Crown copyright courtesy Microbiological Research Establishment)

Examples Parker[9] described the role of Thiobacilli in the corrosion of concrete sewer pipes; evolution of hydrogen sulphide from the sewage leads to the deposition of inorganic sulphur compounds on the roof of the pipe, and these are oxidised to sulphuric acid by the bacteria, causing a characteristic corrosion pattern in which the roof of the pipe becomes decayed. Iron pipes carrying polluted effluents, and concrete manhole covers examined at the National Chemical Laboratory have been corroded for similar reasons; corrosion of Mouchel cooling towers[10] had a like origin, as had corrosion of buried iron gas mains in south London[11]. Corrosion of statues in France has been partly attributed to Thiobacilli[12], corrosion of buildings and stonework probably has a similar origin, the source of sulphur for the bacteria being atmospheric pollution. Occasionally, in the experience of the National Chemical Laboratory, ornamental cements containing sulphur have been used for facing buildings, and these form ideal substrates for Thiobacilli. Materials containing sulphur have been used for jointing water mains and Thiobacilli may cause corrosion by forming acid from them[13]. Instances in which vulcanised rubber has been corroded by these bacteria are known[14]. Corrosion by Thiobacilli is probably more widespread than the published examples would indicate.

Prevention Methods of prevention already summarised may be used singly or in combination. Elimination of sulphur and certain of its compounds are most effective but more recently more resistant materials such as polythene or asbestos are used to replace iron and concrete where acid corrosion in severe.

Corrosion by Ferrobacillus

F. ferro-oxidants[15] is capable of accelerating the oxidation of pyritic (FeS_2) deposits at acid pH values. It is usually found in association with Thiobacillus and was known as Thiobacillus ferroxidans before the distinction between the two organisms was appreciated. It is responsible for pollution problems arising from acid waters in gold and bituminous coal mines[16, 17]; such waters are corrosive to pumping machinery[18] and mining installations (see Fig. 2.20).

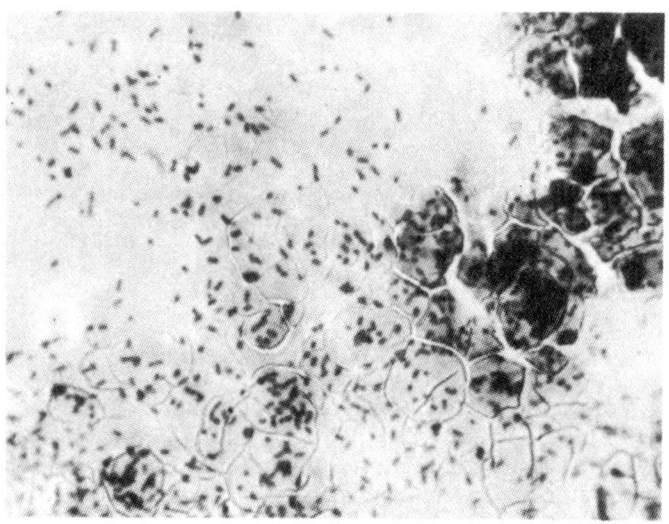

Fig. 2.20 *Ferrobacillus ferro-oxidans* (NCIB 8 451), bacteria and encrustations of ferric oxides; the proportion of bacteria was much increased by filtering and centrifugation, × 260. (Crown copyright courtesy Microbiological Research Establishment)

Detection Corrosive waters formed by these bacteria have a pH in the region of 2 to 3, show a brown deposit of basic ferric sulphate, and contain free sulphuric acid.

Prepare and sterilise by heating a medium of $(NH_4)_2SO_4$, 0.15 g; KCl, 0.05 g; $MgSO_4 7H_2O$, 0.5 g; KH_2PO_4, 0.05 g; $Ca(NO_3)_2$, 0.01 g in 1000 ml of tap water. Prepare a 10% solution of $FeSO_4 7H_2O$ and sterilise by filtration. Add 1 ml $FeSO_4$ solution to 100 ml of medium in a 250 ml conical flask, inoculate with the material being examined, check that the pH is approximately 3.5, and incubate in air at room temperature. Growth of F. ferro-oxidans is indicated by formation of a brown precipitate as compared with slow, slight browning in an uninoculated culture; the organism grows very slowly and may take up to a month to show unequivocal results; the responsible bacteria are straight rods, often difficult to see microscopically amid the debris.

Prevention Neutralisation of acid waters with lime has been recommended but the resulting sulphate-laden water may present a disposal problem. Use of acid-resistant machinery and pipes is more satisfactory.

Mechanism and sulphur oxidation Apart from its intrinsic interest the economic importance of acid corrosion and more lately interest in ore leaching, has stimulated considerable work on the oxidation of sulphur, Fe^{2+} and Mn^{2+}. It must be stressed that the Thiobacilli are obligate aerobes, i.e. that depend on molecular oxygen as a terminal electron acceptor. Possible reactions for the oxidation of sulphur are[19]:

$$4S^- \rightarrow 2S_2O_3^- \rightarrow S_4O_6^- \rightarrow SO_3^- + S_3O_6 \rightarrow 4SO_3^- \rightarrow 4SO_4^{2-}$$

The central role of SO_3^- is apparent.

Reactions leading to SO_4^{2-} formation from pyrites are possibly:

$$2FeS_2 + 7O_2 + 2H_2O \rightarrow 2FeSO_4 + 2H_2O \qquad \ldots(2.8)$$

$$4FeSO_4 + O_2 + H_2SO_4 \rightarrow 2Fe_2(SO_4)_3 + 2H_2O \qquad \ldots(2.9)$$

$$FeS_2 + Fe_2(SO_4)_3 \rightarrow 3FeSO_4 + 2S \qquad \ldots(2.10)$$

$$2S + 3O_2 \rightarrow 2H_2O + H_2SO_4 \qquad \ldots(2.11)$$

some of these reactions, e.g. 2.8 and 2.10, take place slowly in the absence of bacteria but are accelerated in their presence.

Other Acid Corrosion

Less well studied than the effects of the Thiobacilli are corrosion reactions due to the formation of acids from the oxidation of organic materials. These may include the products of microbial attack on protective coatings such as hesssian sacking and bitumen coatings used for iron pipes initiated by the cellulose-decomposing bacteria. Paper and synthetic rubber coatings for insulation cables may also be attacked. Under strongly aerobic conditions CO_2 is the end product of the oxidation of the organic material, and lead carbonate has been detected as a corrosion product of lead-coated underground cables. Under semi-anaerobic conditions organic acids accumulate and these may lead to simple acid corrosion or alternatively may accelerate corrosion by chelation of passive layers on metal. Besides bacteria, moulds and yeasts may accumulate organic acids even under aerobic conditions and in some cases may synthesise complex 'secondary metabolites' some of which, although only weakly acid, are powerful chelating agents. These may be of special significance when microbial slimes accumulate on metal surfaces, as relatively high concentrations of potentially corrosive products may be trapped in them, and corrosion pits result. It is possible that massive pitting in aluminium fuel tanks in aircraft may originate in this way[20].

Microbial-accelerated Cathodic Depolarisation of Ferrous Metals

Corrosion of iron and steel, especially in anaerobic conditions such as waterlogged soils, is usually caused by sulphate-reducing bacteria of which the genus Desulphovibrio is the most commonly occuring. The presence of organic materials such as acetate often stimulates these organisms' reducing power whereby sulphate is reduced to sulphide, but some at least of this genus appear to grow as essentially as chemolithotrophs, and reduce sulphate as follows:

$$SO_4^{2-} + 4H_2 \rightarrow S^{2-} + 4H_2O \qquad \ldots(2.12)$$

Simple corrosion by H_2S should yield exclusively FeS:

$$Fe + H_2S \rightarrow FeS + H_2 \qquad \ldots(2.13)$$

whereas both hydroxide and sulphide would be expected if the cathodic reaction was due to the mechanism proposed by von Wolzogen Kuhr and van der Vlugt[21].

$$4Fe + 4H_2O + SO_4^{2-} \rightarrow 3Fe(OH)_2 + FeS + 2OH^- \quad \ldots(2.14)$$

It can be seen from equation 2.14 that the ratio of iron corroded to iron in the form of sulphide should be 4:1, but values from 0.9 to 48 are commonly obtained experimentally. Subsequently it was shown by Booth[22] and his co-workers that the ratios of the corrosion products were dependent on the particular strain of Desulphovibrio and on their rates of growth. Later the activity of the enzyme hydrogenase which bring about the reaction:

$$H_2 \rightleftharpoons 2e^- + 2H^+ \qquad \ldots(2.15)$$

was correlated directly to the ratios of corrosion products. Further work, especially with methods of continuous or semi-continuous growth of pure cultures coupled with appropriate enzymic and electrochemical measurement, largely confirms the important role of hydrogenase in cathodic depolarisation[23], but also suggests rates are affected by precipitated ferrous sulphide which can, in the presence of excess ferrous ions, form films on the metal with the possibility of setting up local concentration cells. The role of ferrous sulphide and ferrous salts is as yet unexplained, but the concentration of ferrous ions appears to have an effect even on the anaerobic corrosion of buried pipes in the field as well as in the laboratory. The proposed mechanisms emphasise the role of environmental conditions on anaerobic microbial corrosion.

It is noteworthy that rates of corrosion in the field are often much higher than those in the laboratory with pure cultures. This emphasises the complexity of 'natural' eco-systems[24].

In considering this, two aspects concern engineers, firstly in any new engineering venture involving buried iron or steel and concrete structures. The soil conditions must be evaluated by boreholes and pits with in situ and laboratory testing. Any evaluation must take into account seasonal and exceptional water regimes in the soil. Off site changes in these must also take account of the changes that are likely to occur by reason of engineering works which are expected to alter soil structure and consequently microbial populations and activities.

The term aggressive is often used to imply some approximately quantitative estimate of the likelihood of corrosion and depends on measuring factors such as soil water (resistivity), pH, redox potential, salt concentrations and bacterial populations in order to establish criteria for the prediction of corrosion rates[25]. Similar measurements for predicting corrosion in rivers and bottom deposits have been described. Confirmation of these prognostic tests was made on buried or immersed metal plates with periodic measurements of metal/soil or metal/water potentials with reference to copper/copper sulphate electrodes; weight losses were determined after burial for up to three years. Aggressive soils are characterised by resistivity of less than 2000 cm or a redox potential of less than 400 mV (see Table 2.21 for evaluation of E_h); when a soil is 'borderline' and its resistivity and redox potential approximate to the values above, the water content provides a further criterion and it is regarded as aggressive when the water content is greater than 20%. These criteria have been confirmed to be valid for mild steel with a few exceptions, and have been found to be fairly satisfactory for lead and zinc but invalid for aluminium. The metal/soil potential gave little information of value.

Table 2.21 Redox potentials* of soils in relation to corrosiveness

Range of E_h	Classification of corrosiveness
100 mV	Severe
100–200 mV	Moderate
200–400 mV	Slight
400 mV	Non-corrosive

* The redox potential is determined with a probe consisting of a platinum electrode and a Hg/Hg_2Cl_2—Cl or Ag/Ag Cl—Cl reference electrode. If E_p is the potential of the platinum probe, E_r the potential of the reference electrode and E_h the redox potential of the soil (in mV on the hydrogen scale) then

$$E_h = E_p + E_r + 60(pH-7)$$

where pH is the pH of the soil

Detection of Anaerobic Corrosion

This is immediately recognised in smooth pitting with a black corrosion product and smell of hydrogen sulphide when the metal object is first exposed; cast iron shows graphitisation. The iron sulphide corrosion product oxidises rapidly on exposure to air and should be examined quickly; in doubtful cases acidification on exposure is confirmatory. Isolation and counting the bacteria depends on establishing strict anaerobiosis. Methods for examining soil and water are given below.

Sampling

Soil samples from the levels in which structure or pipes are to be laid are filled to the top of screw capped bottles, and bacteriological tests are made within 24 hours.

Isolation and Enumeration

Methods have recently been evaluated by Mara and Williams[26] and for most purposes the modified ISA medium is suitable: iron sulphite agar (oxoid) 23 g/l; $FeSO_4 7H_2O$, 0.5 g/l; 7% sodium lactate 5.0 g/l; $MgSO_4 7H_2O$, 2.0 g/l. The medium is adjusted to pH 7.5 with sterile NaOH after autoclaving at 121°C for 15 min. For liquid cultures, screw-cap test tubes are filled to the brim, for enumeration sterile test tubes (150 × 16 mm) are filled to within 5 mm of the top with inoculated agar medium, covered with a cap of sterile 1.5% w/v agar and closed with a polypropylene cap. Incubation is at 30°C and colonies are counted until the maximum number has developed. These methods are prone to give false positives and large errors, and should only be attempted where adequate microbiological backing is available.

Sulphate-reducing bacteria are present in virtually all soils and the qualitative procedure is valuable because it works only when relatively large numbers are present. Hence a positive result with this test is a rough indication of a particularly aggressive soil, though a negative result does not necessarily mean that the soil is innocuous. Quantitative procedures for enumerating sulphate-reducing bacteria in soils and water have been available only for the last few years and data on populations in aggressive and non-aggressive soils are therefore scantly. Highly polluted waters and soil are known to contain 10^5 and 10^6 viable sulphate-reducing bacteria per millilitre or gram; waters with less than 10^2 bacteria per millilitre are usually innocuous from the pollution point of view. It likely that the aggressiveness of soils would follow broadly the population of viable sulphate-reducing bacteria within these limits.

As already stated, aggression will also depend on soil E_H according to the scale given in Table 2.21. Temperature is also a controlling factor and both psychrophylic (cold) and thermophilic (hot) forms are known, e.g. in electrical transformers, hot water systems.

Hydrogenase Determination

This enzyme is of wide occurrence in bacteria where it is concerned with the reduction of nitrate and CO_2 as well as sulphur. Methods for its estimation depend on measuring some activity of hydrogenase by (a) dye reduction (benzyl viologen or methylene blue), (b) isotopic exchange and (c) evolution of molecular hydrogen. Interpretation of quantitative results is difficult due to the complex relationship between the enzyme cell structure and the particular method selected[27].

Sulphate Reducing Bacteria

The range of bacteria species capable of sulphate reduction has been greatly expanded since the studies of Pfennig and Widdel[28]. By using a wide range of growth media and physical conditions well over twelve new species have been added to this genus, including Desulfosarcina, Desulfobacter and Desulfococcus etc. As their names suggest these differ in their morphology but in addition their growth requirements and conditions also differ. One characteristic of these new species is that their growth rates are much slower than Desulfovibrio. In addition to these true bacteria Stetter (1987) has isolated an Archaebacterium (Archaeglobus) able to grow and reduce sulphate at 65 to 80°C. This newly characterised group are thought to have merged with the eubacteria and evolved as a distinct family including Methanogens and Halophiles which may occur in consortia associated with metal corrosion. Most 'biosulphur' deposits originated in Permian and Jurrasic times, periods in which there was a burst of sulphur reduction as yet unexplained[29].

Corrosion Due to Microbes Other Than Sulphur Metabolisers

In recent years it has become apparent that widespread microbial infections of materials in the manufacturing industries can lead to corrosion for the reason briefly outlined above. Examples include the instant rusting of machined parts, corrosion of machine tools, aircraft fuel tanks, hydraulic systems, strip steel etc.

The precise role of any specific organism in these instances is difficult to determine and will probably remain the province of specialists, largely because of the ubiquitous occurrence of microbes. However, a number of simplified tests have recently been devised to assist engineers and chemists in diagnosing whether or not a particular corrosion is biological in origin. These are based on (1) direct microscopy, (2) measurements of microbial metabolism (oxygen uptake, dye reduction, extracellular enzyme activity) and (3) direct enumeration of specific species of selective media. These are briefly discussed below:

1. Microscopy[30]. A binocular phase-contrast microscope equipped with low-power lens and an achromatic × 40 objective and × 10 compensating eyepiece. An oil-immersion lens (× 100) is required for stained material. Dark ground illumination is also probably worthwhile. Size measurements can be made with an image splitting eyepiece (Vickers Instruments, York). The significance of microscopal examination depends on familiarity with the material and long experience of its microbial populations. The scanning electron microscope is increasingly used for examining the association of microbes and metals under corrosion attack. When coupled with electron probe analysis it becomes a powerful analytical tool.
2. Two simple metabolic tests may be used for evaluating infections by aerobic and anaerobic organisms.
    (a) Red spot test[31]. This relies on the reduction of a soluble colourless tetrazolium dye to an insoluble coloured formazan by respiring microbes. This can be done on plates (Oxoid Limited), ampoules or slides (BDH Limited). The test can be calibrated by estimating by eye the extent of red formazan deposition and correlating it with elaborate counting methods or merely by taking account of any particular situation. Anaerobic bacteria give rapid responses, slower or little response is given by most moulds and yeasts[32]. Anaerobes[33] can be enumerated in the solidifying media for Desulphovibrio (p. 2.93).

(b) A test based on an enzyme released by microbial growth is illustrated by the Avtur Test (BDH Limited) for jet fuel infections. This is based on an approximate assay of the enzyme acid phosphatase.

3. The medium already described for enumerating Thiobacilli (*p.* 2.88) is a typical selective media and depends on selecting for a mixed population an organism with a specific growth requirement sufficiently different from most of its fellows in that environment. Further selectivity can be designed into a medium by the addition of suitable growth inhibitors, e.g. bile salts (to select coliforms) penicillin (to select Gram negatives), etc. Temperature and other physical parameters can also be used for selection pressure on a mixed community. The use of media enriched with biocide is particularly important for estimating the likelihood of resistant species occurring during treatment of an industrial plant to stop microbial corrosion. Selection media are available for isolating and enumerating fungi (Corn metal agar; Oxoid Limited) and yeasts (Tryptone soya broth; Oxoid Limited) containing 10% w/v sucrose, which inhibits the growth of most bacteria and fungi[34]. Sampling can be very important especially if false positive results are to be avoided. The usual practice is to collect directly into sterilised bottles, e.g. plastic (Sterilin Limited), from appropriate sections of a system assuming that, for instance, in multi-phase systems representative samples of each phase are collected, e.g. oil and water lubricating lines. Both liquids should be run to waste for at least 30s before collecting. Samples are assayed as quickly as possible or stored at 2-4°C in the dark. After testing, samples are disposed by autoclaving or immersion in a suitable disinfectant overnight.

The advent of Biotechnology now mainly directed at medical diagnostics and more recently to the food industry is likely to yield more rapid and simple tests for measuring microbial mass, enzymes etc. and these, e.g. a clip slide measuring ATP, adapted to corrosion diagnosis[35].

Biofilms

Most laboratory studies on microbial corrosion have been made in growth chambers such as chemostats with pure cultures; loss of metal from strips immersed in such 'homogeneous' systems has been followed by a variety of methods. However, it is apparent that the natural systems under which corrosion occurs are much more complex than this. In soil, for instance, the microbial population is complex and far from uniform as is the supporting soil structure. Corrosion is a surface phenomenon and it is those microbes at the surface with which the name 'biofilm' is now applied[36]. Even though such films may be 10-20 micron in depth, ingenious studies with computer driven micro-probes shows that condition through the film may vary from oxygen saturation at the outer surface to complete anaerobiosis at the surface of the metal[37]. This is reflected by a layered composition of microbes each type selected by their responses to differences in environmental conditions. Growth and metabolism and their chemical effects, not least corrosion rates will be affected. There is no doubt that the ability to form such films in a controlled manner and to investigate their properties will increasingly give many incites to the mechanisms involved in the field. This applies especially to the low rates of corrosion found in model systems compared to those in nature.

It is worthwhile drawing attention to health hazards associated with film infected water systems which also cause corrosion. Two of the most common are Legionnaires disease and so called 'humidifier fever'. Because of strong adhesion of biofilms and diffusion rates through the film treatment based on cleaners and chemical sterilisers such as chlorine often fail; similar considerations apply to other systems in industry, e.g. food, paint, oil and gas are examples where biofilm activities have given massive problems.

To conclude it must be stressed that recent work has directed attention to the interplay between different microbial species in most of the corrosion effects described. Microbial corrosion is therefore one special instance of the rapidly developing field of Microbial Ecology[38].

D. E. HUGHES

REFERENCES

1. Hughes, D. E. and Hill, E. C., Mining and Metallurgical Congress Paper No. 24 (1969)
2. Hill, E. C. in *Microbial Aspects of Corrosion*, Ed. Miller, J.D.A., Medical and Technical Publishing, Aylesbury (1971)
3. Miller, J. D. A. and Tiller, A., ibid.
4. Le Roux, N. W., ibid.
5. Stafford, D. A. and Callely, A. G., Symposium on Effluent Treatment, B.C.R.A., Chesterfield, to be published.
6. Butlin, K. R. and Postgate, J. R., in 'Autotrophic Micro-organisms', *Symp. Soc. Gen. Microbiol.*, Cambridge University Press (1954)
7. Fletcher, A. W., in *Microbial Aspects of Corrosion*, Ed. Miller, J.D.A., Medical and Technical Publishing, Aylesbury (1971)
8. Williams, A. R., Stafford, D. A., Callely, A. G. and Hughes, D. E., *J. Bact.*, **33**, 656 (1970)
9. Parker, C. D., *Aust. J. Exp. Biol. Med. Sci.*, **23**, 81, 91 (1945)
10. Taylor, C. B. and Hutchinson, G. H., *J. Soc. Chem. Ind.*, London, **66**, 54 (1947)
11. Chemistry Research 1956, H.M.S.O., London, 16, 61 (1956)
12. Pochon, J., Coppier, O. and Tchan, Y. T., *Chim. et Industr.*, **65**, 496 (1951)
13. Frederick, L. R. and Starkey, R. L., *J. Amer. Wat. Wks. Ass.*, **40**, 729 (1948)
14. Thaysen, A. C., Bunker, H. J. and Adams, M. E., *Nature*, London, **155**, 322 (1945)
15. Leathen, W. W., Kinsel, N. A. and Braley, S. R., *J. Bact.*, **72**, 700 (1956)
16. Braley, S. A., *Min. Engng.*, N. Y., **8**, 314 (1956)
17. Braley, S. A., *Min. Engng.*, N. Y., **9**, 76 (1957)
18. Butlin, K. R. and Vernon, W.H.J., *J. Inst. Water Engineers*, **3**, 627–637 (1949)
19. Trudinger, P. A., *Advances in Microbial Physiology*, **3**, 111 (1969)
20. Elphick, J., in *Microbial Aspects of Metallurgy*, Ed. Miller, J.D.A., Medical and Technical Publishing, Aylesbury (1970)
21. von Wolzogen Kuhr, C.A.H. and van der Vlugt, L. S., *Water*, Den Haag, **18**, 147 (1934)
22. Booth, G. H., *Microbiological Corrosion*, M & B Monographs, Mills and Booth Ltd., London (1971)
23. Panhania, I. P., Moosavia, A. N., Hamilton, W. A., *J. Gen. Microbiol.*, **132**, 3357
24. Hamilton, W. A., *Ann. Rev. Microbiol.*, **39**, 195
25. Starkey, R. L. and Wight, K. M., *Anaerobic Corrosion of Iron in Soil*, Amer. Gas. Ass., New York (1945)
26. Mara, D. D. and Williams, D.J.A., *J. Appl. Bact.*, **33**, 543, (1970)
27. San Pietro, A., *Methods in Enzymology*, Vol. II, Academic Press, N.Y. 861

28. Pfennig, N., Widdel Trupor, H. G., *The Prokaryotes*, Ed. Starr, M. P., Springer-Verlag, Berlin (1981)
29. Postgate, J. R., *The Sulphate Reducing Bacteria*, Cambridge University Press, 2nd Ed. (1984)
30. Barer, R., *The Microscope*, Blackwell, Oxford (1959)
31. Hill, E. C. and Pemberthy, I., *Metals and Materials*, **2**, 359 (1968)
32. Hill, E. C. and Gibbon, O., *Metal Finishing*, **15**, 395 (1969)
33. Hill, E. C., *Aircraft Engineering*, 24 July (1970)
34. Callely, A. G., *Process Biochemistry*, **3**, 11 (1967)
35. Biologically Induced Corrosion. Proc. Int. Conf., Gaitlersbury, Nat. Ass. Corrosion Eng., University of Delaware (1983)
36. Hamilton, W. A., *Ecology of Microbial Communities*, 41 S.G.M., Symp., Cambridge University Press (1984)
37. Wimpenny, J. W. T., Lovitt, R. W., Coombes, J. P., *Microbes In Their Natural Environment*, Cambridge University Press (1983)
38. *Experimental Microbiol Ecology*, Ed. Burns, R. G., Slater, J. H., Blackwell, Edinburgh (1982)

2.7 Chemicals

For the purposes of this section a *chemical* may be defined as a substance useful technologically and containing over 95% of the principal chemical.

In general most textbooks on corrosion control, including this one, give data on the physical and mechanical properties of a particular metal or alloy and then outline its suitability in various environments, rather than considering a particular chemical in relation to its effect on various materials of construction. This approach may be difficult to appreciate, since the design engineer will be given, at first, the environment he has to work with and then is required to choose the most suitable materials. To date this information is grossly inadequate because of the enormity of collating such a large amount of data. For example, if some 400 chemicals are identified as being handled and processed on a large scale and if there are 10 suitable materials, then 4 000 systems would have to be considered. Since temperature, concentration and solution velocity are important in determining corrosion rate, and if only five levels of each of the three variables are considered, then the number of experiments to be carried out would be $4\,000 \times 5^3 = 500\,000$.

There are in addition several other factors that accelerate corrosion and must be taken into account; these include crevices, galvanic coupling, tensile stress, aeration, presence of impurities, surface finish, etc. If these were also taken into consideration then several million experiments would have to be performed to compile such data. There are many instances where two or more chemicals exert a marked synergistic action such that low dissolution rates obtained in either environment become much greater in the presence of both. Further, the corrosiveness of a chemical will be affected by the presence of certain impurities, which may act as either accelerators or inhibitors. To take all these factors into account would add to an already impossible task and as Evans has remarked[1], 'There are not enough trained investigators in the world to obtain the empirical information to cover all combinations of conditions likely to arise'. Unfortunately corrosion science has not yet reached the stage where prediction, based on a few well established laws, allows selection of materials to be made without recourse to a vast amount of data.

Sources of Information

In the chemical industry, many processes are required to contain and handle solutions of complex composition containing many aggressive ingredients, but there exists a readily available supply of materials that can be used in the most aggressive environments. The purpose of this section is to indicate the procedure that an engineer could adopt if he was called upon to design plant items for chemicals for which he had little or no prior knowledge or experience. This is a situation that exists within many engineering industries that do not utilise metallurgists or materials scientists and are generally unaware of the existence of the specialist corrosion engineer or scientist and his importance at the planning stage. The most important sources are:

1. The 'ideal' source book for designers, which is the one in which the individual chemicals are listed together with the corrosion rates for a variety of materials under different conditions of temperature, pressure, velocity, etc.
2. Corrosion resistance data lists on specific materials offered for sale. However the engineer must also consider mechanical and physical properties and, last but not least, the cost of the material, its fabrication and protection. Thus reference must be made to books, journals and data that provide this information.
3. Information based on experience, which includes national standards, specifications and codes of practice, and also the technical and scientific literature reviews and reports. Corrosion engineers, technologists and scientists employed by specialist organisations or as consultants form the most important source of information, and their advice should be sought wherever possible.

Selection Based on Chemical Environment

The design engineer who requires full information will be disappointed because such books, tables and monographs do not exist. In the present situation the engineer will turn to official standards, etc. However, they are relatively few in number and, though an important source of information, suffer from the lack of reliable data for many popular materials in a wide range of environments that may lead to pitting, stress-corrosion cracking, crevice corrosion and corrosion fatigue.

For the engineer who cannot find a suitable specification for a particular chemical environment various sources of information exist. Perhaps the most comprehensive text presently available is the *Corrosion Guide*[2]. In this book a great deal of information has been gleaned from the world's literature in an attempt to fulfil the engineers' requirements. Because of the magnitude of the problem, three or four levels of corrosivity are given, but only a small but detailed amount of data are reported for certain chemicals with regard to temperature, concentration and velocity. The data given may be useful for making an initial survey of materials prior to further detailed planning or testing. Many of the 800 or so chemicals mentioned do not corrode all of the materials (metals, ceramics and plastics) and this book could be considered as the most useful source of preliminary information. The guide gives corrosive effects of a large number of individual inorganic and organic chemicals and also some important groupings such as fatty acids, cements, concrete, mortar, plaster, oil and varnishes. However, it must again be emphasised that the Corrosion Guide provides no information other than corrosion behaviour, and that this alone is insufficient in a rational approach to materials selection. The corrosion resistance of mild steel is poor in most environments, but this is frequently outweighed by the fact that it has good mechanical properties, and is readily available, cheap and readily fabricated by welding, etc.

When the information is available, the Corrosion Guide provides detailed corrosion data on the preparation of various chemicals. For example, in the section on sulphuric acid, the corrosion rates for several alloys are given when used at various stages of an actual process involving that acid. In the section on phosphoric acid, cognisance of the method of production shows also the influence of the minor constituents as well as the major chemical on the corrosion of various materials. Nevertheless, it must be emphasised that even a book as comprehensive as the Corrosion Guide can only cover a limited number of all the possible chemicals used in practice.

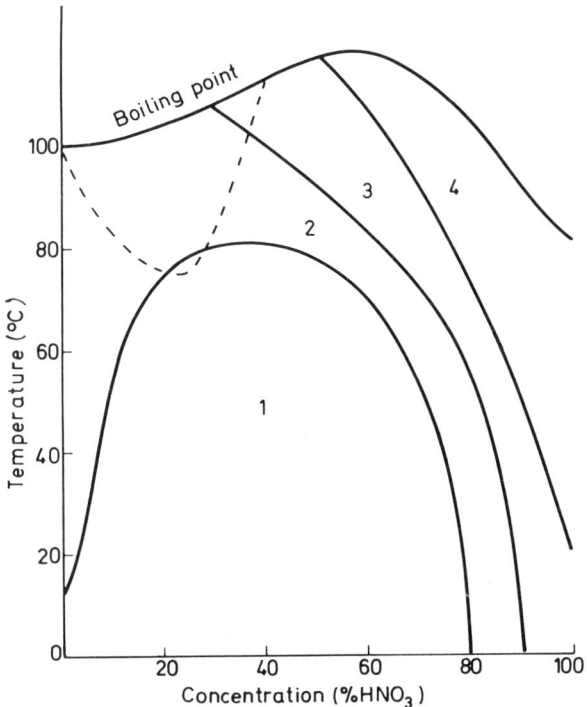

Fig. 2.21 Safe operating conditions for various steels in nitric acid solutions. Iso-corrosion lines at $0 \cdot 1 \text{ gm}^{-2}\text{h}^{-1}$

Area 1: Fe-13Cr
Areas 1 + 2: Fe-17Cr
Areas 1 + 2 + 3: Fe-18Cr-8Ni
Areas 1 + 2 + 3 + 4, but excluding area within broken line: Fe-15Si
(after Berg[3])

Where a large collection of data exists then it may be effectively condensed in the form of diagrams. A popular method is the use of 'iso-corrosion rates' plotted on co-ordinates of temperature and concentration for one material and one chemical. Because of the large amount of data on the common acids there are many examples of this type of diagram, e.g. the work of Berg[3] who has chosen metals and alloys that are readily available. He has excluded many metals and alloys on the grounds that they are either 'Non-resistant or can be substituted by cheaper materials....'

From these diagrams an initial selection can be made for materials to be used for the chemicals considered, and although temperature and concentration are included, the effects of flow and the effect of crevices, galvanic coupling or stresses are not indicated. Thus for sulphuric acid, 'chemical lead' and 'silicon iron' appear to cover a wide range of temperature and concentration; however, the mechanical weakness of the former and the brittleness and difficulties in fabrication of the latter, might mitigate against their use in say pumps, valves and unsupported pipes. For these purposes the more expensive nickel-base alloys have also been listed which may be preferred on various grounds which might include ease of fabrication and tolerance to thermal cycling.

An interesting interpretation of these temperature–concentration diagrams has been given by Nelson[4]. He has combined information of all the materials on the one diagram so that the best materials for the whole range of conditions are seen together. The diagrams taken from Berg[3] have been combined in Fig. 2.21 to show the value of this method of presentation.

Selection Based on the Properties of the Material

The more usual method of presenting data on the corrosiveness of various chemicals is by reference to a specific class of metal or material. Thus in the present book, the sections devoted to individual classes of materials contains lists of chemicals and in some cases details of their behaviour under various conditions of concentration, flow and temperature (see in particular the sections devoted to metals and alloys).

Manufacturers[5] and specialist materials development associations publish extensive corrosion data in the form of monographs, and this form of presentation is also used in national standards[6]. The most recent comprehensive text in this category is perhaps the publication by the Zinc Development Association[7]. The work is important in that the section on chemicals also deals with common, though complex, chemical formulations, e.g. fire-extinguisher fluids, soaps and syndets, agricultural chemicals such as pesticides and fertilisers. This publication also demonstrates the mammoth task of recording all the available data for just one material. A comparable book for mild steel would probably be much larger, whereas for many other materials the information has not yet been determined. Thus at best, only very incomplete data are available in this form.

Selection Based on Experience

The scientific and technical literature[8] abounds with much information, adding slowly to the massive factual data required for the design engineer who relies solely on his own literature search.

The material development associations and manufacturers have, by their own research and development, accumulated a great deal of information about their own product and this is transmitted directly to potential users of their materials.

Selection Based on Scientific Principles

Ideally the design engineer requires an equation which condenses all this information and from which he can calculate the effect of a particular chemical upon a range of materials, and the limiting conditions of say temperature, concentration and velocity. To achieve this objective he needs to know which of the properties of the chemical and the material are the most important in determining the interaction leading to corrosion.

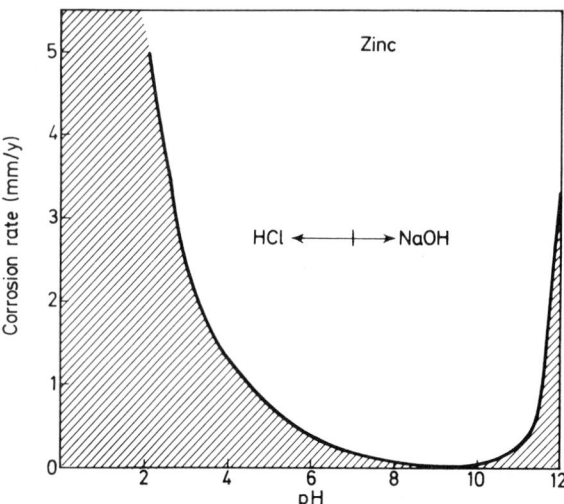

Fig. 2.22 Effect of pH on metals relying on passive films for protection, e.g. zinc

The majority of metals and alloys available depend for their resistance to corrosion on the properties of an oxide film or corrosion product which is formed initially by the corrosion process. In many cases the protectiveness of the oxide film is determined by its stability in aqueous solutions in a specific pH range, either chemically dissolving to form aquocations at lower pH values or complex anions (aluminate, ferroate, plumbate, zincate, etc.) at higher pH values (Fig. 2.22). An important property of the chemical is therefore the pH value that it develops when dissolved in water. For many materials and many chemicals this is the overriding factor and in many cases the oxide film is found to dissolve uniformly by interaction with hydrogen ions, the rate varying with concentration in accordance with equations such as:

$$\text{Rate} = \text{constant} \times (\text{conc. of } H^+ \text{ ions})^n$$

As the film dissolves more oxide film is formed, i.e. the metal/oxide interface progresses into the metal, and the overall rate may be low enough to be acceptable for a particular process. In other cases, the corrosion products precipitate on the surface of the oxide and either accelerate the overall rate by enhancing diffusion of ions through the porous outer layers or, when less porous layers are formed, access of hydrogen ions to the inner oxide surface is reduced thus decreasing the rate.

The ability to form a second barrier film when the conditions are such that the oxide film is dissolving rapidly, is another important property of the chemical, i.e. its ability to form highly insoluble corrosion products which would allow the physical blockage of the surface. There are, however, a certain number of chemicals that have the ability to cause non-uniform dissolution of oxide films and the more serious form of pitting corrosion occurs. Certain anions have the ability to attack the oxide film in small areas and allow corrosion at high rates confined to these areas. This often leads to deep penetration of the metal although overall weight-loss is very low. Many chemicals appear to have this ability but the most insidious are the halides, particularly chloride, which occurs as impurity in many chemicals, raw materials and water supplies. Many chemicals have a reputation of being difficult to handle but this is often due to the presence of these types of impurity. For example, many chlorinated hydrocarbons in the presence of water hydrolyse to form hydrochloric acid which is responsible for the corrosiveness of these non-ionic substances and the possibility of attacks on ships' cargo compartments.

Rationalisation of Data

For this purpose a chemical may be said to either (*a*) dissolve a material uniformly, the rate depending on pH or (*b*) non-uniformly leading to pitting corrosion. There are, of course, examples of intermediate behaviour but this simple division leads to a method of assessing the probable behaviour of a chemical. Only (*a*) will be discussed in detail since pitting corrosion is dealt with elsewhere in this book. (See Section 1.6 and Chapter 21.)

Uniform Corrosion

The corrosion rate of a metal, which depends for its protection on a passive oxide film, may be predicted from a simple empirical adsorption law (Freundlich):

$$C_R = AC^n$$

where C_R = the corrosion rate,
A = the specific rate constant,

C = the concentration of hydrogen ions and
n = an integer,

which may be modified by taking logarithms and substituting pH for C, when the equation becomes

$$\log C_R = K - n(\text{pH})$$

where $K = \log A$.

Reliable pH data and activities of ions in strong electrolytes are not readily available. For this reason calculation of corrosion rate has been made using weight-loss data (of which a great deal is available in the literature) and concentration of the chemical in solution, expressed as a percentage on a weight of chemical/volume of solution basis. Because the concentration instead of the activity has been used, the equations are empirical; nevertheless useful predictions of corrosion rate may be made using the equations.

In strong acid solutions many common structural materials dissolve uniformly and this assumption is reasonable in many real situations. The data given in the monograph by Berg[3] are used in order to demonstrate the universal application of the technique. Four main types of behaviour may be identified for metals and alloys in various acids at different temperatures and concentrations.

Type 1. *Increasing corrosion rate with increasing concentration and temperature* In this case the equation obeyed is

$$\log C_R = n \log C - \frac{A}{T} + K,$$

where A, K and n are constants specific to a particular material and chemical. This relationship predicts that corrosion rate C_R increases continuously with temperature and concentration.

Example 1. Hard lead (antimoniacal) can be used in sulphuric acid to quite high concentration but it displays an increasing corrosion rate with increasing temperature and concentration. Relationships are complex, but the general form of the equation may be used:

$$\log C_R = 4\log C - \frac{2\,500}{T} - 1\cdot 25$$

The prediction is not very accurate, but in Fig. 2.23 comparison between the actual and calculated value shows the effectiveness of the empirical equations especially at higher temperatures. The iso-corrosion line obtained by substituting $0\cdot 1 \text{ gm}^{-2}\text{h}^{-1}$ in the equation gives a reasonable guide to the temperature and concentration limitations of this material. The activation energy, which may be obtained from the temperature coefficient, indicates an activation-controlled reaction, so that flowing solutions should have a small effect on corrosion rates. Providing that the designer allows for the low mechanical strength of lead, the material could be considered for situations where high flow is involved, but abrasion of the protective film of corrosion products must be avoided.

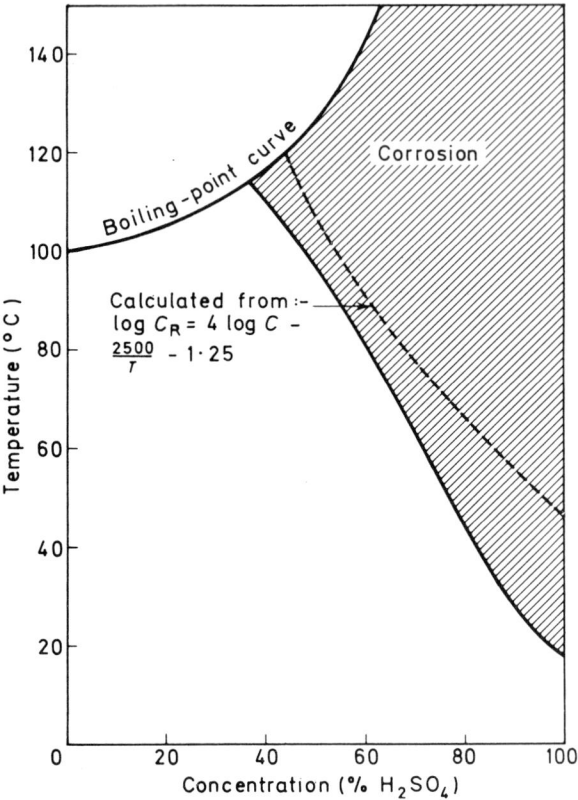

Fig. 2.23 Type 1 behaviour; increasing corrosion rate with increasing temperature and concentration, e.g. lead in sulphuric acid. Iso-corrosion lines at $0 \cdot 1 \, \text{gm}^{-2}\text{h}^{-1}$

Type 2. *Decreasing corrosion rate with increasing concentration and temperature* Metals and alloys that show this behaviour are important because they can be used for more concentrated acids. Mild steel is a good example in that it resists concentrated H_2SO_4 and HNO_3. This behaviour is associated with true passivity in the case of HNO_3, but in H_2SO_4 there is also a precipitation of corrosion products which appear to block the surface and prevent further attack. For higher temperatures and less concentrated solutions, mild steel is not resistant but other ferrous alloys have been developed that retain this characteristic over a much wider range of temperature and concentration. The empirical equations may be applied, and although an increase in temperature increases the rate, the form of the equation must be changed to allow for the decrease in corrosion rate with increasing concentration:

$$\log C_R = K - n\log C - \frac{A}{T}$$

From the temperature factor A, an activation energy may again be calculated which gives a useful indication of the influence of flow-rate on corrosion rate.

Example 2. The corrosion rate of silicon iron (Fe–15Si) in static H_2SO_4 may be predicted from:

$$\log C_R = 6\cdot 54 - 1\cdot 71 \log C - \frac{1\,540}{T}$$

The calculated and experimental iso-corrosion lines are given in Fig. 2.24 and show reasonable agreement with experimental values when the corrosion rate was $0\cdot 1\,\text{gm}^{-2}\text{h}^{-1}$. However, at a concentration of 50% there is a sudden decrease in rate and the material is useful at all temperatures beyond this concentration. The activation energy has been calculated as approximately 30kJ/m and suggests that the corrosion rate will increase with increasing flow-rate, which could increase considerably the corrosion rates for static solutions quoted above. A diffusion-controlled reaction would be in accord with a salt passivation or blocking-type mechanism on a surface otherwise protected by SiO_2. The material is, however, used for pumps, stirrers, etc. in high concentrations of sulphuric acid where a thick anodic film is formed.

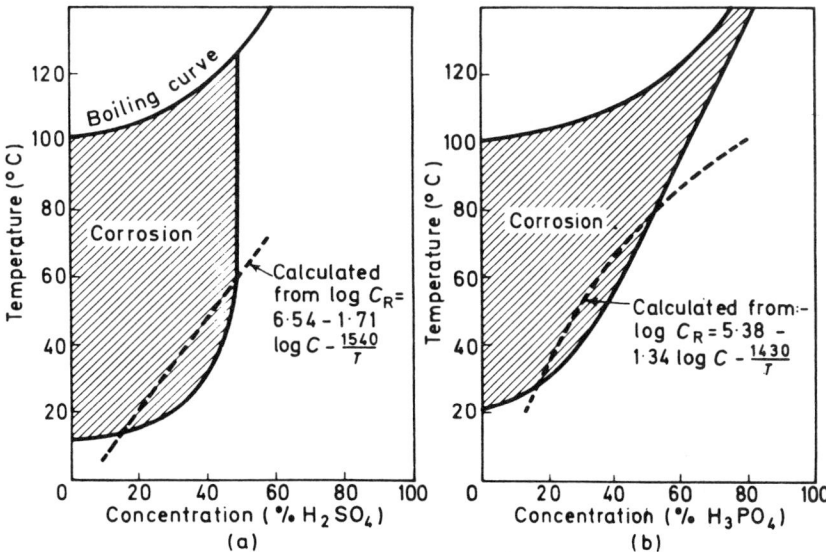

Fig. 2.24 Type 2 behaviour; decreasing corrosion with increasing temperature and concentration, e.g. (*a*) Fe–15Si in H_2SO_4 and (*b*) aluminium bronzes in H_3PO_4. Iso-corrosion lines at $0\cdot 1\,\text{gm}^{-2}\text{h}^{-1}$

Example 3. Aluminium bronze (Cu–7Al–3Fe) in H_3PO_4 follows the equation:

$$\log C_R = 5\cdot 38 - 1\cdot 34 \log C - \frac{1\,430}{T}$$

This alloy is suitable for high concentrations of de-aerated phosphoric acid providing the concentration is above a certain maximum, which varies with temperature in accordance with the equation. The calculated activation energy suggests that diffusion control may be operating and that flowing solutions should influence corrosion rate to a marked extent. The corrosion rate is influenced by aeration which is perhaps responsible for the observed diffusion control.

Type 3. *Chemicals that show a maximum rate at a certain concentration*
A great many metals and alloys are unattacked in dilute solution but the rate increases with increase in concentration up to a maximum and then decreases with further increase in concentration. Thus the rates in the very dilute and highly concentrated acid may be acceptable. This behaviour has been observed for many metals and alloys in acid, alkali and neutral solutions of many salts. Brasher[9] *et al.* have shown that for many chemicals the maximum rate occurs at an acceptably low concentration of chemical followed by a low rate with further increase in concentration. Therefore, chemicals may be classed as 'aggressive' or 'inhibitive', depending on the position of the maximum corrosion rate in relation to anion concentration. The falling off in corrosion rate arises from the formation of an anodic film that is not affected by an increasing concentration of hydrogen ions. This has often been identified as a crystalline metal salt or oxide associated with the low solubility of the corrosion product. The observed behaviour is therefore a combination of both Types 1 and 2 manifesting itself over the range of temperatures and concentrations.

Example 4.   *Nickel-chromium steel* (Fe–18Cr–10Ni–2Mo) *in* H_2SO_4. In the range where the rate increases with concentration the relationship is:

$$\log C_R = 8\cdot 2 + 1\cdot 67 \log C - \frac{3\,245}{T}$$

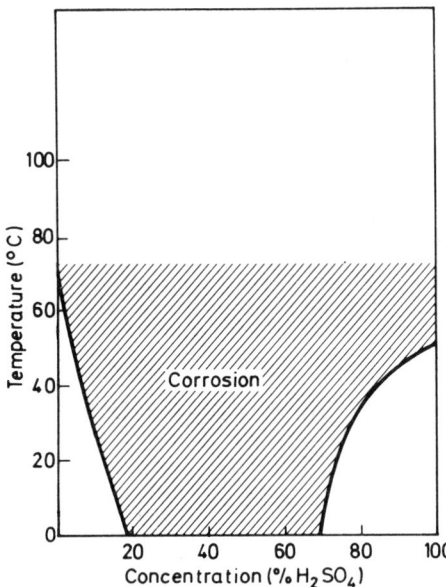

Fig. 2.25 Type 3 behaviour; maximum corrosion rate with change in concentration, e.g. nickel and chromium steel in H_2SO_4. Iso-corrosion lines at $0\cdot 1\,\text{gm}^{-2}\text{h}^{-1}$

when C is in the range 0–40% and T in the range 20–60°C. For the range where there is a decreasing rate:

$$\log C_R = 117 \cdot 5 - 45 \cdot 2 \log C - \frac{10\,000}{T}$$

when C is in the range 70–100% and T is in the range 10°C–100°C.

In the region of ascending rate the reaction is clearly governed by a diffusion process and is susceptible to flow-rate, whereas at high concentrations temperature is of greater importance than solution velocity which would have very little effect on the corrosion rate. At the high rates of corrosion (above about $10\,\text{gm}^{-2}\text{h}^{-1}$) these relationships do not apply between 40 and 60% H_2SO_4 because of the transition between the two types of behaviour. The use of the equations is most effective when some acceptably low rate is chosen, say $0 \cdot 1\,\text{gm}^{-2}\text{h}^{-1}$ and the iso-corrosion line is calculated over a range of temperatures and concentrations (see Fig. 2.25).

Type 4. *Decrease in corrosion rate with increase in temperature* An increase in temperature generally increases reaction rates, and the previous examples show that this applies to corrosion rates. However, at certain high rates of corrosion a decrease in rate can occur when the solubility of certain anodic products is exceeded, owing to surface coverage by such films. Because an increase in temperature leads to an increase in corrosion then it should be possible to reduce corrosion when the dissolution is greater than a certain high rate. This type of behaviour has been found and gives an insight into the seemingly uncharacteristic behaviour of some materials in chemical processing plant.

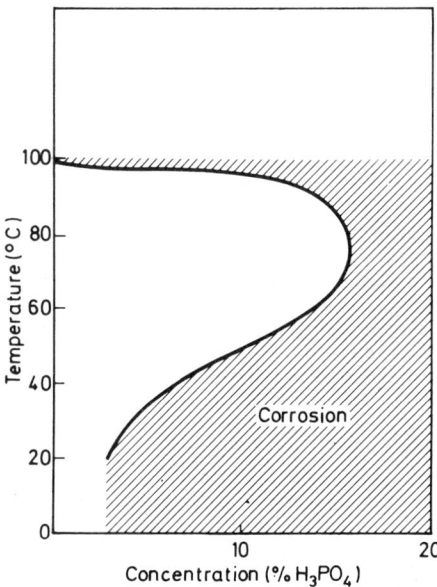

Fig. 2.26 Type 4 behaviour; decreasing corrosion rate with increase in temperature, e.g. Fe–17Cr in H_3PO_4. Iso-corrosion lines at $0 \cdot 1\,\text{gm}^{-2}\text{h}^{-1}$

Example 5. Chromium steel (Fe–17Cr) in phosphoric acid at low concentrations shows a decreasing rate with increasing temperature (see Fig. 2.26) presumably due to surface coverage by metal phosphates.

It is apparent that each metal/chemical system should conform with one or more of the four types of behaviour already mentioned, and the behaviour should therefore be capable of prediction by means of one of the following:

(i) $\log C_R = K + n\log C - \dfrac{A}{T}$ (iii) $\log C_R = K + n\log C + \dfrac{A}{T}$

(ii) $\log C_R = K - n\log C - \dfrac{A}{T}$ (iv) $\log C_R = K - n\log C + \dfrac{A}{T}$

Example 6. Hastalloy B (Ni–26Mo–4Fe) in formic acid shows all the previous types of behaviour and includes the falling off in rate with increase in temperature (see Fig. 2.27).

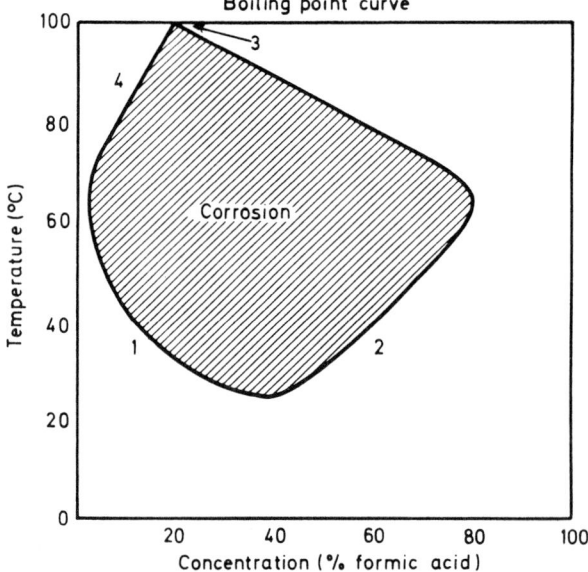

Fig. 2.27 Example of Types 1, 2, 3 and 4 behaviour. Hastalloy B in formic acid
1. Increasing corrosion rate with increasing concentration and temperature
2. Decreasing corrosion rate with increasing concentration and temperature
3. Maximum corrosion rate at a certain concentration
4. Decrease in corrosion rate with increase in temperature

In the main there exists, for each system of a chemical in contact with those metals and alloys that rely on a passive film, the possibility of an increase in corrosion rate with increasing concentration but reaching a maximum and followed by a decrease in rate. If the concentration when this maximum is reached is low, then the chemical is 'inhibitive'. The effect of temperature on corrosion is dependent on the position of the maximum concentration. For many chemical/metal systems this maximum may be at a temperature beyond the boiling point of the solution, in which case only part of the behaviour described in Types 1, 2 and 3 is observed.

The boiling point line is an artificial barrier, since constant temperature prevails only at constant pressure and concentration. However, many chemical processes are carried out under pressure and the onset of the behaviour described for Type 4 will manifest itself for many other metals and alloys. Thus serious corrosion at higher temperatures resulting from higher pressures may not necessarily occur.

Non-uniform Corrosion

The arbitrary division of behaviour has been made because of the extreme behaviour of some chemicals that initiate small areas of attack on a well-passivated metal surface. The form of attack may manifest itself as stress-corrosion cracking, crevice attack or pitting. At certain temperatures and pressures, minute quantities of certain chemicals can result in this form of attack. Chloride ions, in particular, are responsible for many of the failures observed, and it can be present as an impurity in a large number of raw materials. This has led to the development of metals and alloys that can withstand pitting and crevice corrosion, but on the whole these are comparatively expensive. It has become important, therefore, to be able to predict the conditions where more conventional materials may be used. The effect of an increase in concentration on pitting corrosion follows a similar relationship to the Freundlich equation where

$$C_R = KC^n,$$

but concentrations below which pitting does not occur are generally very low. The effect of temperature may again be represented by the Arrhenius equation:

$$C_R = B\exp(-A/T))$$

where A and B are constants. The corrosion rate C_R is meaningful when expressed as $1/\gamma$, where γ is the time required for the onset of pitting. Conditions may therefore be chosen when γ is large.

In other sections of this book the anions and other chemicals which enhance pitting, crevice and stress corrosion are discussed in greater detail.

Equilibrium Potential-pH Diagrams with Anions at Various Temperatures

There are now many diagrams available for metals and alloys which have been calculated not only for metal—H_2O systems, but also for metal–H_2O-anion equilibria (Sections 1.4 and 7.6).

Many authors have now produced diagrams in which the effects of temperature have been calculated usually over the range 20°C to 200°C. In many cases, experiments have been over the range carried out to verify the predictions made in the diagrams.

The following list is a guide to the availability of some of these diagrams:

Anion effects

Fe-S-H_2O 10, 11	25°C to 300°C
Fe-Cl-S-H_2O 12, 13	25°C to 300°C
FeO-Cl-H_2O 12	
Fe-CO_3-H_2O 14	
Fe-MoO_4-H_2O 15	
Ni-S-H_2O 12	25°C to 250°C
Ni-P-H_2O 16	25°C to 90°C
Cr-S-H_2O 17	25°C to 250°C
Cr-Cl-H_2O 17	
Cu-S-H_2O 18	25°C to 200°C
Cu-CO_3-H_2O 14	25°C
Pb-CO_3-H_2O 14	25°C
Zn-CO_3-H_2O 14	25°C
Sn-Cl-H_2O 19	25°C to 300°C

Temperature

Mn-H_2O 20	25°C to 300°C
Mo-H_2O 21	25°C to 300°C
Pt-H_2O 21	25°C to 300°C
Ti-H_2O 21	25°C to 300°C
Al-H_2O 22	25°C to 300°C
Fe-H_2O 23	
Ag-H_2O 23	
Cr-H_2O 21, 24	

Alloys

70%Cu 30%Zn-Cl-H_2O 25
$MgZn_2$-Cl-H_2O 26
2%Mg, 6·4%Zn, Cu-Cl-H_2 26

P. J. BODEN

REFERENCES

1. Evans, U. R., *The Corrosion and Oxidation of Metals*, Arnold, London (1960)
2. Rabald, E., *Corrosion Guide*, Elsevier, London, 2nd edn (1968)
3. Berg, F. F., *Corrosion Diagrams*, V.D.I.-Verlag, G.m.b.H., Düsseldorf (1965)
4. Nelson, G. A., *Corrosion Data Survey*, Shell Development Co., Emeryville, Calif. (1960)
5. Examples are: *Wiggin Alloys 100*, Henry Wiggin Co. Ltd., Hereford; *Copper in Chemical Plant*, Copper Development Association, London (1960); *Corrosion Resistance of Stainless Steel*, Publication 112/1, Firth Vickers Stainless Steel Ltd., Sheffield; *Corrosion Resistance of Titanium*, Imperial Metals Industries (Kynoch) Ltd., Birmingham
6. British Standard Code of Practice C.D. 3 003: *Linings of Vessels and Equipment for Chemical Processes*. Part 1: Rubber, Part 2: Glass Enamel, Part 3: Lead, Part 4: P.V.C., Part 5: Epoxy Resins, Part 6: Phenolic Resin, Part 7: Corrosion and Heat Resistant Materials, Part 8: Precious Metals, Part 9: Titanium and Part 10: Brick and Tile
7. Slunder, C. J. and Boyd, W. K., *Zinc: Its Corrosion Resistance*, Zinc Development Association, London (1971)
8. Examples are: *Conference on Materials Selection*, Gen. Ed. Verink, E. D., Jr., Gordon Beach, Met. Soc., **40**, London (1966). See particularly, Gachenbach, J. E., 'Material Selec-

tion in Chemical Industry' and Koelbl, H. and Schulze, J., *Selection of Materials for Chemical Industry*, Proceedings 4th International Conference on Metallic Corrosion, Amsterdam, N.A.C.E., New York (1972)
9. Brasher, D. M., Reichenberg, D. and Mercer, A. D., *Brit. Corr. J.*, **3**, 144 (1968)
10. Barry, T. I. in *Diagrams of Chemical and Electrochemical Equilibria*, CEBELCOR, **142**, Brussels (Aug. 1982)
11. Biernat, R. J. and Robins, R. C., *Electrochimica Acta*, **17**, 1261 (1972)
12. Macdonald, D. D., Syrett, B. C., *Corrosion*, **35**, 10, 471 (1979)
13. Yang, X. Z., *E.P.R.I.*, Palo Alto, California (1981)
14. Masuko, N., Inque, T., Kodama, T., *Proc. Int. Cong. Met. Corr.*, **2**, 280, Nat. Res. Comm. Canada. Toronto (1984)
15. Kodama, T., Ambrose, J. R., *Corrosion*, **33**, 155 (1977)
16. Gool, A., Boden, P. J., Harris, S. J., *Trans IMF*, **66**, 67 (1988)
17. Macdonald, D. D., Syrett, B. C., Wing, S. S., *Corrosion*, **35**, 1, 1, (1979)
18. Kwok, O. J., Robins, R. G., *Int. Symp. Hydrometallurgy*, AIMMPE, Chicago (1973)
19. House, C. I., Kelsall, G. H., *Electrochimica Acta*, **29**, 10, 1459 (1984)
20. Macdonald, D. D., *Corr. Sci.*, **16**, 461 (1976)
21. Lee, J. B., *Corrosion*, **37**, 467 (1981)
22. Macdonald, D. D., Butler, P., *Corr. Sci.*, **13**, 259 (1973)
23. Pound, B. G., Macdonald, D. D., Tomlinson, J. W., *Electrochimica Acta*, **24**, 929 (1979)
24. Radhakrishnamurty, P. *et al.*, *Corr. Sci.*, **22**, 753 (1982)
25. Verink, Jr., E. D. in *Electrochemical Techniques for Corrosion Studies*, 43 N.A.C.E., Houston, Texas, (1977)
26. Ugiansky, M., Kruger, J., Staehle, R., *Proc. 7th Congr. Met. Corr.*, ABRACO, Brazil, 605 (1979)

2.8 Corrosion by Foodstuffs

Foodstuffs, like other chemical substances, are responsible in all phases of processing (including packaging) for corrosive effects of different kinds on constructional materials. These effects are, in addition, influenced by environmental conditions of processing, i.e. by the temperatures involved, by the rates of flow (in particular, erosion is associated with very high rates of flow), and by alternating stresses which may be present in component parts of process machinery (this effect is specifically called *stress* corrosion). The presence or absence of oxygen is generally important in relation to the extent of corrosion produced.

The corrosive effects to be considered (mainly simple corrosion of metals) are, as would be expected from the edible nature of foodstuffs which are not excessively either acidic or basic but which may contain sulphur, less severe than those often encountered with inedible materials containing reactive substances. The importance of corrosive effects where foodstuffs are concerned lies not so much in the action of the foodstuffs on the metal involved as in the resultant metal contamination of the foodstuff itself, which may give rise to off-flavours, in the acceleration of other undesirable changes (by the Maillard reaction[1] for example), and in the possible formation of toxic metallic salts. Metal ions generally have threshold values of content for incipient taste effect in different liquid foodstuffs. Except in the case of the manufacture of fruit juices and pickles, process plant failure through corrosion must be rare. Nevertheless all foodstuffs, particularly liquid ones, should be regarded as potentially corrosive and capable of metal pick-up which may be undesirable.

Construction Materials

The most suitable construction material should usually be selected on the basis of prior experience or direct tests. An exact analysis of the constituents of foodstuffs is not always possible, since they are often of complex composition, but it is usually the organic acids present which are the important corrosive agents. Corrosion rates not in excess of 10–$100\,\text{gm}^{-2}\,\text{d}^{-1}$ are generally to be sought in food process plant. It is also necessary to bear in mind that in many cases the plant will have to be regularly disinfected by cleaning and sterilising solutions. These should either be non-corrosive or contain inhibiting additives. As always, the choice of construction material is determined in practice in relation to ease of fabrication and cost, which depend upon complexity of construction (cf. jacketed pans with a h.t.s.t. pasteuriser). Electrolytic couples should be avoided. Since bacteria may be pathogenic or, like certain metals, cause off-flavours, the constructional material for foodstuffs plant should be of such a nature that its surfaces can easily be kept clear and free of bacterial lodgment.

In selection of construction materials, the tendency is therefore to play safe; stainless steel is on this account generally specified[2] and is always used in cases of doubt. The suitability of other materials should not, however, be overlooked[3]. Prior to the general advent of stainless steel in 1924, tinned copper was much in vogue. Today titanium, although expensive, represents a possible alternative[4].

The precise effect of different quantities of various trace metals should be considered in relation to the specific type and variety of foodstuff in question. Differing quantities of inhibiting substances may be present in different varieties of the same foodstuff. Thus for example different quantity levels of trace metals are reported as causing off-flavours in various fruit juices, and again the quantity level of trace metals causing rancidity in edible oils depends upon whether the oils are crude or refined. The possible toxicity of these trace metals in different foodstuffs has been investigated, recommendations[4] by the Food Standards Committee of the Ministry of Agriculture, Fisheries and Food are made from time to time, and there are statutory regulations in force in different countries[5] for some foodstuffs. To state the position briefly, aluminium and tin in the quantities normally encountered are considered as non-toxic. Tin, however, in excess is considered undesirable, and may be responsible for gastro-intestinal upsets, so that the accepted limit for canned foods is taken at 250 p.p.m.

Particular groups of foodstuffs which create corrosive environments, and the processing of these foodstuffs, will now be briefly discussed. Indications will also be given of construction materials recommended to meet these situations.

Foodstuffs

Liquid Foodstuffs

As these demand a considerable degree of processing, they are probably the most important group. Within this group, there are certain foodstuffs which are slightly acid in reaction, such as instant coffee extracts and the fruit juices (for example, lemon juice at a pH of 4, and blackcurrant at 2·5). The processing[6,7] of these fruits involves milling, pressing, concentration by evaporation as necessary (temperatures are of the order of 40 to 50°C), pasteurisation (h.t.s.t.), subsequent storage and canning. The common metals, together with aluminium, will be attacked; the degree to which they are susceptible to corrosion is determined by oxygen content and the presence of hydrogen acceptors. Some published information on rates of attack is available[8-10], e.g. boiling lemon juice and tomato juice give figures of 1 400 and 180 gm^{-2} d^{-1} respectively on aluminium. The use of sulphur dioxide for the protection of fruit juices against moulds and bacteria presents an additional corrosion hazard.

The important commercial feature of these juices, especially significant with blackcurrant and tomato juices, is their ascorbic acid (or vitamin C) content, of which loss by oxidation is known to be accelerated both by heat and by metal (particularly copper) contamination. The effect of copper has been carefully investigated for pure ascorbic acid[11], and more recently ascorbic acid in blackcurrant juice and model systems[12]. There are, however, oxidation inhibitors of different kinds (which may themselves be heat-sensitive) present in various fruits, which give differing results. The presence of metals will also affect flavours[13], may cause discoloration, and may give rise to clouding effects, as in apple juice[14].

The first choice of construction material for processing fruit juices is stainless steel (BS En 58A or En 58B — the B type being similar to the A type but providing in addition for resistance to weld decay — or En 58J). Where sulphur dioxide preservation is conducted, high-molybdenum stainless steel (type En 58J) is used. Both mild steel and copper should be avoided. There are, however, other materials available, e.g. tinned copper (provided that the coating is satisfactory, and that it is shown to be otherwise suitable), enamelled metal (less popular than formerly on account of the danger of chipping), and Pyrex glass for piping and linings. Gun-metal and aluminium bronze are useful metals for various parts of pressing equipment. Plastic materials are suitable for storage-tank linings[15], provided that the surfaces can be sterilised. Nickel and Monel have certain applications; Monel for example offers good resistance to erosion. Aluminium, though known to give fair resistance to organic acids, is not generally used for normal-temperature work; certain alloys may be found more suitable than the pure metal.

For canning, conventional tin cans are available[16,17], but it is generally recommended that the coating be protected by acid-resisting lacquers, and, when sulphur dioxide has been used[18], by an acid- and sulphur-resisting lacquer. The effect of nitrates in water causing detinning is of further importance.

Milk is subjected to the process operations[6] of pasteurisation (h.t.s.t., $71 \cdot 1$–$72 \cdot 2°C$ for 15 s), evaporation (temperatures of the order of 40–50°C), homogenisation, sterilisation and drying. In addition, milk is processed into other dairy products such as butter, cream and cheese by what are essentially normal-temperature operations. Milk is approximately neutral in reaction, although lactic acid is present; the lactic acid content is increased by natural souring or by the artificial souring necessary for cheese and butter manufacture. This is perhaps the only constituent of milk which is responsible for any metal attack. Protective films of precipitated colloids may be formed by heating. Milk is sensitive in flavour to the presence of such metals as copper and iron, the chief result being a 'castor oil' or 'tallowy' taste which probably arises from fat or lecithin turning rancid. Tin is regarded as not affecting the flavour of milk. Since the flavour of milk is also influenced to a great extent by bacterial action, cleansing and sterilising solutions are regularly used in milk process plant.

Stainless steel is now generally recommended for all phases of processing (type En 58J is not normally necessary except in cases where the process liquid is specially acid)[19]. Aluminium is widely used but suffers from the defect of pitting[20] at higher temperatures. Tinned copper is suitable, where the construction permits its use. Nickel is satisfactory for cold milk and for milk that is being heated up, but is stated to be unsuitable for milk which is cooling down. Glass or enamelled steel is suitable for storage materials and for pipe-lines. Where rapid temperature fluctuations in process operations are involved, however, glass is at a disadvantage. Plastics are used for lining milk-storage vessels and for piping up to temperatures of say 50°C. There are certain reservations; with p.v.c., certain plasticisers used may impart an objectionable taste, and certain brands become opaque. Alkaline cleansing agents (up to strengths of 5% caustic soda solution equivalent) do not affect p.v.c., whereas phenolformaldehyde linings are affected. There are, however, other agents available for such linings.

The edible oils and the margarine emulsions derived from them are large tonnage foodstuffs. High- (i.e. 190°C in deodorisation) and medium-temperature operation, and low-temperature storage are involved. The refining of crude oils is carried out in the presence of caustic soda at 60°C. Refined oils would not be expected to attack the common metals, but in fact edible oils, particularly in the crude state, contain varying amounts of fatty acids, and margarine emulsions contain salt, both of which are capable of attacking metals. Here again, the effect of the metal contaminant on the foodstuff is more important than the corrosive effect of the foodstuff on the metal. It is now common knowledge that copper and iron impurities act as catalysts in the oxidation of oils, to cause slight or marked rancidity. Various oils, however, contain different inhibiting substances. Stainless steel is the prescribed material (En 58A, En 58B; or En 58J for emulsions containing salt). Deodorising plant handling refined oils may be fabricated in mild steel. Glass-lined mild-steel tanks are available for storage.

The manufacture by fermentation and/or distillation of alcoholic beverages containing a wide range of organic materials including acids, whose effect it would be difficult to assess, is traditionally carried out in copper plant, and this has found full consumer acceptance.

Sugar Products

The foodstuff sugar is relatively inert as a chemical, though when it is processsed as an aqueous solution slightly acid conditions may be present and boiling temperatures may be involved. There are no deleterious effects due to trace metals. Mild steel is generally recommended for sugar processing and for the handling of aqueous solutions in allied industries. The processing of fruits into jams and purées, which consists essentially in boiling in open pans, is closely related. The environmental factors here are the natural acidity of the fruits and the possible presence of sulphur dioxide from stored fruit pulps. Traditionally, copper plant, clad on cast iron, uncoated, tinned or even silvered, is used, but stainless steel is now widely adopted.

In the manufacture of sugar confectionery, including chocolate, the main ingredients are sugar and glucose, milk (including condensed milk), and cocoa fat, and the essential operations those of boiling and compounding. Cocoa fat, like other edible fats, is liable to oxidative rancidity. The modern choice of constructional material is again stainless steel.

Vegetable Processing

In pickle and sauce manufacture, vinegar (an approximately 5% solution of acetic acid), and salt at temperatures up to boiling point are the important corrosive agents. Sodium chloride, particularly in the presence of oxygen and acids, is known to cause rapid attack. Metal contamination (particularly iron) is likely to cause darkening by reaction with tannin present, which is either leached from wooden vessels or derived from certain spices. In pickle and sauce manufacture, stainless steel (En 58J) is recommended; it is interesting to note, however, that it is claimed that the flavour of piccalilli liquor is more readily brought out in the presence of iron[21]. This belief finds expression in the practice of using cast iron process vessels.

Meat and Fish Products

Despite attention to hygiene and cleanliness, corrosion of base metals by meat juices and deterioration of meat and fish owing to metal contamination is liable to occur. Stainless steel is recommended in soup and paste manufacture, and aluminium has a certain application. Mild steel is however used in the corned beef industry, for meat pre-cooking.

Other Food Products

Many foodstuffs are in the form of solids or processed powders, and do not offer serious corrosion problems, though mild steel equipment in infrequent use, or after washing down, can develop slightly rusted surfaces. This material is usually undesirable if it finds its way into food products. Scouring batches of dry foodstuffs is one solution to the problem if stainless steel is not used or affordable. Hygiene and cleanliness are, however, dominant factors.

Meat and fish are very liable to bacterial putrefaction; in this connection an interesting innovation is the increasing use of easily cleaned aluminium fish boxes. It is also possible that copper should be avoided in contact with herrings, which have a high fat content.

Mention should also be made of glutamic acid and invert sugar which are used in foodstuffs and demand the use of hydrochloric acid-resistant material in manufacture, and of the essential flavouring oils which should preferably be stored and prepared in stainless steel and aluminium equipment.

R. J. CLARKE

REFERENCES

1. Reynolds, T. M., 'The Chemistry of Non-enzymic Browning', *Advances in Food Science*, **14**, 168 (1965)
2. Gilroy, P. E., *Food*, **21**, 255 (1952)
3. Various authors, *Ann. Falsif.*, **213** (1950)
4. O'Keefe, J., *Bell's Food and Drugs*, Butterworths, London, 14th edn. (1968)

5. Reith, J. F., *Ann. Bromatologia*, **8**, 145 (1956)
6. Clarke, R. J., *Process Engineering in the Food Industries*, Butterworths, London (1957)
7. Tressler, D. K. and Joslyn, M. A., *Fruit and Vegetable Juice Processing Technology*, Avi. Westport (1968)
8. Blount, A. L. and Bailey, H. S., *Trans. Amer. Inst. Chem. Eng.*, **18**, 139 (1926)
9. Gilroy, P. E. and Champion, F. A., *J. Soc. Chem. Ind.*, **67**, 407 (1948)
10. McKay, R. J. and Worthington, R., *Corrosion Resistance of Metals and Alloys*, Reinhold, New York (1936)
11. Joslyn, M. A. and Miller, J., *Food Res.*, **14**, 325 (1949)
12. Timberlake, C. F., *J.S.F.A.*, **8**, 159 (1957) and **2**, 268 (1960)
13. Schrader, J. H. and Johnson, A. H., *Industr. Eng. Chem. (Industr.)*, **26**, 179 (1934)
14. Kieser, M. E. and Timberlake, C. F., *J.S.F.A.*, **8**, 151 (1957)
15. Docherty, A. C. and Hughes, H., *Chem. Ind.*, 1 171 (1959)
16. Hartwell, R. R., *Advances in Food Research*, **3**, 327 (1950)
17. Alderson, M. G., *Food Manuf.*, **67** (1970)
18. Board, P. W., Holland, R. V. and Elborne, R. G., *Journal Sci. Food. Agric.*, **18**, 232 (1967)
19. Bottom, G. H., *J. Soc. Dairy Tech.*, **6**, 179 (1953)
20. Aziz, P. H. and Goddard, H. P., *Industr. Eng. Chem. (Industr.)*, **44**, 1 791 (1952)
21. Smith, E., *Corrosion Resisting Steels—Application in the Pickle and Sauces Industry*, B.F.M.I.R.A., Scientific and Technical Survey, December (1950)

2.9 Mechanisms of Liquid-metal Corrosion*

The corrosion of metals and alloys by liquid metals generally follows the pattern of the formation of metallic alloys, i.e. solution and intermetallic compound formation and the corrosion process is often one of simple dissolution in the liquid metal. In some special cases electron-transfer processes — involving reducible impurities in the liquid metal — may modify or even override the simple dissolution process. This is especially true when, as with liquid alkali metals, the solubility of structural metals is very low. Very often under isothermal conditions equilibrium between an alloy and liquid can be approached. Continued corrosive attack is then possible only if the equilibrium is disturbed by removing in some way the dissolved corrosion product from the system. Thus the nature of liquid-metal corrosion varies depending on whether the fluid is static or is moving relative to its container, on whether the temperature is constant or varying throughout the system, and on whether the container is a single metal or a composite of two or more metals. Most processes, however, involve solution as a first step.

No adequate theory is available to explain the variation in the solubilities of metals in molten metals. Both Kerridge[1] and Strachan and Harris[1] noted that plotting the solubilities of metals in a number of solvent metals showed a periodic variation with the *solute* and not the solvent (see Fig. 2.28), i.e. a given metal such as manganese showed a consistently high solubility in molten magnesium, tin, bismuth and copper, compared with iron or chromium. Kerridge correlated this variation with the solute lattice energy and hence with the latent heat of fusion. This correlation is useful in making qualitative predictions.

In the more practical sense solution may be uniform or localised.

Preferential solution can take two forms:

1. Leaching — one component of an alloy is preferentially dissolved, an example being nickel which is leached from stainless steels by molten lithium or bismuth, sometimes to such an extent that voids are left in the steel.
2. Intergranular attack — the liquid penetrates along the grain boundaries, owing either to the accumulation of soluble impurities in the boundaries or to the development at the junction of a grain boundary with the metal surface of a low dihedral angle to satisfy surface-energy relationships.

---

* Testing procedures for liquid-metal corrosion are given in Chapter 19.

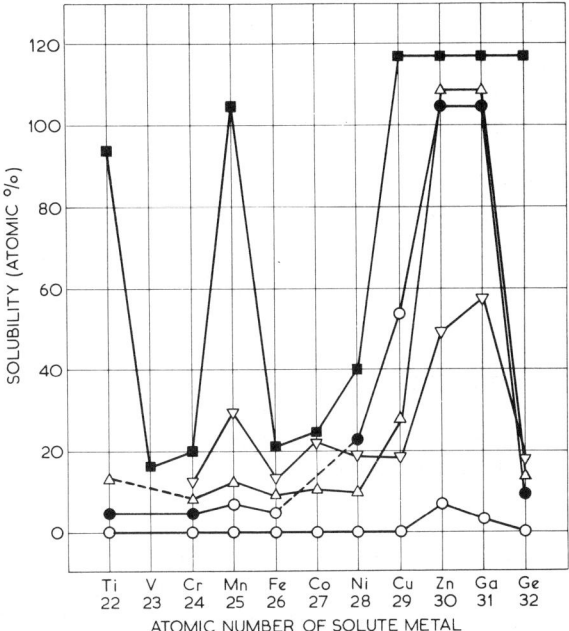

Fig. 2.28 Solubilities of the first row of transition elements in five liquid metals. (To avoid overlap of points, the graphs have been set at different positions on the solubility axis)

	Solvent metal	Temperature (°C)	For true solubility, deduct from axis reading:
○	Hg	20	—
●	Mg	700	4%
△	Sn	450	8%
▽	Bi	400	12%
■	Cu	1 200	16%

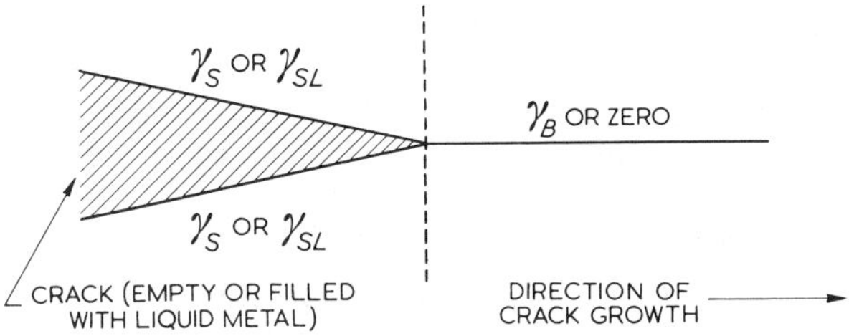

Fig. 2.29 Energies involved in the growth of a crack

When this process is accompanied by stress, catastrophic failure can occur, a classical example being the action of mercury on brass.

The situation may be described in terms of the surface-energy changes when a crack propagates through a solid metal as shown in Fig. 2.29 where γ_S is the solid–gas interfacial energy, γ_{SL} is the solid–liquid interfacial energy and γ_B is the grain-boundary energy.

Tabulated below are the energy changes involved for different cracking modes, with numerical values for the case of copper in contact with liquid lead with a dihedral angle at a copper grain boundary of 90° and $\gamma_S = 1\cdot 8 \text{ J/m}^2$, $\gamma_B = 0\cdot 6 \text{ J/m}^2$ and $\gamma_{SL} = 0\cdot 4 \text{ J/m}^2$:

Transgranular:	$2\gamma_S = 3\cdot 6 \text{ J/m}^2$
Grain-boundary cracking:	$2\gamma_S - \gamma_B = 3\cdot 0 \text{ J/m}^2$
Grain-boundary cracking in the presence of liquid metal.	$2\gamma_{SL} - \gamma_B = 0\cdot 2 \text{ J/m}^2$
Transgranular cracking with the crack filled with liquid metal:	$2\gamma_{SL} = 0\cdot 8 \text{ J/m}^2$

It is seen that the presence of the liquid metal greatly lowers the surface-energy change for grain-boundary cracking[3].

Temperature Gradient

The solubility of metals, S, in molten metals generally varies with temperature according to the relationship

$$\log S = A - (B/T)$$

where A and B are constants for a given system. It is therefore possible for more material to dissolve from a container at its highest-temperature end than at the low-temperature end, and if the melt flows round the container by natural or forced convection the liquid arriving at the cold region will be supersaturated and will precipitate solute until equilibrium is attained. If it is then recycled to the hot end it dissolves more metal until saturated and then returns to the cold end to precipitate this excess. This process is termed 'thermal gradient mass transfer'. It can best be illustrated by circulating a corrosive metal such as bismuth round a thermal convection loop of the type shown in Fig. 2.30. After a prolonged period of operation, thinning of the inner wall of the hot section and a precipitate on the wall of the cold section can be clearly observed. (The latter is often termed a *plug* since it eventually blocks the pipe to liquid flow.) This process has been analysed in some detail and the various stages are detailed in Fig. 2.31. The overall rate-controlling step has been shown by Ward and Taylor[4] to be diffusion through the boundary film of solute atoms into the flowing stream.

To be precise, they found that the solution of solid copper in liquid lead and bismuth obeyed the following equations.

Under static conditions, at temperature T

$$n_t = n_0[1 - \exp(-KS/V)]$$

where n_t = concentration of solute after time t, n_0 = saturation concentration of solute, S = surface area of solid exposed to liquid of volume V, $K = K_0 \exp[-\Delta E^+/RT]$ (ΔE^+ = activation energy for solution).

Under flowing conditions

$$\frac{dn_t}{dt} = K(S/V)(n_0 - n_t)$$

If, therefore, the solute atoms can be prevented from entering the boundary film from the solid the process will be halted. A method for doing this

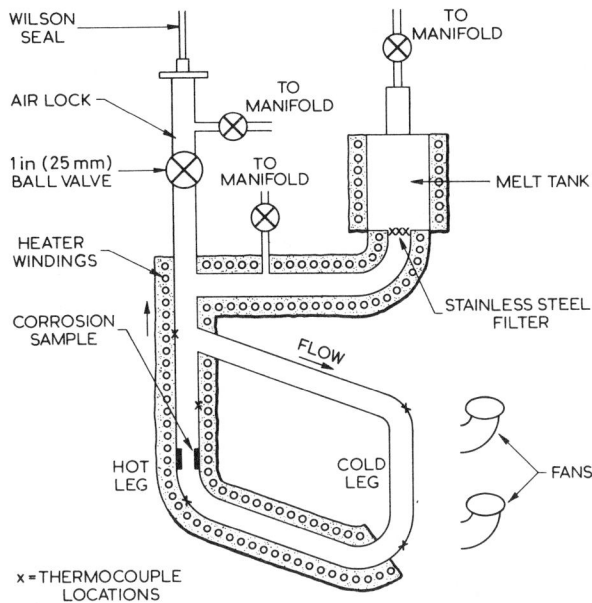

Fig. 2.30 Thermal convection loop

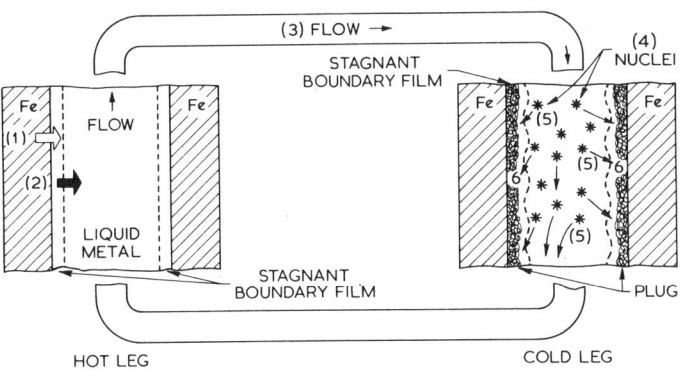

Fig. 2.31 Stages in thermal gradient mass transfer

1. Solution
2. Diffusion
3. Transport of dissolved metal
4. Nucleation
5. Transport of crystallites
6. Crystal growth and sintering (plug formation)

(after Brookhaven National Laboratory)

was discovered by workers at the U.S. General Electric Company some time ago when developing a mercury boiler for electric power generation[5]. They found that small quantities of dissolved titanium, zirconium, chromium, nickel and aluminium were effective as inhibitors of the corrosion of steels by hot mercury, the first two being particularly so. Later interest in the use of liquid bismuth as a carrier of uranium in a liquid-metal-fuelled reactor led to the extension of the use of zirconium inhibitor to bismuth in steel circuits and to an elucidation of the inhibiting mechanism[6,7]. The zirconium reacts with the nitrogen, which is always present in steel to the extent of about 100 p.p.m., to form a surface layer of ZrN which is thermodynamically a very stable compound and is an effective diffusion barrier. Furthermore, as long as there is residual zirconium in solution in the bismuth (or mercury) and dissolved nitrogen in the steel, the film is self-healing. Mercury boilers have operated successfully for thousands of hours relying on this principle.

In recent years there has been a continued interest in the use of alkali metals, notably sodium and lithium, as heat exchange media in nuclear reactors and fusion systems respectively and as chemical reactants in fuel cells. This interest is reflected in the proceedings of several major conferences which are referenced in the bibliography (see p. 2.109).

Generally speaking corrosion processes in liquid alkali metals are either concerned with dissolution of the component (general or selective), chemical reaction between the component and non-metallic impurities O_2, C, N_2, H_2, which are soluble in liquid sodium at the ppm level, or a combination of both processes where dissolution is followed by chemical reaction in the liquid phase. Solubilities of constructional materials—refractory metals and the components of iron and nickel base alloys—in liquid alkali metals are much less than in more noble metals, mercury and bismuth, and solubilities in liquid sodium at 650°C can range from a fraction of a ppm (refractory metals) to 1-10 ppm for metals like Fe, Cr and Ni. Elements such as Ni, Cu and the precious metals have appreciable solubilities in liquid lithium and it is generally considered that alloys of high nickel content have limited use in lithium systems operating with a temperature gradient.

In heat exchange systems corrosion processes are more concerned with dynamic, not static, liquid alkali-metal environments, consequently the previously quoted corrosion rate equation, is more applicable in this type of system. Resistances to mass transport may be present in either the hot or cooler parts of the circuit and the kinetics of mass transfer may be rate controlled either by dissolution or deposition at the solid surface, by transport of material through the adjacent laminar sub-layer or by phase boundary reactions in the solid metal. In diffusion-controlled mass-transfer situations involving turbulent fluids Epstein[8] has suggested that mass transfer equations can be derived from heat transfer analogies and expressions relating corrosion rate to the dimensionless groups. Reynolds No. (Re) and Schmidt No. (Sc) have been found to have some application where corrosion rates are sensitive to changes in flow velocity or diffusivity in the liquid phase. The equation suggested by Epstein to meet this situation is of the form:

$$\text{Rate} = 0.023 \, (D/d) \, C_W \, (\text{Re})^{0.8} \, \text{Sc}^{0.33}$$

in which C_W is the concentration of the dissolved species at the wall, D is the liquid phase diffusivity for the soluble species and d is the pipework diameter. The velocity term is incorporated in the Re No. expression.

Non-metallic impurities in liquid alkali metals play a major role in the corrosion of materials either by affecting metal solubilities, forming spallible corrosion products on the metal surface, promoting liquid metal embrittlement or bulk embrittlement of the surface or by sensitising the structure for further attack by other impurities e.g. O_2. As in other corrosive environments the direction and magnitude of these impurity reactions are dictated by free energy and solubility relationships for both solid and liquid phases.

In some metal components it is possible to form oxides and carbides, and in others, especially those with a relatively wide solid solubility range, to partition the impurity between the solid and the liquid metal to provide an equilibrium distribution of impurities around the circuit. Typical examples of how thermodynamic affinities affect corrosion processes are seen in the way oxygen affects the corrosion behaviour of stainless steels in sodium and lithium environments. In sodium systems oxygen has a pronounced effect on corrosion behaviour[9] whereas in liquid lithium it appears to have less of an effect compared with other impurities such as C and N_2. According to Casteels[10] Li can also penetrate the surface of steels, react with interstitials to form low density compounds which then deform the surface by bulging. For further details see non-metal transfer.

One important and perhaps unique feature of corrosion in the alkali metals is the formation of corrosion products based on complex ternary oxides. Horsley[11] has shown that oxides of the type $(Na_2O)_2 FeO$ can form when iron is corroded in sodium under conditions where the standard state binary oxides FeO, Fe_2O_3 and Fe_3O_4 are thermodynamically unstable. Weekes[12] in his analysis of those factors which affect corrosion behaviour, has also suggested this type of oxide may play a role in the corrosion of stainless steels in sodium. Addison[13] has shown that many transition metals form complex oxides with Na_2O and corrosion products based on the ternary oxide Na_3NbO_4 have been identified on the surfaces of niobium after exposure to the sodium containing oxygen[14]. This type of oxide is mechanically unstable in flowing sodium environments and therefore it is relatively easy to promote fresh surfaces for further attack by oxygen impurities. Complex oxides of the type $NaCrO_2$ also feature in the corrosion of stainless steels in alkali metals. $NaCrO_2$ for example may exist either as an oxide film on the surface of the steel during the initial stages of corrosion or it may, under more adverse conditions, penetrate the grain boundaries and become a precursor for grain detachment.

Impurity reactions can be controlled or eliminated by adequate purification of the liquid metals and in pumped loop systems this can be achieved by using techniques known as cold trapping and hot trapping. Cold trapping involves taking a small percentage of the main loop flow and by-passing it through a container which is cooled to the required temperature to precipitate out the impurities. Hot trapping on the other hand involves removal of impurities by chemical reaction between the soluble species and a material which has a higher thermodynamic affinity for the impurity than the liquid metal or its containment. In sodium systems cold trapping can remove oxygen impurities down to the 1–3 ppm level whereas hot trapping using Zr heated to 600°C can take the levels down to <1 ppm. Corrosion losses recorded on stainless steels exposed to flowing sodium at these low levels are very small[9] (Fig. 2.32). In certain systems where metal solubilities are relatively high (e.g. Ni in liquid lithium) the use of cold traps can encourage thermal gradient mass-transfer, consequently under these circumstances other methods of purification may be required.

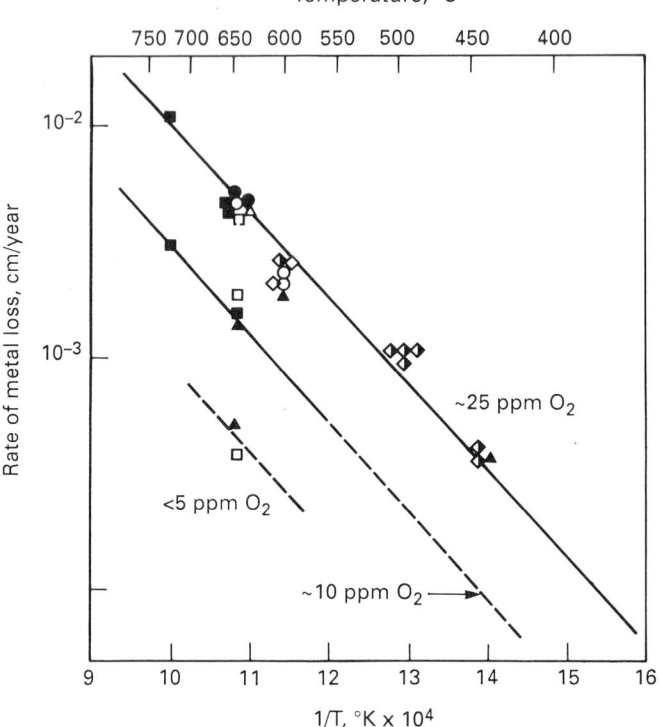

Fig. 2.32 Corrosion losses for stainless steels exposed to flowing sodium at differing oxygen levels

Chemical Gradient

In a container or a 'loop' in which a molten metal is flowing but where there is not necessarily a temperature gradient, a type of corrosion can occur by transfer of material along an activity gradient ('activity' being used here in the thermodynamic sense — see Section 20.1). Again it is convenient to discuss this phenomenon under two headings:

Metal Transfer

This is often referred to as 'isothermal mass transfer' since it does not require a thermal gradient. When two different metals are separated by a third, liquid, metal and one or both solid metals dissolve in the liquid, an alloying action may be observed. The liquid metal itself may form an intermetallic compound on the surface of one or both of the solid metals, in which case a diffusion barrier will build up and the reaction will gradually slow down. If, however, this does not occur, an alloy of the two solid metals will be formed either in the melt or on the surface of one of the solids. The most common effect is the formation of a layer at the surface of the less soluble of the two solid metals, this layer being made up of all or some of the compounds which are formed between the two solids. The rate-controlling step is the diffusion of the less soluble metal through this surface layer to react with the other species of solute atom. Some good examples have been provided by the studies by Covington and Woolf[15], who have investigated, among other systems, aluminium and molybdenum in molten lead, tin and bismuth. In bismuth, three of the four intermetallic compounds were identified in the layer on the molybdenum in the following order from the molybdenum surface: Al_3Mo, Al_5Mo and $Al_{12}Mo$. In tin only Al_3Mo was observed, while in lead two compounds were formed, the controlling process probably being the rate of diffusion through the layers of the different compounds and the rate of solution of the more soluble metal (aluminium) in the liquid metal.

Non-metal Transfer

The elements of primary importance in this context are oxygen, nitrogen, carbon and hydrogen. In the technology of the liquid alkali metals they play a predominant rôle. Their origin is associated with leakages in the circuit, impurities remaining after construction or residual impurities in the liquid metal. It is convenient to discuss these four elements separately.

Oxygen In cold-trapped sodium containing only a few p.p.m. of oxide, refractory metals such as Nb, V, Ta and Zr can suffer severe oxidation. Their behaviour is much the same as in oxidising gases at high temperature, forming non-adherent oxide films. Oxygen can also diffuse into the metal to produce solid solutions with consequent embrittlement. To prevent this occurring the oxygen level in the sodium is brought down below the p.p.m. level with a 'hot-trap'.

Nitrogen Nitrogen is virtually insoluble in pure sodium (10^{-4} p.p.m. at 500°C and 1 MN/m^2). It is, however, quite soluble in molten lithium (2 000 p.p.m. at 300°C and 100 kN/m^2) and can accelerate the already corrosive action of lithium, particularly on steels. Zirconium and yttrium have been used to 'getter' nitrogen.

Carbon The solubility of carbon in sodium has been measured; it is considered lower than the corresponding value for oxygen (2 p.p.m. of carbon at 520°C) but is sufficiently high to give rise to undesirable effects. Carburisation of refractory metals and of austenitic stainless steels has been observed in sodium contaminated with carbon; e.g. oil, grease or a low alloy ferritic steel the source of which can be either decomposed organic material, e.g. oil, or a ferritic steel of low or zero alloy content. The latter is an example of chemical-gradient transfer against the temperature gradient since the activity of carbon in Fe–18Cr–12Ni, possibly stabilised with titanium or niobium, is clearly lower than that in a plain carbon steel and there is, therefore, a driving force for carbon transfer. A deterioration in properties of both steels occurs, the austenitic becoming embrittled and the ferritic softened. The effect can be minimised if the carbon activity in the ferritic steel is reduced to that in the stainless steel by the incorporation in the steel of a 'carbon stabiliser' such as titanium. Hot trapping with zirconium removes carbon as well as residual oxygen, but generally carbon sources should be kept from liquid metal circuits containing materials sensitive to carburisation effects.

Hydrogen Sodium and the other alkali metals react with moisture to produce the oxide or hydroxide plus hydrogen. The latter dissolves to some extent in sodium and may transfer to refractory metals which dissolve hydrogen, e.g. zirconium, thereby causing embrittlement. It is probable that this will be serious only in systems running at fairly low temperatures, since hydrides are generally unstable at high temperatures where hydrogen diffuses very rapidly.

In summary for non-metal transfer situations chemical thermodynamics is a useful guide to probable behaviour. The transfer of a non-metal, X, dissolved in a molten metal, M', to another metal, M'', will depend on the relative free energies of formation of $M'X$ and $M''X$ (see Section 7.6). Thus sodium will give up oxygen to Zr, Nb, Ti and U, as the free energy of oxide formation of these metals is greater than that for sodium; on the other hand, sodium will remove oxygen from oxides of Fe, Mo and Cu unless double oxides are formed.

Summary

Corrosion by liquid metals can take the form of simple solution, alloying between the liquid metal and solid metal, intergranular penetration, impurity reactions, temperature gradient mass-transfer and dissimilar metal or chemical activity gradient mass-transfer. Experience has shown that in systems operating with a temperature or chemical activity gradient, mass-transfer can occur from one point in the system to another to produce metal loss, deposition of corroded material and changes in metallurgical structure. From an engineering stand-point these processes can affect the structural integrity of components and the thermal-hydraulic performance of heat exchange systems by reducing load carrying capability (corrosion), by fouling and blocking of tubes (deposition) and by degrading mechanical properties (structural changes).

G. LONG
A. W. THORLEY

REFERENCES

1. Kerridge, D. H., *Reactor Technol.*, **1**, 215 (1961)
2. Strachan, J. F. and Harris, N. L., *J. Inst. Met.*, **85**, 17 (1956–7)

3. Rostoker, W., McCaughey, J. M. and Markus, H., *Embrittlement by Liquid Metals*, Reinhold, New York (1960)
4. Ward, A. G. and Taylor, J. W., *J. Inst. Met.*, **85**, 145 (1956-7) and **86**, 36 (1957-8)
5. Reid, R. C., Amer. Soc. Mech. Engrs., Tech. Paper No. 51-S-13 (1951)
6. Frost, B. R. T. et al., *Proceedings of the Second Conference on the Peaceful Uses of Atomic Energy*, United Nations, Geneva, **7**, Paper P.270 (1958)
7. Klamut, C. J. et al., ibid., paper P.2406 (1958)
8. Epstein, L. F. 'Liquid Metal Technology', *Chemical Eng. Progress*, Symposium Series No. 20, **53**, 67
9. Thorley, A. W. and Tyzack, C., *Symposium on Alkali Metal Coolants, Vienna 1966*, International Atomic Energy Agency, Vienna, 97 (1967)
10. Casteels, F. et al., *Proc of 3rd Int Conf on Liquid Metal Engineering and Technology*, **3**, 73 (1984)
11. Horsley, G. W., *J. Iron St. Inst*, **182**, 43 (1956)
12. Weekes, J. R., Klamut, C. J. and Gurinsky, D. H., *Symposium on Alkali Metal Coolants, Vienna 1966*, International Atomic Energy Agency, Vienna, 97 (1967)
13. Addison, C. C. et al., *Int. Symposium on the Alkali Metals, Nottingham, 1966*, Spec Pub. No. 22, The Chemical Society London, (1967)
14. Thorley, A. W. and Tyzack, C.,*5th European Congress of Corrosion, Paris 1973*, Paper 56, (1973)
15. Covington, A. K. and Woolf, A. A., *Reactor Technol.*, **35** (1956) and 3 (1961)

BIBLIOGRAPHY

Various authors, *International Symposium on the Alkali Metals, Nottingham, 1966*. Special Publication No. 22, The Chemical Society, London (1967)

Various authors, *Symposium on Alkali Metal Coolants, Vienna, 1966*. International Atomic Energy Agency, Vienna (1967)

Various authors, *International Conference on Sodium Technology and Large Fast Reactor Design*, Argonne, Ill., U.S.A. ANL 7520, Pts. 1 and 2, N.B.S., U.S. Dept. of Commerce, Springfield, Va. (1968)

Various authors, *International Conference on Liquid Alkali Metals*, British Nuclear Energy Soc., Nottingham Univ., April (1973)

Various authors. *Proceeding of International Conference on Liquid Metal Technology in Energy Production*, Champion, Pennsylvania, Conf 760503-P1-P2 (1976)

Various authors. *Second International Conference on Liquid Metal Technology in Energy Production*, Richland, Washington, Conf. 800401-P1-P2 (1980)

Various authors. *Liquid Metal Engineering and Technology. Proceedings of 3rd International Conference*, Oxford, England, **1**, **2** and **3** (1984)

2.10 Corrosion in Fused Salts

Interest in the use of fused salts in industrial processes is continually increasing and these media are gradually becoming accepted as a normal field of chemical engineering. The change is being accelerated by the increasing demand for the newer refractory metals—often produced by processes involving fused salts—and also by the novel chemical engineering techniques which have been developed in the nuclear-energy industry. For example, a nuclear reactor using molten fluorides as a fluid fuel has operated, and this has involved the use of pumps, heat exchangers and similar equipment to circulate the high-temperature melt[1].

In certain applications it has not always been easy to find suitable metallic container materials, particularly in the nuclear-energy industry, where, for certain applications, corrosion resistance of the same order as that required by the fine chemical industry has to be achieved in order to prevent contamination of the process stream. Such difficulties have stimulated the study of corrosion in fused salts and have led to a fairly high degree of understanding of corrosion reactions in these media.

The subject is also closely related to fuel-ash corrosion which in most cases is caused by a layer of fused salts such as sulphates and chlorides[2,3]. Attention has been focused on the electrochemistry of this type of corrosion[4,5] and the relevant thermodynamic data summarised in the form of diagrams[6-8]. Fluxing and descaling reactions also resemble in some respects reactions occurring during the corrosion of metals in fused salts. A review of some of the more basic concepts underlying corrosion by fused salts (such as acid-base concepts and corrosion diagrams) has appeared[8].

Scope of the Present Treatment

There are two cases in which a metal can be attacked by a salt melt: if it is soluble in the melt, or if it is oxidised to metal ions. In the first case, attack occurs by direct dissolution without oxidation of the metal and the mechanism is likely to be closely similar to attack by liquid metals (Section 2.9). If the solubility is appreciable, excessive corrosion can be expected, but with few exceptions metals appear to be appreciably soluble only in their own salts. Most of the metals of the first and second groups of the periodic table are soluble in their own halides, and in certain cases there is complete miscibility at high temperatures[9-11].

The present survey will be confined to corrosion arising as a result of oxidation of the metal to ions, since little information on corrosion involving only metal-solubility effects is available.

General Principles

Electrochemically, the system metal/molten salt is somewhat similar to the system metal/aqueous solution, although there are important differences, arising largely from differences in temperature and in electrical conductivity. Most fused salts are predominantly ionic, but contain a proportion of molecular constituents, while pure water is predominantly molecular, containing very low activities of hydrogen and hydroxyl ions. Since the aqueous system has been extensively studied, it may be instructive to point out some analogues in fused-salt systems.

Setting aside for the moment effects due to passivity, a metal can be oxidised (corroded) only if it is in contact with something it can reduce; in an aqueous electrolyte, either hydrogen ions are displaced to form hydrogen gas, or some other solute (often dissolved oxygen) is reduced; in a salt melt, either one of the metal cations (corresponding to hydrogen ions in the aqueous solution) is displaced to produce metal, or some molecular species (such as dissolved chlorine or oxygen) or an ion (such as nitrate or ferric ion) is reduced. In aqueous electrolytes, therefore, metals which are too noble to displace hydrogen corrode only if dissolved oxygen or some other reducible substance is present, and when the reducible material has been used up, corrosion ceases; the metal has become 'immune', i.e. has reached electrochemical equilibrium with its environment. The position is similar in fused salt corrosion, and the relative nobilities of the salt melt and the metal are important. A noble metal in contact with a pure melt of a base-metal cation can react only to a very limited extent provided the anion is not reducible (but see References 4–8); for example, nickel cannot displace sodium from a sodium chloride melt to any appreciable extent and hence nickel will not corrode in molten sodium chloride unless some other reducible impurity is present.

The displacement reaction between a metal, M, and a salt melt, AB, can be written:

$$\underset{\text{metal}}{M} + \underset{\substack{\text{salt} \\ \text{melt}}}{AB} \rightleftharpoons \underset{\substack{\text{metal} \\ \text{corrosion} \\ \text{product}}}{MB} + \underset{\substack{\text{displaced} \\ \text{metal}}}{A} \qquad \ldots(2.16)$$

for which there is an equilibrium constant*

$$K = \frac{a_{MB} a_A}{a_M a_{AB}} \qquad \ldots(2.17)$$

The metal M and the salt AB, being pure substances, have an activity of one, while the corrosion product MB and the displaced metal A will in most cases

---
* See Sections 1.4 and 9.1.

be in solution in the melt at an activity of less than one. Where the metal M is noble and the cation A of the salt is base, the equilibrium constant (2.17) has a very small value, so that the activities of displaced metal A and corrosion product MB remain small. Provided there are no other factors present which might remove or combine with MB or A, the system will reach equilibrium after a small amount of reaction (corrosion) has occurred and the system will then be 'immune' against further corrosion. Distillation of the displaced metal A out of the melt is one of the factors which can prevent equilibrium being reached, and under such conditions corrosion can continue indefinitely. Such factors have been considered in some detail for chloride melts[12]

It is usual to choose a container metal for fused salts sufficiently noble for the displacement reaction (2.16) to be negligible, and the most important aspects of corrosion are, as in aqueous solutions, those which involve reducible impurities, although in a salt melt there is also the additional possibility of a reducible anion (see above). All such factors can be described as controlling the 'oxidising power' of the melt, which can be defined in terms of a 'redox potential' just as in aqueous solutions[13]. The redox potential is expressed by relationships of the form

$$E = E^\ominus + \frac{RT}{zF} \ln \frac{a_{\text{ox}}}{a_{\text{red}}} \qquad \ldots(2.18)$$

For example, the redox potential of a system containing oxide ions and oxygen* could be expressed as:

$$E = E^\ominus_{O_2/O^{2-}} + \frac{RT}{2F} \ln \frac{a_{O_2}^{\frac{1}{2}}}{a_{O^{2-}}} \qquad \ldots(2.19)$$

Similar equations can be formulated for other equilibria, e.g. NO_3^-/NO_2^-, Na^+/Na, etc.

The redox potential of a melt is a measure of its aggressiveness towards metals, and a metal in contact with the melt will react with it until its potential becomes the same as the redox potential of the melt[14]. The potential of the metal will depend on the activity of its ions in solution in the melt

$$E = E^\circ_M + \frac{RT}{zF} \ln a_{M^{z+}} \qquad \ldots(2.20)$$

so that the higher the redox potential (oxidising power) of the melt, the more will the metal have to oxidise (corrode) — giving M^{z+} ions — before equilibrium can be reached. (Oxidation of the metal will also result in reduction of the oxidising species present in the melt, which under certain conditions can cause the redox potential of the melt to fall during the course of the reaction, in which case equilibrium will occur at a potential rather lower than the initial redox potential of the melt[14].)

---

*Peroxide and superoxide ions have recently been identified in molten systems. These ions are formed by the oxidation of oxide ions by oxygen or oxy-anions such as nitrate (see Reference 11).

Conditions Under Which Corrosion Occurs in Practice

It follows from what has been said that the conditions under which metals are exposed to fused salts will in practice be of three types:

1. Conditions under which equilibrium between metal and salt melt can be reached and maintained.
2. Conditions under which equilibrium cannot be reached.
3. Conditions under which passivity is possible.

Item 3 has not been considered yet, so its importance in melts will be discussed first.

In metal/melt systems, just as in aqueous systems, one possible way of ensuring adequate corrosion resistance is to choose conditions such that the metal is passive, which requires that it should become covered with an adherent, compact, insoluble film or deposit, preventing direct contact of the metal with its environment. Any melt which reacts with a metal to give a corrosion product insoluble in the melt is in principle capable of passivating the metal, e.g. passivity can be expected to occur in oxidising salts in which metal oxides are sparingly soluble. Thus, iron is highly resistant to alkali nitrate melts because it becomes passive, and passivity has also been observed by electrode potential measurements of an iron electrode in chloride melts containing nitrates[14], although in this case the oxide corrosion product is not particularly protective. In general, fused salts are 'good' solvents for inorganic compounds so that passivity is not likely to be a widely encountered phenomenon.

In cases where passivity is impossible, corrosion can be prevented if the metal can reach equilibrium with the melt (case 1). The system usually undergoes some corrosion initially, when traces of oxidising impurities are reduced and the redox potential of the melt falls (Fig. 2.33). Finally, after a certain amount of corrosion has occurred, the metal becomes immune and corrosion ceases. In Fig. 2.33 complete equilibrium between metal and melt was still not quite reached even after several hundred hours exposure.

The equilibrium of metals with molten sodium hydroxide[15] and with fused alkali chlorides[14] has been studied in detail. Williams, Grand and Miller[15] studied the reaction of molten sodium hydroxide with nickel,

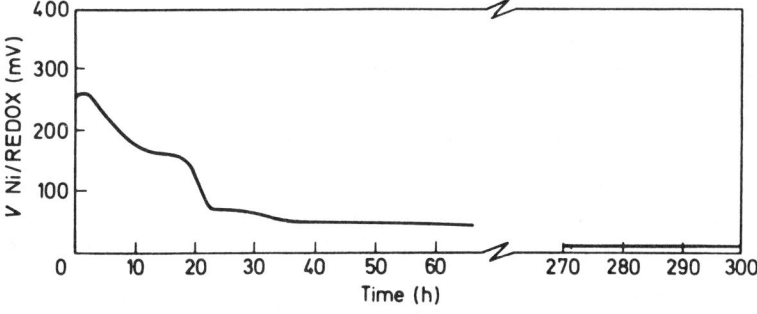

Fig. 2.33 Potential difference V between a redox electrode and a nickel electrode immersed in an alkali chloride melt; 700°C, argon atmosphere[14]

copper, Monel, gold, chromium, titanium, tantalum, beryllium, manganese, silver, cobalt, iron and ferrous alloys. The primary reaction can be represented as:

$$\text{NaOH} + \text{metal} \rightleftharpoons \text{Na}_2\text{O} + \text{metal oxide} + \text{H}_2$$

and for each metal at each temperature, equilibrium was reached at some characteristic and reproducible pressure. Fig. 2.34 shows the rate of approach to equilibrium for nickel at 800, 900 and 1000°C*.

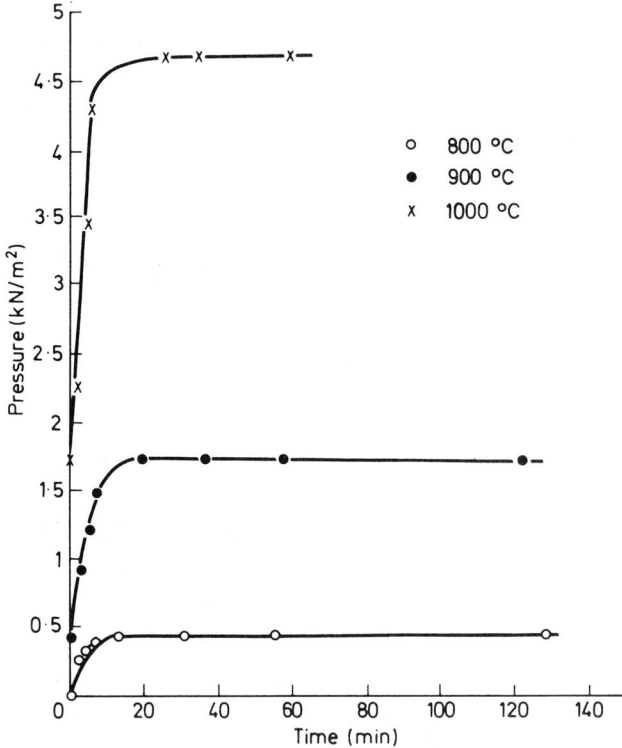

Fig. 2.34 Rate of pressure development over NaOH/Ni system[15]

The reaction can be inhibited by applying a hydrogen pressure at the beginning of the reaction, or alternatively can be driven to the right by sweeping away the hydrogen formed, e.g. by purging with a gas stream. Secondary reactions such as

$$\text{Na}_2\text{O} + \text{metal} \rightleftharpoons \text{metal oxide} + 2\text{Na}$$

can also be driven to completion by continually removing one of the reaction products, in this case by distillation of sodium metal. The course of equilibration of nickel with fused alkali chlorides was followed electro-

---

* Mathews and Kruh[15(a)], in similar experiments, were unable to obtain steady-state pressures, but this was probably due to the larger size of vessel used, to temperature gradients, and to difficulties due to 'creeping' of the fused sodium hydroxide up the vessel walls.

chemically by Littlewood and Argent[14], who showed the importance of geometric factors. The time taken for the metal to come to equilibrium with the melt depends on the amount of oxidising impurities present and on the ratio 'surface area of metal/volume of melt', being longer for a small surface area of metal in contact with a large volume of melt. Littlewood and Argent suggested that the kinetics were determined largely by diffusion processes in the melt. Equilibration was hindered when the chloride melts were in contact with glass containers which are not inert at high temperature.

Under certain conditions, it will be impossible for the metal and the melt to come to equilibrium and continuous corrosion will occur (case 2); this is often the case when metals are in contact with molten salts in practice. There are two main possibilities: first, the redox potential of the melt may be prevented from falling, either because it is in contact with an external oxidising environment (such as an air atmosphere) or because the conditions cause the products of its reduction to be continually removed (e.g. distillation of metallic sodium and condensation on to a colder part of the system); second, the electrode potential of the metal may be prevented from rising (for instance, if the corrosion product of the metal is volatile). In addition, equilibrium may not be possible when there is a temperature gradient in the system or when alloys are involved, but these cases will be considered in detail later. Rates of corrosion under conditions where equilibrium cannot be reached are controlled by diffusion and interphase mass transfer of oxidising species and/or corrosion products; geometry of the system will be a determining factor.

Special Features of Fused-salt Corrosion

'Electrochemical' corrosion of metals Since the aggressiveness of salt melts is governed by redox equilibria, and is often controlled by composition of the external atmosphere[16], effects analogous to 'electrochemical' or 'oxygen-concentration' corrosion in aqueous systems can occur in salt melts. Tomashov and Tugarinov[17] determined cathodic polarisation curves in fused chlorides and concluded that the cathodic reactions of impurities could be represented as:

$$O_2 + 4e \rightarrow 2O^{2-}$$
$$CO_2 \rightleftharpoons CO + O; \quad O + 2e \rightarrow O^{2-}$$
$$H_2O \rightleftharpoons H^+ + OH^-; \quad H^+ + e \rightarrow \tfrac{1}{2}H_2$$
$$SO_4^{2-} + 8e \rightarrow S^{2-} + 4O^{2-}$$
$$2NaNO_3 + 6e \rightarrow Na_2O + 2NO + 3O^{2-}$$
$$2NaNO_2 + 2e \rightarrow Na_2O + 2NO + O^{2-}$$

The anodic reaction was:

$$M - e \rightarrow M^+$$

Since metals are electronic conductors, the anodic and cathodic reactions will not necessarily occur at the same site, and 'anodic' and 'cathodic' areas can develop as in aqueous solutions. For example, 'wash-line' attack is often a feature of corrosion by fused salts in contact with air.

Thermal mass transfer (see also Section 2.9) When a temperature gradient exists in a system containing metal in contact with molten salt, thermal potentials are set up, causing removal of metal at high-temperature points and deposition of metal at cooler places[18,19]. This mass transfer is essentially different in nature from that met in liquid-metal corrosion, which is simply a temperature-solubility effect. In fused salts, both the corrosion and deposition reactions are electrolytic, and it has been shown that an electrical path is necessary between the hot and cold regions of the metal. Edeleanu and Gibson suggest that this type of mass transfer be called *'Faradaic mass transfer'* to indicate that it requires an electrolytic current[19a].

Mass-transfer deposits can lead to blockages in non-isothermal circulating systems, as in the case of liquid-metal corrosion. In fused salts, the effect can be reduced by keeping contamination of the melt by metal ions to a minimum; e.g. by eliminating oxidising impurities or by maintaining reducing conditions over the melt[19].

Selective attack Corrosion of alloys at high temperatures is complicated by effects due to diffusion, particularly where the alloy components have different affinities for the environment, and corrosion of an alloy in a fused salt at high temperature often exhibits features similar to those of internal oxidation. Selective removal of the less noble component occurs, and as it diffuses outwards, vacancies move inwards and segregate to form visible voids (Kirkendall effect). Since diffusion rates are faster at grain boundaries than in the grains, voids tend to form at the grain boundaries and specimens often have the appearance of having undergone ordinary intercrystalline corrosion. More careful examination has shown, however, that in the case of Fe-18Cr-8Ni corroding in a fused 50-50 NaCl/KCl melt at 800°C in the presence of air, the attack is not continuous at the boundaries, and the voids formed are not in communication with each other[20]. In high-nickel alloys, a greater proportion of voids are formed within the grains and the appearance of intercrystalline attack is less marked[20-24]. When Inconel is exposed to fused sodium hydroxide, a two-phase corrosion-product layer is formed, resulting from growth of the reaction product — a mixture of oxides and oxysalts — into the network of channels[22].

Selective removal of the less noble constituent has been demonstrated by chemical analysis in the case of nickel-rich alloys in fused caustic soda[21-23] or fused fluorides[24], and by etching effects and X-ray microanalysis for Fe-18Cr-8Ni steels in fused alkali chlorides[20]. This type of excessive damage can occur with quite small total amounts of corrosion, and in this sense its effect on the mechanical properties of the alloy is comparable with the notorious effect of intercrystalline disintegration in the stainless steels.

Practical Determinations of 'Corrosion Rates'

From what has been said already, it is clear that determinations of 'corrosion rates' from small-scale experiments must be treated with great caution. If the metal cannot passivate, it will corrode until it becomes immune, at which point the corrosion rate will fall to zero; between initial exposure and the attainment of immunity the corrosion rate will be continually changing. If, on the other hand, it is impossible for the metal to come to equilibrium with the melt, then the rate of corrosion, although probably constant, will be primarily controlled by diffusion and interphase mass-transfer rates, and the geometry of the system will be an overriding factor. For this reason, it is not always possible to correlate the results of different workers under apparently similar conditions, nor can such results be expected to correspond particularly closely to the amount of corrosion encountered in larger-scale apparatus.

It is not worth while, therefore, to give a digest of experimentally determined 'corrosion rates', but Table 2.21 indicates some sources of such data and their nature. (Some references to data on compatibility of fused salts with non-metallic materials have been included for the sake of completeness.) It should be remembered, that in the case of alloys, failure usually arises from selective attack which causes porosity of the container, even though the wall may appear on casual inspection to be quite sound[26,27].

Table 2.21 Corrosion data in fused salts (see also Reference 37)

Melt	Metal	Remarks	Reference
NaOH	Ni, Cu, Monel, Au, Ag, Cr, Fe, Ti, Be, Ta, Mn	System allowed to reach equilibrium	15
NaOH	Inconel		22
NaOH	High-Ni alloys		21
NaOH	Ni	Including thermal mass transfer	19
NaOH, LiOH, KOH	Ni, Cu, Fe, stainless steel		32
Fused alkalis	Various		25
$LiNO_3$, $NaNO_3$, KNO_3	Ni, Cu, Duralumin		30
$BaCl_2$/KCl/NaCl	Fe–Ni–Cr alloys and mild steel	With reference to salt-bath heat treatment pots	26, 27
NaCl, KCl, $CaCl_2$	Fe		17
KCl, NaCl, LiCl	Ni, Cu and 3 steels		28
NaCl/KCl and LiCl/KCl mixtures	Fe		29
$BaCl_2$/KCl, $CaCl_2$/NaCl and $MgCl_2$/KCl eutectic mixtures	Fe		31
Carnallite ($MgCl_2 \cdot KCl \cdot 6H_2O$)	Fe	Especially effect of dehydration	33
KCl, NaCl and K_2TaF_7 mixture	Inconel		23
Fluoride mixtures	High-Ni alloys		24
Various	Various	Includes glasses and ceramics	34
Various	Various	Includes glasses	35

One interesting feature of comparative experiments with a series of salts having a common anion is that the aggressiveness of the salts towards metals is dependent on the nature of the cation. The aggressiveness of chloride melts in contact with air is in the order[17,28]:

$$LiCl \sim MgCl_2 \sim CaCl_2 \gg NaCl > KCl$$

In the case of $CaCl_2$ and NaCl, the order corresponds with the corrosion behaviour expected from cathodic polarisation curves[17]. The order of aggressiveness of chlorides can also be explained on the basis of redox potentials of the melts, calculated on thermodynamic grounds from the free energies of formation of the appropriate oxides and chlorides[36]. The order of aggressiveness of nitrates is complicated by passivity effects[30], while that of alkalis in contact with air is [32]:

$$KOH > NaOH > LiOH$$

This is the reverse order to the aggressiveness of chlorides and indicates that the mechanism of corrosion in the two systems is different, i.e. in the latter case it involves the discharge of hydrogen as in acid aqueous solutions.

Measures to Reduce Corrosion

The obvious method of reducing corrosion in fused salts is to choose a system in which either the metal can come to equilibrium with the melt, or else truly protective passivity can be attained. In most cases in industry neither of these alternatives is used. In fact, fused salt baths are usually operated in air atmosphere, and the problem is the prevention of *excessive* corrosion. This can be done in two ways, *(a)* by reducing rates of ingress of oxidising species (mainly O_2 and H_2O) from the atmosphere, and rates of their diffusion in the melts, and *(b)* by keeping the oxidising power (redox potential) of the melt low by making periodic additions to the bath.

Access of air to the melt can be reduced by fitting a cover over it, or by floating a layer of powdered graphite on the melt surface. Diffusion rates in the melt can be reduced only by lowering the temperature, and this invariably reduces the amount of corrosion (except in certain cases in nitrate melts where passivity effects occur[30]).

Additions made to salt baths to lower the oxidising power of the melt are known as 'regenerators'. In 'neutral' chloride salt baths (i.e. those causing neither carburisation nor decarburisation of steel), regenerators fall into two classes, *viz.* alkalis and reactive metals. Alkali additions raise the O^{2-} ion activity in the melt, and since the oxygen partial pressure over the melt is constant, this results in a fall in redox potential (equation 2.19). Reactive metals combine with oxidising impurities in the melt, and in general the more base the metal the greater its effectiveness. The baser metals, such as sodium or potassium, are the most reactive, but would be inconvenient in practice. Magnesium is sometimes used, but slightly more noble metals with volatile chlorides, such as silicon, are commoner. Evaporation of the $SiCl_4$ as it is formed drives the reaction to completion. Besides reducing the aggressiveness of salt baths, certain regenerators also affect the carburising tendency of salt baths, but the subject is outside the scope of this discussion.

Recent Developments

Helpful surveys and reviews can now be found in the literature and readers are invited to consult them. Janz and Tomkins gave a critical data collection[37] from pre-79 publications (458 ref.). Rahmel[38] and later, Numata et al.[39] have dealt with the general principles of corrosion and have given some examples concerning alkali halides and oxysalts which are the most interesting ones with regard to their occurring in several applications. More recently, Rapp[40] and Rameau et al.[41] have discussed metal corrosion by gases in the presence of condensates ('Hot Corrosion'). So, only some additional informations relating to publications of the past decade will be given here.

Corrosion of Metals and Alloys

Molten halides are largely used for electrowinning of metals, alloys, and gases, and in high-temperature batteries. Feng et al.[42] have shown that Fe, Co, Ni, Cu and Mo are considerably less corroded in molten LiCl-KCl eutectic when this melt contains lithium oxide which is due to oxide film formation. Recently, corrosion of Cu and Mg was investigated in HF-KF mixtures[43-44] because of their use as conducting busbars in fluorine electrowinning. Copper busbars are preferred in low acidic mixtures while magnesium is a more corrosion resistant material in high acidic and low temperature mixtures. *Sulphate melts* have been investigated in relation with their role in the 'Hot Corrosion', in the presence of aluminium ions[45] or ferric ions[46] which make the melts more aggressive, e.g. a deleterious effect on iron corrosion due to an increased Fe^{3+} content of the salt was observed for low concentrations of this ion ($<12\,mol\%$)[46]. The thermochemical stability of the molten *Nitrate-Nitrite mixtures* and corrosion of iron and stainless steels by these melts were extensively studied (as a function of temperature and oxoacidity) in relation with their use as a coolant and storage fluid in solar thermal electric power plants[47-49]. In particular, passivation of iron is observed only in a narrow acidity domain where $NaFeO_2$ can be formed. It was also demonstrated that a nitriding process appears only as a consequence of the oxidation process[48]. Hsu et al. have investigated the corrosion resistance of chromium[50] and iron[51] in presence of *Carbonate melts* ($Li_2CO_3-K_2CO_3$) at 650°C, in connection with their use as current collector materials in molten carbonate fuel cells. Thermochemical calculations were carried out to establish phase-stability diagrams for the (Fe and Cr)−Li−K−C−O systems, and good agreement was observed between the predicted (deduced from calculated phase stability diagrams) and the experimentally observed corrosion products.

Corrosion of Non-oxide Ceramics

In view of their importance, corrosion of non-oxide ceramics were studied in chloride, sulphate and carbonate melts. Tressler et al. have[52] studied the corrosion resistance at 1000°C of commercially available SiC and Si_3N_4 ceramics against pure Na_2SO_4 and NaCl and their eutectic mixture. The corrosiveness of these salts decreases in the following sequence: $Na_2SO_4 > Na_2SO_4 + NaCl \gg NaCl$, showing the dependence of the corrosion effects on the free-oxide ion activity. Si_3N_4 is much less reactive than SiC. It was also reported that in addition to corrosive etching, contaminant ion (Na, Cl and S) penetration was noticeable and hence mechanical properties might be altered. The severe attack on SiC by Na_2SO_4 (and also by Na_2CO_3) was later confirmed by Jacobson and Smialek[53].

Stress Corrosion Cracking

Very few studies concern the stress corrosion cracking phenomena in molten salts. We can point out the works of Blackburn *et al.*[54] concerning the behaviour of the Ti-8Al-1Mo-1V alloy in molten LiCl-KCl at 350°C, and that of Conseil[55] who has studied the corrosion fatigue of 304L, 316L and 321 stainless steels in molten nitrates. Recently, Atmani and Rameau[56,57] have described a tensile apparatus suitable for corrosion tests in molten salts. The behaviour of 304L stainless steel was studied in molten $NaCl$-$CaCl_2$ at 570°C using either a constant strain rate or a constant load technique. Intergranular corrosion fracture was shown and the role of $M_{23}C_6$ precipitation in the crack propagation was evidenced.

Concluding Remarks

'Material/Molten Salt/Gas' System Chemistry

As demonstrated by the recent works on corrosion, it is worthwhile to have at our disposal (or to acquire) a perfect knowledge of the whole environmental conditions i.e. of the considered molten salt chemistry and its dependence on the nature of the cover gas. This allows us to predict, from a thermodynamic point of view, the operating conditions leading to reduced corrosion effects, the consistency of which is being further experimentally tested.

How to Lower the Corrosion Effects?

Besides the measures described in the main part of this paper, significant improvement of the corrosion resistance of a material can be obtained by use of inhibitors. In this case, good results seem to be achieved with rare earths[58-60] employed in low amounts as alloying metals or as protective dense oxide coatings. The use of coatings often means the research into the best compromise between mechanical properties and corrosion resistance of the material.

D. INMAN
G. PICARD

REFERENCES

1. Bettis, E. S., *et al.*, *Nuclear Sci. Engng.*, **2**, 804, 841 (1957)
2. Shirley, H. T., *J. Iron St. Inst.*, **182**, 144 (1956)

3. Jarvis, W. R., *Chem. & Ind. (Rev.)*, 1 687 (1961)
4. Burrows, B. W. and Hills, G. J., *J. Inst. Fuel*, **39**, 168 (1966)
5. Cutler, A. J. B., Hart, A. B., Fountain, M. J. and Holland, N. H., A.S.M.E. Publication 67-WA/CD-4 (1967)
6. Rahmel, A., *Electrochim. Acta*, **13**, 495 (1968)
7. Bombara, G., Baude, G. and Tamba, A., *Corrosion Sci.*, **8**, 393 (1968)
8. Inman, D. and Wrench, N. S., *Brit. Corros. J.*, **1**, 246 (1966)
9. Blander, M. (Ed.), *Molten Salt Chemistry*, Interscience, New York (1964)
10. Sundheim, B. R. (Ed.), *Fused Salts*, McGraw-Hill, New York (1964)
11. Mamantov, G. (Ed.), *Molten Salts*, Marcel Dekker, New York (1969)
12. Edeleanu, C. and Littlewood, R. *Electrochim. Acta*, **3**, 195 (1960)
13. Littlewood, R. and Edeleanu, C., *Silicates Industr.*, **26**, 447 (1961)
14. Littlewood, R. and Argent, E. J., *Electrochim. Acta*, **4**, 155 (1961)
15. Williams, D. D., Grand, J. A. and Miller, R. R., *J. Amer. Chem. Soc.*, **78**, 5 150 (1956)
15a. Mathews, D. M. and Kruh, R. F., *Industr. Engng. Chem.*, **49**, 55 (1957)
16. Littlewood, R. and Argent, E. J., *Electrochim. Acta*, **4**, 114 (1961)
17. Tomashov, N. D. and Tugarinov, N. I., *Zh. Prikl. Khim.*, **30**, 1 681 (1957)
18. Smith, G. P., Steidlitz, M. E. and Hoffman, E. E., *Corrosion*, **13**, 56lt (1957)
19. Lad, R. A. and Simon, S. L., *Corrosion*, **10**, 435 (1954)
19a. Edeleanu, C. and Gibson, J. G., *J. Inst. Met.*, **88**, 321 (1960)
20. Edeleanu, C., Gibson, J. G. and Meredith, J. E., *J. Iron St. Inst.*, **196**, 59 (1960)
21. Smith, G. P. and Hoffman, E. E., *Corrosion*, **13**, 627t (1957)
22. Smith, G. P., Steidlitz, M. E. and Hoffman, E. E., *Corrosion*, **14**, 47t (1958)
23. Bakish, R. and Kern, F., *Corrosion*, **16**, 533t (1960)
24. Manly, W. D. *et al.*, *Proceedings of the Second Conference on the Peaceful Uses of Atomic Energy*, United Nations, Geneva, 1958, Pub. No. A/CONF/15/P/1990
25. Evans, U. R., *The Corrosion and Oxidation of metals*, Arnold, London, 356–357 (1960)
26. Jackson, J. H., *Alloy Cast. Bull.*, No. 16, 1 (1952)
27. Jackson, J. H. and LaChance, M. H., *Trans. Amer. Soc. Metals*, **46**, 157 (1954)
28. Gurovich, E. I., *Zh. Prikl. Khim.*, **27**, 395 (1954)
29. Kochergin, V. P. and Stolyarova, G. I., *Zh. Prikl. Khim.*, **29**, 730 (1956)
30. Gurovich, E. I., *Zh. Prikl. Khim.*, **29**, 1 358 (1956)
31. Kochergin, V. P., Garpinenko, M. S., Skornyakova, O. N. and Minullina, M. S., *Zh. Priki. Khim.*, **29**, 621 (1956)
32. Gurovich, E. I., *Zh. Prikl. Khim.*, **32**, 836 (1959)
33. Kochergin, V. P., Kabirov, A. V. and Skornyakova, O. N., *Zh. Prikl. Khim.*, **27**, 883 (1954)
34. Mackenzie, J. D., In *Physicochemical Measurements at High Temperatures*, Eds. J. O'M. Bockris, J. L. White and J. D. Mackenzie, Butterworths, London, 334 (1959)
35. Lunden, A., *Corrosion Sci.*, **1**, 62 (1961)
36. Littlewood, R., *J. Electrochem. Soc.*, **109**, 525 (1962)
37. Janz, G. J. and Tomkins, R. P. T., *Corrosion*, Houston, **35**, 485 (1979)
38. Rahmel, A., in *Molten Salt Technology*, Lovering, D. G., (Ed.) Plenum Press, New York and London, 266 (1982)
39. Numata, H., Nishikata, A. and Haruyama, S., *Boshoku Gijutsu*, **33**, 103 (1984)
40. Rapp, R. A., *Corros. Sci.*, **42**, 568 (1986)
41. Rameau, J. J., Duret. C., Morbioli, R. and Steinmetz, P., in '*Corrosion des Matériaux à Haute Température*', Béranger, G., Colson, J. C., Dabosi, F., (Eds) Les Editions de Physique, France, 527 (1985)
42. Feng, X. K. and Melendres, C. A., *J. Electrochem. Soc.*, **129**, 1245 (1982)
43. Rouquette, S., Ferry, D. and Picard, G., in *Molten Salts*, C. L. Hussey, S. N. Flengas, J. S. Wilkes and Y. Ito (Eds) The Electrochemical Society softbound series, Vol 90–17, p. 492; Rouquette, S. and Picard G., *Electrochim. Acta*, **38**, 487 (1993)
44. Germanaz, P., Nicolas, F., Rouquette, S., Lamirault, S., Ferry, D. and Picard, G., *Materials Science and Engineering*, A120, 329 (1989)
45. Griffiths, T. R., King, K. and Marchant, S. H., *Proc 4th Int. Symp. on Molten Salts*, The Electrochem. Soc. ed., **84–2**, 451 (1984)
46. Numata, H., *Corros. Sci.*, **44**, 724 (1988)
47. Picard, G., Flament, T., Trémillon, B., Saint-Paul, P. and Spiteri, P., Fr. Patent # 82 16201, **2 533 578** (1982)

48. Picard, G. S., Flament, T. A., and Trémillon, B. L., *J. Electrochem. Soc.*, **132**, 863 (1985); Spiteri, P., Saint-Paul, P., Picard, G., Lefebvre, H. and Trémillon, B., *Materials Science and Engineering*, **87**, 369 (1987); Picard, G. S., Lefebvre, H. M. and Trémillon, B. L., *J. Electrochem Soc.*, **134**, 52 (1987); *Proc. the joint Int. Symp. on Molten Salts*, The Electrochem. Soc. ed., **87-7**. 1028 (1987); Spiteri, P., *Procèsverbal EDF-DER D 499/T 41*, **28-12** (1981); D 554/T 40, **1-6** (1984)
49. Fernandez, H., Carter, J. M. and Osteryoung, R. A., *Proc. 4th Int. Symp. on Molten Salts*, The Electrochem. Soc. ed., **84-2**, 468 (1984)
50. Hsu, H. S. and De Van, J. H., *J. Electrochem. Soc.*, **133**, 2077 (1986)
51. Hsu, H. S, De Van, J. H. and Howell, M., *J. Electrochem. Soc.*, **134**, 3038 (1987)
52. Tressler, R. E., Meiser, M. D. and Yonushonis, T., *J. Am. Ceram. Soc.*, **59**, 278 (1976)
53. Jacobson, N. S. and Smialek, J. L., *J. Electrochem. Soc.*, **133**, 2615 (1986)
54. Blackburn, M. J., Feeney, J. A. and Beck, T. R., *Advances in Corrosion Science and Technology*, Fontana, M. G. and Staehle, (eds.) **3**, Plenum Press, New York (1973)
55. Conseil, B., Thesis, Poitiers—France (1981)
56. Atmani, H. and Rameau, J. J., *Corros. Sci.*, **24**, 279 (1984); **27**, 35 (1987); *Materials Science and Engineering*, **88**, 221 (1987)
57. Rameau, J. J. and Atmani, H., *Materials Science and Engineering*, **88**, 247 (1987)
58. Okanda, Y., Fukusumi, M., Nenno, S. and Newkirk, J. B., *Metall. Trans.*, **14A**, 2131 (1983)
59. Okanda, Y., Fukusumi, M., Mizuuchi, K. and Nenno, S., *Trans. Jpn. Inst. Met.*, **27**, 680 (1986)
60. Séon, F. M., *J. Less-Common Met.*, **148**, 73 (1989)

2.11 Corrosion Prevention in Lubricant Systems

Lubricants are not generally regarded as being corrosive, and in order to appreciate how corrosion can occur in lubricant systems it is necessary to understand something of the nature of lubricants. Once, lubricants were almost exclusively animal or vegetable oils or fats, but modern requirements in the way of volume and special properties have made petroleum the main source of supply. In volume, lubricants now represent about 2% of all petroleum products; in value, considerably more.

There are many hundreds of different varieties of lubricants, many of them tailored to meet particular requirements. Lubricating greases are solid or semi-solid lubricants made by thickening lubricating oils with soaps, clays, silica gel or other thickening agents. Synthetic lubricants, which will operate over a very wide range of temperature, have been developed mainly for aviation gas-turbine engines. These are generally carboxylic esters and are very expensive products.

The main function of most lubricants is to reduce friction and wear between moving surfaces and to abstract heat. They also have to remove debris from the contact area, e.g. combustion products in an engine cylinder, swarf in metal-cutting operations. Sometimes they have to protect the lubricated or adjacent parts against corrosion, but this is not a prime function of most lubricants. On the other hand, many lubricants do contain corrosion inhibitors and some lubricating oils, greases, mineral fluids and compounds are specially formulated to prevent the corrosion of machinery or machine parts, particularly when these components are in storage or transit. These temporary protectives are described in Section 17.3.

The Manufacture of Lubricating Oils

Mineral lubricants may be distillates or residues derived from the vacuum distillation of a primary distillate with a boiling point range above that of gas oil[1,2,3]. They are mixtures of hydrocarbons containing more than about 20 carbon atoms per molecule, and range from thin, easily flowing 'spindle' oils to thick 'cylinder' oils. For hydrocarbons having the same number of carbon atoms per molecule, the higher the proportion of carbon to hydrogen, the more viscous the oil and the lower the viscosity index.

Distillate lubricating oils can be conveniently divided into three groups — low viscosity index oils (LVI oils), medium viscosity index oils (MVI oils) and high viscosity index oils (HVI oils). LVI oils are made from naphthenic distillates, with low wax contents so that costly dewaxing is not required. MVI oils are produced from both naphthenic and paraffinic distillates; the paraffinic distillates have to be dewaxed. HVI oils are prepared by the solvent extraction and dewaxing of paraffinic distillates. Solvent extraction is a physical process which removes the undesirable constituents, thereby improving viscosity index and the oxidation and colour stability. White oils are obtained by the more drastic refining of low viscosity lubricating oil distillates to remove unsaturated compounds and constituents that impart colour, odour and taste. They are usually solvent extracted and then repeatedly treated with strong sulphuric acid or oleum and alkali, and finally 'clay'-treated to remove surface-active compounds. Acid and clay treating is expensive and is being superseded by hydrofinishing, a catalytic hydrogenation treatment.

The residues from the vacuum distillation can also be refined to provide very viscous lubricants. The residues from paraffinic base oils are generally solvent extracted and dewaxed. The main use of these products (bright stocks) is as blending components for heavy lubricants. Thus residues from naphthenic base oils, which are also used as blending components for heavy lubricants, are normally not extracted.

The performance characteristics of a lubricating oil depend on its origin and on the refining processes employed, and in order to ensure consistent properties these are varied as little as possible. Some aero-engine builders insist on a complete re-evaluation of a lubricant, costing many thousands of pounds, whenever there is a change of source (crude) or refining process.

Corrosion and Corrosion Prevention in Lubricant Systems

Deterioration of lubricants in use Lubricating oils deteriorate in service in two ways, *viz.* they become contaminated and they undergo physical and chemical changes due to oxidation[4]. In engines the common contaminants are airborne dust and wear products, unburnt fuel, fuel combustion products and water. The oxidation products are mainly acidic materials and asphaltenes. Asphaltenes in association with fuel contaminants and water form sludges and lacquers. The acidic materials resulting from oxidation are generally weak organic acids although in extreme cases strong mineral acids may be produced. However, contamination by fuel combustion products is the source of almost all strong-acid contamination in lubricants[5]. Thus, distillate diesel fuels can contain up to 1% sulphur (by weight) and residual diesel fuels up to 3% sulphur. This sulphur is oxidised to sulphur acids, and sulphuric acid condensate may be encountered on the cooler surfaces[6]. In gasolines the sulphur content is very low but halogen compounds are added to the gasolines as scavengers for the lead-based anti-knock compounds. These halogen compounds can give rise to halogen acids[7].

Lubricants with a high aromatic content (LVI and some MVI oils) tend to oxidise to give sludge-forming compounds, although some naphthenic oils give organic acids. Paraffinic oils (HVI and some MVI oils) oxidise more slowly to give weak acids. In a plentiful supply of oxygen, oxidation proceeds at a significant rate at temperatures above about 60–130°C depending on the composition of the lubricant[8]. This oxidation is an extremely complicated process which involves the formation of organic peroxides as intermediates[9]. It is catalysed by the presence of metals, particularly copper, iron and lead[10].

The rôle of additives The progressive development of engines and general machinery resulting in more arduous operating conditions, and particularly the use of longer oil-change periods, means that neither straight mineral oils nor compounded oils (mineral oils to which a proportion of an animal or vegetable oil has been added) are adequate for modern service requirements.

Despite the introduction of new, improved methods of refining it has been necessary to enhance the performance of lubricants by the use of additives, either to reinforce existing qualities or to confer additional properties. Once additives were regarded with some suspicion—an oil that needed an additive was necessarily an inferior oil; today they are an accepted feature of lubricants. Almost all quality lubricants on sale today contain one or more additives. An enormous range of additives are available for use in lubricants[1,11-13], some produced by the oil companies and others provided by specialist manufacturers. Additives are usually named after their particular function, but many additives are multifunctional. Thus, an anti-wear additive may also protect a surface against corrosion. The main types of additives that can enhance the anti-corrosion behaviour of lubricants are listed in Table 2.22.

Extreme pressure additives Many additives, essential to the performance of the lubricant, provide no corrosion protection and some additives may become corrosive in certain circumstances. Thus EP (Extreme Pressure) additives contain chemical groups which are designed to react chemically with metal surfaces when normal lubrication fails, forming easily sheared layers of metal oxides, sulphides, chlorides or phosphates, thereby preventing catastrophic wear and seizure. Reaction between EP compounds and metal surfaces should only occur at local hot spots and the layers formed are extremely thin. However, if the operating conditions are very severe these layers are continually generated and removed as they fulfil their antiwear function. A process of this nature is sometimes called 'chemical' wear, and if sliding surfaces operate continually under these conditions loss of metal from the rubbing surfaces can ultimately result in failure. Alternatively, all the EP additive may be used up (depleted) and then failure by seizure will occur. EP agents are intended to cater for the occasional overload condition and it must be emphasised that machinery should be designed so that it does not require the continual action of EP agents to function satisfactorily.

Obviously the selection of an EP additive requires great care; if it is too active it may give rise to excessive metal removal under normal operating conditions (see the section on corrosion by sulphur additives). Also, if a component is prone to fatigue pitting in service the presence of an overactive EP agent may result in corrosion fatigue[14].

Table 2.22 Additives that can enhance the anti-corrosion behaviour of lubricants

Additive	Function	Typical chemical types
Oxidation inhibitor (anti-oxidant)	To increase oxidation reistance of lubricants by interfering with the chain reactions which give rise to acid and asphaltene formation	Oil-soluble amine and phenol derivatives for temperatures up to about 120°C. For higher temperatures dialkyldithiophosphates and other compounds listed below as metal deactivators; also P_2S_5-olefin and -terpene reaction products
Metal deactivator	To form inactive protective films on metal surfaces which otherwise might catalyse oxidation and corrosion reactions	Trialkyl and triaryl phosphites, organic dihydroxyphosphines, some active sulphur compounds, diamines; in lubricating greases, mercaptobenzothiazole and phosphites
Corrosion inhibitor	To protect metal surfaces, particularly bearing surfaces (copper, silver, lead) against corrosion	Zinc dialkyldithiophosphates, sulphurised olefins, sulphurised terpenes
Rust inhibitor	To eliminate rusting in presence of moisture	Polar compounds, such as metallic soaps, esters and derivatives of dibasic acids; barium and calcium sulphonates
Water repellant	To impart water-resistant properties, particularly in greases	Aliphatic amines, hydroxy fatty acids and some organic silicone polymers
Basic compound	To neutralise strong mineral acids (from fuels)	Barium or basic calcium salts of sulphonic acids and alkylsalicylic acid. 'Superbasic' and 'hyperbasic' additives are basic salts with a large excess of base
Dispersants	To keep insoluble combustion and oxidation products in suspension and dispersed	Salts of phenolic derivatives; polymers containing barium, sulphur and phosphorus; calcium or barium soaps of petroleum sulphonic acids

Additive interaction Modern high-performance lubricants contain a number of additives, each with a particular, special function. Thus a turbine lubricant may contain an oxidation inhibitor, a rust inhibitor, an extreme pressure (EP) agent and an anti-foam compound. A high-grade diesel-engine lubricant may contain a VI improver, a dispersant, an anti-oxidant, a corrosion inhibitor, a basic compound, a pour-point depressant and an anti-foam compound. Sometimes these additives may have undesirable side effects or interact adversely[1]; in a turbine oil the rust inhibitor may act as an emulsifier interfering with demulsification; in a diesel lubricant the dispersant may promote oil oxidation. Frequently anti-corrosion additives may not be able to exert their maximum effect because they are competing for sites on metal surfaces. The development of a successful new lubricating oil requires much skill and experience and always necessitates considerable laboratory and field testing in order to strike the right balance between the various additives.

Additives in greases Oil additives may be incorporated in the oil phase in a grease. However, with greases there is also the opportunity to employ oil-insoluble anti-corrosion agents, since these can be incorporated in the grease along with other thickening agents, e.g. sodium nitrite is incorporated in greases in this way.

Additive depletion Additives are consumed or 'depleted' as they fulfil their respective functions. It is for this reason, more than any other, that it is necessary to change engine oils regularly.

Corrosion by sulphur additives Sulphur compounds occur naturally in most lubricants and many oil additives contain sulphur. In a properly formulated lubricant these sulphur compounds should be inactive at ambient temperature. At elevated temperatures they may decompose to give more active materials which can stain and corrode metals, particularly silver and copper. However, these same sulphur compounds have many beneficial qualities[1,4,10]; this is why they are not removed completely in refining and why they are used as additives.

Thus, sulphur compounds in lubricants generally act as anti-oxidants, preventing acid and sludge formation. They also form very thin films on metal surfaces protecting them from acid or peroxide attack. In addition, sulphur compounds are often used as EP agents. The oil chemist must try to strike a balance—the activity of the sulphur must be high enough for it to exert a beneficial effect and yet not so high as to stimulate corrosion.

All too frequently lubricants containing sulphur are exposed to more severe operating conditions than intended, and staining and corrosion results. This has given sulphur compounds and sulphur-containing additives, particularly those of the dithiophosohate type, a bad name. Silver bearings are still used in certain diesel and aero-engines and if the lubricants for these engines contain sulphur compounds with too much chemical activity, severe corrosion ensues[5]. A more widespread problem is the corrosion of phosphor-bronze alloys[5,15] (containing about 10% tin) particularly in little-end bushes in diesel engines where temperature can exceed 200°C. Some engine builders and operators hold sulphur additives entirely responsible, but this opinion cannot be substantiated, for corrosion can occur with lubricants containing only natural sulphur compounds[5]. Two important metallurgical factors, affecting the corrosion resistance of phosphor-bronze alloys are the amount of alloying element in solution in the copper-rich phase and the porosity of the alloy. For example, if the amount of tin in solution can be increased by special casting techniques or heat treatments the corrosion resistance is greatly increased. Similarly, zinc or silicon in solution also increases the resistance of copper to sulphur corrosion. If the alloy is porous the lubricant is drawn into the pores where it stagnates, and, at high temperatures, becomes very corrosive, e.g. copper catalyses oil oxidation with the consequent formation of corrosive sulphur compounds.

The most satisfactory solution to this problem is to employ a corrosion-resistant alloy, and alloys of the gun-metal type, containing 2-4% zinc, have proved completely satisfactory. The substitution of zinc for phosphorus gives sounder castings and improves the corrosion resistance of the copper-rich matrix.

Corrosion in Specific Lubricant Systems

Lubricants are specially formulated to meet all manner of requirements and the major corrosion problems associated with important classes of lubricants will now be outlined.

Engine lubricants Engine lubricants are exposed to severe operating conditions, being subjected to high temperatures, the products of combustion and a plentiful supply of oxygen. Consequently, unless the oil is changed at appropriate intervals strong mineral acids and weak organic acids may accumulate. In addition, droplets of water may be formed and these can contain strong mineral acids derived from the fuel combustion gases. These droplets sometimes give rise to emulsions which deposit in the colder portions of an engine, e.g. on the rocker-box covers; ferrous surfaces are most affected by this condensed moisture. A special dynamic corrosion test has been developed to study corrosion in these two-phase (water-in-oil) systems[16]. Problems associated with the retention of water in engine lubricants are likely to become more acute as anti-pollution devices are fitted to engines. The harmful effects are best countered by anti-rust and basic additives, and in diesel engines burning high-sulphur fuels, e.g. marine diesel engines, very high levels of basicity are required.

Cast or sintered copper-lead or lead-bronze alloys are widely used for engine bearings. The lead phase in such bearings is readily attacked by weak organic acids and almost all the lead can be leached out unless preventive measures are employed[17]. The lead may be protected by a precision electro-deposited overlay of a lead-tin or lead-indium alloy[18]. About 3% tin or 5% indium in lead will render the lead resistant to attack by oil-oxidation acids[5,19]. One reason why leaded bearings are protected by an overlay, and not by incorporating the protective alloying elements in the underlying lead, is that both tin and indium dissolve preferentially in copper. In a cast or sintered bearing, therefore, any tin or indium will be found in solution in the copper-rich phase, leaving the lead-rich phase susceptible to attack.

In overlay bearings operating above about 140°C, the tin or indium in the overlay diffuses towards, and alloys with, the underlying copper, depleting the overlay and reducing its resistance to corrosion[5]. This depletion by diffusion can be combatted by the use of a diffusion barrier or 'dam', e.g. a nickel-rich layer between the bearing alloy and, the overlay[20].

In gasoline engines lead halides accumulate in the lubricant; occasionally these give rise to the corrosion of aluminium-alloy pistons[7] and very rarely to corrosion on aluminium-tin bearings[5].

Steam-turbine lubricants Lubricants in steam turbines are not exposed to such arduous conditions as those in engines. The main requirement is for high oxidation stability. However, they may be exposed to aqueous condensate or, in the case of marine installations, to sea water contamination, so they have to be able to separate from water easily and to form a rust-preventing film on ferrous surfaces, and it is usual to employ rust inhibitors. The problem of tin oxide formation on white-metal bearings is associated with the presence of electrically conducting water in lubricants and can be over-come by keeping the lubricant dry[21,22].

Gear lubricants In addition to the usual oxidation and corrosion inhibitors, lubricants for heavily loaded gears almost always contain EP additives containing sulphur, chlorine or phosphorus. In order to function, these additives must react locally with the metal surfaces, and yet the extent of the reaction should not be such that it could be described as corrosive, or promote fatigue pitting[1]. These EP additives may be quite safe with ferrous metal surfaces, but may cause severe corrosion on copper alloys, e.g. on bronze worm wheels[12] if for any reason excessive temperatures arise.

Some turbine lubricants have to lubricate the turbine gears as well as the turbine; in these circumstances any EP additives employed should not be corrosive in the presence of moisture[12].

Metal-working lubricants Metal-working lubricants can be divided into two categories, *viz.* cutting lubricants and forming lubricants. Three types of cutting fluids are widely used: soluble oils, water-base fluids and straight cutting oils. Soluble oils are low-viscosity mineral oils which when added to water form stable oil-in-water emulsions. Their main function is to cool rather than to lubricate, but they should contain rust inhibitors to prevent corrosion of both the workpiece and the machine. Water-base fluids contain no oil and employ the normal corrosion inhibitors used in coolants, e.g. sodium nitrite, sodium benzoate and triethanolamine phosphate. Straight cutting oils are used in severe operations where good lubrication is essential, and generally contain EP agents. These EP agents must be carefully selected so that they do not stain the workpiece or the machine tool[12]. For example, very active sulphur compounds should not be employed with copper alloys[23]. Chlorinated additives should be stable in moist air to avoid the risk of hydrochloric acid formation.

Forming lubricants include lubricants for rolling, drawing, extruding and forging. A vast range of compounds including fatty oils and compounded oils are used in these operations, and a major requirement is that they should not stain the workpiece during the forming operation or during subsequent annealing or in storage. Consequently, all oils and additives employed should be completely volatile, in addition to affording protection against rusting.

Lubricants for other applications Specially formulated lubricants are required for steam engines, compressors and exhausters, refrigerators, hydraulic equipment, textile machinery, transformers and switchgear, nuclear power plants and many other diverse applications. Most of these lubricants will contain a carefully balanced set of additives, including some to prevent corrosion either directly by protecting the metal surface, or indirectly by preventing deterioration of the lubricant and combatting the action of contaminants. When corrosion is encountered with a specialised lubricant, the cause is not likely to be any weakness on the part of the lubricant; it is probable that the lubricant is being exposed to operating conditions far more extreme than those for which it was designed.

Recent Developments in Lubricant Base Stocks and Additives

There have been no fundamental changes in the nature of lubricating oil base stocks or in the types of additives used to improve their performance. This enable existing sections to be updated by general references[24-39] so providing the opportunity to introduce topics that have assumed greater importance in recent years. Improvements in the performance of additive packages stem primarily from modifications to the molecular structure of existing additive types and from a better understanding of the interaction of additives with each other and with polar species in base stocks[24, 28b, 28e, 33, 34, 37, 38, 39]. Thus, effectiveness of basic (alkaline) additives has been greatly improved by increasing their solubility in base stocks, by exploiting synergistic action between two similar additive types and by the use of a third additive to enhance (catalyse) the performance of the other two. Similar principles have been used to augment the performance of oxidation inhibitors.

Deterioration of Lubricants in Services

Much greater attention is now given to oil condition monitoring[28e]. Lubricants are expensive and users have no wish to change them until necessary. On the other hand, great damage can be caused by continuing to use a lubricant that no longer meets specification. Methods for the testing and analysis of lubricants are the subject of many Company, National and International Standards[41-45]. Typical tests for used diesel engine oils include viscosity, fuel dilution, flash point, water content, ash level, insolubles and neutralisation value. Neutralisation value is a measure of the acidity TAN (Total Acid Number), or alkalinity, TBN (Total Base Number) of the oil. Laboratory examination of used oil samples and reporting results takes time. Results can be obtained more quickly by using do-it-yourself kits[46, 47]. Regardless of who carries out the analyses the lubricant supplier should be prepared to give guidance on analytical methods, interpretation of results and if an oil change is required.

Oil condition monitoring is often extended to include techniques that give advance warning of the deterioration or impending break-down of machinery, whether or not this is associated with loss of lubricant performance. The first, and still most widely used method is SOA[48] (spectrometric oil analysis) which can measure the concentrations of up to 21 metallic elements. Generally, sudden changes in amounts of elements present are considered more significant than absolute levels. Other monitoring techniques[28a, 49, 50, 51, 52] include Ferrography, particle counting and magnetic plugs, all of which can monitor the number, size and morphology of extraneous material. On-stream versions of all techniques except SOA are now available.

Health and Safety

The solvent action of mineral oil base stocks can cause skin problems and prolonged exposure may have been the origin of a few skin cancers[28d, 53]. The use of additives that might be in any way harmful to health, e.g. ortho-tricresyl phosphate (anti-wear) and sodium mercaptobenzothiazole (anti-corrosion) has been discontinued where skin contact is likely.

The presence of fungi and bacteria in water-base or water-contaminated lubricants such as machining fluids and marine diesel crank case lubricants may promote corrosion, but have not been shown to be harmful to man[54-58].

Corrosive Wear in Piston Engines

Modern, highly-efficient engines, requiring minimum maintenance, able to accept a wide range of fuels and meeting strict exhaust emission composition rules present special lubrication problems. In particular it is essential that the lubricant can prevent corrosion from fuel combustion products retained in the engine. Corrosive wear can be prevented by alkaline additives which neutralise the acids responsible, by additives that absorb them and prevent them reaching metal surfaces and by anti-rust additives that are adsorbed on metal surfaces and prevent the access of water and oxygen[34, 35, 59, 60, 61]. Often all three additive types will be used in association although their relative concentrations will depend on the nature of the corrosive agents. In spark-ignition engines the use of liquid methane, liquified petroleum gas (propane and butane), methanol, ethanol and lead-free gasoline can give specific corrosion problems[62-65]. Broadly, they generate organic acids of shorter chain length and greater corrosivity than normal leaded gasolines so higher additive levels are required.

Diesel-fuel sulphur levels can range from about 0.2% to 5.0%. The higher levels only occur in low and medium speed engines operating on residual fuels, but because of the high combustion pressures sulphuric acid condensate forms at temperatures up to 200°C. On cross-head engines, which have independent cylinder lubrication systems the neutralising power of the fresh oil depends on both oil feed rate and alkalinity[61]. The alkalinity is measured as TBN[41, 42] and lubricants with a TBN of 100 are now available. This means that the lubricant has an acid neutralising capacity equivalent to a 10% aqueous solution of KOH in oil-soluble form. Experience shows that when fuels with sulphur contents of 2.0% or more are used the TBN of the cylinder drainings should not fall below 10 if excessive wear rates are to be avoided. However, the use of highly alkaline cylinder oils with low sulphur fuels is not recommended; not only is it needlessly expensive, it can give rise to high wear rates[35, 61].

In trunk piston engines running on high sulphur fuels the copious quantities of oil splashed up from the crank case provide a greater reservoir of alkalinity and an initial TBN of 25-30 is generally adequate. Nevertheless, the TBN of the oil should not fall below about three times the sulphur content of the fuel.

Water Base Lubricants

In some applications, e.g. mining machinery, non-flammable lubricants are specified and water-base or water-containing fluids are used. Standard corrosion inhibitors are used to combat corrosion but even a small amount of water in a mineral oil lubricant can adversely affect the fatigue life of components such as rolling bearings and gears[66,67]. Fire resistant water-base fluids can reduce the fatigue life of a rolling contact bearing by 90%[68]. These premature failures are usually attributed to hydrogen embrittlement. However, additives used to restrain hydrogen embrittlement in acid pickling and electroplating operations do not help. Worthwhile improvements in fatigue life can be achieved by using additives of a completely different nature[69] and by special heat-treatment techniques[70].

Cavitation and Erosion

Cavitation and erosion are increasingly the cause of failure in plain bearings and hydraulic systems[71]. Two types of cavity can form in lubricants, vaporous and gaseous. Only vaporous cavities can form and collapse rapidly and the very high pressure associated with their collapse can cause mechanical (impact) damage on surfaces. Common causes of cavitation are pressure fluctuations associated with the flow of the liquid and the vibration of a surface in contact with it. Unlike other kinds of damage, vaporous cavitation is generally encountered on the unloaded areas of bearings[71-72]. When the damage is due solely to cavitation, the damaged surfaces are rough: when foreign particles are present they are smooth (cavitation erosion).

Vaporous cavitation can remove protective films, such as oxides, from metals and so initiate corrosion[71]. In addition, the very high local pressures and temperatures associated with the final stage of cavity collapse can induce chemical reactions that would not normally occur. Thus certain additives are damaged by cavitation and their decomposition products can be corrosive.

R. W. WILSON

REFERENCES

1. O'Connor, J. J. (Ed.), *Standard Handbook of Lubrication Engineering*, McGraw-Hill Book Company, New York (1968)
2. Cameron, A. (Ed.), *Principles of lubrication*, Longmans Green, London (1966)
3. *Modern Petroleum Technology*, Institute of Petroleum, London, 3rd edn (1962)
4. Zuidnor, H. H., *The Performance of Lubricating Oils*, Reinhold Publishing Corporation, New York, Chapman and Hall, London, 2nd edn. (1959)
5. Wilson, R. W. and Shone, E. B., *Joint Course on Tribology*, Paper 4, Institution of Metallurgists, London (1968)
6. Broeze, J. J. and Wilson, A., *Gas and Oil Power*, **44**, 386 (1949)
7. Graham, R., *Shell Aviation News*, June, 18 (1950)

8. Fowle, T. I., *Conference on Lubrication and Wear*, Paper 19, Institution of Mechanical Engineers, London, 568 (1968)
9. Garner, F. H. and Wilson, B. S., *J. Inst. Petrol*, **37**, 225 (1951)
10. Denison, G. H., *Ind. Eng. Chem.*, **36**, 477 (1944)
11. Ford, J. F., *J. Inst. Petrol*, **54**, 198 (1968)
12. Braithwaite, E. R. (Ed.), *Lubrication and Lubricants*, Elsevier Publishing Company, London (1967)
13. *Lubrication*, **54**, 85 (1968)
14. Galvin, G. D. and Naylor, H., *Lubrication and Wear*, Third Convention, Paper 6, Inst. Mech. Engrs., 56 (1965)
15. Quayle, J. P., *Copper*, **3** No. 5, 12 (1969)
16. Hughes, R. I., *Corrosion Science*, **9**, 535 (1969)
17. *Lubrication*, **8**, 145 (1953)
18. Forrester, P. G., *Met. Rev.*, **5**, 507 (1960)
19. Wilson, R. W. and Shone, E. B., *Anti-Corrosion Methods and Materials*, **17** No. 8, 8 (1970)
20. Forrester, P. G., *Modern Materials*, **4**, Academic Press, New York and London, 173 (1964)
21. Bryce, J. B. and Roehner, T. G., *Trans. Inst. Mar. Engrs.*, **73**, 377 (1961)
22. Lloyd, K. A. and Wilson, R. W., *Lubrication and Wear*, Seventh Convention, Paper 10, Inst. Mech. Engrs., 76 (1969)
23. *Lubrication*, **55** No. 7, 65 (1969)
24. *The Petroleum Handbook*, Elsevier Science Publishers B. V., Amsterdam, 6th Edition (1983)
25. Hobson, G. D. and Pohl, W., (Ed.), *Modern Petroleum Technology*, Applied Science Publishers, Barking, Essex, England (1973)
26. Ranney, M. W., *Lubricant Additives*, Noyes Data Corporation, New Jersey (1978)
27. Vincentz, C. R., *Additives for Lubricants* Bartz, W. J., Verlay, (Ed.) Hanover, FRG (1985)
28. Jones, M. H. and Scott, D., (Ed.), *Industrial Tribology*, Elsevier Scientific Publishing Company, Amsterdam (1983)
    (*a*) Wilson, R. W. and Shone, E. B., Chap. 5, *The Diagnosis of Plain Bearing Failures*
    (*b*) Lansdown, A. R., Chap. 9, *Selection of Lubricants*
    (*c*) Soul, D. M., Chap. 10, *Lubricant Additives, their Application, Performance and Limitations*
    (*d*) Eyres, A. R., Chap. 12, *Health and Safety Aspects of Lubricants*
    (*e*) Collacott, R. A., Chap. 18, *On Condition Maintenance*
29. Boner, C. J., *Modern Lubricating Greases*, Scientific Publications (G.B.) Ltd. (1976)
30. Documentation Tribology: *Wear, Friction and Lubrication*, BAM (Federal Inst. for Testing Materials), Berlin (biennial)
31. Tribology: Wear, *Friction and Lubrication*, Documentation of the R & D Programme of the Federal Ministry of R & D (BMFT) 8 vols. to date, Springer-Verlag, Berlin (ongoing)
32. Booser, E. R., (Ed.) *Handbook of Lubrication: Theory and Practice of Tribology*, **1**, Application and Maintenance, CRC Press, Boca Raton, USA (1982)
33. Helm, J. L., 'The Changing Lube Market', *Lubr. Eng.*, **36**, 81 (1980)
34. Brook, J.H.T., *Corrosion Control by Lubricating Oils*, Proc. Symp. Practical Aspects of Corrosion Inhibition, Soc. Chem. Ind., London (1980)
35. Wilson, R. W., *Corrosion Prevention in Lubricant Systems*, 6th Int. Cong. Metallic Corrosion, **2**, 1479, Australasian Corrosion Association (1981)
36. Tourret, R., & Wright, E. P., (Ed.) *Performance and Testing of Gear Oils and Transmission Fluids*, Institute of Petroleum, London (1981)
37. Murray, D. W., McDonald, J. M. and Wright, P. G., 'Effect of Basestock Composition on Lubricant Oxidation Performance', *Pet. Rev.* **36**, 36 (1982)
38. Marsh, J. F., *Colloidal Lubricant Additives*, Chem and Ind., No. 14, 470 (1987)
39. Barcroft, F. T. and Park, D., 'Interactions on Heated Metal Surfaces Between Zinc D. P. and Other Lubricating Oil Additives', *Wear*, **108**, 213 (1986)
40. Hiley, R. W., 'Corrosion of Tin-Base Babbitt Bearings to Form Tin Oxides', *Trans. Inst. Mar. Engrs.*, **91**, 52 (1979)
41. *ASTM Standards on Petroleum Products and Lubricants*, American Society for Testing and Materials (Annual Publication)
42. *IP Standard for Petroleum and Its Products*, The Institute of Petroleum, U.K. (Annual Publication)

43. *Significance of ASTM Tests for Petroleum Products*, ASTM Special Publication C (1977)
44. Klamann, D., *Lubricants and Related Products (Synthesis, Properties, Applications, International Standards)*, Verlag Chemie GmbH, Weinheim, FRG (1984)
45. Pohl, W., *Diesel Engine Reference Book*, (Ed.) Lilly, L.C.R. Chap. 16, 'Lubrication and Lubricating Oils, Butterworths, London (1984)
46. Williams, W. T., *Lubr. Eng.*, **33**, 191 (1977)
47. *Lubrication and Fuels in Ships*, Shell ADC-V Oil Print Analyser; Shell Minilab., Shell International Trading Company (1978)
48. Waggoner, C.A., *Materials Report 71-A*, Defence Research Establishment, Canada (1971)
49. Beerbouer, A., *Lubr. Eng.*, **32**, 285 (1976)
50. Yardley, E. D., *Wear*, **56**, 213 (1979)
51. Pocock, G., *Introduction to Ferrography*, British Inst. Non-Destructive Testing, London (1979)
52. Kwon, O. K., Kong, H. S., Kim, C. H. and Oh, P. K., *Trib. Int.*, **20**, 153 (1987)
53. Warne, T. M., and Halder, C. A., *Lubr. Eng.*, **42**, 97 (1986)
54. 'Code of Practice for Metal-Working Fluids', *Inst. of Petr.*, Heydon & Son Ltd., London (1978).
55. Bennett, E. O., *Trib. Int.*, **16**, 133 (1983)
56. Bennett, E. O. and Bennett, D. L., *ibid* **18**, 169 (1985)
57. Hill, E. C., *ibid.*, **16**, 136 (1983)
58. Hill, E. C. and Genner, C., *Proc. 3rd Int. Tribology Congr.*, 3A, 92-120, Tech. Univ. Radon, Poland (1981)
59. Cartwright, S. J. and Carey, L. R., Society of Automotive Engineers, *SAE Special Publication SP 473*, (1980)
60. Thomas, F. J. and Hold, G. E., American Society of Mechanical Engineers, *ASME Paper No. 80-DGP-40*, ASME Energy Technology Conf., New Orleans (1980)
61. Golothan, D. W., *Trans. Inst. Mar. Engrs.*, **90**, Series A, Part 3, 137 (1978)
62. Marbach, H. W., Frame, E. A., Owens, E. C. and Naegeli, D. W., Society of Automotive Engineers, *SAE Preprint 811199 for Oct. Meeting*, (1981).
63. Ryan, T. W., Naegeli, D. W., Owens, E. C., Marbach, H. W. and Barbee, J. G., *ibid*, Preprint 811200.
64. Thring, R. H., *Automative Eng.*, **92**, 60 (1984)
65. Wederpohl, E., *Institut fur Erdolforschung, Tatigkeitsbericht*, Jahrg, 12 (1980)
66. Grunberg, L., Scott, D. and Jamieson, D. T., *Phil. Mag.*, **93**, 1553 (1963)
67. Scott, D., *Treatise on Materials Science and Technology*, **13**, 121, Academic Press, N.Y. (1979)
68. *Guidelines for Derating the Life of Ball-Bearings when Used with Fire Resistant Fluids*, Ass. Hydraulic Equipment Manufacturers (1977)
69. Murphy, W. F., Polk, C. J. and Rowe, C. N., *ASLE Conf., Preprint No. 76-LC-48-1*, Boston, Oct. (1976)
70. Hollox, G. E., Hobbs, R. A. and Hampshire, J. M., *Wear*, **68**, 229 (1981)
71. Dowson, D., Godet, M. and Taylor, C. M. (Eds) *Cavitation and Related Phenomena in Lubrication*, 1st Leeds-Lyon Symp. on Tribology, Section 7, 177, Inst. Mech. Eng. (1975)
72. James, R. D., Garner, D. R. and Warriner, J. F., *13th Congr. Int. des Mach. a Combust (CIMAC)*, Vienna, **1**, 78.1, Publ. CIMAC Brit. Nat. Committee, London (1979)

2.12 Corrosion in the Oral Cavity

Introduction

Dentistry is concerned with preventing and treating oral disease, eliminating pain, restoring the oral apparatus to function and improving aesthetics. The oral cavity is a complex environment, in which hard and soft tissues (teeth, palate and gums) are continuously bathed in saliva and are subjected to varying loads during mastication and deglutition (swallowing). Saliva varies in composition and pH from person to person, and from hour to hour for a given individual; it comprises a mixture of inorganic anions (predominantly chlorides and phosphates), organic acids, enzymes, bacteria and gastric secretions such as mucin[1-3]. The salinity of saliva approaches that of seawater and tends to be highly corrosive to most non-noble metals. The total forces exerted during mastication vary with the age, musculature and other physiological factors for the patient and the existing dentition, typically being ca. 60 kg for the molar teeth in a healthy dentate patient, giving rise to stress values in the range $3-17 \, MN/m^2$, depending upon the food being chewed[3,4]. Clearly, the mouth is a hostile environment and dental materials are required to be mechanically strong and resistant to degradation caused by stresses and the oral environment.

Decayed teeth are treated by removing the decay and then mechanically shaping the cavity to provide optimum retention of a restoration or 'filling'. Anterior (front of the mouth) teeth are restored with aesthetic materials such as composite resins which are modified epoxies or other glassy polymers containing inorganic fillers, such as quartz or glass. Badly decayed or broken anterior teeth are often given porcelain crowns. Posterior teeth, which perform the masticatory functions of crushing and chewing food, require mechanically stronger materials. Teeth with one or more cusps in situ are restored with dental amalgam while those requiring more extensive restoration are provided with metallic crowns covering all or part of the tooth. These crowns may be covered by porcelain for aesthetic reasons. Missing teeth may be replaced by pontics which are attached, usually by soldered joints, to crowns on the adjacent or abutment teeth to form a bridgework or fixed partial denture (FPD).

Larger numbers of lost teeth are replaced by a removable partial denture (RPD), a framework or base to which are attached the replacement teeth. RPDs are fabricated from a variety of noble and base metal alloys as well as polymers such as acrylic resin. Complete dentures, provided when all the teeth are lost, are usually made of acrylic resin but sometimes metals. If there has been significant loss of the bony support, metallic or ceramic implants may be placed in the remaining bone with posts projecting through the gums into the oral cavity, and dentures are attached to the posts. Orthodontics, the specialty dealing with tooth realignment and positioning, commonly uses stainless steel brackets, bands and wires attached to the teeth to effect tooth movement.

Clearly a variety of metals and alloys are used in dentistry. Corrosion occurring in the oral cavity generally does not result in catastrophic failure but rather discolouration of restorations, staining of the teeth and, in some cases, allergic reactions due to the release of metallic ions into the biosystem. Corrosion can also cause disruption of restorations, leading to leakage at the restoration-tooth margins and to secondary dental decay. There has also been a report of corrosion of implanted plates and screws following oral surgery that delayed healing of a fracture and resulted in osteomyelitis of the mandible[5].

Dental Amalgam

Traditionally, the most widely used restorative material was dental amalgam, some 80% of all restorations comprising this material wholly or in part, and even today, dental amalgams constitute a high percentage of posterior restorations. Dental amalgam is prepared by triturating or alloying mercury with filings or spheres of a silver-base alloy to form a plastic mass which is inserted into the prepared tooth cavity where it sets to a hard mass. For decades, the composition of the conventional amalgam alloy powder remained largely unchanged from 65-70% Ag, 25-29% Sn, 0-6% Cu and 0-2% Zn, the principal component being the silver-tin eutectic, Ag_3Sn, known as the gamma phase. When this reacts with mercury, two new phases, gamma-1, Ag_2Hg_3, and gamma-2, $Sn_{7-8}Hg$, are formed so that the set mass contains residual gamma particles, gamma-1 and gamma-2[4,6]. The gamma-2 phase is mechanically weak and susceptible to corrosion.

The corrosion mechanism of dental amalgam has been studied extensively and it is accepted that the gamma-2 phase is anodic to the other phases present[4,6-10]. Corrosion of gamma-2 releases mercury which may react with residual gamma or the gamma-1 phase, the former reaction forming more gamma-2 and gamma-1 phases, while the latter forms mercury-rich gamma-1. The formation of additional gamma-1 and gamma-2 phases ensures continued corrosion of the Sn-rich phase throughout the body of the restoration. The mercury-rich gamma-1 phase is cathodic to gamma-1[7,8] but the reaction is slow and polarises rapidly. Zero resistance ammetry studies[11] have shown that amalgam is susceptible to crevice corrosion and readily establishes galvanic cells, notably differential pH cells, cells between polished and corroded areas on the surface as well as those arising from compositional differences within a restoration. Large galvanic currents are generated when amalgam is in contact with gold (or aluminium) and the magnitude of the current is affected by the type of alloy used to prepare the amalgam, but saliva exerts an inhibitory effect on these galvanic currents[12]. While there is good correlation between *in vitro* corrosion and fracture of restoration margins[12,13], *in vitro* studies correlates poorly with the *in vivo* corrosion of amalgam, which is lower than predicted. This low *in vivo* corrosion rate has been ascribed to the formation of various salts such as the oxides, hydroxides, phosphates and oxychlorides of tin, copper and zinc present in the amalgam and the cathodic inhibiting action of the CO_3^{2-}-HCO_3^- system in saliva[2,10,14]. Amalgam corrosion, however, does have the benefit that the corrosion products seal the margins between tooth and restoration.

The poor mechanical properties, corrosion and other problems associated with gamma-2 containing (conventional) dental amalgams led to the development of the high copper, reduced-tin content, amalgam alloys which are strengthened by the presence of Ag-Cu particles. When triturated with mercury, the gamma-2 phase reacts with Ag-Cu to form the eta phase, Cu_6Sn_5 and more gamma-1. The reduced gamma-2 content of these amalgams results in greater strength, better clinical behaviour and lower corrosion rates than the conventional amalgams[4,6].

There have been numerous reports of possible allergic reactions to mercury and mercury salts and to the mercury, silver and copper in dental amalgam as well as to amalgam corrosion products[15-20]. Studies of the release of mercury by amalgams into distilled water, saline and artificial saliva tend to be conflicting and contradictory but, overall, the data indicate that mercury release drops with time due to film formation[9,21] and is less than the acceptable daily intake for mercury in food[22]. Further, while metallic mercury can sensitise, sensitisation of patients to mercury by dental amalgam appears to be a rare occurrence[18]. Nevertheless, there is a growing trend to develop polymer-based posterior restorative materials in order to eliminate the use of mercury in dentistry.

Noble Metal Alloys

Traditionally, full or partial coverage crowns, bridgework, RPDs, porcelain-fused-to-metal restorations (PFMs) and even complete denture bases were cast from high carat gold alloys. These alloys contain >75% Au and are based on the ternary Au-Ag-Cu alloys with the mechanical properties improved through additions of Pt, Pd and Ir. These high gold alloys possess near ideal casting characteristics as well as excellent corrosion resistance.

In recent years, however, the increasing price of gold has resulted in greater use of alloys containing less gold and greater amounts of silver and palladium as well as various base metal alloys (see later) for dental castings. Lower gold content alloys have good mechanical properties and can be accurately cast but they do not possess the oral corrosion resistance of high gold alloys. Further, these alloys are often metallurgically heterogeneous or at least less homogeneous than the high noble-metal alloys, which will affect their corrosion susceptibility due to galvanic coupling effects[23].

Cast restorations often exhibit little corrosion but rather an unaesthetic tarnishing or discolouration. There appears in fact to be an inverse relationship between oral corrosion and tarnish for many low gold and copper-rich alloys in potential ranges where the corrosion rate is limited by the rate of the cathodic reaction[23,24]. This effect is possibly due to the uniform distribution of anodic and cathodic sites in single phase alloys with no preferential deposition of corrosion products and consequently little tarnish. In contrast, the anodic and cathodic reactions in multiphase alloys are separated and the different nobilities of these phases may give rise to numerous bimetallic galvanic cells. Other systems, typically Ag–Pd alloys, do appear to show a correlation between tarnish and corrosion[25].

Most cast dental restorations are subjected to some form of subsequent heat treatment such as annealing, hardening or soldering. This often induces changes in the structural state or in the phases present and may establish local galvanic cells. Potentio-dynamic polarisation studies[26] have shown that high gold alloys are unaffected by their thermal history but the corrosion susceptibility of low golds (containing <60% Au) and Ag–Pd alloys are markedly effected by heat treatment. As-cast and age-hardened structures were found to be more susceptible to corrosion than those annealed at 100° below the solidus temperature. Other workers[24,27] also have reported that as-cast low golds are more prone to *in vivo* and *in vitro* tarnish than alloys solution heat-treated at 700°C. This does not appear to be the case for high Pd–Cu alloys used for porcelain bonding which show polarisation resistance values independent of the metal pre-treatment[29].

Multiphase gold or palladium-based alloys never show dissolution of Au or Pd but often exhibit progressive surface ennoblement due to selective dissolution of copper or silver from the outer 2–3 atomic layers[29-32]. Heat treatment often decomposes multicomponent alloys into a Pd–Cu rich compound and an Ag-rich matrix with corrosion of the latter phase in deaerated artificial saliva and S^{2-}-containing media[33]. Au–Cu-rich lamellae have similarly been observed, again with preferential attack on Ag-rich phases or matrix. These effects presumably arise from the ability of the noble alloy phases to catalyse the cathodic reduction of oxygen[27-30].

Generally, a greater Pd level in Pd-based single phase alloys results in increased corrosion resistance in the potential range and Cl^- level typical of the mouth[25,28,34,35], as might be expected. Corrosion of low gold alloys occurs in Cl^- media, typically with attack occurring on the Ag-rich matrix, but phosphate and bicarbonate ions have an inhibiting effect while the isothiocyanate ion can induce passivity[33,36]. Various amino acids, typical of body fluids, can induce passivity but it appears that solution pH is the controlling factor in corrosion for high Ag-content alloys, passivation being enhanced by basic solutions[37].

Base Metal Alloys

Approximately 90% of all RPDs are now cast from base metal alloys containing principally chromium, cobalt and nickel, with chromium being the element present in all such alloys. Commonly, these cast chromium alloys contain various alloying elements, typically ≤5% Mo, ≤1% Fe, 25–30% Cr and the balance Co although there are some widely used alloys containing

15–20% Cr and ≤5% each of Al, Mn and Mo with the balance Ni. The physical properties of these base metal alloys are controlled by the minor alloying elements present, ie. C, Mo, Be, W and Al. The upper Cr level is limited to 30% to avoid casting difficulties and the formation of the brittle sigma phase while the lower level is 15–20% to maintain corrosion resistance. There has also been a limited application of stainless steel for wrought RPDs. Oral implants are usually fabricated from Ti, Ta and Cr–Co alloys and most orthodontic appliances utilise stainless steel brackets and wires although there is increasing application of Co–Cr, beta-titanium and Ni–Ti alloys for orthodontic spring wires.

There has been comparatively little published on the corrosion behaviour of cast chromium alloys or on their *in vivo* corrosion products. The available data are somewhat contradictory but it appears that these alloys tend to passivate in synthetic saliva solutions and in the potential range and chloride level corresponding to that of the oral cavity[38-42]. It has been reported that *in vitro* polarisation results for Co–Cr alloys in isotonic (0.9%) saline closely correlate with *in vivo* polarisation in rabbits[43] and also that there are no significant effects on corrosion behaviour arising from mucin or the amino acids present in the proteins found in the mouth[40,41]. The corrosion rates of stainless steel in isotonic saline were also found to be comparable to or less than those measured *in vitro*[44]. Other work indicates, however, that the corrosion rate of Cr–Co is several times greater in the mouth than that found *in vitro*[45] and proteins (from calf serum) were found to reduce corrosion rates and metal ion release from cast chromium alloys and a Ni–Cr–Be alloy[46,47]. There is, however, no evidence of corrosion of Be when present in Ni-base alloys[40]. There are no indications of pitting attack or tarnishing of cast chromium alloys in 5% sulphide solution or in the mouth. Crevice corrosion of Co–Cr alloys does not occur in ferric chloride solution[38] but it can occur when Co–Cr is in contact with gold in phosphate-buffered saline[48]. Ni-base alloys, however, are susceptible to crevice corrosion[38] and galvanic coupling with gold increases the metal dissolution rate[47].

Metal ion release from cast chromium alloys and stainless steel occurs *in vivo* as a result of corrosion and/or dissolution of passive films. *In vitro* studies[39] indicates that the release of metals into artificial saliva is less than 0.2 ppm over a two-month period but a recent *in vivo* study[49] shows that release of Cr and Co is readily detected in a short time (5 min) in the saliva of patients wearing Co–Cr dentures. Although metal release decreased with the age of dentures, presumably due to passivation effects, it still occurred over a long period, paralleling the results of other workers with Co–Cr and particularly Ni–Cr alloys over 35 week test periods[48]. This long-term metal ion release would account for the many reports of mucosal and systemic allergic reactions to cast chromium alloys[50,52]. The allergens in such cases are thought to be Ni and Cr but patients allergic to Ni are sometimes also allergic to Co. Although the literature on oral reactions to Cr and Co is sparse, allergic responses to Ni and Co were found in five out of ten Ni-sensitive patients patch-tested with Ni-containing Co–Cr alloys[53]. *In vivo* release of Ni and Co through corrosion has been shown to occur with implanted alloys. The released metal localises in tissue near the implant and to a lesser extent in various organs and is thought to result in allergic reactions[54,57]. Additionally, Ni–Cr casting alloy powders have been shown to have cytotoxic potential[58].

There is, however, considerable debate over whether metal release from dental prostheses can cause sensitisation and/or allergic reactions and some consider it to be unlikely[18,42]. Overall there are relatively few reported cases of local or systemic hypersensitivity resulting from base metal dentures and it has been reported that the incidence of nickel allergy with intra-oral exposure to Ni alloys is only 4% compared to 6% in those without exposure[59]. *In vivo* studies with guinea pigs[60] indicate that exposure to Ni and Cr from dietary intake or intra-oral alloys can induce a state of partial tolerance to both metals. The authors suggested that individuals not hypersensitive to Ni or Cr may become partially tolerant as a result of intra-oral exposure to the metals. Further, they suggested that this could have application as a preventive protocol for those at high risk of sensitisation[60].

There has been growing clinical interest in magnetic retention of overdentures. Typically a permanent samarium-cobalt magnet, encapsulated in stainless steel or a high palladium alloy for strength and corrosion resistance, is fitted into the denture and a ferromagnetic alloy is cemented into the residual tooth root. The ferromagnetic alloys are principally Pd–Co alloys with small ($\leq 2.5\%$) additions of Ga and ($\leq 2\%$)P. These alloys appear to have good corrosion resistance in the oral potential range (-100 to $300\,\text{mV}$, vs S.C.E.) based on electrochemical studies[61] but there are, as yet, few reports on these materials which will probably be of increasing clinical importance in the future.

Galvanic Effects in the Mouth

Corrosion resulting from intra-oral mixed metal and other galvanic cells has been mentioned previously and dentists may often observe blackening of restorations and other effects but not always recognise their source. Many serious oral conditions such as lichen planus, leukoplakia and oral cancer have been ascribed to galvanic cells in the mouth[17,62-64]. These couples may arise from different metals being in contact, electrical circuits being established between a gold crown and an amalgam core separated by a film of dental cement or even a soldered cast prosthesis with separations in the solder joints. Patients with such galvanic couples commonly have either a subjective complaint of a metallic taste or sensation or an objective complaint such as chronic inflammation of the mucosa and possibly neurologic complaints. Intra-oral measurements of electrode potentials and the properties of patients' saliva did not establish a relationship between the measured potentials of individual metallic restorations and orofacial complaints[65,66]. Intra-oral potential measurements made on patients with mixed metals in contact also indicated that large potential differences could exist between similar metals[67]. These authors indicated that a potential difference greater than $50\,\text{mV}$ between metals was harmful. While the pathological effects arising from oral galvanic cells varied with individual patients and was not always proportional to the observed electrode potentials, regression and often disappearance of oral lesions occurred when harmfully high potential differences were eliminated[63, 67].

Conclusions

Oral corrosion of metallic restorations does not, *per se*, generally result in serious damage to the structure. Corrosion can result, however, in various local and systemic effects, notably the hypersensitivity and allergic reactions reported by many workers. Galvanic cells created by mixed metal couples can delay fracture healing and induce oral lesions and cancer.

J. A. von FRAUNHOFER

REFERENCES

1. Jenkins, G. M., *The Physiology of the Mouth*, 3rd. ed., Blackwell, Oxford (1973)
2. Palaghias, G. 'The Role of Phosphate and Carbonic Acid-bicarbonate Buffers in the Corrosion Processes of the Oral Cavity', *Dental Materials*, **1**, 139-144 (1985)
3. von Fraunhofer, J. A. 'Oral Implants', in *Scientific Aspects of Dental Materials*, (Ed.) J. A. von Fraunhofer), Butterworths, London (1975)
4. Craig, R. G., *Restorative Dental Materials*, 7th ed., C. V. Mosby, St Louis (1985)
5. Steiner, M., von Fraunhofer, J. A. and Mascaro, J. 'The Role of Corrosion in Inhibiting the Healing of a Mandibular Fracture', *Journal of Oral Surgery*, **39**, 140-143 (1981)
6. Wing, G., 'Dental Amalgam', in *Scientific Aspects of Dental Materials*, (Ed.) J. A. von Fraunhofer), Butterworths, London (1975)
7. von Fraunhofer, J. A. and Staheli, P. J. 'Corrosion of Amalgam Restorations: A New Explanation', *British Dental Journal* **130**, 522-524 (1971)
8. von Fraunhofer J. A. and Staheli, P. J. 'Corrosion of Dental Amalgam', *Nature*, **240**, 304-306 (1972)
9. Okabe, T. 'Mercury in the Structure of Dental Amalgam', *Dental Materials*, **3**, 1-8 (1987)
10. Do Duc, H., Meyer R. M. and Tissot, P. 'Electrochemical Behaviour of Sn_8Hg (gamma-2) and Dental Amalgam in a Phosphate Buffer Solution', *Electrochimica Acta*, **25**, 851-6 (1980)
11. von Fraunhofer, J. A. and Staheli, P. J. 'The Measurement of Galvanic Corrosion Currents in Dental Amalgam', *Corrosion Science*, **12**, 767-773 (1972)
12. von Fraunhofer, J. A. and Staheli, P. J. 'Gold-Amalgam Galvanic Cells: The Measurement of Corrosion Currents', *British Dental Journal*, **132**, 357-362 (1972)
13. Sarkar, N. K., Osborne, J. W. and Leinfelder, K. F. '*In Vitro* Corrosion and *In Vivo* Marginal Fracture of Dental Amalgams', *Journal of Dental Research*, **61**, 1262-1268 (1982)
14. Awad, S. A. and Kamel, K. M. 'Behaviour of Tin as Metal-metal Phosphate Electrode and Mechanism of Promotion and Inhibition of its Corrosion by Phosphate Ions', *Journal of Electroanalytical Chemistry*, **24**, 217-25 (1969)
15. Finne, K., Goransson, K. and Winckler, L. 'Oral Lichen Planus and Contact Allergy to Mercury', *International Journal of Oral Surgery*, **11**, 236-39 (1982)
16. Frykholm, K. O., Frithiof, L. Fernstrom, A. I. B., Moberger, G., Blohm, S. G. and Bjorn, E. 'Allergy to Copper Derived from Dental Alloys as a Possible Cause of Oral Lesions of Lichen Planus' *Acta Dermatovenerol*, **49**, 268-81 (1969)
17. Lundstrom, I. M. C. 'Allergy and Corrosion of Dental Materials in Patients with Oral Lichen Planus', *International Journal of Oral Surgery*, **12**, 1-9 (1982)
18. Burrows, D. 'Hypersensitivity to Mercury, Nickel and Chromium in Relation to Dental Materials', *International Dental Journal*, **36**, 30-34 (1986)
19. Catsakis, L. H. and Sulica, V. I. 'Allergy to Silver Amalgams', *Oral Surgery, Oral Medicine and Oral Pathology*, **46**, 371-375 (1978)
20. Craig, R. G. 'Biocompatibility of Mercury Derivatives', *Dental Materials* **2**, 91-96 (1986)
21. Ferracane, J. L., Mafiana, P., Cooper, C. and Okabe, T. 'Time-dependent Dissolution of Amalgams into Saline Solution', *Journal of Dental Research*, **66**, 1331-1335 (1987)

22. Olstad, M. L., Holland, R. I., Wandel, N. and Hensten Pettersen, A. 'Correlation between Amalgam Restorations and Mercury Concentrations in Urine', *Journal of Dental Research*, **66**, 1179-1182 (1987)
23. Holland, R. I., Jørgensen, R. B. and Herø, H. 'Corrosion and Structure of a low-gold Dental Alloy', *Dental Materials*, **2**, 143-146 (1986)
24. Herø, H. and Jørgensen, R. B. 'Tarnishing of a Low-gold Dental Alloy', *Journal of Dental Research*, **62**, 371-376 (1983)
25. Ishizaki, N. 'Corrosion Resistance of Ag-Pd Alloy System in Artificial Saliva: An Electrochemical Study', *Journal of the Osaka Dental University*, **3**, 121-133 (1969)
26. Bessing, C., Bergman, M. and Thoren, A. 'Potentiodynamic Polarization Analysis of Low-gold and Silver-Palladium Alloys in Three Different media, *Dental Materials*, **3**, 153-159 (1987)
27. Herø, H. and Valderhaug, J. 'Tarnishing *in vivo* and *in vitro* of a Low Gold Alloy Related to its Structure', *Journal of Dental Research*, **64**, 139-143 (1985)
28. Metzger, P. R., Vrijhoef, M. M. A. and Greener, E. H. 'Corrosion Resistance of Three High-Palladium Alloys', *Dental Materials*, **1**, 177-179 (1985)
29. Niemi, L., Minni, E. and Ivaska, A. 'An Electrochemical and Multispectroscopic Study of Corrosion of Ag-Pd-Cu-Au Alloys', *Journal of Dental Research*, **65**, 888-891 (1986)
30. Hultquist, G. and Herø, H. 'Surface Enoblement by Dissolution of Cu, Ag and Zn from Single Phase Gold Alloys', *Corrosion Science*, **24**, 789-805 (1984)
31. German, R. M. 'The Role of Microstructure in the Tarnish of Low-gold Alloys', *Metallography*, **14**, 253-266 (1981)
32. Gniewek, J., Pezy, J., Baker, B. G. and Bockris, J. 'The Effect of Noble Metal Additions upon Corrosion: An Auger Spectroscopic Study', *Journal of the Electrochemical Society*, **125**, 17-23 (1978)
33. Niemi, L. and Herø, H. 'Structure, Corrosion and Tarnishing of Ag-Pd-Cu Alloys', *Journal of Dental Research*, **64**, 1163-1169 (1985)
34. Vaidyanathan, T. K. and Prasad, A. '*In vitro* Corrosion and Tarnish Analysis of the Ag-Pd Binary System', *Journal of Dental Research*, **60**, 707-715 (1981)
35. Sastri, S., Vaidyanathan, T. K. and Mukherjee, K. 'Potentiodynamic Polarization of Silver-Palladium Alloys in Chloride Solutions', *Metal Transactions A*, **13A**, 313-317 (1982)
36. Sarkar, N. K., Graves, R. A., Park, J. R. and Usha, M. G. 'Corrosion of Gold Casting Alloys in Selected Biological Media', *Journal of Dental Research*, **66**, 205 (Abstr. 792) (1987)
37. Sarkar, N. K., Graves, R. A., Park, J. R. and Usha, M. G. 'Effects of Selected Amino Acids on the Chloride Corrosion of Gold Casting Alloys', *Journal of Dental Research*, **66**, 206 (Abstr. 794) (1987)
38. Hodges, R. J. 'The Corrosion Resistance of Gold and Base Metal Alloys', in *Alternatives to Gold Alloys in Dentistry* (Ed.) T. M. Valega, Sr., DHEW Publication No. (NIH) 77-1227 106-138 (1987)
39. Espevik, S. 'Corrosion of Base Metal Alloys *in vitro*', *Acta Odontologia Scandanavia*, **35**, 113-117 (1978)
40. Sarkar, N. K., Graves, R. A., and Park, J. R. 'Corrosion of Nickel-Chromium Casting Alloys in Selected Biological Media', *Journal of Dental Research*, **66**, 206 (Abstr. 793) (1987)
41. Sarkar, N. K., Graves, R. A. and Park, J. R. (b) 'Electrochemical Behaviour of Ni-Cr Alloys in Chloride Solutions containing Amino Acids', *Journal of Dental Research*, **66**, 206 (Abstr. 795) (1987)
42. Anusavice, K. J. 'Council on Dental Materials, Instruments and Equipment: Report on Base Metal Alloys for Crown and Bridge Applications: Benefits and Risks', *Journal of the American Dental Association*, **111**, 479-483 (1985)
43. Buchanan, R. A. and Lemons, J. E. '*In Vivo* Corrosion-Polarisation Behavior of Titanium-base and Cobalt-base Surgical Alloys', *Transactions of the Eighth Annual Meeting of the Society for Biomaterials*, Orlando, Florida (1982)
44. Revie, R. W. and Greene, N. D. 'Comparison of the *In Vivo* and *In Vitro* Corrosion of 18-8 Stainless Steel and Titanium', *Journal of Biomedical Materials Research*, **3**, 465-470 (1969)
45. Gettleman, L., Cocks, F. H., Darmiento, L. A., Levine, P. A., Wright, S. and Nathanson, D. 'Measurement of *In Vivo* Corrosion Rates in Baboons and Correlation with *In Vitro* Tests', *Journal of Dental Research*, **59**, 689-707 (1980)

46. Hensten-Pettersen, A. and Jacobsen, N. 'Nickel Corrosion of Non-precious Casting Alloys and the Cytotoxic Effect of Nickel *In Vitro*', *Journal of Bioengineering*, **2**, 419–425 (1978)
47. Herø, H., Valderhaug, J. and Jørgensen, R. B. 'Corrosion *in vivo* and *in vitro* of a commercial NiCrBe alloy', *Dental Materials*, **3**, 125–130 (1987)
48. Moberg, L. E. 'Long-term Corrosion Studies *In Vitro* of Gold, Cobalt-Chromium and Nickel-Chromium Alloys in Contact' *Acta Odontologia Scandanavia*, **43**, 215–222 (1985)
49. de Melo, J. F., Gjerdet, N. R. and Erichsen, E. S. 'Metal Release from Cobalt-Chromium Partial Dentures in the Mouth', *Acta Odontologia Scandanavia*, **41**, 71–74 (1983)
50. Wood, J. F. L. 'Mucosal Reaction to Cobalt Chromium Alloy' *British Dental Journal*, **136**, 423–424 (1974)
51. Levantine, A. V. 'Sensitivity to Metal Dental Plate' *Proceedings of the Royal Society of Medicine*, **67**, 1007 (1974)
52. Brendlinger, D. L. and Tarsitano, J. J. 'Generalized Dermatitis due to Sensitivity to a Chrome-Cobalt Removable Partial Denture', *Journal of the American Dental Association*, **81**, 392–394 (1970)
53. Magnusson, B., Bergman, M., Bergman, B. and Söremark, R. 'Nickel Allergy and Nickel Containing Dental Alloys', *Scandinavian Journal of Dental Research*, **90**, 163–167 (1982)
54. Samitz, M. H. and Katz, S. A. 'Nickel Dermatitis Hazards from Prostheses', *British Journal of Dermatology*, **92**, 287–290 (1975)
55. Bergman, M., Bergman, B and Söremark, R. 'Tissue Accummulation of Nickel Released due to Electrochemical Corrosion of Non-precious Dental Casting Alloys', *Journal of Oral Rehabilitation* **7**, 325–300 (1980)
56. Brune, D., Kjaerheim, A., Hensten-Pettersen, A. and Marion, L. 'Corrosion of Dental Alloys Studied by Implantation and Nuclear Tracer Technique', *Acta Odontologia Scandanavia*, **41**, 129–134 (1983)
57. Stenberg, T. and Bergman, B. 'Release and Uptake of Cobalt from Cobalt-Chromium Alloy Implants', *Acta Odontologia Scandanavia* **41**, 149–154 (1983)
58. Woody, R. D., Huget, E. F. and Horton, J. E. 'Apparent Cytotoxicity of Base Metal Casting Alloys', *Journal of Dental Research*, **56**, 739–743 (1977)
59. Moffa, J. P., Ellison, J. F. and Hamilton, J. C. 'Incidence of Nickel Sensitivity in Dental Patients', *Journal of Dental Research*, **62**, 199 (Abstr. 271) (1983)
60. Vreeburg, K. J. J., de Groot, K., von Blomberg, M. and Scheper, R. J. 'Induction of Immunological Tolerance by Oral Administration of Nickel and Chromium', *Journal of Dental Research*, **63**, 124–128 (1984)
61. Vrijhoef, M. M. A., Mezger, P. R., Van der Zel, J. M. and Greener, E. H. 'Corrosion of Ferromagnetic Alloys used for Magnetic Retention of Overdentures', *Journal of Dental Research*, **66**, 1456–1459 (1987)
62. Lain, E. S. 'Electrogalvanic Lesions of the Oral Cavity Produced by Metallic Dentures', *Journal of the American Medical Association*, **100**, 717–720 (1933)
63. Banoczy, J., Roed-Petersen, B. Pindborg, J. J. and Inovay, J. 'Clinical and Histologic Studies on Electrogalvanically Induced Oral White Lesions' *Oral Surgery, Oral Medicine and Oral Pathology*, **48**, 319–323 (1979)
64. Solomon, H. A. and Reinhard, M. C. 'Electric Currents from Dental Metals', *American Journal of Cancer*, **22**, 606–610 (1934)
65. Hakansson, B., Yontchev, E., Vannberg, N.-G. and Hedegard, B. 'An Examination of the Surface Corrosion State of Dental Fillings and Constructions. I. A Laboratory Investigation of the Corrosion Behaviour of Dental Alloys in Natural Saliva and Saline Solutions', *Journal of Oral Rehabilitation*, **13**, 235–246 (1986)
66. Yontchev, E., Hakansson, B., Hedegard, B. and Vannberg, N.-G. 'An Examination of the Surface Corrosion State of Dental Fillings and Constructions. II. A Clinical Study on Patients with Orofacial Complaints', *Journal of Oral Rehabilitation*, **13**, 365–382 (1986)
67. Inovay, J. and Banoczy, J. 'The Role of Electrical Potential Differences in the Etiology of Chronic Diseases of the Oral Mucosa', *Journal of Dental Research*, **40**, 884–890 (1961)

2.13 Surgical Implants

Introduction

Metallic devices have been used to repair and replace parts of the human body for centuries. Archaeological evidence clearly indicates that surgical procedures were performed in several ancient civilisations. The use of surgical metal implants in humans was first recorded in 1562 when a gold prosthesis was used to close a defect in a cleft palate[1].

Progress in surgery, however, was slow and mixed liberally with superstition until the latter part of the nineteenth century. Pasteur's and Lister's aseptic surgical techniques, developed around 1883, and shortly thereafter Roentgen's discovery of X-rays in 1895, added a new dimension to orthopaedic surgery[2]. As the occurrence of infection was brought under control, the relationship between material properties and the success of implant surgery became more clearly apparent. Tissue compatibility, corrosion resistance and strength were the critical characteristics found to be necessary. The noble metals, gold and silver, met the first two criteria but lacked strength for applications with high stress. Metals such as brass, copper and steel had adequate strength for many applications but exhibited poor corrosion behaviour and tissue compatibility.

In the beginning of the twentieth century, surgical techniques were developed for the fixation of bone fractures with a plate and screw combination. Sherman-type bone plates were fabricated from the best available alloy at the time, vanadium steel. By the 1920s the use of vanadium steel became questionable because of poor tissue compatibility. At that time however, no other alloy was available with high strength and good corrosion resistant properties.

During the 1930s, stainless steels containing 18% chromium and 8% nickel were first used for surgical implants. This material had far superior corrosion resistant properties than anything that had been available up to that time and immediately attracted the interest of orthopaedic surgeons. Bone plates, screws and other fixation devices were fabricated and used as surgical implants. Although the material performed better than anything else available, it still showed some susceptibility to attack in the saline environment of the human body. In 1926, when Strauss patented the 18-8 SMo stainless steel, containing 2 to 4% molybdenum and a reduced carbon content of 0.08%, a material was created which promised improved resistance to acid and chloride containing environments. This material formed the basis for the type 316L alloy in common use today. Also in the 1930s, a cobalt-chromium-molybdenum casting alloy previously used for dental appliances began to be used for surgical implants.

When titanium became commercially developed in the late 1940s, it was very soon evaluated as a suitable surgical implant material. The metal possessed a good combination of mechanical and corrosion resistant properties and also demonstrated outstanding tissue compatibility characteristics. Although a few internal fixation devices were also used in the United States in the 1950s and 1960s, the most extensive clinical use of titanium alloys was in England.

Interest in the Ti-6Al-4V alloy and Extra Low Interstitial (ELI) versions of this alloy for total joint prostheses spurted in the United States in the late 1970s. This alloy now finds wide application in orthopaedic surgery[3].

In the early 1980s a high-nitrogen austenitic stainless steel was introduced for use in orthopaedic surgery. The starting point for the development of this strong, highly corrosion resistant steel was the well-known Cr-Ni-Mo austenitic stainless steel type 316S16. Solid solution strengthening was achieved by increasing chromium and nitrogen contents and supplemented by a small addition of niobium. These alloying additions also served to improve corrosion resistance substantially while the use of a high nitrogen content allowed the attainment of an austenitic structure without increase of the use of the expensive element nickel, despite the presence of a high chromium plus molybdenum content. The balance of composition had been carefully selected to allow the realisation of substantially improved mechanical and corrosion resistant properties without excessive increase in cost or handling characteristics[4,5]. In its cold-worked state this material exhibits very high yield and fatigue limiting strengths.

Significant advances have also been made by forging titanium alloy (Ti-6Al-4V) and cobalt chromium alloys; cold working multiphase cobalt based alloys and by hot isostatically pressing cobalt chromium alloy powders. The property values claimed by the manufacturers are far in excess of the minimum values specified in the British, American and International Standards[6].

Non-metallic Materials

Most modern joint replacement systems rely on a metal-plastic bearing arrangement. Plastics had not been in use in orthopaedic surgery much before 1958 when Charnley[7,8] introduced polytetrafluoroethylene (Teflon) for an acetabular prosthesis which was cemented into place with cold-curing polymethylmethacrylate. When the Teflon showed an unacceptable degree of wear, Charnley introduced high-density polyethylene as an alternative material. He subsequently went on to use ultra high molecular weight polyethelene. This is currently the main plastic which is used in orthopaedic surgery.

Several other polymers, non-metals and ceramics, are currently being implanted for applications outside orthopaedic surgery, typical examples being polypropylene, polyacetal, polyethylene glycol terephthalate, polysulphone, carbon fibre, zirconia and aluminium oxide ceramics.

The mechanisms by which polymers undergo degradation in the human body are not yet completely understood. Examples of breakdown of these materials are illustrated by the embrittlement and excessive wear of polyester sockets exposed to the mechanical, biochemical and thermal stresses of the physiological milieu, as well as by the fatigue fractures, excessive wear and additional cross-linking (embrittlement) that have been observed in polyethylene sockets.

Material Selection

There are a number of ways to reduce corrosion[9]. Altering the environment is a common method used for industrial applications. Changing the electrolyte, concentration, reducing the temperature, or adding chemical inhibitors are all are used to decrease corrosion. Unfortunately, these techniques cannot be used for surgical implants since the environment is fixed and cannot be altered without destructive biological effects. Coatings are also widely used to retard corrosion. They have only limited use for protecting implants since many are subjected to abrasion and wear, especially orthopaedic devices. Applied current protection methods (anodic and cathodic protection) are too clumsy and impractical for use with surgically implanted devices. The only generally useful way to reduce the corrosion of surgical implants is by using an appropriate alloy and/or other materials[10].

The materials which are currently specified in the National and International Standards for use as implants together with their associated mechanical properties are shown in Table 2.23.

These standards also outline requirements for surface finish, grain size, heat treatment, metallurgical cleanness, absence of delta ferrite and alloy segregation to ensure that besides having a well balanced chemistry the alloys shall be in the proper metallurgical condition to yield optimum mechanical and corrosion resistant properties.

The Environment

It is widely appreciated that the deterioration of metal and plastic implant materials within the body is one of the most important aspects of implant surgery. This particular application of materials places an almost unique demand on the resistance to deterioration. The reasons are basically twofold, for not only may the environmental effects alter the structure and properties of the material, which may itself affect the function of the implant and hence the well-being of the patient, but also the by-products of any structural change may have harmful effects on the patient[16].

Ideally, a metallic implant should be completely inert in the body. However, that is rarely the case. The body environment is extremely hostile to all foreign materials and therefore, the effect of the environment on

Material	Type and condition	Typical standard for implant application	Ultimate tensile strength MPa min	0.2% tensile yield stress MPa min	Young's modulus $\times 10^3$ MPa	Elongation at fracture % min	Compressive strength MPa	Vickers hardness	Fatigue strength (10^8 cycles) MPa
Stainless steel	316 L annealed	BS 7252/1 Comp. D ISO 5832/1 ASTM F138	465	170	200	40	—	183	245–300
Cobalt-chromium molybdenum alloy	as cast	BS 7252/4 ISO 5832/4 ASTM F75	665	450	200	8	—	300	235–340
Titanium pure	annealed	BS 7252/2 type 1 ISO 5832/2 ASTM F67	240	170	127	24	—	240	250–280
Stainless steel	316 L cold worked	BS 7252/1 Comp. D ISO 5832/1 ASTM F138	505–605	195–295	200	35	—	320	300
Cobalt-chromium tungsten-nickel alloy	'wrought' heat treated	BS 7252/5 ISO 5832/5 ASTM F90	860	310	230	30	—	450	480
Cobalt-nickel chromium molybdenum alloy	medium hard	BS 7252/6 ISO 5832/6 ASTM F562	1000	650	230	20	—	350	400–450
Titanium alloy	Ti-6Al-4V annealed	BS 7252/3 ISO 5832/3 ASTM F136	860	780	111	10	—	350	400–440
High nitrogen stainless steel	annealed	BS 7252/9 ISO 5832/9	740	430	200	30	—	269	460
High nitrogen stainless steel	cold worked	BS 7253/2 ISO 9569 DP	1150	810	200	15	—	365	640
Aluminium oxide ceramic	Al_2O_3	BS 3531/8 ISO 6474 ASTM F603	270	N/A	380	0	4000	2000	N/A
Polyethylene	ultra high molecular weight RCH 1000	BS 7253/5 ISO 5834/2 ASTM F648	35	21	0.5	350	20	—	N/A
Polymethylmethacrylate	bone cement	BS 7253/1 ISO 5833/1 ASTM F451	25	N/A	2	5	70	—	<14
Cortical bone	wet	—	80–160	N/A	20	1–3	130–280	20–30	30

N/A = Not available

the implant and the effect of the implant on its host tissue are of primary concern.

A surgical implant is constantly bathed in extracellular tissue fluid. Basically water, this fluid contains electrolytes, complex compounds, oxygen and carbon dioxide. Electrolytes present in the largest amounts are sodium (Na^+) and chloride (Cl^-) ions. Most of the fluids existing in the body (such as blood, plasma and lymph) have a chloride content (and pH) somewhat similar to that of sea water (about 5 to 20 g/l and pH about 8)[17].

A 0·9% salt solution is considered to be isotonic with blood. Other electrolytes present include bicarbonate ions (HCO_3^-) and small amounts of potassium, calcium, magnesium, phosphate, sulphate and organic acid ions. Included among the complex compounds and present in smaller amounts are phospholipids, cholesterols, natural fats, proteins, glucose and amino acids. Under normal conditions the extracellular body fluid is slightly alkaline with a pH of 7·4[2].

In 1965 Zapffe[18] found that after fracture trauma local pH values decreased to 5.3–5.6. As healing took place the pH increased gradually to the normal 7·4. Another source of low local pH *in vivo* is the presence of crevices between components of a fixation device. A restriction of the oxygen supply to these locations can lead to pH values of about unity[19].

With internal fixation, the disturbance of the blood supply to the bones is often accompanied by severe pathalogical changes that may affect healing and variation in the equilibrium state electrochemically[20,21]. The ionic species also perform numerous functions that include maintenance of the body pH and participation in the oxidation-reduction (electron-transfer) reactions[22]. Normal imbalance occurs in the fluid compartment and different transport of ions and non-uniform changes normally accompany disease states. For example, during intensive care, after accidents or surgical operations, the fluid compartments are often disturbed. From an electrochemical viewpoint, the acceleration of corrosion can be due to the differential conditions existing along the implant surface. These conditions may be responsible for the formation of electrochemical cells accompanied by active metal dissolution at favoured localised spots at the implant-body fluid interface[10].

Reaction of the host tissue to metallic implants is affected by many factors including shape and size of the implant, movement between the implant and tissue, extent of corrosion attack, general degradation of the implant, and the biological activity of the resulting by-products of corrosion or degradation.

When an implant is surgically inserted into the human body, the internal environment is greatly disturbed. Haematomas are likely to collect around the implant, resulting in a lowered pH. Laing[23] observed pH values as low as 4·0 in healing wounds. The low pH usually persists until the haematomas are reabsorbed after several weeks[2].

From what has been said it is obvious that the environmental conditions within the human body are quite hostile and vary according to the degree of interaction with the body and the implant and the degree of trauma and infection associated with the implantation procedure.

There are a series of other factors which can result in altering the local environmental conditions and lead to various forms of corrosion and/or failure of the implant. These are discussed in the next section.

Corrosion Phenomena

In-Vivo and In-Vitro Testing

A number of studies have been carried out on the difficult task of measuring corrosion parameters *in-vivo*. These have involved measurement of corrosion potential[24-26] or corrosion rate using polarisation methods[27-30]. Only one of these studies has involved a functional loading situation. Brown and Simpson[33] measured corrosion potential versus time for a 316L stainless steel bone screw/plate combination in a sheep tibia *in-vivo*. A potential shift toward more active potentials was observed as the sheep increased the loading of the device by walking on a treadmill. Thull *et al.*[32] have performed corrosion experiments in dogs using telemetric methods. Corrosion potential versus time measurements were made for various implant alloys. The results were interpreted in terms of a situation where implant/tissue relative movement established a fretting corrosion situation. No information on stress levels or corrosion rates were obtained in either of these studies.

In-vitro fretting corrosion experiments have been performed using weight loss[31], polarisation[29] and corrosion potential measurements. Thull and Schaldach[32] have performed corrosion potential versus time measurements *in-vitro* in a joint simulator with different values of applied load. Brown and Simpson[33] have performed similar studies with screw/plate fretting. Both studies found larger shifts towards active potentials with larger loads[34].

One of the main problems of corrosion testing *in-vivo* and the interpretation of the mechanisms of corrosion which have taken place *in-vivo* after the implant has been removed from a patient is well illustrated in Figure 2.35 used by Semlitsch and Willet[14] to illustrate the types of corrosion which are associated with a total hip joint replacement. This figure shows that many corrosion mechanisms could be taking place simultaneously in a system of this nature.

To understand each of the corrosion mechanisms which could take place *in-vivo* and because of the difficulties of conducting *in-vivo* experiments most of the experimental work has been carried out *in-vitro* under somewhat simplified experimental conditions.

Anodic polarisation curves for the basic materials used for implants (e.g. stainless steel type 316S12, Titanium-6Al-4V and a Co–Cr–Mo casting alloy) have been obtained under varying conditions of surface finish, environment, chloride ion concentration, pH, aerated and de-aerated solutions at various temperatures and are well documented in the literature[36-38]. Typical results are shown in Tables 2.24 and 2.25. These results show that the basic materials used for implantation exhibit good corrosion resistant properties in simulated body fluid environments (Hanks's physiological solution) and in much more aggressive solutions such as 0.23 M $[Cl^-]$ acidified to pH 1.5, which simulate deep crevice and infection conditions. The materials' rest potentials are well below their respective breakdown potentials and are therefore stable under these conditions. These results only give an indication of how the materials behave under non-stressed, non-crevice and non-fretting conditions.

Ogundele and White[39] carried out a series of polarisation studies on surgical grade stainless steels in Hanks's physiological solution. Under

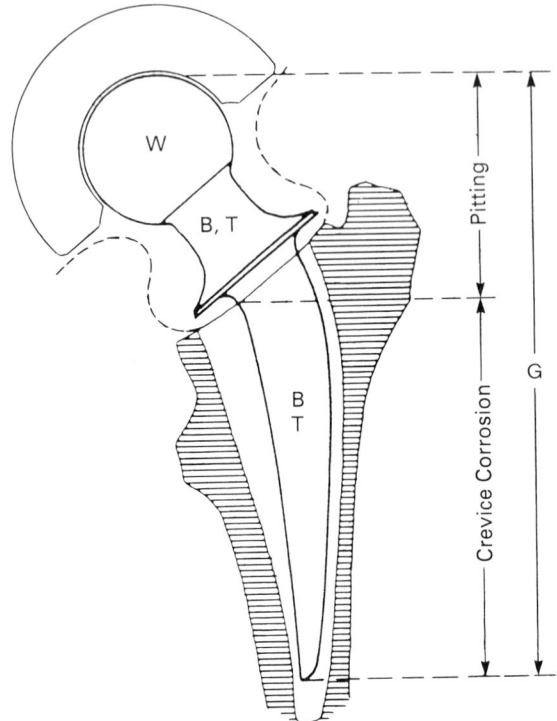

Fig. 2.35 Mechanical stresses and corrosive effects to which a joint endoprosthesis is exposed in the human body; W-wear (mechanical and/or depassivation), B-bending (corrosion fatigue), T-torsion, G-general and localised attack.

Table 2.24 Breakdown potentials (mV) for 316 stainless steel, titanium and cobalt-chromium-molybdenum alloy in oxygen-free 0.17 M NaCl solution at 37°C using a silver/silver chloride reference electrode.

Material	After Hoar and Mears[24] Hanks's solution	After Bultitude and Morris[35] Ringer's solution
316 Stainless steel	400–480	160–650
Titanium	9000	800[a]
Cr-Cr-Mo (cast) Alloy	870	790[b]

[a] Only slight increase in current produced on raising the potential to +1·8 volts.
[b] Breakdown potential difficult to define. The current increases more rapidly at +0·79 volts.

varying conditions of temperature, [Cl$^-$], [HCO$_3^-$] and pH and are summarised as follows:

1. Increasing concentrations of chloride ions adversely affected corrosion resistance in lowering passivation breakdown potentials, marginally lowering corrosion potentials, lowering passivation potentials and generally increasing the propensity to pitting attack as evidenced

Table 2.25 Breakdown potentials for 316S12 stainless steel (cold worked), high nitrogen stainless steel (cold worked), titanium-6Al-4V and cast-cobalt-chromium-molybdenum alloy in continuously aerated aqueous acidified chloride solution 0.23 M [Cl$^-$] pH 1.5 at 25°C*.

Material	British standard	Condition	Breakdown potential mV(s.c.e.)	Rest potential mV(s.c.e.)
316 S12 Stainless steel	BS7252/1 Comp. D	cold worked	180	−100
High nitrogen stainless steel	BS7252/9	cold worked	850	300
Titanium-6Al-4V	BS7252/3	annealed	1700	150
Co–Cr–Mo Alloy	BS7252/4	cast	850	155

* Results provided by De Puy International Ltd. Leeds.

by the increase in size of hysteresis loops from potentiodynamic cycling tests.
2. Increasing concentrations of bicarbonate tended to raise the breakdown potentials but also increased the corrosion potentials. This, in combination with a high chloride concentration, high bicarbonate concentrations may raise the corrosion potentials such that they border on passivation breakdown. The increase in hysteresis loop size on potentiodynamic cycles with increasing bicarbonate concentration shows a lowered resistance to pitting attack and crevice corrosion.
3. Variations in pH promoted increases in corrosion potentials from acid pH levels to neutral pH; thereafter, however, corrosion potentials were lowered in alkaline solutions to more active values. Decreasing pH caused a lowering of breakdown potentials in the presence of Cl$^-$ and an increase in the current densities for passivation.
4. Increasing temperature tended to raise corrosion potentials, lower breakdown potentials, thereby generally lowering corrosion resistance. Ogundele and White[39] re-emphasised that 316L type stainless steels are generally corrosion resistant in biological fluids and are suited to application in orthopaedic implants. They support their argument by the fact that there is a low incidence of implant failures (<5% from corrosion and corrosion-related mechanisms or both)[40].

Pitting Corrosion (Sections 1.5 and 1.6)

The literature contains a number of studies on the susceptibility of the cobalt-based alloys to pitting corrosion. *In-vitro* studies conducted by Mueller and Greener[38], involving static conditions, revealed no evidence of pitting having occurred. Syrett and Wing[41], utilising cyclic polarisation analyses, observed that neither as-cast nor annealed Co–Cr–Mo alloy demonstrated hysteresis loops in their cyclic polarisation curves. They concluded that this substantiated the fact that neither should be susceptible to pitting or crevice corrosion. Other investigators[42-48] studying the susceptibility of the cobalt alloy to pitting corrosion under *in-vitro* static conditions, have shown the alloy not to be susceptible. However, when the alloy is either severely cold worked or subjected to fretting or cyclic loading conditions[49-50] pitting corrosion has been observed *in-vitro*[25].

The addition of a minimum of 2% molybdenum content in type 316 stainless steel has been shown to reduce the tendency for pitting-type corrosion in chloride environments. Hoar and Mears[24] postulated that chloride ions accelerate the corrosion of stainless steel by penetrating the oxide film. The chloride-contaminated film then loses its passivating quality and a local attack on the metal follows, creating a pit. The exact mechanism by which molybdenum strengthens the oxide film is not clearly understood[2].

Stress Corrosion Cracking (Sections 1.5, 1.6 and Chapter 8)

Pitting and stress corrosion cracking, although usually associated with stainless steels in chloride media, have not been observed on recovered surgical implants. Implants often exhibit cracks and surface pitting, but these are most likely the result of improper manufacture rather than corrosion[51-52].

Sheehan *et al.*[53] studied stress corrosion cracking (SCC) of type 316L stainless steel *in-vitro*. Tests were conducted using specimens with electropolished surfaces in slow strain-rate tension tests and Schneider intramedullary nails in static bend tests. They concluded, as previous investigators had done, that SCC is not a mode of failure of type 316L stainless steel implants *in-vitro* and found no indication that SCC of this material would occur *in-vivo*. Bundy and Desai[54] studied SCC of type 316L stainless steel and ELI Ti-6Al-4V using fracture mechanics and measuring crack propagation velocity versus stress intensity in environments of $MgCl_2$, HCl and Ringer's solution. Crack propagation occurred in precracked type 316L stainless steel in Ringer's solution held at a potential that disrupted the passive film. The conclusion from this investigation was that SCC of type 316L stainless steel could occur *in-vivo* if these conditions existed.

Edwards *et al.*[55] carried out controlled potential, slow strain-rate tests on 'Zimaloy' (a cobalt-chromium-molybdenum implant alloy) in Ringer's solution at 37°C and showed that hydrogen absorption may degrade the mechanical properties of the alloy. Potentials were controlled so that the tensile sample was either cathodic or anodic with respect to the metal's free corrosion potential. Hydrogen was generated on the sample surface when the specimen was cathodic, and dissolution of the sample was encouraged when the sample was anodic. The results of these controlled potential tests showed no susceptibility of this alloy to SCC at anodic potentials.

Corrosion Fatigue (Section 8.6)

The corrosion fatigue failure of total joint prostheses has been widely studied. It can be concluded from an extensive literature survey carried out by Leclerc[6] that provided the alloy used in the manufacture of the joint prosthesis conforms with the current national and international standards and that it is in the correct metallurgical condition the contribution of corrosion in the corrosion fatigue failures of joint prostheses is marginal. The corrosion contribution increases with the time that the prosthesis is present in the patient.

It has also been shown by various testing laboratories that the *in-vivo* failures can be simulated using *in-vitro* simulators without the presence of a corrosive medium and provided that a torsional component of load is applied to the prosthesis.

Bechtol[56] in reviewing his clinical experience with 1087 cemented Charnley and Bechtol total hip replacements concluded that prosthetic stem failure is actually the final sequence of previous failures of bony and cement support. His suggested sequence of events leading to prosthetic stem failure have been substantiated in the literature and are as follows:

1. Poor blood supply in the area of the medial femoral neck (as the consequence of a pre-existing pathological condition or excessive early weight bearing) leads to initial loosening of the cement bond;
2. Resorption of bone in the medial neck places the forces of continued weight bearing on the bone cement in that area and may lead to medial and/or distal migration of the femoral cement and prosthetic component;
3. Fracture of cement transversely near the distal end of the prosthetic stem or in the area of the medial neck results in a slight downward or medial shift of stem position and further increases the forces upon the stem;
4. Repeated flexing of the proximal stem in a fatigue mode over a period of months or years results in a widening of the separation between the metal and cement laterally and ultimately in an actual deformity or fracture of the metal stem.

Hughes, Jordan and Orman[19] reported on some corrosion fatigue tests that they carried out on an Amsler Vibrophore with specimens immersed in saline environments with a pH range 2 to 7·4. They found that titanium showed a slight decrease in fatigue properties over the pH range whereas stainless steel showed a drastic decline in fatigue strength at pH 4 and below this level became inferior to the titanium. Rollins *et al.*[57] tested steels in chloride solutions of varying pH and found that corrosion in the range pH 4–10 was independent of pH since the process was controlled by oxygen diffusion to cathodic areas, i.e. it depended on the solubility of oxygen in the solution. Below pH 4 there was an increase in corrosion rate associated with cathodic hydrogen evolution. Although there was insufficient data to analyse fully the effects of changes of pH, it is clear that the increased corrosion rate reported below pH 4 was making a contribution to the reduction of the fatigue strength of the stainless steels.

Piehler *et al.*[58] reported on corrosion fatigue of hip nails and emphasised the importance of materials selection and design. Corrosion-fatigue tests of nail plates were conducted on a flexure fatigue machine to compare devices of identical design but different material. Large plate designs had superior corrosion-fatigue performance over small plates. Devices made of

Ti-6Al-4V were superior in corrosion-fatigue performance to the type 316L devices even though some fretting and wear occurred in the countersinks.

Fretting and Crevice Corrosion (Section 1.6 and 8.7)

One of the most serious corrosion problems associated with type 316 stainless steel is its susceptibility to crevice corrosion. The incidence and extent of this type of corrosion in surgical implants was stressed by Scales et al.[59] who reported the presence of crevice corrosion in 24% of type 316L bone plates and screws examined after removal from patients. This record however compared favourably with the presence of crevice corrosion in 51% of 18-8 stainless plates, demonstrating the superiority of the molybdenum-containing grade.

Further improvements in corrosion resistance have been made by minimising the non-metallic inclusion content of type 316 stainless steel. A routine examination of over 100 type 316LVM surgical implants, removed for various reasons, did not show evidence of pitting corrosion or general attack but have revealed small areas of crevice corrosion in bone plate/screw interfaces. A crevice is usually formed by screw head/bone plate counter contiguity, initiating the corrosion process. Constant minute flexural movements of the highly stressed bone plate aggravates the corrosion damage through fretting of the contacting surfaces[2].

Syrett and Davis[42] conducted *in-vivo* studies wherein they implanted crevice corrosion specimens of Co-Cr-Mo in dogs and rhesus monkeys for up to two years. Their results indicated the alloy was not susceptible to crevice corrosion. Galante and Rostoker[60] implanted crevice-type specimens of Co-Cr-Mo and Ti-6Al-4V in the back of rabbits for 12 months. Although no evidence of severe corrosion was found in any of the specimens, several of the titanium and cobalt specimens did show signs of 'single pits' in the crevice regions.

From tests conducted in chloride solutions at temperatures greater than 100°C, Griess[61] concluded that for titanium to corrode in crevice regions, the solution within the crevice must become and remain significantly acid. However, it should be emphasised that the temperature in the physiological environment is much lower (37°C). In the physiological corrosion test performed by Fraker et al.[62] the pH of one of their test solutions was 1·9; they reported no significant effect on the passive behaviour of titanium or Ti-6Al-4V. Solar et al.[63] also found that the pH had no significant effect in the range of 1·5 to 9·0.

Galvanic Corrosion (Section 1.6 and 1.7)

The specimen design used in the study by Rostoker et al.[48] was such that it simulated both galvanic coupling and crevice conditions. Specimens were immersed in a 1% saline solution at 37°C, and examined by optical microscopy after exposures of a few to 100 days. No corrosion was observed on Ti-6Al-4V when the alloy was either uncoupled, coupled with itself (simple crevice), or coupled with type 316L stainless steel, cast Co-Cr-Mo alloy or graphite. In contrast, type 316L stainless steel exhibited multiple pitting in all tests. Levine and Staehle[64] reported similar findings, and concluded that type 316L stainless steel is the most susceptible to crevice attack compared with wrought Co-Cr-Mo, commercial purity titanium and Ti-6Al-4V. They also concluded that 'problems can occur with any of the widely used metals, however, especially if metals are mixed'. The use of dissimilar metals in surgery has recently been reviewed by Mears[65] who suggests that titanium may be used with a number of less passive metals without fear of increased corrosion of the latter.

The successful clinical use of titanium and cobalt-chromium alloy combinations has been reported[66]. Lucas et al.[25] also investigated this combination using electrochemical studies based on mixed potential and protection potential theories. Verification of these studies was made by direct coupling experiments. The electrochemical studies predicted coupled corrosion potentials of -0.22 V and low coupled corrosion rates of $0.02\,\mu$ A/cm^2. Direct coupling experiments verified these results. The cobalt-titanium interfaces on the implants were macroscopically examined and no instances of extensive corrosion were found. Overall, the *in-vitro* corrosion studies and the examination of retrieved prostheses predicted no exaggerated *in-vivo* corrosion due to the coupling of these cobalt and titanium alloys.

Sensitisation (Section 3.3)

The solubility of carbon in austenitic stainless steels is much greater at normal softening treatment temperatures (1000-1100°C) than at lower temperatures and the formation of $Cr_{23}C_6$ carbides, largely at grain boundaries, during slow cooling through or on subsequent reheating within the approximate temperature range 500-750°C can lead to serious depletion of chromium with a decrease in propensity to become passive at or near the grain boundaries, so that intergranular corrosion may occur when the steel is in contact with a corrodent[67].

Intercrystalline corrosion was a serious problem with the austenitic stainless steels early in their development since carbon contents then were relatively high, e.g. En58J contained up to 0.12% C, 316 type stainless steel contained up to 0.08% C. The problem in relation to surgical implants has been reported by Scales et al.[59] and as a result of this and several other reports the British, American and International Standards specified the use of a 316S12 type austenitic stainless steel which contains 0.03% C max. The use of the lower carbon content stainless steels as specified in the various standards has now eliminated the problem of sensitisation of implants. If manufacturers do use the 0.08% C versions they have to be very careful with the forging temperatures or anneal the prostheses afterwards.

Conclusions

The basic materials which are called for in the National and International Standards, provided that they are in their correct metallurgical condition, have adequate strength and corrosion resistance for most orthopaedic applications.

As the surgical techniques advance, the average age of the patient subjected to total joint replacement is decreasing and therefore the life expectancy of the implant is increasing. This obviously places more emphasis on the fatigue and corrosion resistant properties of the materials which are used in orthopaedic surgery.

The modern materials technologists and surgeons are experimenting with have different types of surface coatings on the prostheses in an effort to induce bony in-growth, enhance fixation and/or eliminate the use of bone cement. Typical coatings involve plasma spraying, sintering alloy powders or diffusion bonding meshes onto the surfaces and/or coating the surfaces with hydroxyapatite. Whilst these coatings are resolving some orthopaedic problems they might generate a new series of difficulties which the manufacturer will have to overcome, especially as far as fatigue life and crevice corrosion resistance is concerned.

M. F. LECLERC

REFERENCES

1. Wickstrom, J. K., Surgical Implants—The Mechanical and Environmental Problems, *Journal of Materials*, **1**, 366–372 (1966)
2. Bardos, D. I., Stainless Steels in Medical Devices, in *Handbook of Stainless Steels*, (Eds) D. Peckner and I. M. Bernstein, 42·1–42·10. McGraw-Hill, New York (1977)
3. Luckey, H. A., Kubli, Jr. F., 'Editors Introduction', in *Titanium Alloys in Surgical Implants*, ASTM Publication STP 796, 1–3, Philadelphia (1983)
4. Truman, J. E., 'Stainless Steels-a Survey Part 2: Special Grades', *The Metallurgist and Materials Technologist*, 75–78 (1980)
5. Truman, J. E., 'An Austenitic Stainless Steel of Improved Strength and Corrosion Resistance', *Stainless Steel Industry*, **6**, 21–32 (1978)
6. Leclerc, M. F., 'A Review of the Factors Influencing the Mechanical Failure of the Femoral Component used in Total Hip Replacement', Proceedings: *Engineering in Orthopaedic Surgery and Rehabilitation*, 36–48, published by Bioengineering Unit, Princess Margaret Rose Hospital, Edinburgh (1982)
7. Charnley, J., 'Total Prosthetic Replacement of the Hip Joint Using a Socket of High Density Polyethylene', *Wrightington Hospital Publication No. 1 Centre for Hip Surgery*, (1966)
8. Charnley, J., (Ed.) *Low Friction Arthroplasty of the Hip*, Springer-Verlag, Berlin, Heidelberg, New York (1979)
9. Fontana, M. G., and Green, N. D., (Eds) *Corrosion Engineering*, 2nd Edition, 194–222, McGraw-Hill, New York (1978)
10. Greene, N. D., 'Corrosion of Surgical Implant Alloys: A Few Basic Ideas', in *Corrosion and Degradation of Implant Materials*, Second Symposium, (Eds) A. C. Fraker and C. D. Griffin, 5–10 ASTM Publication STP 859, Philadelphia (1985)
11. British Standards Institution *Surgical Implants*, Specification Numbers BS 7251 to BS 7254, London (1990–93)
12. International Organisation for Standardisation *Implants for Surgery—Metallic Materials—Wrought HIgh Nitrogen Stainless Steel*, Specification Number ISO 5832 Part 9, Geneva (1992)
13. Swanson, S. A., Freeman, M. A. R., Vernon-Roberts, B., and Weightman, B., 'The Scientific Basis of Joint Replacement—Properties of Materials', (Eds) S. A. Swanson and M. A. R. Freeman, 1–17, Pitman Medical Publishing Co. Ltd (1977)
14. Semlitsch, M., and Willert, H. G., 'Properties of Implant alloys for Artificial Hip Joints', *Medical and Biological Engineering and Computing*, 18, 511–520 (1980)
15. Chas. F. Thackray Ltd. *Ortron 90*, Technical Publication 2655, Leeds, England (1981)

16. Williams, D. F., The Deterioration of Materials in Use, in *Implants in Surgery*, (Eds) D. F. Williams and R. Roaf, 137–183. W. B. Saunders, Philadelphia (1973)
17. Pourbaix, M., 'Electrochemical Corrosion of Metallic Biomaterials', *Biomaterials*, **5**, 122–134 (1984)
18. Zapffe, C. A., 'Human Body fluids Affect Stainless Steel', *Metal Progress*, **67**, 95–98, (1965)
19. Hughes, A. N., Jordan, B. A. and Orman, S., 'The Corrosion Fatigue Properties of Surgical Implant Materials. Third Progress Report – May 1973', *Engineering in Medicine*, **7**, 135–141 (1978)
20. Walker G. D., *Journal of Biomedical Materials Research* (Symposium), **5**, 11–26 (1974)
21. Jacobson, B., and Webster, J. B., Chapter 10. Surgery, in *Medicine and Clinical Engineering*, (Eds) Jacobson and Webster 525–532 Prentice-Hall Inc., New Jersey (1977)
22. Guyton, A. C., Chapter 33, 'Partition of the Body Fluids Osmotic Equilibria between Extracellular and Intracellular Fluids', in *Textbook of Medical Physiology* (5th Edition), (Ed.) Guyton A. C., 424–437, W. B. Saunders, Philadelphia (1976)
23. Laing, P. G., Compatibility of Biomaterials, *Orthopaedic Clinics of North America*, **4**, 249–273 (1973)
24. Hoar, T. P., and Mears, D. C., 'Corrosion Resistant Alloys in Chloride Solutions', *Proceedings of the Royal Society – Series A*, **294**, 486–510 (1966)
25. Lucas, L. C., Buchanan, R. A., Lemons, J. E., and Griffin, C. D., 'Susceptibility of Surgical Cobalt-Base Alloy to Pitting Corrosion', *Journal of Biomedical Materials Research*, **16**, 799–810 (1982)
26. Svare, C. W., Belton, G., and Korostoffe, E., 'The Role of Organics in Metallic Passivation', *Journal of Biomedical Materials Research*, **4**, 457–467 (1970)
27. Revie, R. and Greene, N. D., 'Comparison of the In-Vivo and In-Vitro Corrosion of 18-8 Stainless Steel and Titanium', *Journal of Biomedical Materials Research*, **3**, 465–470 (1969)
28. Colangelo, V. T., Green, N. D., Kettelkamp, D. B., Alexander, H., and Campbell, C. J., 'Corrosion Rate Measurements In-Vivo', *Journal of Biomedical Materials Research*, **1**, 405–414 (1967)
29. Steinemann, S. G., 'Corrosion of Surgical Implants – In-Vivo and In-Vitro Tests', in *Evaluation of Biomaterials*, (Eds) G. D. Winter, J. L. Leray and K. deGroot, *Advanced Biomaterials*, **1**, 1–34 (1980)
30. Buchanan, R. A. and Lemons, J. E., 'In-vivo corrosion – Polarization Behaviour of Titanium-Base and Cobalt-Base Surgical Alloys', *Transactions of the 8th Annual Meeting of the Society of Biomaterials*, **5**, 110 (1982)
31. Brown, S. A. and Merritt, K., 'Fretting Corrosion in Saline and Serum', *Journal of Biomedical Materials Research*, **15**, 479–488 (1981)
32. Thull, R. and Schaldach, M., 'Corrosion of Highly Stressed Orthopaedic Joint Replacements', in *Engineering in Medicine: Advances in Artificial Hip and Knee Joint Technology*, (Eds) M. Schaldach and D. Hohmann, 242–256. Springer Verlag, Berlin, Heidelberg, New York (1976)
33. Brown, S. A. and Simpson, J. P., 'Crevice and Fretting Corrosion of Stainless-Steel Plates and Screws', *Journal of Biomedical Materials Research*, **15**, 867–878 (1981)
34. Bundy, K. J., Vogelbaum, M. A. and Desai, V. H., 'The Influence of Static Stress on the Corrosion Behaviour of 316L Stainless Steel in Ringer's Solution', *Journal of Biomedical Materials Research*, **20**, 493–505 (1986)
35. Bultitude, F. W. and Morris, J. R., *Laboratory Study of the Corrosion of Implants*, Atomic Weapons Research Establishment, Aldermaston, England, Report GRO 44/83/29 (1969)
36. Revie, R. W. and Greene, N. D., 'Corrosion Behaviour of Surgical Implant Materials 1: Effects of Sterilisation', *Corrosion Science*, **9**, 755–761 (1969)
37. Revie, R. W. and Greene, N. D., 'Corrosion Behaviour of Surgical Implant Materials 2: Effects of Surface Preparation', *Corrosion Science*, **9**, 763–770 (1969)
38. Mueller, H. J. and Greener, E. H., 'Polarization Studies of Surgical Materials in Ringer's Solution', *Journal of Biomedical Materials Research*, **4**, 29–41 (1970)
39. Ogundele, G. I. and White, W. E., 'Polarization Studies on Surgical-Grade Stainless Steels in Hanks' Physiological Solution', in *Corrosion and Degradation of Implant Materials*, second symposium, (Eds) A. C. Fraker and C. D. Griffin, 117–135 ASTM Publication STP 859, Philadelphia (1985)
40. White, W. E., Postlethwaite, J. and LeMay, I., in *Microstructural Science* (Vol. 4) 145–157, American Elsevier, New York (1976)

41. Syrett, B. C. and Wing, S. S., 'Pitting Resistance of New and Conventional Orthopaedic Implant Materials—Effect of Metallurgical Condition', *Corrosion*, 34A, 138–145 (1978)
42. Syrett, B. C. and Davis, E. E., 'Crevice Corrosion of Implant Alloys—A comparison of In-Vitro and In-Vivo Studies', Presented at Spring meeting of the ASTM, Kansas City, MO. (1978)
43. Sury, P., Corrosion Behaviour of Case and Forged Implant Materials for Artificial Joints, Particularly with Respect to Compound Designs. Research and Development Department, Sulzer Brothers Ltd., CH-8401, Winterthur, Switzerland.
44. Cohen, J. and Wulff, J., 'Clinical Failure Caused by Corrosion of a Vitallium Plate', *Journal of Bone and Joint Surgery*, 54A, 617–628 (1972)
45. Rose, R. M., Schiller, A. L. and Radin, E. L., 'Corrosion-Accelerated Mechanical Failure of a Vitallium Nail-Plate', *Journal of Bone and Joint Surgery*, 54A, 854–862 (1972)
46. Cahoon, J. R., Bandyopadhya, R. and Tennese, L., 'The Concept of Protection Potential Applied to the Corrosion of Metallic Orthopaedic Implants', *Journal of Biomedical Materials Research*, 9, 259–264 (1975)
47. Morral, F. R., 'Crevice Corrosion of Cobalt Based Surgical Alloys', *Journal of Materials*, 1, 384 (1966)
48. Rostoker, W., Pretzel, C. W. and Galante, J. O., 'Couple Corrosion Among Alloys for Skeletal Prostheses', *Journal of Biomedical Materials Research*, 8, 407–419 (1974)
49. Syrett, B. C. and Wing, S S., 'An Electrochemical Investigation of Fretting Corrosion of Surgical Implant Materials', *Corrosion*, 34, 379–386 (1978)
50. Cohen, J., 'Corrosion Testing of Orthopaedic Implants', *Journal of Bone and Joint Surgery*, 44A, 307–316 (1962)
51. Colangelo, V. J. and Greene, N. D., 'Corrosion and Fracture of Type 316 SMO Orthopaedic Implants', *Journal of Biomedical Materials Research*, 3, 247–265 (1969)
52. Cahoon, J. R. and Paxton, H. W., 'Metallurgical Analyses of Failed Orthopaedic Implants', *Journal of Biomedical Materials Research*, 2, 1–22 (1968)
53. Sheehan, J. P., Morin, C. R. and Packer, K. F., 'Study of Stress Corrosion Cracking Susceptibility of Type 316L Stainless Steel In-Vitro', in *Corrosion and Degradation of Implant Materials*, Second Symposium', (Eds) A. C. Fraker and C. D. Griffin, 57–72 ASTM Publication STP 859, Philadelphia (1985)
54. Bundy, K. J. and Desai, V. H., 'Studies of Stress-Corrosion Cracking Behaviour of Surgical Implant Materials using a Fracture Mechanics Approach', in *Corrosion and Degradation of Implant Materials*, second symposium', (Eds) A. C. Fraker and C. D. Griffin, 73–90, ASTM Publication STP 859, Philadelphia (1985)
55. Edwards, B. J., Louthan, M. R. and Sisson, R. D., 'Hydrogen Embrittlement of Zimaloy: A Cobalt-Chromium-Molybdenum Orthopaedic Implant Alloy', in *Corrosion and Degradation of Implant Materials*, Second Symposium, (Eds) A. C. Fraker and C. D. Griffin, 11–29 ASTM Publication STP 859, Philadelphia (1985)
56. Bechtol, C. O., 'Failure of Femoral Implant Components in Total Hip Replacement Operations', *Orthopaedic Review*, 4, 23–29 (1975)
57. Rollins, V., Arnold, B. and Lardner, E., 'Corrosion Fatigue in High Carbon Steel', *British Corrosion Journal*, 5, 33 (1970)
58. Piehler, H. R., Portnoff, M. A., Sloter, L. E., Vegdahl, E. J., Gilbert, J. L. and Weber, M. J., 'Corrosion-Fatigue Performance of Hip Nails: The Influence of Materials Selection and Design', in *Corrosion and Degradation of Implant Materials*, Second Symposium, (Eds) A. C. Fraker and C. D. Griffin, 93–104, ASTM Publication STP 859, Philadelphia (1985)
59. Scales, J. T., Winter, G. D. and Shirley, H. T., 'Corrosion of Orthopaedic Implants, *Journal of Bone and Joint Surgery*, 41B, 810–820 (1959)
60. Galante, J. and Rostoker, W., 'Corrosion: Related Failures in Metallic Implants and Experimental Study', *Clinical Orthopaedics and Related Research*, 86, 237–244 (1972)
61. Greiss, J. C., 'Crevice Corrosion of Titanium in Aqueous Salt Solutions: *Corrosion*, 24, 96 (1968)
62. Fraker, A. C., Ruff, A. W. and Yeager, M. P., *Proceedings of the Second International Titanium Conference*. (Eds) Joffee, R. I. and Burke, H. M., Plenum, New York (1973)
63. Solar R. J., Pollack, S. R. and Korostoffe, E., 'In-vitro Corrosion Testing of Titanium Surgical Implant Alloys: An Approach to Understanding Titanium Release from Implants', *Journal of Biomedical Materials Research*, 13, 217–250 (1979)
64. Levine, D. L. and Staehle, R. W., 'Crevice Corrosion in Orthopaedic Implant Metals, *Journal of Biomedical Materials Research*, 11, 553 (1977)

65. Mears, D. C., 'The Use of Dissimilar Metals in Surgery', *Journal of Biomedical Materials Research*, **6**, 133–148 (1975)
66. Jackson-Burrows, H., Wilson, J. N. and Scales, J. T., 'Excision of Tumors of Humerus and Femur with Restoration of Internal Prostheses', *Journal of Bone and Joint Surgery*, **57B**, 148 (1975)
67. Truman, J. E., 'Stainless Steels – A survey Part 1: Development of Types', *The Metallurgist and Materials Technologist*, 15–23 (1980)

Glossary of Medical Terms

Cholesterol: A fatlike, pearly substance, a monatomical alcohol, $C_{27}H_{45}OH$, crystallising in the form of acicular crystals and found in all animal fats and oils, in bile, blood, brain tissue, milk, yolk of egg, the medullated sheaths of nerve fibres, the liver, kidneys and adrenal glands.

Distal: Away from the body or origin, e.g. the free end of a limb.
Note: Contrast with proximal.

Endoprosthesis: A permanent prosthesis used wholly within the body, e.g. as a replacement for a bone, a joint, a tendon or a ligament.

Extracellular: Outside of a cell or cells.

Haematoma: The collection of blood in tissues due to haemorrhage.

Intramedullary nail: An implant for introduction into the marrow cavity of long bones forming an internal splint. The nail may be straight or curved and of V, clover-leaf (e.g. Kuntscher nail) or circular cross-section.

In-vitro: Literally, 'in glass', e.g. in laboratory apparatus.

In-vivo: In a living organism, e.g. a human being.

Lateral: Away from the median plane of the body.
Note: Contrast with medial.

Lymph: A transparent, slightly yellow liquid of alkaline reaction, found in the lymphatic vessels.

Medial: Towards the median plane of the body.
Note: Contrast with lateral.

Median: The longitudinal anteroposterior plane dividing the body into apparently similar halves.

Pathology: The study of disease.

Phospholipids: A lipid containing phosphorus which on hydrolysis yields fatty acids, glycerin, and a nitrogenous compound. Lecithin cephalin and sphingomyelin are the best known examples.

Physiological environment: The study of the function of the body.

Plasma: Anything formed or moulded. The fluid portion of the blood in which the corpuscles are suspended.

Prosthesis: Any device that replaces an anatomical part or deficiency.

Proximal: Nearer to the body or origin, e.g. the root of a limb.
Note: Contrast with distal.

Resorption: The loss of substance through physiological or patholigical means, such as loss of dentin and cementum of a tooth.

Telemetric: The making of measurements at a distance from the subject, the measurable evidence of the phenomena under investigation being transmitted by radio signals.

Total joint replacement: The replacement of a joint by an endoprosthesis.

Trauma: A wound or injury.

3 FERROUS METALS AND ALLOYS

3.1	Iron and Steel	**3**:3
3.2	Low-alloy Steels	**3**:23
3.3	Stainless Steels	**3**:34
3.4	Corrosion Resistance of Maraging Steels	**3**:78
3.5	Nickel–Iron Alloys	**3**:92
3.6	Cast Iron	**3**:101
3.7	High-Nickel Cast Irons	**3**:115
3.8	High-Chromium Cast Irons	**3**:128
3.9	Silicon–Iron Alloys	**3**:136
3.10	Amorphous (Ferrous and Non-Ferrous) Alloys	**3**:146

3.1 Iron and Steel

Introduction

Bare iron and steel are liable to rust in most environments but the extent of the corrosion depends upon a number of factors, the most important of which are the composition and surface condition of the metal, the corrosive medium itself and the local conditions.

With regard to the effect of composition, ferrous metals fall into three broad categories:

1. The ordinary cast irons, wrought irons and steels, to which no alloying elements are added, and which are vulnerable to corrosion.
2. Low-alloy steels, which contain about 2-3% of alloying elements, commonly copper, chromium and nickel. These steels still rust, but under certain conditions in the atmosphere, the rust formed becomes adherent and protective so that the corrosion rate becomes several times less rapid than with the ordinary steels mentioned above. These steels are often termed *weathering* steels.
3. Stainless steels, which contain high percentages of alloying elements, e.g. 18% chromium, 8% nickel and 3% molybdenum. Steels of this type are practically non-corrodible in appropriate circumstances.

The discussion that follows will be concerned mainly with materials in the first of these categories. Moreover, since the corrosion of cast iron is discussed elsewhere (see Section 3.6), and since little wrought iron is produced nowadays, the subject matter will virtually resolve itself into the corrosion of ordinary carbon steels, as used in mass for general purposes. The corrosion of low-alloy steels and that of stainless steels are considered in Sections 3.2 and 3.3 respectively.

The treatment will begin with a brief consideration of the mechanism of rusting and of the influence of variations in the steel itself. It will be completed by short surveys of present knowledge of the rusting of ordinary mild steel in the three natural media: air, water and soil.

Mechanism of Rusting

In pure dry air at normal temperatures a thin protective oxide film forms on the surface of polished mild steel. Unlike that formed on stainless steels it is not protective in the presence of electrolytes and usually breaks down in air, water and soil. The anodic reaction is:

$$Fe \rightarrow Fe^{2+} + 2e^-$$

In de-aerated solutions, the cathodic reaction is

$$2H^+ + 2e^- \rightarrow H_2$$

This occurs fairly rapidly in acids but very slowly in alkaline and neutral solutions. In the presence of oxygen the following reaction occurs in slightly alkaline and neutral solutions:

$$O_2 + 2H_2O + 4e^- \rightarrow 4OH^-$$

This is the common form of cathodic reaction in most environments. The OH^- ions react with Fe^{2+} ions to form ferrous hydroxide:

$$Fe^{2+} + 2OH^- \rightarrow Fe(OH)_2$$

This is oxidised to ferric hydroxide $Fe(OH)_3$, which is a simple form of rust. The final product is the familiar reddish brown rust $Fe_2O_3 \cdot H_2O$, of which there are a number of varieties, the most common being the α form (goethite) and the γ form (lepidocrocite). In situations where the supply of oxygen is restricted, Fe_3O_4 (magnetite) or $\gamma\, Fe_2O_3$ may be formed.

This is a simplified treatment but it serves to illustrate the electrochemical nature of rusting and the essential parts played by moisture and oxygen. The kinetics of the process are influenced by a number of factors, which will be discussed later. Although the presence of oxygen is usually essential, severe corrosion may occur under anaerobic conditions in the presence of sulphate-reducing bacteria (*Desulphovibrio desulphuricans*) which are present in soils and water. The anodic reaction is the same, i.e. the formation of ferrous ions. The cathodic reaction is complex but it results in the reduction of inorganic sulphates to sulphides and the eventual formation of rust and ferrous sulphide (FeS).

Effect of Variations in the Steel

Ordinary steels are essentially alloys of iron and carbon with small additions of elements such as manganese and silicon added to provide the requisite mechanical properties. The steels are manufactured from a mixture of pig iron and scrap, which is treated in the molten state to remove excess carbon and other impurities. The steel may be continuously cast into strands or cast into individual ingots. The final product is then produced by rolling, drawing or forging. During hot rolling and forging the steel surface is oxidised by the air and the scale produced, usually termed *millscale*, may have an important influence on the corrosion of the steel, as will be discussed later.

The structure of millscale consists of three superimposed layers of iron oxides in progressively higher states of oxidation from the metal side outwards, viz. ferrous oxide (FeO) on the inside, magnetite (Fe_3O_4) in the middle and ferric oxide (Fe_2O_3) on the outside. The relative portions of the three oxides vary with the rolling temperatures. A typical millscale on 9.5 mm mild steel plate would be about 50 μm thick, and contain approximately 70% FeO, 20% Fe_3O_4 and 10% Fe_2O_3.

If millscale was perfectly adherent, continuous and impermeable, it would form a good protective coating, but in practice millscale soon cracks and flakes off in places. In air, the presence of millscale on the steel may reduce the corrosion rate over comparatively short periods, but over longer periods the rate tends to rise. In water, severe pitting of the steel may occur if large amounts of millscale are present on the surface.

Generally, the process of manufacture has no appreciable effect on the corrosion characteristics of steel. Slight variations in composition that inevitably occur from batch to batch in steels of the same quality have little effect with the exception of copper, the effect of which is discussed more fully in Section 3.2.

Briefly, the addition of about 0·2% of copper results in a two to three-fold reduction in the corrosion rate in air compared with a copper-free steel[1,2].

Variations in the other elements in ordinary steels affect the corrosion rate to a marginal degree, the tendency being for the rate to decrease with increasing content of carbon, manganese and silicon. For example, in the open air a steel containing 0·2% of silicon rusts about 10% less rapidly than an otherwise similar steel containing 0·02% of silicon.

Wrought iron, which has been replaced by mild steel, contained appreciable quantities of slag and corroded at a rate about 30% less than mild steel when exposed in air. However, artificially produced Aston-Byers iron corrodes at about the same rate as mild steel.

The corrosion rates of wrought iron and mild steel when immersed in seawater or buried in soil are not significantly different when the copper contents are similar.

Investigations of the effects on corrosion of mechanical working of steels indicate some influence on pitting but little on general corrosion.

Sea-water immersion tests were carried out to determine the effects of rolling direction and tensile stress on the corrosion of a steel containing 0·14C, 0·47Mn and 0·04Si[41]. Specimens were cut from plates parallel to and perpendicular to the rolling direction. There was little difference in general corrosion performance, although pitting was worse on the plate cut parallel to rolling. Under anaerobic conditions in harbour sediments the effects on pitting were more marked.

A study of pitting on the internal surfaces of radiators has shown the incidence of pitting to be related to the state of deformation in tension of the steels. The pitting increased with the degree of deformation[42].

Rusting in Air

Controlling Factors

The rusting of bare steel in the atmosphere is controlled by the climatic conditions at the exposure site. The main factors are the availability of moisture, and the extent to which the air is polluted, but other less important ones, such as temperature, must also be considered. (See also Section 2.2.)

Moisture: the critical humidity Moisture can reach a steel surface directly in liquid form as a result of precipitation processes, i.e. rain and dew, but the water vapour that is always present in air can also, under certain conditions, cause steel to rust at relative humidities well below saturation.

This important fact was first demonstrated by Vernon[3] in a series of classical experiments, some of which are summarised graphically in Fig. 3.1. He showed that rusting is minimal in pure air of less than 100% relative humidity but that in the presence of minute concentrations of impurities, such as sulphur dioxide, serious rusting can occur without visible precipitation of moisture once the relative humidity of the air rises above a critical and comparatively low value. This value depends to some extent upon the nature of the atmospheric pollution, but, when sulphur dioxide is present, it is in the region of 70-80%. Below the critical humidity, rusting is inappreciable, even in polluted air.

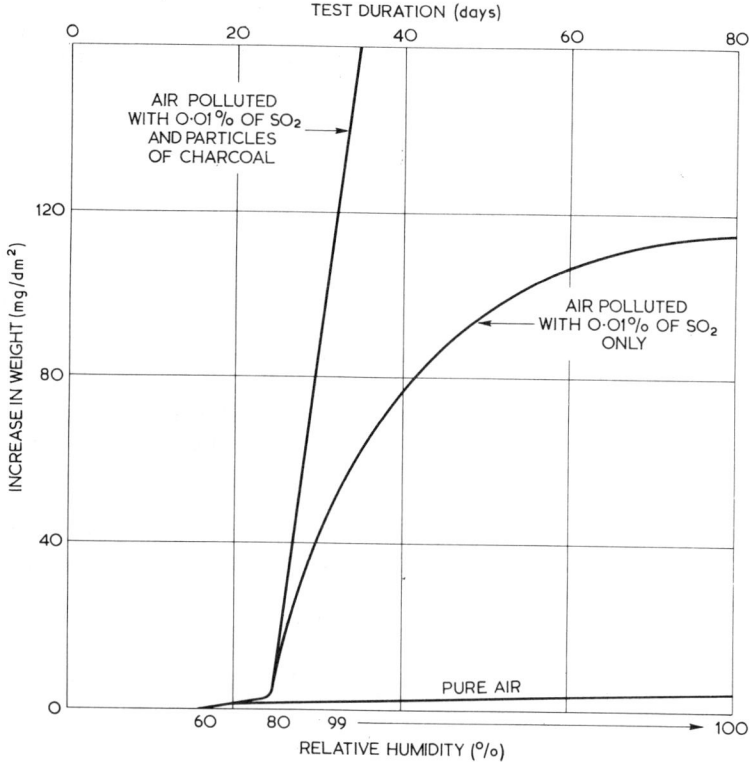

Fig. 3.1 Effect of relative humidity and atmospheric pollution on the rusting of iron (after Vernon[3])

The presence of moisture on steel above the critical humidity but below the saturation point may be caused by an adsorption mechanism or by the presence of particles of deliquescent salts on the surface. Once rusting has begun, the composition of the rust already formed will influence the relative humidity at which further rusting will occur, because rusts formed in polluted atmospheres contain hygroscopic salts. The method by which moisture reaches the surface is probably less important, however, than the length of time during which the steel remains wet. Estimates of the amount of water (in gm^2) on the surface have been made[45]. These indicate that the presence of water is the critical factor not the total amount as shown below.

Critical relative humidity	0·01
100% relative humidity	1
Covered by dew	10
Wet from rain	100

Pollution Although humidity plays a vital part, the impurities in the air are decisive in determining the rate of rusting in atmospheres of the requisite humidity; in their absence rusting is not serious even in highly humid air. The most important impurity in industrial atmospheres is sulphur dioxide, although chlorides and ammonium salt[4] also may have an effect. Generally, near the coast chlorides have the most pronounced effect. In the presence of chlorides, rusting can continue at humidities as low as 40%[5]. As a rule, the chloride concentration in the air falls off rapidly with distance inland, but steel rusts at almost incredible rates on surf beaches in the tropics where it is exposed to a continuous spray of sea salts in the surf. This is shown by the results of some tests made in Nigeria that are given in Table 3.1.

Table 3.1 Effect of sea salts on the rate of rusting*
(Tests made in Nigeria on behalf of BISRA by the former Tropical Testing Establishment, Ministry of Supply)

Distance from surf	Salt content of air†	Rate of rusting (mm/y)
50yd	11·1	0·950
200yd	3·1	0·380
400yd	0·8	0·055
1 300yd	0·2	0·040
25 miles	—	0·048

*The specimens were of ingot iron and were exposed for one year.
†The salt content of the air was determined by exposing wet cloths, and is expressed as mg NaCl d^{-1} (100 cm^2)$^{-1}$.

In most districts, however, sulphur dioxide and dust particles are the main corrosive pollutants. It has been demonstrated that there is a direct relationship between sulphur dioxide in the atmosphere and the corrosion of steel exposed to it (see Fig. 3.2). In a series of tests carried out in the Sheffield area, sulphur dioxide accounted for about 50% of the variations in corrosion rate at the different sites[6].

Various methods can be used to determine the concentration of sulphur dioxide in the vicinity of steel test specimens. They all provide suitable information on the relationship between sulphur dioxide and corrosion. However, the actual amount of sulphur dioxide in contact with the steel surface is more important than the concentration, as shown in the work reported by Walton et al.[43]

The significance of the amount of sulphur dioxide rather than the concentration has been demonstrated by other workers who have studied the effects of atmospheric flow rate. An increase in steel corrosion with increase in atmospheric flow rate at a constant volume concentration of sulphur

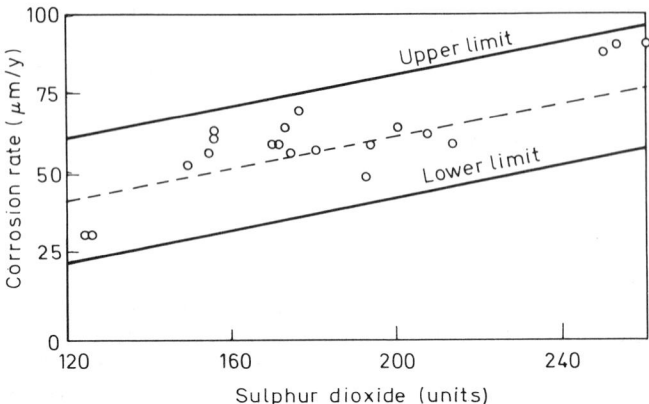

Fig. 3.2 Relationship between sulphur dioxide and corrosion (after Chandler and Kilcullen[6]). Note that 1 unit is the average daily reading in μg of SO_2 per m^3 of air

dioxide has been observed[44]. The effect of flow has also been shown by comparisons of measurements of sulphur dioxide using the lead dioxide method, where the amount of SO_2 in the air measured over comparatively small areas was influenced by the siting of the 'candle'.

Sulphur dioxide in the air originates from the combustion of fuel and influences rusting in a number of ways. For example, Russian workers consider that it acts as a cathodic depolariser[7], which is far more effective than dissolved oxygen in stimulating the corrosion rate. However, it is the series of anodic reactions culminating in the formation of ferrous sulphate that are generally considered to be of particular importance. Sulphur dioxide in the air is oxidised to sulphur trioxide, which reacts with moisture to form sulphuric acid, and this in turn reacts with the steel to form ferrous sulphate. Examination of rust films formed in industrial atmospheres have shown that 5% or more of the rust is present in the form of iron sulphates[8] and $FeSO_4 \cdot 4H_2O$ has been identified in shallow pits[9].

Workers who have studied atmospheric corrosion processes are generally agreed that the loss of iron as a sulphate accounts in only a small measure for the effects of sulphur dioxide[10-13]. There is not complete agreement on the detailed mechanism but once ferrous sulphate has been formed it is able to promote further rusting. This has been demonstrated by allowing steel to rust in an atmosphere containing sulphur dioxide and then transferring it to a clean atmosphere, where the corrosion is continued at an enhanced rate, at least for a time[14].

There is a cyclic variation in the amount of sulphate found in rust formed on steel exposed at different times of the year (Fig. 3.3), and the amount depends on the month of the year rather than on the period of exposure, at least for periods of up to two years[8]. Consequently, the month of exposure will have an important influence on the corrosion rate up to a year or so (see Table 3.2). In one test, specimens exposed for two months from September corroded at 0·035 mm/y compared with 0.0136 mm/y for specimens exposed from May, and the amount of rust formed on steel follows this cyclic pattern over periods of up to two years and possibly for

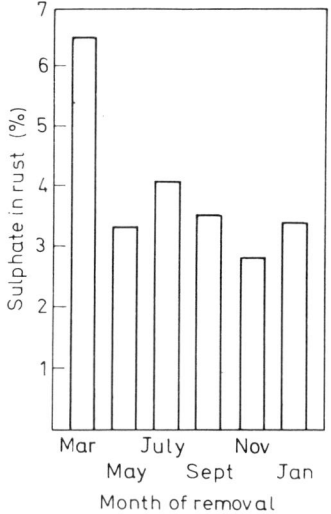

Fig. 3.3 Sulphate in adherent rust on steel exposed at Battersea in January (after Chandler and Stanners[8])

Table 3.2 Variations in rate of rusting of steel exposed at different months of the year

Month test begun	Rate of rusting (mm/y)
January	0·081
March	0·064
May	0·076
July	0·070
September	0·094
November	0·089

longer periods. However, the rust retained on steel is a surprisingly small proportion of the total rust formed, e.g. over a period of one year only about one-tenth of all the rust produced was actually retained on the steel.

The effects of ammonium contamination have been studied and some workers consider that it has a significant influence on the corrosion of steel[15,16].

Dust particles also play an important rôle (Fig. 3.1). They act as nuclei for the initial corrosion attack and as some particles are hygroscopic their presence tends to increase the periods of wetness of the steel surface.

Temperature Although ambient temperature would be expected to have an influence on the rate of rusting, its effect is not clearly defined despite the efforts of workers to establish a relationship[17,18]. It has an effect on relative humidity and consequently an indirect effect on corrosion. However, fluctuations in temperature may be more important than average temperatures because they influence condensation and the rate of drying of moisture in contact with steel.

Rates of rusting in different climates Some representative figures to show how the rate of rusting of bare steel in the open air varies in different parts of the world are given in Table 3.3. They relate to 100 × 50 × 3.2 mm (4 in × 2 in × $\frac{1}{8}$ in) pieces of ingot ironly freely exposed in a vertical position for one year.

Table 3.3 Rates of rusting of mild steel outdoors in different climates (BISRA)

Type of atmosphere	Site	Corrosion rate (mm/y)
Great Britain		
Rural or suburban	Godalming	0·048
	Llanwrtyd Wells	0·069
	Teddington	0·070
Marine	Brixham	0·053
	Calshot	0·079
Industrial	Motherwell	0·095
	Woolwich	0·102
	Sheffield	0·135
	Frodingham	0·160
	Derby	0·170
Overseas		
Rural or suburban	Khartoum	0·003
	Abisko, North Sweden	0·005
	Delhi	0·008
	Basrah	0·015
	State College, Pa, U.S.A.	0·043
	Berlin-Dahlem	0·053
Marine	Singapore	0·015
	Apapa, Nigeria	0·028
	Sandy Hook, N.J., U.S.A.	0·084
Marine/industrial	Congella, South Africa	0·114
Industrial	Pittsburgh, Pa, U.S.A.	0·108
Marine, surf beach	Lagos	0·615

The results follow the expected pattern, the highest rate being that observed on a surf beach at Lagos. The magnitude of this rate (0·6 mm/y) will be better appreciated by the statement that an ingot iron specimen as above, 3·2 mm thick, would be completely destroyed by rusting from both sides in about $2\frac{1}{2}$ years. As previously stated, rusting in Great Britain is roughly proportional to the atmospheric pollution; in accordance with this, industrial sites are several times more corrosive than rural, suburban or marine ones.

The low rates of rusting observed at Khartoum, Abisko, Delhi, Basrah and Singapore are primarily associated with the absence of serious pollution. Moreover, at most of them the relative humidity is low, e.g. at Khartoum the relative humidity lies below the critical value for rusting throughout the whole year.

Despite these results it should not be assumed that corrosion rates of steel will necessarily be low in all comparatively non-polluted desert environments. In regions such as the Arabian Gulf, considerable variations in corrosion rates may occur between inland and coastal sites. This arises not only from the salt content of the air but also from sand which is blown on to the steel. Although temperatures are high during the day, condensation may occur at night. The moisture produced can react with the salt to provide corrosive conditions.

The effects of different types of sand on the corrosion of mild steel have been studied in the laboratory[46]. It was concluded that fine sand has a higher salt content and is more corrosive than coarse sand within the partical size range < 0·25–2·4 mm.

In a survey of atmospheric corrosion in the Canadian arctic and sub-arctic regions rates as low as 2–5 μm/y were recorded at inland sites[49]. Within 1 km of the sea, rates of 21–34 μm/y were measured.

Effect of the exposure conditions The absolute values for the rate of rusting given in Table 3.3 would be affected by the mass of the specimen itself and by other factors such as the orientation of the steel, the climatic conditions prevailing at the time of exposure, and the duration of exposure.

The *orientation* of the steel influences the rate of rusting through its effect on the amounts of moisture and pollutants that can reach the surface. The groundward side of a horizontal surface is protected from rain but it is also shielded from the drying action of the sun and often of the wind, so that dew tends to remain in contact with the steel there for longer periods; moreover, harmful solid particles and soluble salts are not leached away. Consequently, the groundward side may corrode more rapidly than the skyward side. The same considerations apply to steel exposed obliquely. The relative corrosion of the opposite faces of a vertical steel plate will largely depend on the direction of the prevailing wind.

In agreement with these considerations, the results of tests at Derby[19] have shown that specimens exposed at 45° corroded 10–20% more than vertical specimens, and that 54% of the total loss was on the underside. In American tests[20] on specimens exposed at 30° to the horizontal, 62% of the loss was on the underside.

This influence of the sheltering and the orientation of steel on the corrosion rate has been further demonstrated in tests carried out 228 m from the sea at Kure Beach, North Carolina[47]. In these tests the corrosion rates over a 4-year period varied by a factor of five depending on the orientation and degree of sheltering. Generally, the east-facing specimens exposed at 30° from the horizontal corroded at the highest rate and west-facing specimens exposed at the same angle at the lowest rate.

The influence of height above the ground on the corrosion of steel has been demonstrated in two series of American tests. In tests carried out at Cape Kennedy[48], specimens exposed at 9·1 m above the ground and 55 m from the sea corroded at just over a third of the rate of those at ground level. Increasing the height of exposure to 18·2 m resulted in a slightly lower rate of corrosion.

Tests at Kure Beach[47] approximately 24 m from the sea showed a similar reduction at elevations of about 44 m but an increase in corrosion at levels up to 7·6 m. Clearly the relationship between height of exposure and corrosion is influenced by the topography of the site. In the same series of tests carried out 244 m from the sea, the corrosion rate was about half that at ground level for steel exposed at higher elevations.

Effect of mass The rate of rusting of steel in the atmosphere is affected to some extent by the mass of the part concerned, because this determines the speed at which the surface temperature adjusts itself to fluctuations in the ambient temperature, the amount of condensation during humid periods, and the time during which dew or rain remains in contact with the steel. For example, in a test over 12 months at the National Chemical Laboratory under sheltered conditions outdoors, thick steel plates rusted more than thin ones as is shown below.

Plate thickness (mm)	55	28	12·5	5
Average general penetration (mm)	0·038	0·033	0·031	0·030

Evidently, therefore, correct design can play an important part in reducing the corrosion of steel structures through lessening the danger of local attack due to mass metal effects and condensation.

Effect of surface condition As previously noted, millscale on steel may decrease the corrosion rate over short periods. However, over longer periods the surface condition is not usually a determining factor as can be seen from Table 3.4 which shows the results of 5-year tests at Sheffield.

Table 3.4 Effect of the surface condition at the time of exposure on the atmospheric corrosion of mild steel (BISRA) (5 years outdoor exposure at Sheffield)

Surface condition	Average penetration (mm)
As-rolled	0·545
Pickled	0·545
Sandblasted	0·532
Machined	0·534
Polished	0·532

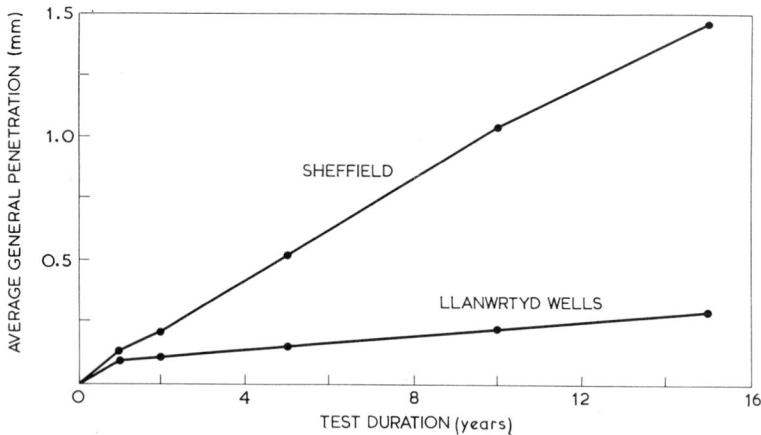

Fig. 3.4 Rusting of mild steel (0·02% Cu) in the open air at Sheffield and Llanwrtyd Wells (after Reference 2)

It is important to know whether the rate of rusting of steel outdoors increases, remains constant or decreases with increasing *duration of exposure*. The position over a long period is adequately illustrated by the curves drawn in Fig. 3.4 showing the results of tests carried out for 15 years in a rural environment at Llanwrtyd Wells and in an industrial atmosphere at Sheffield[2]. At both sites the rate of rusting fell considerably after the first year or two, but then remained fairly constant for the rest of the test period. In actual figures the average rate of rusting in mm/y over the first two years was 0·058 for Llanwrtyd Wells and 0·118 for Sheffield; the corresponding rates over the last 10 years were 0·015 and 0·025, respectively.

Similar observations have been made in the United States, where a marked decrease in the rate of rusting has generally been observed after the early stages. For example, in one test, the total corrosion over the first two years was about 0·1 mm but the additional corrosion during the next 10 years was only about 0·06 mm. These are slow rates of rusting, however, and it is possible that the rust formed was more protective than usual, because of the relatively non-aggressive test environment.

The corrosion curves in Fig. 3.4 were obtained some years ago. Corrosion is markedly influenced by the pattern of pollution, which is changing in the United Kingdom, and consequently the long-term corrosion rates may change. There is some evidence based on more recent tests to indicate that in many industrial environments the corrosion rate of steel over periods of 15 years will drop to a greater extent than is shown in Fig. 3.4.

It should be made clear that all the rates of rusting in the atmosphere just quoted, relate to average general penetration and take no account of pitting. Serious pitting of steel exposed to atmospheric corrosion is uncommon on simple test plates, but it may be necessary to allow for this in some practical cases, where local attack may be occasioned by faulty design and other factors.

Rusting indoors and in enclosed spaces Although much of what has been written above is of general application, the experimental data apply specifically to exposure outdoors. The conditions of exposure indoors are somewhat different, because the steel is not exposed to rain and direct sunlight. Rusting depends on the condensation of moisture, which may evaporate from the surface much more slowly under enclosed conditions than outdoors; moreover, the rust tends to remain on the surface and may build up in time to a thick scaly layer.

The rates of rusting vary so much with the particular conditions of exposure that it is difficult to generalise, but useful information regarding rusting in enclosed spaces will be found in papers by Holzworth, Thompson and Boegehold[21] and by Stanners[22].

The first three authors were concerned with the rusting of motorcar bodies. They found that the rusts formed on steel under sheltered and exposed conditions, respectively, differed markedly in chemical composition, structure and protective properties. The second paper gives the results of exposure tests in many different indoor atmospheres, from which the following representative rates of rusting over one year are taken.

Type of atmosphere	Rate of rusting (mm/y)
Domestic kitchens and bathrooms	0·0025–0·010
Laundry	0·0075
Bleach house	0·043
Sulphuric acid plant	0·048
Paper mill	0·068
Locomotive shed	0·080
Steelworks' pickling plant	>0.45

Rusting in Water

Introduction

The corrosion of steel by natural and industrial waters is a complicated and many-sided phenomenon, which cannot be dealt with completely within the narrow compass available here. The difficulty arises from the fact that, of the three main considerations involved — the composition and surface condition of the steel, the quality of the water, and the operating conditions — the last is generally the most important. Moreover, in industry the operating conditions vary widely, and many of them need individual study. Consequently, all that will be attempted here is to state a few general facts and principles. For more detailed treatment the reader is referred to the recognised textbooks such as those of Evans[10], Hasse[23] and Butler and Ison[24]. A brochure published by the Iron and Steel Institute[19] may also prove helpful.

Effect of the Metal Composition

All ordinary ferrous structural materials, mild steels, low-alloy steels and wrought irons corrode at virtually the same rate when totally immersed in natural waters. Wrought iron may be slightly more resistant than mild steel; in a test in sea-water at Gosport, Scottish wrought-iron specimens lost about 15% less weight after 12 months immersion than specimens of ordinary mild steel. As shown in Table 3.5, the process of manufacture and the composition of mild steel do not affect its corrosion rate appreciably[25].

Although small copper additions generally have no influence on the corrosion rate of steel immersed in water, under certain boiler conditions small amounts of copper of up to 0·3% may have a marked effect on reducing the depth of pitting. This is attributed to the catalytic action of metallic copper on the formation of a protective layer of magnetite which diverts the attack sideways producing general rather than pitting corrosion[23].

Generally, over 3% of alloying additions such as chromium are necessary to obtain any marked improvement in the corrosion-resistance of steel in waters.

Table 3.5 Rates of rusting of mild steels in sea-water (BISRA)
(total immersion for 203 days at Plymouth)

Type of steel*	Analysis (%)				Average general penetration (mm/y)
	C	Mn	P	S	
1. Basic Bessemer, rimming					
Ordinary	0·05	0·64	0·06	0·02	0·143
High phosphorus	0·03	0·31	0·14	0·04	0·143
High phosphorus and sulphur	0·03	0·30	0·10	0·07	0·148
2. Open-hearth, rimming					
Ordinary	0·13	0·33	0·03	0·03	0·143
From haematite pig	0·06	0·32	0·01	0·03	0·140
3. Open-hearth, killed					
Ordinary	0·10	0·35	0·03	0·02	0·140
From haematite pig	0·11	0·34	0·01	0·03	0·136
4. Open-hearth, killed					
Ordinary	0·22	0·71	0·03	0·03	0·143
From haematite pig	0·21	0·58	0·02	0·03	0·158

*The copper contents of the steels, which were supplied through the courtesy of l'Office Technique pour l'Utilisation de l'Acier (France), varied from 0·03 to 0·11%. The killed steels contained about 0·04% Al and 0·1% Si.

Effect of the Surface Condition

The surface condition of the steel at the time of exposure is of great importance however. This is because many natural waters are good electrolytes, so that there is ample opportunity for electrolytic corrosion when steel is permanently in contact with them. The presence of millscale on the surface is more dangerous, for example, when steel is immersed in sea-water than when it is exposed to air, for the galvanic cell formed by millscale and bare steel can operate much more freely under the former conditions. This may lead to rapid pitting; pits up to 1·25 mm deep were found on as-rolled steel specimens after six months immersion in sea-water at Gosport[26].

It follows that for many practical purposes where steel is exposed to water without a protective coating — boiler tubes are a good example — it is desirable to remove millscale before putting the parts into use.

Effects of Welds

Serious pitting may occur in the area of welds, particularly in sea-water. Corrosion rates of up to 10 mm/y have been reported in weld joints of ice-breakers. The severe corrosion has been attributed to galvanic effects between the weld metal and the steel plate. The use of more noble electrodes for welding are reported to overcome this problem[27].

Effect of the Water Composition

Saline and acid waters are particularly aggressive to mild steel, so the composition of the water is clearly important in determining the rate of rusting of steel exposed to it. Some of the main factors here are the nature and amount of the dissolved solids, which influence the electrical conductivity, pH value and hardness of the water, the carbon dioxide and oxygen contents, and the presence of organic matter.

Dissolved solids The effect of dissolved solids is complex. The presence of inorganic salts, notably of chlorides and sulphates, should promote corrosion, because they increase the conductivity of the water, thereby facilitating the electrochemical rusting process; moreover, chlorides at least may be detrimental to the development of protective films (see below). Alkaline waters tend to be less aggressive than acid or neutral waters, and rusting can be repressed entirely by making the water strongly alkaline. Unfortunately, at pH values just insufficient to give complete passivation, there is a grave danger of severe pitting, even though the total corrosion is reduced, and this for many purposes is a greater evil.

The most important property of the dissolved solids in fresh waters is whether or not they are such as to lead to the deposition of a protective film on the steel that will impede rusting. This is determined mainly by the amount of carbon dioxide dissolved in the water, so that the equilibrium between calcium carbonate, calcium bicarbonate and carbon dioxide, which has been studied by Tillmans and Heublein[28] and others, is of fundamental significance. Since hard waters are more likely to deposit a protective calcareous scale than soft waters, they tend as a class to be less aggressive than these; indeed, soft waters can often be rendered less corrosive by the simple expedient of treating them with lime (Section 2.3).

Dissolved gases Oxygen and carbon dioxide are the most important dissolved gases in water. Oxygen is an effective cathodic depolariser and the cathodic reaction in water is generally oxygen reduction (see Sections 1.4 and 9.1). At ordinary temperatures in neutral or near neutral water, dissolved oxygen is necessary for any appreciable corrosion of steel. Increasing the oxygen concentration results in an acceleration of the corrosion of steel up to a certain concentration but beyond this the rate of corrosion is reduced. In slowly-moving distilled water, it has been found[29] that the critical concentration is 12 m/l. This value increases with temperature and in the presence of certain dissolved salts; it decreases with high velocities.

Carbon dioxide affects the acidity of the water and, as already noted, influences the formation of protective carbonate scales.

Presence of organic matter Another important factor is that most natural waters are far from being sterile. They contain greater or lesser amounts of organic matter, both living and dead. Some of the dead organic matter, e.g. peat residues, may render the water corrosive by making it acid, but in most cases the living organisms probably exert the greater influence. In natural sea-water fouling occurs, and in fresh waters algae may grow. Moreover, there are a number of strains of bacteria, such as the sulphate-reducing bacteria (see Section 2.6), that can influence the rusting process under immersed conditions.

Effect of the operating conditions The operating conditions have an important influence. Generally the factors involved are complex; they include temperature, rate of flow, design features and stray currents.

The temperature of the water affects the rate of rusting in several ways. First, the corrosion process shares the general tendency of chemical reactions to increase in speed with rising temperature. More important, however, are the effects of temperature on the nature and solubility of the corrosion products. For example, a rise in temperature will often throw down a carbonate scale; moreover, it increases the rate of diffusion of oxygen through water but decreases the solubility of this gas. Some of these effects are conflicting, with the result that under certain laboratory conditions at least the rate of rusting/temperature curve for steel immersed in water passes through a maximum before the boiling point is reached – at about 80°C in experiments made by Friend[30].

The rate of water flow is also most important. This determines the supply of oxygen to the rusting surface, and may remove corrosion products that would otherwise stifle further rusting. A plentiful oxygen supply to the cathodic areas will stimulate corrosion, but so may smaller supplies at a slow rate of flow, if this leads to the formation of differential aeration cells (see Section 1.6).

At sufficiently high rates of flow in natural waters enough oxygen may reach the surface to cause partial passivity, in which case the corrosion rate may decrease[31]. In sea-water, owing to the high concentration of chloride ions, the corrosion rate increases with velocity. In one series of tests, corrosion under static conditions was 0·125 mm/y, 0·50 mm/y at 5 ft/s and 0·83 mm/y at 15 ft/s.

Design must also be considered. Sharp changes in the direction of flow, as in a badly designed water box, may lead to severe local damage by impingement attack. Severe galvanic corrosion can result from the injudicious juxtaposition of steel with non-ferrous metals, such as copper or bronze. Even when the ferrous and non-ferrous metals are not in direct contact, local corrosion cells can be set up round small particles of non-ferrous metal that have been dissolved by the water and redeposited on to the steel. Corrosion cells can also be formed when steel is in contact with solutions of different saline contents; for example, 'long line' currents have been observed in a lock gate that was in contact with a layer of fresh river water flowing out over a layer of sea-water.

Rates of rusting in natural waters It will be evident from the preceding remarks that the results of laboratory experiments or field tests on corrosion by natural waters can be applied to practical cases only with considerable reserve. Yet the reader may well wish to gain some general idea of the rates of rusting involved. For this purpose a few representative experimental results are given in Table 3.6; they apply to ordinary low-carbon structural steel tested under the conditions stated. The figures are for the average general penetration over the whole test areas. As an indication of the rate of pitting, it may be noted that in the sea-water tests of the Institution of Civil Engineers[32] the maximum depth of pitting for descaled mild steel after 15 years immersion was about 2·3 mm; when the steel had been immersed in the as-rolled condition with its millscale, a figure as high as 7·6 mm was observed. Under half tide immersion conditions the corrosion rate of steel may be increased by a factor of 2 to 5 compared with the results for total immersion quoted above.

Table 3.6 Rates of rusting of mild steel in natural waters
(total immersion)

Type of water	Test authority	Test site	Test duration (years)	Average general penetration (mm/y)
Sea-water	Institution of Civil Engineers[32]	Halifax, Nova Scotia	15	0·108
		Plymouth	15	0·065
	BISRA[33]	Emsworth	5	0·065
Fresh water	Institution of Civil Engineers[32]	Plymouth: fresh-water reservoir	15	0·043
River water	Office Technique pour l'Utilisation de l'Acier[34]	La Cadène: granite bed, very pure water	5	0·068
		Dôle: highly calcareous water	5	0·010

Rusting in Soil

Introduction

The practical importance of soil corrosion to Great Britain was well brought out some years ago by a report of a Departmental Committee of the Ministry of Health[35] which will repay study. It is even more difficult to generalise about this problem than about rusting in water. With regard to the bare metals, the problem is of little more than academic interest, because in good practice iron and steel are not buried in corrosive soils without an adequate protective coating, often supplemented by an effective scheme of cathodic protection. The operating conditions are again of extreme importance, e.g. corrosive long-line currents may originate where sections of a pipeline run through adjacent dissimilar soils, or, more frequently, stray currents from adjacent electrical installations may enter the pipe; in both cases severe corrosion damage may be caused where the currents leave the pipe. The presence of stones, tree roots or attack by rodents may perforate the protective coating.

Although all these complicating factors cannot be reproduced in small-scale tests, it will be of value to summarise the main knowledge that has been gained from long-period burial trials conducted in the United States[36] and in Great Britain[37]. The subject will be considered under two heads: effect of metal composition and effect of the soil.

Effect of the Metal Composition

The corrosion of iron and steel in soil is generally electrochemical in character but the conditions are such that the corrosion products usually remain in contact with the metal. Moreover, the rate of oxygen supply is often low in comparison with that in air or in water. This is probably the main reason why, considered broadly, there are no major differences in the general corrosion rates of ordinary steels and cast irons when buried in most types of soil. These rates of general attack are comparatively low, say 0·038 mm/y over a 10-year period for burial in a clay soil, as compared with 0·1–0·13 mm/y for complete exposure outdoors in an industrial atmosphere.

Local corrosion or pitting is more important for practical purposes than the rate of general corrosion, and may proceed 10 times or so more rapidly than this. Inasmuch as certain types of cast iron are liable to suffer *graphitic corrosion**, whereas steel does not, steel might theoretically be expected to show to some advantage when used for buried pipelines. In practice, however, a cast-iron pipe has to be of stouter wall than a steel pipe for equal strength, and it is doubtful whether any distinction between the rust resistance of the two materials in the soil is justified.

It is also doubtful whether the surface condition of the bare metal at the time of burial has much significance. Some authorities consider that the casting skin on cast iron is protective, but the evidence on this point is conflicting. On the other hand, it is desirable to remove the millscale from steel when, as is usual, a protective coating is applied. If the millscale was left on, it might cause the coating to spall.

The effect of stray currents arising from a d.c. source or from cathodic protection of an adjacent structure are considered in Sections 11.5 and 11.6.

Effect of the Soil

Soils vary greatly in corrosiveness, and the type of soil affects the corrosion rate much more than any variation in the ferrous material or in its method of manufacture. Although it is difficult to assess the corrosiveness of a particular soil beforehand, much useful information can be obtained from a well-conducted soil survey on the site.

In general, dry, sandy or calcareous soils, with a high electrical resistance, are the least corrosive. At the other end of the scale are the heavy clays and the highly saline soils, whose electrical conductivity is high. The depth of the water table is also important; much depends on whether the buried iron or steel is permanently above or below this, or even more perhaps on whether it is alternately 'wet' or 'dry'. The variation in corrosion rate with depth of burial is illustrated by the results given in Table 3.7, which also serve to indicate the rates of average general penetration in typical British soils. It will be noted that the depth of burial had no consistent effect, which is not surprising since the average depth of the water table and the seasonal fluctuations in this varied from one site to another.

The maximum general corrosion rate reported in tests carried out by the U.S. National Bureau of Standards[36,38] is 0·068 mm/y, the maximum rates obtained in tests carried out in the United Kingdom by BISRA[39] and the National Physical Laboratory[40] are 0·035 mm/y and 0·050 mm/y. However, the pitting rate was much greater, maximum pits of 0·25 mm/y have been reported from American and 0·30 mm/y from British tests.

---

*Graphitic corrosion is associated with the presence of graphite flakes in the iron matrix, and results in the local replacement of iron by 'plugs' of graphite and corrosion products.

Table 3.7 Effect of depth of burial on the rusting of mild steel flats (BISRA)
(test duration 5 years)

Site	Type of soil	Average general penetration (mm/y)	
		1·37 m	0·61 m
Benfleet	London clay	0·0185	0·0361
Gotham	Keuper marl	0·0132	0·0094
Pitsea	Alluvium	0·0353	0·0284
Rothamsted	Clay with flints	0·0201	0·0213

Bacterial activity often plays a major part in determining the corrosion of buried steel. This is particularly so in waterlogged clays and similar soils, where no atmospheric oxygen is present as such. If these soils contain sulphates, e.g. gypsum and the necessary traces of nutrients, corrosion can occur under anaerobic conditions in the presence of sulphate-reducing bacteria. One of the final products is iron sulphide, and the presence of this is characteristic of attack by sulphate-reducing bacteria, which are frequently present (see Section 2.6).

Effect of the Duration of Burial

Finally, it should be added that the extensive field tests made in the United States[38] indicate that buried steel rusts less and less rapidly as time goes on, both as regards general attack and pitting. This can be illustrated by the typical results shown in Fig. 3.5. Field tests made in British soils by BISRA have not, however, exhibited the same tendency; in these rusting has been roughly proportional to the duration of burial.

K. A. CHANDLER
J. C. HUDSON

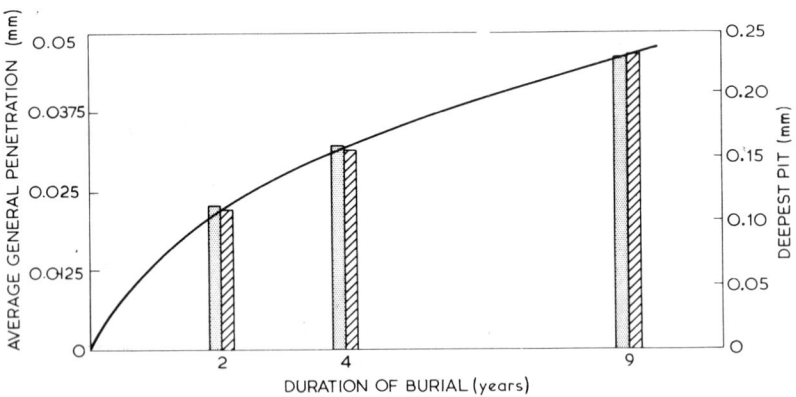

Fig. 3.5 Effect of duration of burial on the corrosion of mild steel (after Romanoff[38]). *Left:* average general penetration deduced from loss in weight, *right:* deepest pit. Figures are averages for 16 soils

REFERENCES

1. Larrabee, C. P. and Coburn, S. K., *Proc 1st Int. Congr. Met. Corros.*, 1961, 276, London, Butterworths (1962)
2. Sixth Report of the Corrosion Committee, Spec. Rep. No. 66, Iron and Steel Institute, London (1959)
3. Vernon, W. H. J., *Trans. Faraday Soc.*, **31**, 1 668 (1935)
4. Ross, T. K. and Callaghan, B. G., *Corr. Sci.*, **6**, 337 (1966)
5. Chandler, K. A., *Br. Corros. J.*, **1**, July (1966)
6. Chandler, K. A. and Kilcullen, M. B., *Br. Corr. J.*, **3**, March (1968)
7. Rosenfel'd, I. L., *Proc. 1st Congr. Met. Corros.*, 1961, 243, London, Butterworths (1962)
8. Chandler, K. A. and Stanners, J. F., *Proc. 2nd Int. Congr. Met. Corr.* (N.A.C.E. Houston), 325 (1963)
9. Tanner, A. R., *Chemistry & Industry*, 1 027 (1964)
10. Evans, U. R., *The Corrosion & Oxidation of Metals*, Edward Arnold, London (1960)
11. Schikorr, G., *Werkstoffe Korros.*, **14** No. 2, 69 (1963)
12. Schwartz, H., *Werkstoffe Korros.*, **16** No. 2, 93 (1965)
13. Schwartz, H., *Werkstoffe Korros.*, **16** No. 3, 208 (1965)
14. Schikorr, G., *Korrosion und Metallschutz*, **17**, 305-313 (1941)
15. Harrison, J. B. and Tickle, T. C. K., *J. Oil Colour Chem. Assoc.*, **45** No. 8, 571 (1962)
16. Ross, T. K. and Callaghan, B. G., *Nature*, London, **211**, 25 (1966)
17. Golubev, A. Y. and Kadyrov, M. Kh., Repr. No. 3-67-487/13, Gosinti, Moscow (1967)
18. Sereda, P. J., *Ind. Eng. Chem.*, **52**, 157 (1960)
19. Dearden, J., *J.I.S.I.*, **159**, 241 (1948)
20. Larrabee, C. P., *Trans. Electrochem. Soc.*, **85**, 297 (1944)
21. Holzworth, J., Thompson, R. F. and Boegehold, A. L., *Trans. Soc. Auto. Engrs.*, **64**, 221 (1956)
22. Stanners, J. F., *J. Appl. Chem.*, **10**, 461 (1960)
23. Haase, L. W., *Werkstoffzerstorüng und Schutzschich bildung Im Wasserfach*, Verlag Chemie Weinheim Bergstr. (1951)
24. Butler, G. and Ison, H. C. K., *Corrosion and its Prevention in Waters*, Leonard Hill, London, 74 (1966)
25. Hudson, J. C., *J.I.S.I.*, **166**, 123 (1950)
26. Hudson, J. C., The Corrosion of Iron & Steel, Chapman and Hall, London, 61 (1940)
27. Uusitalo, *Proc. 2nd Int. Congr. Met. Corr.*, N.A.C.E., Houston, 812 (1963)
28. Tillmans, J. and Heublein, O., *Gesundheitsing*, **35**, 669 (1912)
29. Uhlig, H., Triadis, D. and Stern, M., *J. Electrochem. Soc.*, **102**, 59 (1955)
30. Friend, J. N., *Carnegie Schol. Mem.*, **11**, 113 (1922)
31. Uhlig, H. H., *Corrosion Handbook*, John Wiley, 391 (1948)
32. Friend, J. N., *18th Report of the Committee of the Institution of Civil Engineers on the Deterioration of Structures of Timber, Metal and Concrete Exposed to the Action of Sea Water*, London (1940)
33. Hudson, J. C. and Stanners J. F., *J.I.S.I.*, **180**, 271 (1955)
34. Baudot, H. and Chaudron, G., *Rev. Met.*, **43**, 1 (1946)
35. Ministry of Health, Interim Report of the Departmental Committee on the Deterioration of Cast Iron and Spun Iron Pipes, H.M.S.O., London (1950)
36. Romanoff, M., *Underground Corrosion*, National Bureau of Standards, Circular 579, Washington (1957)
37. Hudson, J. C. and Acock, J. P., *Symposium on the Corrosion of Buried Metals*, The Iron & Steel Inst., Special Report No. 45, London (1952)
38. Romanoff M., *J. Res. Nat. Bur. Stand.*, **660**, 223-224 (1962)
39. Hudson, J. C. and Watkins, K. O., BISRA Open Report No. MG/B/3/68
40. Booth, C. N. et al., *Br. Cor. J.*, **2** No. 3, 104-118 (1967)
41. Mor, E. D., Travesc, E. and Ventora, G., *Br. Corros. J.*, **11**, 40, January (1976)
42. Jelinek, J., Neufeld, P. and Pickup, G. A., *Br. Corros. J.*, **13**, 112, March (1978)
43. Walton, J. R., Johnson, J. B. and Wood, G. C., *Br. Corros. J.*, **17**, 59, February (1982)
44. Vannerberg, N. G., Electrochem Soc, Pittsburg 78-2, Oct (1978) (Extended abstract p. 314)

45. Barton, K., Bartonova, S. and Beranek, E., *Werkstoffe U Korrosion*, **25**, 659 (1974)
46. Awad, G. H., Abdel Halim, F. M. and El Arabi, R. M., *Br. Corros. J.*, **15**, 140, March (1980)
47. Laque, F. L., *Materials Performance*, **21**, 17, April (1982)
48. *ASTM Special Technical Publication*, No. 435 (1968)
49. Biefer, G. J., *Materials Performance*, **20**, January (1981)

3.2 Low-alloy Steels

Physical and Mechanical Properties

The mechanical properties of low- or medium-carbon structural steels can be improved considerably by small alloy additions. For example, 1% of chromium will raise the yield point of 0.2% carbon steel from about 280 MN/m^2 to 390 MN/m^2. This has led to the development of a range of so-called 'low-alloy steels' with high tensile properties. A typical example is grade 817M40 (En 24), which contains 0.4% C, 0.2% Si, 0.6%, Mn, 1.2%, Cr, 0.3% Mo and 1.5% Ni.

Although, originally at least, the main object was to increase the strength of the steel, improvements in the mechanical properties of unalloyed steels have resulted in a considerable overlap in properties between the two classes. In some cases, though by no means all, low-alloy additions, besides making further improvements in properties possible, may also enhance resistance to corrosion. It is to such steels that this chapter specifically refers. As a class they are by no means uncorrodible but under favourable conditions, such as when they are freely exposed outdoors, some of them rust several times less rapidly than unalloyed mild steel. Methods of fabrication differ little in principle from those already described. The low-alloy steels specifically designed to be slow rusting are commonly called weathering steels, and to optimise this corrosion resistance the alloying elements most commonly used are chromium, nickel and copper. To maintain rust resistance and uniform appearance, matching welding rods made from suitable low-alloy steels have to be used.

Corrosion Behaviour in Aqueous Environments

Theoretical Considerations

The improvement in rust resistance achieved through low-alloy additions obviously depends on the nature and amounts of the alloying elements — incidentally their effects are not additive — and to an even greater degree on the nature of the corrosive environment. To make a broad generalisation, weathering steels show to maximum advantage when they are freely exposed to the open air in industrial environments but, even then, their performance varies with compass orientation, prevailing wind direction and degree of shelter. This is illustrated in Fig. 3.6, which compares the relative rates of corrosion of a weathering steel and mild steel at a U.K. industrial location after nine years exposure[1]. The greatest rate of corrosion was on panels facing the north-westerly direction, which is wettest for the longest period of time. However, this is where the low-alloy steel showed the greatest advantage such that the corrosion losses over a complete structure become more uniform.

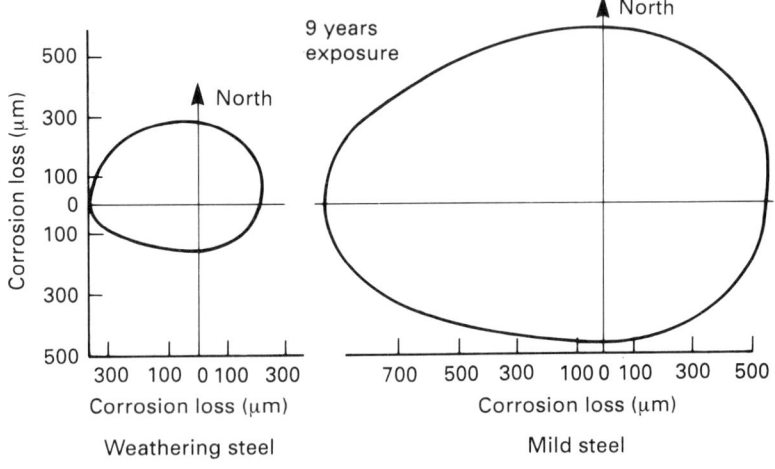

Fig. 3.6 Corrosion losses of steels exposed vertically, facing different compass directions

Cathodic additions (such as copper and chromium) to low-alloy steels influence the rate of rusting by raising the potential of the surface to more noble values so encouraging passivation[2]. Electrochemical measurements certainly seem to bear this out and they have been used in attempts to develop improved compositions[3,4].

Initially, a weathering steel appears to rust like mild steel and quickly assumes a fine, sandy appearance. However, unlike mild steel whose oxide repeatedly spalls off, the surface rust layer stabilises with time, provided that the exposure conditions allow the steel to dry out periodically. The rust then becomes darker, granular and tightly adherent whilst any pores or cracks become filled with insoluble salts. This protects the underlying steel by reducing the permeability of the oxide layer to water and air[5], both of which must be present simultaneously at the metal surface for rusting to continue. Because of the need for intermittent drying to stabilise the oxide film it is doubtful, from the corrosion aspect, whether the use of weathering steels is worthwhile where immersion in natural waters, or burial in soil, is involved.

Characteristic Features of Corrosion Behaviour

When low-alloy steels are exposed outdoors, the rust formed on them is generally darker in colour and much finer in grain than that formed on ordinary steel. Moreover, the slowing down in rusting rate with time (cf. Section 3.1, p. 3:13) seems to be more marked for low-alloy steels than for ordinary steels. This can be illustrated by the BISRA[6] figures given in Table 3.8.

Table 3.8 Variation of rate of rusting with time (BISRA)[6]

Steel	Rate of rusting (mm/y)		Ratio, B/A
	A 1st and 2nd years	B 6th to 15th year	
Ordinary mild steel (Cu 0.02%)	0.129	0.094	0.73
Low-ally steel (Cr 1.0%: Cu 0.6%)	0.077	0.025	0.33

The distinguishing feature of the behaviour of the slow-rusting low-alloy steels is the formation of this protective rust layer. Corrosion in conditions where it cannot form is little different from that of unalloyed steel, although the particular alloying elements present will have some influence on the actual rate at which corrosion occurs.

Statements of this kind can give only a general idea of the improvements to be gained by the use of low-alloy steels in the atmosphere, because so much depends upon the conditions of exposure. In particular, the beneficial effects observed in the open air do not generally extend to conditions where the steel is enclosed and sheltered from the rain. Thus, the rates of rusting of bare steels, exposed for five years in Dove Holes Tunnel, England, were as follows[7]:

Copper content of steel (%)	0.02	0.2	0.5
Rate of rusting (mm/y)	0.061	0.060	0.059

As a further example of the ineffectiveness of low-alloy additions in slowing down rusting under sheltered conditions, tests by BISRA in indoor atmospheres failed to reveal any substantial difference in the rusting of a chromium-copper steel and of an ordinary mild steel in most of them[8]. The test sites covered a wide range of domestic and industrial conditions, from bathrooms to locomotive sheds.

Corrosion in Natural Environments

In the Atmosphere

The general effect of low-alloy additions on the rusting of structural steel in the open air is illustrated in Fig. 3.7, which shows the results of tests conducted by BSC for ten years in an industrial district of Sheffield[1]. Figure

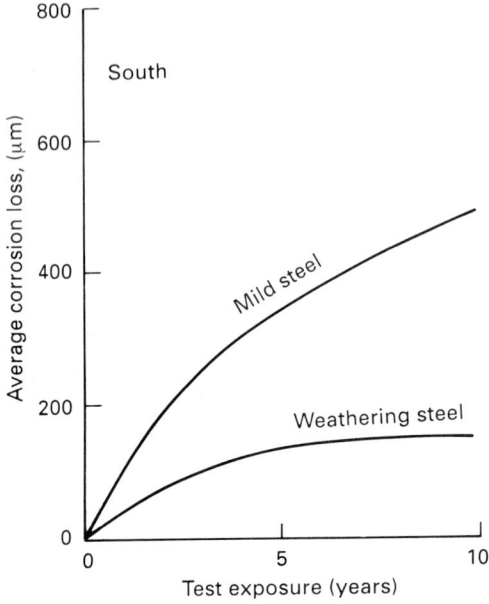

Fig. 3.7 Effect of low-alloy additions on the corrosion of steel outdoors[1]

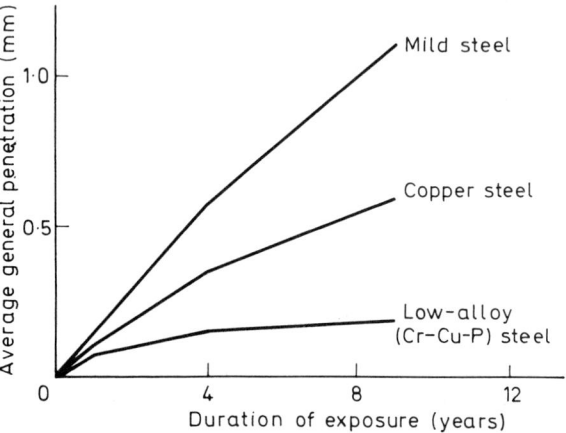

Fig. 3.8 Effect of low-alloy additions on the corrosion of steel outdoors (after Edwards[9])

3.8 refers to similar trials of nine years duration at Rotherham[9]. Similar curves based on American tests[10], reproduced in Fig. 3.9, show substantially the same features. The American specimens were exposed obliquely, so that the rates of rusting of the skyward and groundward faces differed; the figures given are averages for both. The fact that the rates of rusting are markedly slower in the American than in the British tests is mainly due to differences in the corrosiveness of the test atmospheres.

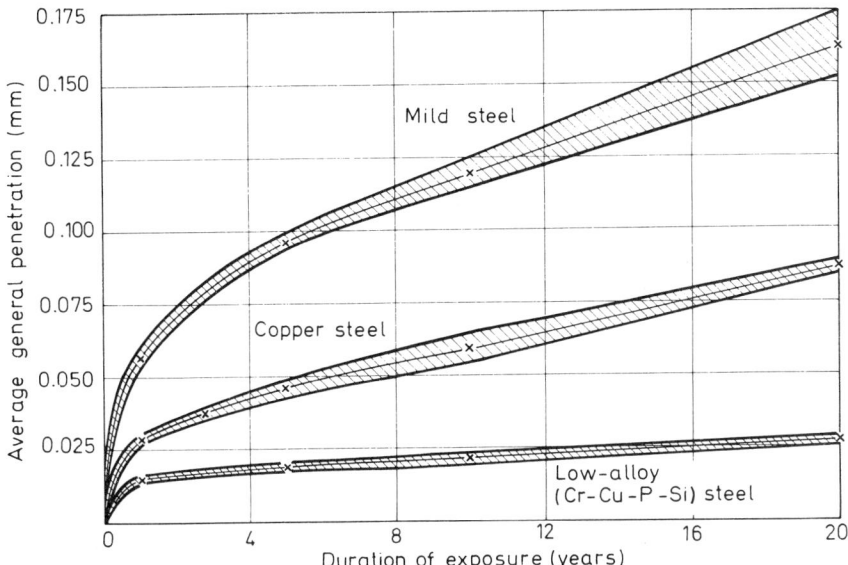

Fig. 3.9 Effect of low-alloy additions on the corrosion of steel outdoors at Kearny, N.J. (after Larrabee and Coburn[10])

The effects of the various alloying elements are not additive. Bearing this in mind, the practical effect of individual elements can be summarised as follows.

1. *Copper* additions up to about 0.4% give a marked improvement, but further additions make little difference.
2. *Phosphorus*, at least when combined with copper, is also highly beneficial. However, in practice, levels above about 0.10% adversely affect mechanical properties.
3. *Chromium*, in fractional percentages, has a significant influence on corrosion rates. While it appears to be beneficial, some conflicting results have been reported and its contribution to the reduced corrosion of complex low-alloy steels containing copper and phosphorus is not large.
4. *Nickel*, while reducing corrosion rates a little, is not as important in its effect as the aforementioned elements.
5. *Manganese* may have a particular value in chloride-contaminated environments, but its contribution is little understood.
6. *Silicon* is in a similar position to manganese, with conflicting evidence as to its value.
7. *Molybdenum* has been little used in low-alloy steels, but may be as effective as copper and is worthy of further study.

The effects of these and other alloying additions and of their combinations have been extensively reviewed[11] (over 180 references). The data in Table 3.9 are not consistent but give some indication of the extent of the variations that can occur.

Table 3.9 Effect of individual low-alloy additions on the rate of rusting of low-carbon steels in the open air

Basis steel composition (%)	Alloying elements, %* and rusting rates†																
	Al	Rusting rate	Cr	Rusting rate	Cu	Rusting rate	Mn	Rusting rate	Mo	Rusting rate	Ni	Rusting rate	P	Rusting rate	Si	Rusting rate
Tests at Kearny, N.J.[10]‡																	
0·01-0·02 Cu 0·25-0·40 Mn ≤0·01 P, ≤0·10 Si					0·24	21							0·10	26	0·61	049
0·01-0·02 Cu 0·25-0·40 Mn 0·06 P, 0·2 Si			1·3	100	0·21	52					1·0	44			0·53	100
0·05 Cu 0·25-0·40 Mn ≤0·01 P, ≤0·10 Si					0·24	69							0·10	59	0·56	76
0·05 Cu 0·25-0·40 Mn 0·06 P, 0·2 Si			1·3	91	0·21	77					1·0	64			0·50	92
Tests at Sheffield[12]§																	
0·01-0·02 Cu 0·5-0·6 Mn 0·02-0·06 P 0·1-0·2 Si			2·6	38	0·23	52	1·8	79	0·55	60	3·1	30				
0·06 Cu 0·5 Mn 0·05-0·06 P 0·2 Si	1·6	72	1·0	67	0·23	72							0·16	87	0·81	93

*Total amount, including that in basis steel. †Taking the rate of rusting of the unalloyed basis as 100. ‡Duration of test 15·5 years. §Average values for 1, 2 and 5 years.

Immersed

The main elements that alter the rate of rusting of low-alloy steels when immersed in natural waters are aluminium, copper, chromium, molybdenum and nickel, but other additions, e.g. manganese, silicon, phosphorus and sulphur, may have minor roles. The action of some alloying elements can be beneficial, neutral or detrimental, depending upon whether localised or uniform corrosion is being considered and whether the steel is fully, partially or intermittently immersed. After a large programme of collaborative work between a number of research laboratories in Europe, Table 3.10 was drawn up to summarise the findings[12]. In using the table to select

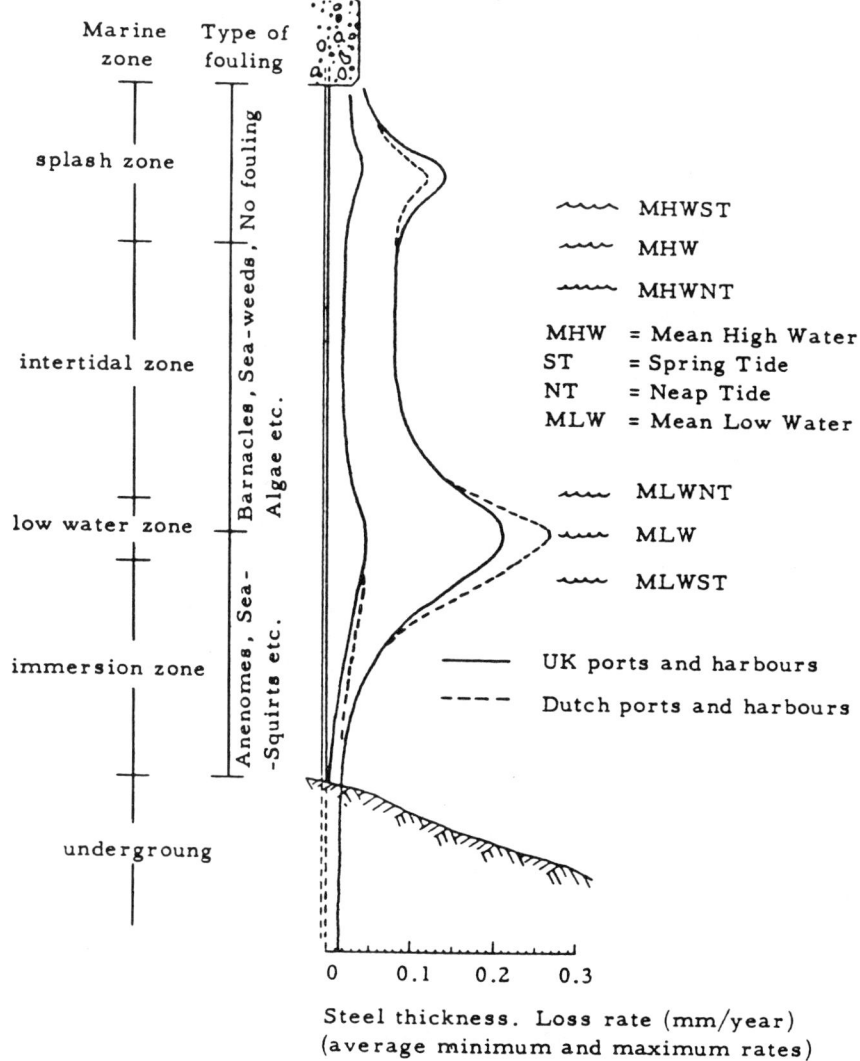

Distribution of steel pile corrosion as a function of marine exposure zone and marine fouling

Table 3.10 Effect of alloying elements on marine corrosion resistance

Corrosion type	Environment	Favourable	Neutral	Unfavourable
Uniform corrosion	Immersion	Mn Si Al Mo ($t > 4$ yrs) Cr ($t \leqslant 4$ yrs)	Ni	P S Cu Mo ($t \leqslant 4$ yrs) Cr ($t > 4$ yrs)
	Tidal and splash zone	P	Cu, Cr, Ni	
	Marine atmosphere	P, Si, Mn, Cu, Cr, Ni (Mo, V, Ti)		
Local corrosion* (esp. pitting)	Immersion		Cu, Cr	Ni
	Tidal and splash zone	Cu	Ni	Cr

t = exposure time
* No SCC has been observed for $R_{p0.2} < 850$ MPa

materials for partial immersion, a further complicating factor is that for unprotected steel the corrosion rates are always highest, irrespective of steel or water composition, at the water lines, especially in tidal and wave conditions. This is illustrated in the diagram on p 3:29, which shows the average variation in corrosion depth of steel piles that were measured in a survey of U.K. ports and harbours[13].

From a consideration of Table 3.10, a copper–phosphorus steel might be chosen for its resistance to corrosion in the critical tidal and splash zone. However, the variation of corrosion resistance is much greater than the difference between various alloy steels so it is improbable that low-alloy steels will corrode more slowly than mild steel in most practical environments. This conclusion is supported by Forgeson *et al.* who concluded from extensive tests in fresh and salt waters of the Panama Canal Zone that: 'Proprietary low-alloy steels were not in general more resistant to underwater corrosion than the mild unalloyed carbon steel'[14].

Underground

In tests by BISRA, made over three years in a heavy clay soil at Binfield, additions of chromium and copper had no beneficial effect on the rusting of buried mild steel, as is shown in the following figures[15]:

Alloying elements (%)	Nil	Cu 0.5	Cr 1.0	Cr 0.6, Cu 0.5
Rate of rusting (mm/y)	0.026	0.033	0.031	0.026

The apparent differences have little significance since the experimental error was in the region of 10%.

Additional evidence is available from tests made by the National Bureau of Standards on ten varieties of steel, which were buried in 15 typical American soils from 1937 to 1950[16]. The results showed that, with few

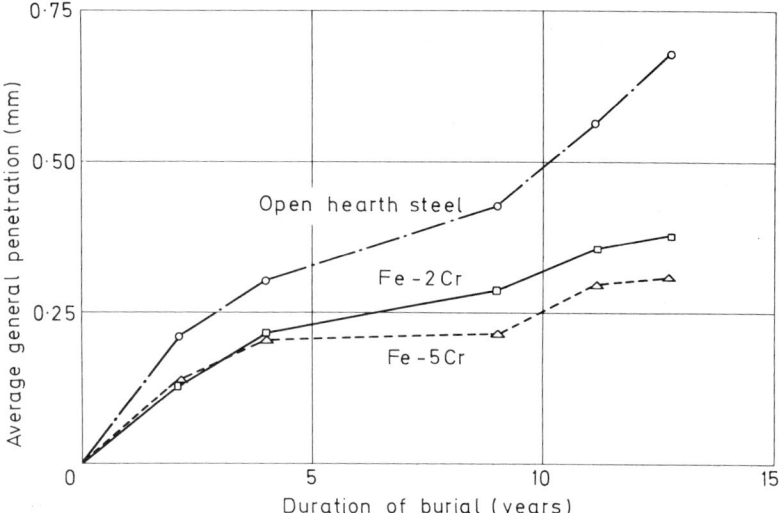

Fig. 3.10 Effect of chromium content on the corrosion of buried steel (after Romanoff[16])

exceptions, the corrosion of low-alloy steels containing coppers nickel and molybdenum in various combinations did not differ by more than 20%, from that of ordinary carbon steel. Additions of 2 or 5% of chromium did increase the corrosion resistance, however, as is indicated in Fig. 3.10*.

Applications in Industry

Most structural steelwork that is exposed to the atmosphere is given a protective coating of some kind. If this coating is continuously maintained in perfect condition, so that no rusting of the steel takes place, there is no advantage from the corrosion aspect in using a low-alloy steel instead of mild steel. If, on the other hand, it is probable that the protective coating will be damaged or allowed to deteriorate, the use of a low-alloy steel should be considered. The more compact rust film formed on these steels will be less likely to cause the coating adjacent to the corroded areas to spall off, and the rate at which breakdown of the coating spreads will be reduced. Several investigators have reported better performance and durability of painting schemes on low-alloy steels than on ordinary steel. For example, Copson and Larrabee have written[17]: 'Both field tests and service experience have shown that paint coatings are more durable on high-strength low-alloy steels than on carbon steel or on copper steel. Any rust which forms at breaks or holidays or underneath the paint film is less voluminous on the low-alloy steels. Owing to the smaller volume of rust there is less rupturing of the paint film and, hence, less moisture reaches the steel to promote further corrosion.'

---

*The points shown in these graphs have been calculated by the present authors from the original data: they are averages for the three soils Nos. 58, 63 and 64 in the American paper. There were only single samples of carbon steel and of 2% Cr steel, but there were three 5% Cr steels, the results of which have been averaged. Some of the chromium steels also contained about 0.5% Mo.

In conformity with this, low-alloy steels could be used with advantage for such purposes as the production of agricultural machinery, on which the coating is liable to get knocked about, and which is frequently left in the open fields for long periods.

Other obvious applications of low-alloy steels are for roofing sheets, wire ropes and chain-link fencing. The advantages here, though appreciable, are not so great as would appear at first sight, because a high proportion of such products is at least galvanised, if not plastic-coated as well, and the effect of the low-alloy additions begins to operate only after the zinc coating has failed.

Weathering steels have also been used for the bodies of railway wagons in many countries including America, Japan, South Africa and the U.K. The results of trials on coal gondolas in service on an American railroad[18] indicate that sheets made from a commercial low-alloy steel (0.14% P, 0.02% S, 0.8% Si, 0.4% Cu, 1.0% Cr) would last one-and-a-half to twice as long as copper steel (0.3% Cu) sheets, if undue local attack at laps and ledges (due to the accumulation of wet coal dust) was prevented*. Similar results have been obtained in England. In a service trial[19] of 15 years duration on floor-plates of coal wagons, three low-alloy steels were found to give about 50% greater life expectancy than mild steel. The three steels contained 0.35% Cu, 0.44% Cu with 1.34% Mn, and 0.33% Cr with 0.90% Cu. There was no significant difference in performance between them.

In another service trial[20] on the Western Region of British Rail, sleepers of copper steel (Cu 0.5%) were found to lose up to 50% less weight by rusting than ordinary steel sleepers when laid in the open track. In agreement with what has been stated above, there was, however, little difference between the performances of the two types of sleepers when they were laid inside the Severn Tunnel.

However, the most widespread use of weathering steels in the U.K., and elsewhere, has been for buildings and bridges, especially where maintenance painting is particularly difficult, dangerous, inconvenient or expensive[21]. Bridges over land, rivers, railways, roads and estuaries fall into this category, although in the last two cases care should be taken with respect to airborne salinity. Road bridges can be affected by salt-laden atmospheres or water, produced as a consequence of winter ice and snow clearing with deicing salt and grit. The chloride can be in the form of an airborne spray thrown up by passing vehicles or as a result of leaks in the bridge deck. In the presence of salt-water many materials, including steel, paint, reinforced concrete, aluminium, etc., deteriorate at an accelerated rate. Weathering steels are no exception, and higher than normal corrosion rates should be expected if they are exposed to saline waters or frequent spraying with salt. The important criterion is design. Many bridges have been built successfully from weathering steels but at the design stage it is important to consider the possible effects of road salt in order to obtain the maximum maintenance-free life.

---

*This proviso draws attention to the fact already mentioned that the enhanced corrosion resistance of low-alloy steels is not usually obtained in wet conditions. Presumably the copper steel sheets, in turn would have outlasted sheets of ordinary steel.

In order to obtain a uniform colour, it is essential to remove all mill-scale and residual grease or oil stains, preferably by blasting. The detailing of all sections should be such as to avoid pockets, crevices and any location which will collect and retain moisture and dirt for long periods. Any such locations, as well as faying surfaces, should be painted for corrosion protection. The paint requirements for weathering steels are exactly the same as for carbon steel and the slow rusting nature of the weathering steel will result in all paint systems having an extended life before maintenance is required.

An important aspect of design is to predict the lines of run-off of surface water. This is because the water will contain minute particles of brown rust, especially in the pre-stabilisation period, that will stain some surfaces. Matt, porous surfaces stain particularly easily and run-off should not be over concrete, stucco, galvanised steel, unglazed brick or stone.

J. C. HUDSON
J. F. STANNERS
R. A. E. HOOPER

REFERENCES

1. Hooper, R. A. E. and Lee, B. V., *Proc. 12th International COR-TEN Conference*, Florida, 1985, United States Steel, Pittsburgh (1985)
2. Tomashov, N. D. and Lokotilov, A. A., *Annals-Korriziya i Zaschita Stallei*, **171**, Mashgiz, Moscow (1959)
3. Becker, G., *Arch. Eisenhüttenwes*, **36**, 489 (1965)
4. Pourbaix, M., *Rapport Technique No. 160*, Cebelcor, Brussels (1969)
5. Skorchelletti, V. V. and Tukachinsky, S. E., *Zh. Prikl. Khim., Leningr.*, **26** No. 1, 30 (1953)
6. *Sixth Report of the Corrosion Committee*, The Iron and Steel Institute, Special Report No. 66, London (1959)
7. *Fourth Report of the Corrosion Committee*, The Iron and Steel Institute, Special Report No. 13, London (1936)
8. Stanners, J. F., *J. Appl. Chem.*, **10**, 461 (1960)
9. Edwards, A. M., *Proc. Symp. on Developments in Methods of Prevention and Control of Corrosion in Buildings*, British Iron and Steel Federation, London (1966)
10. Larrabee, C. P. and Coburn, S. K., *Proc. First International Congress on Metallic Corrosion*, London, 1961, Butterworths, London, 276 (1962)
11. Chandler, K. A. and Kilcullen, M. B., *Brit. Corros. J.*, **5** No. 1,24 (1970)
12. Songa, T., *International Conference on Steel in Marine Structures*, Paris, 1981, ECSC, Luxembourg (1981)
13. Morley, J. and Bruce, D. W., *ECSC Technical Report*, EUR 8492, EN (1983)
14. Forgeson, B. W., Southwell, C. R. and Alexander, A. L., *Corrosion*, **16**, 105t (1960)
15. Hudson, J. C., Banfield, T. A. and Holden, H.A., *J. Iron St. Inst.*, **146**, 107 (1942)
16. Romanoff, M., Underground Corrosion, *National Bureau of Standards Circular 579*, US Government Printing Office, Washington (1957)
17. Copson, H. R. and Larrabee, C. P., *Bull. Amer. Soc. Test Mater.*, No. 242, 68 (1959)
18. Kelly, B. J., *Corrosion*, **7**, 196 (1951)
19. Hudson, J. C., *Iron St. Inst.*, **194**, 45 (1960)
20. Hudson, J. C., *Iron St. Inst.*, **169**, 13 (1951)
21. *Weathering Steel in Bridgework*, British Steel Corporation, London (1971)

3.3 Stainless Steels

Introduction

The scope of the term stainless steel has not been precisely defined, but for general purposes it may be considered to include alloys whose main constituent is iron but which also contain not less than 10% Cr. As with low-alloy steels, a distinction between low or medium carbon grades and high carbon grades must also be drawn, the latter being more in the nature of alloy cast irons. These are used mainly for oxidation resistance at high temperatures and for applications where abrasion resistance allied to a certain amount of corrosion resistance is required, and will not be considered in this section.

The term *stainless* steel is not, of course, a strictly accurate description, but the difference in behaviour between chromium-bearing steels and carbon steels in many environments is so marked as to justify the adoption of the name.

H. Brearley was the first to see clearly the commercial possibilities of the 12–13% Cr martensitic steel, introduced to the public in the form of stainless cutlery over the period 1912–1915. As a result, he has been very widely considered the 'inventor' of stainless steel. In fact, a remarkable amount of study had been made of such alloys over the previous decade. For example, Guillet (1902–1906) and Portevin (1909–1911) in France, and Giesen (1907–1909) in England had worked on the metallurgy and physical properties. They noted etching difficulties, but made no special study of corrosion resistance. On the other hand, Monnartz in Germany had, in 1908–1909, given detailed consideration to this aspect. He showed a remarkable appreciation of passivity and its relation to oxidation and potential, the rôle of chromium content in resistance to nitric acid, and the effect of carbon. He discussed carbide stabilisation by chromium and the use of molybdenum additions for improved corrosion resistance. He also carried out tests in the atmosphere, tap water and sea-water. In spite of this, it was left to Brearley to visualise the full practical possibilities of such a steel, given the optimum combination of composition, heat treatment and surface finish.

The austenitic steels with the well known 18Cr–8Ni basis came into use in the 1920s.

As with the chromium-iron alloys, the study of the records of early development reveal a complicated story, again leading to the pioneering work of Guillet in France and Giesen in England during the first decade of this century. Maurer and Strauss, working for Krupp in Germany, were prominent in the development over the next 10 years, but the earlier Krupp steel was of higher chromium and carbon content, as indicated by Strauss, who in 1924 gave 0·25C-20Cr-7Ni as typical (A.S.T.M. 1924). Practical development of this steel in Sheffield led to lowering of the chromium and carbon contents and by 1924 the familiar 18-8 composition with 0·10-0·15% C was established.

The Krupp work had shown interesting improvements in acid resistance resulting from molybdenum and copper additions, and the use of 2-3% Mo for more difficult acid conditions was soon established. Other early additions were made to overcome susceptibility to intercrystalline corrosion, culminating in the general use, by the early thirties, of titanium additions for carbide stabilisation, followed shortly after by the alternative use of niobium.

Classification of Stainless Steels

The various requirements of not only the designer and user, but also of the fabricator and steel producer, have led to considerable development in these steels. Not only have new grades been devised but also the traditional grades have, in some instances, been slightly modified, in some cases repeatedly to facilitate production by modern methods. This has resulted in a situation which may appear bewildering with numerous specifications and proprietary grades, the objectives of which may not be immediately clear. The more commonly used grades are covered by national standards and these are illustrated here by reference to the current British Standards, BS 970: Part 1: 1983 and BS 1449: Part 2: 1983. Not all the grades therein are included here but the more widely used are dealt with. There are very close equivalents to these in the national specifications of most industrial countries and in the I.S.O. standards. A number of specialised grades are not covered (the situation varies from country to country) and some of the more specialised types will be illustrated by representative steels. It should be noted that these are only illustrative, and that there are often other quite similar types of equal merit.

The basic corrosion behaviour of stainless steels is dependent upon the type and quantity of alloying. Chromium is the universally present element but nickel, molybdenum, copper, nitrogen, vanadium, tungsten, titanium and niobium are also used for a variety of reasons. However, all elements can affect metallurgy, and thus mechanical and physical properties, so sometimes desirable corrosion resisting aspects may involve acceptance of less than ideal mechanical properties and vice versa.

The way that alloying affects structure is illustrated in Fig. 3.11. Which is a modification of the diagram first proposed by Schaeffler. The diagram is for steel cooled rapidly from 1050°C to room temperature. Fig. 3.11 is for a ternary (Fe-Cr-Ni) system with weight percentage of chromium shown on the horizontal axis and of nickel on the vertical. As can be seen, the diagram is divided with four major areas according to structure: martensite, austenite, ferrite and austenite plus ferrite. These four structures provide a convenient basis for classification. All other elements behave as either chromium or nickel in affecting structure, and these have to be taken into

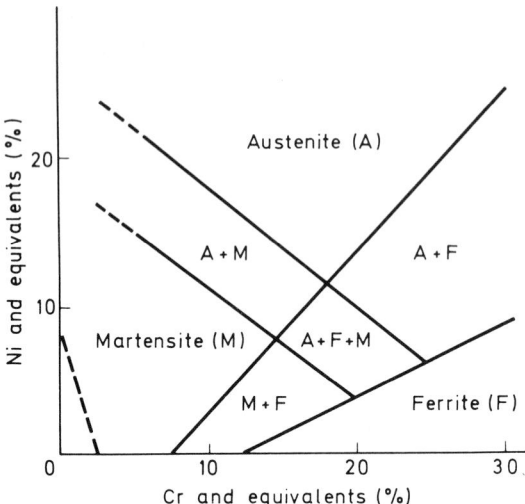

Fig. 3.11 Structure of steels after cooling from elevated temperature as determined by composition. *A* austenite, *M* martensite and *F* δ-ferrite (after Schaeffler)

account when predicting structure for a given composition. The degree of effect relative to chromium or nickel varies and equivalence factors have been proposed for all the commonly used elements. For further details References 1 to 3 should be consulted.

Martensitic Steels

A plain carbon steel, when heated to elevated temperature, has a face-centred cubic structure (austenite), but when this is cooled to normal room temperature at any but a rapid rate, it tranforms to the body-centred cubic α-ferrite with the formation of iron carbides. This ferritic structure is soft and ductile. If cooling is sufficiently rapid however, transformation is to martensite (with much of the carbon retained in solid solution) which is hard and consequently more brittle, the actual hardness depending upon the carbon content. One consequence of alloying with chromium is to decrease progressively the cooling rate necessary to give the martensitic transformation. Thus steels falling in the martensitic area of Fig. 3.11 which contains sufficient chromium to be 'stainless' have a martensitic structure on cooling in air from elevated temperature except in large section form. With many grades a ferritic structure can be obtained by retarding the cooling rate, for instance by furnace cooling. While desirable for some purposes, the high hardness of a martensitic structure is not so for others and this has transformed hardness of rapidly cooled steel may be reduced by further sub-critical heat treatment, i.e. by tempering, this resulting in the formation of ferrite and chromium-bearing carbides. The tempering process is described in detailed in Reference 2.

Most of the martensitic stainless steels are based on 11·5 to 13·5% chromium and those from BS 970 and BS 1449 are shown in Table 3.11.

STAINLESS STEELS

Table 3.11 Martensitic steels

Steel	C	Si	Mn	S	P	Cr	Ni	Mo	Others
410S21[a]	0·09–0·15	1·0	1·0	0·030	0·040	11·5–13·5	1·00		
416S21[a]	0·09–0·15	1·0	1·5	0·15–0·35	0·060	11·5–13·5	1·00	0·60	
416S29[a]	0·14–0·20	1·0	1·5	0·15–0·35	0·060	11·5–13·5	1·00	0·60	
416S37[a]	0·20–0·28	1·0	1·5	0·15–0·35	0·060	12–14	1·00	0·60	
416S41[a]	0·09–0·15	1·0	1·5	0·06	0·040	11·5–13·5	1·00	0·60	Se 0·15–0·35
420S29[a]	0·14–0·20	1·0	1·0	0·030	0·040	11·5–13·5	1·00		
420S37[a]	0·20–0·28	1·0	1·0	0·030	0·040	12·0–14·0	1·00		
431S29[a]	0·12–0·20	1·0	1·0	0·030	0·040	15·0–18·0	2·06–3·0		
Alloyed 12% Cr[b] steel	0·12	0·3	0·6	0·015	0·015	12·25	0·80	0·60	V 0·18
Alloyed 12·1 Cr[b] steel	0·12	0·4	0·8	0·015	0·015	11·5	2·3	1·4	V 0·15, Nb 0·3
Precipitation hardening steel[b]	0·04	0·4	0·7	0·015	0·015	14·0	5·5	1·6	Cu 1·8, Nb 0·25
Controlled transformation steel[b]	0·05	0·4	1·3	0·015	0·015	15·8	5·5	1·6	Cu 2·0, Ti 0·08

[a] BS970: Part 1:1983. Values maxima unless stated.
[b] Proprietary grades. Values typical.

The grades with the 410 or 420 numerals are the basic 13% chromium type with varied carbon content. The additions of sulphur or selenium (possibly with phosphorus) to some grades (416 group) is to improve machinability. 431S29 has increased chromium content to improve corrosion resistance, but reference to Fig. 3.11 shows that such addition alone would lead to a mixed martensite–δ-ferrite structure with certain disadvantages to mechanical properties. The nickel addition is to limit ferrite content.

Developments of the 410S21 type have involved alloying with molybdenum, vanadium, tungsten, niobium and nickel, either singly or in various combinations, with the objective of modifying the tempering characteristic for reasons explained later. Two examples of this group of steels are given in Table 3.11.

Other more highly alloyed types, of which a typical example is given in Table 3.11, have the designation of precipitation hardening martensitic. Relative to the simple 13% chromium types they have a substantial nickel content and low carbon with additions from molybdenum, copper, aluminium, titanium and niobium. These offer improved corrosion resistance, strength, toughness, weldability and fabrication properties, but not always together.

A group analytically quite similar to the precipitation hardening types, but metallurgically different, is the 'controlled transformation' or 'semi-austenitic' group, a representative example of which is in Table 3.11. These offer further advantages in fabrication.

For further details of the special martensitic steels, References 1 to 4 should be consulted.

Ferritic Steels

From Fig. 3.11, it can be seen that by increasing the chromium content while maintaining a limited amount of nickel-equivalent elements, first mixed martensite–ferrite structures are produced and then fully ferritic. This is δ-ferrite, that is a body-centred cubic structure stable at all temperatures. Relative to martensite it is soft, but it is also usually brittle. For this latter reason, usage has in the main been in small section form. This and some other disadvantages are offset for some purposes by attractive corrosion resistance or physical properties.

The British Standard ferritic grades are in Table 3.12. Three of these (403S17, 405S17 and 409S17) have, in fact, considerable amounts of martensite as cooled from elevated temperature, this being softened by treatment in the 750–850°C temperature range.

In the last decade there has been increased interest in the ferritic steels stimulated originally by the availability of new steel-making processes which gave hope that the brittleness problem could be solved by suitable control of carbon and nitrogen contents. This hope has only been partially realised, but as a result a number of new grades have been marketed which do represent useful additions to the range. These have became known collectively as 'Super Ferrities'. Some examples are in Table 3.12. A substantial amount of relevant information was presented at the conference indicated in Reference 5.

Table 3.12 Ferritic steels

Steel	C	Si	Mn	S	P	Cr	Ni	Mo	Others
403S17[a]	0·08	1·0	1·0	0·030	0·040	12·0–14·0	1·00		
405S17[a]	0·08	1·0	1·0	0·030	0·040	12·0–14·0	1·00		Al 0·10–0·30
409S17[a]	0·08	1·0	1·0	0·030	0·040	10·5–12·5	1·00		Ti 6 × C–1·0
430S17[a]	0·08	1·0	1·0	0·030	0·040	16·0–18·0	1·00		
434S17[a]	0·08	1·0	1·0	0·030	0·040	16·0–18·0	1·00	0·90–1·30	
18-2[b]	0·020	0·5	0·4	0·004	0·020	18·5	0·1	2·0	N 0·012, Ti 0·13, Nb 0·4
26-1[b]	0·002	0·3	0·1	0·015	0·010	26	0·1	1·0	N 0·006
[b]	0·01	0·25	0·20	0·004		25·5	2·0	3·0	N 0·015, Ti 0·5
[b]	0·012	0·22	0·2	0·002		29	0·8	4·0	N 0·026, Ti 0·6
[b]	0·01	0·25	0·2	0·004		25·5	2·0	3·0	N 0·016, Ti 0·5

[a] BS 1449 (1983). Values maxima unless stated.
[b] Proprietary 'super ferrities'. Typical analyses.

Austenitic Steels

All elements, to differing degrees, depress the temperature range over which austenitic transforms to martensite and, by suitable alloying, this can be depressed to below ambient so that an austenitic structure is retained. Being soft, such a structure has the advantages of very high ductility and toughness, while the combinations of chromium and nickel most frequently be used to give this structure (based on the original famous '18/8') offer good resistance to a wide range of corrodents. The British Standard Grades are listed in Table 3.13A. The functions of the various additions other than chromium and nickel will be discussed later. As with the other generic groups, there are produced a large number of alloys not covered by the standards, examples of which are listed in Table 3.13B, and these are intended to give enhanced strength, and/or corrosion resistance. Again it is stressed that these are merely illustrative of the types available.

Duplex Steels

The name implies a two-phase structure and thus could be applied to any two-phase region of Fig. 3.11, but in fact is normally taken to imply the austenitic plus ferrite field with deliberate control to give a substantial content of both phases. The idea is not new—such alloys having been used for many years to some degree, notably in France and Sweden—but there has recently been enhanced interest in the type. Improvement in strength and in some aspects of corrosion resistance relative to the standard austenitics are available. A large amount of information regarding these steels was made available at the symposium recorded in Reference 6. Some anlayses are given in Table 3.14.

Physical Properties

Typical physical properties for a selection of steels are given in Table 3.15.

Table 3.13A Austenitic steels ex BS970 or BS1449

Steel	C	Si	Mn	S	P	Cr	Ni	Mo	Others
303S31	0·12	1·0	2·0	0·15–0·35	0·06	17·0–19·0	8·0–10·0	1·0	
303S42	0·12	1·0	2·0	0·06	0·06	17·0–19·0	8·0–10·0	1·0	Se 0·15–0·35
304S11	0·030	1·0	2·0	0·030	0·045	17·0–19·0	9·0–12·0		
304S15	0·06	1·0	2·0	0·030	0·045	17·5–19·0	8·0–11·0		
315S16	0·07	1·0	2·0	0·030	0·045	16·5–18·5	9·0–11·0	1·25–1·75	
316S11	0·030	1·0	2·0	0·030	0·045	16·5–18·5	11·0–14·0	2·0–2·5	
316S13	0·030	1·0	2·0	0·030	0·045	16·5–18·5	11·5–14·5	2·5–3·0	
316S31	0·07	1·0	2·0	0·030	0·045	16·5–18·5	10·5–13·5	2·0–2·5	
316S33	0·07	1·0	2·0	0·030	0·045	16·5–18·5	11·0–14·0	2·5–3·0	
317S12	0·030	1·0	2·0	0·030	0·045	17·5–19·5	14·0–17·0	3·0–4·0	
317S16	0·06	1·0	2·0	0·030	0·045	17·5–19·5	12·0–15·0	3·0–4·0	
321S31	0·08	1·0	2·0	0·030	0·045	17·0–19·0	9·0–12·0	Ti 5 × C–0·80	
347S31	0·08	1·0	2·0	0·030	0·045	17·0–19·0	9·0–12·0	Nb 10 × C–1·00	

All values maxima unless otherwise indicated.

Table 3.13B Typical analyses of some proprietary, highly alloyed austenitic steels

Steel	C	Si	Mn	Cr	Ni	Mo	Others
Solution-hardened austenitic	0·04	0·4	3·75	21·5	9·5	2·75	N 0·4, Nb 0·3
'Super austenitic'	0·02	0·4	0·4	19·0	33·0	2·2	Cu 3·0, Nb 0·5
'Super austenitic'	0·015	0·4	0·5	20·0	18·0	6·1	Cu 0·8, N 0·2
'Super austenitic'	0·015	0·5	1·5	20·5	25·0	4·7	Cu 1·5
'Super austenitic'	0·02	0·4	1·5	20·0	25·0	6·5	
'Super austenitic'	0·015	0·4	0·4	27·0	31·0	3·5	Cu 1·0

Table 3.14 Typical analyses of some duplex steels

C	Cr	Ni	Mo	Cu	W	N
0·02	18·5	4·9	2·7			0·07
0·05	22·5	8	2·4	1·5		
0·02[a]	22·0	5·5	3·0			0·18
0·02	25·0	5·5	3·0	1·8		0·18
0·02	25·0	6·5	2·5	0·5		0·16
0·02	25·0	7·0	3·5	0·8	0·8	0·22

[a] W. 1·4462.

Mechanical Properties

Martensitic Steels

These steels are normally used in the hardened and tempered conditions and the effect of tempering temperature on the strength of two grades is shown in Fig. 3.12. The lower carbon grades are most frequently used in a relatively soft state and typical properties are shown in Table 3.16. The toughness indicated is excellent, but it should be remembered that these steels show a sharp ductile/brittle transition at a critical testing temperature that varies according to strength and other features (such as grain size). Low toughness values at ambient temperature must be expected with treatments giving U.T.S. values above approx. 950 MN/m^2. Special proprietary grades with superior transition temperatures at higher strength levels have been developed (Table 3.16). Typical curves showing how toughness is related to temperature are given in Fig. 3.13 for 13% Cr steel and a modified type.

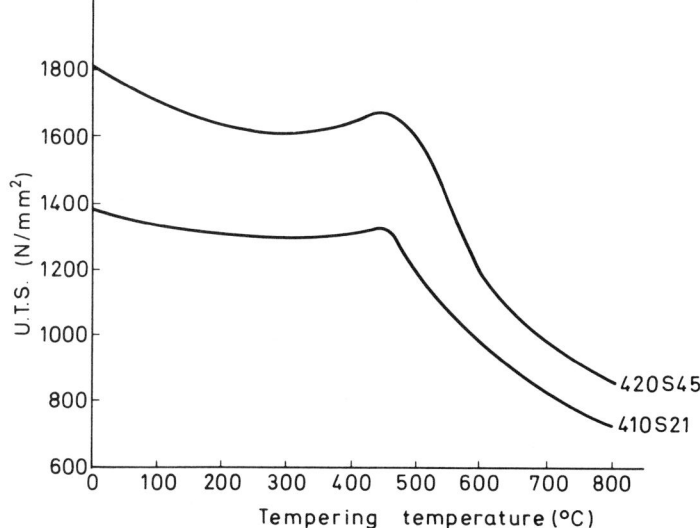

Fig. 3.12 Strength (U.T.S.) of two 13% Cr, martensitic steels as affected by tempering at various temperatures for 1 h following air cooling from 980°C

Table 3.15 Physical properties of some stainless steels

	410S21 (fully tempered)	420S37 (cutlery temper)	431S29 (fully tempered)	Precipitation hardening 550°C overaged	430S17 (softened)	304S16 (softened)	321S31 (softened)	347S31 (softened)	316S31 (softened)
Specific gravity (g/cm³)	7·73	7·74	7·70	7·70	7·70	7·90	7·90	7·93	7·96
Specific heat (J/g°C at 20°C)	0·48	—	0·48	—	—	0·50	0·50	0·50	0·50
Specific resistance (μΩ cm at 20°C)	56	60	72	77	62	73	71	72	73
Thermal conductivity (W/m°C at 20°C)	24·70	20·93	18·84	17·17	21·77	15·91	15·91	15·91	16·33
Coefficient of thermal expansion (20–100°C)	0·000011	0·000011	0·000010	0·000011	0·000010	0·000016	0·000016	0·000017	0·000016
Magnetic permeability	750	750	190	110	290	1·005–2·03	1·5–2·0	1·005–1·03	1·05

Table 3.16 Mechanical properties of some martensitic steels

	410S21 (750°C temper)	420S29 (750°C temper)	420S37 (cutlery temper, 200°C)	Alloyed to 12% Cr steel 2 (650°C temper)
0·2% P.S.(MN/m^2)	350	480	1 380	825
U.T.S. (MN/m^2)	605	710	1 530	1 020
Elongation (% 4$\sqrt{A}$)	33	27	17	22
Reduction of area (%)	70	60	47	61
Fatigue limit (10^8 cycles, MN/m^2)	±230	±310	—	—
Impact strength, Izod (J)	133	108	22	79

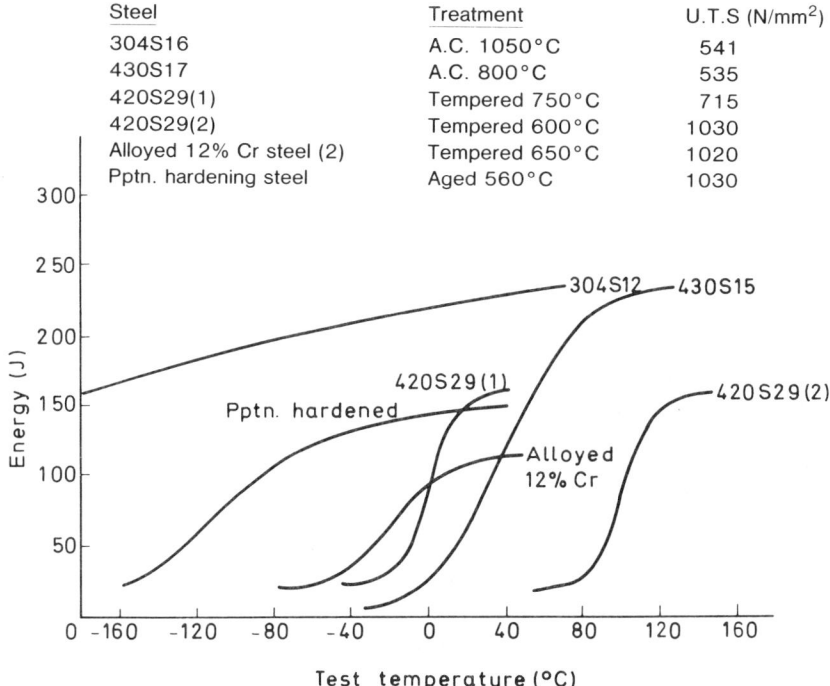

Steel	Treatment	U.T.S (N/mm^2)
304S16	A.C. 1050°C	541
430S17	A.C. 800°C	535
420S29(1)	Tempered 750°C	715
420S29(2)	Tempered 600°C	1030
Alloyed 12% Cr steel (2)	Tempered 650°C	1020
Pptn. hardening steel	Aged 560°C	1030

Fig. 3.13 Impact strengths (Charpy 'V'-notch) at various temperatures for several stainless steels

Where high hardness values are required with steels such as 420S37, it is customary to use low temperature tempering in order to obtain optimum corrosion resistance. Such treatments are usually confined to lighter sections as the toughness is limited. Typical properties of 420S37 are given in Table 3.16.

With the precipitation hardening types, high strengths can be obtained with good toughness. A feature of these steels is that the ductile–brittle transition is less sharp, although low impact values are obtained at very low temperatures. Properties for a typical example are shown in Tables 3.17 and Fig. 3.13.

Table 3.17 Mechanical properties of typical precipitation hardening and semi-austenitic stainless steels

	Precipitation hardening*			Semi-austenitic*†	
	Fully overaged[1]	550°C overaged[2]	Peak hardened[3]	Softened[4]	Peak hardened[5]
0·2% P.S. (MN/m^2)	550	990	1 130	310	1 080
U.T.S. (MN/m^2)	930	1 040	1 210	800	1 160
Elongation (% 4$\sqrt{A}$)	27	23	22	18	10
(% 50 mm)	–	–	–	30	30
Reduction of area (%)	66	65	59	–	–
Fatigue limit (10^8 cycles, MN/m^2)	±650	±590	±475	–	–
Impact strength, Izod (J)	115	100	68	–	–

Notes: * See Table 3.11
 † Sheet specimens
 1. A.C. 1 050°C A.C. + 2 h 750°C + 2 h 620°C
 2. A.C. 1 050°C A.C. + 2 h 750°C + 2 h 550°C
 3. A.C. 1 050°C A.C. + 2 h 850°C + 4 h 450°C
 4. A.C. 1 050°C A.C.
 5. A.C. 1 050°C A.C. + 2 h 750°C + 4 h 450°C

Ferritic Steels

The ferritic steels also show sharp ductile–brittle transition but the transition temperatures, at least for those grades in British Standards, are normally above ambient for all conditions of heat treatment. A transition curve for 430S17 is included in Fig. 3.12. Because of this limitation in toughness, use of these steels is usually restricted to small sections where brittle fracture hazards are not relevant. They are normally used in the annealed state when yield strength and U.T.S. are typically of the order of 270 and 540 M/mm^2, respectively. Use of modern melting methods to limit the amount of carbon and nitrogen to very low levels can depress the transition temperature to below ambient provided due care is taken with other features such as grain size, but the special grades illustrated in Table 3.12 are generally used in thin section for their corrosion-resisting properties rather than improved mechanical properties.

Austenitic Steels

The austenitic steels included in the British Standards have excellent toughness, even at very low temperatures, as illustrated typically by one grade in Fig. 3.13. They cannot be hardened by heat treatment and have a very low limit of proportionality in the annealed state. A high work hardening rate gives relatively high U.T.S. and high uniform elongation before reaching the U.T.S. These characteristices can be very useful in cold forming operations. The properties to be expected from the whole group of '300' steels in Table 3.13A are shown in Table 3.18.

The austenitic structure can be strengthened by cold work, and high strengths, still with adequate ductility, can be obtained by cold rolling, drawing or stretching[7]. The power requirement for such strengthening

Table 3.18 Mechanical properties of austenitic steels

0·2% P.S. (MN/m^2) (varies according to section and method of working)	150–240
U.T.S. (MN/m^2) (varies according to section and method of working)	500–650
Elongation (% $4\sqrt{A}$) (varies according to grade)	45– 70
Reduction of area (%) (varies according to grade)	55– 80
Izod (J) (varies according to grade)	110–162
Fatigue limit (10^8 cycles, MN/m^2)	±275

limits its application to thinner sections, but the flow stress is reduced markedly by increasing the working temperature and, provided the recrystallisation temperature is not exceeded, strengthening can still be obtained. Strengthening by such 'warm working' can be applied to plate and certain forged items.

An alternative method of strengthening is solution hardening by further alloying. Several elements commonly used in steel (e.g. silicon, chromium, molybdenum) give some increase in proof strength proportional to their contents, but the most effective is the interstitial element nitrogen which can be added to the '18/8' types up to about 0.2% and can increase the 0.2% proof stress value to 300 N/mm^2 without detriment to corrosion resistance. More recently grades with up to 0.4% nitrogen (e.g. see Table 3.14B) have become available which give 0.2% proof stress values up to 460 N/mm^2, often with improved corrosion resistance.

Duplex Steels

These steels have higher 0.2% proof and U.T.S. values than do the standard austenitic grades. The reason is not entirely clear and is probably due to a combination of effects. Strength is higher for alloys of higher alloying element content and higher ferrite content. An example of the properties is given for W.1.4462 steel (see Table 3.14) which has a 0.2% proof strength of 400 N/mm^2, U.T.S. 770 N/mm^2 and elongation 34% in the ideally heat-treated state. The Izod impact value of 160 J is well maintained with reducing temperature, the 50% energy transition temperature for 65 mm/ diameter bar being −95°C. Ductility and especially toughness can decrease with slower cooling rates following annealing, and so lower values are to be expected from bigger sections. The extent of loss depends upon the alloy involved. For details Reference 6 should be consulted.

Methods of Fabrication

All the standard stainless steels may be hot forged by hammering, pressing, drop stamping or extrusion, without difficulty although they are 'stiffer', especially the nickel-bearing types, than the low alloy and carbon steels. Working ranges of 1 100–900°C and 1 200–900°C are common for the iron-chromium and iron-chromium-nickel types, respectively. It is normal to heat treat after forging in order to develop optimum mechanical properties and, in some instances, corrosion resistance. With the martensitic types, normalising and tempering (air cooling from the austenitic range followed by reheating to some temperature below the austenitic range), annealing (furnace cooling from the austenitic range) or simply tempering, may be adopted. For the ferritic steels, heating at 750–800°C and air cooling is common, while for the austenitics, heating to 1 000–1 100°C followed by air cooling or quenching depending on steel and section, is normally carried out. Duplex steels are usually heated to 1 050–1 100°C and then cooled rapidly in water to minimise sigma-phase formation, although small sections can be cooled in air. In larger sections, severe thermal shock must be avoided, both when heating and cooling with the ferritic types and with the martensitic types in the hardened condition, or cracking may occur. The austenitic types are very crack resistant but distortion can occur and internal stresses be generated if severe thermal gradients are caused.

Shaping of sheets, bars, plates, tubes, etc. may be carried out hot or cold in most instances. The ductility of all the steels in the softened state is good in the cold, and it is normally only necessary to heat if insufficient power is available to allow cold forming. As low a temperature as possible is normally adopted to minimise oxidation, but the metallurgical considerations already noted or to be discussed later must be borne in mind when selecting the temperature. While ductility is high in all cases, the rates of work hardening of the softened martensitics and ferritics on the one hand, and the austenitics on the other are quite different. With the former it is low and so these steels are ideal for cold heading and deep drawing, but can only be stretch formed to a moderate degree. The austenitics work harden rapidly and while this is a drawback for some forming methods, it gives the steels a very high uniform elongation value in tension and they are admirably suited to stretch forming.

All the steels can be joined by mechanical methods, but the possibility of and the consequences of crevice corrosion must be considered. The various methods can be utilised successfully in many circumstances, although hot rivetting should not be adopted as oxidation between surfaces makes crevice corrosion very probable. Welding by all the normal processes is feasible with the austenitic types, which, being non-hardening and tough, require few precautions to prevent weld cracking. Flame welding using hydrocarbons is not ideal because of the possibility of carburisation. The 'heat treatment' caused locally by the welding heat can reduce corrosion resistance (i.e. lead to 'weld decay') in some circumstances, but this phenomenon is well understood and the standard grades given in Table 3.12 are either very resistant or virtually immune. The duplex steels are also readily weldable and, provided there is suitable selection of composition of material to be welded and of consumables, there is no need of post-welding heat treatment. The martensitic steels, being air hardening, require special care with pre- and post-welding heating to avoid weld or heat-affected zone cracking, and the problem is greater with higher carbon levels. The precipitation hardening types are much easier to weld. Ferritic steels are also liable to cracking except in thin section.

The need for heat treatment after forming or welding is a complex topic and can only be mentioned briefly here. Generally speaking, however, unless the forming operation has been very severe it is not necessary to heat treat to restore mechanical properties of austenitic types, although in special cases it may be advisable to do so to relieve stresses. With the martensitic steels it is imperative to heat treat after welding to soften the weld and with some of the ferritics to restore corrosion resistance.

All the stainless steels can be machined in the softened states, but they may present some problems unless the correct techniques are adopted. This is especially so with the austenitic grades where the extreme ductility minimises chip breaking and the work hardening may cause difficulties unless modest cuts are made. The 'free-cutting grades' (those with high sulphur contents or selenium additions) are much easier to machine, but it must be remembered that they have somewhat reduced corrosion resistance, ductility and weldability compared to their normal counterparts. Detailed machining instructions are readily available from steel suppliers.

Corrosion Behaviour

Theoretical Considerations

The useful corrosion resistance of a stainless steel is due almost entirely to the fact that it exhibits passivity in a wide range of environments. The nature of passivity has been discussed extensively in Sections 1.4 and 1.5 and so will not be considered in detail here. The working hypothesis adopted by many concerned with stainless steel is that the passive film is essentially an oxide, and this has proved a reasonable assumption for practical purposes.

The general form of the anodic polarisation curve of the stainless steels in acid solutions as determined potentiostatically or potentiodynamically[8] is shown in Fig. 3.14, curve *ABCDE*. If the cathodic curve of the system *PQ* intersects this curve at *P* between *B* and *C* only, the steel is passive and the film should heal even if damaged. This, then, represents a condition in which the steel can be used with safety. If, however, the cathodic curve *P'Q'* also intersects *ED* the passivity is unstable and any break in the film would lead to rapid metal solution, since the potential is now in the active region and the intersection at *Q'* gives the stable corrosion potential and corrosion current.

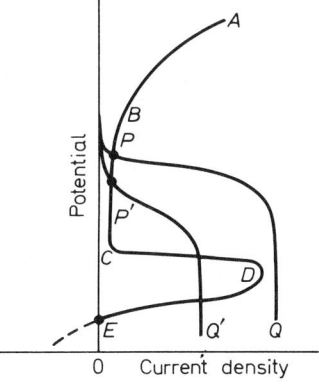

Fig. 3.14 Idealised form of a potentiostatic anodic polarisation curve *ABCDE* for stainless steels as determined in sulphuric acid solution. *PQ* and *P'Q'* are two cathodic polarisation curves that lead to passivity and corrosion, respectively

Reducing acids or acid mixtures tend to give cathodic curves at low potentials, thus favouring corrosion, but oxidising solutions give curves of high potential and so favour passivity unless the curve is at such a high level that it intersects the anodic curve at BA, the region of transpassivity where the film is rendered unstable by oxidation. Besides providing an insight into the passivity breakdown mechanism, the determination of curves of this type provides a convenient method for studying the effects of steel composition, heat treatment and microstructure on corrosion resistance. The actual values of points A, B, C and D depend, of course, on the solution and temperature used, but standardisation of test technique allows useful comparisons to be made. The relative effects of some of the alloying elements utilised may be gauged from the figures for several standard steels shown in Table 3.19. Increasing chromium content depresses potential C, therefore usefully extending the range of stable passivity, but it also increases the critical current at point D so that higher corrosion rates are obtained in the absence of passivity. Nickel also depresses potential C but markedly reduces the critical current. Molybdenum also has beneficial effects on both breakdown potential and critical current. Other alloying elements also have beneficial effects (while some are harmful), but, with the exception of copper and silicon which are utilised in some proprietary alloys, these are the only alloying elements which are used with the objective of improving the basic corrosion resistance of the steel to acids.

Table 3.19 Some critical values from anodic polarisation curves determined potentiodynamically in 20% sulphuric acid at 27°C (see Fig. 3.13)

Steel	Potential* C (V)	Potential* E (V)	Critical current density D (mA/cm^2)
410S21	+0·17	−0·42	50
431S29	−0·02	−0·42	12
430S17	−0·20	−0·55	100
434S17	−0·25	−0·55	15
304S15	−0·25	−0·37	0·15
316S31	No breakdown under test conditions used		

*Versus S.C.E.

In near neutral or alkaline salt solutions, anodic polarisation curves of the type shown in Fig. 3.15 are obtained in which the active loop CDE of Fig. 3.14 is no longer apparent. At high potentials curve $B'A'$ will represent oxygen evolution, but at low potentials (curve BA) in the presence of halide ions it will be indicative of pitting corrosion (Section 1.6). This form of attack represents a considerable hazard for some applications. The depassivation potential, often referred to as the breakthrough potential or the pitting potential, is not an absolute value for a given steel. It can vary as a result of differing test procedures and also according to the electrolyte involved.

The aggressivity of halides varies, with bromide and chloride being most aggressive. Increasing concentration of the halide also depresses the pitting potential as demonstrated for two steels in Fig. 3.16. Certain ions in solution act as inhibitors (e.g. nitrate) raising the pitting potential while others depress it (e.g. sulphide). Temperature and pH also have effects as illustrated

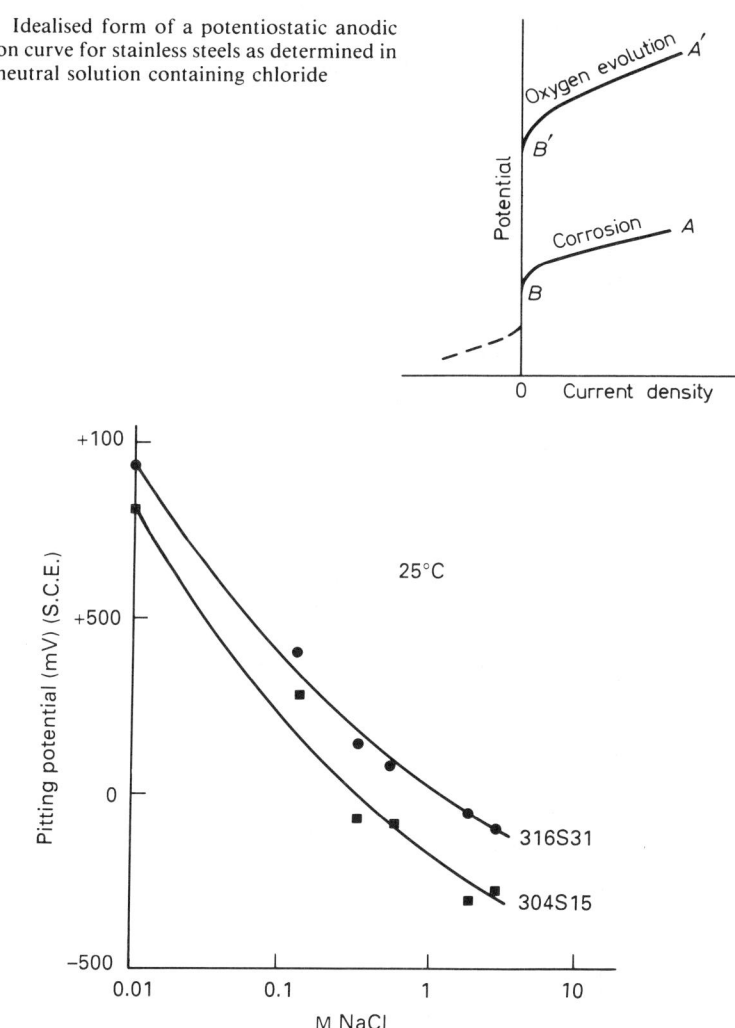

Fig. 3.15 Idealised form of a potentiostatic anodic polarisation curve for stainless steels as determined in a neutral solution containing chloride

Fig. 3.16 Pitting potential versus NaCl content

in Figs. 3.17 and 3.18. In practice, without polarisation from an external source, a steel will corrode if it attains a potential above the critical value relevant to the solution with which it is in contact. Thus the nature and amount of any oxidising agents (of which dissolved oxygen is the most common) is important. Thus halide salts that are also oxidants (e.g. ferric chloride, cupric chloride) are especially dangerous. Obviously anodic protection is not possible in pitting solutions, but cathodic protection is.

While the pitting potential of a steel is not an absolute value, provided that test conditions are strictly standardised, it can be used to compare relative resistances to pitting corrosion initiation of steels and to determine the effects

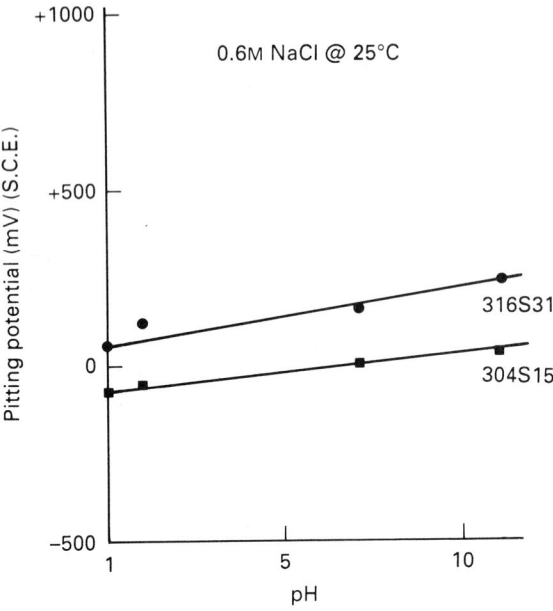

Fig. 3.17 Pitting potential versus pH

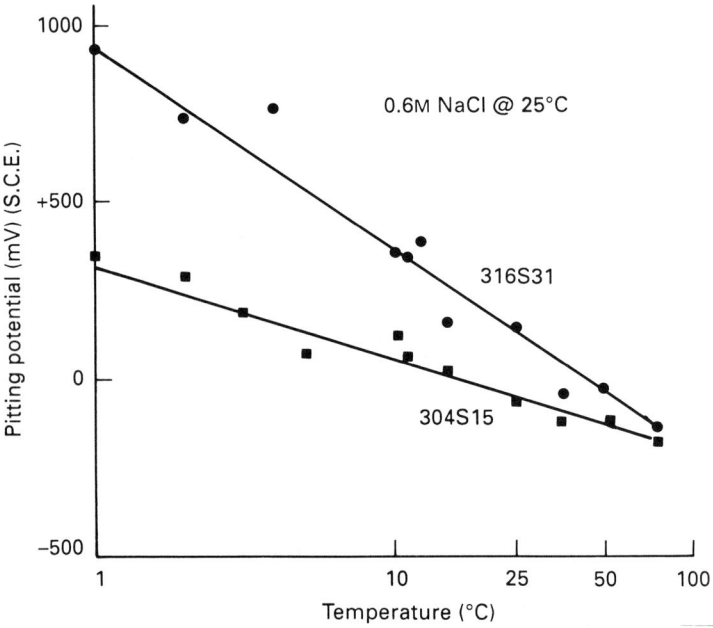

Fig. 3.18 Pitting potential versus temperature

of different types of alloying or of varying heat treatment. This procedure is now widely used as is that of critical pitting temperature determination. In the latter approach a sample is maintained at a constant potential in a standard solution while the temperature is increased in a step-wise manner until pitting commences (anodic current increases).

The effects of a number of alloying elements on pitting resistance can be seen from the pitting potentials of various standard, proprietary and experimental steels given in Table 3.20. The test technique was a potentiodynamic

Table 3.20 Pitting potentials for various stainless steels[a]

B.S. Number (where applicable)	Cr	Mn	Ni	Mo	N	S	Others	Pitting potential (mV) (S.C.E)
410S21	13·2							−200
416S21	13·3	1·5		0·40		0·27		−300
431S29	16·6		2·50					−100
430S17	16·0							−100
304S16	18·8		10·0					−70
303S31	17·9	1·2	9·7	0·40		0·25		−220
315S16	18·65		9·62	1·25				0
316S33	17·5		11·4	2·7				140
317S16	17·5		13·1	4·1				420
321S31	18·25		9·0				0·43 Ti	−45
347S31	18·28		9·95				0·77 Nb	−80
Precipitation hardening martensitic	13·7		5·5	1·63			1·67 Cu, 0·29 Nb	−80
Solution hardened austenitic	21·7	3·9	9·3	2·6	0·40		0·30 Nb	>850
	28·2		1·2		0·12			210
'Super ferritic'	26·3		2·2	3·1				>850
'Super austenitic'	20·1		18·2	6·1	·20		Cu ·8	>850
Duplex	22·0		5·4	2·9	·19			850
Experimental	22·6	3·8	20·0				Nb ·3	−30
Experimental	22·5	4·2	20·4		·39		Nb ·3	50
Experimental	22·1	4·2	20·3	1·0			Nb ·3	−10
Experimental	22·7	3·8	19·9	1·0	·41		Nb ·3	200
Experimental	22·8	4·3	20·0	2·6			Nb ·3	125
Experimental	22·3	4·3	20·4	2·7	·31		Nb ·3	>850

[a] Determined potentiodynamically in 0·6 M NaCl + 0·1 M NaHCO$_3$ at 25°C (part B in Fig. 3.15)

one using a solution of 0.6 M sodium chloride and 0.1 M sodium bicarbonate at 25°C. The pitting potential was that to give a current of 10^{-5} A/cm^2. The beneficial effects of chromium, molybdenum and nitrogen are clearly apparent. Not so obvious from these data are the effects of nickel which is slightly beneficial, and manganese which is somewhat detrimental. The markedly adverse effect of substantial sulphur additions (normal range as impurity is 0.005–0.02%) is clearly shown. Other elements also have beneficial or detrimental effects, but only chromium, molybdenum and nitrogen are used extensively to promote resistance to localised corrosion. It is becoming increasingly common to describe the pitting corrosion resistance of stainless steels in terms of these elements using formulae, of which the following is, perhaps, the most popular:

% Cr + 3.3 × % Mo + 16 × % N

There clearly must be effects from other elements present in the steel incidentally or functionally, but a reasonable relationship between this factor and the measured pitting potential is shown for a large number of steels in Fig. 3.19. Of interest, but not yet explained, is the shape of the curve as discussed elsewhere[10].

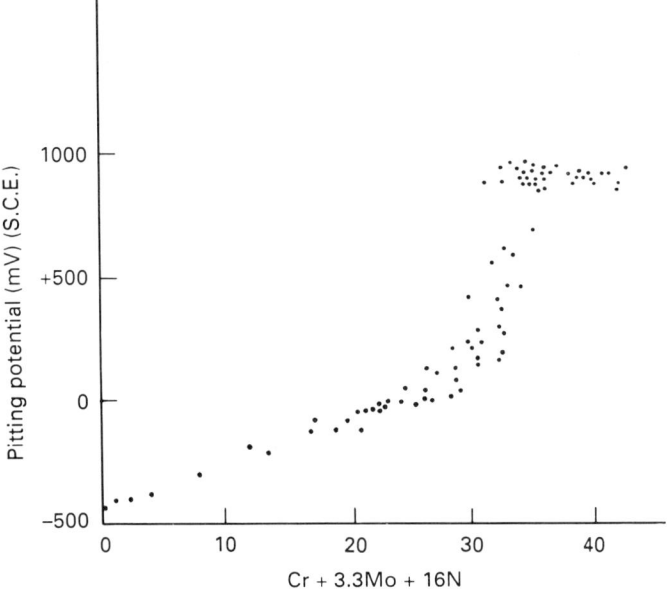

Fig. 3.19 Pitting potential (0·6M NaCl at 25°C) versus a composition factor. Cr, 0·021-28·5; Mo, 0·01-4·23; N, 0·01-0·455; Ni, 0·22-20·39; Mn, 0·22-4·53; Si, 0·13-0·71

Metallurgical Considerations and Forms of Corrosion

The corrosion resistance of any of the stainless steels is at its best when the material is single phase and in a homogeneous state. In acid solutions corrosion is then usually essentially uniform in nature except for grain orientation effects and possibly etch pitting. In very strongly oxidising solutions (i.e. conditions of transpassivity) however, apparently homogeneous austenitic steels may exhibit intercrystalline corrosion, which is believed to be associated with segregation of certain interstitial elements at grain boundaries, but is of little practical importance since stainless steels are not suitable for use in such solutions. Pitting or localised corrosion is rare in correctly treated steels in acid solutions, but can be obtained in some circumstances. In near neutral or alkaline halide-bearing solutions, however, any corrosion is by pitting and the consequences may be serious. The mechanism of pitting has been discussed in Sections 1.5 and 1.6 and will not be considered further here.

In practice, of course, few alloys are homogeneous in composition, the segregation produced during solidification being reduced but not eliminated during subsequent working and heat treatment. This is the case with stainless steels, and while microsegregation of the major alloying elements does not usually lead to any significant effect on overall corrosion resistance, some differential corrosion rates and etching effects of various surfaces can be noted under environmental conditions causing significant corrosion.

The presence of second phases also implies partitioning of the elements with consequent alloying element segregation. In some martensitic steels there can be retained austenite following cooling from the hardening temperature while conversely, with some austenitic steels, there can be some martensite present in the softened state or, more commonly, induced by cold deformation. Since the austenite–martensite transformation is rapid and occurs at low temperatures, there is little chance of alloying element segregation while the crystallographic state itself has little effect on corrosion resistance. Thus the presence of austenite in martensite and vice versa has little practical effect, although substantial acid attack can reveal a difference. When austenite is formed in a martensitic steel from prolonged heating below the AC_3 temperature it does have a different composition from the now tempered martensite matrix. This situation can apply with the precipitation hardening grades when overaged, but the practical effect on corrosion resistance is very small. Austenite may form in some, but not all, ferrite steels (depending upon composition) during high-temperature heat treatment. In the case of the less highly alloyed types (e.g. 430S17) this transforms to martensite on cooling, while with the more highly alloyed types it remains as austenite. In both cases there is reversion to ferrite and homogenisation with the correct heat treatment but even if such treatment is omitted, the effect on corrosion resistance is small.

The most likely second phase in martensitic and in austenitic steels is delta ferrite although, usually because of other considerations, care is taken to balance the composition so that it is avoided. In the cases of castings and weld metal, however, a small amount of ferrite is usually present in the interests of soundness. The ferrite phase is higher in chromium and molybdenum but lower in nickel, nitrogen and copper (for instance) than the austenitic phase but the differentials between the ferrite and the martensite or austenite in the near equilibrium (i.e. correctly heat-treated) state are not great, so that both the phases in a two-phase structure have quite similar corrosion resistances. There are differences, however, and the phases can be revealed on etching, while in certain media there can be selective attack in practice. There is little practical hazard except in the case of weldments in the as-deposited state. The ferrite content at very high temperature is greater than after normal heat treatment and the rapid cooling following welding can lead to retention of an increased amount of ferrite. This leads to more marked element partitioning with a more noticeable effect an corrosion resistance, especially if the ferrite content is such as to form continuous paths in the matrix. This danger is avoided by control of composition or by post-weld heat treatment. However, as-deposited weld metal proves perfectly satisfactory for many applications.

In duplex steels, ferrite is a major intentional constituent. The features noted above are relevant; the two phases do have differing analyses but not to such an extent as to cause serious corrosion problems. A further structural feature, theoretically possible in other grades but usually only of any consequence in the duplex steels, is the formation of sigma or chi phases. These are rich in chromium and molybdenum and form on heating in the 550–950°C temperature range. The upper temperature limit for their formation and the time–temperature relationship necessary to produce a significant amount depends on steel composition and for most steels there is no practical danger (unless prolonged periods at elevated temperature are expected in service). Chi and sigma formation can be much more rapid with the duplex steels and can occur during cooling following heat treatment. The steels more highly alloyed with chromium and molybdenum are most at risk and quenching after heat treatment is advisable. Even with quenching the problem can arise with larger sections. Both toughness and corrosion resistance can be affected. Sigma and chi phases themselves are very resistant to corrosion but their formation leads to depletion from the adjacent material of chromium and molybdenum (the phases form preferentially in ferrite). Thus the full corrosion resistance of the alloy is diminished. While due care and attention must be paid to this aspect, it does not normally represent a serious limitation with this group of steels.

Precipitation of Carbides

The most marked structural effect on corrosion resistance is that associated with chromium-bearing carbides, which can occur in all three groups of steels. The martensitic types of hypo-eutectic composition have essentially all carbon in solid solution in the as-transformed (i.e. the hardened) condition. Tempering reduces strength as shown in Fig. 3.11 and causes the precipitation of carbides; at lower temperatures the carbides are essentially of iron, but at higher temperatures they are chromium-rich. The diffusion rate of carbon which is an interstitial element is much greater than that of chromium, and as a result chromium gradients are set up adjacent to the growing carbide particles. If the carbide particle distribution is such that the chromium-reduced regions can overlap, a continuous low-chromium path can be formed which may lead to selective attack. Such attack can occur along prior austenite boundaries and also along martensite lath boundaries. The treatments giving susceptibility to such attack are indicated in Fig. 3.20; at low temperatures there is insufficient chromium in the carbides to have an effect, while at higher temperatures diffusion of chromium is rapid enough to prevent severe gradients, although the overall effective chromium content is reduced somewhat, as is the corrosion resistance as a result. Obviously the effects are more pronounced with higher carbon steels. As the effects are associated with a precipitation reaction, the time of tempering has an effect as well as temperature*. For discussion of the tempering processes see References 2 and 3.

*Martensitic stainless steels are usually used in the softened (tempered at or above 650°C) or in the fully hardened condition (tempered at or below 250°C) so that there is no substantial reduction in corrosion resistance resulting from carbide precipitation. However, the hard soldering of knife blades can result in carbide precipitation and pitting of the blade at the area adjacent to the handle, and care must be taken in the soldering process to avoid this danger.

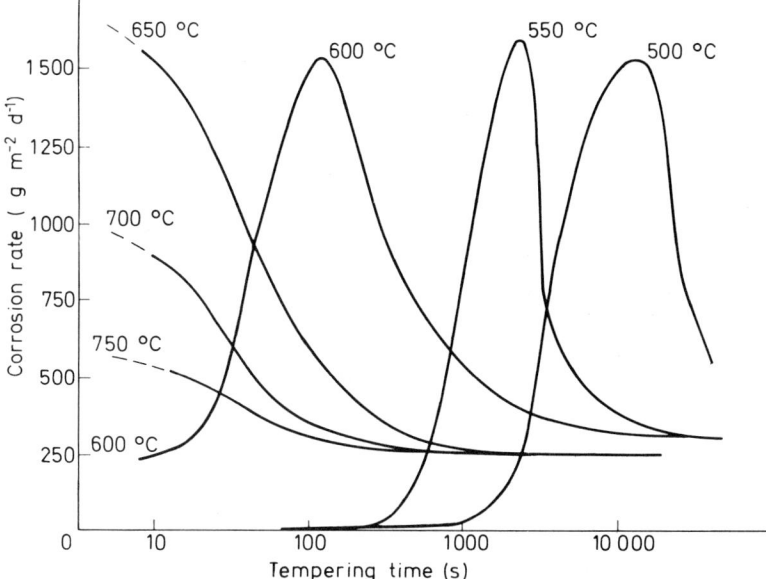

Fig. 3.20 Effect of tempering treatment on corrosion resistance of 420S45 (air cooled, 980°C). Corrosion tests in 10% nitric acid solution at 20°C

Austenitic steels of the 304S15 type are normally heat treated at 1 050°C and cooled at a fairly rapid rate to remove the effects of cold or hot working, and in this state much of the carbon is in supersaturated solid solution. Reheating to temperatures below the solution treatment temperature leads to the formation of chromium-rich $M_{23}C_6$ precipitates predominantly at the grain boundaries with the production of chromium gradients and reduced corrosion resistance as is the case with the martensitic steels. Any attack is

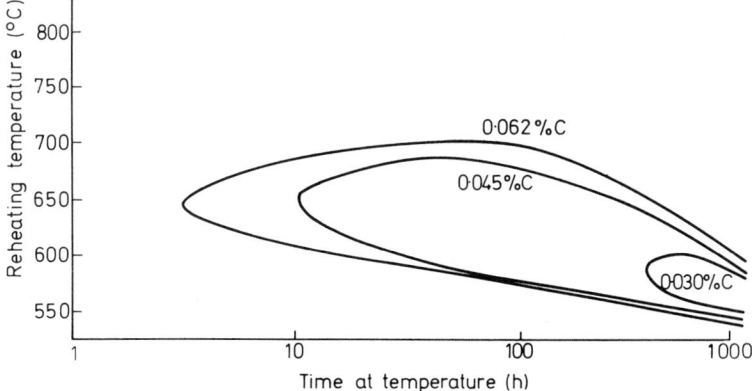

Fig. 3.21 Temperature-time-sensitisation diagrams for three austenitic Cr–Ni steels solution treated at 1 050°C. The curves enclose the treatments causing susceptibility to intercrystalline corrosion in a boiling $CuSO_4 + H_2SO_4$ test reagent

intergranular in this instance. The effect of temperature and the time of reheating on establishing whether steels are susceptible to this form of attack or not, is shown in Fig. 3.21. The corrosive medium used was that specified in ASTM A262E and it should be noted that the precise position and area of the curves are relevant to this test medium only, although the general form is similar for other test media. Much has been written regarding the relative merits of various accelerated tests, but it is generally accepted in this country that the copper sulphate-sulphuric test gives an adequate guide to steel condition for most purposes. For other test methods ASTM 262 should be consulted. It should also be noted that not all corrosive media produce intergranular attack on 'susceptible' steels.

The metallurgical condition leading to susceptibility to intercrystalline corrosion can be caused by cooling from the solution treatment temperature at a sufficiently slow rate as well as by isothermal treatment. Large sections in certain grades may thus require quenching in oil or water to ensure optimum corrosion resistance. The heating introduced by welding can also produce the undesirable metallurgical condition in bands a short distance on each side of the weld. This may lead to the localised attack known as *weld decay* (see Section 9.5).

Since the susceptibility to intercrystalline attack is due to carbide formation it follows that low carbon steels are more resistant to the phenomenon, as illustrated in Fig. 3.21. Thus the alloy with a carbon content suitable for the thermal treatment to be encountered in fabrication should be selected. An alternative method for ensuring resistance to intercrystalline corrosion is to 'stabilise' the carbon, that is to restrict its solubility at heat treatment temperature by incorporating a strong carbide-forming element in the steel, and titanium or niobium are utilised for this purpose (321S31 and 347S31). Whether low-carbon grades or stabilised grades are selected for a given application where intercrystalline corrosion may be a hazard depends on a variety of factors. The stabilised grades are especially useful where service may involve periodic heating into the critical temperature range followed by contact with some aqueous corrodent. As shown in Fig. 3.21, even the low carbon grades can be sensitised in time. Low carbon grades may be preferred where welding is followed by heating into the critical range for a relatively short time for the purpose of stress relief. The zone immediately adjacent to the weld is subjected to a very high temperature in welding which causes solution of the titanium or niobium carbides so that chromium-rich carbides may form on reheating. Thus a narrow band immediately adjacent to the weld may corrode in service. This is known as *knife-line* attack to distinguish it from *weld decay* where the corrosion zones are some slight distance ($\simeq 2$ mm) from the weld.

The ferritic steels may also undergo intercrystalline corrosion as a result of grain boundary carbide formation. In the normal softened state (treated $\simeq 800°C$) the carbon is largely precipitated and the ferrite composition homogenised so that further heating at lower temperatures has no adverse effect. During solution treatment above $950°C$, however, carbon is redissolved. Sensitisation can then occur at lower temperatures but the rate is so rapid that it can only be suppressed by very rapid cooling which is not practically feasible. Thus weld decay is very possible in service unless a remedial heat treatment (∿800°) is applied. Unlike the austenitics, the weld decay band is the higher temperature zone immediately adjacent to the fusion zone and, of course, the fusion zone itself if matching filler is used. This rapid sensitisation behaviour is due to the lower solubility of carbon and higher difussion rates in ferrite.

T.T.S curves as those in Fig. 3.21 can be developed for material cooled very rapidly following solution treatment but the 'C' curve range is ∿400–550°C and the 'nose' is at very short times. Freedom from sensitisation in welding can be obtained by ensuring extremely low carbon (and nitrogen) but such levels are not commercially feasible. Stabilisation by niobium and titanium is feasible, but higher ratios are needed than with austenitic steels. With most of the 'super ferritic' group a combination of a practical low carbon level and titanium addition is used.

Precipitation of Nitrides

Nitrogen can dissolve at elevated temperatures and precipitate as chromium nitrids at lower temperatures with similar effects to those described in the previous section, although there are some differences in detail.

Nitrogen is not an alloying addition with most of the martensitic group but is present, usually at about 0.02%, as an incidental. It can be a deliberate addition (up to 0.05%) with some specials and higher values are being proposed for some creep-resisting steels. Similar effects as with carbon may be anticipated but adverse effects in corrosion behaviour should not arise from good heat treatment practice avoiding tempering between 350 and 650°C.

Nitrogen is also present as an incidental in austenitic steels, usually at about 0.03%, but the only practical consequence is that allowance for its presence should be made when calculating the amount of titanium added for stabilisation. Nitrogen is being used to an increasing extent as a deliberate addition because of the improvement in strength[12] and resistance to some forms of corrosion. Modest additions to the 18/10 types actually retard carbide sensitisation and at 0.2% nitrogen the T.T.S. curves for low-carbon steels are very similar to those for low-nitrogen equivalents. Further alloying with nitrogen is possible by melting and freezing under increased pressure or, for modified steel analyses[14], at atmospheric pressure. At these higher nitrogen levels sensitation due to C_2N formation is possible under adverse heat treatment conditions. This aspect is discussed in Reference 13.

In the ferritic steels the effects of nitrogen and carbon are indistinguishable one from the other and the normal incidental level is sufficient to cause weld decay susceptibility. Thus in the 'super ferritic' group both carbon and nitrogen are controlled to a low practicable level and sufficient titanium is added to stabilise both elements.

Stress-accelerating Effects

It is worthy of note that the simultaneous presence of a sustained tensile stress and a corrodent with a sensitised steel may result in rapid cracking rather than relatively slow, general intergranular penetration. Perhaps stress-accelerated intercrystalline corrosion in a better term for this than stress-corrosion cracking. In its worst manifestations, the presence of stress may cause failure of steel at a degree of sensitisation that would have been insufficient to give problems in the environment without stress.

Corrosion in Natural Environments

Atmospheric Corrosion

One of the major assets of stainless steels has proved to be their resistance to discoloration in the atmosphere and even the least corrosion-resistant alloys have been used indoors with success. Knives, for instance, must be made from the 13% chromium martensitic types to give the required hardness yet they retain their bright attractive appearance without special cleaning other than ordinary washing. Many other domestic and kitchen articles have been made from the simple ferritic and austenitic steels with similar success. Outdoor service is more arduous, however, and the martensitic steels are not used where appearance is important. The ferritic steels 430S17 and 434S17 have been and are used for motor car trim (the extent varying according to fashion) and retain their appearence well. In one way this is surprising as simple exposure tests out of doors show loss of reflectivity, especially with 430S17. It is thus presumed that thin films of wax and grease applied in the normal coarse of operation and cleaning have a beneficial effect.

The mechanism of atmospheric corrosion of stainless steel has not been widely explored. Good service indoors, even of the lesser alloyed grades, may be attributed at least in part to the predominantly low relative humidity in heated buildings, but short time tests indoors in chambers giving high humidity have shown at least delayed initiation relative to outdoor exposure. Since the gaseous composition of air indoors must be similar to that outdoors it may be presumed that the species causing the difference are transported to the surface as airborne solids, as airborne free liquid droplets or in rain. The last of these is improbable since samples sheltered from the rain outdoors invariably corrode faster than samples exposed alongside but not sheltered.

Exposure tests have shown that steel corrodes more rapidly at coastal and industrial sites than at rural, inland sites. Thus, it is likely that airborne chlorides and sulphurous gases are major causes; some laboratory tests have confirmed the adverse effects of both these acting separately and together. From such testing there is also evidence that deposited carbon is detrimental. Any corrosion is in the form of very fine pitting, the degree being reflected in pit density rather than size. The pits cannot usually be resolved with the naked eye, and any degradation is perceived as a loss in reflectivity or a 'misting' of the surface. Unless the steel is cleaned regularly, a more immediate visual effect on a corroding surface is rust staining.

The effects of some alloying elements on relative behaviour in an industrial atmosphere (Sheffield, U.K.) are shown in Table 3.21A. For comparison, data for simultaneous tests on carbon steel and some non-ferrous material are given. Results are as weight loss over a five-year period and data from two series are given, these being at the same site but with 24 years separating them. The wide differences between results in the two series are attributed to changes in degree and possibly type of atmospheric pollution. As with pitting under immersed conditions, a markedly beneficial influence of molybdenum is obvious, and this is further demonstrated by the more recent data of Table 3.21B in which testing was for 18 years at a 'heavy industrial' site and assessment was by pit density and pit depth measurements. Table 3.21C shows the effect of geographical location, the sites being classified as 'severe marine, heavy industrial, semi-industrial and rural'[15]. Other prolonged tests on various steels at various sites have been described elsewhere. In these tests assessment was by appearance[16]. In all the test programmes it was found that smooth surfaces give better results than rough ones and that the quality of abrasive used can be of significance for abraded surfaces.

Table 3.21A Atmospheric exposure test, Sheffield, 5 years. Results: loss weight (g/m^2); means of multiple samples

Material	1982/1943 series	1962/1967 series
Carbon steel	3 700	
410S21	270	
430S17	135	45
434S17	—	15·5
18Cr–8Mn	165	
12Cr–12Ni	160	
304S15	85	4·5
25Cr–20Ni	80	
321S31	70	
30% Cr	70	
316S31	1	·4
Al	165	
Cu	185	
Phosphor	200	
Zn	385	
54 Cu–44Ni	440	

Table 3.21B Effect of molybdenum content on atmospheric corrosion. Heavy industrial site. 18 years

Material	Mo content (%)	Pit density (number/cm^2)	Pit depth (μm)
304S15	0·31	3 870	81
315S16	1·44	1 000	52
316S33	2·70	625	35·5
317S16	3·45	290	17·5

Table 3.21C Effect of environment type on atmospheric corrosion. 18 years

Site	Pit density (number/cm^2)		Pit depth (μm)	
	304S15	316S33	304S15	316S33
Rural	2030	225	20	17·5
Semi-industrial	3030	420	21	18
Heavy industrial	3870	625	81	35·5
Marine	3160	355	85	24

Where retention of appearance is of prime importance, the molybdenum-bearing steels are used almost exclusively in the U.K. Different opinions have been expressed as to whether regular cleaning is necessary to obtain the best results. Data have now become available which show that the absence of such clearing is of no detriment far fully exposed steel (although clearing obviously removes any staining) but can be of value where steel is sheltered from rain. In Table 3.22A are given results from a series of tests in Sheffield, U.K. (1962–1968), the values being total area of pitting as a percentage as approved to number of fits. Table 3.22B contains results for 316S33 steel exposed to a marine atmosphere either sheltered or unsheltered.[15]

Table 3.22A Effect of periodic working on the behaviour of several stainless steels exposed to an industrial atmosphere (Sheffield, 1962–1968)

Frequency of washing	Area corroded (%)			
	430S17	434S17	304S16	316S11
Weekly	23·8	3·66	3·60	<·01
Monthly	27·9	4·20	4·19	<·01
Quarterly	47·3	4·74	4·59	<·01
Not washed	26·5	3·60	2·0	<·01

Table 3.22B Effect of periodic washing (every six months) and shelter from rain on 316S33 samples exposed to a marine atmosphere (Reference 15)

	Number of pits/cm^2	Pit depth (μm)
Fully exposed	355	23
Fully exposed and washed	355	24
Sheltered	452	71
Sheltered and washed	420	46

Corrosion in Natural Waters

One of the main factors which establishes the corrosivity of water to stainless steel is the chloride content. Also significant are oxygen content and pH, and it is also probable that other features such as hardness and the nature and concentration of other anions and cations have effects. Water temperature and flow velocity can also be important. Any corrosion takes the form of pitting or, if crevices are available, larger areas of attack within the crevice.

The 13% chromium, martensitic steels are unsuitable for use in sea-water. Type 431S29, some precipation hardening martensitics, the 304 group and, especially, the 316 group have proved successful for many applications, but it must be appreciated that presence or absence of corrosion with these and with other grades in sea-water depends on the details of use. Generally speaking, the greater the chromium, molybdenum and nitrogen content, as discussed earlier, the better the chances of success in service.

Of great significance is the velocity of the sea-water, and the initiation and propagation of pits is more likely under stagnant conditions. Obviously stagnant conditions will apply permanently within crevices to which seaater has access whether these be of an engineering nature, due to the growth of marine organisms, or caused by deposits of sand or silt. Within such crevices, the pH is reduced by hydrolysis of metal ions introduced by diffusion through the passive film so reducing the potential for the onset of pitting[17]. With flowing seawater, marine organisms will not grow and deposits are less likely.

Thus it will be seen that design and the way in which an artefact is used will largely influence which steel may prove successful. Simple precautions such as the flushing out of pumps and pipework with fresh water after use can mean the difference between satisfactory behaviour and limited life. The test data in Table 3.23[18] illustrate the marked effect of water velocity. Those applications where prolonged contact with static sea-water cannot be avoided have provided some of the impetus for the development of the more highly alloyed austenitic, ferritic and duplex steels which are finding increasing favour for arduous sea-water service.

Table 3.23 Effect of seawater velocity on pitting corrosion of 316S31. Tests for 3½ years (Reference 18)

	Velocity 1·2 m/s	*Velocity zero*
Number of pits	0	87
Maximum pit depth (mm)	0	1·98
Average pit depth (mm)	0	0·96

Contact of brass, bronze, copper or the more resistant stainless steels with the 13% Cr steels in sea-water can lead to accelerated corrosion of the latter. Galvanic contact effects on metals coupled to the austenitic types are only slight with brass, bronze and copper, but with cadmium, zinc, aluminium and magnesium alloys, insulation or protective measures are necessary to avoid serious attack on the non-ferrous material. Mild steel and the 13% chromium types are also liable to accelerated attack from contact with the chromium-nickel grades. The austenitic materials do not themselves suffer anodic attack in sea-water from contact with any of the usual materials of construction.

With river waters and other less saline waters the possibility of corrosion is reduced, but the same general principles apply.

Potable waters when cold naturally represent less of a hazard than many of the natural waters, but on the other hand they are often used at higher temperatures. The admirable service over many years of kitchen and catering utensils and cutlery show how resistant all the grades are, but prolonged contact at elevated temperatures may introduce special problems. For relatively high chloride waters (say >200 p.p.m.) only the more corrosion resistant types should be used for hot water. Heat transfer from steel to water can lead to corrosion at very high heat fluxes and also at lower heat fluxes with hard waters due to the concentration of chlorides underneath water scale deposits. Similar concentrations can occur in crevices in heat transfer surfaces, and can, in both instances, lead either to pitting, or, in extreme cases with the austenitic types, to stress-corrosion cracking.

Soil Corrosion

Stainless steels have not been widely used in applications where they are buried in soil, but some applications have involved underground service. Various stainless steels from the 13% Cr to the molybdenum-bearing austenitic types were included in the comprehensive series of tests in a variety of soils reported by Romanoff[19]. High-chloride poorly-aerated soils proved most aggressive, but even here the austenitic types proved superior to the other metals commonly used unprotected. Of special interest is the fact that though corrosion was by pitting there was little or no increase in pit depth after the first few years.

Corrosion in Chemical Environments

Acids

The value of stainless steels may be illustrated by consideration of some of the stronger acids.

It should be noted that the data presented were obtained under laboratory conditions in still solutions of pure acids. Only a selection of the steels covered by British Standards are included. Type 410S21 may be taken as typical of the alloyed 12–13% chromium martensitic group. The more highly alloyed ferrite types available as proprietary grades were not developed with resistance to acids in mind, but passivity can apply over a wide range of conditions, although corrosion rates in the absence of passivity are not usually attractive. The same comments are relevant to the duplex steels, but the more highly alloyed, proprietary austenitics offer both improved passivity and reduced kinetics, especially those grades with increased chromium and molybdenum contents.

Sulphuric acid Stable passivity is possible under a range of conditions for all the stainless types, the transition to the active state being favoured by an increase in acid concentration and temperature. With the 13% Cr types, passivity is only possible in very dilute solutions and corrosion in the active region is very rapid; the range of acid strengths over which steels are passive is extended somewhat with the higher chromium ferritics. The nickel, and in some instances, molybdenum content in the austenitic grades and in the precipitation-hardening martensitic types, not only enhances passivity but also retards corrosion in the active state so that in some instances materials can give a useful life while corroding. Copper is also beneficial. Titanium and niobium present in some grades do have a slightly beneficial effect, but are not added primarily for this reason, and for practical purposes types 321S31 and 347S31 may be considered similar to 304S15. Corrosion rates for various grades determined in static solutions, not deliberately aerated, at 20°C are shown in Fig. 3.22. The effect of acid temperature on the behaviour of the more resistant grades is illustrated in Fig. 3.23.

The presence of reducing agents or oxidising agents in the acid, even in small quantity, can reduce or extend the range of passivity by depressing or raising the cathodic curve in the diagram illustrated in Fig. 3.14. Practical

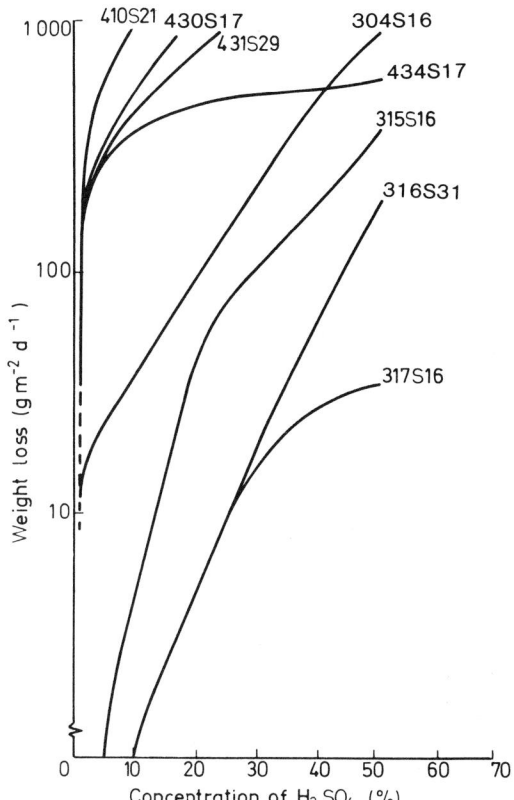

Fig. 3.22 Corrosion rates of various stainless steels in different concentrations of sulphuric acid at 20°C

use may be made of this by utilising small additions of an oxidant as a cathodic reactant. Copper sulphate or nitric acid are often used for this purpose; the marked effect of the latter is illustrated in Fig. 3.24.

Polarisation from an external source may also affect the range of passivity. Cathodic polarisation may depress the potential from the passive to the active region (see Fig. 3.14) and thus care should be taken to avoid contact with any other corroding metal. Anodic polarisation, on the other hand, can stabilise passivity provided that the potential is not increased into the range of transpassivity (see Fig. 3.14) and anodic protection is quite feasible.

The 'super austenitics' with high contents of nickel and molybdenum with some copper have much enhanced resistance to sulphuric acid solutions.

Hydrochloric acid Most of the general remarks in the last section dealing with sulphuric acid are also applicable to hydrochloric acid, but the chloride ion makes this acid a very aggressive one and the application of stainless

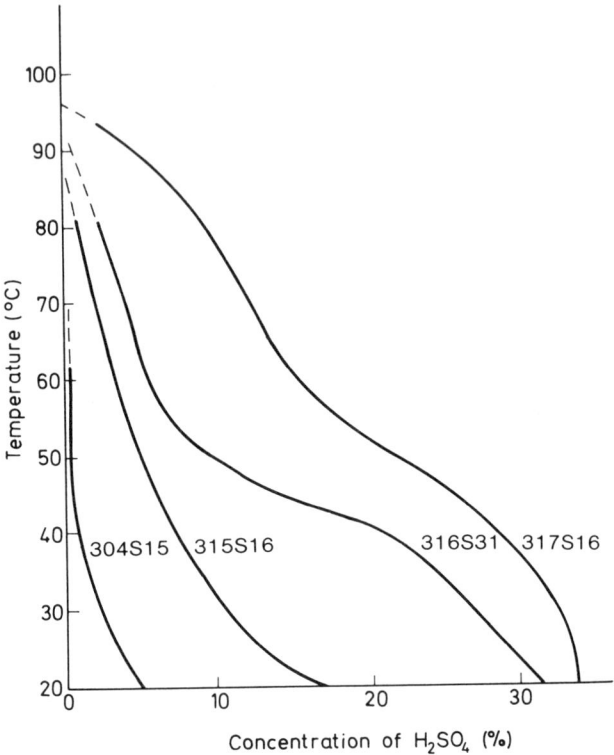

Fig. 3.23 Strength and temperature of sulphuric acid solutions and type of stainless steel to give a corrosion rate of $2 \cdot 5 \text{ g m}^{-2} \text{ d}^{-1}$

steels is confined to low concentrations. Corrosion behaviour is illustrated in Figs. 3.25 and 3.26. While passivity is possible with this acid as was noted for sulphuric acid, anodic protection either by impressed current or addition of an oxidant is not practicable and, indeed, could be harmful.

Nitric acid Of the alloying elements utilised, chromium and nickel are beneficial in reducing attack in nitric acid. The effect of acid strength on corrosion rates of three types of steels is shown in Fig. 3.27. At the 13% Cr level, enhanced attack occurs at both low and high acid strength but at 17% Cr the attack is less at lower acid strengths. Nitric acid is especially likely to produce more rapid attack at grain boundaries and thus for long term service, especially if welded plant is involved, the extra-low-carbon austenitic grades (<0·03% C) or the stabilised grades are preferred. Of the latter, the niobium-bearing variety withstands the more onerous conditions better than the titanium-bearing type. From laboratory and plant experience, the top limits of concentration tolerable appears to be those summarised in Table 3.24. It should be noted that some corrosion occurs even below these limits and some contamination of acid must be expected. High concentrations of nitric acid raise the potential of the steel into the transpassive region with a consequent increase in the corrosion rate.

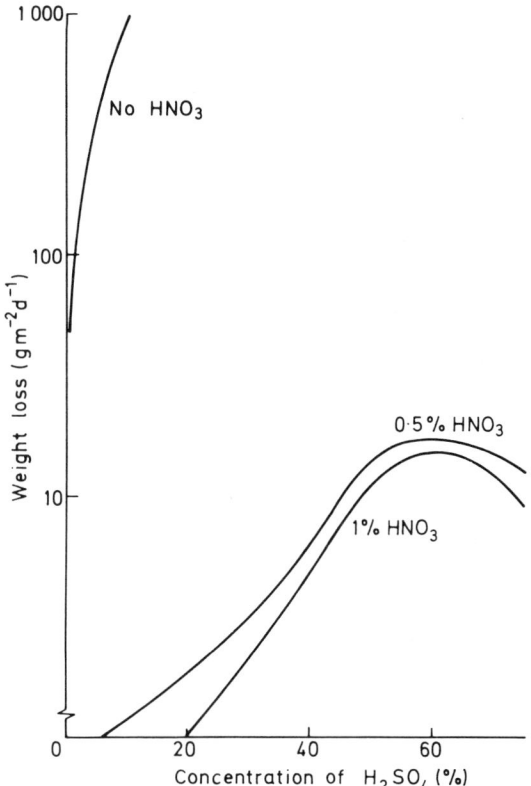

Fig. 3.24 Inhibiting effect of nitric acid on corrosion of 304S15 steel in sulphuric acid at 100°C

For special applications special silicon-bearing austenitic steels are produced by a few manufacturers. Silicon contents are 4–5·3% and there is a corresponding increase in nickel and very low (<0·015%) carbon. In the welded state these are giving good service in nitric acid of over 90% strength up to 75°C.

Phosphoric acid The austenitic grades are resistant to all strengths up to 80°C but are limited to 30–40% concentration at boiling point, the molybdenum-bearing types having the best corrosion resistance. Some test data for various types is shown in Fig. 3.28. Industrially, this acid is often encountered in an impure state with appreciable amounts of sulphuric and hydrofluoric acid present so that process testing is likely to be particularly necessary. The 'super-austenitic' steels have enhanced resistance to phosphoric acid solutions.

Acetic acid Acetic acid may be taken as being typical of the stronger organic acids. Considerable variability between tests is shown in the 'borderline' region at boiling point, and attack on plain austenitic types may rise to approximately 1 mm/y in the more concentrated solutions. 316S16 steel is

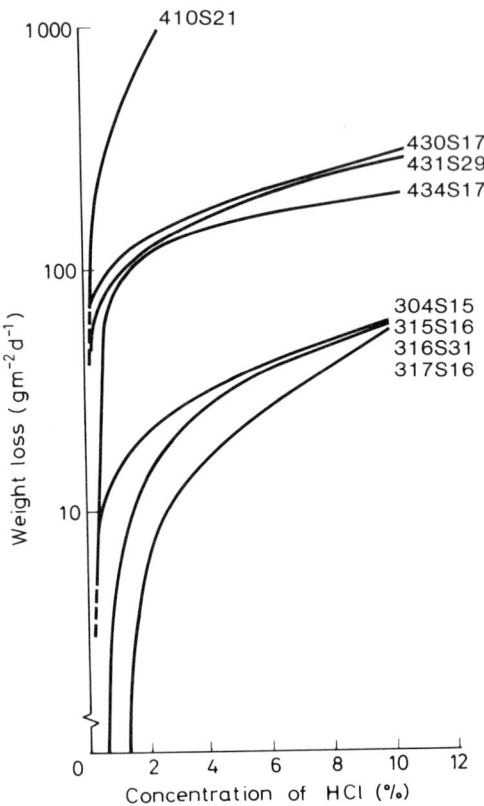

Fig. 3.25 Corrosion rates of various stainless steels in different concentrations of hydrochloric acid at 20°C

superior to non-molybdenum steels, with losses of up to 0·1 mm/y with boiling 80% acid.

With stainless steels optimum resistance is dependent on aeration, and due account must be taken of this in considering operating conditions where previous experience has been with non-ferrous alloys which require an absence of oxygen for full resistance. There is therefore no need to run off the early distillate from a batch to avoid impurity during initial de-aeration. On the other hand, complete exclusion of oxygen for the benefit of non-ferrous metals in a mixed plant have very serious consequences for the stainless steel, and this must be kept in mind at the design stage.

Other acids Formic acid is more liable to cause pitting than is acetic acid, and it must be considered with particular care at higher temperatures. Citric, oxalic, lactic and sebacic acids are also liable to cause appreciable attack in strong solutions at the higher temperatures, and for the borderline conditions molybdenum is a useful addition to the steel. On the other hand, many organic acids are less corrosive than acetic, and non-molybdenum austenitic steels may be used with them without trouble even at 100°C. Such acids

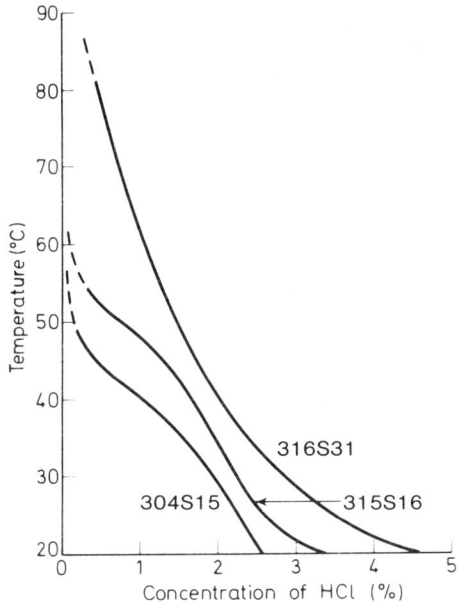

Fig. 3.26 Strength and temperature of hydrochloric acid solutions and type of stainless steel to give a corrosion rate of $2 \cdot 5 \text{ g m}^{-2} \text{ d}^{-1}$

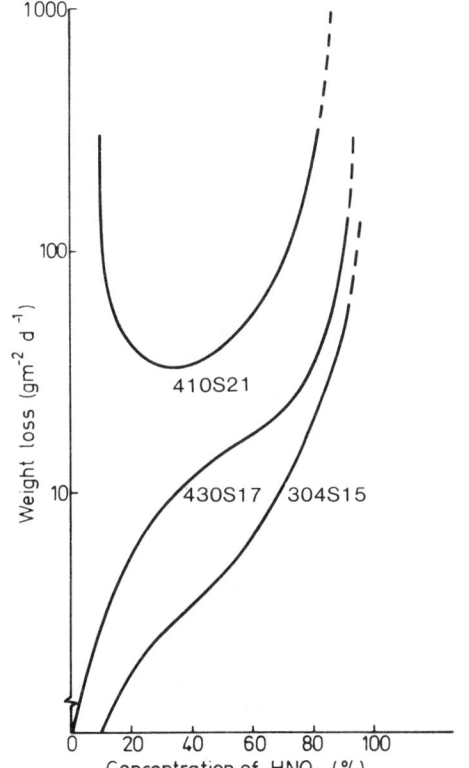

Fig. 3.27 Corrosion rates of various stainless steels in nitric acid solutions at boiling point

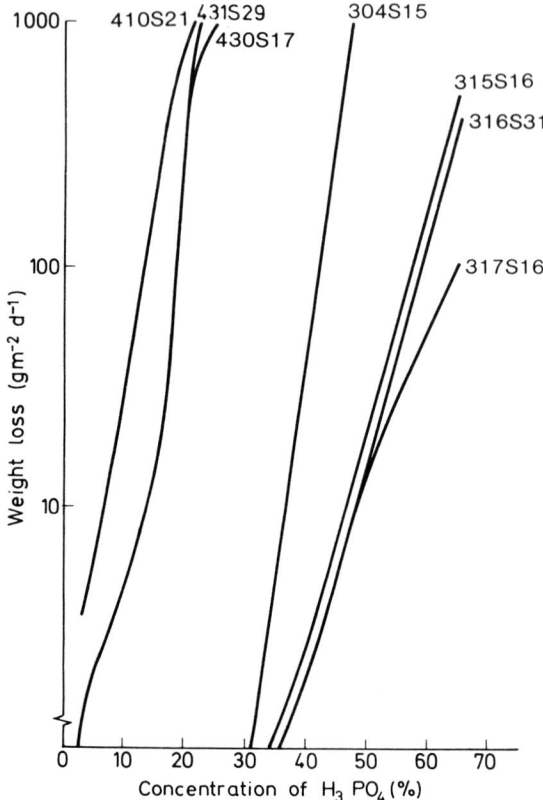

Fig. 3.28 Corrosion rates of various stainless steels in phosphoric acid solutions at boiling point

Table 3.24 Maximum nitric acid concentration that can be tolerated by austenitic stainless steels

Temperature (°C)	Stabilising element	Nitric acid concentration (%)
20°C	Ti	70
	Nb*	99
40°C	Ti	70
	Nb*	95
60°C	Ti	70
	Nb*	80
80°C	Ti	70
	Nb	70

*Or extra low carbon (<0·03%) grade.

include benzoic, carbolic, cresylic, malic, picric, pyrogallic and tartaric. The higher fatty acids such as oleic and stearic give no attack up to processing temperatures far above 100°C.

Of the inorganic acids, boric and sulphurous acids are fully resisted under many conditions, but the tendency of the latter to oxidise to sulphuric must be appreciated. This can lead to a dangerous sulphurous/sulphuric mixture so that the molybdenum-bearing austenitic steel may be required.

Hydrofluoric acid and hydrochloric acid have similar effects, and even the austenitic steels are only suitable for use with cold dilute solutions.

Alkalis

There is no attack by ammonia, and caustic alkalis can be used in contact with the austenitic types up to moderately high temperatures, but in strong solutions of caustic alkalis at temperatures approaching boiling point some attack will occur.

Salts

Resistance to solutions of most salts is good unless the salt can dissociate to give a critical concentration of the relevant acid. Exceptions are halides, of which bromides are potentially the most dangerous, closely followed by chlorides. Whether corrosion (usually of the pitting type) of a particular grade occurs depends upon the concentration of the salt, the temperature, the pH and the potential attained. Salts producing a high potential (e.g. ferric chloride, cupric chloride) are especially dangerous. Similarly, mixtures of salts can be dangerous for the same reason (e.g. sodium chloride plus copper sulphate). Resistance of the steel to such condition is dependent, in the main, on chromium, molybdenum and nitrogen content.

Halogens

In the presence of moisture all the halogens are corrosive to stainless steel, although dilute iodine solutions can be used, e.g. under hospital conditions, for contact for short duration.

Non-acid Organic Compounds

Resistance to non-acid organic materials is usually very good. It is necessary to bear in mind that some halogen derivatives are liable to either break down or to hydrolyse in contact with moisture, giving the corresponding halogen acid, which may cause attack.

Domestic and Catering

Resistance of most grades to foodstuffs and beverages in preparation and cooking in excellent, although the free-cutting grades are best avoided. For knives and cutting equipment the hardenable, but less corrosion-resistant, martensitic grades are needed and are adequate if heat treated correctly and properly maintained. They should not be left for prolonged periods with salt or food acids or 'soaked' for long periods. The austenitic types are widely used for cooking utensils and tableware because of advantages in fabrication, but the ferritic 430S17 has also been used successfully. Similarly, the austenitic types are used predominantly for preparation surfaces. Where there is prolonged contact with saline solutions or with salt, the molybdenum-bearing types are preferred, but in any event washing to remove the salts should be undertaken as soon as possible after use. Stainless steels are readily sterilised, this being a substantial advantage over some other materials, but it should be remembered that some sterilants, e.g. hypochlorites, are very corrosive and contact time should be minimised.

Ease of fabrication is the reason that 304S14 is preferred for most other domestic items including kitchen cabinets, food storage cabinets, refrigerator parts and trim, washing machine parts and trim, vacuum cleaner parts and trim, decorative ware, serving dishes etc, and, of course, sinks and drainers. For many of these 430S17 is satisfactory, but the need for heat treatment after fabrication by welding is a disadvantage in some instances.

Conjoint Action of Static Stress and Corrosion

The simple martensitic grades are most widely used in the softened state in which they do not appear to be susceptible to stress-corrosion cracking. If tempered so that carbide-induced selective corrosion is possible (see Fig. 3.16) the application of a stress may cause rapid penetration by cracking in many environments. Tempering at temperatures below that at which chromium-rich carbides form (~350°C) leaves the steel in a strong condition (strength is proportional to carbon content). Then stress-corrosion cracking of the hydrogen-induced type can occur, the probability increasing with steel strength, applied stress and solution aggressivity. In the latter context, reducing pH, increasing chloride content and the presence of cathode poisoners, such as sulphides, are very significant factors. If increased strength is achieved by other means, such as alloying to retard softening on tempering or the use of a precipitation hardening process, cracking can also occur but the relative susceptibilities of the steels vary. This topic has been discussed in detail elsewhere[20].

Stress-corrosion cracking of the austenitic steels can also occur, and the consensus, if not universal, view is that the mechanism is one of the disruption of the passive film by strain at the advancing crack tip but not on the crack sides. Thus the corrodent involved must be capable of localised depassivation, i.e. of the type giving a pitting tendency. In practice the presence of the chloride ion has proved to be the critical factor with the exception of caustic cracking which can occur with these steels in hot, concentrated caustic solutions. Since the presence of chloride is quite widespread, it is, at first sight, surprising that stress-corrosion cracking is not more prevalent. However, practical experience and laboratory testing have shown that, even in the presence of stress, pitting corrosion is more likely. The critical feature appears to be temperature, with most examples of failure being at elevated temperature, i.e. above 80°C. Laboratory studies have confirmed the effect of temperature, although cracking has been induced at 40°C under most salt deposits and at room temperature in acid solutions. The stress involved in practical failures has usually proved to be incidental, e.g. arising from fabrication, rather than functional. Stress relieving presents difficulties with such steels but is applied in special cases. All the steels in the '18/8' group have generally similar resistance to cracking. Various elements in solid solution can have either detrimental or beneficial effects, but the only one of practical significance is nickel of which substantial additions are beneficial. For further data References 21-23 should be consulted.

Plain chromium, ferritic steels are much more resistant and for a time were considered virtually immune to stress-corrosion cracking. It is now known that failure can be caused, especially if the steels contain addition of copper, cobalt or nickel. Even so, resistance is superior to that of the standard austenitics, and ferritics are used where stress-corrosion cracking of the austenitic grades could be a possibility[20].

While not immune to cracking, the duplex steels are also much more resistant than are the austenitics and are increasingly being used for this reason. Their characteristics have been reviewed elsewhere[20].

Conjoint Action of Cyclic Stress and Corrosion

Stainless steels are subject to fatigue failure under 'dry' conditions as are all metallic materials, having distinct fatigue limits where level is dependent on steel type and heat treatment. The limits can be depressed by the simultaneous action of a corrodent, the degree depending upon the nature of the corrodent. Under severe conditions the limit can be displaced to very low values and it is customary to describe resistance by an endurance limit, that is the cyclic stress to give rupture at a specific number of cycles when in contact with a specific corrodent. Some comparative data are in Table 3.25.

Applications

Martensitic Types

By far the greatest use has been for engineering purposes where the corrosion-resisting requirements are not high and applications are legion. Typical are steam and water valves, pump, and turbine and compressor components and shafting. Where the simple 13% Cr steel has proved inadequate in corrosion resistance, the higher chromium type 431S29 has been used and, more latterly, the precipitation hardening types. In the hardened state, the steels are used where abrasion resistance is required. 13% Cr steels are widely used for steam turbine blading and, with the increasing size of turbines, higher strength, tough versions have been developed e.g. alloyed 12% Cr steel 2 of Table 3.11.

A major use of the 13% Cr higher carbon type is for the manufacture of corrosion-resistant cutting edges. Type 420S45 has been the standard for

Table 3.25 Rotating bend fatigue test results for a number of stainless steels

Steel	C	Cr	Ni	Mo	Others	RP·2 (MPa)	Rm (MPa)	Fatigue[b] Limit[a] (±MPa)	Endurance Limit[b] (±MPa)
410S21	0·09	13·12				440	614	340	130
12% Cr Mo	0·09	12·30		0·65		555	722	415	140
431S29	0·15	16·79	2·55			739	989	370	246
Pptr. Hardening martensitic	0·04	13·67	5·44	1·65	Cu 1·66, Nb 0·30	1000	1060	550	230
430S17	0·06	16·40				205	402	200	110
434	0·04	16·01		1·03		326	511	240	140
304S16	0·035	18·48	10·73			230	591	260	230
321S31	0·06	18·09	9·75		Ti 0·44	229	604	260	215
347S31	0·055	18·26	9·45		Nb 0·80	243	638	280	215
315S16	0·05	17·07	10·94	1·80		285	582	270	230
316S33	0·04	17·43	13·53	2·84		223	287	270	260
317S16	0·04	19·36	14·35	3·47		249	630	270	230
304 + N	0·05	18·95	9·51		0·19	380	739	300	260
316 + N	0·03	17·42	10·89	2·80	0·195	359	738	310	290
'Super Austenitic'	0·03	21·24	40·95	3·15	Cu 1·98	285	671	270	280
Solution hardened austenitic	0·04	21·53	9·4	2·75	Mn 3·76, N 0·39, Nb 0·30	484	858	385	370
Duplex	0·02	22·04	5·37	2·88	N 0·19	475	761	450	355

[a] Tests in dry air at room temperature.
[b] Tests with sample moistened with 3% sodium chloride solutions at room temperature.

domestic knives for many years. The weldable type 409S19 is finding increasing use for the manufacture of long-life exhaust systems for automobiles.

Ferritic Types

Major uses of the ferritic steels have been on motor vehicles as trim and in domestic equipment such as cutlery and hollow ware, but use has also been made in refrigerators, washing machines and on sinks and similar fittings. Some types would no doubt find much wider application in the chemical field and other fields where their superior corrosion resistance would be a considerable advantage if it was not for the fact that the austenitic types have advantages (sometimes considerable) in fabrication. However, the availability of the low interstitial weldable types and the 'super ferritics' is increasing in scope.

Austenitic Types

Architectural Considerable use has been made of austenitic steels for decorative aspects such as shop fronts, fittings, door and window frames and, in some cases, extensive wall cladding. Many of these applications may be termed functional as well as decorative since although initial expenditure may be more than for competitive materials, the absence of the necessity for maintenance is a considerable attraction. In the UK, at least, use of the molybdenum-bearing 316S31 type is now almost universal although molybdenum-free grades have been used with success in the past. The use of steels for fully functional purposes is also growing and the molybdenum-bearing grade may not be necessary. A recent example is the use of 316S31 for topside cladding on offshore oil-rigs. Roofing in stainless steel (even 430S17) is being used to an increasing extent on the continent, while load-bearing members such as masonry fixings are being made in the austenitic grades.

Marine engineering Austenitic steels are used extensively for the manufacture of marine propeller shafts, the molybdenum-bearing steels being most popular. Where extremely high torques are operative, the precipitation hardening martensitics may be used as long as there is not prolonged mooring in static seawater. The high yield-point austenitic steels are also used for high torque shafts and have seawater resistance superior to that of the 316 group. Austenitic steels are also used for rudder components and propellers, and the duplex steels with improved strength are also finding such service. Austenitic steels are also widely used for wire ropes, yacht rigging fittings, deck fittings and other items exposed to spray.

Corrosive conditions are, of course, much less onerous when contact with fresh water only is involved and less resistant grades may be used in some instances.

Domestic and catering applications Austenitic steels are ideal for most food preparation, and in this connection are used for table tops, sinks and cooking vessels, including steam-heated pans for hotels and canteens. Other household uses include washing machines, steam irons, mixing machines and cutlery. The importance of a polished non-porous surface in promoting ease of cleaning and general hygiene is obvious, and this type of finish is readily available with these steels.

Food and drink production and distribution

Dairy work Applications include vessels for milk storage and sterilisation, cooling units, cream separators and cheese and butter-making equipment, as well as general dairy fittings, bottle-washing machinery and tankers for bulk milk transport. Extensive use is made of these steels in equipment used in production of ice cream and dried milk.

Brewing Austenitic steels are used in fermentation vessels, storage tanks and attemperators, in yeast processing, and for transport tanks and barrels.

Preserves and fruit juices Much use is made of sulphur dioxide to preserve fruit pulps and juices, and for these applications molybdenum-bearing steels should be specified.

Soups and sauces These products may involve highly corrosive mixtures, in this case from combined chlorides and acidity. For this reason it is often necessary to use the molybdenum-bearing type of steel.

Baking Ease of cleaning is again important, and austenitic steels are very suitable for making items such as mixing machinery and working tables.

Surgery and hospital uses Surgical instruments with cutting edges are made from the hardenable martensitic types and sometimes from special, extra-high carbon grades to ensure good cutting edges. Other instruments are generally made from the 304 group but where 'springiness' is needed (i.e. surgical forceps) the hardenable martensitics are used. The austenitic types are used for many items of theatre equipment and utensils and the molybdenum-bearing grades are preferred where severe sterilisation is involved.

The low carbon 316S11 or 316S13 grades are used for bone nails, plates, screws, fixing wires and joint prostheses, but the more highly alloyed nitrogen-bearing types are finding increasing favour for the last of these applications.

Photographic processing Developing gives no trouble with the Cr-Ni steels, but for contact with acid fixing solutions 316S16 steel should be specified.

Textile applications Much use has been made of non-molybdenum Cr-Ni steels in dyeing. On the other hand, many dyeing solutions contain chlorides or other corrosive substances such as formic acid, and here it is often wiser to use the 316 group.

For bleaching with alkaline peroxide, the 304 group is satisfactory but with moist sulphur dioxide or hypochlorites the 316 group should be used. In the case of hypochlorites, contact should be short or confined to alkaline sodium hypochlorite.

Chemical engineering

Acids One of the most extensive uses of austenitic steels has been for synthetic nitric acid plant, including oxidation coolers, absorption towers, storage vessels and pipelines, as well as transport tanks. For storage of cold 98% acid or for near-boiling solutions above 60% strength, 347S31 or 304S11 should be used.

Phosphoric acid processing is usually associated with onerous conditions, involving hot strong acid, often with appreciable contents of impurities such as sulphuric or hydrofluoric acids. Limited use is made of the austenitic steels, especially 316 group.

Sulphurous acid occurs in bisulphite manufacture, as well as in various processes such as cellulose pulp digestion and oil refining. Since some oxidation to sulphuric acid, which makes the conditions much more corrosive, is often involved, the use of molybdenum-bearing types is generally preferred.

The 316 group steels are used for vinegar distillation and other high-temperature acetic acid applications, the non-molybdenum types being used at lower temperatures. With the higher fatty acids such as oleic and stearic, the non-molybdenum types are resistant even up to boiling point. Lactic, citric and oxalic acids are used with these steels over more limited ranges, the 316 group being necessary in many cases.

Processes involving less corrosive organic acid such as carbolic, cresylic, salicylic and many others, are satisfactorily catered for by the non-molybdenum steels. This also applies to the less corrosive inorganic acids such as boric, nitrous, and carbonic and silicic.

Fertilisers For nitrates, non-molybdenum austenitic steels are satisfactory, but in the manufacture of ammonium sulphate some free acid is often present, so that evaporators and centrifugal dryer baskets in this case are generally made from molybdenum-bearing steels. For super-phosphates this has limited application.

Cellulose products There is little scope for the use of austenitic steels in the alkali processes for the digestion of wood pulp, but both molybdenum-free and molybdenum-bearing steels are extensively used for sulphite pulp digestion, choice depending on concentration, temperature and working experience.

The presence of free sulphuric acid in rayon-spinning baths limits application of the austenitic steels, but they are used for acetylation of cellulose in the acetate process. They are also used for dissolving and spinning solutions in the cuprammonium processes.

The austenitic steels are extensively used for nitrating cellulose, owing to their good resistance to many sulphuric-nitric acid mixtures.

Fats and soaps Their excellent resistance to the higher fatty acids makes the austenitic steels valuable constructional materials for plant dealing with hydrogenation or other treatment of oleic, stearic, and similar acids.

In soap-making some use is made of the molybdenum-bearing type for dealing with glycerine recovery from soap liquors containing sodium chloride.

Oil refining and by-products Various acid conditions, e.g. sulphuric acid and sulphur dioxide, are encountered in refining, and for such conditions 316 group steel have proved useful, with the most resistant higher-nickel Cu–Mo types for the more difficult applications.

Fine chemicals There are many applications of the austenitic steels in the manufacture and storage of fine chemicals and pharmaceutical products. These include storage tanks, pipelines, valves, stills, steam-jacketed pans, mixing vessels, filters and tableting machinery. Considerable use has been made of these steels in penicillin production.

High-pressure synthesis Here austenitic steels are applied to autoclaves and pressure-vessel linings, pipelines, valves and other equipment dealing with a variety of products, often at quite high temperatures.

Atomic power production The necessity for avoiding contamination of operative liquids, together with other requirements which must be met in selecting constructional materials in this highly specialised field has resulted in the choice of austenitic steels for applications in heat exchangers, pressure vessels, pipelines and fuel processing.

J. E. TRUMAN

REFERENCES

1. *The Handbook of Stainless Steels*, (Ed. Peckner, D. and Bernstein, I. M.), McGraw-Hill (1977)
2. Irvine, K. J., Crowe, D. J. and Pickering, F. B., *J.I.S.I.*, **195**, p. 386 (1960)
3. Pickering, F. B., *Physical Metallurgy of Stainless Steel Development*, International Metals Reviews, p. 227, December (1976)
4. Truman, J. E., *Stainless Steels-A Survey*, Met. and Mat. Tech., Pt. 1, p. 15, January (1980), Pt. 2, p. 75, February (1980)
5. *Proc. Conf. Stainless Steels 77*, (Ed. Barr, R. Q.), Climax Molybdenum Co. (1977)
6. *Proc. Conf. Duplex Stainless Steels.* (Ed. Lula, R. A.), ASM (1983)
7. Llewellyn, D. T. and Murray, J. D., *I.S.I. Special Report No. 86*, 197 (1964)
8. Edeleanu, C., *J.I.S.I.*, **188**, p. 122 (1958)
9. Truman, J. E., Coleman, M. J. and Pirt, K. R., *Brit. Corr. J.*, **12**, p. 236 (1977)
10. Truman, J. E., *Proc. Conf. U.K. Corrosion '87*, I.C.S.T. (1987)
11. Hochmann, J., Desestret, A., Jolly, P and Mayoud, R., *Proc. Conf. Stress Corrosion Cracking and Hydrogen Embrittlement of Iron Base Alloys*, (Ed. Stahle, R. W., Hochmann, J., McCright, R. D., and Slater, J. E.), N.A.C.E. (1977)
12. Kendal, A., Truman, J. E. and Lomax, K., *Proc. Conf. High Nitrogen Stress 88*, Lille (1988), Inst. Metals (1988)
13. Truman, J. E., *Ibid*.
14. Truman, J. E. and Lomax, K., *Ibid*.
15. Needham, N. G., Freeman, P. F., Wilkinson, J. and Chapman, J., *Proc. Conf. Stainless Steel '87*. Int Metals, p. 215 (1988)
16. Chandler, K. A., *I.S.I. Pub. No. 117*, p. 127 (1969)
17. Oldfield, J. W., and Sutton, W. H., 'Crevice Corrosion of Stainless Steels', *European Congress on Corrosion*, London (1977)
18. Muller, G. E., *Corrosion 75*, paper 120, N.A.C.E. (1975)
19. Romanoff, M., Nat. Bureau of Standards, *Circular 579*, 222, April (1957)
20. Truman, J. E., International Metals Reviews, Vol. 26, No. 6 (1981)
21. Latanision, R. M. and Staehle, R. W., *Fundamental Aspects of Stress Corrosion Cracking*, N.A.C.E., p. 214 (1969)
22. Theus, G. J. and Staehle, R. W., as ref. 11, 845
23. Truman, J. E., as ref. 11, 111

BIBLIOGRAPHY

Monypenny, J. H. G., *Stainless Iron and Steel*, Chapman and Hall, Vol. 1 (1951) and Vol. 2 (1954)
Colombier, L. and Hochmann, J., *Stainless and Heat Resistory Steels*, Arnold (1967)
Handbook of Stainless Steels, (Ed. Peckner, D. and Bernstein, I. M.), McGraw-Hill (1977)
Sedricks, A. J., *Corrosion of Stainless Steels* J. Wiley and Sons (1979)
The Metallurgical Evolution of Stainless Steels, (Ed. Pickering F. B.), The Metals Soc. (1979)

3.4 Corrosion Resistance of Maraging Steels

Composition

Maraging steels are a class of high strength steels of very low carbon content. Strengthening is achieved by the use of substitutional elements to produce age hardening in the martensitic iron-nickel matrix. The term *maraging* was thus coined from the words 'martensite' and 'age hardening'.

The development of maraging steels began in the late 1950s, on steels containing 20% and 25% Ni using a combination of aluminium, titanium and niobium as age-hardening elements. The characteristics of these steels are contained in Reference 1. Later work[2] revealed the important synergistic age-hardening effect of cobalt plus molybdenum and led to the development of the 18% Ni maraging steels. Using titanium as a supplementary hardening element, and with appropriate balancing of cobalt and molybdenum, nominal yield strengths in the range of 1 370–2 400 MNm^{-2} can be achieved. Table 3.26 lists the nominal composition of the 18% Ni maraging steels. Other types of maraging steel include an alloy (17% Ni) developed for use as a casting, and a 12Ni–5Cr–3Mo alloy. Stainless-type alloys have also been developed but are outside the scope of this section.

Table 3.26 Nominal composition of 18% Ni maraging steels

Maraging steel type	Nominal composition (%)					Nominal yield strength (MN m^{-2})
	Ni	Co	Mo	Ti	Fe	
18% Ni 200	18	8·5	3	0·2	Balance	1 380
18% Ni 250	18	8	5	0·4	Balance	1 720
18% Ni 300	18	9	5	0·6	Balance	2 050
18% Ni 350	17·5	12·5	3·75	1·7	Balance	2 390
Cast alloy	17	10	4·6	0·3	Balance	1 580

Data for the cast and the Cr-bearing alloys are contained in References 36 and 5 respectively. Discussion in this article is restricted to the 18% Ni maraging steels.

Structural Features

On cooling to room temperature after annealing, maraging steels transform completely to martensite. The as-annealed structure consists of packets of parallel lath-like martensite platelets arranged within a network of prior-austenite grain boundaries. The platelets have a high dislocation density but are not twinned.

On heat treating at 485°C a very rapid age-hardening reaction takes place and greatly strengthens the material. Although the nature of the precipitates formed is still uncertain, the consensus of opinion is that ageing for several hours at 485°C results in a Ni_3Mo phase while longer times produce an Fe_2Mo phase. There may also be a titanium precipitate, η-Ni_3Ti or $Ni_3(Mo, Ti)$[3]. Ageing at higher temperatures or longer times results in some reversion to austenite, which may be stable at room temperature (depending on time and temperature of ageing), and a lower strength.

Physical and Mechanical Properties

Table 3.27 Summary of physical properties for the 18% Ni 200 to 18% Ni 350 alloys

Density $8 \cdot 0$–$8 \cdot 1$ g/cm^3
Crystal structure Martensite (body centred cubic); austenite (face centred cubic)
Lattice parameter Martensite $2 \cdot 856$–$2 \cdot 862$Å at room temperature; austenite (retained) $3 \cdot 58$Å
Thermal conductivity $19 \cdot 68$–$20 \cdot 93$ kW/m^2 °C (20–100°C)
Electrical resistivity 60–70 $\mu\Omega$ cm when solution annealed at 815°C; 35–50 $\mu\Omega$ cm when maraged at 485°C for 3 h
Melting temperature 1 430–1 445°C
Transformation temperature M_s 145–200°C; M_f 77–145°C; A_s 445°C
Nominal length change $-0 \cdot 06\%$ to $-0 \cdot 10\%$ during maraging

Table 3.28 Summary of nominal mechanical properties for the 18% Ni maraging alloys

Property	18% Ni 200	18% Ni 250	18% Ni 300	18% Ni 350
Yield strength—0·2% offset (MNm^{-2})	1 310–1 550	1 650–1 820	1 780–2 060	2 270–2 480
Ultimate tensile strength (MNm^{-2})	1 340–1 580	1 680–1 860	1 820–2 100	2 300–2 510
Elongation in 25·4 mm (1 in), (%)	11–15	10–12	7–11	6–10
Reduction in area (%)	35–67	35–60	30–50	25–45
Modulus of elasticity E (kNm^{-2})	$18 \cdot 0 \times 10^7$	$18 \cdot 5 \times 10^7$	$18 \cdot 9 \times 10^7$	$19 \cdot 3 \times 10^7$
Hardness (Rockwell C)	44–48	48–50	51–55	56–59
Impact Charpy V-notch (J)	35–68	24–45	16–26	7–14
Notch tensile				
0·0128 m bar $K_t = 10$ (MNm^{-2})	2 390	2 350–2 510	2 900–3 100	–
0·00762 m bar $K_t = 10$ (MNm^{-2})	–	2 560–2 660	–	1 360–1 490
Fracture toughness (K_{Ic}) (MNm$^{-2}\sqrt{m}$)	101–176	98–165	88–143	44–82
Endurance limit (MNm^{-2})				
Smooth bar 10^8 cycles	620–795	620–760	760–900	690
Notched bar 10^8 cycles ($K_t = 2 \cdot 2$)	275–345	275–380	275–415	–
($K_t = 2 \cdot 8$)	–	–	–	352

Note. In all cases treatment was for 1 h at 815°C plus 3 h at 485°C.

Summaries of the physical and mechanical properties of the 18% Ni maraging steels are given in Tables 3.27 and 3.28. The mechanical properties are

Table 3.29 Comparison of toughness of maraging steels and other high strength alloys*

Alloy	Yield strength (MNm^{-2})	K_{1c} (MNm$^{-2}\sqrt{m}$)
18% Ni 200	1 380	110-176
18% Ni 250	1 720	98-165
18% Ni 300	1 930	88-143
D6 AC	1 380	88-99
H-11	1 790	66-71
AISI 4340	1 790	61-66
AMS 6430	1 510	61-71
Ti-16V-2·5Al	1 170	49-55
Aluminium 7075.T6	415-485	39-66

*Data after Reference 9.
Note. K_{1c} is the plane strain fracture toughness (see Section 8.9).

highlighted by good ductility, toughness and a lack of notch sensitivity. The plane strain fracture toughness of maraging steels is superior to other alloys at comparable strength levels (Table 3.29).

Fabrication

Maraging steels have been produced both by air and vacuum melting. Small amounts of impurities can decrease toughness significantly, sulphur in particular is detrimental and should be kept as low as possible. Silicon and manganese also have a detrimental effect on toughness and should be maintained below a combined level of 0·20%. Such elements as C, P, Bi, O_2, N_2 and H_2 are kept at the lowest levels practicable.

The maraging steels are readily hot worked by conventional rolling and forging operations. A preliminary homogenisation at 1 210°C–1 260°C is normally used prior to hot working at that temperature. During subsequent hot working, extended times at, or slow cooling through, temperatures from 760°C–1 100°C should be avoided since they produce embrittlement[4]. Maraging steels can be cold worked up to 85% before requiring intermediate annealing (because of low work-hardening characteristics) but are usually annealed after smaller reductions.

Heat treatments are relatively simple and normally consist of annealing for 1 h at 815°C followed by ageing for 3 h at 485°C. Recently, double annealing treatments have grown in favour. Machining or fabrication is easily performed in the as-annealed condition. Subsequent age hardening generally introduces small and predictable dimensional changes.

Theoretical Considerations and Characteristics of Corrosion Behaviour

Corrosion in Natural Environments

In atmospheric exposure 18% Ni maraging steel corrodes in a uniform manner[5], and becomes completely rust covered. Pit depths tend to be more shallow than for the low-alloy high-strength steels[6]. Atmospheric corrosion rates[7] in industrial (Bayonne, New Jersey) and marine (Kute Beach, North Carolina) atmospheres are compared with those for low-alloy steel in Figs. 3.29, 3.30 and 3.31. The corrosion rates drop substantially after the first year or two and in all cases the rates for maraging steel are about half the corrosion rate for HY80 and AISI 4340 steels.

The corrosion rates for both maraging steel and the low alloy steels in seawater are similar initially, but from about 1 year onwards the maraging steels tend to corrode more slowly as indicated in Fig. 3.32[7]. The corrosion rates for both low alloy and maraging steel increase with water velocity[5]. During sea-water exposure the initial attack was confined to local anodic areas, whereas other areas (cathodic) remained almost free from attack; the latter were covered with a calcareous deposit typical of cathodic areas in sea-water exposure. In time, the anodic rust areas covered the entire surface[6].

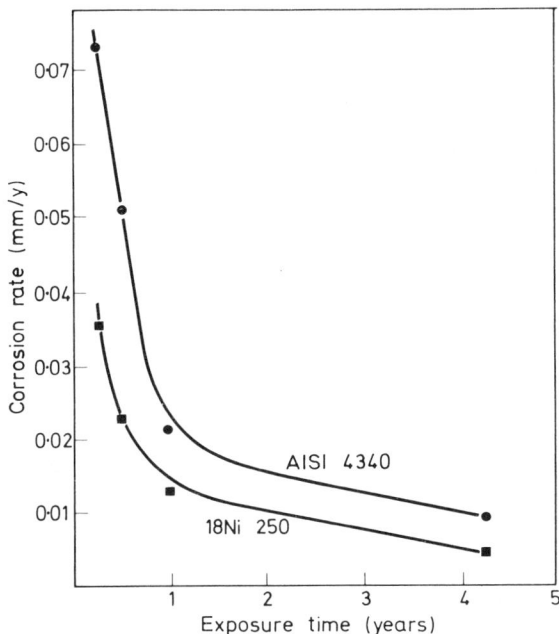

Fig. 3.29 Corrosion rates of maraging and low alloy steels in an industrial atmosphere at Bayonne, N.J. (after Kenyon, Kirk and van Rooyen, *Corrosion*, **27**, 390 (1971)[7])

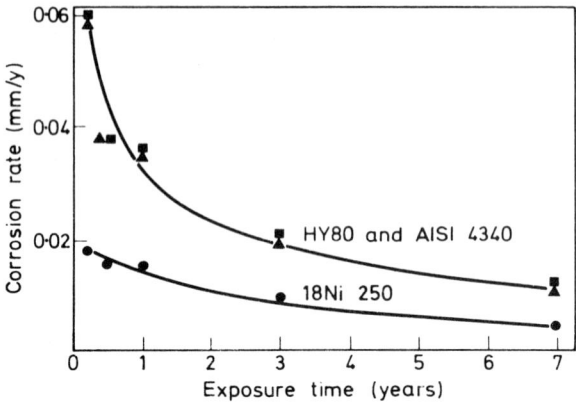

Fig. 3.30 Corrosion rates of maraging and low alloy steels 24·4 m from the sea at Kure Beach, N. C. (after Kenyon, Kirk and van Rooyen, *Corrosion*, **27**, 390 (1971)[7])

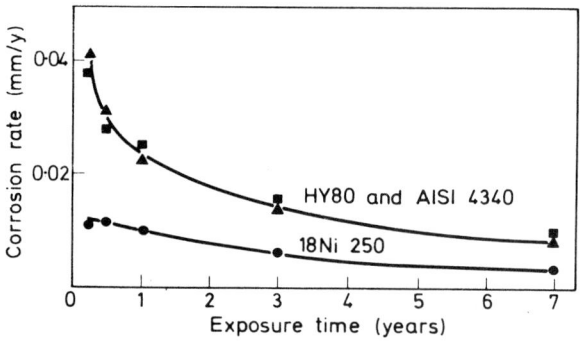

Fig. 3.31 Corrosion rates of maraging and low alloy steels 244 m from the sea at Kure Beach, N. C. (after Kenyon, Kirk and van Rooyen, *Corrosion*, **27**, 390 (1971)[7])

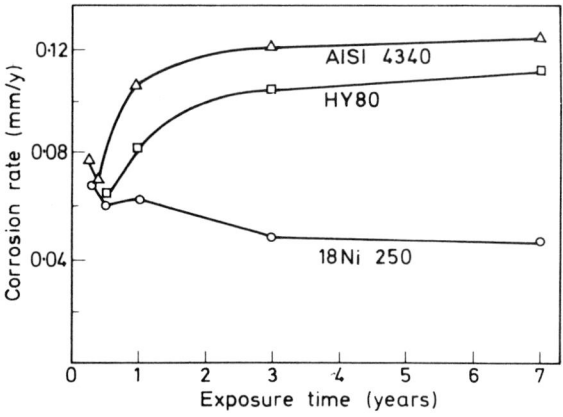

Fig. 3.32 Corrosion rates of maraging and low alloy steels in sea-water flowing at 0·6 m s (after Kenyon, Kirk and van Rooyen, *Corrosion*, **27**, 390 (1971)[7])

Polarisation tests[8] indicate that maraging steel does not exhibit passive behaviour in 3% NaCl, and that the polarisation curves are unaffected by changes in heat treatment.

Corrosion in Chemicals

The corrosion rates of maraging steels in acid solutions such as sulphuric, hydrochloric, formic and stearic acids are substantial although lower than those of the low-alloy high-strength steels[9]. Polarisation studies[10] indicate that maraging steels exhibit active-passive behaviour in 1N and 0·1N sulphuric acid. The corrosion potential, critical current density, primary passivation potential and passive current density are all affected by variations in ageing treatment. The critical and passive current densities increase as the structure is varied from fully annealed to fully aged. The normal heat treatment produces critical and passive current densities of $0·4 mA/cm^2$ and $0·2 mA/cm^2$ respectively ($0·1 mA/cm^2 \approx 1·2 mm/y$[11]). Stress corrosion of maraging steel may also occur in acid environments.

Stress Corrosion Cracking

The 18% Ni maraging steels do not display passivity and normally undergo uniform surface attack in the common environments. Of more serious consequence, however, for all high strength steels, is the degree of susceptibility to stress corrosion cracking (s.c.c.).

For high strength steels in general, the s.c.c. process in aqueous media is characterised by delayed failure, which consists of an incubation period followed by slow and, at times, intermittent crack growth. Failure can occur on loading to some fraction of the yield stress, or through the action of residual stresses, often in environments as mild as humid air[12]. The degree of susceptibility depends on the mode of loading and is highest in cases of plane strain loading (triaxial stresses), with tensile loading and plane stress bending, in that order, representing less severe loading conditions[13] (tensile stresses of course are present in all the above modes of loading). In general, susceptibility to s.c.c. increases with increasing yield strength. Different alloy types however, vary in their degree of resistance, and at comparable strength levels maraging steels compare favourably in cracking resistance to other high strength steels.

There has been some controversy as to whether s.c.c. occurs by active path corrosion or by hydrogen embrittlement. Lack of space does not permit a full treatment of this subject here. References 14 and 15 are recent reviews on the s.c.c. of high strength steels and deal with the mechanism of cracking (see also Section 8.4). It is appropriate to discuss briefly some of the latest work which appears to provide pertinent information on the cracking mechanism. It should be noted, however, that cracking in all alloy systems may not be by the same mechanism, and that evidence from one alloy system need not constitute valid support for the same cracking mechanism in another.

Recently-developed techniques[16,17] have enabled studies to be made of the electrochemical conditions near the tip of a growing stress corrosion crack. It was found that under freely corroding conditions the pH of the solution near the crack tip was about 3·8 for all eleven steels (including maraging steels) studied[18]. Furthermore, potential measurements indicated that thermodynamic conditions were satisfied for hydrogen ion reduction. Further potential-pH measurements were made on AISI 4340 steel exposed to 0·6N NaCl solutions of different pH and polarised to potentials both negative and positive with respect to the corrosion potential[17]. It was found that the pH of the solution in the crack was determined solely by electrochemical reactions at the crack tip irrespective of starting pH. It was also apparent that regardless of the impressed potential, the electrochemical conditions in the crack satisfied the thermodynamic requirements for the production of hydrogen. These results indicate that it is not necessary to invoke an active path mechanism and that hydrogen is available even during anodic polarisation of AISI 4340 steel.

Activation energy measurements for s.c.c. of H-11 (38 kJ/g atom)[19] and AISI 4340 (36 kJ/g atom)[20] steels in water and moist air suggest that the cracking rate in these alloys is controlled by the diffusion of hydrogen, since these activation energies are in close agreement with the value of 39 kJ/g atom obtained for the diffusion of hydrogen through AISI 4340 membranes[21]. Activation energies have not been determined for cracking in maraging steels, nor have they been determined for any steel in situations of anodic polarisation.

There is little question that at strongly negative potentials cracking occurs by hydrogen embrittlement. Also, the activation energy studies of H-11 and AISI 4340 strongly suggest that hydrogen embrittlement is the controlling mechanism in these cases. For all other cases there does not appear to be a strong basis for favouring one mechanism over the other or to discard the possibility that each mechanism shares in the control of the cracking process. The latter possibility for maraging steel has been supported by a recent study of cracking response to polarisation, fractographic studies and pH determinations of the corrodent at the tip of the crack[22].

Testing for Stress Corrosion Cracking

The stress corrosion resistance of maraging steel has been evaluated both by the use of smooth specimens loaded to some fraction of the yield strength and taking the time to failure as an indication of resistance, and by the fracture mechanics approach[23] which involves the use of specimens with a pre-existing crack. Using the latter approach it is possible to obtain crack propagation rates at known stress intensity factors (K) and to determine critical stress intensity factors (K_{ISCC}) below which a crack will not propagate (see Section 8.9).

Any test (several such tests are used) in which time to failure of smooth specimens is determined is an overall measure of the incubation period to initiate a crack, the ability to resist the propagation of a stress corrosion crack and the ability to resist final mechanical fracture. Since this test does not indicate the relative merits of an alloy in each individual aspect of the cracking process it is probably of less benefit to a design engineer. The use of precracked specimens in the fracture mechanics approach follows from the philosophy that structures are likely to contain crack-like defects. The use of precracked specimens promotes a rapid change in chemistry of the solution at the crack tip and shortens or eliminates the initiation time for crack propagation. Data from both types of tests are available and are reviewed here.

Cracking Resistance in Smooth Materials

Maraging steel in the strength range 1 240–1 720 MNm^{-2} tested as 'U' bends in sea-water, displayed good resistance, as it did not fracture in periods up to 2–3 years although there was considerable general corrosion and fouling. However, microcracks were observed after 6 months. Similar behaviour[5,6] of 'U'-bends and bent beam specimens can be expected in industrial or marine atmospheres although general corrosion is less severe. By comparison AISI 4340 at strength levels of 1 660 MNm^{-2} failed in about 1 week in both sea-water and atmospheric tests[5,6]. Maraging steel of yield strength at or above 2 060 MNm^{-2} was not resistant and failed rapidly.

Maraging steel welds are somewhat less resistant than base plate. 'U'-bend exposure of 1 240 MNm^{-2} strength welds survived for up to 2 years in sea-water while at 1 380 MNm^{-2} failures occurred in 2–18 months[7].

It is possible to provide cathodic protection to base plate up to 1 720 MNm^{-2} yield strength, by coupling to mild steel or possibly to zinc[5,6], but zinc and metals more active than zinc tend to induce hydrogen embrittlement. Welds up to 1 380 MNm^{-2} may be cathodically protected by zinc, but at impressed potentials of $-1 \cdot 25$ V (S.C.E.) both 1 240 and 1 380 MNm^{-2} welds fail rapidly due to hydrogen embrittlement[7]. Neither mild steel nor zinc couples protect AISI 4340 steel[5].

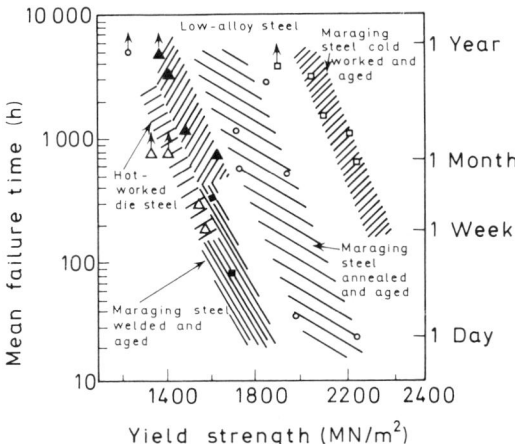

Fig. 3.33 Bent-beam test results in aerated distilled water. These specimens were exposed to the environment at a stress of 70% of yield (after Setterlund, *Materials Protection*, **4** No. 12, 27 (1965)[24])

Although tests on smooth specimens indicate that cathodic protection of maraging steel is possible, tests on specimens with pre-existing cracks indicate a greater sensitivity to hydrogen embrittlement during cathodic polarisation[7]. The use of cathodic protection on actual structures must therefore be applied with caution, and the application of less negative potentials than are indicated to be feasible in smooth specimen tests is to be recommended if it is assumed that structures contain crack-like defects.

Further evidence of the relative resistance of maraging steel is reproduced in Fig. 3.33 from Reference 24. Maraging steel is shown to be superior to a die steel and low alloy steel (both unidentified) in bent beam tests stressed at 75% of the yield strength in distilled water. Also shown is the beneficial effect in smooth surface tests of cold rolling. Shot peening has a similar beneficial effect[7].

Resistance to Propagation of Pre-existing Cracks

Critical Stress Intensity Factor It has become common to use K_{ISCC}, the critical stress intensity factor, as a measure of the resistance of an alloy to s.c.c. Tests are performed on specimens which are precracked by a fatigue machine and must be of sufficient dimensions to ensure plane strain conditions. Recommendations on precracking and dimensions are given elsewhere[25, 26].

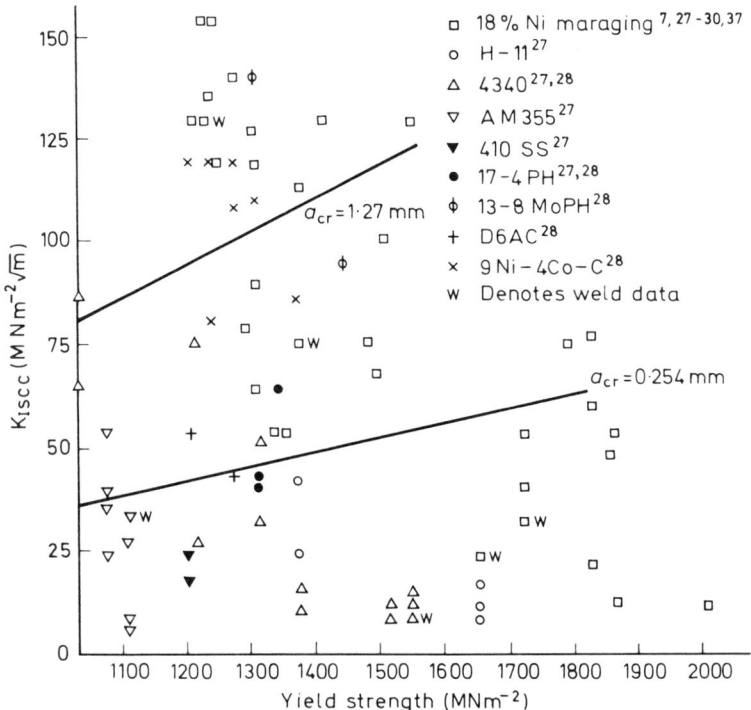

Fig. 3.34 Comparison of K_{ISCC} as a function of yield strength of 18% Ni maraging and other high strength steels

Figure 3.29 presents a summary comparing K_{ISCC} values of maraging steel, with values for H-11, AISI 4340, AM 355*, 410 stainless steel, 17-4 PH†, 13-8 Mo PH†, D6AC‡ and 9Ni-4Co-C§ steels. The data have been taken from a number of sources[7, 27-30, 37] and are for exposures in aqueous environments, with and without NaCl. No attempt has been made to distinguish between different environments since they do not affect the results appreciably. Plotted points marked W refer to data for welds. Also included in Fig. 3.34 are lines of critical crack depth a_{cr}. The region below a line of specified critical crack depth corresponds to combinations of strength and K_{ISCC} for which a long crack of the specified depth will propagate when stressed to the yield point, whereas the region above the line corresponds to the strength and K_{ISCC} combinations for which the crack will not propagate. The critical crack depth (assuming yield point stresses) for cracks whose length greatly exceeds their depth is given by

$$a_{cr} = 0 \cdot 2 \left(\frac{K_{ISCC}}{\sigma_y} \right)^2$$

where σ_y is the yield stress[23].

In general it is clear that maraging steels compare favourably with other high strength steels, and offer comparatively high K_{ISCC} values over a wide range of strength. It is also clear that maraging steels can withstand a greater crack depth without crack propagation.

A further estimation of the corrosion resistance of maraging steel can be obtained from data on the rate of crack propagation. Although the rate of crack propagation has been found to be a function of stress intensity in some alloys, for many alloys and heat treatments there is a range of stress

Table 3.30 Crack propagation rates for a number of high strength steels*

Alloy	Yield strength (MNm^{-2})	Crack velocity (mm/s)
Maraging 250	1 570	$1 \cdot 31 \times 10^{-5}$
Maraging 250	1 690	$1 \cdot 40 \times 10^{-5}$
Maraging 300	1 950	$2 \cdot 75 \times 10^{-4}$
Modified maraging 300†	2 180	$1 \cdot 18 \times 10^{-4}$
Modified maraging 300† (underaged)	1 810	$0 \cdot 72 \times 10^{-3}$
Modified maraging 300† (overaged)	1 760	$2 \cdot 88 \times 10^{-5}$
4340	1 430	$1 \cdot 65 \times 10^{-3}$
D6AC	1 540	$1 \cdot 74 \times 10^{-4}$
H-11	1 420	$1 \cdot 19 \times 10^{-5}$
HP 9-4-25	1 330	$2 \cdot 54 \times 10^{-5}$
HP 9-4-45 (bainitic)	1 460	$1 \cdot 06 \times 10^{-5}$

* Data after Carter[31].
† No details on the modification were available.

* Registered Trademark of Allegheny Ludlum Industries
† Registered Trademark of Armco Steel Corporation
‡ Registered Trademark of Vasco-Teledyne Company
§ Registered Trademark of Republic Steel Corporation

intensity above K_{ISCC} and approaching K_{IC} where the rate of crack propagation is independent of stress intensity[31]. Table 3.30, summarised from Reference 31, presents a list of crack growth rates for a number of high strength steels whose cracking rates were independent of stress intensity. The cracking rate for maraging steel is seen to be slower than for 4340 and D6AC and equivalent to H-11 and HP 9Ni–Co–C steels, at about 10^{-5} mm/s. The fact that cracks propagate very slowly in maraging steels has an important consequence related to their stress corrosion testing. In determining K_{ISCC} by dead-weight-loaded cantilever-beam tests, it has been recommended[32] that maraging steel should withstand 1 000 h without failure to ensure that the applied stress intensity is at, or lower than, K_{ISCC}. 100 h was considered adequate for the more rapidly cracking low-alloy steels[32].

Effect of Metallurgical Variables on s.c.c.

It is notable that while it is possible to produce maraging steels with consistently uniform mechanical properties, the stress corrosion properties are subject to scatter, as indicated in Fig. 3.34. To a large extent this scatter is an indication of the greater sensitivity of s.c.c. resistance to metallurgical variables. Although the variation in cracking resistance is not well understood*, and the reaction to certain treatments not always consistent, certain observations may be used to indicate guidelines for improved properties.

Cracking has generally been observed to be intergranular[6,33-35] with isolated cases of transgranular cracking[29]. Both Ti_2S[29] and $Ti(C, N)$[29,35] on prior austenite boundaries have been suggested to be related to greater susceptibility. Since prolonged time in the temperature range 760–1 100°C favours the precipitation of Ti_2S and $Ti(C, N)$ on prior austenite boundaries, such prolonged exposures should be avoided both in processing and in annealing. It is preferable that annealing should be done at the minimum temperatures required to develop desired properties.

Studies of the effect of ageing temperature on cracking behaviour have shown rather marked effects. Underageing at temperatures of 455°C or lower was found to increase greatly the rate of crack propagation without affecting the mode of crack propagation (intergranular) or K_{ISCC}[31,33,34]. Table 3.30 lists crack growth rates for a modified 18% Ni 300 steel as a function of ageing treatment. Other work[29] found that underageing not only increased the crack growth rate, but also decreased the resistance to cracking (K_{ISCC}) and resulted in a change from a transgranular cracking mode to an intergranular mode. Electron microscopy was unable to detect any grain boundary phases. No satisfactory explanation for this behaviour was apparent. Overageing appears to offer slightly slower crack growth rates (Table 3.30), however it did not significantly improve K_{ISCC}[29]. The best combination of properties is obtained with the normal ageing treatment at 485°C.

Studies on welds showed that cracking resistance as indicated in 'U'-bend tests[7] was substantially increased by a post weld anneal (1 h at 815°C) prior

---

*This aspect and the general stress corrosion and hydrogen embrittlement behaviour are considered in detail in Reference 37.

to the normal ageing treatment. Material aged in the as-welded condition was less resistant. The most significant structural difference resulting from the two heat treatments was a finer dispersion of austenite ribbons in material annealed before ageing.

High Temperature Corrosion

Little data are available on hot corrosion behaviour. Figure 3.35 indicates maraging steel to have better resistance to air exposure at 535°C than a 5% Cr tool steel[9]. Metallographic examination indicates that exposure to

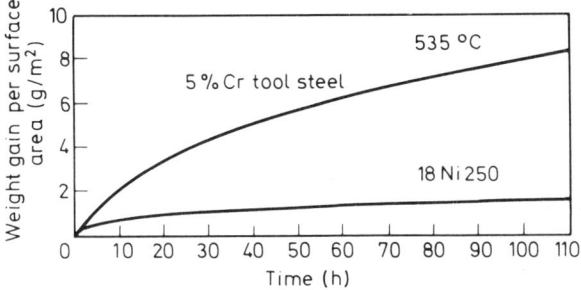

Fig. 3.35 Oxidation rate at 535°C of 18Ni250 maraging steel compared with a generally-available tool steel. These tests were performed on $\frac{1}{4}$ in (6·35 mm) cubes placed in refractory cubicles and exposed to still air for total times of 5, 25 and 100 h. The weight gain includes the scale formed during heating and cooling (after Reference 9)

air at elevated temperatures results in reaction with both oxygen and nitrogen, forming both oxides and titanium carbo-nitrides. Under some conditions of reduced oxygen partial pressure, selective sub-surface oxidation of iron can occur.

Applications

Maraging steels have found varied uses in the aerospace and aircraft industries. These uses have included rocket motor cases, landing gear components, aircraft forgings and fasteners. Other areas of usage include machine tool and die applications, and extrusion hardware. Marine uses include hydrofoil foil systems and aircraft arrester hooks.

Recent Developments

During the 1970s, the price of cobalt increased enormously and then declined. These price fluctuations, plus the concern about the future supply of cobalt, caused serious declines in the use of the cobalt-containing grades of maraging steel. In response to this market change, a new cobalt-free grade of maraging steel has been developed[38]. In the United States, this alloy is presently being manufactured by Teledyne Vasco under the trade name VascoMax T-250.

The composition of this steel is given is 18·5Ni, 3·0Mo, 1·4Ti, 0·1Al. Its strength is 1 720 MPa and its mechanical properties approximate those of the 18% Ni 240 maraging steel. The stress-corrosion behaviour of this alloy has not been studied in detail. A preliminary study[39] of the crack growth behaviour of 12·7 mm thick WOL specimens tested freely corroding in a 3·5% NaCl solution at room temperature showed a threshold stress-intensity of 27 MPa$\sqrt{m}$. This value is somewhat less than the threshold values usually observed in the 18% Ni 250 steel. At higher stress intensities, the crack growth data showed a plateau value of approximately 20×10^{-2} mm/h over a range of stress intensities from 36 to 63 MPa$\sqrt{m}$. Comparable crack velocity plateaux have been seen in the older grades of maraging steel.

Concerning the mechanism of stress corrosion in maraging steels, the common view in recent years has been that cracking in chloride solutions was due to a hydrogen-embrittlement mechanism. A recent study by Craig and Parkins[40], however, has shown good evidence that cracking can proceed by hydrogen embrittlement at low potentials, by anodic dissolution at high potentials, and possibly by combinations of both mechanisms at intermediate potentials. The effects of precracking or pitting of smooth specimens was also examined. In many instances the local changes in chemistry of the test solution in the cracks or pits was more important than the stress concentrations at these locations.

D. P. DAUTOVICH

REFERENCES

1. Bieber, C. G., *Metal Progress*, **78**, 99 (1960)
2. Decker, R, F., Eash, J. T. and Goldman, A. J., *Trans. Quart. A.S.M.*, **55**, 58 (1962)
3. Floreen, S., *Met. Rev.*, **13**, 115 (1968)
4. Novak, C. J., D.M.I.C. Memo 196 (1964)
5. Kirk, W. W., Covert, R. A. and May, T. P., *Met. Eng. Quart.*, **8**, 31 (1968)
6. Dean, S. W. and Copson, H. R., *Corrosion*, **21**, 95 (1965)
7. Kenyon, N., Kirk, W. W. and van Rooyen, D., *Corrosion*, **27**, 390 (1971)
8. Stavros, A. J. and Paxton, H. W., Paper presented at NACE Meeting, Chicago, Illinois (1971)
9. *Data Bulletin on 18% Ni Maraging Steels*, The International Nickel Company, Inc. (1964)
10. Nam Bui, Pieraggi, B. and Dabosi, F., *Mem. Sci. Rev. Met.*, **68**, 223 (1971)
11. France, W. D., Jr. and Mazzatenta, E. D., *Materials Engineering*, **62**, Sept. (1970)
12. Johnson, H. H., *Proceedings of Conference of Fundamental Aspects of Stress Corrosion Cracking*, National Association of Corrosion Engineers, 439 (1969)
13. Hayden, H. W. and Floreen, S., *Corrosion*, **27**, 429 (1971)
14. Phelps, E. H., *Proceedings of the Conference of Fundamental Aspects of Stress Corrosion Cracking*, National Association of Corrosion Engineers, 398 (1969)
15. Kennedy, J. W. and Whittaker, J. A., *Corrosion Science*, **8**, 359 (1968)
16. Brown, B. F., Fujii, C. T. and Dalhberg, E. P., *J. Electrochem. Soc.*, **116**, 218 (1969)
17. Smith, J. A., Peterson, M. H. and Brown, B. F., *Corrosion*, **26**, 539 (1970)
18. Brown, B. F., Extended Abstract of paper presented at Fourth International Congress on Metallic Corrosion, Amsterdam, Netherlands (1969)
19. Johnson, H. H. and Willner, A. M., *Appl. Mat. Res.*, **4**, 34 (1965)
20. van der Sluys, W. A., Presented at First National Symposium on Fracture Mechanics, Lehigh University, USA (1967)

21. Beck, W., Bockris, J. O'M., McBreen, J. and Nanis, L., *Proc. Roy. Soc.*, (London), **A290**, 221 (1966)
22. Syrett, B. C., *Corrosion*, **27**, 270 (1971)
23. Brown, B. F., *Met. Rev.*, **13**, 171 (1968)
24. Setterlund, R. B., *Mat. Protection*, **4** No. 12, 27 (1965)
25. American Society for Testing and Materials, pub. No. STP 381 (1965)
26. American Society for Testing and Materials, pub. No. STP 410 (1966)
27. Freedman, A. H., *J. Mater.*, **5** No. 12, 431 (1970)
28. Brown, B. F., Naval Research Laboratory Report 7168 (1970)
29. Dautovich, D. P., The International Nickel Company, Inc., unpublished work
30. Steigerwald, E. A., presented at ASM Conference on Stress Corrosion Cracking, Philadelphia, Pa., Aug. (1970) Original work by Carter, C. S., Boeing Co. Report D6-19770, Nov. (1967)
31. Carter, C. S., *Corrosion*, **27**, 471 (1971)
32. Brown, B. F., paper presented at ASM Conference on Fracture Control, Philadelphia, Pa., Jan. (1970). Referred to by Carter, C. S., (Ref. 33)
33. Carter, C. S., *Met. Trans.*, **1**, 1 551 (1970)
34. Stavros, A. J. and Paxton, H. W., *Met. Trans.*, **1**, 3 049 (1970)
35. Parkins, R. N. and Haney, E. G., *Trans. Met. Soc.*, **242**, 1 943 (1968)
36. Sadowski, E. P. and Koppi, W. A., *Trans. Am. Foundrymen's Soc.*, **75**, 294 (1967)
37. Dautovich, D. P. and Floreen, S., Paper presented at the International Conference on Stress Corrosion Cracking and Hydrogen Embrittlement of Iron Base Alloys, Unieux-Firminy, France, June (1973)
38. Floreen, S., U.S. Patent 4, 443, 254, April 17 (1984)
39. Floreen, S., unpublished work
40. Craig, I. H. and Parkins, R. N., *Brit. Corros. J.*, **19**, 3-16 (1984)

3.5 Nickel–Iron Alloys

Use of Nickel–Iron Alloys

Nickel–iron alloys have a number of important applications that are derived from such special physical properties as their unique magnetic characteristics in the regions of 35, 50 and 80% nickel and from their abnormally low thermal expansion in the region of 36–50% nickel. Although not specifically used as corrosion-resistant materials, their high resistance to attack from many common environments is of benefit in their specialised applications.

Electrochemical Characteristics

The potentiodynamically-derived polarisation curves of Beauchamp[1] (Fig. 3.36) demonstrate the effect of increasing nickel content on the anodic behaviour of iron–nickel alloys in 1N H_2SO_4. The maximum current in the active region is reduced and the potentials moved to more noble values; the current in the passive region is increased and some evidence of secondary passivity appears. The greater nobility with increasing nickel content in the active region is of importance in acid environments where hydrogen evolution is the major cathodic reaction, and results in significantly lower rates of corrosion. In neutral environments, the protection provided by a layer of insoluble corrosion products is of greater significance.

Atmospheric Corrosion

The addition of small amounts of nickel to iron improves its resistance to corrosion in industrial atmospheres due to the formation of a protective layer of corrosion products. Larger additions of nickel, e.g. 36% or 42%, are not quite so beneficial with respect to overall corrosion since the rust formed is powdery, loose and non-protective, leading to a linear rate of attack as measured by weight loss. Figure 3.37 of Pettibone[2] illustrates the results obtained.

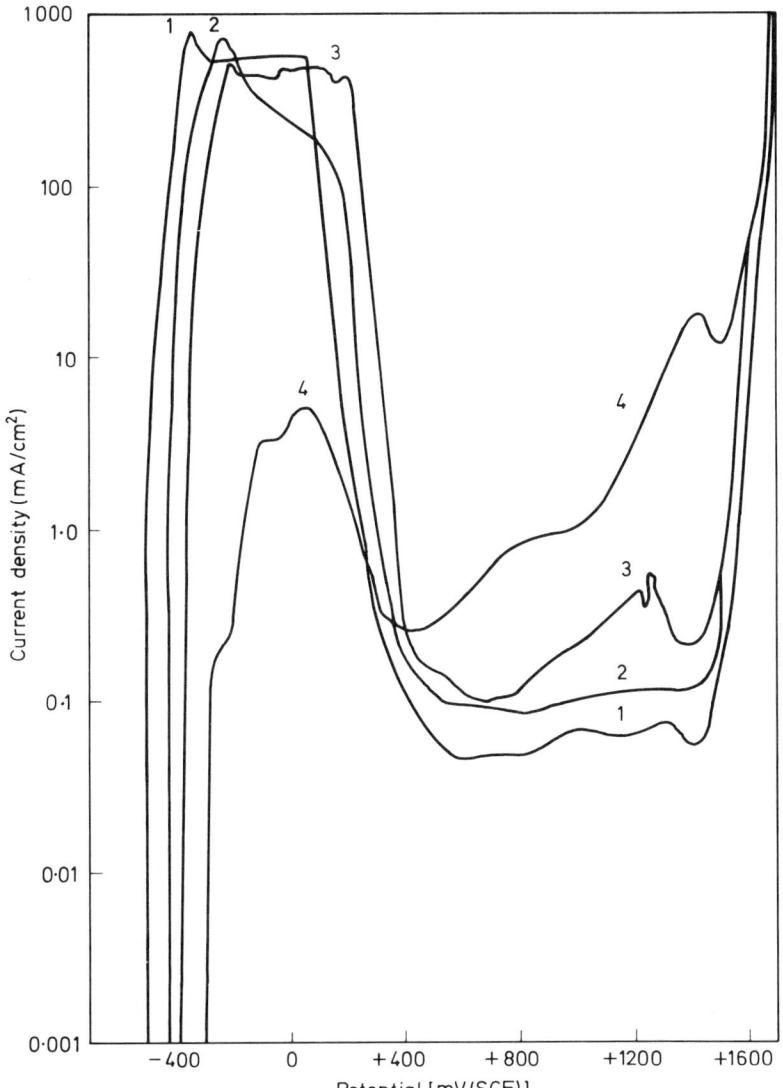

Fig. 3.36 Effect of nickel on the anodic behaviour of iron alloys in 1 N H_2SO_4 at 25°C. Curve 1 Fe; curve 2 Fe-10Ni; Curve 4 Ni (after Beauchamp[1])

With respect to resistance to pitting corrosion, there is an increasing advantage to be obtained by increasing the nickel content up to 50%. There is little distinction between the Fe-50Ni alloy and pure nickel. Data on the corrosion of Fe-36Ni alloy at an industrial site in the USA are reported by La Que and Copson[3] and at a European site by Evans[4].

In marine atmospheres the overall rates of corrosion are reduced progressively with increase in nickel content up to about 35%, but with small improvement thereafter. The rates of corrosion at various sites, reported by

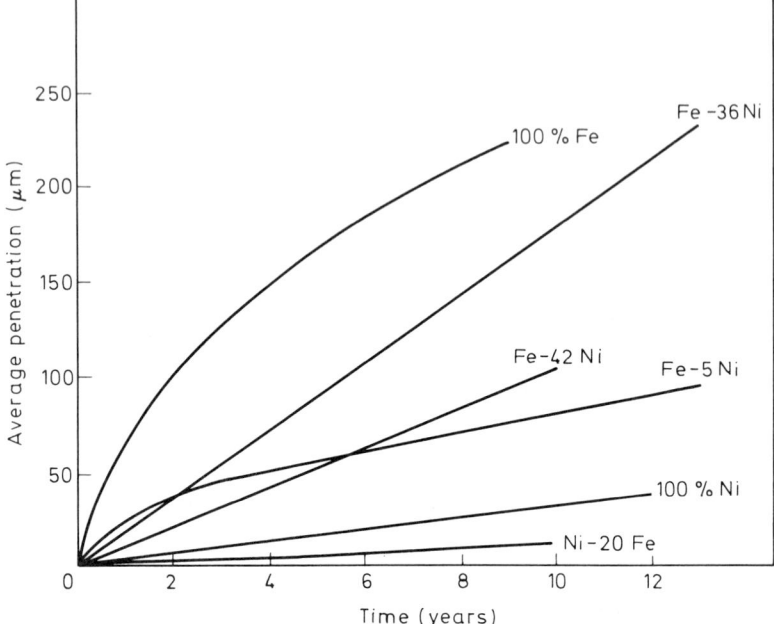

Fig. 3.37 Resistance of nickel-iron alloys to corrosion by an industrial atmosphere (Bayonne N. J., USA) (after Pettibone[2])

Table 3.31 Resistance of Fe-36 Ni and mild steel to corrosion in marine atmospheres[5]

Location	Corrosion after exposure for 15 years			
	Average (gm^{-2} d^{-1})		Localised (max. pit depth, mm)	
	Fe-36Ni	Mild steel	Fe-36 Ni	Mild steel
Colombo, Ceylon	0·08	5·5	0	—
Auckland, New Zealand	0·02	1·5	0	2·43
Halifax, Nova Scotia	0·03	0·65	0·1	1·64
Plymouth, England	0·08	3·5	0·19	1·09

Friend[5], show the superiority of Fe-36Ni over mild steel with respect to both average and localised corrosion (Table 3.31)

Sea-water

The average rates of corrosion of Fe-36Ni alloy exposed to alternate immersion in sea-water are appreciably greater than those that occur when the alloy is exposed to marine atmospheres. Although the rates of corrosion are significantly below those observed for mild steel (Table 3.32) the superiority over mild steel in not so great with respect to pitting attack.

NICKEL–IRON ALLOYS

Table 3.32 Resistance of Fe–36Ni and mild steel to corrosion during alternate immersion in sea-water[5]

Location	Corrosion after exposure for 15 years			
	Average (gm^{-2} d^{-1})		Localised (max. pit depth, mm)	
	Fe–36Ni	Mild steel	Fe–36Ni	Mild steel
Colombo, Ceylon	0·64	3·4	1·0	2·55
Auckland, New Zealand	0·09	0·32	0·24	0·36
Halifax, Nova Scotia	0·24	1·5	2·59	2·15
Plymouth, England	0·36	1·4	0·25	1·58

Nickel–iron alloys fully immersed in sea-water may suffer localised corrosion which can be severe under conditions where oxygen is constantly renewed at the surface and the formation of protective corrosion products is hindered, e.g. in fully-aerated flowing sea-water. In quieter, less oxygenated conditions, average corrosion rates of Fe–36Ni are low and well below those for mild steel, as exemplified in the data given in Table 3.33[5]. However the resistance to localised attack is not improved to the same extent.

Table 3.33 Resistance of Fe–36Ni alloy and mild steel to corrosion in sea-water[5]

Location	Corrosion after exposure for 15 years			
	Average (gm^{-2} d^{-1})		Localised (max. pit depth, mm)	
	Fe–36Ni	Mild steel	Fe–36Ni	Mild steel
Colombo, Ceylon	0·8	2·0	2·5	6·5
Auckland, New Zealand	0·5	2·0	1·08	2·59
Halifax, Nova Scotia	1·3	2·2	3·49	1·23
Plymouth, England	0·8	1·5	1·82	2·75

Fresh Water

Nickel–iron alloys suffer significantly less corrosion than mild steel when exposed to a soft, fresh water but Friend[5] found that the resistance to pitting is only slightly greater (Table 3.34).

Table 3.34 Resistance of Fe–36Ni and mild steel to corrosion in fresh water[5]

Corrosion after exposure for 15 years			
Average (gm^{-2}d^{-1})		Localised (max. pit depth, mm)	
Fe–36Ni	Mild steel	Fe36Ni	Mild steel
0·07	0·94	2·00	2·20

Resistance to Acids

Much of the information available on resistance of nickel-iron alloys to corrosion by mineral acids is summarised by Marsh[6]. In general, corrosion rates decrease sharply as the nickel content is increased from 0 to 30-40%, with little further improvement above this level. The value of the nickel addition is most pronounced in conditions where hydrogen evolution is the major cathodic reaction, i.e. under conditions of low aeration and agitation. Results reported by Hatfield[7,8] show that the rates of attack of Fe-25Ni alloy in sulphuric and hydrochloric acid solutions, although much lower than those of mild steel, are still appreciable (Tables 3.35 and 3.36). In solutions of nitric acid, nickel-iron alloys show very high rates of corrosion.

Table 3.35 Resistance of Fe-25Ni and carbon steel to corrosion by sulphuric acid

Alloy	Corrosion rate (mm/year)					
	5% H_2SO_4		25% H_2SO_4		50% H_2SO_4	
	15°C	40°C	15°C	40°C	15°C	40°C
Carbon steel	62	183	93	378	0·95	2·8
Fe-25Ni	0·45	1·4	0·45	1·8	0·9	1·8

Table 3.36 Resistance of Fe-25Ni and carbon steel to corrosion by hydrochloric acid

Alloy	Corrosion rate (mm/year)					
	5% HCl			25% HCl		
	15°C	40°C	60°C	15°C	40°C	60°C
Carbon steel	23	40	41	63	188	185
Fe-25Ni	0·45	2·3	5·4	0·9	0·9	19

Resistance to Alkalis

Addition of nickel improves the resistance of iron and steel to corrosion by alkaline solutions. The beneficial effect is most pronounced in hot, strong caustic solutions as illustrated by the results[9] on nickel cast irons in Table 3.37.

Table 3.37 Resistance of nickel cast irons to corrosion by hot caustic soda*

% Nickel in cast iron	Corrosion rate (mm/year)	% Nickel in cast iron	Corrosion rate (mm/year)
0	1·8 to 2·3	15	0·8
3·5	1·2	20	0·09
5	1·2	30	0·01

*Specimens immersed for 54 d in evaporator concentrating caustic soda from 50 to 60% under 88 kN/m^2 vacuum.

Resistance to Salt Solutions

Nickel–iron alloys are more resistant than iron to attack by solutions of various salts. In alternate immersion tests in 5% sodium chloride solution Fink and De Croly[10] determined values of $2 \cdot 8$, $0 \cdot 25$ and $0 \cdot 5 \, \text{gm}^{-2} \text{d}^{-1}$ for alloys containing 37, 80 and 100% nickel compared with $46 \, \text{gm}^{-2} \text{d}^{-1}$ for iron. Corrosion rates of about $0 \cdot 4 \, \text{gm}^{-2} \text{d}^{-1}$ are reported by Hatfield[7] for Fe–30Ni alloy exposed to solutions containing respectively 5% magnesium sulphate, 10% magnesium chloride and 10% sodium sulphate; the same alloy corroded at a rate of about $1 \cdot 2 \, \text{gm}^{-2} \text{d}^{-1}$ in 5% ammonium chloride.

In a study of the corrosion of ten binary nickel–iron alloys in 3% sodium chloride solution, Schwerdtfeger[11] found the average corrosion rate to decrease from $1 \cdot 4 - 1 \cdot 6 \, \text{gm}^{-2} \text{d}^{-1}$ for alloys containing 0–16% nickel alloy to $0 \cdot 1 \, \text{gm}^{-2} \text{d}^{-1}$ for a 57% nickel–iron alloy. There was little further reduction in rate of weight loss for the higher nickel alloys. However, the alloys showed increasing tendency to suffer from pitting and crevice corrosion with increasing nickel content.

Stress Corrosion

In tests lasting for 14 days, Copson[12] found that the susceptibility of steel to stress-corrosion cracking in hot caustic soda solutions increased with increase in nickel content up to at least $8 \cdot 5\%$. Alloys containing 28% and more of nickel did not fail in this period. In boiling 42% magnesium chloride the 9% nickel–iron alloy was the most susceptible of those tested to cracking (Table 3.38). Alloys containing 28 and 42% nickel did not fail within 7 days.

Table 3.38 Resistance of iron–nickel alloys to stress corrosion cracking in boiling 42% magnesium chloride[12]

| Composition of alloy | | | | | Hardness | Time to cracking (days) | Comments |
Ni	C	Mn	Si	Fe			
nil	0·19	1·65	0·20	bal.	89 Rb	No cracking after 11 days	
2·02	0·19	0·46	0·18	bal.	77 Rb	7	Few shallow cracks
4·96	0·15	0·51	nil	bal.	96 Rb	<3	Profuse deep cracks
8·67	0·10	0·76	0·23	bal.	24 Rc	<3	Cracked in two
27·88	0·03	0·18	0·06	bal.	81 Rb	No cracking after 7 days	
41·79	0·02	0·18	0·08	bal.	77 Rb	No cracking after 7 days	
99·41	0·10	0·24	0·02	0·13	20 Rc	No cracking after 7 days	

Couper[13] reports cracking of an Fe–36 Ni alloy in 10–55 days in this medium. Radd et al.[14] have noted cracking of Fe–36 Ni alloys at ambient temperatures in an unspecified environment, but this possibly may have been residual traces of acid copper chloride etching solution.

Bimetallic Corrosion

Bimetallic corrosion of nickel–iron alloys may be of significance in welding operations. Ni–45 Fe alloys are used as filler materials in the welding of cast irons but the favourable area relationship of weld metal to base plate mitigates the effect of the more noble characteristics of the nickel-iron alloy. Thus their application in corrosive environments is rarely of concern.

Of more serious practical significance is iron contamination of nickel-clad steel welds. Tables 3.39 and 3.40 show the increase in corrosion of various nickel-iron alloys which may occur when coupled to nickel in

Table 3.39 Bimetallic corrosion between nickel and nickel-iron alloys in 16% calcium chloride solution[2]

Alloy	Corrosion rate* (mm/y)	
	Coupled	Uncoupled
100 Ni	—	0·02
Ni-5 Fe	0·045	0·045
Ni-10 Fe	0·7	0·055
Ni-20 Fe	1·0	0·044

* Room temperature test of duration 120 days, solution agitated. Cathode: anode area = 100:1.

Table 3.40 Bimetallic corrosion between nickel-iron alloys in sodium hydroxide[2]

Galvanic couple	Corrosion rate* (mm/y)				
	23% NaOH		50% NaOH	75% NaOH	
	Coupled	Uncoupled	Coupled	Coupled	Uncoupled
Ni-5 Fe	0·04	0·035	0·06	0·0250	0·02
Nickel	0·015	0·01	0·02	0·04	0·04
Ni-10 Fe	0·07	0·03	0·48	0·38	0·035
Nickel	0·01	0·01	0·015	0·02	0·04
Ni-20 Fe	0·095		0·04	0·05	0·015
Nickel	0·015		0·01	0·03	0·04
Ni-30 Fe	0·04	—	0·2	0·31	—
Nickel	0·02	0·01	0·01	0·02	0·04
Ni-40 Fe	0·04	—	0·21	0·34	—
Nickel	0·02	0·01	0·01	0·02	0·04

* Area ratio Ni:Ni-Fe = 10:1.

calcium chloride or sodium hydroxide solutions. It is evident that in the calcium chloride solution, 5% iron contamination of weld metal can be tolerated whilst in the sodium hydroxide solution bimetallic corrosion would not become significant until contamination had exceeded 20%.

Recent Developments

In recent years some additional data on the corrosion resistance of the Ni-Fe alloys has been generated, but it is fairly limited and the foregoing material summarises the majority of the available information. The additional data falls into the categories, electrochemical, resistance to acids and resistance to hydrogen cracking.

Electrochemical Characteristics

This work has been carried out by Marcus and his co-workers[15-18] and deals with the influence of sulphur on the passivation of Ni-Fe alloys. For sulphur-containing Ni-Fe alloys, sulphur segregates on the surface during anodic dissolution. Above a critical sulphur content a non-protective thin sulphide film is formed on the surface instead of the passive oxide film.

Resistance to acids

It is in this area that most work has been carried out, particularly in relation to corrosion resistance in sulphuric acid solutions[19-23]. Bourelier et al.[19] and Raicheff et al.[20] investigated the inhibitive effect of chloride ions on corrosion in sulphuric acid. The inhibition efficiency was found to depend on the alloy composition, alloy surface and chloride concentration. The more aggressive the environment, the greater the inhibition efficiency. Yagupol'skaya et al.[21] studied the effect of iodine additions to sulphuric acid on the corrosion resistance of Ni and Ni-Fe alloys. Again there was an inhibitive effect caused by the halide ion.

Uto et al.[22] studied Ni-Fe-Si alloys, adding Si to improve the castability of the simple Ni-Fe materials. They found that alloys containing 0-70% Fe and 5-10% Si, with the balance being Ni, are usable in 0-85% sulphuric acid at temperatures up to 80°C and in 0-10% hydrochloric acid at temperatures up to 40°C.

Cid et al.[23] studied the corrosion resistance of Ni, 5% Fe-Ni and 10% Fe-Ni alloys in the trans-passive region in sulphuric acid. For a given acid concentration the addition of iron reduced the corrosion rate. It was concluded that the addition of small percentages of Fe was doubly beneficial, decreasing both general and intergranular corrosion.

Resistance to Hydrogen Cracking

Marquez et al.[24] studied the effect of cold rolling on the resistance of Ni-Fe alloys to hydrogen cracking. It was found that low carbon, 10-19% Ni-Fe alloys become considerably more resistant to hydrogen cracking after severe cold rolling. The observed resistance decreased with increasing carbon content and the improvement was directional, the optimum effect applying to specimens stressed in the longitudinal direction.

G. N. FLINT
J. W. OLDFIELD

REFERENCES

1. Beauchamp, R. L., Dissertion, Ohio State University, Figure 27 (1966)
2. La Que, F. L. and Copson, H. R., (Ed.), *Corrosion Resistance of Metals and Alloys*, Reinhold Publishing Corporation, New York, Van Nostrand Reinhold Ltd, London, 2nd edn., 458 (1963)
3. *Symposium on Atmospheric Corrosion of Non-Ferrous Metals*, Amer. Soc. Test. Mat., 58th Annual Meeting, June 29 (1955). Spec. Tech. Publn. No. 175, 141-158

4. Evans, T. E., *4th International Congress on Corrosion*, Amsterdam (1969)
5. Friend, J. N., *Deterioration of Structures of Timber, Metal and Concrete Exposed to the Action of Sea Water*, London, 18th Report of the Committee of the Institution of Civil Engineers (1940)
6. Marsh, J. S., *The Alloys of Iron and Nickel: Vol. I Special-purpose Alloys*, McGraw-Hill, New York and London, 495 et seq. (1938)
7. Hatfield, W. H., *Engineer*, **134**, 639 (1922)
8. Hatfield, W. H., *J. Iron and Steel Institute*, **108**, 103 (1923)
9. *Corrosion Engineering Bulletin No. 2*, International Nickel Co.
10. Fink, C. G. and De Croly, C. M., *Trans. Amer. Electrochem. Soc.*, **56**, 239 (1929)
11. Schwerdtfeger, W. J., *J. of Research*, **70C**, 187 (1966)
12. Rhodin, T. H., (Ed.), *Physical Metallurgy of Stress Corrosion Fracture*, Interscience, New York, 259-262 (1959)
13. Couper, A. S., *Materials Production*, **8** No. 10, 17 (1969)
14. Radd, F. J., Wolfe, L. H. and Crowder, L. H. and Crowder, L. H., The World Petroleum Conference (1967)
15. Marcus, P., Olefjord, I. and Oudar, J., *Corr. Sci.*, **24**, 259 (1984)
16. Marcus, P., Olefjord, I. and Oudar, J., *Corr. Sci.*, **24**, 269 (1984)
17. Marcus, P. and Oudar, J., *Proc. Conf. 'Passivity of Metals and Semiconductors'*, Bombannes, France, June 1983, (Ed. Froment, M.), Elsevier 119 (1983)
18. Marcus, P., Bournet, A. and Oudar, J., *Mem. Sci. Rev. Metall.*, **78**, 509 (1981)
19. Bourelier, F. and Vu Quang, K., *Proc. Conf. 10th Int'l Congr. Metallic Corr*, Madras, India, November 1988, p. 2813 (1987)
20. Raicheff, R., Aroyo, M. and Aropadjan, S., *Werkst. und Korr.*, **33**, 25 (1982)
21. Yagupol'skaya, L. N. Lavrenko, V. A. and Kozachenko, E. V., *Zasch. Metal.*, **109**, 291 (1974)
22. Uto, Y., Kitajima, H. and Kai, T., *Nippon KInzoku Gakkaishi*, **27**, 18 (1963)
23. Cid, M., Pelerin, J. and Petit, M. C., *Mem. Sci. Rev. Metall.*, **77**, 951 (1980)
24. Marquez, J. A., Matshusima, I. and Unlig H. H., *Corrosion*, **26**, 216 (1970)

3.6 Cast Iron

Composition and Structure

Cast iron is the term applied to a wide range of ferrous alloys, whose principal distinguishing feature is a carbon content in excess of 1·7%. The relatively low melting point of these alloys compared with that of steels and their tendency to expand slightly on solidification, which make them admirably suited for the production of components by casting, result from this feature of their composition.

Four main types of iron are commonly encountered, viz. *white* iron in which all the carbon is in solid solution; *grey* iron in which most of the carbon is present as graphite flakes; *nodular* or *ductile* iron in which most of the carbon is present as graphite nodules, produced during solidification of the casting; and finally *malleable* iron in which most of the carbon is present as graphite nodules, produced subsequent to solidification by heat-treatment of the casting. These classes can be further sub-divided by consideration of their matrices. White iron has a matrix of pearlite containing amounts of free carbide which depend on the carbon content of the iron. Grey iron usually has a pearlitic matrix; this is the form usually meant by the term *cast* iron and is the most common alloy of the family. Nodular

Table 3.41 Composition ranges of cast iron alloys

Type of iron	Microstructure		Total carbon (%)	C.C† (%)	Si (%)	Mn (%)	S (%)	P (%)
	Graphite form	Matrix						
White iron	None	Pearlite + carbide	1·7–3·0	All	0·8–1·3	0·4	<0·15	<0·5
Grey iron	Flake	Pearlite	2·7–4·0	<0·9	0·5–3·3	0·3–1·0	<0·15	<1·4
Nodular graphite iron*	Nodules	Pearlite or ferrite	3·3–3·9	<0·9	1·6–2·5	0·4	<0·01	<0·1
Blackheart malleable	Nodules	Ferrite	2·0–2·7	None	0·8–1·2	0·1–0·6	<0·15	<0·2
Whiteheart malleable	Nodules	Pearlite	3·3–3·9	0–1·2	0·3–0·8	0·1–0·5	<0·4	<0·1

*Nickel 0–1·5%, magnesium 0·04–0·10%
†Combined carbon.

graphite irons solidify with a pearlitic matrix, but in order to develop the full ductility of the iron, the castings are often subsequently annealed to give a ferritic matrix. Malleable irons are produced by two different processes which result in either ferritic or pearlitic matrices depending on the process adopted, but even the pearlitic iron is usually produced with a surface layer of ferrite.

The figures quoted in Table 3.41, while not authoritative in indicating upper and lower limits, give some idea of the range of analysis to be expected for each type of iron. Because of this variation in composition, cast irons are usually specified in terms of their mechanical properties rather than on an analytical basis.

Effect of Structure and Composition on Corrosion Resistance

Structure

An essential difference may be observed between the behaviours of steel and cast iron components immersed in an environment in which rust is precipitated at some distance from the corroding surface. The steel will waste away at a steady rate and its overall dimensions will steadily diminish, whereas cursory examination of the cast iron may suggest that it has not corroded at all, since its dimensions appear to be substantially unchanged. This difference arises from the fact that the cast iron contains in its microstructure several more or less corrosion-resistant components which are largely or completely absent from the microstructure of a steel. The most important of these corrosion-resistant micro-constituents are graphite, phosphide eutectic, and, to a lesser extent, carbide. When the cast iron corrodes in such a way that the corrosion product is deposited at some distance from the corroding surface, a skeleton is left behind comprising graphite flakes stiffened, in the case of phosphoric irons, by phosphide-eutectic cells and plugged by the carbonaceous debris resulting from the decomposition of the pearlite, silicic acid derived from the oxidation of the silicon dissolved in the iron, and whatever rust is precipitated relatively close to the metal (see Fig. 3.38). This skeleton, although much inferior in strength to the original iron, nevertheless retains sufficient strength to resist moderate erosion and may preserve the original contour of the component quite well.

The amount of graphitic residue retained on a corroding surface depends partly on the morphology of the graphite and partly on the corrosivity of the medium. In general, a coarse flake graphite tends to give a more permeable and less strong residue than a finer graphite, while nodules produce an even weaker residue. However, these differences are only clear at very high corrosion rates. For example, while flake graphite iron retains virtually all the graphitic residue when corroded in 0·5% sulphuric acid, the residue from a nodular graphite iron is largely detached; in 0·05% sulphuric acid, however, there is little difference in the amounts retained by the two irons[1]. The effect of the graphite in the iron on the corrosion process depends on the residue the residue thickness — thus in 0·5% acid, graphite stimulates attack on the nodular graphite iron because of its ability to act as an efficient cathode for

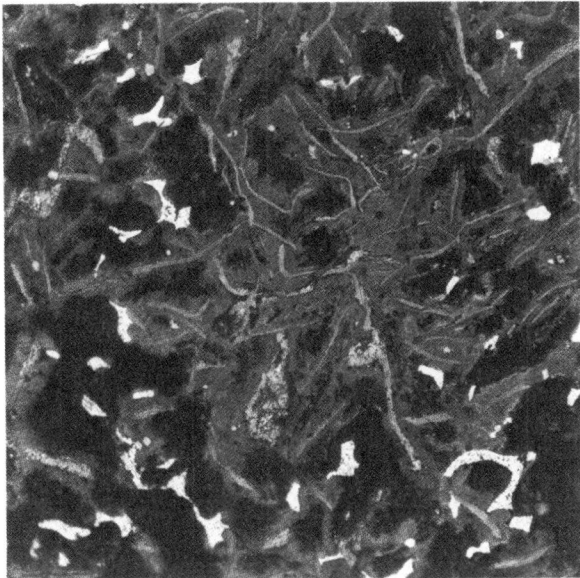

Fig. 3.38 Typical graphitic residue, containing graphite flakes and some unattacked phosphide eutectic

hydrogen evolution, but in 0·05% acid, once the residue thickness reaches about 0·1 mm the attack tends to become stifled[1]. An interesting sidelight on this aspect is that flake graphite irons exposed in hot 70% sulphuric acid often do not evolve hydrogen, which surprisingly is apparently adsorbed by the graphite, presumably because of the high surface tension forces involved in bubble formation in this medium.

The general increase in the use of nodular graphite iron for the production of castings has led to work on the effect of graphite morphology on the corrosion resistance of the iron. In neutral environments the effect of graphite form seems small, but Russian work has reported that irons with finely-dispersed or nodular graphite have a much higher resistance to sulphuric acid solutions than normal flake graphite iron[2], while pipes produced by centrifugal casting processes, which necessarily have fine structures, are claimed to have better general corrosion resistances, whether they are grey or ductile iron, than pipes produced by casting in non-rotating moulds[3].

Berenson and Wranglen[4] have suggested that the presence of magnesium sulphide inclusions in ductile cast irons may act as corrosion-initiating sites and may therefore make ductile iron more liable to corrosion than grey iron but this view has been contested by Collins[5].

Composition

Small variations in the composition of cast irons, or even the addition of small amounts of alloying elements, generally have little effect on the corrosion resistance.

Graham et al.[6], however, working on the corrosive wear of automobile cylinders and piston rings exposed to high sulphur fuels, showed that irons exposed to 70% sulphuric acid at 130°C are attacked at rates dependent on the silicon content of the iron, the rate being relatively low at below 1% Si but rising to a peak at about 2% Si (Fig. 3.39). The effect seems to be associated with the development of a corrosion stifling film on the lower silicon irons. The laboratory work was broadly substantiated by piston-ring wear results and tests in boiler-combustion simulation rigs.

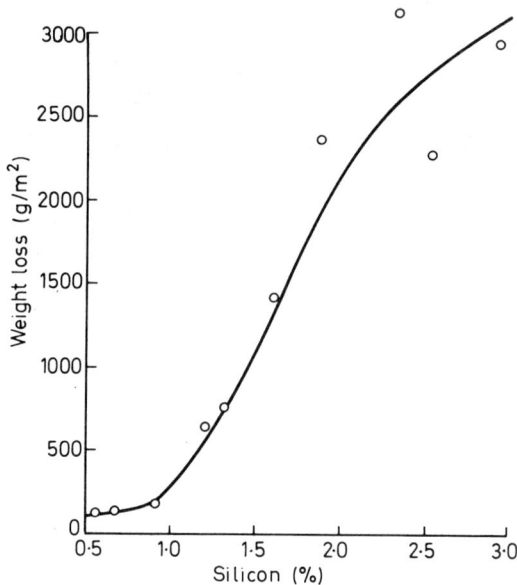

Fig. 3.39 Weight loss for cast iron exposed in 70% H_2SO_4 at 130°C for 16 h (after Graham et al.[2])

Whittaker and Brandes[7] followed up this work, carrying out tests on irons containing small amounts of copper and nickel. The addition of 0·6% Cu to irons containing 2% Si gave a significant improvement in corrosion resistance, but the addition to irons containing less than 1·5% Si decreased the corrosion resistance, which nevertheless remained greater than that for the higher silicon irons; nickel additions produced a smaller effect.

Copper additions appear to have the particular effect of reducing the corrosion stimulating effect of the sulphur content of an iron exposed to acid and the effect is thus less marked in low sulphur irons. Because sulphur can stimulate corrosion in acidic environments, it is usually kept as low as possible in irons to be used under these conditions. A low sulphur content is in any case metallurgically desirable.

Nickel has often been added to castings to be used with alkalis, either molten or in solution. There is, however, no conclusive evidence to support any belief in its efficacy.

Corrosion in Natural Environments

Atmospheric Corrosion

Because cast iron components are normally very heavy in section, the relatively low rates of attack associated with atmospheric corrosion do not constitute a problem and little work has been carried out on the phenomenon. A summary of some of the data available is given in Table 3.42. The most extensive work in this field was initiated by the A.S.T.M.[8] in 1958 and some of the results produced by these studies are quoted in Table 3.43. It will be noted that there is a marked fall in corrosion rate with time for all the metals tested.

Corrosion by Waters

The corrosivity of a natural water depends on the concentration and type of impurity dissolved in it and especially on its oxygen content. Waters of similar oxygen content have generally similar corrosivities, e.g. well-aerated quiescent sea-water corrodes cast iron at rates of 0·05–0·1 mm/y while most well-aerated quiescent 'fresh' waters corrode iron at 0·01–0·1 mm/y.

Those waters in which the carbon dioxide content is in excess of that required as bicarbonate ion to balance the bases present are among the most aggressive of the 'fresh' waters. Hard waters usually, though not invariably, deposit a carbonate scale and are generally not appreciably corrosive to cast iron, corrosion rates of less than 0·02 mm/y being frequently encountered. Water-softening processes do not increase the corrosivity of the water provided that the process does not result in the development of an excess of dissolved carbon dioxide.

Waters of pH less than 6 may be expected to be corrosive, but, because any weak acids present in the solution may not be fully ionised, it does not follow that water of pH greater than 7 will not be corrosive. Mine waters are particularly corrosive to cast iron, often to such an extent as to preclude its use with them, because of their relatively high acid content, derived from the hydrolysis of ferric salts of the strong acids, mainly sulphate, and because the ferric ion can act as a powerful cathodic depolariser.

While well-aerated near-neutral waters are normally much more corrosive than poorly-aerated waters, waters with near zero oxygen contents may cause high rates of corrosion if active sulphate-reducing bacteria, which can act as very efficient depolarising agents, are present. A corrosion rate of 1·5 mm/y has been observed on cast iron exposed to such a water.

The presence of active sulphate-reducing bacteria usually results in 'graphitic corrosion' and this has led to a useful method of diagnosing this cause of corrosion. The leaching out of iron from the graphitic residue which is responsible for the characteristic appearance of this type of corrosion leads to an enriched carbon, silicon and phosphorus content in the residue as compared with the original content of these elements in the cast iron. Sulphur is usually lost to some extent but when active sulphate-reducing bacteria are present, this loss is offset by the accumulation of ferrous sulphide in the residue with a consequent increase in the sulphur content of the residue out

Table 3.42 Corrosion rate of steels and irons in the atmosphere (g m^{-2} d^{-1})

Environment	Rural			Urban			Industrial			Marine		
Source	A.S.T.M.[8]	Roll[9]	Friend[10]	Nekrytyi[11]	Roll[9]	A.S.T.M.[8]	Dearden[12]	Friend[10]	Roll[9]	LaQue[13]	A.S.T.M.[8]
Exposure period years	12	2	6	1	2	12	1	6	2	—	12
Material:											
Steel	0·23			1·2		0·27	3·4	2·4–3·2		3·6	1·38
Grey iron			1·4–2·1				3·2	1·1–1·2		0·6	2·0
White iron			0·1–0·3				1·3				
Malleable iron											
(a) <0·1% S	0·11	3·3		2·1	4·9–6·0	0·14			3·3		0·43
(b) >0·1% S	0·15					0·20			3·6		0·75
Nodular iron											
(a) Pearlitic	0·10					0·15				0·9	0·37
(b) Ferritic	0·11					0·17					0·72

Table 3.43 A.S.T.M. atmospheric corrosion data (g m^{-2} d^{-1})

Location		State College, Pa. (rural)			Kure Beach, N.C. (marine)			Newark, N.J. (industrial)		
Duration (years)		1	3	12	1	3	12	1	3	12
Metal	Condition									
Ferritic ductile iron	As cast	0·90	0·36	0·11	1·51	0·85	0·72	1·29	0·51	0·17
	Machined	0·56	0·31	0·09	0·90	0·63	0·60	0·88	0·36	0·12
Pearlitic ductile iron	As cast	0·62	0·30	0·10	0·96	0·53	0·37	0·15	0·43	0·15
	Machined	0·50	0·22	0·07	0·82	0·47	0·27	0·70	0·30	0·10
Malleable iron >0·1% S	As cast	0·75	0·40	0·15	1·41	1·11	0·75	1·53	0·70	0·20
Mild steel	Rolled	0·97	0·52	0·23	3·02	2·01	1·38	1·75	0·81	0·27

of all proportion to the content of the other elements present. The figures quoted in Table 3.44 illustrate the point.

Table 3.44 Increase in sulphur content of graphitic corrosion residue due to the presence of active sulphate-reducing bacteria

	Total carbon (%)	Si (%)	P (%)	S (%)
Metal	3·39	2·31	1·17	0·10
Graphitic corrosion residue (G.C.R.)	10·3	7·4	3·1	4·4
Ratio G.C.R. metal	3·0	3·2	2·7	44·0

Velocity of water As the velocity of flow of a water over cast iron increases, the supply of oxygen to the corroding surface is increased. Eventually, the supply of oxygen to the surface will reach a level sufficient to permit the formation of a strongly adherent, impermeable rust scale on the corroding surface. When this level is reached, the corrosion rate either ceases to change with increasing water velocity (Fig. 3.40[14]) or decreases, sometimes to negligible values.

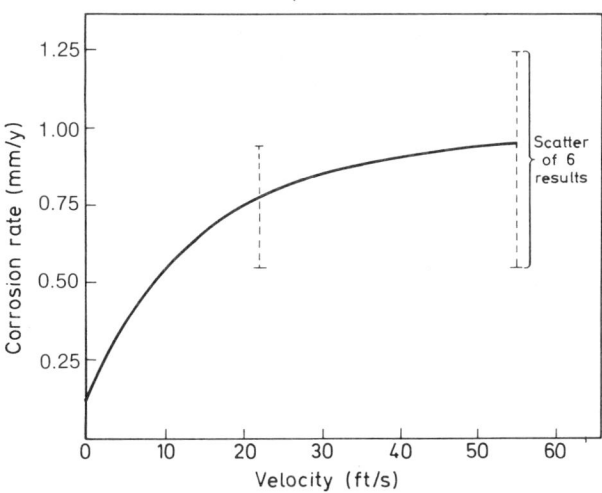

Fig. 3.40 Effect of velocity upon cast iron in sea-water; Rotor test at a water temperature of 25°C (after Higgins[14])

Very rapid and highly localised pitting is sometimes observed on components exposed to very turbulent flow conditions leading to cavitation in the stream. In general, these conditions appear to induce corrosion rather than erosion on cast iron surfaces, in contradistinction to what usually happens with other metals, apparently because the erosive component of the liquid flow scours away corrosion-stifling films and allows the development of very active electrochemical cells on the exposed metal surfaces[15].

Soil Corrosion

Since buried pipes for water, sewage and gas are a major use of cast iron, the corrosion of buried iron structures needs special consideration in any study of the corrosion properties of cast iron. It is also a very complex topic that is not fully understood.

Because pipes are used to contain fluids, failure occurs when there is leakage. With grey iron pipes, leakage usually arises because the pipe fractures as a result of soil stresses or top-loading exceeding the hoop strength of the pipe. With ductile iron pipes, leakage is usually through small corrosion pits. For this reason grey iron pipes are susceptible to both uniform and pitting corrosion while ductile iron pipes are only seriously affected by pitting.

A characteristic of the corrosion on buried ferrous metals is that the attack is usually mostly in the form of pitting, especially with the cast irons. This raises a problem in measuring the extent of corrosion in burial trials. Usually both the weight loss, measuring the average loss of section, and the deepest pit, measuring the maximum loss of section, are reported. For assessing the severity of the attack on buried pipes, the second parameter is clearly the most important.

The rate of attack on buried iron will of course depend on the corrosivity of the soil in question. The best guide to this at present is the work carried out by the National Bureau of Standards in the United States between 1922 and 1955[16]. In this very detailed and extensive study, the chemical characteristics of more than 100 soils drawn from all over the US were examined and compared in relation to the corrosion behaviour of small steel and iron specimens that had been buried in them for up to about 17 years. Detailed study of the published results suggests, however, that the only parameter to give any useful correlation with the severity of the corrosion is the electrical resistivity of the soil. Even this is only valid up to about 2000 ohm cm with a correlation coefficient of about -0.5[17]. However electrical resistivity is the factor most often used at present for dividing soils into corrosivity classes. In the UK it is usually assumed that all soils having a resistivity greater than 4000 ohm cm (measured at pipe depth by the four-pin Wenner method) are not aggressive to bare iron pipes, while soils of lower resistivity will require the pipes to be given some protection. The setting of the resistivity value greatly above that suggested by the NBS data reflects the risk of inaccuracy in measuring the soil resistivity and the relatively poor correlation between resistivity and severity of corrosion. However, the application of protection in accordance with this rule has worked well in practice and Booth[18] has confirmed its general utility in a survey he carried out on 89 soils in the UK, although he has suggested that a better basis for classification would be:

	Aggressive	Non-aggressive
Resistivity and/or	<2 000 ohm cm	>2 000 ohm cm
Redox potential at pH 7	<0·40 V (N.H.E.) or <0·43 V if in clay or	>0·40 V (N.H.E.) or >0·43 V if in clay
Borderline cases resolved by water content	>20% w/w	<20% w/w

Much more detailed schemes have been proposed in the USA[19] and in Germany[20] but examination of these shows that they also depend essentially on the resistivity measurement[21].

Although resistivity surveys can decide which soils along a projected pipeline are probably aggressive to buried ferrous metals and which are not, it has been found that when unprotected pipes fail in service the local soil resistivity does not correlate to any significant extent with the corrosion rate at the pitted site. Indeed, Stokes[22] showed from a survey of statistics of ductile iron pipe failures in the UK that the most likely pitting rate for an aggressive soil is about 1 mm/y with a range of about 0·3 to 3 mm/y. Since the soil electrolyte in most of these aggressive soils is likely to be a dilute salt solution of low dissolved oxygen content and pH about 8·5, such a consistent corrosion rate seems logical. Collins[23] has suggested that the tendency for low resistivity soils to be aggressive is due to the fact that for the most part they are poor draining and poorly aerated heavy clays of low flowability and compactability and thus more liable to create differential concentration cells on the pipe surface than free-flowing soils such as sands. He argues on this basis that, as the soil resistivity falls, the probability of serious pitting increases but the pitting rate remains substantially constant. They would explain the poor correlation between the chemical parameters of the soil and its corrosivity. It also suggests that the physical characteristics of the soil should be better indicators of its corrosivity.

It is common experience that corrosive soils tend to be the heavy clays, especially if they have been subjected to working by, for example, heavy earth-moving machinery. Lighter soils are usually only corrosive if they have been contaminated by industrial debris, especially ashes, ferrogenous slags and carbonaceous material such as cinders.

It should be noted that it is extremely difficult to predict service lives of buried pipelines from the results of controlled trials with small specimens, whether in the laboratory or in the field. For example a study on the comparative corrosion resistances of ductile and grey iron pipes carried out jointly by European pipemakers in 1964–1973[24,25] indicated a mean pitting rate of 0·35 mm/y for uncoated ductile iron pipe exposed in a typical heavy Essex clay of 500–900 ohm cm resistivity for 9 years. This is clearly at odds with the rate of 1 mm/y normally found on a corroded service pipe from such a soil. The discrepancy appears to be due to the use of specimens that were only a third of a pipe length each and were buried separately. It may reflect the contribution of the total surface area of the pipe as a cathode to the corrosion current at the anodic area at the pitting site.

Although iron pipes suffer from the same corrosion risk as steel pipelines, associated with the generation of a galvanic cell with a small anode and a large cathode, the risk is mitigated for iron pipelines because the electrical continuity is broken at every pipe joint. For this reason long-line currents are uncommon in iron lines and cathodic protection is rarely necessary. It also accounts for the ability to protect iron lines by the application of non-adherent polyethylene sleeving[26].

Corrosion in Industrial Environments

Corrosion by Acids

In general, unalloyed grey or white cast irons possess no useful resistance to dilute mineral acids. In very dilute acids the presence of air, or other oxidising agents such as ferric salts, appreciably increases the corrosion rate. If corrosion rates are to be held below 0·25 mm/y in moderately aerated solutions, it is unwise to exceed a total acid concentration of 0·001N, irrespective of the acid concerned.

Mineral acids Unalloyed cast iron possesses no useful resistance to hydrochloric acid at any concentration or temperature. Dilute sulphuric, nitric and phosphoric acids are also very aggressive, corrosion rates amounting to several centimetres per year in some cases. Owing to the insolubility of surface films of ferrous sulphate in strong sulphuric acid, however, concentrations of this acid in excess of 65% can generally be withstood by cast iron at room temperature, corrosion rates being less than 0·12 mm/y. Concentrations of sulphuric acid greater than 97% at temperatures up to 300°C do not corrode cast iron at more than 0·12 mm/y. Free SO_3 (oleum) does not increase the corrosion rate. Any influence operating against the formation or retention of the insoluble sulphate film, for example high liquid velocities or abrasion of the metal surfaces, may lead to excessive corrosion rates, even in acid concentrations greater than 65%. Collins[27] has described some grey iron castings which had suffered deterioration when handling oleum or sulphur trioxide due to penetration of the sulphur trioxide into the iron, apparently along graphite flakes, which generated an accumulation of internal stress and eventually caused blistering or even fracture of the casting. In no cases, however, was there any marked corrosion of the iron.

While concentrated nitric acid passivates steel, the phenomenon is too unreliable to permit cast irons to be used with confidence, even for strong nitric acids. The evidence available[28,29] in relation to mixed nitrating acids,

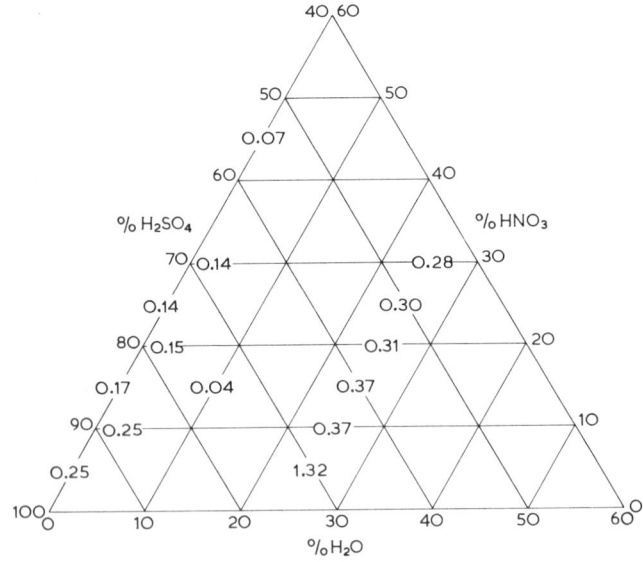

Fig. 3.41 Corrosion rates (mm/y) for cast iron in contact with mixed acids (H_2SO_4, HNO_3, H_2O). Figures indicate corrosion rates at room temperature in acid of the composition indicated

however, suggests that the corrosion rates may be lower here than in solutions of pure nitric or sulphuric acids of comparable dilution. Typical results are shown in Fig. 3.41.

Organic acids Dilute solutions of organic acids, especially if well aerated, attack cast iron at uneconomical rates. Temperature and velocity are also accelerating factors.

Corrosion by Alkalis

Dilute alkali solutions do not corrode cast iron at any temperature, but hot solutions exceeding about 30% concentration will attack it, with an accompanying evolution of hydrogen, to form a ferrite. Broadly speaking, if corrosion rates are to be held below 0·2 mm/y the temperature should not exceed 80°C. Corrosion rates may range as high as 1·25-2·5 mm/y in boiling solutions of more than 50% concentration. Molten caustic soda (650°C) may attack cast iron initially at rates around 20 mm/y, but it is probable that this figure decreases somewhat after a few days. In spite of these high corrosion rates, caustic-concentration and fusion pots are made from cast iron, since the material is relatively cheap, and the thick wall necessary for mechanical strength also gives the pot long life. It is of the utmost importance to ensure that concentration and fusion pots are cast in sound metal, as any unsoundness, particularly at the bottom of the pan, will lead to pitting attack and premature failure. The addition of 1 to 3% nickel to metal for these castings is said to be beneficial.

Corrosion by Salt Solutions

The corrosivity of a salt solution depends upon the nature of the ions present in the solution. Those salts which give an alkaline reaction will retard the corrosion of the iron as compared with the action of pure water, and those which give a neutral reaction will not normally accelerate the corrosion rate appreciably except in so far as the increased conductivity of the solution in comparison with water permits galvanic effects to assume greater importance. Chlorides are dangerous because of the ability of the anions to penetrate otherwise impervious barriers of corrosion products.

Those salts which hydrolyse to give an acid solution, e.g. the strong acid salts of aluminium, iron and, to a lesser extent, calcium, give solutions which may be very corrosive to cast iron, particularly if they are well aerated. When oxidising salts are also present in these acid solutions, a particularly dangerous system may be created. It is owing to this combination of oxidising and acidic character that mine waters are so corrosive.

If oxidising salts are present in neutral solutions they may reduce corrosion of the iron by the establishment of thin protective films on the metal surface. Their effectiveness is considerably diminished if much chloride ion is also present in solution.

Ammonium salts are more corrosive than those of the alkaline metals, the difference increasing directly with concentration and temperature. If free ammonia is also present, however, the pH value may be raised to a point at which little corrosion can occur. If ammonium salt solutions are exposed to locally high temperatures, it is possible that ammonia may be lost from the salts with a corresponding increase in the free acid content of the solution, and the development of a more corrosive character.

Corrosion under Stress

Irons exposed under conditions of cyclic stress are liable to corrosion fatigue. Some data on this effect have been given by Collins and Smith[30], but a more extensive study has been reported by Palmer[31]. By its nature, it is difficult to give absolute values for resistance to corrosion fatigue since this will be affected by the duration and frequency of cycling and the corrosivity of the environment. Palmer exposed specimens in a Wöhler fatigue machine to sprays of demineralised water, 3% sodium chloride solution and demineralised water containing various inhibitors, using a cycle frequency of about 3 000 cycles/min and a duration of 50 000 000 or 100 000 000 cycles. The results for the uninhibited solutions are summarised in Table 3.45. Palmer found that corrosion fatigue due to water spray could be eliminated or mitigated by some of the inhibitor systems examined, but it was

Table 3.45 Corrosion fatigue limiting strengths (MN/m^2)

Environment	Air*	Water*	3% NaCl Solution†
Pearlitic grey iron	126	100	39
Ferritic grey iron	93	77	23
Pearlitic ductile iron	270	224	46
Ferritic ductile iron	208	178	46

* Fatigue strength based on 50×10^6 cycles.
† Fatigue strength based on 100×10^6 cycles.

apparently easier to inhibit damage on grey irons that on ductile irons which could only be inhibited by 0·25% potassium chromate solution. He suggested this was due to an inability of the other systems to maintain a continuous passive film on the iron; grey iron is less sensitive to a notch effect in fatigue so that the presence of local sites of attack would be less important to this metal provided that the overall corrosivity of the solution were depressed.

Brown et al.[32] have reported that there have been reported that there have been several incidents of cracking of steel components carrying and storing reformed or coal gas. They suggest this was due to a stress-corrosion cracking mechanism associated with the partial inhibition of the steel surface by carbon monoxide absorbed from the gaseous contents. The phenomenon was more fully investigated by Kowaka and Nagata[33] who showed that the presence of carbon monoxide, carbon dioxide and water was essential for the cracking to take place. Similar cracking was subsequently observed on a ductile iron pipeline carrying a wet mixture of carbon monoxide and carbon dioxide. Studies to define the conditions which permit this corrosion mechanism indicated that a partial pressure of carbon monoxide in excess of about 0·5 bars is necessary to induce cracking in ductile iron stressed to about 300 MN/m^2. Higher CO partial pressures may, however, induce cracking at somewhat lower tensile stresses.

Harkness[34] has described a mode of stress accelerated corrosion of flake graphite iron pipes that he refers to as 'fissure corrosion'. This is produced on pipes buried in corrosive soils under significant bending stress and is manifested by deep narrow fissures, often largely filled with graphitic corrosion residue, oriented at right angles to the direction of stress. The nature of the attack and the factors affecting its incidence have been described in more detail by Palmer[35]. He showed that under very corrosive conditions, fissure corrosion could be produced in flake graphite irons by bending stresses of the order of 120 MN/m^2. It is not known whether there is a threshold stress below which the attack does not occur. Nodular graphite irons require a bending stress of the order of 250 MN/m^2 before fissuring occurs, and in this material the fissures are much less deleterious than in flake graphite irons since they are more shallow and less notch-like in character. Flake graphite irons exposed to fissures are liable to fail mechanically after relatively minor penetration of the pipe wall by the fissures, but no nodular graphite iron pipe has yet been caused to fail by fissure penetration alone.

H. H. COLLINS

REFERENCES

1. Collins, H. H., *BCIRA J.*, **10**, 543 (1962)
2. Fetisov, N. M., Marinchenko, B. V. and Volodin, V. V., *Russian Castings Protection*, 408 (1973)
3. Balakin, V. G. et al., *Liteinoe Proizvodstvo*, 31 (1975)
4. Berenson, J. and Wranglen, G., *Corrosion Science*, **20**, 937
5. Collins, H. H., *Corros. Sci.*, **21**, 259 (1981)
6. Graham, R., Prado, O. S., Collins, M. H., Brandes, E. A. and Farmery, H. K., *Proc. Inst. Mech. Engs.* **174**, 617, (1960)
7. Whittaker, J. A. and Brandes, E. A., *Foundry*, 70 (1962)
8. Mannweiler, G. B., *Proc. Amer. Soc. Testing. Mats.*, **72**, Appendix 1, 42 (1972)
9. Roll, F., *Werkstoffe u. Korrosion*, **12**, 209 (1962)
10. Friend, J. N., *Carnegie Scholarship Memoirs*, **18**, Iron and Steel Institute, 61 (1929)
11. Nekryti, S. S. and Karpov, V. T., *Russian Castings Production*, 369 (1961)
12. Dearden, J. and Swindale, J. P., *J. Iron St. Instn.*, **185**, 227 (1957)
13. Laque, F. L., *Corrosion*, **14**, 485t (1958)
14. Higgins, R. I., unpublished data
15. Collins, H. H., *Pitting of Diesel Cylinder Liners*, Diesel Engineers' and Users' Association (1961)
16. Romanoff, M., *Underground Corrosion*, US Nat. Bur. Stnd., Washington DC, Circular No. 579 (1957)
17. Schwerdtfeger, W. J., *J. Res. Bur. Stnd.*, **69C**, 71 (1985)
18. Booth, G. H., *Brit. Corr. J.*, **2**, 109 (1967)
19. Smith Harry, W., *Cast Iron Pipe News*, **38**, 3, winter (1971)
20. *Deutscher Verein des Gas und Wasserfaches*, Technisches Regel GW9, March (1986)
21. Winkler, A., *Gas und Wasserfach/Erdgas*, **120**, 335 (1979)
22. Stokes, R. F., *Chemistry and Industry*, 659, September (1983)
23. Collins, H. H., *Conf. Corrosion and the Water Industry*, UMIST, Manchester (1986)
24. Collins, H. H., Fuller, A. G. and Harrison, J. T., *12th World Gas Conference*, Nice, Report 19 U/D (1973)
25. De Rosa, P. J. and Parkinson, R. W., *Water Research Engineering Technical Report*, **241**, 22 (1986)

26. Collins, H. H. *BCIRA J.*, 157 (1982)
27. Collins, H. H., *Metallurgia*, 71, 177, April (1985)
28. Bate, S. C., *Chem. Age*, London, **15**, 419 (1926)
29. Chapman, F. F., *Trans. Amer. Inst. Chem. Engs.*, **18**, 7 (1926)
30. Collins, B. L. and Smith, J. O., *Proc. ASTM*, **42**, 639 (1942)
31. Palmer, K. B., *Proceedings of Conference on Engineering Properties and Performance of Modern Iron Castings*, BCIRA, 110 (1972)
32. Brown, A., Harrison, J. T. and Wilkins, R., *Corros. Sci.*, **10**, 547 (1970)
33. Kowaka, M. and Nagata, S., *Corrosion*, **32**, 395 (1976)
34. Hardness, I. D., *1st Int. Conf. on Internal and External Protection of Pipes*, Durham, X35 (1975)
35. Palmer, K. B. *BCIRA J.*, **30**, 1257 (1982)

3.7 High-nickel Cast Irons

Composition and Properties

The addition of about 20% nickel to cast iron produces materials with a stable austenitic structure; these materials are sometimes known as austenitic cast irons but are more often referred to commercially as *Ni-Resist* cast irons. The austenitic matrix of these irons gives rise to very different mechanical and physical properties to those obtained with the nickel-free grey cast irons. The austenitic matrix is more noble than the matrix of unalloyed grey irons and it was shown in the early work of Vanick and Merica[1] that the corrosion resistance of cast iron increases with increasing nickel content up to about 20% (Fig. 3.42).

Although the Ni-Resist irons, due to their austenitic matrix, are tougher and more shock resistant than the nickel-free grey irons, those in which the carbon is present in the flake graphite form (F.G. irons) still exhibit certain disadvantages due to the graphite structure. Much better strength and impact properties can be obtained by treating the iron with a small quantity of magnesium sufficient to give a residual content of 0·05-0·1%, which converts the graphite to a spheroidal form (S.G. irons). The Ni-Resist irons are available in both flake and spheroidal graphite forms and typical structures, consisting of flake or spheroidal particles dispersed throughout the austenitic matrix, are shown in Figs. 3.43 and 3.44. The matrix also contains small amounts of carbides, the amounts of which increase with increasing chromium content.

The first alloys in the Ni-Resist series, containing about 20% nickel, were introduced in the 1930s and soon became established in both corrosion and heat resistance applications. The range of alloys has been extended over the years and a total of twenty grades of austenitic irons have been developed with nickel contents varying from 13 to 35%. Each material has somewhat different characteristics so that the most appropriate grade must be selected to obtain the most advantageous properties for any particular application. The compositions and mechanical properties of the principal grades of austenitic cast irons are summarised in Table 3.46. There are six basic grades of flake graphite austenitic iron and five basic grades of spheroidal graphite austenitic iron. There is no spheroidal austenitic iron corresponding to Type 1 Ni-Resist since it is difficult to obtain a good spheroidal graphite structure in an austenitic iron containing more than 2% copper. A considerable number of modified grades also exist which differ in composition and properties from these basic grades. Specifications for eight commonly used grades of austenitic cast irons are given in BS 3468:1962.

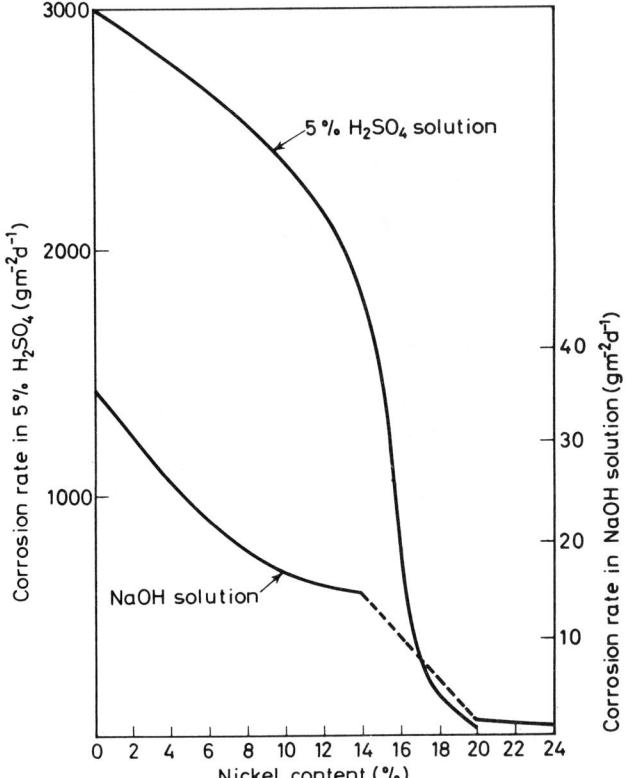

Fig. 3.42 Effect of nickel content on corrosion resistance

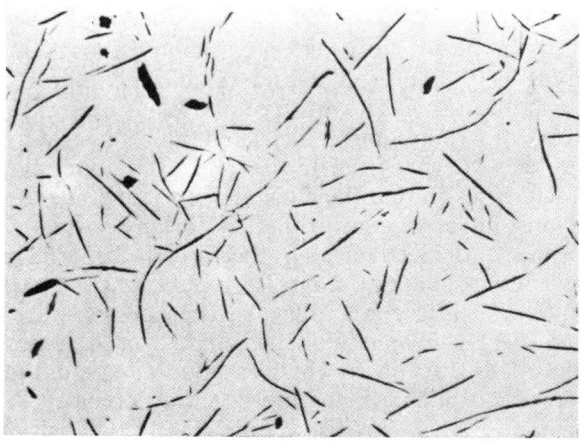

Fig. 3.43 Structure of typical flake graphite austenitic iron

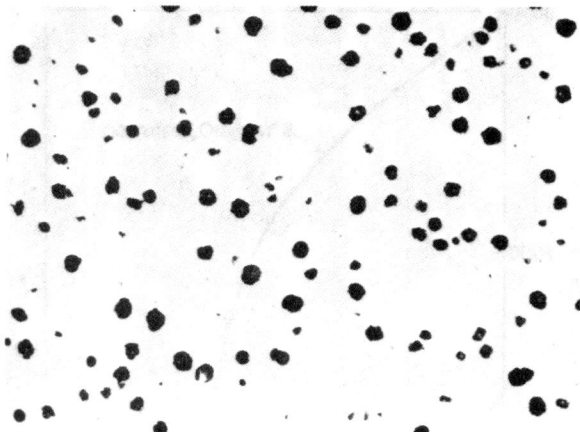

Fig. 3.44 Structure of typical spheroidal graphite austenitic iron

The tensile strengths of the spheroidal graphite irons are generally about twice those of their flake graphite equivalents and can be further improved, by about $8 \times 10^7 \, \text{Nm}^{-2}$, by quenching the iron in oil or water from temperatures of 925–1 000°C. This treatment is even more effective when applied to chill castings but the ductility of these is lower because of the increased amount of carbide formed as a result of chilling. The impact resistance of the spheroidal graphite grades is much better than that of the equivalent flake graphite irons and elongation values as high as 40% can be attained with the S.G. irons. The mechanical properties of the austenitic irons are also good at low temperatures which can be useful in a number of chemical plant and cryogenic applications.

The austenitic irons show excellent casting properties and good machinability, which, in combination with the good mechanical properties and good corrosion resistance, ensures wide use of these materials in many applications.

Aqueous Corrosion Behaviour

The austenitic cast irons show better corrosion resistance than the ferritic irons primarily due to the nickel content of the austenitic matrix.

Potential-current density (E–i) curves, which have been determined[2] for a number of the austenitic cast irons and also for the nickel-free ferritic irons, indicate that in general the austenitic cast irons show more favourable corrosion characteristics than the ferritic irons in both the active and passive states.

In de-aerated 10% sulphuric acid (Fig. 3.45) the active dissolution of the austenitic irons occurs at more noble potentials than that of the ferritic irons due to the ennobling effect of nickel in the matrix. This indicates that the austenitic irons should show lower rates of attack when corroding in the active state such as in dilute mineral acids. The current density maximum in the active region, i.e. the critical current density (i_{crit}) for the austenitic irons tends to decrease with increasing chromium and silicon content. Also the current densities in the passive region are lower for the austenitic irons

Table 3.46 Composition and properties of principal grades of flake and spheroidal graphite austenitic case irons

BS 3468 designation	Ni-Resist type	Composition (wt. %)						Minimum tensile strength ($\times 10^7$ N/m^2)	Brinell hardness
		C	Si	Mn	Ni	Cr	Cu		
AUS101A	Type 1	3·0	1–2·8	1–1·5	13·5–17·5	1·75–2·5	5·5–2·5	17	130–170
AUS102A	Type 2	3·0	1–2·8	0·8–1·5	18–22	1·75–2·5	0·5 max.	17	125–170
AUS105	Type 3	2·6	1–2	0·4–0·8	28–32	2·5–3·5	0·5 max.	17	120–160
—	Type 4	2·6	5–6	0·4–0·8	29–32	4·5–5·5	0·5 max.	17	150–210
—	Type 5	2·4	1–2	0·4–0·8	34–36	0·1 max.	0·5 max.	14	100–125
AUS104	—	1·6–2·2	4·5–5·5	1–1·5	18–22	1·8–4·5	0·5 max.	19	248 max.
AUS202A	Type D-2	3·0	1·75–3	0·7–1	18–22	1·75–2·5	—	37	140–200
AUS205	Type D-3	2·6	1·5–2·8	0·5 max.	28–32	2·5–3·5	—	37	140–200
—	Type D-4	2·6	5–6	0·5 max.	29–32	4·5–5·5	—	42	170–240
—	Type D-5	2·4	1·5–2·8	0·5 max.	34–36	0·1 max.	—	37	130–180
AUS204	—	3·0	4·5–5·5	1–1·5	18–22	1–2·5	—	37	230 max.

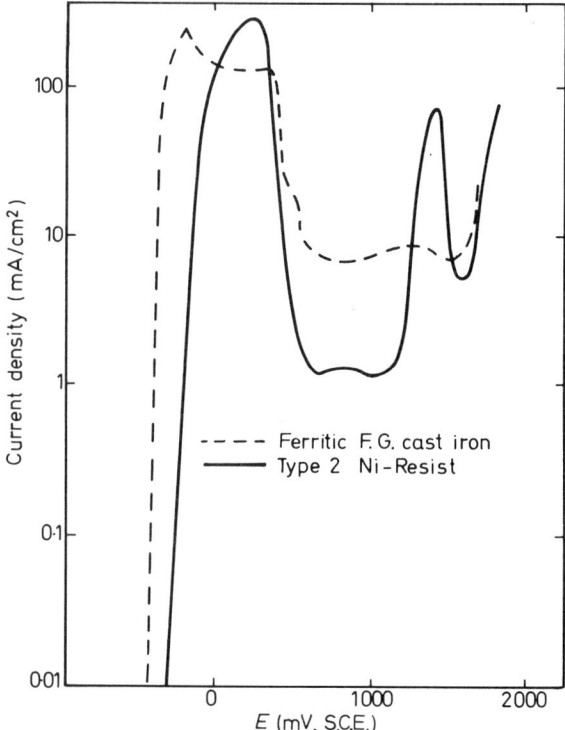

Fig. 3.45 Potential-current density curves in 10% sulphuric acid solution at 25°C

than for the ferritic. These observations indicate that the austenitic irons, particularly those of higher chromium and silicon content, should show superior passivating properties to the ferritic irons.

Similar curves determined in 50% sodium hydroxide solution at 60°C show (Fig. 3.46) that the austenitic irons exhibit more noble active dissolution and also lower current densities in the active and passive regions than the ferritic irons; the current densities in both regions decrease markedly with increasing nickel content (Fig. 3.47).

In 3% sodium chloride solution at 60°C the austenitic irons again show superior characteristics to the ferritic. The breakdown potentials determined in this environment, which provide a relative measure of the resistance to attack in neutral chloride solutions, are generally more noble for the austenitic irons than for the ferritic (Table 3.47). This indicates that the austenitic irons should show better corrosion resistance in such environments.

The more favourable electrochemical characteristics exhibited by the austenitic irons in this range of environments are reflected in the corrosion behaviour of the alloys discussed below.

One of the outstanding properties of the austenitic irons is their resistance to graphitic corrosion or 'graphitisation'. In some environments ferritic cast irons corrode in such a manner that the surface becomes covered with a layer of graphite. This compact graphite layer, being more noble than the matrix, markedly increases the rate of attack. The austenitic irons rarely form this

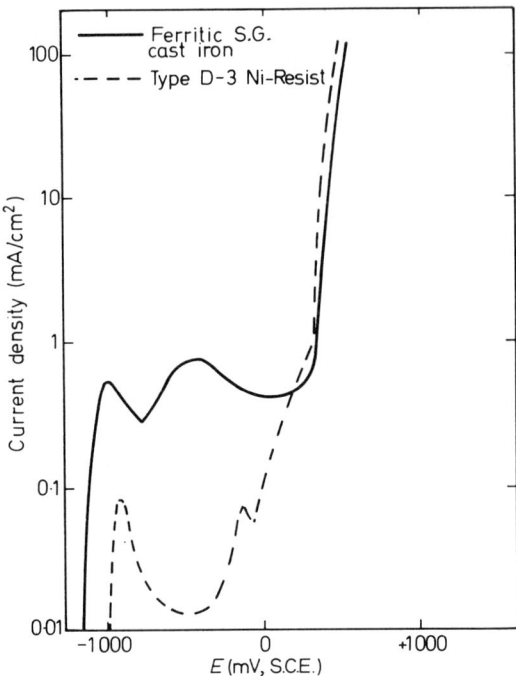

Fig. 3.46 Potential-current density curves in 50% sodium hydroxide solution at 60°C

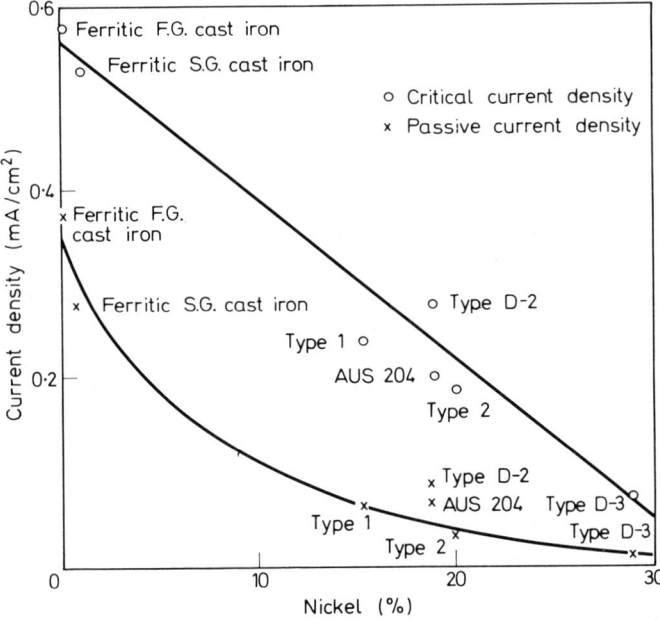

Fig. 3.47 Current densities in active and passive regions for ferrite and austenitic cast irons in 50% NaOH

Table 3.47 Breakdown potentials evaluated from E-i curves in 3% NaCl at 60°C

Alloy	Breakdown potential (mV vs S.C.E.)
AUS101A	−470
AUS102A	−520
AUS202A	−570
AUS205	−620
AUS204	−620
Ferritic S.G. cast iron	−670
Ferritic F.G. cast iron	−720
Mild steel	−620

graphite layer and consequently, in environments where graphitic corrosion is a problem, perform much better than low alloy cast irons.

Practical experience indicates that the corrosion resistance of the flake and spheroidal graphite irons is similar in many environments; however, the spheroidal graphite irons have shown superior corrosion resistance to the equivalent flake graphite grades in a number of cases[3].

Atmospheric Corrosion

Although the Ni-Resist irons will not remain rust-free when exposed to the atmosphere their corrosion resistance is much better than that of plain cast iron or mild steel. The results of a 7.5 year exposure trial carried out in a

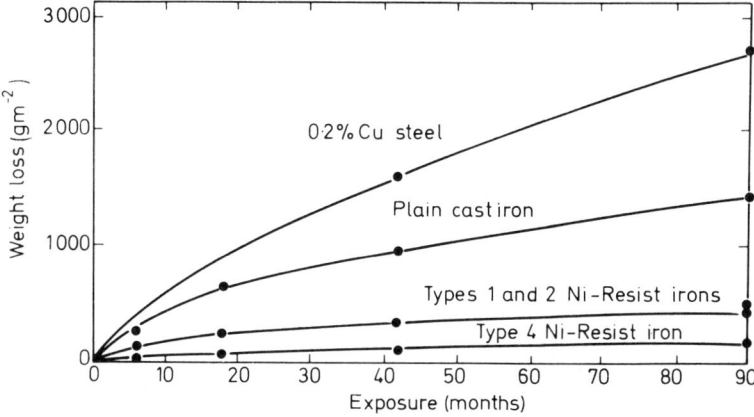

Fig. 3.48 Results of a 7.5 year exposure test programme on 150 × 100 mm panels at Kure Beach, N.C.

marine environment at Kure Beach, North Carolina, USA are shown in Fig. 3.48. The corrosion rates derived from the curves after 7.5 years exposure are given in Table 3.48.

Table 3.48 Corrosion rates after exposure for 7·5 years at Kure beach

Alloy	Corrosion rate (mm y^{-1})
0·2% Cu steel	0·020
Cast iron	0·010
Types 1 and 2 Ni-Resist	<0·003
Type 4 Ni-Resist	<0·003

Natural Waters

Water which is used for cooling purposes in refineries and chemical plant can cause severe problems of corrosion and erosion. Ordinary cast irons usually fail in this type of environment due to graphitic corrosion or corrosion/erosion. Ni-Resist irons however show better corrosion resistance, due to the nobility of the austenitic matrix, and are preferred for use in the more aggressive environments such as those containing appreciable amounts of carbon dioxide or polluted with chemical wastes or sea-water.

The austenitic irons have also been shown to exhibit better corrosion resistance than the ferritic irons in sea-water. Tests over long periods of time have shown that Ni-Resist irons of Types 1, 2 and 3 corrode at rates of 0·020 to 0·058 mm y^{-1} in relatively quiet sea-water. Under similar conditions low alloy cast irons have shown corrosion rates ranging from 0·066 to 0·53 mm y$^{-1(4)}$. The Ni-Resist irons maintain this superiority over a wide variety of conditions (Figs. 3.49 and 3.50) both in stationary and flowing sea-water. In a test lasting 740 days in sea-water moving at 1·5 m/s low

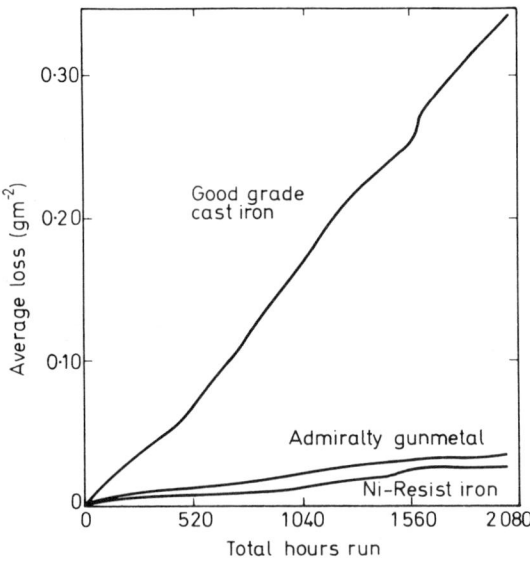

Fig. 3.49 Relative corrosion of Ni-Resist iron, cast iron and gunmetal in aerated sea-water

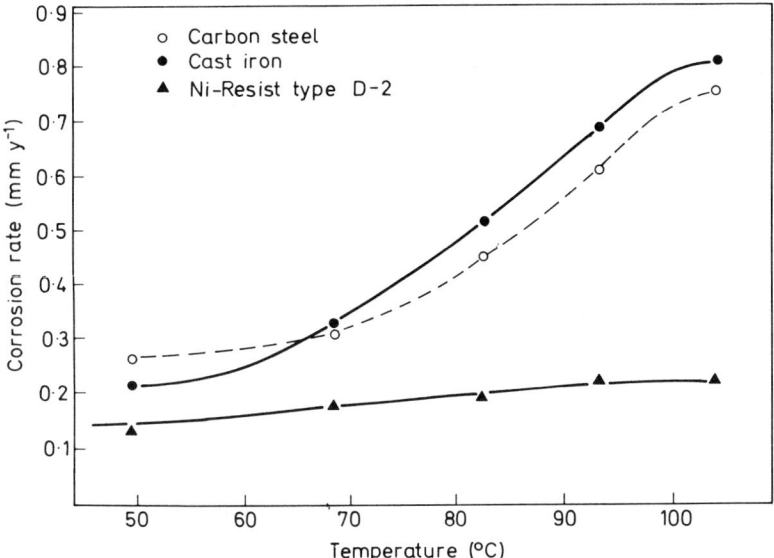

Fig. 3.50 Corrosion rate versus temperature in de-aerated sea-water; continuous test exposure for 156 d

Table 3.49 Sea-water corrosion/erosion test carried out at 8 m/s at 28°C for 60 d

Alloy	Average corrosion rate (mm y^{-1})
Cast iron	6·9
2% Ni cast iron	6·1
Type 1 Ni-Resist	0·74
Type 2 Ni-Resist	0·79
Type 3 Ni-Resist	0·53

alloy cast iron showed a corrosion rate of 1·3 mmy^{-1} compared to 0·050 mmy^{-1} for Type 2 Ni-Resist. In tests carried out at controlled temperature at a higher velocity of 8 m/s (Table 3.49) the Ni-Resist irons again showed better properties than low alloy cast irons or mild steel.

Acids

Under certain conditions of temperature and concentration the austenitic cast irons show useful resistance to hydrogen-evolving mineral acids.

The austenitic irons can be usefully applied in handling very dilute solutions of sulphuric acid at ambient or moderately elevated temperatures under conditions which can be very corrosive to ordinary cast iron and carbon steel. Austenitic irons have also given satisfactory service in handling concentrated sulphuric acids, but although they show low corrosion rates in such environments they are not markedly superior to the unalloyed cast irons. Type 1 Ni-Resist and the high silicon grades AUS104 and AUS204 are the types most generally used in sulphuric acid environments.

The austenitic irons are superior to ordinary cast iron in their resistance to corrosion by a wide range of concentrations of hydrochloric acid at room temperature (Table 3.50). However, for practical uses where such factors as velocity, aeration and elevated temperatures have to be considered, the austenitic irons are mostly used in environments where the hydrochloric acid concentration is less than 0·5%. Such environments occur in process streams encountered in the production and handling of chlorinated hydrocarbons, organic chlorides and chlorinated rubbers.

Table 3.50 Corrosion of Type 1 Ni-Resist, cast iron and carbon steel in unaerated hydrochloric acid solutions at room temperature

Acid concentration (%)	Corrosion rate (mmy^{-1})		
	Ni-Resist	Cast iron	Carbon steel
1·8	0·13	23	15
3·6	0·38	30	36
5·0	0·46	38	46
10·0	0·41	30	48
20·0	1·1	32	69
27·0	3·0	30	60
36·0	9·4	28	30

The austenitic irons are also useful in some circumstances for handling organic acids such as dilute acetic, formic and oxalic acids, fatty acids and tar acids. They are more resistant to organic acids than unalloyed cast irons, e.g. in acetic acid the austenitic irons show corrosion rates 20–40 times lower than the ferritic iron (Table 3.51).

Table 3.51 Corrosion of Type 1 Ni-Resist and ferritic cast iron in acetic acid in laboratory tests at 15°C

Acid concentration (%)	Corrosion rate (mmy^{-1})	
	Cast iron	Ni-Resist
5	17	1·0
10	22	0·5
25	20	0·5
50	16	2·0

The austenitic irons show poor resistance to solutions of nitric acid even when dilute and at low temperatures.

Alkalis

Austenitic cast irons show particularly good corrosion resistance in alkaline environments, even better than that shown by low alloy cast irons. The resistance to corrosion improves with increasing nickel content (Fig. 3.51),

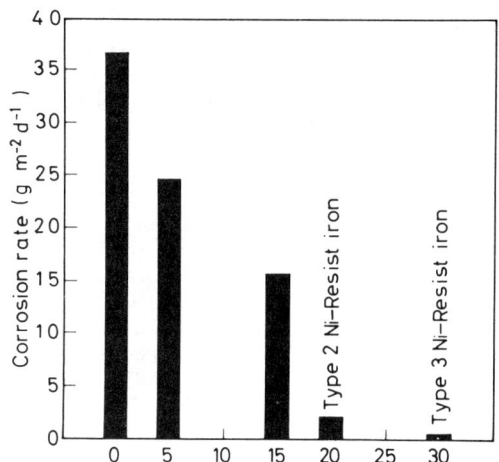

Fig. 3.51 Effect of nickel additions to cast iron in reducing corrosion by caustic alkalis

Table 3.52 Corrosion of Ni-Resist irons, cast iron and carbon steel in caustic soda solutions

Alloy	Corrosion rate (mmy^{-1})	
	14% NaOH (88°C)	74% NaOH (127°C)
Cast iron	0·2	1·93
Carbon steel	0·2	0·38
Ni-Resist Type 1	0·07	—
Ni-Resist Type 2	—	0·15
Ni-Resist Type 3	—	0·06
Ni-Resist Type D-2	—	0·13

and the irons containing about 30% nickel, such as Type 3 Ni-Resist, show the best resistance. This high corrosion resistance enables Ni-Resist cast irons to be used in caustic soda processes where both high temperature and concentrations are encountered (Table 3.52). One of the most important uses for the austenitic irons has for many years been in caustic soda production plant where the materials find use in many components. Molten caustic soda is very much more aggressive than aqueous solutions and a corrosion rate of 6·6 mmy^{-1} has been recorded[5] for austenitic iron in molten sodium hydroxide at 670°C.

The austenitic irons also show good corrosion resistance in caustic alkalis containing sulphides and mercaptans and have therefore proved useful materials for the construction of pumps, valves and piping in caustic soda regenerators in oil refineries.

Salts

The corrosion rates of the Ni-Resist irons in salt solutions depend upon the chemistry of the salt. Solutions which are alkaline or neutral in reaction are not generally corrosive to high nickel irons and even brines containing calcium and magnesium chloride can be safely handled by the austenitic irons (Table 3.53). Those salts which hydrolyse to give an acidic solution are more corrosive to high-nickel irons although the corrosion rate is still less than for unalloyed cast iron in the same medium. The corrosion rates of Type 1 Ni-Resist and cast iron given in Table 3.54 demonstrate that the austenitic iron shows better resistance than the nickel-free cast iron in a wide range of salt solutions.

Table 3.53 Corrosion of Type 1 Ni-Resist and cast iron in brine solutions

Environment	Temperature (°C)	Corrosion rate (mmy^{-1})	
		Ni-Resist	Cast iron
14% NaCl + 16·7% CaCl$_2$ + 3·4% MgCl$_2$. pH ≈ 6	69	0·08	0·53
Saturated NaCl	93	0·12	1·85

Table 3.54 Corrosion of Type 1 Ni-Resist and cast iron in various inorganic salt solutions

Salt solution	Concentration (%)	Temperature (°C)	Corrosion rate (mmy^{-1})	
			Ni-Resist	Cast iron
Aluminium sulphate	5	16	0·41	1·0
Aluminium chloride	5	16	0·08	1·3
Aluminium chloride	5	93	0·15	4·8
Aluminium sulphate	5	16	0·15	0·76
Aluminium citrate	30	Room	1·5	550
Aluminium thiocyanate	50	27	0·38	3·3
Potassium aluminium sulphate	5	16	0·25	0·76
Zinc chloride	30	Boiling	2·0	16
Ammonium nitrate	5	Room	0·23	0·76
Manganese chloride	10	77	0·038	0·79
Ammonium chloride	20	93	0·25	5·8

Applications

The individual characteristics and uses of the basic grades of the austenitic irons are given in Table 3.55. The major uses for these materials occur in the handling of fluids in the chemical and petroleum industries and also in the power industry and in many marine applications. The austenitic irons are also used in the food, soap and plastics industries where low corrosion rates are essential in order to avoid contamination of the product. Ni-Resist grades Type 2, 3 or 4 are generally used for such applications but the highly alloyed Type 4 Ni-Resist is preferred where low product contamination is of prime importance.

Table 3.55 Characteristics and uses of basic grades of austenitic cast irons

BS 3468 Designation	Ni-Resist type	Characteristics	Uses
AUS101A	Type 1	Least expensive austenitic iron; good corrosion resistance particularly in acidic media	Pumps, valves, furnace components
AUS102A AUS202A	Type 2 Type D-2	Good corrosion resistance; better than Type 1 in alkaline environments	As for Type 1 but preferable for alkaline solutions; used in soap and plastic industries
AUS105 AUS205	Type 3 Type D-3	Good thermal shock resistance; high resistance to erosion particularly in alkaline media	Pumps, valves, pressure vessels, filter parts, exhaust gas manifolds
— —	Type 4 Type D-4	Best corrosion resistance and erosion resistance of the austenitic irons	Castings for industrial furnaces; used in food industry for low contamination of product
— —	Type 5 Type D-5	Very low thermal expansion; good dimensional stability	Scientific instruments, glass moulds
AUS104 AUS204	— —	Good resistance to high temperature oxidation; good corrosion resistance in sulphuric acid	Pumps and valves

The modified grades of Ni-Resist are often sufficiently different to the basic grades to be used in additional applications. Ni-Resist Types 1B, 2B and D-2B have a higher chromium content than the corresponding basic grades which increases the erosion resistance of the materials; these grades are less expensive than the other grades with good erosion resistance, i.e. Ni-Resist Types 3 and 4. A chromium-free grade of austenitic iron, Ni-Resist Type D-2C, has particularly good ductility and shows good mechanical properties down to −100°C. Even better low temperature properties are, however, obtained with Ni-Resist Type D-2M, a chromium-free 4% manganese grade, which was specially developed for cryogenic applications such as the separation of aromatic hydrocarbons and the production of ethylene. Ni-Resist Type D-4A is a recently developed grade of austenitic iron which has particularly good resistance to high temperature oxidation and better ductility than the standard Type D-4 grade[6].

A. C. HART

REFERENCES

1. Vanick, J. S. and Merica, P. D., *Trans. Amer. Soc. Steel Treat.*, **18**, 923 (1930)
2. Evans, T. E. and Hart, A. C., unpublished results
3. Swales, G. L., *The Industrial Chemist*, Jan.-March (1963)
4. Uhlig, H. H., (Editor), *The Corrosion Handbook*, John Wiley, New York, 385–386 (1948)
5. La Que, F. L., *Cast Iron in the Chemical and Process Industries*, Grey Iron Founders' Society Inc., Cleveland, Ohio (1945)
6. Cox, G. J., *British Foundryman*, 1–14, Jan. (1970)

3.8 High-chromium Cast Irons

Composition

There is no clear demarcation between high-chromium steels and high-chromium cast irons other than the fact that components are fabricated from the steels, and cast in the irons. In practice, however, the irons are usually found to have carbon contents of between 0·6 and 3%, while most of the steels contain less than 0·3% carbon.

Of the high-chromium irons, those used for components requiring a high degree of corrosion resistance normally contain 25-35% chromium, although it has been suggested by Küttner[1] that much higher contents may sometimes be necessary to give adequate resistance in certain environments. It is commonly agreed[2] that a useful formula for the minimum chromium content of a corrosion-resistant iron is:

$$\%Cr = (\%C \times 10) + 12$$

i.e., a 1·5% carbon alloy should contain not less than 27% chromium. This is supported by work carried out by Küttner[1] with various solutions, although his own interpretation of his results is regarded as faulty.

The limitations imposed by this formula, together with the fact that the alloys in practice rarely contain more than 35% chromium, suggest that the maximum carbon content of the irons should be 2·3%, and in fact the irons normally contain between 1·0% and 2·0% carbon, unless some property other than corrosion resistance is the most important.

Silicon may be present in high-chromium irons in amounts varying between 0·5 and 2·5%. Its effect is to increase fluidity in the foundry and improve the surface quality of castings. Further effects are to refine the eutectic carbides in the iron, to produce a more uniform structure and to raise the temperature at which the matrix transforms from ferrite to austenite with consequent dimensional changes. Additions above 2·5%[3,4] have an embrittling effect.

Structure

Irons of the compositions indicated above all have structures similar to that shown in Fig. 3.52, that is, a uniform dispersion of chromium-iron complex carbides in a matrix of chromium-containing ferrite. The chromium content of the ferrite is not known, although it is assumed to be about 10-13%. The

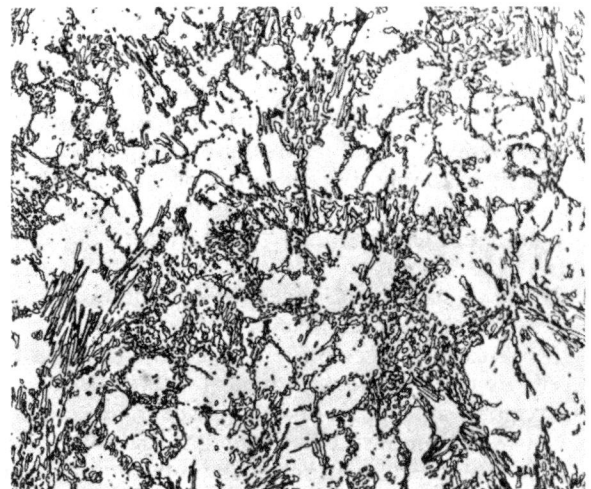

Fig. 3.52　Microstructure of 30% chromium iron. Analysis: total C 1·6, Si 1·8, Cr 31. Etched in Murakami's reagent, total magnification × 150

carbides are probably mixtures of the types Cr_7C_3 and $Cr_{23}C_6$, in which some of the chromium has been replaced by iron[5].

Mechanical Properties

Broadly speaking, the high-chromium irons are hard but not completely unmachinable. Typical properties for irons of the compositions described above in the as-cast state are:

Brinell hardness	320 HB
Transverse strength	695 MN/m^2
Tensile strength	463 MN/m^2

The hardness of the alloys makes them particularly useful in environments where abrasion or wear resistance may be important.

Production

As already indicated, these irons are used for the production of components by casting.

The principal difficulties in the production of castings in this alloy are its high shrinkage, which entails some tendency to the development of porosity, and the ready formation of oxide skin, which may cause cold laps in the casting. Castings must in consequence be produced by methods similar to those employed for steel castings and care must be taken to avoid the introduction of oxide into the mould.

Corrosion Resistance

The high-chromium irons undoubtedly owe their corrosion-resistant properties to the development on the surface of the alloys of an impervious and highly tenacious film, probably consisting of a complex mixture of chromium and iron oxides. Since the chromium oxide will be derived from the chromium present in the matrix and not from that combined with the carbide, it follows that a stainless iron will be produced only when an adequate excess (probably not less than 12%[1,2]) of chromium over the amount required to form carbides is present. It is commonly held, and with some theoretical backing, that carbon combines with ten times its own weight of chromium to produce carbides[1]. It has been said that an increase in the silicon content increases the corrosion resistance of the iron[1]; this result is probably achieved because the silicon refines the carbides and so aids the development of a more continuous oxide film over the metal surface. It seems likely that the addition of molybdenum has a similar effect, although it is possible that the molybdenum displaces some chromium from combination with the carbon and therefore increases the chromium content of the ferrite.

The irons are most useful in environments containing a plentiful supply of oxygen or oxidising agents; anaerobic or reducing conditions may lead to rapid corrosion. Physical effects such as abrasion or sudden dimensional changes induced by temperature fluctuations may rupture the film and allow corrosion to take place. The iron will also be subject to corrosion by solutions containing anions, such as those of the halides, which can penetrate surface films relatively readily.

Atmospheric Corrosion

Provided there is a suitable excess of chromium over carbon in the alloy, the irons will not rust when exposed to the atmosphere in the as-cast state. Alloys which have been found to tarnish in the as-cast state because of an inadequate excess of chromium may be found to be completely stainless in the machined and polished state, presumably because a thin film is more likely to be continuous on a smooth surface than on a rough one.

Natural and Industrial Waters

Because of its mechanical properties and the difficulties associated with its production, high-chromium iron is mostly used in environments which are particularly aggressive to other cast alloys. It is most useful for handling acid waters containing oxidising agents, for example mine waters and industrial effluents. Because many of these waters tend to contain solid matter in suspension, which can lead to abrasion of metals exposed to them, the very hard high-chromium iron is often the most suitable material for pumps handling these solutions. There is always a possibility that abrasive slurries may damage components made from this material by breaking down the oxide film and allowing corrosion to take place, but provided there is a plentiful supply of oxygen or oxidising agent at the metal surface this danger is reduced by the rapid healing of the film.

Acids

The most comprehensive data available about the corrosion behaviour of high-chromium irons in a wide variety of acid, basic and saline solutions are those issued by Bergische Stahl-Industrie of Remscheid, Germany[6], and this has formed the basis of all the following comments. Additional data given by Küttner[1] and Houdremont and Wasmuht[7] are generally, but not completely, in agreement with this.

Figure 3.53 has been derived from data[6] for corrosion by nitric acid solutions of an alloy nominally containing 29% chromium and 0·8% carbon. It

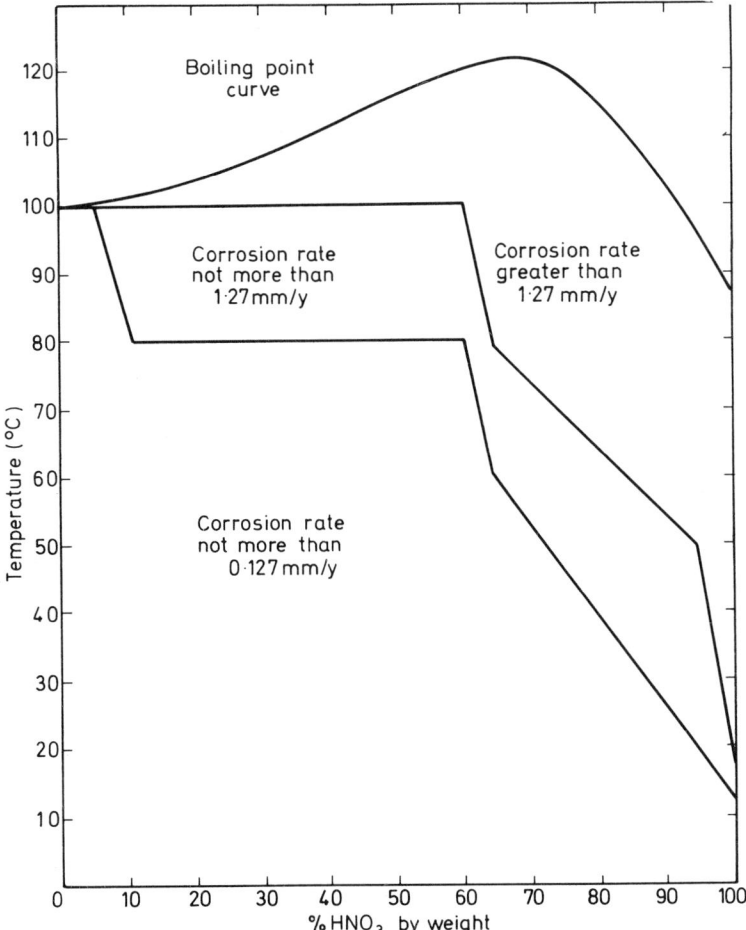

Fig. 3.53 Resistance of high-chromium iron to nitric acid solutions

is interesting to note that the very useful resistance afforded by this alloy in these solutions is roughly complementary to that of the high-silicon irons, the high-chromium iron being more suitable for dilute solutions and the high-silicon iron for concentrated solutions. Because of the lack of information about the method of obtaining the results on which this and the subsequent diagram are based, it must be understood that the diagram gives only an indication of corrosion rates likely to be encountered and should not be considered as authoritative.

The data available[6] suggest that high-chromium irons have no useful resistance to sulphuric acid of more than 10% concentration at any temperature. At temperatures above 20°C corrosion rates in excess of 1·27 mm/y are probable even for acid of less than 10% concentration. The addition of 2% molybdenum appears to produce an appreciable increase in the resistance to this acid at very low and very high concentrations (Fig. 3.54).

It is doubtful whether the irons have any useful resistance to hydrochloric acid solutions at any concentration or temperature. It has, however, been claimed that the molybdenum-containing alloy is attacked by 1% acid at

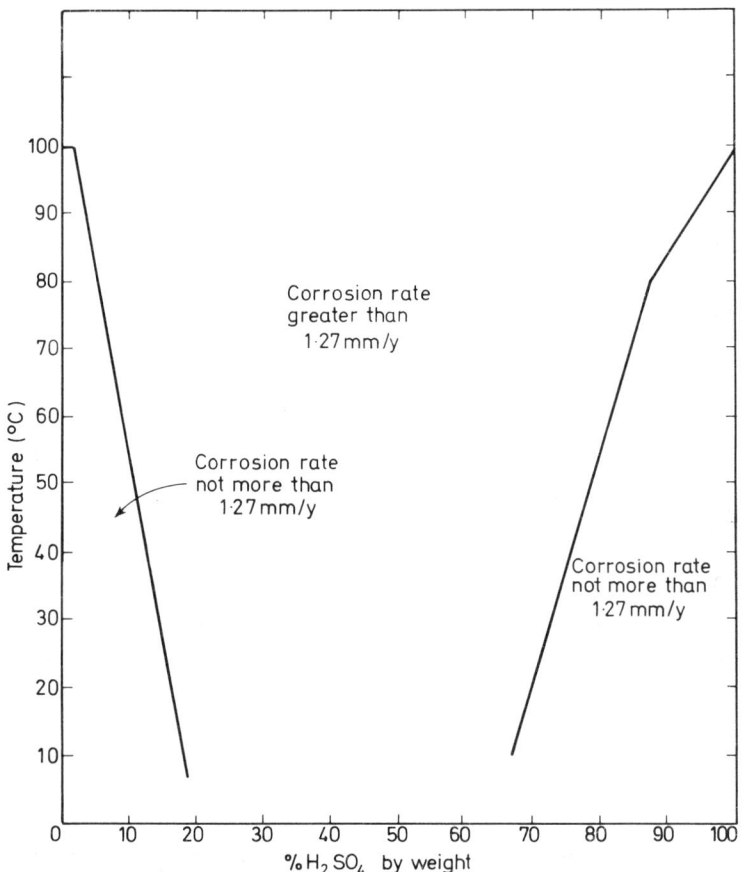

Fig. 3.54 Resistance of high-chromium iron containing 2% molybdenum to sulphuric acid solutions

20°C at rates not exceeding 1·27 mm/y; 2% acid, however, attacks this alloy at rates greater than 1·27 mm/y.

Although the irons have such poor resistance to sulphuric acid solutions, they are reported to be much more resistant to sulphuric acid/nitric acid mixtures which rarely cause corrosion rates of more than 1·27 mm/y. Aqua regia is corrosive to the alloys, although Küttner[1] has reported that an increase in the chromium content of the alloys, apparently according to the formula

$$\%Cr = (\%C \times 5) + 36$$

is effective in producing a resistance to this solution.

Other acids are completely resisted by the irons at room temperature, although high corrosion rates can sometimes develop at elevated temperatures.

Alkalis

The data available suggest that the high-chromium irons do not offer any better resistance to alkalis than unalloyed grey iron, which would normally be preferred in view of its lower cost and its mechanical properties.

Salt Solutions

Many salts which are corrosive towards unalloyed iron because of their tendency to hydrolyse to release acid, e.g. calcium and zinc chlorides, are not dangerous to high-chromium irons. The more corrosive salts, typified by aluminium sulphate and ferric chloride, are, however, corrosive to high-chromium irons. Hot aluminium sulphate solutions can give corrosion rates greater than 1·27 mm/y although cold solutions corrode the alloys at rates not exceeding 0·127 mm/y.

Ferric chloride solutions are particularly aggressive to high-chromium irons. Rates of attack greater than 12 mm/y have been recorded for a 25% solution at 20°C. The useful resistance of the alloys to mine waters which contain this salt is probably because the concentration involved is very much lower than this.

Küttner[1] has indicated that irons of the higher chromium content which he reports as being able to resist aqua regia are also resistant (corrosion rate not more than 0·127 mm/y, to a cold 30% solution of ferric chloride.

Resistance to High Temperature Corrosion

Irons containing around 12% chromium[8] may be used for service up to 900°C whilst those in the range 30–35% chromium, 1–2% carbon and 1–2% silicon have a scaling resistance up to 1 050°C[9,10], being employed for furnace hearths, kiln furniture, heat exchangers, etc. The composition range represents a compromise between ease of founding, avoiding ferrite/

austenite changes on thermal cycling and maximum scaling resistance consistent with the avoidance of the long term embrittling effect of σ-phase formation in the temperature range 600–800°C[11].

Corrosion-Erosion Resistant High Chromium Alloy Iron[12]

High-chromium alloys with carbon contents in the range 0·5–2·0% afford a useful compromise between resistance to corrosion and resistance to abrasion. As the carbon content is increased, the resistance to abrasion improves, but corrosion resistance is reduced. The matrix structure of this range of high-chromium alloy irons can be largely austenitic or it can be transformed to martensite by heat treatment. There has been increased interest in this series of alloy irons in recent years because they would seem to offer a cost-effective solution to problems encountered in handling abrasive slurries arising from gas scrubber installations in coal-fired power stations. They are also seen as candidate materials for the high-speed high-pressure pumps necessary in coal liquefaction projects, since they are able to resist abrasion at temperatures at which many abrasion-resistant steels would soften.

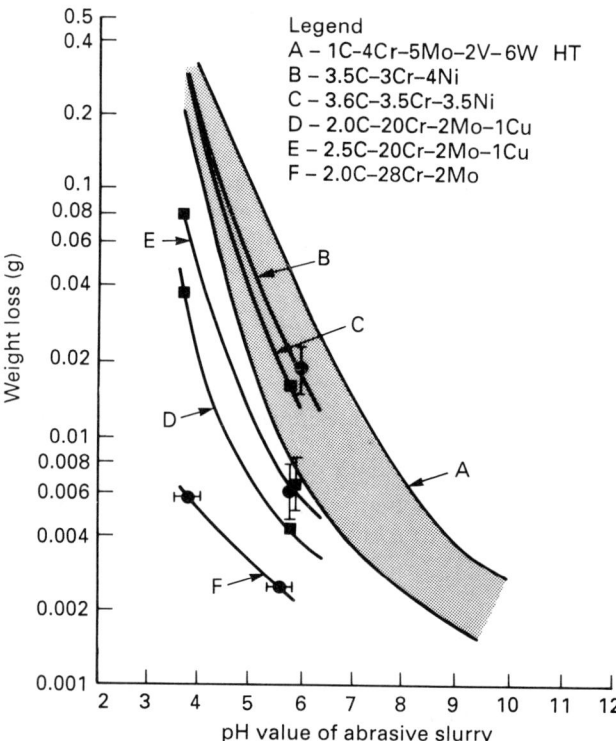

Fig. 3.55 Influence of pH value on weight loss in the slurry abrasion test

Tests carried out on the slurries encountered in gas scrubber installations in power-generating plants have shown that corrosion is generally less damaging than abrasion. Thus, it has been possible to utilise irons of higher carbon contents than the traditional corrosion-resistant chromium irons. Development of alloy irons to resist abrasion in somewhat corrosive environments has benefited from the data generated in recent years on the effect of alloying elements such as nickel, copper, manganese and molybdenum, on the hardenability of the high-chromium irons and on the stability and properties of high-chromium austenites.

Figure 3.55 shows that the high-chromium irons containing 2–2·5% carbon and 20–28% chromium have relatively good resistance to slurry abrasion at pH values down to 4. In more acidic environments, metal loss rates accelerate rapidly. Alloys with lower carbon content or those with higher molybdenum and nickel contents have been developed for service in more aggressive environments. Potentiodynamic polarisation curves generated for three high-chromium irons[13] show that the lower carbon material exhibits passivation behaviour in all four test solutions at current densities and potentials that show some promise of significant corrosion resistance (Figs. 3·56–3·59). This has been borne out in field trials on erosion–corrosion resistant

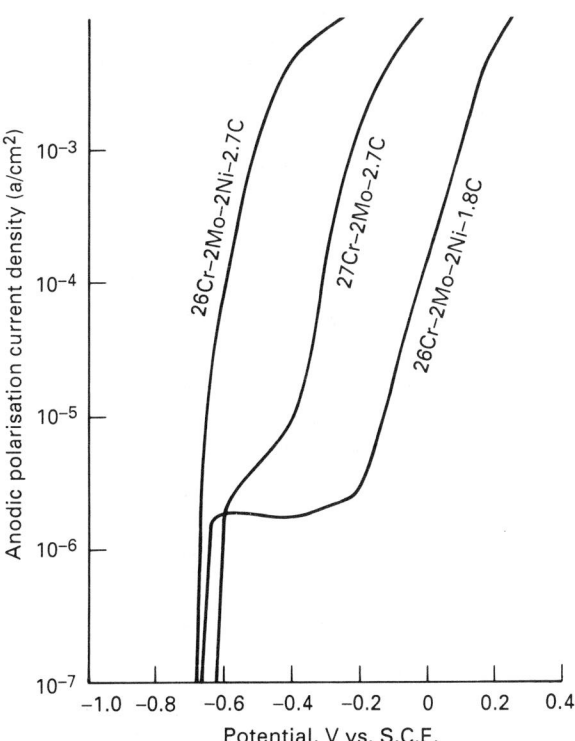

Fig. 3.56 Potentiodynamic polarisation curves for high-chromium white irons in nitrogen-saturated solution containing 800mg/l Cl^-, pH 3.5, 25°C

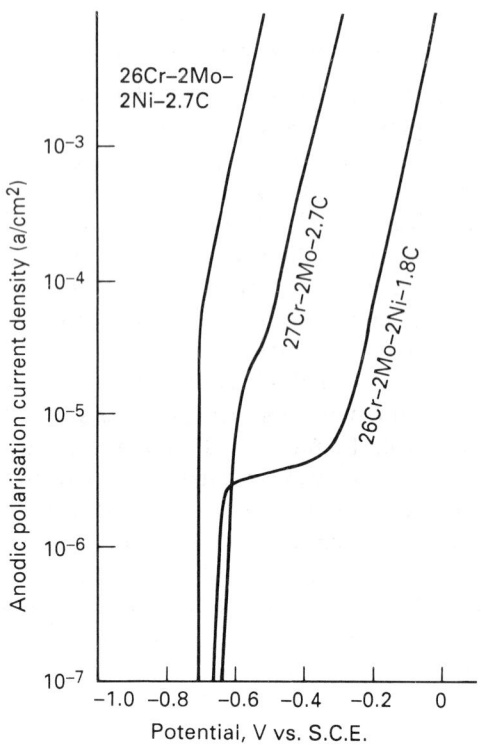

Fig. 3.57 Potentiodynamic polarisation curves for high-chromium white irons in nitrogen-saturated solution containing 4 000 mg/l Cl$^-$, pH 3.5, 25°C

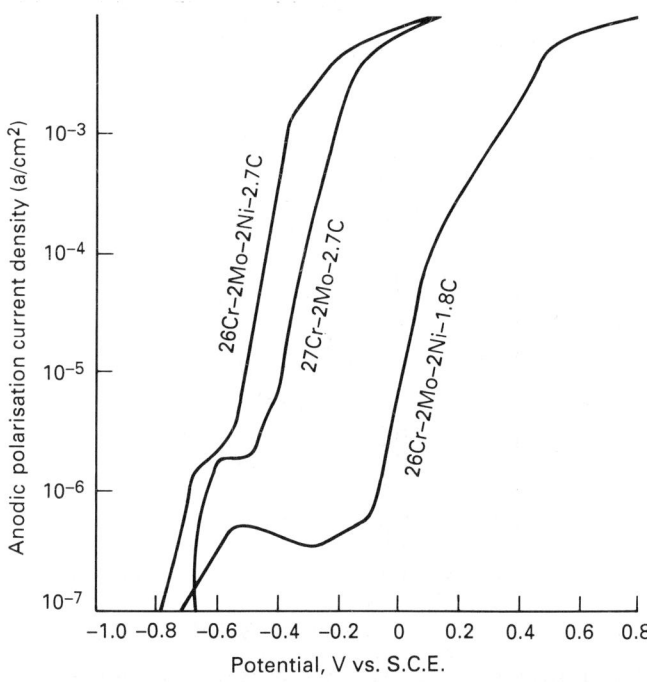

Fig. 3.58 Potentiodynamic polarisation curves for high-chromium white irons in nitrogen-saturated solution containing 800 mg/l Cl$^-$, pH 7, 25°C

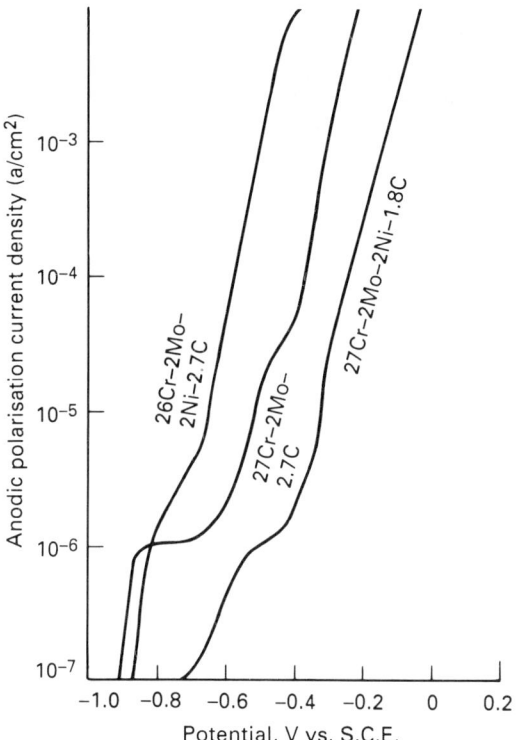

Fig. 3.59 Potentiodynamic polarisation curves for high-chromium white irons in nitrogen-saturated solution containing 4 000 mg/l Cl⁻, pH 7, 25°C

pump alloys providing the corrosion conditions were not severe. Although the high-chromium high-carbon alloy irons are clearly not suitable for service in highly corrosive conditions, they offer a cost-effective solution to severe abrasion in many important applications, particularly when measures are taken to control pH.

J. DODD

REFERENCES

1. Küttner, C., *Tech. Mitt. Krup.*, No. 1, March 17 (1933)
2. Kinzel, A. B. and Franks, R., *Alloys of Iron and Chromium*, Vol. 2: *High Chromium Alloys*, McGraw-Hill, New York (1940). See especially pp. 228-260
3. Valenta, E., *I.S.I. Carnegie Schol. Mem.*, **19**, 79-165 (1930)
4. Valenta, E. and Poboril, F., *Chemie et Industrie*, June, 633-648 (1933)
5. Jackson, R. S., *J.I.S.I.*, **208**, 163-167 (1970)
6. *Corrodur Bergit*, Bergische Stahl-Industrie, Remscheid, Germany (1957). See especially pp. 19-34
7. Houdremont, E. and Wasmuht, R., *Krupp. Mh.*, No. 12, 331 (1931). See also *Metals and Alloys*, **4**, 13, Feb. (1933)
8. Hallett, M. M., *J.I.S.I.*, **170**, 321-329 (1952)
9. Colton, W. J. *Brit. Foundryman*, **56**, 237-261 (1963)
10. Dixon, R. H. T. and Cumberland, J., *Foundry Trade Journal*, **116**, 721-726 and 785-791 (1964)

11. Boyes, J., *B.C.I.R.A.*, 715-731 and 461-475 (1963)
12. Lizlov, G., *Corrosion Resistance of 28% Chromium 2% Molybdenum White Iron*, unpublished report, Climax Molybdenum Co. Laboratory
13. Dodd, J., 'Recent Developments in Abrasion Resistant High Chromium-Molybdenum Iron, Low-Alloy Manganese Steels and Alloyed Nodular Iron of Importance in the Extraction and Utilization of Energy Resources', *J. Materials for Energy Systems, A.S.M*, **2**, 65-76, September (1980)

3.9 Silicon–Iron Alloys

Composition and Structure of Silicon Irons

The normal silicon content of cast iron (up to about 3%) has little effect upon the corrosion resistance of the alloys. In alloys containing much greater amounts of silicon, however, the silicon is responsible for the development of a marked increase in chemical resistance. These alloys can be divided into two types: those containing 4 to 10% silicon, which are used in applications requiring an iron with good resistance to oxidation at high temperatures, and those containing 12 to 18% silicon, which are used in applications requiring an iron with very high resistance to acid attack. These last are commonly referred to as the 'high-silicon irons'.

The alloys containing less than 11% silicon have resistances to low-temperature corrosion not substantially different from those of low-silicon irons containing similar amounts of other alloying elements, and will not be further discussed in this section.

Table 3.56 Analysis of typical silicon-iron alloys

Name	Total carbon (%)	Si (%)	Mn (%)	S (%)	P (%)	Ni (%)	Cr (%)	Mo (%)
Grey cast iron	3·5	2·0	0·5	0·1	0·1	—	—	—
Hypersilid 14/16	0·65	14·5	0·5	0·02	0·15	—	—	—
Duriron	0·85	14·5	0·6	<0·05	<0·1	—	—	—
Durichlor 51	1·00	14·5	0·6	<0·05	<0·1	—	5·0	1·0*
Hypersilid 16/18	0·35	17·0	0·5	0·02	0·1	—	—	—

Table 3.57 Mechanical properties of silicon-iron alloys

Name	Ultimate tensile strength (MN/m^2)	Elongation 2 in (50 mm) gauge length	Brinell hardness no.
Grey cast iron	23	2%	180
14·5% silicon iron	129	Nil	540
Durichlor	108	Nil	450

The analyses of typical high-silicon irons are given in Table 3.56 and their typical mechanical properties in Table 3.57.

All these alloys are characterised by high hardness values and low resistance to impact. In this they are probably more similar to stoneware than to other metals but they are superior to stoneware in thermal conductivity and in their resistance to thermal shock, which, however, is poor compared with that of other metals. Moreover, it is usually easier to make castings of silicon iron than to fabricate required parts from stoneware.

The microstructure of the high-silicon irons containing less than 15·2% silicon consists of a matrix of α silico-ferrite[1] in which the majority of the carbon present in the alloy is distributed as fine graphite flakes. A typical microstructure of Fe–14·5 Si is shown in Fig. 3.60. The use of chromium or molybdenum in the alloy will result in some carbides appearing in the microstructure. Since the amount of free graphite in the microstructure is critical to the performance of the alloy in service it is necessary to increase the total carbon content to compensate for that lost as carbides. The irons containing more than 15·2% silicon some η phase is also present[1]. The hardness and brittleness of the silicon iron is due to the nature of the silico-ferrite. An attempt[2] has been made to produce high-silicon iron with a nodular graphite structure, with the intention of improving the mechanical properties of the alloy, but because the low strength of these alloys is due to the matrix rather than to the graphite form, the nodular graphite irons have proved to be little, if at all, superior in strength to the conventional alloys.

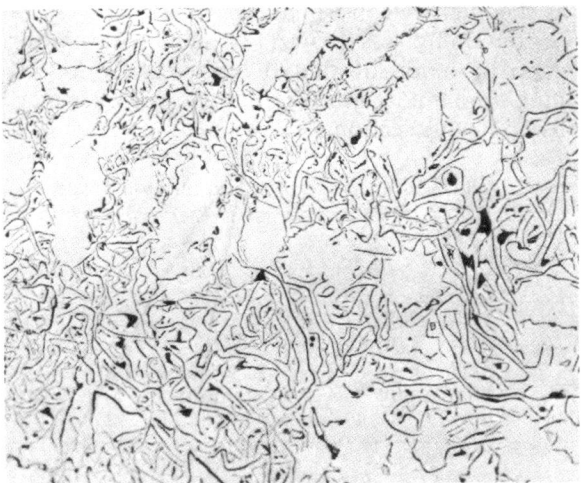

Fig. 3.60 Typical structure of 14·5% silicon iron

All the components made in silicon irons are produced by casting. The high-silicon irons are best melted in crucible or electric furnaces and their production tends to be concentrated in the hands of a few highly specialised foundries. Because the mechanical properties of these high-silicon irons preclude any machining other than grinding, it is necessary to design the castings made from these alloys in such a way that the necessity for subsequent treatment is kept to the minimum; for this reason the closest liaison between designer and foundry is essential.

In view of the poor mechanical properties of the high-silicon irons, the development of any stresses in the castings during solidification is very dangerous, since they may cause the casting to crack in subsequent service. To overcome this risk, it is often desirable to strip the castings from the moulds while they are still red hot and to anneal them at 850°C for 4-5 h, followed by slow cooling[3].

General Corrosion Behaviour

The outstanding resistance to corrosion exhibited by the high-silicon alloys is believed to be due to the development of a corrosion-resistant film containing a large proportion of silica. The full protective value of the film does not develop for most applications until at least 14·25% silicon is present in the alloy (Fig. 3.61). Increase in the content of silicon above 14·5% does not have a dramatic effect upon the corrosion resistance of the alloy (Fig. 3.61), although the further increase in film density is of service if the casting is to be exposed to solutions containing halide ions, especially hydrochloric acid.

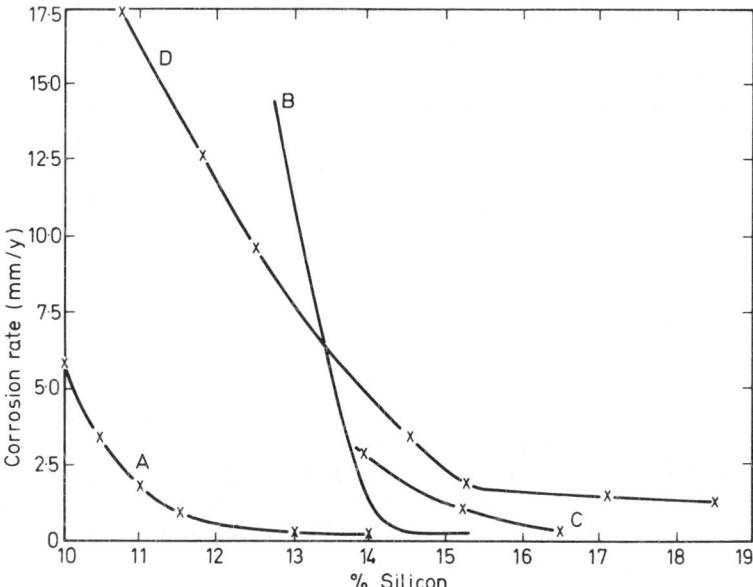

Fig. 3.61 Dependence of corrosion resistance of high-silicon iron on silicon content. A 70% HNO_3, B 20% H_2SO_4 at b.p., C 10% HNO_3 at b.p., D 10% HCl at 80°C

Since the formation of the silica film does not depend on any particular property of the corrosive environment, high-silicon irons can resist attack by a very wide range of environments. Solutions which are capable of dissolving silica, even in a small degree, are, however, inimical to silicon irons, and there are also a few ions capable of penetrating the silica film, which can cause relatively serious corrosion of the metal. The presence of chromium

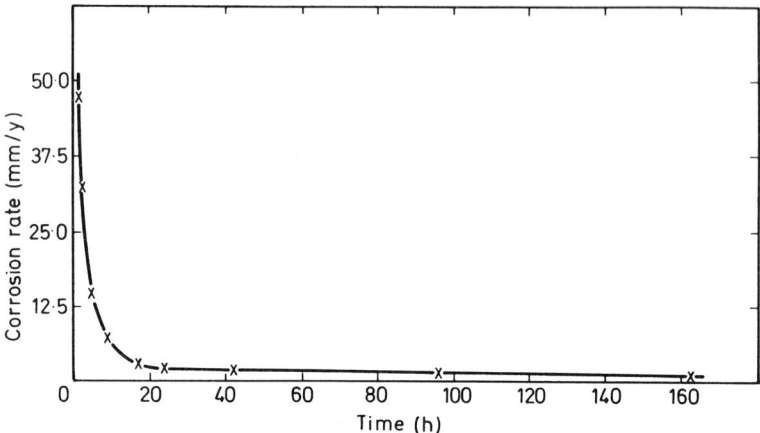

Fig. 3.62 Variation of corrosion rate of 14·5 Si iron in boiling 30% H_2SO_4 with time (Duriron Co. data)

in the alloy has been found to greatly strengthen the protective film and thus prevent the selective penetration by these corrosive ions. This is especially true of the chloride ion which is so commonly encountered in chemical applications.

Before the silica film can form, some corrosion of the metal must necessarily take place, and it follows that initial corrosion rates are high. Fig. 3.62 illustrates this point and suggests that uniform rates of corrosion are not reached until at least 100 h after the onset of the attack. As a result, useful data on the corrosion of high silicon irons can be obtained only from tests of at least this duration.

Corrosion in Natural Environments

Although the high-silicon irons are often used in circumstances which expose them to atmospheric, water or soil corrosion, they are rarely installed specifically to resist these agencies. Their corrosion resistance is such, however, that in fact no normally occurring environment ever causes serious attack. This is not to say that these irons can be regarded as stainless, and in fact alloys containing less than 14·7% silicon have been reported as becoming rusty in a moist atmosphere[3].

Corrosion in Industrial Environments

Acids

Since the corrosion resistance of the high-silicon alloys depends upon the permanence and impermeability of a thin silica film on the surface of the metal, it is obvious that any reagent which can damage the film will cause accelerated corrosion of the metal. For this reason all solutions containing hydrofluoric acid must be regarded as incompatible with the alloys.

As silica is not attacked by any acid other than hydrofluoric it might be expected to act as an effective barrier to attack by any other acid solutions, but in fact, while the high-silicon iron is resistant to attack by most acids, it is corroded relatively severely by hydrochloric, hydrobromic and sulphurous acids. The aggressive character of the two halogen acids may be ascribed to the readiness with which their relatively small anions can penetrate a passive film.

Attempts made to produce an alloy more resistant to hydrochloric acid have resulted in alloys containing 17–18% silicon or 14·5% silicon and chromium plus 3% molybdenum. The first is produced in Britain, and the second in the United States. The reason for the increase in resistance to hydrochloric acid of the Fe–18 Si alloy is thought to lie primarily in the increased density of the silica-rich film left on the metal by initial corrosion. The addition of 6% chromium with some molybdenum to Fe–14·5 Si causes the formation of extremely stable complex carbides with the consequent complete elimination of graphite[1] plus the formation of a more penetration-resistant silica film, probably containing chromium in substantial quantity.

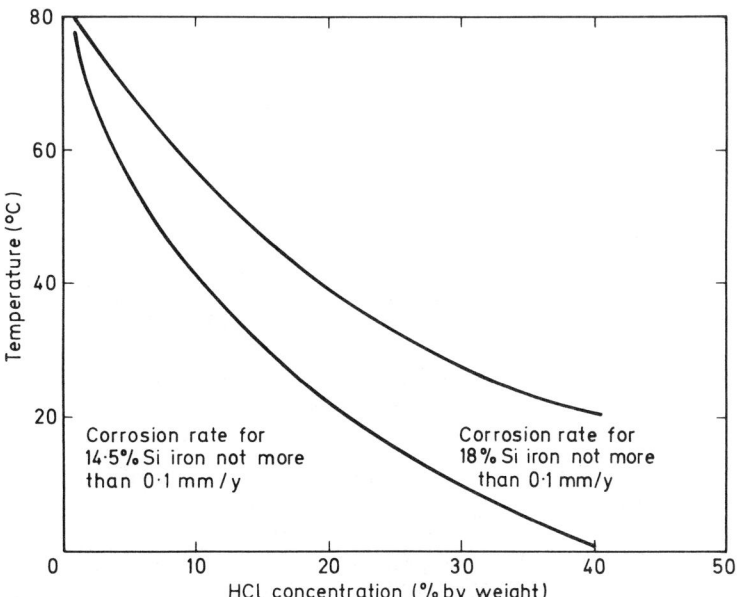

Fig. 3.63 Limits of use of silicon irons for handling hydrochloric acid solutions (former Ministry of Supply data) for Fe–18Si and Fe–14·5Si alloys

Figure 3.63 was prepared by a Silicon Iron Advisory Panel to the former Ministry of Supply between 1940 and 1945 as a guide to the ranges of hydrochloric acid concentration and temperature in which the two types of high-silicon iron could be safely used. The limits to the ranges are set by those conditions outside which the corrosion rate exceeds 0·1 mm/y. It should be emphasised that these curves give a guide to operating conditions only, and are not always in complete agreement with other published work. Borderline cases should always be checked by corrosion tests, and in the interpretation of these the large variations in corrosion rate which may arise because of subtle differences between laboratory and plant conditions must always be borne in mind.

The evidence at present available concerning the corrosion of high-silicon irons by sulphurous acid is insufficient to allow the formation of any theory about the mechanism by which the silica film barrier is broken down in the presence of this acid. As far as is known, this acid is corrosive to all types of high-silicon iron.

Probably the most useful characteristic of the high-silicon irons is their ability to withstand sulphuric acid at all temperatures and concentrations. The maximum rate of corrosion which can develop has been reported to be 0·482 mm/y in 30% sulphuric acid at boiling point[4], and this falls to a minimum rate of 0·025 mm/y when the acid concentration exceeds 60% and the temperature is at boiling point (Fig. 3.64). The former Ministry of Supply

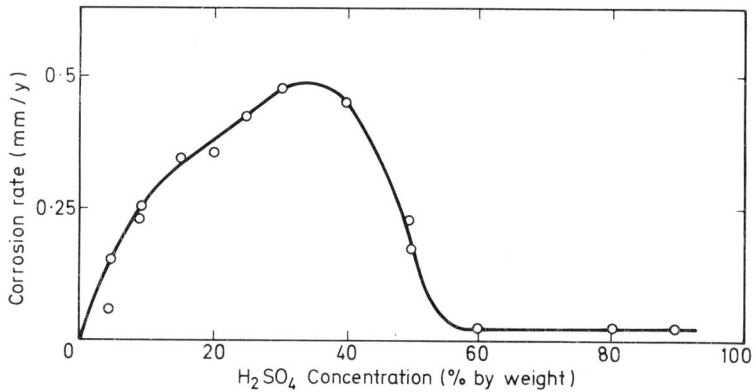

Fig. 3.64 Variation of corrosion rate with concentration of boiling sulphuric acid (Duriron Co. data)

Advisory Panel prepared a chart to indicate the regions of temperature and acid concentration in which the corrosion rate of Fe–14·5 Si was less than 0·1 mm/y (Fig. 3.65). This curve is at variance with the results from which Fig. 3.64 is derived, which suggest that the dotted curve in Fig. 3.65 may represent the true state of affairs. At best, Fig. 3.65 can be regarded only as a guide and the remarks already made about the interpretation of Fig. 3.63 are also applicable here. Even a 12% silicon-containing alloy provides satisfactory corrosion resistance to boiling sulphuric acid and has been applied on occasion in the commercial concentration of this acid. The lower silicon alloy has improved mechanical properties which allows it to resist better the rigours of a higher temperature service.

It has been reported[5] that oleum is corrosive to high-silicon iron and it is not normally recommended for withstanding this acid.

Nitric acid is also withstood by high-silicon iron. The concentrated acid is believed to reinforce the silica film by the formation of a passive iron oxide

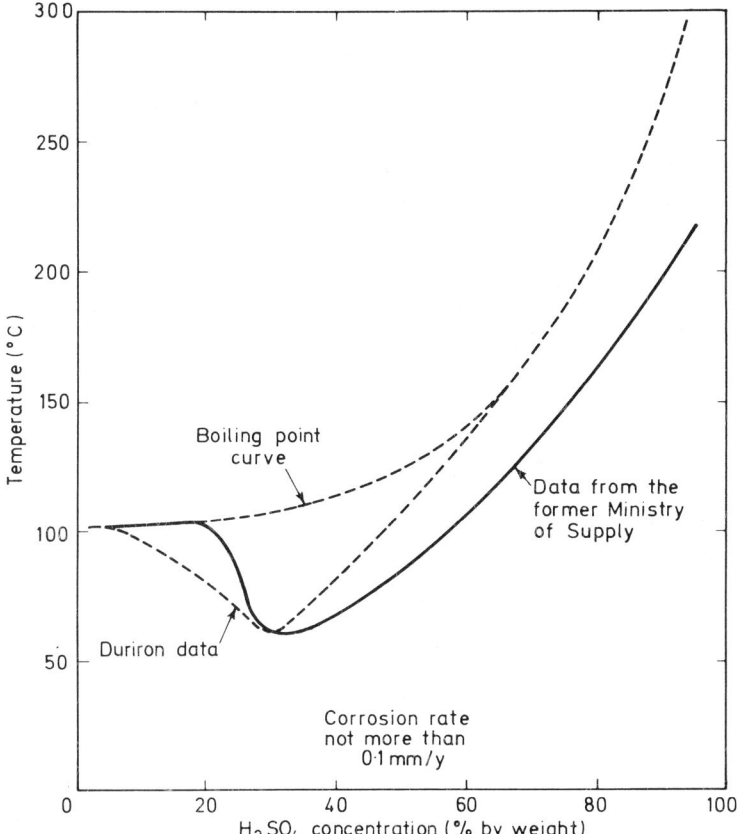

Fig. 3.65 Limits of use of silicon iron for handling sulphuric acid solutions

film, and this assumption is supported by the fact that the highest rates of attack are associated with hot dilute solutions. The curve prepared by the former Ministry of Supply Advisory Panel to indicate the regions of acid concentration and temperature in which the corrosion rate of Fe–14·5 Si is less than 0·1 mm/y is shown in Fig. 3.66. Fontana[6] has published a similar curve which is at variance with that produced in the UK for concentrations of less than 30%, as shown by the dotted line in Fig. 3.66.

The results published in the literature suggest that solutions of orthophosphoric acid at all concentrations and temperatures are not corrosive to high-silicon irons. Kosting and Heins[7] quote figures given by Duriron Co. together with figures obtained in their own tests of 24 hours duration; these are given in Table 3.58.

It will be observed that (probably mainly because of the short test period they used) their results indicate significantly higher corrosion rates than those given by Duriron Co. but neither source suggests that the rates are dangerously high. It seems probable that solutions of pyrophosphoric or tetraphosphoric acid are liable to give rise to excessive corrosion, although there is evidence to suggest that pyrophosphoric acid may be handled safely

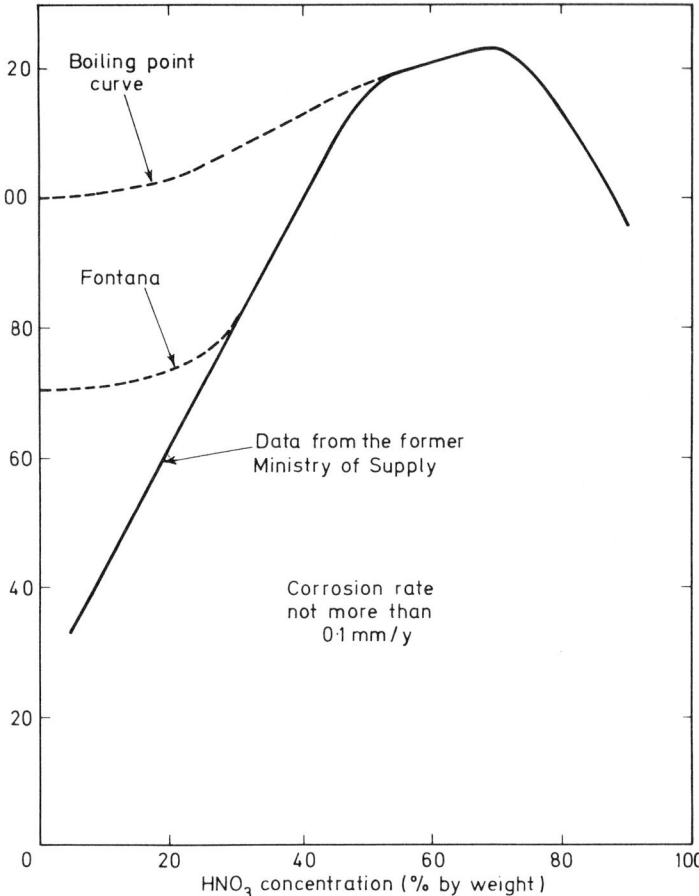

Fig. 3.66 Limits of use of silicon iron for handling nitric acid solutions

Table 3.58 Corrosion of high-silicon irons by ortho-phosphoric solutions

Acid concentration (%)	Temperature (°C)	Corrosion rate (mm/y)	Origin of data
10	82–88	0·007	Duriron Co.
	98	0·147	Kosting & Heins[7]
25	82–88	0·010	Duriron Co.
	98	0·038	Kosting & Heins[7]
50	98	0·185	Kosting & Heins[7]
87	82–88	0·010	Duriron Co.

in the cold and even possibly up to 80 to 90°C. Most crude phosphoric acid contains appreciable amounts of fluoride which lead to excessive corrosion rates.

High silicon iron offers excellent resistance to attack by all concentrations of nitric-sulphuric acid mixtures. The mixed acid corrodes the iron at rates never greater and often lower than the individual acids of comparable concentration.

All organic and other weak acids commonly encountered in industry are satisfactorily handled by high-silicon irons.

Alkalis

High-silicon irons are inferior to unalloyed grey irons in their resistance to attack by alkalis. For example, boiling 20% caustic soda solution or 50% caustic soda at 80°C will attack high-silicon irons at rates of the order of 1·27 mm/y (grey iron would be attacked at rates not greater than 0·25 mm/y).

The irons are not recommended even for so weak a base as ammonium hydroxide, if the liquid temperature is greater than 20°C. The alternate handling of acids (for which the alloy is normally resistant) and alkalis may also prove troublesome since the alkali will normally prevent the formation of the protective silica film on which its acid resistance depends.

Salts

Salts giving an alkaline reaction may be corrosive to the irons, and while neutral solutions can be handled safely there is usually little point in using high-silicon irons for these relatively innocuous solutions. The irons are useful in handling acidic solutions, subject to the restrictions already referred to regarding the halide, sulphite and phosphate ions.

Resistance to High-temperature Corrosion

High-silicon irons may be used at elevated temperatures if the process requires it. For example, 20-in diameter (0·5 m) pipe has been used for acid concentrations handling boiling 95% sulphuric acid at approximately 285°C where the products of combustion reach temperatures of the order of 590°C. The principal limitation on their use is imposed by their relatively low thermal conductivities and susceptibility to cracking from thermal shock; this demands that the rate of application or removal of heat should not be rapid.

Cathodic Protection

The use of high-silicon irons as anodes for impressed-current cathodic-protection systems is described in Sections 10.3 and 10.4.

J. DODD
W. A. LUCE

REFERENCES

1. Hurst, J. E. and Riley, R. V., *J. Iron St. Inst.*, **155**, 172 (1947)
2. Dumitrescu, T., Medeleanu, V., Nicolaid, M. and Dinu, I., *Rev. Métall. (Roumania)*, **3** No. 2, 19 (1958)
3. Hurst, J. E., *Proc. Inst. Brit. Foundrym.*, **37**, B46 (1943-44)
4. Duriron Co. (published literature)
5. Luce, W. A., *Chem. Engng.*, **61**, 246 (1954)
6. Fontana, M. G., *Industr. Engng. Chem. (Industr.)*, **45**, 91A, 92A, 94A, May (1953)
7. Kosting, P. R. and Heins, C., *Industr. Engng. Chem. (Industr.)*, **23**, 140-150, Feb. (1931)
8. Bryan, W. T. and Uhlig, H. H. (Ed.), *The Corrosion Handbook, Wiley*, New York; Chapman and Hall, London, 201-207 (1948)

3.10 Amorphous (Ferrous and Non-ferrous) Alloys

Structural Characteristics

Amorphous alloys stable at ambient and higher temperatures consist of at least two components without any long-range atomic order. They are produced by a variety of constituents from the gas, liquid and aqueous phases. Vitrification of metal surfaces is also caused by destruction of the long-range atomic order in the surfaces of solid metals.

Amorphous alloys are free of defects associated with the crystalline state, such as grain boundaries, dislocations and stacking faults. Because the formation of the structure without long-range atomic order is based on the prevention of solid state diffusion during solidification, amorphous alloys are free of compositional fluctuations resulting from solid state diffusion such as second phases, precipitates and segregates. The amorphous alloys are, therefore, regarded as chemically ideal homogeneous alloys which are thermodynamically metastable single-phase solid solutions supersaturated with alloy constituents. The formation of the single-phase supersaturated solid solution is quite suitable for the production of new alloys possessing specific properties by alloying with necessary amounts of the appropriate elements.

Corrosion Resistance

The corrosion behaviour of amorphous alloys has received particular attention since the extraordinarily high corrosion resistance of amorphous iron–chromium–metalloid alloys was reported[1]. The majority of amorphous ferrous alloys contain large amounts of metalloids. The corrosion rate of amorphous iron-metalloid alloys decreases with the addition of most second metallic elements such as titanium, zirconium, vanadium, niobium, tantalum, chromium, molybdenum, tungsten, cobalt, nickel, copper, ruthenium, rhodium, palladium, iridium and platinum[1-9]. The addition of chromium is particularly effective. For instance amorphous Fe-8Cr-13P-7C alloy passivates spontaneously even in 2 N HCl at ambient temperature[10]. (The number denoting the concentration of an alloy element in the amorphous alloy formulae is the atomic percent unless otherwise stated.)

Amorphous Fe–3Cr–13P–7C alloys containing 2 at% molybdenum, tungsten or other metallic elements are passivated by anodic polarisation in 1 N HCl at ambient temperature[11]. Chromium addition is also effective in improving the corrosion resistance of amorphous cobalt-metalloid[12] and nickel-metalloid[13,14] alloys (Fig. 3.67). The combined addition of chromium and molybdenum is further effective. Some amorphous Fe–Cr–Mo-metalloid alloys passivate spontaneously even in 12 N HCl at 60°C. Critical concentrations of chromium and molybdenum necessary for spontaneous passivation of amorphous Fe–Cr–Mo–13P–7C and Fe–Cr–Mo–18C alloys in hydrochloric acids of various concentrations and different temperatures are shown in Fig. 3.68[15].

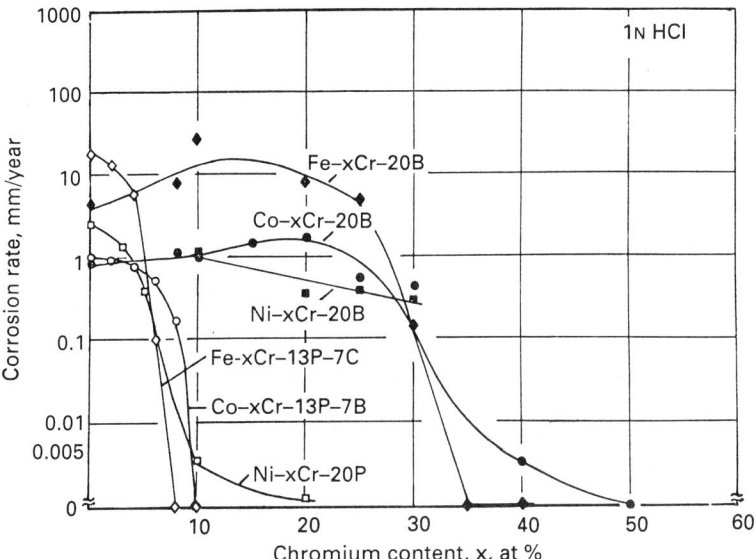

Fig. 3.67 Changes in corrosion rates of amorphous Fe-, Co- and Ni-base alloys measured in 1 N HCl at room temperature as a function of alloy chromium content[13,14]

Some metal–metal alloys also form an amorphous structure. From a corrosion point of view, amorphous metal–metal alloys containing valve metals are particularly useful. In strong acids with high oxidising power, such as boiling nitric acid, the alloys whose corrosion resistance is based mostly on the presence of chromium are corroded, but amorphous alloys containing valve metals, such as tantalum, show very high corrosion resistance which is much higher than that of the crystalline metal (Fig. 3.69)[16]. Some of these alloys are spontaneously passive even in azeotropic boiling 6 N HCl[17]. Phosphoric acid does not contain a particularly aggressive anion, but because of their high boiling points, boiling concentrated phosphoric acids are quite corrosive. As shown in Fig. 3.70[18], amorphous Ni–Ta alloys are more corrosion resistant than crystalline tantalum metal in hot concentrated phosphoric acid, and various amorphous alloys have corrosion resistance comparable with crystalline tantalum metal.

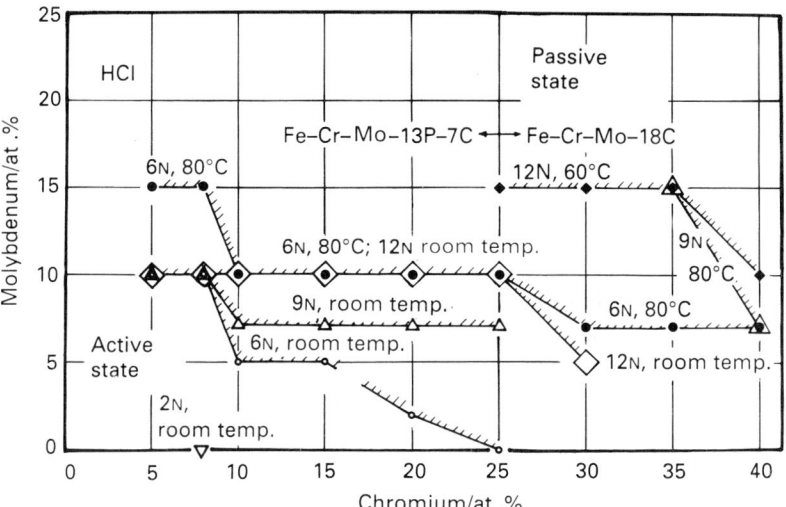

Fig. 3.68 Critical concentrations of chromium and molybdenum necessary for spontaneous passivation of amorphous Fe-Cr-Mo-13P-7C and Fe-Cr-Mo-18C alloys in hydrochloric acid of various concentrations and temperatures[15]

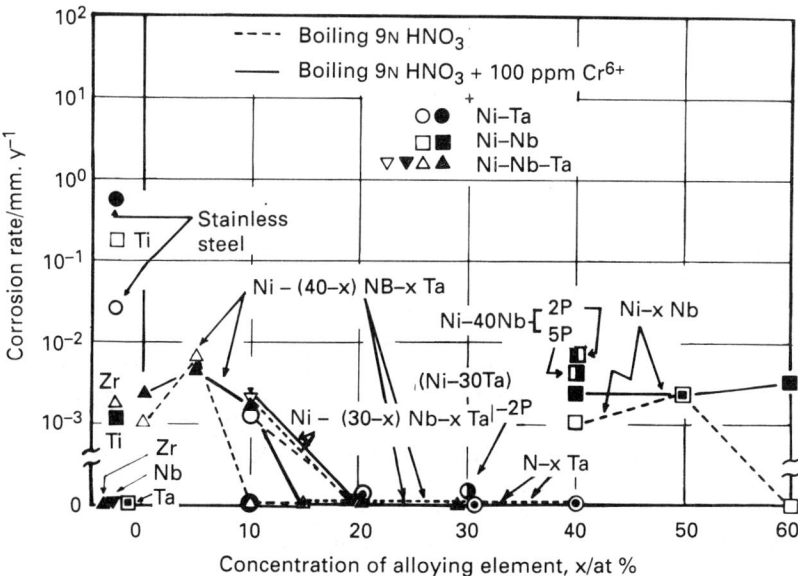

Fig. 3.69 Corrosion rates of amorphous Ni-Nb, Ni-Ta and Ni-Nb-Ta alloys in boiling 9 N HNO_3 with and without 100 ppm Cr^{6+} ion[16]

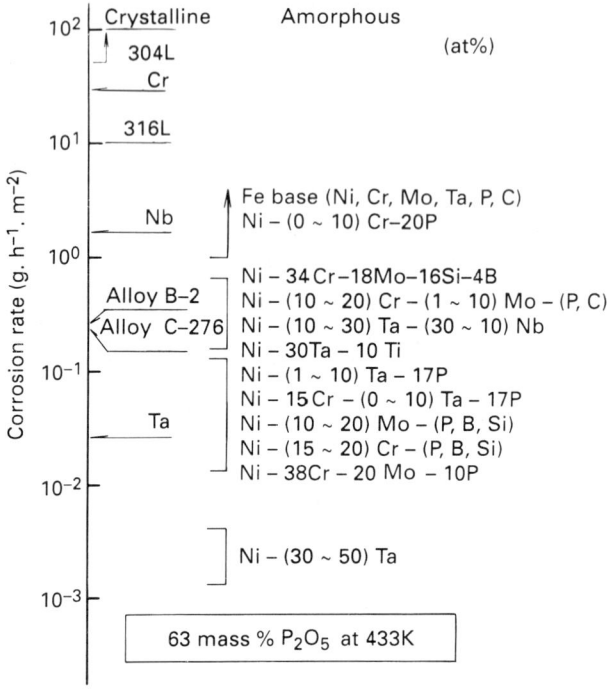

Fig. 3.70 Corrosion rates of amorphous and crystalline alloys measured in 87 wt% H_3PO_4 at 160°C[18]

Passive Film

X-ray photoelectron spectroscopic study[19] of the spontaneously passive amorphous Fe-10Cr-13P-7C alloy in 1 N HCl revealed that the passive film consists of Cr^{3+}, O^{2-}, OH^- and H_2O, and hence the passive film has been called a passive hydrated chromium oxyhydroxide film $(CrO_x(OH)_{3-2x}nH_2O)$. Subsequent investigations have revealed that chromium enrichment occurs not only in passive films formed on amorphous alloys[13-15, 20-22], but also in those on crystalline[23-25] alloys whose corrosion resistance is owing to the presence of chromium. It has been shown[26] that the resistance against passivity breakdown is higher when the chromium content of the passive film is higher. Thus, the higher the ability of an alloy to concentrate chromic ion in the passive film, the higher is the corrosion resistance of the alloy. The concentration of chromic ion in passive films formed on amorphous alloys is far higher than the concentration in films on crystalline alloys (Table 3.59).

The passive films formed by the addition of sufficient amounts of valve metals to amorphous nickel–valve-metal alloys are exclusively composed of valve-metal oxyhydroxides or oxides such as $TaO_2(OH)$[16], $NbO_2(OH)$[16] or Ta_2O_5[18]. Consequently, amorphous alloys containing strongly passivating elements, such as chromium, niobium and tantalum, have a very high ability

Table 3.59 Concentration of chromic ion in passive films formed on amorphous alloys and stainless steels in 1 N HCl at ambient temperature

	Cr^{3+}/Total metallic ions	Passivation	Reference
Amorphous alloys			
Fe–10Cr–13P–7C	0.97	Spontaneous	19
Fe–3Cr–2Mo–13P–7C	0.57	Anodic polarisation	11
Co–10Cr–20P	0.95	Spontaneous	21
Ni–10Cr–20P	0.87	Spontaneous	22
Stainless steel			
Fe–30Cr–(2Mo)	0.75	Anodic polarisation	24
Fe–19Cr–(2Mo)	0.58	Anodic polarisation	25

to concentrate beneficial ions in their passive films and have a high corrosion resistance resulting from spontaneous passivation.

Chemical Homogeneity

The high corrosion resistance of amorphous alloys disappears on heat treatment that produces crystallisation[27-32]. Figure 3.71 shows an example of the

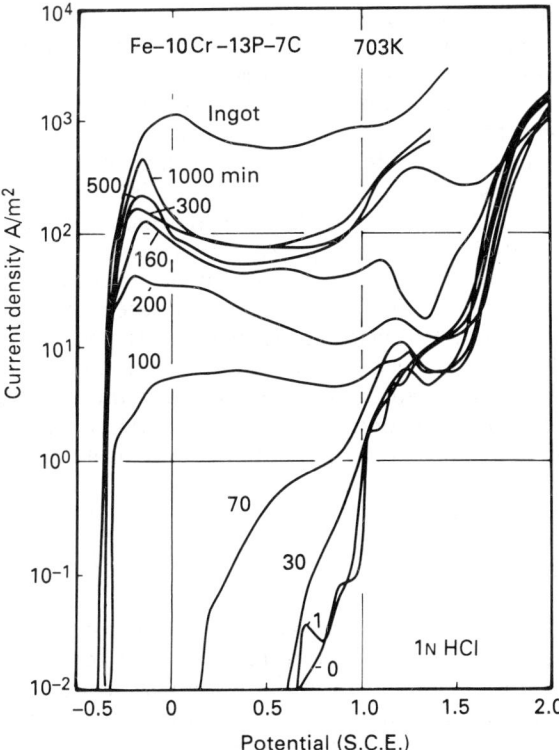

Fig. 3.71 Change in polarisation curve of amorphous Fe–10Cr–13P–7C alloy in 1 N HCl with the time of heat treatment at 723 K. The time of heat treatment is expressed in the figure in minutes[30]

effect of heat treatment in the case of Fe–10Cr–13P–7C[30]. A microcrystalline metastable phase is formed in the amorphous matrix by heat treatment at 703 K for 100 min. The alloy becomes no longer spontaneously passive in 1 N HCl as soon as the microcrystalline phase appears in the amorphous matrix, and the anodic dissolution current continues to increase with increasing time of heat treatment. This occurs because of the introduction of chemical heterogeneity into the homogeneous single phase of the amorphous alloy. Rapidly solidified microcrystalline stainless steels also have high pitting corrosion resistance[33,34], and detrimental defects on which a stable passive film does not form are mostly precipitates and segregates of impurities[34]. The chemically homogeneous single-phase nature of amorphous alloys which are free of defects resulting in the formation of a uniform passive film is responsible for the high corrosion resistance of these alloys.

Fast Passivation

When the chromium-enriched passive film is formed on amorphous and crystalline iron–chromium alloys, containing no noble metals such as nickel, the composition of the alloy surface just under the chromium-enriched passive film is almost the same as that of the bulk alloy[24]. Hence, the formation of a chromium-enriched passive film results from selective dissolution of alloy constituents unnecessary for passive film formation. When an alloy is able to passivate, fast active dissolution of the alloy results in rapid enrichment with beneficial ions. The passivating ability is, therefore, closely related to the activity of the alloy[14]. The thermodynamically metastable nature of amorphous alloys is responsible for their high reactivity when they are not covered by a passive film, and hence is responsible for the fast passivation by the formation of the film in which the beneficial ions are highly concentrated. As shown in Fig. 3.67 for iron-, cobalt- and nickel-based alloys, when the alloy chromium content is not high enough to cause spontaneous passivation, the more active iron-based alloys dissolve rapidly and the more noble nickel-based alloys dissolve slowly. The fast dissolution in iron-based alloys is effective in concentrating the chromic ions, so that iron-based alloys passivate spontaneously with the addition of a small amount of chromium. In contrast, the slowly dissolving noble nickel-based alloys require the addition of larger amounts of chromium for spontaneous passivation.

Metastable Nature

Amorphous alloys are in a thermodynamically metastable state, and hence essentially they are chemically more reactive than corresponding thermodynamically stable crystalline alloy[1,27,35]. If an amorphous alloy crystallises to a single phase having the same composition as the amorphous phase, crystallisation results in a decrease in the activity of the alloy related to the active dissolution rate of the alloy[35].

Since amorphous alloys can be regarded as metallic solids with a frozen-in melt structure, the liquid structure freezes at different temperatures depending upon quenching conditions with the consequent formation of different amorphous states. Accordingly, even for amorphous alloys of the same composition, anodic dissolution currents are not always identical owing to different structural relaxation intensities[36-38].

Effect of Metalloids

As can be seen in Fig. 3.67, the corrosion resistance of amorphous alloys changes with the addition of metalloids, and the beneficial effect of a metalloid in enhancing corrosion resistance based on passivation decreases in the order phosphorus, carbon, silicon, boron[39] (Fig. 3.72). This is attributed partly to the difference in the speed of accumulation of passivating elements due to active dissolution prior to passivation[40].

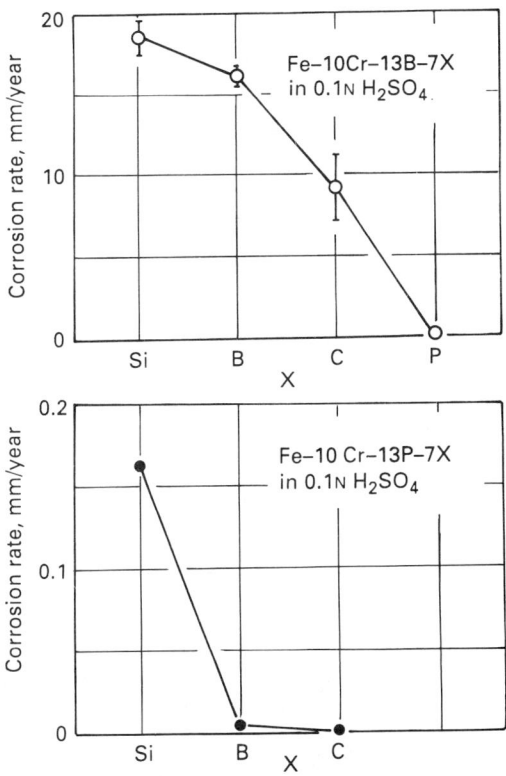

Fig. 3.72 Average corrosion rates of amorphous Fe-10Cr-10B-7X and Fe-10Cr-13P-7X alloys in 0.1 N H_2SO_4 at 30°C, where X is one of the metalloids indicated on the abscissa[39]

The effect of metalloids on the corrosion resistance of alloys also varies with the stability of polyoxyanions contained in their films. Phosphorus and carbon contained in iron-chromium-metalloid alloys do not produce passive films of phosphate and carbonate in strong acids, and so do not interfere with the formation of the passive hydrated chromium oxyhydroxide film[20,40]. In contrast, boron-containing alloys require the addition of large amounts of chromium to increase the passivating ability by concentrating the chromium oxyhydroxide in the surface films because the films contain chromium borate[21,40].

The thickness of the passive films discussed above is up to 3-4 nm. In contrast, the surface film on an amorphous Cu-40Zr alloy continues to grow to over 100 nm in 1 N H_2SO_4, but the addition of only 2 at% phosphorus is effective in depressing the film growth to a few tens of nanometres[41,42]. The addition of a few atomic percent of phosphorus to amorphous Ni-30Ta alloys results in a decrease in the corrosion rate in boiling 6 N HCl by about four orders of magnitude[18].

The corrosion rate of amorphous Ni-P alloys in 1 N HCl is lower than those of crystalline nickel metal and amorphous Fe-P-C alloy by factors of about 5 and 250, respectively, and is further decreased by the addition of various elements[43] (Fig. 3.73).

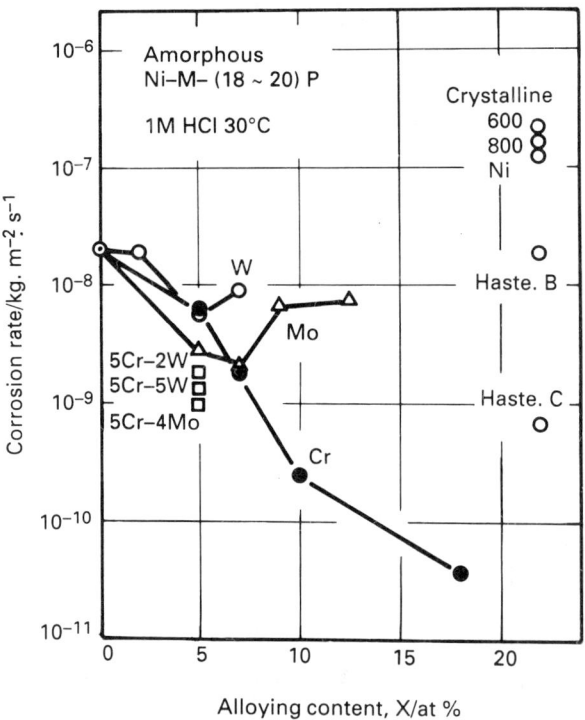

Fig. 3.73 Average corrosion rates of amorphous Ni-P alloys measured in 1 N HCl at 30°C. Included are average corrosion rates of crystalline nickel and nickel-base alloys[43]

Silicon-containing amorphous metal-metalloid alloys form surface films. Sputter-deposited Fe-Si alloys containing 25 at.% or more silicon are passivated by anodic polarisation in dilute sulphuric acid owing to the formation of a SiO_2 film[44]. Melt-spun amorphous Fe-39Ni-10B-12Si alloy is more resistant against pitting corrosion than the amorphous Fe-40Ni-20B alloy owing to the formation of a silicon-enriched surface film[45]. An increase in the silicon content of amorphous Fe-B-Si alloys extends the passive potential range[46]. Increasing the silicon content of amorphous Fe-10Cr-5Mo-B-Si alloys leads to a decrease in current densities in both the active and passive regions in 6 N HCl at 25°C without changing the open circuit corrosion potential owing to the formation of a SiO_2-like substance along with the hydrated oxyhydroxide film[47].

Passivity Breakdown

The chemically homogeneous amorphous alloys with high passivating ability form uniform passive films in which beneficial ions are highly concentrated. Passivity breakdown occurs in the form of general corrosion only when the whole film is dissolved under very aggressive conditions[48]. The high passivating ability also provides high resistance against crevice corrosion[49,50]. The crevice corrosion potentials and protection potentials of these alloys are very high.

Active Path Corrosion

Stress-corrosion cracking based on active-path corrosion of amorphous alloys has so far only been found when alloys of very low corrosion resistance are corroded under very high applied stresses[51,52]. However, when the corrosion resistance is sufficiently high, plastic deformation does not affect the passive current density or the pitting potential[53], and hence amorphous alloys are immune from stress-corrosion cracking.

Hydrogen Embrittlement

Amorphous alloys are capable of absorbing far higher amounts of hydrogen than conventional crystalline steels[54]. Thus, some amorphous alloys fail by hydrogen embrittlement when they are corroded under tensile-stressed conditions. However, increasing corrosion resistance by alloy modifications, such as increasing the chromium and/or molybdenum contents of amorphous iron-based alloys, reduces hydrogen absorption and hence hydrogen embrittlement[55].

Oxidation

The oxidation behaviour of amorphous alloys studied below their crystallisation temperature is not greatly different from that of crystalline metals, although the presence of large amounts of metalloids complicates the situation[56-58].

The amorphous structure favours internal oxidation unless a protective oxide film is formed as, for example, under low oxygen partial pressures[59].

Production Methods

The thickness of amorphous alloys is dependent upon production methods. Rapid quenching from the liquid state, which is the most widely used method, produces generally thin amorphous alloy sheets of 10–30 μm thickness. This has been called melt spinning or the rotating wheel method. Amorphous alloy powder and wire are also produced by modifications of the melt spinning method. The corrosion behaviour of amorphous alloys has been studied mostly using melt-spun specimens.

Laser and electron beam processing are effective methods for preparing amorphous surface alloys covering conventional crystalline bulk metals[60,61].

Sputter deposition is capable of producing thick alloys. The corrosion behaviour of amorphous sputter deposits is similar to that of their melt-spun amorphous counterparts[62,63]. However, sputter deposits prepared using conventional sputtering apparatus have never been defect-free, and hence the substrate metals are corroded in aggressive environments[64]. Technological improvements to the sputter deposition process have enabled the preparation of defect-free sputter deposits. Sputtering is particularly suitable for the production of special amorphous alloys such as Cu–Ta[65], Al–Nb[66] and Al–Ta[66], which cannot be prepared even in crystalline phase mixtures by conventional methods, e.g. melting, because the boiling points of copper and aluminium are far lower than the melting points of the valve metals. These alloys containing tantalum or niobium have very high corrosion resistance.

Various amorphous alloys can be prepared by plating[67]. Plating is particularly suitable for the preparation of thinner amorphous alloys than is possible by melt spinning, e.g. <1 μm, although production of defect-free alloys is difficult.

Ion implantation and ion mixing produce amorphous alloys as thin as only several tens of nanometres. Implantation of metalloids such as phosphorus in austenitic stainless steel has been known to produce amorphous surface alloys having high corrosion resistance[68,69].

K. HASHIMOTO

REFERENCES

1. Naka, M., Hashimoto, K. and Masumoto, T., *J. Japan Inst. Metals*, **38**, 835 (1974)
2. Hashimoto, K., Naka, M. and Masumoto, T., *Sci. Rep. Res. Inst. Tohoku University*, **A-24**, 48 (1976)
3. Naka, M., Hashimoto, K. and Masumoto, T., *Sci. Rep. Res. Inst. Tohoku University*, **A-26**, 283 (1977)
4. Naka, M., Hashimoto, K. and Masumoto, T., *J. Non-Cryst. Solids*, **29**, 61 (1978)
5. Hashimoto, K., Asami, K., Naka, M. and Masumoto, T., *Sci. Rep. Res. Inst. Tohoku University*, **A-27**, 237 (1979)
6. Masumoto, T., Hashimoto, K. and Naka, M., *Proc. 3rd Int. Conf. Rapidly Quenched Metals*, The Metals Society, London, 435 (1978)
7. Naka, M., Hashimoto, K. Inoue, A. and Masumoto, T., *J. Non-Cryst. Solids*, **31**, 347 (1979)

8. Cadet, P., Keddam, M. and Takenouti, H., *Proc. 4th Int. Conf. Rapidly Quenched Metals*, The Japan Institute of Metals, Sendai, **2**, 1477 (1982)
9. Kovacs, K., Farkas, J., Kiss, L., Lovas, A. and Tompa, K., 'Rapidly Quenched Metals', *Proc. 4th Int. Conf. Rapidly Quenched Metals*, The Japan Institute of Metals, Sendai, **2**, 1471 (1982)
10. Kobayashi, K., Hashimoto, K. and Masumoto, T., *Sci. Rep. Res. Inst. Tohoku University*, **A-29**, 284 (1978)
11. Hashimoto, K., Naka, M., Noguchi, J., Asami, K. and Masumoto, T., in *Passivity of Metals*, (Frankenthal, R. P. and Kruger, J., eds.), The Electrochemical Society, Princeton, N.J., 156 (1978)
12. Naka, M., Hashimoto, K. and Masumoto, T., *Proc. 3rd Int. Conf. Rapidly Quenched Metals*, The Metals Society, London, 449 (1978)
13. Hashimoto, K., Kasaya, M., Asami, K. and Masumoto, T., *Corros. Engng. (Boshoku Gijutsu)*, **26**, 442 (1977)
14. Naka, M., Hashimoto, K. and Masumoto, T., *J. Non-Cryst. Solids*, **34**, 257 (1979)
15. Hashimoto, K., Kobayashi, K., Asami, K. and Masumoto, T., *Proc. 8th Int. Cong. Metallic Corrosion*, DECHEMA, Frankfurt/Main, **1**, 70 (1981)
16. Kawashima, A., Shimamura, K., Chiba, S., Matsunaga, T., Asami, K. and Hashimoto, K., *Proc. 4th Asian-Pacific Corrosion Control Conference*, Tokyo, **2**, 1042 (1985)
17. Shimamura, K., Kawashima, A., Asami K. and Hashimoto, K., *Sci. Rep. Inst. Tohoku Univ.*, **A-33**, 196 (1986)
18. Mitsuhashi, A., Asami, K., Kawashima, A. and Hashimoto, K., *Corros. Sci.*, **27**, 957 (1987)
19. Hashimoto, K., Masumoto, T. and Shimodaira, S., in *Passivity and Its Breakdown on Iron and Iron-Base Alloys*, (Staehle, R. W. and Okada, H., eds.), NACE, Houston, 34 (1975)
20. Asami, K., Hashimoto, K., Masumoto, T. and Shimodaira, S., *Corros. Sci.*, **16**, 909 (1976)
21. Hashimoto, K., Asami, K., Naka, M. and Masumoto, T., *Corros. Engng. (Boshoku Gijutsu)*, **28**, 271 (1979)
22. Kawashima, A., Asami, K. and Hashimoto, K., *Corros. Sci.*, **24**, 807 (1984)
23. Asami, K., Hashimoto, K. and Shimodaira, S., *Corros. Sci.*, **18**, 151 (1978)
24. Hashimoto, K., Asami, K. and Teramoto, K., *Corros. Sci.*, **19**, 3 (1979)
25. Hashimoto, K. and Asami, K., *Corros. Sci.*, **19**, 251 (1979)
26. Asami, K. and Hashimoto, K., *Corros. Sci.*, **19**, 1007 (1979)
27. Hashimoto, K., Osada, K., Masumoto, T. and Shimodaira, S., *Corros. Sci.*, **16**, 71 (1976)
28. Diegle, R. B. and Slater, J. E., *Corrosion*, **32**, 155 (1976)
29. Kulik, T., Baszkiewicz, J., Kaminski, M., Latuszkiewicz, J. and Matyja, H., *Corros. Sci.*, **19**, 1001 (1979)
30. Naka, M., Hashimoto, K. and Masumoto, T., *Corrosion*, **36**, 679 (1980)
31. Kapusta, S. and Heusler, K. E., *Z. Metallkd.*, **72**, 785 (1981)
32. Diegle, R. B., *Proc. 4th Int. Conf. on Rapidly Quenched Metals*, The Japan Institute of Metals, Sendai, **2**, 1457 (1982)
33. Tsuru, T. and Latanision, R. M. *J. Electrochem. Soc.*, **129**, 1402 (1982)
34. Kawashima, A. and Hashimoto, K., *Corros. Sci.*, **26**, (1982)
35. Huerta, D. and Heusler, K. E., *Proc. 9th Int. Cong. Metallic Corrosion*, National Research Council Canada, Ottawa, 222 (1984)
36. Masumoto, Y., Inoue, A., Kawashima, A., Hashimoto, K., Tsuai, A. and Masumoto, T., *J. Non-Cryst. Solids*, **86**, 121 (1986)
37. Nagarkar, P. V., Searson, P. C. and Latanision, R. M., *Proc. Symp. on Corrosion, Electrochemistry and Catalysis of Metallic Glasses*, (Diegle, R. B. and Hashimoto, K., eds.), the Electrochemical Society, Pennington, 118 (1988)
38. Habazaki, H., Ding, S.-Q., Kawashima, A., Asami, K., Hashimoto, K., Inoue, A. and Masumoto, T., *Corros, Sci.*, **29**, (1989)
39. Naka, M., Hashimoto, K. and Masumoto, T., *J. Non-Cryst. Solids*, **28**, 403 (1978)
40. Hashimoto, K., Naka, M., Asami, K. and Masumoto, T., *Corros, Engng. (Boshoku Gijutsu)*, **27**, 279 (1978)
41. Burleigh, T. D. and Latanision, R. M., in *Passivity of Metals and Semiconductors*, (Froment, M., ed.), Elsevier, Amsterdam, 317 (1983)
42. Burleigh, T. D. and Latanision, R. M., *Proc. 9th Int. Cong. Metallic Corrosion*, National Research Council of Canada, Ottawa, **2**, 645 (1984)
43. Kawashima, A., Asami, K. and Hashimoto, K., *J. Non-Cryst. Solids*, **70**, 69 (1985)

44. Brusic, V., MacInnes, R. D., and Aboaf, J., in *Passivity of Metals*, (Frankenthal, R. P. and Kruger, J., eds.), The Electrochemical Society, Princeton, N.J., 170 (1978)
45. Janik-Czachor, M., *Werk. u. Korr.*, **34**, 47 (1983)
46. Janik-Czachor, M., *Werk. u. Korr.*, **34**, 451 (1983)
47. Hashimoto, K., Asami, K. and Kawashima, A., *Proc. 9th Int. Cong. Metallic Corrosion*, National Research Council of Canada, Ottawa, **1**, 208 (1984)
48. Hashimoto, K., in *Passivity of Metals and Semiconductors*, (Froment, M., ed.), Elsevier, Amsterdam, 1983, 235 (1983)
49. Diegle, R. B., *Corrosion*, **35**, 250 (1979)
50. Diegle, R. B., *Corrosion*, **36**, 362 (1980)
51. Pampillo, C. A., *J. Mater. Sci.*, **10**, 1194 (1975)
52. Archer, M. D. and McKim, R. J., *Corrosion*, **39**, 91 (1983)
53. Devine, T. M., *J. Electrochem. Soc.*, 124, 38 (1977)
54. Kawashima, A., Hashimoto, K. and Masumoto, T., *Corros. Sci.*, **16**, 935 (1976)
55. Kawashima, A., Hashimoto, K. and Masumoto, T., *Corrosion*, **36**, 577 (1980)
56. Hunderi, O. and Bergerson, R., *Corros. Sci.*, **22**, 135 (1982)
57. Thomas, M. T. and Bear, D. R., *Proc. 4th Int. Conf. Rapidly Quenched Metals*, The Japan Institute of Metals, Sendai, **2**, 1453 (1982)
58. Ley, L. and Riley, J. D., *Proc. 7th Int. Vacuum Cong.*, 2031 (1977)
59. Bigot, J. Calvayrac, Y., Harmeline, H., Chevalier, J-P. and Quivy, A., *Proc. 4th Int. Conf. Rapidly Quenched Metals*, The Japan Institute of Metals, Sendai, **2**, 1463 (1982)
60. Yoshioka, H., Asami, K., Kawashima, A. and Hashimoto, K., *Corros. Sci.*, **27**, 981 (1987)
61. Kumagai, N., Samata, Y., Jikihara, S., Kawashima, A., Asami, K. and Hashimoto, K., *Mater. Sci. Engng*, **99**, 489 (1988)
62. Wang, R., *J. Non-Cryst. Solids*, **61** and **62**, 613 (1984)
63. Diegle, R. B. and Merz, M. M., *J. Electrochem. Soc.*, **127**, 2030 (1983)
64. Anderson, R. A., Dobisz, E. A., Perepezko, J. H., Thomas, R. E. and Wiley, J. D., in *Chemistry and Physics of Rapidly Solidified Materials*, (Berkowitz, B. J. and Scattergood, R. O., eds.), the Metallurgical Society of AIME, Warrendale, 111 (1983)
65. Shimamura, K., Miura, K. Kawashima, A., Asami, K. and Hashimoto, K., *Proc. Symp. on Corrosion, Electrochemistry and Catalysis of Metallic Glasses*, (Diegle, R. B. and Hashimoto, K., eds.), the Electrochemical Society, Pennington, 232 (1980)
66. Yoshioka, H., K. Kawashima, A., Asami, K. and Hashimoto, K., *Proc. Symp. on Corrosion, Electrochemistry and Catalysis of Metallic Glasses*, (Diegle, R. B. and Hashimoto, K, eds.), the Electrochemical Society, Pennington, 242 (1988)
67. Watanabe, T. and Tanabe, Y., *J. Metal Finishing Soc. Japan*, **32**, 600 (1981)
68. Grant, W. A., *Nuclear Instruments and Methods*, **182/183**, 809 (1981)
69. Clayton, C. R., Wang, Y-F. and Hubler, G. K., in *Passivity of Metals and Semiconductors*, (Froment, M., ed), Elsevier, Amsterdam, 235 (1983)

4 NON-FERROUS METALS AND ALLOYS

4.1 Aluminium and Aluminium Alloys	**4**:3
4.2 Copper and Copper Alloys	**4**:38
4.3 Lead and Lead Alloys	**4**:77
4.4 Magnesium and Magnesium Alloys	**4**:99
4.5 Nickel and Nickel Alloys	**4**:117
4.6 Tin and Tin Alloys	**4**:158
4.7 Zinc and Zinc Alloys	**4**:169

4.1 Aluminium and Aluminium Alloys

Aluminium and the aluminium alloys lend themselves to many engineering applications because of their combination of lightness with strength, their high corrosion resistance, their thermal and electrical conductivity and heat and light reflectivity, and their hygienic and non-toxic qualities. The variety of forms in which they are available also enhances their utility.

Composition and Mechanical Properties

Pure aluminium has good working and forming properties, high resistance to corrosion, low mechanical strength, and high ductility. The diverse and exacting technical demands made on aluminium alloys in different applications are met by the considerable range of alloys available for general and specific engineering purposes (BS 1470-75, 1490), each of which has been designed and tested to provide various combinations of useful properties. These include strength/weight ratio, corrosion resistance, workability, castability, or high-temperature properties, to mention but a few. The compositions and properties of these standard alloys are given in Tables 4.1 to 4.4.

More specialised alloys are covered by the *DTD* and *L* series for aircraft applications and include the high strength Al-Zn-Mg alloys. The medium strength weldable Al-Zn-Mg compositions are finding increasing utility in engineering and a national specification may be anticipated in the near future.

Where free machining characteristics are required, this may be achieved by additions of cadmium, antimony, tin or lead (e.g. BS 4300/5). Materials for electrical use are of special composition (BS 2627, 3988), while bearings are manufactured from Al-Sn alloys.

Composites of aluminium alloy with a thin cladding on one or both surfaces of a more anodic aluminium alloy or pure aluminium, enable sheet, plate and tube to be produced with special combinations of strength and corrosion resistance appropriate to service conditions. Although originally applied to high strength aircraft alloys, this principle of cladding is now utilised in several important industrial applications.

The I.S.O. designations may be correlated directly with the British Standard General Engineering series and partially with the American Aluminium Association designations. The nearest equivalents for the three systems are given in Table 4.5, although differences in alloying practice in America

Table 4.1 Some wrought British Standard aluminium alloys for general engineering purposes (non-heat treatable alloys)

Material designation		Major alloying constituents (nominal %)	Tensile strength range* (N/mm²)	Resistance to atmospheric attack	Cold forming(2)	Machining	Suitability for			Protective anodising
BS 1470-75	I.S.O.						Fusion welding(3)		Resistance spot welding	
							Oxy-gas	Inert gas shielded arc		
1	Al 99·99	99·99 Al	100 max.	V	V	P	V	V	F	V
1A	Al 99·8	99·8 Al	125 max.	V	V	P	V	V	F	V
1B	Al 99·5	99·5 Al	55–135	V	V	F	V	V	G	V
1C	Al 99·0	99 Al	60–140	V	V	F	V	V	G	V
N3	Al Mn1	1·25 Mg	90–175	V	V	F	V	V	V	G
N4	Al Mg2	2·25 Mg	160–225	V	G	G	G	G	V	V
N5	Al Mg3·5	3·5 Mg	215–275	V	G	G	F	G	V	V
N6(1)	Al Mg5	5·0 Mg	250 min.	G	G	G	F	G	V	G
N8	Al Mg4·5 Mn	4·5 Mg	275–345	V	G	G	F	G	V	G

Notes: 1. Rivet and screw stock only.
2. Ratings are for material in the optimum condition for forming.
3. Ratings are given for correct technique and filler rod and take into account the properties of material after welding.
V = very good, G = good, F = fair, P = poor and U = unsuitable.
* Where strength varies with temper the specified minima are quoted. Similarly where properties are influenced by the fabrication process, the lowest minima are given.

Table 4.2 Heat treatable alloys in the solution treated condition, but naturally aged where applicable

Material designation		Major alloying constituents (nominal %)	Tensile strength* (N/mm²)	Resistance to atmospheric attack	Suitability for					
					Cold forming(2)	Machining	Fusion welding(3)		Resistance spot welding	Protective anodising
BS 1470–75	I.S.O.						Oxy-gas	Inert-gas-shielded arc		
H9	Al Mg Si	0·7 Mg, 0·5 Si	125 min	V	G	G	G	G	G	V
H12	Al Cu2 Nil Mg Fe Si	2·3 Cu, 1·0 Ni, 0·9 Mg, 0·9 Fe, 0·9 Si	310 min	F	F	G	U	G	U	F
H15	Al Cu4 Si Mg	4·5 Cu, 0·8 Si, 0·5 mg, 0·8 Mn	370 min	F	F	G	U	G	V	F
HC15	Al Cu4 Si Mg	As H15, clad with 1B	375 min	G	F	G	U	F	V	V
H20	Al Mgl Si Cu	1·0 mg, 0·6 Si, 0·25 Cu (0·5 Mn or 0·25 Cr)	215 min	G	G	G	F	G	G	G
H30	Al Si Mg Mn	1·0 Si, 0·8 Mg, 0·7 Mn	200 min	G	G	G	F	G	G	G

Notes: 2. Ratings are for material in the optimum condition for forming.
3. Ratings are given for correct technique and filler rod and take into account the properties of material after welding.
* Where strength varies with temper the specified minima are quoted. Similarly where properties are influenced by the fabrication process, the lowest minima are given.

Table 4.3 Heat treatable alloys in the fully heat treated condition

Material designation		Major alloying constituents (nominal %)	Tensile strength* (N/mm²)	Resistance to atmospheric attack	Cold forming(2)	Machining	Suitability for			Protective anodising
BS 1470–75	I.S.O.						Fusion welding(3)		Resistance spot welding	
							Oxy-gas	Inert-gas-shielded arc		
H9	Al Mg Si	0·7 Mg, 0·5 Si	150 min	G	G	V	G	G	G	V
H12	Al Cu2 Nil	2·3 Cu, 1·0 Ni, 0·9 Mg, 0·9 Fe, 0·9 Si	385 min	F	P	V	U	G	U	F
H15	Al Cu4 Si Mg	4·5 Cu, 0·8 Si, 0·5 Mg, 0·8 Mn	400 min	P	P	V	U	G	V	F
HC15	Al Cu4 Si Mg	As H15, clad with 1B	400 min	G	P	G	U	F	V	V
H16	Al Cu2 Mg1·5 Fe1 Nil	2·3 Cu, 1·5 Mg, 1·2 Fe, 1·1 Ni	430 min	P	P	V	U	G	U	F
H20	Al Mg Si Cu	1·0 Mg, 0·6 Si, 0·25 Cu (0·5 Mn or 0·25 Cr)	280 min	G	F	V	F	G	G	G
H30	Al Si Mg Mn	1·0 Si, 0·8 Mg, 0·7 Mn	280 min	G	F	V	F	G	G	G

Notes: 2. Ratings are for material in the optimum condition for forming.
3. Ratings are given for correct technique and filler rod and take into account the properties of material after welding.
* Where strength varies with temper the specified minima are quoted. Similarly where properties are influenced by the fabrication process, the lowest minima are given.

ALUMINIUM AND ALUMINIUM ALLOYS

Table 4.4 Cast British Standard aluminium alloys for general and special purposes

BS 1490	Major alloying constituents (nominal %)	Condition	Min. tesile strength* (N/mm²)		Fluidity	Resistance to hot tearing	Pressure tightness	Machin-ability	Resistance to corrosion	Protective anodising	Comments
			Sand cast	Chill cast							
					GENERAL PURPOSE ALLOYS						
LM2	10Si, 1·5 Cu	M	–	150	G	V	V	G	G	P	General purpose die casting, particularly for thin sections
LM4	5 Si, 3 CU	M	140	160	G	G	V	G	F	P	Widely used, general purpose alloy
		TF	230	280							
LM6	12 Si	M	160	190	V	V	V	F	V	P	Excellent castability for thin sections and intricate shapes, good corrosion resistance
LM20	12 Si	M	–	190	V	V	V	G	G	P	Excellent castability, but less corrosion resistant than LM6
LM24	8·5 Si, 3·5 Cu	M	–	180	G	V	V	G	F	P	General purpose die casting
LM25	7 Si, 0·3 Mg	M	130	160	G	V	V	G	V	P	High strength with good corrosion resistance
		TF	230	280							
LM27	7 Si, 2 CU, 0·4 Mn	M	140	160	G	V	V	G	G	P	General purpose alloy

Note: * Where strength varies with temper the specified minima are quoted. Similarly where properties are influenced by the fabrication process, the lowest minima are given.

4:8 ALUMINIUM AND ALUMINIUM ALLOYS

Table 4.4 (continued)

SPECIAL PURPOSE ALLOYS

Alloy	Composition	Condition								Applications
LM0	99·5 Al min.	M	—	—	F	F	F	F	V	V	Suitable for corrosive environments and electrical applications
LM5	4·5 Mg, 0·4 Mn	M	140	170	F	F	P	G	V	V	Suitable for corrosive environments and for decorative applications
LM9	12 Si, 0·4 Mg, 0·4 Mn	M	—	190	G	V	G	F	G	P	Good combination of castability, corrosion resistance and strength
			240	295						
LM10	10 Mg	TB	280	310	F	G	P	G	V	F	Combines toughness with shock resistance
LM12	10 Cu, 0·3 Mg	M	—	170	F	G	G	V	P	P	Specially suitable for hydraulic equipment
LM13	11 Si, 1 Cu, 1 Mg	TF	170	280	G	V	F	F	F	P	Used for pistons
		TF7	140	200						
LM16	5 Si, 1·3 Cu, 0·5 Mg	TB	170	230	G	G	G	G	F	P	Elevated temperature applications
		TF	230	280						
LM18	5 Si	M	120	140	G	V	V	F	V	P	Used for food handling equipment
LM21	6 Si, 4 Cu, 0·4 Mn, 0·2 Mg	M	150	170	G	G	G	G	P	P	High strength with good engineering characteristics
LM22	5 Si, 3 Cu, 0·4 Mn	TB	—	245	G	G	G	G	P	P	High shock resistance, used for structural applications
LM26	9·5 Si, 3 Cu, 1 Mg	TE	—	210	G	V	V	F	P	P	Used for pistons
		TE	—	170						
LM28	18 Si, 1·5 Cu, 1 Mg, 1 Ni	TE	120	190	F	G	F	P	G	P	Special piston applications
LM29	23 Si, 1 Cu, 1 Mg, 1 Ni	TE	120	190	F	G	F	P	G	P	Special piston applications
		TF	120	190						
		M	—	130						
LM30	17 Si, 4·5 Cu, 0·5 Mg	TS	—	160	G	F	F	F	P	P	Automobile cylinder blocks

Note: V = very good, G = good, F = fair and P = poor.

Table 4.5 British Standard, I.S.O. and comparable aluminium association designations

BS 1470-75	I.S.O.	Nearest AA designation	Major alloy constituents (nominal %)
1	Al 99·99	1099	99·99 Al
1A	Al 99·8	1080	99·8 Al
1B	Al 99·5	1050	99·5 Al
1C	Al 99·0	1110	99 Al
N3	Al Mn1	3003	1·25 Mn
N4	Al Mg2	5052	2·25 Mg
N5	Al Mg3·5	5154	3·5 Mg
N6	Al Mg5	5456	5.0 Mg
N8	Al Mg4·5 Mn	5083	4·5 Mg
H9	Al Mg Si	6063	0·7 Mg, 0·5 Si
H12	Al Cu2 Nil Mg Fe Si	–	2·3 Cu, 1·0 Ni, 0·9 Mg, 0·9 Mg, 0·9 Fe, 0·9 Si
H15	Al Cu4 Si Mg	2014	4·5 Cu, 0·8 Si, 0·5 Mg, 0·8 Mn
HC15	Al Cu4 Si Mg	Al clad 2024	As H15, clad with 1B
H16	Al Cu2 Mg1·5 Fe1 Nil	–	2·3 Cu, 1·5 Mg, 1·2 Fe, 1·1 Ni
H20	Al Mg1 Si Cu	6061	1·0 Mg, 0·6 Si, 0·25 Cu (0·5 Cu (0·5 Mn or 0·25 Cr)
H30	Al Si Mg Mn	6351	1·0 Si, 0·8 Mg, 0·7 Mn

make a direct correlation impossible in several cases. Notably, where European practice utilises a minor addition of manganese, a similar effect is frequently achieved by a chromium addition in American practice; also, there is a greater tendency to use small copper additions in American alloys than in European alloys.

The British Standard alloys use a systematic letter notation to indicate the form and heat treatment of the material. Details are given in Table 4.6. For example, the strongest condition is H8 for a non-heat-treatable alloy, TB or TD for a single heat-treatment alloy, and TE, TF and TH for a double heat-treatment alloy.

The non-heat-treatable alloys (prefixed *N*) are hardened by cold work and attain the desired properties by a combination of annealing and cold work. The hard material has markedly increased strength with only slightly reduced corrosion resistance.

The heat-treatable alloys (prefixed *H*) — notably the Al-Cu-Mg and the Al-Mg-Si types — can be heated at 480–535°C for a period between 20 min and some hours to obtain solution of the alloying elements, and then rapidly quenched. This solution treatment gives increased strength, and may also give slightly increased corrosion resistance. Further strengthening of certain alloys is achieved by an additional lower temperature heat-treatment for longer periods (1–20 h or more, according to the alloy) which promotes precipitation of the alloying elements within the metal crystal structure. With some alloys this ageing treatment takes place at room temperature. The ageing or precipitation treatment slightly reduces the corrosion resistance of most alloys.

Comprehensive details of alloy properties and characteristics are provided in the publications of the major aluminium companies and independent organisations[1].

Table 4.6 Temper and heat treatment symbols for aluminium alloys-suffixes

British Standard designation	Meaning
\multicolumn{2}{c}{WROUGHT MATERIALS}	
M	As manufactured. Material which acquires some temper from shaping processes in which there is no special control over thermal treatment or amount of strain hardening
O	Annealed. Material which is fully annealed to obtain the lowest-strength condition
H1, H2 H3, H2 H5, H6 H7, H8	Strain hardened. Material subjected to the application of cold work after annealing (or hot forming) or to a combination of cold work and partial annealing/stabilising in order to secure the specified mechanical properties. The designations 1-8 are in ascending order of tensile strength
TB	Solution heat treated and naturally aged. Material which receives no cold work after solution heat treatment except as may be required to flatten or straighten it. Properties of some alloys in this temper are unstable
TD	Solution heat treated, cold worked and naturally aged
TE	Cooled from an elevated temperature shaping process and precipitation treated
TF	Solution heat treated and precipitation treated
TH	Solution heat treated, cold worked and then precipitation treated
CAST MATERIALS	
M	As cast
TS	Stress relieved only
TE	Precipitation treated
TB	Solution treated
TB7	Solution treated and stabilised
TF	Solution treated and precipitation treated
TF7	Full heat treatment plus stabilisation

Physical Properties

Some of the more useful physical and mechanical properties of aluminium are given in Tables 4.7 and 4.8. The common wrought forms are rolled plate (prefixed P), clad plate (PC), sheet and strip (S), clad sheet and strip (C), bars, rods, and sections (E), extruded round tube and hollow sections (V), drawn tubes (T), wire (G), rivet stock (R), bolt and screw stock (B), and forgings and forging stock (F). Castings are made in sand moulds or in metal moulds known as dies; the most widely used methods involve casting either under gravity or under pressure. Aluminium and aluminium alloys are fabricated into products such as roll plate, sheet, extruded sections, drawn tube, etc. by all the familiar processes, with modifications appropriate to the temper or condition of the material. Joining may be carried out by mechanical methods (such as riveting and bolting), brazing, soldering, adhesive bonding, or welding. The argon-shielded arc welding methods (mig and tig) are particularly appropriate where corrosion resistance of welded joints is of importance[2].

ALUMINIUM AND ALUMINIUM ALLOYS

Table 4.7 Physical properties of aluminium

Atomic number	13
Atomic volume	10·0
Atomic weight	26·97
Valency	3
Crystal structure	Face-centred cubic
Interatomic distance	2·863 Å
Electrochemical equivalent (g/Ah)	0·3354
Density at 293 K (kg/m^3)	2 700

m.p. (K)	Thermal					Electrical			
	Sp. heat at 293 K (J/kg K)	Mean sp. heat (293–931 K) (J/kg K)	Latent heat of fusion (kJ/kg)	Coeff. of linear exp. (293–393 K) (m/K)	Thermal conductivity at 273 K (W/m K)	Elec. Vol. resistivity at 293 K ($\mu\Omega$ cm)	Elec. vol. conductivity at 293 K (% I.A.C.S.)	Temp. coeff. of elec. resistance per K for 293 K	Thermoelectric power vs platinum (mV/100 K)
931	896	1 047	387	0·61 × 10^{-6}	214	2·7–3·0	63–57	0·0041	+0·41

Table 4.8 Mechanical properties of aluminium

Young's modulus (MN/m^2)	Torsion modulus (MN/m^2)	Poisson's ratio	Compressibility ($dv/v_0 dp$)	
			at 293 K	at 400 K
59×10^3	24×10^3	0·34	$1·45 \times 10^{-6}$	$1·70 \times 10^{-6}$

Selection of Purity or Alloy Type

It will be noted that the materials covered by the BS specifications fall into several distinct groups, sometimes with apparently small differences within the group. Characteristics which could influence the selection of the most appropriate material for a specific application are tabulated in Table 4.1 for wrought products, but some elaboration is desirable since the successful utilisation of aluminium begins with the selection of alloy. Additionally, mention should be made of materials not covered by the BS General Engineering series.

Pure Aluminium

Within the BS series the corrosion resistance of unalloyed aluminium increases with increasing metal purity.

The use of the 99·8% and 99·99% grades is usually confined to those applications where very high corrosion resistance or ductility is required. The chemical industry can advantageously use these purities for handling some products, but because of their low mechanical strength they are sometimes used as a cladding material for a stronger substrate.

Decreasing the purity results in increased strength for the 99% and 99·5% grades, which still retain a high resistance to corrosion. The 99% pure metal may be considered the more useful general purpose metal for lightly stressed applications such as panelling and cooking utensils.

Aluminium-Manganese Alloy

The alloy N3 is the sole BS alloy of this binary system. In sheet form, the combination of good corrosion resistance with adequate mechanical strength results in large tonnages being used in building, cooking utensils and sundry general applications.

Aluminium-Magnesium Alloys

For general use, the Al–Mg system is represented by N4, N5 and N8 with increasing magnesium content respectively. The corrosion resistance of all these alloys is extremely good, while the level of mechanical properties obtainable makes them ideally suited for structural use in aggressive conditions.

The characteristics of these alloys make them ideal for boat and ship-building, for which a long history of satisfactory performance is on record for the higher magnesium alloys. Where strength is less critical the lower magnesium alloys may be used with similar success and are recommended for aqueous conditions.

Elevated temperatures should be avoided with N5 and N8, since the precipitation of Mg_2Al_3 over a period of time can lead to serious structural corrosion. In case of doubt regarding this aspect, the manufacturer should be consulted.

Aluminium-Magnesium-Silicon Alloys

The heat-treatable Al–Mg–Si alloys H9, H20 and H30 are predominantly structural materials, all of which have a high resistance to corrosion.

The low Mg + Si content of H9 facilitates the production of complex extrusions with a good surface finish making H9 a natural choice for glazing sections and other architectural features. Higher mechanical properties are obtainable with the H20 and H30 compositions, which are therefore more suitable for load bearing structures.

The corrosion resistance of the Al–Mg–Si alloys is slightly inferior to that of the Al–Mg alloys, but where maximum obtainable strength is required then a fully heat-treated Al–Mg–Si alloy would generally be preferable to an Al–Mg alloy with comparable properties obtained by cold working.

Aluminium-Copper Alloys

The composition of these alloys extends beyond the binary system and they may be categorised as the Duralumin type H15 and the complex types H12 and H16.

The mechanical properties and characteristics of H15 cause it to be used for those applications where high strength is the prime criterion, outweighing its poor resistance to corrosion. Protection by anodising or painting is desirable, when satisfactory performance may be expected except in the most severe conditions. Alternatively, the HC15 clad version has a corrosion resistance similar to its pure cladding, provided that repeated heat treatments have not caused excessive copper diffusion into the pure cladding. The use of H15 for machined components is fairly common, but cannot be recommended where the service conditions will be aggressive.

Where the retention of strength at elevated temperatures is required, then the alloys H12 and H16 should be considered. Because of their copper content the corrosion resistance is mediocre and for service in aggressive environments the Al–1Zn clad version to DTD 5070 would generally be preferred to the unclad metal.

Aluminium-Zinc-Magnesium Alloys

The Al–Zn–Mg alloy system provides a range of commercial compositions, primarily for those areas where strength is a major consideration. None of these alloys are yet included in the BS General Engineering series although their use in Europe and America is quite well established. Essentially the range of compositions may be conveniently divided into two categories.

The high strength alloys contain a Zn + Mg content well in excess of 6% and are used in specialist structures such as aircraft. The risk of stress corrosion cracking in these alloys may be accentuated by incorrect heat treatment or composition and they cannot be recommended for general use (Section 8.5).

The other group of alloys are those with a Zn + Mg content not exceeding 6%. These have been used for general engineering, when natural ageing after welding can be utilised to permit the fabrication of strong welded structures. In particular, these medium strength Al-Zn-Mg alloys have been successfully used for transport applications and it seems probable that this will increase in the near future. With correct manufacturing procedures the risk of stress corrosion with these alloys is negligible and the resistance to unstressed corrosion is only slightly inferior to the Al-Mg-Si structural alloys.

Corrosion Behaviour in Aqueous Environments

Theoretical Considerations of Corrosion Behaviour

Aluminium is a very reactive metal with a high affinity for oxygen. The metal is nevertheless highly resistant to most atmospheres and to a great variety of chemical agents. This resistance is due to the inert and protective character of the aluminium oxide film which forms on the metal surface (Section 1.5). In most environments, therefore, the rate of corrosion of aluminium decreases rapidly with time. In only a few cases, e.g. in caustic soda, does the corrosion rate approximate to the linear. A corrosion rate increasing with time is rarely encountered with aluminium, except in aqueous solutions at high temperatures and pressures.

The corrosion resistance of aluminium and its alloys is largely due to the protective oxide film which within seconds attains a thickness of about 10 Å on freshly exposed metal; continuation of growth is markedly influenced by the environment, being accelerated by increasing temperature and humidity. Immersion in water results in rapid oxide thickening. The behaviour of the oxide may be modified by impurities or alloying additions. In aluminium-magnesium alloys the presence of magnesia in the oxide imparts a characteristic bloom to metal stored under humid conditions. The possible effects of minor impurities or additions is well illustrated in the case of tin, whose modifying effect upon the oxide[3] is utilised in obtaining a highly electronegative potential in aluminium sacrificial anodes (see Section 10.2).

The oxidation of aluminium at room temperature is reported to conform to an inverse logarithmic equation for growth periods up to 5 years duration[4]. At elevated temperatures, oxidation studies over shorter periods illustrate conformity to parabolic, linear and logarithmic relationships according to time and temperature. These kinetic variations are attributed to different mechanisms of film formation[5,6].

The various equilibria of the $Al-H_2O$ system have been collated by

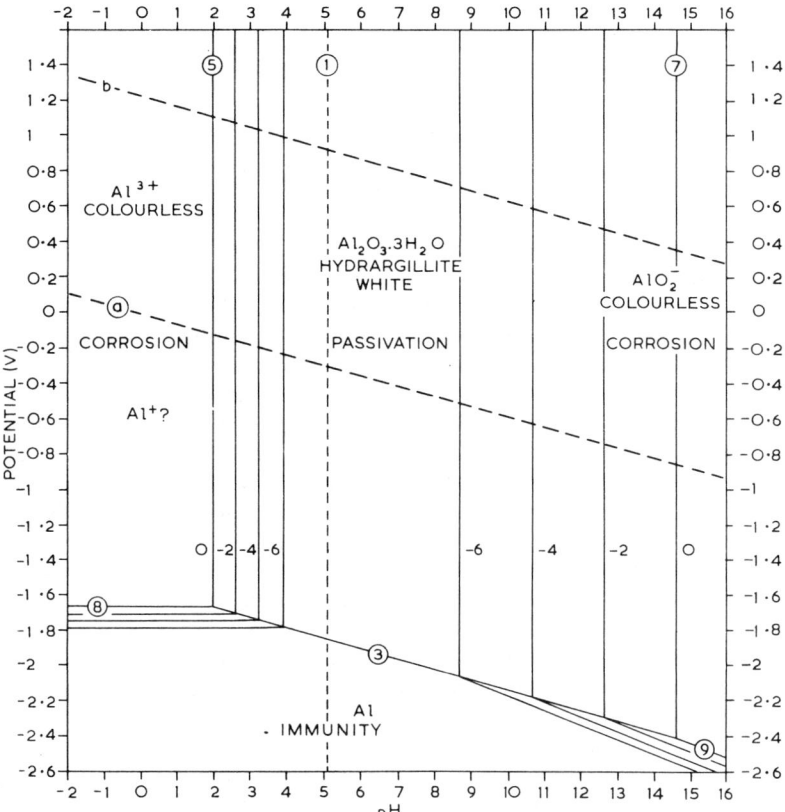

Fig. 4.1 Potential versus pH diagram for Al/H$_2$O system at 25°C (after *Corrosion*, **14**, 496t (1958))

Pourbaix *et al.* in a potential versus pH diagram (Fig. 4.1). This diagram indicates the theoretical circumstances in which aluminium should show corrosion (forming Al^{3+} at low pH values and AlO$_2^-$ at high pH values), passivity due to hydrargillite, i.e. Al$_2$O$_3 \cdot$3H$_2$O (at near-neutral pH values) and immunity (at high negative potentials). The nature of the oxide actually varies according to temperature, and above about 75°C, boehmite (Al$_2$O$_3 \cdot$H$_2$O) is the stable form. It should be noted that the potential-pH diagram does not indicate one of the most important properties of aluminium, i.e. its ability to become passive in strongly acid solutions of high redox potential such as concentrated nitric acid (see also Section 1.4).

Characteristic Features of Corrosion Behaviour

General Dissolution

This occurs in strongly acid or strongly alkaline solutions, but there are specific exceptions. Thus in concentrated nitric acid the metal is passive and the kinetics of the process are controlled by ionic transport through the oxide film, while inhibitors such as silicates permit the use of some alkaline solutions up to pH 11·5 to be used with aluminium. Even where corrosion may occur to a limited extent aluminium is often preferred to other metals because its corrosion products are colourless.

Pitting

This is the most commonly encountered form of aluminium corrosion. In certain near-neutral aqueous solutions a pit once initiated will continue to propagate owing to the fact that the solution within the pit becomes acid, and the alumina is no longer able to form a protective film close to the metal. When the aluminium ions migrate away from the areas of low pH, alumina precipitates as a membrane, further isolating and intensifying local acidity, and pitting of the metal results (see Section 1.6). Solutions containing chlorides are very harmful, particularly when they are associated with local galvanic cells, which can be formed for example by the deposition of copper from solution or by particles such as iron unintentionally embedded in the metal surface. In alkaline media pitting may occur at mechanical defects in the oxide. Pits usually have no crystallographic shape although structurally indicative etch pits can be produced on aluminium.

Where perforation is the criterion of failure, statistical analysis may be judiciously applied to the distribution and depth of pits. Aziz[7] shows that the maximum pit depth on comparatively small test pieces can be related linearly with the maximum depths to be expected in service on large areas over the same period of time. This involves the use of special probability paper (graph paper ruled in such a way that data involving random probabilities may be plotted to give straight-line relationships). Other work from the same laboratory indicates that the use of a small size of panel or of an insufficient number of panels may invalidate pitting test results. Media which are capable of causing pitting may produce no attack when the panels are too small or may attack only a percentage of the panels.

Intercrystalline Corrosion

This is also electrochemical in nature, the galvanic cell being formed because of some heterogeneity in the alloy structure, which may arise from major or trace alloying additions or from minor elements present. In the aluminium-copper type alloys, precipitation of $CuAl_2$ particles at the grain boundaries leaves the adjacent solid solution anodic and prone to corrosion[8]. With aluminium-magnesium alloys the opposite situation occurs, since the precipitated phase Mg_2Al_3 is less noble than the solid solution. However, serious intercrystalline attack in these two alloys is not usual, provided that correct manufacturing and heat treatment conditions are observed.

In the case of the aluminium-magnesium system, most commercial alloys are usually supersaturated, so that elevated service temperatures and inexpert heat treatment are inadvisable, since any resultant grain boundary precipitation may induce susceptibility to intercrystalline attack. The extent of this susceptibility may be approximately deduced from the continuity of

Mg_2Al_3 at the boundaries, continuous or nearly continuous films being extremely detrimental and discrete widely spaced particles being relatively harmless.

Trace elements which adversely affect intercrystalline attack are normally controlled at a safe level. Copper is particularly pertinent in this respect since relatively small additions can cause a marked increase in intercrystalline attack in some alloy systems (Sections 1.3 and 1.7).

Stress Corrosion

This form of corrosion is of limited occurrence with only a few aluminium alloys[9], in particular the higher strength materials such as the Al–Zn–Mg–Cu type and some of the Al–Mg alloys, wrought and cast, with the higher magnesium contents, notably after specific low-temperature heat treatments such as occur during stove enamelling. Stress corrosion is intergranular on aluminium alloys (see Section 8.5).

Filiform Corrosion

This appears as a random non-branching white tunnel of corrosion product either on the surface of non-protected metal or beneath thin surface coatings. It is a structurally insensitive form of corrosion which is more often detrimental to appearance than strength, although thin foil may be perforated and attack of thin clad sheet (as used in aircraft construction) may expose the less corrosion resistant aluminium alloy core. Filiform corrosion is not commonly experienced with aluminium, as reflected by the insignificance afforded it in reviews on the phenomena[10] (Section 1.6).

Layer Corrosion

This may occur on material which has a marked fibrous structure caused by rolling or extrusion. The attack is rapid and very selective, forming partly detached layers of relatively uncorroded material. It is regarded by some authorities as a form of stress corrosion, the stress being either inherent in the metal or produced through the pressure of the larger volume of the corrosion product. It is rare, occurring mainly in copper-bearing alloys, but can occur in a number of environments, including some regarded as only mildly corrosive. Suitable adjustments of ageing treatments and copper content may largely overcome the effect in the higher-strength Al–Cu type alloys[11] (Section 1.3).

Effect of Composition

Few general statements can be made regarding the effect on corrosion resistance of alloying elements or impurities. A useful summary of the information has been prepared by Whitaker[12]. Copper is usually harmful causing increased susceptibility to intercrystalline or general attack, so that alloys containing copper should be regarded as less corrosion resistant than copper-free materials. There are however exceptions to this generalisation, such as an improved stress corrosion resistance in Al–Zn–Mg alloys obtained by a small copper addition[13,14]. Alternatively, the presence of copper may be utilised to delay perforation at the expense of increased general corrosion.

With increasing purity of aluminium, greater resistance to corrosion is developed. On high-purity materials, however, any pits which develop are likely to be deeper though fewer in number than those formed in more impure metal. In some special applications, notably in contact with ammonia solutions or pure water at elevated temperatures and pressures, the iron and silicon present in commercial-purity metal are beneficial and retard corrosion. Up to about 5% magnesium improves the corrosion resistance to sea-water.

Bimetallic Corrosion (Section 1.7)

Aluminium is anodic to many other metals and when it is joined to them in a suitable electrolyte—which may even be a damp porous solid—the potential difference causes a current to flow and considerable corrosion can result. Corrosion is most severe when the resistance of the electrolyte is low, e.g. sea-water. In some cases surface moisture on structures exposed to an aggressive atmosphere can give rise to galvanic corrosion. In practice, copper, brasses, and bronzes in marine conditions cause the most trouble. The danger from copper and its alloys is enhanced by the slight solubility of copper in many solutions and its subsequent redeposition on the aluminium to set up active local cells. This can occur even when the copper and aluminium are not originally in contact, e.g. when water running over cuprous surfaces subsequently comes into contact with aluminium. Similarly, water washings from lead can cause pitting of aluminium. The controlling factor with lead and cuprous washings is the solvency of the water, so that soft waters are the most damaging in this respect. The successful utilisation of these metals in close proximity to aluminium, e.g. in plumbing and roofing, therefore requires careful design to avoid the transfer of a harmful solute to the aluminium.

Contact with steel, though less harmful, may accelerate attack on aluminium, but in some natural waters and other special cases aluminium can be protected at the expense of ferrous materials. Stainless steels may increase attack on aluminium, notably in sea-water or marine atmospheres, but the high electrical resistance of the two surface oxide films minimises bimetallic effects in less aggressive environments. Titanium appears to behave in a similar manner to steel. Aluminium–zinc alloys are used as sacrificial anodes for steel structures, usually with trace additions of tin, indium or mercury to enhance dissolution characteristics and render the operating potential more electronegative.

Aluminium may accelerate attack on zinc alloys; this is particularly noticeable when there is an unfavourable area ratio, as with galvanised fittings in aluminium sheets. In alkaline solutions, however, the aluminium may be preferentially attacked.

The copper-bearing aluminium alloys are more noble than most other aluminium alloys and can accelerate attack on these, notably in sea-water. Mercury and all the precious metals are harmful to aluminium.

Mechanical and Design Factors

Stress below the proof stress does not normally affect corrosion rates. Cyclic stresses in combination with a corrosive environment (corrosion fatigue) can produce failure at below the ordinary fatigue limit. Alloys susceptible to intergranular attack may corrode faster when stressed (see Section 8.5).

Soldered or brazed joints will usually have lower corrosion resistance than the parent metal, but sound welded joints with resistance to attack equal to that of the parent metal can be obtained in most alloys[2]. Many assemblies contain angles, pockets or crevices which attract moisture originating either from extenal sources or from condensation. The corrosion so caused could often be avoided by slight redesign of the assembly, the provision of drain holes of at least 5 mm dia., and the avoidance of horizontal surfaces being among the more important features. Crevices may be filled with jointing compounds. In static assemblies these compounds may be of the setting variety, but in assemblies subject to vibration or movement, as on ships, it is essential that the mastic used should not become too rigid as it might crack in service. It is advisable to incorporate chromates in jointing compounds to inhibit attack by any moisture that may penetrate.

Corrosion in Natural Environments

Atmospheric

The aluminium alloys as a group weather outdoors to a pleasant grey colour, which deepens to black in industrial atmospheres. Superficial pitting occurs initially but gradually ceases, being least marked on high-purity aluminium. With some alloys, including the copper-bearing alloys and the medium-strength Al–Zn–Mg alloys, additional protection, e.g. painting, is desirable in the more aggressive atmospheres to avoid any risk of intercrystalline corrosion.

Gases such as hydrogen sulphide and carbon dioxide do not increase the corrosivity of the atmosphere towards aluminium[15]. Service experience extends over 70 years and includes such well-known examples as Eros, Piccadilly Circus, London, which is in excellent condition, although cast in a low purity (98%) aluminium, and a cupola of San Gioacchino, a church in Rome which was covered in 1897 with sheet 1·25 mm thick and now shows attack to a depth of less than 0·13 mm. Twenty-year tests at selected marine, industrial and rural sites in the U.S.A.[16] have shown that the greater part of the attack takes place in the first year or two and that thereafter the rate of attack maintains a low value. Results from typical environments are shown in Fig. 4.2, and it is apparent that clad alloys give the best results. The relatively high percentage strength losses are due to the extremely thin test specimens. After 20 years the average measured depth of attack for an aluminium–copper alloy at a sea coast test site did not exceed 0·15 mm. The falling-off in rate of pitting with time is in sharp contrast to the behaviour of the older-established structural metals which have a fairly uniform corrosion rate throughout their life, and indicates that the relative merit of aluminium increases with scheduled life.

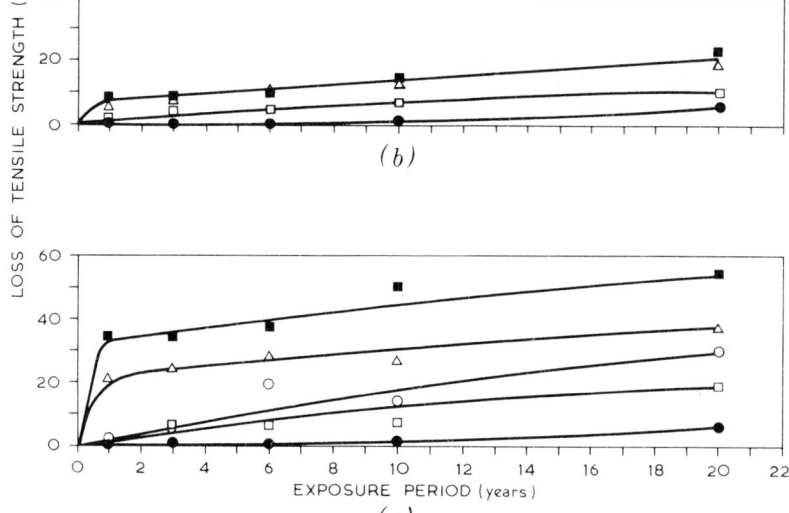

Fig. 4.2 A.S.T.M. 20-year corrosion tests:
- ■ · 2107-T3
- □ 3003-H14
- △ 6051-T4
- ○ 1100-H14
- ● Alclad 2017-T3

(a) State College, Pa. (rural). Premachined tension specimens 0·89 mm thick. Curves for 1100-H14, 3003-HI4 and Alclad 2017-T3 fall below curve shown; (b) New York, N.J. (industrial); (c) La Jolla, Calif. (seacoast) (after A.S.T.M. Symposium on Atmospheric Corrosion of Non-ferrous Metals, 27 (1955))

Aggressive environments include marine conditions and particularly industrial atmospheres containing high concentrations of acid gases such as sulphur dioxide; rain washing is beneficial in both environments, while dampness and condensation alone can accentuate the rate of attack in the presence of chlorides and acidic sulphates.

The relative aggressivity of industrial, marine and rural conditions has been clearly demonstrated by the results of seven year tests in the U.S.A. and British Isles[17], and in this work the benefit from rain washing was especially manifest for the industrial sites in the British Isles (Fig. 4.3).

The combination of acidic sulphates and condensation experienced in some industrial conditions, can cause a particularly voluminous loose corrosion product on some alloys, such as NS3. Where this is likely to be troublesome, cladding with high purity metal is recommended and has been successfully employed, for example on the underside of aluminium-roofed industrial buildings.

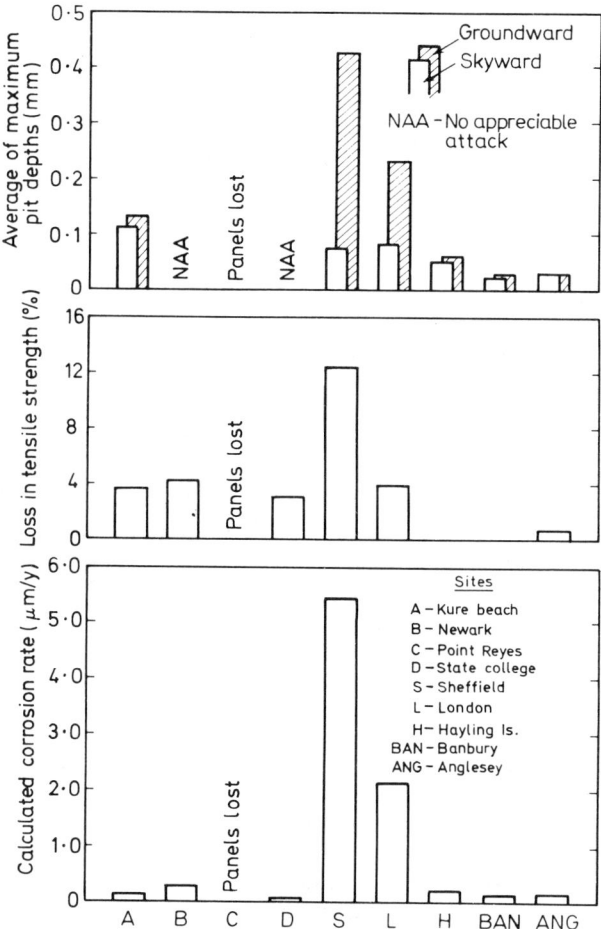

Fig. 4.3 Comparison of weathering effects at United States and English sites — 1199-H18 (and Sl-H) alloy — 7 year data (after *Metal Corrosion in the Atmosphere*, A.S.T.M. Pub. STP435, 151 (1967))

While the continual removal of atmospheric pollution by rain washing is beneficial, the removal of the protective corrosion product is obviously undesirable. The retention of the weathered surface is therefore usually preferred unless aesthetic considerations are of major importance, in which case abrasive or specialist chemical cleaning are effective.

In urban areas, atmospheric fall-out of carbon from partially burned fuel can cause severe localised pitting by galvanic action, although this is not commonly encountered.

Indoors, aluminium retains its appearance well, and even after prolonged periods may show no more than slight dulling or on aluminium–magnesium alloys a slight bloom. This superficial deterioration can be accelerated by the presence of moist conditions and condensation which in extreme cases may lead to staining.

The presence of condensation in confined spaces, such as the turns of a coil or stacked sheets, can cause a more severe staining accompanied by a thick bloom. Oiling or the use of interleaving is sometimes successful in preventing this damage in marginal cases, but improved storage conditions or the elimination of crevice conditions are preferable.

Natural Waters

Immersed aluminium and its alloys have excellent resistance to attack by distilled or pure condensate water, and are used in industry in condensing equipment and in containers for both distilled and deionised water, as well as in steam-heating systems[18].

Of the more important British Standard alloys, only those which contain copper as a major alloying constituent are likely to corrode in unpolluted sea-water, but pollution of the sea-water may cause localised pitting attack to occur on other aluminium alloys. The Al–Mg alloys containing up to about 4·5% magnesium offer particularly good combinations of corrosion resistance and strength. Fouling collects readily on aluminium alloys, as on other materials, and where it may be necessary to use paints containing cuprous oxide for anti-fouling purposes the risk of bimetallic corrosion can be substantially inhibited by a chemical pretreatment of the aluminium followed by a chromate priming paint. Mercury-containing anti-fouling compositions must never be used, as serious bimetallic corrosion will result.

The behaviour of aluminium in natural fresh water and tap waters may vary as these waters differ widely in their dissolved solid content. No corrosion occurs immediately on immersion of aluminium and its alloys in these near-neutral waters, and aluminium gives satisfactory service with all types of tap water provided regular cleaning and drying can take place, as occurs with aluminium hollow-ware. In some waters, black or brown stains which are largely due to optical effects associated with the oxide film on the metal surface, occur. Although somewhat unsightly, the film is quite harmless and can be removed by simple methods such as boiling of fruit (e.g. rhubarb). Alternatively, preliminary boiling with pure water provides some protection against the staining, but can hardly be considered justifiable in most cases.

The combination of carbonate, chloride and copper is more damaging than if they are present singly or if one of them is absent[19,20], so that some supply waters are naturally more aggressive than others. The role of copper is of particular relevance, since as little as 0·02 parts per million can initiate pitting in hard waters[21], although more is required in soft waters which are otherwise less aggressive. In this context however it must be remembered that soft waters are inherently more cupro-solvent than hard waters; consequently the conjoint use of aluminium and copper fittings is rarely advisable irrespective of the necessity for avoiding galvanic interaction when the two metals are in direct contact.

Once pitting has started it may continue in solutions which would themselves be incapable of initiating corrosion. In waters of all types, the rate of increase in the depth of pitting falls off rapidly with time. Water movement (of the order of 0·3 m/s or more) will reduce pitting or prevent its initiation. A rise in temperature tends to lead to higher corrosion rates at existing pits, but even with the most aggressive hard waters, above about 50°C the oxide-forming mechanisms act to prevent the initiation of pitting, as shown by the long and satisfactory service given by aluminium hollow-ware which is assisted in some waters by scale formation.

Where aluminium is to be used in direct contact with cold natural waters with no possibility of regular cleaning, clad aluminium alloys are the preferred materials. An Al–1·2 Mn alloy clad with Al–1·2 Zn is suitable. The cladding is anodic to the core and corrosion is therefore restricted to the surface cladding, thus obviating the risk of perforation. Cladding with super-purity aluminium is preferable where it is important to have the minimal degree of total corrosion, but in this case the potential relationship with the core is more critical and in some circumstances the cladding can actually become cathodic. Sacrificial protection may also be obtained from sprayed coatings of appropriate composition which can be applied to extrusions and castings as well as to sheet, rod, plate and tubes. In practice, unclad aluminium–manganese alloys have been used for piping soft waters in this country and, more widely, in the USA.

Underground Corrosion by Soils

This is largely related to the presence of moisture which can leach out soluble constituents from the soil. As is the case with natural waters, the nature of the corrosive environment is a more important factor than the alloy used, provided that copper-bearing alloys are avoided. At the present time it is impossible to produce a satisfactory classification of soils in respect of their aggressive action on aluminium alloys. Made-up ground, particularly when it includes cinders, is usually extremely corrosive, while neutral clays are often least corrosive. It is desirable that protection should be given to all aluminium materials buried in soils[22,23] except where there is previous experience of satisfactory service from aluminium in a given soil. Pipe wrappings based upon bitumen or chromates are effective, while for cable sheathing a continuous plastic coating provides both electrical and corrosion protection. Cathodic protection has been utilised for pipelines[24,25] but is not widely practised; close control is necessary since over-protection can result in alkali attack. Potentials in the region of $-1 \cdot 0$ V vs saturated Cu/CuSO$_4$ are favoured, although some divergence of opinion exists in this respect.

Corrosion in Chemical Environments

Detailed information about behaviour of specific chemicals is given in several works of reference[25-30].

Acids

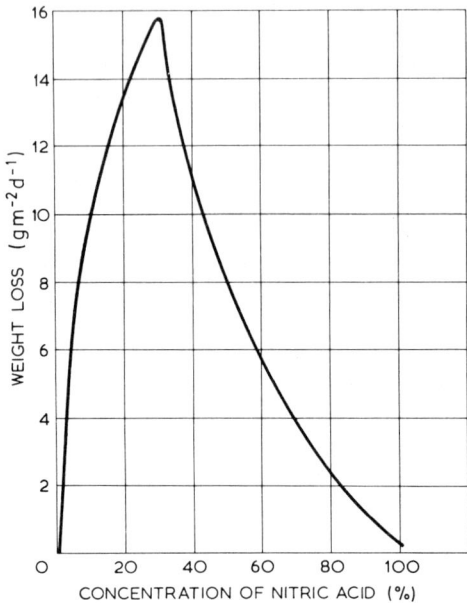

Fig. 4.4 Action of nitric acid of various concentrations on commercial-purity aluminium at 20°C (after Reference 26)

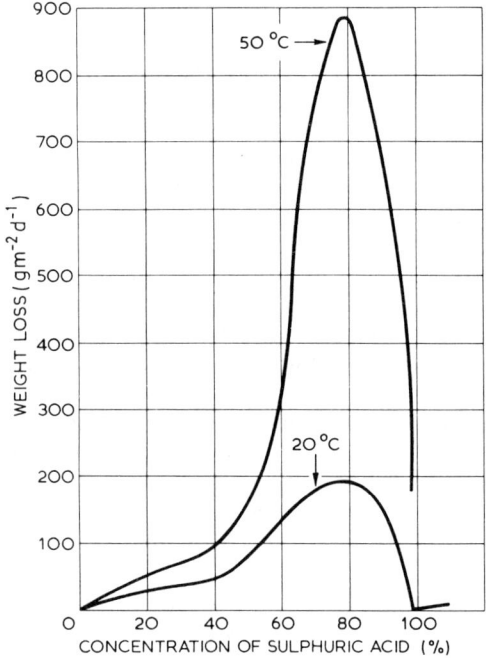

Fig. 4.5 Action of sulphuric acid of various concentrations on commercial-purity aluminium (after Reference 26)

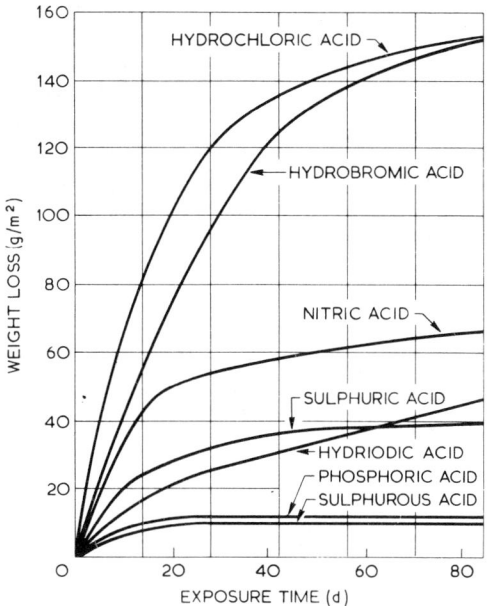

Fig. 4.6 Action of dilute (0·1 N) solutions of inorganic acids on commercial-purity aluminium at 25°C (after Reference 26)

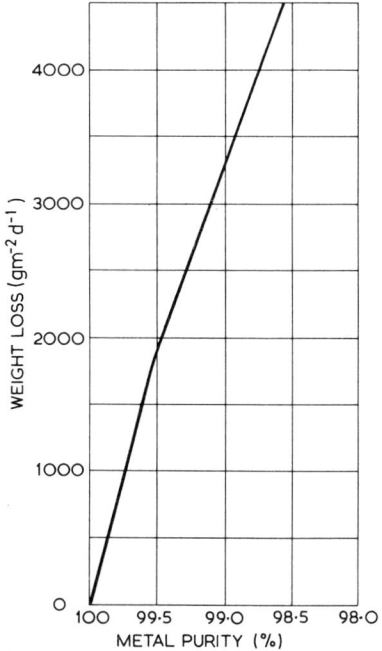

Fig. 4.7 Action of 40% hydrochloric acid on aluminium of various purities at 20°C (after Reference 26)

Most acids are corrosive to aluminium-base materials. The oxidising action of nitric acid at concentrations above about 80%, however, causes passivation of aluminium. Very dilute and very concentrated sulphuric acid dissolves aluminium only slowly. Figures 4.4 and 4.5 give corrosion data at various concentrations for these two acids. The corrosion rates of aluminium in other inorganic acids in dilute solution are shown in Fig. 4.6. Boric acid also exerts little attack on aluminium, while a mixture of chromic and phosphoric acids can be used for the quantitative removal of corrosion products from aluminium without attacking the metal.

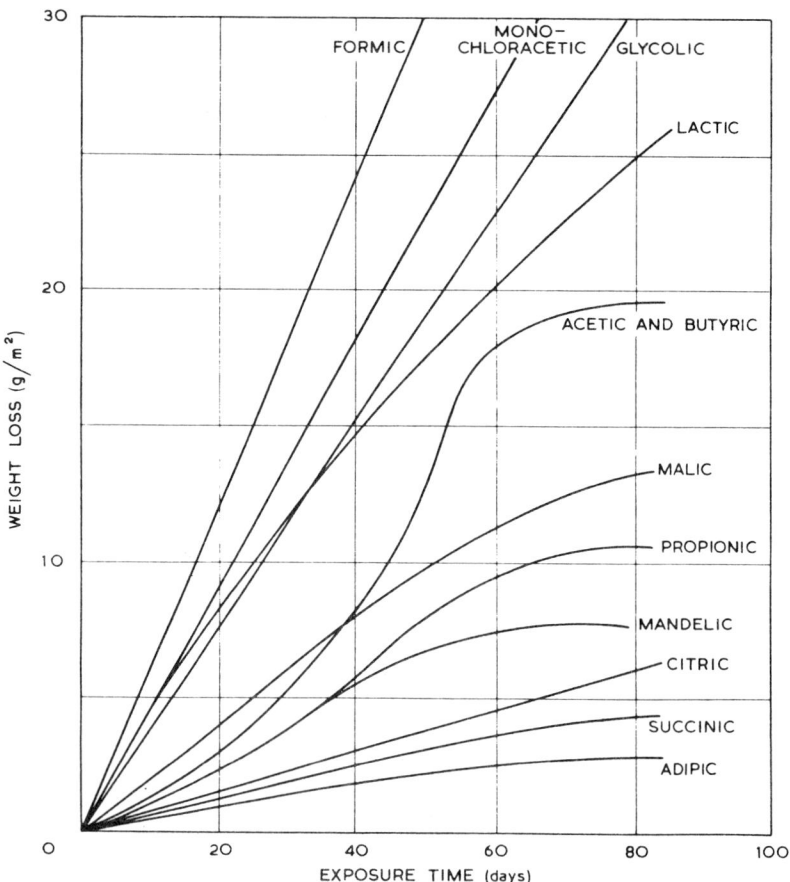

Fig. 4.8 Action of dilute(0·1 N) solutions of organic acids on commercial-purity aluminium at 25°C (after Reference 26)

The effect of commercial metal purity (impurities mainly iron and silicon) on corrosion by 40% hydrochloric acid is shown in Fig. 4.7. This curve is typical of that obtained with many acids.

Organic acids usually have low rates of attack on aluminium, notable exceptions to this generalisation being formic acid, oxalic acid and some chloride-containing acids such as trichloroacetic acid. Corrosion rates for dilute organic acid solutions are given in Fig. 4.8. Glacial acetic acid (pH 3) has no significant corrosive effect on aluminium but the rate of attack increases rapidly with decreasing concentration or in the absence of the traces of water normally present. The rate of corrosion in an acid solution rises rapidly with temperature, often doubling or more with a 10°C rise.

Alkalis

Alkalis are generally corrosive to aluminium; caustic soda is in fact used for chemical milling of aluminium. 99·0% aluminium is, however, resistant to ammonium hydroxide, even at pH 13, while the action of more dilute caustic alkalis can be inhibited by the use of silicates. Mild alkalis such as sodium carbonate are moderately corrosive and are not recommended for washing aluminium hollow-ware. Synthetic detergents, in general, give satisfactory service in cleaning aluminium, but those containing uninhibited sodium carbonate may give some surface roughening. Inhibitors such as silicates can prevent attack by the more dilute solutions.

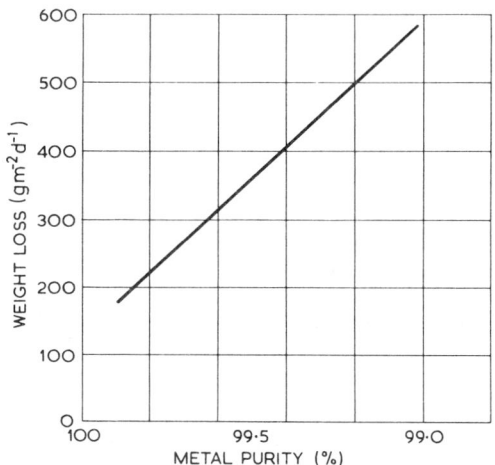

Fig. 4.9 Action of 5·6% potassium hydroxide solution on aluminium of various purities at 20°C (after Reference 26)

Alloys of aluminium with magnesium or magnesium and silicon are generally more resistant than other alloys to alkaline media. The corrosion rate in potassium and sodium hydroxide solutions decreases with increasing purity of the metal (Fig. 4.9), but with ammonium hydroxide the reverse occurs.

Inorganic Salts

Most simple inorganic salt solutions cause virtually no attack on aluminium-base alloys, unless they possess the qualities required for pitting corrosion, which have been considered previously, or hydrolyse in solution to give acid or alkaline reactions, as do, for example, aluminium, ferric and zinc chlorides. With salts of heavy metals—notably copper, silver, and gold—the heavy metal deposits on to the aluminium, where it subsequently causes serious bimetallic corrosion.

Some salts, notably chromates, dichromates, silicates, borates and cinnamates, have marked inhibitive power and are very effective in closed-circuit water systems. Care must be taken to ensure that a sufficient quantity of such anodic inhibitors as chromates is added, as otherwise attack, though occurring at fewer points, may be more severe at these points. Chromates and dichromates have little inhibitive power in strongly acid solutions.

Aluminium is used in hydrogen peroxide (H.T.P.) processing and storage equipment partly because of its high corrosion resistance but also because it does not cause degradation of the peroxide.

Organic Compounds

With many organic compounds, aluminium shows high corrosion resistance either in the presence or absence of water. The lower alcohols and phenols are corrosive when they are completely anhydrous—rarely encountered in practice—since repair of breaks in the natural protective oxide film on aluminium cannot take place in the absence of water. Amines generally cause little attack unless very alkaline.

Processing and storage equipment for many chemicals, including acetaldehyde, formaldehyde, nylon salt, methyl methacrylate, carbon tetrachloride, glycerol, triacetin, proprionic acid, acetic acid and acetic anhydride, is manufactured from aluminium alloys, primarily because of their excellent corrosion resistance.

Antifreeze solutions based on ethylene glycol additions to water have been standardised (BS 3150–3152), the standard differing in the type of corrosion inhibitor present. Inhibition of antifreeze with sodium mercaptobenzothiazole and triethylamine phosphate (NaMBT + TEP) has been used for many years with complete success in contact with aluminium, e.g. in aero-engines, but difficulties with graphitisation of case iron engine components in the solution have led to the introduction of two other types of inhibitors: (*a*) benzoate plus nitrite, and (*b*) borax, usually with soluble oils. Service experience has indicated that corrosion of aluminium components in these inhibited solutions occasionally takes place, though most trials give satisfactory results.

In refrigerating systems, halogen derivatives of methane and ethane marketed under the trade names of Arctons and Freons are without action on pure aluminium and its copper-free alloys in dry conditions, but in wet conditions monochlorodi-, dichloromono-and trichloromonofluoromethanes can hydrolyse to produce slight attack on the aluminium.

Aluminium has good resistance to petroleum products, and an Al–2Mg alloy is used for tank heating coils in crude-oil carriers. Caked-on deposits must be removed from the coils by hot sea-water cleaning in order to maintain effective heat transfer and prevent corrosion. Aluminium is also used in the petroleum industry for sheathing for towers, heat exchangers, transport and storage tanks and scrubbers. Many industries use aluminium alloys for heat exchangers, clad alloys being used where pitting corrosion is liable to be initiated by one of the contacting materials. Heat exchangers in the gas industry have utilised duplex tubes, with aluminium on the water side and steel on the gas side in cases where aluminium is unsuitable owing to the presence of catechol which can attack it.

Aluminium does not become brittle at low temperatures and for this reason (and because of its corrosion resistance) it has been adopted for the carriage and storage of liquefied methane.

High-temperature Corrosion

Dry Atmospheres

When exposed at high temperatures in dry atmospheres aluminium is highly resistant to corrosion by most of the common gases, other than the halogens or their compounds.

High-temperature Aqueous Systems

When aluminium corrodes at temperatures below 90°C in aqueous systems, attack is usually by pitting. At temperatures between 90 and 250°C (for the attainment of which considerable pressures are needed) uniform attack is the commonest form of aqueous corrosion. Above about 250°C, uniform attack is merely the prelude to highly destructive intergranular attack. The corrosion products from the uniform attack form a film which includes a barrier layer and a bulk film analogous to those formed during anodising (Section 15.1); it is the bulk film which controls the corrosion rate, which is not significantly affected by most common dissolved ions[31]. The onset of intergranular attack occurs at about the same time as the crystallisation of the amorphous barrier layer oxide. Kinetic studies indicate that over the temperature range from 100 to 363°C the oxidation rate law is successively inverse logarithmic, parabolic and linear[32].

The requirements of nuclear energy application fostered an interest in special alloys for service in high temperature aqueous environments, but their utilisation has not been widespread. Encouraging results have been reported for alloys of 2Ni–0·5Fe[33] and 1·2Ni–1·8Fe[34].

It has been suggested that the role of nickel (as $NiAl_3$) is to provide sites of low hydrogen overvoltage, where cathodically liberated hydrogen may be liberated without disrupting the protective oxide[35]. The distribution of such sites is apparently critical however, since high corrosion resistance is associated with a fine dispersion of the second phase, while the electronic conductivity of the film is probably also important[36].

Steam forms a protective white film at temperatures up to about 250°C, but above this temperature steam can, under some conditions, react with aluminium progressively to form aluminium oxide and hydrogen. Sintered aluminium powder (S.A.P.) has relatively good resistance to steam at 500°C, but at about 300°C an addition of 1% nickel to the S.A.P. is needed to prevent rapid disintegration.

Molten Salts and Metals

Aluminium-base alloys resist the action of many molten salts which are nearly neutral in reaction. Molten sodium nitrate or mixtures of sodium nitrate and potassium nitrate are used for salt bath heat treatment of some aluminium alloys.

Molten metals generally attack aluminium alloys. Both zinc and tin form alloys by dissolution of aluminium, although the aluminium does not melt. Molten lead is inert to aluminium, and molten lead baths can be used for heating aluminium alloys. Mercury, molten at room temperature, amalgamates readily with aluminium alloys if their naturally formed oxide films are temporarily removed by scratching, and rapid corrosion occurs on subsequent exposure to moist air or water. Under stressed conditions cracking will frequently result, since mercury penetrates into the aluminium alloy selectively at grain boundaries. Contact of aluminium with mercury is extremely dangerous and severe corrosion can occur with a very small amount of mercury.

Aluminium in Contact with other Materials

The aluminium alloys recommended for building purposes (not including the high strength alloys containing copper) have good resistance to concretes, mortars, plasters and asbestos cement products. When freshly mixed, some of these materials release traces of alkaline products which may be sufficient to stain aluminium or to etch it slightly. As soon as the mixture is set, however, the attack ceases and even after many years' service, attack on embedded aluminium is found to be negligible[37,38].

With cement and concrete under continually wet conditions, there may be some surface attack. This decreases rapidly with time, and the strength of components is not significantly affected. Under embedment conditions, bituminous protection is advisable, to avoid risk of cracking of the concrete due to stresses set up owing to the bulk of the corrosion product. Plasters are generally even less aggressive than Portland cement. In damp environments, some corrosion of aluminium may arise in contact with the more open-grained building stones and brickwork, but the hard stones, such as granite, are inert. With building stone and brickwork, as with soils (Section 9.3), it is the nature of the products which can be leached out which will determine whether aluminium corrodes. Unprotected aluminium, in the form of nails for example, can be satisfactorily used in contact with precast concrete blocks, which are usually non-corrosive to aluminium. Magnesium oxy-chloride compositions (used for flooring), on the other hand, stimulate corrosion of aluminium under moist conditions, as will many insulation materials based on magnesium and calcium silicates.

Plastics are generally without action on aluminium and are widely used to provide insulation between other metals and aluminium, while the use of aluminium/plastic laminates is increasing. Rubber has no effect upon aluminium.

A few acid woods, such as oak, chestnut and Western red cedar, accelerate surface weathering of aluminium, but do not usually give rise to serious attack[39]. Timber preservatives containing soluble copper compounds should be avoided; creosote and zinc napthenate are satisfactory preservatives for wood in contact with aluminium.

Common packaging materials are a potential source of aggressive substance[40], and careful selection is recommended to avoid surface deterioration. Where paper is in contact with aluminium, the chloride content should be below 0·05%, sulphate content below 0·25%, copper content below 0·01% and the pH of aqueous extracts in the range pH 5·5-7·5, in order to avoid corrosion in damp conditions. Papers and felts used in building applications should also conform to this specification as a minimum requirement and be of the highest quality, since metallic copper found in materials of inferior origin can result in severe local galvanic attack of aluminium.

Tarpaulins are sometimes treated with copper-containing preservatives and water leached from these has been found to cause corrosion of underlying aluminium sheets.

Fibreglass insulation produced from soda glass can cause pitting in conditions where leaching of alkali occurs, for example, by condensation: the use of Fibreglass produced from Pyrexglass is therefore preferred. Common putties of whiting/linseed oil composition do not attack aluminium; adhesion is obtained by allowing the metal surface to weather, or by applying an etch primer treatment to the metal. Both thermosetting adhesives (e.g. the phenolic types) and thermoplastic adhesives (such as paraffin and microcrystalline waxes, or bitumen) are non-corrosive to aluminium. In general, adhesives applied to aluminium should not contain chlorides in excess of 0·05% (as NaCl) of the solid content, and should be free from copper- or mercury- containing anti-fungicides. The presence in the adhesive of borax or sodium silicate is beneficial when one of the adhesive components is of an acid character.

Recent Literature Surveys

The corrosion resistance of aluminium in a variety of media has been reported. It has been observed that mono-chloroacetic acid has no corrosive effect and di-chloroacetic acid has negligible effects up to 5 m but tri-chloroacetic acid produces a vigorous reaction[41]. The effects of some transition metal and heavy metal cation on dissolution of aluminium in neutral and acid chloride containing solutions has been reported by Khedr and Lashien[42] while the corrosion rate of aluminum conductors in integrated circuits has received the attention of Lerner and Eldredge[43]. An experimental pH-potential diagram for aluminium in seawater is available[44]. The environmental chemistry factors affecting surface film destruction have received some discussion[45] and the energy transfer in aluminium dissolution is represented by a potential energy surface diagram[46]. The cathodic protection of aluminium in seawater in considered by Gundersen and Nisancioglu[47,48].

The sources of characteristic emission generated in aluminium alloys exposed to various environments is presented by Arora[49] while Graver and Wiedmer have undertaken an electrochemical investigation into Al-rich intermetallics[50], Larsen-Basse has reported on the corrosion of aluminium alloys in ocean thermal energy conversion seawaters[51], Ahmed on corrosion and its prevention in de-salination plants[52], Lashermes on marine environment effects[53], similarly Huppartz and Krajewski[54], Kunze on the corrosivity of various foodstuffs to aluminium packaging materials[55], Rogozhina *et al.* in the corrosion of a range of aluminium alloys used in agricultural enclosures[56], while reactions with nitric acid have been covered by Singh *et al.*[57] and Horn[58]. Singh has shown that surface roughness has an important effect on the well-water staining of aluminium and its alloys[59].

The addition of magnesium to Al-Li-Cu-Zr-Ge alloy results in a more consistent corrosion behaviour[60]. Overageing increases the exfoliation corrosion resistance as well as the resistance to electrochemical corrosion. On the other hand, the addition of germanium to Al-Li results in the underaged alloy being more stable in terms of breakdown and repassivation compared to pure aluminium or the binary alloy[61] although ageing can produce a grain boundary structure and associated precipitation effects that reduce the corrosion resistance[62]. Mauret and Lacaze studied the water corrosion of Al-Mg and Al-Cu-Mg using gas chromatography of hydrogen[63], Huppartz and Wieser report on the electrochemical behaviour of Al-Mg-Mn and Al-Mg-Si in seawater and hard water with the relevant E-Z diagrams[64] and Shirkhanzadeh and Thompson provide information concerning the corrosion of Al-Ga in alkaline solutions[65].

As seen above Moran *et al.*, has commented upon the exfoliation corrosion of Al-Li-Cu-Zr-Ge[60]. The mechanism has been investigated by Reboul and Bouvaist[66] and a mathematical model suggested by Robinson[67]. The influence of alloy elements in Al-Zn-Mg has been reported by Reboul[68] and it has been shown that exfoliation corrosion of Al-Mg-Si in irrigation water is also governed by alloy and impurity concentration[69].

A number of studies have been undertaken concerning the suitability of various inhibitors including p-quinone and acetic acid[70], the former being of little value in the case of AA1060 alloy in aqueous potassium nitrate, carboxyliacids[71], Diamines[72], Complexons such as zinc phosphate[73], oxalates[74], and morpholine and thiosemi-carbozide derivatives[75]. An XPS investigation of dichsomate and molybdate in chloride ion-containing solutions showed that, under the conditions used, chromium exists on the surface primarily as Cr(III) whereas molybdenum exists as Mo(IV) and Mo(VI)[76]. Zanzuichi and Thomas report on the use of inhibitor for aluminium films in integrated circuits[77].

Pitting corrosion always remains a worthy subject for study, particularly with reference to mechanism, and the problem conveniently divides into aspects of initiation and growth. For 6061 alloy in synthetic seawater, given sufficient time, pit initiation and growth will occur at potentials at or slightly above the repassivition potential[78]. In an electrochemical study, it was found that chloride ions attack the passive layer as a chemical reaction partner so that the initiation process becomes one of cooperative chemical and electrochemical effects[79].

A focused laser beam was used by Alkine and Feldman[80] to create local depassivation thus providing a novel and specific approach to initiation of value in fundamental studies; yet Bonora et al.[81] found that pitting resistance may be improved if the entire surface area be irradiated with such a beam; it was argued that a chemically inert surface was produced by the irradiation treatment. Thompson et al.[82] report on the involvement of preexisting flaws or weak spots in the surface film and Fokin and Koteneu[83] describe an ellipsometric study or pit formation and repassivation.

Increasing the hydrostatic pressure can increase pitting susceptibility and decrease the passivation range as a result of the decreased thickness and increased number of defects in the oxide layer[84]. Hunkeler and Bohni[85] quantitatively examined pit growth as a function of time, potential and electrolyte conductivity. They considered that growth occurred as a result of a primary change in the properties of the surface area of the pit caused by adsorption of chloride ions while growth is ohmically controlled which, under ideal conditions, results in a square root growth law[86]. Alwitt et al.[87] found what they consider is a unique form of pitting corrosion during the anodic dissolution of aluminium in hot chloride ion containing solution. A high density of fine etch tunnels were produced extending along the ⟨100⟩ directions and evolving from cubic etch pits when all but one wall of a pit becomes passivated; dissolution rate is high from this active pit surface.

Computer simulation of etch pit morphology provided good agreement with experiment for Idemoto and Koura[48], morphology in chloride ion and nitrate ion containing solution also having been investigated by Klinger and Feller[89] while both Mansfield et al.[90] and Sharland and Tasker[91,92] have been involved with mathematical modelling of pitting corrosion. Furthermore reports are available concerning the pitting of aluminium foils[93], the effect of molybdenum[94], rapid solidification processing[95], both of the latter being beneficial, alternating current[96], brazing[97] and in Al-Zn-Mg-Cu alloy[98] and 5083/6061 alloys[99].

Hitzig et al. have produced a simplified model of the aluminium oxide layer(s) to explain impedance data of specimens prepared under different layer formation and sealing conditions[100]. The model also gives consideration to the formation of active and passive pits in the oxide layer. Shaw et al. have shown that it is possible to electrochemically incorporate molybdenum into the passive film which, as previously noted[101], improves the pitting resistance.

Interest has been aroused in connection with the formation of electrochemical films on aluminium covered with a thermally grown film[102-4]. Both the thermally and the anodically grown film are amorphous normally but growing ananodic film on top of a thermal film results in the anodic film being crystalline. Less charge seems to be required compared with the anodising of a 'clean' aluminium surface and so presumably the crystallised film can withstand a higher electric field than the amorphous film. Fundamental studies elucidating the growth and properties of barrier-type films have been reported by Skeldon et al.[106], Csanady et al.[107], Ebihara et al.[108], Fukuda and Fukushima[109], Menezes et al.[110], Wittberg et al.[111], and Thompson et al.[112].

Strazzi has reviewed methods of sealing oxide films[113] and Omata et al. find that adhesion of paint films to anodised layers depends on penetration of the paint into the micro-pores of the anodic layer[114]. Faller has made a comparison of a number of anodising processes and process parameters after weathering the anodised specimens for five years in industrial and marine atmospheres[115]. Elevated temperature oxidation behaviour of Al-Mg-Li[116] and Al-Zn-Mg[117] have also been reported.

Film dissolution in acetic acid-acetate buffers has been investigated by Valand using the potential step method[118] in neutral and acid solutions using high potential cathodic polarization[119] and dissolution in KF solution has been shown to follow an empirical relation incorporating a film thickness parameter[120]. A change in dissolution rate occurred indicating a duplex oxide with the inner layer dissolving more easily than the outer layer.

The volume of reported work concerning the environmentally assisted cracking of aluminium alloys, particularly the Al-Zn-Mg type, is quite phenomenal and cannot adequately be reviewed in this general update.

Surface reactions and their relation to environmentally assisted cracking of Al-Mg has been reported by Ford[121] and Pathania and Trumaris[122] while Lee and Pyum[123] have undertaken an electrochemical study of Al-Cu-Mg showing that its SCC rate is affected by prior metallurgical history. Dietzer et al.[124] determined Kiscc of 2024 in 3·5% sodium chloride solution finding that the value was little affected by the three different loading methods used in their study. With the Al-Li-Cu alloy fatigue bonded in 3·5% sodium chloride solution, pitting was found to be important to crack initiation but this was dependent on the strain rate range involved[125]. For the Al-Li-Cu-Mg[126-8] K_{iscc} decreased with increasing ageing time, no doubt a result of the precipitation of S-(Al-Li) at high angle boundaries and associated PFZ (precipitate-free-zone) formation.

The Al-Zn-Mg alloy[122, 129-142] studies have been mainly concerned with Kiscc determination for different bonding modes, environmental conditions and sample metallurgical history. Failure is very much a function of grain size, grain boundary precipitation and formation of PFZS; thus, work that attempts to improve the situation by alloy/microstructure modification is prominent. This involves compositional changes and heat treatment designed to affect segregation phenomena and precipitate type, morphology and location as a result of ageing. There are two generally accepted mechanisms for SCC; film rupture with anodic dissolution and hydrogen-assisted cracking. Which of these occurs appears to be dependent on environmental chemistry and the potential of the alloy in the environment[131, 140] although bonding mode can also be important[133]. A number of studies have addressed the role of hydrogen in SCC of aluminium alloys[122, 131, 139, 143, 145]. Although there is still no general consensus of opinion, it does seem that hydrogen affects the plastic deformation properties of the aluminium matrix in the crack tip zones.

The related alloy system, Al-Zn-Mg-Cn is also well documented[146-152]. Overageing is reported to be beneficial since modification of the grain boundary precipitate aspect ratio occurs[148].

Bucci[113] has produced a useful and extensive report of value for the selection of suitable aluminium alloys to resist both SCC and corrosion fatigue while Khobaib[154] discusses a range of beneficial inhibitors suitable under conditions of corrosion fatigue.

Finally, reports are available on the durability of adhesively bonded aluminium joints[155,156].

J. C. BAILEY
F. C. PORTER
A. W. PEARSON
R.A. JARMAN

REFERENCES

1. *The Properties of Aluminium and its Alloys*, 6th edn, The Aluminium Federation, Birmingham (1968)
2. Blewett, R. V. and Skerry, E. W., *Metallurgia*, **71**, 73 (1965)
3. Keir, D. S., Pryor, M. J. and Sperrey, P. R., *J. Electrochem. Soc.*, **114**, 777 (1967)
4. Godard, H. P., *J. Electrochem. Soc.*, **114**, 354 (1967)
5. Aylmore, D. W., Gregg, S. J. and Jepson, W. B., *J. Inst. Met.*, **88**, 205 (1959-60)
6. Bartlett, R. W., *J. Electrochem. Soc.*, **111**, 903 (1964)
7. Aziz, P. M. and Godard, H. P., *Corrosion*, **12**, 495t (1956)
8. Hunter, M. S., Frank, G. R. and Robinson, D. L., 2nd *International Congress on Metallic Corrosion*, 66 (1963)
9. Champion, F. A., *J. Inst. Met.*, **83**, 385 (1954-55)
10. Barton, J. F., *Paint Manufacture*, Nov., 53 (1964) and Dec., 47 (1964)
11. Bell, W. A. and Campbell, H. S., *J. Inst. Met.*, **89**, 464 (1960-61)
12. Whitaker, M. E., *Metal Ind.*, 80, 183, 207, 227, 247, 263, 288, 303, 331, 346, 387 (1952)
13. Chadwick, R., Muir, N. B. and Grainger, H. B., *J. Inst. Met.*, **85**, 161 (1956-57)
14. Bushy, J., Cleave, J. F. and Cudd, R. L., *J. Inst. Met.*, **99**, 41 (1971)
15. Aziz, P. M. and Godard, H. P., *Corrosion*, **15**, 529t (1959)
16. *Symposium on Atmospheric Corrosion of Non-Ferrous Metals*, Amer. Soc. Test. Mat., Special Technical Publication No. 175 (1956)
17. *Metal Corrosion in the Atmosphere*, Amer. Soc. Test. Mat., Special Technical Publication No. 435 (1968). Papers by McGeary, *et al.*, p. 141 and Ailor, J. R., p. 285
18. Symposium on Corrosion by High Purity Water, *Corrosion*, **13**, 151t (1957)
19. Davies, D. E., *J. Appl. Chem.*, **9**, 651 (1959)
20. Rowe, L. C. and Walker, M. S., *Corrosion*, **17**, 353t (1961)
21. Porter, F. C. and Hadden, S. E., *J. Appl. Chem.*, **3**, 385 (1953)
22. Gilbert, P. T. and Porter, F. C., *Iron and Steel Inst.*, Special Report No. 45, 55-74 (1951)
23. Raine, P. A., *Chem. and Ind.* (Rev.), 1102, 1196 (1956)
24. Sprowls, D. O. and Carlisle, M. E., *Corrosion*, **17**, 125t (1961)
25. *Day Chemische Verhalten von Aluminium*, Aluminium-Verlag GmbH, Dusseldorf (1955)
26. *Aluminium in the Chemical and Food Industries*, British Aluminium Co. Ltd., London (1959)
27. *Aluminium with Food and Chemicals*, Alcan Booth Industries Ltd. (1966)
28. Ritter, F., *Korrosionstabelien Metallischer Werkstoffe,* Springer-Verlag, Vienna (1944)
29. *Aluminium with Food and Chemicals*, The Aluminium Association, New York (1967)
30. *Process Industries Applications of Alcoa Aluminium*, Alcoa, Pittsburgh, U.S.A.
31. Troutner, V. H., *Corrosion*, **15**, 9t (1959)
32. Dillon, R. L., *Corrosion*, **15**, 13t (1959)
33. Perryman, E. C. W., *J. Inst. Met.*, **88**, 62 (1959)
34. Dillon, R. L. and Bowen, H. C., *Corrosion*, **18**, 406t (1962)
35. Draley, J. E. and Ruther, W. E., *Corrosion*, **12**, 480t (1956)
36. Greenblatt, J. H. and Macmillan, A. F., *Corrosion*, **19**, 146t (1963)
37. Porter, F. C., *Metallurgia*, **65**, 65 (1962)
38. Jones, F. E. and Tarleton, R. D., *Effect of Embedding Aluminium and Aluminium Alloys in Building Materials*, National Building Studies Research Paper No. 36, H.M.S.O., London
39. Farmer, R. H. and Porter, F. C., *Metallurgia*, **68**, 161 (1963)
40. Scott, D. J. and Skerrey, E. W., *Br. Corros. J.*, **5**, 239 (1970)
41. Mansour H. *et al.*, *Bull. of Electrochemistry*, **2**, 449-451 (1986)
42. Khedr M. G. A. and Lashien, A. M. S., *J. Electrochem. Soc.*, **136**, 968-72 (1989)

43. Lerner, I. and Eldridge, J. M., *ibid.*, **129**, 2270-73 (1982)
44. Gimenez P. *et al.*, *Rev. Aluminium*, 518, 261-72 (1982)
45. Godard, H. P., *Materials Performance*, **20**, 9-15 (1981)
46. Foley, R. T. and Nyuyen, T. H., *J. Electrolem. Soc.*, **129**, 464-7 (1982)
47. Gundersen, R. and Nisancioglu K., *Corrosion*, **46**, 279-85 (1990)
48. Nisancioglu, K., Lunder, O. and Holtan, H., *ibid.*, **41**, 247-57 (1985)
49. Arora, A., *ibid.*, **40**, 459-65 (1984)
50. Graver, R. and Wiedmer, E., *Wekstoffe Korros.*, **31**, 550-5 (1980)
51. Larsen-Basse, J., *Materials Performance*, **23**, 16-21 (1984)
52. Ahmed, Z., *Anti-Coros. Wleth, Mat.*, **28**, 4-10 (1981)
53. Lashermes, M., *Rev. Aluminium*, 523, 505-11 (1982)
54. Huppatz, W. and Krajewski, H., *Weistroffe Korros.*, **30**, 673-84 (1979)
55. Kunze, E., *Aluminium*, **52**, 296-301 (1976)
56. Rogzhina, E. P., Koltunova, G. A., Pashkova, O. A. and Goluber, A. I., *Zashch, Met.*, **25**, 120-24 (1989)
57. Singh, D. D. N., *et al.*, *J. Electrochem. Soc.*, **129**, 1869-74 (1982)
58. Born, E-M., *Werkstoffe Korros.*, **41**, 32-3 (1990)
59. Singe, T., *Aluminium*, **57**, 187-9 (1981)
60. Moran, J. P., *et al.*, *Corrosion*, **43**, 374-82 (1987)
61. Colvin, E. L., Cahen Jr., G. L., Stoner, G. E. and Starke, E. A., *Corrosion*, **42**, 416-21 (1986)
62. Kumai, C., Kusinski, J., Thomas, G. and Devine, T. M., *ibid.*, **45**, 294-302 (1989)
63. Mauret, P. and Laraze, P., *Corros Sci.*, **22**, 321-9 (1982)
64. Huppartz, W. and Wieser, D., *Werkstoffe Korros*, **40**, 57-62 (1989)
65. Shirkhanzadeh, M. and Thompson, G. E., *Electrochim. Acta*, **33**, 939-40 (1989)
66. Reboul, M. G. and Bouvaist, J., *Werkstoffe Korros.*, **30**, 700-12 (1979)
67. Robinson, M. J., *Corros. Sci.*, **22**, 775-90 (1982)
68. Reboul, M. G. and Bouvaist, J., *Rev., Aluminium*, 491, 41-55 (1980)
69. Zahavi, J. and Yahalom, J., *J. Electrochem. Soc.*, **129**, 1181-5 (1982)
70. Onuchukwa, I. and Oppong, F. W., *Corros. Sci.*, **26**, 919-26 (1986)
71. Moussa, M. N. and El-Togoury, M. M., *AntiCorros, Meth. Mat.*, **37**, 4-8 (1990)
72. Al-Suhybani, A. A., *Corros. Prev. Control*, **37**, 11-16 (1990)
73. Kuznetsov, Yu. I. and Bardasheva, T. I., *Zashch. Met.*, **24**, 234-40 (1988)
74. Wilhelmsen, W. and Grande, A. P., *Electrochim. Acta*, **33**, 927-32 (1988)
75. Anon., *Anti-corros. Meth. Mat.*, **35**, 4-8 (1988)
76. Bairamow, A. K., *Corros. Sci.*, **25**, 69-73 (1985)
77. Zanzuichi, P. J. and Thomas III, J. H., *J. Electrochem. Soc.*, **135**, 1370-1376 (1988)
78. Aylor, D. M. and Moron, P. J., *ibid.*, **133**, 868-72 (1986)
79. Tomisanyi, L., Varga, K. and Bartik, I., *Electrochim. Acta*, **34**, 855-9 (1989)
80. Alkine, R. and Feldman, M., *J. Electrochem. Soc.*, **135**, 1850-51 (1988)
81. Bonora, P. L., *et al.*, *Thin Solid Films*, Lausanne, **81**, 339-45 (1981)
82. Thompson, G. E., *et al.*, *J. Electrochem. Soc.*, **129**, 1515-17 (1982)
83. Fokin, M. N. and Kotenev, V. A., *Zashch. Met.*, **24**, 111-4 (1988)
84. Beccaria, A. M. and Poggi, G., *Corrosion*, **42**, 470-75 (1986)
85. Hunkeler, F. and Bohrii, H., *Werkstoffe Korros.*, **34**, 593-603 (1983)
86. *Idem.*, *Corrosion*, **40**, 534-40 (1984)
87. Alwitt, R. S., *et al.*, *J. Electrochem. Soc.*, **131**, 13-7 (1984)
88. Idemoto, Y. and Koura, N., *J. Metal Finishing Soc*, Japan, **37**, 30-5 (1986)
89. Klinger, R. and Feller, H. G., *Aluminium*, **57**, 224-7 (1981)
90. Mansfield, F., Lin, S., Khim, S. and Shih, H., *J. Electrochem. Soc.*, **137**, 78-82 (1990)
91. Sharland, S. M. and Tasker, D. W., *Corros. Sci.*, **28**, 603-20 (1988)
92. Sharland, S. M., *ibid.*, **28**, 621-30 (1988)
93. Aylor, D. M. and Moran, P. J., *J. Electrochem. Soc.*, **133**, 949-51 (1986)
94. Mosher, W. C., *et al.*, *ibid.*, **133**, 1063-4 (1986)
95. Yoshioku, H., *et al.*, *Corros. Sci.*, **26**, 795-812 (1986)
96. Vu Quang, K., *et al.*, *J. Electrochem. Soc.*, **130**, 1248-52 (1983)
97. Hattori, T. and Sakamota, A., *Welding J.*, **61**, 3395-425 (1982)
98. Maitra, S. and English, G. C., *Met. Trans.*, **13A**, 161-6 (1982)
99. El-Boujclaini, M., Ghali, E. and Galibois, A., *J. Appl. Electrochem.*, **18**, 257-64 (1988)
100. Hitzig, J., *et al.*, *J. Electrochem. Soc.*, **133**, 887-92 (1986)
101. Shaw, B. A., Davis, G. D., Fritz, T. L. and Olver, K. A., *ibid.*, **137**, 359-60 (1990)

102. Kobayashi, K., et al., ibid, **133**, 140-1 (1986)
103. Crevecour, C. and de Wit, H. J., ibid., **134**, 808-16 (1987)
104. Partridge, P. G. and Chadbourne, N. C., J. Mater. Sci., **24**, 2765-74 (1988)
105. Skeldon, P., et al., Thin and Olid Films, **123**, 127-133 (1985)
106. Xu, Y., et al., Corros Sci., **27**, 83-102 (1987)
107. Csanady, A., et al., Corros Sci., **24**, 237-248 (1984)
108. Ebinhara, K., et al., J. Metal Finishing Soc., Japan, **33**, 156-64 (1982)
109. Fukuda, Y. and Fukushima, T., Electrochim. Acta., **28**, 47-56 (1983)
110. Menezes, S., Haak, R., Hagen, G., Kendig, M., J. Electrochem Soc., **136**, 1884-6 (1989)
111. Wittberg, T. N., Wolf, J. D. and Wang, P. S., J. Mater, Sci., **23**, 1745-7 (1988)
112. Thompson, G. E., et al., Trans. Inst. Metal Finishing, **58**, 21-5 (1980)
113. Strazzi, E., Alluminio, **50**, 496-9, 520-5 (1981)
114. Omata, K., et al., Aluminium, **57**, 811-3 (1981)
115. Faller, F. E., ibid., **58**, E23-5 (1982)
116. Csanady, C. and Kurthy, J., Mat. Sci., **16**, 2919-22 (1981)
117. anon., Corros. Sci., **22**, 689-703 (1982)
118. Valard, T., Electrochim. Acta, **25**, 287-92 (1980)
119. Cabot, P. L., et al., Corros. Sci., **26**, 357-9 (1986)
120. Abou-Romia, M. M. and El-Basiouny, M. S., Corrosion, **42**, 324-8 (1986)
121. Ford, P., et al., J. Electrochem. Soc., **127**, 1325-31 (1980)
122. Pathania, R. S. and Trumans, D., Met. Trns. **12A**, 607-12 (1981)
123. Lee, K. W. and Pyum, S. I., Metell., **36**, 280-3 (1982)
124. Dietzel, W. D., Schwalbe, K. H. and Wu, D., Fatigue Fract. Eng. Mater. Struct., **12.**, 495-510 (1989)
125. Rebiere, M. and Magnin, T., Mater. Sci. Eng., **A128**, 99-106 (1990)
126. Dorward, R. C. and Hasse, K. R., Corrosion, **43**, 408-13 (1987)
127. Dorward, R. C., ibid., **46**, 348-52 (1990)
128. Ahmad, M., Mater. Sci. Eng. **A125**, 1-14 (1990)
129. Lunarska, E. and Szklarska-Smialowski, Z., Corrosion, **43**, 414-24 (1987)
130. Trumans, D., ibid., **42**, 601-8 (1986)
131. Kim, Y. S. and Pyum, S. I., Brit. Corros. J., **18**, 71-5 (1983)
132. Mankowski, G. and Dabosi, F., Corrosion, **40**, 552-8 (1984)
133. Mudlee, M. P., Thompson, A. W. and Bernstein, I. M., ibid., **41**, 127-36 (1985)
134. Rajan, K., et al., J. Mat. Sci., **17**, 2817-24 (1982)
135. Richter, J. and Kaesche, H., Werkstoffe Korros., **32**, 289-95 (1981)
136. Scamans, G. M., Aluminium, **57**, 268-74 (1981)
137. Holroyde, N. J. H. and Hardie, D., Met. Technol., **9**, 229-34 (1982)
138. Rahman, M. S., et al., Z. Metallkunde, **73**, 589-93 (1982)
139. Christocloulou, L. and Flowers, H. M., Acta Metall., **28**, 481-7 (1980)
140. Lotto, C. A. and Cottis, R. A., Corrosion, **45**, 136-41 (1989)
141. Ratke, L., Z. Metallkd., **81**, 144-8 (1990)
142. Onoro, J., Moreno, A. and Ranninger, C., N. Mater. Sci., **24**, 3888-91 (1989)
143. Bond, G. M., Robertson, I. M. and Birnbaum, H. K., Acta Metell., **36**, 2193-7 (1988)
144. Zeides, F., Mater. Sci. Eng., **A125**, 2-30 (1990)
145. Watson, J. W., Shen, Y. Z. and Meshi, M., Met Trans. **19A**, 2299-304 (1988)
146. Hasse, K. R. and Dorward, R. C., Corrosion, **41**, 663-9 (1986)
147. Hermann, J., J. Mat. Sci., **16**, 2381-6 (1981)
148. Narasimha Rao, B. V., Met Trans., **12A**, 1356-9 (1981)
149. Cordier, H., et al., Metall., **36**, 33-40 (1982)
150. Dorward, R. S. and Hasse, K. R., Corros. Sci., **22**, 251-7 (1982)
151. Swanson, R. E., et al., Scripta Metell., **16**, 321-3 (1982)
152. Sarker, B. et al., Met Trans., **12A**, 1939-43 (1981)
153. Bucci, R. J., Eng. Fruct. Mechanics, **12**, 407-44 (1979)
154. Khobaib, M., Lynch, C. T. and Vahlcliek, F. W., Corrosion, **37**, 285-92 (1981)
155. Minford, J. D., Int. J. Adhesion Adhesives, **2**, 25-8 (1982)
156. Cotter, J. C. and Kohler, R., ibid., **1**, 23-8 (1980)

4.2 Copper and Copper Alloys

Copper and copper alloys are amongst the earliest metals known to man, having been used from prehistoric times, and their present-day importance is greater than ever before. Their widespread use depends on a combination of good corrosion resistance in a variety of environments, excellent workability, high thermal and electrical conductivities, and attractive mechanical properties at low, normal and moderately elevated temperatures.

A wide range of cast and wrought alloys is available. For detailed expositions of properties and uses the reader is referred to publications on many specialised aspects obtainable from the Copper Development Association offices in various countries. Relevant publications of the British Standards Institution include BS 1400, *Copper Alloy Ingots and Castings*[1] and BS 2870-5, *Copper and Copper Alloy Wrought Products*[2]. All standards of the American Society for Testing and Materials relating to copper and copper alloys are included in a volume published annually[3].

Composition and Properties

The mechanical properties of wrought alloys[4] depend on composition and metallurgical condition. At the extremes, annealed pure copper has a tensile strength of 180 MN m^{-2} and a hardness of 40 H_V, and heat-treated beryllium copper can have a tensile strength of 1 300 MN m^{-2} and a hardness of 390 H_V. Summaries of typical properties of some of the more important wrought and cast copper alloys are given in Tables 4.9 and 4.10.

Coppers The purest grade of copper commercially available, and that with the highest electrical conductivity, is oxygen-free high-conductivity copper. The minimum copper content required by some specifications is 99·99%, and the method of manufacture is such that no residual deoxidant is present. Oxygen itself has very little effect on conductivity, and the 'tough pitch' coppers (either electrolytic or fire-refined), containing about 0·04% oxygen, are high-conductivity materials.

One disadvantage of tough pitch coppers is the embrittlement that is liable to occur when they are heated in atmospheres containing hydrogen. For many purposes, therefore, and particularly where fabrication is involved, deoxidised coppers are preferred. The usual deoxidising agent is phosphorus, and specifications require residual phosphorus contents of between

Table 4.9 Typical properties of wrought alloys

Alloy	Density (g/cm³)	Melting pt. (liquidus, °C)	Coefficient of expansion × 10⁶	Electrical conductivity % I.A.C.S.	Thermal conductivity (W/m°K)	Tensile strength (MN/m²)	Elongation %	Vickers hardness
H.C. copper	8·94	1 083	18	103	390	180–340	10–60	40–110
Deoxidised non-arsenical copper	8·93	1 682	18	80	340	180–340	10–55	40–120
Arsenical copper	8·93	1 080	17	45	175	220–360	10–55	40–125
Tellurium copper	8·93	1 075	18	96	360	230–320	10–50	40–110
Beryllium copper	8·2	955	18	23	85	500–1 300	2–40	110–390
85/15 brass	8·74	1025	19	35	155	280–540	8–70	65–170
70/30 brass	8·53	955	20	27	125	280–600	5–75	55–180
60/40 brass	8·38	905	21	29	125	370–600	5–45	75–180
Aluminium–brass	8·33	980	19	23	100	320–700	6–75	70–200
Naval brass	8·41	890	21	25	110	370–620	5–45	75–180
H.T. brass	8·35	890	21	23	105	520–770	8–35	90–220
15% nickel silver	8·69	1 060	16	7	35	350–700	4–55	70–220
30% nickel silver	8·87	1 190	17	5	20	390–700	4–50	90–220
5% tin–bronze	8·89	1 050	18	17	80	340–740	5–70	70–220
12% tin–bronze	8·70	995	19	8	50	460–830	5–65	110–250
Silicon–bronze	8·52	1 030	18	8	40	260–630	5–75	60–200
7% aluminium–bronze	7·95	1 050	18	15	80	430–770	4–65	80–210
90/10 cupro–nickel	8·91	1 150	16	10	50	310–620	8–55	90–200
70/30 cupro–nickel	8·94	1 240	16	5	30	370–700	5–55	95–210

Table 4.10 Properties of cast alloys

Alloy	Min. tensile strength (MN/m^2)	Min. elongation (%)
10Sn–0·2P	230	7
10Sn–10Pb	190	5
10Sn–2Zn	260	15
5Sn–5Zn–5Pb	200	15
9·5Al–2Fe	490	20
Silicon-bronze	310	15
30Zn–2Pb	190	12
Naval brass	310	20
H.T. brass (up to 2·5Al)	460	20
H.T. brass (up to 5·0Al)	590	15
H.T. β brass	740	12

0·004 and 0·05%. Phosphorus-deoxidised coppers commonly have electrical conductivities about 80% of those of pure copper.

Arsenical coppers containing about 0·4% arsenic (tough pitch or deoxidised) are used where increased strength at elevated temperatures is required. Additions of cadmium (1·0%), chromium (0·5%), and silver (0·1%) also give improved high-temperature properties, but without any serious loss of electrical conductivity. Tellurium (1·0%) gives improved machinability. An addition of about 2% beryllium gives a heat-treatable alloy that can develop extremely high strength.

Brasses Brasses are basically alloys of copper and zinc, containing between about 10 and 45% Zn, but many other additions are made and the resulting alloys are the most complicated of all the copper-base series. The single-phase (α) brasses, containing up to about 37% Zn in the binary alloys, may have additions of 1% Sn (Admiralty brass), 2% Al (aluminium-brass), or 1–2% Pb for ease of machining. Duplex (α-β) brasses containing more than 37% Zn, may have additions of 1% Sn (Naval brass), or 1–3% Pb to assist machining. Both α and α-β brasses, with and without lead, are used in the cast as well as the wrought form.

High-tensile brasses are α-β (or, occasionally, β) alloys containing up to 5% Al and 1–2% of one or more of the following: Sn, Pb, Fe, Mn. These alloys also are used in both wrought and cast form.

Copper/nickel alloys Alloys containing 5–30% Ni, used mostly in the wrought condition[5], have a very good combination of properties. For optimum corrosion resistance, additions of 0·5–2·0% each of Fe and Mn are made.

Tin-bronzes and gunmetals Alloys containing 3·0–12·5% Sn and 0·02–0·04% P, known as *phosphor-bronzes*, are widely employed. Cast as well as wrought alloys are used, and cast leaded bronzes are also available.

Gunmetals are alloys of copper, tin and zinc, with or without lead, used in the cast condition. Commonly used alloys are (*a*) 10Sn–2Zn, and (*b*) 5Sn–5Zn–5Pb.

Aluminium-bronzes Aluminium-bronzes usually contain 5–10% Al, the structure being duplex when more than about 8% Al is present. Plain Cu–Al alloys are sometimes used, but wrought (single-phase) alloys may have additions of about 0·25-2% of one or more of the following: Ni, Fe, Mn, Ag, Sn, As. Cast alloys of high strength and complex structure usually contain about 10% Al and additions of Fe, Mn and Ni.

Silicon-bronzes Silicon-bronzes usually contain 1·5-3% Si and 0·5-1% Mn[6]. They are used in wrought or cast form, though the cast alloys may also contain some Zn and Fe.

Nickel silvers These wrought alloys consist essentially of Cu, Zn and Ni, with Ni in the range 10-30%. Leaded nickel brasses are also used, usually where some machining is involved.

General Considerations of Corrosion Behaviour

Copper is the first member of Group IB of the periodic table, having atomic number 29 and electronic configuration 2.8.18.1. Loss of the outermost electron gives the cuprous ion Cu^+, and a second electron may be lost in the formation of the cupric ion Cu^{2+}.

Copper occurs in the uncombined state in nature and is relatively easily obtained by the reduction of its compounds. It is not very active chemically and oxidises only very slowly in air at ordinary temperatures.

In the electrochemical series of elements, copper is near the noble end and will not normally displace hydrogen, even from acid solutions. Indeed, if hydrogen is bubbled through a solution of copper salts, copper is slowly deposited (more rapidly if the process is carried out under pressure). (See Section 1.2 for thermodynamic considerations.)

As copper is not an inherently reactive element, it is not surprising that the rate of corrosion, even if unhindered by films of insoluble corrosion products, is usually low. Nevertheless, although the breakdown of a protective oxide film on copper is not likely to lead to such rapid attack as with a more reactive metal such as, say, aluminium, in practice the good behaviour of copper (and more particularly of some of its alloys) often depends to a considerable extent on the maintenance of a protective film of oxide or other insoluble corrosion product.

Many of the alloys of copper are more resistant to corrosion than is copper itself, owing to the incorporation either of relatively corrosion-resistant metals such as nickel or tin, or of metals such as aluminium or beryllium that would be expected to assist in the formation of protective oxide films. Several of the copper alloys are liable to undergo a selective type of corrosion in certain circumstances, the most notable example being the *dezincification* of brasses. Some alloys again are liable to suffer stress corrosion by the combined effects of internal or applied stresses and the corrosive effects of certain specific environments. The most widely known example of this is the *season cracking* of brasses. In general brasses are the least corrosion-resistant of the commonly used copper-base alloys.

The various grades of copper available do not differ to any marked extent in their corrosion resistance, and a choice is usually based on other grounds. Subsequent references to the corrosion behaviour of copper may therefore be taken to apply broadly to all types of copper.

The choice of alloy for any particular application is determined by the desired physical, mechanical and metallurgical properties. Within these limits, however, a range of materials is usually available. It is essential that at the very earliest stage the choice of materials and the details of design of the installation should be considered from the point of view of corrosion, if the best performance is to be obtained in service. This is particularly true of copper alloys, where protective measures are not normally applied.

Several books contain general summaries of the corrosion behaviour of copper and its alloy[7-15] and the formation of copper corrosion products and methods for their identification have been described in a number of papers[16].

Electrode Potential Relationships

The standard potentials for the equilibria

$$Cu^{2+} + 2e \rightleftharpoons Cu \quad \quad \quad \quad \ldots(4.1)$$
$$Cu^{+} + e \rightleftharpoons Cu \quad \quad \quad \quad \ldots(4.2)$$
$$Cu^{2+} + e \rightleftharpoons Cu^{+} \quad \quad \quad \quad \ldots(4.3)$$

are $+0\cdot34$ V, $+0\cdot52$ V and $+0\cdot17$ V respectively, based on values in the book by W. M. Latimer[17]. For the equilibrium

$$2Cu^{+} \rightleftharpoons Cu^{2+} + Cu$$
$$K = a_{Cu^{2+}}/(a_{Cu^{+}}^{2})^{2} \quad \quad \quad \quad \ldots(4.4)$$

K has the value of about 1×10^{6} at 298 K, and in solutions of copper ions in equilibrium with metallic copper, cupric ions therefore greatly predominate (except in very dilute solutions) over cuprous ions. Cupric ions are therefore normally stable and become unstable only when the cuprous ion concentration is very low. A very low concentration of cuprous ions may be produced, in the presence of a suitable anion, by the formation of either an insoluble cuprous salt or a very stable complex cuprous ion. Cuprous salts can therefore exist in contact with water only if they are very sparingly soluble (e.g. cuprous chloride) or are combined in a complex, e.g. $[Cu(CN)_2]^-$, $[Cu(NH_3)_2]^+$. Cuprous sulphate can be prepared in non-aqueous conditions, but because it is not sparingly soluble in water it is immediately decomposed by water to copper and cupric sulphate.

The equilibrium between copper and cuprous and cupric ions is disturbed by the presence of oxygen in solution, since the reaction shown in equation 4.3 is facilitated, the oxygen acting as an electron acceptor.

Behaviour of Copper Electrodes

The electrode potential behaviour of copper in various solutions has been investigated and discussed in considerable detail by Gatty and Spooner[18]. According to these workers a large part of the surface of copper electrodes in aerated aqueous solutions is normally covered with a film of cuprous oxide and the electrode potential is usually close to the potential of these film-covered areas. The filmed metal simulates a reversible oxygen electrode at the existing oxygen concentration and pH, less an overvoltage determined by the existing current density. The principal factors that affect the electrode potential are thus the nature of the solution and the way in which this influences the area of oxide film, and the supply of oxygen to the metal surfaces.

In solutions containing chloride there is a tendency for the establishment of the Cu/CuCl/Cl⁻ electrode potential, so that the activity of chloride ions is an important factor in determining the electrode behaviour. From a knowledge of the solubility products of cuprous chloride and cuprous oxide it is possible to predict under what conditions chloride or hydroxyl ions are the potential-determining ions. According to Gatty and Spooner, chloride determines the potential if $a_{OH^-} < 10^{-8.1} \times a_{Cl^-}$ and hydroxyl if $a_{OH^-} > 10^{-8.1} \times a_{Cl^-}$. This will not hold in concentrated solutions, however, since complex $[CuCl_2]^-$ ions as well as simple ions will be present. A further factor to be considered is the ready formation of insoluble basic compounds. In solutions not containing chloride (e.g. sulphate or nitrate solutions), corrosion rates are usually lower and the electrode potential is more steady over

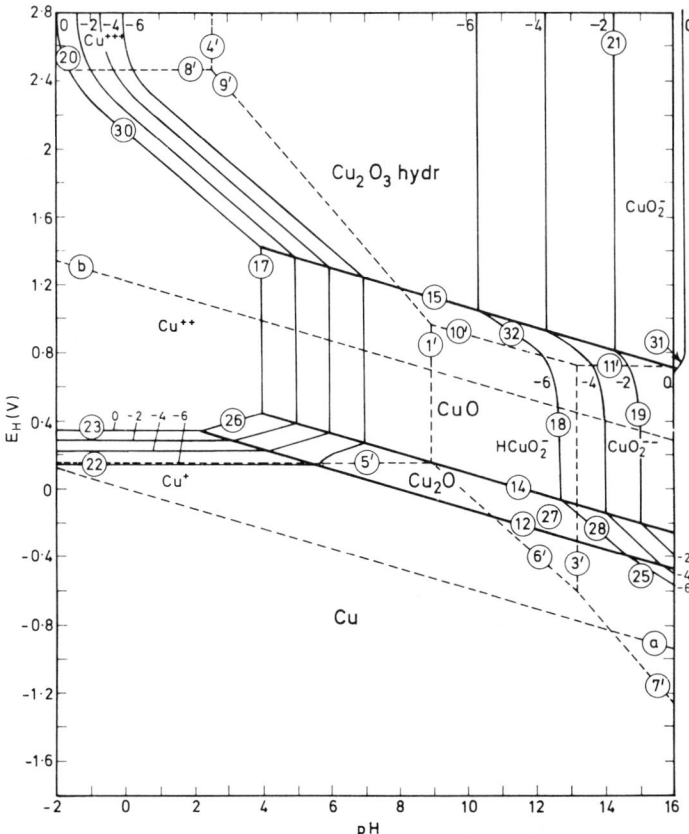

Fig. 4.10 Potential-pH equilibrium for the system copper-water at 25°C (courtesy M. J. N. Pourbaix of Centre Beige d'Étude de la Corrosion, after Delhez, R., Depommier, C. and van Muylder, J., Report RT 100, July (1962))

a wide range of conditions. In this case Gatty and Spooner consider that the rate of corrosion is probably determined by the rate at which metal ions can escape through pores in the protective oxide film, and this is supported by the results of experiments on the anodic and cathodic polarisation of copper.

The potential-pH equilibrium diagrams devised by Pourbaix[19, 20] are considered in detail in Section 1.4, and one of the diagrams for the Cu–H_2O system is shown in Fig. 4.10. Such diagrams are of considerable assistance in discussing many problems related to the chemistry, electrochemistry, electrodeposition and corrosion of copper. It is well recognised, of course, that the thermodynamic approach has limitations, the most important of which is that though predictions can be made about the possibility of a given reaction proceeding in certain circumstances no information can be gained about the rate at which it will proceed.

A method of representing the behaviour of copper in dilute aqueous solutions by means of corrosion-current/pH diagrams has been given by Rubinić and Marković[21].

A study of the behaviour of copper when anodically polarised has been made by Hickling and Taylor[22], using an oscillographic method that records the variation of potential with quantity of electricity passed. In alkaline solutions the main stages of polarisation were (*a*) the charging of a double layer, and (*b*) the formation of a film of cuprous oxide which was almost at once oxidised to cupric oxide. In 0·1 N NaOH the film was about four molecules thick when oxygen evolution first commenced. In buffer solutions of decreasing pH, the formation of sparingly soluble salts preceded or accompanied the formation of the oxide film and in acid solutions giving soluble copper salts no passivity developed, the anodic process being merely dissolution of copper. Other workers have also studied the anodic behaviour of copper or copper alloys in alkaline[23] and in acid[24] solutions.

Atmospheric Corrosion

Copper has a high degree of resistance to atmospheric corrosion, and is widely used for roofing sheets, flashings, gutters and conductor wires, as well as for statues and plaques. The resistance of copper and its alloys is due to the development of protective layers of corrosion products, which reduce the rate of attack. The formation, in the course of time, of the typical green 'patina' gives copper roofs a pleasing appearance; indeed efforts have been made to produce it artificially or to accelerate its formation[25, 26]. The nature of the corrosion products formed on copper exposed to the atmosphere was exhaustively studied by Vernon and Whitby[25, 27]. In the early periods of exposure the deposit contains sulphide, oxide and soot. By the action of sulphuric acid and by the oxidation of sulphide, copper sulphate is formed, and this hydrolyses and forms a coherent and adherent basic copper sulphate. This approximates initially to $CuSO_4 \cdot Cu(OH)_2$, but gradually increases in basicity until after 70 years or so it becomes $CuSO_4 \cdot 3Cu(OH)_2$ and is identical with the mineral brochantite. In some cases small quantities of basic carbonate, $CuCO_3 \cdot Cu(OH)_2$ (malachite), are also present, and near the sea coast basic chloride $CuCl_2 \cdot 3Cu(OH)_2$, (atacamite) is produced.

Even very near the sea coast, however, sulphate usually predominates over chloride. The presence of atmospheric pollution is thus an essential factor in the development of green patina.

In laboratory tests Vernon[28] showed that the relative humidity and the presence of sulphur dioxide have a profound effect on the rate of corrosion of copper, as of many other metals. When the relative humidity was less than 63%, there was little attack even in the presence of much sulphur dioxide, but when the relative humidity was raised to 75%, corrosion became severe and increased with the concentration of sulphur dioxide present.

By exposing specimens to the atmosphere at different times of the year, Vernon found that the rate of attack on copper was determined by the conditions prevailing at the time of first exposure. For specimens first exposed in winter there was a linear relationship between increase in weight and time of exposure, indicating that the layer of corrosion product formed under these conditions was non-protective. For specimens first exposed in summer the square of the increase in weight was proportional to the time of exposure, indicating that the coating formed in summer, when the atmospheric pollution was relatively low, was protective. The parabolic law holds when the corrosion product layer obstructs the access of the corrosive agent to the metal, the rate of attack then being inversely proportional to the thickness of the layer. The protective character of the layer persisted through subsequent periods when the pollution was relatively high.

Copper tarnishes rapidly when exposed to atmospheres containing hydrogen sulphide[29] and the reaction is not dependent on the presence of moisture.

Atmospheric corrosion tests on copper and several copper alloys were carried out by Hudson[30] at a number of sites in Great Britain. Corrosion damage was assessed by one or more of the following methods: gain in weight, loss of weight after cleaning, loss of electrical conductivity, and loss of tensile strength. Hudson found that the resistance to atmospheric corrosion was high and that the rate of attack tended to decrease with time of exposure. Little difference was found between the behaviour of arsenical copper and high-conductivity copper, and most of the alloys tested behaved very similarly except for the brasses, which deteriorated more rapidly owing to dezincification.

Several series of atmospheric exposure tests have been carried out since Hudson's work, and the loss in weight data obtained in five of the most important investigations are summarised in Table 4.11. In all cases, losses in tensile strength were also determined, and the results from the two methods were, in general, in good agreement. However, for alloys suffering selective attack (such as dezincification of brasses), change in mechanical properties usually provided a more reliable indication of deterioration than weight loss. Some other findings common to all the tests were that (a) corrosion rates decreased with time, (b) least attack occurred at rural sites and most in urban and industrial atmospheres, (c) corrosion was uniform and with few exceptions there was no significant pitting.

Tracy, Thompson and Freeman[31] exposed specimens of 11 different grades of copper in the form of sheet and wire to rural, marine and industrial atmospheres in the USA for periods up to 20 years. The differences in the behaviour of the materials were small and of little, if any, practical significance. Very similar results for various types of copper were found by

Mattsson and Holm[34] in Sweden and Scholes and Jacob[35] in the UK (see Table 4.11).

The results of tests on copper alloys have been given by Tracy[32], Thompson[33], Mattsson and Holm[34] and Scholes and Jacob[35], the first two of these investigations being made under the aegis of the American Society for Testing and Materials. The tests of Tracy, and Scholes and Jacob were both for periods up to 20 years; in those of Thompson, and Mattsson and Holm specimens have been removed after 2 years and 7 years and further specimens remain exposed for removal after 20 years. The numbers of materials tested are given in Table 4.11; they included brasses, nickel silvers, cupro-nickels, beryllium coppers and various bronzes. Mattsson and Holm tested 14 alloys in the form of rod in addition to the sheet materials, the results for which are given in Table 4.11.

In the tests described by Tracy, a high-tensile brass suffered severe dezincification (Table 4.11). The loss in tensile strength for this material was 100% and for a non-arsenical 70/30 brass 54%; no other material lost more than 23% during 20 years' exposure. In Mattsson and Holm's tests the highest corrosion rates were shown by some of the brasses. Dezincification caused losses of tensile strength of up to 32% for a β brass and up to 12% for some of the α-β brasses; no other materials lost more than 5% in 7 years. Dezincification, but to a lesser degree, occurred also in the α brasses tested, even in a material with as high a copper content as 92%. Incorporation of arsenic in the α brasses consistently prevented dezincification only in marine atmospheres.

In the tests described by Thompson, the alloy showing the lowest rate of attack at all sites was a bronze containing 7Al–2Si. Relatively high corrosion rates were shown by Cu–5Sn–0·2P at a marine site, and Cu–2·5Co–05Be in the industrial environment. The beryllium-copper alloys were the only materials to show measurable pitting, the deepest attack being 0·06 mm after 7 years. In Scholes and Jacob's tests some pitting, intergranular or transgranular penetration, or selective attack occurred on some of the alloys. The maximum depth of attack exceeded 0·2 mm in 20 years on 6 of the 21 materials (three brasses, two nickel silvers and Cu–20Ni–20Mn), but exceeded 0·5 mm in 20 years only on Cu–20Ni–20Mn and 60/40 brass. These two latter alloys lost up to 73% and 13% respectively of their tensile strength; no other alloys lost more than 10% in 20 years.

From the work described and other investigations[199], it is evident that copper and most copper alloys are highly resistant to atmospheric corrosion. In general, copper itself is as good as, or better than, any of the alloys. Some of the brasses are liable to suffer rather severe dezincification and it is unwise to expose these to the more corrosive atmospheres without applying some protection.

When unusually rapid corrosion of copper and its alloys occurs during atmospheric exposure, it is likely to be for one of the following reasons:

1. Extreme local pollution by products of combustion.
2. Bad design or construction, e.g. the presence of crevices where moisture may lodge for long periods.
3. Constant dripping of rain water contaminated by atmospheric pollution (e.g. from near-by chimney stacks) or by organic acids from lichens, etc.

Table 4.11 Atmospheric corrosion tests on copper and copper alloys

	No. of types of copper	No. of different alloys	No. of sites	Period of exposure (years)	Average rates of attack from weight losses (mm/y × 10⁴)		
					Rural sites	Marine sites	Urban/industrial sites
Tracy, Thompson and Freeman[31]	11	—	4	20	3·6-4·3	6·9-9·4	8·6-12
Tracy[32]	2	9	7	20	0·5-7·6	1·3-23*	13-30*
Thompson[33]	1	17	4	7	3·3-10	4·3-25	13-27
Mattsson and Holm[34]	{4	—	3	7	5-6	7-8	10-12
	{—	18	3	7	2-5	6-11	9-22
Scholes and Jacob[35]	{4	—	2	20	—	6-10	11-20
	{—	17	2	20	—	8-26	14-38

*Rates of attack for a high-tensile brass were 45 × 10⁻⁴ to 115 × 10⁻⁴ mm/y.

4. Corrosion fatigue due to inadequate allowance for expansion and contraction with consequent buckling as the temperature fluctuates.

Most of these can be avoided by attention to design.

Corrosion in tropical environments has been the subject of several papers[36], some of which deal with corrosion at bimetallic contacts[37].

Soil Corrosion

Several extensive series of soil-corrosion tests have been carried out by the National Bureau of Standards in the United States, and the results have been summarised by Romanoff[38]. In one series two types of copper and ten copper alloys were exposed in fourteen different soils for periods up to 14 years. The results for the copper specimens are summarised in Table 4.12.

The behaviour of the phosphorus-deoxidised and tough pitch coppers was in general very similar. At the less corrosive sites, copper was, with few exceptions, the best material, but most of the alloys lost not more than about twice as much weight, with maximum depths of attack usually not more than two or three times as great as with copper. At the other sites copper was also usually rather better than the alloys, but some of the alloys were occasionally superior.

The three most corrosive sites were rifle peat (pH 2·6), cinders (pH 7·6) and tidal marsh (pH 6·9). Corrosion of some of the alloys was particularly severe in the cinders. The behaviour of the brasses tested, particularly those high in zinc, was rather different from that of the other materials. In most cases dezincification occurred and the brasses were the worst materials; in

Table 4.12 Soil-corrosion tests on copper by National Bureau of Standards and British Non-ferrous Metals Research Association

	Period of exposure (years)	Average rate of attack from loss in weight (mm/y $\times 10^4$)	Maximum rate of pitting (mm/y)
BNFMRA 1st series:			
5 least corrosive soils	10	0·5–2·5	Nil
BNFMRA 2nd series:			
4 least corrosive soils	5	5·0–25	0·040
Nat. Bur. Standards:			
9 least corrosive soils	14	4·0–25	0·043
Nat. Bur. Standards:			
2 next most corrosive soils	14	25–130	0·033
BNFMRA 1st series:			
acid clay and acid peat	10	53–66	0·046
BNFMRA 2nd series:			
cinders	5	66	0·32
Nat. Bur. Standards:			
3 most corrosive soils:			
rifle peat, cinders, tidal			
marsh	14	160–355	0·115

the cinders, for instance, several brass specimens were completely destroyed by dezincification. In some of the soils rich in sulphides, however, the brasses were the best materials.

The British Non-Ferrous Metals Research Association carried out two series of tests, the results of which have been given by Gilbert[39] and Gilbert and Porter[40]; these are summarised in Table 4.12. In the first series[37] tough pitch copper tubes were exposed at seven sites for periods of up to 10 years. The two most corrosive soils were a wet acid peat (pH 4·2) and a moist acid clay (pH 4·6). In these two soils there was no evidence that the rate of corrosion was decreasing with duration of exposure. In the second series[40] phosphorus-deoxidised copper tube and sheet was exposed at five sites for five years. Severe corrosion occurred only in cinders (pH 7·1). In these tests sulphides were found in the corrosion products on some specimens and the presence of sulphate-reducing bacteria at some sites was proved. It is not clear, however, to what extent the activity of these bacteria is a factor accelerating corrosion of copper.

Cinders and acid peaty soils are obviously among the soils most corrosive toward copper. There is, however, no direct relationship between the rate of corrosion and any single feature of the soil composition or constitution[41]. For instance, in the American tests corrosion in several soils with either low pH or high conductivity was not particularly severe, while the British tests show that high chloride or sulphate contents are not necessarily harmful.

The above-mentioned tests show that bare copper can safely be buried in a wide range of soils without fear of excessive corrosion. Experience of the behaviour of copper water service pipes, which are used widely, confirms this. Trouble is confined to 'made-up' ground containing cinders, etc. and a few other aggressive soils, and in these circumstances it is necessary to apply protection such as bitumen-impregnated wrappings or plastic coatings. Tin coatings cannot be recommended since experience shows that accelerated attack is liable to occur at pores and scratches in the coating, leading to premature failure. Copper water pipes have been known to fail by the action of stray electric currents but this is not a common cause of trouble.

There is also agreement between the soil-corrosion tests carried out by the National Bureau of Standards and practical experience of the behaviour of hot-pressed brass fittings used for joining copper water service pipes. These duplex-structure brass fittings are liable to suffer attack by dezincification in many soils in which copper behaves satisfactorily, and for burial underground fittings of copper or gun metal are to be preferred. In general, it may be said that unless there is some special reason for using a copper alloy, it is preferable to choose copper for applications involving service underground.

Copper and Copper Alloys in Natural Waters

Copper and copper alloy pipes and tubes are used in large quantities both for conveying fresh and salt waters and in condensers and heat exchangers where fresh or salt waters are used for cooling. Pumps, screens, valves and other ancillary equipment may also be largely constructed of copper alloys. Large tonnages of these materials are therefore used in power stations, on board ship, in sugar factories and in oil refineries, as well as in hot-and cold-water circuits and heating and cooling systems in hospitals, hotels, factories and homes.

Corrosion problems that arise are frequently discussed under the headings (a) sea-water, and (b) fresh waters, but there is, in fact, no sharp dividing line, since some harbour, estuarine and brackish well waters are mixtures of sea-water and fresh water and are often variable in composition. In the past, corrosion problems were serious, particularly in sea-water service, but resistant alloys have been developed and although trouble still occasionally arises this is more frequently due to poor design or operation rather than to lack of materials suitable for the application.

There are several distinctive types of corrosion that copper and copper alloys may suffer, particularly in sea-water, but also on occasion in fresh waters. The more important of these are discussed briefly below.

Impingement attack When moving water flows over copper or copper alloys the turbulence may be sufficient to cause breakdown of the surface film. This is particularly likely to happen if air bubbles entrained in the water break as they hit the metal surface. The resulting corrosion is characteristic, producing clean-swept pits, often of a 'horseshoe' shape as shown in Fig. 4.11. This type of attack was first described by Bengough and May[42,43]. The action can be very rapid, the local anodes being depolarised by the continuous removal of metal ions and corrosion product, and the local cathodes by the dissolved oxygen in the rapidly moving well-aerated water stream.

Factors that tend to increase the severity of impingement attack are increase of water speed and particularly of local turbulence, pollution of the water, and, within certain limits, increase in the size and the amount of entangled air bubbles (see also Section 1.6).

A laboratory test designed to simulate the conditions occurring in condenser tubes in practice was devised by May[43] and newer versions of this 'jet impingement apparatus' have been described[44], as has some of the testing equipment in use in the USA[45]. Use of the jet impingement apparatus has been an important factor in the development of alloys resistant to impingement attack, but it has to be borne in mind that the results obtained when the water is recirculated may be different from those obtained when it is passed once through the apparatus, as shown by Gilbert and LaQue[46].

Fig. 4.11 Typical impingement attack on Admiralty brass condenser tube. Magnification × 2

Dezincification of brasses When dezincification occurs, regions of the brass become replaced by a porous mass of copper which, though retaining the shape of the original article, has virtually no strength. There has long been discussion as to whether there is selective corrosion of the zinc in the brass, which leaves the copper behind, or whether complete dissolution of the brass occurs, followed by re-deposition of copper. Possibly both processes occur in different circumstances. The mechanism has been investigated and discussed by Evans[7], Fink[47], Lucey[48], Feller[49] and Heidersbach[50], and is referred to in many other papers[51].

With a single-phase brass the whole of the metal in the corroded areas is affected. Dezincification may proceed fairly uniformly over the surface, and this 'layer type' takes much longer to cause perforation than the localised 'plug type' that more often occurs[52]. With a two-phase brass the zinc-rich β phase is preferentially attacked as shown in Fig. 4.12. Eventually the α phase may be attacked as well. The zinc corrosion products that accompany dezincification may be swept away, or in some conditions may form voluminous deposits on the surface which may lead to blockages, e.g. in fittings.

In general, the rate of dezincification increases as the zinc content rises, and great care needs to be exercised in making brazed joints with copper/zinc brazing alloys, particularly if they are to be exposed to sea-water. Under these conditions, a properly designed capillary joint may last for some time, but it is preferable to use corrosion-resistant jointing alloys such as silver solders (e.g. BS 1845, Type *AG1* or *AG5*)[53].

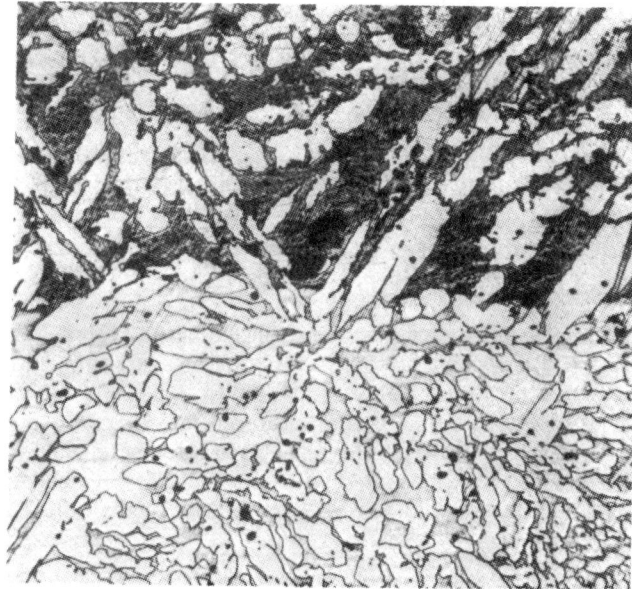

Fig. 4.12 Dezincification of two-phase brass showing preferential attack of the β phase (upper half of photomicrograph). Magnification × 133

Factors that cause increased rates of dezincification are high temperature, high chloride content of water and low water speed. Dezincification is also likely to occur preferentially beneath deposits of, for instance, sand or silt on the metal surface, or in crevices where there is a low degree of aeration.

Addition of about 0·04% arsenic will inhibit dezincification of α brasses[42] in most circumstances and arsenical α brasses can be considered immune to dezincification for most practical purposes[54]. There are conditions of exposure in which dezincification of these materials has been observed, e.g. when exposed outdoors well away from the sea[34], or when immersed in pure water at high temperature and pressure, but trouble of this type rarely arises in practice. In other conditions, e.g. in polluted sea-water, corrosion can occur with copper redeposition away from the site of initial attack, but this is not truly dezincification, which, by definition, requires the metallic copper to be produced *in situ*. The work of Lucey[48] goes far in explaining the mechanism by which arsenic prevents dezincification in α brasses, but not in α-β brasses (see also Section 1.6). An interesting observation is that the presence of a small impurity content of magnesium will prevent arsenic in α brass from having its usual inhibiting effect[55].

Additions of antimony or phosphorus, in amounts similar to arsenic, are claimed to be also capable of preventing dezincification of α brasses. Most manufacturers use arsenic, however, and it certainly appears desirable to avoid phosphorus, since Bem[56] has shown that this element can, in some circumstances, lead to an undesirable susceptibility to intercrystalline corrosion. The same appears to be true of excessive amounts of arsenic (over about 0·05%).

No reliable method of inhibiting dezincification of two-phase brasses has been discovered. Various additions, including arsenic, have been advocated from time to time, but nothing is known that will render α-β brasses immune to dezincification under all conditions of exposure. The addition of 1% tin can markedly reduce the rate of dezincification, and naval brass (61Cu–38Zn–1Sn) is attacked considerably more slowly than 60/40 brass in sea-water, though there may be virtually no difference in most fresh waters. Some of the cast complex high-strength two-phase brasses containing tin, aluminium, iron and manganese appear to have relatively good resistance to dezincification, but they are by no means immune to it.

Selective attack in other alloys Selective attack analogous to dezincification can occur in other copper alloys, particularly aluminium bronzes and less frequently tin bronzes[57], cupro-nickels[58], etc. In recent years de-aluminification of aluminium bronzes has been studied extensively[59] and the results indicate that whilst α-phase alloys suffer such attack comparatively rarely, alloys of higher aluminium content can be more susceptible. The electrochemical relationships are such that preferential corrosion of the second phase is liable to occur in α-γ_2 alloys, but α-β alloys are relatively resistant to attack. Retention of β phase is favoured by rapid cooling after casting or after high-temperature heat treatment, and also by incorporating manganese in the alloy.

Deposit attack and pitting When water speeds are low and deposits settle on the surface (particularly at water speeds below about 1 m/s), pitting of copper and copper alloys is liable to occur by differential aeration effects.

In sea-water systems such attack may occur under dead barnacles or shellfish, the decomposing organic matter assisting corrosion. Pitting is most likely to occur in polluted in-shore waters, particularly when hydrogen sulphide is present. In such contaminated waters non-protective sulphide scales are formed and these tend to stimulate attack.

Corrosion of condenser tubes and related equipment There have been many surveys[60] of the problems of corrosion of condenser and heat-exchanger tubes and related components in marine service and others[61] dealing with oil refinery service.

Corrosion of condenser tubes was a problem of great magnitude during the first quarter of this century. Its solution was based on research carried out for the Institute of Metals by Bengough *et al.*[42,52,62], one of whom, May[43,54], remained associated with the research when it was transferred to the auspices of the British Non-Ferrous Metals Research Association in 1930. A history of condenser tubes up to 1950 has been published[206].

In early times 70/30 brass condenser tubes failed by dezincification and Admiralty brass (70Cu–29Zn–1Sn) was brought into use. This proved little better, but some time later the addition of arsenic was found to inhibit dezincification. Failures of Admiralty brass by impingement attack became a serious problem, particularly as cooling water speeds increased with the development of the steam turbine. The introduction of alloys resistant to this type of attack was a great step forward and immediately reduced the incidences of failure.

The alloys in most common use today are aluminium brass (76Cu–22Zn–2Al–0·04As) and cupro-nickels containing appropriate iron and manganese additions[15,63]. Three cupro-nickel alloys are in widespread use containing (approximately) (i) 30Ni–0·7Fe–0·7Mn, (ii) 30Ni–2Fe–2Mn and (iii) 10Ni–1·5Fe–1Mn. These materials are extensively and successfully used in ships, power stations, oil refineries, etc., in condensers and heat exchangers, with nominal water speeds through the tubes of up to about 3 m/s, sometimes with much entangled air present. At the highest water speeds there is a rather greater factor of safety with 70/30 cupro-nickel, and this alloy is also usually to be preferred under most polluted water conditions, although occasionally other alloys are as good or even better. There is evidence to indicate that when the operating conditions involve relatively high temperatures, aluminium-brass or 90/10 cupro-nickel is to be preferred to 70/30 cupro-nickel[58]. Admiralty brass is no longer considered a suitable alloy for sea-water service, except possibly where water speeds are very low, i.e. not more than about 1 m/s. In some oil refineries, for instance, Admiralty brass is preferred because it has good resistance to corrosion by oil products, and in these installations the heat exchangers are designed to have low cooling-water speeds, so that corrosion from the sea-water side is kept within reasonable bounds.

Tin–bronze containing about 12% Sn has been shown to have good resistance to impingement attack[64], to attack by acid cooling waters, and to abrasion in cooling waters containing suspended solid particles, but the alloy has so far only been used on a limited scale. The alloy most commonly used when suspended abrasive matter is a problem is the 30% Ni alloy containing 2% each of Fe and Mn. Aluminium–bronze tubes have sometimes given good results, but their use has been limited because of their susceptibility to pitting attack.

The occasional failures that still occur in condenser tubes are usually due to one (or sometimes several) of the following factors:

1. Localised attack or pitting in badly contaminated waters.
2. Pitting under decaying barnacles, shell fish or other deposits.
3. Impingement attack due to high local water velocities, e.g. at partial obstructions in a tube such as pieces of coke, shell fish, etc.
4. Erosion due to sand or other abrasive particles suspended in the water[65].
5. Use of tubes of the wrong alloy, or of incorrect composition, or containing manufacturing defects.

A difficult condenser-tube corrosion problem arises from the use of polluted cooling waters from harbours and estuaries that may be severely contaminated. All condenser-tube materials are liable to suffer corrosion in these circumstances, and the choice of materials is made difficult by the fact that different orders of merit apply at different locations and even at the same location at different times. The state of the water when the tubes first enter service may well determine whether or not a satisfactory life will be obtained[66]. The most corrosive waters are those containing free hydrogen sulphide produced by the action of sulphate-reducing bacteria. Waters may also be rendered abnormally corrosive by the presence of small amounts of organic sulphur compounds produced by bacterial action, as shown by Rogers[67].

Corrosion of power-station condenser tubes by polluted waters[68] has been particularly troublesome in Japan and efforts have been made to study the problem by electrochemical methods[69] and by exposing model condensers at a variety of power station sites[70]. Improved results have been reported using tin brasses[71] or special tin bronzes[72]. Pretreatment with sodium dimethyldithiocarbamate is reported to give protective films that will withstand the action of polluted waters[73], though the method would be economic only in special circumstances.

Electrochemical studies, including the determination of polarisation curves, have been carried out in recent years by many authors[74] in endeavours to understand more fully the mechanism of protective film formation on copper alloys in sea-water. Other authors have described experiences with condenser tubes in fresh or brackish waters[75]. Methods of maintaining tube cleanliness include chlorination[76], use of high molecular weight water-soluble polymers[77] and use of the Taprogge system of circulating sponge-rubber balls[78,79].

Condenser tube-plates of Naval brass usually undergo some dezincification in sea-water service, but this is normally not serious in view of the thickness of metal involved. Attack can, however, be more serious with 60/40 brass (Muntz metal) and such plates may have to be renewed during the life of the condenser. Increasing use is now being made of tube-plates of more resistant materials such as aluminium-brass, silicon-bronze, aluminium-bronze or cupro-nickel. Plates that are too large to be rolled in one piece can be fabricated by welding together two or more pieces. In some special applications the tubes are fusion welded or explosively welded to the tube-plates.

Fusion welding operations are rather more difficult with brasses than with other copper alloys (because of evolution of zinc fume from brasses).

Condenser water-boxes were hitherto usually made of unprotected (or poorly protected) cast iron and these afforded a measure of cathodic protection to the tube-plates and tube ends. This beneficial effect has been lost with the general adoption of water-boxes completely coated with rubber or some other impervious layer, or of water-boxes made from resistant materials such as gunmetal, aluminium-bronze or cupro-nickel, or steel clad with cupro-nickel or Monel. To prevent attack on tube-plates and tube-ends in these circumstances, it is highly desirable to install either a suitable applied-current cathodic-protection systems[80], or sacrificial soft-iron or mild-steel anodes. Ferrous wastage plates have the additional advantage that the iron corrosion products introduced into the cooling water assist in the development of good protective films throughout the length of the tubes. This is particularly important in the case of aluminium-brass tubes; indeed, with such tubes it may be desirable, as an additional preventive measure, to add a suitable soluble iron salt (such as ferrous sulphate) regularly to the cooling water. Cases of the success of such treatment in power station condensers have been described by Bostwick[81] and Lockhart[82], and other workers[79,83] have since studied the effects of ferrous sulphate treatment on tube behaviour.

As it has become increasingly necessary to supplement natural sources of fresh water in various parts of the world, processes for producing fresh water from sea-water have been intensively studied and the literature dealing with the subject is very extensive. Distillation is currently the process most widely used and during recent years increasing numbers of multi-stage flash-distillation plants have been installed in various countries, many of the larger units being capable of producing several millions of gallons of fresh water per day. In these plants, sea-water passes through horizontally-disposed tubes and steam 'flashed' from the brine condenses on the outside. In some parts of the plant the conditions are similar to those in steam condensers, but in other parts the sea-water has been treated to remove dissolved gases and is at much higher temperatures. In another distillation process receiving considerable study, films of sea-water fall down the inner walls of comparatively large-diameter vertically-disposed tubes, usually of fluted configuration, and evaporation takes place due to the heating effect of steam condensing externally. Copper-alloy tubes are used in large numbers for the heat-exchange units in distillation plants, mainly aluminium-brass and the various cupro-nickel alloys, and the factors affecting choice of materials have been considered in several papers[84].

For ships' cooling-water trunking and salt-water services in the engine room and elsewhere, including fire mains, plumbing and air-conditioning systems, more resistant alloys are taking the place of copper or galvanised steel, which were formerly extensively used, but which do not have adequate resistance to attack by sea-water. Both aluminium-brass and the Cu-10Ni-1·5Fe alloy are widely used and, being highly resistant to impingement attack, normally give excellent service. In some special naval applications pipelines of 70/30 cupro-nickel are used. It is important that correct fabrication and installation techniques are used. Carbonaceous residues from fillers used in bending operations must be avoided or pitting corrosion may occur in service. Jointing materials of low corrosion resistance should not be used, silver brazing or appropriate welding methods being the correct techniques. Residual stresses, if present, can cause stress-corrosion failures of aluminium-brass pipelines in service.

Copper alloys in wrought or cast form are used for other purposes in ships and other marine installations, such as for propellers[85], bearings, valves and pumps. One widespread application of aluminium-brass is its use for heating coils in tankers carrying crude oil or petroleum products. Some corrosion problems encountered in this and other applications on board ship have been described by Gilbert and Jenner[86].

Fresh Waters

Fresh waters are, in general, less corrosive towards copper than is sea-water, and copper is widely and satisfactorily used for distributing cold and hot waters in domestic and industrial installations[15,87]. Copper and copper alloys are used for pipes, hot-water cylinders, fire-back boilers, ball floats, ball valves, taps, fittings, heater sheaths, etc. In condensers and heat exchangers using fresh water for cooling, tubes of 70/30 brass or Admiralty brass are usually used, and corrosion is rarely a problem.

Joints in copper components may be a source of trouble. Copper/zinc brazing alloys may dezincify and consequently give rise to leaks[90]. In some waters, soft solders are preferentially attacked unless in a proper capillary joint. Copper/phosphorus, copper/silver/phosphorus, and silver brazing alloys are normally satisfactory jointing materials. Excessive corrosion of copper is sometimes produced by condensates containing dissolved oxygen and carbon dioxide. Rather severe corrosion sometimes occurs on the fire side of fire-back boilers and on electric heater element sheaths under scales deposited from hard waters[91].

Dezincification of brasses This may occur, particularly in stagnant or slowly-moving warm or hot waters relatively high in chloride and containing little carbonate hardness[88]. Dezincification of α brasses is inhibited by the usual arsenic addition (see Fig. 4.12), but two-phase brasses are liable to severe attack in some waters[89]. In such waters the use of duplex-structure brass fittings should be avoided.

Impingement attack Copper may occasionally suffer this form of attack in systems where the speed of water flow is unusually high and the water is one that does not form a protective scale, e.g. a soft water containing appreciable quantities of free carbon dioxide[92]. Ball valve seatings may also suffer an erosive type of attack. The corrosion of ball valves, including the effect of chlorination of the water, has been studied by several workers[93].

Dissolution Some waters continuously dissolve appreciable amounts of copper[94]. Factors that favour this action are high free carbon dioxide, chloride and sulphate contents, low hardness, and increase of temperature. The trouble is therefore most prevalent in hot, soft, acid waters. The corrosion is general and the resulting thinning is so slight that the useful life of the pipe or component is virtually unaffected (unless impingement attack occurs). Trouble is usually confined to (a) stimulation of the corrosion of components of zinc-coated steel[95], light alloys[96], and sometimes bare steel with which the water subsequently comes into contact; and (b) the formation of green stains in baths, sinks, etc. owing to the combination of copper with soaps.

In de-aerated conditions, for instance in most central heating systems, little if any attack on copper occurs[92,97]. As far as drinking waters are concerned, copper is not classified as a toxic substance or hazardous to health. To avoid any difficulties due to unpalatability, the maximum continuous copper content should not exceed 1·0 p.p.m., with a limit of 3 p.p.m. in water after standing overnight in copper pipes. A review of the subject by Grunau[98] makes reference to 394 published papers.

Pitting Occasionally copper water pipes fail prematurely by pitting. This most often occurs in cold waters originating from deep wells and boreholes and has been shown by Campbell[99] to be associated with residues of carbon produced in the bore of the tubes during bright annealing, as a result of decomposition of residual drawing lubricant. It is therefore necessary for manufacturers to take steps to avoid these harmful residues. This trouble has occurred in many different countries[100].

Failures of this type are confined to certain districts, and Campbell has shown that in many supply waters in Great Britain pitting cannot proceed, even in tubes containing dangerous cathodic films, owing to the presence in the water of small amounts of a naturally-occurring inhibitor, probably an organic colloidal material, that stifles pitting of copper. Trouble therefore only occurs in waters that contain little or no inhibitor.

Pitting failures also occasionally occur in copper water cylinders[101] and as a result of a study of this problem Lucey[102] has made suggestions about the mechanism of pitting of copper in supply waters.

In hot-water pipes, failures sometimes occur in certain areas supplied with soft waters from moorland gathering grounds. The waters concerned contain a few hundredths of a part per million of manganese, and in the course of several years' exposure, a deposit rich in manganese dioxide is laid down in the hottest parts of the system. This may cause pitting and eventually lead to failure. Hot-water pitting of another type is sometimes experienced in soft waters having a high sulphate content in relation to the carbonate hardness and a relatively low pH value[103].

Behaviour in Chemical Environments

Detailed information on the action of a wide range of chemicals on copper and copper alloys is given in a number of publications, particularly those listed under References 7-12, 104 and 105.

When contemplating the use of copper-base materials for industrial purposes it is necessary to bear in mind that even though a satisfactory life of the component may be obtained, trouble can arise from other causes:

1. Copper compounds can be tolerated only in small amounts in potable waters or substances that are to be consumed.

2. Copper compounds are highly coloured, and a very small amount of corrosion may lead to staining and discoloration of products.
3. Stimulation of the corrosion of vital parts made of more anodic metals may occur if they are connected to copper.
4. Very small amounts of copper taken into solution may cause considerable corrosion of more anodic metals elsewhere in the system, particularly zinc[95], aluminium[96], and sometimes steel[106]. Small particles of copper deposited from solution set up local cells that cause rapid pitting.

Despite these qualifications copper and its alloys are used extensively and successfully in much chemical equipment. Uses include condensers and evaporators, pipelines, pumps, fans, vacuum pans, fractionating columns, etc. Tin-bronzes, aluminium-bronzes and silicon-bronzes are used in some circumstances because they present better corrosion resistance than copper or brasses.

Acid solutions Copper does not normally displace hydrogen even from acid solutions, and it is therefore virtually unattacked in non-oxidising conditions. Most solutions that have to be handled contain dissolved air, however, and this will cause cathodic depolarisation and enable some corrosion to take place[107]. It is difficult, therefore, to lay down any general recommendations for the use of copper in acid solutions, since the rate of attack depends so greatly on the particular circumstances. Under fairly mild conditions copper or copper alloys are successfully used for handling solutions of hydrofluoric[104,108], hydrochloric[8-10,104,109], sulphuric[8-10,110], phosphoric[8,9,104,111] and acetic and other fatty acids[8-10,104,112]. Rates of corrosion, in general, increase with concentration of acid, temperature, amount of aeration[113] and speed of flow[114]. Tin-bronzes[8], aluminium-bronzes[7,115], silicon-bronzes[7,8] and cupro-nickels[8] are among the copper alloys most resistant to acids. Brasses should not normally be used. All copper-base materials are attacked rapidly by oxidising acids such as nitric, strong sulphuric, etc.

The dissolution of copper and of brasses in acid solutions has been studied by several authors[116]. Various substances have inhibiting effects on the rate of attack of copper or brasses in nitric acid[117] and in hydrochloric acid[118].

Neutral and alkaline solutions Copper-base materials are resistant to alkaline solutions[8,9,119] over a wide range of conditions but may be appreciably attacked by strong solutions, particularly if hot. Copper/nickel alloys usually give the best results in alkaline solutions. Copper and copper alloys should be avoided if ammonia[8,9,120] is present, owing to the danger of both general corrosion and, if components are under stress, stress corrosion.

Copper is satisfactory for handling solutions of most neutral salts[7-9,104,121] unless aeration and turbulence are excessive. An exception is provided by those salts that form complexes with copper, such as cyanides, and solutions containing oxidising agents, such as ferric or stannic compounds[8].

Other chemicals Copper and copper alloys are unsuitable for handling hydrogen peroxide[104,122] or molten sulphur[104,123]. Hydrogen sulphide accelerates corrosion of most copper-base materials. In its presence brasses high in zinc are usually found to behave better than other copper alloys[8].

Halogens have little action on copper at room temperature when dry, but are corrosive when wet. Hypochlorite solutions are corrosive[124]. Most organic compounds are without appreciable action[8,104]. Copper and copper alloys are extensively and successfully used in refrigeration systems employing organic refrigerants such as CCl_2F_2. Attack can, however, occur if halogenated compounds hydrolyse in the presence of moisture to give traces of hydrochloric acid. Copper alloys are widely used for handling hydrocarbon oils, though if sulphur compounds are present attack can be serious[7]. The effects of synthetic detergents on copper have been investigated[125], and several authors[9,104,126] have discussed various aspects of the behaviour of copper and its alloys in the food-processing industry.

Oxidation and Scaling

Several authors[7,8,127-129] have reviewed the literature on the oxidation and scaling of metals, including copper.

Copper

The volume ratio (see Section 1.9) for cuprous oxide on copper is 1·7, so that an initially protective film is to be expected. Such a film must grow by a diffusion process and should obey a parabolic law. This has been found to apply for copper in many conditions, but other relationships have been noted. Thus in the very early stages of oxidation a linear growth law has been observed (e.g. at 1 000°C)[130].

At 180–290°C it was found[131] that the parabolic law first applied but subsequently changed to a logarithmic relationship of the type

$$y = K \log B(t + 1/B)$$

B being a constant. Other workers have reported a cubic relationship under some conditions.

Evans[132] has shown how the effect of internal stresses in growing films may have various effects that can lead to any one of the first three growth laws referred to above.

At medium and high temperatures[133,134] copper ultimately follows the parabolic law[128,135]. It has been shown[136] using radioactive tracers that the diffusion of copper ions in cuprous oxide is the rate-determining step at 800–1 000°C, and there is considerable evidence in favour of the view that metal moves outwards through the film by means of vacant sites in the oxide lattice[133].

When oxidation is a diffusion process the oxidation rate should be related to the temperature by the Arrhenius equation

$$K = A \exp[-Q/RT]$$

where K is the rate constant, A a constant, R the gas constant, T the absolute temperature, and Q the activation energy. Values that have been obtained for A and Q are summarised by Tylecote[128], Pilling and Bedworth[137], Feitknecht[138] and others give values of Q of about 0·17 MJ for temperatures of 700–1 000°C, while at lower temperatures (up to 500°C) Vernon[139] and others obtained values of about 85 kJ. These values are in agreement with calculations by Valensi[140] based on the assumption that at the high temperatures the oxidation proceeds by the reaction of oxygen with metallic copper to produce cuprous oxide while at lower temperatures the rate is determined by the reaction between oxygen and cuprous oxide to form cupric oxide.

At low temperatures (e.g. 300°C) the film consists almost entirely of CuO. As the temperature increases the film consists of a layer of Cu_2O beneath a layer of CuO, the proportion of Cu_2O increasing until at high temperature the film is almost entirely Cu_2O. The precise composition of the film depends, however, on a number of factors, including temperature, time, oxygen concentration in the atmosphere, etc. Tylecote[141] has investigated the composition, properties and adherence of scales formed on various types of copper at temperatures between 400 and 900°C. At the higher temperatures the scales formed on coppers containing phosphorus were more brittle and less adherent than those formed on coppers containing no phosphorus.

Studies have been carried out of the effects at high temperatures of sulphur[142] and of atmospheres containing hydrogen sulphide[143], steam[144, 145], sulphur dioxide[145] and hydrogen chloride[145].

Copper Alloys

With copper alloys containing more noble metals the oxide will be substantially pure copper oxide since the oxides of the noble metals have higher dissociation pressures than the copper oxides. With alloys containing baser metals, however, the alloying element will appear as an oxide in the scale, often in greater concentration than in the alloy itself, and sometimes to the exclusion of copper oxides. The dissociation pressures of many oxides have been calculated by Lustman[146].

Whether the rate of oxidation of an alloy of copper with a baser metal is less or more than that of copper will depend on the concentration of the alloying element and the relative diffusion velocities of metal atoms or ions in the oxide layers. There is extensive literature on the oxidation behaviour of copper alloys[128, 129, 147]. According to Wagner's theory[148] the rate of oxidation will be largely influenced by the electrical conductivity of the film, and the theory is therefore supported by the fact that the alloying elements giving maximum oxidation resistance, i.e. beryllium, aluminium and magnesium, form oxides having very low conductivities, as shown by Price and Thomas[149]. Wagner calculated that when sufficient aluminium was present in copper to cause the formation of an alumina film the oxidation rate should be decreased by a factor of more than 80 000. Experiment showed a factor of only 36, but when Price and Thomas carried out initial oxidation under very slightly oxidising conditions, producing only a film of alumina, the oxidation rate on subsequent exposure to full oxidising conditions was decreased by a factor of about 240 000. Hallowes and Voce[145] found that selective oxidation of a 95Cu–5Al alloy by this method gave protection from atmospheric oxidation up to 800°C unless the film was scratched or otherwise damaged, or the atmosphere contained sulphur dioxide or hydrogen chloride.

The effects on oxidation resistance of copper as a result of adding varying amounts of one or more of aluminium, beryllium, chromium, manganese, silicon, zirconium are described in a number of papers[147]. Other authors have investigated the oxidation of copper-zinc[150,151] and copper-nickel alloys[151,152], the oxidation of copper and copper-gold alloys in carbon dioxide at 1 000°C[153] and the internal oxidation of various alloys[154].

Copper alloys have been used extensively in high-pressure feed-water heaters in power generating plant. Experience has shown that when such heaters are operated intermittently, 70Cu–30Ni or 80Cu–20Ni alloy tubes suffer fairly rapid and severe steam-side (external) oxidation with the formation of exfoliating scales. This corrosion, which may be associated with ingress of air during shutdowns, has been the subject of several published papers[155]. The behaviour of other alloys for feed-water heater service has also been discussed[156].

Stress Corrosion (Chapter 8)

Failure of copper alloys may occur by cracking due to the combined influence of tensile stress and exposure to a corrosive environment. When the stresses are produced in components during manufacture the trouble is usually known as *season cracking* and failures of brass components due to this form of stress corrosion have been known for many years[157-160].

Only certain specific environments appear to produce stress corrosion of copper alloys, notably ammonia or ammonium compounds or related compounds such as amines. Mercury or solutions of mercury salts (which cause deposition of mercury) or other molten metals will also cause cracking, but the mechanism is undoubtedly different[161]. Cracks produced by mercury are always intercrystalline, but ammonia may produce cracks that are transcrystalline or intercrystalline, or a mixture of both, according to circumstances. As an illustration of this, Edmunds[162] found that mercury would not produce cracking in a stressed single crystal of brass, but ammonia did.

Stress Corrosion of Brasses

Alloys containing only a few per cent of zinc may fail if the stresses are high and the environment sufficiently corrosive. Most types of brass, besides the plain copper/zinc alloys, appear to be susceptible to stress corrosion. An extensive investigation of the effect of additions to 70/30 brass was carried out by Wilson, Edmunds, Anderson and Peirce[163], who found that about 1% Si was markedly beneficial. Other additions were beneficial under some circumstances and none of the 36 additions tested accelerated stress-corrosion cracking. Further results are given in later papers[164]

In general, the susceptibility to stress corrosion appears to increase with increase in zinc content, but in some circumstances alloys containing 64–65% Cu were found to be rather more affected than those containing 60% Cu[165].

Many workers[157,159,166] have investigated the residual stresses introduced by different working processes in brasses of various compositions and the annealing treatments necessary to remove these stresses or reduce them to a safe level. A 'stress-relief anneal' at about 300°C will usually lower internal stresses to comparatively small values without much effect on the hardness of the material.

Specifications for brass products customarily include provision for carrying out a mercurous nitrate test[157] to ensure that unduly high residual tensile stresses are not present, but a satisfactory result in this test does not guarantee freedom from cracking in environments containing ammonia. More searching tests involving exposure to ammonia have therefore been devised. The standardisation of stress-corrosion cracking tests and their correlation with service experience have been described in several papers[167]. Other authors[168] have described practical cases of stress-corrosion cracking, usually involving tensile stresses applied in service. Two possible preventive measures are the use of coatings[169] or inhibitors[170].

The behaviour of a wide range of α, α-β and β brasses in various corrosive environments was studied by Voce and Bailey and the results summarised by Whitaker[171]. Penetration by mercury and by molten solder was intercrystalline in all three types of brass. In moist ammoniacal atmospheres the penetration of unstressed brasses of all types was intercrystalline. Internal or applied stresses accelerated the intercrystalline penetration of α brasses and initiated some transcrystalline cracking, and also caused severe transcrystalline cracking of β alloys and transcrystalline cracking across the β regions in the two-phase brasses. Immersion in ammonia solution, however, caused intercrystalline cracking of stressed β brasses.

β brasses containing 3% or more aluminium failed with an intercrystalline fracture when stressed at about the 0·1% proof stress in air. The behaviour of alloys of this type was subsequently studied by Perryman[172], and by Bailey[173], who has shown that cracking in air occurs only when moisture is present. It has been confirmed that β brasses are prone to crack in service[174].

High-strength α-β brasses containing up to about 5% Al (with small amounts of Fe, Mn, Sn, etc.) used for propellers, parts of pumps, nuts and bolts, etc. usually give good service but occasionally suffer intercrystalline failure, for instance in contact with sea-water. Examination of such failures usually reveals thin dezincified layers along the cracks, but it is difficult to decide whether the crack or the dezincification occurred first.

The theoretical aspects of stress-corrosion cracking have attracted much attention in recent years. Amongst the copper alloys, the behaviour of brasses in ammoniacal environments has been most studied. Whilst cracking has been shown to be possible in contact with some other corrosive agents, ammonia has the most powerful effect. Evans[132] suggests that this is because ammonia is a feeble corrodent that produces little attack except at regions such as grain boundaries or other lattice imperfections and because it prevents accumulation of copper ions in the crevices formed owing to the formation of stable complexes, $Cu[(NH_3)_4]^{2+}$. The mode of cracking (intercrystalline or transcrystalline) can be affected by changes in composition of the brass or by changes in the nature of the environment[175]. Mattsson[176] found that on immersion in ammoniacal solutions of different pH values, stressed brasses cracked most rapidly at pH 7·1–7·3 and that in this range black surface films formed on the metal. The rôle of tarnish films has been further studied subsequently[177]. Many authors have studied electrochemical[178] or metallurgical[179] aspects of the stress corrosion of copper alloys and discussed theories of the mechanism. Papers on the subject have been included in several symposia or conferences on stress corrosion of metals[159, 180, 181]. The stress cracking of brasses was reviewed by Bailey[182] and subsequent reviews of stress corrosion[183] contain references to the subject.

Stress Corrosion of Other Copper Alloys

Evidence indicates that pure copper is not liable to undergo stress corrosion[184-186] but instances are known of the failure by stress corrosion of copper containing about 0·4% As[187] or 0·02% P[184]. Failure can also occur with copper-beryllium[188], copper-manganese[189], aluminium-bronzes[190], tin-bronzes, silicon-bronzes, nickel silvers[191] and cupro-nickels[191, 192]. Most of these alloys are much less susceptible to cracking than brasses[185, 193]. Under some conditions, however, aluminium-bronzes can be very prone to cracking[194]. In ammoniacal environments the cracks tend to be transcrystalline, and in steam atmospheres intercrystalline[195]. Additions of 0·35% Sn or Ag are claimed to be effective in preventing intercrystalline cracking[196, 197]. An aluminium-bronze containing 2% Ni and 0·5–0·75% Si is claimed to have good resistance to stress corrosion[196, 198].

Thompson and Tracy[184] carried out tests in a moist ammoniacal atmosphere on stressed binary copper alloys containing zinc, phosphorus, arsenic, antimony, silicon, nickel or aluminium. All these elements gave alloys susceptible to stress corrosion. In the case of zinc the breaking time decreased steadily with increase of zinc content, but with most of the other elements there was a minimum in the curve of content of alloying elements against breaking time. In tests carried out at almost 70 MN/m^2 these minima occurred with about 0·2% P, 0·2% As, 1% Si, 5% Ni and 1% Al. In most cases cracks were intercrystalline.

Protective Measures

The good behaviour of copper and copper alloys is dependent upon correct choice of material, good design of equipment, and proper methods of operation. If proper attention is given to these factors there will usually be no need for protective measures. In special cases, however, e.g. to prevent the dissolution of small amounts of copper or to maintain a high-grade finish, metallic coatings of one or more of the following metals may be applied: tin, lead, nickel, silver, chromium, rhodium, gold. In other cases painting, varnishing or lacquering may be desirable, or if the conditions are very severe, as in some corrosive soils, heavier protection such as bituminous or plastic coatings may be necessary. Brasses that are liable to suffer dezincification or stress corrosion may need protection where other copper alloys would be satisfactory unprotected. Sometimes use is made of the principles of cathodic protection, e.g. steel 'protector blocks' in condenser water-boxes.

In some circumstances, use of inhibitors may be a desirable remedial measure. For instance, benzotriazole has been found of considerable value for preventing staining and tarnishing of copper products[200]. Sodium diethyldithiocarbamate also has useful inhibiting properties[201]. Other types of inhibitors can be of value in condensate systems[202] and in acid solutions[117,118]. Reviews have been given of corrosion inhibitors for copper[203] and brasses[204].

The danger of accelerated attack on copper-base materials due to coupling with other metals is small since copper is usually the cathodic member of the couple, but precautions are often necessary to prevent excessive corrosion of the anodic member. Surveys of the behaviour of couples involving copper and copper alloys have been given[11,205]. One material that has been found capable of accelerating attack on copper in practice is graphite; hence graphitic paints are undesirable. Occasionally the action between different copper-base materials may be appreciable, e.g. gunmetal may stimulate the corrosion of copper or brass in sea-water.

Copper Alloys in Marine Environments

Much attention continues to be devoted to the corrosion behaviour of copper alloys in an increasing range of marine applications[207,244,245,246].

Many publications have dealt with the long-established uses of copper alloys for condensers and heat exchangers and there have been several reviews of the selection of materials for these applications[208] and others discussing the factors that may lead to corrosion problems[209]. Specific aspects covered include the effects of velocity[210], ferrous sulphate treatment[211,212,213] and sponge ball cleaning[212,213,214]; the latter may lead to greater corrosion, though ferrous sulphate treatment can offset the effect. At low seawater temperatures[215] alloys appear to be more susceptible to attack and the beneficial effects of ferrous sulphate additions to the seawater are reduced. Chlorination[212,216,244] can cause increased attack in some circumstances, aluminium brass being more susceptible than cupronickels. Presence of sulphide pollution[217,244] causes serious corrosion if polluted and aerated conditions alternate, or if oxygen and sulphides are simultaneously present.

Attention has been focussed on the cupronickels[218,245,246], which have been shown to have extremely low long-term corrosion rates in quiet or slowly-moving seawater[219]. The 90/10 alloy[220,221,246] is of particular interest: it is well established for pipelines on ships[222,245] and has become widely used for piping systems on offshore platforms[221,223,224,245]. In addition to its good corrosion resistance, 90/10 cupronickel is resistant to marine macrofouling[219,225,244,245] (providing it is not cathodically protected). This has led to proposals for uses such as construction of ships' hulls[223,226,227,228,244,245], fish cages for aquaculture[233,227,229] and cladding of steel offshore structures in the tidal/splash zones[244,245]. Large-scale use of copper alloys in desalination plant (particularly multi-stage flash units) has continued and much information has appeared on the selection of materials and their performance in service[230,244].

Investigations into the effects of arsenic and phosphorus in single-phase brasses on their susceptibility to intergranular attack and stress-corrosion cracking in seawater[231] have shown that the normal addition of arsenic to inhibit dezincification (about 0.04%) has no significant adverse effect[232]. Other problems investigated have included de-alloying of aluminium bronzes[233, 244], effects at bimetallic contacts[234] and influence of siphonic effects[235].

Major efforts have been made to understand the nature of films formed on copper alloys in seawater. These films have quite different characteristics[236, 244, 245] to those formed in sodium chloride solution[237], the differences being associated with the presence of organic material in the natural environment. Protective films often have a duplex structure, with an outer layer rich in iron providing impingement resistance and an inner layer giving chemical/electrochemical protection. With aluminium brass a colloidal mixed hydroxide inner layer provides a buffering action; with cupronickel, however, there is a chloride-rich layer which strongly inhibits the cathodic reaction. However, the structure of the films is affected by variables such as water velocity, temperature and oxygen content. The polarisation resistance technique has been used to evaluate the films formed on condenser tubes in service[238].

Other Topics

Methods of avoiding pitting failures in copper cold-water tubes have been further studied[239]. Many hot-forged brass water fittings are now made from modified alloys that have an $\alpha\beta$ structure during forging and are then heat-treated to a dezincification-resistant α structure[240]. The corrosion resistance of β aluminium brasses (shape memory effect alloys) has been studied[241].

Stress corrosion of brasses continues to form the subject of much research[242], as does the effect of inhibitors in various circumstances[243].

Microbiologically-induced corrosion of copper water pipes in institutional buildings has been reported from several countries. The results of research, leading to remedial measures, have been summarised by Geesey[247].

P. T. GILBERT

Acknowledgement This section is based on the article 'Chemical Properties and Corrosion Resistance of Copper and Copper Alloys' which formed Chapter XVIII of the American Chemical Society Monograph No. 122, *Copper: The Science and Technology of the Metal its Alloys and Compounds,* edited by Professor Allison Butts, and published by the Reinhold Publishing Corporation, New York, in 1954. Acknowledgement is hereby made to the Reinhold Publishing Corporation for permission to use the above-mentioned section as a basis for the present chapter.

REFERENCES

1. *Copper Alloy Ingots and Copper and Copper Alloy Castings,* BS 1400 (1973)
2. Specifications for Copper and Copper Alloys:
   BS 2870, *Sheet, Strip and Foil* (1980); BS 2871, *Tubes* (1971); BS 2872, *Forging Stock and Forgings* (1989); BS 2873, *Wire* (1969); BS 2874, *Rods and Sections* (1986); BS 2875, *Plate* (1969)

3. *Copper and Copper Alloys*, American Society for Testing and Materials, Philadelphia (1992)
4. *Standard Handbook — Copper, Brass, Bronze, Wrought Mill Products*, 8th edn, Copper Development Association Inc., New York (1985); 'Properties and Applications of Wrought Coppers and Copper Alloys', *Metal Prog.*, **98**, 85 (1970)
5. 'Wrought Cupro-Nickels', *Mater. in Des. Engng.*, **49**, 127 (1959); Katz, W., *Werkst. u. Korrosion*, **15**, 977 (1964)
6. 'Silicon Bronzes', *Mater. in Des. Engng.*, **50**, 112 (1959)
7. Evans, U. R., *Metallic Corrosion Passivity and Protection*, Edward Arnold, London (1946); Evans, U. R., *The Corrosion and Oxidation of Metals*, Edward Arnold, London (1960)
8. Uhlig, H. H. (Ed.), *Corrosion Handbook*, Wiley, New York and Chapman and Hall, London (1948)
9. LaQue, F. L. and Copson, H. R., *Corrosion Resistance of Metals and Alloys*, 2nd edn, Reinhold, New York (1963)
10. Speller, F. N., *Corrosion, Causes and Prevention*, McGraw-Hill, New York (1951)
11. Burghoff, H. L., *Corrosion of Metals*, Amer. Soc. Metals Monograph, Cleveland, Ohio, 100–130 (1946)
12. *Metals Handbook*, Amer. Soc. Metals, Metals Park, Ohio, **1**, 983–1005 (1961)
13. Rogers, T. Howard, *Marine Corrosion Handbook*, McGraw-Hill Co. of Canada Ltd., Toronto (1960)
14. Rogers, T. Howard, *Marine Corrosion*, Butterworths, London (1968)
15. Cairns, J. H. and Gilbert, P. T., *The Technology of Heavy Non-Ferrous Metals and Alloys*, Butterworths, London (1967)
16. Anon., *Corrosion*, **15**, 199t(1959); **16**, 131t(1960); Lasko, W. R. and Tice, W. K., *Corrosion*, **18**, 116t (1962); Erdös, E., *Werkst. u. Korrosion*, **19**, 385 (1968)
17. Latimer, W. M., *Oxidation Potentials*, Prentice-Hall, New Jersey (1952)
18. Gatty, O. and Spooner, E. C. R., *Electrode Potential Behaviour of Corroding Metals in Aqueous Solutions*, Clarendon Press, Oxford (1938)
19. Pourbaix, M. J. N., *Thermodynamics of Dilute Aqueous Solutions* (Trans. by J. N. Agar), Edward Arnold, London (1949)
20. Pourbaix, M. J. N., *Atlas of Electrochemical Equilibria in Aqueous Solutions*, Pergamon, Oxford (1966)
21. Rubinić, L. and Marković, T., *Werkst. u. Korrosion*, **10**, 666 (1959)
22. Hickling, A. and Taylor, D., *Trans. Faraday Soc.*, **44**, 262 (1948)
23. Feitknecht, W. and Lehnel, H. W., *Helv. Chim. Acta.*, **27**, 775 (1944); Wilde, B. E. and Teterin, G. A., *Brit. Corrosion J.*, **2**, 125 (1967)
24. Bonhoeffer, K. F. and Gerischer, J., *Z. Elektrochem.*, **52**, 149 (1948); Stolica, N. D. and Uhlig, H. H., *J. Electrochem. Soc.*, **110**, 1215 (1963); Mansfield, F. and Uhlig, H. H., *J. Electrochem. Soc.*, **117**, 427 (1970); Varenko, E. S. et al., *Zashchita Metallov.*, **6**, 103 (1970)
25. Vernon, W. H. J., *J. Inst. Met.*, **49**, 153 (1932)
26. Freeman, J. R. and Kirby, P. H., *Metals and Alloys*, **5**, 67 (1934)
27. Vernon, W. H. J. and Whitby, L., *J. Inst. Met.*, **42**, 181 (1929); **44**, 389 (1930); **49**, 153 (1932); Vernon, W. H. J., *J. Chem. Soc.*, 1853 (1934)
28. Vernon, W. H. J., *Trans. Faraday Soc.*, **27**, 255, 582 (1931)
29. Evans, U. R., *Trans. Electrochem. Soc.*, **46**, 247 (1924)
30. Hudson, J. C., *Trans. Faraday Soc.*, **25**, 177 (1929); *J. Inst. Met.*, **44**, 409 (1930); *J. Birmingham Met. Soc.*, **14**, 331 (1934); *Metal Ind.*, **44**, 415 (1934); *J. Inst. Met.*, **56**, 91 (1935)
31. Tracy, A. W., Thompson, D. H. and Freeman, J. R., Special Technical Publication No. 175, A.S.T.M., 77–87 (1955)
32. Tracy, A. W., STP No. 175, A.S.T.M., 67–76 (1955)
33. Thompson, D. H., *Metal Corrosion in the Atmosphere*, STP 435, A.S.T.M., 129 (1968)
34. Mattsson, E. and Holm, R., *Metal Corrosion in the Atmosphere*, STP 435, A.S.T.M., 187 (1968)
35. Scholes, I. R. and Jacob, W. R., *J. Inst. Met.*, **98**, 272 (1970)
36. Compton, K. G., *Trans. Electrochem. Soc.*, **91**, 705 (1947); Ambler, H. R. and Bain, A. A. J., *J. Appl. Chem.*, **5**, 437 (1955); Forgeson, B. W. et al., *Corrosion*, **74**, 73t (1958); Hummer, C. W., Jr., Scuthwell, C. R. and Alexander, A. L., *Mater. Protection*, **7** No. 1, 41 (1968)

37. Cole, H. G., R.A.E. Report MET82 (1954); Compton, K. G., Mendizza, A. and Bradley, W. W., *Corrosion*, **11**, 383t (1955)
38. Romanoff, M., *Underground Corrosion*, Nat. Bur. Stand. Circ. 579, Supt. of Documents, Washington, D.C. (1957)
39. Gilbert, P. T., *J. Inst. Met.*, **73**, 139 (1947)
40. Gilbert, P. T. and Porter, F. C., Iron and Steel Inst. Special Report No. 45, 55-74, 127-134 (1952)
41. Marković, T., Sevdić, M. and Rubinić, L., *Werkst. u. Korrosion*, **11**, 87 (1960)
42. Bengough, G. D. and May, R., *J. Inst. Met.*, **32**, 81 (1924)
43. May, R., *J. Inst. Met.*, **40**, 141 (1928)
44. May, R. and de V. Stacpoole, R. W., *J. Inst. Met.*, **77**, 331 (1950)
45. LaQue, F. L. and Stewart, W. C., Milaux et Corros., **23**, 147 (1948); LaQue, F. L., *Proc. Amer. Soc. Test. Mater.*, **52**, 1 (1952)
46. Gilbert, P. T. and LaQue, F. L., *J. Electrochem. Soc.*, **101**, 448 (1954)
47. Fink, F. W., *Trans. Electrochem. Soc.*, **75**, 441 (1939); Evans, U. R., *ibid.*, 446
48. Lucey, V. F., *Brit. Corrosion J.*, **1**, 9 and 53 (1965)
49. Feller, H-G., *Z. Metallkunde*, **58**, 875 (1967)
50. Heidersbach, R., *Corrosion*, **24**, 38 (1968)
51. Polushkin, E. P. and Shuldener, H. L., *Trans. Amer. Inst. Min. (Metall.) Engrs.*, **161**, 214 (1945); Kleinberger, R., Okuzumi, H. and Perio, P., *Métaux (Corrosion-Ind.)*, No. 413, 40-43 (1960); Kenworthy, L. and O'Driscoll, W. G., *Corrosion Tech.*, **2**, 247 (1955); Piatti, L. and Grauer, R., *Werkst. u. Korrosion*, **14**, 551 (1963); Hashimoto, K., Ogawa, S. and Shimodaira, S., *Trans. Japan Inst. Met.*, **4**,42 (1963); Robinson, F.P.A. and Shalit, M., Corrosion Tech., **11**, 11 (1964); Rowlands, J. C., *Proc. 2nd Internat. Cong. Met. Corrosion*, New York, 1963, N.A.C.E., Houston, 795 (1966); Frade, G. and Lacombe, P., *Mém. Sci. Rev. Métall.*, **63**, 649 (1966); Sugawara, H. and Ebiko, H., *Corrosion Sci.*, **7**, 513 (1967); Joseph, G. and Arce, M. T., *Corrosion Sci.*, **7**, 597 (1967); Langenegger, E. E. and Robinson, F.P.A., *Corrosion*, **24**, 411 (1968) and **25**, 59 (1969); Horton, Ralph M., *Corrosion*, **26**, 160 (1970); Rothenbacher, P., *Corrosion Sci.*, **10**, 391 (1970); Pötzl, R. and Lieser, K. H., *Z. Metallkunde*, **61**, 527 (1970)
52. Bengough, G. D., Jones, R. M. and Pirret, R., *J. Inst. Met.*, **23**, 65 (1920)
53. Upton, B., Brit. *Corros. J.*, **1**, 134 (1966)
54. May, R., *Trans. Inst. Mar. Engrs.*, **49**, 171 (1937); Sherborne, H. F. (p. 76), Bailey, G. L. (p. 78). Discussion of Bradbury, E. J. and Johnson, L. W., *Trans. Inst. Mar. Engrs.*, **63**, 59 (1951)
55. Breckon, C. and Gilbert, P. T., *Chem. and Ind.*, Jan. 4, 35 (1964)
56. Bem, R. S., *The Engineer*, **206**, 756 (1958)
57. Clark, W. D., *J. Inst. Met.*, **73**, 263 (1947)
58. Breckon, C. and Gilbert, P. T., *Proc. 1st Internat. Cong. on Met. Corrosion*, London, 1961. Butterworths, London, 624 (1962)
59. Gleekman, L. W. and Swanby, R. K., *Corrosion*, **17**, 144t (1961); Rowlands, J. C., *Corrosion Sci.*, **2**, 89 (1962); Smith, A. A., *Corrosion Prev. and Control*, **10**, 29 (1963); Upton, B., *Corrosion*, **19**, 204t (1963); Piatti, L. and Grauer, R., *Werkst. u. Korrosion*, **14**, 551 (1963); Neiderberger, R. B., *Modern Castings*, **45** No. 3, 115 (1964); Gaillard, F. and Weill, A. R., *Mem. Sci. Rev. Met.*, **61**, 437 (1964); Maersch, R. E. and Ciesleiwicz, J. M., *Mater. Protection*, **3** No. 7, 54 (1964); Shinoda, G. and Amano, Y., *Trans. Japan Inst. Met.*, **4**, 231 (1963); Tanabe, Z., *Corrosion Sci.*, **4**, 413 (1964); Arnaud, D. *et al.*, *Fonderie*, No. 226, 403 (1964); Arnaud, D., *Fonderie*, No. 275, 88 and No. 281, 355 (1969); Shibard, P. R. and Balachandra, J., *Anti Corrosion Methods Mater.*, **14** No. 2, 10 (1967)
60. Gilbert, P. T. and May, R., *Trans. Inst. Mar. Engrs.*, **62**, 291 (1950); Gilbert, P. T., *Trans. Inst. Mar. Engrs.*, **66**, 1 (1954); Breckon, C. and Baines, J. R. T., *Trans. Inst. Mar. Engrs.*, **67**, 1 (1955); Bradbury, E. J. and Johnson, L. W., *Trans. Inst. Mar. Engrs.*, **63**, 59 (1951); Slater, I. G., Kenworthy, L. and May, R., *J. Inst. Met.*, **77**, 309 (1950); Bethon, H. E., *Corrosion*, **4**, 457 (1948); Eichhorn, K., *Werkst. u. Korrosion*, **8**, 657 (1957) and **21**, 535 (1970); Nothing, F. W., *Metaill*, **10**, 520 (1956) and **16**, 1089, 1196 (1962); Maurin, A. J., *Corros. and Anti-Corros.*, **5**, 275, 383 (1957) and **6**, 15 (1958); Todhunter, H. A., *Corrosion*, **11**, 221t (1955) and *Power*, **100**, 85 (1956); Gilbert, P. T., *Chem. and Ind.*, July 11, 888 (1959); Nowlan, N. V., *Corrosion Tech.*, **7**, 397 (1960); Otsu, T., *Sumitomo Light Metal Tech. Rep.*, **1** No. 1, 62 (1960); Gilbert, P. T., *Inst. Mar. Engrs. Materials Section Symposium*, March, 14 (1968); Gilbert, P. T., *Trans. Inst. Mar. Engrs.*, **82** No. 7,

6 (1970); Kingerley, D. G., *Brit. Chem. Engng.*, **6** No. 1, 20 (1961); Sisson, A. B., *Corrosion*, **17**,18 (1961); Hall, B. N., *Corrosion Prev. and Control*, **10**, 49 (1963); Malcolm, R. R., *Australasian Corros. Engng.*, **7** No. 3, 17 and **7** No. 10, 25 (1963); Serre, J. and Laureys, J., *Corrosion et Anti-Corrosion*, **11**, 305, 360 (1963) and *Corrosion Sci.*, **5**, 135 (1965); Kenworthy, L., *Trans. Inst. Mar. Engrs.*, **77**, 149 (1965); Page, G. C., *Anti-Corrosion Methods Mater.*, **14** No. 5, 13 (1967); Yandushkin, K. N., *Zashchita Metallov.*, **6**, 46 (1970)
61. Tracy, A. W., *Corrosion*, **1**, 103 (1945); Mitchell, N. W., *Corrosion*, **3**, 243 (1947); Van der Baan Sj., *Corrosion*, **6**, 14 (1950); Mason, J. F., *Corrosion*, **12**, 199t (1956); Rust, A. D., *Corrosion Tech.*, **3**,185 (1956); Gilbert, P. T., *Soc. Chem. Ind. Monograph*, No. 10, 111-120 (1960); Bird, D. B. and Moore, K. L., *Mater. Protection*, **1** No. 10, 70 (1962)
62. Bengough, G. D., *J. Inst. Met.*, **5**, 28 (1911); Bengough, G. D. and Jones, R. M., *J. Inst. Met.*, **10**, 13 (1913); Gibbs, W. E., Smith, R. H. and Bengough, G. D., *J. Inst. Met.*, **15**, 37 (1916); Bengough, G. D. and Hudson, O. F., *J. Inst. Met.*, **21**, 167 (1919)
63. Bailey, G. L., *J. Inst. Met.*, **79**, 243 (1951); Tracy, A. W. and Hungerford, R. L., *Proc. Amer. Soc. Test. Mater.*, **45**, 591 (1945); LaQue, F. L. and Mason, J. F., *Proc. Amer. Petrol Inst.*, **30M III**, 103 (1950); Todhunter, H. A., *Corrosion*, **16**, 226t (1960) and *Mater. Protection*, **6** No. 7, 45 (1967); LaQue, F. L. and Stewart, W. C., *Corrosion*, **8**, 259 (1952); Krafack, K. and Franke, E., *Werkst. u. Korrosion*, **4**, 310 (1953); Se Ui Yu and Turkovsaya, A. V., *Tsvetnaya Metall.*, **4**, 145 (1961); May, T. P. and Weldon, B. A., *Rev. Nickel*, No. 3, 183 and No. 4, 219 (1966)
64. Chapman, J. and Cuthbertson, J. W., *J. Soc. Chem. Ind.*, **58**, 100, 330 (1939); Cuthbertson, J. W., *J. Inst. Met.*, **72**, 317 (1946)
65. Tanabe, Z., Sumitomo Light Metal Tech. Rep., **9** No. 3, 167 (1968)
66. Baker, L., *Trans. Inst. Mar. Engrs.*, **65**, (1953)
67. Rogers, T. Howard, *J. Inst. Met.*, **75**, 19 (1948-49)
68. Sato, S., *Sumitomo Light Metal Tech. Rep.*, **6** No. 1, 42 (1965); Tanaka, R., Sumitomo Light Metal Tech. Rep., **3** No. 3, 55 and **3** No. 4, 1 (1962) and **6** No. 3, 152 (1965); Changarnier, J., *Corros. Anti-Corros.*, **1**, 8 (1953)
69. Shimodaira, S., Sugawara, H. and Sato, S., *Sumitomo Light Metal Tech. Rep.*, **4** No. 1, 31 (1963); Tanabe, Z., *Sumitomo Light Metal Tech. Rep.*, **5** No. 1, 16 (1964)
70. Otsu, T. and Sato, S., *Sumitomo Light Metal Tech. Rep.*, **2** No. 2, 23 and **2** No. 4, 27 (1961); Otsu, T. and Okawa, M., *Sumitomo Light Metal Tech. Rep.*, **3** No. 3, 35 (1962); Otsu, T., Sato, S. and Watanabe, T., *Sumitomo Light Metal Tech. Rep.*, **4** No. 2, 21 (1963); Tanaka, R. and Tanabe, Z., *Sumitomo Light Metal Tech. Rep.*, **5** No. 1, 9 (1964); Tanaka, R., *Sumitomo Light Metal Tech. Rep.*, **5** No. 3, 188 (1964) and **6** No. 1, 71 (1965); Sato, S. and Sagiska, K., *Sumitamo Light Metal Tech. Rep.*, **11** No. 2, 1 (1970)
71. Tanabe, Z., *Sumitomo Light Metal Tech. Rep.*, **6** No. 2, 119 (1965)
72. Sato, S., Proc. 4th Int. Cong. Met. Corrosion Amsterdam (1969). Nat. Assoc. Corrosion Eng., Houston, 795 (1972)
73. Rowlands, J. C., *J. Appl Chem.*, **15**, 57 (1965)
74. Grubitsch, H., Hilbert, F. and Sammer, R., *Werkst. u. Korrosion*, **17**, 760 (1966); Tanabe, Z., *Sumitomo Light Metal Tech. Rep.*, **8** No. 1, 10 (1967); Meany, J. J., Jr., *Mater. Protection*, **8** No. 10, 27 (1969); Mor, E., Scotto, V. and Trevis, A., *Corrosion (Paris)*, **18** No. 2, 67 (1970); Baudo, G. and Giuliani, L., *Werkst. u. Korrosion*, **21**, 332 (1970); North, R.F. and Pryor, M. J., *Corrosion Sci.*, **9**, 509 (1969) and **10**, 297 (1970); Giuliani, L. and Bombara, G., *Brit. Corrosion J.*, **5**, 179 (1970)
75. McAllister, R. A., *et al.*, *Corrosion*, **17**, 579t (1961); Erdös, E., *Schweizer Archiv.*, **30**, 251 (1964); Mifflin, R. C. and Bird, D. B., *Mater. Protection*, **8** No. 9, 72 (1969)
76. Sato, S., *Sumitomo Light Metal Tech. Rep.*, **3** No. 3, 106 (1962)
77. Sherry, A. and Gill, E. R., *Chem. and Ind.*, Jan. 18, 102 (1964); Edwards, B. C., *Corrosion Sci.*, **9**, 395 (1969)
78. Gilbert, P. T., Chem. and Ind., July 11, 888 (1959)
79. Sato, S., Nagata, K. and Ogiso, A., *Sumitomo Light Metal Tech. Rep.*, **11** No. 3, 1 (1970)
80. Peplow, D. B., *Brit. Power Engineering*, **1** No. 5, 51 (1960); Attwood, P. G. and Richards, N. G., *Corrosion*, **17**, 8t (1960); Crennel, J. T. and Sawyer, L. J. E., *J. Appl. Chem.*, **12**, 170 (1962); Page, G. G., *Proc. 2nd Intermat. Cong. Met. Corrosion*, New York (1963). *Nat. Assoc. Corrosion Eng.*, Houston, 275 (1966)
81. Bostwick, T. W., *Corrosion*, **17**, 12 (1961)
82. Lockhart, A. M., *Proc. Inst. Mech. Engrs.*, **179**, 495 (1964-65)

83. North, R. F., *Corrosion Sci.*, **8**, 149 (1968); Gasparini, R., Della Rocca, C. and Ioannilli, E., *Corrosion Sci.*, **10**, 157 (1970)
84. *Proceedings of Conference on the Role of Copper and its Alloys in Desalination Equipment*, London, December (1966), Copper Development Association. See Stewart, J. M., p. 21, Gilbert, P. T., p. 31, Weldon, B. A. and Tuthill, A. H., p. 39; Tuthill, A. H. and Sudrabin, D. A., *Metals Engng. Quart.*, **7** No. 3, 10 (1967); Fink, F. W., Tech. Rep. 704/6, Copper Development Association, New York (1966) and *Mater. Protection*, **6** No. 5, 40 (1967); Schoraten, A., *Metall.*, **22**, 1153 (1968); Cohen, A. and Rice, L., *Mater. Protection*, **8** No. 12, 67 (1969) and **9** No. 11, 29 (1970); Bom, P. R., *Brit. Corrosion J.*, **5**, 258 (1970)
85. Campbell, H. S. and Carter, V. E., *J. Inst. Metals*, **90**, 362 (1962); Murphy, T. J. and Jack, J. B., *Shipping World and Shipbuilder*, Jan. 21, 282 (1965)
86. Gilbert, P. T. and Jenner, B. J., *Inst. Marine Engrs. International Marine and Shipping Conference*, London, June (1969)
87. Campbell, H. S., *Chem. and Ind.*, 692 (1955); Hatch, G. B., *J. Amer. Waterworks Assoc.*, **53**, 1417 (1961); N.A.C.E. Tech. Rep. 60-11 and *Corrosion*, **16**, 453t (1960); Schafer, G. J., *New Zealand J. Sci.*, **5**, 475 (1962)
88. Tumer, M. E. D., *Proc. Soc. Water Treatm. Exam.*, **10**, 162 (1961) and **14**, 81 (1965)
89. Baldwin, A. B. and Campbell, H. S., *Brit. Waterworks Assoc. J.*, **43**, 13 (1961); Schafer, G. J. and Dall, R. A., *Australasian Corrosion Engng.*, **10** No. 3, 9 (1966); Ladeburg, H., *Metall.*, **20**, 33 (1966); Simmonds, M. A. and Huxley, W. G. S., *Australasian Corrosion Engng.*, **11** No. 11, 9 (1967)
90. Schafer, G. J., Foster, P. K. and Marshall, T., *New Zealand J. Sci.*, **4**, 194 (1961)
91. Schafer, G. J. and Dall, R. A., *Brit. Corrosion J.*, **3**, 12 (1968); Harrison, P. S., *Electrical Times*, **153**, 219 (1968)
92. Obrecht, M. F., *Corrosion*, **18**, 189t (1962)
93. Ingleson, H., Sage, A. M. and Wilkinson, R., *J. Inst. Wat. Engrs.*, **3**, 81 (1949); Wormwell, F. and Nurse, T. J., *J. Appl. Chem.*, **2**, 685 (1952); Solelev, A., *J. Inst. Wat. Engrs.*, **9**, 208 (1955)
94. Tronstad, L. and Veimo, R., *J. Inst. Met.*, **66**, 17 (1940); Kenworthy, L., *J. Instn. Heat. Vent. Engrs.*, **8**, 15 (1940); Gilbert, P. T., *Proc. Soc. Water Treatm. Exam.*, **15**, 165 (1966)
95. Kenworthy, L., *J. Inst. Met.*, **69**, 67 (1943)
96. Porter, F. C. and Hadden, S. E., *J. Appl. Chem.*, **3**, 385 (1953)
97. Davenport, W. H., Nole, V. F. and Robertson, W. D., *J. Electrochem. Soc.*, **106**, 1005 (1959); Ives, D. J. G. and Rawson, A. E., *J. Electrochem. Soc.*, **109**, 447 (1962); Obrecht, M. F. and Pourbaix, M., *J. Amer. Waterworks Assoc.*, **59**, 977 (1967)
98. Grunau, E. B., Städtehygiene, No. 7, 153 (1967)
99. Campbell, H. S., *J. Inst. Met.*, **77**, 345 (1950); *J. Appl Chem.*, **4**, 633 (1954); *Proc. Soc. Wat. Treatm. Exam.*, **3**, 100 (1954) and *Proc. 2nd. Internat. Cong. Met. Corrosion*, New York (1963), N.A.C.E., Houston, 237 (1966)
100. Schafer, G. J., *Australasian Corrosion Engng.*, **6** No. 8, 15 (1962) and *New Zealand Plumbing Rev.*, **1** No. 9, 10 (1964); Rambow, C. A. and Holmgren, R. S., Jr., *J. Amer. Waterworks Assoc.*, **58**, 347 (1966); Pourbaix, M., Van Muylder, J. and Van Laer, P., *Corrosion Sci.*, **7**, 795 (1967); von Franqué, O., *Werkst. u. Korrosion*, **19**, 377 (1968); Lihl, F. and Klamet, H., *Werkst. u. Korrosion*, **20**, 108 (1969); Walker, I. K. and Page, G. G., *Australasian Corrosion Engng.*, **13** No. 4, 13 (1969); Kennett, A., *Australasian Corrosion Engng.*, **13** No. 4, 5 (1969); Gilbert, P. T., *Australasian Corrosion Engng.*, **13** No. 5, 13 (1969)
101. Schafer, G. J. and Dall, R. A., *Australasian Corrosion Engng.*, **7** No. 10, 33 (1963)
102. Lucey, V. F., *Brit. Corrosion J.*, **2**, 175 (1967)
103. Mattsson, E. and Fredriksson, A-M., *Brit. Corrosion J.*, **3**, 246 (1968)
104. Lee, J. A., *Materials of Construction for Chemical Process Industries*, McGraw-Hill, New York (1950)
105. Rabald, E., *Corrosion Guide*, Elsevier Publishing Co., New York (1951); Carmenisch, K. P., *Pro-Metal*, **13** No. 73, 288 (1960); Ritter, F., *Korrosionstabellen Metallischer Werkstoffe*, Springer-Verlag, Vienna (1958); LaQue, F. L., *Corrosion*, **10**, 391 (1954); Heim, A. T., *Industr. Engng. Chem.*, **49**, 63A, 64A, 66A (1957); Baker, S., *Corrosion Tech.*, **8**, 8 (1961); Tracy, A. W., *Chem. Engng.*, **69**, 130, 152 (1962); Anon., *Industr. Engng. Chem.*, **40**,1827 (1948), **43**, 2218 (1951) and **49**, 63A (1957)
106. Gould, A. J. and Evans, U. R., *J. Iron St. Inst.*, **155**, 195 (1947)

107. Lacan, M., Markovic, T. and Rubinic, L., *Werkst. u. Korrosion*, **10**, 767 (1959)
108. Holmberg, M. E. and Prange, F. A., *Industr. Engng. Chem.*, **37**, 1030 (1945); Anon., *Industr. Chem.*, **30**, 609 (1954); Lingnau, E., *Werkst. u. Korrosion*, **8**, 216 (1957)
109. Fontana, M. G., *Industr. Engng. Chem.*, **42**, 69A (1950)
110. Groth, V. J. and Hafsten, R. J., *Corrosion*, **10**, 368 (1954)
111. Bulow, C. L., *Chem. Engng.*, **53**, 210 (1946)
112. Friend, W. Z. and Mason, J. F., *Corrosion* **5**, 355 (1949), N.A.C.E. Report and *Corrosion*, **13**, 757t (1957)
113. Russell, R. P. and White, A., *Industr. Engng. Chem.*, **19**, 116 (1927); Damon, G. H. and Cross, R. C., *Industr. Engng. Chem.*, **28**, 231 (1936)
114. Cornet, I., Barrington, E. A. and Behrsing, G. U., *J. Electrochem. Soc.*, **108**, 947 (1961)
115. Caney, R. J. T., *Aust. Engr.*, **64**, 54 (1954) and U.K. Pat 718, 987; Zitter, H. and Kraxner, G., *Werkst. u. Korrosion*, **14**, 80 (1963); Piatti, L. and Fot, E., *Werkst. u. Korrosion*, **15**, 27 (1964)
116. Gregory, D. P. and Riddiford, A. C., *J. Electrochem. Soc.*, **107**, 950 (1960); Talati, J. D., Desai, M. N. and Trivedi, A. M., *Werkst. u. Korrosion*, **12**, 422 (1961); Bumbulis, J. and Graydon, W. F., *J. Electrochem. Soc.*, **109**, 1130 (1962); Kagetsu, T. J. and Graydon, W. F., *J. Electrochem. Soc.*, **110**, 856 (1963); Feller, H-G, *Corrosion Sci.*, **8**, 259 (1968); Otsuka, R. and Uda, M., *Corrosion Sci.*, **9**, 703 (1969)
117. Rana, S. S. and Desai, M. N., *Indian J. Technol.*, **5**, 393 (1967); Desai, M. N. and Shah, Y. C., *Anti-Corrosion Methods Mater.*, **15** No. 12, 9 (1968); Desai, M. N., Shah, Y. C. and Gandhi, M. H., *Corrosion Sci.*, **9**, 65 (1969); Padma, D. K. and Rama Char, T. L., *Anti-Corrosion Methods Mater.*, **16** No. 4 (1969); Desai, M. N., Shah, Y. C. and Punjani, B. K., *Brit. Corrosion J.*, **4**, 309 (1969)
118. Ammar, I. A. and Riad, S., *Corrosion Sci.*, **9**, 423 (1969)
119. Anon., *Proc. Amer. Soc. Test. Mater.*, **35**, 161 (1935); Desai, M. N. and Rana, S. S., *Werkst. u. Korrosion*, **17**, 870 (1966)
120. Radley, J. A., Stanley, J. S. and Moss, G. E., *Corrosion Tech.*, **6**, 229 (1959); Schaefer, B. A., *Corrosion Sci.*, **8**, 623 (1968); Bartoniček, R., Holinka, M. and Lukašovská, M., *Werkst. u. Korrosion*, **19**, 1032 (1968); Green, J. A. S., Mengelberg, H. D. and Yolken, H. T., *J. Electrochem. Soc.*, **117**, 433 (1970); Jenkins, L. H. and Durham, R. B., *J. Electrochem. Soc.*, **117**, 768 (1970)
121. Dubrisay, R. and Chesse, G., *Compt. Rend. Acad. Sci. Paris*, **220**, 707 (1945)
122. Reichert, J. S. and Pete, R. H., *Chem-Engng.*, **54**, 218 (1947)
123. West, J. R., *Chem. Engng.*, **58**, 281 (1951)
124. Botham, G. H. and Dummett, G. A., *J. Dairy Res.*, **16**, 23 (1949)
125. Holness, H. and Ross, T. K., *J. Appl. Chem.*, **1**, 158 (1951); Bukowiecki, A., *Schweizer Archiv. Angew. Wiss.*, **24**, 355 (1958)
126. Mason, J. F., *Corrosion*, **4**, 305 (1948); Inglesent, H. and Storrow, J. A., *J. Soc. Chem. Ind.*, **64**, 233 (1945); Clendenning, K. A., *Canad. J. Res. F (Technology)*, **26**, 277 (1948)
127. *Review of Oxidation and Scaling of Heated Metal Solids*, D.S.I.R., H.M.S.O., London (1935); Vernon, W. H. J., *Chem. and Ind. (Rev.)*, **59**, 87 (1940)
128. Tylecote, R. F., *J. Inst. Met.*, **78**, 259 (1950–51); Cabrera, N. and Mott, N. F., *Rep. Progr. Phys.*, **12**, 163 (1948–49); Rönnquist, A. and Fischmeister, H., *J. Inst. Met.*, **89**, 65 (1960–61)
129. Kubaschewski, O. and Hopkins, B. E., *Oxidation of Metals and Alloys*, Butterworths, London (1953); Hauffe, K., *Oxydation von Metallen und Legierungen*, Springer-Verlag, Berlin (1956)
130. Wagner, C. and Grunewald, K., *Z. Phys. Chem.*, **40**, 455 (1938B)
131. Dighton, A. L. and Miley, H. A., *Trans. Electrochem. Soc.*, **81**, 321 (1942)
132. Evans, U. R., *Symposium on Internal Stresses in Metals and Alloys*, Inst. Metals, London, 291–310 (1947), *Trans. Electrochem. Soc.*, **91**, 547 (1947) and Research, London, **6**, 130 (1953)
133. Mott, N. F., *Trans. Faraday Soc.*, **35**, 1175 (1939); **36**, 472 (1940); **43**, 429 (1947); *Nature, London*, **145**, 996 (1940); *J. Chim. Phys.*, **44**, 172 (1947); *Research*, London, **2**, 162 (1949); Price, L. E., *Chem. and Ind. (Rev.)*, **56**, 769 (1937)
134. de Carli, F. and Collari, N., *Chim. et Industr.*, **33**, 77 (1951); McKewan, W. and Fassell, W. M., *J. Metals. N. Y.*, **51**, 1127 (1953); Paidassi, J., *Acta Metallurgica*, **6**, 216 (1958); Lohberg, K. and Wolstein, F., *Z. Metallk.*, **46**, 734 (1955); Baur, J. P., Bridges, D. W. and Fassell, W. M., *J. Electrochem. Soc.*, **103**, 273 (1956); Gulbrausen, E. A., Copan, T. P. and Andrew, K. F., *J. Electrochem. Soc.*, **108**, 119 (1961); Rönnquist, A., *J. Inst.*

Met., **91**, 89 (1962); Yoda, E. and Siegel, B. M., *J. Appl. Phys.*, **34**, 1512 (1963); Wallwork, G. R. and Smeltzer, W. W., *Corrosion Sci.*, **9**, 561 (1969)
135. Tylecote, R. F., *J. Inst. Met.*, **78**, 327 (1950-51) and **81**, 681 (1952-53)
136. Bardeen, J., Brattain, W. H. and Shockley, W., *J. Chem. Phys.*, **14**, 714 (1946); Castellan, G. W. and Moore, W. J., *J. Chem. Phys.*, **17**, 41 (1949)
137. Pilling, N. B. and Bedworth, R. E., *J. Inst. Met.*, **29**, 529 (1923)
138. Feitknecht, W., *Z. Elektrochem.*, **35**, 142, 500 (1929)
139. Vernon, W. H. J., *J. Chem. Soc.*, 2273 (1926)
140. Valensi, G., *Pittsburgh International Conference on Surface Reactions*, Corrosion Publishing Co., Pittsburgh, 156-165 (1948)
141. Tylecote, R. F., *J. Inst. Met.*, **78**, 301 (1950-51)
142. Oudar, J., *Métaux*, **35**, 397, 445 (1960)
143. Dyess, J. B. and Miley, H. A., *Trans. Amer. Inst. Min. (Metall.) Engrs.*, **133**, 239 (1939); Vernon, W. H. J., *Trans. Faraday Soc.*, **19**, 839 (1924)
144. Preston, G. D. and Bircumshaw, L. L., *Phil. Mag.*, **20**, 706 (1935)
145. Hallowes, A. P. C. and Voce, E., *Metallurgia, Manchr.*, **34**, 95 (1946)
146. Lustman, B., *Metal. Prog.*, **50**, 850 (1946)
147. Dennison, J. P. and Preece, A., *J. Inst. Met.*, **81**, 229 (1952-53); Spinedi, P., *Metallurg. Ital.*, **45**, 457 (1953); Collari, N. and Spinedi, P., *Metallurg. Ital.*, **46**, 403 (1954); Blade, J. C. and Preece, A., *J. Inst. Met.*, **88**, 427 (1959-60); Maak, F. and Wagner, C., *Werkst. u. Korrosion*, **12**, 273 (1961); Wallbaum, H. J., *Werkst. u. Korrosion*, **12**, 417 (1961); Maak, F., *Z. Metallkunde*, **52**, 538 (1961); Zwicker, U., *Metall.*, **16**, 1110 (1962); Kapteijn, J., Couperus S. A. and Meijering, J. L., *Acta Metall.*, **17**, 1311 (1969); Sanderson, M. D. and Scully, J. C., *Corrosion Sci.*, **10**, 165 (1970)
148. Dunwald, H. and Wagner, C., *Z. Phys. Chem.*, **22**, 212 (1933B); Wagner, C., *Z. Phys. Chem.*, **21**, 25 (1933B), Pittsburgh International Conference on Surface Reactions, Corrosion Publishing Co., Pittsburgh, 77-82 (1948); Hoar, T. P. and Price, L. E., *Trans. Faraday Soc.*, **34**, 867 (1938)
149. Price, L. E. and Thomas, G. J., *J. Inst. Med.*, **63**, 21 (1938)
150. Schückher, F. and Lampe, V., *Pro-Metal.*, No. 105, 192 (1965)
151. Wood, G. C. and Chattopadhyay, B., *J. Inst. Met.*, **98**, 117 (1970)
152. Whittle, D. P. and Wood, G. C., *J. Inst. Met.*, **96**, 115 (1968) and *Corrosion Sci.*, **8**, 295 (1968)
153. Swaroop, B. and Wagner, J. B., Jr., *J. Electrochem. Soc.*, **114**, 685 (1967)
154. Rhines, F. N., *Corros. Mat. Prot.*, **4**, 15 (1947); Ashby, M. F. and Smith, G. C., *J. Inst. Met.*, **91**, 182 (1963); Bolsaitis, P. and Kahlweit, M., *Acta Metall.*, **15**, 765 (1967); Pötschke, J., Mathew, P. M. and Frohberg, M. G., *Z. Metallkunde*, **61**,152 (1970)
155. Moore, C. and Bindley, D., *Proc. 2nd Internat. Cong. Met. Corros.*, New York (1963), Nat. Assoc. Corros. Eng., Houston, 391 (1966); Castle, J. E., Harrison, J. T. and Masterson, H. G., *ibid*, 822; Hopkinson, B. E., *A.S.M.E.*, Paper No. 62-WA-274 (1962); Wiedersum, G. C. and Tice, E. A., *A.S.M.E.*, Paper No. 64-WA/CT-3 (1964); Castle, J. E., Harrison, J. T. and Masterson, H. G., *Brit. Corros. J.*, **1**, 143 (1966)
156. Otsu, T. and Sato, S., *Trans. Jap. Inst. Met.*, **2**, 153 (1961); Sato, S., *Sumitomo Light Metal Tech. Rep.*, **5** No. 1, 2 (1964); **5** No. 2, 27 (1964); **5** No. 3, 231 (1964); **5** No. 4, 290 (1964); Brush, E. G. and Pearl, W. L., *Corrosion*, **25**, 99 (1969)
157. Moore, H., Beckinsale, S. and Mallinson, C. E., *J. Inst. Met.*, **25**, 35 (1921); Moore, H. and Beckinsale, S., *J. Inst. Met.*, **23**, 225 (1920)
158. *Bibliography on Season Cracking*, Proc. Amer. Soc. Test. Mater., **41**, 918 (1941)
159. *Symposium on Stress-corrosion Cracking in Metals*, Amer. Soc. Test. Mater.–Amer. Inst. Min. (Metall.) Engrs., Philadelphia (1944)
160. Nelson, G. A., *Bull. Amer. Soc. Test. Mat.*, No. 240, 39 (1959)
161. Robertson, W. D., Trans. Amer. Inst. Min. (Metall.) Engrs., **191**, 1190 (1951)
162. Edmunds, G., Ref. 159, pp. 67-89
163. Wilson, T. C., Edmunds, G., Anderson, E. A. and Peirce, W. M., Ref. 159, pp. 173-193
164. de Jager, W. G. R., *Metall.*, **4**, 138, 185 (1950); Steinle, H., *Metall.*, **9**, 492 (1955); Kamath, K. V., *Freiberger Forschungshefte*, B56 (1961); Lihl, F. and Hutter, H., *Metall.*, **21**, 884 (1967); Sato, S. and Nosetani, T., *Sumitomo Light Metal Tech. Rep.*, **10** No. 2, 83 (1969); Syrett, B. C. and Parkins, R. N., *Corrosion Sci.*, **10**, 197 (1970)
165. Ref. 8, pp. 78-79
166. Sato, S., *Sumitomo Light Metal Tech. Rep.*, **1** No. 3, 45 (1960); Kamath, K. V. and

Erdmann-Jesnitzer, F., *Metall.*, **14**, 1061 (1960); Thompson, D. H., *Chem. Engng.*, **68** No. 3, 130 (1961); Laub, H., *Metall.*, **20**, 1174 (1966); Adamson, K., *Corrosion Sci.*, **7**, 537 (1967); Erdmann-Jesnitzer, F. and Kaeslingk, N., *Werkst. u. Korrosion*, **20**, 493 (1969); Sabbadini, L., *Metallurgia Italiana*, **62**, 228 (1970)
167. Edmunds, G., Anderson, E. A. and Waring, R. K., Ref. 159, pp. 7-18; Bulow, C. L., Ref. 159, pp. 19-35; Jamieson, A. L. and Rosenthal, H., Ref. 159, pp. 36-46; Hellsing, S., Lissner, O., Rask, S. and Ström, B., *Werkst. u. Korrosion*, **8**, 569 (1957); Aebi, F., *Z. Metaelk.*, **49**, 63 (1958); Thompson, D. H., *Mater. Res. Standards*, **1**, 108 (1961); Szabo, E., *Werkst. u. Korrosion*, **14**, 162 (1963); Mattsson, E., Lindgren, S., Rask, S. and Wennström, G., *Current Corrosion Research in Scandinavia*, Kemian Keskusliitto, Helsinki, 171 (1965)
168. Breckon, C. and Gilbert, P. T., *Metal Ind.*, **93**, 89, 114 (1958); Sinclair, N. A. and Albert, H. J., *Mater. Protection*, **1** No. 3, 35 (1962); Fox, D. K., *Modern Castings*, **42** No. 6, 51 (1962); Baumann, G., *Werkst. u. Korrosion*, **13**, 737 (1962); Sato, S., *Sumitomo Light Metal Tech. Rep.*, **4** No. 1, 48 (1963); Uhlig, H. H. and Sansone, J., *Mater. Protection*, **3** No. 2, 21 (1964); Peters, B. F., Carson, J. A. H. and Barer, R. D, *Mater. Protection*, **4** No. 5, 24 (1965); Logan, H. L. and Ugiansky, G. M., *Mater. Protection*, **4** No. 5, 79 (1965); Serre, J., *Corrosion et Anti-corrosion*, **14** No. 1, 9 (1966)
169. Laub, H., *Metall.*, **20**, 597 (1966) and **21**, 173 (1967); Laub, H., *Metalloberfläche*, **20**, 413, 453, 493 (1966)
170. Sato, S. and Nosetani, T., *Sumitomo Light Metal Tech. Rep.*, **10** No. 3, 175 (1969)
171. Whitaker, M. E., *Metallurgia*, Manchr., **39**, 21, 66 (1948)
172. Perryman, E. C. W., *J. Inst. Met.*, **83**, 369 (1954-55); Perryman, E. C. W. and Goodwin, R. J., *J. Inst. Met.*, **83**, 378 (1954-55)
173. Bailey, A. R., *Metal Ind.*, **80**, 519 (1952) and *J. Inst. Met.*, **87**, 380 (1959)
174. Sheehan, T. L. and Dickerman, H. E., *J. Amer. Soc. Nav. Engrs.*, **58**, 586 (1946)
175. Pugh, E. N., Craig, J. V. and Montague, W. G., *A.S.M. Trans. Quart.*, **61**, 468 (1968)
176. Mattsson, E., *Electrochimica Acta*, **3**, 279 (1961)
177. Forty, A. J. and Humble, P., *Phil. Mag.*, **8** No. 86, 247 (1963); Proc. 2nd Internat. Cong. Met. Corrosion, New York (1963), N.A.C.E., Houston, 80 (1966); McEvily, A. J., Jr. and Bond, A. P., *J. Electrochem. Soc.*, **112**, 131 (1965)
178. Dix, E. H., Jr., *Proc. Amer. Soc. Test. Mater.*, **41**, 928 (1941); Read, T. A., Reed, J. B. and Rosenthal, H., Ref. 159, 90-110; Chaston, J. C., *Sheet Metal Ind.*, **24**, 1395 (1947); Zürrer, T., *Pro-Metal.*, **13**, 307 (1960); Forty, A. J., *Metal Progr.*, **75**, 154 (1959); Graf, L. and Budki, J., *Z. Metallk.*, **46**, 378 (1955); Graf, L., *Corros. Anti-Corros.*, **6**, 151 (1958); Graf, L. and Lacour, H. R., *Z. Metallk.*, **51**, 152 (1960) and **53**, 764 (1962); Graf, L. and Richter, W., *Z. Metallk.*, **52**, 834 (1961); Aebi, F., *Z. Metallk.*, **46**, 547 (1955) and **47**, 421 (1956); Logan, H. L., *J. Res. Nat. Bur. Stand.*, **48**, 99 (1952) and **56**, 159 (1956); Bakish, R. and Robertson, W. D., *J. Electrochem. Soc.*, **103**, 320 (1956); Edeleanu, C. and Forty, A. J., *Phil. Mag.*, **5** No. 58, 1029 (1960); Graf, L., Proc. 2nd Internat. Cong. Met. Corrosion, New York (1963), N.A.C.E., Houston, 89 (1966) and *Metall.*, **18**, 1163, 1287 (1964); Lynes, W., *Corrosion*, **21**, 125 (1965); Ohtani, N. and Dodd, R. A., *Corrosion*, **21**, 161 (1965); Pugh, E. N. and Westwood, A. R. C., *Phil. Mag.*, **13**, 167 (1966); Pugh, E. N., Montague, W. G. and Westwood, A. R. C., *A.S.M. Trans. Quart.*, **58**, 665 (1965); Hoar, T. P. and Booker, C. J. L., *Corrosion Sci.*, **5**, 821 (1965); Fairman, L., *Corrosion Sci.*, **6**, 37 (1966); Tanabe, Z., *Sumitomo Light Metal Tech. Rep.*, **7** No. 3, 137 (1966); Takano, M. and Shimodaira, S., *Trans. Japan Inst. Met.*, **8**, 239 (1967) and *Corrosion Sci.*, **8**, 55 (1968); Murakami, Y. and Ikai, Y., *Trans. Japan Inst. Met.*, **8**, 246 (1967); Lahiri, A. K., *Brit. Corrosion J.*, **3**, 289 (1968); Lahiri, A. K. and Banerjee, T., *Corrosion Sci.*, **8**, 895 (1968); Hoar, T. P. and Rothwell, G. P., *Electrochimica Acta.*, **15**, 1037 (1970)
179. Swann, P. R. and Nutting, J., *J. Inst. Met.*, **88**, 478 (1960); Swann, P. R., *Corrosion*, **19**, 102t (1963); Swann, P. R. and Pickering, H. W., *Corrosion*, **19**, 369t, 373t (1963); Tromans, D. and Nutting, J., *Corrosion*, **21**, 143 (1965); Rönnquist, A., *Jernkontorets Ann.*, **149**, 604 (1965); Brown, B. F., *Met. Mater.*, **2** No. 12, 171 (1968); Graf, L., *Werkst. u. Korrosion*, **20**, 408 (1969)
180. *Symposium on Internal Stresses in Metals and Alloys* (1947), Institute of Metals, London (1948)
181. *Stress Corrosion Cracking and Embrittlement* (Electrochem. Soc. Symposium), Ed. Robertson, W. D., Wiley, New York (1956); *Physical Metallurgy of Stress-Corrosion Fracture* (A.I.M.E. Symposium), Ed. Rhodin, T. N., Interscience, New York (1959);

Conference on Fundamental Aspects of Stress Corrosion Cracking, Ohio State Univ. (1967), N.A.C.E., Houston (1969)
182. Bailey, A. R., *Metall. Rev.*, **6** No. 21, 101 (1961)
183. Logan, H. L., *Met. Engng. Quart.*, **5**, 32 (1965); Parkins, R. N., *Metall. Rev.*, **9** No. 35, 201 (1964); Engell, H-J. and Speidal, M. O., *Werkst. u. Korrosion*, **20**, 281 (1969)
184. Thompson, D. H. and Tracy, A. W., *J. Metals*, N.Y., **1**, 100 (1949)
185. Cook, M., Ref. 180, p. 73
186. Pugh, E. N., Montague, W. G. and Westwood, A. R. C., *Corrosion Sci.*, **6**, 345 (1966); Uhlig, H. H. and Duqette, D. J., *Corrosion Sci.*, **9**, 557 (1969)
187. White, L. F. and Blazey, C., *Metal Ind.*, **75**, 92 (1949)
188. Sylwestrowicz, W. D., *Corrosion*, **25**, 168, 405 (1969) and **26**, 160 (1970)
189. Lahiri, A. K. and Banerjee, T., *Corrosion Sci.*, **5**, 731 (1965); Chatterjee, U. K., Sircar, S. C. and Banerjee, T., *Corrosion*, **26**, 141 (1970); Chatterjee, U. K. and Sircar, S. C., *Brit. Corrosion J.*, **5**, 128 (1970)
190. Blackwood, A. W. and Stoloff, N. S., *A.S.M. Trans. Quart.*, **62**, 677 (1969)
191. Helliwell, B. J. and Williams, K. J., *Metallurgia*, **81**, 131 (1970)
192. Graf, L., *et al., Z. Metallk.*, **54**, 406 (1963)
193. Thompson, D. H., *Corrosion*, **15**, 433t (1959)
194. Marshall, T. and Hugill, A. J., *Corrosion*, **13**, 329t (1957)
195. Klement, J. F., Maersch, R. E. and Tully, P. A., *Corrosion*, **15**, 295t (1959)
196. Norden, R. B., *Chem. Engng.*, **65**, 194, 196 (1958)
197. Klement, J. F., Maersch, R. E. and Tully, P. A., *Metal Prog.*, **75**, 82 (1959), *Corrosion*, **16**, 519t (1960) and U.S. Pat. 2 829 972
198. Robertson, W. D., Grenier, E. G., Davenport, W. H. and Nole, V. F., *Metal Prog.*, **75**, 152 (1959) and U.K. Pat. 802 044
199. Wiederholt, W., *Werkst. u. Korrosion*, **15**, 633 (1964); Laub, H., *Metall.*, **22**, 1116 (1968)
200. Dugdale, I. and Cotton, J. B., *Corrosion Sci.*, **3**, 69 (1963); Cotton, J. B., *Proc. 2nd Internat. Cong. Met. Corrosion*, New York (1963), N.A.C.E., Houston, 590 (1966); Walker, R., *Anti-Corrosion Methods and Mater.*, **17** No. 9, 9 (1970); Cotton, J. B. and Scholes, I. R., *Conf. on the Protection of Metal in Storage and in Transit*, Brintex Exhibitions Ltd., London (1970); Poling, G. W., *Corrosion Sci.*, **10**, 359 (1970)
201. Bhatt, I. M., Soni, K. P. and Trivedi, A. M., *Werkst. u. Korrosion*, **18**, 968 (1967)
202. Tinley, W. H., *Chem. and Ind.*, Dec. 12, 2036 (1964); Obrecht, M. F., *Proc. 2nd Internat. Cong. Met. Corrosion*, New York (1963), N.A.C.E., Houston, 624 (1966)
203. Desai, M. N., Rana, S. S. and Gandhi, M. H., *Anti-Corrosion Methods Mater.*, **17** No. 6, 17 (1970)
204. Desai, M. N., Shah, Y. C. and Gandhi, M. H., *Australasian Corros. Engng.*, **12** No. 3, 3 (1968); Desai, M. N. and Shah, Y. C., *Werkst. u. Korrosion*, **21**, 712 (1970)
205. Rance, V. E. and Evans, U. R., *Corrosion and its Prevention at Bimetallic Contacts*, H.M.S.O., London (1956).
206. Gilbert, P. T., Historical Metallurgy Soc. Conference, Birmingham, Paper Copper 4 (1984)
207. LaQue, F. L., *Marine Corrosion, Causes and Prevention*, John Wiley & Sons, New York (1975); Schumacher, M., *Sea Water Corrosion Handbook*, Noyes Data Corp., Park Ridge, N.J. (1979); Gilbert, P. T., *Mater. Performance*, **21** (2), 47 (1982)
208. Gilbert, P. T., *Proc. 6th Internat. Cong. Met. Corrosion*, Sydney, Australia (1975); Joncheray, D. Guegan, F., and Groix, F., Metaux Corrosion Ind., No. 644,140 (1979); Toscer, G., Metaux Corrosion Ind., No. 606,68 (1976); Sato, S. and Nagata, K., Sumitomo Light Metal Tech. Rep., **19** (3, 4), 83 (1978)
209. Popplewell, J. M., N.A.C.E. Nat. Conference, Houston, Paper No. 21 (1978); Syrett, B. C., *Corrosion*, **32**, 242 (1976); Syrett, B.C., and Coit, R.L., *Mater. Performance*, **22** (2), 44 (1983)
210. Henrikson, S., and Knuttson, L., *Brit. Corrosion J.*, **10**, 128 (1975); Efird, K. D., *Corrosion*, **33**, 3 (1977); Lush, P. A., Hutton, S. P., Rowlands, J. C. and Angell, B., *Proc. 6th European Cong. Met. Corrosion*, London, p. 137 (1977); Lush, P. A., Hutton, S. P., Rowlands, J. C. and Angell, B. *Proc. 5th Internat. Cong. Marine Corrosion and Fouling*, Barcelona, p. 200 (1980)
211. Effertz, P. H. and Fichte, W., *Der Maschinenschaden*, **49**, 163 (1976); VGB Kraftwerkstechnik, **57**, 116 (1977); Sato, S., and Okawa, M., *Sumitomo Light Metal Tech. Rep.,* **17**, (1, 2), 17 (1976); Sato, S., Nosetani, T., Yamaguchi, Y. and Onda, K., *Sumitomo Light Metal Tech. Rep.*, **16** (1, 2), 23 (1975); Gilbert, P. T., Chem. and Ind.

Supplement 2 No. 13, 37 (1977); Hack, H. P., and Gudas, J. P., *Mater. Performance*, **18** (3), 25 (1979); **19** (4), 49 (1980); Heaton, W. B., *Brit. Corrosion J.*, **12**, 15 (1977); **13**, 57 (1978); Henrikson, S., Asberg, M. and Holm, R., *Proc. 8th Scandinavian Corrosion Conference*, Helsinki (1978)
212. Kawake, A., Ikushima, Y., Iijuma, S., Sato, S. and Nagata, K., *Sumitomo Light Metal Tech. Rep.*, **18** (3, 4), 1 (1977)
213. Elmer, K., Edison Electrical Institute Power Station Sub-Committee Conference (April 1975)
214. Lo, B. K., *Electrotechnik*, **53**, 831 (1979)
215. Francis, R., *Brit. Corrosion J.*, **18**, 35 (1983)
216. Francis, R., *Mater. Performance*, **21** (8), 44 (1982)
217. Niederberger, R. B., Gudas, J. P. and Danek, G. J., N.A.C.E. Nat. Conference, Houston, Paper No. 76 (1976); MacDonald, D. D., Syrett, B. C. and Wing, S. S., *Corrosion*, **35**, 367, 409 (1979); Syrett, B. C. and Wing, S. S., N.A.C.E. Nat. Conference Chicago, Paper No. 33 (1980); Efird, K. D. and Lee, T. S., *Corrosion*, **35**, 79 (1979); Gudas, J. P., and Hack, H. P., *Corrosion*, **35**, 67 (1979); Gudas, J. P., and Taylor, D. W., *Corrosion*, **35**, 259 (1979); Efird, K. D. and Lee, T. S., N.A.C.E. Nat. Conference, Houston, Paper No. 24 (1978); De Sanchez, S.R. and Schiffrin, D. J., *Corrosion Sci.*, **22**, 585 (1982); Schiffrin, D. J. and De Sanchez, S. R., *Corrosion*, **41**, 31 (1985)
218. Richter, H., *Werkstoffe Korrosion*, **28**, 671 (1977); Drolenga, L.J.P., Ilsseling, F. P. and Kolster, B. H., *Werkstoffe Korrosion*, **34**, 167 (1983)
219. Efird, K. D. and Anderson, D. B., Mater. Performance, **14** (11), 37 (1975)
220. Gilbert, P. T., *Brit. Corrosion J.*, **14**, 20 (1979)
221. Ijsseling, F. P., Krougman, J. M. and Drolenga, L. J. P., *Proc. 5th Internat. Cong. Marine Corrosion and Fouling*, Barcelona, p. 146 (1980)
222. Vreeland, D. C., *Mater. Performance*, **15** (10), 38 (1976); Gilbert, P. T. and North, W., *Trans. Inst. Mar. Engrs.*, **84**, 9 (Mater. Section Symposium) (1972)
223. Nicholson, R. B. and Todd, B., *Metallurgist Mater. Tech.*, **12**, 302 (1980)
224. Gilbert, P. T., *Metallurgist Mater. Tech.*, **10**, 316 (1978); Gilbert, P. T., *Proc. Internat. Symposium Corrosion and Protection Offshore*, Cefracor, Paris (1979); Copper Development Assoc. (UK), *Copper Alloys for Offshore Tech.*, Publication CM-L39 (1976); Lim, L. H., Conference Offshore Europe '77, Aberdeen, Paper No. 36680 (1977)
225. Efird, K. D., *Mater. Performance*, **15** (4), 16 (1976); Chandler, H. E., *Metals Progress*, **115**, 47 (1979)
226. Manzolillo, J. L., Thiele, E. W. and Tuthill, A. H., Soc. Naval Arch. and Mar. Engrs. Conference, New York (1976); Obrzut, J. J., *Iron Age*, **220**, 36 (1977); **222**, 35 (1979); Anon., *Metal Construction*, **11**, 181 (1979)
227. Moreton, B. B. and Glover, T. J., *Proc. 5th Internat. Cong. Marine Corrosion and Fouling Barcelona*, p. 267 (Biology) (1980)
228. Prager, M. and Thiele, E. W., *Welding J.*, p. 17 (July 1979); Middleton, L. G., R. Inst. Naval Arch/Copper Development Assoc. Symposium (Jan. 1980), London; Schorsch, E., Bicicchi, R. T. and Fu, J. W., *Trans. Soc. Naval Arch. & Marine Engrs.*, **86** (1978); Moreton, B. B., *Metallurgist Mater. Tech.*, **13**, 247 (1981)
229. Internat. Copper Research Assoc. Newsletter No. 9, p. 1 (1979)
230. Desalination Materials Manual, Dow Chem. Co. for US Office of Water Research & Tech. (1975); Materials Failure Identification Manual for Sea Water Desalination Plants, Aqua-Chem Inc. for US Office of Water Research & Tech., O.R.N.L. (1976); Todd, B., *Middle East Water & Sewage J.* (Oct./Nov. 1977); Temperley, T., *Desalination*, **33**, 99 (1980); Hill, K., *Desalination*, **25**, 111 (1978); Sato, S. *Trans. Japan Inst. Metals*, **19**, 575 (1978); Sato, S. and Nagata, K., *Sumitomo Light Metal Tech. Rep.*, **18** (1, 2), 11 (1977); Schrieber, C. F., Boyce, T. D., Oakes, B. D. and Coley, F. H., *Mater. Performance*, **14** (2), 9 (1975); Oakes, B. D., *Mater. Performance*, **15** (1), 44 (1976); Ross, R. W. and Anderson, D. B., *Mater. Performance*, **14** (9), 27 (1975); Ross, R. W., *Mater. Performance*, **18** (7), 15 (1979)
231. Sato, S. and Nagata, K., *Sumitomo Light Metal Tech. Rep.*, **15**, 174 (1974); Bianchi, G., Mazza, F., Sivieri, E. and Torchio, S., *Proc. 6th European Cong. Met. Corrosion*, London, p. 271 (1977); Torchio, S., *Corrosion Sci.*, **21**, 59, 425 (1981)
232. Campbell, H. S., *Brit. Corrosion J.*, **18**, 206 (1983)
233. Rowlands, J. C. and Brown, T. R. H. M., *Proc. 4th Internat. Cong. Marine Corrosion & Fouling*, Juan-les-Pins, p. 475 (1976); Culpan, E. A. and Rose, G., *Brit. Corrosion J.*, **14**, 160 (1979); Ferrara, R. J. and Caton, T. E., *Mater. Performance*, **21** (2), 30 (1982)

234. Gilbert, P. T., *Proc. 5th Internat. Cong. Marine Corrosion & Fouling*, Barcelona, p. 210 (1980); Scholes, I. R., Astley, D. J. and Rowlands, J. C., *Proc. 6th European Cong. Met Corrosion*, London, p. 161 (1977)
235. Page, G. G., *New Zealand J. Sci.*, **26**, 415 (1983)
236. Ijsseling, F. P., Krougman, J. M. and Drolenga, L.J., *Proc. 5th Internat. Cong. Marine Corrosion & Fouling*, Barcelona, p. 146 (1980); Efird, K. D., *Corrosion*, **31**, 77 (1975); Efird, K. D., *Corrosion*, **33**, 347 (1977); MacDonald, D. D., Syrett, B. C. and Wing, S. S., *Corrosion*, **34**, 289 (1978); Ijsseling, F. P. and Krougman, J. M., *Proc. 6th European Cong. Met. Corrosion*, London, p. 181 (1977); Ilsseling, F. P. and Krougman, J. M., *Proc. 4th Internat. Cong. Marine Corrosion & Fouling*, Juan-les-Pins (1976); Castle, J. E., *Corrosion Sci.*, **16**, 3 (1976); Epler, D. C. and Castle, J. E., *Corrosion*, **35**, 451 (1979); Shone, E. B., Brit. Corrosion J., **10**, 33 (1974)
237. Kato, C., Ateya, B. G., Castle, J. E. and Pickering, H. W., *J. Electrochem. Soc.*, **127**, 1890, 1897 (1980); Kato, C. and Pickering H. W., *J. Electrochem. Soc.*, **131**, 1225 (1984)
238. Parker, J. G. and Roscow, J. A., *Brit. Corrosion J.*, **16**, 107 (1981)
239. *Proc. Internat. Symposium Corrosion of Copper & Copper Alloys in Building*, Tokyo, Japanese Copper Development Assoc., 12 papers (1982); Cornwall, F. J., Wildsmith, G. and Gilbert, P. T., *Brit. Corrosion J.*, **8**, 202 (1973); A.S.T.M. Spec. Tech. Pub. 576 p. 155 (1976)
240. Bowers, J. E., Oseland, P. W. and Davies, G. C., *Brit. Corrosion J.*, **13**, 177 (1978)
241. Terwinghe, F., Celis, J. P. and Roos, J. R., *Brit. Corrosion J.*, **19**, 115 (1984)
242. Sparks, J. M. and Scully, J. C., *Corrosion Sci.*, **16**, 619 (1974); Kermani, M. and Scully, J. C., *Corrosion Sci.*, **18**, 833 (1978); **19**, 89, 489 (1979); Scully, J. C., *Metal Sci.*, **12**, 290 (1978); *Corrosion Sci.*, **20**, 297 (1980); Takano, M. and Staehle, R. W., *Trans. Japan Inst. Metals*, **19**, 1 (1978); Takano, M., *Trans. Japan Inst. Metals*, **18**, 787 (1977); *Corrosion*, **30**, 441 (1974); Kawashima, A., Agrawal, A. K. and Staehle, R. W., A.S.T.M. Spec. Tech. Pub., **665**, p. 266 (1979); Linder, M. and Mattsson, E., *Proc. 6th European Cong. Met. Corrosion*, London (1977); Uhlig, H., Gupta, K. and Liang, W., *J. Electrochem. Soc.*, **122**, 343 (1975); Holroyd, N. J. H., Hardie, D. and Pollock, W. J., *Brit. Corrosion J.*, **17**, 103 (1982)
243. Gupta, P., Chaudhary, R. S. and Prakash, B., *Brit. Corrosion J.*, **18**, 98 (1983); Walker, R., *Corrosion*, **20**, 290 (1973); Lewis, G., *Brit. Corrosion J.*, **16**, 169 (1981); Subramanyan, N. C., Sheshadri, B. S. and Mayanna, S. M., *Brit. Corrosion J.*, **19** (4), 177 (1984)
244. Conference 'Copper Alloys in Marine Environments' Birmingham UK, April 1985. Copper Development Association. 20 papers
245. Conference 'Marine Engineering with Copper-Nickel' London April 1988. Institute of Metals. 12 papers
246. Parvizi, M. S., Aladjem, A. and Castle, J. E., *Internat. Materials Reviews*, **33** (4), 169 (1988)
247. Geesey, G. G., Lewandowski, Z. and Fleming, H-C., *Biofouling/Biocorrosion in Industrial Water Systems*. Lewis Publishers Inc. Chelsea, Michigan (1993) or 4)

4.3 Lead and Lead Alloys

Introduction

Lead forms a series of relatively insoluble compounds, many of which are strongly adherent to the metal surface. In conditions where a stable continuous film can form, further reaction is often prevented or greatly reduced. Thus the general good corrosion resistance of lead results from the formation of relatively thick protective films of corrosion product.

The major uses of lead in the UK are in batteries, and in sheet and pipe of which the vast majority is sheet for building purposes. These applications account for about one third each of lead used. This situation is unique, since in all other countries batteries account for most of the lead market. A small but very important application is sheet and pipe for the chemical industry. Lead is no longer installed for water services. Lead cable sheathing which accounts for 5% is in general decline, but is valued in niche applications such as on oil rigs where resistance to hydrocarbons is important. The use of lead for anodes accounts for a very small tonnage, but is still of great importance to the industries which use them.

Lead sheet is used in the building industry throughout continental Europe and to a lesser extent Australia, but hardly at all in the USA. Other aspects of lead consumption follow the same general trends worldwide.

Composition and Mechanical Properties

Lead

While lead of purity in excess of 99.99% is commercially available, it is very rarely used owing to its susceptibility to grain growth and fatigue failure by intercrystalline cracking, and indifferent mechanical properties. Because of its generally superior corrosion resistance, pure lead to BS 334:1982 type A, shown in Table 4.13, is occasionally used in chemical plant, but only if there is no suitable alternative.

Lead Alloys

Besides Type A lead, nine lead alloys are specified in British Standards for various purposes. Their compositions and impurity limits are given in Table 4.13. In addition, alloys for batteries and for anodes are of importance. In due course it is likely that European standards will supersede the current national ones.

Of the elements commonly found in lead alloys, zinc and bismuth aggravate corrosion in most circumstances, while additions of copper, tellurium, antimony, nickel, silver, tin, arsenic and calcium may reduce corrosion resistance only slightly, or even improve it depending on the service conditions. Alloying elements that are of increasing importance are calcium, especially in maintenance-free battery alloys and selenium, or sulphur combined with copper as nucleants in low antimony battery alloys. Other elements of interest are indium in anodes[1,2], aluminium in batteries[3], and selenium in chemical lead as a grain refiner[108].

BS 334:1982 Compositional limits of chemical lead defines the composition of five grade of lead (Types A, B1, B2, B3 and C) and also gives guidance on selection, a method for the determination of creep strength and an empirical test for corrosion resistance, the flash test*, which, however, does not guarantee compliance with the standard. Chemical analysis is always to be preferred. It has been shown that flash points rise slightly with increasing copper and tellurium, remain constant with small additions of silver and fall with bismuth, zinc, tin and antimony[4]. Type A lead should only be used in a vibration-free environment and where the superior corrosion resistance is of paramount importance. For general chemical plant use, type B1 copper lead is to be preferred on account of its much greater structural stability, especially at elevated temperatures. Its mechanical properties are also significantly better. Type B2 copper tellurium lead has extremely good fatigue resistance which is retained to a greater extent at elevated temperatures than type B1. The main effect of tellurium is to form a fine-grained uniform grain structure, to enhance work hardening, and to delay recrystallisation. The silver content in type B3 also delays recrystallisation and promotes a large-grained stable structure which is creep-resistant[5,6]. Type C antimonial lead is used for valves, pump bodies and fatigue-resistant applications, but is not suitable for use at temperatures above 60°C owing to a rapid increase in creep rate, or in sulphuric acid concentrations above 60%.

BS 801:1984 Composition of lead and lead alloy sheaths for electric cables gives compositional requirements for lead and three alloys, B, E and 1/2 C. The impurity limits for lead are more relaxed than for type A lead, but lead to this grade can also be prone to intercrystalline cracking, which has been observed in the transport of cables as well as in service. Alloy E contains tin and antimony, alloy B 0.85% antimony, and 1/2 C tin and cadmium. Alloy B is suitable for use in environments where severe vibration is

---

* A test for evaluating the chemical quality of grades A, B1, B2 and B3 by resistance to H_2SO_4. A specimen is placed in 95–96% H_2SO_4 and the temperature raised to 300°C in 7 min. The 'flashing' is due to a sudden increase in the rate of formation of $PbSO_4$ and should not happen below 285°C, or 300°C for lead for use at elevated temperatures.

Table 4.13 Composition of types of lead and lead alloys commonly used in the United Kingdom

Specification	Use	Alloying element[1] or impurities (%)											
		Sn	Sb	Ag	Cu	Ni	Fe	Bi	Cd	Zn	Te	As	S	Pb
BS 334:1982 Type A	Specialised chemical applications	0·001	0·002	0·002	0·003	0·001	0·003	0·005	Trace[2]	0·002	—	Trace	Trace	99·99 min
BS 334 Type B1 Copper–Lead	General chemical applications	0·001	0·002	0·002	0·050 0·070	0·001	0·003	0·005	Trace	0·002	—	Trace	Trace	Balance
BS 334 Type B2 Copper–Tellurium–Lead	Specialised chemical applications	0·001	0·002	0·002	0·050 0·070	0·001	0·003	0·005	Trace	0·002	0·020 0·050	Trace	Trace	Balance
BS 334 Type B3 Copper–Silver–Lead	Specialised chemical applications	0·001	0·002	0·003 0·005	0·003 0·005	0·001	0·003	0·005	Trace	0·002	—	Trace	Trace	Balance
BB 334 Type CL	Specialised chemical applications	0·001	2·5 11	0·01	0·01	0·005	0·003	0·015	Trace	0·002	—	0·01	Trace	Balance

Specification	Application											
BS 801:1984 Lead	Cable sheathing	0·35	0·15	0·005	0·06	—	—	0·05	0·02	0·002	0·005	0·005	—	Balance[3]
BS 801:1984 Alloy B	Cable sheathing	0·01	0·80 0·95	0·005	0·06	—	—	0·05	0·02	0·002	0·005	0·005	—	Balance[3]
BS 801:1984 Alloy E	Cable sheathing	0·35 0·45	0·15 0·25	0·005	0·06	—	—	0·05	0·02	0·002	0·005	0·005	—	Balance[3]
BS 801:1984 Alloy 1/2C	Cable sheathing	0·18 0·22	0·005	0·005	0·06	—	—	0·05	0·06 0·09	0·002	0·005	0·005	—	Balance[3]
BS 1178:1982	Sheet for building	0·005	0·01	0·01	0·03 0·06	—	—	0·05	—	0·05	—	—	—	Balance[3]

1. One figure indicates a maximum impurity level. Two figures show maximum and minimum content of alloying additions.
2. Trace is defined as less than 0·005%
3. Total other elements 0·01% max.

expected, alloy E is somewhat resistant to vibration and, as previously mentioned, unalloyed lead is not at all resistant. These materials are less corrosion resistant than chemical lead, but their performance is adequate in underground or marine environments.

BS 1178:1982 Milled lead sheet for building purposes lays down requirements for composition, structure, thickness, freedom from defects, width and length, and marking. The specified copper content stabilises the structure of the material, conferring resistance to thermal fatigue cracking caused by grain growth and thermal cycling.

Lead acid batteries currently use antimonial alloys of a range of compositions or lead–calcium–(tin) alloys depending on the application. They are proof against the comparatively weak acid used and offer good resistance to oxygen evolved during charging, but have a variety of advantages and disadvantages which are covered later.

Corrosion Behaviour

The standard electrode potential, $E^0_{Pb^{2+}/Pb} = -0.126\,V$[7,8], shows that lead is thermodynamically unstable in acid solutions but stable in neutral solutions. The exchange current for the hydrogen evolution reaction on lead is very small ($\sim 10^{-13} - 10^{-11}\,A\,cm^{-2}$), but control of corrosion is usually due to mechanical passivation of the local anodes of the corrosion cells as the majority of lead salts are insoluble and frequently form protective films or coatings.

Anodic Behaviour

Lead is characterised by a series of anodic corrosion products which give a film or coating that effectively insulates the metal mechanically from the electrolyte (e.g. $PbSO_4$, $PbCl_2$, Pb_3O_4, $PbCrO_4$, PbO, PbO_2, $2PbCO_3.Pb(OH)_2$), of which $PbSO_4$ and PbO_2 are the most important, since they play a part in batteries and anodes. Lead sulphate is important also in atmospheric passivation and chemical industry applications.

In an aqueous electrolyte, the anodic behaviour of lead varies greatly depending on the conditions prevailing. Extensive reviews of the anodic behaviour of lead have been produced[10-12, 19, 20, 21]. Under certain conditions, the passive film may be converted to lead dioxide which has an electronic resistivity of $1 - 4 \times 10^{-4}$ ohm cm. Two polymorphs of PbO_2 exist, α and β. Both are non-stoichiometric and on a lead substrate there is always an oxygen deficiency[29, 30] An upper limit of n in PbO_n has been given as $1\cdot99$ and various lower limits of $1\cdot938$, $1\cdot875$ and below have been given (25, 30, 137, 138). In practice there is a variation in oxygen content owing to resistance to diffusion of 0 and 0^- species through the film[20, 25]. The structure of lead oxide films can be very complex and detailed studies have been undertaken[6, 9-11, 15-17, 25, 27, 28, 39], often using recently developed techniques[26]. Figure 4.13 shows the regions of thermodynamic stability for the compounds which can form in the $Pb-H_2SO_4-H_2O$ system.

PbO_2 has a low overpotential for the liberation of oxygen from H_2SO_4[32] and KOH[28] solutions, and for chlorine[33].

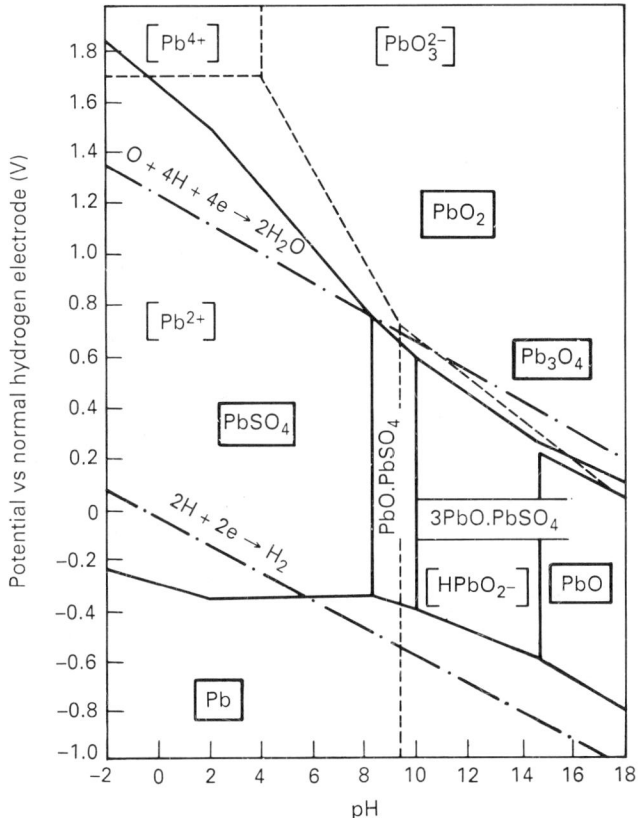

Fig. 4.13 Potential-pH diagram of lead in the presence of sulphate ions at unit activity and 25°C (after Ref. 139), reproduced by permission of Pergamon Press.

Cathodic Behaviour

Cathodic disintegration can occur with lead, observable as a grey cloud of fine metal particles. Hydrogen evolved on the surface of the lead can be absorbed if the current density is sufficiently high[34,35]. Above this level, 'avalanche penetration' can occur, leading to the formation of lead hydride, which leads to disintegration in the manner described[37]. Electrochemical implantation of alkali metals can also lead to disintegration[36].

Thermodynamics of the Pb-H₂O, Pb-H₂O-X systems

Pourbaix et al.[38] have studied the $Pb-H_2O$, $Pb-H_2O-X$ systems where X is a non-metal, and have established the domains of thermodynamic stability of lead, lead cations and anions, and insoluble compounds of lead. Figure 4.15 shows the Pb-H_2O system; it can be seen that in the region of high and low pH, corrosion occurs owing to the amphoteric nature of lead (cf. Zn, Al, Sn). This is a significant factor in the behaviour of lead in actual environments.

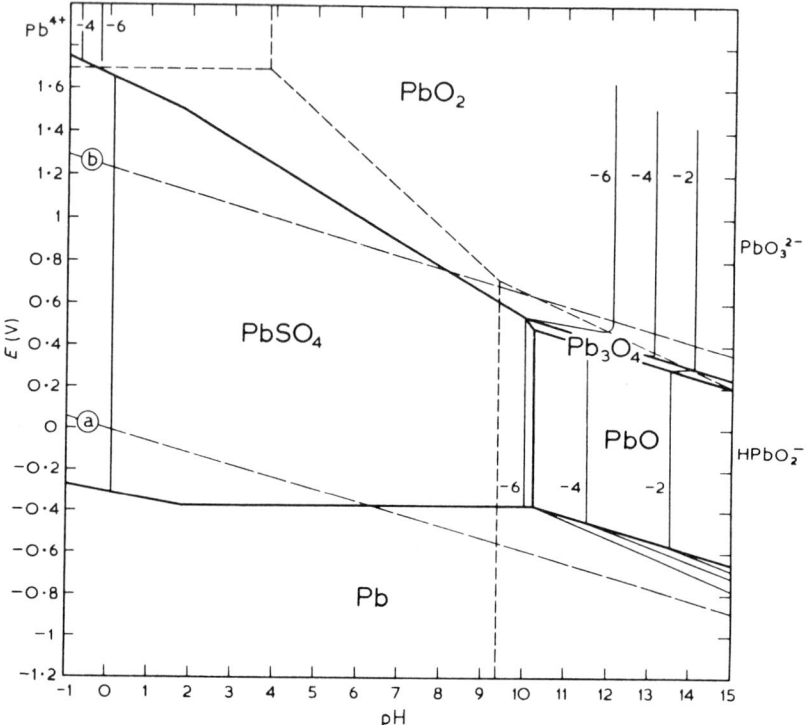

Fig. 4.14 Potential/pH diagram for the Pb-H_2O system. The area between ⓐ and ⓑ corresponds to the thermodynamic stability of water. Light lines represent equilibrium conditions between a solid phase and an ion at activities 1, 10^{-2}, 10^{-4} and 10^{-6}. Heavy lines represent equilibrium conditions between two solid phases. Broken lines represent equilibrium conditions between two ions for a ratio of these ions equal to unity (after Delahay, Pourbaix and van Rysselberghe[38])

In contrast with the Pb-H_2O system, it can be seen in Figure 4.13 that in the presence of SO_4^{2-} the corrosion zone in the region of low pH no longer exists, owing to the thermodynamic stability of $PbSO_4$. The Pb-H_2O-CO_2 system has been expressed in a similar pH/potential diagram[38] in which account has been taken of insoluble carbonates and basic carbonates of lead.

The predictions of the pH/potential diagram are generally fulfilled, but in very concentrated acid solutions, attack may diminish, owing to the relative insolubility of the relevant salt in the acid. Thus, lead nitrate, although soluble in water, has (owing to common ion effect) only slight solubility in concentrated nitric acid, and the corrosion rate is reduced. Similarly, lead chloride is less soluble in moderately concentrated hydrochloric acid than it is in water. In concentrated hydrochloric acid, however, the converse is the case, since in high chloride-ion concentrations lead forms soluble complex anions.

Inspection of products of corrosion and correlation with thermodynamic data frequently gives an indication of the cause of corrosion. Thus from Figure 4.14 it will be seen that a potential of about 1V (neglecting the influence of other ions which would be unlikely to decrease this value) is required for the formation of lead dioxide in neutral solution. This is somewhat higher than that likely to be generated by galvanic action between dissimilar metals commonly used in civil engineering, and the presence of lead dioxide among the corrosion products is usually taken as an indication that an impressed current is responsible for corrosion.

From an inspection of the more common compounds of lead, it will be seen that, in many environments, the corrosion product will be relatively insoluble (Table 4.14). Often, however, compact protective films are prevented from forming on the surface of the metal. The nature of the film is influenced by the mode of crystallisation, and in the case of the lower oxides for example, frequently little protection is afforded. Lead dioxide often forms a good adherent film, especially when it is produced from a sulphate film or other adherent compounds during anodic oxidation. When it is formed away from the surface by chemical reaction it gives no protection. It is a strong oxidising agent and, unlike the lower oxides, is not affected by most acids. Concentrated hydrochloric acid gradually dissolves it to form hexachloroplumbic acid, and with alkalis, plumbates are formed.

Table 4.14 Compounds of lead

Compounds of lead	Formula	Solubility at 25°C	Product colour
Acetate	$(CH_3COO)_2Pb$	soluble (55 g/100 ml)	white
Bromide	$PbBr_2$	$5 \cdot 7 \times 10^{-6}$	white
Carbonate	$PbCO_3$	$3 \cdot 3 \times 10^{-14}$	white
Basic carbonate	$2PbCO_3 \cdot Pb(OH)_2$	as above	white
Chloride	$PbCl_2$	1×10^{-4}	white
Chromate	$PbCrO_4$	$1 \cdot 8 \times 10^{-4}$	orange
Dioxide	PbO_2	insoluble	black or dark brown
Fluoride	PbF_2	$3 \cdot 7 \times 10^{-8}$	white
Formate	$Pb(CHO_2)_2$	soluble (1·6 g/100 ml)	white
Hydroxide	$Pb(OH)_2$	4×10^{-15}	white
Iodide	PbI_2	$1 \cdot 4 \times 10^{-8}$	yellow
Monoxide	PbO	insoluble	yellow-red
Nitrate	$Pb(NO_3)_2$	soluble (60 g/100 ml)	white
Phosphate	$Pb_3(PO_4)_2$	10^{-55}	white
Sulphate	$PbSO_4$	1×10^{-8}	white
Sulphide	PbS	$3 \cdot 4 \times 10^{-28}$	black
Sulphite	$PbSO_3$	insoluble	white
Triplumbic-tetroxide	Pb_3O_4	insoluble	red

With sparingly soluble salts of lead, the compactness of the deposits may be strongly influenced by the concentration of the relevant anion. Very low concentrations frequently resulting in imperfect coatings.

Atmospheric Corrosion

Lead is used for roofing, gutters, flashings, downspouts, etc. and exhibits excellent resistance to air (dry or humid). The sequence of patina formation is orthorhombic PbO → basic lead carbonate → normal lead carbonate → normal lead sulphite → normal lead sulphate[40]. The oxide is initially converted to plumbonacrite ($6PbCO_3.3Pb(OH)_2.PbO$) and hydrocerrusite ($2PbCO_3.Pb(OH)_2$)[41]. While these have an extremely low solubility, they can produce a white flocculant 'run-off' in wet weather, which can stain surrounding surfaces in the very early stages of exposure[42,43,75]. In marine environments, the initial film reacts with sodium chloride when wet to produce basic lead chloride and sodium hydroxide. This may result in corrosion of adjacent materials such as aluminium[42-43]. The lead patina stabilises, but takes approximately twice as long as in other atmospheric environments to do so. A common treatment for new lead is a resin-based patination oil which suppresses the formation of basic carbonates allowing the slow controlled growth of a strongly adherent normal carbonate patina from the outset[42,43,75]. Galvanic corrosion is not normally significant because the corrosion films formed are electrically insulating, although an isolated instance of severe galvanic corrosion of lead in contact with stainless steel in the presence of lime mortar has been reported.

Severe corrosion can be caused by organic acid fumes such as acetic or formic acids. These can be liberated by new wood, especially oak, and also by varnishes, glues, urea formaldehyde, plastics, fabrics and drying-oil paints, which can liberate fumes for a considerable time after application[47].

Water

Distilled Water

In distilled water free from dissolved gases, corrosion is slight though significant. The rate is increased by the presence of oxygen. With oxygen together with very small concentrations of carbon dioxide, very rapid corrosion takes place, with basic carbonates forming a white turbidity. At moderate CO_2 concentrations, a degree of passivation of the lead surface occurs, but corrosion is still significant. At high CO_2 contents, corrosion is increased due to the formation of soluble bicarbonate[48]. Lead is therefore not suitable for distilled water containers. It is, however, used for steam heating coils, but if the condensate is not recycled without access to air, rapid failure is likely.

Condensation corrosion is also a common cause of failure in lead-work on buildings. Trapped water is evaporated from and condensed on the underside of the lead during thermal cycling in the environment. This repeated condensation causes the production of lead oxide and lead hydroxide which is soluble and migrates away from the surface, leaving it unpassivated. Subsequent reaction with CO_2 in the atmosphere produces copious quantities of basic lead carbonate, resulting in blistering, perforation, and finally disintegration of the lead[43,126,127]. Adequate ventilation and adherence to codes of practice are essential to prevent this[128]. Water can be admitted through cracks caused by thermal fatigue, which is a consequence of overfixing the lead, using sheets which are too large, or of using lead containing insufficient copper.

Natural Waters

Because of the long life of lead pipework, water may be conveyed through existing lead pipes for some years to come. This may not be hazardous if the waters contain sufficient carbonate, sulphate or silicate (*see* Section 2.3), and are alkaline. The presence of 'aggressive carbonic acid'[49] or organic acids will render relatively hard waters plumbosolvent. Waters from peaty moorland frequently contain quinic acid from the roots of bilberry and heather which increases attack, as do aggressive agents such as nitrates and carbon dioxide in stagnant water. Rain water is also frequently plumbosolvent. Treatments given to water include deacidification with milk of lime, whiting, or by limestone bed, and removal of organic material with alkalis and aluminium sulphate. These treatments will encourage the formation of a protective carbonate scale[52,53]. Zinc orthophosphate treatment is also reported to control the dissolution of lead from pipes[44]. The permitted lead content of tap water in the EC is currently 50 $\mu g/l$[54] whilst in the USA it is 15 $\mu g/l$[46].

Lead usually has excellent resistance to seawater owing to the formation of a passive film of basic carbonate and carbonate-chloride double salts[55,70], which should be compared with its behaviour in solutions of alkali chlorides (see *salts* p. 4:87).

Underground Corrosion

Stray-current Corrosion

Stray currents are a source of damage to buried metal structures (see Section 10.5) and lead pipes and cable sheaths are particularly susceptible to it. Although lead can corrode under cathodic (alkaline) conditions, it is generally the anodic sites on the pipe or cable sheath which corrode. Lead is considered to be endangered if the current density is more than 25 mA m^{-2}[56]. This is influenced by the conductivity of the soil, which is largely determined by the moisture content, but may be affected by salting of roads in winter. The limit of corrosion may be considered as 100 metres from the current source[56]. Non-metallic links in pipework may break electrical continuity. This will produce more numerous corrosion sites, but they are frequently less intense. The surface will normally be covered with a mostly whitish corrosion deposit associated with either a smooth pitted surface or a more general rough etched appearance[57]. The corrosion product may comprise oxides, carbonates, hydroxides and chlorides[58]. Glassy watery crystals containing $PbCl_2.Pb(OH)_2$ and $PbCl_2.6PbO.2H_2O$ have been identified. The use of lead pipes for earths for alternating currents has also resulted in serious corrosion[58,59]. No protective coating is fully effective, but some give good protection[56,50,58,60-63].

Electrolytic corrosion may also occur on the inside of cable sheaths by the passage of current from the cable sheath to the wire[56,58].

Electrochemical Corrosion

Corrosion cells are established by inhomogeneity of the lead[69] or its environment, although severe corrosion due to metal composition is not common. 'Geological cells' formed between soils which differ in water content, degree of aeration, or the presence of various chemicals or bacteria can give rise to the passage of large corrosion currents at an e.m.f. of up to 1.5 V. Extensive long-term tests have been conducted on lead in soils[71-73, 76, 78]. The worst combination of soils is wet clay and cinders. The carbon in the cinder acts as an efficient cathode and severe anodic corrosion takes place in the clay environment. Moisture held in the clay permits the passage of relatively high currents. Anodic corrosion can occur when cables are in contact with dissimilar metals such as steel support racks or copper bonding ribbon. A new (clean) section of cable may also become anodic to an old (passivated) cable and can corrode.

Soils of high permeability are less aggressive since water tends to be mobile, so reducing concentration cells[77] and frequently drains readily to allow free movement of oxygen, thus reducing the effect of aeration cells. Both mechanisms give rise to pitting corrosion. Where oxygen circulates freely, a stable patina is often formed which is similar to that formed in air. Sandy soils tend to be among the best. Very large grained soils are normally good for the reasons given above, but under certain conditions severe localised pitting can be caused due to aeration cells[79]. Clays and silts tend to be worst. Cables are often laid in sand or crushed chalk.

Sulphates, silicates, carbonates, colloids and certain organic compounds[80] act as inhibitors if evenly distributed, and sodium silicate has been used as such in certain media. Nitrates tend to promote corrosion, especially in acid soil waters, due to cathodic de-polarisation and to the formation of soluble nitrates. Alkaline soils can cause serious corrosion with the formation of alkali plumbites which decompose to give (red) lead monoxide. Organic acids and carbon dioxide from rotting vegetable matter or manure also have a strong corrosive action. This is probably the explanation of 'phenol corrosion', which is not caused by phenol[81], but thought to be caused by decomposition of jute or hessian in applied protective layers[82-85].

Calcium hydroxide leached from incompletely cured concrete causes serious corrosion of lead (see Section 9.3). This is because carbon dioxide reacts with the lime solution to form calcium carbonate, which is practically insoluble. Carbonate ions are therefore not available to form a passive film on the surface of the lead[86]. Typically, thick layers of PbO are formed, which may show seasonal rings of litharge (tetragonal PbO) and massicot (orthorhombic PbO)[87, 88].

To prevent underground corrosion, lead is frequently protected with coatings of tar, bitumen, resin, etc., which are only effective if they completely insulate the metal from corrosive agents and stray currents. No coating is fully effective, but some give good protection[50, 56, 58, 60-63]. The most successful method used is cathodic protection[64] which for impressed currents, if correctly applied, can protect indefinitely (see Chapter 10). It is effective at a potential of $E^\circ = -0.8$ V[65] or about 0.1 V more negative than its equilibrium potential in the soil in question[66]. Both impressed currents[67] and sacrificial anodes have been used[68,72]. An excessively negative potential can increase the pH of the environment, thus causing corrosion. Caustic soda has also been observed from electrolysis of de-icing salt.

Patches of conductive lead sulphide can be formed on lead in the presence of sewage. This can result in the flow of a large corrosion current[89]. Sulphate-reducing bacteria in soils can produce metal sulphides and H_2S, which results in the formation of deep pits containing a black mass of lead sulphide[90]. Other micro-organisms may also be involved in the corrosion of lead in soil[64,91].

Cables are frequently laid in ducting for protection, but are still susceptible to corrosion by aeration cells set up between the cables and the duct walls, and to attack by corrosive solutions, especially from concrete ducts. They are also prone to corrosion by organic acids from wooden ducting, and to galvanic corrosion with iron supports. Damage by insects and animals may also occur[92].

Chemicals

Corrosion data reported as weight losses can be misleading because of the high density of lead; volume losses or yearly penetration figures are to be preferred for this metal. It should also be remembered that in chemical applications the thickness of lead used is usually greater than that of other metals, and higher corrosion rates, by themselves, are therefore not so serious.

Since lead is protected by relatively thick films of corrosion products, short-term tests can be misleading, as once the film has formed there will be a significant decrease in the corrosion rate.

Several sources of corrosion data are available, which should be consulted for specific information on corrosion resistance[93-97,129].

Gases

Lead will resist chlorine up to about 100°C[97], is used for dry bromine at lower temperatures[98] and is fairly resistant to fluorine[94]. Hydrofluoric acid does not passivate lead, so lead should not be used in this environment. Lead is very resistant to sulphur dioxide and fairly resistant to sulphur trioxide, wet or dry, over a wide temperature range[94].

Acids

Mineral Acids

Sulphuric acid is frequently made, stored and conveyed in lead. The corrosion resistance is excellent (*see* Figure 4.15) provided that the sulphate film is not broken in non-passivating conditions. Rupture of the film may be caused by erosion by high velocity liquids and gases containing acid spray.

In such an environment an inner lining of acid-resistant brick is often used. Thermal cycling may also disrupt the film. Acid of more than 85% concentration tends to dissolve the lead sulphate film, although lead has been used in cold quiescent conditions with concentrations of over 90%.

Nitrosylsulphuric acid, and nitrosyl chloride formed as a result of chloride in the water, can cause corrosion in sulphuric acid and lead-chamber plants. Alloying is not generally beneficial in this instance[99] and some elements (such as copper) can increase the corrosion rate.

Nitric acid readily attacks lead if dilute and the metal should not be used for handling nitrate or nitrite radicals except at extreme dilutions and preferably with a passivating reagent such as a sulphate, which will confer some protection. An example of this is the wash water from cellulose nitrate units. Corrosion decreases to a minimum at 65–70% HNO_3 and lead has been used for storage of nitric acid in the cold at this concentration[94, 101]. Resistance to a mixture of 98·85% H_2SO_4 and nitric acid of 1·50–1·52 S.G. can be excellent[101].

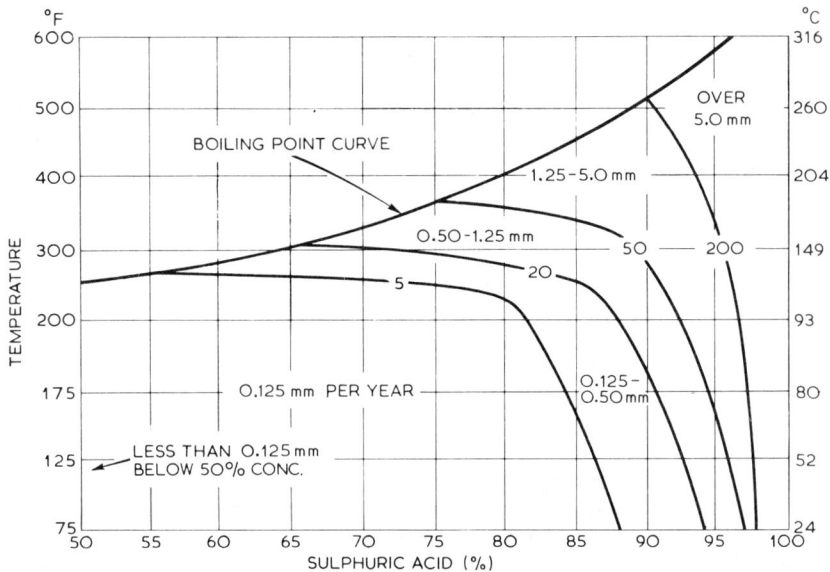

Fig. 4.15 Corrosion of lead by sulphuric acid as a function of temperature. Concentrations below 50% are not shown because resistance of lead is very good even at temperatures including boiling (after Fontana[100])

Hydrochloric acid should be regarded as aggressive to lead and its use cannot generally be recommended, although a satisfactory life has been obtained with acid of up to 30% concentration at ambient temperature and 20% concentration at 100°C. Antimonial lead is markedly more resistant[93, 97]. Resistance of lead to corrosion by HCl is presumably due to the formation of a protective film of lead chloride which is only slightly soluble at these concentrations combined with the rate-limiting effect due to the high hydrogen evolution overpotential of lead. With mixed hydrochloric and hydrofluoric acids for pickling steel, the behaviour of lead is uncertain. The life can be increased, however, by adding the hydrofluoric acid first to passivate the surface. The presence of aluminium fluoride can prevent the formation of a protective film and severe corrosion may result[102].

Phosphoric acid and chromic acid   Lead has good corrosion resistance to these acids. Its resistance to chrome-plating solutions will be discussed later.

Organic Acids

Acetic acid   Lead is attacked by most weak organic acids which produce water soluble lead salts, in the presence of air or organic oxidants. Lead is resistant to cold glacial acetic acid and is used for making storage vessels[93]. Aqueous acetic acid, solutions containing acetates, and acetic acid vapour rapidly corrode lead. The lead oxide protective film is dissolved, yielding salts which are carbonated in the presence of CO_2 and water to form basic lead carbonate, which in these circumstance does not form a passive film. In the absence of oxygen, corrosion in dilute solutions (0·01M or less) is slight[103]. Formic acid behaves in a manner similar to acetic acid. Lead is resistant to oxalic, tartaric and fatty acids only in the absence of oxygen.

Dilute (0·1N–0·001N) acetic, propionic, butyric, succinic and lactic acids all corrode lead to about the same extent. Pyruvic acid appears to inhibit corrosion after a short period of attack. In most cases the corrosion products are x $PbCO_3$, y $Pb(OH)_2$, the ratio of x to y being 2:1, and corrosion is intergranular[104].

Lead in building can be corroded by organic acids from new wood, decaying wood and lichens (*see* Sections 9.3 and 18.10). This is a common phenomenon with run-off from lichens which grow on tiles and slates. Where this occurs, a sacrificial strip of lead has been advocated[126].

Fuel oil   Organic acids are thought to be responsible for corrosion of lead by fuel oil. Formerly, mercaptans were held to be the cause, but now it is believed that naphthenic acids are responsible.

Lubricating oils   Bearings do not normally fail due to corrosion, but where this has occurred it has been associated with the acidity of white oils, the peroxide content and the presence of air. Peroxides are the controlling factor, but corrosion is reduced in the absence of air. The corrosion product consists of a basic lead salt of two or more organic acids[105] (*see* Section 2.11).

Alkalis

Lead is not particularly resistant to alkalis, but in some cases the corrosive action of sodium hydroxide and potassium hydroxide can be tolerated (KOH to 50% and up to 60°C, NaOH to 30% and 25°C, 10% and 90°C)[93]. The rapid attack of lime solutions is discussed earlier (*also see* Section 19.3).

Salts

Lead is not generally attacked rapidly by salt solutions (especially the salts of the acids to which it is resistant). The action of nitrates and salts such as potassium and sodium chloride may be rapid. In potassium chloride the corrosion rate increases with concentration to a maximum in 0.05M solution, decreases with a higher concentration, and increases again in 2M solution. Only loosely adherent deposits are formed. In potassium bromide adherent deposits are formed, and the corrosion rate increases with concentration. The attack in potassium iodide is slow in concentrations up to 0.1M but in concentrated solutions rapid attack occurs, probably owing to the formation of soluble $KPbI_3$. In dilute potassium nitrate solutions (0.001M and below) the corrosion product is yellow and is probably a mixture of $Pb(OH)_2$ and PbO, which is poorly adherent. At higher concentrations the corrosion product is more adherent and corrosion is somewhat reduced[106].

Details of the corrosion behaviour of lead in various solutions of salts are given in Figure 4.16.

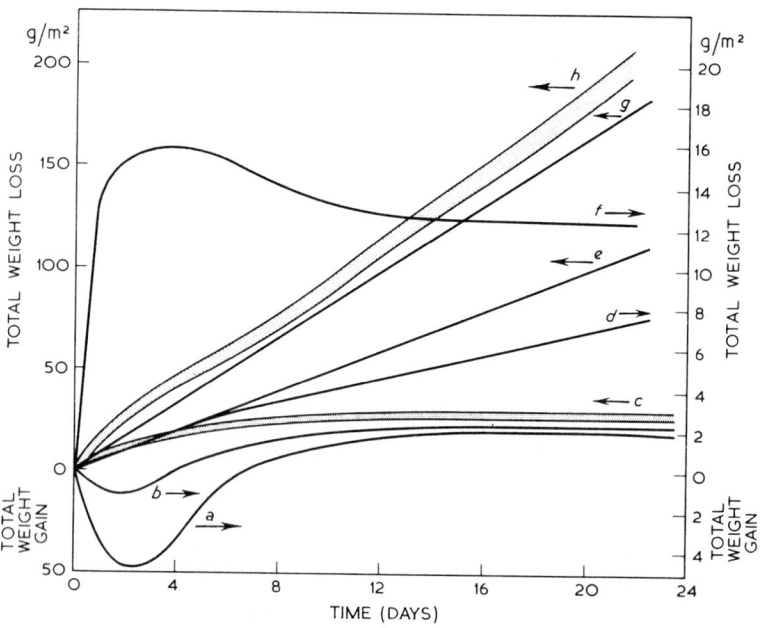

Fig. 4.16 Relationship between weight loss or weight gain and time for lead immersed in various environments (selected from Reference 107).

LEGEND

a 0·5 N $(NH_4)_2SO_4$; pH 2·9
b 0·5 N Na_2SO_4; pH 5·4
c 0·01 N NaOH; pH 12·8
   0·01 N $Ba(OH)_2$; pH 11·9
d 0·5 N NaCl; pH 4·9
e 0·5 N CH_3COONH_4 + NaOH; pH 8·9
f 0·25 N NaF; pH 6·3
g 0·5 N CH_3COONH_4 + CH_3COOH; pH 4·6
h 0·04 N $Ca(OH)_2$; pH 13
   0·5 N CH_3COONa; pH 16
   0·5 N $Ba(OH)_2$; pH > 13

Lead Anodes

Anodes for electroplating and for electrolysis of brine are frequently made of lead and lead alloys. Despite the formation of a passive film of lead dioxide (see anodic behaviour), there is generally a very slow continued corrosion which leads to thickening of the PbO_2 film. The tensional forces produced can cause growth of the anode. The film may also crack, releasing PbO_2 particles. Alloying elements are frequently added for strength or to stabilise the film. Rolled or extruded structures are generally more resistant to corrosion than cast.

In seawater, lead anodes with 1 or 2% silver may be used for cathodic protection of ships[13,110] at current densities of up to 120 A m^{-2} [112]. Lead with 6% antimony and 1% silver has also been recommended. It is thought that silver might provide small stable nucleation sites for PbO_2 formation[13,111] in a manner similar to the Pb/Pt bi-electrode[24,111] (see Section 11.3), which is serviceable at 250 A m^{-2}. A lead, 1% Ag, 0.5% Bi or 0.5% Te alloy with a platinum micro-electrode will perform well at 500 A m^{-2}.

In environments containing sulphuric acid, the introduction of cobalt ions into solution reduces the corrosion rate of pure lead markedly[115]. There is disagreement over the effectiveness of cobalt with antimonial alloys[113,114,118]. It is important to keep a PbO_2 electrode well above the $PbO_2/PbSO_4$ potential or rapid corrosion will occur[116]. Anodes for the electrowinning of copper have traditionally been made of antimonial lead, but new high purity processes have necessitated a change to lead-calcium-tin alloys[114,117]. The less porous nature of the PbO_2 layer reduces the amount of lead transferred to the cathode to 2-3 ppm. In the electrolytic recovery of zinc, the traditional anode has been made of an arsenical hypoeutectic lead-silver alloy known as Tainton lead. Lead-thallium has been reported to show good resistance[119], and interest in the excellent behaviour of lead-calcium-silver[120] has recently been revived[121]. Other alloys used are lead-tin and lead-silver-tin. Anodes made by sintering powdered lead, alone and with additions of cobalt and of silver have been reported to have good corrosion resistance[122]. More exotic approaches involving embedded catalytic particles[123] and catalytic particles in a semiconducting polymer coating[124] have also been suggested.

Chrome plating anodes, tanks and pipelines are normally made of lead containing 6% antimony despite occasional and sometimes spectacular failures. The electrolyte is usually chromic and sulphuric acids, with fluosilicic acid in the case of mixed catalyst baths. Tanks are sometimes treated anodically when new to produce a PbO_2 coating. If this is subsequently damaged, severe local corrosion can occur. Flame treatment of the surface of tank linings has been found to be beneficial, and treatment of anodes has been advocated. Rolled or extruded anodes are generally preferred. Corrosion is controlled by the sulphate concentration of the bath, with maximum corrosion at 4-9 g/l depending upon temperature. It has been found that 0·5 g/l of magnesium fluosilicate supresses corrosion without affecting the plating process[125]. With high efficiency baths, all alloys suffer rapid and severe pitting corrosion when no current is passing. Relatively little attack occurs with BS 334 type A, Type B2, type C (8%) and 7% Sn/Pb in mixed catalyst baths, whereas with simple sulphate-catalyst baths, type B2 and tin-lead are attacked at room temperature and at 40°C, and types A and C are attacked at room temperature only.

Battery Corrosion

The complex nature of the lead-acid battery is dealt with in several excellent sources[11, 14, 20, 45, 131]. Lead acid batteries typically consist of lead alloy supports which carry an electrochemically active mass, the composition of which differs between positive and negative plates, and with the state of charge of the battery. Failure normally occurs in the positive grids of a battery. The main cause of failure is loss of contact between the grid and the active mass due to 'grid growth' which is caused by the change in volume of the active material during the charge/discharge cycle, and by corrosion of the metal surface, which can be accelerated by stress.

The process of grid growth is restricted by utilising a sufficiently strong grid. This is achieved by a combination of appropriate alloy composition and physical characteristics of the grid such as dimensions of the grid wires. Grid growth is rapidly accelerated in the event of intergranular corrosion occurring.

The corrosion rate is greatest close to the reversible $PbO_2/PbSO_4$ potential as a result of a solid state reaction between PbO_2 and the underlying lead surface[116]. This corresponds to the rest or open circuit condition.

Passivation at the metal/active mass interface, or of the active mass itself can also lead to failure. Detrimental changes in the morphology of the active mass and microstructural changes in the grid material can also occur.

Traditionally battery grids have been made from lead with 6-14% antimony with a small amount of arsenic. While alloys in the region of 5-6% antimony are still used in some industrial, deep discharge and traction applications, high antimony contents have been largely replaced in automotive batteries by complex low antimony or lead-calcium-(tin) alloys. The reduction in antimony content has been made possible by the introduction of nucleants. It is increasingly common to find different alloys used in the positive and negative grids. Also expanded grids from wrought alloys are now widely used.

High antimony alloys exhibit high strength, good castability and give good deep cycling performance. The latter requires that the active mass has good adhesion to the metal, is structurally stable during cycling and does not passivate. Recent work has confirmed that antimony reduces shedding of active material[134], produces a surface film of greater porosity which becomes more porous during cycling, that it promotes stability of the active mass, and has shown that $PbSO_4$ is more reluctant to nucleate on antimonial lead[132]. Although corrosion rates may appear quite high, attack is normally of a general nature which allows a satisfactory service life. This is because the eutectic is preferentially corroded, which reduces intergranular corrosion. Antimony reduces the oxygen overpotential on the positive grid. Sb^{5+} ions can migrate from the positive grid to the negative and be reduced to metallic antimony[135]. This reduces the hydrogen overpotential, leading to excessive gassing, thus consuming water from the electrolyte, reducing charge efficiency and liberating stibine. During overcharge, antimony increases the rate of formation of the inner corrosion layer on the positive grid.

Low antimony alloys typically contain less than 3% antimony, with some alloys containing as little as 0.6%. The most commonly used alloys are 1.3-1.8% Sb. They always contain As to assist hardening, and a nucleating agent such as Se or S with Cu. These are necessary because the coarse dendritic structure is prone to porosity and hot cracking during casting. The addition of nucleating agents gives a fine grained structure with good corrosion resistance. Tin is often added to increase fluidity in casting alloys. The reduced antimony content allows the production of low-maintenance batteries which require the addition of water infrequently in the second half of their service life; corrosion is also reduced. There is still enough antimony present, however, to prevent premature failure by passivation of the positive grids.

Lead-calcium-(tin) alloys are used in maintenance-free automotive starting lighting and ignition (SLI) batteries, in stationary batteries[140, 142] and have been suggested for applications such as negative grids in some traction batteries[140]. It is essential that a correct calcium content and a suitable calcium-tin ratio is used. In the binary lead-calcium alloys, a fine grained structure with serrated grain boundaries is produced by a discontinuous precipitation reaction. It is thought that the serrated nature of the grain boundaries reduces the severity of intergranular corrosion. The addition of tin changes the nature of the precipitation reactions to give two areas of stability. One is with high calcium-low tin and the other is in the region below 1.8% tin and less than 0.07% calcium.

Outside these areas, a secondary precipitation reaction occurs which eventually gives a structure which is very susceptible to corrosion[136]; this is in the form of both deep penetrating corrosion and general corrosion. The secondary precipitation reaction takes place at the grain boundaries and can be initiated by a brief high temperature excursion after a considerable service life. Failure can subsequently occur in weeks. In the regions of stability, these alloys are very stable and exhibit high corrosion resistance. The behaviour of wrought lead-calcium-(tin) alloys is completely different but with the appropriate composition and processing conditions the final structure is extremely corrosion resistant[141].

Lead-calcium alloys containing tin are generally less corrosion resistant, but less prone to passivation. The formation of semiconducting SnO_2 in the film on the positive grid tends to reduce passivation caused by deep discharge. It is common, therefore, to use the ternary alloy for positive grids and binary for the negative, which saves on the considerable cost of tin. Despite the inferior deep-discharge performance and reduced stability of active mass structure, acceptable life can now be achieved in SLI batteries, and use in other areas is increasing. Batteries made from these alloys have a much reduced rate of self-discharge compared with antimonial alloys, thus giving a longer shelf-life, and maintain a high discharge voltage throughout their life.

P. C. FROST
E. LITTAUER
H. C. WESSON

REFERENCES

1. Hine, F., Ogata, Y., Yasuda, M. 'Consumption of Lead-silver Alloy Anodes in Sulfuric Acid, *B. Electrochem.*, **4**, 61-65 (1988)
2. Sumitoma Metal Industries Ltd Japan Pat. 5928598 (1984)
3. Prengaman, R. E., 'Structure Control of Non-Antimonial Lead Alloys via Alloy Additions, Heat Treatment and Cold Working, Pb80, *Ed. Proc. 7th Int. Lead Conf.*, Madrid, Lead Development Association, London (1983)
4. Kawabata, R., Miyase S. and Tagaya, M. 'Effects of Alloying Elements on Flash Points of Lead', *Trans. J.I.M.*, **5**, 85 (1964)
5. Rutter, J. W. and Aust, K. T., *Trans AIME*, **218**, 682 (1960)
6. Heubner, U. and Reinert, M., 'Effect of Small Silver Contents on the Characteristics of Lead and its Alloys', Pb80, *Seventh International Lead Conference*, Lead Development Association, London
7. Lingane, J. I., *Amer. Chem. Soc.*, **60**, 724 (1938)
8. Landolt-Bornstein, *Physikalisch-Chemische Tabellen and Erganzungsbande*, Berlin, Springer 1923, 1927, 1931, 1935, 1936
9. Bullock, K. R., 'Electrochemical and Spectroscopic Method of Characterising Lead Corrosion Films', *J. Electroanal. Chem.*, **222**, 347-366 (1987)
10. Von Fraunhofer, J. A. *Anti-Corros.* Nov. (1968) and Dec. (1968)
11. Kuhn, A. T. ed. *The Electrochemistry of Lead*, Academic Press, London (1979)
12. Pavlov, D., in *Procs. of the Symposium on Advances in Lead-Acid Batteries*, Bullock K. R. and Pavlov, D., (Eds), Electrochem. Soc. 110, (1984)
13. Shreir, L. L. and Hayfield, P. C. S., in *Cathodic Protection, Theory and Practice*, Ashworth, V. and Booker, C. J. L. Ellis Horwood, Chichester, 108 (1986)
14. Dasoyan, M. A. and Aguf, I. A., *Current Theory of Lead Acid Batteries*, Technology Ltd, Stonehouse, with ILZRO Inc. N.Y. 46 (1979)
15. Bullock, K. R., Trischan, G. M. and Burrow, R. G., 'Photoelectrochemical and Microprobe Laser Raman Studies of Lead Corrosion in Sulphuric Acid', *J. Electrochem. Soc.*, **130**, 1283 (1983)
16. Bullock, K. R. and Butler, M. A., 'Corrosion of Lead in Sulphuric Acid at High Potentials', *J. Electrochem. Soc.*, **133**, 1085 (1986)
17. Bullock, K. R. *J. Electrochem. Soc.* **127**, 662 (1980)
18. Caulder, S. M. and Simon A. C. in Refs. 11, 43.
19. Dawson, J. L. in Ref. 11, pp. 309
20. Pavlov, D. in McNicol, B. D. and Rand, D. A. J. eds, *Power Sources for Electric Vehicles*, Elsevier, Amsterdam 142 (1984)
21. Burbank, J., Simon A. C. and Willihnganz E. in Tobias C. W. ed., *Advances in Electrochemistry and Electrochemical Engineering,* **8**, Wiley Interscience, New York 157 (1971)
22. Thirsk, H. R. and Wynne-Jones, W. F. K. *Trans. Inst. Metal Finish*, **29**, 35 (1957)
23. Jones, P., Thirsk, H. R. and Wynne-Jones, W. F. K. *Trans. Faraday Soc.*, **52**, 1003 (1956)
24. Littauer, E. L. and Shreir, L. L. *Proceedings of the First International Congress on Metallic Corrosion, London (1961)*, Butterworths, London 374 (1962); Shreir L. L. *Corrosion*, **17**, 118t (1961)
25. Pavlov, D. and Rogachev, T. 'Dependence of the Phase Composition of The Anodic Layer on Oxygen Evolution and Anodic Corrosion of Lead Electrode in Lead Dioxide Potential Region', *Electrochim. Acta.*, **23**, 1237 (1978)
26. Bullock, K. R. 'Electrochemical and Spectroscopic Methods of Characterising Lead Corrosion Films, *J. Electroanal. Chem.*, **222**, 347 (1987)
27. Pavlov, D. and Popova, R. *Electrochim-Acta*, **15**, 1483 (1970)
28. Ruetschi, P. *J. Electrochem. Soc.*, **120**, 331 (1973)
29. Dini, J. W. and Helm, J. R. *Metal Finish*, **67**, (8/1969)
30. Beck, F. in Ref. 11, pp. 70
31. Hampson, N. A. in Refs. 11, 30
32. Jones, P. Lind, R. and Wynne-Jones, W. F. K. Trans. Faraday Soc., **50**, 972 (1954)
33. Littauer, E. L., PhD Thesis University of London, (1961).
34. Ives, D. G. and Smith, F. R. *Trans. Faraday Soc.*, **63**, 217 (1967)
35. Smith, F. R. *Disc. Faraday Soc.*, **56**, 113 (1973)
36. Hayes, M. and Kuhn, A. T. in Ref. 11, pp. 207

37. Salzberg, H. W. *J. Electrochem. Soc.*, **100**, 588 (1953)
38. Delahay, P., Pourbaix, M. and Van Rysselbergh, P. *J. Electrochem. Soc.*, **98**, 57 (1951)
39. Carr, J. P. and Hampson, N. A. 'The Lead Dioxide Electrode', *Chem. Rev.*, **72**, 6, 679–703 (1972)
40. Tranter, G. C. 'Patination of Lead: An Infra-red Spectroscopic Study', *Br. Corros. J.* **11**, 4, 222 (1976)
41. Olby, J. K. *J. Inorg, Nucl. Chem.*, **28**, 2507 (1966)
42. Hill, R. H., Frost, P. C. and Smith, R. 'Corrosion of Aluminium in Contact with Lead in Atmospheric Environments', in: *Pb80, Ed. Proc. 7th Int. Lead, Conf.*, Lead Development Association, London 194–203 (1983)
43. Hill, R. H., Frost, P. C. and Smith, R. 'Various Aspects of Weathering and Corrosion of Lead in Building Applications', in: *PB83 Ed. Proc. 8th Int. Lead Conf.*, Lead Development Association, London, 103 (1985)
44. Boffardi, B. P. and Sherbondi, A. M. 'Control of Lead Corrosion by Chemical Treatment, Paper 445, Corrosion '91 NACE Conf., Cinncinnati, Mar. 11-12 (1991)
45. Brown, H. E. *Lead Oxide Properties and Applications*, ILZRO, New York 194–234 (1985)
46. US Govt. 'Safe Drinking Water Act', (1974), Amendment (June 1991)
47. Brill, R. H. Ed. *Science and Archaeology*, MIT Press 91–99 (1971)
48. Hofmann, W. *Lead and Lead Alloys Properties and Technology*, Springer-Verlag, English Trans. by Lead Development Association, London 302 (1970)
49. Wesson, H. C. *Corrosion Prevention and Control,* **6**, 9, 12 (1959)
50. Evans, U. R. *An Introduction to Metallic Corrosion*, 2nd Ed. Arnold, London (1963)
51. Heap, J. H. *J. Soc. Chem. Ind.*, **32**, 771 (1913)
52. Hawkes, C. A. *Chem. Ind.*, **264**, (1944)
53. Miles, G. *J. Soc. Chem Ind.*, **67**, 10 (1948)
54. E.E.C. Directive 80/778/EEC, on the Quality of Running Water Intended for Human Consumption, July 1980.
55. Beccaria A. M. *et al.*, 'Corrosion of Lead in Sea Water', *Br. Corros. J.* **17**, 2, 87 (1982)
56. Haehnel, O. ETZ 60, 713 (1939)
57. Ref. 48, p. 315
58. Glander, F. and Glander, W. *Z. Metallk*, **44**, 97 (1953)
59. Amy, L. and Moujnos, C. *Rev. Gen. Elect.* **66**, 187 (1957)
60. Reiner, S. *Z. Metallkde*, **30**, 277 (1938)
61. Borel, I. *Bull. Ass. Suisse Electr.*, **28**, 54 (1937)
62. Gosden, J. H. *Chem. and Ind.*, 1069 (1956)
63. Radley, W. G. *Electr. Engng.* **57**, 168 (1938)
64. Uhlig, H. H. *The Corrosion Handbook*, 5th Ed., New York/London (1955)
65. Hornung, R. *Techn. Mitt. PTT.*, **31**, 265, 318 (1953)
66. Compton, K. G. *Corrosion*, (Houston), **12**, 37 (1956)
67. Doyle, E. J. *Corrosion*, (Houston), **11**, 17 (1955)
68. Robinson, H. A. and Featherly, R. L. *Corrosion*, (Houston), **3**, 349 (1947)
69. Ref. 48, p. 313
70. Beccaria, A. M. *et al.*, 'Investigation on Lead Corrosion Products in Sea Water and In Neutral Saline Solutions', *Werkstoffe und Korros.*, **33**, 416-420 (1982)
71. Romanoff, M. *Underground Corrosion*, NBS 579, National Bureau of Standards, April 227 (1957)
72. Robson, W. W. and Taylor, A. R. *Some Experiments in The Mechanism of Corrosion of Lead Pipes in Soils*, Report MM/19/54, Associated Lead Manufacturers Ltd. (1954)
73. Denison, I. A. and Romanoff, M. J. *J. Res. Nat. Bur. Stand*, **44**, 259 (1950)
74. Beavers, J. A., Koch, G. H. and Berry, W. E. *Corrosion of Metals in Marine Environments*, MCIC Report MCIC-86-50 (1982)
75. Cook, A. R. and Smith, R. 'Atmospheric Corrosion of Lead and Its Alloys', in Ailor, W. H. *Atmospheric Corrosion*, Wiley Interscience, New York (1982)
76. Logan, K. R. *Underground Corrosion*, Nat. Bur.-Stand, Circular C450 (1945)
77. Haase, L. W. *Werkst. U. Korros.* **2**, 90 (1951)
78. Burns, R. M. *Bell Syst. Tech. J.*, **15**, 603 (1936)
79. Burns, R. M. and Salley, W. J. *Indust. Eng. Chem.*, **22**, 293 (1930)
80. Ref. 48, p. 310
81. Ref. 48, p. 293
82. Radley, W. G. and Richards, C. E. *J. Inst. Elect. Engng.* **35**, 685 (1939)

83. Senez, J. C. and Pichinoty, F. *F. Corros. Anticorros*, **5**, 203 (1957)
84. Cole, E. L. and Davies, R. L. *Chem. and Ind.* (Rev.) **39**, 1030 (1956)
85. Bonde, G. and Lunn, B. *Ingen. Intern. Edit.* **2**, 103 (1958)
86. Ref. 48, p. 294
87. Brčić, B. S. and Šiftar, J. *Corrosion et Anticorrosion*, **6**, 342 (1958)
88. Wolff, E. F. and Bonilla, C. F. *Trans. Electrochem. Soc.*, **79**, 307 (1941)
89. Schmelling, E. L. and Roschenbleck, B. *Werkst. U. Korrosion*, Mannheim, **9**, 529 (1958)
90. Reinitz, B. B. *Corrosion*, (Houston), **9**, 425 (1953)
91. Wolzogen-Kuhr, C. V. **Water, 40**, 281 (1956)
92. Ref. 48, p. 317.
93. *Lead for Corrosion Resistant Applications*, Lead Industries Association Inc., New York (1974)
94. *Corrosion of Lead*, Lead Development Association, London (1971)
95. *Corrosion Resistance of Lead and Lead Alloys*, Chem. Eng. Feb (1953)
96. Rabald, E. *Resistance of Lead to Corrosives, from Corrosion Guide*, (1951)
97. *Lead in Modern Industry*, Lead Industries Association, New York (1952)
98. Frost, P. C. *The Corrosion of Lead and Lead Alloys in Bromine*, Report MM/4/81, Lead Industries Group Ltd., (1981)
99. Wickert, K. *Korros-Metallsch. Beih.* **20**, 147 (1944)
100. Fontana, M. G. *Industr. Engng. Chem.*, **43**, 9, 105A (1951)
101. Ref. 48, p. 291
102. Camuil, J. *Sci. Industr. Res.* **20**, 114 (1944)
103. Turnbull, D. and Frey, D. R. *J. Phys. Colloid Chem.*, **51**, 681 (1947)
104. Coles, E. L., Gibson, J. G. and Hinde, R. M. *J. Appl. Chem.* **8**, 341 (1958)
105. Wilson, B. S. and Garner, F. H. *J. Inst. Petrol.*, **37**, 225 (1951)
106. Vaivads, A. and Liepina, L. *Latv. P.S.R. Zinat. Akad. Vestis*, **8**, 119 (1954)
107. Katz, W. *Metalloberfläche*, **11**, A.161 (1953)
108. Heubner, U. and Reinert, M. *Development of Improved Lead Materials for Chemical Plant*, Pb80 7th Int. Lead Conf., Madrid, Lead Development Association, London 204–214 (1983)
109. Heubner, U. and Reinert, M. *Effect of Small Silver Contents on the Characteristics of Lead and its Alloys*, Pb80 7th Int. Lead Conf. Madrid, Lead Development Association, London 130–137 (1983)
110. Barnard, K. N., Christie, G. L. and Gage, D. G. 'Service Experience with Lead Silver Alloy Anodes in Cathodic Protection of Ships', *Corrosion*, **15**, 11, 581–586 (1959)
111. Peplow, D. B. and Shreir, L. L. 'Lead/Platinum Electrodes for Marine Applications', *Corr. Tech.* Apr. (1984)
112. Morgan, J. H. 'Lead Alloy Anode for Cathodic Protection', *Corr. Tech.* 10/12. 348–352 (1958)
113. Koch, D. F. A. *Electrochimica Acta*, **1**, 32 (1959).
114. Eggett, G. and Naden, D. 'Developments in Anodes for Pure Copper Electrowinning from Solvent Extraction Produced Electrolytes', *Hydrometallurgy*, Elsevier, Amsterdam, **1**, 123–137 (1975)
115. Lander, J. J. *J. Electrochem. Soc.*, **99**, 467 (1954)
116. Lander, J. J. *J. Electrochem. Soc.*, **103**, 1 (1956)
117. Prengaman, R. D. *Wrought Lead-Calcium-Tin Anodes for Electrowinning*, AIME Conference, Los Angeles, 28/2/84.
118. Gendron, A. S., Ettel, V. A. and Abe, S. 'Effect of Cobalt Added to Electrolyte on Corrosion Rate of Pb-Sb Anodes in Copper Electrowinning', *Canad. Met. Quart.*, **14**, 1, 59–61 (1975)
119. Ref. 48, p. 298.
120. Hanley, H. R., Clayton, C. Y. and Walsh, D. F. *Trans. AIME*, Yearbook 91, 275 (1930)
121. UK Pat. App. GB 2149424A (1983)
122. Kir' yakov, G. Z. and Dunaev, Yu. D. *Izvest. Akad. Nauk, Kazakh. S.S.R. Ser. Khim*, **2**, 32 (1957)
123. UK Pat. App. GB 2085031A (1980)
124. UK Pat. App. GB 2096643A (1981)
125. Carter, V. E. and Campbell, H. S. *J. Met. Finish.*, **8**, 103 (1962)
126. Corrosion of Lead Roofing, Interim Report for Ecclesiastical Architects; and Surveyors' Association (1986)
127. Joyce, S. J. Thesis, Brighton Polytechnic (1983)

128. Murdoch, R. 'The Lead Sheet Manual', Vols 1, 2, and 3, Lead Sheet Association, London, (1990, 1992, 1993)
129. Heubner, U. *et al.*, Metallgesellschaft 'Lead Handbook' (1983)
130. Ref. 48, pp. 347
131. Sharpe, T. F. in Bard, A. J. *Encyclopedia of Electrochemistry of the Elements*, **1**, 235 (1974)
132. Webster, S., Mitchell, P. J., Hampson, N. A. and Dyson, J. I. 'The Cycle Life of Various Lead Alloys in 5M H_2SO_4, *J. Electrochem. Soc.*, **1**, 133 (1986)
133. Peters, K. and Young, N. R. 'Some Aspects of Corrosion in Lead Acid Batteries, *I. Chem. E. Symposium Series*, **98**, 185 (1980)
134. Gibson, I. K., Peters, K. and Wilson, F. in Thompson, J. Ed., *Power Sources*, **8**, Academic Press, London 565 (1980)
135. Dawson, J. L., Wilkinson, J. and Gillibrand, M. I. in Collins, D. H. Ed., *Power Sources*, **3**, Oriel Press, Newcastle-upon-Tyne, 1-9 (1970)
136. Prengaman, R. D. 'Structural Control of Non-Antimonial Lead Alloys Via Alloy Additions, Heat Treatment and Cold Working', Pb80 Ed., *Proc. 7th Int. Lead Conf.*, Lead Development Association, London 34 (1983)
137. Bystrom, J. *Ark. Kemi, Mineral Geol.*, 20A, 11, (1945)
138. Katz, T. *Ann. Chim.*, (Paris), **5**, 5 (1950)
139. Barnes S. C. and Mathieson, R. T. in Simon, A. C. *Batteries* (2), Collins, D. H. Ed., Pergamon Press, New York 41 (1965)
140. Ref. 20 p. 212
141. St. Joe Minerals Corporation, U.K. Patent 1 338 823 (1970)
142. Prengaman, R. D. *Improvements in Alloys, Oxides and Expanders for Lead Batteries*, Lead Development Association, London 3 (1984)

4.4 Magnesium and Magnesium Alloys

Magnesium is a divalent metal and is silvery white in appearance. It is the eighth most abundant element and sixth most abundant metal. The atomic weight is 24·32 and the specific gravity of the pure metal 1·738 at 20°C. The structure is close packed hexagonal. The melting point is 650°C and the boiling point 1 107°C. The specific heat at 20°C is 1·030 kJ/kg °C and the thermal conductivity at 20°C is 157·5 W/m°C; the electrochemical equivalent is 0·126 mg/C. The standard electrode potential $E^{\circ}_{Mg^{2+}/Mg} = -2\cdot 37$ V, but in 3% sodium chloride the electrode potential is $-1\cdot 63$ V (vs S.C.E.), i.e. $-1\cdot 38$ V (vs S.H.E.).

For engineering purposes magnesium is rarely used in the unalloyed condition. Small percentages of aluminium, zinc, etc. as indicated in Table 4.15 are added to improve mechanical and other properties. Magnesium is itself used for alloying with other metals.

Applications of magnesium alloys are many and varied but their light weight — about two-thirds that of the aluminium alloys — and high strength-to-weight ratio have made them of particular interest to the aircraft and guided-weapons industries. In the transport industry too, their light weight is attractive. Other features for which they are noted are their high stiffness-to-weight ratio, their great ease of machinability, good casting qualities and high damping capacity.

In the nuclear engineering field special magnesium-base alloys are extensively used as canning materials for uranium in gas-cooled reactors.

The compositions of the more common magnesium-rich alloys used in Great Britain are given in Table 4.15. Similar alloys are in use in the USA and elsewhere and their American designation together with the equivalent British alloys are given in Table 4.16. Many of the casting alloys are given various simple heat treatments to improve their properties, while the wrought alloys can be obtained in a number of tempers.

As with other electronegative metals, the absence of serious corrosion of these alloys in ordinary industrial atmospheres is largely a result of the formation of protective films which inhibit further attack. Similarly, when serious corrosion does occur, or when it occurs after a period of successful use, it can usually be traced to a change in conditions of such a nature that protective films already formed have suffered dissolution or break-down. No alloying ingredients are known which effect any substantial improvement

Table 4.15 Nominal composition of magnesium-rich alloys

Designation		Al	Zn	Zr	Th	Rare earth metals	Fractionated rare earth metals	Be	Mn	Ag
Casting alloys	Z5Z	—	4·5	0·7	—	—	—	—	—	—
	RZ5	—	4·0	0·7	—	1·2	—	—	—	—
	TZ6	—	5·5	0·7	1·8	—	—	—	—	—
	MSR A	—	—	0·6	—	—	1·7	—	—	2·5
	MSR B	—	—	0·6	—	—	2·5	—	—	2·5
	ZRE1	—	2·2	0·6	—	2·7	—	—	—	—
	ZT1	—	2·2	0·7	3·0	—	—	—	—	—
	MTZ	—	—	0·7	3·0	—	—	—	—	—
	A8	8·0	0·5	—	—	—	—	—	0·3	—
	AZ91	9·5	0·5	—	—	—	—	—	0·3	—
	AZ91X	9·4	0·4	—	—	—	—	—	0·3	—
	C	7·5–9·5	0·3–1·5	—	—	—	—	0·0015	0·15 (min)	—
	AZG	6·0	3·0	—	—	—	—	—	0·3	—
	ZE63	—	6·0	0·6	—	2·5	—	—	—	—
Wrought alloys	ZW3	—	3·0	0·6	—	—	—	—	—	—
	ZW1	—	1·3	0·6	—	—	—	—	—	—
	ZW6	—	5·5	0·6	—	—	—	—	—	—
	ZTY	—	0·5	0·6	0·75	—	—	—	—	—
	AM503	—	—	—	—	—	—	—	1·5	—
	AZ31	3·0	1·0	—	—	—	—	—	0·3	—
	AZM	6·0	1·0	—	—	—	—	—	0·3	—
	AZ855	8·0	0·4	—	—	—	—	—	0·3	—
	ZM21	—	2·0	—	—	—	—	—	1·0	—
	ZM61	—	6·0	—	—	—	—	—	1·0	—

% Composition (remainder magnesium)

Table 4.16 Equivalent British and American designations for magnesium alloys

Wrought alloys		Cast alloys	
British designation	American designation (ASTM)	British designation	American designation (ASTM)
ZW6	ZK60A	Z5Z	ZK51A
AM503	M1A	RZ5	ZE41A
AZM	AZ61A	TZ6	ZH62A
AZ855	AZ80A	MSR	QE22A
ZM21	ZM21	ZRE1	EZ33A
ZM61	ZM61	ZT1	HZ32A
		MTZ	HK31A
		A8	AZ81A
		AZ91	AZ91
		ZE63	ZE63

in the general corrosion behaviour of the magnesium alloys; though manganese is usually regarded as beneficial this is probably because it offsets the deleterious effects of iron and other cathodic metals which may be present rather than because it makes any positive improvement in the resistance of the magnesium itself. All metals added to produce the various alloys are intended simply for the improvement of the physical and mechanical properties and not for any effect they may have on the corrosion behaviour. On the other hand, none of these additions has any very marked deleterious effect on the corrosion resistance of the alloys.

Since corrosion resistance depends on film formation, it follows that the behaviour of the alloys will vary considerably with the medium to which they are exposed. The corrosion of the metal is governed largely by the solubility and other characteristics of the film. Thus, magnesium fluoride is very insoluble in hydrofluoric acid and as a consequence magnesium does not dissolve in this acid. Initial attack forms a film of magnesium fluoride and even though the film is not impervious to other corrosive influences it effectively seals the metal against further reaction. In dilute aqueous hydrofluoric acid, attack may take place, and if so it will be of a pitting type similar to that which takes place in tap water. In fact the corrosion is due to the water itself and not to the acid. Magnesium sulphate on the other hand is readily soluble in dilute sulphuric acid and no protective film is formed when magnesium reacts with this acid; attack is rapid and continuous with evolution of hydrogen. It should be noted, however, that magnesium sulphate is only slightly soluble in *concentrated* sulphuric acid. When, therefore, magnesium is immersed in strong sulphuric acid, initial attack produces a film of magnesium sulphate which quickly saturates the acid at the interface, and the reaction is reduced to a vanishingly low rate. As long as water is excluded no further attack takes place.

In considering the corrosion behaviour of magnesium alloys, therefore, it is of the utmost importance to know the nature of the medium to which the metal is to be exposed. In general, atmospheric attack in damp conditions is largely superficial; aqueous solutions bring about attack which varies not only with the solute but with the volume, movement and temperature

MAGNESIUM AND MAGNESIUM ALLOYS

Table 4.17 Standard electrode potentials (S.H.E.)

Metal	Potential V
Magnesium	−2·37
Aluminium	−1·66
Zinc	−0·76
Iron	−0·44

Table 4.18 Steady-state potential of several metals in 1 M solutions of different electrolytes (vs 0·1 N Calomel, $E_H = 0·336$ V)

Metal	Potential (V)								
	Sodium chloride	Sodium sulphate	Sodium chromate	Hydrochloric acid	Nitric acid	Sodium hydroxide	Ammonium hydroxide	Calcium hydroxide (sat'd)	Barium hydroxide (sat'd)
Magnesium	−1·72	−1·75	−0·96	−1·68	−1·49	−1·47	−1·43	−0·95	−0·88
Aluminium	−0·86	−0·50	−0·71	−0·80	−0·49	−1·50	−0·80	−1·54	−1·53
Zinc	−1·15	−1·19	−0·67	−1·14	−1·06	−1·51	−1·50	−1·40	−1·49
Iron	−0·72	−0·76	−0·16	−0·66	−0·58	−1·22	−0·18	−0·30	−0·25

of the liquid. Many organic liquids are quite inert to magnesium, but some of those which contain reactive polar groups, as might be expected, are reactive in some degree towards the metal.

Galvanic corrosion of magnesium, i.e. the enhanced corrosion to which the anodic member of a pair of metals in contact is subject to when both are in contact with a common electrolyte, is of considerable practical importance, since magnesium is anodic to all other structural metals in most electrolytes.

The standard electrode potential of magnesium is given, along with the potentials of other metals, in Table 4.17 and the steady-state potentials of magnesium in various solutions are listed in Table 4.18[1].

Corrosion of Magnesium Alloys in Atmospheric Conditions

In clean, dry atmospheres (with r.h. below about 60%) uncontaminated magnesium alloys retain a lustrous surface almost indefinitely. If the atmosphere is clean but not dry and the humidity approaches 100% a scattered pattern of corrosion spots appears after a period, but considerable areas of unaffected surface remain for a very long time. (This effect is probably attributable to galvanic corrosion on a micro-scale, each spot representing corrosion of the magnesium adjoining a cathodic particle contained in the surface as an impurity.) It is quite otherwise if the surface has been contaminated either by corrosive dusts or by cathodic particles introduced by abrasive treatments of various kinds. For example, if a piece of magnesium which has been shot blasted is exposed to damp conditions, the whole surface rapidly becomes covered with a greyish layer of corrosion product. In drier atmospheres the layer may resemble a patina and develop very slowly.

In many ways the corrosion of magnesium alloys in normal atmospheric conditions is a close approximation to the *initial* formation of rust on mild

Table 4.19 Corrosion rates of magnesium-rich alloys (g m^{-2} d^{-1}) in three different environments

Alloy	Type of test		
	Immersion for 30 days in 3% NaCl solution	Sea-water spray 3 times per day for 6 months	Atmospheric exposure for 2 years
A8	3·5–133·0 av. 72·7	1·5–2·5 av. 2·0	0·22–0·26 av. 0·24
AZ91	1·3–144·0 av. 52·6	0·5–2·3 av. 1·7	0·19–0·26 av. 0·23
Z5Z	5·5–9·3 av. 8·0	0·7–0·9 av. 0·8	0·34–0·35 av. 0·34
ZRE1	20·5–32·7 av. 26·6	1·2–2·0 av. 1·6	0·36–0·37 av. 0·36
RZ5	12·0–40·2 av. of 26 tests 32·6	1·1–1·6 av. of 12 tests 1·3	—
TZ6	26·8–67·6 av. of 19 tests 53·4	0·9–1·2 av. of 4 tests 1·1	0·40–0·44 av. of 4 tests 0·41
High-purity A8	3·0	0·4	0·28

steel when similarly exposed and is just as superficial. Study of the cross-sections of a number of test bars of different alloys which were exposed to the weather for three years at Clifton Junction, Manchester, showed that the loss of section was remarkably small and the deterioration of mechanical properties was not greater than could be attributed to this and to the roughening of the surface which had occurred.

Corrosion rates in normal industrial atmospheres measured as loss of weight over a period are extremely uniform among the various alloys. Table 4.19, last column, gives the corrosion rates (in $g\ m^{-2}\ d^{-1}$) for a number of alloys determined at Clifton Junction in recent years. The highest value recorded ($0\cdot 4\ g\ m^{-2}\ d^{-1}$) is equivalent to a rate of penetration of $0\cdot 076$ mm/y, which is appreciably less than that of mild steel.

Composition of Corrosion Product

The compound produced by the interaction of magnesium with moist air appears to be essentially magnesium hydroxide ($S_p = 10^{-11\cdot 05}$). Magnesium oxide reacts slowly with water to form hydroxide so that films of oxide formed at high temperatures, during hot-forming operations for example, eventually become hydrated. The final composition of the corrosion product undoubtedly depends upon the nature of the atmosphere and the gases it contains. Thus the natural air-formed corrosion product in industrial areas always consists largely of carbonate and sulphate in addition to hydroxide.

In marine atmospheres magnesium chloride is formed and eventually oxychloride by reaction with magnesium hydroxide formed at the same time. Since the chloride is hygroscopic, moisture is attracted and the corrosive effect is hence much worse than that of water alone.

Marine Atmospheres

If by a 'marine atmosphere' one means the conditions at a site within a few metres of high-water mark in an otherwise clean country atmosphere, then corrosion of unprotected magnesium alloys is remarkably small. The presence of actual liquid spray is exceptional, and during periods of high wind, when such spray is generally produced, the humidity is usually low and evaporation of the droplets is rapid. This explains why crystals of salt may be found on exposed metal yet little evidence of severe corrosion may be seen. Salt solution is corrosive but dry salt particles are almost without effect. Two other factors which operate in such regions are: (*a*) the washing effect of clean rain followed by rapid evaporation; and (*b*) the scouring by blown sand particles. Unpainted magnesium items exposed near the sea coast invariably exhibit a dull, slightly rough surface, with no corrosion product evident. This appearance is in sharp contrast to that of metal which has been exposed to conditions of persistent dampness; in such cases the corrosion product, considerably more voluminous than the metal from which it is formed, is retained on the surface as a greyish white powder. The glossy surface of painted specimens and the polish of newly prepared bare metal specimens exposed near the sea coast is also quickly lost. Corrosion rates of painted specimens which have been exposed so long that the paint has perished are often higher than those of bare metal. The cause of this is probably the retention of water by the sponge-like texture of the deteriorated paint*. In a similar way corrosion may be most in evidence at junctions and narrow gaps. This is due not so much to true 'crevice corrosion' as to the retention of moisture. Surface moisture evaporates quickly, and the incipient corrosion to which it gives rise quickly comes to an end. In a narrow recess, however, trapped moisture, unable to evaporate, continues the corrosion process over long periods. True crevice corrosion on the other hand is a phenomenon which owes its nature to the development of anodic areas within the crevice, caused by exclusion of oxygen for example, and does not appear to occur with the magnesium alloys.

In conditions where sea-water spray may be deposited regularly on magnesium articles with no alleviating mechanism for its removal, or where breaking waves may drench the components, the effect is quite different. Corrosion of bare metal will be heavy and will be intensified at junctions with other more noble metals. Unless magnesium alloys can be adequately protected in such combinations it is better to avoid their use. This matter is dealt with under the section on protection.

In an experiment similar to that referred to on p. 4.100, tensile test bars were exposed at Clifton Junction, Manchester, for six months, during which time they were sprayed three times daily with sea-water. Whereas exposure to industrial atmosphere alone had little effect, bars of the same alloys were much more heavily attacked by sea-water spray.

High-purity Alloys and Galvanic Corrosion

In addition to the alloying ingredients which are added, certain other metals are usually present in small amounts. In the alloys which contain aluminium, for example, iron usually amounts to about $0 \cdot 02-0 \cdot 05\%$. By special techniques and care in melting this can be reduced to about one-tenth of the above figure. Many workers have shown that such high-purity alloys have a markedly better resistance to salt water than those of normal purity, but their behaviour towards industrial atmospheres is not greatly different. Furthermore, the practical value of the higher resistance to corrosion is largely offset when components are used in electrical contact with other more cathodic metals. The effect of a steel bolt for example, even when it has been zinc or cadmium plated, is much greater at the point of contact than that of the excess of local cathodes in the impure alloys. Galvanic corrosion at joints with other metals therefore is not markedly less in the case of the high-purity alloys. Nevertheless, such alloys have their place, and when they can be used without other metal attachments provide better intrinsic resistance to corrosion by sea-water than the alloys of normal purity.

Alloys containing zirconium as a grain-refining agent have the iron content automatically reduced to about $0 \cdot 004\%$ by settling out of impurities during the alloying procedure.

---

*Private communication from A. P. Fenn, Esq., Birmingham Aluminium Casting (1903) Co. Ltd.

Reference to Table 4.19 will show that greatly superior corrosion rates of the high-purity alloys are only in evidence in the more severe conditions of test by immersion in salt water, and that in less drastic conditions, and especially in industrial atmospheric exposure, there is little to choose between the alloys.

Figure 4.17 illustrates the corrosion occurring on high-purity AZ31 and ZW3 in contact with steel bolts. Tested alone in sea-water, the corrosion rate of the former is much the lower. It is evident from the illustration, however, that the governing factor in galvanic corrosion is the type of electrolyte present rather than the composition of the alloy.

Nature of the Corroded Surface

When corrosion occurs on a smooth machined magnesium alloy surface, this surface is roughened by the chemical action, and after the initial attack the degree of roughness does not change appreciably. In the usual industrial atmospheric conditions the attack is uniform, but in immersed conditions, including corrosion under pools of condensate, attack may be, and usually is, irregular; some areas become anodic to other areas and, as corrosion proceeds at the former, a pitted effect results. Even in atmospheric attack the roughening is really a microscopic form of pitting.

There is a noticeable difference between the appearance of the aluminium-containing magnesium-rich alloys on the one hand and the zinc/zirconium-containing magnesium alloys on the other. In the former the microscopic pits in the surface which has been exposed to the weather tend to be narrow and relatively deep, while in the latter they are wider and tend to overlap, leading to a slightly wavy appearance.

The unequal attack which occurs in tap water, condensate and other mild electrolytes may lead to perforations of thin-gauge sheet and even to deep pitting of castings. In stronger electrolytes the effect is variable. In chloride solutions such as sea-water, attack on the metal usually results in the pitting of some areas only, but where the metal surface has been rendered reactive, as by shot blasting, attack may be so rapid that uniform dissolution over the whole surface may occur. In either case magnesium-base alloys are not usually suitable for use in aqueous liquids since they are not intrinsically resistant to these electrolytes.

Methods of Corrosion Testing

In considering the corrosion of magnesium and its alloys it is important to examine the methods available for assessing corrosion tendencies and particularly those known as accelerated tests. Tests carried out by immersion in salt water or by spraying specimens regularly with sea-water are worthless as a means of determining the resistance of magnesium alloys under any other than the particular test conditions. Extrapolation to less corrosive conditions is not valid and even the assessment of the value of protective measures by such means is hardly possible. The reason is to be found in the fact that corrosion behaviour is directly related to the formation of insoluble

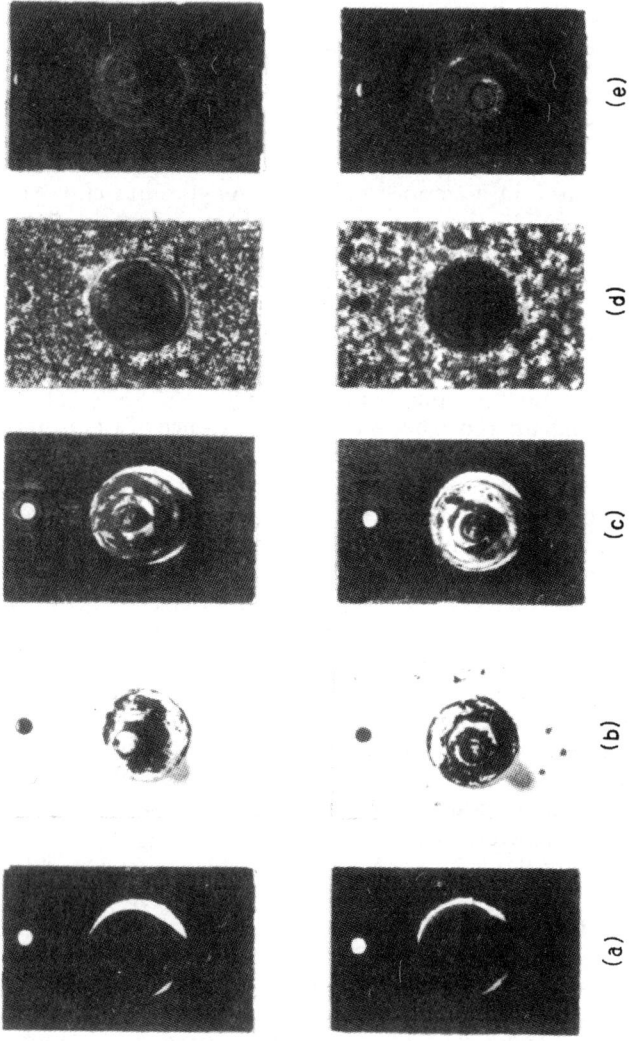

Fig. 4.17 Samples of high-purity AZ31 (upper photographs) and ZW3 (lower photographs) magnesium-base alloys, fitted with mild-steel nuts and bolts and exposed to a variety of corrosion conditions. (*a*) 4·5 hours' immersion in 3% salt soln., (*b*) 180 days' immersion in distilled water, (*c*) 4 days' immersion in borehole water, (*d*) 180 days' in humidity cabinet sea-water spray and (*e*) 180 days' atmospheric exposure

films. In chloride solutions there is no stable insoluble film formed from the solution itself and no previously formed film (by chemical reaction) is impermeable to the chloride ion. Even existing protective films are penetrated relatively easily by chloride ions, and organic films of paint or varnish are subjected to osmosis and to swelling in conditions which are quite unlike normal experience. Except for the specific purpose of determining the behaviour of the materials in dilute chloride solutions, accelerated methods of test of this nature are inadmissible and the results misleading. Corrosion in the atmosphere is usually a continuous, relatively gentle, but persistent, process. Since it is continuous the time scale cannot be shortened and to attempt to obtain results merely by increasing the severity of the conditions is illogical. Such tests can only give information about the behaviour in the chosen conditions. The only true indication of behaviour is obtained by exposing samples to the conditions to be experienced, if these are known or can be assessed. If the severity is of an intermittent nature then it is usually permissible to accelerate the time basis by omitting or reducing the intervals of less severe exposure. Even in this case, however, the effects of recovery by drying out, etc. should not be overlooked. In particular, the samples exposed should take account of the effects of joints with both similar and dissimilar materials; acid vapours from wood and plastics and the electric stress introduced by coupling to other metals in the presence of an electrolyte can vitiate completely any deductions concerning the protective value of paints based on tests on isolated pieces of metal.

Intergranular Corrosion (Sections 1.3 and 1.7)

True intergranular corrosion of magnesium alloys does not occur for the reason that the grain-boundary constituent is invariably cathodic to the grain body. It follows, therefore, that corrosion will be principally concentrated on the grains and the grain-boundary constituent will not only be more resistant to attack but will in some measure receive cathodic protection from the corrosion of the neighbouring grain.

Examples are occasionally quoted which purport to show attack at the grain boundaries, but this is not intergranular corrosion properly so called; indeed it is the opposite and might better be called granular attack, for it is the grain and not the boundary which is preferentially attacked. Because the grain boundary is cathodic to the grain proper, attack is concentrated on the area of the anodic grain adjoining the boundary until eventually the grain may be undercut and fall out of the matrix. The important difference from true intergranular corrosion is quite clearly that attack can proceed only grain by grain and cannot make its way through the body of the material following a grain-boundary path.

As usually cleaned for microscopical examination, a corroded specimen has invariably lost the delicate tracery of intergranular material in the cleaning process and thus may present something of the appearance associated with intergranular attack. If special steps are taken, however, it is possible to mount a corroded specimen of magnesium alloy with the grain boundaries still intact, showing where some grains have entirely disappeared while others are in the process of dissolution round the edges where the cathodic

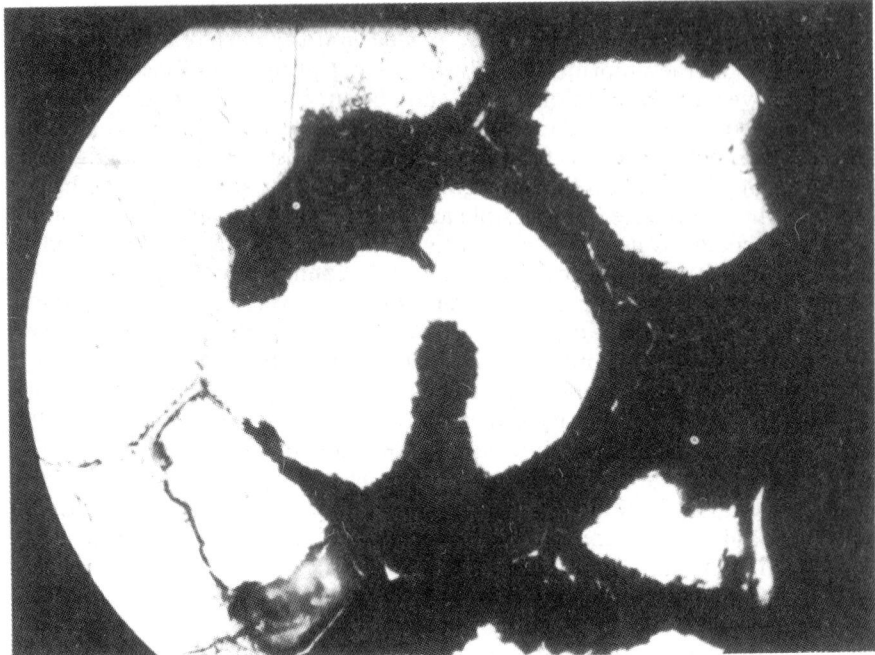

Fig. 4.18 Corroded grain in ZRE1 showing the grain boundaries still intact. Attack occurs on the periphery of the grains and thus is not intergranular

effect of the grain-boundary material is strongest. This is illustrated in Fig. 4.18 which shows a sample of corroded ZRE1 with the intergranular constituent still intact in many areas and with one grain almost etched free from the containing network.

Protection of Magnesium Alloys

The proneness or otherwise to corrosion is essentially the same in all the magnesium-base alloys and it is important to note that the requirements of protection therefore do not vary for the magnesium alloy under consideration. If conditions are such that any one of the alloys can be used satisfactorily without protection, then any other of the alloys can be so used. On the other hand, if a given protective scheme is found necessary for a particular alloy, then the same protective scheme will be found necessary (and will be equally effective) with any other magnesium-base alloy.

It will be realised that since the tendency of magnesium to corrode is governed by the nature of the environment to which it is exposed the degree of protection necessary is also controlled by the same factor.

The methods of protection available are of two basic kinds: chemical or electrochemical treatments which oxidise the metal and produce a film which is more stable than the metal itself, and coating methods which rely upon the application of some extraneous material to provide a more or less impervious coating and thus restrict access of corrosive influences. Generally the two methods are complementary and for best effects, or for protection under the more drastic conditions, they are used in conjunction with each other.

The first class referred to above includes various simple treatments in which the metal is dipped in acidic or near-neutral solutions, usually containing chromate in the hexavalent condition, which has a strong passivating action on many reactive metals. The resulting chemical film which is formed on the magnesium-rich alloys consists essentially of magnesium oxide but also contains chromium compounds. This class of treatment also includes methods of anodic oxidation achieved either by galvanic or electrical means. In general the films resulting from such electrical methods are thicker and often harder than those produced by the simple immersion methods, and usually form more effective barriers against corrosion. Both types, however, should be regarded as suitable for withstanding normal atmospheric conditions for relatively short periods only, since they are all porous or at least permeable to water to some degree. They are, in fact, to be considered as foundation treatments for coating with more impermeable organic materials such as paint and enamel. (See also Section 15.3.)

The second class of protective measures includes all those processes generally known as painting, as well as 'temporary' protective treatments such as greasing and oiling. It is conventional to apply paint films over one or other of the chromate-containing oxide films (produced in one of the immersion-type baths), but paints can be applied equally well to the electrolytically developed coatings. It is not usually wise to apply paint directly to the bare metal. The reason for this is twofold. In the first place, a chromated surface, especially a freshly chromated surface, is not as likely to have deteriorated in storage as a bare metal surface and consequently will provide a better basis for paint. Secondly, and more important, the natural surface of magnesium alloys in contact with damp air is alkaline, because of the presence of the naturally formed oxide and hydroxide, and this may lead to the rapid deterioration of paint films. This is especially true of the oil-based and some of the synthetic air-drying paints which are sensitive to alkali.

In all cases a priming coat of paint containing some form of hexavalent chromium such as zinc or strontium chromate should be used.

Table 4.20 lists a number of the better-known processes for producing protective films on magnesium alloys by chemical and electrochemical processes.

High Temperature Stoved Epoxy Resins

In recent years use has been made of the strong adhesion, toughness and water impermeability of some of the epoxy resins to secure greatly improved surface protection of magnesium alloys. By this means it has been possible to employ these alloys even in situations where they are drenched repeatedly with sea-water.

Not all the epoxy resins are equally efficacious and all have to be stoved at a relatively high temperature (180–220°C) in order to develop the requisite

Table 4.20 The more usual cleaning and chromating processes for magnesium-base alloys*

Process	D.T.D Spec.	United States Spec.	Purpose	Composition (% by wt.)	Method of use	Resulting appearance	Suitable container	Notes
Pickling and cleaning								
Nitric acid pickle	911C, Sec. 2.1.1	MIL-M-3171C, Sec. 3.2.4	Cleaning rough castings	5–10% conc. nitric acid in water	Repeated brief dips until clean, followed by thorough rinse, preferably with hose	Clean and bright but may be loose black smut on Mg-Zr-Zn alloys	Glass, polythene, rubber and earthenware	Metal removal
Sulphuric acid pickle	911C, Sec. 2.1.1	MIL-M-3171C, Sec. 3.2.4	Cleaning rough castings	2–5% conc. sulphuric acid in water	Repeated brief dips until clean, followed by thorough rinse, preferably with hose	Clean and bright but may be loose black smut on Mg-Zr-Zn alloys	Glass, polythene, rubber and earthenware	Metal removal
Fluoride anodising treatment	911C, Sec. 2.1.3	MIL-M-3171C, Sec. Type VII, 3.9	Super cleaning for maximum corrosion	10–30% ammonium bifluoride in water	Application of 120 V a.c. for 5–30 min	Smooth white matt	Rubber, ebonite and polythene	No dimensional change M.E.L.B. Pat. 721 445
Caustic soda clean	—	MIL-M-3171C, Sec. 3.2.3	Cleaning finished parts or greasy components generally Removal of paint	2–5% caustic soda in water	Immerse and boil for 15–30 min; wash thoroughly	Clean, showing no water 'break' when wetted with cold water	Steel	No dimensional change Proprietary brands of alkaline cleaners may be used provided these do not attack metal Soda ash with washing soda and soap are suitable for preparing alkaline cleaning baths
Chromic acid bath	911C, Sec. 2.1.4	MIL-M-3171C, Sec. 3.4.1	Removal of old chromate and fluoride films; also removal of corrosion product and oxide films	10–15% chromic anhydride (CrO_3) in water	Immerse and boil for 15 s to 30 min or as necessary; wash thoroughly	Clean grey appearance	Glass and steel	No dimensional change unless much impurity is present For small delicate parts 0·1–0·5% silver chromate may be added to reduce attack caused by impurities such as chloride in corrosion product

MAGNESIUM AND MAGNESIUM ALLOYS

Process	Specification	Purpose	Composition	Operating conditions	Appearance	Masking materials	Remarks
Chromating							
Acid chromate bath (I.G. bath) (chrome pickle)	911C, App. 11, Bath iv	MIL-M-3171C, Type I, Sec. 3.5	Protection in storage Paint foundation Repair of chromate film	15% sodium or potassium dichromate and 20–25% nitric acid (s.g. 1·42)	Immerse for 10 s to 2 min, drain for 5 s, wash in cold or warm water	Golden bronze often with iridescent colours	Glass, earthenware, slate, aluminium and stainless steel	Metal removal Cannot be used on parts to fine tolerances Useful for rough castings
Hot half-hour bath (R.A.E. bath) (black bath)	911C, App. 11, Bath iii	MIL-M-3171C, Sec. 6.4.1	Good paint foundation Protection in storage	3% ammonium sulphate, 1·5% potassium or sodium dichromate and 0·5–0·75% 0·880 ammonia solution	Immerse in bath and simmer for 30 min; wash	Usually black; light brown on D.T.D. 118	Glass, steel and aluminium	No dimensional change Used chiefly for finished work
Chrome-manganese bath (M.E.L. black bath)	911C, App. 11, Bath v	MIL-M-3171C, sec. 6.4.1	Good paint foundation Protection in storage	10% sodium or potassium dichromate, 5% magnesium sulphate and 5% manganese sulphate	Immerse in bath for 2 h at 20°C; proportionately less if bath is heated	Usually black or dark brown; light brown on D.T.D. 118	Glass, steel and aluminium	No dimensional change Used chiefly for finished work where heating may be undesirable where inserts are present
Special surface treatments							
H.A.E. process	—	MIL-M-45202B (ORD)	Corrosion and abrasion resistance	12% potassium hydroxide, 1% aluminium (high purity) 3·5% tri-sodium phosphate (crystals), 3·5% potassium fluoride (anhydrous) and 2·2% potassium manganate	Application of up to 90 V a.c. for up to 2 h	Dark brown, ceramic-like coating, brittle	Glass, rubber and steel	Increase in dimensions of 0·025 to 0·050 mm
Dow 17 process	—	MIL-M-45202B	Corrosion and abrasion resistance	24% ammonium bifluoride, 10% sodium dichromate and 8·6% orthophosphoric acid (85%)	Application of up to 110 V a.c. at 50 to 500 A/m², temp. 70–80°C; time 2–30 min	Grey-green to dark green dependent on thickness	Steel, rubber and vinyl-based materials	Increase in dimensions of 0·005 to 0·038 mm

*Table reproduced by permission of Magnesium Elektron Ltd.

adhesion to the metal and the necessary water resistance in the resin. Furthermore, a technique of baking the metal before application of the resin, followed by dipping of the still warm metal, is essential in order to fill subcutaneous blemishes. In the absence of such a procedure trouble can still arise from 'pockets' which retain moisture and which are only bridged by a normally applied resin film.

The Effect of Surface Finishing on the Corrosion Behaviour of Magnesium Alloys

It is probable that all corrosion of magnesium alloy surfaces exposed to a damp atmosphere, or still more immersed in an electrolyte, is largely galvanic in origin and much influenced by the presence of exposed cathodic particles. Some of these are present in the alloys as unavoidable impurities, and nearly all foreign metallic particles not in solution are cathodic to magnesium. Various methods of chemical pickling or etching as well as mechanical means of metal treatment may remove a proportion of these, but unfortunately they will usually expose a further number in the lower layers. Furthermore, some methods of mechanical abrasion may increase the number of foreign cathodic particles in the surface by entrapment, and even pickling solutions are not exempt from causing the deposition of more noble metals in solution by displacement. Partly used pickling baths which have been in use for some time, in particular, become enriched in cations of other metals and may redeposit these metals by displacement. In practice, pickling baths based on nitric acid are less likely to give rise to this effect though they may not be able to remove foreign particles already *in situ*.

Chief among the processes which bring about harmful effects on the corrosion resistance of magnesium alloy surfaces are shot and grit blasting and the use of emery cloths and papers. Various blasting operations are used for the removal of adhering foundry sand from sand castings but it should be recognised that they lower very considerably the natural resistance of the surface and reduce its ability to form protective films. Metal particles either of the abrasive itself or of material scoured from the equipment, are lodged in the magnesium surfaces by blasting operations, while splinters of emery (which form quite effective cathodes) are picked up by the use of such materials. On the other hand, glass papers are usually harmless; the splinters of glass are non-conductive and therefore incapable of acting as cathodes. Despite the fact that silicon carbide is a conductor, the use of Carborundum paper and belts does not usually lead to any serious deterioration in the corrosion resistance of magnesium alloys.

By the use of many commercial abrasive processes, the corrosion resistance of magnesium alloys can be reduced to such an extent that samples of metal that may lie quiescent in salt water for many hours will, after shot blasting, evolve hydrogen vigorously, and the corrosion rate, as measured by loss of weight, will be found to have increased many hundred-fold. The effect in normal atmospheres is naturally much less, yet the activation of the surface is an added hazard and is the opposite of passivation which is essential if later-applied paint finishes are to have proper durability.

The use of chromating baths and acid pickles is powerless to remove all the evil effects of such treatments, but one of the electrolytic processes, namely that of Fluoride Anodising at a high voltage in a solution of ammonium bifluoride, is very effective in removing cathodic foreign metals. In this process the magnesium surface itself is quickly converted to insoluble and non-conductive magnesium fluoride and this reaction thereafter terminates. The current is thereupon automatically directed to and concentrated on the local metal cathodes which are conductive, and these are either dissolved or dislodged from the surface. Carbon, in the form of graphite resulting from the use of die lubricants in forming and in pressure diecasting, may also be an active cathode. It is not dissolved by the electrolysis but the film is undercut and insulated from the metal surface by a layer of magnesium fluoride. In this condition it is less harmful than when in direct contact with the metal; furthermore, it can more readily be removed by treatment in chromic acid or in hot caustic soda solution, processes which, in the absence of prior fluoride anodising, are not completely effective.

Recent Developments

The detrimental effects of 'heavy metal' impurities and surface contamination on the corrosion performance of magnesium alloys have been described (4.101). Speciality high purity alloys, with extremely good corrosion resistance, were developed for use in the nuclear industry and this concept has recently been applied to the most commercially used magnesium alloy-AZ91. Quantitative studies[2,3] have determined the threshold levels for Fe, Ni and Cu impurities in this alloy system below which a 50 to 100 fold improvement in salt fog corrosion resistance is obtained. The attainment and control of these low impurity levels for both high pressure die and sand castings has been demonstrated[2,4] resulting in the ASTM designation of the high purity alloys AZ91D and AZ91E respectively.

Die cast AZ91D components have passed the stringent corrosion tests and field trials[5,6] of several automotive manufacturers and items, such as grilles, clutch housings, aircleaners, valve covers and wheels have already given several years trouble free service. AZ91E sand castings have recently been specified for aerospace applications as replacements for existing 'normal purity' AZ91C components.

The high purity concept is being extended to the die casting alloy AM60B[7] and to other alloys in the Mg-Al system. In the Mg-Zr system alloys containing yttrium and rare earth additions have been developed[8,9], WE54 (Mg-5% Y, 4% RE-Zr) WE43 (Mg-4% Y, 3% RE-Zr). These alloys are available in both wrought and sand cast forms and possess high strength at both ambient and elevated temperatures. They also exhibit a high level of corrosion resistance. Fe and to a lesser extent Ni impurities are naturally controlled[10] to low levels in the presence of Zr. In the absence of Zn or other active[11] alloying constituent good corrosion resistance, similar to AZ 91 E, is obtained.

Corrosion rates of 0.1–0.2 mg/cm^2/day under ASTMB117 salt fog conditions are typical for WE54 and the high purity AZ91 alloys which are comparable with those of some aluminium alloys[8]. This considerable improvement in corrosion resistance does not however protect against galvanic corrosion. In corrosive environments the standard techniques to reduce or eliminate the effect of galvanic couples must still be employed[12,13].

Rapid solidification technology has been applied to several magnesium alloy systems[14] and extruded material of some of these systems have exhibited excellent corrosion resistance.

Fluxless melting[15] techniques, employing protective atmospheres of air, carbon dioxide and sulphur hexafluoride (SF_6), are now being used by many foundries. Flux inclusions in castings, particularly pressure die castings, have in the past contributed to magnesium's poor corrosion reputation. By employing fluxless techniques the risk of deleterious flux inclusions, due to improper melt handling, is avoided.

Chromate conversion coatings[13] are still the most widely used pretreatments prior to painting. With the increased emphasis of the hazards associated with hexavalent chromium, several chromate free treatments[13,16,17] have been used on magnesium. These treatments are not as effective as chromating and consequently should be restricted for use in mild environments only. The NH35 chromate treatment[18], with its significantly reduced chromium addition, has been developed for use on high purity AZ91 pressure die castings.

Epoxy based primer systems remain the best suited for the corrosion protection of magnesium. Cathodic epoxy electrophoretic paints[6], chromate inhibited epoxy-polyamide primers[19] and high temperature stoving epoxy sealers[20] are used to provide protection up to 180°C. For higher temperature applications up to 300°C, epoxy silicone or polyimide[21] based systems can be used.

The following checklist is given as a general guide[22] to minimise the corrosion of magnesium components in service:

1. Design—good design to minimise exposed dissimilar metal couples, radius sharp edges and avoid water traps. Allow for protection of mating faces.
2. Specify good quality castings, forgings and extrusions.
3. Select protective scheme to suit operational environment—for new applications err on the side of overprotection until performance experience has been obtained.
4. Ensure a clean metal surface free from cathodic contaminants.
5. Apply good quality conversion coatings.
6. Ensure correct organic protection scheme application as soon as possible after conversion coating.
7. Observe 'wet assembly' procedures on exposed galvanic couples.
8. Inspect and maintain protection.

K. G. ADAMSON
D.S. TAWIL

REFERENCES

1. Mears, R. B. and Brown, C. D., *Corrosion*, **1**, 113 (1945); Hanawalt, J. D., Nelson, C. E. and Peloubet, J. A., *Trans. Amer. Inst. Min. (Metall.) Engrs.*, **147**, 275 (1942)
2. Reichek, Clark, Hillis, *Controlling the Salt Water Corrosion Performance of Magnesium AZ91 Alloy*, SAE Paper 850417

3. Hillis, *The Effects of Heavy Metal Contamination on Magnesium Corrosion Performance*, SAE Paper 830523
4. Clark, AZ91E Magnesium Sand, Casting Alloy. The Standard for Excellent Corrosion Performance, Proc. IMA World Magnesium Conference, Los Angeles, June (1983)
5. Kaumle, Toemmeraas, Bolstad; *The Second Generation Magnesium Road Wheel*, SAE Paper 850420
6. *Product Design and Development for Magnesium Die Castings*, Dow Chemical Co. publication
7. Hillis, Reichek; *High Purity AM60 Magnesium Alloy*, SAE Detroit Congress Feb. 26 (1986)
8. Unsworth; 'Developments in Magnesium Alloys for Casting Applications', *Metals and Materials*, 83–86, February (1988)
9. King, Fowler, Lyon; Light-weight Alloys for Aerospace Applications II, *Proc. TMS Meeting*, New Orleans, USA, Feb 17–21, 1991, pp. 423–437
10. Emley; *Principles of Magnesium Technology*, Pergamon Press, 176–190, 685 (1966)
11. Hanawalt, Nelson, Peloubet; 'Corrosion Studies of Magnesium and its Alloys', *Trans Am. Inst. Mining Met. Eng.* 147, 273–299 (1942)
12. Hawke; *Galvanic Corrosion of Magnesium*, 14th International Die Casting Congress, Toronto, May (1987)
13. *ASM Metals Handbook*, 9th Ed, **13**, 740–754 (1987)
14. Das, Chang, Raybould; 'High Performance Magnesium Alloys by Rapid Solidification Processing', *Light Metal Age*, Dec. 5–8 1986
15. 'Use of Air/CO_2/SF_6 Mixtures for the Improved Protection of Molten Magnesium-Couling', *Proc. IMA World Magnesium Conference*, Oslo, June (1979)
16. *Corrosion and Protection of Magnesium*, AMAX Magnesium Publication, (1984)
17. *Magnesium: Designing Around Corrosion*, Dow Chemical Co. Publication, (1982)
18. *The NH35 Chromating of Magnesium Pressure Die Castings*, Norsk Hydro Publication, (1985)
19. Robinson; 'Evaluation of Various Magnesium Finishing Systems', *Proc. IMA World Magnesium Conference*, New York (1985)
20. *Clear Baking Resin for Surface Sealing Magnesium*, U.K. Specification DTD5562, HMSO
21. *Improved Protection of Magnesium Alloys Against Synthetic Aviation Lubrications at Elevated Temperatures*, Rendu, Tawil; SAE Paper 880869
22. *Surface Treatments for Magnesium Alloys in Aerospace and Defence*, Magnesium Elektron Publication

BIBLIOGRAPHY

Metals Handbook, **1**, 8th edn, American Society for Metals, Chicago (1961)

Pearlstein, F. and Teilell, L., 'Corrosion and Corrosion Prevention of Light-Metal Alloys' Paper No. 114, *Corrosion*, 73, Anaheim, March (1973)

Emley, E. F., *Principles of Magnesium Technology*, Pergamon Press, Chapter xx (1966)

Adamson, K. G., King, J. F. and Unsworth, W., *Evaluation of the Dow 17 Treatment for Magnesium Alloys*, Ministry of Defence D.Mat. Report No. 192, February (1973)

King, J. F., Adamson, K. G. and Unsworth, W., *Impregnation of Anodic Films for the Protection of Magnesium Alloys*, Ministry of Defence D.Mat. Report No. 193, February (1973)

Adamson, K. G., King, J. F. and Unsworth, W., *Evaluation of High Temperature Resistant Coatings for the Protection of Magnesium Alloys*, Ministry of Detence D.Mat. Report No. 196, July (1973)

4.5 Nickel and Nickel Alloys

Physical and Mechanical Properties

Composition of Metal and Alloys

Commercially pure nickel has good mechanical properties and good resistance to many corrosive environments and therefore finds application where this combination of properties is required. Of more importance, however, is the fact that nickel forms a wide range of alloys having desirable engineering and corrosion-resistant properties. With regard to corrosion resistance to aqueous solutions, among the most important of these alloying elements are Cr, Fe, Cu, Mo and Si. Since the range of corrosion-resistant nickel alloys includes some that owe their corrosion resistance to passivity and others that are resistant because they are sufficiently noble not to displace hydrogen from acidic solutions, the corrosive environments in which nickel alloys can be successfully used are very varied, embracing acids, salts and alkalis (both oxidising and non-oxidising in character) sea-water, natural waters and the atmosphere and combinations of these encountered industrially.

In addition to nickel alloys, nickel also forms an important alloying element in stainless steels and in cast irons, in both of which it confers additional corrosion resistance and improved mechanical and engineering properties, and in Fe–Ni alloys for obtaining controlled physical and magnetic properties (see Chapter 3). With non-ferrous metals nickel also forms important types of alloys, especially with copper, i.e. cupro-nickels and nickel silvers; these are dealt with in Section 4.2.

Nickel is also widely used as an electrodeposited underlay to chromium on 'chromium-plated' articles, reinforcing the protection against corrosion provided by the thin chromium surface layer. Additionally the production of articles of complex shape to close dimensional tolerances in nickel by electroforming – a high-speed electrodeposition process – has attracted considerable interest. Electrodeposition of nickel and the properties of electrodeposited coatings containing nickel are dealt with in greater detail in Section 14.7.

The nominal compositions of commercially pure wrought nickel and the main types of modern corrosion-resistant nickel alloys are given in Table 4.21; some of these supersede earlier variants no longer in production. Applications of nickel alloys are not confined to those where corrosion resistance to aqueous solutions is a prime requirement, and the complete range of nickel alloys that are available commercially for other specialised uses, notably those involving service at high temperatures, is therefore much greater than indicated by Table 4.21. The corrosion and oxidation resistance of nickel alloys at elevated temperature, is described in Section 7.5.

In general, the alloys listed in Table 4.21 are confined to those in which nickel is the principal alloying element, but it should be noted that highly alloyed stainless steels containing 20-30% Cr, and 20-30% Ni with additions of molybdenum and copper have some features in common with the Ni-Cr-Fe-Mo-Cu alloys given in the table.

In addition to the alloys in Table 4.21, Ni-Sn and Ni-Ti alloys also possess useful corrosion resistance. Ni-Sn alloys are extremely brittle and, because of this, are used only as electrodeposited coatings. Ni-Ti alloys over a wide range of compositions have been studied, of which perhaps the intermetallic compound NiTi (55·06Ni-44·94Ti) has attracted the most interest.

Structural Features and Physical and Mechanical Properties

Nickel normally crystallises in the f.c.c. structure; it undergoes a magnetic transformation at 357°C and is ferromagnetic below that temperature. In all the alloys shown in Table 4.21 the f.c.c. (austenitic) structure is substantially retained, and in consequence most of the alloys possess the combination of properties required of materials for widespread industrial acceptability, i.e. tensile strength, ductility, impact strength, hardness, hot and cold workability, machinability and fabrication.

Table 4.22 gives the physical properties for nickel and a range of nickel alloys; Table 4.23 shows the mechanical properties. The data given in these tables are those published by manufacturers. It is seen that, compared with nickel, the alloys have considerably lower thermal conductivity and much higher electrical resistivity. As with nickel, some of the alloys undergo magnetic transformation; e.g. the Ni-Cu alloy 400 has a transformation temperature close to 0°C. The mechanical properties in Table 4.23 are generally those of wrought material in the annealed condition those of materials in other conditions and of cast alloys may differ appreciably. In all cases alloying considerably increases the proof stress and tensile strength. The elongation values of wrought alloys are generally only slightly below those of nickel. The hardness values of wrought alloys are generally below 200 H_V for annealed material.

As with stainless steels, some nickel alloys have a propensity to form intergranular precipitates of carbides and intermetallic phases during heat treatment and sometimes during welding. The presence of such intergranular precipitates may render the materials susceptible to intergranular attack in certain corrosive environments. To minimise this possibility, the content of carbon and, in some cases, other alloying elements is carefully controlled. The subject of intergranular corrosion of specific nickel alloys and of methods of avoiding it is dealt with in greater detail later in this chapter.

Table 4.21 Nominal compositions of corrosion-resistant nickel alloys

Type and designation	C	Ni + Co	Cr	Fe	Mo	Cu	Ti	Al	Nb	W	Other
Ni											
200	0·08	99·6		0·2		0·1					
201	0·01	99·6		0·2		0·1					
Ni-Cr-Fe											
600	0·08	Bal.	15·5	8·0		0·2					
600L	0·025(−)	Bal.	15·5	8·0							
601	0·05	Bal.	23·0	14·1		0·5		1·4			
690	0·03	Bal.	30·0	9·5							
800	0·05	32·5	21·0	Bal.		0·4	0·4	0·4			Ti:C 15(+)
800L	0·03(−)	32·0	21	Bal.			0·4	0·4			
Ni-Cr-Fe-Mo											
718	0·03	52·5	19·0	Bal.	3·05		1·0	0·6	5·0		Zr 0·02(−)
H-9M	0·03(−)	Bal.	22	19	9					2	
Ni-Cr-Fe-Mo-Cu											
G 3	0·007	Bal.	22·2	19·5	7·0	2·0			0·25 X	0·75	
G 30	0·03(−)	Bal.	29·5	15·0	5·0	1·7			0·7 X	2·5	
825	0·03	42·0	21·5	Bal.	3·0	2·2	0·9	0·1			Ti:C 30(+)
825 h Mo	0·025(−)	42·5	21·0	Bal.	6·2	2·2	0·8	0·2			
925	0·02	42·0	21·0	Bal.	3·0	2·2	2·1	0·3			
20	0·02(−)	37·2	20·0	Bal.	2·5	3·5					Nb:C 8(+)
28	0·015(−)	31·0	27·0	Bal.	3·5	1·2					N 0·05
Ni-Cr-Mo											
C 276	0·005	Bal.	15·5	5·5	16·0					3·7	V 0·35(−)
C 4	0·015(−)	Bal.	16·0	3·0(−)	15·5		0·7(−)				
C 22	0·01(−)	Bal.	22·0	3	13·0				0·2(−)	3·0	V 0·35(−)
625	0·05	Bal.	21·5	2·5	9·0		0·2	0·2	3·6 X		
Ni-Mo											
B2	0·01(−)	Bal.	1·0(−)	2·0(−)	28·0						
Ni-Cu											
400	0·2	Bal.		1·2		31·5					
K 500	0·1	Bal.		1·0		29·5	0·6	2·7			

(−) maximum (+) minimum Xincl.Ta.

Bal. – substantially the balance of the alloy composition, although other elements such as deoxidants and impurities in small amounts are included in the balance.

Table 4.22 Physical properties of corrosion-resistant nickel alloys

Type and designation	Melting range (°C)	Density (kg m^{-3})	Specific heat (J kg^{-1} K^{-1})	Mean coefficient of thermal expansion (K^{-1})	Thermal conductivity (W m^{-1} K^{-1})	Electrical resistivity (Ω m)	Modulus of elasticity (G Pa)
Ni							
200	1435–1445	8.89 × 10^3	456	13.3 × 10^{-6}	74.9	0.09 × 10^{-6}	214
201	1435–1445	8.89	456	13.3	79.2	0.08	207
Ni-Cr-Fe							
600	1370–1425	8.42	461	13.3	14.9	1.03	214
600 L		8.45	460	14.0	14.8	1.05	214
601	1300–1370	8.05	448	13.75	11.2	1.22	206.5
690		8.14	450	14.5	13.9	1.15	210
800	1355–1385	8.02	460	14.2	11.7	0.99	196
800 L		8.0	550	15.9	11.5	0.97	200
Ni-Cr-Fe-Mo							
718		8.2	430	14.2	11.4	1.24	204
H-9M							
Ni-Cr-Fe-Mo-Cu							
G 3		8.30	453	14.6	10.0	1.13	199
G 30		8.22		12.8	10.2	1.16	202
825	1370–1400	8.14	441	13.9	11.1	1.00	198
825 h Mo		8.3	500	15.0	12	1.10	200
925	1310–1365	8.14	435	13.2		1.17	201
20	1370–1425	8.05	500	14.9	11.7	1.03	195
28	1330–1370	8.0	442	15.0	10.8	0.99	195
Ni-Cr-Mo							
C 276	1325–1370	8.89	427	11.2	9.4	1.30	205
C 4		8.64	406	10.8	10.1	1.25	211
C 22	1355–1400	8.69	414	12.4	10.1	1.14	206
625	1290–1350	8.44	410	12.8	9.8	1.29	208
Ni-Mo							
B2		9.22	373	10.3	11.1	1.37	217
Ni-Cu							
400	1300–1350	8.83	419	14.1	21.8	0.51	179
K 500	1315–1350	8.46	419	13.7	17.5	0.62	179

Table 4.23 Typical mechanical properties of corrosion-resistant nickel alloys

Type and designation	Form of material	0.2% proof stress (M Pa)	Tensile Strength (M Pa)	Elongation (%)	Hardness (HV)
Ni					
200	Annealed sheet	157	450	44	
201	Annealed	103	403	50	100 max.
Ni-Cr-Fe					
600	Annealed sheet	269	629	42	180 max.
600 L	Annealed	180 min.	550 min.	30 min.	
601	Annealed sheet	292	675	46	151 max.
690	Annealed	300 min.	600 min.	45 min.	
800	Annealed	249	592	30 min.	179
800 L	Annealed	180 min.	450 min.	35 min.	
Ni-Cr-Fe-Mo					
718	Solution annealed, precipitation hardened	1035 min.	1240 min.	12 min.	
H-9M	Annealed sheet	372	730	57	
Ni-Cr-Fe-Mo-Cu					
G 3	Solution heat-treated plate	311	692	58	172
G 30	Solution heat-treated plate	324	689	56	
825	Annealed sheet	317	672	42	
825 h Mo	Annealed	240 min.	550 min.	25 min.	
925	Annealed rounds	356	769	49	176
20	Annealed	240 min.	550 min.	30 min.	220 max.
28	Solution heat-treated sheet	220 min.	500 min.	35 min.	225 max.
Ni-Cr-Mo					
C 276	Solution heat-treated sheet	355	792	61	192
C 4	Annealed sheet	421	801	54	200
C 22	Solution heat-treated sheet	407	800	57	205
625	Annealed sheet	414 min.	827 min.	30 min.	247 max.
Ni-Mo					
B2	Annealed sheet and plate	412	894	61	215
Ni-Cu					
400	Annealed	216	542	51.5	
K 500	Annealed sheet	275 min.	620 min.	25 min.	170 max.

Methods of Fabrication

Nickel and wrought nickel alloys may be fabricated by welding or, less commonly, by brazing or silver soldering. In order to minimise the deleterious effects that may result from integranular precipitation, either low-heat-input welding procedures employing flux-coated electrodes, or the MIG, TIG or plasma arc procedures, are recommended. Thick sections may be welded using the submerged arc process and a relatively restricted heat input. Oxyacetylene welding is rarely used because of the high heat input and the danger of carbon transfer into the metal.

Corrosion Behaviour in Aqueous Environments

Theoretical Considerations

Nickel occupies an intermediate position in the electrochemical series; $E^{\ominus}_{Ni^{2+}/Ni} = -0\cdot227$ V, so that it is more noble than Zn and Fe but less noble than Sn, Pb and Cu. Figure 4.21 shows a revised potential–pH equilibrium (Pourbaix) diagram for the Ni–H_2O system at 25°C[1]. The existence of the higher anhydrous oxides Ni_3O_4, Ni_2O_3 and NiO_2 shown in an earlier diagram[2] appears doubtful in aqueous systems in the absence of positive identification of such species. It is seen that:

1. Nickel is thermodynamically stable in neutral and moderately alkaline solutions although not in acidic or strongly alkaline solutions.
2. The metal would be expected to dissolve in acidic solutions forming Ni^{2+} ions with liberation of H_2.
3. The metal should be capable of passivation by forming a surface layer of $Ni(OH)_2$ and perhaps NiO (see later) of nickel in neutral and moderately alkaline solutions.
4. The metal may be unstable in strongly alkaline solutions, dissolving to form $Ni(OH)_3^-$ ions.
5. In strongly oxidising neutral and alkaline conditions passivation should be possible through formation of a film of NiOOH.

On the basis of these data, nickel is considered to be a slightly noble metal, although in practice, as will be seen below, it is considerably more corrosion resistant in both acidic and alkaline solutions than would be predicted from Fig. 4.19.

Several complications are involved in the calculation of potential–pH equilibrium diagrams for temperatures other than 25°C[3,4,5], including the fact that the pH scale itself varies with temperature; thus, diagrams in which the pH scale refers to the temperature for which the equilibria are calculated are probably preferable for most purposes[5]. The most notable consequence of increasing temperature on the equilibria appears to be a widening of the pH range within which the hydroxide $Ni(OH)_2$ is thermodynamically stable.

Anodic Behaviour of Nickel

Many investigators have studied the anodic behaviour of nickel. A complete discussion of the reactions occurring during anodic dissolution and passivation of the metal is outside the scope of this chapter, which is confined to a brief summary of the main features of practical significance.

Anodic E–i curves for nickel obtained by potentiostatic, potentiokinetic or, in earlier days, galvanostatic techniques, have been published by many workers. Unfortunately, good agreement is not always found between data from different sources. The principal reasons for the discrepancies appear to lie in the nature and amount of impurities in the metal[6,7,8] or in the solution[9,10], both of which may have a profound effect on the shape of the curve, and in variations in experimental procedure[11-17].

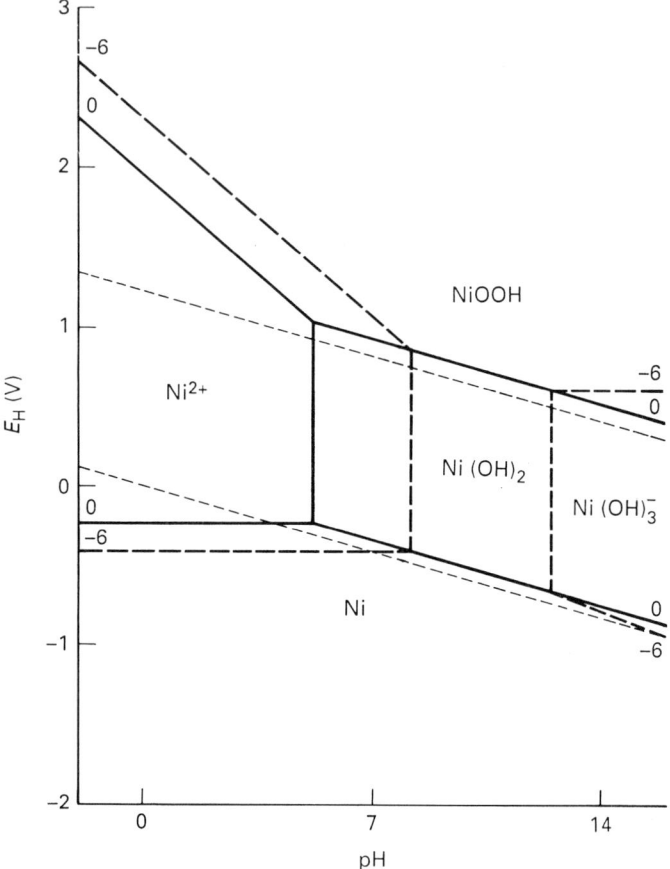

Fig. 4.19 Potential-pH equilibrium diagram for the Ni-H$_2$O system at 25°C (after Silverman[1])

Figure 4.20 shows a curve for nickel in 0.5 M H$_2$SO$_4$[18] which illustrates the main features of the anodic behaviour of the metal that are of interest with regard to its corrosion resistance. It is seen that in acidic solution nickel is capable of passivation and that the extent of the passive range (*DE*) is considerable, ≈0·5 V. The passivation of nickel in acidic solution is a feature not predicted by the potential–pH equilibrium diagram (see Section 2.1) and is one reason why, in practice, the corrosion resistance of the metal in acidic solutions is better than that indicated from consideration of thermodynamic equilibria. A second, perhaps more important, reason lies in the fact that in the active range (*ABC*) the anodic overpotential is substantial because the exchange current density for nickel dissolution is small (Table 21.17). This, coupled with the fact that in the electrochemical series nickel is only moderately negative with respect to hydrogen, $E^{\theta}_{Ni^{2+}} = -0\cdot 227$ V, equilibrium, means that in practice the rate of dissolution of nickel in acidic solutions is slow in the absence of oxidants more powerful than H$^+$ or of substances capable of making the anodic reaction kinetically easy. The anodic dissolution current density of nickel in the active state as a function of potential does, however, depend in a critical manner on the rate at which the measurements are made[13,17,19] and on pH[13]. To explain this, Sato and Okamoto[13] proposed that in acidic solution anodic dissolution of nickel is catalysed by OH$^-$ and proceeds by way of the following reaction sequence:

$$Ni + OH^- \rightarrow NiOH(ads.) + e^-$$
$$NiOH(ads.) \rightarrow NiOH^+ + e^-$$
$$NiOH^+ \rightarrow Ni^{2+} + OH^-$$

the overall rate of reaction being controlled by the concentration of OH$^-$ ions. Burstein and Wright[17] consider that the first stage in the sequence, i.e. formation of NiO(ads.) to form a pre-passive layer is the rate determining step. This mechanism appears to provide a basis for explaining the sluggish anodic dissolution of nickel in acidic solution and also to account in part for the variations in the anodic behaviour reported from different sources. In solutions containing high concentration of Ni^{2+} and SO$_4^{2-}$ Vilche and Arvia[20] consider that dissolution of Ni to Ni^{2+} and formation of Ni(OH)$_2$ are competing processes in the pre-passive region.

The anodic dissolution of nickel is also dependent on the amount of cold work in the metal[19,21], and in the active region the anodic current density of cold worked material at a given potential is up to one order of magnitude greater than that of annealed material.

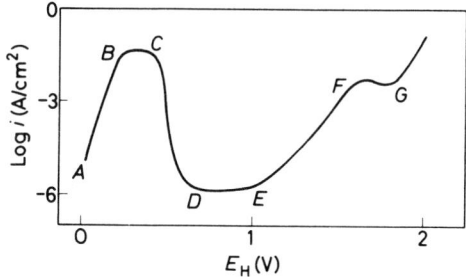

Fig. 4.20 Potentiostatic E-log i curve for nickel, anodically polarised in 0·05 M H$_2$SO$_4$ saturated with N$_2$ at 25°C (after Sato and Okamoto[18])

At high potentials in acidic solution nickel becomes transpassive (EF), and in this region corrosion occurs preferentially at grain boundaries[21], as with stainless steels. In the passive and transpassive states anodic dissolution results in the formation of Ni^{2+} ions in solution[21]. At still higher potentials nickel exhibits secondary passivity (FG), and although the anodic current is several orders of magnitude greater than in the passive region (DE) it is not localised at grain boundaries[21]. At potentials above the range of secondary passivity the anodic current density increases and dissolution proceeds through an oxide film, probably NiOOH and is accompanied by evolution of O$_2$. In this region grain boundaries are preferentially attacked again[21]. The corrosion behaviour of nickel in acidic solutions in the regions of transpassivity, secondary passivity and beyond, is of limited practical significance, since these potentials are beyond the range of the redox potentials of most aqueous solutions.

The influence of temperature on the anodic behaviour of nickel has been studied[3,8], and in acidic and neutral solutions the active-passive transition is not observed at temperatures greater than about 100°C (Fig. 4.21).

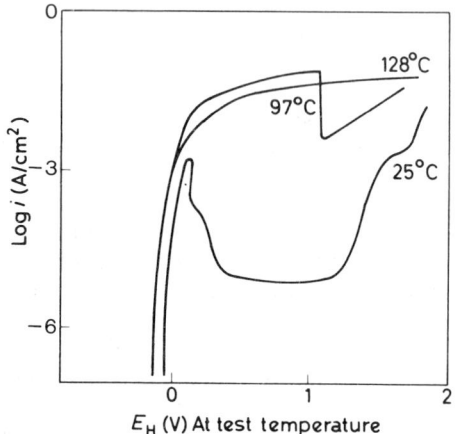

Fig. 4.21 Effect of temperature on the anodic behaviour of nickel in 0·025 M H_2SO_4 + 0·025 M K_2SO_4 (pH 1·3) de-aerated with H_2. The curves were determined potentiokinetically at a scan rate of 2 V/h and proceeding from negative to positive (after Cowan and Staehle[3])

As with most other metals, the anodic behaviour of nickel is influenced by the composition of the solution in which measurements are made, particularly if the solution is acidic. Acidic solutions containing Cl^- ions[22-29] or certain sulphur compounds[9] in particular have a pronounced influence both in increasing the rate of anodic dissolution in the active range and in preventing passivation, and in stimulating localised corrosion[30]. Thiourea and some of its derivatives have a complex effect, acting either as anodic stimulators or inhibitors, depending on their concentration[30].

In alkaline solutions, except possibly in high concentrations at elevated temperatures, nickel is normally passive.

Passivity of Nickel

In many aqueous solutions nickel has the ability to become passive over a wide range of pH values. The mechanism of passivation of nickel and the properties of passive nickel have been studied extensively — perhaps more widely than for any other element, except possibly iron. In recent years the use of optical and surface analytical techniques has done much to clarify the situation[31,32]. Early studies on the passivation of nickel were stimulated by the use of nickel anodes in alkaline batteries and in consequence were conducted in the main in alkaline media. More recently, however, attention has been directed to the passivation of nickel in acidic and neutral as well as alkaline solutions.

Most authorities nowadays accept the view that passivity of nickel, as of most other metals, is due to the formation of a film of oxide or hydrated oxide. Ellipsometric measurements, both in alkaline solution[33] and in acidic solution[34], support the existence of surface oxide films on passive nickel several nanometres thick, although impedance measurements[35] suggest that, in acidic solutions at least, the passive layer is electrically complex and is not an ideal dielectric.

In acidic solutions the film has been reported to be hydrated nickel oxyhydroxide, $NiO_y(OH)_{2-2y} \cdot MH_2O$ in which y is greater in the passive film than in the pre-passive film formed in the active region[36]. In neutral solutions films consisting of NiO[37] and $Ni(OH)_2$[32] possibly with some NiO[31] have been described. In alkaline solutions $Ni(OH)_2$ has been reported[38,39].

In alkaline solutions, galvanostatic measurements[40] suggest that passivation of nickel is due to formation of a monolayer of $Ni(OH)_2$. This probably forms by a solid state process involving nucleation and growth, according to the general model for such growth proposed by Armstrong, Harrison and Thirsk. In some alkaline conditions, particularly concentrated solutions at higher temperatures, thicker films are undoubtedly formed.

As indicated when discussing anodic behaviour the mechanism of film formation is complex, involving adsorption of OH^- ions to form a pre-passive layer followed by either dissolution or film formation as alternative processes.

In certain concentrated acidic solutions, e.g. H_2SO_4, nickel, whilst not truly passive, may exhibit 'pseudo-passivity' owing to crystallisation of a layer of nickel salt (in conc. H_2SO_4 probably β-$NiSO_4 \cdot 6H_2O$) on the surface[41].

Influence of Alloying on Anodic Behaviour of Nickel

During recent years a considerable amount of information has been published on the anodic behaviour of nickel alloys. The data include studies both of binary alloy systems in which nickel forms the major alloying component and of more complex commercially produced nickel alloys. The data are sufficiently numerous to permit a rational and fairly complete interpretation of many of the corrosion-resistant properties of nickel alloys on the basis of their anodic behaviour.

Potential/anodic current density curves illustrating the influence of binary alloying additions to nickel are shown as follows: Cr, Fig. 4.22; Fe, Fig. 4.23; Cu, Fig. 4.26; Mo, Fig. 4.28 (curve for Alloy B); Si, Fig. 4.29; Sn, Fig. 4.30; Ti, Fig. 4.31; Al, Fig. 4.32; and Mn, Fig. 4.33. The deductions that may be drawn from the data about the influence of these alloying elements on the anodic behaviour of nickel are summarised in Table 4.24.

It should be noted that the data refer mostly to the behaviour of the alloys in H_2SO_4. Passivity is, however, influenced by the composition of the solution as well as that of the metal and for this reason the influence of alloying additions may be different in solutions containing other ions. In particular, Cl^- and other similarly aggressive ions have a large influence and may prevent passivation, either completely or partially. If passivity cannot be maintained over the entire surface of the metal, pitting develops, and this is considered later.

Broadly speaking the binary alloying additions fall into two categories: (1) those that improve passivity of Ni, viz. Cr, Si, Sn, Ti, Al and (2) those

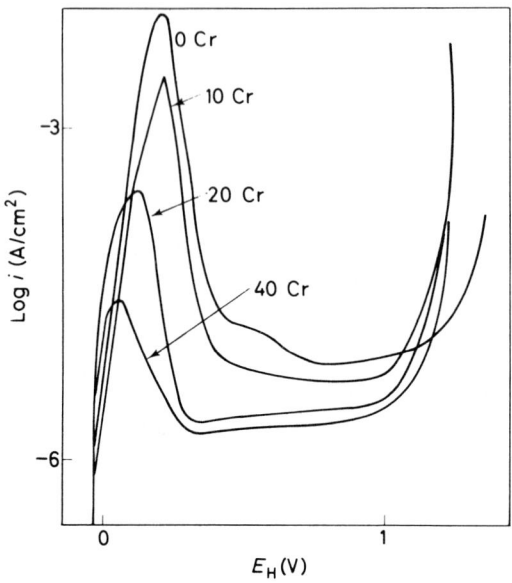

Fig. 4.22 Effect of chromium content on the anodic behaviour of Ni–Cr alloys in 0·5 M H_2SO_4 (de-aerated with H_2) at 25°C; the potential was increased incrementally by 0·025 V every 3 min (after Hodge and Wilde[42])

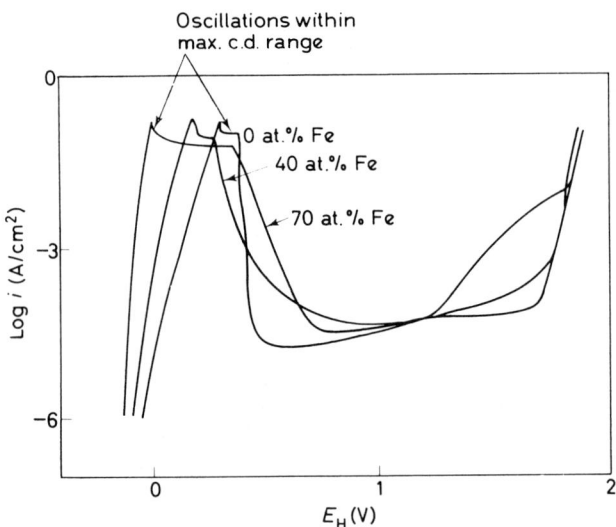

Fig. 4.23 E-log i relationship for the anodic behaviour of Ni–Fe alloys in 0·5 M H_2SO_4 (de-aerated with H_2) at 25°C (after Economy, et al.[43]; see also References 44-46

Table 4.24 Influence of alloying on anodic behaviour of nickel

Alloy addition (%)	Fig. no	Influence on anodic behaviour				
		Active region	Max. c.d. prior to passivation, i_{crit}	Potential of passivation, E_p	c.d. in passive region, i_p	Passive region
0–40Cr	4·22	Potential range reduced	Large decrease	Less noble	Considerable decrease	Potential range increased
0–70 at. % Fe	4·23	Potential range increased	Little effect on magnitude, but potential range of max. c.d. increases. Oscillations often observed within max. c.d. range	More noble	Little effect	Potential range reduced
0–70Cu	4·26	Potential range increased	Increase	More noble, no passivation above ≈50% Cu	Large increase	Potential range reduced, eliminated above ≈50% Cu
0–28Mo	4·28	Potential range moved to more noble potentials	No passivation	No passivation	No passivation	No passivation
0–16·5Si	4·29	Potential range moved to less noble potentials	Decrease	Less noble	Large decrease	Potential range increased
Electrodeposited Sn–35Ni	4·30	Potential range moved to much less noble potentials		Much less noble	Large decrease	Potential range much increased
0–100Ti	4·31	Disappears		Much less noble?	Large decrease	Potential range much increased
0–10Al	4·32	Potential range increased	Decrease	Less noble	0–7% Al considerable decrease	Potential range increased
0–62 at % Mn	4·33	Potential range increased to much more noble potentials	Large increase	More noble	Large increase	Potential range greatly decreased

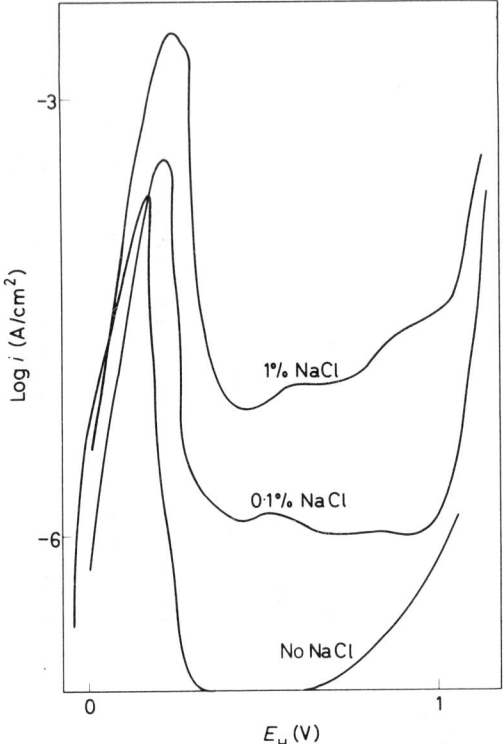

Fig. 4.24 Anodic behaviour of Alloy 600 in 0·5 M H_2SO_4 (de-aerated with N_2) at 24°C containing different concentrations of Cl^- ions (after Piron, et al.[47])

that ennoble Ni, viz. Cu and Mo. Iron and manganese do not belong in either category. Although Ni-Fe alloys can be passivated, their passivity is less than that of nickel and they are also less noble than nickel. In the presence of chromium, however, iron has a considerably beneficial influence on passivity, as may be seen by comparing the curve for the Ni-15Cr-8Fe Alloy 600 in Fig. 4.24 with the curves for binary Ni-Cr and Ni-Fe alloys in Figs. 4.22 and 4.23 respectively.

Alloying elements which enhance the passivity of nickel are expected to improve the corrosion resistance to oxidising media, in particular acidic solutions containing oxidants. Generally, this is found to be so in practice, although it should be noted that strongly oxidising acids, e.g. HNO_3 and H_2CrO_4, or other acidic solutions containing powerful oxidants may render such alloys transpassive, in which condition the corrosion resistance may be impaired. In less oxidising media, particularly in acidic solutions where hydrogen evolution is the cathodic process, not all alloying elements which improve passivity are beneficial, although some are. In these circumstances chromium, silicon and probably aluminium are unhelpful and might be expected to confer little benefit, because the passivation potential, although displaced to slightly more negative values, is not displaced sufficiently to permit passivity to develop in hydrogen-evolving acidic solutions. In contrast, alloying additions of titanium and tin (in the electrodeposited Sn-35Ni

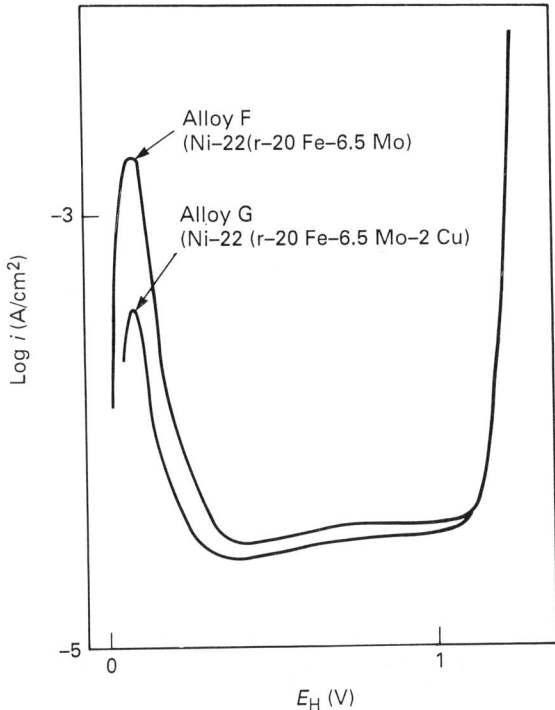

Fig. 4.25 Anodic behaviour of Alloy F and Alloy G in boiling 10% H_2SO_4 de-aerated with H_2 (the potential was increased incrementally every 3 min;. after Leonard[48]; see also Reference 49)

alloy) are undoubtedly beneficial because the active/passive transition is displaced to sufficiently negative potentials that enable passivity to be maintained in non-oxidising acidic solutions.

In fact silicon is also often beneficial, especially in H_2SO_4. In dilute H_2SO_4, Ni-Si alloys containing about 10% Si do not passivate spontaneously, but the rate of anodic dissolution rapidly falls to a low value owing to the formation of a silicon-rich surface layer. In concentrated H_2SO_4 such Ni-Si alloys are passive, whilst in H_2SO_4 of intermediate concentration the corrosion behaviour is complex, being governed by the nature of the cathodic process, which changes as corrosion proceeds[53].

The alloying elements molybdenum and copper do not, by themselves, enhance passivity of nickel in acid solutions, but instead ennoble the metal. This means that, in practice, these alloying elements confer benefit in precisely those circumstances where chromium does not, viz. hydrogen-evolving acidic solutions, by reducing the rate of anodic dissolution. In more oxidising media the anodic activity increases, and, since binary Ni-Mo and Ni-Cu alloys do not passivate in acidic solutions, they are generally unsuitable in such media.

Relatively small amounts of molybdenum in Ni-Cr-Fe alloys, as in stainless steels, render passivation much easier and it may be seen from Fig. 4.25

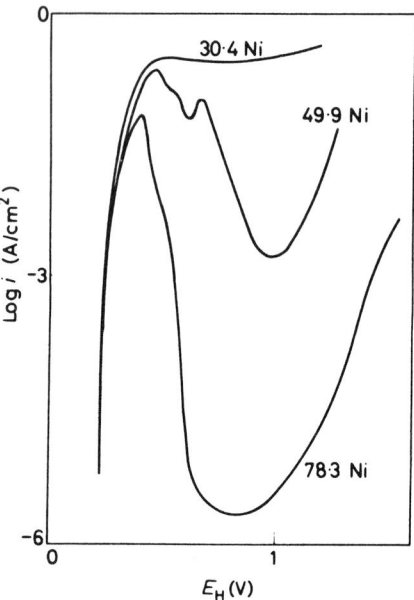

Fig. 4.26 Anodic behaviour of Ni–Cu alloys in 0·5 M H_2SO_4 (de-aerated with N_2) at 25°C; the curve was determined potentiokinetically at 0·4 V/h for the 78·3 and 49·9% Ni alloys and at 3 V/h for the 30·4% Ni alloy proceeding from more positive to more negative (after Osterwald and Uhlig[50])

that the further addition of 2% Cu enhances the effect. The major effect is to reduce the maximum current density prior to passivation, $i_{crit.}$, although the current density in the passive range, i_p, is also reduced. Potential/anodic current density curves of Ni–Cr–Fe–Mo and Ni–Cr–Fe–Mo–Cu alloys plotted in the conventional way do not show these effects clearly, but they may be illustrated by employing fast scan rates[49] or elevated temperatures (see Fig. 4.25). Because of the effect on $i_{crit.}$, Ni–Cr–Fe–Mo and Ni–Cr–Fe–Mo–Cu alloys have good corrosion resistance to acidic solutions both in oxidising conditions and when corrosion is accompanied by hydrogen evolution.

The addition of chromium to Ni–Mo alloys containing about 15% Mo confers passivity, as may be seen by comparing the curves in Fig. 4.28 for Alloy B (Ni–28Mo), Alloy N (Ni–16·5Mo–7Cr) and Alloy C (Ni–16Mo–15·5Cr). Chromium, however, displaces the active region in these alloys to more negative potentials, so that whilst the chromium-containing alloys are more corrosion resistant than the chromium-free alloy in oxidising acidic media, they are less resistant in most hydrogen-evolving acidic solutions.

An interesting illustration of the effect that quite small alloying additions may sometimes have on anodic behaviour is seen in Fig. 4.27[51] from a comparison of the Ni–30Cu alloy Alloy 400 with its age-hardening variant Alloy K500, which contains 2·7% Al and 0·6% Ti. The presence of these elements in the latter alloy is responsible for a well-defined passive region, whereas the former alloy shows only a slight tendency to passivate in acidic

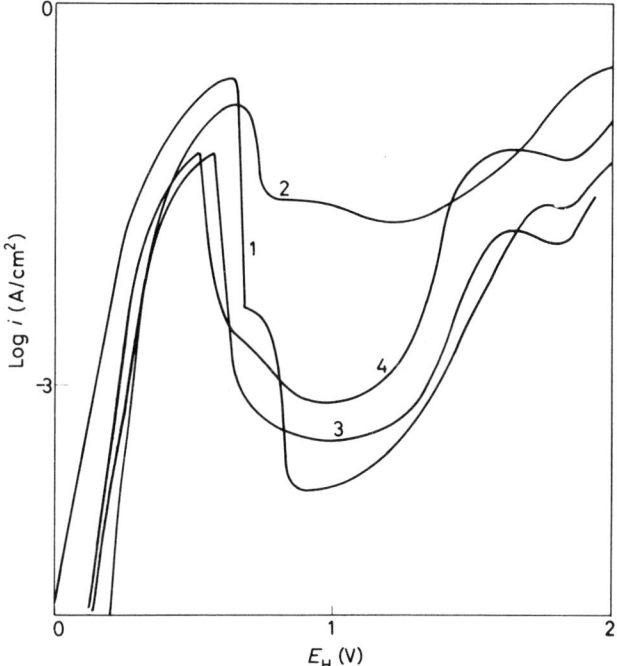

Fig. 4.27 Anodic behaviour of Ni-Cu alloys in 10% H_2SO_4 at ambient temperature. 1, Ni; 2, Alloy 400; 3, Alloy K500 solution treated; 4, Alloy K500 aged (after Flint and Barker[51])

solutions. Furthermore, a clear distinction may be seen between the passivity of the age-hardening alloy in the solution-treated condition, where aluminium and titanium are substantially in solid solution, and in the aged condition, where the alloy is strengthened by precipitates of Ni_3Al and Ni_3Ti. In addition to the removal of most of the aluminium and titanium from solid solution, precipitation also increases the effective copper content of the matrix. Both of these effects may be responsible for the reduction in passivity of the aged material.

Another indication of the influence of precipitated phases on anodic behaviour may be seen in the curve for Alloy C in Fig. 4.28, where the small peak in the middle of the passive range is probably attributable to anodic dissolution of an intermetallic phase (μ) and M_6C carbide[58].

The influence of minor alloying elements and the effect of formation of other phases on the anodic behaviour of nickel alloys are thus not negligible and should not be ignored.

Pitting (Section 1.6)

Pitting of nickel and nickel alloys, as of other metals and alloys, occurs when passivity breaks down at local points on the surface exposed to the corrosive environment, at which points anodic dissolution then proceeds whilst the

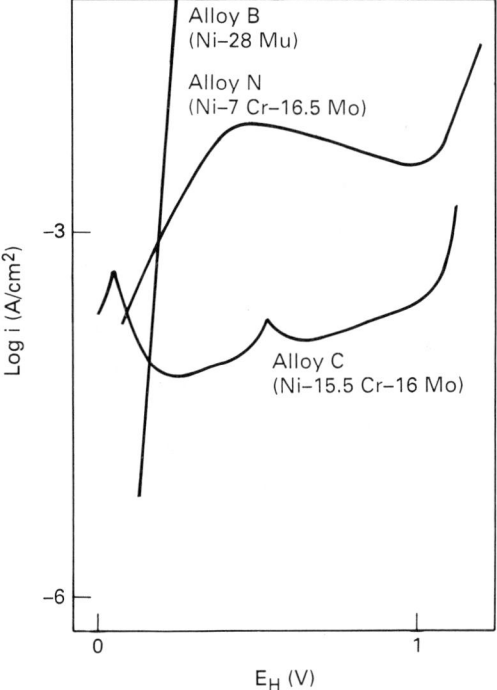

Fig. 4.28 Anodic behaviour of Alloys B, C and N in boiling 10% H_2SO_4 de-aerated with H_2; the potential was increased incrementally (after Leonard[48])

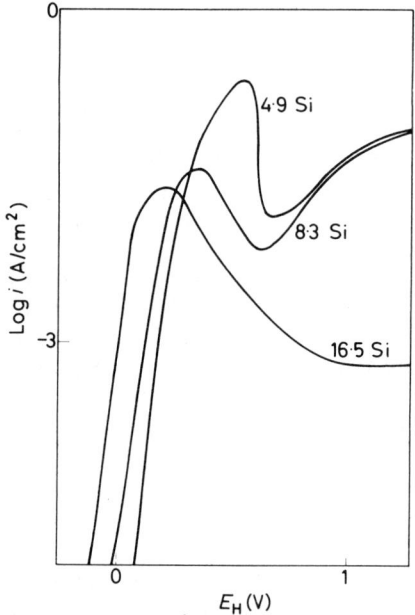

Fig. 4.29 Anodic behaviour of Ni-Si alloys in 25% H_2SO_4 (de-aerated with N_2) at ambient temperature (after Barker and Evans[52])

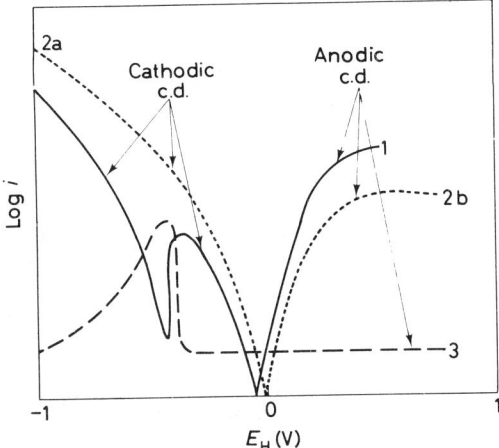

Fig. 4.30 Suggested anodic behaviour of electrodeposited Sn-35Ni alloy; 1, 'observed' curve; 2a, H_2 evolution; 2b, H_2 oxidation; 3, 'true' anodic curve (after Clarke and Elbourne[54])

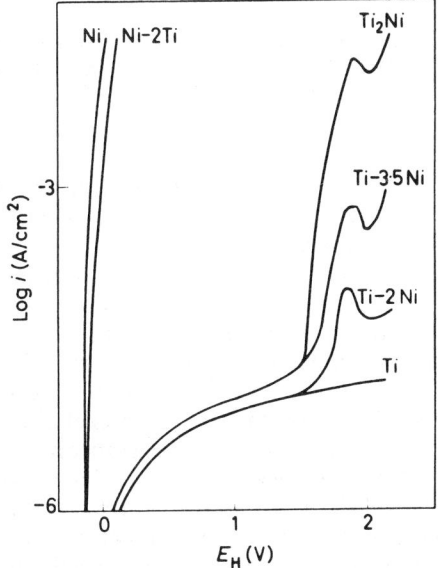

Fig. 4.31 Anodic behaviour of Ni-Ti alloys in HCl + 3·5% NaCl(pH 1), de-aerated with argon, at 22·2°C; the potential was increased by 0·02 V every minute (after Sedriks, et al.[55])

major part of the surface remains passive. Since most (sometimes all) of the cathodic reaction accompanying corrosion is distributed over the passive surface, it follows that the fewer the number of sites of breakdown the more intense is the anodic dissolution at each site, i.e. the fewer the pits the faster they grow, at least in the early stages. Pitting of nickel has been shown to develop preferentially near structural features in the metal, such as grain boundaries, and also at imperfections in the surface, such as scratches[59].

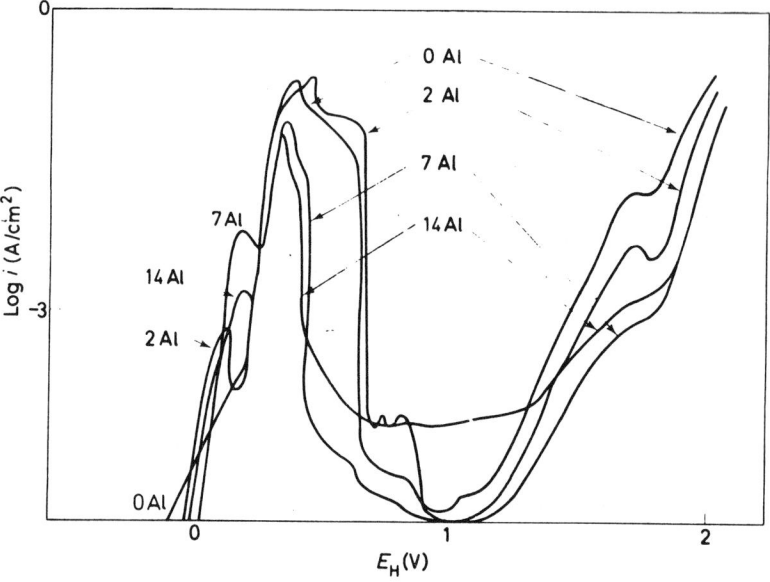

Fig. 4.32 Anodic behaviour of Ni–Al alloys in 0·5 M H_2SO_4, de-aerated with H_2, at 22°C; the potential was increased by 0·01 or 0·02 V every 3 min in the active range and by 0·04 V in the passive range (after Crow, et al.[56])

Electropolishing appears to be helpful in reducing the tendency of pits to develop at surface imperfections, but not necessarily at sites associated with structural features of the metal.

In practice, pitting of nickel and nickel alloys may be encountered if the corrosive environment contains chloride or other aggressive ions and is more liable to develop in acidic than in neutral or alkaline solutions[24]. In acidic solutions containing high concentrations of chloride, however, passivity is likely to break down completely and corrosion to proceed more or less uniformly over the surface. For this reason nickel and those nickel alloys which rely on passivity for their corrosion resistance are not resistant to HCl.

Figure 4.34 illustrates, by means of potential/anodic current density curves, the influence of pH and Cl^- ions on the pitting of nickel[22]. The tendency to pit is associated with the potential at which a sudden increase in anodic current density is observed within the normally passive range (E_B on Curve 1 in Fig. 4.34). It can be seen that in neutral 0·05 M Na_2SO_4 containing 0·02 M Cl^- (Curve 1) E_B has a value of approximately 0·4 V E_H. When pitting develops, the solution in the pits becomes acidic owing to hydrolysis of the corrosion product (see Section 1.6) and when this occurs the anodic current density increases by at least two orders of magnitude and tends to follow the curve obtained in 0·05 M H_2SO_4 + 0·02 M NaCl (Curve 2). Comparison of Curves 2 and 3 illustrates the influence of Cl^- ions on the pitting process.

Owing to the hydrolysis reaction, pit development is an autocatalytic process and often there is an induction period before pit growth attains

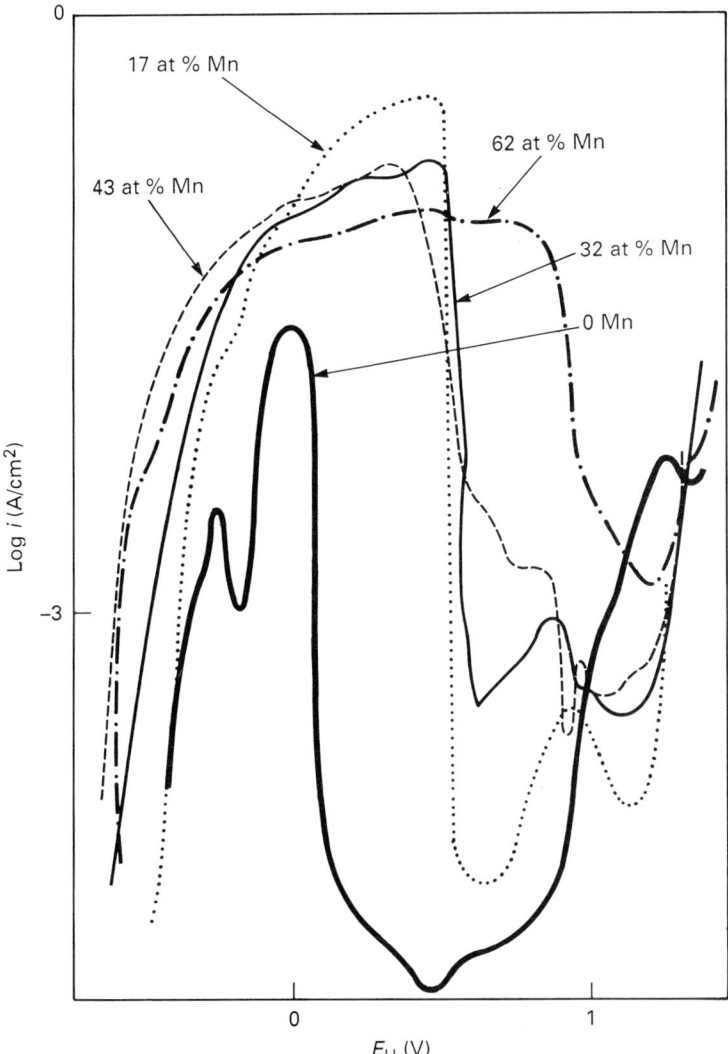

Fig. 4.33 Anodic behaviour of Ni–Mn alloys in 0·5 M H_2SO_4 saturated with H_2 at 20°C (after Horton et al.[57])

observable proportions. In some circumstances, e.g. neutral and alkaline solutions, the induction period may be very long in practice. As with other passive metals and alloys, development of pitting in nickel may be inhibited in flowing solutions[60].

Figure 4.35[61] illustrates the effect of temperature on the rate of development of pitting, measured as a corrosion current in an acidic solution containing Cl^-; it is seen that quite small increments in temperature have large effects. The influence of temperature is of considerable significance when metals and alloys act as heat transfer surfaces and are hotter than the corrosive environment with which they are in contact. In these circumstances,

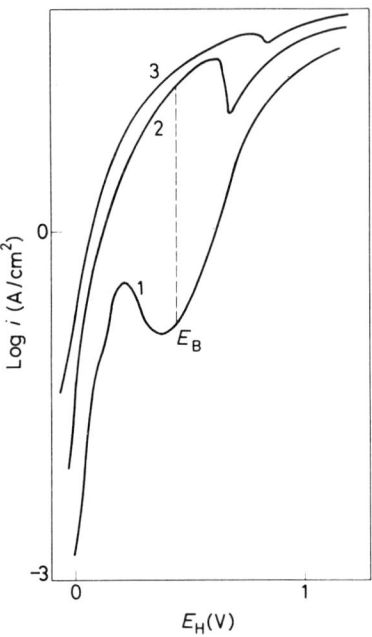

Fig. 4.34 Influence of pH and Cl^- ions on the anodic behaviour of nickel in $SO_4^{2-} + Cl^-$ ion solutions at 20°C (potentiokinetic polarisation at 0·05 V/min). 1, 0·05 M Na_2SO_4 + 0·02 M NaCl; 2, 0·05 M H_2SO_4 + 0·02 M NaCl; 3, 0·05 M H_2SO_4 + 0·05 M NaCl (after Szklarska-Smialowska[22])

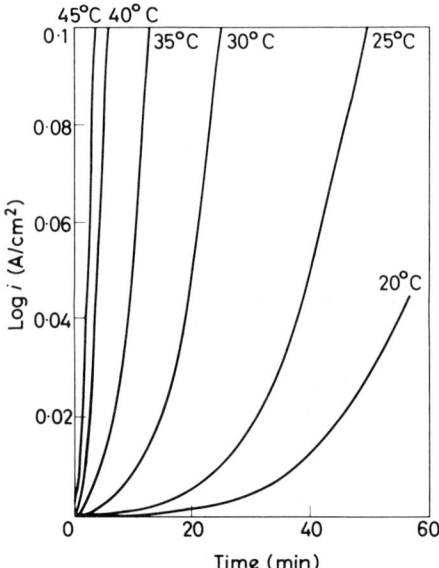

Fig. 4.35 Influence of temperature on breakdown of passivity of nickel in H_2SO_4 + Na_2SO_4 solution (pH 0·4) containing 0·05 M Cl^- (after Gressmann[61])

deep pointed pits may develop rather than the shallower rounded pits usually found when there is no thermal gradient. A possible explanation is that anodic dissolution becomes concentrated at the base of the growing pit in preference to its sides under the influence of the thermal gradient in the metal.

Figure 4.36 shows the influence of pH on the breakdown potential of nickel in alkaline solutions containing Cl^- ions, and it is apparent that the breakdown potential becomes more positive as the pH increases, i.e. breakdown is unlikely unless the solution has a very high redox potential.

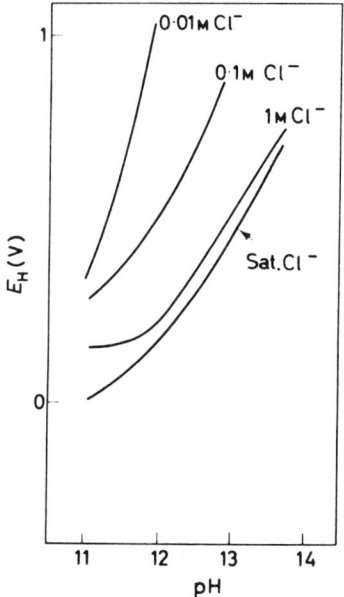

Fig. 4.36 Influence of pH and Cl^- ion on the breakdown potential of commercial nickel in alkaline solutions (0·001-5 M NaOH) de-aerated with N_2 (after Postlethwaite[62])

Alloying nickel with other elements has a marked influence on the susceptibility to pitting. Figure 4.37 shows the variation of the breakdown potential with chromium concentration for binary Ni-Cr alloys[63], and it is seen that breakdown becomes significantly less probable as the chromium increases above 10%. Alloying with iron in addition to chromium yields a further improvement, as may be seen from Fig. 4.24, which shows that the Ni-15Cr-8Fe alloy Alloy 600 exhibits little tendency to breakdown even in an acidic solution containing 1% NaCl. In practice, Ni-Cr-Fe alloys exhibit a high degree of pitting resistance and, as with stainless steels, the addition of a few per cent molybdenum improves their resistance even further.

Nickel alloys which rely on nobility for their corrosion resistance, viz. Ni-Cu and Ni-Mo alloys in acidic solution, do not usually pit in these circumstances. It should be noted, however, that the Ni-Cu alloy Monel 400 normally forms a protective oxide film in neutral and alkaline solutions, and this is of particular significance with regard to its corrosion resistance to

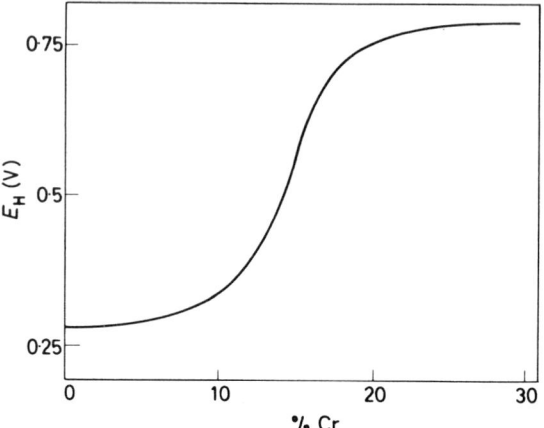

Fig. 4.37 Influence of the chromium content of Ni–Cr alloys on the breakdown potential in 0·1 M NaCl at 25°C de-aerated with N_2 (after Horvath and Uhlig[63])

sea-water. In circumstances where the supply of O_2 is insufficient to maintain the film in good repair, as in stagnant conditions, pitting may develop.

Crevice Corrison (Section 1.6)

In recent years crevice corrosion has received increased attention owing to the serious hazards that develop if this type of localised attack is overlooked or ignored. Crevice corrosion can be an especially serious problem with passive metals and alloys because breakdown of passivity in the deoxygenated solution that develops in crevices leads to anodic dissolution. Of the nickel alloys those containing molybdenum and, to a lesser extent, copper offer the best-resistance to this form of attack. Ni–Cr–Mo alloys are among the most resistant of metallic materials to crevice corrosion, although their resistance may be impaired if intergranular precipitates of molybdenum-rich M_6C carbide are allowed to form[64] (see Intergranular Corrosion). In cast materials at least, solution heat-treatment of the 625 type of alloy is beneficial and if such a heat-treatment is given, reduction of the niobium content of the alloy may be cost-effective[65].

Intergranular Corrosion (Sections 1.3 and 1.7)

As with most other metal and alloys systems, nickel and certain of its alloys may suffer intergranular corrosion in some circumstances. In practice, intergranular corrosion of nickel alloys is usually confined to the vicinity of welds as a result of the effects produced by the welding operation on the structure of the material in those regions. Alloys that are subjected to other similarly unfavourable thermal treatments may also become susceptible. The compositions of most commercial nickel alloys that are marketed today are, however, carefully controlled to minimise the possibility of intergranular corrosion developing in welded material during service.

Intergranular corrosion of nickel and its alloys is nearly always associated with grain boundary precipitates. In certain commercial grades of nickel, which contain carbon as an impurity, lengthy exposure to high temperatures may result in the formation of a grain boundary film of graphite which in some circumstances renders the material susceptible to intergranular corrosion on subsequent exposure in an environment to which the material is otherwise well suited, viz. caustic alkalis; with nickel this form of corrosion may be intensified by stress in the metal. For these reasons, the low-carbon grade of commercial nickel, Nickel 201, is, in practice, preferred where this form of attack is a possibility. With material of higher carbon content the possibility of intergranular corrosion developing to a serious extent may be minimised by applying a stress-relieving heat treatment after fabrication. The presence of other elements in nickel, notably sulphur, may also render the metal liable to intergranular penetration and embrittlement.

The types of chromium-containing nickel alloys that owe their corrosion resistance to passivity, viz. Ni–Cr–Fe, Ni–Cr–Fe–Mo and Ni–Cr–Fe–Mo–Cu alloys, may become susceptible to intergranular corrosion in circumstances broadly similar to those that produce susceptibility in stainless steels[66, 67]. In these materials, preferential attack by the corrosive environment occurs at zones immediately adjacent to grain boundaries at which precipitates of the chromium-rich carbides $M_{23}C_6$ or possibly M_7C_3 have formed, the attack being concentrated on the chromium-depleted zones adjacent to the precipitate, since these zones cannot become passivated[68, 69, 70]. As with stainless steels, the appropriate preventative measures are to minimise carbide formation by controlling the carbon content of the material to levels as low as practicable — nowadays 0·02% C max. is attainable — to increase the chromium content and to add elements such as titanium and niobium to form carbides more stable than $M_{23}C_6$ with the residual carbon and thus prevent chromium-depletion. It should be noted, however, that owing to the higher activity of carbon in nickel-rich alloys than in stainless steels, a greater proportion of stabilising element such as titanium is needed in the former materials than in the latter[71]. Intergranular corrosion of stainless steels and Ni–Cr–Fe alloys has been observed to occur in the absence of grain boundary carbide precipitates in the alloy during laboratory tests in highly oxidising acidic solutions such as HNO_3 containing chromates or dichromates[72, 73], and is associated with segregation of P and Si to grain boundaries. A review of intergranular corrosion of alloys in the Fe–Ni–Cr system, including stainless steels and nickel alloys, is available[72].

Another type of nickel alloy with which problems of intergranular corrosion may be encountered is that based on Ni–Cr–Mo containing about 15% Cr and 15% Mo. In this type of alloy the nature of the grain boundary precipitation responsible for the phenomenon is more complex than in Ni–Cr–Fe alloys, and the precipitates that may form during unfavourable heat treatment are not confined to carbides but include at least one intermetallic phase in addition. The phenomenon has been extensively studied in recent years[58, 64, 74-79]. The grain boundary precipitates responsible are molybdenum-rich M_6C carbide and non-stoichiometric intermetallic μ

phase (Ni, Fe, Co)$_3$ (W, Mo, Cr)$_2$[58]. Depending on the nature of the corrosive environment attack in this type of alloy may be either at depleted zones adjacent to grain boundaries or on the grain boundary precipitates themselves. Thus two different mechanisms of intergranular corrosion operate in this type of alloy, one involving attack on the depleted regions being observed in HCl (and perhaps other hydrogen-evolving acidic solutions), the other, in which the precipitates themselves are preferentially attacked, being observed in more highly oxidising acidic media. An observation of significance made some years ago was that limiting the silicon content of this type of alloy to very low levels reduced the tendency for formation of the intermetallic phase during welding[75] and this led to the introduction of improved commercial alloys of the C276. More recently a composition possessing even greater thermal stability, i.e. C4, has been developed, in which iron and tungsten present in the earlier alloys have been largely replaced by nickel[79] and further alloys have been introduced with higher Cr and lower Mo contents, e.g. C22 and 625 (see Table 4.21).

Ni-Mo alloys containing about 28% Mo are a third category of nickel alloy liable to intergranular corrosion in the welded condition. In these alloys preferential corrosion may develop at zones adjacent to welds exposed to HCll and other hydrogen-evolving acids in which this type of alloy is used. Corrosion is preferentially concentrated on molybdenum-depleted zones adjacent to grain boundaries in which molybdenum-rich M_6C carbide has precipitated. The susceptibility of this type of alloy to intergranular corrosion is reduced by controlling the carbon and iron content to levels as low as is practicable and also by addition of about 2% V[80] or 3·5-5% W[81]. Niobium may also be a beneficial addition[80], but titanium and zirconium accelerate intergranular corrosion of this type of alloy[81].

Bimetallic Corrosion

Owing to their intermediate position in the galvanic series, nickel and nickel alloys may stimulate corrosion of metals less noble to themselves when in bimetallic contact and thus receive cathodic protection or suffer intensified corrosion from contact with more noble metals and graphite. In general, in mild environments such as unpolluted atmospheric conditions, nickel and nickel alloys are compatible with a fairly wide range of other metals and alloys, but in strong electrolytes such as sea-water and marine atmospheres the range of compatible couples is less. Table 4.25 gives guidance in very general terms, but should not be assumed to apply in every circumstance, since other factors may influence the issue. The relative surface areas of the two metals in contact plays a large part in determining whether bimetallic corrosion is serious or not, and the combination of a small area of the more negative (less noble) metal or alloy in contact with a large area of the more noble material is usually the most dangerous situation (see Section 1.7). Protection of the less noble metal by painting or other means, if properly carried out, is usually effective in minimising bimetallic corrosion. In aggressive environments nickel and the different types of nickel alloy are not necessarily wholly compatible one with another.

Corrosion in Natural Environments

The Atmosphere

Nickel and nickel alloys possess a high degree of resistance to corrosion when exposed to the atmosphere, much higher than carbon and low-alloy steels, although not as high as stainless steels. Corrosion by the atmosphere is, therefore, rarely if ever a factor limiting the life of nickel and nickel alloy structures when exposed to that environment.

Table 4.25 Bimetallic corrosion effects of nickel and nickel alloys
(General guidance only; other factors, including relative surface areas, often exert an important influence)

Corrosive environment	Corrosion of nickel or nickel alloy is stimulated by bimetallic contact with:	Bimetallic contact with nickel or nickel alloy has little or uncertain influence*:	Bimetallic contact with nickel or nickel alloy stimulates corrosion of:
Most atmospheric conditions except marine atmospheres	Au Pt	Rh Pd C (graphite) Ti Cu and Cu alloys Stainless steels Cr plate C steel† Al and Al alloys† Mg and Mg alloys†	Pb Sn Soft solders Cd Zn Galvanised steel Al clad Carbon steel‡ Al and Al alloys‡ Mg and Mg alloys‡
Sea-water and marine atmospheres	C (graphite) Graphitised cast iron Au Pt	Ag and Ag brazing alloys Cu and Cu alloys Pb Sn Soft solders Other Ni alloys	Austenitic cast iron§ Low-alloy steels Cast iron (ungraphitised) Wrought iron Carbon steel Cd Al and Al alloys Zn Mg and Mg alloys

* Little effect in most atmospheres, except marine. Effects in sea-water and marine atmospheres depend on surface area relationships.
† If properly painted. ‡ If unpainted or improperly painted. § Contact with small area of Alloy 400 has little effect.

The appearance of bright nickel is, however, impaired by exposure to moist, polluted atmospheres owing to the phenomenon known as *fogging*. Vernon showed more than 60 years ago that for 'fogging' of nickel to occur, a high humidity—greater than about 70% r.h.—and the presence of SO_2 were both necessary[82]. Fogging is due to the catalytic oxidation of SO_2 in polluted atmospheres by the nickel surface and subsequent corrosion of the nickel by the liquid film of H_2SO_4 thus formed on the surface, the corrosion product—a basic nickel sulphate—being responsible for fogging. In the early stages the film can be readily removed by wiping with a cloth, but once the surface has become fogged the bright appearance cannot be restored merely by wiping and mild abrasion is needed. Some nickel alloys, viz. the

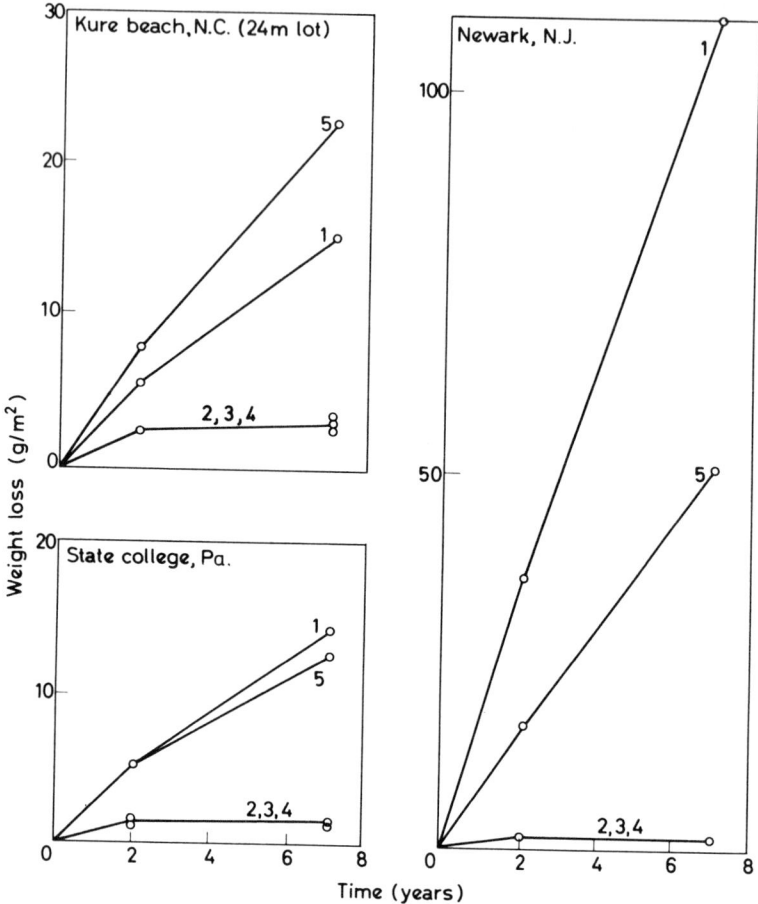

Fig. 4.38 Atmospheric corrosion of nickel and nickel alloys during exposure tests at sites in the USA. 1, Nickel 200; 2, Alloy 600; 3, Alloy 800; 4, Alloy 825; 5, Alloy 400 (after van Rooyen and Copson[85])

Ni-30 Cu Alloy 400 also undergo fogging, but alloys containing 15% Cr or more do not exhibit this phenomenon. Fogging is prevented by a very thin film of chromium deposited on the surface—a fact which forms the basis for the bright appearance of decorative chromium-nickel plate (see Sections 13.7 and 13.8).

Nickel and nickel alloys do not form thick layers of corrosion products when freely exposed to outdoor atmospheres in circumstances where the surface is periodically washed by rain, but such deposits may form on sheltered surfaces. Quantitative data on the rate of loss of metal and of pitting of nickel and nickel alloys exposed to outdoor atmospheres are available[83-86]. Figure 4.38 shows results obtained at three sites in the USA over a 7 year period[85] and Fig. 4.39 gives results from a 10 year test at Birmingham[86]. In both series of tests, Ni-Cr-Fe alloys gave lower weight losses than nickel itself or Ni-Cu alloys and the American results bring out the

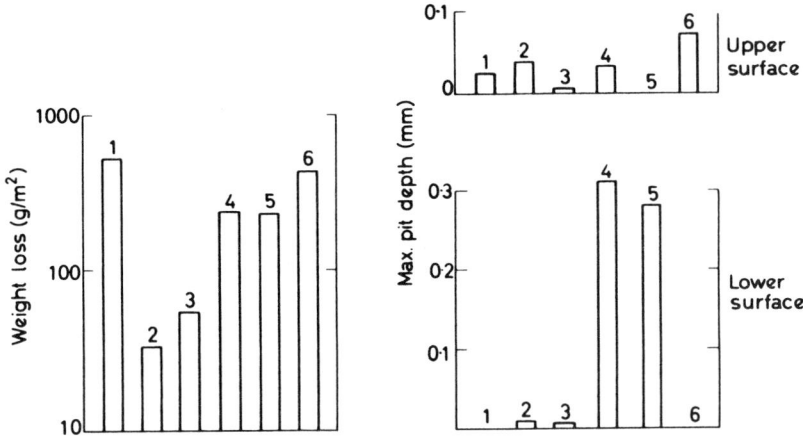

Fig. 4.39 Atmospheric corrosion of nickel and nickel alloys at Birmingham, England, during exposure tests of 10 year duration. 1, Nickel 200; 2, Alloy 600; 3, Alloy DS (Fe-37Ni-18Cr-2Si); 4, Alloy 400; 5, Alloy K500; 6, Ni-28Mo (after Evans[86])

point that over long periods the corrosion rate of Ni–Cr–Fe alloys in the atmosphere declines to a very low value whilst that of nickel and Ni–Cu alloys remains approximately linear. Comparison of the data in Figs. 4.38 and 4.39 shows that corrosion at the UK site was several times greater than that at the most aggressive American site. This has also been observed with stainless steels in a test programme where a direct comparison was made between identical test samples[87].

Fresh Water

Nickel and nickel alloys are normally resistant to fresh water and natural waters at temperatures up to normal boiling point; there may, however, sometimes be a risk of pitting in waters of high acidity or high salinity in stagnant conditions. In flowing conditions, oxygen dissolved in the water is normally sufficient to maintain passivity. Aerobic bacteria appear to have little influence, but corrosion may become severe in the presence of bacteria-induced decay products.[88] Steam condensates containing O_2 and CO_2 may, however, be aggressive towards nickel and Ni–Cu alloys, in which circumstances Ni–Cr–Fe alloys are more resistant.

Sea-water

Nickel and nickel alloys possess good resistance to sea-water in conditions where the protective properties of the passive film are fully maintained. As pointed out above, Ni–30 Cu Alloy 400, in contrast to its behaviour in acidic solution, normally forms a protective film in neutral and alkaline environments, including sea-water; this alloy and its age hardening variant

Alloy K500 is widely used in sea-water. A particularly valuable feature of the behaviour of nickel and its alloys in sea-water is the ability of the protective surface film to remain in good repair in highly turbulent and erosive conditions. Because of this the alloys are used extensively in pumps and valves and other similar equipment in contact with sea-water flowing at high velocity. The protective film on nickel, Ni-Cu and Ni-Cr alloys is normally kept in good repair providing the effective sea-water velocity is greater than approximately 2 m/s[89] and in these circumstances overall corrosion rates are normally of the order of 0·01 mm/y.

In sea-water flowing at slower velocities and more especially in stagnant conditions, pitting and crevice corrosion may develop, particularly beneath deposits and marine growths at the surface of the metal. Some data for the Ni-30 Cu Alloy 400[90] are shown in Fig. 4.40; the corrosion was mostly pitting.

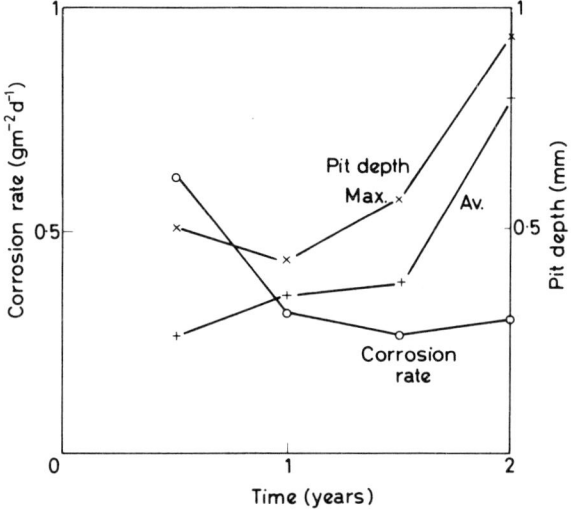

Fig. 4.40 Corrosion of Monel 400 in sea-water at Port Hueneme Harbour, Cal., USA (after Brouillette[90])

Ni-Cr-Fe alloys are liable to suffer more intensive pitting than Ni-Cu alloys and nickel itself in low velocity sea-water[89], but the addition of a few per cent of molybdenum to Ni-Cr-Fe alloys greatly improves the resistance to pitting and crevice corrosion. Table 4.26 shows some data reported by Niederberger, Ferrara and Plummer[91] which illustrate the magnitude of the improvement. It will be seen from Table 4.26 that Ni-Cr-Mo alloys possess the best resistance to corrosion and pitting in sea-water. A specimen of Ni-Cr-Mo alloy has been reported to be immune to corrosion in sea-water over a 10 year period, having suffered no weight loss and no pitting[92]. With cast material annealing may improve resistance to crevice corrosion if niobium is also present in the alloy[65]. When immersed in deep sea-water nickel and nickel alloys undergo less corrosion than in shallow conditions[93].

Table 4.26 Corrosion of nickel alloys in quiet and in slow moving sea-water (after Niederberger, et al[91])

Alloy composition	Quiet sea-water			Slow moving sea-water		
	Wt. loss (g)	Range of pit depth (mm)		Wt. loss (g)	Range of pit depth (mm)	
		Exposed	Crevice		Exposed	Crevice
Ni-35Cu	20·40	0-0·5	0·75-0·8	21·10	0·25-0·38	0·6-0·68
Ni-30Cu-3Al	19·50	0·55-0·65	0·55-0·58	24·40	0·8-1·0	0·5-0·7
Ni-16Cr-7Fe	11·85	3·25	3·25	12·55	3·25	0·73-3·25
Ni-35Cr-2Fe	9·32	3·0	0·15-0·7	7·62	3·0	0·38-1·65
Ni-47Fe-20Cr	15·72	3·4	3·63	24·50	3·63	3·63
Ni-27Mo	42·80	0·5-0·53	0·33-0·35	54·80	0·25-0·35	0·2-0·4
Ni-16Mo-7Cr-4Fe	0·50	Nil	0-0·05	1·45	Nil	0·08-0·1
Ni-30Fe-21Cr-3Mo	0·25	0-0·03	0-0·23	0·20	0·08-0·15	0·03-0·65
Ni-22Cr-9Mo-2Fe	Nil	Nil	Nil	0·25	Nil	Nil
Ni-20Cr-5Mo-6Fe	0·20	Nil	0·35-0·63	0·15	0-0·05	Nil
Ni-16Cr-16Mo-4W	Nil	Nil	Nil	0·10	Nil	Nil

Conditions of exposure: Quiet sea-water—suspended from raft.
　　　　　　　　　　　Slow moving sea-water—velocity 0·3-0·6 m/s.
Location: Harbor Island, Wrightsville, Beach, N.C., USA.
Duration of test: 2 years; panel size 12 × 3 in (0·305 × 0·076 m).
　　　　　　　　　crevice area 1 × 1 in (0·025 × 0·025 m).

Underground

As other cheaper materials usually give satisfactory performance, nickel and nickel alloys are not normally required for applications involving resistance to corrosion underground. Data on their behaviour in these circumstances are therefore sparse; in particular, whether micro-organisms responsible for the accelerated corrosion of ferrous and other metals in certain anaerobic soils have any influence on nickel and its alloys, is uncertain.

Corrosion in Chemical Environments

Acids

The wide range of corrosion-resistant nickel alloys that are produced commercially is capable in practice of handling most types of acid. Since the nickel-alloy range includes some that are corrosion resistant by virtue of their relative nobility and others that owe their resistance to passivity, alloys suitable both for hydrogen-evolving acids and for more oxidising acids are available. Table 4.27 contains a summary of data mainly derived from laboratory corrosion tests to illustrate the behaviour of individual alloys in some common mineral and organic acids.

The data in Table 4.27 refer to solutions of pure acids; in practice the presence of impurities often has a large influence and modifies the corrosion resistance to a greater or lesser extent. Oxygen from the air stimulates corrosion of alloys of the relatively noble type, including nickel itself, Ni–Cu and Ni–Mo alloys, but may be helpful in maintaining passivity of the other type, viz. alloys containing 15% Cr or more. Other oxidants such as Fe^{3+} or Cu^{2+} (which may sometimes in practice be present through corrosion of ferrous and copper-base alloys also in contact with the acidic environment) usually have a similar influence. The presence of halide ions, especially Cl^- and F^-, in H_2SO_4, H_3PO_4 and HNO_3 is usually highly detrimental to the corrosion resistance of both the noble and passive types of alloys. It should be noted, however, that Alloy 690 possesses sufficient resistance to HNO_3-HF mixed acids to be practically useful and that a Co-20Cr-15W-10Ni alloy is similarly resistant to certain HNO_3-HCl mixtures.

In addition to impurities, other factors such as fluid flow and heat transfer often exert an important influence in practice. Fluid flow accentuates the effects of impurities by increasing their rate of transport to the corroding surface and may in some cases hinder the formation of (or even remove) protective films, e.g. nickel in HF. In conditions of heat transfer the rate of corrosion is more likely to be governed by the effective temperature of the metal surface than by that of the solution. When the metal is hotter than the acidic solution corrosion is likely to be greater than that experienced by a similar combination under isothermal conditions. The increase in corrosion that may arise through the heat transfer effect can be particularly serious with any metal or alloy that owes its corrosion resistance to passivity, since it appears that passivity breaks down rather suddenly above a critical temperature, which, however, in turn depends on the composition and concentration of the acid. If the breakdown of passivity is only partial, pitting may develop or corrosion may become localised at hot spots; if, however, passivity fails completely, more or less uniform corrosion is likely to occur.

Table 4.27 provides a basis for selecting the nickel alloy type likely to be suitable for service in particular acids. The Ni-Cr-Fe-Mo-Cu and Ni-Cr-Fe-Mo alloys, both wrought and cast, are the types most often selected for H_2SO_4, and they possess the additional advantage that their resistance is not greatly affected by the presence of SO_2. Cast Ni-Si alloys containing 9% Si or more and, preferably, alloying additions of copper, titanium and molybdenum[94, 95] are also used for H_2SO_4. Most nickel alloys have good resistance to pure H_3PO_4, but the presence of halide ion impurities reduces the resistance in the higher concentration range. Alloy 690 possesses good resistance to HNO_3 and is one of the few metals able to withstand the combination of HNO_3 plus HF; it is not, however, resistant to mixtures of HNO_3 and HCl. Among metallic materials Alloy B2 is one of the most suitable for handling HCl, particularly in the absence of air and other oxidants; in oxidising conditions Ni-Cr-Mo alloys are usually more suitable. Nickel itself and Alloy 400 both possess good resistance to HF; in practice Alloy 400 is used for aqueous and anhydrous HF, but precautions are necessary against stress-corrosion cracking. In practice, nickel and nickel alloys have good resistance to most organic acids.

Alkalis

Nickel and its alloys are among the most resistant metallic materials to caustic alkalis. Nickel itself possesses outstanding resistance to NaOH and KOH and is used to contain these substances over the entire concentration and temperature ranges that are of practical interest, viz. 0–100% and up to 350°C. At the higher concentrations and temperatures KOH is significantly more corrosive than NaOH towards nickel and in these circumstances the metal is sometimes cathodically protected.

At temperatures above 300°C, low-carbon nickel (0·02% C) is preferred to avoid the possibility of intergranular attack developing after long exposure; if material of higher carbon is employed it should be annealed after fabrication and before exposure to caustic alkalis to prevent stress-assisted intergranular corrosion.

The corrosion rate of nickel in sodium hydroxide is adversely affected by heat transfer by small amounts of oxidisable alkaline sulphur-containing salts, e.g. Na_2SO_3, $Na_2S_2O_3$, Na_2S and, at high temperatures, by alkaline oxidising agents, viz. $NaClO_3$ and Na_2O_2. In the former circumstance Alloy 600 is more resistant than nickel, but not in the latter. When Alloy 600 is used for service in caustic alkalis, it should be stress relieved after fabrication to minimise the possibility of stress-corrosion cracking.

Other nickel alloys, notably Alloy 600, also possess good resistance to caustic alkalis.

Salts

Nickel and nickel alloys generally possess good corrosion resistance to acidic, neutral and alkaline salts, including halides, that are not oxidising in character. Oxidising salts are usually corrosive towards Ni, Ni–Cu and Ni–Mo alloys, but not to Ni–Cr and Ni–Cr–Fe–Mo–Cu alloys unless they contain appreciable quantities of both oxidiser and halide ions, e.g. $FeCl_3$, $CuCl_2$, NaOCl. Ni–Cr–Mo alloys are among the few metallic materials that are resistant to oxidising halide salts.

Wet and Dry Gases

Nickel and its alloys are usually resistant to dry gases, including NH_3, SO_2, F_2, Cl_2, HCl and HF even at high temperatures, and are often the preferred materials for handling such gases. Nickel and Alloy 600 are used in service at elevated temperatures with dry Cl_2, HCl, F_2 and HF. Alloy 600 and certain other nickel alloys are resistant to dry SO_2 and dry NH_3. When moist or under dew-point conditions these gases are in many instances appreciably more corrosive towards nickel and most nickel alloys, with some exceptions. Ni–Cr–Mo alloys do, however, possess good resistance to condensates containing SO_2 and Cl at temperatures well in excess of 100°C, and also to solutions containing NH_3 and its salts. Ni–Cr–Mo alloys are among the most resistant metallic materials to moist halogens.

Table 4.27 Corrosion resistance of nickel and nickel alloys to acidic solutions

Alloy	H_2SO_4	H_3PO_4	HNO_3	HCl	HF	Other acids
Nickel 200 and 201	A, 0–20%[1], RT B, 0–50%[1], 70°C	A, 0–85%[1], RT	U	A, 0–10%[1]	B, 0–70%[1], RT A, 0–90%[1], RT	Acetic, B, 0–100%, BP Formic, A, 0–90%[1], RT
Alloy 400	A, 0–85%[1], 30°C B, 0–60%[1], 95°C B, 25–85%[2], 30°C	A, 0–90%[1], 100°C	U	A, 0–8%[1], RT	B, 0–60%[1], RT A, >90%[1], RT	Acetic, A, 0–100%[1], RT B, 0–40%[2], RT B, 65–100%[2], RT
Alloy B2	A, 0–96%[1], 65°C A, 0–50%, BP	A, 0–85%[1], 65°C A, 0–50%, BP	U	A, 0–37%[1], RT B, 0–25%[1], 65°C B, 0–2%, BP	A, 0–45%[1], RT	Acetic, A, 0–100%, BP Formic, A, 0–90%, RT, BP B, 0–60%, 65°C
Alloy C276 C4 C22	A, 0–96%, 65°C B, 0–10%, BP	A, 0–85%, 65°C A, 0–50%, BP	A, 0–70%, RT A, 0–30%, 65°C B, 30–70%, 65°C B, 0–10%, BP	A, 0–37%, RT A, 0–2%, 65°C B, 2–37%, 65°C A, 0–1%, BP	A, 0–5%, RT B, 5–45%, RT	Acetic, A, 0–100%, BP Formic, A, 0–90%, 65°C B, 0–90%, 65°C
Alloy G	B, 0–10%, BP	A, 0–30%, BP B, 30–85%, BP			A, 0–45%, RT	

(continued opposite)

Table 4.27 (continued)

Alloy	H_2SO_4	H_3PO_4	HNO_3	HCl	HF	Other acids
Alloy 825	A, 0-5%, 80°C B, 0-45%, BP B, 40-80%, 100°C	A, 0-20%, BP B, 20-80%, BP A, 0-80%, 75°C	A, 0-30%, BP B, 30-70%, BP A, 0-70%, 75°C		B 38-70%, 60°C	Acetic, A, 0-100%, BP Formic, A, 0-100%, BP Oxalic, A, 0-10%, BP B, 10-50%, BP
Alloy 600	A, 0-70%, RT[3]	A, 0-80%, RT[3]		A, 0-2%, RT B, 2-15%, RT[3]	S	Acetic, A, 0-100%, RT

The data show corrosion resistance as a function of acid composition, concentration and temperature. Since the data are mostly derived from laboratory corrosion tests in pure solutions, they should not be taken as a firm indication of performance in service.

A = < 0·1 mm/year
B = 0·1-0·5 mm/year
S = no data, but often suitable in service
U = unsuitable

% = concentration w/w
RT = room temperature
BP = boiling point

(1) = air-free solutions; aeration increases corrosion
(3) = saturated with air
(3) = not resistant at high temperatures

Organic Compounds

Nickel and nickel alloys are resistant to many organic compounds and are often suitable for handling organic acids, alcohols and halogenated hydrocarbons. It should be borne in mind, however, that halogenated organic compounds may undergo hydrolysis in the presence of water or steam and release appreciable quantities of the corresponding halogen hydracid, and this will often dictate the choice of the alloy. Detailed information should be sought concerning the suitability of alloys for particular circumstances.

Water and Steam at High Temperatures

The corrosion rates of nickel and nickel alloys in pure water and steam at elevated temperatures are generally extremely low, typically of the order of 1 μm/year. The metal and its alloys are therefore often selected for service in these environments in circumstances where contamination of the water by metal ions is to be avoided. It should be noted, however, that the possibility of stress corrosion may need to be taken into account in certain circumstances (see below). Additionally where phosphate water treatment has been used in PWR secondary heat exchangers, severe localised corrosion has occurred when alkaline phosphates have been permitted to accumulate[98, 99].

Conjoint Action of Stress and Corrosion (Chapter 8)

As with alloys of other metals, nickel alloys may suffer stress-corrosion cracking in certain corrosive environments, although the number of alloy environment combinations in which nickel alloys have been reported to undergo cracking is relatively small. In addition, intergranular attack due to grain boundary precipitates may be intensified by tensile stress in the metal in certain environments and develop into cracking. Table 4.28 lists the major circumstances in which stress corrosion or stress-assisted corrosion of nickel and its alloys have been recorded in service and also shows the preventive and remedial measures that have been adopted, usually with success, in each case.

With regard to stress-corrosion cracking in the Ni-Cr-Fe system, including both nickel-base alloys and stainless steels, a vast number of papers has been published. A detailed review of work published before 1969 is available[96] and the authors have since published additional data[97].

The susceptibility of nickel alloys, principally Alloys 600 and 800 to stress-corrosion in water-cooled nuclear reactor heat-exchanger circuits has received much attention. The influence of both metallurgical variables (e.g. alloy composition, heat-treatment) and water chemistry (additives, inhibitors) have been extensively studied and reviewed.[99-102]

In recent years several new Ni-Cr-Fe-Mo and Ni-Cr-Fe-Mo-Cu have been introduced with improved resistance to sulphide stress cracking in sour oil and gas environments.[103-104]

Table 4.28 Stress corrosion of nickel and nickel alloys

Alloy type	Environment	Type of cracking	Preventive or remedial measures
Nickel 200	Caustic alkalis, high concentrations and high temperature	IG	1. Stress-relieve after fabrication 2. Use low-carbon nickel (0·02% C max.)
Alloy 600 Alloy 400	Caustic alkalis, high concentrations and high temperature	Usually IG	Anneal after fabrication
Ni–Cr–Fe and Ni–Cr–Fe–Mo–Cu (Ni approx. <40%)	Chloride solutions at elevated temperature	TG	1. Use alloys with higher nickel content[96] 2. Control amounts of minor alloying elements and impurities[96] 3. Remove or reduce Cl^- in environment if possible
Alloy 600 Alloy 800	Water-cooled nuclear reactor circuits	Usually IG	1. Control alloy composition and processing 2. Control water chemistry 3. Stress relieve after fabrication
Alloy 800	Polythionic acids	IG	Stabilise material against intergranular corrosion
Alloy 400	Hydrofluoric acid	Usually IG	1. Stress relieve after fabrication 2. Control composition of weld electrode/filler 3. Control residual O_2 and other oxidising agents
Ni–Cr–Fe–Mo and Ni–Cr–Fe–Mo–Cu	Sour oil and gas	IG	1. Select appropriate alloy and strength level

IG = Intergranular; TG = Transgranular.

High-temperature Corrosion and Oxidation

Nickel and Ni–Cr alloys are among the most resistant metallic materials to corrosion and oxidation at high temperatures and are widely used to resist corrosion by gases and molten salts at elevated temperatures (see Sections 7.1 and 7.5).

Applications in Industry

As will have become apparent, nickel and corrosion-resistant nickel alloys have wide ranges of application, particularly in industries where strongly acidic, strongly alkaline or strongly saline environments are encountered. Table 4.29 lists some of the more important applications in those industries where these conditions most frequently arise, i.e. in the chemical, petrochemical, oil and gas, nuclear and conventional power generating, textile, paper, marine, desalination and food processing industries. The list is by no means exhaustive and there are many other applications of a similar nature in these and other industries. The table should, nevertheless, serve

Table 4.29 Applications of Ni and corrosion-resistant nickel alloys

Alloy type	Industry						
	Chemical	Petrochemical	Nuclear and power	Marine and desalination	Textile and paper	Food	Other
Ni	Production of caustic alkalis Production of phenol Salt evaporators	Chlorination processes Oxy-chlorination processes Production of phenolic resins Production of synthetic rubber, plastics Production of chlorinated hydrocarbons	Fluorination processes		Production of viscose rayon	Processes where freedom from metallic contamination is necessary	Production of fine chemicals
Ni-Cr			Fuel element processing in HNO_3 + fluorides				
Ni-Cr-Fe (approx. 75% Ni)	Production of caustic alkalis especially in the presence of S	Production of halogenated hydrocarbons and phenolic resins	Heat exchangers in water-cooled reactors Fluorination processes		Paper pulp production (alkaline processes) Steam-heated dryers Textile rolls	Processes and storage where freedom from contamination is necessary	
Ni-Cr-Fe (approx. 32% Ni)	HNO_3 condensers	Reforming processes	Superheater tubes				
Ni-Cr-Fe-Mo	Production of fertilisers	Handling acid sludges	Fuel element processing		Wood pulp digesters SO_2 strippers	Processes where freedom from stress corrosion is necessary	Pickling of steel Handling waste gases containing SO_2
Ni-Cr-Fe-Mo-Cu	Sulphonation processes	Hydrocarbon processes involving H_3PO_4			Synthetic fibre manufacture		

(continued opposite)

Table 4.29 (continued)

Alloy type	Industry						
	Chemical	Petrochemical	Nuclear and power	Marine and desalination	Textile and paper	Food	Other
Ni-Cr-Mo	Chlorination processes			Applications where no pitting and no loss of reflectivity are necessary	Bleaching operations		
Ni-Mo	Processes involving HCl and non-oxidising acidic chlorides	Distillation columns containing acidic chlorides					
Ni-Cu	Production and recovery of HF	HF alkylation processes. Sulphur stripping columns. Distillation columns containing acidic chlorides. Handling acid sludges	Fluorination processes	Valves, impellers, propeller shafts, fasteners. Demisters in desalination plants		Handling refrigerating brines. Salt production. Evaporators	Pickling of steel
Ni-Si	Processes involving conc. H_2SO_4						

to indicate the wide range of corrosive conditions in which nickel and its alloys are used in practice.

A major feature to note is that in many applications the requirement is not only for a corrosion-resistant material but also for one that will not contaminate the product. This is particularly so in the food processing industry, where freedom from metallic contamination is important both from the point of view of preserving the product in good condition and also to avoid rendering it toxic. Nickel is advantageous in both respects. Similar considerations of product purity arise in the production of certain chemical intermediates, notably phenol that is to be used for production of synthetic fibres. Another application where, in addition to a very high degree of integrity of the material, minimum release of corrosion product is required is in the heat exchangers of water-cooled nuclear reactors. Alloys 600 and 800 have been widely used in the heat exchangers of pressurised water-reactors.

Increasing concern over environmental pollution from sulphur-bearing flue gases has led to the development of new and improved nickel alloys for flue gas scrubber systems; (e.g. the Ni-Cr-Fe-Mo alloys in Table 4.21).

Nickel and most nickel alloys are available in the usual wrought forms — plate, sheet, bar, tube, etc. — and also in some cases as clad steel plate. Sheet material may be employed as corrosion-resistant liners in process vessels and some of the alloys may also be used as weld overlays to provide a corrosion-resistant surface. In applications where a higher strength than that normally available is required, high-strength variants of some of the materials, i.e. Ni, Ni-Cu and Ni-Cr alloys, are available. These materials are strengthened by precipitation hardening, and heat treatment is necessary to develop the full strength. Nickel and most of the nickel alloy types are available in cast form, and it should be noted that the Ni-Si alloy and some of the Ni-Cr-Fe-Mo-Cu alloys with higher than normal silicon are only available as castings. These materials are used primarily for pumps and valves. Certain wrought nickel, Ni-Cr-Fe and Ni-Cu alloys can be supplied and fabricated to meet the statutory requirements for pressure-vessel construction.

T. E. EVANS

REFERENCES

1. Silverman, D. C., *Corrosion*, **37**, 546 (1981)
2. Pourbaix, M., *Atlas of Electrochemical Equilibria in Aqueous Solutions*, Pergamon Press, 333 (1966)
3. Cowan, R. L. and Staehle, R. W., *J. Electrochem. Soc.*, **118**, 557 (1971)
4. Brook, P. A., *Corr. Sci.*, **12**, 297 (1972)
5. Ashworth, V. and Boden, P. J., *J. Electrochem. Soc.*, **119**, 720 (1972)
6. DiBari, G. A. and Petrocelli, J. V., *J. Electrochem. Soc.*, **112**, 99 (1965)
7. Hart, A. C., *Metal Fin. J.*, **19**, 332 (1973)
8. Marshall, G. W. and Jones, M. T., *Corr. Sci.*, **14**, 15 (1974)
9. de Gromoboy, T. S. and Shreir, L. L., *Electrochim. Acta.*, **11**, 895 (1966)
10. Kesten, M., *Corrosion*, **32**, 94 (1976)
11. Chatfield, C. J. and Shreir, L. L., *Corr. Sci.*, **12**, 563 (1972)
12. Ijzermans, A. B., *Corr. Sci.*, **10**, 113 (1970)
13. Sato, N. and Okamoto, G., *J. Electrochem. Soc.*, **111**, 897 (1964)

14. Pigeaud, A., *J. Electrochem. Soc.*, **122**, 80 (1975)
15. Kesten, M., *J. Electrochem. Soc.*, **119**, 722 (1972)
16. Felloni, L., Palomboarini, G., Cammarota, G. P. and Lanzoni, E., *Br. Corr. J.*, **16**, 156 (1981)
17. Burstein, G. T. and Wright, G. A., *Electrochem. Acta*, **20**, 95 (1975)
18. Sato, N. and Okamoto, G., *J. Electrochem. Soc.*, **110**, 605 (1963)
19. Hoar, T. P., *Trans. Inst. Metal Fin.*, **39**, 166 (1962)
20. Vilche, J. R. and Arvia, A. J. *Corr. Sci.*, **18**, 441 (1978)
21. Kunze, E. and Schwabe, K., *Corr. Sci.*, **4**,109 (1964)
22. Szklarska-Smialowska, Z., *Corr. Sci.*, **12**, 527 (1972)
23. Vilche, J. R. and Arvia, A. J., *Corr. Sci.*, **15**, 419 (1975)
24. Sussek, G. and Kesten, M., *Corr. Sci.*, **15**, 225 (1975)
25. Kesten, M., *Corr. Sci.*, **14**, 665 (1974)
26. Tokuda, T. and Ives, M. B., *J. Electrochem. Soc.*, **118**, 1404 (1971)
27. Zamir, M. and Ives, M. B., *Corrosion*, **29**, 319 (1973)
28. Zamir, M. and Ives, M. B., *J. Electrochem. Soc.*, **121**, 1141 (1974)
29. Rätzer-Scheibe, H. J. and Feller, H. G., *Z. Metallkunde*, **63**, 351 (1972)
30. Clark, P. N., Jackson, E. and Robinson, M., *Br. Corr. J.*, **14**, 33 (1979)
31. Hummel, R. E., Smith, R. J. and Verink, E. D., *Corr. Sci.*, **27**, 803 (1987)
32. Smith, R. J., Hummell, R. E. and Ambrose, J. R., *Corr. Sci.*, **27**, 815 (1987)
33. Hayfield, P., in *Advances in Corrosion Science and Technology*, Ed. by M. G. Fontana and R. W. Staehle, **2**, Plenum Press, 96 (1972); (quoting results of Tronstad)
34. Bockris, J. O'M., Reddy, A. K. N. and Rao, B., *J. Electrochem. Soc.*, **113**, 1133 (1966)
35. Lovrecek, B. and Lipanovic, S., *Corr. Sci.*, **10**, 865 (1970)
36. Kawashima, A., Asami, K. and Hashimoto, K., *Corr. Sci.*, **25**, 1103 (1985)
37. Macdougall, B., Mitchell, D. F., and Graham, M. J., *Corrosion*, **38**, 85 (1982)
38. Visscher, W. and Dmjonovic, A., 25th ISE Meeting Paper 138, Zürich (1976)
39. Bode, H., Dehmelt, K. and Witte, J., *Electrochem. Acta*, **11**, 1079 (1966)
40. Davies, D. E. and Barker, W., *Corrosion*, **20**, 47t (1964)
41. Gilli, G., Borea, P., Zucchi, F. and Trabanelli, G., *Corr. Sci.*, **9**, 673 (1969)
42. Hodge, F. G. and Wilde, B. E., *Corrosion*, **26**, 146 (1970)
43. Economy, G., Speiser, R., Beck, F. H. and Fontana, M. G., *J. Electrochem. Soc.*, **108**, 337 (1961)
44. Shiobara, K., Sawada, Y. and Morioka, *Trans. Inst. Met. Japan*, **5**, 97 (1965)
45. Sayano, R. R. and Ken Nobe, *Corrosion*, **25**, 260 (1969)
46. Condit, D. O., *Corr. Sci.*, **12**, 451 (1972)
47. Piron, D. L., Koutsoukos, E. P. and Ken Nobe, *Corrosion*, **25**, 151 (1969)
48. Leonard, R. B., *Corrosion*, **24**, 301 (1968)
49. Morris, P. E. and Scarberry, R. C., *Corrosion*, **26**, 169 (1970)
50. Osterwald, J. and Uhlig, H. H., *J. Electrochem. Soc.*, **108**, 515 (1961)
51. Flint, G. N. and Barker, W., *Proc. Chem. Engng. Group (S.C.L)*, **43**, FB1 (1964)
52. Barker, W. and Evans, T. E., unpublished data
53. Evans, T. E. and Hart, A. C., *Electrochim. Acta.*, **16**, 1955 (1971)
54. Clarke, M. and Elbourne, R. G. P., *Corr. Sci.*, **8**, 29 (1968)
55. Sedriks, A. J., Green, J. A. S. and Novak, D. L., *Corrosion*, **28**, 137 (1972)
56. Crow, W. B., Myers, J. R. and Marvin, B. D., *Corrosion*, **27**, 459 (1971)
57. Horton, R. M. and Kwangchula Kim, *Corrosion*, **30**, 13 (1974)
58. Hodge, F. G., *Corrosion*, **29**, 375 (1973)
59. Tokuda, T. and Ives, M. B., *Corr. Sci.*, **11**, 297 (1971)
60. Postlethwaite, J., Huber, B. and Makepeace, D., *Corrosion*, **42**, 646, (1986)
61. Gressmann, R., *Corr. Sci.*, **8**, 325 (1968)
62. Postlethwaite, J., *Electrochim. Acta.*, **12**, 333 (1967)
63. Horvath, J. and Uhlig, H. H., *J. Electrochem. Soc.*, **115**, 791 (1968)
64. Streicher, M. A., *Corrosion*, **32**, 79, 1976
65. Davies-Smith, L. R., Lane, J. D. and Riley, T., *Br. Corr. J.*, **22**, 90 (1987)
66. Henthorne, M. and DeBold, T. A., *Corrosion*, **27**, 255 (1971)
67. Brown, M. H., *Corrosion*, **25**, 438 (1969)
68. Tedmon, C. S. and Vermilyea, D. A., *Corrosion*, **27**, 377 (1971)
69. Raymond, E. L., *Corrosion*, **24**, 180 (1968)
70. Briant, C. L. and Hall, E. L., *Corrosion*, **43**, 539 (1987)
71. Copson, H. R., Hopkinson, B. E. and Lang, F. S., *Proc. A.S.T.M.*, **61**, 879 (1961)

72. Cowan, R. L. and Tedmon, C. S., in *Advances in Corrosion Science and Technology*, Ed. by M. G. Fontana and R. W. Staehle, **2**, Plenum Press (reviewing results of several authors), 293 (1973)
73. Vermilyea, D. A., Tedmon, C. S. and Broecker, D. E., *Corrosion*, **31**, 222 (1975)
74. Gräfen, H. and Böhm, G., *Z. Metalikunde*, **51**, 245 (1960)
75. Class, I., Gräfen, H. and Scheil, E., *Z. Metallkunde*, **53**, 283 (1962)
76. Streichart, M. A., *Corrosion*, **19**, 272t (1963)
77. Samans, C. H., Meyer, A. R. and Tisinai, G. F., *Corrosion*, **22**, 336 (1966)
78. Leonard, R. B., *Corrosion*, **25**, 222 (1969)
79. Hodge, F. G., *Corrosion*, **29**, 375 (1973)
80. Flint, G. N., *J. Inst. Met.*, **87**, 303 (1959)
81. Pavlov, S. S. and Svistunova, T. V., *Metallov. Term. Obrab. Metallov.*, **10**, 20 (1970)
82. Vernon, W. H. J., *J. Inst. Met.*, **48**, 121 (1932)
83. Copson, H. R., *A.S.T.M. Spec. Tech. Pubin. No.* 175, 141 (1955)
84. Copson, H. R. and Tice, E. A., *Werkst. u. Korrosion*, **15**, 645 (1964)
85. van Rooyen, D. and Copson, H. R., *A.S.T.M. Spec. Tech. Publn. No.* 435, 175 (1968)
86. Evans, T. E., *Proc. 4th Int. Congr. on Met. Corr.*, 408 (1969)
87. Evans, T. E., *Werkst. u. Korrosion*, **15**, 797 (1964)
88. Guillaume, I., Grionaudeau, J., Valensi, G. and Brisou, J., *Corr. Sci.*, **13**, 97 (1973)
89. Tuthill, A. H. and Schillmoller, C. M., paper presented at 'The Ocean Science and Engineering Conference', Washington D.C. (1965)
90. Brouillette, C. V., *Corrosion*, **14**, 352t (1958)
91. Niederberger, R. B., Ferrara, R. J. and Plummer, F. A., *Mats. Perf.*, **9**, 18 (1970)
92. Weisert, E. D., *Corrosion*, **13**, 659t (1957)
93. Reinhart, F. M., *U.S. Naval Civil Engng. Laboratory Tech. Note N*-1023
94. Barker, W., Evans, T. E. and Williams, K. J., *Brit. Corr. J.*, **5**, 76 (1970)
95. Barker, W., Williams, K. J. and Evans, T. E., *Chem. Proc. Engng.*, **51**, 57 (1970)
96. Latinision, R. M. and Staehle, R. W., *Proc. Conf. Fundamental Aspects of Stress Corrosion Cracking*, 214, NACE (1969)
97. Staehle, R. W., Royuela, J. J., Reredon, T. L., Serrate, E., Morin, C. R. and Farrar, R. V., *Corrosion*, **26**, 451 (1970)
98. Pessall, N., Dunlap, A. B. and Feldman, D. W., *Corrosion*, **33**, 130 (1977)
99. Weeks, J. R., *Corrosion Problems in Energy Conversion and Generation*, ed. C. S. Tedmon Jr., *The Electrochem. Soc.*, P322 (1974)
100. van Rooyen, D., *Corrosion*, **30**, 73 (1974)
101. van Rooyen, D., *Corrosion*, **31**, 327 (1975)
102. Patharia, R. S., *Corrosion*, **34**, 149 (1978)
103. Harris, J. A. and Lamke, T. F., Corrosion 82 Conference, Houston, Texas, Paper 137 (1982)
104. Harris, J. A., Lemke, T. F., Smith, D. F. and Moeller, R. H., Corrosion 84 Conference, New Orleans, Louisiana, Paper 310 (1984)

4.6 Tin and Tin Alloys

The most important forms in which tin is used[1] are as follows:

1. Coatings for other metals.
2. Tin of more than 99% purity.
3. Tin hardened by additions of 1–2% Cu or Sb.
4. Pewter with 90–95% Sn, 4–8% Sb and 1–2% Cu.
5. Soft solder with tin and lead in all proportions.
6. Bearing metal with a large range of proportions of tin, antimony, copper and lead, or with tin (5–30%) in aluminium.
7. Diecasting alloys containing 70–80% Sn with antimony, copper and lead, either singly or combined.

Corrosion behaviour of the above-listed forms of tin can be considered as being basically similar, except in the case of solders and bearing metals, where the wide composition range and special duties of the materials necessitate additional comment. The behaviour of massive tin is a guide to the performance of coatings (Section 13.5).

Physical Properties

The use of unalloyed tin is restricted by its low melting point (232°C) and by its low tensile strength (15 MN/m^2). On the other hand, its melting point, and its ability to wet other metals, facilitates its use as solder and as a coating, while its softness and high ductility make it suitable for cold working and for bearing applications.

Of the elements normally present in tin-rich alloys, lead forms a simple eutectic system with a eutectic composition at 63% Sn, and copper and antimony have a small solid solubility and form the intermetallic compounds Cu_6Sn_5 and SbSn respectively[1].

Tin recrystallises readily at room temperatures so that effects of mechanical working are slight, arising from differing grain size and not from stress. Corrosion in sodium carbonate solution was found[2,3] to be less for finer grained material, but there is some doubt as to the general application of this finding.

The normal crystal form of tin is body-centred tetragonal, but a low-temperature allotrope, 'grey tin', is cubic. The transformation temperature is 13.2°C but, unless there is inoculation with the low-temperature form, the change rarely occurs even at very low temperatures, as the quantities of impurity normally present in commercial tin act as inhibitors. Since the transformation produces 'warts' of grey product, its effects are not distinguishable by the casual observer from those of local corrosion; usually, products suspected to be grey tin prove on investigation to be corrosion product.

The impurities likely to be present in nominally pure tin are unlikely to affect its corrosion resistance, except for minor effects on the rate of oxidation in air. Small aluminium contents, however, may result in a severely embrittling intercrystalline attack by water. The addition of antimony counteracts this effect. Although 0·1% magnesium appears to be tolerable, larger amounts produce effects similar to those of aluminium.

Theoretical Consideration of Corrosion Behaviour

The Pourbaix diagram[4] for tin (Fig. 4.41) refers only to solutions in which formation of soluble tin complexes or protective layers of insoluble salts does not occur. There are few instances of the formation of protective layers other than oxide on tin, and although the formation of soluble complexes is more common, the diagram provides a useful general indication of the

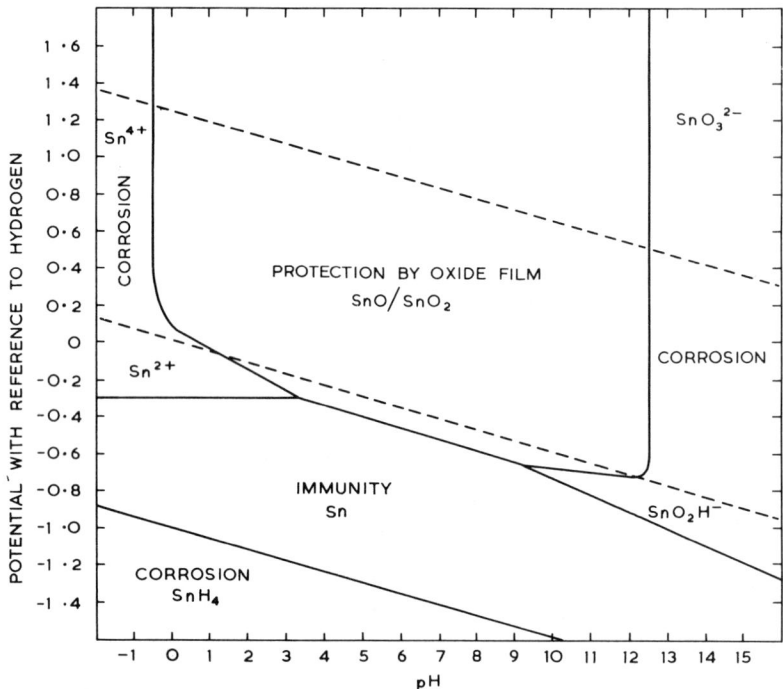

Fig. 4.41 Potential-pH diagram for tin. The full lines enclose the zones where the stability of various compounds or ions makes the indicated ion possible. The broken lines indicate the limits of stability of water at a pressure of 1 atm (after Reference 4 and Cebelcor)

conditions in which the corrosion of tin is possible. The wide field of stability of the two oxides, some of it extending below the field of stability of water, indicates easy passivation over a large range of pH, but, in either strongly acid or strongly alkaline solutions, tin may be dissolved. The region at depressed potentials where the possibility of the formation of SnH_4 exists has no known practical importance, but the action has been observed in the laboratory[5].

In zones where, according to the diagram, the dissolution of tin is possible, the rate of corrosion may be very slow. The hydrogen overpotential of tin is high. (In a range of concentrations of H_2SO_4[6], the Tafel constant a was 1·29 and b was 0·118, the exchange current for the hydrogen evolution reaction was 10^{-11} A cm^{-2} with a transfer coefficient of 0·50; in KOH solutions[7], the Tafel constant a varied between 1·36 and 1·53 with concentration, b was 0·120 and the exchange current was 3×10^{-6} A cm^{-2}.) Consequently, in alkaline or moderately acid solutions free from oxygen or oxidising agents, corrosion of pure tin may be barely detectable, unless the tin is in contact with another metal of low hydrogen overpotential. Also, tin forms complex ions with many acids including those commonly found in fruits (citric, oxalic, malic). This action has important effects on the galvanic behaviour of tin. The normal electrode potential of tin is −0·136 V, but the reduction of stannous ion activity resulting from formation of complexes may depress the corrosion potential to make tin anodic to iron[8, 8(a)].

The stability of tin over the middle pH range (approximately 3·5-9), its solubility in acids or alkalis (modified by the high hydrogen overpotential), and the formation of complex ions are the basis of its general corrosion behaviour. Other properties which have influenced the selection of tin for particular purposes are the non-toxicity of tin salts and the absence of catalytic promotion of oxidation processes that may cause changes in oils or other neutral media affecting their quality or producing corrosive acids.

Atmospheric Corrosion

Oxidation of tin in dry air is slow; the metal remains bright and interference colours are not developed below about 180°C. On a rolled tin surface heated in air[9, 10], the thickness of the oxide grew according to a logarithmic law at temperatures up to about 160°C and according to a parabolic law at higher temperatures, for which the oxide was identified as SnO. For electropolished tin heated in oxygen at pressures of 130 N/m^2 and above, three stages of oxidation were observed at temperatures up to 220°C[11-13]:

1. An initial sigmoid growth curve during nucleation.
2. A logarithmic growth curve when cavities acted as diffusion barriers.
3. Erratic behaviour caused by random film rupture.

The oxide formed was identified as α-SnO at temperatures down to 75°C. However, a mixture of SnO and SnO_2 is formed in dry air at lower temperatures and, in humid air, at temperatures up to at least 100°C[39].

Small additions, e.g. 0·1%, of indium, zinc or phosphorus, reduce the rate of oxidation[9] and an addition of antimony, thallium or bismuth accelerates it[14].

Humidity of the air, in the absence of polluting gases or dusts, increases the rate of oxidation[14(a)] and may cause yellow discoloration. In an ordinary uncleaned atmosphere, some corrosion product may be formed in due course; tin products are not hygroscopic and tin is not attacked at relative humidities below 100% unless the dust falling on the surface is hygroscopic, or impurities in the metal are able to form a hygroscopic product. Indoors, in a laboratory atmosphere without special pollution, a grey film, increasing in weight linearly with time ($0 \cdot 004$ g m^{-2}d^{-1}), is formed[15]. The reflectivity of the surface is slowly lost if it is left untouched, but may be preserved by regular washing; in one experiment, when the surface was washed at intervals of three weeks, a water wash was adequate for six weeks and although the use of soap was necessary thereafter almost complete preservation of reflectivity was achieved[16]. In a Stevenson Screen out of doors, i.e. with exposure to all atmospheric influences except rain, the corrosion rate declined somewhat with time[15].

The impurities ordinarily present in the atmosphere do not appreciably affect the character of corrosion. No tarnishing effect is exerted by hydrogen sulphide, sulphur dioxide and other acids in low concentrations, including formic, acetic and other organic acids which when evolved from wood or insulating materials are so often destructive to metals other than tin in the confined spaces of electrical equipment or of packages. Chlorides accelerate corrosion and tend to form a white corrosion product containing oxychloride. The presence of some impurities, notably zinc, in the metal may cause tarnishing and loss of brightness in atmospheres containing SO_2[17].

The atmospheric pollution prevailing in special industrial or laboratory locations may induce more severe corrosion, e.g. the vapours from concentrated hydrochloric or acetic acid will etch tin, and moist sulphur dioxide will produce a sulphide tarnish, as will hydrogen sulphide at temperatures above about 100°C. The halogens attack tin readily. The commonly used volatile corrosion inhibitors are without adverse action although the benefit derived from their use is doubtful.

When tin is fully exposed out of doors, corrosion is uniform, and the rate falls only slightly with time. The metal becomes dull and accumulates a compact layer of pale grey product, mainly stannous oxide. Rates observed during exposures in the USA for periods of up to 20 years were as follows[18]:

Industrial	$0 \cdot 0013$–$0 \cdot 0018$ mm/y
Sea coast	$0 \cdot 0018$–$0 \cdot 0028$ mm/y
Rural	$0 \cdot 0005$ mm/y

Corrosion in Near-neutral Aqueous Media

Pure tin is completely resistant to distilled water, hot or cold. Local corrosion occurs in salt solutions which do not form insoluble compounds with stannous ions (e.g. chloride, bromide, sulphate, nitrate) but is unlikely in solutions giving stable precipitates (e.g. borate, mono-hydrogen phosphate, bicarbonate, iodide)[19]. In all solutions, oxide film growth occurs and the potential of the metal rises. Any local dissolution may not begin for several days but, once it has begun, it will continue, its presence being manifested at first by small black spots and later by small pits. Movement of the solution may prevent pitting; stagnation, especially in crevices where the tin touches another solid surface, favours its progress. Contact with a more noble metal such as copper or nickel increases the number and intensity of pits; contact with metals such as aluminium and zinc gives cathodic protection.

As indicated above, the bicarbonate ion inhibits the process, which does not occur, therefore, in many supply waters; attack is most likely in waters which by nature or as a result of treatment have a low bicarbonate content and relatively high chloride, sulphate or nitrate content. The number of points of attack increases with the concentration of aggressive anions and ultimately slow general corrosion may occur. During exposure of 99·75% tin to sea-water for 4 years, a corrosion rate of 0·0023 mm/y was observed[20]. Corrosion in soil usually produces slow general corrosion with the production of crusts of oxides and basic salts; this has no industrial importance but is occasionally of interest in archaeological work.

Corrosion by Acids

Since the high overvoltage restricts hydrogen evolution, corrosion in organic acids or dilute non-oxidising mineral acids is controlled by the rate of supply of oxygen. In solutions of acid open to air, with specimens of size 50 × 20 mm completely immersed, corrosion rates were, in a range of 0·1 N organic acids, 400–500 g m^{-2}d^{-1} and, in 0·1 N hydrochloric or sulphuric acids, 600 g m^{-2}d^{-1}. In absence of oxygen, the rates in the two mineral acids were 100–150 g m^{-2}d^{-} and in the organic acids, negligible. Phosphoric acid forms a protective layer and, even in the presence of air, the corrosion rate is only about 20 g m^{-2}d^{-1}. Nitric acid by reason of its oxidising action attacks tin freely even in the absence of oxygen, but chromic acid forms a protective film. This film, which contains chromic oxide and tin oxides, will, after withdrawal from the acid, give some degree of protection against mildly corrosive conditions; hot solutions of chromic acid, alone or mixed with phosphoric acid, are used as passivating solutions.

The compounds that inhibit acid attack upon steel exert little or no restraint on acid attack upon tin.

Corrosion by Alkalis

The Pourbaix diagram indicates the possibility of attack by solutions of pH values above about 9·5, but the position of this limit is influenced by temperature, by the constitution of the solution, and by the surface condition of the metal. Corrosion will ensue if the surface oxide can be dissolved; this will invariably take place if the pH exceeds 12, and may occur even at pH values between 10 and 12.

Once corrosion begins, its rate is governed by the oxygen supply and temperature and is not greatly affected by the character of the alkali. Rates of attack for specimens completely immersed in still solutions open to air are about 600 g m^{-2}d^{-1} at 30°C and 1000 g m^{-2}d^{-1} at 70°C. In intermittent immersion such as is experienced in the cleaning of tinned ware by alkaline detergents, however, the rate of corrosion is affected by the nature and concentration of the solution, since these affect the time required for removal of the oxide film at each fresh immersion[21].

Saturated ammonia solutions do not attack tin, possibly because of the negligible oxygen content, but more dilute solutions behave like those of other alkalis of comparable pH.

The removal of oxygen from an alkaline solution, as by the addition of sodium sulphite, can prevent corrosion unless the tin is in contact with another metal, such as steel, from which hydrogen can be evolved. Additions of oxidising agents in small amounts stimulate corrosion but sufficiently large additions produce passivity. Alkaline chromate solutions in the passivating range produce a film containing chromium oxide, which has some protective value[9,22].

Corrosion by Other Liquid Media

Sulphide solutions, sulphurous acid and some foodstuffs containing organic sulphur compounds, produce stains of sulphide, but the rate of loss of metal is low. Milk and milk products are usually without action, although local corrosion has been known to occur in dairy equipment.

Beer initially dissolves a trace of tin and this may be sufficient to cause a haze in the liquor, but the action is usually checked after a period of contact.

In general, near-neutral aqueous products are without action except for possible sulphide staining or, when there are dissolved salts present, some local corrosion. The slight acidity which may develop in solutions of some organic compounds such as formaldehyde or alcohols can be tolerated.

Many organic liquids, including oils (essential, animal, vegetable or mineral), alcohols, fatty acids, chlorinated hydrocarbons and aliphatic esters, are without action. The absence of any catalytic action of tin on oxidative changes is helpful in this respect. When, however, mineral acidity can arise, as with the chlorinated hydrocarbons containing water, there may be some corrosion, especially at elevated temperature.

Tin as an Anode

Tin when made anodic shows passive behaviour as surface films are built up but slow dissolution of tin may persist in some solutions and transpassive dissolution may occur in strongly alkaline solutions. Some details have been published for phosphoric acid[23] with readily obtained passivity, and sulphuric acid[24] for which activity is more persistent, but most interest has been shown in the effects in alkaline solutions. For galvanostatic polarisation in sodium borate[25] and in sodium carbonate solutions[26] at 1×10^{-6}–50×10^{-6} A/cm^2, simultaneous dissolution of tin as stannite ions and formation of a layer of SnO occurs until a critical potential is reached, at which a different oxide or hydroxide (possibly SnO$_2$) is formed and dissolution ceases. Finally oxygen is evolved from the passive metal. The nature of the surface films formed in KOH solutions[27] up to 7 M and other alkaline solutions[28] has also been examined.

If the applied current density is reduced when a tin anode has been made passive in alkaline solution with the formation of a brown film and evolution of oxygen, the surface film changes to one of yellow colour and dissolution of tin as stannite ions proceeds freely[29]. This effect is exploited in the electrodeposition of tin from sodium or potassium stannate solutions.

It is possible to obtain conditions[30] in which the anodic film continues to grow to form a blue or black layer, and this, although not exceptionally protective, has uses in the treatment of baking pans. A typical anodising solution contains 100 g/l $Na_2HPO_4 \cdot 12H_2O$ and 20 ml/l phosphoric acid, and is used at 350 A/m^2 at 60–90°C for about 10 min.

Solders

By the nature of their use as jointing material, solders are usually presented to a corrosive environment as a small area within a much larger area of another metal. Thus, if the solder is anodic to the metal it joins, and if the corroding medium has good electrical conductivity, damaging corrosion is possible.

Solders are anodic to copper, but soldered joints in copper pipes are widely used without trouble for cold supply waters; possibly corrosion is restricted by the deposition of cathodic carbonate scales and the formation of insoluble lead compounds. Hot supply waters tend to be more aggressive and, where these are involved, it is wise to tin any copper which has a soldered joint. Electrolytes of high conductivity such as sea-water will also attack soldered joints in copper.

Soldered brass seldom gives trouble. In radiators, antifreeze solutions have been alleged to cause corrosion, possibly because materials such as ethylene glycol sometimes detach protective deposits. Sodium nitrite, valuable as a corrosion inhibitor for other metals in a radiator, tends to attack solders, but sodium benzoate is safe and, in addition, protects the soldered joint against the action of nitrites[31]. In an investigation[32] of other inhibitors in ethylene glycol solutions, 1% borax, either alone or in combination with 0·1% mercaptobenzothiazole, appeared to be satisfactory.

Solders are cathodic to steel, zinc and cadmium, and anodic to Monel metal. Although tin or tin-coated metals may be used safely in contact with aluminium when they are not fused with it, a joint in aluminium made with a tin-lead solder is liable to destructive corrosion. The formation, on fusion, of the grain-boundary state, which, as already mentioned, makes aluminium so dangerous an impurity in tin, is responsible. Tin-zinc solders may be used; the zinc gives a useful degree of protection.

For environments in which tin is less readily corroded than lead, corrosion resistance of the alloy decreases as the lead content increases; the decrease may, in some circumstances, be sharp at a particular composition. In the more corrosive media, such as nitrite solution, a sharp increase of corrosion rate is observed as the lead content increases beyond 30%. In waters with low contents of dissolved salts, the corrosion rate increases slowly with lead content up to about 70% and then rises more steeply, but in the general run of supply waters the ability of lead to form protective insoluble anodic products is helpful to the durability of solder. Selective dissolution of tin has been reported to occur in prolonged contact of solders with solutions of anionic surface active agents[33]. This is consistent with some observations of the behaviour of pure tin in such solution[34].

Soldered joints, especially those to be used in a static environment, are, if insufficient care is taken, liable to corrosion by residues of flux, which by their nature as oxide removers are potentially corrosive. It is, however, possible to select fluxes which are active when hot but give non-corrosive residues when cold.

If it is necessary to use more vigorous materials, such as zinc chloride, any residues must be rinsed away; owing to the tendency of zinc chloride to yield a screen of basic chloride in a neutral rinse, a preliminary rinse in water acidified with hydrochloric acid is advisable.

Bearing Metals

There are three main types of Babbitt alloy for bearings: high-tin alloys (substantially lead-free), intermediate-tin alloys containing some lead, and high-lead alloys. The high- and intermediate-tin alloys have the resistance of tin to the weak acids and sulphur compounds which may be present in used oils; the lead-rich alloys are rather less resistant[35]. The anti-oxidant action of tin is useful in limiting acidity. Rare instances of corrosion of tin-rich bearings have occurred, with the formation of hard crusts of tin oxide in which the intermetallic compounds of tin with copper and with antimony may remain embedded. Although the causes are not wholly established, some access of water to the bearing is an essential factor and accumulation of breakdown products of constituents of the lubricating oil may contribute[36]. The nobility of intermetallic compounds of tin relative to tin itself has been demonstrated[37].

When free access of salt water to a bearing is possible, the Babbitt alloys are not suitable since they are cathodic to steel shafts. For underwater bearings, alloys with 70% Sn, 1·5% Cu and the balance Zn, are used; the possible dissolution of zinc gives cathodic protection to the shaft, although the more easily replaced bearing suffers some corrosion.

Uses of Tin

Apart from the special uses in solders and bearings metal referred to already, and as coatings, tin and its alloys find employment where advantage can be taken of their physical properties and their fair resistance to tarnish and corrosion in near-neutral environments. Tin pipe is used to condense steam for high-purity distilled water, as a conveyor of beer and soft drinks, especially in coils through cooling media, and, in a larger size, as organ pipes. Some pharmaceutical and food products are packed in tin collapsible tubes and tinfoil coverings are used on cork wads for jar and bottle closures. Pewter is most valued for the decorative forms into which it is easily worked or cast, but it is also used for drinking vessels and dishes.

Atmospheric Corrosion

The corrosion of tin in various atmospheres has been extensively monitored recently using XPS and AES techniques[38-41]. While it is difficult to resolve the peaks from the tin oxides and establish their degree of surface hydration, there is general agreement that both SnO and SnO_2 may be present depending on the temperature of exposure to oxygen.

Corrosion by Acids

The corrosion of tin by nitric acid[42] and its inhibition by n-alkylamines has been reported[43]. The action of perchloric acid on tin has been studied[44] and sulphuric acid corrosion inhibition by aniline, pyridine and their derivatives[45] as well as sulphones, sulphoxides and sulphides[46] described. Attack of tin by oxalic, citric and tartaric acids was found to be under the anodic control of the Sn^{2+} salts in solution in oxygen free conditions[47]. In a study of tin contaminated by up to 1200 ppm Sb, it was demonstrated[48] that the modified surface chemistry catalysed the hydrogen evolution reaction in deaerated citric acid solution.

Corrosion by Alkalis

The deliberate growth of tin oxide or mixed oxide films in both acid and alkaline solutions has been reviewed recently[49]. The effect of additions to alkaline solutions on their attack on tin has been considered[50] as has the tendency for pitting corrosion to occur in solutions containing chloride ions[51,52]. In connection with this, potential pH diagrams for $Sn-H_2O-Cl$ systems have been published[53].

Tin as an Anode

The use of alkaline tin plating baths continues to stimulate research into the anodic dissolution of tin into sodium hydroxide solution[54,55]. It has been shown[41] that in the prepassive region, tin (II) oxide and hydroxide exist on the surface of the metal. At the onset of passivity, these change to tin (IV) oxide and hydroxide and the extent of surface hydration changes rapidly[56].

Solders

There is an accelerating trend away from the use of lead-containing solders in contact with potable water. The effects of galvanic corrosion of one of the substitute alloys (Sn3%Ag) in contact with a number of other metals including copper have therefore been studied[57]. The corrosion of tin/lead alloys in different electrolytes including nitrates,[58] nitric and acetic acids,[59] and citric acid[60] over the pH range 2-6 were reported. The specific alloy Pb/15%Sn was studied[61] in contact with aqueous solutions in the pH range

2-6. Higher acidity caused greater corrosion but contamination by sulphur dioxide or carbon dioxide inhibited attack. By contrast, chloride ions were found to have a mild aggressive effect.

The effect of O_2, SO_2, NO_2, H_2S, Cl_2, CO and NH_3 on Sn/50%Pb in atmospheres of different relative humidity were investigated[62] but only SO_2 and NO_2 were active at low concentrations (<100 ppm). An XPS study of Sn/50%Pb solder exposed to O_2, H_2O and NO_2 was conducted to establish both the surface species formed and the ratio of the concentration of each metal in the surface. Previous XPS studies had only considered the interaction of tin/lead solder with the air[40,63,64].

Corrosion of solders used in the electronics industry is usually a function of the presence of residues from various manufacturing and assembly operations. Corrosion in heat exchangers, particularly in automobiles[65,66] is a more significant problem and a test methodology has been described as well as various factors controlling the corrosion of tin-lead alloys in radiators[67].

Bearing Metals

The corrosion of tin-rich white metal bearings is rare and consequently detailed studies of the phenomenon have not been carried out. However, useful contributions to the understanding of the factors affecting bearing corrosion have been made recently[12,15].

S. C. BRITTON

REFERENCES

1. Hedges, E. S. et al., *Tin and its Alloys*, Edward Arnold, London (1960)
2. Derge, G., *Trans-Electrochem. Soc.*, **75**, 449 (1939)
3. Derge, G. and Marcus, H., *Trans. Amer. Inst. Min. (Metall.) Engrs.*, **143**, 198 (1941)
4. Deltombe, E., de Zoubov, N. and Pourbaix, M., Technical Report No. 25, Centre Belge d'Étude de la Corrosion (1955)
5. Salzburg, H. W. and Mies, F., *J. Electrochem. Soc.*, **105**, 64 (1958)
6. Quintin, M. and Hagymas, G., *J. Chim. Physique*, 541 (1964)
7. Ross, T. K. and Firoiu, C., *Electrochim. Acta.*, **8**, 877 (1963)
8. Hoar, T. P., *Trans. Faraday Soc.*, **30**, 472 (1934)
8a. Willey, A. R., *Br. Corros. J.*, **7**, 29 (1972)
9. Britton, S. C. and Bright, K., *Metallurgia, Manchr.*, **56**, 163 (1957)
10. Trillat, J. J., Tertian, L. and Britton, S. C., *Métaux-Corros.-Ind.*, **32**, 475 (1957)
11. Boggs, W. E., Kachik, R. H. and Pellissier, G. E., *J. Electrochem. Soc.*, **108**, 6 (1961)
12. Boggs, W. E., Trozzo, P. S. and Pellissier, G. E., *J. Electrochem. Soc.*, **108**, 13 (1961)
13. Boggs, W. E., *J. Electrochem. Soc.*, **108**, 124 (1961)
14. Boggs, W. E., Kachik, R. H. and Pellissier, G. E., *J. Electrochem. Soc.*, **110**, 4 (1963)
14a. Britton, S. C. and Sherlock, J. C., *Br. Corros. J.*, **9**, 96 (1974)
15. Kenworthy, L., *Trans. Faraday Soc.*, **31**, 1331 (1935)
16. Kenworthy, L. and Waldram, J. M., *J. Inst. Met.*, **55**, 247 (1934)
17. Britton, S. C. and Clarke, M., *Trans. I.M.F.*, **42**, 195 (1964)
18. Hiers, G. O. and Minarcik, E. J., Symposium on the Atmospheric Corrosion of Nonferrous Metals, Spec. Tech. Pub. No. 175, Amer. Soc. Test. Mat., 135 (1956)
19. Hoar, T. P., *Trans. Faraday Soc.*, **33**, 1152 (1937)
20. Friend, J. N., *J. Inst. Met.*, **39**, 111 (1928)
21. Britton, S. C. and Michael, D. G., *J. Appl. Chem.*, **5**, 402 (1955)
22. Britton, S.C. and Angles, R. M., *J. Appl. Chem.*, **4**, 351 (1954)
23. Ragheb, A. and Kamel, L. A., *Corrosion*, **18**, 153t (1962)

24. Machu, W., Azzam, A. R. and Halashi, G. M., *Metallöberflache,* **9A**, 53 and 73 (1955)
25. Shah, S. N. and Davies, D. E., *Electrochim. Acta.*, **8**, 663 (1963)
26. Davies, D. E. and Shah, S. N., *Electrochim. Acta.*, **8**, 703 (1963)
27. Hampson, N. A. and Spencer, N. E., *Br. Corros. J.*, **3**, 1 (1968)
28. El Wakkad, S. E. S., El Din, A. M. and Sayed, J. K., *J. Chem. Soc.*, 3103 (1954); El Din, A. M. and El Wahab, F. M. A., *Electrochim. Acta.*, **9**, 883 (1964); Barbulescu, F. and Roller, L., *Rev. Roumaine Chim.*, **10**, 491 (1965)
29. Kerr, R., *J. Soc. Chem. Ind.*, **57**, 405 (1938); Bianchi, G., *Chim. e Ind.*, **29**, 295 (1947)
30. Kerr, R. and Macnaughtan, D. J., *J. Electrodes. Tech. Soc.*, **12**, 19 (1937)
31. Wormwell, F. and Mercer, A. D., *J. Appl. Chem.*, **3**, 22 (1953)
32. Levy, M., *Ind Engng. Chem.*, **51**, 209 (1959)
33. Watts, C., *Engineering Materials and Design*, **4**, 740 (1961)
34. Ross, T. K., *J. AppL Chem.*, **5**, 10 (1955); Ross, T. K. and Harris, W., *J. Appl. Chem.*, **10**, 24 (1960)
35. Symposium on Lead-base Babbitt Alloys, *Metal Progr.*, **69**, 174 (1956)
36. Bryce, J. B. and Roehner, T. G., *Trans. Inst. Mar. Engrs.*, **73**, 377 (1961)
37. Clarke, M. and Britton, S. C., *Corrosion Sci.*, **3**, 207 (1963)
38. Lau, C. L. and Wertheim, *J. Vac. Sci. Technol.*, **15**, 2, 622 (1978)
39. Okamoto, Y., Carter, W. J. and Hercules, D. M., *Appl. Spectroscopy*, **33**, 3, 287 (1979)
40. Farrell, T., *Metal Science*, March, 87 (1976)
41. Ansell, R. O., Dickinson, T., Povey, A. F. and Sherwood, P. M. A., *J. Electrochem. Soc.*, **124**, 9, 1360 (1977)
42. Abd el Haleem, S. M., Khedo, M. G. A. and El Kot, A. M., *Corr. Prev. and Control*, **28**, 2, 5 (1981)
43. Baraka, A., Ibrahim, M. E. and Al-Abdullah, M. M., *Metallöberfleche*, **35**, 7, 263 (1981)
44. Stirrup, B. N. and Hampson, N. A., *Surface Technology*, **4**, 73 (1976)
45. Abd Aal, M. S. and Assaf, F. H., *Trans. of the SAEST*, **15**, 2, 107 (1980)
46. Abd Aal, M. S., Abd Wahab, A. A. and Assaf, F. H., *Metallöberfleche*, **34**, 8, 323 (1980)
47. Gouda, V. K., Rizkalla, E. N., Abd-el-Wahab, S., and Ibrahim, E. M. *Corr. Sci.*, **21**, 1 (1981)
48. Leidheiser, H., Rauch, A. F., Ibrahim, E. M. and Granata, R. D., *J. Electrochem. Soc.*, **129**, 8, 165 (1982)
49. Gabe, D. R., *Surface Technology*, **5**, 6, 463 (1977)
50. Costa, J. M. and Cullere, J. R., *Corr. Sci.*, **16**, 587 (1976)
51. Abd. Aal, M. S. and Assaf, F. H., *J. Electrochem. Soc. India*, **30**, 1, 38 (1981)
52. Sinicki, C., *Compte. Rend.*, **290**, 14, 255 (1980)
53. House, C. I. and Kelsall, G. H., *Electrochimica Acta*, **29**, 10, 1459 (1984)
54. Dickinson, T. and Lofti, S., *Electrochimica Acta*, **23**, 513 (1978)
55. Kapusta, S. D. and Hackerman, N., *Electrochimica Acta*, **25**, 1625 (1980)
56. Kapusta, S. D. and Hackerman, N., *J. Electrochem. Soc.*, **129**, 9, 1886 (1982)
57. Linder, M. and Mattson, F., *Proc. 7th Scand. Corr. Congr.*, 19, Trondheim, 1975
58. Marshall, A., Pierey, R. and Hampson, N. A., *Corr. Sci.*, **15**, 23 (1975)
59. Fawzy, M. A., Sedahmed, G. H. and Mohamed, A. A., *Surface Technology*, **14**, 3, 257 (1981)
60. Radovici, O. and Popescu, B., *Revue Roumaine de Chimie*, **15**, 12, 1799 (1970)
61. Bullock, J. S., *Union Carbide Corp., Report No. Y2132*, (1978)
62. Tompkins, H. G., *J. Electrochem. Soc.*, **120**, 5, 651 (1973)
63. Bird, R. J., *Metal Science Journal*, **7**, 109 (1973)
64. Lin, A. W. C., Armstrong, N. R. and Kuwana, T., *Anal. Chem.*, **49**, 8, 1228 (1977)
65. Walker, M. S., *Materials Performance*, **18**, 4, 9 (1979)
66. Falke, W. L., Schwaneke, A. E. and Lee, A. Y., *Welding Jnl. Suppl.*, **52**, 10, 460 (1973)
67. Beal, R., *Engine Coolant Testing: State of the Art, ASTM STP 705, American Soc. for Testing and Materials*, 327 (1980)
68. Dunford, J. J., *Jnl. of Naval Eng.*, **23**, 1, 54 (1976)
69. Lloyd, K. A. and Wilson, R. W., *Proc. Tribology Conv., Inst., Mech. Engineers*, 76, 28-30 May, Gothenberg, (1969)

4.7 Zinc and Zinc Alloys

The excellent resistance of zinc to corrosion under natural conditions is largely responsible for the many and varied applications of the metal. In fact nearly half the world consumption of zinc is in the form of coatings for the prevention of corrosion of steel fabrications exposed to the atmosphere and to water. For its varied applications zinc is obtainable in a number of grades. Ordinary commercial (G.O.B.) zinc contains up to about 1·5% total of lead, cadmium and iron. Electrolytic zinc has a minimum zinc content of 99·95% and contains small amounts of the same impurities. Special high-purity zinc has a minimum of 99·99% zinc. Even purer zincs are commercially available.

The special high-purity zinc (99·99%) is used mainly for the production of diecasting alloys containing 4% aluminium and 0·04% magnesium and some-times 1% copper, as shown in Table 4.30, which gives the composition of the two alloys laid down by BS 1004: 1972, and of some newer zinc alloys.

Physical and Mechanical Properties

The bluish-white form of the unalloyed metal is familiar. It is moderately hard and quite malleable at normal temperatures. The tensile strength and impact resistance of the unalloyed metal are low, so zinc is not to be regarded as a structural metal.

The physical properties of zinc and the diecasting alloys are given in Table 4.31.

The good mechanical properties of the diecasting alloys depend on the fine-grain structure resulting from rapid chilling in the water-cooled die. The impact strength of the pressure-diecast alloys at normal temperatures is higher than that of cast aluminium alloys and grey cast iron, and although cooling below normal temperatures produces a sudden drop in the impact strength, this is still within the range given by many other cast metals at ordinary temperatures. (Variation of mechanical properties of the alloys with temperatures is given in Table 4.32.) Other factors, such as the design of castings, also influence mechanical properties.

Its malleability and ductility make it possible to produce zinc in sheet, strip and plate form by rolling and as rod and wire by extrusion; continuous casting is also used. Rolled zinc in sheet and strip form is a well-established

Table 4.30 Composition of ingot zinc alloys

Alloy†	Alloying elements (%)					Impurities % (max.)*					
	Al	Cu	Mg	Ti	Other	Fe	Cu	Pb	Cd	Sn	Zn
3BS 1004A	3·9–4·3	—	0·04–0·06	—	—	0·075	0·03	0·003	0·003	0·001	Balance
5BS 1004B	3·9–4·3	0·75–1·25	0·04–0·06	—	—	0·075	—	0·003	0·003	0·001	Balance
2	3·9–4·3	2·6–3·9	0·025–0·050	—	—	0·03	—	0·004	0·003	0·002	Balance
7	3·9–4·3	—	0·0–0·02	—	0·005–0·020Ni	0·075	0·10	0·0020	0·002	0·001	Balance
8	8·0–8·8	0·8–1·3	0·015–0·030	—	—	0·10	—	0·004	0·003	0·002	Balance
Ilzro 16	0·01–0·04	1·0–1·5	—	0·15–0·25	0·10–0·20 Cr 0·30–0·40Ti+Cr	0·075	—	0·003	0·003	0·001	Balance
12†	10·5–11·5	0·5–1·25	0·01–0·03	—	—	0·075	—	0·004	0·003	0·002	Balance
27	25·0–28·0	2·0–2·5	—	—	—	0.075	—	0·004	0·003	0·002	Balance
Alzen 305	30	5	—	—	—	0·03	—	0·003	0·003	0·001	Balance

*The limits for impurities indicate the need for using special high-purity zinc in making zinc casting alloys even when the aluminium content is very low. In the latter case, the precaution is necessary to avoid dangerous contamination of the other alloys if mixing should occur. Indium, thallium, nickel and manganese are also restricted in BS 1004 alloys.

† ZA12 is primarily an alloy for gravity casting in metallic or non-metallic moulds. Ilzro 8 and 16 is for pressure-die casting components which need better creep resistance than the BS 1004 alloys provide; No. 16 is recommended for high-temperature applications. Alloy No. 2 can be pressure die cast and is sometimes chosen for sliding components because of its hardness. It is also used for casting sheet-metal-forming dies and plastic moulds, but cannot be employed for moulding chlorine compounds (e.g. p.v.c.), some rubbers or polystyrene expanded with steam, because of its corrosion under such conditions. Alloy No. 7 is for pressure die casting and is claimed to display improved fluidity because of the reduced magnesium content. Alzen 305 is a bearing alloy used mainly as gravity cast shells, and castings and continuously cast or extruded rod. It should not be used in contact with salt or alkaline solutions nor at temperatures above 120°C. ZA8 and ZA27 are for casting by various processes.

Table 4.31 Physical properties of zinc and cast zinc alloys

Material	Atomic weight	Specific gravity (g/cm³)	Melting point (°C)	Electrical conductivity (Cu = 100)	Thermal conductivity (Cu = 100)	Coefficient of expansion per °C
Zinc	65·39	7·14	419·4	28·3	29·3	$3·95 \times 10^{-5}$
Alloy 3	—	6·7	382–387	26·6	30	$3·7 \times 10^{-5}$
Alloy 5	—	6·7	379–388	25·9	29	$2·7 \times 10^{-5}$
Alzen 305	—	4·8	480–580	31	28	$2·2 \times 10^{-6}$

Table 4.32 Variation with temperature of mechanical properties of diecast test bars of alloys 3 and 5, BS 1004

Temperature (°C)	Tensile strength (N/mm²)		Elongation % on 50 mm × 0·3 mm		Impact strength	
	3	5	3	5	3	5
95	200	240	30	23	58	—
40	250	290	16	13	57	62
20	280	350	11	8	57	60
10	—	—	—	—	42	56
0	295	370	9	8	10	53
−10	—	—	—	—	5	24
−20	—	—	—	—	3·5	5
−40	320	370	4·5	3	3	3·5

roofing material, particularly in parts of France and northern Europe. Diecastings can be made readily on account of the low melting points and the good flow properties of the diecasting alloys. The other major uses are in brass manufacture, coatings for steel and in sacrificial anodes.

Corrosion Properties

Both zinc and zinc alloys have excellent resistance to corrosion in the atmosphere and in most natural waters. The property which gives zinc this valuable corrosion resistance is its ability to form a protective layer consisting of zinc oxide and hydroxide, or of various basic salts, depending on the nature of the environment. When the protective layers have formed and completely cover the surface of the metal, the corrosion proceeds at a greatly reduced rate.

Alloys containing up to 1% copper and about 0·1% titanium have been used in strip form for roofing on a large scale in France, Germany, the UK and elsewhere since about 1960, but no details are available on how their corrosion resistance compares with that of the traditional unalloyed rolled zinc. Initially they are lighter in colour and they remain bright a little longer.

Composition of the Protective Films

In dry air, a film of zinc oxide is initially formed by the influence of the atmospheric oxygen, but this is soon converted to zinc hydroxide, basic zinc carbonate and other basic salts by water, carbon dioxide and chemical impurities present in the atmosphere.

Below about 200°C, the film grows very slowly and is invisible and very adherent. It is often said that the first layer formed governs the corrosion resistance of the zinc throughout its life. If the film becomes too thick, it is liable to break away or become porous, when, of course, it ceases to provide protection. Moreover, zinc oxide occupies a larger volume than the zinc from which it originated and, as the layer thickens, strains are set up which lead to the production of fissures.

Vernon[1] claims that in outdoor atmospheres the corrosion product consists largely of zinc oxide, hydroxide and combined water, but also contains zinc sulphide, zinc sulphate and carbonate. The following table[2] gives the composition of typical films formed in an industrial atmosphere.

Table 4.33 Composition of films formed in industrial atmosphere

Compound	%
ZnO	37·0
Included water	20·0
Sulphates:	
$\quad ZnSO_4$	5·2
$\quad PbSO_4$	9·0
$\quad Na_2SO_4$	1·8
$ZnCl_2$	0·2
Carbonates:	
$\quad CaCO_3$	1·1
$\quad MgCO_3$	0·4
$\quad ZnCO_3$	17·4
Fe_2O_3	2·5
Sand	4·5

The formation of the protective layers is governed largely by the pH of the environment. Since zinc forms an amphoteric oxide, both acid and alkaline conditions adversely affect its corrosion behaviour by interfering with the formation of the protective layers. Figure 4.42 shows how the corrosion rate of zinc varies with the pH[3], and it will be seen from this that the attack is most severe at pH values below 6 and above 12·5, while within this range the corrosion is very slow.

The electrochemical properties of zinc also have a large bearing on its corrosion behaviour. Zinc* is negative to $E^{\ominus}_{H^+/H_2}$ and magnesium and aluminium excepted, to most other metals commonly encountered, including those found in the less pure forms of zinc. This means that when zinc is in contact with these metals sacrificial electrochemical action can take place, with zinc forming the anode. Contact with other metals and impurities can

---

*$E^{\ominus}_{Zn^{2+}/Zn} = -0\cdot76$ V (SHE) and the corrosion potential of Zn approximates to this value in sea water, but is more positive in fresh waters, particularly at elevated temperatures.

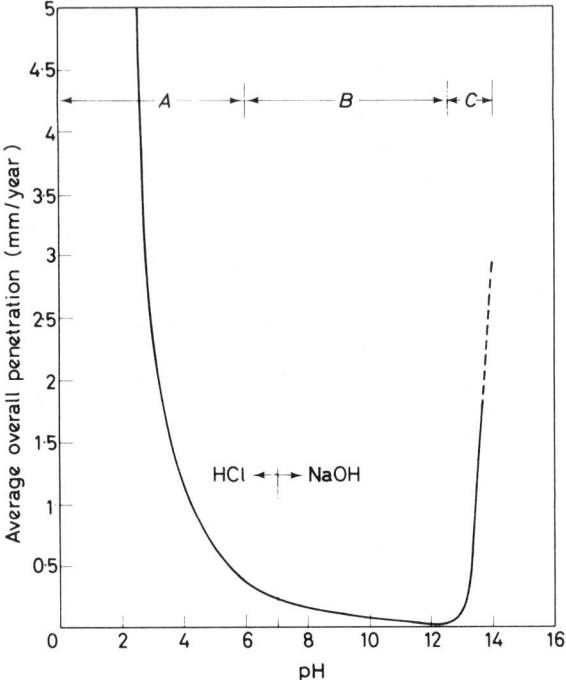

Fig. 4.42 Effect of pH on the rate of corrosion of zinc, A, rapid corrosion; B, stable film — low corrosion rate; C, rapid corrosion (after Roetheli, et al.[3])

therefore cause zinc to corrode rapidly, and it is important to avoid intermetallic contacts with metals such as copper, tin or nickel, under immersion in water. On the other hand the overvoltage for hydrogen evolution on zinc is high, and this can, in some circumstances, lead to a reduction in the corrosion rate, although under normal atmospheric conditions oxygen is generally present in sufficient concentration to act as a cathode reactant.

The Corrosion of Zinc in the Atmosphere

In dry air the stability of zinc is remarkable. Once the protective layer of zinc oxide formed initially is complete, the attack ceases. Even under under normal urban conditions, such as those in London, zinc sheet 0·8 mm thick has been found to have an effective life of 40 years or more when used as a roof covering and no repair has been needed except for mechanical damage. The presence of water does, of course, increase the rate of corrosion; when water is present the initial corrosion product is zinc hydroxide, which is then converted by the action of carbon dioxide to a basic zinc carbonate, probably of composition similar to $ZnCO_3 \cdot 3Zn(OH)_2$[4]. In very damp conditions unprotected zinc sometimes forms a loose and more conspicuous form of corrosion product known as 'wet storage stain' or 'white rust' (see p. 4.171).

Industrial atmospheres usually accelerate the corrosion of zinc. When heavy mists and dews occur in these areas, they are contaminated with considerable amounts of acid substances such as sulphur dioxide, and the film of moisture covering the metal can be quite acid and can have a pH as low as 3. Under these conditions the zinc is dissolved but, as the corrosion proceeds, the pH rises, and when it has reached a sufficiently high level basic salts are once more formed and provide further protection for the metal. These are usually the basic carbonate but may sometimes be a basic sulphate. As soon as the pH of the moisture film falls again, owing to the solution of acid gases, the protective film dissolves and renewed attack on the metal occurs. Hudson and Stanners[5] conducted tests at various locations in order to determine the effect of atmospheric pollution on the rate of corrosion of steel and zinc. Their figures for zinc are given in Table 4.34 and clearly show the effect which industrial contamination has on the corrosion rate.

In areas near the sea coast the rates of corrosion may be increased somewhat by the sea spray containing soluble chlorides, but the rates are still much lower than those prevailing in heavily polluted industrial areas. The white corrosion product which is sometimes found under these conditions probably consists of the basic chloride Zn_2OCl_2[(4)].

Anderson[6], reporting on a series of 20-year exposure tests carried out by the ASTM, quotes the average rates of corrosion for various types of environment which are given in Table 4.35.

These results clearly demonstrate (*a*) the effect of moisture in increasing the corrosion rate six times in the rural areas compared with the arid areas

Table 4.34 Atmospheric corrosion of zinc in various part of the UK

Site	Degree of pollution	Standard pollution*	Corrosion rate (mm/y)	
			Zinc	Steel
Godalming	Slight	0·23	0·00107	0·0425
Teddington		0·82	0·00211	0·0633
Hornchurch	Moderate	0·92	0·00324	0·0627
London (Victoria Park)		1·28	0·00445	0·0756
Barking (Greatfield Park)	Heavy	1·60	0·00843	0·0750
Salford (Ladywell)		2·31	0·00843	0·105
Sheffield (Hunshelf Bank)	Severe	3·58	0·0129	0·107
Billingham (Council Offices)		5·24	0·0125	0·180

* Calculated on weight (mg) of SO_3 absorbed per day by 1 dm^2 of lead dioxide

Table 4.35 Further examples of atmospheric corrosion rates for zinc[6]

Atmosphere	Corrosion rate (mm/y)
Industrial	0·0064
Sea coast	0·0015
Rural	0·0011
Arid	0·00018

and (b) the considerable increase of attack in the industrial areas. Anderson considers the principal features which control the rate of corrosion of zinc are (a) the frequency of rainfall and dew fall, (b) the acidity of the atmosphere and (c) the rate of drying. In situations where the drying of zinc of retarded corrosion is found to be most severe, for example the bottom strand of a galvanised wire fence which is shielded from the sun by grass and weeds.

Ambler[7] has attempted to find a relationship between the corrosion of zinc and iron and atmospheric salinity in the UK. This followed previous tests in Nigeria[8], when it was concluded that the governing factor in the corrosion of steel and zinc was airborne salt and that there was a relationship between corrosion and the distance from the sea. In the UK, however, no such relationship was found to exist, and the governing factor in the corrosion of zinc in the atmosphere is confirmed to be the amount of sulphur dioxide pollution.

White rust If a fresh zinc surface is allowed to stand with large drops of dew on it, as may easily happen if it is stored in a closed place in which the temperature varies periodically, it is attacked by the oxygen dissolved in the water, owing to differential aeration between the edges and the centres of the drops. A porous form of zinc oxide builds up away from the surface and quickly takes up carbon dioxide from the air to form the basic carbonate known as *white rust* or *wet storage stain*.

This type of corrosion can take place on any new surface of zinc and is best prevented by storing the metal in a dry, airy place until a protective layer has been formed. Zinc which has been properly aged in this way is safe against white-rust formation. Various methods are employed to prevent white rust. A chromate treatment is widely used for zinc-plated articles and for galvanised sheet, and occasionally for zinc die castings. Fatty substances, such as oils or lanolin, are sometimes used to protect larger items.

Corrosion of Zinc in Water

When zinc is immersed in distilled water containing dissolved oxygen, a protective film, probably consisting of zinc hydroxide, is slowly formed, and this extends over nearly the whole surface. At certain points the metal seems to remain uncovered and local attack continues, resulting in pitting. This may be quite serious and can lead to the rapid penetration of a zinc sheet. The attack does not spread over the whole surface of the zinc but is confined to these small local areas, although similar pits are liable to appear at points in a vertical line directly below the seat of the original corrosion. It has been suggested that the areas attacked occur where particles of corrosion product have fallen and are resting on the surface, shielding it from oxygen. The point thus shielded becomes anodic and suffers corrosion. If oxygen replenishment is made uniform over the whole surface, e.g. by 'whirling'[9] a zinc plate, pitting is avoided and the protective film of zinc hydroxide is found over the entire surface.

Natural waters all contain dissolved salts to a certain extent and these tend to form a scale on the surface of the metal. Since such scales have a protective effect it is to be expected that corrosion will be less severe in hard water. This is in fact true, and it is often found that distilled water is more corrosive than natural waters.

The effect of pH on the corrosion of zinc has already been mentioned (p. 4.170). In the range of pH values from 5·5 to 12, zinc is quite stable, and since most natural waters come within this range little difficulty is encountered in respect of pH. The pH does, however, affect the scale-forming properties of hard water (see Section 2.3 for a discussion of the Langelier index). If the pH is below the value at which the water is in equilibrium with calcium carbonate, the calcium carbonate will tend to dissolve rather than form a scale. The same effect is produced in the presence of considerable amounts of carbon dioxide, which also favours the dissolution of calcium carbonate. In addition, it is important to note that small amounts of metallic impurities (particularly copper) in the water can cause quite severe corrosion, and as little as 0·05 p.p.m. of copper in a domestic water system can be a source of considerable trouble with galvanised tanks and pipes.

Sea-water contains considerable amounts of soluble salts, particularly sodium chloride, which is present in concentrations from 1 to 25%. The North Sea, for example, contains about 3% sodium chloride, 0·47% magnesium sulphate, 0·2% magnesium chloride and 0·1% calcium chloride. The carbon dioxide content is about 0·0005 to 0·01% and the pH is between 7·6 and 8·1. The high chloride content would tend to increase the rate of corrosion, and this usually takes the form of pitting under these conditions. The corrosive influence of the chloride ions is, however, inhibited by the presence of magnesium and calcium ions by virtue of the formation of a protective layer of magnesium and calcium salts (calcareous scale).

The effect of the magnesium salts has been clearly demonstrated by Schikorr[10]. Zinc immersed in a solution containing 30 g/l of sodium chloride showed a weight loss of 198 g/m^2 after 14 days. When the solution contained in addition 12 g/l of magnesium chloride, the loss in weight was only 4 g/m^2 after the same period. Artificial sea-water was also tested and gave a weight loss of 5 g/m^2.

Zinc roofs are quite satisfactory at the coast, where they receive a large amount of salt-water spray, and many British piers have been covered with sheet zinc which has lasted 50 years and more. Most of the zinc actually immersed in sea-water is in the form of zinc coatings, the behaviour of which is discussed in a later section of this book (see Section 14.4). Experience with these coatings has proved the value of zinc in sea-water compared with many other metals in this environment[11].

The effect of temperature on the corrosion of zinc in water It is found that the temperature has a marked effect on the rate at which zinc corrodes in water. The corrosion rate in distilled water reaches a maximum in the temperature range 65–75°C. This variation in the corrosion rate with temperature is attributed to changes in the nature of the protective film. At lower temperatures the film is found to be very adherent and gelatinous, while at temperatures around 70°C it becomes distinctly granular in character and much less adherent. Above 75°C it again tends to become more adherent and assumes a very compact and dense form. It is believed that the granular coating formed at temperatures around 70°C is more porous

Table 4.36 Dependence of corrosion rate of zinc on temperature

Temperature (°C)	Corrosion rate (mm/y)
20	0·020
50	0·071
55	0·38
65	3·05
75	2·34
95	0·31
100	0·12

than the others and permits greater access of the dissolved oxygen to the metal.

Experiments in which rolled high-grade zinc was immersed in distilled water for 15 days at various temperatures are described by Cox[12]. The figures which he obtained are given in Table 4.36.

Soil Corrosion

The factors influencing the corrosion of metals in soil are more numerous than those prevailing in air or water, and the electrochemical effects are more pronounced. Moreover, soils vary widely in their composition and behaviour even over very short distances. It is difficult therefore to obtain reliable data. It is evident, however, that zinc has considerable resistance to corrosion when buried, and the greatest attack is caused by soils which are acid or contain large amounts of soluble salts.

In tests carried out by the National Bureau of Standards in the USA[13] specimens of copper alloys, lead, zinc and zinc alloys were buried at a number of different sites for periods varying from 11 to 14 years. The soils tested covered a pH range from 2·6 to 9·4 and resistivities ranged from 62 to 17 800 Ω cm. The weight losses and maximum depths of pitting were recorded, and the results indicated that the most severe corrosion occurred in soils of poor aeration having high acid and soluble-salt contents.

Although there was no significant difference in weight loss between rolled zinc and zinc alloys, the maximum penetration of the alloys definitely appeared to be greater than that of rolled zinc.

BISRA tests on galvanised steel pipe buried for five years at five different sites are described by Hudson and Acock[14]. The galvanised pipes resisted corrosion rather better than steel at all sites. Galvanised pipes of small diameter are frequently used to provide underground water services in farms and similar establishments, and little trouble is experienced.

Corrosion Resistance of Zinc in Chemical Environments

Acids and alkalis Zinc dissolves in liquids whose pH is below about 5 and above 12·5. Table 4.37 shows the approximate pH value of 0·01 and 0·1 N solutions of some common acids and alkalis, from which it is seen that all of the acids included would attack zinc, while only the stronger solutions of most of the alkalis would cause any corrosion. Zinc dissolves in caustic soda solution to form sodium zincate, i.e. Na_2ZnO_2.

Table 4.37 pH values of common acids and alkalis

Acids	Concentration	pH
Hydrochloric	0·1 N	1·1
Hydrochloric	0·01 N	2·0
Sulphuric	0·1 N	1·2
Sulphuric	0·01 N	2·1
Oxalic	0·1 N	1·6
Formic	0·1 N	2·3
Acetic	0·1 N	2·9
Acetic	0·01 N	3·4
Bases		
Sodium hydroxide	0·1 N	13·0
Sodium hydroxide	0·01 N	12·0
Potassium hydroxide	0·1 N	13·0
Potassium hydroxide	0·01 N	12·0
Sodium metasilicate	0·1 N	12·6
Trisodium phosphate	0·1 N	12·0
Sodium carbonate	0·1 N	11·6
Borax	0·1 N	9·2
Sodium bicarbonate	0·1 N	8·4

Owing to the high overpotential for hydrogen evolution pure zinc is attacked only very slowly by dilute sulphuric and hydrochloric acids, but impure zinc dissolves quite readily, evolving hydrogen. Thus when ordinary commercial zinc is placed in a dilute acid it immediately starts to corrode slowly. As corrosion proceeds the impurities present are reprecipitated as a black metallic sponge which may further accelerate the reaction owing to the increased surface area of low overpotential cathodic sites, e.g. copper. Some metallic impurities, notably aluminium, retard the corrosion of zinc by acids. This is due to the production of highly protective films. Other metals, such as tin and lead, retard the corrosion initially, but when after a few hours they are deposited as a metallic sponge the corrosion is as fast as with ordinary zinc.

Salt solutions When a zinc sheet is immersed in a solution of a salt, such as potassium chloride or potassium sulphate, corrosion usually starts at a number of points on the surface of the metal, probably where there are defects or impurities present. From these it spreads downwards in streams, if the plate is vertical. Corrosion will start at a scratch or abrasion made on the surface but it is observed that it does not necessarily occur at all such places. In the case of potassium chloride (or sodium chloride) the corrosion spreads downwards and outwards to cover a parabolic area. Evans[15] explains this in terms of the dissolution of the protective layer of zinc oxide by zinc chloride to form a basic zinc chloride which remains in solution.

Feitknecht[16] has examined the corrosion products of zinc in sodium chloride solutions in detail. The compound on the inactive areas was found to be mainly zinc oxide. When the concentration of sodium chloride was greater than 0·1 M, basic zinc chlorides were found on the corroded parts. At lower concentrations a loose powdery form of a crystalline zinc hydroxide appeared. A close examination of the corroded areas revealed craters which appeared to contain alternate layers and concentric rings of basic chlorides and hydroxides. Two basic zinc chlorides were identified, namely $6Zn(OH)_2 \cdot ZnCl_2$ and $4Zn(OH)_2 \cdot ZnCl_2$. These basic salts, and the crystalline zinc hydroxides, were found to have layer structures similar in general to the layer structure attributed to the basic zinc carbonate which forms dense adherent films and appears to play such an important rôle in the corrosion resistance of zinc against the atmosphere. The presence of different reaction products in the actual corroded areas leads to the view that, in addition to action between the major anodic and cathodic areas as a whole, there is also a local interaction between smaller anodic and cathodic elements.

Organic chemicals Most organic liquids, other than the stronger acids, only attack zinc slowly. Zinc is, therefore, suitable for storage tanks for liquid hydrocarbons such as motor fuels, for phenols and for trichlorethylene degreasers. Zinc or zinc-coated vessels are not recommended for use in contact with acid foodstuffs, but are regularly used for dry foods. Zinc in small quantities is beneficial in the human diet.

Cathodic Protection by Zinc Anodes

Zinc should give a potential of $-1·05$ V vs. $Cu/CuSO_4$ and should have a driving potential of about $-0·25$ V with respect to cathodically protected steel. Zinc is therefore sufficiently negative to act as a sacrificial anode, and its first use for such purposes was on the copper-sheathed hulls of warships more than a century ago. The first attempts to fit zinc anodes to steel hulls, however, were a complete failure, for the sole reason that it had not been realised that the purity of the zinc was of paramount importance. The presence of even small amounts of certain impurities leads to the formation of dense adherent films, which cause the anodes to become inactive.

The major harmful impurity is iron, and by keeping the iron content to less than 15 p.p.m. it became possible to produce perfectly satisfactory anodes of zinc[17,18]. Alternatively the effect of the iron can be neutralised by alloying the zinc with certain metals, among which aluminium and silicon or cadmium have been found to be particularly effective. The presence of cadmium causes the corrosion product to fall away evenly, leaving an active surface (see Section 10.2).

Zinc alloy anodes are generally very efficient, owing to their non-polarising characteristics and to the absence of parasitic reactions when buried or immersed. As zinc's 'self-corrosion' is, therefore, very small, the anode's high efficiency of 85–95% holds throughout the current-density range, whereas the efficiency of magnesium, about 50–55% at high current densities, may fall to only 30% at low current densities. The use of these zinc alloys for reference electrodes to monitor some hull-protection schemes or as permanent reference cells located along pipelines, underlines their non-polarising characteristics. The electrochemical equivalent of zinc indicates that 10·5 kg should give 1 A/year, but in practice this amount of charge is given by 11–12 kg of zinc compared with 7·7–9 kg of magnesium. As the cost of zinc is low, zinc anodes are found to be very economical. Moreover, since zinc is fairly dense, the volume consumption (1·6 litre A^{-1} year^{-1}) is much lower than that for magnesium (5 litre A^{-1} year^{-1}) and the dimensions of the zinc anode are correspondingly smaller.

The driving potential of zinc to cathodically protected steel is 200–250 mV and this is considerably lower than the 700 mV given by magnesium. While this value is ideal in sea-water or other electrolytes of low resistivity, the use of zinc is not always practicable in environments of higher electrical resistance. For example, it is not likely to be of much use for protecting large underground systems in a high resistivity soil, but on the other hand it proves to be of value for smaller underground units, such as storage tanks, situated in soils of resistivity below about 3 000 Ω cm. Olive[19], for instance, in the USA, discussed the application of zinc anodes for protecting underground equipment at petrol-filling stations. Larger installations which employ a considerable number of zinc anodes are the protective schemes for the steel gas mains in Houston and New Orleans[20]. Of a total of some 1 200 galvanic anodes in New Orleans about 1 000 are of zinc. This is a good example of a case where a system of zinc anodes can be used to protect a large underground installation under the right soil conditions. Zinc is used quite widely for the protection of bare service pipes of small diameter, and is receiving increased acceptance for the protection of large-diameter coated pipes in built-up areas in order to minimise interference on adjacent mains. It is also used for protecting galvanised cold-water tanks.

Where the use of zinc anodes is practicable, the low driving potential is a great advantage since the resistance of the steel to be cathodically protected is the controlling resistance, and the current output of the anode varies with the requirements of the cathode. Thus it can be said that zinc anodes are largely self-governing.

By far the largest use of zinc anodes is in sea-water for the protection of ships' hulls and North Sea pipelines and drilling rigs. The high conductivity of the sea-water and the excellent natural resistance of zinc to corrosion make it very effective in this application. Many examples of trials with zinc anodes on ships can be quoted, and much work has been done in this respect by the US Navy. One example is a paper by Carson[18], describing zinc anodes attached directly to the hull and found to have a life of 8 to 10 years. This paper also gives the results of tests carried out at the Pacific Naval Laboratory, showing the effect of various impurities in the zinc. Carson looks into the economics of zinc alloy anodes and concludes that for small and medium hull sizes, e.g. 1 400 m^2 wetted hull area, zinc alloy anodes are more economical than the most competitive alternative systems available and have the additional advantages of more even current distribution, minimum risk of paint-stripping, and maintenance-free long life. For larger hull areas zinc alloy anodes are at least as economic as many alternative systems.

Zinc anodes are made in a variety of shapes and sizes, ranging in weight from 2·25 to 11 kg and in shape from cylindrical rods to rectangular bars.

When used underground they are usually placed in a backfill, consisting of gypsum, sodium sulphate and clay, which may be added loose, shipped in a bag around the anode, or obtained in cast form.

Behaviour of the ZA alloys in aerated water from pH 2.0 to 13.0

As part of an International Lead Zinc Research Organisation Programme, the corrosion behaviour of the ZA alloys has been studied at the Noranda Research Centre in Canada[21]. Air was bubbled continuously through distilled water flowing through the corrosion cells, its pH being controlled by additions of hydrochloric acid or sodium hydroxide. The temperature was 22 ± 2°C. Pure zinc sheet was included for comparison. The results are based on immersion times of 4 to 15 days and show that ZA27 undergoes little attack between pH6 and 11.5, like pure zinc. At pH 4.0 there was preferential attack of the decomposed β-phase, whereas at pH 12.8 there was selective dissolution of the aluminium-rich primary dendrites leaving a sponge-like surface. The samples were gravity cast plates.

Behaviour of ZA alloys in neutral salt spray

The test followed ASTM B-117. Pure zinc and the 4% aluminium diecasting alloy were corroded slightly more than ZA8 and ZA12 during up to 600 hours exposure; after longer periods the difference diminished. ZA27 corroded at about one-third the rate of the other alloys, probably because attack occurs on the zinc rich matrix, while ZA27 contains a significantly larger amount of the aluminium-rich phase and behaves like the aluminium diecasting alloy 380 also included in the test[21].

Behaviour of the ZA alloys at a waste water plant

Gravity die cast ZA alloy test plates and 99.99% pure rolled zinc samples were exposed at a waste water treatment plant in Detroit, Michigan, USA. The results after one year are summarised in Table 4.38[21].

Recent Developments

In the 1970s and 1980s alloys with higher contents of aluminium have become more important, especially for casting using metal or graphite moulds. Table 4.30 has been modified to cover these alloys. A super-plastic alloy containing about 22% aluminium has also been developed, but only has very limited use.

The physical properties of the newer alloys have been well documented (see bibliography). Essentially, they lie between those of Alloy 5 (Alloy B) and of Alzen 305 in Table 4.31, broadly in relation to their aluminium content. The mechanical properties vary considerably according to the method

Table 4.38

Exposure site	Sample	Corrosion rate $gm^{-2} y^{-1}$
Atmospheric–Indoor		
Bar screen (High H_2S concentration	Zinc	1
and humidity)	ZA8	2
	ZA12	1
	ZA27	1
Atmospheric–Outdoor		
Primary sedimentation tank	Zinc	8–12
(Relatively high H_2S and	ZA8	9
humidity)	ZA12	9
	ZA27	4
Secondary sedimentation tank	Zinc	9
(Relatively high H_2S and	ZA8	10
humidity)	ZA12	10
	ZA27	9
Immersed		
Primary sedimentation tank	Zinc	1,239–1,601
(Waste water with low O_2 containing	ZA8	625
phosphorus precipitants and	ZA12	2,181
coagulants)	ZA27	1,209
Secondary sedimentation tank	Zinc	857–1,174
(Oxygenated waste water	ZA8	324
containing microorganisms	ZA12	352
	ZA27	253

Note A corrosion rate of 300 g m^{-2} y^{-1} corresponds to a uniform depth of penetration between 50 and 60 μm for the ZA alloys.
Visual examination of the immersed samples showed that ZA8 suffered moderate, uniform corrosion in both tanks. ZA12 showed extensive localised corrosion in the primary tank but slight uniform corrosion with localised corrosion at several spots in the secondary tank. ZA 27 suffered extensive localised corrosion over most of the surface in the primary tank but only slight uniform corrosion with several very small pits dispersed all over the surface in the secondary tank.

of casting. The five major alloys are broadly compared in Table 4.39. Up to about 15% aluminium the materials are non-sparking and can be used in mines.

Table 4.39 Typical Mechanical Properties of Zinc Alloys

Property	Alloy				
	3	5	8	12	27
Tensile strength (N/mm^2)					
Sand or gravity cast			250	300	420
Pressure die cast	283	328	372	400	440
Elongation (% on 5.65 $\sqrt{\text{area}}$)					
Sand or gravity cast			1	1	4
Pressure die cast	10	7	8	5	5
Hardness (Brinell)					
Sand or gravity cast			90	95	115
Pressure die cast	82	91	105	105	115
Shear strength (N/mm^2)					
Sand or gravity cast			240	250	290
Pressure die cast	214	262	276	325	
Fatigue strength (5 $\times$ 10^8 cycles)					
Sand or gravity cast			52	104	172
Pressure die cast	48	57			
Impact strength (J)					
Sand gravity			20	25	48
Pressure die cast	58	65	40	28	13

The corrosion behaviour of the zinc-aluminium casting alloys can, for practical purposes, be considered as similar to pure zinc — even a factor of 2 in corrosion rate would be of little significance in a solid product. This is to be contrasted with both the behaviour and significance of zinc-aluminium alloys when produced as coatings (Chapter 14.4) — in this latter case, the production process is designed to give inherently corrosion-resistant structures and only limited corrosion attack can be tolerated because the coating is thin. With castings, the only corrosion criterion is that harmful impurities such as lead, tin and cadmium, which could cause inter-crystalline corrosion, shall be below specified and low levels; the difference in levels of general corrosion is unlikely to affect the choice of alloy.

The general corrosion rate of zinc and zinc alloys in practice often have been shown to be much less than in simulated conditions; this is because many naturally occurring substances act as inhibitors. Figure 4.42 is a good example of this. The diagram is valuable for the qualitative relationship between acid, neutral and alkaline conditions but, in practice, the corrosion rates are usually very much lower than indicated by the pH because of the effect of other dissolved constituents and the barrier effect of corrosion products. Seawater around the British Isles is much less corrosive to zinc than tropical seawater.

The atmospheric corrosion data in Table 4.34 (and also Table 13.8) is related to historic environments. Current use in the industrial areas listed with acidic pollution would show much lower corrosion rates as the corrosion of zinc in the atmosphere is essentially related to the SO_2 content (and the time of wetness) and in many countries the sulphurous pollution has been greatly reduced in the past 20 years. Zinc also benefits from rainwater washing to remove corrosive poultices: thus, although initial corrosion rates are usually not very different on upper and lower surfaces, the latter tend — with time — to become encrusted with corrosion products and deposits and these are not always protective.

The bibliography is expanded from that given in the previous edition.

A. R. L. CHIVERS
F. C. PORTER

REFERENCES

1. Vernon, W. H. J., Second Experimental Report to the Atmospheric Corrosion Committee, B.N.F.M.R.A., *Trans. Faraday Soc.*, **22**, 113 (1927)
2. Bablik, H., Galvanising Hot Dip, Spon, London, 333 (1950)
3. Roetheli, B. E., Cox, G. L. and Littreal, W. B., *Metals and Alloys*, **3**, 73 (1932)
4. Morriset, P., *Zinc et Alliages*, **20**, 15 (1959)
5. Hudson, J. C. and Stanners, J. F., *J. Appl. Chem.*, **3**, 86 (1953)
6. Anderson, E. A., Amer. Soc. Test. Mater. Special Publication No. 175, June (1955)
7. Ambler, H. R., *J. Appl. Chem.*, **10**, 213 (1960)
8. Ambler, H. R. and Bain, A. A., *J. Appl. Chem.*, **5**, 437 (1955)
9. Evans, U. R. and Davies, D. E., *J. Chem. Soc.*, 2607 (1951)
10. Schikorr, G., *Z. Metallk.*, **32**, 314 (1940)
11. Hudson, J. C. and Banfield, T. A., *J. Iron St. Inst.*, **154**, 229 (1946)
12. Cox, G. L., *Ind. Engng. Chem.*, **23**, 902 (1931)
13. Denison, I. A. and Romanoff, M., *J. Res. Nat. Bur. Stand.*, **44**, 259 (1950)
14. Hudson, J. C. and Acock, G. P., *Tests on the Corrosion of Buried Iron and Steel Pipes*, Iron and Steel Inst. Special Report No. 45 (1953)

15. Evans, U. R., *The Corrosion and Oxidation of Metals*, Edward Arnold, London (1960)
16. Feitknecht, W., *Chem. and Ind. (Rev.)*, 1102 (1959)
17. Crennell, J. J. and Wheeler, W. C. G., *J. Appl. Chem.*, **8**, 571 (1958); **6**, 415 (1956)
18. Carson, J. A. H., *Corrosion*, **16**, 99 (1960)
19. Olive, M. J., *Corrosion*, **16**, 9 (1960)
20. Trouard, S. E., *Corrosion*, **13**, 21 (1957)
21. Progress reports Nos 9 and 10, Project ZM-287 International Lead Zinc Research Organization, 292 Madison Avenue, New York 10017, USA

BIBLIOGRAPHY

ILZRO *Engineering Properties of Zinc Alloys*, 2nd edition, 116, ILZRO (1981)
Johnen, H. J. *Zink-Taschenbuch*, 347, Metall-Verlag GmbH, Berlin (1981)
BSI PD 6484 *Corrosion at bimetallic contacts and its alleviation*, BSI, London, 32, (1984)
ASTM *Symposia on corrosion behaviour ASTM STP 175, 290, 435 AND 646* ASTM, Philadelphia (1956, 1959, 1968, and 1978)
Slunder, C. J. and Boyd, W. K. *Zinc: its corrosion resistance*, 2nd edition, ILZRO Res, Triangle Park, Ill, 250, (1983)
Schikorr, G. *Atmospheric Corrosion Resistance of Zinc*, English edition 1965, ZDA, London, 4, (1964)
Wiederholt, W. *Behaviour of Zinc in Water*, Metall-Verlag GmbH, Berlin, 162, English typescript translation available in ZDA library (1965)
Wiederholt, W. *Behaviour of Zinc with Chemicals*, Korrosionverhalten von Zinc, Band 3 (in German but identifiable data), 110 (1976)
Witt, C. A. *Corrosion Behaviour of Zinc in Contact with Bitumen*, Korrosionverhalten von Zink, Band, **4**, 47, (in German) (1980)
Desai, M. N., Rana, S. S. and Gandhi, M. H. *Corrosion inhibitors for zinc*, Typescript copies with ILZRO and ZDA (1969)
ILZRO *Zinc dust and powder*, ILZRO, Res Triangle Park, NC, 118, (1982)
Porter, F. *Zinc Handbook*, Marcel Dekker Inc. New York 629 (1992)

5 RARER METALS

5.1 Beryllium   **5**:3
5.2 Molybdenum   **5**:10
5.3 Niobium   **5**:24
5.4 Titanium and Zirconium   **5**:36
5.5 Tantalum   **5**:62
5.6 Uranium   **5**:78
5.7 Tungsten   **5**:87

5.1 Beryllium

Introduction

As a light, strong metal, beryllium holds considerable promise as a useful engineering material, but because of an inherent directional brittleness, a really significant commercial use, e.g. in the aircraft industry, has not proved possible. It has been used to a limited extent in aerospace applications, and it was employed as heat shields for the Project Mercury space capsule. It has also found use in precision guidance systems when fairly pure environmental conditions can be assured.

In common with magnesium and zirconium the metal has little tendency to capture neutrons, and there was promise that a significant industrial use would arise in nuclear engineering, but after extensive trial in gas cooled reactors, its ultimate commercial employment in that context was also deemed inappropriate.

For many years, however, beryllium has been used as a minor alloying addition to other metals, particularly copper, and although its main attribute lies in improving mechanical properties, it generally also improves the corrosion resistance of the parent metal. For example, in the case of copper, beryllium addition has been found to increase oxidation resistance[1], as well as the wet corrosion resistance, particularly in corrosion fatigue[2] where the strengthening effect is advantageous. By ensuring conditions under which preferential oxidation of the beryllium constituent occurs—the principle of 'selective oxidation'[3]—the oxidation resistance of copper-beryllium can be much improved relative to a normal surface. A similar effect can be produced on silver, where deposition of beryllium oxide, e.g. by cathodic deposition from ammoniacal beryllium sulphate or nitrate, effects very considerable improvement in the resistance to tarnishing. Beryllium has been used for the purpose of improving oxidation resistance in magnesium, to produce the Magnox series of alloys used for fuel sheathing in carbon-dioxide-cooled reactors[4]. The particular composition used in the Calder Hall reactor, for example, contains about 0·01% beryllium and 0·8% aluminium. In accord with the improved resistance to oxidation, the incendivity of Magnox alloys is also reduced relative to that of pure magnesium. This had been found with magnesium some years earlier[5] when as little as 0·001% beryllium raised the air ignition temperature of a magnesium-aluminium-zinc alloy from 580°C to over 800°C.

Metallurgy

General

Beryllium is a light metal (s.g. 1·85) with a hexagonal close-packed structure (axial ratio 1·568). The most notable of its mechanical properties is its low ductility at room temperature. Deformation at room temperature is restricted to slip on the basal plane, which takes place only to a very limited extent. Consequently, at room temperature beryllium is by normal standards a brittle metal, exhibiting only about 2 to 4% tensile elongation. Mechanical deformation increases this by the development of preferred orientation, but only in the direction of working and at the expense of ductility in other directions. Ductility also increases very markedly at temperatures above about 300°C with alternative slip on the 1010 prismatic planes. In consequence, all mechanical working of beryllium is carried out at elevated temperatures. It has not yet been resolved whether the brittleness of beryllium is fundamental or results from small amounts of impurities. Beryllium is a very poor solvent for other metals and, to date, it has not been possible to overcome the brittleness problem by alloying.

As already indicated, the brittleness of beryllium has so far been the main determining feature in the technology, and because of the mechanical anisotropy, the most widely practised method of fabrication is via powder metallurgy.

Two other factors are noteworthy: the deleterious effects on chemical and mechanical properties of small amounts of impurities residual from extraction of the metal, and its toxicity. The first of these factors is obviated by vacuum melting the raw metal (for purification) as an essential prerequisite to further processing. The toxicity of beryllium is essentially a pulmonary problem and great care must be taken in handling the finely divided metal or its compounds. In practice, this type of activity is usually carried out under well-ventilated conditions. Certain tolerance levels for atmospheric beryllium are now internationally accepted[6] and merit careful study before work on beryllium is embarked upon.

Extraction and Fabrication

Beryllium is extracted from the main source mineral, the alumino-silicate *beryl*, by conversion to the hydroxide and then through either the fluoride or the chloride to the final metal. If the fluoride is used, it is reduced to beryllium by magnesium by a Kroll-type reaction. The raw metal takes the form of 'pebble' and contains much residual halides and magnesium. With the chloride on the other hand, the pure metal is extracted by electrolysis of a mixture of fused beryllium chloride and sodium chloride. The raw beryllium is now dendritic in character, but still contains residual chloride.

Before further processing, the raw metal must be purified. Various methods of leaching have been tried on a laboratory scale, but in practice the method usually adopted is vacuum induction melting. The ingot so produced is then converted to powder by swarfing, followed by grinding. On a laboratory scale the grinding is generally done by ball milling, while in production, milling is carried out between beryllium-faced plates.

The particle size of powder most often used for consolidation is −200 mesh (74 μm sieve aperture), and the most widely practised method of consolidation is hot pressing *in vacuo*.

Setting aside the necessity for hot working and the toxicity problem, the fabrication of consolidated beryllium generally follows normal lines; rolling, extrusion, drawing, forging, etc. have all been successfully carried out. It is interesting to note that because of the high chemical activity of beryllium, allied to the method of consolidation from powder, the usual grade of metal contains about 1 to 2% of beryllium oxide. It could therefore be considered almost a mild cermet rather than a conventional pure metal. Specifications for chemical composition are detailed in Table 5.1.

Table 5.1 Chemical composition of various forms of beryllium

Form	Analysis								Notes
	Be Assay % min.	BeO % max.	Fe % max	Al % max.	C % max.	Mg % max.	Si % max.	Ni % max.	
Pebble	96·0	—	0·15	0·14	—	—	0·1	0·04	Remainder slag
Ingot	99·0	1·0	0·15	0·14	0·14	0·08	0·1	0·04	
Powder	98·0	2·0	0·18	0·16	0·12	0·08	0·12	—	Any other metallic impurity 0·04% max.

Corrosion Behaviour

Aqueous Environments

In its general corrosion behaviour, beryllium exhibits characteristics very similar to those of aluminium. Like aluminium, the film-free metal is highly active and readily attacked in many environments. Beryllium oxide, however, like alumina, is, a very stable compound (standard free energy of formation = −579 kJ/mol), with a bulk density of 3·025 g/cm^3 as compared with 1·85 g/cm^3 for the pure metal, and with a high electronic resistivity of about 10^{10} Ω cm at 0°C. In fact, when formed, the oxide confers the same type of 'spurious nobility' on beryllium as is found, for example, with aluminium, titanium and zirconium.

It is often difficult to correlate or categorise the corrosion behaviour of beryllium because of past inconsistencies in the quality of metal studied and differences between the chemical composition and metallurgical history of current material and of past grades. For example, although vacuum melting now produces a fairly consistent grade of beryllium (often designated QMV), this method of purification was not introduced intil the late 1940s. Prior to this, powder was produced by grinding 'pebble' without purification and hence small amounts of residual halide slag frequently persisted through to the fabricated product.

Furthermore, even though a consistent quality of beryllium is now produced, the chemical composition falls far short of the standards found for instance in aluminium; generally, the main impurities consist of about 1% of beryllia at grain boundaries, about 0·15% of iron and 0·05-0·1% of other elements such as silicon, aluminium and carbon.

Beryllium is readily attacked by most acids and, being amphoteric, is slowly attacked by caustic alkalis with the evolution of hydrogen. As might be anticipated, in view of the controlling influence of the surface film of beryllia on corrosion behaviour, concentrated nitric acid has little effect on beryllium[7], while the dilute acid results in slow attack. Hot acid is much more reactive. Nitric acid is in fact often used to pickle-off residual mild steel from hot-extruded clad beryllium.

Anything which will affect the characteristics of the surface film of beryllia can be expected to affect the corrosion resistance of the metal. For example, dissolved fluoride or chloride ions give rise to increased pitting attack, analogous to the attack on aluminium in similar circumstances. At beryllium carbide inclusions in the metal, hydrated oxide or beryllium hydroxide is formed in moist air by hydrolysis of the carbide[8]; depending on the size of the inclusion, it would be expected that pitting of the underlying metal might ensue. The presence of cations which will deposit heavy metals and set up local cathodic areas has a pronounced adverse effect on corrosion resistance. Cupric ions in particular, at a concentration of less than 1 p.p.m., have been found to cause serious pitting in 0·005 M hydrogen peroxide aqueous solution at 85°C; ferric ions also appear to increase the rate of corrosion, although apparently not as seriously as copper.

Conversely to the above, any factor which tends to maintain the protective character of the beryllia film will obviously increase corrosion resistance, and, in this respect, the presence of anodic inhibitors such as sodium dichromate, up to about 40 p.p.m., will effectively suppress pitting of beryllium in water.

Most of the controlled corrosion studies on beryllium have been carried out in the USA in simulated reactor coolants. The latter have usually been water, aerated and de-aerated, containing small amounts of hydrogen peroxide and at temperatures up to 300-350°C. Many variables have been examined, covering surface condition, chemical composition, temperature, pH, galvanic effects and mechanical stress[8].

Surface condition Machined, abraded and pickled surfaces all exhibit much the same behaviour in water, and after exposure of up to about one year at temperatures less than 100°C average attack measures 0·0025-0·0050 mm/y. Almost always, however, corrosion of beryllium in water is accompanied by pitting and, on machined surfaces, pits of as much as 0·25 mm have been observed in 0·0005 M hydrogen peroxide at 85°C. Under similar conditions, annealed material has been found to be somewhat less resistant to attack than either machined or pickled surfaces.

Chemical composition The effect of carbide inclusions in aggravating attack in moist air has already been referred to, but in testing in 0·0005 M hydrogen peroxide, pH 6, at 85°C, the presence of carbide up to 0·26% by weight, iron up to 0·4% by weight, silicon up to 0·2% by weight, and aluminium up to 1·05% by weight, in both vacuum-cast and hot-pressed beryllium, did not appear to affect average corrosion rate; after tests of 13 months, the corrosion rate measured only 0·0025 mm/y. Pitting in these tests appeared to be worst in the aluminium-containing specimens, but even in those it did not exceed 0·050 mm.

Temperature Results of tests carried out in the temperature range 250–350°C are not consistent. In one series for example, on extruded beryllium for seven days in distilled water at 270°C, average attack varied between 0·015 and 0·53 mm/y. In contrast, other work on hot-pressed material gave an average attack of less than 0·0025 mm/y.

Variation of pH Between 4 and 8, variation of pH does not appear to have a very significant effect on corrosion rate in de-aerated water.

Galvanic effects Galvanic effects would be expected with such metals as copper, but even with stainless steel (US grade 347) tests in water containing small amounts of hydrogen peroxide at 85°C showed definite evidence of accelerated attack. Extruded beryllium was used in both static and dynamic conditions, and attack on the beryllium test specimens was some 3 to 5 times greater than that on uncoupled controls coupled to the steel. Galvanic effects with various aluminium alloys, however, were not so definite. This is perhaps not surprising in view of the close similarities between the two metals.

Effect of stress The first reported work on the stress-corrosion cracking of beryllium related to its use in water containing 0·005 M hydrogen peroxide at pH 6–6·5 and at about 90°C[9]. Although in these circumstances some pitting occurred, there was no evidence of stress corrosion even though the tests employed extruded metal stressed at up to 90% of the yield stress. Subsequently, stress corrosion has been reported on exposure to synthetic sea-water[10]. In this environment stress-corrosion failure was reported in a 40 h test at 70% of the 0·2% yield strength, the crack habit being transgranular in character.

Gaseous Environments

The possible employment of beryllium in nuclear engineering and in the aircraft industry has encouraged considerable investigation into its oxidation characteristics. In particular, behaviour in carbon dioxide up to temperatures of 1 000°C has been extensively studied[11-17], and it has been shown that up to a temperature of 600°C the formation of beryllium oxide follows a parabolic law but with continued exposure 'break-away' oxidation occurs in a similar fashion to that described for zirconium. The presence of moisture in the carbon dioxide enhances the 'break-away' reaction[15-16]. It has been suggested that film growth proceeds by cation diffusion and that oxidation takes place at the oxide/air interface[18].

It is reported that beryllium powder, of unspecified particle size, will burn in air at 1 200°C and react with nitrogen at 500°C[5]. Fluorine appears to attack beryllium at room temperature, and the other halogens, nitrogen dioxide and hydrogen sulphide are said to attack it at elevated temperatures[5].

Dry hydrogen chloride gas readily attacks solid beryllium above about 500°C with the formation of volatile beryllium chloride. Beryllium carbide and nitride are similarly attacked, but not beryllium oxide; this behaviour is of use in one method for the determination of beryllium-oxide in metallic beryllium.

There is a reaction between beryllium and nitrogen that starts at about 750°C and is appreciable at 850°C, beryllium nitride being formed[11]. The reaction with oxygen is less sluggish and at 900°C in oxygen oxidation proceeds at about twice the rate of nitride formation. Thus when beryllium is heated in air, beryllium nitride forms only a small proportion of the total scale—about 0·75% after 1 h at 1 000°C.

Protective Measures

From what has already been indicated, it will be apparent that use of beryllium in almost any commercial environment involving moisture may require some form of surface protection and it should be recognised that apart from the hazard of pitting corrosion, precautions must usually be taken against adverse galvanic coupling with heavy metals.

Protective measures range from chemical conversion coatings and anodising to the application of more substantial protective layers, e.g. enamels. For a more detailed treatment of the subject, the reader is advised to consult References 19, 20, 21 and 22.

Recent Developments

The early promise of wide applications for beryllium has not materialised, despite improvements in purity and more efficient means of consolidation such as isostatic hot pressing, because of the metal's toxicity, brittleness and cost. It is now chiefly of interest in the specialised fields of aerospace and nuclear applications. BrushWellman[23] is currently the sole commercial primary producer of beryllium metal in the West.

Jepson[24], presented a well documented review of the earlier research literature on its corrosion behaviour for the anticipated applications. Relatively little has been published since the present review on the behaviour of the metal in less severe or normal atmospheric environments apart from that by Mueller and Adolphson[25], which has been adapted as the chapter 'Corrosion of Beryllium', in the *ASM Metals Handbook*. This includes some 30 new references, half of which are private communications, reflecting the extent of unpublished work in this field. Procedures for handling, cleaning, protecting and packaging for shipment or storage to guard against these problems are given.

Additional work on protection includes painting[26] and anodizing. Paine and Stonehouse[27] used a high temperature silicone base paint on beryllium heat-sinks for aircraft brakes, together with manganese plating on the mating steel parts to give cathodic protection in fresh and salt water solutions. Further work on stress corrosion in aerated synthetic sea water[28] includes cathodic protection by an impressed galvanic potential[29].

J. B. COTTON
E. G. KING

REFERENCES

1. Fröhlich, K. W., *Z. Metallk.*, **28**, 368 (1936)
2. Gough, H. J. and Sopwith, D. G., *J. Inst. Met.*, **60**, 143 (1937)
3. Price, L. E. and Thomas, G. J., *J. Inst. Met.*, **63**, 21 (1938)
4. Brooks, P. E. et al., *J. Inst. Met.*, **88**, 500 (1959-60)
5. Kroll, W. J., in *The Corrosion Handbook* (Ed. Uhlig, H. H.), Wiley, New York; Chapman and Hall, London (1948)
6. Eisenbud, M., in *The Metal Beryllium* (Eds. White, D. W., Jr. and Burke, J. E.), Amer. Soc. Metals, Cleveland, Ohio, 620 (1955)
7. Kjellgren, B. R. F., *Rare Metals Handbook*, Reinhold, New York, 31 (1954)
8. English, J. L., in *The Metal Beryllium* (Eds. White, D. W., Jr. and Burke, J. E.), Amer. Soc. Metals, Cleveland, Ohio, 530 (1955)
9. Logan, H. L. and Hessing, H., *Summarising Report on Stress Corrosion of Beryllium*, NBS-6, Nat. Bur. Stand., Washington, Dec. (1955)
10. Miller, R. A. et al., *Corrosion*, **23**, 11-14 (1967)
11. Gulbransen, E. A. and Andrew, K. F., *J. Electrochem. Soc.*, **97**, 383 (1950)
12. Williams, J. and Munro, W., A.E.R.E. Report M/M108 (1956)
13. Livey, D. T. and Williams, J., *Proceedings of the Second Conference on the Peaceful Uses of Atomic Energy*, United Nations, Geneva (1958) Paper A/CONF/15/P/319
14. Werner, W. J. and Inouje, H., Conference on the Metallurgy of Beryllium, Institute of Metals, Paper 32, Oct. (1961)
15. Pennah, P. J. et al., Conference on the Metallurgy of Beryllium, Institute of Metals, Paper 10, Oct. (1961)
16. Higgins, J. K. and Antill, J. E., *J. Nuclear Mat.*, **5**, No. 1, 67 (1962)
17. Smith, R. et al., Conference on the Metallurgy of Beryllium, Institute of Metals, Paper 30, Oct. (1961)
18. Kerr, I. S. and Wilman, H., *J. Inst. Met.*, **84**, 379 (1955-56)
19. Miller, P. D. and Boyd, W. K., *Mat. Eng.*, **68**, 33-36 (1968)
20. Mackay, T. L. and Gilpin, C. B., *Electrochem. Tech.*, **6** (1968)
21. Stonehouse, A. J. and Beaver, W. W., *Mat. Protect.*, **4**, 24-28 (1965)
22. Booker, J. and Stonehouse, A. J., *Mat. Protect.*, **8**, 43-47 (1969)
23. BrushWellman Engineered Materials, Beryllium/Mining Division, South River Road, Elmore, Ohio 43416, USA
24. Jepson, W. B., *Corrosion of Light Metals*, 221-56 (1967)
25. Mueller, J. J. and Adolphson, D. R., *Beryllium Science and Technology*, **2**, 417-33 (1979)
26. Terlo, G. Ya., AD 628185, Dept of Navy (1966)
27. Paine, R. M. and Stonehouse, A. J., Corrosion/77, Paper No 26 (1977)
28. Prochko, R. J. et al. *Mat. Protect.* **5**, No. 12, 39-42 (1966)
29. King, T. T. and Myers, J. R., *Corrosion*, **25**, No. 8, 349-51 (1969)

5.2 Molybdenum

Molybdenum, although once considered a rather unusual metal with a few highly specialised applications, e.g. as support wires in tungsten-filament lamps and in thermionic valves as anode and grid materials, today finds extensive use in the missile and aerospace industry, in the production of high-temperature hydrogen and inert-atmosphere furnaces and furniture, in the production of solid-state electronic devices and in electrodes and bushings used in glass-making processes. Molybdenum and its alloys are now also finding increasing use in chemical and petrochemical plant. Historically, molybdenum has been employed in high-temperature sulphuric acid service and molybdenum-lined equipment has been used in organic chlorination processes. The Mo–30W alloy, which is a complete solid solution, has demonstrated its superiority as a material of construction for liquid metal applications such as the pumping of molten zinc.

It is not subjected to hydrogen embrittlement as is tantalum, niobium and nickel alloys, and thus is able to sustain thermal and mechanical shock after exposure to gaseous hydrogen at high temperatures.

It is produced in the form of a powder and is consolidated either by powder metallurgy techniques, by arc casting or, more recently, by electron-beam melting. Most commercially available molybdenum is however made by powder metallurgy, and it is only when very large pieces are required that arc melting is used. Electron-beam melting gives rise to large crystals in the metal which limits its application.

Physical and Mechanical Properties

Unalloyed molybdenum is available commercially in almost all forms, from forging billets and plate to seamless tubing, foil and wire. In addition, the following molybdenum-base alloys can be considered as established engineering materials: Mo–0·5Ti–0·8Zr (designated *TZM*) and Mo–30W. A Mo–30Ta alloy has also been produced for specific applications where high corrosion resistance is required.

At ambient temperatures the strength of molybdenum is comparable to that of normalised low-alloy steel and moderately higher than that of the austentic stainless steels. However, whereas the low-alloy steels are limited to use at service temperatures of about 550°C and stainless steels to about

Table 5.2 Physical properties of molybdenum*

Atomic number	Atomic weight	Density at 20°C (g/cm³)	Melting point (°C)[1]	Boiling point (°C)[1]	Linear coeff. of expansion (°C)[2,3]	Sp. heat (J/g°C)[2]	Thermal				Electrical			Thermal neutron cross section (barns/atoms)[5]
							Thermal conductivity (W/cm°C)[1,2]	Vapour pressure (mm Hg)[4]	Total radiation (W/cm²)	Total Emissivity[2,4]	Conductivity % IACS	Resistivity (μΩ cm)[1,2]	Temp. coeff. of electrical resistivity (°C)	
42	95·95	10·22	2 610	5 660	6·65×10⁻⁶ at 20–1 593°C	0·276 at 20°C 0·239 at 538°C	1·42 at 20°C 1·26 at 204°C	1 at 3 297°C 760 at 4 827°C	0·55 at 730°C, 6·3 at 1 330°C and 19·2 at 1 730°C	0·07 at 100°C 0·13 at 1 000°C	36	5·5 at 20°C 23·9 at 727°C	4·7×10⁻³ at 10–100°C	2·7

*Note. Superscripts outside parentheses denote reference numbers at the end of the section.

870°C, unalloyed molybdenum retains useful strength up to 1 200°C, and the *TZM* alloy has strength up to 1 650°C.

At elevated temperatures the thermal conductivity is far superior to that of Fe-18Cr-8Ni and the Ni and Cu binary alloy Monel, two other alloys widely used in severe chemical service.

The mechanical properties reported in the literature for molybdenum and its alloys are frequently at variance. That this should be so is not surprising as the properties of molybdenum and its alloys are greatly affected by the prior history of the material, both thermal and mechanical. Far too often values are used without reference to the sources of the material, various states of heat treatment, etc. When mechanical properties are an important feature of the design application, advice should always be sought on the suitability as only the manufacturer has the complete data on the history of his own product. Physical and some typical mechanical properties given for general guidance are shown in Tables 5.2 and 5.3.

Table 5.3 Mechanical properties of molybdenum

$UTS^{2,6}$ (MN/m^2)	Yield stress^2 (MN/m^2)	Modulus of elasticity^6 (GN/m^2)	Hardness			Recrystal- lisation temp.^{6}	Stress relieving temp.^{6}
			Hard rolled (VPN)	Annealed (VPN)	Sintered material^2 (Rockwell)		
317–871 (0·25 mm sheet) 517–1 206 (sintered material) 827–1 379 at 20°C 137–206 at 1 000°C	432 on 100 mm extrusion, to 586 on 1·4 mm sheet at 20°C	317 at 20°C 282 at 500°C 268 at 1 000°C	280	250	95B–27C	900– 1 200°C	800°C

Methods of Fabrication

Forming The fabrication of molybdenum is largely dictated by the ductile-brittle transition temperature. Most operations, except those on thin sheet or wire, are carried out warm and it is often necessary to heat not only the workpiece but also the die.

Molybdenum can be spun, flow turned and deep drawn as well as pressed. The temperature required to produce components satisfactorily varies with the thickness of the molybdenum sheet.

Specific fabrication temperatures are given by Czarnecki *et al.*[7] for various forming processes, including roll forming, hydropressing, joggle and die forming on molybdenum-titanium alloy sheet.

Forging Ingots of molybdenum produced either by sintering or by arc-melting are readily forged after heating to about 1 500°C. The heating of large ingots can be performed without protection, but thinner sections are heated in a hydrogen atmosphere to limit oxidation.

Rolling Sheet rolling is performed cold with intermediate anneals. Warming of the sheet and tools improves the rate of working; thicker material can be rolled at still higher temperatures.

Swaging   Swaging is performed hot, the temperature commencing at about 1 300°C and falling as the ductility of the metal improves to about 700°C.

Drawing and spinning   Spinning and cupping are facilitated if molybdenum is gently heated; suitable temperatures can be found in Reference 7.

Machining   Molybdenum can be machined by any of the standard methods such as milling, turning, drilling, boring, grinding, shaping, threading and tapping. The low coefficient of expansion of molybdenum makes it necessary to keep the tool cool when drilling in order to prevent seizing of the tool and possible cracking of the metal. Drilling is normally carried out dry but swarf can be carried away with a soluble cutting oil. When milling, cutters must be kept sharp and copious coolant provided; 1.1.1-trichlorethane is recommended for the coolant.

Welding   It is possible to weld molybdenum using a TIG-shielded arc-welding process. A heat-affected zone is unavoidable and grain growth must be anticipated.

These characteristics always give a weld which has less ductility than the parent material and no method of welding is entirely satisfactory particularly if the joint is to be stressed.

Resistance welding   Spot and seam welding are used to join molybdenum for electronic use; this technique is not satisfactory for large-scale work.

Electron-beam welding   Molybdenum can be electron-beam welded but the technique has only limited use.

Plasma welding   This method of welding is comparatively new and has not yet found much application.

Brazing   Satisfactory brazed joints in molybdenum have been made using oxyacetylene torch and furnace brazing techniques.

Soldering   Conventional soft solders can be used, but it is first essential to nickel- or copper-plate the molybdenum.

Corrosion Behaviour

Theory

The theoretical aspects of molybdenum's corrosion behaviour are complex and there is as yet no clear cut, generally applicable picture. There are, however, a large number of literature references which include data on polarisation, passivation and potential of molybdenum under widely assorted conditions. The electrode potential of molybdenum depends on its surface condition. For example, some tests showed an E_H of $+0.66$ V when the molybdenum was passivated by treatment with concentrated chromic acid and -0.74 V after activation by cathodic treatment in sodium hydroxide.

The potential in aqueous solutions depends on the pH of the solution as demonstrated by Masing and Roth[8], Shatolov and Marshakov[9], and others.

As an example of the magnitude of this effect, Shatolov and Marshakov[9] in one series of tests in buffered chloride solution found that the irreversible potential of molybdenum could be expressed as

$$E = 0 \cdot 37 - 0 \cdot 045 \mathrm{pH}$$

Molybdenum is widely used as an alloying addition to stainless steels to facilitate the formation of the passive film and to improve resistance to pitting attack (see Section 3.3).

Corrosion in Natural Environments

Atmospheric Molybdenum begins to oxidise in air at 300°C and oxidation becomes rapid at 500°C and the rate of attack is very rapid by 1200°C[10]. Below 500°C oxidation proceeds according to a parabolic law, indicating some degree of protection. The oxidation proceeds by a two-step process with molybdenum dioxide (MoO_2) as the inner oxide layer and molybdenum trioxide (MoO_3) as the outer layer. Above 500°C, MoO_3 begins to volatilise, and at 600°C the rate of evaporation of MoO_3 becomes significant. At about 770°C the rate of evaporation of MoO_3 equals its rate of formation, and as the temperature increases, the volatilisation rate becomes extremely rapid. The ultimate oxidation rate is linear, with the rate-determining step being the oxidation of the dioxide to the trioxide.

The formation of molten MoO_3 above 815°C results in a catastrophically accelerating effect due to the following factors: (*a*) the liquid oxide flows off the metal surface, (*b*) the rate of diffusion of oxygen through the liquid phase is high, and (*c*) the molten oxide can also act as a flux.

The crystal structures observed during the oxidation of molybdenum consist of stable molybdenum dioxide in contact with the metal throughout the range 300–700°C. As the film thickens in the low-temperature range, the trioxide predominates on the surface. At 400°C, molybdenum trioxide is no longer observed and molybdenum dioxide is the only oxide observed.

The inability of molybdenum to withstand oxidising conditions at even moderate temperatures has led to investigations of means of reducing the oxidation rate. Alloying has been examined and rejected since attainment of good oxidation resistance was at the expense of high temperature strength, and protective coatings appear to be the most effective. A number of coatings have been developed for molybdenum; for temperatures up to 1 200°C Mo-base alloys applied as cladding or sprayed coatings, and chromium–nickel electroplates are effective.

For high temperatures several coatings based on siliconising are available. One of the most successful, designated *W3*, has been used for long periods of say 500 h at 1 200°C and even at 1 700°C for short term (1 h) applications (Fig. 5.1). This coating, which is basically $MoSi_2$, has several advantages: it is relatively thin (about 0·05 mm) and it can be applied over rivets and other fasteners. Bolts and nuts have been successfully coated and used. The coating is applied by either a pack diffusion technique similar to chromising or by an electrochemical technique, and coating procedures have been evolved for the protection of complex assemblies.

Immersed Molybdenum has good resistance to synthetic sea water, the rate of attack up to 60°C being less than 0·1 mm/y[11] and it is only slightly corroded when exposed to synthetic sea water spray at 60°C for periods of 10, 20 and 30 days.

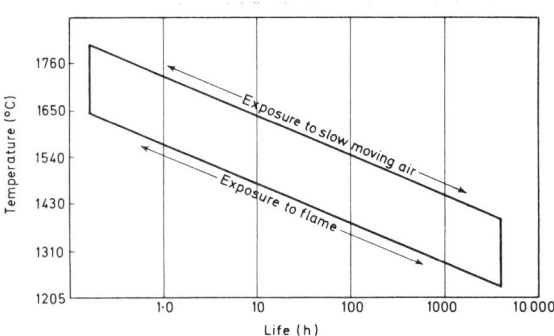

Fig. 5.1 Protection offered by three commercial silicide coatings overlap broadly in this performance band (after *Prod. Engng.*, June 25, 63 (1962))

Corrosion in Chemical Environments

Acids Molybdenum exhibits good resistance to hydrochloric, hydrofluoric, phosphoric and sulphuric acids (see Figs. 5.2, 5.3 and 5.4). In static tests[12] in these acids at Ohio State University it was found that the corrosion rate for relatively long test periods (670 h) was about ten times that experienced in 47 h tests.

Oxidising conditions severely reduce molybdenum's corrosion resistance, and aeration of the above acids causes a pronounced increase in the corrosion rate. It is rapidly attacked by oxidising acids such as nitric acid, and by reducing acids containing oxidisers such as HNO_3, $FeCl_3$, etc. It is less resistant at 100°C, particularly in 10% acetic acid (the corrosion rate being 0·33 mm/y), 10% formic acid (0·2 mm/y) and 0·25% benzoic acid (0·25 mm/y).

Molybdenum dissolves rapidly in 70% sulphuric plus 20% nitric acids at 35°C, but exhibits good resistance to 30% hydrochloric plus 7% sulphuric acid and 10% citric plus 1% sulphuric acids at 100°C. It corrodes three times as fast in 10% acetic acid containing 5% sulphuric acid at 100°C than it does in pure 10% acetic (1 mm/y against 0·33 mm/y).

The presence of 2% formic acid in acetic acid has relatively little effect on the corrosion of the metal. Among the impurities added at the 0·2% level to 10% acetic acid, only mercuric chloride caused an appreciable increase in corrosion rate (0·71 mm/y), and all the other additions appeared to inhibit corrosion.

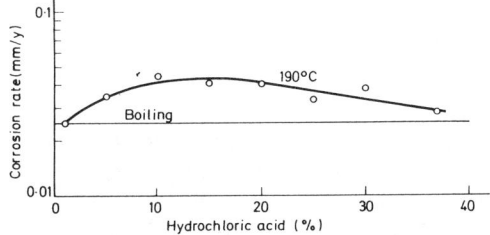

Fig. 5.2 Molybdenum corrosion tests in hydrochloric acid (after Bishop, C. R., *Corrosion*, **19**, Sept., 308t (1963))

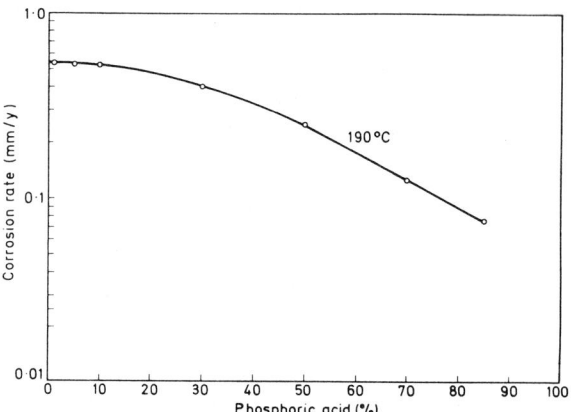

Fig. 5.3 Molybdenum corrosion tests in phosphoric acid (after Bishop, C. R., *Corrosion*, **19**, Sept., 308t (1963))

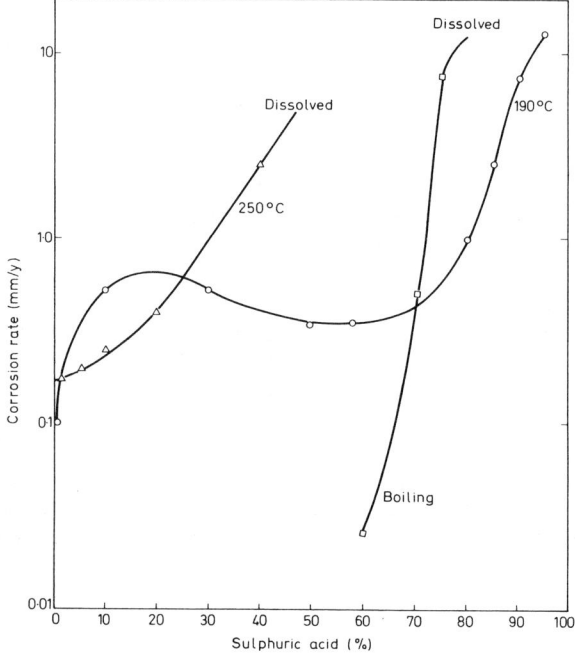

Fig. 5.4 Molybdenum corrosion tests in sulphuric acid (after Bishop, C. R., *Corrosion*, **19**, Sept., 308t (1963))

When small amounts of impurities were added to concentrated phosphoric acid, the corrosion rate was not significantly affected. The addition of only 0·007% ferric ion to concentrated hydrochloric acid caused only a slight increase in corrosion at 35°C, but at 100°C the corrosion rate increased to 6·35 mm/y.

Alkalis Molybdenum is moderately resistant to aerated solutions of ammonium hydroxide and is inert when oxygen is excluded. It has only fair resistance in aerated 1% sodium hydroxide at 35°C and 60°C but its resistance is better in a 10% solution at both these temperatures. It is severely corroded in sodium hypochlorite solutions (pH 11 or higher) at 35°C (Table 5.4).

Table 5.4 Corrosion in alkaline solutions[11] (duration of test 6 days)

Solution and concentration	Corrosion rate (mm y^{-1})			
	35°C		60°C	
	Air	Helium	Air	Helium
Conc. sodium hypochlorite (5-6% available Cl_2)	11·4	11·9	—	—
Dilute sodium hypochlorite (10:1)	2·44	2·42	—	—
Ammonium hydroxide				
0·35% NH_3	0·14[1]	0[1]	—	—
15% NH_3	0·24	0	—	—
Conc. 29% NH_3	0·0025[2]	—	—	—
	0·02[3]	—	—	—
Sodium hydroxide				
1%	0·38	0	0·49	0·069
10%	0·10	0·0025	0·18	0

Notes. 1. Average of 4 specimens.
2. Non-aerated 6-day test.
3. Non-aerated 28-day test continuation of 6-day test.

Table 5.5 Corrosion in inorganic salt solutions[11] (duration of test 6 days)

Solution and concentration	Corrosion rate mm y^{-1})					
	35°C		60°C		100°C	
	Air	Helium	Air	Helium	Air	Helium
Sodium chloride (3%)	0·01[1]	0[1]	—	—	0·061	0·012
Aluminium chloride (10%)	0·0025	0	—	—	0·0305	0·005
Ammonium chloride (10%)	0·005[3]	0[3]	—	—	0·0585	0
Ferric chloride (20%)	40·6	34·5	—	—	—	—
Cupric chloride (20%)	19·3	6·35	—	—	—	—
Mercuric chloride						
0·5%	0·485	0·485	—	—	—	—
1·0%	0·61[2]	0·43[2]	0·84	0·81	—	—
5·0%	0·305[3]	0·33[3]	1·72	1·9	—	—
	0·38[2]	0·203[2]	1·5	1·5	—	—

Notes. 1. 30-day test.
2. 15-day test average of 4 specimens.
3. Average of 4 specimens.

Inorganic salt solutions Molybdenum has excellent resistance to 3% sodium chloride, 10% aluminium chloride and 10% ammonium chloride at temperatures up to 100°C. It is severely corroded by 20% solutions of ferric and cupric chlorides at 35°C and is subject to pinhole-type pitting in mercuric chloride solutions (Table 5.5).

Table 5.6 Corrosion by molten metals[13]

Metal	Temperature (°C)	Resistance	Comments
Aluminium	660	Poor	Rapid attack
Bismuth	up to 1 430	Excellent	
Bi–32Pb–16Sn	800	Excellent	
Cerium	800	Good	Poor above 800°C
Caesium	up to 870	Excellent	Poor above 870°C
Copper	up to 1 300	Good	0·14% Cu soluble in Mo
Europium	–	Good	Used as crucibles
Gallium	450	Good to poor	Good to 400°C
Gold	–	Good	Used as crucibles
Lead	1 200	Excellent	Solubility <0·005 at. %
Pb–55·5Bi	900	Excellent	} Arc cast material
	1 090	Excellent	
Lithium	1 425	Excellent	Solubility less than 10^{-4} at %
Magnesium	1 000	Excellent	Solubility less than 2×10^{-4} at. %
Mercury	up to 600	Excellent	
	600 to 700	Limited	
Plutonium	–	Good	
Potassium	1 200	Excellent	
Rubidium	1 035	Excellent	
Samarium	–	Good	Used as containers
Scandium	1 625	Poor	
Silver		Good	Used as crucibles
Silver plus 1–15% titanium	1 020	Satisfactory	Used as crucibles
Sodium	1 020	Excellent	
Sodium containing 0·5% oxygen	700	Fair	
Tin	>520	Poor	Rapid attack

Industrial-type Environments

Molten metals Molybdenum has good resistance to many molten metals (see Table 5.6) and molybdenum and its alloys are finding application in metal producing and fabricating operations. Its high strength to weight ratio, corrosion resistance and resistance to attack by liquid metals makes it readily suited for advanced heat transfer systems, and these factors coupled with its low thermal neutron cross section have led to it being considered as a canning material in nuclear reactors. The alloy Mo–30W also has unique corrosion properties in certain media, being for all practical purposes completely resistant to attack by molten zinc at normal casting temperatures.

Molybdenum is extremely resistant to attack by molten glass up to 1 000°C if sulphur is present, and up to 1 400°C in its absence. In addition, any reaction products formed by the two materials are colourless. Consequently molybdenum is used in the production of optical and soda lime glasses. Molybdenum is, however, corroded by lead glasses which lose their gloss as a result of the interaction between the two materials[14].

Fused salts Molybdenum has excellent resistance to a wide range of fused salts and has been used in the fused salt electrolysis of magnesium, platinum, thorium and uranium. In the production of pure magnesium, molybdenum is used to couple graphite electrodes electrically. Molybdenum cathodes are used with other materials in the recovery and purification of spent nuclear fuels from fused salt mixtures where contamination of the electrolyte must be prevented.

Gases Molybdenum has a fair resistance to chlorine, bromine and iodine but not to fluorine and it has excellent resistance to a wide range of other gases (see Table 5.7).

Refractories Molybdenum is compatible with a wide range of refractory materials as can be seen from Table 5.8.

Galvanic Corrosion

Work by the US Bureau of Mines[11] involving galvanic couple experiments showed that the normally low corrosion rates of molybdenum were reduced further by contact with aluminium, SAE 1 430 steel or magnesium in aerated solutions of synthetic sea water or 3% sodium chloride.

Table 5.7 Resistance of molybdenum to gases [16, 17]

Gas	Temperature (°C)	Resistance	Comments
Air	370	Good	Above 370°C poor
Bromine (dry)	up to 455	Excellent	Over 455°C poor
Dichlorodifluoro- methane +1% sulphur hexafluoride	up to 315 650	Good Poor	
Hydrocarbons	up to 1 100 above 1 100	Good Fair	Some carburisation may occur
Nitric oxide	1 125 over 1 125	Good Fair	
Nitrogen	up to 2 400 over 2 400	Excellent Good	Nitride case forms
Nitrogen tetroxide	1 155	Poor	
Oxygen	up to 370	Good	Over 370°C poor
Steam	650	Fairly good	Over 650°C poor
Sulphur dioxide	up to 650 over 650	Excellent Fair	Attack rapid at 1 165°C

Table 5.8 Resistance to refractories [17]

Refractory material	Temperature (°C)	Resistance	Comments
Alumina	up to 1 900	Excellent	
Carbon	up to 1 100 over 1 100	Good Fair	Superficial brittle carbide case forms
	over 1 370	Poor	Molybdenum carburises rapidly and completely
Glass	up to 1 400	Excellent	No sulphur, lead or tin present
Silica	650 1 900–2 000	Excellent Very good	Helium atmosphere
Zirconia	up to 2 000	Excellent	
Zirconium oxide +8% yttrium oxide	2 350	Good	

Contact with copper in sodium chloride solutions reduces the corrosion of molybdenum, but in synthetic sea water the corrosion rate is somewhat higher. In 1% sodium hydroxide, molybdenum corrodes slightly faster when coupled with SAE 1 430 steel with or without the presence of air. It is protected by contact with copper in aerated 0·46% sulphuric acid. It is not significantly affected by contact with 316 or Carpenter 20 stainless steels in sulphuric acid solutions.

Molybdenum tends to be protected by vanadium in aerated 7·1% hydrochloric acid and it receives a high degree of protection when coupled with copper in this medium. Molybdenum corrodes somewhat faster than normal in 3·1% nitric acid when coupled with tungsten. It is not affected by contact with titanium in 3·1% nitric acid. It is protected by aluminium and copper in aerated 10% formic acid and by aluminium in air-aerated 9% oxalic acid. In the latter solution, copper had only a slight protective effect when coupled with molybdenum.

Behaviour of Molybdenum Alloys

The US Bureau of Mines[15] found the chemical and galvanic corrosion behaviour of both the *TZM* and Mo–30W alloy to be generally equal or superior to that of unalloyed molybdenum in many aqueous solutions of acids, bases and salts. Notable exceptions occurred in 6·1% nitric acid where both alloys corroded appreciably faster than molybdenum. In mercuric chloride solutions the *TZM* alloy was susceptible to a type of crevice corrosion which was not due to differential aeration. The alloys were usually not adversely affected by contact with dissimilar metals in galvanic couple experiments, but the dissimilar metals sometimes corroded galvanically. Both alloys were resistant to synthetic sea water spray at 60°C.

Other alloys of molybdenum which have been investigated for their corrosion resistance contain 10–50% Ta and were found to have excellent resistance to hydrochloric acid. Ti–Mo alloys were found to resist chemicals that attack titanium and Ti–Pd alloys, notably strong reducing acids such as hot concentrated hydrochloric, sulphuric, phosphoric, oxalic, formic and trichloroacetic. For example, a Ti–30Mo alloy has the following corrosion rates: in boiling 20% hydrochloric acid, 0·127–0·254 mm/y; in 10% oxalic acid at 100°C, 0·038 mm/y, which compares favourably with the respective rates of 19·5 and 122 mm/y for the Ti–0·2Pd alloy.

Applications in Industry

Molybdenum has always been used in quite large quantities in the electronics industry, and the use of molybdenum windings, protected by hydrogen, for high-temperature furnaces is wide-spread. Molybdenum windings and radiation screens are being used in increasing numbers in electric furnaces for the bright annealing of stainless steels and similar materials. The availability of the metal in massive form has opened up new fields for molybdenum and one of the first applications for larger components was in the glass industry where electrodes and stirrers used in the melting of glass are constructed from molybdenum. Molybdenum, or more usually the Mo–30W alloy, is finding increasing use in applications involving molten metals such as the die casting of molten zinc[18,19]. Molybdenum crucibles are used in the production and handling of rare earth metals and compounds such as lithium platinide, and boats of molybdenum are used extensively in metal vapour deposition processes. Molybdenum as electrodes and other components finds a use in molten halide salt electrolysis, one of the principal applications of which is in the recovery of nuclear materials.

In the chemical process industry molybdenum has found use as washers and bolts to patch glass-lined vessels used in sulphuric acid and acid environments where nascent hydrogen is produced. Molybdenum thermocouples and valves have also been used in sulphuric acid applications, and molybdenum alloys have been used as reactor linings in plant used for the production of *n*-butyl chloride by reactions involving hydrochloric and sulphuric acids at temperatures in excess of 170°C. Miscellaneous applications where molybdenum has been used include the liquid phase Zircex hydrochlorination process, the Van Arkel Iodide process for zirconium production and the Metal Hydrides process for the production of super-pure thorium from thorium iodide.

In the polyacrylic synthetic fibre industry, carbonitrided molybdenum guides have been used in place of chromium plated steel because of their resistance to corrosion and erosion. Chemicals that attack molybdenum are listed in Table 5.9.

Table 5.9 Chemicals that attack molybdenum [16,17]

Reagent	Concentration (%)	Temperature (°C)	Resistance
Potassium carbonate	100	260	Poor (attack greater than $1 \cdot 3$ mmy^{-1})
Potassium hydroxide	50	100	Good
	100	Fused	Poor
Sodium carbonate	100	315	Poor (attack greater than $1 \cdot 3$ mmy^{-1})
Sodium hydroxide	50	100	Good
	100	Fused	Poor
Ferric chloride	50	Room	Poor
Lead nitrate	–	Fused	No reaction
Lead oxide	–	Fused	Poor (violent reaction)
Potassium chlorate	–	Fused	Poor (violent reaction)
Potassium nitrate	–	Fused	Poor (violent reaction)
Potassium nitrite	–	Fused	Poor (violent reaction)
Sodium peroxide	–	Fused	Poor (violent reaction)

Although molybdenum is resistant to molten glass, except leaded, molybdenum components not coated with glass but exposed to the oxidising furnace atmosphere corrode rapidly due to volatilisation of molybdenum oxide above 370°C. To overcome this, stirrers etc. for use in glass plant are physically clad with platinum sheet in vulnerable areas. Modern plating techniques have enabled dense platinum coatings to be put onto the surface of the molybdenum and it is expected that this technique will be exploited further in the near future.

Molybdenum shows good corrosion resistance to zirconia up to 2000°C but above about 1200°C, zirconia becomes electrically conductive and thus care must be taken in the design of high temperature furnaces using zirconia refractories, especially 3-phase, as at the operating temperature the refractories carry most of the current.

Arc cast molybdenum appears to have a greater corrosion resistance than the powder metallurgical product, especially in large section. This is thought to be due to the greater density of the cast variety. Electron beam melted molybdenum, although dense, has an exceptionally large grain size and this presents problems in fabrication, extrusion appearing to be the most satisfactory method.

Flame sprayed molybdenum articles have poor corrosion resistance, no doubt owing to the porosity of the coating. However, modern plasma spraying techniques produce a dense coating and this should lead to more widespread use of clad materials such as molybdenum clad steel where the clad product should have the same corrosion resistance as the solid material.

Several coating techniques are now available to overcome the oxidation problems with molybdenum above 300°C. One of these, based on molybdenum disilicide, is finding increased usage in flame breakout shields for aero-engines where tests have shown (unpublished work) that the coated material can withstand a high pressure torching type flame attack at temperatures in excess of 2000°C.

Molybdenum is used for high energy laser mirrors which require water cooling. Corrosive action of the circulating cooling water can be prevented by coating the waterways with a thin film of tungsten by chemical vapour deposition. US Pat Application 308976 (1982).

Molybdenum glass melting electrodes have recently found new usage in reclamation of contaminated land. Holes are drilled into the ground on an 11–18 ft grid pattern 50 ft deep and glass melting electrodes inserted. The area between the electrodes on the surface of the ground is covered with graphite and a fume extraction hood with associated scrubbing equipment.

Current is passed into the electrodes causing the ground temperature to rise. Volatiles are vaporised off and collected in the scrubbing systems for safe disposal. The current is increased up to 4000 amps over a period of several days causing the soil to melt, incorporating non volatile elements into the fused mass. Other volatiles and organics migrate to the surface where they combust in the presence of air. The system is then moved to the next area for decontamination whilst the fused mass is allowed to solidify and cool. Back filling to allow for shrinkage enables the land to be re-used for agriculture or building.

J. BENTLEY
C. E. D. ROWE

REFERENCES

1. Lyman, T., *A.S.M. Metals Handbook* (1961)*
2. Hampel, C. A., *Rare Metals Handbook*, 2nd edn, Reinhold (1961)*
3. Krikorian, O. H., *Thermal Expansion of High Temperature Materials*, VCRL 6 132, Sept. (1960)*
4. Bockris, J. O'M., White, J. L. and Mackenzie, J. D,, *Physical and Chemical Measurements at High Temperatures*, Butterworths, 369–371 (1959)*
5. MacDonald, R. N. and Boucon, H. H., *Nucleonics*, **20**, August 8th, 158 (1962)*

*These references can be found collectively in *Properties of Refractory Materials*, by S. J. Burnett, U.K.A.E.A. Research Report No. R4657, H.M.S.O. (1969)

6. Chelius, J., *Machine Design*, March 1st (1962)*
7. Czarnecki, E. G., Stacey, J. T. and Zimmerman, D. K., *Refractory Metal for Glide and Re-entry Vehicles*, Met. Soc. Conference, **17** and *Refractory Metals and Alloys*, **2** (Edited by Semchyshen and Perlmutter), Interscience Publishers, New York (1963)
8. Masing, G. and Roth, C., 'Behaviour of Molybdenum and Nickel and Some Molybdenum-nickel Alloys in Acid Electrolytes', *Werkstoffe und Korrosion,* **3**, 176-186, 253-263 (1952)
9. Shatalov, A. Ya. and Marshakov, I. A., 'Electrode Potentials and Corrosion of Molybdenum and Tungsten', *Zhurnal Fizicheskoy Khimii*, **28** No. 1, 42-50 (1954)
10. Jaffee, R. J., 'Proceedings on an International Symposium on High Temperature Technology', Asilomar, California, 1959, McGraw-Hill, New York, 61 (1960)
11. Acherman, W. L., Carter, J. P., Kenahan, C. B. and Schlan, D., *Corrosion Properties of Molybdenum, Tungsten, Vanadium and some Vanadium Alloys*, Report of Investigations No. 6 715, US Bureau of Mines (1966)
12. *Corrosion Tests on Metallic Molybdenum*, Engineering Experimental Station, Ohio State University, Project 142 (1957-1959)
13. Barto, R. L. and Hard, D. T., 'Refractory Metals in Liquid Metal Handling', *Research/Development*, Nov. 26 (1966)
14. 'Corrosion of Molybdenum by Molten Glass', *Kazus Ocku Yogyo Kyrokai Shi*, **72**, 108-113 (1964)
15. Ackerman, W. L., Carter, J. P. and Schlain, D., 'Corrosion Properties of the TZM and Mo-30W Alloy, US Bureau of Mines Report No. 7 196, Aug. (1969)
16. *Corrosion Properties of Special Alloys and Rare Metals*, Jacob and Korres G.m.b.H, 473 Ahlen, W. G.
17. Nair, F. B. and Briggs, J. Z., *Corrosion Resistance of Molybdenum and Molybdenum Base Alloys*, Molybdenum Metal
18. Gilbert, R. W., 'New Mo-30W Alloy Replaces Tool Steels for Die Casting Nozzles', *Die Casting Engineer*, March (1964)
19. Burman, R. W. and Litchfield, G., 'Severe Molten Zinc Corrosion is Reduced by Improved Molybdenum Tungsten Alloy', *Engineering and Mining Journals*, **164** No. 4, April, 88 (1963)
20. Fitzpatrick, V. F. Timmerman; Buelt, J. L., 'In situ vitrification – A candidate process for in situ destruction of hazardous waste' Pacific West Laboratory, PNL-SA-14065 (1986)

BIBLIOGRAPHY

Cox, F. G., 'Joining Molybdenum', *Welding and Metal Fabrication*, Sept. (1961)
Nair, F. B., 'Molybdenum Resists Corrosion at High Temperatures', *Design Engineering*, Oct., 104-105 (1964)
Hargrave, D. P., 'Molybdenum and Tungsten', Paper 5 of conference on New Engineering Materials, *Proceedings of Institution of Mechanical Engineers*, **180**, Part 3D (1955-1956)
Northcott, L., *Molybdenum*, Butterworths, London (1956)
Harwood, J. J. (Ed.), *The Metal Molybdenum* (the protection of molybdenum against high temperature oxidation), American Society for Metals, Cleveland, Ohio (1958)
Technical Notes, Climax Molybdenum Co., Jan. (1964)

5.3 Niobium

Niobium is always found in nature associated with tantalum and it closely resembles tantalum in its chemical and mechanical properties. It is a soft ductile metal which, like tantalum, work hardens more slowly than most metals. It will in fact absorb over 90% cold work before annealing becomes necessary, and it is easily formed at room temperature. In addition, welds of high quality can be produced in the metal. In appearance the metal is somewhat similar to stainless steel; it has a density slightly higher than stainless steel and a thermal conductivity similar to 1% carbon steel.

It is somewhat less corrosion resistant than tantalum, and like tantalum suffers from hydrogen embrittlement if it is made cathodic by a galvanic couple or an external e.m.f., or is exposed to hot hydrogen gas. The metal anodises in acid electrolytes to form an anodic oxide film which has a high dielectric constant, and a high anodic breakdown potential. This latter property coupled with good electrical conductivity has led to the use of niobium as a substrate for platinum-group metals in impressed-current cathodic-protection anodes.

The mechanical properties of niobium are dependent on the previous history of the material and the manufacturer should be consulted if these properties are likely to be critical. Physical and some typical mechanical properties are set out in Tables 5.10 and 5.11.

Methods of Fabrication

Niobium possesses excellent room temperature fabrication characteristics compatible with all conventional production practices. Large reductions (up to 90%) of recrystallised material can be made without intermediate process annealing. Secondary fabrication operations such as stamping, drawing or forming into completed shapes can be performed cold. Intermediate anneals are dependent on the amount of work involved. In tube drawing or deep drawing, annealed niobium should be used. Reductions of 60 to 80% with multiple draws are customary before re-annealing, but the initial draw should have a depth not greater than 40 to 50% of the diameter.

It machines in a similar manner to soft copper, and high-speed-steel tools with high cutting speeds are most satisfactory. Trichloroethane is recommended as a cutting medium and the work must be kept well flooded at all

Table 5.10 Physical properties of niobium

Melting point[1] (°C)	Boiling point[1] (°C)	Density[1] (g/cm³)	Specific heat[2] (J/g°C)	Thermal conductivity[55] (W/cm°C)	Resistivity[1,3] (μΩ/cm at 20°C)	Coefficient of thermal expansion[2,4] (×10⁻⁶/°C)	Cross section thermal neutrons[5] (barns/atom)
2 468	4 927	8·57	0·268 at 15°C 0·320 at 1 227°C	0·523 at 0°C 0·691 at 1 873°C	15	7·1 at 20°C 7·39 at 0–400°C	1·16

Table 5.11 Mechanical properties of niobium

Yield stress[2,6] (MN/m²)	UTS[2,7] (MN/m²)	Modulus of elasticity[2,7] (GN/m²)	Poisson's ratio	Hardness (VPN)	Resistance to thermal shock	Workability (ductile to brittle trans. temp.)(°C)	Recrystallisation temperature[7]	Stress relieving temperature[7]
288 at 20°C 68 at 1 050°C	517–1 034 at 20°C 241 at 500°C 89–117 at 1 000°C	188 at 204°C 113 at 600°C 108 at 900°C	0·38	77–173	Good	−150 (based on rupture tests)	900–1 300°C	800°C

times. If it is not possible to use this coolant, satisfactory results can be obtained with water-soluble oil coolants.

It can be welded by resistance, tungsten-inert gas (TIG), plasma arc and electron beam techniques. To protect the metal from attack by air, resistance welding is carried out under water and the TIG method is best performed in a chamber of argon. The latter three methods produce ductile welds that equal the base metal in most of its characteristics.

Niobium closely resembles tantalum in its mechanical properties and for more detailed information relating to the fabrication of niobium see Section 5.5 on tantalum.

Corrosion Resistance

Niobium like tantalum relies for its corrosion resistance on a highly adherent passive oxide film; it is however not as resistant as tantalum in the more aggressive media. In no case reported in the literature is niobium inert to corrosives that attack tantalum. Niobium has not therefore been used extensively for corrosion resistant applications and little information is available on its performance in service conditions. It is more susceptible than tantalum to embrittlement by hydrogen and to corrosion by many aqueous corrodants. Although it is possible to prevent hydrogen embrittlement of niobium under some conditions by contacting it with platinum the method does not seem to be broadly effective. Niobium is attacked at room temperature by hydrofluoric acid and at 100°C by concentrated hydrochloric, sulphuric and phosphoric acids. It is embrittled by sodium hydroxide presumably as the result of hydrogen absorption[8] and it is not suited for use with sodium sulphide.

Atmospheric Niobium like several other refractory metals is extremely reactive with atmospheric oxygen. It will in fact react with air at temperatures as low as 200°C[9] although reaction does not become rapid until temperatures above red heat (about 500°C) are reached; at 980°C the rate is 0·05 mm/h[10] and at 1 200°C the rate is 300 mm/h[11]. It is not attacked by oxygen at 100°C but the attack is catastrophic at 390°C. At lower temperatures a thin adherent oxide film is formed on the surface of the metal, but at higher temperatures, above red heat, the oxide diffuses rapidly throughout the metal with consequent embrittlement. At elevated temperatures the metal reacts with all the common gases including nitrogen (300–400°C), water vapour (300°C), carbon dioxide, carbon monoxide and hydrogen (250°C).

Protection of niobium and its alloys from oxidation in air is accomplished by coating, e.g. with zinc deposited by holding in zinc vapour at 865°C[12] or coating with a layer of chemically stable oxide, nitride or silicide. Silicide coatings applied by pack cementation, fused slurry[13] or by electrolytic methods have been found to be one of the most effective means of preventing oxidation of the metal.

Water The corrosion resistance of pure niobium in water and steam at elevated temperatures is not sufficient to allow its use as a canning material in water-cooled nuclear reactors. Alloys of niobium with molybdenum, titanium, vanadium and zirconium however have improved resistance and have possibilities in this application. Whilst the Nb–10Ti–10Mo alloy offers the best corrosion resistance the Nb-7V alloy seems more practical on the basis of weldability. It also has good high-temperature strength properties.

Acids[10, 14-16] Niobium is resistant to most organic acids and to mineral acids especially under oxidising conditions. Figs. 5.5 to 5.7 show the corrosion behaviour of niobium in laboratory tests in various concentrations of sulphuric and hydrochloric acids at the boiling point and at 190°C and 250°C, and in phosphoric acid at the boiling point. It has excellent resistance to nitric acid, the rate of attack in 70% acid at 250°C being only 0·25 mm/y. In dilute sulphurous acid at 100°C the corrosion rate is 0·0125 mm/y, but in concentrated acid at the same temperature it is greater than 0·25 mm/y.

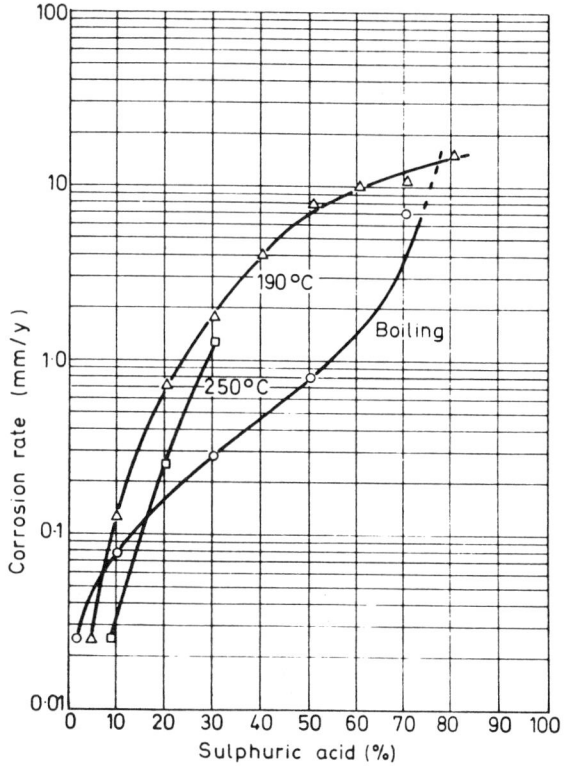

Fig. 5.5 Niobium corrosion in sulphuric acid[15]

Alkalis[8, 10, 14] Though niobium is not attacked by most alkalis at room temperature it is seriously attacked at 98°C, and severe embrittlement is obtained in concentrated alkali at room temperature and virtually all alkalis at 98°C. See Table 5.12 for detailed corrosion rates and embrittlement ratings.

There is evidence to show that the corrosion product when niobium is attacked by sodium hydroxide is $Na_8Nb_6O_{19}.18H_2O$.

Salts[10, 17, 18] Tests on niobium have only been carried out in a limited number of salt solutions; however, in the main niobium exhibited similar resistance to tantalum in most salt solutions, and like tantalum it is attacked by salts that hydrolyse to form alkalis.

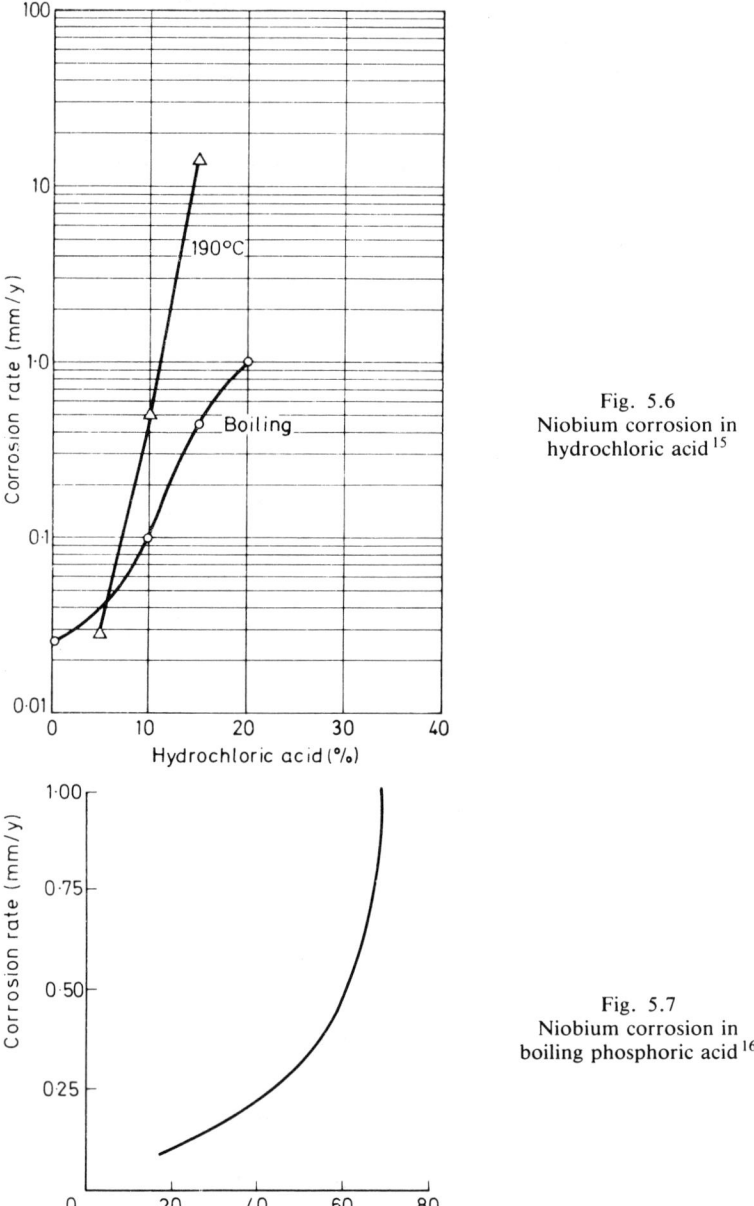

Fig. 5.6 Niobium corrosion in hydrochloric acid[15]

Fig. 5.7 Niobium corrosion in boiling phosphoric acid[16]

Gases[9, 17, 19, 20] It is unattacked by most common gases, e.g. nitrogen, hydrogen, oxygen, carbon dioxide, carbon monoxide and sulphur dioxide (wet or dry) up to 100°C, and it is inert to chlorine and bromine (both wet and dry) to 100°C. It is, however, attacked by nitrogen at 300–400°C, hydrogen at 250°C, oxygen at 200°C, carbon and carbon-bearing gases at 1 200–1 400°C and by chlorine at 200–250°C.

Table 5.12 Corrosion in alkalis[8, 10, 14]

Reagent	Concentration (%)	Temperature (°C)	Corrosion rate (mm/y)	Embrittlement rating
NaOH	1	98	0·74	B
	5	98	1·17	C
	10	98	2·00	C
KOH	1	98	0·60	A
	5	98	2·56	C
Na_2CO_3	10	98	1·63	B
	20	98	1·60	C
K_2CO_3	10	98	2·90	C
	20	98	2·46	C
Na_2PO_4	25	98	1·34	B
Na_2S	9	98	0·09	A
NaOH and Na_2S	9	98	4·15	B
NaOH		Molten		Severe attack at 535°C
KOH		Molten		Dissolves metal at 360°C

Note. Embrittlement ratings were obtained by bending a wire corrosion-test-specimen and the following notations were used:
A wire unbroken when sharply bent.
B wire broken when sharply bent.
C wire broken when handled or slightly bent.

Liquid Metals[21]

Bismuth Niobium is resistant to bismuth at temperatures up to 560°C[22] but is attacked at higher temperatures[23-25] and is therefore not considered a suitable container for handling liquid bismuth even under oxygen-free conditions[26]. Furthermore, the stress-rupture properties of niobium are significantly lowered when the metal is tested in molten bismuth at 815°C[24, 27].

Gallium It is slightly less resistant than tantalum to gallium, showing good resistance at temperatures up to 400°C but poor resistance above 450°C[28-30].

Lead Although subject to slight penetration at 980°C it shows no detrimental effects in stress rupture tests when tested in molten lead at this temperature[27] or at 815°C[24]. It is highly resistant to mass transfer in liquid lead as indicated by data obtained in tests at 800°C with a thermal gradient of 300°C[31].

Lithium Niobium has good resistance to molten lithium at temperatures up to 1 000°C[28, 32].

Mercury In static tests niobium shows good resistance to mercury at temperatures up to 600°C[28].

Sodium, potassium and sodium-potassium alloys Liquid sodium, potassium or alloys of these elements have little effect on niobium at temperatures up to 1 000°C[28, 33, 34], but oxygen contamination of sodium causes an increase in corrosion[35, 36]. Sodium does not alloy with niobium[37]. In mass transfer tests, niobium exposed to sodium at 600°C exhibited a corrosion rate of approximately $1 \text{ mg cm}^{-2}\text{d}^{-1}$. However, in hot trapped sodium at 550°C no change of any kind was observed after 1 070 h[38].

Thorium-magnesium In static tests the thorium–magnesium eutectic had no appreciable effect on niobium at 850°C[39].

Uranium Short-term tests indicate that the practical upper limit for niobium as a container material for uranium is about 1 400°C[40]. Niobium is dissolved in a uranium-bismuth alloy in less than 100 h at a temperature of 800°C[41]. Uranium eutectics with iron, manganese or nickel, corroded niobium at 800°C[42] and 1 000°C[43]. It is significantly attacked by uranium–chromium at 1 000°C[44].

Zinc Molten zinc is reported to attack niobium at a significant rate at temperatures above 450°C[45]. It is attacked by zinc at 600°C and shows increasing solubility with temperature up to 850°C.

Galvanic effects If niobium is cathodic in a galvanic couple the results can prove disastrous because of hydrogen embrittlement. If niobium is the anode in such a couple it anodises so readily that no damage occurs and the galvanic current drops to a very low value due to the formation of an anodic oxide film.

Anodic oxide formation Lakhiani and Shreir[46] have studied the anodic oxidation of niobium in various electrolytes, and have observed that temperature and current density have a marked effect on the anodising characteristics. The plateau on the voltage/time curve has been shown by electron microscopy to correspond with the crystallisation of the oxide and rupture of the previously formed oxide. It would appear that this is a further example of 'field recrystallisation' — a phenomenon which has been observed previously during anodisation of tantalum[47]. No significant data on the galvanic behaviour of niobium are available; however, its behaviour can be expected to be similar to tantalum.

Alloys of Niobium

Niobium-Tantalum Niobium and tantalum form solid-solution alloys which are resistant to many corrosive media and possess all the valuable properties of the pure metals. This could have great practical value since in a number of branches of technology it might permit the replacement of pure tantalum by a cheaper alloy of niobium and tantalum. Miller[48] and Argent[49] reported data on the resistance of the niobium-tantalum system, but the tests were only carried out under mild conditions and the data have only limited significance. However, Gulyaev and Georgieva[16] and Kieffer, Bach and Slempkowski[50] carried out tests at elevated temperatures and their work indicated that the corrosion rates of the alloys are substantially that of tantalum provided the niobium content does not exceed 50%.

Niobium-Zirconium Nb-0·75Zr has excellent mechanical properties and similar corrosion resistance to pure niobium; higher zirconium concentrations reduce the corrosion resistance.

Niobium-Titanium Nb-8Ti exhibits unusual behaviour: although the corrosion resistance is slightly lower than the pure metal it shows no sign of embrittlement in sulphuric and hydrochloric acids. The higher the titanium content the lower the corrosion resistance.

Niobium-Zirconium-Titanium Niobium alloys containing zirconium and titanium have improved resistance to high-temperature water[51] and have been evaluated for use in pressurised-water nuclear reactors.

Niobium-Vanadium The presence of vanadium reduces niobium's corrosion resistance to most media. The alloy containing 12·6 at. %V however has excellent resistance to high-temperature water and steam, and this property and the alloy's relatively low neutron cross section give it considerable potential for nuclear applications.

Niobium-Molybdenum The addition of molybdenum to niobium within the solid solution range gives improved corrosion resistance to hydrochloric and sulphuric acids.

Industrial Applications of Niobium

Nuclear Niobium finds use in some nuclear reactors on account of its compatibility with uranium and liquid sodium/potassium at fast reactor temperatures.

Impressed-current cathodic-protection anodes Niobium has, a high anodic breakdown potential (100 V in sea water), a good electrical conductivity (13% that of Cu), good mechanical properties and anodises readily forming an adherent passive oxide film. These properties have led to it being used as a substrate for platinum in impressed-current cathodic-protection anodes for use in high-resistivity waters and other situations where high driving potentials are required to obtain good current spread. Niobium has the advantage over tantalum in that it is less costly, and its cost can be decreased by using a composite electrode with a copper core that also increases the conductivity of the anode[52].

Capacitors Niobium's electrical properties have also led to its investigation as a capacitor material; however, as far as is known there has been no significant commercial application of the material in this field.

Chemical plant It has been reported from some plants producing hydrochloric acid that tantalum condensers are being replaced by ones of niobium, and in certain petroleum plant niobium is being specified for its corrosion resistance and mechanical properties.

Electrical Niobium is finding growing use in components for high-pressure sodium lamps.

Miscellaneous Niobium also finds use in satellite launch vehicles and spacecraft and one of the principal applications for niobium-base alloys is in the production of super-conducting devices.

Recent Developments

Corrosion and other properties of niobium have been reviewed in Symposia held in 1981[53] and 1982[54], and in updated suppliers' literature[55,56].

Wider application would result from improving the oxidation resistance of niobium and its alloys at elevated temperature. Oxidation kinetics are parabolic to paralinear over the range 400 to 600°C and linear from 700 to 900°C[57], although oxidation rate is irregular in the range 600 to 810°C[58]. In contrast the kinetics of nitriding are parabolic over the range 400°C to 1100°C[57]. At room temperature the oxidation of niobium in oxygen is logarithmic and pressure dependent[59].

The oxidation rate of niobium in air from 800°C to above 1000°C can be decreased by alloying e.g. with hafnium, zirconium, tungsten, molybdenum, titanium or tantalum[60,61]. However, the preferred fabricable alloys still require further protection by coating[60]. Ion implantation improves thermal oxidation resistance of niobium in oxygen below 500°C[62].

In hydrochloric acid at temperatures up to 100°C, the corrosion rate decreases with time and ferric iron concentration[63]. The presence of air does not affect the general corrosion rate but in 10N acid it promotes pitting attack, which also arises in chloride-containing methanolic solutions in the absence of sufficient water to effect passivation[64]. Alloying niobium with 2.5% or more of tantalum significantly decreases corrosion rates in hydrochloric acid[65].

Niobium is resistant to pitting and general corrosion in hydrobromic acid up to the azeotropic concentration of 47 wt% and 124°C; the presence of free bromine enhances passivity[66].

Anodic oxide film properties depend upon ion concentration in acid chloride[67] and in alkaline[68] solutions; films are more compact and crack-free in acid solution[69]. Alloying with more than 47% of nickel gives good resistance to hydrogen embrittlement in potassium hydroxide solution[70].

Cathodic protection applications in fresh water include use of ferrite-coated niobium[71], and the more usual platinum-coated niobium[72]. Platinised niobium anodes have been used in seawater, underground[73] and in deep wells[73,74] and niobium connectors have been used for joining current leads[75]. Excellent service has been reported in open-seawater, where anodic potentials of up to 120V are not deleterious, but crevice corrosion can occur at 20 to 40V due to local surface damage, impurities such as copper and iron, and under deposits or in mud[75].

Recent information on the behaviour of niobium in molten salts is sparse and confined to a few specific, mixed-salt environments[76].

J. BENTLEY
I. R. SCHOLES

REFERENCES

1. Lyman, T., *Metals Handbook*, A.S.M. (1961)*
2. Hampel, C. A., *Rare Metals Handbook*, Reinhold, 2nd edn (1961)*
3. Quarrel, A. G., *Niobium, Tantalum, Molybdenum and Tungsten*, Report of Conference, University of Sheffield (1960)*

---

*These references can be found collectively in *Properties of Refractory Metals*, by S. J. Burnett, U.K.A.E.A. Research Report No. AERE R4 657, H.M.S.O. (1969).

NIOBIUM 5:33

4. Krikorian, O. H., *Thermal Expansion of High Temperature Materials*, UCRL 6 132, Sept. (1960)*
5. MacDonald, R. N. and Boucom, H. H., *Nucleonics*, **20** No. 8, Aug., 158 (1962)*
6. Mordike, B. L., *J. Inst. Metals*, **88** No. 6, 272 (1960)*
7. Chelius, J., *Machine Design*, March 1 (1962)*
8. Tingley, I. I. and Rogers, R. R., 'Corrosion of Niobium and Tantalum in Alkaline Media', *Corrosion*, **21** No. 2, 132 and 136, April (1965)
9. Balke, C. W., in *Corrosion Handbook* (Ed. H. H. Uhlig), 620-621 and 720-722, Wiley (1948)
10. Macleary, D. L., 'Testing of Niobium and Niobium Alloys', *Corrosion*, **18**, 67t-69t, Feb. (1962)
11. Jaffee, R. J., *Proceedings on an International Symposium on High Temperature Technology*, Asilomar, California, 1959, McGraw Hill, New York, 61 (1960)
12. Wehrmann, *Corrosionomics*, Fansteel Metallurgical Corporation, Sept. (1956)
13. Priceman, C. and Soma, L., *Development of Fused Slurry Silicide Coatings for the Elevated Temperature Oxidation Protection of Niobium and Tantalum Alloys*, Report AFML-TR-68-210, Sylvania Electric Products Inc., Dec. (1968)
14. Cox, F. G., *Niobium Welding and Metal Fabrication*, 352-358, Oct. (1965)
15. Bishop, C. R., 'Corrosion Tests at Elevated Temperatures and Pressures', *Corrosion*, **19** No. 9, Sept., 308t-314t (1963)
16. Gulyaev, A. P. and Georgieva, I. Ya., *Zashchita Metallov*, **1** No. 6, 652-657, Nov.-Dec. (1965)
17. *Technical Data on Fansteel Niobium*, Bulletin TDB, Fansteel Metallurgical Corporation
18. *Corrosion Tables of Special Materials and Rare Metals*, Jacob and Korves GmbH
19. Meyll Shpeydel, *Collection Nioby i Tañtal* (Nb and Ta), edited by O. P. Kolchina (1960)
20. Gulbransen, E. A. and Andrew, K. F., *Trans. A.I.M.E.*, **188**, 586-599 (1950)
21. Barto, R. L. and Hurd, D. T., *Research and Development*, 26-30, Nov. (1966)
22. Stoughton, L. D. and Sheehan, T. V., *Mechanical Eng.*, **78**, 699-702 (1956)†
23. Lloyd, E. D. in *Plansee Proceedings* 1958—*High Melting Metals*, Metallwork Plansee AG, Reutte Tyrol, 249-256 (1959)†
24. Parkman, R. and Shepard, O. C., US Atomic Energy Commission Publication No. ORO 45, June 11 (1951)†
25. Frost, B. R. T. and Addison, C. C., et al., in *Proceedings of Second United Nations International Conference on the Peaceful Uses of Atomic Energy*, Geneva, 1958, **7**, Reactor Technology, 139-165†
26. Parr, G. W. and Graham, L. W., *Bull. Inst. Metals*, **4**, 125-126, Dec. (1958)†
27. Grassi, R. C., Bainbridge, D. W. and Harman, J. W., US Atomic Energy Commission Publication No. AECU 2 201, July 31 (1952)†
28. Miller, E. C., Chapter 4 of *Liquid Metals Handbook*, US Atomic Energy Commission, Navy Dept., Washington, D.C., 144-183 (1952)†
29. Wilkinson, W. D., US Atomic Energy Commission Publication No. ANL-5027, Aug. (1953)
30. Jaffee, R. I., Evans, R. M., Fromm, E. A. and Gonser, B. W., US Atomic Energy Commission, Publication No. AECD-3 317 (1949) and *Metal Abstracts*, **20**, 241 (1952)†
31. Cathcart, J. V. and Manly, W. D., *Corrosion*, **12**, 87t-91t (1956)†
32. Cunningham, J. E., US Atomic Energy Commission Publication No. ORNL-CF-51-7-135, 78, July 23 (1951)†
33. Reed, E. L., *J. Am. Ceram. Soc.*, **37**, 146-153 (1954)†
34. Cottrell, W. B. and Mann, L. A., *Nucleonics*, **12**, 22-25, Dec. (1954)†
35. Wyatt, L. M. and Dickinson, F. S., *Welding and Metal Fabrication*, **25**, 378-385, 396 (1957)†
36. Raines, G. E., Weaver, C. V. and Stang, J. H., US Atomic Energy Commission, Publication No. BMI-1 284, Aug. (1958)†
37. Adams, R. M. and Sittig, M., in *Liquid Metals Handbook, Sodium-NAK Supplement*, US Atomic Energy Commission, Navy Dept., Washington, D.C., 14 (1955)†
38. Eichelberger, R. L., US Atomic Energy Commission Publication No. BNL-489; Proceedings of the French-American Conference on Graphite Reactors, 168-173, Nov. 12-13 (1957)†

† These references can be found collectively in *Columbium and Tantalum*, edited by F. T. Sisco and E. Epremian, Chapt. 8, Wiley (1963).

39. US Atomic Energy Commission Publication No. ISC-1 118, Semi-annual Summary, Research Report in Engineering for July-December, Ames Laboratory, Ames, Iowa (1958)†
40. McIntosh, A. B. and Bagley, K., *J. Inst. Metals*, **84**, 251-270 (1955-1956)†
41. US Atomic Energy Commission Publication No. ISC-978, Semi-annual Research in Engineering for July-December, Ames Laboratory, Ames, Iowa (1957)†
42. US Atomic Energy Commission Publication No. ISC-607, Quarterly Summary Research Report in Metallurgy for January, February and March, Ames Laboratory, Ames, Iowa (1955)†
43. US Atomic Energy Commission Publication No. ISC-506, Quarterly Summary Research Report in Metallurgy for April, May and June, Ames Laboratory, Ames, Iowa (1954)†
44. US Atomic Energy Commission Publication No. ISC-423, Quarterly Research Report in Metallurgy for July, August and September, Ames Laboratory, Ames, Iowa (1953)†
45. Hodge, W., Evans, R. M. and Haskins, A. F., *J. of Metals*, **7**, 824-832 (1955)†
46. Lakhiani, D. M. and Shreir, L. L., *Nature, Lond.*, **188**, 4 744 (1960)
47. Vermilyea, D. A., *J. Electrochem. Soc.*, **102**, 207 (1955)
48. Miller, G. L., *Tantalum and Niobium*, Academic Press, 328 (1959)
49. Argent, B. B., *J. Inst. Metals*, **85**, 547-551 (1957)
50. Kieffer, R. von, Bach, H. and Slempkowski, I., *Werkstoffe und Korrosion*, No. 9, 782-784, Sept. (1967)
51. Dayton, R. W. and Tipton, C. R., Battelle Memorial Inst. Progress Reports: BMI 1 324, March 1 (1959), BMI 1 340, March 1 (1959) and *Nuclear Science Abs.*, **13** Nos. 18 089 and 18 090 (1959)†
52. 'Niobond' and 'Tibond', Marston Excelsior, Wolverhampton, England (1974)
53. Lupton, D., Aldinger, F. and Schulze, K., Niobium in Corrosive Environments, *Niobium 81, Proceedings of the International Symposium*, San Francisco, edited by Harry Stuart, The Metallurgical Society of AIME, 533-560, (1981)
54. *Refractory Metals and Their Industrial Applications*, symposium sponsored by ASTM Committee B10 New Orleans, ASTM Special Technical Publication 849 (1982)
55. Columbium (Niobium), Teledyne Wah Chang Albany, PO Box 460, Albany, Oregon.
56. Columbium (Niobium), KBI Division of Cabot Corporation, PO Box 1462, Reading, PA.
57. Strafford, K. N., *Corrosion Science*, **19**, 49-62 (1979)
58. Clenny, J. T. and Rosa, C. J., *Met Trans*, **11A**, 1385-1389 (1980)
59. Grundner, M. and Halbritter, J. *Surface Science*, **136**, 144-154 (1984)
60. Inouye, H., in *Reference 1*, 615-636
61. Babitzke, H. R., Siemens, R. E., Asai, G. and Kato, H., Development of Columbium and Tantalum Alloys for Elevated-Temperature Service, *Bureau of Mines Report of Investigations 6558*, US Department of the Interior, (1964)
62. Pons, M., Caillet, M. and Galerie, A., *Mater. Chem. Phys*, **15**, 45-60 (1986)
63. Covino, B. S. Jr, Carter, J. P. and Cramer, S. D., *Corrosion*, **36**, 554-558 (1980)
64. Palit, G. C. and Elayaperumal, K., *Corrosion Science*, **18**, 169-179 (1978)
65. Krehl, M., Schulze, K., Olzi, E. and Petzow, G., *Z. Metallkde*, **74**, 358-363 (1983)
66. Uehara, I., Sakai, T., Ishikawa, H., Ishii, E. and Nakane, M. *Corrosion*, **42**, 492-499 (1986)
67. El-Basiouny, M. S., Bekheet, A. M. and Gad Allah, A. G. *Corrosion*, **40**, 116-119 (1984)
68. Bulhões, L. O. S. and D'Alkaine, C. V., In *Proceedings of the 8th International Congress on Metallic Corrosion* (Mainz 1981) Frankfurt-am-Main; Dechema (1981)
69. Vijayan, C. P., Claessens, P. L. and Piron, D. L. *Corrosion*, **37**, 170-174 (1981)
70. Sugimoto, K., Belanger, G. and Piron, D. L. *Corrosion*, **36**, 437-441 (1980)
71. Kumar, A., Segan, E. G. and Bukowski, J. *Materials Performance*, **23**, 24-28 (1984)
72. Baboian, R., *Materials Performance*, **22**, 15-18 (1983)
73. Baboian, R., *Materials Performance*, **18**, 9-15 (1979)
74. Toncre, A. C., *Materials Performance*, **19**, 38-40 (1980)
75. Hayfield, P. C. S. *Materials Performance*, **20**, 9-15 (1981)
76. Janz, G. J. and Tomkins, R. P. T., Corrosion in Molten Salts; An Annotated Bibliography, *Corrosion*, **35**, 485-504 (1979)

BIBLIOGRAPHY

Meyll Shpeydel, *Collection Niobi i Tantal*, Ed. O. P. Kolchina, 565–571 (1960)
Miller, G. L., *Tantalum and Niobium*, Butterworths, London (1959) *Columbium and Tantalum*, Ed. F. T. Sisco and E. Epremian, Wiley, New York and London (1963)
Materials and Design, 74–78, Jan. (1968)
Tantalum, Niobium and Zirconium, Murex Ltd., Rainham, Essex

5.4 Titanium and Zirconium

Titanium and zirconium both belong to Group IVa of the Periodic Table and have many similarities in their metallurgical and chemical properties. They both have a very strong affinity for oxygen, and their excellent corrosion resistance results from the presence at the metal surface of a tenacious compact film of oxide. At temperatures in excess of 1 000°C both metals readily take up oxygen, nitrogen, hydrogen and carbon, the product being so brittle that it cannot easily be worked. Thus it was not until the advent of novel methods of reduction from the chlorides, followed by techniques of purification and consolidation by remelting under high vacuum or an inert atmosphere, that the metals were rendered sufficiently ductile to become attractive to the engineer.

The development of titanium resulted from the ever-increasing demands of the aircraft and aero-engine designer for materials of high strength but low density. The demand for zirconium arose from the requirement in nuclear reactors for a material having a reasonable transparency to neutrons coupled with moderate strength and good oxidation resistance in high-temperature water and in hot carbon dioxide.

Thus, while titanium and zirconium were primarily developed for specific purposes, their increasing availability in commercial quantities made it economically possible for the chemical industry to take advantage of their exceptionally good corrosion resistance.

TITANIUM

Physical Properties

In its physical properties titanium shows several interesting differences from the commonly employed structural metals. It undergoes a crystal transformation at 882°C; above this temperature it has a b.c.c. structure, designated a β phase, while below it, it has a c.p.hex. structure, known as the α phase. This α phase has a $c:a$ ratio of 1·587 — significantly lower than that for other hexagonal metals such as magnesium, zinc and cadmium. There are thus more slip planes available for deformation, and high-purity titanium is therefore a relatively ductile metal at room temperature, e.g. it can be deformed 95% or more between anneals. In many alloys the phase transformation can be used to obtain a moderate increase in strength, but at the expense of ductility. Thus the commercially pure metal is soft enough to be

Table 5.13 Physical properties of commercially pure titanium

Atomic number	Atomic weight	Crystal structure		Melting point (°C)	Density (g/cm³)	Thermal conductivity (W/m⁻¹K⁻¹) at 20°C	Electrical resistivity (μΩ/cm) at 20°C	Specific heat (J kg⁻¹K⁻¹) 0-100°C	Thermal expansion (20-100°C) (°K)	Magnetic susceptibility (10⁻⁶ cgs units⁻¹)	Electrode potential w.r.t. sat calomel (V)
		Below 882°C	Above 882°C								
22	47.9	a = 2.9504 Å b = 4.683 Å c/a = 1.587	a = 3.3065 Å	1660 ± 10	4.51	17	48.2	528	7.6×10^{-6}	+3.43	−1.75

readily cold formed, and the higher-strength alloys are easily forged. Machining is carried out with conventional machine tools, but at slower speeds than is usual for most metals and alloys, while titanium and most of its alloys may be welded by argon-arc techniques with argon shielding on both sides of the weld. Physical properties of titanium are given in Table 5.13.

Mechanical Properties

The mechanical properties of titanium are affected by small amounts of oxygen and nitrogen in solid solution. It is therefore possible to produce a number of grades of commercially pure titanium, with mechanical properties suited to the different applications for which the material is intended.

Table 5.14 Mechanical properties of commercially pure titanium

0·2% Proof stress (MPa)	Tensile strength (MPa)	Elongation on 50 mm (%)	Fatigue limit (% of T.S.)	Bend radius	Young's modulus (GPa)	Density (g/cm^3)
130–460	270–740	30–15	50	$1t - 2\frac{1}{2}t$	105–125	4·51

The figures of Table 5.14 indicate the range of values normally attained with the commercially available metal.

Corrosion Behaviour

General Titanium is intrinsically very reactive, so that whenever the metal surface is exposed to air, or to any environment containing available oxygen, a thin tenacious surface film of oxide is formed. This oxide, which is present on fabricated titanium surfaces at normal or slightly elevated temperatures, has been identified as *rutile*, a tetragonal form of titanium dioxide, and it is the presence of this surface film which confers upon titanium excellent corrosion resistance in a wide range of corrosive media.

The fundamental mechanism involved in the binding of the surface film to the substrate is not known with any certainty. According to the Pilling-Bedworth principle, the oxide formed at the surface of titanium should occupy a larger volume than the metal from which it was formed, and hence it will be compressively stressed. It is argued that thin compact films resulting from inward movement of oxygen, such as those formed on titanium, are more protective than surface films formed by outward movement of cations. It has been pointed out[1,5] that outward movement of cations may well leave vacancies at the metal/oxide interface, leading to a weakening of the bond between oxide and metal.

Nakayama[2], however, has suggested that, for rutile, which is tetragonal in structure, the strong bond between metal and oxide results from the favourable spacing between titanium ions in the rutile lattice and those in the metal structure. This explanation, however, does not account for the fact that other oxides of titanium, such as *brookite*, which is orthorhombic, and *anatase*, which is tetragonal, are also protective[3].

It is perhaps a significant observation that protective films on titanium appear usually to be formed when the metal surface has access to water, even though this may be present only in trace quantity and in vapour form. Thus, if titanium is exposed to some vigorously oxidising environments in the complete absence of moisture, surface films produced may not provide protection and oxidation in depth may take place, often in the form of a violent exothermic reaction. Examples of this pyrophoric behaviour are to be found in reactions, which may be initiated at room temperature, between titanium and dry oxygen-rich atmospheres[4] at pressures above 345 kN/m^2 between titanium and dry chlorine[6], and between titanium and dry nitric acid containing nitrogen dioxide[7]. In such reactions the concentration of oxidising agent is clearly a determining factor, but the presence of moisture plays a significant part in inhibiting the attack. Thus, in oxygen-rich atmospheres the limiting oxygen concentration, below which the exothermic reaction does not take place, is of the order of 35%, the limiting concentration of nitrogen dioxide below which the reaction does not occur in dry, red, fuming nitric acid is about 1%, while in gaseous chlorine the presence of 0·013% of water vapour is sufficient to prevent significant attack. It is perhaps pertinent to record that one proposed mechanism of passivation of the metal surface suggests that it could well result from direct reaction between the metal and hydroxyl ions[8].

Chemical environments There is ample evidence to suggest that in aqueous liquid environments the presence of an oxidising species, possibly also coupled with the presence of hydroxyl ions, results in the formation of passive films at the surface of titanium. Thus, titanium is very resistant to corrosion in nitric acid at room temperature and at boiling point. While titanium has a significant corrosion rate in acids that normally produce hydrogen on reaction with metal, e.g. sulphuric acid, hydrochloric acid, the addition of small amounts of oxidising reagents results in the formation of passive films. Hence titanium is resistant to attack in mixtures of strong sulphuric and nitric acids, in mixtures of strong hydrochloric and nitric acids, in strong hydrochloric acid containing free chlorine and even in sulphuric and hydrochloric acids containing small amounts of cations such as ferric and cupric salts capable of producing an oxidising reaction[9,10] (see Sections 1.4 and 1.5).

Titanium is almost invariably resistant towards neutral salts, particularly halides, at temperatures up to 100°C, and in respect of the latter environments it is significantly more resistant than stainless steel. In strong solutions of caustic alkalis, on the other hand, titanium tends to form soluble titanates, and it is not as resistant as say, nickel. While at low or moderate concentrations of alkali there is no significant attack, the metal has appreciable solubility in concentrated or molten caustic alkali. Titanium is however resistant to attack by aqueous ammonia at all concentrations and temperatures and to anhydrous ammonia[10].

The overall pattern of behaviour of titanium in aqueous environments is perhaps best understood by consideration of the electrochemical characteristics of the metal/oxide and oxide-electrolyte system. The thermodynamic stability of oxides is dependent upon the electrical potential between the metal and the solution and the pH (see Section 1.4). The Ti/H_2O system has been considered by Pourbaix[11]. The thermodynamic stability of an insoluble phase does not mean that it will form a protective film which will isolate the metal surface from the environment, as this depends on the physical properties of the film, and these cannot be predicted from thermodynamic data.

In general, however, for titanium immersed in acid solutions, potentials above zero on the saturated calomel scale are conducive to the formation of protective oxide, while at certain negative potentials hydride films, which also confer some protection, can be formed[12]. Between the potential at which a continuous hydride film is formed and that at which protective oxide films appear, soluble titanium ions are produced and rapid corrosion ensues.

The concept of a 'protective potential' region explains why the addition of oxidising substances or oxidising metallic ions to a non-oxidising acid often prevents the corrosion of titanium. For example, it has been shown that the addition of cupric ions to 3N hydrochloric acid raises the metal/electrolyte potential into the passive region and thus results in the formation of protective films[9,13]. The same concept of a protective potential region also explains why the addition of chlorine to strong hydrochloric acid causes titanium to become passive, and why the presence or absence of traces of dissolved oxygen in formic acid leads to the existence of 'borderline passivity' in that acid. Thus in the passivation of titanium the redox potential of its solution is of particular significance.

The deliberate raising of the electrical potential of titanium, either by the attachment of discrete particles of a noble metal, such as platinum or palladium, at the surface, or by the application of positive direct current to force the formation of a protective film, is dealt with at a later point. The electrochemical aspect of the corrosion of titanium is comprehensively treated in a number of papers[3,12,14-16].

The behaviour of titanium in a wide range of chemical environments is well summarised in the literature produced by the prime producers of titanium[10] and the reader is advised to consult it for detailed information on corrosion rates. In general, titanium is only resistant to pure 'non-oxidising' acids such as sulphuric and hydrochloric, at dilutions of the order of 2-5% and at moderate temperatures, e.g. 60°C, although, as already indicated, the presence of air and oxidising agents can improve this situation. It has excellent resistance to corrosion in most organic acids, but there are certain exceptions in that in formic acid a condition of borderline corrosion exists at concentrations above about 10%, while there is considerable attack by oxalic acid even in dilute concentrations. In citric acid there is significant corrosion in 50% strength at 100°C, and the metal is not resistant to trichloroacetic acid. Stress corrosion occurs in dry methanol, as discussed later.

In its resistance to liquid metals, titanium shows variable behaviour, the rate of attack often depending upon temperature and increasing with rise in temperature. By thickening the surface film of oxide, resistance to attack is enhanced, and, for example, repeated repair of the surface film renders titanium resistant, on a limited-time basis, to molten zinc in galvanising baths. A surface-oxide thickening technique also enables titanium to be employed in contact with molten aluminium. Titanium equipment is also used in applications involving lead-tin solders, and it is resistant to mercury, at least up to 150°C.

The behaviour of titanium in a wide range of chemical environments is summarised in Table 5.15. In practice, however, whatever the intrinsic reaction between a metal and its environment, account must often be taken of

Table 5.15 Resistance of titanium and zirconium to chemical reagents

Reagent*	Concentration by weight (%)	Temperature (°C)	Ratings† Titanium	Zirconium
Acetic acid, aerated	5, 25, 50, 75, 99·5	boiling	A	A
Acetic acid, non-aerated	99·5	boiling	A	A
Acetic anhydride	99	room	A	A
	99·5	140		A
Aluminium chloride	5, 10	100	A	A
	25	100	C	A
Aluminium sulphate	10, 30	boiling		A
Ammonia, anhydrous (1·4 MN/m^2)	100	40	A	
Ammonium chloride	1, 10, saturated	100	A	A
Ammonium hydroxide	28	room, 60, 100	A	A
Ammonium sulphate	saturated	50		A
Aniline hydrochloride	5, 20	100	A	A
Aqua regia (1HNO$_3$:3HCl)	—	room	A	C
Barium chloride	5, 20	100	A	A
Benzene	—	room	A	A
Benzoic acid	saturated	room, 60	A	A
Bromine	liquid	30	C	
	moist liquid	room	C	
	vapour	30	A	
Bromine-saturated water	—	room, 60	A	
Calcium chloride	5, 10, 25	100	A	A
	62	154	AC	AC
	73	177	B	C
Calcium hypochlorite	2, 6	100	A	A
	18–20	21–24	A	C
Carbon tetrachloride	100	room, 50	A	A
Chlorine gas, dry (less than 0·005% H$_2$O)	—	30	C	A
Chlorine gas, wet (more than 0·013% H$_2$O)	—	75	A	C
Chlorine-saturated water	—	room	A	A
Chloroacetic acid	100	100	A	A
Chromic acid	10, 50	boiling	A	A
	36·5	90	A	A
	10, 20, 30	20, 50, 100	A	A
Citric acid, aerated	10, 25, 50	100	A	A
Citric acid, non-aerated	50	boiling	B	A
Cupric chloride	1	35	A	A
	2·5, 5, 7·5, 10	35	A	B
	12·5, 15, 20	35	A	C
	1	60	A	A
	2·5	60	A	B
	5, 7·5, 10, 12·5, 15, 20	60	A	C
	1, 2·5, 5, 7·5,	100	A	B

TITANIUM AND ZIRCONIUM

Table 5.15 (continued)

Reagent*	Concentration by weight (%)	Temperature (°C)	Ratings† Titanium	Ratings† Zirconium
Cupric chloride (continued)	10, 12·5, 15, 20	100	A	C
	40	boiling	A	C
	55	118	A	C
Dichloroacetic acid	100	100	A	B
Ethyl alcohol	95	boiling	A	A
Ferric chloride	1, 2.5, 5, 10	35	A	A
	15, 20	35	A	B
	25, 30	35	A	C
	1, 2·5, 5	60	A	A
	10, 15	60	A	B
	17·5–30	60	A	C
	1, 2·5	100	A	A
	5	100	A	B
	10, 30	100	A	C
	50	113, 150	A	C
Formaldehyde	37	boiling	A	A
Formic acid, aerated	10, 25, 50, 90	100	A	A
Formic acid, non-aerated	10	boiling	A	A
	25, 50, 90	boiling	C	A
Furfuryl alcohol	—	170		A
Hydrochloric acid, aerated	0·5	35	A	A‡
	1·3	60	A	A‡
	1	100	B	A‡
	2, 3	100	C	A‡
	4, 5	60	B	A‡
	5	room	A	A‡
	7·5, 10	35	B	A‡
	15, 37	35	C	A‡
	20	room	B	A‡
Hydrochloric acid, non-aerated	1	35	A	A‡
	1	70	A	A‡
	1	boiling	C	A‡
	3, 5	room	A	A‡
	3, 5	70	B	A‡
	3, 5	boiling	C	A‡
	5	50	C	A‡
	10, 18	room	B	A‡
	10	70	C	A‡
	20, 37	room	C	A‡
Hydrochloric acid/ nitric acid mixtures	1:3, 2:1, 3:1 4:1	room	A	C
	7:1, 20:1	room	A	
	5:1	boiling	B	
Hydrofluoric acid	1	room	C	C
Hydrofluoric acid, anhydrous	100	room	C	C
Hydrogen peroxide	3,6,30	room	AB	
	10	50		A
Hydrogen-sulphide-saturated water	—	70	A	
Lactic acid, aerated	10, 25, 50, 85	100	A	A
Lactic acid, non-aerated	10, 25, 50, 85, 100	boiling	A	A
Magnesium chloride	5, 20, 42	boiling	A	
	42	35, 60, 100		A

TITANIUM AND ZIRCONIUM

Table 5.15 (continued)

Reagent*	Concentration by weight (%)	Temperature (°C)	Ratings† Titanium	Ratings† Zirconium
Manganous chloride	5, 20	100	A	A
Mercuric chloride	1, 5, 10, saturated	100	A	A
Monochloroacetic acid	30	80	A	A
	100	boiling	A	A
Nickel chloride	5, 10	100	A	A
	20	100		A
Nitric acid, aerated	5, 10, 30, 40, 50, 60,	100	A	A
	69·5			
	20	290	B	
	40	200	B	A
	65	175	A	A
	70	270	B	
Nitric acid	65	boiling	A	A
non-aerated	90	room	A	A
	red fuming	room, 50, 70	§	§
Nitric acid/hydrofluoric acid mixture	15:1	room	C	
Oxalic acid, aerated	0·5, 1, 5, 10	35	A	A
	0·5, 1, 5, 10, 25	60, 100	C	A
Phenol	25, 50, 100	20, 50 boiling		A
Phosphoric acid	5, 10, 20, 30	35	A	A
	35–85	35	B	A
	5–35	60	B	A
	5	100	B	A
	10	80	C	A
Potassium hydroxide	10	boiling	A	A
	50	27	A	A
Propionic acid	vapour	190	C	
Propylene oxide	100	34		A
Silver nitrate	20	room, 60	A	A
Sodium bisulphate	saturated	room	A	
Sodium carbonate	20	boiling	A	
Sodium chlorate	saturated	room	A	
Sodium chloride	saturated	room, 111	A	A
Sodium cyanide	saturated	room	A	
Sodium dichromate	saturated	room	A	
Sodium hydroxide	5–10	21	A	A
	10	boiling	A	A
	50	38–57	A	A
	73	113–129	AB	A
	caustic fusion, 73% to anhydrous NaOH	121–538	C	B
Sodium hypochlorite, 10% w/v av. chlorine	—	boiling	A	
	40	80	A	
Sodium nitrate	saturated	room	A	
Sodium phosphate	saturated	room	A	A
Sodium silicate (75°Tw.)	—	room, 60	A	

Table 5.15 (continued)

Reagent*	Concentration by weight (%)	Temperature (°C)	Ratings† Titanium	Zirconium
Sodium sulphate	saturated	room, 60	A	
Sodium sulphide	saturated	room	A	
Sodium sulphite	saturated	room	A	
Stannic chloride	5, 24	100	A	A
Sulphur, molten	100	240	A	
Sulphur dioxide, dry	100	room, 60	A	
SO$_2$ gas saturated with water vapour (see also sulphurous acid)	—	room, 60, 70	A	
Sulphuric acid	1, 3, 5	35	A	A
	10	35	B¶	A
	20–50	35	C¶	A
	60–70	35	B¶	A
	above 70	35	C	
	above 80	35	C	C
	1, 5	boiling	C	A
	10	50	C	A
Sulphurous acid (see also SO$_2$/H$_2$O)	6	room	A	A
	6	100		A
Tannic acid	25	100	A	A
Tartaric acid	10, 25, 50	100	A	A
Trichlorethylene (unstabilised)	—	boiling	A	A
				A
Trichloroacetic acid	100	100	C	C
Trisodium phosphate	5–20	35–100		A
Uranyl sulphate	U$_1$, 4g/l	250		A
Zinc chloride	20, 50, 75	150	A	
	75	200	B	
	80	173, 200	C	
	75–90	150–250	C	

* Aqueous solutions, not aerated unless otherwise indicated.
† A. Attack <0·125 mm/y. The material is regarded as suitable for use where little dimensional change can be tolerated.
  B. Attack between 0·125 and 1·25 mm/y. The material may be suitable for use provided that some corrosion is permissible.
  C. Attack >1·25 mm/y. The material is ordinarily considered unsuitable.
‡ See text, p. 5.54.
§ Not recommended. See pp. 5.47 and 5.53.
¶ See p. 5.46.

the existence of special circumstances, usually localised in character, which can give rise to corrosion even with an apparently innocuous medium. Thus, the presence of bimetallic couples, the existence of deep crevices, the presence of abrasive particles in a liquid stream, the incidence of local tensile stress, or the application of reciprocating stresses, are all features which demand consideration when a specific material of construction is to be selected.

Resistance to erosion Titanium has outstanding resistance to erosion resulting from the presence of abrasive particles entrained in cooling water and in process liquors[17-20]. In one practical trial as a steam condenser tube, under circumstances known to result in rapid erosion of conventional condenser-tube alloys, titanium was virtually unmarked after more than 15 years service. As a turbine-blade material subject to impingement by water droplets moving at very high speeds, titanium has been shown to be superior to the conventional Fe–13Cr, Fe–18Cr–8Ni +Ti and to Monel. Under such circumstances, the harder the blade the better resistance it offers to erosion, and titanium alloys of the Ti–6Al–4V type give even better service than commercially pure titanium[18], particularly for large turbine generators.

Resistance to crevice corrosion Titanium is more resistant to crevice corrosion than most conventional metals and alloys, particularly where differential aeration is involved, e.g. it is very resistant to crevice attack in sea water at normal temperatures. This form of corrosion becomes more severe when acidity develops in a crevice and this is more prone to occur under conditions of heat transfer[21,79,85]. Under these circumstances, especially in the presence of halide, even titanium may suffer attack, and the metal should not be employed in strong aqueous halides at temperatures in excess of 130°C. This limiting temperature can be raised to 180°C by use of the Ti–0·15Pd alloy[21-23] or by coating with noble metals[86]. (See also Sections 1.4 and 1.6.)

Some crevice attack upon titanium can also occur in the presence of gaseous chlorine gas at temperatures below 100°C, but this is mainly confined to crevices formed between titanium and organic sealing compounds. Here again, the Ti–0·15Pd alloy is less prone to attack.

Titanium in contact with other metals In most environments the potentials of passive titanium, Monel and stainless steel, are similar, so that galvanic effects are not likely to occur when these metals are connected. On the other hand, titanium usually functions as an efficient cathode, and thus while contact with dissimilar metals is not likely to lead to any significant attack upon titanium, there may well be adverse galvanic effects upon the other metal. The extent and degree of such galvanic attack will depend upon the relative areas of the titanium and the other metal; where the area of the second metal is small in relation to that of titanium severe corrosion of the former will occur, while less corrosion will be evident where the proportions are reversed[17]. Metals such as stainless steel, which, like titanium, polarise easily, are much less affected in these circumstances than copper-base alloys and mild steel.

In acid solutions, the behaviour of titanium/dissimilar-metal couples may differ from that just described, and on occasion titanium may be anodic to stainless steel and even to aluminium[24]. In chemical-plant environments, therefore, it is usual to take the precaution of insulating titanium from adjacent components constructed from other metals.

Resistance to stress-corrosion cracking Commercially pure titanium is very resistant to stress-corrosion cracking in those aqueous environments that usually constitute a hazard for this form of failure, and with one or two exceptions, detailed below, the hazard only becomes significant when titanium is alloyed, for example, with aluminium. This latter aspect is discussed in Section 8.5 under titanium alloys.

For commercially pure titanium, the specific environments to be avoided are pure methanol and red, fuming nitric acid[25-28,65], although in both environments the presence of 2% of water will inhibit cracking. On the other hand, the presence of either bromine or iodine in methanol aggravates the effect. When it does occur, stress-corrosion cracking of commercially pure titanium is usually intergranular in habit.

Resistance to fatigue and corrosion fatigue The resistance of titanium to fracture by fatigue, induced by imposition of rapidly reversing stresses, compares favourably with that of the more conventional metals and alloys. Commercially pure titanium has a definite fatigue limit, in air, at about half its tensile strength, and at this figure fracture may take place at between 10^7 and 10^8 reversals. In this respect the commercially pure metal resembles steel rather than the non-ferrous alloys. Reversed stresses at a figure below the limit indicated are not likely to result in fatigue failure, irrespective of the number of reversals applied.

For many metals, the presence of corrosive environments coupled with reversing stress results in fracture by corrosion fatigue at a stress level well below that of the normal fatigue limit. While, given the appropriate environment, titanium is not immune to this effect, its generally good resistance to corrosion renders corrosion fatigue a comparatively rare event. Thus, the fatigue limit for titanium wetted with sea water is very similar to the figure obtained in air[29]. It is, therefore, not surprising that valve springs and valve plates of titanium alloys, used in gas compressors, give better performance than the conventional alloy steels.

High-temperature behaviour Commercially pure titanium is an established material for use at the moderately elevated temperatures attained in aircraft exhaust shrouds and firewalls, but neither titanium nor titanium alloys are suitable for use at really high temperatures. The tensile strength of commercially pure titanium shows a steady fall with increase in temperature, the tensile strength at 350°C being approximately half that at room temperature. The creep strength is improved by suitable alloying and an alloy containing 8% Al, 4·5% Sn, 4% Zr and smaller quantities of Nb, Mo, Si has a high creep strength at temperatures up to 600°C.

Above about 600°C penetration of oxygen and nitrogen occurs. It has already been indicated that the presence of these elements renders the titanium brittle, and this feature must be taken into account in considering the use of titanium at elevated temperatures. Titanium has nevertheless been successfully employed as an autoclave lining in steam atmospheres at a temperature of 400°C and a pressure of 10 Mpa.

Examples of the Use of Titanium in Chemical Plant

As with all fairly expensive materials of construction, economy in the use of titanium can often be achieved by using it in the form of thin linings upon a thicker load-bearing support. The soundest and most economical method of achieving this is by explosively bonding thin titanium sheet to thick steel plate. This technique, coupled with that of welding the duplex plate into large reaction vessels, is well established[30-31,66]. The decision as to whether to use solid titanium or to employ explosively clad steel, depends upon the size and wall thickness of the construction and is largely decided upon economic grounds. The reader is advised to consult the specialist suppliers and fabricators before deciding which form to employ.

For some chemical plant applications the iron content of the titanium employed can influence its behaviour, e.g. in some strengths of nitric acid, and in chlorine dioxide, preferential weld attack may occur if the iron content of the titanium is above a certain critical level. This effect is only encountered in a few specific environments[66], but where these are involved, it is recommended that titanium with an iron content of less than 0·05% is specified.

There are also occasions, particularly in hydrogen-containing atmospheres, when surface contamination of the titanium with iron can result in localised corrosion and embrittlement. This effect can be countered by avoidance of undue contamination with iron during fabrication, by post-fabrication cleaning and by post-fabrication anodising[10,16,67]. It should be emphasized, however, that in general use in the marine and chemical industries discussed below, iron levels up to 0·2% do not adversely affect corrosion resistance.

Examples of the use of titanium in the chemical industry are briefly summarised below. For more detailed treatment the reader is advised to consult Reference 32.

Titanium is being employed in the bleaching industry[36] where the good corrosion resistance of the material makes it particularly suitable for equipment in both textile and paper pulp bleaching processes. In the dye-stuffs industry, the inertness of titanium eliminates any products of corrosion which might cause discoloration of the products. A similar situation can also exist in areas like the plastics, pharmaceuticals, and food-stuffs industries.[33]

Equipment lined with titanium has been employed for organic reaction vessels used in contact with nitric acid at elevated temperatures and pressures[34], in a large rotary ammonium chloride dryer[35], for emulsion pans holding photographic solutions[35], and for general service in corrosive liquors. Shell and tube heat exchangers have been used to handle hydrochloric acid containing free chlorine[37] and chromic acid[38], while titanium pumps have been found to be useful for organic chlorides containing hydrochloric acid and free chlorine[40].

Important applications for titanium have been developed in processes involving acetic acid, malic acid, amines, urea, terephthalic acid, vinyl acetate, and ethylene dichloride. Some of these represent large scale use of the material in the form of pipework, heat exchangers, pumps, valves, and vessels of solid, loose lined, or explosion clad construction. In many of these the requirement for titanium is because of corrosion problems arising from the organic chemicals in the process, the use of seawater or polluted cooling waters, or from complex aggressive catalysts in the reaction.

Titanium is the only one of the more common structural metals which is not attacked by wet chlorine gas and it is thus widely used as a heat exchange material for cooling the gas after the electrolysis stage. Preheating of sodium chloride brine is carried out in titanium plate heat exchangers, while titanium butterfly valves, demisters, and precipitators handle the chlorine gas produced in the cell. The most important use of titanium in chlorine production is as anodes in place of graphite in the electrolytic process. This is covered in more detail later.

The resistance of titanium in nitric acid is good at most concentrations and at temperatures up to boiling[32,33]. Thus tubular heat exchangers are used in ammonium nitrate production for preheating the acid prior to its introduction into the reactor via titanium sparge pipes. In explosives manufacture, concentrated nitric acid is cooled in titanium coils and titanium tanks are used in the reprocessing of spent nuclear fuel elements by dissolution in nitric acid.

The excellent corrosion resistance of titanium in sea water has led to one of the largest present and potential future uses for the material outside the aerospace industries. To all intents and purposes, commercially pure titanium is completely unattacked by seawater at ambient and moderately elevated temperatures. This has resulted in titanium becoming firmly established as a heat exchange material for power station condensers, for desalination plant, and in on-shore and off-shore oil installations[39] where seawater or other polluted waters are used as the cooling medium. Titanium fittings are replacing stainless steel for racing yachts[90] and the material is also being used for many applications in Naval vessels, particularly minesweepers where its non-magnetic properties are an advantage.

A rapidly growing use in the medical field[91] is for surgical implants as either bone plates and screws, joint replacements, or for the repair of cranial injuries. Here, titanium and its alloys have the advantages of complete compatibility with body fluids, low density, and low modulus. Applications also exist in dentistry.

Titanium impellers have been used in pumps employed for the conveyance of corrosive and erosive ore slurries, for organic chlorides containing hydrochloric acid and free chlorine[40], for handling moist chlorine gas, and in the wood-pulp and the textile-bleaching industry, particularly with sodium hypochlorite[36].

In the electroplating industry, the use of titanium as hooks[41] and as heating and cooling coils for temperature control of certain acidic liquors has improved the control of plating baths[38, 42]. Perhaps the most significant advance has been in the nickel plating industry where solid nickel anodes have been largely replaced by titanium baskets holding nickel shot and nickel shapes. In this development, the titanium itself is anodically passivated, but at the same time the passive film allows electron transfer to occur between contacting surfaces of titanium and nickel so that the latter is anodically dissolved.

Anodic passivation also allows titanium to be employed as a jig for aluminium anodising baths[43], because the protective anodic film formed on titanium allows passage of electronic current to the metal contact while virtually suppressing flow of ionic current through the anodically-formed surface film. This aspect is discussed in more detail in relation to special applications.

In the field of electrowinning and electrorefining of metals, titanium has an advantage as a cathode, upon which copper particularly can be deposited with finely balanced adhesion that allows the electrodeposited metal to strip easily when required. Titanium anodes are also being employed as a replacement for lead or graphite in the production of electrolytic manganese dioxide.

In the field of nuclear energy, titanium has been used for processing of fuel elements, where this demands use of nitric acid or aqua regia[44-45], and for control-rod mechanism, in which the short half-life of irradiated titanium is of advantage.

Environmental considerations in recent years have dictated that sulphur bearing compounds are removed from the exhaust gases of coal burning power stations in order to reduce the incidence of 'acid rain'. The flue gas desulphurisation process (FGD), while removing the sulphur, also changes the character of the waste gases and makes them more corrosive to the materials from which power station chimneys are normally constructed. Considerable service experience has now demonstrated that titanium is resistant to the conditions and this represents one of the most promising uses of the material for the future.

Special Applications

Anodic Passivation

It has already been indicated that titanium is not particularly resistant to corrosion in hot, strong acids of the type that usually generate hydrogen upon reaction with metals—acids such as sulphuric or hydrochloric. In contact with such acids, corroding titanium assumes a negative electrical potential (approximately -0.7 V, S.C.E.).

If this negative potential is artificially raised by slow increments, a critical potential level may be attained, at which corrosion dramatically ceases and the metal acquires a protective film. The potential level at which this occurs usually lies between -0.5 V and -0.2 V (S.C.E.), and it is evident that at this potential level the metal/electrolyte interface has attained a thermodynamic state conducive to the formation of a stable, insoluble titanium dioxide[46]. This protective surface film has been shown to consist mainly of *anatase*, a tetragonal form of the oxide, and if the potential is further raised, the film thickens, giving a series of interference tints until it reaches a dark purple colour at a maximum limiting thickness of about 2×10^{-7} m. Once this film has become established, there is very little further passage of current into the electrolyte and corrosion virtually ceases as long as the 'protective' potential is applied. The most obvious means of attaining this potential is by application of an anodic direct current from an external source, but the same effect can also be attained to some extent by coupling the titanium to a more noble element such as carbon, or one of the platinum group metals. The latter method is, however, of limited application because it is dependent upon the potential level attained by the noble element, and this may not be sufficiently high to provide a mixed potential which is above the critical value for film formation (see Sections 1.4 and 1.5). Nevertheless, Stern and his associates[47-49] have shown that the addition of 0.2% palladium to titanium produces a discrete dispersion of palladium particles at the surface, which permits the combination to offer an adequate resistance to corrosion in 5% boiling sulphuric and hydrochloric acids. Cotton[3,50,51] has shown that application of a d.c. potential of about 2 V between titanium and a suitable cathode can prevent corrosion in a wide range of strong non-oxidising acids at concentrations and temperatures which present considerable handling difficulties with most metallic materials of construction[81,84].

One full-scale practical application of this principle of anodic passivation is found in titanium heat exchangers handling 8% sulphuric acid containing hydrogen sulphide and carbon disulphide employed in viscose rayon processing[9,52]. It is conservatively estimated that each of these anodically passivated units performs the duty previously undertaken by three graphite heat exchangers. In doing this they require a current of only 1.5 A supplied at 15 V.

Use as Anodes

As indicated above, when a positive direct current is impressed upon a piece of titanium immersed in an electrolyte, the consequent rise in potential induces the formation of a protective surface film, which is resistant to passage of any further appreciable quantity of current into the electrolyte. The upper potential limit that can be attained without breakdown of the surface film will depend upon the nature of the electrolyte. Thus, in strong sulphuric acid the metal/oxide system will sustain voltages of between 80 and 100 V before a spark-type dielectric rupture ensues, while in sodium chloride solutions or in sea water film rupture takes place when the voltage across the oxide film reaches a value of about 12 to 14 V. Above the critical voltage, anodic dissolution takes place at weak spots in the surface film and appreciable current passes into the electrolyte, presumably by an initial mechanism involving the formation of soluble titanium ions.

Thus titanium by itself cannot function as an efficient anode for the passage of positive direct current into an electrolyte. The surface film of oxide formed upon the titanium has, however, a most useful property: while it will not pass positive direct current into an electrolyte (more correctly, while it will not accept electrons from negatively charged ions in solution), it will accept electrons from, or pass positive current to, another metal pressed on to it. Hence a piece of titanium which has pressed on to its surface a small piece of platinum will pass positive direct current into brine and into many electrolytes, at a high current density, via the platinum, without undue potential rise, and without breakdown of the supporting titanium[53,54].

Platinised titanium anodes (titanium carrying a thin surface film of platinum, of the order of 0·0025 mm thick) have proved successful in cathodic-protection systems employing impressed-current techniques, as electrodes for electrodialysis of brackish water, and in many applications where established anode materials suffer significant corrosion. Platinum-coated titanium anodes can operate without breakdown at very high current densities, of the order of 5 000 A/m^2, in sea water, as although the very thin platinum coating may be porous the underlying titanium exposed at the pores will become anodically passivated[55].

In aqueous chloride where it is necessary to use platinised titanium anodes coated over only part of their surface, e.g. titanium rod tipped with a thin platinum film, it may be necessary to limit the applied voltage to 12 V.

The development of platinised titanium has been extended to include the replacement of platinum by deposits of other forms of corrosion-resistant conducting surfaces, such as platinum-iridium and ruthenium oxide. Apart from their corrosion resistance, these surfaces have the ability of operating electrochemically at lower overvoltages than plated platinum or graphite. Thus, for a range of electrochemical cells used in the chlor-alkali industry for the production of chlorine and sodium chlorate, etc. there is a significant advantage in using them compared with graphite[82], and they are now the preferred choice for such applications.

By a mechanism similar to that discussed in relation to platinum coating, titanium can function as a conducting jig to support aluminium components and assemblies in conventional anodising baths. In this application the exposed titanium acquires the insulating film, but allows current to pass to the aluminium at the points of contact[56].

Pyrophoric Tendency

If titanium is exposed to certain vigorously oxidising environments, oxidation does not cease at the surface, and a rapid exothermic reaction in depth ensues. The fundamental reason for this remarkable change in the character of the oxidation is not known with any certainty, but it is significant that in almost every instance the presence of a small quantity of water, sometimes in trace amounts, prevents this rapid oxidation in depth.

Investigation into the effect has been mainly devoted to reactions with red fuming nitric acid[57]. It seems that in red fuming nitric acid a preliminary reaction results in the formation of a surface deposit of finely divided metallic titanium; ignition or pyrophoricity can then be initiated by any slight impact or friction. The tendency to pyrophoricity increases as the nitrogen dioxide content of the nitric acid rises from zero to maximum solubility at about 20%, but decreases as the water content rises, the effect being nearly completely stifled at about 2% water.

Other media in which titanium is subject to pyrophoricity are anhydrous liquid or gaseous chlorine[58], liquid bromine, hot gaseous fluorine, or oxygen-enriched atmospheres at moderately low pressures.

Titanium Alloys

Mechanical properties of various titanium alloys are given in Table 5.16. In general the corrosion behaviour of those titanium alloys developed for the aircraft industry is very similar to that of unalloyed titanium[59]. The addition of some alloying elements may increase resistance to one medium, but decrease it to others[60].

Additions of zirconium confer a significant increase in corrosion resistance, particularly in sulphuric and hydrochloric acids[59,61]. At alloying additions of the order of 50% Zr, however, there can be a significant diminution in resistance to oxidation[62] and the welding of titanium to zirconium is not advisable, because within the welded zone the proportion of titanium to zirconium will almost inevitably fall within the sensitive composition range.

The addition of 0·2% palladium to titanium decreases the corrosion rate in boiling 5% sulphuric acid by a factor of 500, and in boiling 5% hydrochloric acid by a factor of 1 500, in relation to the rates obtained with unalloyed titanium. The addition of palladium in these quantities thus provides an adequate measure of resistance to relatively weak concentrations of the acids mentioned[48].

From the corrosion-resistance aspect, one of the most effective additions to titanium is that of molybdenum. According to Yoshida and his colleagues[63-64], the addition of 15% Mo produces an alloy fully resistant to virtually all concentrations of sulphuric and hydrochloric acid at room temperatures, while with 30% Mo, the alloy is resistant to all strengths of boiling sulphuric acid up to a concentration of 40% by weight, and to 10% boiling hydrochloric acid.

The stress-corrosion cracking hazard for titanium alloys containing aluminium is significantly higher than that obtaining for commercially pure titanium, and in addition to stress-corrosion cracking in methanol and red

Table 5.16 Mechanical properties of some titanium alloys

Nominal composition in weight % and characteristics		0.2% proof stress (min) (MPa)	Tensile strength (MPa)	Elongation (min) (%)	Young's modulus (typical) (GPa)	Fatigue limit (% of T.S.)	Bend radius on 2 mm	Density (g/cm³)	Stress for 0.1% total plastic strain in 100 h (MPa)	Production range
Ductile medium strength Ti-2·5Cu alloy, weldable and age hardened BSTA 21–24, 52–55, 58	Annealed	400	540–770	16	105–120	60–65	2t	4·56		S, B, W, E
	Solution treated and aged	525	650–880	10	105–120	60–65		4·56		S, B, W, E
Small additions of Pd giving improved resistance to non-oxidising acids		170	330–420	25	105–120	50	1t	4·51		S
Medium strength Ti-6Al-4V alloy BS TA10–13, 28, 56	Sheet	900	960–1270	8	105–120	55–60	5t	4·42		S
	Rod	830	900–1160	8	105–120	55–60		4·42		B, W, E
High strength Ti-4Al-4Mo-2Sn 0·5Si alloy. Creep resistant up to 400°C BSTA 45–51, 57		960	1100–1280	9	110–130	50–60		4·60	465 at 400°C	B, E
Very high strength Ti-4Al-4Sn-4Mo-0·5Si alloy BSTA 38–42		1095	1250–1420	8	110–130	40–50		4·62		B, E
High strength Ti-11Sn-5Zr-2·25Al-1Mo alloy, creep resistant up to 450°C BSTA 18–20, 25–27		970	1110–1340	8	105–110	55–60		4·84	385 at 450°C	B, E

Table 5.16 (continued)

Medium strength Ti–6Al–5Zr–0·5Mo–0·2Si alloy, weldable and creep resistant up to 520°C BSTA 43, 44	Room	850	990–1140	6	125	50	4·45	B, E
	520°C	480	620–780	9	125	50	4·45	300 B, E
Medium strength Ti–5·5Al–3·5Sn–3Zr–1Nb–0·3Mo–0·3Si alloy weldable and creep resistant up to 550°C	Room	820	950 min	10	120	50	4·51	B, E
	540°C	460	590 min	12	120	50	4·51	300 B, E
Medium strength Ti–6Al–7Nb alloy for surgical implant applications		800	900–1200	10	105	55–60	4·52	B
Medium strength Ti–5·8Al–4Sn–3·52r–0·7Nb–0·5Mo–0·35Si–0.06C alloy, weldable and creep resistant up to 600°C	Room	910	1030	6	120	60	4·51	B
	600°C	450	585	9				
High strength Ti–15Mo–3Nb–3Al–0·25i alloy, oxidation resistant		965	1035–1350	4	96		4·92	St
High strength Ti–3·5Al–8V–6Cr–4Mo–4Zr alloy, deep hardenable and corrosion resistant		1180	1250	11	106		4·82	B, E

Table 5.17 Physical properties of unalloyed zirconium

Atomic number	Atomic weight	Crystal structure		Melting point (°C)	Density at 20°C (g/cm³)	Thermal conductivity (W/m °K)	Electrical resistivity (at 20°C) (μΩ/cm)	Temperature coefficient of resistivity (°C)	Specific heat (J/g°C)	Thermal expansion per °C	Standard electrode potential (V)	Thermal neutron absorption cross-section. Reactor grade	
		Below 865°C	Above 865°C									Microscopic	Macroscopic
40	91·2	c.p. hex at 25°C $a = 3\cdot232$ Å $c = 5\cdot15$ Å	b.c.c. at 900°C $a = 3\cdot61$ Å	1845	6·490	22	39·7	44	0·276	$5\cdot89 \times 10^{-6}$	$-1\cdot53$	0·180 barn/atom	0·08 mm

fuming nitric acid, cracking has been observed in salt solution, in hot solid sodium chloride and in uninhibited chlorinated hydrocarbons. Because of the importance of these alloys to the aircraft industry there has been considerable laboratory investigation of the effect and the reader is advised to consult References 65 and 66 and the literature for a comprehensive treatment of the subject. (see also Section 8.5)

Viewed in perspective, evidence of failure in service has been rare and the practical hazard is certainly very much lower than would appear from the results of laboratory tests. In chlorinated hydrocarbons the effect can be controlled by the addition of inhibitors, and, for example, the appropriate commercial degreasants containing these inhibitors are specified in a British detence standard*.

ZIRCONIUM

The growth of nuclear engineering with its specialised demands for materials having a low neutron absorption coupled with adequate strength and corrosion resistance at elevated temperatures, has necessitated the production of zirconium in relatively large commercial quantities. This specific demand has resulted in development of specially purified zirconium, and certain zirconium alloys, for use in particular types of nuclear reactor.

In its natural state, zirconium is associated with hafnium, and for use in nuclear reactors it is essential to separate the two because hafnium readily absorbs neutrons. This situation gives rise to bulk production of two forms of raw zirconium metal, a hafnium free reactor grade and a commercially pure hafnium bearing quality (ASTM designations R60001 and R60702 respectively). A number of different zirconium alloys are also commercially available including one containing tin, iron, chromium, and nickel additions (R60802) and a similar material (R60804) but without the nickel. Both of these are used in water cooled nuclear reactors. A zirconium $2\frac{1}{2}$% niobium alloy (R60901) provides a heat treatment capability, while in the chemical industry a similar alloy (R60705) offers good corrosion resistance and better strength than commercially pure zirconium.

Generally, for the chemical engineer not particularly associated with atomic energy, unalloyed zirconium containing hafnium is an appropriate choice for those occasions which require the special corrosion resistant properties exhibited by the metal.

Physical and Mechanical Properties

The physical properties of unalloyed zirconium are recorded in Table 5.17.

Mechanical properties of these grades of zirconium depend to a large extent upon the purity of zirconium sponge used for melting. Hardness and tensile strength increase rapidly with rise in impurity content, notably oxygen, nitrogen and iron. Typical mechanical properties of chemical grades of zirconium are listed in Table 5.18.

* *The Cleaning and Preparation of Metal Surfaces*, Defence Standard 03-2/1 (1970), obtainable from the Ministry of Defence, First Avenue House, High Holborn, London, W.C.1.

Table 5.18 Mechanical properties of chemical grades of zirconium

ASTM designation	0·2% proof stress (MPa)	Tensile strength (MPa)	Elongation (%)	Bend radius
R 60702	207 (min)	379 (min)	16 (min)	5*t*
R 69705	379 (min)	552 (min)	16 (min)	3*t*

Table 5.19 gives the physical properties of Zr-Sn-Cr-Ni alloy.

Table 5.19 Physical properties of Zr-Sn-Cr-Ni alloy

Alloy nominal composition (%)	Density at 20°C (g/cm^2)	Electrical resistivity at 21°C ($\mu\Omega$ cm)	Coefficient of linear thermal expansion (°C)		Thermal neutron absorption cross-section	
			20–700°C	25–600°C	Microscopic	Macroscopic
Zr-1·5 Sn-0·1 Cr-0·12 Fe -0·05 Ni	6·57	74	6·5 × 10^{-6}	—	0·22-0·24 barn*/atom	—

*1 barn = 10^{-24} cm^2.

The mechanical properties of the alloys will vary slightly according to the purity of sponge, and also with heat treatment.

Table 5.20 Minimum mechanical properties of nuclear grade zirconium alloys

ASTM designation	Condition	Direction of test	0·2% proof stress (MPa)	Tensile strength (MPa)	Elongation (%)
R 60001	Annealed	Transverse	207	296	18
R 60802	Annealed	Transverse	303	386	25
R 60804	Annealed	Transverse	303	386	25
R 60901	Annealed	Transverse	344	448	20
R 60901	Cold worked	Transverse	385	510	15

Behaviour of Commercially Pure Zirconium in Aqueous Environments

Zirconium, like titanium, depends upon the integrity of a surface film, usually of oxide, for its corrosion resistance, but there are differences in behaviour between the two metals when they are exposed to aggressive aqueous environments.

In general, zirconium does not equal titanium in resistance to certain oxidising media, but it is superior in non-oxidising acids, and in caustic alkalis. The presence of certain impurities in zirconium influences the corrosion behaviour, and while small amounts of hafnium are not deleterious, carbon in amounts greater than 0·06% lessens resistance to hot concentrated hydrochloric acid by a factor of several hundreds[68,69]. The contrast in behaviour between titanium and zirconium in a wide range of media is illustrated in detail in Table 5.15.

To summarise, zirconium performs well in nitric acid at all concentrations up to 70% and temperatures up to 200°C[68], but it will react pyrophorically in a fashion similar to titanium in concentrated nitric acid containing free nitrogen dioxide. If there are appreciable amounts of hydrochloric acid present together with nitric acid, there may be severe attack, and, in contrast to titanium, zirconium is not resistant to aqua regia containing three parts nitric to one part hydrochloric acid. Towards chromic acid, zirconium is resistant at least up to a strength of 50% at a temperature of 90°C. In saturated chlorine water the corrosion rate of zirconium is virtually nil, but unlike titanium it is attacked in moist gaseous chlorine and not in dry chlorine at room temperature[70]. In solutions of metal chlorides, behaviour appears to depend upon whether the chloride solution tends to be oxidising or reducing, and in general zirconium is not as resistant as titanium. Thus it is not resistant to boiling ferric or cupric chlorides at strengths greater than 10%, but it is resistant to mercuric, stannic, manganous, nickel, ammonium, zinc, magnesium, barium and sodium chlorides, and to sea water. Behaviour in aluminium chloride is worth noting, for zirconium is resistant to boiling 25% aluminium chloride, while titanium is attacked. Both metals corrode in boiling 62% calcium chloride.

In resistance to hydrochloric and sulphuric-acids zirconium shows a significant advantage over titanium[68]. With pure hydrochloric acid at 100°C the corrosion rate is negligible up to the constant boiling strength, i.e. 20% w/w at atmospheric pressure, but at 200°C under pressure there is appreciable attack at acid strengths greater than 18% by weight. The presence of traces of copper and iron in the hydrochloric acid can result in a significantly increased rate of attack, and, for example, in boiling 20% acid 1 000 p.p.m. of iron or copper raises the rate of attack from less than 0·0075 mm/y to the barely acceptable level of 0·5 mm/y. In sulphuric acid where traces of metal ions do not appear to be unduly troublesome, there is no appreciable corrosion up to 66% w/w at boiling point; the rate of attack, however, increases rapidly in boiling 70% acid, and at 200°C under pressure there is significant uniform corrosion at about 40% w/w. The presence of chlorine in sulphuric acid can seriously increase the rate of corrosion.

With phosphoric acid the performance of zirconium is again distinctly superior to that of titanium, for while, in general, use of titanium is limited to strengths less than 30% w/w, for zirconium there is no appreciable corrosion at room temperature up to 80% strength. As temperature rises there is an inflection in the corrosion-rate curve, an unacceptable rate being reached in boiling acid at 50% strength. As temperature rises beyond this, the corrosion rate again decreases[68], until at 200°C, under pressure, there is again negligible attack in 80% acid.

Neither titanium nor zirconium is recommended for use in hydrofluoric acid.

Zirconium is also resistant to attack in a wide range of organic acids, one useful difference from titanium being that it is not corroded in boiling deaerated formic acid at concentrations of 25% and upwards, in which titanium exhibits borderline passivity. In strong chlorinated organic acids, however, there may be some attack at elevated temperatures.

It is in its behaviour to caustic alkalis that zirconium shows itself to be superior to those other elements of Groups IV and V whose resistance to corrosion results primarily from an ability to form surface films. Thus, in contrast to tantalum, niobium and titanium, zirconium is virtually completely resistant to concentrated caustic solutions at high temperatures, and it is only slightly attacked in fused alkalis. Resistance to liquid sodium is good. Zirconium is thus an excellent material of construction for sections of chemical plant demanding alternate contact with hot strong acids and hot strong alkalis—a unique and valuable attribute.

Because of its good performance in mineral acids, there is little need or incentive to invoke anodic passivation techniques for zirconium. The metal can be anodised in sulphuric acid, but, again in contrast to the behaviour of titanium, it does not form a stable anodic film in chloride solutions, and even in neutral sodium chloride, zirconium rapidly corrodes if an anodic potential of 2 V is applied.

Applications in Industry

The chemical industry now provides a major area for the use of zirconium equipment. The material is employed in the form of heat exchangers, stripper columns, reactor vessels, pumps, valves, and piping for a wide variety of chemical processes. These include hydrogen peroxide production, rayon manufacture, and the handling of phosphoric and sulphuric acids and ethyl benzene. Gas scrubbers, pickling tanks, resin plants, and coal gasification reactors are some of the applications where the good corrosion resistance of zirconium towards organic acids is utilised. A particularly useful attribute is the ability of the material to withstand environments with alternating acidity and alkalinity.

Special Applications

It has already been indicated that the principal use for zirconium is in the field of nuclear engineering. The very nature of this application demands the lowest possible corrosion rate, and this has necessitated a great deal of investigation into the oxidation rate of zirconium, when exposed to hot water, steam and carbon dioxide.

When zirconium oxidises in these environments at elevated temperatures the reaction kinetics follow a law which can be formulated as

$$w = Kt^n$$

where w = weight gain, t = time, and K and n are constants at a constant temperature. Initially n has a value of between $\frac{1}{3}$ and $\frac{1}{2}$, and the rate of oxidation decreases with time. However, when a certain thickness in the surface film is attained, the value of n may change and become equal to or greater than unity. The corrosion rate will then become constant or will increase. This type of behaviour, which can occur with several metals or alloys, has been called 'breakaway' corrosion. Within the period at which the value of n remains below unity, the monoclinic oxide film produced on zirconium is hard, glossy, adherent and usually black or dark coloured. When the kinetic change takes place the character of the film changes, and continued oxidation may lead to heavy surface spalling.

Unalloyed zirconium produced from Kroll sponge quickly reaches the breakaway point when exposed to steam or hot water at reactor temperatures. Early investigation in the United States established that this behaviour resulted from the almost inevitable presence of nitrogen, but that the deleterious effect could be countered by an addition of tin[71], and the alloy known as Zircaloy 2, containing about 1·5% Sn, 0·1% Fe, 0·1% Cr, 0·05% Ni was developed for use in water-cooled reactors. Even with this alloy, metallurgical treatment during fabrication is known to affect performance, and a rigorous scheme of corrosion testing is employed[72,73] to ensure that the semi-fabricated material and finished product conform to a high degree of corrosion resistance. This test involves the autoclaving of carefully prepared coupons for fourteen days in pure steam at a temperature of 400°C and a pressure of 10 Mpa. At the conclusion of the test, satisfactory material has a weight gain of $28 \pm 10 \, \text{mg/dm}^2$, and is covered with a glossy black lustrous film. Defective material manifests itself by high weight gains (up to as much as $100 \, \text{mg/dm}^2$) and the appearance in the surface film of white corrosion product.

Most of the considerable volume of published work on the behaviour of zirconium relates to its use in nuclear reactors in contact with water or steam, e.g. in pressurised steam the control of oxidation by use of boric acid has been reported[75]. The reader is advised to consult the reviews on this important aspect of the subject cited under References 76 and 77.

It should be noted that swarf from a zirconium-titanium alloy containing approximately 50% by weight of each element is prone to pyrophoricity in air. It has also been reported[62] that when zirconium is welded to titanium, the welded zone is much more sensitive to corrosion than either of the parent metals. If, therefore, it is proposed to use any construction in which zirconium is welded to titanium, caution should be observed in the machining of welds, and the corrosion behaviour of the weld should be checked by prior testing in the environment with which the construction will be employed.

The pyrophoric tendency of zirconium in contact with red fuming nitric acid has already been mentioned.

There is some evidence that the increase in corrosion recorded when zirconium is exposed to hydrochloric acid at 200°C under pressure results from intergranular penetration[78].

Finally, perhaps, it should be pointed out that because the behaviour of zirconium is often adversely influenced by the presence of impurities in corrosive environments, corrosion testing prior to use should be carried out in actual plant liquors rather than in purer synthetic solutions.

J. B. COTTON
B. H. HANSON

REFERENCES

1. Evans, U. R., *The Corrosion and Oxidation of Metals*, Arnold, London, 39–48 (1960)
2. Nakayama, Castings Research Laboratory Report, No. 5, Waseda University, 57–59 (1956)
3. Cotton, J. B., *Werkst. u. Korrosion*, Weinheim, **2** No. 3, 152 (1960)
4. Adamson, G. M., Jr. et al., *Proceedings of the Second U.N. International Conference on the Peaceful Uses of Atomic Energy*, Paper P/1 993, United Nations, Geneva (1958)
5. Jackson, J. D., *Mat. Prot.*, **4** No. 1, 30–33 (1965)
6. Millaway, E. E. and Kleinman, M. H., *Corrosion*, **23**, 88–97 (1967)
7. Wullner, R. L., *Mat. Prot.*, **4** No. 1, 55–56 (1965)
8. Mueller, W. A., *J. Electrochem. Soc.*, **107**, 157 (1960)
9. Cotton, J. B., *Chemical Engineering Progress*, **66** No. 10, 57–62 (1970)
10. *Corrosion Resistance of Titanium*, IMI Titanium Ltd., P.O. Box 704, Witton, Birmingham
11. Pourbaix, M., Rapport No. 21, Centre Belge d'Étude de la Corrosion, Brussels (1953)
12. Fischer, W. R., *Werkst. u Korrosion*, Weinheim, **10**, 243 (1959)
13. Schlain, D. and Smetko, J. S., *J. Electrochem. Soc.*, **99**, 417 (1952)
14. T.M.L. Report No. 57, Titanium Metallurgical Laboratory, Battelle Memorial, 116–153, Oct. 29 (1956)
15. Stern, M. and Wissenberg, H., *J. Electrochem. Soc.*, **106**, 754 (1959)
16. Thomas, N. T. and Nobe, K., *J. Electrochem. Soc.*, **116**, 1 748 (1969)
17. Cotton, J. B. and Bradley, H., *Chem. and Ind.* (Rev.), 643 (1958)
18. Cotton, J. B. and Downing, B. P., *Trans. Inst. Mar. Engrs.*, **69**, 311 (1957)
19. Feige, N. G. and Kane, R. L., *Metals Engr. Quart.*, **7**, 27–29 (1967)
20. Sims, M. H., *Power*, **112**, 890–96 (1968)
21. Greiss, J. C., *Corrosion*, **24**, 96–109 (1968)
22. France, W. D. and Greene, N. D., *Corrosion*, **24**, 247–51 (1968)
23. Takamura, A., *Corrosion*, **23**, 306–13 (1967)
24. Schlain, D., US Bureau of Mines, Report No. 4 965, April (1953)
25. Sedriks, A. J., *A.S.M. Trans. Quart.*, **61**, 625–27 (1968)
26. Harey, E. G. and Wearmouth, W. R., *Corrosion*, **25**, 87–91 (1969)
27. Sedriks, A. J., *Corrosion*, **25**, 325–28 (1969)
28. Rittenhouse, J. B. et al., *Trans. Amer. Soc. Metals.*, **51**, 871, 895 (1959)
29. Inglis, N. P., *Chem. and Ind.* (Rev.), 180 (1957)
30. *Engineering*, Jan. 6 (1967)
31. Obrig, H. and Ehle, J. C., *Chem. Process. Engrg.*, **50** (1969)
32. Hanson, B. H., *The Chemical Engineer*, 276–79, April (1978)
33. Barron, L. J., *Light Metal Age*, **14** Nos. 3 and 4, 16 (1956)
34. *Industr. Engng. Chem.*, **50**, 934 (1958)
35. Connolly, B. J., *Chem. Proc. Engng.*, **39**, 247 (1958)
36. Bomberger, H. B., *Industr. Engng. Chem.*, **49**, 1 658 (1957)
37. Carmichael, M. L., *Battelle Techn. Rev.*, **5** No. 12, 9 (1956)
38. *Steel*, **143** No. 26, 62 (1958)
39. Ishii, Y. and Hoskino, Y., *Chem. Engng.*, Tokyo, **21**, 559 (1957)
40. Frazer, G. T. et al., *Mat. and Meth.*, **43**, 112 (1956)
41. *Light Metal Age*, **17**, 27, Oct. (1959)
42. *Corrosion*, **15**, 82 (1959)
43. Hames, W. T., *Aircraft Prod.*, Lond., **20**, 369 (1958)
44. Savolainen, J. E. and Dlanco, R. E., *Chem. Engng. Prog.*, **53**, 78F (1957)
45. Peterson, C. L. et al., *Industr. Engng. Chem.*, **51**, 32 (1959)
46. Schmets, J. and Pourbaix, M., *Proceedings of the 6th Meeting of the International Committee for Electrochemical Thermodynamics and Kinetics*, Poitiers, 1954, Butterworths, London (1955)
47. Stern, M. and Wissenberg, H., *J. Electrochem. Soc.*, **106**, 755 (1959)
48. Stem, M. and Wissenberg, H., *J. Electrochem. Soc.*, **106**, 759 (1959)
49. Stem, M. and Bishop, C. R., *Amer. Soc. Met.*, Preprint No. 165 (1959)
50. Cotton, J. B., *Chem. and Ind.* (Rev.), 68 (1958)
51. Cotton, J. B., *Werkst. u Korrosion*, **11**, March 3 (1960)
52. Evans, L. S., Hayfield, P. C. S. and Morris, M. C., *Proc. 4th Intern. Congress on Metallic Corrosion*
53. Cotton, J. B., *Chem. and Ind.* (Rev.), 492 (1958)
54. Cotton, J. B., *Platinum Metals Rev.*, **2**, 45 (1958)

55. Shreir, L. L., *Platinum Metals Rev.*, **4**, 15 (1960)
56. Jones, J. C., *Prod. Finish.*, Lond., **12** No. 12, 81 (1959)
57. Rittenhouse, J. B. *et al.*, *Trans. Amer. Soc. Metals*, **51**, 871, 895 (1959)
58. Millaway, E. E. and Kleinman, M. H., *Corrosion*, **23**, 88-97 (1967)
59. Golden, L. B. *et al.*, *Trans. Amer. Soc. Metals*, **51**, 871, 895 (1959)
60. Schlain, D. and Kenahan, C. B., *Corrosion*, **14**, 405t (1958)
61. Andreeva, V. V. and Gluklova, *J. Appl. Chem.*, **11**, 390 (1961)
62. Cotton, J. B., *Chem. and Ind.*, 357-358 (1962)
63. Yoshida, S. *et al.*, *J. Govt. Mech. Lab.*, Tokyo, **10**, 2-21 (1956)
64. Yoshida, S. *et al.*, *J. Jap. Inst. Metals*, **21** No. 3, 183 (1957)
65. S.T.P. 397, A.S.T.M., 1916 Race St., Philadelphia, USA
66. Jackson, J. D. and Boyd, W. K., *The Science, Technology and Application of Titanium*, Pergamon, 267-281 (1966)
67. Imperial Chemical Industries, Ltd., Brit. Pat. 1 187 771 (15.4.70)
68. Kuhn, W. E., *More Zirconium Facts*, **1** No. 2, Carborundum Metals Company, 4 (1957)
69. Kuhn, W. E., *Chem. Engng.*, 156 (1960)
70. Gegner, P. J. and Wilson, W. L., *Corrosion*, **15**, 341t and 350t (1959)
71. Thomas, D. E., Proceedings of the First UN Conference on Atomic Energy, Geneva, 1955, **9**, Paper P/537, 407, United Nations, Geneva
72. Cotton, J. B. and Gallant, P. E., *Proceedings of the First International Congress on Metallic Corrosion,* London, April, 1961, Butterworths, London, 458 (1962)
73. Kass, S., *Corrosion*, **16**, 137 (1960)
74. O'Driscoll, W. G., Tyzack, C. and Raine, T., *Proceedings of the Second Conference on the Peaceful Uses of Atomic Energy*, Geneva, 1958, **5**, Paper P. 1 450, Geneva, 75 (1958)
75. Britton, C. F., *J. Nuc. Mat.*, **15** No. 4, 263-277 (1965)
76. Coleman, C. E. and Hardie, D., *J. Less Common Metals*, **11**, 168-85 (1966)
77. Rosa, C. J., *J. Less Common Metals*, **16**, 173-201 (1968)
78. Kuhn, W. E., *Corrosion*, **16**, 141t (1960)
79. Cotton, J. B., Localised Corrosion 676 NACE International Corrosion Conference series NACE-3 (eds R. W. Stahle, B. F. Brown, J. Kruger and A. Agrawal) (1974)
80. Schutz, R. W., Grauman, J. S., and Hall, J. A., 5th International Conference on Titanium, Munich, 2617-24 (1984)
81. Hayfield, P. C. S. and Hanson, B. H. *Chemical Processing*, **16** (5), 52 May (1970)
82. Coulter, M. O., *Modern Chlor-Alkali Technology*, S.C.I. 5.71 (eds J. H. Collins and J. H. Entwhistle) (1980)
83. Cotton, J. B. and Scholes, I. R., *Trans. Inst. Mar. Eng.*, **84**, 16. 538
84. Satoh *et al.*, 5th International Conference on Titanium, (eds G. Lutjering, U. Zwicker and W. Bunk) Munich, 1165-71 (1984)
85. Kobayashi *et al.*, 4th International Conference on Titanium, (eds H. Kimura and O. Izumi) Kyoto, 2613-22 (1980)
86. Fukuzuka *et al.*, 4th International Conference on Titanium, (eds H. Kimura and O. Izumi) Kyoto, 2631-38 (1980)
87. Hanson, B. H., 2nd International Conference on Titanium, (eds R. I. Jaffee and H. M. Burte) Boston, 2419-29 (1972)
88. Gehring, G. A. and Kyle, R. J. Paper 60, *Corrosion* (1982)
89. Brettle, J., *Metals and Materials*, 442-51 Oct. (1972)
90. Hanson, B. H., Seahorse, **97**, 46-7 Nov./Dec. (1986)
91. Hanson, B. H., *Materials and Design*, Vol vii, No. 6, 301-7 Nov./Dec. (1986)

5.5 Tantalum

General

Tantalum is one of the most versatile corrosion-resistant metals. Its corrosion behaviour can be compared with that of glass in most environments. This behaviour is attributed to the stable passive film of Ta_2O_5 produced on the surface during exposure.

The pure metal has a very high melting point (2996°C) and is blue-grey and like lead in appearance. It has a density of about twice that of carbon steel (16.6 g/cm^3) and a similar thermal conductivity. It is one of the refractory metals and suitable for high temperature application under protective conditions.

It can be readily cold worked, but hot working, however, must be avoided as the metal reacts with gases such as oxygen, nitrogen and carbon dioxide with resultant embrittlement. It can be machined, although care is necessary to obtain a good surface finish. The high strength, good workability and excellent corrosion resistance permit the use of very thin walled components, a commonly employed thickness in chemical plant being about 0.3 mm.

These properties, coupled with the metal's ability to promote bubble-type vapour formation on the surface when heating liquids, and dropwise condensation when condensing vapours, make the metal an ideal constructional material for heat-transfer equipment for use with strong acids.

The absence of corrosion, coupled with the fact that scale and other deposits appear to be dislocated by thermal cycling, result in a finish on tantalum heating surfaces that is as good as the original, even after 20 or 30 years in service, and also ensure that good heat-transfer properties are maintained throughout the life of the equipment. The use of tantalum for process equipment also ensures freedom from contaminations of the product.

The mechanical properties of tantalum are dependent on the previous history of the material and the manufacturer should be consulted if these properties are likely to be critical. The physical and some typical mechanical properties are listed in Tables 5.21 and 5.22. The effect of the temperature on the strength and elongation of tantalum sheet in vacuum is shown in Figs. 5.8 and 5.9.

Table 5.21 Physical properties of tantalum

			Thermal				Electrical	
Density (g/cm^3)	Melting point[1] (°C)	Boiling point[1] (°C)	Thermal neutron absorption cross section[3] (barn/atom)	Linear coeff. of expansion[2] (°C)	Thermal conductivity[2] (W/cm°C)	Specific heat[2] (J/g°C)	Electrical resistivity[1,2] (μΩ/cm)	Temperature coeff. of resistivity (°C)
16·6	2 996	5 425	21	20–500°C 6·6 × 10^{-6} 20–1 500°C: 9·0 × 10^{-6}	5·44 at 20°C 7·52 at 1 106°C	0·142 at 0°C 0·161 at 1 227°C	12·43 at 20°C 54·8 at 1 000°C	3·82 × 10^{-3} at 0–100°C 3 × 10^{-3} at 0–1 000°C

Table 5.22 Mechanical properties of tantalum

Modulus of elasticity[2,4] (GN/m^2)	Poisson's ratio	Yield stress (MN/m^2)	UTS[2,4] (MN/m^2)	Ductile to brittle transition temperature[2] (°C)	Recrystallisation temperature[4] (°C)	Stress relieving temperature[4] (°C)	Hardness[2] (VPN)		Stability[2] (% creep rate/h, min.)
							Annealed	Hard worked	
186 at 20°C 151 at 1 000°C	0·35	179–1 060 at 27°C and 44–310 at 500°C	689–1 034 at 20°C and 103–138 at 1 000°C	None detected down to −196	1 050–1 500	900	80–100	180	0·113 at 750°C, 96 GN/m^2

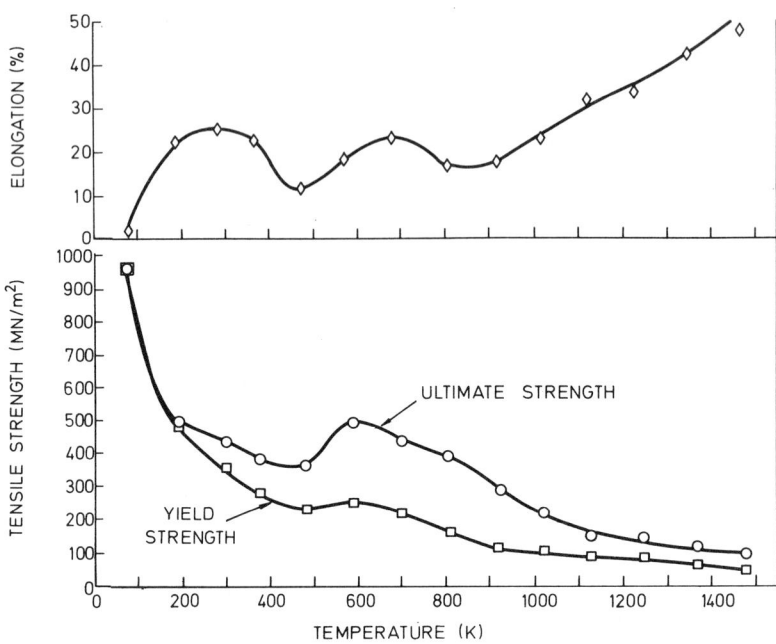

Fig. 5.8 Effect of temperature on the tensile strength and elongation of tantalum

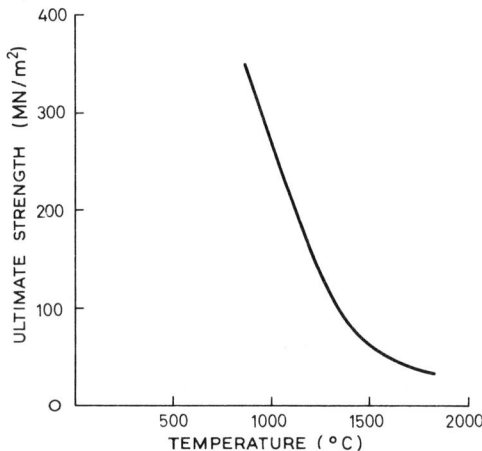

Fig. 5.9 Effect of temperature on the ultimate strength of tantalum

Methods of Fabrication

The high melting point and reactivity of tantalum with the permanent gases at high temperatures prevents conventional consolidation by melting and casting in air. The metal is in fact consolidated by vacuum sintering, vacuum-arc melting and electron-beam melting of powder compacts. Vacuum sintering yields metal of fine grain, whereas electron-beam melting yields softer coarse-grained metal which requires cold forging prior to rolling. Metal produced by all three techniques will absorb considerable cold work before annealing is necessary.

Rolling and swaging Vacuum-sintered bar can be cold rolled, and reductions up to 90% between anneals are possible. Arc-cast and electron-beam-melted material is generally forged at room temperature prior to rolling and swaging.

Drawing Tantalum has a tendency to gall and is normally anodised to provide a surface which will carry a drawing lubricant. Seamless tube is produced by cupping followed by drawing or by hollow shells.

Spinning Tantalum can be formed by all conventional spinning techniques, provided a lubricant such as tallow is employed, and can be spun into configurations which cannot be produced by other forming methods.

Machining Tantalum is readily machined using high-speed-steel tools, provided a lubricant such as trichloroethane is employed.

Blanking and cutting Tantalum can be blanked, cut and sheared using similar equipment and techniques to those used for austenitic stainless steel.

Joining Tantalum can be joined by riveting, brazing and welding; however, due to the good properties of welded joints the former techniques are seldom used.

Welding Because of the reactivity of the hot metal with the permanent gases, conventional welding techniques cannot be used. In general the practical methods are restricted to tungsten-electrode inert-gas (TIG), resistance, electron-beam (EB) and plasma-arc welding. To ensure statisfactory TIG welds, welding should be done in an inert-gas-filled chamber. Material thinner than 0·5 mm cannot readily be TIG welded and resistance welding has to be used. Spot welding can be carried out in air and under water. EB welding gives a contamination-free narrow weld and heat-affected zone, irrespective of material thickness, and plasma-arc welding has been used in 0·05–1·0 mm sheet and gives a weld with similar properties to EB welds.

Economical Considerations

The relatively high cost of tantalum has been a limiting factor in its use. Fabrication techniques, in which thin linings of tantalum are used, result in equipment at a much lower cost than an all-tantalum construction.

The long life and reliability of tantalum equipment in severe-corrosion applications often more than offsets its higher initial costs. Therefore, a new situation has been created for utilising the benefits of tantalum products. When tantalum is properly applied, it can often be justified not only on a field replacement basis but also on initial installation.

Corrosion Resistance

Tantalum's corrosion resistance is due to the presence of a thin continous surface film of tantalum pentoxide (Ta_2O_5). Thus the metal is passive and approaches the inertness of gold and platinum in a large number of very aggressive environments. The metal itself in the active state lies below zinc in the thermodynamic nobility table presented by Pourbaix[5]. In the passive state, its oxide film, however, puts it just below rhodium and above gold in the Pourbaix practical nobility table. The oxide film adheres well and appears to be free from porosity. At elevated temperatures a suboxide layer develops between the metal and the upper oxide film (Ta_2O_5) interface. This suboxide layer is not stable at temperatures higher than 425°C. When heated above this temperature only the stable pentoxide exists and the internal stress set up by the metal during oxide conversion causes the protective oxide film to flake and spall. Owing to this phenomenon, high temperature application of tantalum is limited in atmospheric environments under oxidative conditions.

Available reports indicate that tantalum is an effective passive metal in most of the chemical environments, at ambient temperature and up to about 100°C. There are only a few environments in which tantalum corrodes in a rate higher than 1mm/y, at temperatures up to about 100°C.

Fluorine and Fluoride Environments

Tantalum is severely attacked at ambient temperatures and up to about 100°C in aqueous atmospheric environments in the presence of fluorine and hydrofluoric acids. Flourine, hydrofluoric acid and fluoride salt solutions represent typical aggressive environments in which tantalum corrodes at ambient temperatures. Under exposure to these environments the protective Ta_2O_5 oxide film is attacked and the metal is transformed from a passive to an active state. The corrosion mechanism of tantalum in these environments is mainly based on dissolution reactions to give fluoro complexes. The composition depends markedly on the conditions. The existence of oxidizing agents such as sulphur trioxide or peroxides in aqueous fluoride environments enhance the corrosion rate of tantalum owing to rapid formation of oxofluoro complexes.

Hydrogen Embrittlement

Tantalum has a high solubility for hydrogen, forming two internal hydrides, but the exact mechanism of their formation is not precisely known.

There is evidence that embrittlement can occur at temperatures below 370°C. Clauss and Forestier[6] in fact reported that embrittlement can occur when tantalum is deformed in contact with hydrogen at room temperature. Examination of the literature indicates that one of the few defects in the resistance of tantalum to corrosion in aqueous media lies in its susceptibility to hydrogen embrittlement. Although it is inert in concentrated hydrochloric at temperatures as high as 110°C, some reaction occurs at appreciably higher temperatures and sufficient hydrogen may be absorbed to cause embrittlement. Since it becomes cathodic in galvanic cell circuits with virtually all constructional metals, it must be electrically insulated from other metals with which it could come into contact in a common electrolyte, in order to prevent hydrogen discharge and entry into the metal. Anodising the tantalum, or addition of selected oxidising agents to the environment[19] are proposed to reduce hydrogen embrittlement.

Reactions with Gases: Hydrogen, Nitrogen, Oxygen

Tantalum and tantalum alloys react with hydrogen, nitrogen and oxygen at temperatures above 300°C. Hydrogen is dissolved in the metallic matrix above 350°C[8] and evolved at higher temperatures of about 800°C[9,10]. The dissolved hydrogen embrittles the tantalum and its alloys. This effect can be used to prepare tantalum powder.

The reaction with small amounts of nitrogen results in an increased hardness, tensile strength and electrical resistivity. Tantalum is embrittled by higher amounts of nitrogen. The reaction takes place at temperature above 400°C[8]. Nitrides among other phases form at the surface, but at higher temperatures these decompose and all the nitrogen is liberated at 2100°C[11].

Generally, the most important reaction is that of tantalum with oxygen, since it tends to form oxides when heated in air. Reaction starts above 300°C and becomes rapid above 600°C[19]. The scale is not adherent, and if the oxidised material is heated above 1000°C oxygen will diffuse into the bulk of the material and embrittle it. At 1200°C catastrophic oxidation attack takes place at a rate of about 150 mm/h[13]. Oxygen is not driven off by heating alone, but in vacuum above 2300°C it is removed as a suboxide. The first step of the conversion mechanism of tantalum into oxide was shown to occur by the nucleation and growth of small plates along the $\{100\}$ planes of the BCC metal[21,22].

The presence of a few atomic percent of oxygen in tantalum increases electrical resistivity, hardness, tensile strength, and modulus of elasticity, but decreases elongation and reduction of area, magnetic susceptibility, and corrosion resistance to HF[23].

The main protective method against atmospheric catastrophic attack is surface coatings of silicides, and aluminides[24].

Atmospheric Conditions

Tantalum has a high resistance to general outdoor atmospheres. Tantalum and the Ta-10W alloy are virtually immune to sea water at ambient conditions and tantalum is only tarnished in oxygenated sea water at 26°C.

Acid Media

Tantalum is practically inert to nitric acid at all concentrations and temperatures. The corrosion rate in 70% acid at 270°C is about 0·1 mm/y. It also

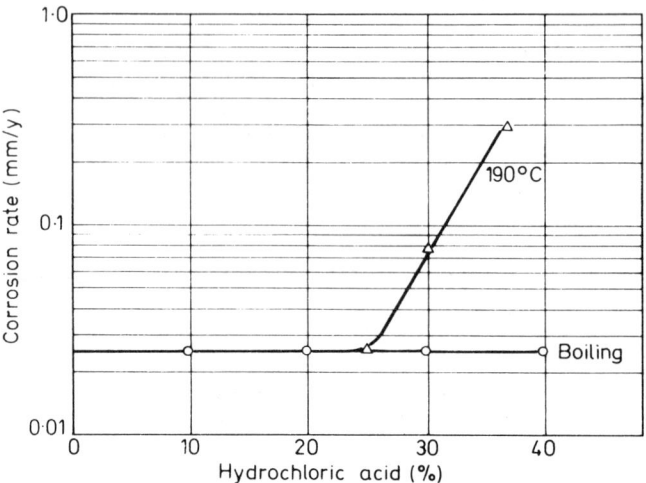

Fig. 5.10 Corrosion of tantalum by hydrochloric acid[18]

resists fuming nitric acids up to at least 150°C and hydrochloric acid at all concentrations up to 190°C though above 25% the corrosion rate rises rapidly and, in addition, the entry of hydrogen caused embrittlement (Fig. 5.10). Tantalum is completely inert to hydrochloric acid mixtures even in the presence of sulphuric acid and its salts in all proportions and concentrations up to boiling point. It is not corroded by phosphoric acid at concentrations up to 85% and temperatures up to 200°C, provided flouride ions, often found in commercial acid, do not exceed 5 p.p.m. It is practically inert to perchloric acid, chromic acid, hypochlorous acid, hydrobromic acid, hydriodic acid and most organic acids provided they do not contain flourides, flourine or free sulphur trioxide. One exception to flouride attack appears to be in certain chromium plating baths in which fluoride is used as the catalyst, the corrosion rate in 40% CrO_3 plus 0·5% F at 55-60°C being 0·0005 mm/y.

It is completely inert to 98% sulphuric acid to at least 160°C and to even higher temperatures at lower concentrations. Practically, it may be used to 200°C in all concentrations and to 225-250°C at concentrations between 80% and 90%. Fuming sulphuric acid containing sulphur trioxide attacks tantalum at room temperature as do hydrofluoric and fluorosilicic acids.

Specific information is given in Figs. 5.10, 5.11 and Table 5.23.

Alkali Media

Sodium hydroxide (NaOH) and potassium hydroxide (KOH) solutions do not dissolve tantalum, but tend to destroy the metal by formation of successive layers of surface scale. The rate of the destruction increases with concentration and temperature. Damage to tantalum equipment has been experienced unexpectedly when strong alkaline solutions are used during cleaning and maintenance.

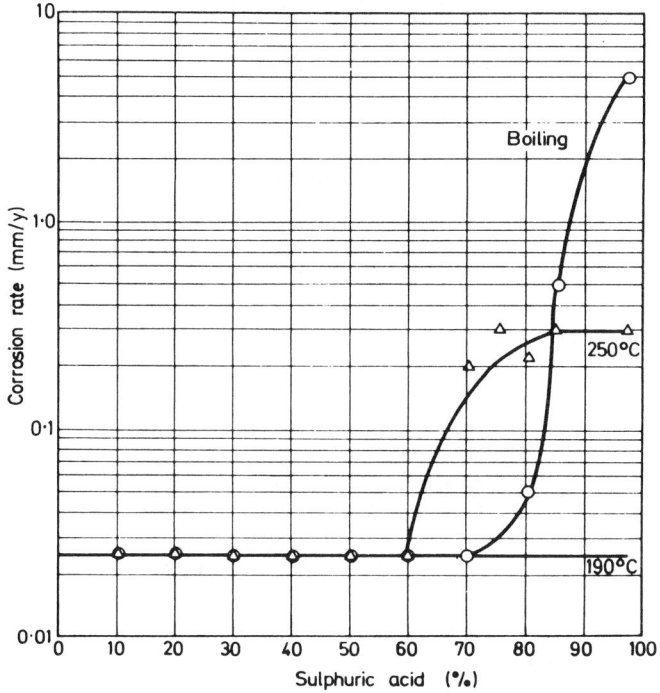

Fig. 5.11 Corrosion of tantalum by sulphuric acid[18]

Table 5.23 Corrosion by miscellaneous acids[14, 15, 17]

Acid	Concentration (%)	Temperature (°C)	Corrosion rate (mm/y)
Chromic	All concentrations	100	<0·08
	10–50	Boiling	<0·12
	36·5	90	<0·12
	70	25	<0·12
	70	100	0·12–0·9
Phosphoric (air free)	10–85	50–250	<0·12
	96	215–220	0·033
	96	225–230	0·094

Tantalum is attacked, even at room temperature, by concentrated alkaline solutions. However, tantalum is fairly resistant to dilute alkaline solutions. In one long-term exposure test in a paper mill, tantalum suffered no attack in a solution with a pH of 10. Attack by alkaline solutions is summarised in Table 5.24.

It suffers mild attack in fused sodium hydroxide at about 315°C, but is severely attacked at 535°C; it is also strongly attacked by fused potassium hydroxide at 360°C and sodium carbonate at 850°C.

Table 5.24 Corrosion by alkalis [14-17, 25]

Alkali	Concentration (%)	Temperature (°C)	Corrosion rate (mm/y)	
Sodium hydroxide	10	Room	0·00025	No noticeable embrittlement
	5	98	0·0033	
	10	98	0·0082	
	50	38–57	0·9–1·25	
	50	120	>1·3	
	73	113–129	0·9–1·25	
Potassium hydroxide	5	100	<0·08	
	10	Boiling	<0·12	
	50	27	0·9–1·25	
	40	100	>1·25	

Salts in Aqueous Solutions

Tantalum has excellent resistance to virtually all salts including chlorides (especially cupric and ferric chloride), sulphates, nitrates and salts of organic acids, provided (a) they do not contain fluorides, fluorine and free sulphur trioxide, or (b) hydrolyse to produce strong alkalis.

Other Environments

Halogens Although tantalum is severely attacked by flourine at room temperature it does not react with liquid chlorine, bromine and iodine up to 150°C and the metal suffers no appreciable attack in wet or dry bromine, chlorine and iodine below 250°C. It is virtually uncorroded by hydrogen bromide and hydrogen chloride below 370°C, attack starting at about 375 and 410°C respectively.

Liquid metals The corrosion resistance of tantalum depends on the metallurgical interaction between the liquid metal and tantalum. Generally good resistance is observed in low melting point liquid metals.

Bismuth Liquid bismuth has little action on tantalum at temperatures below 1000°C[26, 27-29], the rate of attack at 870°C being less than 0.13 mm/y, and exerts not detrimental effects on the stress rupture properties of tantalum at 815°C, but is causes some intergranular attack at 1000°C[31].

Calcium Tantalum is only slightly attacked by calcium at 1200°C, the rate of attack being 0.37 mm/y[32].

Gallium The resistance of tantalum to molten gallium is considered to be good up to 450°C, but poor above 600°C, the rate of attack at 482°C being less than 0.13 mm/y.

Lead Tantalum is highly resistant to liquid lead at temperatures up to 1000°C[26] with a rate of attack of less than 0.025 mm/y[33]. It exhibits no detrimental effects when stress rupture tests are conducted in molten lead at 815°C[33].

Lithium Tantalum possesses good resistance to molten lithium up to 1000°C[26,36] the rate of attack at this temperature being less than 0·13 mm/y.

Mercury In static tests tantalum showed good resistance to mercury at temperatures up to 600°C[26], the rate of attack at 590°C being less than 0·13 mm/y.

Silver Tantalum is only slightly attacked by silver at 1200°C, the rate of attack being 0·25 mm/y[32].

Sodium, potassium and sodium-potassium alloys Liquid sodium, potassium or alloys of these elements have little effect on tantalum at temperatures up to 1000°C[26,31,35], the rate of attack of 1200°C being less than 0·13 mm/y, but oxygen contamination of sodium causes an increase in corrosion[36,37]. In fact, if oxygen is present, attack may commence as low as 250°C. Sodium does not alloy with tantalum[30].

Thorium-magnesium In static tests the thorium-magnesium eutectic had no appreciable effect on tantalum at 1000°C[38], the corrosion rate being less than 0·13 mm/y.

Tin Static tests in liquid tin at 1740°C for 1 h showed that some tantalum dissolved (0·33% Ta was found in the tin); the corrosion rate at 260°C is 1·3 mm/y.

Uranium Short-term tests indicate the practical upper limits for tantalum as a container material for uranium to be 1450°C[40]. However, attack below these temperatures is significant, since a tantalum crucible with a wall 1·52 mm thick was completely corroded in 50 h in 1275°C.

Zinc Molten zinc attacks tantalum at a significant rate above 450°C[42], the rate of attack at 370°C being less than 0·13 mm/y. At 750°C the attack becomes appreciable[43].

Aluminium, magnesium and cadmium Tantalum is attacked at a corrosion rate higher than 1 mm/y by these molten metals. The related temperatures are: aluminium 660°C, magnesium 650°C, and cadmium 370°C.

Organic compounds In general, tantalum is completely resistant to organic compounds and is used in heat exchangers, spargers, and reaction vessels[60].

Fine chemicals, foods, and pharmaceuticals The immunity of tantalum to corrosion also ensures product purity and undesired side reactions in the processing of fine chemicals, foods, and pharmaceuticals.

Body fluids and tissues Tantalum is a very stable passive metal and completely inert to body fluids and tissues. Bone and tissue do not recede from tantalum, which makes it attractive as an implant material for the human body[45,60].

Carbon, Boron, and Silicon Tantalum reacts at elevated temperatures (not stated) directly with carbon, boron, and silicon to form Ta_2C and TaC, TaB and TaB_2, and $TaSi_2$, respectively[45].

Phosphorous Tantalum phosphides, TaP and TaP_2, are formed by heating tantalum filings in phosphorous vapour at 750 to 950°C[45].

Sulphur Tantalum reacts with sulphur or H₂S at red heat to form tantalum sulphide Ta_2S_4.

Sulphur Trioxide Tantalum is attacked by sulphur trioxide at ambient conditions at rates higher than 1 mm/y.

Selenium and Tellerium Tantalum is attacked by selenium and tellurium vapours at temperatures higher than 80°C. Only slight attack is observed on the metal by liquid selenides and tellurides of ytirum, the rare earths, and uranium at temperatures of 1300 to 2100°C, and tantalum is considered to be a satisfactory material in which to handle these intermetallic compounds.

Oxidizing gases Tantalum reacts with oxidising gases at elevated temperatures.

Galvanic Effects

As indicated previously, a galvanic couple in which tantalum is the cathode can prove disastrous because of embrittlement. On the other hand, if tantalum is the anode it passivates so readily in most environments that no damage occurs, and the galvanic current drops to a very low value. Haissinksky[46] studied couples of tantalum with platinum, silver, copper, bismuth, antimony, molybdenum, nickel, lead, tin, zinc and almiunium in 0·1 N H_2SO_4. Except when tantalum was coupled to zinc or aluminium it was the more negative member (anode) of the couple. However, the galvanic current rapidly decreased as the tantalum passivated. In hydrofluoric acid tantalum was again more positive than zinc and aluminium, but more negative than platinum, silver, copper, antimony, nickel and lead. The latter six couples result in high steady-state currents, because tantalum corrodes rather than passivates in fluoride solutions. The apparently anomalous behaviour of tantalum coupled with bismuth or iron in hydrofluoric acid is explained by Haissinksky as being due to the formation of insoluble fluorides on the surfaces of the bismuth or iron electrodes.

Time is an important factor in determining whether tantalum will be damaged by galvanic effects[47]. This is illustrated in Table 5.25. Aluminium becomes negative in dilute NaCl, HCl and NaOH after 1 h, and the polarity of other metals coupled with tantalum changes with time. The details of these tests are not available, and it is diffcult to assess the significance of the actual potential values. In practice it would be dangerous to depend on laboratory tests to provide information as to whether tantalum is negative in a given galvanic couple situation.

Alloys of Tantalum

Tantalum-Niobium Alloying tantalum usually decreases the corrosion resistance of the metal due to metallic contamination of the Ta_2O_5 passive film. The corrosion rates in HCl and H_2SO_4 environments increase roughly

Table 5.25 Potential of tantalum in tantalum-metal couples[47]

Time (min)	Coupled metal	Potential of tantalum (mV)		
		3% NaCl electrolyte	1% HCl electrolyte	1% NaOH electrolyte
0·5	Hastelloy B	−34·3	−30·5	−104
60	Hastelloy B	+2·6	+1·0	−4·3
0·5	Hastelloy C	−50·0	−72·0	−136
60	Hastelloy C	−2·8	−2·7	−4·4
0·5	Nickel	−65·1	−36·0	−180
60	Nickel	−2.5	+0·1	−4·4
0·5	Lead	−1·6	+51·0	−10
60	Lead	+60·1	+52·9	+29·8
0·5	Aluminium	+128	+275	+621
60	Aluminium	+230	+208	+509
0·5	304 stainless steel	−52·0	−	−
60	304 stainless steel	−2·9	−	−

in proportion to the niobium content in the alloy. Only tantalum contents higher than 60% appears promising for boiling 70% sulphuric acid. Additions of zirconium, hafnium, chromium or vanadium to binary tantalum–niobium alloy enhance the corrosion rate of the binary alloys[48, 54].

Tantalum–Molybdenum Schumb, Radtke and Bever[55] studied the corrosion resistance of tantalum–molybdenum alloys that form a continuous series of solid solutions. The results of tests of up to 500 hours duration (Table 5.26) indicate the corrosion resistance of the alloy to be substantially that of tantalum, provided its concentration exceeds 50%.

Table 5.26 The corrosion behaviour of tantalum–molybdenum alloys in concentrated sulphuric and hydrochloric acids at 55°C; solutions saturated with oxygen

Tantalum (atomic %)	Average corrosion rate (mg dm^{-2}d^{-1})	
	Conc. H$_2$SO$_4$	Conc. HCl
0·	0·8	1·8
10·1	0·9	1·7
20·1	0·8	1·8
30·0	1·0	0·9
40·0	0·9	1·0
50·0	0·0	1·0
61·2	0·0	0·0
71·5	0·0	0·0
82·8	0·0	0·0
91·4	0·0	0·0
100	0·0	0·0

Tantalum–Tungsten Braun, Sedlatschek and Kieffer[56] examined tantalum–tungsten alloys in 50% potassium hydroxide up to 80°C and in 20% hydrochloric acid at 20°C. In the alkaline solution the corrosion rate was a maximum when the tantalum was over 60 at.%. In hydrofluoric acid the alloy system exhibited the relatively low corrosion rates associated with tungsten until the tantalum concentration exceeded 80 at.%.

Staples and Galloway[58] examined the corrosion resistance of the Ta-10W alloy in hydrochloric, sulphuric and nitric acids and in sodium hydroxide solution and found that there was virtually no difference in corrosion rate between the alloy and pure tantulum in 10-30% hydrochloric acid to 175°C, 70-90% sulphuric acid to 205°C and 60% nitric acid to 190°C. In addition, in 5% sodium hydroxide the alloy had a lower corrosion rate than pure tantalum. This alloy also has yield and ultimate tensile strengths approximately double those of pure tantulum but it is considerably more difficult to work and fabricate.

Tantalum-Titanium Bishop[57] examined the corrosion resistance of this alloy system in hydrochloric, sulphuric, phosphoric and oxalic acids and found that alloys containing up to about 50% titanium retained much of the superlative corrosion resistance of tantalum. Under more severe conditions, a titanium content of below 30% appears advisable from the standpoint of both corrosion resistance and hydrogen embrittlement, although contacting or alloying the material with noble metals greatly decreases the latter type of attack. Tantalum-titanium alloys cost less than tantalum because titanium is much cheaper than tantalum, and because the alloys are appreciably lower in density. These alloys are amenable to hot and cold work and appear to have sufficient ductility to allow fabrication.

Other Tantalum alloys It has been observed that the presence of a small amount of iron or nickel, for example, in a tantalum weld makes that site subject to about the same acid attack as would be experienced by iron or nickel alone. Galvanic action, as well as simple chemical attack, is undoubtedly involved.

Industrial Applications of Tantalum

About one half of the tantalum used in the chemical industry is employed on sulphuric acid duties, mainly in the form of multiple-tube bayonet heaters.

A further quarter is operating in media containing chlorine or its derivatives and it is in this field and that of the bromine industry that the first tantalum chemical plant was put on steam in the early 1930s. A large variety of tantalum equipment is in fact used in the synthesis and handling of hydrochloric acid and tantalum heat-exchanger units are employed in a wide range of processes involving nitric acid.

Tantalum, because of its negligible corrosion rate, finds use in the pharmaceutical and food manufacturing industries where even the smallest amount of metallic impurity cannot be tolerated in many products.

By virtue of the high breakdown potential of the oxide film (approximately 155 V in sea water and 280 V in low conductivity water of pH = 7) tantalum has found use as a substrate for platinum in impressed-current cathodic-protection anodes, which can be used at high impressed voltages (50 V) and high current densities. However, because of its lower cost, niobium is preferred for systems that have to operate at high voltages

Reconcentration of sulphuric acid A very large amount of tantalum heater surface has been installed in plants for the reconcentration of diluted sulphuric acid arising from metal pickling, oil refinery operations and from petrochemical processes producing alcohols and ketones. Since reconcentration provides a means of overcoming a waste of disposal problem, the use of such plants is expanding[17].

Plants producing and handling halogens and halogen compounds Tantalum finds extensive use in the production and handling of hydrochloric and hydrobromic acid, chlorine and bromine and many of their derivatives. Absorbers, coolers and heaters which show considerable advantages in terms of heat-flux capabilities and corrosion resistance have been used on hydrochloric acid duties for over 40 years and condensers have been used in bromine plants for at least the same period. Typical applications of tantalum in the bromine and chlorine industries are listed in Table 5.27[60,61,62].

Table 5.27 Typical applications of tantalum equipment in bromine and chlorine industry

Product	Operation	Equipment
Benzyl chloride	Chlorination of toluene	HCl absorbers
Bromine, crude	Steam and bromine condensation	Condensers
Bromine, pure	Purification from chlorine and organics	Boilers, condensers, complete purification plants
Chlorine	Brine cooling	Heat exchangers
Chloroacetic acid	HCl absorption	HCl absorbers
Chlorobenzene, also monochlorobenzene and paradichlorobenzene	Chlorinator operation HCl absorption	Condensers, HCl absorbers
Ethyl bromide	Alcohol bromination	Special HBr reactor: anhydrous HBr plant
Halogens (except fluorine)	Chlorine, bromine, iodine generators and recovery systems	Bayonet Heaters, coils, condensers, regulator parts, thermometer wells
Hydrobromic acid	Generation	Bayonet heaters, coils, condensers, synthetic HBr plants
Hydrochloric acid	Production, purification, recovery, processing	Bayonet heaters, heat exchangers, coils, condensers, HCl absorbers, synthetic HCl plants, acid coolers, gas coolers, chlorine burners, strippers, thermometer wells
Hydrochloric acid C.P.	Distillation	Bayonet heaters, coils, condensers, complete stills
Hydrochloric acid, anhydrous	Production	Absorbers, coolers, strippers, chlorine burners, complete plants
Hydriodic acid	Generation and recovery	Bayonet heaters, coils, condensers
Hydrogen chloride	Production	Absorbers, coolers, strippers, chlorine burners, complete plants
Iodine	Recovery from sour brines	Coils

Plants handling aqua regia Aqua regia is used extensively in the extraction and refining of the precious metals, and tantalum, as one of the few metals resistant to this medium, is used for dissolution/evaporation pans, reactor lids and all immersed ancillary equipment.

Nitric acid plants Tantalum heat exchangers and sparge pipes find extensive use in plants producing high purity nitric acid, ammonium nitrate and terephthalic acid.

J. BENTLEY

REFERENCES

1. Lyman, T., *Metals Handbook*, A.S.M. (1961)
2. Hampel, C.A., *Rare Metals Handbook*, 2nd ed., Reinhold (1961)
3. MacDonald, R.N. and Boucom, H.H., *Nucleonics*, **20**, 158, Aug. 8th (1962)
4. Chelius, J., *Machine Design*, March 1st (1962)
5. Pourbaix, M., *Atlas of Electrochemical Equilibria in Aqueous Solutions*, Pergamon Press (1966)
6. Clauss, A. and Forestier, *Compt. Rend. Acad. Sci.*, Paris, **246**, 3241-3243 (1958)
7. Bishop, C.R. and Stern, M., *J. Metals*, N.Y., **13** 144 (1961)
8. Gulbransen, F.A. and Andrew, K.F., *J. Metals*, New York, **180**, 586 (1950) and *Trans. A.I.M.E.*, **188**, 586-599 (1950)
9. Sieverts, A. and Berger, E., *Ver. Deutsch. Chem. Ges.*, **44**, 2394-2402 (1911)
10. Sieverts, A. and Bruning, H., *Z. Physik. Chemie.*, **A174**, 365-369 (1935)
11. Wright, D.A., *Nature*, Lond., **142**, 794 (1938)
12. Schmidt, F.F. and Ogden, H.R., *The Engineering Properties of Tantalum and Tantalum Alloys*, DMIC Report No. 189, DMIC Batelle Memorial Institute, Sept. 13th (1963)
13. Jaffee, R.J., *Proceedings of an International Symposium on High Temperature Technology*, Asilomer, California, 1959, McGraw-Hill, New York, **61** (1960)
14. Takamara Akira, *Corrosion Engineering*, **16**, No. 3, March (1967)
15. *Corrosion Tables of Special Materials and Rare Metals*, Jacob and Korves GmbH, Ahlen
16. Barto, R.L. and Hurd, D.T., *Research/Development*, Nov., 26-30 (1966)
17. *Corrosion Data Survey*, compiled by Nelson, G.A., National Association of Corrosion Engineers (1967)
18. Bishop, C.R., *Corrosion*, **19**, Sept. 308 (1963)
19. Stern, M. and Bishop, C.R., Corrosion and Electrochemical Behavior, in *Columbium and Tantalum*, Sisco, F.T. and Epremian, E., Ed., John Wiley & Sons (1963)
20. Dreyman, E.W., Precious Metal Anodes: State of the Art, *Mater. Protect. Perform.*, **II** No. 9, Sept 1972, pp. 17-20, John Wiley & Sons (1963)
21. Gebhardt, E. and Seghezzi, H.D., *Z. Metallkd.*, **50**, 248 (1959)
22. Bakish, R., *J. Electrochem. Soc.*, **105**, 71 (1958)
23. Schmidt, F.F., Klopp, W.D., Albrecht, W.M., Holden, F.C., Ogden, H.R. and Jafee, R.J., Technical Report WADD-TR-59-13, United States Air Force (1959)
24. Hallowell, J.B., Maykuth, D.J. and Ogden, H.R., Silicide Coatings for Tantalum and Tantalum-Base Alloys, in *Refractory Metals and Alloys III: Applied Aspects*, **30**, Part 2, American Institute of Mining, Metallurgical and Petroleum Engineers, Dec. (1963)
25. Tingley, I.I. and Rogers, R.R., 'Corrosion of Niobium and Tantalum in Alkaline Media', *Corrosion*, **21**, April, 132 (1965)
26. Miller, E.C., Chapter 4 of *Liquid Metals Handbook*, Atomic Energy Commission, Navy Department, Washington, D.C., 144-183 (1952)
27. Stoughton, L.D. and Sheetan, T.V., *Mechanical Engineering*, **78**, 699-702 (1956)
28. Seifert, J.W., *Nuclear Sci. Abstracts*, **13**, No. 19667, 2635, 2636, (1959)
29. Lloyd, E.D., '1958 High Melting Metals', *Plansee Proceedings*, Metallwerk Plansee A.G., Reutte, Tyrol, 249-256 (1959)
30. Parkman, R. and Shepard, O.C., US Atomic Energy Commission Pub. ORD-45, June 11th (1951)
31. Reed, E.L., *J. Am. Ceramic Society*, **37**, 146-153 (1954)
32. US Atomic Energy Commission Pub. No. ISC-607, Quarterly Summary Research Report in Metallurgy for Jan. to March, Ames Laboratory, Ames, Iowa (1955)

33. Wilkinson, W.D., Hoyt, E.W. and Rhode, H.V., US Atomic Energy Commission Pub. ANL-5449 (1955)
34. Cunningham, J.E., US Atomic Energy Commission Pub. No. ORNL-CF-51-7-135, July 23rd (1951)
35. Cottrell, W.B. and Mann, L.A., Nucleonics, 12, Dec., 22-25 (1954)
36. Wyatt, L.M. and Dickinson, F.S., *Welding and Metal Fabrication*, 25, 378-385, 396 (1957)
37. Raines, C.E., Weaver, C.V. and Strang, J.H., US Atomic Energy Commission Pub. No. BMI-1284, Aug. 21 (1958)
38. Adams, R.M. and Sittig, M., in *Liquid Metals Handbook*, Sodium NAK Supplement, Atomic Energy Commission, Navy Dept., Washington, D.C., 14, (1955)
39. US Atomic Energy Commission Pub. ISC-396, Quarterly Summary Research Report in Metallurgy for April to June, Ames Laboratory, Ames, Iowa (1953)
40. McIntosh, A.B. and Bagley, K.Q., *J. Inst. Met.*, 84, 251-270 (1955-1956)
41. US Atomic Energy Commission Pub. No. ANL-5560, Chemical Engineering Division, Summary Report, Jan. to March, Argon National Laboratories, Argon, III (1956)
42. Hodge, W., Evans, R.M. and Haskins, A.F., *J. of Inst. Metals*, 7, 824-832 (1955)
43. Dekany, J.P., Lavendal, H.W., Winsch, I.O. and Pavlik, J., US Atomic Energy Commission Pub. No. ANL-5996, Summary Report, Jan. to March (1959)
44. Taylor, D.E., Tantalum and Tantalum Compounds, in *Encyclopaedia of Chemical Technology*, 19, 2nd ed., John Wiley & Sons, 630-652 (1969)
45. Hampel, C.A., Ed., *Tantalum in Rare Metals Handbook*, 2nd ed., Reinhold (1967)
46. Haissinsky, M., *Metaux et Corrosion*, 23, 15-18 (1948)
47. Wehrmann, R., *Corrosionomics*, Fansteel Metallurgical Corp., Sept. (1956)
48. Miller, G.L., *Tantalum and Niobium*, Butterworth, London (1959)
49. Argent, B.B., *J. Inst. Met.*, 85, 547-551 (1957)
50. Kieffer, von R., Bach, H. and Stempkowski, I., *Werkst. u. Korrosion*, Sept. 782-784 (1967)
51. Miller, G.L., *Tantalum and Columbium*, Academic Press (1959)
52. Mosolov, A.V., Nefedova, I.D. and Klinov, I.Y., Corrosion and Electrochemical Behaviour of Niobium-Tantalum Alloys in HCl at Elevated Temperatures and Pressures, *Z. Metallov.*, (in English), 4, No. 3, May-June 248-251 (1968)
53. Gulyaev, A.P. and Georgieva, I.Y., Corrosion Resistance of Binary Niobium Alloys, *Z. Metallov.*, (in English) 1 No. 6, Nov.-Dec. 652-657 (1965)
54. Lupton, D. and Aldinger, F., Possible Substitutes for Tantalum in Chemical Plant Handling Mineral Acids, in *Trends in Refractory Metals, Hard Metals, and Special Materials and Their Technology*, Proceedings of the 10th Plansee Seminar, Reutte, Austria 101-130 (1981)
55. Schumb, W.C., Radtke, S.F. and Bever, M.B., *Ind. Eng. Chem.*, 42, 826-829 (1950)
56. Braun, H., Sedlatschek, K. and Keiffer, R., *J. of the Less Common Metals*, 1, No. 6, Dec., 413-419 (1959)
57. Staples, B.G. and Galloway, W.S., Jr., *Materials Protection*, July, 34-39 (1968)
58. Bishop, C.R. and Powell, R.L., *Corrosion*, 18, No. 6, 205t-210t, June (1962)
59. Bishop, C.R. and Stern, M., *J. Metals*, N.Y., 13, 144 (1961) and *Corrosion*, 17, Aug., 379 (1961)
60. Metal Handbook, 13, 725 (1987)
61. Itzhak, D., Internal reports related to the environmental behaviour of tantalum in the bromine company production plants, Bromine Comp. Ltd., Israel (1990)
62. Straze, H. and Itzhak, D. 'The Corrosion Resistance of Ti, Ta and Nb Under Induced Potential in Halide Environments', 11th Int. Con. Cong., Florence, Italy, April (1990)

BIBLIOGRAPHY

Bachman, W.T., 'Do Not Overlook Refractory Metals for Corrosive Environments', *Materials in Design Engineering*, 106, Nov. (1966)
Charles, R., 'A Champion for Tantalum', *Metal Bulletin*, Dec. 22nd (1970)
Cheiius, J., 'Understanding the Refractory Metals', *Chemical Engineering*, Dec. 10, 178 (1962)
Cooper, M.J. and Mannox, D.J., 'Tantalum and its Alloys as Engineering Materials', *Proc. Mech. Eng.*, 180, Part 3D (1965-1966)
Sisco, F.T. and Epremian, E., *Columbium and Tantalum*, Wiley (1963)
Vorhis, F.H., 'Exotic Metals for Process Equipment', *Chemical Engineering*, Oct 25, 182 (1965)

5.6 Uranium

For many years the corrosion of uranium has been of major interest in atomic energy programmes. The environments of importance are mainly those which could come into contact with the metal at high temperatures during the malfunction of reactors, *viz.* water, carbon dioxide, carbon monoxide, air and steam. In all instances the corrosion is favoured by large free energy and heat terms for the formation of uranium oxides. The major use of the metal in reactors cooled by carbon dioxide has resulted in considerable emphasis on the behaviour in this gas and to a lesser extent in carbon monoxide and air.

Water

Uranium reacts with water to form uranium dioxide, hydrogen and uranium hydride. However, the hydride usually has only an ephemeral existence and reacts itself with water to form uranium dioxide and hydrogen. Reaction rates decrease with pH below 2 and it has been suggested that the solid products form by the inward diffusion of hydroxyl ions through the oxide[1]. The oxide is produced mainly as a non-adherent powder, and a linear rate law is obeyed. Autoclave tests have demonstrated that the rate constant increases markedly with temperature up to at least 300°C (Figure 5.12*)[2]. The corrosion rate is also influenced by dissolved gases in the water. In particular the presence of oxygen decreases substantially the reaction rate[2] but makes the metal susceptible to crevice corrosion and pitting attack. The inhibition is most marked at the lower temperatures at which the oxygen solubility is highest and the hydrogen product is not sufficient to reduce the oxygen content locally. The oxygen could exert its influence by being adsorbed preferentially on the oxide[3] or by removing the disruptive influence of the formation of uranium hydride. An alternative view of such 'hydrogen effects' relates them to changes in the electrical properties of the oxide detected by impedance measurements during corrosion[4]. Mechanical

---

* Corrosion rates have been given as rates of weight gain because they are the basis of most measurements. Penetrations calculated from the results are dependent upon the relative amounts of different products, and hence require information on the relative abundance of products, which is generally not known quantitatively and can depend upon experimental conditions such as temperature.

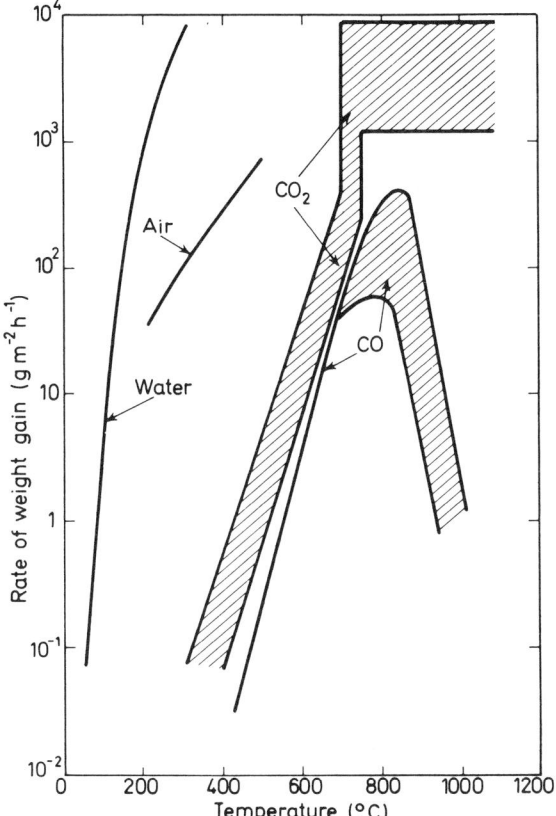

Fig. 5.12 Oxidation rates of uranium in distilled water, air, CO_2 and CO as a function of temperature

failure which is probably associated with stress corrosion as distinct from hydrogen pick-up, has been obtained for specimens of uranium[5] and various γ-quenched alloys. For at least the alloys it is enhanced by chloride ions[6] and is inhibited by cathodic polarisation[7]. At high temperatures the only resistant metallic materials based upon uranium are alloys containing appreciable amounts of the γ-phase stabilising elements, molybdenum, niobium and zirconium[2], the silicide U_3Si[8] and the aluminides. However, the alloys, as distinct from the intermetallic compounds, fail in a sudden disintegrating manner after long periods of exposure, due to, it is believed, hydride formation within the metal.

Atmospheric Corrosion

Uranium tarnishes readily in the atmosphere at room temperature. Electropolishing inhibits the process whilst etching in nitric acid activates the surface. Uranium dioxide and hydrated UO_3 are the principal solid products, although uranium hydride may have a transitory existence. The corrosion is enhanced by water vapour and hence is governed by the humidity conditions[9]. However, the presence of oxygen markedly inhibits attack by water vapour[3]. The view has been expressed that the corrosion is electrochemical in nature with the hydrated UO_3 being formed at cathodic areas[10].

Gases at High Temperatures

Carbon Dioxide and Carbon Monoxide[11]

The solid corrosion products in carbon dioxide and carbon monoxide are uranium dioxide, uranium carbides and carbon. The major reaction with carbon dioxide results in the formation of carbon monoxide:

$$U + 2CO_2 \rightarrow UO_2 + 2CO \quad \ldots(5.1)$$

and in the formation of carbide the carbon transfer from the gas phase occurs mainly as a consequence of subsequent reactions of carbon monoxide with the metal through pores of the oxide scale. Reaction with carbon monoxide must, of necessity, involve the transfer from the gas phase of a carbon atom with each oxygen atom. This transfer could occur by direct reaction or via the thermal decomposition of the gas to produce carbon dioxide:

$$2CO \rightarrow C + CO_2 \quad \ldots(5.2)$$

which can react subsequently by reaction 5.1. Diffusion studies on $UO_2^{(12)}$ demonstrate that oxygen ions are the major diffusing species in the oxide, and therefore that the oxide grows by the inward passage of oxygen. Volume changes associated with the oxidation lead to large stresses in the oxide and subsequently to the formation of solid products in the form of powders and/or cracked porous scales.

At the temperatures of interest, a linear rate law is quickly established, and the rate determining step is believed, in general, to be diffusion of oxygen ions through a thin layer of adherent oxide of constant average thickness at a given temperature. For carbon dioxide, the rate constant (Fig. 5.12) increases steadily with temperature until there is a sudden increase in rate, together with some self-heating from the high heat of reaction, at or near the β-γ phase transition in the metal (780°C); at higher temperatures there is little or no dependence of reaction rate on temperature up to 1 000°C. The majority of the oxide forms as either a non-adherent powder (the particle size of which increases with temperature), or, at the highest temperatures, as a cracked adherent scale. The formation of scale accounts for the lack of temperature dependence for the rate constant at the highest temperatures and could result either from sintering of the oxide, or, more probably, from the growth stresses being relieved by plastic deformation of the oxide and/or the underlying metal, rather than rupture of the oxide. At the highest temperatures the rate determining step might be gaseous diffusion through the porous scale[13]. The presence of small amounts of water vapour (greater than 100 p.p.m.) and oxygen (greater than 10 p.p.m.) enhances attack significantly at the lower temperatures of 400-500°C[11], and under these conditions preferential attack is frequently observed at carbide inclusions in the metal[14]. Major alloy additions of silicon (greater than 3·8%) impair oxidation resistance at all temperatures whilst additions of copper (1%), titanium (5-10%) and molybdenum (10-15%) decrease rates, but only at the higher temperatures[15]. The beneficial influence of copper is associated with the enrichment of the element at the metal/oxide interface as the UCu_5 phase[16]. The mechanisms responsible for the effects of the other alloying elements have not been firmly established but can be attributed to changes in the physical defects and thicknesses of the adherent oxide, and hence in its protective properties.

The oxidation rates in carbon monoxide (Fig. 5.12) are less than those for carbon dioxide. They increase steadily with temperature up to 800°C but then decrease markedly by a factor of 100 up to 1 000°C. The decrease in rate can be attributed to the beneficial factors operating at the higher temperatures with carbon dioxide and to the unfavourable thermodynamics for reaction 5.2 resulting in low equilibrium partial pressures of carbon dioxide.

Air

The high oxygen potential of air leads to the formation of higher oxides of uranium than uranium dioxide. U_3O_8 in the form of a powder is the main product after long times at temperatures above 200-300°C, although the lower oxides may be formed as intermediaries and at lower temperatures. The volume change associated with its formation is greater than for uranium dioxide, and as with carbon dioxide leads to a linear rate law being established. This volume change combined with the high oxygen potential gives fast reaction rates[17] by comparison with the behaviour in carbon dioxide (Fig. 5.12). The fast rates and the large heat change lead to an appreciable self-heating of specimens and an associated increase in reaction rate. In this situation the supply of oxygen from the air can be rate determining. The susceptibility of the metal to self-heating increases with the surface area/volume ratio of components and powders may ignite at room temperature[18]. However, large blocks of metal are not normally liable to self-heating until 400-600°C and then need an external heat source to maintain the situation. Alloy additions of molybdenum (5-15%) markedly reduce reaction rates[15,19]. Minor additions of other elements may either enhance or reduce the susceptibility to ignition[20] (e.g. aluminium or copper, respectively).

Steam

The mechanisms of corrosion by steam are similar to those for water up to 450°C, but at higher temperatures are more closely related to the behaviour in carbon dioxide. Studies at 100°C have demonstrated that uranium hydride is produced during direct reaction of the water vapour with the metal and not by a secondary reaction with the hydrogen product[1]. Also at 100°C it has been shown that the hydride is more resistant than the metal[1]. Inhibition with oxygen reduces the evolution of hydrogen and does not involve reaction of the oxygen with the uranium[3]. Above 450°C the hydride is not stable and hydrogen is released directly to the gas phase. Also, the cohesion and protection provided by the uranium dioxide increase with the formation of a dense scale, at least for short times at the highest temperatures. As a consequence the extent of reaction after a period greater than 100 min is at a maximum at 300–400°C with a rate of weight gain of $10^3 \mathrm{g\,m^{-2}h^{-1}}$. The reaction rate increases only slowly with temperature from 500 to 1 200°C, although one piece of work reported a marked increase in attack near the β–γ transition in the metal[21], analogous to that obtained in carbon dioxide. A parabolic rate law has been found to apply at temperatures of 500–1 200°C for periods of 30 min–6 h[22], although other work indicates that this law is established for only 1–2 h above 880°C and that a linear law generally applies[21]. It follows that for a period of 1–2 h the extent of corrosion in steam straddles that for carbon dioxide, being greater at temperatures less than 700°C and comparable or less at higher temperatures.

Behaviour of Irradiated Uranium

The behaviour of irradiated uranium has been studied mainly with respect to the release of fission products during oxidation at high temperatures[23]. The fission products most readily released to the gas phase are krypton, xenon, iodine, tellurium and ruthenium. The release can approach 80–100%. For ruthenium it is dependent upon the environment and only significant in the presence of oxygen to form volatile oxides of ruthenium.

Studies of the influence of irradiation on the kinetics of oxidation have been confined to post-irradiation work. In general, prior irradiation increases reactivity, although there are considerable inconsistencies in the enhancements obtained[11,23,24]. The effects can be derived from an increased surface area associated with the swelling voids produced in the metal by the irradiation, and can also probably arise to a lesser extent from chemical effects of the fission products.

Coatings

Metallic and organic-based coatings have been developed to protect uranium from oxidation at low and high temperatures. The work on metallic coatings has covered intermetallic compounds and solid solutions of uranium with aluminium, zirconium, copper, niobium, nickel and chromium[26]. Aluminium- and zirconium-based coating[25-27] can be particularly effective in reducing attack over an extended range of temperature. Also, nickel plates can provide protection against atmospheric corrosion[28]. Preliminary work on organic coatings demonstrated that they can enhance atmospheric corrosion[29,30]. This behaviour is attributed to the loss of the inhibiting action of oxygen due to water vapour permeating the coatings more readily than oxygen. However, a special coating composed of a styrenebutadiene copolymer containing powdered aluminium has been developed with a low permeability for water vapour which will reduce the corrosion in humid air by a factor of at least 10 at temperatures up to 40°C[29,30].

Conclusions

The behaviour of uranium has been well characterised for a variety of environments of importance in the nuclear industry. The corrosion is governed by the constitution and physical character of the solid reaction products which in turn are determined mainly by the oxygen potential of the environment, the temperature and the presence of water. The mechanisms of attack are known in broad outline. A major area in need of more detailed study is the influence of irradiation both prior to and during oxidation.

Recent Developments

Extensive work into the corrosion and oxidation of uranium and its alloys has been undertaken over the past decade but much of this is in the form of Ministry and industrial reports which are not necessarily readily available. The present review concentrates on the work published in the normal scientific and technical press.

The effect of alpha-radiolysis of water on the corrosion potential of UO_2 has been measured where the cell construction allowed the source to be brought within 30 μm of the UO_2 electrode[31]. Oxidising conditions were provided at the UO_2 surface and over a period of 30 h the main process occurring appeared to be the catalytic decomposition of H_2O_2 to H_2O and $\frac{1}{2}O_2$ with some oxidation of the UO_2. The oxidation of uranium by water was studied by Fuller et al.[32] using infrared and sorption analysis. Oxidation occurred cyclically forming laminar layers of oxide which spall off because of the strain at the oxide-metal interface. The reaction rate is directly proportional to the amount of adsorbed water on the oxide product and transport was rapid through the open hydrous product. Dehydration of the hydrous oxide irreversibly forms a more inert oxide which cannot be rehydrated to the degree that prevails in the original hydrous product by uranium oxidation with water. An anomalous temperature dependence was observed for the oxidation of oxycarbide layers on the surface of uranium metal[33]. Normally, oxidation or corrosion reactions are expected to proceed more rapidly as water temperature increases but the removal of the outer-most atomic layers of carbon from uranium oxycarbide by oxygen reproducibly proceeds at a much faster rate at 25°C than at 280°C.

The post irradiation examination of two uranium-aluminide fuel plates exposed to aqueous corrosion showed failure due to pinhole corrosion during irradiation to about 76% of the maximum burn-up limit[34]. It was thought that the cladding failed by pitting corrosion initiated at pre-existing pits at a hot spot. Fuel plate material was washed out through these pinholes due to aqueous corrosion and erosion. Egert[35] also looked at corrosion of coated uranium finding corrosion occurring at defects present in the coating. An equation was derived to predict the extent of substrate corrosion given the number of defects per unit area and knowledge of the corrosion kinetics of the substrate.

Galvanic corrosion reports have emerged from two sources. In the first[36], the chemical compatibility of uranium carbides and Cr-Fe-Ni alloys was discussed. Evaluation was by thermodynamic modelling and experimental phase studies. Two reaction temperatures (700 and 1 000°C) were used to simulate normal and over temperature operation of advanced liquid metal fast breeder reactor fuel-cladding couples. In the second, McIntyre et al.[37] coupled depleted uranium-$\frac{3}{4}$% titanium to aluminium, magnesium or mild steel in synthetic seawater. The galvanic current was monitored with time. Gravimetric measurements, polarisation resistance measurements and the concept of mixed potential theory were used to calculate corrosion rates. The galvanic currents must be monitored over extended periods of time to detect changes in galvanic corrosion behaviour. Good agreement was obtained for corrosion rates calculated using the concepts of mixed potential theory and those obtained from gravimetric methods.

Balooch et al.[38] has investigated the reaction of water vapour and oxygen with liquid uranium using molecular beam mass spectrometric methods. For a clean liquid uranium surface, a water reaction probability of about 0·4 and for oxygen about 0·6 were found. These were temperature independent. The values decreased as the coverage of the surfaces by islands of oxides increased and, for water, approached a value of 0·08 for the surface when completely covered with an oxide layer of about 50 nm. The room temperature kinetics of oxygen accumulation on sputtered cleaned U–0·1% Cr alloy surface was followed using AES and direct recoil spectrometry[39] and a mechanism proposed of island growth of 1·5 nm thickness spreading over the surface.

Uranium alloys are very susceptible to stress corrosion cracking and a knowledge of the surface stresses involved are essential. In an uncoated U-Ti alloy these have been found to be relatively large and compressive at 365 MPa but the presence of nickel or zinc coatings on the rod surface led to much smaller compressive stress[40]. The stress corrosion behaviour of U–$7\frac{1}{2}$Nb–$2\frac{1}{2}$Zr in oxygen and hydrogen gases over a temperature range of -20 to 100°C under pressures varying from $0·3 \times 10^{-6}$ to 0·15 MPa have been analysed using a fracture mechanics approach[41]. Stress corrosion cracking mapping and cracking kinetics were determined as functions of stress intensity factor, temperature and pressure. It was found that the mechanism responsible for SCC varied with the experimental conditions used.

Powell and his co-workers have continued to explore the hydrogen embrittlement problem associated with uranium alloys[42-44]. Looking at the internal hydrogen embrittlement of gamma-stabilised uranium alloys (containing molybdenum, niobium or Nb + Zr) they found that the tensile ductility decreased only slightly with increasing hydrogen content up to a critical hydrogen concentration above which the tensile ductility drops to nearly zero[42]. The only alloy of those studied not displaying this sharp drop was one containing $7\frac{1}{2}$%Nb and $2\frac{1}{2}$%Zr. This was probably because it was not possible to achieve a sufficiently high hydrogen content under the experimental conditions used.

The critical hydrogen content for the ductility loss increased with increasing hydrogen solubility in the alloy. The fracture surfaces were not characteristic of those found under conditions of SCC. In terms of hydrogen and deuterium solubility in a similar series of bcc alloys, the equilibrium constants were determined at infinite dilution as a function of temperature[43]. The free energy function was expressed in terms of the bound-proton model.

The best value for the number of hydrogen sites per metal atom in these alloys was found to be three; the zero-point enthalpy of formation of hydrogen in these alloys was very nearly a linear function of alloy composition and the Einstein temperature of the bound-proton (1680 ± 80 K) was not a function of alloy composition. The internal hydrogen embrittlement of alpha-prime uranium-5·7% niobium alloy shows an enhanced microvoid coalescence fracture mode with loss of tensile ductility for hydrogen concentration less than 23 μg per g, the alloy having been solution annealed at 800°C and water quenched[44]. Specimens with 36 μg per g of hydrogen had much lower ductilities and exhibited a new, possibly hydride, phase which was associated with brittle transgranular fracture when this phase has a lenticular morphology extending well across the parent metal grains.

Finally, a book has recently been published covering corrosion problems related to nuclear waste disposal[45]. It discusses a variety of subjects including corrosion behaviour and SCC of copper, carbon steels and high alloy steels under conditions related to nuclear waste disposal. Special attention is paid to pitting and problems associated with hydrogen gas generation from corrosion processes.

J. E. ANTILL
R. A. JARMAN

REFERENCES

1. Baker, M. McD., Less, L. N. and Orman, S., *Trans. Faraday Soc.*, **62**, 2 513 (1966)
2. Wanklyn, J. N. and Jones, P. J., *J. Nucl. Mat.*, **6**, 291 (1962)
3. Baker, M. McD., Less, L. N. and Orman, S., *Trans. Faraday Soc.*, **62** 2 525 (1966)
4. Leach, J. S. L., *J. Inst. Met.*, **88**, 24 (1959)
5. Hughes, A. N., Orman, S. and Pictor, G., *Corrosion Science*, **10** 239 (1970)
6. Magnani, N. J., *J. Nucl. Mat.*, **42**, 271 (1972)
7. McLaughlin, B. D., *J. Nucl. Mat.*, **43**, 343 (1972)
8. Bourns, W. T., *Corrosion Testing of Uranium Silicide Fuel Specimens*, AECL-2 718 (1968)
9. Waber, J. T., *A Review of Recent Data on the Corrosion Behaviour of Uranium* LA 2 035 (1959)
10. Kato, T. R. and Vonderbrink, V. O., *Atmospheric Corrosion of Uranium*, NLCO-860 (1962)
11. Pearce, R. J., Whittle, I. and Hilton, D. A., *J. Nucl. Mat.*, **33**, 1 (1969)
12. Belle, J., *J. Nucl. Mat.*, **30**, 3 (1969)
13. Pearce, R. J., *J. Nucl. Mat.*, **34**, 332 (1970)
14. Hayfield, P. C. S., Graham, R. L. and Ramshaw, G., 'Localised Oxidation of Uranium in Carbon Dioxide and Carbon Monoxide in the Range 200-500°C', Paper No. 5 of Inst. of Met. Sym. on Uranium and Graphite (1962)
15. Antill, J. E. and Peakall, K. A., *Less Common Metals*, **3**, 239 (1961)
16. Antill, J. E. and Peakall, K. A., *J. Electrochem. Soc.*, **110**, 1 146 (1963)
17. Baker, L. and Bingle, J. D., *J. Nucl. Mat.*, **20** 11 (1966)
18. Baker, L., Schnizlein, J. G. and Bingle, J. D., *J. Nucl. Mat.*, **20**, 22 (1966)
19. Isaacs, J. W. and Wanklyn, J. N., *The Reaction of Uranium with Air at High Temperatures*, AERE R-3 559 (1960)
20. Schnizlein, J. G., Baker, L. and Bingle, J. D., *J. Nucl. Mat.*, **20**, 39 (1966)
21. Hopkinson, B. E., *J. Electrochem. Soc.*, **106**, 102 (1959)
22. Wilson, R. E. and Martin, P., *Isothermal Studies of the Uranium Steam Reaction by the Volumetric Method*, **148**, ANL 6 569 (1962)
23. Parker, G. W. *et al.*, Out-of-Pile Studies of Fission Product Release from Overheated Reactor Fuels at ORNL, 1955-1965, ORNL-3 981 (1967)
24. Fischer, D. F. and Schnizlein, J. G., *J. Nucl. Mat.*, **28**, 124 (1968)
25. Baque, P., Koch, P., Dominget, R. and Darras, R., CEA-R-3 638

26. Buddery, J. H., Clark, M. E., Pearce, R. J. and Stobbs, J. J., *J. Nucl. Mat.*, **13**, 169 (1964)
27. Pearce, R. J., Giles, R. D. and Tavender, L. E., *J. Nucl. Mat.*, **24**, 129 (1967)
28. Orman, S., Owen, L. W. and Picton, G., *Corrosion Science*, **12**, 35 (1972)
29. Orman, S., *Atom*, No. 150, 93 (1969)
30. Orman, S. and Walker, P., *J. Oil and Colour Chemists' Association*, **48**, 233 (1965)
31. Bailey, M. G. *et al.*, *Corros. Sci.*, **25**, 233 (1985)
32. Fuller, E. C. *et al.*, *J. Nucl. Mat.*, **120**, 174 (1984)
33. Ellis, W. P., *Surface Sci.*, **109**, L567 (1981)
34. Vinjami, K. and Hobbins, R. R., *Nucl. Technology*, **62**, 145 (1983)
35. Egert, C. M., *Corrosion*, **44**, 36 (1988)
36. Beahm, E. C. and Culpepper, C. A., *Nucl. Technology*, **54**, 215 (1981)
37. McIntyre, J. F., Lefeave, E. P. and Musselman, K. A., *Corrosion*, **44**, 502 (1988)
38. Balooch, M., Olander, D. R. and Siekaus, W. J., *Oxid. Met.*, **28**, 195 (1987)
39. Swissa, E., Bloch, J., Atzmony, U. and Mintz, M. H., *Surf. Sci.*, **214**, 323 (1989)
40. Sha, W. and Wang, Y.-H., *J. Less Common Met.*, **146**, 179 (1989)
41. Lepoutre D., Nomine, A. M. and Miannay, D., ibid., **121**, 521 (1986)
42. Powell, G. L., Kroger, J. W. and Bennet, R. K., *Corrosion*, **32**, 9 (1976)
43. Powell, G. L., *J. Phys. Chem.*, **83**, 605 (1979)
44. Powell, G. L. and Northcutt, W. G., *J. Nucl. Mat.*, **132**, 47 (1985)
45. 'Corrosion problems related to nuclear waste disposal', The Institute of Materials, London, (1992)

5.7 Tungsten

Occurrence

Tungsten is 26th in the table of abundance in the earth's crust at 69 gms/tonne. (C.f. copper 70 gms/tonne and nickel 80 gms/tonne.) The common ores of tungsten are wolframite $(Fe.Mn)WO_4$ containing about 76% WO_3 and scheelite $CaWO_4$ containing 80.53% WO_3. Ores containing 0.25–2.5% WO_3 are economic to extract, and after concentration by floatation, are pulverised, mixed with alkaline compounds such as soda lime and calcined in rotary kilns to produce sodium tungstate. This is dissolved and treated with mineral acid to yield tungstic acid H_2WO_4 which can be reduced in hydrogen to produce reasonable purity tungsten powder or dissolved in ammonia to produce ammonium paratungstate (APT). Controlled crystallisation yields high purity APT crystals which are calcined to produce pure tungsten oxide. This is reduced in continuous furnaces with a flow of hydrogen at 800–1000°C to yield tungsten powder suitable for processing to solid material.

Because of the high melting point of tungsten 3410°C no refractories are available to contain molten tungsten, so fabrication by direct casting cannot be carried out. Tungsten can be arc cast or electron beam melted, but, because of the coarse grain size, further fabrication is only possible by extrusion. The vast majority of tungsten is fabricated by powder metallurgical techniques where the powder is pressed in a tool steel die or hydrostatically into billets to approx 60% of the theoretical density. The billets are sintered by direct resistance heating at 3000°C or indirect heating at 2500°C to increase the density above 90% theoretical, the minimum density suitable for further processing by rolling, swaging, forging or extrusion. Mechanical working commences with the sintered bar heated to about 1700°C. As the section is reduced, the processing temperature is also reduced so that thin sheet and wires can be rolled or drawn at room temperature.

Fabrication

Tungsten sheets or rods can readily be fabricated by bending, spinning, flow turning, punching, stamping and riveting, provided that work is carried out above the ductile/brittle transition temperature but below the recrystallisation temperature. These temperatures depend on the material thickness and are around room temperature for thin sheets and wires. Typically 1 mm thick sheet should be fabricated between 250°C and 850°C.

Uses

50-55% of tungsten is used in the hard metal industry.
18-20% is used in tool steels.
6% is used in superalloys.
5% is used as tungsten chemicals.
15% is used as wrought products.
Wrought tungsten has a variety of uses including wire lamp and valve filaments, rocket nozzles, contacts, heating elements, evaporating coils and boats, X-ray targets, susceptors and radiation screens in vacuum furnaces. It is these applications where tungsten may be subjected to some form of corrosion. Other uses of tungsten include semi-conductors and gamma-ray radiation shielding. However, these latter uses are beyond the scope of this review.

Tungsten, both pure, and containing up to 2% thoria is used in high intensity arc lamps where its low vapour pressure, good conductivity, high melting point and resistance to halogen gases are advantageous. Tungsten pins are used in electron tubes where its thermal expansion characteristics are similar to glass. These pins are slightly oxidised and the oxide dissolves in the glass forming a good seal. In the aerospace industry tungsten satisfies the requirements for a material that can withstand temperatures above 3000°C at pressures of several hundred atmospheres with high thermal stresses and exposure to highly corrosive gases.

Difficulties in fabrication of tungsten tend to preclude its use in the chemical industry except as a last resort.

Other uses are in thin film technology where coatings are applied by vacuum deposition. Tungsten boats, or coils fabricated from wire are heated by direct resistance heating and used to evaporate Ag, Al, Au, B, Ba, Ce, Cr, Fe, In, Mg, Mn, Ni, Pa, Pt, SiO, Te, V, Zn and Zr.

Tungsten alloys

The use of tungsten carbide alloyed with cobalt, well known as Hard Metal is used for cutting tools because of its excellent wear resistance but finds little use as a corrosion resistant material.

Alloys with thoria (ThO_2) are used for TIG (Tungsten Inert Gas) welding electrodes and in electronic applications where its increased electron emission properties and high temperature strength prove advantageous.

Alloys with rhenium, another high melting point metal (3180°C) exhibit outstanding high temperature properties insofar as they have a higher recrystallisation temperature than pure tungsten and are still ductile in the recrystallised condition. Common alloys with rhenium contain 3%, 5% or 26% rhenium. The 3% and 5% alloys combine ductility with reasonable weldability whilst the 26% alloy finds usage in wire form as a thermocouple material usually in combination with the 5% tungsten/rhenium alloy. Tungsten/rhenium alloys are not widely used because of the very high cost of the rhenium used in the alloying. Corrosion resistance of tungsten/rhenium alloys is similar to that of pure tungsten.

Cleaning

Because of the high ductile/brittle transition temperature of tungsten, it is necessary to carry out fabrication, especially of heavy sectioned material, at temperatures in excess of 1000°C. The surface of the tungsten becomes oxidised, the degree and tenacity of the oxide depending on the temperature used for the fabrication operation. Heavy sectioned, heavily oxidised tungsten should be cleaned in a molten mixture of 90% sodium hydroxide and 10% sodium nitrite at 400°C. This mixture attacks the material violently and great care should be taken during this process. The tungsten must be dry before putting it into the molten medium as dangerous sputtering could occur. After, cleaning, which only takes a few seconds, the component should be removed from the medium and allowed to cool naturally in air to room temperature before washing free of the caustic medium in hot water. This cooling stage is necessary because the reaction to the caustic medium is so violent, the temperature of the component rises rapidly to white heat and sudden cooling, even in boiling water would cause cracking of the component due to thermal shock. Alternatively, providing the component is of sufficient strength, the oxide layer can be removed by a sand-blasting operation. Oxide can also be removed by immersion in one of the following mixed acid solutions using concentrated acids:

Nitric acid + hydrofluoric acid + water (Ratio 1:1:1 by volume)

Nitric acid 45 vol% + hydrofluoric acid 27.5 vol% + glacial acetic acid 27.5 vol%

Perchloric acid + phosphoric acid + water (Ratio 1:1:2)

Lightly oxidised tungsten can be cleaned by heating in 10% caustic soda solution at about 80°C followed by rinsing in water and then dilute hydrochloric acid solution (10%) to neutralise remaining caustic before finally rinsing in water and drying. Alkaline potassium ferricyanide solution can also be used for removing light oxide films – a mixture of equal parts of 10% potassium hydroxide and 10% $K_3[Fe(CN)_6]$ followed by a dilute hydrochloric acid rinse as above.

Corrosion properties

The resistance of tungsten to common reagents is summarised in Table 5.28.

Table 5.28 Resistance of tungsten to various reagents

Material	Conditions	Reaction
Acids		
Acetic acid Glacial	100°C	Slight attack
Aqua regia	hot or cold	Rapid attack
Chromic acid 10%	100°C	Attacked
HF + HNO_3	hot or cold	Rapid attack
Hydrochloric acid	cold conc.	No attack
Hydrochloric acid	hot, dil or conc	Slight attack
Hydrofluoric acid, pure	100°C	No attack
NaOCl	conc., cold	Attacked
Nitric acid	cold, dil. or conc.	No attack
Nitric acid	hot, dil. or conc.	Slight attack
Oxalic acid	100°C	Slight attack
Phosphoric acid	cold	No attack
Phosphoric acid	hot	Slight attack
Sulphuric acid	cold, dil. or conc.	No attack
Sulphuric acid	hot dil. or conc.	Slight attack
Alkalis		
Ammonia + H_2O_2	cold	Rapid attack
Ammonia + H_2O_2	hot	Rapid attack
Ammonia solution	hot or cold	No attack
Molten NaOH or KOH with KNO_2 KNO_3 or $KClO_4$	molten	Violent attack
Sodium hydroxide	molten	Rapid attack
Sodium or potassium hydroxide solution	hot or cold	No attack
Halogens		
Bromine	above 250°C	Attacked
Chlorine	above 250°C	Attacked
Fluorine	hot or cold	Attacked
Iodine	above 800°C	Attacked
Other gases		
Carbon dioxide	above 1200°C	Oxidises
Carbon monoxide	above 1000°C	Carburises
Hydrocarbons	above 1000°C	Carburises
Hydrogen	all temperatures	No attack
Hydrogen Chloride	below 600°C	No attack
Hydrogen Sulphide	above 400°C	Slight attack
Inert gases	all temperatures	No attack
Nitrogen	above 2300°C	Nitrides form
Nitrogen	below 2000°C	No attack
NO_2, NO, N_2O	above 400°C	Oxidises
Oxygen or air	above 400°C	Oxidises
Sulphur dioxide	above 700°C	Oxidises
Water vapour	above 800°C	Oxidation/reduction reaction occurs
Water vapour	dew point −25°C max all temperatures	No attack
Water vapour	dew point −25 to +20°C below 750°C above 750°C	No attack Oxidises

Table 5.28 Continued

Material	Conditions	Reaction
Vacuum	10^{-4} torr	Evaporation above 2800°C
Molten Metals		
Aluminium	below 700°C	No attack
Bismuth	below 980°C	No attack
Gallium	below 800°C	No attack
Lithium	below 1620°C	No attack
Magnesium	below 600°C	No attack
Mercury	below 600°C	No attack
Potassium		No attack
Sodium	below 900°C	No attack
Sodium/potassium alloy	below 900°C	No attack
Zinc	below 750°C	No attack
Refractories		
Al_2O_3	below 1900°C	No attack
BeO	below 1980°C	No attack
Fire brick	below 1600°C	No attack
Graphite	above 1400°C	Carburises
MgO	below 1980°C	No attack
Quartz and glass	molten	No attack
ThO_2	below 2200°C	No attack
Uranium Oxide	below 3000°C	No attack
ZrO_2	below 1600°C	No attack

Future prospects

Because of the special techniques necessary to fabricate complex shapes in tungsten, it is only used as a last resort where all other metals are unsuitable and its high melting point and strength render it useful. Increased usage in the glass and ceramic fibre industry is envisaged. Chemical and physical vapour deposition techniques now being exploited enable tubes, e.g. copper, to be lined with dense tungsten and this may offer new uses in the chemical industry. Similarly, plasma sprayed tungsten powder has been used to line pots for containing molten lead.

For the chemical engineer unfamiliar with tungsten, or indeed, any other of the refractory metals, the *Corrosion Brochure*[1] published by *Processing Magazine* is a useful starting point for assessing the suitability of refractory metals for a given application. The sections on tungsten, molybdenum, tantalum and niobium are updated annually by the author of this review. However, for more specific information, contact should be made, at a early stage of the project, with manufacturers of these materials so that up to date information can be obtained.

C. E. D. ROWE

REFERENCES

1. *Corrosion Brochure*, published annually by Techpress Publications (Northside House, 69 Tweedy Road, Bromley, Kent UK)

6 THE NOBLE METALS

6 The Noble Metals

Introduction

The outstanding characteristics of the noble metals are their exceptional resistance to corrosive attack by a wide range of liquid and gaseous substances, and their stability at high temperatures under conditions where base metals would be rapidly oxidised. This resistance to chemical and oxidative attack arises principally from the inherently high thermodynamic stability of the noble metals, but in aqueous media under oxidising or anodic conditions a very thin film of adsorbed oxygen or oxide may be formed which can contribute to their corrosion resistance[1]. An exception to this rule, however, is the passivation of silver and silver alloys in hydrochloric or hydrobromic acids by the formation of relatively thick halide films.

Their high initial cost, combined with a mechanical strength which is generally inferior to that of base metals, results in the noble metals being used as sheaths, linings, or other thin coatings on strong supporting structures. They are also widely used as electrodeposits, as described in Section 14.9. Silver and, to a lesser extent, platinum and its alloys are sometimes employed in more massive section for the construction of chemical plant, as for example distillation stills and bushings for the glass industry. The chemical, synthetic-fibre, glass and metallurgical industries make very wide use of the noble metals and their alloys wherever equipment must be protected from chemical attack, or a product protected from contamination by the products of corrosion or erosion.

Physical and Mechanical Properties

Although in the majority of their applications the choice of noble metals is determined by their chemical rather than by their physical and mechanical properties, some consideration of the latter is necessary. The relevant information for the noble metals as a whole is given in Tables 6.1 and 6.2, and details relating to the individual metals will be found in the following paragraphs.

Silver

Silver in the fully annealed state is a soft, ductile metal which is easily fabricated into the very wide range of forms employed in industry by the

Table 6.1 The physical properties of the noble metals

Metal	Atomic number	Atomic weight	Lattice structure	Density at 20°C (g/cm³)	Melting point (°C)	Thermal conductivity at 0–100°C (W/m°C)	Specific heat at 0°C (J/kg°C)	Coefficient of linear expansion at 20–100°C $\times 10^6$	Thermal neutron cross-section (barns) (10^{-28} m²)	Resistivity at 0°C (μΩ cm)	Temperature coefficient of resistance 0–100°C $\times 10^3$
Platinum	78	195·09	fcc	21·45	1 769	73	131·2	9·1	9	9·85	3·9
Iridium	77	192·22	fcc	22·65	2 443	148	128·4	6·8	430	4·71	4·3
Osmium	76	190·2	cph	22·61	3 050	87	129·3	6·1	15	8·12	4·2
Palladium	46	106·4	fcc	12·02	1 552	76	244·3	11·1	6	9·93	3·8
Rhodium	45	102·91	fcc	12·41	1 960	150	246·4	8·3	150	4·33	4·6
Ruthenium	44	101·07	cph	12·45	2 310	105	230·5	9·1	3	6·80	4·2
Silver	47	107·87	fcc	10·49	961	419	234	19·7	63	1·59	4·1
Gold	79	196·97	fcc	19·32	1 063	311	128	14·4	99	2·06	3·9

Table 6.2 The mechanical properties of the noble metals

Metal	Tensile strength (annealed) (MN/m²)	Elongation (annealed) (%)	Modulus of elasticity (E) (GN/m²)	Hardness (annealed) (HV)
Platinum	124	40	172	40
Iridium	1 100	–	516	220
Osmium	–	–	556	350
Palladium	172	40	117	40
Rhodium	688	5	316	100
Ruthenium	496	10	417	240
Silver	139	60	72	26
Gold	108	70	72	20

normal metal-working techniques such as drawing, spinning, rolling, etc. Silver work-hardens appreciably during fabrication. The mechanical strength of silver is markedly affected by an increase in temperature, and falls to about 25% of the initial value for cold, hard-worked silver when the metal is heated to just over 200°C.

Silver, with a thermal conductivity of 419 W/m°C is a somewhat better conductor of heat than copper, and this property is often utilised in the construction of heat exchangers, evaporator linings, etc.

Gold

Gold is an extremely soft and ductile metal, and exhibits very little work-hardening during deformation. Very severe cold rolling can increase its hardness from 20 HV in the fully annealed state to 70 to 75 HV, but this produces only a slight increase in its strength. The ultimate tensile strength of pure gold is too low to allow self-supporting structures of any size to be constructed of this metal—its use is almost entirely restricted to thin linings or electrodeposits on base-metal equipment.

Platinum and its Principal Alloys

Pure platinum is soft, ductile and easily fabricated. Its mechanical properties are profoundly affected by the degree of cold working to which it has been subjected and by the presence of slight impurities or alloying constituents. In its applications it is frequently alloyed with another metal of its group—the melting points of its alloys with Rh, Ir, Os and Ru being higher than the parent metal, those with Pd being lower. In most cases the strength, rigidity, hardness and resistance to corrosion are improved by alloying. Addition of certain base metals, however, can lead to embrittlement and failure of platinum and its alloys even if the base metal is present only in trace quantities.

Platinum work-hardens at approximately the same rate as palladium or copper, a 25% cold reduction in thickness raising the hardness from about 40 HV in the fully annealed state to 75 HV. Cast platinum is usually slightly harder than the same grade of metal in the wrought and annealed state.

Alloys with rhodium Rhodium alloys readily with platinum in all proportions, although the workability of the resulting alloy decreases rapidly with increasing rhodium content. Alloys containing up to about 40% rhodium, however, are workable and find numerous applications. The principal physical and mechanical properties of rhodium-platinum alloys are listed in Table 6.3.

The resistance of rhodium-platinum alloys to corrosion is about the same as or slightly better than that of pure platinum, but they are much more stable at high temperatures. They have excellent resistance to creep above 1 000°C, a factor which largely determines their extensive use in the glass industry, where continuous temperatures sometimes exceeding 1 500°C are encountered. Rhodium additions to platinum reduce appreciably the volatilisation of pure platinum at high temperatures.

Alloys containing more than 40% rhodium, while very difficult to fabricate, are almost immune from attack by aqua regia. The Pt−10Rh alloy is particularly resistant to attack by free wet chlorine such as that produced by the combustion of halogenated organic vapours.

Table 6.3 Physical and mechanical properties of rhodium–platinum alloys

Rhodium content (wt%)	Melting point (°C)	Hardness (annealed) (HV)	Hardness (hard) (HV)	Tensile strength (hard) (MN/m^2)
0	1 769	40	100	124
5	1 825	55	130	206
10	1 850	65	160	310
20	1 900	82	210	480
30	1 920	95	250	550
40	1 940	100	290	620

Alloys with iridium Iridium alloys with platinum in all proportions, and alloys containing up to about 40% iridium are workable, although considerably harder than pure platinum. The creep resistance of iridium–platinum alloys is better than that of rhodium–platinum alloys at temperatures below 500°C. Their stability at high temperatures, however, is substantially lower, owing to the higher rate of formation of a volatile iridium oxide.

Alloys with ruthenium Additions of ruthenium have a most marked effect upon the hardness of platinum, but the limit of workability is reached at about 15% ruthenium, owing to the fact that ruthenium belongs to a crystallographic system different from that of platinum. Apart from a somewhat greater tendency to oxide formation at temperatures above 800°C, the resistance to corrosion of ruthenium–platinum alloys is comparable to that of iridium–platinum alloys of similar composition.

Chemical Properties

The factors leading to the high resistance of the noble metals to chemical attack, i.e. their thermodynamic stability over a wide range of conditions and the possibility of the formation of very thin protective films under oxidising conditions, have already been mentioned. A factor tending to reduce corrosion resistance in aqueous solutions is the tendency of these metals to form complexes with some anions.

In the absence of complexing agents the noble metals are extremely resistant to corrosion by aqueous solutions of alkalis and salts, and to dilute acids. The resistance of gold and the platinum metals to corrosion by concentrated acids is summarised in Table 6.4, and to halogens in Table 6.5[2]. The resistance of silver to oxidising acids is generally lower than that of the other noble metals, while in halogen acids it forms a protective film of insoluble halide. Silver also differs from the other noble metals in forming a sulphide tarnish film in industrial atmospheres, due to the action of sulphur compounds.

In reviewing the corrosion resistance of the individual metals in aqueous solutions it is useful to compare the observed behaviour with that predicted

Table 6.4 Corrosion ratings of noble metals in acid

Acid	Temp. (°C)	Au	Ir	Os	Pd	Pt	Rh	Ru
Aqua regia	Room	D	A	D	D	D	A	A
	100	D	A	D	D	D	A	A
HI, 60%	Room	A	A	B	D	A	A	A
	100	—	A	C	—	D	A	A
HBr, 62%	Room	A	A	A	D	B	B	A
	100	—	A	C	D	D	C	A
HCl, 36%	Room	A	A	A	A	A	A	A
	100	A	A	C	B	B	A	A
HF, 40%	Room	A	A	A	A	A	A	A
Nitric, 70%	Room	A	A	C	D	A	A	A
Nitric, 95%	Room	B	A	D	D	A	A	A
	100	—	A	D	D	A	A	A
Perchloric, s.g. 1·6	Room	A	—	—	A	A	—	—
	100	A	—	—	C	A	A	A
Phosphoric, 100 g/l	Room	—	A	—	A	A	A	A
	100	A	A	D	B	A	A	A
Selenic, 42%	Room	A	—	—	C	A	—	—
	100	A	—	—	D	C	—	—
Sulphuric, 98%	Room	A	A	A	A	A	A	A
	100	A	A	A	C	A	B	A

A No appreciable corrosion
B Some attack, but not enough to preclude use
C Attacked enough to preclude use
D Rapid attack

Table 6.5 Corrosion ratings of noble metals in halogens

Halogen	Temp. (°C)	Au	Ir	Os	Pd	Pt	Rh	Ru
Sat. Cl water	Room	D	A	—	A	A	A	B
	100	—	—	—	D	A	—	—
Moist Cl	Room	D	A	C	D	B	A	A
Dry Cl	Room	B	A	A	C	B	A	A
Sat. Br water	Room	D	A	—	B	A	A	B
Moist Br	Room	C	A	B	D	C	A	A
Dry Br	Room	D	A	D	D	C	A	A
I in alcohol	Room	B	A	—	B	A	B	B
Moist I	Room	A	A	A	B	A	B	A
Dry I	Room	A	A	B	A	A	A	A

A No appreciable corrosion
B Some attack, but not enough to preclude use
C Attacked enough to preclude use
D Rapid attack

by the pH–potential diagrams developed by Pourbaix and his co-workers[3]. On the following diagrams the limits of stability of water are indicated by the dotted lines *a* and *b*.

Silver

The behaviour of silver in different environments is determined by three principal factors: (*a*) the high nobility of the metal, (*b*) the formation of passive protective films and (*c*) the tendency of the metal to form complex ions in solution.

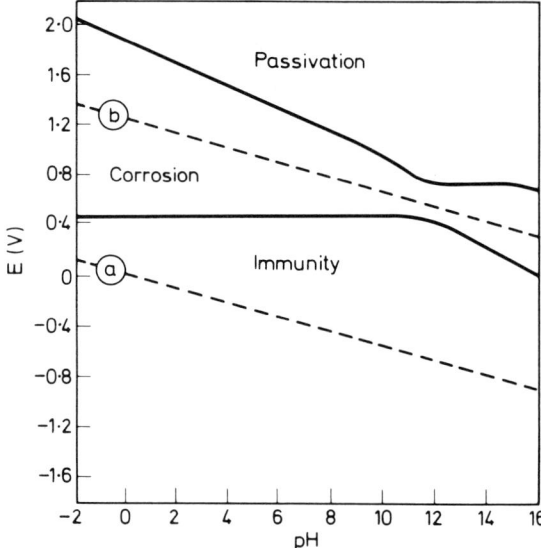

Fig. 6.1 Theoretical domains of corrosion, immunity and passivation of silver, at 25°C

Silver, with a standard electrode potential $E_{Ag^+/Ag} = 0.79$ V, is exceeded in nobility only by gold and the platinum-group metals. The thermodynamics of the behaviour of silver in reducing acids, in the presence or absence of oxygen or complexing agents, has been considered in Section 1.4. Reference to the Ag/H$_2$O diagram[3] (Fig. 6.1) shows that at low potentials —below about 0·4 V—the metal remains immune to attack over almost the whole pH range. This diagram, of course, is of limited value, and is considerably modified in the presence, for example, of CN$^-$ or Cl$^-$ ions. The presence of halides, with the exception of fluoride, substantially increases the zone of passivity, owing to the very low solubility products[4] of silver halides. (Solubility products at 25°C: AgCl = 1.7×10^{-10}; AgBr = 5.0×10^{-13}; AgI = 8.5×10^{-17}; AgF—very soluble.)

On the other hand, the presence of CN$^-$ ions greatly increases the zone of corrosion, owing to the formation of complex ions. Silver, therefore, is thermodynamically stable in reducing acids, e.g. hydrochloric acid, acetic acid, phosphoric acid, provided oxidising substances are absent.

Oxidising acids, e.g. nitric acid, hot sulphuric acid at concentrations exceeding 80% and reducing acids containing oxidising agents, will be corrosive to silver, and the diagram shows that an extensive zone of corrosion occurs at elevated potentials in the acid region.

When silver is passivated by a halide film, as is formed for example in hydrochloric acid, the film is tenacious, self-healing and highly insoluble. It may be reduced to the metal, however, by coupling the silver with such metals as zinc, aluminium, and, in the case of chemical plant, Hastelloy. In such instances the silver will continuously corrode.

As may be seen from the diagram, silver in highly alkaline solution corrodes only within a narrow region of potential, provided complexants are absent. It is widely employed to handle aqueous solutions of sodium or potassium hydroxides at all concentrations; it is also unaffected by fused alkalis, but is rapidly attacked by fused peroxides, which are powerful oxidising agents and result in the formation of the AgO^+ ion[5]. Table 6.6 gives the standard electrode potentials of silver systems.

Silver is attacked by most compounds of sulphur, becoming covered with a yellow, brown or black sulphide film.

Table 6.6 Standard electrode potentials of some silver systems[5]

System	Potential (V)
Ag^+/Ag	0·799
Ag^{2+}/Ag^+	1·98
AgO^+/Ag^{2+}	2·1
Ag_2O/Ag	0·344
AgO/Ag_2O	0·57
Ag_2O_3/AgO	0·74

Table 6.7 Standard electrode potentials of some gold systems

System	Redox potential (V)
Au^+/Au	1·68
Au^{2+}/Au^+	<1·29
Au^3/Au^{2+}	<1·29
Au^{3+}/Au^+	1·41
Au^{3+}/Au	1·50
$AuBr_2^-/Au$	0·96
$AuCl_4^-/Au$	1·0
$H_2AuO_3^-/Au$	0·7

Gold

The high resistance of gold to attack by a very wide range of corrosive media results from its very high nobility. It may be seen from Fig. 6.2 that gold is immune to attack over the whole range of pH values at redox potentials below about 0·4 V, and that the zone of immunity extends to higher potentials at the lower end of the pH range. Gold, however, is easily complexed, and its solubility in hydrochloric acid containing an oxidising agent (e.g. nitric acid) results from a combination of high redox potential and the formation of chlor–auro complex ions[6], $AuCl_4^-$. The unstable Au^+ ion and the easily reducible Au^{3+} ion also readily form stable complexes. The Au^{2+} form is not stable in aqueous solutions, existing only in the solid, insoluble sulphide form. Standard electrode potentials of gold are given in Table 6.7.

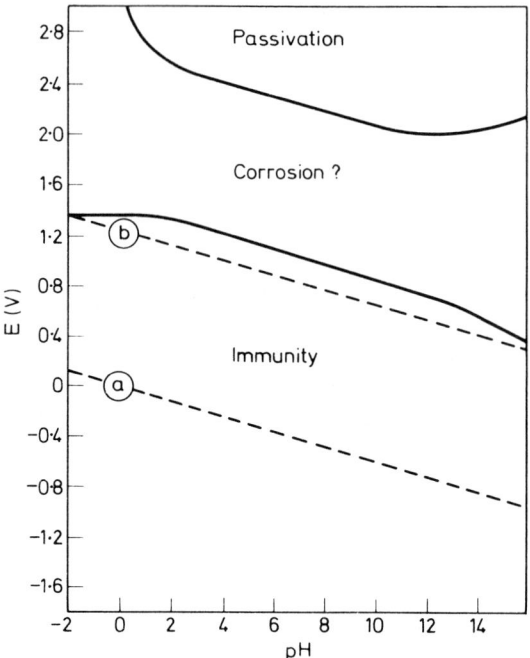

Fig. 6.2 Theoretical conditions of corrosion, immunity and passivation of gold, at 25°C

Gold is unaffected in alkaline solutions, but in the presence of cyanides the $Au(CN)_2^-$ ion is readily formed:

$$Au(CN)_2^- + e^- \rightarrow Au + 2CN^-; \quad E = 0 \cdot 6 \text{ V}$$

The dissociation constant for the reaction

$$Au(CN)_2^- \rightarrow Au^+ + 2CN^-$$

is $K = 5 \times 10^{-39}$. Thus in the presence of cyanide ions gold is a powerful reducing agent and is readily oxidised by oxygen. This reaction forms the basis for the extraction of gold from its ores on an industrial scale.

Platinum-group Metals

All the six platinum-group metals are highly resistant to corrosion by most acids, alkalis, and other chemicals. Their high nobility is the main factor determining their chemical resistance, and the formation of complex ions in solution is principally responsible for their dissolution under certain conditions.

As may be seen from the potential–pH diagram[2,14] (Fig. 6.3) platinum is immune from attack at almost all pH levels. Only in very concentrated acid solutions at high redox potentials (i.e. under oxidising conditions) is there a zone of corrosion. This accounts for the solubility of platinum in aqua regia. Platinum is also prone to complex-ion formation, and this can lead

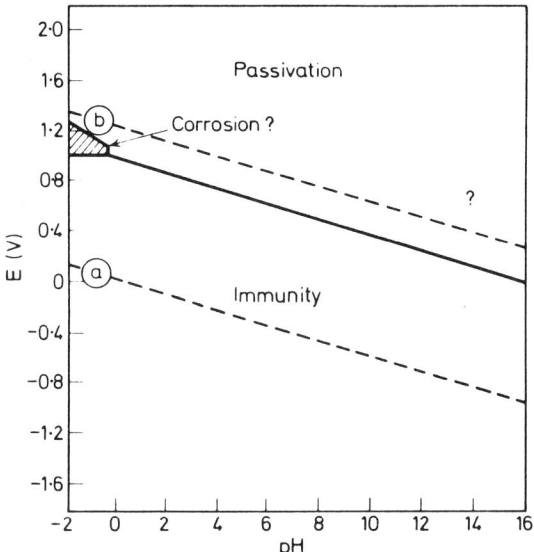

Fig. 6.3 Theoretical domains of corrosion, immunity and passivation of platinum, at 25°C

to difficulties in interpreting the pH–potential diagrams when practical aqueous solutions are evaluated.

Platinum is unaffected by most organic compounds*, although some compounds may catalytically decompose or become oxidised on a platinum surface at elevated temperatures, resulting in an etched appearance of the metal. Carbon and sulphur do not attack platinum at any temperature up to its melting point. Molten platinum may dissolve carbon, but the solubility of the latter in solid solution is virtually zero.

The resistance of rhodium to chemical attack is remarkable, and surpasses that of platinum. Its domain of stability (as seen from Fig. 6.4) is extremely wide, and in the absence of complexing agents it is stable in aqueous solutions of all pH values. In the massive form it is unattacked by caustic alkalis, acids and oxidising agents, including aqua regia. When finely divided, however, it is attacked by concentrated sulphuric acid and aqua regia.

The behaviour of iridium is closely analogous to that of rhodium; its corrosion diagram is very similar and it is, with rhodium, one of the least corrodible of metals. It is unattacked by alkalis, acids or oxidising agents in aqueous solution, although a fused mixture of caustic potash and potassium nitrate will attack it. The metal has an excellent resistance to fused lead oxide, silicates, molten copper and iron at temperatures up to 1 500°C. Additions of iridium to platinum considerably raise the corrosion resistance of the latter to a very wide range of reagents.

Compared with platinum, rhodium and iridium, palladium has much less resistance to chemical attack. Its theoretical corrosion diagram is depicted in Fig. 6.5, from which it may be seen that the metal is stable in the presence

---

*Certain organic compounds form complexes with platinum, and this accounts for the fact that the thin coating of platinum on titanium is rapidly corroded when platinised titanium is used as an anode in plating baths containing organic addition agents.

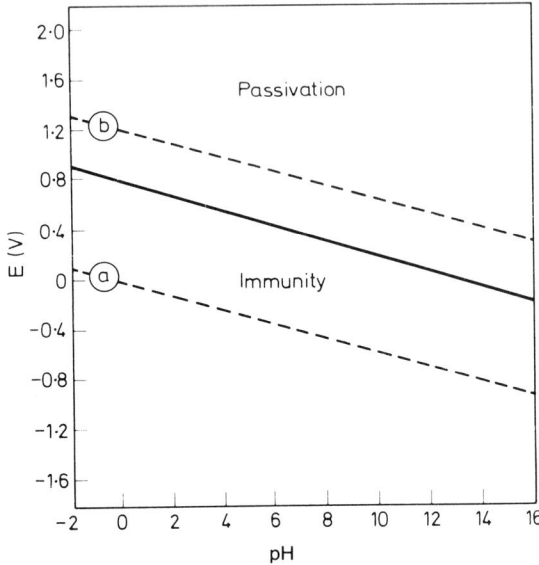

Fig. 6.4 Theoretical domains of corrosion, immunity and passivation of rhodium, at 25°C

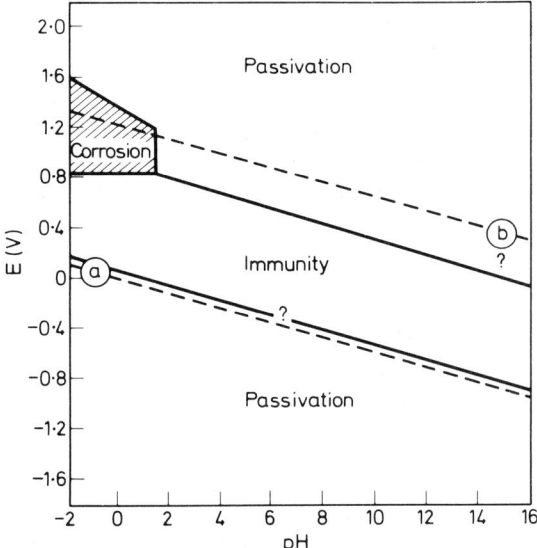

Fig. 6.5 Theoretical domains of corrosion, immunity and passivation of palladium, at 25°C

of aqueous solutions of all pH values with the exception of strong oxidising agents and complexing substances. In practice, palladium is not attacked by chlorine water, except at high temperatures, and is not tarnished in moist air. Non-oxidising acids, e.g. acetic, oxalic, hydrofluoric and sulphuric acids, have no effect on the metal at ordinary temperatures. Strongly oxidising acids, however, e.g. hydrochloric acid containing nitric acid, rapidly attack palladium. Dilute nitric acid attacks palladium only slowly, but the metal is rapidly corroded by the concentrated acid. Alloys of palladium with platinum, however, retain most of the corrosion resistance of platinum. In ordinary atmospheres palladium is resistant to tarnish, but some discoloration due to sulphide-film formation may take place in industrial atmospheres containing sulphur dioxide. Alkaline solutions, even in the presence of oxidising agents, are without effect on the metal. This may be due to a thin passivating film of PdO which is more stable than $Pd(OH)_2$ and which prevents further attack.

Ruthenium and osmium are decidedly less noble than the other four metals of the platinum group. Both exist in numerous valency states and very readily form complexes. Ruthenium is not attacked by water or non-complexing acids, but is easily corroded by oxidising alkaline solutions, such as peroxides and alkaline hypochlorites.

High-temperature Properties of the Noble Metals[7]

Silver and Gold

Owing to their relatively low melting points and mechanical strengths, silver and gold find very few applications at elevated temperatures. Silver below its melting point has considerable resistance to oxide-film formation, but molten silver dissolves appreciable quantities of oxygen, which comes out of solution as silver oxide or bubbles dispersed throughout the metal when the metal solidifies. Gold is not subject to oxide-film formation at any temperature up to its melting point, but may be covered by a thin adsorbed layer of oxygen. The absence of an oxide film enables gold to be pressure-welded to itself at room temperatures.

Platinum-group Metals

The excellent resistance of platinum, rhodium and iridium to oxidation at high temperatures finds numerous applications in technology, in particular in the form of platinum-based alloys. Osmium and ruthenium form volatile oxides which may be isolated (OsO_4 and Ru_2O_3), and they are not widely used.

Platinum, while it does not form a measurable oxide film, is covered by a tenaciously held layer of adsorbed oxygen[8]. It volatilises at an increasing rate as its temperature rises above 1 000°C, and in the presence of moving oxygen or air the rate of volatilisation is considerably increased. In the presence of oxygen the volatilisation proceeds via the formation of a volatile, unstable oxide which has not to date been identified[9]. Rhodium, iridium and palladium exhibit oxide-film formation, the last named at temperatures as low as 600°C[10]. Palladium oxide dissociates again at temperatures above 870°C, the metal appearing bright up to its melting point. Absorption of oxygen without film formation occurs, however, and the palladium increases in weight. Platinum is more volatile than rhodium and iridium in the range 900–1 200°C, but their volatilities are about equal at temperatures around 1 300°C. Below 1 100°C the alloys of platinum with rhodium and palladium are less volatile than pure platinum, but palladium–platinum alloys are not employed at these temperatures owing to the oxygen absorption of the palladium already referred to[11]. Rhodium–platinum alloys at high temperatures show no preferential loss of either metal, and are widely used. Iridium–platinum alloys show greater loss of weight on heating in air, because of the greater rate of oxidation of iridium and the higher volatility of the oxide of this metal. Iridium is thus lost preferentially from iridium–platinum alloys.

Volatilisation of platinum and its alloys when subjected to continuous high operating temperatures may be substantially reduced by avoiding contact with oxygen, especially oxygen which is in movement or circulation owing to convection currents. This is achieved by completely embedding the metal in a high-grade pure alumina refractory cement or block—flame-sprayed coatings, for example, are effective in preventing free circulation of air over the metal. Only very high-grade alumina, free from silica or other oxides more easily reduced under reducing conditions can be employed, otherwise contamination and embrittlement of the platinum may result from partial reduction of such oxides.

Grain-growth of platinum and its alloys when operating continuously at high temperatures is often responsible for failure of the metal, resulting from weaknesses developed by large inter-crystal boundaries. This defect may be obviated to a considerable extent by the use of sintered metal produced by powder-metallurgical techniques[12], or by the incorporation of a small amount of a refractory oxide, carbide or nitride in powder form in the body of the metal. The latter method provides a source of nuclei for crystal formation, and the resulting pure metal or alloy has considerably higher strength and longer life at high temperatures[13]

Applications of the Chemical Properties of the Noble Metals

Protective Linings and Sheaths

Many chemical-plant vessels constructed of mild steel, copper, or other base metals may be protected from corrosive attack by their contents by means of a lining of one of the noble metals. This form of construction utilises the higher strength of the base metal—as, for example, in the walls of pressure vessels—and at the same time a minimum amount of a much more expensive noble metal is used. The use of noble metals in plant is frequently determined by the desire to avoid contamination of valuable reactants or substances by the products of corrosion of base metals.

Silver linings are generally 0·5 mm to 1 mm thick, although occasionally greater thicknesses are employed. The interior of the vessel to be lined must be free from roughness, projections, etc. and maximum radii of corners must be ensured. Exit pipes are lined similarly, and continuous fusion-welded seams ensure that no contact between base-metal vessel and corrosive contents takes place. A very large range of chemical plant may be lined with silver, but a solid silver construction is sometimes employed for condenser coils, distillation heads, etc. In such instances the silver is generally 1 mm to 2·5 mm thick. Stirrers, thermocouple pockets, inlet tubes, etc. are either constructed of solid silver or sheathed with this metal to prevent contact of the base with corrosive materials.

Linings in silver may be of two types, viz. 'loose' and 'bonded'. Loose linings provide good contact between vessel and lining — adequate for good heat transfer — but the establishment of a partial or full vacuum in the vessel would cause the air gap between lining and vessel to expand, resulting in collapse and failure of the lining.

Bonded silver linings are fabricated for mild steel or copper vessels. They are soldered *in situ* to the walls of the vessel by means of a special tin-silver solder. The melting point of this solder is approximately 280°C, and 200°C is recommended as the maximum continuous operating temperature for linings bonded with it. Since the whole of the silver is firmly adherent to the vessel, bonded linings are suitable for operation under vacuum conditions, and provide excellent heat-transfer characteristics.

Platinum and rhodium–platinum and iridium–platinum alloys are frequently employed to line and sheath autoclaves, reactor vessels and tubes, and a wide range of equipment. Linings are generally 0·13 mm to 0·38 mm thick, and for certain applications co-extruded platinum-lined Inconel or other metal reactor or cooling tubes are fabricated. In such cases the platinum is bonded to the base metal, but in all other instances platinum linings are of the 'loose' type.

Bursting Discs

Bursting discs are the simplest and most certain form of protection against the effect of over-pressure in a closed system. They cannot fail to operate,

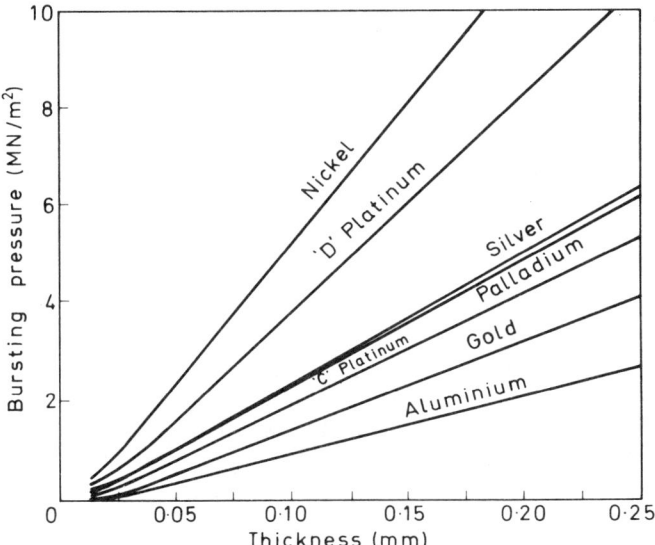

Fig. 6.6 Relationship between bursting pressure and original thickness for seven disc materials in a standard 1 in (25·4 mm) orifice

they have low inertia, they open immediately and they provide a full-bore passage for the discharge of pressure. Bursting discs are completely leak-proof until they burst, and can be made to withstand full vacuum. A most valuable aspect of their behaviour is their inherent tendency always to 'fail safe'.

The protection of vessels containing corrosive materials presents a special problem for the selection of bursting discs—a rapid rate of corrosion can lead to a high frequency of failures. In addition, the 'creep' of a metal disc when under tension at elevated temperatures would tend to weaken it and result in premature failure.

Bursting discs may be fabricated of gold, silver, platinum or palladium. The recommended maximum temperatures for continuous use are 80°C for gold, 150°C for silver, 300°C for palladium and 450°C for platinum. Figure 6.6 gives bursting pressure/disc thickness data for these metals and for aluminium and nickel.

Spinnerets

The spinning of viscose rayon for the production of yarn, tyre cord and staple fibre involves the extrusion of an alkaline solution of cellulose into an acid bath. The orifices through which individual fibres are extruded are often extremely small—down to 30 μm or less in diameter, and their dimensional accuracy must be maintained to a very high degree for long periods while operating in two highly corrosive media simultaneously. Often many thousands of such orifices are contained in a single spinneret.

Platinum–gold alloys are widely employed as materials of construction for rayon spinnerets, in particular the 30Pt–70Au containing 0·5 Rh as a grain-refining additive. This alloy, capable of being hardened substantially by suitable heat treatment, allows the orifices to be pierced at a hardness of about 120 HV and after treatment provides a finished spinneret with a hardness of approximately 220 HV. This greater hardness permits a high polish to be produced in a scratch-resistant exit face. Small grain size ensures that the holes produced have a high grade of uniform circularity. Other alloys for spinneret construction include rhodium–platinum, iridium–platinum, iridium–rhodium–platinum, ruthenium–platinum and ruthenium–palladium.

Noble Metal Solders

Joining one metal to itself or to another has frequently presented many metallurgical and chemical problems. For a number of years palladium-containing brazing alloys have found widespread applications, as they offer several advantages over other conventionally used materials:

(a) They possess considerable strength at high temperatures.
(b) They have low vapour pressures, and are thus suitable for applications in vacuum systems.
(c) They permit brazing of metals such as cobalt-based alloys, Nimonics, tungsten, zirconium, molybdenum and beryllium which could not otherwise be joined.

(d) They have little erosive effect on parent metals, permitting thin sections to be joined.
(e) They permit ceramics to be joined directly to low-expansion alloys.
(f) They have a high resistance to corrosion.

Silver–copper–palladium alloys with liquidus temperatures of 800–1 000°C have very low vapour pressures combined with good wetting and flow characteristics and are widely employed in vacuum work. They exhibit a lower tendency to stress corrosion than silver–copper, and do not form brittle alloys with other metals.

Silver–palladium–manganese brazes possess excellent creep characteristics and have been developed for high-temperature applications involving the use of cobalt or nickel-based alloys, heat-resistant steels, molybdenum and tungsten. Their liquidus temperatures lie in the range 1 100–1 250°C.

Nickel–manganese–palladium brazes are resistant to attack by molten alkali metals and find applications in sodium-cooled turbine constructions. Their freedom from silver and other elements of high thermal neutron-capture cross-section allows them to be used in liquid-metal-cooled nuclear reactors.

Copper–palladium–nickel–manganese brazes give very low erosion of the parent metals to be brazed, and are therefore used to join thin sections of stainless steels and high-nickel alloys.

Gold–copper alloys exhibit exceptional resistance to corrosion, and have very low vapour pressures. Gold–nickel alloys, with similar low vapour pressure, are somewhat stronger than gold–copper at high temperatures. Both series of alloys are widely employed in vacuum systems.

The addition of beryllium and silicon to nickel–palladium alloys gives very good high-temperature brazes, especially for alloys containing aluminium and titanium.

Pure ruthenium powder or mixed ruthenium–molybdenum powders have been found able to effect good joints between molybdenum and tungsten. A eutectic melting above 1 900°C is formed, and joints produced in hydrogen atmospheres at 2 100°C operate satisfactorily at 1 500°C. A cobalt–palladium–gold alloy has also been reported to be useful in brazing molybdenum.

Silver-based low-temperature brazing alloys have been known for many years and are widely employed throughout the engineering and chemical industries. Their output is measured in many tonnes per year, and their utility in joining almost all the commonly used materials of construction has made them indispensable for the production of a very wide range of plant, equipment and structures of all kinds. They exhibit a high resistance to corrosion by industrial atmospheres and possess excellent strengths at temperatures considerably higher than may be employed for lead or tin-based solders (see Section 10.5).

Anodic Behaviour of the Noble Metals

As with the chemical behaviour of the noble metals in aqueous solutions, their anodic behaviour closely follows the predictions of the Pourbaix diagrams[2,14], if due allowance is made for the formation of complexes.

Silver generally corrodes anodically below the reversible oxygen potential, unless an insoluble oxide or salt is formed. When silver is used as an anode in sulphuric acid solutions, its behaviour shows an analogy with that of lead. Silver sulphate, Ag_2SO_4, is first formed, and this acts as a passive film[15]. When the potential is raised the sulphate is oxidised to AgO, which may be cathodically reduced back to Ag_2SO_4 at a potential lower than that required for its initial formation. When made anodic in nitrate solutions, silver generally dissolves quantitatively as Ag^+, and this forms the basis of the electrorefining techniques widely used in industry. Similar considerations apply to the anodic behaviour in cyanide solutions where silver forms a complex cyanide. In chloride solutions silver anodes become covered with a layer of silver chloride. When made anodic in alkaline solutions, silver becomes covered with a thin film of Ag_2O silver oxide, which protects it from further attack.

In chloride solutions, gold, when made anodic, initially dissolves in accordance with the equation $Au + 4Cl^- \rightarrow AuCl_4^- + 3e$. The dissolution gradually decreases, and the gold surface becomes passive owing to an adsorbed layer of oxygen[16]. Gold anodes in perchloric acid are passivated by a film of oxide, Au_2O_3, if used at potentials above 1·4 V. If the potential is raised above 1·95 V, the oxide film becomes flaky, and the gold is severely attacked[17]. In sulphuric acid an initial period of dissolution exists in accordance with the equation $Au \rightarrow Au^+$, followed by passivation resulting from a film of hydroxide, $Au(OH)_3$[18]. Ozone may be produced at sufficiently high current densities and the $AU(OH)_3$ is deposited with an accompanying attack on the gold below it to form an unstable peroxide. Continued production of oxide on the surface produces increasing passivity.

The anodic behaviour of platinum and certain of its alloys is of considerable technical importance, since they can be employed under a wide range of conditions without appreciable corrosion, and often in circumstances where no other metal can be used. Their use industrially has recently been extended by applying them as thin coatings to a substrate of a passive metal such as tantalum or, more commonly nowadays, titanium, to reduce the cost. Platinised titanium anodes are discussed in detail in Section 11.3.

Platinum can be used as an anode for the evolution of oxygen without corrosion over the whole pH range. At one time it was considered that the platinum was inert, so that oxygen could be evolved reversibly in the thermodynamic sense. However, it has been found that the reversible oxygen potential is not normally attained at a platinum surface, and that adsorbed oxygen or an oxide film play a part in the reaction. The nature of the surface film on platinum and some of the other noble metals has been the subject of considerable study[19]. Its exact nature is still in dispute, and probably varies for the different metals and under different conditions. The important point for technical applications is that it is very thin and is conductive, so that platinum anodes can operate at high current densities without excessive polarisation. However, the overpotential for the evolution of oxygen or chlorine on the surface of the thin oxide film is appreciably greater than on the bare metal surface which is considered to be a serious limitation in relation to the successful use of platinum in fuel cells[20]. Comparatively thick oxide films are only produced on platinum by direct currents with a considerable superimposed a.c. component or by alternating current under some conditions[21].

For chlorine evolution, platinum generally shows very low rates of attack, and platinum anodes can be used for cathodic protection in sea-water[22], and for chlorine production. However, in strongly acid solutions at low current densities platinum can be attacked; due to the complexing action of the chloride ion, the area of the potential-pH diagram (Fig. 6.3) in which corrosion occurs is somewhat enlarged, so that an anode operating at low overvoltage may be at a potential within this region. At higher current densities the overvoltage rises and passivation occurs.

The rate of corrosion of platinum anodes in chloride solutions increases if a considerable a.c. component is superimposed on the d.c. supply[23].

Rhodium and iridium have a resistance to anodic corrosion comparable with that of platinum, and are more resistant to the influence of alternating currents. A platinum-iridium alloy, in the form of a coating on titanium, is preferred to pure platinum for the production of chlorine from brine[24], due to its improved corrosion resistance and lower overvoltage.

Palladium is considerably less resistant to anodic corrosion than platinum, though it may be used for evolution of oxygen from alkaline solutions. It is attacked rapidly when used as an anode in sea-water, and dissolves quantitatively in acid chloride solutions.

Ruthenium dissolves anodically in alkaline solutions, as predicted by Pourbaix[3], but its corrosion resistance when made anodic in acid solutions is variable. Under some conditions the volatile and toxic tetroxide is evolved. Osmium is even more reactive anodically than ruthenium.

Recently it has been shown that the oxides of the platinum metals can have a higher corrosion resistance than the metals themselves[25], and have sufficient conductivity to be used as coatings for anodes, e.g. with titanium cores. Anodes with a coating of ruthenium dioxide are being developed for use in mercury cells for the electrolysis of brine to produce chlorine[26], since they are resistant to attack if in contact with the sodium-mercury amalgam.

The use of small platinum pins to promote the formation of lead dioxide on lead anodes is described in Section 11.3.

Alloying Additions of Noble Metals to Facilitate Passivity

The addition of a small percentage of a noble metal to a base metal such as stainless steel or titanium can provide sites of low overvoltage for the cathodic reduction of dissolved oxygen or hydrogen ions. This permits larger currents and hence more positive potentials to be obtained at the anodic region, and promotes passivation under some circumstances[27]. This effect has been demonstrated for stainless steels[2] but has not been adopted in practice, since under other conditions the noble metal addition accelerates corrosion[27].

The method is more useful with titanium, and the effect of alloying titanium with a small amount of palladium is described in Section 5.4. The use of platinum in the prevention of hydrogen embrittlement in tantalum, considered in Sections 5.5 and 1.7, also depends on the low hydrogen overvoltage of the added element.

High-temperature Applications of the Platinum-group Metals

The principal applications of the outstanding stability of platinum and its alloys at high temperatures lie in their use as materials of construction for equipment to handle molten glass and as electrical resistance windings for high-temperature furnaces.

In handling molten glasses at temperatures which sometimes reach 1 500°C, and frequently lie around 1 400°C, platinum alloys fulfil several functions:

(a) They protect refractory bricks and equipment from the erosive effect of flowing glass, thus greatly prolonging the life of those parts which would otherwise have to be replaced frequently.
(b) They protect the glass from contamination either by particles of refractory broken off by erosion or by the dissolution of refractory bricks or parts.
(c) They enable very high dimensional accuracy to be maintained when used to line or construct equipment designed to meter, control, or form molten glass.

The most usual forms of platinum or, more frequently, of the platinum alloys containing 5 or 10% of rhodium, are thin protective sheaths contoured to conform with the shape of the underlying refractory. Sheaths of this type, generally 0·25 mm to 0·65 mm thick, are widely employed to protect tank lips, skimmer blocks, stirrers, thermocouple pockets, etc. More substantial—thicknesses-up to 2·5 mm thick—are used to protect orifices whose dimensional accuracy must be maintained to a high degree.

Bushings for the manufacture of glass fibre contain a large number of small, accurately shaped orifices which must maintain their dimensional accuracy to a very high degree. In addition, such bushings must remain free from distortion and act as electrical resistance heaters providing an even temperature profile to give uniform rates of flow through individual orifices. Operating continuously at temperatures around 1 400°C, such bushings have to give long life under very arduous conditions. They are generally made of Pt–10Rh and supported by accurately shaped refractory bricks.

Rhodium in rhodium–platinum alloys imparts a slight grey colour to glasses, especially to those of high lead content, so that pure platinum is utilised for the construction of crucibles or other equipment to handle highquality colour-free optical glasses. Even platinum can impart traces of colour if the glass is high in alkali content, but the effect is reduced in oxidising atmospheres.

Investigations on the wettability of platinum alloys by glasses have established that this property, as measured by the contact angle between glass and metal, is strongly dependent on the surrounding atmosphere. It rises sharply as the atmosphere changes from oxidising to reducing, and is also dependent on the affinity for oxygen of the alloying constituent, which is generally rhodium but may also be gold, beryllium, or iridium.

Rhodium–platinum alloys containing up to 40% Rh are used in the form of wire or ribbon in electrical resistance windings for furnaces to operate continuously at temperatures up to 1 750°C. Such windings are usually completely embedded in a layer of high-grade alumina cement or flame-sprayed alumina to prevent volatilisation losses from the metal due to the free circulation of air over its surface. Furnaces of this type are widely employed for steel analysis, ash fusions and other high-temperature analytical procedures.

Working with Noble Metals

Silver

Silver is available in three principal grades for industrial purposes — fine silver, the normal commercial product containing a minimum of 99·95% silver which is used in most plant and equipment; chemically pure silver containing a minimum of 99·99% silver, used for catalytic and special purposes where the presence of certain trace impurities may adversely affect its resistance to corrosion, and oxygen-free silver. The latter grade is as pure as fine silver and is used for equipment which may be heated in the presence of hydrogen without danger of embrittlement resulting from combination of the hydrogen with oxygen dissolved in the metal.

Silver is available in many standard forms — sheet, strip, foil of thicknesses down to 0·013 mm, rod, wire down to 0·013 mm diameter, gauze, tubes, bi-metal as silver-clad copper or phosphor-bronze and many others. It is easily fabricated by the normal techniques of rolling, spinning, drawing, etc. and readily joins to itself by fusion-welding using argon-arc welding. Welding by oxy-hydrogen flame may also be employed, but the resulting weld is not as satisfactory owing to the possibility of oxygen absorption while the metal is molten, followed by embrittlement by hydrogen. Fine silver filler rods may be used, and hammering the weld fillet down to the contour of the surrounding metal produces a very strong joint. A wide range of silver-based or soft (tin-based) solders may also be used.

Gold

Gold is available for industrial purposes in a grade containing a minimum of 99·9% gold in a wide range of forms — sheet, foil, tube, wire, etc. It is easily fabricated, and when it is being joined to itself may be fusion-welded with an oxy-hydrogen flame or hammer-welded at temperatures well below the melting point. Soft-soldering is not recommended.

The Platinum-group Metals

Platinum is capable of being refined to an extremely high degree of purity, resulting in a metal of remarkably consistent and reproducible properties. Certain traces of impurities, however, have profound effects on some of its properties, and the correct grade of platinum must always be selected for any particular application. Among the applications for which special grades of platinum have been developed may be listed the following:

Chemical and catalytic. This grade of platinum is for conversion to catalysts, gauzes and chemical compounds. Spectrographic analysis is employed to control the presence of trace impurities harmful in these applications.

High-temperature. Depending on the intended continuous operating temperature, specific impurities known to affect the mechanical properties of platinum at these temperatures are excluded. Grain-stabilised platinum containing a refractory additive to reduce grain growth to a minimum and to increase its resistance to creep and mechanical strength is also available.

Equipment. The impurities in this grade of platinum, intended for all normal purposes in chemical plant, electrodes, crucibles, etc. other than for use above 1 000°C, consist only of traces of the other platinum metals.

Platinum in the forms detailed above and in its more usual alloys with other noble metals is available as sheet, foil down to 0·0064 mm thick, tube, rod, wire down to 0·0064 mm diameter, Wollaston wire down to 0·001 mm diameter, and clad on thin sections of base metals, e.g. copper, nickel, Inconel, etc.

Platinum, palladium and the normal alloys of platinum used in industry are easily workable by the normal techniques of spinning, drawing, rolling, etc. To present a chemically clean surface of platinum and its alloys after fabrication, they may be pickled in hot concentrated hydrochloric acid to remove traces of iron and other contaminants—this is important for certain catalytic and high-temperature applications. In rolling or drawing thin sections of platinum, care must be taken to ensure that no dirt or other particles are worked into the metal, as these may later be chemically or electrolytically removed, leaving defects in the platinum.

When platinum or its alloys are being joined, properties of the weld or solder must be such that it is no less corrosion or oxidation-resistant for the application in question than the parent metal. Platinum and its alloys are readily joined to themselves and to certain base metals, e.g. iron, nickel, copper. The principal methods for joining platinum are as follows:

(*a*) Fusion welding, using a platinum or alloy filler rod of the same composition as the parent metal and a shielded electric arc or an oxy-hydrogen flame (an oxy-acetylene flame may cause carbon pick-up by the molten metal). The weld fillets are then cold hammered to the contours of the surrounding metal to provide a strong joint.

(*b*) Hammer-welding. Platinum and rhodium-platinum alloys when cleaned are readily hammer-welded to themselves and to each other at temperatures in the range 800–1 000°C. The welds so produced are completely homogeneous.

(*c*) Soldering. Fine gold, copper, silver–palladium or platinum–palladium–gold–copper alloys may be used to solder platinum to itself, to its alloys, or to steels, nickel, etc. No fluxes are used, and soft solders should not be employed.

Economies Gained by Using Noble Metals

In spite of their high initial cost, the noble metals in industry offer substantial advantages and economies in use. They avoid the frequent and expensive replacement of plant on account of corrosion, they protect the product from contamination, allowing a higher quality with its attendant premium on value to be produced, and they frequently permit the carrying out of operations which would not otherwise be possible.

When they have served their purpose or become damaged the noble metals will realise a very high proportion of their initial cost. No matter in what form they are utilised, very efficient processes are available for effecting their complete recovery. This factor often makes noble metals the most economic in use in the chemical and engineering industries. For many applications, no other metal or group of metals can fulfil their function as efficiently, combined with such a low net cost to their user.

Recent Developments

Investigations into the electrochemical corrosion of the noble metals, particularly gold and platinum, have been reported widely over the past decade. Both gold and silver have a high dissolution rate in acidic thiourea solutions containing iron ions as an oxidant; this rate depends on thiourea and iron concentration[9]. The anodic dissolution of gold in potassium cyanide solution is of interest to a number of industries. In aqueous alkaline cyanide, a study as a function of potential within the range -0.9 to $+0.4$ V (SCE) showed three maxima at -0.65, $+0.04$ and $+0.38$ V. The decrease in the dissolution current between these peaks were ascribed to film formation[30-32]. A single electron transfer reaction was predicted over the potential range -0.9 to $+0.3$ V.

James and Hagee[33] have reported a study of the high temperature vaporization chemistry in the Au-Ce system including the formation of vapour complex species of gold and silver with copper and iron.

The selective dissolution of Ag-Au alloys showed that the anodic dissolution is controlled by non-steady state diffusion of silver from the bulk alloy to the surface at low potentials[34]. This effect increases with increase in potential and caused by the formation of vacancies in the surface layer, the concentration of which depends on potential. Above a certain critical potential, characterisation of each of the alloys studied, the dissolution increased because of the destruction of the gold-enriched surface layer. In a similar vein, Forty[35] has looked at the TEM of surface morphology changes during selective dissolution of silver in 50Au50Ag alloy by nitric acid. It was revealed that surface diffusion of gold plays an important part in depletion gilding. As the surface silver atoms dissolve, residual gold atoms re-form into gold-rich islands so that fresh silver atoms are continuously exposed to the corrosive agent, layer by layer.

Duncan and Frankenthal[36] report on the effect of pH on the corrosion rate of gold in sulphate solutions in terms of the polarization curves. It was found that the rate of anodic dissolution is independent of pH in such solutions and that the rate controlling mechanism for anodic film formation and oxygen evolution are the same. For the open circuit behaviour of ferric oxide films on a gold substrate in sodium chloride solutions containing low iron concentration it is found that the film oxide is readily transformed to a lower oxidation state with a Fe^{2+}/Fe ratio corresponding to that of magnetite[37].

Nicol has reported extensively on the anodic oxidation of gold[38-40]. For acidic solutions[40], a potential pH diagram is shown delineating the zones of compound stability in aqueous solutions and a graph depicts the standard reduction potentials of gold complexes with cyanide, thiosulphate, iodide, sulphite, bromite and chloride ions as well as with thiourea and water. A table summarises the reactions with the principal Zigadns, describes anodic products, states gold yields and indicates passivation states by oxides. For reactions in alkaline solutions[39], particularly alkaline cyanides, cyanide appears to be the species in which gold has the most stable anodic reaction at high pH values. It is concluded that gold–cyanide reactions have, so far, not been explained adequately.

Gold electro-deposits on stainless steel and nickel-based alloys have been assessed in a variety of laboratory and industrial test atmospheres, and the extent of tarnishing determined from contact resistance measurements[41]. Angerer and Ibl[42] report on the electrodeposition of cobalt hardened gold from an acid citrate both in terms of current efficiency, deposit carbon and cobalt content, porosity and morphology for a variety of current densities and electrolyte flow rates. The process became mass transport limited at $50\,\text{mAcm}^{-2}$. Cobalt additions decrease the current efficiency caused by a decrease in the Rydropen over potential and an increase in the over potential of gold deposition. The properties of gold deposits produced by a pulsed current density and asymmetric a.c. plating instead of d.c. for a cyanide-citrote–sulphate electrolyte without alloy addition is considered by Dini and Johnson[43].

Several gold based alloys used for electrical contacts have been evaluated by exposure to oxidation, H_2S or SO_2[44] whilst the corrosion failure mechanism associated with gold metallisation in electronic circuits have been reported[45,46]. Growth of gold shorts from a cathodic conductor occurs if chloride ions are present whilst a voluminous reaction product of $Au(OH)_3$ is produced by the anodisation of an anodically biased conductor. No electronic circuit using gold conducting paths, is totally immune to corrosion failure.

The stress corrosion cracking of low carat gold alloys during manufacture and storage is a perennial problem and has been reviewed by Dugmore and DesForges[47].

The importance of gold dental alloys is apparent from the many papers regularly being published on the subject. Cracks in gold crowns cemented on amalgam restorations are seen as evidence of stress corrosion[48] with the possibility of mercury diffusion into the crown and electrochemical dissolution of the grain boundaries. Laser welded dental gold alloys show less corrosion than soldered alloys and the different solidification conditions have been held responsible[49]. Similarly, the micro-segregation of silver- and/or copper-rich phases are thought to be the cause of the low chloride resistance of 45% and 65% gold dental alloys[50]. Laub and Stanford discusses the position and future testing of new gold dental alloys with reference to their tarnish resistance and corrosion behaviour[51] whilst Prasad[52] summarises the papers presented at the second annual conference of the International Precious Metals Institute covering both the physical and chemical metallurgy aspects of gold dental alloys. Finally, whilst considering gold, Chaston[53] provided an extensive review of the industrial uses of gold.

The thermodynamic behaviour of silver and solubilities of silver and its compounds have been computed in an electrochemical study of silver in potassium hydroxide solutions at high temperature[54].

Platinum has also had its share of attention in recent years. The effect of phosphoric acid concentration on the oxygen evolution reaction kinetics at a platinum electrode using 0·7 m–17·5 m phosphoric acid at 25°C has been studied with a rotating disc electrode[55]. The characteristics of the ORR are very dependent on phosphoric acid concentration and H_2O_2 is formed as an intermediate reaction. Also, platinum dissolution in concentrated phosphoric acid at 176 and 196°C at potentials up to 0·9 (SHE) has been reported[56].

A study of the corrosion of platinum in alkaline solutions at 20 and 70°C show the rate to be higher in the presence of oxygen than that in inert atmosphere the difference decreasing as the potential varies from 0·9 to 1·2 V[57]. The dissolution appears to involve the intermediate formation of surface oxides either by discharge of water molecules or by interaction with molecular oxygen. Biss[58] has reported on the oxygen evolution at platinum electrodes in alkaline solution whilst the temperature and concentration effects on the electrochemical reduction of oxygen in potassium hydroxide solutions has also been considered[59]. An interesting approach to anodic dissolution has been provided by using both radioisotopes and electrochemical methods[60]. The electrolyte employed was alkaline gallate and corrosion was under conditions simulating the electrorefining of gallium metal. The corrosion loss of platinum increases with gallium content of the electrolyte; SEM showed that dissolution may be localised near non-metallic inclusions.

Turning now to the acidic situation, a report on the electrochemical behaviour of platinum exposed to 0·1 M sodium bicarbonate containing oxygen up to 3970 kPa and at temperatures of 162 and 238°C is available[61]. Anodic and cathodic polarisation curves and Tafel slopes are presented whilst limiting current densities, exchange current densities and reversible electrode potentials are tabulated. In weak acid and neutral solutions containing chloride ions, the passivity of platinum is always associated with the presence of adsorbed oxygen or oxide layer on the surface[62]. In concentrated hydrochloric acid solutions, the possible retardation of dissolution is more likely because of an adsorbed layer of atomic chlorine[62].

A number of bi-electrodes have been studied for application as insoluble anodes in electroplating[63]; platinised titanium, Ti-Pt, Ti-Cu and Ti-Ag. Anodic polarisation measurements in various copper, nickel, chromium and tin plating solutions together with passivation current densities are used to discuss performance and suitability.

The oxidation kinetics of both gold and palladium alloys as a function of temperature and pressure have been reported by Opara et al[64] and the behaviour of palladium and palladium oxide when heated in gaseous hydrochloride acid at 20–1000° by Ivashentsev and Ryumin[65].

A potential pH diagram for the Pd-H_2O-Cl system is presented and described[66] whilst the rest potentials of palladium electrodes in oxygen-saturated solutions of different pH values have been determined[67]. From the latter, $dEr/dpH = -65$ mV in acid and -35 mV in alkaline solutions, and $dEr/dlgP_{O_2} = 30$ mV for both pH ranges.

The weight change with time for pure rhodium annealed at 1400°C in vacuum and subject to oxygen, nitrogen or air at 1400–1800°C and pressures of 0·0001–760 tarr is used to discuss the metal's resistance to oxidation and sublimation by the formation of volatile oxides[68]. The stability of rhodium is found to be sensitive to temperature. Air and nitrogen atmospheres decrease the sublimation of 1800°C.

Finally, reviews of the oxidation reactions of the platinum metals[69] and new metallurgical developments in the field of precious metals[70] have been published.

G. W. WALKIDEN
R. A. JARMAN

REFERENCES

1. Anson, F. C. and Lingane, J. T., *J. Amer. Chem. Soc.*, 79, 4 901 (1957)
2. La Que, F. L. and Copson, H. R., *Corrosion Resistance of Metals and Alloys*, Chapman & Hall, London
3. Pourbaix, M. J. N., *Atlas of Electrochemical Equilibria*, Pergamon Press, Oxford (1966)
4. Latimer, W. L., *Oxidation Potentials*, Prentice Hall, New York, 191 (1961)
5. Latimer, W. L., *Oxidation Potentials*, Prentice Hall, New York, 197 (1961)
6. Latimer, W. L., *Oxidation Potentials*, Prentice Hall, New York, 195 (1961)
7. Powell, A. R., *Platinum Metals Rev.*, No. 2, 95 (1958)
8. Andreeva, V. V. and Shishakov, N. A., *J. Appl. Chem.*, 11, 388 (1961)
9. Alcock, C. B. and Hooper, G. W., *Proc. Roy. Soc.*, (A)254, 551 (1960)
10. Raub, E. and Plate, W. Z., *Metallk.*, 48, 529 (1957)
11. Betteridge, W. and Rhys, D. W., *Proc. First International Congress on Metallic Corrosion*, London, 1961. Butterworths, London, 186 (1962)
12. Unpublished, Johnson, Matthey & Co. Ltd., Research Laboratories, London
13. Johnson, Matthey & Co. Ltd., UK Pat. 830 628 (1958)
14. Pourbaix, M. J. N., Van Muylder, J. and de Zoubov, N., *Platinum Metals Rev.*, 3 No. 47, 100 (1959)
15. Jones, P. and Thirsk, H. R., *Trans. Faraday Soc.*, 50, 732 (1954)
16. Armstrong, G. and Butler, J. A. V., *Trans. Faraday Soc.*, 30, 1 173 (1934)
17. Jirsa, F., *Coll. Czech. Chem. Comm.*, 13, 505 (1948)
18. Jirsa, F. and Buryanek, O. Z., *Electrochem.*, 29, 126 (1923)
19. Hoare, J. P., *Advances in Electrochemistry and Electrochemical Engineering*, 6, Interscience Publishers, New York, 201–288 (1967)
20. Bockris, J. O'M. and Reddy, A. K. N., *Modern Electrochemistry*, Vol. 2, Macdonald, London, 1 350 (1969)
21. Altmann, S. and Busch, R. H., *Trans. Faraday Soc.*, 45, 720 (1949)
22. Walkiden, G. W., *Corrosion Technology*, Jan.-Feb. (1962)
23. Juchniewicz, R. and Hayfield, P. C. S., *Proc. 3rd Int. Congr. on Metallic Corrosion*, Moscow, 1966. Vol. III, English Ed., 73–82
24. Anon., *Platinum Metals Rev.*, 13 No. 3, July, 103 (1969)
25. Beer, H. B., UK Pat. 1 147 442 (1969)
26. Anon., *Platinum Metals Rev.*, 14 No. 3, July, 92 (1970)
27. Hoar, T. P., *Platinum Metals Rev.*, 2, 117 (1958)
28. Tomashov and Chernova, *Dokl. Akad. Nauk SSSR*, 89, 121 (1953)
29. Won, C. W. and Cho, T. R., *J. Korean Inst. Metals*, 25, 495 (1985)
30. Kirk, D. W. *et al.*, *J. Electrochem. Soc.*, 125, 1436 (1978)
31. Kirk, D. W. *et al.*, Ibid., 126, 2287 (1979)
32. Tan, T. P. and Wan, C. C., *Chem. Tech. Biotechnol.*, 29, 427, (1979)
33. James, S. E. and Hager, J. P., *Metall. Trans.*, 9B sol. (1978)
34. Vyazovikina, N. V. and Marshakov, I. I., *Zasch. Met.*, 15, 656 (1979)
35. Forty, A. J., *Nature*, 282, 597 (1979)
36. Duncan, B. S. and Frankenthal, R. P., *J. Electrochem. Soc.*, 126, 95 (1979)
37. Formavo, L., *Borros. Sci.*, 20, 1251 (1980)
38. Nicol, M. J., *Gold Bull.*, 13, 46 & 105 (1980)
39. Nicol, M. J., Ibid., 14, 42 (1982)

40. Nicol, M. J., Ibid., **13**, 105 (1980)
41. Mayer, U., *Galvano*, **504**, 203 (1980)
42. Angerer, H. and Ibl, N., *J. Appl. Electrochem.*, **9**, 219 (1979)
43. Dini, J. W. and Johnson, H. R., *Gold Bull.*, **13**, 31 (1980)
44. Schiff, K. L. and Harmsen, N., *Metall.*, **30**, 620 (1976)
45. Frankenthal, R. P. and Becker, W. H., *J. Electrochem. Soc.*, **126**, 1718 (1979)
46. Frankenthal, R. P. and Kruger, J., *Gold Bull.*, **18**, 46 (1985)
47. Dugmore, J. M. M. and Des Forges, C. D., Ibid., **12**, 140 (1979)
48. Oden, A. and Tullberg, M., *Acta Odontol. Scand.*, **43**, 15 (1985)
49. van Benthem, H. and Vahl, J., *Dtsh. Zahnartzi*, **40**, 286 (1985)
50. Sarkar, N. K. et al., Dent, F., *Res.*, **58**, 568 (1979)
51. Laub, L. W. and Stanford, J. W., *Gold Bull.*, **14**, 13 (1981)
52. Prasad, A., Ibid., **11**, 134 (1978)
53. Chaston, J. C., *Int. Met. Rev.*, **22**, 25 (1977)
54. Pound, B. G. et al., *Electrochim. Acta*, **24**, 929 (1979)
55. Hsueh, K. L. et al., *J. Electrochem. Soc.*, **131**, 823 (1984)
56. Bindra P. et al., Ibid., **126**, 1631 (1979)
57. Khrushchuaue, E. I. et al., *Prot. Met.*, **15**, 560 (1979)
58. Biss, V. I., *J. Electrochem. Soc.*, **133**, 1621 (1986)
59. Park, S. M. et al., Ibid., **133**, 1641 (1986)
60. Mironov, Yu. M. et al., *Prot. Met.*, **12**, 532 (1976)
61. Francis, P. E. and Shawski, S., *Brit. Corros. J.*, **12**, 230 (1977)
62. Markov, S. S. et al., *Prot. Met.*, **13**, 89 (1977)
63. Kruger, M. and Gabe, D. R., *Trans. Inst. Met. Fin.*, **54**, 127 (1976)
64. Opara, B. K. et al., *Prot. Met.*, **15**, 398 & 494 (1979)
65. Ivashentsev, Ya I. and Ryumin, A. I., *Sov. Non-Ferrous Met. Research*, **6**, 205 (1978)
66. Kammel, R. et al., *Metall.*, **30**, 414 (1976)
67. Vracar, Lj M. et al., *J. Electrochem. Soc.*, **134**, 1695 (1987)
68. Ismail, M., *Platinum Met. Rev.*, **23**, 22 (1979)
69. Raub, C. J., *Metall.*, **32**, 802 (1978)
70. Raub, C. J., Ibid., **30**, 638 (1976)

7 HIGH-TEMPERATURE CORROSION

7.1	Environments	**7**:3
7.2	The Oxidation Resistance of Low-Alloy Steels	**7**:16
7.3	High-Temperature Corrosion of Cast Iron	**7**:53
7.4	High-Alloy Steels	**7**:67
7.5	Nickel and its Alloys	**7**:91
7.6	Thermodynamics and Kinetics of Gas-Metal Systems	**7**:145

7.1 Environments

Introduction

The terms 'hot corrosion' or 'dry corrosion'* are normally taken to apply to the reactions taking place between metals and gases at temperatures above 100°C; i.e. temperatures at which the presence of liquid water is unusual. The obvious cases of wet corrosion at temperatures above 100°C, i.e. in pressurised boilers or autoclaves, are not considered here. In practice, of course, common metals and alloys used at temperatures above normal do not suffer appreciable attack in the atmosphere until the temperature is considerably above 100°C. Thus iron and low-alloy steels form only the thinnest of interference oxide films at about 200°C, copper shows the first evidence of tarnishing at about 180°C, and while aluminium forms a thin oxide film at room temperature, the rate of growth is extremely slow even near the melting point.

Consideration will also be given to attack arising from contact with solids such as refractories, and with molten materials such as salts, glasses, and lower-melting-point metals and alloys. On a fundamental basis, the distinction between some of these latter reactions and normal-temperature aqueous corrosion is not always clear, since galvanic effects may be of significance in both cases, but for practical purposes a distinction can be made on the basis of the temperature involved.

A few cases occur in which hot-corrosion and wet corrosion are interdependent, the wet corrosion arising from the condensation of liquids generated during a period at elevated temperatures. The formation of condensates of hydrobromic acid in engines burning anti-knock fuels containing ethylene dibromide is important in this context. Such cases are properly considered as aqueous corrosion.

Most hot-corrosion phenomena of practical significance are controlled by the kinetics of the reactions proceeding, rather than by the thermodynamic stability of the reactants or products involved. It must, however, be borne in mind that reaction rates determined under simplified laboratory conditions are frequently inapplicable to the more complicated conditions experienced in practice. Factors of major importance in this context are stress and thermal cycling.

*The terms 'wet' and 'dry' in relation to corrosion are discussed in Section 1.1.

Stress This can be of importance in two ways. An effect analogous to aqueous stress corrosion, in which attack is accelerated at grain boundaries, may be produced; this is particularly noticeable and very well known in cases of attack by other molten metals. Stress may also accelerate general attack when the metal is suffering progressive deformation, as in creep or fatigue; the mechanical properties of the metal and of the scale may differ sufficiently from one another for the scale to crack periodically as deformation of the metal increases, and thus produce corresponding increases in the rate of corrosive attack. Gulbransen and Andrew[1] have reported effects of this nature on the rate of oxidation of nickel–chromium alloys, although the test specimens were cooled to normal temperature for the strain to be imposed. It was shown that the magnitude of the change in the rate of oxidation depended critically on the silicon content of the alloy. Abnormal oxidation effects during high-temperature fatigue tests on Nimonic alloys have been described by Betteridge[2], accelerated oxidation occurring at the point of fatigue cracking.

Thermal cycling This can lead to cracking and flaking of scale layers, either by stresses due to differential expansion of scale and underlying metal, or by dimensional changes associated with phase changes in the metal or in the scale. The latter phenomenon has proved particularly troublesome in efforts to produce a protective scale on molybdenum-base alloys[3]. Scale flaking, whatever the cause, leads to acceleration of the corrosion rate. Although some laboratory tests simulate the effects of thermal cycling it must be remembered that the size and shape of test sample is significant. Bruce and Hancock[4] have developed a vibrational technique for studying the onset of scale cracking during oxidation and have used it to study the behaviour of scales on nickel and iron in the temperature range 570–1 000°C.

The more fundamental aspects of the oxidation of metals have been fully described in Chapter 1, and by Kubaschewski and Hopkins[5], while practical considerations, particularly in relation to commercially important alloys, have been dealt with by Hessenbruch[6].

Air and Oxygen

Although corrosion by oxidation in air is undoubtedly the type of attack of greatest practical importance it is remarkable that relatively few cases are entirely free from the complications arising from contamination of the reacting atmosphere. Except in the case of heating by electrical means, contamination arises from the combustion products, in gaseous, liquid or solid form, while with electrical heating the oxidation is often complicated by the presence of fumes or vapour arising from the materials being processed.

Perhaps the closest approach to pure oxidation in everyday conditions arises in domestic electric heating appliances where the elements are exposed to the air. At some points the elements are necessarily in contact with supporting refractories, and if these are not of adequate purity, accelerated corrosion leading to early failure can occur. In a similar way the sheathed radiant-type elements of electric cookers usually fail owing to the corrosive effects of contaminants such as animal fats or salts from spilled liquids.

Wires for these applications are commonly assessed by a standardised test such as that described in ASTM B76-81, which involves electrical heating for periods of 2 min to a temperature of 1 177°C (or other temperature depending on the composition of the wire) until failure by burn-out occurs. The test is a satisfactory comparative one which simulates service conditions apart from the acceleration resulting from the considerably increased temperature, but it is not suitable for the determination of basic data. Accurate setting of temperature and its maintenance at a uniform level along the length of a wire are difficult. Furthermore, the steep temperature gradient from the wire into the surrounding cool atmosphere produces disturbances due to thermal diffusion effects. These are of particular importance in the case of wires giving rise to volatile oxides (such as nickel-chromium wires forming chromic anhydride) since the heavy oxide molecules are rapidly removed by thermal action from the vicinity of the wire. This phenomenon is probably responsible for the fact that the appearance of an electrically heated wire, particularly in relation to scale adhesion, is frequently different from that of a wire heated in a furnace under similar conditions.

In the case of alloys having one constituent considerably more reactive to oxygen than the others, conditions of temperature, pressure and atmosphere may be selected in which the reactive element is preferentially oxidised. Price and Thomas[7] used this technique to develop films of the oxides of beryllium, aluminium, etc. on silver-base alloys, and thereby to confer improved tarnish resistance on these alloys. If conditions are so selected that the inward diffusion of oxygen is faster than outward diffusion of the reactive element, the oxide will be formed as small dispersed particles beneath the surface of the alloy. The phenomenon is known as *internal oxidation* and is of quite common occurrence, usually in association with a continuous surface layer of oxides of the major constituents of the alloy.

Kubaschewski and Hopkins[5] consider the conditions of the gaseous phase which influence the rate of corrosion of metals; apart from major variations of composition, they refer also to the effects of minor impurities, gas pressure, flow rate and ionisation.

Impurities

The corrosion rate can be influenced considerably by the presence of small proportions of contaminants, particularly if these are adsorbed on to the metal surface and thus influence the nature of the initially formed film of corrosion product. The presence of water vapour in air is, of course, of greatest practical importance, and marked effects from this cause on the scaling rate of steel have been reported[8]; the direction of the observed effect has not been constant and it is possible that the influence of hydrogen, formed by dissociation of the water, in decarburising the steel is a factor of importance. Similarly, it has been reported that humidity influences the life of nickel–chromium wires for electrical heating elements subjected to the ASTM life test referred to earlier, a change from dry air to air saturated with water vapour at normal temperature reducing the life by approximately one-half[9]. The nitrogen in air can sometimes play a significant rôle in corrosion, particularly in relation to steels containing chromium or aluminium which have high affinities for nitrogen; even in cases in which the nitrogen does not enter into the chemical processes of corrosion it may influence the rate of attack by oxidation, presumably by being adsorbed on to the metal surface. Thus Kubaschewski and Hopkins[5] point out that different reaction rates result from using on the one hand dry air and on the other pure oxygen at the equivalent partial pressure of the oxygen in the air.

Pressure

Variation in the pressure of the reacting gas can affect corrosion processes in two ways. In the cases more usually met with in practice, in which the corrosion rate is controlled by diffusion processes in the surface film of corrosion product, the influence of gas pressure on corrosion rate is slight. If, however, the dissociation pressure of the oxide or of a constituent of the scale lies within the range involved, the stability of the corrosion product will be critically dependent on the pressure. The effect of temperature is, however, far more critical and thus, in practical cases, pressure variations rarely decide the stability of corrosion products.

Rate of Gas Flow

The rate of flow of the gases over the metal surface is generally of little significance. In some special cases, however, it has a very marked influence on the rate of attack, which is high in stagnant atmospheres but low if adequate ventilation is provided; alloys containing appreciable proportions of molybdenum show this effect to a marked degree; it is associated with the volatility of the oxide, MoO_3, and its catalytic action in promoting further oxidation of the alloy.

Ionisation

Little effect is exerted by ionisation of the gas phase on corrosion rate.

Steam

Corrosion by essentially pure steam arises principally in connection with power generating plant. Temperatures up to about 600°C in association with pressures up to about 15 MN/m^2 are involved, although in the most advanced super-critical installations being planned for the future, temperatures of 650°C and pressures of 40 MN/m^2 are under consideration. The highest temperatures occur in the superheater tubes, but it is probable that the severest corrosion conditions on these components will arise from the presence of fuel ashes on the outside of the tubes rather than from the steam on the inside. Perhaps the most critical components in such installations will be the main steam pipes which are more highly stressed than the superheater tubes. The corrosion resulting from the action of high-temperature steam generally leads to the formation of oxides, although hydroxides may be formed. The reaction depends on the dissociation pressure of the oxide in relation to the partial pressure of the oxygen in the water vapour.

The importance of the dissociated hydrogen in contributing to the decarburisation of carbon steels heated in water vapour has already been mentioned.

Carbon and Oxides of Carbon

The importance of these agents in leading to hot corrosion has increased considerably in recent years as a result of the development of graphite-moderated nuclear reactors using carbon dioxide as coolant. Reaction between carbon dioxide and the graphite leads to the presence of a small content of carbon monoxide (which increases with increasing temperature) in the atmosphere. The formation of oxide films results from the presence of dissociated oxygen, but at the same time carburisation may occur with steels or alloys having constituents which form stable carbides. Indeed, carbon monoxide, sometimes with additions of hydrocarbons, is a very effective gas-carburising agent for steels, as the partial pressure of the oxygen is insufficient to lead to oxidation of iron. In the case of materials forming more stable oxides, e.g. chromium-containing steels or alloys, oxidation may proceed in parallel with carburisation.

Sulphur-containing Gases

Sulphur-containing atmospheres are of importance mainly in the effect which they exert when present in small proportions in combustion products, but studies have been made of the effects of pure sulphur vapour, hydrogen sulphide and sulphur dioxide on various metals. The effect of pure sulphur is usually straightforward, leading to the formation of a sulphide film which may be more or less protective according to its structure and properties. Some sulphides have low melting points, or form low-melting-point eutectics with the metal, and in these cases attack is rapid above the critical temperature, usually taking the form of intergranular penetration. For example, in the nickel–sulphur system a eutectic forms at 645°C, whereas in the iron–sulphur system the minimum melting point is 988°C. Hence it is normally found that iron-base alloys are more satisfactory than nickel-base alloys in sulphurous atmospheres at temperatures within this range. Attack by hydrogen sulphide bears some similarity to that by sulphur, the hydrogen having a negligible effect in most cases. Sulphur dioxide, however, frequently behaves very differently, as the formation of oxide films apparently largely prevents the absorption of sulphur. Useful details of the effects of hydrogen sulphide and sulphur dioxide on various metals are provided by Farber and Ehrenberg[10].

Hydrogen

Attack upon metals by hydrogen rarely results in the formation of surface films of compounds since the gas is usually readily soluble in the metal, with a high rate of diffusion, and hydrides, if formed, are generally rather unstable. The effect of hydrogen usually follows from its reducing action on certain constituents in the alloys; probably the most familiar examples of this are the decarburisation of steels by the conversion of carbides to volatile hydrocarbons, and the embrittlement of tough-pitch copper by the reduction of intergranular oxide. Hydrogen occurs as a major constituent of the atmosphere in furnaces used for bright annealing or for brazing, and in such cases the temperatures involved may rise to as high as 1 200°C. In the chemical engineering field many processes of synthesis or hydrogenation are carried out with hydrogen-rich atmospheres at temperatures up to about 600°C and at pressures rising to 1 000 atm. Much of the understanding of the attack on steels by hydrogen has been obtained from work on problems arising in the chemical industry.

Nitrogen

Although nitrogen is the major constituent of air, it plays a minor rôle in the scaling of most metals on heating in air, its effect being completely overshadowed by that of oxygen. This follows from the fact that the dissociation pressures of the nitrides of the commoner metals are greater than the partial pressure of the nitrogen. Chromium, aluminium and some of the less common metals such as titanium, molybdenum and tungsten, however, form stable nitrides which can result from heating materials containing these elements in air. Reaction between nitrogen and metals is of principal importance, however, in reducing atmospheres, particularly in cases in which absorption of nitrogen is deliberately sought, i.e. in the surface-hardening of steels by nitriding. The atmosphere in such cases is anhydrous ammonia, dissociation of which occurs at the surface of the steel, although only a small proportion of the resultant nitrogen is absorbed. Materials resistant to nitrogen and not acting as dissociation catalysts for ammonia are required as containers for such treatments, and stainless steels of the 18/8 or 25/20 types, or Inconel (76Ni-15Cr-8Fe) have been widely used.

Combustion Products

The various gas mixtures arising from the combustion of fuels form the major group of atmospheres leading to the hot corrosion of metals. Almost all natural and processed or synthetic fuels consist predominantly of hydrocarbons, carbohydrates, or more complex carbon compounds, so that the principal constituents of the combustion products are the oxides of carbon and water vapour, together with the residual nitrogen from the air. In conditions of incomplete combustion the gases may contain hydrogen, hydrocarbons and carbon monoxide in appreciable proportions, with only small proportions of oxygen, while when complete combustion is effected with

Table 7.1 Composition of waste gases after complete combustion of fuels

Fuel	Waste gases (% by vol.)		
	CO_2	H_2O	N_2
Bituminous coal	17–19	4–9	73–77
Anthracite	18–19	3–4	77–78
Lignite, peat, wood	17–18	10–16	66–72
Coke, charcoal	19–20	1–4	77–78
Petroleum products	13–14	11–14	73–75
Coal gas	8–11	20–22	69–71
Blast-furnace gas	22–25	0–2	74–77
Coke-oven gas	8–10	20–22	70–72
Producer gas	14–19	6–14	72–76
Water gas	13–16	15–18	67–71
Oil gas	9–11	18–22	68–72

excess air these reducing gases will be absent, but there will be larger amounts of oxygen. An indication of the range of compositions to be encountered is given by the values quoted for different fuels in Table 7.1 (see References 11 and 12). In most fuels some sulphur-containing compounds are present and the combustion gases then contain sulphur dioxide, sulphur trioxide or sometimes hydrogen sulphide. It is thus evident that under ordinary working conditions it is not uncommon for equipment to be subjected to atmospheres whose compositions vary from time to time, and this is one factor which can lead to abnormally accelerated corrosion rates. In particular, when the atmosphere can vary from oxidising to reducing in character, the equilibrium constitution of the scale formed on an alloy may change, with resultant cracking or porosity in the layer. The protective action of a coherent scale layer is thus much reduced.

A particular type of corrosion observed with nickel–chromium and nickel–iron–chromium alloys and involving both carburisation and oxidation, takes place at temperatures around 1 000°C in carbon-containing atmospheres. The attack, which is termed *green-rot* (see Section 7.5 (p. 7:129)) on account of the colour of fractured surfaces of the attacked alloy, occurs particularly in atmospheres of the bright-annealing type when they are contaminated with carbon-containing gases or vapours, but can also arise with incompletely burnt hydrocarbon fuels. This type of attack has been reported to be considerably accelerated by cyclic variations of the atmosphere from oxidising to reducing.

Ashes

Apart from the direct corrosive action of the gaseous products from burnt fuels, an important effect frequently arises from the deposition of ashes on the metal surfaces involved. Ashes normally consist of complex mixtures or compounds of oxides, the compositions varying very widely and the precise state of combination being uncertain. Ranges of ash compositions reported for different samples of fuels of three broad classes are given in Table 7.2 (see

Table 7.2 Composition of fuel ashes

Ash composition (%)	Fuel		
	British coals*	Peat	Fuel oil
SiO_2	25–50	20–50	2–20
Al_2O_3	20–40	7–15	2–20
Fe_2O_3	0–0	0–20	1–60
TiO_2	0–3	–	–
CaO	1–10	5–30	0–10
MgO	0.5–5	5–25	–
Na_2O	1–6	0–6	1–30
V_2O_5	–	–	5–60
SO_3	1–12	5–20	7–30
Cl_2	–	0–1	–
P_2O_5	–	1–3	–

* P. J. Jackson (private communication) considers that present-day compositions of ashes from British coals would range as follows:

SiO_2	Al_2O_3	CaO	MgO	Na_2O	K_2O	Cl_2	P_2O_5
25–60	15–40	1–20	0–5	0·2–3	0·2–5	0–5	0–1

References 13, 14, 15 and 16), from which it will be noted that the ranges are so wide that in any specific case the actual ash analysis must be obtained before consideration can be given to probable corrosion reactions. The corrosive attack by an ash is frequently critically dependent on the temperature, accelerating steeply when the melting point of a particular ash constituent is exceeded. This is particularly noticeable in the case of the vanadium-containing ashes from crude or residual mineral oils which seriously attack most heat-resisting alloys when the temperature is above about 650°C, the melting point of vanadium pentoxide. The rapid attack follows fluxing of the normally protective oxide scale by the ash, and the attack on the underlying metal may be in the form of further oxidation or may include attack by sulphur compounds or other active constituents of the ash. Various investigators have compared the rates of attack on different alloys by selected real or synthetic ashes and somewhat contradictory results have been reported. Thus, whereas Evans[16] stated that low-nickel or nickel-free iron-base alloys were more resistant to vanadium corrosion at 732°C than nickel or cobalt-base alloys, Harris and his associates[17] concluded that the latter type of alloys should be reasonably resistant if they contain less than 30% iron, less than 2% vanadium, and more than 16% chromium. Betteridge, Sachs and Lewis[18] also found that nickel–chromium alloys were much more resistant to vanadium attack than were austenitic steels. It is probable that the different conclusions arose from the difference in character of the attack which follows initial penetration of the protective scale, on account of variations in composition of the selected ashes.

The deposits formed in internal combustion engines by high-octane petrols may be classed as ashes; they consist of mixtures of lead oxides, bromides and sulphates derived from the anti-knock additives and, of course, exert their main corrosive effect on the parts operating at the highest temperature: the exhaust valves and the sparking-plug electrodes.

The formation of an ash is not always deleterious, since a layer of a less reactive material may serve to restrict access to the alloy of more reactive agents either in the ash itself or in the atmosphere. The erosive action of ash also has to be considered in addition to the purely corrosive action. With high-speed gas streams, as in gas turbines, erosion is frequently the more important effect.

Halogens and Halogen Compounds

The halogens are particularly important in hot environments because many of the metallic halides have relatively high vapour pressures, and the corrosion products are therefore vaporised instead of forming protective surface layers. This is particularly true in the case of chromium and aluminium, both important constituents of heat-resisting alloys. However, the attack is mainly on the metals rather than on the oxides, so that an initially-formed oxide film confers appreciable protection and the action of halogens in a predominantly oxidising atmosphere is much less than in a reducing atmosphere. Small traces of acid gases such as hydrochloric acid can, however, considerably accelerate oxidation rates in air, particularly of copper-base alloys.

Molten Salts

A wide variety of molten salt baths is used in industrial practice and there exists a correspondingly wide range of possible chemical reactions with metallic parts being heated in the baths, or with the containers holding the baths. Baths consist largely of mixtures of nitrates, carbonates or halides of the alkaline or alkaline-earth metals, and these generally have a ready solubility for oxides of other metals. Hence, although the salt baths may be sufficiently oxidising to react with the alloys concerned to form oxide, this product is taken into solution in the bath and does not form a protective scale; attack therefore progresses steadily. In some cases absorption of components of the salt mixture into the alloy may occur, with either deleterious or beneficial results. Thus, with baths containing borates, reduction of boric oxide may occur by the action of reactive constituents such as aluminium, resulting in the absorption of boron into the alloy; with nickel-base alloys a low-melting-point eutectic may then be formed with serious effects on the properties of the alloy. On the other hand, alkali-cyanide baths are used for the heat-treatment of steel parts, absorption of carbon and nitrogen into the steel giving surface hardening.

Molten-salt environments are fully treated in Section 2.10.

Refractories, Ceramics, Glasses, etc.

The possible attack on metals and alloys by materials of this type may be considered in a manner similar to attack by molten salts. If a molten phase exists in the material at the temperature under consideration it is possible that solution of the oxides formed by the alloy will occur, thus preventing the formation of a protective film. Molten glasses act in this manner, and refractories with traces of impurities, perhaps in localised regions, can act very similarly. Minerals such as mica and asbestos frequently give rise to troubles due to impurities which either lower the melting point or impair the electrical insulation at elevated temperatures. These materials, which are silicates of magnesium and calcium, often contain small quantities of iron oxides or sulphides and other impurities which make them unsuitable for use at temperatures above about 500 or 600°C, particularly if electrical insulation characteristics are required.

A possibility which must be considered is that of reaction between a refractory and an alloy occurring in consequence of the reduction of a constituent of the refractory by a metal in the alloy. Thus alumina has a higher free energy of formation than silica, and therefore aluminium is capable of reducing silica. At a high enough temperature, if oxygen is not available in the atmosphere to combine with the aluminium, this reaction is likely to take place. In order to avoid reactions of this type, only pure and very stable refractories, such as alumina, magnesia, zirconia or beryllia, should be used in contact with reactive metals at high temperatures.

Molten Metals

If the major constituents of a solid alloy in contact with a liquid alloy are highly soluble in the latter without formation of compounds, progressive attack by solution is to be expected. If, on the other hand, a stable intermetallic compound is formed, having a melting point above the temperature of reaction, a layer of this compound will form at the interface and reduce the rate of attack to a level controlled by diffusion processes in the solid state. By far the most serious attack, however, occurs in the presence of stresses, since in this case the liquid alloy, or a product of its reaction with the solid alloy, may penetrate along the grain boundaries, with resultant embrittlement and serious loss of strength.

Molten metals as a corrosive environment are discussed more fully in Section 2.9.

Recent Developments

The different processes and their material requirements are reviewed in References 19 and 20, while annual conferences have been held under the auspices of the U.S. Bureau of Standards and other interested bodies since 1976[21]. The processes involved embrace combustion, gasification and liquefaction, each of which presents characteristically different corrosive environments.

Combustion of coal may take place in conventional fixed beds using lump coal and in which temperatures up to 1 300°C may be reached; by entrained flow in which pulverised coal is injected into the combustion zone with the air, reaching temperatures up to 1 500°C; or in the more recently developed systems of fluidised-bed combustion, again using pulverised coal but with temperatures restricted to below about 900°C[22]. While most processes operate at near to normal pressure, some fluidised-bed processes are designed for pressures up to about 20 atm. The environments involved are basically those referred to under 'Combustion Products' but the corrosive attack on metallic components is to a major extent due to alkali metal sulphates formed from constituents in the coal and deposited in a molten condition on the metals. For components immersed in the bed of a fluidised-bed system the lower temperatures lead to less volatilisation of the alkali sulphates, but low partial pressures of oxygen may reduce the effect of oxide scales in minimising sulphur attack.

Coal gasification processes also use fixed-bed, fluidised-bed and entrained-flow methods, but involve the supply of restricted amounts of air or oxygen to burn some of the coal to provide the heat source, and steam to yield the required gaseous product. The latter may be mixtures of hydrogen and carbon monoxide as a substitute for natural gas or as industrial fuel gas, or methane for chemical feedstock. The various processes are all carried out at elevated pressures and at temperatures ranging from 500°C to over 1 800°C. While the major metallic components of the plant, such as the reaction vessel, may be refractory lined to minimise the temperature reached and to protect them to some extent from the environment, other critical components such as heat-exchanger tubes may be exposed. The atmosphere is highly reducing with low oxygen and high sulphur partial pressures and a high carbon activity. The behaviour of constructional materials in gasification plant has been reviewed in References 22 and 23.

Liquefaction of coal involves the extraction of carbon by solvents at high pressures and at temperatures up to about 500°C, followed by separation of the extract, which is then hydrogenated in the presence of a catalyst to yield hydrocarbon oils. The corrosion conditions are not regarded as severe.

In most processes of coal conversion the corrosive action of the gaseous environment may be aggravated to some degree by the erosive action of the entrained solids, the ashes or partly-burned fuel termed char.

The high temperatures and high gas velocities arising in gas-turbine engines have led to much concern at the threat of corrosive attack on critical components. Turbine entry temperatures may range between 700°C and 1 300°C and, while for aero engines refined fuels with low sulphur contents are used, oxidation may be a significant problem and corrosion may be aggravated by contaminants such as sulphates and chlorides introduced either in minor amounts in the fuel or, more seriously, in the ingested atmosphere. Aircraft operating in marine environments such as from aircraft carriers, or helicopters involved in sea rescues or oil-rig servicing, are particularly vulnerable to hot corrosion of turbine components. For the higher gas temperatures air-cooled turbine blades are used, restricting the metal temperature to below about 1 150°C, the maximum permissible with the most advanced materials on grounds of creep and fracture resistance. Even then protective coatings are widely used both to provide increased corrosion resistance and to give thermal insulation. Coatings of a number of different types have been developed, embracing intermetallic compounds, usually aluminides, formed by diffusion or overlay techniques, and oxide ceramic coatings, primarily based on zirconia, applied by spraying[25,26]. Erosion of turbine components may also pose serious difficulties to aircraft operating in sandy locations, while imperfect combustion may lead to erosion by carbonaceous residues from the fuel.

Industrial gas turbines generally operate at lower temperatures than aircraft jet engines but may be more susceptible to hot corrosion because they may use less-refined fuels, even crude or residual oils containing relatively high proportions of sulphur or vanadium. Coatings of the type referred to above are increasingly being used to provide protection.

A further environment in which interest has been shown in recent years arises in the use of helium as the primary coolant in high-temperature nuclear reactors. The gas is circulated continuously around the reactor core and a secondary heat exchanger, and becomes contaminated, mainly with hydrogen and carbon monoxide from the graphite moderator. Tests in the temperature range 650–800°C have shown that surface and intergranular oxidation and carburisation can occur, affecting the long-time creep rupture properties of materials[27].

W. BETTERIDGE

REFERENCES

1. Gulbransen, E. A. and Andrew, K. F., Amer. Soc. Test. Mat., Sp. Techn. Publ. No. 171, 35–48 (1955)
2. Betteridge, W., *The Nimonic Alloy*, 1st edn, Arnold, London, 218 (1959)
3. Gleiser, M., Larsen, W. L., Spiser, R. and Spretnak, J. W., Amer. Soc. Test. Mat., Sp. Techn. Publ. No. 171, 65–88 (1955)
4. Bruce, D. and Hancock, P., *J. Inst. Mets.*, **97**, 140 (1969)
5. Kubaschewski, O. and Hopkins, B. E., *Oxidation of Metals and Alloys*, 2nd edn, Butterworths, London (1962)
6. Hessenbruch, W., *Zunderfesterlegierungen*, Springer, Berlin (1940)
7. Price, L. E. and Thomas, G. J., *J. Inst. Met.*, **63**, 21 (1938)
8. Siebert, C. A. and Donnelly, H. G., *Trans. Amer. Soc. Metals*, **28**, 372 (1940)
9. Brasunas, A. de S. and Uhlig, H. H., *Bull. Amer. Soc. Test. Mat.*, No. 182, 71 (1952)
10. Farber, M. and Ehrenberg, D. M., *J. Electrochem. Soc.*, **99**, 427 (1952)
11. Spiers, H. M., *Technical Data on Fuel*, Publn. Brit. Natl. Ctee., World Power Conf., 6th edn (1962)
12. *Efficient Use of Fuel*, H.M.S.O., London (1944)
13. Crossley, H. E., *External Boiler Deposits*, Inst. of Fuel Special Study of the Ash and Clinker Industry, Paper No. 4 (1952)
14. Waddams, J. A. and Wright, J. C., *J. Inst. Fuel*, **32**, 246 (1959)
15. Sykes, C. and Shirley, H. T., Iron and Steel Inst. Sp. Rep. No. 43, 153–169 (1952)
16. Evans, C. T., Amer. Soc. Test. Mat., Sp. Techn. Publ. No. 108, 59–105 (1951)
17. Harris, G. T., Child, H. C. and Kerr, J. A., *J. Iron Steel Inst.*, **179**, 241 (1955)
18. Betteridge, W., Sachs, K. and Lewis, H., *J. Inst. Petrol.*, **41**, 170 (1955)
19. Marriott, J. B., Van de Voorde, M. and Betteridge, W., *Coal Conversion Processes and their Materials Requirements*, EUR 9182 (1984)
20. Meadowcraft, D. B. and Manning, M. I. (eds), *Corrosion Resistant Materials for Coal Conversion Systems*, Applied Science Publishers, London, (1983)
21. *Materials for Coal Conversion and Utilization*, Annual Conferences 1976–1982, NBS, Gaithersburg, Maryland
22. Howard, J. R. (ed.), *Fluidized Beds: Combustion and Applications*, Applied Science Publishers, London (1983)
23. Hill, V. L. and Black, H. L., (eds), *Properties and Performance of Materials in Coal Gasification Environments*, ASM, Ohio (1981)
24. Norton, J. F. (ed), *High Temperature Materials Corrosion in Coal Gasification Atmospheres*, Applied Science Publishers, London (1984)
25. Restall, J. E., 'Surface Degradation and Protective Treatments.' In *Developments in Gas Turbine Materials – I*, Meetham, G. W. (ed), Applied Science Publishers, London, ch. 10 (1981)

26. Lang, E. (ed), *Coatings for High Temperature Applications*, Applied Science Publishers, London (1983)
27. Bates, H. G. A., Betteridge, W., Cook, R. H., Graham, L. W. and Lupton, D. F., 'The Behaviour of Metals in High-Temperature Reactor Helium for Steam Generators,' *Nuclear Technology*, **28**, 424–440 (Mar. 1976)

7.2 The Oxidation Resistance of Low-Alloy Steels

Introduction

Low alloy steels are generally considered to comprise plain carbon steels and steels with a total alloying content of up to 12%. As such, they are much cheaper than more highly alloyed materials and are often used in large quantities in heavy engineering industries. Whilst these materials are not generally selected for resistance to high temperature corrosion (the material choice is largely dictated by cost, ease of fabrication and mechanical properties) they are often required to operate in high temperature aggressive environments. For instance, the power generation, refuse incineration and chemical process industries use many miles of low alloy steel heat exchanger tubes. Hence, the high temperature oxidation properties of low alloy steels are often important in determining component life.

The extent to which low alloy steels react to high temperature corrosive environments is the subject of this chapter. In view of the commercial importance of these steels, the published literature on this topic is extensive and is being continually enlarged. The reader is encouraged to refer to the many excellent papers and current issues of the journals, referenced at the end of the chapter, for more detailed and contemporary information on the topic.

Factors Governing Oxidation Behaviour

In the absence of stress, the survival of a given component is largely determined by the extent to which the material of construction reacts with the environment in which it resides. This can largely be categorised by the extent to which the following two criteria are satisfied[1].

1. Is the material thermodynamically stable in the environment?
2. If not, will the reaction rate between the environment and the material be slow enough to give an acceptable life?

In the vast majority of cases at elevated temperatures, the answer to the first question is no. With the exception of gold and platinum, which are generally too expensive for large-scale industrial use, are in short supply and do not have the required mechanical properties, most materials will react to some degree with their environment. Most highly alloyed steels and superalloys satisfy the second criterion for a wide range of high temperature environments. However, these materials may be far too expensive for large-scale industrial applications. The oxidation behaviour of most metals and alloys in a high temperature environment is, therefore, governed by the degree of protection afforded by any oxide scale that forms[1]. This depends upon the oxide melting point, its mechanical integrity and the rate of diffusion of elemental species, present within both the environment and the alloy, through the oxide scale.

In certain cases (e.g. Cr above 1 000°C) the CrO_3 oxide which forms is volatile, and clearly affords no protection to the substrate[1]. In other cases a solid oxide scale forms which may, or may not, be continuous. The extent to which a solid surface scale protects the metal depends upon complete surface coverage. This behaviour can largely be categorised according to whether the volume of oxide produced is less or greater than the volume of metal consumed during the oxidation reaction. This principle was originally advanced by Pilling and Bedworth[2], with the oxide:metal volume ratio being known as the Pilling Bedworth Ratio (PBR), where:

$$PBR = \frac{\text{Volume of oxide produced}}{\text{Volume of metal consumed}}$$

If the PBR is less than unity, the oxide will be non-protective and oxidation will follow a linear rate law, governed by surface reaction kinetics. However, if the PBR is greater than unity, then a protective oxide scale may form and oxidation will follow a reaction rate law governed by the speed of transport of metal or environmental species through the scale. Then the degree of conversion of metal to oxide will be dependent upon the time for which the reaction is allowed to proceed. For a diffusion-controlled process, integration of Fick's First Law of Diffusion with respect to time yields the classic Tammann relationship[3], commonly referred to as the Parabolic Rate Law:

$$x^2 = Kt \qquad (7.1)$$

where x is a measure parameter, K is the rate constant and t is time. The progress of oxidation (x) is generally measured by weight gain, weight loss, scale thickness, or retreat of the metal surface. However, other parameters, such as loss of oxidant in the environment, may be used. The time interval may be measured in seconds, hours, 1 000s of hours and years. As a consequence, many different units of the measured oxidation parameter and the time period have been reported in the literature, leading to many differing units for the quoted rate constant.

In some circumstances, the reaction rates may not be exactly parabolic, and even initially parabolic rates may be influenced by changes within the oxide scale with time. As an oxide scale grows, the build-up of inherent growth stresses, externally applied strains and chemical changes to either oxide scale or metal may all compromise the initial protection offered by the scale, leading to scale breakdown and ultimately partial or complete loss of protection; paralinear, or linear kinetics may ensue. In other circumstances, as will be seen later in this chapter, very small additions of contaminants to

7:18 THE OXIDATION RESISTANCE OF LOW-ALLOY STEELS

the environment may radically modify the oxidation response, by either favouring non-protective scales from the outset, or causing rapid breakaway corrosion following an initially protective period. Hence, there are many hidden dangers in extrapolating short-term oxidation kinetic data to long times, and great care is required when utilising published oxidation data to assess potential corrosion rates and component lives in industrial environments.

The temperature dependence of a diffusion controlled reaction is typically described by the Arrhenius relationship:

$$K = a \exp \frac{(-Q)}{RT} \quad (7.2)$$

where a is a constant, Q is the activation energy, R is the gas constant, and T is the temperature (K). Hence the rate constant (K) will be influenced by the temperature of exposure, due to the increased ion mobility at higher

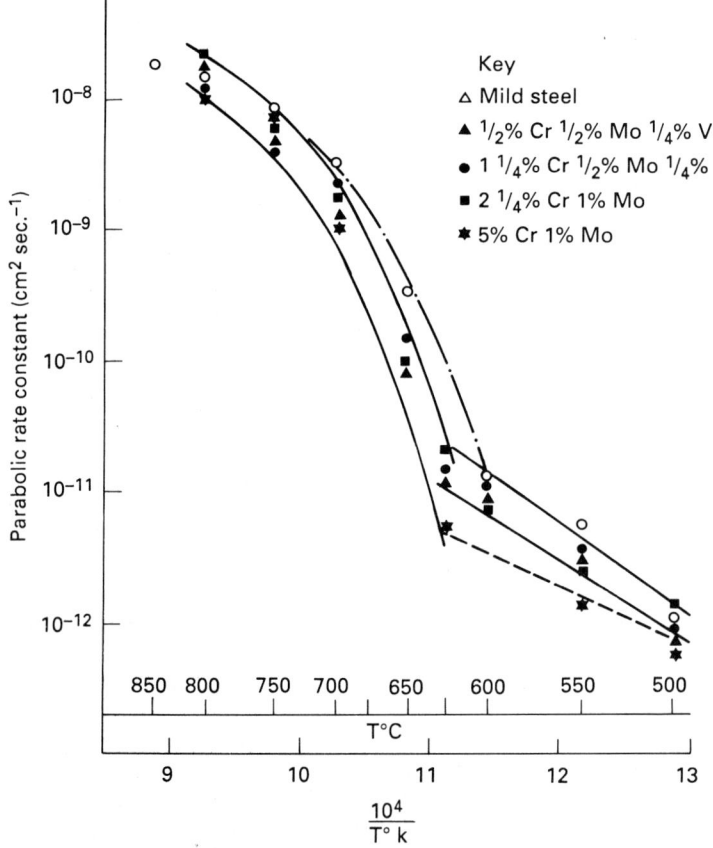

Fig. 7.1 Arrhenius plot for the oxidation of mild steel and low-alloy steels in air showing a sharp break in the slope and curvature due to the appearance of FeO in the scale above 570°C

temperatures, via the activation energy term. Again great care is required when extrapolating data beyond the temperature range of the initial experimental data. Sharp changes in the Arrhenius slope may occur with the appearance of a new, thermodynamically favoured, species within the scale. In addition, an Arrhenius plot of the temperature dependence of the reaction rate may reveal curvature, due to the progressive change from one rate determining process to another. Both features are observed in the Arrhenius plot for low-alloy steels (Fig. 7.1).

The Oxidation of Iron

For iron in most oxidising environments, the PBR is approximately 2.2 and the scale formed is protective. The oxidation reaction forms a compact, adherent scale, the inner and outer surfaces of which are in thermodynamic equilibrium with the metal substrate and the environment respectively, and ion mobility through the scale is diffusion controlled.

The rate of oxidation of iron is then governed by the stabilities of the various phases, which are in turn a function of the temperature and oxygen partial pressure of the environment. Examination of the $Fe-O_2$ phase diagram (Fig. 7.2) reveals that the principal solid oxide phases below 570°C will be Fe_3O_4 (magnetite) and Fe_2O_3 (haematite). Above 570°C, FeO (wustite) appears as a third phase within the scale. These phases are present within the scale as individual layers, with the layer sequence dictated by the equilibrium oxygen partial pressure (pO_2) for phase stability prevailing at the given temperature. Hence, the oxide phase stable at the lowest pO_2 (FeO at $>570°C$) is found closest to the metal substrate, whereas the phase stable at the highest pO_2 (Fe_2O_3) is found closest to the oxidising environment. If, however, the pO_2 of the environment is low enough, only FeO will be formed. At intermediate values of the pO_2, Fe_3O_4 will also form and, for most industrial environments, the pO_2 is sufficient for the formation of an outer layer of Fe_2O_3. Whilst these observations are true for bulk scales, FeO has also been found to be stable in very thin films at temperatures down to 400°C[4] and within narrow cracks at 500°C[5].

In the initial stages of oxidation, nucleation of oxide occurs at favoured crystallographic sites, followed by preferential lateral spread from these nuclei to form a continuous thin film over the iron surface. At low temperatures ($<200°C$) thin films, 100s of nanometres thick, form exceedingly rapidly. Ion mobility is driven by space charges between gas/oxide and oxide/metal interfaces. As the scale thickens, the initial rapid growth rate is superseded by an extremely slow, logarithmic or inverse logarithmic growth law.

For iron oxidising at temperatures between 350 and 500°C, Fe_3O_4 nucleates first and grows laterally over the surface[6]. Once complete coverage is achieved, the Fe_3O_4 thickens parabolically. Ultimately, nucleation and lateral overgrowth by Fe_2O_3 occur, slowing the growth rate of the Fe_3O_4 appreciably, due to the lower effective pO_2 at the Fe_3O_4 surface, i.e. the pO_2 falls from that of the environment to that governed by the equilibrium between Fe_3O_4 and Fe_2O_3 at the prevailing temperature. Both FeO and Fe_3O_4 nucleate first and grow out of the surface of pure Fe and

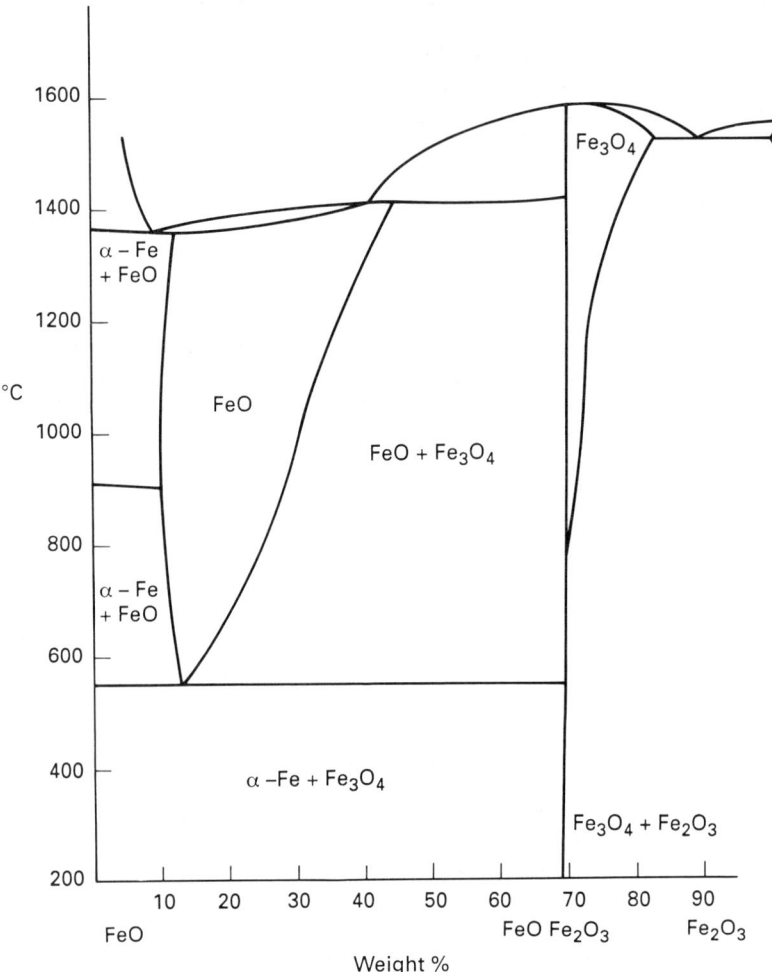

Fig. 7.2 Fe–O phase diagram showing the principal solid phases stable on iron and wide stoichiometry limits of FeO (after White[182])

Fe–3%Cr alloy at 700 to 800°C in low pressure O_2 (10^{-3} Pa).[7,8]. Whilst Fe_2O_3 is thermodynamically stable, it can only form once the oxygen available for Fe_2O_3 growth has increased sufficiently. Initially the available oxygen is continually depleted by the formation of Fe_3O_4. The length of this induction period prior to Fe_2O_3 growth was a direct function of the prevailing pO_2, being 3 h at 40 Pa, but only 1 min at 101 kPa[7].

Above 570°C, a distinct break occurs in the Arrhenius plot for iron, corresponding to the appearance of FeO in the scale. The Arrhenius plot is then non-linear at higher temperatures. This curvature is due to the wide stoichiometry limits of FeO; limits which diverge progressively with increasing temperature. Diffraction studies have shown that complex clusters of vacancies exist in $Fe_{(1-x)}O$[9,10]. Such defect clustering is more prevalent in oxides that demonstrate a high degree of non-stoichiometry; with vacancy-hole complexes possibly being present at the higher deviations[11]. Single defects only occur when these oxides are closest to their stoichiometric composition. These considerations lead to a complex dependence of the cation diffusion coefficient on oxide stoichiometry[12], and hence temperature.

Mass transport measurements have shown that cation transport predominates in FeO (Fe^{2+}) and Fe_3O_4 (Fe^{2+}, Fe^{3+}), whereas anion transport predominates in Fe_2O_3 (O^{2-}). This leads to the well-accepted growth scheme for multi-layered scale growth on iron shown in Fig. 7.3, with the governing equations for individual layer growth being:

$$Fe_3O_4 + Fe = 4FeO \qquad \text{above } 570°C$$
$$4Fe_2O_3 + Fe = 3Fe_3O_4$$
$$2Fe_3O_4 + O = 3Fe_2O_3$$

It was originally considered that the growth of these layers was largely controlled by lattice diffusion, following the theories expounded by Wagner[13,14]. However, experimental work has shown that, for Fe_3O_4 below 600°C and Fe_2O_3 at all temperatures, theoretical calculations of oxide growth using lattice diffusion coefficients are unable to predict the growth rates observed in practice[15,16]. Tracer studies during the growth of Fe_3O_4 on Fe at 500°C have shown that Fe_3O_4 growth is dominated by outward diffusion of Fe ions along short-circuit paths in the oxide[17]. These short-circuit paths are largely considered to be the oxide grain boundaries[15]. For Fe_3O_4 above 600°C, the observed parabolic growth rate constants are within an order of magnitude of those calculated from tracer diffusion of Fe in

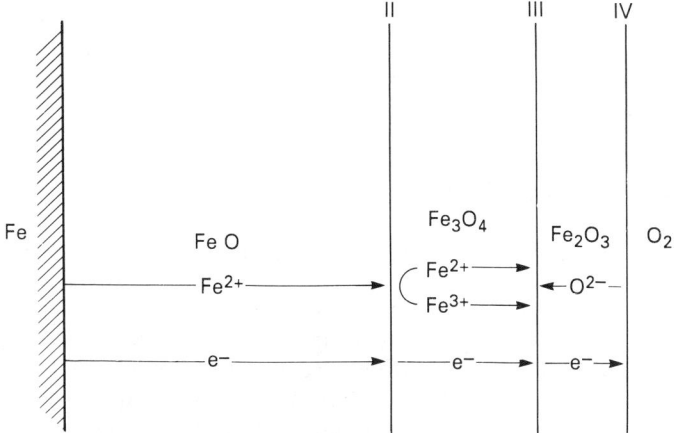

Fig. 7.3 Simplified scheme for the diffusion-controlled growth of multilayered scales on pure iron and mild steel above 570°C

Fe_3O_4[18]. A similar correspondence between theory and practice has been found for growth of Fe_3O_4 by the solid state reactions from FeO and Fe_2O_3 between 600 and 1 200°C[18]. The growth rate of FeO is within 10% of the theoretical rate expected from Fe lattice diffusion, calculated according to the Wagner theory[19].

The measured activation energy for Fe_2O_3 growth is of the order of 169 to 222 kJ/mol[20]. This activation energy is much lower than that expected from measurements of the tracer diffusion of Fe or O^{21} and the measured rate constant for Fe_2O_3 growth is approximately ten times greater than that calculated according to Wagner[13]. O^{18} tracer diffusion studies on the growth of Fe_2O_3 during the oxidation of Fe at 823 K have shown that, whilst the primary diffusion species through the crystal lattice is Fe ions, fast, inward O_2 diffusion occurs down cracks in the scale[22].

The oxidation rate of pure Fe in O_2 has been shown to be affected by specimen shape and the original surface profile[23]. For iron oxidising in O_2, 50:50 O_2 + H_2O and O_2 + CO_2 rough surfaces have been found to oxidise more slowly than smooth surfaces, since surface irregularities hinder the oxide flow[23,24]. The oxide is unable to deform sufficiently to maintain intimate contact with the metal surface such that porous scales are formed[25].

For the oxidation of pure Fe at 500°C, the early reaction rate is more rapid on cold-rolled than annealed surfaces. On cold-worked surfaces, the scale tends to be rough and poorly adherent and contains less Fe_3O_4[26]. Cold-work enhances the oxidation rate by mopping up vacancies so that pore-free scales are formed which maintain contact with the substrate[27-31]. The author[32] has found an effect of stored cold-work during the oxidation of fracture surfaces at 600°C, which disappears with increasing oxidation due to the effect of surface annealing and consumption of the cold-worked layer by oxidation. At long oxidation times (>200 h), the oxidation rate of a fracture surface becomes indistinguishable from that of an annealed emery-ground surface[32]. At temperatures above approximately 625°C, there is no effect of cold-work[23].

General Alloying Effects on Oxidation

The general requirements for higher temperature alloys is that they should be cheap, possess adequate mechanical strength, be resistant to chemical degradation by the environment and be easily fabricated. These requirements are often conflicting, and a compromise is required for alloy design. Simple iron alloys form oxides which are not normally protective enough at temperatures above about 550°C[33]. Therefore other elements are needed to form a more protective scale. These elements usually comprise chromium, aluminium and/or silicon. For these elements to confer adequate protection, the scale that they develop should be stoichiometric (to minimise ionic transport rates), free of gross defects, stress free at temperature, resist spalling and not be volatile by further reaction with the environment[34].

It has already been shown that bulk lattice diffusion is not generally considered to be the rate-controlling process for the oxidation of iron in most real situations. Hence the classical Wagner treatment, whereby the valency of the alloying element increases or decreases the number of lattice defects, provides little clue as to the overall effects of alloying elements on the oxidation of iron alloys. Rather, studies have shown that the effects of alloying elements on the oxidation of iron alloys are largely brought about by their mobility (or lack of mobility) in the iron oxide lattice, and the effect this has on scale morphology, phase structure and oxide plasticity. The addition of even small quantities of alloying elements can profoundly influence the scale morphology produced and hence the subsequent oxidation behaviour.

The Fe_3O_4 unit cell comprises an inverse spinel structure, with Fe^{2+} and Fe^{3+} cations occupying octahedral and tetrahedral interstitial sites within a close-packed oxygen lattice. The occupation of an interstitial site is accompanied by a specific site energy and particular cations show a preference for occupation of that site in which they sit most comfortably, i.e. the lowest energy site (Fig. 7.4). The difference between the potential energy of a cation in a preferred site to that in a non-preferred site is termed the crystal field preference energy[35] (CFPE) and is principally determined by the number of electrons on the d-shell. The direction of travel of a cation through the magnetite lattice is via alternate octahedral-tetrahedral-octahedral site transfers[36]. These interstitial sites are interconnected by saddle points within the lattice and, in order to transfer from one interstitial site to another, the diffusing ion must acquire enough energy to surmount this energy barrier and pass through the saddle point. If the direction of movement of the cation is from a favoured to a non-favoured site, additional energy is required to overcome the CFPE. If, however, the direction of cation movement is in the opposite direction, there is a net reduction in the total lattice energy equivalent to the CFPE. It is apparent that, if the

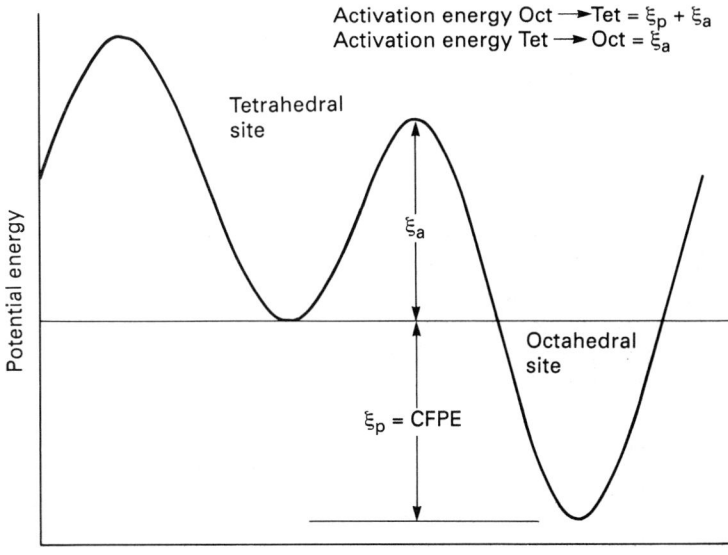

Fig. 7.4 Potential energy/lattice position diagram for occupation of interstitial sites in Fe_3O_4 lattice alloying element cations

cation shows a high CFPE between interstitial sites, then cation transfer is much easier in one direction than the other and diffusion of the cation through the lattice will be severely hindered. The CFPE for Cr, Mo and V in octahedral sites in magnetite is high, whereas that for Mn is virtually identical to that of Fe, both of which are very low. Hence Cr, Mo and V are virtually immobile in the magnetite lattice, whilst Mn is found throughout the scale at atom ratios to the iron similar to those found in the original metal[37,38].

As a result of this lack of mobility of certain alloying elements, a new phase layer appears in the scale. This new layer grows inward from the original metal surface, is in intimate contact with the substrate alloy[39] and has an alloying element composition approximately 1.5 times that of the original metal[37]. Moreau[40] identified this inward growing layer on Fe–Cr steels as $FeCr_2O_4$ globules within a wustite matrix. Further, Rahmel[39] determined that the inner layer of four layered scales grown on iron alloys containing Cr, Mo, V and Si consisted of FeO containing $(Fe, X)_3O_4$ where X is Cr, Mo, V or Si. The oxygen transport through the inner layer is thought to take place via pores within the inner layer, since solid-state diffusion of oxygen through the magnetite lattice is five orders of magnitude too low[41], and grain boundary diffusion is also too low, to account for the observed growth rates[42]. The oxygen pathways are now thought to comprise grain boundary triple points and transient microvoids, continuously created and rearranged by creep[43].

At temperatures below approximately 600°C, numerous studies have shown that the thickness of the inward growing layer approximates to that of the outwardly growing magnetite + haematite scale, with the location of the spinel/magnetite interface approximately located at the original metal surface[34,37,44-46]. Hence, the new spinel formed approximately balances the volume of metal consumed, implying equal amounts of oxygen and cation transport in duplex film growth[34] and also implying that the rate of growth of the two scales are linked to one another. Moreover, it is now considered that the overall growth rate is mainly determined by the rate of cation diffusion outward through the outer layer, since the cations must first diffuse out to provide space for inner layer growth. Hence the extent of inner layer growth is dictated by the recession of the oxide metal interface.

These observations have been rationalised by Robertson and Manning[43] who suggest that, for single-layer growth, the metal ions are transported as metal vacancies to the scale/metal interface. This interface consists of incomplete planes of metal atoms, or ledges, on which there are incomplete rows or jogs, analogous to interfacial edge dislocations. A metal vacancy in the oxide, on arrival at the interface, aligns with a jog atom on the metal surface. Oxidation then occurs when the jog atom moves into the oxide and annihilates the vacancy. Repetition of this process causes continual stripping of the metal ledge and allows oxide to fall into the space created (analogous to the climb of an edge dislocation). The surrounding oxide lattice then relaxes, maintaining adhesion. Since Cr is much less mobile in Fe_3O_4 than Fe, a Cr ion tends to remain at the site where it first enters the oxide and inhibits dislocation movement[47]. This prevents the scale falling into the space created by the outward diffusion of Fe ions and microvoids (a few atom sites) are formed at the scale/metal interface. This initiates duplex growth

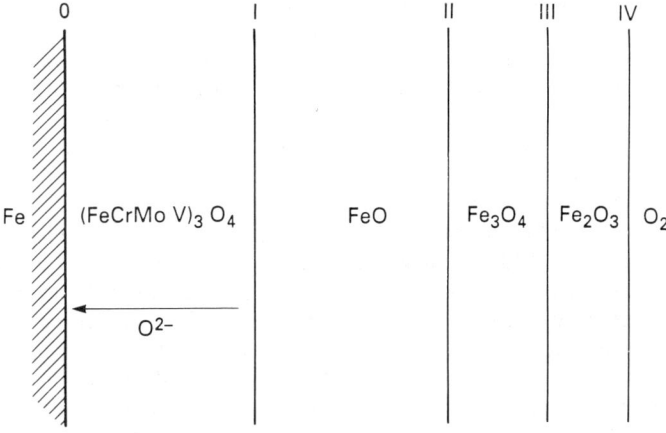

At 0 $3(Fe, Cr, Mo, V) + 4O \rightleftharpoons (Fe, Cr, Mo, V)_3O_4$
I. $FeO \rightleftharpoons Fe + O$
II III & IV as mild steel (Figure 7.12)

Fig. 7.5 Simplified scheme for the diffusion-controlled growth of multilayered scales containing spinel layers on low-alloy steels

by creating the space required for inner layer growth, without compromising adhesion.

For many low alloy steels, therefore, the scale phase sequence is as shown in Fig. 7.5 and the governing equations for individual layer growth are similar to those for pure iron, with the addition of:

$$Fe_3O_4 = 2Fe^{2+} + Fe^{3+} + 4O^{2-}$$

at the spinel/magnetite interface and

$$3(FeX) + 4O^{2-} = (FeX)_3O_4$$

at the metal/spinel interface. The Fe^{2+} and Fe^{3+} ions diffuse through the overlying magnetite (or wustite) to promote higher oxide reduction as before. The O^{2-} ions percolate through pores and microfissures in the underlaying spinel to promote growth of the spinel layer.

With some alloying elements, e.g. Si and Al, the concentration of element required in the lattice to form their own oxide is so low that preferential formation of pure Al_2O_3 or SiO_2, which are both highly protective, may occur. For the reaction

$$AO + B(\text{in alloy}) = A(\text{in alloy}) + BO$$

the limiting mole fraction (N_A) of A required for protection is given by

$$\frac{N_A}{1 - N_A} = \exp\frac{1}{RT}(\Delta G_{AO} - \Delta G_{BO}) \qquad (7.3)$$

where ΔG_{AO} and ΔG_{BO} are the free energies of formation of the oxides AO and BO, respectively.

Provided the mole fraction of A does not fall below N_A, then the oxide AO will be formed exclusively. The important criterion is the ratio of the oxidation parabolic rate constant to that of the diffusion coefficient of A[48]. For Al in Fe, the parabolic rate constant is very low, whilst the diffusion coefficient is relatively high, whereas the diffusion coefficient of Cr is much lower. Hence, the bulk alloy composition of Al in iron required for the exclusive formation of Al_2O_3 at any given temperature is lower than the Cr concentration required for the exclusive formation of Cr_2O_3.

By adding a third element (C) it is possible to increase the likelihood of forming compound BO, at a lower concentration than would be required in the pure binary alloy $A-B$, provided that the third element C has an affinity for O intermediate between that of A and B[49]. If the mobility of B in the ternary alloy is high enough and element C has sufficient thermodynamic activity in the alloy to form CO, then the potential of O may be lowered to such an extent that BO forms more readily, i.e. element C acts as a getter for element B. Such behaviour is characteristic of, for instance, Fe-Al-Si and Fe-Cr-Si alloys.

Effects of Specific Alloying Elements on the Oxidation of Iron

Carbon

The addition of carbon to iron principally affects the subsequent oxidation response via the oxidation of the carbon in the steel to form CO and CO_2. Runk and Kim[50,51] and Nosek and Werber[52] have all reported similar observations on the oxidation of Fe-C alloys at temperatures up to 400°C. It appears that magnetite nucleates first over the ferrite, the growth reaction obeying two-stage logarithmic kinetics with electron transfer through the scale being the rate-controlling process. Due to the formation of a kinetic barrier of CO and CO_2 at the carbide/oxide interface, only a thin film (15 nm) of randomly oriented crystallites of γ-Fe_2O_3 of approximately 7 nm mean grain diameter form over the cementite. Continued oxidation results in grain growth in the γ-Fe_2O_3 due to strain-induced grain boundary migration and a transition from γ- to α-Fe_2O_3. This latter transformation is accompanied by a 2.3% volume decrease which causes cracking and allows the carbon oxides to escape. Ultimately, lateral magnetite overgrowth of the cementite occurs and, once Fe_3O_4 has spread over the whole surface, protective kinetics prevail, since the magnetite is impermeable to the carbon oxides.

Bohnenkamp and Engell[53] and Caplan et al.[54,55] have also reported rapid carbon loss from the steel during the initial stage of oxidation at higher temperatures (circa 850°C) followed by a much lower, or zero, loss of carbon later. Caplan et al. measured the CO_2 evolution by infrared gas analysis and reported that the percentage of carbon loss from 0.1%C, 0.4%C, 0.8%C and 1.2%C steels was overall very small and may be redistributed in the metal. There was no carbon loss detected at 700°C. The overall oxidation rates were all found to be parabolic at 850°C and less than the oxidation rate of pure Fe. In the Fe-C alloys the individual phases of FeO, Fe_3O_4 and

Fe_2O_3 were found to be less regular than those of pure Fe and were often highly porous.

Malik[56] reported that, at temperatures between 600 and 850°C, in 101 kPa oxygen, the oxidation rate of Fe–5%M–C steels (where M was Si, Ti, V, Nb, Ta, Cr, W or Ni) fell as the carbide stability increased. The oxidation of all of the alloys obeyed parabolic kinetics, although some breakaway occurred following an incubation period. This breakaway was attributed to scale disruption, as a result of CO_2 evolution, with the carbon loss being most rapid during the first 5 min. Whilst the amount of carbon loss increased with the carbon content of the alloy, as did the oxidation rate constant, the total carbon loss was very much lower than that available. Those alloys forming a pure carbide phase were found to have a lower oxidation rate than those alloys comprising a solid solution phase or cementite. All of the binary Fe–5%M alloys displayed a similar reaction rate, which was approximately one order of magnitude lower than that of pure Fe, due to the formation of mixed oxides or spinels in the scale. The Fe–5%M–C alloys always showed two-layered scales, with an inner mixed oxide or spinel overlaid with Fe_2O_3. The scales formed on the high-carbon alloys were generally more compact and adherent (following initial scale disruption by C loss) due to carbide dispersion improving scale integrity. Malik argued that the carbide-forming elements retard the diffusion of carbon in austenite, reducing the overall scaling rate. In non-carbide-forming alloys, such as Ni and Si, the oxidation rate was greater, due to a higher carbon mobility in the steel.

At 850°C and 1.2% C the oxidation rate was found to be in the order Fe–Cr–C > Fe–C > Fe–Ni–C > Fe–Ti–C > Fe–Ta–C > Fe–Nb–C > Fe–V–C > Fe–W–C, which is nearly, but not exactly, the sequence of carbide stabilities.

Aluminium

On oxidation, aluminium forms the highly refractory, and hence protective, Al_2O_3. However, addition of aluminium to steels can cause embrittlement problems. There is a need, therefore, to realise the protective benefit of aluminium at as low a concentration as possible within the steel. At aluminium concentrations below 2.4 wt%, bulky stratified scales, comprising Fe_2O_3 and Fe_3O_4, with an inner layer of Al_2O_3 or $FeAl_2O_4$ are formed at 800°C[57]. At 2.5% aluminium, large areas of Al_2O_3 were always observed with iron oxide nodules. Formation of these iron oxide nodules is only suppressed once the aluminium content exceeds approximately 7 wt%[57-59].

Ahmed and Smeltzer[60], Pons et al.[61] and Smith et al.[62] have all found that, at 1173 K, iron alloys containing around 5–6% aluminium initially form a rapidly growing duplex scale comprising an outer α-Fe_2O_3 layer overlaying an inner $(FeAl)_3O_4$ layer. However, with prolonged oxidation, an Al_2O_3 layer eventually forms at the oxide/metal interface and as precipitates within the alloy. The reaction rate gradually decreases as the Al_2O_3 at, or near, the surface coalesces eventually to form a continuous film which virtually stops the outward diffusion of Fe ions. Electron Backscattered Mossbauer Spectroscopy studies have shown that the outer Fe_2O_3 contains approximately 10% Al^{3+} and the inner layer comprises Al_2O_3 with some

Fe^{3+} [63]. The critical Al content for the exclusive formation of Al_2O_3 has been found to be raised by the presence of Ti and B[64].

The beneficial effects of adding both aluminium and chromium to steels has been demonstrated by Tomaszewicz and Wallwork[64] during oxidation studies on Fe–Al–Cr alloys at 800°C in pure oxygen at 26.6 kPa. They showed that Al acts as a primary getter for oxygen, nucleating Al_2O_3, with Cr acting as a secondary getter, nucleating Cr_2O_3. If there was sufficient Cr and Al in the alloy, no iron oxides were formed. The total Al + Cr content to suppress nodule formation was found to be in the range of 7–8%, with 7% Al required at 0% Cr, but only 3% Al required at 5% Cr.

The diffusion coefficient for S in Al_2O_3[65] at 950°C is approximately 100 times lower than that in Cr_2O_3[66]. Hence, for high-temperature applications in S environments, aluminium confers a much greater degree of protection than that afforded by chromium.

Silicon

The addition of silicon to iron has been reported by many authors to confer significant corrosion protection. Rahmel and Tobolski[67] found that the addition of up to 4% Si to Fe, exposed to pure oxygen, $O_2 + H_2O$ or CO_2 in the temperature range 750–1 050°C, gave a limiting corrosion rate due to the formation of an iron–silicate layer. However, Robertson and Manning[68] indicate that Fe–Si oxides are generally immiscible, and several authors[69–72] have reported that the corrosion protection arises from the formation of a SiO_2 healing layer, beneath the magnetite, which acts as a barrier to outward transport of metal ions. At the low oxygen potentials at the base of the scale, and for very thin films, charged effects in the amorphous SiO_2 network and electronic carriers control the growth of the SiO_2[42, 73].

Adachi and Meier[74] studied the oxidation of Fe–Si alloys under isothermal and cyclic oxidation conditions, in air, between 900°C and 1 100°C. They found that the air oxidation rate decreased with the silicon content, such that at 10 wt% Si the oxidation rate was lower than that conferred by Cr, due to the formation of a continuous film of SiO_2. The oxidation kinetics of these steels was found to be linear, due to the diffusion of Fe through a film of SiO_2 of constant thickness. This Fe then caused dilution of the SiO_2 to form Fe_2SiO_4 and produced the Fe_2O_3 as an outer layer. At greater than 10% Si, SiO_2, overlaid with Fe_2O_3, comprised the total scale.

At the low oxygen potentials found in CO/CO_2 environments, the critical concentration of Si for the selective formation of a continuous film of SiO_2 is only approximately 0.05%[75]. However, at higher oxygen potentials, the higher growth rate of transient iron oxides suppresses the growth of this continuous SiO_2 scale. Even at 5% Si in the steel, no continuous SiO_2 layer formed during oxidation in air, whereas such a film did form during oxidation in Ar at a pO_2 of 10^{-4}.

Logani and Smeltzer[76–78] have observed that, for Fe–1.5%Si at 1 000°C in CO/CO_2, the initial slow reaction rate was followed by regions of linear behaviour due to the amorphous SiO_2 film being consumed by the growth of wustite–fayalite nodules during the early stages. These wustite–fayalite nodules were nucleated at alloy grain boundaries and then grew laterally to inundate the SiO_2 films over alloy grains.

During the oxidation of high Si content steels in high-pressure CO_2, the oxidation reaction can suddenly switch to a highly protective mode which proceeds extremely slowly after an initial incubation period[68]. This follows the formation of a healing layer, comprising a line of amorphous silicon-rich oxide along the oxide/metal interface with Cr enrichment to 30–40 at% just above the healing layer and Cr depletion in adjacent metal sites. Slowing of the oxidation reaction by the SiO_2 permits selective oxidation of Cr. Cr and Si are then synergistic, with less Si being required in Cr-containing steels than in straight Fe–Si steels, due to secondary gettering. For Fe–Si, then, 2.5–3% Si is required for healing layer formation, irrespective of the temperature. For Fe–Cr–Si, however, the critical Si content decreases with increasing Cr content and temperature.

Manganese

Since Mn is both soluble in iron oxides and mobile to the same extent as Fe, the addition of Mn to steels has little effect on the overall scaling rate in air or oxygen. Jackson and Wallwork[79] have shown that between 20% and 40% manganese must be added to steel before the iron oxides are replaced by manganese oxides. However, Mn supresses breakaway oxidation in CO/CO_2 possibly by reducing the coalescence of pores in the oxide scale.

The addition of up to 15% Mn to pure Fe, under sulphidising conditions at 1073 K, leads to a small increase in the scaling rate[82]. At 2% Mn, MnS forms as stringers in the subscale, but these do not form a coherent layer even at a concentration of 15% Mn. The increase in scaling rate is possibly due to increased short-circuit diffusion, since metal diffusion in MnS is much slower than in FeS[83].

Sulphur

The presence of small quantities of S in steels has little effect on the initial scaling rates in air, but may be detrimental to long-term scale adhesion. Sulphur has, however, been shown to be detrimental to breakaway oxidation in CO/CO_2 environments[81]. However, sulphur has been shown to reduce the total uptake of carbon in the steel under CO/CO_2[84] and reduce the scale thickening rate. In this context, free-cutting steels were found to oxidise at a significantly lower rate, as did steels subjected to pretreatment in H_2S.

Phosphorus

Like sulphur, phosphorus appears to have little effect on the overall scaling of iron alloys in air. It may, however, play a role in suppressing breakaway oxidation in carbon steels in CO/CO_2 environments. Donati and Garaud[80] found that the tendency for breakaway was lower over ferrite, where P segregates. To confirm this, the authors doped pure Fe with P and found that the breakaway rate slowed down at 350 ppm phophorus and was totally supressed at 900 ppm phosphorus. A similar benefit has been reported by Dewanckel et al.[81].

Nickel

The addition of Ni to Fe dramatically reduces the oxidation rate due to the virtual insolubility of NiO in FeO[85]. With time, the Ni concentrates at the FeO/substrate interface, reducing the oxidation rate by reducing the Fe activity at the base of the FeO and thus reducing the stability of the FeO. Menzies and Lubkiewicz[86] found that the oxidation of an Fe-12%Ni alloy in O_2 obeyed parabolic kinetics at all temperatures between 700°C and 1 000°C. At 700°C, the Ni suppressed the formation of FeO, with the scale comprising only Fe_3O_4 and Fe_2O_3. Progressive enrichment of Ni occurred in the alloy as the substrate was consumed by oxidation. When the concentration of Ni reached 50-60%, the Ni entered the spinel phase, leading to the formation of $Ni_{(x)}Fe_{(3-x)}O_4$ with x approximately equal to 0.24 near the alloy surface and less than 0.01 near the Fe_2O_3. At 900°C to 1 000°C, the Ni entered the spinel in the early stages of oxidation, with x values of 0.4 at 900°C. This led to a reduction in the parabolic rate constant. At 1 000°C FeO is the stable oxide phase formed on Fe-Ni alloys, even up to 80% Ni, since FeO is more stable than NiO[87].

Chromium

Of all of the alloying elements added to steels, Cr has been the most used for improving the corrosion properties. In terms of high-temperature oxidation, steels containing approximately 10% Cr are capable of forming a continuous, highly protective film of Cr_2O_3[33]. Significant reductions in the oxidation rate are realised at lower Cr concentrations due to the formation of FeCr spinels and the suppression of FeO formation to temperatures in excess of 570°C[88]. At 700°C FeO exhibits a very narrow stability range on the Fe-Cr-O phase diagram (Fig. 7.6) for Cr up to 6%[89]. The wustite stability range is almost negligible on Fe-0.5%Cr due to the high reactivity of Cr towards O_2 at 1 000°C. However, whilst Cr_2O_3 is normally protective on steels in air or O_2 up to approximately 900°C, volatile CrO_3 may form at higher temperatures[33]. Whilst the addition of Cr to steels is normally considered beneficial, Dewanckel et al.[81] found that low quantities (< 500 ppm Cr) were detrimental to the breakaway performance of low-alloy and carbon steels.

Many studies have shown that surface pretreatment of Fe-Cr alloys has a strong effect on the scale morphology and subsequent oxidation rate[27,90-92]. For instance, Caplan[27] indicated that several Fe-Cr alloys show improvement in the corrosion resistance due to cold work, with greater than 16% Cr required to show the optimum benefit. Khanna and Gnanamoorthy[90] examined the effect of cold work on 2.25%Cr-1%Mo steels at temperatures between 400°C and 950°C over 4 h in 1 atm O_2. They found that up to 90% reduction by cold rolling had a negligible effect on the oxidation rate up to 700°C. However, above 700°C there was a general reduction in the kinetics

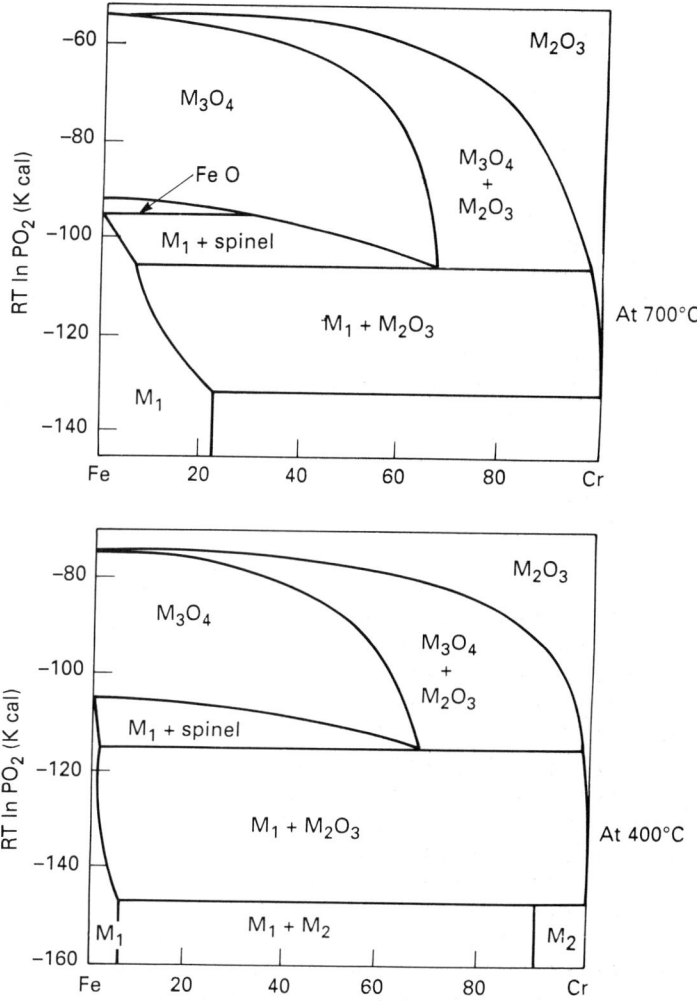

Fig. 7.6 Fe–Cr–O phase diagram for Fe–Cr alloys at 400°C and 700°C

due to enhanced Cr diffusion in the alloy, leading to the formation of a Cr-rich spinel. In addition, they found that breakaway corrosion occurred at 900°C on annealed samples, but this did not occur on samples which had been cold-worked. For Fe–10%Cr at 600°C in air, Hossain found that the oxidation resistance increased with the degree of cold work[91].

Molybdenum

In the temperature range 500–1 000°C, the addition of between 0.5% and 5.6% Mo decreases the oxidation rate of iron by a factor of almost ten, with the maximum benefit being obtained at a concentration of approximately 2%[93]. Like many alloying elements, Mo promotes the formation of a duplex spinel/magnetite scale, with the Mo retained within the inner spinel layer[91].

Others

Small additions of Ce have been shown to have a favourable influence on oxide growth of several Fe-Cr alloys by improving scale adherence and acting as nucleation sites for Cr_2O_3[94]. Levels of Ce as low as 0.024% reduce the carbon uptake of steels in carbonaceous atmospheres by several orders of magnitude. Trace concentrations of As and Sn have been found to improve the breakaway properties of mild and low alloy steels in CO/CO_2, whereas Cu has been found to be detrimental[81].

Stress Effects

Whilst alloying elements may initially determine the protective nature of an oxide scale, the response of this oxide scale to stress is often crucial in determining the long-term oxidation performance of an alloy in an industrial environment[95]. The stresses in an oxide scale may arise both internally (due to growth stresses) and externally (due to applied stresses)[96]. In general, oxide scales do not possess sufficient slip systems for plasticity and thus rely upon diffusion controlled creep for plastic deformation and stress relief[97]. At temperatures below approximately 600°C only elastic deformation is possible and stress relief by cracking is likely to occur. If the scale/metal interface is strong, and the bulk oxide is weak, through-scale cracking results when the combined stresses exceed a critical value[98]. If the interface has a lower strength than the oxide bulk strength, decohesion occurs first. A minimum energy is required for cracking or decohesion[99,100]. Evans and Lobb[98] have calculated that cracking occurs when the strain energy per unit volume (w^*) of oxide contained in layer thickness (t) equals the work required for internal cracking (G_c) or decohesion (G_d). For internal cracking:

$$w^* = \frac{8PG_c}{fL} \sim \frac{4G_c}{fL} \quad (7.4)$$

where L = side length of unit volume, f = fraction of stored energy in the oxide, G_c = energy for unit area of cracked oxide surface, and P = geometric parameter (~ 0.5). For decohesion:

$$tw^* = \frac{G_d}{f} \quad (7.5)$$

where f = energy/unit area of fresh metal at the oxide/metal interface.

Through-scale cracking may not necessarily be detrimental to an alloy if rapid scale healing can occur[101], and scale delamination can reduce the total scaling rate if the scale remains adjacent to the alloy surface[96,102]. In this instance, scale separation confounds the transport of metal ions across the interface into the oxide. However, if through-scale cracking occurs in conjunction with delamination, scale spalling may follow[98], with an attendant loss of protection. For pure metals which oxidise in a parabolic manner, repetitive scale spalling may produce a much enhanced total metal loss over time, since the reaction kinetics continually revert to the initial rapid period of the oxidation curve and paralinear kinetics ensue.

The majority of heat-resisting alloys contain at least one element which is selectively oxidised (e.g. Cr, Si, Al). The resulting scale is highly protective, but depletion of the secondary element may occur if the diffusion rate of this element in the alloy is low. Then repetitious loss of the scale can be profoundly detrimental to long-term performance, since depletion of the alloying element near the surface may occur to such an extent that its concentration falls below the critical level for the exclusive reformation of the highly protective scale[103, 104].

Growth Stresses

As indicated earlier, protective oxide scales typically have a PBR greater than unity and are, therefore, less dense than the metal from which they have formed. As a result, the formation of protective oxides invariably results in a local volume increase, or a stress-free oxidation strain[105, 106]. If lateral growth occurs, then compressive stresses can build up, and these are intensified at convex and reduced at concave interfaces by the radial displacement of the scale due to outward cation diffusion (Fig. 7.7)[48, 106, 107].

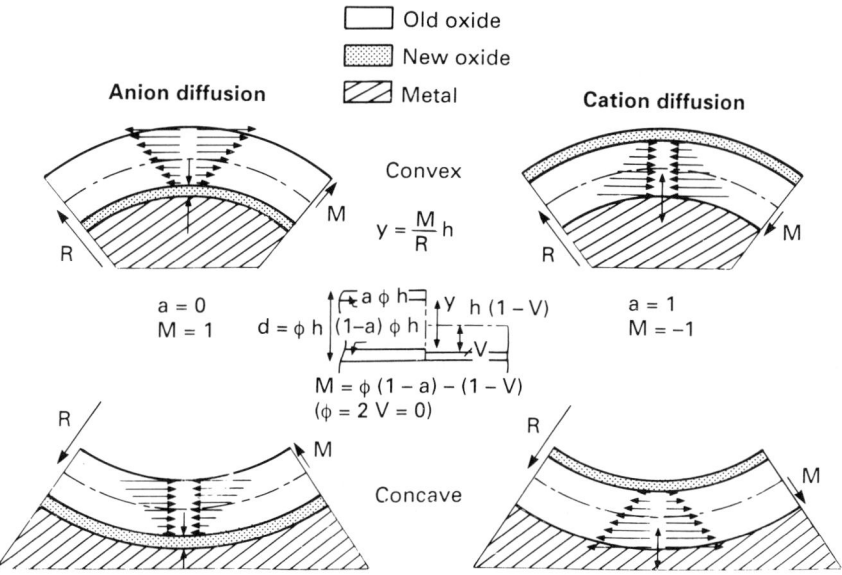

Fig. 7.7 Oxide growth stresses on curved surfaces (after Hsueh and Evans[105])

The reduced oxidation near sample corners is related to these stress effects, either by retarded diffusion or modified interfacial reactions[105]. Manning[108] described these stresses in terms of the conformational strain and distinguished between anion and cation diffusion, and concave and convex surfaces. He defined a radial vector M, describing the direction and extent of displacement of the oxide layer in order to remain in contact with the retreating metal surface, where:

$$M = \frac{V_{ox}}{V_{me}}(1 - a)1(1 - v) \tag{7.6}$$

and V_{ox}/V_{me} is the PBR, a is the amount of oxide formed at the surface and v is the volume gain due to vacancy injection into the metal. If v is zero, the sign of M depends upon the predominance of anion ($a = 0$) or cation ($a = 1$) diffusion. The enforced radial displacement (δy) of the oxide layer results in a tangential or radial reaction stress in both the oxide and the metal. The strain rate ($\delta e/\delta t$) of the oxide layer then depends upon the radius of curvature (R) and the rate of metal loss ($\delta h/\delta t$) as:

$$\frac{\delta e}{\delta t} = \frac{M(\delta h)}{R(\delta t)} \tag{7.7}$$

This leads to high strains on regions of high curvature and zero strain on a flat surface. If new growth occurs stress free, then the oldest region of oxide will be the most highly stressed, this being located at the oxide surface for anion-controlled growth and at the oxide/metal interface for cation-controlled growth. Hsueh and Evans[105] carried out a similar analysis and calculated the distribution of radial and tangential stress within the oxide scale as a function of scale thickness and specimen geometry.

Whisker growth is a process often reported for the relief of compressive growth stresses during oxidation[109-111]. This whisker formation, and scale buckling of thin haematite layers, has been linked to the stresses arising from oxide formation at the magnetite/haematite interface as a result of the countercurrent of cation and anion diffusion in both oxides[107]. Acoustic emission studies of 2.25%Cr–1%Mo steel at 900°C have shown that, as the oxide thickness increases, growth stress builds up in the scale and is relieved by scale cracking[112].

System-applied Stresses

The tensile failure strain of oxides grown on EN2 steel between 600°C and 900°C lies in the range 1×10^{-4} to 2.5×10^{-4} [113]. Components in service may be stressed beyond these failure strains, leading to scale cracking.

Ward et al.[114] have shown that, under cyclical loading, the oxidation rate of steels is similar to that under unstressed isothermal conditions, provided the fatigue stress is below the stress required to exceed the scale failure strain. If, however, the failure strain is exceeded, the oxidation rate is accelerated due to repetitive scale failure, and linear kinetics are observed.

Low cycle fatigue loading of 9.5%Cr steel at 650°C in air has been shown to enhance uniform scale formation as well as promote nodular scale formation at cracks[101]. However, no cracks were found in the scale after exposure, indicating that any cracks that form must heal very quickly. This healing of cracks was attributed to the overgrowth of the chromium-rich oxide by an iron rich oxide.

Thermal Stresses

Under thermal cycling conditions, the principal source of stress within the oxide scale is the temperature change[98]. Christl et al.[102] have noted that, when cooling 2.25%Cr-1%Mo steel from 600°C in air, compressive stresses build up in the haematite, whilst tensile stresses build up in the magnetite and spinel layers. This arises because the thermal expansion coefficients of the individual oxide layers increase in the order: α metal $<$ α spinel $<$ α magnetite $<$ α haematite[96].

Multilaminated scales have been reported following thermal cycling[115], and Rolls and Nematollahi[116] have studied the influence of thermal cycling on the oxidation of 1%Cr-0.5%Mo low-carbon steel. The oxidation kinetics were found to be mostly parabolic, with thin scales (10 μm) more prone to spalling than thick scales (20 μm). The authors reported that the higher the temperature drop on cycling, the greater was the degree of scale disruption. They derived a qualitative relationship to describe this behaviour:

$$Y = \frac{f(NO)}{x} \quad (7.8)$$

where Y = degree of scale detachment, N = number of cycles, x = thickness, and O = cooling rate. This implies that thicker scales cooled slowly show less detachment than thinner scales cooled rapidly. The oxide scale comprised spinel, magnetite and wustite, with only the spinel layer remaining adherent during thermal cycling. From this the authors concluded that the bulk scale cohesive strength was less than the inner spinel adhesive strength. Scale detachment was observed at voids and microcracks produced at the outer/spinel interface during isothermal oxidation. Hence, the number of scale layers that became detached during thermal cycling was governed by the number of parallel rows of voids in the scale and not by the number of thermal cycles.

Commercial Low-alloy Steels in Air or Oxygen

Simms and Little[117] have examined the early stages of scale growth on 2.25%Cr-1%Mo steel at 600°C in dry flowing oxygen. Between 1 and 22 h they found that a thick oxide layer spreads laterally over a thin oxide layer. After 50 h, no thin areas of oxide layer were left. Whiskers gradually developed on the outer surface and these were well defined after 100 h. Fracture sections of the oxide revealed that the thin scales were duplex whereas the thicker scale was triplex, with a middle layer of Fe_3O_4 which spread laterally with time. The authors concluded that the first phases to develop were fine equiaxed α-Fe_2O_3 overlying a doped spinel. Later, nucleation and

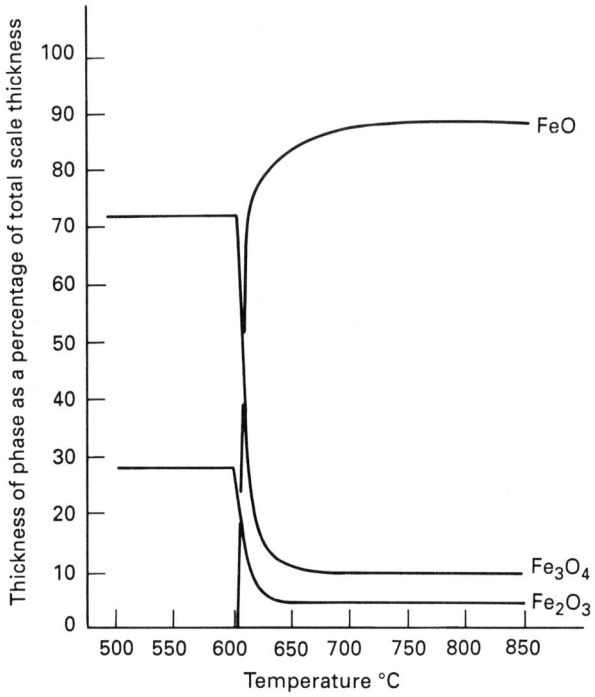

Fig. 7.8 Relative thickness of wustite, magnetite and haematite on mild steel as a function of formation temperature

lateral spread of coarse columnar grains of Fe_3O_4 occurred between the two first-formed layers. The Fe_2O_3 appears to reach a limiting thickness at which time Fe_3O_4 nucleates.

The author[118] has carried out exposures of a range of low-alloy steels (up to 5% Cr) in laboratory air between 500°C and 850°C. Plain carbon steel displayed a duplex scale, comprising 75% Fe_3O_4, 25% Fe_2O_3 at all temperatures below 600°C (Fig. 7.8). Above 600°C, a coarse, columnar-grained layer of FeO, occupying approximately 90% of the scale thickness was observed. For the Cr-containing steels, a distinct spinel phase was observed at temperatures in excess of around 615°C, with all of the Cr, Mo and V incorporated within this layer, beneath the Fe_3O_4 and Fe_2O_3. FeO was not observed in the scale until temperatures exceeded 650°C, the FeO appearing as a fourth layer between the spinel and Fe_3O_4 layers. Parabolic oxidation kinetics were observed for all steels at all temperatures. Below 600°C there was little difference in the oxidation kinetics between the chromium steels until the Cr level reached around 5%, this difference mainly arising due to an increasing activation energy for Fe_3O_4 growth with increasing Cr content. There was no significant difference in the activation energy for Fe_2O_3 growth, irrespective of the Cr content of the steel (Fig. 7.9).

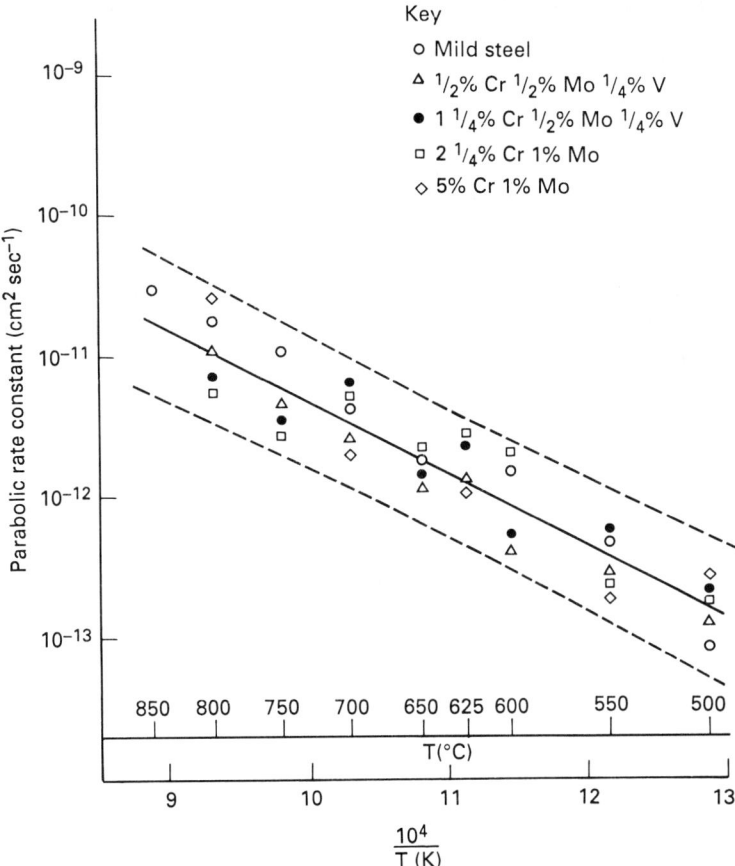

Fig. 7.9 Arrhenius plot for the haematite thickening rate on mild steel and low-chromium alloy steels in air

Industrial Environments

Steam

A number of authors have reported studies of the oxidation of low-alloy steel in steam. The general observations indicate strong similarities with oxidation in air, the kinetics being typically parabolic[119,120] and the scales typically comprising Fe_2O_3, Fe_3O_4, FeO and spinel phases, dependent upon alloy composition, temperature and oxygen partial pressure of the environment[121-123].

For Fe in steam, water vapour or CO_2 below 570°C, a two-layered Fe_3O_4 layer is observed, the inner layer growing by O_2 diffusion inwards. Similarly, Potter and Mann[124] reported the formation of a duplex Fe_3O_4 layer during the oxidation of mild steel in steam between 300°C and 550°C.

Inward growing spinel layers, containing all of the alloying elements are observed on low alloy steels[121], with outer layer growth being controlled by outward diffusion of iron via grain boundaries[46,125]. Since the solid state diffusion of O_2 is five orders of magnitude too low to account for the observed oxidation rate[41], the inner layer is assumed to be porous. Mayer and Manolescu[126] have confirmed the presence of the required network of interconnecting pores in the inner layer.

Rahmel and Tobolski[127] have shown that the iron core below triplex scales of $FeO/Fe_3O_4/Fe_2O_3$ grown in O_2 + water vapour at 950°C become enriched in H, indicating that water vapour can penetrate relatively thick wustite scales. Cracks are a valid path for such diffusion but penetration can occur even when the oxide can deform plastically and there is no evidence of cracking[128].

Cory and Herrington[120] measured the location of H_2 formation in 9Cr steel at temperatures between 501°C and 552°C and 2.25Cr, 2.25CrNb and Fe-C steels at 501°C. The slope of all plots of H_2 evolution indicated parabolic behaviour. For the 9Cr steel, the H_2 evolution was found to be

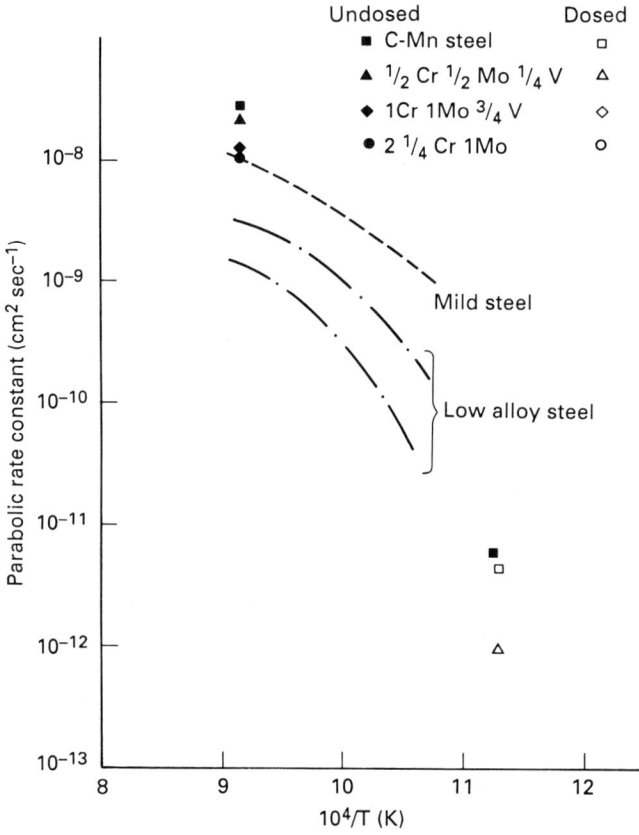

Fig. 7.10 Kinetics of wustite growth on mild steel and low-chromium alloy steels in air and steam (after Wiles[119])

independent of the steam partial pressure, Fe diffusion in the Fe_3O_4 being the rate controlling step. All alloys below the 9Cr level were found to oxidise at two to three times the rate of the 9Cr steel.

At temperatures in excess of 750°C the addition of water vapour accelerates the rate of growth of FeO (Fig. 7.10) by producing large pores in the FeO[119,129]. At much higher temperatures (1 200°C) Sheasby et al.[130] have found that addition of steam to O_2-N_2, O_2-H_2O-N_2 and H_2O-N_2, in a simulated reheating atmosphere furnace, caused increases in scale growth due to improved adhesion at the scale/metal interface. They concluded that water vapour enhances scale creep as previously reported by Tuck et al.[131].

Griskin et al.[132] reported that there is no apparent effect of steam pressure on the rate of oxidation of Cr-Ni steels at temperatures between 600°C and 650°C at 10.1-20.2 MPa. Similar observations for Cr-Mo and Cr-Mo-V steels between 500°C and 600°C have been made by Wiles[119]. She compared low-alloy steel samples exposed to 101 kPa steam with power plant components that had operated for up to 150 000 h in steam at 17.25 MPa and found no significant difference in the oxidation rates (Fig. 7.11).

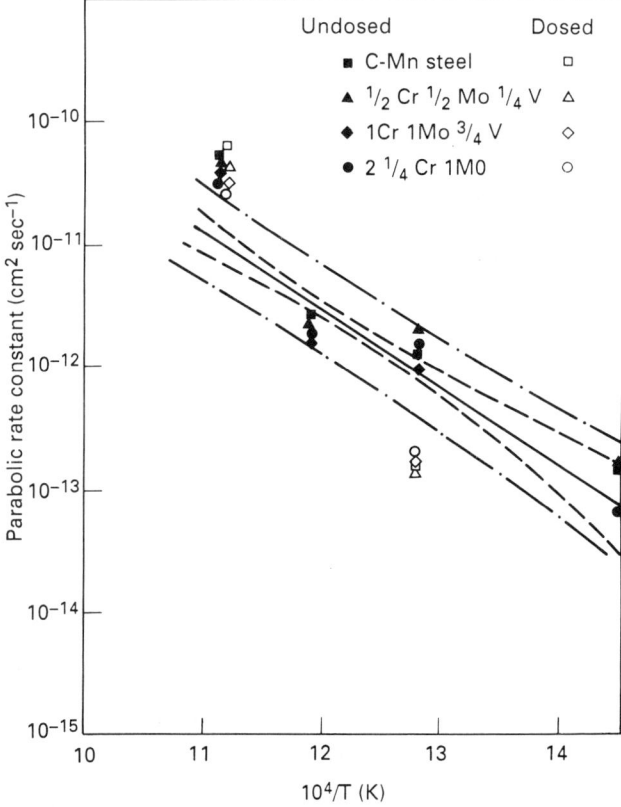

Fig. 7.11 Comparison of overall scaling kinetics of low-alloy steels in laboratory air and power station steam at 17·25 MPa (after Wiles[119])

Whilst total pressure appears to have little effect on the overall scaling kinetics, the oxygen partial pressure of the environment does influence the overall scale morphology. At a low enough pO_2, 9%Cr steels in steam produce a layered scale of Fe_3O_4 and spinel, with no Fe_2O_3 overlay[125]. Wiles[119] also noted that the proportion of Fe_2O_3 in the scales grown on low-alloy steel was directly dependent upon the oxygen partial pressure of the environment.

This variation in the amount of Fe_2O_3 in the outer portion of the scale has been found to be of importance in industrial applications. The oxide scales formed on the steam side of superheater and reheater tubes in power generation contain varying amounts of Fe_2O_3[133]. It has been shown that the proportion of Fe_2O_3 in the scale dictates the spalling propensity, the greater the Fe_2O_3, the greater the tendency to scale spalling during thermal cycling[96,102,133]. Spalling of the bore oxide scale can then lead to potential problems with turbine erosion and tube blockage[96,134].

Combustion Gases

A considerable amount of work has been carried out into the corrosion of steels in the gases produced during the combustion of fossil fuel due to extensive use of low alloy steels as heat exchanger tubes in power generation. Combustion gases contain many species, such as CO, CO_2, SO_2, SO_3, H_2S and HCl, arising from elements within the fuel. The many different combinations of operating temperature and chemical stoichiometry of combustion reactions lead to many possible complex corrosion reactions.

In coal-fired power stations, severe corrosion of the steam-generating tubes in the furnace walls has been attributed to reducing conditions with high pCO, low pO_2, high pHCl, and high local heat fluxes and flame impingement[135,136]. In areas of high corrosion, large concentrations of CO are almost always present, and thick Fe_3O_4 scales, with islands and bands of FeS are observed[135,136]. Under the most severe reducing conditions, thick columnar scales of almost pure FeS have been observed[136]. The high corrosion rate areas are also subject to a significant flux of sulphur-bearing carbonaceous material in the flame envelope[135] which may locally exacerbate reducing conditions and raise the pH_2S[136]. Outside the high corrosion rate areas, the furnace atmospheres are found to be relatively O_2-rich and the corrosion scales comprise protective Fe_3O_4[135].

Lees and Whitehead[136] have shown that differences in boiler design lead to differences in furnace atmospheres, which are subsequently reflected in differences in scale morphology and corrosion performance. Hence they report that there is no unique scale morphology which is characteristic of furnace wall corrosion. They also warn that the scale that is examined during an investigation may not be an exact reflection of the scale on the tube surfaces during operation due to the possible hydrolysis of the scale on cooling (when hot flue gas is replaced by moist air) and the redistribution of phases in the scale due to the loss of the incident heat flux.

There is extensive experience within the UK of the combustion of high-chlorine coals because of the unusually high chlorine contents found in indigenous British coals[137]. These Cl levels average 0.25%, with some seams as high as 0.8% or more Cl. This Cl is present in the coal in a single basic form, probably as weakly bonded ions, and is released from coal very rapidly during combustion as HCl[138].

At medium to high corrosion rates (>100 nm/h) the corrosion scale found on the surface of corroding furnace wall tubes in the UK is invariably separated along 50% of the fireside circumference by a Cl-rich layer of uncertain composition. The stoichiometry suggests that this layer lies between $FeCl_2$ and FeOCl compositions[136]. However, high furnace wall corrosion rates may occur at low Cl concentrations if the conditions are severely reducing and may also arise if the concentrations of SO_2 and HCl in the flue gas are high. Under these conditions the scales comprise typically Fe_3O_4 with FeS islands and lamellae.

The influence of HCl on the corrosion rates and mechanisms of furnace wall corrosion has still not been finalised. UK data from low-pressure plant (Fig. 7.12) has shown that:

$$R = 1380[Cl] - 290 \qquad (7.9)$$

where R is the corrosion rate in nm/h and [Cl] is the percentage Cl in the coal[136]. However, very high corrosion rates can occur at low chlorine contents and sudden upsurges in corrosion rates have been correlated with long periods of sustained heavy load running in the absence of large changes to the coal chlorine content. Laboratory studies have investigated the role of HCl in the fireside corrosion mechanism, with HCl favouring a transition

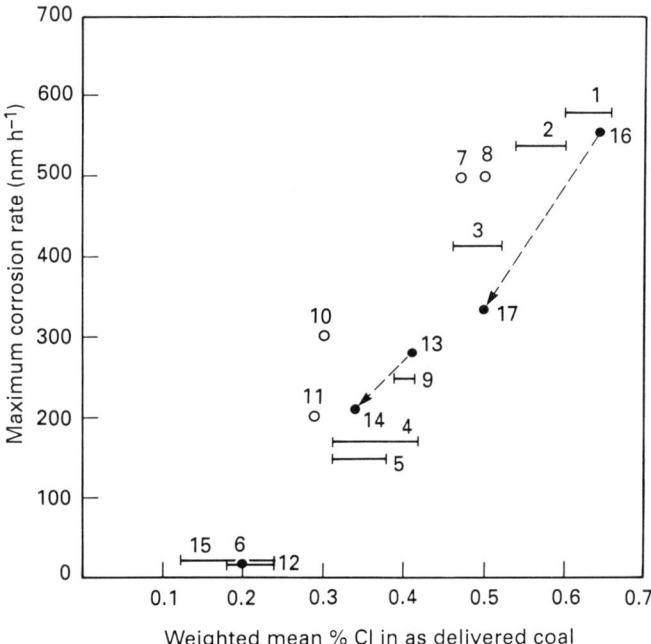

Fig. 7.12 Influence of coal chlorine content on the furnace wall corrosion rates of mild steel tubes in low-pressure coal-fired power plant (Lees, C.E.G.B., private communication)

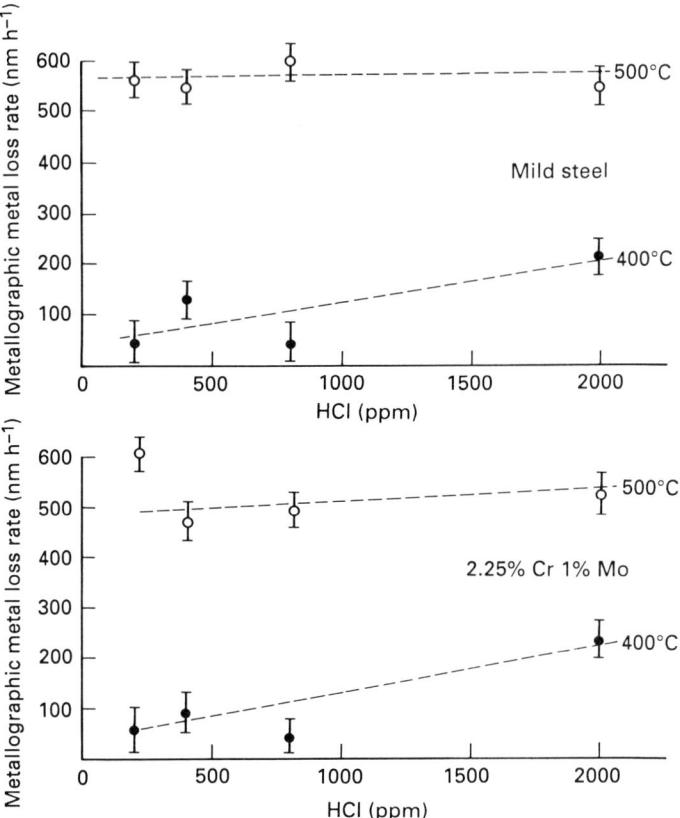

Fig. 7.13 Influence of coal chlorine content on the corrosion rates of low-alloy steels exposed to laboratory simulation of furnace wall corrosion (Brooks, C.E.G.B., private communication)

from parabolic to linear kinetics above 0.2% Cl in the coal, but no increase in corrosion rate with increasing coal Cl content above this threshold (Fig. 7.13) has been demonstrated[139]. The current understanding is that reducing conditions, coal chlorine content and high heat fluxes most probably act synergistically to produce the intense corrosion conditions often observed.

Superheater/reheater corrosion in fossil-fuel-fired boilers is caused by the deposition of alkali sulphates on to the tube surface[140]. The corrosion rates increase rapidly at temperatures above 600°C as the sulphates become molten. These molten sulphates contain free SO_3 which dissolves the protective oxide to form Fe-based sulphates. The corrosivity of the molten sulphates depends strongly upon their melting points, which are themselves strongly dependent upon the ratio of Na and K in the deposits.

Higher than design strain rates in service ($>10^{-9}\,s^{-1}$) can also cause disruption of the corrosion scale, leading to enhanced metal loss and corrosion-assisted cracking of the substrate[141]. Laboratory exposures show parabolic rates with and without Cl additions to the gas. There appears to be a direct effect of HCl on the scale integrity. Low Cl (0.2% HCl) produces porous and discontinuous Fe_2O_3, whilst high Cl (0.8% HCl), causes complete disintegration of the Fe_2O_3 and an irregular and porous magnetite scale. It was originally thought that the direct relationship between Cl content and corrosion rate was due to such a direct influence of HCl on scale integrity. It is now known, however, that HCl aids the release of potassium from the coal, which increases the overall corrosion rate[142].

Fluidised-bed systems produce higher combustion intensities at lower temperatures than combustion of pulverised fuel in conventional fossil-fuel-fired boilers. The mineral matter for corrosion does not form fused salts and is not expected to release corrosive species. Fluidised bed combustors can, therefore, burn lower grade, cheaper fuel in smaller plant with better pollution control than traditional boilers[143].

Minchener et al.[144] report that the bubble phase of atmospheric fluidised bed combustion has a pO_2 in the range 2×10^{-1} to 2×10^{-2}. Combustion in the dense phase is sub-stoichiometric, with the pO_2 as low as 10^{-13}, and SO_2 and SO_3 present in the range 500–5 000 ppm. Low Cr–Mo steels show heavy scaling in these conditions, whereas 9–12% Cr steels show good resistance to sulphidation up to 650°C. Roberts et al.[145], however, report that for pressurised fluidised-bed combustion, ferritic steels at or below 9% Cr show heavy general corrosion above 540–560°C.

Chemical Environments

The oil industry frequently uses stainless steels or exotic bonded alloys for the processing of crude oil in the temperature range 200–600°C. These materials are very expensive and there is a strong economic incentive for finding cheaper alloys which are resistant to H_2S and some gaseous organic sulphides arising from the S content of the crude oil[146].

Metal sulphides show the same type of predominant defects as metal oxides, i.e. cations in $Fe_{(1-x)}S$, $Cr_{(2+y)}S_3$. The defect concentration in most sulphides is much higher than those in the corresponding oxides, but the defect mobilities are only slightly higher. Thus the higher diffusivities and growth rates are determined by the higher defect concentrations. Cr and Al only slightly reduce the corrosion rate, and much higher Al is needed than that required for oxidation protection. Very protective scales are only formed at a S pressure lower than that for formation of the base-metal sulphide[147].

Mrowec et al.[148] examined the resistance to high-temperature corrosion of Fe alloys with Cr contents between 0.35 and 74 at% Cr in 101 kPa S vapour. They found that the corrosion was parabolic, irrespective of the temperature or alloy composition, and noted that sulphidation takes place at a rate five orders of magnitude greater than oxidation at equivalent temperatures. At less than 2% Cr, the alloys formed $Fe_{(1-x)}S$ growing by outward diffusion of Fe ions, with traces of $FeCr_2S_4$ near the metal core.

Narita and Nishida[149] examined the sulphidation of low Cr–Fe alloys at 700–900°C in 101 kPa of pure S. They found that the addition of small quantities of Cr significantly decreased the corrosion rate due to the formation of

$FeCr_2S_4$ in the inner reaches of the scale. The scale comprised an outer FeS layer, with an inner layer of FeS, $FeCr_2S_4$ and Cr_3S_4 in varying amounts depending upon the Cr content. The corners of specimens corroded more rapidly than flat faces due to breakaway conditions. At low Cr contents the rate was increased, but above 4-6% Cr the parabolic rate constant decreased. Above 7.4% Cr an intermediate layer, containing $FeCr_2S_4$ and varying amounts of Cr_3S_4, the proportion of the latter increasing with increasing Cr content, formed between the inner and outer layers.

In view of this potentially rapid degradation of Cr-containing steels by high-temperature sulphidation in petrochemical and coal gasification reactors, Al is much used in the Fe alloys for these applications[150-153]. Al in Fe reduces the sulphidation rate in S_2 vapour by up to two orders of magnitude[154] due to the high thermodynamic stability of aluminium sulphide relative to iron sulphide, the low rate of sulphidation of Al compared with pure Fe and the large PB ratio of Al_2S_3 (3.7). The addition of 5% Al in Fe in 101 kPa S_2 vapour between 500°C and 700°C resulted in the rate of reaction decreasing by a factor of ten. Paralinear kinetics were observed, with the inner layer of a duplex FeS scale containing a finely dispersed Al_2S_3 phase which acts as a diffusion barrier to Fe^{2+} migration. Increasing the temperature to 800°C resulted in a rapid take off of the corrosion rate, with catastrophic corrosion rates above 800°C due to the large volume of Al_2S_3 causing an increase in scale porosity.

Condit et al.[155] examined the sulphidation of several Fe-Cr-Al alloys under a variety of sulphidising conditions. They noted that, in the early stages of sulphidation, a thin compact inner layer forms which is high in Cr and Al. Subsequently, a thicker microcrystalline outer layer forms with a uniform Fe, Cr and Al composition. Formation of the outer compact layer was favoured by increasing pS_2 and decreasing temperature, with the layer forming much more rapidly in H_2S than in pure S_2. The sublayer disappeared more or less rapidly dependent upon alloy composition. The authors propose three stages for scale development. First a thin compact layer forms due to the penetration of S into the alloy with preferential formation of sulphide from those metals with the highest affinity for S. $Fe_{(1-x)}S$ also forms due to the abundance of Fe in the alloy.

The outer layer then dissociates to release sulphur which dissolves in grain boundaries of the alloy to form Cr and Al sulphides. The Fe released by this dissociation sulphidises again at the interface between the two layers. The volume increase associated with the conversion of metal to sulphide generates mechanical stress which causes the outer layer to break up and permits permeation of S. This initiates a second stage, where growth of the scale is linear and comprises a porous outer layer, with FeS, Cr_2S_3 and Al_2S_3 evenly distributed, possibly as $FeCr_2S_4$, $FeAl_2S_4$ and $FeCr_{(x)}Al_{(2-x)}S_4$. S then diffuses through the pores to the scale/metal interface. The third stage comprises the formation of an outer compact layer of $Fe_{(1-x)}S$ and continued thickening of the inner layer.

Addition of Cr to Fe-Al alloys aids the formation of Cr sulphides and Al_2S_3 which together markedly reduce the sulphidation rate[146]. Between 2% and 5% Cr then, more than 3% Al is required to obtain protection. At 9% Cr, however, only 1% Al is needed to give protection since the Cr is sufficiently active to lower the S potential seen by Al (secondary gettering). Thus all Fe–Cr–Al alloys initially form FeS, Cr sulphides and Al_2S_3. The Cr and Al are then exposed to a much lower pS_2 at the scale/metal interface and Cr sulphides and Al_2S_3 grow preferentially if their activity is high enough. Ultimately, a protective layer of Al_2S_3 or Al_2S_3 + Cr sulphides develops at the scale/metal interface and the reaction rate decreases substantially.

Karlson et al.[156] found, from on-site experience of cement-producing plant, that corrosion of Fe surfaces may occur in gases containing O_2, SO_2 and alkali chlorides such as NaCl and KCl between 300°C and 500°C. They reported that the corrosion rates may be extraordinarily high (5–10 mm/month) implying liquid-phase corrosion. Laboratory simulation of the plant conditions demonstrated the need for both SO_2 and the alkali chloride in the environment. The principle corrosion reaction was found to be:

$$2Fe_2O_3 + 12[K, Na]Cl(s, l) + 12SO_2(g) + 9O_2(g) \rightarrow$$
$$4[K, Na]_3Fe(SO_4)_3(s, l) + 6Cl_2(g)$$

A thermodynamic evaluation of this equation indicated that the reaction could proceed with SO_2 levels as low as 100 ppm.

CO/CO_2

Failures of mild steel components in Magnox reactors in the UK and Italy after approximately 5 years of operation alerted the world to the potential for breakaway oxidation of low-alloy steels in CO/CO_2 environments[80]. The CO_2, 1% CO, 300 vppm CH_4, 250 vppm H_2O, 100 vppm H_2 environment used in CAGRs was selected on the need to minimise oxidation of the graphite reactor core and deposition of C from the coolant gas[157]. Corrosion rate tests of 15 000–20 000 h, in the limited range of conditions anticipated by the designers, showed that the maximum reduction in corrosion rate of ferritic steels in CO_2 at 600°C is realised at around the 9% Cr level[158–160]. Therefore 9Cr1Mo steel was chosen for the evaporator and primary superheater sections of the CAGR[158]. However, in the late 1960s, Taylor (reported in Reference 158) identified evidence for a significant change in the corrosion mechanism for 9% Cr steels at around 550°C. This change could lead to rate inversion with increasing temperature in steels containing 0.7–0.8% Si, or breakaway in 0.4–0.5% Si steels.

Because of their importance to the nuclear power generation industry, these observations initiated a vast amount of research into the oxidation of low-alloy steels in CO/CO_2 environments. It is now clear that low-alloy steels exhibit three types of behaviour when exposed to CO/CO_2, i.e. protective, transitional and linear-breakaway (Fig. 7.14), with the time to breakaway and the breakaway rate being of crucial importance in determining component life.

For mild and low-alloy steels in CO_2 the first scale to form is a compact coarse columnar layer of Fe_3O_4[161–163]. Growth of this layer is controlled by outward grain boundary diffusion of Fe ions[17, 163, 164]. The inward countercurrent of vacancies is initially annihilated at the metal surface, but eventually vacancy condensation at the scale/metal interface gives decohesion, the scale develops microporosity, an inner layer grows within the space created by the departing metal ions and a duplex scale forms[163–165]. Then

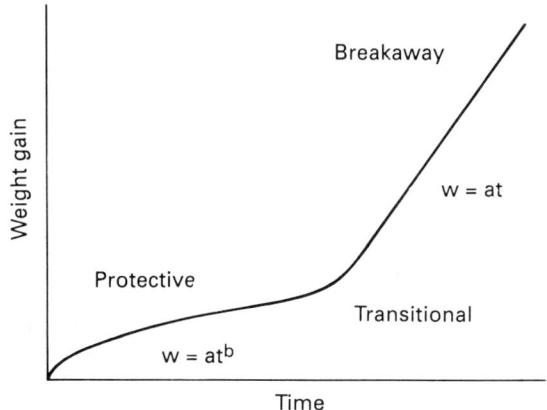

Fig. 7.14 Schematic diagram of stages of low-alloy steel oxidation in CO/CO_2

for Cr-containing steels, the protective scale comprises an outer, coarse columnar-grained magnetite layer and an inner, slightly porous, Cr-rich spinel of fine (0.1 μm) equiaxed grains[68, 162, 166]. As with low-alloy steels in other environments, the Cr is not mobile in the scale, but is oxidised *in situ*[163, 166]. Studies have shown that the M_3O_4 spinel nucleates at asperities on the surface and duplex growth is also known to be favoured in the vicinity of inclusions and specimen corners[163]. Once initiated, the inner layer grows by inward diffusion of O, probably as CO_2, down microfissures and micropores[17, 113, 167], in both the lateral and vertical directions, until a complete layer is obtained[164]. The growth of this layer subsequently follows the parabolic rate law.

During this protective stage, decreasing the water and CO content of the gas appears to decrease the rate constant of Cr-containing steels[157] but has little effect on the rate constant of carbon steels[161, 168]. The rate constant of all steels has been found to decrease with decreasing temperature and increasing Si content[103, 157, 161, 169, 170]. Ferguson et al.[168] have also reported that increasing the S content of carbon steels reduces the parabolic rate constant.

An increase in the Si content of the steel has been reported to give a significant increase in the duration of the protective regime for carbon steel[161, 168, 169]. A similar benefit has also been reported for Cr-containing steels[157, 158, 163, 171]. Increases in the duration of the protective regime are also realised with reductions in the CO and H_2O contents of the gas[168], in the temperatures[158, 161, 168] and, for carbon steels, in the surface roughness[161] or, for 9Cr steels, with increased surface cold work[158]. Robertson and Manning[43] found that breakaway may also be delayed by some S-containing gases.

Following the initial protective period, under certain conditions of temperature, alloy and gas composition[158], the oxidation goes through a transitional stage into breakaway. Several authors have reported that breakaway oxidation is initiated once the scale reaches a critical thickness[157, 158] or weight gain[68] and only occurs below an initially protective duplex layer[43, 163].

Both single-layer protective and duplex-layer protective scales have a PBR of 2.1, whereas breakaway scales show a PBR of approximately 2.7, containing around 23% porosity[163], and graphitic carbon grains (up to 6% by weight) between the oxide grains[161, 168].

During this transitional period, and once full breakaway is established, growth occurs throughout the oxide and close to the scale/metal interface[172]. Growth below the oxidant surface requires oxidant transport through the scale, but solid-state lattice diffusion is much too slow to account for the rates observed[17]. Therefore, a porous scale is required. Mechanisms for pore growth have been postulated by several authors[165, 173-176] who have suggested that the transition from single-layer to duplex growth is due to the initiation of pores within the scale. Then breakaway was thought to be caused by increased porosity giving unlimited access of oxidant to metal surface. However Atkinson and Smart[47] have shown single layer scales to be slightly porous. Robertson and Manning[43] have proposed that oxidant access is always available and the type of scale which develops is dependent upon conditions at the scale/metal interface. Breakaway occurs if space is created by continuous scale deformation or creep[177].

The inward penetration of CO_2 via cracks and micropores gives:

$$3Fe + 4CO_2 = Fe_3O_4 + 4CO$$
$$2CO = O_2 + 2C \quad \text{via the Boudouard reaction}$$

at the scale/metal interface[113, 167].

The carbon initially diffuses into the steel but ultimately, the steel may become saturated with C[177]. There is no detectable solubility of C in FeO, Fe_3O_4, MnO or Cr_2O_3[178], and C can only permeate through pores or faulty scales[94]. Thus, the C then deposits at the scale/metal interface and prevents the formation of a coherent protective layer once it reaches a critical activity[177]. Carbides in the steel have been found to be the preferred sites for breakaway. These are either pre-existing carbides or carbides precipitated by C injection during oxidation[163]. Precarburised or graphite-painted steel breaks away rapidly, as do thin foils, due to the smaller C sink available[163]. Pritchard et al.[84] reported that the proportion of C in the scale increased with scale thickness and water content of the gas, and was higher in breakaway oxide. For 9% Cr steel, breakaway oxidation is associated with heavy carburisation of the metal and C deposition within the oxide, with preferential breakaway occurring at corners and edges[166].

For Fe in CO/CO_2 at atmospheric pressure, Surman[179] found that if

$$10^{-3} < \frac{pCO}{pCO + pCO_2} < 0.3$$

then the oxidation is parabolic and very little C deposition occurs. He concluded that magnetite is not sufficiently catalytic to promote the Boudouard reaction unless CO > 10% and moisture is present and surmised that H_2O promotes the formation of a Boudouard catalyst. If the CO is greater than 0.4 in the above expression, oxidation and carbon deposition occur simultaneously at a linear rate[180].

German and Littlejohn[161] have observed that increasing the Si content of carbon steel reduces the linear rate constant during breakaway and Banks and Lorimer[177] have shown that Cr and Si reduce the creep rate of Fe_3O_4, thus reducing the post-breakaway rate of Cr-containing steels. However, Si has been reported as having no effect on the post-breakaway rate on 9Cr steels[157,171]. Small changes to the CO and H_2O content of the environment and the temperature have no significant effect on the breakaway rate constant, but a large reduction in the CO can cause reversion to protective kinetics[180], as does reducing the CO_2 pressure to atmospheric[181]. Increasing the CO switches breakaway on again[180].

L. W. PINDER

REFERENCES

1. Holmes, D. R. and Stringer, J. in *Corrosion of Steels in CO_2*, Ed. Holmes, D. R., Hill, R. B. and Wyatt, L. M., British Nuclear Energy Society, London, 165 (1974)
2. Pilling, N. B. and Bedworth, R. E., *J. Inst. Met.*, **29**, 529 (1923)
3. Tammann, G., *Z. Anorg. Allgem. Chem.*, **111**, 78 (1920)
4. Gulbransen, E. A. and Ruka, R., *Trans. AIME*, **188**, 1500 (1950)
5. Pinder, L. W., C.E.G.B. Unclassified Report, MID/SSD/80/0050/R, August 1980
6. Howe, C. I., McEnany, B. and Scott, V. D., *Corr. Sci.*, **25** No. 3, 195 (1985)
7. Goursat, A. G. and Smeltzer, W. W., *Oxid. Met.*, **6** No. 2, 101 (1973)
8. Kuroda, K., Labun, P. A., Welsh, G. and Mitchell, T. E., *Oxid. Met.*, **19** Nos 3/4, 117 (1983)
9. Koch, F. and Cohen, J. B., *Acta Crystallography*, **B 25**, 275 (1969)
10. Cheetham, A. K., Fender, B. E. G. and Taylor, R. I., *J. Phys. C.*, **V**, 2160 (1971)
11. Catlow, C. R. A., Mackrodt, W. C., Norgett, M. J. and Stoneham, A. M., *Phil Mag A*, **40** No. 2, 161 (1979)
12. Chen, W. K. and Peterson, N. L., *J. Phys. Chem. Solids*, **36**, 1097 (1975)
13. Wagner C., *Atom Movements*, ASM, Cleveland, 153 (1951)
14. Wagner, C., *Z. Phys. Chem.*, **B 621**, 25 (1933)
15. Atkinson, A., *Rev. Mod. Phys.*, **57**, 437 (1985)
16. Atkinson, A. and Taylor, R. I., *J. Phys. Chem. Solids*, **46**, 469 (1985)
17. Atkinson, A. and Taylor, R. I., *High Temperature – High Pressure*, **14**, 571 (1982)
18. Dieckmann, R. and Kohne, M., *Ber. Bunsenges Phys. Chem.*, **87**, 495 (1983)
19. Garnaud, G. and Rapp, R. A., *Oxid. Met.*, **11**, 193 (1977)
20. Channing, D. A. and Graham, M. J., *Corr. Sci.*, **12**, 271 (1972)
21. Channing, D. A. Dickerson, S. M. and Graham, M. J., *Corr. Sci.*, **13**, 933 (1973)
22. Francis, R. and Lees, D. G., *Corr. Sci.*, **16**, 847 (1976)
23. Rahmel, A., *Werkstoffe und Korrosion*, **16** No. 10, 837 (1965)
24. Rahmel, A., *Korrosion*, **18**, 41 (1966)
25. Eubanks, K. G., Moore, D. G. and Pennington, W. A., *J. Electrochem. Soc.*, **109**, 382 (1962)
26. Svedung, I., Hammar, B., and Vannerberg, N. G., *Oxid. Met.*, **6** No. 1, 21 (1973)
27. Caplan, D., *Corr. Sci.*, **6**, 509 (1966)
28. von Fraunhofer, J. A. and Pickup, G. A., *Corr. Sci.*, **10**, 253 (1970)
29. Janssen, S. and Lehtinen, B., *Metallurgie*, **7**, 61 (1967)
30. Caplan, D. and Cohen, M., *Corr. Sci.*, **6**, 321 (1966)
31. Price, W. R., *Corr. Sci.*, **7**, 473 (1967)
32. Pinder, L. W., C.E.G.B. Unclassified Report, MID/SSD/80/0057/R, August 1980
33. Stott, F. H., *Mat. Sci. Tech.*, **5** No. 8, 734 (1989)
34. Atkinson, A., *Mat. Sci. Tech.*, **4** No. 12, 1046 (1988)
35. Dunitz, J. D. and Orgel, L. E., *J. Phys. Chem. Solids*, **3**, 318 (1957)
36. Azaroff, L. V., *J. Appl. Phys.*, **32** Part 9, 1658 (1969)
37. Cox, M. G. C., MacEnany, B. and Scott, V. D., *Phil. Mag.*, **26**, 839 (1972)
38. Hodge, J. D., *J. Electrochem. Soc.*, **125** No 2, 55c (1978)
39. Rahmel, A. Z., *Electrochem.*, **66** No 4, 363 (1962)
40. Moreau, J., *Compte Rendu*, **236**, 85 (1953)
41. Surman, P. L. and Castle, J. E., *Corr. Sci.*, **9**, 771 (1969)
42. Atkinson, A. in *Oxidation of Metals and Associated Mass Transport*, Ed. Dayananda, M. A. *et al.* Warrendale, P., The Metallurgical Society of the AIME, 29 (1987)

43. Robertson, J. and Manning, M. I., *Mat. Sci. Tech.*, **4**, 1064 (1988)
44. Bruckman, A., Emmerich, R. and Mrowec, S., *Oxid. Met.*, **5** No. 2, 137 (1972)
45. Fromhold Jr., A. and Sato, N., *Oxid. Met.*, **16** Nos. 3/4, 203 (1981)
46. Harrison, P. L., *Oxid. Met.*, **22** Nos. 1/2, 35 (1984)
47. Atkinson A. *Phil Mag. B*, **55**, 637 (1987)
48. Whittle, D. P., Evans, D. J., Scully, D. B. and Wood, G. C., *Acta Met.*, **15**, 1421 (1967)
49. Wagner, C., *Corr. Sci.*, **5**, 751 (1965)
50. Runk, R. B. and Kim, H. J., *Oxid. Met.*, **2** No. 3, 285 (1970)
51. Kim, H. J. and Runk, R. B., *Oxid. Met.*, **2** No. 3, 307 (1970)
52. Nosek, E. and Werber, T., *Oxid. Met.*, **25** Nos. 3/4, 121 (1986)
53. Bohnenkamp, V. J. and Engell, H. J., *Arch. Eisenhuettenw.*, **33**, 359 (1962)
54. Caplan, D., Sproule, G. I., Hussey, R. J. and Graham, M. J., *Oxid. Met.*, **13**, 255 (1979)
55. Caplan, D., Sproule, G. I., Hussey, R. J. and Graham, M. J., *Oxid. Met.*, **12**, 67 (1978)
56. Malik, A. U., *Oxid. Met.*, **25** Nos. 5/6, 233 (1985)
57. Tomaszewicz, P. and Wallwork, G. R., *Oxid. Met.*, **19** Nos. 5/6, 165 (1983)
58. Tomaszewicz, P. and Wallwork, G. R., in *High Temperature Corrosion*, Ed. Rapp, R. A., NACE, Houston (1983)
59. Boggs, W. E., *J. Electrochem. Soc.*, **118**, 906 (1971)
60. Ahmed, H. A. and Smeltzer, W. W., *J. Electrochem. Soc.*, **133**, 212 (1986)
61. Pons, M., Caillet, N. and Galerie, A., *Corr. Sci.*, **22**, 239 (1982)
62. Smith, P. J., Beauprie, R. M., Smeltzer, W. W., Stevanovic, D. V. and Thompson, D. A., *Oxid. Met.* **28** Nos. 5/6, 259 (1987)
63. Ahmed, H. A., Underhill, R. P., Smeltzer, W. W., Brett, M. E. and Graham, M. J., *Oxid. Met.*, **28** Nos. 5/6, 347 (1987)
64. Tomaszewicz, P. and Wallwork, G. R., *Oxid. Met.*, **19** Nos. 3/4, 75 (1983)
65. Wagner, J. B., in *Defects and Transport in Oxides*, Ed. Smeltzer, M. S. and Jaffe, R. I., Plenum Press, New York, 283 (1974)
66. Seybolt, A. U., *Trans. AIME*, **242**, 752 (1968)
67. Rahmel, A. and Tobolski, J., *Werkstoffe und Korrosion,* **16** No. 8, 662 (1965)
68. Robertson, J. and Manning, M. I., *Mat. Sci. Tech.*, **5**, 741 (1989)
69. Darken, L. S., *Trans. AIME*, **150**, 157 (1942)
70. Tuck, C. W., *Corr. Sci.*, **5**, 631 (1965)
71. Svedung, I. and Vannenberg, N. G., *Corr. Sci.*, **14**, 391 (1974)
72. Wood, G. C., Richardson, J. A., Hobby, M. G. and Banstead, J., *Corr. Sci.*, **11**, 659 (1971)
73. Rochet, F., Rigo, S., Froment, M., d'Anterroches, C., Maillot, C., Roulet, H. and Dufour, G., *Phil. Mag. B*, **55**, 309 (1987)
74. Adachi, T and Meier, G. H., *Oxid. Met.*, **27** Nos. 5/6, 347 (1987)
75. Atkinson, A., *Corr. Sci.*, **22**, 87 (1982)
76. Logani, R. C. and Smeltzer, W. W., *Oxid. Met.*, **3** No. 3, 279 (1971)
77. Logani, R. C. and Smeltzer, W. W., *Oxid. Met.*, **1** No. 3, 3 (1969)
78. Logani, R. C. and Smeltzer, W. W., *Oxid. Met.*, **3** No. 1, 15 (1971)
79. Jackson, P. R. S. and Wallwork, G. R., *Oxid. Met.*, **20** Nos. 1/2, 1 (1983)
80. Donati, and Garaud, *Corrosion of Steels in CO_2*, Ed. Holmes, D. R., Hill, R. B. and Wyatt, L. M., British Nuclear Energy Society, London, 28 (1974)
81. Dewanckel, B., Leclercq, D. and Dixmier, J., *Ibid*, 42
82. McAdam, G. and Young, D. J., *Oxid. Met.*, **28** Nos. 3/4, 165 (1987)
83. Nishida, K., Narita, T., Tani, T. and Sasaki, G., *Oxid. Met.*, **14**, 65 (1980)
84. Pritchard, A. M., Antill, J. E., Cottell, K. R. J., Peakall, K. A. and Truswell, A. E., in *Corrosion of Steels in CO_2*, Ed. Holmes, D. R., Hill, R. B. and Wyatt, L. M., British Nuclear Energy Society, London, 73 (1974)
85. Menzies, L. A. and Tomlinson, W. J., *JISI.*, **204**, 1239 (1958)
86. Menzies, I. A. and Lubkiewicz, J., *Oxid. Met.*, **3** No. 1, 41 (1971)
87. Dalvi, A. D. and Coates, D. E., *Oxid. Met.*, **5** No. 2, 113, 135 (1972)
88. Yearian, H. J., Randell, E. C. and Longo, T. A., *Corrosion*, **12**, 515 (1956)
89. Douglas, D. L., Gesmundo, F. and de Asmundis, C., *Oxid. Met.*, **25** Nos. 3/4, 235 (1986)
90. Khanna, A. S. and Gnanamoorthy, J. B., *Oxid. Met.*, **23** Nos. 1/2, 17 (1985)
91. Hossain, M. K., *Corr. Sci.*, **19**, 1031 (1979)
92. Caplan, D. and Sproule, G., *Oxid. Met.*, **9**, 459 (1975)
93. Rahmel, A., Jaeger, W. and Becker, K., *Arch. Eissenhuttenw.*, **30**, 351 (1959)
94. Wolfe, I., Grabke, H. J. and Schmidt, P., *Oxid. Met.* **29** Nos. 3/4, 289 (1988)

95. Baxter, D. J. and Natesan, K., *Rev. High Temp. Materials*, **3/4**, 149 (1983)
96. Schutze, M., *Mat. Sci. Tech.*, **4**, 407 (1988)
97. Stringer, J., *Corr. Sci.*, **10**, 513 (1970)
98. Evans, H. E. and Lobb, R. C., *Corr. Sci.*, **24** No. 3, 209 (1984)
99. Kubaschewski, O. and Hopkins, B., *Oxidation of Metals and Alloys*, Butterworth, London (1969)
100. Manning, M. I. and Metcalfe, E., *Proc. Sixth European Congress on Metallic Corrosion*, London, 121 (1977)
101. Barbehon, J., Rahmel, A. and Schutze, M., *Oxid. Met.*, **30** Nos. 1/2, 85 (1988)
102. Christl, W., Rahmel, A. and Schutze, M., *Oxid. Met.*, **31** Nos. 1/2, 1 (1989)
103. Whittle, D. P., *Oxid. Met.*, **4** No. 3, 171 (1972)
104. Deadmore, D. L. and Lowel, C. E., *Oxid. Met.*, **11** No. 2, 91 (1977)
105. Hsueh, C. H. and Evans, A. G., *J. Appl. Phys.*, **54**, 6672 (1983)
106. Douglass, D. L., *Oxidation of Metals and Alloys*, American Society of Metals, Metals Park, Ohio, 137 (1971)
107. Mitchell, T. E., Voss, D. A. and Butler, E. P., *J. Mat. Sci.*, **17**, 1825 (1982)
108. Manning, M. I., *Corr. Sci.*, **21**, 301 (1981)
109. Evans, A. G., Crumley, G. B. and Demaray, R. E., *Oxid. Met.*, **20** Nos. 5/6, 193 (1983)
110. Norin, A., *Oxid. Met.*, **9** No. 3, 259 (1975)
111. Appleby, W. K. and Tylecoate, R. F., *Corr. Sci.*, **10**, 325 (1970)
112. Jha, B. B., Raj, B. and Khanna, A. S., *Oxid. Met.*, **26** Nos. 3/4, 213 (1986)
113. Hancock, P. and Hurst, R. C., in *Corrosion of Steels in CO_2*, Ed. Holmes, D. R., Hill, R. B. and Wyatt, L. M., British Nuclear Energy Society, London, 320 (1974)
114. Ward, G., Hockenhull, B. S. and Hancock, P., *Met. Trans.*, **5**, 1451 (1974)
115. Forrest, J. E. and Bell, P. S., *Corrosion and Mechanical Stress at High Temperatures*, Applied Science Publishers, London, 339 (1981)
116. Rolls, R. and Nematollahi, M., *Oxid. Met.*, **20** Nos. 1/2, 19 (1983)
117. Simms, N. J. and Little, J. A., *Mat. Sci. Tech.*, **4**, 1133 (1988)
118. Pinder, L. W., C.E.G.B. Unclassified Report SSD/MID/R58/77, November 1977.
119. Wiles, C., PowerGen, private communication.
120. Cory, N. J. and Herrington, T. M., *Oxid. Met.*, **28** Nos. 5/6, 237 (1987)
121. Cory, N. J. and Herrington, T. M., *Oxid. Met.*, **29** Nos. 1/2, 135 (1988)
122. Hauffe, K., *Oxid. Met.*, 285 (1965)
123. Effertz, P. H. and Miesel, H., *Machinenshaden*, **55**, 14 (1971)
124. Potter, E. C. and Mann, G. M. W, *Proc. NACE 2nd International Congress on Metallic Corrosion*, New York, 872 and 878 (1963)
125. Hurst, P. and Cowen, H. C., *Proc. Conf. Ferritic Steels for Fast Reactor Steam Generators*, British Nuclear Energy Society, London, (1977)
126. Mayer, P. and Manolescu, A. V., *High Temperature Corrosion*, Ed. Rapp, R. A., NACE, Houston, Texas, 368 (1983)
127. Rahmel, A. and Tobolski, J., *Corr. Sci.*, **5**, 333 (1965)
128. Kofstad, P., *Oxid. Met.*, **24** Nos. 5/6, 265 (1985)
129. Bruckman, A. and Mrowec, S., *Corr. Sci.*, **7**, 173 (1973)
130. Sheasby, J. S., Boggs, W. E. and Turkdogan, E. T., *Met. Sci.*, **18**, 127 (1984)
131. Tuck, C. W., Odgers, M. and Sachs, K., *Corr. Sci.*, **9**, 271 (1969)
132. Griskin, A. M., Perkov, V. G., Sentyurev, V. P. and Yaschenko, Ya Y., *Thermal Engineering*, **16**, 121 (1969)
133. Armitt, J., Holmes, D. R., Manning, M. I. and Meadowcroft, D. B., *The Spalling of Steam Grown Oxides from Superheater and Reheater Tube Steels*, EPRI-FP-686,TPS 76-655 Final Report (February 1978).
134. Lux, J. A., American Power Conference, Chicago, Illinois, 29 April to May 1 (1974)
135. Clarke, F. and Morris, C. W., in *Corrosion Resistant Materials for Coal Combustion Systems*, Ed. Meadowcroft, D. B. and Manning, M. I., Applied Science Publishers, London, 47 (1983)
136. Lees, D. J. and Whitehead, M. E., *Ibid.*, 63
137. Latham, E. P., Meadowcroft, D. B. and Pinder, L. W., CRSC-EPRI Int. Conf. on Chlorine in Coal, Chicago, October, 1989 (proceedings to be published)
138. Gibb, W. H., in *Corrosion Resistant Materials for Coal Combustion Systems*, Ed. Meadowcroft, D. B. and Manning, M. I., Applied Science Publishers, 25 (1983)
139. Brooks, S. and Meadowcroft, D. B., *Ibid.*, 105

140. Laxton, J. W., Meadowcroft, D. B., Clarke, F., Flatley, T., King, C. W. and Morris, C. W., C.E.G.B, private communication.
141. Mayer, P. and Manolescu, A. V., in *Corrosion Resistant Materials for Coal Combustion Systems*, Ed. Meadowcroft, D. B. and Manning, M. I., Applied Science Publishers, London, 87 (1983)
142. Cutler, A. J. B. and Reask, E., *Corr. Sci.*, 21, 789 (1981)
143. Perkins, R. A., in *Corrosion Resistant Materials for Coal Combustion Systems*, Ed. Meadowcroft, D. B. and Manning, M. I., Applied Science Publishers, London, 219 (1983)
144. Minchener, A. J., Lloyd, D. M. and Stringer, J., *Ibid.*, 299
145. Roberts, A. G., Raven, P., Lane, G. and Stringer, J., *Ibid.*, 323
146. Zelanko, P. D. and Simkovich, G., *Oxid. Met.*, 8 No. 5, 343 (1974)
147. Mrowec, S. and Przybylski, K., *Oxid. Met.*, 23 Nos. 3/4, 107 (1985)
148. Mrowec, S., Walec, T. and Weber, T., *Oxid. Met.*, 1 No. 1, 93 (1969)
149. Narita, T. and Nishida, K., *Oxid. Met.*, 6 No. 3, 181 (1973)
150. Sutherland, R. B. and Prescott, G. R., *Corrosion*, 18, 277t (1961)
151. Backensto, E. B., Prior, J. E., Sjooberg, J. W. and Manuel, R. W., *Corrosion*, 18, 253t (1962)
152. Malinowski, E., *Metal*, 94 No. 4, (1962)
153. Burns, F. J., *Corrosion*, 25, 119 (1969)
154. Strafford, K. N. and Manifold, R., *Oxid. Met.*, 1, 229 (1969)
155. Condit, R. H., Hobbins, R. R. and Birchenall, C. E., *Oxid. Met.*, 8 No. 6, 409 (1974)
156. Karlsson, A., Moller, P. J. and Johansen, V., *Corr. Sci.*, 30, 153 (1990)
157. Rowlands, P. C., Garrett, J. C. P., Hicks, F. G., Lister, S. K., Lloyd, B. and Twelves, J. A., in *Corrosion of Steels in CO_2*, Ed. Holmes, D. R., Hill, R. B. and Wyatt, L. M., British Nuclear Energy Society, London, 247 (1974)
158. Holmes, D. R., Mortimer, D. and Newell, J., *Ibid.*, 151
159. Newell, J. E., *Nucl. Energy Int.*, 17, 637 (1972)
160. Taylor, J. W. and Trotsenberg, P. V., in *Corrosion of Steels in CO_2*, Ed. Holmes, D. R., Hill, R. B. and Wyatt, L. M., British Nuclear Energy Society, London, 180 (1974)
161. German, P. A. and Littlejohn, A. C., *Ibid.*, 1
162. Hussey, R. J., Sproule, G. I., Caplan, D. and Graham, M. J., *Oxid. Met.*, 11, 65 (1977)
163. Gibbs, G. B., Pendlebury, R. E. and Wooton, M. R., in *Corrosion of Steels in CO_2*, Ed. Holmes, D. R., Hill, R. B. and Wyatt, L. M., British Nuclear Energy Society, London, 59 (1974)
164. Cox, M. G. C., McEnaney, B. and Scott, V. D., *Ibid.*, 247
165. Gibbs, G. B., *Oxid. Met.*, 7, 173 (1973)
166. Harrison, P. L., Dooley, R. B., Lister, S. K., Meadowcroft, D. B., Nolan, P. J., Pendlebury, R. E., Surman, P. L. and Wooton, M. R., in *Corrosion of Steels in CO_2*, Ed. Holmes, R. D., Hill, R. B. and Wyatt, L. M., British Nuclear Energy Society, London, 220 (1974)
167. Harrison, P. L., Dooley, P. B., Lister, S. K., Meadowcroft, D. B., Nolan, P. J., Pendlebury, R. E., Surman, P. L. and Wooton, M. R., *Ibid.*, 220
168. Ferguson, J. M., Garrett, J. C. P. and Lloyd, B., *Ibid.*, 15
169. Rowlands, P. C., Garrett, J. C. P., Popple, L. A., Whittaker, A. and Hoaksey, A., *Nucl. Energy*, 25 No. 5, 267 (1986)
170. Camona, G. A., Imbergamo, M. and Ronchetti, C., in *Corrosion of Steels in CO_2*, ed. Holmes, D. R., Hill, R. B. and Wyatt, L. M., British Nuclear Energy Society, London, 45 (1974)
171. Grandison, N. O. and Facer, R. I., *Ibid.*, 208
172. Gleave, C., Calvert, J. M., Lees, D. G. and Rowlands, P. C., *Proc. Roy. Soc.* [*A*], 379, 409 (1982)
173. Mrowec, S., *Corr. Sci.*, 7, 563 (1967)
174. Gibbs, G. B. and Hales, R., *Corr. Sci.*, 17, 487 (1977)
175. Evans, A. G., Rajdev, D. and Douglass, D. L., *Oxid. Met.*, 4, 151 (1972)
176. Kofstad, P., *Oxid. Met.*, 24, 265 (1985)
177. Banks, P. and Lorimer, G. W., in *Materials to Supply the Energy Demand*, Ed. Hawbolt, E. B. and Mitchell, A., American Institute of Mechanical Engineers, New York, 231 (1982)
178. Wolfe, I. and Grabke, H. J., *Solid State Comm.*, 54, 5 (1985)
179. Surman, P. L., *Corr Sci.*, 13, 825 (1973)

180. Surman, P. L. and Brown, A. M., in *Corrosion of Steels in CO_2*, Ed. Holmes, D. R., Hill, R. B. and Wyatt, L. M., British Nuclear Energy Society, London, 85 (1974)
181. Goodison, D., Harris, R. J. and Goldenbaum, P., British Joint Corrosion Group Symposium on Metal–Gas Reactions in Atmospheres Containing CO_2, London, March, (1967)
182. White, J. CEGB Private Communication

7.3 High-temperature Corrosion of Cast Iron

Introduction

When cast iron is exposed to high temperatures under oxidising conditions, oxidation of the metal results, with the formation of a surface scale. In addition, the dimensions of the component become distorted. Although such dimensional changes can occur also in inert atmospheres or in vacuum, the evidence available suggests that this 'growth' is frequently associated with oxidation, and accordingly it is appropriate to consider it as an aspect of the corrosion of the iron.

The composition of the atmosphere to which components at high temperature may be exposed varies very widely, and most work on these aspects has accordingly been carried out in clean air. The aggressiveness of air is considerably enhanced by the presence of trace amounts of other reactive gases such as steam, carbon dioxide and sulphur dioxide; thus the figures subsequently quoted may in fact be appreciably lower than those encountered in specific atmospheres. The data presented should, however, prove an adequate guide to the order of the effect to be expected.

Growth

Components designed for high-temperature duty may either remain at a steady high temperature for their entire life, or, as more commonly happens, may undergo cyclic variation between a minimum temperature, often room temperature, and a maximum temperature. The maximum temperature involved may be either above or below the critical temperature range of the iron. This is the range within which the transformation between ferrite or pearlite and austenite occurs and for the majority of unalloyed irons it may be regarded as being 700–850°C. (See Section 9.2.)

Conditions of cyclic reheating are more severe than conditions of steady high temperature, and cyclic reheating through the critical range is particularly liable to cause excessive growth of the iron.

Generally, the studies which have been carried out suggest that growth of up to 40% by volume can occur within the first 40 h of cyclic reheating to 900°C with a frequency of 1–4 h/cycle, while subsequent cycling produces growth at a rate rarely exceeding 20% increase in volume in 100 h. The rate of growth which develops increases with increase in temperature and possibly also with increase in frequency of cycling. Although the fact that a 60% increase in volume may occur after only 140 h of cyclic reheating suggests that unalloyed iron is totally unsuited for such applications, iron is in fact extensively used under such conditions, e.g. as furnace doors and fire bars. This may be partly because these applications involve lower cycling frequencies than those which cause the very high rates of growth mentioned, but undoubtedly a major factor determining the use of unalloyed cast iron for such duties is its cheapness, which outweighs the superior growth resistance of more highly alloyed and more expensive irons.

At temperatures below the critical range, much less growth occurs, rarely exceeding 3% for 100 h of cyclic reheating. Here too the rate of growth depends on the temperature and the frequency of cycling. At temperatures below 400°C growth becomes negligible for most irons while below 350°C it is negligible for all irons. This threshold is probably related to the marked decrease in strength which occurs when irons are heated above 400°C, which results in the component being more easily distorted by the development of internal volume changes. Clearly, unalloyed irons have a very considerable usefulness up to about 700°C, and even in steam plant, where dimensional stability is important, there is a case for the use of unalloyed iron at temperatures up to 400°C.

At temperatures below the critical range, an important cause of growth is graphitisation, i.e. the decomposition of the carbide constituent of pearlite to give ferrite and graphite. Unalloyed irons usually contain up to 0·8% combined carbon and complete graphitisation of this can theoretically result in a volume increase of 1·6%[2]. This value has been confirmed by Gilbert and White[1] who have shown that ferritising a fully pearlitic iron gives a linear growth of up to 0·7% (i.e. about 2·1% volume increase). Clearly, the rate of growth due to this mechanism will be controlled by the stability of the carbide in the pearlite and this will vary with the composition of the iron. The presence of certain elements, notably silicon, decreases the stability of the carbide, while it is stabilised by the presence of other elements, notably chromium. An iron with a low silicon content and containing some chromium may thus be expected to have good growth resistance, but since excessive carbide stability can lead to a hard, brittle alloy, there is a limit to the benefit which can be derived from such stability. It should be emphasised that unless large amounts of carbide-stabilising elements are present in the iron, all that will be achieved is a slower *rate* of growth; there will not be a decrease in the total growth possible.

Phosphorus appears to have a beneficial effect on the growth rate. At sub-critical temperatures it helps to stabilise the carbide, while at temperatures up to about 900°C the presence of the hard phosphide eutectic network restricts the deformation to which the much more ductile matrix would otherwise be subject. Since the phosphide eutectic melts at about 950°C, irons containing appreciable amounts of this constituent should clearly not be exposed to this temperature.

Another cause of growth which is of equal importance with graphitisation is the penetration of oxides into the metal along the graphite flakes. This presumably takes place because oxidising gases can be adsorbed on to the graphite and so allowed access to the metal/graphite interface. Since the oxides are more bulky than the metal from which they are derived, internal stresses are set up and growth results. As might be expected, the amount of growth due to internal oxidation increases as the graphite content increases (Fig. 7.15) and also as the section size increases, since this leads to a coarsening of the graphite. On the other hand, a white iron which contains no graphite is very growth-resistant since it does not readily graphitise, nor is it easily penetrated by oxidation. For similar reasons, nodular graphite irons are resistant to growth.

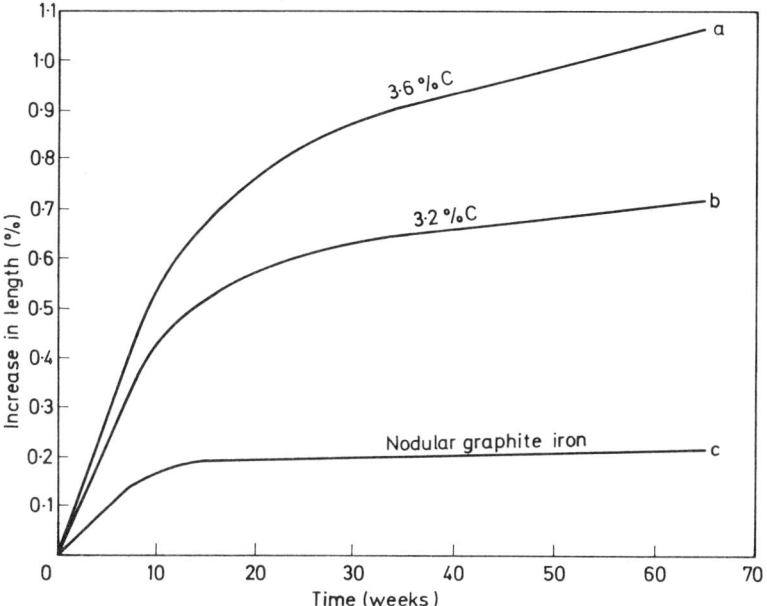

Fig. 7.15 Effect of amount and distribution of graphite on growth in air at 500°C. Curve *a*: Iron 6 (Fe-3·61C_{total}-1·63Si-0·76Mn-0·094S-0·28P). Curve *b*: Iron 1 (Fe-3·25C_{total}-1·58Si-0·65Mn-0·107S-0·25P). Curve *c*: Iron 21 (Fe-3·39C_{total}-1·73Si-0·41Mn-0·013S-0·05P-0·67Ni-0·075Mg-0·004Ce) (after Gilbert[4])

At temperatures above the critical range, the maximum amount of growth due to graphitisation may account for less than 10% of the total growth observed. Undoubtedly a large contribution to the total growth is made by the oxidation of the iron, since the stresses set up in the oxide layers by the differences between the expansion of the oxides and the iron during the alternate heating and cooling cycles generate cracks in the scale, which prevent the reaction from ever becoming self-stifling. The increase in oxidation rate due to the temperature rise does not, however, satisfactorily account for the marked increase in growth rates when the critical range is exceeded. Benedicks and Löfquist[2] have given an interpretation of some dilatometer curves produced by Kikuta[3] which explains the sudden increase in growth at these higher temperatures in terms of the ferrite-austenite transformation.

This explanation implies that at each complete cycle through the critical range there is a net expansion which is due to the fact that the expansion involved in the austenite to ferrite change does not balance the contraction involved in the ferrite-austenite change. For an iron containing 0·7% combined carbon initially, the net growth per cycle involved may be up to 0·5% by length (1·5% by volume). The net growth per cycle decreases with the number of cycles but the possibility that each cycle, at least initially, can contribute this amount of growth suggests that the mechanism can give rise to very high growth rates. As with sub-critical growth, the duration of each stage of the cycle and the rate of heating and cooling will largely determine the rate of growth achieved, very slow and very rapid cycling being probably the least dangerous in this context.

All the remarks so far made have been concerned with conditions of cyclic reheating. When an alloy is held at a steady temperature above the critical range, some growth will arise from graphitisation, partly offset by the contraction involved in the ferrite-austenite transformation, but most of the growth will be due to oxide penetration.

Work carried out by Gilbert[4] on irons maintained at 500°C for 64 weeks (Fig. 7.15) has shown that in ordinary unalloyed flake irons graphitisation and oxidation cause roughly equal amounts of growth, and that as the carbon content increases the effect of oxidation becomes more important and the overall rate of growth increases. Nodular graphite irons grow very slowly under these conditions.

Irons designed specifically for good oxidation- and growth-resistance have highly oxidation-resistant matrices, containing either no carbides at all or very stable carbides, and have critical temperatures either below room temperature or above the maximum temperature anticipated. The alloys most commonly used are Silal, Niresist, Nicrosilal and Fe–30Cr. Details of these irons and their properties are given in Table 7.3. The extremely fine graphite structure present in Silal probably makes a major contribution to its good heat resistance. However, when Silal is produced with nodular graphite, its heat-resistance is further enhanced.

Two other alloys which have been used for their good oxidation- and growth-resistance are Cralfer (Fe–7Al–0·75Cr) and Fe–14·5Si. The production of the former, however, entails considerable difficulties while the latter has poor mechanical properties and poor resistance to thermal shock, with the result that neither is extensively used for this purpose today.

Scaling

When an iron is exposed to an oxidising atmosphere, it develops a scale which consists of a series of layers of oxides of varying composition. The thickness of the scale naturally depends on the temperature and the duration of oxidation (t). The scale does not, however, thicken at a uniform rate with time since its very presence reduces the accessibility of the metal surface to the oxidising gases. Ideally, the thickness of the scale should increase as $t^{\frac{1}{2}}$, but in practice cracks develop in the scale, and these allow the gases to reach the metal surface somewhat more readily than is postulated by this relationship. Cracking will always tend to occur as the film

Table 7.3 Heat-resisting irons

Name	Composition					Mechanical properties				Critical temperature (°C)	Growth characteristics
	Total carbon	Si	Ni	Cr	Cu	Structure	Ultimate tensile strength (MN/m²)	Elongation	Hardness (H_B)		
Silal	2.5	6.0	—	—	—	Fine graphite in silico-ferrite matrix	154	Nil	280	>920	Nil after 80 × 1½ h at 870°C (White and Elsea[5])
*Nicrosilal	2.0	5.0	20.0	2.0	—	Fine graphite in austenitic matrix with some complex carbides	216	2%	140	<20	<1% by length after 240 h at 950°C (Hallett[6])
*Niresist	3.0 3.0	2.0 2.0	20.0 14.0	2.0 2.0	— 7.0	Fine graphite in austenitic matrix with some complex carbides	232	2%	150	<20	<1% by length after 240 h at 950°C (Hallett[6])
High-chromium iron	1.5	1.0	—	30	—	Complex carbides in chromium-ferrite matrix	463	Nil	320	>1100	<1% by length after 240 h at 1050°C (Hallett[6])

*These irons may sometimes contain up to 5% Cr which improves the heat resistance slightly and makes the alloy stronger and harder.

thickness increases, but if the iron is subjected to cyclic variations in temperature, the variations in thermal expansion and contraction of the individual constituents of the scale and the metal may be expected to increase the extent to which cracking occurs. Under such conditions the scale may become completely detached from the metal, thus allowing the high initial rates of oxidation to be reproduced. With flake graphite cast iron, however, since there is a tendency for oxide penetration to occur along the graphite flakes, the scale is keyed to the metal surface rather more securely than is the case with steels and therefore becomes detached rather less readily.

The atmosphere most commonly encountered by irons at elevated temperatures is probably air contaminated with traces of carbon dioxide, sulphur dioxide and steam. Because of the difficulty of defining such an atmosphere, most of the work on scaling has been carried out with relatively

Table 7.4 Scaling of unalloyed cast iron

Reference	Composition				Details of test		Results reported	Scaling rate $(\mathrm{mg\,dm^{-2}d^{-1}})$
	Total carbon	Si	Mn	P	Duration	Temperature (°C)		
Hallett[6]	2·98	1·14	1·07	0·18	240 h	750	$9 \times 10^3 \mathrm{mg/dm^2}$	900
					240 h	850	$21·6 \times 10^3 \mathrm{mg/dm^2}$	2 160
					240 h	950	$45·6 \times 10^3 \mathrm{mg/dm^2}$	4 560
					60 h	1 050	$47 \times 10^3 \mathrm{mg/dm^2}$	18 800
Cameron[8]	3·01	2·46	0·81	—	2 000 h	705	0·188 in/y	449*
	3·43†	2·47	0·45	—	2 000 h	705	0·053 in/y	127*
White, Rice and Elsea[7]	3·42	2·61	0·73	0·18	240 h	870	$0·32 \mathrm{g/cm^2}$	3 200
Timmerbeil[9]	3·15	1·70	0·80	—	96 h	700	$70 \mathrm{g/m^2 d^{-1}}$	700
					96 h	800	$230 \mathrm{g/m^2 d^{-1}}$	2 300
Timmerbeil[10]	3·12	1·99	0·42	0·27	96 h	600	$10 \mathrm{g/m^2 d^{-1}}$	100
					96 h	700	$40 \mathrm{g/m^2 d^{-1}}$	400
					96 h	800	$230 \mathrm{g/m^2 d^{-1}}$	2 300
					96 h	900	$460 \mathrm{g/m^2 d^{-1}}$	4 600
Gilbert[4.17]	3·04	1·23	0·73	0·29	6 y	400	$3·6 \mathrm{mg/cm^2}$	0·17
					64 wk	500	$8·4 \mathrm{mg/cm^2}$	1·9
Gilbert and White[1]	3·29	1·43	0·55	0·24	32 wk	600	$4·9 \mathrm{mg/cm^2}$	22
					32 wk	650	$128·7 \mathrm{mg/cm^2}$	57·5
Gilbert[4.17]	3·39†	1·73	0·41	0·05	6 y	400	$2·0 \mathrm{mg/cm^2}$	0·09
					64 wk	500	$5·27 \mathrm{mg/cm^2}$	1·18
Gilbert and White[1]	3·57†	1·98	0·35	0·03	32 wk	600	$43·3 \mathrm{mg/cm^2}$	19·4
					32 wk	650	$138·8 \mathrm{mg/cm^2}$	61·9
Maitland[13]	3·35†	1·2	0·33	0·06	456 h	650	$2·53 \times 10^3 \mathrm{g/dm^2}$	133
					504 h	700	$5·14 \times 10^3 \mathrm{g/dm^2}$	245
					456 h	800	$20·6 \times 10^3 \mathrm{g/dm^2}$	1 095
					456 h	900	$42·1 \times 10^3 \mathrm{g/dm^2}$	2 216
	3·06†	2·0	0·33	0·06	456 h	650	$2·17 \times 10^3 \mathrm{g/dm^2}$	114
					504 h	700	$4·74 \times 10^3 \mathrm{g/dm^2}$	226
					456 h	800	$23·3 \times 10^3 \mathrm{g/dm^2}$	1 226
					456 h	900	$48·7 \times 10^3 \mathrm{g/dm^2}$	2 563

* Weight increase due to Fe_2O_3.
† Nodular graphite irons containing about 1% Ni.

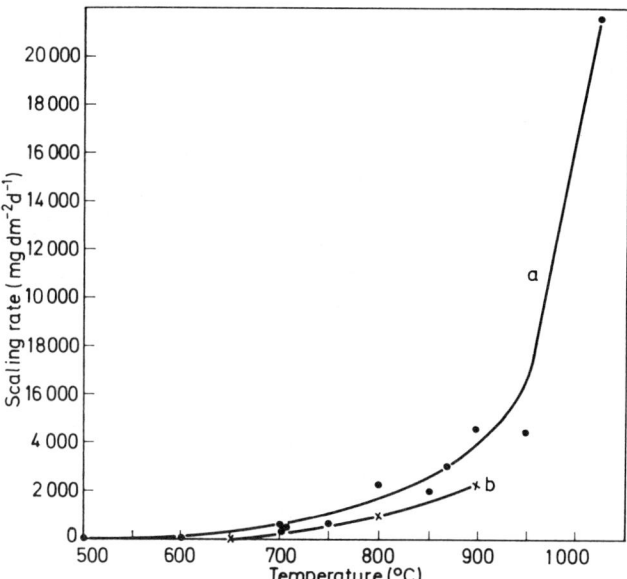

Fig. 7.16 Dependence of scaling rate on temperature (see Table 7.4). Curve *a*: flake graphite irons. Curve *b*: nodular graphite irons

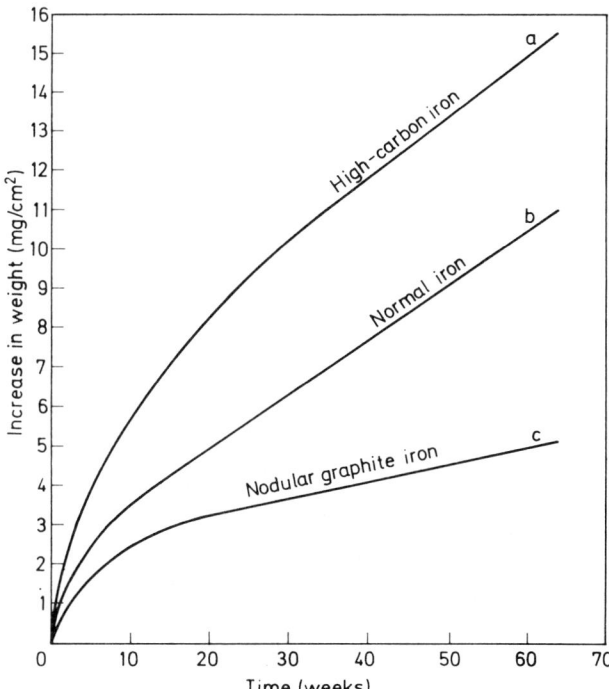

Fig. 7.17 Effect of amount and distribution of graphite on scaling in air at 500°C. Composition as given in Fig. 7.15 (after Gilbert[4])

pure air. Some figures quoted by various authorities for the scaling rates of unalloyed irons in clean air at various temperatures are shown in Fig. 7.16 which has been compiled from data given in Table 7.4.

The most important factor influencing the rate of scaling in a low-alloy iron is the amount and distribution of the graphite. Gilbert[4] has shown that at 500°C an iron with 3·6% total carbon scales and grows very much more than an iron with 3·2% total carbon while nodular graphite irons are even more resistant to scaling (Figs. 7.15 and 7.17). Gilbert and Barnes[14] showed that in a flake graphite iron, for constant carbon content, scaling decreases as the section size decreases, i.e. as the graphite flakes become finer.

Some data given by Hatfield[11] for the oxidation of cast iron in pure dry CO_2, O_2, SO_2 and water vapour at various temperatures are shown in Fig. 7.18. It will be noted that although these atmospheres are more oxidising than air the scaling rates are only two or three times those produced in clean air. Work carried out by Nicholson and Kwasney[12] on the other hand, which was generally similar to that of Hatfield but used relatively impure moist gases, gave scaling rates in oxygen roughly 200 times those observed by Hatfield. This is in line with the general observation that the

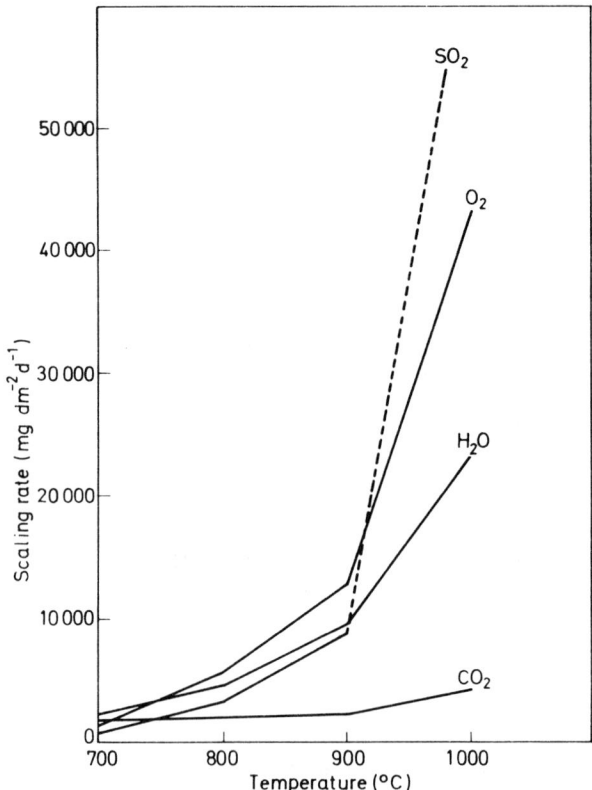

Fig. 7.18 Scaling rates for cast iron in pure gases (Fe-3·20C_{total}-1·13Si-0·72Mn-0·125S-0·58P) (after Hatfield[11])

presence of traces of gases such as steam, carbon dioxide and sulphur dioxide in air can appreciably increase the scaling rate. No final explanation of this effect has yet been produced, but it is possible that it is connected with the development of a more porous scale. Such a scale might be expected to result if the first product of the reaction of the atmosphere with the metal were a compound, e.g. a hydroxide or carbonate, which subsequently decomposed to give the oxide.

The more highly alloyed materials, already mentioned as having useful resistances to growth, have also good resistances to scaling. Figure 7.19 shows the relative resistance of some of these alloys to oxidation. It will be noted that the increase in the chromium content of the austenitic irons from 2 to 5% has had a beneficial effect upon the scaling resistance of these irons. There is some evidence to suggest that the scaling resistance of these irons is further improved if they contain nodular instead of flake graphite[15].

Cox and Gilbert[16] have given a detailed account of the behaviour of the austenitic irons when exposed for 32 weeks to temperatures ranging from 320°C to 800°C, with particular reference to the effect of this exposure on

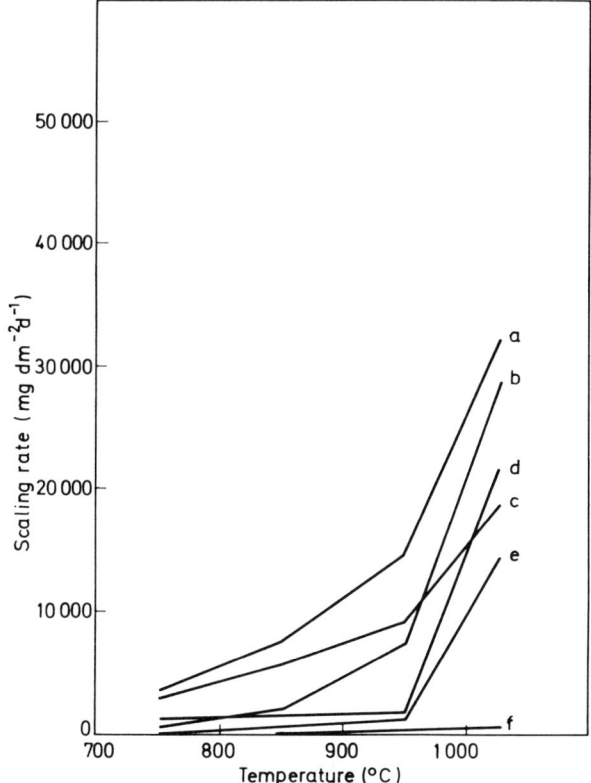

Fig. 7.19 Scaling resistance of high-alloy irons (after Hallett[6]). Curve a: Niresist (Fe-14·6Ni-2·1Cr-7·2Cu). Curve b: Silal (Fe-5·7Si). Curve c: Niresist (Fe-14·4Ni-4·8Cr-7·4Cu). Curve d: Nicrosilal (Fe-4·6Si-22·5Ni-2·5Cr). Curve e: Nicrosilal (Fe-4·9Si-22·8Ni-4·6Cr). Curve f: (Fe-1·3C-33Cr)

Table 7.5 Recommendations for the use of austenitic cast irons at 650–800°C (after Cox and Gilbert[16]). Suitabilities indicated by graded code: 1 = very poor, 5 = excellent

Reference No. (BS 3468:1974)	Typical composition (%)					Heat resistance		Effect on mechanical properties				Overall suitability
	C_{total}	Si	Ni	Cr	Cu	Growth	Scaling	Strength	Elongation	Impact	Hardness	
Flake graphite												
L–Ni Cu Cr 15 6 2	2.5	2.0	15	2.0	7.5	1	2	None	None	Increased	None	1
L–Ni Cr 20 2	2.5	2.0	20	2.0	–	1	2	None	None	Increased	None	1
L–Ni Si Cr 20 5 3	2.0	5.5	20	3.0	–	4	4	None	None	Slightly increased	Increased	4
L–Ni Cr 30 3	2.0	2.0	30	2.5	–	3	3	None	None	Slightly increased	Slightly increased	3
Nodular graphite												
S–Ni Cr 20 2	2.5	2.0	20	2.0	–	2	1	None	None	Increased	None	1
S–Ni 22	2.5	2.0	22	–	–	2	1	None	None	Increased	None	1
S–Ni Si Cr 20 5 2	2.5	5.5	20	2.0	–	5	4	None	None	Slightly increased	Increased	5
S–Ni Cr 30 3	2.0	2.0	30	2.5	–	4	3	None	None	Slightly increased	None	4

their mechanical properties. From this work they have formulated a series of recommendations that are dependent upon the service temperature range. The recommendations for the range 650–800°C are given in Table 7.5.

Long-term Growth and Scaling at 350°C and 400°C

Growth and scaling tests of 21 years duration for a wide range of engineering cast irons have shown[18] that no significant growth will occur in flake graphite cast irons to BS 1452:1977 when they are exposed in air at 350°C, and in nodular cast irons to BS 2789:1973 and malleable cast irons to BS 310:1972 when exposed in air at 400°C. The scaling properties of nodular cast irons are also superior to those of flake graphite cast irons. In applications where it is necessary to maintain very close dimensional stability throughout the service life of a casting, a safe working temperature in air for engineering flake graphite cast irons is generally 350°C and for nodular and blackheart malleable irons, 400°C.

Growth measured as percentage increase in length becomes significant in engineering flake graphite cast irons after about 4 years exposure at 400°C. Once growth commences it continues at an approximately constant rate and this rate depends on the chemical composition and graphite structure of the iron. In flake-graphite irons growth at 400°C, as at higher temperatures, increases as the quantity and coarseness of the graphite increases with increase in carbon content. The effect of carbon content is shown by the results for irons *a* and *b* in Fig. 7.20. Results for the same irons exposed at 500°C are given in Fig. 7.15. These effects are consistent with growth resulting from structural breakdown of the matrix due to graphitisation and growth due to surface oxidation of the graphite both increasing as the carbon content increases. When flake graphite appears in a relatively coarse form but widely dispersed in the matrix structure, as may occur when a medium-carbon iron is cast into a relatively large section size, the growth resistance of the iron may be increased as shown for iron *d*. This iron has a similar chemical composition to that of iron *b* cast into a smaller section size. This behaviour is consistent with graphitisation being reduced when the graphite is more widely dispersed to increase the length of the diffusion paths to the graphite. The alloying elements chromium and molybdenum in amounts up to 0·5%, either singly or in combination, have a marked effect in improving the growth resistance of a flake graphite cast iron as shown for iron *e*. At 400°C, growth resistance for such an iron compares favourably with that of a nodular cast iron.

Growth in nodular cast irons at 400°C in times up to 21 years is generally insignificant and the use of small alloy additions is then unnecessary. For best growth resistance, a nodular iron to BS 2789 should be produced from recarburised steel charges or charges based on refined pig irons since these contain trace amounts of the normal pearlite-stabilising elements. Irons produced from relatively pure pig-iron charges should be avoided because any pearlite present in the matrix will break down more readily and result in slightly greater growth. In nodular cast irons, growth is determined by the stability of the matrix structure. There is no significant growth at sub-critical temperatures due to surface oxidation because this cannot penetrate below

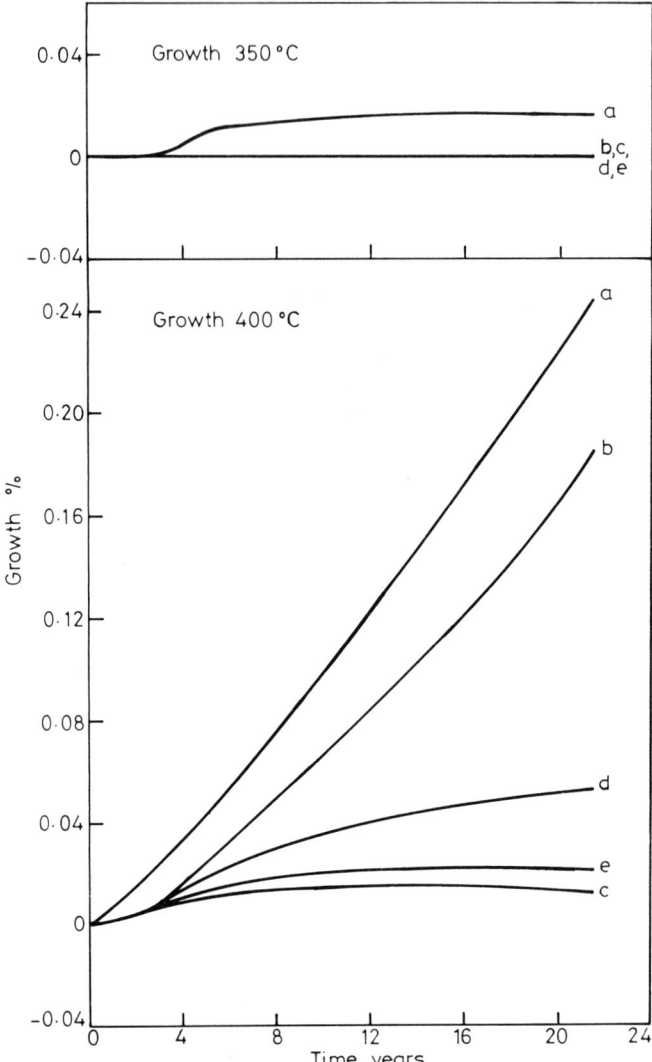

Fig. 7.20 Long-term growth (increase in length %) at temperatures of 350°C and 400°C in flake graphite and nodular cast irons. *a*, high carbon (3·7%) grey iron. *b*: engineering grey iron, 3·2% carbon. *c*: nodular graphite iron. *d*: engineering grey iron, 3·15% carbon; large section casting with coarse but widely dispersed flake graphite. *e*: engineering grey iron alloyed with up to 0·5% molybdenum and/or up to 0·5% chromium

the immediate surface when graphite is present in the nodular form. Fully annealed ferritic nodular irons and ferritic blackheart malleable irons have very good growth resistance because graphitisation is complete at the start and no further structural breakdown can occur during high-temperature service.

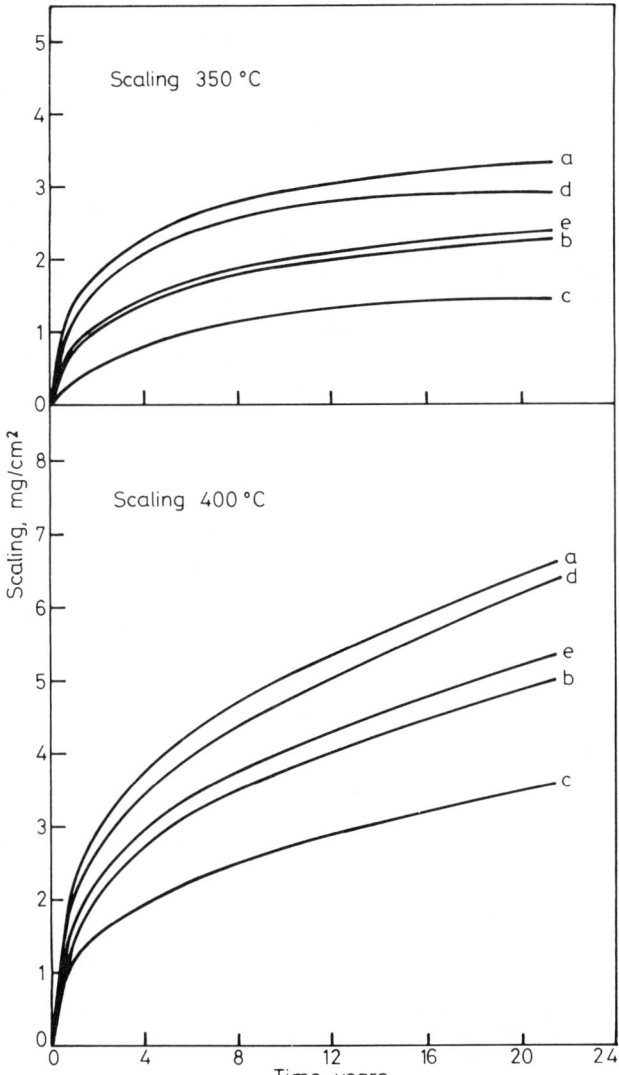

Fig. 7.21 Long-term scaling (increase in weight related to surface area) at temperatures of 350°C and 400°C in flake graphite and nodular cast irons. *a*: high carbon (3·7%) grey iron. *b*: engineering grey iron, 3·2% carbon. *c*: nodular graphite iron. *d*: engineering grey iron, 3·15% carbon; large section casting with coarse but widely dispersed flake graphite. *e*: engineering grey iron alloyed with up to 0·5% molybdenum and/or up to 0·5% chromium

Scaling results for the flake graphite cast irons *a*, *b*, *d* and *e* in Fig. 7.21 show that the weight increase related to surface area due to scaling rises as the quantity of graphite increases with increase in carbon content (iron *a*), and as the graphite becomes more widely dispersed in the matrix (iron *d*). Apart from carbon content, changes in the base metal composition and the use of small alloy additions of nickel, molybdenum and chromium have little, if any, significant effect on the scaling properties of flake graphite irons. Small additions of molybdenum and/or chromium are shown in Fig. 7.21 (iron *e*) to result in no increase in scaling resistance although the same alloy additions have a marked effect in improving growth properties (Fig. 7.20).

When graphite is present in the nodular form, sub-surface oxidation, similar to that which occurs along graphite flakes in a flake graphite cast iron, cannot occur and, consequently, the scaling resistance of a nodular iron is always superior to that of a flake graphite iron. Scaling resistance in nodular irons is not significantly affected by changes in metal composition and small alloy additions and is similar in irons with pearlitic or ferritic matrix structures. The scaling curves shown in Fig. 7.21 for a nodular iron are typical of irons to BS 2789.

When the long-term tests at 400°C and 350°C are considered in relation to shorter-term test data obtained in the temperature range 450–650°C, it is clear that growth and scaling in cast irons are temperature- and time-dependent processes as in other metals. Short-time test results at relatively high temperatures can be used to assess relative growth and scaling resistance of the same irons exposed for longer times at lower temperatures.

H. H. COLLINS
G. N. J. GILBERT

REFERENCES

1. Gilbert, G. N. J. and White, D. G., *J. Res. Brit. Iron Res. Assn.*, **11**, 295 (1963)
2. Benedicks, C. and Löfquist, H., *J. Iron St. Inst.*, **115**, 603 (1927)
3. Kikuta, T., *Sci. Rep. Tohoku Univ.*, **11**, 1, April (1922)
4. Gilbert, G. N. J., *J. Res. Brit. Cast Iron Res. Assn.*, **7**, 478 (1959)
5. White, W. H. and Elsea, A. R., *Trans. Amer. Foundrym. Soc.*, **57**, 459 (1949)
6. Hallett, M. M., *J. Iron St. Inst.*, **170**, 321 (1952)
7. White, W. H., Rice, L. P. and Elsea, A. R., *Trans. Amer. Foundrym. Soc.*, **59**, 337 (1951)
8. Cameron, J. A., *Corrosion*, **10**, 295 (1954)
9. Timmerbeil. H., *Giesserei*, **38**, 25 (1951)
10. Timmerbeil. H., Clas, G. and Mattern, O., *Giesserei*, **38**, 476 (1951)
11. Hatfield, W. H., *J. Iron St. Inst.*, **115**, 483 (1927)
12. Nicholson, J. H. and Kwasney, E. J., *Trans. Electrochem. Soc.*, **91**, 681 (1947)
13. Maitland, R. J. and Hughes, I. C. H., *J. Res. Brit. Cast Iron Res. Assn.*, **7**, 203 (1958)
14. Gilbert, G. N. J. and Barnes, G. M., *J. Res. Brit. Cast Iron Res. Assn.*, **14**, 9 (1966)
15. Sefing, F. J., *Trans. Amer. Foundrym, Soc.*, **63**, 638 (1955)
16. Cox. G. J. and Gilbert, G. N. J., *J. Res. Brit. Cast Iron Res. Assn.*, **14**, 423 (1966)
17. Gilbert, G. N. J., *J. Res. Brit. Cast Iron Res. Assn.*, **14**, 555 (1966)
18. Gilbert, G. N. J. and Pidgeon, C. L., unpublished work at BCIRA

7.4 High-Alloy Steels

Introduction

The basic importance of chromium in relation to passivity in the resistance of steels to aqueous corrosion is parallelled by its rôle as the most important alloying element in practice in securing the high-temperature oxidation resistance of heat-resisting steels. It is not surprising, therefore, that these two aspects of corrosion resistance developed together, and that early investigators of the chromium-iron alloys showed interest in the potential usefulness of these materials for high-temperature oxidation resistance. Thus Chevenard[1], discussing early studies in France, has indicated that the scaling resistance of Fe–Ni–Cr alloys was appreciated at least as early as 1917.

In the UK the rapidly growing interest in high-alloy steels after World War I coincided with expanding needs of industry for materials of improved strength and oxidation resistance at elevated temperatures. Aero-engine developments, rising steam temperatures and pressures in power production, and the increasing importance of high-temperature, high-pressure processes in oil cracking and synthetic chemistry, were all contributory factors in the development of heat-resisting steels in the decade from 1920.

Aitchison[2], writing in 1919 on valve steels for internal combustion engines, gave scaling curves indicating the relative resistance of Fe–12Cr steel, rising to a factor of 10 over ordinary steel at 950°C. In 1922 Dickenson[3] included a section on scaling resistance in a paper dealing with the behaviour of steels at low red heat. He tested various steels in gas and electric muffles at temperatures up to 1 100°C, and the curves produced by him showed considerable improvement over ordinary steel for Fe–15Cr and especially Ni–23Fe–12Cr.

In America, the 1924 A.S.T.M. Symposium on Corrosion and Heat-resisting Alloys contained a particularly interesting paper by Fahrenwald[4]. Tests using gas-fired muffles with excess air over the temperature range 820–1 260°C were described; the results showed clearly the advantages of chromium and chromium–nickel high-alloy steels, including Fe–13Cr, Fe–18Cr, Fe–25Cr and Fe–36Ni–17Cr. The particular value of Fe–25Cr for sulphur roasting plants was pointed out, with a warning against the use of nickel steels for this purpose and an indication that under oxidising conditions attack by alkali salts from the fluxing of the protective chromium oxide scale on the high-chromium steels might be expected.

This symposium also introduced a paper by Johnson and Christiansen[5] on steels for internal-combustion-engine valves, which discussed the merits of Fe-14Cr, Fe-9Cr-3·5Si, Fe-19Ni-14Cr, Ni-27Fe-18Cr and Fe-25Ni-18Cr-3Si.

A third paper in the 1924 symposium, by Vanick[6], dealt with deterioration resulting from attack by hot reducing ammonia gases, with particular reference to the rôle of hydrogen in producing intergranular fissuring.

The first systematic study of the effects of individual and combined combustion gases on pure metals and a wide range of alloy steels was carried out by Hatfield[7] and published in 1927. The gases included oxygen, steam, carbon dioxide, sulphur dioxide and a complex gas which simulated the mixture obtained from combustion of ordinary fuels. A selection from the data is given in Table 7.6 to illustrate the main points. They show the major importance of chromium, with added effect from silicon. It was noted that the presence of the nickel in Fe-18Cr-8Ni did not lead to breakdown in sulphur dioxide in spite of the very severe attack by this gas on pure nickel. The various gases behaved very similarly with regard to extent of attack on Fe-19Cr and Fe-18Cr-8Ni.

The general pattern of alloy composition in relation to scaling resistance was well established by the end of the 1920s, and later development has been largely on the lines of modification to provide the best combinations of strength, ease of fabrication and oxidation resistance to meet the expanding requirements of industry and power generation. In particular, the increasing importance of high strength and low deformation at high temperatures has resulted in the development of a large number of complex alloys combining high creep strength with adequate oxidation resistance.

Typical Heat-resisting Steels

In that they show a superior resistance to oxidation at elevated temperatures over low-alloy types, all the stainless steels listed in BS970, Part 1:1983 (Wrought Steels for Mechanical and Allied Engineering Purposes. Ferritic, Martensitic and Austenitic Stainless and Heat Resisting Steels) and BS1449:1983 (Steel Plate, Sheet and Strip, Part 2: Stainless and Heat Resisting Steels) may be termed heat-resisting steels. Many of them are indeed used for this purpose, for example those listed in Table 7.7, only two of which were developed with oxidation resistance primarily in mind. Two grades in this table have now been withdrawn from British Standards, and the identities given relate to the earlier BS970:1970 and BS1449:1967. They are included because of interesting properties, and equivalent grades are available to other national specifications.

The metallurgy of the three groups (martensitic, ferritic and austenitic) has been discussed in Section 3.3 and will not therefore be dealt with any further here. Only one of the martensitic group is listed since the resistance to oxidation of the rest included in the standards is essentially similar. While having useful properties, applications of this group are limited because of fabrication difficulties, notably in welding, for which reason the lower carbon grades are preferred. While the 11-13% Cr imparts an attractive oxidation resistance, the engineering value of any material is also deter-

Table 7.6 Effect of various gases on the oxidation of metals and alloy steels* (gains in $gm^{-2}d^{-1}$)

Material	O_2			H_2O			CO_2			SO_2			Complex gas†
	700°C	800°C	900°C	700°C	800°C	900°C	700°C	800°C	900°C	700°C	800°C	900°C	900°C
Iron	510	600	1240	620	850	580‡	590	720	1130	350	790	7400	1140
Nickel	10	10	30	3	4	10	4	9	40	920	3660§	840‖	50
Chromium	5	10	20	1	4	10	3	3	10	2	4	30	20
Fe-37Ni	40	100	140	30	130	360	30	100	250	200	1410	1900	300
Fe-8Cr-3Si	3	20	20	3	5	9	2	2	4	5	10	50	10
Fe-13Cr	2	20	30	5	6	150	8	9	160	4	10	100	150
Fe-19Cr	6	7	20	2	6	20	2	8	40	5	5	8	10
Fe-18Cr-8Ni	10	10	30	2	5	6	4	10	30	10	20	20	30

*Data from Hatfield[7].
†Composition 73% N_2, 12% CO_2, 10% H_2O, 5% O_2 and 0.05% SO_2.
‡Scale disintegrating.
§Some scale lost
‖Scale firmly adherent.

Table 7.7 Specifications of some steels used for high-temperature applications (BS970:1983)

	C	Si	Mn	Cr	Ni	Mo	Ti	Nb	Type
410S21	0.09–0.15	1.0 max	1.0 max	11.5–13.5	1.00 max				Martensitic
430S17	0.08 max	1.0 max	1.0 max	16.0–18.0	0.5 max				Ferritic
442S29[a]	0.10 max	0.80 max	1.0 max	18.0–22.0					Ferritic
304S15	0.06 max	1.0 max	2.0 max	17.5–19.0	8.0–11.0				Austenitic
321S31	0.08 max	1.0 max	2.0 max	17.0–19.0	9.0–12.0		$5 \times$ C–0.80		Austenitic
347S31	0.08 max	1.0 max	2.0 max	17.0–19.0	9.0–12.0			$10 \times$ C–1.0 max	Austenitic
316S31	0.07 max	1.0 max	2.0 max	16.5–18.5	10.5–13.5	2.00–2.50			Austenitic
309S24[a]	0.15 max	0.20–1.00	0.5–2.0	22.0–25.0	13.0–16.0				Austenitic
310S31	0.15 max	1.5 max	2.0 max	24.0–26.0	19.0–22.0				Austenitic

[a] BS970:1970

mined by its strength and the simple 410S21 type alloy has been used as the basis for development of a series of higher strength heat-resisting steels, data for some of which will be given later.

The ferritic steel 430S17 has enhanced oxidation resistance and finds some applications in sheet form, but its strength at elevated temperature is low. The higher chromium (20-30%) ferritic types show excellent oxidation resistance, but have poor elevated-temperature strength and, being difficult to produce and fabricate, are not used in large quantity. Cast versions of 27-30% Cr are quite widely used, especially where oxidation resistance, coupled with abrasion resistance, is required when high carbon contents are utilised. Such alloys are brittle.

The 18% Cr austenitic steels are quite widely used because of their ease of production and fabrication as well as their high-temperature properties. Those listed in Table 7.7 are most commonly used. The stabilised types (321S31 and 347S31) are preferred where heating to temperatures likely to cause susceptibility to intercrystalline corrosion (see p. 3.54) is interspersed with conditions where aqueous corrosion could be a hazard. Very low carbon grades are not usually used owing to their somewhat inferior strength. The higher chromium alloys 309S24 and 310S31 were developed to provide alloys with oxidation resistance superior to that of the 18% Cr type. Where strength at very high temperatures (>800°C) is required a high-carbon cast version of type 310S24 may be used where other requirements allow it. This type of alloy will not be discussed here, but information is given in Reference 8.

Physical Properties

Data for a variety of alloys have already been given in Section 3.3, page 3.35.

Mechanical Properties

The martensitic steels are used in the hardened and tempered conditions, the tempering temperatures used obviously being in excess of the proposed service temperature. Alloying elements used to improve the creep strength include molybdenum, vanadium, niobium, cobalt and tungsten and these also have the effect of increasing the resistance to softening or tempering so that the proof strengths of the creep-resisting variants are substantially higher than those for the simple Fe-13Cr alloy. 0·2% proof strengths at various temperatures for several martensitic types are shown in Fig. 7.22. Creep and long-term rupture strengths are shown in Figs. 7.23 to 7.25 as stress to give 0·1% total plastic strain in 10 000 h and as stress to give rupture in 10 000 and 100 000 h plotted as a function of temperature.

The ferritic steels rapidly lose strength at elevated temperature as shown in Fig. 7.26 and are of little value for load-bearing applications.

The austenitic grades, used mainly in the solution treated (softened) state, have low strength at ambient temperature but maintain strength at elevated temperatures much better than the martensitics and the ferritics. As can be seen from figs. 7.23 to 7.25, creep and rupture strengths are far superior

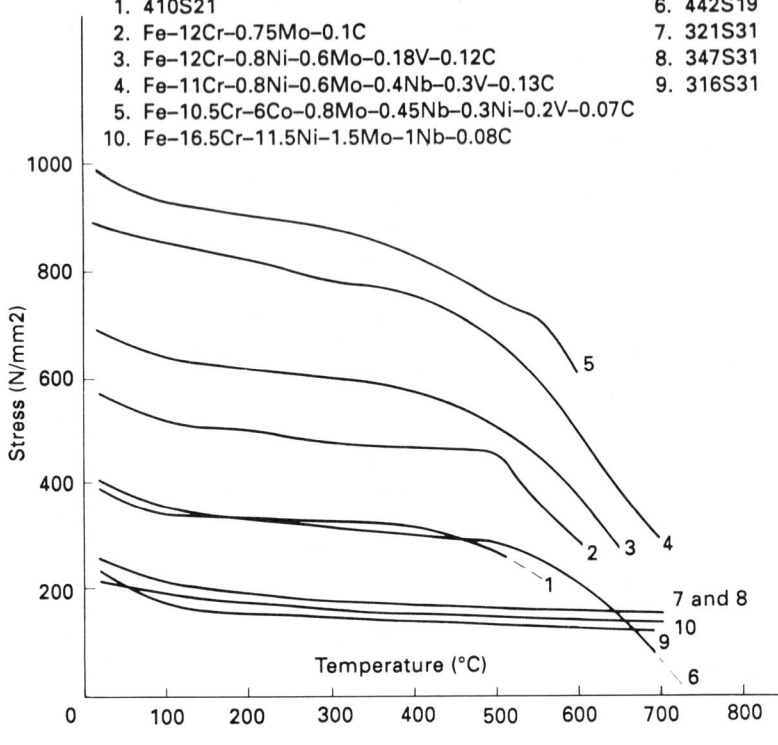

Fig. 7.22 0·2% proof stress values for various steels

above 600°C. Below this temperature proof strength is the limiting factor, but this can be improved somewhat in certain grades by alloying with nitrogen.

No more than a brief outline of the mechanical properties can be given here, for detailed information Reference 9 should be consulted. It should be noted that while steels used for creep resisting purposes may conform to the standard specifications, sometimes specially limited composition ranges within these specifications are used in the interests of strength, structural stability or resistance to embrittlement.

Fabrication

Forming and fabrication characteristics are described in Section 3.3 on stainless steels. Creep-resisting steels are, of course, intended to resist deformation at elevated temperatures, but in fact the mechanical power required for deformation at the forging temperature is little greater than that required for the stainless steels.

Creep-resisting steels often have to be used in thicker sections than is the case with stainless types and this can lead to the need for special techniques for forming and welding.

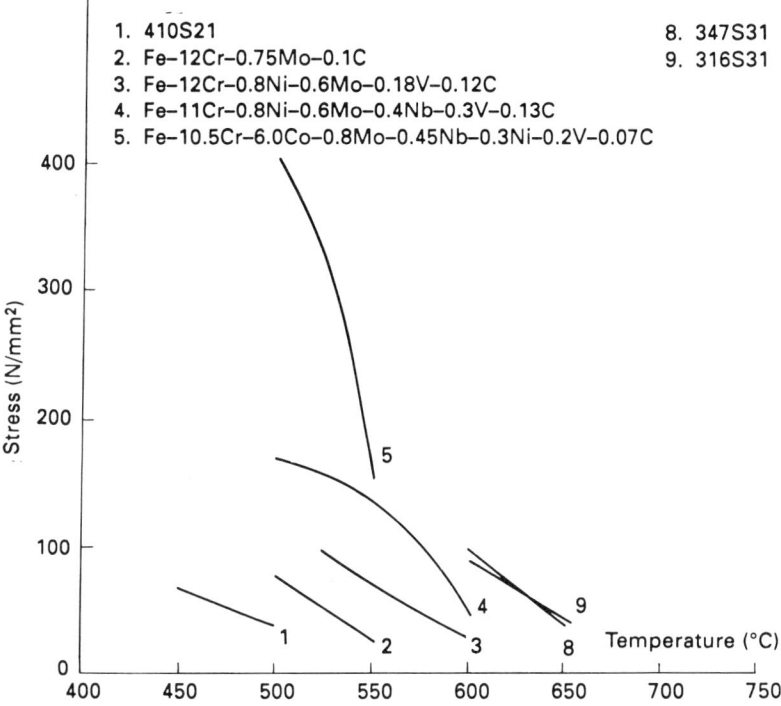

Fig. 7.23 Stresses to give 0·1% total plastic strain at 10 000 h for various steels

High-temperature Corrosion

Scale Structure and Oxidation Rates

Since the paper by Pilling and Bedworth[10] in 1923 much has been written about the mechanism and laws of growth of oxides on metals. These studies have greatly assisted the understanding of high-temperature oxidation, and the mathematical rate 'laws' deduced in some cases make possible useful quantitative predictions. With alloy steels the oxide scales have a complex structure: chromium steels owe much of their oxidation resistance to the presence of chromium oxide in the inner scale layer. Other elements can act in the same way, but it is their chromium content which in the main establishes the oxidation resistance of most heat-resisting steels.

In 1929 Pfeil[11] published a most interesting account of the way layered structures form and the manner in which they influence oxidation rates. From detailed studies of the growth and composition of scales he was able to show clearly how the formation of barrier layers reduced scale formation by hindering outward diffusion of iron through the scale. Naturally, this work had to be largely based on the study of scales of sufficient thickness so that the mechanism of the early stages of oxidation could not be studied in this way. Pfeil analysed the outer, middle and inner layers of scales formed

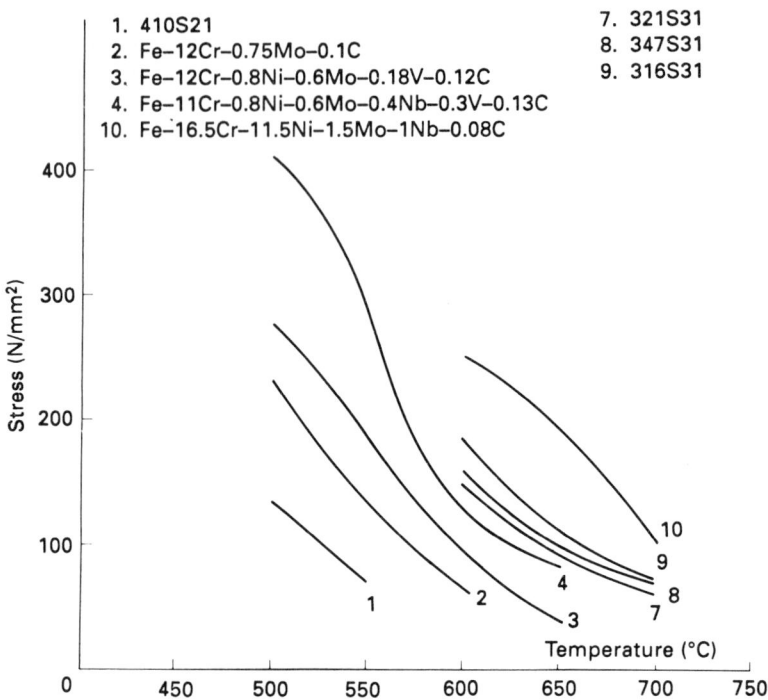

Fig. 7.24 Stresses to give rupture at 10 000 h for various steels

on steels containing various alloying elements and thus was able to demonstrate, for example, that the inner layer of scale formed on a 13% Cr steel in air at 1 000°C contained 34% Cr_2O_3, the middle layer 1·64% and the outer layer 0·89%.

The development of the electron-probe microanalyser has given research workers a powerful tool with which to determine composition variations of scale layers and also of underlying metal. Wood and his co-workers have used this instrument to great advantage to help explain the behaviour of iron–chromium alloys[12-14]. He found that on oxidation at high temperature (1 100°C) protective scales are largely Cr_2O_3, containing small proportions of iron oxide. The formation of the Cr_2O_3 causes depletion of the subjacent alloy in chromium, and the chromium content of the alloy at the interface may be as low as 5·3% without apparent transformation of the Cr_2O_3 to a spinel. Failure of the protective Cr_2O_3 scale with time is considered due to scale lifting and cracking followed by rapid oxidation of the depleted alloy. As the catastrophic break-through progresses the content of the inner layer is diluted to a limiting value of 20-25% Cr, while an outer, virtually pure iron oxide layer develops and the depleted subjacent alloy is almost entirely eliminated. These findings largely explain the observed behaviour of chromium-rich steels at temperature in oxidising atmospheres. Oxidation rates increase only relatively little with increasing temperature until, above some temperature, they increase rapidly. The exact value of this 'breakdown'

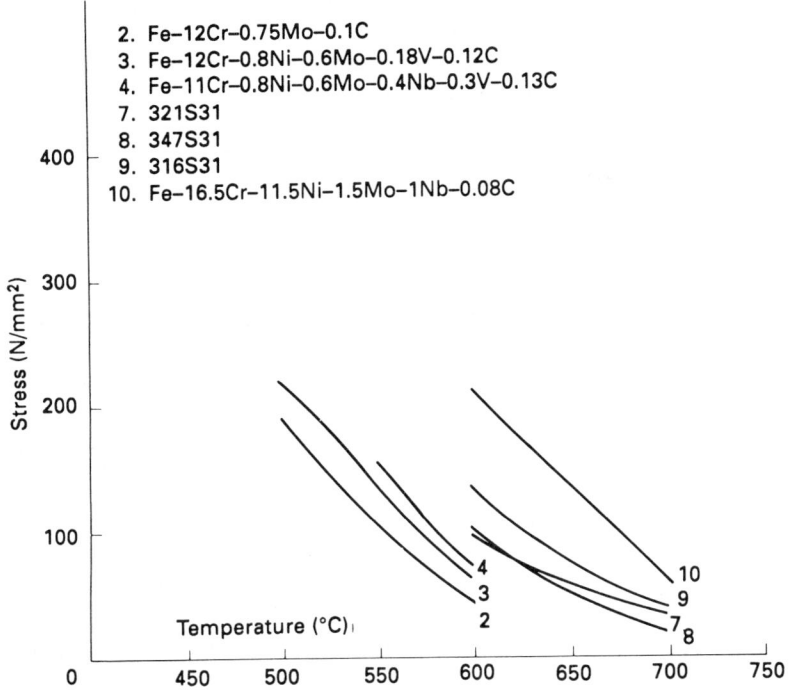

Fig. 7.25 Stresses to give rupture at 100 000 h for various steels

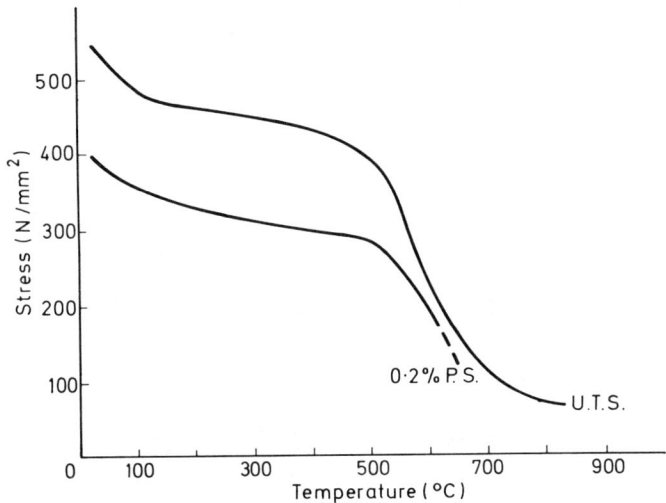

Fig. 7.26 Tensile strength of ferritic stainless steel 442S29

temperature, depends on alloy content but can also be affected by other features external to the alloy. For a detailed review of the mechanism of oxidation of chromium-bearing steels with a comprehensive bibliography Reference 15 should be consulted.

Attack by Gases

Flue Gases

There is naturally a desire with corrosion by oxidation to produce in the laboratory quantitative data which can be used for design purposes. This can be a more questionable procedure in the field of oxidation than is even the case with 'wet' corrosion since so many features can affect the results obtained. Apart from variations in atmosphere composition, gas velocity, rate of heating or cooling, frequency of thermal cycling, method of sample preparation, sample geometry and time of test can all have marked effects on corrosion rate measured under some circumstances. Most laboratory tests are useful mainly in allowing comparison of alloys and a general assessment of the range of temperatures over which an alloy may be useful, although it must be recognised that, ideally, the behaviour of an alloy selected for service should be further checked under conditions closely simulating service, especially if envisaged service is near the temperature ceiling for the alloy indicated by the laboratory tests.

A test procedure which has proved very useful was first described by Hatfield[16]. The samples are cylinders 32 × 12·5 mm in diameter with a standard abraded finish which are supported on open-ended refractory boats in a tubular furnace. In the original test the atmosphere, which was produced by burning towns gas with a 50% excess of air, was passed over the specimens at a standard velocity after first preheating to test temperature over refractory packing in a separate furnace chamber. More latterly, natural gas has been used with suitable modification of air:gas ratio to give

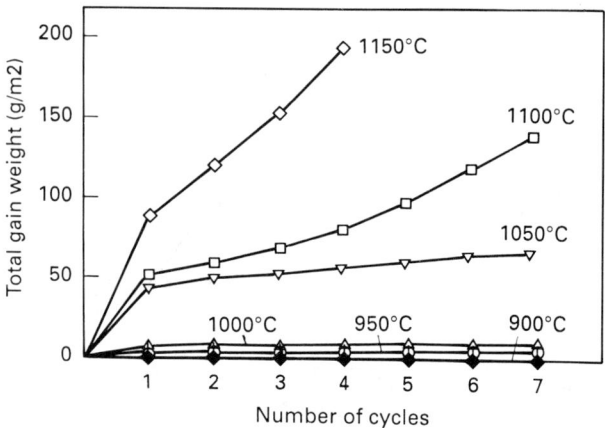

Fig. 7.27 Total weight gain versus number of 6-h test cycles (20·14% Cr)

a generally similar test atmosphere. No significant variation in results has been noted. Test times are short, but thermal cycling is incorporated so that any disruptive effects of differential expansion and contraction of metal and oxide may operate. The specimen is heated to temperature for seven 6-h cycles with intermediate cooling to room temperature and weighing, together with any loose scale shed during cooling. Before each test cycle the specimen is lightly brushed and reweighed. Gain in weight versus cycle plots for a number of temperatures determined for a ferritic steel with 20·14% chromium are shown in Fig. 7.27. The behaviour shown is typical of all steels although the temperatures above which rapid oxidation occurs differ. At lower temperatures, the oxidation rate falls with time (cycles) as a protective scale grows, but the gain in weight before the near-protective behaviour is established increases with temperature. Above some critical temperature there is marked progressive oxidation usually with periodic scale shedding. The change from protective to semi-protective behaviour can sometimes occur during the seven cycles of a test (i.e. breakaway oxidation). Obviously it could also occur after some longer time, although experience has shown that a temperature 50°C below that at which rapid oxidation appears in this text is a reasonable choice for maximum service temperature.

The total gain in weight over the seven 6-h cycles is designated the scaling index and this value is plotted against test temperature for a series of steels of varying chromium content in Fig. 7.28. These were laboratory-produced

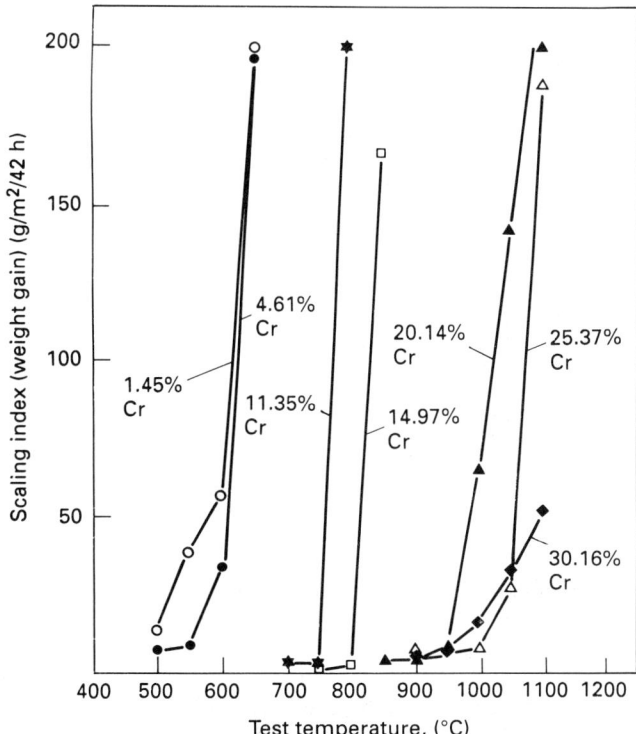

Fig. 7.28 Scaling index versus temperature for several chromium steels

steels containing about 0·05% carbon, 0·5% silicon and 0·5% manganese, but no other alloying except for the inevitable small amounts of impurities. The sharp change from protective to limited protective behaviour with increasing temperature can be clearly seen. From such plots a simpler method for presenting results can be derived, that is the temperatures at which certain scaling indices could be obtained. The temperatures to give indices of 10, 50 and 100 g/m^2 are convenient values and are referred to as the SI_{10}, SI_{50} and SI_{100} temperatures.

SI temperatures are given for a number of tests, including some carbon and low-alloy types for comparison, in Table 7.8. As well as the types listed in Table 7.7, a selection of creep-resisting grades is included. In addition some of the special stainless steels (see Section 3·3) are also included to demonstrate the effects of some other alloying elements.

Table 7.8 Oxidation resistance of a number of steels as shown by a short time, cyclic test

C	Si	Mn	Cr	Ni	Mo	V	Nb	Others	Grade	SI (°C)		
										SI_{10}	SI_{20}	SI_{100}
0.55	0.25	0.82	0.09	0.09						< 400	600	620
0.24	0.26	0.55	3.12	0.34	0.59					< 500	600	640
0.38	0.22	0.61	3.11	0.94	0.94	0.20				< 500	600	615
0.53	3.37	0.50	8.49	0.26						950	960	1010
0.10	0.19	0.48	9.04	0.36	0.92					< 500	640	695
0.07	0.23	0.64	9.09	9.08						< 500	605	620
0.08	0.21	0.37	12.86	0.25						800	820	835
0.28	0.31	0.31	12.80	0.50						790	800	825
0.14	0.29	0.41	16.61	2.50						750	790	850
0.09	0.19	0.46	12.28	0.32	0.68					800	810	820
0.14	0.42	0.86	10.89	0.58	0.85	0.28				800	805	820
0.11	0.43	1.06	10.59	0.81	0.62	0.18	0.39			800	810	815
0.11	0.48	0.73	11.43	2.62	1.33	0.13	0.21			750	760	770
0.07	0.48	0.89	10.70	0.60	0.76	0.17	0.33	6.01 Co		800	810	825
0.045	0.55	0.26	15.94	4.08			0.29	3.26 Cu		780	807	818
0.05	0.32	0.78	13.86	5.50	1.61		0.37	1.74 Cu		750	770	830
0.05	0.42	0.37	16.01	0.25						850	860	875
0.05	0.39	0.73	20.48	0.22						900	1000	1075
0.05	1.49	1.34	28.86	1.82				0.16N		900	1120	1160
0.075	0.55	0.85	13.12	0.09				4.01Al		1060	1100	1125
0.05	0.51	1.04	18.68	10.03						860	880	915
0.07	0.76	0.80	18.25	8.95				0.43 Ti		820	860	890
0.05	0.40	0.76	18.28	9.95			0.77			820	860	890
0.05	0.19	1.69	16.38	10.50	2.52					700	820	845
0.11	0.47	1.17	22.00	14.38						980	1060	1070
0.07	0.48	1.49	25.20	20.24						1030	1090	1180
0.045	0.60	1.12	20.32	33.50				0.25 Ti, 0.33Al		970	1010	1140

The major beneficial effect on oxidation resistance comes from alloying with chromium, silicon and aluminium. Chromium represents the basic alloying addition for most oxidation-resisting steels and can be accommodated up to about 14% with a martensitic structure and 30% (practically) in a ferritic structure. Suitable alloying with nickel allows austenitic structures also with high chromium contents, 25% chromium being the highest value used currently (310S31). While silicon and aluminium both strongly complement the beneficial effects of chromium, they are both strong ferrite formers, which limits their use. Aluminium is used in some very resistant steels but these are ferritic and so can only be used at the high temperatures available from an oxidation resistance point of view when stressing is relatively low. Unlike aluminium, silicon is present in small quantities (0·2-0·5%) in most commercially produced steels (see Table 7.8) and there is evidence that even such small amounts contribute substantially to the behaviour of chromium steels.

In Fig. 7.29, SI_{100} temperatures are plotted against silicon content for a series of martensitic steels with 10·89-13·14% chromium. Nickel is much used to control structure. It can have a slightly adverse effect on the oxidation resistance of martensitic steels but is beneficial in the larger amounts relevant to the austenitic types. The other commonly used alloying elements have little effect (at least in the quantities used) although manganese in substantial amounts is somewhat detrimental and molybdenum can be harmful if service conditions are such that the volatile MoO_3 can attain significant levels in the gases adjacent to the steel. Rare earths in small quantities can be beneficial, as they can in other alloy systems.

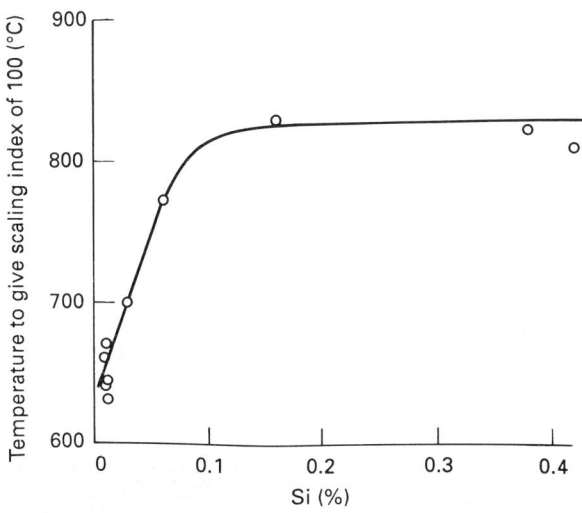

Fig. 7.29 Temperature to give a scaling index of 100 versus silicon content (10·89-13·14% Cr)

Although behaviour does not show any great variation over quite a wide range of oxidising atmospheres, the air:fuel ratio can exert some effect. This was shown by tests on an Fe-18Cr-8Ni alloy at 850°C[17]. The steel was submitted to a test of the type just described, but in atmospheres produced by catalytic burning of 2:1, 4:1 and 6:1 air:towns gas (pre-natural gas), the 4:1 ratio corresponding to stoichiometric combustion. Mean oxidation gains expressed as g/m^2 were:

6:1 air:gas   32
4:1 air:gas   12
2:1 air:gas   4

At temperatures below 850°C this effect of varying air:gas ratio tended to disappear and it was not apparent in tests at 750°C. From the data shown in Table 7.8 it can be seen that 850°C is about the temperature of transition from protective to semi-protective behaviour, so atmospheric effects would be at their greatest.

The effect of sulphur from the gas phase is critically dependent on the effectiveness of fuel combustion. With good combustion to the limit of the available oxygen, and even down to 50% air deficiency, no serious effect was found from high-sulphur fuel in tests with 321S12 steel up to the usual limit of service temperature at 850°C, as shown in Table 7.9. With 310S24 steel at 1 100°C some effect from high sulphur content was found in 2:1 air:gas with effective combustion, but none in 4:1 and 6:1 mixtures.

Table 7.9 Comparison of high- and low-sulphur fuel in tests at 850°C with 321S12 steel[17]

Air:gas ratio	Sulphur in gas*	Oxidation gain in seven 6-h test cycles (g/m^2)
6:1	Low	26
	High	27
4:1	Low	17
	High	21
2:1	Low	10
	High	12

*The towns gas contained 460 mg of S/m^3. High sulphur addition was made in the form of H$_2$S equivalent to 0·5% SO$_2$ in the 6:1 combustion products, i.e. about three times as much as from ordinary high-sulphur fuels.

With poor combustion, on the other hand, very severe acceleration of attack, dependent upon the formation of sulphide in the scale, can occur. This destroys the protective action of the scale, and results in sulphide penetration of the metal in advance of oxidation. The effect is illustrated by tests

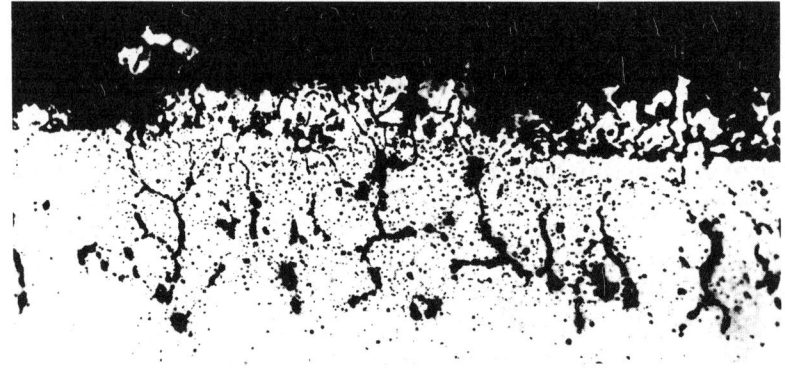

Fig. 7.30 Unetched section through Fe–25Cr–21Ni after attack in 4:1 air:gas + H$_2$S at 1 100°C with poor combustion, showing sulphide penetration; × 150

with 310S24 steel in stoichiometric 4:1 air:gas at 1 100°C, in which the burner was modified to give incomplete combustion. The tests were for two 6-h cycles, and oxidation gains were (a) good combustion, high-sulphur fuel, 16 g/m^2; (b) bad combustion, low-sulphur fuel, 12 g/m^2; (c) bad combustion, high-sulphur fuel, 319 g/m^2. Sulphide penetration into the metal under condition (c) is illustrated in Fig. 7.30.

Air

Air tends to be less aggressive than the flue gas used for the standard test described earlier, but the useful range of temperatures for each steel is effectively similar. Edwards and Nicholson[18] reported some long-term testing of four austenitic grades in air saturated with water (at room temperature) at temperatures of 650°C–875°C for up to 10 000 h. They make the point, and show convincingly, that in any long-term assessment, metal wastage must not be based on scaling alone but that the effect of subsurface penetration must also be considered (this applies also, of course, to testing in other gases). Thus their values (Fig. 7.31) are compounded from surface loss and subsurface penetration, and their work is especially valuable in that an assessment of the effect of long-term heating on mechanical properties was also made.

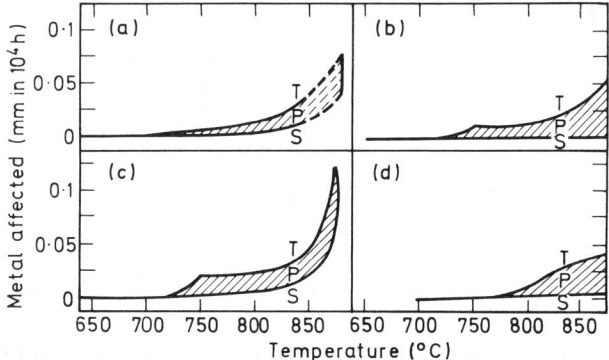

Fig. 7.31 Metal wastage of several steels due to oxidation in air (saturated with water at room temperature) for 10 000 h at various temperatures. (a) Type 302S31, (b) type 321S31, (c) type 316S31 and (d) type 310S31. $T_{total} = S_{surface} + P_{penetration}$ (after Edwards and Nicholson[18])

Nitrogen can be absorbed from air during prolonged heating, but with steel in the unstressed or lightly stressed state the rate is very slow except for temperatures above 1 050°C. Considerable absorption can occur at lower temperatures during creep, however. This fact is presumably due to the exposure of oxide-free surface during creep, and it has been noted that nitrogen absorption is especially marked at cracks. The following nitrogen contents have been reported[19] for the 0·75 mm surface layers of creep specimens in 347S31 steel (original nitrogen content 0·053%) after creep failure:

650°C (life 10 970 h under 108 MN/m^2), 0·077% N$_2$
700°C (life 37 958 h under 46 MN/m^2), 0·65% N$_2$
800°C (life 16 629 h under 1·5 MN/m^2), 0·90% N$_2$.

Steam

Modern boiler developments involving increased steam temperatures and pressures have made it increasingly important to consider the behaviour of high-alloy steels under conditions typical for superheater tubes and steam pipes. For satisfactory service the steels must, of course, possess adequate mechanical properties, especially creep resistance, but they must also be sufficiently resistant to oxidation to ensure long life. Short-term laboratory tests are of value in yielding comparative data for different steels and, in fact, results generally similar to those for the flue gas test already described are obtained, but prolonged tests approximating more nearly to service conditions are desirable.

Rohrig, van Duzer and Fellows[20] exposed samples in an experimental superheater fed with steam at 2·6 MN/m^2 from a power plant. Some 42 materials were tested for periods of up to 16 000 h, attack being estimated after test by weight loss following descaling. It was concluded that at 593°C attack continues at a high rate on carbon steel, whereas the rate for most alloy steels decreases with time (Table 7.10).

Table 7.10 Losses from exposure for 7 461 h in steam at 2·62 MN/m^2 and 593°C *

Steel	Calculated penetration in 10 000 h (mm)
Mild steel	0·107
A.I.S.I. 403 (Fe–12Cr)	0·015
A.I.S.I. 347 (Fe–18Cr–8Ni + Nb)	0·002
A.I.S.I. 309 (Fe–25Cr–12Ni)	0·002
A.I.S.I. 310 (Fe–25Cr–20Ni)	0·003
Fe–35Ni–15Cr	0·002

* Data after Rohring, van Duzer and Fellows[20].

Eberle, Ely and Dillon[21] tested commercial tubes in a small superheater receiving plant steam at 14 MN/m^2 and superheating it from 538° to 677°C. Penetration was estimated from scale thickness measurements after 6 950 h and comparison was made between the attack by steam on the inside of the tubes and that by flue gas from pulverised coal firing on the outside (Table 7.11).

A collaborative test programme covering low-alloy and high-alloy steels was carried out by the Central Electricity Generating Board and various steelmakers. Samples were exposed in specially constructed chambers held at 566°C, 593°C and 621°C fed with power-station steam at a pressure of 3·45 MN/m^2 for times of up to 16 286 h. In the assessment of the results both metal lost from the surface and subsurface penetration were measured. The results have been reported by King, Robinson, Howarth and Perry in a C.E.G.B. report. Selected data are shown in Fig. 7.32, in which the broken lines have been obtained by extrapolation of the experimental results.

Table 7.11 Comparison of internal and external scaling of superheater tubes after 6 950 h in steam at 13·8 MN/m^2 and 500-670°C*

Steel	Estimated penetration (mm/y)	
	Steam	Flue gas
A.I.S.I. 304 (Fe-18Cr-8Ni)	0·038	0·021
A.I.S.I. 321 (Fe-18Cr-8Ni+Ti)	0·038	0·023
A.I.S.I. 347 (Fe-18Cr-8Ni+Nb)	0·010	0·023
A.I.S.I. 318 (Fe-16Cr-13Ni-3Mo+Nb)	0·013	0·029

*After Eberle, Ely and Dillon[21] and relating to tests in a superheater raising steam from 540°C to 670°C.

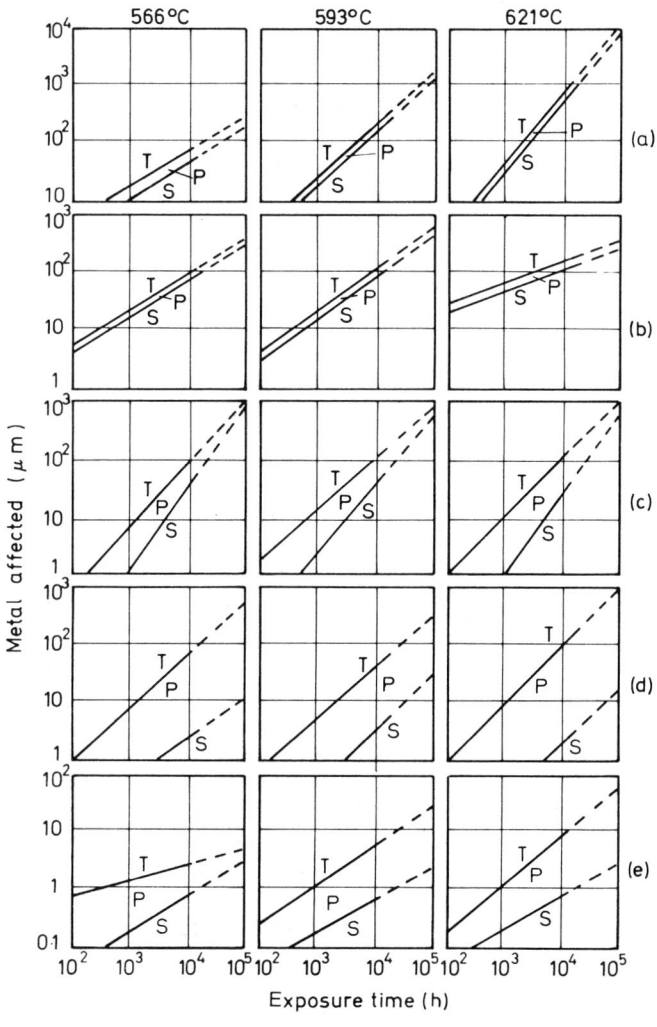

Fig. 7.32 Metal wastage of several steels due to oxidation in steam at various temperatures. (a) Mild steel. (b) Fe-2Cr-0·25Mo, (c) Fe-12Cr + Mo + V, (d) A.I.S.I. 316 and (e) Fe-18Cr-12Ni-1Nb. $T_{total} = S_{surface} + P_{penetration}$ (after King et al.)

Other Industrial Gases

All oxidising gases can lead to oxide formation on chromium steels at elevated temperatures and in some instances this can be associated with absorption of some other substance in the steel. Carbonaceous gases are a good example and whereas high-alloy steels successfully resist flue gases even under conditions of considerable air deficiency, reduction of oxygen content eventually leads to conditions under which at a sufficiently high temperature considerable carburisation of the metal occurs. An example is the endothermic gases used as protective atmospheres for other metals which, at elevated temperature, can rapidly cause embrittlement of high-alloy steel.

The absorption of nitrogen from air has been mentioned and similar effects can occur with nitrogen under similar circumstances. However, nitriding is much more likely in ammonia or in gases containing ammonia, as indicated by the following figures for the nitrogen contents of the outer 0·25 mm layers of samples of 310S31 steel (initial N content 0·06%) after 250 h in ammonia: 500°C, 0·25% N_2; 900°C, 0·55% N_2; 1 000°C, 0·92% N_2; 1 050°C, 1·19% N_2. Nitriding leads to serious embrittlement.

Hydrogen at high pressure and temperatures above 400°C has a considerable adverse effect on carbon steel, dissolving in the steels and combining with carbides to produce methane and so causing fissuring and considerable embrittlement. However, chromium stabilises the carbides and stainless steel may be safely used in hydrogen at dull red heat[22].

Ash Attack

The degree of oxidation in a gaseous environment can be modified greatly by the deposition of even small amounts of certain fuel ashes. The topic of ash corrosion has been reviewed with extensive bibliography[23], but some consideration of high-alloy steels will be given here. Any substance which can form a low melting point mixture with the normally protective oxide scale (i.e. 'flux' the oxide) formed on high-alloy steels, is potentially dangerous. While such substances are not common in fuels the danger should be borne in mind where high-alloy steels are used as containment vessels for high-temperature processes.

Sulphates, which form part of the ash from the combustion of many fuels, are not harmful to high-alloy steels, but can become so if reduction to sulphide occurs. This leads to the formation of low melting point oxide–sulphide mixtures and to sulphide penetration of the metal. Such reduction is particularly easy if the sulphate can form a mixture of low melting point with some other substance. Reduction can be brought about by bad combustion, as demonstrated by Sykes and Shirley[17], and it is obviously important to avoid contact with inefficiently burnt fuels when sulphate deposits may be present. Reduction can also be brought about in atmospheres other than reducing ones and the presence of chlorides or vanadium pentoxide has been shown to be sufficient to initiate the reaction. It has also been shown[24] that it can be initiated by prior cathodic polarisation in fused sodium sulphate. The effect of even small amounts of chloride on oxidation in the presence of sulphate is illustrated in Fig. 7.33[17].

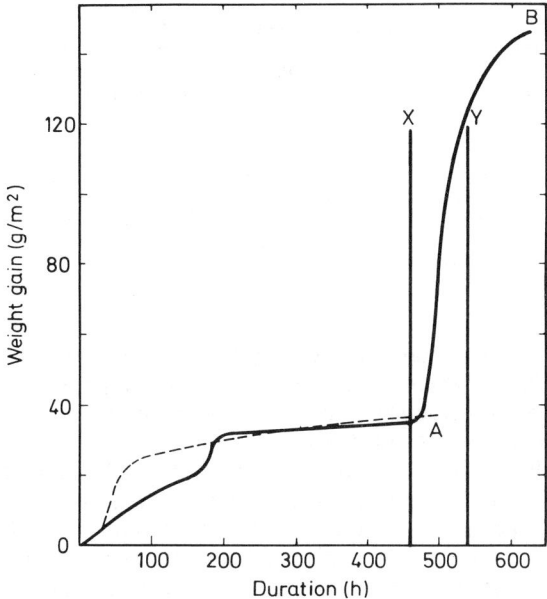

Fig. 7.33 Gains in weight due to oxidation of type 347S31 steel in air at 750°C while in contact with Na₂SO₄. Curve A plain, and curve B containing 0·3% NaCl (period X-Y) (after Sykes and Shirley[17])

Not all sulphates are as readily reduced as sodium sulphate, for instance, calcium sulphate does not usually lead to sulphide penetration, although the presence of other substances with calcium sulphate may lead to accelerated oxidation for other reasons. The results for laboratory tests on a series of metals and alloys in sodium sulphate + sodium chloride and calcium sulphate + calcium chloride mixtures are shown in Table 7.12[25]. In many cases sulphide peneration could be noted with the sodium salts but not with the calcium salts.

Table 7.12 Effect of 90:10 sulphate:chloride mixtures on various metals at 750°C (tests for 6 h in air)*

Material	Loss of metal (g/m²)		
	No mixture	Sodium salts	Calcium salts
Cr	7	30	110
Ni	20	60	110
Mild steel	490	1 430	1 120
Fe-20Cr	4	50	1 350
Fe-28Cr-2Ni	3	60	1 160
Fe-22Ni-14Cr	2	40	950
Fe-18Cr-12Ni+Nb	9	50	470
Ni-13Cr	2	2 700	20
Ni-20Cr+Ti+Al	4	3 700	20

* Data after Shirley[25].

Table 7.13 Corrosion tests in air with specimen half immersed in sodium chloride

Test temperature (°C)	Weight loss after descaling following a 24 h test (g/m^2)				
	Steel				
	430	304	321	347	310
550	100	30	20	20	30
650	320	270	100	210	190
750	1 050	660	650	400	750

It has been suggested that corrosion by sulphates can occur by the formation of pyrosulphates which melt at relatively low temperatures. Accelerated corrosion due to the presence of pyrosulphates has been demonstrated for ferrous alloys including stainless steel[26].

The rôle of chlorides in the presence of sulphates has already been mentioned, but these can also have a serious effect in the absence of other contaminants. The presence of chloride not only leads to considerable acceleration of oxidation rate but can also give substantial subsurface intergranular penetration of the steel. Corrosion test results for several steels are shown in Table 7.13. The attack noted on the calcium sulphate + calcium chloride mixtures indicated in Table 7.12 can possibly be attributed solely to the presence of chloride. Very small amounts of chloride are sufficient to cause serious acceleration, as illustrated in Fig. 7.34. Samples of 347S31 steel were heated in air to 650°C for 20-h cycles. Between cycles the samples were cooled to room temperature and weighed together with any loose scale,

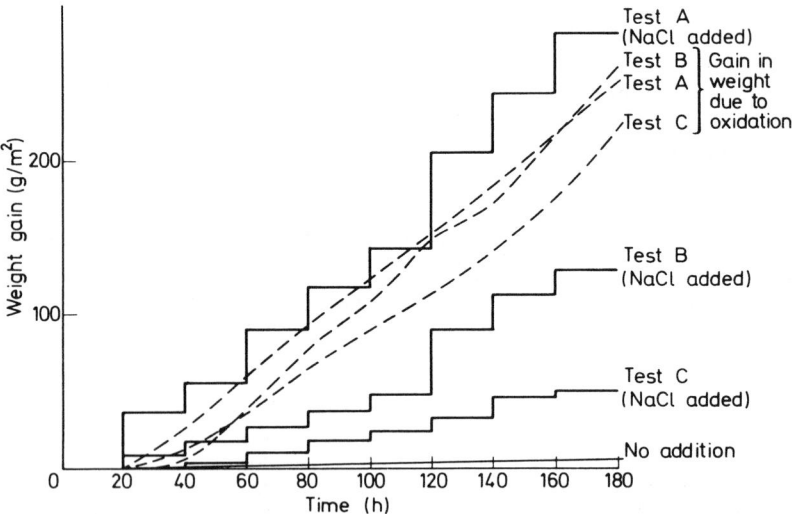

Fig. 7.34 Gains in weight of 347S31 steel in air at 650°C with various periodic additions of sodium chloride. Stepped curves show quantity of NaCl added, broken curves show gains in weight attributable to oxidation

brushed, weighed again and then dipped in a dilute sodium chloride solution, removed, dried and weighed yet again to give the quantity of salt added to the surface. Gains in weight attributable to chloride and to oxidation are shown; it should be noted that the latter are probably low since no account was taken in this method of assessment of any losses of volatile substances. In tests of this sort, when the addition of chloride is discontinued, the oxidation rate slowly returns to the very low value to be expected in air at this temperature. While most interest has been in effects of chlorides, there is evidence that other halides can have similar effects. The subject of corrosion in steam-raising plant burning chloride-bearing coal was considered in the conference reported in Reference 27.

In the case of oil ash, the most serious damage is associated with vanadium compounds. Organic vanadium compounds which cannot be economically removed from the residual oils form vanadium pentoxide in combustion which can have a considerable corrosive effect on heat-resisting steels. Vanadium pentoxide itself has a melting point in the neighbourhood of 660°C and forms compounds which have even lower melting points. Prior to the development of the gas turbine the problem was not of such overriding importance, since components, such as superheater tubes in steam boilers, which had to be thin, were at a low enough temperature to escape serious action. Tube supports in superheater and oil-cracking furnaces fired with residual oil often suffered severe attack, but as they were cast products, relatively inexpensive and replaceable, they were regarded as being of necessity expendable even though losses of metal of thicknesses $\geqslant 10$ mm/year from Fe-24Cr-21Ni and Fe-30Cr steel supports at 800-900°C could occur. With the advent of the gas turbine and the possibility of higher steam temperatures an entirely new order of assessment was necessary as the very much smaller rates of attack occurring in the range 650-800°C involved quite prohibitive expense in relation to costly creep-resisting high-alloy steel blades or tubes. No complete solution has emerged, but useful measures to reduce the seriousness of the attack in some cases have been evolved, and the limits within which vanadium-bearing oils may reasonably be used have been defined.

The oil ash problem is complicated and intensified by the presence of alkali oxides and sulphates which form a variety of low-melting complexes with vanadium oxide. An idea of the complications occurring in practice can be obtained from the account of extensive field and laboratory tests carried out for a committee of the Council of British Manufacturers of Petroleum Equipment[28], with particular reference to superheater supports and metal temperatures in the range 600-850°C. The field tests were carried out in three boiler installations, one marine and two land-based. Materials tested included Fe-26Cr, Fe-18Cr-8Ni+Nb, Fe-24Cr-22Ni and Fe-32Ni-18Cr+Ti, as well as a number of nickel-base alloys. It was concluded that deposits in the superheater zones of land-based boilers tend to have a higher V:Na ratio than is present in the original oil, but that with installations operating at sea, contamination by sea-water may reverse this relation. Deposits of high V:Na ratio accelerated oxidation primarily through scale fluxing, and were most corrosive to steel, while deposits of high sodium sulphate content accelerated oxidation through sulphide action and were most damaging to alloys of high nickel content. A ferritic 26% Cr steel was the only one which showed good resistance to deposits containing sodium and vanadium in all proportions; this type of alloy unfortunately has relatively poor stress-carrying capacity at high temperature.

Since the main action of vanadium is related to fluxing, much attention has been given to the inhibition of this action by formation, through suitable conditions, of higher-melting compounds; calcium and magnesium compounds are the most generally favoured of such additives[29]. Preliminary washing of the oil to reduce sodium to very low limits has also been advocated[30]. The stability of calcium and magnesium vanadates is, unfortunately, not great enough to prevent their substantial conversion to sulphates by the sulphur oxides normally present in the flue gases, so that with additions of alkaline earth oxides to, say, two or three times the stoichiometric equivalent of the V_2O_5, stabilisation as vanadates is incomplete, and the improvement only partial. For critical components the only safe procedure is usually either to limit the temperature at which steels are used to a maximum of about 600°C, or to use more expensive distillate-oil fuel free from vanadium.

The whole subject of vanadium attack has been reviewed by Sachs[31].

An interesting case of 'ash' attack is encountered with valves in engines powered by high octane fuels containing lead compounds. These compounds are deposited from the gases as mixtures of lead oxide, sulphate and bromide, and can cause serious scale-fluxing effects with high-alloy valve steels.

Molten Salts

Molten salt baths are widely used in heat treatment and for steel carburising. As in the case of ash attack, danger to alloy steel containers arises mainly from enhanced oxidation brought about by scale fluxing. Such oxidation can only proceed where oxidising conditions obtain, so that while alkali chlorides form useful heat-treatment baths for steel, they produce severe attack, even with heat-resisting steel containers, at the surface of the bath, with formation of chromates and ferrites. Austenitic Fe–23Cr–12Ni and Fe–35Ni–15Cr steels are used for cyanide-hardening-bath containers, but conversion of cyanide to carbonate in use brings danger of fluxing attack.

Molten alkali hydroxides are particularly dangerous, not only because of scale fluxing, but also because they induce stress corrosion where stress is a serious factor.

Molten Metals

Heat-resisting steels have limited uses in contact with molten metals. They are not recommended for use with molten zinc, cadmium, aluminium, antimony or copper, because of excessive attack and embrittlement effects. In brazing and silver soldering, contact between the molten non-ferrous alloy and the steel occurs for only a very limited period of time.

With molten lead or tin, limited use of high-alloy steels is possible. In the case of containers for lead baths, it is important to avoid the combination of lead oxide and air at the bath surface because of fluxing action, but with mechanical removal of lead oxide and use of carbonaceous coverings, Fe–35Ni–15Cr and Fe–25Cr steels are successfully used. With tin, behaviour depends considerably on temperature, slight action taking place at 300°C and considerable attack at 600°C.

The alkali metals have acquired special interest through their suitability for use as heat-exchanging fluids in atomic reactors. They are generally satisfactorily resisted by the heat-resisting steels, although detailed studies have shown effects on prolonged contact at high temperatures. Thus, Brasunas[32] describes the leaching of nickel by lithium at 1 000°C from Fe–18Cr–10Ni steel, with transformation of the nickel-impoverished surface layer to a ferritic structure and ultimate production of subsurface cavities. He also indicates that there is some penetration by sodium and lead at this temperature with precipitation of intermetallic compounds within the steel.

Applications

The heat-resisting steels are used for a wide range of general engineering and chemical engineering applications where the corrosion resistance, and in some instances strength, of the lower-alloy steels is inadequate. The martensitic steels, because of their lower oxidation resistance, are normally used for the less onerous conditions, and certain limitations in ease of fabrication generally precludes their use for large structures and containment vessels. Their combination of moderate corrosion resistance plus strength at modest temperatures has led to widespread use as turbine discs and blades, bolts and similar parts. They have also been used on steam plant for the less onerous conditions.

The ferritic steels are limited in scope because of lack of hot strength, but the cheaper types such as 430 are used in sheet form for the fabrication of parts such as heat exchangers. The higher chromium varieties are of importance in being much more resistant to sulphur attack than the nickel-bearing types and so are widely used as superheater supports or in sulphide-roasting furnaces, mainly as castings. Strength limitations and brittleness call for care in design.

The austenitic steels combine good oxidation resistance with ease of fabrication and thus are most widely used. In addition, while being quite weak at room temperature, they are among the strongest materials in the 550–750°C range and are thus widely used for this purpose. Typical applications are furnace parts, heat exchangers, gas turbine parts, steam superheaters and piping, and chemical plant equipment for containing reactions and products at elevated temperatures.

J. E. TRUMAN

REFERENCES

1. Hatfield, W. H., *J.I.S.I.*, **115**, 517 (1927)
2. Aitchison, L., *Engineering*, **108**, 799 (1919)
3. Dickenson, J. H. S., *J.I.S.I.*, **106**, 103 (1922)
4. Farenwald, F. A., *Proc. Amer. Soc. Test. Mat.*, **24**, 310 (1924)

5. Johnson, J. B. and Christiansen, S. A., *Proc. Amer. Soc. Test. Mat.*, **24**, 383 (1924)
6. Vanick, J. S., *Proc. Amer. Soc. Test. Mat.*, **24**, 348 (1924)
7. Hatfield, W. H., *J.I.S.I.*, **115**, 483 (1927)
8. Edeleanu, C. and Estruch, B., *I.S.I. Special Report No. 86*, 220 (1964)
9. *High Temperature Properties of Steels*, I.S.I. Publication No. 97 (1967)
10. Pilling, N. E. and Bedworth, R. E., *J. Inst. Met.*, **29**, 529 (1923)
11. Pfeil, L. B., *J.I.S.I.*, **119**, 501 (1929)
12. Wood, G. C. and Whittle, D. P., *Corrosion Science*, **4** No. 3, 263–269 (1964)
13. Wood, G. C. and Whittle, D. P., *Corrosion Science*, **4** No. 3, 293–315 (1964)
14. Wood, G. C., *Corrosion Science*, **2**, 255–269 (1962)
15. Wood, G. C., *Corrosion Science*, **2**, 173–192 (1962)
16. Hatfield, W. H., *J. Inst. Fuel*, **11**, 245 (1938)
17. Sykes, C. and Shirley, H. T., *I.S.I. Special Report No. 43*, 153 (1951)
18. Edwards, A. M. and Nicholson, A., *I.S.I. Publication No. 117*, 149 (1969)
19. Kirkby, H. W. and Truman, R. J., *I.S.I. Special Report No. 64*, 244 (1959)
20. Rohrig, I. A., van Duzer, R. M. and Fellows, C. H., *Trans. Amer. Soc. Mech. Eng.*, **66**, 277 (1944)
21. Eberle, F., Ely, F. A. and Dillon, J. A., *Trans. Amer. Soc. Mech. Eng.*, **76**, 665 (1954)
22. Inglis, N. P. and Andrews, W., *J.I.S.I.*, **128**, 383 (1933)
23. Hancock, P., *Corrosion of Alloys at High Temperatures in Atmospheres Consisting of Fuel Combustion Products and Associated Impurities*, H.M.S.O. (1968)
24. Simons, E. L., Browning, G. V. and Liebhafeky, H. A., *Corrosion*, **11**, 505t (1955)
25. Shirley, H. T., *J.I.S.I.*, **182**, 144 (1956)
26. Jonakin, J., *The Mechanism of Corrosion by Fuel Impurities*, 648 and 649, Butterworths, London (1963)
27. Meadowcroft, D. B. and Manning, M. I. (eds), *Proc. Conf. on Corrosion Resistant Materials for Coal Conversion Systems*, Applied Science Publishers (1982)
28. *British Petroleum Equipment News*, **7** No. 4, 54 and **7** No. 5, 48 (1959–1960)
29. Buckland, B. O., Gardner, G. M. and Sanders, D. G., *Residual Fuel Oil Ash Corrosion*, A.S.M.E. Paper A-52-161 Preprint (1952)
30. Buckland, B. O. and Sanders, D. G., *Modified Residual Fuel for Gas Turbines*, A.S.M.E.. Paper 54-A-246 Preprint (1954)
31. Sachs, K., *Metallurgia*, **57**, 123, 167, 224 (1958)
32. Brasunas, A. de S., *Corrosion*, **9**, 78 (1953)

7.5 Nickel and its Alloys

Oxidation

Pure Nickel

Although the oxidation of nickel has been extensively studied it is only recently that the process has been clearly understood. The relative simplicity of the system in which only a single-phase layer of oxide, NiO, forms has encouraged research, and a further simplification is that the expansion coefficients of the oxide and metal are similar, (17.1 and 17.6 $\times 10^{-6}\,°C^{-1}$, respectively,) so that the effects of thermal cycles can be largely neglected.

Nickel, in comparison with metals such as iron, cobalt and copper, has a relatively good resistance to oxidation at high temperatures. The growth of the oxide generally follows a parabolic law, but deviations are observed depending on the surface preparation, alloy purity and microstructure. Figure 7.35 shows a comparison of the parabolic rate constant for the oxidation of high-purity nickel with the tracer lattice diffusion coefficient of nickel in NiO, and it can be seen that it is only at temperatures in excess of about 1 200°C that the activation energies of the two processes become similar (230–250 kJ mol^{-1})[1]. At lower temperatures the rate of oxidation is increasingly greater than would be predicted by assuming that the process is controlled by bulk diffusion in the oxide lattice, with activation energies being reported in the range 155–170 kJ mol^{-1}. The effect of prior cold-work in the nickel is to increase the oxidation rate, but the observed rate law is usually less than parabolic. Both these observations suggest that the rate of oxidation is controlled by grain boundary diffusion in the oxide; the less than parabolic rate observed in the cold-worked material occurs because the initially fine-grained oxide coarsens during the oxidation process thereby eliminating some short-circuit diffusion paths. Models have been developed to describe the oxidation reaction where the rate is controlled by dual-lattice and grain-boundary diffusion, in which the effective diffusion coefficient is given by,

$$D_{\text{eff}} = D_l + 2(D'\delta)/g \qquad (7.10)$$

where g is the grain size normal to the growth direction, δ is the grain boundary width, and D' and D_l are the diffusion coefficients of the boundary

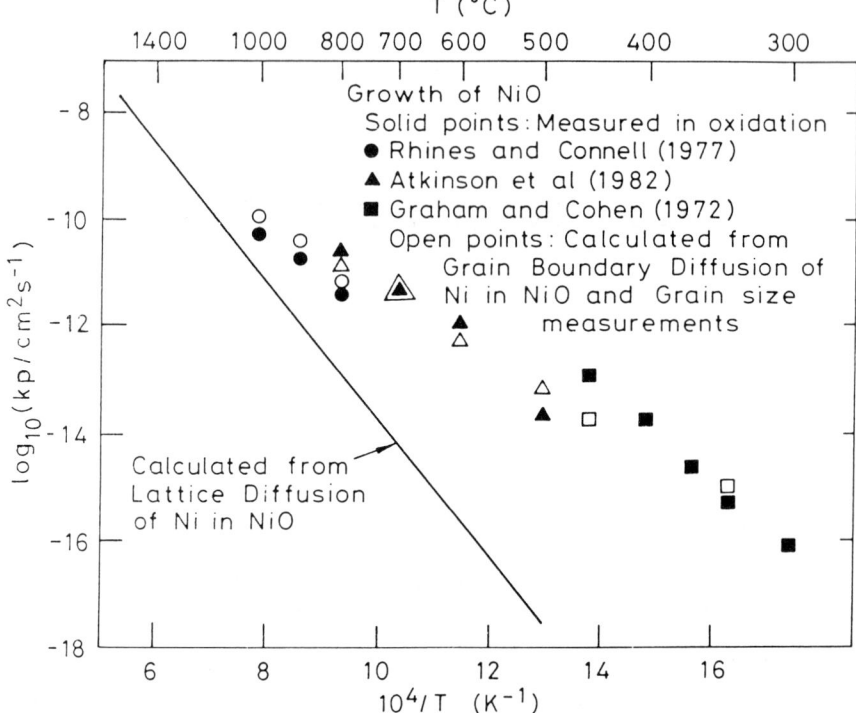

Fig. 7.35 Arrhenius plot of the parabolic rate constant for the oxidation of Ni to NiO (after Atkinson[1])

and the lattice, respectively. This model has been used by Atkinson[1] to calculate parabolic rate constants for nickel where grain boundary diffusion dominates, and there was good agreement between the calculated values and those obtained experimentally, as shown in Fig. 7.35.

A single layer of nickel oxide forms during the early stages of growth of the oxide, but as the layer thickens a duplex structure develops; this consists of an inner region of equiaxed, fine-grained crystallites and an outer region of large columnar crystals. The inner layer is generally more pronounced on less pure material, and is believed to be due to the presence of impurities segregating to the grain boundaries thereby inhibiting grain growth. Tracer diffusion studies have shown that the outer layer grows by movement of nickel vacancies along the grain boundaries, and the inner layer by *molecular oxygen* penetration along microcracks and fissures which are present in the outer layer due to the build-up of stress in that layer. In the case of nickel oxide, compressive stresses result because of the constraints imposed on the oxide layer by the receding metal.

Dilute Nickel Alloys

The resistance of nickel to oxidation may be modified considerably by alloying, although the rate of oxidation still in general obeys a parabolic rate

NICKEL AND ITS ALLOYS

law, the rate constant increasing exponentially with temperature. In general the rate constant increases linearly from the value for nickel with increasing additions of a second element, but above a given level, which depends on the solute, the change becomes slower and for some elements the rate constant then decreases. Results obtained by Horn[2] are illustrated in Fig. 7.36, which shows that beryllium, silicon and chromium in particular can pro-

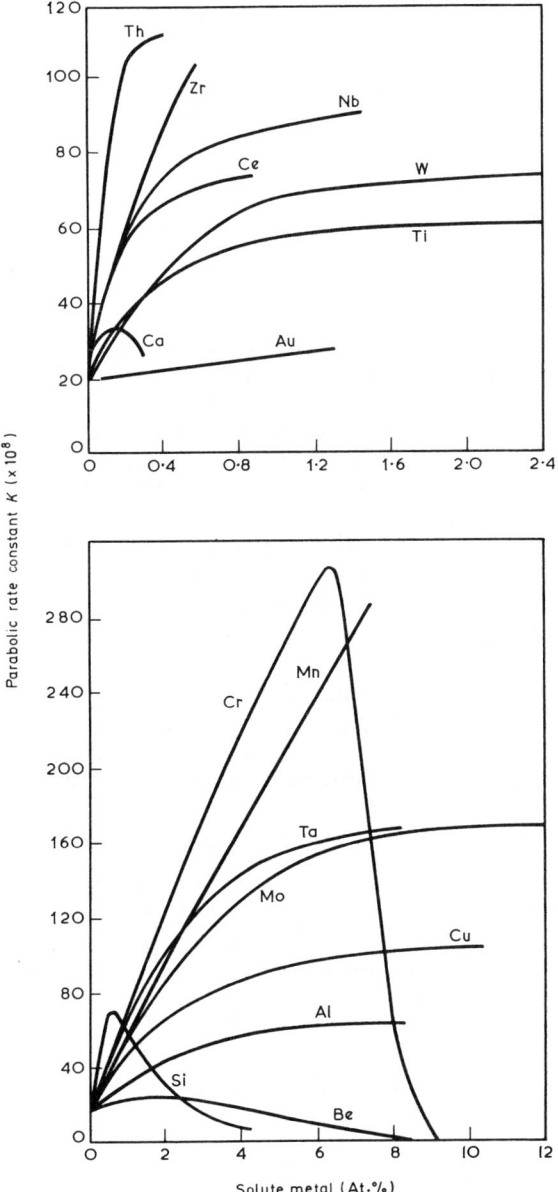

Fig. 7.36 Effect of alloying on the rate constant for oxidation of nickel at 900°C[2]

Table 7.14 Relative specific scaling rate constant for binary nickel alloys at 900°C[2]

Element	$K_{sp}* \times 10^8$
Be	17
Ca	216
Al	15·8
Si	102
Ti	78·7
Zr	204
Ce	275
Th	383
Cr	48·2
Mo	26·5
W	66·5
Mn	35·5
Cu	24·4
Au	7·3

*$K_{sp} = (K_{solute} - K_{nickel})$/atomic % solute, where K_{nickel} is approximately 20×10^{-8}.

duce enhanced oxidation resistance. From the linear portions of the rate constant/concentration curves, Horn calculated the difference in rate constant from that of nickel produced by 1 atomic % of each of the solutes investigated and obtained the values given in Table 7.14. The concentration at which the curves deviate from linearity is associated with that at which the oxide of the solute begins to form a complex oxide with NiO, e.g. $NiCr_2O_4$, or a discrete second phase. At the lower concentrations the effect had been attributed to lattice distortion due to the solute ion; but it is now generally believed to be associated with either Wagner Hauffe doping of the lattice (Section 1.8), where there is significant solubility in the oxide lattice, or with modification of the grain boundary structure where segregation is an important effect.

It would appear that the effects of impurities at the grain boundary must be either (a) to increase the diffusion rates or (b) to influence the microstructure and increase the number of short-circuit paths. However, theoretical modelling of the grain boundary structure by Duffy and Tasker[3] and

Table 7.15 'Alloying factor of oxidation' for nickel alloys*

Element	F† for stated addition (Wt. %)					
	5	10	20	30	40	50
Chromium	4	$1\frac{1}{2}$	$\frac{1}{3}$	$\frac{1}{4}$	$\frac{1}{6}$	$\frac{1}{5}$
Manganese	3	4	—	—	—	—
Tantalum	3	2	$\frac{1}{2}$	$\frac{1}{2}$	$\frac{1}{2}$	1
Molybdenum	$2\frac{1}{2}$	3	—	—	—	—
Copper	$1\frac{1}{2}$	$1\frac{1}{2}$	2	$2\frac{1}{2}$	9	—
Niobium	1	$\frac{1}{3}$	$\frac{1}{3}$	1	—	—
Platinum	1	—	—	$\frac{3}{4}$	—	$\frac{1}{2}$

*Derived from curves of Reference 5.
† Approximate oxidation rate of alloy relative to that of pure nickel.

*In the tables and figures $K \times 10^x$ indicates that the actual values given are $K \times 10^{-x}$.

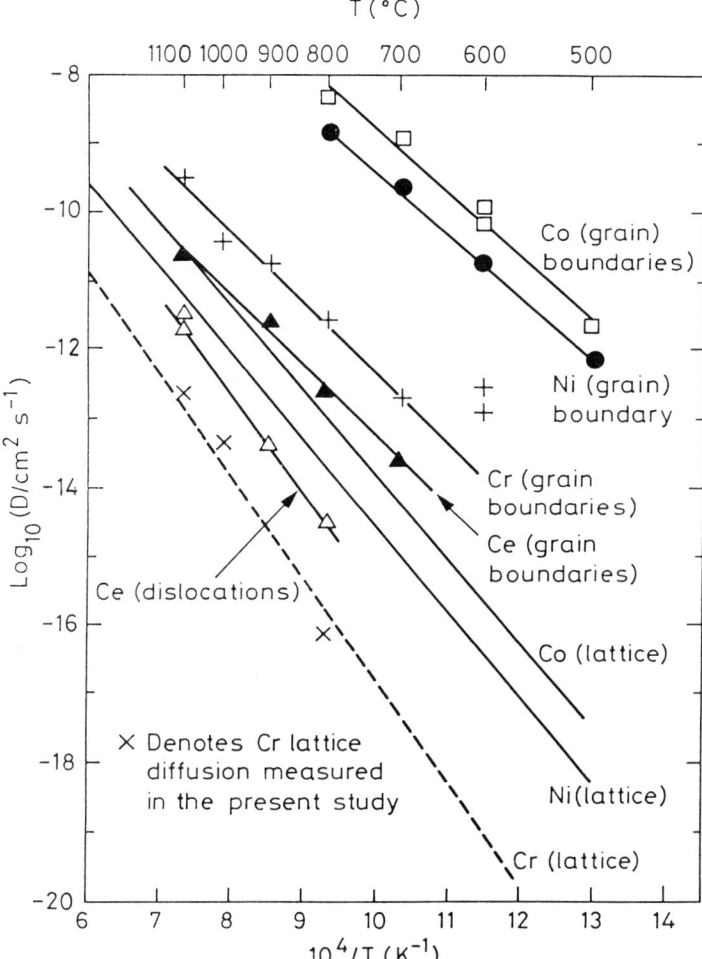

Fig. 7.37 Diffusion coefficients for some impurities in NiO grain boundaries compared with the corresponding lattice diffusivities (the grain boundary width is assumed to be 1 nm) (after Atkinson and Taylor[4])

experimental measurements of grain boundary diffusion rates[4] indicate that impurities often decrease grain boundary diffusion rates (Fig. 7.37). Thus it would appear that the effects of impurities in increasing the oxidation rate of nickel most probably result from a reduction in the oxide grain size with a consequent increase in the number of short-circuit diffusion paths.

A semi-quantitative indication of the effects of different elements on the resistance to oxidation of nickel is given in Table 7.15 which lists values for the relative oxidation rate with respect to that of nickel for different concentrations of solute element. These values are approximately valid over quite wide ranges of time and temperature[5].

Nickel-Chromium Alloys

Isothermal Oxidation The alloys based on the nickel-chromium system are of paramount importance in the field of high-temperature alloys. As shown in Tables 7.16 and 7.17, addition of chromium to nickel has a complex effect on the oxidation behaviour; small additions are deleterious, the isothermal oxidation rate increasing with chromium content to a maximum at about 7% chromium. With less than about 9% Cr, internal oxidation occurs and the chromium content of the matrix is sufficiently reduced for the alloys to appear magnetic (Table 7.17), although before oxidation only alloys with less than 7% Cr are magnetic. A progressive improvement in oxidation resistance results from further additions of chromium up to a chromium level reported variously as about 20% and 40-90%, depending on temperature, and these alloys are more resistant than either of the constituent metals although pure crack-free chromium gives oxidation values very similar to those for commercial Ni-20Cr alloys[8].

Barrett and his colleagues[9], and Kosak[10] have summarised existing information on the scales formed on nickel-chromium alloys. Up to about 10% Cr, the thick black scale is composed of a double layer, the outer layer being nickel oxide and the inner porous layer a mixture of nickel oxide with small amounts of the spinel $NiO \cdot Cr_2O_3$. Internal oxidation causes the formation of a subscale consisting of chromium oxide particles embedded in the nickel-rich matrix. At 10-20% Cr the scale is thinner and grey coloured and consists of chromium oxide and spinel with the possible presence of some nickel oxide. At about 25-30% Cr a predominantly chromium oxide scale is

Table 7.16 Oxidation data for nickel-chromium alloys[6]

Chromium content (%)	Temperature (°C)	pO_2 (atm)	Rate constant (K_p) ($g^2 m^{-4} s^{-1} \times 10^2$)
0·0	900	air	0·28
1·97	900	air	4·9
4·12	900	air	5·8
5·89	900	air	8·2
8·0	900	air	0·0
0·0	1 000	1	3·48
0·3	1 000	1	15·0
1·0	1 000	1	25·8
3·0	1 000	1	28·3
10·0	1 000	1	5·55
0·0	1 096	1	5·48
0·32	1 096	1	23·6
0·92	1 096	1	29·7
2·0	1 096	1	39·6
3·45	1 096	1	46·8
5·67	1 096	1	58·5
7·64	1 096	1	67·8
8·71	1 096	1	30·8
11·1	1 096	1	3·79
14·9	1 096	1	0·35
20·0	1 096	1	0·07

Table 7.17 Results of oxidation tests on nickel–chromium alloys[7]

Material composition (%)		4 days at 1 038 °C			Magnetic response†	8 days at 954 °C	
Ni	Cr	Weight gain (gm^{-2}s$^{-1} \times 10^4$)	Oxide thickness (mm) Outer scale	Inner zone		Mass gain (gm^{-2}s$^{-1} \times 10^4$)	Maximum penetration (mm)
100	—	6·6	0·127	0.0	M	—	—
97	3	10	0·127	0·076	M	—	—
96	4	—	—	—	—	0·61	0·056
94	6	8	0·254	0.076	M	—	—
92	8	—	—	—	—	0·50	0·061
91	9	3·6	0·152*	0·076*	M	—	—
91	9	0·64	0·178*	0·127*	W	—	—
88	12	—	Thin	Nil	N	—	—
86	14	—	—	—	—	0·03	<0·025
85	15	0·28	Thin	Nil	N	—	—
82	18	0·22	Thin	Nil	N	—	—
80	20	—	—	—	—	0·01	<0·025

* Irregular attack; † and M, magnetic; W, weakly magnetic; N, non-magnetic.

observed and Barrett associates optimum scaling resistance with the minimum chromium concentration necessary for the exclusive formation of this oxide rather than the spinel. Other workers have reported results very similar to those described above. Some workers[11] have also reported the presence of considerable amounts of nickel oxide in the scale on an 80Ni–20Cr alloy after oxidation for relatively short periods (100 h) in air. The nickel oxide results from the initial transient oxidation stage where both nickel and chromium oxide crystals nucleate simultaneously and a period of time is required to establish a complete healing layer of Cr_2O_3[12]. However, Pfeiffer[13] considers that the spinel formed initially is reduced to chronic oxide by a displacement reaction, while Pfeiffer[14] and Douglas and Armijo[15], have associated the protective effect with the presence of the spinel constituent. The varying reports on the exact nature of the scale formed can be explained by the marked dependence of the oxidation behaviour on experimental conditions.

The binary Ni–20Cr alloys find their main application in the resistance heater field for both domestic and industrial purposes. Ternary additions, e.g. silicon or manganese at a level of the 3 wt%, can further improve the good oxidation resistance of these materials due to the formation of layers of silica or of manganese spinel, respectively. Wei and Stott[16] showed that addition of 1% Al to Ni–29Cr prevented the formation of $NiCr_2O_4$ during the transient oxidation stage, probably because of the more rapid nucleation of Cr_2O_3. It is suggested that this is due to the presence of an oxygen acceptor (aluminium) which reduces diffusion of oxygen into the matrix and thereby promotes the establishment of the Cr_2O_3 layer. However, there is some doubt about the validity of this mechanism, and an alternative possibility is that the additional nuclei of alumina are suitable sites for the nucleation and growth of the scale. Stott and Wood[17] found that addition of 4% Al to Ni–29Cr was sufficient to promote exclusive formation of alumina.

Other more complex alloys based on the nickel–chromium system are the high-temperature creep-resistant or super-alloys, whose rapid development was closely associated with that of the aircraft gas turbine. The nickel–chromium base was selected originally on the grounds of good oxidation resistance at high temperatures, but subsequent development has been directed largely towards enhanced high-temperature strength. The main strengthening elements used almost universally are titanium and aluminium and although Fig. 7.36 shows that small additions of either of these metals reduce the oxidation resistance of nickel, the effect particularly of aluminium in nickel–chromium alloys is beneficial[18]. Whilst the basic constituents of the scale on the nickel–chromium alloys with small ternary additions are generally unchanged, the compositional changes near the surfaces of such alloys during oxidation indicate that aluminium and titanium must be involved in the process[19]. The presence of alumina has been detected in the oxide formed on Udimet 500, a Ni–19Cr–3Al alloy and alumina together with Ni–Al spinel and chromium oxide on Udimet 700 (Ni–15Cr–4Al)[20].

Other compositional modifications of the nickel–chromium base to promote high-temperature strength have involved additions of many elements, notably carbon, cobalt, molybdenum and/or tungsten, and niobium. Each of these has its own effect on oxidation resistance, and although relevant ternary alloys have been investigated, the variations of composition are now so numerous that measurements of oxidation resistance have in general been made on selected established compositions (Tables 7.18 and 7.19), and no systematic attempt has been made to follow the detailed interaction effects of the separate elements. A useful summary of results published on various nickel-base superalloys is included in a paper by Pettit and Meier[24]. They draw attention to the generally excellent resistance to oxidation of such alloys, but again emphasise the beneficial role of aluminium in enhancing this property. Figure 7.38 is a diagram showing the type of scale that would form on a number of nickel-base alloys as a function of composition when exposed in air at 1 100°C (see Table 7.20 for alloy composition).

Indeed, for operation at temperatures above 1 000°C (now required of advanced aircraft turbine blading materials) it appears that reliance must be placed on alumina as the protective layer, partly because in high velocity oxidising gas streams volatility of CrO_3 leads to appreciable loss of scale by the oxidation of the Cr_2O_3 layer, and partly because alumina is inherently a much better diffusion barrier.

Particular mention should also be made of molybdenum which is a constituent, usually up to 5%, of a large proportion of the materials intended for service at the very highest temperatures. The extremely poor oxidation resistance of molybdenum itself and of its alloys is well known, and is attributed to the volatility of the oxide MoO_3 which prevents the formation of a stable protective scale. Table 7.15 indicates a deleterious effect of molybdenum on the oxidation resistance of nickel, although Preece and Lucas[25] report only a slight increase in oxidation rate caused by addition of molybdenum to nickel, and state that no volatilisation of MoO_3 was noticed, probably because of the moving atmosphere used. Whilst for the commercial superalloys containing about 5% Mo acceptable oxidation resistance can be maintained at temperatures within the normal operating range, a deleterious effect of such additions is noticed during heat treatment at temperatures above about 1 100°C.

Table 7.18 Oxidation data for nickel-base high-temperature alloys[21]

Alloy	Composition (%)					Rate constant K ($g^2 m^{-4} s^{-1}$)						
	Ni	Cr	Mo	Al	Others	816°C	871°C	927°C	928°C	1093°C	1149°C	1204°C
Al-modified Nichrome V	76	19	—	4	—	—	—	—	0.97–2.5	3.1–3.8	6.2–9.2	22
Nichrome V	80	20	—	—	—	—	—	—	8.9–25.3	82–83	161–184	36
Nb-modified Nichrome V	79	20	—	—	1 Nb	—	—	—	4.7–5.6	66–80	89–124	378–417
Inconel 702	Balance	15	—	3	0.5Ti	0.83–0.97	1.8–3.6	2.8–4.4	16.9–19.2	—	—	—
Hastelloy 235	Balance	16	7	1.3	1.9Ti	3.6–7.2	11–12	37–50	132–179	—	—	—
Hastelloy W	Balance	5	20	—	—	0.27	—	—	1135–1253	—	—	—
Inconel X	Balance	15	—	1	7Fe -2.4Ti -1Nb	0.69–0.72	1.1–3.9	—	—	—	—	—
Inconel	Balance	15	—	—	7Fe	0.15–0.24	1.1–2.2	—	—	—	—	—

Table 7.19 Oxidation data for commercial nickel-chromium-base alloys

Alloy	Composition (%)							Weight change (g/m^2)						
								Loss after 100h at temperature, and descaling[22]				Gain in 60h micro balance[23]		
	Ni	Cr	Co	Mo	Fe	Ti	Al	800°C	900°C	950°C	1000°C	1100°C	800°C	950°C	1100°C
Nimonic 75 NC20T	Balance	20	—	—	—	0.4	—	5.5	11.8	40.0	66.6	89.2			
Nimonic 80A NC20TA	Balance	20	—	—	—	2.4	1.2	6.4	26.2	39.6	59.6	112.6		11	20
Nimonic 90 NCK20TA	Balance	20	16	—	—	2.4	1.2	4.6	25.2	55.0	104	112.3	3.8	13	40
Nimonic 100 NCKD20TA	Balance	11	20	5	—	1.5	5.0	0.8	2.7	8.7	16.1	116.1		6	60

Note: 80 g/m^2 corresponds to a depth of about 0.01 mm assuming uniform metal loss over the whole specimen surface.

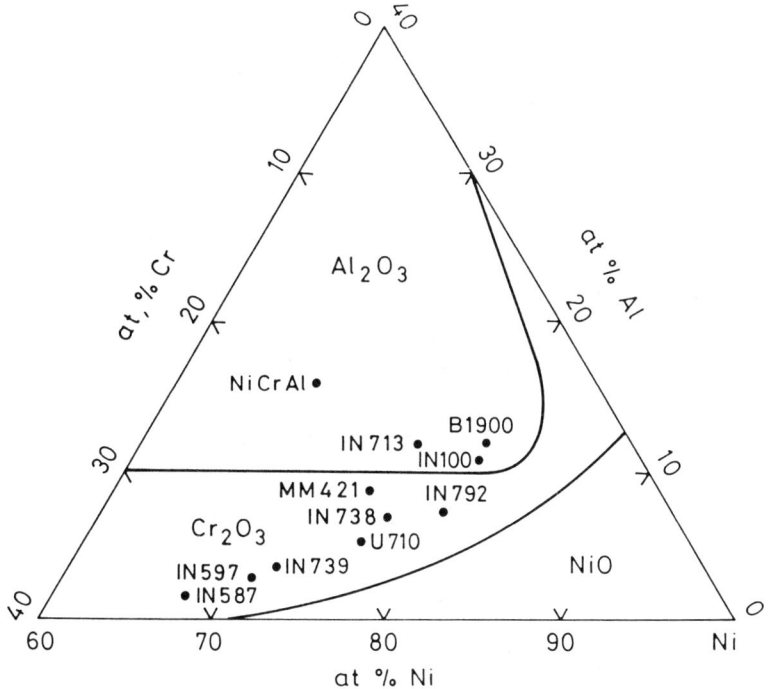

Fig. 7.38 Diagram showing the type of oxide scales formed as continuous layers upon a number of superalloys at temperatures of about 1 100°C (after Pettit and Meier[24])

Table 7.18 shows that the rate constants at 928°C for the Hastelloy alloys are considerably higher than those for molybdenum-free compositions, although the very low chromium content of Hastelloy W is doubtless a significant factor in this connection. It is noteworthy that the molybdenum-containing low-chromium alloy listed in Table 7.19 is generally superior to the others but this high resistance to oxidation is associated with its relatively high aluminium content.

Niobium appears to have a slightly beneficial effect on the oxidation resistance of nickel–chromium-base alloys, although at 1 200°C the addition of niobium to Nichrome (Table 7.18) produces a much higher rate constant. Titanium on the other hand appears to have a slightly deleterious effect on oxidation resistance at low temperatures.

Cyclic Oxidation In many industrial applications it is particularly important for the component to be resistant to thermal shock; for example, resistance-heating wires or blading for gas turbines. Chromia, and especially alumina, scales that form on nickel-base alloys are prone to spalling when thermally cycled as a result of the stress build-up arising from the mismatch in the thermal expansion coefficients of the oxide and the alloy as well as that derived from the growth process. A very useful compilation of data on the cyclic oxidation of about 40 superalloys in the temperature range 1 000–1 150°C has been made by Barrett et al.[26].

Table 7.20 Nominal compositions[a] of some alloys discussed in this chapter

Alloy	C	Mn	Si	Cr	Ni	Co	Mo	W	Nb	Ti	Al	B	Zr	Fe	Other
B-1900	0.1	0.2	0.25	8	Bal	10	6	0.1	0.1	1.0	6	0.105	0.08	0.35	4.3Ta
IN 738	0.17	0.2	0.3	16	Bal	8.5	1.75	2.6	0.9	3.4	3.4	0.01	0.10	0.5	1.75Ta
Mar M 200	0.15	—	—	9	Bal	10	—	12.5	1.0	2.0	5.0	0.015	0.05	—	—
IN 100	0.18	—	—	10.0	Bal	15	3.0	—	—	4.7	5.5	0.014	0.06	—	1.0V
IN 713	0.12	—	—	12.5	Bal	—	4.2	—	2.0	0.8	6.1	0.012	0.1	—	—
IN 792	0.21	—	—	12.7	Bal	9	2	3.9	—	4.2	3.2	0.02	0.05	—	3.9Ta
IN 597	0.05	—	—	24.5	Bal	20.0	1.5	—	1.0	3.0	1.5	0.012	0.05	—	0.02Mg
IN 587	0.05	—	—	28.5	Bal	20.0	—	—	0.7	2.3	1.2	0.003	—	—	—
U 710	0.07	0.1	0.2	18	Bal	15	3	1.5	—	5.0	2.5	0.02	0.05	0.5	—
MM 421	0.15	0.2	0.2	15.5	Bal	10	1.75	3.5	1.75	1.75	4.25	0.015	—	1.0	—
Hast X	0.15	1.0	1.0	21.8	Bal	2.5	9.0	0.6	—	—	—	—	—	18.5	—
TD NiCr	0.1	—	—	21	Bal	—	—	—	—	—	—	—	—	—	2.7ThO$_2$
Cabot Alloy 214	0.04	—	—	16	Bal	—	—	—	—	—	4.5	—	—	4.0	0.1Y
Cabot Alloy 600	0.08	0.8	0.5	21	32.5	—	—	—	—	0.4	0.4	—	—	46	0.4Cu
310SS	0.08	2.0	1.5	25	21.5	—	—	—	—	—	—	—	—	Bal	—
Incoloy 800	0.08	1.0	0.5	16	Bal	—	—	—	—	0.3	0.35	—	—	8	0.5Cu

[a] All compositions are given in weight percent

Small additions of 'active elements' (i.e. elements with a high affinity toward oxygen) and notably the rare earths are known to be very effective in promoting the formation of an adherent oxide layer that is resistant to thermal cycles. The active element can be added in elemental form or as an oxide dispersoid. In the latter case a novel series of alloys produced by mechanical alloying, the so-called oxide dispersion strengthened (ODS) materials that have been developed primarily for enhanced high temperature strength, also show good oxidation resistance[27]. The alloy MA6000 (Ni–15Cr–4·5Al–4W–2·5Ti–2Mo–2Ta–1·1Y_2O_3) is now being used in some gas turbine applications. Early versions of this type of material, e.g. thoria dispersed (TD) alloys, were evaluated some time ago and Stringer et al.[28] reported that TD–Ni20Cr had excellent resistance to spalling and reduced oxidation rates compared with the simple binary alloy.

As well as improving adhesion of the scale, the active element addition also reduced the growth rate and the concentration of chromium or aluminium required for preferential formation of the scale, particularly for the chromia-forming alloys. Whittle and Stringer[29] reviewed the various theories that have been proposed to account for this effect; these include enhanced scale plasticity, formation of a graded seal, modification to the oxide growth process, stronger chemical bonding at the interface, elimination of voids by inert oxide particles acting as vacancy sinks, and oxide protrusions into the alloy which act as 'pegs' to improve adhesion. More recently, Luthra and Briant[30], Smeggil et al.[31] and Lees[32] have proposed that segregation of sulphur to the scale/alloy interface is responsible for the poor adhesion of the oxide, and that the effect of the 'active element' is to scavenge the sulphur present in the alloy and so restore the intrinsically strong bond between the oxide and the substrate. Luthra and Briant[33] have been unable to confirm this effect experimentally, but both Funkenbusch et al.[34] and Smialek[35] have reported results that indicate that, for high purity alloys with sulphur contents of less than about 10 ppm, adherent oxide scales were formed on alumina-forming nickel-base alloys without rare earth alloying additions.

A lively debate continues on this topic and the protagonists of the various mechanisms discussed their respective positions in a recent issue of *Oxidation of Metals*[36]. The effect of active elements was also thoroughly reviewed at a recent conference[37] where it was suggested that, for the alumina-forming alloys, the sulphur effect was important as was the effect on the plasticity of the scale. For chromia-forming alloys, however, improved adhesion was claimed to result from modification to the oxide growth process by the action of the rare earth elements segregating to the oxide grain boundary and thereby altering diffusion processes. It was suggested that in the latter case complete blocking of grain boundary diffusion might occur so that lattice diffusion could then control the overall growth rate. It is clear, however, that no one theory can satisfactorily explain all the experimental observations.

Intergranular Oxidation Intergranular penetration of oxide can be a serious problem particularly when thin-walled components are used in load-bearing applications such as cooled turbine blades. Mass change data often do not adequately reflect the extent of this type of attack, and examination of polished cross-sections is required. This latter method is not straightforward since frequently the extent of attack is quite variable around the circumference of the specimen or component. Nicholls and Hancock[38] have advocated the use of statistical methods to achieve a more complete description of the corrosion process and in the case of internal penetration, where the maximum depth of attack must be determined, use of extreme value statistics increases the confidence of the measurements. The depth of intergranular penetration generally increases parabolically with time as would be expected for a process controlled by inward diffusion of oxygen and severe attack has been reported for alloys which have relatively low concentrations of alloying addition that are insufficient to promote external oxide formation.

Nickel–Iron and Nickel–Chromium–Iron Alloys

There are no significant high-temperature applications for alloys of nickel with iron. The scales formed in air consist of nickel oxide and iron oxide and the latter is usually present in the form of the spinel, $NiO \cdot Fe_2O_3$[39]. In the case of the more dilute nickel alloys, internal oxidation of nickel was observed[40]. Substitution of a substantial proportion of nickel by iron results in a deterioration in the oxidation resistance of nickel–chromium

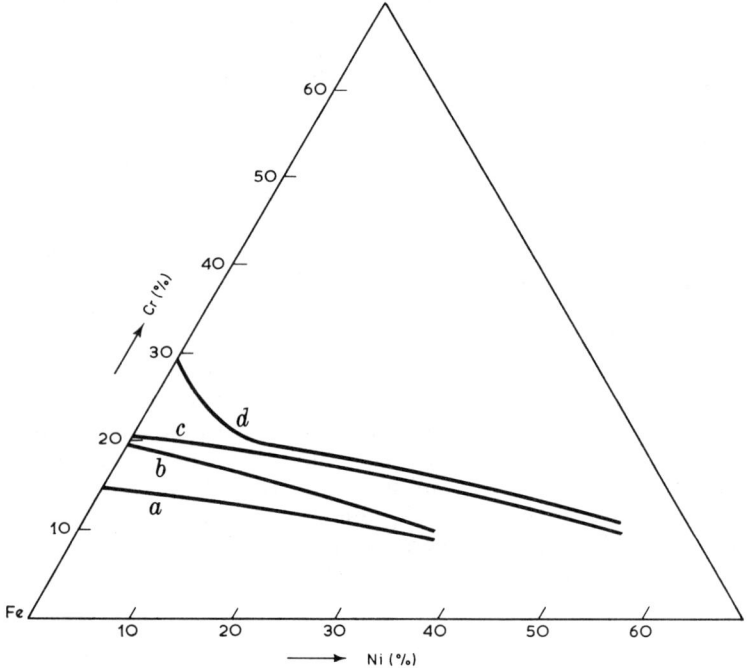

Fig. 7.39 Iso-corrosion lines for Ni–Cr–Fe alloys heated in air[41]
    a. 2·5 mm/y, 871°C   *c*. 2·5 mm/y, 1 093°C
    b. 2·5 mm/y, 982°C   *d*. 3·8 mm/y, 1 204°C

alloys with chromium contents below about 16%. At higher chromium levels, however, the effect of iron is less significant, except perhaps at very high temperatures. Figure 7.39, which illustrates the results of investigations on a range of cast nickel–chromium–iron alloys, shows that, as iron replaces nickel, a higher chromium content is required to ensure satisfactory corrosion resistance in air.

For economic reasons, and because of their advantages in more complex atmospheres, which will be discussed later, nickel–chromium–iron alloys have been developed for many kinds of industrial heating applications. Typical compositions contain about 20 or 40% Fe with 15–20% Cr, the balance being substantially Ni. Under fluctuating heating conditions, the depletion of nickel content results in reduced resistance to scaling, particularly at higher temperatures. Cyclic heating tests on a range of materials of varying iron, nickel and chromium contents have been made by Eiselstein and Skinner[42] (Table 7.21) who concluded that the minimum nickel and chromium contents to promote adequate scaling resistance were 10Ni–14Cr at 760°C, 15Ni–18Cr at 871°C, and 35Ni–20Cr at 982°C. Trueman and Pirt[43] carried out a detailed study on the cyclic oxidation of numerous commercial austenitic steels which confirmed the pronounced beneficial effects of chromium, aluminium and silicon. Nickel was also beneficial for alloys with chromium contents greater than 20%, but contrary to Eiseltein and Skinner[42], it was found[43] that it was detrimental at chromium contents of 11–16%. Manganese and titanium were also detrimental, while additions of molybdenum, vanadium, tungsten and niobium had no apparent effect on behaviour. Table 7.22 lists the main results obtained in terms of temperatures for breakdown of protective behaviour and temperatures for a given weight gain in 42 h.

Table 7.21 Oxidation of nickel-base alloys and nickel–chromium steels in cyclic temperature tests of 100 h duration[42]
(heating cycle: 15 min at temperature, air cool for 5 min)

Material type and composition	Weight change at test temperature (%)		
	760°C	871°C	982°C
Nickel		3·6	
Ni–20Cr	−0·016	−0·016	−0·398
Nimonic 75 (Ni–20Cr–0·5Ti)		0·225	0·433
Inconel (Ni–15Cr–7Fe)	−0·225	0·0	−0·173
Ni–15Cr–22Fe	−0·066	−0·527	−0·546
Incoloy (Fe–42Ni–13Cr)		0·151	−1·22
Type 310 stainless steel (Fe–20Ni–25Cr)	−0·024	−0·283	−20·7
Type 309 stainless steel (Fe–13Ni–23Cr)	0·023	0·086	−31·5
Type 347 stainless steel (Fe–11Ni–18Cr)		−37·6	
Type 329 stainless steel (Fe–4Ni–23Cr)	0·320	−26·1	

Nickel-Aluminium Alloys

The transition from non-protective internal oxidation to the formation of a protective external alumina layer on nickel aluminium alloys at 1 000–1 300°C was studied by Hindam and Smeltzer[44]. Addition of 2% Al led to an increase in the oxidation rate compared with pure nickel, and the development of a duplex scale of aluminium-doped nickel oxide and the nickel aluminate spinel with rod-like internal oxide of alumina. During the early stages of oxidation of a 6% Al alloy somewhat irreproducible behaviour was observed while the α-alumina layer developed by the coalescence of the rod-like internal precipitates and lateral diffusion of aluminium. At a lower temperature (800°C) Stott and Wood[45] observed that the rate of oxidation was reduced by the addition of 0·5–4% Al which they attributed to the blocking action of internal precipitates accumulating at the scale/alloy interface. At higher temperatures up to 1 200°C, however, an increase in the oxidation rate was observed due to aluminium doping of the nickel oxide and the inability to establish a healing layer of alumina.

The discovery of ductile intermetallics, by addition of small amounts of boron for example, combined with their good oxidation resistance, has stimulated interest in their use as structural materials, and the development of Ni_3Al and $NiAl$ is now being actively pursued[46]. These intermetallics have been used for many years as coatings for gas turbine blades to improve oxidation resistance[47]. Despite the technical importance of the system there have been relatively few detailed studies of its oxidation behaviour. However Hindam and Smeltzer[48] investigated the formation of α-alumina on β-NiAl, which was shown to grow initially by inward diffusion of oxygen, but in the fully developed layer counter-current diffusion of aluminium and oxygen in the grain boundaries and aluminium diffusion in the oxide lattice was observed; the oxide contained 0·5% Ni in solution.

Nickel-Copper Alloys

The literature on the oxidation of nickel–copper alloys is not extensive and emphasis tends to be placed on the copper-rich materials. The nickel-rich alloys oxidise according to a parabolic law and at a rate similar to that for nickel; Corronil (Ni-30Cu) exhibited a parabolic rate behaviour below 850°C but a more complex behaviour involving two parabolic stages above 900°C. Electron diffraction examination of the oxide films formed on a range of nickel–copper alloys showed the structures of the films to be the same as for the bulk oxides of the component metals and on all the alloys examined only copper oxide was formed below 500°C and only nickel oxide above 700°C[49].

Sulphidation

Pure Nickel

The behaviour of nickel is very dependent upon the type of sulphidising gas that is used, and the effects of sulphur vapour, hydrogen sulphide,

Table 7.22 Commercial austenitic stainless steels. Analysis, breakdown temperature range and temperatures for scaling indices 1, 5, and 10[43]

Steel No.	C	Si	Mn	S	P	Cr	Ni	Mo	Ti	Others	Specn. no.	Breakdown temperature range (°C)	SI_1[a] (°C)	SI_5[b] (°C)	SI_{10}[c] (°C)
41	0.05	0.51	1.04	0.021	0.025	18.68	10.03				304S15	850–900	860	880	915
42	0.07	0.76	0.80	0.008	0.020	18.25	8.95		0.43		321S12	800–850	820	860	890
43	0.05	0.40	0.76	0.013	0.017	18.28	9.95			0.77Nb	347S17	800–850	820	860	890
44	0.07	0.49	1.57	0.008	0.030	18.65	9.42	1.25				800–850	800	810	820
45	0.05	0.19	1.69			16.38	10.50	2.52				700–800	700	820	845
46	0.05	0.43	1.56	0.015	0.024	17.47	11.41	2.68			316S16	800–850	850	820	875
47	0.07	0.51	1.50	0.284	0.028	18.00	9.10	0.33			303S21	800–850	950	960	970
48	0.09	0.33	1.54	0.267	0.022	17.40	9.60	0.40	0.70		325S21	950–1000	860	900	955
49	0.03	0.59	2.85	0.014	0.014	18.90	10.50			$0.196N_2$		850–900	800	830	860
50	0.03	0.40	2.85	0.017	0.015	18.40	12.10			$0.192N_2$		800–850	810	870	910
51	0.06	0.34	0.75	0.021	0.031	12.55	12.52	2.7				650–700	660	710	755
52	0.06	0.41	8.65	0.013	0.010	17.70	4.23			$0.165N_2$	284S19	650–700	660	740	800
53	0.09	0.27	1.09	0.005	0.017	16.82	11.58	1.31		0.90Nb		700–750	700	810	870
54	0.17	0.53	0.98	0.005	0.018	14.20	9.80	2.18	0.78	2.50Cu		700–750	750	775	810
55	0.17	0.40	0.96			13.85	16.98	2.33	0.63	2.53Cu		800–850	845	910	970
56	0.11	0.47	1.17	0.010	0.020	22.00	14.38					1050–1100	980	1060	1070
57	0.07	0.48	1.49	0.006	0.021	25.20	20.24				310S24	1050–1100	1030	1090	1180
58	0.05	0.34	0.63	0.004	0.017	23.42	17.90		0.43			1000–1050	960	1070	1110
59	0.09	0.55	0.80	0.016	0.027	23.03	17.28		0.71			1000–1050	940	1030	1075
60*	0.15	1.04	0.96			23.68	10.56			0.72W		1000–1050	1005	1060	1075
61*	0.18	0.93	0.76			22.82	11.24			2.8W		1000–1050	1005	1050	1050
62	0.29	1.12	0.88	0.055	0.021	20.42	7.83			2.16W		850–900	860	955	950
63	0.52	0.10	3.94	0.006	0.013	20.50	3.94			$0.432N_2$	349S52	700–750	740	840	890
64	0.045	0.60	1.12	0.003	0.021	20.32	33.50		0.25	0.33Al		1000–1050	970	1010	1140
65*	0.16	2.16	0.82	0.017	0.020	18.62	29.78		0.24	7.60Co		1000–1050	1005	1085	1130
66*	0.35	0.48	1.20	0.018	0.019	18.95	38.88					950–1000	995	1115	1135

[a] SI_1 = temperature in °C for mass gain of 1 mg cm^{-2} in 42 h
[b] SI_5 = temperature in °C for mass gain of 5 mg cm^{-2} in 42 h
[c] SI_{10} = temperature in °C for mass gain of 10 mg cm^{-2} in 42 h
* Casting

sulphur dioxide and sulphur trioxide, will be considered in separate sections.

Sulphur Vapour/Hydrogen Sulphide The rapid attack due to sulphidation of metals forming low-melting-point sulphides has been referred to in Section 7.1; nickel is a classic example of such metals. The binary nickel-sulphur phase diagram shows that the sulphide Ni_3S_2 forms a eutectic with nickel which melts at 635°C, and the maximum solubility of sulphur in nickel is only about 0·005%. At temperatures well below the eutectic temperature, contamination of nickel by sulphur leads to serious intercrystalline attack and resultant embrittlement, the rate of penetration being greatest at temperatures in the region of 550–650°C. Sulphur attack on nickel during heating is produced by even the smallest amounts of sulphur-bearing contaminant and is virulent under reducing; neutral or oxidising conditions. Since oxidation too can proceed intergranularly, a method of identifying the nature of the attack is often useful. Etching the affected material in an aqueous solution containing 20% sodium nitrate will stain sulphur-affected boundaries but leave oxidised boundaries unattacked.

The attack upon nickel by sulphur is so rapid at elevated temperatures and is produced by such small traces of contaminant that the most stringent precautions and cleaning schedules are required prior to the heating of nickel in service and during processing; storage time should be minimised and the metal protected to avoid atmospheric contamination, and handling should be with clean gloves, all possible contact with grease or oil being eliminated. Even with such precautions the use of a thorough cleaning process is to be recommended immediately before heating[50]. At low partial pressures of sulphur, intercrystalline attack is not observed either below or above the eutectic temperature of 635°C and the observed embrittlement under such conditions has been attributed to hardening of the grains by sulphur in solid solution in the matrix. At a pressure of about 130 Pa of sulphur vapour, however, sulphides are readily observed in the grain boundaries.

Pfeiffer[51] contends that, in undeoxidised nickel at least, the low sulphur contents normally found are insufficient to cause grain-boundary embrittlement and that the latter is, generally, due to intergranular oxides. In the presence of sulphur-containing gases, however, the level of sulphur required,

Table 7.23 Comparison of properties of Ni, NiO and NiS[52,53]

	Sulphidation k_p ($g^2cm^{-4}s^{-1}$)	Oxidation k_p ($g^2cm^{-4}s^{-1}$)		
Ni	1.1×10^{-6} (620°C)	9.1×10^{-11} (1000°C)		
	Self-diffusion coeff ($D_m cm^2 s^{-1}$)	Deviation from stoichiometry	ΔG (1000°C) $kJ\ mol^{-1}$	mp (°C)
NiO	1.1×10^{-11} (1000°C)	0.0001 (1000°C)	−127	1984
NiS	1.4×10^{-8} (800°C)	0.080 (700°C)	−88	810 (Ni-Ni_3S_2 635)

i.e. about 0·1%, is readily attained and deposits of sulphide are observed. Rates of sulphide formation are very much greater than those found for oxides due to the more defective sulphide structure. Mrowec[52] and Kofstad[53] have compared the relevant properties of nickel sulphide and oxide (see Table 7.23). The mobile species in nickel sulphide is believed to be the doubly-charged nickel vacancy, V''_{Fe}, and using classical point defect theory it can be shown that the parabolic rate constant, k_p, should be proportional to $p(S_2)^{1/2}$. However, the observed dependence is $p(S_2)^{1/30}$. Mrowec[52] has suggested that, because of the high defect concentration, extended defects form which invalidate assumptions based on the classical theory of point defects. At temperatures between 480°C and 620°C Tideswell[54] using X-ray and electron diffractometry, identified a thick inner layer of Ni_3S_2 with a thin outer layer of NiS. It has been suggested[55] that differences in behaviour were observed when nickel was exposed in H_2/H_2S mixtures or in sulphur vapour, but it is now realised that similar rates of attack are found in both cases[56].

Sulphur Dioxide An excellent account of the reaction of nickel with SO_2 and SO_3 ($SO_2 + O_2$) has been given by Kofstad[53] and is summarised below.

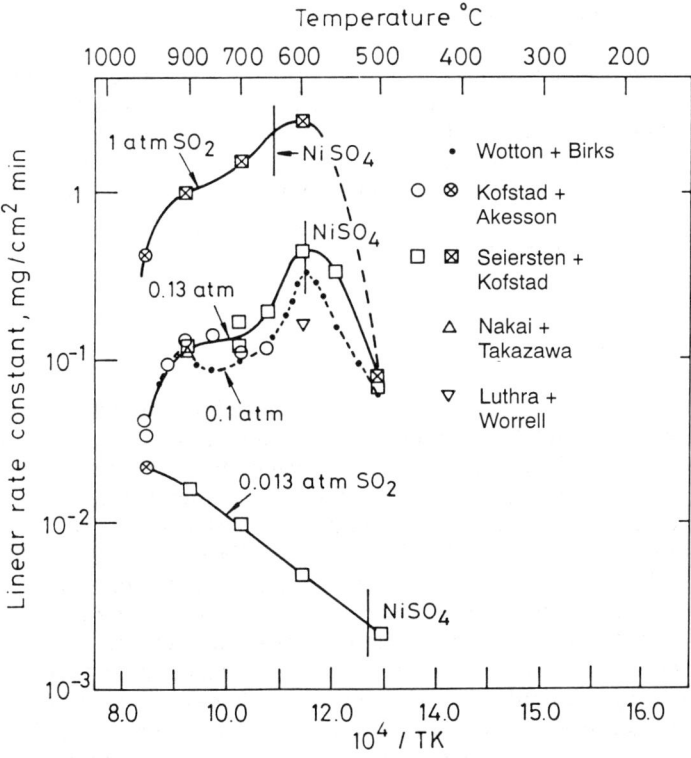

Fig. 7.40 The temperature dependence of the linear rate constant for reaction of nickel with SO_2 at temperatures in the range 500–1 000°C and at pressures from 0·013 to 1 atm (after Kofstad[53], Wotton and Birks[58], Kofstad and Akesson[59], Luthra and Worrell[60], Nakai and Takasawi[61] and Seiersten and Kofstad[62])

In sulphur dioxide linear kinetics are generally observed due to control by phase boundary reactions, i.e. adsorption of SO_2. Rahmel[57] suggested that this is one of the conditions which favours simultaneous nucleation of sulphide and oxide at the gas/scale interface. The main reaction products are NiO, Ni_3S_2, $Ni-S_{liq}$ and $NiSO_4$, depending on the temperature and gas pressure; for example, according to the following reaction:

$$7Ni + 2SO_2 \rightarrow 4NiO + Ni_3S_2$$

The sulphide usually forms an interconnected network of particles within a matrix of oxide and thus provides paths for rapid diffusion of nickel to the interface with the gas. At high temperatures, when the liquid Ni-S phase is stable, a duplex scale forms with an inner region of sulphide and an outer porous NiO layer. The temperature dependence of the reaction is complex and is a function of gas pressure as indicated in Fig. 7.40[58-62]. A strong dependence on gas pressure is observed and, at the higher partial pressures, a maximum in the rate occurs at about 600°C corresponding to the point at which $NiSO_4$ becomes unstable. Further increases in temperature lead to the exclusive formation of NiO and a large decrease in the rate of the reaction, due to the fact that Ni_3S_2 becomes unstable above about 806°C.

Sulphur Trioxide ($SO_2 + O_2$) Linear reaction rates are observed due to phase boundary control by adsorption of the reactant, SO_3. Maximum rates of reaction occur at a SO_2/O_2 ratio of 2:1 where the SO_3 partial pressure is also at a maximum. With increasing $O_2:SO_2$ ratio the kinetics change from linear to parabolic and ultimately, of course, approach the behaviour of the Ni/NiO system. At constant gas composition and pressure, the reaction also reaches a maximum with increasing temperature due to the decreasing SO_3 partial pressure with increasing temperature, so that $NiSO_4$ formation is no longer possible and the reaction rate falls.

Dilute Nickel Alloys

Dilute binary alloys of nickel with elements such as aluminium, beryllium and manganese which form more stable sulphides than does nickel, are more resistant to attack by sulphur than nickel itself. Pfeiffer[63] measured the rate of attack in sulphur vapour (13 Pa) at 620°C. Values around $0 \cdot 15 \text{g m}^{-2}\text{s}^{-1}$ were reported for Ni and Ni-0·5Fe, compared with about $0 \cdot 07-0 \cdot 1 \text{g m}^{-2}\text{s}^{-1}$ for dilute alloys with 0·05% Be, 0·5% Al or 1-5% Mn. In such alloys a parabolic rate law is obeyed; the rate-determining factor is most probably the diffusion of nickel ions, which is impeded by the formation of very thin surface layers of the more stable sulphides of the solute elements. Iron additions have little effect on the resistance to attack of nickel as both metals have similar affinities for sulphur. Alloying with other elements, of which silver is an example, produced decreased resistance to sulphur attack. In the case of dilute chromium additions Mrowec[52] reported that at low levels (<2%) rates of attack were increased, whereas at a level of 4% a reduction in the parabolic rate constant was observed. The increased rates were attributed to Wagner doping effects, while the reduction was believed to result from the formation of a heterophase duplex scale in which a thiospinel formed and reduced rates of diffusion.

Despite their improved resistance to general corrosion by sulphur, the dilute alloys with many elements are sensitive to intergranular attack and embrittlement, and, at temperatures above 635°C, to eutectic formation.

In general, greatly reduced rates of attack are observed for impure or dilute nickel alloys compared with pure nickel when exposed to $SO_2 + O_2$ atmospheres. Haflan et al.[64] have attributed this to the segregation of impurities at the sulphide/oxide interface causing breakup of the sulphide network. For example in the case of silicon additions, it has been shown that silicates form and it has been proposed that these alter the wetting characteristics of the sulphide and prevent the establishment of an interconnected sulphide network.

Nickel-Chromium Alloys

Sulphur Vapour/Hydrogen Sulphide At very low partial pressures of sulphur its great affinity for chromium results in an increased rate of attack as the chromium content of nickel–chromium alloys is raised[65], at higher sulphur potentials (e.g. 1 atm) and after longer times even small additions of the order of 1% chromium produce increased resistance to attack, while larger additions modify considerably the behaviour of nickel in sulphur-containing atmospheres. In general chromium is much less attacked than nickel, which suggests that the sulphide is, to some extent at least, protective[66,67]. Chandler and McQueen[68] have noted the higher activation energy for sulphidation of chromium (147 kJ mol^{-1}) than that for nickel (92 kJ mol^{-1}) and suggest that this reflects the protective nature of the sulphide on chromium compared to that on nickel. In nickel–chromium alloys with chromium contents upwards of 10%, the greater affinity of sulphur for chromium results in the formation of chromium rather than nickel sulphide, at the metal surface or internally. This compound, which does not form a low-melting-point eutectic, has a melting point of 1 550°C. Furthermore, while attack upon the nickel–chromium alloys is partially intergranular, the sulphides form as isolated globules rather than as continuous films. At low sulphur potentials a duplex, compact and adherent scale is formed, kinetic measurement[69] indicating that these two layers grow according to a parabolic law, the rate constant for the outer nickel-rich scale being some four times greater than that for the chromium sulphide inner layer. Whilst initially inward migration of sulphur probably occurs beneath the protective scales, subsequent growth of both layers is apparently by diffusion of metal ions outwards[70]. Variation of sulphur potential at sub-atmospheric values has little effect on the rate of attack at 700°C but considerable variation is reported for the stoichiometry of both scale layers from CrS to Cr_2S_3 and from NiS to Ni_3S_2, and Nowak et al.[71] showed that the thiospinel, $NiCr_2S_4$, could form on Ni–23Cr and Ni–33Cr in H_2S or S_2.

Mixed Oxidants (sulphidising and oxidising gas mixtures) Many industrial environments consist of gas mixtures containing both sulphurous and oxidising components so that, as shown in the previous section, there is competition between oxide and sulphide formation. In the case of Ni–Cr alloys, Cr_2O_3 can form in 'reducing' atmospheres where H_2S is stable. An example application of this process that has been extensively studied recently is coal gasification plant operating at about 900–1 000°C, where typically gas mixtures (coal gasification atmospheres – CGA) have oxygen partial pressures of about 10^{-13} Pa and sulphur partial pressures of about 10^{-2} Pa. Under these conditions iron and nickel oxides would not be stable but Cr_2O_3, Al_2O_3 and both nickel and iron sulphide are stable (see Fig. 7.41)[72]. Many workers have studied the transition between the formation of protective oxide or of non-protective sulphide, and it has been demonstrated that the transition does not correspond to the conditions predicted from purely thermodynamic considerations. Kinetic factors, such as the diffusivity of the different alloying elements and of the reactive species as well as the morphology of the scale, can determine the reaction path. Natesan[73] has termed this transition the kinetic boundary, and for alloys of nickel with about 20% Cr, the protective oxide formation is observed at a pO_2 about two to three orders of magnitude greater than would be predicted from the phase stability diagram. A 46% chromium–nickel binary alloy is reported to form a very stable chromia layer in these conditions[74].

Another factor that determines the long-term stability of the protective oxide layer is its ability to prevent sulphur penetration which would lead to the eventual formation of chromium sulphide beneath the external oxide layer. With most commercial nickel chromium alloys internal sulphidation

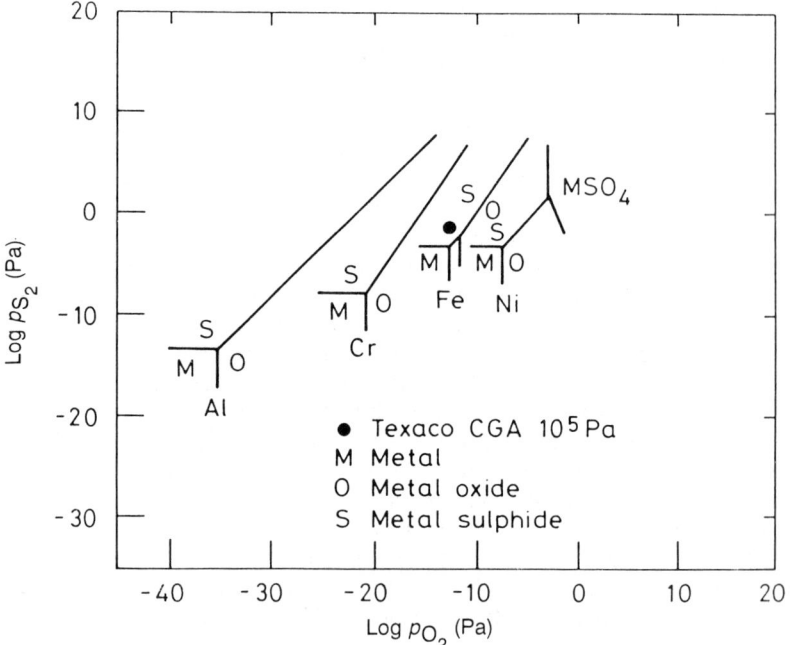

Fig. 7.41 A typical coal gasifier atmosphere at 871°C plotted in the M-O-S thermodynamic stability diagram for elements commonly present in stainless steels and high temperature alloys (after Perkins[72])

is observed and this usually heralds breakaway where rates of attack are greatly increased, with the tendency for this to occur increasing with increasing pS_2 and decreasing pO_2. It is proposed that sulphur can penetrate the scale either by lattice or grain boundary diffusion, or as molecular species through pores and microcracks in the oxide layer. The consensus is that penetration is largely as the molecular species[75]. If sulphur diffusion does occur then, as shown by Benbyamani et al.[76], grain boundary transport is the dominant process at temperatures up to 1 000°C.

Extensive studies have been carried out by Giggins and Pettit[77] and by Vasantasree and Hocking[78] on a range of nickel chromium alloys with up to 50% alloying addition. Generally the principles outlined above can be used to interpret the experimental observations, where the thermodynamics of the reaction are a major factor determining the rate of attack, depending upon whether oxide or sulphide is the stable phase.

Continued exposure of the nickel–chromium alloy to more severely sulphurising and reducing atmospheres results in local depletion of chromium to such an extent that nickel sulphide and the eutectic are formed internally. The latter constituents are not often observed in service failures, but the relative instability of nickel sulphide in the presence of chromium sulphide can result in its reduction to nickel during slow cooling on shut down. That nickel sulphide is formed is suggested by the frequent occurrence of blisters, associated with the formation of molten eutectic on the surface of sulphur-attacked specimens[67].

The detailed appearance of an attacked alloy can vary markedly, depending on the conditions of exposure. In practice the conditions are frequently intermittently oxidising and reducing. The mechanism of attack then involved appears to be that the initial formation of chromium sulphide results in a gradual impoverishment of the matrix with respect to chromium and a consequent loss of oxidation resistance. Rapid internal oxidation of the chromium remaining in the matrix then occurs, to give the intimate mixture of oxide and metallic nickel characteristic of attacked regions. The loss of chromium from the surface regions also raises the Curie temperature towards the value for nickel and attacked specimens therefore appear magnetic. Oxidation of a pre-sulphidised Ni–15Cr alloy was shown by Spengler and Viswanathan[79] to result in further penetration of the alloy by sulphur with consequent nucleation of fresh chromium sulphide particles in advance of the oxidation front, supporting the self-regeneration mechanism proposed by some earlier workers.

Although iron sulphide also forms a eutectic with the metal this melts at 988°C, and at temperatures in the region of 700 to 800°C alloys with substantial proportions of nickel replaced by iron and a chromium level maintained at about 20% show advantages over nickel–chromium-base alloys in resistance to sulphur attack.

The successful application of nickel–chromium–iron alloys as structural components of industrial furnaces and as chambers and containers in chemical processing under conditions of exposure involving sulphur substantiates their good resistance to this form of corrosion. These materials are used for service temperatures in the range 750–1 200°C, the upper limit of serviceability being determined largely by the chromium content of a particular alloy. Results of corrosion tests (Table 7.24)[80] on cast nickel-

Table 7.24 Corrosion resistance of nickel-chromium-iron alloys in oxidisng and reducing flue-gas atmospheres of varying sulphur content[80]

Material composition (%)			Metal loss (mm/y)									
			Reducing atmosphere						Oxidising atmosphere			
			982°C					1093°C	982°C			1093°C
			Sulphur content (g m^{-3})					Sulphur content (g m^{-3})	Sulphur content (g m^{-3})			Sulphur content (g m^{-3})
Ni	Fe	Cr	0	0.115	2.3	6.9	11.5	0.115	11.5	0.115	2.3	6.9	0.115	2.3
70	20	10	0.25	0.5	0.5	25	≥10	–	–	1–2	2	<0.5	<1	<0.5
70	15	15	<0.25	<0.5	0.5	10	≥10	–	–	<0.5	0.5	<0.5	–	–
70	10	20	<0.25	<0.5	<0.5	2	≥10	–	–	<0.5	<0.5	<0.5	–	–
40	50	10	1	2	15	>25	–	2	>75	3	3	2	8	1
40	45	15	0.25	0.5	2	>25	–	1	50	0.5	0.6	<0.5	1.5	–
40	40	20	<0.25	<0.5	<0.5	>25	–	0.5	2	<0.5	<0.5	–	<1	–

chromium–iron alloys in oxidising and reducing flue-gas atmospheres illustrate the beneficial effect of increased chromium content and the deleterious effects of increased temperature and sulphur content of the atmosphere. A very large testing programme on alloys for possible use in

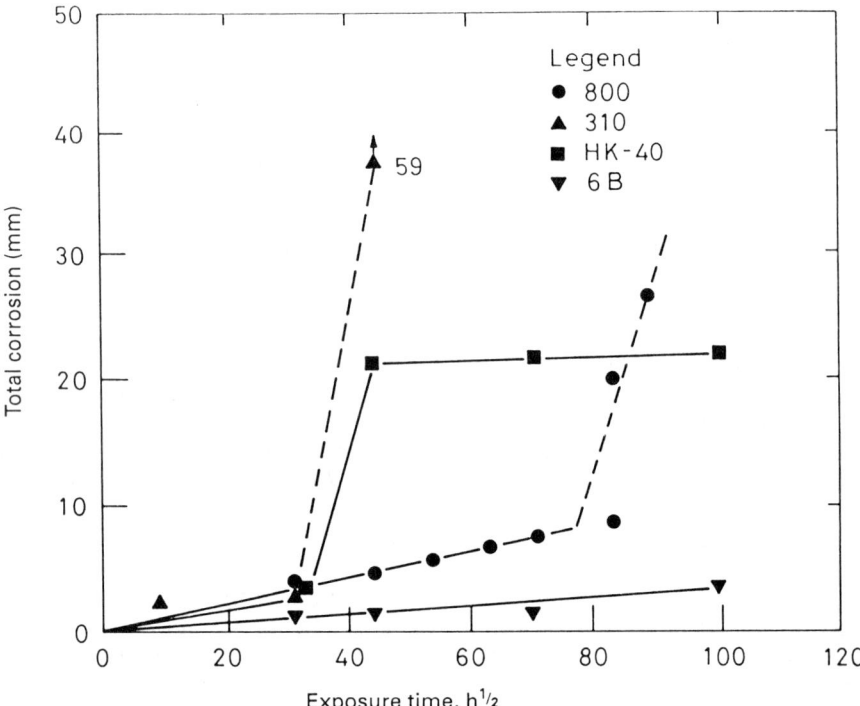

Fig. 7.42 Long-term exposure of selected high temperature alloys at 932°C in a coal gasification atmosphere containing 0·5% H$_2$S (after Hill and Meyer[81])

coal gasification plants was carried out in the U.S.A. involving exposures for periods of up to 10 000 h at temperatures up to about 1 000°C. The results showed that, while many iron chromium–nickel alloys showed protective behaviour in the early stages, breakaway was frequently initated after about 1 000 h exposure as the results shown in Fig. 7.42 illustrate[81]. At about 900°C the transition to breakaway for nickel-base alloys was relatively insensitive to the chromium content of the alloy compared with the behaviour of iron- and cobalt-base alloys, but at about 1 000°C an increase in the chromium content from about 25% to 30% resulted in a significant increase in the time to breakaway in these materials. The breakaway reaction is believed to occur due to the presence of liquid sulphide phases on top of the protective chromia scale. The chromia layer has often been described as 'leaky' since it is an imperfect barrier to diffusion of cations so that iron and nickel from the underlying alloy can diffuse to the surface and react to form the molten sulphide.

It is widely recognised that improved resistance to sulphur attack can be obtained in commercial alloys by small additions of certain elements, notably manganese, silicon and aluminium. Figure 7.43 shows how the depth of metal converted to scale, and more particularly the total penetration by sulphur, is reduced by increasing silicon content in nickel–chromium–iron

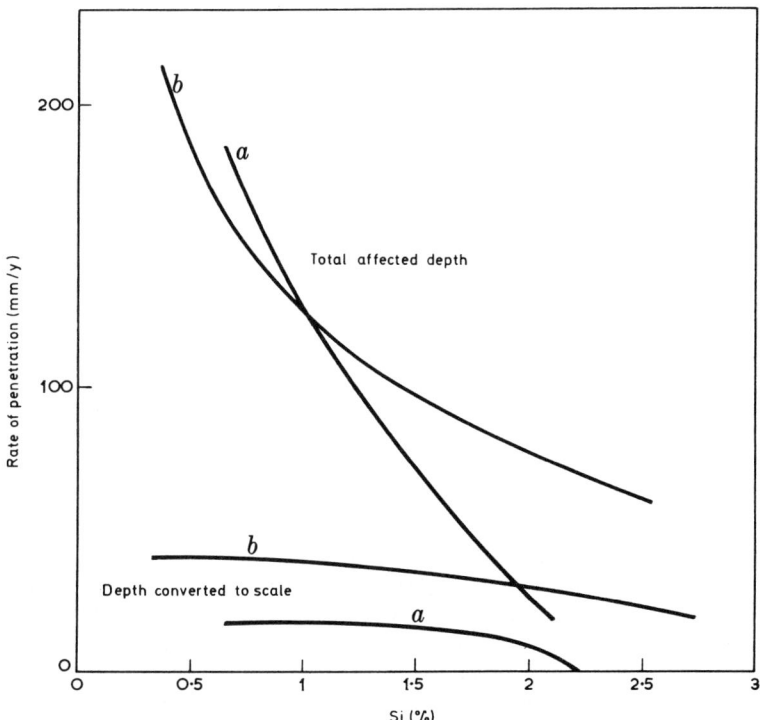

Fig. 7.43 Effect of silicon content on corrosion of Ni–Cr–Fe alloys in a reducing flue-gas atmosphere containing $6 \cdot 9$ g m^{-3} S (100-h tests at 982°C)[80]
  a. Ni–28Fe–12Cr   b. Ni–50Fe–15Cr

alloys. A similar marked effect of aluminium is observed for an addition of about 1% with a further slight gain at 4·5% Al (Table 7.25). Grabke et al.[75] showed that addition of 2-4% aluminium to Incoloy 800H and Type 310 stainless steel was beneficial in conditions where an alumina layer formed under the chromia scale and thus prevented outward diffusion of iron, nickel and manganese. Lai[82], however, reported that the nickel-chromium-aluminium alloy, Haynes 214, was not resistant to sulphidation in the temperature range 750-1 000°C.

Table 7.25 Effect of aluminium on the corrosion at 982°C of Fe-35Ni-15Cr alloy in a reducing flue-gas atmosphere containing 2-3 g m^{-3} of sulphur (100-h test period)[80]

Aluminium content (%)	Metal loss (mm/y)	Total penetration (mm/y)
0	5	65
0·25	5·9	45
0·5	7·2	70
1·1	0·22	4·5
2·5	0·22	10
4·25	0·1	9

Hot-salt Corrosion

Reactions of contaminants in the fuel or air in the combustion zone can result in the formation of compounds which can condense as molten salts onto cooler components in the system. This type of process can occur when fuels containing sulphur or vanadium are burnt. In the case of sulphur contaminants, alkali sulphates form by reactions with sodium which may also be present in the fuel or in the combustion air, and for vanadium-containing fuels low-melting-point sodium vanadates or vanadium pentoxide are produced, particularly when burning residual oils high in vanadium. Attack by molten salts has many features in common which will be illustrated for the alkali-sulphate-induced attack, but which will be subsequently shown to be relevant to the case of vanadate attack.

Alkali Sulphate-induced Attack

Sulphur attack on nickel-chromium alloys and nickel-chromium-iron alloys can arise from contamination by deposits resulting from the combustion of solid fuels, notably high-sulphur coals and peat. This type of corrosion, which has been observed on components of aircraft, marine and industrial gas turbines and air heaters, has been associated with the presence of metal-sulphate and particularly sodium sulphate[83,84] arising directly from the fuel or perhaps by reaction between sodium chloride from the environment with sulphur in the fuel. Since such fuels are burned with an excess of air, corrosion occurs under conditions that are nominally oxidising although the deposits themselves may produce locally reducing conditions.

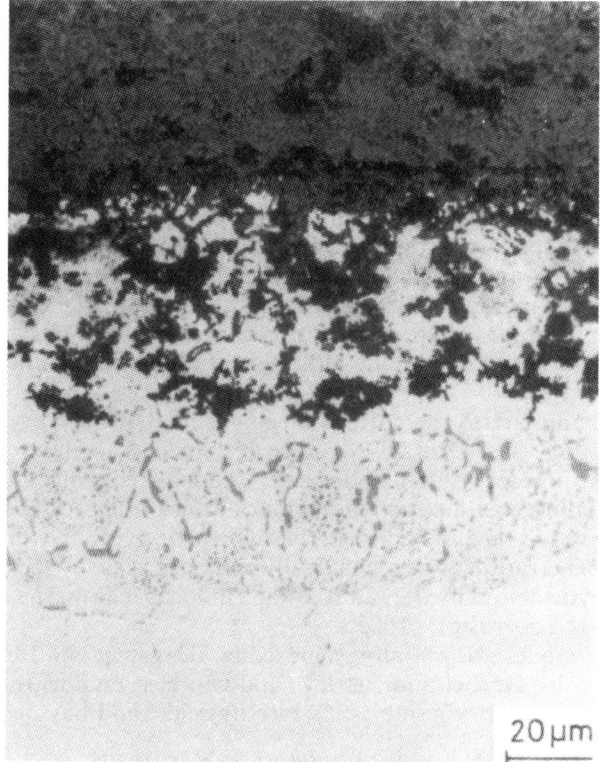

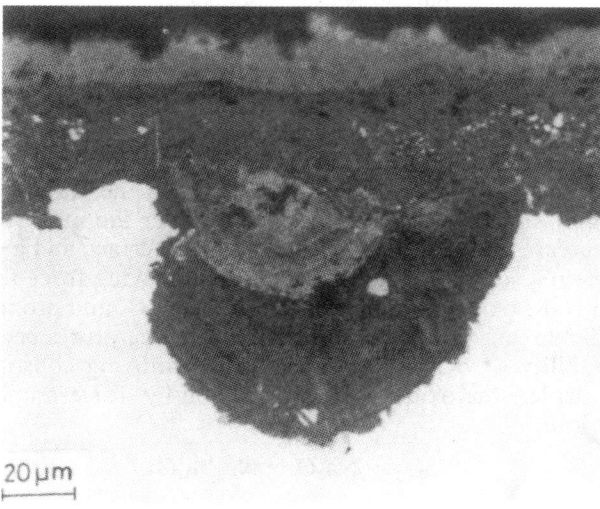

Fig. 7.44 Optical micrographs of Nimonic 105 turbine blades after (a) 1 000 h at 830°C and (b) 1 000 h at 675–750°C in an engine trial with 0·01 ppm NaCl in the intake air (after Saunders and Nicholls[85])

Examination of a component attacked under such conditions indicates that the corrosion has some similarities to gaseous sulphidation discussed in the previous section, in that in one form of the attack internal sulphides form in the chromium-containing alloys. Two types of attack have been observed which have come to be known as 'type I' (high temperature hot corrosion – HTHC) and 'type II' (low temperature hot corrosion – LTHC). Type I attack is usually found in the temperature range 800–950C, and type II at lower temperatures with a maximum in the rate of attack occurring at about 700°C. Typical micrographs of these types of corrosion are shown in Fig. 7.44[85]. In type I attack on nickel alloys, an outer porous oxide layer is formed over a region containing mixed metal and oxide and within the alloy internal sulphides are present. For type II, corrosion pits are observed with little or no internal sulphide although sulphides are found within the pit.

Stringer[86] summarised the various mechanisms that have been proposed to account for this type of attack which he described as accelerated oxidation induced by the presence of the molten salt. There is frequently an incubation period before the transition to higher rates of attack (the propagation stage), which is believed to be due to the salt dissolving the oxide layer by a 'fluxing' process, the penetration of a mechanically damaged scale by the molten salt, or an electrolytic action by the salt to establish local cells on the surface and so accelerate the corrosion process.

The fluxing model was initially proposed by Bornstein and DeCrescente[87] and developed by Goebel and Pettit[88], and can best be illustrated by considering the decomposition of Na_2SO_4 into its acidic and basic components:

$$Na_2SO_4 \rightarrow Na_2O \ (base) + SO_3 \ (acid)$$

So that with basic fluxing of nickel oxide the following reaction occurs

$$NiO + Na_2O \rightarrow Na_2NiO_2$$

and for acidic fluxing

$$NiO + SO_3 \rightarrow NiSO_4$$

The processes involved in basic fluxing are schematically represented for the case of nickel in Fig. 7.45. It can be seen that basic fluxing would be favoured by processes that increase the oxide ion activity in the Na_2SO_4 melt, as is shown in Fig. 7.46. The formation of oxide under the molten salt would result in a decrease in the pO_2 at the melt oxide/interface and a consequent increase in the pS_2 so that there is now sufficient driving force for diffusion of sulphur into the oxide, thereby decreasing the pSO_3 and promoting basic fluxing. Acidic fluxing will be favoured by low temperatures because of the increasing stability of SO_3, or by reactions of alloying constituents with Na_2SO_4 that deplete the oxide ion concentration by, for example, complex oxide formation:

$$MoO_3 + Na_2O \rightarrow Na_2MoO_4$$

Rapp and Goto[89] pointed out that in order to sustain the dissolution reaction a solubility gradient must be present in the molten salt layer. Thus oxide can dissolve at the oxide/melt interface, migrate down a concentration gradient to a site of lower solubility where precipitation occurs; this is

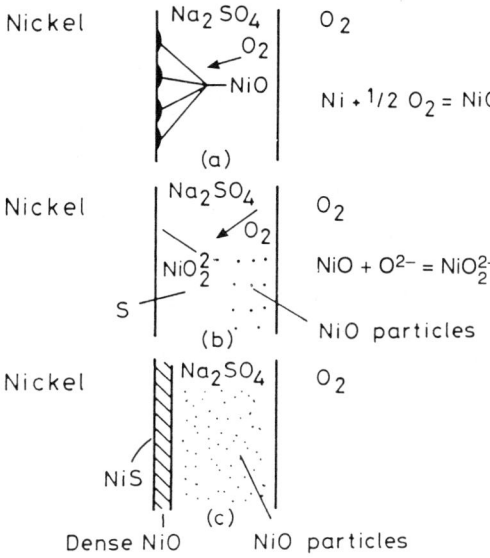

Fig. 7.45 Model of the basic fluxing process (after Goebel and Pettit[88])

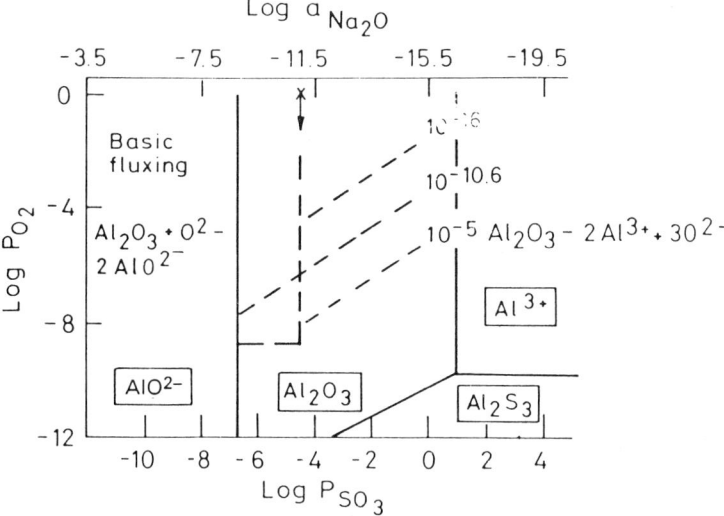

Fig. 7.46 The stable phases in the Al-O-S system in molten Na_2SO_4 at 1 000°C as a function of acidity and oxygen activity (after Goebel and Pettit[88])

believed to be the origin of the external porous oxide layer found on samples exposed to this form of attack. Type I hot-corrosion is generally believed to be the result of basic fluxing or of direct attack on the substrate by the melt through a cracked oxide layer, and type II is the result of acidic fluxing by low melting eutectics of, for example, 38% $NiSO_4 \cdot Na_2SO_4$ (m.p. 671°C) which are stabilised by a high pSO_3[90].

The solubility of oxides relevant to the hot-salt corrosion of superalloys has been measured by Rapp and co-workers, and Fig. 7.47 is taken from a recent review of their work[91]. It is noteworthy that there is a difference of about six orders of magnitude in basicity between the solubility minima for the most basic oxides and the protective alumina, chromia and silica which are all acidic oxides. Chromia scales are resistant to acid fluxing because the solubility minimum corresponds to the conditions in the gas turbine. It is also suggested that chromia scales are able to support a buffering action and resist large changes in the basicity of the melt. The solubility data do not immediately suggest the cause of the superiority of chromia relative to alumina, but Rapp[91] suggested that this might be the result of the faster growth of chromia. Another possibility is that the alumina scale is more vulnerable to mechanical damage which would permit direct attack on the alloy.

From the foregoing discussion it is evident that chromium additions are particularly beneficial in conferring resistance to both type I and type II forms of attack. Generally coatings are used to protect superalloys and the M-Cr-Al-Y type with chromium levels of up to 40% with relatively low aluminium (down to about 5% in some cases) are advocated for resistance to type II attack[92,93], and about 25% Cr and 12% Al for type I. Usually nickel is preferred to cobalt as the base material of the coatings resistant to type II attack because of the reduced stability of the $NiSO_4 \cdot Na_2SO_4$ eutectic compared with the corresponding cobalt-based salt, and coatings of this type are less prone to cracking due to greater ductility of the Ni-Cr-Al-Y system[47].

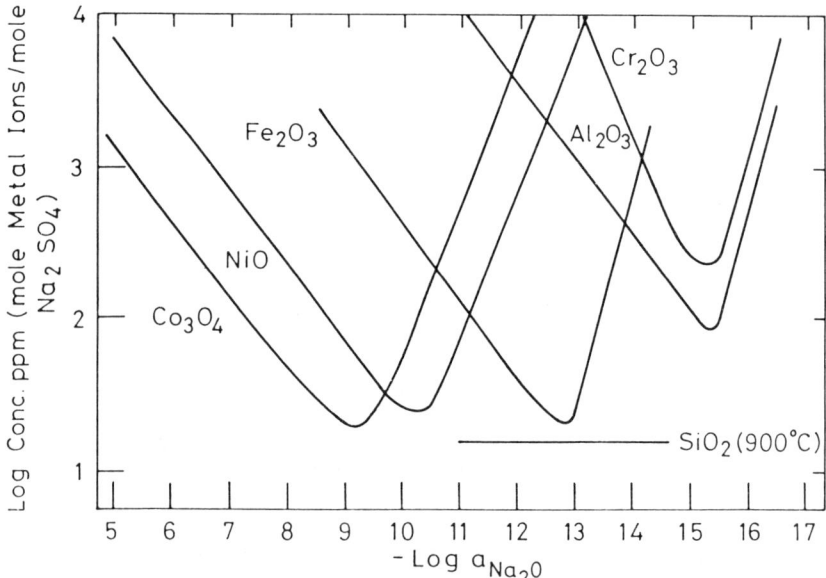

Fig. 7.47 Measured oxide solubilities in fused Na_2SO_4 at 1 200 K and $pO_2 = 1 \cdot 01 \times 10^5$ Pa (after Rapp[91])

Oxidation tests on Nimonic 90A, in which sodium chloride was introduced into the atmosphere, showed that this constituent produces a significant deterioration in the protective nature of the normally adherent film[94]. Although under certain service conditions the presence of sodium chloride is likely, this is not always so, and thus the general applicability of the results of laboratory tests in sodium sulphate and mixtures involving sodium chloride may be questioned. Test procedures for hot-salt corrosion have been reviewed by Saunders and Nicholls[85] who concluded that burner rig testing is the most appropriate procedure provided contaminant flux rates similar to those found in an operating turbine are used in the rig.

Vanadate-induced Attack

One of the most important impurities, found particularly in the residual oils, is vanadium, occurring in the resultant ash as vanadium pentoxide. The fluxing action of the ash with which corrosion is normally associated[95-97], and the marked difference in severity of attack at temperatures above and below the melting point of the ash again suggest molten salt induced attack.

Complex sodium vanadates form with melting points as low as 535°C, but Kvernes and Seiersten[98] reported that in the presence of Na_2SO_4 melting points of about 400°C can be observed. Luthra and Spacil[99] have suggested that the composition of the melt is determined by the following equilibrium:

$$Na_2SO_4 + V_2O_5 \rightarrow 2NaVO_3 + SO_3$$

so that pSO_3 will be an important parameter that determines the course of the reaction.

These melts are very acidic so that basic oxides such as NiO and CoO would be strongly attacked, and as Rapp[100] showed, addition of $NaVO_3$ to Na_2SO_4 shifts the solubility minimum to more basic values while raising its absolute value. Thus the acidic oxides Cr_2O_3, Al_2O_3 and particularly SiO_2 would be expected to be more resistant to this type of attack.

Another important factor controlling corrosion is the nature of the corrosion product which may also have low melting points, e.g. some vanadates and vanadium bronzes (V_2O_5–MO–V_2O_4, where M is metal) whereas nickel orthovanadate [$Ni_3(VO_4)_2$] has a melting point of 1 220°C. In some cases where Al_2O_3 or TiO_2, which are glass modifiers, are the corrosion products, glassy compounds with low melting points form by reaction with V_2O_5 (a glass former)[101].

The general principles outlined above can usually explain the behaviour of materials in laboratory tests and in service in operating plant. For example, Bornstein and DeCrescente[101] found that pure nickel and nickel–chromium alloys were not greatly affected by the presence of V_2O_5, whereas a nickel aluminide or an alumina-forming superalloy B 1900 suffered catastrophic attack due to the formation of low-melting glassy phases. Similar results were also obtained by Betteridge et al.[102] who also noted that silicon additions were beneficial, as might be expected, but iron-base alloys were seriously attacked (Table 7.26).

Table 7.26 Results of tests on nickel and nickel alloys heated at 850°C in contact with vanadium pentoxide[102]

Alloy composition	100 h			250 h			284 h		
	Mass of V_2O_5 (g)	Mass loss (g)	Loss due to V_2O_5 (g)	Mass of V_2O_5 (g)	Mass loss (g)	Loss due to V_2O_5 (g)	Mass of† V_2O_5 (g)	Mass loss (g)	Loss due to V_2O_5 (g)
Nickel	0	0·32	—	0	0·79	—	0	0·88*	—
	0·005	0·39	0·07	0·005	0·57	—	0·01	0·58	—
	0·010	0·42	0·10	0·011	0·46	—			
Ni-20Cr	0	0·020	—	0	0·022	—	0	0·02*	—
	0·006	0·035	0·015	0·005	0·030	0·008	0·01	0·19	0·17
	0·010	0·038	0·017	0·009	0·052	0·030	—	—	—
Ni-20Cr-4Al	0	0·017	—	0	0·016	—	0	0·018*	—
	0·005	0·069	0·052	0·005	0·152	0·136	0·01	0·28	0·26
	0·010	0·115	0·098	0·010	0·163	0·147			
Ni-20Cr-2·5Ti	0	0·148	—	0	0·308	—	0	0·088*	—
	0·006	0·222	0·073	0·005	0·246	—	0·01	0·29	0·20
	0·010	0·185	0·036	0·010	0·257	—			
Ni-20Cr-20Co-2·5Ti	0	0·048	—	0	0·054	—	0	0·06*	—
	0·005	0·077	0·029	0·004	0·128	0·073	0·01	0·236	0·175
	0·009	0·098	0·050	0·009	0·108	0·054			
Ni-14Cr-6Fe	0	0·015	—	0	0·014	—	0	0·015*	—
	0·006	0·029	0·013	0·006	0·023	0·009	0·01	0·186	0·171
	0·010	0·035	0·020	0·010	0·036	0·011			
Ni-10Si-3Cu							0	0·416	—
							0·01	0·435	0·019
Ni-15Cr-5Si							0	0·055	—
							0·01	0·165	0·110
Ni-14Cr-10Si							0	0·02	—
							0·01	0·152	0·132
Ni-15Cr-15Si							0	0·065	—
							0·01	0·208	0·143
Fe-37Ni-18Cr-2Si	0	0·039	—	0	0·036	—	0	0·04*	—
	0·005	0·314	0·275	0·006	0·342	0·305	0·01	1·05	1·01
	0·010	0·426	0·387	0·010	0·418	0·382			

* Estimated from values for 250-h test.
† 0·01 g added every 24 h.

The ash deposits resulting from the combustion of solid and oil fuels often contain appreciable quantities of other corrodants in addition to vanadium pentoxide. One of the more important of these is sodium sulphate, and the effects of this constituent in producing sulphur attack have been mentioned. The contents of sodium sulphate and vanadium pentoxide present in fuel oil ash can vary markedly and the relative merits of different materials depend to a great extent upon the proportions of these constituents. Exposure of heat-resisting alloys of varying nickel, chromium and iron contents to ash deposition in the super-heater zones of oil-fired boilers indicated a behaviour pattern depending on the composition of the alloy and of the ash concerned[103]. In a land-based boiler in which the vanadium pentoxide: sodium sulphate ratio was high, corrosion was due primarily to the fluxing action on the oxide scale, and the alloys containing more than 60% nickel showed greater resistance than those of low nickel content. On the other hand, in a marine boiler in which the ratio of vanadium to sodium sulphate was low, oxidation was accelerated owing to sulphur contamination, and these deposits were most corrosive to the high-nickel alloys. The alloys of intermediate nickel content (30–60%) were the most susceptible to deposits containing vanadium and sodium sulphate in all the proportions investigated. Figure 7.48 summarises the results of this investigation.

However, under more realistic test conditions Hancock and Islam[104] showed that in burner rig tests with contaminant flux rates greater than about 0.1 mg cm^{-2}h^{-1} the corrosion rate of nickel- and cobalt-base superalloys was largely independent of alloy composition in the temperature range 700–850°C. However, in burner rig tests at 600°C, simulating diesel engine combustion, Saunders et al.[105] reported that Nimonic 80A (20% Cr) had superior resistance to Stellite 6 (Co-28%Cr) and EN 52 (Fe-8%Cr-3%Si).

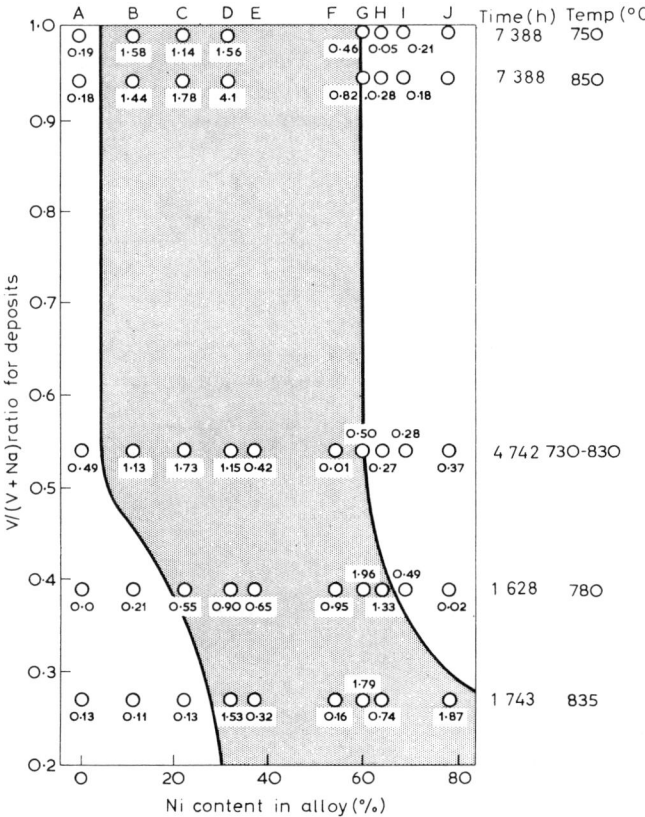

Fig. 7.48 Effect of composition of alloy and deposit on corrosion, shown as weight loss (g cm^{-2}) against plotted points. (Shaded area shows most attack, i.e. >0.5 g cm^{-2})[103]

Seiersten and Kofstad[106] point out that in simple laboratory tests at 650-800°C using sodium vanadate deposits the rate of attack on pure nickel was increased by a factor of about four to five while a plasma-sprayed Ni-Cr-Al-Y coating was even more severely corroded, and the corresponding Fe-Cr-Al-Y coating reacted even more rapidly. However, in dynamic burner rig testing, where continuous replenishment of the deposit occurs, the Fe-Cr-Al-Y coating exhibited corrosion resistance superior to that of Ni-Cr-Al-Y. These results were interpreted in terms of the ability of Ni-base materials to form a stable vanadate [$Ni_3(VO_4)_2$] so that in tests involving only one application of salt an increase in the Na_2O concentration of the melt occurred making it more basic and hence less agressive to NiO. The corresponding iron vanadate is less stable. However, in the burner rig where continuous replenishment of salt occurs this limitation does not arise and the overall corrosion rate is controlled by the relative solubilities of the oxides in the molten vanadate.

A survey of a broad range of Ni-Cr-Fe-Co alloys immersed in an 80% V_2O_5 + 20% Na_2SO_4 mixture over the temperature range 700-1100°C revealed a major effect of chromium content in determining corrosion resistance, the level of chromium required increasing with increasing test temperature[95]. Again at this ratio of vanadium:sodium, attack was largely by the fluxing effect on the protective film of molten vanadates. Results for binary alloys are given in Fig. 7.49. The provision of the high (50-60%) chromium-nickel alloys in usable forms represents perhaps the most promising metallurgical approach to a solution of this particular corrosion problem so far[107-109].

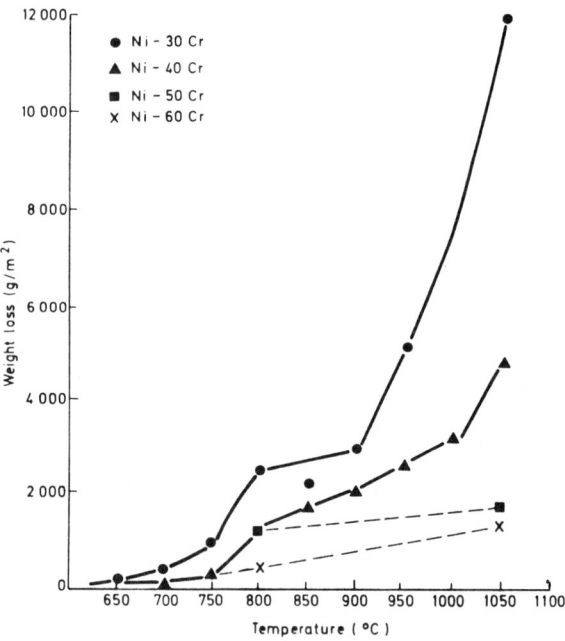

Fig. 7.49 Effect of temperature on corrosion of nickel-chromium alloys exposed to an 80% V_2O_5 + 20% Na_2SO_4 salt mixture for 120 h[95]

Very recently Nicholls and Stephenson[110] confirmed the beneficial effects of chromium additions in a comprehensive series of laboratory tests using synthetic ash deposits in which 75 different alloys were examined. They concluded that nickel- or iron-based material with chromium contents in excess of 25% offered the best resistance to attack in their tests.

Cutler et al.[111] investigated the corrosion resistance of ferritic and austenitic steels in oil-buring power stations. There was a marked superiority of the ferritic materials particularly at gas temperatures of 1 150°C (see Fig. 7.50).

The addition of lead compounds to petroleum fuels leads to attack by the combustion products on components, particularly exhaust valves and sparking-plug electrodes, of piston engines. Deposition of lead sulphate results in both sulphur attack of the type described in another section and corrosion more directly associated with lead itself. For the spark-plug electrodes, nickel alloys with manganese and silicon have proved very satisfactory from the corrosion point of view, while the use of a protective nickel–chromium alloy coating applied to steel exhaust valves by a welding technique has been established for many years. For high-performance engines, the valves themselves have been manufactured from a high-strength nickel–chromium-base alloy, but with increased operating temperatures further corrosion resistance has been required and the application of an aluminised coating has been found effective.

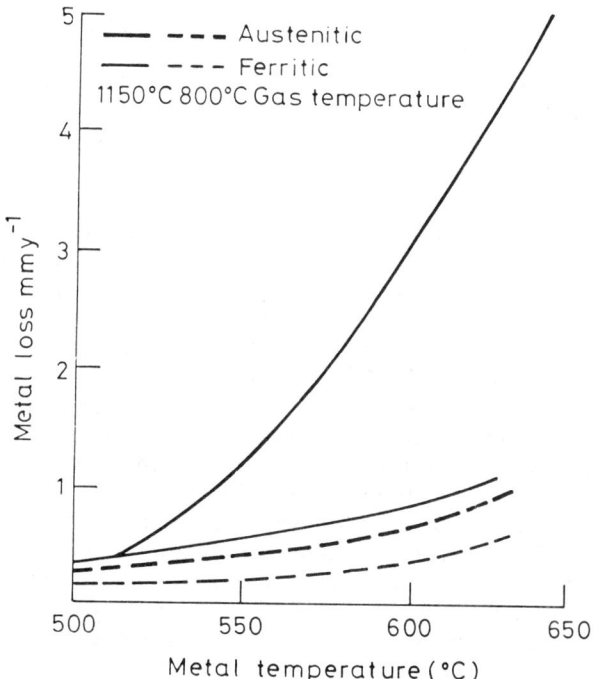

Fig. 7.50 Metal loss of austenitic and ferritic steels after exposure in oil-burning power stations as a function of metal and gas temperature (after Cutler et al.[111])

Carburisation and Attack by Carbon-containing Gases

While carburisation itself is not a normal corrosion process*, in that there is no metal wastage, absorption and diffusion of carbon can lead to significant changes in the mechanical properties of the affected material and in particular to marked embrittlement. Furthermore, initial carburisation can produce an acceleration of the normal oxidation process, a phenomenon that is notable in nickel–chromium alloys.

The question of the compatibility of metals and alloys with carbon and carbonaceous gases has assumed considerable importance in connection with the development of the gas-cooled nuclear reactor in which graphite is used as a moderator and a constituent of the fuel element, and carbon dioxide as the coolant. Tests of up to 1 000 h on a series of metals and nickel-containing alloys under pressure contact with graphite at 1 010°C[112] showed that only copper was more resistant than nickel to diffusion of carbon and that the high-nickel alloys were superior to those of lower nickel content. The more complex nickel–chromium alloys containing titanium, niobium and aluminium were better than the basic nickel–chromium materials.

Tests on a wide range of alloys at temperatures varying from 704 to 927°C have been made by Bernsen et al.[113] to determine the temperature limits beyond which engineering materials carburise when held in contact with graphite. Table 7.27 lists the maximum penetrations of the carburised zones; while nickel in general showed no visible evidence of carburisation the associated hardness measurements indicated solution of carbon even at 704°C. At this temperature the chromium-containing alloys showed little tendency to carburisation, but at 816°C carburisation leading to the formation of chromium carbide was rapid.

Of the carbon oxide gases, carbon dioxide is the less corrosive, leading normally to mild oxidation only and under certain conditions, e.g. low partial pressures of CO_2 or in the presence of steam, to decarburisation of nickel alloys. Tests on nickel and nickel alloys in carbon dioxide at 1.38×10^7 Pa pressure at 704 and 816°C showed that very little increase in mass occurred in up to 1 000 h even at the higher temperature (Table 7.28). The scales formed were similar in constitution to those obtained in air and the rate of scaling approximated to the parabolic law. In addition to the general scale formation, nickel exhibited intergranular oxidation while Inconel (Ni-16Cr-7Fe) and Nichrome V (Ni-20Cr) showed localised pitting. Later work[114] on Nimonic 75 (Ni-20Cr-0·4Ti-0·1C) at pressures of 1 atm or less has shown that oxidation predominates and was generally protective, but at low partial pressure the oxide film was less coherent and uniform, and mass losses due to decarburisation and evaporation of chromium from the specimen surface were recorded at 900–1 000°C.

In the gas-cooled reactor, reaction between the coolant and the moderator results in formation of a proportion of carbon monoxide in the atmosphere. This gas can be carburising to nickel-base alloys but the results of tests[115] in which CO_2 was allowed to react with graphite in the furnace indicate that the attack on high-nickel alloys is slight, even at moderately high temperatures and is still mainly due to simple oxidation.

---

* See discussion of 'corrosion' in Section 1.1.

Table 7.27 Effect of time and temperature on the visible penetration of carburisation in nickel alloys under contact pressure (1–79 N/mm^2) with graphite[113]

Material type and composition	704°C Time (s × 10^{-6})	704°C Maximum penetration (mm)	816°C Time (s × 10^{-6})	816°C Maximum penetration (mm)	927°C Time (s × 10^{-6})	927°C Maximum penetration (mm)	Remarks
Nickel	1·08(1)	<0·0025	1·25	<0·0025	3·6	<0·0025	(1) Grain-boundary attack 0·038 mm
	4·45(2)	>0·75	2·55	<0·0025			(2) Fine intergranular precipitate
			6·29	<0·0025			Hardness increase at each temperature
Nichrome V (Ni–20Cr)	1·08	<0·0025	1·25	0·43			Hardness increase in 350 h at 816°C but not at 704°C
	4·45	0·330	2·55	0·41			
			6·29	0·53			
Inconel (Ni–16Cr–7Fe)	1·08	0·0025	2·55	0·76			Hardness increase in 350 h at 816°C but not at 704°C
	4·45	0·279	2·67	0·76			
Inconel X (Ni–15Cr–7Fe–2·5Ti)	1·08	<0·0025	1·25	0·20			
	4·45	0·175	2·55	0·33			
			6·29	<0·0025			
Inconel 702 (Ni–16Cr–3·5Al)	3·34	0·545	2·67	0·15	3·6	>0·76	
Incoloy 901 (Ni–35Fe–13Cr–2·5Ti)	3·34	<0·0025	2·67	0·76	3·6	>0·76	

Table 7.28 Mass changes during exposure of nickel and nickel alloys to carbon dioxide[113]

Material type and composition	Mass gain (g m^{-2})*			
	0.36×10^6s		3.6×10^6s	
	704°C	816°C	704°C	816°C
Nickel	13.0	20.0	13.0	30.0
Nichrome V (Ni-20Cr)	0.8	1.3	1.1	2.5
Nimonic 80 (Ni-20Cr-2.5Ti-1.5Al)	1.1	6.0	2.5	15.0
Inconel (Ni-16Cr-7Fe)	0.9	2.0	1.0	5.0
Inconel X (Ni-15Cr-7Fe-2.5Ti)	1.8	7.5	2.5	15.0
Hastelloy B (Ni-28Mo-5Fe)	12.5	15.0	15.0	40.0

*Approximate values.

Industrial atmospheres of the so-called 'controlled' type are frequently used for operations such as bright annealing, carburising and nitriding, and are selected as being reducing to the component being treated. Common examples are cracked ammonia and the products of combustion of coal gas or natural gas, the latter normally being desulphurised. Table 7.29 summarises the satisfactory service experience with nickel and Inconel (Ni-16Cr-7Fe) thermocouple protection tubes in such atmospheres. More vigorous carburising conditions may be achieved deliberately by the addition of hydrocarbons into the gas stream as in many petrochemical processes such as ethylene production and steam reforming, or accidently, by the introduction of oily components into the furnace. Under such conditions nickel-base alloys are susceptible to two forms of severe carburisation sometimes referred to as 'green-rot' and metal dusting. In the former, internal chromium carbides are formed during carburisation which often embrittle the alloy, and in process plant a subsequent oxidising cycle would result in oxidation of the carbides at the grain boundaries. If the component then fractures a

Table 7.29 Service experience with thermocouple protection tubes in reducing gases and other media[116]

Material	Temperature (°C)	Atmosphere composition (%)						Other media	Life (years)
		CO	CO$_2$	N$_2$	H$_2$	H$_2$O	S		
Nickel	815–1 070	5	8	64	5	18	0·000 8		1–2
Inconel (Ni-16Cr-7Fe)	980–1 260	5	8	64	5	18	0·000 8		1–2
Inconel	980–1 200	2	11	64	2	21	0–000 8		1–2
Inconel	370–980							Lead	3–5
Inconel	790–870							Carburising mixture (cyanide compound)	3–5

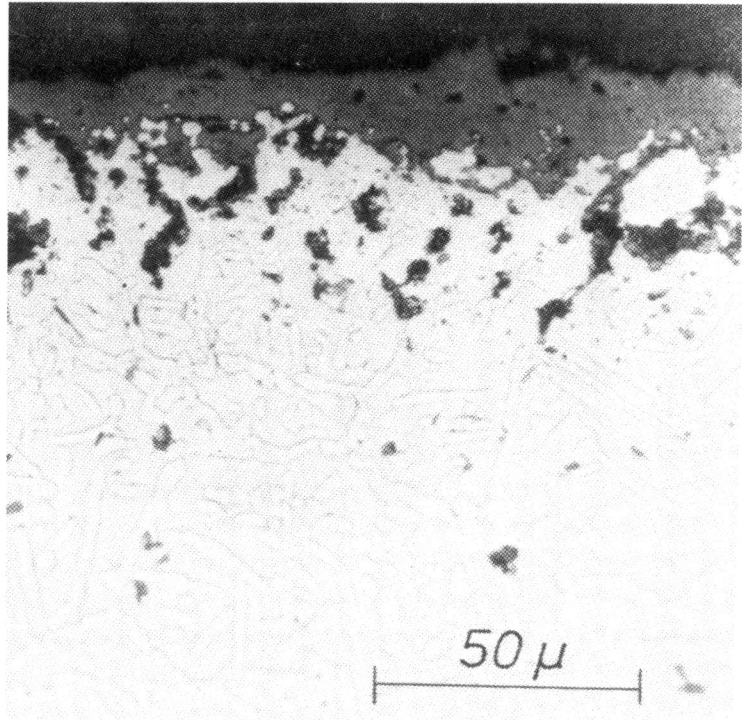

Fig. 7.51 Optical micrograph of Incoloy 800H after 50 h at 1 000°C in a CH_4-H_2 mixture ($a_c \approx 1$) followed by 300 h at 1 000°C in air (after Grabke and Schnaas[117])

characteristic green colour is observed on the fracture surface because of the presence of chromium oxide. Figure 7.51 is a typical cross-section showing this type of attack[117].

Metal dusting usually occurs in high carbon activity environments combined with a low oxygen partial pressure where carburisation and graphitisation occur. Usually pits develop which contain a mixture of carbon, carbides, oxide and metal (Fig. 7.52). Hochmann[118] proposed that dusting occurs as the result of metastable carbide formation in the high carbon activity gas mixture which subsequently breaks down into metal plus free carbon. The dependence of the corrosion resistance of these nickel alloys on the protective oxide film has been described; accelerated or internal oxidation occurs only under conditions that either prevent the formation, or lead to the disruption, of this film. In many petrochemical applications the pO_2 is too low to permit chromia formation (ethylene furnaces for example) so that additions of silicon[119] or aluminium[120] are commonly made to alloys to improve carburisation resistance (Fig. 7.53).

Hall *et al.*[121] pointed out that carburisation is controlled by three independent processes, i.e. carbon deposition, carbon ingress (through the protective scale) and carbon diffusion through the matrix. Carbon deposition usually occurs by decomposition of CH_4 adsorbed on the surface[122] or the catalytic decomposition of CO (Boudouard reaction). Hydrogen

Fig. 7.52 Optical micrograph of a 5% Cr steel after service in a petrochemical plant showing typical 'metal dusting' behaviour (after Hochmann[118])

sulphide additions have been shown to be very effective in controlling carbon deposition by preferential adsorption to exclude CH_4 or CO[123]. Alternatively, the removal of catalysts for the Boudouard reaction (for example Fe, Ni Fe_2O_3 or Cr_2O_3) by use of alloys or coatings that form Al_2O_3 or SiO_2 has also been shown to be effective.

The second stage in the carburisation process, that of carbon ingress through the protective oxide layer, is suppressed by the development of alumina or silica layers as already discussed and in some cases protective chromia scales can also form. Diffusion and solubility of carbon in the matrix has been shown by Schnaas et al.[124] to be a minimum for binary Fe-Ni alloys with a nickel content of about 80%, and Hall[125] has shown that increasing the nickel content for the nickel-iron-25%-chromium system resulted in lower rates of carburisation (Fig. 7.54).

Since the depletion of the matrix with respect to chromium appears to be the basic factor involved, it would be expected that additions of elements having greater affinity than chromium for carbon would increase resistance to attack and such effects have been demonstrated by several investigations[7,118]; where the beneficial effects of additions of titanium,

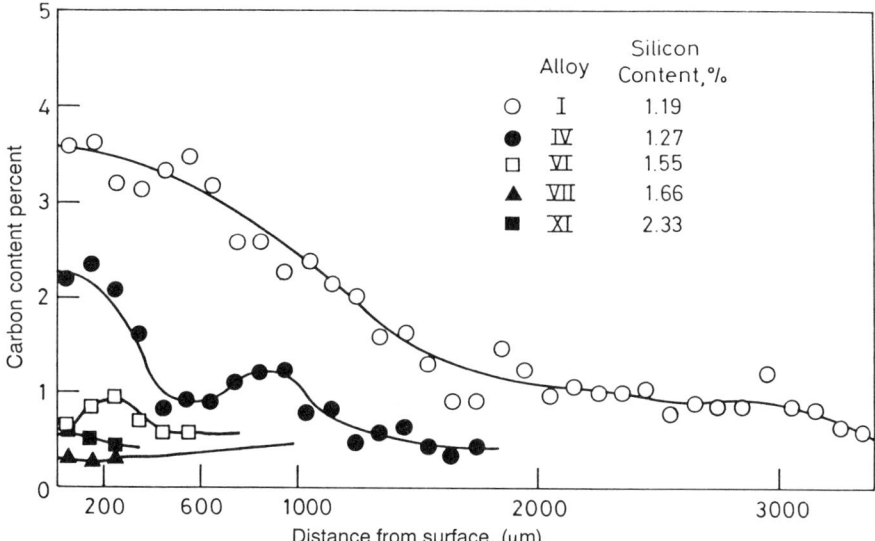

Fig. 7.53 Carbon concentration profiles of various centrifugally cast steels of differing silicon content, after 100 h at 1 093°C in a gas mixture with the following composition at the furnace inlet $H_2/34\%$ $CH_4/30\%$ H_2O (after Kane[119])

tungsten and niobium and to a lesser extent of molybdenum. The complex high-strength superalloys generally contain one or more of these carbide-forming additions and these alloys are relatively resistant to such attack; Nimonic 90 (Ni–20Cr–16Co–2·4Ti–1·2Al) for instance was unattacked after 1 000 hours exposure to carbon monoxide at 950°C.

Nitrogen

Nickel-rich alloys are in general relatively resistant to nitriding and are therefore suitable materials for service in nitrogen-containing atmospheres, e.g. controlled atmospheres utilising the products of cracked ammonia. Inconel (Ni–16Cr–7Fe) is widely used for the fabrication of nitriding boxes while nickel thermocouple-protection tubes give from 5 to 10 years service life at 677–788°C in a bright-annealing atmosphere consisting of 90% nitrogen and 10% hydrogen.

Cyclic heating tests on electrically heated nickel–chromium and nickel–chromium–iron wires in nitrogen[126] showed that the lives to burn-out were at least as good as those obtained in air. The same was true in ammonia at 1 050°C, but at 1 150°C and above the life was less in ammonia than in air (Table 7.30) indicating, as would be expected, that it is atomic nitrogen which is most effectively absorbed.

Table 7.31 lists data for the corrosion of nickel and some nickel alloys by ammonia at 500°C. At higher temperatures the more complex hardened nickel–chromium-base alloys are more resistant than the binary alloys and

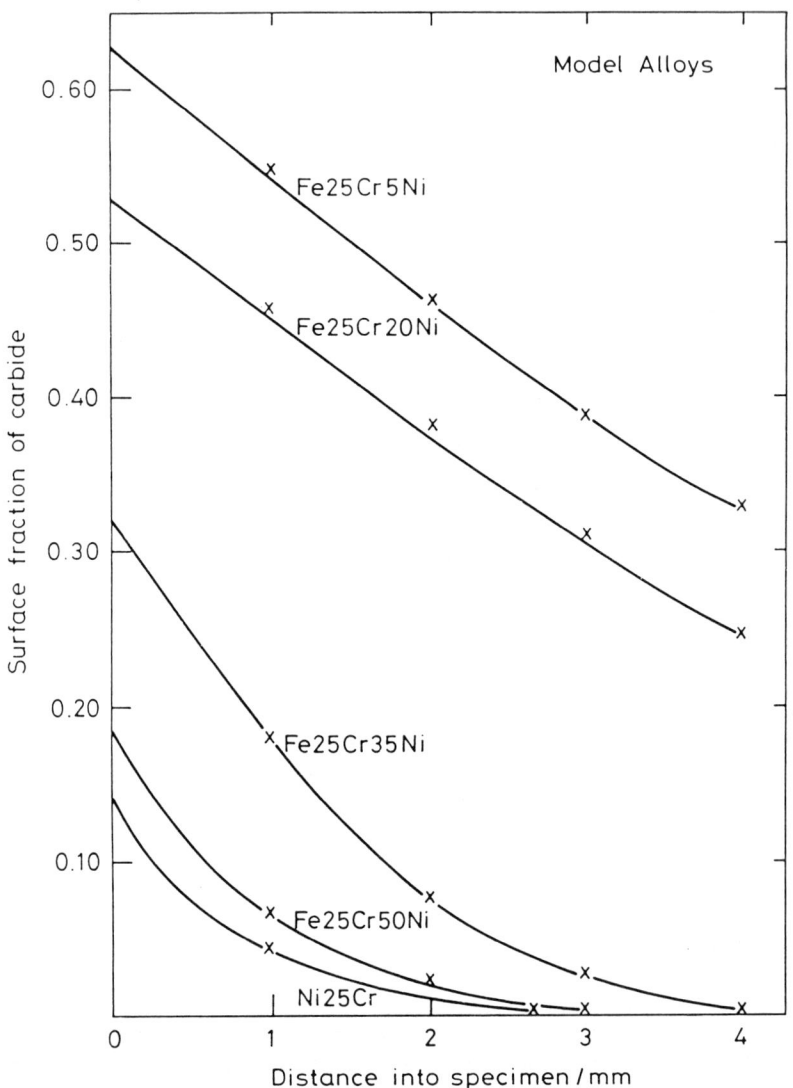

Fig. 7.54 Carbide area profiles for various Fe-Cr-Ni alloys after nine carburising and oxidising exposures for 120 h and 24 h, respectively, at 1 050°C with three intermittent thermal cycles to RT. Carburising gas, 3% CH_4, 20% CO, 40% H_2, balance N_2; oxidising gas, air (after Hall[125])

although the former may be nitrided, little or no surface hardening results. Samples of Nimonic 80A (Ni-20Cr-2·5Ti-1·5Al) exposed in wet cracked ammonia for nearly 5 000 h at 816°C showed no evidence of intergranular attack and no changes of microstructure other than those expected due to heating in an inert atmosphere[128]. In creep tests at temperatures above about 800°C where extensive intergranular cracking occurs in the tertiary stage prior to rupture, internal formation of titanium nitride has been

Table 7.30 Results of cyclic heating tests in various atmospheres[126]

Alloy composition (%)			Maximum cycle temperature (°C)	Life to burn-out ($s \times 10^{-3}$)		
Ni	Fe	Cr		Air	Ammonia	75% H_2:25% N_2
80	—	20	1 050	2 268	3 240	
			1 150	522	166	
			1 200	320	43	1 645
30	50	20	1 150	382	212	1 832
			1 200	194	11	342

Table 7.31 Corrosion of nickel and nickel alloys in anhydrous ammonia at 500°C (1 540-h tests)[127]

Material composition (%)			Maximum penetration rate (mm/y)	
Ni	Fe	Cr	Polished	Oxidised
100	—	—	2·00	2·49
80	—	20	0·18	0·13
59	26	15	0·18	0·18
35	46	19	0·43	0·28
78	8	14	0·15	0·25

observed near the cracks; this is associated with an increased resistance to deformation, and a further secondary stage of creep occurs.

Attack by Halogen Gases

Although there is seldom an industrial requirement to expose metallic components to pure halogenic atmospheres, these gases are frequently present in minor quantities in many environments in energy conversion and process industries. Perhaps the best examples are: (a) in waste incineration where PVC in waste can lead to combustion atmospheres containing up to about 0·2% HCl, depending on waste sorting and blending procedures[129], and (b) in the combustion of chlorine-containing coal which, notably in the UK, has caused severe corrosion of boiler tubes[130]. Recent work on gasification of coal has shown that chloride can also be a problem in these reducing environments, particularly if dew-point corrosion occurs during cyclic operation of the plant[131].

The main characteristic of attack by halogens at elevated temperatures is that most reaction products are volatile compared with the solid products that form in all cases considered hitherto in this chapter. Thus, in cases where metals are exposed to pure halogen gases large mass losses are usually reported with very little external scale formation. Li and Rapp[132] showed that internal chloridation occurred when nickel–chromium alloys were exposed to Ni + $NiCl_2$ powders at 700-900°C. However, where oxide scales can also form, as in combustion gases, the oxide layer was usually highly

Table 7.32 Material degradation data for specimens exposed to Air–2.13% Cl_2 at 900°C[135]

Alloy	Exposure time (h)	Weight change (mg mm^{-2})	Scale thickness (μm)	Metal lost per side (μm)	Depth of internal penetration (μm)
Ni–27%Cr	48	−1.179	N/A	160	N/A
Ni–27%Cr–4% Al	48	+3 × 10^{-3}	N/A	50	N/A
Ni–27%Cr (preoxidised)	48	−0.593	N/A	120	N/A
Ni–27%Cr–4%Al (preoxidised)	48	+1 × 10^{-3}	N/A	40	N/A
Molybdenum	4	−2.8	—	250	complete consumption
Tungsten	75	−3.6	—	280	complete consumption
Hastelloy C176	48	−0.92	N/A	N/A	N/A
Haynes 230	48	−0.1	N/A	N/A	N/A
Alloy 263	48	−0.08	N/A	N/A	N/A
Alloy 214	48	+5 × 10^{-3}	N/A	N/A	N/A
310 Stainless	1	+0.019	45	20	100
	6	−0.161	180	110	100
	24	−0.177	260	140	125
	48	−0.181	265	150	130

Material	Time (h)				
Inconel 600	1	$+3 \times 10^{-3}$	5	5	—
	6	−0.091	10	40	55
	24	−0.23	10	155	140
	48	−0.352	11	175	250
Inconel 671	1	+0.188	190	15	200
	6	+0.355	250	15	300
	24	+0.90	450	50	complete penetration
	48	−0.66	450	completely corroded	
Fecralloy	1	−0.077	100	20	
	6	−0.163	95	100	
	24	completely corroded			
310 stainless (preoxidised)	1	-5×10^{-3}	45	20	
Inconel 600 (preoxidised)	1	−0.01	10	30	
Inconel 671 (preoxidised)	1	+0.281	200	50	
Fecralloy (preoxidised)	1	+0.154	130	20	

fractured and porous, which is believed to be related to the formation of volatile products beneath the oxide layer causing scale cracking. The mechanism for halogen penetration of the scale is not known, but Kofstad[133] suggested that migration down grain boundaries or microchannels was the most likely process. Kofstad also proposed that at the very low pO_2 values that exist at the alloy/oxide interface chlorine, for example, might react directly with the protective oxide converting it into a volatile chloride, but if the chlorine were also in contact with the alloy, direct chloridation would be the more favoured reaction.

It is clear, therefore, that when considering the behaviour of alloys their overall corrosion rate will be controlled by the relative volatility of the halide or oxyhalide, and generally rates of attack are dependent on the flow rate of the atmosphere, with diffusion through a stagnant layer the thickness of which is dependent upon the Reynolds and Schmidt numbers. Thus steady-state linear rates of loss of material are to be expected. Nickel-base alloys would be expected to have superior resistance to halogen attack compared with iron-base materials because of the lower volatility of the nickel reaction products compared with iron halides, as has indeed been found by many workers[134,135]. A useful compilation of results of exposure to air-chloride mixtures for a number of nickel- and iron-base alloys and some refractory materials is given in Table 7.32. It can be seen that addition of aluminium is particularly beneficial because of the low vapour pressure of AlOCl ($7 \cdot 3 \times 10^{-10}$ Pa at 900°C in equilibrium with air, 2% Cl_2 and Al_2O_3)[135]. Daniel and Rapp proposed that at partial pressures greater than about 10 Pa volatilisation can become significant. In the case of pure nickel, attack in fluorine proceeds at much lower rates than in either chlorine or bromine due to the lower vapour pressures of the fluorides[136]. For a comprehensive review of attack by halogens the reader is referred to Daniel and Rapp[136] which gives very valuable basic information on the mechanism of attack in these environments.

Recent industrial experience for alloys in waste incineration plants has indicated the superiority of nickel-base alloys compared with iron-base and iron-containing alloys as would be expected from the previous discussion of the volatility of chlorides. Nickel-base alloys with no addition of iron and relatively high chromium contents have significantly improved performance in these applications (see Tables 7.33 and 7.34.)[137].

Corrosion by Molten Metals and Salts

The physical characteristics of liquid metals and salts and particularly their high heat capacities and thermal conductivities have long established them as materials for use in applications where rapid heat transfer is required at relatively high temperatures. Unfortunately these media are also highly corrosive to metals generally. Corrosion of metallic materials by liquid metals is usually associated with solution of the former in the latter though another important form involves intergranular penetration by the liquid into the solid. Modification of these basic types occurs under conditions involving a temperature gradient, when solution of the solid in the hot region is accompanied by deposition in the cold region, or of a concentration gradient, when

Table 7.33 Corrosion rack tests of nickel- versus iron-base alloys used in waste incineration plants

Alloy	Mass loss[a] (%)	Depth of subsurface attack[b] (mm)
Akron municipal incinerator		
Test temperatures: 611-732°C		
Exposure time: 1 104 h		
625	1.65	0.075
617	7.50	0.102
690	10.10	0.381
020	44.3	0.381
601	46.6	0.381
800H	55.6	0.383
825	60.7	1.016
600	100.0	Destroyed
Saugus municipal incinerator		
Test temperatures (average): low 730°C, high 869°C		
Exposure time: 552 h		
020	<0.02	0.127
617	<0.02	0.051
690	<0.02	0.102
625	0.22	0.305
800H	15.0	0.635
825	43.0	2.234
601	100.0	Destroyed
600	100.0	Destroyed

[a] Mass loss (%) = [ML(i) − ML(f)/ML(i)] × 100
[b] Internal attack exlusive of corrosion mass losses

a metal from one part of the system is dissolved and redeposited in another region initially deficient in this element. Relatively small amounts of impurities such as oxygen or nitrogen may affect the rate of such corrosion by incompletely understood mechanisms, which may, however, involve such factors as the limit of solubility or the wettability of the solid component, or the formation of a barrier to the reaction.

Nickel and nickel-rich alloys must be considered as having generally poor resistance to molten metals. Eldred[138] has made a systematic investigation of the attack of liquid metals on solid metals and alloys, and his results for nickel, and nickel–chromium and nickel–copper alloys, are summarised in Table 7.35. These are for tests at up to 500°C and apart from potassium and sodium all the low-melting-point metals investigated produced moderate to severe attack on the nickel-rich materials. Furthermore, the values for many of the combinations given in the table indicate a marked tendency to preferential intergranular attack.

Tests in thermal convection loops with a maximum temperature of 800°C[139] showed that nickel and nickel-rich alloys with chromium and other additions were relatively very susceptible to the mass-transfer type of attack by lead. On the other hand nickel-rich alloys with molybdenum were among the better of the alloys tested (Table 7.35).

As in other types of corrosion, attack by molten metals depends to a great extent on the detailed conditions of corrosion. Thus satisfactory service life has been reported for Inconel (Ni–16Cr–7Fe) thermocouple protection tubes

Table 7.34 Nominal composition of alloys (weight percent)

Alloy		Fe	Ni	Cr	Mo	Al	Ti	Si	Mn	C	Others
INCONEL[a] alloy	600	7.2	Bal	16.0	—	—	—	0.2	0.2	0.04	
	601	14.1	Bal	23.0	—	1.4	—	0.2	0.5	0.05	
	617	14.1	Bal	22.0	9.0	1.0	—	—	—	0.07	12.5 Co
	625	2.5	Bal	21.5	9.0	0.2	0.2	0.3	0.3	0.05	3.65 NG
	690	9.5	Bal	30.0	—	—	—	—	—	0.03	
INCOLOY[a] alloy	800	Bal	32.0	20.5	—	0.3	0.3	0.4	0.8	0.04	
	825	Bal	42.0	21.5	3.0	0.2	0.09	0.4	0.7	0.03	
INCO[a] alloy	020	Bal	34.0	20.5	2.5	—	—	—	—	—	3.4 Cu, 0.5 NG

[a]INCONEL, INCOLOY and INCO are registered trademarks of the Inco family of companies

in molten lead at temperatures up to 982°C (Table 7.29). For such service, however, the Inconel must be in the solution-annealed condition as the presence of residual stress greatly reduces the service life. Cases of failure in service of nickel–chromium-base alloy tubing due to attack by residual traces of low-melting-point alloys used during the tube manufacture provide further evidence of the susceptibility of these materials to a form of stress corrosion. Russian workers[140] demonstrated a marked reduction in stress-rupture strength of a nickel–chromium alloy in the presence of a lead–bismuth alloy and also the rapid destruction of Ni–20Cr and Ni–27Cr–5Al in the presence of molten tin[141].

The relatively good resistance of nickel-rich materials to attack by molten sodium and potassium has been reported by several investigators. In static tests, nickel and Ni–20Cr alloys showed no evidence of interaction with metallic sodium[142] and Inconel also had good resistance to attack. In dynamic thermal loop tests, stainless steels and Inconel were reported to be the best available structural materials up to 650°C, although at higher temperatures the stainless steel became superior to the nickel alloy on account of preferential leaching of nickel from the latter. Cheng and Ruther's loop tests of nickel in liquid sodium at 550–715°C[143] indicated linear rate relationships at each temperature investigated, an important

Table 7.35 Attack of liquid metals on nickel and nickel-rich alloys[138]

Material type and composition	Liquid metal	Test condition		Category of attack*	Mass change (g)	Change in diameter† cm			
		Temp. (°C)	Time (s × 10⁻³)			A	B	C	D
Nickel	Bismuth	500	4.5	S	−1.65	0.086	0.094	0.166	0.172
Brightray (80Ni–20Cr)		500	360	M	+0.77	—	0.046	0.011	0.053
Monel		500	4.5	S	+0.81	—	0.156	0.012	0.118
Nickel	Cadmium	500	36	S	—	0.015	0.111	0.299	0.353
Brightray		500	360	S	+0.21	0.002	0.116	0.008	0.104
Monel		500	0.9	S	−0.20	0.014	0.058	0.046	0.082
Nickel	Mercury	500	360	M	—	ND§	0.005	—	—
Nickel	Potassium‡	400	1 080	N	—	ND	—	—	—
Nickel	Sodium‡	400	1 080	N	—	ND	—	—	—
Brightray		275	360	N	—	ND	—	—	—
Monel		275	360	N	—	ND	—	—	—
Nickel	Lead	500	360	S	+0.21	0.002	0.050	—	—
Brightray		500	360	M	+0.28	0.0045	0.005	—	—
Monel		500	360	M	+0.22	+0.0005	0.03	—	—
Nickel	Tin	500	36	S	+0.31	0.003	0.053	0.032	0.119
Monel		500	0.9	M	−0.47	0.022	0.034	0.034	0.047
Nickel	Zinc	500	360	S	+12.2	+0.158	0.06	+0.146	0.08
Brightray		500	360	S	+56.7	+1.1	0.35	—	—
Monel		500	3.6	M	−0.28	+0.004	0.007	—	0.027

* S, Severe; M, Moderate; N, Very slight or no attack.
† Negative unless stated otherwise; A. Mean change at middle of specimen; B. Including internal penetration at middle of specimen; C. Maximum change in diameter; D. Including internal penetration at position of C.
‡ Oxide contamination of liquid metal.
§ ND. Not determined.

Table 7.36 Results in thermal convection loop tests of material in contact with molten lead[139]
(Maximum loop temperature 800°C, minimum 500°C)

Material type and composition	Time to complete plugging of loop ($s \times 10^{-3}$)	Remarks
Niobium	1 962*	No mass transfer
Molybdenum	1 800*	No mass transfer
Hastelloy B (Ni-28Mo-5Fe)	1 814*	Little mass transfer
Ni-25Mo	2 419*	Little mass transfer
Nickel	7.2	Heavy mass transfer
Fe-30Ni	990	Heavy mass transfer
Inconel (Ni-16Cr-7Fe)	324	Heavy mass transfer
Nichrome V (Ni-20Cr)	43	Heavy mass transfer

* Test discontinued.

increase with temperature (activation energy for corrosion of 122 kJ mol^{-1} rather greater than that required for simple solution) and negligible effects of oxygen content of the sodium, or neutron flux. The rates of corrosion, however, are some 10–20 times those obtained for stainless steel under comparable conditions.

Tests[144] in a Na–K alloy loop at 570–710°C indicated little or no attack in 6 000 h on Inconel alloy 718 (53Ni–19Cr–19Fe base) or Inconel alloy X750 (73Ni–15Cr–7Fe–2·5Ti), a precipitation-hardening material. The surface quality of the latter alloy was reported to influence strongly the corrosion rate and its dependence on temperature[145]. Lithium is generally much more corrosive than sodium, and alloys containing nickel are not considered suitable for prolonged use above 600°C because of the mass transfer problem[146].

Even though not molten, uranium rapidly attacks Nimonic 80A (Ni–20Cr–2·5Ti–1·5Al) at temperatures above 740°C, at which temperature the nickel–uranium eutectic melts.

In general nickel–chromium alloys are not suitable for use in contact with molten aluminium. Grogan[147] investigated the degree of attack in 13½ h at 800°C of a molten Al–8Cu alloy on a series of materials. A nickel–chromium alloy was most severely corroded, followed in order of increasing resistance by Cronite (Ni–23Fe–14Cr), an Fe–27Cr–20Ni–4Al alloy, an Fe–18Cr–7Ni–3Mn alloy, chromium–vanadium steel and grey cast iron. Molten copper and copper-base alloys on the other hand do not seriously attack nickel–chromium alloys, which have been successfully used as dies and die inserts for the gravity diecasting of aluminium bronze at pouring temperatures of 1 150°C.

In contact with molten salts, the nickel-base alloys behave much more satisfactorily than is the general experience with molten metals. For this reason they are considered as structural materials in atomic reactors using fluoride mixtures as coolants and are used as vessels for heat-treatment salt baths, as thermocouple sheaths and in similar applications.

In the United States the alloy Inor 8 or Hastelloy N (Ni–16Mo–7Cr–5Fe) has been developed as a container material for molten fluorides containing uranium. The nickel–chromium–iron alloy originally considered as a suitable material proved unsatisfactory, owing to preferential leaching of chromium, but it was established[148] that a chromium content of less than 8% could be tolerated, while reduction below 6% resulted in inadequate oxidation resistance. In thermal convection loop tests with a maximum temperature of 677°C, attack on the nickel–chromium–iron alloy in 1 000 h penetrated up to 0·1 mm, whereas with Inor 8 only very limited attack up to 0·02 mm was experienced in over 6 000 h[149]. A review of the corrosion of several metals and alloys including nickel and various nickel alloys by fluorides, chlorides and hydroxides, has been given by Shimotake[150].

Nickel thermocouple sheaths have been used satisfactorily in salt baths, e.g. in barium chloride at 1 000–1 100°C and in sodium and potassium chlorides at up to 980°C. If the salt is contaminated with sulphur then an Inconel sheath is preferred to one of nickel. The latter type is also satisfactory in cyanide baths used for case-hardening.

A common constituent of salt baths is boric oxide and the use of such mixtures in contact with nickel-rich alloys containing Ti, Al or other strong reducing elements is to be avoided, owing to the possible formation of a nickel–boron eutectic of low melting point.

While the few examples quoted provide some general guidance as to the behaviour of nickel-rich materials in contact with molten metals and salts, it cannot be over-emphasised that such behaviour can be very considerably modified by the presence of very small amounts of contaminants in the liquid media (see Sections 2.9 and 2.10). The effect of very small contents of sodium chloride on the corrosion of nickel-base alloys by sodium sulphate has been referred to previously and other reported examples involving trace amounts, particularly of gaseous impurities, underline the need for great care in interpretation of experimental results.

R. A. SMITH
S. R. J. SAUNDERS

REFERENCES

1. Atkinson, A., in *Oxidation of Metals and Alloys and Associated Mass Transport*, ed. Dayananda M. A. *et al.*, The Metallurgical Society of the AIME, Warrendale, Pa., 29, (1987)
2. Horn, L., *Z. Metallk.*, **40**, 73 (1949)
3. Duffy, D. M. and Tasker, P. W., Report TP1155, Harwell Laboratory, Oxfordshire, U.K., (1986)
4. Atkinson, A. and Taylor, R. I., *J. Phys. Chem. Solids*, **47**, 315 (1986)
5. Kubaschewski, O. and Hopkins, B. E., *Oxidation of Metals and Alloys*, 2nd. edn., Butterworths (1962)
6. Zima, G. E., *Trans. Amer. Soc. Metals*, **49**, 924 (1957)
7. Copson, H. R. and Lang, F. S., *Corrosion*, **15**, 194t (1959)
8. Evans, E. B., *Corrosion*, **21**, 274 (1965)
9. Barrett, C. A., Evans, E. B. and Baldwin, W. M., *Thermodynamics and Kinetics of Metals and Alloys: High-temperature Scaling of Ni-Cr, Fe-Cr, Cu-Cr and Co-Mn Alloys*, U.S. Army Office of Ordnance Res. Lab. Rept. (1955)
10. Kosak, R., Ph.D. Dissertation, Ohio State Univ. (1969)
11. Yearian, H. J. *et al.*, *Corrosion*, **12**, 561t (1956)
12. Chattopadhay, B. and Wood, G. C., *Oxid. Met.* **2**, 373 (1970)
13. Pfeiffer, I., *Z. Metallk.*, **53**, 309 (1962)
14. Pfeiffer, H., *Elektrowärme Technik.*, **4**, 39, 79 (1955)
15. Douglas, D. L. and Armijo, J. S., *Oxid. Met.*, **2**, 207 (1970)
16. Wei, F. I. and Stott, F. H., *Reactivity of Solids*, **6**, 129 (1988)
17. Stott, F. H. and Wood, G. C., *Corr. Sci.*, **11**, 799 (1971)
18. Ignetov, D. V. and Shamgimova, R. D., NASA Trans. TTF-39 (1960)
19. Malamand, F. and Vidal, G., *Rech. Aéro.*, No. 56, 47 (1957)

20. Wickusik, C. S., AGARD Conf. Proc. No. 52, Paper No. 12 (1970)
21. Richmond, J. C. and Thornton, H. R., Wright Air Development Centre, Tech. Rept. No. 58-164 (1958)
22. Betteridge, W., *The Nimonic Alloys*, Edward Arnold, 234 (1959)
23. Ferré, S., *Rev. Métall,* **56**, 386 (1959)
24. Pettit, F. S. and Meier, G. H., in *Superalloys* 84, ed. Gell, M. *et al.*, The Metallurgical Society of the AIME, Warrendale, Pa., 651 (1984)
25. Preece, A. and Lucas, G., *J. Inst. Met.*, **81**, 219 (1952)
26. Barrett, C. A., Barlick, R. G. and Lowell, C. E., *High Temperature Cyclic Oxidation Data*, NASA TM 83655 (May 1984)
27. Michels, H. T., *Met. Trans. A*, **8A**, 273 (1977)
28. Stringer, J., Wilcox, B. A. and Jaffee, R. I., *Oxid. Met.*, **5**, 11 (1972)
29. Whittle, D. P. and Stringer, J., *Phil. Trans. R. Soc. Lond.*, A295, 309 (1980)
30. Luthra, K. L. and Briant, C. L., Spring Meeting of the Electrochemical Society, Cincinnati, 6-11 May, *Extended Abstracts*, **84-1**, 26 (1984)
31. Smeggil, L. G., Funkenbusch, A. W. and Bornstein, N. S., Spring Meeting of the Electrochemical Society, Cincinnati, 6-11 May, *Extended Abstracts*, **84**, 27 (1984)
32. Lees, D. G., *Oxid. Met.*, **27**, 75 (1987)
33. Luthra, K. L. and Briant, C. L., *Oxid. Met.* **26**, 397 (1986)
34. Funkenbusch, A. W., Smeggil, J. G. and Bornstein, N. S., *Met. Trans. A*, **16A**, 1164 (1985); and Smeggil, J. G., Funkenbusch, A. W. and Bornstein, N. S. *Met. Trans. A*, **17A** 923 (1986)
35. Smialek, J. L., *Met. Trans. A*, **18A**, 164 (1987)
36. Luthra, K. L., Briant, C. L., Smeggil, J. G., Bornstein, N. S., De Crescente, M. A. and Lees, D. G., *Oxid. Met.*, **30**, 255-67 (1988)
37. J. R. C. Petten, in *The role of active elements in the oxidation behaviour of high-temperature metals and alloys*, The Netherlands, 12-13 Dec. 1988, ed. Lang, E., Elsevier Applied Science Publishers (1989)
38. Nicholls, J. R. and Hancock, P., in *High Temperature Corrosion*, NACE-6, ed. Rapp, R. A., NACE, Houston, 198 (1983)
39. Foley, R. T., *J. Electrochem. Soc.*, **109**, 1202 (1962)
40. Wulf, G. L., Carter, T. J. and Wallwork, G. R., *Corr. Sci.*, **9**, 689 (1969)
41. Brasunas, A. de S., Gow, J. T. and Harder, O. E., *Proc. Amer. Soc. Test. Mat.*, **46**, 870 (1946)
42. Eiselstein, H. H. and Skinner, E. N., A.S.T.M., Special Tech. Pub. No. 165, 162 (1954)
43. Trueman, J. E. and Pirt, K. R., *Br. Corr. J.*, **11**, 188 (1976)
44. Hindam, H. M. and Smeltzer, W. W., *J. Electrochem. Soc.*, **127**, 1622 (1980)
45. Stott, F. H. and Wood, G. C., *Corr. Sci.*, **17**, 647 (1977)
46. Destefani, J. D., *Advanced Materials and Processes*, **2**, 37 (1989)
47. Saunders, S. R. J. and Nicholls, J. R., *Mat. Sci. and Tech.*, **5**, 570 (1989)
48. Hindam, H. M. and Smeltzer, W. W., *J. Electrochem. Soc.*, **127**, 1630 (1980)
49. Hickman, J. W. and Gulbransen, E. A., *Trans. Amer. Inst. Min. (Metall.) Engrs.*, **180**, 519 (1949)
50. *The Heating of Nickel*, Mond Nickel Co. Ltd., Pub. No. 910A (1955)
51. Pfeiffer, I., *Z. Metallk.*, **66**, 516 (1955)
52. Mrowec, St., *Werkstoffe und Korrosion*, **31**, 371 (1980)
53. Kofstad, P., *High Temperature Corrosion*, Elsevier Applied Science Publishers, London, 440 (1988)
54. Tideswell, N. W., *Corrosion*, **28**, 23 (1972)
55. Kramer, L. and Simkovich, G., *Oxid. Met.*, **6**, 91 (1973)
56. Von Bruckman, A. and Mrowec, St., *Werkstoffe und Korrosion*, **25**, 502 (1974)
57. Rahmel, A., *Corr. Sci.*, **13**, 125 (1973)
58. Wotton, M. R. and Birks, N., *Corr. Sci.*, **12**, 829 (1972)
59. Kofstad, P. and Akesson, G., *Oxid. Met.* **12**, 503 (1978)
60. Luthra, K. L. and Worrell, W. L., *Met. Trans. A*, **9**, 1055 (1978) and **10**, 621 (1979)
61. Nakai, H. and Takasawi, H., *Nippon Kinzoku Gakkaeshi*, **40**, 166 (1976)
62. Seiersten, M. and Kofstad, P., *Corr. Sci.*, **22**, 487 (1982)
63. Pfeiffer, I., *Z. Metallk.*, **49**, 267 (1958)
64. Haflan, B. Kofstad, P., *Oxid. Met.*, **25**, 217 (1986) and Anderson, A., Haflan, B., Kofstad, P. and Lillerud, P. K., *Mat. Sci. Eng.*, **87**, 45 (1987)
65. Lifshin, E., Ph.D. Dissertation, Renssalaer Polytechnic Inst., Troy, N.Y., Jan. (1969)
66. Gruber, H., *Festschrift z. 70 Geburtstage v. W. Heraeus*, Hanau, 45 (1930); *Z. Metallk.*, **23**, 151 (1931)
67. Hancock, P., 1st Int. Congress on Metallic Corrosion, London (1961)
68. Chandler, J. and McQueen, H. J., *Corrosion*, **25**, 126 (1969)

69. Romeo, G., Smeltzer, W. W. and Kirkaldy, J. S., *J. Electrochem. Soc.*, **118**, 740 (1971)
70. Romeo, G., Smeltzer, W. W. and Kirkaldy, J. S., *J. Electrochem. Soc.*, **118**, 1336 (1971)
71. Nowak, J. F., Lambertin, M. and Colson, J. C., *Corr. Sci.*, **18**, 971 (1978)
72. Perkins, R. A., *Proc. Conf. 'Environmental Degradation of High Temperature Materials'*, ed. Denner, S. G., The Institution of Metallurgists, London, 5/1-17 (1980)
73. Natesan, K., in *High Temperature Corrosion*, NACE-6, ed. Rapp, R. A., NACE, Houston, 336 (1983)
74. Stroosnijder, M. J. and Quadakkers, W. J., *High Temp. Tech.*, **4**, 141 (1986)
75. Grabke, H. J., Norton, J. F. and Casteels, F. G., *High Temperature Alloys for Gas Turbines and Other Applications* 1986, ed. Betz, W. *et al.*, D Reidel Publishing Co., Dordrecht, 245 (1986)
76. Benbyamani, M., Ajersch, F. and Kennedy, G., *J. Electrochem. Soc.*, **136**, 843 (1989)
77. Giggins, C. S. and Pettit, F. S., *Oxid. Met.*, **14**, 363 (1980)
78. Vasantasree, V. and Hocking, M. G., *Corr. Sci.*, **16**, 261, 279 (1976)
79. Spengler, C. J. and Viswanathan, R., *Met. Trans.*, **3**, 161 (1972)
80. Jackson, J. H., 4ème Congrès Int. du Chauffage Industriel, Oct. Group 1, Section 16, Paper No. 144 (1952)
81. Hill, V. L. and Meyer, H. S., in *High Temperature Corrosion in Energy Systems*, ed. Rothman, M. F., The Metallurgical Society of the AIME, Warrendale, Pa., 29 (1985)
82. Lai, G. Y., in *High Temperature Corrosion in Energy Systems*, ed. Rothman, M. F., The Metallurgical Society of the AIME, Warrendale, Pa., 227 (1985)
83. Avery, H. S. and Matthews, N. A., *Trans. Amer. Soc. Met.*, **38**, 975 (1947)
84. Seybolt, A. V., *Corr. Sci.*, **11**, 751 (1971)
85. Saunders, S. R. J. and Nicholls, J. R., *Thin Solid films*, **119**, 247 (1984)
86. Stringer, J., in *High Temperature Corrosion in Energy Systems*, ed. Rothman, M. F., The Metallurgical Society of the AIME, Warrendale, Pa., 3 (1985)
87. Bornstein, N. S. and DeCrescente, M. A., *Trans. AIME*, **245**, 1947 (1969)
88. Goebel, J. A. and Pettit, F. S., *Met. Trans.*, **1**, 1943, 3421 (1970)
89. Rapp, R. A. and Goto, K. S., in *The Hot Corrosion of Metals By Molten Salts*, ed. Braunstein, J. *et al.*, Electrochemical Soc., Pennington, N.J., 81 (1981)
90. Luthra, K. L. and Shores, D. A., *J. Electrochem. Soc.*, **127**, 2202 (1980)
91. Rapp, R. A., *Mat. Sci. Eng.*, **88**, 319 (1987)
92. Luthra, K. L., *J. Electrochem. Soc.*, **132**, 1293 (1985)
93. Goebel, J. A., Hecht, R. J. and Vargas, J. R., *Proc. 'Fourth Conf. on Gas Turbine Materials in a Marine Environment'*, U.S. Naval Sea Systems Command, Washington DC, 635 (1979)
94. Hurst, R. C., Johnson, J. B., Davies, M. and Hancock, P., in *Deposition and Corrosion in Gas Turbines*, ed. Hard, A. B. and Cutler, A. J. B., Applied Science Publishers, Barking, 143 (1973)
95. Fontaine, P. I. and Richards, E. G., A.S.T.M. Spec. Tech. Pub. No. 421, 269 (1967)
96. Schlapfer, P., Amgwerd, P. and Preis, H., *Schweiz. Arch. Angew. Wiss.*, **15**, 291 (1949)
97. Sachs, K., *Metallurgia*, **57**, 123, 167, 224 (1958)
98. Kvernes, I. and Seiersten, M., in *High Temperature Corrosion*, ed. Rapp, R. A., National Association of Corrosion Engineers, Houston, Tex., 615 (1983)
99. Luthra, K. L. and Spacil, H. S., *J. Electrochem. Soc.*, **129**, 649 (1982)
100. Rapp, R. A., *Mat. Sci. Eng.*, **87**, 315 (1987)
101. Bornstein, N. S. and DeCrescente, M. A., in *Properties of High Temperature Alloys with Emphasis on Environmental Effects*, ed. Foroulis, Z. A. and Pettit, F. S., The Electrochemical Soc., Princetown, N.J., 626 (1976)
102. Betteridge, W., Sachs, K. and Lewis, H., *J. Inst. Petrol.*, **41**, 170 (1955)
103. Part 1 of Report of Council of British Manufacturers of Petroleum Equipment, Corrosion I. Cttee, *Brit. Petrol. Equipm. News*, **7**, 54 (1959)
104. Hancock, P. and Islam, M., *Proc. Conf. on High Temperature Alloys*, ed. Kirman, I. *et al.*, Metals Society, London, 1979
105. Saunders, S. R. J., Spencer, S. J. and Nicholls, J. R., *Proc. Conf. on Diesel Engine Combustion Chamber Materials for Heavy Fuel Operation*, 26-27 Oct. 1979, Inst. of Marine Engineers, London, in press.
106. Seiersten, M. and Kofstad, P., cited in Kofstad, P., *High Temperature Corrosion*, Elsevier Applied Science, London, 497 (1988)
107. Penrice, P. J., Stapley, A. J. and Towers, J. A., *J. Inst. Fuel*, **39**, 300 (1966)
108. Parry, P. J., Bridges, P. J. and Taylor, B., *J. Inst. Met.*, **97**, 373 (1969)
109. Ennis, P. J. and Bridges, P. J., *J. Inst. Met.*, **100**, 346 (1972)
110. Nicholls, J. R. and Stephenson, D. J., *Proc. Conf. on Diesel Engine Combustion Chamber Materials for Heavy Fuel Operation*, 26-27 Oct. 1989, Inst. of Marine Engineers, London, in press.

111. Cutler, A. J. B., Flateley, T. and Hay, K. A., *CEGB Research*, Oct., 13 (1978)
112. Gerds, A. F. and Mallett, M. W., U.S. Atomic Energy Commission Rept. BMI-1 261 (1958)
113. Bernsen, S. A. *et al.*, U.S. Atomic Energy Commission Rept. TID-7 564, 243 (1958)
114. Kalvenes, O., Piene, K. and Kofstad, P., *Corr. Sci.*, **4**, 211 (1964)
115. International Nickel Ltd., unpublished data
116. Kline, E. M. and Hall, A. M., *Metals and Alloys*, **21**, 401 (1945)
117. Grabke, H. K. and Schnaas, A., in *Alloy 800*, ed. Betteridge, W. *et al.*, North-Holland Publishing Co., Amsterdam, 195 (1978)
118. Hochmann, R. F., in *Properties of High Temperature Alloys with Emphasis on Environmental Effects*, ed. Foroulis, Z. A. and Pettit, F. S., The Electrochemical Soc., Princetown, N.J., 715 (1976)
119. Kane, R. H., *Corrosion*, **37**, 187 (1981)
120. Lai, G. Y. and Rothman, M. F., *Proc. Conf. 'Corrosion '84'*, paper 11, National Association of Corrosion Engineers, Houston, Tex., (April 1984)
121. Hall, D. J., Hossain, M. K. and Jones, J. J., *Material Performance*, **24**, 25 (1985)
122. Grabke, H. J., Paulitschke, W., Tauber, G. and Veifhaus, H., *Surface Science*, **63**, 377 (1977)
123. Grabke, H. J., Moller, R. and Schnaas, A., *Werkstoffe und Korrosion*, **30**, 794 (1979)
124. Schnaas, A. and Grabke, H. J., *Oxid. Met.*, **12**, 387 (1978)
125. Hall, D. J., National Physical Laboratory, 1989, unpublished work
126. Pfeiffer, H., *Arch. Eisenhüttenw.*, **29**, 575 (1958)
127. Ihrig, H. K., *Industr. Engng. Chem.*, **41**, 2 516 (1949)
128. Pray, H. A., Peoples, R. S. and Ericson, G. L., U.S. Atomic Energy Commission Rept. BMI-269 (1953)
129. Mayer, P. and Manolescu, A. V., *Corrosion*, **36**, 369 (1980)
130. Brooks, S. and Meadowcroft, D. B., in *Corrosion Resistant Materials for Coal Conversion Systems*, ed. Meadowcroft, D. B. and Manning, M. I., Applied Science Publishers, London, 105 (1982)
131. Perkins, R. A. and Bakker, W. T., *Proc. Conf. Materials for Coal Gasification*, ed. Bakker, W. T. *et al.*, 10–15 Oct., Cincinnati, Ohio, ASM International, 85 (1987)
132. Li, Y. K. and Rapp, R. A., *Trans. Jpn. Inst. Met.*, **24**, 687 (1983)
133. Kofstad, P., *High Temperature Corrosion*, Elsevier Applied Science, London, 504 (1988)
134. Hossain, M. K., Rhoades-Brown, J. E., Saunders, S. R. J. and Ball, K., *Proc. Conf. 'UK Corrosion '83'*, Birmingham, Nov., Inst. of Corr. Sci. and Technology, 61 (1983)
135. Stott, F. H., Prescott, R., Elliott, P. and Al'Atia, M. H. J. H., *High Temp. Tech.*, **6**, 115 (1988)
136. Daniel, P. L. and Rapp, R. A., in *Advances in Corrosion Science and Technology*, ed. Fontana, M. G. and Staehle, R. W., **5**, 55 (1976)
137. Ganesan, P. and Smith, G. D., *Proc. Conf. 'High Temperature Materials in Fluidised Bed Combustion Systems and Process Industries'*, ed. Ganesan, P. and Bradley, R. A., Cincinnati Ohio, Oct., ASM International, 167 (1987)
138. Eldred, V. W., U.K. Atomic Energy Research Est. Report No. X/R1 806 (1955)
139. Cathcart, J. V. and Manly, W. D., *Corrosion*, **12**, 87t (1956)
140. Kishkin, S. T. and Nikolenko, V. V., *C.R. Acad. Sci. U.R.S.S.*, **110**, 1 018 (1956)
141. Petrova, A. R., Kaufman, V. G. and Vdovina, L. M., *Metallov term Obrab Metallov*, No. 10, 73 (1969)
142. Hoffman, E. E. and Manly, W. D., Nuclear Engng. Sci. Congr., Cleveland (1955), Preprint No. 74
143. Cheng, C. F. and Ruther, W. E., *Corrosion*, **28**, 20 (1972)
144. Svedberg, R. C., Keeton, A. and Amman, R. L., Westinghouse Astronuclear Lab. Rep. TME 1 857 (1968)
145. Borgstedt, H. V., Drechsler, G. and Frees, G., *Proc. Sym. Alkali Metal Coolants*, Vienna, 119 (1966)
146. Champeix, L., *Energie Nucleaire*, **8**, 471 (1966)
147. Grogan, J. D., *J. Inst. Met.*, **44**, 279 (1930)
148. Badger, F. S., *Chem. Engng.*, **66**, 162, 164, 166 (1959)
149. MacPherson, H. G., U.S. Atomic Energy Commission Rept. ORNL-2 551 (1958)
150. Shimotake, H. and Hessen, J. C., *Adv. Chem. Service*, No. 64, 149 (1967)

7.6 Thermodynamics and Kinetics of Gas–Metal Systems

The theoretical treatment of gas–metal systems can become very involved because of the large number of variables which may have to be considered simultaneously. These can arise not only from the nature of the atmospheres, temperature (or temperature gradient across the interaction surface layers) and partial pressures of active gases, but also from the composition of the metal in terms of its purity, alloying elements and surface treatment, stresses, vibration, irradiation etc. Fortunately, some of these factors have very little influence on the equilibria, and need to be considered in specific cases only. Any full theoretical treatment of these systems invariably must involve both the disciplines of thermodynamics and kinetics. In real systems these two can be directly interconnected[1] when dealing with gas–metal surface reactions which result in the formation of two or more surface phase layers. Thus, at the same temperature for a given gas–metal system one can observe drastically different rates of interaction (linear, parabolic or any other rate) when the gaseous composition of the atmosphere is altered and the relative partial pressures of the gaseous components are different. This drastic effect on the kinetics may result from the formation of different surface-phase layers which, in turn, may be protective or non-protective. Ideally, one would like to be able to predict theoretically the composition and the physical properties of any surface layer which could be formed in a given gas–metal system as a function of temperature and the partial pressures of the active gases.

A brief literature survey shows that over the years it has been found that the simplest and most convenient way of predicting the composition of the reaction products in these systems may be achieved by means of graphical thermodynamics. This, at a glance, gives a wealth of information which normally would require extensive calculation or pages of tables. Therefore, in this section it is intended to show the construction and use of the three main types of diagrams which have evolved over the years, namely (1) Ellingham-type diagrams, (2) Pourbaix and subsequent thermodynamic phase stability diagrams, and (3) integral free energy of mixing versus concentration diagrams.

Each of these diagrams can be used with advantage for solving specific types of problems but it is only with experience that one is able to judge which of these three types of diagram leads to the most direct solution of a problem.

The Ellingham – $\Delta G^0/T$ or $\Delta \mu/T$ – diagrams show better than the other two types of diagrams the preference for a reaction between a large number of different metals reacting with a particular gas (e.g. O_2, CO, CO_2, H_2, NH_3, Cl_2, HCl, H_2S etc.) as a function of temperature. Just by looking at the relative position of the relevant lines it is possible to assess which metal will preferentially react with a given gas and how this preference with respect to the other metals may be affected by changes in the temperature and the composition and pressure of the atmosphere.

The thermodynamic phase stability diagrams appear to be preferred by corrosion scientists and technologists for the evaluation of gas–metal systems where the chemical composition of the gaseous phase consisting of a single gas or mixture of gases has a critical influence on the formation of surface reaction products which, in turn, may either stifle or accelerate the rate of corrosion. Also, they are used to analyse or predict the reason for the sequence of formation of the phases in a multi-layered surface reaction product on a metal or alloy.

The integral free energy–concentration diagrams have proved essential for the assessment of systems that are sensitive to stress and/or small changes in chemical potentials of the gaseous phase, prediction of the composition of non-stoichiometric compounds and the possibility of a composition different to that of the phase shown by the equilibrium phase diagram when growing as a surface layer. In these particular problems this graphical method has proved to be superior to the preceding methods as it is based on the simultaneous fulfilment of two conditions – the lowest free energy change and a common chemical potential of a given component in all the phases in contact with each other.

Free Energy – Temperature Diagrams (Ellingham Diagrams)

The thermodynamic functions which are required to be manipulated for studying high-temperature equilibria are relatively few in number and the mathematical equations involved can be reduced to about six. The fact that consideration will be given here only to high-temperature processes also leads to a considerable simplification of the algebra. For example, the standard free-energy change when an ordered compound MX is formed between metal M and non-metal X is given at temperature T by the equation

$$\Delta G_T^\ominus = \Delta H_0^\ominus + \int_0^T \Delta Cp \, dT - T \int_0^T \Delta Cp \, d\ln T$$

where $\Delta H_0^\ominus$ is the heat of the reaction at the absolute zero and ΔCp is the difference in heat capacities between reactants and products. Above 298 K a good approximation for the above equation is

$$\Delta G_T^\ominus = \Delta H_{298}^\ominus - T\Delta S_{298}^\ominus$$

where $\Delta H_{298}^\ominus$ and $\Delta S_{298}^\ominus$ are the standard heat and entropy change in the reaction at 298 K. Thus the free-energy change is a simple linear function of temperature and diagrams can be constructed which show this relationship when compounds are formed between say, metals and oxygen[2] (Fig. 7.55)*. These $\Delta G/T$ relationships are straight lines having changes in slope at temperatures where either the metal or oxide melts, boils or undergoes a crystallographic transformation. These changes in slope can be taken into account by adding or subtracting, depending on whether the metal or the metal oxide is involved, the free-energy change for a phase change. Thus for a transformation t, the following correction is required:

$$\Delta G_t = \Delta H_t - T\Delta S_t$$

$$= \Delta H_t - \frac{T\Delta H_t}{T_t}$$

where T_t is the transformation temperature.

The standard free-energy change for a reaction $M + X \rightarrow MX$ is also related to the equilibrium constant for the corresponding reaction

$$K = \frac{a_{MX}}{a_M a_X} \quad (7.11)$$

by the equation

$$-\Delta G^\ominus = RT \ln K = 19 \cdot 15 T \log K \quad (7.12)$$

Substituting for K in equation 7.12 from equation 7.11

$$-\Delta G^\ominus = RT \ln a_{MX} - RT \ln a_M - RT \ln a_X$$

which can be written as

$$-\Delta G^\ominus = \Delta \mu_{MX} - \Delta \mu_M - \Delta \mu_X$$

where $\Delta \mu_i = RT \ln a_i$ is the chemical potential of i.

Thus, as a simple example, when a pure metal M is in contact with the pure metal compound MX, then $a_{MX} = a_M = 1$ and

$$-\Delta G^\ominus = -\Delta \mu_X$$

Furthermore, if an alloy of M is in contact with pure MX,

$$-\Delta G^\ominus = -\Delta \mu_M - \Delta \mu_X$$

and finally when pure M is in contact with a solid or liquid solution containing MX

$$-\Delta G^\ominus = \Delta \mu_{MX} - \Delta \mu_X$$

A chemical reaction can occur only if $-\Delta G > 0$, i.e. if $-\Delta G$ is positive; in addition the value $a = 1$ is by definition the maximum activity for a condensed component where the pure phase is taken as standard state, thus $\Delta \mu$ is always negative. This discussion will be restricted to gases where $p \leqslant 1$; taking $p = 1$ atm (101·325 kN/m^2) as the standard state for the gas, X, it is evident that $\Delta \mu_X$ is always a negative quantity or zero.

---

*In this section there are a number of $\Delta G^\ominus - T$ diagrams in which $\Delta G^\ominus$ is expressed in the original unit of *kilocalories*, rather than in the SI units of *joules*. This has been done in order to avoid the difficult redrawing and recalculating of the diagrams involved.

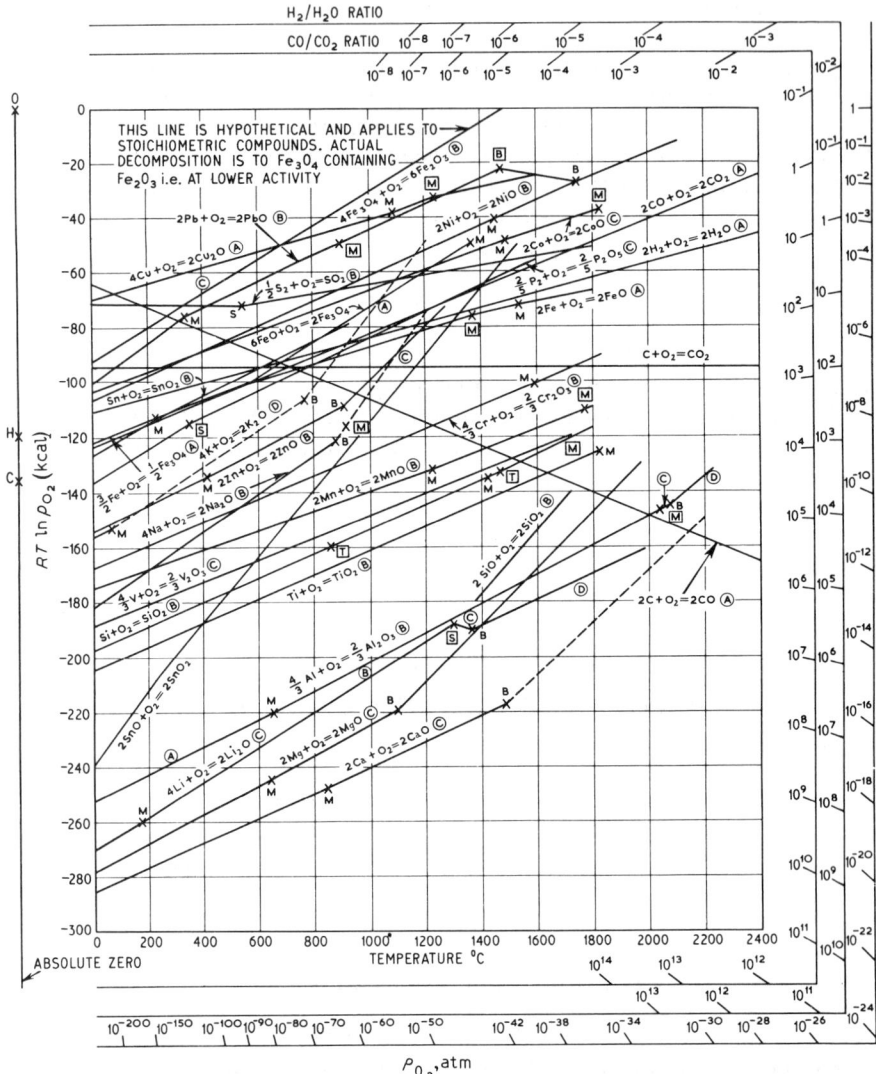

Fig. 7.55 Free energy of formation of oxides (after Richardson and Jeffes[2])

KEY				
Ⓐ		± 1 kcal	M, M̄	: melting point, metal, oxide, resp.
Ⓑ	suggested	± 3 kcal	B, B̄	: boiling point, metal, oxide, resp.
Ⓒ	accuracies	± 10 kcal	S, S̄	: sublimation point, metal, oxide, resp.
Ⓓ		± >10 kcal	T, T̄	: transition point, metal, oxide, resp.

Note: 1 kcal ≈ 4.2 kJ

The pressure of oxygen developed by an intimate mixture of a pure metal and its oxide is conventionally referred to as the dissociation pressure (p_{O_2}) and it can be seen that at a given temperature T

$$-\Delta G_T^\ominus = -\Delta \mu_{O_2} = -19 \cdot 15 T \log p_{O_2}$$

Thus the sequence of the free-energy lines in the diagram for a series of metal oxides shows those with the highest dissociation pressures at the top (smallest negative values of $\Delta\mu_{O_2}$) and those with the lowest dissociation pressures at the bottom, i.e. in the sequence of their stabilities. Clearly if a pure metal is exposed to an oxygen-bearing atmosphere where the pressure of oxygen is greater than the dissociation pressure of the oxide at the operating temperature, the metal oxide will be formed; if the gas contains

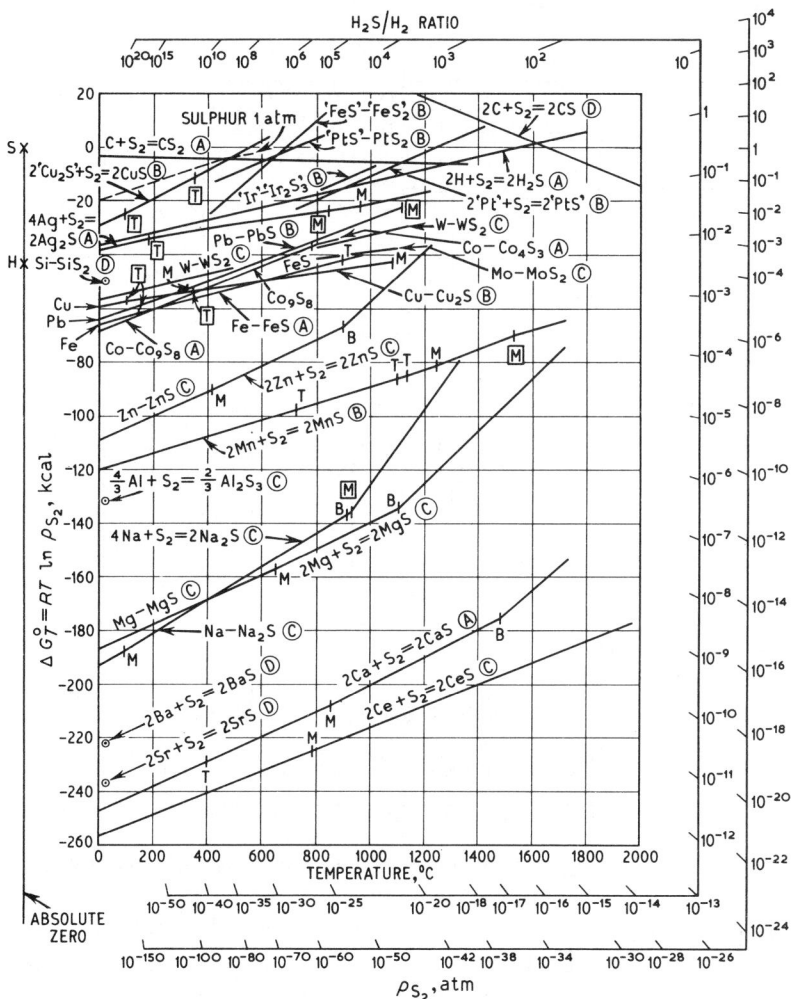

Fig. 7.56 Free energy of formation of sulphides (after Richardson and Jeffes[3])

KEY	Ⓐ		± 1 kcal	M, M̄	: melting point, metal, sulphide, resp.
	Ⓑ	suggested	± 3 kcal	B, B̄	: boiling point, metal, sulphide, resp.
	Ⓒ	accuracies	± 10 kcal	S, S̄	: sublimation point, metal, sulphide, resp.
	Ⓓ		± >10 kcal	T, T̄	: transition point, metal, sulphide, resp.

Note: 1 kcal ≈ 4.2 kJ

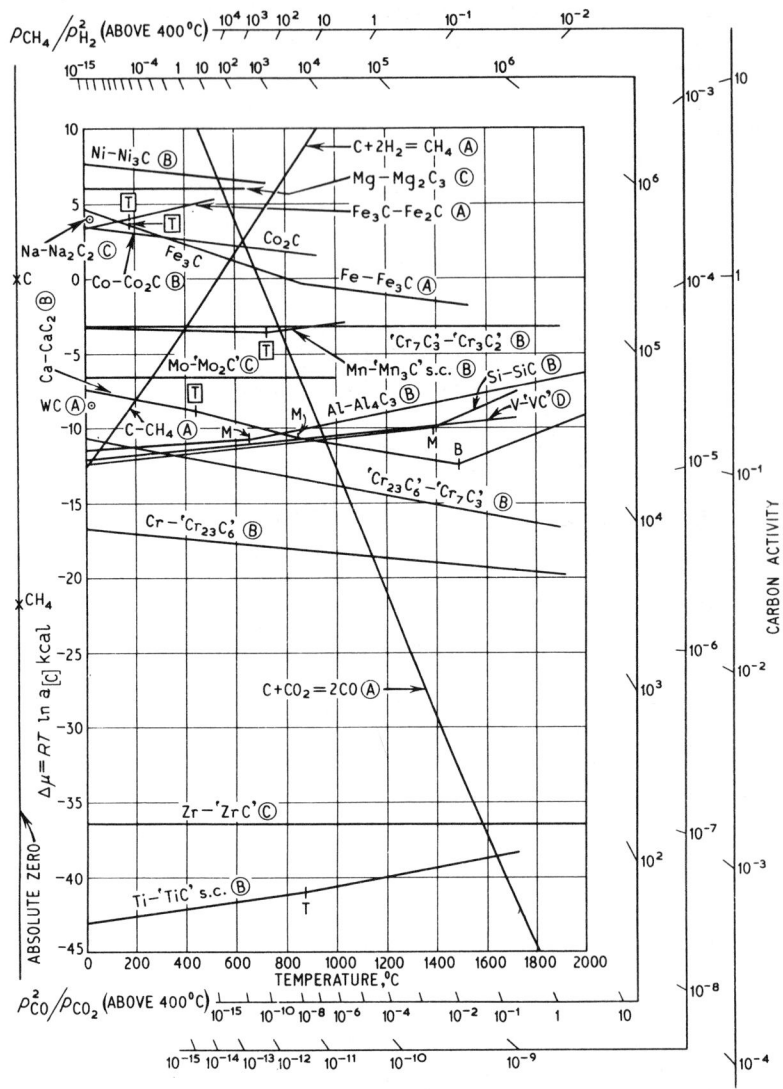

Fig. 7.57 Free energy of formation of carbides (after Richardson and Jeffes[4])

KEY Ⓐ ± 1 kcal
Ⓑ suggested ± 3 kcal M, M̄ : melting point, element, carbide, resp.
Ⓒ accuracies ± 10 kcal T, T̄ : transition point, element, carbide, resp.
Ⓓ ± > 10 kcal B, boiling point, element.

Note: 1 kcal ≈ 4.2 kJ

a lower oxygen pressure than the dissociation pressure, the oxide cannot be formed as a pure phase. It also follows from the equations given above that an alloy of a metal will require a higher oxygen pressure to form the pure oxide, and conversely a pure metal can form an oxide solid solution or liquid slag at a lower pressure than that required for pure oxide formation.

Similar free-energy diagrams, which can be interpreted in exactly the same way, have been constructed for sulphides[3], carbides[4] and nitrides[5] (Figs. 7.56 to 7.58).

It is unnecessary to go to the lengths of *calculating* the oxygen or sulphur potentials of gas phases in order to use these diagrams in certain simple cases. Consider the oxidation of a metal by a hydrogen/water-vapour atmosphere.

The reaction involved here is

$$M + H_2O \rightarrow MO + H_2$$

Therefore

$$-\Delta G^\theta = (\Delta\mu_{MO} - \Delta\mu_M) - RT \ln p_{H_2O}/p_{H_2}$$

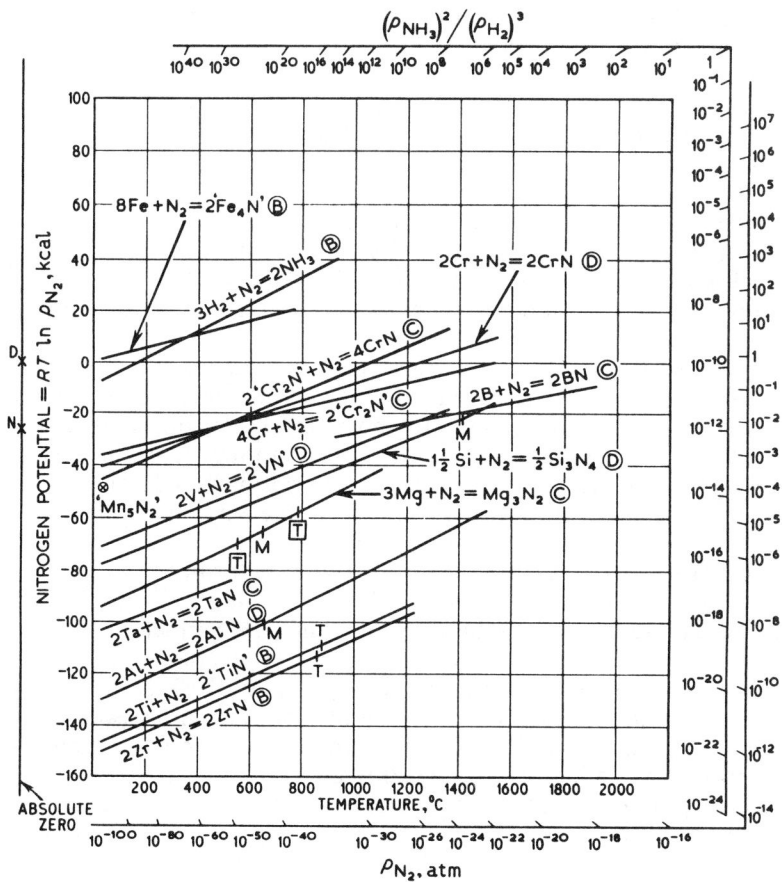

Fig. 7.58 Free energy of formation of nitrides (after Pearson and Ende[5])

KEY	B		± 3 kcal	M : melting point, element.
	C	suggested accuracies	± 10 kcal	T, [T] : transition point, element, nitride, resp.
	D		± >10 kcal	boiling point, element.

Note: 1 kcal ≈ 4.2 kJ

Thus when the oxidation of pure metal to pure oxide is being considered

$$-\Delta G^{\ominus} = -RT \ln p_{H_2O}/p_{H_2} = -\Delta\mu_{H_2O} + \Delta\mu_{H_2}$$

and the oxygen dissociation pressure of the metal/metal oxide is the same as that of the hydrogen/water-vapour mixture. It is thus a practical advantage to have a scale around the edge of the diagram showing values of $\Delta\mu_{O_2}$ (oxygen potential) for various ratios of p_{H_2O}/p_{H_2}.

For a mixture of hydrogen and steam at equal partial pressures, the oxygen potential will be equal to the standard free energy of formation of water vapour from hydrogen and oxygen at all temperatures. Therefore, the extrapolation of the standard free-energy line for the system $2H_2 + O_2 \rightarrow 2H_2O$ intersects the H_2/H_2O scale at the right-hand side of the diagram at the ratio 1/1. The point marked H on the left-hand side of the diagrams (Figs. 7.55 and 7.56) is the extrapolation of the same line to the absolute zero, and is thus equal to the standard heat of formation of water vapour from hydrogen and oxygen at 298 K.

When the hydrogen/water vapour ratio is 100/1, the point on the H_2/H_2O ratio scale representing this ratio is obtained by subtracting the chemical potential for a product molecule (i.e. H_2O in the reaction $H_2 + \frac{1}{2}O_2 \rightarrow H_2O$) at an activity of 1/100 from the standard free-energy line and extrapolating the resulting line to meet the scale at the point marked '$10^2/1$'. It should be observed that the value of the chemical potential for any substance at an activity of 0·01 is the same as that for oxygen at a pressure of 10^{-2} atm, and hence can be obtained from the diagram by using the oxygen potential scale.

From these examples the construction of this scale is apparent, and, as a corollary, it should be noted that the oxygen potentials of CO/CO_2 mixtures can be obtained by joining the point marked C on the left-hand side of the diagram, at the absolute zero of temperature, with the appropriate CO/CO_2 ratios marked on the scale at the right-hand side of the diagram.

A similar scale for p_{H_2S}/p_{H_2}, is attached to the sulphur diagram, and one for $p_{CH_4}/p_{H_2}^2$ to the carbon diagram, etc.

As examples, it can be seen from Fig. 7.55, by projecting the line which connects O and the MnO curve at 1 000°C to the p_{O_2} scale, that the dissociation pressure of manganese oxide (MnO) in contact with pure manganese is 10^{-24} atm pressure of oxygen at a temperature of 1 000°C. Similarly, it can be shown from the diagram that MnO is reduced to pure manganese in an atmosphere consisting of hydrogen and steam in the proportions 10^4:1 above 1 000°C, and in an atmosphere of carbon monoxide/carbon dioxide in proportions 10^5:1 above 1 000°C.

Referring to Fig. 7.56 it can be shown that the dissociation sulphur pressure, as S_2 molecules, of a mixture of copper and copper sulphide is 10^{-8} atm at about 900°C, and sulphide is formed on copper in an atmosphere of H_2/H_2S in the proportions 10^3:1 at all temperatures below 720°C.

Figure 7.59 shows the standard free energies of formation of metal chlorides as a function of temperature[6].

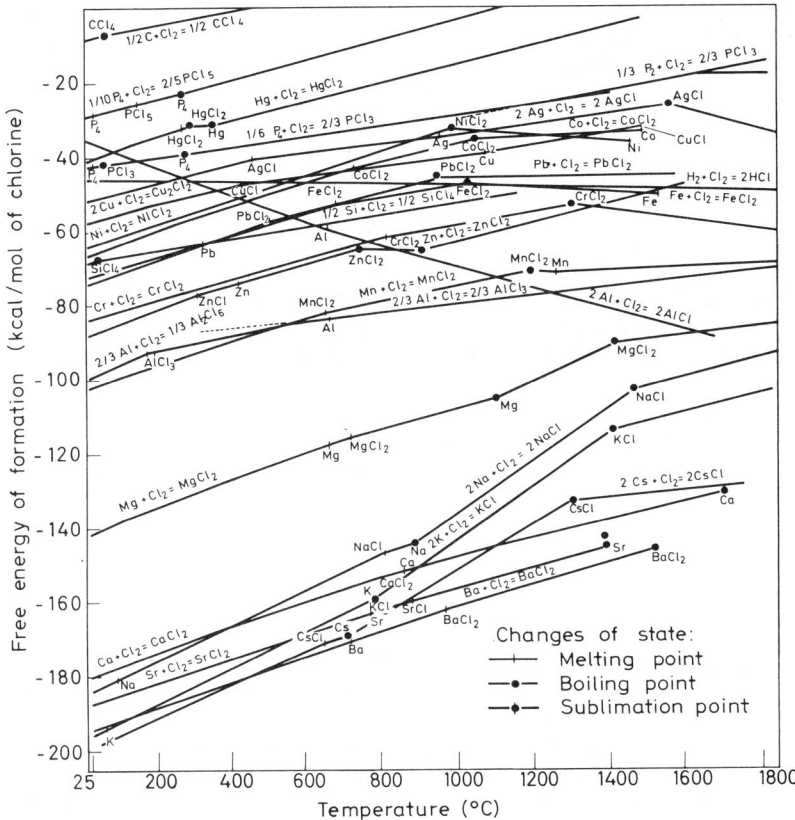

Fig. 7.59 Standard free energies of formation of metal chlorides as a function of temperature (after Villa[6]). Note: 1 kcal ≈ 4·2 kJ

Dilute Metallic Solutions

The metallurgist is concerned with the formation of homogeneous solutions of small amounts of impurities in metals as well as with the formation of compounds. The limit of solubility of impurities is frequently very small, less than one atomic per cent in concentration, and in these dilute solutions Henry's law is applicable, i.e. the activity of a dilute solute is proportional to the concentration of solute in the solution. Consider a dilute solution of an element A which has a high vapour pressure in the pure state at the temperature T, the vapour being monatomic, in solution in element B which has an immeasurably low vapour pressure at the same temperature. Then if the pressure of A could be measured unambiguously for a range of dilute solutions it would be found that

$$p_A = kx_A = k'(A)* \qquad (7.13)$$

*The value of k' obviously depends on whether atomic or weight per cent is used for expressing the concentration of A.

where x_A is the mole fraction of A in solution and (A) is the atomic, or alternatively the weight per cent of A in solution. If the vapour species of A were di-atomic instead of monatomic, then

$$p_{A_2} = kx_A^2 = k'(A)^2 \qquad (7.14)$$

which is Sievert's law for the solution of diatomic gases in metals. Similarly for a triatomic gas, e.g. SO_2

$$p_{SO_2} = kx_S x_O^2 = k'(S)(O)^2 \qquad (7.15)$$

In such a binary solution, the chemical potential of the solute $\Delta\mu_A$ and that of the solvent $\Delta\mu_B$ are related to the *integral free energy* of formation of the solution, ΔG^S, per mole, containing a mole fraction x_A of component A, and x_B for component B, by the expression

$$\Delta G^S = x_A \Delta\mu_A + x_B \Delta\mu_B$$

Corresponding to the integral heat and entropy of formation of the solution are the partial molar heats ΔH_i and entropies ΔS_i of solution of the components where

$$\Delta\mu_i = \Delta H_i - T\Delta S_i$$
$$\Delta H^S = x_A \Delta H_A + x_B \Delta H_B$$
$$\Delta S^S = x_A \Delta S_A + x_B \Delta S_B$$

From the algebraic form of these equations it can be seen that the partial and integral values of a thermodynamic function for a solution are interrelated simply. Figure 7.60 shows the integral value of a function ΔZ for a binary solution, as a function of x_B. At any given mole fraction of each component, the relevant values of the partial properties can be obtained by drawing a tangent to the integral curve at the given composition of the solution; ΔZ_A and ΔZ_B are the intercepts of this tangent with the A-rich and B-rich sides respectively. It also follows that ΔZ_A is the change in the value of the function Z for the component A when 1 mol of A is transferred from the standard state, usually the pure substance, to an infinite volume of the solution of the given concentration, so that the concentration of each species in the solution remains unchanged during the operation.

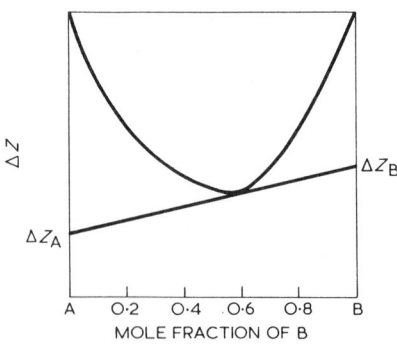

Fig. 7.60 Relationship between partial and integral properties

In the dilute range of concentration the partial heat of solution, ΔH_A, of A would be constant and the partial entropy of solution would be given by

$$\Delta S_A = a - R \ln x_A \qquad (7.16)$$

where a is a constant, and the standard state for A is pure A.

At higher concentrations, Henry's law no longer applies and activities must be substituted for the concentration terms. This statement implies that in the region of solution where Henry's law is valid, the activity coefficient γ of A, defined by

$$a_A = \gamma_A x_A = \gamma_A'(A)$$

is constant. Further, if we choose pure A as the standard state for A then γ_A has a constant value whose magnitude depends on the chemical identity of A, whereas if the standard state is the infinitely dilute solution of A then γ_A' has the constant value of unity in the Henry's law region. The equilibrium constant for, say, a dilute solution of sulphur in iron in equilibrium with an H_2/H_2S atmosphere can be simply expressed[7] as

$$K = (p_{H_2}/p_{H_2S}) \times (\%S)_{Fe}$$

when an infinitely dilute solution of sulphur in iron is taken as the standard state.

Solutions in Solid Iron

The dilute solutions of elements in solid iron are, at present, the only system for which the thermodynamics has been reasonably well worked out experimentally. The remainder of this section will therefore be devoted to the diagrammatic representation of data for these systems which have been evolved by Richardson[4].

The heat and entropy of solution of a dilute constituent remain constant when the infinitely dilute solution is taken as the standard state, provided that the solute obeys Henry's law and that no crystallographic change or change of state of the solvent occurs in the temperature range under consideration[8]. Thus within a given range of temperature in which the solvent remains unchanged, the partial free energy of solution of the solute may be represented on a free-energy/temperature diagram by a straight line. The intercept of this line with the free-energy ordinate at the absolute zero equals the heat of solution, and the slope gives the partial entropy of solution. However, when the stability of a dilute solution of a substance in iron is being compared with the stabilities of compounds, it is preferable to use the pure substance as standard state, in which case the slope of the free-energy line for the dilute solution of given concentration is given by equation 7.16. At the temperature at which the solvent undergoes a change in crystal structure there will be a discontinuous break in the line and, in the new structure, the free-energy line will have a different intercept at the absolute zero, indicating a change in the heat of solution, and a different slope which indicates a change in the constant a in equation 7.16 for the partial entropy of solution.

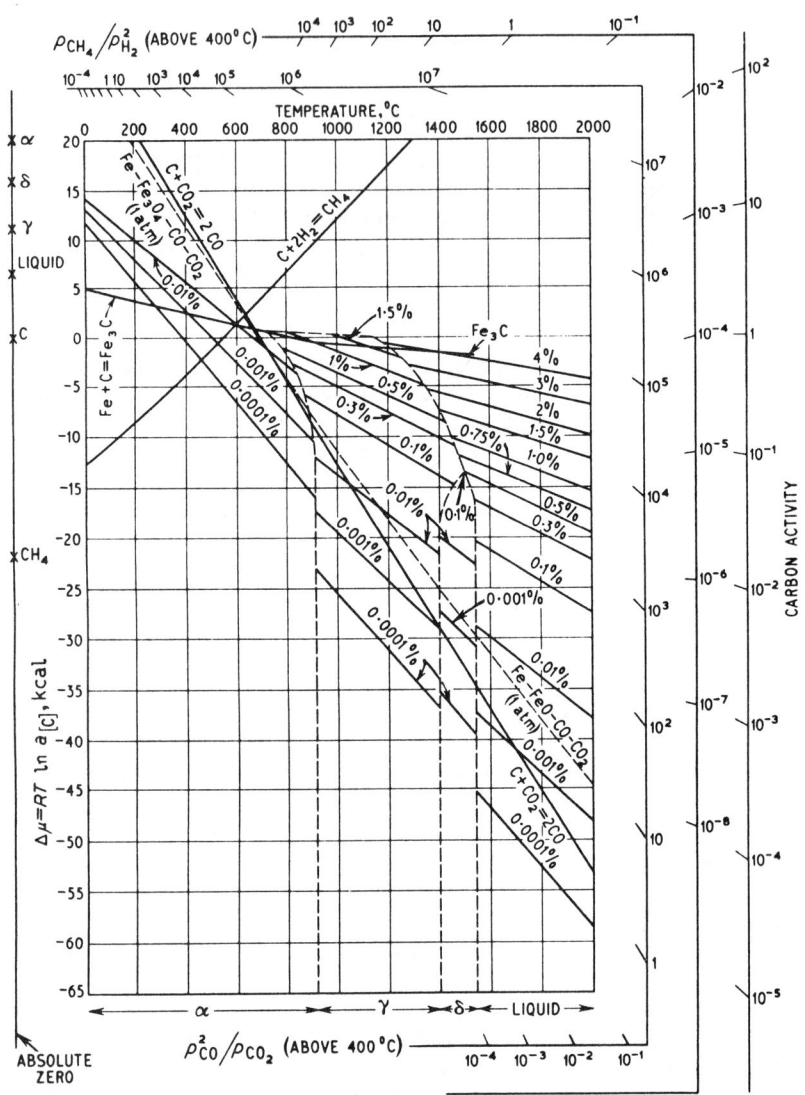

Fig. 7.61 Carbon potentials in iron (after Richardson and Jeffes[4]). Suggested accuracy of all lines ±1 kcal

By making use of experimental data for solutions of carbon in iron, Richardson has constructed a free-energy diagram showing the partial free energy of solution of carbon in α, γ and δ, and liquid iron (Fig. 7.61) which is similar to the earlier Ellingham diagram[9]. The figure is divided by the broken lines into areas of constant crystal structure of the iron solvent, and each area is traversed by lines (dotted in Fig. 7.62) showing the partial free energy of solution of carbon, at a concentration shown on each line, as a

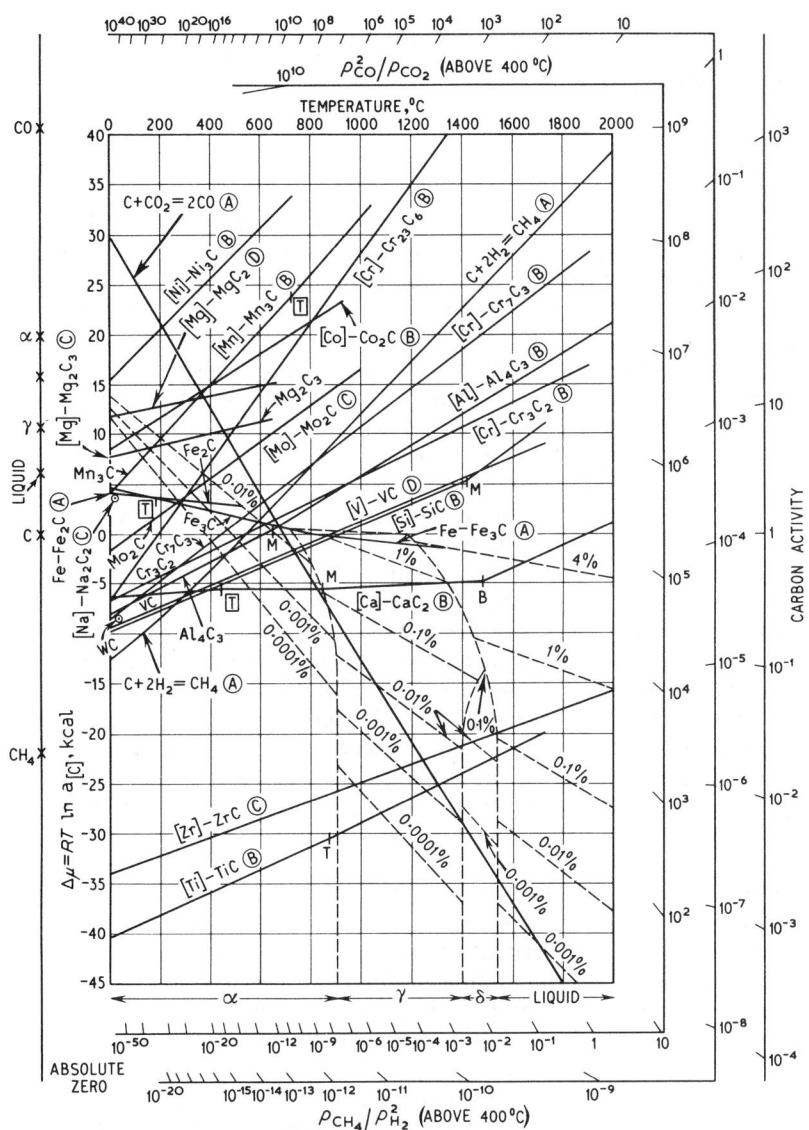

Fig. 7.62 Superposition of Figs. 7.57 and 7.61 (after Richardson and Jeffes[4])

KEY Ⓐ
　　　Ⓑ　suggested
　　　Ⓒ　accuracies
　　　Ⓓ

± 1 kcal
± 3 kcal
± 10 kcal
± > 10 kcal

M: melting point, element.
B: boiling point, element.
T, T̄ : transition point, element, carbide, resp.

Note: 1 kcal ≈ 4.2 kJ

function of temperature. The intercepts at the absolute zero, marked α, γ, δ, and liquid, give values for the partial molar heats of solution of carbon in each crystallographic form of the solvent.

Segregation of Carbides

This diagram may be usefully combined with the standard free-energy diagram for the carbides (Fig. 7.57) to indicate the equilibrium conditions under which carbide particles will segregate from a given solution consisting of an alloying element and carbon in solid iron. Figure 7.62 shows how the diagram for carbon in pure iron and the diagram for the formation of carbides by metals at an activity of 0·01 can be superimposed. The combined diagram can be used to calculate, for example, the temperature below which particles of vanadium carbide can be expected to begin separating from an iron alloy containing vanadium at an activity of 0·01, with respect to pure vanadium as standard state, and carbon at a concentration of 0·1 wt%. Since only a small amount of vanadium is present in the alloy, the activity of carbon at this concentration will be the same as that in pure iron to a good approximation. Thus the solution diagram needs no amendment. However, the carbide diagram gives the standard free energies of formation of carbide from the pure substances and the alloy contains vanadium at an activity of 0·01. It is necessary, therefore, to draw a straight line joining that for vanadium carbide at the absolute zero, and spaced a distance equal to $19 \cdot 15 T \log 0 \cdot 01$ *above* this line, across the diagram. (The values of this function can be read by joining the cross marked C on the middle left-hand side of the diagram with 'carbon activity' $= 10^{-2}$ on the right-hand side of the diagram.) When this line is drawn, it can be seen that the line for

$$[V]_{Fe} + C \rightarrow VC$$

lies above that for 0·01% C in pure iron at temperatures above 840°C but is below it at temperatures lower than 840°C. Clearly then, carbide particles can begin to form as a separate phase only below 840°C in this alloy. At higher concentrations of vanadium in iron the carbide can form, with 0·01 wt% of carbon, at higher temperatures, but not significantly higher until a large proportion of vanadium, raising the activity of vanadium by an order of magnitude, is present.

Similar diagrams for sulphides and nitrides can be constructed from the data given here and the work of Rosenquist and Dunicz[10], and Darken and Gurry[11].

Effects of Large Amounts of Alloying Elements

The diagrams which have just been described are of only limited value because the presence of an alloying element in solution in the iron influences the thermodynamic behaviour of the solute. Thus it is well known that the solubility of gases in metals at constant pressure is changed by addition of alloying elements[12], and since this is only another way of saying that the activity coefficient of the gas atoms in the solution has been changed, we might expect this effect to pervade the whole field of alloy-solution thermodynamics.

The direction in which a given alloying addition will change the activity coefficient of another dilute solute can be predicted semi-quantitatively along the following lines[13]. Let us consider the simple case of a random solution of atoms A, B and S, where S is the dilute solute, in which the energy of binding of a given S atom in the solution may be obtained by assigning a fixed value to the energy of interaction between the A-S and B-S pairs which are formed and the number of A-B pairs which are broken when 1 mol of S is introduced into a large amount of the alloy, under the restriction that the alloy remains at constant composition during the process. It can be shown that

$$\Delta H_S^{A-B} = x_A \Delta H_S^A + x_B \Delta H_S^B - x_A \Delta H_A^{A-B} - x_B \Delta H_B^{A-B}$$

the solution being regarded as so dilute that S-S pairs are not formed. We may then approximate

$$\Delta H_S = RT \ln \gamma_S$$

in which case

$$\ln \gamma_S^{A-B} = x_A \ln \gamma_S^A + x_B \ln \gamma_S^B - \frac{\Delta G_{AB}^{XS}}{RT}$$

where γ_S^{A-B} is the activity coefficient of S in the A-B alloy of atom fraction x_A of A and x_B of B, γ_S^A is the activity coefficient of S in solution in pure A, γ_S^B is that for S in solution in pure B, and ΔG_{AB}^{XS} is the excess free energy of mixing of the A-B alloy at this composition. The derivation of this equation has not been attempted here, and the interested reader should consult Reference 13 for further details.

It can now be seen that if the A-B alloy is ideal, i.e. $\Delta G_{A-B}^{XS} = 0$, then

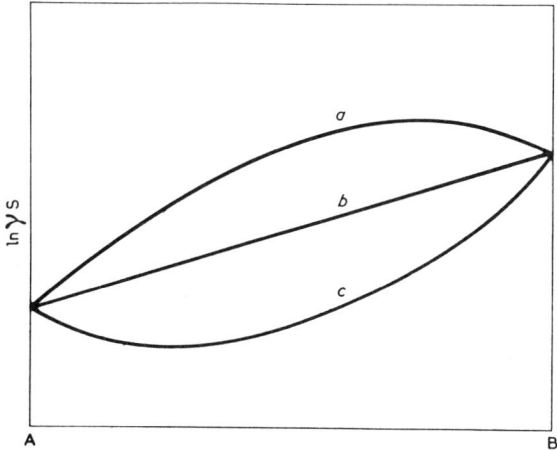

Fig. 7.63 Variation of log γ_S in A-B alloy. Curve a, A-B solvent ideal; b, A-B solvent with ΔG^{XS} negative; c, A-B solvent with ΔG^{XS} positive

the activity coefficient of S will be *decreased* by addition of B if B interacts more strongly with S than does A, while it will be *increased* if B interacts less strongly than A with S. If the excess free energy of mixing of the A-B alloy is positive, or, as an approximation, if the heat of formation of the A-B alloy is positive, then γ_S will be *decreased* by the addition of B even when A and B have interactions of equal strength with S, and, conversely, γ_S will be *increased* if the heat of mixing is negative in these circumstances. The effects are shown diagrammatically in Fig. 7.63.

This simple expression can be used to obtain only a semi-quantitative idea of the effect of an alloying element because the assumptions of randomness and a constant pairwise energy of interaction between atoms are only approximations to the truth in most systems.

For quaternary and more complex alloys a suggestion of Chipman and Sherman[14] might be used. Chipman's school have made use of the symbol ε_C^X for the rate of change of $\ln \gamma$ of the dilute solute, C, with small additions of alloying elements, X. Thus for the solution of carbon in iron:

$$\varepsilon_C^X = \frac{\partial \ln \gamma_C^{Fe}}{\partial x_X}$$

and it has been suggested that for small additions of several alloying elements to iron, the effect on the activity coefficient of a solute, such as carbon, can be obtained from the expression[15]

$$\varepsilon_C^{X-Y-Z} = x_X \varepsilon_C^X + x_Y \varepsilon_C^Y + x_Z \varepsilon_C^Z$$

Values of ε_C^X which have so far been obtained experimentally[16] are shown below.

Alloying element	Si	Cr	Mn	Mo
$\varepsilon_C^X = \partial \ln \gamma_C / \partial x_X$	10	$-4 \cdot 3$	$-0 \cdot 5$	$-0 \cdot 8$

These apply to *liquid* iron as the solvent.

The Segregation of Carbides from Stainless Steel Containing Small Amounts of Carbon

As an example of the way in which these data could be used, the temperatures at which carbides separate from an 18/8 stainless steel are calculated for carbon contents of $0 \cdot 1$, $0 \cdot 01$, $0 \cdot 001$ and $0 \cdot 0001$ wt%. These calculations, which of necessity involve several approximations due to our present lack of knowledge, demonstrate the value of the thermodynamic approach to problems involving the precipitation of phases which may have a pronounced effect on the corrosion behaviour of the alloy (see Section 3.3).

The steel will be considered to be an ideal ternary solution, and therefore at all temperatures $a_{Cr} = 0 \cdot 18$, $a_{Ni} = 0 \cdot 08$ and $a_{Fe} = 0 \cdot 74$. Owing to the γ-phase stabilisation of iron by the nickel addition it will be assumed that the steel, at equilibrium, is austenitic at all temperatures, and the thermodynamics of dilute solutions of carbon in γ iron only are considered.

The effect of nickel on the activity coefficient of carbon will be neglected and the effect of chromium will be taken from the value in the liquid state. From the values quoted above, $\varepsilon_{Cr}^C = -4 \cdot 3$ at $1\,600°C$, and assuming that the effect of chromium is simply to change the *heat* of solution of carbon in iron then the point γ on the left-hand side of Fig. 7.61 must be depressed by an amount

$$\partial \Delta H_{C(\gamma)} = -4 \cdot 576 \times 1873 \times 0 \cdot 335 = -12 \cdot 02 \text{ kJ}$$

the last term being the change in log γ_C at 1 600°C when $x_{Cr} = 0 \cdot 18$.

Now in the case of chromium carbide separation from the steel, three possible crystal structures may be taken up, those of Cr_4C, (or $Cr_{23}C_6$), Cr_7C_3 and Cr_3C_2. It is necessary first to calculate the free energies of formation of the compounds from pure chromium and carbon. The results are:

$$23/6 \, Cr + C \rightarrow 1/6 \, Cr_{23}C_6 \quad \Delta G^\theta = -16\,380 - 1 \cdot 54 T \text{cal} \quad A$$

$$7/3 \, Cr + C \rightarrow 1/3 \, Cr_7C_3 \quad \Delta G^\theta = -13\,900 - 2 \cdot 05 T \text{cal} \quad B$$

$$3/2 \, Cr + C \rightarrow 1/2 \, Cr_3C_2 \quad \Delta G^\theta = -10\,000 - 1 \cdot 39 T \text{cal} \quad C$$

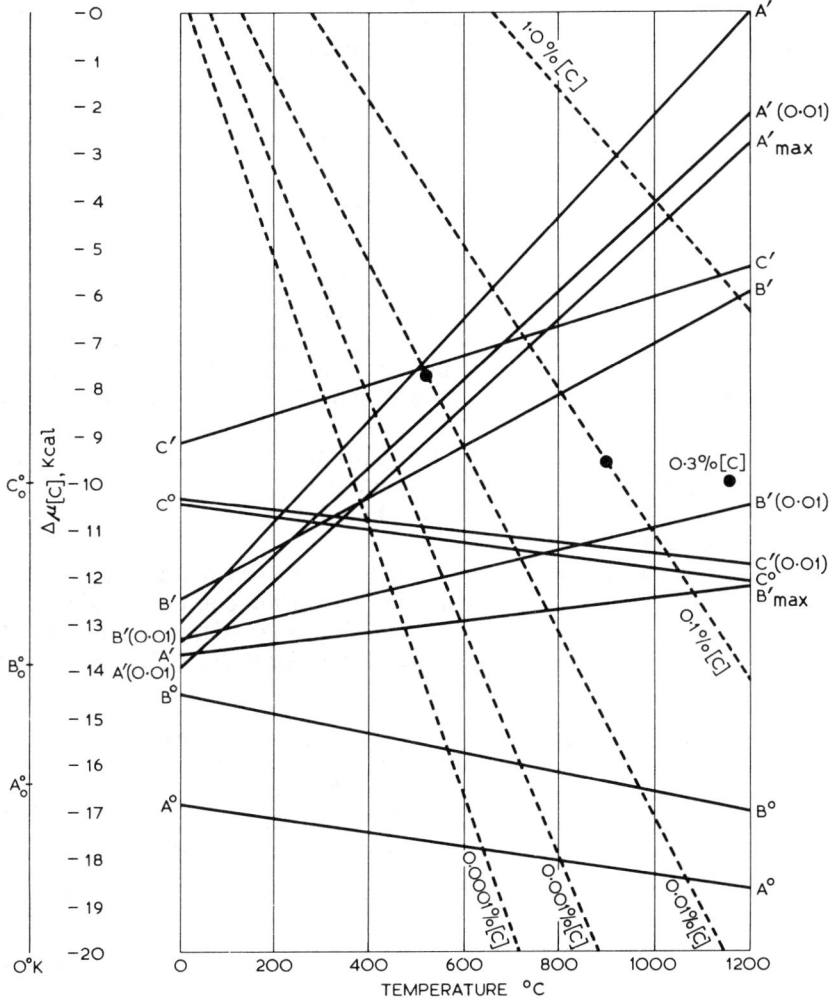

Fig. 7.64 Carbide formation free-energy diagram for Fe-18Cr-8Ni

Since the chromium activity is 0·18 for the formation of carbide in the steel, each of the standard free-energy lines (A°, B°, C°) derived for the carbides must be corrected (moved upwards) for this lower activity (A′, B′, C′).

Finally, the lines for the formation of these carbides at activities 0·01* [e.g. A′ (0·01), B′ (0·01)] are shown in Fig. 7.64 and a summary of the results of the calculation is given in Table 7.36.

It can be seen from the diagram that a phase consisting of Cr_7C_3 at an activity of 0·01 [B′(0·01)] would segregate at 1 000, 720, 550 and 440°C from steels containing 0·1, 0·01, 0·001 and 0·000 1% carbon respectively. Phases containing $Cr_{23}C_6$ would be unstable with respect to this segregation at the temperatures stated, but would separate at an activity of 0·01 at temperatures of 710, 540, 440 and 370°C. A Cr_3C_2 phase might appear at 0·01 activity at the highest carbon content and above 830°C, and below 1 030°C.

Now to complete the solution of the problem one would need to know the solution laws for iron, and a small amount of nickel, in each of these carbide phases, since equilibrium requires that a_{Fe} and a_{Ni} in the segregated carbide must be 0·74 and 0·08 respectively as well as a_{Cr} being 0·18. At present nothing is known about these laws except that the metal atoms might well be randomly distributed in the carbide phase, in which case, as an example,

$$a_{Cr_7C_3} = x^7{}_{Cr_7C_3}$$

in the ideal case.

It is known that $Cr_{23}C_6$ can contain up to 25% iron on the metal atom sites, and Cr_7C_3 up to about 60% iron[17]. Therefore the minimum activities of $Cr_{23}C_6$ and Cr_7C_3 have been calculated for these phases of *maximum* iron content using the ideal laws to calculate activities from carbide composition. The free-energy lines which were thus obtained are shown as $A'_{(max.)}$ and $B'_{(max.)}$.

The picture which emerges from this extremely simplified calculation is that a Cr_7C_3 phase should always precede a $Cr_{23}C_6$ phase in segregation from stainless steel and that the latter should appear at a temperature of 780°C for the carbon content of 0·1 wt%.

According to the phase diagram which has gained acceptance for this system, the Cr_7C_3 phase never appears at low carbon contents, and a cubic phase of the $Cr_{23}C_6$ type separates at 900°C for 0·1% [C], and about 500°C for 0·01% [C]. These points, together with one for 0·3% [C] are shown by

Table 7.36 Values of $\Delta\mu_C$ for chromium carbides at 1 000 K

Product	$\Delta G^\ominus$	$\Delta G^\ominus - n\Delta\mu_{Cr}$ ($x_{Cr} = 0\cdot 18$)	$\Delta G^\ominus - n\Delta\mu_{Cr} + m\Delta\mu_{carbide}$ ($a = 0\cdot 01$)
$Cr_{23}C_6$; $n = 23/6$, $m = 1/6$	−17 940	−17 940 + 13 000 = −4 940	−4 940 − 1 520 = −4 460 cal
Cr_7C_3; $n = 7/3$, $m = 1/3$	−15 950	−15 950 + 7 900 = −8 050	−8 050 − 3 060 = −11 110 cal
Cr_3C_2; $n = 3/2$, $m = 1/2$	−11 400	−11 400 + 5 060 = −6 340	−6 340 − 4 580 = −10 920 cal

the black dots on the diagram. An agreement between the calculated and measured temperatures and compositions for carbide segregation could thus be achieved only by strong departures from the ideal laws in the carbide phases. Alternatively it is possible that the separation of a Cr_7C_3 phase has not so far been observed because of rapid transformation in the solid state to the $Cr_{23}C_6$ phase which is stable at lower temperatures. Such a transformation has been observed in the α Fe-Cr alloys.

Concentrated Ternary Solutions

When both solutes are present in large amounts, i.e. greater than about 1 at. % of each, no simple theoretical treatment is available to predict their mutual effects on thermodynamic properties. In this case, recourse must be made to the various solutions of the ternary Gibbs-Duhem relation

$$x_1 \, d\mu_1 + x_2 \, d\mu_2 + x_3 \, d\mu_3 = 0$$

In order to make any practical use of this equation, a good deal of experimental data are usually required for a ternary system, and it will be found that, at present, such data are seldom available in the literature. The methods of evaluation of such data are fully described in the works of Chipman and Elliott[18] and of Schuhmann[19].

Thermodynamic Phase Stability Diagrams

Pourbaix's pioneering work[20] on the graphical presentation of gas-metal equilibria and the concept of stability zones and their boundaries between the various stable compounds lead to the second type of diagrams. Figure 7.65 shows a Pourbaix plot of the log p_{O_2} of a system against the reciprocal of the absolute temperature for the Zn-O-C systems[20]. The stability zones under varying conditions of temperature, pressure and atmosphere composition are more completely defined than in the Ellingham diagram. However, the diagrams are considered to be more complex and therefore the object of this presentation is defeated unless the scale is greatly enlarged.

Over the years, Pourbaix and his co-workers in the CEBELCOR Institute, founded under his direction, extended these diagrams by including lines for metastable compounds[21]. Figure 7.66 illustrates such a presentation for the Fe-O system over the temperature range 830-2200 K. Pourbaix used these diagrams as a basis for a discussion of the stability of metallic iron (solid, liquid and vapour phases), the oxides of iron as a function of oxygen pressure and temperature from which he explained the protection of iron at high temperature by immunity and passivation. He also pointed out the

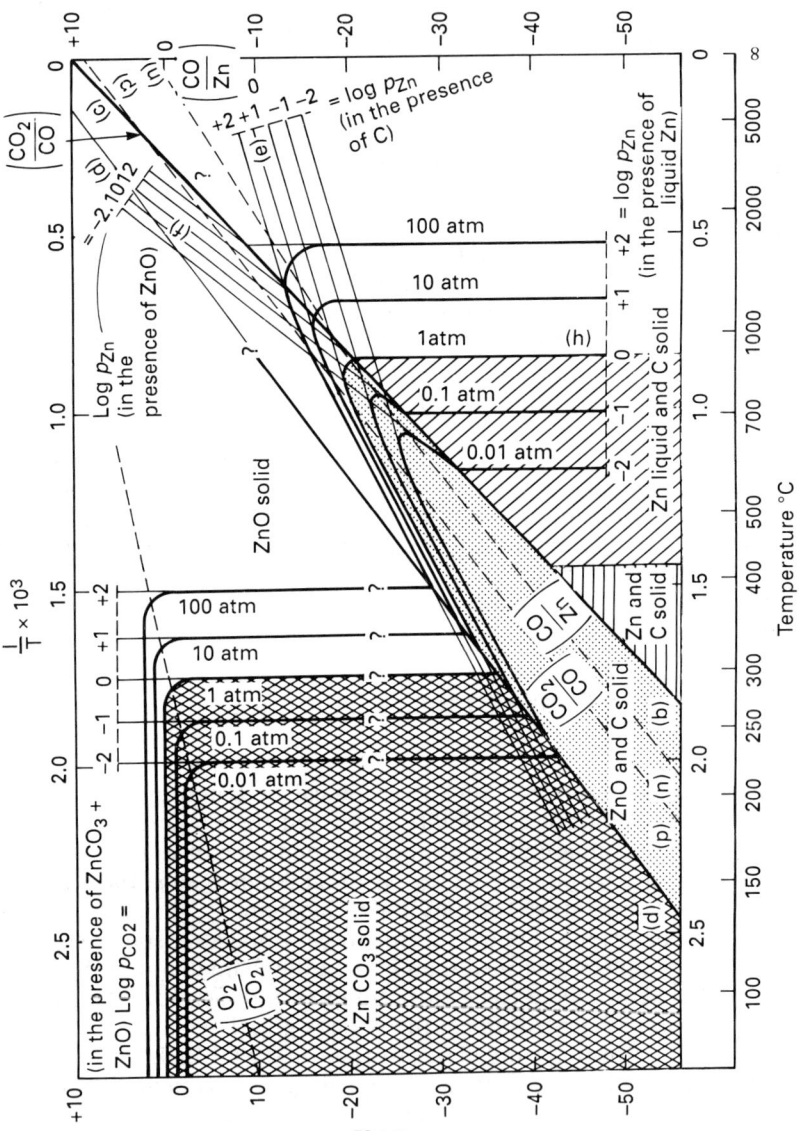

Fig. 7.65 Equilibrium in the Zn-O-C system (after Pourbaix[20])

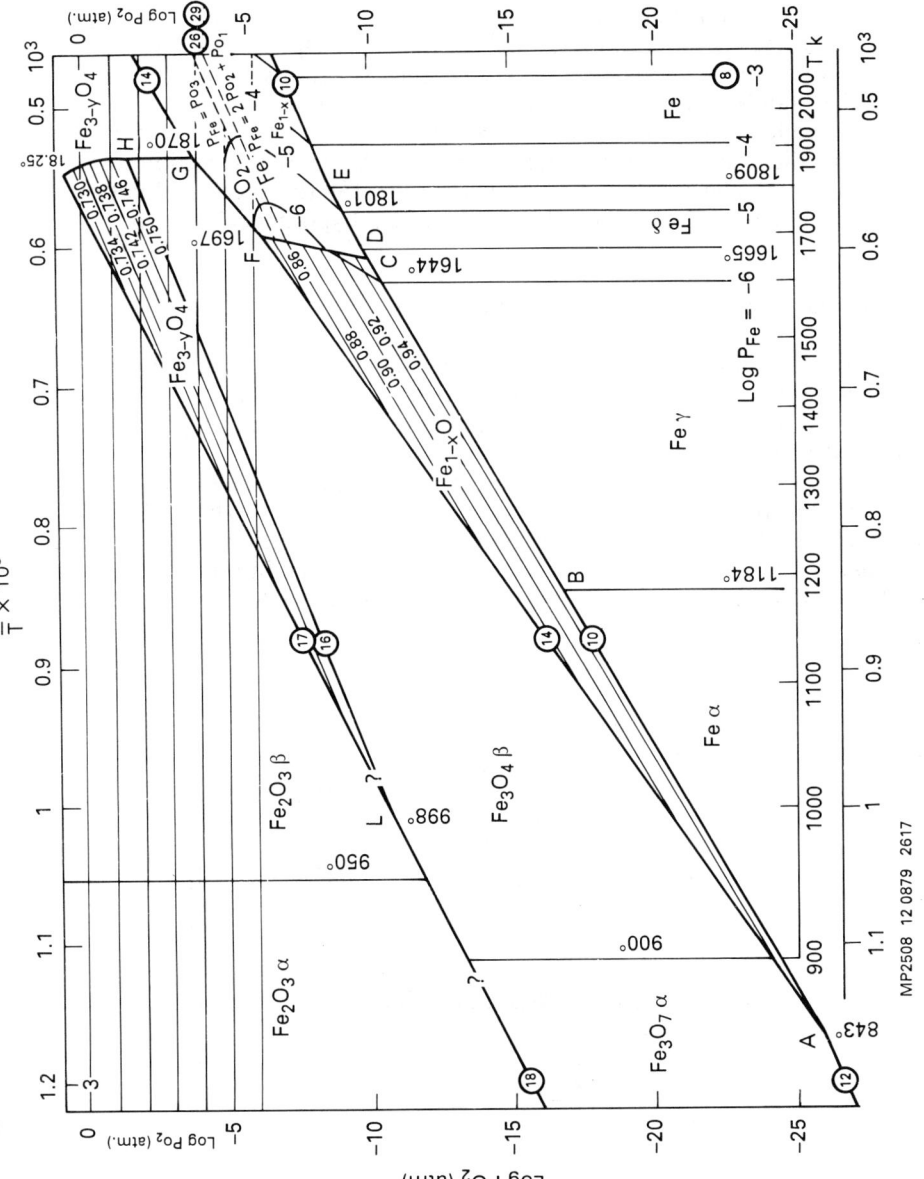

Fig. 7.66 Equilibrium diagram for log $p_{O_2} = f(1/T)$ for the Fe-O system (between 830 and 2 200 K) (after Pourbaix[21])

value of these diagrams in the fields of geology, metallurgy, corrosion and catalysis among others.

A convenient way of representing the thermodynamic information for a given system is by means of isothermal phase stability diagrams which show the ranges of gas compositions over which a condensed phase can exist either singly or in equilibrium with another condensed phase or phases. Kellog and Basu[22] for Pb-S-O and Ingraham[23] for the Ni-S-O, Fe-S-O, Cu-S-O and Co-S-O systems, pioneered the use of such diagrams by considering the relative stability of condensed phases in these systems.

A detailed explanation of the construction of thermodynamic phase stability diagrams may be found in References 22-25. In this section the basic principles of construction and interpretation for the specific situation of gas-metal equilibria will be addressed using a hypothetical system.

Construction of Phase Stability Diagrams

The method of construction of this type of diagram will be illustrated using the general case of the three component system metal-sulphur-oxygen (M-S-O) whose values of ΔG_T^0 for the reactions between the various condensed phases are given in Table 7.37 on page 7:191.

Assume that at the isothermal temperature of interest the following stable condensed phases (solid or liquid) can be formed: M, MO, MS, MSO_4. From the Phase Rule it is clear that the maximum number of condensed phases in contact with each other can be three, in addition to the gaseous phase (SO_2 and O_2). Following the suggestion of Kellog and Basu[22], the

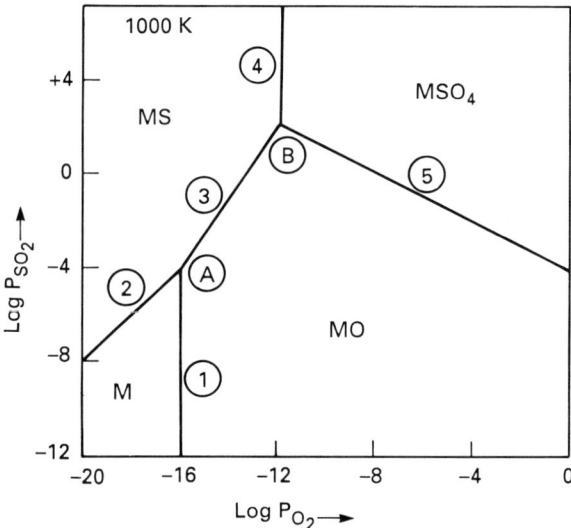

Fig. 7.67 Phase stability diagram for a metal-sulphur-oxygen (M-S-O) system at 1 000 K. (For the thermodynamic data $\Delta G_{1\,000}^0$ for the various across-boundary reactions, see Table 7.37)

Table 7.37 Data for the construction of thermodynamic phase stability diagram M-S-O at 1 000 K

Boundary Between phases	Reaction Considered	Slope of boundary	ΔG^0_{1000} (kJ)	Calculated data from ΔG^0_{1000} and the appropriate equations				
				$-\log K$	$\log P_{SO_2}$	$\log P_{O_2}$	Line no.	Remarks
M/MO	$M + \tfrac{1}{2}O_2 \rightleftarrows MO$	parallel to Y axis	-153.2	-16	NA	-16	1	independent of P_{SO_2}
M/MS	$M + SO_2 \rightleftarrows MS + O_2$	1	$+229.8$	$+12$	-4	-16	2	—
MS/MO	$MS + \tfrac{3}{2}O_2 \rightleftarrows MO + SO_2$	$\tfrac{3}{2}$	-38.3	-20	-4	-16	3	—
MS/MSO$_4$	$MS + 2O_2 \rightleftarrows MSO_4$	parallel to Y-axis	-459.6	-24	NA	-12	4	independent of P_{SO_2}
MO/MSO$_4$	$MO + SO_2 + \tfrac{1}{2}O_2 = MSO_4$	$-\tfrac{1}{2}$	-76.6	-4	$+2$	-12	5	—

*Assumed data of ΔG^0_{1000} for the purpose of illustrating the calculation of the position of the boundary lines and triple points A and B see Fig. 7.67

phase stability diagram may be constructed by plotting the sulphiding potential $\log p_{SO_2}$ along the vertical axis and the oxidising potential $\log p_{O_2}$ along the horizontal axis as in Fig. 7.67. The position and the slopes of the boundary lines between the areas of stability for the condensed phases of the system are then calculated from the appropriate chemical equations describing the reactions which take place when one condensed stable phase reacts to form the other phase. The only thermodynamic data required are either the standard Gibb's free energy change at the chosen temperature (ΔG_T^0) or the equilibrium constant for the reaction at the given temperature (K_T).

Usually, the most convenient boundary to calculate first is that between the pure metal (M) and the metal oxide (MO), i.e. the M/MO boundary since it will be parallel to the y-axis. Using the balanced reaction for the formation of the oxide

$$M + \tfrac{1}{2}O_2 \rightleftharpoons MO \qquad \text{at } 1\,000 \text{ K}$$

and the relationship between ΔG_T^0 and K_T gives:

$$\Delta G_{1\,000}^0 = -RT \ln K_{1\,000} = -RT \ln \frac{a_{MO}}{a_M p_{O_2}^{\frac{1}{2}}}$$

For pure M and MO by definition $a_M = a_{MO} = 1$.

Converting $\ln p_{O_2}$ to $\log p_{O_2}$, rearranging terms and substituting values for $\Delta G_{1\,000}^0$ (Table 7.38), R and T, the following is obtained:

$$-153 \cdot 2 \times 10^3 = 19 \cdot 15 \times 10^3 \times \tfrac{1}{2} \log p_{O_2}$$

or

$$\log p_{O_2} = -16$$

Since p_{SO_2} does not take part in the reaction, the boundary line between M and MO is independent of $\log p_{SO_2}$ and so given by a straight vertical line at $\log p_{O_2} = -16$, parallel to the y-axis (line 1 in Fig. 7.67). It should be noticed that stability areas across the boundary follow the sequence of condensed phases shown in the equation, i.e. on the left-hand side of the boundary pure metal is the stable phase and on the right-hand side the pure metal oxide.

To determine the position of the boundary between M and MS the following chemical reaction is used:

$$M + SO_2 \rightleftharpoons MS + O_2$$

For pure M and MS $a_M = a_{MS} = 1$
Using equation 7.12:

$$\Delta G_{1\,000}^0 = -19 \cdot 15 T \log \frac{p_{O_2}}{p_{SO_2}} = 19 \cdot 15 T (\log p_{SO_2} - \log p_{O_2})$$

Rearranging:

$$\log p_{SO_2} = \log p_{O_2} + \frac{\Delta G_{1\,000}^0}{19 \cdot 15 T}$$

This is the equation of a straight line of the form $y = mx + c$ when $\log p_{SO_2}$ is plotted against $\log p_{O_2}$, where $y = \log p_{SO_2}$, m is the slope which here is

THERMODYNAMICS AND KINETICS OF GAS-METAL SYSTEMS 7:169

unity, $x = \log p_{O_2}$ and c is a constant which in this case is $\Delta G^0_{1\,000}/19 \cdot 15T$. Substituting values for $\Delta G^0_{MS} = 229 \cdot 8 \times 10^3$ and $\log p_{O_2} = 10^{-16}$ (where the previous line and the new line intersect) gives:

$$\log p_{SO_2} = \frac{229 \cdot 8 \times 10^3}{19 \cdot 15 \times 10^3} - 16 = -4$$

Thus, the coordinates of the point of intersection (A in Fig. 7.67) of the two boundary lines M/MS and M/MO are $\log p_{SO_2} = -4$ and $\log p_{O_2} = -16$.

These calculated data are now sufficient to draw the boundary line between the stability areas of M and MS. This is constructed by drawing a straight line having a slope of $+1$ from the point of intersection A. Next, the position of the boundary between MS and MO can be calculated from the reaction:

$$MS + \tfrac{3}{2}O_2 \rightleftharpoons MO + SO_2$$

Thus,

$$\Delta G^0_{1\,000} = -19 \cdot 15 T \log \frac{p_{SO_2}}{p_{O_2}^{\frac{3}{2}}}$$

which, on rearranging, gives

$$\log p_{SO_2} = \tfrac{3}{2} \log p_{O_2} - \frac{\Delta G^0_{1\,000}}{19 \cdot 15 T}$$

The slope of the line is $+\tfrac{3}{2}$ and the line is drawn from the point A to the MS/MSO_4 boundary to be determined next.

The boundary between the MS and MSO_4 stability areas is calculated from the reaction:

$$MS + 2O_2 \rightleftharpoons MSO_4$$

Thus

$$\Delta G^0_{1\,000} = -19 \cdot 15 T \log \frac{1}{p_{O_2}^2} = +19 \cdot 15 T 2 \log p_{O_2}$$

or

$$\log p_{O_2} = \frac{-459 \times 10^3}{19 \cdot 15 \times 10^3 \times 2} = -12$$

Since SO_2 does not take part in the above reaction, the boundary between MS and MSO_4 stability areas is independent of $\log p_{SO_2}$ and is given by a vertical line (4) at $\log p_{O_2} = -12$ parallel to the $\log p_{SO_2}$ axis. The intersection point of the MS/MSO_4 line (4) with that of MO/MS boundary line (3) at point B of Fig. 7.67 completes the stability area of the MS phase (lines 2, 3 and 4).

Finally, the boundary between MO/MSO_4 is calculated from the reaction:

$$MO + SO_2 + \tfrac{1}{2}O_2 \rightleftharpoons MSO_4$$

Thus

$$\Delta G^0_{1\,000} = -19\cdot 15 T \log \frac{1}{p_{SO_2} p_{O_2}^{\frac{1}{2}}}$$

or

$$\log p_{SO_2} = \frac{\Delta G^0_{1\,000}}{19\cdot 15 \times 10^3} - \frac{1}{2}\log p_{O_2}$$

Since this line will start at point B of Fig. 7.67, i.e. when the values of $p_{O_2} = -12$ and $\Delta G^0_{1\,000} = -76\cdot 6\,\text{kJ}$, one can calculate the value of $\log p_{SO_2}$ for the point B by substituting these values in the previous equation:

$$\log p_{SO_2} = \frac{-76\cdot 6 \times 10^3}{19\cdot 15 \times 10^3} - \frac{1}{2}(-12) = +2$$

Therefore, the calculated coordinates of the triple point for the coexistence of MO, MS and MSO_4 are $\log p_{SO_2} = +2$ and $\log p_{O_2} = -12$ and the slope of the MO/MSO_4 boundary is $-\frac{1}{2}$. The straight line from point B having slope $-\frac{1}{2}$ gives the boundary line (5) between the stability areas of MO and MSO_4. This completes the construction of the phase stability diagram for M–S–O at 1 000 K.

The stability phase diagrams contain a wealth of information. Using some selected examples from the literature it is intended to show their range of application in the field of corrosion.

Control of Gas Composition for Surface Stability

Many industrial applications of materials at elevated temperature involve their exposure to complex gas mixtures. Usually it is assumed that the main oxidising species control reaction rates[26] by forming protective oxides, whereas the formation of sulphides, chlorides etc, which may be solid or liquid, can be detrimental to the performance of the material. In practice, using steel as an example, the partial pressures of oxygen (p_{O_2}), sulphur (p_{S_2}), halogenic gases and the activity of carbon (a_C) are controlled by establishing the relevant safe gas equilibria to prevent sulphidation and carburisation of the steel. It is relatively simple to obtain graphically from $\Delta G^0/T$ diagrams and their nomographic scales the ratios of binary equilibrium gas mixtures H_2/HCl, H_2/H_2O, CO/CO_2 and CH_4/H_2 in contact with a particular metal or condensed phase[27]. However, in multicomponent atmospheres it may be necessary to take advantage of specialised computer data banks and the iterative routines such as MTDATA in use at NPL[28] and their facilities for automatically plotting the phase stability diagrams for metals and alloys relevant to the temperature and gaseous conditions of interest. Many such centres are available to outside contract[50].

In Fig. 7.68 the oxidising and sulphiding potentials of four different atmospheric environments, i.e. conventional coal combustion (A), fluidised bed combustion (B), conventional coal gasification (C) and coal gasification using nuclear heat (D), are shown on the thermochemical phase stability

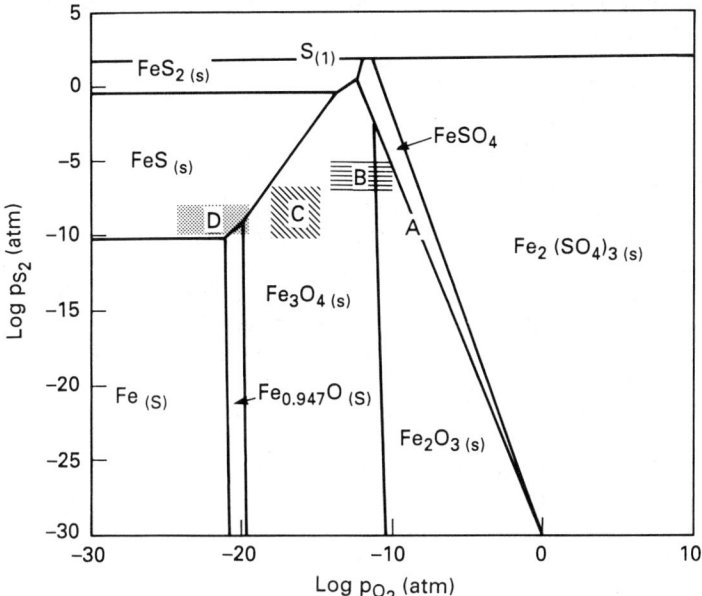

Fig. 7.68 Thermochemical stability diagram for the system Fe–S–O at 1 000 K showing the relative corrosion potentials of the atmospheres in conventional coal combustion (A), fluidised bed combustion (B), conventional coal gasification (C) and coal gasification using nuclear heat (D) (after Gray and Starr[29])

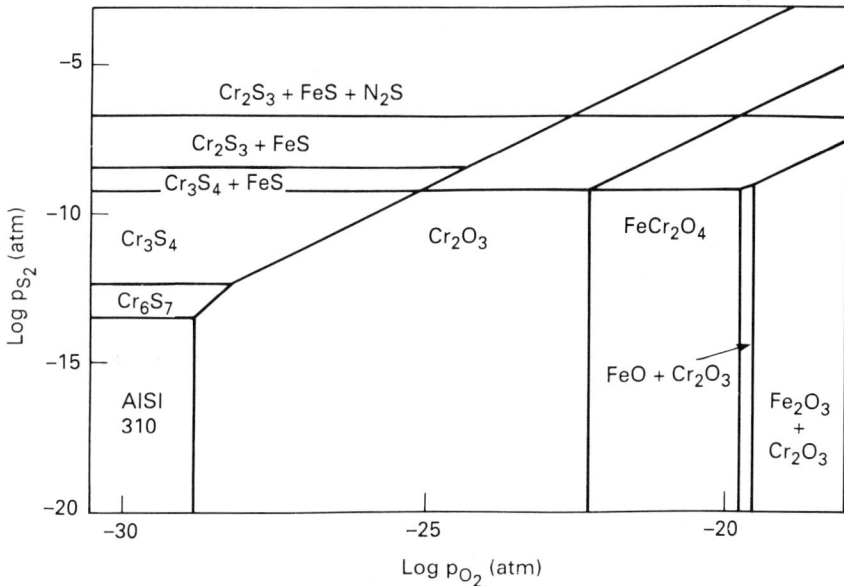

Fig. 7.69 Thermochemical stability diagram for the 310 stainless-steel–S–O system at 750°C (from Gray and Starr[29] after Natesan and Chopra[30])

diagram for the Fe-S-O system at 1 000 K[29]. From the position of the four areas (A, B, C, D) shown on the diagram, the aggressiveness of the environment can be seen as it increases with decreasing p_{O_2} and increasing p_{S_2}. Thus the corrosivity of the different atmospheres increases from process A to D. However, in such an evaluation of the aggressiveness of atmospheres it is necessary to take into account the differences in the feedstock and other process conditions employed[29].

Figure 7.69 and 7.70 show the phase stability diagrams at 1 023 K for the 310 stainless-steel-S-O system and that for the Cr-O-S system relevant to Incoloy 800H alloy, respectively. Comparing these diagrams it is apparent that the boundary between steel AISI 310 and Cr_2O_3 is at a slightly lower oxygen potential (about $-27 \cdot 5$) than that between Incoloy 800H which has the same boundary at $-26 \cdot 5$ oxygen potential. However both these diagrams illustrate how convenient it is to obtain the composition of the atmosphere at the isothermal temperature within which each particular phase may be formed. Obviously, this information is particularly useful when assessing the critical gas composition at which the protective Cr_2O_3 oxide can be expected to be stable. However, it has been observed that at low oxygen pontentials the gas compositions must be made more oxidising by about two to three orders of magnitude than those predicted by the equilibrium values; this is possibly because of kinetic effects[31]. Figure 7.70 gives the additional kinetic phase boundary separating the stability area of Cr_2O_3 and the adjoining stability areas. This observation was confirmed from XPS

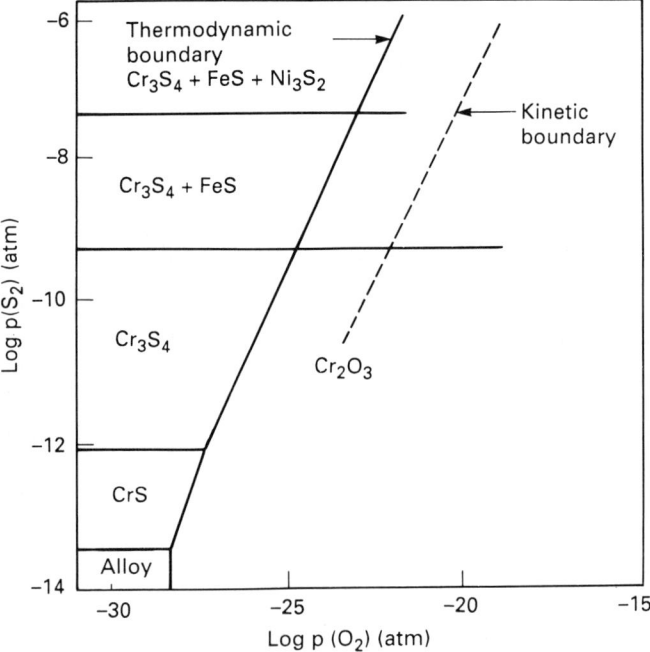

Fig. 7.70 Phase stability diagram for the Cr-O-S system on Incoloy 800H at 1 023 K showing thermodynamic and kinetic boundaries (after Natason[31])

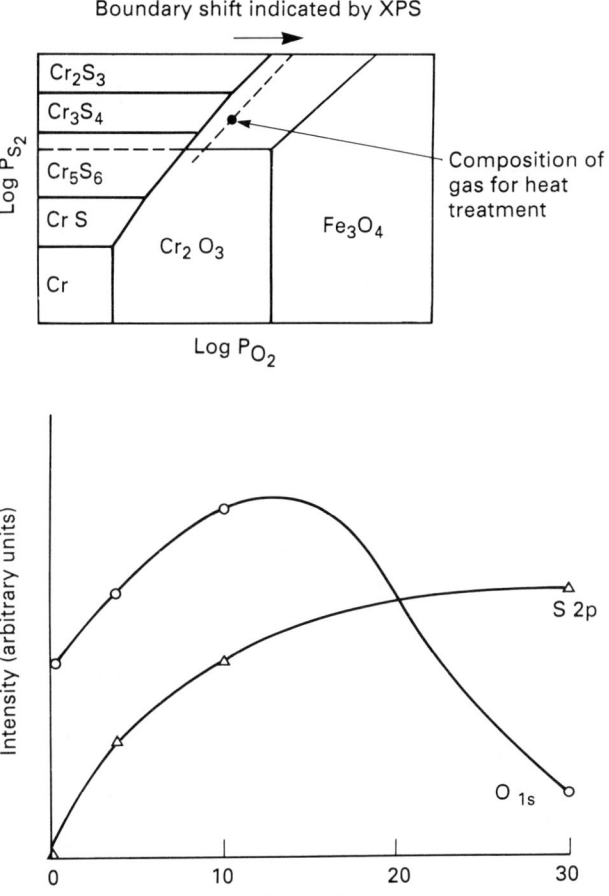

Fig. 7.71 An activity diagram showing the competing formation of sulphides and oxides on chromium. The XPS data (*lower*) show how sulphide replaced oxide as the surface anion when oxide samples were heated in the gas composition marked on the O–S diagram, implying that the boundary should be moved. (Reprinted with permission from Pergamon Press; after Huang[33])

work where it was established analytically[32] that there was a definite boundary shift to a higher oxygen potential (Fig. 7.71).

Phase Stability Diagrams with a Liquid Phase

There are many examples of these diagrams being used to predict or assess the disruptive effect of a liquid phase at different oxygen potentials on the protective properties of oxides. In Fig. 7.72 the stability of Al_2O_3 is shown in equilibrium with liquid Na_2SO_4 as a function of the oxygen activity and the acidity of the liquid salt, at 1 273 K. This stability phase diagram shows that the oxygen potential boundary between basic fluxing and the stable

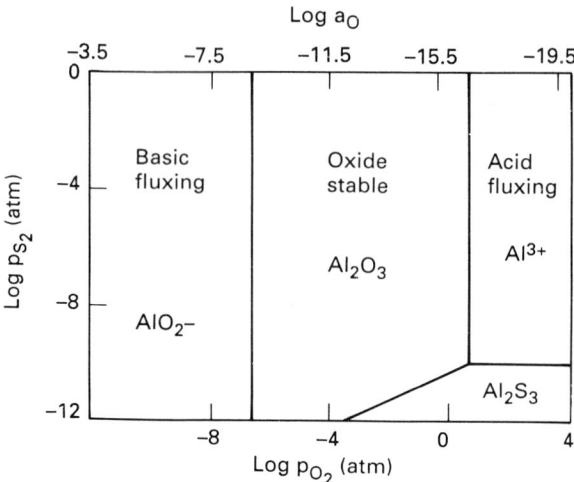

Fig. 7.72 A thermodynamic phase stability diagram for Al-O-S species in the eqilibrium with liquid Na₂SO₄ at 1 000°C as a function of the oxygen activity and the acidity of the salt (after Stringer[34])

oxide is $\log a_O = -8$ approximately, whilst acid fluxing and the stable oxide is $\log a_O = 17$ approximately.

In recent work[35] phase stability diagrams were used to evaluate the effect of molten Na_2SO_4 on the kinetics of corrosion of pure iron between 600°C and 800°C by drawing a series of superimposed stability diagrams for Na-O-S and Fe-O-S at 600°C, 700°C and 800°C and thus to account for the differences in the corrosion behaviour as a function of temperature.

Phase Stability Diagrams and the Formation of Volatile Halides

Another problem in high-temperature corrosion can be the effect of the formation of volatile metallic halides which can, in turn, disrupt the integrity of a protective surface oxide. Figure 7.73 shows that in the Ti-O-Cl system at very low oxygen potentials, volatile $TiCl_3$ can be formed directly from TiO and Ti, whereas from Fig. 7.74 it is clear that in the system U-O-Cl at 450°C the volatile chloride cannot be formed directly from the oxides.

Phase Stability Diagrams for Two or More Metals

These isothermal diagrams can be used to consider the phase stability areas for more than one metal in contact with a common atmosphere and thus to assess the condensed phases which can be stable under the prevailing conditions. Figure 7.75 shows a stability diagram having phase areas for Co-S-O (*solid lines*) and for Cu-S-O system (*broken lines*). From this diagram[23] it can be seen clearly that at 950 K at certain gas mixtures, pure metals Co and

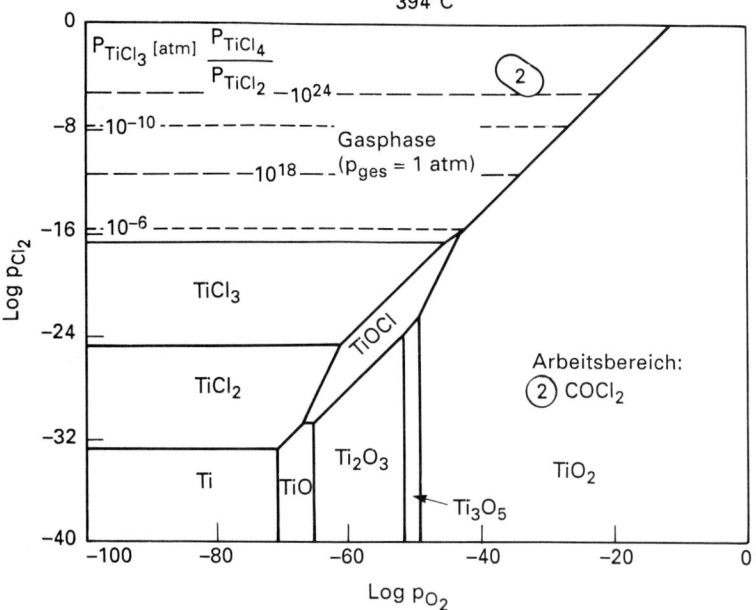

Fig. 7.73 Composition ranges in the Ti–O–Cl system at 394°C (after Knacke[36])

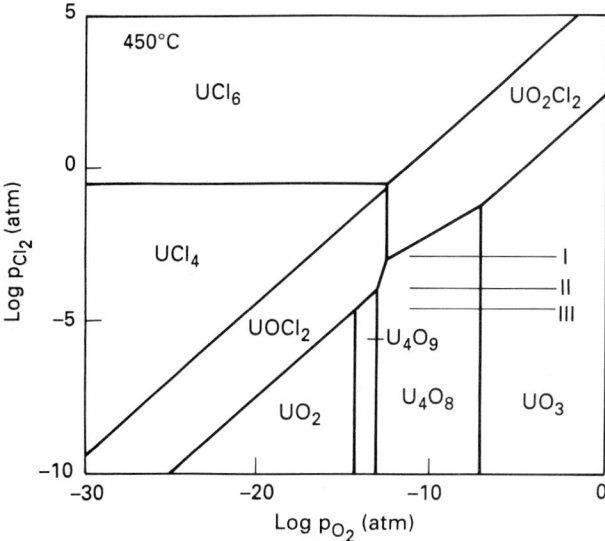

Fig. 7.74 Composition ranges in the U–O–Cl system at 450°C (after Knacke[36])

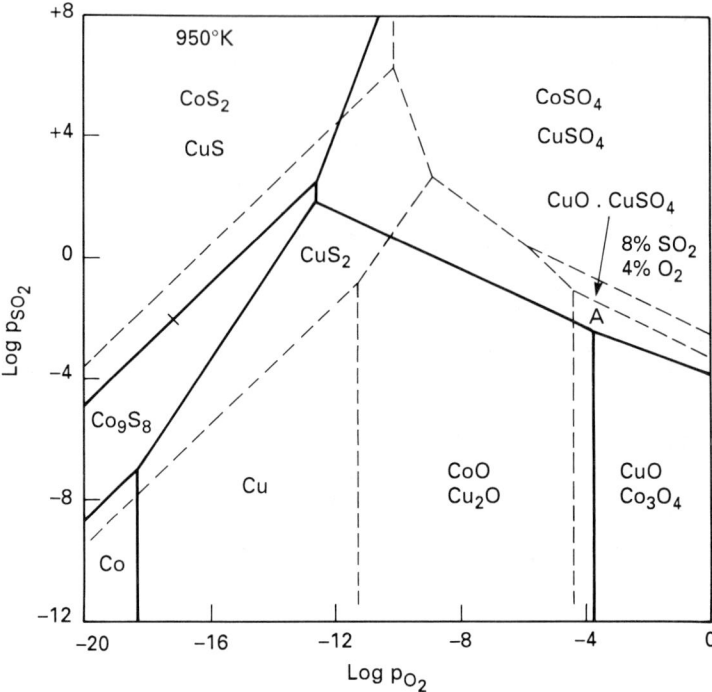

Fig. 7.75 Superimposed predominance area diagrams at 950 K for the Co-S-O system (*solid lines*) and the Cu-S-O system (*broken lines*). Within the area A, $CoSO_4$ and CuO are the stable phases (after Ingraham[23])

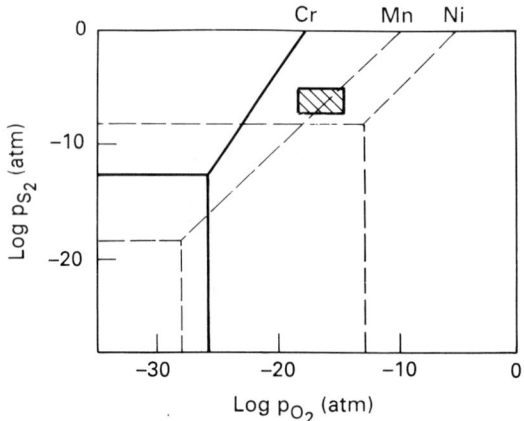

Fig. 7.76 Superimposed simplified thermodynamic stability diagrams for three elements with oxygen and sulphur at 871°C. The shaded rectangle indicates possible activity ranges in coal gasification atmospheres (after Stringer[34])

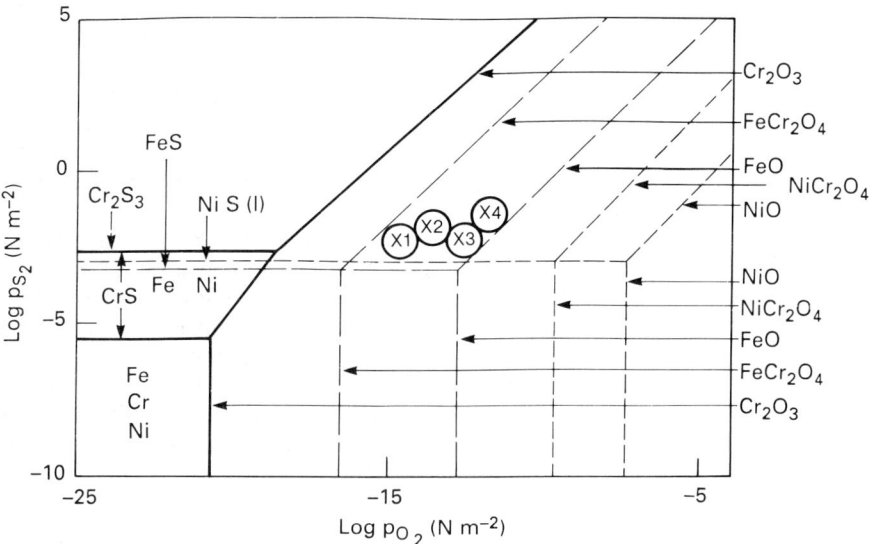

Fig. 7.77 Thermodynamic stability diagram for the Fe–Ni–Cr system at 1 143 K, assuming metal activities to be unity. ----, phase boundaries involving Fe; ----, phase boundaries involving Ni; ——, phase boundaries involving Cr. The location of environments 1, 2, 3, and 4 are indicated by X (after Stott and Smith[37])

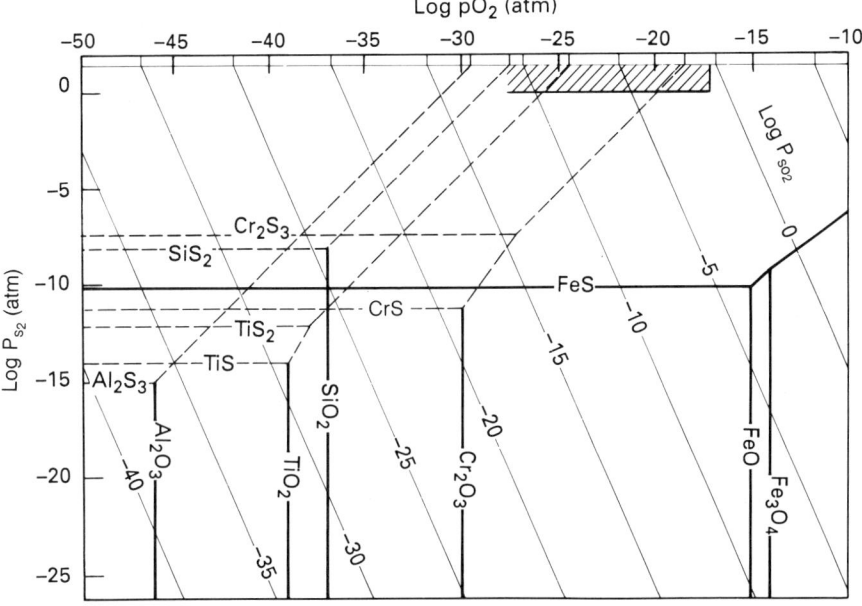

Fig. 7.78 Thermodynamic stability diagram for some oxides and sulphides at 1 000 K (after Lions[38])

Cu will be unaffected by the atmosphere, whilst at other atmospheric compositions the pure oxides will be stable. Figure 7.76 illustrates a simplified diagram[34] at 871°C for three metallic elements Cr/Mn/Ni–S–O in a heat-resisting alloy; the range for coal gasification is also included. It is clear that Cr_2O_3 is stable in all these atmospheres, but NiS will be stable under these atmospheric conditions above 620°C in the form of a eutectic liquid with Ni. Thus, an alloy of Cr and Ni may produce either of these phases or their mixtures leading to corrosion problems.

Figure 7.77 shows a diagram for the three metals Fe, Cr, Ni as a function of sulphide potential against oxygen potential. This diagram has been used to select atmospheres in the study of high-temperature corrosion in which relatively small changes in oxygen and sulphur have a marked effect on the kinetics of corrosion, scale morphology and scale composition of 34Fe–39Ni–27Cr alloy ingots. The atmospheres selected for the study are shown in Fig. 7.77 as $X1$, $X2$, $X3$ and $X4$. Figure 7.78 shows the stability diagram at 1 000 K for Al, Ti, Si, Cr and Fe sulphides at oxidising potentials between, $\log p_{O_2}$, -10 and -50 and suphiding potentials, $\log p_{S_2}$, between 0 and

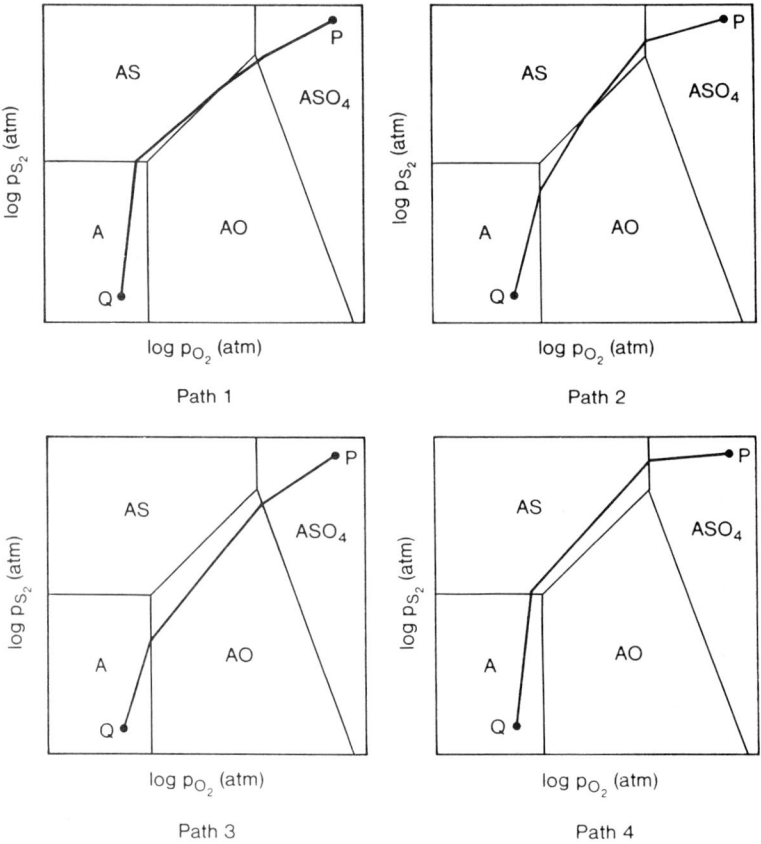

Fig. 7.79 Five possible reaction paths on a schematic thermodynamic phase stability diagram, and the corresponding distribution of phases in the reaction systems (after Stringer[34])

−25. Isobaric lines for $\log p_{SO_2}$ ranging from 0 to −40 are also included in the diagram as straight lines[38]. The diagram was produced for a study aimed at finding improved materials which would be immune to sulphur corrosion and lead to the increased efficiency of thermal and nuclear power stations. From Fig. 7.78 it is clear that the oxides are unstable under high sulphur pressure and very low oxygen pressure. It is also clear that the formation of SO_2 has to be taken into consideration as the reaction between sulphur and oxygen significantly lowers the oxygen activity; under high sulphur pressure and low SO_2 pressure only some oxides are stable (Al_2O_3, TiO_2 and SiO_2), and the oxides of Fe, Co. Ni and perhaps Cr are decomposed.

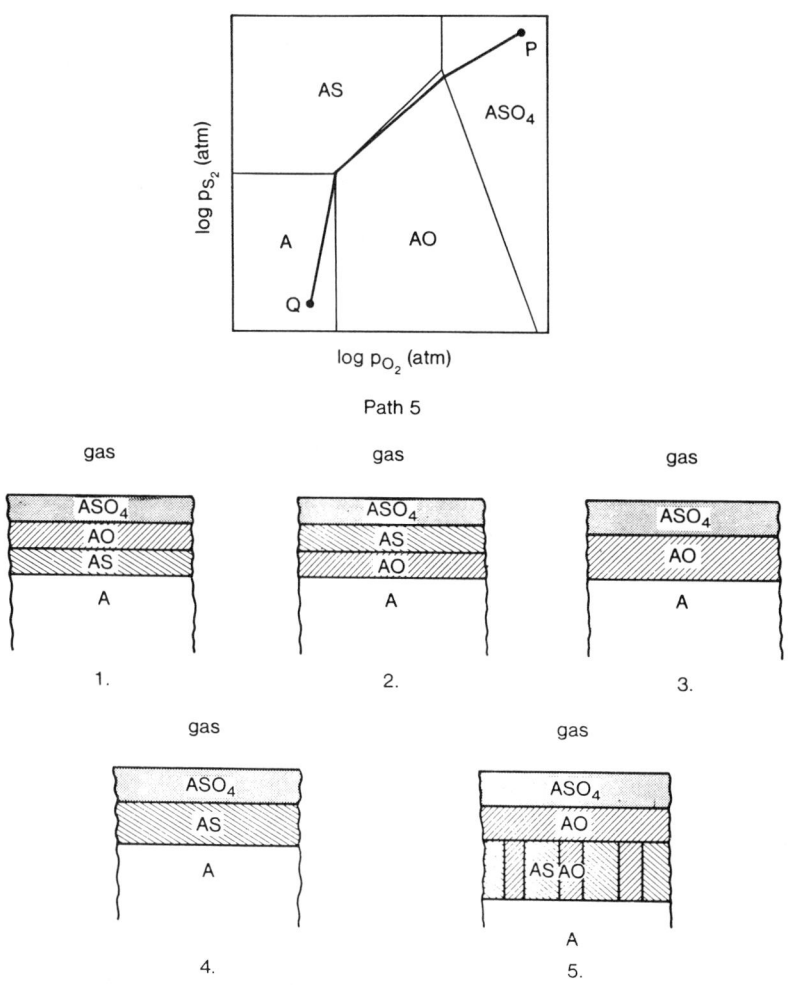

Fig. 7.79 (continued)

Phase Stability Diagrams and the Sequence of Phases in Surface Interaction Layers (Reaction Paths)

A number of authors[39,40] have indicated how the stability diagrams may be used to suggest or explain the possible sequence of phases in surface interaction layers formed during gas–metal interactions. Stringer and Whittle[41] suggested the concept of the 'reaction path' on the stability diagrams which enables prediction of the sequence of phases in surface layers formed when the activity of the oxidant follows a certain path through the various stability areas. Figure 7.79 shows five possible paths in a stability diagram for the A–S–O system. Stringer[34], using the reaction path graphical method in the A–S–O system and the five possible paths shown in Fig. 7.79, accounted for the following sequence of interaction phase layers: $ASO_4/AO/AS/A$, $ASO4/AS/AO/A$, $ASO_4/AO/A$ or $ASO_4/AS/A$ and $ASO_4/AO/AS_\gamma$-AO/A.

The slopes of the reaction path lines, between the point P (giving oxygen and sulphur activities in the gaseous atmosphere) and Q (initial oxygen and sulphur activities in the metal A), are determined by the relative diffusion rates of the species in the phases. From the diagram it can be seen that small differences in slopes can result in significantly different distributions of phases. Stringer[34] points out that only the lack of precise knowledge of diffusion coefficients prevents accurate calculation of reaction paths, and therefore these diagrams, at the moment, are more useful for the interpretation of oxygen and sulphur potential from the observed phase distributions than for predictive purposes.

It is also clear that small changes in the position of points P and Q can have a significant effect on the phase distribution in the surface layers. From the diagrams it is also seen that, when the metal A is saturated with oxygen and sulphur, and therefore the point Q is located at the corner of the rectangle giving the stability area of the metal A, then the innermost phase layer will consist of a mixed sulphide and oxide layer.

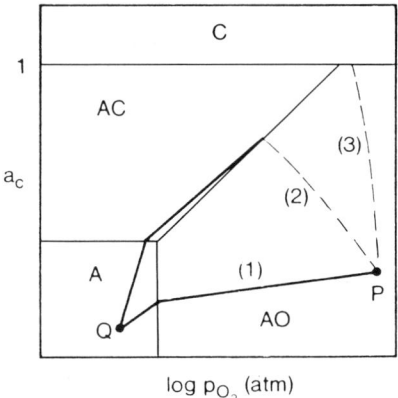

Fig. 7.80 A schematic thermodynamic phase stability diagram for the A–C–O system, showing three reaction paths. Paths 2 and 3 are only possible if gaseous diffusion in pores in the oxide product results in a carbon activity increase through the scale, as shown in Fig. 7.81 (after Stringer[34])

THERMODYNAMICS AND KINETICS OF GAS-METAL SYSTEMS 7:181

Figures 7.80 and 7.81 illustrate the use of a reaction path graphical method for systems where the surface oxide is porous. Figure 7.80 shows the phase stability diagram for a metal-carbon-oxygen (A-C-O) system. Considering reaction path *1* between P and Q it follows that only the metal oxide AO could be formed. However, if the oxide is porous, gaseous molecules of CO_2 can now penetrate to the metal surface (see Fig. 7.81) and thus, following the reaction $CO_2 + A \rightarrow AO + CO$, there must be a gradual increase in the CO partial pressure towards the metal/metal oxide interface within the porous oxide. Figure 7.80 shows the effect of the change of the CO/CO_2 ratio on the carbon and oxygen activities. If the carbon activity rises high enough (see reaction path *2*) carburisation may be possible, or even carbon deposition if a_C exceeds unity, as shown in path *3* Fig. 7.80.

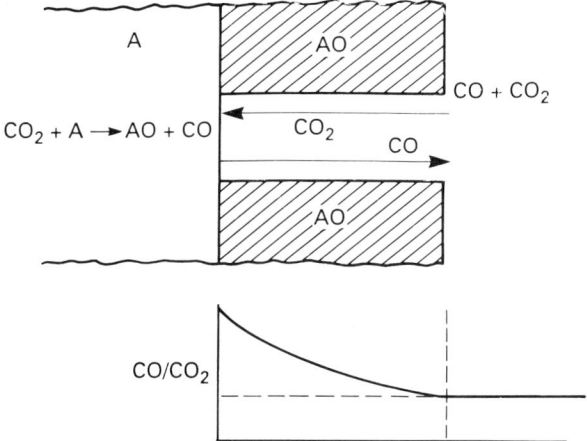

Fig. 7.81 A sketch illustrating how gaseous diffusion processes in pores within an oxide layer can result in an increase in the CO/CO_2 ratio, and hence the carbon activity, through the layer (after Stringer[34])

Phase Stability Diagrams and the Effect of Temperature

Figure 7.82 shows a three-dimensional phase stability diagram for the Fe-S-O system between 800 and 1 000 K. These diagrams are obtained from a knowledge of the variation of ΔG_T^0 for the different reactions which describe the appropriate phase boundary. In general, changes in temperature may have a significant effect on the areas of stability. These may become larger or smaller as the temperature is increased. A detailed description of the method for their production and interpretation may be found elsewhere[23, 42, 43].

Integral Free Energy-Concentration Diagrams

As mentioned earlier, this type of diagram may be useful for the quantitative thermodynamic assessment of gas-metal systems which form non-stoichio-

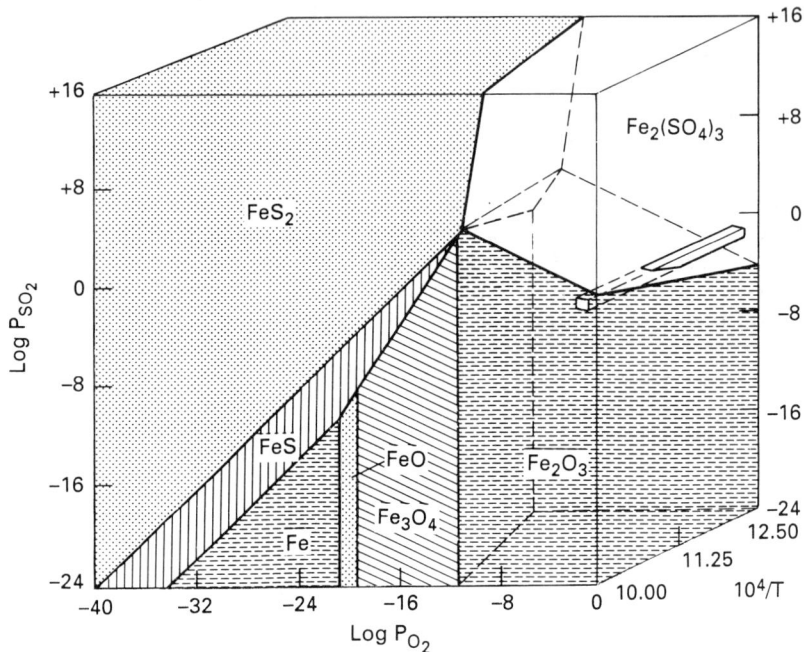

Fig. 7.82 Predominance volume diagram for the Fe-S-O system for the temperature range 800-1 000 K (after Ingraham[23])

metric condensed phases and which are sensitive to relatively small stresses in the condensed phases or small changes in chemical potentials[44], in the gaseous phase and other physical or chemical effects that can be expressed in terms of energy[45] (e.g. irradiation, vibrations, gravity, grain size, penetrating liquids, variation of surface tension, magnetic effects[46] etc.). It will be shown that any effect that slightly alters the relative position of integral free energy-concentration curves may have a drastic effect on the equilibrium and disturb the stability or the composition of the condensed phases.

In this sub-section it is intended first to outline the theoretical basis of these diagrams by considering a simple metal-A-gas-B binary system followed by a quantitative treatment of a hypothetical metal M (at. wt. 50) and oxygen binary system. Finally the application of these diagrams will be illustrated using the Ti-C, $Fe_{(s)}$-$Zn_{(v)}$ and $Fe_{(s)}$-$Zn_{(l)}$ systems.

The key to an efficient use of these diagrams is the understanding of the properties of a common tangent to two or more free energy-concentration curves and in particular the information which may be obtained from the so-called 'tangency rule'. It is therefore intended to develop the tangency rule using a simple isothermal binary system of a pure solid metal A and a pure gas B at 1 atm pressure having two non-stoichiometric solid compounds denoted by phase I and phase II. Figure 7.83 shows the ΔG_m-concentration diagram for this system. The vertical axis y, represents the isothermal free energy changes (ΔG_m) - which are obtained when one mole of a mixture

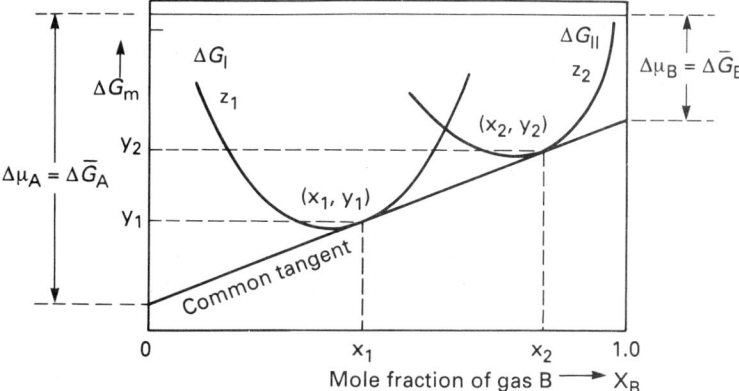

Fig. 7.83 The graphical method for obtaining equilibrium compositions from free energy vs. composition curves at a given temperature. Points of contact give equilibrium compositions X_1 for phase I and X_2 for phase II

having a given composition is formed. The horizontal axis gives the composition of the mixture in mole fraction, of the gas B. On the left-hand side of the scale we have $X_A = 1$ (i.e. pure metal) and the concentration of the gas $X_B = 0$, whereas on the right-hand side of the scale $X_A = 0$ and $X_B = 1$ (i.e. pure gas B at 1 atm pressure) since $X_A + X_B = 1$. The integral free energy changes with concentration for phases I and II are shown as ΔG_I and ΔG_{II} curves, respectively.

Using the diagram it is now possible to predict quantitatively the equilibrium composition of the two condensed phases when in contact with each other, i.e. when growing as solid layers on the surface of the metal A. According to Gibbs[47], in a two-component system any condensed phases at equilibrium will have to satisfy simultaneously two energy conditions, namely (1) the mixture of the phases will acquire the lowest overall free energy, and (2) the chemical potential (or partial molal free energy) of a particular component must be the same in all the phases that are in contact with each other. These two conditions are now sufficient to predict the exact composition of condensed phases at equilibrium with each other.

Gibbs' definition of the chemical potential[47] of the gaseous component B in a mixture at constant temperature T and pressure p is given by

$$\left(\frac{\partial \Delta G_m}{\partial X_B}\right)_{Tp} = (\Delta \mu_B)_{T,p} = (\Delta \overline{G}_B)_{T,p}$$

Therefore this partial differential represents mathematically the tangent to any ΔG–concentration curve. In our case for a common tangent to ΔG_I and ΔG_{II} we must have not only a common slope

$$\frac{dy_1}{dx_1} = \frac{dy_2}{dx_2} = m$$

but also a common intercept on the ΔG_m axis. Thus the equation for a common slope and intercept in Fig. 7.83 must be of the form

$y_1 - y_2 = m(x_1 - x_2)$. It can be shown[48] that the intercept of this tangent on the y axis where $X_B = 1$ (pure gas B at 1 atm pressure) gives the numerical value of $\Delta\mu_B = \Delta\bar{G}_B$ for 1 mol of B dissolved in a mixture of AB having a composition of the point of contact of the common tangent to any ΔG_m-concentration curve. Similarly, the intercept of the same tangent on the y-axis at $X_A = 1$ (pure metal) gives the numerical value of $\Delta\mu_A = \Delta\bar{G}_A$, i.e. the chemical potential (or partial molal free energy) of 1 mol of metal A dissolved in the phase having the composition given by the point of contact of the tangent to the ΔG-concentration curve.

In Fig. 7.83 using the common tangent construction the equilibrium compositions of phase I–phase II at their boundary are found, from the points of contact, to be respectively $X_B = X_1$ and $X_B = X_2$.

Figure 7.84 shows a more complicated isothermal, binary system consisting of metal M and gaseous oxygen O_2 at temperature of 1 000 K. The free-energy–concentration diagram for the system shows four condensed phases, the first being a solid solution of oxygen in the metal, followed by non-stoichiometric condensed phases of nominal compositions M_2O^x, MO^x and MO_2^x. Using the tangency rule, it will be shown that each of these oxides must have a region of homogeneity over a range of composition in which it will be the sole stable phase. In Fig. 7.84 the vertical axis represents the isothermal free-energy change associated with the formation of one mole of the M–O mixture of a given composition, expressed in mole fractions of the metal (X_M) and oxygen (X_O), shown along the horizontal axis.

When a pure solid metal M is in contact with a gas containing oxygen, at first a solid solution of oxygen in the metal is formed. The Gibbs free energy of mixing (ΔG_m) for the corresponding concentrations of oxygen in the solid solution are shown by the curve a–b–c. Note the section a–b of the curve is the only part which can be determined experimentally, whereas the section b–c, representing a supersaturated solution is either extrapolated from the a–b section or calculated theoretically. The values of the free energies of mixing, producing any possible phase, can now be calculated using computer techniques[49] in conjunction with the appropriate thermodynamic data coupled with the relevant phase diagram. It is worth noting that there are already a number of powerful programs[50] which, in conjunction with stored thermodynamic data, can be used to calculate theoretically these curves for an ever-increasing number of binary, ternary and even quaternary systems.

Once the solid solution is produced a surface layer of M_2O^x oxide phase will be formed, having an excess of the metal. This new phase has a separate $\Delta G_{M_2O^x}$-concentration curve shown in Fig. 7.84 by d–e–f–g. This curve is followed by that of ΔG_{MO^x} phase shown by h–i–j–k. Finally a layer of MO_2^x will be formed, and its free-energy–composition curve is shown by l–m–n–o. Applying now the tangency rule, by drawing common tangents to neighbouring ΔG_m-concentration curves, the range of stability of the oxides is determined. As the chemical potential of the oxygen $\Delta\mu_O$ or $\Delta\bar{G}_O$ is fixed by the oxygen pressure in the atmosphere, the equilibrium composition of the MO_2^x oxide layer exposed to the atmosphere is obtained by drawing a tangent to the $\Delta G_{MO_2^x}$ curve with an intercept on the ΔG_m axis when $X_O = 1$ equal to the value of the oxygen chemical potential of the atmosphere (i.e. in our case $= \frac{1}{2}RT\ln p_{O_2}$ where p_{O_2} is the oxygen pressure in the

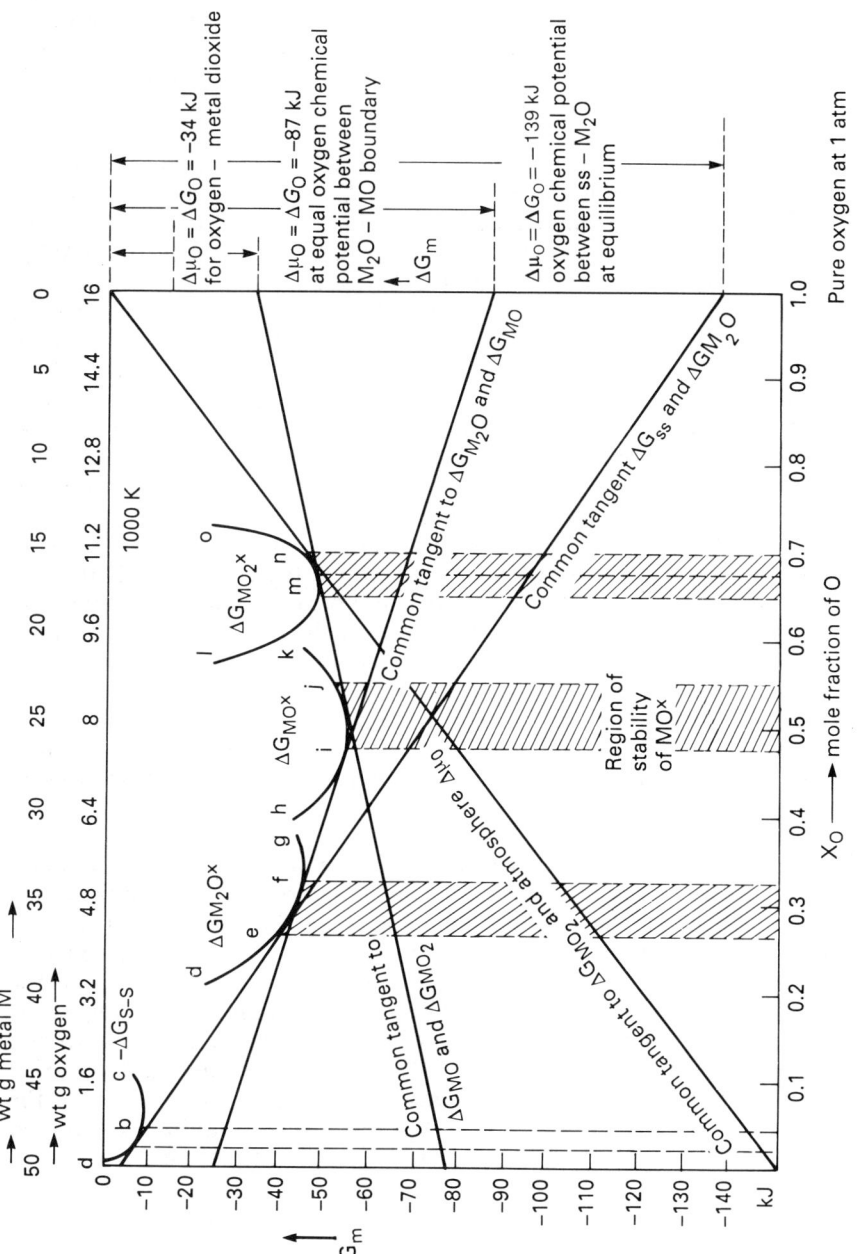

Fig. 7.84 Free energy diagram for a binary system consisting of metal M and gaseous oxygen O_2 at a temperature of 1 000 K

atmosphere – here 1 atm and pure O_2, therefore $\Delta\mu_O = 0$). Thus a tangent from the point $\Delta G_O = 0$ to $\Delta G_{MO_2^x}$ gives the equilibrium composition. The mole fraction of oxygen in the MO_2^x can now be read off the composition axis. The point of contact, n, gives the oxygen equilibrium composition $X_O = 0.7$ in MnO_2^x. The mole fraction of the metal M is $X_M = 1 - X_O = 0.3$. Therefore the equilibrium composition of the outermost oxide layer will be $M_{0.3}O_{0.7}$ with respect to an atmosphere consisting of oxygen at 1 atm pressure.

The common tangent between $\Delta G_{MO_2^x}$ and ΔG_{MO^x} gives points of contact for MO_2^x $X_O = 0.675$ and $X_M = 0.325$ and $X_O = 0.55$ and $X_M = 0.45$ for MO^x (see Fig. 7.84). Similarly the equilibrium compositions on the MO^x and M_2O^x boundary are found by drawing a common tangent to the ΔG_{MO^x} and $\Delta G_{M_2O^x}$ curves. The points i (on the ΔG_{MO^x} curve) and f (on the $\Delta G_{M_2O^x}$ curve) give the equilibrium compositions $M_{0.525}O_{0.475}$ and $M_{0.675}O_{0.325}$ for the two oxides at their boundary. Finally the equilibrium compositions at the boundary between the solid solution and the adjacent oxide layer (M_2O^x) are found by drawing the common tangent to $\Delta G_{M_2O^x}$ and ΔG_{ss}. Points b (on the ΔG_{ss} curve) and point e (on the $\Delta G_{M_2O^x}$ curve) give the respective equilibrium compositions. Thus, the maximum solubility of oxygen in the metal is found from the point b to be $X_O = 0.045$ and the composition of the M_2O^x in contact with the saturated solid solution is $M_{0.735}O_{0.265}$.

Closer examination of Fig. 7.84 shows that each of the non-stoichiometric oxides has a region of homogeneity over which the compound is the sole stable phase. It has been observed, from a number of gas–metal systems, that the lower oxides (here M_2O^x and MO^x) usually show a wider region of non-stoichiometric behaviour than the higher oxides (here MO_2^x).

Regions of Homogeneity of Non-stoichiometric Oxides in the Surface Interaction Layers and the Effect of Oxygen Pressure on their Range of Stability

In Fig. 7.84 each oxide has two points of contact produced by common tangents. These two points predict the range of composition within which each of the oxides is the sole stable phase. Thus the composition of MO_2^x oxide will vary from an oxygen mole fraction $X_O = 0.7$ on the surface of the oxide exposed to the oxygen atmosphere to $X_O = 0.675$ at the MO_2^x/MO^x boundary. It is also clear from the diagram that as long as the oxygen chemical potential remains between $\Delta\mu_O = 0$ (i.e. $p_{O_2} = 1$ atm) and $\Delta\mu_O = -34$ kJ (i.e. $p_{O_2} = 2.8 \times 10^{-4}$ atm) the outermost surface oxide layer will consist of an MO_2^x oxide phase. However, as the pressure of the oxygen is lowered to between 1 and 2.8×10^{-4} atm the equilibrium oxygen contents in the MO_2^x surface layer decrease predictably from $X_O = 0.7$ to $X_O = 0.675$.

The exact equilibrium concentration of oxygen in the MO_2^x oxide in contact with the gas phase can be obtained by first calculating the oxygen chemical potential in the atmosphere, using the relationship $\Delta\mu_O = \frac{1}{2}RT\ln p_{O_2}$, and then drawing a tangent from that point on the $X_O = 1$ axis to the $\Delta G_{MO_2^x}$ curve. The point of contact with $\Delta G_{MO_2^x}$ curve will give the composition of the MO_2^x in contact with the atmosphere at the p_{O_2}. It is also

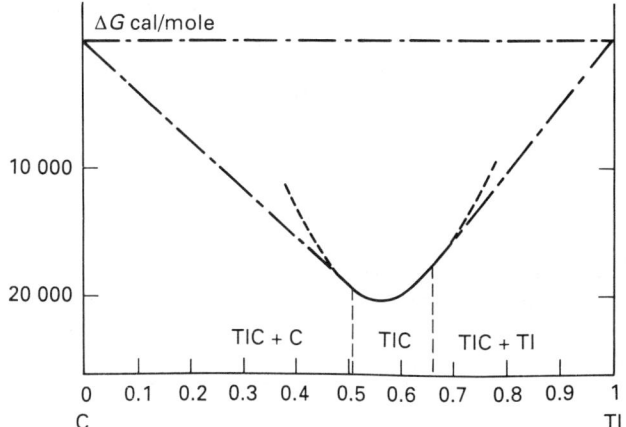

Fig. 7.85 Common tangent method applied to $\Delta G^{fcc}_{(Ti_xC_{1-x})}$ ref $\mid ^{Ti\ bcc}_{C\ graphite}$. The value of the free energy of formation of stoichiometric TiC (ref.: Ti b.c.c, C graphite) calculated from this equation (after Teyssandier et al.[51])

evident from Fig. 7.84 that, once the pressure of oxygen falls below the critical $p_{O_2} = 2 \cdot 8 \times 10^{-4}$ atm, the MO_2^x oxide phase cannot form since no common tangent can be drawn from the new oxygen potential point to $\Delta G^x_{MO_2}$ curve without intersecting other ΔG_m-concentration curves. It is also worth noting that any change in the oxygen pressure in the atmosphere from $p_{O_2} = 1$ atm to the vicinity of $p_{O_2} = 2 \cdot 8 \times 10^{-4}$ atm will have no effect on the compositions of the MO^x, M_2O^x and solid solution phases.

Between oxygen pressures of $2 \cdot 8 \times 10^{-4}$ and 8×10^{-10} atm the outermost oxide layer will consist of the MO^x phase. Its exact surface composition can be predicted by using the common tangent in the same manner as described for MO_2^x. From Fig. 7.84 it is clear that, as the oxygen pressure in the atmosphere is reduced the composition of the surface oxide layer will vary in a predictable manner from $X_O = 0 \cdot 55$ to $X_O = 0 \cdot 475$.

Once the oxygen pressure falls to between 8×10^{-10} and $3 \cdot 04 \times 10^{-15}$ atm, the M_2O^x oxide phase will be the only stable phase whose outermost surface layer composition will change from $X_O = 0 \cdot 325$ to $X_O = 0 \cdot 275$. Below an oxygen pressure of $3 \cdot 04 \times 10^{-15}$ atm no oxide will be formed and the equilibrium solubility of oxygen in the solid solution for a particular oxygen pressure can be predicted once again using the tangency rule.

ΔG_m–Concentration Diagram and the Range of Stability of TiC at 1 900 K

Free-energy–concentration diagrams have been used in the study of the thermodynamic influence on the non-stoichiometry of the solid titanium carbide deposited from H_2–CH_4–$TiCl_4$ gas mixtures at 1 900 K[51]. The authors show how, from the partial pressure measurements of Ti vapour over a range of solid non-stoichiometric $Ti_xC_{(1-x)}$ compositions, the free-energy–concentration curves were calculated as a function of their composition. Figure 7.85 shows one of the plots obtained for $\Delta G_{Ti_xC_{(1-x)}}$ together with two tangents drawn to the free-energy–concentration curve. One tangent from the point $\Delta G = 0$ (on the left-hand side vertical axis, i.e. for $Ti_xC_{(1-x)}$ at equilibrium with C-graphite), and the other tangent from $\Delta G = 0$ (right-hand side vertical axis, i.e. for $Ti_xC_{(1-x)}$ at equilibrium with solid b.c.c. pure titanium metal). The compositions at the two points of contact obtained from the tangency rule predicted the equilibrium compositions of titanium carbide. On the left-hand side of the diagram, i.e. on the $C/Ti_xC_{(1-x)}$ boundary, the equilibrium composition was found to be $Ti_{0.51}C_{0.49}$ and that on the right-hand side, i.e. on the $Ti_xC_{(1-x)}/Ti$ boundary, was found to be $Ti_{0.66}C_{0.34}$. Thus the thermodynamic range of homogeneity of the f.c.c. $Ti_xC_{(1-x)}$ non-stoichiometric TiC was found to be between $X_{Ti} = 0.51$ and $X_{Ti} = 0.66$. In practice these predicted values were found to be correct at the outer boundaries of the deposited titanium carbide. It is interesting to note that, using the thermodynamic data for $\Delta G_{Ti_xC_{(1-x)}}$ and then applying the common tangent method, computer calculated limits of the range of stability of the non-stoichiometric phase were: $X_{Ti} = 0.49$–0.67, fitting well with the observed limits of $X_{Ti} = 0.51$–0.66.

ΔG_m–Concentration Diagrams and the Effect of Physical and Chemical Factors on the Composition and Stability of Surface Interaction Layers ($Fe_{(s)}$–$Zn_{(l)}$ and $Fe_{(s)}$–$Zn_{(v)}$) Systems

There are a number of examples in the literature where, during gas–metal or gas–liquid–metal-surface reactions, certain phases, shown in the relevant equilibrium phase diagrams, do not form. In other cases the composition of these phases may be different from those expected from the normal equilibrium phase diagram. In all these cases neither ΔG^0–T diagrams nor the phase stability diagram proved to be of much use, and therefore attempts have been made to apply the ΔG_m–concentration diagrams to analyse thermodynamically the reason for the differences between the phases obtained under laboratory experimental conditions, such as in the study of equilibrium phase diagrams, and those encountered on a large industrial scale, during which the phases were formed as surface layers. For example, in the $Fe_{(s)}$–$Zn_{(l)}$ system the protective outermost ζ-phase layer, which according to the equilibrium phase diagram should be a stable phase up to 530°C, does not form during galvanising above about 495°C resulting in a rapid linear rate of attack of the steel and an unacceptable quality of galvanising.

Because of the financial importance of this process to steel producers (about one-third of all the steel produced in the world is subsequently galvanised) a great deal of research has been carried out throughout the world to establish the 'true equilibrium phase boundaries' in the $Fe_{(s)}$–$Zn_{(l)}$ system and the critical temperature of stability of the ζ phase. Since the ΔG^0–T diagrams or the phase stability diagrams could not account for these discrepancies in this system, ΔG_m–concentration curves were used for

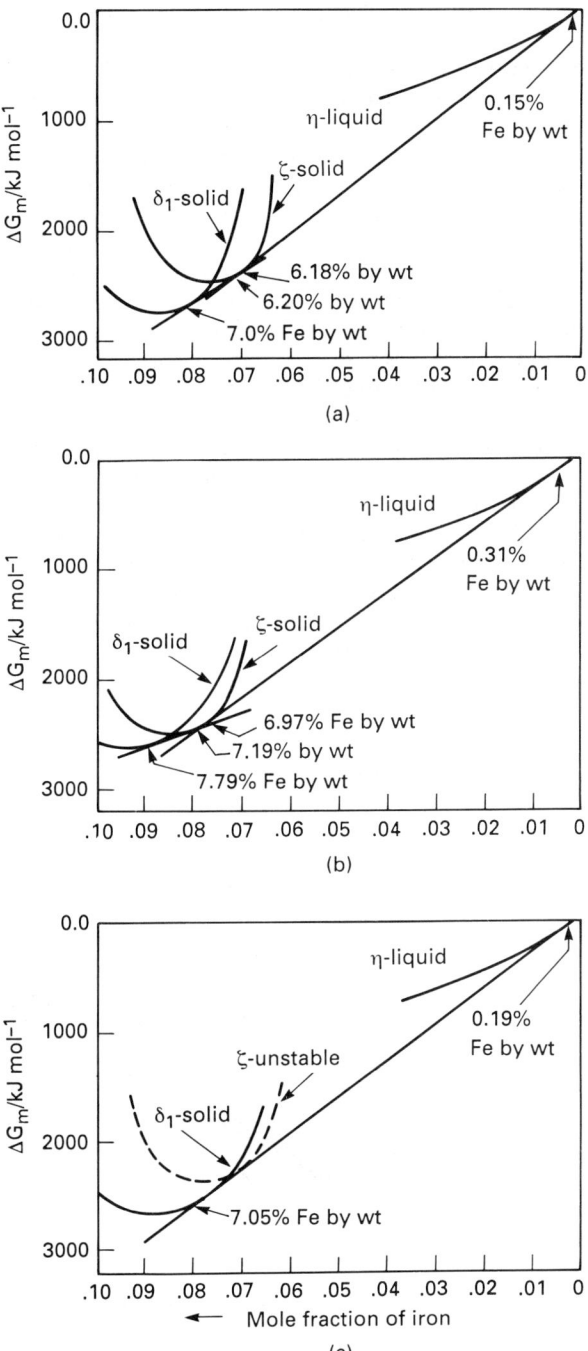

Fig. 7.86 Free-energy-concentration curves for $Fe_{(s)}$–$Zn_{(l)}$ at 505°C under (a) equilibrium conditions, (b) pressure conditions and (c) galvanising conditions (after Mackowiak[55])

both the $Fe_{(s)}$–$Zn_{(l)}$ and $Fe_{(s)}$–$Zn_{(v)}$ systems shown in Figs. 7.86 and 7.87, respectively.

The ΔG_m-concentration curves for $Fe_{(s)}$–$Zn_{(l)}$ at 505°C were calculated for the liquid phase η, and for the condensed phases from experimental results described elsewhere[52]. In the construction of these diagrams it was assumed that the position of the ΔG_m-concentration curve for the liquid η phase remained unaffected by pressures and stresses and therefore was the same in all three diagrams Fig. 7.86A, B and C. The ΔG_m-concentration curves for all condensed phases (Γ, Γ_1, δ_1, ζ) were then fixed with respect to the ΔG_{liquid} curve by the constraint imposed by the use of the common tangent rule. For clarity three separate diagrams were produced. Figure 7.86a shows the situation under equilibrium conditions, as during the study of the equilibrium phase diagram[53]. Figure 7.86b shows equilibria under compressive pressure[54], and Fig. 7.86c shows curves under galvanising conditions where the tensile stress in the ζ-phase layer has altered the ΔG_m-concentration curve for the ζ phase in the upward direction.

Comparing Fig. 7.86a and b, it is clear that the reason for the dramatic increase in the solubility of iron at 505°C from the normal equilibrium value of 0·15% to 0·31% iron by weight under pressure resulted from the slight upward movement of the ΔG-concentration curve for the ζ phase, whereas the position of the liquid η phase remained the same in both cases. It is also worth noting that the composition of the ζ layer at the ζ/liquid boundary is 6·97% iron by weight which is higher than that shown in the equilibrium phase diagram (6·18% iron). There is also an increase in the contents of iron

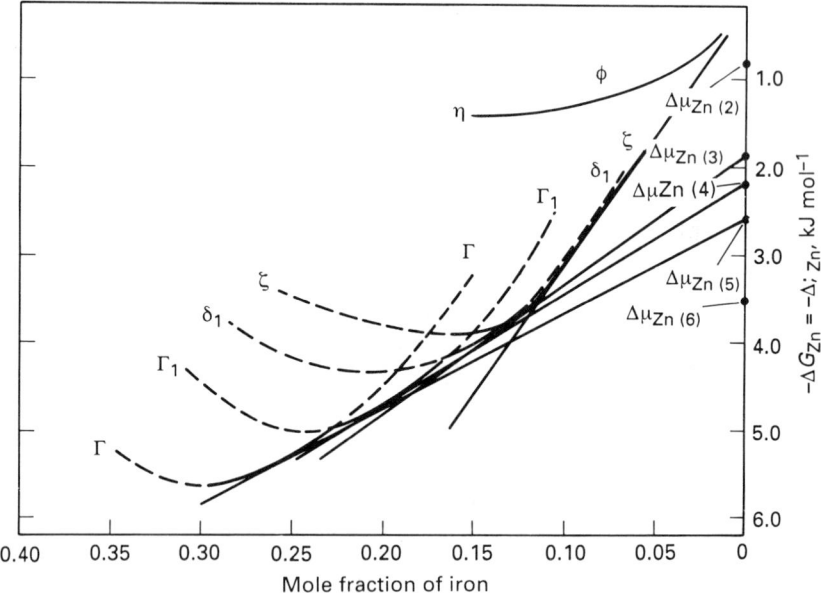

Fig. 7.87 Construction showing position of free-energy-conservation curves of Γ, Γ_1, δ_1, ζ, and η phases for the $Fe_{(s)}$–$Zn_{(v)}$ system at 793 K (after Mackowiack[55])

in both the δ_1 phase and the ζ phase at the δ_1/ζ boundary. Figure 7.86c shows that under galvanising conditions the ΔG_m-concentration curve for the ζ-phase (shown by a broken line) is above the common tangent between the ΔG_m for the η liquid phase and δ_1 solid phase and therefore the ζ phase is unstable under these conditions.

These three diagrams show clearly how a very small change in the relative position of the ΔG-concentration curves for the different phases can have a dramatic effect on the composition and stability of the solid phases in the $Fe_{(s)}$-$Zn_{(l)}$ system. A detailed discussion of these diagrams can be found elsewhere[55]. In addition, Fig. 7.86 shows that any physical effects in the solid phases, which can be expressed in terms of energy, can be added to the ΔG_m-concentration curves. Thus the curves can be corrected for these effects and the resulting new equilibria from the shift of the ΔG-concentration curves (i.e. change in composition of phase, their stability, etc.) can be predicted simply by the use of the tangency rule.

Figure 7.87 shows a ΔG_m-concentration diagram for $Fe_{(s)}$-$Zn_{(v)}$. It was constructed from the experimental data shown in Table 7.37. The method of construction is described elsewhere[44]. Figure 7.87 can now be used, by applying the constraints imposed by the tangency rule, to explain why in Fig. 7.88a and b, where the chemical potentials (shown in the diagram) of zinc vapour varied between 0 and $-1\cdot 81$ kJ mol^{-1}, the total interaction surface layer consisted of Γ, Γ_1, δ_1 and ζ layers; in Fig. 7.88c at a chemical potential only slightly lower ($-2\cdot 11$ kJ mol^{-1}) only Γ and Γ_1 layers were present whilst at $-2\cdot 55$ kJ mol^{-1} only a Γ outermost layer was formed.

The micrographs in Fig. 7.88 show clearly how from a knowledge of the ΔG_m-concentration diagrams it is possible to select the exact reaction conditions for the production of tailor-made outermost surface phase layers of the most desired composition and thus of the optimum physical and chemical properties for a given system. In addition it shows that according to thermodynamics, there can be predictable differences in the composition of the same outermost phase layer prepared at the same conditions of temperature but under slightly different vapour pressures.

Similar results, to the Fe-Zn system were obtained in the $Ti_{(s)}$-$Al_{(l)}$ and $Ti_{(s)}$-$Al_{(v)}$ system where, in the solid-liquid couples some of the expected surface layer phases were not formed, whereas in the solid-vapour system it was possible to obtain all the phases[56] and predict from the ΔG_m-concentration curves the compositions at the different layer phase boundaries.

In the literature some basic relationships have been derived correlating the physical effects on phase and their influence on the value of ΔG_m-concentration curves[45]. These mathematical relationships may be used for 'correcting' the ΔG_m-concentration diagrams. Thus Castleman[58] used the Gibbs' principle (tangency rule) to calculate the equilibrium of the metallic phases growing in Al-U and Ni-Al systems under different hydrostatic pressures. He derived an equation for the change in free energy of a phase on compression. De Boer[59] considered the thermodynamic effect on the dissolution of solids under a simple pressure system (homogeneous and isotropic). Other research workers[47,60,61,62,63] attempted to correlate the effect of stresses and strains, using mechanical theories, with the thermodynamic consideration of equilibria.

The effect of grain size on the value of molal free energy change was also

Table 7.38 Details of specimen preparation

Sample No.	Temp. Fe, K	Temp. Zn, K	ΔT, K	P_{Zn} in system, Pa	$P°_{Zn}$ at 793 K, Pa	$-\Delta \bar{G}_{Zn}$, $\Delta\mu_{Zn}$, kJ mol^{-1}	Phases present in final layer
1	793	792	1	294.9	299.9	0.11	$\Gamma:\Gamma_1:\delta_1:\zeta$
2	793	788	5	268.5	299.9	0.73	$\Gamma:\Gamma_1:\delta_1:\zeta$
3	793	781	12	228.0	299.9	1.81	$\Gamma:\Gamma_1:\delta_1:\zeta$
4	793	779	14	217.9	299.9	2.11	$\Gamma:\Gamma_1$
5	793	776	17	203.7	299.9	2.55	Γ
6	793	770	23	176.3	299.9	3.50	Γ

Composition at interfaces of individual phase layers at 739 K

Sample No.	Composition, at.−%Fe				
	Γ Fe/Γ−Γ/Γ_1	Γ_1 Γ/Γ_1−Γ_1/δ_1	δ_1 Γ_1/δ_1−δ_1/ζ	ζ δ_1/ζ−$\zeta/Zn(v)$
1	32.1−24.2	22.4−16.1	14.7−12.5	11.8−11.5
2	31.9−23.6	21.6−15.4	14.8−13.4	12.2−12.1
3	32.3−25.1	21.8−16.6	14.7−13.6	12.9−12.7
4	32.8−25.4	22.2−20.7	np	np
5	33.0−26.5	np	np	np
6	32.8−27.5	np	np	np

np = not present in total layer

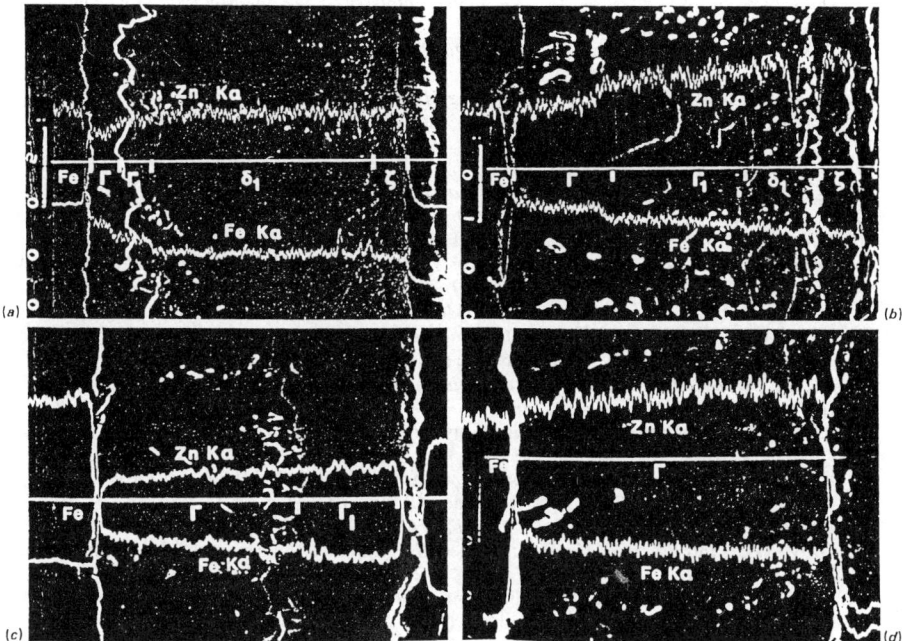

Fig. 7.88 Scanning electron micrographs of cross-sections through interaction layers with superimposed Fe and Zn K_a line scans across the layers (*a*) Sample 2 ($\times$ 210); (*b*) sample 3 ($\times$ 550); (*c*) sample 4 ($\times$ 760); (*d*) sample 6 ($\times$ 1 000) (after Mackowiak and Short[44])

considered by de Boer who derived an expression for the difference in molal free energy as a function of its crystal size[40]. It is beyond this short section to present a full account of all the relationships which may be found in the literature and which may be used to correct the ΔG_m-concentration curves for effects such as vibrations, irradiation, acceleration, capillarity and any others which can be expressed in terms of energy.

J. C. B. ALCOCK
E. EASTERBROOK

REFERENCES

1. Behaviour of High Temperature Alloys in Aggressive Environments, in *Proc. Petten Int. Conf.*, ed. Kirman, I. *et al.*, The Metals Society, 1050 (1980)
2. Richardson, F. D. and Jeffes, J. H. E., *J.I.S.I.*, **160,** 261 (1948)
3. Richardson, F. D. and Jeffes, J. H. E., *J.I.S.I.*, **171,** 165 (1952)
4. Richardson, F. D. and Jeffes, J. H. E., *J.I.S.I.*, **175,** 33 (1953)
5. Pearson, J. and Ende, U., *J.I.S.I.*, **175,** 52 (1953)
6. Villa, H., 'Thermodynamic Data of the Metallic Chlorides', *J. Soc. Chem. Ind.*, No. 1 (supplementary issue), S9–S18 (1950)
7. Sherman, C. W., Elvander, H. I. and Chipman, J., *J. Metals*, **2,** 234 **(1950)**
8. **Lumsden, J.,** *Thermodynamics of Alloys*, Inst. Metals, London (1952)

9. Ellingham, H. J. T., *J. Soc. Chem. Ind., Lond.*, **63**, 125 (1944)
10. Rosenquist, T. and Dunicz, B. L., *J. Metals*, **4**, 604 (1952)
11. Darken, L. S. and Gurry, R. W., *Physical Chemistry of Metals*, McGraw Hill, New York (1953)
12. Sieverts, A. and Krumbhaar, H., *Ber. Dtsch. Chem. Ges.*, **43**, 893 (1910)
13. Alcock, C. B. and Richardson, F. D., *Acta Metallurg.*, **6**, 385 (1958)
14. Chipman, J. and Sherman, C. W., *J. Metals*, **4**, 597 (1952)
15. Wagner, C., *Thermodynamics of Alloys*, Addison-Wesley, New York (1952)
16. Chipman, J., *J.I.S.I.*, **180**, 97 (1955)
17. Goldschmidt, H., *J.I.S.I.*, **160**, 345 (1948)
18. Elliott, J. F. and Chipman, J., *Trans. Faraday Soc.*, **47**, 138 (1951)
19. Schuhmann, R., *Acta Metallurg.*, **3**, 219 (1955)
20. Pourbaix, M. S. N., *Disc. Farad. Soc.*, **4**, 139 and 223 (1947)
21. Pourbaix, M. S. N., The Industrial Use of Thermochemical Data, in *Special Publ. No. 34*, The Chemical Society, London, 55 (1980)
22. Kellogg H. H. and Basu, S. K., *Trans. Mat. Soc. AIME*, **218**, 76 (1960)
23. Ingraham, T. R., in *Applications of Fundamental Thermodynamics to Metallurgical Processes*, ed. Fitterer, G. R. Gordon & Breach, New York London, 179 (1967)
24. Alcock, C. B., *Principles of Pyrometallurgy*, Academic Press, London, New York, 6 (1976)
25. Rao, Y. K., Stoichiometry and Thermodynamics of Metallurgical Processes, Cambridge University Press, London 626 (1985)
26. Rhys-Jones, T. N. (ed.), *Surface Stability*, Inst. of Met., 113 (1989)
27. As Reference 25, pp 379-381; as Reference 48, pp 170-173
28. Davies, R. H. and Barry, T. I., *MTDATA Handbook*, N.P.L. (1989)
29. Gray, J. A. and Starr, F., in Proc. Petten Int. Conf., ed. Kirman, I. et al., The Metals Society, 3 (1980)
30. Natason, K. and Chopra, O. K., First Int. Conf. on Materials for Coal Conversion and Utilisation, Gaithersburg, Maryland, 11 (1976)
31. Natason, K., *High Temperature Corrosion*, ed. Rapp, R. A., N.A.C.E., Houston, 336 (1983)
32. Castle, J. E., *Surface and Interface Analysis*, **9**, 345 (1986)
33. Huang, T. T. et al., *Corr. Sci.*, **24** 167 (1984)
34. Stringer, J., in Proc. Patten Int. Conf., ed. Kirman, I. et al., The Metals Society, 739 (1980). Figure 7.72 after Parkins, R. A. and Voule, S. J., Annual Report to EPRI Project No RP 979-6, 1978
35. Buglia, V. et al., *Corr. Sci.*, **30**, 327 (1990)
36. Knacke, O., in *Metallurgical Chemistry*, ed. Kubaschewski, O., 1972, 549-559, N.P.L., H.M.S.O., London, 549 (1972)
37. Stott, F. H. and Smith, S. in Proc. Patten Int. Conf., ed. Kirman, I. et al., The Metals Society, 781 (1980)
38. Lions, J. et al., in Proc. Patten Int. Conf., ed. Kirman, I. et al., The Metals Society, 769 (1980)
39. Rahmel, A., *Corr. Sci.*, **13**, 125 (1973)
40. Rapp, R. A., *Proc. Workshop on Materials Problems and Research Opportunities in Coal Conversion*, Columbus, Ohio State University, 313 (1974)
41. Stringer, J. and Whittle, D. P., *Proc. First Petten Colloquium on Advanced High Temperature Materials*, **14**, 6 (1977)
42. Rao, Y. K., *Stoichiometry and Thermodynamics of Metallurgical Processes*, Cambridge University Press, London, 631 (1985)
43. Ingraham, R. R., *Trans. Met. Soc. AIME*, **236**, 1064 (1966)
44. Mackowiak, J. and Short, N. R., *Met. Sci.*, **11**, 517 (1977)
45. Mackowiak, J., *Report on Sodium/Steel Interactions*, Sponsored by the Nuclear Installations Inspectorate of the Health and Safety Executive, Ref 98/CS/129/1976 (1977)
46. Miodownik, A. P., *Bulletin of Alloy Phase Diagrams*, **2**, 406 (1982)
47. Gibbs, J. W., *The Scientific Papers*, vol 1, Dover Publication, New York, 65 (1961)
48. Mackowiak, J., *Physical Chemistry for Metallurgists*, George Allen & Unwin, 185 (1966)
49. Kubaschewski, O. et al., *Gases in Metals*, Metals and Metallurgy Trust, ILIFFE Books, London, 18 (1970)
50. Stored data and Software (see Appendix 1)
51. Teyssandier, F. et al., in *Special Publ. No. 34*, The Chemical Society, London, 301 (1980)

52. Brown, W. N., Ph.D. Thesis, University of London (1977)
53. Serebryakova, I. B. and Smirnov, N. S., *Stal*, **25**, 422 (1965)
54. Mackowiak, J. and Short, N. R., *Corr. Sci.*, **16**, 519 (1975)
55. Mackowiak, J., The Industrial Use of Thermochemical Data, in *Special Publ. No. 34*, The Chemical Society, London, 55 (1980) pp 267-279
56. Short, N. R. and Mackowiak, J., *J. Less-Comm. Met.*, 1976, **45**, 301-308. Mackowiak, J. and Shreir, L. L., *J. Less-Comm. Met.*, **1**, 456 (1959)
57. Pascoe, G. and Mackowiak, J., *J.I.M.*, **98**, 253 (1970)
58. Castleman, L. S., *Acta Met.*, **8**, 137-146 (1960)
59. de Boer, R. B., PhD Thesis, University of Utrecht, Holland (1975)
60. MacDonald, G. F. I., *Am. J. Science*, **255**, 226-281 (1957)
61. McLellan, A. G., *Proc. Roy. Soc. A*, **307**, 1-13 (1968)
62. Katchalsky, A. and Curran, P. F., *Nonequilibrium Thermodynamics in Biophysics*, Harvard Univ. Press, Cambridge, Mass., (1967)
63. Kingery, W. D., (ed.) *The Technical Press of MII*, J. Wiley and Sons, N.Y. and Chapman and Hall, London, 187-194 (1959)

Appendix 1

Some Centres Available to Outside Contracts
General Reference:
Metallurgical Thermochemical databases
Contact C. W. Bale and G. Eriksson
Canadian Metallurgical Quarterly
Vol 29 No 2 pp 105-132 (1990)

For details of ThermoCalc Database:
The division of physical metallurgy
KTH S-100 44 Stockholm
Sweden:
Contact Birgitta Jonsson
  Tel: +46 8 790 9140
  Fax: +46 8 100 411
  Email: bosse @ matsc.kth.se
  Datapack (X.25): 24037101046

For details of MTDATA, contact the National Physical Laboratory (NPL)
Queens Road, Teddington TW11 0LW UK
Contact Hugh Davies
  Tel: +44 81 977 3222 ext 6497
  Fax: +44 81 943 2155
  Email: RHD @ UK.Co.NPL.Newton

For details of CHEMSAGE:
GTT mbh Kaiserstrasse 100
5120 Herzogenrath 3
Germany
Contact Bob Fullerton-Batten
  Tel: +49 2407 59533
  Fax: +49 2407 59661

In Australia the CSIRO operate a data base in conjunction with the NPL:
CSIRO Thermochemistry System
PO Box 124
Port Melbourne
Victoria 3207 Australia
Contact A. G. Turnbull
   Tel: +61 3 647 0211
   Fax: +61 3 647 0395
   CSIRONET: (node *MXDIA)

For further information on F*A*C*T:
The Ecole Polytechnique
Box 6079 Station A
Montreal
Quebec
Canada H3C 3A7
Contact A. D. Pelton
   Tel: +1-514-340-4770
   BITNET: J799 @ Polytec1

For general information in the USA:
Alcan Cambridge Technology Center
21 Erie Street
Cambridge Mass 02139 USA
Contact Larry Kaufman
   Tel: +1 617 349 1721
   Fax: +1 617 354 0395

For THERMODATA:
The University of Grenoble
Domaine Universitaire BP 66
38402 Saint-Martin-D'Heres
CEDEX France
Contact B. Cheynet
   Tel: +33 76 427690
   Email: EARN/BITNET: BCHEYNET @ FRGREN81

Additional databases:
THERMOTECH
Surrey Technology Centre
40 Occam Road
The Surrey Research Park
Guildford GU2 5YH
Contact N. Saunders
   Tel: +44 483 502003

8 EFFECT OF MECHANICAL FACTORS ON CORROSION

8.1	Mechanisms of Stress-corrosion Cracking	**8**:3
8.2	Stress-corrosion Cracking of Ferritic Steels	**8**:32
8.3	Stress-corrosion Cracking of Stainless Steels	**8**:52
8.4	Stress-corrosion Cracking of High-tensile Steels	**8**:84
8.5	Stress-corrosion Cracking of Titanium, Magnesium and Aluminium Alloys	**8**:115
8.6	Corrosion Fatigue	**8**:125
8.7	Fretting Corrosion	**8**:165
8.8	Cavitation Damage	**8**:179
8.9	Outline of Fracture Mechanics	**8**:190
8.10	Stress-corrosion Test Methods	**8**:197

8.1 Mechanisms of Stress-corrosion Cracking

The visible manifestations of stress corrosion are cracks that create the impression of inherent brittleness in the material, since the cracks propagate with little attendant macroscopic plastic deformation. In fact, a metal that fails by stress corrosion is usually found to conform to the normal ductility standards for that material and the combination of circumstances that cause a normally ductile metal to behave in this way are the presence of a specific environment, a tensile stress of sufficient magnitude and, usually, a specific metallurgical requirement in terms of the composition and structure of the alloy. The compositions and structures of the alloys and the properties of the environments involved in the various instances of failure indicated in this section are so widely varying as to suggest that rationalisation of all of these experiences in a single explanation would be difficult if not unreal, i.e. it is probable that a number of different mechanisms are involved. This is not to suggest that some systematisation is not possible and indeed the objective of this section is to show that the evidence may be rationalised in a continuous spectrum of mechanisms[1], a concept that has the merit of avoiding the dangers inherent in believing that some highly specific conditions need to be fulfilled for stress corrosion to occur because the mechanism of failure is invariable.

The usual energy balance approach to fracture (see Section 8.9)-by equating the strain energy released to the energy consumed in creating new surface and in achieving plastic deformation–needs modification where corrosion processes are involved to take account of the chemical energy released, and it is the latter that distinguishes stress corrosion from other modes of fracture not involving environmental interaction:

$$\text{Surface energy change} + \text{Plastic work done} = $$
$$\text{Change in initial stored energy} + \text{Electrochemical energy released}$$
(8.1)

Since the surface energy term will usually be negligible by comparison with the plastic work term in the stress corrosion of ductile materials, it may be neglected. The remaining terms may be derived from fracture mechanics and conventional electrochemical conditions and, for the various boundary conditions indicated by West[2], result in

$$P = K_I^2(1 - \nu^2)/E + (zF\rho\delta/M)\eta \qquad (8.2)$$

where P is an appropriate plastic work term, K_I is the stress-intensity factor, ν is Poisson's ratio, E is Young's modulus, z is the valency of the solvated ions, F is Faraday's constant, ρ is the density, δ is the height of the advancing crack front, M is the molecular mass of the metal and η is the anodic overpotential. This assumes that the mechanism of crack advance involves the localised dissolution of metal, but an equivalent expression could be written involving hydrogen-induced cracking. At the threshold stress-intensity K_{Iscc}, i.e. the minimum value of K_I for stress-corrosion cracking, equation 8.2 yields.

$$K_{Iscc} = \{(E/1 - \nu^2)(P - zF\rho\delta\eta/M)_{min}\}^{\frac{1}{2}} \qquad (8.3)$$

Clearly the variables that may influence K_{Iscc}, or the threshold stress for initially plain specimens, and hence the susceptibility to stress-corrosion cracking, are P and η, i.e.

$$K_{Iscc} = \{k_i(P - k_2\eta)_{min}\}^{\frac{1}{2}} \qquad (8.4)$$

Lowering of the plastic work term P will result from an increase in the effective yield stress or an increase in the work hardening rate in the crack-tip region, either or which, for constant η, will lower K_{Iscc} and hence increase the susceptibility to stress corrosion. An increase in the anodic overpotential η_a (the potential of the metal is made more positive) relative to the plastic work term will also increase susceptibility to stress corrosion. The anodic overpotential will be some function of the electrochemical conditions within the crack that control the active to passive transition which determines whether or not cracking occurs, i.e. it will be some function of pH, anion activity, metal composition and electrode potential. The interdependence of these terms upon the structure and composition of the metal, upon the details of the electrochemical conditions at the crack tip in terms of local cell action and film formation, and upon the response of the metal to the presence of stress in creating new metal at the crack tip, make the quantification of the argument extremely difficult and its relation to the detailed mechanisms of crack propagation virtually impossible, as West[2] indicates. However, the recognition of the need for a critical balance between a number of variables if stress corrosion is to occur, and the fact that this balance may be achieved in a number of ways, is important, and not least in relation to the diagnosis and the prevention of stress-corrosion failures.

Stress-corrosion Crack Propagation Models

The implication of the foregoing equations, that stress-corrosion cracking will occur if a mechanism exists for concentrating the electrochemical energy release rate at the crack tip or if the environment in some way serves to embrittle the metal, is a convenient introduction to a consideration of the mechanistic models of stress corrosion. In so far as the occurrence of stress corrosion in a susceptible material requires the conjoint action of a tensile stress and a dissolution process, it follows that the boundary conditions within which stress corrosion occurs will be those defined by failure under a stress in the absence of corrosion, and failure by corrosion in the absence of stress. Between these extremes, wherein stress corrosion occurs, it is necessary to consider how corrosion processes may be influenced by the application of stress to a metal and how fracture may be facilitated by corrosion.

When stress corrosion involves very localised dissolution, with the geometrical requirements of a crack to be fulfilled, the rate of anodic dissolution may be expressed as a rate of crack propagation

$$CV = \frac{i_a M}{zF\rho} \tag{8.5}$$

where i_a is the anodic current density and the remaining symbols are as defined earlier. Now i_a and hence CV will be dependent upon the nature of the phase being dissolved and also upon the associated cathodic processes that occur elsewhere, and which need to produce a current sufficient to balance that at the crack tip. The chemical natures of the sites for these reactions will therefore be of importance, but how does the imposition of a tensile stress influence this situation to produce stress corrosion? Despic et al.[3] have shown that the dissolution rate of iron in acid solution undergoes a marked rise when the strain passes from the elastic to the plastic condition in dynamic straining experiments and that this result is due to the exposure of high-index planes and of edges at slip steps, as well as increasing surface roughness, as plastic deformation occurs. Stresses in excess of the yield stress may therefore produce locally enhanced activity at surfaces where slip steps emerge, i.e. i_a in equation 8.5 may be increased (by up to an order of magnitude) by stresses promoting plastic strain.

However, there is no difficulty in accounting for the observed rates of crack propagation by stress corrosion for most systems in terms of the currents passed at static bare surfaces without invoking arguments involving markedly enhanced currents resulting from the exposure of slip steps. The real problem is in explaining why the corrosion proceeds along a narrow front to retain the geometry of a crack, implying that most of the exposed surfaces, including the crack sides, must remain relatively inactive. The transition from electrochemically active to relatively inactive behaviour that the sides of a crack must undergo as the tip advances and creates more crack can only be achieved if the environment forms a film, and this implies, in relation to equation 8.5, that the conditions for maximum crack growth rate will be met if i_a is maintained close to the film-free value, i.e. the metal is in the active state and protective films are not allowed to grow over the crack tip or, if this does occur, that the film is repetitively broken. The function of stress, then, in the realistic conditions of film-forming environments will be essentially to prevent, or to fracture, films forming at the crack tip.

In this context, two different circumstances[1] may now be envisaged whereby cracks can propagate by a dissolution-controlled process. The alloy may exhibit structural features, either as a segregate or precipitate, usually at the grain boundaries, that cause a local galvanic cell to be established, i.e. a *pre-existing active path* is involved, as originally suggested by Dix[4]. The precipitate or segregate may act as the anode in the local cell or, by

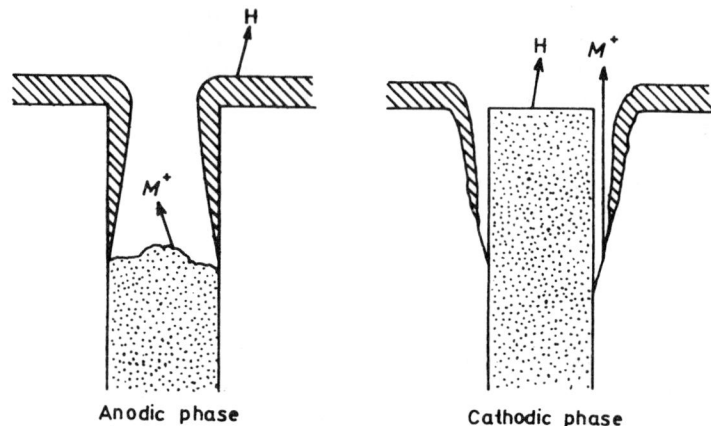

Fig. 8.1 Pre-existing active path mechanisms, in which H represents cathodic hydrogen

acting as an efficient cathode, may cause the dissolution to be localised upon the immediately adjacent matrix (Fig. 8.1). The lattice characteristics in the region of a grain boundary are such that equilibrium segregation of solutes or the nucleation and growth of precipitates are favoured reactions, and so grain boundaries in particular are potential sites for chemical heterogeneity. Where such pre-existing active paths are non-existent, or are inoperative, the disruption of a protective surface film to expose bare metal may result in a second mechanism of crack propagation, as originally suggested by Parr and Straub[5]. Thus stress (or probably more correctly, plastic strain) in the underlying metal may bring about the disruption of the protective film whereby active metal is exposed, in the manner shown schema-

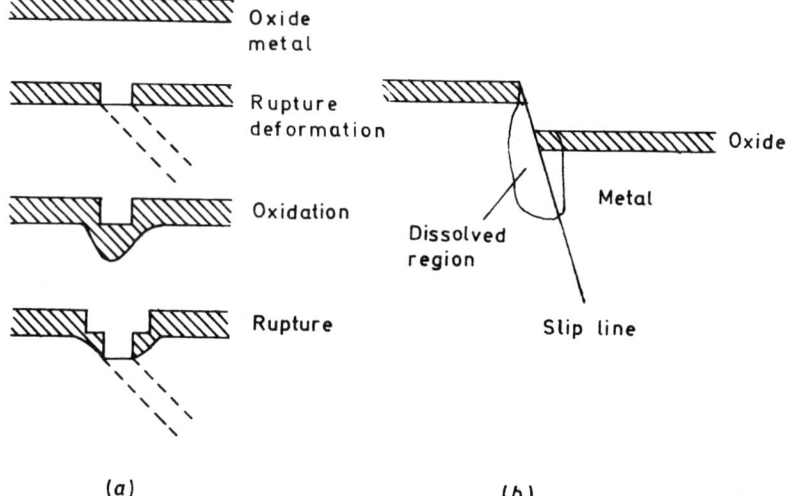

Fig. 8.2 Strain-generated active path mechanisms. (a) Often referred to as the film rupture model and (b) the slip step dissolution model. In both cases growth is by dissolution; film rupture is the rate controlling step, not the mechanism of crack growth

tically in Fig. 8.2. The active path along which the crack propagates is cyclically generated as disruptive strain and film build-up alternate with one another, or propagation is related to the slip characteristics of the underlying metal.

Although somewhat less favoured than a decade or so ago, transgranular cracking by a *strain-generated active path* mechanism remains supported by some workers and a significant body of corroboratory evidence. The localisation of the dissolution in such cases is most often related to slip and particularly that which occurs with some face-centred cubic alloys having a low stacking fault energy or displaying short-range order, where planar groups of dislocations are favoured and cross slip made more difficult[6] (Section 9.2). The observations of Swann and his co-workers[7-9] of dissolution associated with planar dislocation arrays in transmission electron microscopy (TEM) foils led to the suggestion that arrays of fine corrosion tunnels form which subsequently interconnect by the tearing of the remaining ligaments between the tunnels (Fig. 8.3).

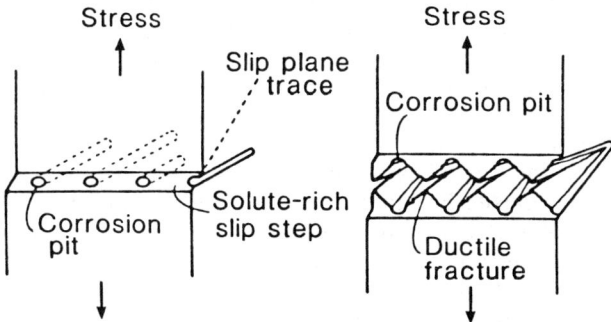

Fig. 8.3 Schematic representation of the stress corrosion cracking mechanism of the pit (after Pickering and Swann[7]). (a) Tubular pits initiated at solute-rich slip step. The pits may, but need not necessarily, follow the slip plane once they are initiated. (b) Ductile tearing along a plane containing the tubular pits. The stress is increased across the plane because of the reduced cross section and the stress raising effect

At about the same time that Swann was developing the tunnelling model, Nielsen[10] was examining, by TEM, the corrosion products removed from stress-corrosion cracks in austenitic stainless steels after exposure to chloride solutions. In general, these take the form of fans showing lamella markings, suggesting that crack growth was discontinuous. It may, of course, be argued that such oxide films are formed after the crack tip has advanced and that they are simply replicating the stress-corrosion fracture surface. However, when Nielsen exposed samples of the steel to $MgCl_2$ solution for only a few minutes oxide-filled corrosion tunnels developed, which were joined by lateral tunnels that increased in number on moving towards the original surface of the specimen, to produce a corrosion-product fan. Somewhat similar effects have been observed[11] in a copper alloy exposed to ammonia vapour and again suggest that localised dissolution processes cannot be ruled out as a contributing factor in the growth of transgranular stress-corrosion cracks in some systems.

The objection that is most frequently raised to such an essentially dissolution-related crack growth model is that it does not appear likely to lead to the matching and interlocking fracture surfaces often observed in the transgranular cracking of face-centred cubic metals[12]. This has resulted in consideration of the possibility that embrittling films formed at the exposed surfaces of metals may play a critical role in stress-corrosion cracking. Edeleanu and Forty[3] observed that the cracking of α-brass single crystals exposed to an ammoniacal solution occurred discontinuously, with short bursts of extremely rapid cracking followed by relatively, long rest periods. It was suggested that truly brittle fracture was associated with the bursts while the rate-controlling periods of non-propagation were concerned with the corrosive processes that established the conditions for further crack bursts. The model requires that α-brass can support a free-running cleavage crack, albeit over short distances of the order of a few microns, and this presented a major difficulty. Thus, while cleavage in a body-centred cubic metal, such as α-iron, has been observed on a microscale, to reflect its well-known tendency for macroscopic cleavage in appropriate conditions, such cleavage of face-centred cubic metals, as in α-brass, has not been demonstrated in like manner. However, relatively recent atomic modelling studies[14, 15] indicate the theoretical possibility of short-range cleavage of ductile metals from an initiating surface film of appropriate characteristics.

Embrittlement of a metal from corrosive reactions, especially whereby hydrogen enters the metal, has often been invoked in the context of stress corrosion. The opposite boundary condition referred to earlier as limiting the regime in which stress corrosion occurs was that of purely mechanical fracture in the absence of corrosive processes. The energy balance of equation 8.1 indicates that, with negligible contribution from dissolution, crack extension will be facilitated by a reduction in the surface energy required to form crack faces or a reduction in the plastic work term by embrittlement of the metal in the crack-tip region. If the environment provides species that are adsorbed at the crack tip to reduce the effective bond strength, then the surface energy is effectively lowered, alternatively the species may diffuse into the metal forming a brittle phase, e.g. a hydride, at the crack tip, or interactions may occur at some region in advance of the crack tip where the stress and/or strain conditions are particularly appropriate for the nucleation of a crack (Fig. 8.4). In the latter case hydrogen is usually regarded as the only species that can diffuse with sufficient speed to account for observed crack propagation rates, and in the present context, hydrogen embrittlement, with the hydrogen derived from corrosion reactions, is considered as a particular instance of stress corrosion. Whilst surface-energy lowering has been suggested[16] as a single mechanism that explains all instances of stress-corrosion cracking, it has particular difficulties in explaining the phenomenon in the more ductile metals. Thus, whilst stress-corrosion cracks propagate without marked macroscopic plastic deformation there is ample evidence to show that localised plastic deformation occurs at the crack tip, and in such circumstances, as indicated in relation to equation 8.1, the surface energy term is negligible in relation to the plastic work term (5 J/m^2 as opposed to 5 kJ/m^2), and so any reduction in surface energy by adsorption will have a negligible effect upon the fracture stress. Moreover, in some

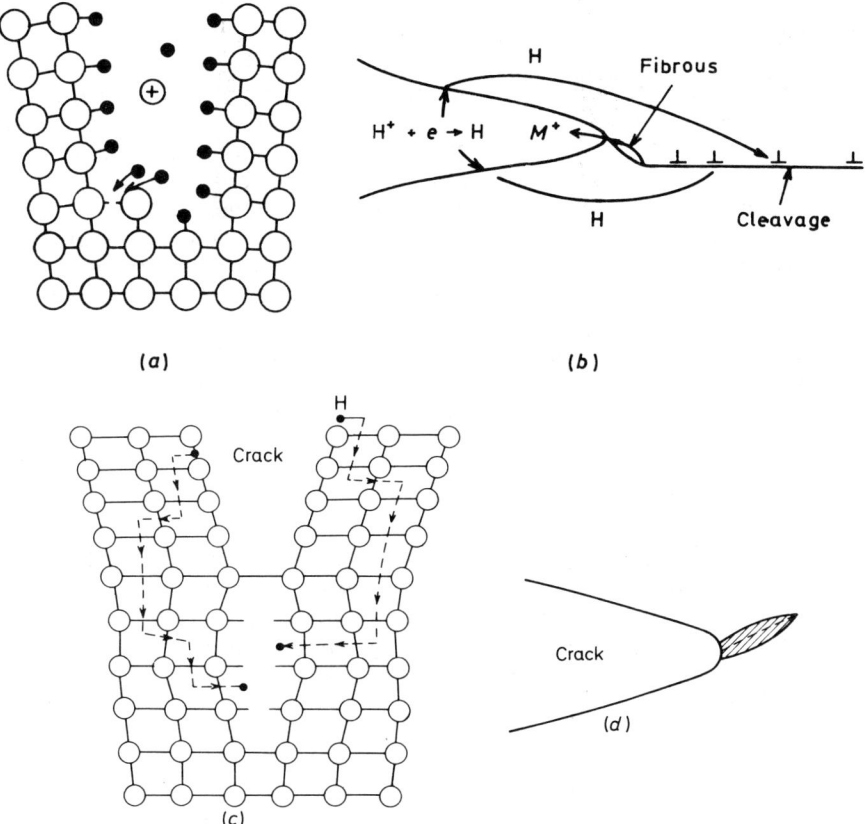

Fig. 8.4 Mechanisms involving embrittlement of the metal. (a) Crack-tip adsorption, (b) hydrogen adsorption, (c) decohesion by hydrogen influx to dilated lattice and (d) crack extension due to brittle hydride particle forming at crack tip

instances of hydrogen-related fracture of metals evidence of the fracture mechanism involving enhanced local plasticity due to the presence of hydrogen[17,18] has become apparent in recent years and, for those cases at least, the mechanism of crack growth is hardly consistent with an approach based on equation 8.1.

In summary then, given the appropriate balance between electrochemical activity and inactivity, localised corrosion may be distributed in a number of ways and result from a number of different mechanisms in promoting stress corrosion. If the structure and composition of the alloy are such that almost continuous paths of segregate or precipitate exist, usually at the grain boundaries, and which are electrochemically different from the matrix, then a latent susceptibility to intergranular corrosion may be activated by the presence of stress. Where pre-existing active paths are inoperative, the strain may generate active paths by rupturing an otherwise protective film and, possibly, activating dissolution at emerging slip steps or initiating micro-cleavage which continues from the film in which it initiates into the underlying metal for some distance before arresting. The more crucial role of stress or strain in the latter case is continued to those alloys that are inherently lacking somewhat in ductility and have a propensity towards brittle fracture, which may be facilitated by a reduction in the energy required for fracture as the result of either adsorption of species or the formation of brittle phases at the crack tip, or of hydrogen in advance of the latter. The suggestion[1] that these different mechanisms of stress corrosion should be considered as occurring within a continuous spectrum, with a gradual transition from one to the others as the dominance of corrosive processes is replaced by stress or strain, leads readily to the notion that alloy composition and structure, electrochemistry and stress may interact in a variety of ways, and that the transformation from one mechanism to another may result from a change in either alloy characteristics or environmental conditions. On the other hand, there are some who consider that all instances of environment sensitive cracking can be explained by a single mechanism of which Galvele[18], with a surface mobility mechanism, is most recent.

The Stress-corrosion Spectrum

Stress corrosion has been the subject of a number of extensive reviews[19-23] resulting from major conferences in recent years, and these, together with the following sections of this volume, avoid any need for a general review of the data. Instead, consideration will be given to some of the implications of the various ideas already referred to in relation to stress corrosion in a variety of systems.

Pre-existing Active Paths

Where cracking is associated with pre-existing active paths, structurally sensitive attack, such as intergranular corrosion, may be expected to be observed upon unstressed specimens, at least in the earlier stages of exposure before secondary reactions such as those that lead to film formation, obscure the metal structure. In Section 8.2 it is indicated that samples of steel polished as for metallographic examination suffer grain boundary attack when immersed for short times under environmental conditions that would lead to cracking of stressed specimens, and that whilst such grain-boundary corrosion does not penetrate to great depths in the absence of stress, it is possible to disintegrate a piece of unstressed mild steel by intergranular corrosion in a boiling nitrate solution with applied anodic current. The observation that carbon steels contain pre-existing susceptible paths for corrosion, the structural distribution of which is related to the paths followed by stress-corrosion cracks, is also apparent in the aluminium-base alloys that undergo stress-corrosion cracking (Section 8.5).

The electrochemical properties of the segregates or precipitates, relative to their associated matrixes, that are involved in these instances of intergranular attack have been the subject of a number of studies following the early work of Dix[4] along these lines. Thus, Doig and Edington[24] used microelectrodes to measure the localised corrosion potentials at grain boundaries in Al–Mg and Al–Cu alloys, and their results correlate well with the cracking propensities of those alloys. The effects of ageing time in these results reflect their effects upon cracking susceptibility. Similarly, heat treatments designed to change the distribution of chemical heterogeneity of grain boundary regions in ferritic and stainless steels and in nickel-base alloys alter the cracking propensities of the latter in ways that support the suggestion that intergranular stress corrosion in these relatively low-strength ductile materials is related to a latent susceptibility to intergranular corrosion.

Of course, an inherent susceptibility to intergranular corrosion is not the only requirement for susceptibility towards intergranular stress corrosion, since for the latter to occur it is necessary for the former to be sustained or enhanced by the application of stress, and there are instances of alloys that are susceptible to intergranular corrosion but not apparently to stress corrosion. The role of stress may be critical in some cases where the material shows a tendency towards intergranular fracture in the absence of corrosive influences, e.g. in some of the high-strength aluminium alloys, or where the structure of the alloy determines whether or not deformation is localised to sustain relatively bare metal in the grain-boundary crack-tip region. In this latter respect it is noteworthy that the α-brasses, which can be caused to fail in an intergranular manner under constant strain conditions in ammoniacal solutions at pH 7.3, show an increased tendency for transgranular cracking under slow-strain-rate conditions[25]. Since the same material, at constant strain, can be made to fail in a transgranular manner by changing the pH of the ammoniacal solution or by small changes in alloy composition, such results simply serve to underline the delicate balance between the factors that promote a particular mechanism of cracking and of the dangers in attempting to rationalise all observations into a single mechanism. Moreover, even within those systems that exhibit intergranular stress corrosion the part played by the response of the metal to the application of stress may be expected to be variable, with an increasing tendency towards a different mechanism, most often resulting in transgranular fracture as the propensity towards intergranular corrosion is reduced and the roles of stress and/or strain become more important.

Strain-generated Active Paths

Many corrosion-resistant alloys owe their electrochemical inactivity to a relatively inert film that forms on the exposed surfaces of the metal, so that the relatively active metal is effectively separated from the environment. If the protective film is disrupted for any reason, such as by plastic strain in the underlying metal, the exposed metal is attacked until the protective film reforms, when further reaction is stifled until the film is again disrupted. With such a mechanism it is claimed[26] that the rate of cracking will be dependent upon the rate of film growth, although the physical characteristics of the film, i.e. its thickness, the extent to which it shows plastic or brittle behaviour and the magnitude of the internal stresses that it contains as a result of its mode of deposition, are also likely to be important. Ellipsometric studies of the rates of film growth on α-brasses in 15 N aqueous ammonia have shown the growth rate to increase with zinc content of the brass, temperature and applied potential, parameters that also increase the stress-corrosion crack-propagation rate, thereby providing support for a film rupture mechanism. The question of whether this would result in intergranular or transgranular cracking is controversial. It is sometimes argued[27] that plastic strain tends to concentrate at grain boundaries, forming dislocation networks in which copious sources exist for plastic flow and hence for promoting intergranular cracking. On the other hand, transgranular cracking will be favoured by planar slip and will therefore be facilitated at low stresses when extensive plastic strain will not result in dislocation networks that would block planar slip. Whilst instances may be cited of transgranular cracking occurring at low stresses and intergranular at high stresses, there is also a considerable amount of evidence to the contrary. The effects[25] of increasing strain rate, and hence of stress, have already been mentioned as resulting in an increased tendency for transgranular cracking, and the effects of increasing amounts of cold work upon the cracking of carbon steels (Section 8.2) are at variance with the expected effects. Preferential oxidation may occur along grain boundaries in the absence of stress, possibly because of equilibrium solute enrichment in such regions, and even where the rupturing of films growing along boundaries is an important part of the stress-corrosion process, as possibly with the α-brasses[28], these are, within the context of the earlier definition, examples of corrosion along a pre-existing active path. The intergranular stress corrosion of the α-brasses therefore constitute a convenient bridge between the pre-existing and strain-generated active path mechanisms.

Where transgranular stress corrosion results from dissolution following repetitive film rupturing it is to be expected that the deformation characteristics of the metal will be important, although these may well be influenced by the presence of the film. It would appear that the height of the slip step formed at a surface must be greater than the thickness of the inactive film if bare metal is to be exposed, and this implies that deformation associated with high slip steps is likely to be more effective in promoting stress-corrosion than when fine slip is operative. However, it should not be assumed that crack initiation will be avoided if only fine slip steps form, since the initiation of cracks in α-brass, for example, in a variety of environments is most often at grain or twin boundaries, despite the subsequent propagation being transgranular[29]. This is probably because the film overlying grain boundaries has different properties from that overlying grain surfaces facilitating crack initiation at the grain boundaries, although if the slip steps are large enough transgranular initiation will also occur. With the face-centred cubic metals, high slip steps are the result of cross-slip being difficult, leading to planar arrays of dislocations, because of low stacking-fault energy or the presence of short range order in the alloy. Swann[6] has shown a relationship between stacking-fault energy and stress-corrosion susceptibility for copper-base alloys and for austenitic stainless steels, indicating also the tendency for transgranular cracking to dominate the more readily planar arrays of dislocations form.

However, whilst the effects of change in alloy composition upon stress-corrosion cracking susceptibility in the present context may be partly due to their effect upon stacking-fault energy, this does not constitute a complete explanation, since alloying may have significant effects upon electrochemical parameters. The effect of the zinc content of brasses upon their filming characteristics has already been mentioned, while in more recent work Sieradzki et al.[30] have shown that the tendency for the dezincification of α-brasses correlates well with the effect of zinc content of the brass upon transgranular cracking. The effect of nickel additions to carbon steels upon cracking in boiling 42% $MgCl_2$ is equally illuminating. Small additions, of the order of a few per cent, have little effect upon the cracking of ferritic steel in boiling nitrate solution and in the absence of nickel such steels will not fail in boiling 42% $MgCl_2$. However, the addition of only 1% Ni will induce a susceptibility to cracking in $MgCl_2$ which follows an increasingly transgranular path as the nickel content is increased, becoming fractographically indistinguishable from the austenitic steel at about 6% Ni. The structure and mechanical behaviour of ferritic steels are not significantly changed by additions of only 1% Ni, yet the change in cracking susceptibility is dramatic and it is difficult to escape the conclusion that this results primarily from changes in electrochemical behaviour[31].

This is not meant to imply that the mechanical behaviour of alloys as reflected in their response to the application of stress is not important in transgranular stress-corrosion, but merely that the relative importance of different parameters can vary from one system to another. The importance of deformation, and hence mechanical behaviour, in transgranular cracking is most crucial in the slip dissolution model. Static dislocations do not usually show evidence of significant chemical activity, unless associated with chemical heterogeneity resulting from solute segregation, but moving dislocations have been suggested as promoting electrochemical activity relevant to stress-corrosion cracking. Hoar and West[32] showed that the currents associated with straining electrodes may be very much greater than those observed at static surfaces, a difference suggested as being the result of yield-assisted depolarisation. The association of such observations with the formation of tunnels from the crack tip and the tearing of the ligaments between the tunnels to produce crack advancement[10,33] is an obvious extension, but Staehle[34] considers neither to be a vital part of the mechanism since they are not relevant in every case of cracking in Fe-Cr-Ni steels. The essential step in the slip dissolution model is that a relatively inactive film is broken by emerging dislocations and a local transient dissolution process ensues. The difference therefore between this and the film rupture model as it is sometimes invoked is largely concerned with the differing emphasis placed upon the acts of film rupture and subsequent metal dissolution as the controlling process.

It has already been mentioned, virtually as an extension of the film rupture model, that a crack initiated in a brittle film may progress into the normally ductile substrate for a small but appreciable distance before being arrested by plastic deformation in the matrix. The attraction of a microcleavage-based mechanism for transgranular stress-corrosion cracking in a number of systems derives from fractographic observations[12] and the emission of discrete acoustic events and electrochemical current transients accompanying crack growth[15]. Thus, stress-corrosion fracture surfaces are characterised by flat, parallel facets separated by steps, opposite fracture surfaces being matching and interlocking. Arrest markings are sometimes observed, suggesting that crack growth is discontinuous, as observed in the experiments of Edeleanu and Forty[13]. Moreover, there is a strong correlation between peak amplitude acoustic emission events and electrochemical current transient peaks during the transgranular cracking of α-brass exposed to NaNO$_2$ solution[15]. Of course, it may be argued that such observations are not unequivocal demonstrations of crack growth by fast cleavage. Thus, arrest markings make no comment upon the processes occurring between successive markings, which simply indicate that the crack stopped. If, as is likely, the crack stops because of plastic deformation and crack yawning, the acoustic emissions and electrochemical current transients could be a consequence of such deformation.

There is a need for measurements aimed at measuring possible cleavage events more directly. Moreover, the expression used in the analytical modelling of cleavage initiated by films[15] appears to involve dislocation–crack interactions which are only likely to be valid under small-scale yielding conditions. Yet the initiation of stress-corrosion cracks in α-brass exposed to NaNO$_2$ solution is associated with the onset of yielding and continues with general yielding[35]. The latter leads to extensive branching, which seems more likely to be related to shear strains being very effective in producing crack growth than to any dynamic effects, not least because crack branching in cleavage-type fracture usually only occurs at very high crack velocities. The debate between the protagonists of the dissolution and the microcleavage mechanisms for transgranular cracking of the more ductile alloys continues, but the concept of localised embrittlement being involved in stress-corrosion cracking in some systems is not in doubt.

Embrittlement of the Metal in the Crack-tip Region

The literature reports many so-called critical experiments that purport to show the operation of a surface energy lowering mechanism of stress corrosion, but the results are frequently equally explicable in terms of some other mechanism. The effect of grain size upon stress-corrosion cracking susceptibility is a typical example, it having frequently been reported that coarse-grained material is more susceptible to cracking than fine-grained material, detailed analysis showing a Petch type of relationship between the grain diameter l, and the stress σ_i, to initiate a stress-corrosion crack, i.e.

$$\sigma_i = \sigma_0 + kl^{-\frac{1}{2}} \quad (8.6)$$

where σ_0 and k are constants, of which k may be related to the surface energy associated with the formation of new surfaces by fracture through

$$k = \left(\frac{6\pi G\gamma}{1-\nu}\right)^{\frac{1}{2}} \quad (8.7)$$

where G is the modulus of rigidity and the other symbols are as defined earlier. Measurement of the dependence of some stress-corrosion fracture stress on grain size therefore allows a surface energy to be obtained, and since Coleman, et al.[36], found the apparent surface energies so determined to be appreciably less than the energy values derived in other circumstances, they concluded that the surface energy associated with crack formation is reduced by the adsorption of some atom or ion species in the stress-corrosion medium. There is, however, an alternative explanation of the grain size dependence of the stress-corrosion behaviour of alloys, and this is concerned with the plastic-flow characteristic of materials as they are influenced by grain size. Thus a relationship of the form of equation 8.6, where σ_i is the flow stress at constant strain and the grain size term arises from the resistance to the formation of a slip band at a grain boundary, can be shown to be relevant to the plastic behaviour of metals, and it follows that the grain-size dependence of stress-corrosion cracking may simply reflect the fact that the latter is related to plastic flow in the material. Results such as those shown in Fig. 8.18 in Section 8.2, indicating similar slopes for the stress-corrosion fracture stress and flow stress plots against grain size, suggest that the effect of grain size in stress-corrosion cracking is as likely to be related to plastic flow effects as it is to surface energy lowering. Similar results are available in relation to the cracking of α-brass in $NaNO_2$ solutions[31].

The specificity of environments that promote stress-corrosion cracking has been adduced in support of a crack-tip adsorption model[16], but such observations do not appear[37] to discount a dissolution mechanism of crack propagation any more than they support an adsorption mechanism. Yet, almost paradoxically, it is from observations on environmental aspects of stress corrosion of high-strength steels that the strongest evidence in support of environmental-induced brittleness in the crack-tip region derives. The solution requirements for cracking in high-strength steels are not highly specific, i.e. failure will occur in a wide range of aqueous and non-aqueous solutions, unlike the situation in relation to the failure of the low-strength ductile alloys, and the common denominator in these environments is hydrogen. The implication is simply that the environment should provide a source of hydrogen, but that species in solution that facilitate the ingress of hydrogen into the metal will enhance cracking, whilst species that lead to the discharge of gaseous hydrogen at the steel surface will retard cracking. In the former category are arsenious salts, which promote hydrogen adsorption and entry, whilst platinum additions to the system may be expected to facilitate hydrogen discharge.

Similarly the effect of increasing cathodic current densities applied in stress corrosion tests may be expected to enhance cracking if hydrogen adsorption is involved in the failure mechanisms. The effects of sodium arsenate and chloroplatinic acid additions to a sodium chloride solution upon the cracking propensity of an 18% Ni maraging steel at various applied cathodic current densities conform with expectations if hydrogen adsorption is the controlling factor in the cracking process[38]. Other observations, such as those involving measurement of the solution pH and the electrode potential at the tip of a propagating stress corrosion crack in a high-strength steel in showing that the conditions there are conducive to hydrogen entry into the steel, are also sometimes adduced in support of a hydrogen-embrittlement mechanism (see Section 1.6 and 8.4). It is worth mentioning, however, that the demonstration of the existence of acid conditions at the crack tip does not exclude the possibility that some crack extension, however small, results from dissolution, which is also likely to be facilitated by the low pH environment at the crack tip. Indeed the production of hydrogen by cathodic reaction requires a balancing anodic reaction, which may occur at the crack tip and result in advancement of the latter.

Whilst discussion continues on the details of hydrogen generation, adsorption and diffusion, and the relative contributions of these to the overall physical mechanism of hydrogen embrittlement, some aspects of the latter have begun to crystallise as the result of experiments conducted in gaseous hydrogen environments[39]. The demonstration that sub-atmospheric pressures of hydrogen gas can readily result in the propagation of cracks in high-strength steels indicates that the mechanism is not likely to involve the diffusion of hydrogen through the metal to voids where a disruptive pressure of gas is generated. This suggests either that hydrogen lowers the surface energy by adsorption or that it accumulates within a few atomic distances from the crack tip, in response to the lowering of its chemical potential by the elastic stress, thereby lowering the cohesive force of the lattice. Oriani[39] prefers the latter explanation because it is the only one that is consistent with the observations of the effect upon crack propagation of small changes in hydrogen gas pressure and the substitution of deuterium for hydrogen. A sufficient reduction in the hydrogen gas pressure surrounding a specimen containing a propagating crack at a given stress intensity caused the crack to stop propagating, but a subsequent increase in pressure, of about $1.6 \, kN/m^2$ from $22 \, kN/m^2$, was sufficient to restart the crack and with a delay time so short that the extra hydrogen entering the lattice as the result of the increased pressure could have diffused no more than a few atom spacings. A similarly rapid response of the crack velocity to small changes in applied cathodic current to a maraging steel immersed in sodium chloride solution has been observed. The effect of deuterium in reducing the response to embrittlement appears not to be related to the difference in transport kinetics of the two isotopes but to their solubilities in the dilated lattice just beyond the crack tip. This again is in agreement with a decohesion model.

An alternative, or possibly an additional, model for hydrogen-induced failure that has received recent support is that based upon the idea originated by Beacham[40], that hydrogen lowers the work for fracture by enhancing localised slip. As Oriani[39] points out, while at first this may seem contradictory, since enhanced plastic deformation would be expected to increase the work of fracture, if the enhanced deformation is directly useful to crack propagation, the difficulty disappears. The most striking illustration of localised decohesion in heavily defined regions at crack tips is due to Birnbaum[17], working with nickel foils strained within a high voltage TEM. It appears that hydrogen causes both localised slip and enhanced decohesion, which receives support from the theoretical modelling of Daw and Baskes[41], showing that the same phenomena that decrease the resistance force for decohesion also decrease the force for shear separation. Although this model is not yet even semiquantified, and some of the experimental observations ambiguous, Oriani[39] regards the decohesion and localised slip models as complementary, rather than competitive.

Nor are these the only models that have acquired strong support, since the formation of brittle hydride phases in the crack-tip region in appropriate metals receives support from the observations of some workers. Thus, Scully and Powell[42] have developed earlier observations on the formation of a hydrides in α-Ti alloys to explain the stress-corrosion cracking of such materials, involving cleavage of the hydride as an important step in the cracking process. Pugh and his co-workers have extended these observations on the importance of hydride formation in the cracking of Ti alloys and have shown that the fracture planes correspond to the habit planes of the hydride, as well as showing that Mg–Al alloys may form hydrides[43]. There are others[44] who believe that stress-corrosion cracking in Ti alloys results from dissolution, but that is not consistent with the effects of immersion in methanolic solutions of HCl prior to straining or the recovery from such exposure in subsequent slow strain-rate tests at very low strain rates[45], which are more readily ascribed to the redistribution of hydrogen.

Environmental Aspects of Stress-corrosion Cracking

It has often been stated that the environmental requirements for stress-corrosion cracking are highly specific, but the list of environments identified as causing cracking in various alloys continues to grow with time and the concept of solution specificity is not so narrow as it was even a decade ago. Nevertheless, it is clear that cracking environments are specific, in the sense that not all possible environments promote cracking, and the electrochemistry of stress corrosion is essentially concerned with explaining this specificity. In very general terms, it is clear that potent solutions will need to promote a critical balance between activity and passivity, since a highly active condition will result in general corrosion or pitting, whilst a completely passive condition cannot, by definition, lead to stress corrosion. Whilst the relative inactivity of all exposed surfaces except the crack tip may be derived from a noble film in the cases of alloys containing sufficiently noble elements, for the great majority of engineering alloys inactivity at exposed surfaces is the result of the presence of oxide films overlaying metal surfaces. It is not surprising therefore to find that the alloys of high inherent corrosion resistance (such as those based upon aluminium or titanium, or the austenitic stainless steels, that readily develop protective films) require an aggressive ion, such as a halide, to promote stress-corrosion cracking. On the other hand, to crack the metals of low inherent corrosion resistance, such as carbon steels or magnesium-base alloys, requires the presence of an environment that is itself partially passivating. Thus, the carbon steels can be made to fail in solutions of anodic inhibitors, such as hydroxides and carbonates, and the cracking of magnesium-base alloys is achieved with an appropriate mixture of CrO_4^{2-} and Cl^- ions, but not with either of these species alone.

The transition from electrochemically active to relatively inactive behaviour that the sides of a crack must undergo as the tip advances by dissolution and creates more crack may be expected to be reflected in the current response of an initially bare surface exposed to the appropriate environment, since dissolution will be associated with the passage of relatively high anodic current densities, but with the passage of time this current will decay if filming occurs. Very rapid rates of current decay are unlikely to permit much dissolution and are not likely therefore to be indicative of conditions conducive to cracking, whilst very slow rates of decay will be more likely to be indicative of pitting than cracking. Intermediate rates of current decay will be those likely to be associated with cracking, and such results are indeed observed[46], but it is not yet possible to predict quantitatively what constitutes a critical rate of decay, although this would be expected to be potential dependent according to the competition between the solvation and filming processes.

A more convenient way of anticipating the range of potentials in which stress-corrosion cracking is likely to occur is available through potentiodynamic polarisation curves. If the potential of an initially film-free surface is rapidly (approximately 1 V/min) changed over an appropriate range, then the currents passed at the surface will indicate ranges of potential in which relatively high anodic activity is likely. The rapid sweep of the potential range has the object of minimising film formation, so that the currents observed relate to relatively film-free or thin-film conditions. If the experiment is now repeated, but with a slow rate of potential change (approximately 10 mV/min) so that time is allowed for filming to occur, comparison of the two curves will indicate any ranges of potential within which high anodic activity in the film-free condition reduces to insignificant activity when the time requirements for film formation are met, and this will indicate the range of potentials within which stress corrosion is likely. Figure 8.5 shows schematic polarisation curves determined under such conditions and indicates the various domains of behaviour expected. The technique correctly anticipates the stress-corrosion cracking of carbon steels in a number of totally different environments[46]. Of course, it is only applicable in those cases where air-formed oxide films can be reductively dissolved so that bare surfaces are created before the potential sweeps; in other cases straining or scraping electrodes must be used to remove the oxide film and the current response of the bared metal then observed potentiostatically at different

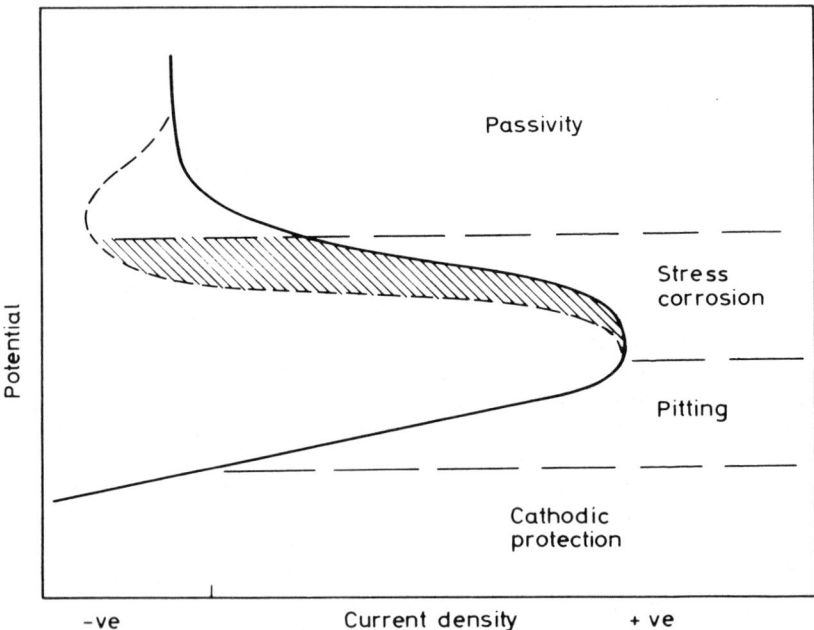

Fig. 8.5 Potentiodynamic polarisation curves and the expected domains of electrochemical behaviour

potentials. However, these different techniques give broadly the same results in any given system.

It is now well established that stress-corrosion cracking only occurs over particular ranges of potential for a given metal–environment combination. Such potential dependence must be related to specific reactions whereby the environmental requirements for cracking are met. Probably the simplest situation in this respect arises with hydrogen-induced cracking, where the hydrogen derives directly from the bulk environment to which the metal is exposed, and in which circumstances the conditions for cracking would be predicted to be met where the potential is below that for hydrogen discharge at the relevant pH. The highest potentials at which hydrogen-induced cracking is observed in various ferritic steels exposed to different solutions lie just below the calculated equilibrium potential for hydrogen discharge as a function of pH[47], so there is reasonable agreement between the predicted and observed behaviours.

Where crack growth is by dissolution associated with filming reactions to retain crack geometry, the potential dependence of cracking should reflect those requirements, again with some pH dependence because of the influences of that quantity upon the potentials at which the various reactions are possible. Where the necessary thermodynamic data are available for the species involved in a particular system, it should be possible to calculate the limits of the cracking domain. This has been done for the cracking of low-strength ferritic steel exposed to phosphate solutions and the agreement between the observed and calculated boundaries of the cracking domain is reasonable[48]. For that system, as with ferritic steels in other environments, the upper boundary of the cracking domain is met when the stable phase becomes $\gamma\text{-}Fe_2O_3$, i.e. at potentials where only the latter forms, cracking does not occur. While the potentials and pH values at which that phase can form will depend upon the phases formed within the cracking domain, it is interesting to consider the location of the potential–pH domains for cracking in various systems involving different ferritic steels in a range of environments at temperatures between 20 and 288°C. Figure 8.6 shows the various cracking domains together with the calculated equilibrium potentials for reactions between Fe_2O_3 and Fe_3O_4 and between Fe_3O_4 and Fe and for hydrogen discharge, all at 90°C as representing an average temperature for the various systems involved[49]. Clearly each cracking domain is associated with the calculated Fe_3O_4/Fe_2O_3 line and indeed in all of these systems, Fe_3O_4 is observed to form under conditions where cracking occurs, although it is frequently associated with other phases, e.g. $FeCO_3$ in the case of cracking by carbonate–bicarbonate solutions and $Fe_3(PO_4)_2$ for cracking by phosphate solutions. Moreover, for most of the systems shown in Fig. 8.6 only ductile failures occur in slow strain-rate tests carried out at potentials high enough to form Fe_2O_3 alone. While it is clear that the anions exert a significant influence upon the location of the cracking domains, the importance of Fe_3O_4 formation within the cracking ranges and Fe_2O_3 formation under conditions associated with ductile fracture appear well established, but the reasons for such less so.

The exceptions in Fig. 8.6 to only ductile failure occurring at potentials high enough to form Fe_2O_3 involve nitrates and high temperature water. In both of those systems cracks grow from pits, and within the pit–crack

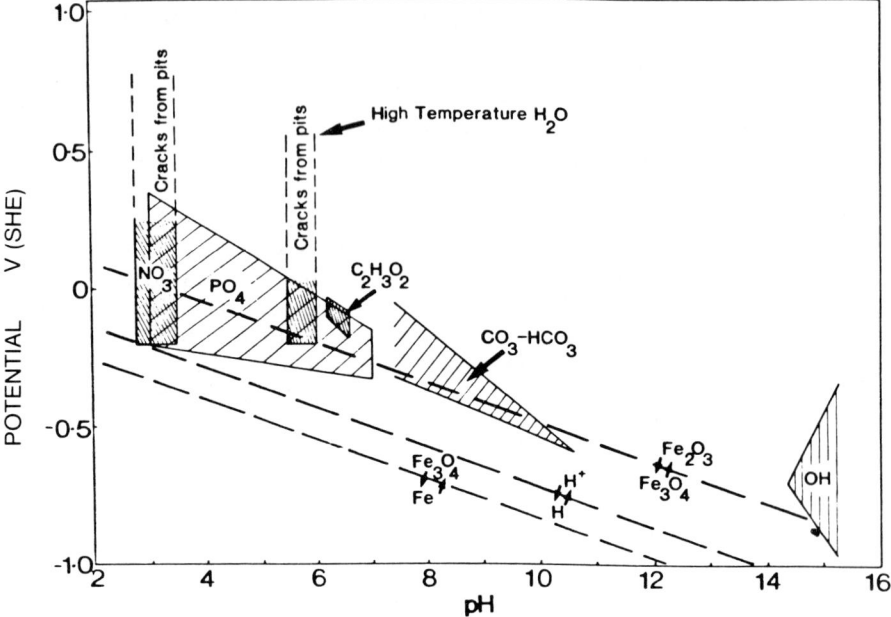

Fig. 8.6 Potential and pH ranges for the stress-corrosion cracking of ferritic steels in various environments, together with the pH-dependent equilibrium potentials for reactions involving $Fe \rightarrow Fe_3O_4$, $H \rightarrow H^+$ and $Fe_3O_4 \rightarrow Fe_2O_3$ (after Congleton et al.[49])

enclaves Fe_3O_4 forms, despite the external surfaces being covered with Fe_2O_3 films. The initiation of stress-corrosion cracks from pits has been observed in a variety of systems and is usually taken as indicative of the local environment within the pit being potent and different from that of the surrounding bulk environment. Where cracking does extend from pits there is usually reasonable correlation of the onset of cracking with the pitting potential. The perturbation of the electrochemical conditions within pits has inevitably led to similar considerations being given to the conditions within crack enclaves and since the early pioneering work of Brown[50] the subject has attracted much attention[51,52]. While there can be no doubt of the existence and importance of localised changes in environment composition and potential within crack enclaves in some systems, it is equally clear that such changes are negligible in other systems. This may be expected to be so where the solution is effectively buffered, the solubility of the solvated species very low and, where the cathodic reaction occurs outside the crack, there is negligible current flow through the crack sides. Such conditions appear to hold for the cracking of ferritic steels in carbonate–bicarbonate and in concentrated hydroxide solutions[53,54].

If significant potential changes exist along cracks then it may be expected that the potential range over which cracking is observed will be a function of whether pre-cracked or initially plain specimens are employed for determining the potential range in which cracking is observed. For a ferritic steel in a carbonate–bicarbonate solution, there are no significant differences in the potential range for cracking for either type of specimen, but this is not so for other systems. Thus, with a maraging steel exposed to NaCl solution at initial pH values of 6 or 11, initially smooth specimens failed in two regimes of potential separated by a region, some 300 mV in extent, in which cracking did not occur[38]. However, pre-cracked specimens did display environment-sensitive crack growth over the whole range of potentials, as indeed did smooth specimens that were pitted before exposure to those conditions that did not promote cracking in unpitted specimens.

This cracking of a maraging steel in two regimes of potential separated by a range of potentials in which cracking did not occur for initially plain specimens is suggestive of cracking by two different mechanisms above and below the range of immune potentials. This has been observed in a number of different systems and has often been interpreted as indicating dissolution-related cracking at the higher potentials and hydrogen-related cracking at the lower potentials. However, the necessity for low potentials to discharge hydrogen has often been queried, more especially where localised acidification of the environment can occur in pits or cracks, thereby raising the potential for hydrogen discharge. This is almost certainly the case with some systems, but it is as well to remember that it cannot be so in all systems for the reasons mentioned earlier, that solution composition changes do not invariably occur in cracks.

If the mechanism by which stress-corrosion cracks propagate involves dissolution at the crack tip, then crack velocities may be expected to be related through Faraday's law to the current density at the crack tip according to equation 8.5. Taking the effective current density as the largest difference between fast and slow sweep rate polarisation curves, or the maximum current densities observed in scraping or straining electrode experiments at appropriate potentials, Fig. 8.7 shows a plot of these current densities against observed crack velocities for a variety of stress-corrosion systems, the line shown being that calculated from equation 8.5. Clearly, for a calculation of this type the agreement between observed and calculated crack velocities is very reasonable, especially since the current density measurements do not take account of the structural dependence of the cracking. An implication of the results shown in Fig. 8.7 is that, for that data, the time during which the crack tip was relatively inactive due to the presence of a film must have been a small proportion of the total time, since otherwise the experimental points would fall well below the calculated line, which assumes continuous dissolution. This is because the crack velocity data in Fig. 8.7 are mostly from slow strain-rate tests which, if conducted at an appropriate strain rate, will prevent filming at the crack-tip. If however the strain rate is less than such values then the crack tips will be inactive for times dependent upon the frequency of film rupture. In such circumstances equation 8.5 needs modification to

$$CV = \frac{Q}{\varepsilon_f} \dot{\varepsilon}_{ct} \frac{M}{zF_l} \qquad (8.8)$$

where Q is the anodic charge (or charge density) passed, ε_f is the strain to rupture the film and $\dot{\varepsilon}_{ct}$ is the crack-tip strain-rate. An expression is available[49] for the crack-tip strain-rate in slow strain rate tests and is of the form

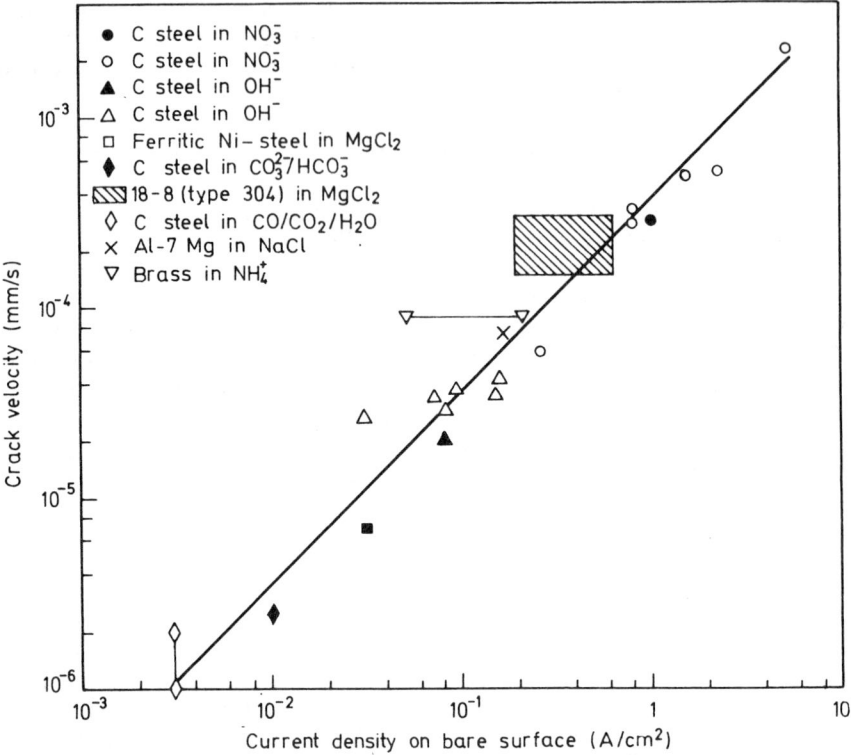

Fig. 8.7 Observed crack velocities and current densities associated with 'bare' surfaces. The line is that calculated from equation 8.5 (after Reference 20)

$$\varepsilon_{ct} = \frac{75}{N}\dot{\varepsilon}_{app} + \frac{CV}{5}\ln\left(\frac{1\,000}{N}\right) \qquad (8.9)$$

where N is the number of cracks along the gauge length and $\dot{\varepsilon}_{app}$ is the applied strain rate. (The constants in equation 8.9 will depend upon the material involved and test specimen size.) Because the first term in equation (8.9) dominates at high $\dot{\varepsilon}_{app}$ and the second term at low $\dot{\varepsilon}_{app}$, there is little effect of crack velocity at high $\dot{\varepsilon}_{app}$; i.e. the crack growth contributes little to the crack-tip strain-rate. However, at low $\dot{\varepsilon}_{app}$, the stress-corrosion crack growth maintains $\dot{\varepsilon}_{ct}$ at values that are appreciably higher than would be obtained if the crack growth had been ignored.

Various workers[55-58] have used equation 8.8, or some modified version thereof, to compare observed with calculated crack velocities as a function of strain rate, but Fig 8.8 shows results[58] from tests on a ferritic steel exposed to a carbonate-bicarbonate solution. The calculated lines move nearer to the experimental data as the number of cracks in equation 8.9 is increased, while the numbers of cracks observed varied with the applied strain rate, being about 100 for $\dot{\varepsilon}_{app} \sim 10^{-6}\,\text{s}^{-1}$, but larger at slower $\dot{\varepsilon}_{app}$ and smaller at higher $\dot{\varepsilon}_{app}$.

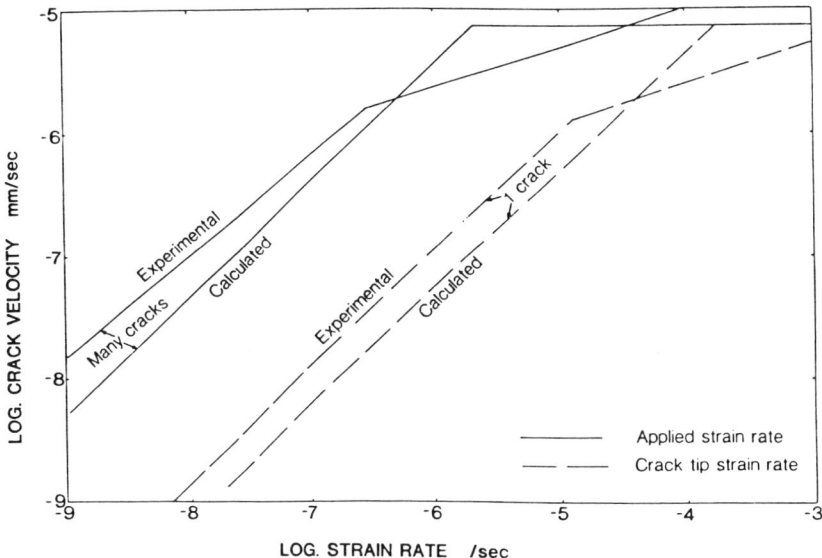

Fig. 8.8 Comparison of calculated and experimental crack velocities as a function of strain rate for a ferritic steel exposed to 1 N Na_2CO_3 + 1 N $NaHCO_3$ at -650 mV(SCE) and 75°C (after Parkins[58])

While equation 8.8 gives reasonable predictions of the crack velocities for small specimens involving relatively small cracks, there is a further factor that must be taken into account with larger specimens and, more importantly, real engineering structures. This concerns the phenomenon of crack merging or coalescence, a matter that is only beginning to be recognised as important even though it is often apparent from the inspection of service failures. The latter almost invariably involve the multiple initiation of cracks, probably over a relatively long period of time, as in laboratory tests. It is probable that most stress-corrosion cracks cease to grow, especially under realistic loading conditions, after relatively small amounts of propagation, possibly because of work hardening in the crack-tip region and a reduction in the crack-tip strain rate. However, with continued crack initiation, some new cracks may form sufficiently near inactive cracks to reactivate the latter. Obviously this requires that the interacting cracks are sufficiently close together for their respective stress fields to interact. Small merged cracks may later cease to propagate, but with continuing nucleation of new cracks they may later be reactivated, these processes continuing until eventually some cracks will reach a size where the stress-intensity factor for relatively rapid crack growth, K_{Iscc}, is reached and the crack velocity will approach that given by equation 8.5. Figure 8.9 indicates schematically these changes in crack velocity with time, which can be quantified in a simple fashion[59] to compare predictions with observed behaviour. Such comparison with service behaviour indicates the importance of crack coalescence, in the absence of which lifetimes would be markedly greater than sometimes experienced and, with a containing vessel, a leak rather than a rupture would more often occur.

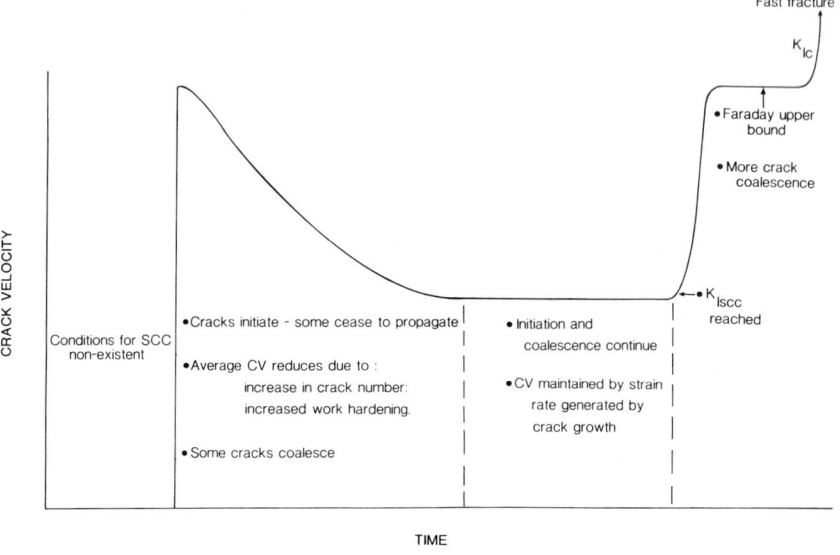

Fig. 8.9 Schematic illustration of the effect of time of exposure upon stress-corrosion crack velocity

The Function of Stress

If crack propagation occurs by dissolution at an active crack tip, with the crack sides rendered inactive by filming, the maintenance of film-free conditions may be dependent not only upon the electrochemical conditions but also upon the rate at which metal is exposed at the crack tip by plastic strain. Thus, it may not be stress, *per se*, but the strain rate that it produces, that is important, as indicated in equation (8.8). Clearly, at sufficiently high strain rates a ductile fracture may be propagated faster than the electrochemical reactions can occur whereby a stress-corrosion crack is propagated, but as the strain rate is decreased so will stress-corrosion crack propagation be facilitated. However, further decreases in strain rate will eventually result in a situation where the rate at which new surface is created by straining does not exceed the rate at which the surface is rendered inactive and hence stress corrosion may effectively cease.

The implications of a significant role for strain rate are wider than the obvious one that stress corrosion should only occur over a restricted range of strain rates. Thus, in constant load tests, since cracks will continue to propagate only if their rate of advancement is sufficient to maintain the crack-tip strain rate above the minimum rate for cracking, it is to be expected that cracks will sometimes stop propagating, particularly below the threshold stress. Such non-propagating cracks are indeed observed below the threshold[60,61]. Moreover, in constant-load or constant-strain tests, the strain rate diminishes with time after loading, by creep exhaustion if the stress remains sensibly constant, and it is found that the stress-corrosion results are sensitive to the relative times at which the stress and electrochemical

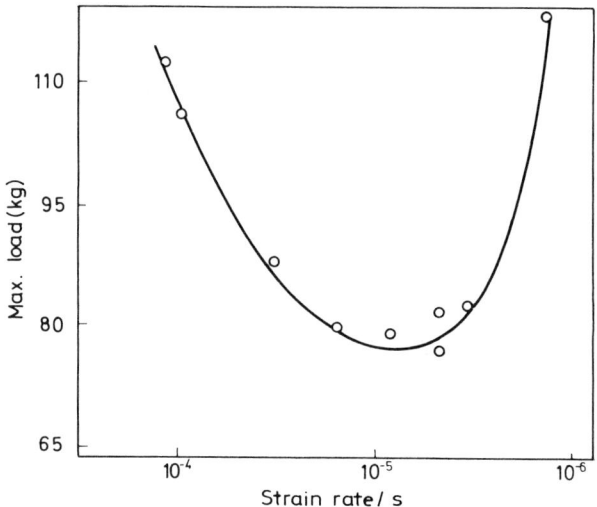

Fig. 8.10 Effect of strain rate upon the cracking propensity of a Mg–Al alloy immersed in a chromate–chloride solution

conditions for cracking are established, i.e. creep at constant load, prior to the establishment of the electrochemical conditions for cracking, delays or prevents cracking[60,61].

However, the most convincing demonstration of the importance of strain rate is obtained from tests in which the strain rate is superimposed, rather than allowed to vary in the inevitable manner of constant load tests. Figure 8.10 shows the effects of various strain rates applied to a Mg–7Al alloy whilst immersed in chromate–chloride solutions, the tests being conducted to total failure and the maximum load achieved being a sensitive measure of whether or not stress-corrosion cracks were produced[60]. If stress-corrosion cracks are not produced then failure is by ductile fracture at the normal UTS for the material, but in the presence of stress-corrosion cracks the maximum nominal stress achieved prior to failure is markedly reduced. It is apparent from Fig. 8.10 that stress-corrosion cracking only occurs within a restricted range of strain rates and that at higher or lower values ductile fracture occurs, as confirmed by fractography.

Experiments on a carbon steel in a carbonate–bicarbonate solution at a cracking potential with the pre-cracked specimens loaded as cantilevers but with the beam displaced at various rates by a device that replaces the conventional load pan, produced the results shown in Fig. 8.11. The changes in net section stress in these tests at various strain rates amounted to less than a few per cent, but the results clearly indicate a lower limiting strain rate below which crack propagation is not observed, followed by a region in which the intergranular stress-corrosion crack velocity is independent of strain rate and then, at relatively high strain rates, a transition to fast transgranular tearing. The strain-rate independent region is to be expected since once the strain-rate is sufficiently high to create bare metal at the crack tip at a faster rate than filming can render the bare metal inactive, the factor controlling the crack velocity will be the rate of metal dissolution which is governed by equation

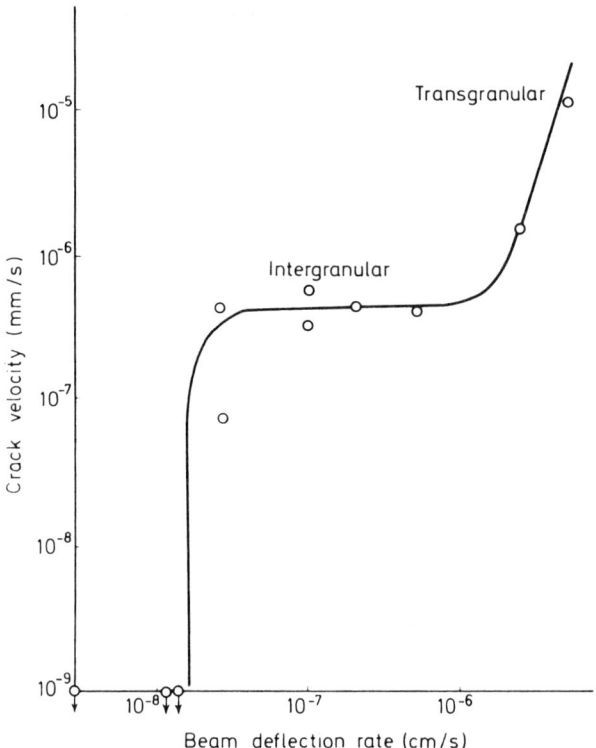

Fig. 8.11 Effect of beam deflection rate of cantilever beam specimens upon stress-corrosion crack velocity of carbon steel in carbonate–bicarbonate solution

8.5. The complementary functions of stress and electrochemistry in this model, involving the creation of bare metal at the crack tip by plastic strain, imply a strong dependence of the strain rate range for cracking, or of the threshold stress in constant load tests, upon the environmental conditions. Such effects are indeed observed, the curve in Fig. 8.10 being capable of a marked shift along the strain-rate scale according to the composition of the environment and whether or not small anodic or cathodic currents are applied, whilst the limiting beam deflection rate below which cracking is not observed in the experiments to which Fig. 8.11 refers can be changed by two or three orders of magnitude by changes in applied potential.

There are indications then that where an active path mechanism is operative, the function of stress in stress-corrosion cracking is to create plastic deformation and therefore that such cracking will be more likely with the lower strength ductile metals. Where the mechanism of cracking involves embrittlement of the metal in the crack-tip region a strain energy argument is involved, and this implies, in relation to equation 8.1, that plastic strain should be minimised and elastic energy maximised for failure, conditions that are most readily met with high yield strength materials. It is well established that the hydrogen embrittlement of steels becomes more marked the higher the yield strength, although changes in structure or composition that result in a change in yield strength, or fracture toughness, may also influence electrochemical reactions, and such parameters as hydrogen diffusivity and these may be as significant as any change in strength in influencing stress-corrosion behaviour. It is also possible that strain rate, as opposed to stress intensity, could be of significance in the stress corrosion of high-strength steels if the environment concerned is one that may lead to filming and the stifling of the reactions that involve the release of hydrogen from the environment or its ingress into the metal. Certainly, there are indications that stress-corrosion cracking in some of the high-strength steels is sensitive to the loading rate and there is a marked similarity to the crack velocity–strain rate curve of Fig. 8.11, and the crack velocity–stress intensity curves[62] obtained from tests on precracked specimens in a typical high-strength steel are shown in Fig. 8.12. Curves similar to the latter have been obtained for high-strength aluminium alloys and for titanium alloys, and the question again arises as to whether it is stress intensity or the strain rate that the latter produces that is important, especially in view of results similar to those shown in Fig. 8.10 for a titanium alloy[42].

Conclusion

The interdependence of the variables in stress corrosion, namely structure, electrochemistry and response to stress, supports the suggestion that these may interact in a variety of ways and if rationalisation of the situation is to be attempted this is more appropriately achieved through the concept of a continuous spectrum of mechanisms rather than a single mechanism. The critical balance between activity and passivity is altered by changes in the structure and composition of the alloy, the response of the latter to the application of stress through changes in mechanical properties and by changes in the environmental conditions. Thus, if the structure and composition of the alloy are such that almost continuous paths of segregate or precipitate exist, usually at the grain boundaries, and which are electrochemically different from the matrix, then a latent susceptibility to intergranular corrosion may be activated by the presence of stress. In the absence of pre-existing active paths, or even in their presence if other conditions hold, the stress may generate active paths by rupturing a protective surface film or by activating dissolution at emerging slip lines. The transformation from a pre-existing to a strain-generated active path mechanism may result not only from physico-metallurgical change in the state of the alloy, but also from changes in the environmental conditions or the crack-tip strain rate.

This greater role of stress or strain in moving away from the pre-existing active path end of the spectrum is continued through to those alloys that undergo local embrittlement of the metal in the crack-tip region. Table 8.1 indicates some of the systems, of metal and environment, that result in stress corrosion, arranged in a series that ranges from those in which the mechanisms are thought to be dominated by dissolution processes to those in which stress, or strain, occupies the more important part of the proposition. While it is very much a matter of opinion as to where specific systems fall in this scheme, perhaps the most significant point about such an arrangement is that it should help to serve as a reminder of the interdependence of the

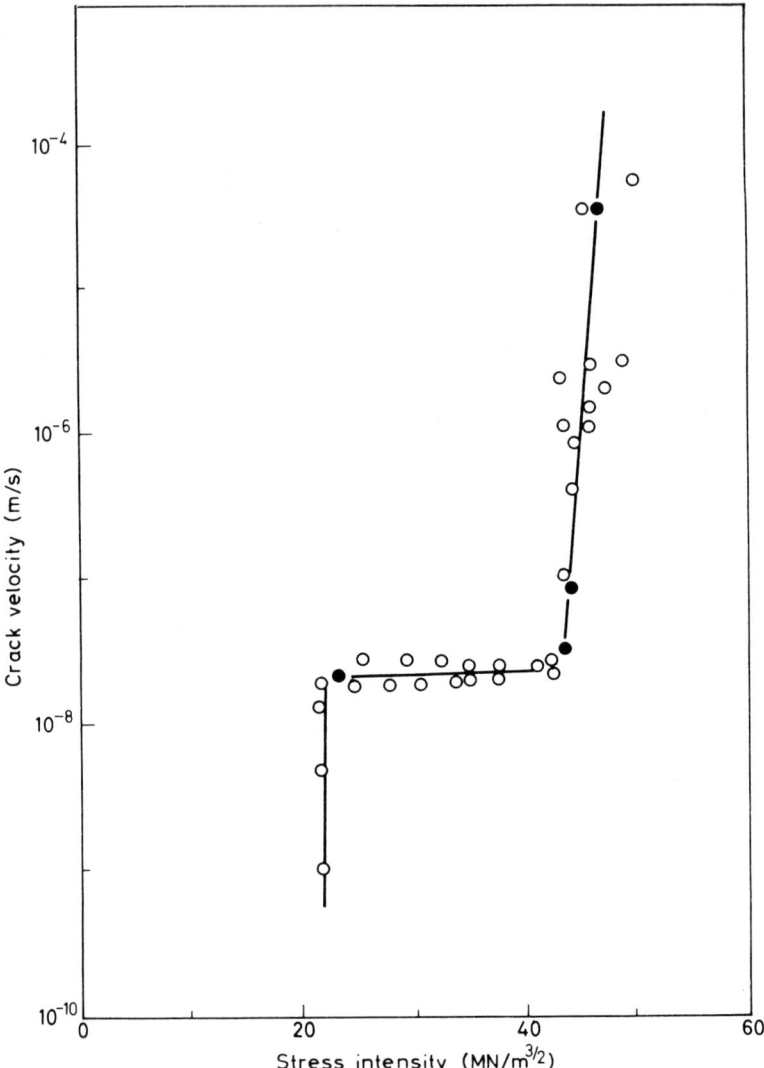

Fig. 8.12 Effect of applied stress intensity upon crack velocity for high-strength (180 GN/m^2 UTS) quenched and tempered steel (AFC 77) in distilled water (after Spiedel[62])

variables and that avoidance of stress-corrosion failure in a specific instance is no guarantee that the preventative action will be equally successful in other circumstances. For instance, whilst nickel additions to a steel are beneficial in relation to caustic cracking, they have little effect upon nitrate cracking and are quite harmful from the viewpoint of cracking in chloride, in that they promote a susceptibility to cracking in the latter not observed in carbon steels. Thus, the avoidance of cracking by a mechanism occupying one part of the spectrum may induce failure in another part by a different mechanism if the interdependence of the variables is ignored.

R. N. PARKINS

Table 8.1 Stress corrosion spectrum

	Corrosion dominated (solution requirements highly specific)			Stress dominated (solution requirements less specific)			
Intergranular corrosion	Carbon steels in NO_3^- solns	Some Al alloys in Cl^- solns, high potentials	Cu–Zn alloys in NH_3^- solns				
	Intergranular fracture along pre-existing paths						
		Fe–Cr–Ni steels in Cl^- solns	Cu–Zn alloys in NO_2^- solns	Mg–Al alloys in CrO_4^{2-} + Cl^- solns	Ti alloys in methanol. Al alloys, low potentials	High strength steels in Cl^- solns	Brittle fracture
		Transgranular fracture along strain-generated paths			Mixed crack paths by adsorption, decohesion or fracture of brittle phase		

REFERENCES

1. Parkins, R. N., *British Corrosion Journal*, **7**, 15 (1972)
2. West, J. M., *Metal Science Journal*, **7**, 169 (1973)
3. Despic, A. R., Raicheff, R. G. and Bockris, J. O'M., *Journal of Chemical Physics*, **49**, 926 (1968)
4. Dix, E. H., *Trans. Amer. Inst. Min. Met. Engrs.*, **137**, 11 (1940)
5. Parr, S. W. and Straub, F. G., *Univ. Ill. Bull.*, **177** (1928)
6. Swann, P. R., *Corrosion*, **19**, 102t (1963)
7. Pickering, H. W. and Swann, P. R., *Corrosion*, **19**, 373t (1963)
8. Swann, P. R., from *The Theory of Stress Corrosion Cracking in Alloys*, Edited J. C. Scully, NATO, Brussels, 113 (1971)
9. Silcock, J. M. and Swann, P. R., from *Environment-Sensitive Fracture of Engineering Materials*, Edited by Z. A. Foroulis, TMS-AIME, Warrendale, Pa, 133 (1979)
10. Nielsen, N. A., *Second International Congress on Metallic Corrosion*
11. Robertson, W. D., Grenier, E. G., Davenport, W. H. and Mole, V. F., from *Physical Metallurgy of Stress-Corrosion Fracture*, Edited by T. N. Rhodin, Interscience, 273 (1959)
12. Pugh, E. N., *Corrosion*, **41**, 517 (1985)
13. Edeleanu, C. and Forty, A. J., *Phil. Mag.*, **5**, 1029 (1960)
14. Sieradzki, K. and Newman, R. C., *Phil. Mag.*, A, **51**, 95 (1985)
15. Sieradzki, K. and Newman, R. C., *J. Phys. and Chem. of Solids*, **48**, p 1101 (1987)
16. Uhlig, H. B., Ref. 19, p 86
17. Birnbaum, H. K., Ref. 23, p733
18. Galvele, J. R., *Corros. Sci.*, **27**, 1, (1987)
19. Staehle, R. W., Forty, A. J. and van Rooyen, D. (eds.), *Proc. Conf. on Fundamental Aspects of Stress Corrosion Cracking*, NACE, Houston (1969)
20. *Proc. Conf. on Stress Corrosion Cracking and Hydrogen Embrittlement of Iron Base Alloys*, NACE (1975)
21. Latanision, R. M. and Pickens, J. R. (eds.), *Atomistics of Fracture*, Plenum Press New York (1983)
22. Sangloff, R. P. and Ives, M. B., (eds.) *Environment-Induced Cracking of Metals*, NACE, Houston (1990)
23. Bruemmer, S. M., Meletis, E. I., Jones, R. H., Gerberich, W. W., Ford, F. P. and Stachle, R. W. (eds.) *Parkins Symposium on Fundamental Aspects of Stress Corrosion Cracking*, TMS-AIME, Warrendale, Pa, (1992)
24. Doig, P. and Edington, J. W., *Brit. Corr. J.*, **9**, 88 (1974)
25. Syrett, B. C. and Parkins, R. N., *Corros. Sci.*, **10**, 197 (1970)
26. Green, J. A. S., Mengelberg, H. D. and Yolken, H. T., *J. Electrochem. Soc.*, **117**, 433 (1970)
27. Vermilyea, D. A., Ref. 20, p 208
28. Pugh, E. N., Ref. 19, p 118
29. Yu, J., Parkins, R. N., Zu, Y., Thompson, G. and Wood, G. C. *Corros. Sci.*, **27**, 141 (1987)
30. Sieradzki, K., Kim, J. S., Cole, A. T. and Newman, R. C., *J. Electrochem. Soc.*, **134**, 1635 (1987)
31. Poulson, B. S. and Parkins, R. N., *Corrosion*, **29**, 414 (1973)
32. Hoar, T. P., and West, J. M., *Proc. R. Soc.*, **A268**, 304 (1962)
33. Swann, P. R. and Embury, J. D., from *High Strength Materials*, Edited by V. F. Zackay, Wiley, New York, p. 327 (1965)
34. Staehle, R. W., Ref. 8, p 233
35. Yu, J., Holroyd, N. J. H., and Parkins, R. N., from *Environment Sensitive Fracture: Evaluation and Comparison of Test Methods*, ASTM STP 821, Edited by S. W. Dean, E. N. Pugh and G. M. Ugianski, p 288 (1984)
36. Coleman, E. G., Weinstein, D. and Restoker, W., *Acta Met.*, **9**, 491 (1961)
37. Parkins, R. N., Ref 20, p 601
38. Craig, I. H., and Parkins, R. N., *Brit Corr. J.*, **19**, 3 (1984)
39. Oriani, R. A., *Corrosion*, **43**, 390 (1987)
40. Beacham, C. D., *Met. Trans.*, **3**, 437 (1972)
41. Daw, M. S. and Baskes, M. I., Sandia Report SAND 86-8863, Sandia Natl. Labs, Albuquerque, N.M.
42. Scully, J. C. and Powell, D. T., *Corros. Sci.*, **10**, 719 (1970)

43. Pugh, E. N., Ref. 21, p 997
44. Beck, T. R., Ref. 8, p 64
45. Ebtehaj, K., Hardie, D. and Parkins, R. N., *Corros. Sci.*, **25**, 415 (1985)
46. Parkins, R. N., *Corros. Sci.*, **20**, 147 (1980)
47. Parkins, R. N., from *The Use of Synthetic Environments for Corrosion Testing*, ASTM STP 970, Edited by P. E. Prancis and T. S. Lee, p 132 (1988)
48. Parkins, R. N., Holroyd, N. J. H. and Fessler, R. R. *Corrosion*, **34**, 253 (1978)
49. Congleton, J., Shoji T. and Parkins, R. N., *Corros. Sci.*, **25**, 633 (1985)
50. Brown, B. F., from *The Theory of Stress Corrosion Cracking in Alloys*, Edited by J. C. Scully, NATO, Brussels, p 186 (1971)
51. Gangloff, R. P., (ed.), *Embrittlement by the Localised Crack Environment*, AIME (1984)
52. Turnbull, A., (ed.), *Corrosion Chemistry within Pits, Crevices and Cracks*, HMSO, London (1987)
53. Parkins, R. N., Craig, I. H. and Congleton, J., *Corros. Sci.*, **24**, 709 (1984)
54. Parkins, R. N., Liu, Y. and Congleton, J., *Corros. Sci.*, **28**, 259 (1988)
55. Gerber, T. L., Garud, Y. S. and Sharma, S. R., *Thermal and Environmental Effects in Fatigue: Research Design Interface*, ASME, PVP Vol. 71, p 155 (1983)
56. Hudak, S. J., Jr., Davidson, D. L. and Page R. A., Ref. 51, p 173
57. Ford, F. P., *Corrosion/86*, Paper No. 327, NACE (1986)
58. Parkins, R. N., *Corrosion*, **43**, p 130 (1987)
59. Parkins, R. N. and Singh, P. M., *Corrosion*, **46**, 485 (1990)
60. Wearmouth, W. R., Dean, G. P. and Parkins, R. N., *Corrosion*, **29**, 251 (1973)
61. Parkins, R. N., *Stress Corrosion Cracking – The Slow Strain Rate Technique*, ASTM STP 665, Edited by G. M., Ugianski and J. H., Payer. ASTM Philadelphia, Pa, p 5 (1979)
62. Spiedel, M. O., *Conference on Hydrogen in Metals*, NACE, 1975

8.2 Stress-corrosion Cracking of Ferritic Steels

The incidences of stress-corrosion failure in ferritic steels, as with most alloys, continues to increase in frequency with the passage of time, probably as the result of the avoidance of general corrosion, the more efficient use of steels, i.e. by employing higher operating stresses, the more extensive use of methods of fabrication that leave relatively high internal stresses in structures, and as diagnostic efficiency has improved. Thus, the cracking of riveted boilers in strong caustic solutions[1] has been experienced for over 80 years, and the failure of evaporating equipment containing ammonium nitrate[2] for little less. The failure of plant used in the cleaning of coal gas[3] and of equipment used in sour oil wells[4] created particularly severe problems 30 to 40 years ago, as did the cracking of anhydrous ammonia storage vessels[5] a decade later. Other environments that have been associated with stress-corrosion failures in ferritic steels have ranged from ferric chloride solution, through acids ranging from fuming sulphuric to hydrocyanic, to a sodium phosphate solution, a fairly comprehensive list having been provided by Logan[6].

More recently the cracking of low-strength ferrite steels by carbonate-bicarbonate environments has become recognised in chemical process plant[7,8] and in high-pressure gas transmission pipelines[9]. Most of these failures have been associated with cracks that followed an intergranular path, but transgranular fractures have been observed in carbon steels in environments including industrially important H_2O–CO–CO_2 mixtures[10]. The compositions and structures of the steels and the properties of the environments involved in these various instances of failure are so widely varying as to suggest that rationalisation of all of these experiences in a single explanation would be difficult if not unreal, i.e. it is probable that a number of different mechanisms are involved (Section 8.1). This is not to suggest that some systematisation is not possible, since from some of the steel environment systems that have been appropriately studied some common trends have emerged.

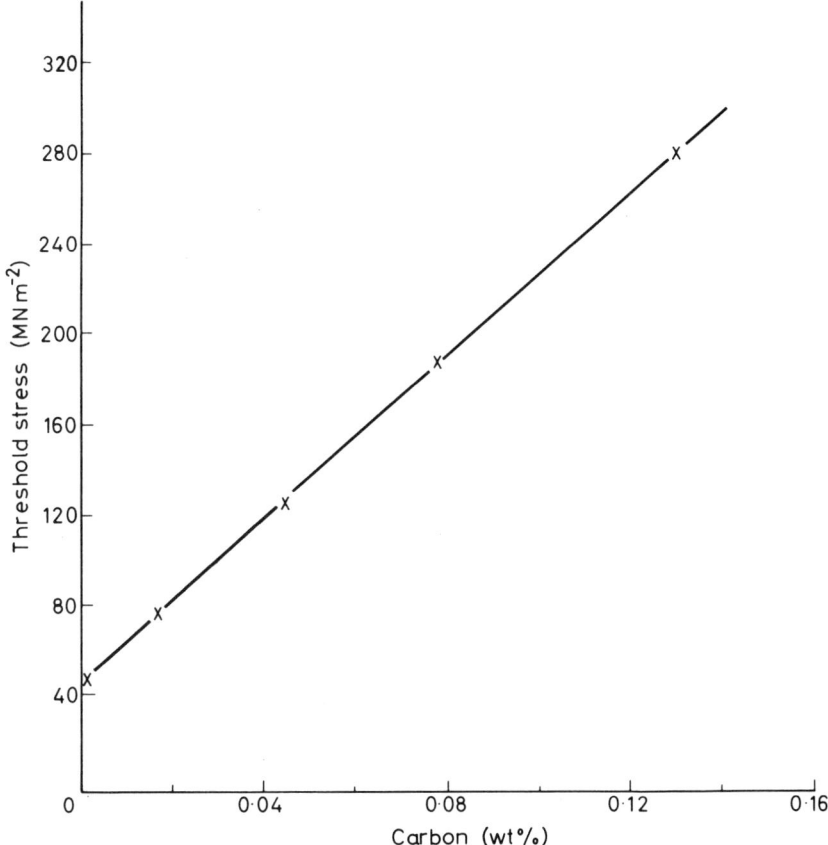

Fig. 8.13 Effect of carbon content of annealed mild steels upon threshold stress for cracking in boiling 4 N NH_4NO_3

Effects of Steel Composition and Structure

Most of the early work carried out in relation to these aspects of the problem used nitrates as the cracking environment where low-strength steels have been the objects of interest. Consequently most of what follows refers to cracking in boiling concentrated nitrate solutions except where otherwise stated. The medium and higher strength steels, such as involved in sour oil well equipment and other applications, are more frequently tested in chloride- or sulphide-containing environments related to service conditions, but the failure of these steels is dealt with elsewhere (see Section 8.4).

For normal commercial-quality mild steels in the annealed or normalised conditions in which they are almost invariably used, various workers have shown that the carbon content of the steel is the major factor determining intergranular cracking susceptibility. Figure 8.13 shows the threshold stresses for a series of commercial mild steels of different carbon contents caused to crack in boiling 4 N NH_4NO_3. The trend of the result suggests

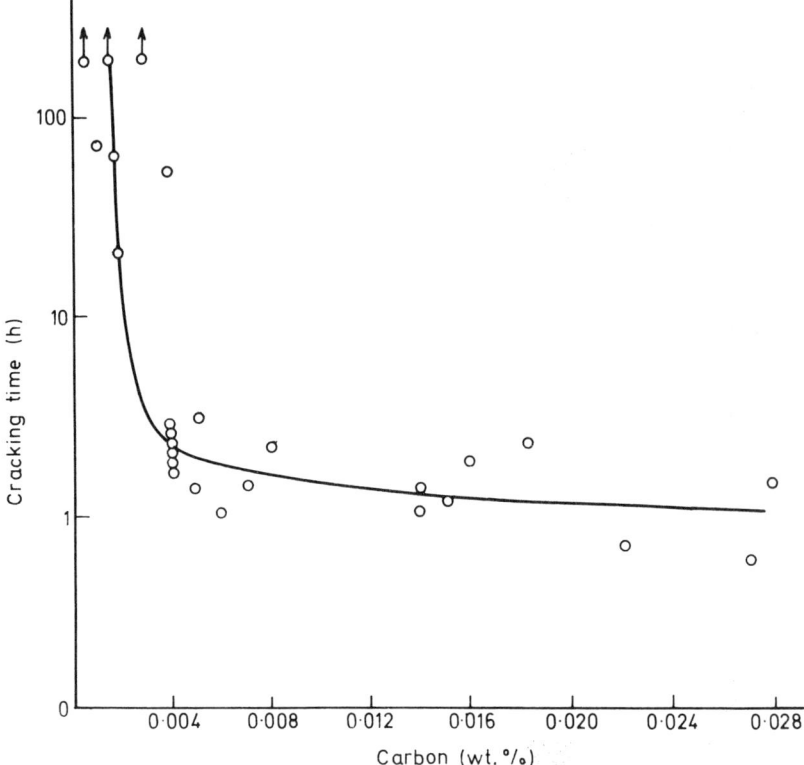

Fig. 8.14 Effect of carbon content of very low carbon steels quenched from 920°C on cracking in a calcium nitrate–ammonium nitrate solution (after Long and Uhlig[11])

that pure iron should be more susceptible to cracking than any steel, but the results[11] shown in Fig. 8.14 indicate that cracking susceptibility goes through a minimum as the carbon is reduced to very low levels, and taken with the results shown in Fig. 8.13 demonstrates the important influence of carbon content in relation to the intergranular cracking of steels in nitrates. Such information as exists indicates that similar trends are observed in relation to the cracking of steels in the carbonate solutions that constitute coal-gas liquors and in strong caustic solutions.

Whilst it appears to be widely accepted that the carbon content of the steel is important, there are differences of opinion as to how the carbon operates. Flis and Scully[12] and Long and Uhlig[11] regard the effect of carbon on the mechanical properties of the steel as the means whereby this element is important in stress corrosion cracking, whereas others[13] regard its electrochemical effects as being more important. Where the cracking is intergranular it appears most probable that segregation of some element or elements to the grain boundaries is likely to be involved and it is well established that carbon can segregate to ferrite grain boundaries. However, other elements can segregate to grain boundaries[14] and it is possible that other species than carbon can promote intergranular cracking of ferritic steels. Thus, it is likely that nitrogen can act in a similar manner to carbon, the most convincing demonstration being that due to Uhlig and Sava[15] who, starting with decarbonised electrolytic iron, which was resistant to cracking, found that the introduction of 0.043% N produced marked susceptibility. Increasing amounts of nitrogen in steel appear to decrease resistance to cracking in nitrates, in the same manner as carbon, while there are results available[16] that suggest essentially similar trends in relation to cracking in boiling hydroxide solutions.

Lea and Hondros[17] prefer phosphorus as the cause of intergranular stress-corrosion cracking of ferritic steels, having defined susceptibility in terms of a 'fragility index' (a product of the propensity of an element to segregate to grain boundaries and its relative harmfulness, atom for atom, once at the grain boundary). From tests upon ingots of mild steel to which different elements were added the data are presented as:

$$\text{Fragility index} = 20\%P + 1.9\%Cu + 1\%Sn + 0.9\%Sb + 0.4\%As \\ + 0.3\%Zn + 0.2\%Ni \, (+ 700\%S + 27\%Ca + 1\%Al)$$
(8.10)

It is claimed that since S, Ca and Al will be present as precipitates they would not in general be detected as grain boundary segregants and their ineffectiveness is indicated by the brackets in equation 8.10. Lea and Hondros do not consider the possible roles of carbon or nitrogen in the cracking of their steels, but from the data obtained phosphorous had the most deleterious effects.

More recently Krautschick et al.[18] measured the cracking responses of iron-phosphorus alloys containing 0.003 to 2 wt% P. They conclude that phosphorus segregation is not necessarily the origin of intergranular stress-corrosion of mild steels in nitrate solutions and that low phosphorus-containing carbon steels could show susceptibility to cracking. On the other hand, studies by Bandyopadhyay and Briant[19] involving the exposure of various low-alloy steels to concentrated sodium hydroxide solution show that phosphorus segregation to the grain boundaries in steels containing up to 0.06 wt% P has a markedly deleterious effect on cracking resistance. However, the same authors indicate that carbon and molybdenum also have deleterious effects in these steels, although within the concentration limits they studied, phosphorus segregation had more deleterious effects.

The results of Lea and Hondros[17] showing a deleterious effect from sulphur in relation to cracking by a nitrate solution is interesting in view of the effects of that element in the transgranular cracking of nuclear reactor pressure vessel steels in high-temperature water[20]. It is thought likely that the sulphur, which exists mostly as manganese sulphide inclusions, creates a localised environment in the region of inclusions, from which cracks are most often initiated. The thought derives support from the adverse effects of the addition of sulphur anions to the bulk environment. Against such effects of sulphur is the report by Bandyopadhyay and Briant[19] of no effect on the intergranular cracking of their low-alloy steels exposed to hydroxide solution.

Where cracks follow intergranular paths due to electrochemical effects related to grain boundary heterogeneity it is likely that selective attack should be observed in such locations by exposure to potent environments even in the absence of applied stress. This has been observed with, low-strength ferritic steels exposed to nitrate, hydroxide or carbonate–bicarbonate solutions. Such attack does not penetrate far along grain boundaries in the absence of stress, but anodic polarisation can cause virtual intergranular disintegration of unstressed steel exposed to a nitrate solution[21]. Bandyopadhyay *et al.*[22] have observed boundary etching in unstressed low-alloy steel, which correlates with phosphorus segregation and cracking propensity as the result of exposure to a sodium hydroxide solution. Such correlations between intergranular stress corrosion cracking propensity and selective attack in grain boundary regions would appear unlikely if the role of grain boundary segregants was solely related to mechanical effects at boundaries, although that is not to imply that such effects are of no consequence in crack growth.

The sometimes contradictory results from different workers in relation to the elements mentioned above extends to other elements[23]. Some of these differences probably arise from variations in test methods, differences in the amounts of alloying additions made, variations in the amounts of other elements in the steel and the differing structural conditions of the latter. Moreover, the tests were mostly conducted at the free corrosion potential, and that can introduce further variability between apparently similar experiments. In an attempt to overcome some of these difficulties, slow strain-rate tests were conducted on some 45 annealed steels at various controlled potentials in three very different cracking environments[23] since, if macroscopic

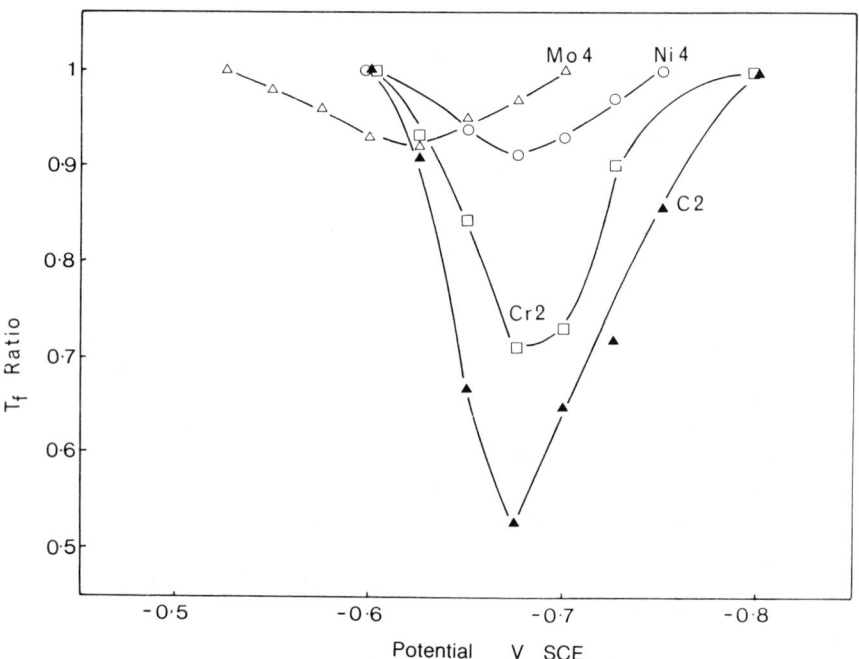

Fig. 8.15 Effects of potential upon the stress-corrosion cracking of various steels in CO_3–HCO_3 solution in slow strain rate tests (after Parkins *et al.*[23])

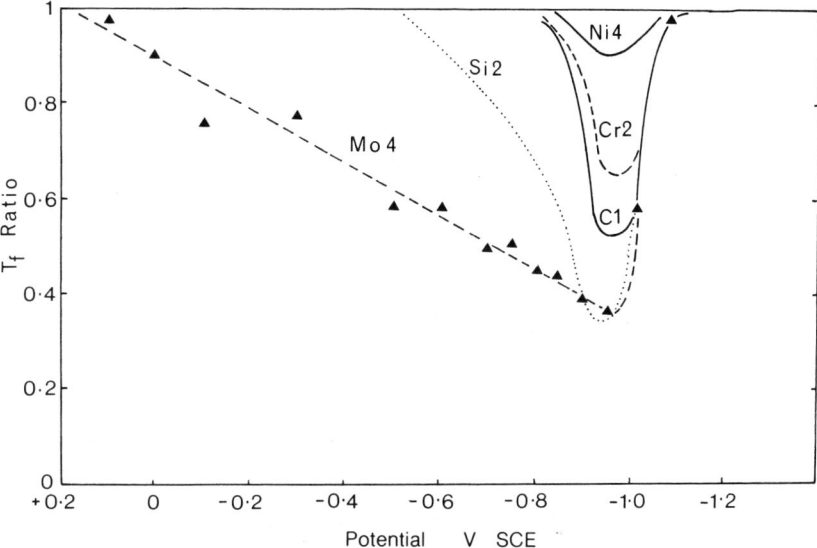

Fig. 8.16 Effects of potential upon the stress-corrosion cracking of various steels in boiling 8.75 N NaOH in slow strain rate tests (after Parkins et al.[23])

electrochemical properties play an important role in determining cracking response, there is no reason to expect that the effects of different alloying elements will be the same irrespective of the environment.

The results are expressed in terms of the effect of applied potential upon the time-to-failure ratio, the latter being derived from the time-to-failure in the test solution divided by the time-to-failure in an inert environment (oil) at the same temperature. A ratio of 1 indicates no susceptibility to cracking and increasing departure of the ratio from 1 indicates increasing susceptibility. Figure 8.15 indicates the beneficial effects that may be derived from additions of specific amounts of chromium, nickel or molybdenum in relation to cracking by a carbonate–bicarbonate solution. Increasing amounts of those elements had the effect of increasing resistance to cracking in that environment, a point returned to below. However, not all alloying additions to ferritic steels are invariably beneficial. Figure 8.16 shows that, while chromium and nickel additions were beneficial in relation to cracking by a sodium hydroxide solution, silicon and molybdenum had quite the reverse effect, particularly in extending the potential range over which cracking was observed. The areas bounded by curves such as those in Figs. 8.15 and 8.16 provide a convenient means of conducting a regression analysis on all of the data. This provides a stress-corrosion index (SCI) that reflects the effects of potential and the severity of cracking, indicating beneficial or deleterious effects according to the direction of change. The regression analysis for the tests in sodium hydroxide gave

$$\text{SCI}_{\text{OH}} = 105 - 45\%\text{C} - 40\%\text{Mn} - 13.7\%\text{Ni} - 12.3\%\text{Cr} - 11\%\text{Ti} + 2.5\%\text{Al} + 87\%\text{Si} + 413\%\text{Mo} \qquad (8.11)$$

reflecting the beneficial effects (negative coefficients) shown in Fig. 8.16 for specific additions of chromium or nickel and the deleterious effects (positive coefficients) from specific additions of silicon and molybdenum. The corresponding equations from tests in the nitrate and the carbonate-bicarbonate solutions were

$$SCI_{NO_3} = 1777 - 996\%C - 390\%Ti - 343\%Al(-132\%Mn) - 111\%Cr - 90\%Mo - 62\%Ni + 292\%Si \quad (8.12)$$

and

$$SCI_{CO_3} = 41 - 17.3\%Ti - 7.8\%Mo - 5.6\%Cr - 4.6\%Ni(-2.9\%Mn)(+1.7\%Si)(+5.6\%Al)(+15\%C) \quad (8.13)$$

When the t ratio (coefficient/standard error of the coefficient) was less than 2 for any element, the latter is bracketed in the regression equation, implying that only the remaining elements should be regarded as having significant effects upon the cracking propensity. The SCI values in the different environments reflect the decreasing potential range for cracking and the decreasing severity of cracking in the order nitrate, hydroxide, carbonate-bicarbonate.

It is probable that several factors are involved in the effects reflected in equations (8.11), (8.12) and (8.13). Thus the coefficients indicate that chromium, manganese and titanium, additions are consistently useful for all three environments, followed by nickel and aluminium in terms of effectiveness, with silicon appearing to be consistently objectionable. This approximate order of merit bears some relationship to the carbide-forming tendencies of the alloying elements. However, it is clear that this is not the only factor that determines the effectiveness of these alloying elements in relation to cracking propensities, since molybdenum is more effective than chromium or manganese as a carbide former, but this is only reflected in the cracking resistance that molybdenum confers in relation to the carbonate-bicarbonate environment, its effect in relation to cracking by hydroxide being deleterious. The electrochemical influences of the alloying elements also appear to be reflected in their effects upon cracking response, with both the dissolution and filming tendencies operative. Thus, there is a general correlation between the effects of the alloying additions upon cracking behaviour and the corrosion of these elements in solutions of similar pH. Aluminium, molybdenum and titanium are well known to show good corrosion resistance in more neutral solutions, with poorer resistance in strongly alkaline or acid environments, except for oxidising acids in the cases of aluminium and titanium. The stress-corrosion results broadly reflect such behaviour. Similarly, nickel is well known to show increased corrosion resistance with increasing pH, while silicon is a very reactive element over a wide range of pH values especially in hot solutions.

Notwithstanding possible explanations of the effects of these various alloying elements, perhaps the most important message arising from the effects reflected in equations (8.11). (8.12) and (8.13) is that the effects of alloying elements vary with the environment. It follows that a steel resistant to cracking in one environment may not be resistant in others, with the effect

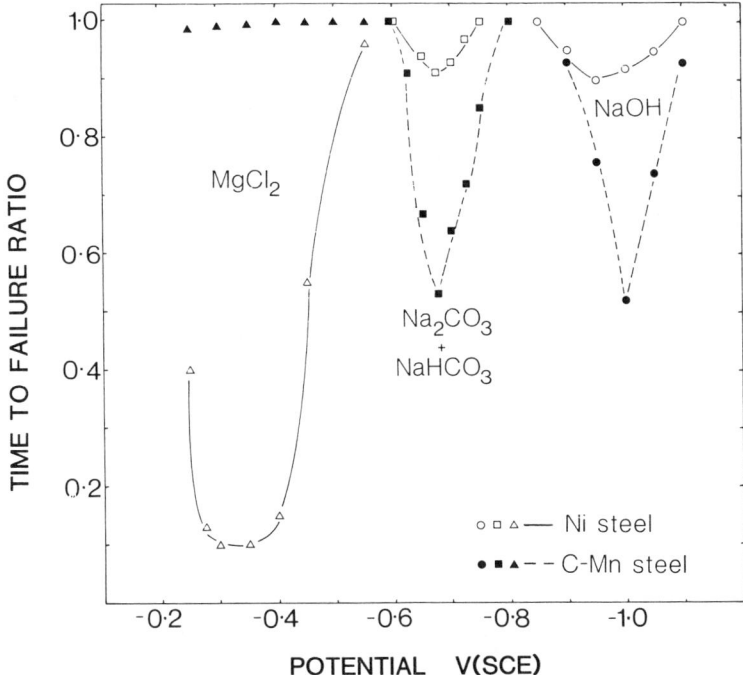

Fig. 8.17 Effects of applied potential upon the time to failure ratio in slow strain rate tests of C-Mn steel, with and without a 6% nickel addition, in boiling 8 M NaOH, 1 M NaHCO$_3$ + 0.5 M Na$_2$CO$_3$ at 75°C, and boiling 4.4 M MgCl$_2$ (after Parkins et al.[23] and Poulson and Parkins[24])

of molybdenum in markedly improving the cracking resistance in carbonate–bicarbonate but being markedly deleterious in relation to hydroxide solutions making that point. Similar effects may be observed in relation to nickel additions. Thus, when almost 6% nickel is present, the resistance to cracking at various potentials by both hydroxide and carbonate–bicarbonate solutions is good, and considerably better than without the presence of nickel, as is apparent from Fig. 8.17. However, in boiling 42% magnesium chloride solutions the nickel-containing steel cracks vary readily[24], yet an unalloyed steel shows no propensity for failure in such a solution. Clearly an alloyed steel developed to have low susceptibility to stress-corrosion cracking in a particular environment will not necessarily show such behaviour in a different environment, a rather obvious point widely recognised in relation to other forms of corrosion but not always recognised in the context of stress corrosion.

The Effects of Heat Treatments

For steels that are most frequently used in the annealed or normalised condition the most important structural parameter that can be influenced by heat treatment is the grain size, although the extensive use of welding as a means of fabricating mild steels means that martensitic and tempered martensitic

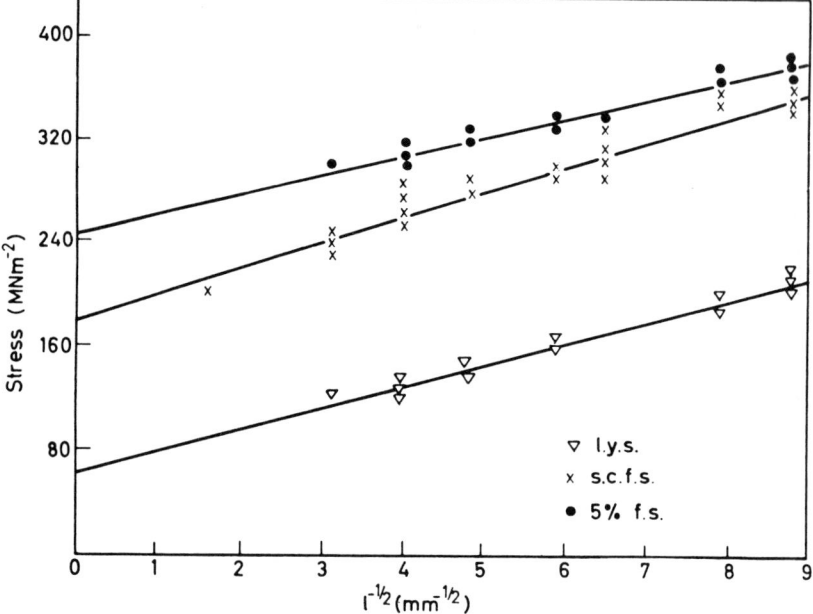

Fig. 8.18 Effects of grain size on lower yield stress, 5% flow stress and stress-corrosion fracture stress for 0.08%C steel in 8 N Ca(NO$_3$)$_2$ (after Henthorne and Parkins[21])

structures may also be encountered. That ferrite grain size has an effect upon stress-corrosion propensity is apparent in the results shown in Fig. 8.18, from which it is clear that coarse-grained steels fracture at appreciably lower stresses than those of smaller grain size. Such results may be interpreted in a number of ways from the more likely saturation of grain boundaries by segregate of limited quantity if the grain size is large, to the effect of grain size on the mechanical properties of steel, which as has already been mentioned are matters of importance in stress-corrosion cracking. The effect of relatively fast rates of cooling from the austenitising temperature is more marked than the grain size effect achieved by differing austenitising temperatures. Thus water quenching from 920°C appears to render steel more susceptible to cracking than oil quenching, and further decreases in the cooling rate through air cooling to furnace cooling further increase cracking resistance. However, it needs to be stressed that these trends are relative and that even with very slow cooling, especially from high austenitising temperatures, many ferritic steels are very susceptible to stress corrosion in certain environments.

The effects upon cracking tendencies of tempering following quenching is, in general, for the marked susceptibility of the water-quenched condition to be mitigated if the tempering temperature is high enough. However, there are some other differences between the results published by various workers. Houndrement, et al.[25] agree with most other workers in showing that tempering above about 300°C increases cracking resistance and that the benefits are maximised when the tempering temperature is 600°C or above. On the other hand, the results of Uhlig and Sava[15] show the full benefits of tempering at temperatures from 250°C upwards with a return to marked susceptibility at 700°C. The latter temperature is dependent upon the time of tempering, marked susceptibility returning after tempering at only 500°C if this is carried out for about 10 h or more. The effects of tempering quoted by Long and Lockington[26] for a 1% Mn alloy with very little carbon are in complete contrast to the results just mentioned. They show that tempering above 200°C increases susceptibility initially and that increasing resistance to cracking begins to be observed when the tempering temperature exceeds about 500°C. The explanation for the variability in these results may lie in the differing carbon contents of the steels used or in the test methods employed, but none of the papers quotes any results from structural studies upon the steels following the various heat treatments.

That the different carbon contents of the steels used by these various workers is a factor in these apparently contradictory results on the effects of tempering quenched carbon-steels is apparent from a study[27] of a range of such steels tested in a nitrate solution. Figure 8.19 shows the threshold stress values as a function of carbon content for the range of steels in the annealed and water-quenched conditions. The deleterious effects of water quenching upon the cracking resistance of the higher carbon steels is readily apparent, whilst with carbon contents below about 0.1% increased resistance to cracking is observed. Those data refer to constant strain tests, but slow strain-rate tests showed the same trends. The implications of the results shown in Fig. 8.19 are that subsequent tempering may be expected to increase susceptibility of the lower carbon steels but decrease that of the higher carbon materials. Figure 8.20 shows that these trends were observed

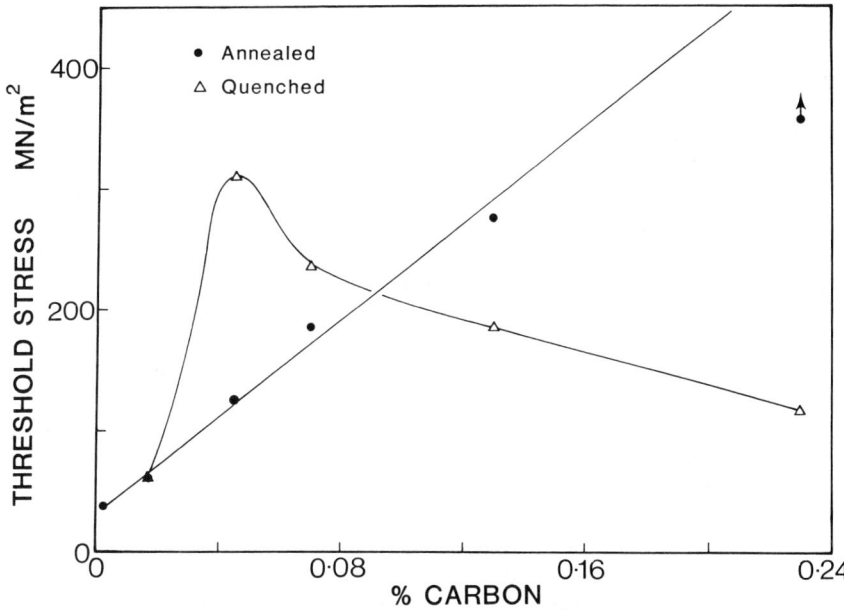

Fig. 8.19 Threshold stresses in a boiling nitrate solution for annealed and quenched steels of different C contents (after Parkins et al.[27])

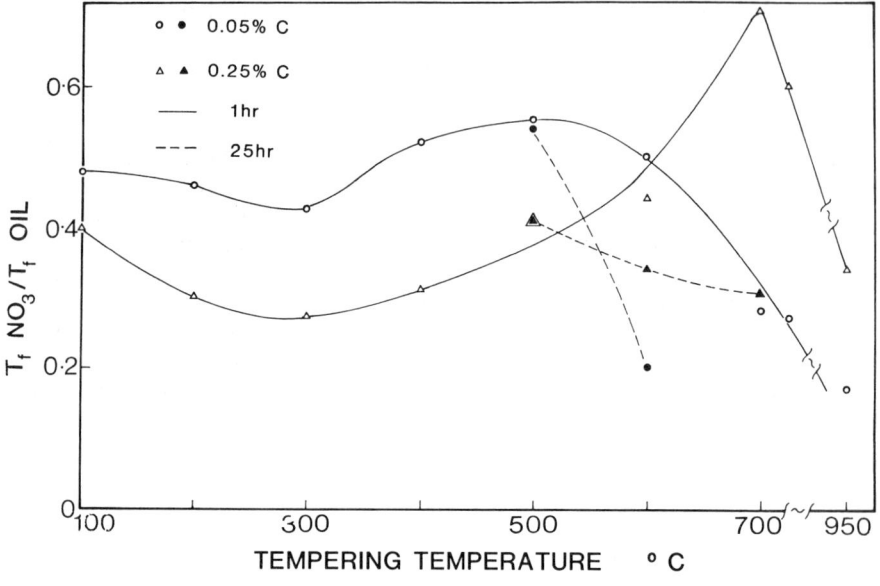

Fig. 8.20 Effects of different tempering times upon the time to failure ratio of two steels tempered at various temperatures (after Parkins et al.[27])

in tests upon tempered specimens of two of the steels, with similar trends shown with the other steels in the series, depending upon their carbon content. Obviously tempering the 0.05%C steel at temperatures in the region of 700°C causes a marked deterioration in cracking resistance, that temperature being lowered with longer tempering times, confirming the trends observed by Uhlig and Sava[15] using a 0.06%C steel. On the other hand, with the 0.15%C steel, tempering at 700°C for 1 hour gave the highest cracking resistance in agreement with the results of Houdrement et al.[25] using a 0.26%C steel.

These various effects of quenching and tempering treatment upon cracking tendency appear to correlate with microstructural changes[27]. High susceptibility is associated with simple, unbranched, crack paths and relatively high crack velocities and occurs at the prior austenite grain boundaries of the higher carbon quenched steels and at the recrystallised ferrite grain boundaries of steels tempered above 500°C for the low carbon contents and above 700°C for the higher carbon materials, for 1 h treatments. High resistance to cracking is associated with lower crack velocities and a marked tendency for multiple branching to develop for short lengths along lath boundaries. These effects are observed in quenched low-carbon steels, where the main cracks follow the prior austenite boundaries and branches develop along the lath boundaries containing auto-tempered carbides, and in the higher carbon steels when tempered to precipitate carbides in the lath boundaries but without recrystallisation occurring. Repeated branching may be expected to reduce the rate of crack propagation, to extend the failure time or increase the chances of a crack ceasing to propagate by the accumulation of corrosion products in the enclave. When the tempering temperature is sufficiently high, and the time sufficiently prolonged, susceptibility increases, in agreement with the observation that prolonged subcritical annealing of pearlite structures to promote carbide spheroidisation at the ferrite grain boundaries increases susceptibility to cracking[28].

The effects upon cracking tendency from tempering higher carbon mild steels mentioned above in relation to cracking by nitrate solutions have also been observed for cracking by carbonate-bicarbonate solutions. In addition, Bandyopadhyay et al.[22] have commented upon the role of preferential attack on large chromium-rich carbides in blunting cracks and reducing crack velocities in low-alloy steels exposed to hydroxide solutions.

Effects of Environment Composition

It has frequently been stated that the environmental requirements for stress corrosion cracking are highly specific, but the relatively extensive list[6] of environments that have been reported as promoting cracking raises queries as to the validity of such statements. It is clear that cracking environments are specific in the sense that not all possible environments promote cracking, but to state that the solution requirements are highly specific may lead to a false sense of security in certain practical situations. It is clear that the propagation of a stress-corrosion crack requires the reactions that occur at the crack tip to proceed at a considerably faster rate than any dissolution processes that take place at the exposed surfaces of the metal, including the crack sides, since otherwise general corrosion or pitting only will be observed. For an inherently reactive metal like mild steel most of the exposed surface will only remain inactive if the surface is passivated, and so environments in which stress corrosion occurs are likely to have considerable oxidising potential. Nitrates and hydroxides, which are of course anodic inhibitors of the corrosion of carbon steels in appropriate circumstances, have such characteristics and are those anions associated with the earliest identified instances of cracking of mild steels.

Other anodic inhibitors of the corrosion of such steels are also capable of promoting stress-corrosion cracking in appropriate circumstances. Thus, carbonate-bicarbonate[29] and phosphate solutions[30] promote dissolution-related cracking in certain potential ranges which, as with nitrates and hydroxides, can be predicted by appropriate electrochemical measurements (Section 8.1). Figure 8.21 shows the current density differences between fast and slow sweep-rate polarisation curves (Fig. 8.5) at various potentials for mild steel immersed in hydroxide, carbonate-bicarbonate and nitrate solutions. Also shown in Fig. 8.21 are the results from controlled potential slow strain-rate tests involving the same solutions, and it is clear that the potential ranges in which cracking occurred are those predicted for each of the three solutions from the electrochemical measurements. Moreover, the severity of cracking, reflected in the time-to-failure ratio, reflects the magnitudes of the current density differences for the three solutions at different potentials, reflecting the trend shown in Fig. 8.7. The potentiodynamic polarisation curves measure the tendencies for the occurrence of dissolution and filming processes, i.e. the combination that is required to promote and retain crack geometry for dissolution-related cracking (Section

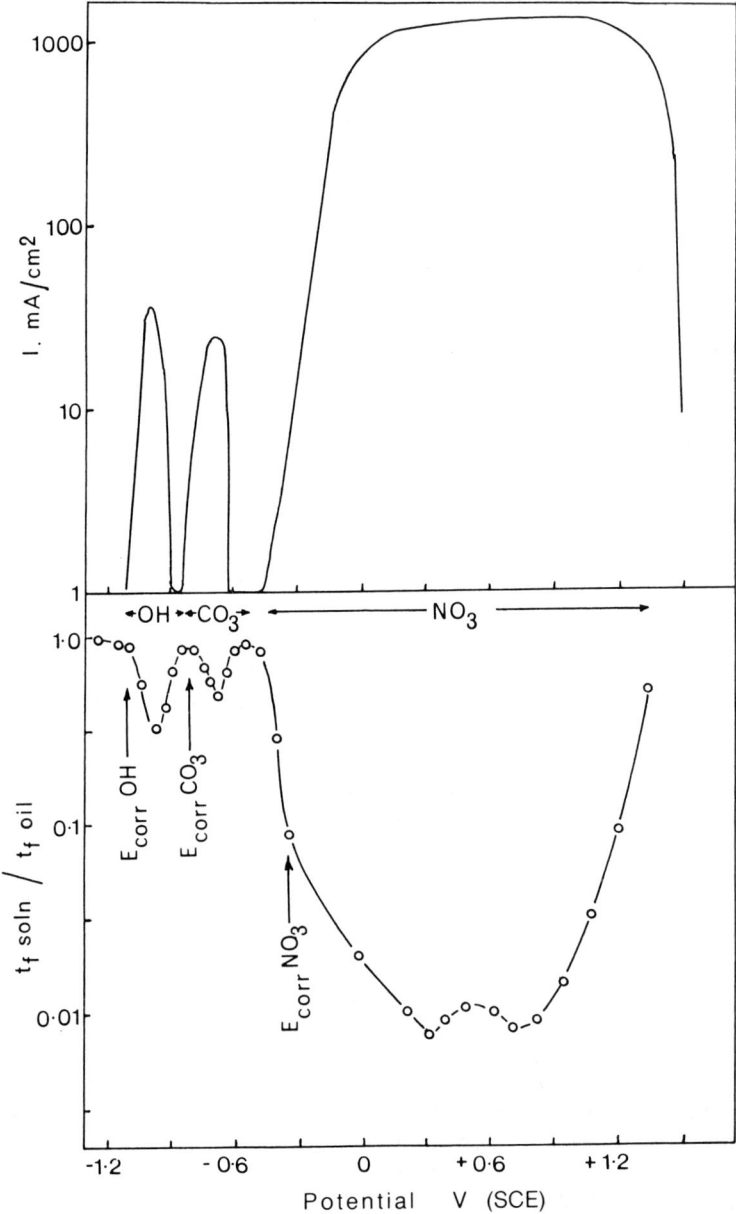

Fig. 8.21 Current density differences between fast and slow sweep rate polarisation curves and stress corrosion cracking suspectiblity as a function of potential for a C-Mn steel in nitrate, hydroxide and carbonate-bicarbonate solutions

8.1), and for low-alloy ferritic steels in a variety of environments they appear to give reasonable indications of cracking propensities, hence they may be useful in assessing the potencies of environments for which cracking data do not exist.

The above mostly refers to intergranular cracking, but there are some media that promote transgranular cracking in ferritic steels. It is likely that the mechanisms of failure in these cases are different from those for intergranular cracking, with strain-generated active paths or localised embrittlement of material in the crack-tip region playing important roles. Perhaps the most topical of these is the cracking of ferritic steels by high temperature water[20]. Although not universally accepted, the evidence tends to support a dissolution-related mechanism of crack growth, rather than one based upon hydrogen ingress. Transgranular cracking of low-carbon steels has been reported as occurring in HCN solutions[32] containing 2.6–3.5 g/l of HCN, and in $FeCl_3$ solutions and chloride-bearing slurries containing ferric oxides and hydrated oxides at 316°C[30]. CO_2–CO–H_2O environments promote transgranular failure[10], increasing quantities of CO, some of which must be present for cracking to be initiated, decreasing the time-to-failure and lowering the stress required[33]. The results may be interpreted in terms of the CO acting as an inhibitor of the attack upon iron by the acidic CO_2–H_2O, a passive film of increasing effectiveness being formed as the proportion of CO is increased, which in turn suggests a film rupture mechanism. Moist H_2S has been reported[34] as causing the failure of cold-drawn high-carbon steel wire when loaded to only 40% of its breaking load in air, although this failure may have resulted from hydrogen-embrittlement in the light of the work that has been carried out in relation to failure in sour oil well equipment[35].

In some environments it appears likely that more than one mechanism of failure may exist, depending upon the potential. Thus ammonium carbonate solutions, which promote intergranular cracking at higher potentials can give rise to transgranular cracking due to a dissolution-related mechanism at somewhat lower potentials, and at much lower potentials to hydrogen-related cracking in slow strain-rate tests. Sodium hydroxide solutions also can promote transgranular hydrogen-induced cracking at potentials appreciably below those that sustain intergranular cracking. Such hydrogen-induced cracking in mild steels in various environments at sufficiently low potentials, although readily produced in slow strain-rate tests, is not normally regarded as a significant problem in industrial situations and so the results mentioned may simply reflect the severity of the slow strain-rate test method. However, it is possible that with cyclic loading, as opposed to the static loading conditions assumed to operate in plant that displays stress corrosion, the environment-sensitive cracking due to hydrogen reflected in slow strain-rate tests may be of relevance. Of course, with high-strength ferritic steels hydrogen-embrittlement is often regarded as the only mechanism of environment-sensitive fracture, for static or dynamic loading. But the two ranges of potentials for cracking, often separated by a regime in which there is insensitivity to cracking, mentioned above in relation to low-strength steels, are sometimes observed with high-strength varieties and suggest to some that more than one cracking mechanism may exist for such materials.

Effects of Additions to Cracking Environments

From the indications already mentioned, that cracking environments render most of the exposed surface inactive whilst allowing dissolution to soluble species at the crack tip when the potential is within the appropriate range, it is possible to categorise the effects of additions to cracking environments as follows. Where cracking occurs at the free corrosion potential, additions to the environment may cause the potential to move outside the cracking range and so prevent failure. Conversely, the free corrosion potential may normally not coincide with the cracking range and certain additions to the environment may result in cracking simply because they cause the free corrosion potential to move into the critical potential range. Alternatively, where the potential remains within the cracking range, additions may influence cracking by their effect upon the passivation or dissolution reactions or both[36]. Such an approach permits rationalisation of the confusing situation that appears to exist from a reading of the literature. Thus, in relation to the cracking of carbon steels in caustic alkalis, laboratory studies have frequently shown that cracking could not be reproduced unless the solution was contaminated with oxidising salts[37] or with oxygen itself[38]. The picture is further complicated by the fact that, while substances such as $KMnO_4$, $NaNO_3$ and Na_2CrO_4 added to NaOH solutions promote cracking at the boiling point, at 250°C they act as inhibitors. Similarly, whilst a little dissolved oxygen apparently promotes cracking, a high concentration prevents failure[38].

The effects of additions in causing the free corrosion potential to move in relation to the potential range for cracking would appear to have considerable practical significance. Thus, the addition of relatively small amounts of nitrates to concentrated NaOH solution, an approach that is sometimes used in treating boiler feed waters in an attempt to avoid caustic cracking, causes the free corrosion potential to become significantly more positive than the cracking potentials in alkali, and failure is no longer observed. However, the same addition will not prevent cracking if the potential is maintained, by potentiostatic control, within the cracking range, so that nitrates should be regarded as unsafe inhibitors of caustic cracking. Na_2SO_4 should also be placed in this category[36] even though the maintenance of a Na_2SO_4/NaOH ratio in boiler feed water in excess of 2.5 is still widely practised as a means of preventing caustic cracking in boilers. In some circumstances additions to environments may promote cracking by causing the free corrosion potential to move into the cracking range. Thus, the free corrosion potential of some mild steels is likely to lie at the boundary of the cracking range in hydroxides so that stress corrosion does not occur. The addition of a small quantity of lead oxide causes the free corrosion potential to move into the cracking range and failure then occurs readily at the free corrosion potential, i.e. without external potentiostatic control. Similarly, the well established effect of oxygen in high temperature pure water in promoting cracking in nuclear reactor pressure vessel steel is probably the result of the oxygen raising the potential into the cracking range[31].

Whilst some additions operate in the manner already indicated it is clear that other substances influence cracking propensity not so much by any effect they may have upon potential, but by modifying the reactions involved in cracking. Tannins and phosphates, and to a lesser extent silicates, prevent cracking in strong NaOH solutions even when the potential is maintained within the cracking range[36]. The reasons for this type of behaviour have not been studied but the relatively high anodic current peak shown by mild steel in 30% NaOH is reduced, by about two orders of magnitude, in the presence of NaH_2PO_4, and to a lesser extent by Na_2SiO_3, which may indicate that these substances operate by hindering the dissolution of ferrite. Tannins have little effect upon the anodic polarisation curve when added to NaOH and their influence in stopping cracking may be related to interference with the cathodic reaction.

Results from various laboratories have shown that additions to nitrate environments appear to influence cracking according to whether they affect the pH of the solution, are oxidising substances or form insoluble products with iron[39]. The effect of the pH upon the cracking potency of nitrates is apparent from studies in which various cations have been incorporated in the solution. Nitrates of acidic cations are the most potent at equivalent strengths and temperatures and this suggests that the cracking capacity of any nitrate solution may be controlled by adjustment of its pH. This is found to be so, small additions of HNO_3 to lower the pH of a given nitrate increasing the cracking tendency, whilst raising the pH with a sufficient quantity of OH^- ions will prevent cracking at the free corrosion potential. This effect of OH^- ions is related to their influence on potential however, since with potentiostatic control in the cracking range, nitrates at a pH of 10 or so will promote failure. Oxidising additions such as $KMnO_4$ and $K_2Cr_2O_7$ all accelerate cracking in nitrates, whilst substances such as H_3PO_4, Na_2HPO_4 and $CO(NH_2)_2$, which may be expected to form insoluble products with iron, retard or prevent cracking.

One of the difficulties associated with much of the data on the effects of additives to potent stress-corrosion cracking environments is that they are the result of *ad hoc* experiments, often without recognition of the importance of potential upon cracking. Free corrosion potentials in laboratory tests can be very different from those that exist in service situations for the same metal-environment combination. Moreover, in view of the balance required between dissolution and passivity for cracking the concentration of an inhibitive substance may be critical. The point is illustrated by the results shown in Fig. 8.22, which indicates the average crack velocities for mild steel specimens tested in a carbonate-bicarbonate solution with various additions of sodium chromate[40]. Cracking is inhibited for all practical purposes by the addition of 0.16 wt% Na_2CrO_4, but Fig. 8.22 shows that at about 0.017% Na_2CrO_4 there is enhanced cracking beyond that observed, with lesser, including zero, additions of chromate. The explanation for such a result, which has been observed with other substances, is that, while sufficient inhibitor will ensure that only passive behaviour is observed, at some intermediate concentrations the grain surface may be effectively passivated but the more active sites, the grain boundaries, are not so protected, thereby increasing the current density in such regions. Clearly it is important to ensure an adequate supply of inhibitor, and Fig. 8.22 also offers a possible explanation for the contradictory statements in the literature of the effects of some additions to cracking environments.

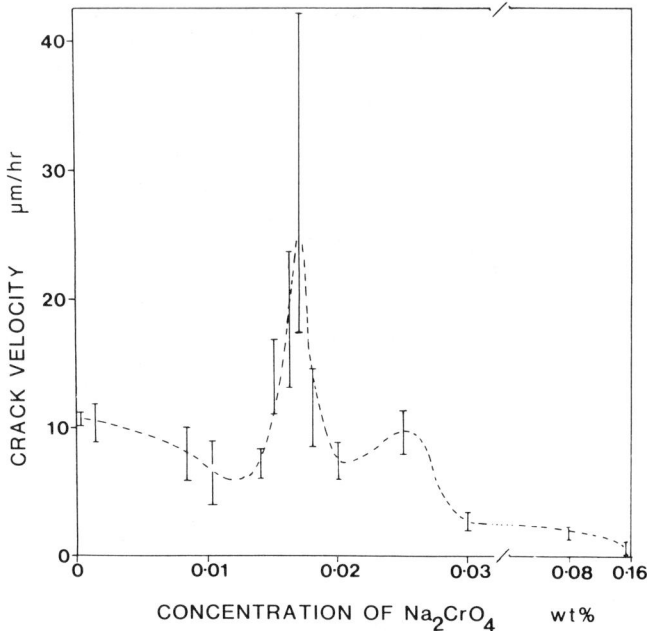

Fig. 8.22 Average crack velocities observed in mild steel specimens tested in 0.5 M Na_2CO_3 + 1 M $NaHCO_3$ at 75°C with various additions of Na_2CrO_4. Results refer to potential of most severe cracking at each chromate concentration; variability in crack velocity in replicate tests shown by lengths of scatter bars (after Tems and Parkins[40])

Methods of Prevention

The incidence of stress-corrosion cracking requires a susceptible alloy to be exposed to a specific environment at stresses above some limiting value, from which it follows that control of the problem may be through manipulation of any or these three parameters[41]. In more detail the choices are:

Metallurgical control —change alloy composition
  —change alloy structure
  —use metallic or conversion coating
Environmental control —apply anodic or cathodic protection
  —remove offending species
  —add inhibitor
  —use organic coating
  —modify temperature
Mechanical control —reduce operating stresses
  —relieve fabrication stresses
  —avoid stress concentrators
  —introduce surface compressive stresses

In a particular service situation, other considerations may preclude some approaches to prevention; for example, weight, strength or economic requirements may dictate the use of an alloy susceptible to cracking. Moreover, the philosophy associated with prevention can differ according to the use of the structure or component involved. Thus, the attitude to slow crack growth in a pressure vessel is likely to be appreciably different from that where the consequences of cracking are no more than the seepage of an innocuous liquid from cracks. Ideally, approaches to prevention should begin at the design stage, but it is not infrequently the case that cracking occurs in an existing plant when such failure had not been anticipated. In such circumstances the approaches to prevention are likely to be restricted and in some situations it may be necessary to accept the continuance of cracking but still attempt to control it by minimising crack growth rates. Some of the implications for prevention or control can be deduced from the factors discussed earlier, but some of the approaches listed above need further brief mention, particularly some of those mentioned under mechanical control.

Many practical instances of stress-corrosion cracking in the lower strength ferritic steels result from the presence in the structure of residual stresses, which are usually in the region of the yield stress, rather than the normally appreciably lower design stresses. Thermal stress relief is likely to be beneficial in such cases, but applying full stress relieving heat treatments to structures can present problems due to inadequate furnace capacity for large fabrications or distortion of the latter at the temperature involved (usually about 650°C for ferritic steels). However, partial stress relief by heating to lower temperatures than that required for full stress relief can be adequate in those cases where the residual and operating stresses can be reduced below the threshold stress[42]. Moreover, these lower temperatures can be achieved with less distortion for furnace annealing or by locally applied heating with large structures.

Since stress corrosion requires the presence of tensile stresses of appropriate magnitude it follows that if compressive stresses are introduced into a surface, cracking should not occur. Shot peening and grit blasting have been shown to be effective in preventing or reducing the incidence of environment-sensitive cracking in ferritic steels and hammer peening, which also leaves surfaces in compression when properly applied, can have similar effects[42]. It is also sometimes possible to leave surfaces at which cracking would otherwise occur in a state of compression by localised heating techniques[43]. These have been developed in relation to the stainless steel pipe cracking problem in boiling water reactors, but the principle should be equally applicable to ferritic steel structures of appropriate geometry.

While plant operating conditions are largely dictated by considerations other than that of control of environment-sensitive cracking, it is worth noting that pressure or temperature fluctuations are likely to induce stress cycles that lower the threshold stress for cracking below that obtained with static loading. Obviously, unnecessary pressure or temperature excursions should be avoided or minimised. While temperature variations leading to thermal stresses may aggravate the problem, especially during plant start-up or shut-down, temperature is an important parameter in another sense in stress corrosion. Thus, the conditions for dissolution-related cracking and the crack velocity are typical thermally activated processes in most instances of cracking in ferritic steels, so that lower temperatures are less likely to result in cracking or be associated with lower crack velocities.

While ideally structures should be designed and fabricated so that environment-sensitive cracking is avoided, in practice it is sometimes necessary to live with the problem. This implies an ability to detect and measure the size of cracks before they reach the critical size that may result in catastrophic failure. Such inspection has important implications for plant design, which should be such as to allow inspection at relevant locations. The latter are regions of high residual stress (welded, bolted or riveted joints) and regions of geometrical discontinuity (notches, crevices, etc.) where stress or environment concentration may occur.

R. N. PARKINS

REFERENCES

1. Stromeyer, C. E., *J. Iron St. Inst.*, **79**, 404 (1909)
2. Jones, J. A., *Trans. Faraday Soc.*, **17**, 102 (1921)
3. *1st Rept. Inst. Gas Engrs.*, Comm. No. 398, Brit. Welding Res. Assoc. F.M.9 Ctee (1951)
4. Fraser, J. P., Eldredge, G. C. and Treseder, R. S., *Corrosion*, **14**, 517t (1958)
5. Phelps, E. H. and Loginow, A. W., *Corrosion*, **18**, 299t (1962)
6. Logan, H. L., *The Stress Corrosion of Metals*, Wiley, New York, pp. 5-7 (1966)
7. Parkins, R. N., Alexandridou, A. and Majumdar, P., *Mats. Perf.*, **25**, 20, (1986)
8. Parkins, R. N. and Foroulis, Z. A., *Mats. Perf.*, **27**, 19 (1988)
9. Parkins, R. N. and Fessler, R. R., *Materials in Engineering Applications*, **1**, 80 (1978)
10. Kowaka, M. and Nagata, S., *Corrosion*, **24**, 427, (1968)
11. Long, L. M. and Uhlig, H. H., *J. Electrochem. Soc.*, **112**, 964, (1965)
12. Flis, J. and Scully, J. C., *Corrosion*, **24**, 326 (1968)
13. Parkins, R. N. and Green, J. A .S., *Corrosion*, **24**, 66,(1968)
14. Hondros, E. D. and Seah, M. P., *Int. Met. Rev.*, **22**, 262, (1977)
15. Uhlig, H. H. and Sava, J., *Trans. A.S.M.*, **56**, 361 (1963)
16. Bohmenkamp, K. *Proc. Conf. on Fundamental Aspects of Stress Corrosion Cracking*, Edited by R. W. Staehle, A. J. Forty and D. van Rooyen, NACE, Houston, p 374, (1969)
17. Lea, C. and Hondros, E. D., *Proc. Roy. Soc.*, **377A**, 477 (1951)
18. Krautschick, H. J., Grabke, J. H. and Diekmann, W., *Corros. Sci.*, **28**, 251 (1988)
19. Bandyopadhyay, N. and Briant, C. L., *Corrosion*, **41**, 274 (1985)
20. Scott, P. M., *Proc. 3rd Symp. on Environmental Degradation of Materials in Nuclear Power Systems – Water Reactors*. Edited by G. J. Theus and J. R. Weeks, AIME – The Metallurgical Society Inc., Warrendale, Pa., p 15, (1987)
21. Henthorne, M. and Parkins, R. N., *Brit. Corrosion J.*, **5**, 186 (1967)
22. Bandyopadhyay, N., Briant, C. L. and Hall, E. L., *Met. Trans. A.*, **16**, 1333, (1985)
23. Parkins, R. N., Slattery, P. W. and Poulson, B. S., *Corrosion*, **37**, 650 (1981)
24. Poulson, B. S. and Parkins, R. N. *Corrosion* **29**, 414 (1973)
25. Houdrement, E., Bennek, H. and Wentrup, H., *Stahl und Eisen*, **60**, 575 and 791 (1940)
26. Long, L. M. and Lockington, N. A., *Corros. Sci.*, **7**, 447 (1967)
27. Parkins, R. N., Slattery, P. W., Middleton, W. R. and Humphries, M. J., *Brit. Corrosion J.*, **8**, 117 (1973)
28. Parkins, R. N., *J. Iron Steel Inst.*, **172**, 149 (1952)
29. Sutcliffe, J. M., Fessler, R. R., Boyd, W. K. and Parkins, R. N., *Corrosion* **28**, 313 (1972)
30. Parkins, R. N., Holroyd, N. J. H. and Fessler, R. R., *Corrosion*, **34**, 253 (1978)
31. Congleton, J., Shoji, T. and Parkins, R. N., *Corros. Sci.*, **25**, 633 (1985)
32. Huckholtz, H. and Pusch, R., *Stahl und Eisen*, **62**, 21 (1942)
33. Brown, A., Harrison, J. T. and Wilkins, R., *Corros. Sci.*, **10**, 547 (1970)
34. Rees, W. P., *Symposium on Internal Stresses in Metals and Alloys*, Inst. of Metals, London, p 333, (1948)
35. Schultz, A. E. and Robertson, W. D., *Corrosion*, **13**, 33 (1957)
36. Humphries, M. J. and Parkins, R. N., *Corros. Sci.*, **5**, 747,(1967)
37. Schroeder, W. E., Berk, A. A. and O'Brien, R. A., *Metals and Alloys*, **8**, 320 (1937)
38. Radeker, W. and Grafen, H., *Stahl und Eisen*, **76**, 1616 (1956)
39. Parkins, R. N., from *Stress Corrosion Cracking and Hydrogen Embrittlement of Iron Base Alloys*, Edited by R. W. Staehle, J. Hochmann, R. D. McCright and J. E. Slater, NACE, Houston, p 601, (1977)

40. Tems, R. D. and Parkins, R. N., *Proc. 5th European Symp. on Corrosion Inhibitors*, Ann. Univ. Ferrara, p 857, (1975)
41. Parkins, R. N., *Mats. Perf.* **24**, (8), 9–20 (1985)
42. Pearson, C. E. and Parkins, R. N., *Welding Research*, **3**, 95t (1949)
43. Danko, J. C., Paper No 162, *CORROSION/84*, NACE, Houston, (1984)

8.3 Stress-corrosion Cracking of Stainless Steels

Intoduction

For the purpose of this section, stainless steels will be assumed to cover the group of alloys which rely mainly on the addition of chromium to iron to impart corrosion resistance. Even within this restricted group there are many alloy types with very different microstructures and mechanical properties and a wide range of susceptibility to stress-corrosion cracking. It is convenient to subdivide these stainless steels into groups under the general headings: martensitic, ferritic, duplex and austenitic, although there are some high-strength grades produced by precipitation hardening heat treatments that could be considered as an additional group. Other groupings are possible, but the above is convenient for discussing resistance to stress-corrosion cracking.

The phenomenon of stress-corrosion cracking can be defined as the occurrence of macroscopic brittle fracture of a normally ductile metal due to the combined action of stress and some specific environment. The environment need not be chemically aggressive in that high general dissolution rates are not required and the phenomenon is complicated by the fact that many different mechanisms can give rise to such cracking. It is often difficult to differentiate between the roles of anodic dissolution and hydrogen absorption on cracking. Also, microstructural changes such as the creation of martensite in the region of the crack tip, and the effect on crack propagation of quite small alternating stresses superimposed on a mean stress, further complicate attempts at complete understanding of stress-corrosion cracking. These aspects are covered in Section 8.1. In this section, the effects of gaseous hydrogen and of deliberate cathodic charging with hydrogen will not be discussed.

Even excluding such instances of hydrogen-assisted cracking, the literature on stress-corrosion cracking of stainless steels is very extensive. A computer search using the key words stress-corrosion cracking, (SCC) and stainless steels generated more than two thousand references in the English language from a single database for the period 1985 to 1990. Fortunately, there have been some excellent reviews published in recent years and the present chapter draws heavily on those as well as on data presented in several recently published books on corrosion of stainless steels.

The good general corrosion resistance of stainless steels is derived from the nature of the oxide generated on iron–chromium alloys containing more than about 12% chromium. Stress-corrosion cracking can occur in these alloys by intergranular cracking, transgranular cracking or as a mixture of both. Intergranular cracking is often promoted by the precipitation of carbides in the grain boundaries of the steel. Carbon diffuses more rapidly through the iron lattice than chromium because of the differences in size of the atoms. Thus, the precipitation of chromium carbide at a grain boundary site can be associated with local denudation of chromium in the adjacent matrix so that preferential corrosion can occur along the low-chromium content grain boundaries (Section 3.3). In contrast, transgranular cracking occurs quite readily in austenitic stainless steels because they tend to have low stacking fault energies so that planar slip is common (Section 20.4). The large slip steps formed at the metal surface can break the protective oxide layer and expose bare metal to the environment. Dissolution occurs rapidly in this region because a small anode is created there that is surrounded by a large cathodic region and dissolution is encouraged to penetrate along the slip planes. Passivation of the flanks of the propagating crack can generate overall crack-like geometry and stress intensification at the crack tip can maintain the creation of bare metal there to sustain the rapid dissolution rate needed for fast crack growth.

The above simple concepts need to be modified and expanded to allow discussion of stress-corrosion cracking of stainless steels in general, especially as the role of hydrogen in assisting crack growth must be accounted for.

Martensitic Stainless Steels

As an approximate guide, the martensitic grades of stainless steels can be defined as those alloys of iron and chromium in which $\%Cr - 17 \times \%C < 12.5$, but which still contain more than 11.5% Cr to give adequate corrosion resistance. On quenching such alloys from high temperatures they will traverse the γ loop in the iron–carbon equilibrium diagram and martensite will be formed to an extent that depends upon the carbon content of the steel. Carbon contents for such steels range from 0.15 to 1.2%, depending upon the strength requirement for the steel. As their name implies, they are used in the quenched and tempered condition for components such as turbine blades, bolts, springs, valve components, cutlery etc. and in the steam generating and chemical industries for many components.

Tempering in the range 400–650°C can be detrimental to the mechanical properties and corrosion resistance (Fig. 8.23)[1]. The latter is considered to occur because of depletion of chromium from the matrix adjacent to the precipitated carbides[2]. In a review of instances of stress-corrosion cracking reported to a supplier of stainless steels for various clients, Truman cites 15 instances of cracking that were all for high hardness material, i.e. 350–650 H_v for quenched and tempered, and 380–430 H_v for precipitation-hardened martensitic types of steel[3]. Typical of this group of stainless steels are the 13Cr turbine blade steels that combine high hardenability, good damping properties, good thermal shock resistance, fatigue resistance and resistance against hot pressurised molecular hydrogen[4].

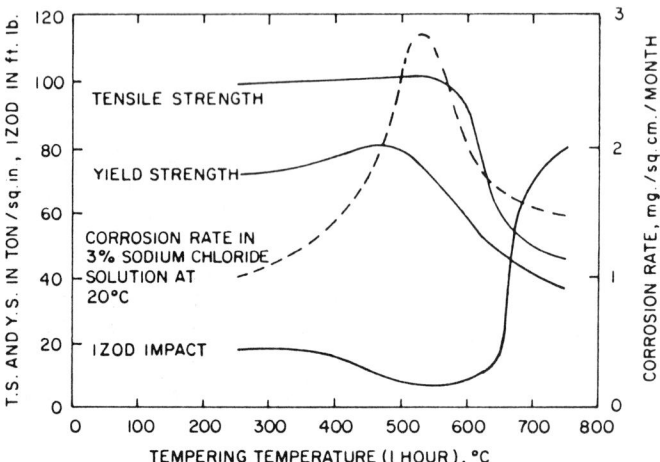

Fig. 8.23 Effect of tempering on the mechanical properties and corrosion resistance of type 420 stainless steel (after Sedriks[5])

The martensitic steels are commonly quoted to the AISI 400 series specifications although that range of numbers also includes some ferritic grades[5,6]. In the as-quenched or as-welded condition the steels are very susceptible to hydrogen cracking and immediate tempering is advisable, especially for the higher carbon content steels. In the tempered state to strength levels typical of many design applications, the steels are susceptible to both stress-corrosion cracking and to hydrogen-assisted cracking. Despite the complication that hydrogen can be generated within cracks and enclaves in acidic solutions and that hydrogen may enter the metal and contribute to the crack propagation mechanism, it is convenient to separate the response of the steel into two categories: (1) cracking under anodic dissolution control and (2) cracking under cathodic charging conditions. The reason for this is that if the anodic dissolution reaction is removed for situations where it is the rate-controlling process then no more hydrogen can be generated within the crack. Also, if anodic dissolution at the crack tip is occurring then some part, if not all, of the crack growth must arise from removal of atoms from the crack tip by corrosion, even if an additional mechanism is needed to account for the observed crack growth rates or to explain the morphology of the fracture surfaces produced. Type 410 (UNS S41000) stainless steel is very susceptible to cracking in 70% NaOH solution and in aqueous chloride solutions[4]. At 1410 MPa yield strength level a USS 12Cr-Mo-V steel cracked in marine and in semi-industrial environments when loaded to 75% of yield[7]. Cathodic polarization decelerates stress-corrosion cracking in NaCl solutions but accelerates it in NH_4Cl solutions, whereas anodic polarisation always accelerates stress-corrosion cracking in all chloride solutions[4].

It would seem easy in principle to separate cracking that proceeds by anodic dissolution from hydrogen-assisted cracking by investigating the effects of polarisation on the crack growth rate, time to failure or some other parameter convenient for assessing susceptibility to stress-corrosion cracking. Making the potential more anodic might be expected to increase susceptibility if cracking were under anodic dissolution control whereas cathodic polarisation should decrease susceptibility. The opposite should be the case for systems under hydrogen-assisted cracking control. The importance of these comments to practical problems is that it would be unwise to apply cathodic protection for corrosion control to a system that is known to be liable to exhibit hydrogen-assisted cracking. Some systems exhibit both types of cracking within different ranges of potential whereas for others there is a virtual overlap, e.g. martensitic steels in boiling NH_4Cl solutions at pH 5.1[4] where it is considered that increased anodic dissolution generates hydrogen that is absorbed into the steel to cause enhanced cracking. It is presumably for this reason that Spaehn adopts the terminologies anodic stress-corrosion cracking and cathodic stress-corrosion cracking so that the mechanism of cracking is not necessarily implied by the description of the cracking.

Tempering of martensitic stainless steels is performed to improve their toughness, but the tempering causes precipitation of carbides at the grain boundaries which has two main effects on the material. First, it can denude the grain boundary regions of chromium making them less resistant to corrosion. This effect can be exacerbated by local galvanic cell conditions set up at the grain boundaries. Second, precipitation at the grain boundaries will alter the mechanical properties of the grain boundaries relative to that of the matrix. Annealing at 500°C gives continuous grain boundary precipitates rather than discrete precipitates and the former are very deleterious and cause intergranular cracking[8].

There are numerous quoted examples of intergranular stress-corrosion cracking in tempered martensitic stainless steels[9-13] and abundant work relating this to the generation of non-equilibrium solute content profiles at the grain boundaries due to intergranular precipitation[14-18]. Thus, for a Super 12Cr–Mo–V stainless steel tested in a boiling 0.01 M NaOH plus 0.1 M NaCl solution, susceptibility was worst when a continuous chromium-depleted concentration profile was produced at the grain boundaries by tempering[19]. Further heat treatment that caused coarsening of the $M_{23}C_6$ precipitates generated overlapping diffusion fields that removed the continuous chromium-depletion zones and removed susceptibility to cracking.

Temperature, electrode potential and solution pH are also important. For a Type 403 stainless steel tested in 0.01 M Na_2SO_4, Bavarian et al. have shown that at a potential chosen to lie within the potential range for cracking at 100°C, cracking was also obtained at 75°C but not at 25°C or 50°C[20].

A large amount of stress-corrosion testing has been performed on pre-cracked specimens (Section 8.9). Crack growth rates of almost 10^{-4} m/s have been recorded for a martensitic stainless steel tempered at 475°C[21] and tested in distilled water ($K = 50 MPa\sqrt{m}$) but the crack growth rate decreased for material tempered at higher temperatures. For a Type 431 steel it was found that K_{Iscc} increased and the crack growth rates decreased by several orders of magnitude, depending upon the applied stress level, when the as-quenched steel was tempered at 650°C (Fig. 8.24)[22]. Spaehn[4] suggests that as the favoured industrial tempering temperature range for such steels

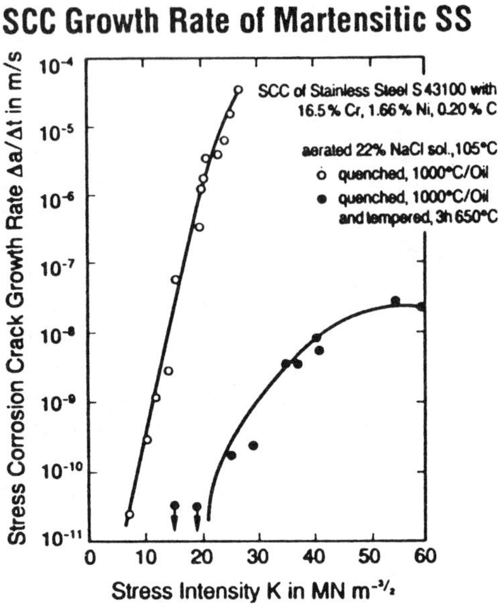

SCC Growth Rate of Martensitic SS

Fig. 8.24 Influence of heat-treatment conditions on the sub-critical stress corrosion growth rate of a nickel-bearing SS as a function of stress intensity. In the as-quenched condition, the steel shows much faster crack grown rates (after Spaehn[4])

is 700–750°C, much higher K_{Iscc} values than the value of about 20 MPa√m quoted by Speidel and much lower crack growth rates than 10^{-8} m/s should be expected even at high K values for properly tempered material.

The occurrence of stress-corrosion cracking in the martensitic steels is very sensitive to the magnitude of the applied stress[4]. For instance, a 13% chromium martensitic steel tested in boiling 35% magnesium chloride solution (125.5°C) indicated times to failure that decreased abruptly from more than 2500 h to less than 0.1 h as the applied stress was increased from 620 MPa to about 650 MPa (Fig. 8.25). However, the effects of stress on time to failure are not always so dramatic. For instance, in the same set of experiments times to failure for a 17Cr–2Ni martensitic steel gradually decreased from more than 800 h to about 8 h as the applied stress was increased from 500 MPa to 800 MPa.

Thus, the dominant parameter controlling the anodic stress-corrosion cracking resistance of martensitic stainless steels is the tempering temperature subsequent to quenching. However, because tempering causes precipitation of carbides and concentration profiles at the grain boundaries which induce intergranular cracking, the susceptibility to stress-corrosion cracking is not simply related to the hardness attained by the steel. The allowable hardness to give freedom from cracking depends upon the type of martensitic steel and the environment to which it is exposed. For instance, a hardness of 350 H_v or less is considered safe for operation in boiler feed water, condensing steam or boiler water, but the maximum allowable hardness may be different in other environments[4]. Indeed, work by Doig *et al.* has shown

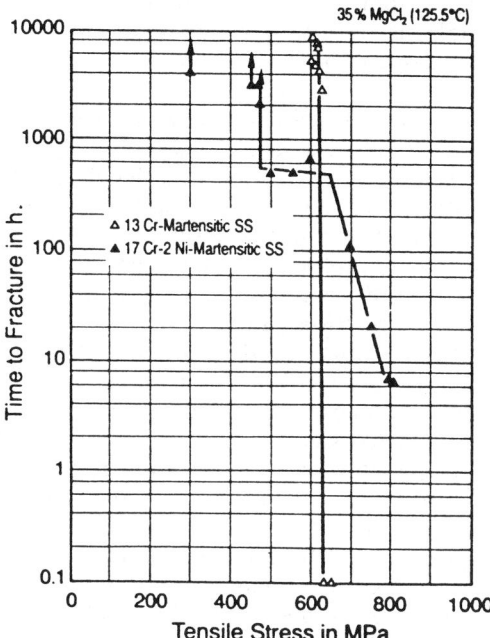

Fig. 8.25 Long-time constant-load tests demonstrating a distinct stress-corrosion cracking threshold stress in the case of a straight 13Cr martensitic SS as opposed to a nickel-bearing SS (after Spaehn[4])

that the allowable hardness of a 12Cr-Mo-V martensitic steel is a function of the tempering temperature for the four-point bend tests they performed in a 0.01 M NaOH/0.1 M NaCl solution at a stress level equal to 90% of yield[19]. Some variation in allowable hardness existed at each tempering temperature, being greater for heat treatments that generated narrow concentration profiles than those that generated wide concentration profiles, although the differences decreased with increasing tempering temperature.

It is not surprising that hardness is important because the mechanical toughness can be expected to decrease with increasing hardness, and the level of residual stress present will also depend on the hardness of the steel, especially for welded components. Thus, the important role of the microstructure in influencing susceptibility to stress-corrosion cracking is consistent with the observation that hardness levels are a good guide to stress-corrosion resistance, but they should not be used universally without due consideration of the specific alloy and the environment in which it is to be used.

Cathodic (Sulphide) Stress-corrosion Cracking

The terminology cathodic (sulphide) stress-corrosion cracking is borrowed from Spaehn's review[4] for the reasons previously mentioned. Generally, cracking of martensitic stainless steels under cathodic conditions will be hydrogen-induced and sulphur or sulphur species, and various other hydrogen recombination poisons, enhance the take-up of hydrogen into the metal, thereby increasing the susceptibility to cracking. Typical of such cracking is that of a pump shaft used in a soot water circuit where the environment was saturated with H_2S and contained solid soot particles[4]. The material was Type 410 stainless steel and had a hardness of >350 H_v. Similarly, a Super 12Cr (X-20Cr-Mo-V 12.1) German steel, which is similar to Type 422 stainless steel but without tungsten present, used in the manufacture of a start-up heater in an ammonia plant failed by hydrogen cracking that was caused by condensing water on the outer tube surface. The crack was in the heat affected zone (HAZ) of a weld where the hardness was >310 H_v.

There is a need for care in welding procedures to prevent hydrogen cracking. The martensitic steels must be allowed to cool to about 80°C after welding to ensure complete transformation to martensite, a mandatory requirement for gaining the necessary toughness on subsequent tempering at 750°C. In the as-quenched state the steel is very susceptible to cathodic stress-corrosion cracking if condensation is allowed to occur on the surface of the steel. Indeed, microsections taken from a similar steel have been known to crack during metallographic observation. Thus, it is extremely important to perform the tempering immediately after the weld has cooled to a sufficiently low temperature to ensure adequate toughness after tempering. The welding and tempering schedule must be well enough defined to avoid both inadequate cooling, and/or excessive delay, prior to tempering. To mitigate the generation of excessively high hardness in welded components, modifications in welding procedures have been developed, such as:

1. austenitic welding by applying preheat to maintain temperatures of about 400°C during welding;
2. martensitic welding with a preheat in the range 100–200°C;
3. partial martensitic welding with a preheat to maintain the temperature in the range 250–400°C during welding.

An alternative to the above is to use more weldable varieties of martensitic steels, such as the low-carbon nickel martensites. These have 12–17% Cr, 3–6% Ni and about 0.05% C. The low-carbon gives better weldability and the high-nickel prevents the formation of delta-ferrite. The development of these steels has been described by Irvine et al.[23,24]. These steels are also susceptible to hydrogen cracking, for example in the petrochemical and gas industries where H_2S in chloride-containing environments may be very damaging. A NACE standard procedure exists that can be used to assess such steels when proposed for operation in very aggressive conditions[25]. Data for a range of steels, including a low-carbon nickel martensitic steel, tested by the NACE procedure indicated no correlation between performance and hardness level[4]. Perhaps of more importance and concern, however, is the observation that cracking occurred at very low stress levels in some cases, e.g. at about 70 MPa sustained stress level in the NACE test. However, if no relevant test data for a specific application are available, it is probable that a choice of the lowest hardness that can be allowed for a particular design situation is likely to provide the greatest resistance to cracking[25–27].

Another aspect of the presence of H_2S in working solutions is that it can reduce the pitting resistance of the steel[28]. The acid conditions within the pits can generate hydrogen. Double tempering treatments are used for some steels but the effects produced are complicated. Whereas resistance to sulphide stress cracking can be increased, the fracture mode changing from intergranular to transgranular, the double tempering impairs pitting resistance and cracks can be initiated at the pits via chloride stress-corrosion cracking.

Ferritic Stainless Steels

The ferritic stainless steels have $Cr\% - 17 \times C\% > 12.5$, so that in cooling from high temperatures they remain completely ferritic, although the formation of austenite is possible in some grades. The main disadvantage of a totally ferritic structure stems from the fact that bcc metals exhibit a ductile to brittle transition with decreasing temperature. The value of the ductile-brittle transition temperature, T_c, is very dependent on the ferrite grain size[29,30]. In welding ferritic stainless steels it is difficult to prevent excessive grain growth in the HAZ adjacent to the welds, and unacceptable low toughness regions can be generated. The ferritic 400 series of alloys tend to have high ductile-brittle transition temperatures[31], i.e. well above room temperature, even before welding. T_c increases with increasing chromium content so that improving the corrosion resistance by increasing the chromium content brings a penalty of increased brittleness[32]. The ferrite grain size alters T_c because the length of a dislocation pile-up possible at a grain boundary increases with increasing grain size. T_c is also dependent upon the flow stress of the material.

Interstitial alloying elements such as carbon and nitrogen will increase the flow stress by locking and by the generation of precipitates. Thus, low carbon and nitrogen levels are advantageous for improving toughness. However, as well as the effect of carbon and nitrogen on toughness, the precipitation of carbides and nitrides in the steels can give sensitization to stress-corrosion cracking because of local depletion of chromium in the matrix adjacent to the precipitates. As precipitation is easier if heterogeneous nucleation is possible, for instance at grain boundaries, the latter can become more prone to chemical attack if they are decorated with precipitates[33-35]. The degree of sensitisation can be reduced by the addition of stabilising elements such as Ti and Nb, but heat treatment at about 800°C is required for complete removal[36] (see Section 1.3).

The advent of vacuum melting and argon-oxygen decarburisation techniques have allowed the production of low-interstitial ferritic stainless steels in recent years. To achieve low ductile-brittle transition temperatures, C + N is kept to less than 100 ppm or to less than 400 ppm if Ti or Nb are present as stabilising elements. The improved mechanical properties are accompanied by better resistance to intergranular corrosion[37]. The low-interstitial Cr-Mo and Cr-Ni-Mo ferritic stainless steels are very resistant to chloride cracking and are used for the manufacture of heat exchangers in the chemical industry[4]. Data exists indicating that the high-strength 18Cr-2Mo steels are resistant to high chloride content oxygenated river water at temperatures up to 130°C[4]. However, cracking in 42% LiCl + thiourea has been reported for high temperature annealed 26Cr-1Mo stainless steel at an applied stress equal to 90% of yield[38]. For material so heat treated, the open circuit potential moved in the active direction relative to mill-annealed material during the test. A prestrain of 5% prior to loading enhanced cracking. The work indicated that susceptibility to stress-corrosion cracking depended upon the inherent corrosion and repassivation rates of the alloy, which could be altered by thermal and by mechanical treatments.

Tests in boiling magnesium chloride solutions have shown that 17Cr ferritic stainless steels exhibit cracking if copper, nickel and/or cobalt are present at greater than certain amounts, and Schmidt and Jarleborg[39] suggest that the total nickel + copper content of 17Cr steels be kept below 0.5%. Small amounts of Ni and Cu can be tolerated by low interstitial ferritic grades in both $MgCl_2$ and $CaCl_2$ solutions, which are both very severe test environments. The data indicate that up to 0.17% Cu and up to 0.6% Ni do not cause susceptibility to cracking in these environments. Between 0.6% Ni and 0.8% Ni the allowable Cu decreases to almost zero for tests in $MgCl_2$ at 140°C but a little Cu can be tolerated and up to 1.2% Ni without cracking occurring in $CaCl_2$ solutions at 130°C[40]. However, as has been pointed out by Staehle[41], misleading indications can sometimes occur from tests in such aggressive environments. The example quoted by Staehle relates to cracking of alloy 600 in high-temperature water, but similar caution is wise with other materials and other environments. Bond and Dundas' data on the effects of nickel, copper and molybdenum on the stress-corrosion cracking of a large series of alloys indicated that cracking occurs when (Ni% + 3 × Cu%) > 1.1%. Molybdenum accentuates the effect of nickel in promoting stress-corrosion cracking. Alloys with up to 5% Mo, but free of nickel, copper or cobalt were highly resistant to stress-corrosion cracking[40,42].

Although cracking can be induced in some ferritic alloys in aggressive solutions, the ferritic grades of stainless steel tend to be more resistant to cracking than the martensitic grades tested in the same solutions. Despite the fact that, as has been already mentioned above, the cracking in Bond and Dundas' work was transgranular in nature, too much emphasis should not be placed on the mode of cracking. For instance, they also refer to work by Streicher indicating that commercial type 430 and 466 stainless steels heat treated at 1095°C then water quenched cracked mainly in a transgranular mode in a $MgCl_2$ solution, but in an intergranular mode in a NaCl solution. The heat treatment had sensitised the steels to intergranular corrosion. Dundas also performed tests on similarly annealed and on welded ferritic steels which indicated susceptibility to intergranular corrosion and cracking at high stress levels in boiling artificial sea water. Annealing at 815°C made the steels immune to cracking.

In any alloy system the four important factors controlling stress-corrosion cracking are: (1) alloy susceptibility; (2) stress level; (3) environment; and (4) electrode potential. The importance of electrode potential should not be overlooked. For instance, Newburg and Uhlig[43] have shown that the susceptibility to cracking of 18Cr alloys of various nickel contents in $MgCl_2$ at 130°C can be altered by polarising the specimen. Cracking was induced by anodic polarisation, the degree of polarisation required depending on the nickel content of the alloy, but the minimum potential for cracking did not vary systematically with increasing nickel content. The experimentally determined minimum potentials for stress-corrosion cracking were more negative than the free corrosion potential for those alloys that cracked under open circuit conditions. An 18Cr–1.5Ni alloy was made to crack in 42% $MgCl_2$ solution at both anodic and cathodic applied potentials, but was immune to cracking in the potential range −650 to 250 mV SHE[44]. Shimoda et al. have demonstrated the adverse effects of increasing the nickel content of a ferritic alloy in the range up to 4% Ni when tested in 42% $MgCl_2$ solution. Thus it was demonstrated that the range of alloys investigated underwent hydrogen cracking after cathodic charging at 25°C and that the cracking of a 1.75Ni alloy could be prevented in slow-strain-rate tests in deaerated 42% $MgCl_2$ at 140°C by cathodic polarisation. Uhlig and coworkers have also demonstrated the susceptibility of ferritic stainless steels to cracking under hydrogen charging conditions[43, 46, 47], with transgranular cracking being the dominant fracture mode,

As has been mentioned previously, deliberate hydrogen charging will not be discussed in this section. The difficulties associated with the interpretation of data when hydrogen cracking versus anodic dissolution are discussed have already been mentioned and are probably of little interest to the practising engineer, although in the fulness of time a detailed understanding of the mechanisms of fracture should assist in alloy development programmes.

The ferritic stainless steels tend to have quite good resistance to pitting. Pits can give stress concentration effects and acidic environments that generate hydrogen as well as the localised corrosion that might in itself be a precursor to cracking by the linking together of pits to form a trench, tunnel etc. If pitting does not occur, some other crack initiation mechanism is required. One possibility is slip step-induced cracking, and for this to occur, the slip step must be large enough to fracture the oxide layer on the steel surface and thereby expose bare metal to the environment. If this is achieved, cracking is likely to be controlled by the magnitude of the local dissolution current density at the exposed bare metal relative to the repassivation rate for the alloy at that location. Locci et al.[38] have obtained slip-step height and stress-corrosion data for a low-interstitial (E-Brite) and a high-interstitial 26Cr–1Mo steel. Slip step heights were similar for both steels but varied significantly with heat treatment and by introducing cold work. The heat-treated states that generated the largest slip-step heights induced greater susceptibility to stress-corrosion cracking. The difference in susceptibility of the two alloys was therefore due to the difference in corrosion and repassivation rates that arose from the different chemical compositions, rather than due to the difference in slip-step heights.

The local dissolution rate, passivation rate, film thickness and mechanical properties of the oxide are obviously important factors when crack initiation is generated by localised plastic deformation. Film-induced cleavage may or may not be an important contributor to the growth of the crack[48] but the nature of the passive film is certain to be of some importance. The increased corrosion resistance of the passive films formed on ferritic stainless steels caused by increasing the chromium content in the alloy arises because there is an increased enhancement of chromium in the film and the films are thinner and more protective. In Ceislak and Duquette's work[49], passivation treatments caused Cr enrichment in the films. At 260°C, thicker films were formed than at lower temperatures. It was observed that chloride ions were not incorporated into the films and it was postulated that the thicker films formed at 260°C might be more defective than those formed at lower temperatures and that chloride ions might attack such defective regions in the film. The thickness of the film might be important if cracking is initiated by localised plastic flow, evidence for which exists for austenitic stainless steels in high temperature water[50, 51].

Another aspect concerning the use of ferritic stainless steels that should be remembered relates to the form of the equilibrium diagram for Fe-Cr alloys and the effect of low temperature annealing treatments (Fig. 8.26). Sigma-phase embrittlement can occur, especially with the higher chromium content alloys. Prolonged use at temperatures greater than 280°C may cause the often referred to 475°C embrittlement, which is due to the decomposition of ferrite into the low Cr α and high Cr α' forms. Providing that such conditioning factors are accounted for, Spaehn[4] claims that the immunity of the low-interstitial Cr-Mo steel to hot chloride and caustic solutions has been proven, as has the immunity of nickel-bearing 'superferritic' steels in industrially important concentrated chloride solutions.

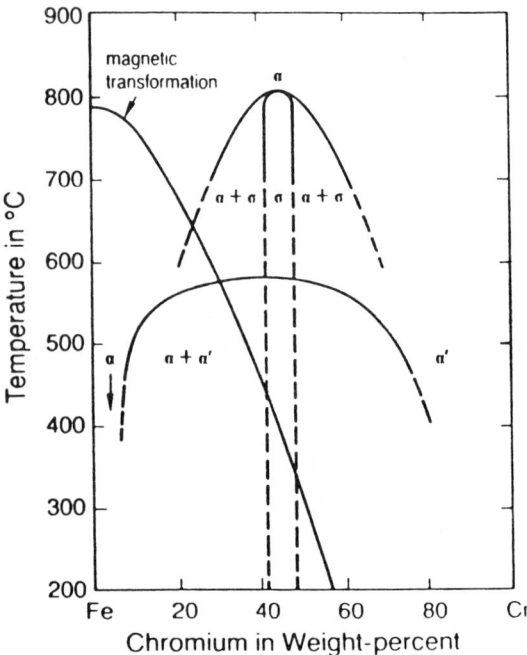

Fig. 8.26 Binary constitution diagram for Fe-Cr alloys (after Spaehn[4])

Duplex Stainless Steels

The development of duplex stainless steels arose from the desire in the chemical industry for alloys with both higher strengths than available with austenitic grades and good corrosion resistance. The alloys developed do have these properties as well as good pitting resistance and high threshold stress values for stress-corrosion cracking. The alloys usually contain between 30 and 70% ferrite. Low ferrite alloys contain 17-21% Cr and 5-12% Ni, whereas the higher ferrite content alloys contain 18.5-26% Cr and 4.5-7% Ni. The proportions of ferrite and austenite present can be altered by heat treatment as well as by changing the chemical composition of the alloy. Moreover, diffusion rates for alloying elements are much higher in ferrite than in austenite. At 700°C, chromium diffuses about 100 times faster in ferrite than in austenite and interstitial alloying elements also diffuse faster in the less close packed bcc lattice than in the close packed austenite. Thus, during heat treatments and during welding, significant partitioning of alloying elements can occur[53]. Below 1 000°C carbides form in the ferrite-austenite grain boundaries and there is a difference in chromium in solid solution for the two phases, the higher chromium content being in the ferrite phase. $M_{23}C_6$ carbides nucleate in the high-chromium ferrite phase and in growing they denude the adjacent area of chromium causing it to transform to austenite. With carbon contents less than 0.03%, there are insufficient carbides precipitated to decorate all of the austenite-ferrite grain boundaries.

Typical alloy compositions are reported by Hochmann et al. (Table 8.2)[52]. Molybdenum or copper is added to some of the alloys.

Edeleanu (54) was the first to record the good stress corrosion resistance of 17Cr ferritic stainless steel compared with that of austenitic stainless steels. He also noted that a duplex stainless steel performed better than an austenitic steel when tested in 42% $MgCl_2$ solution and that cracking of the austenitic stainless steel could be prevented by coupling it to a 17Cr or 20Cr ferritic steel. This was due to the ferritic steel cathodically protecting the austenitic steel. The resistance of duplex steels to stress-corrosion cracking in 200°C water containing 87.5 ppm NaCl increases with increasing ferrite content up to about 40% ferrite[55,56].

The relative merits of duplex steels vis-à-vis austenitic steels depend somewhat upon the test solution. In some cases, little difference exists for tests performed in very concentrated solutions of chlorides, but the duplex stainless steels are generally better than the austenitic grades in more dilute solutions. Data due to Suzuki et al.[57] confirm the maximum resistance to cracking in 42% $MgCl_2$ solution for alloys with about 40% ferrite as does the data of Shimodaira et al.[45]. The data due to the latter authors clearly indicate the relative roles of stress and of the corrosion resistance of the ferrite and austenite phases in stress-corrosion cracking. Austenite grains deform at lower stress levels than ferrite grains. However, as austenite is cathodically protected by the adjacent ferrite, cracks cannot initiate at the slip steps formed in the austenite grains. At higher stress levels, strain incompatibility at the austenite-ferrite grain boundaries can encourage cracks to propagate along those interfaces. At high stress levels, transgranular cracks can propagate through both the ferrite and austenite grains. The plastic

Table 8.2 Typical chemical compositions for duplex stainless steels (after Hochmann et al.[52])

	C	Si	Mn	Ni	Cr	Mo	Cu	% ferrite in the normal use condition
Alloys which are predominantly austenitic:								
URANUS 50 (Creusot-Loire) (AFNOR Z 5 CNDU 21-8)	<0.05	0.5	0.5	7.5	20.5	2.5	1.5	20-35
CF 3A (A C I)	0.03	2.0	1.5	8-12	17-12			0-30
CF 8A (A C I)	0.08	2.0	1.5	8-11	18-21			0-30
0 × 21 H5T (GOST)	<0.08	<0.8	<0.8	5.3	21			~30
0 × 21 H6M2T (GOST)	<0.08	<0.8	<0.8	6.0	21	2.1		30-35
Alloys which are predominantly ferritic								
CD-4M Cu (A C I)	<0.04	<1.0	<1.0	5.5	26	2.0	3.0	65-70
URANUS 55 (Creusot-Loire) (Z 5 CNUD 26-6)	<0.05	0.5	0.5	5.5	26	2.0	3.0	~70
453 S (Avesta-SIS 2324)	0.09			5.0	26	1.5		
3 RE 60 (Sandvik)	<0.03	1.7	1.5	4.7	18.5	2.7		~60
I N 744 × (INCO)	<0.06	0.5	0.5	6.5	26			~50
Remanit 2604 Mo (DEW)	0.1	0.5	1.0	4.5	26	1.5		
Remanit 2604 Mo S (DEW)	<0.06	0.5	1.0	7.0	25	1.5		

deformation of the ferrite phase is quite sensitive to temperature; the yield strength increases markedly with decreasing temperature and twins form more readily the lower the temperature. Twin nucleation requires the formation of partial dislocations which are favoured by low stacking fault energy[58,59] (Section 20.4), so both alloying additions and strain rate can be expected to affect the deformation of the ferrite phase.

In constant-load tests[4] a 22Cr-5Ni-3Mo stainless steel was prevented from cracking by applying a cathodic potential of -130 mV SHE, but the environment for the tests was not specified. Similarly, an 18Cr-1.5Ni steel was prevented from cracking by applying potentials of about -250 mV SHE or less, down to -650 mV, for tests in 42% $MgCl_2$ at 143°C.

As with all stainless steels, precipitation of carbides is likely to impair stress-corrosion resistance. Sigma phase formation by extended heat treatments at 800°C lowered the threshold stress for stress-corrosion cracking of a 22Cr-5Ni-3Mo steel[4] and Spaehn claims that inadequate welding procedures may generate an even greater deterioration in resistance. TIG welding causes a larger reduction in threshold stress than MMA, presumably because of the large difference in heat input for the two processes. The higher heat input for MMA will allow retransformation of delta ferrite into austenite during cooling. Residual stresses, caused either by welding or by subsequent grinding, are also likely to impair resistance to stress-corrosion cracking. However, providing that the welding is performed carefully and the sections are not too large, the resistance to stress-corrosion cracking can be maintained.

Duplex stainless steels can undergo sulphide stress-corrosion cracking, especially if sigma phase is formed and if carbides and/or nitrides are nucleated in the ferrite–austenite grain boundaries. The effect of nitrogen content is complex. Whereas very low nitrogen contents are beneficial, intermediate nitrogen contents cause Cr_2N to precipitate, which can denude the surrounding region of chromium. High-nitrogen steels with about 1.4% nitrogen have good resistance to cracking[60], and this appears to result from an increased resistance of the austenite phase containing dissolved nitrogen.

The resistance to cracking of duplex stainless steels varies with temperature, being less at 60-90°C than at RT or at 150°C[4]. The detailed mechanism of cracking is complex because it involves an interaction between the different mechanical and electrochemical responses of the ferrite and the austenite phases. The relative roles of chloride and hydrogen in cracking are also not yet completely clear[61], although a chloride induced stress-corrosion cracking mechanism for the cracking of a duplex stainless steel in H_2S-saturated NaCl (4.3 M) at temperatures from ambient to 60°C seems to be favoured[61]. However, hydrogen cracking of the ferrite may occur in acid solutions at negative potentials at temperatures not too far removed from room temperature. At elevated temperatures, a critical H_2S level exists above which cracking will occur. No cracking occurred in a 25 Cr duplex stainless steel stressed at 90-95% of yield after 1 000 h exposure to 0.04% H_2S at temperatures up to 204°C. With 1% H_2S present, cracking only occurred at 260°C, whereas cracking occurred in the temperature range 25-204°C with 10% H_2S present, although the specimens exposed at 90°C, 232°C and 260°C did not crack.

The basic guidelines for preventing cracking would seem to be to operate at minimum stress levels, at as low an H_2S concentration as possible and to make sure that welding procedures are adequately specified and followed. Furthermore, extensive periods of operation at temperatures that might cause sigma phase formation should be avoided.

Austenitic Stainless Steels

The austenitic grades of stainless steel of chemical compositions comparable with the AISI 300 series alloys are probably the most extensively used stainless steels. They were originally based on an 18Cr-8Ni alloy which corresponds to the cheapest Cr-Ni alloy that will be almost completely austenitic, nickel being a more expensive metal than chromiumn. This composition corresponds to the upper end of the 301 and the lower end of the 304 specifications. Additional alloying can be made to improve corrosion resistance, mechanical properties, weldability etc (Section 3.3). An excellent summary of the possibilities available is given in Fig. 8.27, due to Sedriks[5]. The Schaeffler diagram[62,63] (Fig. 8.28), though sometimes criticised with respect to its accuracy and general applicability because it was developed for

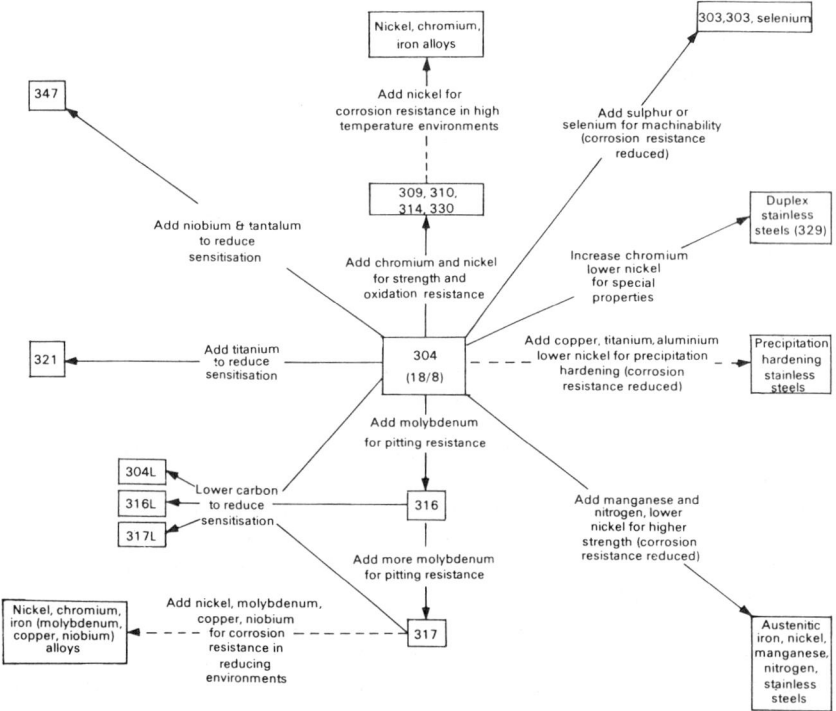

Fig. 8.27 Some compositional modifications of 18/8 austenitic stainless steel to produce special properties. Dashed lines show compositional links to other alloy systems (after Sedriks[5])

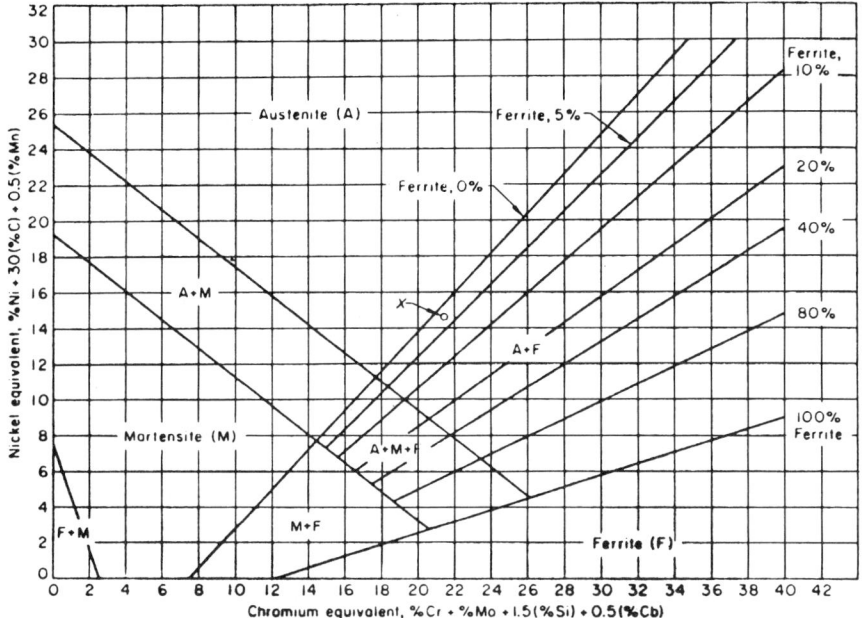

Fig. 8.28 Schaeffler diagram (after Schaeffler[62] and Schneider[63] and Page[64])

use in relation to weld metal, is another good guide to the range of stainless steels available; a modified diagram is available to account for the effect of nitrogen.

The austenitic alloys have high ductilities owing to their fcc structure and low stacking fault energy and good mechanical properties when used for cryogenic applications. As well as delaying the onset of necking by promoting a high work hardening rate, the low stacking fault energy of many of the alloys encourages large slip step generation where the planar slip planes intersect a surface. As corrosion resistance depends upon the existence of a thin, high-chromium content passive oxide layer on the surface of the metal, the large slip step heights formed by plastic deformation are a disadvantage because they disrupt the passive layer.

Two other aspects of the metallurgy of the austenitic stainless steels are important. First, some of the alloys can be unstable when cold worked and transformation from austenite to martensite can occur, though the extent to which this occurs varies from alloy to alloy. Second, chromium is a strong carbide former. Thus, low temperature annealing can cause carbides to precipitate in the matrix and in the grain boundaries. As has been mentioned above, heterogeneous nucleation is easier than homogeneous nucleation, so grain boundaries are the more favoured sites for precipitation. The associated depletion of chromium in the grain boundary regions can be extensive enough to lower the local chromium content to less than the 12% needed to provide good corrosion resistance. Molybdenum can be added to the 18-8 alloy to improve resistance to pitting and general corrosion in halide-containing environments, giving the extensively used type 316 grade of steel.

8:68 STRESS-CORROSION CRACKING OF STAINLESS STEELS

The carbide precipitation that leads to 'sensitised' grain boundary regions can be minimised by reducing the carbon to 0.03% or less but this increases the cost and reduces the strength. The alternative is to add stabilising elements such as titanium or niobium, which are stronger carbide formers than chromium. There are numerous texts that describe the metallurgy of

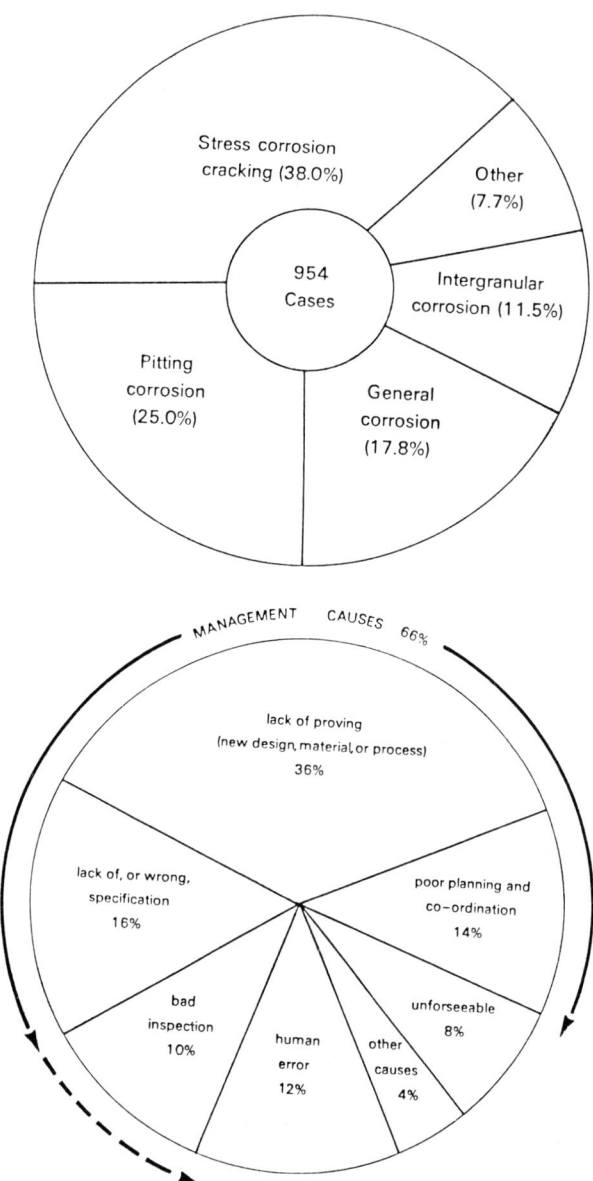

Fig. 8.29 Causes of failure and failure types investigated by a large chemical company in the USA (after Page[65])

the austenitic stainless steels that the reader may consult for a more detailed description of these steels.

With respect to their propensity to stress-corrosion cracking, a useful introduction to the various problems that can arise is given in the monograph by Page[64] and in a subsequent paper[65] on the handling and fabrication of 304 and 316 stainless steels for use in the food industry. Figure 8.29, taken from Reference 65, highlights the regular occurrence of stress-corrosion cracking in such applications.

The fracture mode of stress-corrosion cracks in austenitic stainless steels can be transgranular, intergranular or a mixture of both. One of the earliest environments found to cause problems was solutions containing chlorides or other halides and the data due to Copson (Fig. 8.30) is very informative. The test solution for that data was magnesium chloride at 154°C; the alloys contained 18-20% chromium with various amounts of nickel and the results shown in Fig. 8.30 indicate that the cheapest alloy with a composition of approximately 18Cr-8Ni has the least resistance to cracking in this environment.

It is impossible in this chapter to do justice to the extensive amount of research that has been performed on the stress corrosion of stainless steels, and probably more papers have been written concerning the austenitic grades than on all of the other grades together. Numerous reviews have been published in recent years some of which are listed in references 66-71, but other equally useful reviews are available.

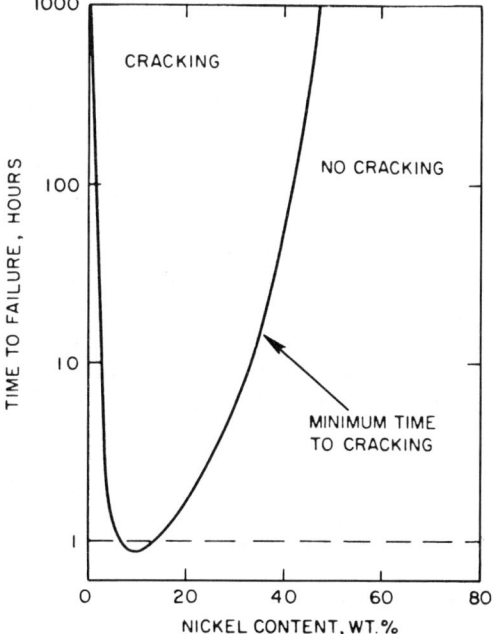

Fig. 8.30 Effect of nickel content on the susceptibility to stress-corrosion cracking of stainless steel wires containing 18-20% chromium in a magnesium chloride solution boiling at 154°C (after Copson[72])

Austenitic stainless steels will exhibit stress-corrosion cracking in hot aqueous chloride solutions, in acid chloride containing solutions at room temperature, in hot caustic solutions and in high-temperature high-pressure oxygenated water.

Copson[72] studied the effect of adding various major alloying elements to a basic 18Cr-8Ni alloy and showed that increased Cr (above 15%) and Mn (from 2 to 4%) were deleterious whereas Ni and Si were beneficial with respect to time to failure in boiling magnesium chloride. The effects of these major alloying elements correlates reasonably well with the way they affect the stacking fault energy[73], which has a minimum at about 18Cr for any given nickel content. However, silicon improves resistance to cracking even though it reduces the stacking fault energy. Silicon is considered to improve resistance by forming a magnesium silicate on the surface of the specimen[69], but there is evidence that silicon is only beneficial with steels containing more than about 0.04% carbon. The effectiveness of silicon disappears if carbide precipitation occurred. Silicon is a ferrite stabiliser, so 3.5-5.5% silicon in an 18-8 steel generates a duplex structure with the improved resistance to stress corrosion for these steels that has been discussed previously[74]. Increasing Cr and Ni give improved resistance to stress-corrosion cracking in caustic solutions[75]. For greater resistance to caustic cracking it is necessary to employ the higher nickel content alloys or even pure nickel[76], but Inconel 600 can be more susceptible than type 316 stainless steel in dilute caustic solutions[77].

Thus, it is difficult, as always, in discussing stress-corrosion cracking, to try to generalise responses. Sedriks[66] has examined the effect of many different alloying elements on the susceptibility of austenitic stainless steels to cracking in chloride-containing solutions (Fig. 8.31) but as is explained in some detail by Hanninen[69], the effects of minor additions are complex and very system dependent. The effect of an alloying addition can vary within the composition range considered and with the environment used for the test. For instance, although molybdenum at the level of 2-4% is generally added to improve pitting and corrosion resistance to chlorides, small additions of molybdenum in the range up to 0.26% markedly reduce the time to failure for U-bend specimens of an 18-8 steel tested in boiling $MgCl_2$ solutions. Also, low phosphorus and nitrogen contents appear beneficial. However, alloys with low nitrogen and low carbon contents cracked in $MgCl_2$ solutions, and the susceptibility to cracking was removed by adding more carbon to the alloy. The no-cracking domain on a log(nitrogen %) versus log (carbon %) plot lay above a line extending from about 0.002% N/0.008% C to about 0.2% N/0.2% C. Of course high-carbon alloys will be more easily sensitised, so the response of the alloys will also depend upon their heat treatment. However, giving a heat treatment in the sensitising temperature range does not necessarily imply increased susceptibility to stress-corrosion cracking. For instance, the resistance of Inconel 600 in caustic solutions was increased by such heat treatments[77,78].

The effects of minor alloying elements such as phosphorus and nitrogen are also complex. Generally, additions of P and N increase susceptibility to stress-corrosion cracking and their effect can be increased by low-temperature ageing[79]. The presence of the alloying elements can influence slip planarity via their effect on stacking fault energy or on short range

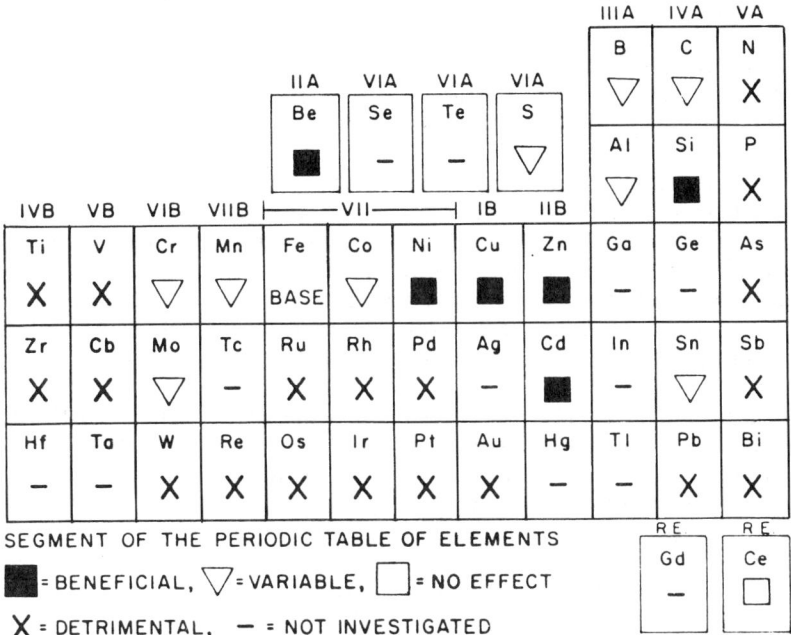

Fig. 8.31 Effect of element shown on resistance of austenitic stainless steels to stress-corrosion cracking in chloride solutions (after Sedriks[5])

order[80], as well as influencing the anodic and cathodic reaction rates in a given solution[81]. Alloying elements present in greater quantities, such as molybdenum, can be beneficial or deleterious depending on the amount present. As has already been stated, Mo is deleterious at the lower composition range in $MgCl_2$[69] and promotes intergranular stress-corrosion cracking[82], but Mo contents greater than 2% have been found to increase the stress-corrosion cracking resistance of a 20Cr–20Ni steel[83]. The stabilising elements Ti and Nb can also decrease resistance to stress-corrosion cracking[84]. Platinum additions have been reported to enhance stress-corrosion cracking, presumably due to enhanced reduction kinetics of hydrogen[85, 86]. The effect of sulphur in the steel is exaggerated if it is present as sulphide-containing inclusions because, on their dissolution, S^-, H_2S and polythionic acid etc. may be formed with ensuing dramatic changes in local chemistry.

The behaviour of austenitic stainless steels in caustic solutions has received less attention than cracking in chloride environments. Transgranular cracking has been reported for low-carbon (< 0.05%) steels in caustic solutions, whereas higher carbon content alloys cracked intergranularly[87]. Wilson and Aspen showed that resistance to cracking was not decreased by sensitisation heat treatments[78]. Type 316 stainless steel has been shown to be more susceptible to cracking in caustic than type 304[88].

It is clear from the above comments that no general conclusions can be drawn concerning the effects on stress-corrosion cracking of various alloying elements added to the steels. Thus, specific data relating to the actual alloy and the environment in which it will be used is almost certainly required in the long term irrespective of which rapid sorting experiments are initially used in the alloy proving and selection stage of design. Nevertheless, some general recommendations have been made in the reviews mentioned above[68,69]. For instance, it has been suggested that alloying elements such as N, P, S, Sb and Sn be minimised, presumably because they tend to segregate to the grain boundaries, and that low-carbon grades will be beneficial for stress-corrosion cracking resistance. Cr concentrations near to 18% should be avoided if possible and if Mo is present the percentage of Cr to be avoided should be reduced by an amount equal to the Mo percentage contained in the alloy. Ni appears beneficial in the range 10–65%, though the high-Ni alloys are not really stainless steels as defined in the context of this section. Si at greater than 4% and Ti at greater than 2% also seem beneficial.

In addition to the alloy compositions being of importance with regard to susceptibility to stress-corrosion cracking, the resistance of the alloy can be altered by microstructural factors. Hanninen has reviewed the available literature quite thoroughly[69] and has concluded that a fine grain size is likely to be beneficial. Strain imposed prior to use tends to be deleterious because deformed material usually acts anodic with respect to unstrained material and because the introduction of plastic deformation may also

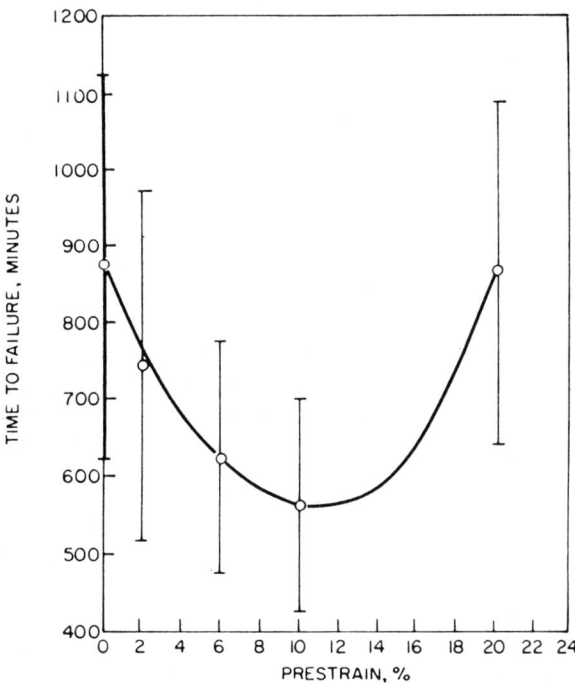

Fig. 8.32 Effect of prestrain on the time to failure of type 310 stainless steel exposed to a magnesium chloride solution boiling at 154°C and stressed at 90% of the yield stress (after Sedriks[5])

introduce high residual stress levels. Cochran and Staehle[89] have shown that the resistance to stress-corrosion cracking of Type 310 stainless steel in $MgCl_2$ at 154°C is least when the material is prestrained about 10% and somewhat similar data exist for Type 321 stainless steel[5] and 316 steel[69] (Figs. 8.32 and 8.33). For 316, Hanninen points out that even small strains can be very dangerous because cracks of about 0.1 mm deep were observed in all of the prestrained specimens that he tested in $MgCl_2$ and because the low dislocation densities associated with small levels of strain favour crack propagation.

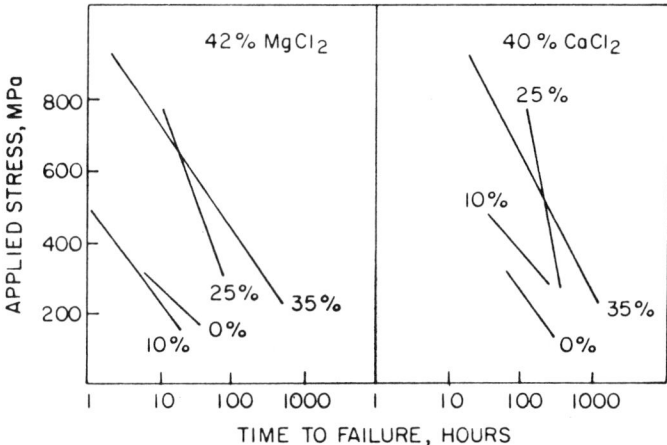

Fig. 8.33 Effect of cold work (%) on the susceptibility to cracking of type 321 stainless steel in boiling magnesium chloride and calcium chloride solutions (after Sedriks[5])

The mode of deformation is important in that dislocation cell structures are less prone to initiate cracks than pile-ups associated with planar slip. Additionally, the tendency for martensite to be formed during deformation affects resistance to stress-corrosion cracking, and this has been used to explain the greater resistance of Type 304 compared with Type 316 steel at prestrains greater than 20%. However, material transformed to martensite can dissolve selectively and a potential difference of about 100 mV has been recorded between martensite and austenite in 1 N H_2SO_4 solution[90]. The selective corrosion can inhibit stress-corrosion cracking because sharp crack morphologies are prevented from forming and such selective dissolution can occur in the absence of tensile stress. Indeed, such selective dissolution has been recorded under compressive loading conditions.

Times to failure for various stainless steels tested in $MgCl_2$ have been shown to increase with increasing proportions of martensite present[91,92]. Perhaps the role of martensite under anodic dissolution conditions is comparable to that of ferrite in duplex stainless steels where the enhanced dissolution of one phase prevents crack initiation in the other. There is, of course, another aspect of martensitic transformation that should be mentioned, i.e. the transformation of austenite to martensite either in the bulk material or at a growing crack tip that can give increased susceptibility to hydrogen-assisted cracking[93]. It appears probable that the susceptibility to cracking in such cases is related to the ease of forming martensite, i.e. the stability of the austenite.

It has also been noticed[94] that thicker corrosion films form on the martensite phase in cold worked steels than on the untransformed matrix, and thicker films can be more brittle and aid crack initiation[50,51,94].

Sensitisation

The largest change in microstructure experienced in austenitic stainless steels is that caused by chromium carbide precipitation in the grain boundaries and the associated chromium depletion in that region giving rise to sensitisation of the steel. This produces both mechanical and electrochemical differences between the matrix and the grain boundaries. The mechanical effect of the carbides is clearly illustrated during low-temperature (less than about 150°C) fracture of sensitised austenitic steels in inert environments when grain boundary separation by microvoid coalescence can be observed. The electrochemical effects are highlighted by the almost featureless intergranular facets formed by intergranular stress corrosion (Fig. 8.34). Significant degrees of chromium depletion occur at the grain boundaries and quantitative models have been developed to estimate the magnitude of the depletion[95] as well as various procedures for experimentally estimating the extent of sensitisation[96-98].

There are four possible contributions to intergranular fracture that can arise from sensitisation. In addition to the previously mentioned effect on

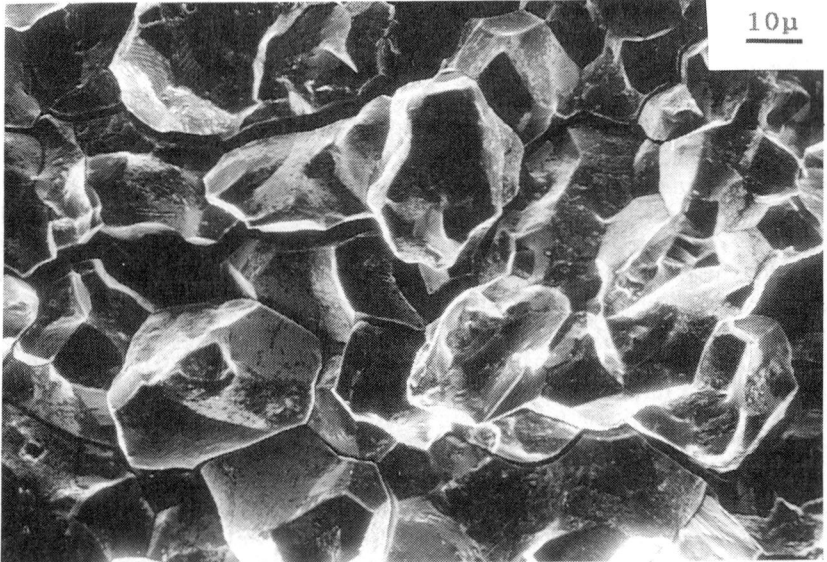

Fig. 8.34 Intergranular cracking of sensitised type 316 stainless steel tested in air-saturated water at 75°C containing 5 ppm chloride

mechanical properties of the grain boundaries compared with the matrix, there are the possibilities that the carbides may be noble with respect to the surrounding material, that the denuded chromium region will be preferentially attacked and that the ageing that caused the carbide precipitation also generated segregation of other impurities to the grain boundaries which may or may not have been incorporated into the precipitates. The situation is further complicated by the fact that steels with discontinuous grain boundary carbides can behave very differently from those with a continuous carbide layer at the grain boundary and also because the chromium depletion at the grain boundary can be partially rectified by extended thermal treatments. This is the basis of the commercially accepted thermal stabilization treatment that was developed for alloy 600 steam generator tubes[99].

Cold work prior to sensitisation heat treatments can enhance the extent of grain boundary corrosion in a given solution although large levels of pre-deformation can induce intragranular precipitation and a corresponding reduction of attack at the grain boundaries. It is important to realise that although an alloy may be sensitised enough to exhibit enhanced grain boundary corrosion in the absence of stress, it may still fail by transgranular cracking in hot chloride solutions[69]. The cracking morphology is influenced by the electrode potential of the metal environment system or by an applied potential.

In addition to the sensitisation that occurs in the temperature range 525°C to 850°C in many stainless steels, it is also possible for sensitisation to develop after long periods of use at lower temperatures. Such low temperature sensitisation is thought to be promoted by pre-seeding of the grain boundaries[100]. Thus, the initial annealing procedure applied to a component may be very important with respect to its service life.

Sensitised structures can undergo intergranular (IG) corrosion in the absence of stress and/or stress-corrosion cracking, and tests aimed at assessing IG attack do not necessarily give a guide to cracking resistance. This is not too surprising considering the important role of stress in stress-corrosion cracking. The effects of carbide precipitation are complicated both by the morphologies of carbides that can be generated and also by the effects due to trace impurities such as phosphorus. Such impurities can be removed from the grain boundaries by dissolution in the carbide[101]. The specific heat treatment schedule or operating temperature can therefore have a major influence on corrosion resistance and stress-corrosion cracking because of partitioning effects and the precise chemical compositions of the inclusions and precipitates present.

Cracking in High Temperature Water

Much of the recent research on stress-corrosion cracking of austenitic stainless steels has been stimulated by their use in nuclear reactor coolant circuits. The occurrence of stress-corrosion cracking in boiling water reactors (BWR) has been documented by Fox[102]. A major cause for concern was the pipe cracking that occurred in the sensitised HAZ of the Type 304 pipework, which is reported to have been responsible for about 3% of all outages of more than 100 h from the period January 1971 to June 1977.

The problem of pipe cracking in high temperature water was soon established as being due to the presence of oxygen in the water which raises the open circuit electrode potential of the steel quite rapidly in the range of dissolved oxygen contents from 10 to 500 ppb[103]. As BWRs typically operate with about 200 ppb oxygen to achieve adequate corrosion control, it was possible for the rest potential of the stainless steel to be high enough to induce cracking. Experimental studies soon demonstrated that high oxygen contents in high-temperature water produced cracking in the sensitised HAZs and that experimentally there was an equivalence between dissolved oxygen in high-temperature water and applied potential for low-oxygen content water in slow-strain-rate tests.

Cracking can be initiated or arrested by appropriate electrochemical potential control as well as by adjusting the level of dissolved oxygen in the water[104,105]. The research has led to some changes in water chemistry control in reactor circuits and to the establishment of potential monitoring so that recorded potentials of less than 0.23 V SHE can be maintained during working, wherever this is possible. The cracking of austenitic stainless steels in high-temperature water has been comprehensively reviewed by Szklarska-Smialowska and Cragnolino[106].

The main concern is the intergranular stress-corrosion cracking that occurs in sensitised material, but transgranular cracking can also occur. For example, Type 304 stainless steel which has been cold worked more than 30% cracks transgranularly in oxygenated water at 288°C[107] and in acid sulphate solutions at anodic polarisation potentials[108,109]. Cold work in the range 5–20% prior to sensitisation heat treatment enhanced susceptibility[107,110,111] whereas deformation in the range 20–30% increased the time to failure.

Various techniques are available for assessing susceptibility to stress-corrosion cracking, e.g. methods based on constant deflection, constant load, cyclic load, continuously increasing load and continuously increasing strain. Each type of test has certain advantages and disadvantages that relate mainly to specimen size and shape requirements and to the time necessary for the test. Times for constant-load and for constant-deflection tests can be very long because cracks need to initiate, and the process is lengthy and characterised by a lot of scatter in the results. The various test methods available do not necessarily give identical rankings for different materials or exactly the same conclusions[112]. For constant-load tests, the yield strength may have to be exceeded by up to 50% or more to achieve acceptably short incubation times for cracking so that the usefulness of the data for life prediction estimates is sometimes debatable.

Similarly, although the slow-strain-rate procedure is now extensively used to investigate stress-corrosion cracking because of the relatively short test times involved and the relative ease of controlling both the environment and the electrode potential during the test, the fact that very large strains are imposed also brings difficulties of interpretation and limits the usefulness of the data for life prediction purposes. The large strains imposed are particularly inconvenient when testing austenitic stainless steels that deform by planar slip because large slip steps form on the specimen surface and mechanically-induced short transgranular cracks are initiated at the slip bands. Such cracks can be distinguished from those involving environment-

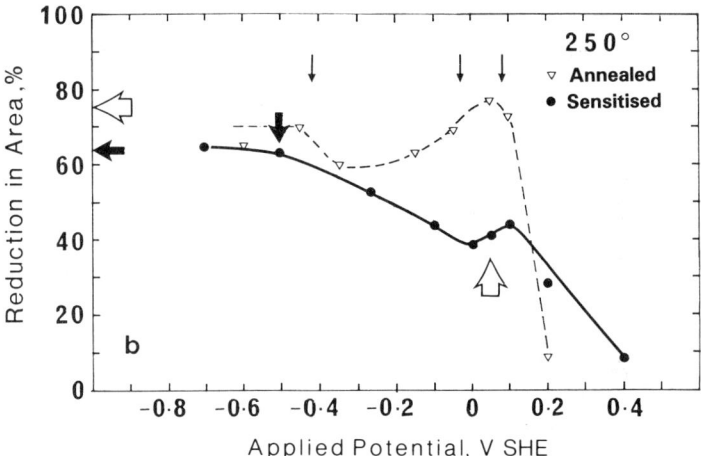

Fig. 8.35 Reduction in area versus applied potential for 316 stainless steel, in 5 ppm chloride content water at 250°C. Horizontal arrows indicate RA% for tests in argon gas; open-annealed, filled-sensitised. Vertical downward pointing arrows on the annealed curve indicate cracking–non-cracking boundaries. Full downward pointing arrow on the sensitised curve indicates commencement of cracking. Open upward pointing arrow on the sensitised curve indicates transition from transgranular to intergranular cracking

assisted cracking, but even the environment-assisted cracks can sometimes be quite short and may only be initiated at quite large strains.

The complicated nature of the response of Type 316 stainless steel to applied potential is illustrated by the data shown in Figs. 8.35 and 8.36, which refer to tests on annealed and on sensitised 316 steel in 5 ppm chloride content water having less than 5 ppb dissolved oxygen but with various applied potentials[50,51]. Figures 8.35 and 8.36 show that the reductions in area for the annealed material were greater at any given potential than for sensitised material. The cracking in the annealed material was transgranular at all potentials but ranges of potential exist where no cracking was observed. The short transgranular cracks that formed at the intermediate potentials required large strains to initiate them and were generated by fracture of the thick oxide layer formed at those potentials. The sensitised material exhibited short transgranular cracking at low potentials, intergranular cracking at high potentials and mixed mode cracking at intermediate potentials.

The strains needed to initiate cracks in both the annealed and the sensitised materials were obtained using tapered slow-strain-rate specimens and the data are given in Fig. 8.36. As can be seen, there is little temperature dependence of the strain needed to initiate cracks in sensitised material whereas the annealed material was most susceptible to cracking at about 250°C. These results indicate the complicated response of Type 316 stainless steel to applied potential and demonstrate that, even though environmentally-assisted cracking may be generated by severe test methods, in this case the slow-strain-rate test, the results obtained must be used with care. For instance, the cracking of the annealed material at low potentials

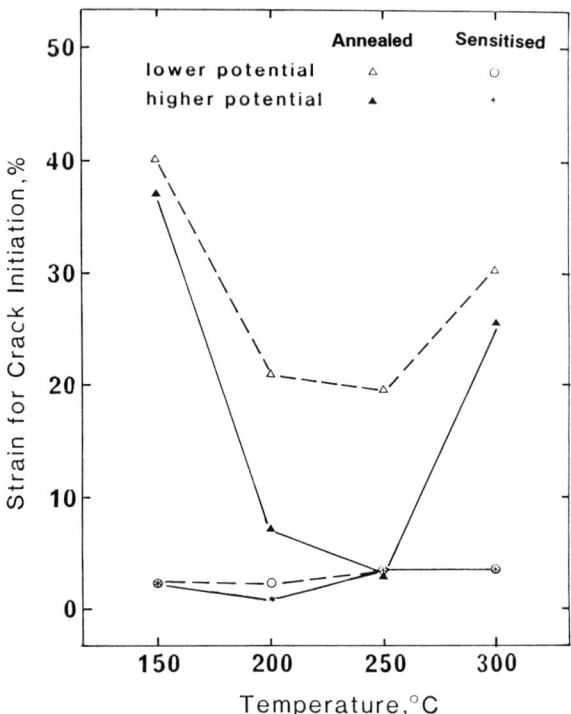

Fig. 8.36 Minimum strains for initiating stress-corrosion cracks in annealed and in sensitised 316 during slow strain rate tests in 5 ppm chloride content water

(implying cracking in very low oxygen content water) only occurs after very large strains are imposed and such large strains will not be experienced by components in high-temperature water systems. Also, the very high potentials used for some of the tests are more positive than the corrosion potentials likely to be achieved in the oxygenated waters experienced by industrial systems. Nevertheless, it is clear that the surface condition of the steel is important in relation to its susceptibility to stress-corrosion cracking.

Steels with thick oxide layers on top of a heavily sensitised surface microstructure have been shown to be very susceptible to cracking in oxygenated water even in the temperature range 50–100°C[113]. The effect of an oxide film promoting easy crack initiation was reported by Berry et al.[114] and rough surfaces or inclusions emerging perpendicular to the surface also facilitate easy crack initiation. Figure 8.37 shows cracking which has initiated in the threaded end of a slow-strain-rate specimen of Type 304L (low-carbon) steel when tested in low-oxygen high-temperature water but under applied potential control. Cracks initiated and grew in the threaded region rather than in the narrower gauge length because sulphur-containing inclusions outcropped at the machined threads for specimens cut with their long axis parallel to the rolling direction. Obviously, dissolution of the inclusion created a crevice and generated a locally very aggressive environment.

Fig. 8.37 Cracks originating at sulphide-containing inclusions that outcrop in threaded end of a slow strain rate specimen tested at 250°C at 0 mV SHE, in lithiated low oxygen content water doped with 100 ppm chloride

Practical Solutions to Stress Corrosion Problems

The requirements for stress-corrosion cracking to occur are a susceptible alloy, the presence of stress, an appropriate environment and a particular range of potentials. Cracking can be prevented by altering one or more of these. The general comments made by Hines[115] are still valid and worth reiterating with some minor supplementation in the light of more recent data.

For aggressive conditions it may be advantageous to use one of the many high-nickel alloys available. The high cost of such alloys may be economically beneficial although some of the alloys are subject to stress-corrosion cracking, but in a more restricted range of environments than the lower nickel alloys that they would replace.

The easiest approach to a solution is likely to be that of altering the environment or to reduce the residual stress levels in the component by appropriate heat treatment providing the latter does not induce deleterious changes in microstructure (sensitisation). However, the aggressive environment may have arisen from a concentration mechanism, and in such cases small changes in the bulk environment are unlikely to be an effective solution to the problem. Care should be taken with welding procedures, or new welding procedures should be devised. An example of the latter is the heat-sink welding that has been successfully applied to 304 tubing in BWR cooling circuits.

As previously mentioned by Hines[115], two points deserve particular mention. First, local concentrations are most likely at 'hot spots' at which the heat transfer conditions are particularly severe, as in blind crevices or gaps in scale in heat exchangers. Excessive scaling is undesirable as partial or complete blockage of tube can lead to particularly severe conditions. The probability of cracking is also reduced by reducing the temperature, particularly if the maximum attainable temperature is reduced. Failures are relatively rare below about 60°C although recent instances of cracking of Type 304 and Type 316 steels in indoor swimming pools have occurred[116], indicating that the 60°C lower limit cannot be universally applied. Also chloride cracking can occur in ambient coastal environments[117,118], with strong indications of the deleterious effect of surface contamination by iron with stress-corrosion cracks initiating at rust spots, although other reasons for crack initiation are possible[119]. This observation leads directly to Hines' second set of general comments, discussed below.

Accidental contamination during erection or repair of steam or hot water systems is potentially dangerous, and careful flushing is desirable. It is also desirable that lagged pipelines be waterproofed if there is any danger of their being wetted by atmospheric moisture or by spillage. Where spillage is a possibility, vessels should not rest directly on concrete. In the last two cases, the use of 'pure' materials for insulation of concrete may reduce the risk of failure, but this precaution is of little value if spillage of process liquors containing chlorides or caustic is possible. It is important to recognise, however, that methods of controlling stress-corrosion cracking that depend on maintaining precise environmental conditions (concentration of chloride, caustic, oxygen or inhibitors, or temperature or pressure) are inherently dangerous, since cracks may be formed very rapidly if control is lost; it is essential therefore to ensure that the control system is effective.

On fully softened material, the time to failure may be considerably increased by a reduction in stress. However, unless a stress relieving or softening treatment is the last stage in fabrication it is probably impossible to avoid the occurrence of some regions where local stresses of yield point magnitude exist, and also of some regions which were cold worked during fabrication, the two classes of region not being necessarily coincident. Such regions occur near welds and in regions where hot or cold forming was necessary; they also arise through accidental dents, stamped identification marks or sheared edges, or through grinding to remove surface defects. Failures have resulted from all these causes. These areas constitute weak spots, which may be particularly dangerous if they coincide with local hot spots, as at the tube/tube-plate joints at the hot end of a heat exchanger.

Stress relief is of little practical value as a means of preventing stress-corrosion cracking in austenitic steels, as cracking occurs at quite low stress levels even in fully softened material and it is difficult to ensure that stresses are reduced to a safe level in a real structure. The technique can however be useful in small items but, even in this case, phase changes which reduce stress-corrosion resistance or have other deleterious effects can occur at the stress relieving temperature.

Cathodic protection has been successfully used in favourable circumstances, but is difficult to apply effectively to complex structures as cracking tends to occur at those points where it is most difficult to ensure that a protective potential is maintained. Anodic protection has been used effectively in low-chloride environments, and this appears a sounder approach if it is certain that pitting conditions will not occur.

J. CONGLETON

REFERENCES

1. Barker, R., *Metallurgia*, **76**, 49 (1967)
2. Truman, J. E., *Brit. Corr. J.*, **11**, 92 (1976)
3. Truman, J. E., *Stress Corrosion Cracking and Hydrogen Embrittlement of Iron Base Alloys*, NACE 5, Houston, Texas, (eds R. W. Staehle, J. Hochmann, R. D. McCright, and J. E. Slater) p. 111 (1977)
4. Spaehn, H., *Environment-Induced Cracking of Metals*, NACE 10, Houston, Texas, (eds R. P. Gangloff and M. B. Ives, p. 449 (1990)
5. Sedriks, A. J., *Corrosion of Stainless Steels*, John Wiley & Sons, New York (1979)
6. Fontana, M. G., *Corrosion Engineering*, 3rd edition, McGraw-Hill, New York (1986)
7. Phelps, E. H., and Loginow, A. W., *Corrosion*, **16**, 325t (1960)
8. El-Sayed, H. A. and Gouda, V. K., *Corrosion Prev. and Control*, **12**, 142 (1986)
9. Truman, J. E., Perry, R. and Chapman, G. N., *Jour. Iron and Steel Inst.*, **202**, 745 (1964)
10. Wilde, B. E., *Corrosion*, **27**, 326 (1971)
11. Phelps, E. H., *Metals Engineering Q.*, **5**, 44 (1973)
12. Hewitt, P. and Hockenhull, B. S., *Corrosion Sci.*, **16**, 47 (1976)
13. Durkin, A. E., *Metal Progr.*, **64**, (1) 72 (1953)
14. Parkins, R. N., *Brit. Corr. J.*, **7**, 15 (1972)
15. Doig, P. and Edington, J. W., *Proc. Roy. Soc.*, **A339**, 37 (1974)
16. Frankenthal, R. P. and Pickering, H. W., *Jour. Electrochem. Soc.*, **120**, 23 (1973)
17. Pande, C. S., Suenaga, M., Vyas, B. and Isaacs, H. S., *Scripta Met.*, **11**, 681 (1977)
18. Poulson, B., *Corrosion Science*, **18**, 371 (1978)
19. Doig, P., Chastell, D. J. and Flewitt, P. E. J., *Metall. Trans. A.*, **13A**, 913 (1982)
20. Bavarian, B., Szklarska-Smialowska, Z. and MacDonald, D. D., *Corrosion*, **38**, 604 (1982)
21. Speidel, M. O., *Corrosion Problems in Energy Conversion and Generation*, (Pennington NJ: The Electrochemical Soc.), 359 (1974)
22. Speidel, M. O., *Metall. Trans. A.*, **12A**, 799 (1981)
23. Irvine, K. J., Llewellyn, D. T., and Pickering, F. B., *Jour. Iron and Steel Inst.*, **192**, 218 (1959)
24. Irvine, K. J., Crowe, D. J. and Pickering, F. B., *Jour. Iron and Steel Inst.*, **195**, 386 (1960)
25. Truman, J. E., *Metals and Materials*, **2**, 208 (1968)
26. Gooch, T. G., *Welding J. Res. Suppl.*, **53**, 287 (1974)
27. Fujii, C. T., *Stress Corrosion-New Approaches*, ASTM STP 610, Philadelphia, PA, 213 (1976)
28. Yoshino, Y. and Ikegaya, A., *Corrosion*, **41**, 105 (1985)
29. Petch, N. J., *Phil. Mag.*, **3** (34), 1089 (1958)
30. Heslop, J. and Petch, N. J., *Phil. Mag.*, **3** (34), 1128 (1958)
31. *Metals Handbook*, Desk Edition, 30. 33. Publ. ASM Metals Park, Ohio (1985)
32. Krivobok, V. N., *Trans ASM*, **23**, 1 (1935)
33. Demo, J. J., *Corrosion* **27**, 531 (1971)
34. Bond, A. P. and Lizlovs, E. A., *Jour. Electrochem. Soc.*, **116**, 1305 (1969)
35. Hodges, R. J., *Corrosion*, **27**, 119 (1971)
36. Henthorne, M., *Intergranular Corrosion of Iron and Nickel Base Alloys, Localised Corrosion-Cause of Metal Failure*, ASTM STP 516, ASTM, 66 (1972)
37. *Stainless Steel '77*, ed. R. Q. Barr, Climax Molybdenum Co. (1977)
38. Locci, I. E., Kown, H. K., Hehemann, R. F. and Troiano, A. R., *Corrosion*, **43**, 465 (1987)
39. Schmidt, W. and Jarleborg, O., Ferritic Stainless Steels with 17% Cr, Climax Molybdenum GmbH
40. Bond, A. P. and Dundas, H. J., *Stress Corrosion Cracking and Hydrogen Embrittlement of Iron Base Alloys*, NACE 5, Houston, Texas, (eds R. W. Staehle, J. Hochmann, R. D. McCright, and J. E. Slater) 1136 (1977)

41. Staehle, R. W., *Environment-Induced Cracking of Metals*, NACE 10, Houston, Texas, (eds R. P. Gangloff and M. B. Ives) p. 561 (1990)
42. Bond, A. P. and Dundas, H. J., *Corrosion*, **24**, 344 (1968)
43. Newberg, R. T. and Uhlig, H. H., *Jour Electrochem. Soc.*, **119**, 981 (1972)
44. Matsushima, S. and Ishihara, T., *Trans. Nat. Res. Inst. Met.*, **17**, 14 (1975)
45. Shimodaira, S., Takano, M., Takizawa, Y. and Kamide, H., *Stress Corrosion Cracking and Hydrogen Embrittlement of Iron Base Alloys*, NACE 5, Houston, Texas, (eds R. W. Staehle, J. Hochmann, R. D. McCright and J. E. Slater) p. 1003 (1977)
46. Uhlig, H. H. and Newberg, R. T., *Corrosion*, **28**, 337 (1972)
47. Marquez, J. A., Matsushima, I. and Uhlig, H. H., *Corrosion*, **26**, 215 (1970)
48. Pugh, E. N., *Corrosion*, **41**, 517 (1985)
49. Cieslak, W. R. and Duquette, D. J., *Corrosion*, **40**, 545 (1984)
50. Congleton, J., Zheng, W. and Hua, H., *Corrosion Science*, **30**, 555 (1990)
51. Congleton, J., Zheng, W. and Hua, H., *Corrosion*, **46**, 621 (1990)
52. Hochmann, J., Desestret, A., Jolly, P. and Mayoud, R., *Stress Corosion Cracking and Hydrogen Embrittlement of Iron Base Alloys*, NACE 5, Houston, Texas, (eds R. W. Staehle, J. Hochmann, R. D. McCright, and J. E. Slater) 956 (1977)
53. Chen, J. and Devine, T. M., Paper 263, *Corrosion 88*, St. Louis, USA (1988)
54. Edeleanu, C., *J. Iron and Steel Inst.*, **173**, 140 (1953)
55. Fontana, M. G., Beck, F. H. and Flowers, J. W., *Metal Progress*, **80**, (Dec), 99 (1961)
56. Flowers, J. W., Beck, F. H. and Fontana, M. G., *Corrosion*, **19**, 186t (1963)
57. Suzuki, T., Hasegawa, H. and Watanabe, M., *Nippon Kunzoku Gakaishi*, **32**, 1171 (1968)
58. Cottrell, A. H., *Dislocations and plastic flow in crystals*, Clarendon Press, Oxford (1953)
59. Mahajan, S. and Williams, D. F., *Int. Met. Rev.*, **18**, 43 (1973)
60. Tsuge, H., Tarutani, Y. and Kudo, T., *Corrosion*, **44**, 305 (1988)
61. Herbsleb, G and Poepperling, R. K., *Corrosion*, **36**, 611 (1980)
62. Schaeffler, A. L., *Metal Progress*, **56**, 680B (1949)
63. Schneider, H., *Foundry Trade Jour.*, **108**, 562 (1960)
64. Page, G. G., *Handling and Fabricating Stainless Steel for the Food Industry*, DSIR Industrial Information Seris No. 4, Publ. Sc. Inf. Publ. Centre, DSIR, Wellington, New Zealand (1984)
65. Page, G. G., *Materials Performance*, **28** (7), 58 (1989)
66. Sedriks, A. J., *J. Inst. Metals.*, **101**, 225 (1973)
67. Theus, G. J. and Staehle, R. W., *Stress Corrosion Cracking and Hydrogen Embrittlement of Iron Base Alloys*, NACE 5, Houston, Texas, (ed R. W. Staehle, J. Hochmann, R. D. McCright and J. E. Slater) p. 845 (1977)
68. Thompson, A. W. and Bernstein, I. M., *Reviews Coat. Corr.*, **2**, 3 (1975)
69. Hanninen, H. E., *International Metallurgical Reviews*, **3**, 85 (1979)
70. Szklarska-Smialowska, S. and Cragnolino, G., *Corrosion*, **36**, 653 (1980)
71. Newman, R. C. and Mehta, A., *Environment-Induced Cracking of Metals*, NACE 10, Houston, Texas, (eds R. P. Gangloff and M. B. Ives) 489 (1990)
72. Copson, H. R., *Physical Metallurgy of Stress Corrosion Cracking*, Interscience, New York, 247 (1959)
73. Neff, D. V., Mitchell, T. E. and Troiano, A. R., *Trans A.S.M.*, **62**, 858 (1969)
74. Wilde, B. E., Stainless Steel Immune to Stress-Corrosion Cracking, US Patent 4002510 (1977)
75. Truman, J. E. and Perry, R., *Brit. Corr. J.*, **1**, 60 (1966)
76. Sedriks, A. J., Floreen, S. and McIlree, A. R., *Corrosion*, **32**, 157 (1976)
77. Berge, P., Donati, J. R., Prieux, B. and Villard, D., *Corrosion*, **33**, 425 (1977)
78. Wilson, I. L. W. and Aspen, R. G., *Stress Corrosion Cracking and Hydrogen Embrittlement of Iron Base Alloys*, NACE 5, Houston, Texas, (eds R. W. Staehle, J. Hochmann, R. D. McCright, and J. E. Slater) 1189 (1977)
79. Eckel, J. F. and Clevinger, G. S., *Corrosion*, **26**, 251 (1970)
80. Swann, P. R., *Corrosion*, **19**, 102t (1963)
81. Murato, T., Sato, E. and Okada, H., *Passivity and its Breakdown on Iron Base Alloys*, (eds H. Okada and R. W. Staehle) NACE, Houston, Texas (1976)
82. Okadam, H., Hosoi, Y. and Abe, S., *Corrosion*, **27**, 424 (1971)
83. Kowaka, M. and Fujikawa, H., *Localized Corrosion* (eds R. W. Staehle, B. F. Brown, J. Kruger and A. Agrawal) NACE, Houston, Texas, 437 (1974)
84. Latanision, R. M. and Staehle, R. W., *Fundamental Aspects of Stress Corrosion Crack-*

ing, NACE 1, (eds R. W. Staehle, A. J. Forty and D. Van Rooyen) NACE, Houston, Texas, 214 (1969)
85. Staehle, R. W., Royuela, J. J., Raredon, T. L., Serrate, E., Morin, C. R. and Farrar, R. V., *Corrosion*, **26**, 451 (1970)
86. Montuelle, J. and da Cunha Belo, M., *Stress Corrosion Cracking and Hydrogen Embrittlement of Iron Base Alloys*, NACE 5, Houston, Texas, (eds R. W. Staehle, J. Hochmann, R. D. McCright, and J. E. Slater) 1125 (1977)
87. Snowden, P. P., *J.I.S.I.*, **194**, 181 (1960)
88. Wilson, I. L. W., Pement, F. W. and Aspen, R. G., *Corrosion*, **30**, 139 (1974)
89. Cochran, R. W. and Staehle, R. W., *Corrosion*, **24**, 369 (1968)
90. Montuelle, J., *Corr. Trait. Prot. Finition*, **16**, 279 (1968)
91. Schreiber, F. and Engell, H. J., *Werkst. Korr.* **23**, 175 (1972)
92. Watanabe, M. and Mukai, Y., *Proc. 4th Int. Cong. on Metal Corrosion*, (ed. N. E. Hamner) NACE, Houston, Texas, 83 (1972)
93. Hardie, D. and Butler, J. J. F., *Mats. Sci. and Tech.*, **6**, 441 (1990)
94. Vermilyea, D. A. and Indig, M. E., *Proc. 5th Int. Congr. on Metal Corrosion*, (ed N. Sato) NACE, Houston, Texas, 866 (1974)
95. Bruemmer, S. M., *Corrosion*, **46**, 698 (1990)
96. *Standard Recommended Practices for Detecting Susceptibility to Intergranular Attack in Stainless Steels*, ANSI/ASTM A262 79 (1979)
97. *British Standard Method for Determination of Resistance to Intergranular Corrosion of Austenitic Stainless Steels: Copper sulphate-Sulphuric acid Method* (Moneypenny Strauss Test) BS 5903 (1980)
98. Clarke, W. L., Cowan, R. L. and Walker, W. L., *Intergranular Corrosion of Stainless Steels*, ASTM STP 656, (Philadelphia, PA, ASTM) 99 (1978)
99. Blanchet, J., Coriou, H., Grall, L., Mahieu, C., Otter, C. and Turluer, G., *Stress Corrosion Cracking and Hydrogen Embrittlement of Iron Base Alloys*, NACE 5, Houston, Texas, (eds R. W. Staehle, J. Hochmann, R. D. McCright, and J. E. Slater) 1149 (1977)
100. Latanision, R. M., *Corrosion of Nickel Base Alloys*, ASM, (ed. R. C. Scarberry) 13 (1985)
101. Banerjee, B. R., Dulis, E. J. and Hauser, J. J., *Trans. ASM*, **61**, 103 (1968)
102. Fox, M., *J. Materials for Energy Systems*, **1**, 3 (1979)
103. Indig, M. E. and McIlree, A. R., *Corrosion*, **35**, 288 (1979)
104. Lin, L. F., Cragnolino, G., Szklarska-Smialowska, Z. and Macdonald, D. D., *Corrosion*, **37**, 616
105. Congleton, J. 'Strain', *The J. of the Brit. Soc. for Strain Meas.*, **26**, 15 (1990)
106. Szklarska-Smialowska, S., and Cragnolino, G., *Corrosion*, **36**, 653 (1980)
107. Clarke, W. L. and Gordon, G. M., *Proc. 5th Int. Congr. on Metal Corrosion*, (ed N. Sato) NACE, Houston, Texas, p. 884 (1974)
108. Vermilyea, D. A., *Corrosion*, **31**, 421 (1975)
109. Vermilyea, D. A., *Jour. Electrochem. Soc.*, **121**, 1190 (1974)
110. Pickett, A. E. and Sim, R. G., *Materials Performance*, **12** (June), 39 (1973)
111. Pedneker, S. and Szklarska-Smialowska, S., *Corrosion*, **36**, 565 (1980)
112. Yang, W., Zhang, M., Zhao, G. and Congleton, J., *Corrosion*, (in press)
113. Congleton, J. and Sui, G., *Corrosion Science*, **13**, 1691 (1992)
114. Berry, W. E., White, E. L. and Boyd, W. K., *Corrosion*, **29**, 451 (1973)
115. Hines, J. G., *Corrosion* (ed. L. L. Shrier) 2nd ed., Oxford, Butterworth-Heinemann, 8:62 (1976)
116. Oldfield, J. W. and Todd, B., *Materials Perform.*, **29**, No. 12, 57 (1990)
117. Kain, R. M., *Materials Perform.*, **29**, No. 12, 60 (1990)
118. Gnanamoorthy, J. B., *Materials Perform.*, **29**, No. 12, 63 (1990)
119. Dillon, C. P., *Materials Perform.*, **29**, 66 (1990)

8.4 Stress-corrosion Cracking of High-tensile Steels

Introduction

There are many applications in which the yield strength of the material of construction is a limiting factor in the design. For example a large proportion of the weight of an offshore oil production platform consists of the structure required to resist wave forces, with a relatively small proportion being the 'topside' equipment which actually does the work. If the strength of the material of the structure could be increased, this would allow the structure itself to be made lighter, with the consequence that the weight of the topside equipment could be increased. Thus materials offering high strength-to-weight or strength-to-volume ratio are in high demand in many fields of engineering. Steels can be produced with yield strengths above 2 000 MPa, and at first sight one might expect these to have widespread application. Unfortunately things are not that simple, and as the yield strength of materials is increased, so other mechanical properties tend to decrease. In particular the materials become more susceptible to brittle fracture, especially when assisted by environmental factors such as hydrogen embrittlement or stress-corrosion cracking. Stress-corrosion cracking failures of high-strength steels have been observed in a wide range of environments, and have been encountered in the chemical, petrochemical, oil, aircraft and nuclear industries. The related problem of hydrogen-induced blister formation or hydrogen-induced cracking tends to be more of a problem in lower strength steels used for crude oil transmission pipelines.

It is now well-established that high-strength steels are, with relatively few exceptions, susceptible to embrittlement as a result of dissolved hydrogen, and the majority of stress-corrosion cracking failures of these materials are attributed to hydrogen embrittlement. This should not be taken to imply that other mechanisms of stress-corrosion cracking do not occur. However, the ease with which hydrogen can be picked up from aqueous environments is such that other stress-corrosion cracking processes are rarely a practical problem for high-strength steels.

Despite considerable developments in the study of stress-corrosion cracking mechanisms in recent years, it remains difficult to draw a clear distinction between those situations which involve hydrogen embrittlement, and those which do not. It is often suggested that one can distinguish between dissolution and hydrogen embrittlement processes by way of the effect of applied polarisation on the cracking. Unfortunately, the electrochemistry of iron is complex, particularly in cracks, crevices and pits, and detailed study shows that it is possible for hydrogen embrittlement to occur with applied anodic polarisation.

A very thorough review of the hydrogen embrittlement of steels was published in 1985[1].

Hydrogen Embrittlement of High-strength Steels

Entry of Hydrogen into Steel

Hydrogen can enter steel either from the gas phase or, by way of the electrochemical reduction of hydrogen-containing species, from the aqueous phase.

Entry from the gas phase Several models have been proposed for the entry of hydrogen from the gas phase, and the details of the process remain uncertain. In general terms however, the reactions involved are the adsorption of molecular hydrogen (or other gas such as hydrogen sulphide), the dissociation of the hydrogen molecule to produce hydrogen atoms adsorbed onto the surface, and the subsequent diffusion of the adsorbed hydrogen atoms into the metal lattice. For perfectly clean and film-free iron surfaces the rate-controlling step in this process may be the dissociation step[2]. In more realistic situations surface films, such as the air-formed oxide film, or the presence of gases, such as oxygen, which adsorb competitively with hydrogen may play an important role.

Entry from the aqueous phase The mechanism of electrochemical production of hydrogen on steel in aqueous solution has received much attention. It is accepted that the reaction occurs in two main stages. The first of these is the initial charge transfer step to produce an adsorbed hydrogen atom. In acid solution this involves the reduction of a hydrogen ion:

$$H_3O^+ + e^- \rightarrow H_{ads} + H_2O$$

In neutral and alkaline solution, where the concentration of hydrogen ions is very low, the reaction switches to the reduction of water molecules:

$$H_2O + e^- \rightarrow H_{ads} + OH^-$$

The second stage of the reaction to produce molecular hydrogen may occur through either of two mechanisms. In the first of these, known as chemical desorption or chemical recombination, two adsorbed hydrogen atoms combine to produce a hydrogen molecule:

$$H_{ads} + H_{ads} \rightarrow H_2$$

Alternatively the adsorbed hydrogen atom may participate in a second electrochemical reaction, known as electrochemical desorption:

$$H_{ads} + H_3O^+ + e^- \rightarrow H_2 + H_2O \quad \text{(acid)}$$

$$H_{ads} + H_2O + e^- \rightarrow H_2 + OH^- \quad \text{(neutral or alkaline)}$$

For iron it is reasonably well established that the reaction goes by way of chemical recombination under most circumstances, although there is some evidence that electrochemical desorption may take over in very alkaline solutions or at large overpotentials.

A third reaction, which goes in parallel with the desorption reaction, is the entry of atomic hydrogen into the steel from the surface adsorbed state:

$$H_{ads} \rightarrow H_{metal}$$

In most circumstances the kinetics of this reaction are controlled by the rate at which the hydrogen can diffuse into the underlying steel, and this reaction is essentially in equilibrium. Consequently it is difficult to study the kinetics of this reaction. A particular situation in which this may be very important relates to the conditions at crack tips, where the hydrogen may be transported into the bulk by dislocation motion, giving rise to very high rates of hydrogen entry.

As the hydrogen entry reaction is generally in equilibrium, the hydrogen concentration just below the entry surface is directly related to the surface concentration or coverage. Consequently the rate of entry of hydrogen into the steel is controlled by the balance between the first and second stages of the hydrogen evolution reaction, since these control the coverage of adsorbed hydrogen. In the case of chemical recombination as the second stage in the reaction, the rate of hydrogen evolution is proportional to $[H_{ads}]^2$ (since two adsorbed hydrogen atoms are involved in the reaction), while the rate of hydrogen entry (or the equilibrium concentration just below the surface, which is usually the controlling factor) is proportional to $[H_{ads}]$. Thus, as the cathodic current increases so the subsurface hydrogen concentration (and hence the rate of hydrogen diffusion through a membrane) increases as the square root of the cathodic current (assuming that

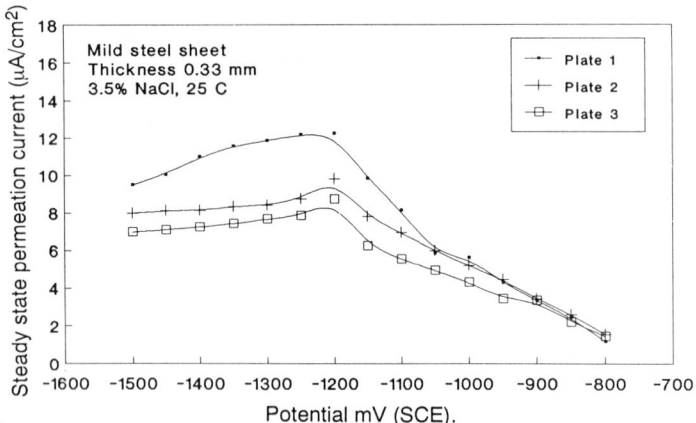

Fig. 8.38 Hydrogen permeation current as a function of applied potential, showing effect of change in reaction mechanism below $-1200\,mV$ (SCE) (after Obuzor[2])

the current due to hydrogen entering the steel is small compared with that due to hydrogen evolution, as is usually the case).

However, when the second stage in the hydrogen evolution reaction is electrochemical desorption, the rate of this reaction is increased as the potential falls, and the adsorbed hydrogen concentration may remain constant or fall, according to the detailed electrochemistry. This results in curves such as that shown in Fig. 8.38 for steel in sodium chloride solution.

Whether the adsorbed hydrogen is produced from the gas phase or from aqueous solution, it appears that the presence of hydrogen atoms distorts the crystal structure of the metal surface[4], and this results in a surface solubility which is higher than that of the bulk. The depth of this distortion is not clear, but it seems possible that the distorted zone may play an important part in initiating brittle-fracture processes.

Location of Hydrogen in Steel

Hydrogen exists in metals in the monatomic form, and is commonly described as atomic hydrogen. In practice the state of charge of the hydrogen atom is not known with any certainty, but it seems probable that it tends to acquire a slight negative charge by attracting electrons from the valence orbitals of the metal lattice. It has been suggested that this results in a weakening of the metal–metal bond which is responsible for hydrogen embrittlement[5].

Hydrogen has a very low solubility in the iron lattice, which makes direct observation of the location of the hydrogen atom in the lattice very difficult. The hydrogen definitely occupies an interstitial site in the bcc iron lattice. Two such sites are normally associated with interstitial solutes in bcc structures, the tetrahedral and the octahedral sites (see Fig. 8.39). Indirect evidence[6] suggests that hydrogen occupies the tetrahedral site.

In addition to interstitial sites in the lattice, hydrogen atoms are also strongly attracted to defect sites in the metal, and these are referred to as 'traps'. Trap sites include vacancies, solute atoms, dislocations, grain boundaries, voids and non-metallic inclusions. Of the various trap types in iron and steel, vacancies are relatively unimportant at ambient temperatures, simply because of their low concentration. Somewhat surprisingly it appears that grain boundaries in pure iron also trap very little hydrogen, although segregation of carbon and other impurity atoms to the grain boundary can increase the tendency for trapping[7]. The more important trap sites in iron and steel appear to be phase boundaries, dislocations, voids and inclusions[8]. Trap types may be classified in various ways, two of the more important being related to the number of hydrogen atoms which can be accommodated in the trap and to the binding energy of the trap. These are summarised in Table 8.3. It should be noted that the terms 'saturable' and 'reversible' are rather loosely defined in relation to hydrogen trapping. Thus some workers take the view that reversible and non-saturable traps are the same (since the equilibrium hydrogen content of the trap will vary with the fugacity). Similarly the term 'reversible' is given a slightly different meaning from that in electrochemical reaction kinetics, where it relates to the activation energy of a reaction, rather than the overall free energy change.

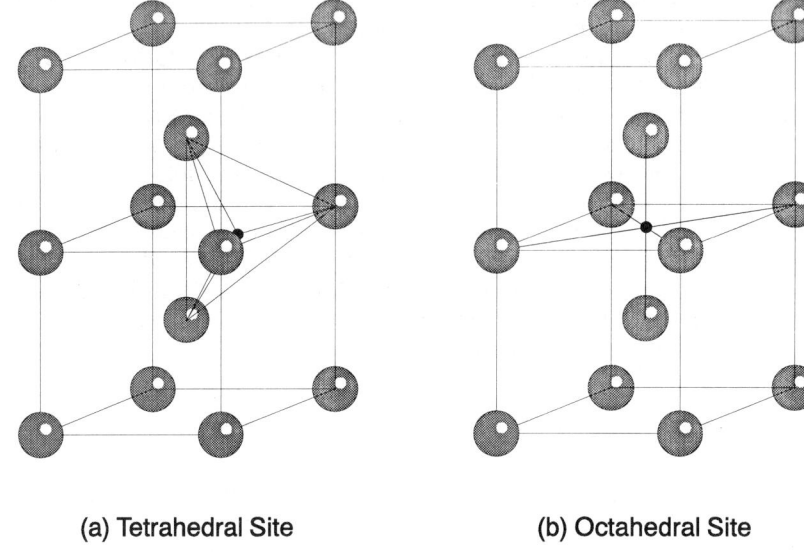

(a) Tetrahedral Site (b) Octahedral Site

Fig. 8.39 Octahedral (a) and tetrahedral (b) sites in the bcc lattice

Table 8.3 Classification of trap types

Classification	Description
Saturable	The number of sites for hydrogen atoms is fixed (e.g. boundaries, dislocations).
Non-saturable	The number of sites for hydrogen atoms in the trap varies according to the fugacity (e.g. voids).
Reversible	The trap binding energy is relatively small, and hydrogen may escape from the trap as well as enter it.
Irreversible	The trap binding energy is large, and hydrogen will not leave the trap at ambient temperature.

There are many ways in which trapping can be studied, but the wide range of trap types and geometries make it difficult to determine the properties of specific trap sites.

Hydrogen trapped in voids consists of adsorbed hydrogen on the walls of the void, together with molecular hydrogen in the void itself. With high fugacities of hydrogen in the steel, such as can be developed in steels in contact with acidic solutions containing hydrogen sulphide, very high pressures may be developed in the void. When combined with the hydrogen embrittlement of the steel around the void, this can lead to the growth of cracks around the void. Such cracks typically develop around non-metallic inclusions which have been flattened by rolling, giving characteristic blisters lying parallel to the rolling direction. This phenomenon is known as *hydrogen-induced cracking*.

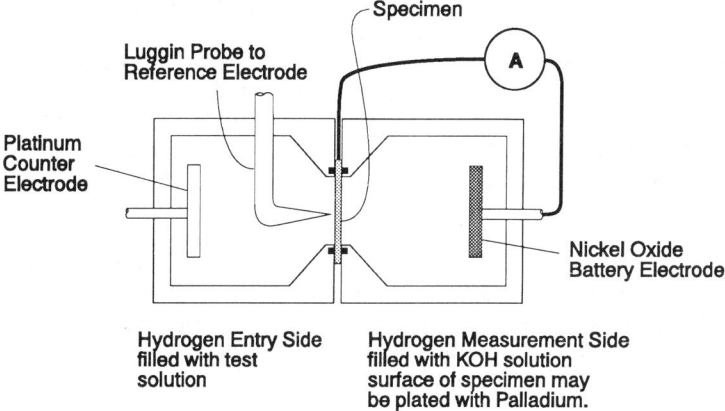

Fig. 8.40 Hydrogen permeation cell (schematic)

Transport of Hydrogen in Iron and Steel

As a small interstitial atom, hydrogen diffuses rapidly in iron, the diffusion rate being of a similar order to that of solutes in aqueous solution.

The study of the transport of hydrogen in steel is commonly undertaken by hydrogen permeation measurements. This involves the permeation of hydrogen through a thin steel membrane (typically less than 1 mm thick). Hydrogen entry may be from the gas phase or from solution, while the flux of hydrogen through the membrane may be determined either by vacuum extraction of the gas to a suitable detector, or, somewhat more simply, by electrochemical oxidation of the hydrogen to hydrogen ions. The latter method forms the basis of the electrochemical hydrogen permeation cell developed originally by Devanathan and Stackurski[9] and illustrated in Fig. 8.40. This has subsequently been developed into a monitoring technique by Berman et al.[10], while Arup[11] has developed a small self-contained sensor using battery technology, shown in Fig. 8.41.

In permeation measurements the first signs of hydrogen diffusing through 1 mm steel membranes can be observed in a few minutes. The practical measurement of diffusion parameters tends to be rather unreproducible,

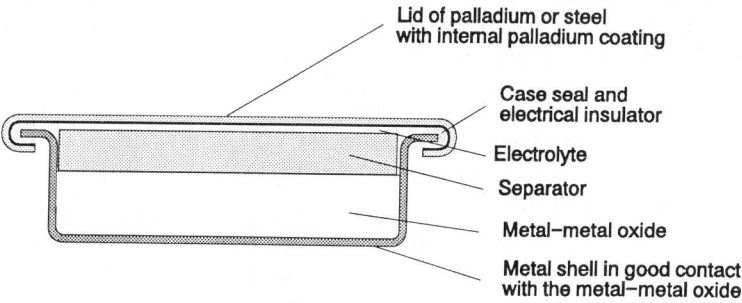

Fig. 8.41 Permeation cell using battery technology (after Arup[11])

because of the role of the various forms of trap, which tend to increase the solubility of hydrogen in the steel, and thereby decrease the apparent diffusion coefficient. For pure iron at temperatures above around 350 K, the activation energy for diffusion is approximately 7.6 kJ/mol, with D_0 being in the region of 1 to 2.5×10^{-7} m^2/s. At ambient temperatures the activation energy ranges from about 4 to 7 kJ/mol, with D_0 ranging from about 0.5 to 1.2×10^{-7} m^2/s. Diffusion coefficients at 298 K are thus about 7×10^{-9} m^2/s.

The above data relate to very pure iron samples with low dislocation densities. In real steels the trapping effects result in much lower apparent diffusivities, which are dependent on the metallurgical state of the steel, as well as its chemical composition. Typical values for the apparent diffusion coefficient of hydrogen in high-strength alloy steel at room temperature are in the region of 10^{-11} m^2/s.

Sources of Hydrogen

The thermodynamics of the reaction of iron with water are such that iron is always thermodynamically capable of displacing hydrogen from water, although the driving force is small at ambient temperatures, except in reasonably strong acids. Thus iron does not suffer from serious corrosion in oxygen-free neutral or alkaline solutions. However, rates of hydrogen evolution which are insufficient to cause significant corrosion may still produce enough hydrogen to create a serious embrittlement problem for high-strength steels. The hydrogen that is produced dissolves readily in the steel, and remarkably small concentrations of hydrogen (less than 1 ppm) can cause embrittlement. In addition to corrosion itself, many of the standard chemical and electrochemical treatments employed to protect steel against corrosion also tend to drive hydrogen into the steel. Some of the major sources of hydrogen are reviewed below.

Welding Hydrogen introduced into welds produces a particularly acute problem, as the weld and the heat-affected zone are inevitably regions of high residual stresses, contain inherent defects and are frequently intrinsically more brittle than the parent material. Thus it is important to minimise the introduction of hydrogen into welds, even for lower strength steels. Gas welding of steels using an oxyacetylene flame will inevitably introduce hydrogen as a result of the hydrogen-containing gases in the flame. In theory electric-arc welding, particularly if the arc is protected from atmospheric moisture by inert gas shielding, will not introduce hydrogen. However, with normal manual metal arc welding using flux-coated electrodes it is possible for the flux coating to absorb moisture from the atmosphere, and this will react with the molten steel to produce hydrogen. For this reason it is good practice to store coated welding electrodes in an oven in order to drive off any moisture (Section 9.5).

Acid Pickling This process is widely used for removing rust and millscale from steel, or for removing internal scales from boilers. The objective of the process is the dissolution of iron oxides or hydroxides or hardness scales, but at the same time the iron will also tend to corrode in the acid, with hydrogen evolution as the cathodic reaction. This leads to hydrogen entry into the steel, which may lead to hydrogen embrittlement or blister formation due to hydrogen-induced cracking.

As dissolution of the steel is normally an undesirable side effect of the pickling process, resulting in loss of metal and wasteful consumption of acid, it is normal practice to add inhibitors (usually referred to as 'pickling restrainers') to the acid. Unfortunately many of these work by interfering with the hydrogen recombination reaction, rather than the initial production of adsorbed hydrogen atoms. This has the effect of increasing the surface coverage by adsorbed hydrogen atoms, and consequently the rate of entry of hydrogen into the steel is increased, even though the overall rate of hydrogen production is reduced. For this reason it is most important that pickling restrainers for use with high-strength steels should be tested for their effects on the uptake of hydrogen as well as their efficiency as inhibitors.

Degreasing and Cleaning Various processes can be applied for the removal of grease and other contaminants prior to painting, electroplating or other surface treatments. Degreasing in organic solvents (e.g. vapour degreasing) is unlikely to generate significant quantities of hydrogen unless the fluid is contaminated with water or other species, such as hydrogen chloride, capable of liberating protons. Cleaning in aqueous alkali without applied polarisation is also unlikely to introduce much hydrogen, although any tendency for pitting corrosion due to chloride contamination could give rise to local problems. Cathodic cleaning, in which hydrogen is deliberately evolved on the steel, is clearly undesirable.

Electroplating Electroplated metal coatings provide a convenient and effective means of protecting steel against atmospheric corrosion, with zinc and cadmium being particularly useful because of their ability to provide sacrificial protection to the steel substrate at breaks in the coating. Unfortunately, all metal deposition processes require the application of potentials at which hydrogen evolution is possible (in theory it would be possible to deposit copper and more noble metals above the hydrogen evolution potential, but as this would also be a potential at which iron would dissolve anodically, this would not produce good deposits). Thus the electrodeposition of coatings almost inevitably introduces hydrogen into the metal. The majority of electrodeposited coatings on iron are much less permeable to hydrogen than iron, the major exceptions to this being other *bcc* metals, with chromium being the only example in widespread use (although even here chromium is generally used with an undercoat of nickel or copper, which will act as a barrier). Consequently most of the hydrogen uptake occurs in the early stages of the deposition process, before a complete coverage of the coating metal has been achieved, and the start of the deposition process is particularly critical in determining the amount of hydrogen absorbed. Clearly the highest possible ratio of metal deposition rate to hydrogen generation rate is required in this period, and this tends to be favoured by high current densities. For this reason (among others) a high current density 'flash' deposit of metal is often used as the first stage of a two-stage plating scheme. Bath composition is also important in controlling the entry of hydrogen into the steel.

Once high-strength steel components have been electroplated, it is possible (and often mandatory) to reduce the damaging effects of the hydrogen by baking at around 200°C. In part this serves to allow hydrogen to diffuse through the coating and out to the atmosphere, but the redistribution of the remaining hydrogen within the steel, reducing damaging local high concentrations, is probably also an important part of this de-embrittlement treatment.

A very important area of use for electroplated high-strength steels is fasteners and other high stress components in aircraft. These are generally protected by cadmium plating, although the toxicity of cadmium and its compounds is giving rise to a search for alternatives. In this application the avoidance of hydrogen embrittlement is clearly essential, and as a result a wide range of standard procedures and tests for cadmium plating of high-strength steels are available as civil or military specifications and codes of practice[12].

Phosphating Phosphating, which is widely used as a pretreatment for steel prior to painting, involves the controlled corrosion of the steel in acid solution, and inevitably leads to the uptake of hydrogen. By limiting the free acid in the phosphating bath, and by introducing oxidising agents to raise the working potential, the uptake of hydrogen may be reduced.

Painting Conventional paints are normally innocuous, but it seems possible that some modern water-based paints, particularly those applied by cathodic electrophoresis, may introduce significant quantities of hydrogen into the steel. Fortunately, these paints are generally stoved after application, and this is effective in removing the hydrogen, which permeates very easily through the paint film[13]. Additionally, these paints are applied primarily to lower-strength steels in motor vehicle bodies and similar applications, and as far as the author is aware no problems have been experienced in service. However, suitable tests would be advisable if it is intended to use these processes with high-strength steels.

Paint strippers may give rise to hydrogen entry into steel, and in critical applications, such as the treatment of aircraft components, commercial paint strippers should be tested before use.

Corrosion in Service Most of the modes of hydrogen entry discussed above involve a single brief charging period, and as a result are treatable, in that the hydrogen can be removed by a suitable de-embrittlement treatment (providing the embrittlement is not so severe that cracks are formed before the de-embrittlement treatment can be applied). In contrast the entry of hydrogen due to corrosion in service is generally continuous. Thus de-embrittlement is not feasible, and the control of hydrogen embrittlement of high-strength steels presents a much more difficult problem.

In general hydrogen will enter steel during any corrosion process involving hydrogen ion or water reduction as one of the cathodic reactions. It is frequently implied that the applied potential must be below the equilibrium potential for hydrogen evolution before hydrogen entry into steel is possible. However, for two related reasons this is not true. First, the equilibrium potential for hydrogen evolution at a given pH is that potential at which protons or water molecules at the metal–solution interface are in equilibrium with molecular hydrogen at a partial pressure of 1 atmosphere. As the partial pressure of hydrogen is reduced, so the equilibrium potential will increase, in accordance with the Nernst equation, and even at potentials 200 mV above the 1 atmosphere equilibrium potential there will be a significant concentration of hydrogen at the metal surface. Secondly, for high-strength steels the matrix concentration of hydrogen required to cause embrittlement is very small, as the hydrogen tends to concentrate at phase boundaries and other trap sites. Thus hydrogen embrittlement has been observed in high-strength steel in contact with gaseous hydrogen at pressures below 0.001 atm.

A second difficulty in predicting the effect of particular corrosion conditions on the rate of hydrogen uptake by steel is the strong influence of local conditions. For passive steel, where the corrosion potential is several hundred millivolts above the 1 atm hydrogen equilibrium potential, one would not expect problems of hydrogen embrittlement, and this is frequently used as an argument against a hydrogen embrittlement mechanism of stress-corrosion cracking in environments such as phosphate and carbonate/bicarbonate. However, the local environment in pits, cracks or crevices may be very different from conditions at the free surface. In particular, acidification may occur due to metal-ion hydrolysis (especially for chromium-containing steels), and the potential in the localised corrosion cavity may be considerably more negative than that measured at the free surface. The net result is that the conditions within the cavity may be favourable for hydrogen evolution, even though the free surface conditions imply that hydrogen embrittlement is very unlikely.

In neutral saline environments, such as seawater, the rate of hydrogen entry is controlled by (among other things) the applied potential. In this case (and probably many others) the response observed, even at a smooth surface, is not as simple as might be expected, largely because of changes in the chemistry of the liquid in immediate contact with the steel. This is indicated in Fig. 8.42 due to Barth and Troiano[14], which shows the rate at which hydrogen permeates through a steel membrane in response to a

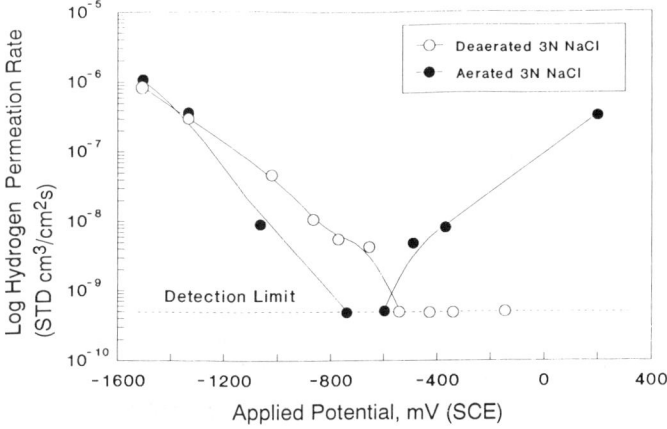

Fig. 8.42 Hydrogen permeation current as a function of applied potential, showing effect of oxygen concentration and applied potential (after Barth and Troiano[14])

range of applied potentials. Similar behaviour has been observed in more recent work which has taken account of the shielding effect of a crack or crevice[15], and this has shown that the effects of applied potential can be explained on the basis of the local potential and pH in relation to the hydrogen equilibrium potential. Figure 8.43 shows typical results obtained in this work, and it can be seen that the hydrogen overpotential and the rate of hydrogen entry into the steel increase with both anodic and cathodic polarisation. Detailed prediction of the effect of applied potentials on the

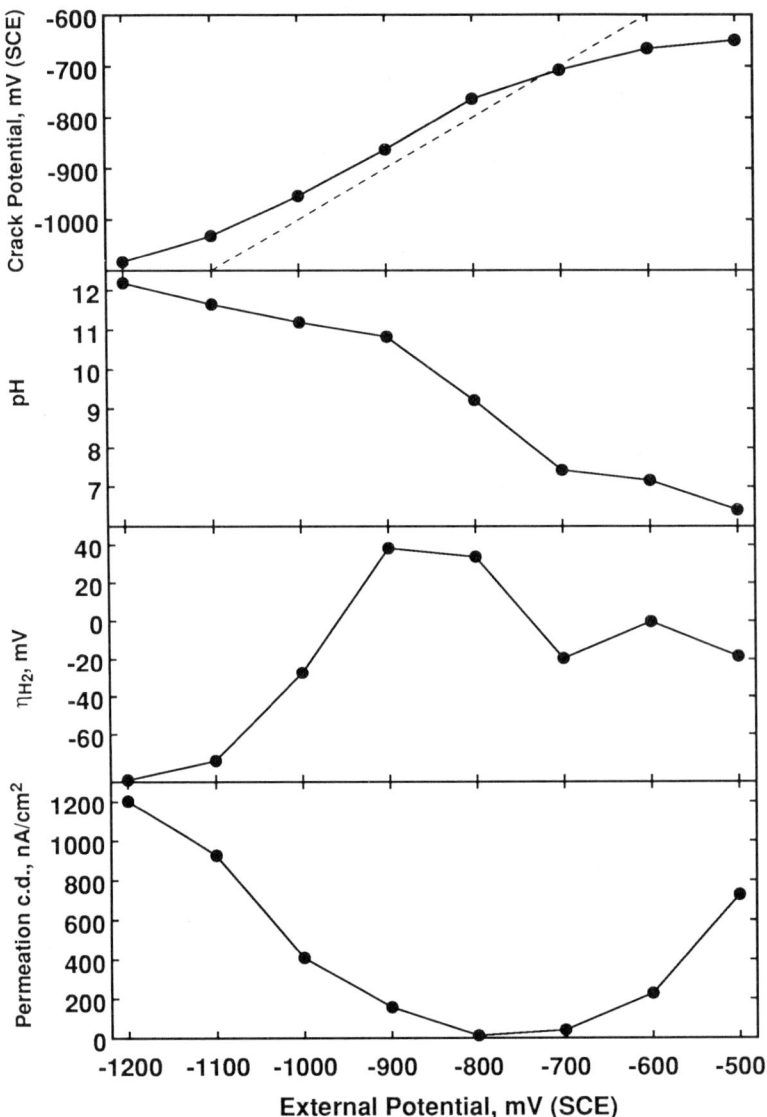

Fig. 8.43 Potential and pH in a simulated crevice, and effect on hydrogen permeation current (after Taqi and Cottis[15])

rate of hydrogen entry require very careful consideration of many factors, including the exact composition of the solution (especially with respect to the oxygen concentration), the geometry of the component (especially the presence of any cracks or crevices) and the mass transport conditions. In general terms, however, when considering the performance of steel in sea-water it seems probable that slight cathodic polarisation to about −750 or −800 mV (SCE) may reduce the rate of hydrogen entry compared with free corrosion conditions, because the increased pH obtained as a result of the cathodic polarisation outweighs the lowered potential. Further decrease in potential increases the rate of hydrogen entry down to about −1200 mV (SCE), when electrochemical desorption takes over as the second stage of the hydrogen evolution reaction, and at more negative potentials the rate of hydrogen entry remains roughly constant.

In deaerated neutral salt solutions anodic polarisation of chromium-free steel tends to lower the rate of hydrogen entry, as the fall in pH due to ferrous ion hydrolysis is relatively limited, and cannot compensate for the increase in potential. However, in aerated solution the ferrous ion can be oxidised to ferric in solution by dissolved oxygen, and as ferric hydroxide is much less soluble than ferrous hydroxide the pH can fall much lower. (This reaction is a serious problem when trying to measure the solution pH in cracks and crevices, and much of the early work which shows very acid pH values in cracks in carbon steels is now known to be incorrect because this reaction was allowed to occur between extracting the crack solution and measuring its pH.) In addition to lowering the surface pH, it seems possible that the precipitated film of ferric hydroxide/oxide also presents an ohmic resistance, and may thereby allow the surface potential to become lower than that measured in the bulk solution. It seems probable that the combination of these two effects was responsible for the increase in permeation current observed for anodic polarisation by Barth and Troiano[14]. Many steels contain small, but significant concentrations of chromium, and it should be appreciated that chromium concentrations of around 1% can markedly increase the tendency for local acidification, owing to the much stronger tendency for chromium ion hydrolysis.

As considerations such as those discussed above have become more widely appreciated, it has become clear that it is quite difficult to specify aqueous environments in which hydrogen entry into steel will occur at a sufficiently low rate that the possibility of hydrogen embrittlement can be discounted. The main candidates for such an environment are strong alkalis and highly oxidising environments, such as nitric acid or nitrates, which are free of species, such as chloride, which predispose to pitting or crevice corrosion, and high temperature environments such as water at 250°C (since the deleterious effects of dissolved hydrogen decrease at high temperatures).

Effect of Hydrogen on Mechanical Properties

Elastic Constants While there is some evidence of small changes in the elastic properties of steel as a result of dissolved hydrogen[16], these changes are small, and of little practical consequence. This is perhaps to be expected in view of the very low solubility of hydrogen in the iron lattice and the small effect on the metal–metal bond strength.

Yield Stress The effect of hydrogen on the yield stress of iron and steels is unpredictable. For very pure iron single crystals and polycrystals the yield stress is frequently found to be decreased by hydrogen, but it may increase or stay the same, depending on the dislocation structure, crystal orientation and purity of the iron[17]. Little information is available for steels.

Plastic Behaviour The effect of hydrogen on the plastic behaviour of iron and steel is somewhat complex, as hydrogen may harden or soften the material, according to its structure and the mode of slip. On the basis of work on single crystals of pure iron Lunarska[17] has concluded that segregation of hydrogen around dislocations decreases their elastic stress fields. At room temperature (where the rate of diffusion of hydrogen is sufficiently high that it can keep up with a moving dislocation) this leads to a softening of single crystals when only one slip system is operative. The hydrogen also suppresses cross-slip of screw dislocations, and this results in increased work hardening when multiple slip systems are active. Lunarska also notes that 'the presence of residuals (even at very low concentration) can drastically change all of these effects'.

It is partly because of the variable effect of hydrogen (giving both softening and hardening, according to the nature of the slip) that the extrapolation of model experiments on very pure iron to predict the behaviour of commercial materials is so difficult. It is further hindered by the ability of dissolved hydrogen to modify the dislocation structure of a straining material.

Brittle Fracture Processes By far the most important impact of dissolved hydrogen on the mechanical properties of steels and particularly high strength steels is the production of brittle fracture where the steel would normally behave in a ductile fashion. If a steel contains dissolved hydrogen this can result in immediate fracture at stresses approaching the fracture stress in the absence of hydrogen, or it may result in delayed failure at lower stresses (Fig. 8.44). The latter behaviour is most pronounced for tensile tests on notched or pre-cracked specimens, as the stress field around the notch or crack creates a high triaxial stress which dilates the metal lattice and tends to attract hydrogen from other parts of the specimen.

Some typical properties of common steels when exposed to environments supplying hydrogen are presented in Table 8.4. This table is presented to illustrate the order of magnitude of hydrogen embrittlement effects, and it is important to appreciate that the heat treatment and mechanical processing of a particular material, as well as the exposure conditions, can markedly affect its resistance to hydrogen embrittlement. Many of the data presented in this table have been extracted from the reviews of Sandoz[18] and McIntyre[19], to which the reader is referred for further information. One useful fact which is indicated by the data in Table 8.4 is the good hydrogen embrittlement resistance of maraging steels when used somewhat below their ultimate capability. These data also illustrate the general result that a given steel will usually have a greater susceptibility to hydrogen embrittlement as it is tempered to a higher strength.

Table 8.4 Typical properties of common high-strength steels in salt solutions

Material	Yield stress (MPa)	Failure time* (h)	K_{Ic} (MPa $\sqrt{m}$)	K_{Iscc} (MPa $\sqrt{m}$)	Plateau velocity (m/s)
HY130	900	no failure	130–170	60–140	5×10^{-8}
300M	1600	40–150	–	115	10^{-8}
4340	1400–1800	–	–	20	10^{-5}–10^{-2}
4340	<1400	–	–	30–80	
4130	1300	–	–	30	10^{-6}–10^{-4}
4130	1050	–	–	120	–
18Ni maraging steel	1900	400	40–100	5–35	–
18Ni maraging steel	1500	10 000	180	40–100	–
18Ni maraging steel	1200	no failure	160	120	10^{-5}

*The failure time presented is the approximate time to failure at an applied stress of 75% of the yield stress of the material.

The Influence of Microstructure and Composition

The influence of the composition and metallurgical structure of steels on their susceptibility to hydrogen embrittlement has been discussed from a theoretical basis by Bernstein and Pressouyre[5]. They considered the nature of the various trap sites in the steel, both with respect to their tendency to accumulate hydrogen, and their sensitivity to fracture in the presence of hydrogen. This approach seems to offer considerable potential in designing alloys for hydrogen resistance without sacrificing other properties, although with our current knowledge the methods are probably more relevant for the rationalisation of results.

As a general rule the damage due to dissolved hydrogen tends to become more severe as the strength of the steel increases. In part this may be associated with the greater resistance of the stronger material to plastic deformation, which facilitates the transition to brittle behaviour. In addition the size of the crack-tip plastic zone, which defines the size of the region to which hydrogen is attracted, is inversely proportional to the yield strength of the material. Hence, for the stronger steels a given amount of dissolved hydrogen will be concentrated into a smaller region, and will therefore have a more damaging effect. However this general effect of strength level does not explain all aspects of hydrogen embrittlement of high strength steel, and steels with the same mechanical properties may be affected by hydrogen in quite different ways as a result of their different microstructures.

A particularly important aspect of the microstructural state of the steel is the condition of the grain boundary. As noted in above, 'pure' grain boundaries do not act as major sites for the trapping of hydrogen, but it is clear that impurity segregation and carbide precipitation at the grain boundary may significantly modify its behaviour. Thus it is now widely recognised that there is a strong link between various forms of temper embrittlement and hydrogen embrittlement, and quite small changes in tempering treatment can give large variations in hydrogen embrittlement resistance[20,21].

On a more positive note, it seems clear that steels can be made more resistant to the effects of hydrogen by incorporating as many strong, finely dispersed traps in the microstructure as is possible, while ensuring that there are no continuous trap sites (such as embrittled grain boundaries).

This may explain the better resistance of steels hardened by cold-working compared with quenched and tempered steels[22]. Whatever the reason, this effect is particularly fortunate in the production of prestressed concrete, where cold-drawn pearlitic (or 'patented') wires give high strength levels (around 1 700 MPa) with good resistance to hydrogen embrittlement. In this particular case it is probably also important that most of the phase boundaries in the structure lie parallel to the tensile axis, allowing them to collect large quantities of hydrogen without serious detriment to the performance of the wire.

Similarly it seems that retained austenite may be beneficial in certain circumstances[23], probably because the austenite acts as a barrier to the diffusion of hydrogen, although in high concentrations (such as those obtained in duplex stainless steels) the austenite can also act as a crack stopper (i.e. a ductile region in the microstructure which blunts and stops the brittle crack).

Theories of Hydrogen Embrittlement

Despite the major technical importance of hydrogen embrittlement, and the wealth of research work on the subject, the mechanism (or perhaps mechanisms) of hydrogen embrittlement remains uncertain. Much of the book edited by Oriani, Hirth and Smialowski[1] is concerned with mechanistic aspects of hydrogen embrittlement, and the reader is referred in particular to the summary by Thomson and Lin[24].

In considering hydrogen embrittlement mechanisms it is important to keep in mind the concentration of hydrogen in the steel, as matrix concentrations are very low, typically of the order of one hydrogen atom for every 10^6 iron atoms. It is very difficult to see how such small amounts of hydrogen can modify fracture properties so markedly, and it must be supposed that hydrogen present in traps or possibly in surface layers (where the solubility may be markedly increased[4]) is the main cause of embrittlement. Thus realistic mechanisms of hydrogen embrittlement will be based on the effect of hydrogen on dislocation behaviour, on the effect of hydrogen at phase boundaries or grain boundaries or on the effect of hydrogen at the metal surface. It might be argued that the role of triaxial stress is to concentrate the hydrogen in the lattice to such a level that it can affect the mechanical behaviour of the steel, but very large concentration factors would appear to be necessary, and in any case the trap sites will presumably still contain much more hydrogen.

Steels are normally ductile at ambient temperatures, although they are often close to brittle behaviour, as is indicated by the ductile-brittle transition temperature. If the conditions at the tip of a sharp crack are considered, it can be seen that brittle fracture will occur if it is easier to break the atomic bond at the tip of the crack than it is to emit a dislocation to blunt the crack (see Thompson and Lin[24]). As dislocation emission is more temperature sensitive than the bond strength it becomes more difficult at low temperatures and brittle fracture occurs. The very severe effects of hydrogen on the performance of steels can be attributed to its role in allowing brittle fracture at ambient temperatures. On the basis of these concepts two possible effects of hydrogen can be identified:

1. Hydrogen can decrease the strength of the metal–metal bond, thereby facilitating brittle fracture. Both the decohesion and surface energy models are based on this premise.
2. Hydrogen can increase the stress required to emit dislocations from the crack tip, thereby making ductile fracture more difficult.

It should be appreciated that these effects are not mutually exclusive.

A third mode of action of hydrogen assumes that the fracture remains inherently ductile, but that the deformation becomes highly localised, thereby markedly reducing the critical strain energy release rate.

In considering the various theories it is also apparent that many of them may be considered as alternative descriptions of essentially the same physical process, or as descriptions of parallel processes which collaborate in the failure. Thus a complete description of hydrogen embrittlement in a given situation will almost inevitably incorporate aspects of several of the following theories.

The Pressure Theory The earliest theory of hydrogen embrittlement was probably the planar pressure theory advanced by Zappfe and Sims[25] in 1941. This essentially proposes that the effect of hydrogen is to create very high pressures of hydrogen gas in voids and other defects within the metal, thereby assisting in the fracture of the steel. While this is an important aspect of the blistering of steel by hydrogen-induced cracking, it cannot of itself explain the hydrogen embrittlement of high-strength steels, where fracture may occur in steel in equilibrium with hydrogen at very low pressures[26]. Some concentration of hydrogen may occur as a result of dislocation transport, but it is difficult to see how significant internal pressures can be generated by hydrogen entering from an external pressure of 0.001 atm.

Decohesion Theories The decohesion models proposed by Troiano, Oriani and others[27-29] suggest that the role of hydrogen is to weaken the interatomic bonds in the steel, thereby facilitating grain boundary separation or cleavage crack growth. In view of the very low hydrogen concentration in the matrix it is necessary for some method to exist by which the hydrogen can be concentrated at the site of the fracture. For cracking along phase or grain boundaries this can be explained in terms of the trapping of hydrogen at the phase boundary. It is a little more difficult to see how transgranular cracking can be explained; processes which have been invoked include the concentration of hydrogen at the region of triaxial tensile stress at the crack tip and local high concentrations of hydrogen being generated by reaction or adsorption at the crack tip.

Surface Energy Theories Surface energy theories were first proposed by Petch and Stables[30]. By lowering the surface energy of the newly-formed crack, the hydrogen reduces the stress intensity required for brittle fracture. As with the decohesion models, surface energy models only seem reasonable for the case of hydrogen derived from surface layers or grain boundaries, since the hydrogen adsorption must occur at the same time as the fracture event in order for the reduction in surface energy to be effective in lowering the energy required for fracture.

Hydride Formation In some systems, notably titanium alloys, hydrogen embrittlement has been attributed to the formation of brittle hydride phases at the crack tip. This has been postulated by Gahr *et al.*[31] as a mechanism of hydrogen embrittlement of niobium. While there is little evidence for hydride formation in steels, it can be argued that hydrides would be unstable and would dissolve as soon as the crack had propagated through them. In view of recent evidence of significant structural rearrangement of steel surface containing chemisorbed hydrogen[4], the possibility should also be considered of the induction of cleavage by a brittle surface film, similar to the film-induced cleavage model of stress-corrosion cracking[32].

Local Plasticity Theories Various workers have suggested that hydrogen acts by reducing the stress required for dislocation motion. This follows observations by several workers (summarised by Morgan and McMahon[33]) of enhanced dislocation motion in thin films exposed to hydrogen. More recently Lynch[34] has proposed that fracture by hydrogen embrittlement involves very fine scale ductility. He found very similar fractography for liquid metal embrittlement and hydrogen embrittlement, and proposed that both processes involve the facilitation of dislocation emission at the crack tip. Thus the crack growth occurs by an extremely localised plastic deformation at the crack tip.

Of the various theories, those based on decohesion appear to give the most satisfactory explanation of the more common case (for high-strength steels) of intergranular hydrogen embrittlement. Transgranular hydrogen embrittlement is somewhat more problematic, with the most critical test probably being the explanation of the effects of hydrogen at relatively low matrix concentrations. A decohesion process may be supported by consideration of such factors as dislocation transport and the concentration of hydrogen at regions of high triaxial tension, but local plasticity and the effects of surface 'hydride' films may also explain the observed behaviour.

Hydrogen Embrittlement Tests

There are several classes of test for hydrogen embrittlement, according to the application. Three general types of mechanical test can be identified, together with chemical and electrochemical tests intended to determine the hydrogen content of steels or the rate of entry of hydrogen from an environment.

Constant Stress Tests The simplest test for hydrogen embrittlement involves applying a constant stress to a specimen. This can be applied both to the testing of samples that already contain hydrogen (e.g. as a result of electroplating) and to the testing of samples in environments causing hydrogen entry. The specimen may take various forms (see the section on tests for stress-corrosion cracking), but is commonly a tensile specimen. Particularly where the test is being used to examine the effect of dissolved hydrogen already present in the steel, the specimen can be notched, in order to develop a region of triaxial stress at the tip of the notch which will tend to concentrate hydrogen.

When a constant stress test of a notched specimen is being used for the evaluation of samples containing hydrogen it is commonly referred to as a sustained load test, although the terms stress rupture and static fatigue are also used in the older literature. The results of a typical sustained load test are shown in Fig. 8.44. At a stress which is frequently (though not necessarily) below the notch tensile strength for hydrogen-free material the specimen will fail instantaneously as a result of the damaging effect of the uniform hydrogen concentration in the steel. At lower stresses delayed failure occurs as hydrogen diffuses to the region of triaxial stress at the tip of the notch. Eventually, as the stress is reduced, the quantity of hydrogen in the neighbourhood of the notch is insufficient to cause failure, and a critical or threshold stress is reached. The behaviour of a steel sample in this test is a function of both the material and the hydrogen concentration, with an increase in hydrogen concentration giving instantaneous failure at lower stresses, shorter times to failure and a lower critical stress. Testing at lower temperatures will give longer failure times, since the rate-controlling process is hydrogen diffusion, but the critical stress will go down, as the intrinsic tendency of the steel to brittle rather than ductile failure is increased. At elevated temperatures hydrogen may be lost by degassing and by trapping, and in this case the critical stress will increase. The failure stress is also strongly influenced by the notch tip radius, with a smaller radius giving shorter times to failure and failure at lower stresses. This can be explained in terms of the size and intensity of the region of triaxial stress at the notch tip, with a sharper notch giving a smaller triaxial region, which can develop a higher hydrogen concentration in a given time. Because of the difficulty of obtaining a reproducible notch tip geometry the sustained load test tends to give rather scattered results. Smooth specimens can be used for sustained load testing, but in this case the defects responsible for local hydrogen concentration will typically be surface or near-surface non-metallic inclusions, and the chance organisation of these will give highly scattered results.

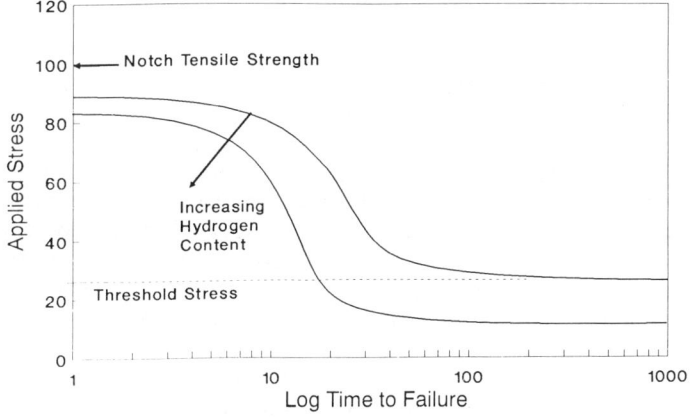

Fig. 8.44 Typical results of static load tests

Smooth specimen results will also tend to be non-conservative, as the majority of real components are liable to contain stress concentrators of one form or another.

In addition to testing for the effects of pre-existing hydrogen in the steel, sustained load tests can also be used to test for the effect of hydrogen entering from the environment. Both smooth and notched specimens can be used. At short times the results obtained will differ from those for steels containing dissolved hydrogen, since time will be required for the hydrogen to enter the steel. However this time can be very short, since hydrogen can diffuse about 10 μm in 1s, and failure times of the order of seconds can be obtained for severe environments and loading conditions. For tests at ambient temperature it is found that the critical stress is reached fairly quickly, and it is generally found that failures will occur within a month of exposure of the samples. This provides something of a problem in relating service behaviour to laboratory tests, since service failures are often observed after much longer exposure times. This may be a result of variations in exposure conditions in service or it may be associated with the need to develop initiating defects by pitting corrosion or similar processes. Whatever the cause, it does lead to the conclusion that sustained load tests cannot be used to predict service life with any accuracy.

Controlled Strain-rate Tests Controlled strain-rate tests were first developed by Parkins (see Ugiansky and Payer[35]) for the study of stress-corrosion cracking. These took the form of constant strain-rate tests (also known, perhaps more accurately, as constant extension-rate tests). Since then alternative forms of test have been developed to modify the conditions under which the specimen is exposed.

The slow strain-rate test is based on the principle that stress-corrosion cracking processes are normally dependent on plastic strain in the material. By extending the specimen very slowly any stress-corrosion cracking phenomena are given every possible opportunity to occur, hence the slow strain-rate test can quickly reveal any tendency to cracking in a given metal–environment combination. A particular advantage of the slow strain-rate test is that failure of the specimen will always occur in a reasonable time, if only by normal ductile fracture processes. Test durations clearly depend on the strain rate and the ductility of the metal, but typical test durations for steels in hydrogen embrittling environments are around one week. In addition to providing a rapid indication of the possibility of stress-corrosion cracking in a particular system, the slow strain-rate test can also be used to study the effect of material and environmental factors on the susceptibility. For example Fig. 8.45 shows the effect of environment composition and tempering temperature on the reduction in area obtained for tests on a quenched and tempered steel in various aqueous solutions. The minimum in the reduction in area which can be seen for tempering temperatures of 350°C provides evidence of the interaction between hydrogen embrittlement and temper embrittlement.

While the conventional slow strain-rate test offers many benefits, it does suffer from a tendency to overstate the susceptibility of materials to hydrogen embrittlement. Thus structural steels of modest strength will fail even under conditions giving relatively low rates of hydrogen entry. This is

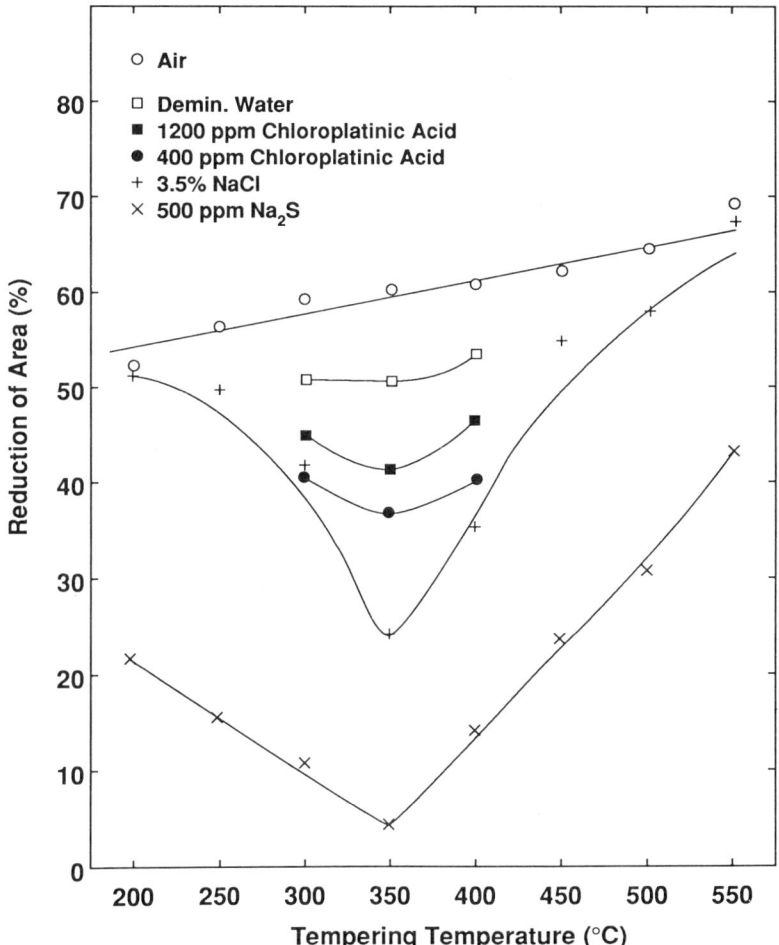

Fig. 8.45 Variation of reduction in area as a function of tempering temperature (after Johnston et al.[21])

because the enforced plastic deformation provides a very severe test condition, which may not be reproduced in service when relatively low stresses are applied to components. This problem is partly addressed by the sustained load test, but this may be unduly mild, as there will be no plastic strain in the specimen once the initial creep strain has died away. In order to provide a test with intermediate severity, Erlings et al.[36] developed a test which exposes the specimen to a quasi-constant strain rate, but at the same time keeps the specimen within the elastic range. This involves prestraining a slow strain-rate specimen up to the yield stress. Then the specimen is exposed to a regular cycle, the stress being reduced (typically) to 90% of the yield stress, slowly strained to 95% of the yield stress, then dropped back to 90%. This test is proposed as a part of a sequence of tests, starting with the low-cost slow strain-rate test, for the qualification of materials for service, as indicated in Fig. 8.46.

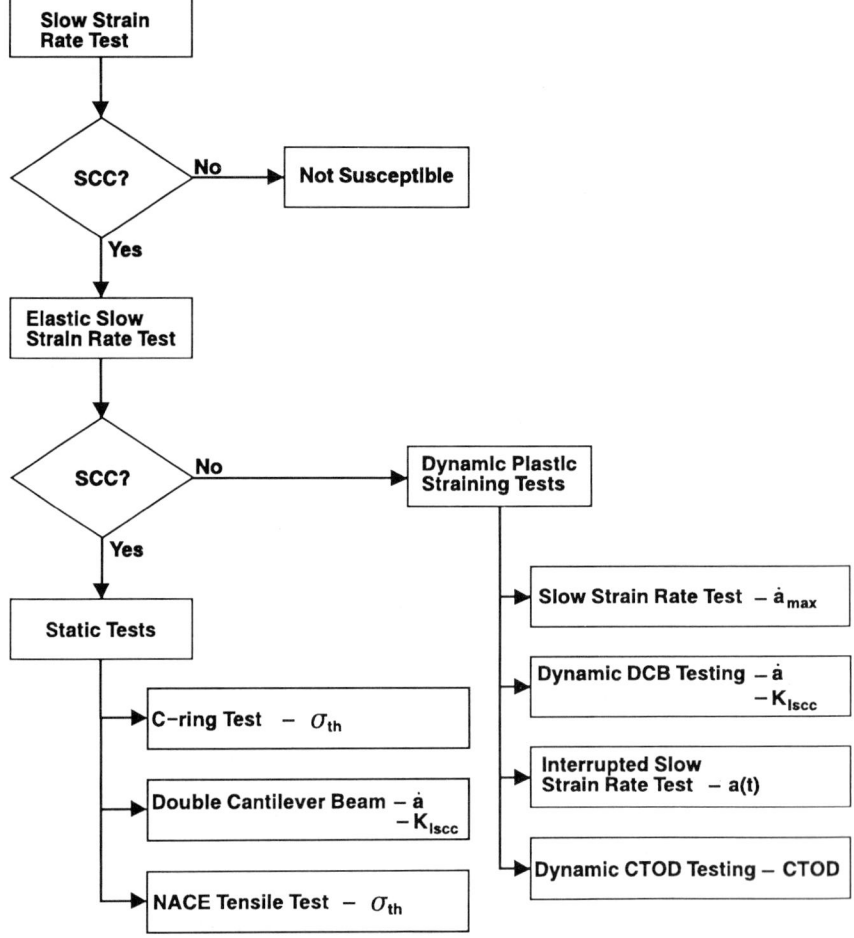

Fig. 8.46 Hydrogen embrittlement test strategy

Another modification to the slow strain-rate test involves the superimposition of a low amplitude sine wave ripple on the slow uniform extension (Fig. 8.47). In effect this produces higher strain rates (which appear to be more damaging for hydrogen embrittlement), while still giving a long test duration, with adequate time for the accumulation of hydrogen in the steel[37].

Fracture Mechanics Tests One problem of both sustained load and slow strain-rate tests is that they do not provide a means of predicting the behaviour of components containing defects (other than the inherent defect associated with the notch in a sustained load test). Fracture mechanics provides a basis for such tests (Section 8.9), and measurements of crack velocity as a function of stress intensity factor, K, are widely used. A typical graph of crack velocity as a function of K is shown in Fig. 8.48. Several regions may be seen on this curve. At low stress intensity factors no crack growth is

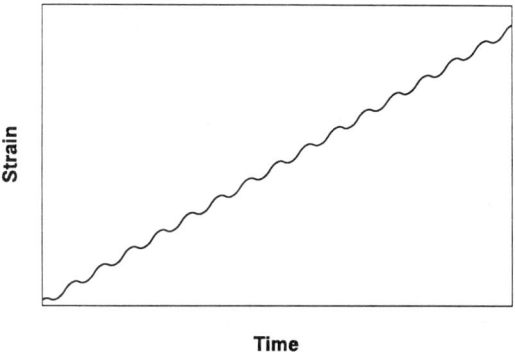

Fig. 8.47 Strain–time history for 'ripple' test (schematic)

observed until a threshold stress intensity factor, K_{Iscc}, is reached. Typically K_{Iscc} is fairly insensitive to the rate of hydrogen entry into the steel (as the crack is not growing at a significant rate, hydrogen will eventually accumulate at a sufficient concentration at the crack tip, whatever the rate of entry). As the stress intensity factor increases above K_{Iscc} the crack growth rate increases rapidly, until a plateau crack velocity is attained. The plateau velocity is a function of the rate of hydrogen entry, and the plateau is usually attributed to the availability of hydrogen at the crack tip being the rate-limiting factor. Finally the crack velocity increases again as the stress intensity factor approaches the critical stress intensity factor for fast fracture, K_{Ic}. In this region the fracture consists partly of ductile tearing and

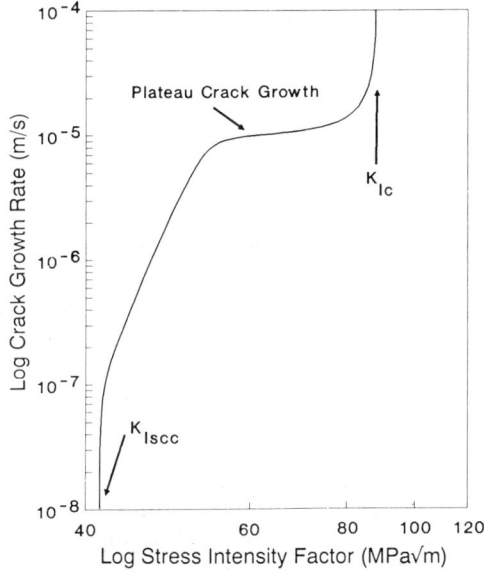

Fig. 8.48 Stress-corrosion crack velocity as a function of applied stress intensity factor

partly of hydrogen embrittlement, with the hydrogen assisting the fracture of the tougher areas in the metal.

The above discussion has assumed that the crack is loaded in mode I (the crack opening mode, with a tensile stress normal to the plane of the crack). Hydrogen has relatively little effect in modes II or III, as these generate shear stresses at the crack tip, rather than tensile stresses, and the shear behaviour of steels is relatively little affected by hydrogen, presumably because dilation of the lattice at the crack tip (which does not occur in modes II and III) is required for hydrogen accumulation.

Controlling Hydrogen Embrittlement in Service

Clearly there are two major ways in which hydrogen embrittlement can be controlled in service; either the hydrogen content of the steel can be reduced (by limiting the rate of entry or by baking to remove existing hydrogen), or the steel can be made more resistant to hydrogen.

Reduction of Hydrogen Content Pre-existing hydrogen in steel, introduced by electroplating or other processing, can be reduced by a suitable de-embrittlement treatment, as discussed above. A more difficult problem is the limitation of hydrogen entry in service. The feasibility of controlling hydrogen embrittlement in this way will clearly depend on the application, and in particular on the source from which the hydrogen is entering the steel. Thus hydrogen pickup from acid pickling can be reduced by the use of appropriate inhibitors, and de-embrittlement treatments can be used after pickling. In contrast hydrogen entry into components in an electrical generator using hydrogen as the cooling gas cannot easily be modified.

In general preventing the entry of hydrogen from the gas phase is liable to be difficult. Inhibition is possible; for example small quantities of oxygen can markedly reduce the rate of hydrogen uptake, presumably as a result of blocking active sites on the metal surface. Similarly ethylene gas will react with adsorbed hydrogen atoms, thereby preventing them from diffusing into the steel. However, there will be few cases in which it is permissible to modify the environment in this way (adding oxygen to hydrogen will certainly cause some concern). Organic coatings will not generally be very effective in reducing hydrogen entry, since these are usually rather permeable to hydrogen. Furthermore, such coatings are usually susceptible to local damage. Metal coatings may offer some reductions in hydrogen entry, as most close-packed metals have a significantly lower diffusion coefficient for hydrogen than iron, although this is partially compensated for by a higher solubility.

In aqueous solutions it becomes somewhat more feasible to modify the entry of hydrogen into the steel. This can be achieved by the addition of inhibitors to the solution, by control of the electrochemical potential of the metal and by coatings.

In situations such as the acid pickling of steel or the use of steel pipes to handle sour oil streams, the use of suitable inhibitors can give a significant reduction in hydrogen entry. In this context it is important to emphasise that the efficiency of an inhibitor in reducing hydrogen entry is not the same as its efficiency in reducing corrosion. Thus arsenic and antimony compounds are effective inhibitors of corrosion of steel in acid, but they achieve this by inhibiting the hydrogen recombination reaction, thereby increasing the rate of hydrogen entry into the steel. In contrast the addition of chloroplatinic acid to a solution will increase the corrosion rate by accelerating the hydrogen evolution reaction. However a part of this acceleration arises because the hydrogen recombination reaction is catalysed by the particles of platinum which are deposited on the surface by an exchange reaction. As a result chloroplatinic acid acts as an inhibitor of hydrogen entry. This can be seen in Fig. 8.45, where the addition of chloroplatinic acid to a 3.5% sodium chloride solution has increased the reduction in area for a given heat treatment. Clearly this is impractical for service use, but several organic inhibitors are available which are effective in reducing both corrosion and hydrogen entry into the steel.

The electrochemical potential has a direct influence on the entry of hydrogen into steel from aqueous solutions, and control of potential represents an obvious way to control hydrogen uptake. Unfortunately the effect of potential on hydrogen entry is complex, and is strongly dependent on solution composition. For example, as has been discussed above, it has been shown[14] that anodic polarisation of steel in salt solution will give an increased rate of hydrogen entry in aerated solution, but negligible entry in deaerated solution. The situation becomes even more complex in passivating environments, especially where pitting or crevice corrosion can occur. As a generalisation, for aerated neutral salt solutions the optimum potential for minimum hydrogen uptake appears to be slightly below typical free corrosion potentials, a typical value for seawater being around −750 to −800 mV (SCE). This potential is also sufficient to reduce corrosion rates, but it gives very little margin for potential differences over the structure being protected. Potentials much below −900 mV will undoubtedly increase the rate of hydrogen entry, and in view of the hazards involved in overprotection, the deliberate application of cathodic protection to high-strength steel structures is probably best avoided.

As in gaseous environments, metal coatings can reduce the rate of hydrogen entry by acting as a low-permeability barrier. In addition, the coating metal may also modify the electrochemical properties at the metal–solution interface, as in the case of traces of platinum derived from chloroplatinic acid. For application in seawater and other neutral salt solutions, zinc and cadmium probably offer the best combinations of properties, having low exchange current densities for hydrogen evolution and relatively low hydrogen permeabilities. In the event that part of the steel substrate is exposed by mechanical damage, a zinc coating has a rather low corrosion potential in seawater, and may be expected to give rapid hydrogen entry. However cadmium gives a potential which is around that at which the minimum rate of hydrogen entry is observed, and is highly recommended as a coating material for high strength steels. Unfortunately, there are moves to phase out the use of cadmium as a result of the toxicity of cadmium compounds. Proposed schemes for 'cadmium replacement' should be examined carefully for their effect on hydrogen embrittlement as well as their corrosion performance.

Organic coatings such as paints are also effective in reducing the corrosion of steel, but as with inhibitors it is not certain that this will also mean that the rate of hydrogen entry is reduced. Organic coatings are typically rather permeable to water, and it is possible for this to be reduced to hydrogen. Whether this hydrogen subsequently enters the steel will depend on the characteristics of the coating-steel interface, and little is known about this. It is recommended that suitable tests should be performed before using organic coatings for protection of high-strength steels.

Increasing Resistance of Steel For a given steel the hardness will play a major part in determining its resistance to hydrogen embrittlement, but other factors are also significant. Thus it is particularly important to obtain an appropriate microstructure, without any temper embrittlement or other deleterious features (see above). Where possible the use of work-hardened ferritic-pearlitic structures will probably give better performance than quenching and tempering, providing the stress is applied parallel to the working direction.

In particularly severe conditions it may be found necessary to use austenitic steels, such as stainless steels from the 300 series. The *fcc* structure of austenite is inherently more ductile than *bcc* ferrite, and it is therefore less liable to switch to brittle modes of fracture in the presence of hydrogen. Additionally the permeability of austenitic steels is less than that of ferritic. The common commercial grades of austenitic stainless steel tend to be rather marginal in respect of the stability of the austenite phase, and they are therefore susceptible to the formation of martensite by strain-induced transformation. This martensite is susceptible to hydrogen embrittlement, and it is therefore important to ensure that steels used for hydrogen service have sufficient austenite-stabilising elements, primarily nickel. This is illustrated in Fig. 8.49, where the effect of nickel content on the residual ductility in hydrogen is presented[38]. Below about 10% Ni there is a strong tendency to strain-induced martensite formation, and low residual ductilities are observed. Once the austenite becomes fully stable, however, there is hardly any loss in ductility due to hydrogen.

Prediction of Behaviour An important aspect of the use of materials in aggressive environments is the prediction of service life, such that planned maintenance procedures can be used. A typical approach that might be suggested for steels in conditions where hydrogen embrittlement can occur is to assume a maximum possible defect in the structure, usually based on the expected capabilities of the inspection system. Using a suitable relationship for the crack velocity as a function of stress intensity factor, the time to failure is calculated by integrating from the initial defect size up to that necessary for fast fracture. While this method is attractive in theory, in practice it is liable to be rather unproductive for high-strength steels. The reason for this is that typical plateau crack velocities for hydrogen embrittlement are very high (between 10^{-6} and 10^{-2} m/s), and the transition from K_{Iscc} to the plateau occurs over a rather small range of stress-intensity factor. Thus, unless the initial defect happens to give a stress-intensity factor fractionally above K_{Iscc}, this calculation will predict either no failure or a very short life (even 10^{-6} m/s corresponds to a crack velocity of 3.6 mm/h). Currently the understanding of the failure of components by hydrogen embrittlement after long periods in service remains somewhat limited, but in many cases it seems probable that pitting corrosion plays an important part in the slow

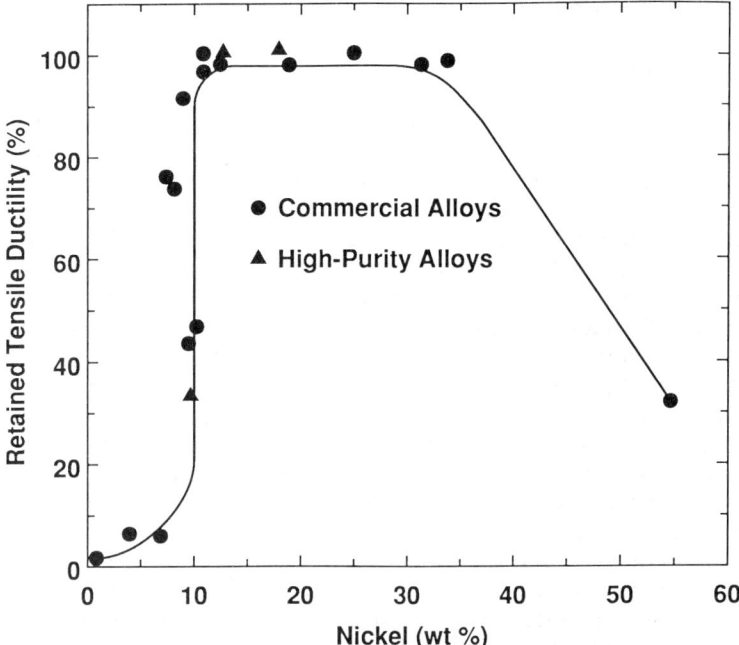

Fig. 8.49 Effect of nickel content on residual ductility of austenitic stainless steels (after Caskey[38])

crack initiation process. Thus prediction of life in hydrogen embrittlement conditions should be concerned with initiation processes in addition to the crack growth above K_{Iscc}, although there is not yet a sufficiently clear understanding of the initiation processes for this to be a productive exercise.

Stress-corrosion Cracking by 'Non-hydrogen' Mechanisms

As indicated by the discussion above, the major stress-corrosion failure mode for high-strength steels is hydrogen embrittlement. To some extent this is probably because of the very marked sensitivity of these materials to hydrogen, which tends to mask any susceptibility to stress-corrosion cracking by other mechanisms. Additionally, other forms of stress-corrosion cracking frequently require more plastic strain than can readily be achieved with these high-strength materials. The reader is therefore referred to the section on the stress-corrosion cracking of lower-strength steels (Section 8.2) for a discussion of environments in which high-strength steels may suffer from stress-corrosion cracking (providing hydrogen does not cause failure first). In assessing the role of hydrogen in a particular stress-corrosion cracking process, it is important to be fully aware of the complexities of the chemistry of iron in aqueous solutions, and in particular the possibility of hydrogen entering the steel in electrochemical conditions which nominally imply an underpotential for hydrogen evolution.

Corrosion Fatigue

Crack Propagation

The large majority of fatigue work in recent years has been concerned with the growth of cracks in the regime where the crack can be described by fracture mechanics criteria such as the stress intensity factor range, ΔK (Sections 8.6 and 8.9). In many ways the corrosion fatigue crack growth behaviour of high-strength steels can be described as a combination of the corrosion fatigue of lower-strength steels superimposed on the hydrogen embrittlement of the higher-strength material. Thus two regimes of behaviour can usually be detected (once above the threshold stress intensity factor range, which is discussed further below).

At values of K_{max} below K_{Iscc} the crack grows by 'pure' corrosion fatigue processes similar to those observed with lower-strength steels. The enhancement of crack growth in this region is usually relatively small compared with tests in air, typically around 3–10 times for aqueous environments, falling to near-zero close to the fatigue threshold. Typical behaviour for HY130 is shown in Fig. 8.50 due to Vosikovsky[39]. This figure also shows the effect of load cycle frequency, and the transition from little effect of hydrogen at 10 Hz to large effects at frequencies of 0.1 Hz and below is typical. The region of frequency-independent growth rate at lower ΔK is somewhat less reproducible. It should be noted that HY130 has particularly good properties for its strength, and K_{Iscc} for the conditions shown in Fig. 8.50 is estimated as greater than 110 MPa $\sqrt{m}$. It should also be noted that Fig. 8.50 shows results for tests at an applied cathodic potential of –1.03 V (SCE), and the effect of hydrogen embrittlement at higher ΔK and low frequency is relatively large. Tests performed at the free corrosion potential (–0.67 V) show a maximum acceleration in growth rate of about four times. For steels with lower values of K_{Iscc} this region of the curve may be entirely swamped by region (b), as shown in Fig. 8.51[40].

As K_{max} approaches K_{Iscc} the crack can also propagate by hydrogen embrittlement processes during the higher load parts of the stress cycle. This forms the basis of various models which have been developed to describe corrosion fatigue, probably the best-known of which are the superposition models due to Wei[41]. In its most recent version this model takes the form:

$$\frac{da}{dN_{total}} = \frac{da}{dN_{air}} + \frac{da}{dN_{scc}} + \frac{da}{dN_{cf}} \quad (8.14)$$

There remains some doubt as to the strict significance of the various terms in the superposition model. This applies particularly to da/dN_{cf}, which can currently only be determined by assuming the superposition model and analysing experimental results accordingly. Similarly, there are questions as to the correct way to sum the various terms (for example, it can be argued that one should consider only the larger of the three terms on the right hand side of the superposition equation). Despite these difficulties, superposition models have proved valuable in analysing the effect of factors such as frequency and temperature on corrosion fatigue.

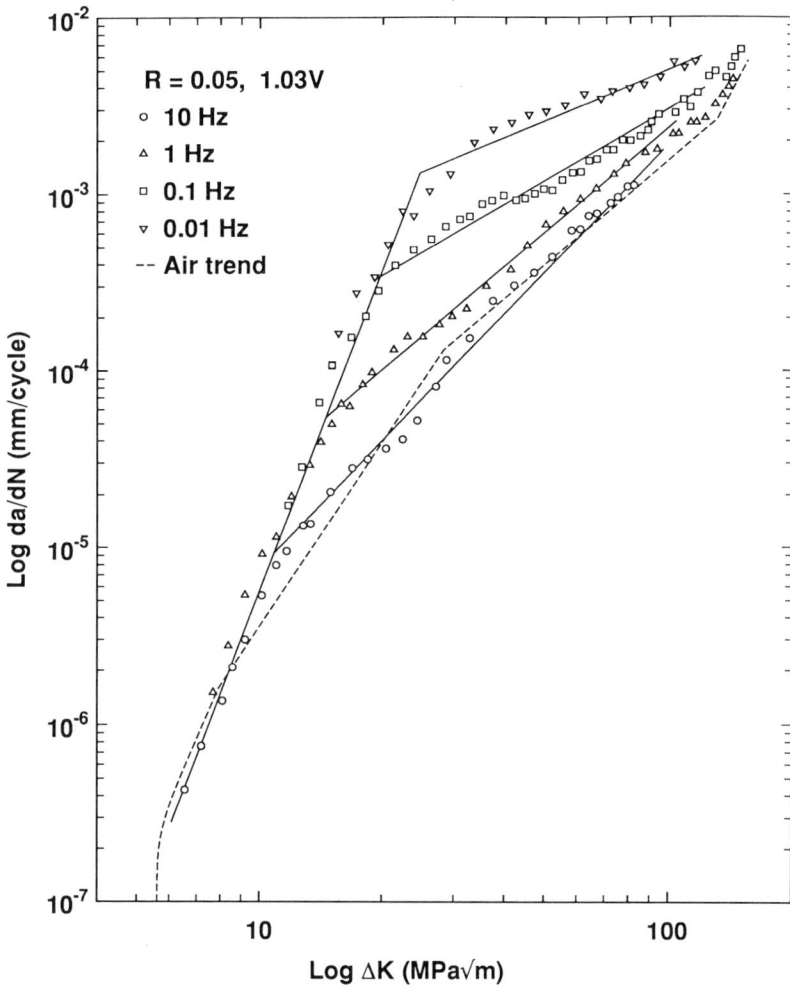

Fig. 8.50 Corrosion fatigue crack velocity as a function of ΔK for HY130 (after Vosikovsky[39])

In crack growth studies on 4130 steel, Gangloff[42] found very large increases in the crack growth rate of small cracks (less than 1 mm in depth), compared to longer cracks in the same nominal conditions (based on ΔK). This was attributed to the difference in the localised conditions at the crack tip[43], which gave a maximum hydrogen activity for the short cracks. Such marked effects have not been observed for lower-strength steels, probably because these are not susceptible to hydrogen embrittlement under static load.

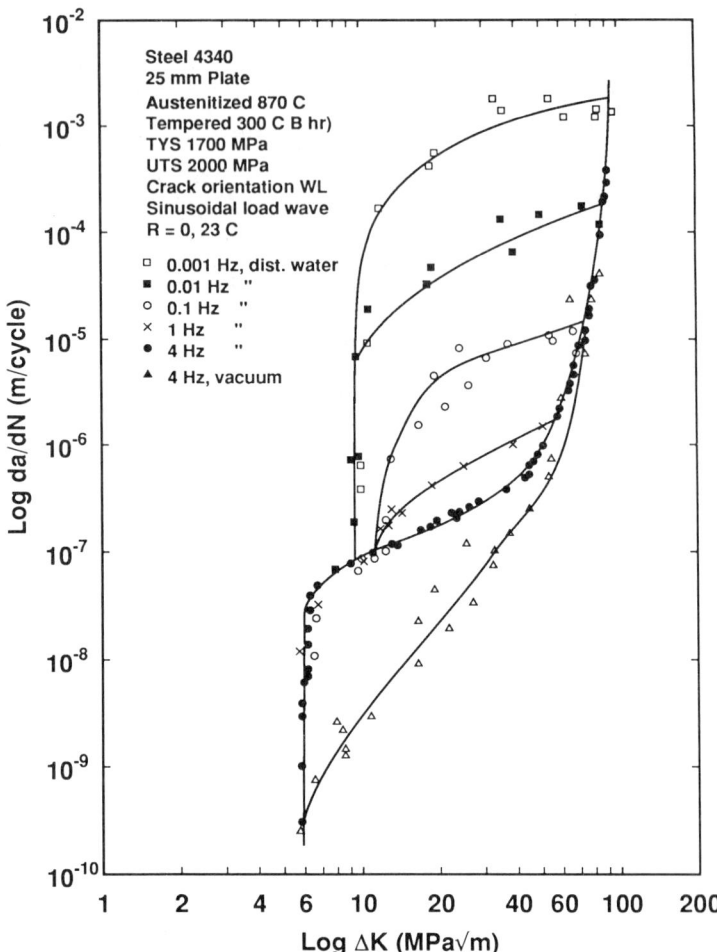

Fig. 8.51 Corrosion fatigue crack velocity as a function of ΔK for 4340 (after Spiedel[40])

Crack Initiation and the Threshold Stress Intensity Factor Range

The development of fracture mechanics and its application to fatigue studies by Paris and Erdogan in the early 1960s[44] has tended to eclipse fatigue tests on smooth specimens, and it is only in recent years that attention has returned to initiation studies. These have started to give an insight into the processes of fatigue crack initiation and the growth of small cracks, but the information available in respect of corrosion fatigue is rather limited.

For tests on smooth specimens it is generally believed that hydrogen will not affect the behaviour for tests in fully reversed bending at high frequencies (since there will be insufficient time within a cycle for the hydrogen to collect at regions of high triaxial stress). For tests with a tensile mean stress, or at low frequencies, there is evidence that dissolved hydrogen can reduce the fatigue life of smooth specimens. In aqueous solution the application of cathodic protection at potentials of around −750 mV (SCE) can be beneficial[45], whereas more negative potentials are damaging. This could be explained on the basis of the effects of potential on hydrogen entry discussed above, but there is also evidence that pitting corrosion around inclusions and corrosion of the subsequent fatigue crack plays a major role in the initiation and early growth of cracks[46].

R. A. COTTIS

REFERENCES

1. Oriani. R. A. Hirth, J. P. and Smialowski, M. (Eds.), *Hydrogen Degradation of Ferrous Alloys* Noyes Publications, Park Ridge, New Jersey (1984)
2. Obuzor, U. W., PhD Thesis, UMIST, Manchester (1989)
3. Pasco, R. W. and Ficalora, P. J., Reference 1, pp. 199-214
4. Imbihl, R., Behm, R. J., Christmann, K., Ertl, G. and Matsushima, T., *Surface Science*, **117**, 257 (1982)
5. Bernstein, I. M. and Pressouyre, G. M., Reference 1, pp. 641-685
6. Puls, M. P., Reference 1, pp. 114-130
7. Asaoka, T., Lapasset, G., Aucouturier and Lacombe, P., *Corrosion*, **34** 39 (1978)
8. Kedzierzawski, P., Reference 1, pp. 271-288
9. Devanathan M.A.V. and Stackursi, Z., *J. Electrochem. Soc.*, **111**(5), 619 (1964)
10. Berman, D. A., Beck W. and Deluccia, J. J., *Hydrogen in Metals*, Ed. Bernstein I. M. and Thompson, A. W., ASM, pp. 575-607 (1974)
11. Arup, H., *Proc. 9th Scandinavian Corrosion Congress*, p. 825 (1984)
12. See, for example, AMS 2401C, ASTM A165, ISO 2082, BS 1706 and BS3382
13. Echizen, Y. and Takada, K., *Kinsoku Hyomen Gijutsu*, **38**(3), 113-115 (1987)
14. Barth, C. F. and Troiano, A. R., *Corrosion*, **28**(7), 259-263 (1972)
15. Taqi, E. A. and Cottis, R. A., *Corrosion Chemistry within Pits, Cracks and Crevices*, Ed. Turnbull, A. NPL, London, pp. 483-494 (1987)
16. Lunarska, E., Zielinski, A. and Smialowski, M., *Acta Met.*, **25**, 305-308 (1977)
17. Lunarska, E., Reference 1, pp. 321-352
18. Sandoz, G., In *Stress Corrosion Cracking in High Strength Steels and in Titanium and Aluminium Alloys*, Ed. B.F. Brown, Naval Research Laboratory, Washington, pp. 79-145, (1972)
19. McIntyre, P., Reference 1, pp. 763-798
20. Hippsley, C. A., *Mater. Sci. Tech.*, **3**(11), 912-922 (1987)
21. Johnston, J. W., Cottis R. A. and Procter, R. P. M., *Electrochem Soc. Extended Abstracts*, **87-2**, 454-455 (1987)
22. Ryder, D. A., Davies T. J. and Strecker, E., *2nd Int. Cong. on Hydrogen in Metals*, Vol. III, Paper 3B2, Pergamon Press, (1977)
23. Kerr, R., Guiterrez-Solona, F., Bernstein I. M. and Thompson, A. W., *Met. Trans. A*, **18A**, 1011-1022 (1987)
24. Thompson R. and Lin, I.-H., Reference 1, pp. 454-511
25. Zappfe C. and Sims, C., *Trans. AIME*, **145**, 225 (1941)
26. Oriani R. A. and Josephic, P. H., *Acta Metall.*, **22**, 1065 (1974)
27. Troiano, A. R., *Trans. A.S.M.*, **52**, 54 (1960)
28. Morlet, J. G., Johnson H. H. and Troiano, A. R., *J. Iron Steel Inst.*, **189**, 37(1958)
29. Oriani, R. A., *Stress Corrosion Cracking and Hydrogen Embrittlement of Iron Base Alloys*, NACE 5, pp. 351-358 (1973)
30. Petch N. J. and Stables, P., *Nature*, **169**, 842 (1952)
31. Gahr, S., Grossbeck M. L. and Birnbaum, H., *Acta Met.*, **25**, 125(1977)
32. Sieradzki K. and Newman, R. C., *Phil. Mag. A*, **51**, 95 (1985)
33. Morgan M. J. and McMahon, C. J., Jr., Reference 1, pp. 608-640
34. Lynch, S., *Scripta Met.*, **13**, 1051 (1979)
35. Ugiansky G. M. and Payer, J. H., (Eds) *Stress Corrosion Cracking-The Slow Strain Rate Technique*, ASTM STP 665 (1979)
36. Erlings, J. G., de Groot H. W. and Nauta, J., *Corros. Sci.*, **27** (10/11), 1153-1167 (1987)

37. Crooker T. W. and Hauser, J. A., II, *A Literature Review on the Influence of Small-Amplitude Cyclic Loading on Stress Corrosion Crack Growth in Alloys*, NRL Memorandum Report 5763, Naval Research Laboratory, Washington DC, April 3 (1988)
38. Caskey, G. R., *Third International Congress on Hydrogen and Materials*, Paris, pp. 611-616 (1982)
39. Vosikovsky, O., *J. Test and Eval.*, **6**, 175-182 (1978)
40. Spiedel, M. O., *Adv. in Fracture Research*, Vol. 6, Ed. Francois, D. Pergamon, Oxford (1982)
41. Wei, R. P., In *Fatigue Mechanisms*, ASTM STP 675, pp. 816-840 (1979)
42. Gangloff, R. P., *Res. Met. Let.*, **1**, 299-306 (1981)
43. Turnbull, A. and Ferriss, D. H., *Corros. Sci.* (1987)
44. Paris P. C. and Erdogan, F., *Basic Eng. Transactions of ASME*, **85D**, 528-534 (1963)
45. Brown, B. F. In *Corrosion Fatigue: Chemistry, Mechanics and Microstructure*, Ed. McEvily A. J. and Staehle, R. W., NACE, pp. 25-28 (1972)
46. Cottis, R. A., Markfield A. and Haritopoulos, P., *Environment Assisted Fatigue*, Ed. Scott, P. M. I. Mech. E. Sheffield (1988)

8.5 Stress-corrosion Cracking of Titanium, Magnesium and Aluminium Alloys

Titanium

Stress-corrosion cracking occurs in titanium alloys in a number of environments, although the number of failures that have occurred under service conditions is very small. Because of the widespread use of titanium alloys in aeroplanes and space vehicles and their increasing use in marine applications it is important that the possibilities of service failures should be removed. As a result a considerable and increasing amount of work has been done on this subject over the last decade as indicated in a recent extensive survey[1].

The two principal forms of stress-corrosion failure are (a) hot salt cracking and (b) room-temperature cracking, the latter occurring in both aqueous and methanolic chloride environments, and in N_2O_4. In addition, environmental failures can occur in alloys in direct contact with some liquid and solid metals, and certain gases.

Hot Salt Cracking[2,3]

Many titanium alloys develop cracks if they are heated under stress while in direct contact with solid chlorides in the presence of oxygen and moisture at temperatures higher than about 250°C, and cracking is predominantly intergranular. Originally the phenomenon was associated with residual traces of NaCl in perspiration[2] and it is sometimes referred to as *fingerprint cracking*. Many other chlorides can cause this type of failure to different degrees. The severity of attach varies[4] with the cationic species, becoming greater in the order $MgCl_2 < SrCl_2 < CaCl_2 < KCl < BaCl_2 < NaCl < LiCl$. Bromides and iodides also cause cracking but fluorides do not.

Mechanism The mechanism of cracking has not been established. Even the corrosion reaction that is responsible for the initial attack has not been determined. Early work[5] led to the suggestion that chlorine gas was generated and could cause fracture by a cyclic process requiring the formation and decomposition of $TiCl_2$:

$$Ti + TiO_2 + 2NaCl + \tfrac{1}{2}O_2 \rightarrow TiCl_2 + Na_2TiO_3$$
$$TiCl_2 + O_2 \rightarrow TiO_2 + Cl_2$$
$$Ti + Cl_2 \rightarrow TiCl_2 \text{ (at the crack tip)}$$

Chlorine has been identified[6] as a reaction product, and furthermore it has been shown[7] that cracking can occur in chlorine gas in the absence of oxygen.

The formation of low-melting-point mixtures has also been considered[8] as a possible step in the cracking process and these have been observed, but the wide range of chlorides that cause cracking suggests that such mixtures may not be necessary.

More recent work has focused attention upon the rôle of moisture. It is not entirely clear that cracking does not occur in its complete absence, but if it is present then it takes part in the corrosion process. Radiotracer techniques have been used[8] to show that hydrogen is retained in areas corroded at 343°C by salt containing tritiated water. Hydrochloric acid is considered to be the important corrosion product which then promotes the entry of hydrogen into the specimen with consequent intergranular separation by a sorption mechanism[8]. To verify this hypothesis, stressed specimens of Ti-8Al-1Mo-1V were tested[8] in glass ampoules containing hydrogen chloride gas at 1 atm and 343°C. Cracking occurred and the time to failure was related to the extent of corrosion which in turn was considered to be dependent upon the moisture content. Cracking was both intergranular and quasi-cleavage, and appeared similar to hot salt cracking.

The production of HCl and hydrogen is thought to arise from the reaction:

$$Ti + 2NaCl + 2H_2O \rightarrow TiCl_2 + 2NaOH + 2H$$

followed by several possible regenerative reactions which would occur at different intervals of time as moisture was absorbed:

$$TiCl_2 + 2H_2O \rightarrow TiO_2 + 2HCl + 2H$$
$$Ti + 2HCl \rightarrow TiCl_2 + 2H$$

The proposed mechanism includes the production of HCl from the pyrohydrolysis of the metal chlorides. Similar reactions are likely for bromides and iodides. Fluorides however are relatively stable and would not be expected to hydrolyse. It was considered that this might account for the inability of fluorides to cause cracking. Hydrogen absorption by titanium alloys exposed to chloride salts at elevated temperatures has been detected[9] and found to be proportional to the amount of moisture participating in the reaction.

Other work on Ti-8Al-1Mo-1V has also indicated[10] that hydrogen absorption occurs. Stressed specimens exposed to hot salt at 454°C for 100 h suffered a marked embrittlement when subsequently tested in air at room temperature. This embrittlement could be removed by vacuum annealing at 649°C for 4 h. The loss of ductility observed was greater as the strain rate of testing was lowered. Both these observations were considered to be indicative that hydrogen was the principal embrittling species. Analysis of the fracture surface of specimens containing 70 p.p.m. of hydrogen that had suffered hot salt cracking has shown[11] that concentrations as high as 12 000 p.p.m. of hydrogen are developed during the cracking process and this would appear to be conclusive evidence that hydrogen plays a major rôle when moisture is present. Cracking of components in dry air with a dewpoint of −84°C was intepreted[12] as showing that only a very low moisture content is required (0.25 p.p.m.). Increasing this by 25 000 fold by water injection did not increase the degree of embrittlement.

Alloy susceptibility Apart from unalloyed titanium all alloys exhibit some degree of susceptibility. An approximate rating derived from laboratory tests[13] indicates three general groups of alloys.

1. Highly susceptible alloys: Ti-5Al-2.5Sn, Ti-12Zr-7Al, Ti-5Al-5Sn-5Zr, Ti-8Al-1Mo-1V, Ti-8Mn.
2. Moderately susceptible alloys: Ti-5Al-5Sn-5Zr-1Mo-1V, Ti-6Al-4V, Ti-6Al-4V-2Sn, Ti-13V-11Cr-3Al.
3. Most resistant alloys: Ti-4Al-3Mo-1V, Ti-11Sn-5Zr-2.25Al-1Mo-0.25Si, Ti-4Mo-4Zr-2Al.

Tests in a $Cl_2 + O_2$ mixture at 427°C have shown[14] that the worst elements for promoting susceptibility are Al, Sn, Cu, V, Cr, Mn, Fe and Ni, while the least harmful are Zr, Ta and Mo. α-phase alloys are generally more susceptible than β-phase alloys. Heat treatment has not been examined extensively, but some heat treatments render some α-alloys more susceptible[3] or change the mode of fracture[6]. The general effect will depend upon the alloy and the heat-treatment cycle. Subsequent cold work can sometimes considerably lower susceptibility[6]. Failure times decrease as either the testing temperature or initial stress value is raised.

It is important to note that a variety of tests have been employed in studies of hot salt cracking and comparison between the results of different workers is not always possible. It is not clear how susceptibility should be defined and, not surprisingly this makes the rating of alloys difficult. It may also be added that at high temperatures of testing, changes may occur within an alloy which make it more or less susceptible than the alloy in the starting condition. The general corrosion rate will increase as the testing temperature is raised and will eventually become unacceptable. The creep rate will also increase and in some alloys this is a more critical criterion than cracking susceptibility.

Preventative methods Cracking can be delayed or possibly prevented by shot peening components (which serves to impose compressive stresses upon their surfaces) and by the prior application of some surface coatings, e.g. nickel plating, or dipped aluminium or zinc coatings[6]. Other work[4] has shown that susceptibility of as-machined specimens is substantially lowered by chemical milling which removes the stressed surface layers. It has also been reported[5] that the amount of corrosion that occurs is reduced and less cracking is observed as the velocity of the gaseous environment in contact with stressed components is increased. This is particularly relevant to components inside aeroplane engines, e.g. compressor blades. Such observations were made at 427°C. Other workers[15] have reported similar observations at 316°C but not at 371°C and most recent work[12] suggests that such effects are extremely small.

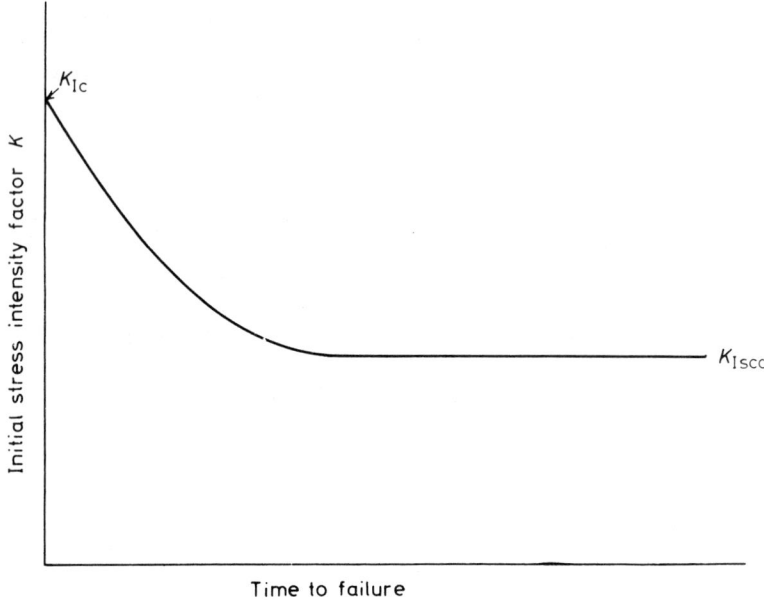

Fig. 8.52 Initial stress intensity factor and time to failure for a susceptible titanium alloy tested in a neutral aqueous environment under plane strain conditions

Room-temperature Cracking

Many titanium alloys are susceptible to stress-corrosion cracking in aqueous and methanolic chloride environments.

Aqueous environments Neutral chloride solutions do not corrode titanium alloys at ambient temperatures, and smooth statically loaded specimens of susceptible alloys do not exhibit failure. In order to nucleate cracking it appears probable that the protective oxide film on alloys must be destroyed and its repair must not occur. If this breakdown occurs then cracking is observed in susceptible alloys. Consequently, the type of test and the type of specimen employed in any selected test are both important considerations, particularly in alloys exhibiting low susceptibility.

Cracking in aqueous solutions was first observed in 1964 when transgranular cracking was found[16] in Ti–7Al–2Nb–1Ta alloy exposed to an NaCl solution in the form of a pre-cracked notched specimen under cantilever load and plane strain conditions, and tested below the plane-strain fracture toughness. Cracking proceeded until the initial stress intensity factor $K (\propto \sigma \sqrt{c})$ reached K_{IC} and tensile overload failure occurred. A plot of initial stress intensity factor K vs. time to failure t_f took the form shown in Fig. 8.52. The value of initial K below which cracking did not occur was designated K_{ISCC}, which represented a threshold value. The ratio K_{ISCC}/K_{IC} was an indication of susceptibility, being small for highly susceptible alloys, e.g. 0.2. Cracking occurs in aqueous solutions only in the presence of Cl^-,

Br⁻ and I⁻, but not F⁻, and failures in distilled water containing no detectable amounts of those species are thought to indicate that the concentrations required are very small. It is possible that the Cl⁻ may come from the metal itself since $TiCl_4$ is formed during the refining of the basic metal.

Transgranular fracture in α-alloys occurs by cleavage on a plane 14–16° to the basal plane[17]. The greatest susceptibility is observed in specimens in which the general cracking plane is parallel to the basal planes, whereas the minimum susceptibility occurs when it is perpendicular to the basal planes[1]. The velocities of crack propagation observed, which range from 10^{-4} to 10 cm/min, are very dependent upon the environmental composition, the specimen's potential, the instantaneous value of K and the mechanical properties of the matrix. Aluminium and oxygen additions produce susceptibility in α-alloys. This has been associated[18] with an increasing tendency to develop co-planar arrays of dislocations which would result in localised regions of high stress concentration. In Ti–8Al alloy heat treated in the $\alpha + \alpha_2$ range, considerable embrittlement is associated with fracture in air along the same transgranular plane[18]. This suggests that the corrosion process serves to initiate mechanical failure during stress-corrosion crack propagation.

Mechanism The mechanism of cracking is not known. The mass–transport–kinetic mechanism[19] attempts to explain the phenomenon as occurring from the result of a high concentration of Cl⁻ ions at the tip of the crack which results in the formation of a layer (or layers) of titanium chloride. This initiates a cleavage crack in the alloy lattice under the influence of the acting tensile component stress. The hydrogen embrittlement mechanism[20] is concerned with the discharge of hydrogen on unfilmed or lightly filmed surfaces at the crack tip. The entry of hydrogen into the deforming volume of metal in front of the tip results in a plastically induced slow-strain-rate hydrogen embrittlement. The consequent loss of ductility is repeated grain by grain as crack propagation continues. Irregular crack propagation has been detected by acoustic emission[21] and is indicated also by fractographic observations[22]. Since the diffusivity of hydrogen is much smaller than the highest velocities observed it has been postulated that cleavage once initiated in regions embrittled by absorbed hydrogen may continue for mechanical reasons to some depth beyond such regions. Consistent with this is the general observation that the highest velocities are observed in the stronger less-ductile alloys.

In neutral solutions the application of cathodic polarisation prevents crack initiation and this could be taken to indicate that hydrogen embrittlement is not the operative mechanism, since the discharge and entry of hydrogen might be expected to fracture the specimen more readily. The beneficial effect of cathodic polarisation has been interpreted[23], however, to result from more rapid film repair in the alkaline catholyte generated by the cathode reaction. The film serves as a barrier to rapid hydrogen entry. Consistent with this is the observation[23] that in an environment of low pH (e.g. 10 N HCl) where film formation would not be expected, cathodic polarisation has no effect upon crack propagation.

Many workers have employed pre-cracked specimens in a number of configurations that permit interchangeable data to be obtained from an analysis derived with fracture mechanics. The pre-crack provides an ideal crevice for corrosion reactions which, as a result of hydrolysis equilibria, lead to low pH conditions (see Section 1.6). The pH at the tip of a crack in a titanium specimen exposed to neutral solutions (pH 7) has been shown[24] to fall to as low as pH 1.8. Such environments will aid the necessary breakdown of the protective film. In alloys that are not very susceptible the testing procedure adopted may affect the results. Thus loading a specimen that is in contact with the environment is a more severe test than one in which the specimen is loaded and the environment then added. The difference arises from the destruction of the passive film at the crack tip caused by the application of the load. Fracture also occurs in flat unnotched specimens dynamically strained over a narrow range of crosshead speeds while in contact with neutral NaCl. Plastic deformation of the surface results in the destruction of the passive film. At high strain rates ductile failure occurs before crack propagation, while at low strain rates repassivation occurs and prevents crack initiation[25].

In pre-cracked specimens the degree of susceptibility decreases with decreasing thickness in statically loaded specimens. It is probable that the explanation for this is not entirely due to mechanical reasons, viz. the departure from plane strain conditions. The transition is dependent upon heat treatment, loading rate, susceptibility and orientation[1]. An additional effect may lie in the greater difficulty of establishing an occluded cell leading to pH changes at a crack tip as the thickness of the specimen is decreased. Such a cell is probably also prevented by many anions. Thus the addition of CrO_4^{2-}, PO_4^{3-} and F^-, and many other species, reduces or prevents cracking. Some cations more noble than titanium have a similar effect, e.g. Cu^{2+}.

Susceptibility to aqueous cracking occurs to different degrees. Some alloys will break in moist air in the pre-cracked conditions, others require immersion in distilled water, while others require immersion in water containing appreciable amounts of dissolved halide. Different heat treatments may produce these different levels of susceptibility in one alloy. The Ti–8Al–1Mo–1V alloy, for example, will fail in laboratory air in the step-cooled condition, but requires immersion in distilled water in the mill-annealed condition and in 0.6 M KCl in the duplex annealed condition[1]. Heat treatment of titanium alloys produces a variety of phase structures, morphology and composition, and the effects upon stress-corrosion susceptibility are complex[26]. Generally, processes increasing the yield stress low K_{IC} and K_{ISCC}, while β processing is beneficial. In α-alloys ageing in the $\alpha + \alpha_2$ field raises susceptibility. In $\alpha + \beta$ alloys untempered martensitic structures are often immune in neutral solutions. Where the β-phase is immune, as has been observed in some alloys containing Mo or V, the grain size, volume fraction and mean free path of the α-phase are all important. The same considerations apply to $\alpha + \beta$ alloys where only the β-phase is susceptible, e.g. Ti–8Mn. In β-alloys transgranular cracking is also observed, e.g. in Ti–13V–11Cr–3Al, and has been described[27] as arising from micro-void formation on (100) planes. Where the β-phase contains a fine Widmanstatten α-phase precipitate and mode of cracking is one of intergranular separation.

Electrochemical aspects of the stress-corrosion behaviour have been investigated, mainly in neutral solutions[1]. The open-circuit potential of Ti–8Al–1Mo–1V is -800 mV (vs. S.C.E.). The crack initiation load reaches a minimum at −500 mV in NaCl and 0 mV in NaI. Cathodic protection occurs below −1 000 mV. Anodic protection is observed in Cl⁻ and Br⁻ solutions over a potential range that is very dependent upon strain rate and heat treatment. A linear relationship is observed between velocity and potential over certain ranges of potential. The velocity is also increased by raising the halide concentration of the testing environment. Raising the pH to high values (13–14) may reduce or inhibit cracking depending upon the alloy, its condition and the type of test employed. Lowering the pH results in the apparent elimination of the threshold stress if the concentration of the hydrochloric acid is high (> 7 M)[1].

Methanolic environments In methanolic environments stress-corrosion fracture of a similar kind is seen in titanium alloys. In α-alloys transgranular cleavage is observed[20] and a wide range of velocities[1]. With additions of HCl, however, an aditional type of fracture is seen in which intergranular separation is seen resulting from a dissolution mechanism accompanied by hydrogen pick-up by the lattice[28]. This aggressive behaviour is therefore different from that observed in neutral aqueous or methanolic environments. It is inhibited by water additions, the amount depending upon the concentration of HCl[29]. Unstressed specimens undergo intergranular attack under open circuit conditions and this is accelerated by the application of anodic polarisation[23]. Solution additions that stimulate the cathodic reaction increase the rate of intergranular attack and thereby shorten times to failure, e.g. Hg^{2+}, Cu^{2+}, Pd^{2+}. The addition of H_2SO_4 and HCOOH also increases attack, as do Br_2 and I_2[30]. Raising the viscosity of the solution by additions of glycerol increases the times to failure[31].

In aggressive methanolic environments no K_{ISCC} is observed. Instead very slow intergranular fracture proceeds at an increasing velocity with increasing values of K until it is superseded by the more rapid cleavage. Where the alloy is not susceptible to cleavage the fracture will be intergranular up to overload failure. The transition in α-alloys depends upon the aluminium and oxygen content, and the degree of cold work[25]. No pre-cracking is required in aggressive environments and in dynamic straining tests fracture occurs at all crosshead speeds below a maximum, since repassivation is not possible[25]. Additions of water will eventually remove this first stage, but not the second unless the alloy is not susceptible to transgranular cleavage in distilled water.

Impressed cathodic currents tend to prevent cracking, the current density required increasing as the water content is lowered. Anodic currents stimulate cracking and there is a linear relationship between velocity and potential up to the pitting potential[1]. Exposure of stressed specimens to aggressive methanolic environments followed by fracture in air results in transgranular fractures similar to stress-corrosion fractures, indicating that some species is absorbed from the environment[22]. This has been postulated to be hydrogen as in aqueous environments[22]. This mechanism has not been firmly established for either environment but there is an increasing amount of evidence to support it. Thus the embrittlement of specimens exposed in the unstressed condition to aggressive methanolic environments can be removed by ageing so that subsequent fracture in air reveals no cleavage[32]. In addition, notched specimens of Ti–Al alloys charged with hydrogen and broken in laboratory air have been shown to cleave along the same plane as the stress-corrosion fracture[33].

Higher alcohols have not been investigated extensively but failures do occur[31]. Other organic liquids cause transgranular fracture, e.g. CCl_4, $C_2H_2I_2$ and a range of commercial Freons which are fully substituted halide compounds[1]. Where these are not aggressive a K_{ISCC} is observed and it is likely that a pre-crack or a dynamic test is required in order to produce stress-corrosion fracture. In those compounds causing fracture which do not appear to contain hydrogen, failure must be the result of residual moisture, providing hydrogen is the species responsible for cracking, but this point has not been established.

In both aqueous and organic environments the crack velocity is related to the instantaneous stress intensity factor, as shown in Fig. 8.53. Three regions may be observed: I, II and III. Regions I and III are not always observed and the specific relationship observed depends upon the alloy composition and heat treatment, the environmental composition and the experimental conditions[1].

Other environments Other environments have been shown to cause stress-corrosion failure although the amount of work done on such failures has not been large. High-purity red N_2O_4 caused failure of a Ti-6Al-4V pressure vessel during testing[34]. Cracking could be prevented by additions of NO or H_2O, but not by additions of NOCl. K_{ISCC} decreases with increasing temperature. Cracking is both intergranular and transgranular by cleavage in Ti-Al alloys and occurs at a slower rate and at lower K values than in neutral NaCl[35]. Cracking occurs at noble potentials and it seems unlikely that hydrogen plays any part in the fracture. Very high crack densities are sometimes observed (25/mm^2). It is thought that fracture is associated with film breakdown and/or the formation of a non-protective film by chemical means:

$$Ti + N_2O_4 \rightarrow TiO(NO_3)_2 + NO^+ + NO_2 + e$$
(unprotective film)
$$2TiO(NO_3)_2 \rightarrow 2TiO_2 + 2N_2O_4 + O_2$$

Commercial titanium and all alloys crack in red fuming HNO_3 containing 20% NO_2. Eliminating NO_2 causes cracking in only some alloys while the addition of 2% H_2O removes susceptibility completely[1]. Molten salts containing halides also cause stress corrosion[36]. Mixed chlorides and bromides at 350°C promote both intergranular and transgranular fracture with maximum velocities as high as 7mm/s. Cracking is very dependent upon both the temperature and the amount of halide present.

Some liquid metals have been observed to cause embrittlement in many titanium alloys. In mercury, for example, Ti-8Al-1Mo-1V exhibits both intergranular and transgranular fracture[36] with velocities as high as 10 cm/s. Heat treatment affects this behaviour in a manner similar to that observed in aqueous and methanolic solutions. Some alloys are embrittled by liquid cadmium and zinc. More surprising, perhaps, is the observed solid metal embrittlement which has been found on titanium alloy components coated with cadmium, silver or zinc[37, 38]. Service failures of cadmium-plated Ti-6Al-4V fasteners have been reported[35], and cracking of this alloy and

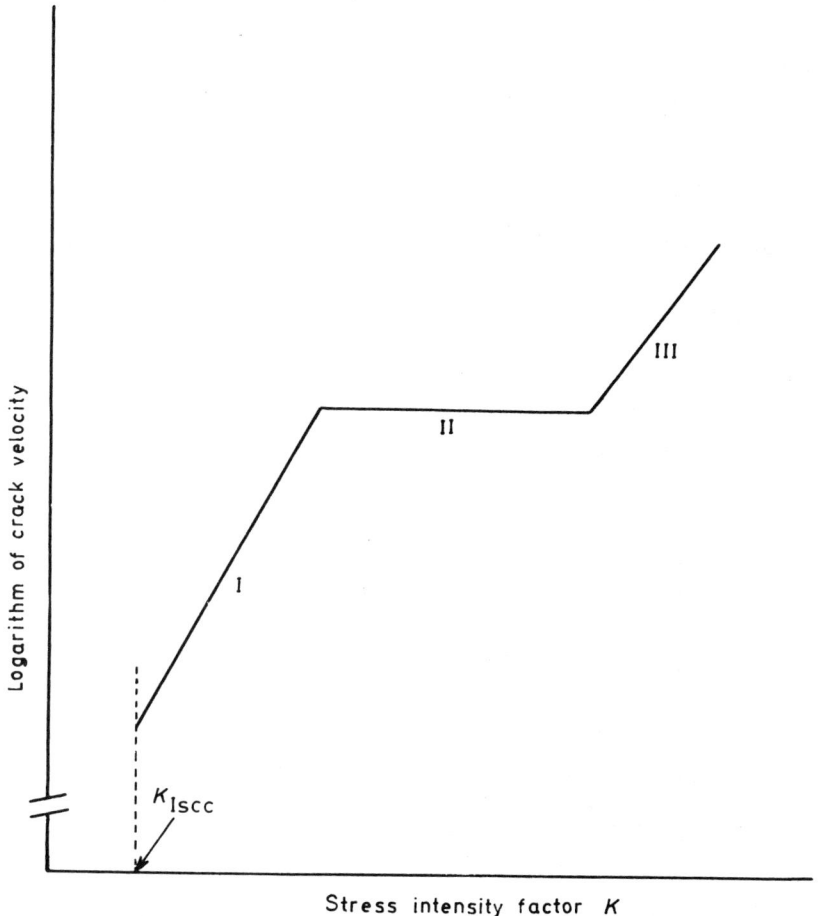

Fig. 8.53 Relationship between instantaneous stress intensity factor and crack velocity for a susceptible titanium alloy in 10 N HCl[26]. Activation energy for Stage I = 113 kJ/mol and for Stage II = 20.9 kJ/mol

of Ti–8Al–1Mo–1V has been produced in laboratory tests on coated specimens in the temperature range 38–316°C[38]. While the mechanism of such failures has not been established, cadmium has been detected on the fracture surface and the fracture process appears similar to that occurring in liquid metal embrittlement. Hydrogen is not thought to be an important factor since such failures are observed in components coated both electrolytically and by vapour deposition.

In addition to the failures in Cl_2 and HCl already referred to, cracking also occurs in hydrogen gas. Stressed Ti–Al alloys bombarded with low-energy protons have failed in a manner similar to that observed in hot salt cracking[9]. Other studies have shown that hydrogen gas can cause slow crack growth in many titanium alloys resulting in fracture surfaces very similar to those resulting from aqueous stress-corrosion failures[39].

Recent Developments

Nearly all the later work on titanium alloys has focused upon the role of absorbed hydrogen in causing stress-corrosion crack propagation. Studies of similarities between stress-corrosion fracture surfaces and fracture surfaces produced in hydrogen, residual hydrogen levels in fracture surfaces and consideration of metallurgical events occurring ahead of the crack tip have all contributed to a clearer understanding of the fracture process. In one study of hot salt cracking, analysis of hydrogen in the fracture surface was made[40] and the conclusion was drawn that it was the species responsible for causing the fracture. There is not universal agreement on this point, however. A study of Ti-8Al-1Mo-1V alloy in a molten salt mixture of highly purified LiCl/KCl at 375°C led to the conclusion that hydrogen was not responsible for the cracking process[41] since it was claimed that there was no source of hydrogen in the mixture. Crack velocities in this mixture were very much higher than those normally reported in hot salt cracking. The observed fractures were similar to those observed in aqueous solutions at lower temperatures and it was suggested that some other undetermined factor was responsible for cracking under the two different exposure conditions[41].

Several studies have contributed support to the hypothesis that absorbed hydrogen causes stress-corrosion cracking at ambient temperatures. Reversible embrittlement experiments on a Ti-O alloy exposed to CH_3OH/HCl solution have shown clearly that an absorbed species is responsible for transgranular cleavage[42]. In dynamic strain-rate stress-corrosion tests additions of Hg^{2+} increased the amount of cleavage occurring in fractures, whereas additions of Pt^{2+} produced less cleavage. These effects were attributed to the increased and decreased amount of hydrogen absorbed, respectively, as a result of the additions. The fracture surfaces of specimens of a Ti-8Al-1Mo-1V alloy after stress-corrosion cracking and those of the same alloy broken by slow strain-rate hydrogen-embrittlement (SSRHE) have been shown to be virtually indistinguishable[43].

Specimens broken in dry ultra-pure argon at various high crosshead speeds have resulted in failures that were completely dimpled but such fractures became increasingly brittle as the crosshead speed was lowered, an effect seen previously[44] and attributed to the presence of a small concentration of internal residual hydrogen. In aqueous solutions and under SSRHE conditions the resulting cleavage-like fractures were much more pronounced and indistinguishable from each other. The discontinuous transgranular fractures were accompanied by periodic acoustic emission signals[45]. Glancing angle electron diffraction revealed the presence of an fcc hydride phase on the fracture surface. The propagation process was considered to consist of the repeated formation and fracture of the hydride phase. Cracking occurred on the $\{10\bar{1}7\}$ plane, as had been previously reported by several workers but, in addition, two specimens with a texture which had this plane tending to lie parallel to the stress axis exhibited $\{100\}$ fractures. Both these planes are hydride habit planes[46, 47].

An analysis of loading mode effects has also provided evidence of the critical role of hydrogen. A stress-intensity factor (K) can be achieved in either a tensile loading mode (mode I) or a shearing mode (mode III) (Section 8.9). Under mode I conditions the volume of metal immediately in

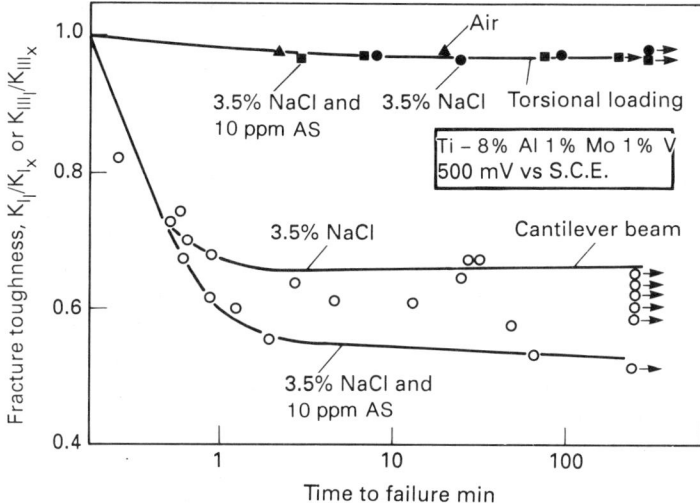

Fig. 8.54 Susceptibility of Ti-8Al-1Mo-1V to stress-corrosion cracking in 3.5% NaCl under both tensile and torsional loading, corresponding to mode I and mode III, respectively. The ordinate consists of the ratio of failure value in solution to failure value in air[50]

front of the crack is subjected to a high triaxial stress. This is not the case for mode III conditions. Dissolved hydrogen atoms accumulate in regions of high triaxial stress[48]. It has been argued[49] that an alloy failing by stress-corrosion cracking mainly as the result of absorbed hydrogen would exhibit a markedly different susceptibility according to the loading mode employed. Such a difference was observed on a high-strength steel, with no failure occurring under mode III conditions[49]. Such an approach was used by Green et al.[50] who examined a Ti-8Al-1Mo-1V alloy in 3% NaCl solution. The results are shown in Fig. 8.54. Under potentiostatic conditions the alloy was not susceptible to stress-corrosion cracking when tested under mode III conditions. Under mode I conditions, however, the value of K in solution was lower than the value in air, with the normalised ratio of the two falling to around 0.7. As a further indication of the important role of hydrogen, the addition of a cathodic poison, As, to the solution lowered the ratio of 0.6. These two effects of loading mode and cathodic poison addition are readily interpretable with respect to a hydrogen-embrittlement model of stress-corrosion crack propagation.

Two additional points concerning the results shown in Fig. 8.54 can be made. First, the authors argued[50] that any cracking process in which the main cause of propagation was the result of some form of anodic dissolution would occur equally under either loading mode. Any difference in loading mode behaviour would only be seen when hydrogen was the embrittling species. In support of this proposed distinction the authors presented results on the intergranular cracking of α-brass specimens in a concentrated ammoniacal solution which is considered to occur as the result of a dissolution mechanism. Specimens were found to be equally susceptible under both loading modes of testing. Secondly, under open circuit conditions the corrosion potential of Ti–8Al–1Mo–1V alloy was about $-800\,mV$ (SCE). If As was added to the NaCl solution under those conditions no increased embrittlement was observed since at that potential AsO_2^- ions will react with water to form AsH_3, arsine, which is a gas. No As atoms will adsorb on the alloy and no effect due to the addition of the poison will be observed. At $-500\,mV$ (SCE) adsorption of As atoms on to the alloy surface will occur and a poisoning effect is therefore observed. It may be added that As additions are inhibitive to the corrosion process, mainly because of it impeding the hydrogen evolution reaction and, possibly, as a result of increasing the solution pH, if added at sufficiently high concentrations.

Strain-rate effects have been examined by several authors. The stress-corrosion crack velocity has been shown to be dependent upon the strain-rate at the crack tip[51]. The K_{Iscc} value has been observed to be dependent upon the loading rate for a Ti–6Al–6V–2Sn alloy in a 3.5% NaCl solution[52]. The value reached a minimum at an intermediate loading rate which was considered to be indicative of hydrogen embrittlement through hydride formation. The calculated strain rate was considered to correspond closely to the theoretical minimum strain rate for hydrogen transport by dislocations[52]. Intergranular fracture observed in region I of the velocity/K curve has been attributed to there being a low density of hydrogen-carrying dislocations which arose preferentially in grain boundary regions and gave rise to grain boundary hydrides[52]. Intergranular fracture in commercial purity Ti in a CH_3OH/HCl mixture was attributed to absorbed hydrogen[53].

The role of hydrogen in the stress-corrosion cracking of β alloys has not been widely investigated. In Ti–13V–11Cr–3Al alloy cracking in aqueous solutions at ambient temperatures occurs as a cleavage process on or close to {100} planes[54] as a discontinuous process[55]. In CH_3OH/HCl solutions reversible embrittlement experiments were interpreted as showing that cleavage was caused by absorbed hydrogen. Consistent with such an analysis, AsO_2^- additions to the solution increased the amount of cleavage observed after stress-corrosion tests, while Pt^{2+} additions resulted in much less cleavage. Hg^{2+} additions caused more cleavage and quinoline additions less. These effects showed that the amount of cleavage observed was not related to the corrosion rate. How hydrogen embrittlement occurs in the bcc β lattice has not been determined. Reversible embrittlement has also been observed in the β-III alloys without the formation of hydrides[56] which was also the conclusion drawn from work on a Ti–28Mo alloy.

The evidence for the role of hydrogen in many cases of stress-corrosion cracking of Ti alloys is strong but it must be remembered that identical fractures can be caused by liquid metal embrittlement in which hydrogen appears to play no role and in other environments which are claimed to have no hydrogen source. Since there is more than one species capable of promoting the transgranular cleavage characteristically observed in α alloys, where such corrosion fractures are observed any analysis must seek to establish which is the most rapidly acting embrittling species.

Magnesium

Stress-corrosion cracking occurs in magnesium alloys in aqueous environments at ambient temperatures. It is found mainly in wrought alloys, although isolated examples of cracking in castings have been reported. The major alloying element responsible for producing susceptibility is aluminium. In practice failure commonly appears to arise from the action of residual stresses.

Mechanism

The mechanism of stress-corrosion cracking in magnesium alloys has not been fully elucidated, but many features have been observed. The rôle of the electrolyte appears to lie in concentrating the corrosion attack in a relatively small number of points resulting in pitting which often precedes cracking. Where little or no pitting occurs cracking is only initiated if the value of the stress is sufficiently high, probably causing rupture of the surface film by plastic deformation of surface metal grains. In polycrystalline specimens the threshold stress below which failure does not occur will be below the stress values below this level, but in single crystals the yield point clearly must be exceeded to produce plastic deformation and this has been observed to be necessary for stress-corrosion cracking to occur[57]. In such environments the capacity to repair ruptured films must not be too great otherwise this process will occur in place of crack initiation. If environments are too aggressive then widespread corrosion may occur in place of pitting and cracking. In laboratory tests mixtures of NaCl plus $Na_2Cr_2O_7$ are commonly used to cause cracking which is preceded by deep pitting; a common mixture consists of 3.5% NaCl plus 2% K_2CrO_4. The aggressive anion is important and cracking is not confined to the Cl^- ion. Thus in 1 N solutions it has been observed[58] that the rate of stress-corrosion cracking decreases according to: $Na_2SO_4 > NaNO_3 > Na_2CO_3 > NaCl + CH_3COONa$, whereas the rate of corrosion of unstressed samples decreases according to: $NaCl > Na_2SO_4 > NaNO_3 > CH_3COONa > Na_2CO_3$. In NaCl plus K_2CrO_4 mixtures the rate of cracking depends upon both the absolute concentration of each species and their ratio. It has been established that KHF_2 solutions also cause stress-corrosion cracking. Little investigation of the action of this electrolyte has been carried out, but since the F^- ion is an inhibitor for the corrosion of magnesium at least part of the electrochemical explanation may lie in inhibited film breakdown and repair kinetics. Thus cracking does not occur in fluoride solutions above a certain limit of concentration. In non-fluoride solutions stress-corrosion crack initiation is inhibited at pH values greater than 10.2[59] and this is also probably related to the greater ease of film formation that occurs in highly alkaline solutions on magnesium surfaces. Under open-circuit conditions or the application of anodic polarisation, only pitting is observed in unstressed specimens under all conditions of heat treatment in environments that cause cracking in stressed specimens.

Cracking is generally observed to be transgranular, but this may change depending upon the heat treatment, the grain size and possibly the environmental pH. Thus in the Mg-6.5Al-1Zn alloy transgranular cracking occurs if specimens are water-quenched[60, 61], a process that produces a very fine transgranular dispersion of FeAl which occurs as platelets[62] that are cathodic to the solid solution matrix. Cracking may include the dissolution of material adjacent to the precipitates which may also be solute depleted. In furnace-cooled material the same alloy exhibits intergranular cracking and this may be associated with the grain boundary precipitate of $Mg_{17}Al_{12}$ which occurs during furnace cooling. This precipitate is also cathodic to the matrix but the potential difference developed is lower than that between FeAl and the matrix. This may be a partial electrochemical explanation for the observation that as the iron content is increased the proportion of transgranular fracture also increases, even in furnace-cooled alloys[63]. Above a grain size of about 0.03 mm cracking is transgranular irrespective of the heat treatment employed. In two-phase alloys intergranular cracking occurs in specimens aged at $\leqslant 150°C$ and transgranular cracking in specimens aged at $>150°C$.

The path of transgranular cracking has not been clearly established and it may vary from one alloy to another. It has been reported as following crystallographic planes[64], possibly the basal plane[60], while others report that no well-defined fracture plane can be discerned in large-grain-size specimens[65], that the crack occurs at a high angle to the basal plane[57], and that fracture occurs as cleavage along (0001), (10$\bar{1}$0) or (10$\bar{1}$1) depending upon which plane is most nearly perpendicular to the operative tensile stress[66].

Cracking is considered to occur by a combination of dissolution and mechanical fracture. It is not necessary therefore to account for the relatively high propagation rates by anodic dissolution since this would require high current densities, e.g. 0.6 mm/min requires 14A/cm². An anodic reaction appears to be occurring during cracking, and cathodic protection prevents crack initiation and arrests crack propagation. During cracking the tip is active and hydrogen is evolved. Potential fluctuations have been detected[67] in Mg-5Al alloys in KHF_2 solutions although not in Mg-7Al alloys in $Cl^-/Cr_2O_7^{2-}$ mixtures[68]. One explanation[68a] has been given for the mechanical part of the fracture which is also indicated by fractographic observations[66, 69, 70] and by irregular specimen extension[70]. This is that cleavage occurs on (3140) planes, as a result of hydrogen absorption. The amount of current required to effect cathodic protection increases with increasing stress on specimens.

A threshold stress below which cracking does not occur is observed in some environments[59, 71]. It is dependent upon the alloy composition and heat treatment, and upon the testing environment. In 3% NaCl solutions it is not well defined and this may arise from the pitting that occurs which can be expected to lower the cross-sectional area locally and gradually raise the effective stress acting across that region. Below the threshold stress, cracks occur which do not propagate to sufficient length to cause complete fracture of specimens[69]. The maximum length of crack observed l, is related to the proof stress of the material σ by the relationship $\sigma^2 l$ = constant. Such cracks have been observed in fatigue and corrosion fatigue[72] and their density and length are very dependent upon the mechanical properties of the alloy[73]. From this analysis emerges a description of alloys undergoing pitting at low stress values, pitting and cracking without complete failure at higher stress values up to the threshold level above which time to failure decreases with increasing value of the initial stress[69].

Alloy Susceptibility

Cracking is observed in commercial Mg-Al-Zn alloys in the range 3-10% Al in which the ratio Al/Zn ≥ 2, and susceptibility increases with increasing Al content. For a given alloy increasing amounts of Fe increase susceptibility[67]. Copper additions also appear to raise susceptibility[58]. It must be emphasised, however, that Mg-Al alloys exhibit stress-corrosion susceptibility even if manufactured from elements of the highest purity available. Most aluminium-free alloys appear to be non-susceptible in most aqueous environments, but KHF_2 solutions appear to be able to cause failure in most alloys including nominally pure magnesium metal in which failure is intergranular. Thus Mg-Mn alloys, for example, are immune in NaCl plus K_2CrO_4 solutions (unless additions of 0.5 Ce are made to the alloys) while they exhibit cracking in KHF_2 solution[67]. Mg-14Li alloys, which have a b.c.c. structure, exhibit intergranular fracture in humid air although this can be prevented by a stabilising treatment consisting of heating for 24 h at 149°C after quenching[70]. The Mg-3Zn-0.7Zr alloy has been reported[67] to fail in distilled water and KHF_2 solution.

Preventative Methods

Stress relieving at low temperatures is commonly used to lower stress-corrosion susceptibility in Mg alloys, e.g. 8 h at 125°C for Mg-6.5Al-1Zn-0.3Mn[74], since higher temperatures tend to lower the yield point. Similar treatments are advisable for welds which can be a source of high residual stresses. Treatments designed to put the surface in a state of compressive stress also tend to prolong stress-corrosion life. Shot peening, surface rolling[71] and abrasion all produce beneficial effects. Surface oxidation followed by anodising is also reported[75] to increase stress-corrosion life. Susceptible alloys can be clad with non-susceptible Mg alloys, but where the edge is exposed, wetting of both the alloy and clad layer is important in order to achieve cathodic protection of the former. The replacement of susceptible alloys with non-susceptible alloys or with alloys exhibiting less susceptibility, can often be undertaken with little loss of mechanical properties. Refining the grain size lowers susceptibility. Heat treatment produces changes in the threshold stress[76] and alterations in crack morphology as already described.

Aluminium

Stress-corrosion cracking occurs in certain aluminium alloys which have been developed for medium and high strength by employing variations in composition, cold work and heat treatment[1, 77, 78]. The main alloys are based upon Al-Mg, Al-Mg and Al-Cu, but stress corrosion also occurs in Al-Ag, Al-Cu-Mg, Al-Mg-Si, Al-Zn and Al-Cu-Mg-Zn alloys. It has not been observed in pure aluminium. In alloys susceptibility appears to increase as the amount of alloying addition that can be taken into a supersaturated solid solution is raised. In ternary and higher element alloys susceptibility is also influenced by the ratio of solute elements[79]. Small additions of Cr, Mn, Zr, Ti, V, Ni and Li can reduce the susceptibility of wrought products in high-purity binary, ternary and quaternary alloys[79, 80]. Stress-corrosion cracking in castings is not common but it is found occasionally[81].

The large majority of failures occur in aqueous environments and therefore most attention has been focused upon them, but failure can occur[1] in N_2O_4, mineral oils, alcohols, hexane and mercury. It has not been established whether failure in these environments is the result of residual moisture. Failures in service arise often from the action of residual stresses which can occur in components as a result of quenching followed by machining. The stress level required to initiate cracking is often very much below the yield stress. Alloys brought to a high-strength condition are particularly susceptible.

Mechanism

The mechanism of cracking in aluminium alloys has not been elucidated, but many factors have been examined. Cracking is nearly always intergranular. Stress-corrosion life is very dependent upon the grain shape and orientation in relation to the acting stress. Stress-corrosion resistance is lowest in the short transverse direction of wrought components since many grain boundaries are then lying orthogonally to the applied stress. Notice of such effects is commonly taken in the design of components. In plane-strain tests a relationship between crack velocity and stress intensity factor is found[1, 82] similar to that shown in Fig. 8.53 for titanium alloys. A large number of alloys exhibit only Stages I and II. Others exhibit Stage III and others two 'plateaus' or Stage II regions[1]. Results similar to those shown schematically in Fig. 8.53 are also obtained. The crack velocity may vary over nine orders of magnitude and determining K_{ISCC} can be difficult since too high a value may be obtained if the velocity-detecting apparatus is not sufficiently sensitive or the length of time of the experiment too short. It has been suggested[1] that K_{ISCC} might be defined as corresponding to a crack velocity of 10^{-8} cm/s.

In equiaxed specimens crack branching is observed in the region of Stage II at a value of K about 1.4 times the K value at the lower end of the stage. Since cracking occurs at low stress values it is not altogether surprising that the precipitation of corrosion products within existing cracks can sometimes exert relatively appreciable stresses which result in crack propagation.

The effect of environmental variables upon the logarithm of velocity *vs.* K relationship has been examined[1] for a few alloys in some conditions of heat treatment. While it cannot be certain that similar results would be obtained with all alloys, the results reported[1, 82] do show interesting features that may have points in common with all alloys. For an Al–Zn–Mg–Cu alloy (7075–T651) the stress-corrosion plateau velocity was a maximum in 5 M KI solution under potentiostatic conditions at -520 mV (*vs.* S.C.E.), reaching about 2×10^{-4} to 5×10^{-4} cm/s, whereas in 3% NaCl under open-circuit conditions the plateau velocity was about 10^{-6} cm/s. Stage I in both cases was the same. The plateau velocity was very sensitive to moisture and depended linearly upon the water vapour of the testing environment[1, 82]. Crack propagation does not occur in argon or hydrogen unless moisture is present[83]. Many alloys exhibit plateau velocities when tested in distilled water that are similar to those observed in moist atmospheres. Additions of Cl^-, Br^- and I^- increase the plateau velocities observed in distilled water by as much as 10^2 times under open-circuit conditions, but many other anions have no effect upon the logarithm of velocity vs. K relationship under open-circuit conditions or over a wide range of potential values. In Cl^-, Br^- and I^- solutions the plateau velocity depends upon the halide concentration for some alloys, e.g. the velocity increases with increasing iodide concentration above a certain minimum for 7079-T651, but this is not universally true. In neutral solutions cathodic polarisation lowers the plateau velocity while anodic polarisation increases it until pitting occurs. In strongly acidic solutions the velocity is less sensitive to potential changes and no cathodic protection is observed[1]. The pH value of the environment does not appear to cause changes in the plateau velocity under open-circuit conditions, but acidic conditions move Stage I to the left in Fig. 8.53 thus giving rise to higher velocities for a given K value. Such effects have been described as indicating that the size of the cathodic zone within the crack where hydrogen is released controls the rate of crack propagation[84]. Generally, lowering the pH shortens the time to failure of specimens, an effect that will include the influence of pH upon the initiation process. Under potentiostatic conditions lowering the pH can cause appreciable increases in the plateau velocity on the cathodic side.

Temperature effects indicate an activation energy of 113 kJ/mol for Stage I and 16 kJ/mol for Stage II in 7079-T651 alloy. Crack velocity in Stage II is lowered as the solution viscosity is increased.

No mechanism for cracking in N_2O_4 has been established[85]. In organic media crack velocities are similar to those obtained in distilled water. Lowering the water content results in lower velocities. Not all authors attribute failures in organic liquids to the residual moisture[86]. Furthermore, part of the fracture may be transgranular[86]. Water additions to methanol increase crack velocities as do halide additions. In oils velocities are similar to those in organic liquids and distilled water.

Much of the extensive work on crack velocity described here has been carried out over a long period by Spiedel[1]. Detailed studies of velocity-dependent and velocity-independent parameters reveal how complex the phenomenon is. The three major alloy systems will now be discussed.

Al-Mg (5000 Series) and Al-Mg-Si (6000 Series) In the binary alloy system strength is obtained mainly by strain hardening. Stress corrosion is thought to be associated with a continuous grain boundary film of Mg_5Al_8 which is anodic to the matrix[87]. Air cooling prevents the immediate formation of such precipitates, but they form slowly at ambient temperatures. Thus only low Mg alloys are non-susceptible (Al-3% Mg). Widespread precipitation arising from plastic deformation[88] with carefully controlled heat-treatment conditions can lower susceptibility. Al-5Mg alloys of relatively low susceptibility are subjected to such treatments. Mn and Cr additions improve the stress-corrosion resistance of Al–6Mg and Al–7Mg alloys[89]. Mn increases the precipitation rate and both elements promote the formation of elongated grains. Al–7Mg alloys are usually very highly susceptible, but a recent development[1] indicates that an alloy of this kind can be produced of low susceptibility.

Al–Mg–Si alloys are strengthened by precipitation hardening in which Mg_2Si is formed. They are not very susceptible to stress-corrosion cracking[77, 90] which only occurs in specimens subjected to a high solution-treatment temperature followed by a slow quench[77]. Ageing such material eliminates susceptibility[77].

Al–Cu and Al–Cu–Mg (2000 Series) These alloys are strengthened by precipitation hardening. Under conditions of natural ageing these alloys are highly susceptible to stress-corrosion cracking. Susceptibility is associated with slow quench rates which also result in grain-boundary corrosion in unstressed specimens, which is thought to arise from electrochemical effects between $CuAl_2$ and solute-depleted zones formed during quenching[91]. Since thick-section material cannot be quenched rapidly quench-rate effects determine the type of component that any particular alloy can be used to make.

During artificial ageing susceptibility passes through a maximum just before peak hardness is achieved. Similar changes occur in the potential difference developed between grains and grain boundaries[92]. After further ageing, precipitation of the equilibrium $CuAl_2$ occurs within the grains and the potential difference between grains and boundaries then disappears. A recent test[93] provides a rapid means of indicating susceptibility to intergranular attack and stress-corrosion cracking. The specimen's potential is measured in a mixture of absolute methyl alcohol and carbon tetrachloride. Corrosion of the grain boundary provides sites for deposition of dissolved copper, whereas an absence of corrosion results in deposits of copper which are non-adherent. The former develops a potential of about -300 mV (vs. S.C.E.) whereas the latter develops a potential of about -1100 mV.

Al–Zn–Mg and Al–Zn–Mg–Cu (7000 Series) These alloys are strengthened by precipitation hardening. Cr, Mn and Zr additions produce elongated grain shapes and inhibit grain growth. High-purity ternary alloys exhibit the highest plateau velocities and although much research has been done on them they are not used in practice. Commercial low-copper alloys are particularly susceptible, and although overageing is generally beneficial the effects of such a treatment are less pronounced with these alloys. Artificial ageing is beneficial but susceptibility in the short transverse remains troublesome[77]. Silver additions to the alloy are reported[94] to improve stress-corrosion resistance. The effect appears to arise from the stimulation of precipitation processes which minimise the width of the precipitate-free zone which arises either from vacancy or solute depletion during quenching. Other workers find that silver gives no improvement either in strength or stress-corrosion resistance[77]. An explanation for this difference appears to lie in experimental procedures[1]. A general conclusion is that ageing temperature and not chemical composition is the most important factor governing short-transverse stress-corrosion resistance in these alloys[95].

Recent developments, particularly in producing alloys suitable for thick sections, is reviewed by Spiedel[1], together with the particular problems of welding.

General

Much discussion of stress-corrosion cracking mechanisms in aluminium alloys has been concerned with the development of anodic areas at grain boundaries. The origin of such areas can be caused by the action of the stress, and susceptible alloys do not necessarily suffer from intergranular corrosion in the absence of stress. Thus in some conditions Al–Mg–Si suffers from intergranular corrosion, but not stress corrosion[96], 7039–T64 suffers from stress corrosion but not intergranular corrosion[97], 7075–T651 suffers from both, 7075–O from neither. The electrochemical effects may arise from solute-depleted zones, precipitates anodic or cathodic to the adjacent matrix, or from the rupture of films at the crack tip by plastic deformation. The effect of relative humidity upon the plateau velocity suggests that there may not be a volume of water at the crack tip[1], a possibility which if established would demand a careful re-examination of possible electrochemical reactions.

From a metallurgical viewpoint the effect of grain shape has been described. On a microstructural level the precipitate–matrix interface properties appear to be important. In alloys of Al–6Zn–3Mg aged to peak hardness, slip occurs in a relatively small number of bands which develop a high density of dislocations. Overageing, which lowers susceptibility, results in plastic deformation occurring in much more diffuse bands of dislocations[98]. Grain-boundary precipitates are important both for electrochemical and mechanical reasons and the precipitate-free zone width (as well as the solute-depleted zone width) may also be important. The precise relative significance of these three micro-structural features has not been fully ascertained and it is a subject of a considerable discussion[99-102]. Much of this centres around the rôle of preferential deformation in the precipitate-free zone resulting in selective dissolution, a process that has not been demonstrated experimentally. Selective corrosion of solute-depleted regions, hydrogen adsorption, tensile-ligament dissolution and general adsorption have also been invoked as major components of mechanistic processes[1]. There is evidence that acidity develops within the region of the crack tip, a pH of 3.5 being observed[103] and the mass-transport-kinetics model[104] appears to explain the plateau velocity as being limited by the kinetics of halide-ion transport to the crack tip. A number of workers[105-107] have provided some evidence that absorbed hydrogen may be at least partly responsible for cracking.

Preventative Methods

The importance of grain shape and the orientation of the applied stress to the short transverse direction has already been pointed out. Overageing also generally lowers strength and stress-corrosion susceptibility. Both the design and manufacture of components are important. Quenched components often have high internal tensile stresses and subsequent machining of such pieces may result in surfaces that readily nucleate cracks. This possibility can be removed by manufacturing components close to the final required size before heat treatment.

Shot peening is a beneficial surface treatment since it puts the surface into a state of compression and generally obscures the grain structure. Subsequent painting of the peened surface is often useful. If pitting occurs then cracking can be expected in susceptible material when the attack penetrates the depth of the compressed surface layer.

Paint coatings can be effective in preventing stress corrosion but it is not always a simple matter to produce and maintain a perfect complete coverage. Galvanic coatings based upon electroplated layers or metal pigmented paints, are commonly used. Such layers do not need to be perfect, but the protection afforded to breaks (or 'holidays') will depend very much upon the localised electrochemical conditions. Metal spraying is also employed and for highly susceptible alloys a thin cladding sheet of aluminium is employed, either on one or both sides. These clad composites are employed for general corrosion resistance and not merely to combat stress corrosion. Anodising is generally not recommended. Cathodic protection is effective but is often not practicable.

Recent Developments

Later work on aluminium alloys has also focused more closely upon the role of hydrogen which had not previously been widely considered as an embrittling species in the stress-corrosion cracking process for these alloys. The idea was not new, however. Reports of intergranular failure under cathodic charging conditions had been made at a much earlier time[108,109]. A reduction in stress-corrosion life and alloy ductility in a high purity Al–5Zn–3Mg alloy had been found in specimens pre-exposed to a 2% NaCl solution[110], an effect that was accentuated if specimens were stressed[111].

In more recent work embrittlement in water vapour-saturated air and in various aqueous solutions has been systematically examined together with the influence of strain rate, alloy composition and loading mode, all in conjunction with various metallographic techniques. The general conclusion is that stress-corrosion crack propagation in aluminium alloys under open circuit conditions is mainly caused by hydrogen embrittlement, but that there is a component of the fracture process that is caused by dissolution. The relative importance of these two processes may well vary between alloys of different composition or even between specimens of an alloy that have been heat treated differently.

The role of humid air has been examined in the embrittlement both of high-purity Al–Zn–Mg alloys and also for a few commercial compositions[112-114]. Loss of ductility in unstressed specimens is a reversible process[110]. Such an effect, when observed, is readily attributable to absorbed hydrogen. In unstressed specimens hydrogen must enter through the unbroken surface film either in an atomic or a protonated form. The thickness, composition and morphology of surface films are all likely to be important factors controlling the rate of hydrogen or hydrogen ion entry. This point was emphasised at an early stage when it was observed that solution heat treatment of a high-purity Al–6Zn–3Mg alloy in the temperature range 450–500°C resulted in an increased sensitisation to pre-exposure embrittlement[112]. The change in sensitisation could be correlated with the formation of crystalline MgO which occurred preferentially at alloy grain boundaries. It appeared that MgO crystallites facilitate the entry of hydrogen into the alloy grain boundaries[112]. Oxide thickening resulted in a reduced rate of embrittlement at room temperature. In much subsequent work a lot of attention has been focused upon the presence of Mg in precipitate-free grain boundaries and its possible role in pre-exposure embrittlement and the stress-corrosion process[115-121]. The retention of this segregation after precipitation has been much discussed[122,123]. High temperature solution heat treatment, slow quenching and overageing may reduce the level of segregated Mg and thereby reduce the hydrogen entry rate. This would account for the beneficial effects that these procedures have upon the stress-corrosion resistance of Mg-containing alloys.

Alloying effects were also examined in this study[112]. Additions of 1.7Cu or 0.14Cr to the high-purity alloy reduced the rate of embrittlement. The chromium-containing alloy and a commercial 7075 alloy both recovered their ductility after exposure to water vapour-saturated air at 20°C unlike the high-purity alloy. The effect of chromium is shown in Fig. 8.55. The high-purity alloy did not recover its ductility in dry air or after storage for 12 h at 68°C in a vacuum of 10^{-7} Torr. The 7075 recovered ductility a little more rapidly than the chromium-containing alloy.

The presence of hydrogen in pre-exposed specimens was revealed by straining specimens in vacuo. Hydrogen evolution occurred in the elastic region of the stress/strain curve, an effect that had been shown to be very much reduced by electropolishing pre-exposed specimens prior to testing[134],

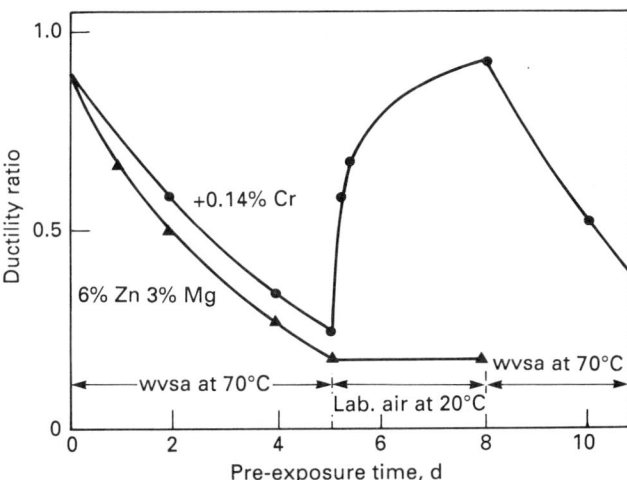

Fig. 8.55 The effect of a 0.14Cr addition on the recovery of ductility of Al–6Zn–3Mg alloys during storage in laboratory air at 20°C after pre-exposure to water vapour-saturated air for 5 days at 70°C. The ductility ratio is the ratio of elongation-to-fracture of specimens broken under the cited conditions and under vacuum conditions[112]

but which had no effect upon the measured ductily. During the plastic straining and at the point of fracture hydrogen evolution from pre-exposed specimens was detected[124].

The embrittlement caused to aluminium alloys by pre-exposure to moist atmospheres or stress-corrosion environments is thought to be due to hydrogen in the atomic form. Intergranular bubbles of hydrogen, formed in association with certain precipitates, have been observed by HV TEM[112,125] and are associated with a lowered degree of embrittlement. Increased resistance to stress-corrosion cracking in Al-Zn-Mg alloys resulted in an increased propensity for hydrogen trapping and a decrease in the permeation rate of hydrogen through unstressed alloy membranes[126]. The ability to trap hydrogen as innocuous bubbles improves the embrittlement resistance of Al alloys[127]. Thus it appears that fracture occurs by hydrogen-induced grain boundary decohesion once all available sites for hydrogen trapping are saturated. High purity Al-Zn-Mg alloys have relatively few sites and therefore embrittle readily. Alloying additions to these alloys and commercial alloys in general result in microstructures that have a much higher density of potential trapping sites for hydrogen.

The mechanism of hydrogen embrittlement of aluminium alloys has not been established. Experiments on a high purity Al-5.6Zn-2.6Mg alloy hydrogenated by exposure to water vapour saturated in air at 70°C indicated that internal hydrogen embrittlement occurs by the formation and rupture of a hydride phase at grain boundaries[128]. Electron diffraction revealed a very thin layer of AlH_3 (~1μm thick), formed probably as the result of a stress-induced mechanism. Two stages of embrittlement were noted: stage I, in which the diffusion of hydrogen into the region ahead of the advancing crack tip was necessary to provide a sufficient concentration of hydrogen to produce AlH_3; and stage II, in which sufficient hydrogen was already present at the grain boundaries to form AlH_3 and hydrogen diffusion during stressing was not therefore required. Such a distinction explains both the strong dependence of stage I upon strain rate, stage I extending as the strain rate is lowered, and the absence of any strain-rate dependency in stage II. In one interrupted stress-corrosion test in a NaCl solution a thin layer on the fracture surface at the intergranular/dimple transition region was observed, although no diffraction pattern was obtained. The authors noted that the stress-corrosion fracture surfaces frequently do not show such a layer. They recognised that stress-corrosion cracks may propagate by a competing and basically different mechanism[128].

Fractographically, failure has been seen to occur discontinuously[129-131], an observation interpreted as being the result of repeated pre-exposure embrittlement. Matched arrest markings have been seen in specimens broken by stress-corrosion in chloride solutions, in water and in some service failures[131]. For the two alloys examined, 7071 and 7179, the average striation spacing was not a strong function of the applied stress intensity factor. Acoustic emission also indicated that cracking was discontinuous[132]. The striation results are in agreement with observations that the effect of stress intensity on stress-corrosion crack propagation in Al-Zn-Mg alloys by the hydrogen-embrittlement mechanism appears to increase the rate of crack jumping rather than to alter the magnitude of the crack advance[132]. If the number of available trap sites for the embrittling hydrogen atoms is fixed by

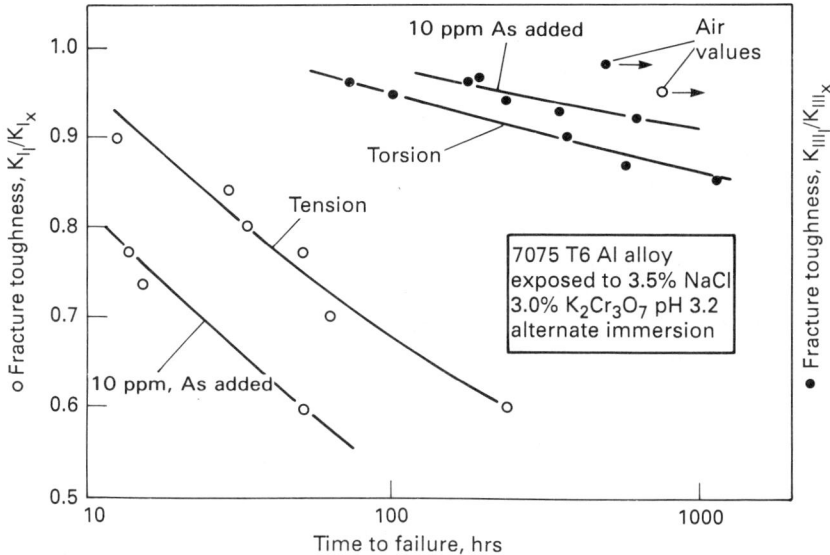

Fig. 8.56 Susceptibility of 7075-T6 Al alloy to stress-corrosion cracking in 3.5% NaCl + 3% $K_2Cr_2O_7$ under both tensile and torsional loading[50]

the alloy composition and its thermal history then the stress operating across the grain boundary must affect the grain boundary diffusion rate such that all the available trap sites are saturated more rapidly and brittle grain boundary failure can be induced by any further accumulation of hydrogen atoms. In this way the kinetics of pre-exposure embrittlement are accelerated by applied stress[111].

The role of loading mode on the stress-corrosion cracking of an Al alloy has been examined with a 7075 alloy in the T6 condition[50] and for 5083[133], with similar results. Figure 8.56 shows results obtained with the 7075 alloy[50]. In the tension test the alloy was more susceptible to cracking than in the torsion test. Unlike Fig. 8.54 for a titanium alloy, however, some cracking did occur under the torsion mode of testing which indicated that cracking occurs both by hydrogen embrittlement and by dissolution with the first factor being more important. In the tension mode the addition of the cathodic poison, As, resulted in more rapid failure, a result entirely consistent with a hydrogen-embrittlement mechanism. The beneficial effect of the As addition in the torsion mode was probably the result of an inhibitive effect upon the dissolution reaction.

The role of the stress in embrittlement and stress-corrosion processes has been examined in some detail by employing the slow strain-rate technique[134, 135]. Specimens of alloy 7179-T651 tested in air or in vacuum after pre-exposure to water at 70°C or in water at various potentials at ambient temperature exhibited a reversible embrittlement in excess of that arising from testing in moist air[134]. The embrittlement was attributed to hydrogen absorption, and recovery was thought to be due to loss of hydrogen (particularly under vacuum) or to diffusion to traps. Potentiostatic tests revealed two potential regions of embrittlement corresponding to one cathodic and one anodic to the open circuit potential. Specimens of alloy 7049-T651 also exhibited a reversible pre-exposure at low strain-rates, with the critical strain-rate decreasing in less aggressive environments[135]. Recovery from pre-exposure embrittlement was only observed when specimens were subsequently strained in an inert environment. In laboratory air or seawater pre-exposure and subsequent strain-rate effects were additive. Potentiostatic tests revealed, as with alloy 7179, that there were two potential regions of embrittlement. Fractography and overageing effects both indicated that the major embrittling species at the free corrosion potential was hydrogen embrittlement. It appeared that hydrogen absorption led first to transgranular fracture and then to intergranular fracture, with the transition occurring at lower local hydrogen concentrations as the strain rate was decreased. Similar results and conclusions were drawn from experiments on Al–6Zn–3Mg, Al–6Zn–3Mg–1.7Cu and Al–6Zn–3Mg–0.14Cr alloys pre-exposed in the solution-treated condition to moist vapour at 115°C[136].

In addition to examining pre-exposure effects, the slow strain-rate testing technique has been used increasingly to examine and compare the stress-corrosion susceptibility of aluminium alloys of various compositions, heat treatments and forms. A recent extensive review[137] draws attention to differences in response to the various groups of commonly employed alloys which are summarised in Fig. 8.57. The most effective test environment was found to be 3% NaCl + 0.3% H_2O_2. The most useful strain rate depends upon the alloy classification.

The susceptibility of Al–Li alloys to stress-corrosion cracking has been

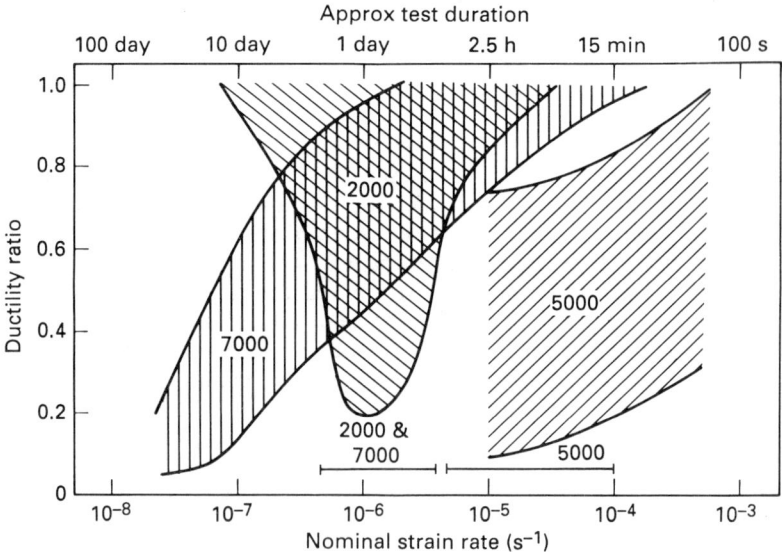

Fig. 8.57 Strain rate regimes for studying stress corrosion cracking of 2 000, 5 000 and 7 000 series alloys[137]. The ductility ratio is the ratio of elongation-to-fracture or reduction in area measured in solution to that measured in a control environment

examined to a limited extent[138]. In alternate immersion crack initiation testing the alloys are less susceptible than the extensively used aerospace alloys. Restricted geometry conditions and thin-film de-aerated electrolyte conditions promote cracking, however, probably as a result of making the necessary development of alkaline crack-tip conditions occur more readily. The pH at the crack tip in these alloys is about 9 and is controlled by the Li^+ concentration[138]. A reversible embrittlement effect has been detected in these alloys thereby suggesting a possible role for absorbed hydrogen in the cracking process. Susceptibility to stress-corrosion cracking is highly dependent upon the Cu content of Al-Li-Mg alloys containing 2-2.5Li and 0-0.6Mg. Crack initiation in plane specimens did not occur in the absence of Cu. Restricted specimen geometry and thin film electrolyte conditions promoted cracking even in Cu-free alloys. In Al-Li and Al-Li-Zr alloys crack initiation did not occur in plane specimens[138]. In DCB specimens, however, crack propagation occurred from notches in a 95% environment at 40°C. Cracking in an Al-2.8Li under alternate immersion conditions has also been reported[139].

J. C. SCULLY

REFERENCES

1. *Stress-Corrosion Cracking in High Strength Steels and in Aluminium and Titanium Alloys* (ed. B. F. Brown), NRL, Washington D.C. (1972)
2. Jackson, J. D. and Boyd, W. K., DMIC Technical Note, Battelle Memorial Institute, Columbus, Ohio (1966)
3. *Stress Corrosion Cracking of Titanium*, ASTM STP 397, ASTM, Philadelphia (1966)
4. Logan, H. L., *Fundamental Aspects of Stress Corrosion Cracking* (ed. R. W. Staehle, A. J. Forty and D. van Rooyen), NACE, Houston, 662 (1969)
5. *TML Report No. 88*, Battelle Memorial Institute, Columbus, Ohio (1957)
6. Peterson, V. C. and Bomberger, H. B., Reference 3, 80 (1966)
7. Kirchner, R. L. and Ripling, E. J., First Interim Report, Materials Research Laboratory, Richton Park, Illinois (1964)
8. Rideout, S. P., Louthan, M. R. Jr., and Selby, C. L., Reference 3, 137 (1966)
9. Ondrejcin, R. S., *Met. Trans.*, **1**, 3031 (1970)
10. Gray, H. R., *Corrosion*, **25**, 337 (1969)
11. Gray, H. R., Aerospace Structural Materials Conference, No. 2 (1969)
12. Gray, H. R. and Johnston, J. R., *Met. Trans.*, **1**, 3101 (1970)
13. Boyd, W. K., Reference 4, 593 (1969)
14. Adams, R. E. and Von Tiesenhausen, E., Reference 4, 691 (1969)
15. Weber, K. E. and Davis, A. D., Lockheed California Co., NASA CR 981 Dec. (1967)
16. Brown, B. F., Lennox, T. J., Jr., Newbegin, R. L., Peterson, M. H., Smith, J. A. and Waldron, L. J., *NRL Memorandum Report 1574*, November (1964)
17. Beck, T. R. and Blackburn, M. J., *J.A.I.A.A.*, **6**, 326 (1968)
18. Blackburn, M. J. and Williams, J. C., Reference 4, 620 (1969)
19. Beck, T. R., Reference 4, 605 (1969)
20. Sanderson, G. and Scully, J. C., *Corros. Sci.*, **8**, 541 (1968)
21. Gerberich, W. W., *2nd International Conference on Fracture*, 919 (1969)
22. Sanderson, G., Powell, D. T. and Scully, J. C., Reference 4, 638 (1969)
23. Powell, D. T. and Scully, J. C., *Corrosion*, **24**, 151 (1968)
24. Brown, B. F., Fujii, C. T. and Dahlberg, E. P., *J. Electrochem. Soc.*, **116**, 201 (1969)
25. Scully, J. C. and Powell, D. T., *Corros. Sci.*, **10**, 719 (1970)
26. Feeney, J. and Blackburn, M. J., *The Theory of Stress Corrosion Cracking in Alloys* (ed. J. C. Scully), N.A.T.O., Brussels, 355 (1971)
27. Fager, D. N. and Spurr, W. F., *Trans. Am. Soc. Metals*, **61**, 283 (1968)
28. Menzies, I. A. and Averill, A. F., *Electrochim. Acta*, **13**, 807 (1968)

29. Mori, K., Takamura, A. and Shimose, T., *Corrosion*, **22**, 29 (1966)
30. Sedriks, A. J., *Corrosion*, **25**, 207 (1969)
31. Sedriks, A. J. and Green, J. A. S., *Corrosion*, **25**, 324 (1969)
32. Spurrier, J. and Scully, J. C., *Corrosion*, **28**, 453 (1972)
33. Mauney, D. A., Starke Jr., E. A. and Hochman, R. F., Reference 11
34. King, E. J., Kappelt, G. K. and Fields, C., *Bell Aerospace Systems Report* (1966)
35. Battelle NASA Report, NASr 100(09) (1969)
36. Beck, T. R., Blackburn, M. J., Smyrl, W. H. and Spiedel, M. O., The Boeing Co. Report, Contract NAS 7-489. No. 14, December (1969)
37. Duttweiler, R. E., Wagner, R. R. and Antony, K. C., Reference 3, 152 (1966)
38. Fager, D. N. and Spurr, W. F., The Boeing Co. Report D6-22691
39. Nelson, H. G., Williams, D. P. and Stein, J. E., *Met. Trans*, **3**, 469 (1972)
40. Binxi, Y., *J. Chinese Society of Corrosion and Protection*, **3**, 41 (1983)
41. Smyrl, W. H. and Blackburn, M. J., *Metall, Trans. A.*, **31**, 370 (1975)
42. Scully, J. C. and Adepoju, T. A., *Corros. Sci.*, **17**, 789 (1977)
43. Koch, G. H., Bursle, A. J. and Pugh, E. N., *Metall. Trans. A*, **9**, 129 (1978)
44. Williams, D. N., *J. Iron Steel Inst.*, **91**, 147 (1962-3)
45. Koch, G. H., Bursle, A. J., Liu, R. and Pugh, E. N., *Metall. Trans. A*, **12**, 1833 (1981)
46. Paton, N. E. and Spurling, R. A., *Metall. Trans. A*, **7**, 1769 (1976)
47. Boyd, J. D., *Trans. A.S.M.*, **62**, 1977 (1969)
48. Wreidt, H. A. and Oriani, R. A., *Acta Met.*, **18**, 753 (1970)
49. St. John, C. and Gerberich, W. W., *Metall. Trans. A*, **4**, 589 (1973)
50. Green, J. A. S., Hayden, H. W. and Montague, W. G., *Effect of Hydrogen on Behavior of Materials*, ed. Thompson, A. W. and Bernstein, I. M., AIME, Warrendale, Pennsylvania, p. 200 (1976)
51. Adepoju, T. A. and Scully, J. C., *Corros. Sci.*, **15**, 415 (1975)
52. Muskowitz, J. A. and Pelloux, R. M., *Metall. Trans. A*, **10**, 509 (1979)
53. Ebtejah, K., Hardie, D. and Parkins, R. N., *Corros. Sci.*, **25**, 415 (1985)
54. Lycett, R. W. and Scully, J. C., *Corros. Sci.*, **19**, 799 (1979)
55. Katz, Y. and Gerberich, W. W., *Int. J. Fract. Mech.*, **6**, 219 (1970)
56. DeLuccia, J. J., Final Report, *NADC076207-30* (June 1976)
57. Meller, F. and Metzger, M., *U.S.N.A.C.A. Tech. Note No. 4019* (1957)
58. Romanov, V. V., *Stress Corrosion Cracking of Metals* (translated from the Russian), 61 (1961)
59. Loose, W. S., *Magnesium*, ASM, Cleveland, Ohio, 173 (1946)
60. Priest, D. K., Beck, F. H. and Fontana, M. G., *Trans. ASM*, **47**, 473 (1955)
61. Priest, D. K., *Stress-Corrosion Cracking and Embrittlement* (ed. W. D. Robertson), Wiley, New York, 81 (1956)
62. Heidenreich, R. D., Gerould, G. H. and McNulty, R. E., *Trans. AIME*, **166**, 15 (1946)
63. Pardue, W. M., Beck, F. H. and Fontana, M. G., *Trans. ASM*, **54**, 539 (1961)
64. George, P. F. and Diehl, H. A., *Met. Prog.*, **62**, 121 (1952)
65. Logan, H. L., *J. Res. Mat. Bur. Stand*, **65C**, 165 (1961)
66. Fairman, L. and West, J. M., *Corros. Sci.*, **5**, 711 (1965)
67. Perryman, E. C. W., *J. Inst. Met.*, **78**, 621 (1950-51)
68. van Rooyen, D., *Corrosion*, **16**, 421t (1960)
68a. Chakrapani, D. G. and Pugh, E. N., *Met. Trans.*, **6A**, 1155 (1975)
69. Wearmouth, W. R., Ph.D. Thesis, University of Newcastle-upon-Tyne (1967)
70. Logan, H. L., *Stress Corrosion Cracking*, Wiley, New York, 217 (1966)
71. Timonova, M. A., *Intercrystalline Corrosion and Corrosion of Metals Under Stress*, Consultants Bureau, New York, 263 (1962)
72. Forrest, P. G., *Fatigue of Metals*, Pergamon, 146 (1962)
73. Brown, B. F. and Beachem, C. D., *Corrosion Sci.*, **5**, 749 (1965)
74. Hunter, M. A., *Metals Handbook*, ASM, Cleveland, Ohio, 234 (1948)
75. Loose, W. S., *The Corrosion Handbook* (ed. H. H. Uhlig), Wiley, New York, 173 (1948)
76. Loose, W. S. and Barbian, H. A., *Stress Corrosion Cracking of Metals*, ASTM/AIME, 273 (1944)
77. Sprowls, D. O. and Brown, R. H., Reference 4, 466 (1969)
78. Graf, L. and Neth, W., *Z. Metallunde*, **60**, 789 and 860 (1969)
79. Petri, H. G., Siebel, G. and Vosskuhler, H., *Aluminium*, **26**, 2 (1944)
80. Engell, H. J., Neth, W. and Suchma, A., *Z. Metallkunde*, **61**, 261 (1970)
81. Rogers, T. H., *Corrosion 1961*, Butterworths, London, 605 (1962)

82. Spiedel, M. O., *The Theory of Stress Corrosion Cracking* (ed. J. C. Scully), N.A.T.O., Brussels, 289 (1971)
83. Watkinson, F. E. and Scully, J. C., *Corros. Sci.*, **11**, 179 (1971)
84. Sedriks, A. J., Green, J. A. S. and Novak, D. L., *Met. Trans.*, **1**, 1815 (1970)
85. Lorenz, P. M., Technical Report AFML-TR-69-99 (1969)
86. Procter, R. P. M. and Paxton, H. W., *ASTM J. of Materials*, **4**, 729 (1969)
87. Binger, W. W., Hollingsworth, E. H. and Sprowls, D. O., *Aluminium* (ed. K. R. van Horn), Vol. I, ASM, 209 (1967)
88. Anderson, W. A., US Patent 3 232 796, February 1 (1966)
89. Niederberger, R. B., Basil, J. L. and Bedford, G. T., *Corrosion*, **22**, 68 (1966)
90. Chadwick, R., Muir, N. B. and Granger, H. B., *J. Inst. Metals*, **82**, 75 (1953-54)
91. Hunter, M. S., Frank, G. R., (Jr.) and Robinson, D. L., *Second International Congress on Metallic Corrosion*, N.A.C.E., Houston, 604 (1966)
92. Mears, R. B., Brown, R. H. and Dix, E. H. (Jr.), *Symposium on Stress Corrosion Cracking of Metals*, ASTM/AIME, 329 (1944)
93. Horst, R. L. (Jr.), Hollingsworth, E. H. and King, W., *Corrosion*, **25**, 199 (1969)
94. Rosenkranz, W., *Aluminium*, **39**, 741 (1963)
95. Staley, J. T., Final Report, Naval Air Systems Command Contract N00019-C8-C-0146 (1969)
96. Gruhl, W., *Metall.*, **19**, 206 (1965)
97. Helfrich, W. J., *Corrosion*, **24**, 423 (1968)
98. Spiedel, M. O., Reference 4, 561 (1969)
99. Sedriks, A. J., Slattery, P. W. and Pugh, E. N., *Trans. ASM*, **62**, 238 (1969)
100. Polmear, I. J., *J. Aust. Inst. Metals*, **89**, 193 (1960)
101. Deardo, A. J. and Townsend, R. D., *Met. Trans.*, **1**, 2573 (1970)
102. Watkinson, F. E. and Scully, J. C., *Corros. Sci.*, **12**, 905 (1972)
103. Brown, B. F., Fujii, C. T. and Dahlberg, E. P., *J. Electrochem. Soc.*, **116**, 218 (1969)
104. Beck, T. R., Blackburn, M. J. and Spiedel, M. O., *Quarterly Progress Report*, No. 11, Contract NAS 7-489 (1969)
105. Gruhl, W., Leichtmetall-Forschunginstitut of Vereinigte Aluminium-Werke AG, Bonn, Report 1970
106. Gest, R. J. and Troiano, A. R., *Corrosion*, **30**, 274 (1974)
107. Montgrain, L. and Swann, P. R., *Hydrogen in Metals* (ed. I. M. Bernstein and A. W. Thompson), A.S.M., Ohio, 575 (1974)
108. Troiano, A. R., *Trans. A.S.M.*, **52**, 54 (1960)
109. Tromans, D. and Pathania, R. S., *The Electrochemical Society: Extended Abstracts N*, 62 (1969)
110. Gruhl, W., *Z. Metallkunde*, **54**, 86 (1963)
111. Gruhl, W. and Brungs, D., *Metall.*, **23**, 1020 (1969)
112. Scamans, G. M., Alani, R. and Swann, P. R., *Corros. Sci.*, **16**, 443 (1976)
113. Alani, R., Scamans, G. M. and Swann, P. R., *Brit. Corros. J.*, **12**, 80 (1977)
114. Scamans, G. M., *J. Mat. Sci.*, **13**, 27 (1978)
115. Vismanadham, R. M., Sun, T. S. and Green, J. A. S., *Corrosion*, **36**, 275 (1980)
116. Vismanadham, R. M., Sun, T. S. and Green, J. A. S., *Metall. Trans. A*, **11**, 85 (1980)
117. Sun, T. S., Chen, J. M., Vismanadham, R. M. and Green, J. A. S., *App. Phys. Letts.*, **31**, 580 (1977)
118. Chen, J. M., Sun, T. S., Vismanadham, R. M. and Green, J. A. S., *Metall. Trans. A*, **8**, 1935 (1977)
119. Scamans, G. M., *Environmental Degradation of Engineering Materials*. ed. Louthan, M. R., Jr., McNitt, R. P. and Sissons, R. D., Jr., Virgina Polytechnic Institute, p. 153 (1981)
120. Scamans, G. M. and Rehal, A., *J. Mat. Sci.*, **14**, 2459 (1979)
121. Malis, T. and Charturvedi, M., *J. Mat. Sci.*, **17**, 1479 (1982)
122. Pickens, J. R., Precht, W. and Westwood, A. R. C., *J. Mat. Sci.*, **18**, 1872 (1983)
123. Holroyd, N. J. H. and Scamans, G. M., *Scripta Met.*, **19**, 915 (1985)
124. Montgrain, L. and Swann, P. R., *Hydrogen in Metals*, A.S.M., Metals Park, p. 575 (1974)
125. Takano, M. and Nagata, T., *Corr. Eng. Japan*, **32**, 456 (1983)
126. Scamans, G. M. and Tuck, C. D. S., *Environment Fracture of Engineering Materials*, ed. Foroulis, Z. A., AIME, Warrendale, Pennsylvania, p. 464 (1974)
127. Christodoulou, L. and Flower, H. M., *Acta Met.*, **18**, 481 (1980)

128. Carialdi, S. W., Nelson, J. L., Yeske, R. A. and Pugh, E. N., *Hydrogen Effects in Metals*, ed. Bernstein, I. M. and Thompson, A. W., AIME, Warrendale, Pennsylvania, p. 437 (1981)
129. Scamans, G. M., *Scripta Met.*, **13**, 245 (1979)
130. Scamans, G. M., *Metall. Trans. A*, **11**, 846 (1980)
131. Scamans, G. M., Reference 36, p. 467.
132. Wood, W. E. and Gerberich, W. W., *Metall. Trans. A*, **5**, 1285 (1974)
133. Pickens, J. R., Gordon, J. R. and Green, J. A. S., *Metall. Trans. A*, **14**, 925 (1983)
134. Hardie, D., Holroyd, N. J. H. and Parkins, R. N., *J. Mat. Sci.*, **14**, 603 (1979)
135. Holroyd, N. J. H. and Hardie, D., *Corros. Sci.*, **21**, 129 (1981)
136. Yuen, L. and Flower, H. M., *Annual Rep.*, Imp. Coll. of Sci. and Tech, London (Sept 1980)
137. Holroyd, N. J. H. and Scamans, G. M., *Slow Strain-Rate Stress Environment-Sensitive Fracture: Evaluation and Comparison of Test Methods*, ed. Dean, S. W., Pugh, E. N. and Ugiansky, G. M., A.S.T.M., Philadelphia, p. 202 (1984)
138. Holroyd, N. J. H., Gray, A., Scamans, G. M. and Hermann, R., *Aluminium–Lithium III*, (eds Baker, C., Gregson, P. J., Harris, S. J. and Peel, C. J.) Institute of Metals, London p 310 (1986)
139. Christodoulou, L., Struble, L. and Pickens, J. R., *Aluminium–Lithium II*, ed. Starke, E. A., Jr. and Sanders, T. H., A.I.M.E., Warrendale, Pennsylvania, p. 561 (1984)

8.6 Corrosion Fatigue*

Introduction

Corrosion fatigue can be defined as a materials failure mechanism which depends on the combined action of repeated cyclic stresses and a chemically reactive environment. The total damage due to corrosion fatigue is usually greater than the sum of the mechanical and chemical components if each were acting in isolation from the other. This simple definition describes a subject of great complexity, combining as it does, many facets of metallurgy, chemistry and mechanical engineering. Numerous laboratory investigations have been carried out emphasising one or more of these aspects, stimulated either by practical requirements for engineering design data, failure analysis, or academic motivations to learn something of the mechanisms of interactions between cyclically deformed materials and their environments. In many respects, there are close parallels with stress-corrosion cracking. However, crack nucleation and self-sustaining growth under the combined action of a constant, not cyclic, tensile stress and a chemically reactive environment is confined to a relatively small number of material-environment combinations. On the other hand, environmental enhancement of a fatigue process can occur with a much wider range of materials and environments because of the ability of the mechanical fatigue process to maintain sharp crack tips in circumstances where non-cyclic stresses could not. Nevertheless, it will also be shown that the dividing line between stress corrosion and corrosion fatigue is not always clear from either a mechanisms or failure analysis viewpoint, in part because it is difficult in practice to be sure that a component, or indeed a test specimen, is subject to a literally constant stress.

Although corrosion fatigue has been recognised and studied for many decades, certainly since World War I, it was not until 1971 that an international conference was held to review the subject[1]. Several reviews are also available from the same period including those of Waterhouse[2] in previous editions of this book and Gilbert[3]. From the time corrosion fatigue was first recognised and described until the 1960s, virtually all experimental investigations used smooth cylindrical specimens which were cyclically stressed until they failed or survived some pre-determined target number of

---

*The work described in this section was undertaken as part of the Underlying Research Programme of the UKAEA.

stress cycles. Data from this type of experiment are typically presented in the form of an S-N curve (Fig. 8.58) which shows the number of cycles to failure, N, as a function of the cyclic stress range, S. The same technique and method of results presentation is still used today, particularly in the context of engineering qualification tests on components and welded connections.

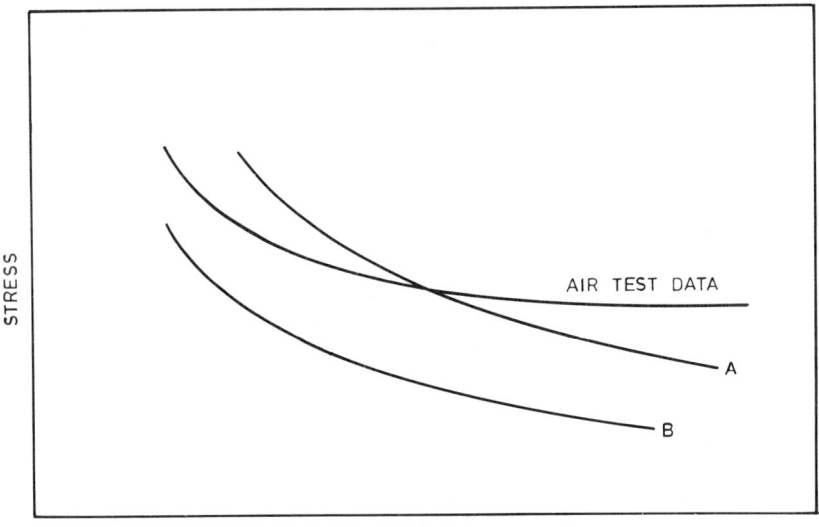

Fig. 8.58 S-N curves for air and corrosion fatigue tests (schematic). A, Corrosion fatigue showing retarded initiation at high stress; B, Corrosion fatigue giving a general lowering of fatigue strength (after Congleton and Craig)

With the development of linear elastic fracture mechanics (Section 8.9) in the 1960s and the recognition that fatigue crack growth rates per cycle, da/dN, could be expressed as a simple function of the cyclic crack-tip stress-intensity value, ΔK, (Fig. 8.59), increasing attention has been focused on measuring the rates of corrosion fatigue crack growth processes. This approach has an important conceptual advantage since it is clear that if a time-dependent process, such as corrosion, is combined with a non-time-sensitive, but stress cycle dependent, fatigue crack nucleation and growth process to give corrosion fatigue, then the cyclic frequency becomes an extremely important variable. It is normally difficult, if not impossible, to investigate fully cyclic frequency effects in an integrated lifetime S-N test. This is because high test frequencies, often considerably greater than 10 Hz, are necessary if the complete S-N curve, including low stress ranges and high cyclic lives greater than 10^6 cycles say, is to be defined in an acceptably short period of time. On the other hand, a test which measures the rate of failure development is not nearly so severely constrained in studying cyclic frequency effects and therefore the time dependent aspects of corrosion fatigue. It should also be noted that the relationship between corrosion fatigue crack growth and S-N data is not necessarily straightforward and will be discussed later on.

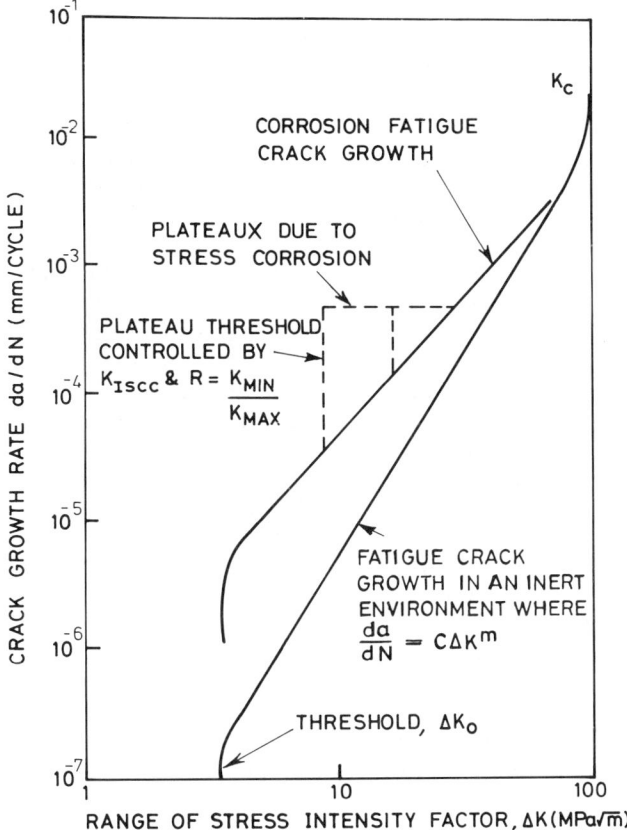

Fig. 8.59 Features of a corrosion fatigue crack growth curve

The diversity of practical corrosion fatigue problems investigated during this century illustrates the range of material–environment combinations which must be considered. For example, during World War I carbon steel towing ropes used in mine sweeping exhibited very short lifetimes which were not improved by increases in steel wire strength. Galvanising did prove to be effective, however. More recent examples associated with the marine environment have concerned the integrity of submarine hulls and of offshore structures for oil and gas production. Aircraft components must also be proved against corrosion fatigue from environments as diverse as water spray affecting undercarriage components to very hot gas environments typical of jet engines. Heat exchangers of all types which can be subjected to water hammer or cyclic thermal stresses associated with their operation also widen the range of materials from steels to nickel, aluminium, copper- or titanium-base alloys and environments from sea or river water to the carefully controlled water chemistries typical of modern boilers and nuclear power reactors.

One method of ordering or categorising this great diversity of materials and environment combinations which will be followed here, is to divide the subject matter up into general groups as follows: (1) gaseous oxidation and adsorption, (2) other non-aqueous environments such as liquid metals, (3) aqueous systems subject to general corrosion and/or pitting, (4) aqueous systems in which the materials are immune to general corrosion, usually by virtue of a barrier coating or cathodic protection, and (5) aqueous systems in which the materials form protective oxide or passive films. Examples will be used below to illustrate corrosion fatigue behaviour of metal–environment combinations falling into each of these categories in order to deduce underlying principles and common themes. It is important to note that, apart from possibly the first and third categories above, an environment need not necessarily be 'corrosive' in the normal sense of the word for it to exert a substantial effect in corrosion fatigue. This arises because local strains associated with the formation and propagation of fatigue cracks can fracture or greatly thin protective films and/or expose highly, chemically reactive, fresh metal surface which can behave chemically in a radically different way to other unstrained surfaces.

In the succeeding sections of this chapter a brief description of the mechanisms of fatigue crack initiation and growth as presently understood is given together with an indication of the various ways in which corrosion may influence these mechanical processes. After that, illustrative examples of corrosion fatigue crack growth and corrosion fatigue endurance in various alloy–environment combinations using the categories given in the previous paragraph are described. The chosen order of presentation of endurance data following crack growth data is done deliberately so that the influence of corrosion on the relative contributions of the crack nucleation and growth phases of failure development in endurance tests can be assessed and thereby linked to the final section on practical applications.

Mechanisms

An important development over the last few decades has been an improved understanding of the mechanisms of how fatigue cracks initiate and grow in metallic materials. An essential first step is the localisation of cyclic, plastic deformation onto favourably orientated slip planes. Any oxidation or adsorption process may prevent slip step reversal and continuing slip on adjacent planes leads to closely spaced groups of slip planes known as persistent slip bands (PSB) (Fig. 8.60). If the surfaces are initially smooth, these slip processes can be shown to be accompanied by intrusions and extrusions of material at the slip band with incipient cracks forming at the intrusion. The irreversible movements of dislocations associated with slip band formation are very complex and vary significantly with metallurgical structure[4,5]. In polyphase materials, the sites for strain localisation may be inclusions, grain boundaries, metallic precipitates or precipitate-free zones (PFZ) or simply a mechanical stress concentration such as a notch or corrosion pit. Following the initial localisation of strain, visible cracks then form on shear planes (stage I) and may propagate in this mode across one or several grains until one dominant crack takes over and propagates perpendicular to the imposed principal tensile stress (stage II). Failure occurs when the remaining ligament breaks by plastic collapse or brittle fracture. Some materials such

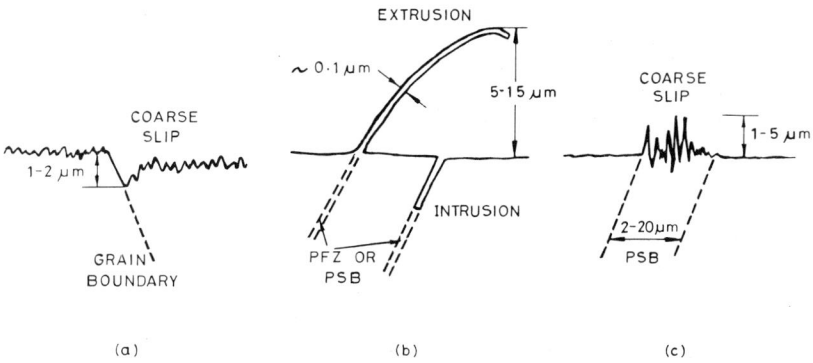

Fig. 8.60 Schematic diagrams showing common surface profiles produced during fatigue: (a) coarse slip and crack initiation adjacent to grain boundaries; (b) extrusions and intrusions; (c) coarse slip within a persistent slip band (after Lynch[7])

as steels exhibit definite fatigue or endurance limits in air or vacuum and at cyclic stress ranges below the limit, fatigue failure does not occur. In such cases the endurance limit stress range coincides with the cyclic stress/strain yield point which is often about half the tensile strength. In other cases a definite fatigue limit is not observed and endurance limits for a suitably large number of cycles, say 10^7 or 10^8, is quoted where the slope of the falling S-N curve is shallow (Fig. 8.58).

It is plain from the description above that the boundary between crack initiation and crack growth is not clear cut. Indeed many would regard the distinction as semantic, or state that most of the fatigue life is spent in crack propagation, however small those cracks might be. Nevertheless, the proportion of cyclic life occupied by the various stages can vary greatly with metallurgical structure, magnitude of the applied cyclic and mean stress, geometry and environment. Only stage II of the growth process (Fig. 8.61) can be properly characterised in terms of the linear elastic parameter, the cyclic range of the crack-tip stress-intensity factor, ΔK. Even this is subject to the conditions that the crack-tip plasticity be contained within an elastic continuum and that the crack is large compared to microstructural dimensions. When these conditions are satisfied and the environment is benign, the familiar Paris equation can characterise the crack growth rate per cycle, da/dN, over a wide range of ΔK (Fig. 8.59).

Stage I crack growth, or stage II growth under high strain conditions, requires more specialised methods of analysis to represent the driving crack-tip stress field. This problem is now known as the 'short crack' problem, defined roughly by cracks 0.01 to 1.0 mm deep dependent on alloy strength and raises unique issues with regard to the influence of chemically reactive environments. Nevertheless, successful quantitative representations of high strain, cyclic endurance by the Coffin–Manson equation[6] predate attempts to characterise fatigue crack growth explicitly and are widely used in low cyclic fatigue design. It is important to note, however, that the shear decohesion processes (Fig. 8.60) associated with fatigue failure in ductile metallic materials are essentially the same throughout all the stages of crack initiation

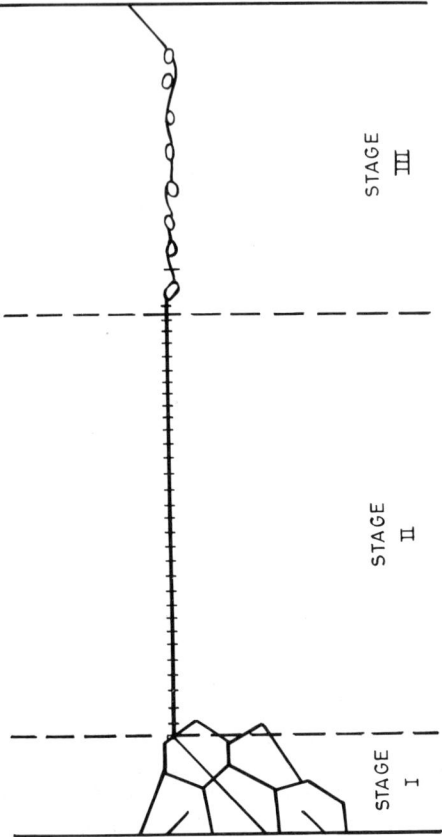

Fig. 8.61 Fatigue crack propagation across specimen section (after Tomkins and Wareing[16])

and growth whether the net section stresses or strains are above or below the elastic limit. Difficulties in characterising fatigue crack growth quantitatively arise only from difficulties in providing adequate descriptions of the crack-tip driving force over the full range of stresses and strains and crack sizes. However, the characteristic crack growth/arrest markings known as striations which are commonly visible on ductile metal fatigue fracture surfaces in benign or mildly oxidising environments are normally associated with stage II growth. Their formation is illustrated in Fig. 8.62.

Before discussing the influence of corrosion on the mechanical deformation processes of fatigue crack initiation and growth in some specific systems, it is useful to have a general mechanistic framework to which specific examples can be related. In the early stages of fatigue crack nucleation, the main effect of corrosion is to accelerate the plastic deformation and slip processes which precede the formation of stage I cracks. These may be broadly classified into four groups: (1) where oxide films interfere with slip reversibility, (2) where adsorbed species influence slip by facilitating the

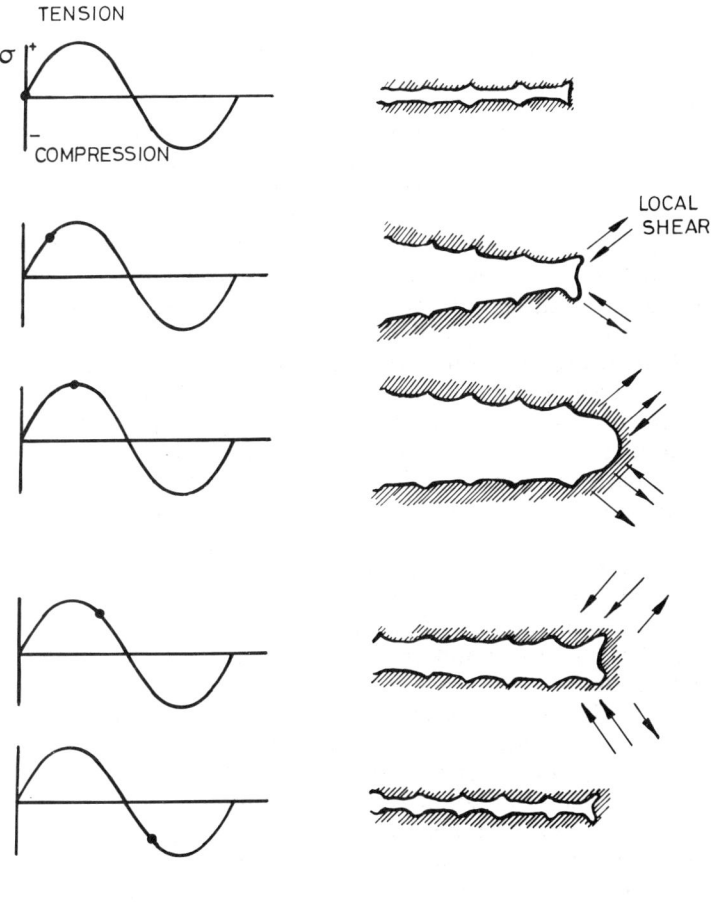

Fig. 8.62 Crack-blunting model of stage II crack progagation (after Laird)[16]

nucleation of dislocations, (3) where corrosion processes result in the injection of an embrittling species such as hydrogen, and (4) where corrosion removes plastically deformed material. The same processes can also influence the later stages of crack growth although two additional considerations come into play. One is the role of oxides or other corrosion products in impeding crack closure and consequent effects on the effective range of the crack-tip stress-intensity. The second is the effect of a long, narrow diffusion path for reactants and products along a macroscopic crack and the possibility of chemical modifications to crack-tip environments relative to the external bulk environment.

Owing to the nature of the fatigue process in ductile metals, it is clear that a constant supply of atomically clean, new surface is presented to interact with any environment. Oxygen and water will normally adsorb strongly and very quickly on these new clean metal surfaces. Any oxidising agent, whether gaseous or aqueous, will react rapidly with fresh metal surface exposed by the fatigue process and the extent of reaction per cycle will then clearly depend on the duration of the cycle period. With gaseous oxidants there is usually little ambiguity about the nature of the processes leading to environmental acceleration or in some cases retardation. Oxygen and air for example at normal ambient temperatures usually generate sufficient oxide to impede any rehealing during the compressive part of the cycle of the new metal surfaces created during crack opening. These processes normally also do not seriously impede crack closure of longer stage II cracks. At higher temperatures, however, the formation of thicker oxide layers can impede crack closure thereby raising apparent crack growth thresholds and reducing crack growth rates. Alternatively, hydrogen and hydrogen sulphide gases are generally thought to accelerate fatigue by supplying hydrogen to cause embrittlement of the plastically deformed metal created by the fatigue process. Thus any processes impeding hydrogen entry can reduce the embrittling effect. Some liquid metal environments can transport carbon to or from the crack-tip zone and thereby alter the rate of crack growth. One author[7] has proposed that liquid metals facilitate the nucleation of dislocations and, on the basis of fractographic evidence, has drawn parallels with many other environments to suggest that the same mechanism operates in other cases.

Water vapour and aqueous solutions are much more difficult to interpret unambiguously. In aqueous systems at the free corrosion potential, anodic processes such as dissolution of persistent slip bands or the crack tip cannot proceed without a corresponding cathodic process. Since dissolved oxygen cannot penetrate far down cracks or crevice geometries, hydrogen evolution is the most likely supporting cathodic reaction in long cracks. Thus, the dominance of dissolution or hydrogen-embrittlement processes in accelerating fatigue crack growth is difficult if not impossible to prove. Even potentiostatically controlled tests can be difficult to interpret when the problems of establishing the crack-tip potential and the chemistry of crack-tip environments are considered. In passing, one may note that the common assertion that acid environments form in stress-corrosion and corrosion fatigue cracks often does not stand up to examination. This is because precisely known preconditions must be satisfied for this to occur; i.e. a potential difference must exist between the crack tip and exterior surfaces which is normally provided by an oxygen concentration cell; the dissolved cation must be hydrolysable; and an acid forming anion must also be present. Persistent slip band or crack-tip dissolution can also act to slow down corrosion fatigue cracking by blunting the crack tip. Clearly, whether dissolution processes lead to crack-tip blunting depends on the kinetics of the dissolution process and the counter effect of stress cycle dependent mechanical sharpening by fatigue. Precipitation processes within cracks can also lead to significant perturbations to crack closure with consequent effects on rates and thresholds as well as effects on diffusion of chemical species into and out of cracks.

The complexity of these chemical and mechanical interactions is such that each metal–environment system must be examined on an individual basis to determine the important processes influencing corrosion fatigue crack nucleation and growth rates. Thus, in the ensuing sections, examples are quoted to illustrate commonly occurring phenomena or establish more general principles with reasonably wide applicability for particular classes of metal/environment combinations. It should be noted, however, that when it becomes necessary to evaluate new metal–environment combinations, there is no unified theory of corrosion fatigue which can avoid the need for experimental data.

Studies of corrosion fatigue in metallic materials have, for the most part, aimed at measuring S–N curves on plain cylindrical specimens or the growth rates of macroscopically large cracks as a function of the cyclic stress-intensity factor, ΔK (Fig. 8.58 and 8.59). One point that requires attention in S–N testing is the diameter of the test section relative to the expected loss of section size by general corrosion. Excessive corrosion could lead to premature failure by simply increasing the effective net section stress which would not then genuinely reflect the behaviour of larger components. Most of the common types of pre-cracked fracture mechanics specimen have been used in investigations of corrosion fatigue crack growth although the compact tension (CT) specimen has been the most popular. The CT specimen has a number of important experimental advantages for this type of work, among them a high mechanical advantage allowing the lowest applied loads of any type of specimen to achieve a given ΔK value and relatively simple indirect monitoring (i.e. non-visual) of crack size by compliance or electrical resistance methods. General experimental techniques for corrosion fatigue tests including guidelines on environment containment and monitoring of environmental chemistry, including most importantly, the specimen corrosion potential in aqueous solutions, have been extensively described[8]. Two aspects of experimental design requiring constant vigilance are unwanted electrochemical effects from containment materials and other fixtures and prevention of interactions between electrical equipment such as potentiostats and electrical resistance crack monitoring devices. However, concerns over possible electrochemical effects arising from the use of d.c. electrical resistance crack measurement techniques appear on present evidence to be unfounded.

An important underlying assumption in much fracture mechanics-based corrosion fatigue testing is the similitude between different geometries given the same environmental conditions in which cracks are assumed to be characterised by one simple parameter, ΔK. This is an assumption which has to be carefully examined, particularly if the results are required for some practical application. For example, will the environmental conditions prevailing in a crack in a CT specimen imitate correctly those in a more realistic semi-elliptical crack shape? In one case for structural steel in seawater, this has been shown to be a reasonable assumption[9]. There is more doubt in the case of pressure vessel steels in simulated light water reactor environments where exchange of dissolved impurities between the crack environment and the bulk has a critical influence on the environmental contribution to cracking[10]. In the case of short cracks, however, there are both mechanical reasons invalidating the representation of crack-tip strains and deformations by ΔK as discussed earlier[11] and theoretical and experimental evidence of the importance of crack size at the millimetre to sub-millimetre level on crack chemistry[12].

Crack Propagation–Non-aqueous Environments

Many fatigue crack propagation experiments are carried out in laboratory air at normal ambient temperatures without any concern that air oxidation or reactions with water vapour might be contributing to the crack extension process. In many cases, but not all, the use of air data as a baseline against which to compare other environmental influences is a reasonable working assumption. Nevertheless, it is well known that in steels, for example, crack growth rates obtained in vacuum are two to three times less than those obtained in air[13]. It may also be noted in passing that fatigue crack growth in many different materials in air can be plotted in a relatively narrow scatter band if expressed as a function of $\Delta K/E$, where E is the elastic modulus, rather than ΔK alone. Crack growth thresholds are also often adversely influenced by air oxidation at ambient temperatures relative to vacuum[13].

At higher temperatures, for example those relevant to gas turbines, air oxidation and other corrosion processes become progressively more important contributors to fatigue crack growth in iron- and nickel-base alloys of commercial interest. An order of magnitude increase in crack growth rates in iron- or nickel-based superalloys for aero engines is not uncommon[8,13]. However, oxide wedging at low ΔK values which reduces crack-tip opening displacements can actually raise the apparent crack growth threshold under constant amplitude fatigue loading. Such a mechanism could be less effective under complex spectrum loading sequences where compressive forces can grind up an accumulating oxide scale.

At even higher temperatures above about one third of the melting temperature of an alloy, creep effects also begin to contribute to the crack extension process as well as air oxidation and the resulting crack growth behaviour can vary in a very complex way with loading and environmental variables[14,15]. Once material failure processes such as creep make a significant contribution to crack growth, non-linear deformation processes occurring in front of the advancing crack invalidate the stress-intensity factor K, or its cyclic range ΔK, as a sensible parameter characterising the near crack-tip stress field. The difficulties inherent in finding an acceptable characterising stress field parameter for crack growth under creep-fatigue conditions have been discussed extensively by Tomkins and co-workers[14,16].

For temperatures below the creep range it has been suggested that oxidation can only accelerate fatigue crack growth to an upper limit defined by half the maximum crack-tip opening per cycle as shown by the examples in Fig. 8.63. This is an important principle to understand because of its potential use in design problems, as are the circumstances under which the principle breaks down. From considerations of the feasible geometry of crack tips, fatigue cracks growing in ductile materials by a shear decohesion process cannot be greater than half the crack-tip opening displacement per cycle. The reason that fatigue cracks usually grow at less than this rate is because in real hardening materials, crack-tip strains are not accommodated on one shear plane emanating from the crack tip but on many planes which spread plastic flow to the crack flanks immediately behind the crack tip. Thus any corrosion process which is indiscriminate in removing material from the crack tip or sides will cause blunting if the resulting combination of environ-

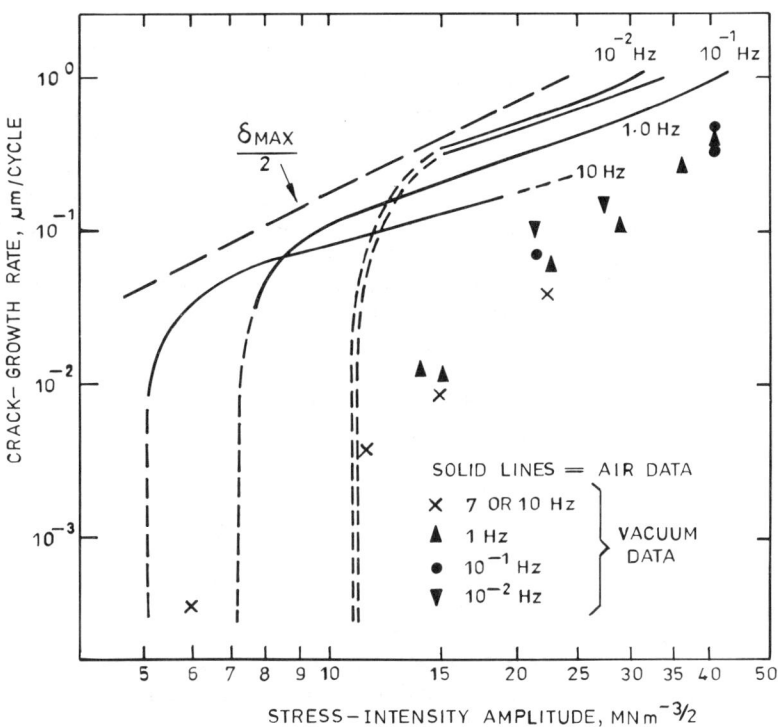

Fig. 8.63 Effect of environment on fatigue crack growth rate in 1% Cr–Mo–V steel at 550°C (after Tomkins and Wareing[16])

mental and mechanical fatigue damage exceeds the theoretical maximum crack-tip opening. Equally, any alternative, potentially self-sustaining failure process such as creep or, as will be seen later, stress-corrosion cracking, that contributes to crack growth will not be contained within this theoretical maximum fatigue crack growth rate defined by the maximum crack-tip opening displacement. At high temperatures, an obvious alternative corrosion process which would cause this generalisation to break down would be hot salt corrosion, particularly sulphidation which occurs in marine gas turbines, and leads to rapid intergranular cracking[13]. It has also been observed that rapid diffusion of oxygen down grain boundaries in some superalloys at high temperatures causes large increases in fatigue crack growth which are unrelated to creep effects[6]. Where creep cavitation occurs, extremely large accelerations in crack growth rate are possible when the cyclic crack-tip opening is of the same order as the cavitation spacing. Such rapid failure processes can be likened to opening a zip fastener through the material.

Some of the most interesting work on the mechanisms of corrosion fatigue crack growth has been done on steels and high strength aluminium alloys in carefully controlled water vapour or hydrogen gas environments. Great care is needed in this type of work to ensure the removal of adsorbed species under vacuum prior to admitting the gaseous environment of interest which must also be very pure[8]. The fracture surfaces of AISI 4340 steel resulting from stress corrosion or corrosion fatigue and the reaction kinetics between water vapour and iron crystals of known orientation have been studied by Auger electron spectroscopy and low energy electron diffraction[17,18]. These results showed that the rate limiting step in the environmental contribution to corrosion fatigue is the reaction between water vapour and iron or possibly iron carbide. Observations of transients in crack growth rates following changes of cyclic frequency strongly suggested that hydrogen produced from the reaction between iron and water vapour is primarily responsible for the environmental enhancement of fatigue crack growth in high strength AISI 4340 steel. The zone of hydrogen damage was also deduced to be somewhat greater (approximately 0.1 to 1.0 mm) than the calculated reversed plastic zone size at the tip of the crack or relevant microstructural dimensions. Similar measurements and deductions have been made for high-strength aluminium alloys[19].

Interest in the role of hydrogen embrittlement in corrosion fatigue, particularly in steels, but also high-strength aluminium alloys and the hydride forming metals such as titanium and zirconium, has prompted much research using hydrogen or hydrogen sulphide gases. In addition there have also been industrial uses and failures of these combinations of materials and environments which have given added impetus to the work.

In steels, the influence of hydrogen on fatigue crack propagation shows close parallels with behaviour in low temperature aqueous environments[20]. For example, research work following a catastrophic failure in 1974 of a $3\frac{1}{2}$ Ni–Cr–Mo–V steel end ring component of a 500 MW generator operating in 5 bar pressure hydrogen established that the high yield stress of the material of 1 250 MPa rendered it susceptible to hydrogen-induced crack growth at constant crack-tip stress intensity. Parallel corrosion fatigue experiments showed the classical above and below K_{Iscc} behaviour seen in high-strength steels in low temperature aqueous chloride solutions (see Fig. 8.59 and next section). Thus any fraction of the cycle period spent with the stress intensity above the static threshold for hydrogen cracking resulted in large increases in corrosion fatigue crack growth rates and the coincident presence of intergranular or brittle facets on the resulting fracture surfaces. At yield strengths below 1 100 MPa these end ring steels were not susceptible to hydrogen cracking under constant loads but there was nevertheless a significant residual frequency-dependent hydrogen-environment effect on fatigue crack growth rates, (Fig. 8.64). Such effects are enhanced by increasing hydrogen pressure (Fig. 8.64) and by the presence of hydrogen sulphide gas, and substantially decreased by air contamination of the hydrogen atmosphere[8,20,21]. This and other work has pointed to the importance of adsorption on fresh metal surfaces created by the fatigue process at the crack tip. The mechanisms by which hydrogen enhances fatigue crack growth once absorbed into the metal remains as much a mystery in this as in other hydrogen-embrittlement research. The reduction in the adverse effect of low-pressure hydrogen gas atmospheres at low cyclic frequencies (Fig. 8.64) is particularly difficult to explain. It may be due to the mismatch between hydrogen and dislocation mobility within the plastic zone, since hydrogen is rather weakly bound to dislocations, or due to minor impurities

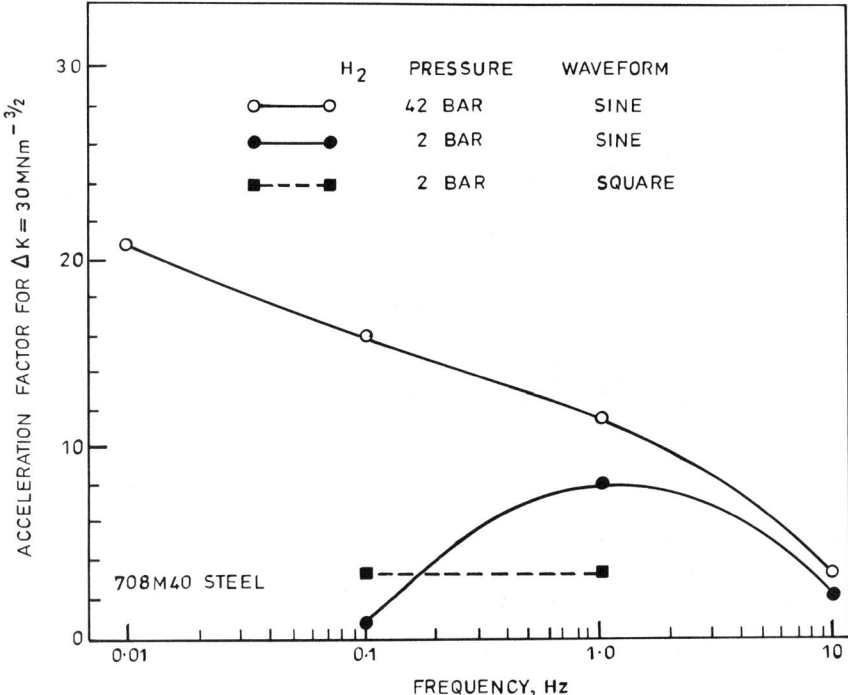

Fig. 8.64 Influence of hydrogen pressure, frequency and waveform on the enhancement of fatigue crack growth in 708M40 steel $\Delta K = 30$ MNm$^{-3/2}$ (after McIntyre[21])

in the gas atmosphere slowly poisoning active metal surface sites for hydrogen adsorption at the crack tip[21].

Corrosion fatigue crack growth in high-strength aluminium alloys is strongly influenced by the presence of water vapour typically between 100 and 10 000 ppm in air or other inert or oxidising gases at normal ambient temperatures[19,22]. The adverse effect of water vapour tends to saturate at the higher partial pressure. The influence of water vapour has been attributed to hydrogen embrittlement. However, although no systematic studies of hydrogen gas atmospheres on fatigue crack growth are available, gaseous hydrogen has not been found to influence appreciably total fatigue life in these alloys. The apparent discrepancy may be due to the extreme reactivity of new aluminium surfaces created at fatigue crack-tips with any oxidising impurity in the gaseous environment and the impervious nature of aluminium oxide films to hydrogen diffusion.

Fatigue crack propagation has been studied extensively in stainless steels over a wide range of temperatures and oxidising environments because of important actual or potential applications in nuclear reactors and steam raising plant[15,23]. Environments such as nitrogen, argon and liquid sodium at temperatures up to 500°C have little influence on fatigue crack growth in ductile stainless steels such as types 304 and 316 compared with vacuum over the same temperature range or room temperature air. Air at 500°C produces more than an order of magnitude increase in propagation rates over a wide frequency range whereas steam at the same temperature causes nearly a further two orders of magnitude increase in crack growth rates. Clearly these are very large and significant effects of environment apparent under conditions where creep interactions are insignificant. Pressurised water at 300°C is not nearly so aggressive, however, at least on solution-annealed stainless steel in the absence of dissolved oxygen (see next section), indicating that strongly thermally activated oxidation processes must operate at higher temperatures.

A considerable technical literature exists on liquid metal embrittlement, but relatively little work has been done on corrosion fatigue crack growth in liquid metals. Most work relates to the influence of sodium at relatively high temperatures around 600°C on stainless steels in the context of core components for fast reactors[15]. In low oxygen (5–10 ppm) sodium, fatigue crack growth rates in type 316 stainless steel are equivalent to those measured in vacuum or inert gases. Carburising or decarburising sodium can enhance these rates by up to a factor of five, however. Other possible contaminants such as lead, tin or zinc may also have very adverse effects. An additional environmental aspect of nuclear reactor components, particularly those in close proximity to or part of the core, is neutron irradiation damage. In general, there does not seem to be a serious adverse effect of irradiation on fatigue crack growth in ferritic or stainless steels until very high doses, say greater than one displacement per atom, are encountered. In these circumstances, helium bubble formation, in particular from n, α reactions with boron within the metal, accompanied by physical swelling occurs and considerable frequency-dependent degradation of fatigue and creep fatigue properties is possible. As in the case of creep cavitation, the most severe effects are observed in stainless steels when the crack-tip opening is of the same order as the helium bubble spacing.

Crack Propagation–Aqueous Environments

Steels

A great deal of experimental work has been carried out using carbon and low-alloy steels in either 3.5% sodium chloride solution or seawater. At the medium-to-low-strength levels, say less than 1 000 MPa yield strength, such materials are not normally susceptible to environmentally-induced cracking (by hydrogen embrittlement) under constant applied loads in aqueous environments unless there are additional sources of hydrogen such as from hydrogen sulphide contamination or excessive cathodic polarisation. By contrast, fatigue crack propagation rates are markedly increased both at the free corrosion potential and at more cathodic potentials consistent with reasonable levels of cathodic protection. The increase in fatigue crack growth rates due to corrosion can be represented by a simple multiplying factor on the corresponding in-air rates like that given earlier in Fig. 8.64 for hydrogen gas environments. Similar observations have been made for quite a wide variety of low-alloy steels freely corroding in 3.5% sodium chloride solution or seawater[8,24,25]. It is seen that aqueous environmental influences on fatigue crack growth are negligible at frequencies of 10 Hz and above and tend to reach a limiting factor at 10^{-2} Hz or lower frequencies. At even lower frequencies, there is evidence that the environmental effect declines due to crack-tip blunting and in combination with low values of ΔK, cracks can actually be arrested because the crack-tip pitting rate is faster than the crack growth rate (on a time base). This has been demonstrated particularly well for intermittent wetting and drying conditions for structural steel in seawater representing splash zone environments on offshore structures[9]. These crack-tip corrosion processes are also thermally activated and an activation energy of about 40 kJ/mole can be deduced from temperature effects on corrosion fatigue crack growth rates around normal ambient temperatures[24].

In addition to the cyclic frequency, the shape of the cyclic waveform also has a marked effect on the environmental contribution to crack growth as originally shown by Barsom[26]. It has been demonstrated in several steel/aqueous environment combinations that the primary environmental contribution to crack extension occurs during the increasing load part of the cycle. Thus, for cycle waveforms of the same period, sine, triangle and positive sawtooth shapes show similar environmental effects, whereas square and negative sawtooth waveforms (i.e. those with a very fast leading edge or rising load) show negligible contributions from the aqueous environment to crack growth when compared with normal laboratory air test results. Clearly, the mechanistic significance is that the environmental influence depends on the length of time in the cycle that new metal surface is being exposed to the chemically reactive solution in the crack enclave.

There is now a considerable body of evidence that points to hydrogen-embrittlement as being primarily responsible for the accelerations in fatigue crack growth seen in steels freely corroding in ambient temperature aqueous environments[28]. For example, as pointed out above, the frequency response of corrosion fatigue in low-alloy steels in hydrogen gas closely resembles that in aqueous environments[27]. In addition, numerous transient effects during changes of experimental conditions seem inexplicable except on the basis of hydrogen embrittlement of a zone of metal just in front of the crack tip. One might anticipate on this basis that cathodic polarisation might increase corrosion fatigue rates with decreasing potential. In fact, a slightly more complex situation arises in which a minimum in the environmental effect is seen at about 100 to 200 mV below the free corrosion potential which then rises as the potential is moved increasingly in the negative direction[8,24]. A possible explanation for the effect of cathodic polarisation has been provided by work on hydrogen permeation rates through low-alloy steel crevices subjected to cathodic polarisation at the crevice mouth[28]. Hydrogen permeation rates at the base of a crevice as a function of externally applied potential exactly match the trend of the environmental contribution to corrosion fatigue rates. This can be readily understood in terms of crack-tip acidification enhancing hydrogen production at the crevice tip at the free corrosion potential, but being reduced at slightly more negative potentials by the accumulation of alkaline cathodic reaction products until finally the rate of hydrogen production increases again as the overpotential for hydrogen evolution becomes greater.

The influence of crack-tip chemistry and electrochemistry on corrosion fatigue crack growth in steels in salt water environments has been extensively reviewed by Turnbull[12]. Corrosion fatigue crack-tip chemistry in large cracks (> 10 mm deep) is surprisingly little disturbed compared with static crevices by mechanical pumping effects, at least at low frequencies such as 0.1 Hz or below. Thus much of the understanding which has developed in recent years concerning crack-tip chemistry in relation to stress-corrosion cracking is also relevant to corrosion fatigue. In ionically conductive solutions such as 3.5% sodium chloride solution or seawater, ohmic drops down cracks or crevices are not large, at least at externally imposed potentials within, say, 500 mV of the free corrosion potential. Thus in seawater, cathodic protection to say -850 mV (versus Ag/AgCl) will give a crack-tip potential of the order of -800 mV (versus Ag/AgCl). This gives rise to a complication for seawater whereby calcareous scale can precipitate both within and outside a crack as a consequence of alkali-forming cathodic reactions. This hard calcareous scale which forms on the crack flanks can have a large effect in reducing the degree of crack opening for a given applied load range with the result that cracks which would otherwise grow at low rates, slow down and even arrest when cathodically polarised in seawater[9]. This phenomenon is most in evidence at crack growth rates approaching the in-air threshold ΔK; i.e. at ΔK values of less than 15 MPa $\sqrt{m}$.

The final major parameter influencing corrosion fatigue crack growth rates in low-alloy ferritic steels in ambient temperature aqueous environments in addition to cyclic frequency, waveform and electrochemical potential is the mean stress level about which the cyclic stress oscillates. It is normal in work on fatigue crack growth to define the mean stress conditions in terms of the stress ratio, R, equal to the ratio of the minimum to the maximum stress or stress intensity in the cycle (Fig. 8.59). The stress ratio has also been found to increase crack propagation rates above the threshold ΔK for crack growth in low-alloy steels in aqueous environments whereas there is little influence of stress ratio in air except on crack growth thresholds themselves. Figure 8.65 shows the combined effects of potential and stress ratio on fatigue crack growth in a structural steel exposed to seawater either at the free corrosion potential or at -1.1 V (versus Ag/AgCl)[9]. Increasing stress ratio increases crack growth rates but the effect apparently saturates between $R = 0.5$ and 0.7. No good mechanistic model has been proposed to explain this effect of R ratio except that in general terms it is clear that a greater proportion of the cyclic crack-tip opening is converted into crack extension at the higher stress ratios in the presence of the aqueous environment.

It will be appreciated from the discussion so far concerning the effect of chemical precipitates in cracks and dissolution rates at crack tips, that when these processes are combined with the influence of R ratio on crack growth thresholds, a rather complex set of interactions is feasible[24,25]. On the whole, higher stress ratios which result in the crack faces being held wider apart than with lower R ratios tend to reduce the influence of crack-tip precipitates and their effect on crack closure. Even in the absence of the complication of precipitates in cracks, a good deal of variability is found in crack growth thresholds in salt water environments relative to those found in air[25,29,30]. It is perhaps not surprising that at low crack growth rates the effect of crack-tip dissolution and any consequential hydrogen-embrittlement

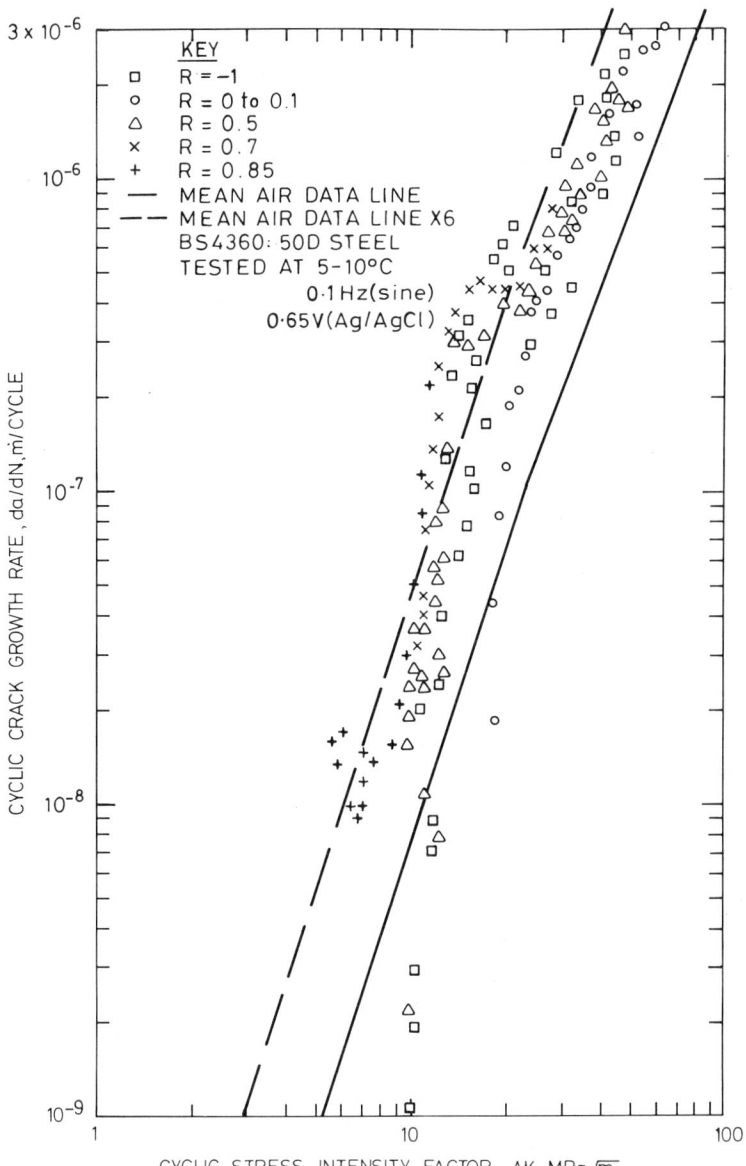

Fig. 8.65 Corrosion fatigue crack growth data for structural steel in seawater at 0.1 Hz, $R = -1$ to 0.85 and -1.10 V (Ag/AgCl) (after Scott[24])

can reduce or increase thresholds dependent on the precise competing kinetics of the electrochemical, mechanical and metallurgical damage processes.

An interesting feature of Fig. 8.65(b), which shows corrosion fatigue crack growth results for a medium strength steel somewhat over-cathodically protected in seawater, is the appearance of a stress-corrosion or plateau-like feature, particularly at high R ratios (cf. Fig. 8.59). In fact there is no evidence that this steel in its as-received condition is at all susceptible to stress-corrosion cracking under these environmental conditions. Nevertheless, a period of crack growth independent of ΔK as in Fig. 8.65(b) is a clear indication of the intervention of a rate-limiting process unrelated to ΔK; in this case most probably the rate of evolution of hydrogen near the crack tip or the rate of diffusion of hydrogen to the crack-tip process zone. Such features are commonly observed in corrosion fatigue tests in alloys which may or may not be susceptible to stress-corrosion cracking[31] and provide a clear indication of how the dividing line between corrosion fatigue and stress corrosion is far from being well defined. It will be seen later when corrosion fatigue systems are discussed in which the breaking and re-healing of passive, protective oxide films are critical to the crack advance mechanism, that the concept of environmental cracking processes dependent on the application of a continuing dynamic strain is not novel. Indeed the so-called slow strain-rate test or constant extension rate test for stress-corrosion susceptibility has been specifically designed to cope with such circumstances and 'windows' of strain rates are commonly found in which environmentally-induced cracking is possible and outside which it is not. Thus we may have environmentally controlled cracking processes in corrosion fatigue dependent on dynamic straining of the crack tip by fatigue forces over a specific range of frequencies but whose rate is not a function of ΔK or any other cyclic crack-tip plasticity parameter because chemical reaction rates or diffusion processes are rate controlling.

In the case of high-strength steels (yield strengths around or greater than about 1 000 MPa), simple models can be employed which superimpose stress-corrosion cracking (by hydrogen embrittlement) on the fatigue process[17,32]. In this case, hydrogen-embrittlement cracking of high-strength steels under constant stress can be well represented as a time-dependent rate, da/dt, as a function of K, the stress-intensity factor, with a well-defined threshold, K_{Iscc}. If a fatigue force is applied, then any fraction of the cyclic ΔK which exceeds K_{Iscc} causes a marked increase in observed corrosion fatigue crack growth rates as illustrated in Fig. 8.59. The fact that this relatively simple model works so well indicates that there is comparatively little strain-rate sensitivity in the constant stress hydrogen cracking process itself, either on K_{Iscc} or on the plateau growth rate.

As indicated earlier, many other metal-environment systems in which mixed fatigue and stress-corrosion-like crack growth processes are possible are not so amenable to such a simple superposition model because the rate of environmental attack is itself strain-rate sensitive. An example in which this has been extensively examined is the case of pressure vessel steels exposed to simulated light water reactor coolants at c. $300°C$[10]. It is known that the rate of crack growth in corrosion fatigue tests on medium-strength reactor pressure vessel steels (A533-B and A508) is very sensitive to the dissolved oxygen concentration between 25 and 100 ppb (which has a strong influence on corrosion potential), the sulphur impurity content of the steel, the sulphur anion concentration in the water and the linear water flow rate. The influence of all these factors has been rationalised on the basis that a sulphur anion rich environment in the crack enclave greatly enhances electrochemical dissolution reaction rates on emerging slip planes at the crack-tip (and consequently, also, enhances nearby cathodic hydrogen evolution reactions). However, these reactions can only take place as the protective oxide film of magnetite (or magnetite plus haematite depending on oxygen concentration) formed rapidly at these high temperatures is broken at the crack tip. This in turn depends on the crack-tip loading rate or strain rate. By representing the environmentally controlled rate of crack growth as a function of crack-tip strain-rate, it has been possible to construct a predictive model which is still basically a superposition model, but one in which the environmental contribution depends not only on the crack-tip stress-intensity exceeding a critical minimum value but also on the effective crack-tip loading rate. Predictions of the influence of frequency and R ratio from the model fit known experimental data very well indeed. The most obvious consequences of this modification of the superposition principle are a dependence of the plateau corrosion fatigue rates on $f^{-\frac{1}{3}}$ rather than f^{-1} of non-strain-rate-sensitive models and the existence of a specific 'window' of cyclic frequencies only within which it is possible to observe any environmental influence on crack growth rates at all.

The study of the growth by fatigue of physically short cracks usually less than 0.1 to 1.0 mm deep is a topic of much current research interest. The study of environmental effects appears to have been confined so far to the influence of high temperature air oxidation of superalloys for aero-engines (see next section for more details) and to steels in salt-water environments. Even in the absence of reactive environments, short cracks grow considerably faster than long ones when expressed as a function of the linear elastic fracture mechanics parameter, ΔK. This can be due to uncontained plasticity at the tip of the crack or microstructurally important features of similar dimensions to the crack size, both of which invalidate the representation of the crack-tip driving force by ΔK. One commonly applied technique to take account of reduced mechanical constraint at a short crack-tip is to plot the crack growth results from both short and long cracks (i.e. conventional fracture mechanics specimens in the second case) as a function of ΔK_{eff} where a correction is made to the nominal ΔK to allow for the minimum stress intensity at which the crack closes. This point is often detected experimentally by electrical potential drop methods. When such corrections are made, short and long crack data are normally self-consistent as a function of ΔK_{eff}. Similar successes of the ΔK_{eff} approach have been achieved in the context of oxide blocking of cracks and pressure effects in viscous liquids. Another older method, though no less successful for low cycle fatigue, has been to express crack growth rates as a power law function of the applied plastic strain range[6].

Crack size effects in corrosion fatigue crack growth have, however, been observed to persist to larger crack sizes than those associated with plasticity and microstructural effects. Notably, increases by up to a factor of 500 in small surface crack growth at depths up to 3 mm compared to longer cracks have been observed in high-strength A4130 low-alloy steel immersed in 3% NaCl solution[11]. At the high steel strength levels used in these tests, short crack effects due to mechanical or metallurgical reasons were only detectable below 0.1 mm. Later experiments on a lower strength HY130 low-alloy steel under the same test conditions showed that short cracks grew two to five times faster than long cracks while a low-strength carbon–manganese steel showed little influence of crack size on growth rates[34]. The effect of environment on crack growth in all these examples was attributed to hydrogen embrittlement (in common with many other similar metal/environment combinations described earlier). It was further argued, supported by some difficult calculations based on necessarily simplified models of corrosion fatigue cracks, that the enhanced environmental effect seen in short cracks was due to the increased availability of hydrogen ions for reduction to embrittling hydrogen atoms. It was suggested that as short crack lengths increased, the rate of hydrogen ion reduction increased to a characteristic maximum whereas oxygen reduction would dominate at or very close to the surface. The decrease after the maximum at even longer crack lengths was thought to be due to transport limitations of the kinetics of the hydrogen ion reduction reaction while the different responses of the three steels was attributed to their inherently differing sensitivities to hydrogen embrittlement. Turnbull has also pointed out that high-strength, low-alloy steels contain significant amounts of chromium which on dissolution and hydrolysis can lower the crack pH much more than is possible from the hydrolysis of ferrous ions[3]. Nevertheless, irrespective of the detailed mechanistic interpretation, the observations reported are an important reminder that the principle of similitude of corrosion fatigue crack growth rates as a function of ΔK cannot always be taken for granted and should always be checked when data are required for practical applications. Another example where this similitude principle may break down was described earlier for pressure vessel steels in high temperature aqueous environments.

Iron–Chromium–Nickel Alloys

Compared with ferritic carbon and low-alloy steels, relatively little information is available in the literature concerning stainless steels or nickel-base alloys. From the preceding section concerning low-alloy steels in high temperature aqueous environments, where environmental effects depend critically on water chemistry and dissolution and repassivation kinetics when protective oxide films are ruptured, it can be anticipated that this factor would be of even more importance for more highly alloyed corrosion-resistant materials.

One steel which has received more attention than most is Type 403 (12% Cr) stainless steel in a medium yield strength condition of 650 MPa[31] because of its importance for turbine blades. For this type of application, cyclic frequencies are relatively high and most of the data relate to frequencies around 30 Hz. At this frequency, distilled water up to the boiling point, steam, seawater and even sulphurous acid environments increase fatigue crack growth rates by up to a factor of five compared to air, with sulphurous acid the most aggressive. As might be anticipated, crack propagation rates observed at lower frequencies and high stress ratios lead to more severe environmental effects. Crack propagation data for distilled water and salt water solutions (0.01 and 1.0 M NaCl) at 100°C show roughly order of magnitude increases in crack growth rates, particularly for ΔK values above 20 MPa $\sqrt{m}$ and frequencies between 40 Hz and 0.1 Hz. Chloride concentration and pH values between 2 and 10 appear to have little influence. However, lower frequencies of 10^{-2} and 10^{-3} Hz can yield many orders of magnitude increase in growth rates even in distilled water. In a hardened condition, Type 403 stainless steel with yield strengths between 1 200 and 1 600 MPa will also suffer from stress corrosion in distilled water at ambient temperature[34].

A comparison between austenitic, austeno-ferritic and ferritic stainless steels in 3% sodium chloride solution[35] has shown distinct differences in the environmental component of crack growth by up to an order of magnitude, even in corrosion fatigue tests at 200 Hz. These environmental effects were shown to be more severe at 0.5 Hz, with the ferritic stainless steel the best of the group and the austenitic stainless steel the worst. Since stainless steels usually depend on oxygen in solution to form protective, passive oxide films, the availability of oxygen down the crack becomes crucial to the interpretation of results such as these. At low cyclic frequencies, oxygen access to the crack tip is unlikely on theoretical grounds[12] and there is some experimental evidence to support this contention[36]. At high frequencies such as 20 or 30 Hz, the importance of pumping becomes more important. Nevertheless, the ranking of these different stainless steels on corrosion fatigue crack growth seems to be more related to their crevice corrosion resistance rather than general corrosion resistance. It appears clear, therefore, that an improved understanding of corrosion fatigue crack growth in these alloys will come about if attention is focused on those factors which affect repassivation rates at crack tips; for example oxygen access and the electrical resistivity of crack enclave solutions and their impact on crack-tip polarisation and dissolution kinetics.

High-frequency experiments do not normally allow enough time for processes more akin to stress-corrosion cracking to appear in corrosion fatigue tests, as made clear in the previous sections on carbon and low-alloy steels. Some evidence that stress corrosion can occur during corrosion fatigue crack growth in stainless steels has been observed in tests at 3 Hz on austenitic stainless steel (type 304) in various halide solutions at ambient temperature where 'plateaux' or periods of constant crack growth rate over specific ranges of ΔK were observed[34]. The influence of pure water environments at temperatures up to 300°C is not large, however, in solution-annealed stainless steels[23]. One particular technological problem worthy of special mention concerns environmentally-induced intergranular cracking in type 304 sensitised stainless steels in Boiling Water Reactor environments, typically at 260 to 290°C. Sensitisation of type 304 steel causes chromium depletion at the grain boundaries in the heat-affected zones of type 304 stainless steel pipe welds. A great deal of work has been done to characterise the mechanism of environmental attack. There is little doubt that intergranular cracking in this material is due to selective dissolution of the chromium-depleted zones at the relatively high corrosion potentials achieved in normal oxygenated (200 ppb) BWR coolants[36]. Further, extensive slow strain-rate stress-corrosion tests have shown that the rate of cracking depends on the imposed strain rate. Similarly, in corrosion fatigue tests, intergranular cracking can also be detected provided both the frequency and the applied ΔK values are low enough. By contrast, no evidence of intergranular cracking is found when either the value of ΔK is too high (> 20 MPa $\sqrt{m}$) or the frequency is too high (> 0.01 Hz).

Relatively little work on corrosion fatigue crack growth in nickel-base alloys has been published[34]. Such alloys are normally selected for their inherently high resistance to corrosion, crevice corrosion and stress-corrosion cracking so that it is not surprising that aqueous environmental effects where measured have not been large. One notable application of a nickel-base alloy, Alloy 600, is for steam generator tubes in pressurised water reactors. Stress-corrosion cracking in Alloy 600 exposed to water environments between 290 and 350°C is exceedingly slow and sensitive to many metallurgical and environmental variables. It can be seen that with stress-corrosion rates typically of the order of 3×10^{-8} mm/s at 325°C and an activation energy of 180 J/mole[37], exceedingly low frequency cycles would be needed to pick up an effect in normal water environments associated with the PWR. However, in concentrated caustic environments (which can accumulate by hide-out mechanisms on the boiler water side of steam generators), stress-corrosion cracking rates are more rapid. As an extreme example, the rate of stress-corrosion crack growth in Alloy 600 in molten caustic soda at 335°C is about 10^{-6} mm/s and significant increases in corrosion fatigue crack growth rates due to this cause are apparent at frequencies of less than 1 Hz as illustrated in Fig. 8.66[34]. This diagram illustrates clearly how careful an investigator must be to conclude that environmental effects on fatigue crack growth are absent or minimal in a particular metal–environment combination. If stress-corrosion rates are very low, as is the case with Alloy 600 in pure water environments even at high temperature, then cyclic frequencies must also be very low to observe an environmental effect in corrosion fatigue. From Fig. 8.66, we can predict that cyclic frequencies less than 10^{-3} Hz would be necessary to observe superposition of stress-corrosion

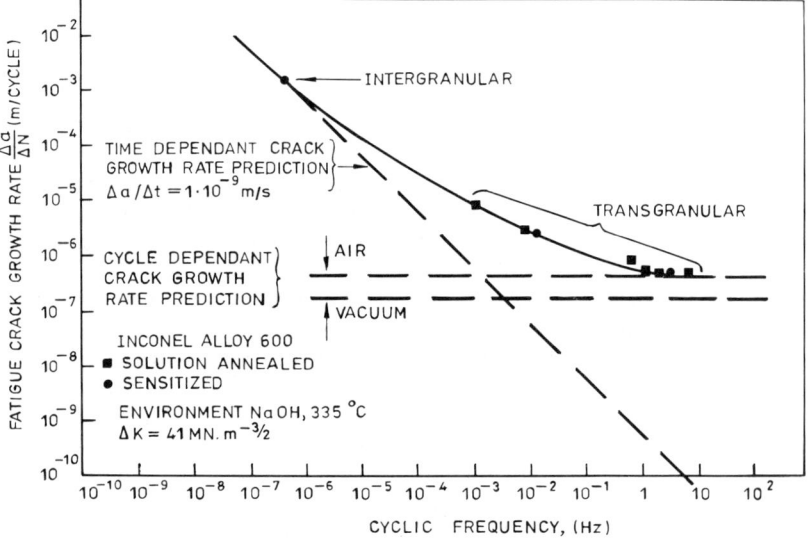

Fig. 8.66 Effect of frequency on the growth rate of corrosion fatigue cracks in alloy IN600 (after Speidel[34])

cracking in Alloy 600 at a rate of 3×10^{-8} mm/s in water at 325°C. Thus, relatively small accelerations of the order of a factor of 2 for crack growth rates reported for sensitised or solution-annealed Alloy 600 in pure deoxygenated water at 288°C are not surprising[38]. Air-saturated water with 7 ppm dissolved oxygen at 288°C was found to be only slightly more aggressive. On the other hand, the high-strength, precipitation hardened version of Alloy 600, Inconel X750, under the same conditions gave very large accelerations in fatigue crack growth which were found to be highly sensitive to heat treatment.

Aluminium Alloys

Three broad classes of aluminium alloys will be considered here; the heat-treatable high-strength aluminium–copper 2000 series and aluminium–zinc–magnesium 7000 series alloys and the non-heat-treatable lower strength aluminium–magnesium 5000 series alloys which are used extensively in marine applications.

In a previous section it has already been observed that high-strength 2000 and 7000 series alloys are sensitive to the presence of water vapour in corrosion fatigue tests. Stress-corrosion susceptibilities of these alloys in low temperature aqueous solutions and the effect of composition and heat treatment have been widely investigated[39]. It is not surprising therefore that when subjected to corrosion fatigue in similar environments, substantial environmental effects can be observed particularly at low frequencies of less than 1 Hz and ΔK values above K_{Iscc}[31,39]. These environmental effects tend to be accompanied by increasing proportions of brittle striations or intergranular cracking when the stress-intensity exceeds the threshold for stress-corrosion cracking, K_{Iscc}. Cyclic waveform at low frequencies does not appear to have a major influence on corrosion fatigue crack growth rate in these cases, probably because the predominant mode cracking is related to stress-corrosion susceptibility which itself is not in this case strongly strain-rate sensitive. Differences between 3.5% sodium chloride solution, natural seawater and simulated seawater and the effect of flow rate for a 7000 series alloy have all been observed to be small or negligible.

Relatively little information on corrosion fatigue crack propagation is available for 5000 series alloys which is surprising in view of their marine applications[31]. At high frequency, 30 Hz, only a slight influence of a seawater environment has been found. For frequencies around 0.1 Hz, a distinct but small effect of seawater on fatigue crack growth has been measured at ΔK values greater than 10 MPa $\sqrt{m}$. This is of a similar order to that found on low- and medium-strength structural steels. Cathodic polarisation and deoxygenation of the environment are also beneficial.

The mechanism of environmental degradation by stress-corrosion cracking or corrosion fatigue has generally been attributed to hydrogen embrittlement[19,22]. However, the reactivity of freshly created aluminium surfaces with any oxidising agent rapidly leads to repassivation. Since the oxide on aluminium is relatively impervious to hydrogen diffusion, and hydrogen diffusion rates are in any case very slow in aluminium, dislocation transport and pumping of the fracture process zone ahead of the crack tip is generally invoked as the mechanistic explanation. Precipitate–matrix interfaces are particularly important sites where separation and crack formation can occur. An alternative explanation has been provided which is based on anodic dissolution of the crack tip and which leads to good quantitative predictions of the influence of aqueous environments on aluminium alloys, even the high-strength ones.[36] Critics point to the adverse effects of water vapour which are similar to those of aqueous environments and where electrochemical explanations are inappropriate. In addition, the adverse effects of aqueous corrosion prior to fatigue tests which can be partially reversed by heat treatment to remove hydrogen are also noted.

Titanium and Zirconium Alloys

These two groups of alloys are discussed together because of their ability to absorb hydrogen and internally precipitate hydrides. Titanium alloys are quite complex from a metallurgical viewpoint and corrosion fatigue crack growth in them is strongly dependent on microstructure[31,40]. Most work appears to have been carried out using a Ti–6Al–4V alloy in various heat treatment conditions leading to varying proportions of α (hexagonal) and β (cubic) phases, although many of the other available titanium alloys have also been investigated from time to time. Nearly all the work on corrosion fatigue crack growth has concentrated on the influence of salt water environments (3.5% sodium chloride or seawater or simulated saline solutions) at normal ambient temperatures.

In common with many of the alloy–environment systems described so far, if the alloy is not susceptible to stress-corrosion cracking under constant stress or stress intensity, then little or no effect of environment on fatigue crack growth is observed. In these cases, frequency, R ratio and potential within the passive or cathodically protected ranges for titanium have no effect on growth rates.

Many high-strength titanium alloys are susceptible to stress corrosion, however, in environments as diverse as aqueous chloride solutions, chloride contaminated methanol and molten salts. The mechanism is generally accepted to be hydrogen embrittlement with the formation of internal hydrides on slip planes, which impede slip and promote cleavage[41]. When tested under corrosion fatigue conditions, those alloys which exhibit stress-corrosion cracking show large environmental effects on fatigue crack propagation when the static stress modes participate. A feature of especial interest in such titanium alloys is the manner in which the apparent threshold for the onset of high crack growth rates (at constant R ratio) varies with cyclic frequency as shown in Fig. 8.67. This behaviour should be contrasted with high-strength steels and aluminium alloys where a single frequency-insensitive threshold parameter, K_{Iscc}, and a constant plateau rate of stress-corrosion crack growth are sufficient to account for the observed cracking rates in corrosion fatigue when superimposed on the inert environment fatigue crack growth rate. Strain rate sensitivity of the stress-corrosion threshold and plateau rate parameters have already been highlighted in connection with lower strength stainless and non-stainless steels under passive conditions. There is evidence too of a frequency dependent threshold to the

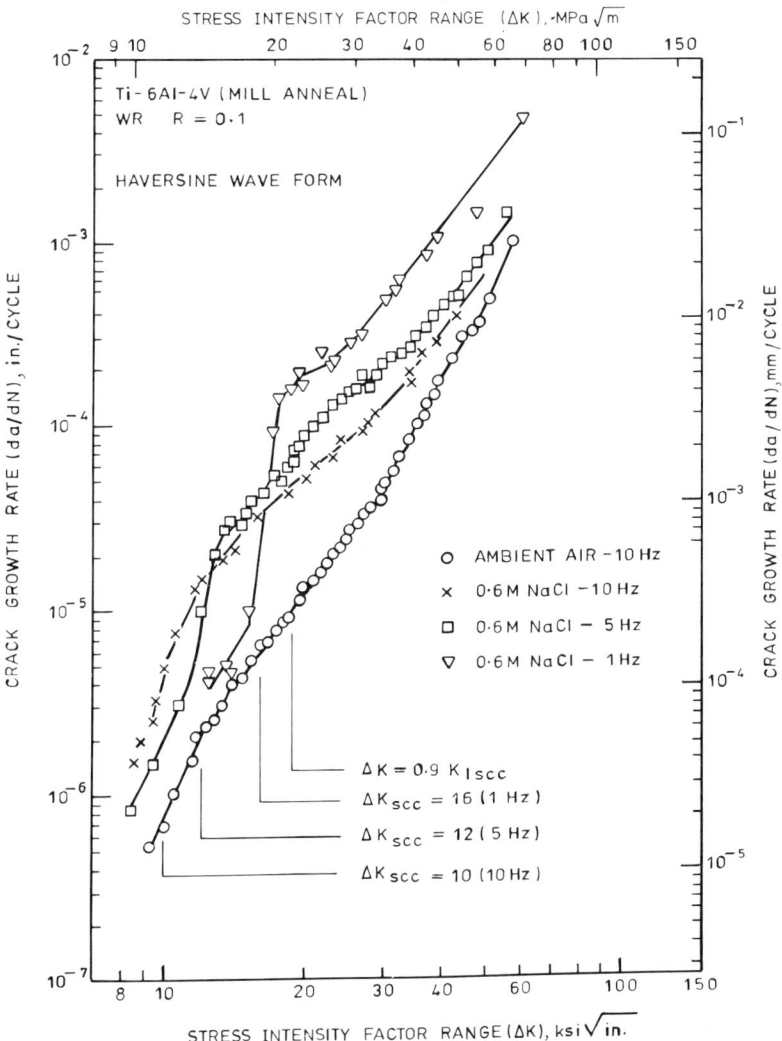

Fig. 8.67 Effect of frequency on corrosion fatigue crack growth behaviour of Ti–6Al–4V in aqueous 0.6 M NaCl (after Pelloux[29])

onset of high plateau corrosion fatigue crack growth rates in medium-strength steels in high temperature aqueous environments. Thus, although the frequency sensitivity of corrosion fatigue crack growth in titanium alloys shown in Fig. 8.67 was regarded as unique when first observed, there is a growing body of evidence for similar effects in other alloy–environment systems.

Zirconium alloys have been much less thoroughly studied than titanium alloys. The main application of interest has been for nuclear reactor components where good corrosion resistance combined with a low neutron capture cross-section has been required. Corrosion fatigue crack growth in these alloys in high temperature (260–290°C) aqueous environments typical of

BWR and PWR coolants have been reviewed[42]. Provided hydride precipitation and thereby stress-corrosion susceptibility is avoided, especially at the normal operating temperatures of water reactors, environmental effects on fatigue crack growth are small. A possible exception arises when zirconium alloys are subjected simultaneously to irradiation, high temperature aqueous corrosion by oxygen-containing BWR coolants and cyclic stresses. Under these circumstances, rather high environmental contributions ($\times 10$) to corrosion fatigue crack growth have been observed. There is clearly a need for further work in this area to sort out the relative importance of dissolved oxygen and irradiation effects both in terms of neutron damage to the material and their effects on oxidising potential.

Copper Alloys

Remarkably little has been published on corrosion fatigue crack propagation in copper and its alloys. In general little or no influence of marine environments has been observed in crack propagation experiments on manganese and nickel–aluminium bronzes although the frequencies employed were quite high (> 2.5 Hz)[31,43].

Corrosion Fatigue Endurance

It has to be stated from the outset in this section that there is rarely a one-to-one correspondence between the effects of environment observed in endurance tests on plain specimens and crack propagation tests on pre-cracked specimens (assuming the same materials, environments and fatigue test variables). In certain circumstances such as welded connections or other components with built-in pre-existing defects, such a correlation is possible. On the other hand, more than 90% of the cyclic life of smooth cylindrical specimens can be spent in propagating a stage I crack across one or two grains in inert environments and little or no relationship exists with standard crack growth test results. The lack of a general correlation shows us that the effects of corrosion on the early stages of crack nucleation and growth are usually different to those observed on macroscopic crack growth. This is despite a general recognition that most of the fatigue life of any artefact, including plain specimens, is taken up in developing a crack, however small, nucleated early in life. Thus in many circumstances, it must be the case that the effects of corrosion on stage I crack nucleation and growth are quite different to those on stage II growth. In addition, it has been noted already that environments themselves can be modified by chemical and diffusion processes set up in long cracks. In view of the above, it is therefore necessary to summarise separately the contents of several detailed reviews of the observations of corrosion fatigue endurance properties of many metal-alloy–environment combinations[2,3,13,15,19,22,25,31,39,44-46].

Gaseous Environments

Early work on the fatigue strength of various metallic alloys including steels, aluminium alloys, copper alloys and nickel-base superalloys in vacuum and in air clearly demonstrated that fatigue performance improved in vacuum[2,3,15,44,45]. At room temperature, these effects are not generally large but in high-strength steels, aluminium, titanium and magnesium alloys, significantly improved fatigue strength or cyclic lives have been observed in dry air compared with moist air[15,31,44]. Other environments such as inert gases and liquid sodium with low partial pressures of oxygen also enhance the fatigue lives of steels compared with air environments as do those which increase bulk material strength such as carburising liquid sodium or neutron irradiation damage[15].

At elevated temperatures, the adverse influence of air oxidation on stainless steels and nickel-base superalloys increases[15,44]. An example is shown in Fig. 8.68 for a nickel-base super alloy where a marked temperature effect on fatigue life was observed in air but which disappeared in vacuum[6]. Such obviously large effects of air oxidation on fatigue life at high temperatures has led to some difficulties in determining the relative importance of oxidation and creep damage in environment–creep–fatigue interactions where the environmental contribution has not been separately investigated. Detailed studies of the frequency dependence of corrosion fatigue lives of superalloys and stainless steels in air at high temperatures has revealed the existence of critical frequencies, typically about 1.0 Hz, above which no effect of air oxidation is found. At lower frequencies, the Coffin–Manson equation can be modified by a frequency-dependent term which successfully correlates all the corrosion fatigue data (Fig. 8.68). This equation in turn can be simply derived by integrating a crack growth power law expressed

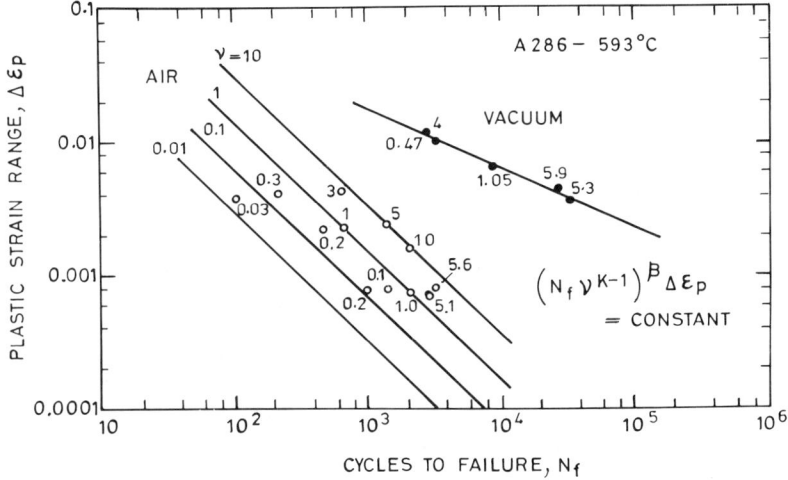

Fig. 8.68 Plastic strain versus fatigue life for A286 in air and vacuum at 593°C. Numbers adjacent to test points indicate frequency, ν, in c.p.m. K and β are material and environment constants (after Coffin[6])

as a function of the plastic strain range with the same frequency-dependent term[6].

Duquette has discussed various hypotheses and supporting observations of mechanisms by which adsorption or oxidation by oxygen and water vapour can influence early slip behaviour, slip band formation and the growth of stage I cracks[44]. At elevated temperature, there is good evidence that intergranular oxidation during preheating of stainless steels and superalloys to the test temperature creates an effective notch for premature crack initiation. At normal ambient temperatures there is much more controversy about how adsorbed species or oxides promote or inhibit slip or rewelding during the compressive part of the cycle and how environments can alter the tensile properties of oxide films. Nevertheless, there is little doubt that water vapour can be a potent cause of hydrogen-embrittlement effects at least in high-strength ferrous, aluminium, magnesium and titanium alloys.

Aqueous Environments

Prior to the modern day preoccupation with the application of fracture mechanics to fatigue and corrosion fatigue crack growth, a very large technical literature of $S-N$ corrosion fatigue results on metal alloys in aqueous environments was published. Gilbert[3] summarised a great number of $S-N$ test results on various alloys in environments such as distilled water, tapwater and seawater. The main effect of corrosion was to decrease by very considerable margins the effective fatigue strengths at any given cyclic life. Sometimes, however, strength was improved at short cyclic lives by very aggressive environments which presumably blunted out incipient fatigue cracks. Examples of typical corrosion fatigue $S-N$ results for carbon steels

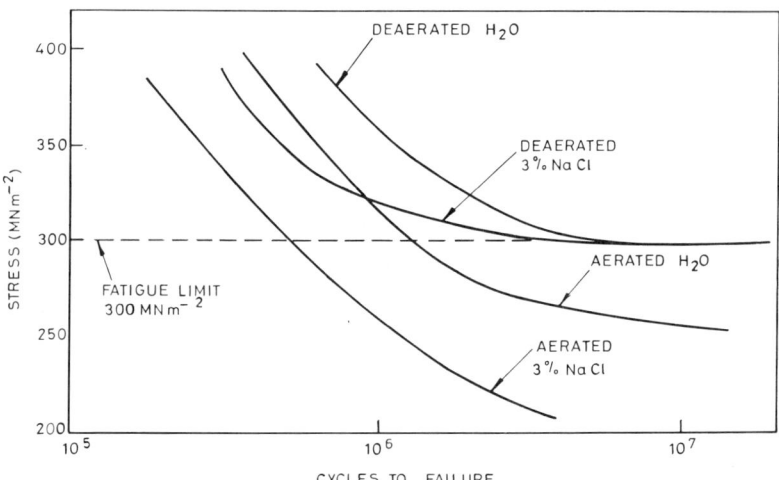

Fig. 8.69 Effect of air and aerated or deaerated distilled water and 3% NaCl solution on fatigue behaviour of steel at 25°C (after Duquette and Uhlig)

are given in Fig. 8.69. It is noteworthy that these large effects of corrosion were observed despite the frequent use of high test frequencies (> 10 Hz) when environmental effects on stage II crack growth rates would be negligible. In some cases no fatigue limit was observed at long cyclic lives. There must be some doubt about the often inferred wide applicability of this observation, however, since specimens were often quite small and general or localised corrosion could reduce the cross-sectional area very significantly in many cases. Endurance limits or fatigue strengths at specific cyclic lives were found to be insensitive to metallurgical condition showing no correlation with tensile strength (in contrast to that observed in air). Corrosion resistance, often specifically pitting resistance, was much more important in determining the endurance limit. Various compilations of fatigue endurance limits as a function of alloy strength have been published but the most recent and most comprehensive due to Speidel[13] are reproduced here in Fig. 8.70 and 8.71. From these figures, it can be concluded that the increased strength of an alloy can only be exploited in corrosion fatigue if first it is resistant to corrosion by the environment. However, it must not be of such a high strength as to be susceptible to hydrogen embrittlement. Speidel has shown how this philosophy has been used to practical advantage in steam turbine blade specifications[39]. Ferritic 12% Cr steels are widely and effectively used in good quality steam but where aggressive condensate is encountered a Ti-6Al-4V alloy is necessary since 12% Cr steels suffer severe losses of fatigue strength under such conditions.

The importance of the prevailing corrosion conditions in determining corrosion fatigue strength is further emphasised by the response of the S-N curve to electrochemical potential and in some instances corrosion

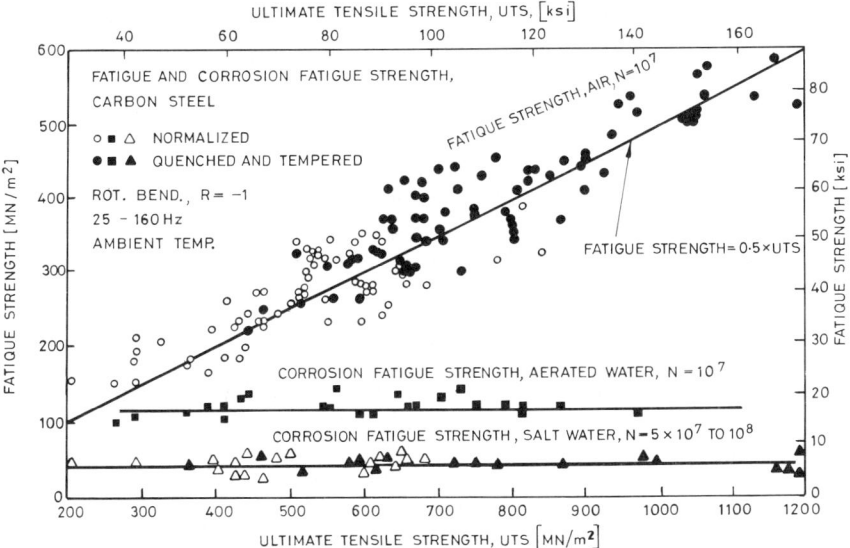

Fig. 8.70 The fatigue strength of carbon steels of varying tensile strengths in air aerated water and seawater (after Speidel[13])

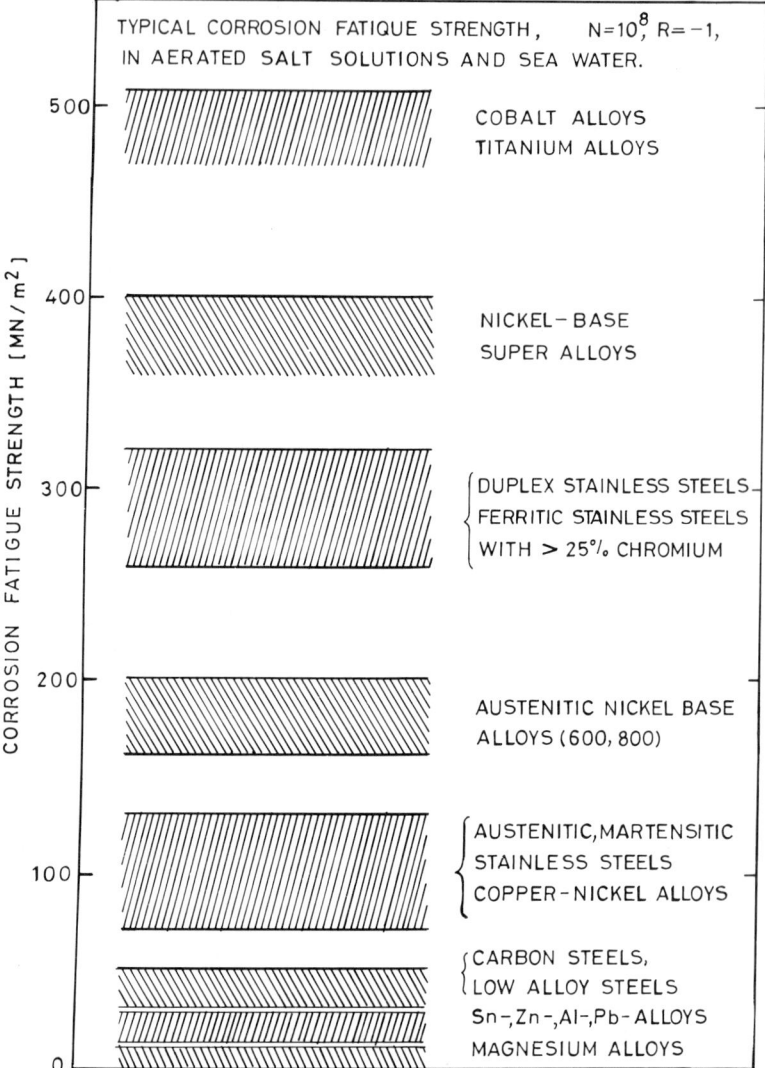

Fig. 8.71 Corrosion fatigue endurance limits for various alloys in aerated salt solutions or seawater (after Speidel[13])

inhibitors. Cathodic protection is very effective at restoring the in-air fatigue endurance limit in carbon steels exposed to distilled water, sodium chloride solution or seawater as is deaeration (Fig. 8.69). In conditions where the metals passivate readily in their environments there can also be a beneficial effect of anodic polarisation. An example is shown in Fig. 8.72 for an austenitic stainless steel in sulphuric acid solution subjected to various imposed potentials[35]. Modest increases in potential in the anodic direction are seen to produce large benefits in corrosion fatigue strength, presumably due to an increase in the kinetics and effectiveness of passivation of emergent

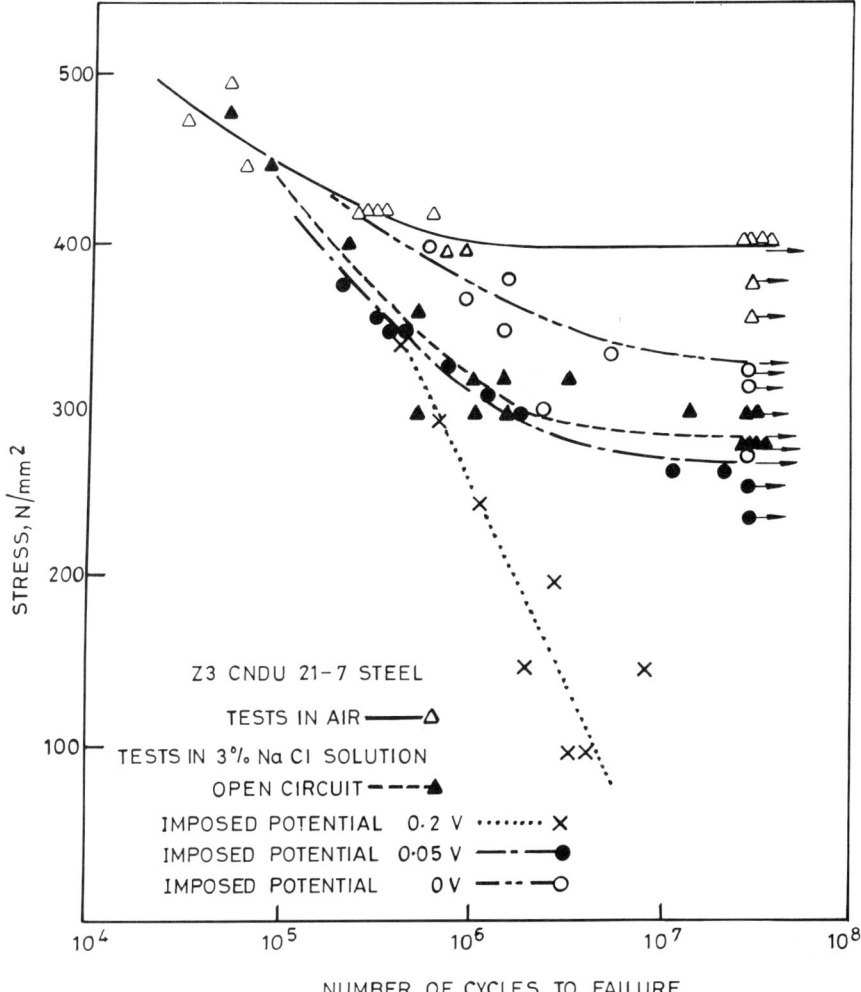

Fig. 8.72 Effect of applied potential on corrosion fatigue behaviour of a ferritic stainless steel in 3% NaCl (after Amzallag et al.[35])

slip steps. If, however, transpassive potentials are applied, then a massive loss in fatigue strength is observed.

On many occasions, microscopic examination of corrosion fatigue failures from initially plain specimens reveals transgranular, slightly branched, cracks apparently emanating from pits. It has not always been clear whether pitting has occurred before or after cracking, but that pitting is certainly detrimental in reducing fatigue strength there is no doubt. An example for an austenitic steel where initially plain specimens were pre-pitted is shown in Fig. 8.73[47]. Similar results have been obtained for low-alloy steels and aluminium alloys. Small hemispherical pits would generate a stress concentration of 2.2 which would only increase to about 3.5 with increasing depth

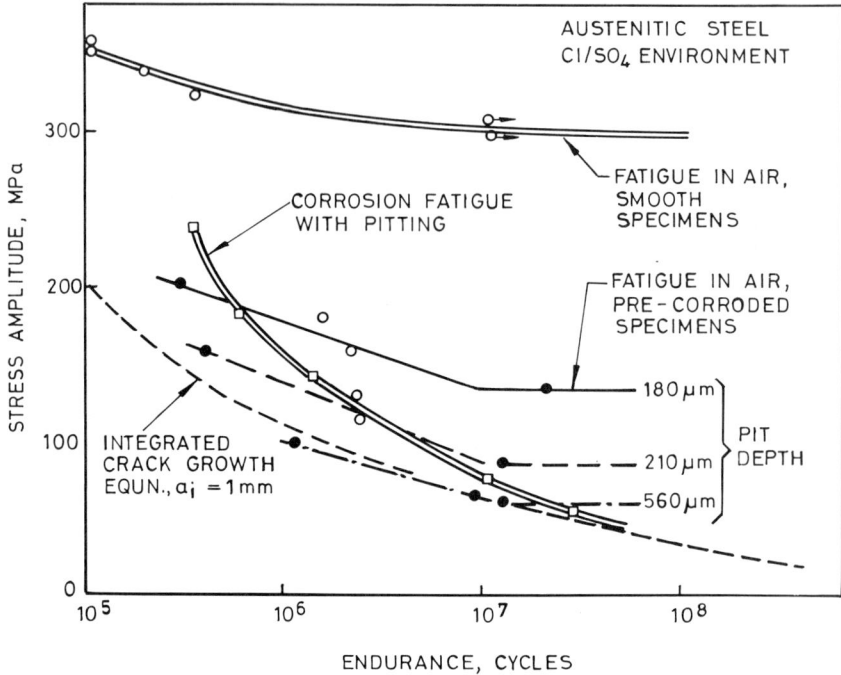

Fig. 8.73 Effect of pitting on the fatigue strength of austenitic steel (after Rust and Swaminathan[47])

and an accompanying change to a hyperbolic shape. Thus, extended pitting exposures do not result in much more severe reductions in corrosion fatigue strength compared with short exposures. The effectiveness of pitting in producing stress-concentrating notches from which cracks initiate has also been demonstrated by comparing the fatigue and corrosion fatigue strengths of notched specimens. The relative effect of corrosion on suitably notched specimens is minor relative to plain specimens[34]. An extreme example of notches in as-fabricated artefacts is that of welding defects introduced into even the best-made welds. These defects such as slag inclusions, undercut etc., are typically 0.5 to 1.0 mm deep. Even low frequency tests on full-penetration fillet welds of structural steel at 0.1 Hz in seawater result in only a modest factor of 2 to 4 reduction in fatigue life relative to air data. This is comparable to the increase in stage II fatigue crack growth rates caused by seawater as observed in crack propagation experiments under linear elastic conditions. Thus the notching effect of pitting is not additive to that introduced by other causes such as in welded fabrications. Further, it is clear that the presence of welding defects or corrosion pitting can reduce fatigue life in the limit to that attributable to crack growth as represented by the Paris equation (Fig. 8.59). Some calculations of fatigue life on this basis are compared with experimental data in Fig. 8.73[48]. Threshold conditions for crack initiation at pits have also been successfully estimated from linear elastic fracture mechanics calculations using the crack growth threshold stress-intensity range and the pit depth treated as though it were a crack depth[49]. Such calculations are limited to pit depths above about 50 to 100 μm depending on alloy strength by linear elastic validity considerations.

Despite the widespread importance and relevance of pitting as the primary reason for severe losses of fatigue strength due to corrosion, it is not the exclusive cause. Some metal–environment combinations have been tested where pitting does not occur or pits are not associated with cracks but where severe environmental effects on fatigue strength have been observed[44,45]. Examples include carbon steels in acid solutions where preferential anodic attack at emergent slip steps without classical pitting is thought to occur, and polycrystalline copper under anodic polarisation where cracks initiate preferentially at dissolved grain boundaries. Hydrogen-embrittlement processes are probably important in high-strength alloys such as 2 000 and 7 000 series aluminium alloys. Nevertheless, dissolution processes of one sort or another whether at persistent slip bands, inclusions or grain boundaries do appear to dominate environmentally-caused losses of fatigue strength.

Observations of the importance of pitting and other localised corrosion processes have led to various criteria being proposed for critical minimum corrosion rates or anodic current densities for an environmental effect on fatigue to be observed[46]. For a cyclic frequency of 30 Hz, the critical minimum anodic current density has been found to be 2–3 μA/cm^2 in steels, 1.2 μA/cm^2 in nickel and 100 μA/cm^2 in copper. A consequence is that in steels in solutions at pH $\geq$ 12, (e.g. alkali-treated 3.5% sodium chloride solution), the corrosion rate falls below the critical minimum and the in-air fatigue strength is restored. Copper and many copper alloys exhibit very low corrosion rates in saltwater environments well below the critical minimum given above and in consequence fatigue strengths are relatively unaffected in such environments.

It is evident from the discussion above that once the corrosion rate is greater than a critical minimum at any given cyclic frequency it is a difficult and complex problem to try and predict its effect on the complete S–N curve without lengthy and laborious experimentation. Waterhouse[2] has reviewed much of the early measurements of corrosion fatigue endurance which aimed to solve this difficult problem and cited particularly the work of McAdam[50] and Endo and Komai[51,52]. A diagram for representing corrosion fatigue data developed by McAdam is shown in Fig. 8.74 where $S_1 \neq S_2 \neq S_3$ etc. are stress range contours and $n_1 \neq n_2 \neq n_3$ etc. are cyclic frequencies typically between a few cycles per week and 100 Hz. McAdam discovered that a wide variety of materials with very different corrosion properties, ranging from mild steel to Monel metal, in different corrosive environments, could be fitted on to the same type of diagram by adjusting the time scales, e.g. mild steels over the time considered were characteristic of the left half of the diagram and Monel of the right. At high frequencies the stress contours cut the constant-frequency lines at right angles, indicating that frequency has little effect on the rate of damage. At low frequencies, i.e. in the bottom right-hand corner of the diagram, the stress contours are crowded closely together and cut the constant-frequency lines at an acute angle. This indicates that the rate of damage is appreciable over a wide range of stress, even at low stresses. The fact that corrosion-resistant materials such as Monel, as well as corrosion-susceptible materials such as mild steel, in mildly corrosive surroundings such as condenser water could suffer considerable damage

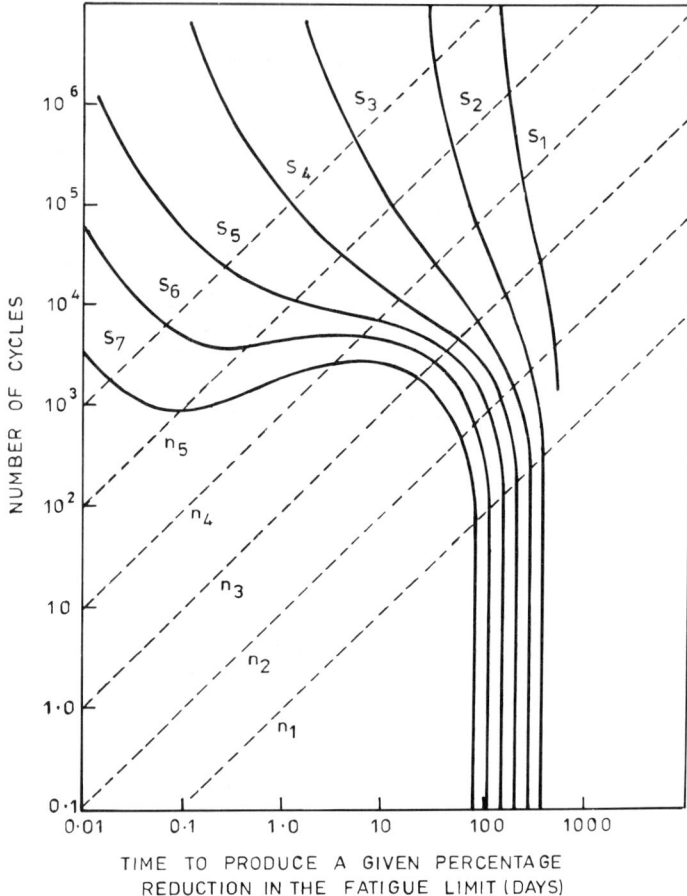

Fig. 8.74 Stress and corrosion time required to produce given percentage reduction (e.g. 15%) of fatigue limit due to corrosion alone (after McAdam[50])

under conditions of a small cyclic load of low frequency, was an important discovery.

The data on which Fig. 8.74 is based are for tests carried out in carbonate well-water. McAdam made the further interesting discovery that if mild steel were tested in condenser water and a similar graph constructed, the set of contours corresponded more closely to the right-hand side of Fig. 8.74, i.e. the behaviour of mild steel in condenser water was similar to that of Monel in carbonate water. The apparent universality of this diagram is an interesting observation, but it has not provoked a basic theory of corrosion fatigue.

Some of the investigations involving electrochemical measurements have been concerned with relating easily determined quanities such as corrosion potential and corrosion current with the behaviour of a material in corrosion fatigue, so that this behaviour can be rapidly assessed without the necessity of the laborious collection of data which was the feature of McAdam's approach. Endo and Komai have derived an expression relating the increase in the corrosion current with the number of cycles of corrosion fatigue[51]. This expression is found to be analogous with an expression relating crack growth with the number of cycles, and hence the increase in corrosion current is found to be a measure of the total crack length. The expression also takes account of the magnitude of the alternating stress, the frequency and the temperature. Since the corrosion current is related to the corrosion potential, the course of corrosion fatigue damage can be followed by potential measurements. In an extension of this work[52], the product of the initial corrosion current density $i_{c,i}$ on the fatigue strained surface and the total life τ_T was shown to be related to the notch sensitivity, η, and the ratio of the fatigue strength in air and in the corrosion medium, k, by the expression

$$\frac{(k-1)}{\eta} = K\sqrt{i_{c,i}\tau_T} \qquad (8.15)$$

where K is a constant. The notch sensitivity is assessed from the stress concentration attributable to pitting. This relation is found to hold for a variety of materials in a particular electrolyte, as illustrated in Figure 8.75.

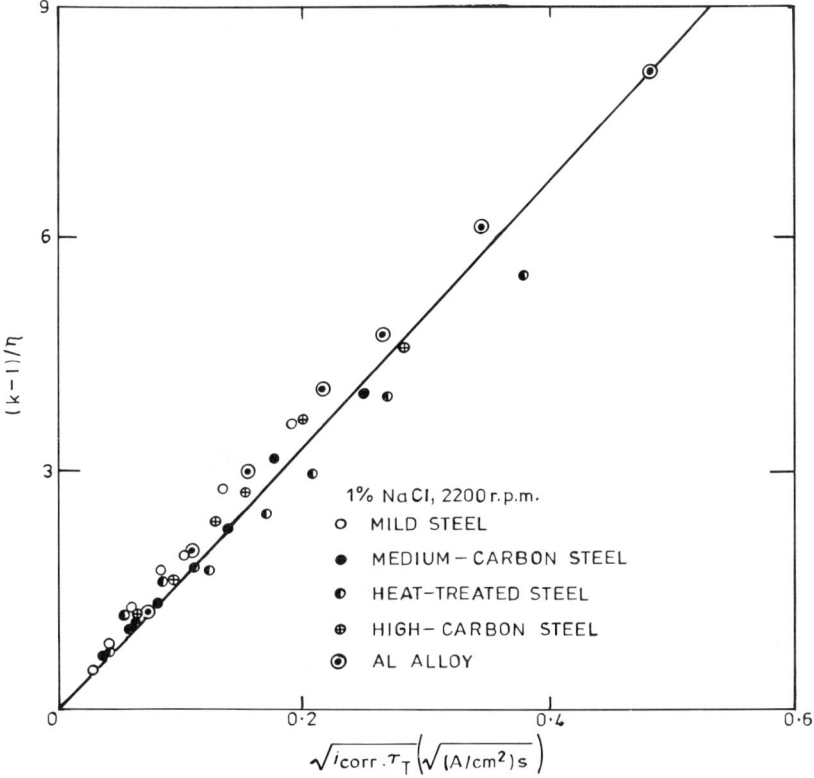

Fig. 8.75 Relation between $(k-1)/\eta$ and $i_{corr}\tau_T$ in 1% NaCl, where k is the ratio of fatigue strength in air to that in a corrosive environment, η the notch sensitivity factor on fatigue strength, i_{corr} the corrosion current density at start of fatigue cycling, and τ_T the total life in seconds (after Endo et al.[52])

Design, Inspection and Prevention

Design against fatigue failure is for the most part still firmly rooted in the cyclic endurance or S–N method; e.g. ASME Section III (1980) for pressurised nuclear components and British Standard 5400 (1980) for other welded fabrications. This is true even though many studies have shown that it is the growth of minor fabrication defects present in all well-made welds which controls cyclic life. In some high technology industries, such as aerospace and nuclear, there is considerable emphasis on defect growth evaluation as part of the design procedure and, in future, designs in some aerospace applications may even be formally based on fatigue crack growth concepts. However, most current interest in fatigue crack growth derives from a need to evaluate remaining life where in-service inspection may find cracks. Thus, when a chemically reactive environment is present, one must consider suitable corrosion protection measures and, where these are not possible or break down, how the influence of corrosion may be introduced into the design or defect evaluation method.

Prevention or minimisation of corrosion fatigue damage has been reviewed by Waterhouse[2]. Firstly good design is important to combat corrosion. In particular, it is necessary to avoid crevices, for example in joints, where stagnation may arise owing to inccessibility to air and the development of an anodic area, or, in the case of a closed circuit, where inhibitors are being used due to the difficulty of replenishing with inhibitor.

The substitution of a more corrosion-resistant material, e.g. Monel metal or stainless steel, is often advocated but this is not necessarily a solution; a 15% chromium steel, for instance, is prone to failure in corrosion fatigue because of the disruption of the normally protective surface film although as noted earlier some other materials are effective, e.g. Ti–6Al–4V alloys[13,39].

The protection of steel against corrosion fatigue has been one of the major preoccupations of corrosion scientists and is often effective if the integrity of the protection system is maintained. The beneficial effect of inducing a compressive stress in the surface of the steel applies to corrosion fatigue as well as normal fatigue provided the surface layer is not penetrated by corrosion. The surface stress can be produced by chemical means such as nitriding or carburising, or by quenching from below the transformation temperature. A third method is by surface rolling or shot peening, both of which leave the surface in a work-hardened condition which probably has the effect of ironing out differences in potential due to local stress concentrations.

Electroplating provides a further possibility for protecting mild steel. Nickel plating is normally very efficient in preventing corrosion in static conditions, but it is known to reduce the fatigue strength of steel because of the tensile stresses in the nickel coating. In corrosion fatigue, nickel plating has little or no effect. Zinc plating, on the other hand, produces a coating in compression, which is in itself effective in causing an increase in the normal fatigue limit. In corrosion fatigue, the performance of zinc-plated material is notably better, because of the added protection given by the sacrificial action of the zinc. The following figures quoted by Waterhouse[2] are for a 0.63% C steel wire in seawater with zero mean stress. The thickness of the plating was 12 μm.

	Fatigue limit dry fatigue (MN/m^2)	Endurance limit corrosion fatigue, MN/m^2 (10^6) cycles
Unplated	620	140
Nickel plated	470	160
Zinc plated	700	530

Some early results for zinc-coated steel show that electroplated coatings of zinc are more efficient in protecting steel against corrosion fatigue than either hot-dipped or sherardised coatings, no doubt on account of the compressive stress present in the electroplated coatings. Zinc has also been used with some success as a pigment in priming paint but this is not so efficient as a plated or galvanised zinc coating because the contact between the zinc particles in the pigment and the steel is not nearly as good as in the metallic coatings. Zinc can also be applied by metal spraying. Another group of materials frequently given corrosion protection by a metal coating is the high-strength aluminium alloys. The protective coating is usually pure aluminium and is applied by spraying in the case of extrusions, or by cladding in the case of sheet.

Examples have been given of the protection that can be afforded by organic coatings, e.g. paints and stoved resin coatings for mild steel, and lanolin impregnation of an anodised coating in the case of aluminium alloys. The beneficial effect on the subsequent fatigue strength in air of coating welds in steel structures with epoxy resins has been demonstrated. With this type of barrier coating, scratches and other blemishes must be avoided.

There are only a few cases where the adverse effects of corrosion on fatigue design S–N curves or fatigue crack propagation evaluation codes have been explicitly included. Two examples are the nuclear and offshore oil production industries where in each case stringent corrosion control measures are implemented. The incorporation of corrosion fatigue considerations into design and inspection codes for these industries has been discussed by Scott, Foreman and Tomkins[53] and by Scott[48]. An important issue is that it is frequently too conservative and uneconomic to neglect in design that part of fatigue life occupied in crack initiation since fabrication and corrosion control measures are taken to avoid serious penalties in that regard. Similarly, crack growth approaches are rarely practicable if based on the maximum feasible rate of crack advance, for example the maximum crack-tip opening displacement per cycle.

In the case of the ASME codes for nuclear pressurised components, the questions of fatigue design and of flaw evaluation are dealt with separately in ASME Section III and Section XI Appendix A, respectively. The design S–N curve for machined butt welds typical of thick section pressurised components is set at a factor of two on stress range or twenty on cyclic life, whichever is more conservative, below the mean of S–N data developed on smooth cylindrical specimens in air. (A somewhat similar design curve obtained by a different method from experimental S–N data for machined butt welds is given in British Standard 5500.) These safety factors are intended to encompass any adverse influence of minor weld defects, size effects, data scatter and environment. As far as environmental effects are concerned it is known that the safety factors are appropriate provided the water has adequately low conductivity and the oxygen content of the water coolant (at 290°C typically) is kept well below air saturation levels. This control of water chemistry is of course mandatory in the operation of such plant. However, in some conventional steam raising plant, particularly components operating at somewhat lower temperatures than 290°C, such as deaerators or feedheaters there is increasing evidence that problems with corrosion fatigue have been, and still are being, encountered particularly where the control of water chemistry has been poor.

The ASME Section XI Appendix A (1980) nuclear pressure vessel code, in contrast to Section III, is concerned with evaluating the consequences for plant safety of any cracks or defects discovered by in-service inspection. In this instance, specific corrosion fatigue crack propagation curves with substantial environmental factors are given for the case where pressure vessel defects are exposed to the primary coolant. Again, water quality is known to affect corrosion fatigue crack propagation in pressure vessel steels but of particular importance is the sulphur inclusion content of the steel. Consequently, much testing is underway internationally to try and improve the data base for this code with the eventual aim of incorporating information such as the importance of steel cleanliness in future editions.

The current design curves for offshore structures contained in the UK Department of Energy Guidance Notes[54] are based on lower bounds, two standard deviations below the mean, of S-N curves for fillet welded joints. Of the many factors influencing fatigue life, the presence of seawater and the effectiveness of corrosion control measures such as cathodic protection are explicitly incorporated in determining the endurance limit and cyclic life at higher stress ranges. However, because of the redundancy common in offshore space frame structures, the safety factors for the design curves are less stringent than for pressure vessels and piping. Nevertheless, there is little doubt that these early attempts at codification of corrosion fatigue data indicate the shape of things to come as materials are exploited to their limits, often in hostile environments.

P. M. SCOTT

REFERENCES

1. Devereux, O. F., McEvily, A. J. and Staehle, R. W., (Eds.) *'Corrosion fatigue: chemistry, mechanics and microstructure'*, NACE-2, National Association of Corrosion Engineers (1972)
2. Waterhouse, R. B., 'Corrosion Fatigue', in Shreir, L. L. (ed.), *Corrosion* 2nd edn, Butterworths, pp. 8:96–8:113 (1976)
3. Gilbert, P. T., 'Corrosion fatigue', *Met. Rev.*, **1**, 379–417 (1956)
4. Grosskreutz, J. C., 'The mechanisms of metal fatigue (II)', Phys. Stat. Sol (b), **47**, 359–96 (1971)
5. Stoloff, N. S. and Duquette, D. J., 'Microstructural effects in the fatigue behaviour of metals and alloys', *CRC Critical Reviews in Solid State Sciences*, **4**, 615–87 (1974)
6. Coffin, L. F., 'Fatigue at high temperature-prediction and interpretation', *Proc. Inst. Mech. Engrs.*, **8**, 109–27 (1974)
7. Lynch, S. P., 'Mechanisms of fatigue and environmentally assisted fatigue', *ASTM STP 675, Fatigue Mechanisms*, Ed. J. T. Fong, pp. 174–203 (1978)

8. *Metals Handbook*, Ninth Edition, Vol. 8, *Mechanical Testing*, American Society for Metals, pp. 361-435 (1985)
9. Thorpe, T. W., Rance, A., Silvester, D. R. V., Scott, P. M. and Morgan, H. G., 'The effect of North Sea service conditions on fatigue crack growth in structural steel', *Fatigue in offshore structural steels*, Thomas Telford Ltd., London, pp. 35-46 (1981)
10. Ford, F. P., 'Overview of collaborative research into mechanisms of environmentally assisted controlled cracking in the low alloy, pressure vessel steel/water system', *IAEA Specialists Meeting on Sub-Critical Crack Growth*, 15-17 May 1985, Sendai, Japan, NUREG/CP-0067 (1986)
11. Gangloff, R. P. and Wei, R. P., 'Small crack-environment interactions – the hydrogen embrittlement perspective', *Proc. AIME Conf. on Short Crack Growth*. AIME (1986)
12. Turnbull, A., 'Review of the electrochemical conditions in cracks with particular reference to corrosion fatigue of structural steel in seawater', *Reviews in Coatings and Corrosion*, **5**, Nos. 1-4, 43-160 (1982)
13. Speidel, M. O., 'Influence of environment on fracture', *Proc. 5th Int. Conf. on Fracture*, Cannes, France, 29 March to 3 Arpril 1981, ICF5 Vol. 6, pp. 2685-704, Pergamon Press (1982)
14. Ashby, M. F. and Tomkins, B., Micromechanisms of fracture and elevated temperature fracture mechanics, Vol. I, ICM 3, Cambridge, UK, Aug. 1979, pp. 47-89, Pergamon Press (1980)
15. Marshall, P., 'The influence of environment on fatigue', in Skelton, R. P. (ed), *Fatigue at High Temperatures*, Applied Science Publishers, pp. 259-303 (1983)
16. Tomkins, B. and Wareing, J., 'Elevated-temperature fatigue interactions in engineering materials', *Metal Science*, **11**, 414-24 (1977)
17. Wei, R. P., 'On understanding environment enhanced fatigue crack growth – a fundamental approach', in Fong, J. T. (ed.), *Fatigue Mechanisms, ASTM STP 675*, pp. 816-40 (1978)
18. Pao, P. S., Wei, W. and Wei, R. P., 'Effect of frequency on fatigue crack growth response of AISI 4340 steel in water vapour', *Proc. of Environment Sensitive Fracture of Engineering Materials*, 24-26 Oct. 1977, Chicago, USA, The Metallurgical Society of the AIME, pp. 565-580 (1977)
19. Duquette, D. J., 'Mechanism of corrosion fatigue of aluminium alloys', *AGARD Conference Proceedings No. 316, Corrosion Fatigue*, Cesme, Turkey, 5-10 April (1981)
20. Stewart, A. T., 'Effect of hydrogen on fatigue crack propagation in steels', *Proc. Int. Conf. on Mechanisms of Environment Sensitive Cracking in Materials*, University of Surrey, 4-7 April 1977, The Metals Society, pp. 400-11 (1977)
21. McIntyre, P., 'Hydrogen-steel interactions during cyclic loading', *Proc. of UK/USSR Seminar on Corrosion Fatigue of Metals*, Lvov, USSR, 19-22 May 1980, The Metals Society pp. 62-73 (1983)
22. Duquette, D. J., 'Mechanisms of crack initiation and propagation in corrosion fatigue', Proc. Int. Conf. on *Mechanisms of Environmental Cracking in Materials*, University of Surrey, 4-7 April 1977, The Metals Society, pp. 305-21 (1977)
23. James, L. A., 'Fatigue crack propagation in Austenitic stainless steel', *Atomic Energy Review* **14**, 37-86 (1976)
24. Scott, P. M., 'Effects of environment on crack propagation', pp. 220-257 in Chell, G. (ed.), *Advances in Fracture Mechanics*, Applied Science Publishers, (1981)
25. Scott, P. M., 'Chemistry effects in corrosion fatigue', in *ASTM STP 801, Corrosion fatigue: mechanics, chemistry and engineering*, American Society for Testing and Materials, pp. 319-345 (1983)
26. Barsom, J. M., 'Effect of cyclic stress waveform on corrosion fatigue crack propagation below K_{ISCC} in high yield steel', *NACE-2*, pp. 424-36 (1972)
27. Atkinson, J. D. and Lindley, T. C., 'The effect of frequency and temperature on environmentally assisted fatigue crack growth below K_{ISCC} in steels', *Proc. The Influence of Environment on Fatigue*, Inst. of Mechanical Engineers, London, 18-19 May (1977)
28. Taqi, E. A. and Cottis, R. A., 'The influence of crevice chemistry on hydrogen uptake by structural steels', in Turnbull, A. (ed.), *Corrosion Chemistry within Pits, Crevices and Cracks*, National Physical Laboratory, London (1984)
29. Pelloux, R. M., 'Corrosion fatigue', *Proc. of Fatigue of Materials and Structures*, Sherbrooke, Canada, pp. 8-26 (July 1978)

30. Ritchie, R. O., 'Near threshold fatigue crack propagation in steels', *International Metals Reviews*, Nos. 5 and 6, pp. 205-230 (1979)
31. Jaske, C. E., Payer, J. H., and Balint, V. S., *Corrosion Fatigue of Metals in Marine Environments*, Springer-Verlag and Batelle Press, (1981)
32. Wei, R. P. and Landes, J. D., 'Correlation between sustained-load and fatigue crack growth in high strength steels', *Materials Research and Standards*, 9, 25-27, 44, 46 (1969)
33. Turnbull, A. and Newman, R. C., 'The influence of crack depth on electrochemistry and fatigue crack growth', *Proc AIME Conf. on Short Crack Growth*, AIME (1986)
34. Speidel, M. O., 'Corrosion fatigue in Fe-Cr-Ni alloys', *Proc. Int. Conf. on Stress Corrosion and Hydrogen Embrittlement of Iron Base Alloys*, Unieux-Firminy, France, 12-16 June 1973. NACE-5, pp. 1071-91 (1976)
35. Amzallag, C., Rabbe, P. and Desestret, A., *Corrosion fatigue behaviour of some special stainless steels*, ASTM STP 642, pp. 117-32 (1977)
36. Ford, F. P., 'Modelling and life prediction of stress corrosion cracking in sensitized stainless steel in high temperature water', *Proc. of ASME Fall Meeting*, 1985
37. Chul Kim, U. R. and van Rooyen, D., 'Strain rate and temperature effects on the stress corrosion cracking of Inconel 600 steam generator tubing in the (PWR) primary water conditions', *Proc. 2nd Int. Conf. on Environmental Degradation of Materials in Nuclear Power Systems-Water Reactors*, Monterey, USA, 9-12 Sept. 1985, American Nuclear Society, pp. 448-55 (1986)
38. Sheeks, C. J., Moshier, W. C., Ballinger, R. G., Latanison, R. M., Pelloux, R. M. N., 'Fatigue crack growth of Alloys X750 and 600 in simulated PWR and BWR environments', *Proc. 1st Int. Conf. on Environmental Degradation of Materials in Nuclear Power Systems-Water Reactors*, Myrtle Beach, USA, 22-25 Aug. 1983, NACR pp. 701-25 (1984)
39. Speidel, M. O., 'Design against environment sensitive fracture', *Proc. 3rd Int. Conf. on Mechanical Behaviour of Materials*, Cambridge, UK, 20-24 Aug. 1979, Vol. 1, pp. 109-37, Pergamon Press (1980)
40. Smith, C. J. E. and Hughes, A. N., 'The corrosion fatigue behaviour of a titanium-6 w/o aluminium-4 w/o vanadium alloy', *Engineering in Medicine*, 7(3), 158-71 (1978)
41. Scully, J. C., 'The role of hydrogen in stress corrosion cracking', *Proc. Int. Conf. on Effect of Hydrogen on Behaviour of Materials*, Moran, Wyoming, USA, 7-11 1975. The American Institute of Mining, Metallurgical and Petroleum Engineers, pp. 129-49 (1976)
42. Gee, C. F., 'Fatigue properties of Zircaloy-2 in a PWR water environment', *Proc. of 1st Int. Conf. on Environmental Degradation of Materials in Nuclear Power Systems-Water Reactors*, Myrtle Beach, South Carolina, USA, 22-25 August 1983, NACE, pp. 687-98 (1984)
43. Mshana, J. S., Vosikovsky, O. and Sahoo, M., 'Corrosion fatigue behaviour of nickel-aluminium bronze alloys', *Canadian Metallurgical Quarterly*, 23, 7-15 (1984)
44. Duquette, D. J., 'Fundamentals of corrosion fatigue behaviour of metals and alloys', in *Proc. Conf. Hydrogen Embrittlement and Stress Corrosion Cracking*, Cleveland, Ohio, 1-3 June 1980, Case Western Reserve University Department of Metallurgy and Materials Science, pp. 249-70 (1980)
45. Congleton, J. and Craig, I. H., 'Corrosion fatigue', in *Corrosion Processes*, Ed. Parkins, R. N., Applied Science Publishers, (1982)
46. Uhlig, H. H. and Winston Revie, R., *Corrosion and Corrosion Control*, 3rd edn, John Wiley and Sons, pp. 148-57 (1985)
47. Rust, T. E. and Swaminathan, V. P., 'Corrosion fatigue testing of steam turbine blading alloys', *EPRI Workshop on Corrosion Fatigue of Steam Turbine Blade Materials*, Palo Alto, California, September 1981, Pergamon Press, (1983)
48. Scott, P. M., 'Design and inspection related applications of corrosion fatigue data', *Mémoire et Etudes Scientifique Revue de Métallurgie*, pp. 651-660 (Nov. 1983)
49. Lindley, T. C., McIntyre, P. and Trant, P. J., 'Fatigue crack initiation at corrosion pits', *Metals Technology* 9, 135-42 (1982)
50. McAdam, D. J., 'Influence of water composition in stress corrosion', *Proc. ASTM*, 31, part II, 259-278 (1931)
51. Endo, K. and Komai, K., 'Electrochemical investigation of the corrosion fatigue of steel in acid solution', *Metalloberfläche*, 22, 378-84 (1968)
52. Endo, K., Komai, K. and Nakamuro, N., 'Estimation of corrosion fatigue strength from corrosion resistance and notch sensitivity of the materials', *Bull. Jap. Soc. Mech. Eng.*, 13, 837-46 (1970)

53. Scott, P. M., Tomkins, B. and Foreman, A. J. E., 'Development of engineering codes of practice for corrosion fatigue', *J. Pressure Vessel Technology*, **105**, 255-62 (1983)
54. Sneddon, N. W., *Background to Proposed New Fatigue Design Rules for Steel Welded Joints in Offshore Structures*, HMSO (May 1981)

8.7 Fretting Corrosion

Definition and Terminology

Fretting or fretting corrosion[1] may be defined as that form of damage which occurs at the interface of two closely fitting surfaces when they are subject to *slight* relative oscillatory slip. The surfaces are often badly pitted and finely divided oxide detritus is formed.

Although the term 'fretting corrosion' implies chemical reaction, it has often been used even when the latter is absent. Campbell[2] has suggested that to avoid confusion the word 'fretting' be used to describe the wear process, and that the expression 'fretting corrosion' be applied in those cases where one or both of the surfaces, or the wear particles from them, react with their environment.

Fretting has been known to engineers by a variety of other names; such terms as 'false brinelling', 'chafing fatigue' and 'cocoa' are in use even today. In Germany, Fink[3] has referred to the phenomenon as 'friction oxidation' (*Reiboxydation*), while the word 'blood' has often been used by engineers of that country. The expression 'false brinelling' was coined primarily to describe the fretting wear process as it occurs in rolling contact bearings, since the damage closely resembles the brinelling of a race which has been subjected to excessive static loading. 'Chafing fatigue' on the other hand, is an expression used in reference to a combined fretting-fatigue failure. Such words as 'cocoa' and 'blood' refer to the reddish brown oxide debris which is often to be seen exuding from fretting ferrous contacts.

Incidence

As almost all materials are susceptible to fretting, its incidence in vibrating machinery is high. Shrink fits, press fits and bolted assemblies, splined couplings, keyed gears, both seatings and tracks of ball and roller races, and even electrical contacts, are all particularly vulnerable. Fretting may not only cause serious dimensional loss of accuracy of closely fitted components, but may also seriously reduce the fatigue strength of a machine component.

Characteristics of Fretting and Factors which Influence the Amount of Damage

Although our knowledge of fretting as a wear problem has been derived mainly from the behaviour of ferrous materials, many of the characteristics observed are common to other metals. The influence of fretting on fatigue, however, must be considered separately, since the magnitude of wear damage, as recorded by volume loss of material, is not a good indicator of possible fatigue damage. The development of fine surface cracks is the important criterion in this situation and this type of damage occurs in its most severe form during the early stages of fretting. Because of this difference *fretting fatigue* is considered subsequently as a separate issue.

The main factors which influence fretting wear can be classified as in the following paragraphs, but as more data become available, it is apparent that a greater measure of interdependence exists than might be indicated by such a simple procedure, and it is becoming less easy to make generalisations. Nevertheless, even with this reservation in mind, it is useful as a first step to list some of the more important factors.

The Atmosphere

In the presence of an inert atmosphere fretting of surfaces still occurs and is generally accompanied by the formation of finely divided debris. In a high vacuum, seizure of metal surfaces may take place.

If fretting of oxidisable metals occurs in air the damage is rather more severe than in an inert gas, but the increase is not usually greater than an order of magnitude. The debris consists mainly of oxide, and in the case of steel it is predominantly α-Fe_2O_3 in a very finely divided form, 0.1–0.01 μm in diameter. The proportion of oxidised to non-oxidised material depends largely upon the hardness of the metal. The most stable form of oxide usually appears as the final product of fretting corrosion, but other forms may occur as intermediate products and non-stoichiometric oxides are often produced. Colour is not a reliable guide to composition and steel may produce debris ranging in colour from red to brown and to black, and in every case the composition of the oxide may be that of α-Fe_2O_3.

The relative humidity of the atmosphere has a large effect on the magnitude of wear[4], but in a direction opposite to that which is encountered in normal corrosion problems. The increasing wear towards lower humidities is accompanied by severe pitting of the surfaces and, under extremely dry conditions, the oxide debris produced from steel surfaces is jet black.

Temperature

This is an important factor because it controls the rate of reaction of the rubbing surfaces with the oxygen in the environment. However, although the presence of oxygen normally accelerates the rate of fretting of oxidisable metals at room temperature, an increase in the reaction rate with oxygen brought about by raising the temperature, has the opposite effect, and reduces the wear. Experiments made with steel and copper surfaces show that once a threshold temperature has been exceeded, the normal type of fretting damage (that is the formation of copious amounts of loose oxide debris) is replaced by the formation of a thick adherent oxide glaze that exhibits low friction and little tendency to generate loose debris[5,6]. In the case of steels, the threshold temperature is around 130–200°C, whereas for copper it is just above room temperature. These glazes remain effective at least up to 300°C and 200°C, respectively, the limits of temperature investigated. Their development is encouraged by improving the surface finish of the rubbing elements, since this probably reduces the risk of breaking up the thin surface oxides initially present on the materials. The glazes are not resistant to an impacting load component.

For a wide range of mild and carbon steels, it is found that the rate of development of this thick oxide glaze formed at elevated temperatures is largely independent of the hardness of the substrate. The level of surface damage represented by the amount of virgin metal converted to oxide is an order of magnitude smaller than the level of wear normally encountered at room temperature and is even less than the damage generated in a nitrogen atmosphere at 20°C.

Load, Amplitude of Slip and the Number of Fretting Cycles

The amount of fretting damage increases in an approximately linear manner with these variables, once the initial stages of fretting are completed[4]. A number of deviations from linearity have been reported, especially with respect to load, where it is often found that there is a tendency for the relationship to become parabolic in form. The superposition of a normal vibratory component of load can cause a very considerable increase in the wear rate[7].

Frequency of Oscillation

Little information exists on the effect of frequency except in the case of steels[8]. With such materials there is a reduction in fretting damage as the frequency is increased to 10 Hz. The form of this relationship is the same at both 20°C and 165°C, these temperatures corresponding to the unglazed and glazed mode of fretting of steel in air, respectively[9]. No frequency effect is observed in nitrogen. The position at higher frequencies is not so well established, but there is some evidence that the wear rate begins to rise again and that this is true for both air and nitrogen environments. The increase in wear rate at lower frequencies is often ascribed to the increased time available for oxidation reactions, but it is also observed that the degree of metal to metal contact is more marked at these frequencies.

Hardness

The effect of hardness is complicated, but like most wear processes an increase in hardness generally leads to a reduction in fretting wear at room temperature. The hardness also controls the form of the debris, softer metals tending to produce a higher proportion of large particles of unoxidised debris. Prior work-hardening of materials, such as iron and mild steel, has no effect on the fretting wear rate.

Lubricants

Oils may act in two ways: they provide a measure of boundary lubrication and they may exclude oxygen from the rubbing zone. However, their effectiveness is not as great as with unidirectional sliding, since there is usually sufficient oxygen present in most cases to allow oxide debris to be generated, and this tends to displace any lubricant film that was initially present between the surfaces.

Mechanism of Fretting Corrosion

Fretting wear, along with most other wear phenomena, is not a process that can be defined in terms of any single mechanism. It consists of a series of events, many of which are common to other wear processes and which may assume greater or lesser significance depending upon the precise nature of the operating conditions, materials and environment.

The overriding difference between fretting and other sliding wear processes lies in the small reciprocating nature of the motion. The damage tends to be of a localised form and any debris which is generated has some difficulty in escaping from the rubbing zone. The oscillatory character of the movement introduces a strong fatigue element into the wear pattern and the reversed shearing of localised material inevitably gives rise to fine surface cracks which may initiate a low stress fatigue failure.

The early stages of fretting with metal specimens bear a strong resemblance to those present in other sliding systems in which adhesion forces play a dominant rôle. In all such cases the initial placing of the two surfaces into contact is accompanied by the formation of localised junctions where physical intimacy of the two surfaces occurs. The material at these junctions yields both plastically and elastically until the area of contact established is sufficient to support the applied load. Oxide or other contaminating films greatly reduce the adhesion across the junctions, but the application of a tangential stress facilitates their dispersal. However, the combined effect of normal and tangential tractions across a junction causes it to grow in size and if it was not for the controlling influence of even physically incomplete intervening low-shear-strength films, this growth might continue unchecked until seizure occurred.

The formation and fracture of these junctions leads to the transfer of small fragments of material from one surface to another and finally, after many such events, the release of a wear particle often in a highly oxidised state. Such a wear process is common to most sliding systems and can undoubtedly occur during fretting movements, especially if the slip amplitude is large. However, many fretting situations occur where the amplitude of slip is very small and perhaps even comparable to the dimensions of a single junction region. In such cases it is not so easy to imagine the transfer mechanism just described, but rather to view fretting action as one in which material within the junction zones is continuously subjected to reversed shear. Such conditions will certainly give rise to junction growth and very strong bonds will rapidly become established. The adhesion usually reaches its maximum value after a few thousand cycles of oscillation[10] and during this period the surface material undergoes considerable plastic deformation. It is thought that the pitted nature of fretted surfaces owes its origin to this early gross deformation period and that subsequent events tend to smooth out the contours. Friction measurements reflect this early growth of bonded surface regions and it is only when these junctions begin to work harden and break up under the combined action of the reversed shear and the corrosive action of an air environment that the friction begins to decrease in magnitude. Metallurgical sections through a fretting zone will usually show the disordered state of the junction material and the presence of oxide structures.

The degree of plastic deformation is much greater for softer materials than those which are hard, and the debris frequently contains a much higher percentage of unoxidised metal. In the case of tool-steel surfaces, fretting debris is entirely oxide and very finely subdivided. The gradual accumulation of oxidised detritus between the rubbing surfaces soon begins to isolate one metal surface from the other and this can be followed by measuring the electrical resistance of the system[11]. Extremely high values of electrical resistance can be recorded when the air is dry, and quite clearly a thick oxide compact has built up between the surfaces and some proportion of the slip is likely to be lost within this layer. The hygroscopic nature of the finely divided oxide compact can be demonstrated by admitting air at 45% r.h. when there is an immediate drop in the resistance. The presence of a water film greatly aids the dispersal of the debris and the fretting damage is more uniform over the surface as well as being of smaller volume. This suggests that soft hydrated oxides may be formed and may act as a lubricant. The large increase in wear rate observed whenever a normal vibratory load is applied to a fretting system is possibly due to the fact that it prevents a thick oxide compact from establishing itself between the surfaces. The early adhesion mode of wear probably persists for a much longer time and may never completely disappear. In practice such conditions must be avoided at all costs.

The manner in which oxide debris is formed when metals fret in air is a subject of considerable controversy and it is clear that no single mechanism can explain all the data. The early theory of Tomlinson[1] was based upon the idea that the surfaces were worn by a process of molecular attrition and that this leads to oxide generation in an oxidising environment. Others considered that fretting was essentially an accelerated oxidation mechanism in which the mechanical removal of oxide prevented the attainment of a stable protective oxide film. Later, Uhlig[8] modified this model to allow some metallic debris to be formed as a result of an adhesion mechanism, but still retaining the corrosion aspect to help explain away the frequency effect[8]. This model encountered difficulties in explaining the decrease in wear with increasing temperature, and Uhlig suggested that perhaps the corrosion aspect could be better represented by a model that involved physical adsorption of oxygen on the steel surface and that the actual formation of the oxide resulted from mechanical activation. More modern theories[12] place greater emphasis on the changing nature of the fretting mechanism, particularly drawing attention to the strong influence of adhesion in the early stages and the significance of corrosion fatigue as a contributory factor in the disintegration of the material making up the junction zones. The later stages of fretting damage are also explained in terms of a microfatigue process rather than one of abrasion by cutting.

The reduction in fretting damage of steel and copper surfaces as a result of increasing the temperature, at least over the temperature range of 20–300°C, has focused attention on the protective rôle that oxidation can play in fretting. It would seem that this arises from a suppression of intermetallic contact during the early adhesion stage of fretting and the improvement provided by a good surface finish substantiates this explanation. Again the presence of a vibratory normal load destroys the opportunity of establishing a protective glaze. It is interesting to note that the frequency effect is of the same form both at room temperature and at slightly elevated temperatures where glazing occurs. It might have been argued that the increased reaction time at the lower frequencies would have enhanced the protective rôle offered by oxidation. These and many other problems still need to be cleared up before an accurate picture of fretting corrosion can emerge.

Preventive Measures

Most of the cases of fretting met with in practice appear to fall into two distinct classes according to whether or not the surfaces involved in the component are intended to undergo some relative motion. If the surfaces are not intended to move, then the first objective should be to prevent slip, either by eliminating the source of vibration, or by increasing the friction between the surfaces. It is believed that the success of certain soft metal electrodeposits in reducing fretting may be due to the improved fit, and hence possibly increased friction, which is obtained from their use. If the displacements cannot be controlled in this way, it may be possible to interpose a thin sheet of an elastic material which can accept the relative movement without slip.

With applications which are intended to undergo relative motion at some stage, once again an attempt should first be made to lower the amplitude of vibration. Should this prove difficult, an improvement in the lubrication conditions will be necessary in order to reduce the amount of intermetallic contact. For example, in the case of rolling bearings the interfacial slip between the rolling elements and the races is accompanied by a relatively large rolling component of motion. If a suitable oil or grease is used, this displacement can be utilised to maintain a film of lubricant on the race.

With plain bearing surfaces of steel, a great improvement can be obtained by phosphating and impregnating the layer with oil. The inherent porosity of the phosphate films provides minute reservoirs for the oil. Should it be necessary to operate a bearing dry, then either a bonded film of MoS_2 or p.t.f.e. (polytetrafluoroethylene) can be used to advantage. Both materials have excellent frictional and wear characteristics.

Fretting Fatigue

Conditions which favour the occurrence of fretting may also be regarded as favourable to metal fatigue and fatigue cracks are often observed to start at points on a surface where fretting has taken place. There is no evidence, however, that the severity of fretting damage, as measured by volume loss of material, is directly related to the magnitude of the reduction in fatigue strength. The significance of fretting is that it allows the formation and rupture of strong adhesive bonds under oscillatory forces and this action generates fine surface cracks which may, or may not, propagate into a major fracture of the component. An understanding of this subsequent crack development is therefore vital if adequate steps are to be taken to avoid an early failure. In order to do this it is useful to examine the normal fatigue behaviour of plain and notched materials subject to reversed loading.

The progress of a normal fatigue failure can be divided into two stages (Stage 1), the formation of slip band cracks at the surface, and secondly (Stage 11), their propagation into the main body of the component and eventual failure (Section 8.6). The first stage of the process is usually marked by the growth of crevices or intrusions along planes aligned in the direction of maximum shear stress and this stage may occupy the major proportion of the final recorded life. However, crack development during this stage eventually ceases, perhaps by the crack meeting an obstacle, such as a grain boundary, and the subsequent propagation of this embryo crack proceeds in accordance with a criterion based upon maximum principal stress, or upon maximum comparative stress in a combined stress situation. This second stage is marked by the appearance of the familiar striations on the crack faces. The stress required for the successful completion of Stage I corresponds to the plain fatigue limit of the material, and is much higher than that which is needed for the subsequent propagation of the surface crack.

The situation is greatly changed, however, if a notch or other stress raiser is present in the surface. If a notch exists, then the stress required to initiate a crack at the root is given by the plain fatigue limit stress divided by the stress concentration factor for the notch, K_t. With some notches the magnitude of K_t may be so large that the nucleation stress is now much lower than the propagation stress and propagation will not occur unless the alternating stress within the main body of the component is of sufficient magnitude. Dormant cracks may thus exist in some engineering components.

In the case of a notched component the propagation stress is both independent of K_t and the notch root radius, and is primarily a function of the combined length l_d of the notch and embryo crack; the criterion for propagation being $\sigma^3 l_d > C$, where C is a material constant and σ the applied cyclic (tensile) stress. Since most non-propagating cracks are short compared with the notch depth we may usually substitute the notch depth for l_d in this expression. Thus assuming that notch root cracks have been formed, a limiting stress exists for propagation, namely $\sigma = (C/l_d)^{\frac{1}{3}}$. If this stress is not reached then dormant cracks will be formed unless an active corrosive environment is present[13].

In many ways the fretting fatigue situation is analogous to the notched fatigue behaviour, not because the fretting damage resembles a mechanical notch, although this view has been held by some workers, but because the fretting action promotes the formation of suitable surface cracks simply by the presence of strong frictional shear stresses. Many of these cracks are observed to be of the order of 100 μm in length and are thus of sufficient length to be propagated by stresses well below the normal plain fatigue limit. Fretting therefore seems to be a remarkably efficient way of achieving the completion of Stage I of the fatigue process.

Once fretting cracks of a suitable length have been formed the life of the component depends upon the rate at which they can be propagated by the Stage 11 mechanism, and this part of the process is subject to the same environmental factors as a normal fatigue failure. The length of the fretting cracks determines the lowest level of stress at which they can be propagated, and hence the new fatigue limit. Once again an active corrosive environment prevents the existence of a fixed fatigue limit and even small 'non-propagating cracks' will not remain dormant for long. Fretting tests carried out in such environments show the same characteristics as corrosion fatigue tests made in the absence of fretting[14].

One important feature about fretting fatigue is the very small amount of physical damage to the surface that is necessary to cause a considerable decrease in fatigue strength of some materials. The damage is often barely discernible to the naked eye. Small slip amplitudes seem to be particularly damaging and this has been attributed to the possibility that larger slip amplitudes and the associated higher wear rate may erase any potentially damaging surface cracks or promote the development of large numbers of very small interacting cracks rather than one or two of greater severity. Only one crack of propagating length is required to cause a component failure. Recent work suggests that the coefficient of adhesion between fretting surfaces in air reaches a maximum value for amplitudes of slip between 30 and 75 μm, and that this level of adhesion is established within a very small number of slip cycles. A few thousand cycles is usually sufficient. Laboratory experiments made with fretting pads clamped to specimens subjected to fatigue stresses indicate that the nominal slip amplitudes which prove the most damaging are those in the range 7.5–14 μm[13]. This range is slightly smaller than the range observed with adhesion experiments, but the sensitivity to slip amplitudes is complex and somewhat higher slip amplitudes prove equally damaging in the presence of a high-tensile mean stress. All the data tends to support the hypothesis that the establishment of strong adhesive bonds between the contacting surface asperities is a major factor in the fretting fatigue process. The strength of such bonds varies in a systematic way with the hardness of the rubbing elements and there is evidence that annealing and over-ageing of age-hardened alloys occurs at the fretting contact. It is claimed that the fretting fatigue strength is related to the adhesion strength of the asperity bridge and that when a critical value of the latter is reached a propagating crack is quickly formed[15].

It is perhaps unfortunate that the sensitivity to fretting is generally greatest in the low slip amplitude regime since this is just the level of slip which is so often experienced with many engineering components. A typical example is the case of a wheel fitted to a shaft. When this combination is subjected to a reversed bending moment localised slip will occur at the edge of the contact area and fretting fatigue cracks may eventually develop there. The establishment of a partially slipped interface is of common occurrence in fitted components and is due to the form of the normal and tangential tractions which have to be transmitted across the interface when the unit is stressed. Slip occurs at all points where the ratio of the tangential to normal traction exceeds the coefficient of friction. Ultimately with sufficient magnitudes of applied stress the whole surface will slip, but this is not usually an acceptable condition. It will be seen therefore that small slip amplitudes are invariably linked with a partially slipped interface and the stress concentrations that exist between the boundary of the slipped and unslipped regions have been suggested as a reason why fretting cracks appear at the boundary. As an extension to this argument it is suggested that the inducement of a high gross slip condition will relieve the stress concentration and hence inhibit the formation of propagating fretting cracks even though this action is likely to lead to greater overall wear.

Mention must be made of the important practical case where a component is subjected to a cyclic stress superimposed upon a high-tensile mean stress. In the case of a simple plain fatigue test, the fatigue strength only falls slowly with rising values of the mean stress (Gerber relationship). Small values of the mean stress have little effect upon the measured fatigue strength because the strength of a plain component is governed by the crack initiation or reversed shear stress stage of the process. However, if cracks have already been developed in the surface as a result of fretting, then the fatigue limit will depend upon the propagating conditions and these will be a function of both the alternating and mean tensile stress[13]. The fatigue strength as a result of applying a mean tensile stress to a cyclically stressed component subjected to fretting shows a steep fall at mean stresses of ca 100 MPa. An alloy steel with UTS 1030 MPa, plain fatigue strength 540 MPa and fretting fatigue strength of 125 MPa with zero mean stress, is reduced to a fretting fatigue strength of 50 MPa with a mean stress of 100 MPa (Fig. 8.76).

Prevention of Fretting Fatigue

Some design aspects have already been mentioned, notably the avoidance of interfacial slip between contacting surfaces or joints. This is best done by avoiding designs in which the interface enters a region of stress concentration. In some cases it may be possible to absorb the movement by the insertion of a resilient layer. A low coefficient of friction can reduce the severity of induced frictional stresses. Resin-bonded P.T.F.E., for instance, is quite effective under conditions of relatively low surface pressure and slip amplitudes. Alternatively, a sacrificial metal coating may be applied, but this method may not prevent a surface crack penetrating the bond interface, especially if this is strong. The influence of the hardness of the rubbing surfaces has already been mentioned and some advantage of this can be taken in certain cases[16]. Surface treatments like case hardening and nitriding can produce a hard and compressively stressed surface layer which inhibits crack propagation. Similar results can be achieved with cold working procedures, such as shot peening.

Reported incidences of fretting have continued to proliferate, in particular in blade/disc fixings in both steam and gas turbines,[17,18] in PWR[19] and AGR[20] nuclear power plants, between the conductors in overhead power

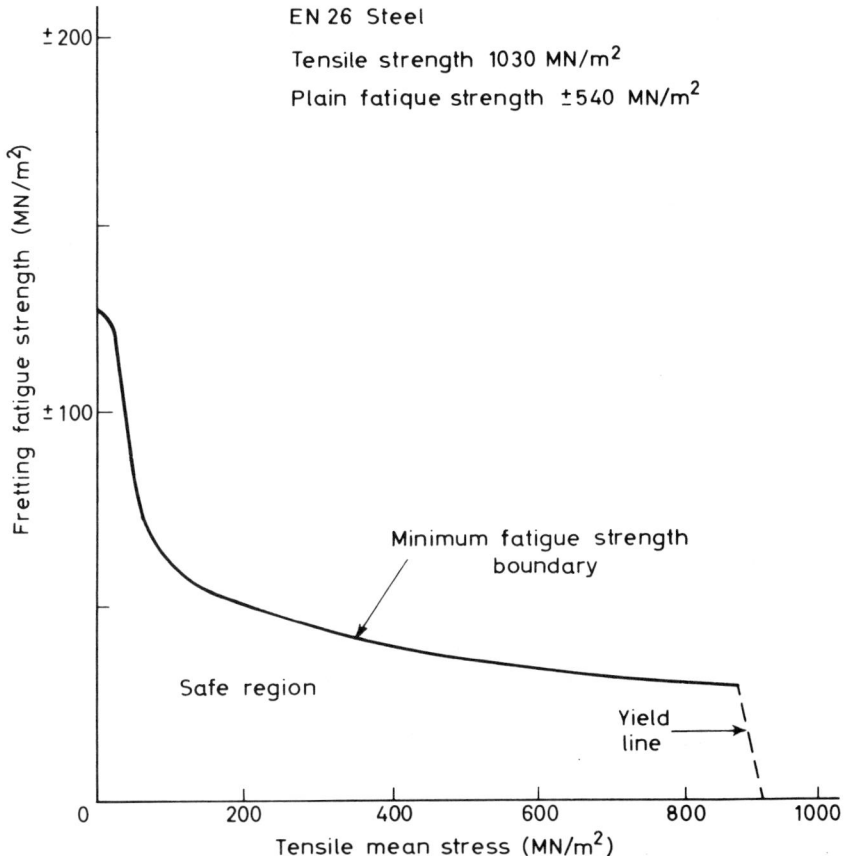

Fig. 8.76 Fretting fatigue strength at various mean stresses

lines[21], in steel ropes operating in marine conditions[22], in orthopaedic implants and fracture fixation devices[23], and in electrical connections subject to vibration[24].

Much of the early work on fretting was confined to mild steel or carbon steel. Although many of the same principles apply to the fretting of other metals such as gold[25], titanium and the superalloys[26], their reaction to the environment may be a more significant factor. In addition, non-metallic materials such as polymers[27], composites[28] and ceramics[29] are becoming widely applied and the principles of contact here are very different from the metallic case.

Recent research has shown the importance of the environment in the fretting process. In high vacuum or non-oxidising gases such as helium or argon, roughening of the surfaces occurs and material may be transferred from one surface to the other, but little loose debris is formed[30] and the initiation of propagating fatigue cracks is more difficult.

Fretting in air-saturated aqueous electrolytes, such as seawater[31] or body fluids[32], produces enhanced removal of material by stimulation of electrochemical reactions, increasing the reaction rates by factors of 10 to 200 compared with air, depending on the frequency. The importance of the chemical factor is emphasised by the fact that the damage can be almost eliminated by cathodic protection[33]. In fretting in artificial body fluid the wear rate is significantly reduced by the presence of proteins[34].

The effect of temperature has been extensively investigated, particularly in relation to materials used in gas turbine aeroengines. The conclusion is that alloys which are capable of forming a spinel-type oxide form the beneficial glaze oxide, with its low friction and low wear properties and its ability to repair itself[35]. Thus the nickel–chromium–iron and nickel–chromium alloys are in this category with the possibility of forming the following spinels: $FeO \cdot Fe_2O_3$, $NiO \cdot Fe_2O_3$, $FeO \cdot Cr_2O_3$, $NiO \cdot Cr_2O_3$. The glaze oxide is only stable at the operating temperature. If the temperature falls to room temperature the film is completely disrupted[36]. Titanium alloys do not behave in this way, possibly because the oxide film is thin due to the solubility of oxygen at temperatures above $800°C$[37]. However, implantation by such species as Ba and Bi has been shown to reduce the coefficient of friction and wear rate to very low values[38]. Whether such films would have the property of self renewal is uncertain.

Fretting at very low temperatures (4 K) has been investigated. The situation appears to be similar to that in high vacuum. An oxide film does not grow at this temperature and so coefficients of friction are high due to adhesion but little wear occurs[39].

The importance of amplitude of slip is emphasised above. In recent years, the area of very low amplitudes and very carefully controlled amplitudes has been investigated. These researches confirm that the specific wear rate (volume removed, per unit distance of sliding per unit applied load) increases dramatically in the region 30–70 μm and then becomes constant, as would be expected in reciprocating or unidirectional sliding[40]. Damage is produced at amplitudes of 1 μm or less[41], but it tends to be characteristic of surface fatigue rather than wear.

Increasing the normal load in some systems will reduce the amplitude of slip and the area over which slip is occurring, but this may result in severe delamination damage[42], and is only to be recommended with considerable caution as a means of reducing fretting damage.

Several investigators have demonstrated the importance of the role of the debris in its effect on the wear rate at large numbers of cycles. Aldham et al.[43] have shown, by using profilometry and contact resistance measurements, that compacted layers of debris on steel support the load and reduce further wear, although metal-to-metal contacts still occur after long periods of time. Escape of loose debris does not affect the process although Colombié et al.[44] claim that removal (blowing out) of the debris results in an increase in the wear rate. Configurations which encourage the escape of debris do increase the frequency of metal-to-metal contact and lead to higher wear rates.[45] It must be remembered that oxide debris is usually abrasive and can cause other problems if it does escape.

The effects of very high frequencies on fretting (up to 20 kHz) have been investigated to see whether fretting tests could be speeded up without impairing the results[46]. At low amplitudes (partial slip regime) surface degradation and fatigue crack initiation were accelerated, but at higher amplitudes (total slip) the higher frequency had little effect on the wear mechanism.

The major advance in fretting fatigue has been the finite element analysis of actual fretting problems, e.g. blade/disc dovetail fixings[47] and the application of fracture mechanisms to the problem[48]. This has led to the concept of the fretting damage factor which is a step forward in the anticipation of fretting problems at the design stage. Its validity has been recently demonstrated in a carefully controlled experiment[49].

The application of fracture mechanisms has shown the acceleration of the crack under fretting conditions in the first 100 μm of its growth[50]. Corrosive conditions, e.g. seawater, further accelerate the initial crack growth rate[51].

Although the effect of normal load shows that there is a limiting pressure of 100 MPa where no further decrease in fatigue strength is observed, it has recently been shown that very high pressure of 9 000 MPa can impede the propagation of the crack, and removal of the load while continuing the fatigue allows the crack to resume propagation[52].

While design is the most satisfactory way of minimising fretting damage, two useful reviews of palliative methods should be noted[53,54]. Shot peening has been shown to be one of the most successful treatments and is widely used on aircraft components. The surface residual compressive stress is the major factor in retarding the propagation of cracks initiated by the fretting[55], but the surface roughening also contributes[56]. Shot peening and the attendant surface roughness also lead to a reduction in the coefficient of friction and wear damage[57,58]. Two extensive reviews have recently been published on fretting fatigue[59] and fretting wear[60].

K. H. R. WRIGHT
R. B. WATERHOUSE

REFERENCES

1. Tomlinson, G. A., Thorpe, P. L. and Gough, H. J., *Proc. Instn. Mech. Engrs., London.*, **141**, 223 (1939)
2. Campbell, W. E., *Symposium on Fretting Corrosion*, Amer. Soc. Test. Mater. Special Technical Publication No. 114, 3 (1953)
3. Fink, M., *Org. Eisenbahnw.*, **84**, 405 (1929)
4. Wright, K. H. R., *Proc. Inst. Mech. Engrs., Lond.* (*B*), **1B**, 556 (1952-53)
5. Wright, K. H. R., Strength of Components, *N.E.L. Report No.* 402, March (1969)
6. Hurricks, P. L., *Wear*, **19**, 207 (1972)
7. de Gee, A. W. J., Commissaris, C. P. L. and Zaat, J. H., *Wear*, **7**, 535 (1964)
8. Uhlig, H. H. *J. Appl. Mech.*, **21**, 401 (1954)
9. Wright, K. H. R., LDR 14/61. National Engineering Laboratory (1961)
10. Bethune, B. and Waterhouse, R. B., *Wear*, **12**, 289 (1968)
11. Fenner, A. J., Wright, K. H. R. and Mann, J. Y., *Proceedings of the International Conference on Fatigue of Metals, Instn. Mech. Engrs.*, London (1957)
12. Hurricks, P. L., 'The Mechanism of Fretting-A Review', *Wear*, **15**, 389 (1970)
13. Field, J. E. and Waters, D. M., NEL Reports Nos. 275 (1967) and 340 (1968)
14. Waterhouse, R. B. and Taylor, D. E., *Wear*, **15**, 449 (1970)
15. Bethune, B. and Waterhouse, R. B., *Wear*, **12**, 369 (1968)
16. Taylor, D. E. and Waterhouse, R. B., *Wear*, **20**, 401 (1972)
17. Ruiz, C., Boddington, P. H. B. and Chen, K. C., *Exp. Mech.*, **24**, 208 (1984)
18. King, R. N. and Lindley, T. C., *Advances in Fracture Research*, vol. 2, Pergamon, Oxford, p. 631 (1982)
19. Hofmann, P. J., Schettler, T. and Steininger, D. A., *Proc. Conf. ASME Pressure Vessels and Piping*, Chicago, Il1 USA, 21-24 July 1986, ASME, New York, 86-PVP-1, 15pp (1986)
20. Jones, D. H., Nehru, A. Y. and Skinner, J., *Wear*, **106**, 139 (1985)

21. Lanteigne, J., Cloutier, L. and Cardou, A, *CEA Report 131-T241*, July 1986, Canada (1986)
22. Pearson, B. R., Brook, P. A. and Waterhouse, R. B., *Wear*, **106**, 225 (1985)
23. Cook, S. D., Giamoli, G. J., Clemow, A. J. T. and Haddad, R. J., *Biomat. Med. Dev. Art. Org.*, **11**, 282 (1983-84)
24. Braunovic, M. *Wear*, **125**, 53 (1988)
25. Antler, M. and Drozdowicz, M. H., *Wear*, **74**, 27 (1981-82)
26. Bill, R. C., *ASLE Trans.*, **16**, 286 (1973)
27. Higham, P. A., Stott, F. H. and Bethune, B., *Corr. Sci.*, **18**, 3 (1978)
28. Friedrich, K., *J. Mater. Sci.*, **21**, 1700 (1986)
29. Horn, D. R., Waterhouse, R. B. and Pearson, B. R., *Wear*, **113**, 225 (1986)
30. Iwabuchi, A., Kato, K. and Kayaba, T., *Wear*, **110**, 205 (1986)
31. Overs, M. P. and Waterhouse, R. B., *Proc. Conf. Wear of Materials*, Reston, Va, USA, 11-14 April 1983, ASME, New York, p. 546 (1983)
32. Williams, R. L. and Brown, S. A., *Trans. Soc. Biomater.*, **7**, 181 (1984)
33. Pearson, B. R. and Waterhouse, R. B., *Proc. Conf. Wear of Materials*, Vancouver, Canada, 14-18 April 1985, ASME, New York, p. 79 (1985)
34. Merritt, K. and Brown, S. A., *J. Biomed. Mat. Res.*, **22**, 111 (1988)
35. Waterhouse, R. B., *Tribology International*, **14**, 203 (1981)
36. Hamdy, M. M. and Waterhouse, R. B., *Proc. Conf. Wear of Materials*, Reston, Va, USA, 11-14 April 1983, ASME, New York, p. 546 (1983)
37. Iwabuchi, A. and Waterhouse, R. B., *Wear*, **106**, 303 (1985)
38. Iwabuchi, A. and Waterhouse, R. B., *Proc. Conf. Wear of Materials*, Vancouver, Canada, 14-18 April 1985, ASME, New York (1985)
39. Iwabuchi, A., Honda, T. and Tani, J., *Cryogenics*, **29**, 124 (1989)
40. Waterhouse, R. B., *Treatise in Materials Science and Technology*, Scott, D. Ed. Academic Press, New York, p. 259 (1979)
41. Kennedy, P. J., Stallings, L. and Peterson, M. B., *ASLE Trans.*, **27**, 305 (1984)
42. Goto, S. and Waterhouse, R. B., Titanium '80 Science and Technology, *Proc. 4th Int. Conf. on Titanium*, vol. 3, Met. Soc. AIME, New York p. 1837 (1980)
43. Aldham, D., Warburton, J. and Pendlebury, R. E., *Wear*, **106**, 177 (1985)
44. Colombié, C., Berthier, Y., Floquet, A., Vincent, L. and Godet, M., *J. Tribology*, **106**, 194 (1984)
45. Kuno, M. and Waterhouse, R. B., *Proc. Conf. Eurotrib.* 1989, Helsinki, Finland, 12-15 June, 1989 **3**, 30 (1989)
46. Söderberg, S., Bryggman, U. and McCullough, T., *Wear*, **110**, 19 (1986)
47. Ruiz, C. and Chen, K. C., *Proc. Conf. Fatigue of Engineering Materials and Structures*, vol. 1, Sheffield, UK, 15-19 Sept., 1986, Inst. Mech. Engrs., London, p. 187 (1986)
48. Hattori, T., Nakamura, M., Sakata, H. and Watanabe, T., *JSME Int. J.*, **31**, 100 (1988)
49. Kuno, M., Waterhouse, R. B., Nowell, D. and Hills, D. A. *Fatigue Fract. Eng. Mater. Struct.*, **12**, 387 (1989)
50. Sato, K., Fujii, H. and Kodama, S., *Wear*, **107**, 245 (1986)
51. Takeuchi, M. and Waterhouse, R. B. *Proc. Conf. Environment Assisted Fatigue*, Sheffield, UK, 12-15 April 1988, Pergamon (in press)
52. Miyagawa, H. and Waterhouse, R. B., *Eurotrib* 1989, *Proc. Conf.*, Helsinki, Finland, 12-15 June 1989 (in press)
53. Chivers, T. C. and Gordelier, S. C., *Wear*, **96**, 153 (1984)
54. Beard, J., *Proc. Conf. Tribology-Friction, Lubrication and Wear Fifty Years On*, vol. 1, London, UK, 1-3 July 1987, Inst. Mech. Engrs., London, p. 311 (1987)
55. Waterhouse, R. B., Noble, B. and Leadbeater, G., *J. Mech. Work. Technol.*, **8**, 147 (1983)
56. Leadbeater, G., Noble, B. and Waterhouse, R. B., *Proc. Conf. Advances in Fracture Research*, vol. 3, Delhi, 4-10 Dec. 1984, Pergamon, Oxford, p. 2125 (1984)
57. Bergman, C. A., Cobb, R. C. and Waterhouse, R. B., *Proc. Conf. Wear of Materials*, vol. 1, Houston, Texas, USA, 5-9 April 1987, ASME, New York, p. 33 (1987)
58. Bilonoga, Yu. L., *Sov. Mater Sci.*, **21**, 282 (1985)
59. Waterhouse, R. B., *Int. Mat. Rev.*, **37**, 77 (1992)
60. Waterhouse, R. B., *ASM Handbook: Friction, Lubrication and Wear Technology*, ASM, Materials Park, Ohio, 242 (1992)

8.8 Cavitation Damage

The subject of cavitation has stimulated the interest of engineers and scientists since the early experiments of Sir Charles Parsons in the late 19th Century, so eloquently described by Burrill[1]. Since that time research effort has been roughly divided between that concerned with the effects of cavitation in fluid flow on the efficiency of hydraulic machines such as turbines, pumps and propellers, and that concerned with the problems of material erosion resulting from the collapse of the vapour bubbles. Several major texts have been published[2-5], and excellent reviews have dealt with specific areas. The latter will be identified in the relevant sections.

Bubble Dynamics

Since the term 'cavitation' was originally suggested by the naval architect R. E. Froude to describe the clouds of bubbles produced by propeller blades the engineering literature usually attributes its occurrence simply to a reduction in liquid pressure to the value of its saturated vapour. More correctly, cavitation occurs when small visible bubbles are formed as a result of pressure reduction to a negative value of about half an atmosphere. The visible bubbles are derived from nuclei which are present in the fluid in the form of minute gas bubbles and solid particles. In addition, nucleation is initiated by gas bubbles trapped in crevices in surfaces containing the fluid. In a survey of nucleating sources in the ocean, Acosta et al.[6] concluded that microbubbles with diameters less than 25 μm were suitable nuclei, but there was insufficient evidence to show that microparticulates in the same diameter range had the same nucleating potential. In engineering pipework systems it is well known that the addition of small quantities of air can 'trigger' cavitation at higher values of the local cavitation index.

The effect of nucleating sources is important when attempting to scale studies with models to full-size components. A common practice in model tests is to expose the fluid to a high static pressure prior to testing in order to drive free gas into solution. In contrast the seeding of water tunnels is considered to improve scaling particularly in high-speed propeller applications[7].

In fluid systems cavitation bubbles form in flow passages where a reduction in pressure occurs usually due to a local increase in velocity. Systems subject to vibration, rather than changes in velocity, may experience cavitation where bubbles are formed in an oscillating-pressure field. Each cycle of oscillation generates and collapses the bubbles within the same volume of liquid.

Bubbles grown from the nuclei identified earlier achieve a maximum radius of perhaps 1 mm then, when the local pressure field increases to a critical value, they collapse violently within a few microseconds. Where the collapse phase occurs close to a solid surface damage is produced which is characterised by the appearance of small pits. The growth and collapse of cavitation bubbles has been extensively studied, and excellent reviews are available[4, 5, 8-11]. Despite much research effort the mechanism by which materials are damaged has not been fully explained. It is, however, clearly related to the conversion of the potential energy associated with the size of individual bubbles prior to collapse into kinetic energy of the surrounding liquid during the collapse phase. Two mechanisms have been proposed. The first involves the pressure of the shock wave radiated towards the surface and the other suggests that damage results from the impact of a high-velocity microjet which emerges from the collapsing bubble.

Models for both damage mechanisms have been proposed by Lush[12] and Grant and Lush[13]. Their analyses take into account effects such as air content and liquid compressibility in calculating collapse shock pressures and it is found that these are comparable with the tensile strength of many commonly used materials. Both Lush and Hammitt acknowledge that the volume of plastic flow required to produce a pit is similar to the work done by a hardness machine indentor, i.e. the stress required is given approximately by the hardness. Lush[14] argues that, since the hardness, when expressed as a stress, is roughly three times the tensile strength together with the probable reduction in collapse pressures to below the maximum calculated values, it is unlikely that the shock wave mechanism would explain pitting except in very soft materials. It would appear that the required magnitude of stress is produced by 'water hammer' pressure developed when a conical tipped microjet having a velocity in the range 100–400 m/s is arrested at the material surface. Despite the magnitude of this pressure, typically 1800 MPa with an ambient pressure of 0.1 MPa, the stress generated may not be sufficient to cause plastic deformation, and damage takes place only after repeated impacts. This implies that the damage mechanism involves a high frequency fatigue stress with the probability that local failure of the material will occur at stresses lower than the tensile strength.

Materials

The intensity of cavitation erosion damage may be extremely severe in fluid flow situations and the consequences of such damage are extensive in terms of cost and loss of operational time. An example in a large civil engineering structure was the collapse of a tunnel discharging water from the Tarbela Reservoir in Northern Pakistan described by Kenn and Garrod[15]. The cause was attributed to the severity of cavitation damage experienced by the concrete walls of the tunnel when high velocity water streams separated from discontinuities such as piers and buttresses. Velocity differentials in excess of 30 m/s were generated and the sheared flow contained vorticity-induced cavitation cavities. In downstream regions where local pressure had recovered, erosion caused penetration of the concrete walls to a depth of 3.4 metres in the period of a few days.

Such examples serve to demonstrate that both designers and operators of hydraulic systems must be aware of the potential consequences of cavitating flows and take steps to minimise their occurrence. Where design and operation are optimised cavitation damage may be further reduced by the use of erosion resistant materials. Figure 8.77 illustrates the results of comparative erosion tests on engineering metallic materials using a laboratory method, described earlier by Godfrey[27], for inducing hydrodynamic cavitation. From cast iron to the Co–Cr–W hard-facing alloys the

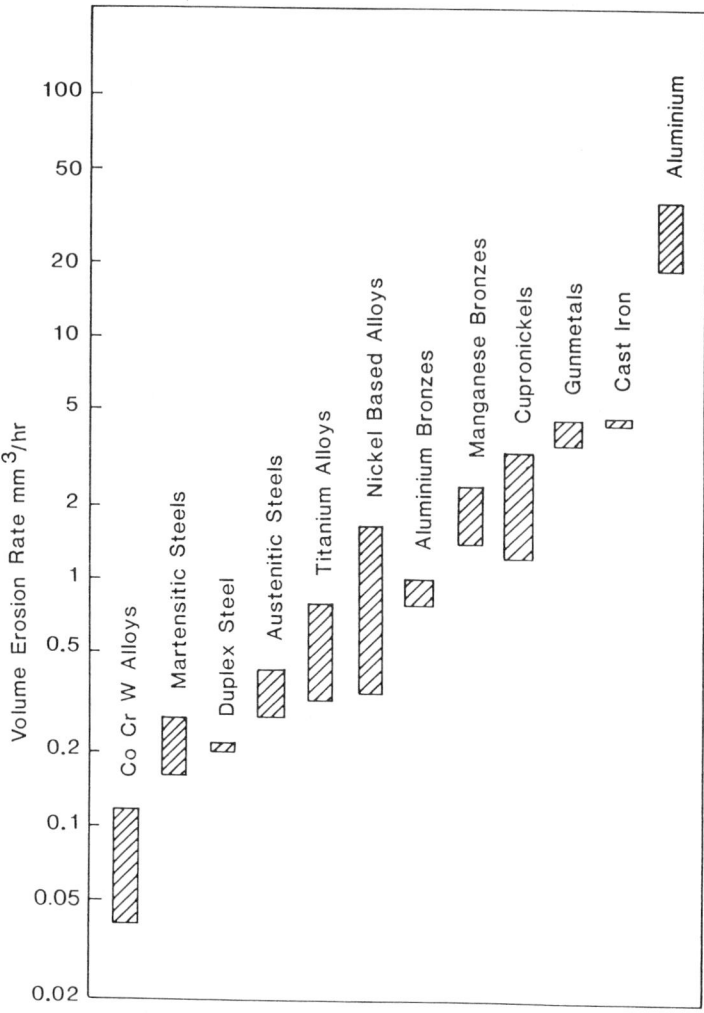

Fig. 8.77 Comparative volume erosion rates for engineering materials

volume erosion rates cover nearly two orders of magnitude representing a considerable difference in erosion resistance. In practice the advantages to be gained, for example in pumps, may be represented by changes in the casing material from gunmetal or cast iron to austenitic steel and the impeller from gunmetal to aluminium bronze, austenitic or duplex steels. Care must be taken in selection where a corrosive fluid is involved and in particular with seawater.

The recent advances in polymer technology have resulted in the increasing use of these materials in engineering applications such as pumps, valves and piping systems particularly for the handling of corrosive fluids. Figure 8.78 gives a ranking of some polymeric materials tested under the same conditions of cavitation intensity as the metallic materials. The test method relies on mass loss to calculate volume erosion rates using the specific gravity and, due to the variability of water absorption of the polymers during the

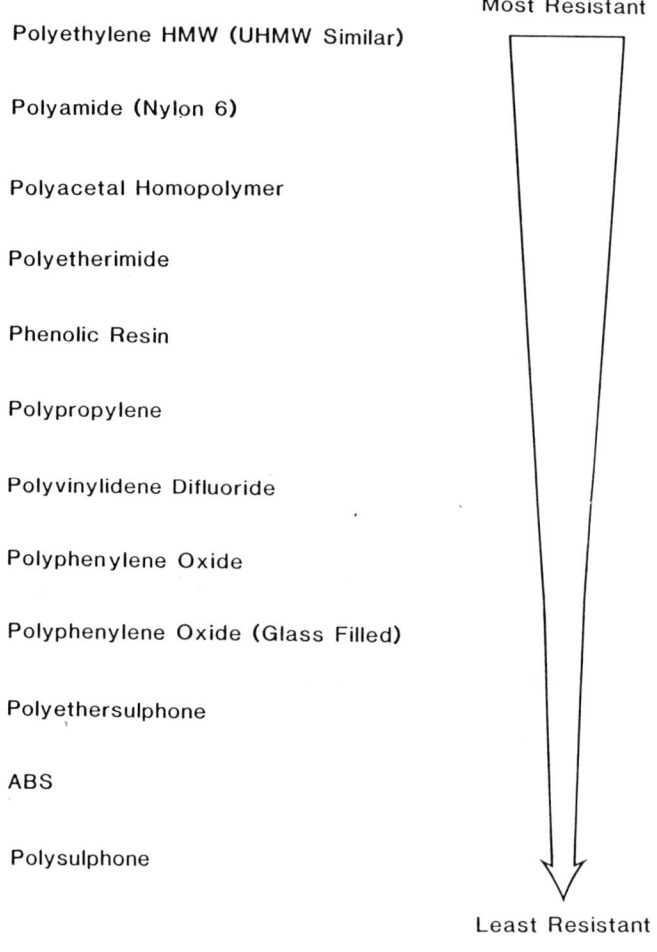

Fig. 8.78 Cavitation erosion resistance ranking of solid polymers

test period, erosion resistance could be expressed only in a qualitative way. It is possible, however, to make reasonably reliable comparisons between some of the more resistant materials. Polyacetal, for example, exhibits similar erosion resistance to nickel aluminium bronze, and the most resistant of the polymers tested, HMW polyethylene, could similarly be compared with the nickel-base and titanium alloys.

The role of coatings as protection against erosion damage has developed as a result, in the case of polymeric materials, of a better understanding of their behaviour when subjected to dynamic loading. This aspect of the use of such materials in naval applications was considered by Angell et al.[16]. Recent developments in elastomeric coatings have produced high-performance polyurethane systems capable of resisting moderate levels of cavitation intensity. A coating thickness of 2–3 mm may be required to ensure resistance against erosion and impact. Provision of conditions to promote the optimum adhesive bond is crucial to the performance of these materials. Ceramic coatings, although brittle, resist low levels of erosion intensity, but porosity reduces long-term performance due to interfacial corrosion particularly in seawater. Improvements in spraying techniques have alleviated this problem to some extent and good erosion resistance has recently been claimed for polymer/ceramic composite systems.

Cavitation Erosion Test Methods

The earlier contribution to this section identified four hydraulic and one vibratory test methods. The latter was subsequently recognised by the American Society for Testing and Materials as a standard method for the determination of cavitation erosion resistance (ANSI/ASTM G32-85). Although possessing many advantages this method has been criticised because the cavitation is not induced hydrodynamically. A useful test has been devised by Lichtarowicz[17] which combines this feature with the simplicity, high erosion intensity and comparatively inexpensive operation offered by the vibratory method. In this test a submerged liquid jet is delivered at high velocity into a chamber, in which the pressure is controlled through a sharp entry parallel bore nozzle of small diameter. Cavitation is induced in the contracted flow region within the nozzle, is stabilised in the parallel section and emerges to impinge upon a disc-shaped test specimen mounted in its path at a known distance from the nozzle. The controlled upstream pressure is generated by a positive displacement pump with modest power requirements. This method may, in the future, become an alternative standard test procedure.

The Erosion Process

The process by which a material becomes eroded when exposed to the forces resulting from the collapse of cavitation bubbles proceeds in a number of stages according to Thiruvengadam and Preiser[18]. In the first of these, the incubation period, no detectable mass loss occurs but damage is visible. In the ductile engineering materials this takes the form of pitting, but brittle materials may also show cracking. During the second phase the erosion rate increases and reaches a maximum value. As the exposure time increases two further stages in the process have been identified which appear to depend upon the test method employed. The third phase is one in which the erosion rate remains more or less constant until the final phase is reached when the rate tends to decrease. These stages were initially defined using a vibratory test method and in venturi type hydrodynamic methods the separation between the stages may be indistinct unless it is possible to remove and weigh the specimen frequently. In practice, erosion rate values obtained from venturi type tests are derived from total volume loss divided by total exposure time as for the data presented in Figs. 8.77 and 8.78.

The erosion rate depends upon the cavitation intensity as defined by the fluid velocity and cavitation number. A general form of the latter is given in the earlier contribution by Godfrey. At constant velocity, with the cavitation number reducing from the value for inception, the erosion rate increases to a maximum and then decreases. At a constant cavitation number, however, the erosion rate varies with powers of velocities ranging from 3 to 10 although a narrower range of 6 to 8 is normally quoted. Harder or more brittle materials tend to have a higher power dependency than ductile materials. Factors which influence erosion rate have been considered in a review by Hutton[19].

Erosion resistance is also dependent upon material properties but, despite the efforts of many researchers, no clear correlation with a single property has emerged. Lush[14], in the light of recent knowledge, analysed the results of Mousson[20] who measured the erosion resistance of many materials in relation to their yield and tensile strengths, ductility and hardness. It appears that, for a particular type of cavitation and intensity, erosion resistance correlates well with hardness for related materials. It is also found that resilient materials of low hardness have a high erosion resistance similar to stainless steel, as indicated earlier, and ductile materials of a given hardness are more resistant than brittle materials of similar hardness. These anomalies and the importance of both ductility and resilience led to a correlating parameter termed the 'ultimate resilience' (UR) given by (tensile strength)/(2 × Young's modulus). Lush concluded that the correlation of erosion resistance with UR is good, but not significantly better than with hardness. A similar conclusion was reached by Hammitt[21], and it would appear that in the range of general engineering alloys a good correlation is obtained where a power of hardness in the range 2.0 to 2.5 is used.

Cavitation in Pumps

Depending on its severity, cavitation in pumps can result in loss of performance, severe erosion, vibration and noise. All these effects may be minimised by attention to design and operation, and by prudent use of erosion-resistant materials. Pumps vary considerably in design and function, and it is convenient to use the centrifugal pump to illustrate cavitation problems because of its common usage in fluid systems.

The cavitation performance depends, to a large extent, on the conditions at the pump inlet or suction. These are defined by the parameter 'net positive suction head' (NPSH):

$$\text{NPSH} = h_a - h_v - h - h_f \tag{8.16}$$

where h_a is the head due to atomspheric pressure, h is the height that the pump is required to lift the fluid, and h_f is the total head loss due to fluid flow in the suction pipework. Pump pressures are conventionally expressed in height of pumped fluid (head) above a datum point. The NPSH, therefore, represents the total head available at the pump suction with reference to the head corresponding to vapour pressure, h_v. To avoid the more serious effects of cavitation the NPSH available at the pump suction must exceed the NPSH required by the pump by the greatest margin possible. NPSH_R is a function of the pump design and increases with increasing flow.

A pump is normally supplied to achieve a specified flow rate by generating sufficient head to overcome the resistance of the system[22]. Figure 8.79a shows typical pump characteristics supplied by a manufacturer in attempting to meet the duty conditions required with optimum efficiency. The rate at which the generated head is reduced with increasing flow rate is determined by the NPSH available. For example, if the initial NPSH of a pump installation is low it rapidly approaches the NPSH required as flow increases until a critical point is reached beyond which cavitation causes the impeller passages to become filled with an air/water mixture (referred to as 'choking') and performance deteriorates rapidly. This point is defined by a measured fall in head (usually 3%) relative to the non-cavitating value. The onset of cavitation occurs at a much higher value of NPSH than the critical value and erosion rate and noise level will have reached a peak and declined before the 3% head drop point is reached. Figure 8.79b shows the zone over which erosion may occur when NPSH is reduced, and illustrates how the problem is exacerbated when the pump is operated away from the design flow rate. This is supported by Fig. 8.79c which shows, for constant pump speed and flow rate the considerable difference in NSPH between cavitation inception and performance breakdown. If freedom from serious erosion is required NPSH should be at least three times the 3% head drop value at the design flow rate (assumed to be the best efficiency condition). For flow rates lower or higher than the design value this should be increased to 6 and 12 times, respectively. It is important to appreciate that severe erosion damage may be occurring without significant loss of performance and that operating conditions should be chosen so that the pump is near to the point of maximum efficiency.

In practice it is expensive, and therefore uneconomic, to produce a pump which operates completely free from cavitation. As a result it is usual for commercial pumps to operate in the NPSH range between inception and a point where erosion damage is unacceptable. The extent of this range may be increased by using impellers made from the more resistant materials shown in Fig. 8.77. The subject of cavitation in pumps has been dealt with extensively by Pearsall[3] and Grist[23].

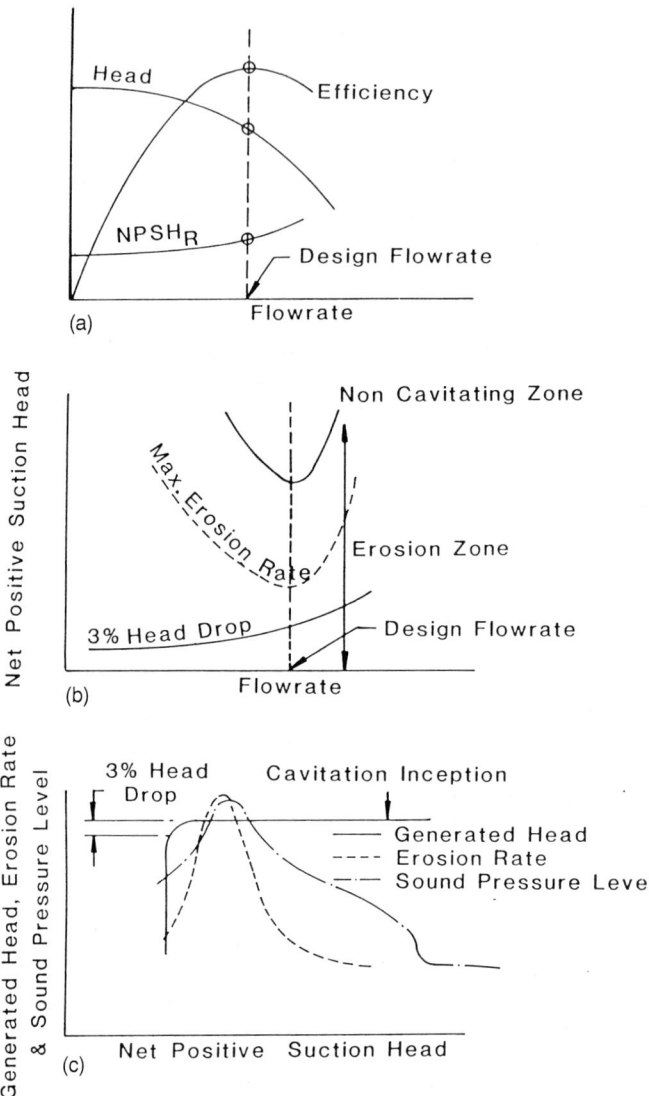

Fig. 8.79 (a) Pump performance characteristics. (b) Cavitation erosion zone. (c) Effect of reducing net positive suction head

Reducing Cavitation in Valves and Orifice Plates

Cavitation in pipe systems is possible wherever there are changes in section or flow direction such as expansions, bends and branches. However, serious erosion problems are normally only associated with components within which flow is severely constricted and consequently accelerated. If pumps are excluded then, in most systems, this situation applies to devices used to control the fluid flow, namely orifice plates and various types of valve.

Lush[24] has proposed cavitation criteria for these components using the empirical data of Tullis and Ball[25] and Boccadoro and Angell[26]. The cavitation index used is based on conditions at the throat of a valve and, correspondingly, the 'vena contracta' of an orifice plate.

$$\sigma_T = \frac{p_T - p_v}{\frac{1}{2}\rho U_T^2} \qquad (8.17)$$

where the subscript T infers throat conditions. Unfortunately neither the pressure p_T nor the velocity U_T can conveniently be measured, but it is found, using Bernoulli's equation, that σ_T is related to the upstream condition denoted by the subscript 0.

$$\sigma_O + 1 = (\sigma_T + 1)(A_O/A_T)^2 \qquad (8.18)$$

A_O and A_T, are the areas of the pipe and valve throat, respectively, at a particular point in the valve opening range. As with the velocity and pressure at the throat, the area A_T cannot usually be measured. However σ_T is also related to the upstream cavitation index σ_O when the flow in the valve is 'choked'. In the valve context this is the condition which exists when further reduction of the downstream pressure fails to increase the flow through the valve. The value of σ_T is given by the expression:

$$\sigma_T = \frac{\sigma_O - \sigma_{Och}}{1 + \sigma_{Och}} \qquad (8.19)$$

Now it appears that the value of σ_{Och} may be estimated by using the loss coefficient K determined at choking provided K is not too small. This is unlikely since in most valves effective flow control occurs at very small throat area when the valve is in the 10–30% open range. The loss coefficient is determined from the pressure loss across the valve and the velocity in the upstream pipe at choking.

$$K_{ch} = \frac{\text{Pressure loss across the valve at choking}}{\frac{1}{2}\rho U_O^2} \qquad (8.20)$$

In the analysis of published data on choking conditions in valves of several different types in the size range 50–250 mm, Lush found that when σ_{Och} was related to K_{ch} by the following expression good agreement was obtained:

$$\sigma_{Och} = 2\sqrt{K_{ch}} + K_{ch} \qquad (8.21)$$

The final estimation of the value of σ_T may appear tedious and several assumptions are made in its derivation, but experimental evidence suggests that it may be used with reasonable accuracy to assess the levels of potentially damaging cavitation erosion. In small valves with nominal bores up to 65 mm cavitation inception occurs in intermittent bursts when the value σ_T is approximately unity. The cavitation becomes continuous and audible as σ_T is reduced to about 0.6, but the risk of damage does not become significant until the value falls below 0.4. As a design criterion the condition of light, steady noise has been described by Tullis as the 'critical' level and is suggested for applications where a limited amount of noise may be tolerated, but unacceptable noise, erosion and vibration are avoided. The critical value of σ_T increases with valve size and the magnitude of this increase, together with that of the inception value, is shown in Fig. 8.80.

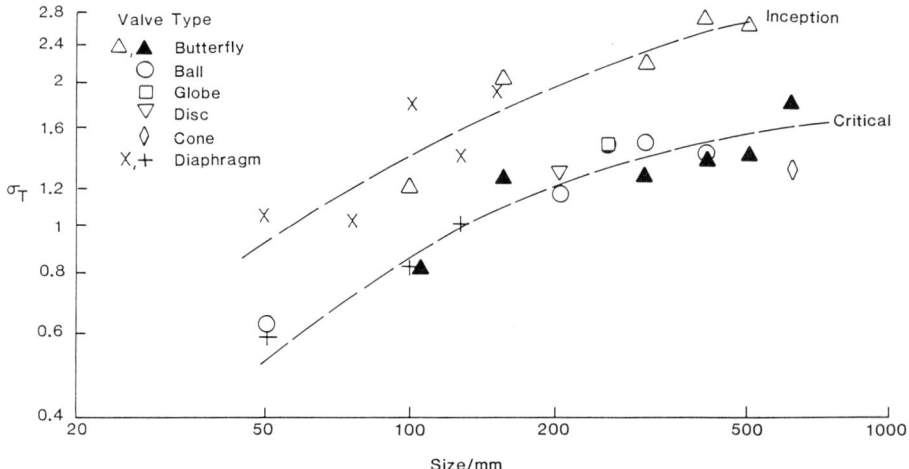

Fig. 8.80 Inception and critical σ_T values for valves including data from Tullis

Orifice plates may be treated in the same way as valves using an appropriate value of σ_T to determine the ratio A_O/A_T which provides an optimum flow control characteristic within acceptable cavitation limits.

B. ANGELL

REFERENCES

1. Burrill L. C., *Trans. Inst. Marine Engrs.*, London, **63** (8) p. 149-167 (1951)
2. Knapp, R. T., Daily, J. W. and Hammitt, F. G., *Cavitation*, McGraw-Hill (1970)
3. Pearsall, I. S., *Cavitation*, Mills and Boon, London (1972)
4. Preece, C. M., *Cavitation Erosion, Treatise on Material Science and Technology*, Vol 16, *Erosion*, Academic Press, London (1979)
5. Hammitt, F. G., *Cavitation and Multiphase Flow Phenomena*. McGraw-Hill, New York (1980)
6. Acosta, A., O'Hern, T. and Katz, J., *International Symposium on Cavitation*, Sendai, Japan p. 1-7 (April 1986)
7. Suhrbier, K. R. and Lecoffre, Y., *ibid*, p. 235-260
8. Arndt, R. E. A., *Annu. Rev. Fluid Mech.*, **13**, p. 273-327 (1981)
9. Plesset, M. S. and Prosperetti, A., *Annu. Rev. Fluid Mech.*, **9**, p. 145-185 (1977)
10. Morch, K. A., *Dynamics of Cavitation Bubbles and Cavitating Liquids, Treatise on Materials Science and Technology*, Vol 16, *Erosion*, Academic Press, London, p. 309-353 (1979)
11. Travena, D. H., *J. Physics D: Applied Physics*, **17**, p. 2139-2164 (1984)
12. Lush, P. A., *J. Fluid Mech.*, **13**, p. 373-387 (1983)
13. Grant, M. McD. and Lush, P. A., *J. Fluid Mech.*, **176**, p. 237-252 (1987)
14. Lush, P. A., *Chartered Mechanical Engineer* (October 1987)
15. Kenn, M. J. and Garrod, A. D., *Proc. Inst. Civil Engrs.*, Part 1, **70**, p. 65-79 (1981)

16. Angell, B., Long, R. F., Weaver, W. R. and Hibbert, J. H., *Proc. 5th Int Conf Erosion by Solid and Liquid Impact*, Cambridge, p. 75-1–75-8 (1979)
17. Lightarowicz, A., *Cavitating Jet Apparatus for Cavitation Erosion Testing, Erosion: Prevention and Useful Applications*, ASTM STP 664, p. 530–549 (1979)
18. Thiruvengadam, A. and Preisher, H. S., *J. Ship Research*, **8**, p. 39–56 (1964)
19. Hutton, S. P., *Proc. ASME Symposium on Cavitation Erosion in Fluid Systems* Boulder, Colorado, USA (1981)
20. Mousson, J. M., *Trans ASME*, **59**, p. 399 (1937)
21. Hammitt, F. G., *Applied Mechanics Reviews*, **32** (6) p. 665–675 (1979)
22. Miller, D. S., *Internal Flow Systems*, BHRA Fluid Engineering (1978)
23. Grist, E., *Proc. Conference on Cavitation*, Institute of Mechanical Engineers, Edinburgh, p. 153–162 (1974)
24. Lush, P. A., *Chartered Mechanical Engineer*, p. 22–24 (September 1987)
25. Tullis, J. P. and Ball, J. W., *Proc. Conference on Cavitation*, Institute of Mechanical Engineers, Edinburgh, p. 55–63 (1974)
26. Boccadoro, Y. and Angell, B., *ibid*, p. 253–259
27. Godfrey, D. J., 'Cavitation Damage', in Shreir, L. L. (ed.), *Corrosion* 2nd edn, Butterworths, pp 8:124–8:132 (1976)

8.9 Outline of Fracture Mechanics

There are a number of fracture modes, the most important of which are ductile overload, which is fairly well understood and can be predicted reasonably accurately, and brittle fracture, which is less predictable from an engineering viewpoint and can cause catastrophic failures due to the speed of the fracture.

The early study of brittle failures, notably those of the Liberty ships, indicated a temperature dependence. This can be illustrated by plotting both fracture stress (σ_f) and yield stress (σ_y) against temperature (Fig. 8.81). Below a certain temperature some materials exhibit a transition from ductile to brittle fracture mode. This temperature is known as the ductile–brittle transition temperature DBTT.

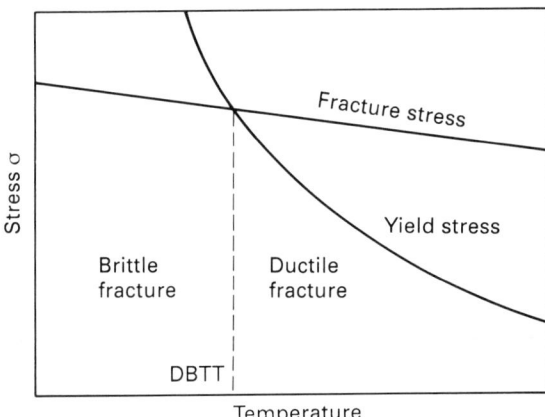

Fig. 8.81 The temperature dependence of fracture stress and yield stress

Materials with a high yield stress tend to go through the ductile to brittle transition at higher temperatures. This property has led to the assumption that true brittle fracture, unlike ductile fracture, is not accompanied by the motion of dislocations. The validity of this assumption is sometimes confirmed by the appearance of brittle fractures, which show essentially no ductility.

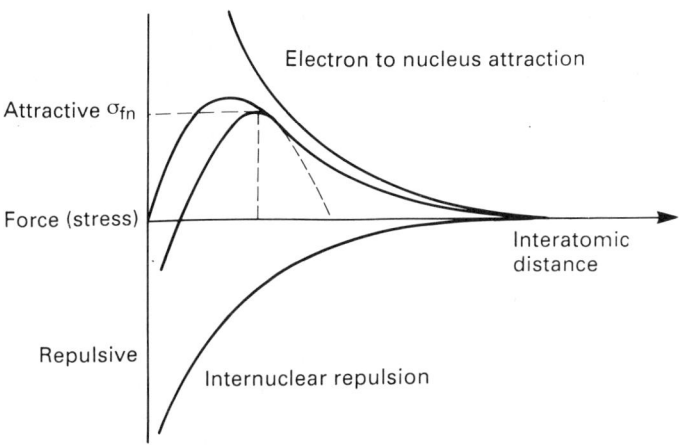

Fig. 8.82 The variation of interatomic force with atomic separation

Brittle fracture may be considered, therefore, as two layers of atoms being pulled apart until the interatomic forces fall below their maximum (Fig. 8.82). Using this information it is possible to calculate the fracture strength of a perfect crystalline solid (σ_{th}), e.g.

$$\sigma_{th} = \sqrt{\frac{E\gamma}{a_0}} \qquad (8.22)$$

where a_0 is the lattice parameter, E is Youngs modulus, and γ is the surface energy. This theoretical fracture strength was found to be approximately two orders of magnitude higher than those measured experimentally in engineering materials. This deviation from the ideal case was attributed to defects within the solid causing localised areas of stress concentrations.

Inglis extended this idea of stress concentrations at defects and derived an equation relating the maximum stress to the size and shape of the defect[1]:

$$\sigma_{max} = \sigma_{app}\left(1 + 2\sqrt{\frac{a}{p}}\right) \qquad (8.23)$$

where σ_{max} is the maximum stress at the crack tip, σ_{app} is the applied stress, a is the half-crack length, and p is the crack-tip radius. This idea was used by Orowan when deriving the fracture strength of a defective solid[2]. He proposed that the sharpest crack would have a crack tip radius of a_0, the lattice parameter. This led to the following equation:

$$\sigma_f = \sqrt{\frac{E\gamma}{4a}} \qquad (8.24)$$

Griffith[3] derived a similar equation using an energy balance approach, equating stored energy with the energy required for crack propagation:

$$\sigma_f = \sqrt{\frac{2E\gamma}{\Pi a}} \qquad (8.25)$$

Equations 8.24 and 8.25 only apply to elastically brittle solids such as glass. However, many engineering materials only break in a truly brittle manner at very low temperature and above these temperatures failures are pseudo-brittle. These have many of the features of brittle fracture but include limited ductility. This plastic work can be included in the above equations, i.e.

$$\sigma_f = \sqrt{\frac{EG}{\Pi a}} \qquad (8.26)$$

where $G = 2\gamma + p$, p being the energy used to cause the plastic deformation during fracture.

This plastic deformation is localised around the crack tip and is present in all stressed engineering materials at normal temperatures. The shape and size of this plastic zone can be calculated using Westergaards analysis[4]. The plastic zone has a characteristic butterfly shape (Fig. 8.83). There are two sizes of plastic zone. One is associated with plane stress conditions, e.g. thin sections of materials, and the other with plane strain conditions in thick sections–this zone is smaller than found under plane stress.

$$r_p^x(\text{plane stress}) = \frac{a}{2}\left[\frac{\sigma_{app}}{\sigma_y}\right]^2 \qquad (8.27)$$

$$r_p^x(\text{plane strain}) = 0.16\frac{a}{2}\left[\frac{\sigma_{app}}{\sigma_y}\right]^2 \qquad (8.28)$$

Where r_p^x is the plastic zone size along the plane of the crack and perpendicular to the applied stress. The above two equations are only a first approximation as the plastic zone contributes to the crack size. The true plastic zone is twice the initial approximation.

The critical crack size is related to the fracture toughness of the material, which can be characterised by the value of the stress-intensity factor, K_{IC}, when the crack begins to propagate. K_{IC} is also known as the plane strain fracture toughness and defines the stress field at a crack tip. Therefore once K_{IC} has been found for a material, the critical crack length, a_{crit}, or the critical fracture strength, σ_{crit} can be found in other geometries provided the equation relating a_{crit}, σ_{crit} and K_{IC} has been defined. The equations usually take the form:

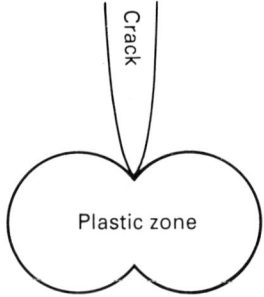

Fig. 8.83 Stressed crack tip with associated plastic zone

$$K_1 = Y\sigma\sqrt{\Pi a} \tag{8.29}$$

where Y is a geometric function of crack length and specimen width.

There are a number of restrictions on the test for K_{IC} for it to be a valid measure of plane strain fracture toughness. Firstly, the plastic zone must not extend through the test piece and secondly the thickness of the material must be such that the test is conducted under plane strain conditions.

The K_{IC} test for fracture toughness is a valuable test for high-strength materials that behave in a brittle manner at ambient temperatures. However, many materials such as mild steel are ductile and as such require very large test specimens. Such test results cannot be used in conjunction with normal engineering structures using reasonable-sized section thickness. To cover this class of materials a branch of fracture mechanics called general yielding fracture mechanics, GYFM, was devised.

One way of looking at the fracture characteristics of a ductile material is by measuring the amount of plasticity at a crack tip prior to crack propagation (Fig. 8.84). One test which measures this is the crack-tip opening displacement (CTOD), δ. Wells[5] has found that δ can be related to the strain energy release rate, G, by the formula:

$$G = \frac{4}{3}\sigma_y\delta \tag{8.30}$$

A similar result was found by Burdekin and Stave[6]. For this test to be valid, the material tested must be of the same thickness as the structure being analysed. This is due to the effect of material thickness on the toughness versus temperature characteristics (Fig. 8.85).

A more universal fracture characteristic for use with ductile materials is the 'J integral'. This is similar to CTOD but relates a volume integral to a surface integral and is independent of the path of the integral[7]; it can be classed as a material property. The J integral can also be used to predict critical stress levels for known crack lengths or vice versa.

Another useful way of predicting the way and rate at which a crack will grow given known conditions is the R curve method used for plane stress studies[8]. A number of tests are required to generate the R curve (Fig. 8.86) which is the graph of G versus crack growth. At low stress, σ_1, no crack growth can occur since the available energy release rate is insufficient. Increasing the stress to σ_2, the available energy equals the critical strain

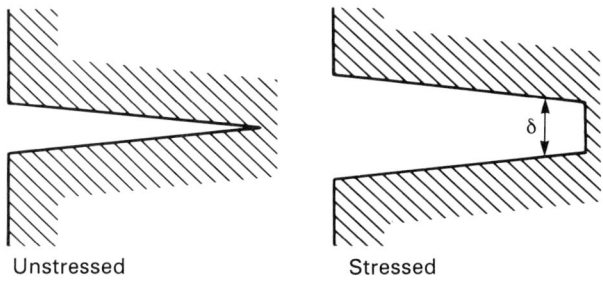

Fig. 8.84 Crack tip opening displacement

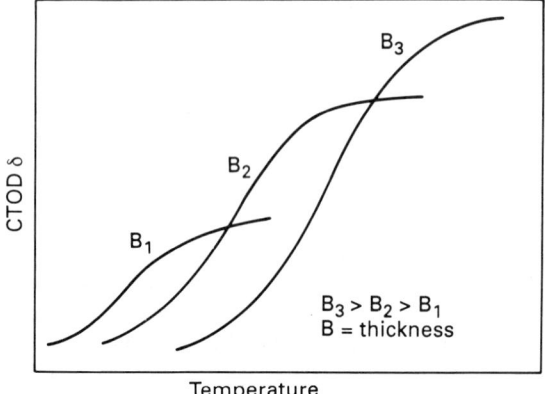

Fig. 8.85 The variation of crack tip opening displacement with temperature

energy release rate and the material fails by fast fracture. However, if the material is in plane stress the crack would not run due to lack of energy but by increasing the stress the crack would grow in a slow stable manner until the applied stress is equal to σ_4 when the crack has grown to critical size and catastrophic failure occurs.

Another important class of fracture is fatigue which is caused by an oscillating stress system. Under these conditions a subcritical crack can grow by discrete steps during each cycle until the crack becomes critical and runs in the normal manner. Three parameters are required to describe the fluctuating stress pattern: the mean stress, the stress range and the period of oscillation. It has been found that some materials exhibit a fatigue limit, i.e. a stress below which no crack growth can occur (Fig. 8.87).

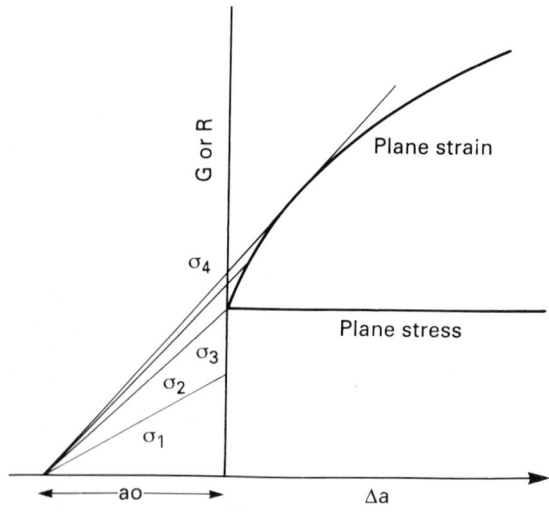

Fig. 8.86 Typical crack resistance R curve

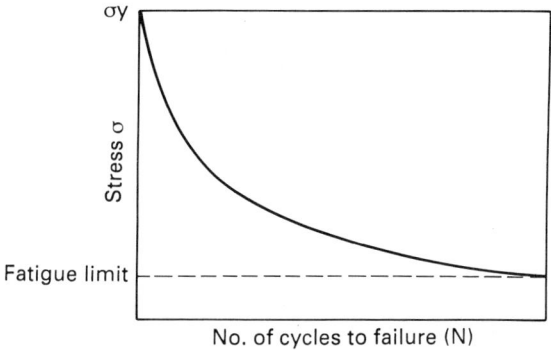

Fig. 8.87 Typical fatigue properties on an S–N curve

An empirical relationship between the rate of crack growth and the stress range has been formulated:

$$\frac{da}{dN} = C(\Delta K)^m \qquad (8.31)$$

where C and m are material constants, $\dfrac{da}{dN}$ = crack growth per cycle and $\Delta K = K_{max} - K_{min}$ where $K_{max} = \sigma_{max}\sqrt{\Pi a}$ and $K_{min} = \sigma_{min}\sqrt{\Pi a}$. Using the above formula, crack growth rates and structure lives can be predicted. If the structure is subjected to a number of different oscillations the effect on the life to failure can be approximated by using the Miner rule:

$$\frac{n_1}{N_1} + \frac{n_2}{N_2} + \ldots \sum \frac{n}{N} = 1 \qquad (8.32)$$

where N_1 is the number of cycles to failure at stress level σ_1 and n_1 is the actual number of cycles endured at that stress.

Materials subjected to high temperatures during their service life are susceptible to another form of fracture which can occur at very low stress levels. This is known as creep failure and is a time dependent mode of fracture and can take many hours to become apparent (Fig. 8.88).

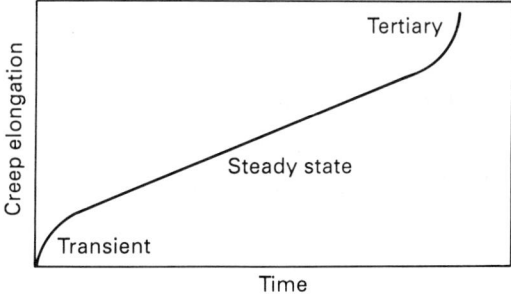

Fig. 8.88 Creep deformation curve

Creep deformation can be split into three separate parts. The first, transient creep, is a short lived phenomenon which gives a high initial rate of deformation but decays according to the expression:

$$\varepsilon_t \propto \beta t^{\frac{1}{2}} \tag{8.33}$$

This is then replaced by a period of steady state creep which, depending on the temperature and stress level, takes up the greatest part of the creep life.

$$\varepsilon_{ss} \propto \gamma t \tag{8.34}$$

β and γ are constants whilst t is time of exposure.

Creep rate ($\dot{\varepsilon}$) during this period can be predicted from the following equations:

$$\dot{\varepsilon} = \text{const. exp}\left[\frac{-Q}{RT}\right] \tag{8.35}$$

for constant stress, and

$$\dot{\varepsilon} = \text{const. exp}\left[-\text{const.}\frac{\sigma}{RT}\right] \tag{8.36}$$

for constant temperature, where Q is the activation energy for the process.

The final stage of creep is known as tertiary creep at which time the creep rate increases rapidly culminating in failure. This acceleration in creep is due mainly to the formation of voids and microcracks in the material which form along the grain boundaries causing the fracture path to be predominantly intercrystalline.

All the above modes of fracture are affected by the environment around the crack tip. This behaviour is typified by the phenomenon of stress-corrosion cracking where a crack, which is subjected to a subcritical stress concentration, will grow in a corrosive environment when $K_1 \geqslant K_{Iscc}$ (K_{Iscc} is the critical stress concentration for stress-corrosion cracking). Therefore, to predict accurately the occurrence of cracking and crack growth rate, not only the materials properties are required but also information on the immediate environmental conditions.

G. N. KING
R. A. JARMAN

REFERENCES

1. Inglis, C. E., *Trans. Inst. Naval Arch., London*, **IV**, 219 (1913)
2. Orowan, E., *Rep. Prog. Phys.*, **12**, 185 (1955)
3. Griffith, A. A., *Phil. Tran. Roy. Soc.*, **A221**, 163 (1921)
4. Westergaard, H. M., *J. Appl. Mech.*, **A49**, 61 (1939)
5. Wells, A. A., *Crack Propagation Symp.*, Cranfield (1961)
6. Burdekin, F. M. and Stave, D. E. W., *J. Strain Anal*, **1**, 2 (1966)
7. Rice, J. R., *J. Appl. Mech. Trans ASME*, 379 (1968)
8. Heyer, R. H., *Fracture toughness evaluation by R-curve methods*. ASTM, STP 527

BIBLIOGRAPHY

Knott, J. F., *Fundamentals of Fracture Mechanics*. Butterworths (1973)
Jayatilaka, Ayal de S., *Fracture of Engineering Brittle Materials*. Applied Science Publishers

8.10 Stress-corrosion Test Methods

Stress-corrosion cracking results from the interactions, in a critically interdependent manner, between an environment, a metal and the response of the latter to the application of an appropriate stress. Recognition of these conjoint requirements has frequently led to the use of tests that attempt to simulate a practical situation, especially with regard to the structure and composition of the material, but less frequently in the manner in which the stress is generated in the testpiece and in achieving representative environmental conditions.

Stressing Systems

Many different methods[1-4] have been used in stressing testpieces, from which it may be reasonably assumed that there is no single method that is markedly superior to all others. Each method may have its peculiar advantages in a given situation, but, ideally, a test method should not be so severe that it leads to the condemnation of a material that would prove adequate for service, or so trifling as to permit the use of materials in circumstances where rapid failure ensues. Methods of stressing testpieces, whether initially plain, notched or precracked, can be conveniently grouped according to whether they involve:

1. a constant total strain or deflection;
2. a constant load;
3. an imposed strain or deflection rate.

Constant deflection tests usually have the attraction of employing simple, and therefore frequently cheap, specimens and straining frames and of simulating the fabrication stresses that are most frequently associated with stress-corrosion failure. Constant load tests may simulate more closely failure from applied or working stresses. Tests involving the application of a constant deflection rate have become fashionable in recent times but their relevance to service failures continues to be debated.

Constant Total-deflection Tests

Prismatic beams stressed by bending offer a simple means of testing sheet or plate material, typical arrangements being shown in Fig. 8.89a to e. Below the elastic limit the stresses may be calculated[1,5] or determined from the response of strain gauges attached to the surface at an appropriate position.

Plastic bending of strip specimens to produce a 'U' bend (Fig. 8.89d and e) will usually allow the use of a lighter restraining system, although some of the effects of the plastic deformation, if not removed by subsequent heat treatment, may be to influence cracking response and the stress obtained in the outer fibres of the specimen is usually less reproducible than with more sophisticated specimens. Tubular material may be tested in the form of 'C' or 'O' rings, the former being stressed by partial closing of the gap (Fig. 8.89c) and the latter by the forced insertion of a plug that is appropriately oversized for the bore. The circumferential stress at the outer surface of a 'C' ring is maximal midway between the bolt holes, but for the 'O' ring it is constant over the periphery, the stresses being readily calculated in terms of measured deflections[1,2].

Constant-deflection tensile tests (Fig. 8.89f and g) are sometimes preferred to bend tests, but for similar cross sections require a more massive restraining frame. In principle this problem may be surmounted by the use of internally stressed specimens containing residual stresses as the result of inhomogeneous deformation. The latter may be introduced by plastic bending, e.g. by producing a bulge in sheet or plate material, or by welding, but such tests provide problems in systematic variation of the initial stress, which will usually be in the region of the yield stress. Moreover, elastic spring-back, in introducing residual stresses by bulging plate or partially flattening tube, may introduce problems, and where welding is involved the structural modifications may raise difficulties unless the test is simulative of a practical situation.

At least as important as the choice of methods of stressing is the realisation of the limitations of the various methods, these having been considered in a review of stress-corrosion test methods[6]. The stiffness of the stressing frame in constant-deflection tests may influence results because of relaxation in the specimen during the initial loading stage and during subsequent crack propagation. Especially in testing ductile materials, the initial elastic strain is converted in part to plastic strain, even if the total deflection remains constant during cracking. This is because as the crack propagates the stress increases on the remaining uncracked portion of the specimen section beyond the crack, eventually reaching the effective yield stress. Yielding will then occur, accompanied by yawning of the crack and frequently with the propagation of a Lüders band that results in a sharp load drop, which is sometimes mistaken as an indication of the crack having advanced by a burst of mechanical fracture. Once load relaxation has been initiated the extent to which it proceeds can vary from specimen to specimen. Thus, Fig. 8.90 shows load relaxation curves for two specimens of the same maraging steel in the same stressing frame, which had a facility for load recording throughout the test. The specimens differed in the extent to which they showed load relaxation prior to sudden fracture, this difference being related to the number of cracks that developed in the specimens. Marked load relaxation

STRESS-CORROSION TEST METHODS

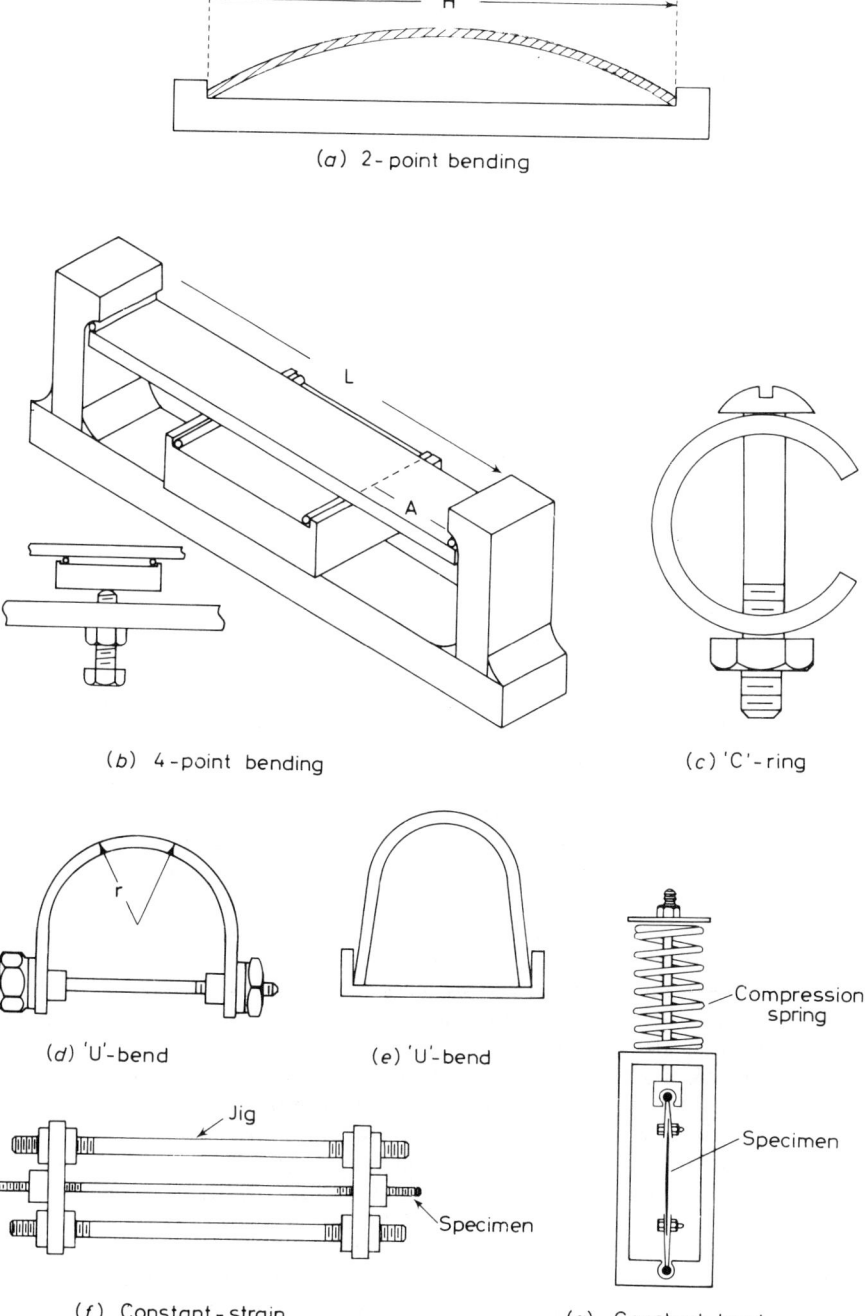

(a) 2-point bending

(b) 4-point bending

(c) 'C'-ring

(d) 'U'-bend

(e) 'U'-bend

(f) Constant-strain tensile test

(g) Constant-load tensile test

Fig. 8.89 Stressing systems for stress-corrosion test specimens; (a)–(f) constant strain, (g) constant load

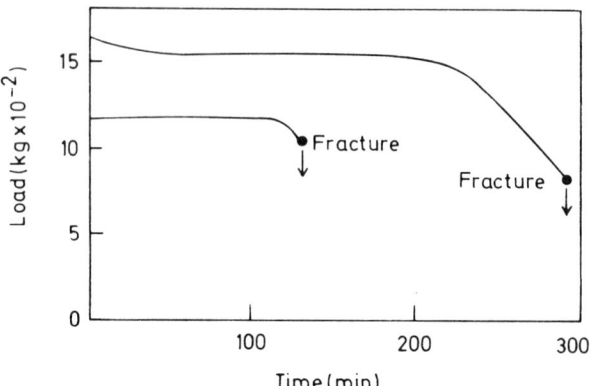

Fig. 8.90 Load relaxation curves for a maraging steel stress corroded in 0.6 M NaCl at pH 2 (after Parkins and Haney[7])

was associated with the development of many cracks in the specimen, and little relaxation with only a few cracks. This can influence the time to failure, as is apparent from Fig. 8.90, where the specimen stressed at the initially higher load took longer to fail than that at the initially lower load. This is because, when only a single stress-corrosion crack develops, it will not need to grow to large dimensions before sudden, final failure occurs, since the applied load remains high, whereas with the marked load relaxation associated with the multi-cracked specimen one of the cracks will need to propagate much further before it reaches the size for sudden fracture at the reduced load. Such an explanation conforms to the observations[7] that the load at fracture is related to the area of stress-corrosion cracking upon the final fracture surface and to the number of cracks initiated.

This type of result will depend upon the nature of the stress-corrosion system being studied, i.e. upon such properties as the fracture toughness of the material and even upon the aggressiveness of the environment employed. It will also vary according to the stiffness of the restraining jig employed, since the stiffer the frame the less the elastic strain that is likely to remain in the specimen after the propagation of a Lüders band, so that a stress-corrosion crack may cease to propagate in some circumstances, especially if the initial stress is in the vicinity of the threshold stress. This indicates some of the dangers inherent in comparing stress-corrosion resistances in terms of times to failure at a given initial stress, an approach that is often practised but which can be misleading. Figure 8.91 shows the results from some tests in which the time to failure of specimens previously cold worked in varying amounts is plotted against initial stress. Comparison of the effects of different amounts of cold work by tests at an initial stress of 280 N/mm^2 or 155 N/mm^2 gives different orders of susceptibility, as shown in Table 8.5. It could be argued that neither of these results is correct because the prior cold work would result in different yield strengths being developed in the three different conditions and that the results should be rationalised by making the comparison as a function of the respective yield strengths. Here

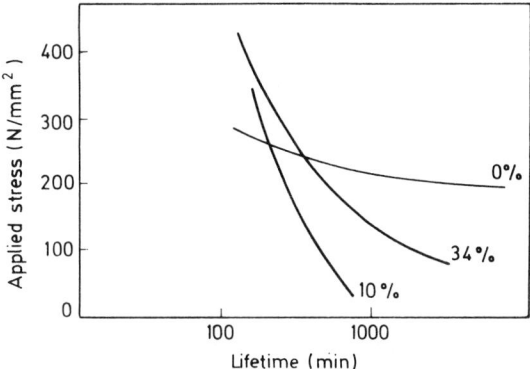

Fig. 8.91 Effects of different amounts of prior cold work (0, 10 and 34%) on the stress corrosion of a 0.07% C steel in boiling 4 M NH_4NO_3

Table 8.5 Relative susceptibilities to cracking of a mild steel in boiling 4 N NH_4NO_3 after various amounts of cold work

Initial stress	Susceptibility of different cold worked conditions		
	Most (%)	Intermediate (%)	Least (%)
280 N/mm^2	0	10	34
155 N/mm^2	10	34	0
100% YS	34	10	0
30% YS	10	34	0

again, however, the order of susceptibility varies according to the rationalised stress at which the comparison is made, as the results in Table 8.5 show. It is difficult to escape the conclusion that a more satisfactory basis of comparison is the threshold, but even the latter may not be a basis for comparison of results obtained using different restraining frames.

The simplicity of the rigs used in the constant-strain tests is an advantage in the application of the corrosive solution. Thus, in the case of two-point bending (Fig. 8.89a) several specimens may be strained in the same rig which can be constructed of plastic and immersed in a tank containing the test solution.

Constant-load Tests

Dead-weight loading (with or without the assistance of levers to reduce the load requirements) of tensile specimens has the advantage of avoiding some of the difficulties already discussed, not the least in allowing accurate determination of the stress if the specimen is uniaxially loaded. The relatively massive machinery usually required for such tests upon specimens of appreciable cross section is sometimes circumvented by the use of a compression spring (Fig. 8.89g) chosen with characteristics that ensure it does not change significantly in length during testing and thereby approximates to a constant-load application. For immersion tests the frame may be coated in PVC and the specimen insulated from the shackles by plastic sleeves and washers to avoid bimetallic effects; alternatively, the specimen may be enclosed in a glass cell containing the test solution. The alternative approach of minimising the size of the loading system by reducing the cross section of the specimen to the dimensions of a wire is dangerous unless failure by stress-corrosion cracking is confirmed by, say, metallography. This is because failure may result from pitting and an attendant increase in the effective stress to the UTS in some stress-corrosion environments. Indeed there is evidence for some systems that before stress-corrosion cracking proper can begin, a pit must form wherein certain chemical or electrochemical conditions are established that permit cracks to be initiated, and in such systems the use of fine wires has obvious pitfalls.

The load relaxation that accompanies some, if not all, constant-deflection tests is replaced in constant-load tests by an increasing stress condition, since the effective cross section of the testpiece is reduced by crack propagation. This suggests that it will be less likely that cracks will cease to propagate once initiated, as may happen with constant deflection tests at initial stresses in the region of the threshold stress, and therefore that threshold stresses are likely to be lower when determined under constant-load conditions than under conditions of constant deflection. Some results due to Brenner and Gruhl[8] for an aluminium alloy (Fig. 8.92) confirm this expectation. These results also show shorter times to failure for the same initial stress with constant load testing and, as already indicated for constant-deflection tests, raise queries as to the significance of time to failure, the parameter so frequently used in assessing cracking susceptibility.

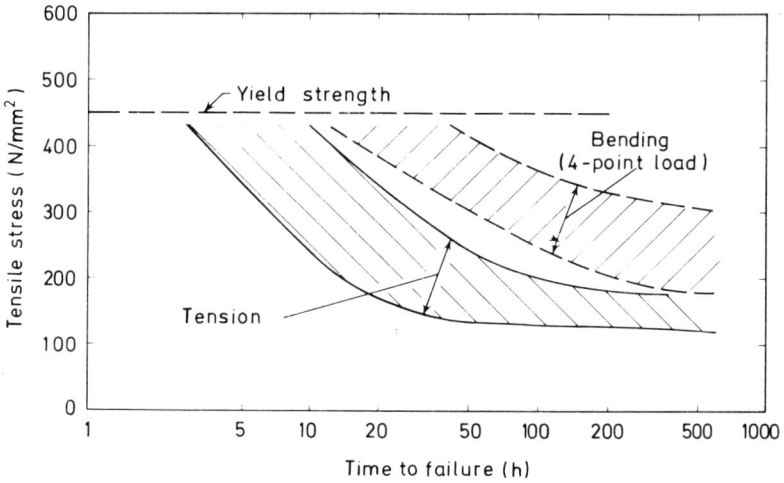

Fig. 8.92 Comparison of test results from bend and tension tests upon Al–Zn–Mg alloy in 3% NaCl plus 0.1% H_2O_2 (after Brenner and Gruhl[8])

Slow Strain-rate Tests

While this method of testing has been in use in some laboratories for two decades or more, and has increased in use considerably in very recent years, there remains some scepticism and unfamiliarity with the method. In essence it involves the application of a relatively slow strain or deflection rate (approx $10^{-6} s^{-1}$) to a specimen[9] subjected to appropriate electrochemical conditions. It should be emphasised that the strain rates employed are very much lower than those involved in straining electrode experiments where the object, the measurement of current transients, is totally different. In slow strain-rate corrosion tests the object is to produce stress-corrosion cracks that, metallographically, are indistinguishable from those produced in constant-load or constant-deflection experiments. The object in all these laboratory tests is normally to obtain data in a relatively short period of time and this is frequently achieved by adopting an approach that increases the severity of the test. In stress-corrosion testing this usually takes the form of increasing the aggressiveness of the environment by changing its composition, temperature or pressure, stimulating the corrosion reactions (galvanostatic or potentiostatic polarisation), increasing the susceptibility of the alloy through changes in structure, or increasing the severity of the stress by the introduction of a notch or precrack. The application of dynamic straining to a stress-corrosion test specimen comes into this last category also, and, like all of the other accelerating approaches, its justification will vary according to the circumstances in which it is used.

Most stress-corrosion crack velocities fall in the range from 10^{-3} to 10^{-6} mm s^{-1}, which implies that failures in laboratory test specimens of usual dimensions occur in not more than a few days. This is found to be so if the system is one in which stress-corrosion cracks are readily initiated, but it is common experience to find that some testpieces do not fail in very extended periods of testing, which are then terminated at some arbitrarily selected time. The consequences are that considerable scatter may be associated with replicate tests and the arbitrary termination of the test leaves an element of doubt concerning what the outcome would have been if it had been allowed to continue to a longer time. Just as the use of precracked specimens assists in stress-corrosion crack initiation, so does the application of slow dynamic strain, which has the further advantage that the test is not terminated after some arbitrary time, since the conclusion is always achieved by the specimen fracturing and the criterion of cracking susceptibility is then related to the mode of fracture. Thus, in the form in which it is normally employed the slow strain-rate method will result in failure in not more than about two days, either by ductile fracture or by stress-corrosion cracking, according to the susceptibility towards the latter, and metallographic or other parameters may then be assigned in assessing the cracking response. The fact that the test concludes in this positive manner in a relatively short period of time constitutes one of its main attractions.

Early use of the test was in providing data whereby the effects of such variables as alloy composition and structure or inhibitive additions to cracking environments could be compared, and also for promoting stress-corrosion cracking in combinations of alloy and environment that could not be caused to fail in the laboratory under conditions of constant load or constant strain. Thus, they constitute a relatively severe type of test in the sense that they frequently promote stress-corrosion failure in the laboratory where other modes of stressing plain specimens do not promote cracking, and in this sense they are in a similar category to tests on precracked specimens. In recent years an understanding of the implications of dynamic strain testing has developed and it now appears that this type of test may have much more relevance and significance than just that of an effective, rapid, sorting test. It may, at first sight, be argued that laboratory tests involving the pulling of specimens to failure at a slow strain rate show little relation to the reality of service failures. In point of fact, in constant-strain and constant-load tests crack propagation also occurs under conditions of slow dynamic strain to a greater or lesser extent depending upon the initial value of stress, the point in time during the test at which a stress-corrosion crack is initiated and various metallurgical parameters that govern creep in the specimen. Moreover, there is an increasing amount of evidence for some systems which suggests that the function of stress in stress-corrosion cracking is to promote a strain-rate which, rather than stress *per se*, is the parameter that really governs crack initiation or propagation. In these cases the minimum creep rate for cracking is as much an engineering design parameter as is the threshold stress or stress-intensity factor obtained from constant-load tests on plain or precracked specimens.

The point may be illustrated by data for a ferritic steel exposed to a carbonate–bicarbonate solution as fatigue precracked cantilever beams subjected to constant loads. Deformation in the plastic zone associated with the precrack is time dependent following load applications and can be measured and the threshold conditions for stress-corrosion cracking defined in terms of a limiting average creep rate over a specific time interval. That limiting creep rate may then be used in subsequent experiments to calculate the threshold stress from creep data determined independently, these calculated threshold stresses then being compared with values determined experimentally. The creep properties of ferritic steels may be varied by prior strain ageing, following different amounts of cold work and Fig. 8.93 shows the observed and calculated threshold stresses from tests on specimens subjected to various strain ageing treatments. Clearly, the general trend of the experimentally determined curve showing the effects of the amount of prior deformation is reflected in the calculated results.

The equipment required for slow strain-rate testing is simply a device that permits a selection of deflection rates whilst being powerful enough to cope with the loads generated. Plain or precracked specimens in tension may be used but if the cross-section of these needs to be large or the loads high for any reason, cantilever bend specimens with the beam deflected at appropriate rates may be used. It is important to appreciate that the same deflection rate does not produce the same response in all systems and that the rate has to be chosen in relation to the particular system studied (see Section 8.1).

The representation of the results from slow strain-rate tests may be through the usual ductility parameters such as reduction in area, the maximum load achieved, the crack velocity or even the time to failure, although as with all tests, metallographic or fractographic examination, whilst not readily quantifiable, should also be involved. Since stress-corrosion failures are usually associated with relatively little plastic deformation, the ductility

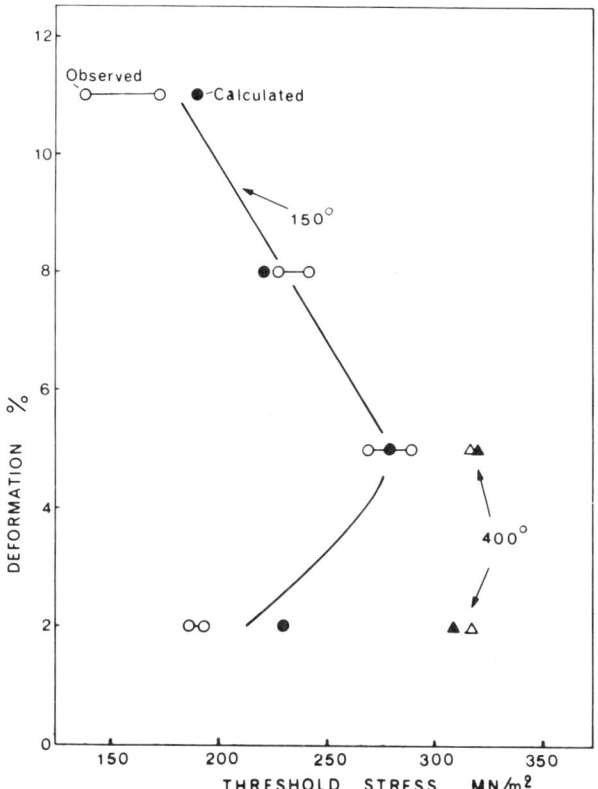

Fig. 8.93 Observed and calculated values of the threshold net section stresses for stress corrosion cracking of a C–Mn steel after various prior deformations and ageing treatments

of specimens will be variable according to the extent to which stress corrosion contributes to the fracture process. This will also influence the shape of the load–extension curve that may be obtained by continuous monitoring of the response of a load cell incorporated in the system, Fig. 8.94 showing the forms of curves obtained with and without attendant stress corrosion. It is apparent from these curves that not only is the extension to fracture dependent upon the presence or otherwise of stress-corrosion cracks, but so also is the maximum load achieved. The latter may be used for expressing cracking susceptibility in some systems, as also may the area bounded by the load–extension curve. However, the variations in maximum load achieved in slow strain-rate tests in circumstances of varying cracking severity are not always large enough for significant distinctions to be made. Even measurements of ductility, such as reduction in area, are not invariably readily made, if only because the final fracture of the specimen does not always follow a simple path and the fitting of the two broken pieces together is not easy. Probably the easiest quantity to measure with reasonable accuracy is the time to failure, which has as much significance in a slow strain-rate test as it does in constant-load or constant-deflection tests. Indeed, the time to failure in slow strain-rate tests is simply related to ductility parameters, a not very

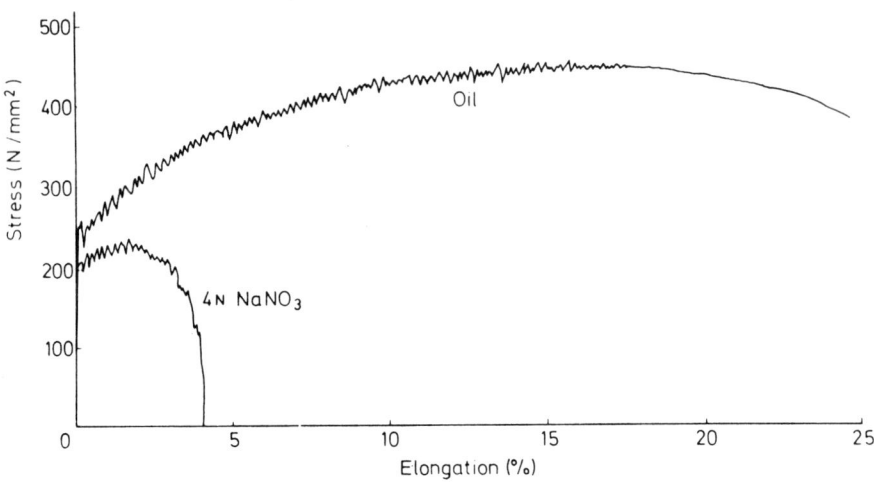

Fig. 8.94 Nominal stress–extension curves for mild steel in oil giving ductile failure, and in 4 N NaNO$_3$ producing stress-corrosion failure, at the same test temperature (104°C)

surprising result when it is remembered that the less the intensity of stress-corrosion cracking the greater will be the ductility to fracture and therefore, the greater the time to failure for a given strain-rate.

Clearly, for slow strain-rate testing to have credence it should give results that are comparable with those obtained by other methods. Figure 8.95 shows some results for tests upon low-alloy ferritic steels in boiling 4 N NH$_4$NO$_3$, the various alloying elements producing a range of cracking susceptibilities as measured by the threshold stresses obtained from constant-strain tests. These results have been normalised by dividing the threshold stress σ_{th} by the lower yield strength σ_y for each steel, whilst the slow strain test results have been normalised by dividing the time to failure in the 4 N NH$_4$NO$_3$ by the time to failure in oil at the same temperature, so that increasing departure from unity indicates increasing cracking susceptibility. The general trend of the results in Fig. 8.95 is clear in indicating reasonable agreement between the two types of test in placing the steels in essentially the same order of merit.

Although slow strain-rate tests are most frequently taken to total failure in order to produce a 'go/no-go' type of result in which threshold stresses are not defined, they can be conducted in a manner that allows such definition. Specimens are preloaded to various initial stresses in the absence of the cracking environment or at a potential that prevents cracking, after which they are allowed to creep until the latter falls below the strain rate to be applied. The applied straining is continued for a sufficient time only to allow cracks to grow to a measurable size. During straining, the stress upon the specimen varies in a manner dependent upon the magnitude of the applied strain rate, hence the importance of restricting the test time to no longer than that necessary to produce measurable cracks. The cracks are probably most conveniently measured by microscopy on longitudinal sections of the gauge lengths, the length of the deepest detectable crack divided by the test time giving an average crack velocity. Figure 8.96 shows some results from tests upon a cast nickel–aluminium bronze exposed to sea-

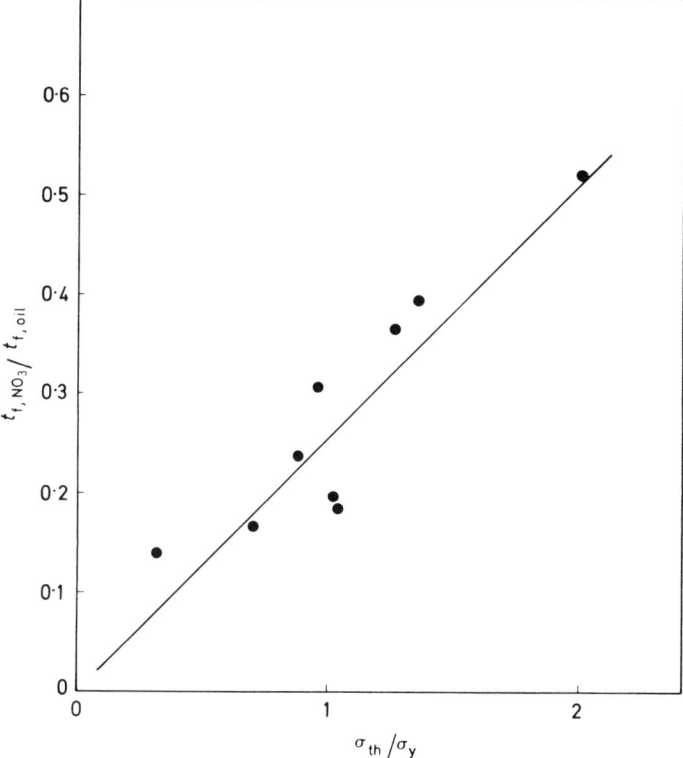

Fig. 8.95 Time to failure ratios from constant-deflection rate tests and normalised threshold stresses σ_{th}/σ_y obtained from constant-strain tests for a series of low-alloy ferritic steels in boiling 4 M NH_4NO_3

water[10] and clearly it is possible to define a threshold stress below which cracking is not observed. However, that threshold stress depends upon the strain rate applied, as is to be expected (see Section 8.1). Another approach to defining threshold stresses in slow strain-rate tests that may sometimes be useful is to use tapered specimens, with the taper angle minimised to avoid complications by resolved components of the tensile load[11]. Applied to the cracking of α-brass exposed to sodium nitrite solutions, a single tapered specimen gave threshold stresses close to those obtained by the use of a number of plain specimens loaded at a given strain rate to various stress levels.

Pre-cracked Testpieces

The literature contains many references to the use of notched, as opposed to pre-cracked or plain, specimens in laboratory studies of stress corrosion, for reasons of improved reproducibility, inability to crack plain specimens under otherwise identical conditions or ease of measuring some parameter such as crack growth rate when the crack location is predetermined. However, the developments in fracture mechanics (see Section 8.9), have resulted in a whole new field of stress-corrosion testing involving the use of specimens

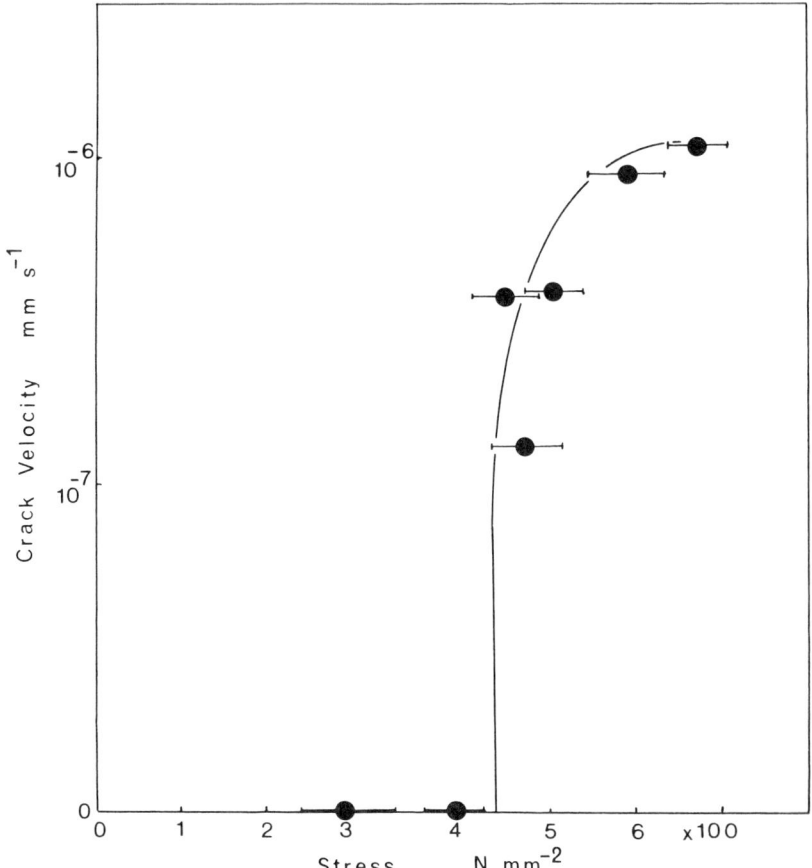

Fig. 8.96 Average stress-corrosion crack velocity from monotonic slow strain rate tests at 1.5×10^{-7} s conducted over various restricted ranges of stress on a cast Ni-Al bronze in seawater at 0.15 V(SCE). The stress range traversed in each test is shown by the length of the bar. (after Parkins and Suzuki[10])

containing a sharp pre-crack, usually produced from a notch by subjecting the specimen to fatigue. The application of fracture mechanics to stress-corrosion cracking is the subject of an admirable review by Brown[12] and various aspects of the method are considered in papers presented at an AGARD conference[13].

The problems associated with the choice of plain specimens for assessing stress-corrosion resistance may, at first sight, appear equally large in relation to precracked specimens in the sense that in the relatively short time during which such tests have been in use a large number of specimen types have been used (Fig. 8.97). However, the differing specimen geometries are rationalised through the stress-intensity factor, with the result that data from different testpieces are comparable, providing appropriate precautions are taken in specimen preparation. The biggest single difficulty is in relation to the large size of specimen that is necessary for highly ductile materials if the concepts of linear elastic analysis are to be applicable. Since it is probable that most service stress-corrosion failures occur in highly ductile

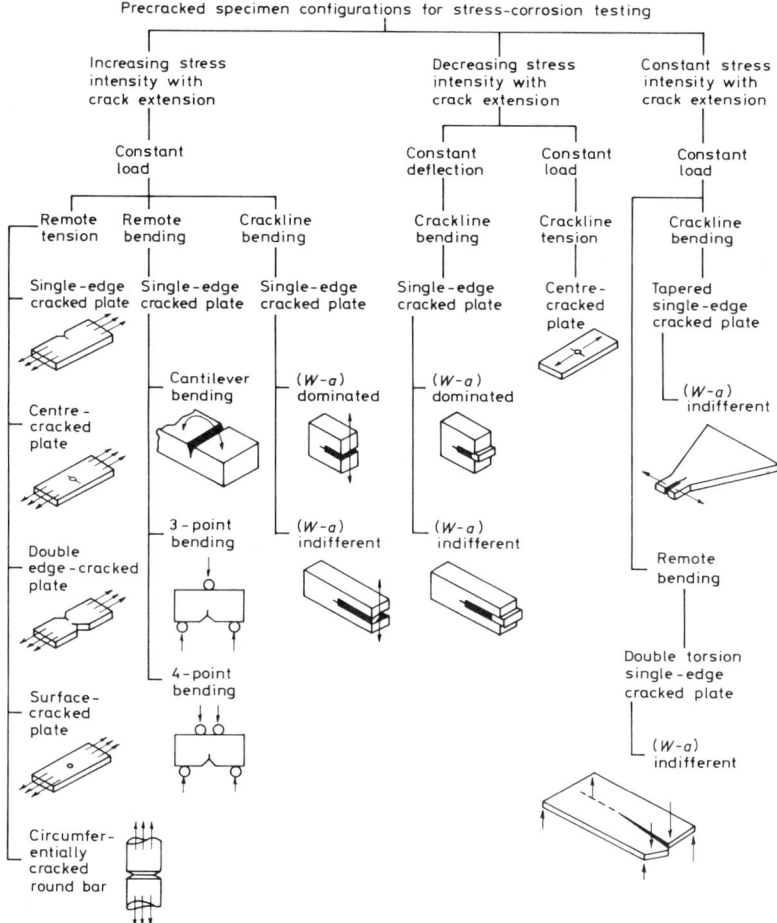

Fig. 8.97 Classification of pre-cracked specimens for stress-corrosion testing (after Smith and Piper[14])

materials in relatively thin sections it is clear that there are problems here, but the use of pre-cracked specimens that dimensionally do not conform to the requirements for linear elastic analysis to be strictly valid is still worthwhile in some instances, and in any case developments in fracture mechanics over the last decade or so allow alternative approaches than that of linear elastic analysis.

Pre-cracked specimens are sometimes useful for other reasons than the analysis that they afford in relation to stress-intensity factors. Such applications may be associated with the simulation of service situations, the relative ease with which stress-corrosion cracks can be initiated at pre-cracks or the advantages that sometimes accrue from the propagation of a single crack. The claim that has sometimes been made of pre-cracked specimen tests–that they circumvent the initiation stage of cracking in plain specimens, erroneously assumed invariably to be related to the creation of a corrosion pit that provides a measure of stress concentration approaching that achieved at the outset with a pre-crack–is rarely entirely valid. Thus, the geometries of a pit, notch or precrack are frequently as important for electrochemical reasons as they are for any reason associated with their influences upon stress distribution. This is because a geometrical discontinuity may be necessary to provide the localised electrochemical conditions, in terms of environment composition or electrode potential, that are necessary for stress-corrosion crack propagation. The objections that have sometimes been made against the use of pre-cracked specimens, e.g. to the validity of introducing a transgranular pre-crack into a specimen that suffers intergranular stress-corrosion cracking, or of the necessity for going to considerable expense to produce a very sharp crack when the introduction of a corrosive environment may blunt the crack by the dissolution, miss the point that such sharp discontinuities do indeed exist in some real materials. Indeed one of the major attractions of pre-cracked specimen testing is that it can provide data that allow the designation of maximum allowable defect sizes in structures for the latter to remain in a safe condition.

In view of the significance of strain rate in stress-corrosion cracking, mentioned earlier, it is as well to remember that its significance is as applicable to pre-cracked specimens as it is to initially plain specimens, in relevant systems. This has a number of implications, not the least of which is the possible influence of time delay between loading pre-cracked specimens and exposing them to the test environment. Moreover, the limiting stress-intensity factor K_{Iscc}, above which cracks grow relatively rapidly (Fig. 8.12) may well depend upon the conditions under which it is determined and it should not be regarded as some property of the material equivalent to, say, a yield stress. There is now a considerable volume of data that show how relatively small fluctuating stresses may reduce the threshold stresses or stress-intensity factors for stress-corrosion cracking and some of these effects are probably related to cyclic loading sustaining creep-related effects. Crack-tip strain rates have consequently become a topic of interest and expressions are available for cyclically loaded pre-cracked specimens[15,16] and also for multi-cracked specimens[17] of the form that initially plain specimens take during slow strain-rate tests.

Comparison of the Results from Plain and Pre-cracked Specimens

It is clear that an initially plain specimen that develops a stress-corrosion crack may, if the geometry is appropriate, conform to the conditions obtaining in an initially pre-cracked specimen. This raises the question, despite the opposing views of the protagonists of the two types of testpiece, as to the comparability of the result from each. Figure 8.98 shows the results[18] obtained from stress-corrosion tests upon a Mg–7Al alloy obtained exposed to a chromate–chloride solution, the cracking susceptibility of the alloy being varied by different heat treatments. The implication of Fig. 8.98 is that the threshold stress σ_{th}, determined upon initially plain specimens of small cross section, is related to the threshold stress intensity K_{Iscc} obtained from pre-cracked specimens of relatively large section. Since K_{Iscc} represents the stress intensity below which an existing crack does not propagate, it would

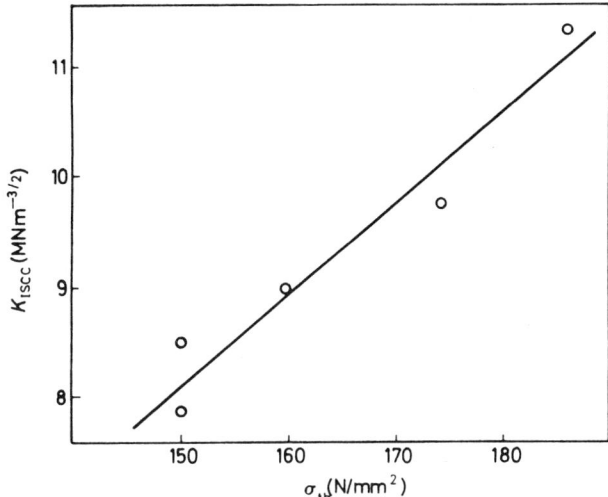

Fig. 8.98 Threshold stress intensities K_{Iscc} from pre-cracked specimen tests, and threshold stresses σ_{th} from plain specimen tests, for a Mg–7Al alloy in various structural conditions tested in chromate–chloride solution (after Wearmouth et al.[18])

appear that the threshold stress given by tests on plain specimens corresponds to values below which cracks do not propagate to give total failure, i.e. the threshold stress for plain specimens is not necessarily the stress below which cracks do not form. Examination of plain specimens stressed below the threshold stress revealed the presence of small stress-corrosion cracks that had ceased to propagate (see Section 8.1) and moreover the maximum sizes of the cracks that did not propagate to total failure were quadratically related to the threshold stress, as would be expected if the concepts of fracture mechanics were applicable to these initially plain specimens. Non-propagating cracks have also been observed at stresses below the threshold stress in other systems, such as low-alloy ferritic steels exposed to various environments, and would therefore support the suggestion that what are being measured in tests upon plain and pre-cracked specimens are not so different as has sometimes been suggested.

Crack Velocity Measurements

In mechanistic studies of stress corrosion and also in the collection of data for remaining-life predictions for plant there is need for stress-corrosion crack velocity measurements to be made. In the simplest way these can be made by microscopic measurement at the conclusion of tests, the assumption being made that the velocity is constant throughout the period of exposure, or, if the crack is visible during the test, *in situ* measurements may be made by visual observation, the difficulty then being that it is assumed that the crack visible at a surface is representative of the behaviour below the surface. Indirect measurements must frequently be resorted to, and these have involved observation of the elongation of the specimen, crack-opening displacement, changes in the electrical resistivity of the specimen and acoustic emissions that sometimes accompany crack extension.

Measurement of the elongation of the specimen is probably the least satisfactory of these, even though it is often the simplest in only requiring a transducer that responds to dimensional change. Multiple cracking of initially plain specimens raises problems in interpreting the data in terms of crack velocities and so the technique has been frequently used for attempting to determine the point in time when cracking was initiated, the preceding time during which the transducer shows no response being equated to an incubation period for cracking. However, such results can be completely misleading because the sensitivity of most transducers is such that they will only detect change when the specimen undergoes some plastic deformation, resulting from the propagation of a crack to the size where the remaining uncracked portion of the section beyond the crack is raised to its yield strength. Consequently, crack propagation can occur during the (so-called) incubation period when the stress is insufficient to cause the propagation of a deformation band. The latter frequently occurs suddenly, producing a sharp response by the transducer, which has sometimes been interpreted as evidence of a burst of fast mechanical fracture but which may in fact be nothing of the sort. It is much more satisfactory to use a crack-opening displacement gauge[19] located across a pre-crack. These gauges usually take the form of two thin cantilever beams to which strain gauges are attached, the beams being located at opposite sides of the extremity of the pre-crack. As crack extension occurs, and the sides of the crack undergo relative displacement the strain gauges respond to the unbending of the beams.

Changes in the electrical resistivity of a specimen containing a propagating crack[20] depend upon applying a high constant direct current at each end of the specimen and measuring the potential difference across electrical leads situated at opposite sides of the crack. The potential field in the region of the crack is disturbed by the presence of the latter and as the crack extends the potential difference between the leads on opposite sides increases, providing that the total current remains constant. This requires a reliable constant current source, and the technique is dependent in some degree upon the exact positioning of the leads and gives less reproducible results if crack branching occurs. The initial thought that the application of d.c. to the specimen may influence the electrochemistry of the stress-corrosion reactions is not sustained in practice and the technique can provide reliable data.

High frequency stress waves are generated when stress-corrosion cracks propagate in some materials, especially the high-strength steels when these undergo hydrogen-induced cracking. The detection of these acoustic signals, which are filtered from lower amplitude background noise, affords a means of studying crack propagation[21]. Whilst the technique involves the use of sophisticated and relatively costly equipment if it is to be correctly practised, it has been suggested that it may also offer a means of distinguishing between active paths and hydrogen-embrittlement mechanisms of cracking[22]. However, that is not universally accepted and the data from acoustic signals need treating with caution[23].

Effects of Surface Finish

It is hardly surprising that the preparation of surfaces of plain specimens for stress-corrosion tests can sometimes exert a marked influence upon results. Heat treatments carried out on specimens after their preparation is otherwise completed can produce barely perceptible changes in surface composition, e.g. decarburisation of steels or dezincification of brasses, that promote quite dramatic changes in stress-corrosion resistance. Similarly, oxide films, especially if formed at high temperatures during heat treatment or working, may influence results, especially through their effects upon the corrosion potential.

However, quite apart from these chemical changes at surfaces occasioned by the method of specimen preparation, physical effects may be important. Paxton and Procter[24] have prepared a review of what little is known about the effects of machining and grinding upon stress-corrosion susceptibility, the most obvious effects being related to surface topography and the introduction of residual stresses into the surface layers. The former is more likely to be important in the higher strength notch-sensitive materials, whilst surface compressive stresses are likely to have the general effect of delaying or preventing failure.

Solution Preparation

Although the list of environments reported as promoting stress-corrosion cracking in any alloy continues to grow with time, the concept of solution specificity remains in that not all corrosive environments will initiate or sustain stress-corrosion cracking in all alloys. Whilst it is inevitable that the environment will always remain as one of the variables that may need to be assessed by stress-corrosion tests, nevertheless certain solutions, by their widespread use over many years, have tended to become standard test solutions for certain types of alloy. Thus, boiling $MgCl_2$ solution for stainless steels, boiling nitrate solutions for carbon steels and 3.5% NaCl for aluminium alloys, to mention but a few, have been extensively used, for example, in comparing the effects of metallurgical variables upon cracking propensities. Such approaches raise two questions, the first concerned with the extent to which 'standard' solutions prepared in different laboratories may be regarded as identical and the second with the extent to which degrees of susceptibility of a range of alloys to cracking in one environment are related to cracking in a different environment.

Whilst the relatively small differences that may be expected to occur between laboratories preparing a solution to the same specification frequently will not influence stress-corrosion test results, there are situations where relatively small changes in environment can promote marked changes in cracking response. Thus, Streicher and Casale[25] have pointed to the potential problems associated with the use of nominally 42% boiling $MgCl_2$ in testing stainless steels. Since the hydrate of $MgCl_2$ is hygroscopic, solution preparation by weighing may lead to appreciable differences in boiling point and hence times to failure in a stress-corrosion test, so that it is preferable to prepare the solution by adding water to the hydrate to achieve a particular boiling point.

Similarly, pH variations resulting from either the initial preparation or from changes during a stress-corrosion test, may exert a marked influence upon results in some systems. Thus, the cracking of carbon steels in nitrates is markedly pH sensitive and, depending upon the volume of solution and the surface area of the specimen exposed, as well as upon the time involved in making the test, significant pH rises can occur and cracking can cease as a result. Moreover, if tests are carried out with anodic stimulation these effects may be aggravated, especially if the counter electrode is immersed in the test cell. In other cases, e.g. the medium and higher strength steels, the initiation and maintenance of cracking frequently requires localised pH changes within the confines of the crack, and these can only occur if the initial conditions of exposure are appropriate.

The oxygen concentration of the solution, as in many instances of corrosion, can also be critical in stress-corrosion cracking tests. Instances are available in the literature that show very markedly different test results according to the oxygen concentration in systems as widely different as austenitic steels immersed in chloride-containing phosphate-treated boiler water[26] and aluminium alloys[27] immersed in 3% NaCl.

The assumption that the relative cracking responses of a series of alloys will be the same irrespective of the environment to which they are exposed can be extremely dangerous (see Section 8.2). Many examples could be quoted of the dangers of drawing conclusions from tests in a given environment and applying these to a different situation, but some results by Lifka and Sprowls[28] will suffice. The results for the relative cracking susceptibilities of three aluminium alloys subjected to different exposure conditions are shown in Fig. 8.99, which indicates that an intermittent spray test using acidified 5% NaCl solution gives the same order of susceptibility for the three alloys as was observed in outdoor exposure tests at three different locations. On the other hand, an alternate immersion test in 3.5% NaCl, widely used for testing aluminium alloys, places the alloys in a completely different order of susceptibility. This single example will suffice to indicate the necessity for simulating service conditions as closely as possible where laboratory data are to be used for selection or design in relation to industrial equipment.

While the dangers inherent in using standardised environments in relation to environment-sensitive fracture are readily indicated by many examples that can be quoted, there remains a problem in relation to alloy development where possible service environments may not always be identifiable at the time of the development programme. In such circumstances it appears desirable that an alloy should be assessed in a range of environments, but even then, it is necessary for realism to be injected into the programme if an excessively large number of test environments are not to be involved. The potential dependence of cracking, with its implications for dissolution and filming reactions or the discharge of hydrogen, suggests that the solution pH is also likely to exert significant influence upon cracking. Plots of cracking domains on potential–pH diagrams sometimes indicate correlations with certain reactions and this may be useful in guiding a testing programme[29],

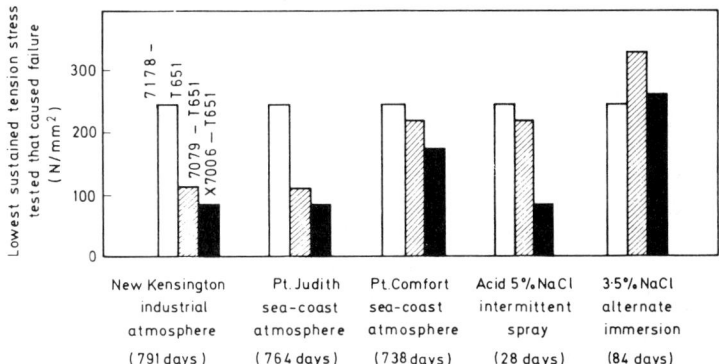

Fig. 8.99 Relative resistance to stress-corrosion cracking of three aluminium alloys subjected to different environments. The stress levels employed corresponded to 75, 50 and 25% of the respective transverse yield strengths (after Lifka and Sprowls[28])

i.e. to involve solutions of different pH values and to explore the cracking propensity as a function of potential.

The importance of potential (see Section 8.1) cannot be overemphasised and some aspects of this part of stress-corrosion testing may be conveniently discussed in the context of Fig. 8.21. This shows the different potential ranges for the cracking of a ferritic steel according to the environment in tests involving potentiostatic control. Also shown are the free corrosion potentials for that steel in the different solutions. These indicate that whilst failure would occur in the nitrate at the free corrosion potentials, this would not be so in the hydroxide or carbonate–bicarbonate solution. This does not mean that carbon steels will never fail by stress corrosion in these two environments at the free corrosion potential, since the latter is, of course, dependent upon the composition of the steel, its surface condition, the composition of the environment and other factors. It is possible, therefore, that as the result of, say, small additions to the environments, added intentionally or present as impurities, the corrosion potential can be caused to be within the cracking range, or that as the result of small additions to a steel, the corrosion potential may fall outside the cracking range. Quite small changes in potential, frequently only a few tens of millivolts, can therefore produce dramatic changes in cracking response and point to the necessity, especially in laboratory tests attempting to simulate a service failure, of reproducing the potential with precision.

Stress-corrosion Test Cells

The cells that contain the specimen and environment for stress-corrosion tests frequently need to be more than a vessel made in some substance, usually glass, that is inert to the environment and which produces no electrical response upon the test specimen. Where cracking is initiated at surfaces through which heat transfer occurs it may be necessary to design a cell in which such an effect is incorporated, since the concentration of substances in solution that may occur at an interface through which heat passes, may play a significant role in promoting cracking, especially if surface deposits allow concentration by evaporation whilst preventing mixing with the bulk or the environment. The cracking of riveted mild-steel boilers and the concentration of carbonate–bicarbonate solutions under pipeline coatings to produce cracking in high-pressure gas transmissions lines, are significant examples. Dana[30] has developed a method for simulating the conditions for cracking of stainless steels in contact with thermal insulating materials, whilst concentration in leaking boiler seams is simulated in the 'embrittlement detector' developed by Schroeder and Berk[31].

Such test cells involve, among other things, a crevice, the essence of which is that the volume of solution that it contains is relatively small compared with the area of exposed metal, a ratio that may influence stress-corrosion test results determined in more conventional cells where crevices do not exist. The experiments of Pugh *et al.*[32], on the stress corrosion of 70–30 brass in ammoniacal solutions of various volumes are particularly instructive in indicating how this ratio may influence results, the time to failure varying by about an order of magnitude for a similar change in solution volume. Changes in the surface area of exposed specimens, apart from the effects already implied, may influence the cracking response for other reasons, as shown by the results of Farmery and Evans[33] for an Al–7Mg alloy immersed in a chloride solution. They found that coupling unstressed to stressed specimens of varying area ratio influences failure times, relatively short times being obtained when the area of unstressed to stressed specimen was large.

Initiation of Stress-corrosion Tests

It may be felt that the initiation of a stress-corrosion test involves no more than bringing the environment into contact with the specimen in which a stress is generated, but the order in which these steps are carried out may influence the results obtained, as may certain other actions at the start of the test. Thus, in outdoor exposure tests the time of the year at which the test is initiated can have a marked effect upon the time to failure[27], as can the orientation of the specimen, i.e. according to whether the tension surface in bend specimens is horizontal upwards or downwards or at some other angle. But even in laboratory tests, the time at which the stress is applied in relation to the time at which the specimen is exposed to the environment may influence results. Figure 8.100 shows the effects of exposure for 3 h at the applied stress before the solution was introduced to the cell, upon the failure of a magnesium alloy immersed in a chromate–chloride solution. Clearly such prior creep extends the lifetime of specimens and raises the threshold stress very considerably and since other metals are known to be strain-rate sensitive in their cracking response, it is likely that the type of result apparent in Fig. 8.100 is more widely applicable.

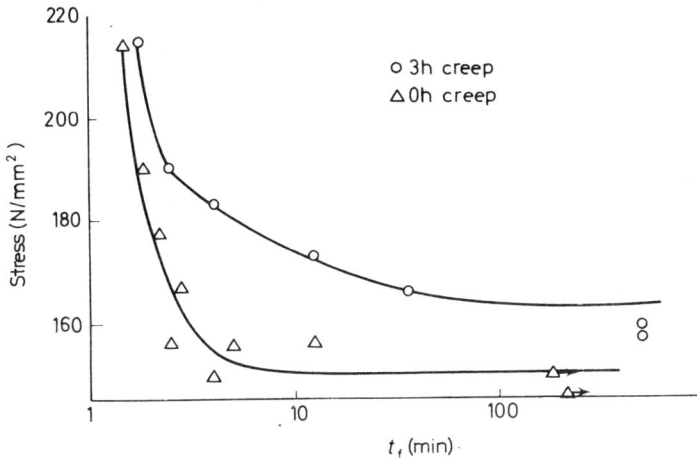

Fig. 8.100 Effect of delay period between application of load and introduction of solution to test cell in the failure of a Mg–7Al alloy exposed to a chromate–chloride solution (after Wearmouth et al.[18])

Hydrogen Embrittlement Tests

The absorption of hydrogen by various materials, including high-strength steels, results in loss of ductility which in turn can result in cracking and fracture when the metal is subjected to a sustained tensile stress (see Section 8.4). Hydrogen may be introduced into these various alloys from the gas phase (during manufacture or welding), or from aqueous solution during surface treatment (pickling, plating, phosphating) or from the environment during a spontaneous corrosion process in which the development of acidity within the crack results in hydrogen evolution and absorption (see Section 1.6). Various test methods[34] may be used to evaluate the effect of hydrogen on the properties of alloys, including some ad hoc tests that were specifically developed for high strength steels.

Although similar constant-load test rigs are used for both active-path corrosion and hydrogen stress cracking there is one fundamental difference in the test procedure. In the case of active-path corrosion testing it is always carried out in the presence of the corrosive environment, but in the case of hydrogen-related cracking, testing may be carried out after hydrogen has been introduced into the alloy either deliberately by gaseous or cathodic charging, or following processes such as welding, pickling or electroplating. However, with pre-charged specimens loss of hydrogen may occur[35] when they are removed from the environment, which results in entry of that substance and so sustained-load tests are also carried out in the presence of an environment (gaseous or aqueous) so that hydrogen is introduced into the testpiece during the application of the tensile stress.

Dynamic Tests

All of the properties evaluated by the conventional tensile test (yield strength, tensile strength, elongation and reduction in area) are affected by the presence of hydrogen, but in the case of the tensile strength and yield strength the effect is significant only when the steel has a very high tensile strength and has been severely embrittled. On the other hand, the reduction in area, and to a lesser extent the elongation, may be used for detecting embrittlement. Hobson and Sykes[36] found that with low-carbon steels there was an almost linear relationship between reduction in area and hydrogen content of the steel. Slow strain-rate tests are sometimes employed in testing materials (and not only steels) after pre-exposure to a source of hydrogen. The strain rate may be critical in that not only can it be too high but if it is too low the hydrogen may diffuse out of the specimen before cracking occurs.

Various types of bend tests have been used to evaluate embrittlement. Beck, Klier and Sachs[37] used thin strip specimens, and determined the decrease in height, Δh, at fracture when the specimen was bent by compressing it at a constant rate in a tensile testing machine (Fig. 8.101). The decrease in height, Δh, gives a measure of the embrittlement, the maximum elongation of the outside fibre of the specimen being calculated from the radius of curvature at maximum bending. In general, the ductility is found to increase with the rate of straining, and for this reason high-strain-rate tests, such as impact tests, are insensitive to hydrogen embrittlement. Where the material is available only in the form of tubing, semicircular specimens may be used in place of flat strip in the compression bend test. The total cross-head travel from the unstressed height along the diameter to the point of fracture gives a measure of embrittlement which may be compared with that obtained from an unembrittled specimen of the same steel.

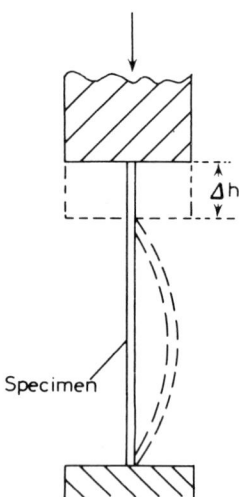

Fig. 8.101 Bend test using a tensile testing machine (after Beck et al.[37])

A constant-rate bend-test machine, which provides an effective method for testing highly embrittled steel wires of high-tensile strength was designed by Zapffe and Haslem[38] (Fig. 8.102). The motor A pulls a chord attached to the travelling arm D that rotates about a pivot pin. The wire specimen G (1.6 × 100 mm) is inserted in D and is supported by the fixed arm F, the arrangement being so designed that tensile or torsional stresses are avoided. The specimen is thus bent around the pivot pan E (radius 1.6 mm) at a constant rate, the angle of bend to cause fracture giving a measure of its ductility. Since ductility increases with rate of straining, the bending rate must be slow and 4°/s is considered to be suitable for detecting embrittlement. A similar machine has been used for studying the embrittlement of spring steel strip after hydrogen has been introduced by cadmium plating.

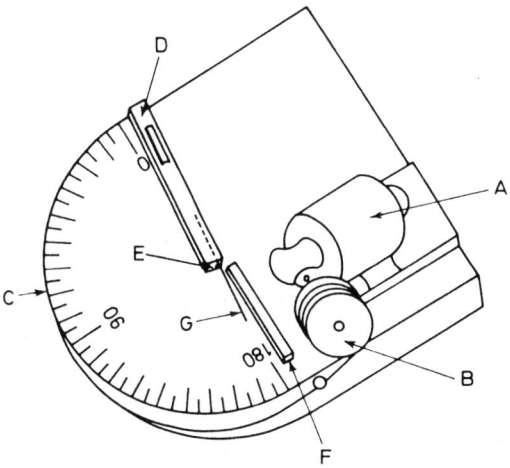

Fig. 8.102 Constant-rate bend test for determining hydrogen embrittlement of wires. A, drive unit; B, pulley; C, semicircular base; D, travelling arm; E, axial pin; F, fixed arm; G, wire specimen (after Zapffe and Haslem[38])

An alternative procedure is the reverse-bend test in which the specimen in the form of wire or strip is bent repeatedly backwards and forwards over a mandrel until it fractures, the number of bends indicating the degree of embrittlement. However, this method is considered to be less sensitive than the single-bend test.

Static Tests

Whereas ductile materials, such as iron and mild steel, are often considered not to crack when charged with hydrogen and subjected to a tensile stress below the yield stress, the position is different with high-strength ferrous alloys where, depending on the strength of the steel and the hydrogen content, failure may occur well below the yield stress. However, the fracture process is not instantaneous and there is a time delay before cracks are initiated; for this reason the phenomenon is sometimes referred to as 'delayed failure'.

In the majority of cases, the tests are conducted using a dead-weight lever-arm stress-rupture rig with an electric timer to determine the moment of fracture, but a variety of test rigs similar to those shown in Fig. 8.89g are also used. The evaluation of embrittlement may be based on a delayed-failure diagram in which the applied nominal stress versus time to failure is plotted (Fig. 8.103) or the specimen may be stressed to a predetermined value (say 75% of the ultimate notched tensile strength) and is considered not to be embrittled if it shows no evidence of cracking within a predetermined time (say 500 h). Troiano[39] considers that the nature of delayed fracture failure can be described by four parameters (see Fig. 8.103):

1. the upper critical stress corresponding to the fracture stress of the unembrittled notched specimen;
2. the lower critical stress, which is the applied stress below which failure does not occur;
3. the incubation period or the time required for the formation of the first crack;
4. the failure time or the time for specimen failure at a given applied stress; in the intermediate stress range this includes a period of relatively slow crack growth.

During the constant-load test it is essential that only axial tensile stresses are applied since any bending stresses that are introduced will result in a higher true stress than that calculated. For this reason the ends of the specimens and the grips must be designed to avoid bending stresses. The ASTM Standard E8-69 specifies that in the case of specimens with threaded ends the grips should be attached to the heads of the testing machine through properly lubricated spherical-seated bearings, and that the distance between the bearings should be as great as is feasible (Fig. 8.104).

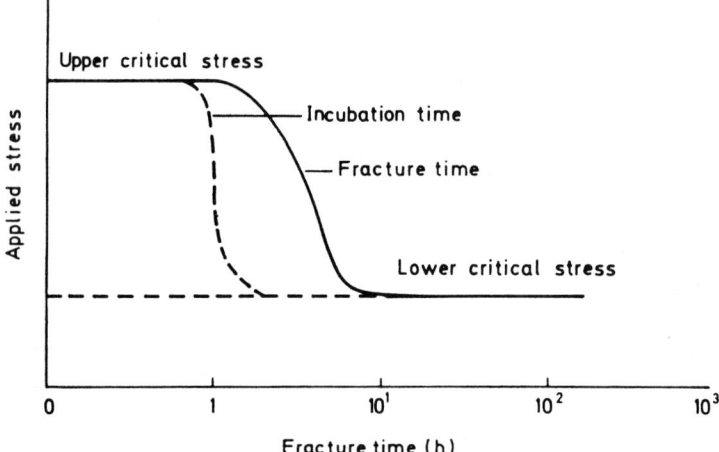

Fig. 8.103 Schematic representation of delayed failure characteristics of a hydrogenated high-strength steel

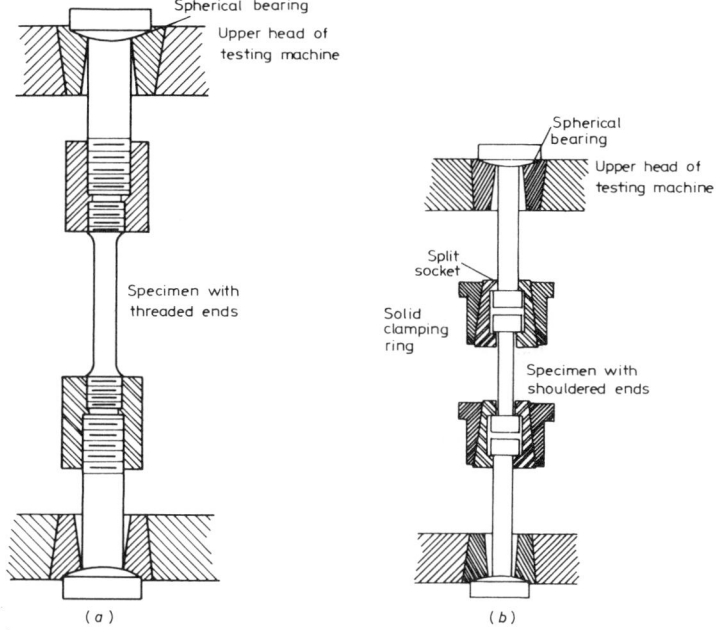

Fig. 8.104 Methods of gripping specimens in order to avoid bending stresses. (a) Device for threaded end specimens, and (b) device for shouldered-end specimens (from *Stress Corrosion Testing*[2])

In order to simplify the test procedure a number of investigators have designed test rigs in which the bulky lever arm is replaced by a loading nut, the stress in the specimen being determined by means of strain gauges; these rigs are similar in principle to that shown in Fig. 8.89g. Figure 8.105 shows a spring-loaded rig that was used by Cavett and van Ness[40] to study the embrittlement produced by hydrogen gas at high pressures, in which the tensile load is applied by compressing a heavy-duty spring.

Raring and Rinebold[41] have devised a method in which the specimen is supported along a diameter of a steel loading ring (Fig. 8.106), and the stress is applied by tightening the bottom nut until the diameter corresponds with the required load. The sudden release of elastic energy stored within the ring when the specimen fractures results in displacement of the tightening nut, and this is used to actuate a microswitch and timer. Williams, Beck and Jankowsky[42] have used notched 'C' rings, the stress being applied by tightening the nut of a calibrated loading bolt which passes through the diameter of the 'C' ring (Fig. 8.107). The strain gauges attached to the bolt form two arms of a Wheatstone bridge circuit and to compensate for temperature changes the other two arms consist of two identical strain gauges attached to a similar unstrained bolt.

R. N. PARKINS

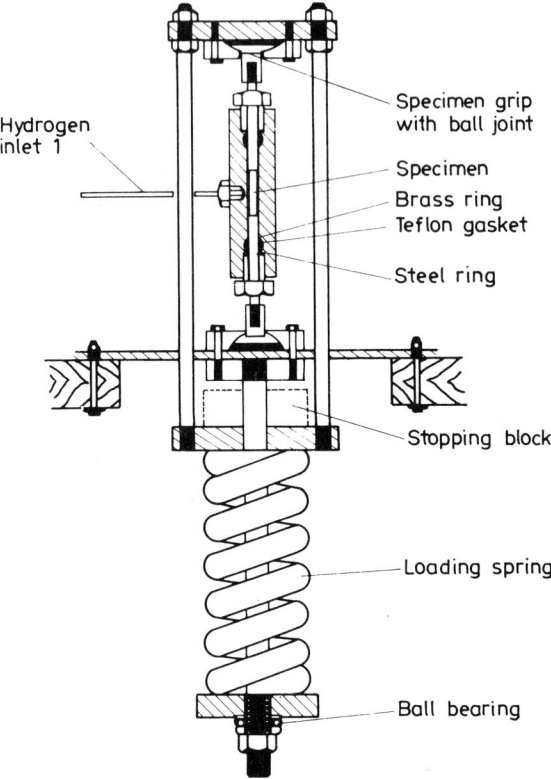

Fig. 8.105 Spring-loaded rig for sustained load testing of a steel specimen in gaseous bydrogen at high pressure (after Cavelt and van Ness[40])

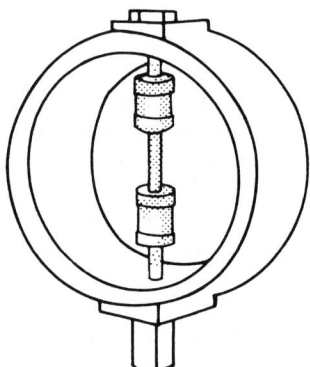

Fig. 8.106 Loading-ring method of stressing a specimen (after Rating and Rinebold[41])

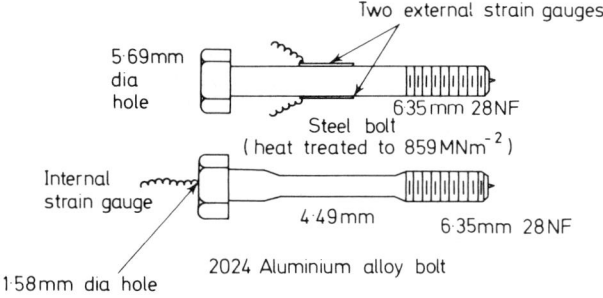

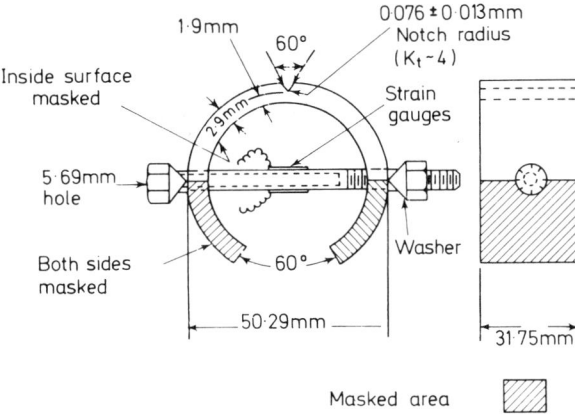

Fig. 8.107 Notched 'C' ring specimen with attached strain gauges

REFERENCES

1. Ailor, W. H. (ed.), *Handbook on Corrosion Testing and Evaluation*, Wiley, New York (1971)
2. *Stress Corrosion Testing*, ASTM STP No. 425 (1967)
3. BS 6980, 'Stress corrosion testing', Parts 1-7, (ISO 7539-1/7) BSI, Milton Keynes
4. Logan, H. L., *The Stress Corrosion of Metals*, Wiley, New York, 273 (1966)
5. Haaijer, G. and Loginow, A. W., *Corrosion*, **21**, 105 (1965)
6. Parkins, R. N., Mazza, F., Royuela, J. J. and Scully, J. C., *Bri. Corr. J.*, **7**, 154 (1972)
7. Parkins, R. N. and Haney, E. G., *Trans. Metall. Soc., AIME*, **242**, 1943 (1968)
8. Brenner, P. and Gruhl, W., *Z. Metall.*, **52**, 599 (1961)
9. Ugianski, G. M. and Payer, J. H. (eds.), *Stress Corrosion Cracking-The Slow Strain Rate Technique*, ASTM STP 665, ASTM, Philadelphia (1975)
10. Parkins, R. N. and Suzuki, Y., *Corros. Sci.*, **23**, 577 (1983)
11. Yu, J., Holroyd, N. J. H. and Parkins, R. N., in *Environment-Sensitive Fracture: Evaluation and Comparison of Test Methods*, ASTM STP 821, Ed. Dean, S. W., Pugh, E. N. and Ugianski, G. M., ASTM, Philadelphia, p. 288, (1984)
12. Brown, B. F., *Met. Rev.*, **13**, 17 (1968)
13. Specialists Meeting on Stress Corrosion Testing Methods, AGARD Conference Proceedings, No. 98, NATO (1972)
14. Smith, H. R. and Piper, D. E., *Stress Corrosion Testing with Precracked Specimens*, The Boeing Co, D6-24872, ARPA 878, June (1970)
15. Lidbury, D. P. G., *Embrittlement by the Localized Crack Environment*, Ed. Gangloff, R. P., AIME, New York, p. 149, (1983)
16. Parkins, R. N., *Corrosion*, **43**, 130 (1987)

17. Congleton, J., Shoji, T. and Parkins, R. N., *Corros, Sci.*, **25**, 633 (1985)
18. Wearmouth, W. R., Dean, G. P. and Parkins, R. N., *Corrosion*, **29**, 251 (1973)
19. Fisher, D. M., Bubsey, R. T. and Srawley, J. E., *Design and Use of a Displacement Gauge for Crack Extension Measurements*, NASA IN-D 3724 (1966)
20. Barnett, W. J. and Troiano, A. R., *Trans. AIME*, **209**, 486 (1957)
21. Gerberich, W. W. and Hartblower, C. E., *Proc. Conf. Fundamental Aspects of Stress Corrosion Cracking*, NACE, p. 420 (1986)
22. Okada, H., Yukawa, K. and Tamura, H., *Corrosion*, **30**, 253 (1974)
23. Pollock, W. J., Hardie, D. and Holroyd, N. J. H., *Br. Corros. J.*, **17**, 103 (1982)
24. Paxton, H. W. and Proctor, R. P. M., Paper No. EM68-520 presented at American Society of Tool and Manufacturing Engineers Symposium on *Surface Integrity in Machining and Grinding*, Pittsburgh (1968)
25. Streicher, M. A. and Casale, I. B., *Proc. Conf. Fundamental Aspects of Stress Corrosion Cracking*, NACE, Houston, p. 305 (1969)
26. Williams, W. I., *Corrosion*, **13**, 539 (1957)
27. Romans, H. B., *Stress Corrosion Testing*, ASTM STP No. 425, p. 182 (1966)
28. Lifka, B. W. and Sprowls, D. O., *Stress Corrosion Testing*, ASTM STP No. 425, p. 342, (1966)
29. Parkins, R. N., in '*The Use of Synthetic Environments for Corrosion Testing*', ASTM STP 970, Ed. Francis, P. E. and Lee, T. S., ASTM, Philadelphia, p.132, (1988)
30. Dana, A. W., *ASTM Bulletin No. 225*, TP 196, p. 46 (1957)
31. Schroeder, W. C. and Berk, A. A., *Bull. US Bur. Mines.*, 443 (1941)
32. Pugh, E. N., Montague, W. G. and Westwood, A. R., *Trans. Am. Soc. Met.*, **58**, 665 (1965)
33. Farmery, H. K. and Evans, U. R., *J. Inst. Met.*, **84**, 413 (1956)
34. Smialowski, M., *Hydrogen in Steel*, Pergamon Press, London (1962)
35. Hardie, D., Holroyd, N. J. H. and Parkins, R. N., *Metal Sci.*, **13**, 603 (1979)
36. Hobson, J. D. and Sykes, C., *J. Iron Steel Inst.*, **169**, 209 (1951)
37. Beck, W., Klier, E. P. and Sachs, G., *Trans. AIME*, **206**, 1263 (1956)
38. Zapffe, C. A. and Haslem, M. E., *Trans. AIME*, **167**, 281 (1946)
39. Troiano, A. R., *Trans. Am. Soc. Met.*, **52**, 54 (1960)
40. Cavett, R. H. and van Ness, H. C., *Welding J. (research supplement)*, **42**, 317 (1967)
41. Raring, R. H. and Rinebold, J. A., *ASTM Bulletin No. 213* (1956)
42. Williams, F. S., Beck, W. and Jankowasky, E. J., *Proc. ASTM*, **60**, 1192 (1960)

8.10A Appendix—Stresses in Bent Specimens

In each of the following equations σ = maximum tensile stress, E = modulus of elasticity and t = specimen thickness.

Two-point bending (Fig. 8.89a)

$$L = \frac{ktE}{\sigma} \sin^{-1}\left(\frac{H\sigma}{ktE}\right)$$

where L = specimen length,
k = constant (1·280) and
H = holder span.

Four-point bending (Fig. 8.89b)

$$\sigma = \frac{12Ety}{3L^2 - 4A^2}$$

where y = maximum deflection,
L = distance between outer supports and
A = distance between inner and outer supports.

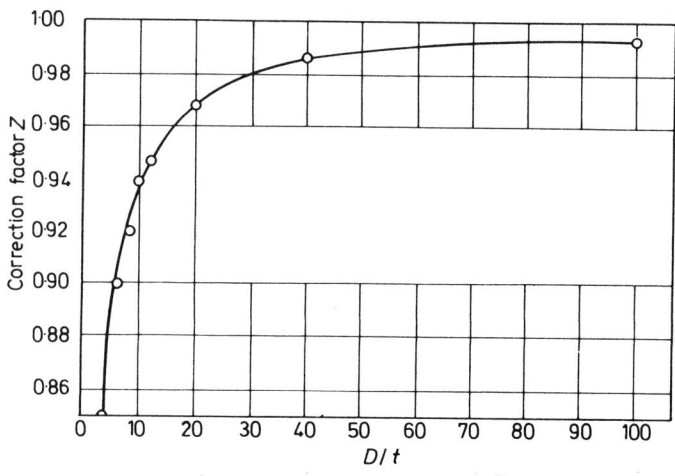

Fig. 8.A1

'*U*'-bends (Figs. 8.89*d* and *e*)

Applied strain $\varepsilon = \dfrac{t}{2r}$, when $t < r$, where r = radius of curvature at section of interest.

'*C*'-rings (Fig. 8.89*c*)

$$\sigma = \frac{4EtZ\Delta}{\pi D^2}$$

where $\Delta = OD_f - OD_i$,
    OD_f = final outside diameter of stressed '*C*'-ring,
    OD_i = initial outside diameter of unstressed '*C*'-ring,
    D = mean diameter, i.e. $(OD - t)$ and
    Z = a correction factor, related to D/t as indicated in Fig. 8.A1.